The Mojave-Sonora Megashear Hypothesis: Development, Assessment, and Alternatives

edited by

Thomas H. Anderson
Department of Geology and Planetary Sciences
University of Pittsburgh
Pittsburgh, Pennsylvania 15260
USA

Jonathan A. Nourse
Department of Geological Sciences
California State Polytechnic University
Pomona, California 91768
USA

James W. McKee
Department of Geology
University of Wisconsin
Oshkosh, Wisconsin 54901
USA

Maureen B. Steiner
Department of Geology and Geophysics
University of Wyoming
Laramie, Wyoming 82071
USA

THE GEOLOGICAL SOCIETY OF AMERICA

Special Paper 393

3300 Penrose Place, P.O. Box 9140 ▪ Boulder, Colorado 80301-9140 USA

2005

Published by The Geological Society of America, Inc.
3300 Penrose Place, P.O. Box 9140, Boulder, Colorado 80301-9140, USA
www.geosociety.org

Printed in U.S.A.

GSA Books Science Editor: Abhijit Basu

Library of Congress Cataloging-in-Publication Data

The Mojave-Sonora megashear hypothesis : development, assessment, and alternatives / edited by Thomas H. Anderson ... [et al.].
p. cm. — (Special paper ; 393)
Includes bibliographic references and index.
ISBN 0-8137-2393-0 (pbk.)
1. Shear zones (Geology)--Mexico--Sonora (State) 2. Geology, Structural--Mexico--Sonora (State) 3. Geology, Stratigraphic—Jurassic. I. Anderson, Thomas H. (Thomas Howard), 1942- II. Special papers (Geological Society of America) ; 393.

QE606.5.M6M65 2005
551.8′0972′17--dc22

2005049457

Cover: Digital elevation model of southwestern North America showing approximate trace of the Mojave-Sonora megashear in red. Left-lateral displacement of ~800–1000 km is implied by apparent correlation of rock assemblages between the Caborca Block and the Inyo Mountains–Death Valley region (white stars). Specific cross-fault ties are proposed and debated throughout this volume. These include Paleoproterozoic basement units (Chs. 1, 4, 5, and 24), Neoproterozoic and Paleozoic marine strata (Chs. 7, 21, and 22), Permo-Triassic plutonic rocks (Chs. 2, 9, and 22), and Lower Mesozoic sedimentary rocks (Chs. 2, 7, 9, 10, and 11). Yellow stars indicate hypothesized cross-fault correlation of Middle Jurassic rhyolite porphyry between Zacatecas and north central Sonora (Chs. 2 and 16). Upper Triassic to Middle(?) Jurassic mélange in Zacatecas (Ch. 16) apparently represents displaced fragments of forearc and trench not present in Sonora. Insets (provided by Patricia Campbell) show examples of mylonite exposed within the axial zone of the type Mojave-Sonora megashear directly southwest of Sonoyta, Sonora. A: Typical outcrop of banded mylonite, with subvertical, northwest-striking foliation and subhorizontal lineation (daypack provides scale). B: Mesoscopic view of mylonitized Triassic granite with left-lateral S-C fabric. Photograph is oriented perpendicular to foliation and parallel to mylonitic lineation. C: Sinistrally sheared muscovite "fish" and recrystallized quartz ribbons that characterize the mylonitic granite (2.5 mm in the long dimension; orientation as in B). Structural and kinematic attributes of these and related rocks, along with detailed geologic maps, are described in Campbell, P.A., and Anderson T.H., 2003, Structure and kinematics along a segment of the Mojave-Sonora megashear: A strike-slip fault that truncates the Jurassic continental magmatic arc of southwestern North America, Tectonics, v. 22, no. 6, p. 16-1–16-21.

10 9 8 7 6 5 4 3 2 1

Contents

Preface

This volume is dedicated to the originator of the controversial Mojave-Sonora megashear hypothesis, Leon T. Silver, professor emeritus at Caltech, in recognition of his seminal contributions to understanding of the tectonic evolution of southwestern North America. Lee Silver studied earth processes in the southwest by careful integration of field observations with geochemical analyses of igneous and metamorphic rocks. He recognized the efficacy of using the age, geologic setting, and geochemical character of igneous rocks to reveal rock patterns that reflect the formation and evolution of continental crust and also to define important structural markers. In this way, Lee was first to recognize two fundamental tectonic processes that occurred in the southwest: left-lateral faulting of Late Jurassic age and westward-directed transport of crust during Cretaceous and Tertiary time.

These two ideas stimulated vigorous debate and generated fertile geologic research that has been pursued for decades. However, the ideas represent only a small part of Lee's broader research that included development and application of U-Th-Pb systematics to the study of zircon, study of the petrology and evolution of circum-Pacific batholiths, and examination of far-flung regions of the Earth and the Moon, meteorites, magmatic processes, and impacts. In all these areas of Lee's interest, he was commonly in the vanguard of current thought.

One of the roots of Lee Silver's scientific productivity is his ability to recognize and formulate fundamental geological questions. An early case in point is his dissertation work in the Johnny Lyon Hills area of Cochise County, Arizona, which provided a new vision of the evolution of Proterozoic crust in the southwest. Lee's work on Precambrian rocks has had wide application in the recognition and solution of Phanerozoic tectonic problems, for example, in evaluating offsets along major lateral faults such as the San Andreas. In that context, Lee attempted to extend knowledge of the Precambrian basement along the trend of the San Andreas to the southeast into Mexico. Lee had made an initial foray into Sonora along the highway parallel to the international border during which he collected a porphyritic granite that yielded an age of 1.4 Ga. Subsequently, he and Anderson, on their return from the 1968 annual meeting of the Geological Society of America at Mexico City, collected several samples of foliated crystalline rocks as they drove northward from Oaxaca through Sonora along the coast highway. Samples also were collected south of Caborca, Sonora, near Rancho Bámori, where Proterozoic ages of basement had been reported. Isotopic analyses of zircon from that sample of rhodacite yielded an interpreted age >1.7 Ga. This result spurred a more extensive reconnaissance that lasted six years and provided the enigmatic results that stimulated the hypothesis of the Mojave-Sonora megashear ("mythical megashear" to some; "Silver's slippery sliver" to others). The megashear hypothesis was originally presented at the national GSA conference in Miami Beach in 1974. It constitutes the unifying theme of this Special Paper.

The ensuing controversy that swirled around the megashear concept was directed most intensely toward the part of the proposed fault that projected into the Mojave Desert of southern California. In this region, the westward protuberance of the San Bernardino Mountains and other ranges to the east and north across the proposed fault extension seemed to preclude its existence. Lee has long argued that the westward protrusion of Proterozoic crystalline rocks mimicked the disrupted isopleths of ages and geochemical signatures shown by granitoids of the Cretaceous batholith. He ardently pursued this idea in the 1980s through his examination of the Rand thrust and his involvement in overseeing the Cajon Pass Deep Drillhole. The "flat" faults above which westward transport occurred may have been detachments or thrusts but the effect is the same—obfuscation of an older (Late Jurassic) fault trace beneath Latest Cretaceous or Tertiary tectonic cover. Presence of a megashear in the Mojave Desert continues to be a contentious issue today.

The papers in this volume examine the Mojave-Sonora megashear—a hypothetical Late Jurassic fault with hundreds of kilometers of left-lateral displacement that is postulated to extend from the Gulf of Mexico northwestward across northern Mexico thence northward along the western margin of North America. Our intents are to: (1) provide the data and insights from which the hypothesis proposed by Silver and Anderson (1974) developed; (2) describe studies of stratigraphy, structure, isotope geochemistry, U-Pb geochronology, rock magnetism, and paleomagnetism that serve to assess the controversial idea; and (3) present works that offer alternative interpretations of the geological relationships in southwestern North America. In order to achieve this goal, we have organized the papers into three groups: development, assessment, and alternatives.

Three papers comprise the development group. An initial "concept" paper by Anderson and Silver presents the U-Pb isotopic results from zircons in crystalline Proterozoic rocks which distinguish the contrasting basement age provinces in Sonora, along with geologic data and observations that reveal the geologic setting of the basement and its pre-Mesozoic cover, and evidence for left-lateral disruption. This concept paper is followed by presentation of the U-Pb isotopic results from zircons in volcanic and plutonic rocks that show the presence of a Middle Jurassic magmatic belt intervening between the Sonoran Proterozoic provinces, the relationship of these magmatic rocks to other Jurassic rocks in the region, and finally how the inferred fault relates to the Middle Jurassic magmatic belt as well as to younger Late Jurassic rocks contemporaneous with postulated fault movements. Anderson, Rodríguez-Castañeda, and Silver are the authors. In the final article of the group, Anderson and Nourse describe numerous Late Jurassic pull-apart basins that distinguish a 200–500-km-wide swath of crust along the northeast side of the megashear and that are interpreted to record a regional transtensional event contemporaneous with megashear faulting.

The assessment papers are arranged roughly geochronologically with respect to the age of the rock units studied. These articles present diverse kinds of new data from northern Mexico and southern Arizona and provide information needed to test the viability of a sinistral Jurassic megashear. Nourse, Premo, Iriondo, and Stahl describe the geology, structure, U-Pb geochronology, and Sm-Nd, Rb-Sr, and U-Pb-Th isotopic characteristics of two contrasting Proterozoic basement blocks that straddle the Sonora-Arizona border near the originally proposed trace of megashear. They consider four alternative paleogeographic scenarios, including a "no-megashear" model, to achieve the present-day distribution of these rocks. Farmer, Bowring, Espinosa Maldonado, Fedo, and Wooden describe the Mojave-like character of Paleoproterozoic basement south of Caborca. They also document the age and petrologic character of distinctive 1.0 Ga plutons restricted to the Caborca block as well as detrital zircons of similar age occurring in nonconformably overlying rocks in Sonora and Nevada. Campos-Enríquez uses magnetic properties of the crust in northern Mexico to delineate domains with contrasting signatures. Stewart reviews the sedimentologic character and ages of Jurassic and older Phanerozoic stratigraphy along the southwestern margin of North America with the goal of comparing facies, thickness, and stratigraphy of exposures in northwestern Mexico with those in California. Centeno-García provides an overview of Upper Paleozoic and Lower Mesozoic stratigraphy in Mexico and provides paleogeographic models for the development of the country. González-León, Stanley, Gehrels, and Centeno-García present new stratigraphic descriptions, Sm-Nd measurements, and U-Pb analyses of detrital zircons from Triassic and Lower Jurassic sedimentary rocks of the El Antimonio Group in the Caborca block. They discuss possible provenances and depositional environments for these rocks in context of a megashear model. Marzolf and Anderson compare the sequence stratigraphy of Triassic and Lower Jurassic sediments near Caborca with similar age strata from other parts of cratonal North America. They argue that the facies patterns and unconformities of the Lower Mesozoic Caborca sequence are out of place and best fit a model in which Caborca originated ~1000 ± 300 km to the northwest. Steiner, Pinos, Lucas, Marzolf, and Estep describe the primary paleomagnetic signature and overprinting recorded in Lower Triassic red beds of the Caborca block. They conclude that these rocks preserve a Triassic pole indicative of paleolatitude significantly farther north. The Caborca rocks were also affected by pronounced clockwise rotation, while undated intrusions into the section preserve a complex magnetic remnance. Haxel, Riggs, Wright, May, and Tosdal provide a new detailed geologic map of the Baboquivari Mountains (southern Arizona), and report new SHRIMP U-Pb analyses that supplement Wright's ca. 1978 data from Jurassic volcanic rocks of this range. Their article offers additional context for separate Middle and Late Jurassic magmatic episodes that may be compared with the Jurassic arc chronology of northern Sonora described in the Anderson et al. paper. Busby, Bassett, Riggs, and Steiner propose a model for strain partitioning within the Jurassic arc of Arizona. They promote an episode of localized left-lateral strike-slip faulting within the Late Jurassic arc, and also postulate that temporal changes in the style and subaerial environment of arc sedimentation was the result of northward translation of Arizona from horse latitudes to mid-latitudes. Bassett and Busby further build upon their intra-arc strike-slip model by documenting the tectonic setting of Glance Conglomerate deposition near the Sawmill Canyon fault in southeastern Arizona. Fastovsky, Hermes, Strater, Bowring, Clark, Montellano, and Hernandez correct an old interpretation that rocks within the Huizachal Canyon reflect Triassic rifting and subsequent opening of the Gulf of Mexico. They show that the Huizachal rocks (La Boca Formation) record Jurassic volcanism and plate convergence. This new interpretation allows the question of eastward transport of those rocks as they lay south of the Mojave-Sonora megashear. Anderson, Jones, and McKee document lithologies and structural relations within a poorly dated (post–early Triassic pre–Early Cretaceous) mélange assemblage in Zacatecas. They speculate that this unit (Taray Formation) represents the missing forearc-trench portion of the Jurassic arc of northern Sonora that has been translated southeastward along the southwest side of the megashear. Steiner documents the paleomagnetic record of Permian plutons and Triassic sediments in the Maya block of Belize. She uses this data to constrain translations and rotations of the Yucatan region that accompanied breakup and subsequent tectonic modification of Pangea. McKee, McKee, and Anderson trace the exposures of Glance Conglomerate from the Mule and Huachuca Mountains of Arizona into Sonora, Mexico. They suggest that fault-bounded basins containing Glance Conglomerate are products of the same stresses that formed the Mojave-Sonora megashear in contrast to being the initial features of an opening rift basin. They also identify large-scale mass-gravity deposits in the Glance Conglomerate and in Cabullona Group; the former are products of localized Jurassic crustal extension, the latter of Late Cretaceous inversion and associated folding. Aranda-Gómez, Housh, Luhr, Henry, and Becker use subtle tectonic features to extend the trace of an older feature (San Marcos fault) that is almost certainly related to the megashear. Such extension of older features—there and elsewhere—helps to delineate the anatomy of the megashear system. Dembosky and Anderson examine the affects of distributed Late Tertiary dextral shear and extension on older paleogeographic elements of southwestern Arizona. Their mid-Miocene palinspastic analysis suggests that pre-Tertiary crust and structures such as the Mojave-Sonora megashear and the Chocolate Mountains anticlinorium have been moved significantly westward and distended by amounts that must combine Late Tertiary transtension with early–mid Miocene extension.

The alternatives group consists of articles that either dispute the existence of the Mojave-Sonora megashear or offer evidence that may not be supportive of the megashear hypothesis. Poole, Perry, Madrid, and Amaya-Martínez argue that the similarities among Paleozoic units in Texas and northern Mexico and those in Sonora are sufficiently strong that they may represent an unbroken sedimentary tract along the margin of southwestern North America. This interpretation contrasts strongly with the megashear hypothesis in which the Sonoran Paleozoic and Neoproterozoic units have been displaced southeastward along the postulated fault. Stevens, Stone, and Miller emphasize the late Paleozoic and early Mesozoic history from which they conclude that: (1) evidence exists for late Paleozoic disruption, and (2) there is no need for Late Jurassic faulting with major displacement. Haenggi and Muehlberger do not dispute the megashear and they suggest that the Chihuahua trough may be a product of a releasing step between Late Jurassic lateral faults. But in contrast to Anderson and Nourse, they argue that the trough is a pull-apart basin formed by right-lateral faulting rather than left lateral. Iriondo, Martínez-Torres, Kunk, Atkinson, Premo, and McIntosh present new $^{40}Ar/^{39}Ar$ ages from white micas in mylonites located along strike with the type locality of the megashear. They argue that these micas are syntectonic and record Laramide thrusting rather than Jurassic sinistral shear. Finally, Housh and McDowell use Sm-Nd, Rb-Sr, and U-Pb isotopic signatures of Late Cretaceous and Tertiary igneous rocks to map the distribution of basement blocks buried beneath the Sierra Madre volcanic cover of Sonora and Chihuahua. They delineate several basement terranes whose distribution is best explained by simple arc accretion mechanisms, and further argue that the trace of the Silver-Anderson megashear is not manifested as a significant crustal discontinuity across Sonora and Chihuahua.

This volume strives to present a balance of arguments for and against existence of the Mojave-Sonora megashear. For the first time, the original U-Pb analyses and field observations that led to development of the megashear concept are published along with their detailed geologic context. The 21 papers included in the assessment and alternatives sections document significant efforts by numerous geoscientists to tackle various aspects of the problem. The area affected by the megashear is vast, and the megashear topic remains controversial; hence, as co-editors, we believe the quantity and length of the articles presented herein are necessary to provide sufficient context and to elucidate the complexity of the region. We hope that the reader finds this collection of articles provides a stimulating base from which to critically evaluate the location, extent, timing, and tectonic significance of the Mojave-Sonora megashear.

Thomas H. Anderson
Jonathan A. Nourse
James W. McKee
Maureen B. Steiner

Geological Society of America
Special Paper 393
2005

The Mojave-Sonora megashear—Field and analytical studies leading to the conception and evolution of the hypothesis

Thomas H. Anderson*
Department of Geology and Planetary Science, University of Pittsburgh, Pittsburgh, Pennsylvania 15260, USA

Leon T. Silver
Division of Geological and Planetary Sciences, California Institute of Technology, Pasadena, California 91125, USA

ABSTRACT

The megashear hypothesis is based upon reconnaissance geologic and geochronologic studies conducted principally from 1968 until 1974 in northwestern Sonora, Mexico. Our research incorporated U-Pb isotopic analyses of more than 70 zircon populations separated from 33 Precambrian rock samples with field relations and maps based upon structural and stratigraphic measurements. The results delineate a region known as the Caborca block and further reveal that the block is a principal element of an unexpected, discordant pattern of Proterozoic basement provinces. The Mojave-Sonora megashear was conceived in an effort to explain: (1) the unexpected pattern of two Proterozoic crystalline provinces with distinct chronologic histories of crust formation (1.8–1.7 Ga, Caborca block versus 1.7–1.6 Ga, Pinal Province); (2) the distribution of contrasting cover rocks overlying these basement blocks, (3) the abrupt northeastern limit of the Caborca block (terrane) against which volcanic and plutonic rocks of mid-Jurassic (mainly 180–160 Ma) age are juxtaposed, and (4) the distribution of Jurassic magmatic units that intervene between the provinces of Proterozoic crust. The similarities that exist between crystalline crust and overlying pre-Jurassic cover in northwestern Sonora, Mexico, and units in the Inyo Mountains–Death Valley region are attributed to the offset of correlative units along a Late Jurassic left-lateral strike-slip fault postulated to extend from the Gulf of Mexico to California and beyond. This large fault or megashear is a principal structure that accommodated 800–1000 km of left-lateral displacement among a set of transforms related to the opening of the Gulf of Mexico. The fault is compatible with Late Jurassic plate motion.

The inferred trace of the Mojave-Sonora megashear is obscured by contractional and extensional deformation and extensive plutonism. These processes, concentrated along the fault, commonly obfuscate and displace fault zone rocks along the inferred trace as well as the rocks adjacent to it. However, the fault zone is exposed in Sierra de Los Tanques near the international boundary between Mexico and the United States, where mylonitic rocks that comprise three aligned, discontinuous, segments crop out

*taco@pitt.edu.

Anderson, T.H., and Silver, L.T., 2005, The Mojave-Sonora megashear—Field and analytical studies leading to the conception and evolution of the hypothesis, *in* Anderson, T.H., Nourse, J.A., McKee, J.W., and Steiner, M.B., eds., The Mojave-Sonora megashear hypothesis: Development, assessment, and alternatives: Geological Society of America Special Paper 393, p. 1–50, doi: 10.1130/2005.2393(01). For permission to copy, contact editing@geosociety.org.

for ~25 km. The zone of mylonitic rocks, which crosses Route 8, 13 km SW of Sonoita, is locally almost 5 km wide and separates Triassic granitoids and Precambrian gneiss from Jurassic volcanic and clastic rocks.

The limited exposure of the fault zone is a principal concern of those who object to the Mojave-Sonora megashear hypothesis. Studies of paleomagnetism, structure, stratigraphy, crustal geochemistry, and detrital zircons do not refute the megashear concept; commonly they reinforce existing evidence in support of the hypothesis.

Keywords: megashear, Sonora, Proterozoic, Jurassic.

INTRODUCTION

Purpose and Context

The Mojave-Sonora megashear (Silver and Anderson, 1974) was conceived in an effort to explain an unexpected pattern of Proterozoic crystalline provinces with distinct chronological histories (Fig. 1; Anderson and Silver, 1981) in Sonora, northwestern Mexico. The purpose of this paper is: (1) to present the field relations and U-Pb interpreted isotopic ages from zircon that led to the delineation of the Proterozoic crystalline rocks distinguishing the Caborca block, a terrane overlain by generally unmetamorphosed sedimentary strata of Late Precambrian to Early Jurassic age (Fig. 2), (2) to elucidate the thinking, based upon information available in 1974, that led to the hypothesis of the Mojave-Sonora megashear, along which the Caborca block is postulated to have been displaced southeastward more than 800 km, (3) to briefly outline subsequent work that helps to delineate the fault, its character and role in the evolution of southwestern North America, and (4) to provide a summary of objections to the Mojave-Sonora megashear hypothesis as well as our responses to these apparent problems.

The paper is subdivided into four main parts: (1) overview and conception of the Mojave-Sonora megashear hypothesis followed by (2) description of field observations with incorporated U-Pb analyses, (3) summary of post-1974 studies that were conducted with the principal objective of assessing the Mojave-Sonora megashear hypothesis, and (4) overview of some objections to the idea of a very big fault, with short rebuttals that highlight our thinking about the apparent problems. A number of other papers in this volume, for example those by Housh and McDowell, Iriondo, Poole et al., and Stevens et al., treat objections to the Mojave-Sonora megashear hypothesis in more detail. The chapter "Jurassic rocks in Sonora, Mexico: Relation to the Mojave-Sonora megashear" (Anderson et al., this volume, Chapter 2) may provide useful additional information about the Mojave-Sonora megashear that complements the Precambrian basement story.

In Sonora the pattern of Proterozoic basement provinces (Fig. 1) emerged from reconnaissance geologic and geochronologic studies conducted principally from 1968 until 1974, while Anderson held a postdoctoral appointment under Silver. Data from a few additional collections that were made after 1974. The yield from the study conducted throughout Sonora, mainly between 1968 and 1978, includes field observations, structural and stratigraphic measurements, maps, and U-Pb isotopic analyses of more than 70 zircon populations separated from 33 Precambrian rock samples (Table 1). Samples and geologic information were gathered during reconnaissance visits to ranges selected on the basis of information from the 1962 1:1,000,000 geologic map of Mexico and published reports by early workers who conducted regional surveys (e.g., Dumble, 1900; Flores, 1927; King, 1939; Salas, 1968). Topography generally was not shown on these geologic maps. We acquired copies of 1:250,000 topographic maps a year or so after we began the reconnaissance. Generally our samples were collected along unmarked tracks to ranches, most of which were not identified on the available maps. Sample locations are based upon mileage from ranches and road intersections and conspicuous physiographic features. A few sites have been revisited and located on 1:50,000 topographic maps; unfortunately many have not and the locations are approximate. Samples were processed and zircon populations extracted and analyzed per methods described in the Appendices. Modifications in the techniques are noted in Anderson and Silver (1977) and Anderson et al. (1980).

The reconnaissance nature of our work in a largely unmapped region of several tens of thousands of square kilometers led to the collection of many samples. We analyzed as many as we could with the intent of identifying and delineating Proterozoic and younger crystalline basement provinces. For many samples, analyses of two zircon fractions were sufficient to identify the province to which the rock belonged. Although the aspiration was to analyze additional zircon fractions for every sample, we were able to do so for only several samples that were unusual or important for geologic or geochronologic reasons. We exert caution with respect to the "ages" assigned to the analyzed rocks through the use of the phrase "interpreted isotopic age." This phrase indicates the tentative nature of the age estimates based upon two analyses and consideration of other factors such as inheritance related to cores in the zircons or unidentified old zircons in the analyzed fraction or overgrowths formed during metamorphism. Despite the problems that may arise from these issues, the conformity of the isotopic results from Sonoran samples with those comprising a large body of existing data from samples previously analyzed from southwestern United States, coupled with new results (Nourse

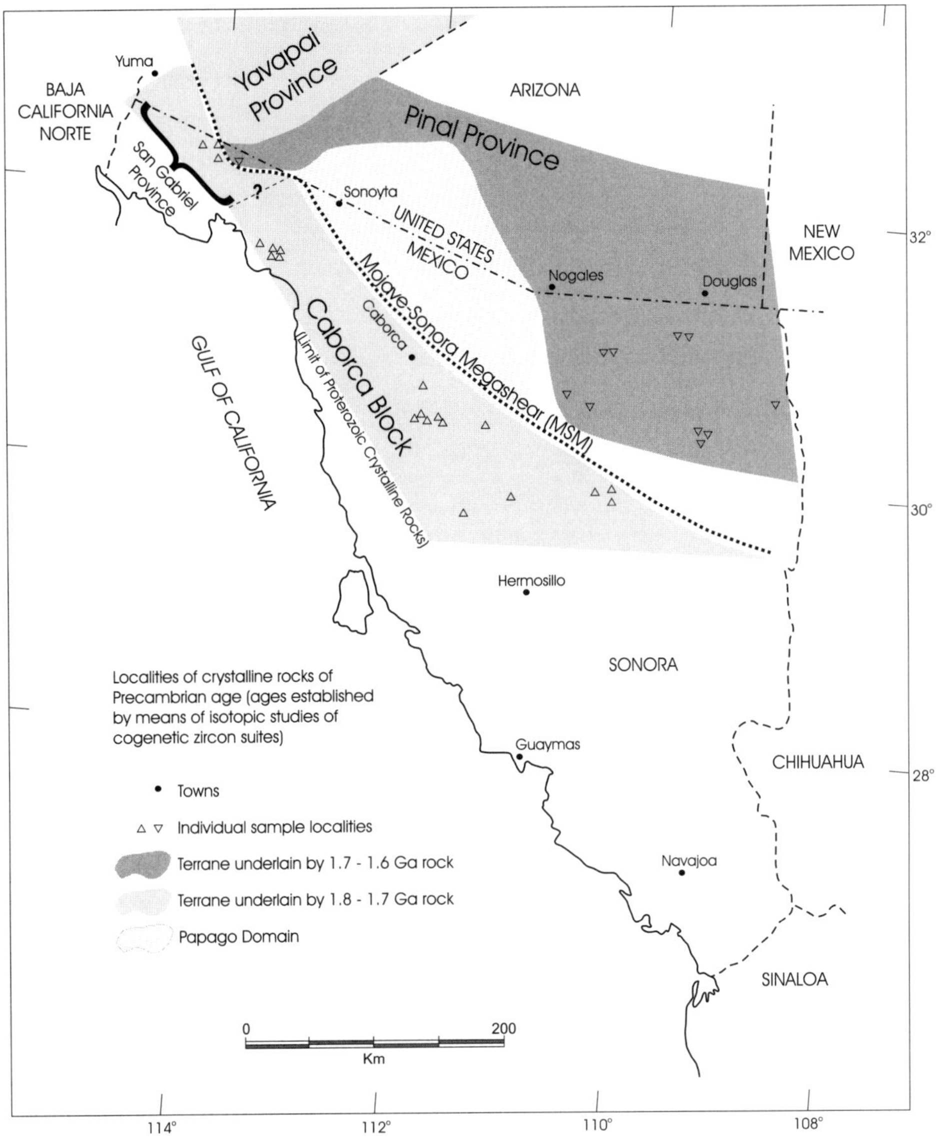

Figure 1. Terranes of Proterozoic crystalline rock, Sonora, Mexico (modified from Figure 1, Anderson and Silver, 1981).

et al., this volume, and other data cited below) indicates that the reconnaissance results serve as valid estimates of the ages of crystallization of zircon either from an igneous melt or from a fluid or vapor containing volatile components mobilized during upper amphibolite or granulite metamorphism and precipitated as a pegmatitic segregation.

Several rock units that we studied have been restudied (e.g., Iriondo et al., 2003; Premo et al., 2003; Nourse et al., this volume; Farmer et al., this volume). If comparisons are available they are noted in the text. Generally, the ages of zircon that we determined overlap with the results from more recent work within the interval of assigned errors, especially for rocks that are slightly or unmetamorphosed. Some ages, which were interpreted from results of analyses in strongly metamorphosed and deformed rocks, are discordant with the age crystallization as suggested by subsequent analyses of single zircons (Nourse et al., this volume; W.R. Premo, 2004, personal commun.). The discordance reflects the inclusion of strongly discordant zircon among the many grains included within our analyzed fractions. The discordant zircon grains are commonly those that are uranium rich. We attempted to isolate suites of zircons with very low uranium contents by means of magnetic separation so that radiation damage would be minimal and susceptibility to lead loss during metamorphism would be low. However, for some samples, we were unable to separate enough of the lowest-uranium grains for analysis and discordant grains were included.

Our most detailed studies, in mountain ranges between Caborca and Bámori Ranch (Fig. 3), included mapping, stratigraphic observations, section measurement, and sample collection. We focused upon this region because: (1) previous studies provided stratigraphic relationships between sedimentary strata and underlying crystalline rocks, and (2) the initial K-Ar and

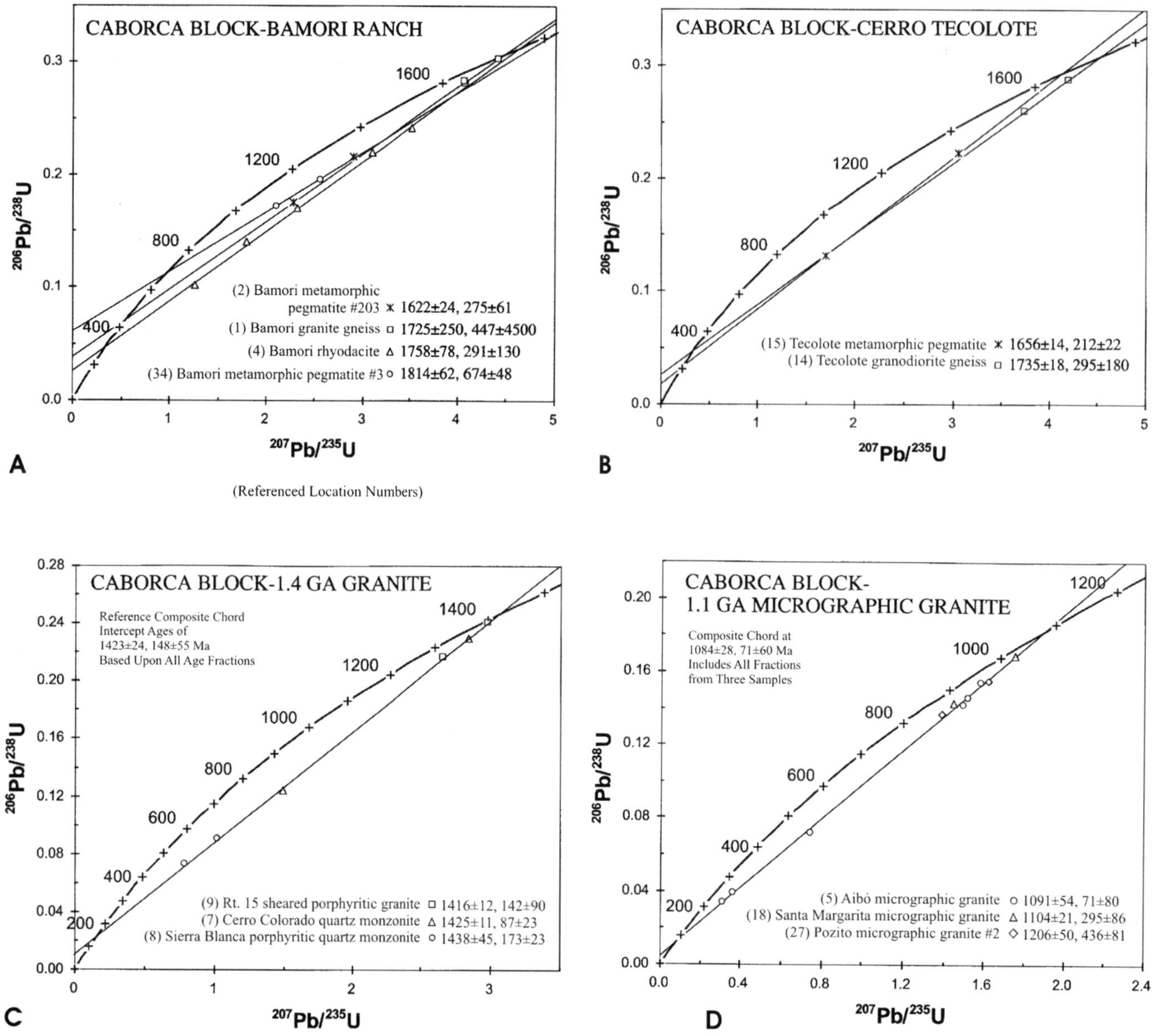

Figure 2. Concordia diagrams showing U-Pb isotopic data for zircons from Proterozoic crystalline rocks in Sonora, Mexico. Numbers in parentheses refer to sample locations shown on Figure 4 and listed in Table 1.

Rb-Sr results from rocks of this region (Damon et al., 1962; Giletti and Damon, 1962) included dates of rocks that were unexpected based upon our knowledge at the time.

The results of our field work (Fig. 4[1]) and U-Pb isotopic analyses (Fig. 2 and Table 1) in the Caborca block are offered below. This information provides the basis for comparison with Precambrian and some younger strata in southwestern Arizona and southern California. Outcrops described in this paper were visited during fast-paced reconnaissance trips. We present brief descriptions of the principal areas, generally where multiple samples of Proterozoic crystalline rocks were collected.

Micrographic granite that yielded crystallization ages of ca. 1.1 Ga is described for the first time (see also Farmer et al., this volume). The unconformable relationship between the granite, intrusive into the older basement units, and overlying miogeoclinal strata constrains the age of sedimentation to be younger than 1.1 Ga (see Stewart et al., 1984, 2002; Stewart and Poole, 1997). The distinctive red to grayish-pink feldspathic granite, unique to the Caborca terrane, also provides a useful structural marker

[1]Figures 4 and 9 are on the CD-ROM and on loose sheets accompanying this volume, and are in the GSA Data Repository as item 2005138, available online at www.geosociety.org/pubs/ft2005.htm, or on request from editing@geosociety.org or Documents Secretary, GSA, P.O. Box 9140, Boulder, CO 80301, USA.

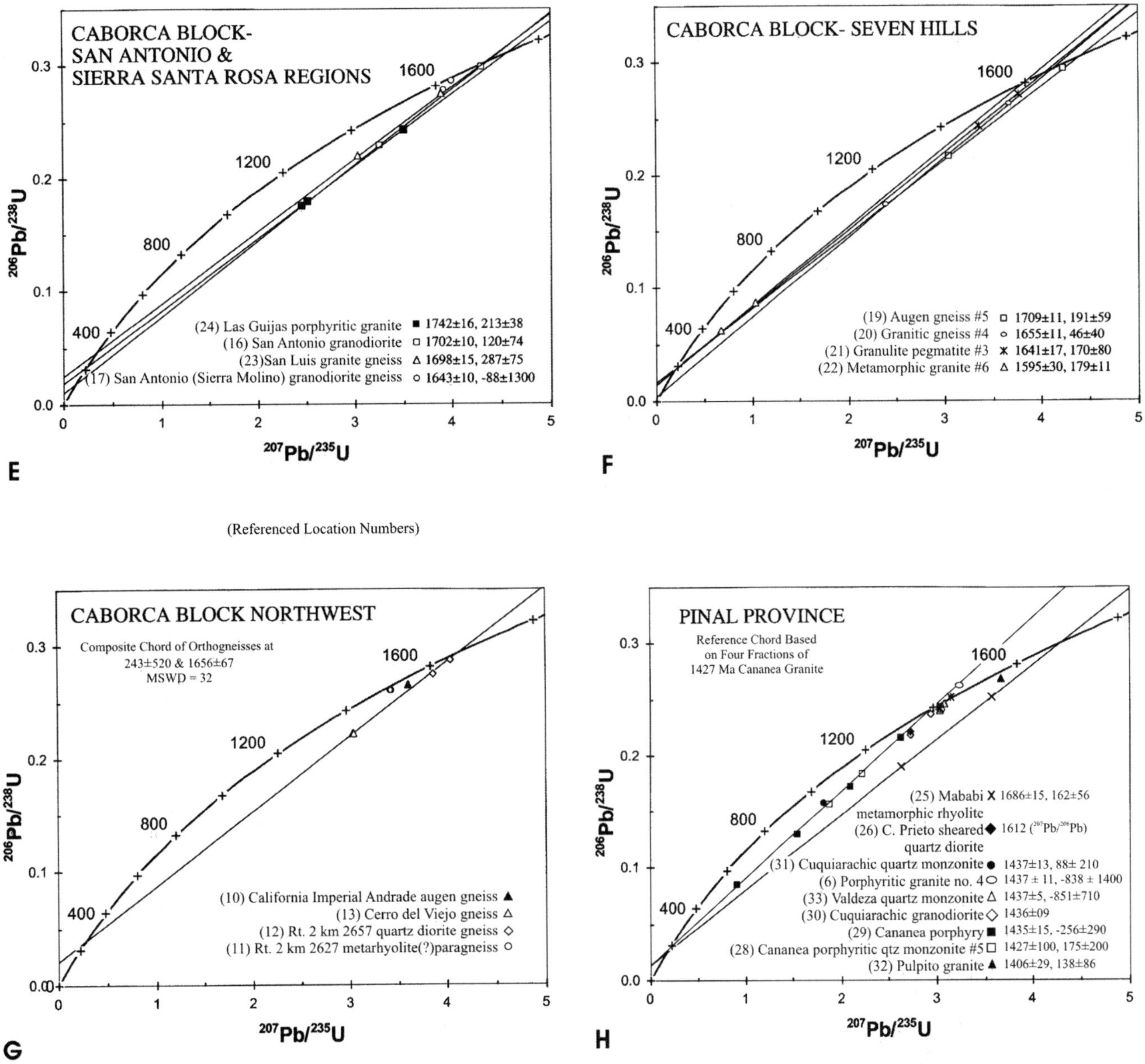

Figure 2 (*continued*).

where incorporated into allochthonous masses thrust across the Mojave-Sonora megashear.

Previous Work and Geologic Overview

The presence of Proterozoic crystalline rocks in Sonora was first indicated by the discovery of an incomplete sequence of Cambrian and older carbonate and clastic rocks near Caborca that rest nonconformably upon crystalline basement (Cooper and Arellano, 1946; Arellano, 1956). Initial radiometric studies were made by Damon et al. (1962) and Gilletti and Damon (1962) on rocks that crop out in ranges between Caborca and Bámori Ranch, ~40 km south (Fig. 3). Between 1962 and 1968 little additional work was conducted on the Precambrian rocks near Caborca aside from brief comments by Fries (1962) who published a geologic summary emphasizing Paleozoic rocks. Concurrent with our work, Merriam (1972) conducted reconnaissance mapping in northwesternmost Sonora, and Eells (1972) mapped and subdivided Late Precambrian and Cambrian sedimentary strata that crop out in ranges west of Bámori Ranch.

TABLE 1. U-PB ANALYTICAL DATA FOR ZIRCONS FROM SAMPLES OF PROTEROZOIC ROCKS MAINLY IN SONORA, MEXICO

CABORCA BLOCK: BAMORI RANCH - CERRO TECOLOTE

Location	Sample Mesh Size Franz Setting	Observed ratios			Radiogenic Pb Atom %			Concentrations (ppm)		Isotope ratios			Apparent ages		
		Pb 206/Pb 204	Pb 207/Pb 204	Pb 208/Pb 204	Pb 206	Pb 207	Pb 208	Pbrad	U	Pb206/U238	Pb207/U235	Pb207/Pb206	Pb206/U238	Pb207/U235	Pb207/Pb206
1	Bamori granite gneiss #204														
	P200 Franz 1.7 NM 1°/25°[b]	4811.6	521.1	314.3	85.98	9.07	4.96	197.7	652.2	0.3032	4.4087	0.10540	1707	1714	1721.31
	P200 Franz 1.7 M 4°/25°[b]	2463.3	272	168.1	86.35	9.05	4.59	224.4	800.6	0.2815	4.0698	0.10490	1599	1648	1712.57
	P200 Franz 1.7 M 8°/25°[b]	2357.5	258.3	165.3	86.34	8.96	4.7	242.4	857.8	0.2838	4.0604	0.10370	1610	1646	1691.38
2	Bamori metamorphic pegmatite #203														
	R200+P200 Franz 1.7 NM 2°/20°[a]	*1954.3	203.6	96.9	88.69	8.61	2.7	470.9	2240.7	0.2168	2.9030	0.09710	1265	1383	1569.15
	R200+P200 Franz 1.7 M 2°/20°[b]	6995	674.2	220.2	89.24	8.42	2.33	538.1	3195.2	0.1749	2.2760	0.09440	1039	1205	1516.14
3	Bamori quartzite #301														
	R200 Franz 1.7 NM 0°/20°[b]	3642.2	485.4	488.8	79.74	10.34	9.92	103.6	300	0.3200	5.7200	0.12960	1790	1934	2092.55
	P200 Franz 1.7 M 3°/25°[b]	2344.6	271.8	254.6	83.11	9.15	7.74	196	996.9	0.1900	2.8858	0.11010	1121	1378	1801.06
4	Bamori rhyodacite #2 (A) + #205 (B)														
	R200+P200 Franz 1.7 NM 5° /20°[b]-B	1283.2	149.4	217.7	80.14	8.48	11.38	85.9	331.4	0.2414	3.5200	0.10580	1394	1532	1728.27
	R200+P200 Franz 1.7 NM 6° /20°[a]-A	345.2	49.2	85.6	80.12	8.22	11.66	85.9	365	0.2192	3.0990	0.10250	1278	1432	1669.89
	R200+P200 Franz 1.7 M 2°/20°[b]-B	1380.2	150.9	253	79.54	7.9	12.56	108.3	588	0.1696	2.3190	0.09930	1010	1218	1611.01
	R200+P200 Franz 1.7 M 6°/20°[a]-A	a363.8	a47.9	a88.4	80.71	7.53	11.76	108.8	727.9	0.1396	1.7970	0.09340	842	1044	1496.02
	R200+P200 Franz 1.7 M 10° /20°[a]-A	b404.7	b50.5	b94.4	81	7.31	11.79	126.2	1165.5	0.1014	1.2630	0.09040	623	829	1434
14	Tecolote granodiorite gneiss #201														
	R200 Franz 1.7 NM 0° /20°[b]	3441.9	376.7	340.8	83.76	8.83	7.41	152.4	515.3	0.2881	4.1894	0.10540	1632	1672	1721.31
	P200 Franz 1.7 M 2.5° /25°[b]	2339.7	257.8	246.9	83.73	8.74	7.53	160	596.2	0.2598	3.7381	0.10433	1489	1580	1702.54
15	Tecolote metamorphic pegmatite #202														
	R200 Franz 1.7 NM 2.5°/20°[a]	4513.4	464.6	301.7	86.32	8.62	5.06	235	1057.2	0.2222	3.0530	0.09884	1293	1421	1602.36
	R100 Franz 1.7 NM 5°/20°[b]	2370.2	235.9	98.4	89.34	8.37	2.29	300.24	2368	0.1318	1.7021	0.09366	798	1009	1501.27
34	Bamori metamorphic pegmatite #3														
	R200+P200 NM 1°/25°[a]	1316.2	123.6	50.95	90.5	8.6	0.9	451	2428.7	0.1955	2.5410	0.09519	1151	1284	1531.84
	R200+P200 M 1°/25°[a]	*820.31	73.14	17.97	90.3	8.02	1.97	589	3598.9	0.1715	2.0910	0.08916	1020	1146	1407.61

* From Pb concentration

[a] Common lead correction Pb 206/Pb 204 = 17.8; Pb 207/Pb 204 =15.6; Pb 208/Pb 204 = 38.0

[b] Common lead correction Pb 206/Pb 204 = 18.2; Pb 207/Pb 204 = 15.6; Pb 208/Pb204 = 38.0

[c] Common lead correction Pb 206/Pb 204 = 18.6; Pb 207/Pb 204 = 15.6; Pb 208/Pb204 = 38.0

A - Bamori rhyodacite #2

B - Bamori rhyodacite #205 (second collection)

(*continued*)

TABLE 1. U-PB ANALYTICAL DATA FOR ZIRCONS FROM SAMPLES OF PROTEROZOIC ROCKS MAINLY IN SONORA, MEXICO (*continued*)

CABORCA BLOCK-1.1 GA MICROGRAPHIC GRANITES AND 1.4 GA GRANITE - U/PB ISOTOPIC RESULTS FROM ZIRCON

Location	Sample Mesh Size Franz Setting	Observed ratios			Radiogenic Pb Atom %			Concentrations (ppm)		Isotope ratios			Apparent ages		
		Pb 206/Pb 204	Pb 207/Pb 204	Pb 208/Pb 204	Pb 206	Pb 207	Pb 208	Pbrad	U	Pb206/U238	Pb207/U235	Pb207/Pb206	Pb206/U238	Pb207/U235	Pb207/Pb206
5	Aibo Micrographic Granite														
	R100 (Franz 1.7 amp) NM 2° / 20° *b	310.80	37.40	1230.40*	78.14a	5.80	16.06	27.172	159.84	0.15436	1.57971	0.07422	925	962	1047
	R200 (Franz 1.7 amp) M 1°/ 20° b	722.10	68.60	191.10	77.35 a	5.82	16.82	34.236	210.90	0.14590	1.51482	0.07529	878	936	1076
	R200 (Franz 1.7 amp) M 2° / 20° a	527.11	53.94	159.58	76.10b	5.73	18.17	39.303	485.80	0.07154	0.74255	0.07528	445	564	1076
	R200 (Franz 0.9 amp) M 10°/ 20° b	638.40	57.40	159.40	79.17 a	5.34	15.50	69.098	1622.32	0.03919	0.36418	0.06740	248	315	850
	R200 (Franz 0.7 amp) M 10°/ 20° b	490.90	46.90	127.50	79.65 a	5.27	15.08	74.272	2015.01	0.03412	0.31152	0.06622	216	275	813
	R200 (Franz 1.7 amp) NM 0.5°/ 20° ad	676.51	66.01	175.39	77.81b	5.96	16.23	28.658	183.29	0.14138	1.49200	0.07653	852	927	1109
7	Cerro Colorado quartz monzonite # 300														
	R200 Franz 1.7 NM 0°/20° b	1274.6	128.5	323.4	75.93	6.82	17.25	103.2	395.4	0.22929	2.8390	0.08986	1331	1366	1423
	P200 Franz 1.7 M 2°/25° b	1058.3	104.8	240.7	78.09	6.7	15.22	156.01	1135	0.12418	1.4992	0.08576	755	930	1333
8	Sierra Blanca porphyritic quartz Monzonite #1														
	R200 Franz 1.7 NM 0°/20° a	358.4	43	57.48	87.74	7.17	5.09	72.9	818.8	0.09038	1.0187	0.08180	558	713	1241
	P200 Franz 1.7 M 1°/25° a	1279.8	99	56	89.2	7	3.81	105	1485.8	0.07291	0.7884	0.07844	454	590	1158
9	Sheared porphyritic granite #100														
	R200 Franz 1.7 NM 1°/20° a	962.4	100.1	124.28	84.69	7.58	7.73	156.8	637.5	0.24102	2.9728	0.08946	1392	1401	1414
	P200 Franz 1.7 M 4°/25° a	6161.2	562.26	562.87	85.15	7.58	7.27	202.1	918.6	0.21671	2.6591	0.08898	1264	1317	1404
18	San Antonio (Santa Margarita) Micrographic Granite														
	R200 (Franz 1.7 amp) NM 1.5°/ 20° b	500.60	52.00	124.40	79.71 a	6.01	14.28	42.196	232.14	0.16840	1.75210	0.07546	1003	1028	1081
	R200 (Franz 1.7 amp) M 2.5°/ 25° b	505.86	51.60	133.40	78.77 a	5.81	15.41	61.070	392.64	0.14239	1.44944	0.07382	858	910	1037
27	Pozito Micrographic Granite														
	R200 (Franz 1.7 amp) NM 1.5°/ 25° c	2752.27	223.69	500.10	80.31c	6.11	13.58	42.696	257.97	0.15445	1.62110	0.07610	926	978	1098
	R200 (Franz 1.7 amp) M 3° / 25° c	1258.49	107.22	276.71	78.96c	5.84	15.20	53.896	361.95	0.13660	1.39190	0.07390	825	886	1039

*Composition calculated using an assumed Pb 206/Pb 208 ratio of 4.86
[a] Common lead correction Pb 206/Pb 204 = 17.8; Pb 207/Pb 204 =15.6; Pb 208/Pb 204 = 38.0
[b] Common lead correction Pb 206/Pb 204 = 18.2; Pb 207/Pb 204 = 15.6; Pb 208/Pb204 = 38.0
[c] Common lead correction Pb 206/Pb 204 = 18.6; Pb 207/Pb 204 = 15.6; Pb 208/Pb204 = 38.0
[d] Using values from comp w/6% correction (data not used for concordia)

(*continued*)

TABLE 1. U-PB ANALYTICAL DATA FOR ZIRCONS FROM SAMPLES OF PROTEROZOIC ROCKS MAINLY IN SONORA, MEXICO (*continued*)

CABORCA BLOCK: SAN ANTONIO REGION & SIERRA SANTA ROSA REGIONS

Location	Sample Mesh Size Franz Setting	Observed ratios			Radiogenic Pb Atom %			Concentrations (ppm)		Isotope ratios			Apparent ages		
		Pb 206/Pb 204	Pb 207/Pb 204	Pb 208/Pb 204	Pb 206	Pb 207	Pb 208	Pbrad	U	Pb206/U238	Pb207/U235	Pb207/Pb206	Pb206/U238	Pb207/U235	Pb207/Pb206
16	San Antonio granodiorite														
	R200 Franz 1.7 NM 0.5°/20° [b]	1352.3	141	160.7	81.76	8.53	9.72	65.49	208.5	0.29864	4.294	0.10427	1685	1692	1701
	P200 Franz 1.7 M 2.5°/25° [b]	463.5	47.8	58.8	81.3	8.38	10.31	76.34	314.87	0.22916	3.259	0.10313	1330	1471	1681
17	Sierra Molino granodiorite gneiss														
	R200 Franz 1.7 NM 2°/20° [c]	3375.1	358.56	283.38	85.08	8.69	6.22	142.23	506.25	0.27800	3.917	0.10218	1581	1617	1664
	P200 Franz 1.7 M 5°/25° [c]	5159.9	537	400	85.34	8.65	6	359	1245.65	0.28605	4	0.10141	1622	1634	1650
23	San Luis granite gneiss														
	R200 Franz 1.7 NM 2°/20° [b]	1844.1	203.9	144.7	86.09	8.88	5.03	157.51	575.71	0.27396	3.896	0.1031	1561	1613	1681
	P200 Franz 1.7 M 5°/25° [b]	2704.4	285.8	167.1	87.06	8.76	4.18	194.29	897.04	0.21934	3.042	0.1006	1278	1418	1635
24	Las Guijas Porphyritic Granite														
	R200 Franz 1.7 NM 1.5° /20° [a]	2829.8	296.3	202.5	85.02	8.9	6.08	136.36	556.51	0.24278	3.505	0.1047	1401	1528	1709
	P200 Franz 1.7 M 6°/25° [a]	2341.6	237.6	148	85.86	8.71	5.43	156.6	867.5	0.17923	2.508	0.1015	1063	1274	1651
	R200 Franz 1.7 M 8°/20° [b]	2417.2	245.2	163.5	85.54	8.68	5.78	155.94	881.48	0.17508	2.449	0.1014	1040	1257	1651

[a] Common lead correction Pb 206/Pb 204 = 17.8; Pb 207/Pb 204 =15.6; Pb 208/Pb 204 = 38.0
[b] Common lead correction Pb 206/Pb 204 = 18.2; Pb 207/Pb 204 = 15.6; Pb 208/Pb204 = 38.0
[c] Common lead correction Pb 206/Pb 204 = 18.6; Pb 207/Pb 204 = 15.6; Pb 208/Pb204 = 38.0

(*continued*)

TABLE 1. U-PB ANALYTICAL DATA FOR ZIRCONS FROM SAMPLES OF PROTEROZOIC ROCKS MAINLY IN SONORA, MEXICO (*continued*)

CABORCA BLOCK: SEVENS HILLS AND NORTHWEST

Location	Sample Mesh Size Franz Setting	Observed Ratios			Radiogenic Pb Atom %			Concentrations (ppm)		Isotope Ratios			Apparent Ages		
		Pb 206/Pb 204	Pb 207/Pb 204	Pb 208/Pb 204	Pb 206	Pb 207	Pb 208	Pbrad	U	Pb206/U238	Pb207/U235	Pb207/Pb206	Pb206/U238	Pb207/U235	Pb207/Pb206
10	Andrade augen gneiss #1														
	R200 Franz 1.7 NM 1°/20° [b,*]	3156	311.6	290.9	83.97	8.29	7.74	103.6	379.5	0.26520	3.6085	0.09870	1516	1551	1600
11	RT.2-KM 2627 metarhyolite (?) gneiss #4														
	R200 Franz 1.7 NM 0°/20° [b]	1663.1	158.3	178.2	83.17	7.92	8.91	156	576.45	0.26039	3.4194	0.09518	1492	1509	1532
	P200 Franz 1.7 M 5°/25° [b]	2373.4	227.5	322.3	81.19	7.78	11.02	392.9	1417.4	0.26033	3.4399	0.09585	1492	1514	1545
12	RT.2-KM 2657 quartz diorite gneiss #														
	R200 Franz 1.7 NM 0°/20° [b]	5000.7	525.1	349	85.86	8.78	5.36	125.3	437.4	0.28692	4.0449	0.10223	1626	1643	1665
	R200 Franz 1.7 M 0.5°/20° [b]	9075.1	926.6	499.5	86.42	8.82	4.76	135.2	492.4	0.27455	3.8634	0.10165	1564	1606	1654
13	Cerro del Viejo augen gneiss #102														
	R200 + P200 Franz 1.7 NM 0°/20° [c]	4452.8	457.6	427.9	84.2	8.39	7.41	164.6	728.6	0.22125	3.041	0.09968	1288	1418	1618
19	Mex-Son 7 Hills Augen gneiss #5														
	R200 (Franz 1.7 amp) NM 0°/20° [a]	2061.3	215.36	152.2	84.87	8.867	6.267	124.00	414.60	0.29371	4.2311	0.10448	1676	1702	1733
	P200 (Franz 1.7 amp) M 2°/25° [a]	2215.6	225.88	160.7	85.09	8.699	6.213	131.70	600.00	0.21606	3.0459	0.10224	1273	1438	1692
20	Mex-Son 7 Hills Granitic gneiss #4														
	P200 (Franz 1.7 amp) NM 0°/25° [a]	2361.2	239.71	147.1	85.92	8.723	5.353	172.69	659.40	0.26186	3.6658	0.10167	1513	1584	1682
	P200 (Franz 1.7 amp) M 6°/25° [a]	1844.1	185.77	114.0	86.02	8.665	5.319	218.97	1261.70	0.17271	2.3990	0.10074	1037	1258	1665
21	Mex-Son 7 Hills Granulite peg #3														
	R200 (Franz 1.7 amp) NM 0°/20° [a]	3713.5	373.3	135.7	87.94	8.84	3.21	237.95	892.70	0.27118	3.7590	0.10052	1561	1605	1660
	R200 (Franz 1.7 amp) M 2°/20° [a]	2480.5	247.79	96.9	87.80	8.771	3.431	260.49	1085.40	0.24377	3.3580	0.0999	1419	1514	1649
22	Mex-Son 7 Hills Metagranite #6														
	R200 (Franz 1.7 amp) NM 1.5°/20° [a]	371.6	32.14	21.2	87.45	7.56	4.99			0.08731	1.0407	0.08649		734	1372
	R200 (Franz 1.7 amp) M 0.5°/20° [a]	404.9	32.05	22.5	88.13	6.977	4.889	72.78	1204.20	0.06163	0.6728	0.07916	389	529	1197

[a] Common lead correction Pb 206/Pb 204 = 17.8; Pb 207/Pb 204 =15.6; Pb 208/Pb 204 = 38.0
[b] Common lead correction Pb 206/Pb 204 = 18.2; Pb 207/Pb 204 = 15.6; Pb 208/Pb204 = 38.0
[c] Common lead correction Pb 206/Pb 204 = 18.6; Pb 207/Pb 204 = 15.6; Pb 208/Pb204 = 38.0
* Imperial County, California

(*continued*)

TABLE 1. U-PB ANALYTICAL DATA FOR ZIRCONS FROM SAMPLES OF PROTEROZOIC ROCKS MAINLY IN SONORA, MEXICO (*continued*)

1.4 GA GRANITE: U/PB ISOTOPIC RESULTS FROM SAMPLES WITHIN SONORAN PINAL PROVINCE CRUST

Location	Sample Mesh Size Franz Setting	Observed ratios			Radiogenic Pb Atom %			Concentrations (ppm)		Isotope ratios			Apparent ages		
		Pb 206/Pb 204	Pb 207/Pb 204	Pb 208/Pb 204	Pb 206	Pb 207	Pb 208	Pbrad	U	Pb206/U238	Pb207/U235	Pb207/Pb206	Pb206/U238	Pb207/U235	Pb207/Pb206
6	Porphyritic granite #4														
	R200 Franz 1.7 NM 0.5°/20° [a]	2538.5	229.3	388.6	80.42	7.26	12.31	116.8	430.95	0.25210	3.1409	0.09030	1449	1443	1432
	P200 Franz 1.7 M 3°/25° [b]	1495.5	136.4	223.9	80.59	7.35	12.07	206.3	517.6	0.24095	3.0305	0.09118	1392	1415	1450
25	Mababi metamorphic rhyolite														
	R200 Franz 1.7 NM 1°/20° [b]	845.7	86.7	150.8	78.07	8	13.92	62.62	224.57	0.2519	3.5586	0.10252	1448	1540	1670
	P200 Franz 1.7M 5°/25° [b]	751.9	75.6	126.7	78.8	7.92	13.28	102	490.52	0.1896	2.628	0.10055	1119	1309	1634
26	Cerro Prieto Sheared Quartz Diorite														
	R200 Franz 1.7 NM 1/2°/20° [c]	8659	874.1	670.1	85.29	8.47	6.24	205.2	759.32	0.268	3.6716	0.09936	1531	1565	1612
29	Cananea porphyritic quartz monz #2 (contact zone)														
	R200 + P200 Franz 1.7 NM 5°/20° [b]	442.8	52.9	108.6	79.74	7	13.26	46.5	234.36	0.18303	2.2156	0.08785	1084	1186	1379
	R200 + P200 Franz 1.7 M 5°/25° [b]	326.6	42.4	87.2	80.23	6.97	12.8	59.27	353.53	0.15562	1.8641	0.0869	932	1068	1358
	aContact zone - hand speciman														
28	Cananea porph quartz monz # 5														
	R200 Franz 1.7 NM 2°/20° [b]	329.7	43.2	92.1	79.22	7.02	13.76	45.89	195.35	0.21532	2.6308	0.0886	1257	1309	1396
	R200 Franz 1.7 M 2°/20° [b]	339.1	44	88.3	80.3	7.11	12.59	49.54	268.74	0.17128	2.0902	0.0885	1019	1146	1393
	P200 Franz 1.7 M 7°/25° [b]	355.2	44.9	88.6	80.84	7.03	12.14	58.33	422.61	0.12908	1.5477	0.08694	783	949.6	1359
	P200 Franz 0.5 M 10°/25° [b]	241.4	33.9	69.8	81.67	6.7	11.64	81.19	953.75	0.08430	0.9099	0.08199	522	657	1245
30	Cuquiriachic granodiorite #3														
	R200 Franz 1.7 NM 0°/20° [b]	513.8	46.7	118	75.73	6.88	17.39	23.83	88.71	0.23538	2.9487	0.09089	1363	1395	1444
	P200 Franz 1.7 M 4.5°/25° [b]	269	24.6	83.1	71.41	6.53	22.06	67.84	258.26	0.21700	2.7363	0.09145	1266	1338	1456
31	Cuquiriachic quartz monzonite #501														
	R200 Franz 1.7 NM 0.5°/20° [b]	579.2	52.4	115.8	77.5	7.01	15.49	59.8	221.22	0.24235	3.0226	0.09047	1399	1413	1435
	P200 Franz 1.7 M 4°/25 [b]	581.1	52.4	108.8	78.28	7.06	14.66	113.9	468.86	0.21991	2.7348	0.09017	1281	1338	1429
32	Pulpito granite #1														
	R200 + P200 Franz 1.7 NM 4.5°/20° [b] *	33.8	16.9	64.8	82.54	6.88	10.58	101.3	615	0.15727	1.8075	0.08333	942	1048	1277
33	Valdeza porph quartz monzonite #4														
	R200 Franz 1.7 NM 0°/20° [c]	3580	340.6	493.1	82.03	7.49	10.48	69.5	270.27	0.24518	3.0852	0.09125	1414	1429	1452
	P200 Franz 1.7 M 3°/25° [c]	861.7	93.5	137	82.65	7.64	9.71	104.5	420.7	0.23856	3.0412	0.0925	1379	1418	1478

[a]Common lead correction Pb 206/Pb 204 = 17.8; Pb 207/Pb 204 =15.6; Pb 208/Pb 204 = 38.0
[b]Common lead correction Pb 206/Pb 204 = 18.2; Pb 207/Pb 204 = 15.6; Pb 208/Pb204 = 38.0
[b]Common lead correction Pb 206/Pb 204 = 18.2; Pb 207/Pb 204 = 15.6; Pb 208/Pb204 = 38.0
[c]Common lead correction Pb 206/Pb 204 = 18.6; Pb 207/Pb 204 = 15.6; Pb 208/Pb204 = 38.0
*From Pb concentration; Pb 206/208 = 8

(*continued*)

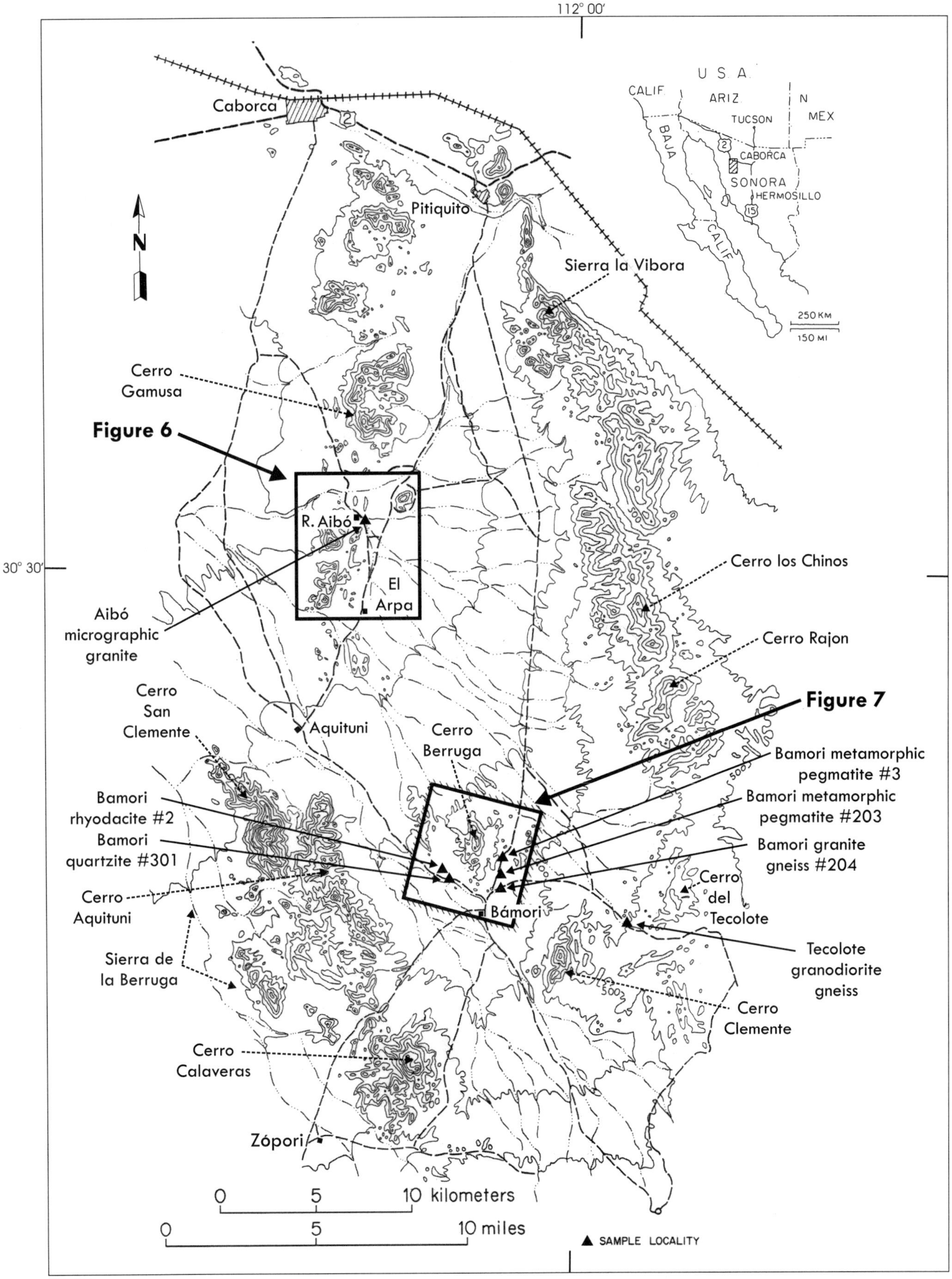

Figure 3. Map of the region between Caborca and Rancho Bámori (Bámuri), showing hills (cerros), ranges (Sierras), and ranches mentioned in the text. Areas of geologic mapping are outlined. Triangles show locations of analyzed samples from the Pitiquito–Cerro Berruga crystalline suite. Contour interval = 100 m.

To the east, in central Sonora, extensive reconnaissance studies by Dumble (1900), Flores (1927), and King (1939), supplemented by paleontologically focused works of Keller (1928) and Imlay (1939), provided useful and informative guides to localities of probable Proterozoic outcrops. In north-central Sonora, mapping of a large area northwest of Santa Ana by Salas (1968) revealed the presence of an extensive assemblage of mainly greenschist-facies rocks comprised of phyllite with subordinate marble, metaconglomerate, quartzite, and metagraywacke. In places, mineral assemblages characteristic of almandine-amphibolite facies occur. Salas concluded that the layered rocks were composed of two suites, both of which he considered to be Precambrian in age, based upon similarities of metamorphic grade to the rocks near Bámori Ranch.

Our work in Mexico stemmed from an extensive geochronologic and geologic survey of Precambrian crystalline rocks in southwestern North America conducted during the 1950s and 1960s (e.g., Cooper and Silver, 1964; Silver, 1965, 1967, 1969, 1978; Silver et al., 1961, 1963; Anderson et al., 1971; Anderson and Silver, 1976). The results led to the recognition of cogenetic suites of volcanic and plutonic rocks representing the crust accreted to southwestern North America (Fig. 5A). Among the exposures of this crust, two parallel regional belts, trending southwesterly, were recognized on the basis of crystallization ages determined using the U-Pb system on zircon (e.g., Silver, 1963; Silver and Deutsch, 1961, 1963). The southeasterly Pinal belt, in southern Arizona and southwestern New Mexico, is distinguished by metasedimentary and volcanic rocks (1.71 Ga and younger), and postkinematic plutons (1.625 Ga and older) (Cooper and Silver, 1964; Silver, 1967, 1969, 1978; Silver and Deutsch, 1963; Conway and Silver, 1989). Northwest of the Pinal Province rocks, older (ca. 1.8 Ga) volcanic rocks and associated plutons (1.75–1.72 Ga), called the Yavapai Province (Silver and Barker, 1968; Anderson et al., 1971; Anderson and Silver, 1976), comprise a parallel belt (Fig. 5A). In Arizona, the contact between the belts of chronologically distinct suites of Proterozoic volcanic, sedimentary, and plutonic rocks occurs within a zone defined by the change in crystallization ages northwestward from those of the younger Pinal Province (ca. 1.7–1.6 Ga) to the Yavapai Province (ca. 1.8–1.7 Ga) (Conway et al., 1987). Among the plutonic and volcanic rocks composing these two chronologically separate provinces no evidence for preexisting basement rocks, such as pre–1.8 Ga zircons in sandstone interbeds, was known (Silver, unpublished data), and the volcano-plutonic suites were considered to be the products of initial crust-forming events. Recent studies in Sonora (e.g., Iriondo, 2001; Iriondo and Premo, 2003) and southern Arizona (Eisele and Isachsen, 2001; Swift and Force, 2001) reveal additional features of the basement provinces but do not negate the effectiveness of the working hypothesis, described below, that we employed at the time.

The contact between the Yavapai and Pinal belts in Arizona projects southwest toward an orthogonal intersection with the

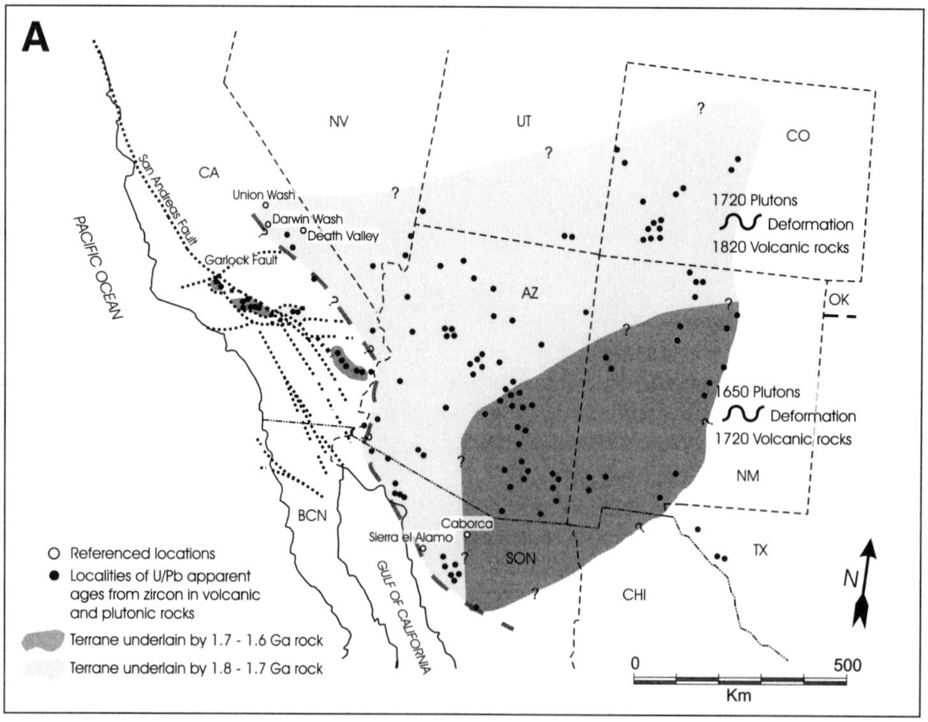

Figure 5. Maps of southwestern North America showing distribution of chronologically distinct suites of crust-forming Proterozoic rocks. Variations among versions A (1969), B (1973), and C (1974) reflect additions of new data, resulting in modifications of the distribution of Precambrian rocks and revised tectonic interpretations. (*Continued on following page.*)

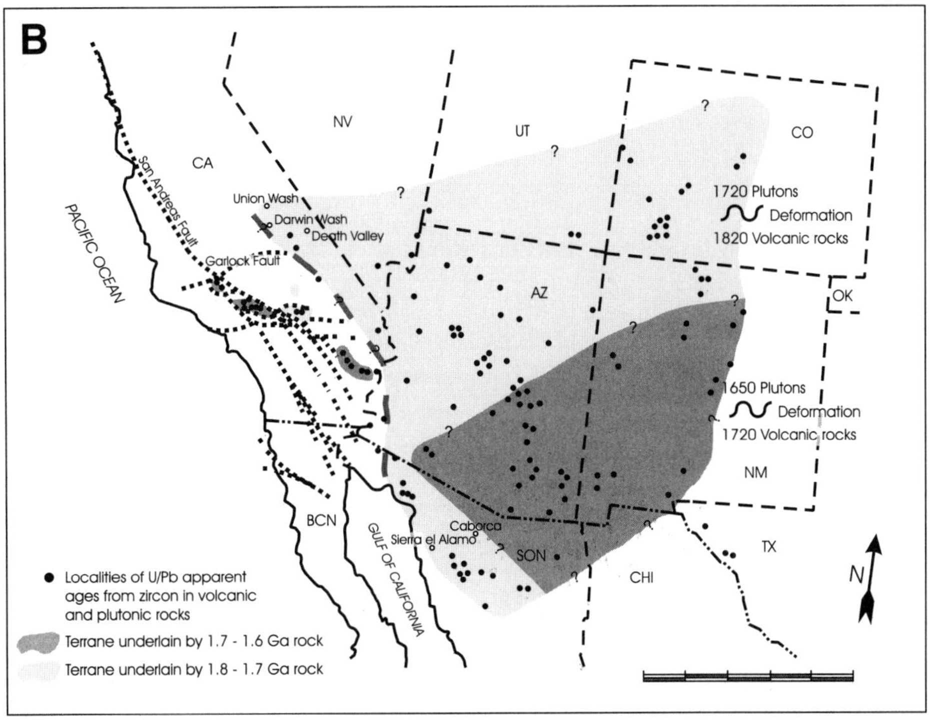

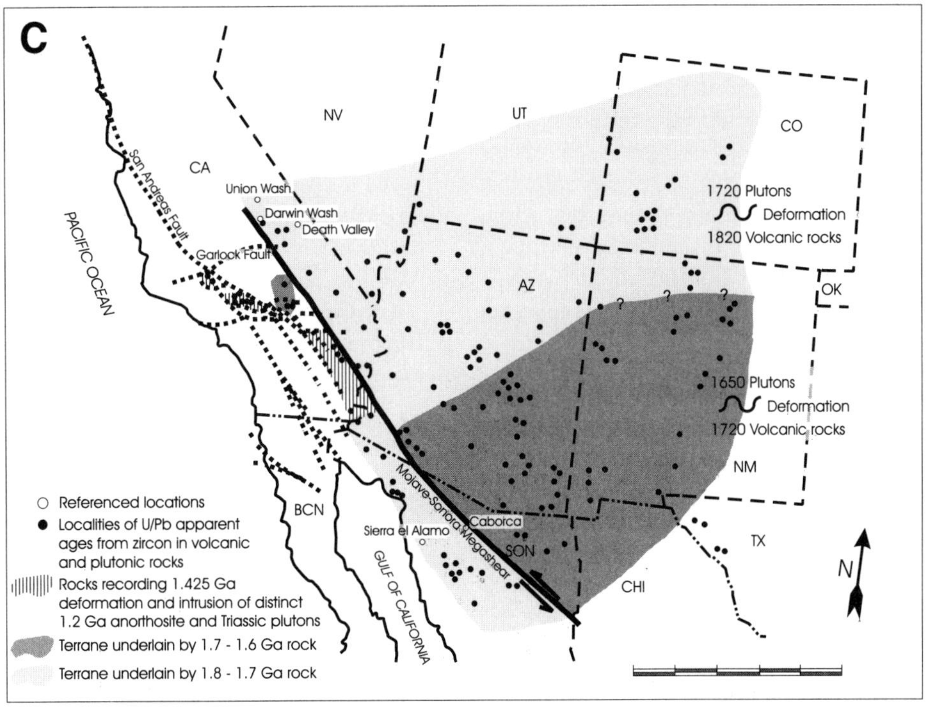

Figure 5 (*continued*).

northwest-striking San Andreas fault north of the head of the Gulf of California (Fig. 5B). In 1968, the significance of the San Andreas fault as a plate boundary along which large displacement may have occurred was newly appreciated and the anticipated truncation of the belts of Proterozoic crust at the San Andreas fault provided an important regional geological context at the initiation of study in 1968. At the beginning of our work, we aspired to: (1) delineate the truncation of the belts of Proterozoic rocks at the San Andreas fault, (2) expand knowledge of the distribution of the distinct provinces of Precambrian crystalline basement in southwestern North America (e.g., Silver, 1978; Conway and Silver, 1989 and references therein) southward into northern Mexico, and (3) identify exposures that could be the source of displaced fragments of ca. 1.2 Ga anorthosite (Silver et al., 1963) and Triassic plutons that crop out west of the San Andreas fault in southwestern California (Fig. 5C).

PROTEROZOIC ROCKS IN SONORA: EMERGENCE OF A PUZZLING PATTERN

The tectonostratigraphic relationships that emerged during the regional study did not support our working hypothesis that: (1) the belts would be truncated at the intersection with the San Andreas fault (Fig. 5A–C), and (2) we would encounter the source(s) of units similar to those in southern California that had been displaced northward along the right-lateral strike-slip San Andreas fault (Fig. 5C). Rather than truncation against the San Andreas fault, the plutonic and volcanic rocks with U-Pb ages that are correlative with those of the Yavapai Province of Arizona appeared to bend sharply southeastward forming a <-shaped pattern open to the southeast. The apparent strong bend in the belt of 1.8–1.7 Ga province rocks into Mexico startled us. The geometry is a pronounced departure from the southwest trend that characterizes the Yavapai belt for hundreds of kilometers across Arizona and New Mexico (Fig. 5A). Nevertheless, in western Sonora, crystalline rocks, the ages of which fall within the same 1.8–1.7 Ga period as those of the Yavapai Province, appeared to bend southeastward and extend more than 500 km southeast across the trend of the younger Pinal belt (Fig. 1).

In Sonora, Mexico, the 1.8–1.7 Ga rocks distinguish a basement domain that we call the Caborca block (Fig. 1). The northeastern limit of the block coincides with the Mojave-Sonora megashear. The town of Caborca at the northeasternmost edge of the Caborca block is underlain by Cambrian strata that face Jurassic volcanic and volcaniclastic strata toward the northeast (Fig. 1). As we learned more about the crystalline basement of the Caborca block, differences between the 1.8–1.7 Ga Caborca crystalline rocks and the nearby Yavapai rocks of southwestern Arizona emerged, despite their similar ages. The siliciclastic metamorphic rocks, conspicuous at Cerro Berruga, generally lack the strong mafic to intermediate volcanic signature common to the closest Yavapai equivalents in southwestern Arizona (cf. Anderson and Silver, 1976; Anderson and Silver, 1981). Further, metaquartzite along the southern margin of Cerro Berruga contains very old detrital zircon (Table 1). The quartzose units in the Caborca block that contrast with the coeval mafic Yavapai rocks of Arizona indicate either an abrupt change of facies or juxtaposition by faulting in order to explain the strong differences between the occurrences of 1.8–1.7 Ga rocks.

The irregular distribution of Nd isotope and other recently determined geochemical values in the Caborca basement rocks (Iriondo and Premo, 2003; Valenzuela-Navarro et al., 2003; Espinoza et al., 2003; Farmer et al., 2003) may also reflect the influence of old crust compatible with the siliciclastic character and the presence of very old detrital zircon in the rocks near Bámori Ranch. Sm-Nd isotopic results show positive, although somewhat depleted, values similar to the island arc–like Yavapai rocks in the eastern part of the Caborca block and along the inferred trace of the Mojave-Sonora megashear and negative values near Bámori Ranch. The values of the Bámori rocks signal the presence of Mojave Province rocks, recognized by Iriondo (2001), that are postulated to contain a substantial Archean source component introduced in the form of sedimentary detritus (Rämö and Calzia, 1998). In the late 1960s the Mojave Province (see Iriondo, 2001, for references and discussion) had not been separated as a distinct subdivision of Proterozoic crust. However, one of us (Silver) recognized the similarities of lithology, texture, and age between the Fenner gneiss in southern California and porphyritic gneiss in the Sierra Santa Rosa (see following), which are now known to have geochemical characteristics common to the Mojave Province.

We believe that the irregularities in the distribution of isotopic values are compatible with displacement along the Mojave-Sonora megashear. Restoration of the proposed 800 km would juxtapose Yavapai-like rocks of central Sonora against the Yavapai of Arizona and the Mojave-like rocks near Bámori Ranch against Mojave Province rocks in California. However, the boundary between the Yavapai and Mojave Provinces may not serve as a good piercing point if it reflects an irregular distribution of strata that were incorporated into the 1.8–1.7 Ga crust and that provided the Archean component to Mojave Province rocks.

The Caborca block is further distinguished by the presence of Neoproterozoic and Cambrian strata that comprise a thick section overlying the Proterozoic crystalline basement. These strata are correlative to those in California, which are interpreted to have accumulated in a miogeoclinal setting (e.g., Stewart, 1970; Stewart et al., 1984, 2002; Stewart, this volume). In Sonora, outcrops of miogeoclinal strata are restricted to the Caborca block (Stewart et al., 2002). No sections of Neoproterozoic and Cambrian miogeoclinal units crop out among exposures of Jurassic igneous rocks that intervene between the two different provinces of Proterozoic crystalline rocks (Haxel et al., 1988). At Cananea (Fig. 4), the closest locality across the Jurassic magmatic zone northeast of the Caborca block where Paleozoic stratigraphic relations with Proterozoic crystalline basement may be observed, nonconformable Cambrian quartzite rests directly upon Mesoproterozoic granite (Valentine, 1936; Anderson and Silver, 1977) and Neoproterozoic miogeoclinal rocks are absent. In southeast Arizona, similar thin Cambrian strata probably accumulated on a platform locally above distinctive units of the Apache Group (older than 1.2 Ga); these are unknown on the Caborca block. The oldest sedimentary strata on the Caborca block that rest nonconformably above crystalline basement locally overlie distinctive bodies of 1.1 Ga micrographic granite. This stratigraphic relationship precludes correlation of the older Apache Group with the Caborca miogeoclinal section.

A transition comparable to that which occurs between the Yavapai and Pinal Provinces in Arizona does not exist in Sonora. During the course of the study it became clear that deformed and undeformed Jurassic volcanic and plutonic rocks (e.g., Corona, 1979; Rodríguez-Castañeda, 1984, 1994) intervene between exposures of rocks of the Caborca block and of the Pinal Province rocks to the northeast (see Anderson et al., this volume, Chapter 2). Many exposures of Jurassic rocks lie within a region designated as the Papago terrane (Fig. 1) that is distinguished by several features; among them is the absence of outcrops of Proterozoic basement and Neoproterozoic and Paleozoic cover (Haxel et al., 1988). Exposures of Papago terrane volcanic and volcaniclastic strata intruded by mid-Jurassic (and younger) plutons are common in the ranges directly northeast of the northeast margin of the Caborca block (Anderson et al., this volume, Chapter 2). Corona (1979) provides descriptions of representative lithologies. They include the following: (1) Volcanic rocks resembling those that commonly yield interpreted U-Pb zircon ages close to 175 Ma consist mainly of quartz-feldspar and feldspar porphyritic tuff locally intercalated with minor amounts of vitric-welded tuff and volcanic breccia. Pink, gray, and brown colors suggest rhyolitic, dacitic, and andesitic composition. (2) Volcaniclastic conglomerate, interbedded with sandstone, siltstone, and rare carbonate rocks. Conglomerate contains subrounded to rounded pebbles, cobbles, and boulders consisting predominantly of andesitic porphyry with fewer fragments of quartz sandstone, sandstone, siltstone, and carbonate. (3) Massive volcaniclastic sandstone, interstratified with thin beds of siltstone and mudstone. (4) Carbonate rocks, although rare, typically occur as small lenses, centimeters thick, interbedded with sandstone and conglomerate.

Plutonic rocks which fall into two (Ko-Vaya and Kitt Peak–Trigo Peaks) of the three superunits described by Tosdal et al. (1989) from the Papago terrane in Arizona are also recognized in Sonora. Ko-Vaya rocks commonly have alkaline tendencies and yield interpreted U-Pb zircon ages between 160 and 145 Ma (Tosdal et al., 1989). Other Sonoran intrusive rocks resemble petrographically distinctive units of the calc-alkaline Kitt Peak–Trigo Peaks supergroup, such as granodiorite with distinctive potassium-feldspar phenocrysts of lavender tinge that commonly yield somewhat older Middle Jurassic ages (see Haxel et al., this volume; Anderson et al., this volume, Chapter 2). The magmatic assemblage is comparable to that formed along an Andean-like convergent margin (Tosdal et al., 1989).

Despite extensive exposures of Jurassic magmatic rocks, the absence of some tectonostratigraphic elements common to convergent margins indicates probable truncation parallel to the belt. Anomalous characteristics of the belt include the following: (1) No mélange or other evidence of an accretionary prism of Jurassic age is known southwest on the magmatic belt or from the Caborca block. (2) No strata of Middle Jurassic age that may be interpreted as forearc to a Mid- to Late Jurassic arc have been discovered on the Caborca block. (3) Local strong ductile deformation of Oxfordian and Kimmeridgian strata adjacent to the northeast side of the fault crosscut Cretaceous plutons. Neither comparable ductile nor contractional deformation is observed in pre-Jurassic rocks of the Caborca terrane.

In summary, a viable geologic history of northern Sonora must incorporate: (1) the distribution of distinct suites of Proterozoic crystalline rocks and the nature of the strata overlying each province, (2) the absence of the record of transition in Paleozoic strata from miogeoclinal to cratonal facies, (3) the presence of volcanic rocks comprising part of a Jurassic arc that intervenes between the Proterozoic provinces, (4) the mainly Late Jurassic deformation, commonly ductile and contractional, which is largely restricted to strata of the intracratonal Jurassic volcanic arc, and (5) the abrupt contact of Yavapai-equivalent rocks that compose the Caborca block against commonly folded and foliated Jurassic volcanic and plutonic rocks.

1974 CONCEPTION OF THE MOJAVE-SONORA MEGASHEAR HYPOTHESIS

In 1974 we presented a synthesis of our field and geochronological studies conducted during the previous 6 years and concluded by proposing truncation of the Pinal and Yavapai Province rocks along a Mesozoic fault with several hundreds of kilometers of left-lateral displacement (Silver and Anderson, 1974). We strove to offer an interpretation of the regional stratigraphic and structural relations compatible with the following: (1) the acute <-shaped bend in the Yavapai Province rocks revealed by the geochronologic studies; (2) the abrupt facies changes in the Yavapai Province that occur from the mafic Arizona rocks to the quartzose Sonoran rocks along the tightly curved Proterozoic margin; (3) the distribution of deformed Jurassic volcanic rocks for which no trench or forearc basin was recognized; (4) the similarities between Upper Precambrian and Cambrian strata of the southern Great Basin and the Caborca region that had been noted as early as 1951 by Eardley and that had been re-emphasized as a result of the stratigraphic work by Stewart (1970); (5) crystalline basement units in the vicinity of Warm Spring Canyon, in southern Death Valley, that not only yielded interpreted U-Pb ages from zircon correlative with the ages between 1.7 and 1.8 Ga (Silver et al., 1961) of the Caborca block but further contained similar quartzose units.

An additional important link to the Death Valley region that greatly influenced our thinking was provided as a result of study of detrital zircon from quartzite horizons in exposures of Pinal and Yavapai Province rocks across Arizona (Silver, unpublished data). Generally, the suites of detrital zircons yield ages that correlate closely with contemporaneous volcanic units, i.e., 1.7 and 1.8 Ga, respectively. Two exceptions to the pattern of coeval ages of detrital zircon in volcanic units and interbedded quartzose clastic units existed in 1974. Proterozoic quartzite beds from southern Death Valley and Bámori Ranch, south of Caborca, contain suites of detrital zircons (multigrain fractions) that yield ages older than 2.0 Ga. The "ages" of the fractions of detrital zircons indicate incorporation of zircon from much older sources than quartzose beds interstratified with the volcanic units in Arizona. These old zircon ages suggest a contribution from Archean sources of which the nearest known occurrences are crust of the Wyoming Province.

The similarities of Proterozoic and younger units near Death Valley to those of the Caborca block spurred additional stratigraphic comparisons among younger Paleozoic and Mesozoic formations. However information about post-Cambrian, pre-Cretaceous units was limited because few fossiliferous exposures were known. The exception is found west-northwest of Caborca at Sierra del Alamo, where Permian rocks are overlain by a thick section of Triassic and Jurassic strata (White and Guiza, 1948; González-León, 1997; Stewart, this volume). Lithostratigraphic equivalents of Late Triassic marine rocks of the Caborca block that had been mapped by Merriam (1963) in Darwin Wash, California (Fig. 5) drew our attention (see Anderson et al., this volume, Chapter 2).

The multiple matching characteristics (lithostratigraphy, age, and thickness) of similar correlative crustal sections as young as Triassic between Caborca and Death Valley, California, suggested to us that lateral displacement after Triassic time, perhaps within the zone of Middle Jurassic volcanic rocks, provided a plausible hypothesis compatible with the geologic relations observed in Sonora. Although no piercing points other than comparable thickness of correlative units were recognized, we concluded that ~800 km of left-lateral displacement is required to restore the Caborca block northwestward opposite the Death Valley region.

The trace of the inferred fault generally corresponds with valleys filled with alluvium. The broad, northwesterly trending valleys in which the inferred trace lies are oblique to the northerly trend of the topographically more prominent late Miocene basins and ranges. Outcrops of fault zone rocks, unknown in 1974, were discovered in 1980. The best exposures occur southwest of Sonoita (Fig. 4), where vertical mylonite with horizontal to subhorizontal lineation records left-lateral strike-slip sense (Campbell and Anderson, 2003). To the southeast, other exposures of the fault that may be overprinted by younger deformation occur northwest of Quitovac (Connors, 1990) and west of Quitovac (Iriondo, 2001) (Fig. 4).

Our initial estimate of the timing of displacement along the Mojave-Sonora megashear was based upon the apparent juxtaposition of Early Jurassic strata, overlying the Caborca block, against mid-Jurassic volcanic rocks north of the block. As we determined the ages of the mid-Jurassic volcanic units, commonly ca. 175 Ma, that were truncated and deformed during Mojave-Sonora megashear faulting, it became evident that the maximum age of faulting was mid-Jurassic or younger. The minimum age was provided by Late Cretaceous plutons that locally intrude deformed Jurassic units.

Contemporaneity and coincidence of this structure with plate motions proposed by Dietz and Holden (1970) to accommodate mid-Atlantic and Gulf of Mexico spreading prompted consideration of the megashear as a transform fault. The magnitude of proposed offset (~800 km) along the fault is sufficient so that the problem of continental overlap, revealed by reconstructions of Pangea such as that proposed by Bullard et al. (1965), is partially relieved. Anderson and Schmidt (1983) later proposed a reconstruction in which offsets restored along a set of three contemporaneous, cogenetic faults, the Mojave-Sonora megashear and two faults to the south, eliminate the overlap of pre-Jurassic continental crust.

PRECAMBRIAN ROCKS OF THE CABORCA BLOCK

The Pitiquito–Cerro Berruga Crystalline Suite

From Caborca southward to Bámori Ranch, ranges containing exposures of the Proterozoic basement and overlying sedimentary cover have been the focus of studies providing characteristics of the units distinguishing the type "Caborca block" (e.g., Cooper and Arellano, 1946; Arellano, 1956; Damon et al., 1962; Fries, 1962; Gilletti and Damon, 1962; Eells, 1972; Anderson and Silver, 1971, 1981). Ranges in which key stratigraphic relationships among Proterozoic units are exposed (Fig. 3) flank the broad valley, which extends from the village of Pitiquito, ~20 km east-southeast of Caborca along Route 2, south for 35 km to Cerro Berruga, north of Bámori Ranch. (Note: Cerro Berruga is distinct from Sierra de la Berruga, nearby to the south, underlain principally by late Neoproterozoic sedimentary strata.) The geologic relationships identified during previous work provided a useful context for our initial studies and we focused considerable attention upon this area where Anderson did initial mapping (Anderson and Silver, 1981) and studied stratigraphic relationships. We used the results of these studies as a principal building block for subsequent interpretations. In light of the importance that we attribute to this initial work it is described here in some detail.

At the north end of the valley, we found that late Precambrian carbonate and clastic beds rest nonconformably upon 1.1 Ga micrographic granite near abandoned Aibó Ranch (Fig. 6), indicating that 1.2 Ga strata of the Apache Group, known to overlie Pinal Province rocks in southern Arizona, are not preserved in the Caborca block. This important stratigraphic relationship, compatible with the presence of the Yavapai-age basement, supported the working hypothesis that the Caborca basement was not a southward extension of the Pinal Province.

Neoproterozoic strata dip away from the axis of the valley, suggesting that the axis of the valley corresponds with the hinge of a breached regional anticline likely modified by normal faults. At the southern end of the valley low hills that rise from the valley floor surround a higher peak called Cerro Berruga, where Proterozoic crystalline rocks were first recognized (Fig. 7; Damon et al., 1962; Giletti and Damon, 1962). A low ridge at the southern

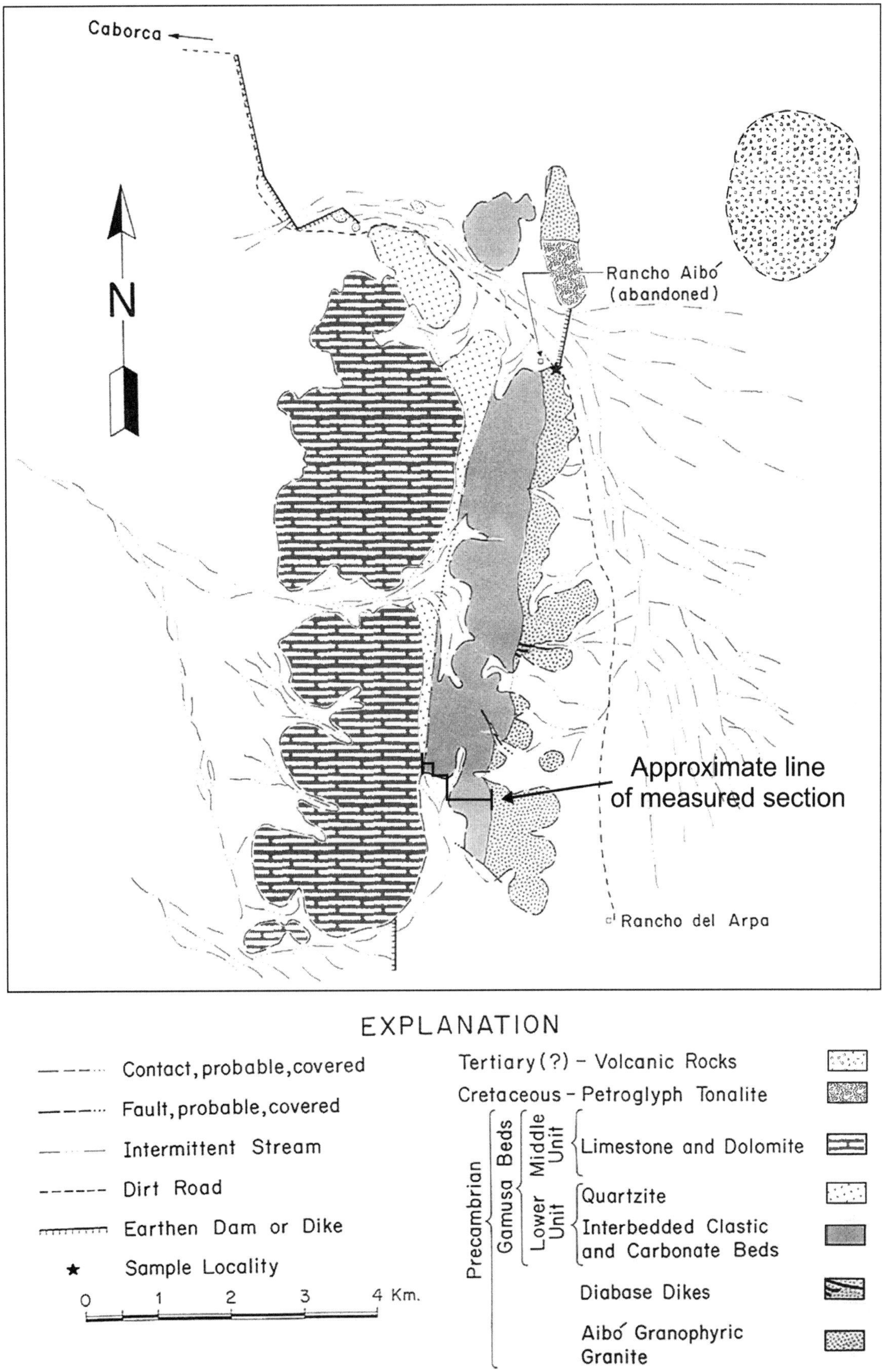

Figure 6. Geologic sketch map of Arpa Hills (Cerros El Arpa) at Rancho Aibó, Sonora, México.

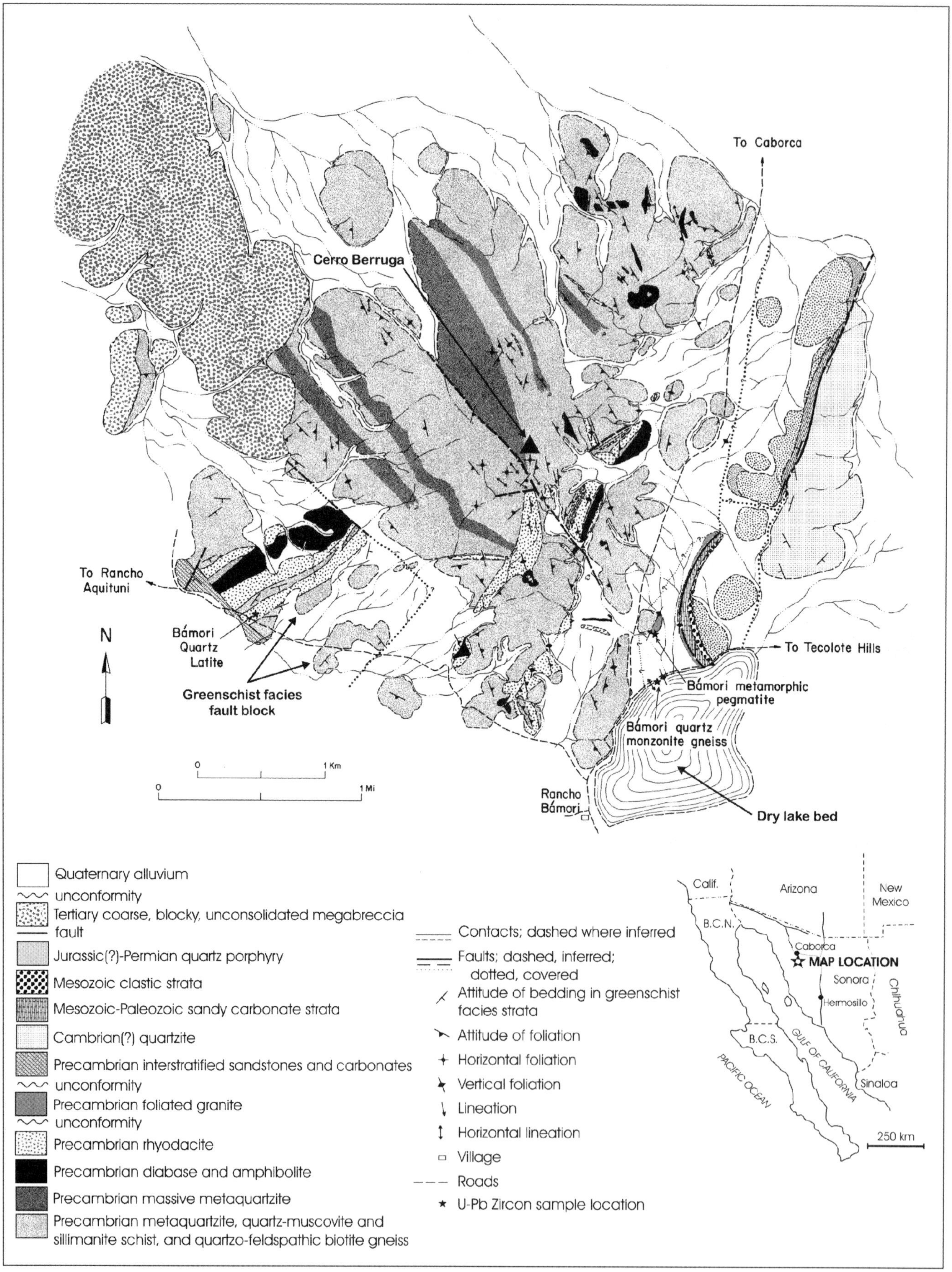

Figure 7. Geologic map of Cerro Berruga and vicinity (modified from Anderson and Silver, 1981).

margin of the hills surrounding Cerro Berruga exposes metamorphic rocks locally overlain nonconformably by southwest-facing sedimentary strata that may mark the southward-plunging nose of the regional anticline. Units composing this thin, partial section are tentatively correlated with the basal beds of the nearest late Precambrian strata to the northeast that crop out below Cerro Rajon, where they rest nonconformably upon granite (Stewart et al., 1984). Samples of gneisses, exposed in the core of the anticlinal structure, yielded the zircons upon which U-Pb isotopic analyses were conducted. The interpreted isotopic ages reveal the stratigraphic framework of the crystalline crust in this area and the broader region encompassed by the Caborca block.

Bámori Ranch (Cerro Berruga): Geology and Geochronology

The low hills surrounding Cerro Berruga are underlain principally by gneiss, metaquartzite, and schist (Fig. 7). Cerro Berruga is remarkable because, at its southwest margin, greenschist-grade metamorphic rocks, within a small area of several square kilometers, reveal stratigraphic relations among layered units. No other exposure of prograde greenschist-facies Proterozoic crust is known within the Caborca block. The southeasterly dipping homoclinal sequence of low-grade metamorphic units is separated by postulated faults from rocks characterized by upper-amphibolite facies comprising the regional terrain. Prior to metamorphism, the greenschist-grade layered sequence consisted mainly of fine- to medium-quartzose clastic strata intruded by sills and concordant lenticular masses of rhyodacite and hornblende diorite (now hornblende amphibolite). Cross-stratification in some metasedimentary beds, contacts with intrusive rocks, and, to a lesser degree, bedding and sedimentary textures are preserved. In places, igneous texture is also well preserved in sills of rhyodacite.

Across the narrow covered intervals that bound the small area of greenschist-grade rocks and that are interpreted as faults, an abrupt increase in metamorphic grade and intensity of deformation occurs. Within this larger region encompassing Cerro Berruga and most of the surrounding hills, quartzite, biotite-muscovite schist, massive quartzo-feldspathic gneiss, and amphibolite sufficiently resemble the overall composition of the less metamorphosed units to suggest correlation. In light of the likely correlation between the low- and high-grade assemblages, we consider the age information from the greenschist rocks to apply to the amphibolite-grade rocks comprising the crust of the Caborca block.

Metasedimentary rocks. The most common lithologies among exposures within the area of greenschist are gray quartzite, muscovite metasandstone and quartz-sericite and muscovite schist. These units commonly weather to shades of brown and show sparkling flakes of white mica along bedding planes. No lithic fragments were seen in the beds neither in the field nor in thin section. Thin sections reveal abundant quartz grains with less common microcline, altered plagioclase and flakes of sericite, muscovite, biotite, and chlorite. The platy minerals are weakly to strongly oriented and define schistosity that is generally parallel to bedding. Quartzose beds are commonly <10 cm thick but in places are tens of centimeters thick. The original thickness of beds of sericitic schist is obscured by schistosity.

More metamorphosed equivalents of the greenschist-grade strata include quartzite, quartz-microcline gneiss (with ubiquitous biotite, chlorite and/or muscovite as accessories), and quartz-biotite schist interlayered with irregularly distributed bands of hornblende-plagioclase amphibolite and plagioclase-quartz-microcline gneiss. Quartzose gneiss is the most common rock type, and massive layers of metaquartzite as thick as 100 m have been used as stratigraphic markers to delineate structure on Figure 7. Rare conglomerate beds crop out.

Two fractions of zircon extracted from moderately cross-bedded quartzite south of the rhyodacite collecting locality (Fig. 3) yield $^{207}Pb/^{206}Pb$ ratios that correspond to ages of 1.8 and 2.1 Ga (Table 1). The ages of these populations of mixed zircons indicate derivation from sources distinctly older than igneous rocks that formed during the generation of the mid-Proterozoic crust. No source from which zircons as old as Archean may be derived is known in the present vicinity of the Caborca block.

Rhyodacite. Sills and lenticular masses of brownish-gray porphyritic rhyodacite intruded the sedimentary rocks as concordant sheets commonly several tens to more than 100 m thick. Contacts with the sedimentary layers are sharp. However, at the margins of the rhyodacite, thin, narrow wedges of sedimentary strata may occur between igneous masses. Locally, shearing may obscure the contact, especially if phenocrysts are sufficiently recrystallized so that textural differences are obfuscated. In these areas the porphyry may resemble mica schist, which is difficult to distinguish from sedimentary schist. The rhyodacite contains plagioclase phenocrysts (as much as 10%) and distinctly fewer quartz phenocrysts (~1%) in an aphanitic matrix. Plagioclase phenocrysts are euhedral to subhedral and are commonly between 1–4 mm long. Almost all grains show moderate to intense sericitization. Quartz phenocrysts, 0.3–0.4 mm in diameter, are commonly round or slightly elongate anhedral individuals or mosaics. The groundmass is composed of common intergrowths of very fine recrystallized quartz and feldspar and common opaque grains. Weakly oriented flakes of biotite and chlorite define schistosity. From the interior of the intrusive masses outward, schistosity becomes better defined and at the margins it may be well developed. The rhyodacite weathers reddish-brown with reddish-orange phenocrysts

Masses of plagioclase-quartz-microcline gneiss (Table 2) among amphibolite facies rocks are interpreted to be correlative with compositionally similar rhyodacite. The gneiss commonly weathers pale brown and may be identified in the field by the lack of stratification, homogeneous texture, and mineralogical composition. Foliation is moderately to well defined by oriented flakes of mica and some samples show lineation defined by elongate to rod-like grains and aggregates. Abundant round or lens-shaped, pale orange porphyroblasts(?) from 1 mm to 5 mm long may consist of single grains of plagioclase with small patches of microcline and/or quartz inclusions or a mosaic of equidimensional grains of plagioclase and microcline (with less abundant quartz). The matrix is recrystallized and consists of anhedral plagioclase, quartz, and microcline with weakly to strongly oriented flakes of mica. Rare samples preserve relict igneous textures with phenocrysts of plagioclase and quartz in a fine-grained matrix. Microscopically, the relict phenocrysts consist of strongly sericitized poikilitic feldspar with numerous inclusions of quartz in a groundmass of feldspar, altered to sericite, epidote minerals, and quartz. The correlation with rhyodacite also is supported by the presence of intercalated amphibolite, plagioclase-quartz-microcline gneiss and quartz-mica schist and gneiss that define coherent, generally concordant belts similar to those of the less metamorphosed rocks.

The U-Pb interpreted age of zircon in the rhyodacite (Fig. 4, #4; Table 1, Bámori rhyodacite #2 and #205) is ca. 1.76 Ga (Figs. 2A and 7). The age is among those characteristic of a magmatic pulse between 1.78 and 1.74 Ga also recognized by Premo et al. (2003). The rhyodacite provides a minimum age for the sedimentary section into which it is intruded.

Hornblende amphibolite. Sills and lenticular masses of hornblende diorite also intrude the metasedimentary rocks at Cerro Berruga. The mafic intrusive rocks crop out among both greenschist- and amphibolite-grade rocks. The best examples are preserved among the greenschist-grade rocks where concordant lenses, in places more than 100 m thick, interfinger with adjacent rhyodacite and quartzose strata. Fresh exposures are greenish-black, whereas weathered surfaces show primary aggregates of black hornblende prisms, commonly more than 1 mm across, in a matrix of small laths of plagioclase and prisms of hornblende. In thin section, anhedral grains of blue-green hornblende are set in a fine-grained matrix of weakly oriented, sericitized plagioclase laths and hornblende prisms with less abundant opaques and rare quartz.

Among amphibolite-grade rocks, amphibolite is common as elongate and irregular masses associated with plagioclase-quartz-microcline gneiss and also as small pods, lenses, and irregular masses in metasedimentary rocks. In fresh exposures it is greenish-black and weathers dark reddish-brown or grayish-orange, depending upon the ratio of hornblende to plagioclase. Weathered surfaces show aggregates and small prisms of hornblende imbedded in lighter-colored groundmass. Elongate plagioclase and hornblende crystals do not define foliation but may define distinctive lineation, although some masses show slight, if any, preferred mineral orientation. In thin section anhedral to subhedral yellowish-green hornblende and plagioclase, altered to white mica and clinozoisite, predominate with less abundant quartz, biotite, opaques, and apatite.

Quartz monzonite gneiss. East of the road between Bámori Ranch and Pitiquito, at the north edge of the dry pond at the ranch, rare exposures of foliated quartz monzonite (adamellite) crop out (Fig. 7; Table 2). Poor outcrops also occur on the east side of the first small mound of quartzite on the east side of the road north of the ranch. A short distance to the east across the dry wash, gneiss, strongly epidotized, underlies a low ridge of gray carbonate. In hand specimen the rock consists of interlayered pink and dark-gray laminae a few millimeters thick, consisting mainly of potassium feldspar and altered mafic minerals, with less conspicuous quartz and plagioclase. At the sample locality, foliation is vertical and strikes N10–20°W, although the strike swings to N25°E northward. Horizontal lineations are defined by undulations on the foliation planes that are parallel with streaks of smeared mineral aggregates. Zircon from the adamellite (Fig. 4, #1; Table 1, Bámori granite gneiss #204) yields an interpreted U-Pb age of ca. 1.73 Ga (Fig. 2A), somewhat younger than the magmatic pulse between 1.78 and 1.74 Ga reported by Premo et al. (2003) but older than their age (1.703 Ga) of the nearby granite at Cerro Rajon.

Microscopically, the gneiss consists of almost-equal amounts of anhedral grains, some of which are elongate, of plagioclase, potassium feldspar, and quartz. Grains are between 0.75 and 2.0 mm in diameter. Moderately oriented chloritized biotite in elongate anhedral to subhedral plates and aggregates including apatite, zircon, and other opaques define planar fabric. Fractured grains and narrow zones distinguished by granulated feldspar and quartz are numerous. Epidote fills thin veins and occurs as irregular patches and as an alteration product of plagioclase. Potassium feldspar is cloudy.

The contact between the gneiss and quartzite is concealed by rubbly float. In places near the contact, the gneiss is transitional into foliated amphibolite. Short dikes and pods of dark, aphanitic rock and feldspathic aplite crop out but record no discernable deformational fabric.

Pegmatites. We distinguish two classes of discordant, quartzo-feldspathic pegmatites: igneous and metamorphic. Igneous pegmatites of simple composition that are commonly several centimeters to more than half a meter wide, consisting mainly of brightly colored orangish-pink potassium feldspar, quartz, and muscovite, cut units within the greenschist terrain. Each of the several pegmatites observed has sharp contacts with country rocks and in places undeformed pegmatite cuts foliated rhyodacite. The long axes of the pegmatite minerals are oriented perpendicular to the walls of the country rock. The second class of pegmatite typically occurs as narrow pods and lenses in strongly metamorphosed masses of paragneiss and orthogneiss. The contacts between this type of pegmatite and country rocks are generally gradational from coarse to medium grains and occur across a zone a few centimeters wide. The pegmatite mineral assemblage is transitional into a quartz-rich core. In thin section, the anhedral grains are commonly 3–4 mm wide, although grains as large as 5–6 mm are present. Some grains are fractured and narrow bands of granulated minerals traverse the pegmatitic rock. Dark, anhedral grains of zircon, commonly metamict, may be visible among the larger crystals. Analyses reveal that the pegmatite zircons are rich in uranium and other radioactive elements probably concentrated during partial melting (Table 1). The mafic-poor mineralogy and gradational contacts between country rocks and pegmatitic bodies are interpreted as an indication that these bodies were generated essentially in place from adjacent rocks during the acme of metamorphism.

Two samples of coarse crystals from unfoliated masses that form generally elongate bodies were collected (Fig. 7). Within the adamellite gneiss, zones 15–20 cm wide, composed mainly of crystals of plagioclase, quartz and potassium feldspar with rare chloritized biotite flakes (Table 2), cut gently across the trend of foliation in slightly more mafic gneiss. Zircon extracted from this metamorphic pegmatite was analyzed in order to determine the age of partial melting during metamorphism. The pegmatite (Fig. 4, #2; Table 1, Bámori metamorphic pegmatite #203) from which zircons were separated crops out ~0.3 km N15°W of the gneiss locality. The zone of pegmatitic segregation trends N15–20°E in gneissic granodiorite containing few layers of amphibolite and quartzo-feldspathic gneiss. Near-vertical foliation in the host gneiss strikes N-S to N15–20°W. The analytical results yield an upper intercept of ca. 1.62 Ga (Table 1; Fig. 2A). A younger age is not precluded by the data.

A second unfoliated pegmatite (Fig. 4, #34; Table 1, Bámori metamorphic pegmatite #3), collected from a block bulldozed from a low road cut along the eastern margin of Cerro Berruga (Fig. 7), is composed of coarse crystals mainly of quartz and feldspar derived from nearby quartzo-feldspathic and amphibole-bearing gneisses. Zircons from the sample are rich in uranium. However, the range in uranium concentration of the strongly discordant zircons is not sufficient to yield a meaningful result (Table 1; Fig. 2A).

Metamorphism. Damon et al. (1962) concluded that initial metamorphism was of the hornblende-hornfels facies and that a later, somewhat lower-grade event resulted in retrogression. Mineral assemblages in thin sections examined during our study support these conclusions.

In the small, fault-bounded area, initial metamorphism was of the middle greenschist facies, as indicated by locally preserved sedimentary and igneous structure, oriented biotite that defines weak foliation in the rhyodacite sills, and the absence of sillimanite or garnet. Among rock samples outside the small fault-bounded area, sillimanite is common, although everywhere it is sheathed by plates or fibrous pseudomorphs of muscovite. Biotite has been moderately to strongly altered to chlorite and muscovite. Plagioclase shows weak to intense alteration, mainly to sericite or less commonly to epidote group minerals. Among the strongly metamorphosed quartzose metasedimentary rocks, a common assemblage includes quartz-microcline-plagioclase-biotite-sillimanite with a totally altered accessory. The relict accessory grains that are interpreted as former garnet occur as irregular, elongate poikiloblastic grains that show ubiquitous alteration to a mixture of fine-grained chlorite and white mica.

The retrograde metamorphism has affected almost all the rocks. Muscovite is a ubiquitous constituent of quartzose rocks that we examined and is probably the latest metamorphic product as suggested by its: (1) presence along the borders of masses of fibrous sillimanite, (2) replacement of biotite, and (3) occurrence as flakes formed in the interior of plagioclase grains athwart fine-grained aggregates of white mica. Commonly, subhedral rectangular plates of muscovite lie at an angle to biotite that defines foliation. These plates may be several millimeters long and enclose shreds of chloritized biotite, quartz inclusions, and opaques. Randomly oriented muscovite is particularly abundant and well developed adjacent to the margins of pegmatites with sharp contacts that cut across foliated rocks of the greenschist-grade terrane.

Structure. The major structure in the amphibolite facies rocks of Cerro Berruga is a northwesterly trending asymmetrical antiform, delineated by massive layers of metaquartzite that plunges gently southeastward (Fig. 7). In the southwestern flank of the antiform, the strike of massive quartzite beds is parallel with the foliation. At other localities intense deformation such as local tight folds recorded by laminated micaceous beds within quartzose gneiss and the lack of stratigraphic markers obfuscates the relationship between structural surfaces and primary stratification. Structural elements such as (1) the strike of compositional layers in gneiss and foliation in micaceous beds, (2) lineation defined by smeared mineral grains, (3) finely rippled surfaces on metaquartzite that may record the intersection of bedding and foliation, (4) small folds that record flowage of metaquartzite, and (5) aligned, mafic minerals on foliation planes, all of which are parallel to the major fold hinge, are interpreted to indicate folding during the acme of Precambrian metamorphism. The nose of the fold has been broken by faults; some are parallel to the hinge, whereas others break across the structure. Cleavage that cuts compositional bands is uncommon, although locally massive quartzite may show spaced fractures parallel to cleavage.

Faults probably isolate the greenschist-grade rocks from an uplifted block of almandine-amphibolite facies rocks as indicated by the discordant structural trends recorded by rocks within each terrane. In contrast to the northwesterly trending fold in the amphibolite facies rocks, bedding in greenschist metasedimentary rocks and foliation in the igneous intrusive rocks strike northeasterly and dip to the southeast. Cross-beds in quartz sandstone indicate that the sequence is a right-side-up homoclinal section. Bedding surfaces of phyllitic units in the northwestern part of the low-grade area show intersecting sets of crenulations, but outcrops that expose these structures are scarce, and too few measurements were made to determine their role in the deformational history.

In the hills east of Cerro Berruga and north of the pond at Bámori Ranch, moderately to shallowly dipping faults separate Jurassic(?) quartz porphyry, underlying sedimentary units and worm-burrowed (scolithus(?) tubes) Cambrian(?) quartzite from orthogneiss. Thousands of meters of Neoproterozoic and Cambrian strata, exposed nearby to the northeast, along the southern flank of Cerro Rajon, have been cut out between the quartzite and the orthogneiss. Additional hundreds or thousands of meters of Paleozoic and Lower Mesozoic strata are missing above the quartzite. The unconformable relation between Lower Jurassic and Cambrian strata north of Cerro Rajon reported by Stewart et al. (1984) suggests that the missing Phanerozoic units have been eroded along a pre-Pliensbachian unconformity that cuts stratigraphically downward from north to south.

In the westernmost part of the Berruga hills, Proterozoic crystalline rocks are overlain by a veneer of unsorted blocks. The blocks, some of which are meters long, are comprised of two main lithologies: rusty weathering quartz sandstone and gray carbonate with abundant algal mats. Small fragments of the underlying crystalline rocks may be found, but are likely derived from outcrops in rare windows through the sheet of debris. The contact between the blocky rubble and underlying basement is horizontal or nearly so. The origin of this debris sheet is unknown. However, its megabreccia-like character and proximity to major, gently dipping faults that cut out strata suggest comparison with early Miocene extensional features known elsewhere in northern Sonora (Nourse et al., 1994).

Cerro del Tecolote Samples

About 10 km east of Bámori Ranch, where the road crosses a low pass on the southwest flank of Cerro del Tecolote (Figs. 3 and 7) massive granodioritic gneiss crops out. The gneiss, strongly recrystallized with a distinct foliation oriented N65°E, 86°S, is interpreted to have been derived from an intrusive body of granodioritic composition (Table 2). Nearby, rocks rich in potassium feldspar are converted to augen gneiss. Common metamorphic veins and pegmatites suggest metamorphism reached high-amphibolite or low-granulite grade. Zircon from the gneiss (Fig. 4, #14; Table 1, Tecolote granodiorite gneiss #201) exposed in a stream cut on the north side of the road (Fig. 3) yields an upper-intercept age of ca. 1.74 Ga (Fig. 2B), slightly older than the orthogneiss near Bámori Ranch. About 12 m west in the same cut, a zoned metamorphic pegmatite (Fig. 4, #15; Table 1, Tecolote metamorphic pegmatite #202) with a quartz core and potassium-rich border zone was collected. Zircon from the body yields an upper-intercept age of ca. 1.66 Ga (Table 1, Figs. 2B and 7).

Postorogenic Proterozoic Plutons (1.4 Ga and 1.1 Ga)

Away from Cerro Berruga, generally unfoliated plutonic rocks comprise additional components of the crystalline basement nonconformably underlying Neoproterozoic strata. The plutons that intrude the 1.7 Ga crystalline basement are texturally and petrographically distinctive and fall within two different age groups: ca. 1.4 and ca. 1.1 Ga. The older intrusives are composed of granite and granodiorite that generally contain large phenocrysts (3–10 cm) of potassium feldspar. The younger plutons are grayish-red micrographic granite.

Porphyritic granite at Cerro Colorado and farther southeast. Representative exposures of porphyritic granite in the Pitiquito–Cerro Berruga suite area occur at the base of the Neoproterozoic section exposed at Cerro Colorado (Figs. 4 and 8). At this locality the granite is composed of pinkish-gray phenocrysts as long as 4 cm in a matrix of quartz, commonly in spherical masses up to 1 cm in diameter, grayish-white plagioclase, as large as 1 cm, and biotite as flakes and aggregates. Opaque grains and allanite are notable accessories. We collected a sample at Cerro Colorado, several kilometers southeast of Cerro Rajon (Figs. 4 and 8), from a richly phenocrystic facies that underlies the low rocky ridge trending ~N80°E. Locally at the margins of the crudely tabular(?) granite body, a transition into more mafic granodiorite or quartz diorite with scattered pink phenocrysts may be observed. Pegmatites also strike easterly. Foliation in the mafic rock strikes northerly. The Cerro Colorado quartz monzonite #300 (Fig. 4, #7) yields an interpreted age of ca. 1.43 Ga. (Table 1; Figs. 8 and 2C).

Constituents of this series are widespread and crop out within and outside of the Caborca block (Fig. 4) as well as at many localities in the southwestern United States (e.g., Anderson and Silver, 1977; Anderson, 1989). Plutons of this generation comprise a volumetrically significant element of Precambrian crust in southwestern North America. Plutonic rocks correlative with this magmatic suite are generally undeformed. However, some masses in Sonora are foliated (e.g., sheared porphyritic granite #100 [Fig. 4, #9] and Sierra Blanca porphyritic quartz monzonite #1 [Fig. 4, #8]). The analyses of characteristically porphyritic Caborca block plutons are consistent with ages in the range of 1415–1440 Ma, as indicated by the ^{207}Pb-^{206}Pb ages (Table 1; Fig. 2C). Further, the age of sheared porphyritic granite #100 (1416 ± 12 Ma) compares well with Iriondo's unpublished result of 1418 ± 5 Ma.

Cerro El Arpa—Undeformed micrographic granite (1.1 Ga) at Aibó Ranch. The second suite of undeformed plutons is known only from Sonora, Mexico, where autochthonous occurrences are restricted to the Caborca block. The exposure at Aibó Ranch is described here in some detail because: (1) the granite is a previously unknown body, and (2) the granite provides a maximum age for nonconformably overlying Neoproterozoic strata. A second large mass of micrographic granite is exposed in the San Antonio region of central Sonora (Fig. 4, Fig. 9 [see footnote 1]; see below). Other masses that are probably allochthonous, having been thrust across the Mojave-Sonora megashear onto Jurassic rocks, crop out north of the San Antonio region (Stephens, 1988; Stephens et al., 1986; Rodríguez-Castañeda, 1990) and west of Sonoita (Campbell and Anderson, 1996, 2003), and north of Quitovac (Iriondo, 2001).

Good outcrops occur south of Pitiquito at the abandoned Aibó Ranch, the ruins of which are at the north end of Arpa Hills (Fig. 6). Arpa Hills, together with Gamuza Hills and several small hogbacks south of El Arpa Ranch, are formed mainly from moderately west-dipping Upper Proterozoic strata nonconformably resting upon red-weathering micrographic granite (Fig. 3). The granite may extend eastward beneath the broad alleviated valley to Sierra La Vibora, where exposures of similar granite are overlain by east-dipping beds (Fig. 3). Thrusts and normal faults obscure some primary stratigraphic relationships among the exposures in Sierra La Vibora (DeJong et al., 1988).

Damon et al. (1962) initially studied the granite during reconnaissance studies at Aibó Ranch. Their work indicated that the granite is Precambrian, although the proximity of the younger crosscutting intrusive bodies may have precluded determining the age of primary crystallization as suggested by the discordance of the Rb/Sr (710 ± 100 Ma) and K/Ar (150 ± 8 Ma) dates.

Field relations and petrographic character. The micrographic granite is cut by a few fine-grained, steeply dipping, altered pyroxene diabase dikes. In places weathered reddish-brown knobs of dike protrude several centimeters above the surface of erosion of the surrounding granite. The present erosion surface is similar to the Proterozoic surface as shown locally by low knobs of dike rock that protrude into exposures of the nonconformably covering Proterozoic sedimentary layers.

A small mass of medium-grained hornblende-biotite quartz diorite also intrudes the granite at the north end of an earthen dam near ruins of Aibó Ranch. In hand specimen, vitreous plates of brownish-black biotite and irregularly formed masses of greenish-black hornblende are set in light-gray to white plagioclase. Generally, the hypidiomorphic-equigranular rock consists of euhedral to subhedral laths of twinned and commonly zoned andesine as individual crystals and inclusions in later-crystallizing constituents. Green hornblende and plates of biotite are subhedral to anhedral. Potash feldspar and quartz are texturally late and occur as scattered anhedral grains of nonuniform size. Accessory minerals include magnetite, apatite, sphene, and zircon (Table 2). The main constituents are little altered and only minor amounts of sericite, calcite, chlorite, and epidote group minerals are present. Zircon from this body yields an interpreted Cretaceous age (Silver, unpublished data).

The medium-grained micrographic granite consists of altered medium grayish-green plagioclase, scattered gray quartz, and finely crystalline aggregates of altered mafic grains in a mosaic of grayish- to pale-red potash feldspar (Table 2). Weathered surfaces show distinctive "brick red" color characteristic of altered, shallow granite intrusives. Samples from other localities in the Arpa Hills are pale to moderate to grayish-red, specked with altered grayish-orange plagioclase crystals, irregular black aggregates of mafic minerals, and less conspicuous gray quartz.

Thin sections reveal anhedral to subhedral quartz and mainly anhedral potash feldspar with less abundant subhedral to euhedral plagioclase in a matrix of intergrown quartz and microperthite. Slender bands of alteration products commonly define exsolution lamellae. Potash feldspar is consistently turbid. Rare quartz grains have hexagonal cross sections but anhedral and subround grains are more common. Micrographic texture is comprised of polygonal, blocky, triangular and vermicular intergrowths, a few tenths of a millimeter in largest diameter, which may be adjacent to and transitional into crystals showing no intergrowth. In places, coarse, elongate, angular to vermicular intergrowths radiate from individual crystals of microperthite. Grains of quartz may be similarly rimmed but generally the enclosing intergrowths show cuneiform texture. Twinned albite exists as subhedral to euhedral crystals, commonly coarser than other constituents. Albite may show alteration to white mica and epidote-clinozoisite. Rarely, microperthite is rimmed with albite. Irregular masses of magnetite and scattered fine-grained aggregates of chlorite, epidote, and muscovite, which delineate outlines of former crystals, perhaps mafic, are in common association with magnetite, commonly rimmed with hematite. Zircon and apatite are common accessories. Among exposures in Arpa hills, lithologic variations include rocks in which all quartz and feldspar are intergrown with vermicular texture. Other varieties display intergrowths and aggregates of anhedral, subhedral, and rare euhedral quartz crudely enclosing larger, anhedral grains of microperthite that commonly contain quartz as single inclusions or enclose quartz in a poikilitic framework.

Isotopic results. Analytical results from six analyzed fractions of Aibó micrographic granite (Fig. 4, #5) are shown on Table 1 (Aibó micrographic granite) and plotted on Figure 2D. The zircon fractions display a wide range in the degree of discordance with $^{206}Pb/^{238}U$ ages within the interval between 218 and 1130 Ma. On the concordia diagram, the six form a linear array, which intersects the concordia curve at 1091 ± 54 Ma (Fig. 2D). Other U-Pb isotopic ages from zircon in the Aibó granite (Iriondo et al., 2003; and Farmer et al., this volume) cluster near 1085 Ma, and our slightly older result may reflect inheritance.

Micrographic granite also crops out in the San Antonio region and as allochthonous masses thrust northward onto Jurassic(?) rocks (Fig. 4; see below). The chord defined by the data from the rocks at all three localities indicates an age of ca. 1.08 Ga (1084 ± 28 Ma, Fig. 2) that coincides with the recent determinations of Iriondo et al. (2003; and Farmer et al., this volume).

Neoproterozoic Strata between Pitiquito and Bámori Ranch

Similarities between the thickness and composition of Upper Precambrian and Cambrian strata of the southern Great Basin and the Caborca region were noted as early as 1951 by Eardley and were re-emphasized by Stewart in 1970. The presence of the 1.1 Ga Aibó granite at the base precludes correlation with either the Pahrump Group in California or Apache Group in central Arizona. In light of the absence of pre–1.2 Ga strata at Aibó and the absence of 1.1 Ga granite in California and Arizona, we were perplexed as to the relationship of the section at Aibó Ranch to other Cordilleran Neoproterozoic strata. A section was measured and described near Aibó Ranch at the north end of the Arpa hills with the objective of characterizing the rocks (Fig. 10) and establishing stratigraphic relations with strata exposed in other ranges (Longoria et al., 1978; Longoria and Perez, 1979; Stewart et al., 1984).

An early problem had to do with the nature of the contact between the granite and overlying strata. Initially, the contact between sedimentary strata and granite was described as intrusive (Cooper and Arellano, 1946); later as a fault (Arellano, 1956; Longoria, 1981). Damon et al. (1962) reported the presence of basal conglomerate resting unconformably upon the Aibó granite. Fragments of granite in basal layers and excellent preservation of the contact where young intrusives and faults are absent corroborate this interpretation (Anderson et al., 1979; Stewart et al., 1984). In the Arpa hills, the contact may be slightly sheared, but large displacement is not suggested by structural or stratigraphic evidence.

Near Aibó Ranch, the incomplete section is 450 m thick with a basal clastic sequence ~20 m thick. The section (Fig. 10) includes a basal bed of medium-bedded, fine to very coarse quartz sandstone and granule conglomerate with clasts of the granite overlain by 2–3 m of gray, pink, or reddish-brown, fine- to coarse-grained thin-bedded quartz sandstone, commonly laminated with interbeds and lenses of dolomite and sandy dolomite. The sandy

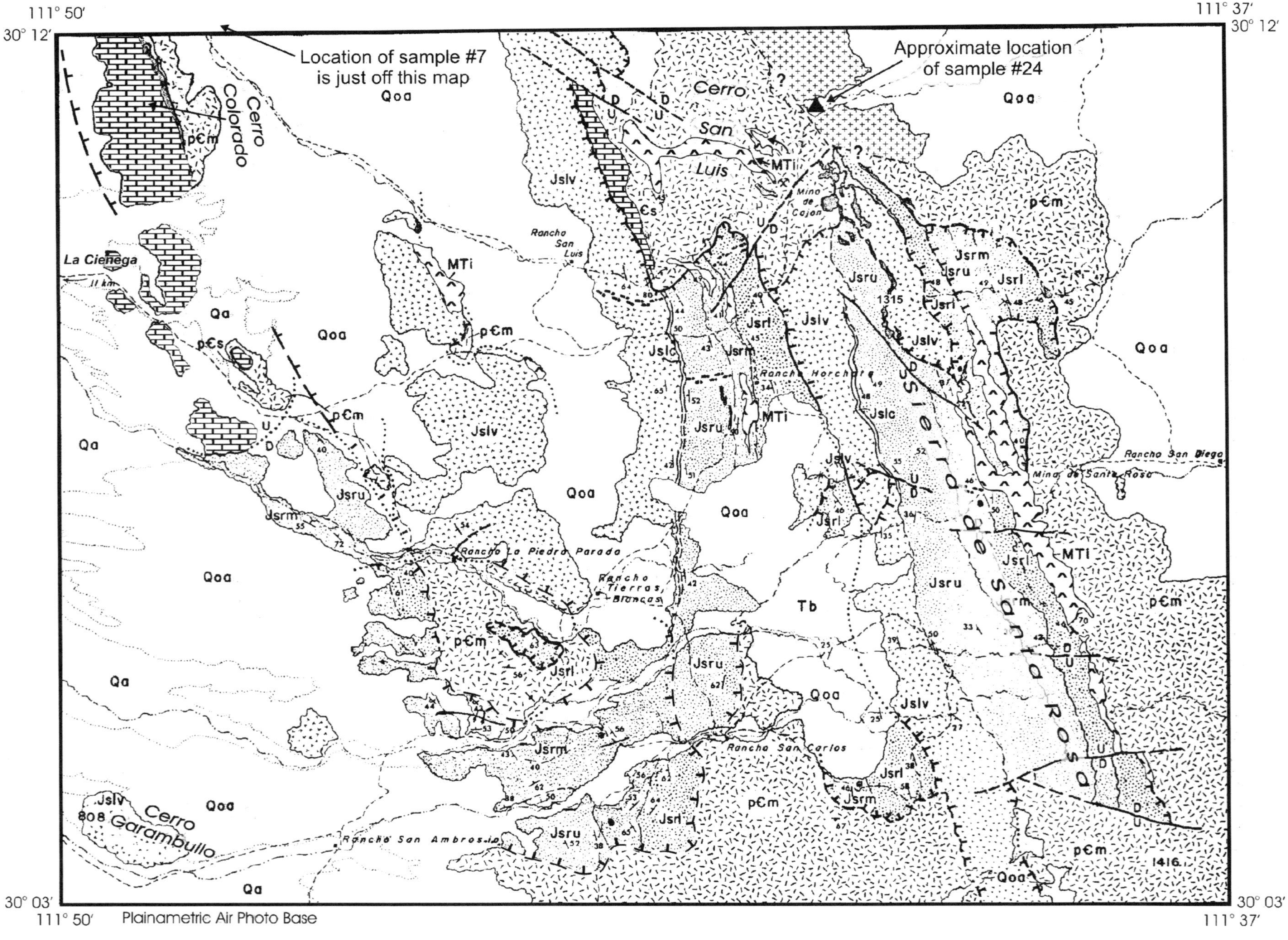

Plainametric Air Photo Base

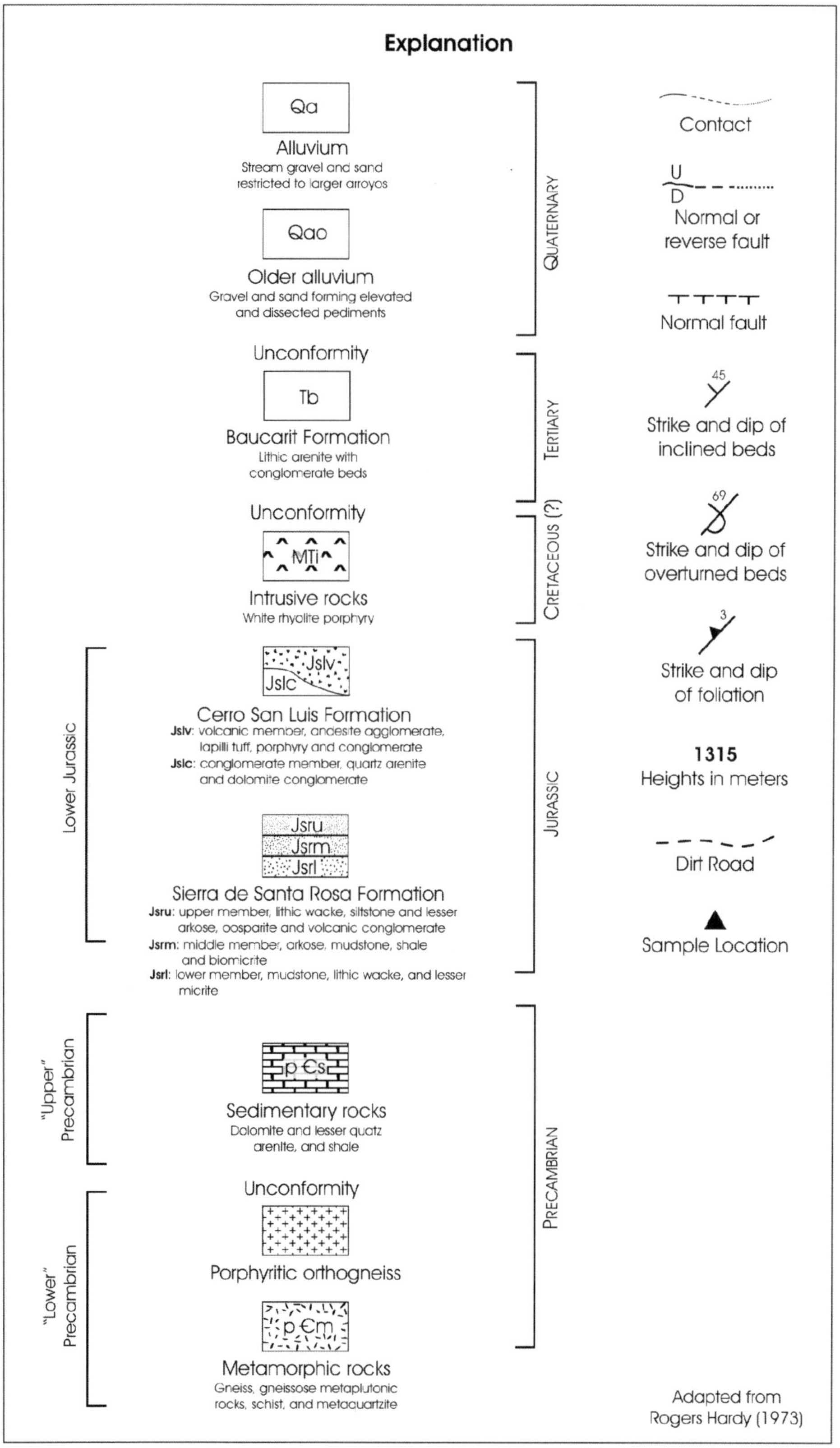

Figure 8 (*on this and previous page*). Geologic map of Sierra Santa Rosa and surrounding area (modified after Hardy, 1973, 1981).

TABLE 2. MODAL ANALYSES OF SOME GRANITIC ROCKS NEAR CERRO BERRUGA

Plagioclase-quartz-microcline gneiss, Cerro Berruga*	1	2	3	4	5	6	Average
Quartz	32	26	32	22	37	36	31
Plagioclase (An12-28)	37	31	36	39	20	30	32
Microcline	17	32	12	20	21	22	21
Chlorite and biotite	12	9	17	14	20	8	13
Muscovite	N.A.	N.A.	N.A.	N.A.	2	2	1
Opaques	1	3	3	3	1	2	2
Apatite	Trace	Trace	Trace	Trace	Trace	Trace	Trace
Zircon	Trace	Trace	Trace	Trace	Trace	Trace	Trace
Epidote	N.A.	N.A.	N.A.	N.A.	N.A.	Trace	N.A.

Bámori granite (Adamellite) gneiss #1 and Bámori metamorphic pegmatite #2†	Adamellite Gneiss	Metamorphic Pegmatite
Quartz	28	32
Plagioclase (An8-38)	28	37
Orthoclase & Microclase	28	30
Biotite and Chlorite	11	<1
Opaque Minerals	1	Trace
Epidote	1	N.A.
Sphene	Trace	N.A.
Apatite	<1	N.A.
Zircon	Trace	Trace
Misc. alteration products e.g. calcite, Ti-oxide	2	<1

Tecolote granodiorite gneiss #14†	
Quartz	19
Altered Plagioclase	40
Perthite	15
Chlorite	22
Opaque Mineral	3
Apatite	Trace
Zircon	Trace
Misc. alteration products (e.g., calcite, Ti-oxide)	Trace

*Based on 1000 points per thin section
†Based on three thin sections - 1000 points per thin section
N.A. = not applicable

beds give way to ~200 m of limestone and dolomite commonly sandy and thin bedded, with interspersed medium to thick-bedded carbonate higher in the section. The basal clastic beds and overlying thin-bedded dolomite are equivalent to the El Arpa Formation, whereas the thicker carbonate beds are similar to the Caborca Formation (Longoria, 1980, 1981; Stewart et al., 1984). Above the carbonate strata, thin-bedded brown dolomite gives way to pale red and grayish-orange siltstone and fine-grained quartzose sandstone overlain by almost 20 m of reddish-orange weathering quartzite that forms a conspicuous ledge. Our measured section stopped at the base of the quartzite horizon. Stewart et al. (1984) named the fine-grained sandstone and siltstone, and overlying strata that include the Johnnie oolite, the Clemente Formation. The maps, measured sections, rock descriptions and proposed correlations in Stewart et al. (1984) and Stewart (this volume) significantly revise the stratigraphy proposed by Longoria and colleagues (Longoria et al., 1978; Longoria and Perez, 1979; Longoria, 1980, 1981; Longoria and González, 1981).

Extent of the Caborca Block as Distinguished by Proterozoic Rocks Temporally Correlative with Yavapai Province Crust

Away from the valley between Pitiquito and Bámori Ranch, outcrops of volcanic and plutonic rocks with ages correlative with ca. 1.7–1.8 Ga Yavapai Province crust or other proxies of Proterozoic crystalline basement such as ca. 1.6 Ga metamorphic pegmatites or ca. 1.4 Ga plutons are considered indications of basement crust that reveal the extent of the Caborca terrane. Three regions that are informally designated: (1) Sierra Santa Rosa, (2) San Antonio, and (3) Seven Hills are described below (Figs. 4, 8, 9, and 11). The regions encompass groups of hills and higher

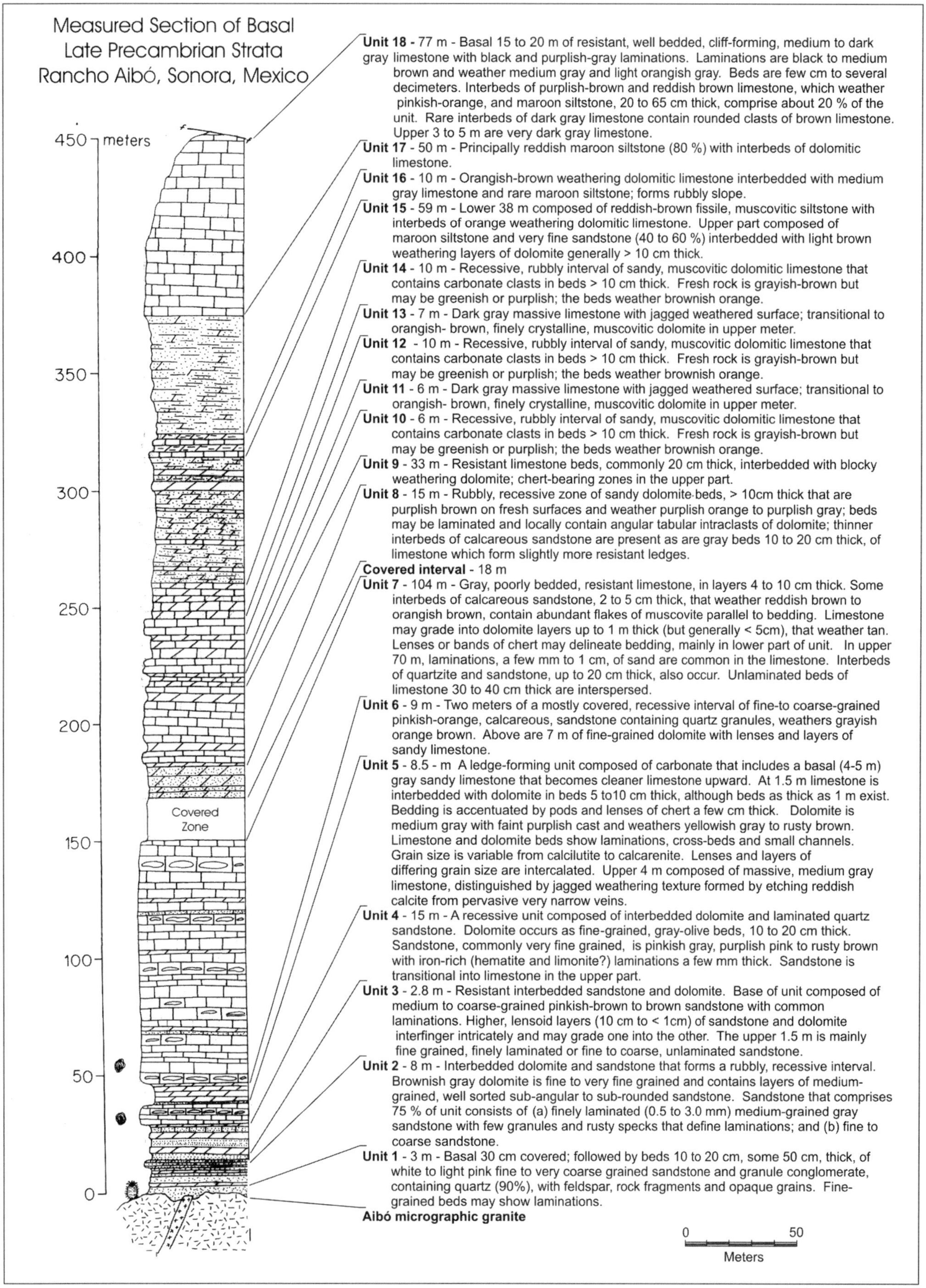

Figure 10. A measured section of Neoproterozoic rocks resting nonconformably upon micrographic granite near Rancho Aibó.

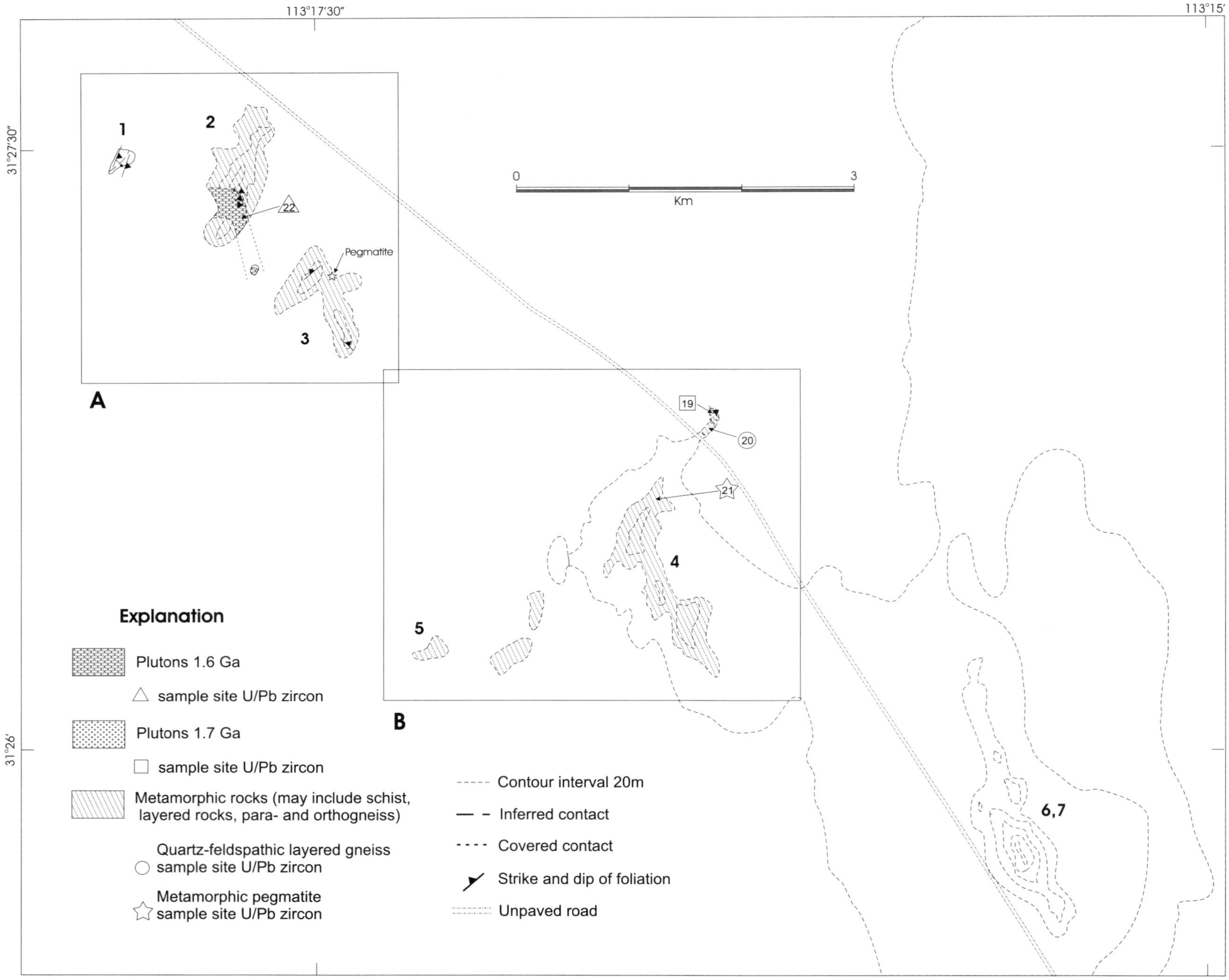
113°17′30″
113°15′
31°27′30″
31°26′
0
3
Km
1
2
3
4
5
6,7
A
B
22
Pegmatite
19
20
21
Explanation
Plutons 1.6 Ga
sample site U/Pb zircon
Plutons 1.7 Ga
sample site U/Pb zircon
Metamorphic rocks (may include schist, layered rocks, para- and orthogneiss)
Quartz-feldspathic layered gneiss sample site U/Pb zircon
Metamorphic pegmatite sample site U/Pb zircon
Contour interval 20m
Inferred contact
Covered contact
Strike and dip of foliation
Unpaved road

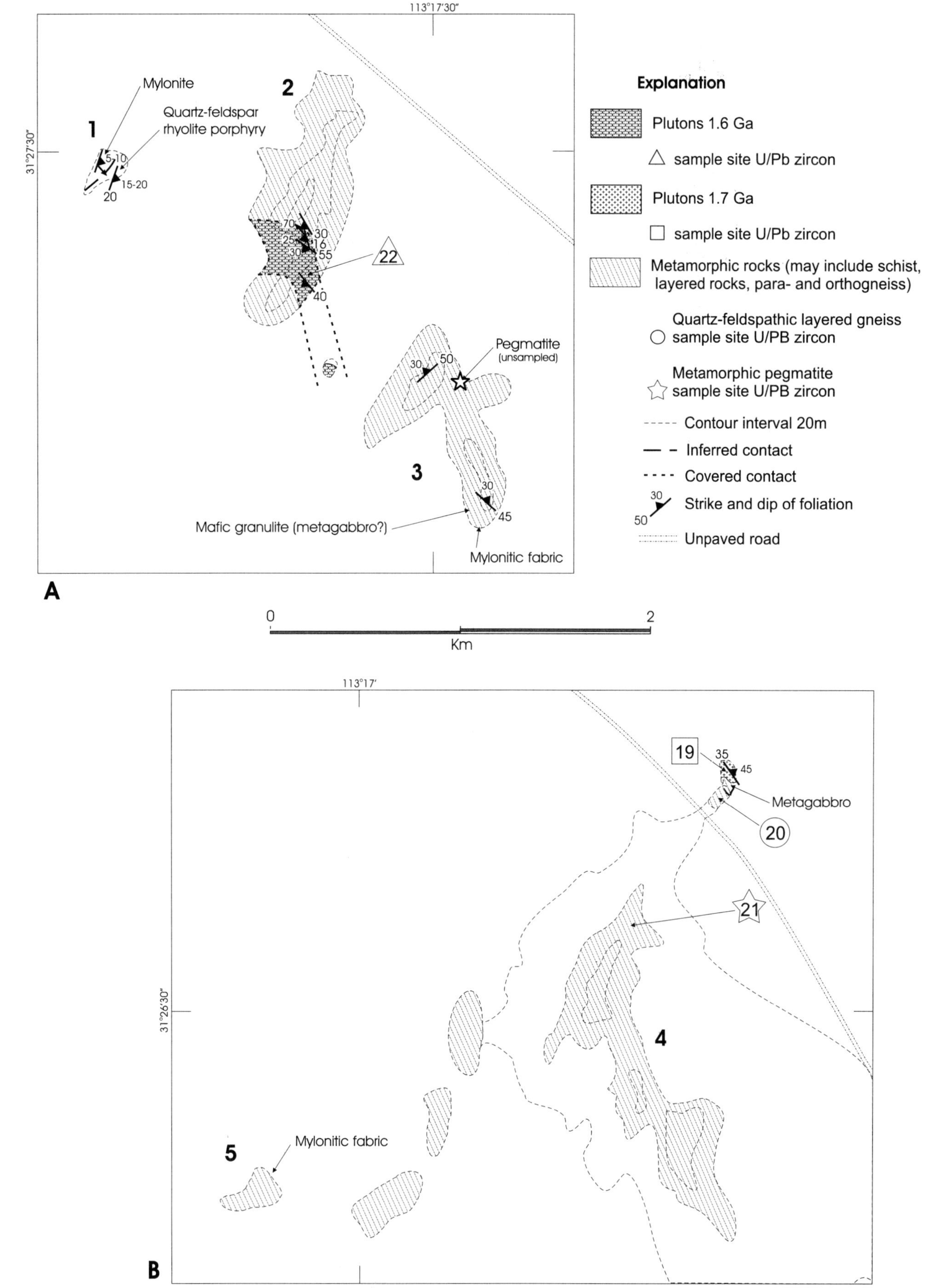

Figure 11 (*on this and previous page*). (A) Geologic sketch map of Seven Hills area, northwestern Sonora. (B) Enlarged map showing: (a) geology of hills 1, 2, and 3; and (b) geology of hills 4 and 5 (Silver, field notes, 1969).

ranges where samples of Proterozoic crystalline rocks correlative with the Yavapai Province were collected. None of these areas was mapped at the time of our visits before 1974. The modified maps of Sierra Santa Rosa and surrounding hills (Hardy, 1973, 1981) and San Antonio region (Rodríguez-Castañeda, 1984, 1988, 1990, 1994) are included to provide an informative context for the sample descriptions. Geological relationships exposed among the Seven Hills are based solely upon our field notes.

Santa Rosa region. About 70 km from Caborca, Sierra Santa Rosa and the surrounding hills, including Cerros Colorado, San Luis, and El Dicacho comprise high ground across a broad north-trending basin east of the Pitiquito-Bámori region (Figs. 4 and 8). Much of the large area dominated by Sierra Santa Rosa was mapped by Hardy (1973, 1981) during his ambitious M.S. thesis work. Although the contacts between major bedrock exposures and alluvium generally trend north, as do the basins intervening between ranges, the strike of principal ridges, bedding and some faults within areas of bedrock is generally northwesterly. Lower Jurassic strata may rest unconformably upon Neoproterozoic or Cambrian strata or directly overlie crystalline basement from which it is separated by low-dipping faults interpreted by Hardy (1973, 1981) as younger on older thrusts. We interpret the dip of bedded units to record tilting of the range toward the southwest. Tilting probably occurred above northwesterly striking listric normal faults. In this perspective, the faults originally mapped as thrusts are gently dipping segments of the listric normal faults above which detachment took place. The detachments with normal slip are probably of early Miocene age, as suggested by common northwest strike of steeper normal faults that is characteristic of this period of extension (cf. Nourse et al., 1994). However, the northwest-striking faults must be transected by a younger set of inferred northerly striking normal faults. The presence of this set of faults is indicated by the topographic expression of the straight western border of the Santa Rosa highlands (i.e., Sierra Santa Rosa, Cerro San Luis, and Sierra San Blas), as well as the narrow basin that intervenes between Sierra Santa Rosa and the elongate, narrow, northerly trending ridge including Cerro Colorado and Garambullo (Fig. 8).

West of Cerro San Luis and at Cerro Colorado, Neoproterozoic strata that rest nonconformably upon basement are overlain by allochthonous Jurassic strata (Fig. 8). In places, as much as 2000 m of Paleozoic and Neoproterozoic strata, normally beneath the Jurassic rocks, are missing in response to the combination of pre-Tertiary erosion and faulting. This stratigraphic relationship is similar to that in the northern part of Cerro Rajon, ~30 km northwest of Santa Rosa, where Lower Jurassic beds unconformably rest upon Middle Cambrian strata (Stewart et al., 1984).

Proterozoic basement in Sierra Santa Rosa is best exposed along the northeast flank of the range. Although Hardy's work was not focused upon the crystalline basement rocks, he recognized quartzose metasediments and foliated gneissic and plutonic units including five principal rock types: (1) light-brown, medium-grained, foliated, muscovite quartzite; (2) dark-gray metaquartzite and black hornfels; (3) medium-grained, gneissic pink granite with gray muscovite schist; (4) "quartzo-feldspathic gneiss porphyry"; and (5) medium- to coarse-grained quartz monzonite (Hardy, 1973, 1981). The description of muscovitic quartzose metasedimentary rocks is comparable to that of the layered suite at Bámori Ranch. Two igneous bodies sampled during our reconnaissance studies are shown on Figure 4. These bodies correspond to Hardy's unit 4, gneiss with feldspar crysts commonly a few centimeters in diameter that may be surrounded by a corona of garnet (Fig. 4, #23; Table 1, San Luis granite gneiss), and unit 5, coarse-grained quartz monzonite with pink orthoclase phenocrysts and aggregates of biotite (Fig. 4, #7; Table 1, Cerro Colorado quartz monzonite #300, discussed here earlier). The interpreted age of the gneiss is ca. 1.7 Ga. (Fig. 2E; Table 1). A third sample (Fig. 4, #24; Table 1, Las Guijas porphyritic granite) from exposures to the east yields an upper-intercept age of ca. 1.74 Ga (Fig. 2E).

The metamorphic overprint and the interpreted age of the San Luis gneiss distinguish this body from the older orthogneisses common to the Caborca block. In 1974, the garnetiferous orthogneiss of the Santa Rosa region was unique in Sonora; the nearest rock with comparable age and lithology of which we had knowledge was the Fenner gneiss of the eastern Mojave Desert (Silver, unpublished data). After 1974, additional study of basement orthogneisses in southern California revealed the presence of similar rocks with sufficiently distinctive isotopic, metamorphic, and chronologic signatures that they have been designated as a separate province, designated as Mojave (Wooden and Miller, 1990; Rämö and Calzia, 1998; Iriondo, 2001, and this paper [Discussion—Mojave rocks]). Additional analysis of the San Luis body (Nourse and Premo, unpublished data) yielded an age of 1.729 ± Ga (Premo, 2004, personal commun.).

The high ground of the Santa Rosa region, distinguished by elevations above 800 m, forms an extensive unnamed upland that is more than 40 km wide. At the northeast corner of the upland, elevations diminish among small (15 × 5 km), crudely elongate basins and ranges, the topographic axes of which trend northwest. Cerro El Dicacho, the highest peak in this area, reaches 955 m. The peak stands above a narrow ridge underlain in part by porphyritic biotite granite cut by shallowly dipping foliation that strikes north. At the sample locality (Fig. 4), 3 km along the ridge southeast of the peak, just east of Las Guijas mine, the granite is homogeneous aside from some narrow mafic-poor zones of granite, parallel to foliation and outcrops of slickensided quartz veins. Nearby to the east (La Virgen mine,) a vein of quartz more than a meter thick cuts well-banded gneiss with intercalations of amphibolite and veins of quartz that record minor folds. The intercept age (1.742 ± 16 Ga) is comparable to the age (1.733 ± 7 Ga) from the same body reported by Iriondo and Premo (unpublished data) and to the ages of plutons to the west (cf. Premo et al., 2003) and to the east (e.g., 1.73 Ga El Creston granite, Valenzuela-Navarro, 2003)

San Antonio region. The San Antonio region (Figs. 4 and 9), named for Rancho San Antonio (one of several in Sonora), is ~70 km southeast of Sierra Santa Rosa and almost 150 km from

Caborca. The area is part of a mountainous block about 40 km wide that lies within the transition from the low, broad valleys surrounding isolated ranges in western Sonora to the high ranges of the volcanic-capped Sierra Madre Occidental. The highest parts of the sierras, which reach almost 2000 m, stand above the narrow, fault-controlled, north-trending valleys of Rios San Miguel (west) and Sonora (east). The principal entrance to the ranch is from the valley of Rio Sonora to the east that follows a narrow arroyo called Cañada Los Jacalitos, which cuts across faulted blocks along the margin of the mountain core. The walls of the arroyo expose massive, unfoliated granite and granodiorite from which samples were collected. Zircon from the undeformed samples (Fig. 4, #18; Table 1, San Antonio (Santa Margarita) micrographic granite; and Fig. 4, #16; Table 1, San Antonio (El Jacalón) granodiorite) yield upper intercept U-Pb ages of ca. 1104 Ma and ca. 1702 Ma, respectively (Table 1; Figs. 2D–E and 9).

The maps by Rodríguez-Castañeda (1984, 1988, 1990, 1994) reveal the stratigraphy and structure of the region (Fig. 9). The plutonic rocks in Cañada Los Jacalitos represent two of three generations of Precambrian intrusives exposed within this up-faulted block. The oldest of the three is the granodiorite (San Antonio sample) mapped as El Jacalón (Castillo-Mendoza, 1987) after a local ranch. The dark-gray hornblende-biotite granodiorite, which weathers light gray, is locally foliated and cut by dikes of microdiorite that are restricted to the body. Granodiorite underlies ~40 km^2 of the western margin of the range.

Micrographic granite, informally designated as Santa Margarita by Rodríguez-Castañeda (1984), is distinguished by the red to pink color of fresh surfaces, feldspathic composition and micrographic texture, observable in thin section and hand specimen. The granite forms the largest body of Precambrian intrusive rock in the San Antonio region, where it underlies some of the highest parts of the ranges. Although the granite may be locally foliated and is fractured and broken by faults, it is generally a massive, unfoliated, highly resistant unit. Rodríguez-Castañeda (1994) notes that Santa Margarita granite intrudes the older El Jacalón granodiorite or may be juxtaposed against the older mass along zones marked by ductile deformation, which he attributes to strike-slip faulting. Jurassic(?) volcanic rocks generally nonconformably overlie the granite, although in places allochthonous masses of detached orthoquartzite rest upon it. Porphyritic granite, called El Tuti, with texture and lithology comparable to the 1.4 Ga Cananea granite (Anderson and Silver, 1977), is exposed in a small window a few tens of kilometers west of the El Jacalón outcrops (Rodríguez-Castañeda, 1990; Fig. 4).

Contacts between country rocks and the Proterozoic plutonic bodies were not observed. Gneiss and schist within a pendant surrounded by Early Tertiary granite is exposed ~3 km southwest of El Jacalón granodiorite. The metamorphic rocks comprise El Alamito complex that includes chlorite-biotite quartzo-feldspathic schist and gneiss cut by mafic and granitic dikes. Gneiss, of intermediate composition, from the complex (Fig. 4, #17; Table 1, Sierra Molino Granodiorite gneiss) yields an upper concordia intercept age of ca. 1.64 Ga (Fig. 2E).

North of the faulted northern margin of the principal mass of Santa Margarita micrographic granite that coincides with the Mojave-Sonora megashear, allochthonous outliers of granite (Rodríguez-Castañeda, 1990) rest upon deformed Oxfordian (Rangin, 1977) strata (Fig. 9). Some of the Jurassic beds record intraformational folds indicative of soft-sediment deformation (M.B. McKee, unpublished mapping). Other allochthonous micrographic granite crops out as much as 40 km north of the Mojave-Sonora megashear (Fig. 4; Stephens et al., 1986). These allochthonous rocks, which are exposed in a northerly trending Tertiary horst, comprise part of La Lamina thrust sheet, which lies above deformed rocks of inferred Late Jurassic age. Rare micrographic granite (Fig. 4, #27; Table 1, Pozito micrographic granite), layered saussurite gabbro and greenschist-grade gneissose leucogranite, granite and subordinate plagioclase-chlorite-hematite schist crop out above the thrust. Two fractions of zircon from the Pozito granite yielded ^{207}Pb-^{206}Pb ages of 1098 and 1039 Ma (Table 1). The micrographic granite and strongly altered gabbroic rocks are highly fractured and locally cleaved but are otherwise undeformed. The metamorphic rocks may be well foliated. Mylonitic rocks are locally preserved within a zone of intense cataclasis and brecciation above the basal thrust. Faults with low to moderate dips as well as anastomosing zones of closely spaced fractures are pervasive.

In addition to the allochthons of micrographic granite, orthoquartzite (principally Los Changos Formation) and rare blocks of gray carbonate may rest upon (Rodríguez-Castañeda, 1990) or occur as exotic masses among (Stephens, 1988) beds of the Oxfordian-Kimmeridgian La Colgada Formation or other Jurassic strata. We believe that the age of emplacement of the allochthons and olistoliths(?) is syn- to postaccumulation of Oxfordian strata. Our interpretation of the age as pre-Cretaceous is based upon: (1) the presence of the masses (olistoliths?) of quartzite among Jurassic strata in the footwall of La Lamina thrust; (2) penecontemporaneous folds and faults in the Oxfordian strata; and (3) the lack of contractional structures in pebbly sandstone bearing Early Cretaceous fossils (Rodríguez-Castañeda, 1984) that locally unconformably overlies the Late Jurassic rocks and provides a minimum age for the deformation of the Oxfordian La Colgada Formation (Rangin, 1977). La Lamina thrust, exposed northwest of Cucurpe along the road to Magdalena, may have been reactivated during Cretaceous contraction (Sosson and Calmus et al., 2000) or Miocene extension (Stewart, 1985, personal commun.).

Seven Hills. Seven Hills is an informal name for an elongate (~S45°E) cluster of small, low, dark-brown hills that protrude from the desert floor southeast of Route 8 ~80 km southwest of Sonoita (Figs. 4 and 11). The hills occur ~16 km along an unpaved road that trends southeast from Route 8 toward Colonia Ortiz Garza. Four of the hills are underlain by metamorphic and plutonic rocks, including layered quartzo-feldspathic gneiss, orthogneiss, and granite. The most northwesterly hill (Fig. 11B, #1) exposes a gently dipping zone of strongly mylonitic rocks that overlie foliated greenschist-grade graywacke and rhyolites of unknown age. The subhorizontal (5–10°SE) mylonite records minor recumbent folds, obscured by strong fracturing, and lineation trending N40°W. The underlying rhyolite(?), which contains porphyroclasts of quartz and feldspar, strikes N20°E and dips 15–20°SE. The most southeasterly hills (Fig. 11A, #6 and #7) expose unfoliated, unmetamorphosed felsic volcanic rocks of probable Cenozoic age.

Hill #2 (Fig. 11B) is composed of folded quartzo-feldspathic gneiss, strongly metamorphosed, as indicated by granoblastic texture of recrystallized grains and common pegmatitic segregations of coarse grains suggesting partial melting. Locally, compositional bands strike ~N30°E. Some reorientation may have taken place above gently dipping zones of breccia that cut the gneiss, which we interpret as faults. The high-grade gneiss has been converted to sericite-biotite phyllonite along zones (N30°W, 70°SW) of ductile deformation that cut the older foliation. Near the southern end of the hill, an equigranular granite dike, roughly 45 m thick, cuts the layered metamorphic sequence along a trace that bends from N70°W to N15°W. The granite is also metamorphosed, likely low-amphibolite(?) or upper-greenschist (?) grade and deformed, as indicated by recrystallized biotite that is aligned along planes of foliation striking N40°W. Despite the overprint, the equigranular, medium-grained granitic texture is still apparent, particularly in the center of the dike where shearing is least developed. A sample (#22, Figs. 11B and 4; Table 1, Mex-Son 7 Hills metagranite #6) collected from a single block in the center of the dike is little modified aside from locally intense epidotization. The dike, which yields an upper intercept age of ca. 1.6 Ga (Fig. 2F) clearly postdates granulite metamorphism, as shown by the lack of layering, absence of pegmatitic segregations, and discordance with foliation in the adjacent gneiss.

Quartzo-feldspathic gneiss locally interlayered with pyroxene granulite, amphibolite and less abundant mafic granulite, which resembles meta-gabbro, underlies hill #3 (Fig. 11B). Foliation swings from N50°E in the north to N45°W at the southeastern tip, where strong shearing has created biotite–muscovite phyllonite and mylonite from gneiss and pegmatitic segregations.

Southwestward the road cuts the extreme northeast corner of hill #4 (Fig. 11B), which is underlain principally by mafic and quartzofeldspathic layered gneiss and less common augen gneiss. The rocks in which excellent granulite texture is preserved show little evidence of retrograde metamorphism. A sample of quartzo-feldspathic gneiss (#20, Figs. 11B and 4; Table 1, Mex-Son 7 Hills granitic gneiss #4), collected from a bulldozed block dislodged from a massive layer with scattered, sparse mafic minerals, yielded an upper intercept age of ca. 1.65 Ga. This age, comparable to the ages of metamorphic pegmatites, suggests the gneisses reached a temperature at which significant lead loss occurred or new zircon formed. A few dozen meters northeast, augen gneiss (#19, Fig. 11B) with closely spaced fractures forms a single ledge that gives the appearance of a crosscutting relationship (although not demonstrable) with the layered units. The augen gneiss and crosscutting gabbroic dikes withstood the granulite metamorphism, as indicated by pegmatitic mafic segregations in the dikes. Feldspar augen, 0.5–1.0 cm, and as much as 25% quartz characterize the gneiss that is foliated (N35°W, 45°NE) but not layered. The gneiss (Table 1, Mex-Son 7 Hills augen gneiss #5; Figure 2F) yields a ^{207}Pb-^{206}Pb age of 1705 Ma from the least-discordant (almost concordant) fraction. Southwest along the main ridge of the hill ~350 m west of the road, segregations in amphibolite form excellent examples of metamorphic pegmatites (#21, Fig. 11B) in which some zircon is visible. A sample was collected in the middle slope between the ridge top and hill base. The pegmatite yields an upper intercept age of ca. 1.64 Ga (Table 1, Mex-Son 7 Hills granulite pegmatite #3; Figure 2F) that is comparable to ages from other pegmatites and some strongly metamorphosed gneisses of the Caborca block. Continuing west to hill #5, low exposures partly obscured by dunes reveal strongly sheared and recrystallized sericitic phyllonite that also underlies the higher ground. Although the original character of the rock is locally unrecognizable, several outcrops contain the transition to granulite gneiss.

The exposures at Seven Hills are the most northwesterly occurrences of rocks typical of 1.7 Ga Caborca block Proterozoic crystalline basement that we identified. Further northwest, post–1.7 Ga plutons become common and high-strain Paleoproterozoic deformation is marked (see Nourse et al., this volume). However, the local zones of northwest-striking, moderately to steeply dipping foliated to mylonitic rock that transect the exposures at Seven Hills also affect the Mesoproterozoic rocks forming Sierra Blanca to the northwest. This range, which stands above basalt flows erupted from the Pinacate volcanic center, contains a highly deformed sequence of porphyritic quartz diorite and amphibolite intruded by a presently deformed porphyritic quartz monzonite. The sparse porphyroclasts of potassium feldspar, which are more common in nearby outcrops, are parallel to foliation oriented N45°W, 60–70°SW. Aplite and pegmatitic dikes are concordant with the foliation, perhaps in response to transposition. Despite the very strong deformation and recrystallization, we infer that the sequence includes metamorphosed volcanics and porphyritic and equigranular plutonic rocks. From the southwest flank of the southeast end of Sierra Blanca, we collected a sample of porphyritic quartz monzonite gneiss (Fig. 4, Locality #8). This rock yielded an upper-intercept age of ca. 1.44 Ga (Table 1; Fig. 2C).

Proterozoic San Gabriel–Like Crust in Northwesternmost Sonora: An Extension of the Caborca Block

In northwesternmost Sonora, west of Sonoita (Fig. 4), crystalline basement and overlying strata of the Caborca block are covered by volcanic rocks erupted from the Pinacate volcano. West of Pinacate (Fig. 4) where crystalline rocks re-emerge, our reconnaissance sampling revealed that orthogneisses record ages younger than those of the Caborca basement to the south (cf. Nourse et al., this volume). Samples collected during the first year of study in 1969 included a quartz diorite gneiss exposed adjacent to Route 2 on the south side of the highway near the northwest end of Sierra Alacranes (Fig. 4, #12; Table 1, Route 2-Km 2657 quartz diorite gneiss). The gneiss, xenoliths, and cross-cutting dikes are foliated and metamorphosed, as shown by the presence of segregations that we interpret as the result of partial melting. Phyllonite and other mylonitic rocks are not present at this locality. This rock and two other porphyritic orthogneisses (Fig. 4, locations #13, #10; Table 1, Cerro del Viejo augen gneiss #102 and Andrade augen gneiss #1) fall along a discordia chord that intersects concordia at ca. 1.66 Ga. At the time of the 1970s work, we compared the ages and petrographic and structural characteristics of the Route 2 rocks with augen gneiss in ranges of the Mojave Desert of southeastern California (e.g., Big Chuckwalla, Little Chuckwalla, and Orocopia Mountains) and San Gabriel Mountains. In California the gneissic rocks and rare intrusive bodies of anorthosite and associated rocks and Triassic plutons were designated informally as San Gabriel terrane (e.g., Silver et al., 1963; Silver, 1971, 1982). As briefly discussed below and by Nourse et al. (this volume), the boundary in Sonora between the Caborca and San Gabriel–like rocks does not separate rocks formed during distinct crust-forming episodes. Rather the boundary probably reflects the overprint of ca. 1.65–1.60 Ga plutonism, deformation, and metamorphism on 1.7+ Ga rocks, some of which are coeval with Caborca/Yavapai rocks.

Our 1974 working hypothesis was that these complex San Gabriel rocks lie southwest of the megashear and are an additional tectonostratigraphic element that, combined with the Caborca block, forms a composite terrane (Fig. 1). U-Pb interpreted ages from igneous rocks in the hills of Sierra Los Tanques, west of Sonoita (Silver, unpublished data; Anderson and Campbell, 1992), combined with recent mapping (Campbell and Anderson, 2003) reveal plutons of Triassic age intruded into gneiss. These data indicate that at least Triassic plutonic rocks, which may be part of the San Gabriel terrane, extend southeastward to Route 8, where they are juxtaposed against Jurassic strata. The absence of the foliation and intrusive bodies of ca. 1.6 Ga orthogneiss, common to San Gabriel terrane, among the older Caborca basement (1.7+ Ga) units at Seven Hills suggests that the position of the contact between Caborca and San Gabriel terranes lies in the vicinity of Route 8, between Sonoita and Puerto Peñasco (Fig. 1). The effect of 1.6 Ga metamorphism is prominent in much of the Caborca block, where it is manifest by metamorphic zircon in pegmatites and paragneisses (e.g., see Seven Hills).

Recent mapping and U-Pb analyses by Nourse et al. (this volume) among ranges flanking Route 2 provides information about relations among the layered gneisses and deformed intrusive bodies of the San Gabriel terrane. Study of single zircons and whole-rock Nd signatures indicates that some augen gneiss in the San Gabriel suite may be derived from plutons similar to those of the 1.7+ Ga Yavapai Province (Premo et al., 2003; Nourse et al., this volume). However, we did not recognize the older plutons. Our sample of Sierra Viejo orthogneiss is lithologically and texturally similar to 1722 ± 19 Ma "augen gneiss" analyzed by Nourse et al. (this volume). Our single analysis of the Andrade sample is from their Pilot Knob locality of Premo et al. (2003) that yielded an age of 1672 ± 11 Ma. As shown in Nourse et al. (this volume) a common denominator of the 1.70–1.73 Ga samples near Route 2 is a strong discordance in the conventional zircon analyses, very similar to the behavior of our samples #12 and #13.

Mesozoic Rocks of the Caborca Block: An Additional Distinguishing Stratigraphic Element

Similarities between Precambrian and Cambrian rocks on the Caborca block and near Death Valley, California, spurred comparisons among stratigraphically higher Paleozoic and Mesozoic formations. However information about post-Cambrian, pre-Cretaceous units was limited because few fossiliferous exposures were known. The exception is found west-northwest of Caborca at Sierra del Alamo where Permian rocks are overlain by a thick section of Triassic and Jurassic strata. Fossiliferous limestone interbedded with shale and coarser clastic rocks had been assigned to the Late Triassic on the basis of ammonites (White and Guiza, 1948). Coal-bearing sandstone-rich units that crop out southeast of Hermosillo (King, 1939) contain plant fossils that are correlative (Alencaster de Cserna, 1961a, 1961b; Pineda 1961). Early Jurassic shale and sandstone that comprise a thick section overlying the Late Triassic strata at Sierra del Alamo had been mapped in ranges (e.g., Sierra Santa Rosa; Hardy, 1973) across the Caborca block to the east based upon the presence of the fossil Weyla. Lithostratigraphic equivalents of Late Triassic marine rocks and overlying Early Jurassic of the Caborca block had been mapped by Merriam (1963) in Darwin Wash.

The Triassic and Jurassic strata that crop out on the Caborca block have since been studied extensively as described in this volume (see contributions by Stewart; Marzolf and Anderson; González-León et al., and Anderson et al., Chapter 2, for latest results, summaries, and references).

PRECAMBRIAN ROCKS OF THE PINAL PROVINCE, NORTHEASTERN SONORA

Exposures of Proterozoic crystalline rocks that distinguish the Caborca block abruptly terminate along the inferred megashear. Across this discontinuity to the northeast, mid-Jurassic volcanic and plutonic rocks that are commonly deformed distinguish a belt of variable width, generally a few tens of kilometers wide. Haxel et al. (1984) noted the absence of exposures of autochthonous pre-Jurassic sedimentary rocks and crystalline basement among exposures of Jurassic igneous and sedimentary rocks and designated the region, which encompasses 20,000 km^2 within south-central Arizona and north-central Sonora, as the Papago terrane (see Anderson et al., this volume, Chapter 2, for discussion of the use of Papago "terrane").

Northeast of the Papago terrane, metamorphic basement crops out as micaceous, schistose rocks generally at the margins of large 1.4 Ga plutons. The metasedimentary rocks resemble units that at the time were grouped under the name Pinal Schist.

In Arizona, U-Pb interpreted ages of units from Pinal Schist and equivalent stratified units indicated accumulation between 1.7 and 1.675 Ga (Silver, 1978; Conway and Silver, 1989). The Pinal rocks are distinctly younger than the stratified section composing part of the crystalline basement of the Caborca block. Two suites of plutons, commonly unfoliated, intrude the Pinal Schist or Pinal Province (Cooper and Silver, 1964; Conway and Silver, 1989). Older bodies yield ages within the interval from 1.64 to 1.615 Ga, whereas younger, regionally widespread plutons and batholiths of porphyritic granite yield ages between 1.45 and 1.425 Ga.

Recent, additional studies reveal a more complex history for the igneous and tectonic evolution of the southern Arizona Precambrian basement (Eisele and Isachsen, 2001; Swift and Force, 2001). Within the Pinal Province, two blocks are distinguished: (1) Pinal block to the northwest, and (2) Cochise block to the southeast. Sedimentation within the Pinal block occurred between 1.68 and 1.65 Ga, whereas volcanic activity in the younger Cochise block is recorded between 1.647 and 1.630 Ga (Eisele and Isachsen, 2001). The boundary between the blocks, which is irregular, has not been traced into Sonora, although the Cochise block appears to extend southward across the international border as shown by Eisele and Isachsen (their Figs. 1 and 7, 2001). This new work does not alter the contrast between ages of volcanism and plutonism between the Pinal and Yavapai Provinces shown by earlier studies (e.g., Silver, 1965, 1969, 1978).

In northern Sonora we sampled an outcrop of volcaniclastic beds exposed in a pass through Sierra Los Ajos, west of Rancho Mababi that includes a rhyolitic tuffaceous layer containing quartz phenocrysts (Fig. 4, #25; Table 1, Mababi metamorphic rhyolite). The layers at this exposure generally strike N50°E with local deviations to a more northerly strike. The dip is moderate, commonly 45°, toward the northwest. The rock is sufficiently fine grained so that hinges of crinkles (0°, N35°E), a lineation (30°, N80°W), and rare kink bands (N10°E, 83°NW) are recorded. Zircons from the rhyolitic layer yield an upper-intercept age of ca. 1.69 Ga (Table 1, Fig. 2H). In light of the recent studies in Arizona that suggest ages of volcanism younger than 1.68 Ga for the Pinal Province, this slightly older age may reflect the presence of older detrital zircon within the tuffaceous layer.

In northeastern Sonora widespread, 1.4 Ga plutons are by far the most abundant intrusive rocks. The extent of exposures of porphyritic granite may indicate the presence of a batholith-size body that contains screens and pendants of schist. The batholith-size mass corresponds with much of the Cananea high (McKee and Anderson, 1998). Exposures at Cananea (Fig. 4, #28 and #29; Anderson and Silver, 1977) yielded an age of ca. 1.43 Ga. (Table 1, Cananea porphyritic quartz monzonite #5 and Cananea porphyritic quartz monzonite #2 [contact zone]; Fig. 2H). Samples from exposures between Cananea and Nacozari (Fig. 4, #30, #31, #33) yield similar ages (Table 1, Cuquiriachic granodiorite #3, Cuquiriachic quartz monzonite #501 and Valdeza porphyritic quartz monzonite #4; Fig. 2H). Farther northeast, the 1.4 Ga granite from Pulpito (Fig. 4, #32; Table 1, Pulpito granite #1) may mark the limit of Precambrian exposures at the western margin of the Sierra Madre Occidental. Subsequent field studies reveal the presence of similar porphyritic granite in Sierras Mesteñas and Cenizos (Fig. 4). The 1.4 Ga porphyritic plutonic rocks do not distinguish Yavapai or Pinal crustal provinces, as they are widespread and occur in both crustal provinces (e.g., Silver et al., 1977; Anderson, 1989). The plutons do however indicate the presence of Proterozoic crust, and our working hypothesis is that in northeastern Sonora they reveal the extent of crust-forming rocks temporally correlative with Pinal Schist.

Plutonic rocks cogenetic with the formation of the Pinal-type crust are rare. Johnny Lyon granodiorite in the Johnny Lyon Hills, Arizona, is typical of this suite (Silver and Deutsch, 1963; Cooper and Silver, 1964). A single, isolated outcrop of this group may occur in Sonora. The small exposure composed of quartz diorite that crops out in a narrow arroyo on the east flank of Cerro Prieto (Fig. 4, #26) that yields a Pb^{207}/Pb^{206} age of 1.612 Ga may be the most southwesterly exposure of Pinal Province rocks in central Sonora. Our interpretation has been that this outcrop delimits the southeastern boundary of the province. However, in light of additional work (e.g., Iriondo, 2001) an alternative explanation is that the quartz diorite is a fragment of Mojave Province. If so, it is probably allochthonous because of its position among thrusts that transported crystalline rocks such as Pozito micrographic granite from the Caborca block northward (see above).

In northwestern Sonora, 1.4 Ga porphyritic granite (Fig. 4, #6) intrudes and metamorphoses siliceous paragneiss along Route 2 (Fig. 4, #11). Both samples yield ages of ca. 1.4 Ga (Table 1, Porphyritic granite #4 and Route 2-Km 2627 metarhyolite(?) gneiss #4; Fig. 2G–H). We originally thought that these units were within the Caborca block because of the proximity to 1.7 Ga granite (Damon, K/Ar, unpublished data) along Route 15. However, Nourse et al. (this volume) include the 1.4 Ga granite and wall rocks as part of the Mazatzal or Pinal Province based upon their recent mapping and radiometric studies (Nourse et al., 2003). Our age (1437 ± 11 Ma) for the granite (Table 1; Fig. 2G) agrees well with the results from the same pluton (1432 ± 6 Ma) presented by Nourse et al. (this volume).

CHRONOLOGIC DEVELOPMENT OF THE MOJAVE-SONORA MEGASHEAR HYPOTHESIS: STUDIES AFTER 1974

Plate Tectonic Analysis

Silver (Silver and Anderson, 1974; Fig. 12) recognized the possibility that a megashear offered a solution to the problem of crustal overlap encountered with the straightforward Pangea reconstruction of Bullard et al. (1965). During the late 1970s Anderson conducted a review of regional geologic studies in Mexico with the hope of recognizing disrupted patterns of pre-Cretaceous rocks in Mexico that might record offset of the magnitude and sense of displacement postulated for the megashear. Earlier workers (e.g., Murray, 1959, 1961; de Cserna, 1970) had noted that Paleozoic rocks, extending southwestward from

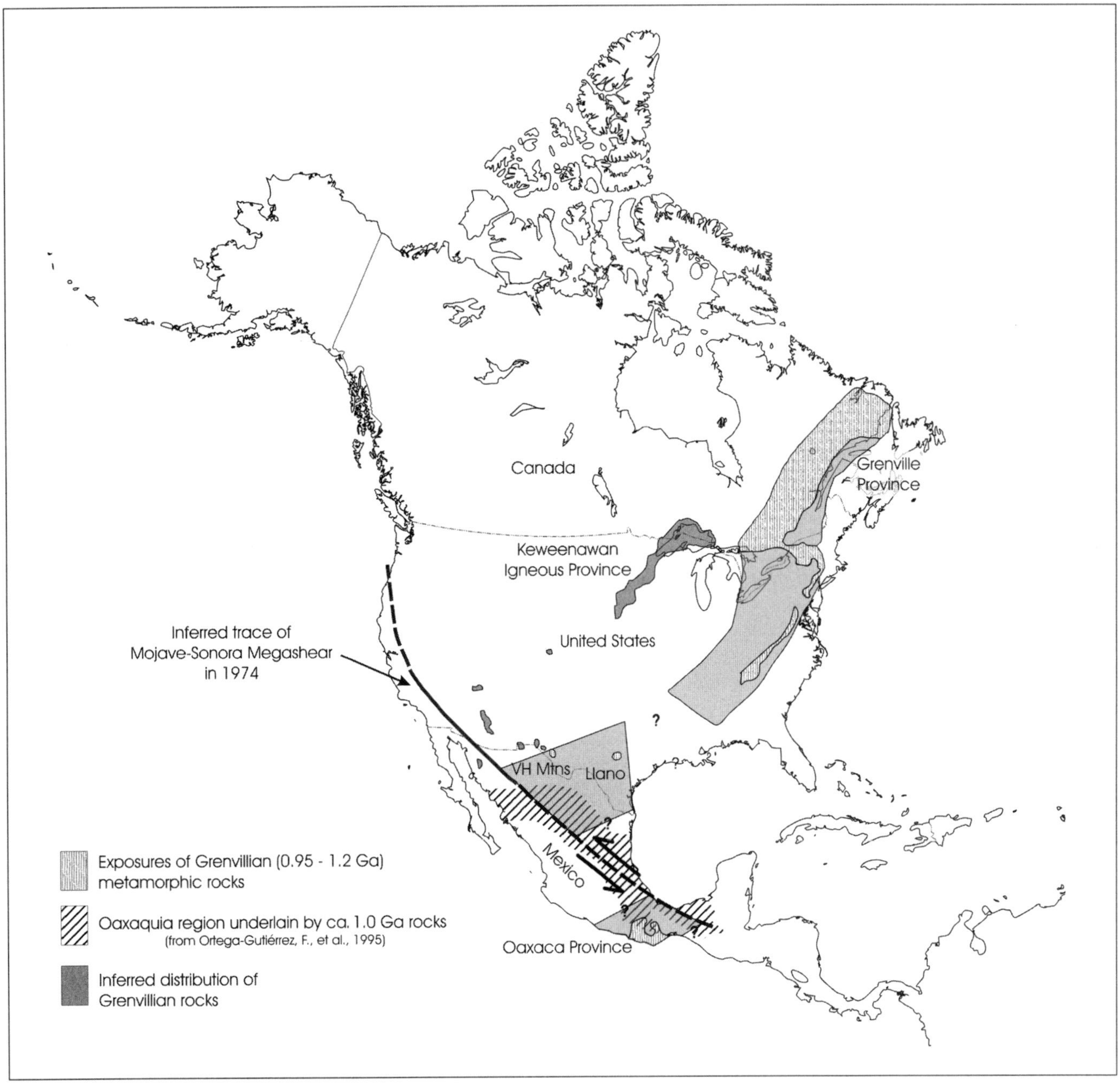

Figure 12. Outline map of southwestern North America showing the initial inferred regional trace of the Mojave-Sonora megashear and the proposed offset of Oaxacan Grenvillian rocks (Silver and Anderson, 1974).

Texas, are not known in central Mexico. The disruption and pattern of pre-Cretaceous rocks following the great bend in the Sierra Madre Oriental, delineated by deeply etched, folded rocks that swing westward south of the Coahuila Peninsula, suggested a feasible eastward inferred path of the Mojave-Sonora megashear.

Anderson and Schmidt (1983) assessed the viability of the proposed eastward extension of the Mojave-Sonora megashear by: (1) determining the pole of rotation and comparing it with the pole for the opening of the Atlantic (Le Pichon and Fox, 1971; Klitgord and Schouten, 1980), and (2) testing the efficacy of restoration of the proposed displacement along the fault with respect to the elimination of the overlap of part of Mexico across northern South America that is revealed in the straightforward Pangea configuration of North and South America (Bullard et al., 1965). Schmidt's construction of oblique Mercator projections from the stage pole of motion for North America (lat. 52°N and long. 79°W) (see Klitgord and Schouten, 1986), calculated for the time interval ca. 160–148 Ma from points along the Mojave-Sonora megashear, reveals two additional linear zones parallel to the Mojave-Sonora megashear that coincide in part with major crustal features (Trans-Mexico volcanic belt; Acapulco trench) in southern Mexico (Anderson and Schmidt, 1983, their Figs. 2 and 3). These zones, interpreted to be transforms related to the megashear set, define three blocks that, if moved to eliminate overlap of the continents, reveal 2400 km of cumulative sinistral displacement.

Field Studies in Eastern Mexico

Anderson and Schmidt (1983) postulated that the megashear curves eastward across northern Mexico, and constrained the inferred fault trace with the distribution of pre–late Jurassic rocks, especially those in the vicinity of the Monterrey salient, distinguished by a great westward swing in the northerly strike of the Sierra Madre Oriental in eastern Mexico. North of the Monterrey salient, Paleozoic rocks that underlie the southern part of the Coahuila block (also described as a peninsula or island; see Figs. 3 and 7 of Anderson and Schmidt, 1983) do not extend southward as shown by outcrops of pre-Cretaceous Mesozoic rocks exposed in the San Julian uplift across the salient. They inferred that the trace of the Mojave-Sonora megashear coincides with the apparent stratigraphic discontinuity between pre-Cretaceous rocks across the salient.

During the late 1970s one of the authors (Anderson) learned of regional mapping being conducted by J. McKee and N. Jones in northeastern Mexico. An outcome of their work in the state of Coahuila is the recognition that initial activity along the regional San Marcos fault, a west-northwest–striking structure bounding the northern margin of the Mesozoic Coahuila Island, was contemporaneous with the age proposed for the megashear (McKee et al., 1990). Regional studies involving McKee, Jones, and Anderson were initiated (e.g., Anderson et al., 1991b; Jones et al., 1995, McKee et al., 1999; Anderson et al., this volume, Chapter 16) to further assess the significance of Late Jurassic faulting by testing the existence of the postulated stratigraphic discontinuity between the Coahuila block and the San Julian uplift to the south.

The initial cooperative study focused upon late Paleozoic, arc-proximal marine strata and related intrusive rocks, named the Las Delicias Formation, that underlie at least 6000 km^2 of the Coahuila block, a term often used for the southern part of the paleogeographic feature known as the Coahuila peninsula or island (McKee et al., 1999, see their figures 25 and 26). The Coahuila block is bounded by faults: to the north, the San Marcos (McKee et al., 1990; Anderson and Nourse, this volume) and to the south, the inferred Mojave-Sonora megashear (Anderson and Schmidt, 1983). The block is appropriately designated as a stratigraphic terrane distinguished by the Las Delicias Formation, the oldest exposed unit.

South of the Delicias terrane exposed in the Coahuila block, across the inferred megashear, no autochthonous Paleozoic or older rocks are known within the San Julian uplift, the principal area of pre-Cretaceous exposures. Among the hills of the uplift, thousands of meters of flows, breccia, laharic conglomerate, air-fall tuff, ash-flow tuff, and tuffaceous siltstone, which are locally strongly deformed, are included within Caopas, Nazas, and Rodeo Formations of probable Middle Jurassic age (Anderson et al., 1991b; Jones et al., 1995).

In the southern part of the San Julian uplift a mudstone-rich body of rock, called Taray Formation, that is characterized by dismembered beds of sandstone and shale and fragments of a great range of sizes in a finer matrix, underlies the volcanogenic rocks (Anderson et al., this volume, Chapter 16). The Taray is interpreted to have formed as part of a Jurassic(?) accretionary prism (López-Infanzón, 1986; Anderson et al., this volume, Chapter 16), although there is no direct evidence of age. The only fossils are Paleozoic ones from exotic (?) clasts of limestone. Its age may be judged by the circumstance of structural and apparent stratigraphic proximity of this to the probable Jurassic volcanic rocks that are logical associates with the mélange absent a commonly intervening forearc basin.

The tectonostratigraphic sequence at this latitude (southwest to northeast) of accretionary prism (Taray), Jurassic(?) arc (Nazas, Rodeo, Caopas Formations), and overlying clastic cover (La Joya) does not crop out northward across the Monterrey salient and must either turn abruptly westward along the southern edge of the Coahuila Island or be truncated and offset along the megashear. Similarities with respect to age, lithology, and chemical character of the volcanogenic strata of northeastern Mexico and those of the Jurassic belt in southwestern North America led Jones et al. (1995) to propose that the northeastern Mexico volcanic units are a fragment of the Jurassic Cordilleran magmatic arc that had been truncated and translated from Sonora southeastward along the megashear.

Although the megashear has been inferred to roughly parallel the Parras basin axis along the Monterrey salient, wells that penetrate Jurassic units (Grajales Nishimura et al., 1992) constrain the subsurface trace to a position north of the basin, just north of the southern edge of the Coahuila block. Analysis and interpretation of gravity data reveal a distinct contrast even further north, under the block (Anderson et al., 1997). The apparent imbrication of the Las Delicias rocks onto the Jurassic arc and the arc onto the accretionary prism at San Julian (Anderson et al., 1991b) may have occurred in response to contraction between the buttress-like Coahuila block and the Jurassic units during southwestward-directed transpression.

Analysis of Microstructures in Steeply Dipping Mylonite along the Postulated Fault Trace

In 1980, we discovered exposures of mylonite southwest of Sonoita, Sonora; pertinent ages were reported by Silver and Anderson (1983). Three exposures of mylonite that crop out along the postulated fault occur within Sierra de Los Tanques, west of the village of Sonoita (Campbell and Anderson, 1996, 2003). These exposures, which extend for 25 km and cross

Mexico Route 8, 13 km southwest of Sonoita, reveal mylonitic crystalline rocks with steep foliations and shallow lineations. Cretaceous plutons (ca. 84 and 70 Ma, U-Pb zircon; unpublished data) intrude the mylonitic rocks. Away from the intrusions, mylonitic rocks have similar metamorphic mineral assemblages and microstructures, indicating that they formed under similar deformation conditions.

The best exposure is in the southeastern part of the study area where a tripartite assemblage consists of: (1) foliated, locally mylonitic Jurassic volcanic and volcaniclastic strata to the northeast, (2) an axial zone of mylonitic rocks, and (3) masses of granitic rocks and compositionally banded quartzo-feldspathic gneiss of Proterozoic(?) and Middle Triassic age. Mylonitic foliation composed of northwest-striking S and C foliations pervades the axial zone; lineations generally plunge shallowly (Campbell and Anderson, 2003). Microstructures record left-lateral shear. Where foliation dips moderately within the northwestern part of the axial zone, restoration of tilted, nonconformably overlying (Miocene?) volcanic layers to horizontal brings the foliation to vertical. Local contractional deformation, recorded by shallow thrusts, folded mylonite, and steep mylonitic foliation with downdip mineral lineation, reveals transpression and development of small-scale flower structures. Thrusts that place Triassic and Precambrian crystalline rocks onto ductilely deformed Jurassic rocks are more numerous and better developed northeast of the principal fault. Along this segment of the megashear, the juxtaposition of weak Jurassic volcanic arc rocks against rigid felsic Precambrian and Triassic crystalline rocks may have facilitated the movement of material northeastward, out of the zone of strike-slip motion.

In the central and northwestern exposures some outcrops of pervasively foliated and lineated rocks are cut by Cretaceous (ca. 84 Ma) granite (Aguajito Springs granite of Haxel et al., 1984) that may be weakly foliated. Some inclusions of mylonite cut by granite record left-lateral shear. The structural fabric of the Triassic and Precambrian mylonitic rocks is interpreted to have formed during left-lateral strike-slip motion and accompanying transpression along the Mojave-Sonora megashear. The northwesterly strike of steep foliation, left-lateral kinematics, and crosscutting relationships with Cretaceous intrusive rocks preclude correlation with younger episodes of Cretaceous contraction and Tertiary extension that are recognized in the southwestern United States and northwestern Mexico.

Paleomagnetic Study

Three Upper Triassic–Lower Jurassic formations in the southwestern United States and northwestern Mexico were investigated paleomagnetically as a means of testing for displacement along the proposed megashear (Cohen et al., 1986). The Luning Formation of west-central Nevada and the Sil Nakya Formation of southern Arizona lie north of the megashear, whereas in northwestern Sonora, the Antimonio Formation lies south of the megashear. Results from the Nazas Formation of presumed Lower Jurassic age from north-central Mexico were incorporated into the interpretations (Nairn, 1976). These four units comprise two sets of correlative formations on opposite sides of the proposed Mojave-Sonora megashear. If Triassic-Jurassic paleomagnetic poles from the Luning and Sil Nakya Formations north of the proposed megashear are compared with poles from the equivalent Antimonio and Nazas Formations south of the megashear, the four poles are statistically indistinguishable from one another at the 95% confidence level, both before and after proposed displacement along the megashear is restored. Nevertheless, the improvement in grouping of the poles after restoration of the displacement according to the model of Anderson and Schmidt (1983) is sufficient to indicate at the 95% confidence level that the two sets of poles are not in fact drawn from the same distribution of poles. The apparent contradiction between the two statistical conclusions may well arise from the very conservative application of Fisher statistics to the formation poles. As an example, the alpha 95 for the Antimonio Formation is based upon use of $N = 3$ (number of localities) rather than upon $N = 19$ (number of sites) or $N = 115$ (number of specimens). These results (Cohen et al., 1986) are interpreted as a strong suggestion that the postulated displacement along the megashear did occur, but not as convincing proof.

Recognition of Transtensional and Transpressional Structures Associated with the Megashear

Widely distributed fault-controlled basins that began to fill with clastic debris in Late Jurassic time are interpreted to be pull-apart basins cogenetic with displacement along the megashear (see Anderson and Nourse, this volume). Folds and thrusts, especially well developed in Jurassic sedimentary and volcanic rocks along the fault record transpression. In Sonora the initial recognition of steep-sided fault blocks attributable to Late Jurassic normal faulting was made by M.B. McKee (1991) who mapped gravity-driven slides of Lower Cretaceous strata along the flank of the Cananea high. She (McKee and Anderson, 1998) (1) postulated that the Cananea high formed before accumulation of the Lower Cretaceous strata and that they moved down the flank of the high in response to Late Cretaceous tectonism, and (2) compared the block-like uplift to similar highlands bounded by steep faults described by Ransome (1904). Nourse (1995) recognized a basin of similar origin as a result of mapping in Sierras El Batamote and El Alamo near Altar, Sonora, Mexico. He attributed the formation of basins in northern Sonora, in which rocks of the Bisbee Group and equivalent strata accumulated, to transtension cogenetic with the Mojave-Sonora megashear.

The transtension idea was first proposed by Harding and Coney (1985) who postulated that a transtensional rift basin, bounded along the southwest margin by the Mojave-Sonora megashear, enclosed the McCoy Mountains Formation and equivalent strata of Apache Wash exposed in Livingston Hills and southern Plomosa, Dome Rock, McCoy, Palen and Coxcomb Mountains in western Arizona and southern California. Anderson and Nourse (this volume) argue that the sedimentary basins are widespread and that they may be recognized by: (1) linked NW- and E-W–striking faults, that commonly delineate partially preserved rhomb-like outlines corresponding to the perimeters of outcrops of conglomerate correlative with Glance and Apache Wash, (2) the Late Jurassic age of strata in the basins as shown by interpreted U-Pb zircon analyses (Fackler-Adams et al., 1997) and diverse fossil assemblages (Lawton and Olmstead, 1995), and (3) the geochemical characteristics of stratigraphically low volcanic flows and tuffs within the basins (e.g., Laubach et al., 1987). The northwest- and east-trending sets of faults and a less-developed third set, composed of northeast-striking right-lateral faults (e.g., Cooper and Silver, 1964; Kluth, 1983), comprise a cogenetic system that is kinematically and chronologically compatible with the Mojave-Sonora megashear. Conglomerate units correlative with basal strata exposed in the McCoy Mountains (McCoy Mountains Formation) and south of Bisbee, Arizona (Glance conglomerate), distinguish major pull-apart basins inboard of the Mojave-Sonora megashear from Texas to California. In addition to the basins mentioned above, other pull-apart structures are recognized at: Plomosa Mountains (Richard et al., 1993), Imuris, Sonora (Nourse, 1995), Chiricahua Mountains (Lawton and Olmstead, 1995), within the Chihuahua Trough (de Cserna, 1970; DeFord and Haenggi, 1970), Mina Plomosas–Placer de Guadalupe (Bridges, 1964), and Valle San Marcos within La Mula Basin (McKee et al., 1990).

Thrusts and folds, some of which are strongly overturned, are well developed in clastic strata of the Jurassic arc assemblage along the trace of the megashear (Corona, 1979; Anderson et al., 1991a; Stephens, 1988; Chepega, 1987; Connors, 1990). These studies show that locally crystalline rocks, some of which correlate lithologically and temporally with Proterozoic units of the Caborca block southwest of the Mojave-Sonora megashear, are imbricated with and thrust over the Jurassic strata to the northeast (Anderson et al., 1984). The most pronounced example of a transpressional thrust underlies allochthonous crystalline rocks at La Lamina Ranch, 40 km from the megashear (Stephens et al., 1986; Fig. 4; also see above, San Antonio region micrographic granite).

Other small allochthons of micrographic granite also occur north of the Mojave-Sonora megashear west of Sonoita where they have been mapped by Campbell and Anderson (2003). These displaced masses may represent erosional remnants of a larger thrust mass (Campbell and Anderson, 1996) that includes the rocks above the Quitobaquito thrust (Haxel et al., 1984). Field relations mapped along the inferred Mojave-Sonora megashear near Route 8 further show outliers south of the fault, suggesting the presence of a positive flower (palm) structure (Campbell and Anderson, 2003).

In Zacatecas, eastern Mexico, Anderson et al. (1991) mapped contractional structures tens of kilometers south of the inferred trace of the Mojave-Sonora megashear that they attributed to transpression. The juxtaposition of Jurassic volcanic rocks (Jones et al., 1995) above mélange (Anderson et al., this volume, Chapter 16) probably occurred in response to shortening during this Late Jurassic deformational event.

Study of the Effects of Extensional (Tertiary) and Contractional (Cretaceous) Overprints on the Paleogeography of the Postulated Fault Trace

Recognition of Tertiary gneiss during early reconnaissance studies (Anderson et al., 1980) and folded and faulted Oxfordian strata (Rangin, 1977) stimulated us to assess the effects of overprinted deformation upon the megashear and related structures. Ductile and brittle deformation of early Miocene age in Sonora is characterized and distinguished by Nourse et al. (1994). A principal conclusion of this work is that Jurassic strata in northwestern Mexico that crop out northeast of the megashear commonly form the footwall of detachment faults. Metamorphism of the footwall rocks diminishes in intensity westward from Imuris to the ranges north of Caborca. Characteristically lineated (N60°E) rocks (Nourse et al., 1994) are not known south of the megashear in western Sonora, although northwest-striking normal faults are prominent locally (DeJong et al., 1988; see also discussion of Santa Rosa region).

Southeast of Quitovac, outcrops of Jurassic volcanic strata form a bulge that protrudes westward, distorting the trace of the inferred megashear. Within these exposures a narrow NW-trending basin that preserves Tertiary strata suggests formation during early Miocene extension (Corona, 1979; Nourse, 2001). Later Miocene extension, manifested as northerly trending basins and ranges, overprints the older structures and locally obscures them. Dembosky and Anderson (this volume) attempt to assess the effects of "basin and range" extension in northwesternmost Sonora and create an early Miocene paleogeography by restoring displacements along cogenetic normal and strike-slip faults.

Cretaceous deformation has been most studied along the megashear between Quitovac and Tuape (De Jong et al., 1988; Jacques-Ayala et al., 1990; Iriondo, 2001; Nourse, 2001; Rodríguez-Castañeda, 1990). Nourse (2001) argued that some strongly folded and foliated rocks are fill that accumulated in a pull-apart basin along the megashear, and he related the deformation to inversion of the basin along reactivated structures. Strong disharmonic folding in Early Cretaceous strata southwest of Cananea is attributed to basinward gravity sliding down the flank of a major Late Jurassic horst (McKee and Anderson, 1998) followed by basin inversion (Rodríguez-Castañeda et al., 1997; Rodríguez-Castañeda, 2000). A full discussion of the effects of Cretaceous deformation of strata within pull-apart basins that formed in Proterozoic crust northeast of the megashear may be found in McKee et al. (this volume).

Additional Studies of Lower Mesozoic Strata Near El Antimonio

In 1974, comparisons between the section on the Caborca block and that exposed near Death Valley focused upon the similarities of lithostratigraphy and thickness of the late Precambrian and early Cambrian units. Younger units are less well known or are poorly exposed. At Sierra del Alamo where the early Mesozoic section is best preserved in northwestern Sonora, stratigraphers had recognized Permian rocks overlain by a thick section of Late Triassic and Early Jurassic strata. John Marzolf (1992, personal commun.) recognized that the absence of Early Triassic formations would be in clear contrast to Lower Mesozoic sections near Death Valley, where Early Triassic strata are well known.

The recognition of possible differences between strata at Sierra del Alamo and Death Valley as well as the desire to collect sufficient data so that detailed comparisons could be made among early Mesozoic units prompted additional stratigraphic and paleontologic work (e.g., Stanley and González-León, 1995; González-León, 1997; González-León and Stanley, 1997 and references therein). A significant result of the new work is the recognition of Early and Middle Triassic strata (Lucas et al., 1997) at Sierra del Alamo and the development of paleogeographic models into which the new data are incorporated (Marzolf, 1997). In summary, new paleontologic (Stanley and González-León, 1995; Stanley, 1997) and stratigraphic information are compatible with the postulated correlation of Caborca block Mesozoic cover with equivalent units near Death Valley, although they do not prove correlation.

Studies of Detrital Zircons on the Caborca Block

Suites of detrital zircons from Proterozoic quartzite exposed south of Cerro Berruga, near Bámori Ranch on the Caborca block yield apparent $^{207}Pb/^{206}Pb$ ages of ca. 1.8 and 2.1 Ga (Table 1). Suites of detrital zircons from quartzite interbedded with local volcanic units of the Mazatzal (ca. 1.7 Ga) and Yavapai (ca. 1.8) Proterozoic crustal provinces in Arizona and New Mexico yield sedimentary ages that correlate closely with the volcanic units (Silver, unpublished data). The isotopic ratios from the suites of mixed populations of detrital zircons from the Caborca block indicate a contribution from Archean sources sufficient to increase the isotopic ratios to unexpected values. No sources of Archean zircons are known in Mexico or in adjacent parts of the United States. In 1974 comparable sedimentary ages from zircon were known only from Proterozoic quartzite beds of the World Beater complex from southern Death Valley, closer to the Archean crust of the Wyoming Province. The presence of Archean detrital zircon was later verified by Wooden and Miller (1990). As noted above, we consider the results from detrital zircons to add a significant additional link between the rocks of Caborca and Death Valley.

Additional studies of detrital zircons from Neoproterozoic and Phanerozoic miogeoclinal and eugeoclinal units in Sonora and Nevada augment our sparse initial data and expand the bases for comparison of widely separated units (Gehrels and Stewart, 1998; Stewart et al., 2001; see also Farmer et al., and González-León et al., this volume). The similarities in the populations of detrital zircons in strata from Sonora and Nevada are compatible with, but do not prove, the hypothesis that the units are offset equivalents. We cannot preclude models involving flow of detritus along the continental margin in a trench or offshore basin, exceptionally strong contour currents or exotic sources that have been subsequently removed from the western North American continental margin (Gehrels and Stewart, 1998). However none of these models solves the problem of rare or absent or distant sources as simply as restoration along the megashear.

Piercing Lines

The initial estimate of ~800 km of offset along the megashear was based upon correlation of suites of rock units with comparable ages, lithologies, and thicknesses (Silver and Anderson, 1974). Additional studies such as those pertaining to unit thicknesses (Stewart, this volume) and the distribution of Jurassic arc rocks along the megashear (Jones et al., 1995) have led to revised estimates. A faulted Triassic intrusive body (Campbell and Anderson, 2003) has excellent potential to provide a firm magnitude if the displaced part of the mass is identified.

An additional estimate of displacement may be provided by lines defined by the intersection of regional stratigraphic units of Triassic and Early Jurassic ages and three truncating erosional surfaces recognized by Marzolf (Marzolf and Anderson, this volume). Restoration of correlative lines, defined by intersections between stratigraphic units of Triassic and Early Jurassic ages and regional erosional surfaces on opposite sides of the fault, suggests ~1000 km of offset. This magnitude compares well with that of Ketner (1986) based upon apparent offset of exposures of Ordovician Eureka Quartzite on opposite sides of the megashear. These estimates are in contrast to that of Stewart et al. (1999) who suggest that ~600 km is sufficient to restore strata in Sierra Agua Verde, east of Hermosillo, Sonora, to a position close to comparable inner-shelf units of the Cordilleran miogeocline exposed in eastern California and southern Nevada.

Age of Faulting

Near Sonoita, Sonora, mylonite derived from Triassic granitic rock and inclusions of mylonite in a Cretaceous pluton restrict the age of lateral faulting to the interval between 233 and 84 Ma (Campbell and Anderson, 1996, 2003). Regional relationships such as: (1) deformation of fossiliferous Oxfordian strata near the fault in central Sonora (Anderson et al., this volume, Chapter 2), (2) 158 Ma hypabyssal rocks displaced along the fault (Jones et al., 1995), (3) emplacement of the undeformed 148 Ma Independence dike swarm near the fault (Chen and Moore, 1979; James, 1989), and (4) crosscutting relationships among deformed and undeformed plutons in the central Mojave (e.g., Miller et al., 1991; Miller and Walker, 2002; Schermer et al., 2001) indicate deformation between 164 and 149 Ma. Synintrusive deformation of a 161 Ma pluton in the Clipper Mountains, southern California (Howard et al., 1995), may have occurred in response to transpression near the fault, comparable to that inferred by Walker et al. (1990) in the Cronese Hills to the north.

Formation of basins (e.g., McCoy, Bisbee, La Mula, San Marcos, and Sabinas, as well as the Chihuahua trough) between Bathonian and Oxfordian time within the borderland of southwestern Mesozoic North America correlates with megashear faulting (Anderson and Nourse, this volume) as shown by the ages of fossils (Lawton and Olmstead, 1995; Olmstead, 1999; Haenggi, 2002) and intercalated volcanic units (162 Ma) (Fackler-Adams et al., 1997). These basins and many smaller ones formed at releasing left-steps within the region of the gentle eastward curve of the megashear (see Anderson and Nourse, this volume). Collectively, indirect and direct evidence of movement along the Mojave-Sonora megashear indicate that between 162 and 148 Ma the fault was most active.

Older displacement along the Mojave-Sonora megashear, such as that suggested by Dickinson and Lawton (2001), requires multiple episodes of displacement because: (1) the Mojave-Sonora megashear is distinguished by juxtaposition of Proterozoic crystalline rocks against Middle Jurassic arc rocks, and (2) Lower Jurassic rocks on the Caborca block record offset along the Mojave-Sonora megashear (e.g., Stewart, this volume).

DISCUSSION OF RELATIONS THAT "REFUTE" THE MEGASHEAR HYPOTHESIS

El Capitan

Outcrops of folded, foliated Paleozoic strata resting upon foliated, locally mylonitic Jurassic granodiorite protrude as small hills from the plain of the Colorado River delta south of Route 2 in northwesternmost Sonora. The highest hill, known as El Capitan, contains beds of late Paleozoic age (Leveille, 1984). The compositions of the sedimentary units and the fossils that they contain correlate with units characteristic of the cratonal cover of North America. The presence of these units, southwest of the inferred trace of the megashear, is considered by Hamilton et al. (1987) to demonstrate a tie to autochthonous cratonal rocks of North America. The beds at El Capitan are not equivalent to the section exposed upon the Caborca block and therefore could not be fragments displaced northward during San Andreas–related faulting. We view the deformed El Capitan exposures as allochthonous masses, displaced southwestward by back thrusting during northeast-directed Cretaceous contraction (see Anderson et al., this volume, Chapter 2). Equivalent structures include the Santa Rosa mylonite of the Peninsular Ranges (Sharp, 1979; Simpson, 1984), the westward-directed thrusts of the Mojave Desert (Silver, 1983), and the deep-seated thrusts proposed by May (1989).

Oaxaquia

Oaxaquia is the name proposed by Ortega-Gutierrez et al. (1995) for Grenville-age granulite facies rocks that share distinctive geologic features such as massif-type anorthosite-charnockite complexes intruded into protoliths rich in sedimentary rocks of shallow-marine platform or continental rift–related facies that record northwesterly striking structural fabric (Fig. 12). Anderson and Schmidt (1983) assumed that: (1) the Grenville-age rocks that form crystalline basement in southern and western Texas extended south into Mexico as far as Oaxaca, and (2) the elongate southward protuberance had been displaced laterally for hundreds of kilometers along the megashear.

Ortega-Gutierrez et al. (1995) argue that the exposures in southern, central, and northeastern Mexico compose a "microcontinent" that may underlie much of central Mexico. According to these authors, fossils in early through middle Paleozoic beds overlying the anorthositic rocks in Oaxaca show sufficient affinities with European and Gondwana biofacies to conclude that Oaxaquia was not part of Laurentia before Mississippian time when the invertebrate faunas converged (Stewart et al., 1993, 1999a). Subsequently, analyses of present-day lead compositions of whole rocks from the ca. 1.0 Ga crystalline complexes in Mexico by Ruiz et al. (1999) reveal differences between the Oaxacan (Guichicovi complex) rocks and the Laurentian Grenvillian rocks farther north. Based upon this work the authors conclude that "Oaxaquia" is composed of at least two pieces of billion-year-old crystalline crust: a southern part with Gondwanan affinities and a northern part akin to Laurentia.

Accretion of the southern part of Oaxaquia, composed of Oaxaca and Guichicove complexes, against the southern margin of Paleozoic North America, probably occurred during convergence beginning in the Mississippian and culminating in the Permian, contemporaneous with emplacement of debris and large slide masses, comprised mainly of carbonate and volcanic rocks, into Las Delicias basin, Coahuila (McKee et al., 1999). The idea that Grenvillian rocks of Oaxaquia were accreted to the southern margin of North America during late Paleozoic orogenesis is supported by recent studies of pre-Mississippian strata in eastern Mexico south of the megashear that commonly contain non–North American faunas (Sanchez-Zavala et al., 1999; Stewart et al. 1999a). The presence of the exotic rocks in central Mexico suggests assemblage during a complex history of tectonism, which ultimately yielded a collage of terranes (Campa and Coney, 1983).

The paucity of exposures on pre-Mesozoic rocks in central Mexico led Anderson and Schmidt (1983) to underestimate their full extent. However, if the megashear lies within the zone of Paleozoic suturing along the southern edge of the Coahuila block between the distinct northern (Laurentian) and southern (Gondwanan) parts of Grenvillian crystalline terrane, then there is no necessary conflict with the concept of Oaxaquia.

Left-Lateral Late Paleozoic Faulting

The change from shallow to deep-water depositional environments as recorded in Permian rocks exposed in California (Stone and Stevens, 1988) and Sonora (Stewart et al., 1997) has been interpreted to indicate development of a subduction-related foreland basin or a basin formed in response to continental fragmentation during strike-slip (transform?) faulting. Hypothetical strike-slip faulting during Permian time that may have fragmented the southwestern margin of North America also has been cited as an explanation for truncation of the Cordilleran miogeocline and the Antler orogenic belt in California and displacement of the Caborca block (e.g., Stevens et al., 1992; this volume). Implicit in these discussions is the idea that displacement of the Caborca block must have occurred during late Paleozoic time. Workers in Sonora (e.g., Poole and Madrid, 1988; Stewart et al., 1990, 1999b; Poole et al., 1995; Stewart et al., 1997) recognize contractional structures, such as thrusts that formed during the Sonoran orogeny of Late Permian to Middle Triassic age (Poole et al., 1995) and accommodated the emplacement of deep-water units onto strata formed on a shelf covered by shallow water. Stewart et al. (1997) note the possibility that the encroaching allochthon may have caused the continental margin to subside and thereby initiated the sedimentologic changes recorded by Permian units during formation of a deep-water foreland basin. However, direct evidence of Permian lateral faulting has not been reported from Sonora.

Snow (1992) proposes a late Paleozoic contractional setting in California that is similar to that described in Sonora. The idea is supported by the work of Stevens et al. (1997) who interpret late Pennsylvanian structures and stratigraphy as recording the eastward emplacement of the Inyo Mountains above the Last Chance thrust in eastern California.

The proposed southeastward displacement of the Caborca block along the Mojave-Sonora megashear is based principally upon the disrupted pattern of Proterozoic basement provinces and was conceived independent of stratigraphic arguments. Variations in the stratigraphic and paleontologic character of correlative formations in Sonora and California and Nevada such as: (1) thickness of Chesterian strata, (2) the presence or absence of unconformities within Pennsylvanian and Permian strata, (3) the timing of changes from shallow to deep-water conditions in Permian rocks (summarized by Stewart et al., 1997, who list references, and Stewart et al., 1999b), (4) distribution and width of Mississippian facies belts (Stevens, 1997), and (5) the number of taxa (18 compared to 33) that occur in Late Triassic beds in Sonora and Nevada, respectively (Stanley, 1997), do not negate the overall striking similarities that we emphasize. We argue that stratigraphic similarities extend through units of Early Jurassic age and that furthermore the extensive deformation of Middle and Late Jurassic units adjacent to the megashear as well as the mylonitization of Late Triassic plutons within the fault zone near Sonoita (Campbell and Anderson, 2003) support our conclusion that the Caborca block was displaced in Late Jurassic time (see Timing above). Jurassic displacement of the Caborca block along the megashear also is supported by regional stratigraphic relationships among Lower Mesozoic strata as noted by Marzolf (1997, p. 48):

"In the Caborca terrane of Sonora, the Moenkopi, Chinle, Dinosaur Canyon, and Glen Canyon sequences are included within the Antimonio Formation and the lower Chinle, Glen Canyon, and possibly the lower San Rafael sequences are included within the Barranca Group. The increasing erosional vacuity southward from the Colorado Plateau at the base of the Chinle sequence and southwestward and westward from the Plateau at the base of the Glen Canyon sequence precludes the Caborca terrane from having occupied its present position in the Triassic and Early Jurassic."

We conclude that stratigraphic and structural relationships along the megashear and on the Caborca block are incompatible with late Paleozoic strike-slip displacements along the megashear as recognized and defined by us (Silver and Anderson, 1974, 1979, 1981).

Interpretation of the Mojave-Sonora megashear as a fault that accommodated major left-lateral slip during Permian-Triassic time (e.g., Dickinson and Lawton, 2001) is unsupported. The relationships do not preclude earlier pre–mid Jurassic lateral faulting and other tectonic processes that affected southwestern North America and resulted in the absence of the presumed source region for volcanic debris in the Chinle Formation of Arizona (Stewart et al., 1986), followed by development of a Jurassic arc across earlier paleogeographic trends. However, postulated pre–mid Jurassic faulting with ~1000 km of inferred sinistral displacement is not related to the Mojave-Sonora megashear as we defined it.

Marathon Connection

Poole and co-authors (Poole and Madrid, 1988; Poole and Perry, 1997) argue that the Caborca block is autochthonous based upon correlation and inferred connection with correlative stratigraphic units in west Texas (Poole et al., this volume). A similar paleogeographic model based upon the idea that Upper Proterozoic and Paleozoic miogeoclinal strata, unbroken by major left-lateral faults, curve around the southwestern corner of North America and across northern Mexico is offered by Stewart et al. (1984, their Figure 18A). The proximity of the margin to northern Mexico is indicated by: (1) debris from marginal Cambrian beds shed northward and incorporated into late Paleozoic strata of Texas (Almazán-Vásquez et al., 1986); and (2) enormous blocks of late Paleozoic volcanic and carbonate rocks transported into the Las Delicias basin, Coahuila, during collision with South America (McKee et al., 1999). We agree with the likelihood that the margin of the southern edge of Paleozoic North America influenced the position of the megashear in east-central Mexico (e.g., Murray, 1986). We believe that as displacement occurred along the megashear, protuberances in western North America such as the Caborca block were trimmed from the margin and left behind.

Paleomagnetic Studies—A Record of Accretion or Lateral Displacement of Exotic Terranes?

Paleomagnetic studies, initiated in order to further test the hypothesis of displacement along the megashear, were conducted by Cohen et al. (1986). They concluded that: (1) the pole from Upper Triassic Antimonio Formation in Sonora and that from the Luning Formation in Nevada may be considered coincident at the 95% confidence level, but the fit is recognizably improved if the Antimonio pole is rotated 16° clockwise as required by displacement along the megashear; and (2) Cretaceous strata from Sierra del Alamo, southwest of the Mojave-Sonora megashear, do not record rotation. In contrast, Molina-Garza and Geissman (1996), who restudied strata of the Antimonio Formation, interpret shallow-inclination, dual-polarity magnetization in the rocks exposed in Sierra del Alamo as a secondary remanence acquired between Middle Jurassic and Early Cretaceous time. Pre-Cretaceous deformation of Antimonio strata is postulated to have occurred during Jurassic accretion of the Antimonio "terrane" with the Caborca block or contraction within an autochthonous Jurassic continental magmatic arc. They further argue that estimates of latitudinal displacement are inconclusive.

However, depositional contacts at the base of the Antimonio "terrane" indicate that the Antimonio Formation accumulated upon Upper Paleozoic strata on the Caborca block rather than as the result of tectonic accretion as postulated by Molina-Garza and Geissman (1996). Stratigraphic studies at Sierra del Alamo (Stanley and González-León, 1995; González-León and Stanley, 1997; González-León et al., 1995) and near Hermosillo (Stewart et al., 1997) that provide the most complete discussion of stratigraphic relations among Upper Triassic and Lower Jurassic strata and with underlying Paleozoic formations constrain the tectonic history of the Antimonio Formation and related rocks. This work postulates correlation of the fossiliferous carbonate beds of northwestern and west-central Sonora with coal-bearing strata of the Barranca Formation. We conclude that the mapping in combination with paleontologic and stratigraphic analyses (1) show that the widespread outcrops of Upper Triassic and Lower Jurassic rocks of Sonora distinguish an extensive region formerly covered by correlative strata, and (2) demonstrate the depositional relationship of the Mesozoic strata to underlying Paleozoic units, including Permian beds, some of which contain unusual giant fusulinids (Dunbar, 1953). Some workers consider these fossils to be exotic to North America (e.g., Ross and Ross, 1983) or at least parautochthonous (e.g., Stevens, 1995). One idea is that the Permian rocks formed away from the margin of western North America, perhaps as part of an island arc (Miller, 1987; Stevens et al., 1990). Subsequently, they must have been accreted and covered by early Mesozoic strata prior to displacement along the Mojave-Sonora megashear. However, the presence of giant fusulinids in North American rocks of West Texas may indicate that the hypothetical island arc was quite close to the North American mainland. The correlations among widespread Triassic and Lower Jurassic units: (1) show that the boundaries of the Mesozoic Antimonio "terrane" as conceived initially by Stanley and González-León (1995) must be expanded to encompass Upper Triassic strata in the region around Hermosillo, and (2) suggest that the hypothesis of a Late Jurassic accreted Antimonio terrane is invalid.

Additional paleomagnetic results from both sides of the megashear in the Caborca region were reported in 1999 by Molina-Garza and Geissman. Two secondary magnetization events are recognized. The younger is associated with Cretaceous magmatism that occurred after postulated megashear activity. An older magnetization in rocks on both sides of the megashear was acquired between 190 and 120 Ma. The mean directions derived from data acquired from rocks on both sides of the megashear are statistically indistinguishable. These data indicate clockwise rotation of the region with respect to poles from strata on the craton of North America. The timing of magnetization is crucial to constraining arguments: if the magnetization occurred between 190 and 180 Ma, most of the volcanic belt north of the Mojave-Sonora megashear had not yet formed; by 140 Ma the Mojave-Sonora megashear was inactive. Magnetization at 160 Ma is necessary if rocks on both sides of the Mojave-Sonora megashear were affected equally and not displaced later.

The Mojave-Sonora Megashear—An Artifact of Late Cretaceous (Laramide) Orogeny?

In Sonora, Late Cretaceous and/or early Tertiary contractional structures, such as folds, thrust faults and associated cleavage, and indications of uplift and cooling, recorded by K/Ar isotopic systems are recorded in rocks along the Mojave-Sonora megashear (De Jong et al., 1988; Jacques-Ayala et al., 1990; Nourse, 2001; Iriondo, 2001, this volume). Along the Mojave-Sonora megashear faults and folds that record Cretaceous deformation are interpreted by some workers (e.g., De Jong et al., 1988; Jacques-Ayala et al., 1990, 1995; Pubellier et al., 1995, Calmus and Sosson, 1995; Iriondo, 2001) to obviate the need for a terrane boundary. The $^{40}Ar/^{39}Ar$ ages between ca. 70 and 39 Ma obtained from white mica in metamorphic rocks near Quitovac (Fig. 4; Iriondo, 2001) and in south-central Arizona (Haxel et al., 1984) are cited in support of the importance of Cretaceous-Tertiary contractional tectonism. Mapping in the region north of Altar by Nourse (2001) corroborates the presence of contractional deformational as part of a protracted tectonic history including: (1) Late Jurassic formation and filling of a transtensional basin along the northeastern side of the Mojave-Sonora megashear, (2) contraction between 71 and 51 Ma during which asymmetric, northeast-verging folds and thrusts and associated foliation formed, and (3) extension during Miocene time.

We conclude that Cretaceous deformation has a diverse expression in Sonora including: (1) folding and faulting of thick sections of Cretaceous Bisbee Group strata during gravity-driven downslope movement (McKee and Anderson, 1998), and (2) subsequent inversion of preexisting Late Jurassic and Early Cretaceous basin sediments accompanied by the development of northwest-trending folds and thrusts (Rodríguez-Castañeda et al., 1997; Nourse, 2001), reactivation of other basin-bounding faults, and a second episode of gravity-driven deformation (see McKee et al., this volume). We do not believe that Cretaceous deformation accounts for the distribution of Proterozoic crust and cover that originally stimulated proposal of the Mojave-Sonora megashear model.

The Mojave-Sonora Megashear in the Mojave Desert—A Mythical Fault Segment?

Our thinking is that in southernmost California the inferred fault trace of the Mojave-Sonora megashear continues northwestward between exposures of the San Gabriel domain and the westernmost occurrences of autochthonous or parautochthonous strata (e.g., Marble Mountains, southeastern California) that overlie crystalline basement formed in the interval between 1.8 and 1.7 Ga. Within the northern part of the eastern Transverse Ranges, Late Cretaceous and/or Paleocene westward-directed thrusts, extensional detachments of Tertiary age, and late Cenozoic left-lateral faults obscure the inferred trace of the Mojave-Sonora megashear (see Anderson et al., this volume, Chapter 2, for details). In parts of this region the surficial trace of the Mojave-Sonora megashear has been carried westward and the deep trace tectonically covered. Despite these complications, we believe that the effects of younger sedimentologic, magmatic, and deformational processes may be recognized and restored and that a plausible route of the northward extension of the Mojave-Sonora megashear fault may reasonably link it with the pre-Cretaceous Melones–Bear Mountain faults extending northwestward along the western margin of the Sierra Nevada.

SUMMARY

We postulate a major left-lateral strike-slip fault to correspond with a regional stratigraphic and structural discontinuity in Sonora, northwestern Mexico. The discontinuity is distinguished by: (1) the peculiar regional distribution and juxtaposition of distinct suites of Proterozoic crystalline rocks (Yavapai and Pinal) and the contrasting nature of the strata overlying each province, (2) the abrupt contact of Yavapa-equivalent rocks that compose the Caborca block against commonly folded and foliated mainly mid-Jurassic volcanic and plutonic rocks comprising part of an Andean-style arc that intervenes between the juxtaposed Proterozoic provinces, (3) outcrops of a major sinistral shear zone, and (4) a narrow belt of deformation, commonly ductile and contractional, restricted to strata of the Jurassic volcanic arc.

The Caborca block is part of a composite terrane as shown by reconnaissance geochronologic studies of gneissic rocks exposed in ranges along Route 2, northwest of Sonoita. Some orthogneisses within this area, which may be porphyritic and intensely deformed, yield interpreted ages younger than most plutons of the Yavapai suite but older than those of Pinal.

The similarities of Proterozoic basement and Neoproterozoic-Cambrian strata near Death Valley to those of the Caborca block spurred additional stratigraphic comparisons among younger Paleozoic and Mesozoic formations. These younger units also show strong similarities. The initial estimate of the timing of displacement along the Mojave-Sonora megashear was based upon the apparent juxtaposition of Lower Jurassic strata, overlying the Caborca block, against mid-Jurassic volcanic rocks north of the block that we assumed were part of a parautochthonous suite that accumulated against the margin of Mesozoic-age southwestern North America. Exposures of the fault near Sonoita, Sonora, restrict the time of faulting to an interval between 233 and 84 Ma. Regional relationships further constrain displacement along the Mojave-Sonora megashear between 166 and 148 Ma. Comparison of thicknesses of Neoproterozoic and Paleozoic strata suggest displacement of at least 800 km, although regional stratigraphic analysis suggests this magnitude to be a minimum. Additional research including plate tectonic analysis, paleomagnetic results, widespread field work, and associated structural studies, stratigraphic analysis, and identification of sediment provenance detrital zircon studies are either supportive or permissive of major displacement along the Mojave-Sonora megashear. Geologic relationships that are apparently incompatible with either the trace or offset of the inferred fault are attributed to strong superposed deformations that have shifted crustal segments during Laramide contraction and Tertiary extension. No geologic relationships in Sonora support displacement during late Paleozoic time. The Mojave-Sonora megashear was likely a transform plate boundary during formation of the Gulf of Mexico.

ACKNOWLEDGMENTS

We benefited from cooperation offered generously by Guillermo Salas, Diego Cordoba, Jose Guererro, and Fernando Ortega-Gutierrez, each of whom served as Director of the Instituto de Geología of Universidad Nacional Autónoma de México. In Sonora, we had constructive interactions with former chiefs of the northwest regional office of the Instituto, including Jesus Najera, Jaime Roldán-Quintana, César Jacques Ayala, and Carlos González-León. Claude Rangin, who conducted extensive mapping principally in northeastern Sonora, generously shared knowledge of stratigraphy and structure gained from study of fossiliferous strata. Anderson was aided in the field by Jaime Alvarez, Arturo Cisneros, Clay Conway, Tom Early, Ken Ludwig, Jay Murray, Robert Powell, David Wenner, and especially Jose Luis Rodríguez-Castañeda. Patti Campbell, Joe Chepega, Karen Cohen, Chris Connors, Franco Corona, John Dembosky, Mary Beth McKee, Jose Luis Rodríguez-Castañeda, and William Stephens, each of whom pursued graduate degrees under Anderson's direction, undertook daunting mapping projects that contributed greatly to our understanding of geological relationships in Sonora. Gordon Haxel and Dick Tosdal aided Anderson in numerous ways, providing logistical support as well as a backboard for ideas. After a field trip to Sonora in 1979, Jim McKee and Norris Jones initiated cooperative work with Anderson in eastern Mexico. For two decades Anderson benefited from knowledge, logistical support, and patient shepherding provided by these two long-suffering gentlemen. Anderson, who began the project with no first-hand knowledge of isotope geochronology, owes a great debt of gratitude to Silver's laboratory staff at Caltech, where the U-Pb isotopic analyses of zircon were performed. The instruction and patient mentoring provided by Jaime Alvarez, chief rock processor, and Gerri Baenteli Silver, lab technician, were essential to any success attributed to Anderson's role in this project. Ken Ludwig and Ed Lidiak were instrumental in the interpretation and presentation of the U-Pb isotopic results. Carlos Aiken, Peter Coney, John Marzolf, Bill Muehlberger, Grover Murray, Jon Nourse, Pete Palmer, Norm Silberling, and Jack Stewart contributed good thinking and good data that they generously shared with Anderson. S.T. Kitinyrere, Bill Muehlberger, Cal Stevens, and Jack Stewart read early versions of the manuscript and offered helpful comments and constructive criticisms. Scott Davidson prepared the illustrations with skill and patient forbearance; Paul Coyle collated the data tables. W.R. Van Schmus and Wayne Premo provided useful reviews of the final draft.

Special thanks to Jon Nourse who reviewed the manuscript multiple times with great care and substantially improved it.

APPENDICES

Analytical Procedure

Fractions of the total zircon population were selected for isotopic analysis on the basis of their uranium content, with the objective of obtaining the greatest range in the degree of discordance. The zircon fractions were fused with about six times their weight of purified sodium-tetraborate flux for varying time periods, depending upon the sample weight. The analytical procedure for extracting lead is the same as that described by Silver et al. (1963). The few modifications made of the proven techniques include the substitution of 10 mL Pyrex beakers for Teflon during the final digestion of the lead dithizonate and a change in loading technique from running PbS on a tantalum filament to $PbNO_3$ in a silica gel–phosphoric acid medium on a rhenium filament (modified from the procedure of Cameron et al., 1969). Uranium was extracted by normal ion exchange techniques.

Isotopic analyses were made on 12 in., single-focusing, solid-source mass spectrometers constructed at the California Institute of Technology. Data obtained on the electron multiplier and simple collector has been corrected for mass discrimination and for very slight nonlinearity in the shunt factors. Mass spectrometer reproducibility is generally ±1% for radiogenic $^{206}Pb/^{204}Pb$ ratios, and ±0.3% for the $^{206}Pb/^{207}Pb$, $^{206}Pb/^{208}Pb$, and $^{235}U/^{238}U$ ratios. Mass spectrometer precision (1σ) for $^{206}Pb/^{207}Pb$ averages about ±0.2%, as does precision for $^{206}Pb/^{208}Pb$ ratios. Uncertainty for $^{206}Pb/^{204}Pb$ is about ±0.5%.

The common lead blank contributed by reagents and glassware is slightly less than those determined by Silver et al. (1963). The use of the silica gel–phosphoric acid loading technique enables us to run 1 μg quantities of lead, which permit significant reduction of sample size. This combined with the smaller amount of flux used (6 rather than 10 times the weight of sample) can reduce the blank significantly. Contribution of common lead from the rest of the procedure has also been reduced so that the range is ~0.06–0.04 μg.

Interpretation of Ages and Uncertainties

Inheritance and lead-loss typically coexist in bulk samples of both volcanic and plutonic zircons. Therefore discordance (indicating inheritance) does not guarantee that the 206/238 age is a minimum, and concordance guarantees neither absence of lead loss nor absence of inheritance (though it does rule out much inheritance of Precambrian zircons). Even if one could apply objective statistics to this reconnaissance data (which in almost all cases here would give very large errors), the underlying model (of only inheritance or only lead-loss) would be unreliable. Even replicate, overlapping concordant analyses are no guarantee of accuracy (at least within the analytical errors), because the hundreds to thousands of grains involved must result in a very high degree of averaging of complexity. Selection by grain size and magnetic properties may not necessarily split out fractions recording the full spectrum of complexity.

The concordia intercept zircon ages and the uncertainties assigned to the ages were calculated and plotted using the algorithms of Ludwig (2001).

Zircon Suites

Separation Procedure

The procedures for liberating and processing cogenetic suites of zircons are the same as those described by Silver and Deutsch (1963). Prior to fusion, the zircon fractions were washed in ~50 mL of hot (80 °C) concentrated HNO_3 for one hour. They were then rinsed three times with quadruply distilled water, followed by three rinses with filtered technical-grade acetone. Preliminary determinations of the uranium concentration of the fractions were made using α-counting techniques. Descriptions of some fractions from selected samples are included below.

Bámori Rhyodacite

The zircon population consists of predominantly stubby anhedral to subhedral individuals, although well-formed crystals are not rare. Almost all grains have suffered corrosion faceting and corrosion rounding, although the intensity of the modification varies greatly among the crystals. Shallow scallops and pits are most common.

Collectively the zircons appear pale red to light-brownish gray, whereas individuals range from nearly colorless or very pale pink to deeper-red shades. In the more radioactive fractions opaque or nearly opaque grains are pale olive and may be brown.

Internal zoning is common, particularly in the deeper-colored grains, where it may be marked by optical discontinuities, by slight color variations or rarely by aligned inclusions. Slender prisms of apatite(?), opaque grains, outlines resembling zircon, equant anhedral forms (probably structural cavities), rare red-brown plates of iron(?) oxide, and irregularly shaped tubes or other forms which in some grains intersect the external outline and are probably cavities, all occur as inclusions. Many fractured grains are present, but it may be difficult to distinguish between fractured grains and anhedral, strongly corroded zircons.

Bámori Quartz Monzonite Gneiss

Collectively the zircons are grayish red, whereas individuals are predominantly transparent pale red. Elongate grains are the rule, although length-to-width ratios rarely exceed 2. Distinctive zircon morphology is still obvious, although all grains have suffered moderate to severe corrosion faceting that has modified the sharp, crisp outlines to rounded, multiply faceted forms. Doubly terminated grains are common, but terminations are rarely sharp. Grains are rarely fractured and internally, they are generally free from zoning and overgrowths. Most grains contain a few inclusions that probably consist predominantly of other small zircons or apatite as suggested by outlines. Less commonly the zircon core is clouded by dusty opaque particles.

Tecolote Granodiorite Gneiss

This suite is grayish, with pale red individuals. Crystals are predominantly elongate, with length-to-width ratios generally ~2:1 or less. Rare grains are slightly more elongate. All grains show the effects of gentle, although extensive, corrosion faceting that has obscured sharp facet intersections. The resulting morphology is subangular and anhedral forms have edges blunted by many tiny facets. Crystals free of inclusions are very rare. The most common inclusions are slender prisms, probably apatite, a few of which occur in almost all grains. Fine opaque particles, red-brown plates, inclusions of probable zircon, and structural cavities were noted as somewhat less abundant types (cf. Silver and Deutsch, 1963, p. 726–728). Rare crystals show poorly defined zones, but generally zoning is absent. Grains with recognizable overgrowths were not observed.

Bámori Metamorphic Pegmatite #203

In the aggregate, pass and retain 200 mesh zircons are light-brownish gray, whereas individuals may be colorless, pale to dark red, dusky yellow or light-olive gray. Larger grains (retain 100 mesh) include rare transparent, moderate-red crystals but predominantly consist of opaque crystals that may be red, pale-yellowish orange, or gray and that range through lighter shade to white. Aside from scattered transparent grains, most small zircons are partially if not completely translucent or have an opaque core with a translucent or transparent rim.

Scattered individuals show excellent euhedral form with maximum length-to-width ratios of ~2; however, more common are stubby almost equidimensional grains that, although defined by individual well-formed facets, are asymmetric and the habits of which are not readily identifiable. Crystal outlines are sharp, and, with exceedingly rare exceptions, grains show no signs of having developed corrosion facets. Numerous broken grains are present, but these likely result from crushing preparatory to mineral separations rather than representing any inherent characteristic of the population.

Members of the suite reveal a variety of internal characteristics including distinct zoning, inclusions, overgrowths, and cores. Many crystals, particularly those that appear grayish-orange in transmitted light, show zoning defined by optical discontinuities and in some grains marked by fine inclusions. The zones generally mirror the external form. Few grayish-pink to moderate red grains reveal zoning. As a rule inclusions are much less abundant in the pegmatite zircons than in zircons of the enclosing gneiss. Slender rods and needles, presumably apatite, are scarce in this population, whereas they are the most abundant type in the zircons derived from the gneiss. Some zircons of this population are free from imperfections, although most contain opaque dust commonly concentrated near the zircon core or less commonly at their terminations. The amount of this material appears to vary with the uranium content and hence the degree of metamictization of the crystal. The characteristic types of inclusions in the zircons are stubby to slender prisms of apatite(?); equant anhedral forms, probable structural cavities; red-brown plates of iron oxide (?); and small fractures.

Some zircons show overgrowths on cores. These are grouped together because the overgrowths consistently surround a core that resembles a zircon derived from the gneiss. These complicated individuals commonly show a pinkish-red zircon surrounded by a narrow symmetrical overgrowth of grayish-orange zircon. Several interesting observations may be made regarding the rimmed zircons. In rare grains within the pink crystal yet a third distinct crystal form exists. The best example shows a distinct doubly terminated pale brownish-red innermost core with a few inclusions and a faint, internal zoning. Zones may be seen commonly in the pale grayish-orange overgrowths and rarely within the surrounded core. The pale red cores may be well formed without distinct corrosion faceting, or less commonly they may show somewhat rounded forms. Some zircons that show strong corrosion faceting exists without overgrowths, however, these crystals are consistently very pale pink to clear and show no zoning or abundant inclusions.

In summary, three categories may be recognized among zircons composing the suite: (1) abundant grayish-orange zircons, generally zoned and likely uranium rich, as suggested by their opacity, that presumably were generated during the formation of the metamorphic pegmatite; (2) zircons that consist of rims around a core, likely derived from the suite of the older gneiss, may show euhedral form, zoning and an inner core (these complicated crystals perhaps preserve some of the primary premetamorphic features of the original igneous zircons, or, where the core shows rounded corrosion-faceted grains, the rim sheaths a grain affected by metamorphism); and (3) very pale to clear, rounded, corrosion-faceted zircons, with few inclusions, that appear to have withstood the full brunt of metamorphism and are now annealed and included in the suite of pegmatite zircons, but which were not sheathed by newly crystallized zircon.

Tecolote Metamorphic Pegmatite

Characteristic of this population are pale red, elongate, poorly formed, anhedral to subhedral crystals. This suite shows marked variations as a function of grain size. Larger grains show deeper colors, are more translucent or opaque, and are less well formed. In the aggregate P 200 grains are uniformly grayish-red, whereas groups of larger grains are lighter as they include more uranium-rich reddish-gray and white grains.

Small subhedral crystals commonly are elongate, with length-to-width ratios of 3, although the most abundant grains are stubby with rounded terminations. Rounding shown by these grains seems to be related to poor primary form rather than due to corrosion faceting and pitting characteristic of metamorphic modifications. Many zircons show distinct, clean lines that define well-formed facets. Most of these grains are transparent with scarce inclusions of which opaque aggregates of dusty appearing material are most common. Other inclusion types include slender rods of apatite(?), red-brown plates of iron oxide(?), cavities and zircon-shaped forms. Fractures are common, particularly in the larger grains. Internal zoning is rare or difficult to identify in the small transparent grains, but is more common in the larger, darker grains. Only one distinct overgrowth was seen, and, like those in the Bámori metamorphic pegmatite, it consisted of a rim of slightly yellowish pale-red zircon surrounding a slightly darker pinkish-red crystal.

Internal characteristics of the large R 100 crystals are obscured by their deep reddish-gray, grayish-red, orangish-gray, or whitish-gray color and by their opacity, presumably a function of radiation damage. External morphology may be nearly euhedral, however, because of fracturing, few grains that show this have been preserved. Anhedral or subhedral, stubby, blunted, rounded forms are common.

Aibó Micrographic Granite

Zircons were separated according to techniques described by Silver (1963) and Silver and Deutsch (1961, 1963). A few grams were recovered from a 57 kg rock sample. Recovered zircons are most abundant in the −50+100 and −100+200 (~90 micron average prism diameter) mesh size range, although significant yield was also retained on 400 mesh. The length-to-width ratio rarely exceeds 3 and averages ~2.

Additional subdivision of the zircon population relative to U-Th content was obtained by splitting the size fractions according to magnetic susceptibility on a Franz isodynamic separator (Silver and Deutsch, 1963). Six fractions, which show among them a 10-fold range in uranium content, were analyzed.

The total population of recovered zircon is grayish-red; however considerable variation exists among different fractions and among individuals. The less radioactive fractions (less than 250 ppm U) consist of transparent and translucent grayish-pink to very pale orange zircons as the main type with less abundant lighter and darker end members. Moderately radioactive (~600 ppm U) fractions are predominantly translucent to transparent grayish-red. Greatest color variation is shown by the fractions with 1000–2000 ppm U. These contain rare translucent light-colored zircons, but for most part consist of opaque white, grayish-orange, gray, grayish-red, reddish-brown and brown zircons.

The zircons are characterized by simple crystal form as shown by scattered euhedral grains. The predominant habit is the tetragonal prism with dipyramidal terminations. In some grains, vicinal faces modify pyramidal terminations. The basal pinacoid is a less common termination. As a whole, the suite is generally subhedral with abundant anhedral grains. The external morphology of many grains is modified by fractures and irregular to smooth concave surface pits, which if strongly developed may obliterate external faces.

Internally, the zircons are characterized by ubiquitous needle-to-rod–shaped inclusions of apatite as well as irregularly shaped cavities and tubes, which in some grains intersect the crystal faces. Also common are opaque, brown, dust-like material; opaque, black, anhedral magnetic grains and reddish-brown material, which commonly exists along fractures. Rare, included euhedral grains have outlines that resemble zircons. Light-colored, transparent zircon with few inclusions may enclose dark, rounded zircon with abundant very small inclusions. However, generally zircons with overgrowths of rims are exceedingly rare. Zoned zircons are not common. In zircons in which zones exist, the zones are distinguished by optical discontinuities, dusty inclusions and/or color variations that mark euhedral to subhedral zone outlines.

REFERENCES CITED

Alencaster de Cserna, G., 1961a, Parte I, Estratigrafia del Triasico Superior de la parte central del Estrada de Sonora: Universidad Nacional Autónoma de México, Instituto de Geologia, Paleontologia Mexicana 11, pt. 1, 18 p.

Alencaster de Cserna, G., 1961b, Parte III, Fauna fosil de la Formacion Santa Clara (Carnico) del Estrada de Sonora: Universidad Nacional Autónoma de México, Instituto de Geologia, Paleontologia Mexicana 11, pt. 3, 38 p.

Almazán-Vásquez, E., Palmer, A.R., and Robison, R.A., 1986, Constraints on the southern margin of North America during the Early Paleozoic: Geological Society of America Abstracts with Programs, v. 18, no. 6, p. 525.

Anderson, C.A., and Silver, L.T., 1976, Yavapai Series—A greenstone belt: Arizona Geological Society Digest, v. 10, p. 13–26.

Anderson, C.A., Blacet, P.M., Silver, L.T., and Stern, T.W., 1971, Revision of Precambrian stratigraphy in the Prescott-Jerome area, Yavapai County, Arizona: U.S. Geological Survey Bulletin 1323-C, p. C1–C16.

Anderson, J.L., 1989, Proterozoic anorogenic granites of the southwestern United States, *in* Jenney, J.P., and Reynolds, S.J., Geologic evolution of Arizona: Arizona Geological Society Digest, v. 17, p. 211–238.

Anderson, T.H., and Campbell, P., 1992, Mylonite at the Mojave-Sonora megashear, northwestern Mexico: Geological Society of America Abstracts with Programs, v. 24, no. 7, p. 147.

Anderson, T.H. and Schmidt, V.A., 1983, A model of the evolution of Middle America and the Gulf of Mexico–Caribbean Sea region during Mesozoic time: Geological Society of America Bulletin, v. 94, p. 941–966, doi: 10.1130/0016-7606(1983)94<941:TEOMAA>2.0.CO;2.

Anderson, T.H., and Silver, L.T., 1971, Preliminary history for Precambrian rocks, Bámori Region, Sonora, Mexico: Geological Society of America Abstracts with Programs, v. 3, no. 2, p. 72–73.

Anderson, T.H., and Silver, L.T., 1977, U-Pb isotope ages of granitic plutons near Cananea, Sonora: Economic Geology and the Bulletin of the Society of Economic Geologists, v. 72, p. 827–836.

Anderson, T.H., and Silver, L.T., 1979, The role of the Mojave-Sonora megashear in the tectonic evolution of northern Sonora, *in* Anderson, T.H., and Roldán-Quintana, J., eds., Geology of northern Sonora Guidebook, Annual Meeting of the Geological Society of America: Estación Noroeste, Instituto de Geología, Universidad Nacional Autónoma de México and the University of Pittsburgh, Pittsburgh, USA, p. 59–68.

Anderson, T.H., and Silver, L.T., 1981, An overview of Precambrian rocks in Sonora: Universidad Nacional Autónoma de México, Instituto de Geología, Revista, v. 5, no. 2, p. 131–139.

Anderson, T.H., Eells, J., and Silver, L.T., 1979, Geology of Precambrian and Paleozoic rocks, Caborca–Bámori region, *in* Anderson, T.H., and Roldán-Quintana, J., eds., Geology of northern Sonora Guidebook, Annual Meeting of the Geological Society of America: Estación Noroeste, Instituto de Geología, Universidad Nacional Autónoma de México and the University of Pittsburgh, Pittsburgh, USA, p. 1–22.

Anderson, T.H., Bajek, D.T., Chepega, J.R., Ichikawa, K.M., Rodríguez- Castañeda, J.L., Stephens, W.E., and Silver, L.T., 1984, Crystalline thrust sheets near the Mojave-Sonora megashear, Sonora, Mexico: Geological Society of America Abstracts with Programs, v. 16, p. 430.

Anderson, T.H., McKee, J.W., Jones, N.W., and Rodríguez-Castañeda, J.L., 1991a, Transpressional structures along the Late Jurassic Mojave-Sonora megashear [abs.]: Geological Society of America Abstracts with Programs, v. 23, no. 5, p. A251.

Anderson, T.H., McKee, J.W., and Jones, N., 1991b, A northwesterly trending, Jurassic fold nappe, northernmost Zacatecas, Mexico: Tectonics, v. 10, p. 383–402.

Anderson, T.H., Aiken, C., Balde, M., De La Fuente, M.F., and Murray, G.E., 1997, The structural grain in northeastern Mexico: Reflection of older tectonic templates: Geological Society of America Abstracts with Programs, v. 29, p. A-1.

Anderson, T.H., Silver, L.T., and Salas, G.A., 1980, Distribution and U-Pb isotope ages of some lineated plutons, northwestern Mexico: Geological Society of America Special Paper 153, p. 269–283.

Arellano, A.R.V., 1956, Relaciónes del Cámbrico de Caborca, especialmente con la base del Paleozoico: Congreso Geológico Internacional, 20th, México, 1956; El Sistema Cambrico, su Paleogeografía y el Problema de su Base–Symposium, v. 2, p. 509–527.

Bridges, L.W., 1964, Stratigraphy of Mina Plomosas–Placer de Guadalupe area *in* Geology of Minas Plomosas–Placer de Guadalupe area, Chihuahua, Mexico, Field trip guide: West Texas Geological Society Publication 64-50, p. 50–59.

Bullard, E., Everett, J.E., and Smith, A.G., 1965, The fit of the continents around the Atlantic: Philosophical Transactions of the Royal Society of London, v. A258, p. 41–51.

Calmus, T., and Sosson, M., 1995, Southwestern extension of the Papago terrane into the Altar Desert region, northwestern Sonora, and its implications, *in* Jacques-Ayala, C., González-León, C.M., and Roldán-Quintana, J., eds., Studies of the Mesozoic of Sonora and adjacent areas: Geological Society of America Special Paper 301, p. 99–109.

Cameron, A.E., Smith, D.H., and Walker, R.L., 1969, Mass spectrometry of nanogram size samples of lead: Analytical Chemistry, v. 41, p. 525–526, doi: 10.1021/ac60272a020.

Campa, M.F., and Coney, P.J., 1983, Tectonostratigraphic terrains and mineral resource distributions in Mexico: Canadian Journal of Earth Sciences, v. 20, p. 1040–1051.

Campbell, P.A., and Anderson, T.H., 1996, Structures and deformation partitioning along a Jurassic plate boundary transform, Sonora, Mexico: Geological Society of America Abstracts with Programs, v. 28, p. A445.

Campbell, P.A., and Anderson, T.H, 2003, Structure and kinematics along a segment of the Mojave-Sonora megashear; a strike-slip fault that truncates the Jurassic continental magmatic arc of southwestern North America: Tectonics, v. 22, no. 6, p. 16:1–21.

Castillo-Mendoza, E., 1987, Geología del área del rancho Ensenada, Municipio de Banámichi, Sonora central, México: La Paz, B.C.S., Universidad de Baja California Sur, tesis professional, 97 p. (inédita).

Chen, J.H., and Moore, J.G., 1979, Late Jurassic Independence dike swarm in eastern California: Geology, v. 7, p. 129–133, doi: 10.1130/0091-7613(1979)7<129:LJIDSI>2.0.CO;2.

Chepega, J.R., 1987, Reconnaissance geology of Tuape, north-central Sonora, Mexico [M.S. thesis]: Pittsburgh, Pennsylvania, University of Pittsburgh, 139 p. (Unpublished).

Cohen, K.K., Anderson, T.H., and Schmidt, V.A., 1986, A paleomagnetic test of the proposed Mojave-Sonora megashear in northwestern Mexico: Tectonophysics, v. 131, p. 23–51, doi: 10.1016/0040-1951(86)90266-0.

Connors, C.D., 1990, Geology of the Quitovac/Sierra La Tonita area, northwestern Sonora, Mexico [M.S. thesis]: Pittsburgh, Pennsylvania, University of Pittsburgh, 75 p. (Unpublished).

Conway, C.M., and Silver, L.T., 1989, Early Proterozoic rocks (1710–1615) in central and southeastern Arizona, *in* Jenney, J.P., and Reynolds, S.J., Geologic evolution of Arizona: Arizona Geological Society Digest, v. 17, p. 211–238.

Conway, C.M., Karlstrom, K.E., Silver, L.T., and Wrucke, C.T., 1987, Tectonic and magmatic contrasts across a two-province Proterozoic boundary in central Arizona, *in* Davis, G.H., and Vanden Dolder, E.M., eds., Geologic diversity of Arizona and its margins: Excursions to choice areas (Geological Society of America Field Trip Guidebook 1987 Annual Meeting, Phoenix): Arizona Bureau of Geology and Mineral Technology Special Paper 5, p. 158–175.

Cooper, G.A., and Arellano, A.R.V., 1946, Stratigraphy near Caborca, northwest Sonora, Mexico: American Association of Petroleum Geologists Bulletin, v. 30, no. 4, p. 606–611.

Cooper, J.R., and Silver, L.T., 1964, Geology and ore deposits of the Dragoon Quadrangle, Cochise County, Arizona: U.S. Geological Survey Professional Paper 416, 196 p.

Corona, F.V., 1979, Preliminary reconnaissance geology of Sierra La Gloria and Cerro Basura, northwestern Sonora, Mexico, *in* Anderson, T.H., and Roldán-Quintana, J., eds., Geology of northern Sonora: Boulder, Colorado, Annual Meeting of the Geological Society of America, Field Trip (#27) Guidebook, p. 32–48.

Damon, P.E., Livingston, D.E., Mauger, R.L., Giletti, B.J., and Alor, J.P., 1962, Edad del Precambrico "Anterior" y de otras rocas del zócalo de la region de Caborca-Altar de la parte noroccidental del Estado de Sonora, México: Universidad Nacional Autónoma de México, Instituto de Geología Boletín, v. 64, p. 11–44.

de Cserna, Z., 1970, Tectonics of northern Mexico (Mesozoic sedimentation, magmatic activity, and deformation in northern Mexico), *in* Seewald, K., and Sundeen, D., eds., The geologic framework of the Chihuahua tectonic belt: West Texas Geological Society Publication 71–59, p. 99–118.

DeFord, R.K., and Haenggi, W.T., 1970, Stratigraphic nomenclature of Cretaceous rocks in northeastern Chihuahua, *in* Seewald, K., and Sundeen, D., eds., The geologic framework of the Chihuahua tectonic belt: Symposium in honor of Professor Ronald K. DeFord: West Texas Geological Society Publication 71–59, p. 175–196.

De Jong, K.A., Escarcega-Escarcega, J.A., and Damon, P.E., 1988, Eastward thrusting, southwestward folding, and westward backsliding in the Sierra la Vibora, Sonora, Mexico: Geology, v. 16, p. 904–907.

Dickinson, W.R., and Lawton, T.F., 2001, Carboniferous to Cretaceous assembly and fragmentation of Mexico: Geological Society of America Bulletin, v. 113, no. 9, p. 1142–1160, doi: 10.1130/0016-7606(2001)113<1142: CTCAAF>2.0.CO;2.

Dietz, R.S., and Holden, C., 1970, The breakup of Pangaea: Scientific American, v. 223, p. 30–41.

Dumble, E.T., 1900, Notes on the geology of Sonora, Mexico: Geological Society of America Bulletin, v. 11, p. 10–14.

Dunbar, C.O., 1953, A giant Permian fusuline from Sonora: Smithsonian Miscellaneous Collection, v. 119, no. 2, p. 14–19.

Eardley, A.J., 1951, Structural geology of North America: New York, Harper and Brothers, 624 p.

Eells, J.L., 1972, Geology of the Sierra de la Berruga, northwestern Sonora, Mexico [M.S. thesis]: San Diego, California State University, 77 p. (Unpublished).

Eisele, J., and Isachsen, C., 2001, Crustal growth in southern Arizona: Geochronologic and Sm-Nd isotopic evidence for addition of the Paleoproterozoic Cochise block to the Mazatzal Province: American Journal of Science, v. 301, p. 773–797.

Espinoza, I., Iriondo, A., Premo, W., and Valencia, M., 2003, Geochemistry and SHRIMP U-Pb zircon geochronology of anorthositic rocks at Sierra El Tecolote in the Caborca block, northwestern Sonora: Geological Society of America Abstracts with Programs, v. 35, no. 5, p. 84.

Fackler-Adams, B.N., Busby, C.J., and Mattinson, J.M., 1997, Jurassic magmatism and sedimentation in the Palen Mountains, southeastern California: Implications for regional tectonic controls on the Mesozoic continental arc: Geological Society of America Bulletin, v. 109, p. 1464–1484, doi: 10.1130/0016-7606(1997)109<1464:JMASIT>2.3.CO;2.

Farmer, G.L., Bowring, S.A., Espinoza, I., and Matzel, J.P., 2003, Isotopic evidence on the nature of Precambrian basement an overlying sedimentary rocks, Caborca, Sonora: Geological Society of America Abstracts with Programs, v. 35, no. 4, p. 67.

Flores, T., 1927, Reconocimiento geológicos en la region central de Estrada de Sonora: Instituto de Geología, Boletín 49, 287 p.

Fries, C., Jr., 1962, Reseña de la geología del Estado de Sonora, con énfasis en el Paleozoic [Review of the geology of the State of Sonora, Mexico, with emphasis on the Paleozoic]: Asociación Mexicana de Geólogos Petroleros Boletín, v. 14, p. 257–273.

Gehrels, G.E., and Stewart, J.H., 1998, Detrital zircon U-Pb geochronology of Cambrian to Triassic miogeoclinal and eugeoclinal strata of Sonora, Mexico: Journal of Geophysical Research, v. 103, no. B2, p. 2471–2487, doi: 10.1029/97JB03251.

Gilletti, J.B., and Damon, P.E., 1962, Rubidium-strontium ages of some basement rocks from Arizona and northwestern Mexico: Geological Society of America Bulletin, v. 72, p. 639–644.

González-León, C.M., 1997, Sequence stratigraphy and paleogeographic setting of the Antimonio Formation (Late Permian–Early Jurassic) Sonora, Mexico, *in* Stanley, G.D., Jr., and Gonzalez-León, C.M., eds., Special issue dedicated to the International Workshop on the Geology of Northwestern Sonora, Mexico: Revista Mexicana de Ciencias Geologicas, v. 14, no. 2, p. 136–148.

González-León, C.M., and Stanley, G.D., Jr., 1997, eds., U.S.-Mexico Cooperative Research–International Workshop on the Geology of Sonora Memoir: Universidad Nacional Autónoma de México, Instituto de Geología, Estación regional Noroeste, Publicaciones Ocasionales 1, 69 p.

González-León, C.M., Taylor, D.G., and Stanley, G.D., Jr., 1995, The Antimonio Formation in Sonora, Mexico and the Triassic-Jurassic boundary: Canadian Journal of Earth Sciences, v. 33, p. 418–428.

Grajales Nishimura, J.M., Terrel, D.J., and Damon, P.E., 1992, Evidencias de la prolongación del arco magmático cordillerano del Triásico Tardío-Jurásico en Chihuahua y Coahuila: Asociación Mexicana de Geólogos Petroleros Boletín, v. 42, p. 1–18.

Haenggi, W.T., 2002, Tectonic history of the Chihuahua trough, Mexico and adjacent USA, Part II: Mesozoic and Cenozoic: Boletín de la Sociedad Geológica Mexicana, tomo LV, p. 38–94.

Hamilton, W., Tosdal, R.M., Stone, P., and Haxel, G.B., 1987, Mesozoic tectonics of southeastern California, *in* Davis, G.H., and VandenDolder, E.N., eds., Geologic diversity of Arizona and its margins: Excursions to choice areas: Arizona Bureau of Geology and Mineral Technology, Geological Survey Branch, Special Paper 5, p. 337–350.

Harding, L.E., and Coney, P.J., 1985, The geology of the McCoy Mountains Formation, southeastern California and southwestern Arizona: Geological Society of America Bulletin, v. 96, p. 755–769, doi: 10.1130/0016-7606(1985)96<755:TGOTMM>2.0.CO;2.

Hardy, L.R., 1973, Geology of an allochthonous Jurassic sequence in the Sierra de Santa Rosa, northwest Sonora, Mexico [M.S. thesis]: San Diego, San Diego State University, 92 p.

Hardy, L.R., 1981, Geology of the central Sierra de Santa Rosa, northwest Sonora, Mexico, *in* Ortlieb, L., and Roldán-Quintana, J., eds., Geology of northwestern Mexico and southern Arizona: Boulder, Colorado, Geological Society of America, Cordilleran Section, Annual Meeting, Field Guides and Papers, p. 75–93.

Haxel, G.B., Tosdal, R.M., May, D.J., and Wright, J.E., 1984, Latest Cretaceous and Early Tertiary orogenesis in south-central Arizona: Thrust faulting, regional metamorphism, and granitic plutonism: Geological Society of America Society Bulletin, v. 95, p. 631–653, doi: 10.1130/0016-7606(1984)95<631:LCAETO>2.0.CO;2.

Haxel, G.B., Anderson, T.H., Riggs, N.R., and Goodwin, L.G., 1988, The Papago terrane: A crustal anomaly in south-central Arizona and north-central Sonora: Geological Society of America Abstracts with Programs, v. 20, no. 3, p. 168.

Howard, K.A., McCaffrey, K.J.W., Wooden, J.L., Foster, D.A., and Shaw, S.E., 1995, Jurassic thrusting of Precambrian basement over Paleozoic cover in the Clipper Mountains, southeastern California, *in* Miller, D.M., and Busby, C., eds., Jurassic magmatism and tectonics of the North American Cordillera: Geologic Society of America Special Paper 299, p. 375–392.

Imlay, R., 1939, Paleogeographic studies in northeastern Sonora: Geological Society of America Bulletin, v. 50, p. 1723–1744.

Iriondo, A., 2001, Proterozoic basements and their Laramide juxtaposition in NW Sonora, Mexico [Ph.D. thesis]: Tectonic constraints of the SW margin of Laurentia: Boulder, University of Colorado, 222 p., 3 plates.

Iriondo, A., and Premo, W., 2003, The Caborca block: An inhomogeneous piece of Paleoproterozoic crust in Sonora: Geological Society of America Abstracts with Programs, v. 35, no. 4, p. 67.

Iriondo, A., Miggins, D., and Premo, W., 2003, The Aibo-type (1.1 Ga) granitic magmatism in NW Sonora, Mexico: Failed continental rifting of Rodinia?: Geological Society of America Abstracts with Programs, v. 35, no. 5, p. 84.

Jacques-Ayala, C., 1995, Paleogeography and provenance of the Lower Cretaceous Bisbee Group in the Caborca–Santa Ana area, northwestern Sonora, *in* Jacques-Ayala, C., Gonzalez-León, C.M., and Roldán-Quintana, J., eds., Studies of the Mesozoic of Sonora and adjacent areas: Geological Society of America Special Paper 301, p. 79–98.

Jacques-Ayala, C., García, Y., Barragán, J.C., and DeJong, K.A., 1990, Caborca-Altar geology: Cretaceous sedimentation and compression, Tertiary uplift and extension *in* Gehrels, G.E., and Spencer, J.E., eds., Geologic excursions through the Sonoran desert region, Arizona and Sonora: Arizona Geological Special Paper 7, p. 183–202.

James, E.W., 1989, Southern extension of the Independence dike swarm of eastern California: Geology, v. 17, p. 587–590, doi: 10.1130/0091-7613(1989)017<0587:SEOTID>2.3.CO;2.

Jones, N.W., McKee, J.W., Anderson, T.H., and Silver, L.T., 1995, Jurassic volcanic rocks in northeastern Mexico: A possible remnant of a Cordilleran magmatic arc [abs.], *in* Jacques-Ayala, C., Gonzalez-León, C.M., and Roldán-Quintana, J., eds., Studies of the Mesozoic of Sonora and adjacent areas: Geological Society of America Special Paper 301, p. 179.

Keller, W.T., 1928, Stratigraphische Beobachtungen in Sonora (nordwest Mexico): Eclogae Geologicae Helvetiae, v. XXI, p. 327–335.

Ketner, K.B., 1986, Eureka Quartzite in Mexico?—Tectonic implications: Geology, v. 14, p. 1027–1030, doi: 10.1130/0091-7613(1986)14<1027: EQIMI>2.0.CO;2.

King, R.E., 1939, Geological reconnaissance in northern Sierra Madre Occidental of Mexico: Geological Society of America Bulletin, v. 50, p. 1625–1722.

Klitgord, K.D., and Schouten, H., 1980, Mesozoic evolution of the Atlantic, Caribbean and Gulf of Mexico [abs.] *in* Pilger, R.H., The Origin of the Gulf of Mexico and the Early Opening History of the Central North Atlantic Symposium Proceedings: Baton Rouge, Louisiana State University, p. 100–101.

Klitgord, K.D., and Schouten, H., 1986, Plate kinematics of the central Atlantic, *in* Vogt, P.R., and Tucholke, B.E., eds., The western North Atlantic region:

Boulder, Geological Society of America, Decade of North America Geology, v. M, p. 351–377.

Kluth, C.F., 1983, Geology of northern Canelo Hills and implications for the Mesozoic tectonics of southeastern Arizona, *in* Reynolds, M.W., and Dolly, E.D., eds., Mesozoic paleogeography of the west-central United States: Society of Economic Paleontologists and Mineralogists, Rocky Mountain Section, Rocky Mountain Paleogeography Symposium 2, p. 159–171.

Laubach, S.E., Reynolds, S.J., and Spencer, J.E., 1987, Mesozoic stratigraphy of the Granite Wash Mountains, west-central Arizona, *in* Dickinson, W.R., and Klute, M.A., eds., Mesozoic geology of southern Arizona and adjacent areas: Arizona Geological Society Digest, v. 18, p. 91–100.

Lawton, T.F., and Olmstead, G.A., 1995, Stratigraphy and structure of the lower part of the Bisbee Group, northeastern Chiricahua Mountains, Arizona, *in* Jacques-Ayala, C., Gonzalez-León, C.M., and Roldán-Quintana, J., eds., Studies of the Mesozoic of Sonora and adjacent areas: Geological Society of America Special Paper 301, p. 21–39.

LePichon, X., and Fox, J.P., 1971, Marginal offsets, fracture zones, and the early opening of the North Atlantic: Journal of Geophysical Research, v. 76, p. 6294–6308.

Leveille, G.P., 1984, Geology of El Capitan, Sonora, Mexico [M.Sc. thesis]: San Diego, California, San Diego State University, 122 p. (Unpublished).

Longoria, J.F., 1980, Estudios y trabajos geológicos en las areas del Estrada de Sonora, donde afloran las rocas del Precambrico: Universidad de Sonora, Departamento de Geología, 158 p.

Longoria, J.F., 1981, Geologic features of northwest Sonora, *in* Longoria, J., ed., Regional geology of northwest Sonora: Geological Society of America, Cordilleran Section Annual Meeting, Hermosillo, Sonora, Mexico, 1981, Proceedings, p. 1–48.

Longoria, J.F., and González, M.A., 1981, Description of the localities to be visited during the GSA field excursion no. 4, *in* Longoria, J., ed., Regional geology of northwest Sonora: Geological Society of America, Cordilleran Section, Annual Meeting, Hermosillo, Sonora, Mexico, 1981, Proceedings, p. 49–64.

Longoria, J.F., and Perez, V.A., 1979, Bosquejo geológico de los cerros Chino y Rajon, cuadrangulo Pitiquito-La Primavera (NW de Sonora): Universidad de Sonora, Boletín Departamento de Geología, v. 1, p. 119–144.

Longoria, J.F., González, M.A., Mendoza, J.J., and Pérez, V.A., 1978, Consideraciones estructurales del Cuadrangulo Pitiquito-La Primavera, NW de Sonora: Universidad de Sonora, Boletín Departamento de Geología, v. 1, no. 1, p. 61–67.

López-Infanzón, M., 1986, Estudio petrogenético de las rocas igneas en las formaiciones Huizachal y Nazas: Boletín Sociedad Geológica Mexicana, v. 47, no. 2, p. 1–42.

Lucas, S.G., Silberling, N.J., and Marzolf, J.E., 1997, Triassic sequence stratigraphy and structural complexity in the Sierra del Alamo Muerto, Sonora, Mexico, *in* Gonzalez-Leon, C.M., and Stanley, G.D., Jr., eds., U.S.-Mexico Cooperative Research–International Workshop on the Geology of Sonora Memoir: Universidad Nacional Autónoma de México, Instituto de Geología, Estación Regional Noroeste, Publicaciones Ocasionales 1, p. 46–47.

Ludwig, K.L., 2001, Isoplot/Ex, version 2.49: A geochronological toolkit for Microsoft Excel: Berkeley Geochronology Center, Special Publication No. 1a, 56 p.

Marzolf, J.E., 1997, Sequence stratigraphy of the Colorado Plateau, early Mesozoic paleogeography, and the Sonoran megashear, *in* Gonzalez-Leon, C.M., and Stanley, G.D., Jr., eds., U.S.-Mexico Cooperative Research–International Workshop on the Geology of Sonora Memoir: Universidad Nacional Autónoma de México, Instituto de Geología, Estación Regional Noroeste, Publicaciones Ocasionales 1, p. 48–52.

May, D.J., 1989, Late Cretaceous intra-arc thrusting in southern California: Tectonics, v. 8, no. 6, p. 1159–1173.

McKee, J.W., Jones, N.W., and Long, L.L., 1990, Stratigraphy and provenance of strata along the San Marcos fault, Central Coahuila, Mexico: Geological Society of America Bulletin, v. 102, p. 593–614, doi: 10.1130/0016-7606(1990)102<0593:SAPOSA>2.3.CO;2.

McKee, J.W., Jones, N.W., and Anderson, T.H., 1999, Late Paleozoic and early Mesozoic history of the Las Delicias terrane, Coahuila, Mexico, *in* Bartolini, C., Wilson, J.L., and Lawton, T.F., eds., Mesozoic sedimentary and tectonic history of north-Central Mexico: Geological Society of America Special Paper 340, p. 161–189.

McKee, M.B., 1991, Deformation and stratigraphic relationships of Mid-Cretaceous to Early Tertiary mass gravity slides in a marine basin in Sonora, Mexico [Ph.D. thesis]: Pittsburgh, Pennsylvania, University of Pittsburgh, 286 p. (Unpublished).

McKee, M.B., and Anderson, T.H., 1998, Mass-gravity deposits and structures in the Lower Cretaceous of Sonora, Mexico: Geological Society of America Bulletin, v. 110, no. 12, p. 1516–1529, doi: 10.1130/0016-7606(1998)110<1516:MGDASI>2.3.CO;2.

Merriam, C.W., 1963, Geology of the Cerro Gordo mining district Inyo County, California: U.S. Geological Survey Professional Paper 408, 83 p.

Merriam, R., 1972, Reconnaissance geologic map of the Sonora Quadrangle, northwest Sonora, Mexico: Geological Society of America Bulletin, v. 83, no. 11, p. 3533–3536.

Miller, J.S., and Walker, J.D., 2002, Mesozoic geologic evolution of Alvord Mountain, central Mojave Desert, California, *in* Glazner, A.F., Walker, J.D., and Bartley, J.M., eds., Geologic evolution of the Mojave Desert and southwestern Basin and Range: Geological Society of America Memoir 195, p. 59–77.

Miller, J.S., Glazner, A.F., Walker, J.D., Martin, M.W., and Harrison, S.E., 1991, Stratigraphy and Mesozoic contractional deformation at Alvord Mountain, central Mojave Desert, California: Geological Society of America, Cordilleran Section, Abstracts with Program, v. 23, p. A79.

Miller, M.M., 1987, Dispersed remnants of a northeast Pacific fringing arc; Upper Paleozoic terranes of Permian McCloud faunal affinity, western U.S: Tectonics, v. 6, p. 807–830.

Molina-Garza, R.S., and Geissman, J.W., 1996, Timing of deformation and accretion of the Antimonio terrane, Sonora, from paleomagnetic data: Geology, v. 24, no. 12, p. 1131–1134, doi: 10.1130/0091-7613(1996)024<1131:TODAAO>2.3.CO;2.

Molina-Garza, R.S., and Geissman, J.W., 1999, Paleomagnetic data from the Caborca terrane, Mexico: Implications for Cordilleran tectonics and the Mojave-Sonora megashear hypothesis: Tectonics, v. 18, no. 2, p. 293–325, doi: 10.1029/1998TC900030.

Murray, G.E., 1959, Introduction and regional geologic summary of field trip area, southeastern Coahuila and Nuevo Leon: South Texas Geological Society, Field Trip Guidebook, p. A1–A4.

Murray, G.E., 1961, Geology of the Atlantic and Gulf Coastal Provinces of North America: New York, Harper & Brothers, 692 p.

Murray, G.E., 1986, Musings about some of the tectonics of the southwestern United States and northern Mexico: The Texas Journal of Science, v. 38, no. 4, p. 301–326.

Nairn, A.E.M., 1976, A paleomagnetic study of certain Mesozoic formations in northern Mexico: Physics of the Earth and Planetary Interiors, v. 13, p. 47–56, doi: 10.1016/0031-9201(76)90106-0.

Nourse, J.A., 1995, Jurassic-Cretaceous paleogeography of the Magdalena region, northern Sonora, and its influence of the positioning of metamorphic core complexes, *in* Jacques-Ayala, C., González-León, C.M., and Roldán-Quintana, J., eds., Studies of the Mesozoic of Sonora and adjacent areas: Geological Society of America Special Paper 301, p. 59–78.

Nourse, J.A., 2001, Tectonic insights from stretched Upper Jurassic–Lower Cretaceous conglomerate, Caborca-Altar region, Sonora, Mexico, *in* Gonzalez, C.M., ed., Geology of northwestern Mexico and adjacent areas: Journal of South American Earth Sciences, v. 14, No. 5, p. 453–474.

Nourse, J.A., Anderson, T.H., and Silver, L.T., 1994, Tertiary metamorphic core complexes in Sonora, northwestern Mexico: Tectonics, v. 13, no. 5, p. 1161–1184, doi: 10.1029/93TC03324.

Nourse, J.A., Premo, W.R., and Iriondo, A., 2003, Gneisses of NW Sonora–SW Arizona border region: Two distinct Paleoproterozoic complexes at the southwest edge of Laurentia: Geological Society of America Abstracts with Programs, v. 35, no. 4, p. 67.

Olmstead, G., 1999, Dinoflagellate cyst biostratigraphy of the La Casita Formation (Upper Jurassic), Cerro Los Panales, Chihuahua, Mexico, *in* Bartolini, C., Wilson, J.L., and Lawton, T.F., eds., Mesozoic sedimentary and tectonic history of north-central Mexico: Geological Society of America Special Paper 340, p. 111–134.

Ortega-Gutierrez, F., Ruiz, J., and Centeno-Garcia, E., 1995, Oaxaquia, a Proterozoic microcontinent accreted to North America during the late Paleozoic: Geology, v. 23, no. 12, p. 1127–1130, doi: 10.1130/0091-7613(1995)023<1127:OAPMAT>2.3.CO;2.

Pineda, A.S., 1961, Parte II, Flora fosil de la Formacion Santa Clara (Carnico) del Estrada de Sonora: Universidad Nacional Autónoma de México, Instituto de Geologia, Paleontologia Mexicana 11, pt. 2, 36 p.

Poole, F.G., and Madrid, R.J., 1988, Comparison of allochthonous Paleozoic eugeoclinal rocks in the Sonoran, Marathon, and Antler orogens: Geological Society of America Abstracts with Programs, v. 20, no. 7, p. A-267.

Poole, F.G., and Perry, W.J., 1997, The late Paleozoic Ouachita-Marathon-Sonora orogenic system along the margin of North America: Geological Society of America Abstracts with Programs, v. 29, no. 2, p. 43.

Poole, F.G., Stewart, J.H., Berry, W.B.N., Harris, A.G., Repetski, J.E., Madrid, R.J., Ketner, K.B., Carter, C., and Morales-Ramirez, J.M., 1995, Ordovician ocean-basin rocks Sonora, Mexico, *in* Cooper, J.D., Droser, M.L., and Finney, S.C., eds., Ordovician odyssey: Society of Economic Paleontologists and Mineralogists, Short Papers for the International Symposium on the Ordovician System, 7th, Las Vegas, Nevada, Pacific Section, book no. 77, p. 277–284.

Premo, W.R., Iriondo, A., and Nourse, J.A., 2003, U-Pb zircon geochronology of Paleoproterozoic basement in northwestern Mexico: Evidence for affinity to SW US provinces: Geological Society of America Abstracts with Programs, v. 35, no. 5, p. 67.

Pubellier, M., Rangin, C., Rascon, B., Chorowicz, J., and Bellon, H., 1995, Cenomanian thrust tectonics in the Sahuaripa region, Sonora: Implications about northwestern Mexico megashears, *in* Jacques-Ayala, C., Gonzalez-León, C.M., and Roldán-Quintana, J., eds., Studies on the Mesozoic of Sonora and adjacent areas: Geological Society of America Special Paper 301, p. 111–120.

Rämö, O.T., and Calzia, J.P., 1998, Nd isotopic composition of cratonic rocks in the southern Death Valley region: Evidence for a substantial Archean source component in Mojavia: Geology, v. 26, no. 10, p. 891–894, doi: 10.1130/0091-7613(1998)026<0891:NICOCR>2.3.CO;2.

Rangin, C., 1977, Sobre la presencia del Jurásico Superior con amonitas en Sonora septentrional: Universidad Nacional Autónoma de México, Instituto de Geología, Revista, v. 1, p. 44–47.

Ransome, F.L., 1904, Geology and ore deposits of the Bisbee Quadrangle Arizona: U.S. Geological Survey Professional Paper 21, 167 p.

Richard, S.M., Spencer, J.E., Tosdal, R.M., and Stone, P., 1993, Preliminary geologic map of the southern Plomosa Mountains, La Paz County, Arizona: Tucson, Arizona, Arizona Geological Survey Open File Report 93-9, 27 p.

Rodríguez-Castañeda, J.L., 1984, Geology of Tuape region, north-central Sonora, Mexico [M.S. Thesis]: Pittsburgh, Pennsylvania, University of Pittsburgh, 157 p.

Rodríguez-Castañeda, J.L., 1988, Estratigrafia de la region de Tuape, Sonora, Mexico: Universidad Nacional Autónoma de México, Instituto de Geología, Revista, v. 7, p. 52–66.

Rodríguez-Castañeda, J.L., 1990, Relaciones estructurales en la parte centro-septentrional del estado de Sonora: Universidad Nacional Autónoma de México, Instituto de Geología, Revista, v. 9, p. 51–61.

Rodríguez-Castañeda, J.L., 1994, Geología del área El Teguachi, Sonora, México: Universidad Nacional Autónoma de México, Instituto de Geología, Revista, v. 11, p. 11–28.

Rodríguez-Castañeda, J.L., 2000, Deformación Sonorense: Tectónica extensional en el área Rancho San Antonio, Sonora norte-central—sus efectos y características en rocas cretácicas y terciarias [Ph.D. thesis]: México, D.F., Universidad Nacional Autónoma de México, 217 p.

Rodríguez-Castañeda, J.L., Garcia y Barragán, J.C., and Anderson, T.H., 1997, Regional faults along the southern margin of the Cananea Uplift—A record of Late Cretaceous inversion of the Sonora basin [abs.]: Resumenes (Sesiones Tematicas)–II Convención Sobre La Evolución Geológica De México Y Recursos Asociados, Pachuca, Hidalgo, Universidad Nacional Autónoma de México, p. 68.

Ross, C.A., and Ross, J.R.P., 1983, Late Paleozoic accreted terranes of western North America, *in* Stevens, C.H., ed., Pre-Jurassic rocks in western North America suspect terranes: Los Angeles, Society of Economic Paleontologists and Mineralogists, Pacific Coast Section, p. 7–22.

Ruiz, J., Tosdal, R.M., Restrepo, P.A., and Murillo-Muneton, G., 1999, Pb isotope evidence for Colombia–southern Mexico connection in the Proterozoic, *in* Ramos, V.A., and Keppie, J.D., eds., Laurentia-Gondwana connections before Pangea: Geological Society of America Special Paper 336, p. 183–197.

Salas, G.A., 1968, Areal geology and petrology of the igneous rocks of the Santa Ana region, northwestern Sonora: Boletín Sociedad Geologica Mexicana, v. XXXI, p. 11–64.

Sanchez-Zavala, J.L., Centeno-Garcia, E., and Ortega-Gutierrez, F., 1999, Review of Paleozoic stratigraphy of Mexico and its role in the Gondwana-Laurentia connections, *in* Ramos, V.A., and Keppie, J.D., eds., Laurentia-Gondwana connections before Pangea: Geological Society of America Special Paper 336, p. 211–226.

Schermer, E.R., Stephens, K.A., and Walker, J.D., 2001, Paleogeographic and tectonic implications of the geology of the Tiefort Mountains, northern Mojave Desert, California: Geological Society of America Bulletin, v. 113, p. 920–938, doi: 10.1130/0016-7606(2001)113<0920:PATIOT>2.0.CO;2.

Sharp, R.V., 1979, Some characteristics of the eastern Peninsular Ranges mylonite zone: Proceedings of Conference VIII "Analysis of actual fault zones in bedrock," convened by R. Speed, R. Sharp, and J. Everden: U.S. Geological Survey Open File Report 79-1239, p. 258–269.

Silver, L.T., 1963, The use of cogenetic uranium-lead isotope systems in zircons in geochronology: Radioactive dating, International Atomic Energy Agency, Athens, Nov. 1992, p. 279–285.

Silver, L.T., 1965, Mazatzal orogeny and tectonic episodicity: Geological Society of America Special Paper 82, p. 185–186.

Silver, L.T., 1967, Apparent age relations in the older Precambrian stratigraphy of Arizona [abs.], *in* Burwash, R.A., and Morton, R.D., eds., Geochronology of Precambrian stratified rocks: Edmonton, University of Alberta, p. 87.

Silver, L.T., 1969, Precambrian batholiths of Arizona: Geological Society of America Special Paper 121, p. 558–559.

Silver, L.T., 1971, Problems of crystalline rocks of the Transverse Ranges: Geological Society of America Abstracts with Programs, v. 3, no. 4, p. 193–194.

Silver, L.T., 1978, Precambrian formations and Precambrian history in Cochise County, southeastern Arizona, *in* Callender, J.F., Wilt, J.C., and Clemons, R.E., eds., Land of Cochise–Southeastern Arizona: New Mexico Geological Society, Guidebook for 29th Field Conference, p. 157–163.

Silver, L.T., 1982, Evolution of crystalline rock terranes of the Transverse Ranges, southern California: Geological Society of America Abstracts with Programs, v. 14, no. 4, p. 234.

Silver, L.T., 1983, Paleogene overthrusting in the tectonic evolution of the Transverse Ranges, Mojave and Salinian Regions: Geological Society of America Abstracts with Programs, v. 15, no. 5, p. 438.

Silver, L.T., and Anderson, T.H., 1974, Possible left-lateral early to middle Mesozoic disruption of the southwestern North American craton margin: Geological Society of America Abstracts with Programs, v. 6, no. 7, p. 955–956.

Silver, L.T., and Anderson, T.H., 1983, Further evidence and analysis of the role of the Mojave-Sonora megashear(s) in Mesozoic Cordilleran tectonics: Geological Society of America Abstracts with Programs, v. 15, no. 5, p. 273.

Silver, L.T., and Barker, F., 1968, Geochronology of Precambrian rocks of the Needle Mountains, southwestern Colorado: Part 1, U-Pb zircon results: Geological Society of America Special Paper 115, p. 204–205.

Silver, L.T., and Deutsch, S., 1961, Uranium-lead method on zircons: New York Academy of Science Annual Reports, v. 91, p. 279–283.

Silver, L.T., and Deutsch, S., 1963, Uranium-lead isotope variations in zircons: A case study: Journal of Geology, v. 71, p. 721–728.

Silver, L.T., McKinney, C.R., and Wright, L.A., 1961, Some Precambrian ages in the Panamint Range, Death Valley, California [abs.]: Geological Society of America Special Paper 68, p. 55.

Silver, L.T., McKinney, C.R., Deutsch, S., and Bolinger, J., 1963, Precambrian age determinations in the western San Gabriel Mountains, California: Journal of Geology, v. 71, p. 196–214.

Silver, L.T., Bickford, M., Van Schmus, W., Anderson, J., Anderson, T.H., and Medaris, L.G., 1977, The 1.4–1.5 b.y. transcontinental anorogenic plutonic perforation of North America: Geological Society of America Abstracts with Programs, v. 9, p. 1176–1177.

Simpson, C., 1984, Borrego-Santa Rosa mylonite zone: A Late Cretaceous west-directed thrust in southern California: Geology, v. 12, p. 8–11, doi: 10.1130/0091-7613(1984)12<8:BSRMZA>2.0.CO;2.

Sosson, M., and Calmus, T., 1990, Early Late Cretaceous overthrusting in northern Sonora responsible for a large-scale displacement of the Nevadian (sic.) Belt, *in* Auboin, J., and Bourgois, J., eds., Tectonics of circum-Pacific continental margins: Proceedings, 28th International Geological Congress, 1989: Utrecht, The Netherlands, VSP, p. 37–50.

Snow, J.K., 1992, Large magnitude Permian shortening and continental margin tectonics in the southern Cordillera: Geological Society of America Bulletin, v. 104, no. 1, p. 80, doi: 10.1130/0016-7606(1992)104<0080:LMPSAC>2.3.CO;2.

Stanley, J.R., Jr., 1997, Upper Triassic fossils from the Antimonio Formation, Sonora, and their implications for paleoecology and paleogeography, *in* González-León, C.M., and Stanley, G.D., Jr., eds., U.S.–Mexico Cooperative Research–International Workshop on the Geology of Sonora Memoir: Universidad Nacional Autónoma de México, Instituto de Geología, Estación Regional Noroeste, Publicaciones Ocasionales 1, p. 62–65.

Stanley, J.R., Jr., and González-León, C.M., 1995, Paleogeographic and tectonic implications of Triassic fossils and strata from the Antimonio Formation, northwestern Sonora, *in* Jacques-Ayala, C., González-León, C.M., and Roldán-Quintana, J., eds., Studies on the Mesozoic of Sonora and adjacent areas: Geological Society of America Special Paper 301, p. 1–16.

Stephens, W.E., 1988, Reconnaissance geology of the Cucurpe region, north-central Sonora, Mexico [M.S. thesis]: Pittsburgh, Pennsylvania, University of Pittsburgh, 317 p.

Stephens, W.E., Anderson, T.H., and Silver, L.T., 1986, La Lamina thrust sheet—A far traveled allochthon of crystalline basement from northwestern Mexico: Geological Society of America Abstracts with Programs, v. 18, p. 762.

Stevens, C.H., 1995, A giant Permian fusulinid from east-central Alaska with comparison of all giant fusulinids in western North America: Journal of Paleontology, v. 69, p. 805–812.

Stevens, C.H., 1997, Mississippian facies belts and the continental truncation hypothesis, *in* González-León, C.M., and Stanley, G.D., Jr., eds., U.S.–Mexico Cooperative Research– International Workshop on the Geology of Sonora Memoir: Universidad Nacional Autónoma de México, Instituto de Geología, Estación Regional Noroeste, Publicaciones Ocasionales 1, p. 65–66.

Stevens, C.H., Yancey, T.E., and Hanger, R.A., 1990, Significance of the provincial signature of Early Permian faunas of the eastern Klamath terrane, *in* Harwood, D.S., and Miller, M.M., eds., Paleozoic and early Mesozoic paleogeographic relations; Sierra Nevada, Klamath Mountains, and related terranes: Geological Society of America Special Paper 255, p. 201–218.

Stevens, C.H., Stone, P., and Kistler, R.W., 1992, A speculative reconstruction of the middle Paleozoic continental margin of southwestern North America: Tectonics, v. 11, no. 2, p. 405–419.

Stevens, C.H., Stone, P., Dunne, G.C., Greene, D.C., Walker, J.D., and Swanson, B.J., 1997, Paleozoic and Mesozoic evolution of east-central California: International Geology Review, v. 39, p. 788–829.

Stewart, J.H., 1970, Upper Precambrian and Lower Cambrian strata in the southern Great Basin, California and Nevada: U.S. Geological Survey Professional Paper 620, 206 p.

Stewart, J.H., and Poole, F.G., 1997, Inventory of Neoproterozoic and Paleozoic strata in Sonora, Mexico: U.S. Geological Survey Open-File Report 02-97, 50 p.

Stewart, J.H., McMenamin, M.A.S., and Morales-Ramirez, J.M., 1984, Upper Proterozoic and Cambrian rocks in the Caborca region, Sonora, Mexico—Physical stratigraphy, biostratigraphy, paleocurrent studies, and regional relations: U.S. Geological Survey Professional Paper 1309, 36 p.

Stewart, J.H., Anderson, T.H., Haxel, G.B., Silver, L.T., and Wright, J.E., 1986, Late Triassic paleogeography of the southern Cordillera—The problem of a source for voluminous volcanic detritus in the Chinle Formation of the Colorado Plateau region: Geology, v. 14, p. 567–570, doi: 10.1130/0091-7613(1986)14<567:LTPOTS>2.0.CO;2.

Stewart, J.H., Poole, F.G., Ketner, K.B., and Madrid, R.J., Roldán-Quintana, J., and Amaya-Martinez, R., 1990, Tectonics and stratigraphy of the Paleozoic and Triassic southern margin of North America, Sonora, Mexico, *in* Gehrels, G.E., and Spencer, J.E., eds., Geologic excursions through the Sonoran desert region, Arizona and Sonora: Arizona Geological Special Paper 7, p. 183–202.

Stewart, J.H., Blodget, R.B., Boucot, A.J., and Carter, J.L., 1993, Middle Paleozoic exotic terrane near Ciudad Victoria, northeastern Mexico, and the southern margin of Paleozoic North America: Proceedings of the First Circum-Pacific and Circum-Atlantic terrane Conference: Universidad Nacional Autónoma de México, Instituto de Geología, p. 147–149.

Stewart, J.H., Amaya-Martínez, R., Stamm, R.G., Wardlaw, B.R., Stanley, G.D., and Stevens, C.H., 1997, Mississippian to Jurassic rocks in Sierra Santa Teresa, Sonora, Mexico: Revista Mexicana de Ciencias Geológicas, v. 14, no. 2, p. 115–135.

Stewart, J.H., Blodget, R.B., Boucot, A.J., and Carter, J.L., and Lopez, R., 1999a, Exotic Paleozoic strata of Gondwana provenance near Ciudad Victoria, Tamaulipas, Mexico, *in* Ramos, V.A., and Keppie, J.D., eds., Laurentia-Gondwana connections before Pangea: Geological Society of America Special Paper 336, p. 227–252.

Stewart, J.H., Poole, F.G., Harris, A.G., Repetski, J.E., Wardlaw, B.R., Mamet, B.L., and Morales-Ramirez, J.M., 1999b, Neoproterozoic (?) to Pennsylvanian inner-shelf, miogeoclinal strata in Sierra Agua Verde, Sonora, Mexico: Revista Mexicana de Ciencias Geológicas, v. 16, no. 1, p. 35–62.

Stewart, J.H., Gehrels, G., Barth, A.P., Link, P.K., Christie-Blick, N., and Wrucke, C.F., 2001, Detrital zircon provenance of Mesoproterozoic to Cambrian arenites in the western United States and northwestern Mexico: Geological Society of America Bulletin, v. 113, no. 10, p. 1343–1356, doi: 10.1130/0016-7606(2001)113<1343:DZPOMT>2.0.CO;2.

Stewart, J.H., Amaya-Martínez, R., and Palmer, A.R., 2002, Neoproterozoic and Cambrian strata of Sonora, Mexico: Rodinian supercontinent to Laurentian Cordilleran margin, *in* Barth, A., ed., Contributions to crustal evolution of the southwestern United States: Geological Society of America Special Paper 365, p. 5–48.

Stone, P., and Stevens, C.H., 1988, An angular unconformity in the Permian section of east-central California: Geological Society of America Bulletin, v. 100, p. 547–551, doi: 10.1130/0016-7606(1988)100<0547:AAUITP>2.3.CO;2.

Swift, P., and Force, E., 2001, Subduction-margin assemblages in the Proterozoic Pinal and Cochise blocks, southeastern Arizona: American Journal of Science, v. 301, p. 755–772.

Tosdal, R.M., Haxel, G.B., and Wright, J.E., 1989, Jurassic geology of the Sonoran Desert region, southern Arizona, southeastern California, and northernmost Sonora: Construction of a continental-margin magmatic arc, *in* Jenny, J.P., and Reynolds, S.J., eds., Geologic evolution of Arizona: Arizona Geological Society Digest, v. 17, p. 397–434.

Valentine, W.G., 1936, Geology of the Cananea Mountains, Sonora, Mexico: Geological Society of America Bulletin, v. 47, p. 53–86.

Valenzuela-Navarro, L.C., Valencia-Moreno, M., Iriondo, A., and Premo, W., 2003, The El Creston granite: A new confirmed Paleoproterozoic locality in the Opodepe area, north-central Sonora: Geological Society of America Abstracts with Programs, v. 35, no. 4, p. 83.

Walker, J.D., Martin, M.W., Bartley, J.M., and Coleman, D.S., 1990, Timing and kinematics of deformation in the Cronese Hills, California, and implications for Mesozoic structure of the southwestern Cordillera: Geology, v. 18, p. 554–557, doi: 10.1130/0091-7613(1990)018<0554:TAKODI>2.3.CO;2.

White, D.E., and Guiza, R., 1948, Antimony deposits of El Antimonio District, Sonora, Mexico: U.S. Geological Survey Bulletin 962-B, p. 81–116.

Wooden, J.L. and Miller, D.M., 1990, Chronologic and isotopic framework for Early Proterozoic crustal evolution in the eastern Mojave Desert region, SE California: Journal of Geophysical Research, v. 95, no. B-12, p. 20,133–20,146.

Manuscript Accepted by the Society 24 March 2005

Geological Society of America
Special Paper 393
2005

Jurassic rocks in Sonora, Mexico: Relations to the Mojave-Sonora megashear and its inferred northwestward extension

Thomas H. Anderson*
Department of Geology and Planetary Sciences, University of Pittsburgh, Pittsburgh, Pennsylvania 15260, USA

José Luis Rodríguez-Castañeda*
Estación Regional del Noroeste, Instituto de Geología, Universidad Nacional Autónoma de México, Apartado Postal 1039, 83000 Hermosillo, Sonora, México

Leon T. Silver
Division of Geological and Planetary Science, California Institute of Technology, Pasadena, California 91125, USA

ABSTRACT

The Mojave-Sonora megashear constitutes a regional boundary between lithologically distinct Jurassic assemblages of different ages. North of the Mojave-Sonora megashear, arc-related volcanic, volcaniclastic, and clastic rocks, intruded by plutons (175–160 Ma) compose part of the Middle Jurassic (commonly ca. 175 Ma) igneous province, previously recognized in Arizona and California. Distinct domains among Jurassic igneous rocks in northern Sonora are: (1) southern Papago, a region where pre-Jurassic rocks are unknown, (2) Nogales-Cananea-Nacozari, where Jurassic rocks are underlain by 1.7–1.4 Ga crystalline basement, and (3) Mojave-Sonora, where strata, including Oxfordian beds, along the north side of the Mojave-Sonora megashear are commonly strongly deformed, as recorded by thrust faults, mylonitic foliation, and recumbent folds. The Mojave-Sonora domain extends across the southwestern margins of the southern Papago and the Nogales-Cananea-Nacozari domains. Strong deformation that distinguishes the zone markedly declines within a few tens of kilometers northward. South of the Mojave-Sonora megashear, in central and southern Sonora, Lower Jurassic clastic and volcaniclastic rocks distinguish the Caborca domain. Upper Jurassic sedimentary rocks, commonly conglomeratic, are abundant north of Mojave-Sonora megashear; a single occurrence is known south of the Mojave-Sonora megashear.

Waning of subduction-related Middle Jurassic magmatism was followed by the abrupt formation, ca. 165 Ma, of Coast Range, Josephine, Great Valley, and Devil's Elbow ophiolites and the Smartville Complex within oceanic pull-aparts west of the margin of the North America plate. The formation of ophiolitic rocks signaled the beginning of transtensional faulting.

*Anderson—taco@pitt.edu; Rodríguez-Castañeda—jlrod@servidor.unam.mx.

Anderson, T.H., Rodríguez-Castañeda, J.L., and Silver, L.T., 2005, Jurassic rocks in Sonora, Mexico: Relations to the Mojave-Sonora megashear and its inferred northwestward extension, *in* Anderson, T.H., Nourse, J.A., McKee, J.W., and Steiner, M.B., eds., The Mojave-Sonora megashear hypothesis: Development, assessment, and alternatives: Geological Society of America Special Paper 393, p. 51–95. doi: 10.1130/2005.2393(02). For permission to copy, contact editing@geosociety.org.

Almost contemporaneously (ca. 163 Ma) the lowest volcanic units and overlying coarse sedimentary beds began to accumulate in fault-bounded continental pull-apart basins such as the McCoy Mountains basin. Other transtensional basins, formed at releasing steps where pull-aparts formed, are well developed within the Papago domain and other parts of southwestern United States and northern Mexico.

From Sonora northward into California the Mojave-Sonora megashear fault zone, developed generally within the Middle Jurassic arc-parallel to the former continental margin, is inferred to link with strands of the Melones and Bear Mountain faults of the Foothills fault system, the Wolf Creek fault, and the Big Bend fault. A protuberance of Proterozoic basement (the Caborca block) that was truncated from the continental margin records ~800–1000 km of left-lateral offset. The displacement of the Caborca block took place south of a major releasing step along the Big Bend fault with the result that a regional pull-apart that coincides with the Great Valley of California developed.

Inboard of the Mojave-Sonora megashear Late Jurassic magmatic rocks crop out near faults at some releasing steps and within floors of some pull-apart structures. The distribution suggests that magma rose along faults and into areas of thin crust. In southern Arizona these igneous rocks are included as part of the Artesa layered sequence and the Ko Vaya plutonic suite.

Oxfordian and younger beds, which crop out north of the Mojave-Sonora megashear may contain exotic blocks and contractional structures that are contemporaneous with the Nevadan orogeny. The variation in the style and intensity of deformation of Middle and Upper Jurassic strata, and Upper Jurassic conglomerate rich in clasts derived from rocks of the Caborca domain, are postulated to record transpression near the Mojave-Sonora megashear that locally overlapped the more widespread transtensional structures in time and space.

The cessation of strike-slip faulting locally began ca. 150 Ma, as shown by undeformed intrusive bodies that cut older deformed Middle Jurassic rocks. By the time that the Independence dikes and correlative rocks were emplaced at 148 Ma, scant evidence of lateral faulting is known.

Intrusions, young volcanic cover, transecting strike-slip faults, and multiple generations of low-angle extensional and contractional faults obscure Jurassic structures in Sonora and southern California. Despite these complications, removal of the effects of superposed structures reveals a viable trace for an inferred Late Jurassic left-lateral fault linking the Mojave-Sonora megashear and more northerly fault segments. The position of this major inferred fault is constrained by distinctive tectonostratigraphic domains.

The Middle and Late Jurassic and earliest Cretaceous plate tectonic history includes (1) subduction (175–165 Ma), (2) coupling (ca. 165 Ma), (3) rifting, transtension, lateral faulting, transpression, and contraction (165–145 Ma), and (4) renewed subduction (ca. 135 Ma) along the western margin of the North America plate and terranes (e.g., Wrangellia) to the west. The structures that record the diverse plate processes and that are preserved best in the overriding North America plate are compatible with a consistently maintained easterly directed maximum compressive stress.

Keywords: Jurassic, megashear, Sonora, sinistral, transform.

INTRODUCTION

Overview

In Sonora the pattern of Jurassic igneous domains (Figs. 1 and 2) emerged from reconnaissance geologic and geochronologic studies conducted principally from 1968 until 1974, while Anderson held a postdoctoral appointment under Silver. Data from a few additional collections that were made after Anderson left Caltech are included. The yield from the study conducted throughout Sonora, mainly between 1968 and 1978, includes field observations, structural and stratigraphic measurements, maps, and U-Pb isotopic analyses of zircon populations separated from 13 Jurassic rock samples (Figs. 1 and A1; Table 1). Samples and geologic information were gathered during reconnaissance visits to ranges selected on the basis of information from the 1962 1:1,000,000 geologic map of Mexico and published reports by early workers who conducted regional surveys (e.g., Dumble, 1900; Flores, 1929; King, 1939; Salas, 1968). No topography was shown on the geologic maps. We acquired copies of 1:250,000 topographic maps a year or so after we began the reconnaissance. Generally our samples were collected along unmarked tracks to ranches, most of which were not identified on the available maps. Sample locations are based upon mileage from ranches and road intersections and conspicuous physiographic features. A few sites have been revisited and located on 1:50,000 topographic maps; unfortunately many have not and the locations are approximate. Samples were processed and zircon populations extracted and analyzed per methods described in the Appendices. Modifications in the techniques are noted in Anderson and Silver (1977) and Anderson et al. (1980). The reconnaissance nature of our work in a largely unmapped region of several tens of thousands of square kilometers led to the collection of many samples. We analyzed as many as we could with the intent of identifying and delineating Proterozoic and younger crystalline provinces. For many samples, analyses of two zircon fractions were sufficient to identify the province to which the rock belonged. Although the aspiration was to analyze additional zircon fractions for every sample, we were able to do so for only several samples that were unusual or important for geologic or geochronologic reasons. We exert caution with respect to the "ages" assigned to the analyzed rocks through the use of the phrase "interpreted isotopic age." This phrase indicates the tentative nature of the age estimates based upon two analyses and consideration of other factors, such as inheritance related to cores in the zircons or unidentified old zircons in the analyzed fraction.

Ages of the Sonoran rocks are assigned based upon interpreted ages derived from U-Pb isotopic results on zircon from igneous rocks (Table 1; Anderson and Silver, 1978, 1979). Interpreted U-Pb ages from volcanic, hypabyssal, and plutonic rocks cluster within the interval 180–145 Ma. Although the results and interpreted ages are not fully amenable to modern error analysis, geologic consistency confirms that the ages are meaningful; furthermore, some fractions are concordant and the data are both internally consistent and consistent with other analyses independently conducted on similar or the same rocks in Arizona.

Lithologically and chronologically distinct assemblages of Jurassic rocks are separated in Sonora, Mexico, by the Mojave-Sonora megashear. North of the Mojave-Sonora megashear, volcanic, volcaniclastic, and clastic rocks that are intruded by plutons together make up part of a Middle Jurassic volcanic province widespread in Arizona and California (e.g., Bassett and Busby, this volume; Haxel et al., this volume). South of the Mojave-Sonora megashear, in central and southern Sonora, fossiliferous Early Jurassic clastic and volcaniclastic rocks and subjacent Paleozoic, Neoproterozoic, and Proterozoic rocks distinguish the Caborca domain, also described as a block or terrane, from the Middle Jurassic arc to the northeast (Anderson and Silver, 1981; this volume; Stewart, this volume). The disparity between the rocks of the Caborca domain and the rocks of northeastern Sonora led to the conclusion that the Mojave-Sonora megashear must be a major fault (Silver and Anderson, 1974).

In Sonora the Mojave-Sonora megashear commonly limits the southwestern margin of a domain of deformed Middle and Late Jurassic rocks. Beds as young as Oxfordian are affected by the deformation that includes thrust faults, mylonitic foliation and folds, and exotic blocks, emplaced penecontemporaneously. The timing of deformation suggests correlation with the Nevadan orogeny of California (Silver and Anderson, 1983).

Restoration of >800 km of left-lateral displacement along the inferred trace of the Mojave-Sonora megashear removes duplicated and offset belts of Middle Jurassic magmatic rocks (Jones et al., 1995) and juxtaposes Early Jurassic, Triassic, and older rocks of the Caborca block (Anderson and Silver, this volume; Stewart, this volume) against lithologically similar and correlative units in California.

Purpose

The purpose of this article is to provide new information about the character, ages, and distribution of Sonoran Jurassic rocks and integrate this information into a regional tectonostratigraphic setting. We attempt to achieve this objective by: (1) Describing the occurrence and character of Jurassic rocks in Sonora, northwestern Mexico (Figs. 1 and 2). (2) Reporting U-Pb interpreted ages from isotopic analyses of zircon mainly from Jurassic volcanic and plutonic rocks (Table 1). (3) Characterizing the sharp discontinuity between the Caborca domain, a region characterized principally by Early Jurassic strata, and domains north of the postulated Mojave-Sonora megashear, where Early Jurassic sedimentary strata are unknown but where Middle and Late Jurassic rocks are common. (4) Proposing that Jurassic arc-rocks in northern Sonora record two distinct processes, namely, (a) magmatism in response to Middle Jurassic subduction beneath the southwestern continental margin of North America, and (b) transection by left-lateral faulting along the Mojave-Sonora megashear (Silver and Anderson, 1974; 1983; Anderson and Silver, 1979). (5) Discussing the hypothesis that lateral faulting was accompanied by local transpressional deformation and more widespread transtension (Anderson and Nourse, this volume) and that local thick sections of Late Jurassic sediments with minor volcanic interbeds accumulated in the transtensional basins. (6) Delineating the northwestward extension of the inferred trace of the hypothetical Mojave-Sonora megashear by describing the rocks and structures that constrain its position and amplifying existing hypotheses (e.g., Anderson and Schmidt, 1983; Harper et al., 1985; Wolf and Saleeby, 1995) by incorporating new ideas and information. (7) And last, by restoring Tertiary dextral displacements among the San Andreas set of faults; westward-directed thrusting in the Mojave region of California, and >800 km of left-lateral displacement along the inferred trace of the Mojave-Sonora megashear to illustrate the probable linkages

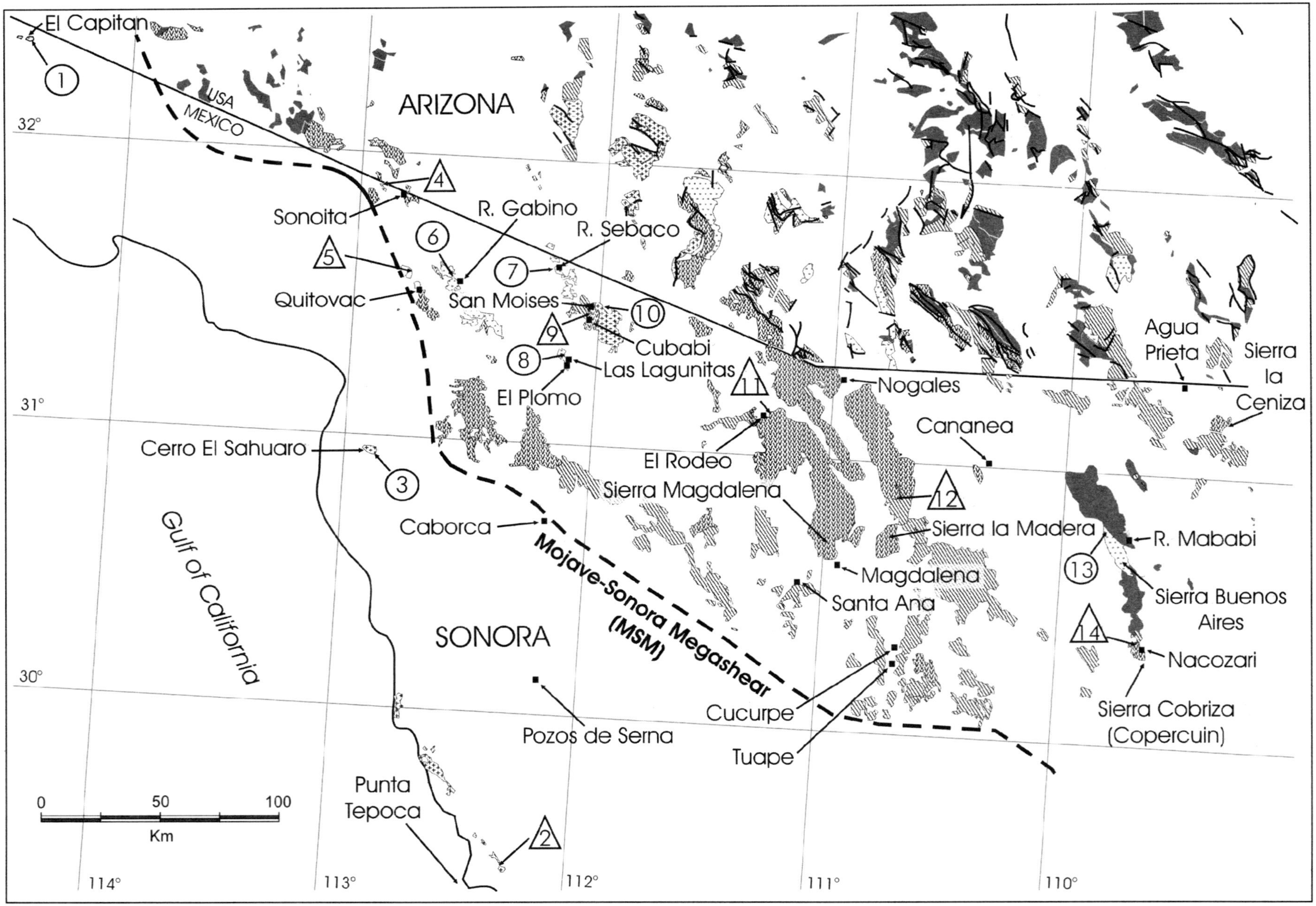

El Capitan
1
USA
MEXICO
ARIZONA
32°
4
Sonoita
R. Gabino
R. Sebaco
6
5
7
Quitovac
San Moises
10
9
Cubabi
8
Las Lagunitas
El Plomo
11
Nogales
Agua
Prieta
Sierra
la
Ceniza
31°
Cananea
Cerro El Sahuaro
3
El Rodeo
Sierra Magdalena
12
Gulf of California
Caborca
Sierra la Madera
R. Mababi
Mojave-Sonora Megashear
(MSM)
13
Magdalena
Santa Ana
Sierra Buenos
Aires
SONORA
14
Nacozari
30°
Cucurpe
Sierra Cobriza
(Copercuin)
Pozos de Serna
Tuape
Punta
Tepoca
0
50
100
Km
2
114°
113°
112°
111°
110°

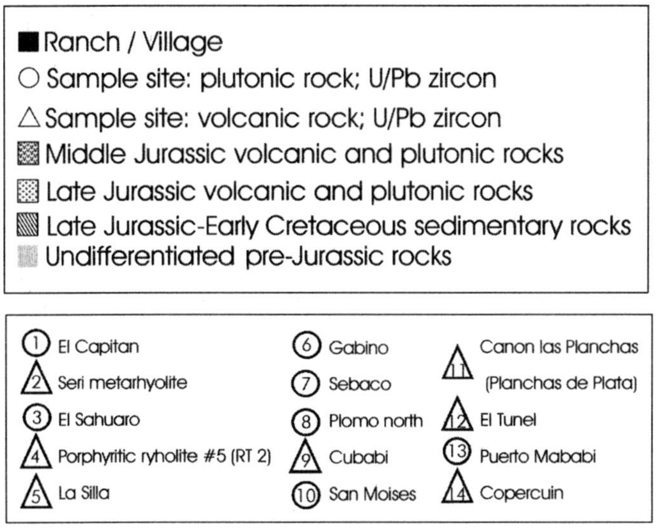

Figure 1. Map showing distribution of Jurassic rocks in northern Sonora and locations of samples from which zircons were extracted and analyzed using U-Pb method.

between the formation of the Gulf of Mexico and major lateral faults along the margin of western North America.

The distribution and character of Early Jurassic rocks in Sonora have been summarized recently by Stewart et al. (1997) and Stewart (this volume) as part of a broader initiative to advance knowledge of the Triassic and Jurassic rocks in Sonora (González-León and Stanley, 1997). We do not intend to duplicate this recent work but to provide complementary data about Sonoran Middle and Late Jurassic rocks and the structures they contain.

Previous Work

Jurassic rocks were first recognized and described in Sonora at Sierra de Santa Rosa and the surrounding region (Keller and Wellings, 1922; Keller, 1928; Jaworski, 1929; Burckhardt, 1930) (Fig. 2, loc. 24). In this region, a section of hundreds of meters of sandstone, shale, and limestone is assigned an Early Jurassic age based upon the occurrence of fossils, including *Weyla mexicana*. Flores (1929) recognized correlative fossiliferous strata at Sierra de Caracahui (Fig. 2, loc. 25) and near San José de Pimas. Later work between Sierra de Santa Rosa and Sierra Caracahui (Ramirez and Acevedo, 1957; Erben, 1956) corroborated these age assignments. To the northwest at Sierra del Alamo (Fig. 2, loc. 10), strata described by White and Guiza (1948) were found to contain Early Jurassic fossils (Imlay, 1943) correlative with those to the southeast. In the south, between Hermosillo and Arivechi, King (1939) mapped clastic beds with some coal horizons as part of the Late Triassic and Early Jurassic Barranca Formation.

Late Jurassic fossiliferous beds are known from only two Sonoran localities: (1) On the Caborca block at Cerros Pozo de Serna (Fig. 2, loc. 28), Beauvais and Stump (1976) described a section, >1000 m thick, composed of shale and lithic sandstone with conglomerate interbeds. Fossils from the shale beds are Late Jurassic (late Oxfordian to late Kimmeridgian). (2) Near Cucurpe, north of the Mojave-Sonora megashear discontinuity, Oxfordian ammonites are preserved in red sandstone and shale (Rangin, 1977). However, extensive exposures of unfossiliferous strata, especially north of the Mojave-Sonora megashear, are also probably of Late Jurassic age (Nourse, 1995, 2001).

In southeastern Arizona, early workers recognized volcanic and volcaniclastic strata they believed to be Jurassic age (e.g., Gilluly, 1956; Cooper and Silver, 1964; Drewes, 1968, 1971a, 1971b, 1971c). However, the stratigraphic position of these beds, above Paleozoic beds and beneath the Late Jurassic to Cretaceous Bisbee Group, constrains the age only within a broad early Mesozoic interval.

Subsequent field and isotopic studies in southern and western Arizona, northern Sonora, and southeastern California, summarized by Tosdal et al. (1989), reveal a Jurassic igneous province composed mainly of rhyodacitic flows, tuffs, volcaniclastic rocks and associated plutons. Additional fieldwork (e.g., Nourse, 1995) and U-Pb interpreted apparent ages from zircon in volcanic strata that yield Middle Jurassic ages (e.g., Asmerom et al., 1990; Riggs et al., 1993; Haxel et al., this volume) corroborate the presence of the regional igneous province. Quartzose sandstone, conglomerate, and rare carbonate beds are at some places present as conspicuous interbeds among pyroclastic horizons. Stratigraphic studies show that some sandstone beds are equivalent to units on the craton (Colorado Plateau) for which Jurassic ages are well established (Drewes, 1971a; Marzolf, 1980, 1982, 1983; Bilodeau and Keith, 1986; Bilodeau et al., 1987; Busby-Spera et al., 1990; Riggs and Haxel, 1990; Riggs et al., 1993).

These supracrustal sequences are intruded by calc-alkaline plutons of late Middle and early Late Jurassic age, and the entire suite of Jurassic rocks is inferred to represent part of a continental-margin magmatic arc (Tosdal et al., 1989). The alkaline character of some volcanic and plutonic rocks within the Jurassic suite, as well as the presence of massive sections of conglomerate and eolian sandstone, have led some workers to conclude that the magmatic arc developed in an extensional tectonic setting (Busby-Spera, 1988; Riggs and Busby-Spera, 1989; Busby-Spera et al., 1990; Haxel et al., this volume). Coarse clastic rocks intercalated with volcanic layers (Davis et al., 1979; Bilodeau et al., 1987) are interpreted to be fill related to Middle Jurassic calderas (Riggs, 1987; Riggs and Busby-Spera, 1990, 1991; Lipman and Hagstrum, 1992). Other lithostratigraphic features related to the calderas are recognized close to the Sonora-Arizona border, and correlative tuff and porphyry extend into Sonora (Segerstrom, 1986, 1987; Nourse, 1990; Riggs and Haxel, 1990; Riggs and Busby-Spera, 1991).

A stratigraphic transition from dominantly volcanic and volcaniclastic units to coarse clastic strata rich in debris derived from Paleozoic cover and crystalline rocks underlying the Jurassic volcanic sections has been recognized in southern Arizona (Briskey et al., 1978; Kluth et al., 1982; Kluth, 1983; Bilodeau et al., 1987), New Mexico (Lawton and Olmstead, 1995), and

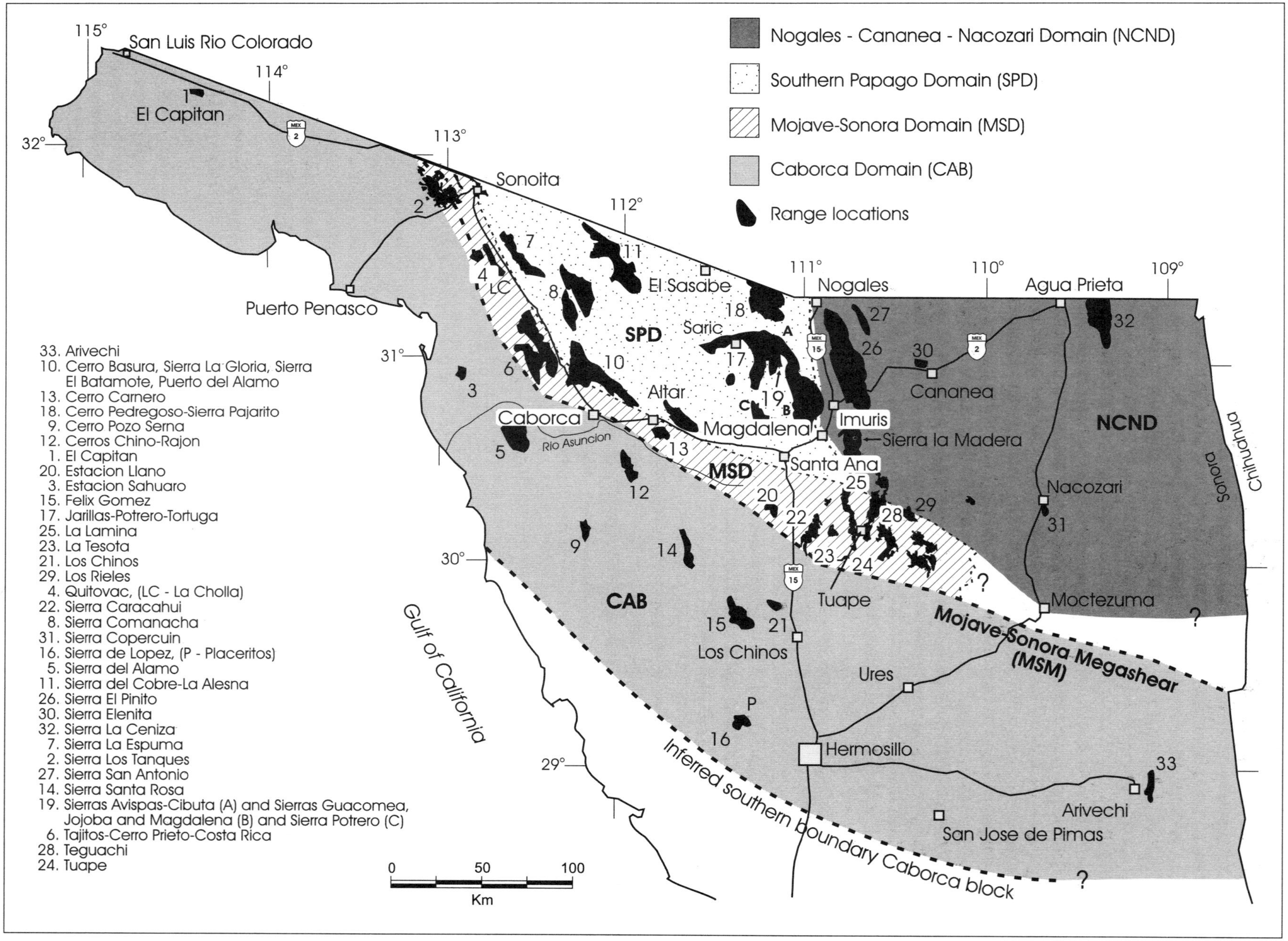

Figure 2. Inferred domains of Jurassic rocks in Sonora and locations of ranges mentioned in text.

north-central Sonora (Rodríguez-Castañeda, 1984, 1986, 1991). Isotopic studies reveal age differences between volcanic units that may be of ash-flow and caldera origin and stratigraphically higher tuff and andesite flow rock that occur as interbeds within the lower parts of thick sections of coarse clastic strata at the base of the sedimentary section.

Whereas pyroclastic rocks of the caldera assemblage yield U-Pb interpreted ages around 170 Ma (Haxel et al., this volume; Asmerom et al., 1990; Riggs et al., 1993), some conglomerate, high in the Jurassic section, may be as young as 151 Ma, as suggested by isotopic analyses of whole-rock Rb-Sr from samples in ash-flow tuffs in the Canelo Hills, Arizona (Kluth et al., 1982). Although neither geochemical (the samples were altered) nor geologic conditions (the samples were from multiple units) were optimal, this age clearly differs from older ages from lower volcanic units. Ash-flow tuff units interbedded within conglomerate south of Canelo Hills yield concordant biotite K-Ar and Rb-Sr whole-rock interpreted ages of 147 ± 6 and 149 ± 11 Ma (Marvin et al., 1978).

Conglomerate, stratigraphically high within the Middle Jurassic volcanic sections, is commonly correlated with conglomerate at the Glance Mine type locality near Bisbee, Arizona. The Glance Conglomerate is at the base of the sedimentary section that continues upward into the Bisbee Group and includes the fossiliferous Mural Limestone of Early Cretaceous age. The conglomerate contains abundant cobbles and boulders of Proterozoic Pinal schist and granite and Paleozoic carbonate that reflect the presence of source rocks along basin margins, which must have been steep and generally fault-controlled. The sedimentologic setting of the Glance Conglomerate has been interpreted as an indication that a change in tectonic setting from arc magmatism to rifting and basin development occurred during Late Jurassic time (Kluth et al., 1982; Bilodeau et al., 1987; Tosdal et al., 1989) in response to regional transtension (Anderson and Nourse, this volume). Additional age constraints for the transition from arc magmatism to transtension come from southern California where clastic rocks of the McCoy Mountains Formation, considered to be an equivalent to Glance Conglomerate, overlie lapilli tuff. U-Pb zircon ages of the lapilli tuff yield an age of 175 ± 8 Ma (Fackler-Adams et al., 1997) consistent with previously published U-Pb isotope ages from Jurassic volcanic rocks in southern Arizona (e.g., Riggs et al., 1993). Stratigraphically higher volcanic units that locally interfinger with siltstone of the lower McCoy Mountains Formation yield younger ages of 155 ± 8 and 162 ± 3 Ma. Based upon these results, Fackler-Adams et al. (1997) conclude that the commonly coarse clastic Late Jurassic strata record the abrupt waning of older Jurassic volcanism beginning in Callovian time.

DOMAINS OF JURASSIC ROCKS IN SONORA

Definitions

In Sonora, Mexico, the Mojave-Sonora megashear separates two regional provinces (Fig. 2). Early Jurassic sedimentary rocks occur only south and southwest of the Mojave-Sonora megashear, whereas Middle Jurassic volcanic, volcaniclastic, clastic, and plutonic rocks are concentrated north of the Mojave-Sonora megashear. The relation of the Middle Jurassic magmatic rocks to Early Jurassic rocks in central Sonora is a matter of debate (e.g., Anderson and Silver, 1979; Rangin, 1982; Stewart et al., 1984; Roure and Sosson, 1986; Pubellier et al., 1995). Late Jurassic rocks are mainly sedimentary and occur principally northeast of the Mojave-Sonora megashear, where they are common (Anderson and Nourse, this volume). However, Late Jurassic strata are not entirely areally restricted, and rare exposures occur south of the Mojave-Sonora megashear. In Sonora south of Hermosillo, little is known of the Jurassic rocks.

We divide the rocks of these two provinces into four domains based on geologic setting and structural characteristics (Fig. 2). In northern Sonora, Middle and Late Jurassic volcanic, volcaniclastic, clastic, and related plutonic rocks have been recognized north of the Mojave-Sonora megashear from Sonoita to Nacozari (Fig. 1). Among the Middle Jurassic igneous rocks and related sedimentary units (Fig. 2) three domains are recognized: the southern Papago domain of northwestern and north-central Sonora and adjacent Arizona (Fig. 3), the Nogales-Cananea-Nacozari domain within north-northeastern Sonora (Fig. 3), and the Mojave-Sonora domain along the Mojave-Sonora megashear itself (Fig. 4).

South of the Mojave-Sonora megashear the Caborca domain (Caborca block) includes Early Jurassic sedimentary strata and other sedimentary rocks, including underlying Triassic, Paleozoic, and Neoproterozoic units that rest unconformably upon Proterozoic (1.8–1.7 Ga and/or 1.4 Ga) crystalline basement. The Caborca domain refers to the same group of rocks as those included within the Caborca terrane (Campa and Coney, 1983) and also encompasses the Caborca block, the informally designated region of Proterozoic crystalline basement south of the Mojave-Sonora megashear, emphasized by Anderson and Silver (1981) (Fig. 2). The differences in age and origin of Jurassic rocks across the Mojave-Sonora megashear in conjunction with the differences of age between Proterozoic crystalline rocks underlying the Caborca block and Proterozoic rocks across the Mojave-Sonora megashear led Silver and Anderson (1974) to conclude that the contact between them is a major, regional, strike-slip fault traversing Sonora, which they called the Mojave-Sonora megashear. The Mojave-Sonora megashear roughly corresponds with Mexico Route 2 from Sonoita to Caborca and from Caborca eastward along Rio Asuncion to the point where the river bends northeastward toward Santa Ana (Fig. 4). From this point the inferred fault continues east toward Estación Llano and Tuape (e.g., Rodríguez-Castañeda, 1990).

In northern Sonora and southern Arizona, the absence of outcrops of Proterozoic crystalline basement beneath common exposures of Middle Jurassic supracrustal rocks led Haxel et al. (1984, 1988) to distinguish this part of the larger Jurassic igneous province of the southwestern United States and northwestern Mexico as the "Papago terrane." The absence of expo-

TABLE 1. U-Pb ANALYTICAL DATA FOR ZIRCONS FROM SAMPLES IN NORTHERN SONORA

Sample—Mesh Size	Observed ratios			Radiogenic Pb			Concentrations		Observed isotope ratios			Apparent ages (Ma)		
	Pb 206/Pb 204	Pb 207/Pb 204	Pb 208/Pb 204	Pb 206	Atom % Pb 207	Pb 208	(ppm) Pb^{rad}	(ppm) U	Pb206/U238	Pb207/U235	Pb207/Pb206	Pb206/U238	Pb207/U235	Pb207/Pb206
Cubabi Rhyolite #1[c]														
R200+P200 Franz 1.7 NM 2°/20°	391.8	34.2	118.7	78.98	3.93	17.08	11.81	409.65	0.0277	0.1901	0.0498	176	177	187
R200+P200 Franz 1.7 M 5°/20°	154.3	22.3	66.6	79.33	3.95	16.72	15.74	588.93	0.0246	0.1692	0.0498	157	159	184
Copercuin Rhyolite #502[b]														
R200 Franz 1.7 NM 0°/20°	183.4	24.4	81.0	76.13	4.06	19.82	13.73	441.29	0.0274	0.2015	0.0533	174	186	340
El Capitan gneiss #1[a]														
R200 Franz 1.7 NM 0°/20°	291.2	31.0	80.6	85.00	4.31	10.69	12.76	475.0	0.0264	0.1848	0.0507*	168	172	228
R200 Franz 1.7 M 0°/20°*	438.7	36.6	1433.8	84.09	4.18	11.73	12.45	470.0	0.0258	0.1765	0.0497	164	165	181
R200 Franz 1.7 M 1°/25°*	447.0	37.3	1524.7	83.93	4.22	11.84	15.29	616.0	0.0241	0.1672	0.0503	154	157	210
*from conc.														
El Sahuaro quartz monzonite[a]														
R200 Franz 1.7 NM $^1/_2$ °/20°	283.4	29.9	4093.9	82.34	4.08	13.59	19.36	765.1	0.0241	0.165	0.0525	153	155	308
R200 Franz 1.7 M 1 °/20°	550.1	42.0	125.8	82.34	4.08	13.59	30.03	1199.0	0.0239	0.1630	0.0495	152	153	173
El Tunel quartz porph #2[c]														
R200 Franz 1.7 NM 1°/20°	634.0	46.1	138.1	82.49	4.09	13.42	16.22	569.1	0.0273	0.1867	0.0496	174	174	174
R200 Franz 1.7 NM 1°/20° [a]	528.4	40.9	121.0	82.49	4.09	13.42	16.20	569.1	0.0273	0.1867	0.0496	174	174	176
[a]Spike before fusion calculation using rad.206/208 = 6.144916 from Pb comp only.														
Gabino porphyritic quartz monzonite[c]														
R200 Franz 1.7 NM 1°/20°	1196.6	74.0	345.9	76.28	3.78	19.94	19.99	664.25	0.0267	0.1823	0.0496	170	170	175
P200 Franz 1.7 M 2°/25°	1400.7	84.1	474.7	73.23	3.63	23.14	28.51	931.0	0.0260	0.1780	0.0496	166	166	174
Canon Las Planchas rhyolite #6[c]														
P200 Franz 1.7 NM 2°/25° comp	368.74	33.0	107.17	80.18	3.97	15.84	17.27	620.43	0.0259	0.1772	0.0496	165	166	174
P200 Franz 1.7 M 3°/25° comp	355.47	32.32	99.79	81.10	4.03	14.87	31.52	1246.83	0.0238	0.1630	0.0496	152	153	178
P200 Franz 1.7 NM 2°/25° [a]	298.5	29.5	93.3	80.18	3.98	15.84	17.30	620.4	0.0259	0.1773	0.0496	165	166	177
P200 Franz 1.7 M 3°/25° [b]	325.5	30.8	94.3	81.10	4.02	14.88	31.50	1246.8	0.0238	0.1629	0.496	152	153	175
[a]Based on sbf using rad. 206/208 = 5.061521 from Pb comp only.														
[b]Based on sbf using rad. 206/208 = 5.452245 from Pb comp only.														
Libertad rhyolite #3														
P200 Franz 1.7 NM 0°/25° [c]	679.3	48.2	113.4	85.95	4.24	9.80	9.73	497.2	0.0196	0.1331	0.0494	125	127	165
P200 Franz 1.7 M 3°/25° [a]	440.5[b]	36.2	105.5	82.75	4.03	13.22	12.60	623.0	0.0195	0.1308	0.0487	124	125	135
[a]SBF conc.														
[b]calculated composition														

(*continued*)

TABLE 1. U-Pb ANALYTICAL DATA FOR ZIRCONS FROM SAMPLES IN NORTHERN SONORA (*continued*)

Sample—Mesh Size	Observed ratios			Radiogenic Pb			Concentrations		Observed isotope ratios			Apparent ages (Ma)		
	Pb 206/Pb 204	Pb 207/Pb 204	Pb 208/Pb 204	Pb 206	Atom % Pb 207	Pb 208	(ppm) Pb[rad]	(ppm) U	Pb206/U238	Pb207/U235	Pb207/Pb206	Pb206/U238	Pb207/U235	Pb207/Pb206
Plomo North porphyritic granite #5[c]														
R200 Franz 1.7 NM 1/2°/20°	1753.1	101.8	562.3	73.96	3.68	22.36	15.83	520.77	0.0261	0.1790	0.0497	166	167	182
P200 Franz 1.7 M 1°/25°	1874.2	108.2	735.2	70.14	3.50	26.36	23.10	721.8	0.0261	0.1793	0.0499	166	167	190
R200 Franz 1.7 NM 1/2°/20° [a]	238.0	26.5	104.3	73.96	3.69	22.35	15.81	520.8	0.0261	0.1793	0.4990	166	167	190
P200 Franz 1.7 M 1°/25° [b]	1493.2	88.9	592.1	70.15	3.49	26.36	23.09	721.8	0.0260	0.1784	0.0497	166	167	180
[a]Sbf calculation using rad.206/208 = 3.308185 from Pb comp only.														
[b]Sbf calculation using rad.206/208 = 2.661497 from Pb comp only.														
Sonora porphyritic rhyolite #5[b]														
R200 Franz 1.7 NM 1/2°/20°	623.8	46.2	196.3	76.22	3.85	19.92	14.76	458.0	0.0284	0.1979	0.0505	181	183	219
P200 Franz 1.7 M 1/2°/25°	834.1	57	261.4	75.50	3.83	20.68	21.68	669.3	0.0283	0.1979	0.0507	180	183	219
Puerto Mababi quartz monzodiorite[c]														
P200 Franz 1.7 NM 0°/25°	513.7	40.7	149.1	78.42	3.98	17.60	15.32	514.21	0.0271	0.1899	0.0508	173	177	230
P200 Franz 1.7M 1/2°/25°	634.3	47.3	206.6	75.46	3.88	20.66	19.14	601.3	0.0279	0.1980	0.0515	177	183	262
P200 Franz 1.7 NM 0°/25° [a]	900.7	59.8	235.9	78.46	3.93	17.61	15.31	514.2	0.0271	0.1878	0.0502	173	175	202
P200 Franz 1.7M 1/2°/25° [b]	787.1	54.5	248.4	75.50	3.83	20.67	19.14	601.3	0.0279	0.1950	0.0507	177	181	226
[a]Based on sbf using rad. 206/208 = 4.456315 from Pb comp only.														
[b]Based on sbf using rad. 206/208 = 3.65304 from Pb comp only														
San Moises granite[c]														
R100+R200+ P200 Franz 1.7NM 10°/20°	228.2	26.0	71.4	82.71	4.12	13.16	10.99	452.5	0.0233	0.1603	0.0498	149	151	188
Rancho Sebaco monzodiorite[c]														
R200 Franz 1.7 NM 0°/20°	200.6	24.8	69.5	81.72	4.13	14.14	6.36	213.0	0.0282	0.1969	0.0506	180	182	221
R200 Franz 1.7M 1 1/2°/20°	1284.8	78.2	248.3	82.27	4.07	13.66	12.91	472.3	0.0258	0.1775	0.0496	164	166	174
Seri metarhyolite #2[a]														
R200+P200 Franz 1.7 NM 3/4°/20°	81.17	19.0	44.5	86.49	4.69	8.82	10.05	403.57	0.0250	0.1873	0.0542	159	174	380
R200+P200 Franz 1.7 NM 3/4/20° [a]	196.48	24.9	57.4	86.15	4,48	9.36	8.93	405.0	0.0221	0.1585	0.0520	142	151	296
R200+P200 Franz 1.7 M 3/4°/20°	175.56	23.92	55.57	85.90	4.53	9.57	10.44	471.09	0.0221	0.1609	0.0527	141	151	318
Silla sheared rhyolite porphyry[c]														
P200 Franz 1.7M 0°/25°	1546.5	91.8	224.4	85.33	4.25	10.41	25.69	1012.8	0.0252	0.1730	0.0498	160	162	188
P200 Franz 1.7M 2°/25°	1569.6	92.8	229.8	85.22	4.24	10.54	33.56	1327.6	0.0250	0.1719	0.0498	159	161	185

Note: common Pb correction: 206/204, 207/204, 208/204

[a] 17.8, 15.6, 38.0

[b] 18.2, 15.6, 38.0

[c] 18.6, 15.6, 38.0

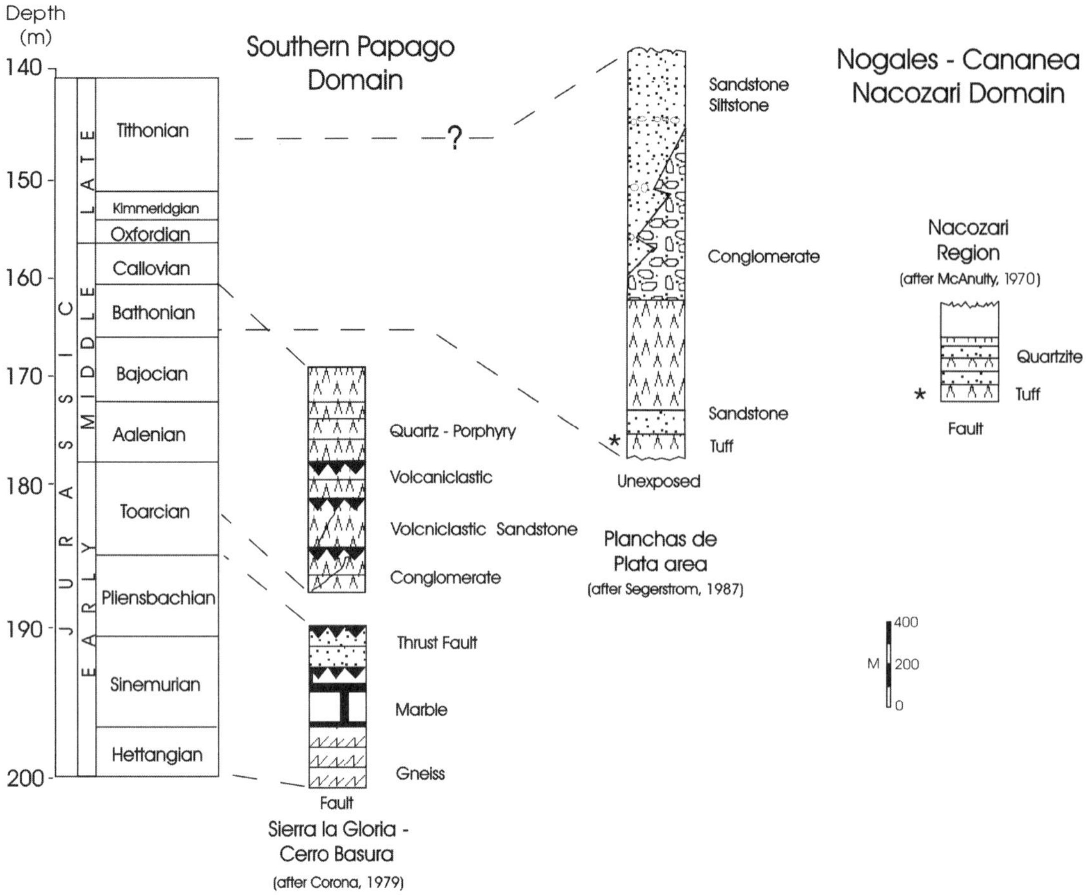

Figure 3. Lithostratigraphic sections from the Caborca and Nogales-Cananea-Nacozari domains. Time scale from Pálfy et al. (2000).

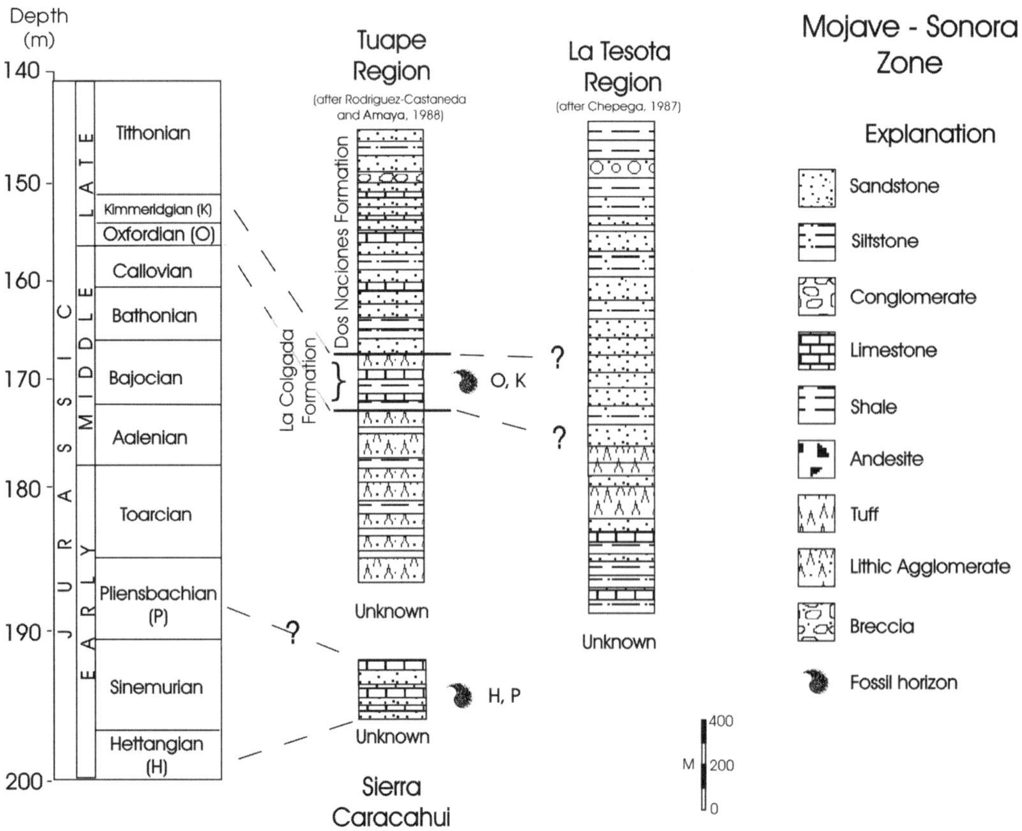

Figure 4. Lithostratigraphic sections from the Mojave-Sonora zone. Time scale from Pálfy et al. (2000).

sures of pre–Middle Jurassic units within the Papago terrane has been interpreted to reflect the development of a Jurassic magmatic arc upon crust extended before development of the Jurassic magmatic rocks (e.g., Stewart et al., 1990). Anderson and Nourse (this volume) suggest that an additional or alternative explanation is that the crust foundered during strong Late Jurassic transtension. This suggestion stems from the extensive exposures of ca. 150 Ma volcanic, hypabyssal and intrusive rocks that formed contemporaneously with transtension. The transtensional processes involved volcanism and development of pull-apart basins that simultaneously spread Late Jurassic volcanic layers across the cover of Middle Jurassic tuff, while extension at releasing steps formed basins that were filled with sediment (Fig. 2).

The use of "terrane" commonly carries the connotation of a fault-bounded entity, however, we know of no evidence indicating that the rocks within the Papago terrane are allochthonous or exotic with respect to adjacent rock bodies. In fact, continuing work within the domain, especially that of Riggs (Riggs and Haxel, 1990; Riggs and Busby-Spera, 1990, 1991; Riggs et al., 1993; Haxel et al., this volume), reveals probable correlation between sandstone horizons interbedded among the Papago Jurassic magmatic rocks and comparable units on the craton to the north. In view of the similarity between the Jurassic rocks in Arizona and Sonora and the absence of exposures of older rocks in both areas we recognize the Papago "terrane" as a distinct area, useful for discussion, but prefer to use the less restrictive "domain." We distinguish the fact that this discussion pertains to Sonoran rocks by means of the qualifier "southern."

The presence of Proterozoic crystalline rocks beneath the Jurassic volcanic rocks characterizes the Nogales-Cananea-Nacozari domain. Proterozoic crystalline crust, comprised of Pinal Schist (ca. 1.7 Ga) and associated 1.4 Ga plutonic rocks, appears east of the Papago domain boundary, which roughly parallels Route 15 between Santa Ana and Nogales. In Arizona, outcrops of Precambrian rocks occur east of the northern extension of this line, near Nogales. However, in Sonora, occurrences of Proterozoic rocks are first exposed farther east, at Cananea (Valentine, 1936; Anderson and Silver, 1977).

Strongly deformed Jurassic rocks, principally along the northeast side of the Mojave-Sonora megashear distinguish the Mojave-Sonora domain. The domain, which extends across the southern Papago and Nogales-Cananea-Nacozari domains, is discussed as a separate entity in order to emphasize the intensity and style of contractional deformation that at some localities is recorded by strata as young as Oxfordian and Kimmeridgian. Entrained olistoliths, recumbent folds, and mylonitic foliation are well developed in some rocks within this domain. Rodríguez-Castañeda and García y Barragán (1994) note that deformation could be as young as earliest Cretaceous, although fossiliferous Early Cretaceous sandstone that unconformably overlies the deformed Oxfordian strata provides the younger limit on the time of deformation. The age of deformation correlates with that called Nevadan in the Sierra Nevada region of California and Nevada.

Southern Papago Domain

The southern Papago domain includes the region from north of the strongly deformed rocks of the Mojave-Sonora domain to the Arizona-Mexico border between Sonoita and Nogales (Fig. 2). Its eastern boundary is located along a line, inferred to mark the eastern limit of exposed Proterozoic crystalline basement, that extends between Nogales and Magdalena. Jurassic units that crop out within at least eleven ranges in the southern Papago domain (Fig. 1) are composed principally of volcanic, volcaniclastic, and plutonic rocks.

Ranges in north-central Sonora including Sierras Cibuta, Avispas, and Pajarito are underlain by volcanic and conglomeratic rocks that are probable equivalents to the Cobre Ridge Tuff and the Topawa Group, which yield interpreted U-Pb ages of ca. 170 Ma (Knight, 1970; Riggs and Haxel, 1990; Riggs and Busby-Spera, 1990, 1991; Riggs et al., 1993; Haxel et al., this volume). The sequence exposed in the vicinity of Cañon Planchas de Plata, which cuts Sierra Las Avispas near the northeastern boundary of the southern Papago domain, may be mapped as two formations, Las Avispas and La Jareta (Segerstrom, 1987). The Las Avispas Formation includes three ash-flow tuff units separated by probable eolian subarkose. Its thickness is at least 500 m. Analysis of zircon from a probable intrusive rhyolite porphyry, from Cañon Las Planchas yields concordant ages from two fractions, the older of which has a $^{206}Pb/^{238}U$ date of 165 Ma (Table 1), providing a best estimate of the age.

La Jareta Formation, a clastic sequence >1300 m thick, disconformably overlies the volcanic rocks with local angularity of ~18° (Segerstrom, 1987). Within La Jareta Formation two members may be differentiated: a lower, predominantly conglomerate member, and an upper member of mainly sandstone and siltstone. The thickness of these members is variable from place to place. Most pebbles in the conglomerate member are similar to the underlying volcanic rocks; much less common are pebbles of reworked subarkose and granite. Locally this unit is as much as 650 m thick.

Segerstrom (1987) notes the likelihood of correlation of Las Avispas Formation with the Cobre Ridge tuff. Rhyolitic volcanic units that crop out as far west as Rancho Cubabi (Fig. 1) along the low eastern flank of Sierra del Cobre (Fig. 2) are also correlative with the volcanic rocks in Arizona, as shown by U-Pb ages. Two fractions from the rhyolite yield concordant dates of 176 and 157 Ma, with the younger age interpreted as resulting from Pb loss (Table 1).

To the south, in Sierras Cibuta and Las Avispas, the effects of early Miocene crustal extension, manifest as foliation, ductile normal faults, and detachments, increase abruptly across the northwest-trending Imuris lineament (Nourse et al., 1994), which extends from northern Sierra La Madera to the north of Sasabe, Sonora (Fig. 2). Nevertheless, work by Nourse (1988, 1989, 1990) and Nourse et al. (1994) in Sierras Magdalena, Guacomea, La Madera, Jarillas, and Potrero, and near Saric reveals a Jurassic(?) section composed of quartz-porphyry with interbeds of quartz-arenite that passes upward into conglomerate with predominant cobbles of volcanic porphyry and quartzite. At Sierra La Madera this section is overlain by the Late Jurassic–Early Cretaceous(?) Cocospera Formation (Gilmont, 1978) that contains pebbles of quartzite, sandstone, carbonate, and volcanic rocks. Sills and plugs of granite porphyry intrude the lower quartzite clast–bearing conglomerate (Nourse, 1990).

The most detailed documentation available for the Middle Jurassic volcanic and sedimentary rocks of the southern Papago domain in Sonora is from Corona (1979, 1980; Corona and Anderson, 1981), who described exposures in Cerro Basura and Sierra La Gloria. Unfortunately, the resistant, well-exposed volcanic rocks in the sierra are commonly in the upper plates of faults. Volcanic rocks consist mainly of quartz-feldspar porphyritic tuff locally intercalated with minor amounts of vitric-welded tuff, volcanic breccia, feldspar-rich dacite and/or andesite. Volcaniclastic conglomerate, interbedded with sandstone, siltstone, and rare limestone, also crop out throughout Sierra La Gloria and Cerro Basura. Conglomerate clasts are subrounded to rounded and include pebbles, cobbles, and boulders and consist predominantly of dark, feldspar porphyry and subordinate quartz sandstone, sandstone, siltstone, and carbonate. Locally, the volcaniclastic conglomerate grades into metaconglomerate that is foliated and metamorphosed to greenschist facies. Conspicuous elongate pebbles trend N60°E. The age of this lineation is indeterminate because both Late Cretaceous contraction, which is recognized in folded strata southeast of these exposures (Nourse, 2001), and early Miocene extension (Nourse et al., 1994) record the same direction of tectonic transport.

Plutonic rocks that crop out within the southern Papago domain are comparable to the petrographically distinctive Kitt Peak–Trigo Peaks superunit (Tosdal et al., 1989). Granodiorite with phenocrysts of distinctive potassium-feldspar of lavender tinge and millimeter-size euhedral sphene is the characterizing lithology of this group. Exposure of this lithology at El Plomo, south of Sierra del Cobre and in the northern Sierra La Espuma, a short distance west of Rancho Gabino, may indicate the presence of a batholith, broken by faults, that underlies hundreds of square kilometers and extends between the two localities (Figs. 1 and 2). The older of two near-concordant fractions of zircon from porphyritic quartz monzonite near Rancho Gabino (Fig. 1) has a $^{206}Pb/^{238}U$ date of 170 Ma, providing the best estimate of the age of intrusion. Two analyzed fractions from porphyritic granite north of El Plomo yield overlapping (but slightly discordant) $^{206}Pb/^{238}U$ dates of 166 Ma, which is likely close to the true age of the granite. Rocks of the same lithology at Kitt Peak, Arizona, yield a concordant U-Pb age of 165 ± 2 Ma (Haxel et al., this volume).

Analyses of two fractions of zircon from monzodiorite near Rancho Sebaco along the flank of Sierra La Alesna (Fig. 1) indicate a Middle Jurassic age based upon: (1) slightly discordant ratios with a $^{206}Pb/^{238}U$ date of 164 Ma and (2) a concordant but low-precision result with a $^{206}Pb/^{238}U$ date of 179 Ma. The common occurrence of 165 Ma plutons in this region suggests that the younger age is more likely and that the older age reflects inheritance of older zircon.

Near Rancho San Moises, close to the border with the Arizona, a single analysis of zircon extracted from granite provides concordant results of 149 Ma, indicating a Late Jurassic age. This pluton is part of a regional suite of alkaline Late Jurassic units composed of volcanic and sedimentary units (Artesa) and associated intrusive rocks (Ko Vaya) (Tosdal et al., 1989). In Arizona, the volcanic rocks of the Artesa sequence are flows, flow breccia, and volcanic conglomerate. Sedimentary strata are commonly argillite, sandstone, pebbly sandstone, and conglomerate. Clasts from cratonal Paleozoic rocks and contemporaneous or older volcanic strata are conspicuous in some conglomerate units. This sequence includes strata high in the Jurassic section, commonly just below the Late Jurassic–Early Cretaceous Glance Conglomerate. Laminated or well-bedded, immature, volcanic litharenite is abundant relative to other sedimentary lithologies.

Artesa sequence rocks extend from southern Arizona southward to the low ranges near Rancho San Moises and Sierras San Manuel, and La Gloria (Anderson, unpublished mapping; Tosdal et al., 1989; Fig. 5). Plutonic rocks of the Ko Vaya supergroup form conspicuous exposures of pink to maroon granite and syenitic rocks that typically weather reddish, pinkish, yellowish or orangish. Ko Vaya rocks are characterized by textural and compositional heterogeneity. An example of the Ko Vaya suite is the perthitic granite that forms the prominent Baboquivari Peak. Plutonic rocks of this group yield interpreted U-Pb ages between 160 and 145 Ma (Tosdal et al., 1989) and commonly are moderately alkaline. Miarolitic cavities, common strong alteration of the plutonic rocks, and the association with hypabyssal porphyry suggest emplacement at shallow depth.

We separate the alkaline igneous rocks that fall within the interval between 160 and 145 from older calc-alkaline rocks with ages that cluster between 175 and 160 Ma. We interpret the composition, ages, paucity of interbedded tuff, and common association with sedimentary strata to indicate a tectonic setting different from that of subduction-related magmatism. We consider the Artesa sequence to correlate with the formation of transtensional basins, beginning ca. 162 Ma.

Nogales-Cananea-Nacozari domain

The Nogales-Cananea-Nacozari domain (Figs. 2 and 3) includes much of northeastern Sonora. It is distinguished from other domains because exposures of Precambrian crystalline basement rocks and Paleozoic strata demonstrate cratonal underpinning. Rocks of known or probable Jurassic age are less extensive than in the southern Papago domain to the west (Fig. 2). Jurassic localities in the Nogales-Cananea-Nacozari domain include: Sierra El Pinito (Nourse, 1990), Cananea area (Ayala-Fontes, 1992; Meinert, 1980), Nacozari region (McAnulty, 1970), Puerto Mababi (this report), and Sierra La Ceniza (Solano-M., 1989).

The most extensive exposures of Jurassic rocks occur in Sierra El Pinito, close to eastern margin of the southern Papago

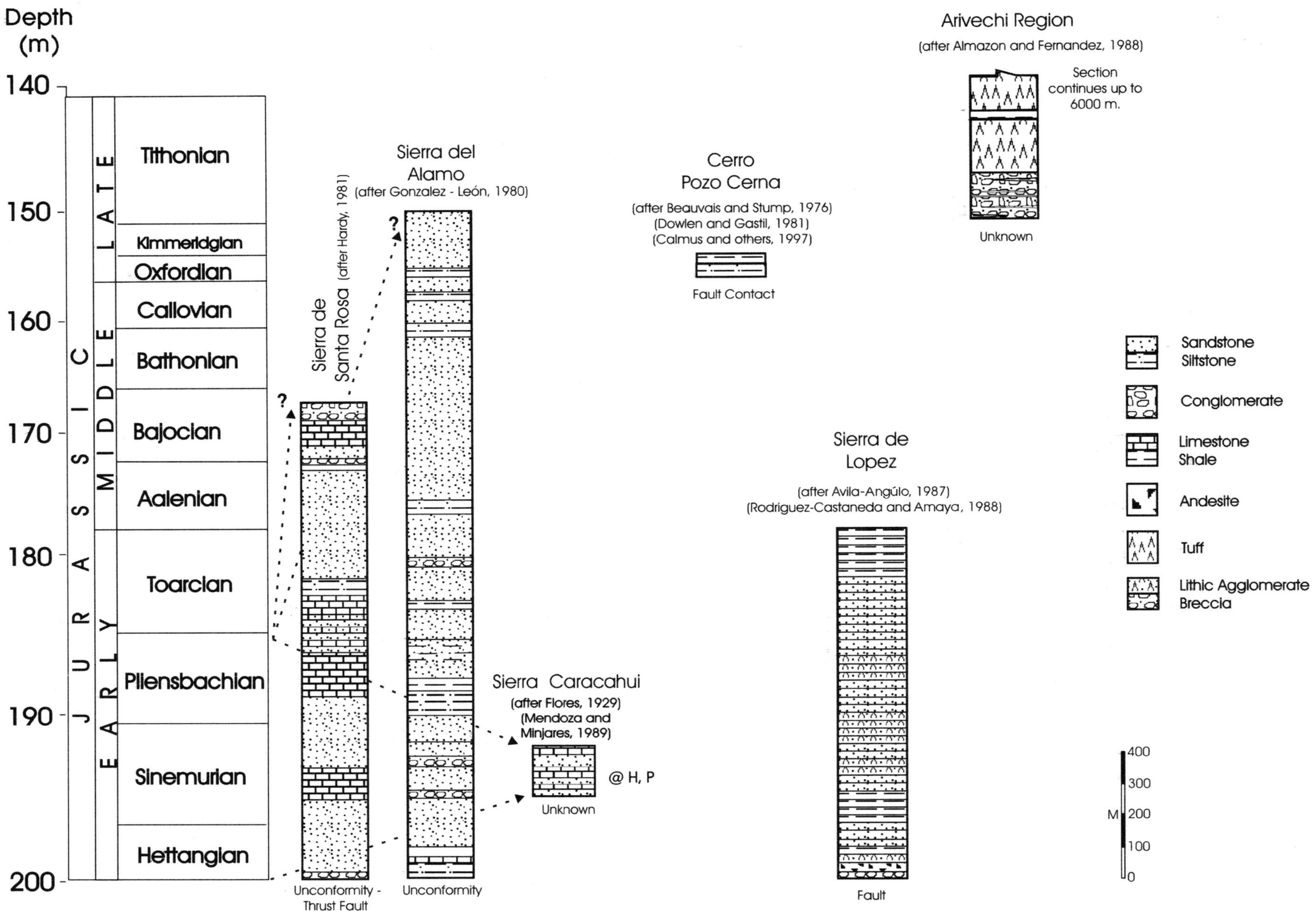

Figure 5. Lithostratigraphic sections within the Caborca block. Time scale from Pálfy et al. (2000).

domain. Although the range is unmapped, interpretation of aerial photographs suggests that a large mass of porphyry underlies its southern part. Outcrops of quartz-porphyry (sample El Tunel, Fig. 1), exposed in road cuts along Route 2, resemble those to the west in the Papago domain and northwest in southern Arizona. One analyzed fraction yields concordant ratios with a $^{206}Pb/^{238}U$ age of 174 Ma, indicating a Middle Jurassic age correlative with those from lithologically similar units of the Cobre Ridge Tuff and Topawa Group.

At Cananea, Ayala-Fontes (1992) noted that rocks described by Valentine (1936) are probably Jurassic. This sequence includes rhyolite and trachyte interbedded with well-sorted quartzite and graywacke (Elenita Formation) and rhyolite, dacite, and latite interbedded with lahar, tuff, and volcaniclastic sandstone (Henrietta Formation). These units are intruded by the El Torre syenite, a pre-Cretaceous pluton (Meinert, 1980).

West of Nacozari, in Sierra Copercuin, McAnulty (1970) mapped a unit of interbedded tuff, quartzite, graywacke, and welded tuff overlain by massive quartzite, 20 m thick, transitional into "calcareous conglomerate" and lenses and beds of impure limestone (Fig. 3). This volcanic-rich unit, ~500 m thick, named Lily Formation, includes tuffaceous horizons from which a geochronology sample was collected. One discordant fraction yielded a $^{206}Pb/^{238}U$ date of 174 Ma, which provides a reasonable upper limit for the age of this unit. The Lily Formation is overlain by carbonate-rich beds of Copercuin Formation, which McAnulty (1970) considered equivalent to units such as the Early Cretaceous Mural Limestone.

Sierra Copercuin lies at the southern end of a long, narrow Tertiary fault block (Figs. 1 and 2), most of which is composed of Proterozoic granite or Paleozoic strata. The Proterozoic rocks are intruded by a biotite granite that forms a high-standing ridge. Analyses of zircon from a sample of granite collected not far west of the pass at Puerto Mababi yield discordant U-Pb ratios (Table 1). The $^{206}Pb/^{238}U$ dates of 173 and 177 Ma suggest a Middle Jurassic age for this pluton. The most northeasterly exposures of probable Jurassic volcanic rocks in Sonora occur at Sierra Ceniza (Solano-M., 1989).

Mojave-Sonora Domain

The Mojave-Sonora domain is composed of Middle Jurassic volcanic and sedimentary strata, as well as fossiliferous Late Jurassic (Oxfordian and Kimmeridgian) beds within a zone, not more than a few tens of kilometers wide, traversing the southern parts of both the southern Papago domain and Nogales-Cananea-Nacozari domain (Fig. 2). Middle and Late Jurassic rocks contain one or more of: (1) mylonitic foliation and microstructures showing left-lateral shear within structural domains of steeply dipping fault planes, (2) ductile and brittle transpressional thrust faults, some of which have mylonitic upper-plate rocks and associated folds, in some places recumbent, (3) allochthonous masses of quartzite, granite, and metamorphic and carbonate rocks, which occur above thrusts or are enclosed among sedimentary units, suggesting gravity-driven penecontemporaneous emplacement.

The southwest margin of the Mojave-Sonora domain coincides with the Mojave-Sonora megashear, which separates the Caborca block from Middle and Late Jurassic igneous and sedimentary units to the northeast. Where the Mojave-Sonora megashear crops out several kilometers southwest of Sonoita, it corresponds with a major zone of mylonite separating Middle Jurassic quartz porphyry from Triassic and Precambrian crystalline rocks (Campbell and Anderson, 1998, 2003). Thrust faults, delineated by upper plates composed of granite, schist, gneiss and/or quartzite, some of which lithologically resemble and in some cases are correlative with units restricted to the Caborca block, are prominent structures within the Mojave-Sonora domain. The hanging walls, of modest areal extent, are emplaced over or imbricated with deformed Jurassic plutonic, volcanic, and volcaniclastic (shale, sandstone, conglomerate) rocks.

We believe that the age of deformation is Late Jurassic because: (1) thrusts along the Mojave-Sonora megashear were emplaced between ca. 180 and ca. 85 Ma as shown by U-Pb ages and (2) in places deformed fossiliferous Oxfordian beds are unconformably overlain by earliest Cretaceous sandstone beds that do not show the earlier deformation. We hold our opinion despite Cretaceous contraction (e.g., Nourse, 2001) and gravity sliding (e.g., McKee, 1991; McKee and Anderson, 1999; Rodríguez-Castañeda, 1997, 2000) and Tertiary detachment faulting or ductile deformation (e.g., Nourse et al., 1994) that overprint the Late Jurassic structures. The age of the deformation within the Mojave-Sonora domain correlates with the Nevadan orogeny (e.g., Schwieckert et al., 1984).

We also believe that the Late Jurassic rocks and the structures that they contain reveal deformation contemporaneous with Mojave-Sonora megashear strike-slip faulting and local transpression. Because of these inferences we describe two major thrusts, the Quitobaquito near Sonoita, and the La Lamina, at Rancho La Lamina, in some detail. We interpret these thrusts, which are generally restricted to areas close to the Mojave-Sonora megashear, to record transpression along the Mojave-Sonora megashear.

Near Sonoita in northwest Sonora (Figs. 1 and 2), the northwesternmost exposures within the Mojave-Sonora domain underlie Sierra Los Tanques (Fig. 2). Near here, Merriam (1972) noted extensive exposures of metarhyolite and metaandesite that he compared to the Santiago Peak volcanics of southern California. In adjacent Arizona, mapping in the Quitobaquito Hills, Puerto Blanco Mountains, and Sonoita Mountains shows Jurassic supracrustal and plutonic rocks in an imbricated stack, including meta-quartz porphyry, greenschist, metaconglomerate, quartzofeldspathic phyllite and semischist, in part calcareous, and variably metamorphosed granite, granodiorite, granite porphyry, and porphyritic granite (Haxel et al., 1984). These rocks are interpreted to be metamorphosed and deformed equivalents of Jurassic volcanic, volcaniclastic, hypabyssal, and clastic rocks, as well as associated intrusive bodies beneath structurally higher gneisses (Haxel et al., 1984; Tosdal and Miller, 1988; Tosdal et al., 1989, 1990). Upper-plate rocks include augen gneiss that yields a zircon U-Pb age of 1749 ± 23 Ma (Premo et al., 2003) correlative with granitic rocks of the Caborca block (Anderson and Silver, this volume).

The ductile Quitobaquito thrust separates the upper-plate gneissic rocks from the Jurassic units and is the principal structure of the area (Haxel et al., 1984). The thrust was interpreted to have formed during Late Cretaceous to early Tertiary orogenesis based upon exposures of weakly foliated Late Cretaceous granite that crop out among the older upper-plate rocks. However, in places this granite intrudes the sequence of foliated, commonly mylonitic rocks in the hanging wall, which leads us to conclude that the Quitobaquito thrust fault is older than Late Cretaceous (Dembosky, 2001).

In Sonora, unfoliated klippen of grayish-red granite crop out between the Mojave-Sonora megashear and Quitobaquito thrust, several kilometers to the north. The granite masses resemble the distinctive 1.1 Ga Aibó granite (Anderson and Silver, this volume), autochthonous exposures of which are known only from the Caborca block. Anderson and Campbell (1994) suggest that klippen of Proterozoic granite and metamorphic rocks that overlie deformed Jurassic units south of the Quitobaquito thrust are remnants of the Quitobaquito hanging wall, which extended ~10 km south to subvertical mylonitic rocks containing microstructures, which record left-lateral displacement sense along the Mojave-Sonora megashear (Campbell and Anderson, 2003).

Low exposures among the topographically higher klippen of granite reveal mylonitic meta-quartz porphyry. Two fractions of zircon (Sonora porphyritic rhyolite #5) from these footwall rocks that crop out in a road cut along Route 2 (Fig. 1; Table 1) yield overlapping but slightly discordant results with $^{206}Pb/^{238}U$ dates of 180 Ma, probably close to the age of this unit. Southward, the footwall rocks include foliated volcanic, volcaniclastic, conglomerate, and sandstone that crop out in the northern part of the Sierra Los Tanques (Figs. 1 and 2) and terminate against the steep, mylonitic rocks that distinguish the Mojave-Sonora megashear (Anderson et al., 1992; Campbell and Anderson, 2003).

Similarly, the upper-plate mylonitic Proterozoic augen gneiss, the fabric of which is atypical of most Caborca block Proterozoic orthogneiss, is most comparable to mylonitic gneiss along the Mojave-Sonora megashear, 10 km to the south. We interpret these rocks and relations to suggest that: (1) the fabrics are megashear age, that is Late Jurassic, rather than Proterozoic and (2) the strongly fractured but generally unfoliated granitic rocks forming the klippen and the mylonitic gneisses, some of which are Proterozoic, were thrust northward from the Caborca block across the Mojave-Sonora megashear above a sole thrust plated with mylonitic rocks (Campbell and Anderson, 1998, 2003).

Cretaceous granitic rocks near Aguajito Spring (ca. 84 Ma, U-Pb zircon, Anderson and Silver, 1979) intrude the mylonitic rocks. The granite is locally moderately foliated (as noted by Haxel et al., 1984); however, the foliation in Sonoran outcrops is not mylonitic (Campbell and Anderson, 2003). Furthermore, at the intrusive contact, the strike of steep mylonitic foliation in host rocks deflects from northwest to west and mylonitic rocks are recrystallized. At places the granite, intruded into the zone of Mojave-Sonora megashear mylonitic rocks, contains inclusions of recrystallized and partially melted laminated gneiss interpreted as formerly having mylonitic fabric (Anderson and Campbell, 1994; Campbell and Anderson, 2003).

In central Sonora, east of Tuape, Late Jurassic deformation is recorded by some Late Jurassic strata, including La Colgada Formation, that are recumbently folded and contain exotic blocks. Uncomformably overlying Cretaceous sedimentary strata are less deformed (Rodríguez-Castañeda, 1984, 1986, 1988) (Fig. 4). La Colgada Formation is a key unit because it contains Oxfordian (Rangin, 1977) and middle Kimmeridgian fossils (Rodríguez-Castañeda, 1991). This formation, which contains folds and intraformational structures indicative of penecontemporaneous deformation (McKee, 1991), is composed of thin to medium, and less commonly thick, beds of slightly metamorphosed shale, sandstone, limestone and crystal-rich tuffaceous layers. The thickness of the formation, though obscured by faulting and folding, is probably as much as 1000 m. Late Jurassic beds also contain blocks or slabs of carbonate strata and quartzite, some hundreds of meters in length (Stephens, 1988; Chepega, 1987). East of Tuape (Fig. 2), adjacent to the inferred trace of the Mojave-Sonora megashear, Rodríguez-Castañeda (1984, 1990) mapped thrusts above which quartzite and distinctive red, micrographic Proterozoic granite (Anderson and Silver, this volume) have been displaced northward onto Jurassic strata. Klippen of similar granite and olistoliths of quartzite, displaced as much as 40 km northward from the Mojave-Sonora megashear, are also recognized west and north of Cucurpe (Anderson et al., 1984; Stephens, 1988; Mendoza-Cordova and Minjarez-Rivera, 1989; Rodríguez-Castañeda and García y Barragán, 1994).

The largest structural outlier, which crops out near Rancho La Lamina along the road between Magdalena and Cucurpe, was mapped by Stephens (1988). At this exposure a prominent, gently dipping, brittle fault, La Lamina thrust (Anderson et al., 1984; Sosson and Calmus, 1990), underlies metamorphic and igneous rocks, including 1.1 Ga micrographic granite (Anderson and Silver, this volume), much like the Aibó granite, known only as an autochthonous unit from the Caborca block.

In the vicinity of the La Lamina thrust at least two periods of Mesozoic contractional deformation can be recognized, one during Late Jurassic time and a second after Aptian-Albian time. We believe that the time of emplacement of Proterozoic granite onto the deformed Late Jurassic footwall strata along the La Lamina thrust must be after, or possibly, contemporaneous with accumulation of the Upper Jurassic beds, as suggested by the presence of the blocks and large slabs of quartzite entrained among tilted Late Jurassic strata of the footwall. Furthermore, La Lamina thrust does not cut Cretaceous beds. Northward-directed Cretaceous thrusts also are inferred deep beneath the footwall of the La Lamina thrust (Sosson and Calmus, 1990). Above the inferred deep Cretaceous thrust, both the footwall and allochthonous granite resting upon the structurally higher La Lamina thrust were transported northward.

Cretaceous strata unconformably resting upon folded Upper Jurassic strata between Cucurpe and Sierra La Lamina record folds and faults with diverse vergences. Near La Lamina, Sosson and Calmus (1990) also mapped south-vergent folds and thrusts. The southward vergence is similar to that described in the San Antonio basin to the southeast (McKee and Anderson, 1999; Rodríguez-Castañeda, 2000), and near Tuape (Rodríguez-Castañeda, 1984) where southwest-directed gravity-driven deformation was concurrent with or followed by Late Cretaceous contraction recorded by north-directed local thrusts (Rodríguez-Castañeda, 1990), gentle folds (Sosson and Calmus, 1990), and basin inversion (Rodríguez-Castañeda, 1997).

Above the deformed fine-grained strata of La Colgada Formation, less deformed rocks include limestone, siltstone, sandstone, and conglomerate of the Dos Naciones Formation, ~1000 m thick. Stratigraphic position and poorly preserved fossils collected from the limestone suggest a late Kimmeridgian to Tithonian age (Jaime Barceló, 1986, oral commun.). The Dos Naciones Formation is overlain by the Bacuchi Formation, ~600 m thick. The age of these beds is only constrained to be between Kimmeridgian and Aptian. The overlying Bacuchi Formation consists of limestone, shale, sandstone, and lenses of conglomerate that are several to as much as 20 m long and as much as than 3 m thick. Clasts consist chiefly of subrounded limestone pebbles and subordinate subrounded to subangular chert pebbles. Rodríguez-Castañeda (1991) considered the Bacuchi Formation to be an equivalent of the Glance Conglomerate.

The nature of the post-Oxfordian strata generally is in contrast to the strongly deformed older volcaniclastic and volcanic strata that crop out at other locations within the Mojave-Sonora domain. Structures within the older units include recumbent folds in sandstone at Teguachi (Rodríguez-Castañeda, 1994), olistoliths of volcanic flow rock within sandstone at La Tesota (Chepega, 1987), and quartzite within shale north of Sierra Caracahui (Mendoza-Cordova and Minjarez-Rivera, 1989) (Fig. 2).

At Estación Llano, the San Francisco gold deposit has stimulated mapping, which also reveals multiple deformation as indicated by faults of two ages (Herrera-Urbina and Perez-Segura, 1990; Calmus et al., 1992; Luna-Barcelo and Gastelum-Morales, 1992). Three principal rock units are recognized: (1) heterogeneous Proterozoic(?) crystalline basement comprising the San Francisco unit; (2) several hundred meters of phyllite, sandstone, and conglomerate with interbeds of tuff comprising the Coyotillo Group of probable Jurassic age (Salas, 1968); and (3) the porphyritic Gauna rhyolite of probable Tertiary age. Well-developed, steep, northwest-striking foliation characterizes the quartzo-feldspathic gneissic basement. Martinez-Araiza (1998) attributes this deformation to shear within a zone several kilometers wide along the inferred trace of the Mojave-Sonora megashear. Two other faults are recognized: an east-striking thrust that separates crystalline basement rocks, which were transported southward over clastic strata of probable Middle Jurassic age, and a listric fault, which strikes N20°W, that separates the post-Jurassic Gauna rhyolite from the subjacent San Francisco basement. The strike of the listric fault and the younger-over-older relationship are consistent with early Miocene extension, known to the northeast (Nourse et al., 1994) and inferred from maps of Sierra Santa Rosa (Hardy, 1981; see also Anderson and Silver, this volume). We infer that the thrust is Late Jurassic and records transpression, during which a sliver of basement was transported southward onto Jurassic strata as part of a series of transpressional structures along the Mojave-Sonora megashear. Despite the similarity of this fault to others near the Mojave-Sonora megashear, no stratigraphic evidence exists that precludes a Cretaceous or Miocene age for this particular structure.

Recumbent folds are conspicuous along Route 2 in road cuts through low hills along the southwestern flank of Cerro Basura (Fig. 2). Folds are recorded in thin- to medium-bedded volcaniclastic sandstone interbedded with siltstone and mudstone (Corona, 1979, 1980; Corona and Anderson, 1981). The sandstone beds along Route 2 were interpreted by Corona (1980) to be part of the deformed Middle Jurassic(?) sequence of Cerro Basura and Sierra La Gloria ranges that lie just north of the inferred trace of Mojave-Sonora megashear (Fig. 3). However, rare fossils from the sandstone are Early Jurassic (Stanley and González-León, 1995), and they probably have been thrust onto units of the southern Papago domain. In either case, they record structures typical of those within the Mojave-Sonora domain that preclude determination of stratigraphic sequence. In the higher parts of the ranges, Corona (1980) mapped thrusts along which the Jurassic rocks are imbricated. Massive volcanic units comprise the cores of individual thrust sheets. Weaker lithologies, such as limestone and volcaniclastic rocks, that are tightly folded and/or strongly foliated crop out adjacent to inferred thrusts. Conformable stratigraphic contacts between units are rare.

To the southeast, around Puerto del Alamo, a sequence of probable Late Jurassic age is composed of sandstone, siltstone, and conglomerate (Willard, 1988; Nourse, 2001). Clasts in the conglomerate are elongate and flattened; other strata generally show cleavage. These rocks are correlative with the "volcaniclastic unit" and "the stretched pebble conglomerate unit" of Corona (1980) in the Sierra La Gloria–Cerro Basura area to the northwest. McComb (1987) reports the presence of Jurassic(?) sequences in Sierra El Batamote, northeast of Caborca, similar to those of the Puerto del Alamo area. Stratigraphic and structural relationships have been clarified by Nourse (2001) (Fig. 3) who recognized that the section of Upper Jurassic quartzitic polymict conglomerate and associated strata are folded about northwest-trending hinges. He attributed folds and foliation in the clastic rocks in Sierra El Batamote to Laramide contraction.

Conglomerate at Sierra El Batamote (8 on Fig. 2) interfingers with basaltic-andesitic flows and monomict-andesite flow breccias, which, in turn, rest upon a rhyolite ignimbrite. The conglomerate fines upward and grades laterally into volcaniclastic sandstone and mudstone interstratified with lacustrine sediments and tuffaceous sandstone. Similar conglomerate underlies Sierra del Alamo, the range immediately northwest of Sierra El Batamote, and exposures occur to the northwest as far as the southeastern part of Sierra La Gloria, where a north-trending fault separates conglomerate from Middle Jurassic volcanic strata. Anderson and Nourse (this volume) postulate that the sedimentary rocks mark the position of the Batamote pull-apart basin that formed at a releasing step along the Mojave-Sonora megashear. Nourse (2001) suggests that the original geometry of the sedimentary basin in which the conglomeratic strata accumulated is poorly preserved due to the strong overprint of Laramide contraction.

Between Cerro Basura and Quitovac to the northwest, strongly faulted volcanic rocks and related sediments underlie low ranges to the west side of Route 2. On the southwest side of this belt among hills and low peaks including Cerro Prieto (Fig. 1), Calmus and Sosson (1995) report a thrust of modest horizontal displacement above which Proterozoic gneiss has been displaced onto a section of probable Jurassic age, consisting of sandstone, conglomerate, quartz porphyry, and andesite. Southeastward transport directions are recorded by asymmetric lenses of quartz and feldspar segregated within metamorphic units and kink folds, and northeastward transport by C and S foliation in the sedimentary section. The age of thrusting is postulated to be Late Cretaceous to Paleocene contraction, but Cretaceous and younger rocks are not involved and the deformation could be older.

At La Cholla Mine, southeast of Quitovac, Iriondo (2001) and Iriondo and Atkinson (2000) attribute ductile thrusts to Eocene contraction based upon $^{40}Ar/^{39}Ar$ release spectra of micas in shear zones. Near Quitovac, crystalline Proterozoic rocks are juxtaposed against Jurassic strata along zones characterized by either steeply dipping mylonitic rocks, as in Sierra San Antonio, or along moderately dipping thrusts where they override Jurassic strata (Connors et al., 1989; Caudillo-Sosa and Oviedo-Lucero, 1990; Caudillo-Sosa et al., 1991; Iriondo, 2001). The footwall rocks at Quitovac comprise foliated Jurassic volcanic and volcaniclastic units with intercalated sandstone and conglomerate. This section includes common folds and faults, dacite with mylonitic fabric, foliated sandstone and rhyodacite. Massive quartz-porphyry is less deformed than other units, except at the contacts, where it is generally foliated (Connors et al., 1989; Caudillo-Sosa and Oviedo-Lucero, 1990; Caudillo-Sosa et al., 1991). Connors et al. (1989) and Connors (1990) describe tectonic slivers of gneiss sandwiched between the dacite and sandstone. These slivers, 60–100 m long, are commonly foliated and altered.

Between Quitovac and Sonoita, exposures of foliated Jurassic rocks are interrupted by intrusions of Cretaceous granite, or are covered. Outcrops of screens, pendants, and wall rocks of deformed Jurassic(?) rocks or megashear mylonite among the Cretaceous intrusions provide additional information. Foliated quartz porphyry exposed in road cuts along Route 2, a few hundred meters east of Puerto La Silla, provide an example of the occurrence of Jurassic rocks in this area. Analyses of two fractions of zircon from porphyry just south of La Silla yield two overlapping but slightly discordant isotopic ratios with $^{206}Pb/^{238}U$ dates of 159–160 Ma, suggesting a Middle Jurassic age (Table 1). The structural relation shown in this region by massive, Cretaceous plutons intrusive into deformed Jurassic units is consistent with our observation that the Cretaceous plutonic rocks do not display the strong, ductile deformation commonly recorded by the intruded country rocks near the inferred trace of the Mojave-Sonora megashear. Locally the intrusions have a weak nonmylonitic foliation that we infer is synchronous with or younger than Cretaceous plutonism.

Caborca Domain (Caborca Block) and its Characteristic Mesozoic Cover

The Caborca terrane is distinguished by a 1.7–1.8 Ga crystalline basement overlain by Proterozoic and Paleozoic sedimentary rocks (Anderson and Silver, 1979, 1981, this volume). The elongate Caborca block lies south of the Mojave-Sonora megashear (Silver and Anderson, 1974; Anderson and Silver, 1979) (Fig. 2) and trends roughly NW-SE, extending from El Capitan in northwesternmost Sonora to the southeast near Arivechi. Most Jurassic rocks of the Caborca block are sedimentary strata bearing fossils older than the Middle Jurassic volcanic units across the Mojave-Sonora megashear to the north. In a few ranges volcanics occur above the Lower Jurassic strata. Some, such as the Tarahumara Formation in east-central Sonora, are known to be Late Cretaceous–early Tertiary age (McDowell et al., 2001). Most other volcanic rocks are undated and they may not be precluded from having Middle Jurassic ages. However, sparse exposures of Late Jurassic volcanic (Fig. 2, Seri locality) and plutonic (Fig. 2, El Sahuaro locality) rocks are interpreted to be part the Middle Jurassic magmatic belt fringing the western margin of the Caborca block. Thc Jurassic igneous rocks along the coast of the Caborca block and correlative rocks in Baja California, discussed below, were displaced along with the Caborca block in Late Jurassic time. If our interpretation that the Jurassic rocks along the coast of the Caborca block are the easternmost exposures of this displaced segment of the Middle Jurassic magmatic belt is correct, then it is unlikely that Middle Jurassic igneous rocks are common on most of the Caborca block east of the coastal exposures.

Triassic and Lower Jurassic rocks of the Caborca block, especially those at Sierra del Alamo (Figs. 2 and 5), have been the focus of considerable study as part of the effort to test proposed correlations between strata in northwestern Mexico and rocks in the Death Valley region as proposed by us (Silver and Anderson, 1974). The important exposure at Sierra del Alamo, west-northwest of Caborca, reveals Permian rocks overlain by a thick section of the Triassic and Jurassic strata (González-León, 1979). Within this section, fossiliferous limestone interbedded with shale and coarser clastic rocks contain Late Triassic ammonites (White and Guiza, 1948). The marine rocks of Sierra del Alamo are correlative with coal-bearing sandstone-rich units containing plant fossils that crop out southeast of Hermosillo (King, 1939; Alencaster de Cserna, 1961a, 1961b; Pineda 1961). Lower Jurassic sandstone, siltstone, and conglomerate, estimated to be more than 2 km thick (Fig. 5), are also part of the El Antimonio Formation (González-León, 1980). Appreciation of the possible differences between strata at Alamo and Death Valley as well as the desire to collect sufficient data so that detailed comparisons could be made among Lower Mesozoic units prompted additional stratigraphic and paleontologic work (e.g., Lucas, 1993; González-León et al., 1995; Stanley and González-León, 1995; González-León, 1997; Stewart et al., 1997; González-León and Stanley, 1997, and references therein). Previously, strata containing the fossil *Weyla* that had been identified in ranges (e.g., Sierras Santa Rosa and Caracahui; Flores, 1929; Jaworski, 1929) to the east, across the Caborca block, were correlated with the Early Jurassic rocks at Sierra del Alamo. Marine rocks are now recognized as far southeast as Santa Teresa, a small range east-southeast of Hermosillo (Fig. 2), where poorly preserved ammonoids occur high in the section (Stewart et al., 1997). Stewart et al. (1997) concluded that Lower Mesozoic strata southwest of the Mojave-Sonora megashear are part of a cohesive domain and provided an overview of localities where early Mesozoic strata crop out, a concise, up-to-date review of recent investigations, and a perspicacious summary of regional stratigraphic relations.

Other occurrences of rocks of known or inferred Early Jurassic age within the borders of the Caborca domain merit mention because of the nature of the stratigraphic sections, their implications for the geologic setting, or deformation that the beds record. These localities from northwest to southeast are: Cerros Chino and Rajon (Longoria and Perez, 1979; Longoria et al., 1978; Imlay, 1980), Sierra de Santa Rosa (Hardy, 1981), Sierra de Caracahui (Mendoza-Cordova and Minjarez-Rivera, 1989), Felix Gomez–Los Chinos area (Vega-Granillo et al., 1991; Chaparro-Mesa, 1992), Sierra de Lopez (Avila-Angúlo, 1987; Rodríguez-Castañeda and Amaya, 1988), and Arivechi (Almazán-Vázquez and Fernández-Aguirre, 1988) (Fig. 2).

At Cerro Chino, Longoria and Perez (1979) described a probable Jurassic section, ~4 km thick of andesite, rhyolite flows and flow breccia, and sandstone. Feldspathic sandstone, siltstone, and conglomerate containing rounded cobbles of quartzite and other resistant lithologies crop out on the pediment surface and southwestern range margin. These rocks resemble some beds of the Lower Jurassic section at Sierra del Alamo (Marzolf, 1985, personal commun.). Imlay (1980) identified ammonites of Early Jurassic (Pliensbachian?) age from shale we collected from low hills in a shallow topographic reentrant at the northern end of Cerro Rajon. Farther south in the range, ~3.5 km of volcaniclastic and volcanic rocks are reported at Cerro Rajon (Longoria and Perez, 1979). Lithologies include sandstone, conglomerate, limestone, volcanic breccia, and rhyolitic tuff. Stewart et al. (1984) (Fig. 3) who proposed significant revision of the earlier work, show an unfaulted contact between Cambrian rocks and undifferentiated overlying Mesozoic units from which we collected the ammonites identified by Imlay (1980). The presence of Early Jurassic rocks overlying Paleozoic strata is typical of the Caborca block. However, the report of volcanic units (Longoria and Perez, 1979) indicates additional significant stratigraphic and structural relationships among Jurassic strata. Our reconnaissance observations are that the volcanic rocks are locally deformed. We suspect from regional relations that the rocks have been thrust southward onto the Caborca block. Further study is needed.

In Sierra de Santa Rosa a sedimentary section of sandstone, limestone, and shale is overlain by tuff, agglomerate, and andesite (Fig. 4) (Hardy, 1981). Paleontologic data from bivalves in the lower part of the sedimentary strata indicate a Early Jurassic (Lias) age (Flores, 1929; Jaworski, 1929). Early Jurassic strata are commonly underlain by faults, probably listric with normal sense of displacement, that separate them from Precambrian basement. The Lower Jurassic section is overlain by conglomerate and andesitic volcanic rocks of Jurassic or younger age (Cerro San Luis Formation) (Hardy, 1981).

In the southern part of Sierra Caracahui (Fig. 2), the Caracahui Formation is composed of sandstone, limestone, shale, and volcanic rocks of intermediate composition, similar to those of the Sierra de Santa Rosa (Fig. 4) (Mendoza-Cordova and Minjarez-Rivera, 1989). Based upon fossil content, Flores (1929) suggests Early Jurassic age for sedimentary units in Sierra Caracahui. The stratigraphic relationship of the Early Jurassic Caracahui Formation to underlying beds is unknown. The base is intruded by granitoids, probably Cretaceous. Conglomerate similar to the Cerro San Luis Formation of the Sierra de Santa Rosa overlies the formation.

Near Félix Gómez–Los Chinos (Fig. 2), volcanic and volcaniclastic rocks, including quartz porphyry and intercalated volcanic breccia, are postulated to be Jurassic (Vega-Granillo et al., 1991). The presumed Jurassic rocks are complexly imbricated with Proterozoic basement (Vega-Granillo et al., 1991), possibly due to Middle Tertiary extension as suggested by common younger-over-older structural relations.

In the Sierra de López, exposures of a sequence of sandstone, shale, and mudstone, bearing fossils of Early Jurassic (Sinemurian) age (Avila-Angúlo, 1987; Rodríguez-Castañeda and Amaya-Martínez, 1988) (Figs. 2 and 5), contain interbeds of andesite and tuff. These Jurassic rocks are in fault contact with Paleozoic rocks. No conformable higher units crop out. This unusual section, the only one on the Caborca block within which Early Jurassic volcanic units are interbedded with fossiliferous strata, may correlate with a more complete sequence near Placeritas. At this locality, B. Poole (2003, personal commun.) reports the presence of strata, commonly conglomeratic, from which Late Triassic plant fossils and Early Jurassic (Sinemurian) bivalves (*Weyla*) have been identified. The Triassic beds rest unconformably upon Devonian and Mississippian strata.

At Cerro Pozo de Serna (Fig. 2), Beauvais and Stump (1976) reported fossiliferous exposures of Oxfordian and Kimmeridgian shale, sandstone, and tuffaceous shale within a fault block flanked by Proterozoic or early Paleozoic dolomite (Dowlen and Gastil, 1981). Subsequent restudy (Calmus et al., 1997) indicates the presence of Early Jurassic ammonoids, bivalves, and gastropods within a section that unconformably overlies Proterozoic and Cambrian strata. The relation of these two sections is unclear.

A volcanosedimentary section near Arivechi is considered to be Jurassic (ca. 150 Ma) by Almazán-Vázquez and Fernández-Aguirre (1988). These strata comprise the Zoropuchi and Tarachi Formation, ~6 km thick, which is composed of volcanic breccia, tuffaceous sandstone, siltstone, shale, rhyolitic to rhyodacitic tuff, and agglomerate (Fig. 5). The process of formation of these units has yet to be explained.

Jurassic Igneous Rocks of the Western Caborca Block

Jurassic igneous rocks are known from three localities in the western part of the Caborca block: (1) granodiorite at El Sahuaro, in west central Sonora; (2) metarhyolite in Sierra Seri, one of the ranges along the coast of the Gulf of California, and (3) granodiorite at El Capitan, in northwestern Sonora (Fig. 1). In the western Caborca block, in hills and ranges that protrude from the plains sloping toward the Gulf of California, we collected two samples, a plutonic rock (Fig. 2, El Sahuaro) and a volcanic rock (Fig. 2, Seri), which yielded Late Jurassic interpreted ages. The relation of these rocks to other Jurassic magmatic rocks in Sonora is unclear because they are younger than the major volcanic units of the igneous belt across the Mojave-Sonora megashear, and they are removed from correlative transtensional igneous rocks of southern Arizona (Anderson and Nourse, this volume). We include them with Middle Jurassic arc rocks outboard of the Caborca block as described below. The sample of plutonic rock was collected from unfoliated granodiorite (initially identified as quartz monzonite, cf. Table 1) that underlies Cerro El Sahuaro, an isolated hill standing above surrounding low plains. Two fractions of zircon yield concordant and overlapping dates with a mean $^{206}Pb/^{238}U$ age of 153 Ma (Late Jurassic).

The volcanic rock crops out among elongate ranges along the coast of western Sonora. The ranges contain metamorphosed rocks preserved as pendants in Cretaceous plutons that include andesite, other volcanic units, and limestone beds within sedimentary sections (Gastil et al., 1976). Zircon from foliated metarhyolite collected from Sierra Seri, east of Punta Tepoca, yields two overlapping but slightly discordant analyses with $^{206}Pb/^{238}U$ dates of 141 Ma, suggesting a Late to Middle Jurassic age.

In northwesternmost Sonora, south of Route 2, foliated strata that rest upon foliated Jurassic granodiorite are exposed among small hills, the highest of which is known as El Capitan. The metasedimentary section above the contact includes phyllite and phyllitic slate (~65 m); carbonate (~12 m); quartzite (~40 m); phyllitic mica schist (~40 m); phyllitic quartzite and quartzite (~25 m); and interbedded limestone and dolomite (~100 m). Cross-beds in some beds indicate that the thick carbonate section is at the top of the section. Subsequent study (Leveille, 1984) indicates that the El Capitan section includes equivalents of Pennsylvanian and Permian Supai Group, the Permian Coconino Sandstone, and the Permian Kaibab Limestone, correlative with cratonal platform rocks of southwestern North America. The beds at El Capitan are not equivalent to the section exposed upon the Caborca block and therefore require either unlikely stratigraphic changes, namely a very abrupt transition from Paleozoic miogeoclinal to cratonal rocks, or complex structural relations such as hinterland directed thrusting, to explain their presence. The presence of these strata that correlate with the upper part of the Grand Canyon section and that crop out southwest of the inferred trace of the Mojave-Sonora megashear has been cited (e.g., Hamilton et al., 1987) as indication that no offset has occurred along the postulated fault or, at least, that it must bend to the southwest around the outcrop.

A concordant fraction from zircon in granodiorite yields a $^{206}Pb/^{238}U$ date of 164 Ma (Table 1), indicating a Middle Jurassic age. At the time the sample was collected we noted that the mass of plutonic rocks, including quartz veins and late pegmatites, is strongly foliated with local gradations into horizons that appear mylonitic. The foliated rocks are strongly fractured. The igneous rocks become highly foliated at the contact with phyllitic quartzite. The contact locally can be placed to within 20 cm.

We suggest that the mylonitic fabric imposed upon the exposed rocks records the effects of simple shear, perhaps in the upper plate of a thrust along which the rocks were tectonically transported southwestward. The interpretation of the deformed El Capitan exposures as part of an allochthonous mass is compatible with regional relations described by May (1989) from which one might infer that the strongly foliated Jurassic granodiorite at El Capitan records deformation close to a deep-seated fault. The inferred fault accommodated southwestward back thrusting during northeast-directed Cretaceous or younger (cf. Silver, 1982, 1983) contraction and links the ductile structures of Maria fold- and thrust-belt (Reynolds et al., 1986) to mylonite zones, such as the Santa Rosa mylonite (Sharp, 1967, 1979; Simpson, 1984) in the Peninsular Ranges to the west.

Neogene closure of the Gulf of California positions the mylonite zones in the eastern Peninsular Ranges closer to and south and southwest of El Capitan (Fig. 6[1]) and south of the west-directed thrusts of the Mojave Desert (Silver, 1982, 1983). The restored position of El Capitan is also southwest of the south- to southwest-directed Cretaceous thrusts that are common within the Maria fold-and-thrust belt (Reynolds et al., 1986; Spencer and Reynolds, 1990) (Fig. 6) extending between the Piute Mountains of California (Fletcher and Karlstrom, 1990) and the Granite Wash and Harquahala Mountains in west-central Arizona (Martinez et al., 1994).

THE DISTRIBUTION OF MIDDLE JURASSIC IGNEOUS ROCKS IN SONORA: IMPLICATIONS OF NEOGENE AND LATE JURASSIC RESTORATIONS

Closure of the Gulf of California also brings bodies of Jurassic orthogneiss, preserved as wall rocks of Cretaceous plutons within the Peninsular Ranges batholith in southern California and northern Baja California, close to the sparse exposures of Jurassic

[1]Figures 6, 7, and 8 are on the CD-ROM and on loose sheets accompanying this volume, and are in the GSA Data Repository as item 2005139, available online at www.geosociety.org/pubs/ft2005.htm, or on request from editing@geosociety.org or Documents Secretary, GSA, P.O. Box 9140, Boulder, CO 80301, USA.

igneous rocks along the coast of Sonora at the western margin of the Caborca block (Fig. 1, 6, and 7 [see footnote 1]). The exposures of Jurassic igneous rocks in western Sonora are thereby revealed to be part of a larger magmatic province including the Cuyamaca Reservoir orthogneiss (161 ± 12 Ma, U-Pb zircon, Thomson and Girty, 1994) in southern California and the Agua Caliente tonalite (164.3 + 2.3 Ma, U-Pb zircon, Schmidt and Paterson, 2002) to the south in Baja California. Together, the Californian and coastal Sonoran exposures of Jurassic igneous rocks reveal an elongate Middle to Late Jurassic magmatic province that lies far west of the principal belt of northern Sonora. According to Schmidt and Paterson (2002), the Baja California belt may underlie the length of the entire Peninsular Ranges batholith.

A domain of Triassic intrusive rocks is also revealed by closure of the Gulf of California. The restoration places Triassic granodiorite gneiss (234 ± 39, U-Pb zircon, Thomson, 1994) in San Diego County closer to Sierra Los Tanques, where the only known exposures of Triassic monzodiorite (interpreted zircon U-Pb isotopic age of ca. 233 Ma [reported in Stewart et al., 1986]) crop out along the southwest side of the Mojave-Sonora megashear near Sonoita. Not only do these Triassic rocks provide an additional link across the Gulf of California but they also correlate closely with orthogneiss in the eastern Sierra Nevada, where they occur in the restoration proposed below.

Neither the coastal exposures nor the northern Sonora outcrops contains a complete assemblage of Middle Jurassic tectonostratigraphic units, such as accretionary prism, forearc basin, and arc, associated with subduction beneath a continental margin. However, each suite of igneous rocks, that is the western Caborca block and the northern Sonora domains, (1) is comparable to an Andean-type magmatic province formed during convergence at a continental margin (e.g., Tosdal et al., 1989), and (2) records the overlap of Jurassic magmatic rocks across a continental margin. The continental margins, one in western Sonora and one in northeastern Sonora, differ with respect to their sedimentary assemblages in that the western Sonora Caborca block is characterized by a thick miogeoclinal section, whereas the Nogales-Cananea-Nacozari domain Jurassic volcanic rocks overlie a cratonal section such as that at Cananea.

The exposures of Jurassic volcanic and plutonic rocks in westernmost Sonora are separated from those in northern Sonora by the intervening Caborca block that underlies a wide gap where no Middle Jurassic igneous or sedimentary rocks have been recognized. Furthermore, exposures of northern Sonoran Middle Jurassic volcanic and plutonic rocks abruptly terminate along the Mojave-Sonora megashear. Field mapping by Rodríguez-Castañeda (1984) (Fig. 6) supports the model of strike-slip faulting subparallel to the trend of the volcanic belt, which would explain the duplication of these Jurassic igneous belts.

Mélange (Anderson et al., this volume) and associated Middle Jurassic igneous rocks in Zacatecas probably represent the missing tectonostratigraphic elements displaced southeastward along the Mojave-Sonora megashear (Jones et al., 1995; Fig. 7). This model is consistent with the distribution of Proterozoic rocks in northern Sonora, southern California, and southern Arizona interpreted by Silver and Anderson (1974) to record major left-lateral displacement. Likewise, Early Jurassic fossiliferous marine strata, restricted in Sonora to the Caborca domain southwest of the Mojave-Sonora megashear, have been displaced from correlative units in west-central Nevada.

INFERRED TRACE OF THE MOJAVE-SONORA MEGASHEAR FROM SONORA TO THE GARLOCK FAULT: A DISCUSSION OF CONSTRAINTS PROVIDED BY DOMAINS AND TERRANES OF MESOZOIC AND PROTEROZOIC ROCKS, AND DISRUPTIONS RELATED TO SUPERPOSED DEFORMATIONS

Introduction

In northern Mexico, the location of the inferred trace of the Mojave-Sonora megashear is interpreted to coincide with abrupt discontinuities within the suite of Middle Jurassic rocks that are interpreted to have been part of an Andean-style margin. The discontinuities bound exposures of the convergent-margin suite that are either incomplete, as indicated by the absence of forearc and trench assemblages in Sonora, or truncated, as suggested by the isolated fragment of Middle Jurassic arc in Zacatecas (Jones et al., 1995). Unfortunately, in many places post-Jurassic intrusions, younger volcanic and sedimentary cover, transecting strike-slip faults and multiple generations of low-angle extensional and contractional faults obscure or disrupt the Mojave-Sonora megashear.

Fault rocks considered to represent the Mojave-Sonora megashear have been described only from a single large group of outcrop southwest of Sonoita, Sonora (Campbell and Anderson, 1998; 2003). At this locality the Mojave-Sonora megashear separates the plutons and diverse volcano-sedimentary rocks that distinguish the Middle Jurassic Andean-type margin of western North America from a composite terrane composed of the Caborca block and the informally designated San Gabriel "terrane" (Fig. 6). The Caborca block, discussed above, includes characteristic Proterozoic and Paleozoic rocks (Anderson and Silver, this volume) overlain by Late Triassic and Early Jurassic strata (Stewart, this volume) and rare Triassic plutonic rocks in the northernmost part (see above). The San Gabriel domain is distinguished by crystalline rocks including anorthosite and related rocks (ca. 1.2 Ga) intrusive into the strongly deformed and metamorphosed 1.8–1.7 Ga crystalline basement (Silver, 1971; Nourse et al., 2003, and this volume; Anderson and Silver, this volume). Triassic plutons, generally with crystallization ages between 230 and 210 Ma, compose an intrusive suite with distinctive variations in composition, gabbro to granite, and texture, uniform grain size to coarsely porphyritic (Barth et al., 1990; 1997). The Triassic plutons locally intrude anorthositic rocks and show a clear association with the San Gabriel Proterozoic rocks. Crystalline rocks of the San Gabriel domain, which lack nonconformable Neoproterozoic and pre-Cretaceous sedimentary cover, underlie part of southwestern Arizona and southeastern California (Powell, 1993; Barth et al., 1997). We believe that the Mojave-Sonora megashear separates the San Gabriel domain from crystalline Proterozoic basement overlain by cratonal Paleozoic cover and intruded by Middle Jurassic plutons to the east.

West of Sonoita, Sonora, exposures of Triassic plutonic rocks along the fault that we define as the Mojave-Sonora megashear are the most southeasterly examples in Sonora. We believe that these Triassic rocks correspond to the southeastern extension of the San Gabriel domain. Further indications of the transition from 1.7 Ga basement rocks typical of the Caborca block to San Gabriel–like rocks occur in outcrops north of the Pinacate volcanic field. Very strong ductile deformation is a common characteristic of former porphyritic granitic rocks that crop out in Sierra Alacranes, north of Pinacate. Field studies reveal the very strong ductile Proterozoic deformation, metamorphism, and plutonism that characterize the rocks of the San Gabriel domain (e.g., Nourse et al., 2003). The characteristic deformation and metamorphism overprint and obscure the slightly older rocks of the Caborca block (Anderson and Silver, this volume; Premo et al., 2003; Nourse et al., this volume). We previously have shown the Caborca block and San Gabriel domains as distinct entities (Anderson and Silver, this volume; Fig. 5), however, the new work indicates the northwestward continuation of the Caborca block Proterozoic crystalline rocks, although they appear to have been strongly modified by ductile deformation, metamorphism, and intrusions of 1.6–1.7 Ga granite, anorthosite, and Triassic granitoids.

Inferred Trace of the Mojave-Sonora Megashear from Northwest Sonora to the Garlock Fault

Our thinking is that in southernmost California the inferred fault trace of the Mojave-Sonora megashear continues northwestward between exposures of the San Gabriel domain and the westernmost occurrences of the autochthonous or parautochthonous cratonal sequence (e.g., Marble Mountains; Fig. 6) overlying crystalline basement initially formed in the interval between 1.8 and 1.7 Ga. Within the northern part of the eastern Transverse Ranges, westward-directed thrusts, younger extensional detachments, and left-lateral faults obscure the inferred trace of the Mojave-Sonora megashear. In this region the surficial trace of the Mojave-Sonora megashear has been carried westward and the deep trace tectonically covered. Despite these complications, we believe that the effects of younger sedimentologic, magmatic, and deformational processes may be recognized and restored and that a plausible route of the northward extension of the Mojave-Sonora megashear fault may reasonably link it with the pre-Cretaceous Melones–Bear Mountain faults extending northwestward along the western margin of the Sierra Nevada (Fig. 6).

The San Gabriel domain is not recognized east of the San Andreas fault and north of the eastern Transverse Ranges of southern California. Triassic plutons that crop out in the San Bernardino, Granite, and Black Mountains, and other ranges within the northern Mojave Desert (Fig. 6) are, in general, slightly older than those among the San Gabriel rocks (Barth et al., 1997). The plutons intrude 1.7–1.8 Ga crust overlain by miogeoclinal strata and Middle Jurassic volcanic units (Miller et al., 1995). These relations show that, at this latitude, Triassic plutons are not one of the distinguishing elements of the San Gabriel terrane west of the Mojave-Sonora megashear as we interpret them to be in southeasternmost California (Fig. 6).

The 1.8–1.7 Ga Proterozoic crystalline basement, overlying Neoproterozoic and Mesozoic strata, and plutons of Triassic, Jurassic, and Early Cretaceous ages comprise parts of the allochthonous masses transported westward during Cretaceous and Tertiary deformation (Silver, 1982, 1983; Martin and Walker, 1992; Glazner et al., 1989, 2002; Walker et al., 2002). The westward tectonic expulsion of rocks within the Mojave Desert disrupted regional paleogeographic patterns, and as a consequence many exposures are difficult to fit into regional syntheses (Glazner et al., 2002).

The post-Jurassic structures, chiefly Cretaceous thrust faults and Miocene extensional faults (and Cretaceous intrusions), obscure the inferred trace of the Mojave-Sonora megashear north of the border between Mexico and the United States. Westward- to southward-directed displacements may have begun between 95 and 85 Ma and continued perhaps until 50 Ma (Wood and Saleeby, 1998). Ductile deformation of this age is recorded in igneous rocks in the southern Sierra Nevada (Wood and Saleeby, 1998), and northern (Fletcher et al., 2002) and southern (May and Walker, 1989) Mojave Desert. Underthrusting of the Rand Schist (87–79 Ma, Silver and Nourse, 1986, 1991) and Pelona-Orocopia Schist (Jacobson et al., 1996; Haxel et al., 2002) continued until Late Cretaceous or Paleocene time. Cretaceous deformation, 85–71 Ma (Fletcher and Karlstrom, 1990; Martinez et al., 1994), much of which is ductile, is also recorded by highly strained sedimentary and crystalline rocks that form the Maria fold-and-thrust belt (Spencer and Reynolds, 1990 and references therein), near the Colorado River. Miocene extension was the final structural event during which the upper crust was fractured above gently dipping faults (Glazner et al., 1989, 2002).

We speculate that masses carried westward above faults inferred to have accommodated westward and southward transport include: (1) the central Mojave block (Martin and Walker, 1992; Schermer et al., 2002) and (2) crystalline rocks of the San Gabriel terrane preserved above the Vincent and Chocolate Mountains faults (e.g., Haxel and Dillon, 1978). The central Mojave block contains crystalline basement (Martin and Walker, 1992); overlying cover of Paleozoic miogeoclinal strata; Middle Jurassic volcanogenic strata; and plutons of diverse ages, including bodies of Triassic, Jurassic, and Cretaceous ages. The rocks crop out in the Shadow, San Bernardino, Granite, Rand, and Black Mountains, among others (Fig. 6). These ranges are or may be above hypothetical Cretaceous thrusts (Silver and Nourse, 1986, 1991; Fletcher et al., 2002). They are also underlain by a Miocene detachment (Glazner et al., 1989) that accommodated 45–60 km of additional westward extension as postulated by Glazner et al. (2002).

The synintrusive conditions and timing of the Cretaceous faulting in the northern Mojave Desert (Fletcher et al., 2002) are comparable with those described by May (1989) in the San Gabriel Mountains to the south. Ductile faults recognized by May in the San Gabriel Mountains probably affected rocks inboard of their position after restoration southward along the San Andreas fault (e.g., Powell, 1993). Mapping to the east of the restored position of the San Gabriel rocks reveals a complex zone of faults separating a hanging wall composed of gneissic rocks from Orocopia Schist in the footwall (Grubensky et al., 1993; Haxel et al., 2002). Displacements of diverse ages and direction are recorded along the zone of principal faulting (e.g., Simpson, 1990). The gneiss, which occurs structurally beneath Middle Jurassic rocks, contains microstructures that record northeast-directed transport (Simpson, 1990). Simpson also points out that conditions under which the foliation in the gneiss formed are indistinguishable from those in the younger, structurally lower Orocopia Schist. In the San Gabriel Mountains, directions from the structurally lower schist are diverse (J. Nourse, 1995, personal commun.). In light of recent interpretations of eastward emplacement of the Orocopia Schist (Jacobson et al., 1996, 2000, 2002) the structural and metamorphic results are enigmatic.

One of us (Silver) has suggested that some of the gneiss east of the Chocolate Mountains, much of which records mylonitic fabric and structurally underlies Middle Jurassic igneous rocks of the southwest North American continental margin, may be derived from Triassic and Jurassic protoliths as illustrated by Haxel et al. (2002). Despite the apparent inconsistencies in tectonic transport and the strong deformation stemming from multiple Tertiary detachments that in places affect the gneiss, we make the following speculations: (1) Westerly directed transport along gently dipping Cretaceous (May, 1989) and Tertiary structures (Haxel et al., 2002) obscures the Mojave-Sonora megashear fault(s). (2) Regionally, the contact between the gneiss and the less deformed Middle Jurassic rocks formerly marked the steep Mojave-Sonora megashear. (3) Most of the Jurassic plutons crop out east of the inferred Mojave-Sonora megashear; some Jurassic plutons in the San Gabriel domain are intrusive (G.B. Haxel, 2004, personal commun.), although among the eastern Transverse Ranges intrusive relations have not been documented (Powell, 2003, personal commun.). Our proposed northward restoration of the San Gabriel domain places it north of the area affected by Middle Jurassic magmatism (Fig. 7) thereby suggesting that either the Jurassic plutons are allochthonous or we have placed the San Gabriel domain too far to the north. (4) Fabric in the gneiss that records northeast-directed transport may have formed before the Cretaceous, perhaps during Late Jurassic transpression.

If these ideas are correct, then in the areas strongly affected by west-directed tectonic transport surficial exposures of the Mojave-Sonora megashear must lie west of Middle Jurassic arc rocks intruded into and erupted onto rocks of the inboard North American cratonal continental margin. In the sections of imbricated units, the continental margin rocks, the Mojave-Sonora megashear fault rocks, and rocks of the San Gabriel domain compose an imbricated allochthon, parts of which were transposed during shear along gently dipping faults (Grubensky et al., 1993; Haxel et al., 2002). The deep parts of the Mojave-Sonora megashear fault, having been truncated by the westward-directed faults, are tectonically buried beneath the allochthons. Some of these deep, older structures may have later been displaced westward along the footwalls of Miocene detachment faults.

North of the detached plate of the central-western Mojave, other Triassic plutons that intrude Paleozoic eugeoclinal units occur within a geographically restricted region including Pilot Knob and Lane Mountain (Fig. 6). The eugeoclinal units and Triassic intrusives are postulated to be allochthonous based upon geochemical arguments (Miller et al., 1995). These authors also argue that the rocks were emplaced after intrusion of Triassic plutons and before intrusion of Jurassic plutons.

The eastern limit of the eugeoclinal rocks trends northerly. The contact could be related to Late Jurassic faulting as suggested by steep foliation with gently plunging lineations of that age that crop out at the contact in Alvord Mountain (Miller and Walker, 2002). However, despite the apparent simple relationship of eugeoclinal rocks separated by a lateral fault from the autochthonous rocks of cratonal North America, complications are apparent. If the lateral fault inferred to be the Mojave-Sonora megashear trace is matched across the Garlock fault to a displaced correlative point, it occurs at the eastern margin of the El Paso Mountains, separating eugeoclinal rocks from Jurassic plutonic rocks. The contact between Jurassic granite and the El Paso Mountains rocks is not mapped as a fault (Walker et al., 2002, Plate 1). In light of the relations of the Jurassic granitoids that appear to intrude across the contact between the allochthonous eugeoclinal rocks and autochthonous miogeoclinal rocks of North America, the surficial trace of the Mojave-Sonora megashear is interpreted to lie west of the allochthonous eugeoclinal rocks apparently stitched by Middle Jurassic plutons above a subduction zone.

Late Jurassic Deformation in the Mojave Desert

Despite the absence of fault zone rocks or other direct evidence of the Mojave-Sonora megashear fault zone separating southwestern North American rocks from displaced terranes to the west, well-developed Middle to Late Jurassic structures in rocks exposed in ranges in the northern Mojave Desert of southern California record lateral faulting, contraction, and extension (Walker et al., 2002). We infer that these structures developed contemporaneously with displacement along the Mojave-Sonora megashear and reflect releasing and constraining bends along its trace(s).

In the northern desert, Walker et al. (1990) recognized the belt-like character of Middle to Late Jurassic contractional deformation that occurs near the western margin of Middle Jurassic plutons. Thrust faults within the Cronese Hills that generally accommodate east-directed emplacement of Jurassic plutonic rocks onto Mesozoic layered rocks between 169 and 154 Ma are correlated with those of the East Sierran thrust system activity in the southern Sierra region (170–148 Ma; Dunne and Walker, 1993) based upon similar timing. Middle to Late Jurassic contractional deformation also has been identified at Alvord Mountain (179–149 Ma; Miller and Walker, 2002), Tiefort Mountains (164–148 Ma; Schermer et al., 2001); Iron Mountain (164–151 Ma; Boettcher and Walker, 1993), and Shadow Mountains (153–148 Ma; Martin et al., 2002). Differences in metamorphic grade among rocks in Tiefort (Schermer et al., 2001) and Alvord (Miller and Walker, 2002) Mountains may reflect: (1) upward movement of fault slices within a fault zone, or (2) proximity to cooling igneous masses, and (3) shuffling during subsequent westward-directed Cretaceous thrusting and Tertiary normal faulting. A comparable east-vergent thrusting event, synchronous with intrusion of the Goldhammer pluton (161 ± 10 Ma), has been mapped in the Clipper Mountains (Howard et al., 1995).

In summary, Middle to Late Jurassic (e.g., Howard et al., 1995) contractional deformation that is prominent in some rocks of the northern Mojave Desert is probably related to transpression during left-lateral faulting (Walker et al., 1990). East-directed thrusts in Tiefort Mountains, Cronese Hills, and Clipper Mountains that contrast with west-directed thrusts in Shadow Mountains may reflect a large positive flower above the inferred zone of lateral faulting.

We interpret Late Jurassic extension, locally recorded in rocks of the central Mojave block (Schermer, 1993; Schermer et al., 2002) and in the eastern Mojave block (e.g., Cowhole Mountains, Wadsworth et al., 1995; Busby et al., 2002) to be transtensional. The timing of normal faulting that occurred between the time of emplacement of the Lower and Upper Sidewinder series in the central Mojave block (163–150 Ma) (Schermer, 1993; Schermer et al., 2002) overlaps the interval during which contraction (169–154 Ma) in the Mojave Desert (Schermer et al., 2002; Walker et al., 2002) is recorded.

Palinspastic Restoration of Post-Jurassic Deformation in the Mojave Desert

As explained above, we postulate that west-directed tectonic transport of rocks above gently dipping faults in parts of the Mojave and Sonoran Deserts has obfuscated the inferred trace of the Mojave-Sonora megashear by displacing the fault zone and the rocks along it westward and covering the roots of the fault with allochthons. Two principal episodes of westward displacement above the faults are recognized: during the Cretaceous and Miocene. We attribute the exposures of North American cratonal and miogeoclinal sections west of the deep parts of the inferred Mojave-Sonora megashear to transport above the subhorizontal faults.

Proposed restorations of displacements are shown on Figures 6 and 7. Some of the displacements of rock bodies are inferred because a fault along which movement is inferred has not been mapped. In the southeast corner of California and northwesternmost Sonora, the Cargo Muchacho and El Capitan sections are presumed to exist in the hanging wall of thrust faults comparable to those manifest as the Santa Rosa mylonite zone. No piercing points are available for these displaced masses and they are minimally restored to a position just east of the inferred trace of the Mojave-Sonora megashear. In southwestern Arizona a modest westward bulge of the inferred trace (Dembosky and Anderson, this volume) is also restored in order to align the inferred trace between likely positions east of Neversweat Ridge and west of Quitobaquito Hills. Neversweat Ridge is an important locality because it exposes the most easterly occurrence of Orocopia Schist. The schist is overlain by gneiss of unproven but possibly San Gabriel domain protolith and yet higher Middle Jurassic rocks. We assume that despite the shuffling along subhorizontal faults, the juxtaposition of likely San Gabriel domain rocks against Jurassic arc rocks indicates the proximity of the Mojave-Sonora megashear. The Quitobaquito Hills are a few kilometers northeast of the Mojave-Sonora megashear fault trace exposed southwest of Sonoita, Sonora. The bulge in the inferred trace is likely related to Tertiary extension but the complex history of Cretaceous and Tertiary low-angle faulting (cf. Haxel et al., 2002) hinders accurate partition of displacements according to age.

In the northern Mojave Desert, the central Mojave block (Walker et al., 2002) and the San Bernardino Mountains to the south contain Proterozoic crystalline basement (Silver, 1971; Coleman et al., 2003) and Phanerozoic sedimentary cover rocks (Stewart et al., 1984) that correlate with characteristic North American stratigraphic sections. The central Mojave block is likely underlain by gently dipping faults of Miocene (Glazner et al., 2002) and Cretaceous ages (Fletcher et al., 2002). Martin et al. (2002) restore displacement along the Tertiary Mojave valley fault that places the Jurassic Sidewinder volcanics (Schermer and Busby, 1994) in the Victorville area south of Cronese Hills. We propose a comparable restoration that takes advantage of the several tens of kilometers postulated by Glazner et al. (2002) and adds modest older westward-directed transport sufficient to: (1) position units in the Shadow and San Bernardino Mountains east of the proposed trace of the Mojave-Sonora megashear, (2) align the swarm of Independence dikes (Carl and Glazner, 2002) among the Jurassic Sidewinder rocks with dikes to the north that are intrusive into Jurassic plutonic rocks and autochthonous (or parautochthonous) with respect to the central Mojave block, (3) position the rocks of the Shadow Mountains, which record Late Jurassic deformation (Martin et al., 2002), close to the inferred trace of the Mojave-Sonora megashear, and (4) align Late Jurassic deformation belts with steep foliation and gently dipping lineation at Iron Mountain with that at Alvord Mountain.

Left-lateral displacement of probable Neogene age along the Garlock fault is restored. Within the west-curving tail of the southern Sierra Nevada, complex patterns of faults (Wood, 1997), Sr_i isopleths (Kistler, 1990), and metamorphic pressures (Wood and Saleeby, 1998; Walker et al., 2002; Fig. 6) record Late Cretaceous–Paleogene deformation (Wood and Saleeby, 1998) involving strong apparent right-lateral faulting along the Garlock zone. The position of the probable major crustal boundary along the Sr_i 0.706 line may also be affected by north-striking right-lateral faulting.

INFERRED TRACE OF THE MOJAVE-SONORA MEGASHEAR NORTH OF THE GARLOCK FAULT

Southern Sierra Nevada

Across the Garlock fault a structural and lithospheric boundary is inferred to correspond with the Sr_i 0.706 line (Kistler and Peterman, 1973; Kistler, 1990; Carl and Glazner, 2002). The boundary commonly separates two regions characterized by distinct rock assemblages. East of the line, largely autochthonous but commonly deformed rocks are composed of Proterozoic basement covered by miogeoclinal Neoproterozoic, Paleozoic, and early Mesozoic strata (Schweickert and Lahren, 1991) and Middle and Upper Jurassic volcanic and clastic units (Dunne and Walker, 1993; Stevens et al., 1998, for summary and refs.). West of the Sr_i 0.706 line, the mainly oceanic Kings sequence is prominent (e.g., Busby-Spera and Saleeby, 1987; Wood, 1997, and references therein). This informally designated sequence (Saleeby et al., 1978) is composed of Upper Triassic to Early Jurassic rocks that crop out in pendants from Dinkey Creek to Tehachapi (Fig. 6). The main lithologic components are: quartzite, quartzose schist, chert-argillite, chert-argillite diamictite, epiclastic slate, and volcanic units, including silicic ash and debris flows and intermediate to mafic lava, tuff, and breccia, all of which accumulated in a marine environment.

The Kings sequence overlies complex Paleozoic basement, much of which is comprised of ophiolitic rocks (Saleeby et al., 1978; Saleeby, 1990). Sedimentary units that may underlie Kings strata include: (1) pre-Devonian Shoo Fly Complex (Girty et al., 1990), Paleozoic ophiolite and mélange, and chert-argillite rocks, including Calaveras Formation (e.g., Saleeby, 1990). The stratigraphic relationship of the Kings rocks to this composite underlying "basement" and the structural and stratigraphic relationships of the basement units to one another and to nearby pre-Mesozoic rocks are a source of debate (Schweickert and Lahren, 1991, 1993). Rocks of the Kings sequence project toward exposures of Shoo Fly basement (Saleeby, 1990, their Fig. 1). However, bodies of Cretaceous granite interrupt the exposures of the older rocks and the relation of Kings sequence to the major exposures of Shoo Fly complex is not transparent.

Kistler (1990) suggests that the Sr_i 0.706 line separates Panthalassan crust, that is, Kings sequence and underlying rocks, from North American lithosphere. The idea is strongly supported by the very distinctive Sm/Nd isotopic signature of the Kings River ophiolite, suggesting derivation from highly depleted mantle domains beneath a Panthalassa spreading ridge (Saleeby, 1990). We postulate that the inferred Mojave-Sonora megashear trace coincides with this boundary, which we expect to lie between pendants containing miogeoclinal and some eugeoclinal rocks to the east and those containing Kings sequence to the west (Fig. 6). The ages and stratigraphic assignments of the Sierra Nevada pendants that are summarized by Saleeby et al. (1978) and Schweickert and Lahren (1991) provide the bases for the trace of the Mojave-Sonora megashear that we infer.

We recognize the significance and possible displacements of probable right-lateral faults (Fig. 7), such as the proto–Kern Canyon fault (Busby-Spera and Saleeby, 1990), Snow Lake fault–Gem Lake–Rosy Finch shear zone (Lahren and Schweickert, 1989; Schweickert and Lahren, 1990; Greene and Schweickert, 1995; Greene et al., 1997; Tikoff and Saint Blanquat, 1997), and Owens Valley faults (Stevens et al., 1995, 1998; Stewart, 1988). However, the strike of these faults is subparallel to the Mojave-Sonora megashear and the overall effects on the position of the Mojave-Sonora megashear are modest. We have restored documented displacements across these faults as well as apparent displacements that offset markers such as the Sr_i 0.706 line.

Eugeoclinal rocks such as those that crop out in the Kern Plateau pendants (Dunne and Suczek, 1991) east of the Sr_i 0.706 line are anomalous. Lithostratigraphic similarities among rocks of the Kern pendants, the El Paso–Pilot Knob–Goldstone–Lane Mountain region, and strata of the Antler allochthon far to the north suggest correlation among these geographically separated sections as implied by Dunne and Suczek (1991) and Greene et al. (1997). However, if (1) the hypothesis by Miller et al. (1995) that the eugeoclinal rocks south of the Garlock fault were emplaced in Early Jurassic time is correct and (2) the allochthonous rocks correlate with the outcrops in the El Paso Mountains north of the Garlock, as suggested by the island of low Sr_i values that the El Paso rocks yield, then the exposures of eugeoclinal rocks on both sides of the Garlock fault may have been emplaced onto or against North American crust during an event later than that recorded by the Antler rocks. The Sr_i values shown by Kistler (1990) not only support probable allochthoneity of the eugeoclinal rocks, they further show that the North American lithosphere underlies the exposures of oceanic rocks as far west as the Sr_i 0.706 line. It is likely that the usefulness of the eugeoclinal exposures to mark the fault trace is compromised by multiple deformational events, both pre- and postmegashear faulting.

We believe that the character of the Kings sequence and its Panthalassan "basement" are clearly distinct from the Caborca block. The Kings sequence, composed of locally deep-water, penecontemporaneously deformed, commonly volcanic beds, contrasts strongly with Antimonio Formation, although they are in part age correlative. The Late Triassic–Early Jurassic, nonvolcanic Antimonio Formation in Sonora that rests conformably upon Permian sedimentary beds compares best to strata exposed in the White-Inyo Mountains in eastern-central California (Fig. 6). Because of the dissimilarities between Antimonio Formation and Kings sequence, we do not believe that Kings' rocks intervene between the Caborca block and comparable exposures in the Snow Lake pendant to the east. We prefer to include the Kings sequence as a northerly continuation of the rocks west of the Mojave-Sonora megashear.

If Kings sequence lies west of the Mojave-Sonora megashear, then restoration of left-lateral displacement places it north of the Mina deflection west or northwest of correlative units within a Late Triassic marine igneous province (Fig. 7). The Caborca block restores to a position west of the Snow Lake pendant (Lahren and Schweickert, 1989). We further postulate that the Mojave-Sonora megashear cuts obliquely across the strike of the Kings rocks toward the southernmost exposure of the Melones fault within the Foothills fault system (Figs. 6 and 7; Clark, 1960).

Bear Mountain, Melones, and Other Steep Faults of the Foothills Set: Extension of the Mojave-Sonora Megashear?

The Foothills fault system (Figs. 6 and 7; Clark, 1960, 1964) comprises several steep, sinuous faults, including Melones and Bear Mountain, that generally separate slivers of Jurassic volcanic and sedimentary sections on the west from older tectonic lenses composed of ultramafic rocks and the Paleozoic Shoo Fly complex on the east (Schweickert and Cowan, 1975; Day et al., 1985; see Edelman et al., 1989b, for additional references). The faults cut rocks as young as Early Kimmeridgian (Clark, 1960). Structural analysis and mapping shows that right steps are common along both sinistral faults, resulting in formation of local transpressional structures (Newton, 1990a, 1990b). We postulate that the Foothills faults represent the northward extension of the Mojave-Sonora megashear, although displacement of hundreds of kilometers that we propose for the Mojave-Sonora megashear cannot be recognized along any single fault strand. The apparent absence of displacement may be the product of (1) the complexity of the fault evolution, i.e., rifting, initiated as early as 163 Ma, accompanied by lateral faulting that overlapped contraction beginning ca. 155 Ma, and (2) subsequent deformation of the faults during Cretaceous contraction. We argue that the distribution and character of (1) regional stratigraphic and tectonostratigraphic units and (2) the expressions of deformation along the Foothills faults support the history we propose despite the constraints on displacements described below. The Gulf of California is a useful analogy, especially if one considers the effects of hypothetical contraction along its eastern margin. There, if contraction were to take place, granite fault blocks (for example Tiburon Island) rifted from the Sonoran mainland might be pushed eastward against the coast thereby closing the existing narrow, shallow, intervening basin. The resulting structures would obfuscate the importance of the rifted margin with its adjacent contemporaneously developed ocean floor to the west

The Melones fault is a regional boundary of Nevadan deformation (Tobisch et al., 1987). In the northern Sierra Nevada, the faults that comprise the inferred northward extension of the Melones fault zone (Clark, 1960), including Downieville, Rich Bar, and Goodyears Creek (Fig. 6), do not truncate and displace well-defined tectonostratigraphic units (Edelman et al., 1989b, their Fig. 1). However, the faults commonly coincide with the margins of tectonostratigraphic units. The Downieville fault marks the western edge of the Shoo Fly Complex that is adjacent to the Feather River peridotite. The Rich Bar and Goodyears Creek faults together form the western boundary of the peridotite that is juxtaposed against the Calaveras Complex. Despite the apparent importance of these faults as suggested by their coincidence with contacts of widespread units along strike, Edelman et al. (1989b) point out that mapping by Wagner et al. (1981) shows that the type Melones fault is continuous with the Dogwood Peak–Gillis Hill faults that lie within the Calveras Complex and associated mélange and ophiolitic rocks. The Gillis Hill fault crosses the Slate Creek thrust without appreciable displacement and therefore must be a relatively minor fault according to Edelman et al. (1989b).

West of the Melones fault another principal fault within the Foothills fault system is the Bear Mountain fault zone (Clark, 1960). Its northward continuation is the Wolf Creek–Big Bend fault (Edelman et al., 1989b) that separates the Jurassic Smartville Complex (Xenophontos and Bond, 1978) from older Mesozoic and Paleozoic oceanic volcanic and sedimentary rocks, mélange and ophiolite to the east. According to Edelman et al. (1989b) displacement along the Wolf Creek–Big Bend fault is constrained by: (1) modest displacement of the intrusive contact between the Smartville and the Yuba River pluton (Bobbitt, 1982); and (2) the presence of correlative early Mesozoic rocks (Clipper Gap Formation) on both sides of the fault.

The Bear Mountain fault shows close geographic relations with swarms of mafic dikes and other intrusives characteristic of the Smartville and Folsom Complexes (Saleeby, 1990). Emplacement of many of the mafic rocks between 164 and 160 Ma was concurrent with transtension along the Bear Mountain fault, as suggested by the presence of dikes along and within the zone of faulting (Edelman et al., 1989a; Saleeby, 1990, and references therein), and consonant with the prominent left-releasing step along the Big Bend segment of the fault. Northwesterly striking dikes with mylonitic texture in the Owens Mountain area along strike of the Bear Mountain fault zone may mark the southward continuation of the fault zone or a cogenetic fault. The syntectonic dikes that are commonly transposed were emplaced between 155 and 148 Ma and deformed mainly between 155 and 151 Ma (Wolf and Saleeby, 1992, 1995). Rootless, mesoscopic folds that show sinistral sense of shear plunge steeply, as do rod-shaped segments formed in mafic layers. Post-tectonic recrystallization has destroyed most microscopic shear indicators.

The steep faults within the Foothills system of the western Sierra Nevada are geographically and temporally associated with contractional deformation. Penetrative cleavage and folds record the deformation that is postulated to have occurred during, as well as after, accumulation of the late Oxfordian–early Kimmeridgian Mariposa Formation. The deformational event is designated as the Nevadan orogeny (see Schweickert et al., 1984; Hill, 1985; Wright and Fahan, 1988; Harper et al., 1994, and Hacker et al., 1995, for overviews and references). Schweickert et al. (1984) distinguish the main phase Nevadan deformation that is composed of cleavage, commonly axial planar, and folds with consistent northwest orientation. A distinguishing characteristic of the main phase of the Nevadan event is its short duration, which, at least locally, may be as brief as 155 ± 3 Ma (Schweickert et al., 1984).

Plutonism, correlative metamorphism, and ductile deformation that are recorded between the Melones and Bear Mountain faults follow so close after the main phase of Nevadan deformation that a genetic link is likely (Saleeby et al., 1989b). Geochronologic studies of plutons along the Bear Mountain fault zone show that ductile deformation accompanied emplacement of plutons as young as 150 Ma and is recorded in masses as young as 146 Ma (Saleeby et al., 1989b). These interpretations are based upon argon release spectra and isochron plots derived from analyses of hornblende and biotite from the plutons and wall rocks. The results fall within the interval between 147 and 137 Ma (Saleeby et al., 1989b). Argon dates suggest that most rocks had cooled sufficiently by 140 Ma to close biotite and hornblende systems to argon loss, although Paterson et al. (1991) argue that fault motion and cleavage development persisted until 123 Ma.

In addition to the short duration of the Nevadan event, the geographic extent of rocks containing contractional structures and associated penetrative fabric is limited to a belt-like zone. The zone of deformed rocks is widest in the northern Sierra Nevada, where main-phase foliations may be recognized within a region ~75 km wide (Schweickert et al., 1984). In the central Sierra, major Nevadan-age fold and fault structures are restricted to rocks adjacent to and west of the Melones fault (Tobisch et al., 1987). If as suggested by Newton (1990a, 1990b), Nevadan structures formed during sinistral transpression along the Melones (and presumably other faults of the Foothills set), then the narrow, regional belt of strong deformation generally close to the faults may be expected.

In Sonora, within the Mojave-Sonora domain, correlative deformation is recorded by blocks of very clean, well-sorted quartzite that are entrained among strata (La Colgada Formation) that locally contain Oxfordian fossils (Stephens, 1988). Similar quartzite that is thrust onto Oxfordian and Kimmeridgian beds is interpreted to record contraction associated with strike-slip faulting along the Mojave-Sonora megashear (Rodríguez-Castañeda, 1990). Early Cretaceous sandstone overlies the tilted Late Jurassic strata (Rodríguez-Castañeda, 1986, 1988). Transpression, if penecontemporaneous with and restricted to the Oxfordian and Kimmeridgian stages, occurred between ca. 157 and 149 Ma, according to the time scale of Pálfy et al. (2000). This period of time overlaps the short contractional pulse (155 ± 3 Ma) attributed to the main phase of the Nevadan orogeny (Schweickert et al., 1984).

West of the Bear Mountain–Wolf Creek fault zone the Smartville Complex initially was thought to overlie a basement terrane composed of ca. 200 Ma rocks that are exposed to the east (Bickford and Day, 1988). However, recent U-Pb ages (Bickford, 2004, personal commun.) support the Sm/Nd data that suggests an igneous age of ca. 178 Ma for Smartville basement (Saleeby et al., 1989a) that is different from the adjacent 200 Ma rocks.

The indications for rifting and juxtaposition of domains with different ages of "basement" along the Bear Mountain fault led Harper et al. (1985; see also Wolf and Saleeby, 1995) to propose the presence of a regional pull-apart basin floored with mafic crust south of the Big Bend fault. This model is compatible with geologic studies (e.g., Blake and Jones, 1974) and seismic and gravity data that is interpreted as showing ophiolitic basement beneath the Great Valley of California (Godfrey et al., 1997). The eastern boundary of the Great Valley ophiolite abuts the lower-velocity rocks of the Sierra Nevada foothills along a steep, abrupt contact that may be traced downward ~20 km into the serpentinized mantle rock. Godfrey et al. (1997) interpret the Great Valley ophiolite as a small piece of oceanic crust, ~70 km wide, that formed in response to backarc spreading. The Great Valley ophiolite was later emplaced above low-velocity rocks of Sierran affinity that extend beneath the Great Valley as a slab almost 20 km thick. Godfrey et al. (1997) point out that extensive Cretaceous plutonism at the base of the ophiolite is an alternative to the thrust emplacement hypothesis as an explanation for the presence of deep, lower-velocity crust.

If the Great Valley ophiolite formed contemporaneously with the Smartville Complex between 165 and 162 Ma, then its age correlates with plate reorganization following cessation of regional Middle Jurassic subduction. Deformation of the ophiolitic rocks by late Kimmeridgian (ca. 152 Ma) is suggested by unconformably overlying fossiliferous beds (Jayko and Blake, 1986, and references therein). Emplacement of the ophiolite onto the Sierran-like crust may have taken place simultaneously with local subduction and closure of the Josephine basin, between ca. 155 and 150 Ma (Harper et al., 1994).

The Mojave-Sonora Megashear–Foothills Fault System, Mariposa and Galice Formations, the Yolly-Bolly Junction, and the Great Valley Pull-Apart Basin

Clark's work (1960, 1964) showed that within and west of the Foothills fault system Late Jurassic marine sedimentary and volcanic rocks accumulated contemporaneously with faulting. The ages assigned to these rock units, including Mariposa flysch and slate and Monte del Oro Formation, are corroborated by subsequent work (e.g., see summary by Saleeby et al., 1991). In the northern part of the Great Valley displacements along the Foothills faults resulted in the formation of an abrupt structural boundary (Godfrey et al., 1997). The boundary separated the newly formed ocean floor of Callovian age (165–162 Ma) from an assemblage to the east comprising a "basement" of oceanic arcs, mélange, and ophiolite (Harper et al., 1985) and overlying Middle Jurassic volcanic units (e.g., Lexington Hill, ca. 170 Ma) intruded by cogenetic plutons (e.g., Scales pluton, ca. 165 Ma) (Fagan et al., 2001, and references therein).

The distribution of the fine-grained, Late Jurassic units coincident with the Foothills system faults suggests that their deposition reflects development of the boundary between oceanic and continental crust coincident with the Foothills faults. The geologic setting of the Late Jurassic flysch-like strata is clear in the western Klamath Mountains where Josephine ophiolite (162–164 Ma) is overlain by hemipelagic units followed by Galice flysch, which accumulated between 157 and 150 Ma (Wyld and Wright, 1988; Harper et al., 1994). Thrusting and metamorphism of the ophiolite and overlying strata that began ca. 155 Ma overlap the interval of flysch accumulation and demonstrate the synchroneity of sedimentation and deformation (Schweickert et al., 1984; Wright and Fahan, 1988; Harper et al., 1994; Hacker et al., 1995).

Galice Formation crops out within a belt, locally narrow, from the northwestern Klamath Mountains southward into northern California (Fig. 6). We speculate that the Upper Jurassic flysch deposits, including Galice Formation and its equivalents, such as Mariposa Formation, mark the position of the truncated Late Jurassic continental margin (e.g., Wyld and Wright, 1988) that we believe corresponded with left-lateral faults. If this speculation is correct, then the trace of a major continental margin truncation structure lies within the Foothills fault system and incorporates the Melones fault at the southeast end of the system with the Bear Mountain fault to the north, continuing along the Wolf Creek fault to the Big Bend fault, where it steps west along a major releasing bend at the northern end of the Great Valley and links to the Bear Wallows–South Fork faults (Fig. 8, see footnote 1). These faults curve northwestward within the Rattlesnake Creek terrane (Wright and Wyld, 1994), which had been part of the zone of convergence margin as late as ca. 164 Ma (Bathonian) (Irwin and Wooden, 1999). The faults extend toward Galice and Rogue Formations and Josephine ophiolite of the western Klamath terrane that probably formed the floor of another pull-apart structure. Younger thrust faults obscure the northwestward trace of the sinistral fault(s) that we infer to have lain within oceanic lithosphere.

We believe that the faults at the Yolly-Bolly junction (Blake and Jones, 1977) reflect an older tectonic template that is the product of a Late Jurassic releasing step (Fig. 8; Harper et al., 1985). We interpret the junction to mark the intersection of the principal lateral fault currently marked by the Bear Wallows–South Fork faults with a left step along a former normal fault with easterly strike that merges eastward with the Big Bend fault. The Bear Wallow and South Fork faults (Wyld and Wright, 1988; Blake et al., 1999) bound narrow panels of slate, metagraywacke, and greenstone that compose the southernmost continuous exposures of Galice Formation. The South Fork fault terminates at the Yolly Bolly junction (Blake and Jones, 1977), where it merges with the Coast Range fault (Jayko and Blake, 1986, their Fig. 3). However, outcrops of tectonic slivers of Galice-like rocks continue southward along the Coast Range fault leading Jayko and Blake (1986) to conclude that the slivers may have been dragged to their position by movement along a fault with, among other possibilities, left-lateral movement.

We suggest that the Coast Range fault corresponds to the southward extension of the former left-lateral fault along the western side of the inferred Great Valley pull-apart. This fault geometry, coupled with the ages of the Smartville Complex, the Coast Range ophiolite, and the probable age of the Great Valley ophiolite, is compatible with the development of late Middle Jurassic (Bathonian) oceanic floor beneath the Great Valley of California as previously suggested by Harper et al. (1985).

Based upon mapping and radiometric and fossil ages of fourteen oceanic or ophiolitic terranes, Murchey and Blake (1993) postulated the ages of spreading ridges and the dates when ocean crust arrived at the continental margin. The analysis led them to introduce the idea of the Klamath plate that formed in the early Mesozoic between 170 Ma and at least 240 Ma. As the buoyant part of the Klamath plate approached the subduction zone, the dip of the slab decreased. As the slab rose and coupled against the overriding plate, subduction waned at ca. 170 Ma. Coupling was followed by rifting and generation of ocean crust in fault-controlled basins. We suggest that the plate tectonic record of subduction, collision, rifting with concurrent sedimentation, lateral faulting, and contraction as discussed above and previously by Harper et al. (1985), are all compatible with convergence continuing throughout Middle and Late Jurassic time (see below). Furthermore, the tectonic events are comparable to those included within the expanded Jurassic orogenesis of Wright and Fahan (1988).

Ramifications of the Great Valley Pull-Apart Model

The hypothesis that the Great Valley ophiolite is largely an autochthonous body formed as the floor of a major pull-apart basin beneath the Great Valley contrasts with earlier models involving northward transport of the ophiolite from near the equator followed by accretion to western North America (e.g., McWilliams and Howell, 1982; Hopson et al., 1992). The interpretation that the ophiolite is far traveled has been based mainly upon paleomagnetic results from ophiolite sequences, including basalt and ultramafic rocks that were interpreted to indicate. their former southerly locations (Kempner, 1977; Luyendyk and Hornafius, 1982; Beebe and Luyendyk, 1983; Beebe, 1986). Subsequent paleomagnetic study of chert above the Coast Range ophiolite at Stanley Mountain indicates that: (1) the earlier interpretations are based upon a secondary magnetization, and (2) the location of the formation of the ophiolite based upon the primary magnetization is within a few degrees of its present position (Hagstrum and Murchey, 1996).

Exposures of Coast Range ophiolite compose outliers west of faults parallel to the San Andreas (McLaughlin et al., 1988) and west of the Coast Range fault and its inferred southward extension (Fig. 8). The ophiolitic sections composing the outliers began to form in Middle Jurassic time, as shown by their ages between 166 and 165 Ma (Mattinson and Hopson, 1992; Pessagno et al., 1993) that are the same as the Great Valley and Klamath ophiolites. Rare outcrops that have older ages (e.g., 173 Ma, Mattinson and Hopson, 1992) may be remnants of the roots of the older arc. The ophiolitic rocks form bodies up to several kilometers long that are incorporated into Franciscan Complex along steep faults (McLaughlin and Ohlin, 1984). The ophiolitic masses are grouped into two distinct suites based upon similarities of lithologic composition, stratigraphy, and age (McLaughlin et al., 1988). The widely separated outliers that are correlated on the basis of similar lithologic character are interpreted by McLaughlin et al. (1988) to record displacements along at least two sets of right-lateral faults with somewhat different northwesterly strikes. One of the distinctive outliers, the Benbow, may have been displaced as much as 320 km northwestward from equivalent rocks near Diablo Mountain (Fig. 8).

We incorporate the displacement of the Benbow outlier with other displacements related to faulting along the San Andreas fault and other right-lateral faults by means of a simple, generalized reconstruction. Our reconstruction is guided by the observation that ophiolitic outliers, especially the Benbow outlier, and granitic rocks, especially those at Point Reyes, are key markers that record the finite strain related to Tertiary and perhaps somewhat older right-lateral displacements (McLaughlin et al., 1988; Powell, 1993, for summary and references) (Fig. 8). Stratigraphic relationships between the oceanic rocks and overlying Tertiary strata suggested to McLaughlin et al. (1988) that important faulting was concurrent with deposition of the cover strata and began between 61 and 52 Ma. However, the ~300 km of Tertiary displacement along the San Andreas fault that is recorded by offset of early Miocene volcanic rocks from the Pinnacles area to the Neenach region (see Powell, 1993, for references) is sufficient to restore the Benbow outlier close to Diablo Mountain. The restoration is most directly achieved along fault strands such as Calaveras, Hayward, Healdsburg–Rogers Creek, and Maacama (McLaughlin et al., 1988, 1996). These faults extend northwest along trend of the San Andreas fault close to the point where they bend westward toward Cape Mendocino (Fig. 8). If 300 km of right-lateral displacement is partitioned among the Calveras, Hayward, and related faults, then restoration returns the Benbow outlier to a position close to the Coast Range fault as shown by McLaughlin et al. (1996). The creation of a viable paleogeography as a result of restoration of Tertiary displacements suggests that older, pre- Miocene right-lateral displacements are either quite modest or somehow obscured.

The objective of the Figure 8 reconstruction, which does not do justice to the far more detailed studies of other workers (e.g., McLaughlin et al., 1996; Powell, 1993, and references therein) is to assess the pull-apart model for the Coast Range–Great Valley ophiolites. If, as we postulate, the ophiolitic rocks formed within the pull-apart structure defined crudely by the present trace of the Coast Range thrust and the northwestward extension of the Mojave-Sonora megashear (Melones, Bear Mountain, Wolf Creek faults), then restoration of right-lateral displacements should bring the correlative ophiolitic rocks closer to the inferred regional Great Valley pull-apart basin.

According to Powell (1993), cumulative displacement along the San Andreas and related faults is ~440 km. Powell partitions the cumulative displacement into 290 km along the central part of the San Andreas and 150 km along parallel faults to the west. South of Tejon Pass partitioning of displacements is complex as shown by Powell (1993). However, comparable magnitude among the set of right-lateral faults is likely as suggested by the 300 km of post–12 Ma displacement recorded during opening of the Gulf of California (Oskin and Stock, 2003).

In order to accommodate the southeastward restoration of the ophiolitic and plutonic markers, the paths of the northerly segments of the San Andreas fault and Rinconada-Reliz and San Gregorio–Hosgri faults are reorganized with respect to the present paths shown by Powell (1993). Inferred earlier paths, different from those shown by Powell (1993) are proposed. (1) Displacement of 300 km recorded along the central segment of the San Andreas fault is partitioned from the northernmost segment of the San Andreas fault onto the Hayward, Healdsburg–Rogers Creek, and Maacama faults that compose probable northwesterly extensions of the pre-present San Andreas fault as inferred by McLaughlin et al. (1988, 1996). These linked faults extend northwestward to the Benbow outlier close to the point where the faults bend westward toward Cape Mendocino (Fig. 8). Restoration of the Benbow outlier, containing Middle to Late Jurassic units, 300 km southeastward brings it to a point opposite lithologically similar arc-related strata near Mount Diablo (McLaughlin et al., 1988, their Fig. 4). The restored position is not far west of the inferred southeastward extension of the Coast Range thrust that we infer to mark the former western margin of the Great Valley pull-apart. (2) The remaining 150 km is partitioned among the Rinconada-Reliz and San Gregorio–Hosgri faults to the west so that the plutonic rocks at Point Reyes are moved closer to Cretaceous batholithic rocks to the south by an additional 150 km. The restoration of 150 km along the linked traces of the San Gregorio, Rinconada, and related faults (San Marcos, Espinosa, Reliz, and King City) coupled with 300 km along the proposed path of the early San Andreas fault results in juxtaposition of Point Reyes plutonic rocks opposite similar granitoids southwest of the Sierra Nevada Batholith as portrayed recently by Wood and Saleeby (1998). In this simple model the same total strain within the set of right-lateral faults is maintained and displacements of 300 km of the Benbow outlier and 450 km of the Point Reyes plutonic rocks are achieved. We have tried to avoid attributing displacements to fault segments along which they are precluded by correlations across a fault. The inferred traces are obviously integrations of displacements throughout Tertiary time. Although they diverge from the present fault configurations based upon the careful, thoughtful analysis by Powell (1993), we hope that the fault paths that we offer are permissive and viable.

The reconstruction restores the rocks at Stanley Mountain southward to a position close to 32°N, compatible with the recent interpretation of paleomagnetic data (Hagstrum and Murchey, 1996; Fig. 8). The positions of the ultramafic rocks and granitic rocks after restoration of right-lateral displacements reveal a bulge westward between the magmatic axes of the Sierra Nevada to the north and the Peninsular Ranges batholith to the south (Fig. 8). The bulge is compatible with geologic relations in the Mojave Desert that, as described above, show apparent westward displacement of the Independence dike set and other less distinct linear markers such as the trend of arc-related Jurassic rocks (Fig. 6). Our inferences about the magnitude, timing, and direction of displacements above gently dipping faults are influenced by the tectonic analyses of Silver (1982, 1983), Wood and Saleeby (1998), and Glazner et al. (1989, 2002), which indicate several tens of kilometers of generally westward displacement. This information guides a second major step in the model during which plutonic and ultramafic rocks that show apparent westward displacement above inferred Late Cretaceous–Paleocene thrusts and Miocene detachments are restored eastward (Fig. 8). The restoration results in: (1) crude alignment of the granitic bodies along the apparent Cretaceous magmatic axis between the Sierra Nevada and Southern California batholiths, (2) positioning the rocks of the Jurassic volcanic belt east of the inferred Mojave-Sonora megashear, and (3) locating the ultramafic rocks of the Coast Range ophiolite closer to the inferred Great Valley pull-apart with which they are cogenetic.

An additional feature of the restoration is that the Peninsular Ranges of southern California lie south of outcrops of late Paleozoic rocks exposed at El Capitan in northwesternmost Sonora (Fig. 8). The folded, foliated El Capitan strata that rest upon locally mylonitic Jurassic granodiorite resemble sedimentary units characteristic of the cratonal cover of North America. Wall rocks of the Peninsular Cretaceous plutons that crop out southwest of the Santa Rosa mylonite zone include deformed carbonate rocks with subordinate quartzite and schist. Powell (1993) interprets these rocks as belonging to the Paleozoic miogeoclinal assemblage. East-northeastward restoration of these wall rocks places them southeast of restored El Capitan at a position that was probably at the southwest margin of North America and may have been within the sedimentary transition from the cratonal cover rocks to the deeper-water units of the miogeocline.

RESTORATION OF POSTULATED LEFT-LATERAL DISPLACEMENT ALONG THE MOJAVE-SONORA MEGASHEAR: RESULTING STRATIGRAPHIC RELATIONS AMONG PRE-CRETACEOUS MESOZOIC ROCKS

We propose that restoration of 890 km left-lateral displacement along the Mojave-Sonora megashear fits best with stratigraphic constraints despite the suggestion that as much as 1000 km (Marzolf and Anderson, this volume) may be recorded. In either case the resulting stratigraphic juxtapositions of pre–Late Jurassic rocks in the Caborca block and those that crop out in California and Nevada (Fig. 7) must be viable. The Antimonio Formation, which crops out in the northern part of the Caborca block, contains important stratigraphic relations that constrain restoration. The formation includes Upper Permian, Triassic, and Early Jurassic fossiliferous, marine strata separated by disconformities at system boundaries (González-León, 1997). The section is essentially continuous from Guadalupian carbonate beds of the Monos Formation to the highest beds of the Antimonio of Pliensbachian(?) age (Lucas et al., 1997). The Antimonio Formation is separated from the underlying Monos by an abrupt lithologic change from massive, cherty carbonate to maroon siltstone and sandstone. The lithologic change occurs within the Guadalupian.

The Antimonio Formation contains no known Triassic volcanic layers. We interpret the absence of volcanic units to indicate that Antimonio Formation must have been positioned south of the line between Saddlebag pendant in the eastern Sierra Nevada, and Yerington, Nevada. At Saddlebag pendant, Triassic rocks including a pluton ca. 233 Ma and volcanics of Koip Formation (Schweickert and Lahren, 1993; Schweickert et al., 1994) correlate with metadiorite (233 Ma) and McConnell Canyon volcanics that crop out at Yerington (Dilles and Wright, 1988). The plutonic ages correlate exactly with those comprising the domain of Triassic intrusive rocks to the southwest (granodiorite gneiss, San Diego County, 234 ± 39 Ma, U-Pb zircon; Thomson, 1994; monzodiorite in Sierra Los Tanques, interpreted zircon U-Pb isotopic age of ca. 233 Ma, reported in Stewart et al., 1986) revealed by closure of the Gulf of California (see above). The precise correlation between the Triassic intrusives in Sonora and those of the Saddlebag-Yerington region is provocative. Restoration of 890 km of left-lateral displacement brings them to the same latitude (Fig. 7). Other Triassic plutons, as young as 218 Ma (Barth et al., 1990), intrusive into San Gabriel basement west of the Mojave-Sonora megashear in southern California and the somewhat younger, 200 Ma units within the Kings complex are restored to a position even further north, west and north of the Saddlebag pendant.

The conformable stratigraphic relationship between the Permian-Triassic Antimonio Formation and Permian Monos Formation is most comparable to that locally exposed between the Union Wash and Conglomerate Mesa Formations in the region of Union Wash, southern Inyo Mountains (e.g., Stevens et al., 1997). This relation holds despite the abrupt changes in stratigraphic relationships and lithologies that occurred in response to fast-paced deformation affecting the Californian strata. Although the strata are conformable at Permian Bluff, south of Union Wash, Stevens et al. (1998) interpret lithologic relationships to indicate a disconformity between the Lower Triassic and Upper Permian strata. This interpretation conflicts with the information from fossils reported by Lucas et al. (1997). The fossil results suggest that the unconformity within the El Antimonio section lies below the Permian-Triassic boundary. A similar position was initially suggested by Stone and Stevens (1987) but discarded following subsequent work (Stevens et al., 1998, their Fig. 5). We believe that stratigraphic similarities between the strata at El Antimonio and in the Union Wash region are strong enough that correlation is likely despite the complex structural relations. Restoration of ~890 km of left-lateral displacement places the Triassic and Early Jurassic beds of the Sonoran Antimonio Formation (González-León, 1997) in the vicinity of Union Wash, California (Fig. 7).

Correlative and lithologically comparable sections of Late Triassic carbonate units and Early Jurassic clastic beds occur northward at Yerington (Dilles and Wright, 1988) and continue to the Late Triassic Hosselkus Limestone and Early to Middle

Jurassic Sailor Canyon sandstone and conglomerate described by Fisher (1990) in the region near Lake Tahoe and to the north.

In our restoration, the Caborca block forms a protuberance of Proterozoic crust that must be rimmed by Middle Jurassic igneous rocks. We infer that the Middle Jurassic magmatic rocks cross southern Sonora in order to link with the correlative belt that occurs along the southwestern North American continental margin in southern California, southern Arizona, and northern Sonora. In the restoration, the inferred belt of Jurassic volcanics, which rim the west margin of Sonora, must cross Sonora and join those of the autochthonous margin near the present Garlock fault. The restoration removes the apparent double belts of Jurassic magmatic rocks that may be inferred from the current exposures in western and northeastern Sonora.

SUMMARY AND INTERPRETATION

Correlation of Jurassic Events Along the Southwest Margin of North America: Orogenic and Plate Tectonic Significance

Hacker et al. (1995) point out that Nevadan and Siskiyou orogenic events may be attributed to either or both collision related to convergence or changes in plate motion. In southern Arizona (Tosdal et al., 1989) and Sonora, Jurassic igneous rocks may be subdivided by age and composition into two groups, which we interpret to reflect change from subduction to transform faulting. An older calc-alkaline suite, the ages of which generally fall between 170 and 160 Ma, is commonly attributed to subduction-related magmatism. A younger suite, generally alkaline, which includes trachyandesite and alkali granite and syenite formed between 162 and 147 Ma, is commonly contained within transtensional basins (see Anderson and Nourse, this volume).

The temporally and compositionally distinct suites of igneous rocks in Arizona, Sonora, and Zacatecas, Mexico, may be correlated with events and processes in California and Oregon. We offer an overview of plate tectonics for the margin of southwestern North America that incorporates the deformational history of the Klamath Mountains (e.g., Harper and Wright, 1984; Hill, 1985; Wright and Fahan, 1988; Harper et al., 1994; Hacker et al., 1995) with the geologic record southeastward along the Mojave-Sonora megashear, a transform that is inferred to have linked plate motions in the Atlantic and Pacific regions (Anderson and Schmidt, 1983). The Jurassic history incorporates subduction followed by collisional events, lateral faulting, and crustal extension and contraction:

1. A contractional event (Siskiyou orogeny) at ca. 175 Ma that is recognized in the Sierra Nevada and northward (Renne and Scott, 1988; Wright and Fahan, 1988) is distinct from the following somewhat younger late Middle Jurassic and Nevadan contractional events. Contractional structures correlative with Siskiyou are not recognized in southern California and Sonora, indicating a restricted geographic extent for the older deformation.

2. Renewed subduction within the interval between ca. 175 and 165 Ma resulted in formation of the western Hayfork volcanoplutonic arc (Wright and Fahan, 1988) and equivalent volcanic units southward along the axis of the Sierra Nevada, across the Mojave Desert, and into Arizona (Cobre Ridge Tuff of southern Arizona; Riggs, 1987; Riggs and Busby-Spera, 1990; Riggs and Haxel, 1990) and northern Mexico (this paper).

3. Blockage, consequent jamming of the subduction zone, and coupling of the subducting and overriding plates led to important changes at the plate margin beginning perhaps as early as 169 Ma.

4. Changes at the plate margin are manifest by generation of new oceanic lithosphere such as Coast Range (167–156 Ma, Saleeby et al., 1991), Josephine (164–162 Ma, Harper, 1980; Harper et al., 1994), Preston Peak (Snoke, 1977), and Devils Elbow (Wyld and Wright, 1988) ophiolites.

5. Much of the Smartville Complex west of the Bear Mountain–Melones fault zones formed during this event. Extension is especially evident at apparent left steps such as at the Big Bend fault (Dilek, 1989) in northern California. At the left step, the floor of the Smartville Complex, which formed ca. 178 Ma, was extended and intruded by dikes and plutons between 164 and 160 Ma (Edelman et al., 1989a; Saleeby et al., 1989a). Further south the Great Valley ophiolite formed the floor of the regional pull-apart structure.

6. Plutons created during subduction and already residing in the crust continued to crystallize after 165 Ma.

7. Pull-apart basins inboard of the Mojave-Sonora megashear along the margin of western North America and especially in northern Sonora, southern Arizona, New Mexico, Chihuahua, and Texas (Anderson and Nourse, this volume) also began to form at this time. In continental basement of southwestern North America, the initial manifestation of the transition from convergence to transtension occurred by 162 Ma, as recorded by tuff that occurs in the lowest units of stratigraphic sections in pull-apart basins such as McCoy Basin (Fackler-Adams et al., 1997). Higher in the stratigraphy of the structural basins volcanic units are rare or absent.

8. Contraction, such as that of late Middle Jurassic age in the Mojave Desert of California (Walker et al., 1990) and recorded by the Goldhammer pluton (ca. 161 Ma) (Howard et al., 1995) to the south, is probably transpressional.

9. Ocean floor formation near the Mojave-Sonora megashear ceased at ca. 157 Ma (Saleeby et al., 1982).

10. Continuing (159–149 Ma), strike-slip faulting accompanied by regional contraction along the Mojave-Sonora megashear is shown by: (a) penecontemporaneous folds and olistoliths in Oxfordian and Kimmeridgian(?) La Colgada Formation within the Mojave-Sonora domain of Sonora, Mexico (Anderson et al., this volume); (b) dikes in the western Sierra Nevada (Wolf and Saleeby, 1992); (c) major thrusts (e.g., Madstone and Orleans; Harper et al., 1990, 1994; Hacker et al., 1995), and (d) deformation in Galice Formation of northern California and Oregon.

11. Although subduction had waned or ceased in many places by ca. 165 Ma, plutons, e.g., Ko Vaya granitoids, that

are generally alkaline and with ages that commonly fall within the interval between 155 and 145 Ma intruded locally along faults bounding pull-apart structures and into crust that had been thinned significantly during extension.

12. The cessation of the Late Jurassic lateral faulting and associated contraction is marked by intrusion of dikes, including the undeformed Independence dike swarm at 148 Ma, inboard of the Mojave-Sonora megashear, the dikes of the Owens Mountain area (Wolf and Saleeby, 1995), and the Ko Vaya granitoids (147± Ma) (Tosdal et al., 1989; Anderson et al., this volume, Chapter 16, Figure 2 and Table 3 therein).

We believe that lateral faulting, "rifting" in major pull-apart (transtensional) structures, local contraction at restraining bends (transpression), and widespread shortening, generally near major lateral faults, are principal tectonic processes related to the evolution of the Mojave-Sonora megashear transform between 165 and 145 Ma. Events during the 20 m.y. interval are compatible with most of the tectonism described by Wright and Fahan (1988) in their expanded view of Jurassic orogenesis. However, as discussed below, the sequence of events isolated by us for discussion is merely a manifestation of an interruption of a much longer interval of eastward directed convergence of proto-Pacific oceanic plates beneath western North America between ca. 175 and 88 Ma (e.g., Beck and Housen, 2003). The Mojave-Sonora megashear extends eastward across northern Mexico between offset correlative segments of the mid-Jurassic magmatic belt (Jones et al., 1995), and it accommodated opening of the Gulf of Mexico (Anderson and Schmidt, 1983).

Interpreted Kinematic Relations Among Convergence, Subduction, Collision, Rifting, Extension, and Transform Faulting

Development of the Middle Jurassic (175–165 Ma) magmatic arc along the southwestern margin of North America records plate convergence (Fig. 9A). Extensional structures with easterly strike, such as dikes and faults that formed soon after subduction ceased ca. 165 Ma (Harper et al., 1985; Wolf and Saleeby, 1995; Anderson and Nourse, this volume), track the orientation of the maximum principal stress in the proto-Pacific Klamath and North America plates (Fig. 9B). If the extensional faults in the formerly overriding plate are parallel to the maximum principal stress during the later strike-slip stress regime and if Middle Jurassic subduction involved orthogonal convergence, then it may be argued that subduction, rifting, and transform faulting and transtension are tectonic manifestations of the deep process(es), perhaps east-oriented asthenospheric flow, that carried the Klamath and North America plates toward one another (Fig. 9C). Below we discuss the plate interactions and attempt to tie them to the style and duration of Middle and Late Jurassic tectonic events along the southwestern margin of North America.

During convergence, stress may or may not be transmitted across the plate boundary. If subduction is efficient (Fig. 9A, 175–165 Ma,) and the subducting slab is consumed with essentially no interaction with the overriding slab, then little stress is transmitted from one plate to another across the plate boundary. If, however, the downgoing slab does not subduct efficiently because it is thick or of low density or both, then collision(s) and jamming (Fig. 9B, ca. 170–167 Ma) may occur. As the thick, buoyant downgoing slab rises and binds against the base of the overriding plate, the upper-plate margin may be dragged toward the plate interior and consequently thickened. If the upper lithosphere thickens sufficiently it may become strong enough to transmit stresses related to the deeper flow into brittle areas within the formerly overriding plate. Within the relatively cool, rigid plate interior the stresses may cause failure that is manifest by brittle deformation caused by fractures forming a systematic pattern of normal faults and complementary strike-slip faults. These faults accommodate, in this case, the shortening imposed by continuing lateral flow of the asthenosphere that drags opposing lithosphere and overlying crustal sections toward one another (167–163 Ma). Soon lateral faulting that develops along the plane of maximum shear stress becomes the most efficient means to accommodate strain across the plate boundary and major throughgoing faults develop (Fig. 9C, 163–148 Ma). Shear strain within the zone of principal displacement may be sufficiently high that segmentation (strain softening) among the highly strained rocks occurs with consequent intrusion of syndeformational plutons (155–148 Ma) into highly attenuated regions. Intrusions into the most strained parts of the crust may weld the transform(s) and anneal the lithosphere with consequent strain hardening. If the lithosphere is sufficiently strengthened, it may again begin to transmit stress across the former transform plate boundary (Fig. 9D, 148–123 Ma). Extension perpendicular to the consistently easterly oriented δ_1 axis is maintained as shown by studies of the Independence dikes (Glazner et al., 1999) and is compatible with the development and deformation of the Owens Mountain dike swarm (Wolf and Saleeby, 1995). Ultimately, if nonsubducting lithosphere is removed by accretion or displacement along lateral faults, subduction of the cooled, now dense lithosphere may resume. The style of deformation in the crust of the overriding slab is a function of the nature of the slab, that is, thin versus thick, cold versus warm, rigid versus viscous.

CONCLUSIONS

U-Pb interpreted ages of zircon from volcanic and plutonic rocks in northern Sonora that are correlative with lithologically similar units in southern Arizona reveal widespread Middle Jurassic magmatism north of the Mojave-Sonora megashear. Exposures of these rocks distinguish the southeasternmost manifestation of a continental margin magmatic belt extending southeast from at least the northern Sierra Nevada of California into northern Mexico. In Mexico volcanism that we relate to subduction began as early as 175 Ma and ended between 165 and 160 Ma. The most extensive volcanic units commonly yield U-Pb interpreted ages between 175 and 170 Ma; ages of widespread exposures of calc-alkaline Kitt Peak–Trigo Peak granitoids cluster at ca. 165 Ma.

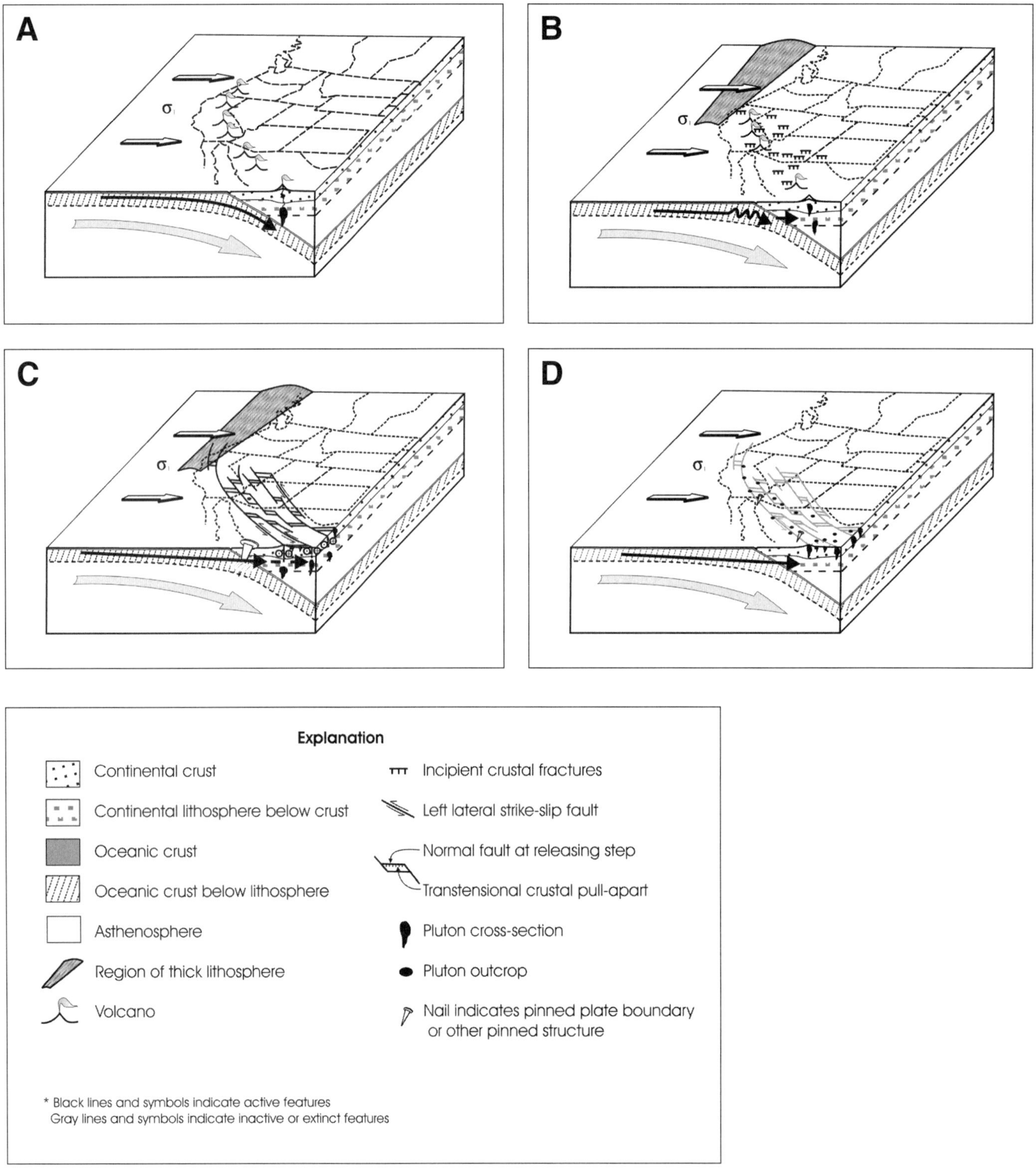

Figure 9. A diagram showing interpreted kinematic relations among Middle to Late Jurassic convergence, subduction, collision, rifting, extension, and transform faulting along the western margin of North America. (A) 175–165 Ma: Plate convergence with unimpeded subduction and consequent volcanism. (B) Ca. 165 Ma: Plate convergence continues although subduction is impeded perhaps by collision of locally thickened lithosphere. As thick subducting slab begins to jam against overriding slab, subduction-related volcanism wanes and transtension begins as recorded by: (1) almost contemporaneous generation of Great Valley, Coast Range, Josephine, and Preston Peak ophiolites in pull-apart basins (167–157 Ma) formed mainly in oceanic lithosphere west of the Mojave-Sonora megashear; (2) fracturing of both overriding and subducting slabs (e.g., dikes in Smartville complex); and (3) development of transtensional basins inboard of the Mojave-Sonora megashear (e.g., McCoy Mountains basin). (C) <162–148 Ma: Plate convergence continues; however, as plates become firmly coupled, a transition from crustal fracturing to transform faulting occurs (e.g., Mojave-Sonora; Melones and Bear Mountain). Strong extensional strain is imposed upon crust within the principal fault zones and major transtensional basins form. Magma intrudes areas of thin crust; the resulting warmed and thickened crust is sufficiently strengthened so that some local ductile deformation recorded by folds and thrusts (Nevadan orogeny) accompanies northerly extension. (D) 148 to ca. 135 Ma: Plate convergence continues; nascent intrusions continue to weld faults and thicken and strengthen crust; stress is transmitted into thickened, warmed crust where contraction may locally occur; by ca. 135 Ma, subduction-related magmatism resumes.

Waning of subduction-related magmatism was followed by the abrupt formation, ca. 165 Ma, of oceanic crust along the western margin of the North America plate. The record of oceanic crust formation is preserved in exposures of the Coast Range, Josephine, Great Valley, and Devil's Elbow ophiolites and the Smartville Complex. The formation of ophiolitic rocks within oceanic pull-aparts such as the Josephine and Great Valley signaled the beginning of transtensional faulting.

Almost contemporaneously (ca. 163 Ma) the lowest volcanic units and overlying coarse sedimentary beds began to accumulate in fault-bounded continental pull-apart basins such as the McCoy Mountains basin. Other transtensional basins, formed at releasing steps where pull-aparts formed, are well developed within the Papago domain and other parts of southwestern United States and northern Mexico.

Late Jurassic magmatic rocks crop out near faults at some releasing steps and within floors of some pull-apart structures. The distribution suggests that magma rose into areas of thin crust and along faults. In southern Arizona these igneous rocks are included as part of the Artesa layered sequence and the Ko Vaya plutonic suite.

The shift from subduction to regional transtension marks the initiation of sinistral strike-slip faulting along the Mojave-Sonora megashear. The Mojave-Sonora megashear, strands of the Melones and Bear Mountain faults of the Foothills fault system, the Wolf Creek fault and the Big Bend fault comprise the zone within which principal displacement took place. The Mojave-Sonora megashear fault zone is generally parallel to the former continental margin except where it transected and displaced a protuberance of Proterozoic basement (the Caborca block). Within the Caborca block, southwest of the Mojave-Sonora megashear, fossiliferous strata consist mainly of Lower Jurassic marine sedimentary rocks. These fossiliferous sections, which lack the common volcanic units of the volcanoplutonic suite, are juxtaposed against distinctly younger but more deformed Middle Jurassic rocks northeast of the Mojave-Sonora megashear. The Lower Jurassic strata are too old to have formed as the forearc assemblage of the Middle Jurassic volcanic belt.

Displacement of the Caborca block records left-lateral sense offset of ~800–1000 km along the main fault. The displacement of the Caborca block took place south of a major releasing step with the result that a regional pull-apart structure, floored with oceanic rocks, developed. This pull-apart coincides with the Great Valley of California. In northern Sonora and southwestern Arizona the Middle Jurassic magmatic arc was transected, truncated, and displaced into northeastern Mexico.

Sinistral lateral faulting continued in places from 163 Ma to ca. 148 Ma. About 155 Ma contractional structures developed during the main phase of the Nevadan orogeny commonly near the main lateral fault(s) and extended into the rocks of the Middle Jurassic magmatic belt. In Sonora, the deformation is recorded by folded and/or imbricated Oxfordian rocks generally consisting of volcanic, volcaniclastic, and clastic strata northeast of the Mojave-Sonora megashear.

The cessation of strike-slip faulting locally began ca. 150 Ma as shown by undeformed intrusive bodies that cut older, deformed Middle Jurassic rocks. By the time that the Independence dikes and correlative rocks were emplaced at 148 Ma, scant evidence of lateral faulting or related transpression is known.

Intrusions, young volcanic cover, transecting strike-slip faults, and multiple generations of low-angle extensional and contractional faults obscure Jurassic structures in southern California. Despite these complications, removal of the effects of superposed structures reveals a viable trace for an inferred Late Jurassic left-lateral fault linking the Mojave-Sonora megashear and Melones fault segments. The position of this major inferred fault is constrained by distinctive tectonostratigraphic domains.

The Middle and Late Jurassic and earliest Cretaceous (175–120 Ma) tectonic history includes subduction, collision, rifting and transtension, lateral faulting, transpression and contraction, and renewed subduction. About 165 Ma, subduction began to wane and rifting became important, as suggested by development of the large areas underlain by new oceanic crust. Between ca. 163 and 145 Ma easterly trending fracturing was very shortly followed by northwesterly striking sinistral faulting along which the extensional faults were linked. By ca. 135 Ma, subduction and related magmatism had resumed.

The progressive deformation recorded extensively in the overriding North America plate mimics a mega-rock deformation experiment. The deformation record includes subduction followed by longitudinal splitting of the plates along their margins. Fracturing that took place after cessation of subduction suggests jamming of the downgoing slab. Continuing plate convergence resulted in transmission of easterly directed maximum compressive stress across the plate boundary between the coupled plates. The orientation of the normal faults accommodating extensional deformation is compatible with an easterly trend for the orientation of the maximum principal stress axis similar to that during earlier subduction. Although fracturing may have briefly preceded shear rupture, deformation at the plate margins was soon partitioned into normal and asymmetric lateral faulting concentrated along the western margin of the North American plate.

Coincidence of the trend of the maximum principal stress axis during subduction and extension and lateral faulting is interpreted as reflecting deep flow beneath the proto-Pacific and North America plates that caused continued convergence between the plates. Extensional deformation at the plate margins suggests that some currents of deep flow were deflected into a relatively narrow channel a few hundred kilometers wide that lay between the converging plates. The channel accommodated mantle flow perpendicular to the maximum principal stress. Longitudinal splitting and shear fracturing were concentrated in the crust above the channel. Continuing, late north-south extension is manifest by dikes of the Independence swarm that were emplaced along northwesterly striking fractures and that opened oblique to their strike. Ductile deformation expressed by folds and foliation was most common in warm rocks of the magmatic belt and within crust attenuated by shear strain and warmed by intrusions. Subduction, collision, rifting and transtension, lateral faulting, regional transpression and contraction may take place under a single stress regime.

ACKNOWLEDGMENTS

Anderson is very grateful to Leon T. Silver, who initiated and supported the first six years of the work reported here. Silver, who served as mentor and colleague, conceived the idea of a great fault to explain the patterns described above. His contributions to the development of the tectonostratigraphy of southwestern North America and to the recognition and interpretation of regional structures have been seminal. Anderson has benefited from the hard work of University of Pittsburgh students, who commonly conducted extensive field mapping projects of largely geologically unknown areas. These stalwarts include D. Bajek, J. Chepega, K. Cohen, C. Connors, F. Corona, J. Dembosky, K. Ichikawa, M.B. McKee, J.L. Rodríguez-Castañeda, and W. Stephens. Bruce Carter, George Davis, Kees De Jong, Gordon Haxel, John Marzolf, Jon Nourse, Bob Powell, Vic Schmidt, Jack Stewart, Dick Tosdal, D.J. Wood and Claude Rangin all have been generous in helping in one way or another with the development of Anderson's ideas with respect to Jurassic rocks in Sonora. Clark Blake pointed out the potential importance of the faults near the Yolly Bolly junction. Geologists of the Instituto de Geología (Universidad Nacional Autónoma de México), the Universidad de Sonora, and the Consejo de Recursos Minerales have been patient, friendly, and very supportive. Jesus Nájera, Jaime Roldán-Quintana, and Guillermo Salas, former directors of the Estación Regional del Noroeste, Instituto de Geología in Hermosillo were most involved in a formal or semi-formal administrative manner. In addition to them, Juan Carlos García y Barragán, Carlos González-León, and César Jacques-Ayala led field trips, talked over data, and facilitated the work. Jim McKee and Mary Beth McKee and Jon Nourse have spared no effort to provide Anderson with a base firmly tied to field relationships for many of the ideas discussed. Research monies from National Science Foundation provided support for much of Anderson's and Rodríguez's work. The Instituto de Geología, Universidad Nacional Autónoma de México (UNAM) provided support to J.L. Rodríguez-Castañeda during his work in cooperation with the Estación Regional del Noroeste. Special thanks to Jose Guerrero and Fernando Ortega-Gutierrez, former Directors of the Instituto de Geología, UNAM, whose encouragement and support were very important during the course of this work. Scott Davidson carefully crafted accurate maps and Paul Coyle collated the data tables. Ken Ludwig generously provided suggestions pertaining to the interpretation of the apparent U-Pb ages of zircon fractions.

Carlos González-León, Karl Karlstrom, S.T. Kitinyrear, Jon Nourse, and Nancy Riggs reviewed preliminary versions of the manuscript, whereas Gordon Haxel, Carl Jacobson, and Jon Nourse reviewed the final version. The constructive assessments included many beneficial suggestions and improvements. Haxel's remarkably thorough reviews were especially helpful, as were the multiple reviews by Nourse. Research grants from the National Science Foundation to Silver funded the early phases of this work.

APPENDIX

Zircon Suites—Separation Procedure

The procedures for liberating and processing cogenetic suites of zircons are the same as those described by Silver (1963) and Silver and Deutsch (1963). Prior to fusion, the zircon fractions were washed in ~50 mL of hot (80 °C) concentrated HNO_3 for 1 hr. They were then rinsed three times with quadruply distilled water, followed by three rinses with filtered technical-grade acetone. Preliminary determinations of the uranium concentration of the fractions were made using α-counting techniques.

Analytical Procedure

Fractions of the total zircon population were selected for isotopic analysis on the basis of their uranium content, with the objective of obtaining the greatest range in the degree of discordance. The zircon fractions were fused with about six times their weight of purified sodium-tetraborat flux for varying time periods, depending upon the sample weight. The analytical procedure for extracting lead is the same as that described by Silver et al. (1963). The few modifications made of the proven techniques include the substitution of 10 mL Pyrex beakers for Teflon during the final digestion of the lead dithizonate and a change in loading technique from running PbS on a tantalum filament to $PbNO_3$ in a silica gel–phosphoric acid medium on a rhenium filament (modified from the procedure of Cameron et al., 1969). Uranium was extracted by normal ion exchange techniques.

Isotopic analyses were made on 12 in, single-focusing, solid-source mass spectrometers constructed at the California Institute of Technology. Data obtained on the electron multiplier and simple collector have been corrected for mass discrimination and for very slight nonlinearity in the shunt factors. Mass spectrometer reproducibility is generally ±1% for radiogenic $^{206}Pb/^{204}Pb$ ratios, and ±0.3% for the $^{206}Pb/^{207}Pb$, $^{206}Pb/^{208}Pb$, and $^{235}U/^{238}U$ ratios. Mass spectrometer precision (1σ) for $^{206}Pb/^{207}Pb$ averages about ±0.2%, as does precision for $^{206}Pb/^{208}Pb$ ratios. Uncertainty for $^{206}Pb/^{204}Pb$ is about ±0.5%.

The common lead blank contributed by reagents and glassware is slightly less than that determined by Silver et al. (1963). The use of the silica gel–phosphoric acid loading technique enabled us to run 1 μg quantities of lead, which permits significant reduction of sample size. This combined with the smaller amount of flux used (6 rather than 10 times the weight of sample) can reduce the blank significantly. Contribution of common lead from the rest of the procedure has also been reduced so that the range is ~0.06–0.04 μg.

Interpretation of Ages and Uncertainties

Inheritance and lead-loss typically coexist in bulk samples of both volcanic and plutonic zircons. Therefore discordance (indicating inheritance) does not guarantee that the 206/238 ages is a minimum, and concordance guarantees neither absence of lead loss nor absence of inheritance (though it does rule out much inheritance of Precambrian zircons). Even if one could apply objective statistics to this reconnaissance data (which in almost all cases here would give very large errors), the underlying model (of only inheritance or only lead-loss) would be unreliable. Even replicate, overlapping concordant analyses

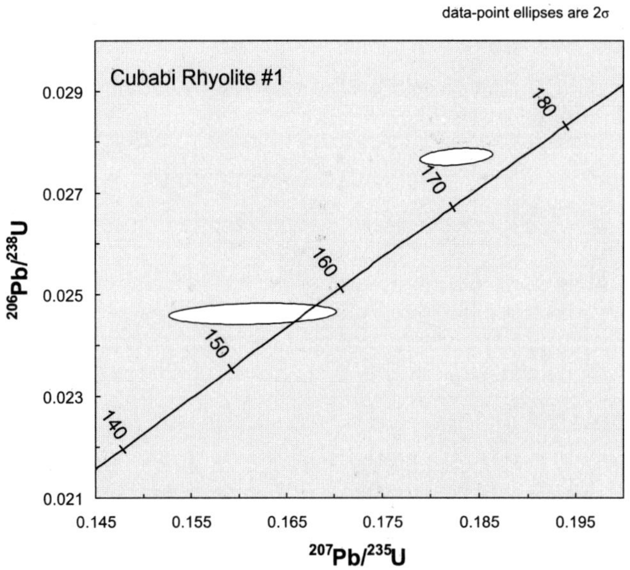

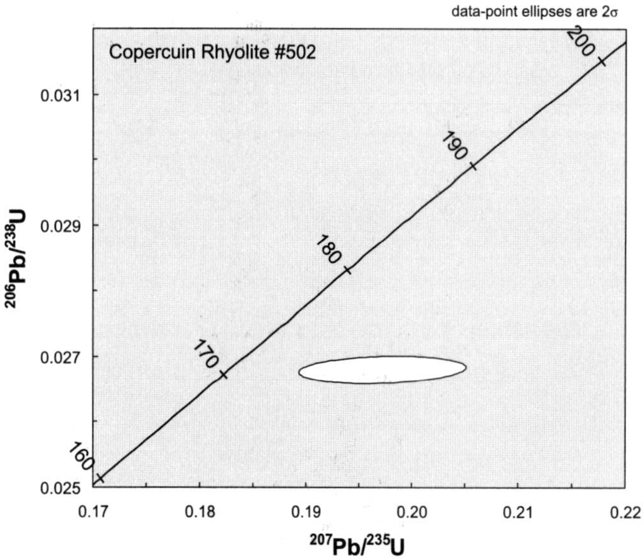

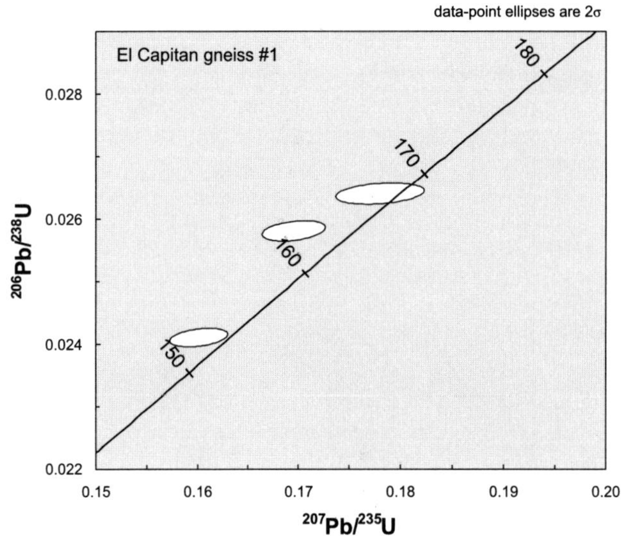

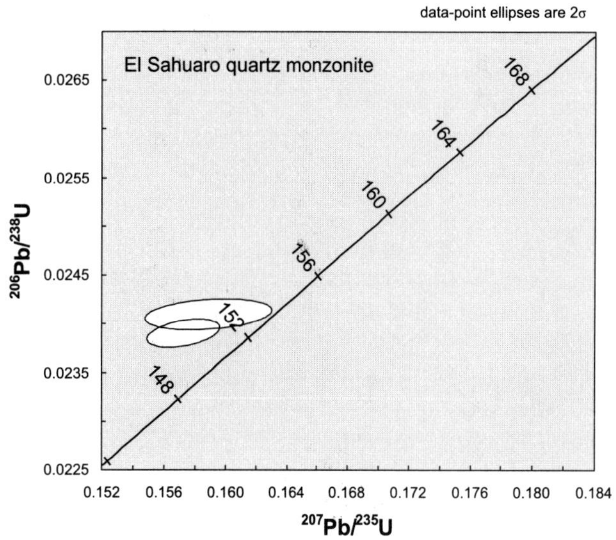

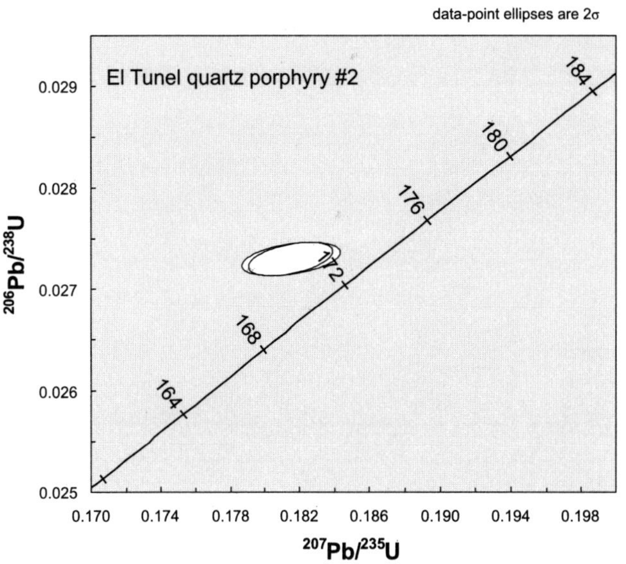

Figure A1 (*continued on following two pages*). Plots of isotopic ratios from fractions of analyzed zircons shown on concordia.

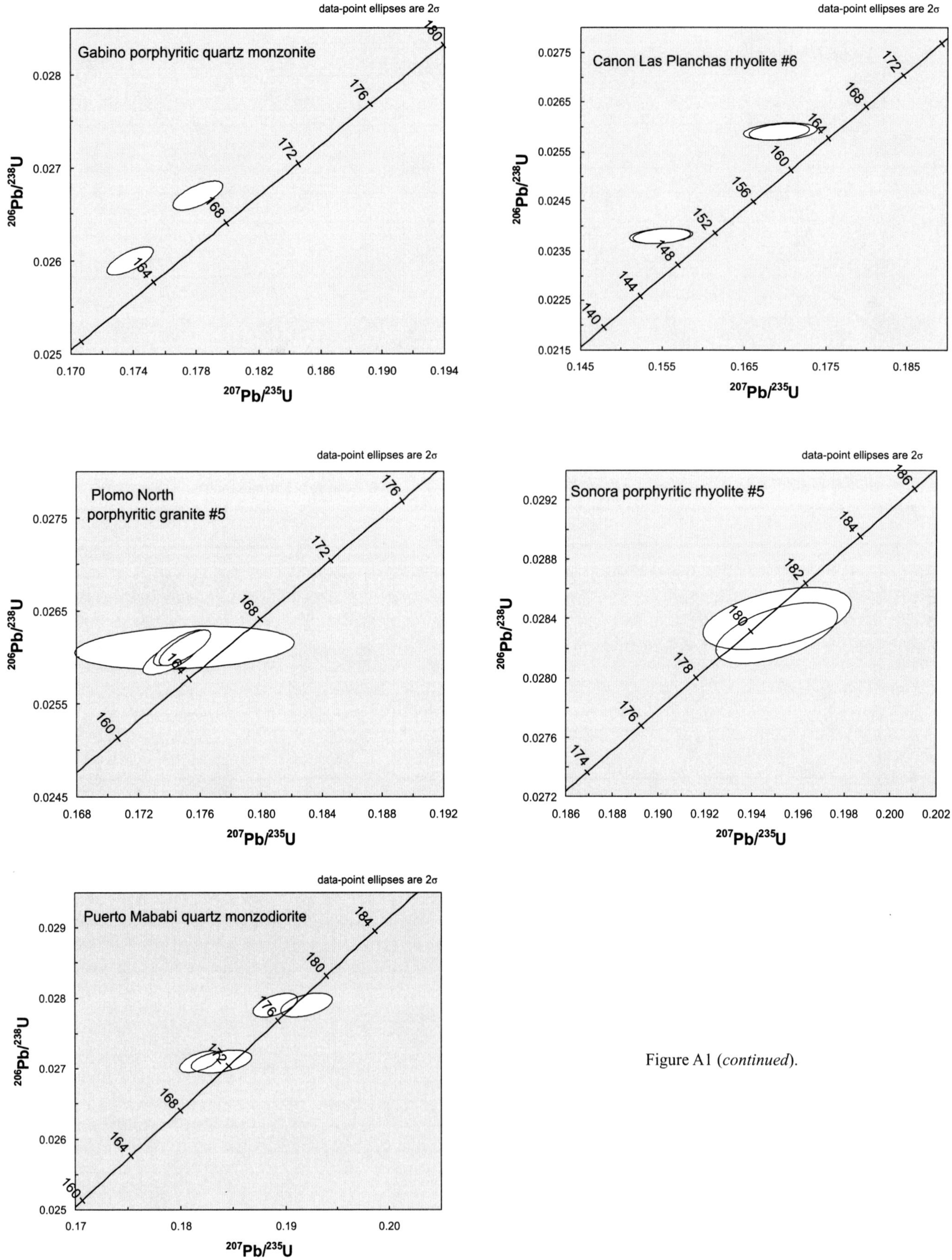

Figure A1 (*continued*).

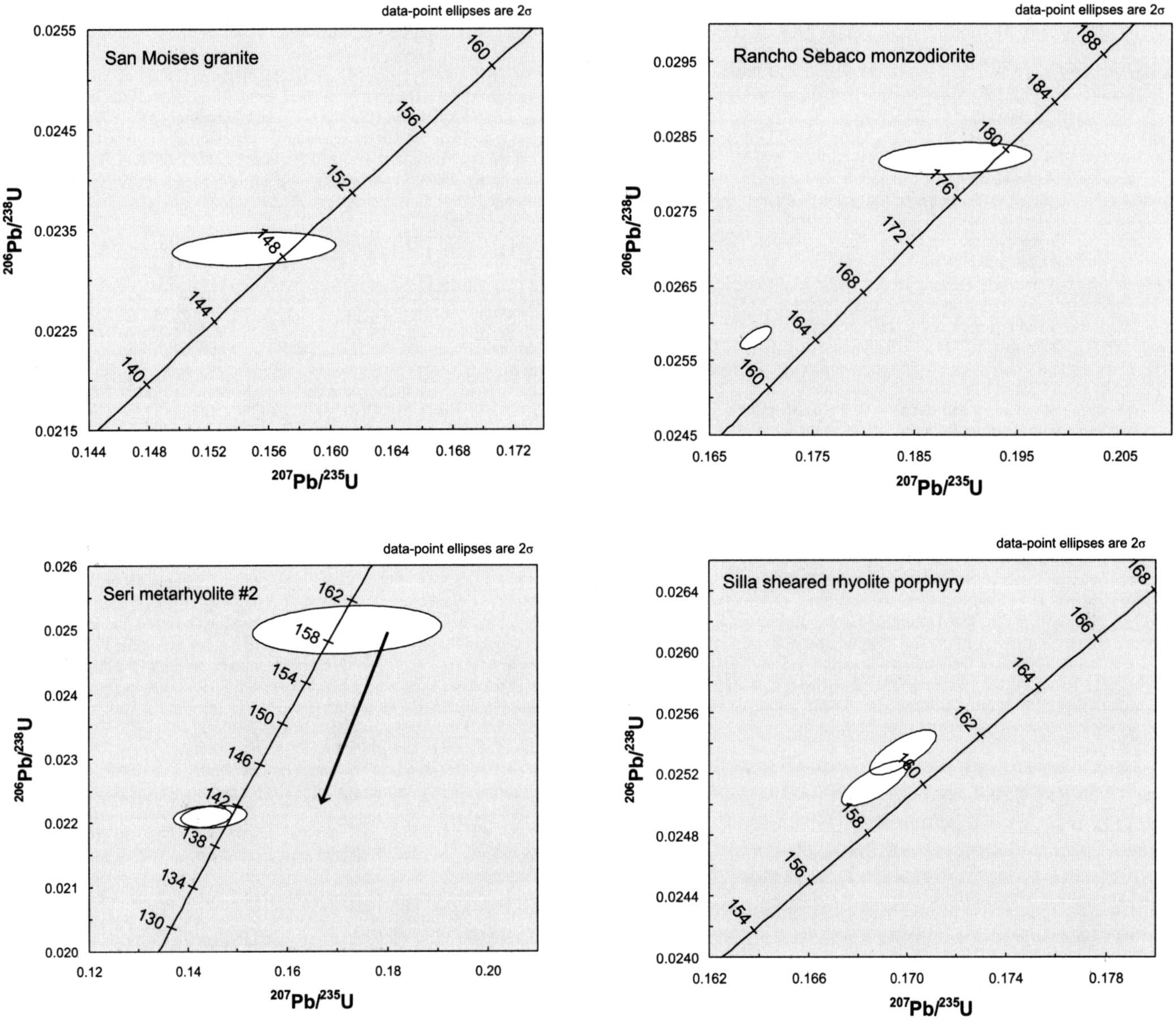

Figure A1 (*continued*).

is in this case no guarantee of accuracy (at least within the analytical errors), because the hundreds to thousands of grains involved must result in a very high degree of averaging of complexity. Selection by grain size and magnetic properties may not necessarily split out fractions recording the full spectrum of complexity.

The concordia intercept zircon ages and the uncertainties assigned to the ages were calculated and plotted using the algorithms of Ludwig (2001).

REFERENCES CITED

Alencaster de Cserna, G., 1961a, Parte I, Estratigrafia del Triásico Superior de la parte central del Estrada de Sonora: Universidad Nacional Autónoma de México, Instituto de Geologia, Paleontologia Mexicana 11, pt. 1, 18 p.

Alencaster de Cserna, G., 1961b, Parte III, Fauna fosil de la Formacion Santa Clara (Carnico) del Estrada de Sonora: Universidad Nacional Autónoma de México, Instituto de Geologia, Paleontologia Mexicana 11, pt. 3, 38 p.

Almazán-Vázquez, E., and Fernández-Aguirre, M.A., 1988, Estratigrafía de la Hoja Arivechi (H12–D56): Segundo Simposio Sobre Geología y Mineria de Sonora, Resúmenes, p. 2–3.

Anderson, T.H., and Campbell, P.A., 1994, The role of the Quitobaquito thrust in Mesozoic deformation Cordillera-Chihuahua corridor: Geological Society of America Abstracts with Programs, v. 26, no. 2, p. 35.

Anderson, T.H., and Schmidt, V.A., 1983, A model of the evolution of Middle America and the Gulf of Mexico–Caribbean Sea Region during Mesozoic time: Geological Society of America Bulletin, v. 94, p. 941–966, doi: 10.1130/0016-7606(1983)94<941:TEOMAA>2.0.CO;2.

Anderson, T.H., and Silver, L.T., 1977, U-Pb isotope ages of granitic plutons near Cananea, Sonora: Economic Geology and the Bulletin of the Society of Economic Geologists, v. 72, p. 827–836.

Anderson, T.H., and Silver, L.T., 1978, Jurassic magmatism in Sonora, Mexico: Geological Society of America Abstracts with Programs, v. 10, no. 7, p. 359.

Anderson, T.H., and Silver, L.T., 1979, The role of the Mojave-Sonora megashear in the tectonic evolution of northern Sonora, *in* Anderson,

T.H., and Roldán-Quintana, J., eds., Geology of northern Sonora: Geological Society of America Field Trip Guidebook, Field trip no. 27, p. 59–68.

Anderson, T.H., and Silver, L.T., 1981, An overview of Precambrian rocks in Sonora: Revista, Universidad Nacional Autónoma de México, Instituto de Geología, v. 5, no. 2, p. 131–139.

Anderson, T.H., Silver, L.T., and Salas, G.A., 1980, Distribution and U-Pb isotope ages of some lineated plutons, northwestern Mexico: Geological Society of America Special Paper 153, p. 269–283.

Anderson, T.H., Bajek, D.T., Chepega, J.R., Ichikawa, K.M., Rodríguez-Castañeda, J.L., Stevens, W.E., and Silver, L.T., 1984, Crystalline thrust sheet near the Mojave-Sonora megashear, Sonora, Mexico: Geological Society of America Abstracts with Programs, v. 16, no. 6, p. 430.

Anderson, T.H., Campbell, P., and Silver, L.T., 1992, Mylonite at the Mojave-Sonora megashear, northwestern Mexico: Geological Society of America Abstracts with Programs, v. 24, no. 7, p. A147.

Asmerom, Y., Zartman, R.E., Damon, P.E., and Shaffiqullah, M., 1990, Zircon U-Th-Pb and whole-rock Rb-Sr age patterns of Lower Mesozoic igneous rocks in the Santa Rita Mountains, southeast Arizona: Implications for Mesozoic magmatism and tectonics in the southern Cordillera: Geological Society of America Bulletin, v. 102, p. 961–968, doi: 10.1130/0016-7606(1990)102<0961:ZUTPAW>2.3.CO;2.

Avila-Angúlo, R., 1987, Consideraciones geológicas y estratigráficas de la porcion NW de Hermosillo [tesis de licenciatura]: Hermosillo, Sonora, Universidad de Sonora, 60 p.

Ayala-Fontes, R., 1992, Geology of the Cananea mining district, *in* Clark, K.F., Roldán-Quintana, J., and Schmidt, R.H., eds., Geology and mineral resources of northern Sierra Madre Occidental, Mexico: El Paso Geological Society, 1992 Field Conference Guidebook, p. 407–412.

Barth, A.P., Tosdal, R.M., and Wooden, J.L., 1990, A petrologic comparison of Triassic plutonism in the San Gabriel and Mule Mountains, southern California: Journal of Geophysical Research, v. 95, p. 20,075–20,096.

Barth, A.P., Tosdal, R.M., Wooden, J.L., and Howard, K.A., 1997, Triassic plutonism in southern California: Southward younging of arc initiation along a truncated continental margin: Tectonics, v. 16, p. 290–304, doi: 10.1029/96TC03596.

Beauvais, L., and Stump, T.E., 1976, Corals, molluscs and paleogeography of Late Jurassic strata of the Cerro Pozo Serna, Sonora, Mexico: Palaeogeography, Palaeoclimatology, Palaeoecology, v. 19, p. 275–301, doi: 10.1016/0031-0182(76)90030-4.

Beebe, W., 1986, A paleomagnetic study of the southern Coast Range ophiolite, California, and tectonic implications [Master's thesis]: Santa Barbara, University of California, 148 p.

Beebe, W., and Luyendyk, B.P., 1983, Preliminary paleomagnetic results from the Llanada remnant of the Coast Range ophiolite, San Benito County, California: Eos (Transactions, American Geophysical Union), v. 64, p. 686.

Beck, M.E., and Housen, B.A., 2003, Absolute velocity of North America during the Mesozoic from paleomagnetic data: Tectonophysics, v. 377, p. 33–54, doi: 10.1016/j.tecto.2003.08.018.

Bickford, M.E., and Day, H.W., 1988, Jurassic ages of arc-ophiolite complexes, northern Sierra Nevada: Implications for the duration of the Nevada "orogeny": Geological Society of America Abstracts with Programs, v. 20, no. 7, p. A274.

Bilodeau, W.L., and Keith, S.B., 1986, Early Jurassic Navajo Aztec-equivalent sandstones in southern Arizona and their paleogeographic significance: American Association of Petroleum Geologists Bulletin, v. 70, p. 690–701.

Bilodeau, W.L., Kluth, C.F., and Vedder, L.K., 1987, Regional stratigraphic, sedimentologic, and tectonic relationships of the Glance Conglomerate in southeastern Arizona, *in* Dickinson, W.R., and Klute, M.A., eds., Mesozoic rocks of southern Arizona and adjacent areas: Arizona Geological Society Digest, v. 18, p. 229–256.

Blake, M.C., Jr., and Jones, D.L., 1974, Origin of Franciscan melanges in northern California, *in* Dott, R.H., Jr., and Shaver, R.H., eds., Modern and ancient geosynclinal sedimentation: Society of Economic Paleontologists and Mineralogists Special Publication 19, p. 345–357.

Blake, M.C., Jr., and Jones, D.L., 1977, Plate tectonic history of the Yolly Bolly junction, northern California: Geological Society of America Field Trip Guidebook, 73rd Cordilleran Section Meeting, Sacramento, California, 17 p.

Blake, M.C., Jr., Harwood, D.S., Helley, E.J., Irwin, W.P., Jayko, A.S., and Jones, D.L., 1999, Geologic map of the Red Bluff 30′ × 60′ Quadrangle, California: U.S. Geological Survey Map I-2542, scale 1:100,000.

Bobbitt, J.B., 1982, Petrology, structure and contact relations of a part of the Yuba Rivers pluton, northwestern Sierra Nevada foothills, California [M.S. thesis]: Davis, University of California, 160 p.

Boettcher, S.S., and Walker, J.D., 1993, Geologic evolution of Iron Mountain, central Mojave Desert, California: Tectonics, v. 12, p. 372–386.

Briskey, J.A., Haxel, G.B., Peterson, J.A., and Theodore, T.G., 1978, Reconnaissance geologic map of the Gu Achi 15′ Quadrangle, Arizona: U.S. Geological Survey Miscellaneous Field Studies Map MF-965, scale 1:62,500.

Burckhardt, C., 1930, Etude synthetique sur le Mesozoique Mexican: Abhandlungen der Schweizerischen Paläontologischen Gesellschaft, Basel, v. 49-50, 280 p.

Busby-Spera, C.J., 1988, Speculative tectonic model for the early Mesozoic arc of the south-west Cordilleran United States: Geology, v. 16, p. 1121–1125, doi: 10.1130/0091-7613(1988)016<1121:STMFTE>2.3.CO;2.

Busby-Spera, C.J., and Saleeby, J.B., 1987, Geologic guide to the Mineral King area, Sequoia National Park, California: Pacific Section, Society of Economic Paleontologists and mineralogists Field Trip Guidebook, 44 p.

Busby-Spera, C.J., and Saleeby, J.B., 1990, Intra-arc strike-slip fault exposed at batholithic levels in the southern Sierra Nevada, California: Geology, v. 18, p. 255–259, doi: 10.1130/0091-7613(1990)018<0255:IASSFE>2.3.CO;2.

Busby-Spera, C.J., Mattinson, J.M., Riggs, N.R., and Schermer, E.R., 1990, The Triassic-Jurassic magmatic arc in the Mojave-Sonoran Deserts and the Sierra Klamath region: Similarities and differences in paleogeographic evolution, *in* Harwood, D., and Miller, M., eds., Late Paleozoic and Mesozoic paleogeographic relations, Klamath-Sierra and adjacent regions: Geological Society of America Special Paper 225, p. 325–337.

Busby, C.J., Schermer, E.R., and Mattinson, J.M., 2002, Extensional are setting and ages of Middle Jurassic eolianites, Cowhole Mountains (eastern Mojave Desert block, California), *in* Glazner, A.F., Walker, J.D., and Bartley, J.M., eds., Geologic evolution of the Mojave Desert and southwestern Basin and Range: Geological Society of America Memoir 195, p. 79–91.

Calmus, T., and Sosson, M., 1995, Southwestern extension of the Papago terrane into the Altar Desert region, northwestern Sonora, and its implications, *in* Jacques-Ayala, C., González-León, C.M., and Roldán-Quintana, J., eds., Studies of the Mesozoic of Sonora and Adjacent Areas: Geological Society of America Special Paper 301, p. 99–109.

Calmus, T., Pérez-Segura, E., and Herrera-Urniba, S., 1992, Geology of Estación Llano (Sonora): A new Proterozoic basement and the San Francisco gold deposit, *in* Clark, K.F., Roldán-Quintana, J., and Schmidt, R.H., eds., Geology and mineral resources of northern Sierra Madre Occidental, México: El Paso Geological Society, 1992 Field Conference Guidebook, p. 313–321.

Calmus, T., Pérez-Segura, E., and Stinnesbeck, W., 1997, La structuracion de la marge Pacifique Nordamericaine et du "terrane" Caborca; apports de la decouverte d'une faune du Jurassique Inferieur et Moyen dans la serie de Pozo de Serna (Sonora, Mexique): Comptes Rendus de l'Academie des Sciences (Paris): Earth and Planetary Sciences, v. 325, p. 257–263.

Cameron, A.E., Smith, D.H., and Walker, R.L., 1969, Mass spectrometry of nanogram-size samples of lead: Analytical Chemistry, v. 41, p. 525–526, doi: 10.1021/ac60272a020.

Campa, M.F., and Coney, P.J., 1983, Tectono-stratigraphic terranes and mineral resource distributions in México: Canadian Journal of Earth Sciences, v. 20, p. 1040–1051.

Campbell, P.A., and Anderson, T.H., 1998, Structure and kinematics along a Jurassic plate boundary transform, the Mojave-Sonora megashear, northwestern Sonora, Mexico, *in* Clark, K.F., ed, Gold deposits of northern Sonora, Mexico: Society of Economic Geologists Guidebook Series, v. 30, p. 177–186.

Campbell, P.A., and Anderson, T.H, 2003, Structure and kinematics along a segment of the Mojave-Sonora megashear; a strike-slip fault that truncates the Jurassic continental magmatic arc of southwestern North America: Tectonics, v. 22, no. 6, p. 16:1–21.

Carl, B.S., and Glazner, A.F., 2002, Extent and significance of the Independence dike swarm, eastern California, *in* Glazner, A.F., Walker, J.D., and Bartley, J.M., eds., Geologic evolution of the Mojave Desert and southwestern Basin and Range: Geological Society of America Memoir 195, p. 117–130.

Caudillo-Sosa, G., and Oviedo-Lucero, L.F., 1990, Geología del área Quitovac, Mpio. de Puerto Peñasco, Sonora [Tesis de Licenciatura]: Hermosillo, Sonora, Universidad de Sonora, 120 p.

Caudillo-Sosa, G., Oviedo-Lucero, L.F., and Rodríguez-Castañeda, J.L., 1991, Falla Quitovac—Resultado de un evento de transpresión del Mojave-Sonora megashear, noroeste de Sonora, México: Universidad Nacional Autónoma de Mexico, Instituto de Geologia, Revista, v. 13, no. 2, p. 140–151.

Chaparro-Mesa., M.E., 1992, Geologia de la carta Los Chinos H12C19, Sonora central [Tesis de Licenciatura]: Hermosillo, Sonora, Universidad de Sonora, 62 p.

Chepega, J.R., 1987, Reconnaissance geology of Tuape, north-central Sonora, Mexico [M.S. thesis]: Pittsburgh, Pennsylvania, University of Pittsburgh, 139 p.

Clark, L.D., 1960, The Foothills fault system, western Sierra Nevada, California: Geological Society of America Bulletin, v. 71, p. 483–498.

Clark, L.D., 1964, Stratigraphy and structure of part of the western Sierra Nevada metamorphic belt, California: U.S. Geological Survey Professional Paper 410, 70 p.

Coleman, D.S., Barth, A.P., and Wooden, J.L., 2003, Metamorphism and orogenesis of Proterozoic rocks in the southwest Mojave province: Geological Society of America Abstracts with Programs, v. 99, no. 5, p. 67.

Connors, C.D., 1990, Geology of the Quitovac-Sierra La Toñita area [M.S. thesis]: Pittsburgh, Pennsylvania, University of Pittsburgh, 75 p.

Connors, C.D., Anderson, T.H., and Silver, L.T., 1989, Expression and structural analysis of the Mojave-Sonora megashear in northwestern Sonora, Mexico: Geological Society of America Abstracts with Programs, v. 21, p. A91.

Cooper, J.R., and Silver, L.T., 1964, Geology and ore deposits of the Dragoon Quadrangle, Cochise County, Arizona: U.S. Geological Survey Professional Paper 416, 196 p.

Corona, F.V., 1979, Preliminary reconnaissance geology of Sierra La Gloria and Cerro Basura, northwestern Sonora, Mexico, *in* Anderson, T.H., and Roldán-Quintana, J., eds., Geology of northern Sonora: Geological Society of America Field Trip Guidebook, Field trip no. 27, p. 32–48.

Corona, F.V., 1980, Reconnaissance geology of Sierra La Gloria and Cerro Basura, northwestern Sonora, Mexico [M.S. thesis]: Pittsburgh, Pennsylvania, University of Pittsburgh, 232 p.

Corona, F.V., and Anderson, T.H., 1981, Stratigraphy and structure of a Jurassic(?) sequence at Sierra La Gloria and Cerro Basura, northwestern Sonora, Mexico: Geological Society of America Abstracts with Programs, v. 13, no. 2, p. 50.

Davis, G.H., Phillips, M.P., Reynolds, S.J., and Varga, R.J., 1979, Origin and provenance of some exotic blocks in Lower Mesozoic red beds basin deposits, southern Arizona: Geological Society of America Bulletin, v. 90, p. 376–384, doi: 10.1130/0016-7606(1979)90<376:OAPOSE>2.0.CO;2.

Day, H.W., Moores, E.M., and Tuminas, A.C., 1985, Structure and tectonics of the northern Sierra Nevada: Geological Society of America Bulletin, v. 96, p. 436–450, doi: 10.1130/0016-7606(1985)96<436:SATOTN>2.0.CO;2.

Dembosky, J.A., Jr., 2001, Paleotectonic restoration of Miocene extensional strain of crystalline rock provinces and terranes in southwestern North America [Ph.D. thesis]: Pittsburgh, Pennsylvania, University of Pittsburgh, 152 p., 12 plates.

Dilek, Y., 1989, Structure and tectonics of an early Mesozoic oceanic basement in the northern Sierra Nevada metamorphic belt, California: Evidence for transform faulting and ensimatic arc evolution: Tectonics, v. 8, p. 999–1014.

Dilles, J.H., and Wright, J.E., 1988, The chronology of early Mesozoic arc magmatism in the Yerington district of western Nevada and its regional implications: Geological Society of America Bulletin, v. 100, p. 644–652, doi: 10.1130/0016-7606(1988)100<0644:TCOEMA>2.3.CO;2.

Dowlen, R.J., and Gastil, R.G., 1981, Reconnaissance geology of Cerro Pozo Serna, western Sonora, Mexico, *in* Ortlieb, L., and Roldán-Quintana, J., eds., Geology of northwestern Mexico and southern Arizona: Geological Society of America, Cordilleran Section, Field Guides and Papers: Instituto de Geología, UNAM, México, p. 431–435.

Drewes, H., 1968, New and revised stratigraphic names in the Mount Wrightson Formation of southeastern Arizona: U.S. Geological Survey Bulletin 1274-C, 15 p.

Drewes, H., 1971a, Mesozoic stratigraphy of the Santa Rita Mountains, southeast of Tucson, Arizona: U.S. Geological Survey Professional Paper 658-C, 81 p.

Drewes, H., 1971b, Geologic map of the Sahuarita Quadrangle, southeast of Tucson, Pima County, Arizona: U.S. Geological Survey Miscellaneous Investigations Map I-613, scale 1:48,000.

Drewes, H., 1971c, Geologic map of the Mount Wrightson Quadrangle, southeast of Tucson, Santa Cruz and Pima Counties, Arizona: U.S. Geological Survey Miscellaneous Investigations Map I-164, scale 1:48,000.

Dumble, E.T., 1900, Notes on the geology of Sonora, Mexico: Geological Society of America Bulletin, v. 11, p. 10–14.

Dunne, G.C., and Suczek, C.A., 1991, Early Paleozoic eugeoclinal strata in the Kern Plateau pendants, southern Sierra Nevada, California, *in* Cooper, J.D., and Stevens, C.H., eds., Paleozoic paleogeography of the western United States, Volume 2: Los Angeles, California, Society of Economic Paleontologists and Mineralogists, Pacific Section, p. 677–692.

Dunne, G.C., and Walker, D., 1993, Age of Jurassic volcanism and tectonism, southern Owens Valley region, east-central California: Geological Society of America Bulletin, v. 105, p. 1223–1230, doi: 10.1130/0016-7606(1993)105<1223:AOJVAT>2.3.CO;2.

Edelman, S.H., Day, H.W., and Bickford, M.E., 1989a, Implications of U-Pb zircon ages for the tectonic settings of the Smartville and Slate Creek Complexes, northern Sierra Nevada, California: Geology, v. 17, p. 1032–1035, doi: 10.1130/0091-7613(1989)017<1032:IOUPZA>2.3.CO;2.

Edelman, S.H., Day, H.W., Moores, E.M., Zihan, S.M., Murphy, T.P., and Hacker, B.R., 1989b, Structure across a Mesozoic ocean-continent suture zone in the northern Sierra Nevada, California: Geological Society of America Special Paper 224, 56 p.

Erben, H.K., 1956, El Jurásico temprano de México: Congreso Geológico Internacional 20, México, D.F.: Sociedad Geológica Mexicana, México, Monografía, p. 393.

Fackler-Adams, B.N., Busby, C.J., and Mattinson, J.M., 1997, Jurassic magmatism and sedimentation in the Palen Mountains, southeastern California: Implications for regional tectonic controls on the Mesozoic continental arc: Geological Society of America Bulletin, v. 109, p. 1464–1484, doi: 10.1130/0016-7606(1997)109<1464:JMASIT>2.3.CO;2.

Fagan, T.J., Day, H.W., and Hacker, B.R., 2001, Timing of arc construction and metamorphism in the Slate Creek Complex, northern Sierra Nevada, California: Geological Society of America Bulletin, v. 113, p. 1105–1118, doi: 10.1130/0016-7606(2001)113<1105:TOACAM>2.0.CO;2.

Fisher, G.R., 1990, Middle Jurassic syntectonic conglomerate in the Mt. Tallac roof pendant, northern Sierra Nevada, California, *in* Harwood, D.S., and Miller, M.M., eds., Paleozoic and early Mesozoic paleogeographic relations: Sierra Nevada, Klamath Mountains, and related terranes: Geological Society of America Special Paper 255, p. 339–350.

Fletcher, J.M., and Karlstrom, K.E., 1990, Late Cretaceous ductile deformation, metamorphism and plutonism in the Piute Mountains, eastern Mojave Desert: Journal of Geophysical Research, v. 95, p. 487–500.

Fletcher, J.M., Miller, J.S., Martin, M.W., Boettcher, S.S., Glazner, A.F., and Bartley, J.M., 2002, Cretaceous arc tectonism in the Mojave block: Profound crustal modification that controlled subsequent tectonic regimes, *in* Glazner, A.F., Walker, J.D., and Bartley, J.M., eds., Geologic evolution of the Mojave Desert and southwestern Basin and Range: Geological Society of America Memoir 195, p. 131–150.

Flores, T., 1929, Reconocimientos geológicos en la región central del Estado de Sonora: Universidad Nacional Autónoma de México, Boletin Instituto Geológico de México, v. 49, 253 p.

Gastil, R.G., and Krummenacher, D., and students, 1976, Reconnaissance geologic map of coastal Sonora between Puerto Lobos and Bahia Kino: Geological Society of America Map and Chart Series MC-16, scale 1:150,000.

Gilluly, J., 1956, General geology of central Cochise County, Arizona, with sections on age and correlation by A.R. Palmer, J.S. Williams, and J.B. Reeside, Jr.: U.S. Geological Survey Professional Paper 281, 169 p.

Gilmont, N.L., 1978, Geology of the Puerto La Bandera area, Sonora, Mexico [Master's thesis]: Flagstaff, Arizona, University of Northern Arizona, 109 p.

Girty, G.H., Gester, K.C., and Turner, J.B., 1990, Pre-Late Devonian geochemical, stratigraphic, sedimentologic, and structural patterns, Shoo Fly Complex, northern Sierra Nevada, California, *in* Harwood, D.S., and Miller, M.M., eds., Paleozoic and early Mesozoic paleogeographic relations: Sierra Nevada, Klamath Mountains, and related terranes: Geological Society of America Special Paper 255, p. 43–56.

Glazner, A.F., Bartley, J.M., and Walker, J.D., 1989, Magnitude and significance of Miocene crustal extension in the central Mojave Desert, California: Geology, v. 17, p. 50–53, doi: 10.1130/0091-7613(1989)017<0050:MASOMC>2.3.CO;2.

Glazner, A.F., Bartley, J.M., and Carl, B.S., 1999, Oblique opening and noncoaxial emplacement of the Independence dike swarm, California: Journal of Structural Geology, v. 21, p. 1275–1283, doi: 10.1016/S0191-8141(99)00090-5.

Glazner, A.F., Walker, J.D., Bartley, J.M., and Fletcher, J.M., 2002, Cenozoic evolution of the Mojave block of southern California, *in* Glazner, A F., Walker, J.D., and Bartley, J.M., eds., Geologic evolution of the Mojave Desert and southwestern Basin and Range: Geological Society of America Memoir 195, p. 19–41.

Godfrey, N.J., Beaudoin, B.C., Klemperer, S.L., and the Mendocino Working Group, 1997, Ophiolitic basement of the Great Valley forearc basin, California, from seismic and gravity data: Implications for crustal growth at the North American continental margin: Geological Society of America Bulletin, v. 108, p. 1536–1562.

González-León, C., 1979, Geology of the Sierra del Alamo, *in* Anderson, T.H., and Roldán-Quintana, J., eds., Geology of northern Sonora: Annual Meeting Geological Society of America Field Trip Guidebook, Field trip no. 27, p. 23–31.

González-León, C., 1980, La Formacion Antimonio (Triásico Superior–Jurasico Inferior) en la Sierra del Alamo, Estado de Sonora: Universidad Nacional Autónoma de México, Instituto de Geologia, Revista, v. 4, p. 13–18.

González-León, C.M., 1997, Sequence stratigraphy and paleogeographic setting of the Antimonio Formation (Late Permian–Early Jurassic) Sonora, Mexico, *in* Stanley, G.D., Jr., and González-León, C.M., Special issue dedicated to the International Workshop on the Geology of Northwestern Sonora, Mexico: Revista Mexicana de Ciencias Geologicas, v. 14, no. 2, p. 136–148.

González-León, C.M., and Stanley, G.D., Jr., 1997, eds., U.S.-Mexico Cooperative Research–International Workshop on the Geology of Sonora Memoir: Universidad Nacional Autónoma de México, Instituto de Geología, Estación Regional Noroeste, Publicaciones Ocasionales 1, 69 p.

González-León, C.M., Taylor, D.G., and Stanley, G.D., Jr., 1995, The Antimonio Formation in Sonora, Mexico, and the Triassic-Jurassic boundary: Canadian Journal of Earth Sciences, v. 33, p. 418–428.

Greene, D.C., and Schweickert, R.A., 1995, The Gem Lake shear zone: Cretaceous dextral transpression in the northern Ritter Range pendant, eastern Sierra Nevada, California: Tectonics, v. 14, p. 945–965, doi: 10.1029/95TC01509.

Greene, D.C., Schweickert, R.A., and Stevens, C.H., 1997, Roberts Mountain allochthon and the western margin of the Cordilleran miogeocline in the northern Ritter Range pendant, eastern Sierra Nevada, California: Geological Society of America Bulletin, v. 109, p. 1294–1305, doi: 10.1130/0016-7606(1997)109<1294:RMAATW>2.3.CO;2.

Grubensky, M.J., Haxel, G.B., and Koch, R.D., 1993, Geologic map of the Castle Dome Mountains: U.S. Geological Survey Miscellaneous Investigations Series Map I-2138, scale 1:62,500.

Hacker, B.R., Donato, M.M., Barnes, C.G., McWilliams, M.O., and Ernst, W.G., 1995, Time scales of orogeny: Jurassic construction of the Klamath Mountains: Tectonics, v. 14, p. 677–703, doi: 10.1029/94TC02454.

Hagstrum, J.T., and Murchey, B.L., 1996, Paleomagnetism of Jurassic radiolarian chert above the Coast Range ophiolite at Stanley Mountain, California, and implications for its paleogeographic origins: Geological Society of America Bulletin, v. 108, p. 643–652.

Hagstrum, J.T., and Lipman, P.W., 1991, Paleomagnetism of three upper Jurassic ash-flows sheets in southeastern Arizona: Implications for regional deformation: Geophysical Research Letters, v. 18, p. 1413–1416.

Hamilton, W., Tosdal, R.M., Stone, P., and Haxel, G.B., 1987, Mesozoic tectonics of southeastern California, *in* Davis, G.H., and Vanden Dolder, E.N., eds., Geologic diversity of Arizona and its margins: Excursions to choice areas: Arizona Bureau of Geology and Mineral Technology, Geological Survey Branch, Special Paper 5, p. 337–350.

Hardy, L.R., 1981, Geology of the central Sierra de Santa Rosa, Sonora, Mexico, *in* Ortlieb, L., and Roldán-Quintana, J., eds., Geology of northwestern Mexico and southern Arizona: Geological Society of America, Cordilleran Section, Field Guides and Papers: Instituto de Geología, UNAM, México, p. 73–78.

Harper, G.D., 1980, The Josephine ophiolite—Remains of a late Jurassic marginal basin in northwestern California: Geology, v. 8, p. 333–337, doi: 10.1130/0091-7613(1980)8<333:TJOOAL>2.0.CO;2.

Harper, G.D., and Wright, J.E., 1984, Middle to Late Jurassic tectonic evolution of the Klamath Mountains, California-Oregon: Tectonics, v. 3, p. 759–772.

Harper, G.D., Saleeby, J.B., and Norman, E.A.S., 1985, Geometry and tectonic setting of sea-floor spreading for the Josephine ophiolite, and implications for Jurassic accretionary events along the California margin, *in* Howell, D.G., ed., Tectonostratigraphic terranes of the circum-Pacific region: Houston, Circum-Pacific Council of Energy and Mineral Resources, Earth Science Series No. 1, p. 239–257.

Harper, G.D., Grady, K., and Wakabayashi, A., 1990, A structural study of a metamorphic sole beneath the Josephine ophiolite, western Klamath terrane, California-Oregon, *in* Harwood, D.S., and Miller, M.M., eds., Paleozoic and early Mesozoic paleogeographic relations: Sierra Nevada, Klamath Mountains and related terranes: Geological Society of America Special Paper 255, p. 379–396.

Harper, G.D., Saleeby, J.B., and Heizler, M., 1994, Formation and emplacement of the Josephine ophiolite and the Nevada orogeny in the Klamath Mountains, California-Oregon: U/Pb and $^{40}Ar/^{39}Ar$ geochronology: Journal of Geophysical Research, v. 99, no. B3, p. 4293–4221.

Haxel, G.B., and Dillon, J.T., 1978, The Pelona-Orocopia Schist and Vincent-Chocolate Mountain thrust system, southern California, *in* Howell, D.G., and McDougall, K., eds., Mesozoic paleogeography of the western United States–Pacific coast paleogeography, Symposium 2: Los Angeles, California, Society of Economic Paleontologists and Mineralogists, Pacific Section, p. 435–469.

Haxel, G.B., Tosdal, R.M., May, D.J., and Wright, J.E., 1984, Latest Cretaceous and early Tertiary orogenesis in south-central Arizona: Thrust faulting, regional metamorphism, and granitic plutonism: Geological Society of America Bulletin, v. 95, p. 631–653, doi: 10.1130/0016-7606(1984)95<631:LCAETO>2.0.CO;2.

Haxel, G.B., Anderson, T.H., Riggs, N.R., and Goodwin, L.B., 1988, The Papago terrane: A crustal anomaly in south-central Arizona and north-central Sonora: Geological Society of America Abstracts with Programs, v. 20, no. 3, p. 168.

Haxel, G.B., Jacobson, C.E., Richard, S.M., Tosdal, R.M., and Grubensky, M.J., 2002, The Orocopia Schist in southwest Arizona: Early Tertiary oceanic rocks trapped or transported far inland, *in* Barth, A., ed., Contributions to crustal evolution of the southwestern United States: Geological Society of America Special Paper 365, p. 99–128.

Herrera-Urbina, S., and Perez-Segura, E., 1990, Facies de metamorfismo en las secuencias de las inmediaciones del poblado de Estacion Llano, Sonora: Boletin del Departmento de Geologia, Universidad de Sonora, Segunda Epoca, v. 7, no. 1–2, p. 7–28.

Hill, L.B., 1985, Metamorphic, deformational and temporal constraints on terrane assembly, northern Klamath Mountains, California, *in* Howell, D.G., ed., Tectonostratigraphic terranes of the circum-Pacific region: Houston, Texas, Circum-Pacific Council for Energy and Mineral Resources, Earth Science Series, p. 173–186.

Hopson, C.A., Mattinson, J.M., Luyendyk, B.P., Beebe, W., Pessagno, E.A., Jr., Hull, D.M., and Blume, C.D., 1992, California Coast Range ophiolite: Jurassic tectonic history: American Association of Petroleum Geologists Bulletin, v. 76, p. 422.

Howard, K.A., McCaffrey, K.J.W., Wooden, J.L., Foster, D.A., and Shaw, S.E., 1995, Jurassic thrusting of Precambrian basement over Paleozoic cover in the Clipper Mountains, southwestern California, *in* Miller, D.M. and Busby, C., eds., Jurassic magmatism and tectonics of the North American Cordillera: Geological Society of America Special Paper 299, p. 375–392.

Imlay, R.W., 1943, Jurassic formations of the Gulf region: American Association of Petroleum Geologists Bulletin, v. 27, p. 1407–1533.

Imlay, R.W., 1980, Jurassic paleobiogeography of the conterminous United States in its continental setting: U.S. Geological Survey Professional Paper 1062, 134 p.

Iriondo, A., 2001, Proterozoic basements and their Laramide juxtaposition in NW Sonora, Mexico: Tectonic constraints of the SW margin of Laurentia [Ph.D. thesis]: University of Colorado, 222 p., 3 plates.

Iriondo, A., and Atkinson, W.W., Jr., 2000, Orogenic gold mineralization along the proposed trace of the Mojave-Sonora megashear: Evidence for the Laramide orogeny in NW Sonora, Mexico: Geological Society of America Abstracts with Programs, v. 32, no. 7, p. 393.

Irwin, W.P., and Wooden, J.L., 1999, Plutons and accretionary episodes of the Klamath Mountains, California and Oregon: U.S. Geological Survey Open-File Report 99-374, 1 plate.

Jacobson, C.E., Oyarzabal, F.R., and Haxel, G.B., 1996, Subduction and exhumation of the Pelona-Orocopia-Rand Schists, southern California: Geology, v. 24, p. 547–550, doi: 10.1130/0091-7613(1996)024<0547:SAEOTP>2.3.CO;2.

Jacobson, C.E., Barth, A.P., and Grove, M., 2000, Late Cretaceous protolith age and provenance of the Pelona and Orocopia Schists, southern California: Implications for evolution of the Cordilleran margin: Geology, v. 28, p. 219–222, doi: 10.1130/0091-7613(2000)028<0219:LCPAAP>2.3.CO;2.

Jacobson, C.E., Grove, M., Stamp, M.M., Vucic, A., Oyarzabal, F.R., Haxel, G.B., Tosdal, R.M., and Sherrod, D.R., 2002, Exhumation history of the Orocopia Schist and related rocks in the Gavilan Hills area of southeasternmost California, *in* Barth, A., ed., Contributions to crustal evolution of the southwestern United States: Geological Society of America Special Paper 365, p. 129–154.

Jaworski, E., 1929, Eine Lias-faune aus Nordwest-Mexiko: Abhandlungen der Schweizerischen Paläontologischen Gesellschaft, Basel, v. 48, 121 p.

Jayko, A.S., and Blake, M.C., 1986, Significance of Klamath rocks between the Franciscan Complex and Coast Range ophiolite: Tectonics, v. 5, p. 1055–1071.

Jones, N.W., McKee, J.W., Anderson, T.H., and Silver, L.T., 1995, Jurassic volcanic rocks in northeastern Mexico: A possible remnant of a Cordilleran magmatic arc, *in* Jacques-Ayala, C., González-León, C.M., and Roldán-Quintana, J., eds., Studies on the Mesozoic of Sonora and adjacent areas: Geological Society of America Special Paper 301, p. 179–190.

Keller, W.T., 1928, Stratigraphische Beobachtungen in Sonora, nordwest Mexico: Eclogae Geologicae Helvetiae, v. 21, p. 327–335.

Keller, W.T. and Wellings, F.E., 1922, Sonora: Compañia Petrolera El Aguila: Geological Report no. 180, 38 p., (unpublished).

Kempner, W.C., 1977, The magnetic properties of the Point Sal ophiolite: A comparison with oceanic crust [Master's thesis]: Santa Barbara, University of California, 134 p.

King, R.E., 1939, Geological reconnaissance of northern Sierra Madre Occidental: Geological Society of America Bulletin, v. 50, p. 1625–1722.

Kistler, R.W., 1990, Two different lithosphere types in the Sierra Nevada, California, *in* Anderson, J.L., ed., The nature and origin of Cordilleran magmatism: Geological Society of America Memoir 174, p. 271–281.

Kistler, R.W., and Peterman, Z.E., 1973, Variations in Sr, Rb, K, Na and initial ^{87}Sr/^{86}Sr in Mesozoic granitic rocks and intruded wall rocks in central California: Geological Society of America Bulletin, v. 84, p. 3489–3512., doi: 10.1130/0016-7606(1973)84<3489:VISRKN>2.0.CO;2.

Kluth, C.F., 1983, Geology of northern Canelo Hills and implications for the Mesozoic tectonics of southeastern Arizona, *in* Reynolds, M.W., and Dolly, E.D., eds., Mesozoic paleogeography of the west-central United States: Society of Economic Paleontologists and Mineralogists, p. 159–171.

Kluth, C.F., Butler, R.F., Harding, L.E., Shafiqullah, M., and Damon, P.E., 1982, Paleomagnetism of Late Jurassic rocks in the northern Canelo Hills, southeastern Arizona: Journal of Geophysical Research, v. 87, p. 7079–7086.

Knight, L.H., 1970, Structure and mineralization of the Oro Blanco mining district, Santa Cruz County, Arizona [Ph.D. thesis]: Tucson, Arizona, University of Arizona, 172 p.

Lahren, M.A., and Schweickert, R.A., 1989, Proterozoic and Lower Cambrian miogeoclinal rocks of Snow Lake pendant, Yosemite-Emigrant Wilderness, Sierra Nevada, California: Evidence for major Early Cretaceous dextral translation: Geology, v. 17, p. 156–160, doi: 10.1130/0091-7613(1989)017<0156:PALCMR>2.3.CO;2.

Lawton, T.F., and Olmstead, G.A., 1995, Stratigraphy and structure of the lower part of the Bisbee Group, northeastern Chiricahua Mountains, Arizona, *in* Jacques-Ayala, C., González-León, C.M., and Roldán-Quintana, J., eds., Studies of the Mesozoic of Sonora and adjacent areas: Geological Society of America Special Paper 301, p. 21–39.

Leveille, G.P., 1984, Geology of El Capitan, Sonora, Mexico [M.S. thesis]: San Diego, California, San Diego State University, 122 p.

Lipman, P.W., and Hagstrum, J.T., 1992, Jurassic ash-flows sheets, calderas, and related intrusions of the Cordillera volcanic in southeastern Arizona: Implications for regional tectonics and ore deposits: Geological Society of America Bulletin, v. 104, p. 32–39, doi: 10.1130/0016-7606(1992)104<0032:JAFSCA>2.3.CO;2.

Longoria, J.F., and Perez, V.A., 1979, Bosquejo geológico de los cerros Chino y Rajon, cuadrangulo Pitiquito-La Primavera (NW de Sonora): Universidad de Sonora, Boletín Departamento de Geología, v. 1, p. 119–144.

Longoria, J.F., González, M.A., Mendoza, J.J., and Pérez, V.A., 1978, Consideraciones estructurales del Cuadrangulo Pitiquito-La Primavera, NW de Sonora: Universidad de Sonora, Boletín Departamento de Geología, v. 1, no. 1, p. 61–67.

Lucas, S.G., 1993, The Triassic-Jurassic boundary section in the Sierra del Alamo, northwestern Sonora: Universidad Nacional Autónoma de México and Universidad de Sonora, III Simposio de la geología de Sonora y áreas adyacentes, Resúmenes, p. 66–68.

Lucas, S.G., Kues, B.S., Estep, J.W., and González-León, C.M., 1997, Permian-Triassic boundary at El Antimonio, Sonora, Mexico: Revista Mexicana de Ciencias Geológicas, v. 14, no. 2, p. 35–40

Ludwig, K.L., 2001, Isoplot/Ex, version 2.49, A geochronological toolkit for Microsoft Excel: Berkeley Geochronology Center, Special Publication No. 1a, 56 p.

Luna-Barcelo, R., and Gastelum-Morales, G., 1992, Geology of the San Francisco Project Estacion Llano, *in* Clark, K.F., Roldán-Quintana, J., and Schmidt, R.H., eds., Geology and mineral resources of northern Sierra Madre Occidental, México: El Paso Geological Society, 1992 Field Conference Guidebook, p. 307–312.

Luyendyk, B.P., and Hornafius, J.S., 1982, Paleolatitude of the Point Sal ophiolite: Geological Society of America Abstracts with Programs, v. 14, p. 182.

Martin, M.W., and Walker, J.D., 1992, Extending the western North American Proterozoic and Paleozoic craton through the Mojave Desert: Geology, v. 20, p. 753–756, doi: 10.1130/0091-7613(1992)020<0753:ETWNAP>2.3.CO;2.

Martin, M.W., Walker, J.D., and Fletcher, J.M., 2002, Timing of Middle to Late Jurassic ductile deformation and implications for paleotectonic setting, Shadow Mountains, western Mojave Desert, California, *in* Glazner, A.F., Walker, J.D., and Bartley, J.M., eds., Geologic evolution of the Mojave Desert and southwestern Basin and Range: Geological Society of America Memoir 195, p. 43–58.

Martinez-Araiza, H., 1998, The San Francisco gold mine, Municipality of Santa Ana, Sonora, Mexico, *in* Clark, K.F., ed., Gold deposits of northern Sonora, Mexico: Society of Economic Geologists Guidebook Series, v. 30, p. 49–58.

Martinez, S.L., Peacock, S.M., and Reynolds, S.J., 1994, Late Cretaceous metamorphism and an inverted metamorphic gradient in the Granite Wash and Harquahala Mountains, west-central Arizona: Geological Society of America Abstracts with Programs, v. 26, no. 2, p. 69.

Marvin, R.F., Naeser, C.W., and Mehnert, H.H., 1978, Tabulation of radiometric ages—including unpublished K-Ar and fission track ages—for rocks in southeastern Arizona and southwestern New Mexico, *in* Callender, J.F., Wilt, J.C., and Clemons, R.E., eds., Land of Cochise, southeastern Arizona: New Mexico Geological Society, Twenty-Ninth Field Conference Guidebook, p. 285–290.

Marzolf, J.E., 1980, The Aztec sandstone and stratigraphically related rocks in the Mojave Desert, *in* Fife, D.L., and Brown, G.R., eds., Geology and mineral wealth of the California desert: Santa Ana, California, South Coast Geological Society, p. 215–220.

Marzolf, J.E., 1982, Paleogeographic implications of early Jurassic(?) Navajo and Aztec sandstones, *in* Frost, E., and Martin, D., eds., Mesozoic-Cenozoic tectonic evolution of the Colorado River region, California, Arizona, and Nevada: San Diego, Cordilleran Publishers, p. 493–501.

Marzolf, J.E., 1983, Early Mesozoic eolian transition from cratonal margin to orogenic-volcanic arc: Utah Geological Society and Mineral Survey Special Studies, v. 60, p. 39–46.

Mattinson, J. M., and Hopson, C.A., 1992, U/Pb ages of the Coast Range ophiolite: A critical evaluation based on new high-precision Pb/Pb ages: American Association of Petroleum Geologists Bulletin, v. 76, p. 425.

May, D.J., 1989, Late Cretaceous intra-arc thrusting in southern California: Tectonics, v. 8, no. 6, p. 1159–1173.

May, D.J., and Walker, N.W., 1989, Late Cretaceous juxtaposition of metamorphic terranes in the southeastern San Gabriel Mountains, California: Geological Society of America Bulletin, v. 101, p. 1246–1267.

McAnulty, W.N., 1970, Geology of the northern Nacozari District, Sonora, Mexico [Ph.D. thesis]: Albuquerque, New Mexico, University of New Mexico, 103 p.

McComb, T.D., 1987, Structural geology of southeastern Sierra El Batamote (northwestern Sonora, Mexico) [M.Sc. thesis]: Cincinnati, Ohio, University of Cincinnati, 111 p. (unpublished).

McDowell, F.W., Roldán-Quintana, J., and Connelly, J.N., 2001, Duration of Late Cretaceous–early Tertiary magmatism in east-central Sonora, Mexico: Geological Society of America Bulletin, v. 113, no. 4, p. 521–531, doi: 10.1130/0016-7606(2001)113<0521:DOLCET>2.0.CO;2.

McKee, M.B., 1991, Deformation and stratigraphic relationships of Middle Cretaceous to early Tertiary mass gravity slides in a marine basin in Sonora, Mexico [Ph.D. thesis]: Pittsburgh, Pennsylvania, University of Pittsburgh, 286 p.

McKee, M.B., and Anderson, T.H., 1999, Mass-gravity deposits and structures in the Lower Cretaceous of Sonora: Mexico: Geological Society of America Bulletin, v. 110, no. 12, p. 1516–1529.

McLaughlin, R.J., and Ohlin, H.N., 1984, Tectonostratigraphic framework of the Geysers–Clear Lake Region, California, *in* Blake, M.C., Jr., ed., Franciscan geology of northern California: Society of Economic Paleontologists and Mineralogists, Pacific Section, v. 43, p. 221–254.

McLaughlin, R.J., Blake, M.C., Jr., Griscom, A., Blome, C.D., and Murchey, B., 1988, Tectonics of foraminifera, translation, and dispersal of the Coast Range ophiolite of California: Tectonics, v. 7, p. 1033–1056.

McLaughlin, R.J., Sliter, W.V., Sorg, D.H., Russell, P.C., and Sarna-Wojcicki, A.M., 1996, Large-scale right-slip displacement on the East San Francisco Bay region fault system, California: Implications for location of late Miocene to Pliocene plate boundary: Tectonics, v. 15, p. 1–18.

McWilliams, M.O., and Howell, D.G., 1982, Exotic terranes of western California: Nature, v. 297, p. 215–217, doi: 10.1038/297215a0.

Meinert, L.D., 1980, Skarn, manto, and breccia pipe formation in sedimentary rocks in the Cananea District, Sonora, Mexico [Ph.D. thesis]: Stanford, California, Stanford University, 232 p.

Mendoza-Cordova, A., and Minjarez-Rivera, V.A., 1989, Cartografia geológica del área Las Peñitas, Sierra Caracahui [Tesis Licenciatura]: Hermosillo, Sonora, Universidad de Sonora, 65 p.

Merriam, R., 1972, Reconnaissance geologic map of the Sonoita Quadrangle, northwestern Sonora, Mexico: Geological Society of America Bulletin, v. 83, p. 3533–3536.

Miller, J.S., and Walker, J.D., 2002, Mesozoic geologic evolution of Alvord Mountain, central Mojave Desert, California, *in* Glazner, A.F., Walker, J.D., and Bartley, J.M., eds., Geologic evolution of the Mojave Desert and southwestern Basin and Range: Geological Society of America Memoir 195, p. 59–77.

Miller, J.S., Glazner, A.F., Walker, J.D., and Martin, M.W., 1995, Geochronologic and isotopic evidence for Triassic-Jurassic emplacement of the eugeoclinal allochthon in the Mojave Desert region, California: Geological Society of America Bulletin, v. 107, no. 12, p. 1441–1457, doi: 10.1130/0016-7606(1995)107<1441:GAIEFT>2.3.CO;2.

Murchey, B.L., and Blake, M.C., Jr., 1993, Evidence for subduction of a major ocean plate along the California margin during the Middle to early Late Jurassic, *in* Dunne, G., and McDougall, K., eds., Mesozoic paleogeography of the western United States, Volume 2: Los Angeles, California, Society of Economic Paleontologists and Mineralogists, Pacific Section, Book 71, p. 227–246.

Newton, M.C., III, 1990a, Tectonostratigraphic history of the southern foothills terrane [Ph.D. dissertation]: Tucson, University of Arizona, 203 p.

Newton, M.C., III, 1990b, Structural control of gold mineralization in the southern mother lode region, *in* Seedorf, E., ed., Geology and ore deposits of the Sierra Nevada and foothills: Geological Society of Nevada, 1990 Fall Field Trip Guidebook, Special Publication 11, p. 107–114.

Nourse, J.A., 1988, Tectonostratigraphic evolution of the Magdalena metamorphic complex, north central Sonora, Mexico: Geological Society of America Abstracts with Programs, v. 20, no. 7, p. A59.

Nourse, J.A., 1989, Geological evolution of two crustal scale shear zones, Part II: The Magdalena metamorphic core complex [Ph.D. thesis]: Pasadena, California, California Institute of Technology, p. 64–396.

Nourse, J.A., 1990, Tectonostratigraphic development and strain history of the Magdalena metamorphic core complex, northern Sonora, *in* Gehrels, G.E., and Spencer, J.E., eds., Geologic excursions through the Sonoran Desert region, Arizona and Sonora: Arizona Geological Survey Special Paper 7, p. 155–164.

Nourse, J.A., 1995, Jurassic-Cretaceous paleogeography of the Magdalena region, northern Sonora, and its influence of the positioning of metamorphic core complexes, *in* Jacques-Ayala, C., González-León, C.M., and Roldán-Quintana, J., eds., Studies of the Mesozoic of Sonora and adjacent areas: Geological Society of America Special Paper 301, p. 59–78.

Nourse, J.A., 2001, Tectonic insights from stretched Late Jurassic–Lower Cretaceous Conglomerate, Caborca-Altar Region, Sonora, Mexico, *in* González-León, C.M., ed., Geology of northwestern Mexico and adjacent areas: Journal of South American Earth Sciences, v. 14, no. 5, p. 453–474.

Nourse, J.A., Anderson, T.H., and Silver, L.T., 1994, Tertiary metamorphic core complexes in Sonora, northwestern Mexico: Tectonics, v. 13, no. 5, p. 1161–1184, doi: 10.1029/93TC03324.

Nourse, J.A., Premo, W.R., and Iriondo, A., 2003, Gneisses of NW Sonora–SW Arizona border region: Two distinct Paleoproterozoic complexes at the southwest edge of Laurentia: Geological Society of America Abstracts with Programs, v. 35, no. 5, p. 67.

Oskin, M., and Stock, J., 2003, Pacific–North America plate motion and opening of the Upper Delfin basin, northern Gulf of California, Mexico: Geological Society of America Bulletin, v. 115, p. 1173–1190, doi: 10.1130/B25154.1.

Pálfy, J., Smith, P.L., and Mortensen, J.K., 2000, A U-Pb and ^{40}Ar/^{39}Ar time scale for the Jurassic: Canadian Journal of Earth Sciences, v. 37, p. 923–944, doi: 10.1139/cjes-37-6-923.

Paterson, S.R., Topics, O.T., and Vernon, R.H., 1991, Emplacement and deformation of granitoids during volcanic arc construction in the foothills terrane, central Sierra Nevada, California: Tectonophysics, v. 191, p. 89–110, doi: 10.1016/0040-1951(91)90234-J.

Pessagno, E.A., Jr., Blome, C.D., Hull, D.M., and Six, W.M., 1993, Jurassic radiolaria from the Josephine ophiolite and overlying strata, Smith River subterrane (Klamath Mountains), northwestern California and southwestern Oregon: Micropaleontology, v. 39, p. 93–166.

Pineda, A.S., 1961, Parte II, Flora fosil de la Formacion Santa Clara (Carnico) del Estrada de Sonora: Universidad Nacional Autónoma de México, Instituto de Geologia, Paleontologia Mexicana 11, pt. 2, 36 p.

Powell, R.E., 1993, Balanced palinspastic reconstruction of pre-late Cenozoic paleogeography, southern California: Geologic and kinematic constraints on the evolution of the San Andreas fault system, *in* Powell, R.E., Weldon, R J., II, and Matti, J.C., eds., The San Andreas fault system: Displacement, palinspastic reconstruction, and geologic evolution: Geological Society of America Memoir 178, p. 1–106.

Premo, W.R., Iriondo, A., and Nourse, J., 2003, U-Pb zircon geochronology of Paleoproterozoic basement in northwestern Sonora, Mexico: Evidence for affinity to SW US provinces: Geological Society of America Abstracts with Programs, v. 35, no. 5, p. 67.

Pubellier, M., Rangin, C., Rascon, B., Chorowicz, J., and Bellon, H., 1995, Cenomanian thrust tectonics in the Sahuaripa region, Sonora: Implications about northwestern Mexico megashears, *in* Jacques-Ayala, C., González-León, C.M., and Roldán-Quintana, J., eds., Studies on the Mesozoic of Sonora and adjacent areas: Geological Society of America Special Paper 301, p. 111–120.

Ramirez, J., and Acevedo, F., 1957, Notas sobre la Geología de Chihuahua: Boletín Asociación Mexicana de Geologos Petroleros, v. 9, p. 583–777.

Rangin, C., 1977, Sobre la presencia del Jurásico Superior con amonitas en Sonora septentrional: Universidad Nacional Autónoma de México, Instituto de Geología, Revista, v. 1, p. 44–47.

Rangin, C., 1982, Contribution a l´etude geologique du System Cordilleran du nord-ouest du Mexique [These de Doctorat d´Etat]: Paris, Universite Pierre et Marie Curie, 588 p.

Renne, P.R., and Scott, G.R., 1988, Structural chronology, oroclinal deformation, and tectonic evolution of the southeastern Klamath Mountains, California: Tectonics, v. 7, p. 1223–1242.

Reynolds, S.J., Spencer, J.E., Richard, S.M., and Laubach, S.E., 1986, Mesozoic structures in west-central Arizona, *in* Beatty, B., and Wilkinson, P.S.K., eds., Frontiers in geology and ore deposits of Arizona and the Southwest: Arizona Geological Society Digest, v. 16, p. 35–51.

Riggs, N.R., 1987, Stratigraphy, structure and geochemistry of Mesozoic rocks in the Pajarito Mountains, Santa Cruz County, Arizona, *in* Dickinson, W.R., and Klute, M.A., eds., Mesozoic geology of southern Arizona and adjacent areas: Arizona Geological Society Digest, v. 18, p. 165–175.

Riggs, N.R., and Busby-Spera, C.J., 1989, Controls of intra-arc subsidence on volcanic and sedimentary facies: Example from the early Mesozoic arc of Arizona and California: 28th International Geologic Congress Abstracts, v. 2, p. 699.

Riggs, N.R., and Busby-Spera, C.J., 1990, Evolution of a multi-vent volcanic complex within a subsiding arc graben depression, Mount Wrightson Formation, Arizona: Geological Society of America Bulletin, v. 102, p. 1114–1135, doi: 10.1130/0016-7606(1990)102<1114:EOAMVV>2.3.CO;2.

Riggs, N.R., and Busby-Spera, C.J., 1991, Facies analysis of an ancient, dismembered large caldera complex and implications for intra-arc subsidence: Middle Jurassic strata of Cobre Ridge, southern Arizona, U.S., *in* Cas, R.A.F., and Busby-Spera, C.J., eds., Volcaniclastic sedimentation: Sedimentary Geology, v. 74, p. 39–67.

Riggs, N.R., and Haxel, G.B., 1990, The Early to Middle Jurassic magmatic arc in southern Arizona: Plutons to sand dunes, *in* Gehrels, G.E., and Spencer, J.E., eds., Geologic excursions through the Sonoran Desert region, Arizona and Sonora: Arizona Geological Survey Special Paper 7, p. 90–103.

Riggs, N.R., Mattinson, J.M., and Busby-Spera, C.J., 1993, Correlation of Jurassic eolian strata between the magmatic arc and the Colorado Plateau: New U-Pb geochronologic data from southern Arizona: Geological Society of America Bulletin, v. 105, p. 1231–1246, doi: 10.1130/0016-7606(1993)105<1231:COJESB>2.3.CO;2.

Rodríguez-Castañeda, J.L., 1984, Geology of Tuape region, north-central Sonora, Mexico [M.S. thesis]: Pittsburgh, Pennsylvania, University of Pittsburgh, 157 p.

Rodríguez-Castañeda, J.L., 1986, Interpretacion del contacto Jurasico-Cretacico en Sonora este-central: Universidad Nacional Autónoma de

México, Instituto de Geología, Nuevas aportaciones a la geología de Sonora, p. 37–48.

Rodríguez-Castañeda, J.L., 1988, Estratigrafia de la region de Tuape, Sonora, Mexico: Universidad Nacional Autónoma de México, Instituto de Geología, Revista, v. 7, p. 52–66.

Rodríguez-Castañeda, J.L., 1990, Relaciones estructurales en la parte centro-septentrional del estado de Sonora: Universidad Nacional Autónoma de México, Instituto de Geología, Revista, v. 9, p. 51–61.

Rodríguez-Castañeda, J.L., 1991, Mesozoic stratigraphy in north-central Sonora: Universidad de Sonora, Boletín del Departamento de Geología, v. 8, no. 1, p. 13–27.

Rodríguez-Castañeda, J.L., 1994, Geología de la región del Teguachi, Estrada de Sonora, Mexico: Universidad Nacional Autónoma de México, Instituto de Geología, Revista Mexicana de Ciencias Geológicas, v. 11, no. 1, p. 11–28.

Rodríguez-Castañeda, J.L., 1997, Gravity sliding structures in Cretaceous–Early Tertiary rocks in north central Sonora, Mexico—Regional significance: Universidad Nacional Autónoma de México, Instituto de Geología, Revista Mexicana de Ciencias Geológicas, v. 14, no. 1, p. 1–12.

Rodríguez-Castañeda, J.L., 2000, Deformación Sonorense: Tectónica extensional en el área Rancho San Antonio, Sonora norte-central—sus efectos y características en rocas cretácicas y terciarias [Ph.D. thesis]: México, D.F., Universidad Nacional Autónoma de México, 217 p.

Rodríguez-Castañeda, J.L., and Amaya-Martínez, R., 1988, Presencia de rocas volcaniclasticas de edad Jurasica en la parte sur de la Sierra de Lopez al NW de Hermosillo, Sonora, Mexico: Sociedad Geológica Mexicana, IX Convencion Geologica Nacional, Memoria, p. 33.

Rodríguez-Castañeda, J.L., and García y Barragán, J.C., 1994, Late Jurassic–Early Cretaceous thrusting in the Altar region: Geological Society of America Abstracts with Programs, Cordilleran Section, v. 26, no. 2, p. 85.

Roure, F., and Sosson, M., 1986, Late Jurassic collision between a composite exotic block and the North American continent: A model for the Cordillera building: Bulletin de la Société Géologique de France, v. 8, p. 945–959.

Salas, G.A., 1968, Areal geology and petrology of the igneous rocks, Santa Ana Quadrangle, Sonora, Mexico [M.S. thesis]: Norman, University of Oklahoma, 110 p.

Saleeby, J.B., 1990, Geochronological and tectonostratigraphic framework of Sierran-Klamath ophiolitic assemblages, *in* Harwood, D.S., and Miller, M.M., eds., Paleozoic and early Mesozoic paleogeographic relations: Sierra Nevada, Klamath Mountains, and related terranes: Geological Society of America Special Paper 255, p. 93–114.

Saleeby, J.B., Goodin, S.E., Sharp, W.D., and Busby, C.J., 1978, Early Mesozoic paleotectonic-paleogeographic reconstruction of the southern Sierra Nevada region, *in* Howell, D.G., and McDougall, K., eds., Mesozoic paleogeography of the western United States, Pacific Coast Paleogeography Symposium 2: Society of Economic Paleontologists and Mineralogists, Pacific Section, p. 311–336.

Saleeby, J.B., Harper, G.D., Sharp, W.D., and Snoke, A.W., 1982, Time relations and structural- stratigraphic patterns in ophiolite accretion, west-central Klamath Mountains, California: Journal of Geophysical Research, v. 87, p. 3831–3848.

Saleeby, J.B., Shaw, H.F., Niemeyer, S., Moores, E.M., and Edelman, S.H., 1989a, U-Pb, Sm/Nd and Rb/Sr geochronological and isotopic study of northern Sierra Nevada ophiolitic assemblages, California: Contributions to Mineralogy and Petrology, v. 102, p. 205–220, doi: 10.1007/BF00375341.

Saleeby, J.B., Geary, E.E., Paterson, S.R., and Tobisch, O.T., 1989b, Isotopic systematics of Pb/U (zircon) and ^{40}Ar/^{39}Ar (biotite-hornblende) from rocks of the central Foothills terrane, Sierra Nevada, California: Geological Society of America Bulletin, v. 101, p. 1481–1492.

Saleeby, J.B., Busby-Spera, C., Oldow, J.S., Dunne, G.C., Wright, J.E., Cowan, D.S., Walker, N.W., and Allmendinger, R.W., 1991, Early Mesozoic tectonic evolution of the western U.S. Cordillera, *in* Burchfiel, B.C., et al., eds., The Cordilleran orogen: Conterminous U.S.: Boulder, Colorado, Geological Society of America, Geology of North America, v. G-3, p. 107–168.

Schermer, E.R., 1993, Mesozoic structural evolution of the west-central Mojave Desert, *in* Dunne, G., and McDougall, K., eds., Mesozoic paleogeography of the western United States, Volume 2: Los Angeles, California, Society of Economic Paleontologists and Mineralogists, Pacific Section, Book 71, p. 307–322.

Schermer, E.R., and Busby, C.J., 1994, Jurassic magmatism in the central Mojave Desert: Implications for arc paleogeography and preservation of continental volcanic sequences: Geological Society of America Bulletin, v. 106, p. 767–790, doi: 10.1130/0016-7606(1994)106<0767:JMITCM>2.3.CO;2.

Schermer, E.R., Stephens, K.A., and Walker, J.D., 2001, Paleogeographic and tectonic implications of the geology of the Tiefort Mountains, northern Mojave Desert, California: Geological Society of America Bulletin, v. 113, p. 920–938, doi: 10.1130/0016-7606(2001)113<0920:PATIOT>2.0.CO;2.

Schermer, E.R., Busby, C.J., and Mattison, J.M., 2002, Paleogeographic and tectonic implications of Jurassic sedimentary and volcanic sequences, *in* Glazner, A.F., Walker, J.D., and Bartley, J.M., eds., Geologic evolution of the Mojave Desert and southwestern Basin and Range: Geological Society of America Memoir 195, p. 93–115.

Schmidt, K.L., and Paterson, S.R., 2002, A doubly vergent fan structure in the Peninsular Ranges batholith: Transpression or local complex flow around a continental margin buttress?: Tectonics, v. 21, no. 5, p. 14-1–14-17.

Schweickert, R.A., and Cowan, D.S., 1975, Early Mesozoic tectonic evolution of the western Sierra Nevada, California: Geological Society of America Bulletin, v. 86, p. 1329–1336, doi: 10.1130/0016-7606(1975)86<1329:EMTEOT>2.0.CO;2.

Schweickert, R.A., and Lahren, M.M., 1990, Speculative reconstruction of a major Early Cretaceous(?) dextral fault zone in the Sierra Nevada: Implications for Paleozoic and Mesozoic orogenesis in the western United States: Tectonics, v. 9, p. 1609–1629.

Schweickert, R.A., and Lahren, M.M., 1991, Age and tectonic significance of metamorphic rocks along the axis of the Sierra Nevada Batholith: A critical reappraisal, *in* Cooper, J.D., and Stevens, C.H., Paleozoic paleogeography of the western United States, Volume 2: Los Angeles, California, Society of Economic Paleontologists and Mineralogists, Pacific Section, p. 653–676.

Schweickert, R.A., and Lahren, M.M., 1993, Triassic-Jurassic magmatic arc in eastern California and western Nevada: Arc evolution, cryptic tectonic breaks, and significance of the Mojave-Snow Lake fault, *in* Dunne, G., and McDougall, K., eds., Mesozoic paleogeography of the western United States, Volume 2: Los Angeles, California, Society of Economic Paleontologists and Mineralogists, Pacific Section, Book 71, p. 227–246.

Schweickert, R.A., Bogen, N.L., Girty, G.H., Hanson, R.E., and Merguerian, C., 1984, Timing and structural expression of the Nevadan orogeny, Sierra Nevada, California: Geological Society of America Bulletin, v. 95, p. 967–979, doi: 10.1130/0016-7606(1984)95<967:TASEOT>2.0.CO;2.

Schweickert, R.A., Lahren, M.M., and Walker, J.D., 1994, New age and structural constraints on volcanism, thrusting and penetrative deformation of rocks in Saddlebag Lake pendant, Yosemite National Park, California: Geological Society of America Abstracts with Programs, v. 26, no. 2, p. 89.

Segerstrom, L., 1986, Geologic setting and silver mineralization in the Planchas de Plata area, northern Sonora, Mexico [M.S. thesis]: Tucson, Arizona, University of Arizona, 167 p.

Segerstrom, L., 1987, Geology of the Planchas de Plata ore, northern Sonora, Mexico, *in* Dickinson, W.R., and Klute, M.A., eds., Mesozoic geology of southern Arizona and adjacent areas: Arizona Geological Society Digest, v. 18, p. 153–164.

Sharp, R.V., 1967, San Jacinto fault zone in the Peninsular Ranges of southern California: Geological Society of America Bulletin, v. 78, p. 705–730.

Sharp, R.V., 1979, Some characteristics of the eastern Peninsular Ranges mylonite zone: Proceedings of Conference VIII, Analysis of actual fault zones in bedrock, convened by R. Speed, R. Sharp, and J. Everden: U.S. Geological Survey Open-File Report 79-1239, p. 258–269.

Silver, L.T., 1963, The use of cogenetic uranium-lead isotope systems in zircons in geochronology: Radioactive dating, International Atomic Energy Agency, Athens, Nov. 1962, p. 279–285.

Silver, L.T., 1971, Problems of crystalline rocks of the Transverse Ranges: Geological Society of America Abstracts with Programs, v. 3, no. 2, p. 193–194.

Silver, L.T., 1982, Evidence and a model for west-directed early to middle Cenozoic basement overthrusting in southern California: Geological Society of America Abstracts with Programs, v. 14, no. 7, p. 617.

Silver, L.T., 1983, Paleogene overthrusting in the tectonic evolution of the Transverse Ranges, Mojave and Salinian regions, California: Geological Society of America Abstracts with Programs, v. 15, no. 5, p. 438.

Silver, L.T., and Anderson, T.H., 1974, Possible left-lateral early to middle Mesozoic disruption of the southwestern North American craton margin: Geological Society of America Abstracts with Programs, v. 6, no. 7, p. 955–956.

Silver, L.T., and Anderson, T.H., 1983, Further evidence and analysis of the role of the Mojave-Sonora megashear(s) in Mesozoic Cordilleran tectonics: Geological Society of America Abstracts with Programs, v. 15, no. 5, p. 273.

Silver, L.T., and Deutsch, S., 1963, Uranium-lead method on zircons: New York Academy of Science Annual, v. 91, p. 279–283.

Silver, L.T., and Nourse, J.A., 1986, The Rand Mountains "thrust" complex in comparison with the Vincent thrust–Pelona Schist relationship, southern California: Geological Society of America Abstracts with Programs, v. 18, no. 2, p. 185.

Silver, L.T., and Nourse, J.A., 1991, Timing of the Rand thrust and its implication for Late Cretaceous tectonics in southern California: Geological Society of America Abstracts with Programs, v. 23, no. 5, p. 480.

Silver, L.T., McKinney, C.R., Deutsch, S., and Bolinger, J., 1963, Precambrian age determinations in the western San Gabriel Mountains, California: Journal of Geology, v. 71, p. 196–214.n T. and Nourse, Jonathan A., the Rand thrust a

Simpson, C., 1984, Borrego–Santa Rosa mylonite zone: A Late Cretaceous west-directed thrust in southern California: Geology, v. 12, p. 8–11, doi: 10.1130/0091-7613(1984)12<8:BSRMZA>2.0.CO;2.

Simpson, C., 1990, Microstructural evidence for northeastward movement on the Chocolate Mountains fault zone, southeastern California: Journal of Geophysical Research, v. 95, p. 529–538.

Snoke, A.W., 1977, A thrust plate of ophiolitic rocks in the Preston Peak area, Klamath Mountains, California: Geological Society of America Bulletin, v. 88, p. 1641–1659, doi: 10.1130/0016-7606(1977)88<1641:ATPOOR>2.0.CO;2.

Solano-M., O., 1989, Estudio geológico de las sierras La Ceniza, El Caloso y Santa Rosa, Municipio de Agua Prieta, Sonora [Tesis Licenciatura]: Hermosillo, Sonora, Universidad de Sonora, 73 p.

Sosson, M., and Calmus, T., 1990, Early Late Cretaceous overthrusting in northern Sonora (Mexico) responsible for a large scale displacement of the Nevadan belt, *in* Aubouin, J., and Bourgois, J., eds., Tectonics of circum-Pacific continental margins, 28th International Geological Congress Proceedings: Utrecht, the Netherlands, VSP, p. 37–50.

Spencer, J.E., and Reynolds, S.J., 1990, Relationship between Mesozoic and Cenozoic tectonic features in west central Arizona and adjacent southeastern California: Journal of Geophysical Research, v. 95, no. B1, p. 539–555.

Stanley, J.R., Jr., and González-León, C.M., 1995, Paleogeographic and tectonic implications of Triassic fossils and strata from the Antimonio Formation, northwestern Sonora, *in* Jacques-Ayala, C., González-León, C.M., and Roldán-Quintana, J., eds., Studies on the Mesozoic of Sonora and adjacent areas: Geological Society of America Special Paper 301, p. 1–16.

Stephens, W.E., 1988, Reconnaissance geology of the Cucurpe region, north-central Sonora, Mexico [M.S. thesis]: Pittsburgh, Pennsylvania, University of Pittsburgh, 317 p.

Stevens, C.H., Pelley, T., and Greene, D.C., 1995, Middle Devonian submarine fans in the eastern Sierra Nevada, California: San Francisco, California, American Association of Petroleum Geologists–Society of Economic Paleontologists and Mineralogists, Pacific Section, Convention Abstracts, p. 46.

Stevens, C.H., Stone, P., Dunne, G.C., Greene, D.C., Walker, J.D., and Swanson, B.J., 1997, Paleozoic and Mesozoic evolution of east-central California: International Geology Review, v. 39, p. 788–829.

Stevens, C.H., Stone, P., Dunne, G.C., Greene, D.C., Walker, J.D., and Swanson, B.J., 1998, Paleozoic and Mesozoic evolution of east-central California, *in* Ernst, W.G., and Nelson, C.A., eds., Integrated earth and environmental evolution of the southwestern United States, The Clarence A. Hall, Jr., Volume: Boulder, Colorado, Geological Society of America, p. 119–160.

Stewart, J.H., 1988, Tectonics of the Walker Lane belt, western Great Basin, Mesozoic and Cenozoic deformation in a zone of shear, *in* Ernst, W.G., ed., Metamorphism and crustal evolution of the western United States (Rubey Volume 7): Englewood Cliffs, New Jersey, Prentice Hall, p. 683–713.

Stewart, J.H., McMenamin, M.A.S., and Morales-Ramírez, J.M., 1984, Upper Proterozoic and Cambrian rocks in the Caborca region, Sonora, México: Physical stratigraphy, biostratigraphy, paleocurrent studies, and regional relations: U.S. Geological Survey Professional Paper 1309, 36 p.

Stewart, J.H., Anderson, T.H., Haxel, G.B., Silver, L.T., and Wright, J.E., 1986, Late Triassic paleogeography of the southern Cordillera—The problem of a source for voluminous volcanic detritus in the Chinle Formation of the Colorado Plateau region: Geology, v. 14, p. 567–570.

Stewart, J.H., Poole, F.G., Keith, K.B., Madrid, R.J., Roldán-Quintana, J., and Amaya-Martinez, R., 1990, Tectonics and stratigraphy of the Paleozoic and Triassic southern margin of North America, Sonora, México, *in* Gehrels, G.E., and Spencer, J.E., eds., Geologic excursions through the Sonoran Desert region, Arizona and Sonora: Arizona Geological Society Special Paper 7, p. 183–202.

Stewart, J.H., Amayo-Martinez, R., Stamm, R.G., Wardlaw, B.R., Stanley, G.D., Jr., and Stevens, C.H., 1997, Stratigraphy and regional significance of Mississippian to Jurassic rocks in Sierra Santa Teresa, Sonora, México, *in* Stanley, G.D., Jr., and González-León, C.M., Special issue dedicated to the International Workshop on the Geology of Northwestern Sonora, Mexico: Revista Mexicana de Ciencias Geologicas, v. 14, no. 2, p. 115–135.

Stone, P., and Stevens, C.H., 1987, Stratigraphy of the Owens Valley Group (Permian), southern Inyo mountains, California: U.S Geological Survey Bulletin, v. 1692, 19 p.

Thomson, C.N., 1994, Tectonic implications of Jurassic magmatism and Early Cretaceous mylonitic deformation, Cuyamaca-Laguna mountains shear zone, southern California [M.S. thesis]: San Diego, San Diego State University, 89 p.

Thomson, C.N., and Girty, G.H., 1994, Early-Cretaceous intra-arc ductile strain in Triassic-Jurassic and Cretaceous continental margin arc rocks, Peninsular Ranges, California: Tectonics, v. 13, p. 1108–1119, doi: 10.1029/94TC01649.

Tikoff, B., and Saint-Blanquat, M., 1997, Transpressional shearing and strike-slip partitioning in the Late Cretaceous Sierra Nevada magmatic arc, California: Tectonics, v. 16, p. 442–459, doi: 10.1029/97TC00720.

Tobisch, O.T., Paterson, S.R., Longiaru, S., and Bhattacharyya, T., 1987, Extent of the Nevadan orogeny, central Sierra Nevada, California: Geology, v. 15, p. 132–135, doi: 10.1130/0091-7613(1987)15<132:EOTNOC>2.0.CO;2.

Tosdal, R.M., and Miller, R.J., 1988, Thermo-tectonic terranes of the Ajo and Lukeville 1° by 2° quadrangle: K-Ar geochronology of early Tertiary and older rocks: U.S. Geological Survey Open-File Report 88-217, p. 5–8.

Tosdal, R.M., Haxel, G.B., and Wright, J.E., 1989, Jurassic geology of the Sonoran Desert region, southern Arizona, southeastern California, and northernmost Sonora: Construction of a continental-margin magmatic arc, *in* Jenny, J.P., and Reynolds, S.J., eds., Geologic evolution of Arizona: Arizona Geological Society Digest, v. 17, p. 397–434.

Tosdal, R.M., Haxel, G.B., Anderson, T.H., Connors, C.D., May, D.J., and Wright, J.E., 1990, Highlights of Jurassic, Late Cretaceous to Early Tertiary tectonics, south-central Arizona and north-central Sonora, *in* Gehrels, G.E., and Spencer, J.E., eds., Geologic excursions through the Sonoran Desert region, Arizona and Sonora: Arizona Geological Survey Special Paper 7, p. 76–88.

Valentine, W.G., 1936, Geology of the Cananea Mountains, Sonora, Mexico: Geological Society of America Bulletin, v. 47, p. 53–86.

Vega-Granillo, R., and Perez-Segura, F., and Chaparro-Mesa, M., 1991, Mantos de corrimiento y cabalgaduras imbricadas en Sonora central: Universidad de Sonora, Boletín del Departamento de Geología, v. 8, p. 37–46.

Wadsworth, W.B., Ferriz, H., and Rhodes, D.D., 1995, Structural and stratigraphic development of the Middle Jurassic magmatic arc in the Cowhole Mountains, central-eastern Mojave Desert, California, *in* Miller, D.M., and Busby, C., eds., Jurassic magmatism and tectonics of the North American Cordillera: Geological Society of America Special Paper 299, p. 327–349.

Wagner, D.L., Jennings, C.W., Bedrossian, T.L., and Bortugno, E.J., 1981, Geologic map of the Sacramento Quadrangle, California: California Division of Mines and Geology, Regional Map Series Map 1A, scale 1:250,000.

Walker, J.D., Martin, M.W., Bartley, J.M., and Coleman, D.S., 1990, Timing and kinematics of deformation in the Cronese Hills, California, and implications for Mesozoic structure of the southern Cordillera: Geology, v. 18, p. 554–557, doi: 10.1130/0091-7613(1990)018<0554:TAKODI>2.3.CO;2.

Walker, J.D., Martin, M.W., and Glazner, A.F., 2002, Late Paleozoic to Mesozoic development of the Mojave Desert and environs, California, *in* Glazner, A.F., Walker, J.D., and Bartley, J.M., eds., Geologic evolution of the Mojave Desert and southwestern Basin and Range: Geological Society of America Memoir 195, p. 1–18.

White, D.E., and Guiza, R., 1948, Antimony deposits of El Antimonio District, Sonora, Mexico: U.S. Geological Survey Bulletin 962-B, p. 81–116.

Willard, J.S., 1988, Geology, sandstone petrography, and provenance of the Jurassic (?)- Cretaceous rocks of the Puerto del Alamo area, northwestern Sonora, Mexico [M.Sc. thesis]: Cincinnati, Ohio, University of Cincinnati, 250 p. (unpublished).

Wolf, M.B., and Saleeby, J.B., 1992, Jurassic Cordilleran dike swarm/shear zones: Implications for the Nevadan orogeny and North American plate motion:

Geology, v. 20, p. 745–748, doi: 10.1130/0091-7613(1992)020<0745:JCDSSZ>2.3.CO;2.

Wolf, M.B., and Saleeby, J.B., 1995, Late Jurassic dike swarms in the southwestern Sierra Nevada Foothills terrane, California: Implications for the Nevada orogeny and North American plate motion, *in* Miller, D.M., and Busby, C., eds., Jurassic magmatism and tectonics of the North American Cordillera: Geological Society of America Special Paper 299, p. 203–228.

Wood, D.J., 1997, Geology of the eastern Tehachapi Mountains and Late Cretaceous–Early Cenozoic tectonics of the southern Sierra Nevada region, Kern County, California [Ph.D. thesis]: Pasadena, California, California Institute of Technology, 287 p., 3 plates.

Wood, D.J., and Saleeby, J.B., 1998, Late Cretaceous–Paleocene extensional collapse and disaggregation of the southernmost Sierra Nevada Batholith, *in* Ernst, W.G., and Nelson, C.A., eds., Integrated earth and environmental evolution of the southwestern United States, The Clarence A. Hall, Jr., Volume: Boulder, Colorado, Geological Society of America, p. 289–325.

Wright, J.E., and Fahan, M.R., 1988, An expanded view of Jurassic orogenesis in the western US Cordillera: Middle Jurassic (pre-Nevadan) regional metamorphism and thrust faulting within an active arc environment: Geological Society of America Bulletin, v. 100, p. 859–876, doi: 10.1130/0016-7606(1988)100<0859:AEVOJO>2.3.CO;2.

Wright, J.E., and Wyld, S.J., 1994, The Rattlesnake Creek terrane, Klamath Mountains, California: An early Mesozoic volcanic arc and its basement of tectonically disrupted oceanic crust: Geological Society of America Bulletin, v. 106, p. 1033–1056, doi: 10.1130/0016-7606(1994)106<1033:TRCTKM>2.3.CO;2.

Wyld, S.J., and Wright, J.E., 1988, The Devils Elbow ophiolite remnant and overlying Galice Formation: New constraints of the Middle to Late Jurassic evolution of the Klamath Mountains, California: Geological Society of America Bulletin, v. 100, p. 29–44, doi: 10.1130/0016-7606(1988)100<0029:TDEORA>2.3.CO;2.

Xenophontos, C., and Bond, G.C., 1978, Petrology, sedimentation, and paleogeography of the Smartville terrane (Jurassic)—Bearing on the genesis of the Smartville ophiolite, *in* Howell, D.G., and McDougall, K., eds., Mesozoic paleogeography of the western United States, Pacific Coast Paleogeography Symposium 2: Los Angeles, California, Society of Economic Paleontologists and Mineralogists, Pacific Section, p. 291–302.

Manuscript Accepted by the Society 24 March 2005

Geological Society of America
Special Paper 393
2005

Pull-apart basins at releasing bends of the sinistral Late Jurassic Mojave-Sonora fault system

Thomas H. Anderson*
Department of Geology and Planetary Science, University of Pittsburgh, Pittsburgh, Pennsylvania 15260, USA

Jonathan A. Nourse
Geological Sciences Department, California State Polytechnic University, Pomona, California 91768, USA

ABSTRACT

A 200–500-km-wide belt along the southwestern margin of cratonic North America is pervaded by northwest- and east-trending faults that flank basins containing thick deposits of locally derived conglomerate and sedimentary breccia. These deposits that crop out mainly in the northern part of mainland Mexico, or southern parts of Arizona and New Mexico are unconformable at their bases, have similar Upper Jurassic and/or Lower Cretaceous stratigraphic ages, and commonly preserve volcanic components in the lower parts of upward-fining sections. We argue that these basins share a common structural origin, based on: (1) the presence of faults, locally preserved, that generally define the basin margins, (2) similar basal units comprised of coarse conglomeratic strata derived from adjacent basement, and (3) locally preserved syntectonic relationships to bounding faults. Fault orientations, and our observation that the faults (and their associated basins) extend south to the inferred trace of the Late Jurassic Mojave-Sonora megashear, suggest that the basins formed in response to transtension associated with sinistral movement along the megashear. Northwest-striking left-lateral strike-slip faults that terminate at east-striking normal faults define releasing left steps at which crustal pull-apart structures formed. These faults, plus a less-developed set of northeast-striking right-lateral faults, appear to comprise a cogenetic system that is kinematically linked with the Mojave-Sonora megashear; that is, the maximum principal stress trends east and the plane containing maximum sinistral shear stress strikes northwesterly.

Late Jurassic faults northeast of the Mojave-Sonora megashear controlled the regional distribution of the pull-apart basins and influenced the orientation and style of many younger structures and intrusions. Most Late Jurassic faults were modified during subsequent episodes of deformation. N60°E-directed contraction during the Late Cretaceous (Laramide) orogeny reactivated older east-striking normal faults as sinistral strike-slip faults; northwest-striking sinistral faults were reactivated as steep reverse faults. Some stratigraphically low units were thrust across basin margins as a result of inversion. Many of the pull-apart basins encompass outcrops of Late Jurassic igneous rocks and/or mineralized Laramide or Tertiary plutons. Some northwesterly

*taco@pitt.edu.

Anderson, T.H., and Nourse, J.A., 2005, Pull-apart basins at releasing bends of the sinistral Late Jurassic Mojave-Sonora fault system, *in* Anderson, T.H., Nourse, J.A., McKee, J.W., and Steiner, M.B., eds., The Mojave-Sonora megashear hypothesis: Development, assessment, and alternatives: Geological Society of America Special Paper 393, p. 97–122. doi: 10.1130/2005.2393(03). For permission to copy, contact editing@geosociety.org.

faults appear to have influenced the position of breakaway zones for early Miocene detachment faults. Despite the common and locally strong structural and magmatic overprinting, remnants of the Late Jurassic faults are recognizable.

Keywords: pull-apart, Late Jurassic, Sonora, sinistral.

INTRODUCTION

Statement of Hypothesis and Objectives

At its type locality near Bisbee, Arizona (Figs. 1, 2, 3; Plate 1[1]), the Upper Jurassic and Lower Cretaceous Glance Conglomerate occurs within an elongate, fault-bounded basin. The long sides of this crudely rhomb-shaped basin coincide with steep northwest-trending faults, and the basin is terminated by east-trending normal faults. Numerous other conglomerate bodies are linked to the Glance by virtue of similarities of texture, depositional environment, and stratigraphic position. These Upper Jurassic–Lower Cretaceous conglomerates lie northeast (Chiricahua Mountains, Arizona), southeast (Mina Plomosas–Placer de Guadalupe and Valle San Marcos, northeastern Mexico), southwest (Imuris and Sierra El Batamote–Sierra del Alamo, northwestern Mexico) and west (McCoy, Palen, and Plomosa Mountains, southern California) of the Glance

[1]Plate 1 is on the CD-ROM accompanying this volume and in the GSA Data Repository as item 2005140, available online at www.geosociety.org/pubs/ft2005.htm, or on request from editing@geosociety.org or Documents Secretary, GSA, P.O. Box 9140, Boulder, CO 80301, USA.

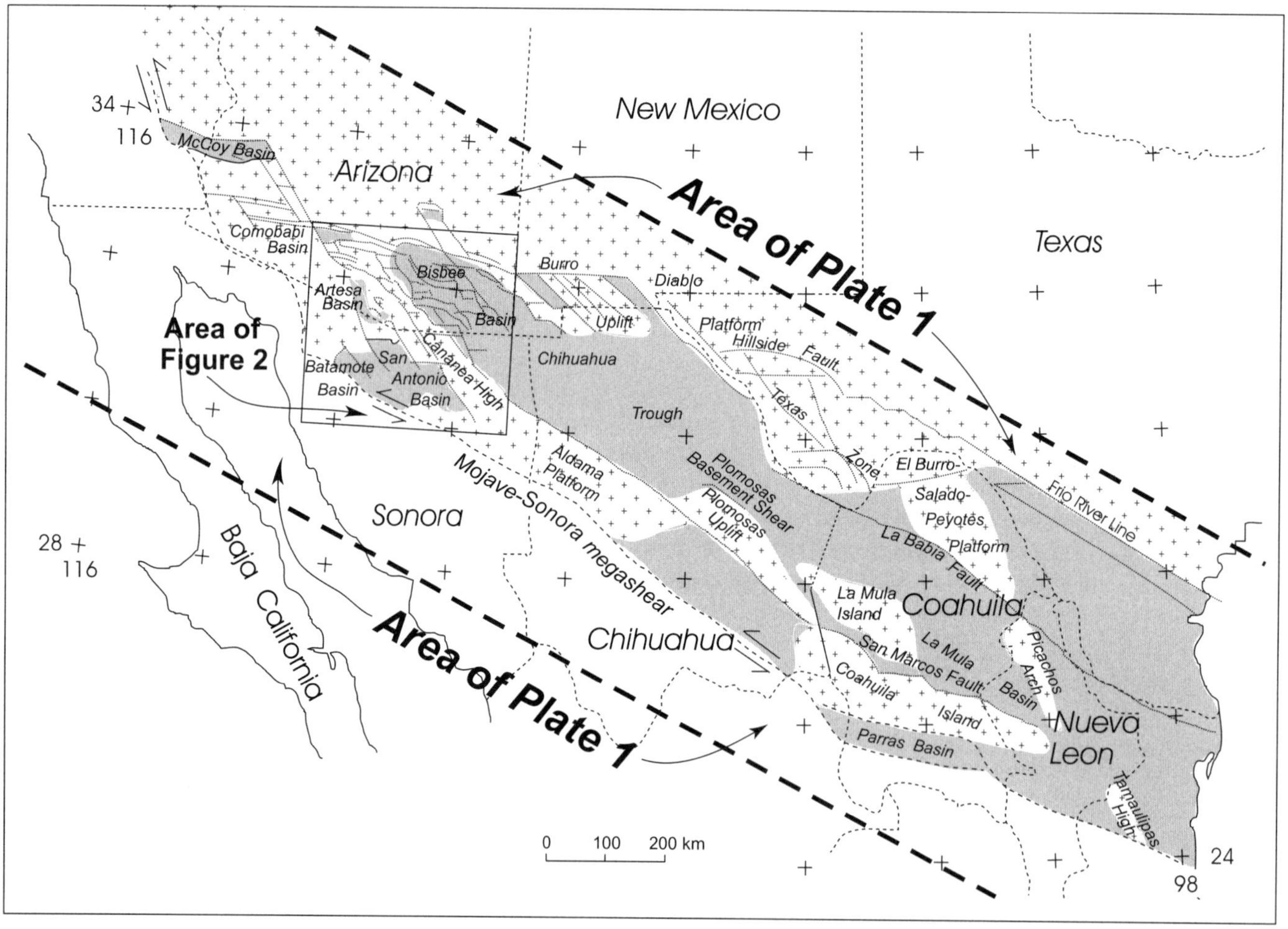

Figure 1. Map of the Mojave-Sonora fault system, showing traces of major lineaments or faults with known or inferred Late Jurassic displacements. Regions of important Late Jurassic–Early Cretaceous uplift or subsidence are highlighted with plus pattern or gray tone, respectively. Note locations of Figure 2 and Plate 1, which detail fault patterns and pre-Cretaceous geology. Late Jurassic regional transtension is implied by linked networks of northwesterly sinistral faults and east-striking normal faults.

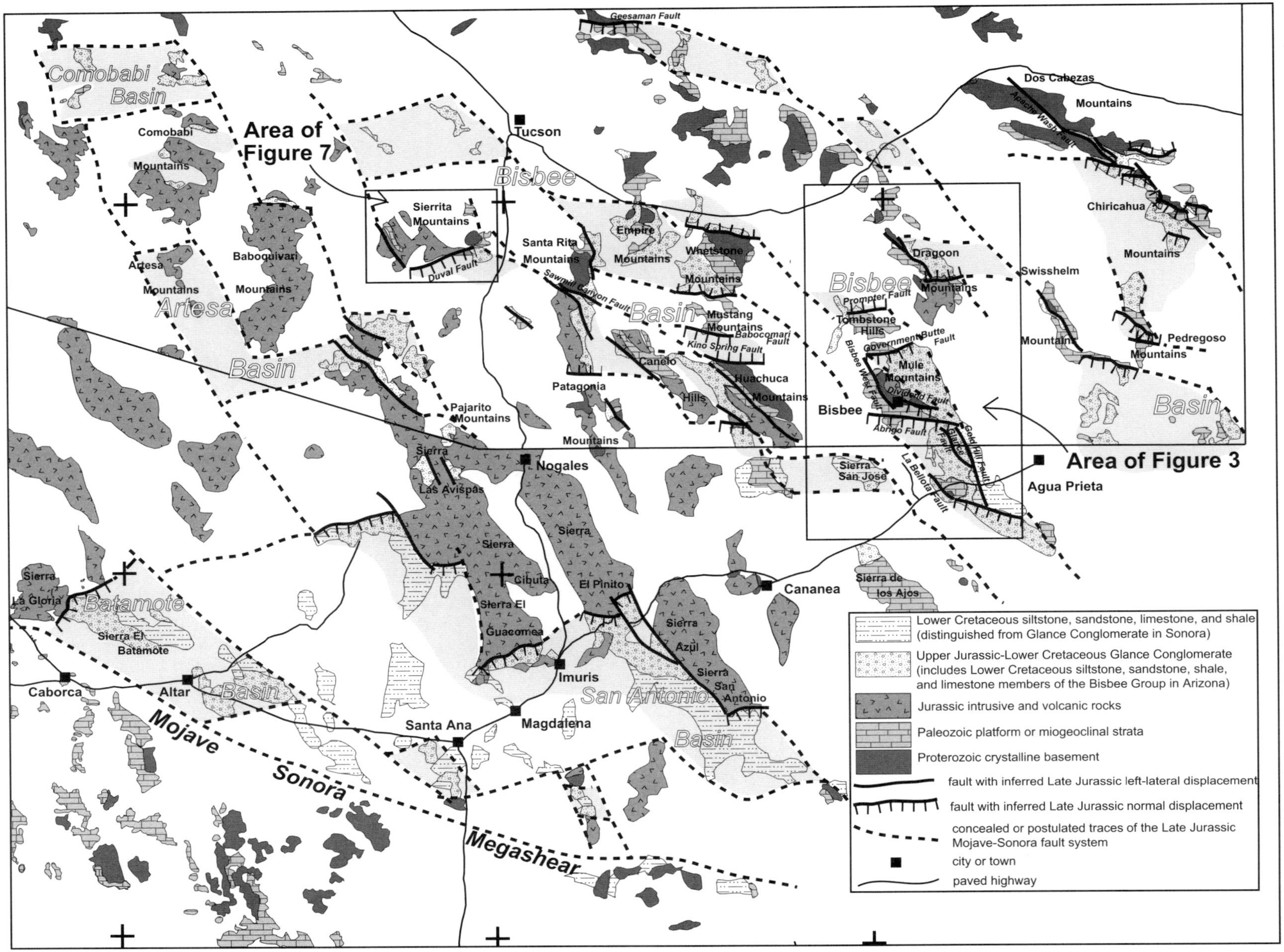

Figure 2. Generalized geologic map of southern Arizona, showing present-day outcrops of Upper Jurassic–Lower Cretaceous sedimentary strata in relation to exposures of Jurassic arc and older rocks. Also highlighted are major faults with known or inferred Late Jurassic displacements, mountain ranges, and towns described in the text. The movement histories of most of the faults shown have been complicated by reactivation during northeast-directed Cretaceous contraction. Data in Arizona digitized from Arizona Geological Survey Map 35 (Richard et al., 2000). Geology in Sonora modified from Nourse (1995, 2001).

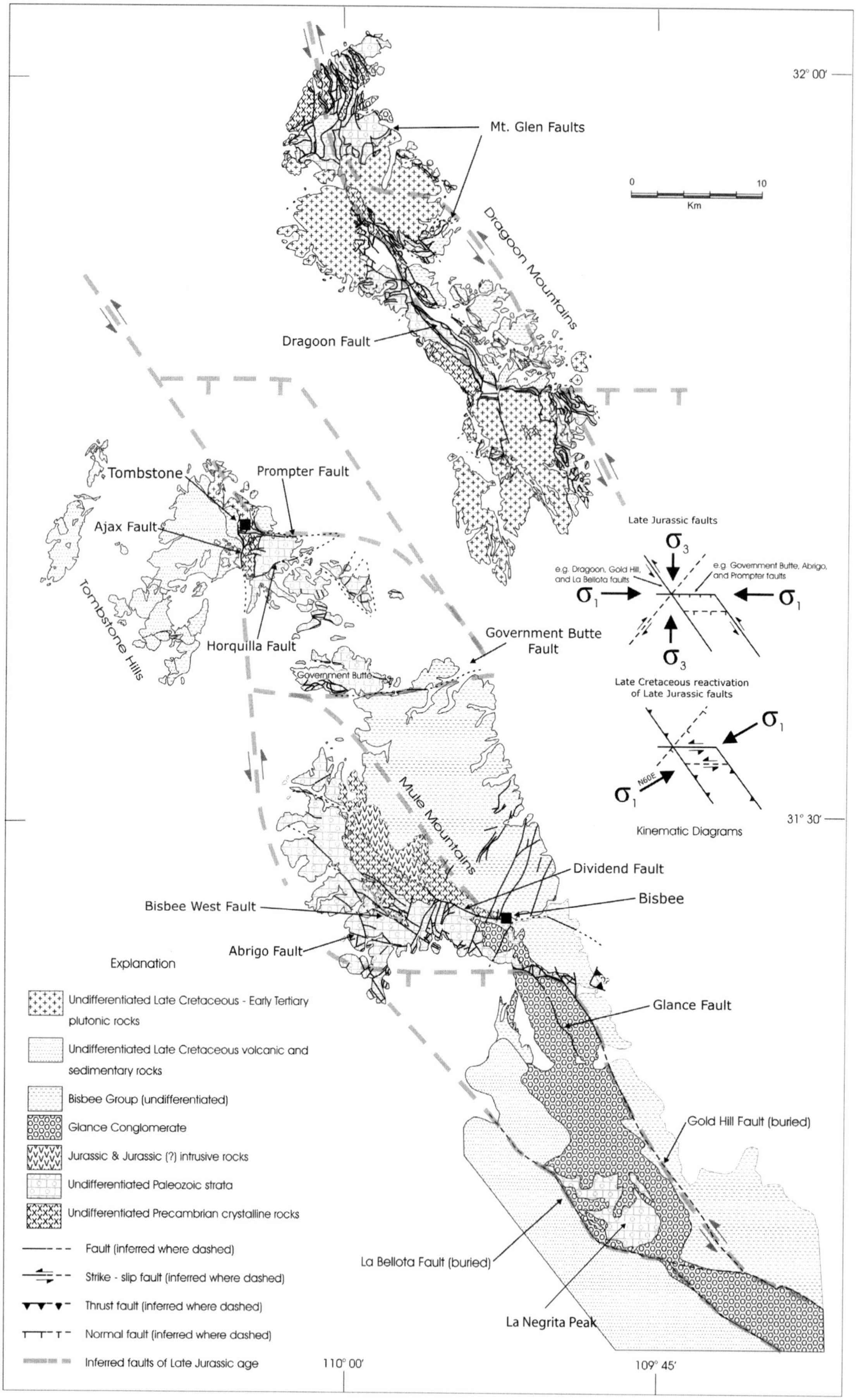

Figure 3. Geologic map of part of southern Arizona and adjacent northern Sonora from Taliaferro (1933), Drewes (1981), González-León and Lawton (1995), and McKee et al. (this volume) that shows postulated pull-apart basins in the Mule Mountains, Dragoon Mountains, and near Tombstone. Kinematic diagram (inset) shows inferred transtensional stress regime for Late Jurassic time and compressional stress regime during Late Cretaceous time.

Conglomerate at its type locality (Figs. 1, 2; Plate 1). All were deposited within a 200–500-km-wide region northeast of the Mojave-Sonora megashear (Silver and Anderson, 1974; Anderson and Schmidt, 1983; Anderson and Silver, 1979, this volume), extending from southeastern California and northwestern Sonora to the Frio River line in southern Texas (Fig. 1). We propose the following hypothesis: The Glance Conglomerate and related coarse clastic sections represent the initial fill of cogenetic Late Jurassic pull-apart basins. These fault-bounded basins formed at ~45° releasing bends or sidesteps within the Mojave-Sonora fault system, and are direct manifestations of distributed brittle shear affecting the southwest North American craton during Late Jurassic sinistral movements on the Mojave-Sonora megashear.

Large pull-apart structures such as the McCoy Basin (Harding and Coney, 1985; Fackler-Adams et al., 1997), and groups of pull-aparts that compose regional basins such as the Bisbee Basin, La Mula Basin, and the Chihuahua Trough, are principal features of the middle Mesozoic crust of southwestern North America. The transtensional basins developed during a transition from Middle Jurassic convergence (with concomitant calc-alkaline volcanism) to Late Jurassic sinistral transform faulting and accompanying alkalic magmatism. In this paper we argue that: (1) the aforementioned Late Jurassic basins display sedimentologic and structural features compatible with a pull-apart model, (2) these basins occur within a belt of broken Proterozoic crust and Middle Jurassic arc basement positioned between the postulated Mojave-Sonora megashear and intact craton, (3) Late Jurassic rupture of continental crust (and attendant formation of narrow depositional basins) was driven by transtension (i.e., wrench faulting, with normal faults linking lateral faults at releasing steps), and (4) the orientation and age of the faulting indicates kinematic relationship to sinistral movement along the Mojave-Sonora megashear.

In many cases we cannot unequivocally prove that sedimentation was synchronous with sinistral and normal fault displacements, but the pattern of faulting that we can observe (or reasonably infer) suggests that the basins are pull-aparts in the classic sense (Mann et al., 1983; Fig. 4). Throughout the study area, Late Jurassic faults and spatially associated syntectonic basins are partly covered by younger deposits, and their original geometry has been further obscured by Laramide and Tertiary tectonism. Some faults along the margins of the eastern basins (Fig. 1; Plate 1) may be reactivated Neoproterozoic or late Paleozoic structures. Despite these complications we believe that the basins are recognizable as distinct entities. For several, we describe geologic evidence bearing on the stratigraphy and sedimentology, and the movement history of boundary faults. We then discuss regional implications of the pull-apart basins in the context of Late Jurassic tectonics of southwestern North America and subsequent structural and magmatic reactivations.

Influential Previous Work

The model of crustal deformation and basin formation offered here arose from ideas that grew from our own work in Sonora south of the type exposures of Glance Conglomerate at Bisbee, Arizona (Anderson et al., 1995; McKee et al., this volume), and near the Mojave-Sonora megashear (Nourse, 1995, 2001), coupled with the indispensable work of Ransome (1904), Bilodeau's (1979, 1982), and Bilodeau et al.'s (1987) effective advancement of knowledge of the Glance Conglomerate. We agree with the conclusion reached by Ransome and Bilodeau that the basins are fault bounded, however, most other workers have employed different models to explain the origin of these rocks. For example, separate articles by Bilodeau (1982) and

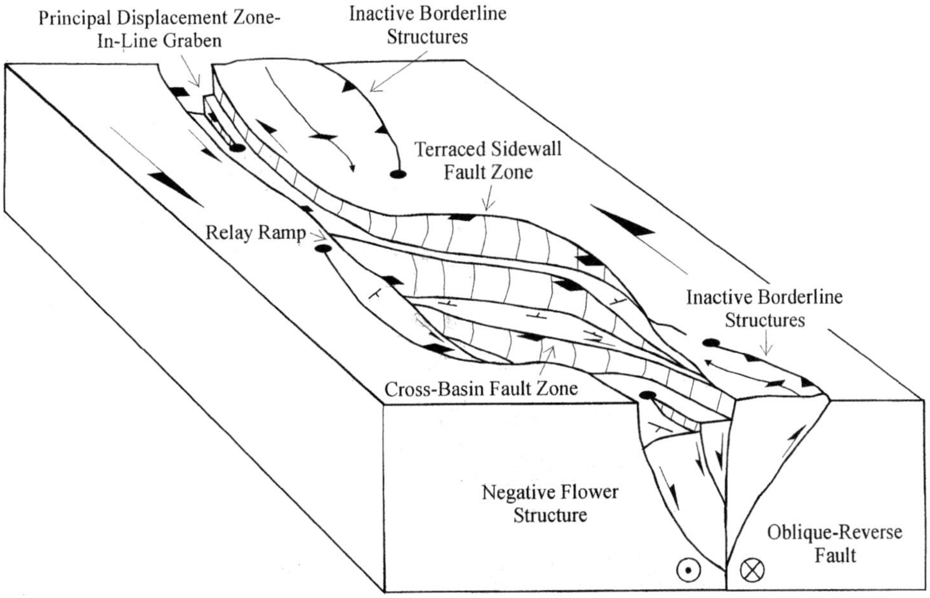

Figure 4. Schematic illustration of a releasing bend between two left-lateral fault strands, highlighting typical paleogeographic features associated with the resulting pull-apart basin. This pattern of northwesterly left-lateral faults linked with east-striking normal faults mimics that observed for the Mojave-Sonora fault system (modified from Aksu et al., 2000).

Dickinson et al. (1989) suggested that the Glance Conglomerate and overlying Bisbee Group accumulated in a back-arc setting or in an aulacogen connected with the Gulf of Mexico along a rift. Lawton and McMillan (1999) and Dickinson and Lawton (2001) developed this concept further in their proposal for a "Borderland rift" system driven by slab rollback. Our paper expands upon Bilodeau's ideas for discrete back-arc rifts but questions the need to lump all of the basins into one region of extension with minimum principal stress oriented orthogonal to the Cordilleran magmatic arc. One argument against the borderland rift model is the presence of a belt of thick, Upper Jurassic conglomerate in northern Sonora (Fig. 1), separate from the Bisbee Basin, suggesting existence of at least two rift systems. Also, the consistent asymmetric geometry of the basins implies that a sinistral shear component accompanied basin formation.

Drewes and Hayes (1983), each of whom have significantly contributed to knowledge of the geology of southern Arizona, disagreed with the idea of Early Cretaceous rifting. They claimed that the dating of sedimentary deposits and faults and the structural complexity of the region left them in doubt "as to whether extensions of his (Bilodeau's) idea to any broadly regional tectonic model are justified" (p. 364). We respect the extensive field mapping conducted by these workers, and we hope that the additional information provided in this report allays their concerns.

Our hypothesis that Late Jurassic transtension pervaded a wide swath of North American crust inboard of the Mojave-Sonora megashear is supported by a variety of independent regional studies. For example, Harding and Coney (1985) proposed that a transtensional rift basin, bounded along the southwest margin by the Mojave-Sonora megashear, enclosed the McCoy Mountains Formation and equivalent strata of southeastern California and southwestern Arizona. Global paleomagnetic analyses by Engebretson et al. (1984) utilized a hotspot reference frame to show that western North America experienced Late Jurassic left-lateral shear coincident with a pronounced drop in normal plate convergence. Finally, recent work on the Late Jurassic Independence dike swarm (Glazner et al., 1999) suggests that many of these northwest-striking dikes were emplaced into left-lateral faults affected by north-south extension rather than tension fractures oriented orthogonal to a northeasterly minimum principal stress direction.

CHARACTERISTICS OF THE LATE JURASSIC PULL-APART BASINS

Mountainous uplifts from southeastern California to west Texas and south into northern Mexico expose thick accumulations of Upper Jurassic conglomerate or breccia gradationally overlain by finer-grained Lower Cretaceous siliciclastic deposits (Bilodeau, 1982; Harding and Coney, 1985; Segerstrom, 1987; Riggs, 1987; Nourse, 1995, 2001; Dickinson et al., 1989; McKee et al., 1990, 1999; Dickinson and Lawton, 2001). A striking characteristic of some conglomerate bodies such as the Glance Conglomerate is the presence of very large clasts. Additional work, described below, shows that coarse clasts in Jurassic conglomerate throughout the region can be attributed to two different depositional settings: (1) they may be Middle Jurassic arc caldera fill deposits, or (2) they may represent the initial fill of steep-walled, Late Jurassic fault-controlled basins. Our work focuses upon the latter occurrences.

Although commonly overprinted by Laramide contraction and/or Miocene extension, the outcrop distributions and fault orientations are consistent with our hypothesis that conglomerate accumulated in basins formed within zones of dilation at releasing bends. Several basins that appear to have pull-apart origins are described below with the intent to provide information about age of formation, regional distribution, and stratigraphic and structural characteristics. Stratigraphic sections from some of these areas are depicted in Figure 5. We provide detailed support for the pull-apart basin model in three ways: (1) We outline the sedimentologic and structural settings of Upper Jurassic clastic and subordinate volcanic deposits that accumulated in basins near the type area of the Glance Conglomerate of southeastern Arizona. (2) We present salient characteristics of rocks and structures near other selected exposures of conglomerate that we correlate with Glance Conglomerate. Key localities are highlighted on Figures 1 and 2 and on Plate 1. (3) In our coverage of each basin's characteristics, we describe local faults that we postulate formed in response to transtension, many of which coincide with basin boundaries and some of which are contemporaneous with deposition of Late Jurassic sediments. We argue that the distribution and pattern of these faults, their association with depositional centers, and their timing are best explained in relation to major sinistral strike-slip displacement along the Mojave-Sonora megashear.

Stratigraphic and Structural Relations in Southern Arizona near the Type Locality of Glance Conglomerate

Stratigraphy, Sedimentology, and Age Control

The Glance Conglomerate is the basal unit of the Bisbee Group, first studied by Ransome (1904) in the Mule Mountains (Figs. 2 and 3). It conformably underlies finer-grained beds of the Morita Formation, Mural Limestone, and Cintura Formation, respectively (Hayes, 1970; Dickinson et al., 1989; Fig. 5). Together, these poorly fossiliferous units provide a practical means of correlating middle Mesozoic sections across southern Arizona, southern New Mexico, and northern Sonora. In southeastern Arizona alone, Glance Conglomerate overlain by finer-grained Lower Cretaceous strata has been recognized in more than 10 ranges and other uplifts (Bilodeau et al., 1987) within the broad region occupied by the "Bisbee Basin" (Dickinson et al., 1986, 1989; Dickinson and Lawton, 2001; Figures 1 and 5).

The only index fossils present in the type section near Bisbee are Aptian-Albian mollusks contained in the Mural Limestone. Throughout the region, Glance Conglomerate contains clasts derived from Proterozoic, Paleozoic, and (commonly) Middle Jurassic sources, and locally is deposited across faults that record

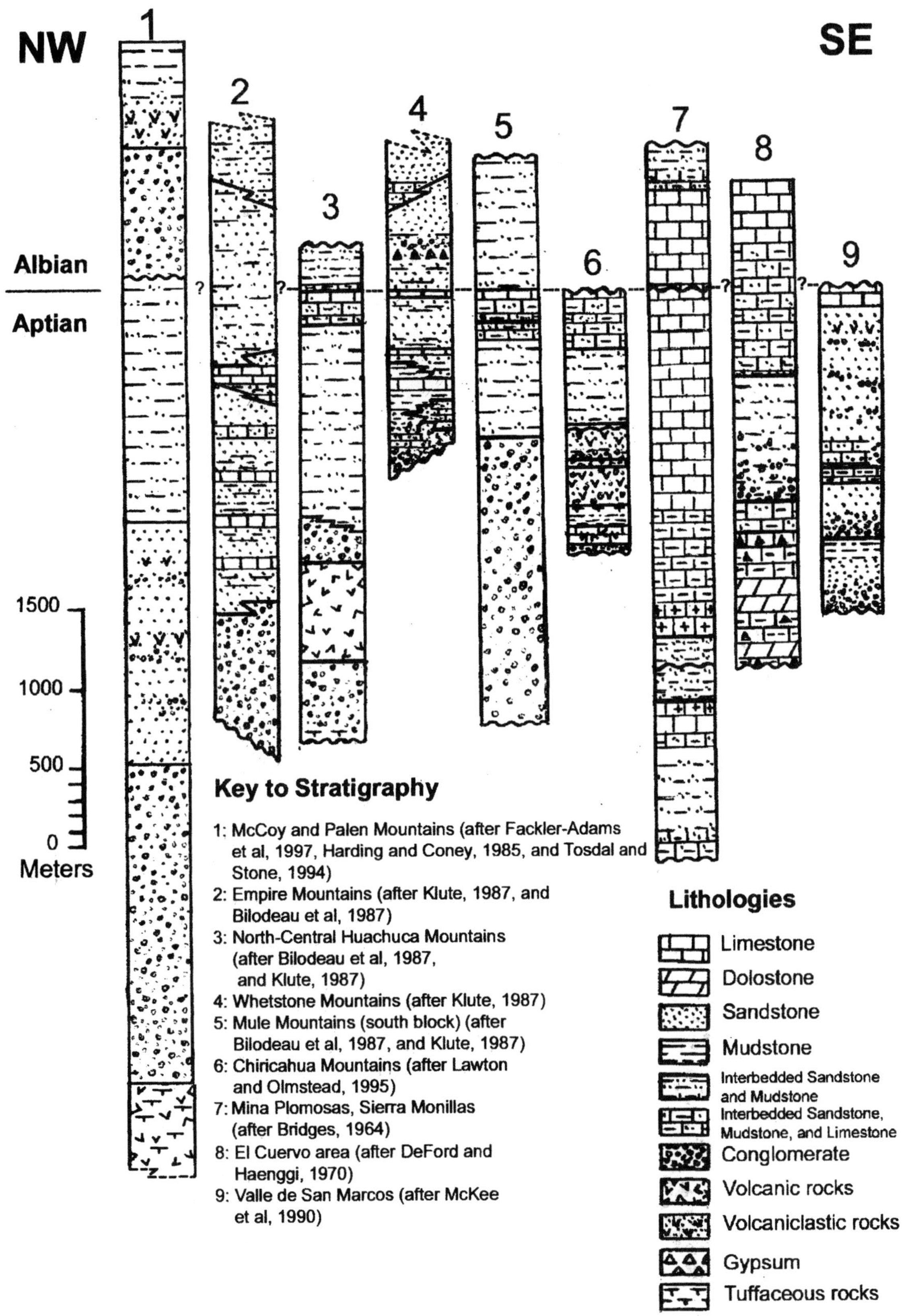

Figure 5. Generalized stratigraphic sections from selected Upper Jurassic–Lower Cretaceous pull-apart basins in Arizona, Chihuahua, Coahuila, and Sonora.

Late Jurassic movements. Thus the stratigraphic age of the type section is crudely constrained between Upper Jurassic and Lower Cretaceous.

Glance Conglomerate was first distinguished near Bisbee, Arizona, south of the steep east-striking Abrigo, Bisbee West, and Dividend normal faults that define the northern margin of a basin (Ransome, 1904; Fig. 3). At the type Glance mine locality, abundant cobbles and boulders of Paleoproterozoic Pinal schist or granite and Paleozoic carbonate reflect the presence of source rocks along the basin margins, which must have been steep. As early as 1904, Ransome recognized that some of these faults record pre-Cretaceous displacement as shown by abrupt thickening of conglomerate across them, from 30 m to more than 2000 m (see also Bilodeau, 1979; Bilodeau et al., 1987). Outcrops of Glance Conglomerate define an elongate body of sedimentary strata bounded by northwest-striking faults (e.g., Gold Hill, Glance faults; Figs. 2 and 3) extending southeastward from the normal faults into Sonora, Mexico. In Sonora, megaboulders and blocks of Paleozoic carbonate, hundreds of meters across, form a cluster that includes the rugged terrain surrounding La Negrita peak (Fig. 3; see also McKee et al., this volume). This jumble of carbonate blocks rests upon a surface, remarkable for its apparent planarity, that separates the megacarbonate blocks from underlying conglomerate and sedimentary breccia composed principally of boulders of Pinal Schist. The southern boundary of the Glance basin is not exposed. The overall pattern of boundary faults mimics that depicted in the pull-apart basin model of Figure 4.

Tens of kilometers west, at Canelo Hills (Kluth, 1982, 1983; Vedder, 1984) and Huachuca Mountains (Hayes and Raup, 1968; Bilodeau, 1979; Vedder, 1984; Fig. 5), volcanic units are interbedded with conglomerate strata residing at stratigraphic positions comparable to the type Glance Conglomerate. Isotopic analyses from volcanic interbeds that crop out in these uplifts provide radiometric age constraints, but also underscore an ongoing controversy about how to distinguish Glance Conglomerate from older volcaniclastic strata associated with eruptive products of the Middle Jurassic magmatic arc.

The existence of the volcanic layers and the presence large carbonate masses that occur as clasts, exotic in relation to the enclosing volcanic rocks in Canelo Hills and Patagonia Mountains (Simons et al., 1966; Davis, 1979; and Kluth, 1982, 1983), Huachuca Mountains (Drewes, 1981), and Pajarito Mountains (Drewes, 1980, 1981; Riggs and Haxel, 1990; Riggs and Busby-Spera, 1991), made recognition of the conglomerate that marks the base of the predominantly sedimentary Bisbee Group difficult. In 1985, Lipman and Sawyer (1985) proposed that some of the coarse breccia units, rich in carbonate debris are parts of Early or Middle Jurassic calderas. As mapping in Canelo Hills progressed (Kluth, 1982; Vedder, 1984), better understanding of stratigraphic relations among major volcanic and sedimentary units led to the recognition of a transition from mainly volcanic units to the overlying sedimentary sequence. Lipman and Hagstrum (1992) reiterated the idea that the sedimentary debris interbedded with volcanic units in Canelo Hills (Kluth et al., 1982; Kluth, 1983) is part of a caldera rock assemblage, based upon additional field work and reinterpretation of paleomagnetic data.

Radiometric studies reveal age differences between volcanic units that may be of ash-flow and caldera origin and stratigraphically higher tuff and andesite interbedded within thick sections of conglomerate. U-Pb isotopic analyses of zircon (e.g., Riggs et al., 1993; Anderson et al., this volume; Haxel et al., this volume) support previous work (Wright et al., 1981; Asmerom et al., 1990) indicating that the caldera rocks and associated conglomerate throughout southern Arizona accumulated during a vigorous burst of volcanic activity principally between 190 Ma and 170 Ma. In contrast, Kluth et al. (1982) argue that Glance Conglomerate is as young as 151 ± 2 Ma, based upon isotopic analyses of whole-rock Rb-Sr from samples of ash-flow tuffs within conglomerate in Canelo Hills. Although neither geologic nor geochemical conditions were optimal for that Late Jurassic age determination, remarkably similar results were obtained from whole-rock Rb-Sr isotopic analyses of volcanic layers from the Temporal Formation in the Santa Rita Mountains to the north (Asmerom et al., 1990). The stratigraphic position of the Temporal, unconformable above tilted Middle Jurassic volcanic units and disconformable below conglomerate mapped as Glance, suggests that this unit records an abrupt syndeformation transition from arc-related volcanism to clastic deposition (Drewes, 1971; Basset and Busby, this volume). Similarly, Marvin et al. (1978) reported concordant biotite K-Ar and interpreted Rb-Sr whole-rock ages of 147 ± 6 and 149 ± 11 Ma, respectively, for ash-flow tuff units interstratified with Glance Conglomerate south of Canelo Hills. These ages clearly contrast with older ages from the Middle Jurassic volcanic substrate, suggesting that the tectonic setting changed from arc magmatism to rifting and basin development during Late Jurassic time (Kluth et al., 1982: Bilodeau et al., 1987).

Boundary Faults

A very striking feature of southern Arizona geology is the spatial coincidence of Upper Jurassic–Lower Cretaceous basin strata with northwest- and east-striking faults (Fig. 2). Literature review (summarized below) reveals that many of these faults record significant movements coeval with sedimentation during Late Jurassic time. Late Cretaceous and/or Tertiary reactivation or deformational overprinting of these faults is ubiquitous. Nevertheless, the overall geometric pattern of faults sets enclosing outcrops of Glance Conglomerate in southern Arizona suggests a regional system of left-stepping northwest-striking sinistral wrench faults linked by east-trending normal faults. This pattern may be extrapolated over a much broader region to the south and southeast (Fig. 1; Plate 1) and supports our thesis that Upper Jurassic "Glance-type" conglomerate accumulated in discrete pull-apart basins in locations that correspond to releasing bends of the Late Jurassic Mojave-Sonora fault system. The salient age constraints and complexities of this fault system in southern Arizona are summarized below.

Northwest-striking discontinuities and faults. In southern Arizona, Titley (1976, p. 74) recognized six northwest-trending regional "zones marking discontinuities" that are distinguished principally by "patterns of rock distribution," in particular, distribution of Paleozoic, Triassic, and Jurassic rocks (Fig. 6; Figs. 6 and 9 of Titley, 1976). Five of these six discontinuities (from southwest to northeast: Comobabi-Nogales, Sawmill Canyon, Silver Bell–Bisbee, Dragoon, and Dos Cabezas) incorporate segments of major mapped faults along which outcrops of Glance Conglomerate and younger Bisbee Group strata end abruptly against older units.

Structural relationships along the faults are complex. Typical examples include the Sawmill Canyon fault (Drewes, 1971; Bassett and Busby, this volume; Figs. 2 and 7) and the Dragoon fault (Keith and Barrett, 1976; Fig. 3). Along the Sawmill Canyon fault, vertical and lateral displacements record diverse movement histories, the earliest of which coincided with deposition of Jurassic rocks (e.g., Drewes, 1971; Basset and Busby, this volume). Where Glance Conglomerate and younger Bisbee Group strata accumulated against steep lateral and normal faults, lithologic discontinuities may mark the presence of buttress unconformities (Figs. 3 and 5 of Bilodeau et al., 1987; Keith and Barrett, 1976). During extensive mapping, Drewes (1981) and colleagues (see references in Drewes, 1981) recognized additional northwesterly trending faults as well as east-trending fault linkages among them. The pattern of rhomb-shaped intersections that emerged from this work (Drewes, 1981) strongly influenced our thinking, as has mapping by McKee et al. (this volume).

Easterly striking faults. Ransome (1904) recognized that easterly trending normal faults such as Abrigo and Dividend (Fig. 3) were fundamental structures that controlled where Glance Conglomerate accumulated. Bilodeau et al. (1987) mapped variations in Glance lithofacies in this area and emphasized the pronounced southward thickening that occurs across the faults. A similar relationship is reported for Glance Conglomerate to the northwest in the Empire Mountains (Fig. 2), where thickness changes occur across an unnamed easterly trending fault zone "that effectively separates two stratigraphically similar but structurally different terranes" (Bilodeau et al., 1987, p. 236). Despite uplift and probable basin inversion during Cretaceous contraction deformation, Glance lithofacies record progressive unroofing of the source area exposed north of the fault zone. The basal units of the conglomerate contain boulders and blocks that may be up to 300 m long.

In the Helmet Peak area of the Sierrita Mountains, Cooper (1971, 1973; Figs. 2 and 7) shows the east-trending "No. 6 thrust" as a principal fault among several north-side-up structures that separate Mesozoic formations from Paleozoic strata to the north and northeast. Although the lowest Mesozoic unit, Rodolfo Formation, south of the faults is considered to be Triassic by Cooper, its age is not well constrained. Sedimentary units,

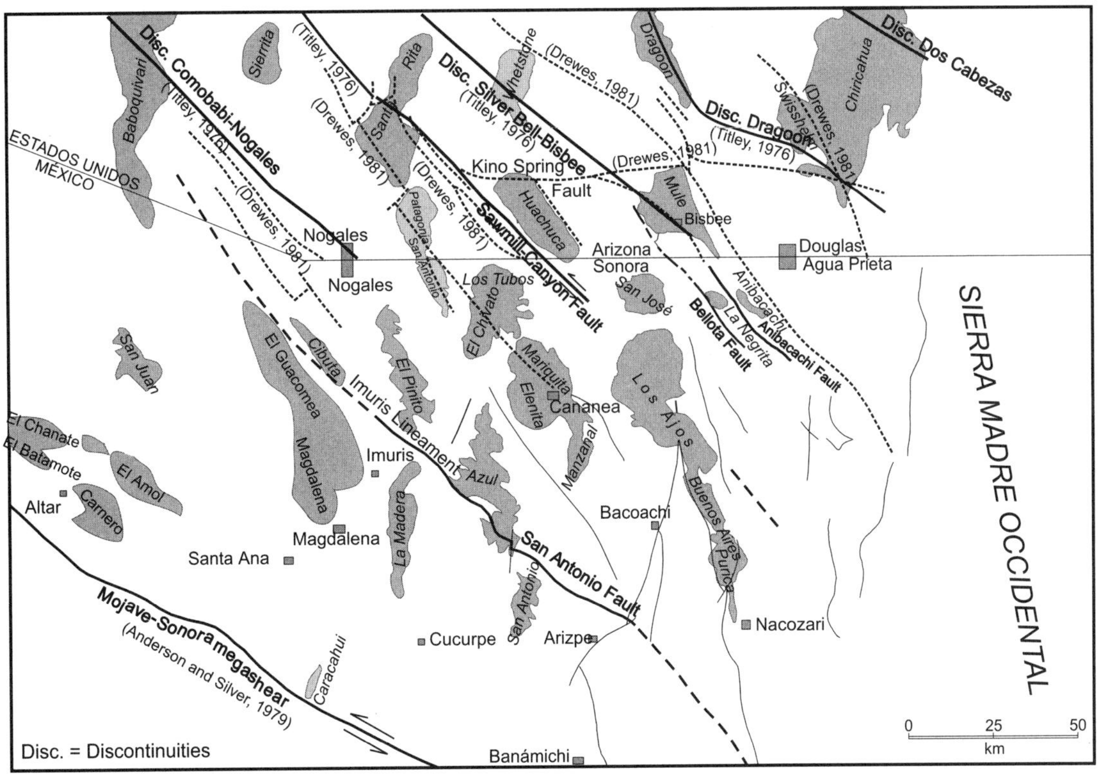

Figure 6. Map of major faults and discontinuities in southern Arizona and northern Sonora, Mexico. Outlines of major ranges are shown in gray. Adapted from Titley (1976), Drewes (1981) and Rodríguez-Castañeda (2000).

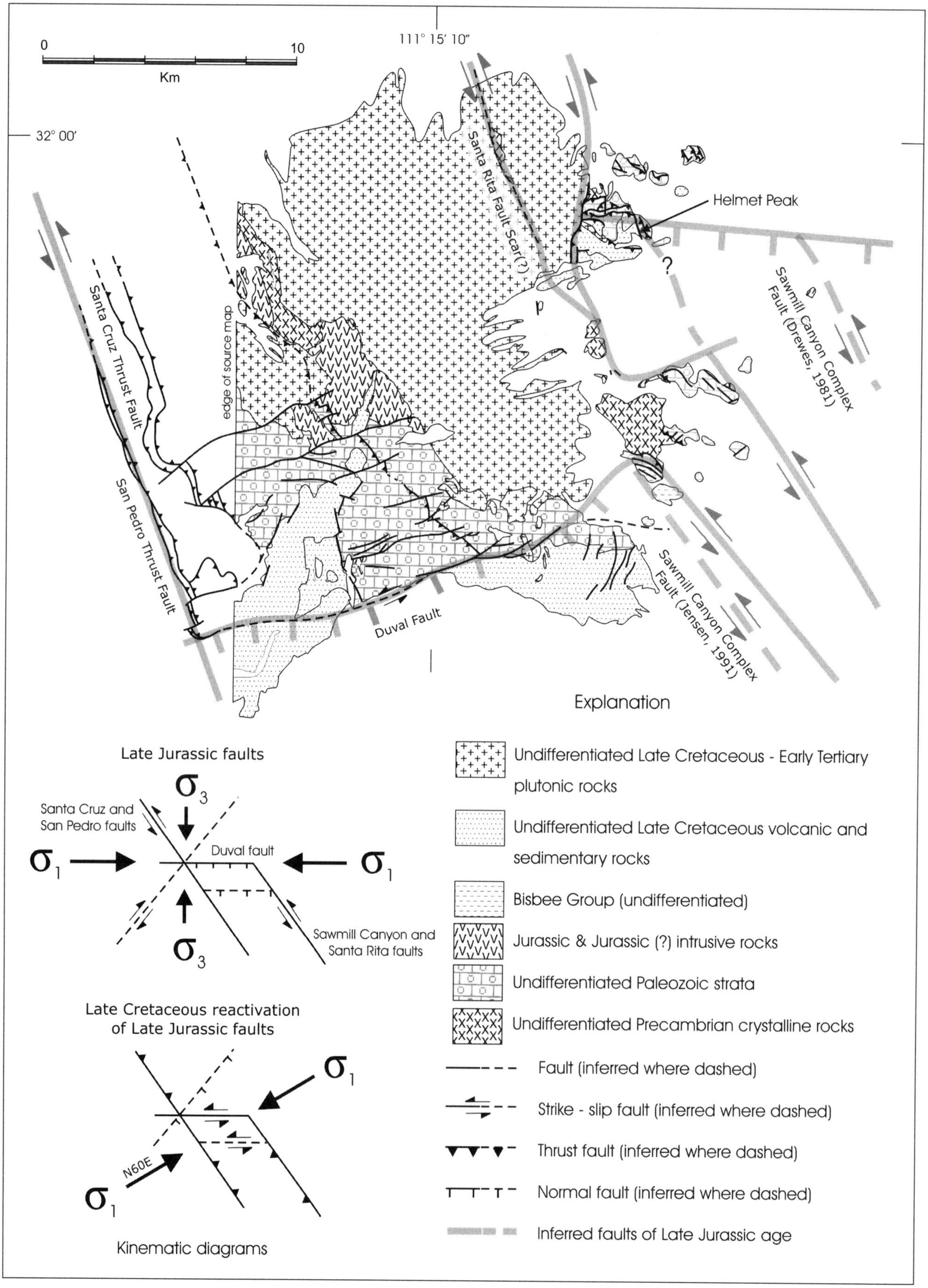

Figure 7. General geologic map of Sierrita Mountains, showing principal Late Jurassic faults reactivated during Cretaceous contraction (after Cooper [1971, 1973] and Drewes [1981]). Kinematic diagram (inset) shows show inferred transtensional stress regime for Late Jurassic time and compressional stress regime during Late Cretaceous time.

consisting of at least 418 m of conglomerate, siltstone, and some sandstone, are overlain by and interfinger with almost 335 m of andesite breccia and flows. The gross features of this unit are comparable to Glance Conglomerate in the Huachuca Mountains (Hayes and Raup, 1968). Of particular note in this region is the interpretation by Drewes (1980) that in the main mountain mass of the Sierritas, lying southwest of the Helmet Peak area (Fig. 7), a second easterly trending fault called Duval bends abruptly northwestward along the parallel, inferred Santa Cruz and San Pedro thrusts. Despite the complex structural relationships in the Sierrita Mountains, the fault geometries and stratigraphic relationships along them suggest to us that the No. 6 thrust and Duval fault, and the San Pedro and Santa Cruz thrusts are reactivated Late Jurassic normal and left-lateral faults, respectively.

To the northeast of Tucson in the Santa Catalina Mountains (Fig. 2), detailed mapping by Janecke (1987) confirms earlier hypotheses (Moore et al., 1941; Pierce, 1958; Suemnicht, 1977) that the east-striking Geesaman fault had an early history as a pre-Cretaceous normal fault. Although strong Cretaceous and Tertiary ductile deformation overprint this fault, Janecke's work reveals significant normal separation along its length. Key evidence is the presence of basal conglomerate of the Bisbee Group, restricted to the hanging wall that contains clasts derived mainly from Upper Paleozoic strata.

In the Patagonia Mountains (Fig. 2), Bussard (1996) mapped megabreccia containing blocks hundreds of meters long, and breccia along the northwesterly striking Harshaw Creek fault. Conglomerate and megabreccia generally occur high in the stratigraphic section where they either rest upon or are interstratified with Middle(?) Jurassic rhyolite lava or tuff. Sills of andesite, meters to tens of meters thick, obscure the upward transition from the silicic volcanic rocks to conglomerate and breccia. Although the andesite bodies are undated, their sedimentologic setting and stratigraphic position suggest correlation with comparable Late Jurassic igneous units of the Sierrita and Huachuca Mountains. Principal exposures of the coarse clastic rocks occur along east-northeasterly trending splays of the Harshaw Creek fault, where they are juxtaposed against blocks of structurally upthrown Paleozoic limestone (Bussard, 1996, his Plate 1). According to Simons (1974), the Harshaw Creek fault records ~4 km of Laramide sinistral separation of Upper Paleozoic carbonate strata. However, Bussard (1996) argues that 2 km of this slip predated deposition of the presumed Upper Jurassic sections.

Drewes (1981) recognized a left-step along the Sawmill Canyon fault where the Babocomari and Kino Spring faults (Fig. 2) splay eastward. The Babocomari fault bounds the south margin of the Mustang Mountains, whereas Kino Spring defines the northern edge of the Huachuca Mountains. We interpret the intervening graben-like structure in which poorly exposed Cretaceous rocks are preserved to be a small pull-apart basin.

Other east-trending faults along which Bisbee Group strata are juxtaposed against older units are recognized in central Cochise County, southeastern Arizona (plates 5 and 6 of Gilluly, 1956). Of these faults the Prompter fault (Fig. 3), which forms the northern boundary of the Ajax Hill horst south of Tombstone, is the best known. Although Gilluly ends the horst a few kilometers south at the Horquilla Peak fault, outcrops of Paleozoic strata distinguish a structural high as far south as the northern margin of the Mule Mountains, where Bisbee Group strata are preserved in a down-dropped block also bounded by an east-striking normal(?) fault reactivated during Cretaceous contraction with left-lateral displacement (Force, 1996).

East of the Tombstone area, in the Dragoon Mountains (Figs. 2 and 3), the principal structure is the complex Dragoon thrust (Gilluly, 1956), which strikes northwest parallel to the axis of the range. Outcrops of Bisbee Group strata delineate a crudely rhomb-shaped area. In the southern Dragoon Mountains, a conspicuous left-step along the Dragoon fault, marked by an easterly trending fault zone along which Bisbee beds to the north are separated from older granite to the south, is interpreted to be a releasing bend.

Northeasterly striking faults. Late Jurassic north- or northeast-trending, right-lateral strike-slip faults are expected complements to a northwesterly striking fault system dominated by sinistral shear (see inset on Fig. 1). Cooper and Silver (1964) mapped numerous faults of this orientation in the Little Dragoon Mountains and Gunnison Hills (Dragoon Quadrangle; their plates 1 and 6) that show dextral displacement of Paleozoic strata. In one place, these faults cut Jurassic volcanic rocks and form escarpments across which Glance conglomerate accumulated (p. 73 of Cooper and Silver, 1964). In Canelo Hills, Kluth (1983) considered steep northwest- and northeast-trending faults to have accommodated vertical offsets during Late Jurassic time.

Other Upper Jurassic Basins with Proposed Pull-Apart Origins

We now turn to numerous other localities of southwestern Arizona, northern Sonora, southern New Mexico, Chihuahua, and Coahuila that preserve thick accumulations of Upper Jurassic conglomerate overlain by Lower Cretaceous strata (Fig. 1; Plate 1). We propose stratigraphic correlation of these coarse clastic sections to previously described areas of southeastern Arizona where the Glance Conglomerate has been documented, and further postulate that these sediments originally accumulated in pull-apart basins. Most of the basins exhibit relationships to adjacent basement and to northwesterly and easterly boundary structures that mimic those described for fault-bounded conglomerate bodies of the Bisbee Basin. Although syntectonic relationships between Upper Jurassic coarse clastic fill and basin-bounding faults are not fully documented, we are intrigued by the repetitive map pattern of northwesterly and easterly faults that encompass exposures of upward-fining conglomerate or breccia. Below we describe the stratigraphy of these basins, some of which preserve their rhomb-like geometry, and summarize what is currently known about their boundary structures. Observations are described in general from west to east and are keyed to geographic and geologic features highlighted on Figures 1, 2, and 7, and Plate 1.

McCoy Basin, Southeastern California and Southwestern Arizona

Sections of Upper Mesozoic sedimentary strata, 7–8 km thick that crop out in ranges from west-central Arizona to southwestern California distinguish the west-northwesterly trending McCoy Basin of Harding and Coney (1985). Rocks within the basin record the interaction of deformation and sedimentation that occurred subsequent to Middle Jurassic volcanism (Richard et al., 1994, and references therein; Fackler-Adams et al., 1997). The McCoy Mountains Formation (Miller, 1944; Harding, 1983; Harding and Coney, 1985), exposed in the ranges of the western McCoy Basin (Fig. 1), provides the best record of Upper Jurassic and Cretaceous deposition. Clastic strata that comprise the formation include siltstone, mudstone, sandstone, conglomerate, and sandy limestone (Tosdal and Stone, 1994; Fig. 5). In the southern Plomosa Mountains of western Arizona (Plate 1), the Apache Wash facies of the lower McCoy Mountains Formation contains units of megabreccia comprised mainly of blocks of pre-Pennsylvania sedimentary units, which may have been emplaced penecontemporaneously as a mass of semicoherent material by gravitational processes and as rock avalanches (Richard et al., 1993).

The age of the lowest units is Late Jurassic as constrained by U-Pb zircon ages (Fackler-Adams et al., 1997; Barth et al., 2004). In the Palen Mountains, zircon from lapilli tuff within the Dome Rock volcanic suite beneath the McCoy Mountains Formation yielded a discordant conventional zircon age of 175 ± 8 Ma (Fackler-Adams et al., 1997). Stratigraphically higher volcanic units that locally interfinger with siltstone of the lowermost McCoy Mountains Formation yielded younger ages of 155 ± 8 and 162 ± 3 Ma. For comparison, recent sensitive high-resolution ion microprobe (SHRIMP) analyses of single zircons from dacitic tuff in the upper Dome Rock sequence of the Palen Mountains indicated a crystallization age of 165 ± 2 Ma (Barth et al., 2004). Based upon their results, Fackler-Adams et al. (1997) concluded that: (1) the formation of the McCoy and Bisbee Basins was synchronous and therefore Glance Conglomerate and the lower McCoy Mountains Formation are correlative, and (2) clastic Upper Jurassic strata record an abrupt waning of volcanism.

SHRIMP analyses of detrital zircons from sandstones in the type section of McCoy Mountains Formation (within the McCoy Mountains) provide additional constraints (Barth et al., 2004), including: (1) the basal sandstone member (625 m thick) is rich in Neoproterozoic-Paleozoic(?) carbonate grains and contains detrital zircons with Proterozoic, Triassic, and Jurassic ages (the youngest of which is 179 Ma), (2) sandstones from the overlying 6.9 km of section contain a detrital zircon assemblage where the youngest component decreases systematically up-section from 116 Ma to 84 Ma, and (3) Late Jurassic detrital zircons (ca. 165 Ma to ca. 145 Ma) are present in all parts of the section except the basal member. From this data set, Barth et al. (2004) conclude that most of the McCoy Mountains Formation accumulated during a protracted period of Early and Middle Cretaceous subsidence. However, the depositional age of the basal member is poorly constrained. Available geochronology indicates that the lower member in the McCoy Mountains Formation must be younger than the age of its youngest detritus (179 Ma), and the lower member in the Palen Mountains is younger than its 165 Ma volcanic substrate. Contrary to the statement of Barth et al. (2004, p. 150) we argue that the data do not preclude a Late Jurassic transtensional origin for the McCoy Basin. Taking into account the above-described field relationships and geochronology, we contend that the basal member records initial development of the McCoy Basin during Late Jurassic at a time when silicic volcanism was still active.

Geochemical data from igneous rocks in the Granite Wash Mountains at the eastern end of the McCoy Basin support the interpretation of a transtensional tectonic setting. Laubach et al. (1987) note the alkalic character of silicic volcanic rocks that occur at the top of the Jurassic volcanic pile where the transition to predominantly sedimentary strata begins. Stratigraphically higher mafic sills and flows that are widespread within overlying clastic strata are also slightly alkalic. Emplacement of the mafic rocks is interpreted to be nearly coeval with sedimentation, as indicated by sedimentary structures at the base of a major sill that indicate intrusion penecontemporaneous with sedimentation (Laubach et al., 1987). Additional geochemical analyses of mafic dikes and sills from the lower McCoy Mountains Formation reveal high-Al basaltic to andesitic compositions indicative of derivation from a mantle source followed by interaction with continental crust, and consistent with emplacement in an extensional setting (Gleason et al., 1999).

The best-preserved contact between McCoy Mountains Formation and underlying volcanic rocks crops out in the Palen Mountains where it trends easterly (Harding and Coney, 1985; Fackler-Adams et al., 1997). The eastern end of the McCoy Basin (Harding and Coney, 1985) is characterized by exposures of limestone-boulder conglomerate in the western Limestone Hills, southern Little Harquahala Mountains, and New Water Mountains. These coarse clastic rocks comprise the lower Apache Wash Formation (= lower McCoy Mountains Formation = Glance Conglomerate), and rest upon Middle Jurassic volcanic units. In places the Apache Wash Formation is separated from Upper Paleozoic beds by faults that have been modified by younger deformation. The complex structural relationships exposed in the ranges at the east end of the McCoy Basin have been carefully studied (Richard et al., 1987; Laubach et al., 1987; Sherrod and Koch, 1987; Reynolds et al., 1987). Results of this mapping indicate that Apache Wash Formation accumulated upon the hanging walls of normal faults as shown in Figure 5 of Richard et al. (1987).

Palinspastic reconstruction of the region for early Tertiary time (Richard et al., 1994) retains the regional easterly trend of the basin. We assume that some contacts between the McCoy Mountains Formation (or equivalent strata) and older rocks originally were buttress unconformities against steep east-striking normal faults. Following the interpretation of Harding and Coney (1985), we infer that the east and west ends of the McCoy Basin were initially bounded by northwest-striking left-lateral faults.

The westernmost of these corresponds to the trace of Mojave-Sonora megashear (Fig. 1; Plate 1).

The Winterhaven Formation (Haxel et al., 1985), a marine sedimentary unit that rests depositionally upon Jurassic rhyodacitic metavolcanic rocks, may be stratigraphically correlative to the McCoy Mountains Formation. Distinctive trachytic volcanic rocks within the lower part of the formation are probably Late Jurassic, based upon correlation with compositionally similar units in other areas (Haxel, 1995, personal commun.). Unlike the Apache Wash facies of McCoy Mountains Formation, however, the Winterhaven Formation does not contain coarse angular fragments. Exposures of this unit are known only from within an elongate, east-trending belt of exposures ~35 × 15 km that lies astride the Colorado River, north of Yuma, Arizona. The Winterhaven has been strongly affected by Cretaceous and Tertiary faulting.

Artesa and Comobabi Basins, South-Central Arizona

Exposures of volcanic, sedimentary, and plutonic units in the Artesa, Quijotoa, and Comobabi Mountains, and Ko Vaya Hills of south-central Arizona, and several hills in northernmost Sonora (Fig. 2) are typical representatives of Late Jurassic layered (Artesa) and plutonic (Ko Vaya) rock suites (Tosdal et al., 1989; Haxel et al., this volume). The margins of two basins are poorly exposed, but crudely rhomb-shaped geometries may be inferred from the distribution of Paleozoic strata and Proterozoic gneiss to the north and northwest, and exposures of Middle Jurassic volcanic rocks to the northeast and southwest (Fig. 2). Basin-bounding faults are probably obscured by Cretaceous and/or Tertiary reactivation. One example is the fault along the west margin of the Baboquivari Mountains that is interpreted by Haxel et al. (1984) to be a shallowly dipping Cretaceous thrust. We speculate that this fault may have initially formed as a steep, Late Jurassic strike-slip fault.

The Artesa Mountains section in includes flows, flow breccia, and volcanic conglomerate as well as argillite, sandstone, pebbly sandstone, and conglomerate. Clasts from cratonal Paleozoic rocks and contemporaneous or older volcanic strata (ca. 170 Ma; see Haxel et al., this volume) are locally conspicuous. In the Comobabi Mountains, conglomerate units within Artesa strata occur just below finer-grained sediments correlated with the Bisbee Group (Dickinson et al., 1989).

Plutonic rocks of the Ko Vaya supergroup in southern Arizona and northern Sonora are characterized by similar textural and compositional variations including those related to grain size, quartz content, color index and/or mafic content, texture, and age. They (1) yield interpreted U-Pb ages between 160 and 145 Ma (Tosdal et al., 1989; Anderson et al., this volume; Haxel et al., this volume), (2) commonly have alkaline tendencies, (3) show strong alteration, (4) may contain miarolitic cavities, and (5) are associated with probable hypabyssal porphyry (Tosdal et al., 1989). Ko Vaya rock representatives include fine- to coarse-grained leucocratic monzogranite and syenogranite, subordinate quartz monzonite, and local granodiorite. They form conspicuous pink to maroon exposures that typically weather red, pink, yellow, or orange. An example is the perthitic granite (147–145 Ma; Tosdal et al., 1989; Anderson et al., this volume) that forms the prominent Baboquivari Peak, as well as spatially associated porphyry. These granitic rocks may show moderate to strong deuteric and (or) hydrothermal alteration. Ko Vaya plutons also include dioritic rocks, such as hornblende-rich quartz monzodiorite and subordinate quartz monzonite and quartz diorite, as well as rare hornblendite. We interpret the age, areal distribution, and shallow setting of these intrusive bodies to record synextensional emplacement into crust broken and thinned at releasing steps. In other words, they are rooted in transtensional basins, most likely along steep normal faults.

Reconnaissance work in Mexico among the Sierras San Manuel, del Cobre, and La Lesna (Anderson, Haxel, and Tosdal, unpublished mapping) reveals volcanic and sedimentary rocks considered by Tosdal et al. (1989) to be part of the Artesa sequence. In Sonora, they note that laminated or well-bedded volcaniclastic sandstone is abundant relative to other lithologies. Tosdal et al. (1989) also recognized rocks similar to those that comprise the Artesa and Ko Vaya units in several Arizona localities outside the Comobabi-Quijotoa area. Limited stratigraphic and geochemical data permits speculation that these isolated volcanic and sedimentary rocks of the Artesa sequence accumulated in basins formed as pull-aparts within the Middle Jurassic magmatic belt.

Sierra El Batamote, Sonora

North of Altar and Caborca, Sonora, Mexico, outcrops of Upper Jurassic conglomerate distinguish an elongate northwest-trending belt between exposures of Jurassic volcanic rocks to the northeast and Paleozoic strata to the southwest (Figs. 1 and 2; Plate 1). Exposures of polymict conglomerate and breccia, locally thicker than 1 km, demarcate fragments of a 60-km-long, Late Jurassic basin bounded on the southwest by the trace of the Mojave-Sonora megashear (Nourse, 2001). Stratigraphic and structural relationships are best documented at Sierra El Batamote (Nourse, 1995, 2001), where the conglomerate and associated strata are folded about northwest-trending hinges parallel to the length of the range. Distinct clasts include Neoproterozoic-Cambrian quartzite and carbonate, Middle Jurassic rhyolite and sandstone, and several varieties of andesite and basalt of presumed Middle or Late Jurassic age. The conglomerate interfingers with basaltic andesite flows and monomict andesite flow breccia or agglomerate, which locally rest upon rhyolite ignimbrite. Subangular boulder-cobble conglomerate fines upward and grades laterally into volcaniclastic sandstone and mudstone interstratified with lacustrine sediments and silicic tuff. Similar conglomerate underlies much of Sierra del Alamo, the range directly northwest of Sierra El Batamote. Fine-grained Lower Cretaceous Bisbee Group beds, recognized in the area by distinctive red and purple colors (Jacques-Ayala and Potter, 1987; Jacques-Ayala, 1995), conformably overlie all of the aforementioned units.

The original geometry of the Batamote Basin and the kinematics of northwest-trending boundary faults are poorly preserved due to the intense overprint of northeast-vergent

Laramide contraction, and local reactivation during southwest-directed mid-Tertiary detachment faulting (Nourse, 2001). A north-striking boundary structure defines the northwest end of the basin at Sierra La Gloria, where exposures of Upper Jurassic conglomerate are juxtaposed against Middle Jurassic quartz-feldspar porphyry. This fault has the appropriate orientation to be a right-lateral conjugate of the Mojave-Sonora megashear (Fig. 2). Alternatively, it may represent a normal fault that was rotated counterclockwise from an original easterly strike as sinistral shear affected the region northeast of the Mojave-Sonora megashear. We speculate that the Mojave-Sonora megashear acted as the master fault that controlled deposition of the Upper Jurassic conglomerate in this region. Other basins of similar age that occur adjacent to the megashear trace to the southeast were originally bounded by secondary northwesterly, northeasterly, and easterly faults, such as those that outline outcrop belts of conglomerate in the San Antonio Basin.

San Antonio Basin, Imuris, Sonora

In north-central Sonora near Imuris, distinct polymict conglomerate occupies a stratigraphic position between Middle Jurassic arc rocks and siliciclastic and carbonate strata of the Bisbee Group (Nourse, 1995; Fig. 2). Along the south-facing flank of Sierra El Pinito, an east-northeast–trending conglomerate belt overlies a sequence of interstratified 174 Ma rhyolite (El Tunel quartz porphyry of Anderson et al., this volume), quartz arenite, and rhyolite-quartz arenite-quartzite cobble conglomerate. Clasts in the younger polymict conglomerate match locally exposed Jurassic rocks and include Paleozoic(?) quartzite cobbles likely derived from sources near Cananea to the northeast (Nourse, 1995). A few kilometers farther east at Sierra Azul, the conglomerate belt makes an abrupt strike change to the southeast. In this area, cobble-pebble conglomerate overlies a small window of Jurassic arc rocks (Fig. 2 of Nourse, 1995) and fines upward to the southwest into a thick section composed of strata equivalent to the Morita, Mural, and Cintura Formations of the Bisbee Group (McKee and Anderson, 1998). Correlative but structurally dismembered Upper Jurassic–Lower Cretaceous rocks are mapped as far southeast as Sierra San Antonio (Rodríguez-Castañeda 1997, 2000; Fig. 2). Ten kilometers west of Imuris, the Upper Jurassic conglomerate ends at a wide northwest-trending valley. Reconnaissance work much farther northwest reveals a separate east-trending belt of conglomerate in a comparable stratigraphic position, i.e., sandwiched between Jurassic arc rocks to the north and Bisbee Group strata to the south.

As described in Nourse (1990, 1995), part of the Jurassic-Cretaceous section near Imuris has been metamorphosed to greenschist facies, and the inferred boundary structures reactivated during middle Tertiary development of the Magdalena-Madera core complex. Nevertheless, the paleogeographic map patterns that emerge upon palinspastic reconstruction are provocative (see Fig. 8 in Nourse, 1995). Great thicknesses of Glance-type conglomerate and overlying Bisbee Group imply the existence of an important Upper Jurassic–Lower Cretaceous basin (or basins) in north-central Sonora. The regional distribution of sediment facies, the coincidence of conglomerate with the northeast and northern basin margins, and linkage of clasts to adjacent Middle Jurassic sources led Nourse (1995) to speculate that sedimentation was controlled by Late Jurassic faults. Near Imuris, the belts of Upper Jurassic conglomerate appear to define the edges of a rhomb-shaped pull-apart, informally designated as the San Antonio basin. A separate basin floored by Upper Jurassic conglomerate may be situated between the Artesa and San Antonio basins (Fig. 2).

Chiricahua Mountains, Arizona

Glance Conglomerate is exposed in the Chiricahua Mountains northeast of Bisbee (Fig. 2) where it forms a relatively thin (25 m) unit basal to 900 m of fossiliferous Upper Jurassic strata (Lawton and Olmstead, 1995; Fig. 5). This section consists of sabkha-type limestone and prodeltaic mudstone or siltstone interbedded with subaqueous basaltic volcaniclastic breccia, basalt pillow lavas, and silicic tuffs, overlain by fluvial arkose, siltstone, and subaerial mafic lava flows (Lawton and Olmstead, 1995). These strata underlie red beds of the Morita Formation, which in turn are overlain by Mural Limestone. Diverse fossil assemblages demonstrate that the clastic and bimodal volcanic strata above the Glance and beneath the Morita Formation accumulated between middle Oxfordian and early Aptian time (Lawton and Olmstead, 1995). The Glance Conglomerate rests unconformably on Permian Concha Limestone, and contains coarse clasts of locally derived carbonate and chert.

Lawton and Olmstead argue that the Glance and overlying strata accumulated in a Late Jurassic–Early Cretaceous fault-controlled rift basin because: (1) the Upper Jurassic strata disappear abruptly north of the Apache Pass fault (Fig. 2), (2) arkose beds suggest erosion from granitic Precambrian basement sources in the north, (3) lacustrine sediments and bimodal volcanic strata support a continental rift setting, (4) the stratigraphy reveals cycles of high-energy fluvial deposition succeeded by rapid subsidence and marine transgression, and (5) there appears to be a regional connection with the Chihuahua Trough, an arm of the actively rifting Gulf of Mexico. The Apache Pass fault is interpreted (correctly, we believe) to be a Late Jurassic normal fault where it bends eastward in the northern Chiricahua Mountains. Principal exposures of Upper Jurassic strata, hundreds of meters thick, crop out south of the series of easterly jogs (left steps) in the Apache Pass fault, recorded by the Wood Mountain fault, Apache Pass fault and subparallel splays (see Figure 1 of Lawton and Olmstead, 1995, p. 36). These authors interpret the abrupt disappearance northward of the section containing mafic flows within a section of siltstone and mudstone containing middle Oxfordian ammonites as indicating "the structural development of a rift basin with a dramatic, fault-bounded northern boundary." We find the rift model of Lawton and Olmstead (1995) intriguing, particularly in light of local fault geometries. The implied syndepositional fault pattern mimics the dog-leg geometries associated with many other pull-apart basins described in this paper.

The Chiricahua Mountains extend southwestward to a point where two smaller ranges, Pedregosa and Swisshelm Mountains, are recognized (Fig. 2). The principal structures exposed in these ranges are a set of westerly striking faults that curve to the northwest at a distinct bend convex toward the southwest (Drewes, 1980; Fig. 20 of Drewes, 1991). The northwesterly faults parallel the narrow, elongate ridge that forms the main topographic element of the Swisshelm Mountains. Here the faults are thrusts along which Precambrian granite and early Paleozoic cover are imbricated and displaced southwestward onto younger Paleozoic strata and overlying undifferentiated Bisbee Group.

To the southeast, near Leslie Canyon, the faults curve eastward (Fig. 1 of Drewes and Thorman, 1978; Fig. 2). Along this segment the faults generally display straight traces indicating steep dips, strike west-northwest, and record down-to-the-south normal and/or left-lateral strike-slip displacements. Bisbee Group and overlying Late Cretaceous volcanic rocks in the Late Jurassic(?) hanging wall are juxtaposed against Paleozoic strata in the footwall across the fault. We propose that these curving faults reflect an original left-step geometry and that thrusting and lateral faulting accommodated N60°E-directed Late Cretaceous contraction, imposed upon a Late Jurassic strike-slip fault that steps eastward to form a releasing bend between the ranges.

Little Hatchet Mountains and Surrounding Ranges, New Mexico

The thickest and most complete section of Late Jurassic and Early Cretaceous strata known in southwestern New Mexico crops out in the Little Hatchet Mountains (Fig. 8) west of the Burro Uplift (Fig. 1). Lucas and Lawton (2000) summarized the stratigraphy of this area, incorporating the thesis work of Harrigan (Harrigan, 1995; Lawton and Harrigan, 1998) with that of earlier workers (Darton, 1922; Lasky, 1947). Darton (1922, 1928) first recognized the Lower Cretaceous strata and compared limestone in the section to the Mural Limestone at Bisbee, Arizona. Lasky (1938) subdivided the thick section of Early Cretaceous strata and introduced the name "Broken Jug" for the lowest unit. Lawton and Harrigan (1998) redefined the Broken Jug Formation and further subdivided it into five informal members, including dolostone, lower conglomerate, fine-grained clastic, upper conglomerate, and basalt. Preliminary studies of fossils collected from Broken Jug "suggest a Late Jurassic age" as indicated by the presence of coral comparable to species in the Oxfordian Smackover Formation (Lucas and Lawton, 2000, p. 189).

In southern New Mexico east-striking faults, exposed as transverse structures in north-trending mountain uplifts, are conspicuous and important geologic features. The best constrained in terms of age and initial offset is the Copper Dick fault in the Little Hatchet Mountains (Fig. 8). As indicated by Lucas and Lawton (2000), movement on the Copper Dick fault was initiated in Late Jurassic time as a down-to-the-south normal fault that restricted the distribution of the contemporaneous Broken Jug Formation. Subsequently, during Laramide contraction the fault was reactivated and accommodated left-lateral (Hodgson, 2000) and dip-slip (Lawton, 2000) displacement.

In light of the similarity of age and tectonic setting of the Broken Jug Formation to other units beneath the Early Cretaceous Formations of the Bisbee Group, we speculate that the Copper Dick fault marks a releasing step originally linked to Late Jurassic northwest-striking sinistral faults. Similar east-striking faults spatially associated with the Bisbee Group crop out in ranges surrounding the Little Hatchet Mountains (Plate 1). Among these are: (1) the Wood Canyon, Goatcamp and Johnny Bull faults in the Peloncillo Mountains (Bayona and Lawton, 2000, and references therein); (2) the South Florida Mountains fault in the Florida Mountains (Clemons, 1998; Amato, 2000, and references therein); (3) the Victorio Mountains and Main Ridge faults that bound a narrow graben-like structure preserving Bisbee Group strata in the Victorio Mountains (McLemore et al., 2000; Kottlowski, 1963; Thorman and Drewes, 1980); (4) faults bounding Atwood Hill in the northern Pyramid Mountains (Thorman and Drewes, 1978; Lasky, 1938); (5) east-striking steep faults in the northern Animas Mountains (Drewes, 1986); and (6) probable faults that control the east-trending folds recorded by Bisbee Group strata in the Brockman Hills (Thorman, 1977; Drewes, 1991).

In southern New Mexico regional lineaments with northwest strike are defined by outcrops of Bisbee Group bounded by thrusts containing Paleozoic strata in the hanging walls. These thrusts are mapped in the Sierra Rica, West Lime Hills of the Tres Hermanas Mountains, and the East Potrillo Mountains (Drewes, 1991). In this region Drewes (1991, his Figs. 13 and 16) shows the pattern of linked northwesterly and easterly striking faults that have been reactivated during Late Cretaceous contraction. Residual gravity anomaly maps of this region (DeAngelo and Keller, 1988) reveal a northwest-trending grain that corresponds to the Burro Uplift (Fig. 1) and certain structures within the fold-and-thrust domains of Drewes (1991). We suggest that this grain primarily reflects crustal blocks initially formed during Late Jurassic transtension.

Chihuahua Trough

The Chihuahua Trough (Plate 1; DeFord, 1964) is an elongate middle Mesozoic basin situated southwest of and roughly parallel to the Rio Grande River between El Paso–Ciudad Juarez and Presidio, Texas. Basal sediments abut the Diablo Platform of west Texas and are bounded on the southwest by the Aldama Platform and Plomosas Uplift (Fig. 1; Gries and Haenggi, 1970). Near Del Rio, Texas, a zone of east-striking faults separates the Chihuahua Trough from the more southeasterly La Mula–Sabinas Basin (discussed below). The Chihuahua Trough has been generally interpreted as a subsiding depocenter connected to the Gulf of Mexico rift. We describe relationships below that suggest initial sedimentation within the trough was controlled by the positions of northwest- and east-striking Late Jurassic faults.

Haenggi (2001; this volume) provides a comprehensive review of work bearing upon understanding of the evolution of the Chihuahua Trough. The oldest sediments are Middle Jurassic evaporites that overlie basement of unknown character.

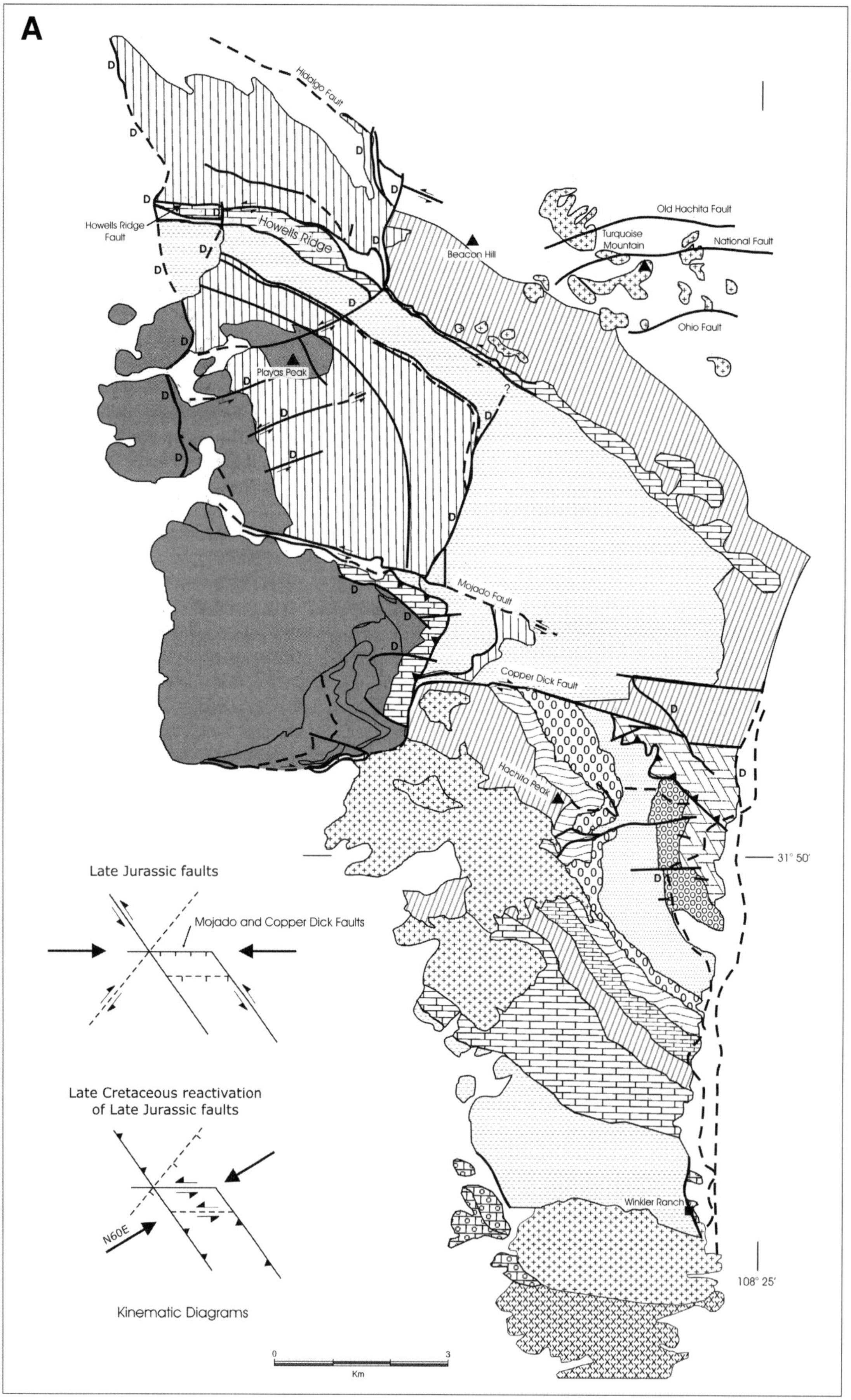
A
Hidalgo Fault
Howells Ridge Fault
Howells Ridge
Beacon Hill
Old Hachita Fault
Turquoise Mountain
National Fault
Ohio Fault
Playas Peak
Mojado Fault
Copper Dick Fault
Hachita Peak
31° 50′
Winkler Ranch
108° 25′
Late Jurassic faults
Mojado and Copper Dick Faults
Late Cretaceous reactivation of Late Jurassic faults
N60E
Kinematic Diagrams
0
3
Km

B

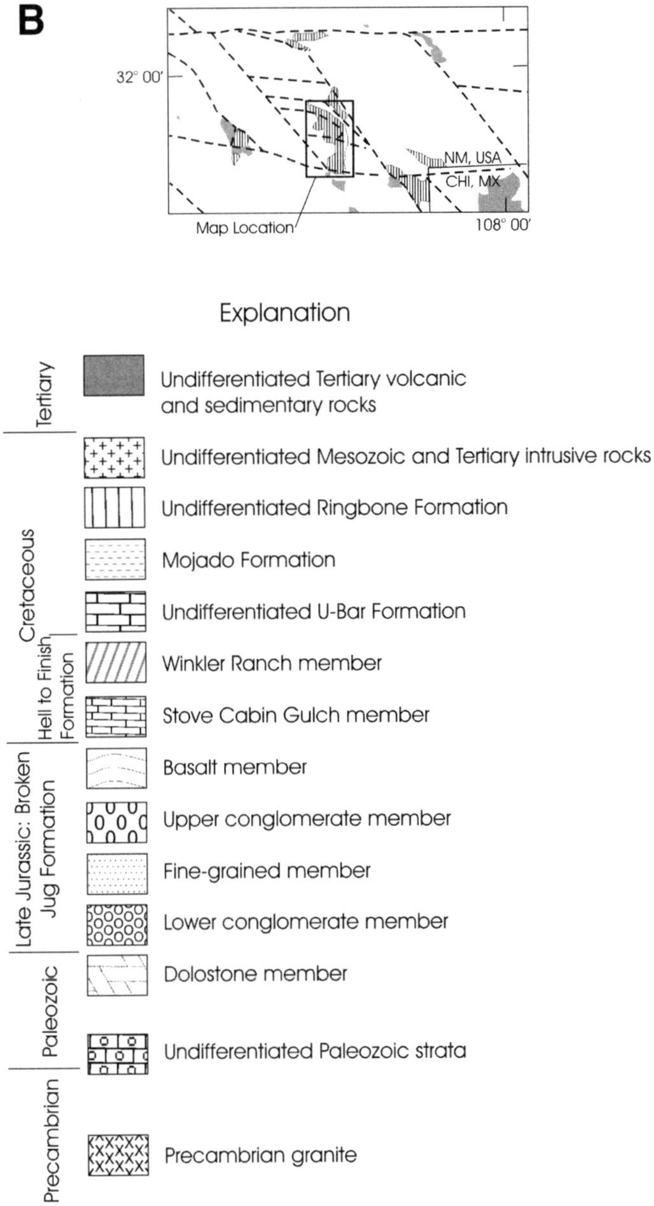

Figure 8 (*on this and previous page*). (A) General geologic map of the Little Hatchet Mountains, showing principal Late Jurassic faults reactivated during Cretaceous contraction (after Lasky [1947], Hodgson [2000], Lucas and Lawton [2000], Basabilvazo [2000], and Channell et al. [2000]). Bold arrows in kinematic diagrams indicate direction of horizontal maximum principle stress, consistent with an inferred transtensional stress regime for Late Jurassic time and compressional stress regime during Late Cretaceous time. (B) Location map and explanation for A.

Thousands of meters of evaporite have been penetrated by drilling on anticlinal structures, but it is assumed that flow into the fold crests exaggerates actual thickness (Haenggi, 2001). Clastic beds interstratified with these evaporites were correlated by Haenggi (1966) to the Jurassic La Casita Formation of Imlay (1952). Early Kimmeridgian to Late Tithonian fossils have been recovered from La Casita outcrops in northeastern Mexico near the Plomosas Uplift (Araujo-Mendieta and Casar-Gonzalez, 1987; Monreal and Longoria, 1999; see description below).

The geometry of the Chihuahua Trough adjacent to west Texas is revealed by the configuration of the overlying Lower Cretaceous Las Vigas lithosome (Haenggi, 1966; DeFord and Haenggi, 1970; Fig. 5). This body of rock thickens abruptly from 0 to more than 1300 m within a few tens of kilometers along a line extending northwesterly along the margin of the Diablo Platform. South of El Paso, the axis of the trough curves strongly westward where it coincides with a depression defined by depths to Precambrian basement that exceed 4400 m (Fig. 5 of Drewes, 1991). An analysis of gravity data integrated with information about basement structure (Jimenez and Keller, 2000) yields more detail about the north end of the trough where two subbasins, El Parabien and Conejo Mendanos, are recognized. A horst-like uplift in northernmost Chihuahua west of Ciudad Juarez–El Paso (Drewes, 1991; Fig. 1) intervenes between the regional Chihuahua Trough and the smaller basins of southern New Mexico. The northeastern flank of the basin is obscured by Late Cretaceous thrusts along which Early Cretaceous hanging wall strata have been displaced northeastward, locally as much as 15 km (Haenggi, 2002).

Part of the southwestern flank of the trough is exposed in the core of a faulted northwest-trending anticlinal structure (Bridges, 1964) called the Plomosas Uplift (Hennings, 1994). The Plomosas Uplift lies at the northeast edge of the broader Aldama block (or platform or regional horst; Fig. 1). Northeast of the uplift, ~3600 m of Upper Jurassic and Lower Cretaceous strata mark the deepest part of the Chihuahua Trough (DeFord and Haenggi, 1970; Fig. 5). The lowest clastic units of these middle Mesozoic strata thin and pinch out against the steep southwestern flank of the Diablo Platform. We postulate that this basin margin coincides with a Late Jurassic sinistral strike-slip fault.

According to Hennings (1994), the Plomosas Uplift was elevated by the combination of regional contraction and left-lateral wrench faulting along a northwest-trending fault named the Plomosas basement shear. Fossiliferous La Casita strata crop out in the flanks of a large anticline near the margin of the uplift adjacent to the postulated basement shear. This 700–1500-m-thick formation is composed principally of clastic rocks that were subdivided into three informal members by Roberts (1989). The lowest member, resting unconformably upon Permian beds, includes sandy to silty mudstone and conglomerate. It is overlain by interbedded shale, marl, and sandstone that contain Kimmeridgian and Tithonian ammonites. The section is interpreted (Roberts, 1989) to record the transition upward from alluvial fan and braid-plain environments to a restricted marine basin filled with prodeltaic turbidites. To the northwest, near Placer de Guadalupe, exposures of correlative strata are as thick as 1500 m. These include shale (locally gypsiferous), shaly limestone, limestone, sandstone and basal conglomerate (Bridges, 1964). The north end of the main La Casita Formation exposures is bounded by an east-trending fault.

The Chihuahua Trough and flanking Diablo Platform (or plateau) terminate in the south against a series of east-striking faults near Del Rio, Texas (Fig. 1). Southeastward across these faults are complementary paleogeographic elements, namely La Mula Basin (Young, 1983; McKee et al., 1990) or Sabinas Gulf (Burkhardt, 1930; Humphrey, 1956; Smith, 1981) and the Tamaulipas Peninsula (Coahuila-Texas craton of Charleston, 1981). The boundary between these paleogeographic elements is the northwest-trending La Babia fault (Charleston, 1981), a regional lineament that coincides with the straight courses of La Babia and Sabinas-Salado River valleys (Fig. 1). The lineament may continue southeastward as far as the town of Mier, Nuevo Leon, where it merges with a similarly trending segment of the Rio Grande, extending toward the Gulf of Mexico.

Haenggi (2002) concludes that: (1) the Chihuahua Trough formed between 163 and 160 Ma, and (2) prominent northerly and northwesterly striking faults record formation of the trough as a right-lateral pull-apart during counterclockwise rotation of the North American plate in response to opening of the Atlantic Ocean. We argue that the geometry of east- and northwest-striking faults, described above, and the relationship of Late Jurassic strata to the faults, is more compatible with pull-apart basin development at releasing steps along left-lateral faults. In our view, the dextral shear sense inferred by Haenggi is based on offset of Precambrian basement that probably developed during Neoproterozoic rifting of Rodinia. This Late Precambrian transform was probably reactivated during Late Jurassic time as a left-lateral structure that controlled sedimentation in the Chihuahua Trough.

La Mula–Sabinas Basin, Coahuila

A conspicuously thick section of Jurassic and Cretaceous strata (~3000 m) crops out in San Marcos valley (Figs. 3 and 7 of McKee et al., 1990; Fig. 5) adjacent to northwesterly striking segments of the San Marcos fault. Basal Late Jurassic units consist of 500 m of polymictic boulder conglomerate interpreted as thick debris flow deposits in a matrix of smaller-scale debris flows and other kinds of sediment gravity-flow deposits. The coarse basal unit is overlain by 500 m of sandstone containing discontinuous beds of debris-flow conglomerate and 300 m of additional conglomerate. The section then fines upward through 600 m to sandstone with minor conglomerate followed by fine sandstone commonly interbedded with shale. The highest 100 m of Jurassic section consists of siltstone and shale without conglomerate. Among these strata calcareous siltstone and shale yield Late Tithonian ammonites. More than 1000 m of Cretaceous beds overlie the Jurassic strata.

Most debris and finer detritus in the Upper Jurassic section was shed northeast across the San Marcos fault from emergent areas of the Coahuila platform (Coahuila Island on Fig. 1). Permian-Triassic intrusive bodies and Upper Paleozoic volcanic and carbonate rocks are the predominant sources. McKee et al. (1990) provisionally accept the assignment of the San Marcos Valley strata to La Casita Formation as assigned by Imlay (1952), although they note lithologic dissimilarity with the type locality.

Exploration wells (Lopez-Ramos, 1980; Eguiluz de Antunano, 2001) reveal a basin ~125 km wide that trends northwesterly between La Mula Island (Jones et al., 1984; Fig. 1) and the Burro-Picachos or Salado platform (Lopez-Ramos, 1980) to the southeast. Commonly, this paleogeographic feature is designated as the Sabinas Basin (e.g., Wilson, 1999). Young (1983) defines the Sabinas Basin as the area distinguished by deposits of Late Cretaceous coal near the city of Sabinas, whereas the northwesterly trending Sabinas Gulf described by Burkhardt (1930) and Humphrey (1956) refers to a marine embayment existing during Late Jurassic and Cretaceous time. McKee et al. (1990) addressed the possible confusion, concluding that "La Mula Basin" should generally refer to the basin in which strata accumulated during Jurassic and Early Cretaceous time.

Northwesterly and west-northwesterly–trending segments of the San Marcos fault zone define the northern boundary of the block-like Coahuila platform for 280 km (Jones et al., 1984; McKee et al., 1990). The gentle westward curve in the San Marcos fault north of Coahuila Island probably resulted in formation of releasing bends and pull-apart basins between it and La Mula Island to the north (e.g., Fig. 25 of McKee et al., 1999). At the village of Sierra Mojada the trace of the fault is lost. However, the alignment of Cretaceous ranges in the area may reflect underlying crustal structure, suggesting that the fault resumes a more northwesterly course striking toward the Plomosas Uplift. We propose that the paleogeographic low area bounded by the Burro-Picachos or Salado platform, La Mula Island, and Coahuila Island formed in response to Late Jurassic transtension.

The oldest rocks in La Mula Basin are granite and metadiorite that yield K-Ar dates between 160 and 164 Ma (Santamaria-O. et al., 1991) and $^{40}Ar/^{39}Ar$ plateau ages of 145 Ma (Garrison and McMillan, 1999). These rocks are from the vicinity of Monclova, where they occur among uplifts that stand above basins commonly trending N75°W. The lowest parts of the basins are floored with Oxfordian deposits, whereas Kimmeridgian-Tithonian strata may mantle "intermediate" blocks (Santamaria-O. et al., 1991). In the southeastern part of La Mula Basin volcanic units including dacite, rhyodacite, andesite, trachyte, basalt, and metamorphosed mafic plutonic and volcanic rocks may be interbedded or intruded into basal units (Santamaria-O. et al., 1991; Garrison and McMillan, 1999). The geochemistry of the igneous rocks is comparable to those produced during rifting of continental crust (Garrison and McMillan, 1999). Haenggi (2002) interpreted the older ages as possibly indicating the presence of arc-related Jurassic rocks north of the Mojave-Sonora megashear, obviating the southeastward displacement of the Cordilleran Jurassic arc offered by Jones et al. (1995) in support of the megashear hypothesis. An alternative hypothesis is that the K-Ar dates record cooling of rift-related igneous rocks emplaced during the transition from Jurassic subduction to transform faulting. Furthermore, comparison of fossil and radiometric ages with the Pálfy et al. (2000) Jurassic time scale suggests that magmatism occurred at two times during the formation of La Mula Basin. Volcanic units low in the section of Callovian and

Oxfordian strata record eruption as early as 161 Ma, during early rifting, whereas the younger ages record Tithonian magmatism through thinned crust. Magmatic pulses at these times are contemporaneous with those of correlative basins in Arizona and California.

DISCUSSION

The Mojave-Sonora Fault System—A Reactivated Zone of Weakness along the Southwest Margin of North America?

Stewart et al. (1984), Almazán-Vázquez et al. (1986), and Poole and Madrid (this volume) postulate that the early Paleozoic continental margin of southwestern North America curved southeastward across northern Mexico. The general geometry of this margin was maintained until late Paleozoic time when formation of Pangea led to convergence in this region as indicated by studies of rocks in the vicinity of Las Delicias, Coahuila (King et al., 1944; McKee et al., 1999, and references therein) that indicate the existence of a continental margin arc that is either parautochthonous or exotic. The southeasternmost segment of the Mojave-Sonora fault system lies along this old continental margin within a zone designated by Murray (1986, 1989) as the California-Tamaulipas geodiscontinuity. He describes the discontinuity (Murray, 1989, p. 211) as "a fundamental, crustal fracture zone characterized by continuing weakness and deformation since the Precambrian…." coinciding with a "zone of tectonic collage" hundreds of kilometers wide extending for ~2500 km from California to the Gulf of Mexico. According to Murray, the tectonic collage is bounded on the north by northwesterly trending lineaments including Walker Lane and the Texas lineament and on the southwest by the Mojave-Sonora megashear and the Torreon-Saltillo-Monterrey fracture zones. Below we discuss the relationship of the most northeasterly lateral faults of the system and some of the linking left-steps to older faults.

Within this region, structures collectively grouped as the "Texas lineament," "Texas zone," or "Texas direction" have been recognized (e.g., Muehlberger, 1965). Ransome (1915) proposed the existence of a major discontinuity near Van Horn that he called the Texas lineament. According to Albritton and Smith (1957, the Texas lineament trends N60°W and separates the Diablo Platform from the Chihuahua Trough. The "Texas direction" of wrench faulting as proposed by Moody and Hill (1956) has a slightly different trend, N70°W, based upon the orientation of the Hillside fault (Fig. 1), the type fault of the Texas direction. Haenggi (2002) argued that evidence along the Hillside fault indicates normal displacements as old as late Paleozoic but does not support strike-slip offset. Geologic maps of the Diablo Plateau northwest of Van Horn (Albritton and Smith, 1957; Wiley and Muehlberger, 1971) reveal three sets of faults (NW-trending, left-lateral strike slip; NE-trending, right-lateral strike-slip, and E-trending normal slip) that cut Precambrian and Paleozoic rocks. We point out that these three fault sets generally coincide with those of the Mojave-Sonora system in orientation and sense of displacement, that is, principal left-lateral strike-slip faults that step at east-trending normal faults. The right-lateral faults are complementary structures. In this model, the Hillside fault represents an extensional structure oriented oblique to the Texas lineament.

As illustrated by Figure 7 in Keith and Swan (1996) the "Texas zone" trends more westerly than the set of Mojave-Sonora left-lateral strike-slip faults. However, integration of the common N50°W-striking left-lateral faults with releasing steps (easterly striking normal faults) produces a more westerly strike for the combined fault sets. Muehlberger provides an apt (1980, p. 113) characterization of the Texas lineament as " a zone of recurrent movement that separates more stable crust of the north from less stable crust on the south. Dip-slip (normal, steep reverse, or thrust) movements are widely demonstrable. Strike-slip movements can be documented for episodes but the amount of slip necessary to produce the observed effects is in miles rather than in hundreds of miles."

The Hillside fault (Moody and Hill, 1956; Fig. 1) is one of several faults that transect the Diablo Platform. East-striking faults distinguish the southern margin of this uplifted block. These faults may also have Paleozoic antecedents as argued by Dickerson (1985), who noted that the Tascotal Mesa fault locally accommodated down-to-the-north movement during late Paleozoic deformation. She also recognized that the fault is the southernmost of four east-striking structures including Chalk Draw–Shafter, Ruidosa, and Candelaria faults (Dickerson, 1980). These structures bound the basement-cored Devils River and Tascotal uplifts and separate them from the Paleozoic Val Verde and Marfa basins to the north (Ewing, 1987). The Chalk Draw and Tascotal Mesa faults record left-lateral strike-slip displacement attributed to Laramide contraction (Calhoun and Webster, 1983) and were probably reactivated again during Tertiary extension.

Structures marking the south margin of the Diablo Platform project east to the Frio River line (Ewing, 1987; Fig. 1), a prominent northwest-trending lineament at the northeast edge of the Mojave-Sonora fault system. This line extends from Corpus Christi to Del Rio, Texas, and separates regions of contrasting structural and stratigraphic history. Of particular relevance is the absence of both Laramide contractional structures and Late Jurassic block faulting northeast of the line. Although the Frio River line shows no evidence of pre-Mesozoic tectonic activity, its orientation and regional extent are suggestive of a significant early history. Thomas (1988) argues that Late Precambrian–early Paleozoic and Mesozoic transforms are similarly oriented in the subsurface of the northern Gulf Coastal Plain. The most southwesterly of these older transforms coincides with the Frio River line.

In southern New Mexico and Arizona, the strike-slip and normal faults of the Mojave-Sonora system are commonly the initial post-Precambrian fault structures recorded by displacements of carbonate-shelf Paleozoic strata that overlie Proterozoic crystalline basement. In general, these do not coincide with the northeast-striking Paleoproterozoic ductile shear zones that transect central Arizona (Karlstrom and Bowring, 1988). Nevertheless, pre-existing features such as the Stockton Pass fault (Swan,

1976) and the Pedregosa Basin (Goetz and Dickerson, 1985) are aligned with faults of the principal strike-slip set of the Mojave-Sonora system, suggesting a long history of deformation along this trend as mused by Murray (1986). In light of these tectonic relationships, it seems probable that the expression and location of the southeast segment of the Mojave-Sonora fault system was strongly influenced by older structures parallel to the former continental margin that served as a plate boundary during late Paleozoic and Late Jurassic time.

The distinctive northwest-trending structural grain of the Mojave-Sonora fault system is commonly obfuscated in this region by northwesterly early Miocene and northerly late Miocene normal faults. However, remnants of fault-related lineaments and structural discontinuities that form the grain may be preserved in Miocene uplifts. Despite earlier fault reactivation during Cretaceous contraction, Late Jurassic structures may be separated from the Miocene and Cretaceous structures in three ways: (1) by age, they were initially active during Jurassic time (Titley, 1976; Drewes, 1981), (2) by orientation, they strike more westerly (N40–60°W) than the early Miocene faults (N30°W), and (3) by structural style, they correspond to zones within which complex lateral and vertical displacements occurred.

Late Jurassic Igneous Rocks Associated with the Pull-Apart Basins

Igneous rocks are penecontemporaneous with the development of Late Jurassic pull-apart basins. For example, alkalic felsic volcanic units in the Canelo Hills have geochemical signatures consistent with formation under conditions in an extensional tectonic setting (Krebs and Ruiz, 1987). Mafic volcanic flows interstratified with Upper Jurassic basin fill of the Chiricahua Mountains include alkali basalt derived from the mantle (Lawton and Olmstead, 1995; Harrigan, 1995) and contaminated by continental crust (Gleason et al., 1999). These igneous units are part of a bimodal suite that first appears stratigraphically high in stacks of arc-related Jurassic volcanic units that generally yield ages of ca. 170 Ma.

Post-arc magmatism is recorded as a regional unit in southern Arizona and northern Sonora that includes the Artesa volcano-sedimentary sequence and intrusive equivalents, composed of the generally shallow Ko Vaya plutons and hypabyssal rocks (Tosdal et al., 1989). The younger rocks of this suite yield crystallization ages younger than 150 Ma (Krebs and Ruiz, 1987; Haxel et al., this volume; Anderson et al., this volume). We speculate that some of the highly felsic granitic plutons of the Ko Vaya type record melts within the arc-heated crust that were locally contaminated by partial melting of diverse older rocks. These bodies intruded fault-controlled basin floors composed of thin crust. Mafic flows and intrusive bodies associated with the Ko Vaya suite or interstratified with clastic sections in pull-apart basins (e.g., Huachuca Mountains, Hayes and Raup, 1968; Chiricahua Mountains, Lawton and Olmstead, 1995) probably were emplaced along steep faults that provided direct conduits through continental crust from deep-seated magma chambers in below-heated mantle lithosphere.

Independence dikes, most of which yield a crystallization age of ca. 148 Ma (Chen and Moore, 1979; James, 1989) may be part of the mafic magmatic suite. However, Hopson (1988) argues that the compositional diversity of the dikes distinguishes them from basaltic dike swarms associated with rifting. The dikes are best known from California where they compose a northwesterly striking swarm extending for 600 km from east-central to southern California. The ages of the dikes, generally ~20 m.y. younger than the principal pulse of Jurassic arc magmatism, and their structural setting (Glazner et al., 1999) are sufficient to disassociate them from convergent margin subduction processes. The dikes record an abrupt transition from dominant sinistral shear along northwest-striking faults to north-south extension (Glazner et al., 1999). In general the dikes lack deformational fabric and we interpret this to indicate the cessation of displacements along the lateral faults as well as the development of pull-apart basins within the Mojave-Sonora fault system.

Structural Inversion and Reactivation of Late Jurassic Faults

Faults of the Mojave-Sonora system provide a crustal template upon which regional N60°E-directed Cretaceous contraction (e.g., Erdlac, 1990) was imposed. Davis (1979) described numerous uplifts along steep faults in southern Arizona. Later publications by Drewes (e.g., 1980, 1981, 1991) and Jensen and Titley (1998) support our hypothesis that steep, pre-Cretaceous, northwest-striking left-lateral faults and east-striking normal faults, have been reactivated consistently as reverse and left-lateral faults, respectively, during Cretaceous contraction.

Near the Mule and Huachuca Mountains, Glance Conglomerate has been thrust over the older bounding faults of pull-apart basins, creating allochthonous masses as mapped by Hayes and Landis (1964) and Hayes and Raup (1968). In southeastern Arizona and southern New Mexico, faults of the Mojave-Sonora system were reactivated and accommodated lateral (Drewes, 1991; Hodgson, 2000) and dip-slip displacement (Lawton, 1996, 2000) during Late Cretaceous contraction. Drewes (1991, his Fig. 13) recognized the pattern of reactivated faults and his structure sections infer the existence of deep detachments that are necessary to accommodate the inversion of the Late Jurassic pull-apart basins.

Dickerson (1985) describes structures, some of which we correlate with Late Jurassic faults, extending from New Mexico into Chihuahua and west Texas. Although two sets of early faults with different orientations are not distinguished, she clearly recognizes the importance of left-lateral slip along pre-existing faults during Late Cretaceous contraction. In west Texas, certain faults record additional Tertiary displacement (Dickerson, 1980; Henry, 1998). Comparable structures are known in the state of Coahuila, Mexico, at the village of Sierra Mojada, where mapping by McKee et al. (1990) along the Late Jurassic San Marcos normal fault revealed local thrusts that appear to be transpressional. Thrusting occurs where a westerly striking segment of the San Marcos fault, reactivated as a left-lateral fault, bends north into a restraining bend.

Laramide Plutonism and Mineralization in the Vicinity of Pull-Apart Structures

Laramide age (i.e., Late Cretaceous–Early Tertiary) plutonic rocks that commonly crop out in the vicinity of pull-apart basins tend to be distributed close to the inferred boundary faults; especially near east-striking normal faults. Exposures of granite along major northwest-striking faults or away from the faulted basin margins are also known. Many of these plutons have been targets of mineral exploration.

We speculate that intrusion into the floors of pull-apart basins occur because the releasing-step normal faults serve as conduits for magma, heat, and fluids to reach areas of extended thin crust in a favorable dilational setting. In some pull-apart basins synextensional Late Jurassic magmatism, such as that described above, may have clogged passageways favorable to magma transport. If older intrusions were not present, then the pull-apart remained vulnerable to heating and perforation from magma upwelling along extensional faults.

The Little Hatchet Mountains of New Mexico (Fig. 8) preserve structural and stratigraphic relationships, summarized recently by Hodgson (2000), Lucas and Lawton (2000), Basabilvazo (2000), and Channell et al. (2000), that are especially suggestive of this process. A Late Jurassic half graben situated between the Copper Dick fault on the north and outcrops of Precambrian(?) Hatchet Gap granite in the hanging wall to the south is the principal pull-apart basin. In addition to Late Jurassic basalt that erupted during filling of the basin, diorite and monzonite of the Cretaceous-Tertiary Sylvanite Intrusive Complex crops out. Further south, at the northern margin of exposures of the Precambrian Hatchet Gap granite, outcrops of the mid-Tertiary Granite Pass granite distinguish an oval-shaped mass, elongate along an east-trending axis (Channell et al., 2000; Fig. 8). North of the Copper Dick fault, exposures of monzonite and diorite among the east-trending Ohio, National, and Old Hachita faults comprise the mid-Tertiary(?) Eureka Intrusive Complex. Spatial correspondence of Laramide plutons in the Little Hatchet Mountains near east-striking faults that record Late Jurassic normal displacements implies preferential emplacement into previously weakened regions of the crust.

Other examples of the coincidence of Late Cretaceous–Tertiary magmatic centers with regions of postulated thin crust, though less well defined than that in the Little Hatchet Mountains, include the following (Fig. 2; Plate 1):

1. The Lordsburg stock. In the northern Pyramid Mountains plutonic rocks crop out south of east-striking faults along which mineralization is localized (Thorman and Drewes, 1978). Thorman and Drewes suggested that similar faults may be guides for exploration for mineral deposits. Although the faults record lateral displacement as part of early Tertiary reactivation, exposures of the granodiorite are limited to the south by east-striking Lower(?) Cretaceous sandstone. We interpret the presence of Early Cretaceous sandstone as indicating the existence of a pull-apart basin in which the sandstone accumulated. The sandstone was uplifted in response to later inversion.

2. The Victorio granite. In the Victorio Mountains Late Cretaceous(?) or Tertiary(?) granite, composed of biotite and muscovite-biotite phases (McLemore et al., 2000), crops out between the east-striking Main Ridge and Victorio Mountains faults (Thorman and Drewes, 1980).

3. The Sierrita Mountains. Porphyry deformed by wrench faulting along an east-striking fault is mineralized (Jensen and Titley, 1998). Esperanza, Sierrita, Pima, Mission, and Twin Buttes mines are located near east-northeast striking faults, including the Duval fault and parallel No. 6 thrust to the northeast, adjacent to a Laramide pluton (Cooper, 1973).

4. Schieffelin granodiorite at Tombstone. The granodiorite intrudes Bisbee Group strata north of the east-striking Prompter fault (Gilluly, 1956).

5. The Stronghold granite. In the northern Dragoon Mountains, the Tertiary Stronghold granite crops out at the north end of an elongate basin filled with Bisbee Group strata.

6. The Turkey Creek caldera. The caldera formed in the basin south of left-steps in the Apache Pass fault (see above) in the Chiricahua Mountains. We interpret the caldera to be an example of a Tertiary igneous body in the midst of a probable pull-apart.

7. Imuris, Sonora. Early and middle Tertiary two-mica granites of the Magdalena-Madera core-complex were emplaced near the contact between the Middle Jurassic arc and Upper Jurassic-Cretaceous sedimentary basins (Nourse, 1995). The east-striking, south-dipping contact delineates the northern boundary of an inferred pull-apart basin (described previously). Interestingly, this belt of two-mica granite appears to mark the breakaway zone of the Magdalena-Madera detachment fault.

8. Cananea, Sonora. Ore deposits occur near Laramide intrusions (Anderson and Silver, 1977) that crop out near the intersection of the faults bounding the northeastern corner of the San Antonio basin.

Some Laramide intrusive rocks emplaced into the regions of thin crust were fractured and mineralized during Late Cretaceous left-lateral reactivation of east-striking faults (e.g., Jensen and Titley, 1998). Late Cretaceous deformation appears to have added to and opened existing fractures, thereby enhancing the porosity and permeability for mineralizing fluids. Additional structural modifications of the Jurassic faults probably occurred during low-angle extensional faulting as suggested by Cooper's mapping (1973).

CONCLUSIONS

Transtension along the Mojave-Sonora megashear with consequent rifting and regional basin formation occurred in concurrence with opening of the Atlantic Ocean and formation of the Gulf of Mexico (Dietz and Holden, 1970; Anderson and Schmidt, 1983). This paper offers an alternative explanation of the development of the rifts and basins recognized by many previous workers (Bilodeau, 1982; Dickinson et al., 1986; Busby-Spera and Kokelaar, 1991; Lawton and Olmstead, 1995; Dickinson and Lawton, 2001; Busby et al., this volume). Our pull-apart basin model is compatible kinematically, temporally, and spatially with the plate motions recognized by Dietz and Holden (1970) and Klitgord and Schouten (1986) but is distinct from previous models, such as those proposing propagation of an aulacogen or thermotectonic basin subsidence during the same time.

The thickest deposits of Upper Jurassic conglomerate accumulated in basins south of generally east-striking normal faults that are linked to regional northwest-striking faults. The normal faults formed at left releasing steps among Late Jurassic (ca. 160–150 Ma) left-lateral faults distributed within a zone a few hundred kilometers wide, extending from the Gulf of Mexico to southern California. The orientation of the basin-bounding strike-slip faults, the demonstrated age of deposits synchronous with faulting, and the presence of basins adjacent to the Mojave-Sonora megashear indicate to us that Late Jurassic opening of the Gulf of Mexico was cogenetic with the Mojave-Sonora transform for which the maximum principal stress trended easterly. The Mojave-Sonora megashear sweeps along the southwestern margin of the Jurassic craton of North America in a position probably influenced in the southeast by a pre-existing boundary between continental and oceanic lithosphere. The pull-apart basins generally demarcate regions of the craton affected by brittle transtension.

Development of the pull-apart basins occurred after formation of calderas and the associated high-energy volcaniclastic deposits between 180 Ma and 165 Ma. Regional correlation among exposures of coarse clastic strata of suspected Jurassic age thus requires distinction of caldera fill or other Middle Jurassic intra-arc sections containing conglomerate and breccia from those that accumulated in Late Jurassic transtensional basins. The principal burst of basin formation that began at ca. 162 Ma correlates with the geologically abrupt Callovian cessation of calc-alkaline volcanism followed by local eruptions of mafic and intermediate volcanic rocks interbedded with conglomerate. We propose that this marked change was plate driven and records the initiation of regional transform faulting. Basins continued to form throughout the interval of most active transform faulting between 158 and 148 Ma. Independence dikes (ca. 148 Ma) show little or no deformation and mark the end of major displacements along lateral faults of the Mojave-Sonora system.

Northeast of the Mojave-Sonora megashear, northwest- and east-striking faults defined a template composed of intersecting, deep-crustal flaws that influenced the style of subsequent Cretaceous and Tertiary deformation throughout the southwestern United States and northern Mexico. Development of a Sevier-like regional fold-and-thrust belt did not occur in the areas where pull-apart basins formed. Instead, Cretaceous contraction was expressed by reactivation of Late Jurassic normal faults as left-lateral strike-slip and Late Jurassic left-lateral strike-slip faults as steep reverse faults.

Areas of thinned crust among the faults of the Mojave-Sonora system influenced the emplacement of igneous rocks and the development of ore deposits during Late Jurassic, Late Cretaceous, and finally Tertiary time. Following regional Late Cretaceous and early Tertiary plutonism, the crust of southwestern North America may have thickened and strengthened sufficiently so that annealing occurred, as is suggested by the prominent sets of Tertiary normal faults (i.e., N30°W "core complex" and N-S "Basin and Range") that commonly break across fault structures of the Mojave-Sonora system.

ACKNOWLEDGMENTS

We acknowledge Lee Silver, teacher, field geologist, petrologist, geochemist, and geochronologist. Lee's knowledge of southwestern North America, based in part upon an enormous data set that he generated, led him to conceive the Mojave-Sonora megashear hypothesis. Jim McKee and Norris Jones introduced Anderson to northeastern Mexico and permitted him to tag along for a decade or so while they studied the pre-Cretaceous geology. Mary Beth Kitz McKee and Jose Luis Rodríguez-Castañeda stuck with thorny thesis problems complicated at times by misdirection from Anderson. Roberto Bernal and Jim and Mary Beth McKee were invaluable colleagues during several years of field work along the Mexico–United States border between Agua Prieta and Cananea, Sonora. Lee Silver supported Nourse's dissertation work (pre-1989) in the ranges surrounding Imuris. Nourse is grateful to E. Stahl, D. Curtis, M. Pratt, M. Magner, B. Kriens, M. Chuang, R. Acosta, and M. Beaumont for cheerful field assistance with several mapping sessions in Sierra El Batamote. The Geological Sciences Department at Cal Poly Pomona provided a field vehicle during the mid 1990's. John Dembosky, Ed Lidiak, and Scott Davidson patiently helped Anderson with figures. Jaime Roldán-Quintana, Carlos González, Cesar Jacques-Ayala, José Luis González-Castañeda, and Juan Carlos Garcia y Barragán, geologists of the Instituto de Geología, Universidad Nacional Autónoma de México, were generous in their support of our research and willingness to discuss the geology of Sonora. Jim McKee commented on early versions of this manuscript. Formal reviews by Gary Gray and Ricardo Presnell provided suggestions for organization, clarity, and illustration that substantially improved the final product.

REFERENCES CITED

Aksu, A.E., Calon, T.J., and Hiscott, R.N., 2000, Anatomy of the North Anatolian fault zone in the Marmara Sea, western Turkey: Extensional basins above a continental transform: GSA Today, v. 10, no. 6, p. 3–7.

Albritton, C.C., and Smith, J.F., 1957, The Texas lineament: 20th International Geological Congress Proceedings, Mexico City, Section 5, p. 501–518.

Almazán-Vázquez, E., Palmer, A.R., and Robison, R.A., 1986, Constraints on the southern margin of North America during the early Paleozoic: Geological Society of America Abstracts with Programs, v. 18, no. 6, p. 525.

Amato, J.M., 2000, Structural relationships in the Florida Mountains, southwestern New Mexico—A review, *in* Lawton, T.F., McMillan, N.J., and McLemore, V.T., eds., Southwest Passage—A trip through the Phanerozoic: New Mexico Geological Society Fifty-First Annual Field Conference, p. 103–108.

Anderson, T.H., and Schmidt, V.A., 1983, The evolution of Middle America and the Gulf of Mexico–Caribbean Sea region during Mesozoic time: Geological Society of America, v. 94, p. 941–966, doi: 10.1130/0016-7606(1983)94<941:TEOMAA>2.0.CO;2.

Anderson, T.H., and Silver, L.T., 1977, U-Pb isotopic ages of granitic plutons near Cananea, Sonora: Economic Geology and the Bulletin of the Society of Economic Geologists, v. 72, p. 827–836.

Anderson, T. H. and Silver, L.T., 1979, The role of the Mojave-Sonora megashear in the evolution of northern Sonora, *in* Anderson, T.H., and Roldán-Quintana, J., eds., Geology of northern Sonora: The Geological Society of America Field Trip Guidebook, Field trip no. 27, p. 59–68.

Anderson, T.H., McKee, J.W., McKee, M.B., and Nourse, J.A., 1995, Coarse clastic deposits of Late Jurassic age in southwestern North America—A record of pull-apart basins: Geological Society of America Abstracts with Programs, v. 27, no. 6, p. A-75.

Araujo-Mendieta, J., and Casar-Gonzalez, R., 1987, Estratigrafia y sedimentologia del Jurassico superior en la cuenca de Chihuahua, norte de Mexico: Revista del Instituto del Petroleo, v. 19, no. 1, p. 6–21.

Asmerom, Y., Zartman, R.E., Damon, P.E., and Shafiqullah, M., 1990, Zircon U-Th-Pb and whole rock Rb-Sr age patterns of Lower Mesozoic igneous rocks in the Santa Rita Mountains, southwest Arizona: Implication for Mesozoic magmatism and tectonics in the southern Cordillera: Geological Society of America Bulletin, v. 102, p. 961–968, doi: 10.1130/0016-7606(1990)102<0961:ZUTPAW>2.3.CO;2.

Barth, A.P., Wooden, J.L., Jacobson, C.J., and Probst, K., 2004, U-Pb geochronology and geochemistry of the McCoy Mountains Formation, southeastern California: A Cretaceous retroarc foreland basin: Geological Society of America Bulletin, v. 116, p. 142–153, doi: 10.1130/B25288.1.

Basabilvazo, G., 2000, The Upper Cretaceous Ringbone Formation, Little Hatchet Mountains, southwestern New Mexico, *in* Lawton, T.F., McMillan, N.J., and McLemore, V.T., eds., Southwest Passage—A trip through the Phanerozoic: New Mexico Geological Society Fifty-First Annual Field Conference, p. 203–210.

Bayona, G., and Lawton, T.F., 2000, Aptian-Albian extensional faulting and subsequent Laramide inversion, central Peloncillo Mountains, southwestern New Mexico, *in* Lawton, T.F., McMillan, N.J., and McLemore, V.T., eds., Southwest Passage—A trip through the Phanerozoic: New Mexico Geological Society Fifty-First Annual Field Conference, p. 85–94.

Bilodeau, W.L., 1979, Early Cretaceous tectonics and deposition of the Glance Conglomerate, southeastern Arizona [Ph.D. dissertation]: Stanford, Stanford University, 145 p.

Bilodeau, W.L., 1982, Tectonic models for early Cretaceous rifting in southeastern Arizona: Geology, v. 10, p. 466–470, doi: 10.1130/0091-7613(1982)10<466:TMFECR>2.0.CO;2.

Bilodeau, W.L., Kluth, C.F., and Veddar, L.K., 1987, Regional stratigraphic, sedimentologic and tectonic relationships of the Glance Conglomerate in southeastern Arizona, *in* Dickinson, W.R., and Klute, M.A., eds., Mesozoic rocks of southern Arizona and adjacent areas: Arizona Geologic Society Digest, v. 18, p. 229–256.

Bridges, L.W., 1964, Stratigraphy of Mina Plomosas–Placer de Guadalupe area *in* Geology of Minas Plomosas–Placer de Guadalupe area, Chihuahua, Mexico: Field Trip Guidebook, West Texas Geological Society Publication 64, p. 50–59.

Burkhardt, C., 1930, Etude synthetique sur le Mesozoique Mexicain: Abhandlungen der Schweizerischen Paläontologischen Gesellschaft, Basel, v. 49-50, 280 p.

Busby-Spera, C.J., and Kokelaar, B.P., 1991, Controls of the Sawmill Canyon fault zone on Jurassic magmatism and extension/transtension in southern Arizona: Geological Society of America Abstracts with Programs, v. 23, no. 5, p. A-250.

Bussard, W.J., 1996, Jurassic Volcanism in the Patagonia Mountains of Southern Arizona [M.S. thesis]: Flagstaff, Northern Arizona University, 92 p.

Calhoun, G.G., and Webster, R.E., 1983, Surface and subsurface expression of the Devils River Uplift: West Texas Geological Society Guidebook, v. 83–77, p. 101–118.

Channell, R., McMillan, N.J., Lawton, T.F., Heizler, M., Esser, R.P., and McLemore, V.T., 2000, Magmatic history of the Little Hatchet Mountains, Hidalgo and Grant Counties, southwestern New Mexico, *in* Lawton, T.F., McMillan, N.J., and McLemore, V.T., eds., Southwest Passage—A trip through the Phanerozoic: New Mexico Geological Society Fifty-First Annual Field Conference, p. 141–148.

Charleston, S., 1981, A summary of the structural geology and tectonics of the state of Coahuila, Mexico, *in* Smith, C.I., and Brown, J.B., eds., Lower Cretaceous stratigraphy and structure, northern Mexico: Field Trip Guidebook, West Texas Geological Society Publication 81–74, p. 28–36.

Chen, J.H., and Moore, J.G., 1979, Late Jurassic Independence dike swarm in eastern California: Geology, v. 7, p. 129–133, doi: 10.1130/0091-7613(1979)7<129:LJIDSI>2.0.CO;2.

Clemons, R.E., 1998, Geology of the Florida Mountains, southwestern New Mexico: New Mexico Bureau of Mines and Mineral Resources Memoir 43, 112 p.

Cooper, J.R., 1971, Mesozoic stratigraphy of the Sierrita Mountains, Pima County, Arizona: U.S. Geological Survey Professional Paper 658-D, 42 p.

Cooper, J.R., 1973, Geologic Map of the Twin Buttes Quadrangle, southwest of Tucson, Pima County, Arizona: U.S. Geological Survey Miscellaneous Geologic Investigations Map I-745, scale 1:48,000.

Cooper, J.R., and Silver, L.T., 1964, Geology and ore deposits of the Dragoon Quadrangle, Cochise County, Arizona: U.S. Geological Survey Professional Paper 416, 196 p.

Darton, N.H., 1922, Geologic structure of parts of New Mexico: U.S. Geological Survey Bulletin, v. 726, p. 173–275.

Darton, N.H., 1928, "Red beds" and associated formations in New Mexico, with an outline of the geology of the state: U.S. Geological Survey Bulletin, v. 794, 356 p.

Davis, G.H., 1979, Laramide folding and faulting in southeastern Arizona: American Journal of Science, v. 279, p. 543–569.

DeAngelo, M.V., and Keller, G.R., 1988, Geophysical anomalies in southwestern New Mexico, *in* Mack, G.H., Lawton, T.F., and Lucas, S.G., eds., Cretaceous and Laramide tectonic evolution of southwestern New Mexico: New Mexico Geological Society Thirty-Ninth Annual Field Conference Guidebook, New Mexico Geological Society, p. 71–75.

DeFord, R.K., 1964, History of geologic exploration in Chihuahua, *in* Geology of Mina Plomosas-Placer de Guadalupe area, Chihuahua, Mexico: Field trip guidebook: West Texas Geological Society, Publication 64-50, p. 116–129.

DeFord, R.K., and Haenggi, W.T., 1970, Stratigraphic nomenclature of Cretaceous rocks in northeastern Chihuahua, *in* Seewald, K., and Sundeen, D., eds., The Geologic framework of the Chihuahua tectonic belt: Symposium in honor of Professor Ronald K. DeFord: West Texas Geological Society, p. 175–196.

Dickerson, P.W., 1980, Structural zones transecting the southern Rio Grande rift—Preliminary observations, *in* Dickerson, P.W., Hoffer, J.M., Callender, J.F. eds., Trans-Pecos region, southeastern New Mexico and west Texas: New Mexico Geological Society Guidebook 31, p. 63–70.

Dickerson, P.W., 1985, Evidence for late Cretaceous–early Tertiary transpression in Trans-Pecos Texas and adjacent Mexico, *in* Dickerson, P.W., and Muehlberger, W.R., eds., Structure and tectonics of Trans-Pecos Texas: West Texas Geological Society Field Conference Publication 85–81, p. 185–194.

Dickinson, W.R., and Lawton, T.F., 2001, Carboniferous to Cretaceous assembly and fragmentation of Mexico: Geological Society of America Bulletin, v. 113, p. 1142–1160, doi: 10.1130/0016-7606(2001)113<1142:CTCAAF>2.0.CO;2.

Dickinson, W.R., Klute, M.A., and Swift, P.A., 1986, The Bisbee basin and its bearing on late Mesozoic paleogeographic and paleotectonic relations between the Cordilleran and Caribbean regions, *in* Abbott, P.L., ed., Cretaceous stratigraphy, western North America: Los Angeles, Pacific Section, Society of Economic Mineralogists and Paleontologists, Book 46, p. 51–62.

Dickinson, W.R., Fiorillo, A.R., Hall, D.L., Monreal, R., Potochnik, A.R., and Swift, P.N., 1989, Cretaceous strata of southern Arizona, *in* Jenney, J.P., and Reynolds, S.J., eds., Geologic evolution of Arizona: Arizona Geological Society Digest, v. 17, p. 447–461.

Dietz, R.S., and Holden, C., 1970, The breakup of Pangaea: Scientific American, v. 223, p. 30–41.

Drewes, H., 1971, Mesozoic stratigraphy of the Santa Rita Mountains, southeast of Tucson, Arizona: U.S. Geological Survey Professional Paper 658-C, 81 p.

Drewes, H., 1980, Tectonic map of southeastern Arizona: U.S. Geological Survey Miscellaneous Geological Investigations Map I-1109, scale 1:125,000.

Drewes, H., 1981, Tectonics of southeastern Arizona: U.S. Geological Survey Professional Paper 1144, 96 p.

Drewes, H., 1986, Geology and structure sections of part of the northern Animas Mountains, Hidalgo County, New Mexico: U.S. Geological Survey Miscellaneous Investigations Series Map I-1686, scale 1:24,000.

Drewes, H., 1991, Description and development of the Cordilleran orogenic belt in the southwestern United States and northern Mexico: U.S. Geological Survey Professional Paper 1512, 92 p.

Drewes, H., and Hayes, P.T., 1983, Comment on "Tectonic models for Early Cretaceous rifting in southeastern Arizona": Geology, v. 11, p. 364–365, doi: 10.1130/0091-7613(1983)11<364:CAROTM>2.0.CO;2.

Drewes, H., and Thorman, C.H., 1978, Major geologic structures between Lordsburg, New Mexico, and Douglas and Tucson, Arizona, *in* Callender, J.F., Wilt, J.C., and Clemons, R.E., eds., Land of Cochise, southeastern Arizona: Twenty-Ninth Field Conference Guidebook, New Mexico Geological Society, p. 291–295.

Eguiluz de Antunano, S., 2001, Geologic evolution and gas resources of the Sabinas Basin in northeastern Mexico, *in* Bartolini, C., Buffler, R.T., and Cantu-Chapa, A., eds., The western Gulf of Mexico basin: Tectonics, sedimentary basins, and petroleum systems: American Association of Petroleum Geologists Memoir 75, p. 241–270.

Engebretson, D.C., Cox, A., and Thompson, G.A., 1984, Plate motions and continental tectonics: Tectonics, v. 3, no. 2, p. 110–124.

Erdlac, R.J., Jr., 1990, A Laramide-age push-up block: The structures and formation of the Terlingua-Solitario structural block, Big Bend region, Texas: Geological Society of America Bulletin, v. 102, p. 1065–1076, doi: 10.1130/0016-7606(1990)102<1065:ALAPUB>2.3.CO;2.

Ewing, T.E., 1987, The Frio River line in south Texas—Transition from cordilleran to northern gulf tectonic regimes, *in* Wilbert, W.P., and Jones, L.P., eds., San Antonio, Texas, Transactions Gulf Coast Association of Geological Societies, v. XXXVII, p. 87–94.

Fackler-Adams, B.N., Busby, C.J., and Mattinson, J.M., 1997, Jurassic magmatism and sedimentation in the Palen Mountains, southeastern California: Implications for regional tectonic controls on the Mesozoic continental arc: Geological Society of America Bulletin, v. 109, p. 1464–1484, doi: 10.1130/0016-7606(1997)109<1464:JMASIT>2.3.CO;2.

Force, E.R., 1996, The Bisbee Group of the Tombstone Hills, southeastern Arizona: Stratigraphy, structure, metamorphism, and mineralization: U.S. Geological Survey Bulletin, v. 2042-B, p. B1–B22.

Garrison, J.M., and McMillan, N.J., 1999, Jurassic continental rift magmatism in northeast Mexico: Allogenic metaigneous blocks in the El Papalote evaporite diapir, La Popa basin, Nuevo Leon, Mexico, *in* Bartolini, C., Wilson, J.L., and Lawton, T.F., eds., Mesozoic sedimentary and tectonic history of north-central Mexico: Geological Society of America Special Paper 340, p. 319–332.

Gilluly, J., 1956, General geology of central Cochise County, Arizona: U.S. Geological Survey Paper 281, 169 p.

Glazner, A.F., Bartley, J.M., and Carl, B.S., 1999, Oblique opening and noncoaxial emplacement of the Independence dike swarm, California: Journal of Structural Geology, v. 21, p. 1275–1283, doi: 10.1016/S0191-8141(99)00090-5.

Gleason, J.D., Spencer, J.E., and Richard, S.M., 1999, Mafic sills in the lower McCoy Mountains Formation, western Arizona: Isotopic evidence for a depleted mantle source and a Late Jurassic link with the Bisbee Basin: Geological Society of America Abstracts with Programs, v. 29, no. 2, p. 10.

Goetz, L.K., and Dickerson, P.W., 1985, A Paleozoic transform margin in Arizona, New Mexico, west Texas and northern Mexico, *in* Dickerson, P.W., and Muehlberger, W.R., eds., Structure and tectonics of Trans-Pecos, Texas: West Texas Geological Society Field Conference Publication 85–81, p. 173–184.

González-León, C.M., and Lawton, T.F., 1995, Stratigraphy, depositional environments, and origin of the Cabullona basin, northeastern Mexico, *in* Jacques-Ayala, C., González-León, C.M., and Roldán-Quintana, J., eds., Studies on the Mesozoic of Sonora and adjacent areas: Geological Society of America Special Paper 301, p. 121–142.

Gries, J.C., and Haenggi, W.T., 1970, Structural evolution of the eastern Chihuahua tectonic belt, *in* Seewald, K., and Sundeen, D., eds., The geologic framework of the Chihuahua tectonic belt: Symposium in honor of Professor Ronald K. DeFord: West Texas Geological Society, p. 119–137.

Haenggi, W.T., 1966, Geology of El Cuervo area, northeastern Chihuahua, Mexico [Ph.D. dissertation]: Austin, University Texas, xxiii + 403 p.

Haenggi, W.T., 2001, Tectonic history of the Chihuahua Trough, Mexico, and adjacent USA, Part I: The pre-Mesozoic setting: Boletin de La Sociedad Geologica Mexicana, Tomo, v. LIV, p. 28–66.

Haenggi, W.T., 2002, Tectonic history of the Chihuahua trough, Mexico and adjacent USA, Part II: Mesozoic And Cenozoic: Boletín de la Sociedad Geológica Mexicana, tomo LV, núm. 1, p. 38–94.

Harding, L.E., 1983, Tectonic significance of the McCoy Mountains Formation, southeastern California and southwestern Arizona [Ph.D. dissertation]: Tucson, University of Arizona, 197 p.

Harding, L.E., and Coney, P.J., 1985, The Geology of the McCoy Mountains Formation, southeastern California and southwestern Arizona: Geological Society of America Bulletin, v. 96, p. 755–769, doi: 10.1130/0016-7606(1985)96<755:TGOTMM>2.0.CO;2.

Harrigan, P.J., 1995, Stratigraphy, depositional environments and tectonic setting of the Broken Jug Formation (Late Jurassic (?)–Early Cretaceous) in the Little Hatchet Mountains, southwestern New Mexico [M.S. thesis]: Las Cruces, New Mexico State University, 101 p.

Haxel, G.B., Tosdal, R.M., May, D.J., and Wright, J.E., 1984, Latest Cretaceous and early Tertiary orogenesis in south-central Arizona: Thrust faulting, regional metamorphism, and granitic plutonism: Geological Society of America Bulletin, v. 95, p. 631–653, doi: 10.1130/0016-7606(1984)95<631:LCAETO>2.0.CO;2.

Haxel, G.B., Tosdal, R.M., and Dillon, J.T., 1985, Tectonic setting and lithology of the Winterhaven Formation; a new Mesozoic stratigraphic unit in southeasternmost California and southwestern Arizona: U.S. Geological Survey Bulletin, v. 1599, 19 p.

Hayes, P.T., 1970, Mesozoic stratigraphy of the Mule and Huachuca Mountains, Arizona: U.S. Geological Survey Professional Paper 658-A, 28 p.

Hayes, P.T., and Landis, E.R., 1964, Geologic map of the southern part of the Mule Mountains, Cochise County, Arizona: U.S. Geological Survey Miscellaneous Geologic Investigations Map I-418, scale 1:48,000.

Hayes, P.T., and Raup, R.B., 1968, Geologic map of the Huachuca and Mustang Mountains, Cochise County, Arizona: U.S. Geological Survey Miscellaneous Geologic Investigations Map I-509, scale 1:48,000.

Hennings, P.H., 1994, Structural transect of the southern Chihuahua fold belt between Ojinaga and Aldama, Chihuahua, Mexico: Tectonics, v. 13, no. 6, p. 1445–1460, doi: 10.1029/94TC00800.

Henry, C.D., 1998, Basement controlled transfer zones in an area of low-magnitude extension, eastern Basin and Range Province, Trans-Pecos, Texas, *in* Faulds, J.E., and Stewart, J.H., eds., The regional segmentation of the Basin and Range Province: Geological Society of America Special Paper 323, p. 75–88.

Hodgson, S.A., 2000, Structural geology and Laramide tectonics of the Little Hatchet Mountains, southwestern New Mexico, *in* Lawton, T.F., McMillan, N J., and McLemore, V.T., eds., Southwest Passage—A trip through the Phanerozoic: New Mexico Geological Society Fifty-First Annual Field Conference, p. 109–116.

Hopson, C.A., 1988, Independence dike swarm: Origin and tectonic significance: Eos (Transactions, American Geophysical Union), v. 69, p. 1479.

Humphrey, W.E., 1956, Tectonic framework of northeast Mexico: Gulf Coast Association of Geological Societies Transactions, v. 6, p. 25–35.

Imlay, R.W., 1952, Correlation of Jurassic formations of North America, exclusive of Canada: Geological Society of America Bulletin, v. 63, p. 953–992.

Jacques-Ayala, C., 1995, Paleogeography and provenance of the Lower Cretaceous Bisbee Group in the Caborca–Santa Ana area, northwestern Sonora, *in* Jacques-Ayala, C., González-León, C.M., and Roldán-Quintana, J., eds., Studies of the Mesozoic of Sonora and adjacent areas: Geological Society of America Special Paper 301, p. 79–98.

Jacques-Ayala, C., and Potter, P.E., 1987, Stratigraphy and paleogeography of Lower Cretaceous rocks, Sierra El Chanate, northwest Sonora, Mexico, *in* Dickinson, W.R., and Klute, M.A., eds., Mesozoic rocks of southern Arizona and adjacent areas: Arizona Geologic Society Digest, v. 18, p. 203–214.

James, E.W., 1989, Southern extension of the Independence dike swarm of eastern California: Geology, v. 17, p. 587–590, doi: 10.1130/0091-7613(1989)017<0587:SEOTID>2.3.CO;2.

Janecke, S.U., 1987, Mesozoic and early Tertiary structural history of the Geesaman Wash Area, northeastern Catalina-Rincon metamorphic core complex, *in* Dickinson, W.R., and Klute, M.A., eds., Mesozoic Rocks of southern Arizona and adjacent areas: Arizona Geological Society Digest, v. 18, p. 189–201.

Jensen, P.W., 1998, A structural and geochemical study of the Sierrita porphyry copper system, Pima County, Arizona [M.S. thesis]:Tucson, University of Arizona, 136 p.

Jensen, P.W., and Titley, S.R., 1998, Wrench fault control of porphyry emplacement at the Sierrita porphyry copper deposit, Arizona: Geological Society of America Abstracts with Programs, v. 30, no. 7, p. A-367.

Jimenez, A.J., and Keller, G.R., 2000, Rift basin structure in the border region of northern Chihuahua, *in* Lawton, T.F., McMillan, N.J., and McLemore, V.T., eds., Southwest Passage—A trip through the Phanerozoic: New Mexico Geological Society Fifty-First Annual Field Conference, p. 79–84.

Jones, N.W., and McKee, J.W., Marquez, B., Tovar, J., Long, L.E., and Laudon, T.S., 1984, The Mesozoic La Mula Island, Coahuila, Mexico: Geological Society of America Bulletin, v. 95, p. 1226–1241, doi: 10.1130/0016-7606(1984)95<1226:TMLMIC>2.0.CO;2.

Jones, N.W., McKee, J.W., Anderson, T.H., and Silver, L.T., 1995, Jurassic volcanic rocks in northeastern Mexico: A possible remnant of a Cordilleran magmatic arc, *in* Jacques-Ayala, C., González-León, C.M., and Roldán-Quintana, J., eds., Studies on the Mesozoic of Sonora and adjacent areas: Geological Society of America Special Paper 301, p. 179–190.

Karlstrom, K.E., and Bowring, S.A., 1988, Early Proterozoic assembly of tectonostratigraphic terranes in southwestern North America: Journal of Geology, v. 96, p. 561–576.

Keith, S.B., and Barrett, L.F., 1976, Tectonics of the central Dragoon Mountains: A new look: Arizona Geological Society Digest, v. 10, p. 169–204.

Keith, S.B., and Swan, M.M., 1996, The great Laramide porphyry copper cluster of Arizona, Sonora, and New Mexico: The tectonic setting, petrology and genesis of a world class porphyry metal cluster, *in* Coyner, A.R., and Fahey, P.L., eds., Geology and ore deposits of the American Cordillera: Geological Society of Nevada Symposium Proceedings, Reno/Sparks, Nevada, April 1995, p. 1667–1747.

King, R.E., Dunbar, C.O., Cloud, P.E., and Miller, A.K., 1944, Geology and paleontology of the Permian area northwest of Las Delicias, southwestern Coahuila, Mexico: Part 1. Geology: Geological Society of America Special Paper 52, 170 p.

Klitgord, K., and Schouten, H., 1986, Plate kinematics of the central Atlantic, *in* Vogt, P.R., and Tucholke, B.E., eds., The Western North Atlantic Region: Geological Society of America, Decade of North American Geology, v. M, p. 351–378.

Kluth, C.F., 1982, Geology and mid-Mesozoic tectonics of the northern Canelo Hills, Santa Cruz County, Arizona [Ph.D. dissertation]: Tucson, University of Arizona, 245 p.

Kluth, C.F., 1983, Geology of the northern Canelo Hills and implications for the Mesozoic tectonics of southeastern Arizona, *in* Reynolds, M.W., and Dolly, E.D., eds., Mesozoic paleogeography of the west-central United States: Tulsa, Oklahoma, Society of Economic Paleontologists and Mineralogists, p. 159–171.

Kluth, C.F., Butler, R.F., Harding, L.E., Shafiqullah, M., and Damon, P.E., 1982, Paleomagnetism of Late Jurassic rocks in the northern Canelo Hills, southeastern Arizona: Journal of Geophysical Research, v. 87, p. 7079–7086.

Kottlowski, F.E., 1963, Paleozoic and Mesozoic strata of southwestern and south-central New Mexico: New Mexico Bureau of Mines and Mineral Resources Bulletin, v. 79, 100 p.

Krebs, C.K., and Ruiz, J., 1987, Geochemistry of the Canelo Hills volcanics and implications for the Jurassic tectonic setting of southeastern Arizona, *in* Dickinson, W.R., and Klute, M.A., eds., Mesozoic Rocks of southern Arizona and adjacent areas: Arizona Geological Society Digest, v. 18, p. 139–151.

Lasky, S.G., 1938, Geology and ore deposits of the Lordsburg mining district, Hidalgo County, New Mexico: U.S. Geological Survey Bulletin, v. 885, 62 p.

Lasky, S.G., 1947, Geology and ore deposits of the Little Hatchet Mountains, Hidalgo and Grant Counties, New Mexico: U.S. Geological Survey Professional Paper 208, 101 p.

Laubach, S.E., Reynolds, S.J., and Spencer, J.E., 1987, Mesozoic stratigraphy of Granite Wash Mountains, west-central Arizona, *in* Dickinson, W.R., and Klute, M.A., eds., Mesozoic Rocks of southern Arizona and adjacent areas: Arizona Geological Society Digest, v. 18, p. 91–100.

Lawton, T.F., 1996, Laramide inversion of Late Jurassic–Early Cretaceous extensional faults, SE Arizona and SW New Mexico: Geological Society of America Abstracts with Programs, v. 28, no. 6, p. 509–510.

Lawton, T.F., 2000, Inversion of Late Jurassic–Early Cretaceous extensional faults of the Bisbee Basin, southeastern Arizona and southwestern New Mexico, *in* Lawton, T.F., McMillan, N.J., and McLemore, V.T., eds., Southwest Passage—A trip through the Phanerozoic: New Mexico Geological Society Fifty-First Annual Field Conference, p. 95–102.

Lawton, T.F., and Harrigan, P.J., 1998, Jurassic strata of the Little Hatchet Mountains, Hidalgo County, New Mexico: New Mexico Geology, v. 19, p. 57.

Lawton, T.F., and McMillan, N.J., 1999, Arc abandonment as a cause for passive continental rifting: Comparison of the Jurassic Borderland rift and the Cenozoic Rio Grande rift: Geology, v. 27, p. 779–782, doi: 10.1130/0091-7613(1999)027<0779:AAAACF>2.3.CO;2.

Lawton, T.F., and Olmstead, G.A., 1995, Stratigraphy and structure of the lower part of the Bisbee Group, northeastern Chiricahua Mountains, Arizona, *in* Jacques-Ayala, C., González-León, C.M., and Roldán-Quintana, J., eds., Studies of the Mesozoic of Sonora and adjacent areas: Geological Society of America Special Paper 301, p. 21–39.

Lipman, P.W., and Hagstrum, J.T., 1992, Jurassic ash-flow sheets, calderas, and related intrusions of the Cordilleran volcanic arc in southeastern Arizona: Implications for regional tectonics and ore deposits: Geological Society of America Bulletin, v. 104, p. 32–39, doi: 10.1130/0016-7606(1992)104<0032:JAFSCA>2.3.CO;2.

Lipman, P.W., and Sawyer, D.A., 1985, Mesozoic ash-flow caldera fragments and their relation to porphyry copper deposits: Geology, v. 13, p. 652–656, doi: 10.1130/0091-7613(1985)13<652:MACFIS>2.0.CO;2.

Lopez-Ramos, E., 1980, Geología de México, Tomo II, 2a. edición: Tesis Resendiz S.A., México, D.F., 454 p.

Lucas, S.G., and Lawton, T.F., 2000, Stratigraphy of the Bisbee Group (Jurassic-Cretaceous), Little Hatchet Mountains, New Mexico, *in* Lawton, T.F., McMillan, N.J., and McLemore, V.T., eds., Southwest Passage—A trip through the Phanerozoic: New Mexico Geological Society Fifty-First Annual Field Conference, p. 175–194.

Mann, P., Hampton, M.R., Bradley, D.C., and Burke, K., 1983, Development of pull-apart basins: Journal of Geology, v. 91, no. 5, p. 529–554.

McKee, J.W., Jones, N.W., and Long, L.E., 1990, Stratigraphy and provenance of strata along the San Marcos fault, central Coahuila, Mexico: Geological Society of America Bulletin, v. 102, p. 593–614, doi: 10.1130/0016-7606(1990)102<0593:SAPOSA>2.3.CO;2.

McKee, J.W., Jones, N.W., and Anderson, T.H., 1999, Late Paleozoic and early Mesozoic history of the Las Delicias terrane, Coahuila, Mexico, *in* Bartolini, C., Wilson, J.L., and Lawton, T.F., eds., Mesozoic sedimentary and tectonic history of north-central Mexico: Geological Society of America Special Paper 340, p. 161–189.

McKee, M.B., and Anderson, T.H., 1998, Mass gravity deposits and structures in the Lower Cretaceous of Sonora, Mexico: Geological Society of America Bulletin, v. 110, no. 12, p. 1516–1529, doi: 10.1130/0016-7606(1998)110<1516:MGDASI>2.3.CO;2.

McLemore, V.T., Dunbar, N., Heitzler, M.T., and Donahue, K., 2000, Geology and mineral deposits of the Victorio mining district, Luna County, New Mexico—Preliminary observations, *in* Lawton, T.F., McMillan, N.J., and McLemore, V.T., eds., Southwest Passage—A trip through the Phanerozoic: New Mexico Geological Society Fifty-First Annual Field Conference, p. 267–278.

Miller, W.J., 1944, Geology of the Palm Springs–Blythe strip, Riverside County, California: California Journal of Mines and Geology, Quarterly Chapter of the State Mineralogist's Report XL, v. 40, no. 1, p. 11–72.

Monreal, R., and Longoria, J.F., 1999, A revision of the Upper Jurassic and Lower Cretaceous stratigraphic nomenclature for the Chihuahua Trough, north-central Mexico: Implications for lithocorrelations, *in* Bartolini, C., Wilson, J.L., and Lawton, T.F., eds., Mesozoic sedimentary and tectonic history of north-central Mexico: Geological Society of America Special Paper 340, p. 69–92.

Moody, J.D., and Hill, M.F., 1956, Wrench fault tectonics: Geological Society of America Bulletin, v. 67, p. 1207–1246.

Moore, B.N., Tolman, C.F., Butler, B.S., and Hernon, R.M., 1941, Geology of the Tucson quadrangle, Arizona: U.S. Geological Survey Open-File Report O-1941, 20 p.

Muehlberger, W.R., 1965, Late Paleozoic movement along the Texas lineament: New York Academy of Sciences Transactions, ser. 11, v. 27, p. 385–392.

Muehlberger, W.R., 1980, Texas lineament revisited: 31st Field Conference Guidebook, Trans-Pecos Region, New Mexico Geological Society, p. 113–121.

Murray, G.E., 1986, Musings about some of the tectonics of the southwestern United States and northern Mexico: The Texas Journal of Science, v. 38, no. 4, p. 301–326.

Murray, G.E., 1989, The California-Tamaulipas geosuture: A review of some facts, interpretations and speculations, *in* Flis, J.E., Price, R.C., and Sarg, J.F., eds. Search for the subtle trap: Hydrocarbon exploration in mature basins: West Texas Geological Society Publication 89–85, p. 211–221.

Nourse, Jonathan A., 1990, Tectonostratigraphic development and strain history of the Magdalena metamorphic core complex, Sonora, Mexico, *in* Gehrels, G., and Spencer, J., eds., Geologic excursions through the Sonoran desert region, Arizona and Mexico: Arizona Geological Society Special Paper 7, p. 155–164.

Nourse, J.A., 1995, Jurassic-Cretaceous paleogeography of the Magdalena region, northern Sonora, and its influence of the positioning of metamorphic core complexes, *in* Jacques-Ayala, C., González-León, C.M., and Roldán-Quintana, J., eds., Studies of the Mesozoic of Sonora and adjacent areas: Geological Society of America Special Paper 301, p. 59–78.

Nourse, J.A., 2001, Tectonic insights from an Upper Jurassic–Lower Cretaceous stretched-clast conglomerate, Caborca-Altar region, Sonora, Mexico, *in* Gonzalez, C.M., ed., Geology of northwestern Mexico and adjacent areas: Journal of South American Earth Sciences, v. 14, no. 5, p. 453–474.

Pálfy, J., Smith, P.L., and Mortensen, J.K., 2000, A U-Pb and $^{40}Ar/^{39}Ar$ time scale for the Jurassic: Canadian Journal of Earth Sciences, v. 37, p. 923–944, doi: 10.1139/cjes-37-6-923.

Pierce, F.L., 1958, Structure and petrology of part of the Santa Catalina Mountains [Ph.D. dissertation]: Tucson, University of Arizona, 86 p.

Ransome, F.L., 1904, The geology and ore deposits of the Bisbee Quadrangle Arizona: U.S. Geological Survey Professional Paper 21, 167 p.

Ransome, F.L., 1915, The Tertiary orogeny in the North American Cordillera and its problems, *in* Rice, W.N., et al., eds., Problems of American geology: New Haven, Yale University Press, p. 287–376.

Reynolds, S.J., Spencer, J.E., and DeWitt, E., 1987, Stratigraphy and U-Th-Pb geochronology of Triassic and Jurassic rocks in west-central Arizona, *in* Dickinson, W.R., and Klute, M.A., eds., Mesozoic rocks of southern Arizona and adjacent areas: Arizona Geological Society Digest, v. 18, p. 65–80.

Richard, S.M., Reynolds, S.J., and Spencer, J.E., 1987, Mesozoic stratigraphy of the Little Harquahala and Harquahala Mountains, west-central Arizona, *in* Dickinson, W.R., and Klute, M.A., eds., Mesozoic rocks of southern Arizona and adjacent areas: Arizona Geological Society Digest, v. 18, p. 101–120.

Richard, S.M., Spencer, J.E., Tosdal, R.M., and Stone, P., 1993, Preliminary geologic map of the southern Plomosa Mountains, La Paz County, Arizona: Arizona Geological Survey Open-File Report 93-9, 27 p.

Richard, S.M., Stanton N., Boettcher, S.S., Hamilton, W.B., Hoisch, T.D., and Tosdal, R.M., 1994, Mesozoic tectonics of the Maria belt, west-central Arizona and southeastern California, *in* McGill, S.F. and Ross, T.M., Geological investigations of an active margin: Geological Society of America, 27th Annual Meeting Cordilleran Section Guidebook, p. 272–291.

Richard, S.M., Reynolds, S.J., Spencer, J.E., and Pearthree, P.A., 2000, Geologic map of Arizona: Arizona Geological Survey Map 35, scale 1:1,000,000.

Riggs, N.R., 1987, Stratigraphy, structure, and geochemistry of Mesozoic rocks in the Pajarito Mountains, Santa Cruz County, Arizona, *in* Dickinson, W.R., and Klute, M.A., eds., Mesozoic rocks of southern Arizona and adjacent areas: Arizona Geological Society Digest, v. 18, p. 165–176.

Riggs, N.R., and Busby-Spera, C.J., 1991, Facies analysis of an ancient, dismembered, large caldera complex and implications for intra-arc subsidence: Middle Jurassic strata of Cobre Ridge, southern Arizona, USA: Sedimentary Geology, v. 74, p. 39–68, doi: 10.1016/0037-0738(91)90034-B.

Riggs, N.R., and Haxel, G.B., 1990, The Early to Middle Jurassic magmatic arc in southern Arizona: Plutons to sand dunes: Arizona Geological Survey Special Paper, v. 7, p. 90–103.

Riggs, N.R., Mattinson, J.M., and Busby, C.J., 1993, Correlation of Jurassic eolian strata between the magmatic arc and the Colorado Plateau: New U-Pb geochronological data from southern Arizona: Geological Society of America Bulletin, v. 105, p. 1231–1246, doi: 10.1130/0016-7606(1993)105<1231:COJESB>2.3.CO;2.

Roberts, D.C., 1989, Sedimentation and tectonics of Cerro Los Pañales area, east-central Chihuahua, Mexico [M.S. thesis]: El Paso, University of Texas, 175 p.

Rodríguez-Castañeda, J.L., 1997, Gravity-sliding structures in Cretaceous–early Tertiary rocks in north-central Sonora, Mexico—Regional significance: Revista Mexicana de Ciencias Geologicas, v. 14, no. 1, p. 1–12.

Rodríguez-Castañeda, J.L., 2000, Deformación Sonorense: Tectónica extensional en el área Rancho San Antonio, Sonora norte-central—sus efectos y características en rocas cretácicas y terciarias [Ph.D. thesis]: México, D.F., Universidad Nacional Autónoma de México, 217 p.

Santamaria-O., D., Ortuno-A., F., and Adatte, T., Ortiz-U., A., Riba-R., A., and Franco-N., S., 1991, Evolucion geodinamica de la Cuenca de Sabinas y sus implicaciones petroleras Estado de Coahuila: Instituto Mexicano del Petroleo, Subdirrecion de Tecnologia de Exploración, Gerencia de investigación Aplicada a la Exploración, Tomo I, 209 p. Segerstrom, L.J., 1987, Geology of the Planchas de Plata area, northern Sonora, Mexico, *in* Dickinson, W.R., and Klute, M.A., eds., Mesozoic rocks of southern Arizona and adjacent areas: Arizona Geological Society Digest, v. 18, p. 153–164.

Sherrod, D.R., and Koch, R.D., 1987, Structure and stratigraphy of Mesozoic sedimentary and volcanic rocks in the New Water Mountains, southwestern Arizona, *in* Dickinson, W.R., and Klute, M.A., eds., Mesozoic rocks of southern Arizona and adjacent areas: Arizona Geological Society Digest, v. 18, p. 81–90.

Silver, L.T., and Anderson, T.H., 1974, Possible left-lateral early to middle Mesozoic disruption of the southwestern North American carton margin: Geological Society of America Abstracts with Programs, v. 6, no. 7, p. 955–956.

Simons, F.S., 1974, Geologic map and sections of the Nogales and Lochiel Quadrangles, Santa Cruz County, Arizona: U.S. Geological Survey Miscellaneous Geological Investigations Map I-762, scale 1:48,000.

Simons, F.S., Raup, R.B., Hayes, P.T., and Drewes, H., 1966, Exotic blocks and course breccias in Mesozoic volcanic rocks of southeastern Arizona, in Geological Survey Research 1966: U.S. Geological Survey Professional Paper 550-D, p. D12–D22.

Smith, C.I., 1981, Review of the geologic setting, stratigraphy, and facies distribution of the Lower Cretaceous in northern Mexico, *in* Smith, C.I., and Brown, J.B., eds., Lower Cretaceous stratigraphy and structure, northern Mexico: Field Trip Guidebook, West Texas Geological Society Publication 81–74, p. 1–27.

Stewart, J.H., McMenamin, M.A.S., and Morales-Ramirez, J.M., 1984, Upper Proterozoic and Cambrian rocks in the Caborca region, Sonora, Mexico—Physical stratigraphy, biostratigraphy, paleocurrent studies, and regional relations: U.S. Geological Survey Professional Paper 1309, 36 p.

Suemnicht, G.A., 1977, The Geology of the Canada del Oro Headwaters, Santa Catalina Mountains [M.S. thesis]: Tucson, University of Arizona, 108 p.

Swan, M.M., 1976, The Stockton Pass fault: An element of the Texas lineament [M.S. thesis]: Tucson, University of Arizona, 119 p.

Taliaferro, N.L., 1933, An occurrence of Upper Cretaceous sediments in northern Sonora, Mexico: Journal of Geology, v. 41, p. 12–37.

Thomas, W.A., 1988, Early Mesozoic faults of the northern Gulf Coastal Plain in the context of opening of the Atlantic Ocean, *in* Manspeizer, W., ed., Triassic-Jurassic rifting, continental breakup and the origin of the Atlantic Ocean and passive margins, Part A: Elsevier, The Netherlands, Developments in Geotectonics 22, p. 463–476.

Thorman, C.H., 1977, Geologic map of the Coyote Peak and Brockman Quadrangles, Hidalgo and Grant Counties, New Mexico: U.S. Geological Survey Miscellaneous Field Studies Map MF-924, scale 1:24,000.

Thorman, C.H., and Drewes, H., 1978, Cretaceous–early Tertiary history of the northern Pyramid Mountains, southwestern New Mexico *in* Callender, J.F., Wilt, J.C., and Clemons, R.E., eds., Land of Cochise, southeastern Arizona: Guidebook Twenty-Ninth Field Conference, New Mexico Geological Society, p. 215–218.

Thorman, C.H., and Drewes, H., 1980, Geologic map of the Victorio Mountains, New Mexico: U.S. Geological Survey Miscellaneous Field Studies Map MF-1175, scale 1:24,000.

Titley, S.R., 1976, Evidence for a Mesozoic linear tectonic pattern in southeastern Arizona: Arizona Geological Society Digest, v. X, p. 71–101.

Tosdal, R.M., and Stone, P., 1994, Stratigraphic relations and U-Pb geochronology of the Upper Cretaceous upper McCoy Mountains Formation, southwest Arizona: Geological Society of America Bulletin, v. 106, p. 476–491, doi: 10.1130/0016-7606(1994)106<0476:SRAUPG>2.3.CO;2.

Tosdal, R.M., Haxel, G.B., and Wright, J.E., 1989, Jurassic geology of the Sonoran desert region, southern Arizona, southeastern California, and northernmost Sonora: Construction of a continental margin magmatic arc, *in* Jenny, J.P., and Reynolds, S.J., eds., Geologic evolution of Arizona: Arizona Geological Society Digest, v. 17, p. 397–434.

Vedder, L.K., 1984, Stratigraphic relationships between the Late Jurassic Canelo Hills volcanics and the Glance Conglomerate, southeastern Arizona [M.S. thesis]: Tucson, University of Arizona, 129 p.

Wiley, M.A., and Muehlberger, W.R., 1971, The Texas lineament, *in* Seewald, K., and Sundeen, D., eds., The geologic framework of the Chihuahua tectonic belt: West Texas Geological Society Publication 71–59, p. 15–23.

Wilson, J.L., 1999, Controls of the wandering path of the Cupido reef trend in northeastern Mexico, *in* Bartolini, C., Wilson, J.L., and Lawton, T.F., eds., Mesozoic sedimentary and tectonic history of north-central Mexico: Geological Society of America Special Paper 340, p. 135–144.

Wright, J.E., Haxel, G.B., and May, D.J., 1981, Early Jurassic U-Pb isotopic ages for Mesozoic supracrustal sequences, Papago Indian Reservation, southern Arizona: Geological Society of America Abstracts with Programs, v. 13, no. 2, p. 115.

Young, K., 1983, Mexico, *in* Moullade, M., and Nairn, A.E., eds., The Phanerozoic geology of the world, Volume II: The Mesozoic, Part B: New York, Elsevier, North-Holland, Inc., p. 61–88.

Manuscript Accepted by the Society 24 March 2005

Printed in the USA

Geological Society of America
Special Paper 393
2005

Contrasting Proterozoic basement complexes near the truncated margin of Laurentia, northwestern Sonora–Arizona international border region

Jonathan A. Nourse*
Department of Geological Sciences, California State Polytechnic University, Pomona, California 91768, USA

Wayne R. Premo
U.S. Geological Survey, Denver Federal Center, Denver, Colorado 80225, USA

Alexander Iriondo*
Department of Geological Sciences, University of Colorado, Boulder, Colorado 80309, USA, and *U.S. Geological Survey, MS 974, Box 25046, Denver Federal Center, Denver, Colorado 80225, USA*

Erin R. Stahl
164 El Camino Way, Claremont, California 91711, USA

ABSTRACT

We utilize new geological mapping, conventional isotope dilution–thermal ionization mass spectrometry (ID-TIMS) and sensitive high-resolution ion microprobe (SHRIMP) U-Pb zircon analyses, and whole-rock radiogenic isotope characteristics to distinguish two contrasting Proterozoic basement complexes in the international border region southeast of Yuma, Arizona. Strategically located near the truncated southwest margin of Laurentia, these Proterozoic exposures are separated by a northwest-striking Late Cretaceous batholith. Although both complexes contain strongly deformed Paleoproterozoic granitoids (augen gneisses) intruded into fine-grained host rocks, our work demonstrates marked differences in age, host rock composition, and structure between the two areas.

The Western Complex reveals a >5-km-thick tilted section of finely banded felsic, intermediate, and mafic orthogneiss interspersed with tabular intrusive bodies of medium-grained leucocratic biotite granite (1696 ± 11 Ma; deepest level), medium-grained hornblende-biotite granodiorite (1722 ± 12 Ma), and coarse-grained porphyritic biotite granite (1725 ± 19 Ma; shallowest level). Penetrative ductile deformation has converted the granites to augen gneisses and caused isoclinal folding and transposition of primary contacts. Exposed in a belt of northwest-trending folds, these rocks preserve southwest-vergent shear fabric annealed during amphibolite facies

*janourse@csupomona.edu. Present address, Iriondo: Centro de Geociencias, UNAM, Campus Juriquilla, C.P. 76230 Juriquilla, Querétaro, México.

Nourse, J.A., Premo, W.R., Iriondo, A., and Stahl, E.R., 2005, Contrasting Proterozoic basement complexes near the truncated margin of Laurentia, northwestern Sonora–Arizona international border region, *in* Anderson, T.H., Nourse, J.A., McKee, J.W., and Steiner, M.B., eds., The Mojave-Sonora megashear hypothesis: Development, assessment, and alternatives: Geological Society of America Special Paper 393, p. 123–182, doi: 10.1130/2005.2393(04). For permission to copy, contact editing@geosociety.org.

metamorphism, when crystalloblastic textures developed. Deformation and regional metamorphism occurred before emplacement of 1.1 Ga(?) mafic dikes.

Throughout the Eastern Complex, meta-arkose, quartzite, biotite schist, and possible felsic metavolcanic rocks comprise the country rocks of strongly foliated medium- and coarse-grained biotite granite augen gneisses that yield mean $^{207}Pb/^{206}Pb$ ages of 1646 ± 10 Ma, 1642 ± 19 Ma, and 1639 ± 15 Ma. Detrital zircons from four samples of host sandstone are isotopically disturbed; nevertheless, the data indicate a restricted provenance (ca. 1665 Ma to 1650 Ma), with two older grains (1697 and 1681 Ma). The pervasively recrystallized Paleoproterozoic map units strike parallel to foliation and are repeated in south-trending folds that are locally refolded about easterly hinges. Southeasterly lineation developed in augen gneiss and host strata becomes penetrative in local domains of L-tectonite. Regional metamorphism associated with this tectonism persisted until ca. 1590 Ma, as recorded by metamorphic growths within some zircon grains. Mesoproterozoic intrusions that crosscut the Paleoproterozoic metasediments and augen gneisses include coarsely porphyritic biotite granite (1432 ± 6 Ma) and diabase dikes (1.1 Ga?). Emplacement of the granite was accompanied by secondary high-U overgrowths, dated at 1433 ± 8 Ma, on some of the Paleoproterozoic detrital zircons, and apparently was also responsible for resetting the whole-rock Pb isotopic systematics (1441 ± 39 Ma) within these Eastern Complex augen gneisses.

Younger plutons emplaced into both Proterozoic basement complexes include medium-grained quartz diorite (73.4 ± 3.3 Ma and 72.8 ± 1.7 Ma), Late Cretaceous hornblende-biotite granodiorite, and Paleogene leucocratic biotite granite. Neogene sedimentary and volcanic strata overlie basement along unconformities that are tilted to the northeast, southeast, or southwest. A brittle normal fault, dipping gently northeast, juxtaposes Tertiary andesite with Paleoproterozoic metasandstone. These relationships suggest that the area shares a common history of mid-Tertiary extension with southwestern Arizona. Later influence of the southern San Andreas fault system is implied by multiple dextral offsets of pre-Tertiary units across northwest-trending valleys.

Our structural, geochronologic, and isotopic data provide new information to constrain pre–750 Ma Rodinia reconstructions involving southwestern Laurentia. Whole-rock U-Th-Pb and Rb-Sr isotopic systematics in both Paleoproterozoic gneiss complexes are disturbed, however, well-behaved Sm-Nd analyses preserve depleted initial ε_{Nd} values (+2 to +4) that are distinct from the Mojave crustal province, but overlapping with the Yavapai and Mazatzal Provinces of Arizona. The Eastern Complex has the appropriate age and Nd isotopic signature to be part of the Mazatzal Province, but records major tectonism and metamorphism at ca. 1.6 Ga that postdates the Mazatzal orogeny. Deformed granitoids of the Western Complex have "Yavapai-type" ages and ε_{Nd} but display structures discordant to the southwesterly Yavapai trend in central Arizona. The Western Complex lies along-strike with similar-age rocks (1.77 Ga to 1.69 Ga) of the "Caborca block" that have only been studied in detail near Quitovac and south of Caborca. Collectively, these rocks form a northwest-trending strip of basement situated at the truncated edge of Laurentia. The present-day basement geography may reflect an original oroclinal bend in the Yavapai orogenic belt. Alternatively, the western Proterozoic belt of Sonora may represent displaced fragments of basement juxtaposed against the Yavapai-Mazatzal Provinces along a younger sinistral transform fault (e.g., the Late Jurassic Mojave-Sonora megashear or the Permian Coahuila transform). Crustal blocks with these specific petrologic, geochronologic, and isotopic characteristics can be found in south-central and northeastern portions of the Australian Proterozoic basement, further supporting a connection between the two continents prior to breakup of the Rodinian supercontinent.

Keywords: Sonora, Proterozoic, Rodinia, SHRIMP, zircon.

INTRODUCTION

Precambrian crystalline rocks in the international border region of northwestern Sonora and southwestern Arizona (Fig. 1) constitute the southwestward limit of Proterozoic basement along the truncated margin of Laurentia near latitude 32°N. They also crop out near a poorly constrained, possibly disrupted intersection between the Mojave, Yavapai, and Mazatzal crustal provinces (Karlstrom and Bowring, 1988; Wooden and Miller, 1990; Wooden and DeWitt, 1991), and the Caborca block (Anderson and Silver, 1979, 1981; Iriondo et al., 2004; Anderson and Silver, this volume). These diverse rocks and structures predate breakup of the Rodinia supercontinent at ca. 750 Ma (Stewart, 1972; Ross et al., 1989; Karlstrom et al., 2000). They occupy a strategic position with regard to paleogeographic reconstructions of the Rodinian and the Laurentian cratons. Integration of our geological mapping, geochronology, and isotopic analyses with recent work on Proterozoic basement at Quitovac (Iriondo, 2001) yields a new data set useful for evaluating which continent, e.g., Antarctica (Moores, 1991), Australia (Karlstrom et al., 1999), Siberia (Sears and Price, 2000), or south China (Li et al., 1995) was attached to southwestern Laurentia in the controversial Rodinia reconstructions. The data also constrain the configuration of certain blocks of Proterozoic crust in Sonora (Fig. 1), and offer a means to assess possible late Paleozoic or Late Jurassic strike-slip displacements of these blocks (Silver and Anderson, 1974; Dickinson, 2000).

We describe new geological mapping, present results from analyses of U-Pb in zircon using conventional isotope dilution–thermal ionization mass spectrometry (ID-TIMS) and sensitive high-resolution ion microprobe (SHRIMP) methods, and report whole-rock Sm-Nd, U-Pb, and Rb-Sr isotopic data from a 2000 km^2 area of crystalline basement (Fig. 2) straddling the Sonora-Arizona border. This area encompasses the cryptic trace of the Mojave-Sonora megashear (Silver and Anderson, 1974; Anderson and Silver, 1981; this volume). Previous workers (Anderson and Silver, 1979) reported the presence of pre–1.67 Ga layered gneisses and 1.45 Ga granite near Mexican Highway 2, but no field relations were described. Our field and laboratory work defines the areal extent of Paleoproterozoic gneiss and Mesoproterozoic granite near the international border, and establishes the stratigraphic sequence and structural chronology. We show that despite gross similarities in fabric development and metamorphic grade, the Paleoproterozoic gneisses form two geographically and compositionally distinct complexes of fine-grained host rocks intruded by granitoids with different emplacement ages of ca. 1.73–1.70 Ga and ca. 1.65 Ga. Mesoproterozoic granite is recognized only in the younger Eastern Complex. Comparison of structural geometry also reveals significant discordance between the two areas.

One exciting result of our study is apparent juxtaposition of "Mazatzal-type" rocks of the Eastern Complex against the older "Yavapai-like" Western Complex. We speculate that the northwest-trending boundary between the two Proterozoic belts (presently obscured by a Cretaceous batholith) represents a Proterozoic suture or a younger strike-slip fault, such as the hypothetical Permian-Triassic "Coahuila transform" (Dickinson and Lawton, 2001) or the Late Jurassic "Mojave-Sonora megashear" (Silver and Anderson, 1974; Anderson and Silver, this volume). To underscore the implications of various terrane juxtaposition models for the configuration of southwest Laurentia, we present several alternative paleogeographic reconstructions.

BASEMENT GEOLOGY AND STRUCTURE

General Overview

Proterozoic crystalline rocks underlie rugged ranges on both sides of Highway 2 in northwestern Sonora and compose several small mountains or isolated hills north of the international border in the Cabeza Prieta region (Fig. 2). These dark-weathering exposures contrast markedly with light-pink Late Cretaceous–early Tertiary biotite ± muscovite granite plutons of the Gunnery Range batholith (Shafiqullah et al., 1980). Conspicuous, tilted nonconformities separate the crystalline rocks from overlying Neogene sections and Quaternary basalt flows that become increasingly abundant from west to east.

The Proterozoic exposures are geographically divided into Western and Eastern Complexes by Late Cretaceous granodiorite of Sierra El Aguila (Anderson and Silver, 1979; Fig. 2). Both complexes contain Paleoproterozoic metaplutonic rocks (augen gneisses) interlayered with fine-grained framework gneisses. Paleoproterozoic gneisses in both areas are similarly recrystallized and record amphibolite facies metamorphism and deformational fabrics. Disparities between the two complexes include: (1) significant discordance in major fold trends, (2) different protolith compositions, and (3) distinctly younger U-Pb zircon ages for augen gneisses of the Eastern Complex.

The Western Complex records intrusion of plutons (1725 ± 12 Ma, 1722 ± 19 Ma, and 1696 ± 11 Ma) into banded gneisses derived from igneous and perhaps immature sedimentary protoliths. In contrast, the oldest rocks in the east are derived from arkose, quartzose sandstone, and possible felsic volcanic protoliths that accumulated no earlier than 1697 Ma. These metasedimentary strata were intruded by two varieties of granite at ca. 1645 Ma. Regional metamorphism and deformation in the Western Complex occurred after 1696 Ma but prior to emplacement of 1.1 Ga(?) mafic dikes. In the Eastern Complex, strong fabrics developed in the ca. 1645 Ma granite augen gneisses are sharply intruded by nonfoliated granite (1432 ± 6 Ma) and diabase (1.1 Ga?).

Within both Proterozoic complexes, the quality of outcrop and degree of geologic coherence between various ranges is excellent, and original mid-crustal Proterozoic structures are remarkably well preserved. Tertiary extension has broken the region into northwest-elongate blocks that have been tilted and probably displaced along detachment faults; nevertheless, Proterozoic stratigraphic and structural features are easily correlated

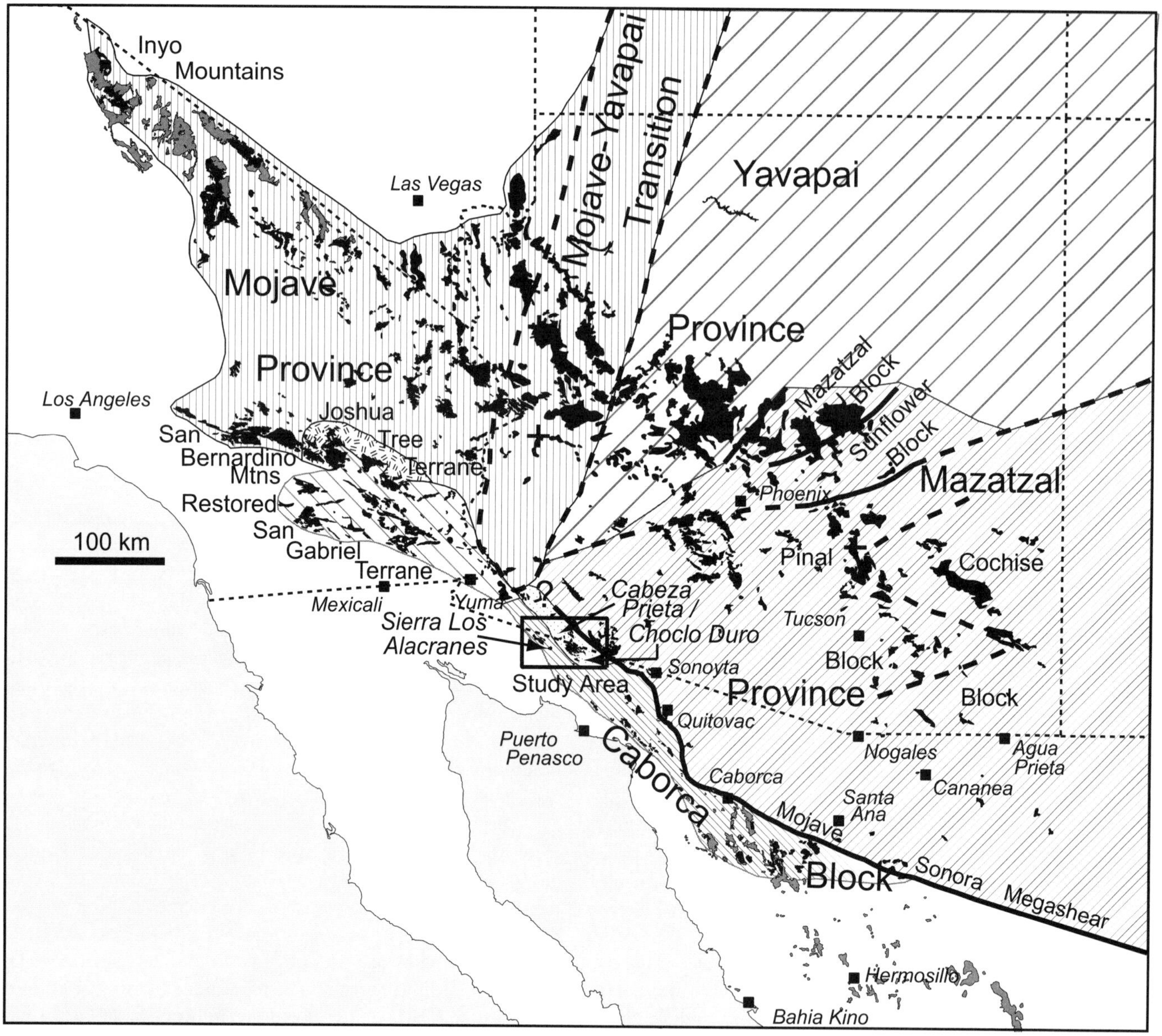

Figure 1. Map showing location of the Western (Sierra Los Alacranes) and Eastern (Cabeza Prieta/Choclo Duro) basement complexes relative to Proterozoic crustal provinces or blocks of southwestern North America. Crustal block or terrane nomenclature from Powell (1982), Anderson and Silver (1981), Karlstrom and Bowring (1988), Wooden and De Witt (1991), Bender et al. (1993), and Eisele and Isachsen (2001). Dashed lines indicate boundaries between Paleoproterozoic crustal provinces, inferred from Pb isotopes (Wooden and DeWitt, 1991). Solid lines indicate structural discontinuities within Mazatzal-age rocks (Karlstrom and Bowring, 1988). The truncated margin of Laurentia corresponds approximately to the western limit of Proterozoic outcrops, shown in black. Neoproterozoic–Lower Paleozoic miogeoclinal strata in the Caborca and Inyo Mountains regions are gray. The San Gabriel terrane is restored ~300 km along the late Cenozoic San Gabriel and San Andreas faults as postulated by Dillon and Ehlig (1993) (Baja California and California borderland are not restored). Trace of the hypothetical Mojave-Sonora megashear is modified from Anderson and Silver (this volume).

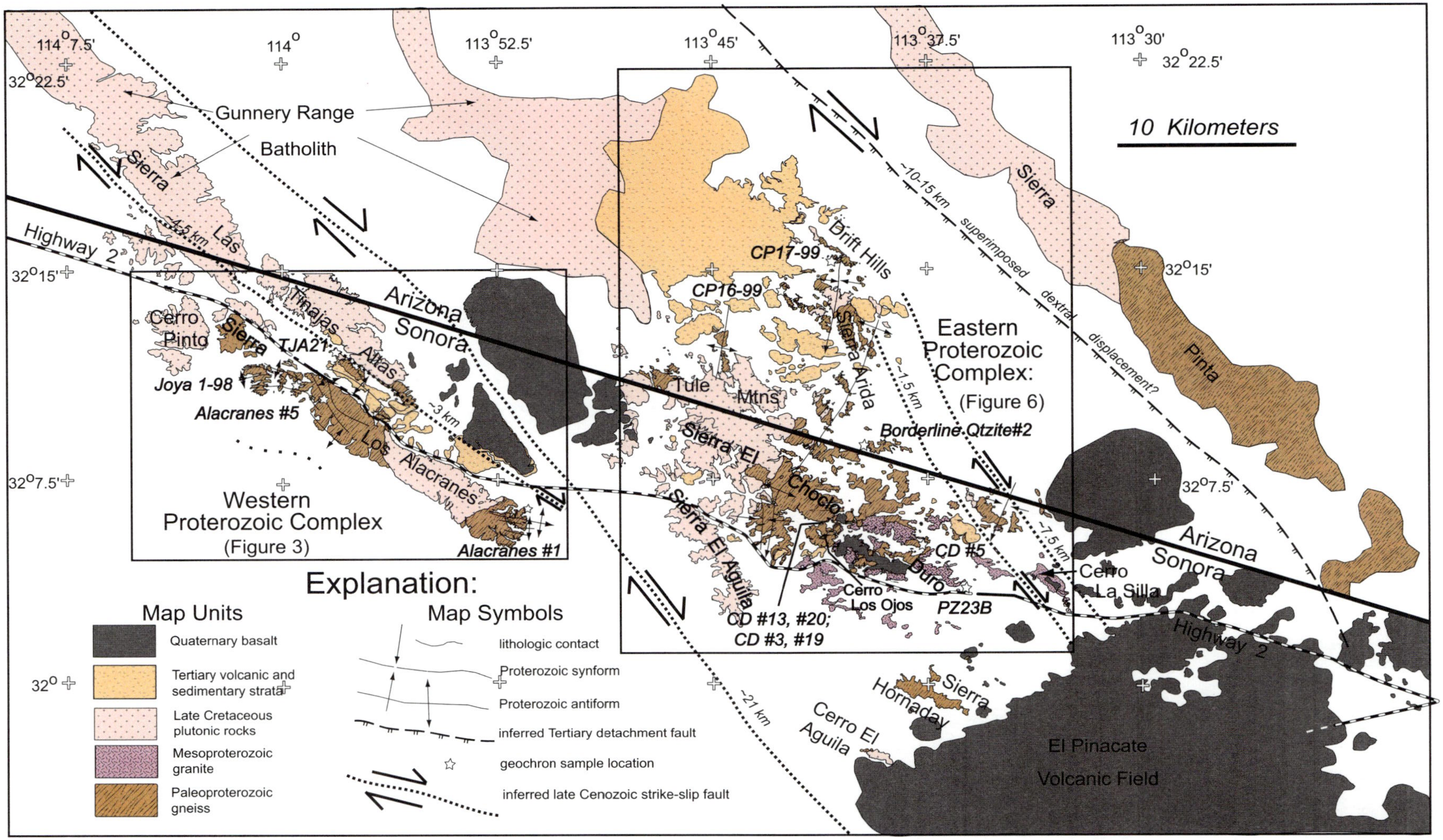

Figure 2. Generalized geologic map of the study area showing sample locations and distribution of Paleoproterozoic gneiss or Mesoproterozoic granite relative to Late Cretaceous–Paleocene plutonic rocks and Quaternary basalts. Outcrops of Neogene volcanic and sedimentary strata define several tilt blocks associated with southwestward displacement on an inferred detachment fault between Sierra Pinta and Sierra Arida. Dotted lines indicate late Cenozoic strike-slip faults, with magnitude of displacement inferred from dextral offsets of distinct basement contacts.

from range to range. Hence, the study area (Fig. 2) contains a powerful new Proterozoic data set ideal for characterizing crust situated at the southwest edge of Laurentia. Field relationships, petrography, and structure are described below from three focus areas where the pertinent outcrops were sampled for geochronologic and isotopic analyses.

Proterozoic Geology and Structure of the Western Complex

Geologic Setting and Petrography of Sierra Los Alacranes and Sierra Las Tinajas Altas

Proterozoic gneisses of the Western Complex underlie much of Sierra Los Alacranes and Sierra Las Tinajas Altas, prominent northwest-trending ranges located adjacent to Highway 2, (Fig. 3). Both mountains contain strongly foliated banded gneiss and Paleoproterozoic granitic gneiss intruded by Late Cretaceous hornblende-biotite quartz diorite and nonfoliated leucocratic biotite granite. The 20-km-long valley separating the two ranges is occupied by a Neogene(?) section composed of (from older to younger): boulder-cobble conglomerate, arkosic sandstone, andesite breccia, and basalt. Sharply exposed basal nonconformities with crystalline basement dip toward the basin axis. Clasts in the basal conglomerate are derived predominantly from basement exposures located 10–50 km to the east. Distinctive boulders of 1.4 Ga granite, Sierra El Aguila granodiorite, and Eastern Complex gneiss are abundant in this unit. Slightly tilted basalt flows that cap the section on Mesas de Malpais (Fig. 3) yielded a whole-rock K-Ar age of 10.49 ± 0.41 Ma (Shafiqullah et al., 1980).

Older banded gneisses. The oldest rocks of the Western Complex are felsic, intermediate, and mafic gneisses characterized by centimeter- to decimeter-scale banding. Felsic and intermediate gneiss (map unit PCgn; Fig. 4A) weather pink or tan to medium gray, and contain variable proportions of biotite, quartz, and feldspar. Crystalloblastic textures range from fine-grained granoblastic to medium-grained porphyroblastic. It is uncertain whether the larger feldspar grains in the porphyroblastic gneisses represent phenocrysts in porphyritic volcanic or hypabyssal protoliths or porphyroclasts in mylonitized plutons. Fine- to medium-grained mafic gneisses and amphibolite (unit PCmgn; Fig. 4B) contain predominantly hornblende, biotite, and plagioclase, with subordinate quartz. Pyroxene may be replaced by hornblende. Bulk compositions and relict plutonic textures in the medium-grained varieties suggest that protoliths included quartz diorite, diorite, and gabbro. Finer-grained varieties may have been derived from andesite or basalt. Part of the banded gneiss is composed of incompletely transposed tabular intrusions that display contacts slightly discordant to host rock banding. Absence of quartzite, marble, and pelitic assemblages in the gneisses precludes the likelihood of mature sedimentary protoliths. We infer that these banded gneisses represent a heterogeneous assemblage of rhyolitic, dacitic, and andesitic flows or porphyries with possible interstratified immature sediments, locally intruded by granite, diorite, and gabbro, i.e., shallow levels of a magmatic arc.

Banded gneiss occurs as xenoliths and screens within all of the Paleoproterozoic plutons described below. Although primary contacts between the gneisses and plutons have been tectonically overprinted, it is still possible to find convincing intrusive contacts near the margins of large plutons. For example, coarse-grained granite dikes (Fig. 4C) emanating from the augen gneiss body of Sierra Las Tinajas Altas (unit PCagn; see below) cut discordantly across compositional layering in the mafic gneiss. Figures 4D and 4E show typical augen gneiss dikes from Sierra Los Alacranes that have been tightly folded with their country rocks.

Paleoproterozoic intrusions. Three distinct Paleoproterozoic granitoids intrude different structural levels of the banded gneiss. All are generally classified as "augen gneisses" because they display a pervasive subsolidus foliation in which eye-shaped feldspars have resulted from tectonic shearing. Foliation in the granitoids and host gneiss is generally concordant. At the scale of Figure 3, variations in foliation orientation define southeast-plunging folds. The Paleoproterozoic augen gneisses are described below from deepest structural levels in the northwest to shallowest levels in the southeast.

Several hills at the northwest end of Sierra Los Alacranes are underlain by pervasively recrystallized, strongly foliated, medium-grained, pinkish-orange, leucocratic biotite alkali granite (unit PCgrgn) that forms concordant sheet-like intrusions into layered felsic and mafic gneiss. Thin sections of the gneissic granite reveal that alkali feldspars have been flattened via recrystallization of strain-free subgrains. Biotite and quartz exhibit similar recrystallization effects, all of which contribute to a macroscopic foliation. Foliation in the granite is concordant with its host gneisses and is isoclinally folded in places. Variably foliated and folded dikes of similar granite observed within structurally higher granodiorite (unit PCgd; described below) suggest that the PCgrgn is the younger of the two plutons. Analyzed granite sample Mina La Joya 1-98 was collected from the hinge region of a prominent synform (Fig. 3).

Medium-grained, slightly porphyritic hornblende-biotite granodiorite (unit PCgd) forms the folded core of northwest Sierra Los Alacranes. This pluton is variably foliated, with the fabric becoming mylonitic near its western margin (Fig. 4F). Xenoliths of felsic and mafic gneiss are abundant near this contact. The granodiorite and gneissic xenoliths display a southwest-vergent deformational fabric (described below) that may penetrate all Paleoproterozoic rocks of the Western Complex. Hornblende exhibits a mottled appearance in hand specimen, and is intergrown with finely recrystallized biotite in thin section. Subhedral sphene is a common accessory mineral. Amphibolite facies metamorphism that accompanied deformation has resulted in epitaxial overgrowths of sphene on magnetite. Analyzed sample Alacranes #5 was collected from the moderately foliated interior of the granodiorite pluton (Fig. 3).

Strongly foliated coarse-grained biotite granite augen gneiss (unit PCagn; Fig. 3) forms two major groups of tabular intrusions

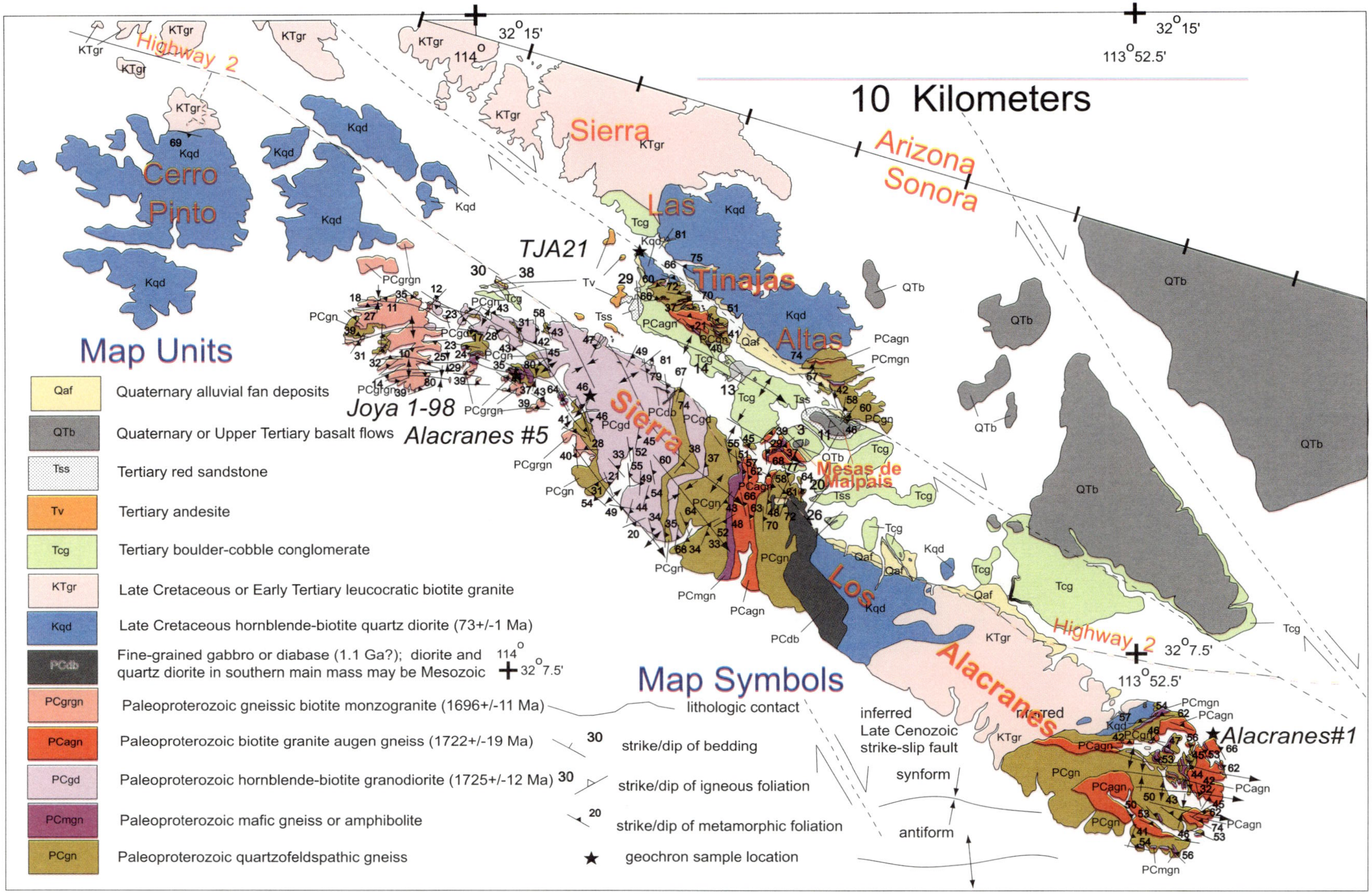

Figure 3. Geologic map of the Western Complex showing locations of analyzed samples.

Figure 4 (*on this and following page*). Photographs showing representative outcrops and field relationships in the Western Complex. (A) Typical outcrop of felsic and intermediate banded gneiss (map unit PCqfgn) with amphibolite layer. Hammer is 40 cm long (B) Mafic gneiss and amphibolite (map unit PCmgn) in northwestern Sierra Los Alacranes. Thin (2–8 cm) layers of quartzofeldspathic gneiss may represent either granitic intrusions or interstratified felsic volcanic rocks. (C) Coarse-grained porphyritic granite dikes near southeastern margin of the biotite granite augen gneiss body of Sierra Las Tinajas Altas (map unit PCagn) intrude mafic gneiss. Note superimposed deformational fabric. (D) Granite dike associated with PCagn body of southeastern Sierra Los Alacranes intrudes mafic gneiss. Note superimposed tight folds with northeast-dipping axial surfaces. View is to the southeast. Hammer handle is 3 cm wide. (E) Close-up view of folded PCagn dike within PCgn unit of southeastern Sierra Los Alacranes. Hammer is 35 cm long. View is to the west; axial surfaces dip gently to the northeast.

Figure 4 (*continued*). (F) Contact between foliated 1722 Ma hornblende-biotite granodiorite (map unit PCgd) and PCgn unit on the west side of northwestern Sierra Los Alacranes. Sample Alacranes #5 was collected from the interior of this granodiorite pluton ~1 km to the northeast. (G) Strongly foliated biotite granite augen gneiss (map unit PCagn) in Sierra Las Tinajas Altas. Largest K-feldspar augen is 5 cm long. (H) 1725 Ma biotite granite augen gneiss in the southeast end of Sierra Los Alacranes. Sample Alacranes #1 was collected from this outcrop.

positioned at structurally high levels of the banded gneiss. This distinct rock (Fig. 4G–H) is characterized by augen-shaped alkali feldspar, the product of flattening, shearing, and recrystallization of phenocrysts that were originally 1–5 cm long. A dark brown, crudely layered appearance results from weathering of 8% to 15% recrystallized biotite that forms wispy stringers around coarse aggregates of feldspar and quartz. Two generations of sphene are present. The younger generation forms epitaxial overgrowths on subhedral sphene cores and on magnetite grains. Foliation is usually concordant with compositional layering in host gneisses. Near pluton margins, 1–10-m-thick sheets of PCagn alternate with slightly discordant screens of banded gneiss. Gneiss xenoliths are less common in the interior of the larger granite bodies. One folded group of augen gneiss intrusions outcrops semicontinuously between the southern Sierra Las Tinajas Altas (Fig. 4G) and the central Sierra Los Alacranes (Fig. 3). Analyzed sample Alacranes #1 (Fig. 4H) was collected from a second group of intrusive sheets exposed at the southeast end of Sierra Los Alacranes.

Late Cretaceous quartz diorite. The Western Complex contains several bodies of weakly foliated to unfoliated medium-grained sphene-hornblende-biotite quartz diorite (Fig. 3; unit Kqd). In contrast to the Paleoproterozoic granodiorite, which is grossly similar in texture and color, hand specimens of Cretaceous quartz diorite contain pristine biotite flakes and subhedral hornblende. Abundant sphene is generally euhedral, and opaque minerals lack the epitaxial sphene overgrowths present in the PCgd and PCagn units. Separate quartz diorite exposures are mapped in the central and southeastern Sierra Los Alacranes, in the southeastern Sierra Las Tinajas Altas, and Cerro Pinto.

Quartz diorite plutons in Sierra Las Tinajas Altas and southern Sierra Los Alacranes display steep intrusive contacts with Paleoproterozoic banded gneiss (Fig. 3). Weak foliation in the quartz diorite is concordant with these contacts. Quartz diorite is sharply intruded by leucocratic porphyritic biotite granite (unit KTgr) in several places, for example, adjacent to the Cerro Pinto microwave tower. That granite is continuous with outcrops of "Gunnery Range granite" that yielded biotite K-Ar ages of 53.1 ± 1.3 Ma and 52.5 ± 1.3 Ma (Shafiqullah et al., 1980). The Tinajas Altas quartz diorite body (analyzed sample TJA 21) appears to be offset ~3–4 km by a northwest-striking dextral strike-slip fault, and probably correlates with the Cerro Pinto quartz diorite (Fig. 3).

Structure of the Western Complex

Pre–1722 Ma tectonic fabric. Wall rocks of the Paleoproterozoic plutons in the Western Complex preserve remnants of deformational fabric that developed prior to emplacement of PCgd. The most common structure is centimeter- to decimeter-scale compositional layering (S_0) in banded gneiss that resulted from transposition of original stratification in volcanic, sedimentary, and sheet-like intrusive protoliths (Fig. 4A). This older fabric was overprinted during the intense deformation described below. Isoclinal folds and relict mylonitic textures may be preserved in xenoliths within less-deformed parts of the 1722 Ma granodiorite (PCgd). Although the original geometry and areal extent of the older fabric are poorly known, protoliths of the banded gneiss experienced significant shearing, folding, and transposition of contacts prior to emplacement of the Paleoproterozoic plutons. Maximum age constraints for this ancient tectonic event await future detailed geochronological study.

Post 1696 Ma–pre 1.1 Ga(?) tectonic fabric. A crystalloblastic, regional metamorphic fabric pervades Paleoproterozoic rocks of the Western Complex (Figs. 5A–D). This fabric is developed throughout the 1725–1696 Ma plutons and their host gneisses but is sharply cut by gabbroic dikes of presumed 1.1 Ga age. We infer that the fabric represents a regional Late Paleoproterozoic event, based on comparison to structures developed in the nearby Yavapai and Mojave crustal provinces. Field observations and map relations indicate the following progressive structural sequence: (1) development of southwest-vergent mylonitic foliation accompanied by transposition and asymmetric folding of intrusive contacts, (2) map-scale closed folding of mylonitic foliation along northwesterly hinges, and (3) regional amphibolite facies metamorphism and recrystallization.

Southwest-vergent noncoaxial fabric (Fig. 5A) is well developed in the 1722 Ma granodiorite (PCgd; Fig. 3) and wall rocks of northwestern Sierra Los Alacranes. Mylonitic "S" and "C" planes define a composite mylonitic foliation (S_1), and west-southwest–oriented stretching lineation (L_1) is conspicuous near the structurally deep western margin of the granodiorite. Asymmetries are commonly preserved despite the presence of a strong crystalloblastic overprint. Abundant mesoscopic "S" and "Z" folds with east- or northeast-dipping axial surfaces (S_2; Fig. 5A) are recorded by mylonitic foliation in the granodiorite and tabular xenoliths of banded gneiss. Similar fold asymmetries are developed in mafic gneisses of central Sierra Los Alacranes and in 1725 Ma granite dikes (Figures 4D and 4E) at the southeast end of this range.

Across the Western Complex, variations in S_0 and S_1 orientation exhibited by map patterns (Fig. 3) and stereonet compilations (Figs. 5B–D) define map-scale closed folds with northwesterly hinges and wavelengths of 0.5 to 2 km. This fold belt generally plunges southeast such that deepest structural levels are exposed at the northwest end of Sierrra Los Alacranes. Because mesoscale asymmetric folds on opposite limbs consistently verge in the same direction (southwest), we argue that the larger folds developed after the noncoaxial shear fabric. Within the limbs of map-scale folds in the southeastern Sierra Los Alacranes and Sierra Las Tinajas Altas, augen gneiss bodies commonly display coaxial foliation with weak downdip lineation. A few southeast-trending lineations (Figs. 5C–D) probably record constriction parallel to major fold hinges.

Grains defining the mesoscopic foliation and lineation are recrystallized throughout the Western Complex. Intense thermal overprint is evident in outcrop and thin section from ubiquitous recrystallization of biotite and the formation of mosaic textures of interlocking subgrains within larger, previously flattened feldspar and quartz phenocrysts. Subtle evidence of thermal activity includes secondary overgrowths of sphene around primary sphene or opaque minerals. We interpret this crystalloblastic fabric to represent the metamorphic culmination of progressive regional contraction recorded by the noncoaxial shear fabric and map-scale folds. Thermal activity outlasted compression, as suggested by poor preservation of linear structures or axial planar cleavage associated with ductile map-scale folding. Folding and attendant metamorphism do not record a Mesozoic event because 1.1 Ga(?) mafic dikes crosscut the ductile fabric and are not folded. No direct timing constraints are currently available, but the regional metamorphism and deformation was intense enough to cause significant isotopic disturbance of some zircons in the augen gneisses (see below).

Proterozoic Geology and Structure of the Eastern Complex

Geologic Setting and Petrography of Sierra Choclo Duro

Important stratigraphic and structural relationships in the Eastern Complex are revealed near Cerro Los Ojos, a rugged hill located directly north of Highway 2 in the southern part of Sierra Choclo Duro (Fig. 6). The prominent Ojos or "eyes" referred to in the name are actually dark patches of diabase intruded into Paleoproterozoic granite. Sierra Choclo Duro and the Cabeza

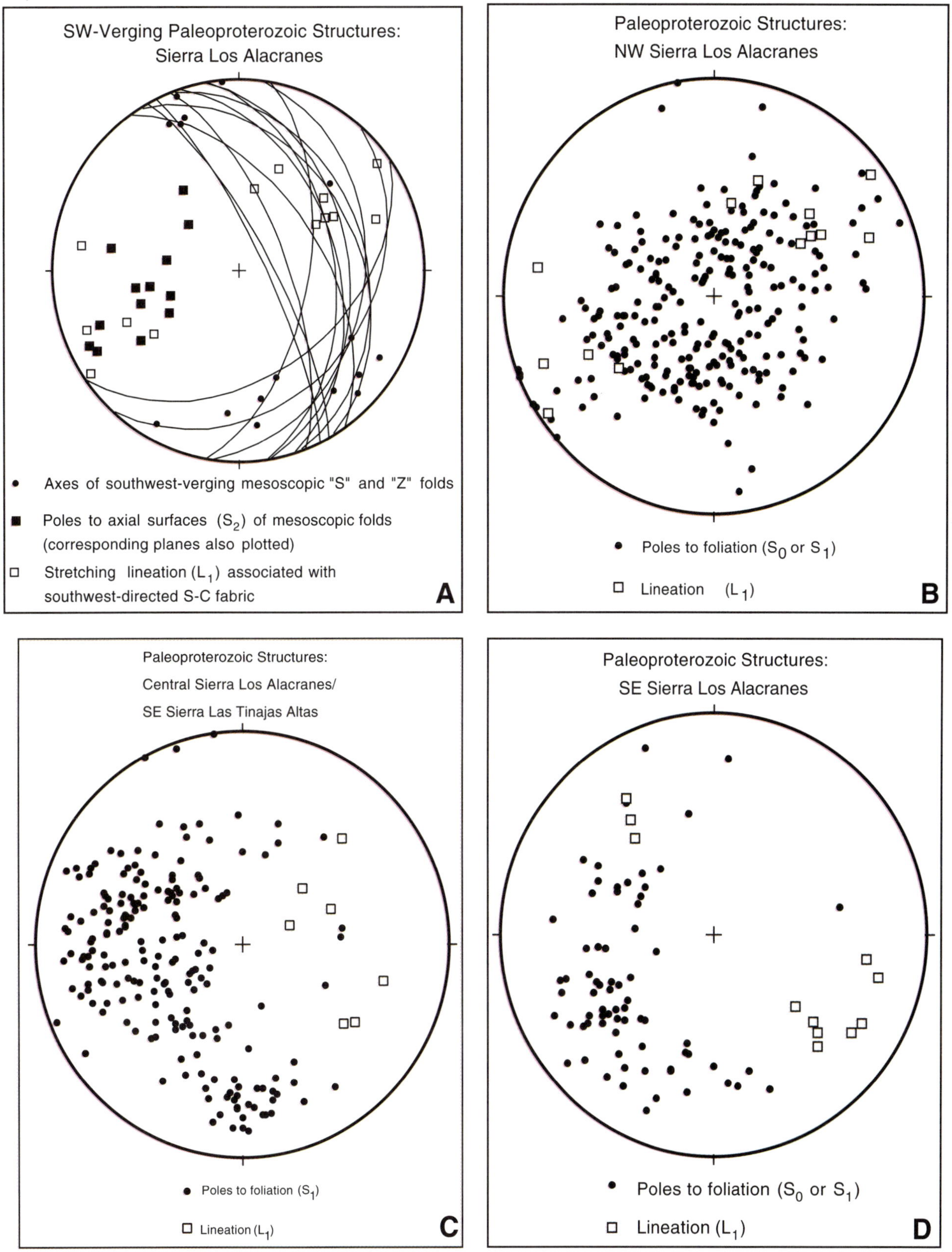

Figure 5. Stereographic projections of structures developed in Paleoproterozoic gneisses of the Western Complex. Stereonet plotting program provided by R. Allmendinger (1995, personal commun.). (A) Mesoscopic structural elements associated with southwest-vergent noncoaxial deformation in Sierra Los Alacranes. This fabric is shared by the 1725–1696 Ma metaplutonic rocks and PCgn host gneisses. (B) Mesoscopic foliation and mineral lineation developed in 1725–1696 Ma metaplutonic rocks and host gneisses in the northwestern Sierra Los Alacranes. 248 foliation poles are plotted. (C) Mesoscopic foliation and mineral lineation developed in 1725 Ma(?) augen gneiss and host gneisses in the southern Sierra Las Tinajas Altas and central Sierra Los Alacranes. 198 foliation poles are plotted. (D) Mesoscopic foliation and mineral lineation developed in 1725 Ma augen gneiss and host gneisses in the southeastern Sierra Los Alacranes. 72 foliation poles are plotted.

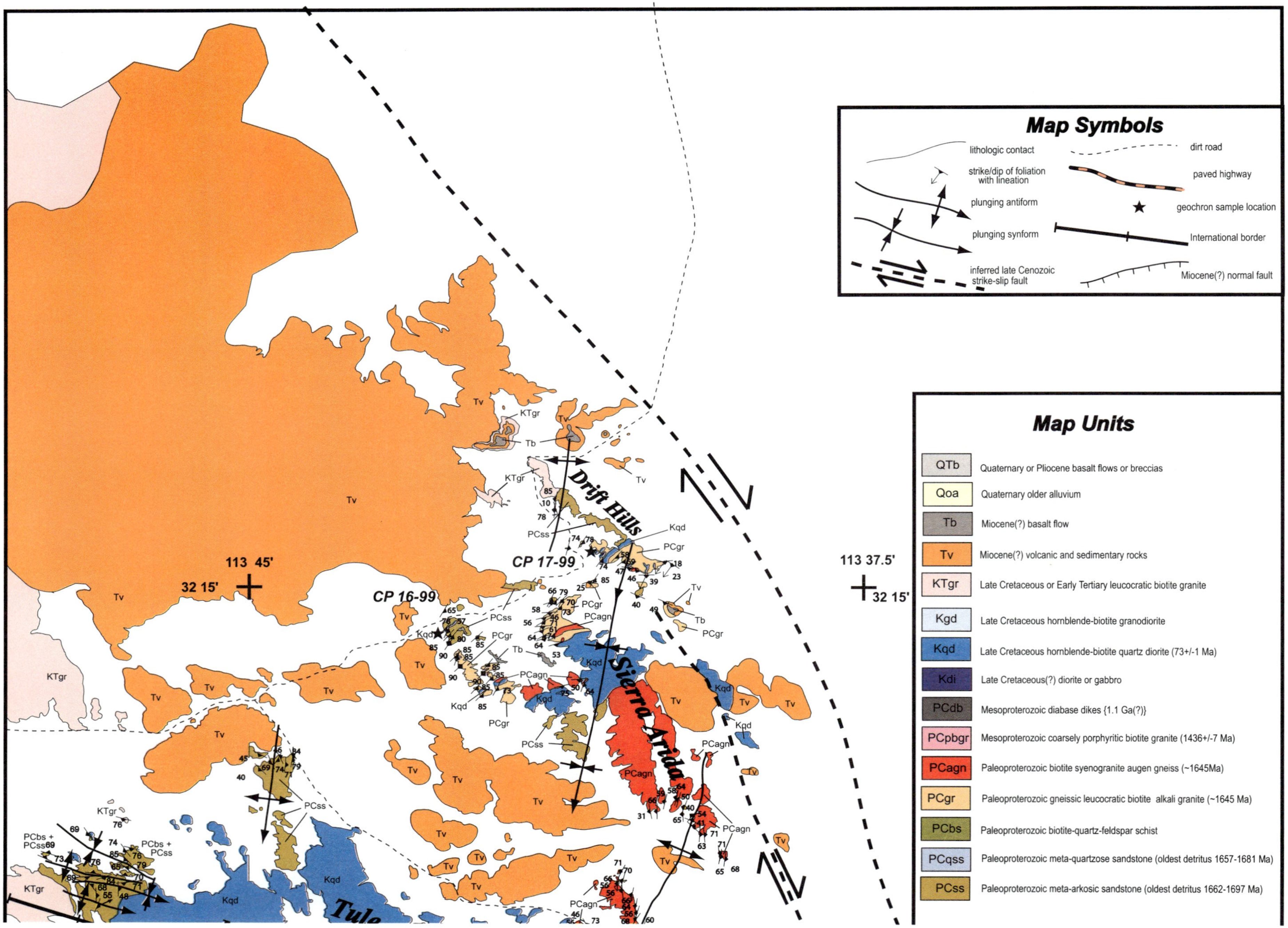
Map Symbols
lithologic contact
strike/dip of foliation with lineation
plunging antiform
plunging synform
inferred late Cenozoic strike-slip fault
dirt road
paved highway
geochron sample location
International border
Miocene(?) normal fault
Map Units
QTb Quaternary or Pliocene basalt flows or breccias
Qoa Quaternary older alluvium
Tb Miocene(?) basalt flow
Tv Miocene(?) volcanic and sedimentary rocks
KTgr Late Cretaceous or Early Tertiary leucocratic biotite granite
Kgd Late Cretaceous hornblende-biotite granodiorite
Kqd Late Cretaceous hornblende-biotite quartz diorite (73+/-1 Ma)
Kdi Late Cretaceous(?) diorite or gabbro
PCdb Mesoproterozoic diabase dikes {1.1 Ga(?)}
PCpbgr Mesoproterozoic coarsely porphyritic biotite granite (1436+/-7 Ma)
PCagn Paleoproterozoic biotite syenogranite augen gneiss (~1645Ma)
PCgr Paleoproterozoic gneissic leucocratic biotite alkali granite (~1645 Ma)
PCbs Paleoproterozoic biotite-quartz-feldspar schist
PCqss Paleoproterozoic meta-quartzose sandstone (oldest detritus 1657-1681 Ma)
PCss Paleoproterozoic meta-arkosic sandstone (oldest detritus 1662-1697 Ma)
Drift Hills
Sierra Arida
Tule
CP 17-99
CP 16-99
113 45'
32 15'
113 37.5'
32 15'

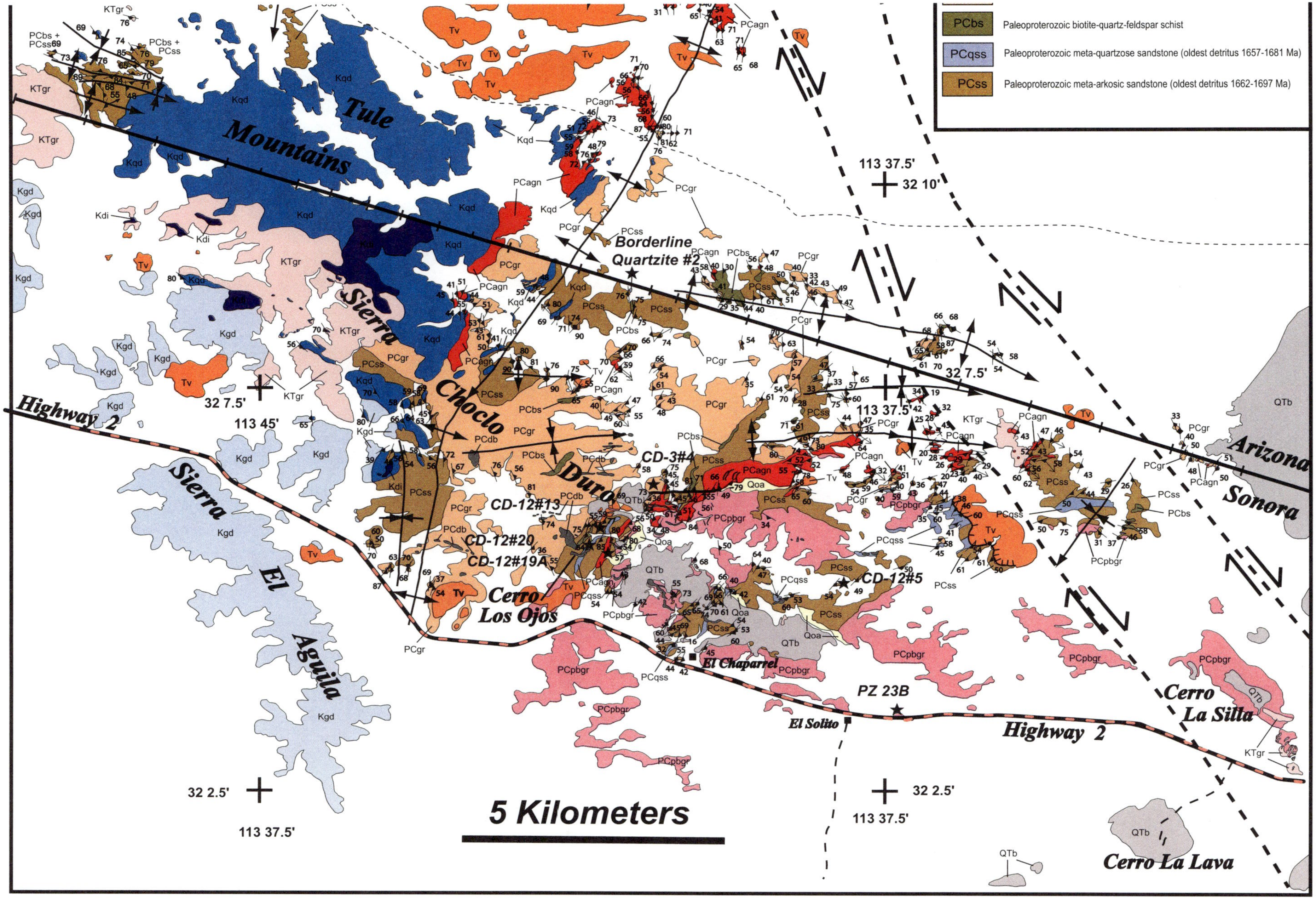

Figure 6. Geologic map of the Eastern Complex, showing the Sierra Choclo Duro and Cabeza Prieta focus areas south and north of the international border, respectively. Sample locations correspond to analyses described in text.

Prieta focus area of southwestern Arizona (Fig. 2; described below) are geographically separated from the Western Complex by a Late Cretaceous plutonic suite that includes the granodiorite of Sierra El Aguila (Anderson and Silver, 1979), quartz diorite similar to our analyzed samples TJA 21 and CP 16–99, and cross-cutting porphyritic monzogranite.

Several smaller hills east of Cerro Los Ojos preserve Pliocene(?) or Quaternary(?) basalt vents erupted through Proterozoic basement (Fig. 7A). Radial dikes emanating from these central pipes and contemporaneous basalt flows that mantle the basement (Fig. 7B) probably are early phases of the Pinacate volcanic field (Lynch and Gutman, 1987), which dominates the landscape to the southeast. The basement stratigraphy and structure revealed in isolated windows through the basalt field is continuous with that mapped in Sierra Choclo Duro and matches the geology north of the international border in the Cabeza Prieta area.

Late Paleoproterozoic paragneisses. Fine-grained, sugary-textured quartzofeldspathic paragneiss constitutes the oldest rock sequence in the Sierra Choclo Duro area (Fig. 6). These foliated strata are mainly derived from interstratified arkosic

Figure 7 (*on this and following page*). Photographs showing representative outcrops and field relationships in the Eastern Complex. (A) Basaltic pipe and lava flow (map unit QTb) erupted through 1432 Ma granite at Cerro La Silla (Fig. 6). View is to the north from Mexican Highway 2. (B) View to the northwest, showing Quaternary(?) basalt vent and associated lava flows draped over 1432 Ma granite and Paleoproterozoic meta-arkose. Note two small windows of meta-arkose beneath the basalt flow. Cerro Los Ojos is the prominent peak to the right. (C) Late Paleoproterozoic coarse-grained biotite granite augen gneiss outcrop from the northeast edge of Sierra Choclo Duro. This map unit (PCagn) continues across the international border into Sierra Arida. Note the inclusion of Paleoproterozoic meta-arkose. Hammer is 40 cm long. (D) Late Paleoproterozoic coarse-grained biotite granite augen gneiss exposure located 1.2 km east of Cerro Los Ojos, along-strike from the location of sample CD-12#19A. To the north, this pluton intrudes the meta-arkose (note screens of fine-grained felsic rock). The strong fabric in this rock body is sharply intruded by 1432 Ma granite.

Figure 7 (*continued*). (E) Xenolith of Paleoproterozoic meta-arkose (map unit PCss) included in nonfoliated 1432 Ma granite of Cerro Los Ojos (map unit PCpbgr). (F) Weakly foliated 1432 Ma porphyritic biotite granite near the northwest margin of the pluton adjacent to Paleoproterozoic host rocks. (G) Northeast-dipping sheets of 1.1 Ga(?) diabase intrude 1432 Ma granite and Paleoproterozoic meta-arkose and granite gneiss on the southeast slope of Cerro Los Ojos. (H) Foliated screen of Paleoproterozoic leucocratic biotite alkali granite (map unit PCgr) within Late Cretaceous quartz diorite (map unit Kqd) in the southern Drift Hills. Sample CP17-99 was collected a few hundred meters west of this location. Quartz diorite sample CP16-99 was collected from a sill intruded into alkali granite 3 km to the south. (I) East-dipping mid-Tertiary(?) volcanic strata (visible on highest peak) overlie Paleoproterozoic granite in the southwestern Sierra Choclo Duro. View is toward the south, with Late Cretaceous granodiorite of Sierra el Aguila in the background.

and quartzose sandstone protoliths. Recrystallized grains range from 1 to 2 mm in diameter. Given the textural uniformity of this unit, and the abundance of quartz, protoliths composed of silicic lava flows and pyroclastic deposits were ruled out near our dated sample localities. However, some felsic gneisses near the international border display coarser grain sizes, intermediate between arkose and granite, and contain more abundant biotite. Some of these felsic rocks with possible porphyritic texture may have been rhyolites and dacites or even hypabyssal intrusives prior to emplacement of Late Paleoproterozoic granite and subsequent metamorphism.

The meta-arkose (unit PCss) is typically composed of 32%–51% microcline and orthoclase, 40%–55% quartz, and 5%–12% plagioclase (Watkins, 2003). Minor components of biotite and muscovite (1%–3%) enhance the foliation. Granoblastic fabric generally obscures primary sedimentary structures, but transposed bedding may be defined by dark laminae in which magnetite and zircon are concentrated. Approximately 5% of the paragneiss is "quartzite" (map unit PCqss), which forms 5–50-m-thick layers oriented parallel to local foliation. Quartz content in this unit ranges from 60% to 85% (Watkins, 2003). The quartzite appears to be restricted to specific stratigraphic horizons within the predominantly feldspathic section (Fig. 6). Subordinate biotite-quartz-feldspar schist layers (unit PCbs), probably derived from siltstone protoliths, weather dark relative to the leucocratic metasandstones and syenogranite. No marbles or calcareous strata occur.

Considering the uniform felsic composition of this metasedimentary sequence, we initially inferred that the sandstones and finer-grained clastic protoliths represent a relatively mature, intracratonal depositional setting. Microcline and quartz grains are abundant, whereas hornblende or volcanic fragments are absent, suggesting erosion of a deeply exhumed granite source. However, a granitic base of the sedimentary sequence is not exposed within the study area. Two samples of meta-arkose (CD-12 #5 and #13) and two samples of quartzite (CD-12 #2 and #20) were collected (Fig. 6) to assess variations in age and provenance recorded by detrital zircons. Preliminary SHRIMP U-Pb analyses (see below) indicate that the sandstones were deposited only a few million years prior to widespread intrusion of granite at ca. 1645 Ma. The detrital zircon data, though isotopically disturbed, suggest that the sediments accumulated in a very dynamic intra-arc setting.

Late Paleoproterozoic granites. Two distinct varieties of foliated Paleoproterozoic granite intrude the metasedimentary sequence and form continuous structural markers that define map-scale folds. The largest body is a medium-grained, slightly porphyritic, leucocratic biotite alkali granite (unit PCgr; Fig. 6), characterized by yellowish stains around recrystallized biotite. Abundant xenoliths and locally discordant contacts indicate an intrusive relationship with the metasandstone. Strong foliation and crystalloblastic fabric pervade this rock, rendering an augen-like appearance to the feldspar. Lineation is locally developed. The granite defines a 1–3-km-thick, south-plunging antiformal sheet that underlies much of Sierra Choclo Duro and extends north of the international border. This sheet is continuous with a synformal body of granite exposed in the Drift Hills of the Cabeza Prieta area. Analyzed sample CD-3 #4 was collected from a representative outcrop of gneissic alkali granite ~1 km northeast of Cerro Los Ojos.

The second Paleoproterozoic intrusion is dark weathering, coarse-grained augen gneiss (unit PCagn; Fig. 7C) derived from a porphyritic biotite syenogranite. This metaplutonic rock contains xenoliths and screens of meta-arkose (Fig. 7D), and forms a 100–500-m-thick folded sill with foliation concordant to that in the adjacent metasedimentary strata. The size and abundance of recrystallized alkali feldspar porphyroclasts locally decreases near the edges of the augen gneiss, suggesting an original porphyritic-aphanitic texture along chilled margins. Close to the main pluton contacts with arkosic wall rocks, 50-cm- to 3-m-thick sills of augen gneiss commonly occur within the metasandstone. These observations demonstrate that the syenogranite does not represent the granitic basement upon which the sediments were deposited; instead, it intruded the sedimentary section prior to deformation.

Age relations between the gneissic alkali granite and syenogranite augen gneiss are equivocal. The two granites are commonly in contact (Fig. 6), but strong foliation masks the original gradational textural relations between the intrusive bodies. Dikes of one unit within the other were not observed. We argue that the augen gneiss (analyzed sample CD-12 #19A) is a coarser-grained border phase of the alkali granite, because it consistently occurs near the upper margin of the larger granite body. Petrographically, the augen gneiss strongly resembles the PCagn unit of the Western Complex. For example, biotite and K-feldspar are recrystallized in a similar fashion, and epitaxial overgrowths of sphene on magnetite are common. However, preliminary U-Pb analyses on zircon (described below) indicate that the syenogranite augen gneiss and gneissic alkali granite of the Eastern Complex are part of a distinctly younger plutonic suite, emplaced at about the same time (within uncertainties) at ca. 1645 Ma.

Mesoproterozoic intrusions. Paleoproterozoic paragneisses and orthogneisses of Sierra Choclo Duro are sharply intruded by 1.4 Ga granite and diabase dikes of presumed 1.1 Ga age. Coarse-grained porphyritic biotite granite (unit PCpbgr; Fig. 6), containing euhedral alkali feldspar phenocrysts as large as 6 cm, forms a large pluton transected by Highway 2 near El Chaparral (Fig. 6). Two thick intrusive sheets extending northeast from the main granite mass display boundaries concordant to broadly folded foliation of the Paleoproterozoic country rocks. Sharp xenolith relations with metasandstone and augen gneiss host rocks are common (Fig. 7E). Biotite is generally recrystallized, but the feldspars are in good shape. Foliation is restricted to rare exposures near pluton margins (Fig. 7F), where feldspar and biotite are aligned parallel to foliation in adjacent Paleoproterozoic gneisses. Analyzed sample PZ 23B was collected from a small hill adjacent to Highway 2 (Fig. 6).

Very dark weathering diabase dikes (unit PCdb; Fig. 7G) with characteristic ophitic texture intrude all Proterozoic units of the Sierra Choclo Duro. These dikes have not been dated, but their composition, texture, and stratigraphic relations suggest correlation to 1.1 Ga diabase dikes described from the Death Valley, Grand Canyon, and central Arizona regions (Silver, 1978; Howard, 1991; Heaman and Grotzinger, 1992). The dikes display irregular margins and vary in thickness from 3 m to 50 m. Contacts and map patterns near Cerro Los Ojos indicate that the principal dike set dips 30° to 50° to the north or northeast.

Late Cretaceous and Paleogene(?) intrusions. The southwestern margin of the Eastern Complex is intruded by a batholith composed of various Late Cretaceous and Paleogene(?) plutons (Fig. 6). The oldest intrusions are weakly foliated bodies of sphene-hornblende biotite quartz diorite (unit Kqd) and poorly foliated diorite. These dark-weathering rocks commonly form inclusions within the biotite granodiorite of Sierra El Aguila (unit Kgd) or the porphyritic biotite monzogranite body (unit KTgr) that extends across the international border into the Gunnery Range batholith. North and northwest of Cerro Los Ojos, several sills of quartz diorite intrude Paleoproterozoic gneiss. These sills are probably part of a larger quartz diorite body that straddles the international border. The quartz diorite resembles our dated sample TJA #21 from Sierra Las Tinajas Altas. Anderson and Silver (1979) reported a U-Pb zircon age of 73 ± 3 Ma for the granodiorite of Sierra El Aguila.

Geologic Setting and Petrography of the Cabeza Prieta Focus Area

Paleoproterozoic gneisses of the Drift Hills, Sierra Arida, and Tule Mountains of Cabeza Prieta (Fig. 6) correlate directly with basement rocks mapped in Sierra Choclo Duro (Figs. 2 and 6). Rocks common to the two areas include the metasedimentary assemblage, foliated leucocratic alkali granite, and biotite syenogranite augen gneiss. Mesoproterozoic granite and diabase dikes have not been recognized, but are known to the east and north (Howard, 1991; Reynolds, 1998). The foliated Paleoproterozoic strata define a south-plunging fold belt, intruded by Mesoproterozoic and Late Cretaceous plutons, that continues south of the border.

Older paragneisses. The oldest rocks in Sierra Arida and the Drift Hills are faintly banded, sugary-textured meta-arkose (unit PCss) interstratified with subordinate biotite-quartz-feldspar schist (unit PCbs), rare quartzite, and fine-grained felsic gneiss probably derived from volcanic protoliths. These rocks are not mapped in detail, but contact relationships with Paleoproterozoic granitoids are similar to those observed directly south in Sierra Choclo Duro. For example, abundant xenoliths of meta-arkose and schist occur within the alkali granite of the Drift Hills, and augen gneiss is commonly interlayered with screens of concordantly foliated felsic paragneiss.

Paleoproterozoic biotite syenogranite augen gneiss. Dark-weathering coarse-grained augen gneiss (unit PCagn) underlies low hills in the eastern part of the Cabeza Prieta study area, where foliation and map patterns define a south-trending antiform-synform pair. This rock unit extends southward across the border, forming a marker horizon that delineates south-plunging folds in the Sierra Choclo Duro area (Fig. 6). It is continuous with the PCagn unit of Cerro Los Ojos, which is intruded by 1.4 Ga granite.

Paleoproterozoic leucocratic biotite alkali granite. Gneissic leucocratic biotite alkali granite (unit PCgr; analyzed sample CP 17–99; Fig. 7H) crops out in the southeastern Drift Hills and continues southward where it structurally underlies unit PCagn. The granite displays a similar recrystallized fabric and yellow halos around biotite that characterize alkali granite of Sierra Choclo Duro. Foliation near the Drift Hills defines a gently south-plunging synform at map scale. In the constricted hinge region, a "swirly" appearance is produced by tight to isoclinal folds of the foliation. Throughout the map area (Fig. 6), the alkali granite consistently underlies augen gneiss. Together these two plutonic phases form a sheet ~2 km thick, intruded into the fine-grained paragneisses. Subsequent east-west shortening has created the synform-antiform pair that is continuous with map-scale folds in the Sierra Choclo Duro area.

Late Cretaceous quartz diorite. In the Drift Hills and Sierra Arida (Fig. 6), several sills of weakly foliated sphene-hornblende-biotite quartz diorite resembling unit Kqd of Sierra Las Tinajas Altas intrude parallel to foliation in the host Paleoproterozoic syenogranite (Fig. 7H). A larger mass of quartz diorite occurs farther south in the Tule Mountains near the international border (Fig. 6), where it is intruded by hornblende-biotite granodiorite and porphyritic biotite granite that underlies northwestern Sierra Choclo Duro. Analyzed sample CP16-99 was collected from a small ridge 3 km southwest of the Drift Hills.

Late Cretaceous–early Tertiary porphyritic monzogranite of Christmas Pass. A small pluton of leucocratic porphyritic biotite monzogranite (unit KTgr) intrudes Paleoproterozoic gneisses in the northwestern Drift Hills. Pegmatite dikes associated with this unfoliated pluton crosscut the gneisses and quartz diorite at consistent orientations of N60W–45NE. Tertiary tilting of the Christmas Pass pluton and its Paleoproterozoic wall rocks is indicated by well-exposed, northeast-dipping unconformities that mark the base of overlying Miocene(?) volcanic and sedimentary strata.

Structure of the Eastern Complex

Post-1645 Ma–pre-1432 Ma fabric. The dominant mesoscopic structure is foliation (S_1) oriented subparallel to compositional boundaries between map units. This foliation is defined by alignment of biotite in all Paleoproterozoic units, and flattened K-feldspar phenocrysts in the granite gneisses. Crystalloblastic foliation displayed by the granitic bodies was probably originally mylonitic, based on top-to-the-northwest S-C fabrics observed at three widely separated locations. Mylonitic lineation (L_1) is intermittently preserved, and tends to be oriented easterly or southeasterly at high angles to major fold trends (Fig. 8A). Near the hinges of map-scale folds, strongly lineated, poorly foliated domains of L-tectonite record constrictional strain. Along fold limbs, most asymmetric microstructures have been obscured by

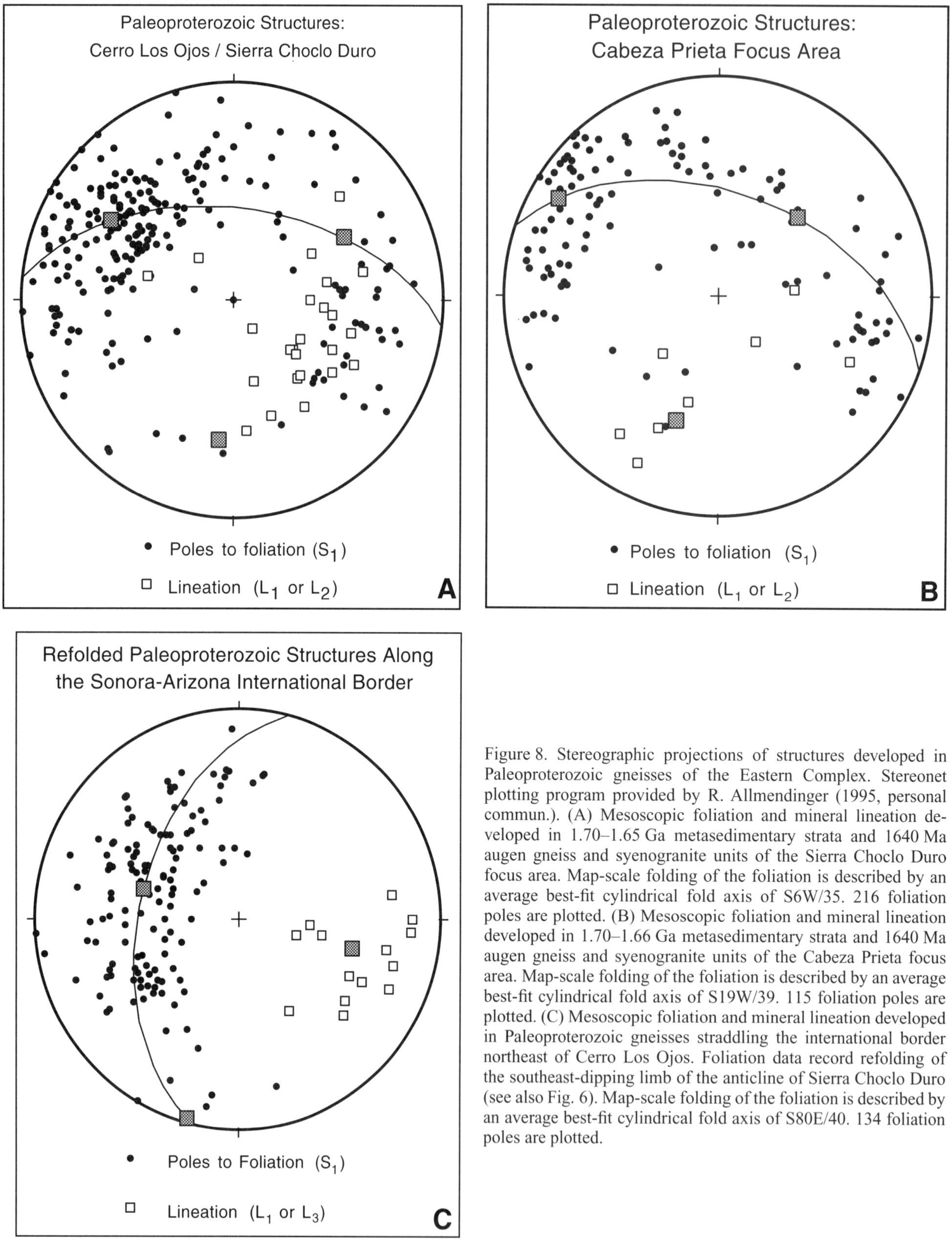

Figure 8. Stereographic projections of structures developed in Paleoproterozoic gneisses of the Eastern Complex. Stereonet plotting program provided by R. Allmendinger (1995, personal commun.). (A) Mesoscopic foliation and mineral lineation developed in 1.70–1.65 Ga metasedimentary strata and 1640 Ma augen gneiss and syenogranite units of the Sierra Choclo Duro focus area. Map-scale folding of the foliation is described by an average best-fit cylindrical fold axis of S6W/35. 216 foliation poles are plotted. (B) Mesoscopic foliation and mineral lineation developed in 1.70–1.66 Ga metasedimentary strata and 1640 Ma augen gneiss and syenogranite units of the Cabeza Prieta focus area. Map-scale folding of the foliation is described by an average best-fit cylindrical fold axis of S19W/39. 115 foliation poles are plotted. (C) Mesoscopic foliation and mineral lineation developed in Paleoproterozoic gneisses straddling the international border northeast of Cerro Los Ojos. Foliation data record refolding of the southeast-dipping limb of the anticline of Sierra Choclo Duro (see also Fig. 6). Map-scale folding of the foliation is described by an average best-fit cylindrical fold axis of S80E/40. 134 foliation poles are plotted.

a coaxial flattening during deformational overprint, and further annealed by pervasive recrystallization.

The map patterns of Figure 6 combined with stereonet compilations of S_1 (Figs. 8A–B) define several folds with 1–2 km wavelengths and hinges that plunge gently south or south-southwest. The most prominent of these is an antiform that extends from the southeastern Sierra Arida through the southwestern Sierra Choclo Duro. Lineations (L_2) with southerly trends (Fig. 8B) are locally developed in the hinge regions, where numerous dip reversals commonly define secondary sets of "M" and "W" folds with wavelengths of a few meters to 200 m. We interpret the south-trending folds to be late-stage manifestations of a progressive post–1645 Ma regional deformation associated with northwest- or west-directed shear.

Variations in S_1 orientation adjacent to the international border northeast of Sierra Choclo Duro delineate a second generation of map-scale folds (Fig. 8C). As indicated on Figure 6, the southeast-dipping limb of the major anticline described above appears to be deformed into a series of east-plunging open folds. Lineations (L_3 superimposed on L_1?) and strongly developed L-tectonites in this area plunge moderately east-southeast. The age and tectonic significance of north-south shortening implied by these structures remains enigmatic.

The pre–1432 Ma structures are thoroughly recrystallized throughout the study area, with amphibolite facies mineral assemblages and statically annealed microstructures preserved in thin section. These textures indicate that thermal activity outlasted the shearing and contraction recorded by the previously described macrostructures. The precise time of deformation and culminating thermal metamorphism is not yet resolved. Our U-Pb zircon analyses (described below) show that most Paleoproterozoic zircons from the Eastern Complex have been isotopically disturbed. The data indicate the likelihood of metamorphic fluid flow and new zircon growth at ca. 1430 Ma, but a separate cryptic event at ca. 1.6 Ga is also suggested.

In summary, Paleoproterozoic rocks of the Eastern Complex are pervaded by crystalloblastic foliation that is openly folded at map scale. This fabric is well developed in ca. 1645 Ma granite gneiss and host strata, but sharply intruded by 1432 Ma porphyritic granite. Field observations indicate the following structural sequence: (1) regional mylonitic foliation development associated with southeasterly lineation and locally preserved northwest-vergent shear fabric, (2) mesoscopic and map-scale folding of foliation about gently south-plunging hinges, (3) local refolding about easterly hinges, and (4) amphibolite facies metamorphism and pervasive recrystallization. Culminating event (4) was probably accompanied by isotopic disturbance of zircons at ca. 1600 Ma.

Structures associated with the 1432 Ma and 1.1 Ga(?) intrusions. Mesoproterozoic porphyritic granite of Cerro Los Ojos is generally intruded parallel to folded S_1 foliation in its Paleoproterozoic host gneisses. The granite has utilized preexisting planar fabric in the country rocks as preferred avenues for emplacement. Northeast of El Chaparral, for example, a granite sill ~1-km-thick follows the trace of folded foliation in its metasedimentary and syenogranite hosts. The Quaternary Pinacate volcanic field covers part of the Mesoproterozoic pluton at the eastern edge of the study area (Figs. 6 and 7A), but map patterns of isolated granite exposures preserve the geometry of a south-plunging synform. Absence of dikes or chilled margins and overall uniformity of texture in the granite suggests crystallization at mid-crustal levels. The granite lacks cleavage, mesoscopic folds, or any systematic ductile structure that might be geometrically associated with the map-scale folding of the Paleoproterozoic strata. It also lacks the crystalloblastic fabric that pervades the folded host gneisses. Thus, intrusion of the 1432 Ma granite postdated a distinct regional tectono-thermal event recorded by recrystallized ductile structures in the ca. 1645 Ma granite gneisses.

Diabase dikes (presumably 1.1 Ga) define a consistently oriented swarm intruded across Paleoproterozoic gneiss and Mesoproterozoic granite (Figs. 6 and 7G). The swarm is highly discordant to regional foliation. These dikes display nonplanar margins that lack chill textures, however, map patterns and field measurements of contacts indicate they were emplaced along fractures that are currently oriented N50-80W/25-45NE. A secondary dike set exhibits northerly strikes. Assuming that this area is part of a northeast-dipping tilt-block domain, removal of horizontal axis rotations associated with late Cenozoic faulting would restore many of the sheet-like diabase intrusions to original subhorizontal orientations. A similar pattern was recognized by Howard (1991) for diabase dikes of Death Valley–central Arizona region.

Late Cretaceous and Cenozoic Tectonic Modifications

Phanerozoic structural overprinting of the Arizona-Sonora border region includes minor foliation development near the margins of Late Cretaceous quartz diorite intrusions, tilting or normal faulting associated with mid-Tertiary extension, and strike-slip translations along presumed splays of the southern San Andreas fault system. Weak foliation that may be synmagmatic in part was observed in some of the Late Cretaceous quartz diorite intrusions. This fabric appears to be restricted to 5–50-m-wide sills emplaced into Paleoproterozoic gneiss or to the margins of larger plutons near contacts with gneiss. In either case, foliation in the quartz diorite is concordant to that in the host gneisses. Crosscutting leucocratic biotite granite and pegmatite do not record this fabric.

Throughout the study area, isolated sections of Neogene conglomerate, arkose, andesite, and basalt overlie the crystalline rocks along nonconformities that are tilted at angles of 10° to 40°. Except for the southwest-dipping section between Sierra Las Tinajas Altas and Highway 2 (Fig. 3), most of exposures dip to the east or northeast (Fig. 7I). Much of the region appears to be broken into a series of tilt blocks, presumably rotated along southwest-dipping normal faults. We suspect that the Eastern Complex is underlain by a detachment fault whose breakaway zone is located in the valley between Sierra Pinta and Sierra Arida (Fig. 2). This hypothetical detachment represents one of a family of early or middle Miocene low-angle normal faults mapped in southwestern Arizona (Richard, 1994; Reynolds, 1998). Antithetic normal faults are also present; for example, a moderately northeast-dipping zone of breccia (Fig. 6) separates Tertiary andesite from a Paleoproterozoic paragneiss footwall. Reconstruction of Proterozoic basement in this region requires restoration of late Cenozoic block rotations and displacements on major normal faults. Given our incomplete Neogene data set, we have not attempted this reconstruction, however it is clear that both Proterozoic complexes have experienced significant northeast-southwest extension; therefore both areas were located farther northeast prior to mid-Tertiary time.

The effects of Late Cenozoic strike-slip faulting are easier to quantify. We have identified several dextral offsets of distinctive rock units that coincide with northwest-trending valleys. As shown on Figures 2, 3, and 6, these inferred faults displace Proterozoic and Late Cretaceous–early Tertiary basement. At Mesas del Malpais (Fig. 3), faults deform a 10 Ma basalt flow and are locally overlain by basalts associated with the Quaternary Pinacate basalt field. The structure in southeastern Sierra Las Tinajas Altas is associated with a mapped breccia zone. Most estimates of displacement are derived from separation of distinct contacts. One major right-lateral fault (Fig. 2) separates the boulder-cobble conglomerate unit of the Western Complex at least 20 km from its bedrock sources in the Eastern Complex. We postulate that this family of right lateral faults represents late Cenozoic splays of the southern San Andreas fault system.

U-PB ZIRCON GEOCHRONOLOGY

Zircon populations were collected from twelve different samples including four recrystallized Paleoproterozoic metasediments, five foliated Paleoproterozoic granitoids, one unfoliated Mesoproterozoic granite, and two Cretaceous quartz diorites (see Figs. 2, 3, and 6 for locations). Zircons were separated using conventional methods (see Appendix A) and fractions for ID-TIMS work are composed of several tens of grains that were handpicked from least magnetic populations. A brief description of each is given in Table 1A, along with the U-Th-Pb analytical data from ID-TIMS work. Analytical methods applied to all U-Th-Pb analyses are also given in Appendix A and correction values in the footnotes of Tables 1A and 1B.

Only two to three fractions per sample were analyzed by the ID-TIMS method in a reconnaissance mode to quickly assess approximate ages and probable U-Pb isotopic complexities. Many of the zircon populations investigated contained grains that showed (1) inherited zircon cores, (2) clear to milky anhedral overgrowths, and (3) discolorations, suggesting significant Pb loss. Even after picking the "very best looking" grains (i.e., those without visible fractures, inclusions, or discolorations, and free of inherited cores), many analyses exhibited complex U-Pb isotopic behavior characterized by significant scatter along concordia to younger ages. This behavior is interpreted as indicating either a Pb-loss event no greater than 500 m.y. after primary crystallization, or the addition of secondary zircon overgrowths, or both. Furthermore, some analyses showed significant Pb loss (up to 50%) during the Mesozoic or younger.

Subsequently, these zircon populations were first imaged and studied under transmitted and reflected light, and cathode luminescence (CL). They were then analyzed using the SHRIMP-RG (reverse geometry) at Stanford University, California, in order to gain access to and measure zones of primary magmatic zircon growth (i.e., to determine crystallization ages). We also analyzed detrital zircons from four Paleoproterozoic metasandstone samples of the Eastern Complex in order to define their provenance(s). The SHRIMP U-Pb isotopic results are given in Table 1B, and the methods used are described in Appendix A and in the footnotes of Table 1B.

Samples from the Western Proterozoic Complex

Paleoproterozoic Metagranitoid Samples

Alacranes #1. As mentioned previously, this sample represents the strongly foliated coarse-grained biotite granite augen gneiss that crops out in the southern and east-central Sierra Los Alacranes and in the southeastern Sierra Las Tinajas Altas. The zircons separated from it are subhedral to euhedral. Most crystals are transparent (colorless to slightly tinted) with some dark randomly oriented inclusions. A few grains are almost opaque. Some inclusions do appear to be inherited zircon cores and these grains were avoided during handpicking.

Three fractions were analyzed by ID-TIMS; the U-Pb isotopic results are illustrated in Figure 9 (open circles). Each of the three analyses produced $^{206}Pb/^{204}Pb$ values in excess of 4500, indicating that essentially all of the Pb is radiogenic, and corrections for common Pb are relatively insignificant. The three fractions do not form a linear array; although two fractions yield the same $^{207}Pb/^{206}Pb$ age of ca. 1645 Ma, and the third fraction [2(82)] plots with a younger $^{207}Pb/^{206}Pb$ age of ca. 1629 Ma (Table 1A). These ages are similar to that reported for a single fraction (1618 Ma) from a sample of this same unit (sample #13, Cerro del Viejo augen gneiss #102; Anderson and Silver, this volume). Because these analyses are not very discordant (less than ~5%), the true age of this pluton was thought to be close to these $^{207}Pb/^{206}Pb$ ages.

Zircons from this sample were then analyzed on the SHRIMP. Twenty-one spot analyses were taken on magmatically zoned areas and eighteen of them (within 7% discordance) yielded a weighted mean $^{207}Pb/^{206}Pb$ age (^{204}Pb-corrected) of 1709.3 ± 8.8 Ma at the 95% confidence level (mean square of weighted deviates [MSWD] = 1.17) with a 28% probability of fit (Fig. 9). Regression of the U-Pb data yielded a similar age: an upper-intercept age of 1705 ± 16 Ma (MSWD = 8.0; not shown). A central cluster of the least discordant analyses (5% or less; N = 10; centers of ellipses indicated by open squares, Fig. 9) yielded a weighted mean $^{207}Pb/^{206}Pb$ age of 1725 ± 12 Ma (MSWD =

TABLE 1A: U-Th-Pb ANALYTICAL DATA FROM ID-TIMS ZIRCON WORK ON SAMPLES FROM EL PINACATE–CABEZA PRIETA

Sample/ Fraction	Fraction Specifics	Sample Wgt (mg)	U (ppm)	Th (ppm)	Pb (ppm)	$^{206}Pb/^{204}Pb$[a]	$^{207}Pb/^{206}Pb$[b]	$^{208}Pb/^{206}Pb$[b]	$^{206}Pb/^{238}U$[c]	$^{207}Pb/^{235}U$[c]	$^{207}Pb/^{206}Pb$[c]	207/206 age (Ma)[#]
ALACRANES #1												
1(82)	16 grains	0.073	465	141	134	4537	0.102857	0.099542	0.2773	3.8661	0.101117	1645
	clear, subhed					(0.40)	(0.193)	(0.518)	(0.166)	(0.171)	(0.040)	(0.74)
2(82)	22 grains	0.048	621	188	176	5633	0.101231	0.098551	0.273848	3.78655	0.100284	1629
	clear, subhed					(0.57)	(0.226)	(0.604)	(0.178)	(0.183)	(0.043)	(0.79)
3(82)	30 grains	0.12	546	167	153	10230	0.1018	0.093572	0.272572	3.80096	0.101137	1645
	clear, subhed					(1.3)	(0.106)	(0.296)	(0.288)	(0.290)	(0.038)	(0.70)
ALACRANES #5												
4(80)	23 grains	0.133	294	224	88.2	7920	0.10435	0.084069	0.293407	4.18408	0.103426	1687
	eu-subhed, clear					(7.1)	(0.083)	(0.259)	(0.169)	(0.218)	(0.123)	(2.3)
5(80)	31 grains	0.159	314	173	95.8	3801	0.106652	0.093032	0.29438	4.20813	0.103676	1691
	eu-subhed, clear					(2.5)	(0.069)	(0.186)	(0.118)	(0.159)	(0.096)	(1.8)
6(80)	41 grains	0.259	294	77.2	88.4	14400	0.104486	0.07951	0.294921	4.22674	0.103944	1696
	more stubby, clear					(6.9)	(0.065)	(0.211)	(0.212)	(0.225)	(0.070)	(1.3)
MINA JOYA #1-98												
5(77)	25 grains	0.074	556	116	150	6589	0.101289	0.069886	0.268003	3.69679	0.100042	1625
	stubby, subhed					(1.00)[@]	(0.123)	(0.328)	(0.065)	(0.117)	(0.097)	(1.8)
6(77)	31 grains	0.057	714	159	198	7489	0.101437	0.07240	0.274338	3.79864	0.100425	1632
	stubby, subhed					(0.85)	(0.091)	(0.310)	(0.116)	(0.125)	(0.047)	(0.87)
TJA #21												
1(80)	31 grains	0.104	260	138	3.83	346.8	0.072145	0.193656	0.01322	0.101664	0.055774	443
	clear, euhedral					(0.31)	(3.65)	(3.34)	(0.411)	(0.561)	(0.344)	(7.6)
2(80)	>50 grains	0.162	355	183	5.11	614.2	0.067741	0.16919	0.013307	0.103261	0.056279	463
	clear, euhedral					(0.21)	(1.80)	(1.76)	(0.810)	(0.830)	(0.171)	(3.8)
3(80)	42 grains	0.29	205	99	3.04	385.09	0.0806	0.210873	0.012861	0.096502	0.054422	389
	clear, stubby					(0.14)	(2.17)	(2.01)	(0.307)	(0.344)	(0.151)	(3.4)
CP-17-99												
4(82)	20 grains	0.033	570	301	124	3442	0.098902	0.199326	0.194167	2.62929	0.098211	1590
	clear, stubby, euhed					(0.30)	(0.512)	(0.632)	(0.210)	(0.218)	(0.055)	(1.0)
5(82)	37 grains	0.074	535	300	113	4371	0.099315	0.192119	0.189569	2.55559	0.097774	1582
	clear, stubby, subhed					(0.31)	(0.250)	(0.325)	(0.115)	(0.123)	(0.043)	(0.81)
6(82)	45 grains	0.108	556	294	116	3686.7	0.100829	0.190262	0.186603	2.52588	0.098173	1590
	clear, stubby, subhed					(0.21)	(0.166)	(0.224)	(0.147)	(0.152)	(0.038)	(0.70)
PZ-23B												
5(75)	16 grains	0.064	330	211	91	2250	0.094634	0.195126	0.24847	3.13825	0.091604	1459
	clear, euhedral					(0.74)	(0.339)	(0.421)	(0.208)	(0.224)	(0.079)	(1.5)
6(75)	28 grains	0.089	333	212	92.5	1916.8	0.095694	0.205351	0.246668	3.08407	0.09068	1440
	clear, subhed					(0.28)	(0.239)	(0.275)	(0.079)	(0.106)	(0.070)	(1.3)
CP-16-99												
1(77)	14 grains	0.049	162	105	2.35	119.12	0.085048	0.28885	0.011657	0.078493	0.048836	140
	clear, euhedral					(0.30)	(11.5)	(8.04)	(1.38)	(2.03)	(1.35)	(32)
2(77)	21 grains	0.041	321	145	5.22	179.8	0.079033	0.207378	0.014126	0.094612	0.048576	127
	clear, euhedral					(0.56)	(6.25)	(5.85)	(0.706)	(1.38)	(1.1)	(26)

[a] Corrected for fractionation only, 0.08 ± 0.03 %/a.m.u. on Faraday Cup, 0.301 ± 0.045 %/a.m.u. using the Daly collector.
[b] Corrected for fractionation and laboratory blank Pb; between 42 and 75 pg total Pb with a measured composition of $^{206}Pb/^{204}Pb$ = 19.05 ± 0.24, $^{207}Pb/^{204}Pb$ = 15.496 ± 0.065, and $^{208}Pb/^{204}Pb$ = 37.87 ± 0.19
[c] Radiogenic ratios; corrected for fractionation, laboratory blank Pb, and initial common Pb using values of Stacey and Kramers (1975) for an approximate age for the sample and second-stage $^{238}U/^{204}Pb$ = 9.74.
[@] Numbers in parentheses are errors (in percent) at the 2-sigma level for the numbers directly above.
[#] Age in millions of years, calculated using Ludwig (1980; 1985), and decay constants of Steiger and Jäger (1977).

TABLE 1B. U-Th-Pb ANALYTICAL DATA FOR SHRIMP SPOT ANALYSES ON ZIRCON FROM EL PINACATE–CABEZA PRIETA SAMPLES

Grains/Spots	Core/ Rim?	U (ppm)	Th (ppm)	Th/U	% comm $^{206}Pb^{*}$	$^{206}Pb/^{238}U$ #	± $^{206}Pb/^{238}U$ @	$^{207}Pb/^{235}U$ #	± $^{207}Pb/^{235}U$ @	$^{207}Pb/^{206}Pb$ #	± $^{207}Pb/^{206}Pb$ @	207/206 AGE	207/206 AGE ±	% DISCORD$
ALACRANES #1														
ALAC1-1	core	298	148	0.50	0.696	0.28949	4.08	4.22375	4.21	0.10548	0.91	1729	19	5
ALAC1-2	core	404	201	0.50	–0.003	0.30528	4.08	4.42609	4.19	0.09873	0.69	1717	17	0
ALAC1-3	core	199	98	0.49	0.514	0.28918	4.11	4.14906	4.33	0.10343	0.80	1698	25	4
ALAC1-4	core	387	80	0.21	1.406	0.26497	5.08	3.85368	5.16	0.10470	0.75	1723	17	12
ALAC1-5	core	1007	834	0.83	–1.071	0.34240	4.09	4.88919	4.19	0.10613	1.32	1689	17	–12
ALAC1-6	core	545	305	0.56	0.556	0.29307	4.64	4.28159	4.71	0.10386	0.51	1731	14	4
ALAC1-7	core	104	53	0.51	0.606	0.29094	4.25	4.25054	4.65	0.10582	1.02	1731	35	5
ALAC1-8	core	857	308	0.36	0.375	0.28993	4.12	4.14699	4.17	0.10406	1.36	1692	12	3
ALAC1-9	core	807	369	0.46	0.531	0.26565	4.07	3.61612	4.13	0.10374	0.63	1600	13	5
ALAC1-10	core	496	251	0.51	0.249	0.30203	4.09	4.42285	4.18	0.10596	1.88	1735	16	2
ALAC1-11	core	304	156	0.51	0.152	0.30146	4.09	4.37632	4.22	0.10517	0.99	1719	19	1
ALAC1-12	core	185	91	0.49	0.791	0.28617	4.11	4.18750	4.32	0.10596	0.77	1734	24	6
ALAC1-13	core	528	187	0.36	0.837	0.27454	4.34	3.91539	4.41	0.10214	1.27	1687	15	7
ALAC1-14	core	312	129	0.41	0.493	0.29130	4.08	4.22428	4.20	0.10568	1.42	1717	18	4
ALAC1-15	core	801	340	0.42	0.767	0.28150	4.11	4.06361	4.18	0.10592	1.03	1709	14	6
ALAC1-16	core	180	75	0.42	0.351	0.29927	4.11	4.36078	4.34	0.10529	1.02	1726	26	2
ALAC1-17	core	1209	612	0.51	0.492	0.28650	4.08	4.10271	4.11	0.10621	0.88	1694	10	4
ALAC1-18	core	334	127	0.38	0.240	0.30124	4.09	4.39924	4.22	0.10515	0.92	1730	19	2
ALAC1-19	core	264	126	0.48	0.183	0.29279	4.10	4.12320	4.29	0.10175	0.85	1663	23	1
ALAC1-20	core	780	373	0.48	–1.021	0.32141	4.13	4.50919	4.22	0.10375	1.07	1656	16	–8
ALAC1-21	core	379	177	0.47	–0.855	0.32332	4.09	4.62504	4.23	0.10357	0.94	1692	20	–7
ALACRANES #5														
ALAC2-1	core	298	87	0.29	0.484	0.28929	1.92	4.15209	2.29	0.10409	1.24	1698	23	4
ALAC-2	core	403	121	0.30	–1.575	0.35532	1.95	5.30095	2.66	0.10820	1.81	1769	33	–11
ALAC-3	core	154	22	0.14	0.106	0.30474	2.45	4.44758	3.06	0.10585	1.84	1729	34	1
ALAC-4	core	273	70	0.26	0.135	0.30207	1.89	4.34536	2.39	0.10433	1.46	1703	27	0
ALAC-5	core	354	87	0.25	1.222	0.26979	2.20	3.91506	2.50	0.10525	1.18	1719	22	10
ALAC-6	core	204	47	0.23	0.829	0.28839	1.97	4.22355	2.60	0.10622	1.71	1736	31	6
ALAC-7	core	238	104	0.44	0.247	0.29864	1.97	4.33382	2.35	0.10525	1.29	1719	24	2
ALAC-8	core	321	126	0.39	–0.551	0.31801	1.89	4.58390	2.21	0.10454	1.16	1706	21	–4
ALAC-9	rim	318	97	0.31	–0.465	0.31738	1.87	4.57566	2.27	0.10456	1.28	1707	24	–4
ALAC-10	core	331	92	0.28	0.570	0.29111	1.90	4.24326	2.28	0.10572	1.27	1727	23	5
ALAC-11	core	321	114	0.36	0.298	0.29618	1.89	4.28674	2.50	0.10497	1.63	1714	30	2
ALAC-12	core	139	47	0.34	–0.166	0.32605	2.03	4.94141	2.74	0.10992	1.85	1798	34	–1
ALAC-13	core	349	82	0.24	0.175	0.30088	1.87	4.36883	2.21	0.10531	1.19	1720	22	1
ALAC-14	core	355	96	0.27	–1.683	0.34944	2.03	5.08130	2.66	0.10546	1.72	1722	32	–12
ALAC-15	core	122	28	0.23	0.627	0.28296	2.44	4.05693	3.51	0.10399	2.52	1696	47	5
ALAC-16	rim	417	123	0.30	0.171	0.30386	1.84	4.44563	2.07	0.10611	0.95	1734	17	1
ALAC-17	core	90	22	0.24	1.094	0.30260	3.29	4.71375	3.92	0.11298	2.14	1848	39	8

(continued)

TABLE 1B. U-Th-Pb ANALYTICAL DATA FOR SHRIMP SPOT ANALYSES ON ZIRCON FROM EL PINACATE–CABEZA PRIETA SAMPLES (*continued*)

Grains/Spots	Core/ Rim?	U (ppm)	Th (ppm)	Th/U	% comm $^{206}Pb^*$	$^{206}Pb/^{238}U$ #	± $^{206}Pb/^{238}U$ @	$^{207}Pb/^{235}U$ #	± $^{207}Pb/^{235}U$ @	$^{207}Pb/^{206}Pb$ #	± $^{207}Pb/^{206}Pb$ @	207/206 AGE	207/206 AGE ±	% DISCORD$
MINA JOYA #1-98														
JOYA-5	core	220	62	0.28	0.000	0.29531	2.89	4.16368	3.06	0.10226	0.98	1666	18	0
JOYA-2	core	539	267	0.50	0.000	0.29498	2.85	4.20009	2.92	0.10327	0.63	1684	12	1
JOYA-3	core	468	136	0.29	0.000	0.29666	2.85	4.24470	2.92	0.10377	0.64	1693	12	1
JOYA-1	core	169	74	0.43	0.000	0.29830	2.94	4.27677	3.13	0.10398	1.07	1696	20	1
JOYA-6	core	237	62	0.26	0.074	0.30407	2.89	4.36871	3.03	0.10420	0.93	1700	17	–1
JOYA-8	core	133	55	0.41	0.000	0.31072	2.94	4.51446	3.19	0.10537	1.23	1721	23	–1
JOYA-7	core	164	49	0.30	0.000	0.29524	2.92	4.29705	3.13	0.10556	1.13	1724	21	3
JOYA-4	core	175	49	0.28	0.000	0.29595	2.91	4.34604	3.01	0.10651	1.06	1740	19	4
CP-17-99														
CP-12	core	468	334	0.71	2.417	0.20125	3.16	2.73073	3.64	0.09841	1.82	1594	34	26
CP-5	core	800	460	0.58	2.793	0.19168	2.21	2.62735	2.44	0.09941	1.02	1613	19	30
CP-14	core	734	620	0.85	1.874	0.22550	2.62	3.10140	2.72	0.09975	0.75	1619	14	19
CP-19	core	279	197	0.71	0.369	0.27530	2.64	3.79334	2.87	0.09994	1.11	1623	21	3
CP-9	core	665	533	0.80	1.483	0.24237	2.64	3.34123	2.76	0.09998	0.80	1624	15	14
CP-7	core	179	221	1.23	–0.045	0.28854	2.68	3.98598	2.95	0.10019	1.23	1628	23	0
CP-10	core	138	135	0.98	1.006	0.26080	2.81	3.60480	3.34	0.10025	1.81	1629	34	8
CP-1	core	2953	733	0.25	1.087	0.25551	2.22	3.54409	2.29	0.10060	0.55	1635	10	10
CP-2	core	360	355	0.99	–0.245	0.29639	2.16	4.11779	2.49	0.10076	1.24	1638	23	–2
CP-17	core	954	559	0.59	0.875	0.26297	2.58	3.65374	2.65	0.10077	0.64	1638	12	8
CP-3	core	799	925	1.16	0.063	0.28762	3.26	3.99702	3.36	0.10079	0.80	1639	15	1
CP-11	core	496	327	0.66	0.945	0.26094	2.62	3.62740	2.76	0.10082	0.88	1639	16	9
CP-18	core	68	60	0.88	1.266	0.25140	3.04	3.50139	3.86	0.10101	2.37	1643	44	12
CP-8	core	246	280	1.14	0.064	0.29011	2.68	4.04109	2.93	0.10103	1.18	1643	22	0
CP-15	core	478	480	1.01	0.985	0.26573	3.40	3.74119	3.51	0.10211	0.84	1663	16	9
CP-20	core	376	369	0.98	0.819	0.26993	2.74	3.80001	2.89	0.10210	0.91	1663	17	7
CP-4	core	146	124	0.85	0.547	0.28065	2.39	3.97555	3.03	0.10274	1.87	1674	35	5
CP-16	core	183	95	0.52	0.290	0.28890	2.71	4.09979	2.98	0.10292	1.24	1678	23	2
CP-13	core	164	139	0.85	0.789	0.27434	3.27	3.89612	3.64	0.10300	1.60	1679	30	7
CD-3 #4														
CD3-4-17	core	95	59	0.62	0.554	0.28842	2.96	3.84101	3.72	0.09659	2.26	1559	42	–5
CD3-4-16	core	113	38	0.34	0.223	0.25881	2.82	3.47374	3.37	0.09735	1.84	1574	34	6
CD3-4-14	core	92	50	0.54	0.574	0.28656	2.81	3.85042	3.45	0.09745	2.00	1576	38	–3
CD3-4-12	core	91	76	0.84	0.154	0.27827	2.89	3.76481	3.59	0.09812	2.12	1589	40	0
CD3-4-5	core	73	47	0.64	1.763	0.27644	3.02	3.77044	6.53	0.09892	5.79	1604	108	2
CD3-4-10	core	89	69	0.78	0.151	0.28868	2.90	3.94307	3.62	0.09906	2.18	1607	41	–2
CD3-4-7	core	92	48	0.52	0.162	0.27285	2.90	3.74128	3.62	0.09945	2.17	1614	40	4
CD3-4-15	core	165	117	0.71	0.263	0.27831	2.74	3.81903	3.18	0.09952	1.62	1615	30	2
CD3-4-19	core	171	140	0.82	0.000	0.30874	2.77	4.24600	2.98	0.09975	1.08	1619	20	–7
CD3-4-13	core	73	42	0.57	0.279	0.30209	2.83	4.18005	3.35	0.10036	1.78	1631	33	–4

(*continued*)

TABLE 1B. U-Th-Pb ANALYTICAL DATA FOR SHRIMP SPOT ANALYSES ON ZIRCON FROM EL PINACATE–CABEZA PRIETA SAMPLES (*continued*)

Grains/Spots	Core/ Rim?	U (ppm)	Th (ppm)	Th/U	% comm ^{206}Pb*	$^{206}Pb/^{238}U$ #	± $^{206}Pb/^{238}U$ @	$^{207}Pb/^{235}U$ #	± $^{207}Pb/^{235}U$ @	$^{207}Pb/^{206}Pb$ #	± $^{207}Pb/^{206}Pb$ @	207/206 AGE	207/206 AGE ±	% DISCORD$
CD3-4-18	core	244	201	0.83	0.058	0.30457	3.55	4.21733	3.66	0.10043	0.90	1632	17	–5
CD3-4-9	core	134	79	0.59	0.000	0.27558	3.19	3.82673	3.57	0.10071	1.61	1637	30	4
CD3-4-6	core	136	78	0.57	0.000	0.28225	2.80	3.93190	3.19	0.10103	1.51	1643	28	3
CD3-4-3	core	151	117	0.78	0.000	0.26768	2.85	3.75957	3.24	0.10187	1.55	1658	29	8
CD3-4-11	core	147	87	0.59	0.089	0.28344	2.96	4.01124	3.39	0.10264	1.64	1672	30	4
CD3-4-1	core	115	60	0.52	0.213	0.27695	2.82	3.92659	3.34	0.10283	1.78	1676	33	6
CD3-4-4	core	69	49	0.71	0.000	0.27301	3.38	3.90051	4.03	0.10362	2.18	1690	40	8
CD3-4-8	core	183	125	0.68	0.000	0.28011	2.77	4.00528	3.07	0.10371	1.34	1692	25	6
CD3-4-21	core	66	35	0.52	0.000	0.28803	2.95	4.14844	3.51	0.10446	1.89	1705	35	4
CD3-4-20	core	88	67	0.76	-0.268	0.29168	2.83	4.22165	3.46	0.10497	2.00	1714	37	4
CD3-4-2	core	96	59	0.61	0.000	0.28086	2.87	4.07304	3.36	0.10518	1.75	1717	32	7
CD-12 #19A														
CD19-9	rim	1217	110	0.09	0.500	0.23006	2.66	2.83726	2.84	0.08945	0.99	1414	19	6
CD19-2	core	54	32	0.59	0.887	0.27860	3.04	3.68858	4.32	0.09602	3.07	1548	58	–2
CD19-8	core	65	39	0.60	0.299	0.28253	3.00	3.78723	4.12	0.09722	2.82	1571	53	–2
CD19-5	core	133	89	0.67	0.259	0.26744	2.85	3.64439	3.65	0.09883	2.27	1602	42	5
CD19-16	core	172	96	0.56	0.000	0.30514	2.74	4.17860	2.94	0.09932	1.08	1611	20	–7
CD19-6	core	60	33	0.54	0.000	0.28278	3.02	3.87674	3.77	0.09943	2.25	1614	42	1
CD19-10	core	129	74	0.57	0.194	0.27666	2.83	3.79986	3.31	0.09961	1.71	1617	32	3
CD19-19	rim	1510	665	0.44	0.090	0.30655	2.66	4.24520	2.69	0.10044	0.38	1632	7	–6
CD19-15	rim	781	115	0.15	0.027	0.32053	2.66	4.44360	2.70	0.10055	0.46	1634	9	–10
CD19-17	core	87	56	0.65	0.373	0.29466	2.81	4.09286	3.49	0.10074	2.07	1638	38	–2
CD19-18	rim	81	49	0.61	0.178	0.28918	2.84	4.02238	3.32	0.10088	1.72	1640	32	0
CD19-1	core	77	50	0.66	0.157	0.30541	2.93	4.27856	3.60	0.10161	2.09	1654	39	–4
CD19-12	core	77	51	0.66	0.077	0.29355	3.04	4.11329	3.74	0.10163	2.18	1654	40	0
CD19-13	core	103	99	0.97	0.275	0.30825	2.87	4.34938	3.27	0.10234	1.57	1667	29	–4
CD19-20	core	70	56	0.79	0.000	0.29030	2.88	4.10047	3.59	0.10245	2.14	1669	40	2
CD19-3	core	127	95	0.75	0.225	0.30318	2.83	4.31430	3.47	0.10321	2.00	1683	37	–1
CD19-11	core	14	12	0.85	0.000	0.29403	3.72	4.21167	5.21	0.10389	3.65	1695	67	2
CD19-4	core	71	60	0.85	0.000	0.27611	3.07	3.98148	3.77	0.10458	2.18	1707	40	8
CD19-7	core	89	44	0.50	0.000	0.28581	3.83	4.17128	4.26	0.10585	1.86	1729	34	6
CD19-14	core	73	41	0.56	–0.039	0.29189	2.86	4.35457	3.33	0.10820	1.70	1769	31	7
CD-12 #2; BORDERLINE QUARTZITE														
CD2-18	core	1889	140	0.07	0.099	0.25695	2.17	3.19500	2.23	0.09018	0.48	1429	9	–3
CD2-5	core	1822	46	0.03	0.020	0.25346	2.23	3.15177	2.25	0.09019	0.36	1430	7	–2
CD2-6	rim	1566	68	0.04	0.005	0.26540	2.18	3.31594	2.21	0.09062	0.38	1439	7	–6
CD2-3	core	2425	66	0.03	0.007	0.26240	2.17	3.27984	2.19	0.09066	0.30	1439	6	–4
CD2-17	rim	1676	42	0.03	0.007	0.25409	2.35	3.17662	2.38	0.09067	0.35	1440	7	–1
CD2-2	core	445	122	0.27	0.000	0.27458	2.22	3.72729	2.30	0.09845	0.62	1595	12	2
CD2-15	core	412	118	0.29	0.089	0.29680	2.21	4.03242	2.29	0.09854	0.62	1597	12	–5

(*continued*)

TABLE 1B. U-Th-Pb ANALYTICAL DATA FOR SHRIMP SPOT ANALYSES ON ZIRCON FROM EL PINACATE–CABEZA PRIETA SAMPLES (*continued*)

Grains/Spots	Core/ Rim?	U (ppm)	Th (ppm)	Th/U	% comm $^{206}Pb^*$	$^{206}Pb/^{238}U$ #	± $^{206}Pb/^{238}U$ @	$^{207}Pb/^{235}U$ #	± $^{207}Pb/^{235}U$ @	$^{207}Pb/^{206}Pb$ #	± $^{207}Pb/^{206}Pb$ @	207/206 AGE	207/206 AGE ±	% DISCORD$
CD2-20	core	1373	46	0.03	0.025	0.25650	2.25	3.49299	2.28	0.09877	0.39	1601	7	8
CD2-1	core	1056	109	0.10	0.008	0.28203	2.25	3.84524	2.28	0.09888	0.39	1603	7	0
CD2-11	core	147	63	0.43	0.127	0.28774	2.28	3.93252	2.56	0.09912	1.16	1608	22	–1
CD2-13	core	694	236	0.34	0.040	0.28496	2.21	3.89550	2.27	0.09915	0.52	1608	10	–1
CD2-9	core	138	100	0.72	0.147	0.28871	2.26	3.96112	2.65	0.09951	1.39	1615	26	–1
CD2-12	core	185	85	0.46	0.106	0.25595	2.57	3.53374	2.81	0.10013	1.14	1627	21	10
CD2-7	core	930	578	0.62	0.033	0.28621	2.18	3.95740	2.23	0.10028	0.43	1629	8	0
CD2-14	core	267	103	0.39	0.000	0.27182	2.24	3.76816	2.43	0.10054	0.93	1634	17	5
CD2-19	core	1359	181	0.13	0.010	0.27014	2.18	3.76018	2.22	0.10095	0.43	1642	8	6
CD2-16	core	212	150	0.71	0.028	0.27323	2.35	3.82240	2.51	0.10146	0.87	1651	16	6
CD2-4	core	244	229	0.94	0.000	0.29602	2.22	4.14602	2.37	0.10158	0.83	1653	15	–1
CD2-8	core	947	624	0.66	0.016	0.29755	2.30	4.17115	2.34	0.10167	0.44	1655	8	–1
CD2-10	core	103	69	0.67	0.000	0.28871	2.28	4.10581	2.59	0.10314	1.22	1681	23	3
CD-12 #5; META-ARKOSE														
CD5-17	rim	1262	47	0.04	0.040	0.24590	2.50	3.03000	2.50	0.08940	0.40	1412	8	0
CD5-22	core	740	111	0.15	0.177	0.23693	2.19	2.98371	2.37	0.09133	0.90	1454	17	6
CD5-15	rim	955	238	0.25	1.790	0.22270	2.70	2.99000	2.70	–0.09750	0.50	1577	9	18
CD5-21	rim	549	205	0.37	0.024	0.29533	2.19	3.99119	2.29	0.09802	0.67	1587	13	-5
CD5-2	core	331	110	0.33	0.127	0.27307	2.21	3.69018	2.38	0.09801	0.88	1587	16	2
CD5-9	core	188	118	0.63	0.086	0.28465	2.25	3.86675	2.48	0.09852	1.03	1596	19	–1
CD5-6	core	84	46	0.54	0.168	0.28916	2.37	3.94563	2.88	0.09896	1.64	1605	31	–2
CD5-7	core	223	89	0.40	0.073	0.28411	2.23	3.89649	2.45	0.09947	1.01	1614	19	0
CD5-18	core	313	196	0.63	–0.230	0.29120	2.50	3.99000	2.60	0.09950	0.70	1614	12	–2
CD5-19	core	158	108	0.68	–0.040	0.29020	2.60	4.00000	2.80	0.10010	1.10	1626	21	–1
CD5-14	core	289	177	0.61	0.034	0.26890	2.22	3.72632	2.39	0.10050	0.89	1633	17	6
CD5-5	core	462	362	0.78	0.030	0.28865	2.22	4.00219	2.31	0.10056	0.64	1634	12	0
CD5-16	rim	648	389	0.60	–0.500	0.30410	2.50	4.22000	2.50	0.10080	0.50	1638	9	–4
CD5-4	core	288	191	0.66	0.000	0.27696	2.23	3.84981	2.37	0.10081	0.81	1639	15	4
CD5-10	core	55	38	0.68	0.286	0.30564	2.40	4.24988	3.05	0.10085	1.88	1640	35	–5
CD5-3	core	221	165	0.75	0.000	0.30136	2.23	4.19500	2.39	0.10096	0.88	1642	16	–3
CD5-8	core	112	71	0.63	0.083	0.29079	2.98	4.06806	3.24	0.10146	1.27	1651	24	0
CD5-11	core	277	216	0.78	0.035	0.27361	2.32	3.83455	2.48	0.10165	0.89	1654	16	6
CD5-13	core	442	240	0.54	0.016	0.28436	2.20	3.99596	2.29	0.10192	0.66	1659	12	3
CD5-1	core	265	142	0.54	0.085	0.28700	2.25	4.03551	2.40	0.10198	0.84	1661	16	2
CD5-20	core	328	208	0.64	0.000	0.29307	2.20	4.13638	2.29	0.10236	0.65	1667	12	1
CD5-12	core	238	116	0.49	0.056	0.29031	2.23	4.16331	2.41	0.10401	0.91	1697	17	3
CD-12 #13; META-ARKOSE														
CLO-13-20	rim	1383	134	0.01	0.012	0.22944	2.21	2.82481	2.25	0.08929	0.40	1410	8	6
CLO-13-9	rim	1843	204	0.11	0.245	0.23793	2.18	2.96786	2.29	0.09047	0.71	1435	14	4
CLO-13-19	rim	1561	366	0.23	0.045	0.27090	2.17	3.59792	2.20	0.09633	0.32	1554	6	1

(*continued*)

TABLE 1B. U-Th-Pb ANALYTICAL DATA FOR SHRIMP SPOT ANALYSES ON ZIRCON FROM EL PINACATE–CABEZA PRIETA SAMPLES (*continued*)

Grains/Spots	Core/ Rim?	U (ppm)	Th (ppm)	Th/U	% comm $^{206}Pb^*$	$^{206}Pb/^{238}U$ #	± $^{206}Pb/^{238}U$ @	$^{207}Pb/^{235}U$ #	± $^{207}Pb/^{235}U$ @	$^{207}Pb/^{206}Pb$ #	± $^{207}Pb/^{206}Pb$ @	207/206 AGE	207/206 AGE ±	% DISCORD$
CLO-13-1	core	229	60	0.26	0.203	0.23628	2.45	3.15462	2.89	0.09683	1.53	1564	29	13
CLO-13-14	core	2588	1812	0.70	0.044	0.28344	2.17	3.79888	2.19	0.09721	0.26	1571	5	–2
CLO-13-18	rim	958	87	0.09	0.036	0.26660	2.30	3.58524	2.34	0.09754	0.45	1578	8	3
CLO-13-10	rim	1504	409	0.27	0.020	0.29769	2.53	4.01797	2.56	0.09789	0.39	1584	7	–6
CLO-13-11	core	484	325	0.67	0.110	0.28294	2.23	3.85400	2.35	0.09879	0.74	1601	14	0
CLO-13-16	rim	958	300	0.31	0.021	0.29523	2.40	4.02481	2.42	0.09888	0.37	1603	7	–4
CLO-13-17	core	1583	800	0.51	0.004	0.31453	2.17	4.31294	2.20	0.09945	0.31	1614	6	–9
CLO-13-6	core	411	306	0.75	0.036	0.30189	2.19	4.15067	2.28	0.09972	0.62	1619	12	–5
CLO-13-13	core	1829	1229	0.67	0.005	0.29837	2.17	4.13170	2.19	0.10043	0.28	1632	5	–3
CLO-13-7	core	291	139	0.48	0.035	0.29044	2.30	4.03163	2.42	0.10068	0.76	1637	14	-0
CLO-13-12	core	501	321	0.64	0.000	0.27899	2.25	3.88250	2.33	0.10093	0.63	1641	12	3
CLO-13-3	core	605	401	0.66	0.027	0.30100	2.24	4.19193	2.30	0.10101	0.51	1643	10	–3
CLO-13-4	core	1176	837	0.71	0.000	0.29920	2.21	4.17128	2.24	0.10111	0.36	1645	7	–3
CLO-13-8	core	146	142	0.98	0.000	0.27662	2.24	3.85786	2.46	0.10115	1.02	1645	19	4
CLO-13-5	core	202	168	0.83	0.029	0.29556	2.22	4.12925	2.38	0.10133	0.87	1649	16	–1
CLO-13-2	core	313	170	0.54	0.058	0.27975	2.21	3.91985	2.41	0.10162	0.97	1654	18	4
CLO-13-15	core	418	288	0.69	0.000	0.29124	2.21	4.09799	2.29	0.10205	0.59	1662	11	1
CD-12 #20; QUARTZITE														
CA-20-21	rim	1835	61	0.03	0.014	0.24317	2.22	3.01839	2.24	0.09003	0.31	1426	6	2
CA-20-18B	rim	457	29	0.06	0.082	0.24180	2.19	3.02949	2.28	0.09087	0.64	1444	12	3
CA-20-24	rim	1471	53	0.04	0.010	0.25857	2.23	3.25060	2.26	0.09118	0.36	1450	7	-2
CA-20-2	core	112	62	0.55	0.215	0.28584	2.29	3.83047	2.74	0.09719	1.50	1571	28	-3
CA-20-10	core	414	162	0.39	0.149	0.26929	2.20	3.65225	2.51	0.09837	1.21	1593	23	4
CA-20-9	core	75	49	0.65	0.227	0.29081	2.31	3.96803	2.75	0.09896	1.49	1605	28	-3
CA-20-15	core	556	300	0.54	0.214	0.24472	2.20	3.34331	2.36	0.09908	0.85	1607	16	12
CA-20-13	core	246	143	0.58	0.112	0.29100	2.21	3.97996	2.40	0.09919	0.93	1609	17	-2
CA-20-20	core	255	165	0.65	0.126	0.28576	2.21	3.91320	2.37	0.09932	0.85	1611	16	-1
CA-20-18	core	115	98	0.85	0.057	0.30609	3.04	4.19375	3.25	0.09937	1.15	1612	21	-7
CA-20-12	core	406	128	0.31	0.596	0.23595	2.20	3.23493	2.68	0.09944	1.53	1614	29	15
CA-20-5	core	184	111	0.60	0.091	0.29480	2.22	4.04964	2.43	0.09963	0.97	1617	18	-3
CA-20-17	core	1337	1202	0.90	0.196	0.27263	2.26	3.74966	2.40	0.09975	0.80	1619	15	4
CA-20-8	core	94	67	0.71	0.122	0.28427	2.28	3.93341	2.65	0.10035	1.35	1631	25	1
CA-20-22	core	128	84	0.65	0.083	0.28622	2.27	3.96353	2.49	0.10043	1.03	1632	19	1
CA-20-6	core	344	192	0.56	0.095	0.31926	2.22	4.42250	2.33	0.10047	0.69	1633	13	-9
CA-20-19	core	140	57	0.40	0.054	0.29991	2.24	4.15535	2.46	0.10049	1.02	1633	19	-4
CA-20-16	core	162	130	0.80	0.132	0.29644	2.24	4.10785	2.50	0.10050	1.12	1633	21	-3
CA-20-1	core	583	331	0.57	0.013	0.23622	2.19	3.29402	2.27	0.10114	0.58	1645	11	17
CA-20-14	core	151	83	0.55	0.041	0.25376	2.26	3.54310	2.49	0.10127	1.04	1648	19	12
CA-20-23	core	195	106	0.55	0.000	0.28194	2.41	3.94400	2.55	0.10146	0.85	1651	16	3
CA-20-4	core	198	184	0.93	0.328	0.29678	2.23	4.15543	2.51	0.10155	1.13	1653	21	–1
CA-20-3	core	568	290	0.51	0.022	0.29702	2.19	4.16807	2.27	0.10178	0.58	1657	11	–1
CA-20-7	core	99	64	0.65	3.363	0.21183	3.93	2.98303	14.58	0.10213	14.04	1663	260	26
CA-20-11	core	105	77	0.74	0.700	0.28918	2.35	4.33960	7.11	0.10884	6.71	1780	122	8

(*continued*)

TABLE 1B. U-Th-Pb ANALYTICAL DATA FOR SHRIMP SPOT ANALYSES ON ZIRCON FROM EL PINACATE–CABEZA PRIETA SAMPLES (*continued*)

Grains/Spots	Core/ Rim?	U (ppm)	Th (ppm)	Th/U	% comm ^{206}Pb*	^{206}Pb/ ^{238}U #	± ^{206}Pb/ ^{238}U @	^{207}Pb/ ^{235}U #	± ^{207}Pb/ ^{235}U @	^{207}Pb/ ^{206}Pb #	± ^{207}Pb/ ^{206}Pb @	207/206 AGE	207/206 AGE ±	% DISCORD$
MESOPROTEROZOIC SAMPLES														
PZ-23B														
PZ23-1		336	111	0.33	0.048	0.24554	0.46	3.07854	0.77	0.09093	0.62	1445	12	2
PZ23-2		386	171	0.44	0.103	0.24463	0.43	3.00490	0.82	0.08909	0.69	1406	13	0
PZ23-3		249	142	0.57	0.064	0.24773	0.52	3.06459	0.91	0.08972	0.74	1420	14	–1
PZ23-4		182	111	0.61	0.159	0.24284	0.63	3.01219	1.37	0.08996	1.22	1425	23	2
PZ23-5		196	77	0.39	0.096	0.24337	0.62	3.05122	1.13	0.09093	0.95	1445	18	3
PZ23-6		226	155	0.69	0.000	0.24848	0.57	3.11670	0.97	0.09097	0.78	1446	15	1
PZ23-7		395	285	0.72	0.042	0.24809	0.43	3.08825	0.78	0.09028	0.65	1432	12	0
PZ23-8		234	142	0.61	–0.015	0.24673	0.56	3.07199	0.90	0.09030	0.70	1432	13	1
PZ23-9		342	208	0.61	0.031	0.25121	0.46	3.09738	0.81	0.08942	0.67	1413	13	–2
PZ23-10		657	323	0.49	0.021	0.24079	0.33	3.02282	0.58	0.09105	0.48	1448	9	4
PZ23-11		543	257	0.47	0.019	0.24826	0.37	3.08505	0.61	0.09013	0.48	1428	9	0
PZ23-12		310	183	0.59	0.051	0.23949	0.51	2.98640	0.86	0.09044	0.69	1435	13	4
PZ23-13		382	186	0.49	0.156	0.24864	0.44	3.08113	0.83	0.08987	0.71	1423	14	–1
PZ23-14		476	260	0.55	–0.003	0.25080	0.39	3.11224	0.63	0.09000	0.49	1426	9	–1
PZ23-15		175	107	0.61	–0.022	0.24648	0.64	3.09781	1.03	0.09115	0.81	1450	15	2

Grains/Spots	Core/ Rim?	U (ppm)	Th (ppm)	Th/U	% comm 206Pb*	207Pb/ 235U #	± 207Pb/ 235U @	238U/ 206Pb #	± 238U/ 206Pb@	207Pb/ 206Pb #	± 207Pb/ 206Pb@	206/238 AGE	206/238 AGE ±	% DISCORD$
CRETACEOUS SAMPLES														
TJA #21														
TJA-1	core	189	66	0.35	3.918	0.03447	88.52	86.42750	4.74	0.02161	88.39	76.59	3.19	103
TJA-2	core	298	133	0.45	0.000	0.07861	6.51	88.04465	3.16	0.05019	5.69	72.56	2.30	64
TJA-3	core	655	405	0.62	3.973	0.11175	17.87	85.74563	3.24	0.06949	17.57	72.83	2.23	92
TJA-4	core	295	307	1.04	0.000	0.07713	7.03	85.14666	3.22	0.04763	6.25	75.26	2.43	7
TJA-5	core	86	48	0.56	0.000	0.07605	11.62	93.00501	4.04	0.05130	10.89	68.61	2.81	73
TJA-6	core	90	50	0.56	0.000	0.07736	14.96	86.33463	4.57	0.04844	14.24	74.15	3.43	39
TJA-7	core	77	46	0.59	0.000	0.07208	14.36	85.93653	4.04	0.04493	13.78	74.82	3.07	225
TJA-8	core	181	101	0.56	5.346			96.41008	4.45			69.41	2.41	
CP-16-99														
CP16-1	core	56	30	0.54	0.000	0.08545	14.13	78.63116	4.41	0.04873	13.43	81.36	3.63	40
CP16-3	core	105	73	0.70	0.000	0.08441	10.11	85.90233	3.83	0.05259	9.36	74.14	2.86	76
CP16-4	core	66	27	0.41	0.000	0.08932	12.60	87.64302	4.31	0.05678	11.84	72.31	3.16	85
CP16-2	core	114	90	0.79	–4.982	0.15071	28.38	82.84063	4.77	0.09055	27.98	73.67	2.81	95

*Common Pb

#Atomic ratios corrected for initial Pb using the amount of ^{204}Pb and corresponding average Earth values from Stacey and Kramers (1975) with the exception of the last two samples that are 207Pb-corrected.

@Errors given are in % at the 1 sigma level

$Degree of discordance, percentage of the distance that the analysis lies along a chord from its extrapolated intersection with concordia (corresponding to its 207/206 age) to the origin at 0 Ma.

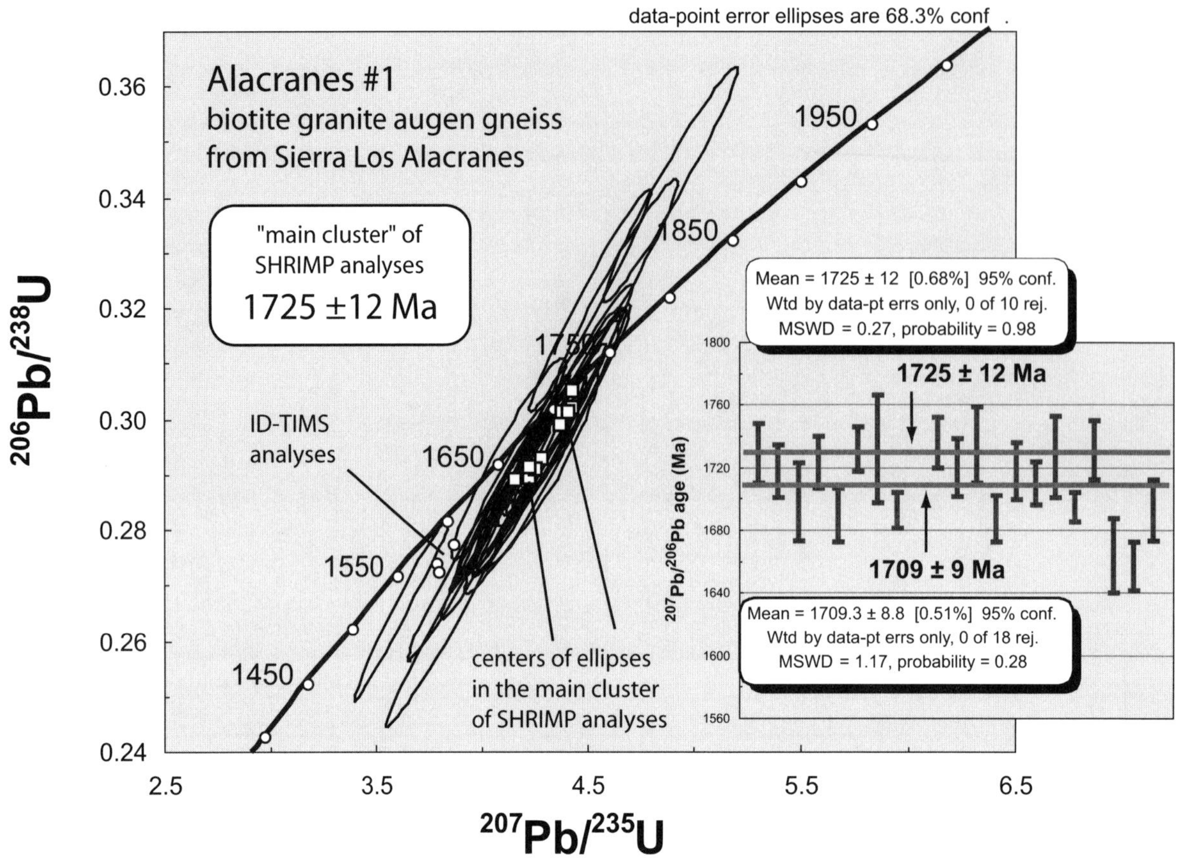

Figure 9. U-Pb zircon concordia diagram showing results of isotope dilution–thermal ionization mass spectrometry (ID-TIMS; circles) and sensitive high-resolution ion microprobe (SHRIMP; ellipses) analyses of zircons from the Western Complex: granite augen gneiss of the southeastern Sierra Los Alacranes (sample #1). Eighteen of twenty-one SHRIMP analyses (within 7% discordance) yielded a weighted mean $^{207}Pb/^{206}Pb$ age (^{204}Pb-corrected) of 1709.3 ± 8.8 Ma at the 95% confidence level (mean quare of weighted deviates [MSWD] = 1.17) with a 28% probability of fit; a main cluster of ten least-discordant SHRIMP analyses (5% or less; open squares) from Alacranes #1 define a weighted mean $^{207}Pb/^{206}Pb$ age of 1725 ± 12 Ma (MSWD = 0.27), which we presently accept as the best estimate for the age of the plutonic protolith. ID-TIMS analyses do not plot with SHRIMP data (see text for explanation). For this diagram and all similar plots, errors for SHRIMP data are 1σ, and ID-TIMS errors are at 2σ, although circles are used to mark ID-TIMS analyses because their ellipses are so small and are always within the size of the circle. See Figure 3 for sample location.

0.27) that we presently accept as the best estimate for the age of the plutonic protolith.

In light of the SHRIMP results, the ID-TIMS analyses clearly exhibit behavior that suggests either Pb loss, probably within 500 m.y. of their crystallization at ca. 1725 Ma, or the presence of secondary overgrowths, such that the analyses represent mixtures of older (ca. 1725 Ma) magmatic portions with younger, yet still Proterozoic overgrowths.

Alacranes #5. This sample is a medium-grained, slightly porphyritic, hornblende-biotite granodiorite, collected from northwestern Sierra Los Alacranes. The pluton is variably foliated and contains xenoliths of felsic and mafic gneisses (Fig. 4F). Zircons separated from it are euhedral to anhedral, clear to yellowish-brown, with variable length:width ratios (1:1–6:1). Most are transparent with no apparent inherited zircon component. Zircon morphologies vary from subhedral, acicular forms to stubbier, multifaceted, "football-shaped" forms; the latter are known to produce unusual isotopic behavior due to secondary zircon growth or overgrowth on primary grains. Fraction 6(80) contained football-shaped grains, in contrast to fraction 4(80) that contained only acicular-shaped grains (Table 1A).

Using the ID-TIMS method, three fractions from this sample composed of different proportions of the two morphologies all yielded $^{206}Pb/^{204}Pb$ values in excess of 3800. The U-Pb isotopic results shown in Figure 10 (open circles) are only slightly discordant and form a quasi-linear array subparallel to the concordia curve. Regression of these analyses yielded concordia upper-intercept ages that are anomalous compared to other ages from southwestern North America. The $^{207}Pb/^{206}Pb$ ages from these analyses range from 1687 Ma to 1696 Ma; the youngest is from the acicular fraction 4(80); the older from the stubbier grains of fraction 6(80), results that are similar to those reported for two

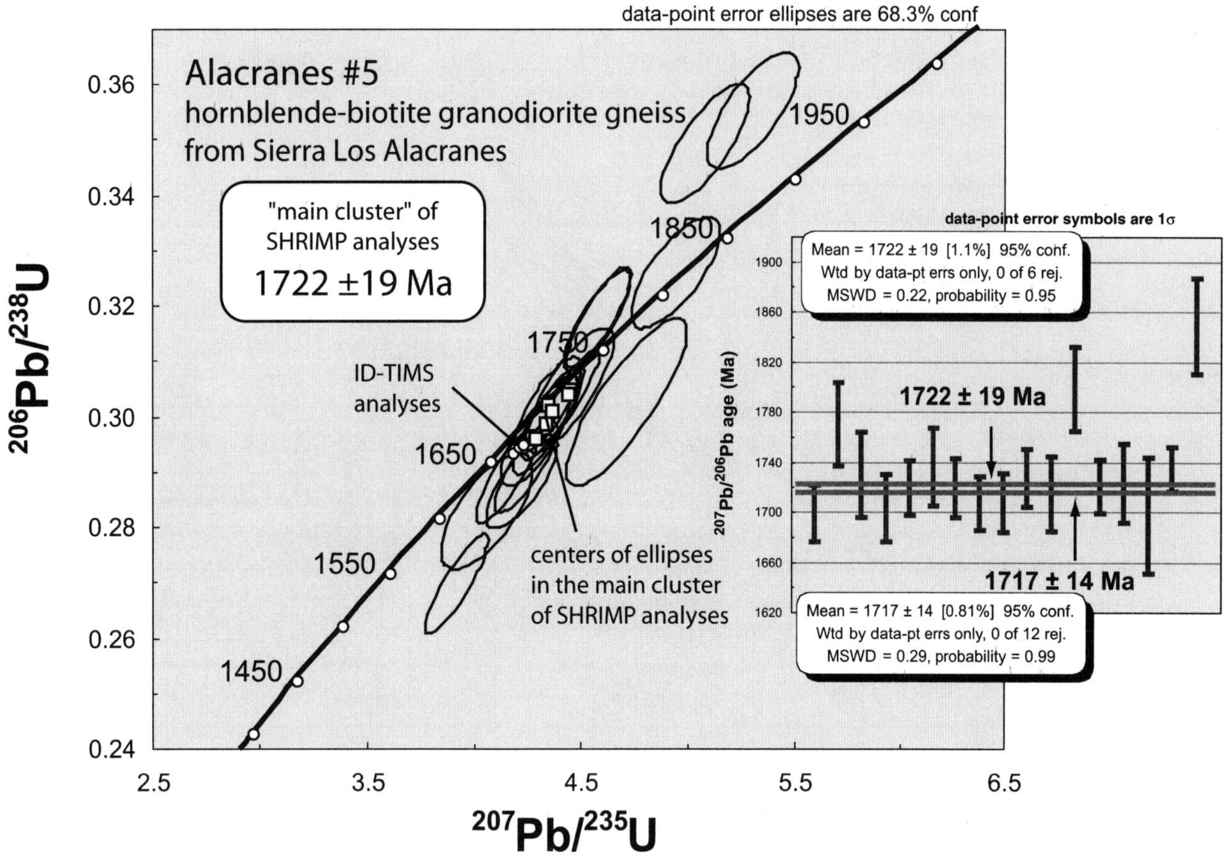

Figure 10. Concordia diagram showing U-Pb zircon analyses from the Western Complex: granodiorite gneiss of the northwestern Sierra Los Alacranes (sample #5). Twelve out of seventeen sensitive high-resolution ion microprobe (SHRIMP) analyses within 6% discordance yielded a weighted mean $^{207}Pb/^{206}Pb$ age of 1717 ± 14 Ma (mean square of weighted deviates [MSWD] = 0.29); a main cluster of six slightly discordant analyses (2% or less; open squares) yielded a weighted mean $^{207}Pb/^{206}Pb$ age of 1722 ± 19 Ma (MSWD = 0.22), which we presently accept as the best estimate for the age of the plutonic protolith. Isotope dilution–thermal ionization mass spectrometry (ID-TIMS) analyses plot near the main cluster of SHRIMP data (see text for explanation). Errors for SHRIMP data are 1σ, ID-TIMS errors are at 2σ, although circles are used to mark ID-TIMS analyses because their ellipses are so small and are always within the size of the circle. See Figure 3 for sample location.

fractions (1654 and 1665 Ma) from a sample of this same rock unit (sample #12, Rt. 2, km 2657, quartz diorite gneiss, Anderson and Silver, this volume). We tentatively interpreted these ID-TIMS results to indicate that the granodiorite was emplaced no earlier than 1687 Ma, and that older zircons (>1696 Ma) were incorporated into the melt (Nourse et al., 2000).

Subsequently, U-Pb isotopic data of zircons from this sample were obtained on the SHRIMP. CL imaging revealed typical magmatic structures and seventeen spot analyses of mainly zircon centers were taken on mainly magmatically zoned areas, but significant scatter was observed in the analyses. Several analyses plot away to the right of the others, suggesting they contain radiogenic Pb due to inheritance of older zircon material. Twelve analyses within 6% discordance yielded a weighted mean $^{207}Pb/^{206}Pb$ age of 1717 ± 14 Ma (MSWD = 0.29; Fig. 10) and a small set (N = 6; centers indicated by open squares) of concordant analyses (2% or less discordance) yielded a weighted mean $^{207}Pb/^{206}Pb$ age of 1722 ± 19 Ma. This age is statistically identical to that measured for the least discordant analyses from Alacranes #1 (Fig. 9), suggesting that both granitoids were contemporaneously emplaced into their host gneisses at ca. 1725 Ma.

The ID-TIMS analyses have slightly younger $^{207}Pb/^{206}Pb$ ages compared to SHRIMP ages, again indicating either Pb loss probably within 500 m.y. of their crystallization at ca. 1725 Ma, or the presence of secondary overgrowths.

Mina La Joya #1-98. This sample is a recrystallized medium-grained, leucocratic, biotite alkali granite collected from the hinge of a synform at Mina La Joya. Zircons separated from it are mostly prismatic, subhedral to euhedral, clear to tan-brown, but also transparent to nearly opaque with variable length:width ratios. Some of the larger grains appeared to contain inherited zircon cores; these were avoided during handpicking.

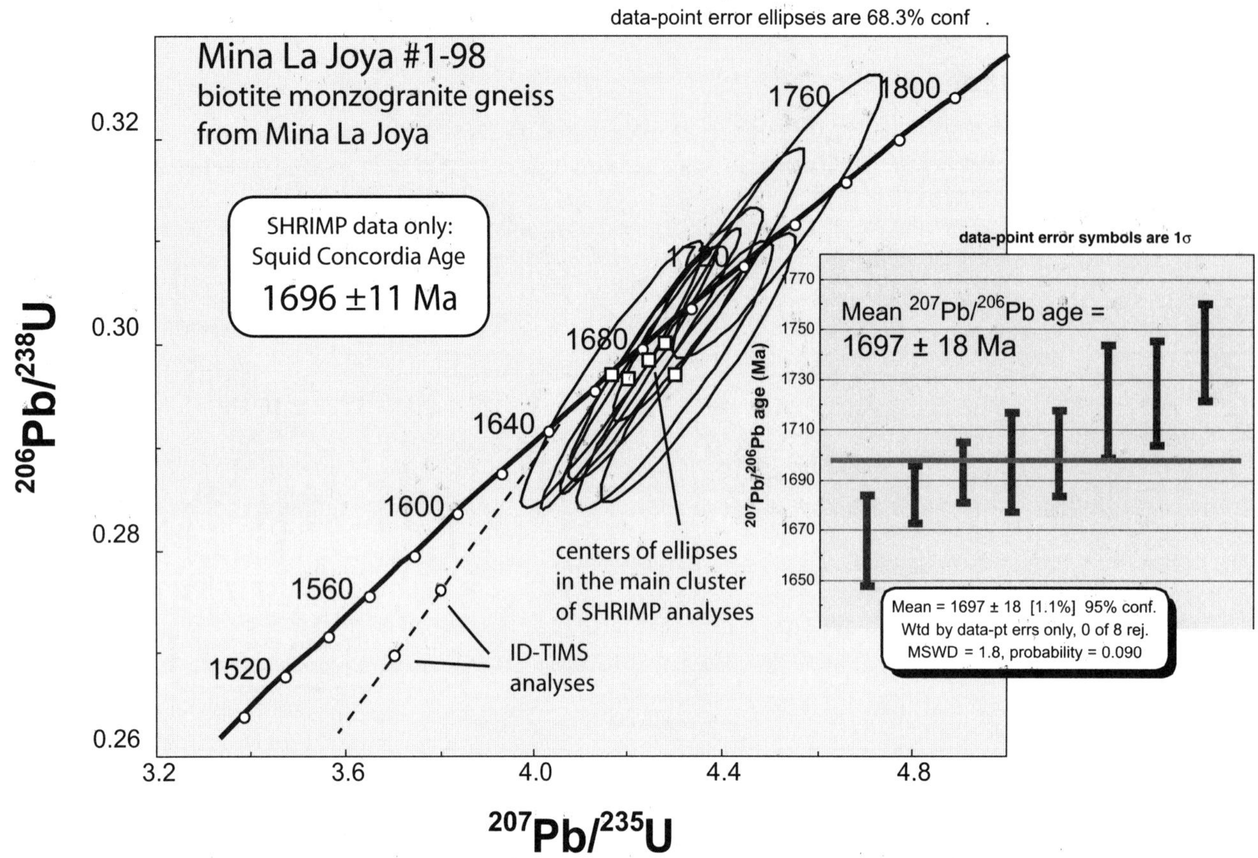

Figure 11. Concordia diagram showing U-Pb zircon analyses from Western Complex: gneissic monzogranite of Mina La Joya (sample #1-98). Five of eight sensitive high-resolution ion microprobe (SHRIMP) analyses (open squares) yielded a squid "concordia age" of 1695.8 ± 10.6 Ma (mean square of weighted deviates [MSWD] = 1.1). Concordia age is calculated using an algorithm that quantitatively tests the assumption of concordance and attempts to find a coherent group from the ^{204}Pb-corrected ^{238}U/^{206}Pb - ^{207}Pb/^{206}Pb data (Ludwig, 2001b). We presently accept this as the best estimate for the age of the plutonic protolith. A weighted mean ^{207}Pb/^{206}Pb age of 1697 ± 18 Ma (MSWD = 1.8) can also be calculated for the eight SHRIMP analyses. Two isotope dilution–thermal ionization mass spectrometry (ID-TIMS) analyses do not plot with the SHRIMP data (see text for explanation). Errors for SHRIMP data are 1σ, ID-TIMS errors are at 2σ, although circles are used to mark ID-TIMS analyses because their ellipses are so small and are always within the size of the circle. See Figure 3 for sample location.

Only two fractions were analyzed by ID-TIMS from this sample; their U-Pb isotopic results shown in Figure 11 (open circles). The two analyses yielded ^{207}Pb/^{206}Pb values in excess of 6500, indicating that all of the Pb is essentially radiogenic, and corrections for common Pb are relatively insignificant. The analyses from this sample define a chord that extrapolated to the concordia curve yielded an upper-intercept age of 1650 ± 7 Ma.

CL images of this zircon population exhibited typical magmatic structures. Using the SHRIMP, eight analyses were obtained from magmatically zoned central areas. All were near-concordant (4% or less) with ^{207}Pb/^{206}Pb ages ranging between ca. 1665 Ma and ca. 1740 Ma. These data yielded a weighted mean ^{207}Pb/^{206}Pb age of 1697 ± 18 Ma (MSWD = 1.8), and a squid "concordia age" of 1696 ± 11 Ma (MSWD = 1.1; Fig. 11), which we interpret to represent the age of emplacement of the Mina La Joya granite.

The ID-TIMS analyses from this sample are significantly more discordant than those analyzed from Alacranes #1 and #5, indicating these zircon grains suffered a significant Pb loss sometime during the Phanerozoic.

Samples from the Eastern Proterozoic Complex

Paleoproterozoic Metagranite Samples

CP-17-99. This sample is a medium-grained, slightly porphyritic, alkali granite gneiss collected from outcrops in the Drift Hills that are intruded by Late Cretaceous quartz diorite dikes (Fig. 7H). It represents part of an extensive sheet of granite intruded into arkosic and quartzose sandstones of the Eastern Complex. Zircons from this sample are small, subhedral to rounded, with color variation from clear to nearly opaque. A few exhibit possible overgrowths. Clear, euhedral forms were handpicked and analyzed by ID-TIMS, with their U-Pb isotopic results shown in Figure 12 (open circles). Three analyses yielded ^{206}Pb/^{204}Pb values in excess of 3400, indicating

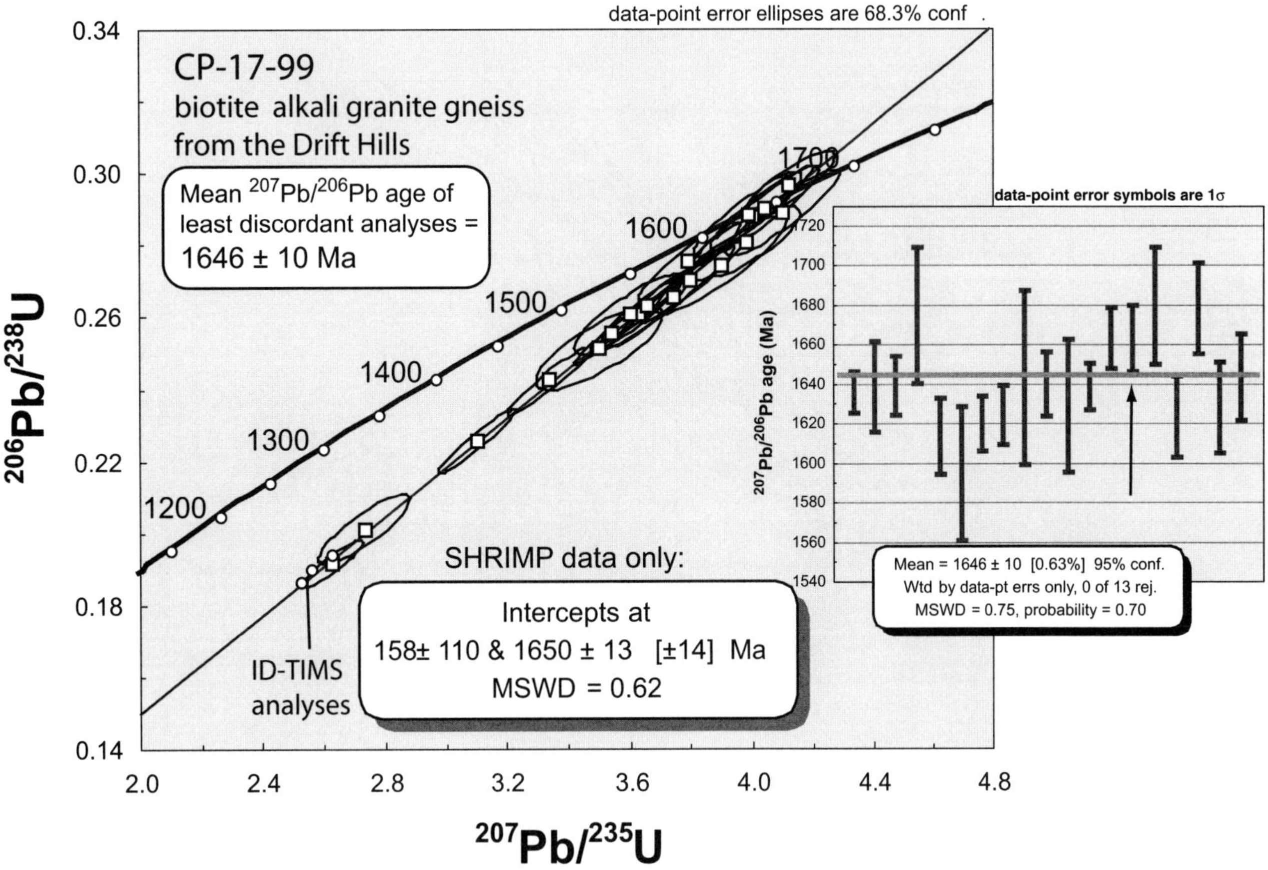

Figure 12. Concordia diagram showing the Pb-loss array defined by three ID-TIMS (isotope dilution–thermal ionization mass spectrometry) and nineteen SHRIMP (sensitive high-resolution ion microprobe) analyses from a alkali granite of the Drift Hills (sample CP-17-99), Eastern Complex. A weighted mean $^{207}Pb/^{206}Pb$ age of thirteen least-discordant (less than 10% discordant) SHRIMP analyses is 1646 ± 10 Ma (mean square of weighted deviates [MSWD] = 0.75). Regression of just the SHRIMP data yielded a concordia upper-intercept age of 1650 ± 13 Ma and lower-intercept age of 158 ± 110 Ma (MSWD = 0.62). SHRIMP errors are shown at the 1σ level, ID-TIMS at the 2σ level, although circles are used to mark ID-TIMS analyses because their ellipses are so small and are always within the size of the circle. See Figure 6 for sample location.

very little common Pb. However, despite modest U contents (~500 ppm; Table 1A), the analyses are quite discordant and close together, leading to a poorly defined concordia upper-intercept age of ca. 1620 Ma.

Zircons from this sample were also measured on the SHRIMP. CL imaging showed mainly typical magmatic structures. The U-Pb results from 19 analyses of zircon centers are shown in Figure 12 (centers of ellipses indicated by open squares). Several of the SHRIMP analyses are considerably discordant (25%–30%), forming an array that extends to the ID-TIMS analyses. This behavior confirms that this zircon population in particular has suffered extensive Pb loss and a regression of the SHRIMP data alone yielded an upper-intercept age of 1650 ± 13 Ma (MSWD = 0.62) and a lower-intercept age of 158 ± 110 Ma, indicating that Pb loss occurred during either the Mesozoic or Tertiary. A weighted mean $^{207}Pb/^{206}Pb$ age of the thirteen least discordant (less than 10% discordant) analyses is 1646 ± 10 Ma (MSWD = 0.75; Fig. 12), which we accept as the best estimate for the age of this augen gneiss, an age certainly much younger (>50 m.y.) than the granitoids of Sierra Los Alacranes.

CD-3 #4. This sample is a strongly foliated and recrystallized augen gneiss derived from a medium-grained, biotite alkali granite protolith similar to sample CP-17-99. Field relationships indicate that it intruded the metasedimentary section prior to metamorphism. A U-Pb age determination was not made using ID-TIMS, but zircons from this sample were analyzed on the SHRIMP. CL images of the zircons showed varied internal structures. Some grains or portions of grains exhibited typical magmatic zonation, however some of the zonations appeared to be distorted or convoluted (after Hoskin and Black, 2000). Other portions of some grains appeared devoid of any structures, but not dark or U-rich as in the case of some overgrowths. Results of 21 SHRIMP spot analyses of zircon centers are shown in Figure 13A. Most of these analyses are slightly discordant (8% or less) and display scatter along concordia with $^{207}Pb/^{206}Pb$ ages ranging between ca. 1560 and 1720 Ma, such that the combined effects of possible

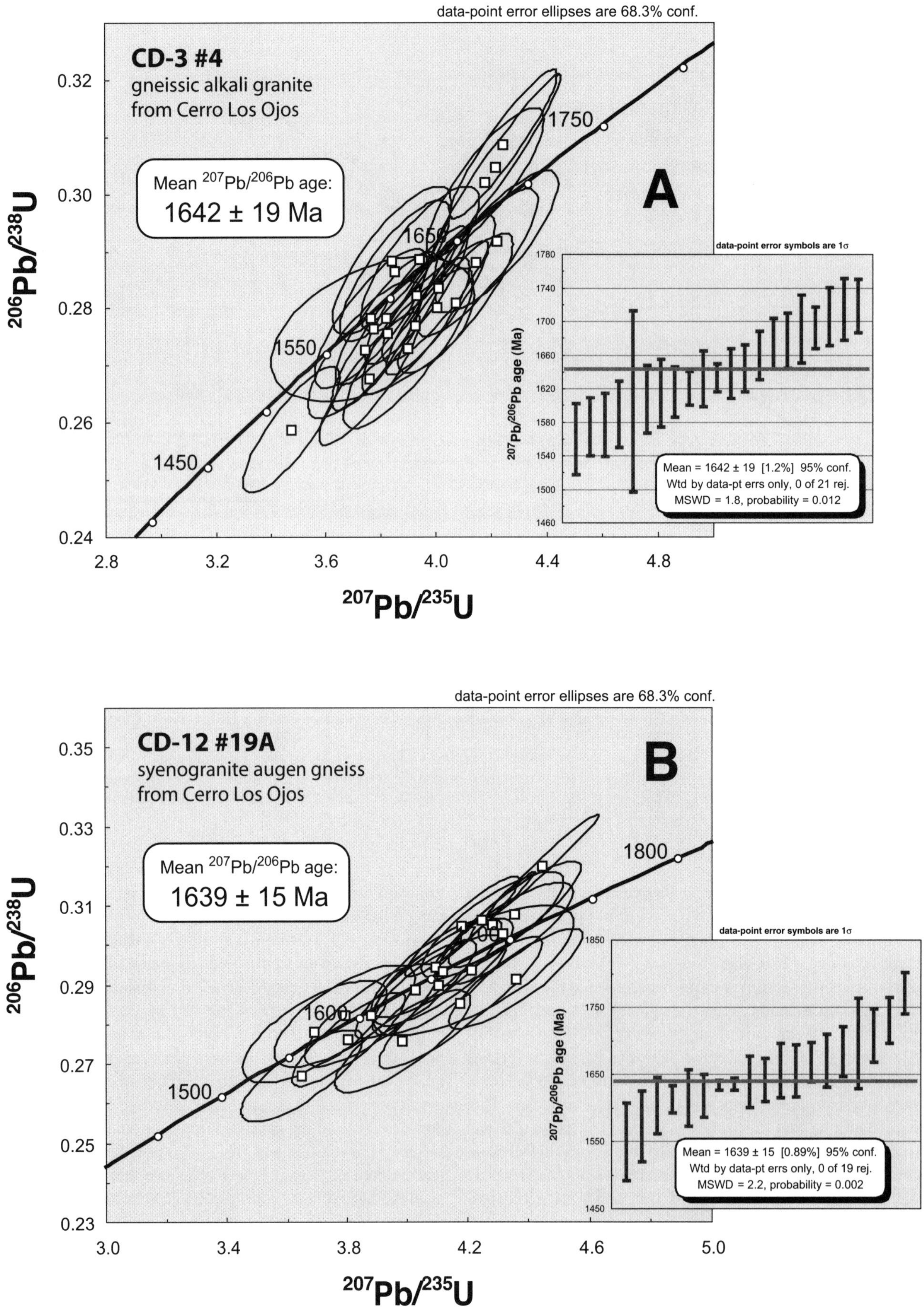

data-point error ellipses are 68.3% conf.
CD-3 #4
gneissic alkali granite
from Cerro Los Ojos
Mean 207Pb/206Pb age:
1642 ± 19 Ma
A
1750
1650
1550
1450
206Pb/238U
207Pb/235U
data-point error symbols are 1σ
207Pb/206Pb age (Ma)
Mean = 1642 ± 19 [1.2%] 95% conf.
Wtd by data-pt errs only, 0 of 21 rej.
MSWD = 1.8, probability = 0.012
data-point error ellipses are 68.3% conf.
CD-12 #19A
syenogranite augen gneiss
from Cerro Los Ojos
Mean 207Pb/206Pb age:
1639 ± 15 Ma
B
1800
1600
1500
206Pb/238U
207Pb/235U
data-point error symbols are 1σ
207Pb/206Pb age (Ma)
Mean = 1639 ± 15 [0.89%] 95% conf.
Wtd by data-pt errs only, 0 of 19 rej.
MSWD = 2.2, probability = 0.002

inheritance, magmatic zircon formation, overprinting, and Pb loss are difficult to resolve. A simple weighted mean $^{207}Pb/^{206}Pb$ age of the data is 1642 ± 19 Ma (MSWD = 1.8), which we tentatively accept as the best estimate for the age of this augen gneiss.

CD-12 #19A. This sample is a biotite-rich augen gneiss derived from coarsely porphyritic syenogranite that forms a border phase of the alkali granite. It shares a strong metamorphic foliation with its sandstone host. We analyzed 19 spots on the SHRIMP (Fig. 13B); 15 central areas and 4 rims. Similar to CD-3 #4, CL imaging of zircons from this sample exhibited a variety of structures. Typical magmatic zoning was apparent in many of the grains; areas we intentionally analyzed in order to obtain a crystallization age. With the exception of one analysis that yielded a near-concordant $^{207}Pb/^{206}Pb$ age of 1414 ± 19 Ma, the data are all within 10% discordance, but exhibit scatter along concordia with $^{207}Pb/^{206}Pb$ ages ranging between ca. 1550 and 1730Ma, behavior very similar to that from sample CD-3 #4, implying multiple isotopic disturbances. It is difficult to resolve a singular concordia age; the crystallization age of the granite protolith is therefore uncertain. A weighted mean $^{207}Pb/^{206}Pb$ age of all the data yielded 1639 ± 15 Ma (MSWD = 2.2).

A common characteristic of SHRIMP spot data from Paleoproterozoic metagranite zircons from the Eastern Complex is the scattering along concordia of essentially concordant analyses with $^{207}Pb/^{206}Pb$ ages between 1550 and 1750 Ma (Table 1B). As stated earlier, it is difficult to resolve upper-intercept ages for these three metagranites. If we compare their weighted mean $^{207}Pb/^{206}Pb$ ages: 1646 ± 10 Ma, 1642 ± 19 Ma, and 1639 ± 15 Ma, they overlap within error. As stated earlier, we interpret these samples as representative of the same suite of granitic material and therefore comagmatic. Based on the ages above, our best estimate for these metagranites from the Eastern Complex is ca. 1645 Ma (weighted mean of the three ages is 1644 ± 8 Ma); however, this estimate should be considered a minimum age allowing for the likelihood that some of the younger spot ages are a result of an isotopic disturbance prior to ca. 1430 Ma (see discussion below).

Figure 13. (A) Concordia diagram showing twenty-one U-Pb SHRIMP (sensitive high-resolution ion microprobe) analyses of zircon cores from alkali granite gneiss of Cerro Los Ojos (sample CD-3 #4), Eastern Complex. Most of these analyses are slightly discordant (8% or less) and display abnormal scatter along concordia, such that the combined effects of possible inheritance, magmatic zircon formation, overprinting, and Pb loss are difficult to resolve. A simple weighted mean $^{207}Pb/^{206}Pb$ age of the data is 1642 ± 19 Ma (mean square of weighted deviates [MSWD] = 1.8). (B) Concordia diagram showing nineteen U-Pb SHRIMP analyses from biotite granite augen gneiss of Cerro Los Ojos (sample CD-12 #19A), Eastern Complex. A weighted mean $^{207}Pb/^{206}Pb$ age of all the data is 1639 ± 15 Ma (MSWD = 2.2). Similar to CD-3 #4, most of these analyses are only slightly discordant (8% or less) and display abnormal scatter along concordia, such that the combined effects of possible inheritance, magmatic zircon formation, overprinting, and Pb loss are difficult to resolve. SHRIMP errors are shown at the 1σ level. See Figure 6 for sample locations.

The timing of profound dynamic-thermal metamorphism is constrained by the age of emplacement of the youngest metagranite at ca. 1640 Ma and intrusion of a 1432 Ma granite (see below) that sharply crosscuts the strong fabric developed in these rocks. The youngest spot data for the metagranites from Cerro Los Ojos indicate that thermal and/or fluid effects may have persisted until ca. 1550 Ma. The metagranite from the Drift Hills, CP-17-99, that is farther away from the 1432 Ma granite, exhibits the least disturbed zircon population and yielded the older age of 1646 ± 10 Ma, which may best represent the true age of these plutons.

Paleoproterozoic Metasandstone Samples

Two samples of meta-arkose and two samples of "quartzite" were collected from different parts of the folded metasedimentary section of the Sierra Choclo Duro focus area (Fig. 6). In general, the detrital zircons are remarkably homogeneous. Many display rough edges and only minor degrees of roundness or abrasion, suggesting they have not traveled far from their source. Twenty or more detrital zircons from each sample were analyzed on the SHRIMP (Table 1B). This SHRIMP reconnaissance of detrital zircons focused on homogeneous regions of grain interiors in an attempt to delineate magmatic age ranges related to provenance as well as evaluate any degree of possible isotopic disturbance. The analytical data yielded similar concordia plots for all four samples (Figs. 14 and 15). Most of the analyses are concordant to slightly discordant (only 13 of 87 analyses are greater than 6% discordant), but scattered along concordia with $^{207}Pb/^{206}Pb$ ages ranging between 1554 Ma and 1697 Ma, similar to the behavior exhibited by recrystallized granite gneiss samples CD-3 #4 and CD-12 #19A. The data scatter probably represents a mixture between older magmatic zircons and portions of zircon that record isotopic disturbances. In CL images, some grains exhibit magmatically produced oscillatory zoning and discrete overgrowths that would indicate complex crystallization histories. Several grains have dark rims or centers in CL images, indicating high-U contents and the probability of metamorphic zircon growth (Fig. 16).

CD-12 #2 (Borderline Quartzite). Although magmatic zonation is evident in some detrital zircon grains from this sample of the Borderline Quartzite, many other grains show featureless centers and a few have distinctive dark rims or outer portions (Fig. 16A). Twenty SHRIMP analyses from mainly central areas (two from rims) are shown in Figure 14A and exhibit a range in $^{207}Pb/^{206}Pb$ ages between 1595 Ma and 1681 Ma (Table 1B; excluding the 1.43 Ga spots), similar to the range exhibited by metagranite samples. Many of these $^{207}Pb/^{206}Pb$ ages are younger than the estimated age of the metagranites (ca. 1645 Ma), indicating the likelihood that these analyses were made on grains that had suffered either Pb loss during an episode not more than a few hundred million years after crystallization or have been metamorphically altered, perhaps recrystallized (e.g., Hoskin and Black, 2000; Hoskin and Schaltegger, 2003). Due to these isotopic disturbances, an accurate age range of provenance cannot be determined, but a probable range of provenance for this quartzite

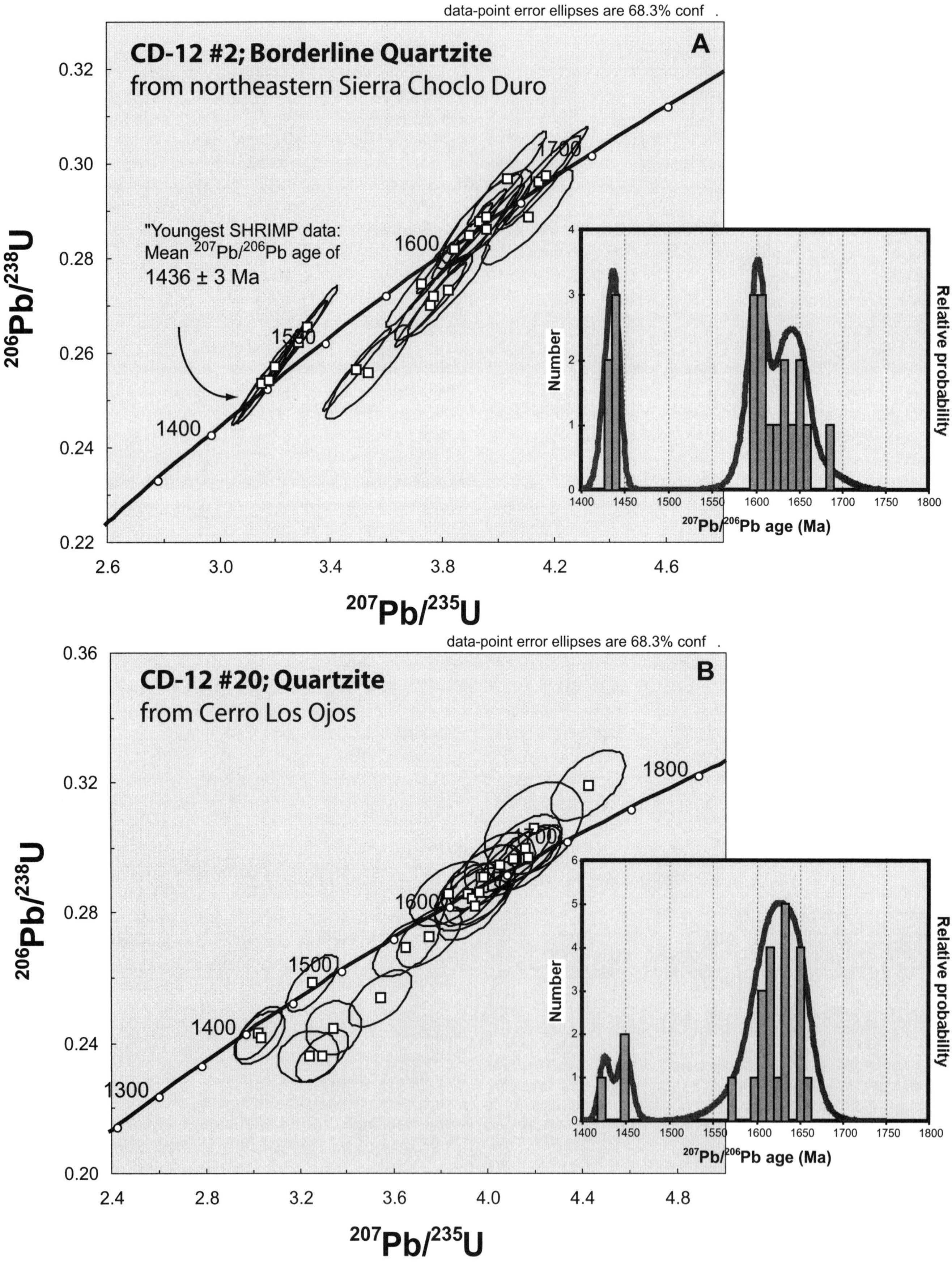
data-point error ellipses are 68.3% conf.
A
CD-12 #2; Borderline Quartzite
from northeastern Sierra Choclo Duro
"Youngest SHRIMP data:
Mean ^{207}Pb/^{206}Pb age of
1436 ± 3 Ma
1400
1500
1600
1700
^{206}Pb/^{238}U
^{207}Pb/^{235}U
Number
Relative probability
^{207}Pb/^{206}Pb age (Ma)
data-point error ellipses are 68.3% conf.
B
CD-12 #20; Quartzite
from Cerro Los Ojos
1300
1400
1500
1600
1700
1800
^{206}Pb/^{238}U
^{207}Pb/^{235}U
Number
Relative probability
^{207}Pb/^{206}Pb age (Ma)

can be estimated at between ca. 1645 and 1681 Ma, most likely excluding sources older than 1700 Ma.

Some younger spot ages were measured from the rims of zircon grains from this sample (Fig. 16A) and include ages between 1429 and 1440 Ma. A weighted mean $^{207}Pb/^{206}Pb$ age of 1436 ± 3 Ma is calculated for these secondary zircon growths (Fig. 14A).

CD-12 #20 (Cerro Los Ojos quartzite). Twenty-five SHRIMP analyses were carried out on mainly center areas of zircon from this quartzite sample from Cerro Los Ojos. CL images of zircons from this sample showed a wide variety of structures, including magmatic zoning often disrupted by streaks of recrystallized zircon and dark rims (Fig. 16B). The U-Pb results show that $^{207}Pb/^{206}Pb$ ages range between 1571 Ma and 1657 Ma (excluding the 1.43 Ga spots; Figure 14B and Table 1B). Invoking the same arguments as those used for the other quartzite sample, many of the $^{207}Pb/^{206}Pb$ ages from this sample are younger than the metagranite mean age of 1644 ± 8 Ma, indicating the likelihood that this zircon population suffered either Pb loss during an episode not more than a few hundred million years after crystallization or metamorphic zircon growth. A probable range of provenance for this quartzite is ca. 1645–1657 Ma, significantly more restricted than the previous quartzite sample.

Three analyses taken on secondary overgrowths (dark high-U rims; Fig. 16B) yielded a weighted mean $^{207}Pb/^{206}Pb$ age of 1437 ± 35 Ma (Fig. 14B).

CD-12 #5 (Cerro Los Ojos meta-arkose). Twenty-two SHRIMP analyses for meta-arkose sample CD-12 #5 were taken on mainly central areas of zircon grains, and the results are shown in Figure 15A. CL imaging revealed a predominance of featureless areas within many of the grains. Although mostly devoid of internal structures, some of the grains showed magmatic zoning but typically toward the outer portions of grains (Fig. 16C). The range in $^{207}Pb/^{206}Pb$ ages for these analyses is between 1577 Ma and 1697 Ma (Table 1B; excluding the 1.43 Ga spots), again similar to the range exhibited by metagranitoid samples. And again, some $^{207}Pb/^{206}Pb$ ages are notably younger than the weighted mean age for the metagranitoids that intrude this metasediment, indicating to us that many of the measured zircon grains had suffered a disturbance to their U-Pb systematics similar to that recognized in sample CD-3 #4. A probable range of provenance for this meta-arkose is ca. 1645–1697 Ma, very similar to that of the Borderline Quartzite, and again very likely excluding sources older than 1700 Ma.

Some younger spot ages were measured from the rims of zircon grains (Fig. 16C) of this sample and include concordant spot ages ca. 1410 Ma and ca. 1585 Ma (Fig. 15A). Interestingly, similar $^{207}Pb/^{206}Pb$ ages are reported for two zircon fractions (1532 and 1545 Ma) from a similar sample (sample #11; Rt. 2, km 2627; metarhyolite gneiss #4) by Anderson and Silver (this volume).

CD-12 #13 (Cerro Los Ojos meta-arkose). CL images of zircons from this sample were similar to the other meta-arkose in that many of the central areas of grains were featureless, although magmatic zonations were found in most grains, sometimes restricted to the outer portions (Fig. 16D). Similar to results from CD-12 #20, twenty SHRIMP analyses on zircon from sample CD-12 #13 exhibit $^{207}Pb/^{206}Pb$ ages between 1554 Ma and 1662 Ma (Fig. 15B; Table 1B). Many of these $^{207}Pb/^{206}Pb$ ages are younger than the mean age of the metagranitoids that intrude this rock unit, indicating once again that many of the analyses were probably taken on grains that had suffered Pb loss either during an episode not more than a few hundred million years after crystallization, or by precipitation of secondary overgrowths. A probable age range of provenance for this meta-arkose is ca. 1645–1662, significantly more restricted in age than those for CD-12 #2 and #5.

Many of the grains from this sample had dark rims (Fig. 16D). Two yielded younger spot ages ranging from 1410 to 1435 Ma, while four others ranged between 1554 and 1603 Ma (Fig. 15B; Table 1B).

Mesoproterozoic Granite Sample (PZ 23B)

This sample is from a coarse-grained, unfoliated, porphyritic biotite granite that intrudes Paleoproterozoic metasandstone and augen gneiss of Cerro Los Ojos in the Eastern Complex. The zircons from it are mostly elongate, prismatic tetrahedrons, with length:width ratios of ~3:1–5:1. Most are clear to pinkish-tan in color. Some grains are dusty with inclusions, and many contain small dark inclusions that suggest inherited material, but no obvious cores were observed. Clear acicular grains were picked to produce two individual fractions for ID-TIMS analyses, the results of which are shown on Figure 17 (solid circles). A lower-intercept age of 1405 ± 10 Ma (not shown) is defined by extrapolation of a chord through the two ID-TIMS analyses that represent a mixture of newly grown magmatic zircon and inheritance of older zircon material. CL images did reveal that many of the grains contained distinct cores of inherited zircon.

Subsequently, zircons from the same sample were analyzed on the SHRIMP, and fifteen spot ages collected from only magmatic zones. The data yielded a weighted mean $^{207}Pb/^{206}Pb$ age of 1432 ± 6 Ma (Fig. 17) that is presently accepted as our best estimate of the time of emplacement of this granite. This age is

Figure 14. (A) Concordia diagram showing twenty U-Pb SHRIMP (sensitive high-resolution ion microprobe) analyses on detrital zircons from Borderline quartzite (sample CD-12 #2), Eastern Complex, exhibiting a range in $^{207}Pb/^{206}Pb$ ages between 1595 Ma and 1681 Ma (excluding the 1.44 Ga spots). A probable range of provenance for this quartzite is 1646–1681 Ma, which notably excludes sources older than 1700 Ma (see text for explanation). Younger spot ages were measured on rims of zircon grains and yielded ages between 1430 and 1440 Ma. A weighted mean $^{207}Pb/^{206}Pb$ age of 1436.1 ± 3.1 Ma is calculated for these secondary zircon growths. (B) Concordia diagram showing twenty-five U-Pb SHRIMP analyses of detrital zircons from Cerro Los Ojos quartzite (sample CD-12 #20), Eastern Complex, with $^{207}Pb/^{206}Pb$ ages ranging between 1571 Ma and 1657 Ma, again excluding sources older than 1700 Ma. A probable range of provenance for this quartzite is 1646–1657 Ma. A weighted mean $^{207}Pb/^{206}Pb$ age for the youngest SHRIMP data is 1437 ± 35 Ma taken on zircon rims. SHRIMP errors are shown at the 1σ level. See Figure 6 for sample locations.

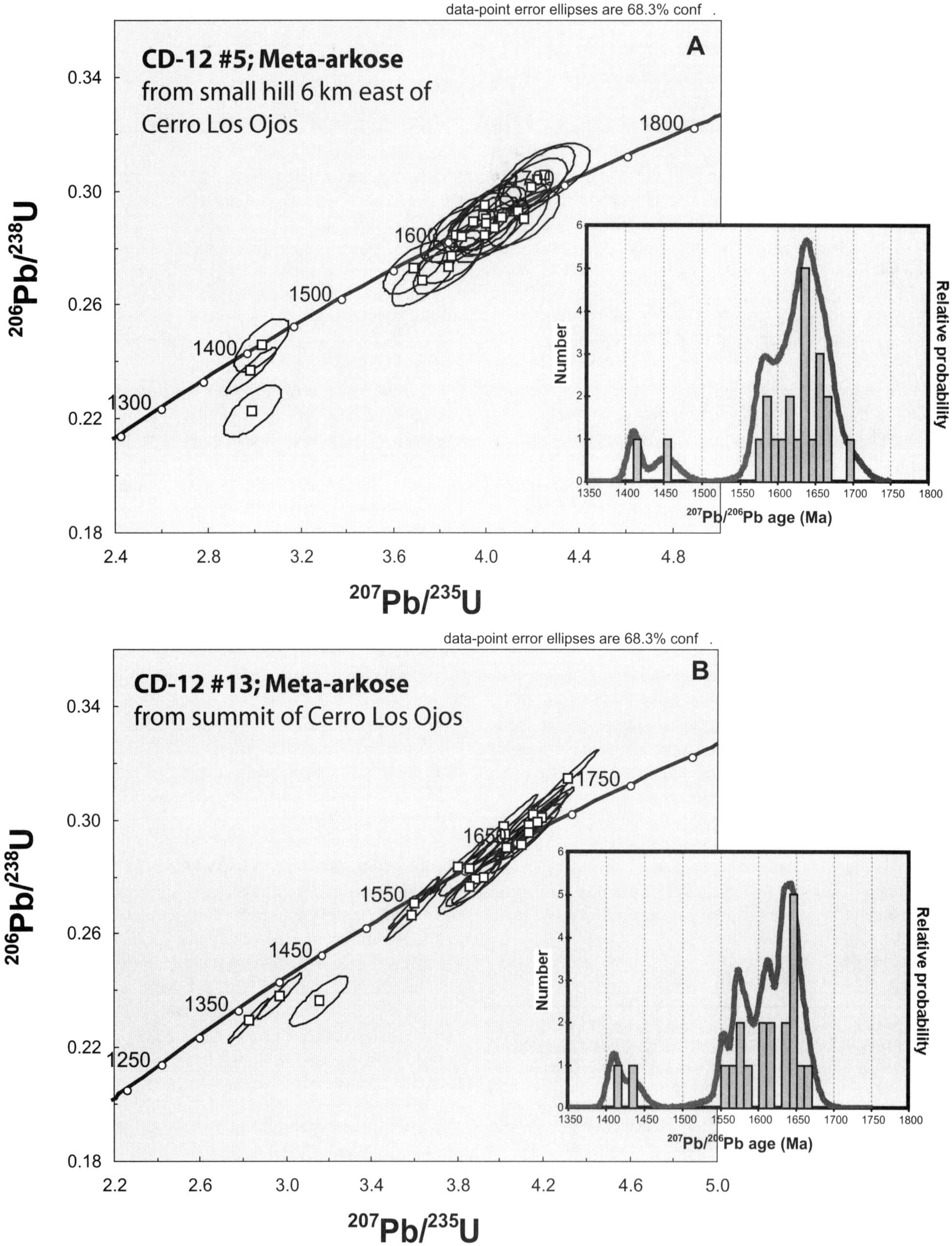

Figure 15. (A) Concordia diagram showing twenty-one U-Pb SHRIMP (sensitive high-resolution ion microprobe) analyses on detrital zircons from Cerro Los Ojos meta-arkose (sample CD-12 #5), Eastern Complex, exhibiting a range in $^{207}Pb/^{206}Pb$ ages between 1577 Ma and 1697 Ma (excluding the 1.44 Ga spots). A probable range of provenance for this quartzite is 1646–1697 Ma, that again excludes sources older than 1700 Ma. Some younger spot ages are measured from the rims of zircon grains and include both ages ca. 1410 Ma and ca. 1585 Ma. (B) Concordia diagram showing twenty U-Pb SHRIMP analyses of detrital zircons from Cerro Los Ojos meta-arkose (CD-12 #13), Eastern Complex, with $^{207}Pb/^{206}Pb$ ages ranging between 1554 Ma and 1662 Ma, again excluding sources older than 1700 Ma. A probable range of provenance for this quartzite is 1646–1662 Ma. Younger spot ages from "disturbed" grains range from 1410 to 1435 Ma, and 1554 to ca. 1619 Ma. SHRIMP errors are shown at the 1σ level. See Figure 6 for sample locations.

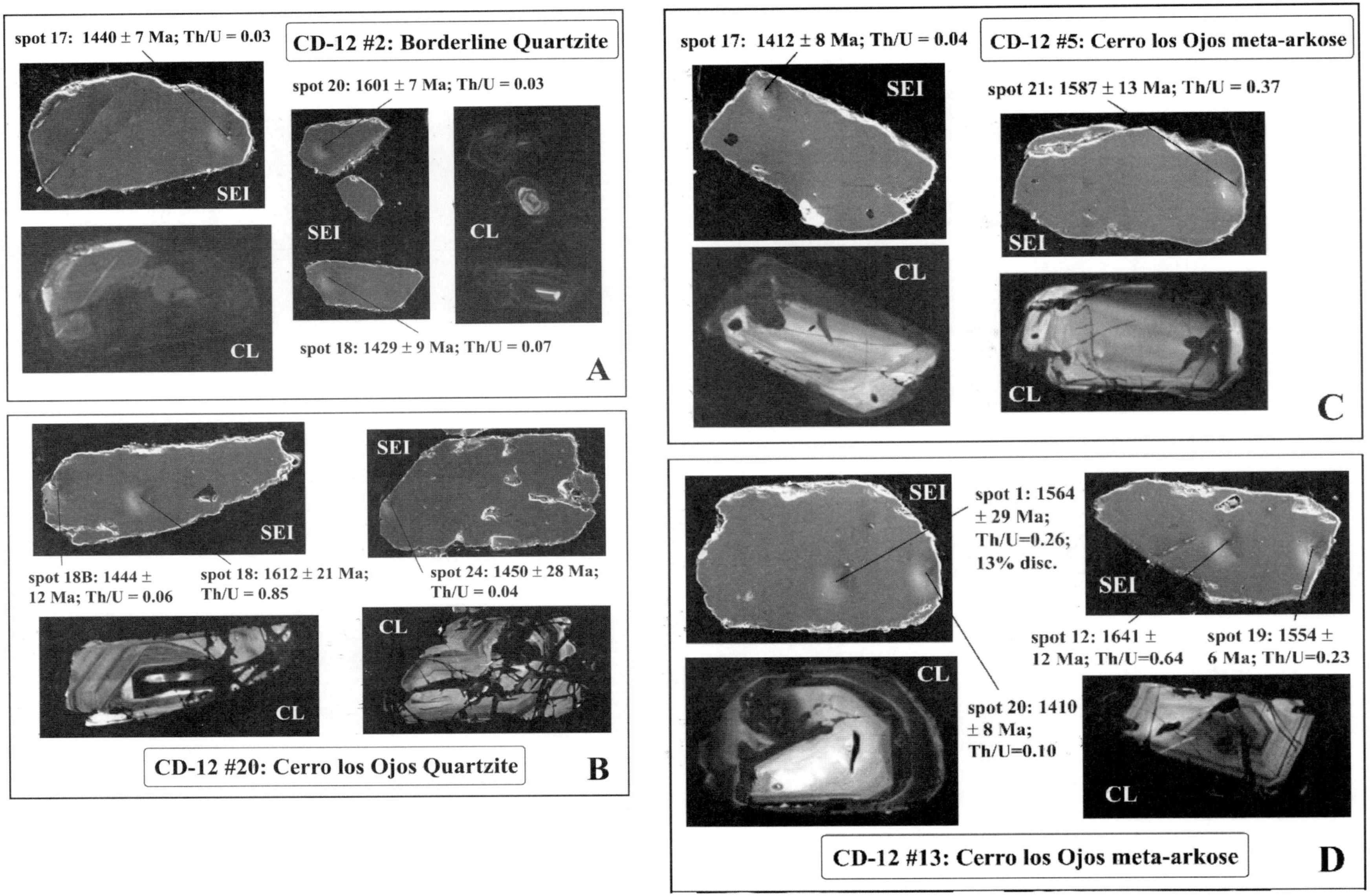

Figure 16. Sensitive high-resolution ion microprobe (SHRIMP) spot locations and ages on representative detrital zircon grains from each of the four metasedimentary samples from the Eastern Complex. Images are shown in both SEI (secondary electron imaging) and CL (cathodoluminescence) at roughly ×200 magnification. Spot ages from rims of these grains are typically younger (most ca. 1.43 Ga) than many of the ages obtained for zircon centers and show reduced Th/U values, less than 0.15 (see text for explanation).

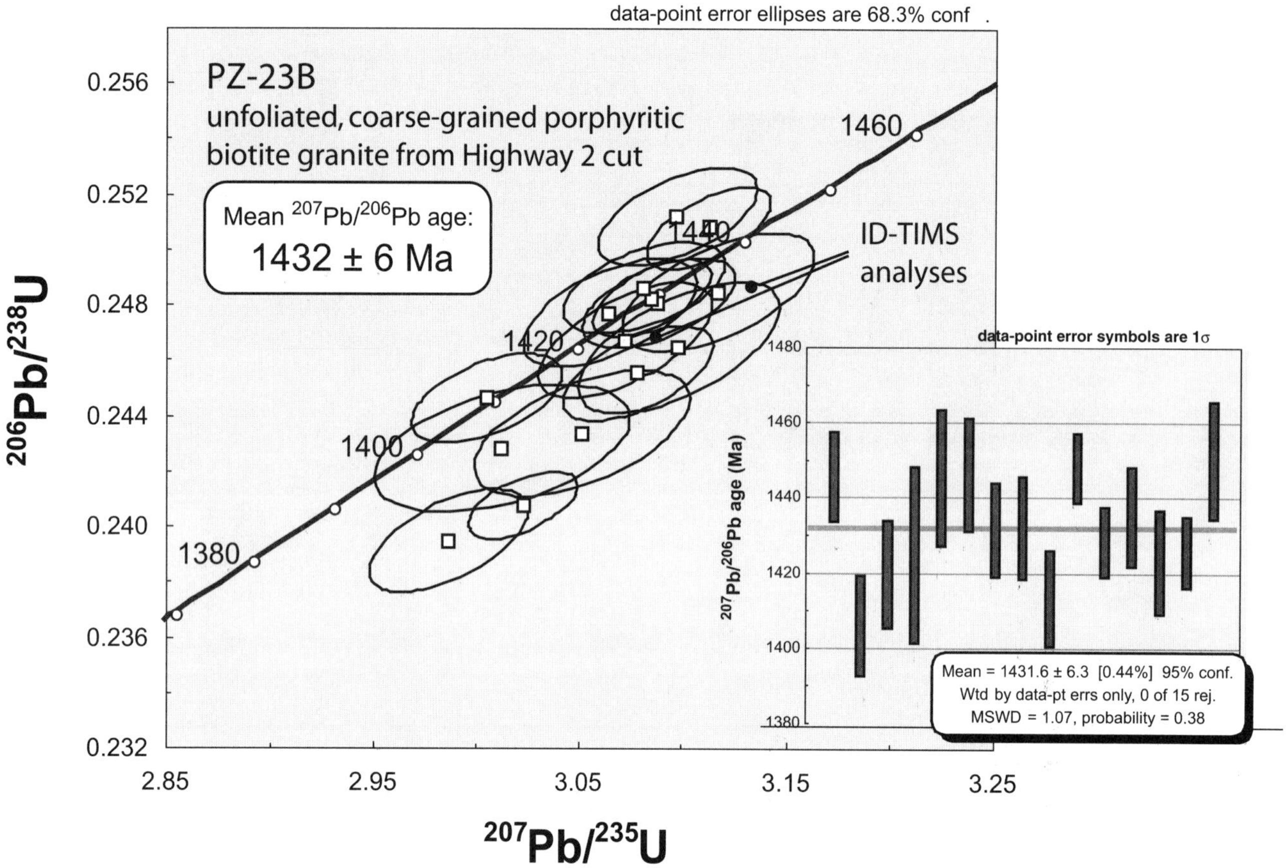

Figure 17. U-Pb concordia diagram showing results of isotope dilution–thermal ionization mass spectrometry (ID-TIMS; solid circles) and sensitive high-resolution ion microprobe (SHRIMP) analyses of zircons from the Mesoproterozoic pluton (sample PZ-23B) of the Eastern Complex. A weighted mean $^{207}Pb/^{206}Pb$ age for the SHRIMP data only is 1431.6 ± 6.3 Ma (mean square of weighted deviates [MSWD] = 1.07). A lower-intercept age that corresponds to the intercept of a two-point chord defined by ID-TIMS analyses is 1405 ± 10 Ma. SHRIMP errors are shown at the 1σ level, ID-TIMS at the 2σ level, although circles are used to mark ID-TIMS analyses because their ellipses are so small and are always within the size of the circle. See Figure 6 for sample location.

indistinguishable from that reported by Anderson and Silver (this volume; sample #8; 1437 ± 5 Ma) for a sample collected from the same granite body, and is a common age for many other "anorogenic" granites known throughout southwestern North America (Anderson, 1989; Bickford and Anderson, 1993).

Cretaceous Quartz Diorite Samples

TJA #21. This sample is a medium-grained, sphene-hornblende-biotite quartz diorite collected from a weakly foliated body that intrudes banded gneiss and augen gneiss in Sierra Las Tinajas Altas within the Western Complex (Fig. 3). Its zircons are euhedral to subhedral dipyramidal prisms, clear to slightly tinted, with variable length:width ratios (1:1–5:1; typically 3:1). The centers of some zircons appear to be inherited cores and are found specifically in the stubby "football" morphologies. Stubby, euhedral, and clear grains were handpicked into three separate fractions and processed using the ID-TIMS method. Their U-Pb isotopic age results are shown in Figure 18A (solid circles) and

Figure 18. Concordia diagrams showing U-Pb analyses from the Late Cretaceous quartz diorites. (A) Concordia diagram showing analyses from the Late Cretaceous quartz diorite of Sierra Las Tinajas Altas (sample TJA #21). Three isotope dilution–thermal ionization mass spectrometry (ID-TIMS) analyses define a chord that intersects concordia at 72.8 ± 1.7 Ma and 1705 ± 160 Ma (mean square of weighted deviates [MSWD] = 0.51). A weighted mean of eight sensitive high-resolution ion microprobe (SHRIMP) $^{206}Pb/^{238}U$ ages is 72.8 ± 1.8 Ma (MSWD = 1.04). We presently accept this as the best estimate for the age of the plutonic protolith. (B) Concordia diagram showing ID-TIMS (circles) and SHRIMP analyses from Late Cretaceous quartz diorite of the Drift Hills Cabeza Prieta focus area (sample CP-16-99). ID-TIMS analyses were difficult to interpret; four SHRIMP analyses also exhibited some scatter; however, three of these data have very similar $^{206}Pb/^{238}U$ ages that yield a weighted mean age of 73.4 ± 3.3 Ma (MSWD = 0.097) that is very similar to the age results from sample, TJA #21; an age we interpret to represent the crystallization age of this quartz diorite of the Cabeza Prieta area.

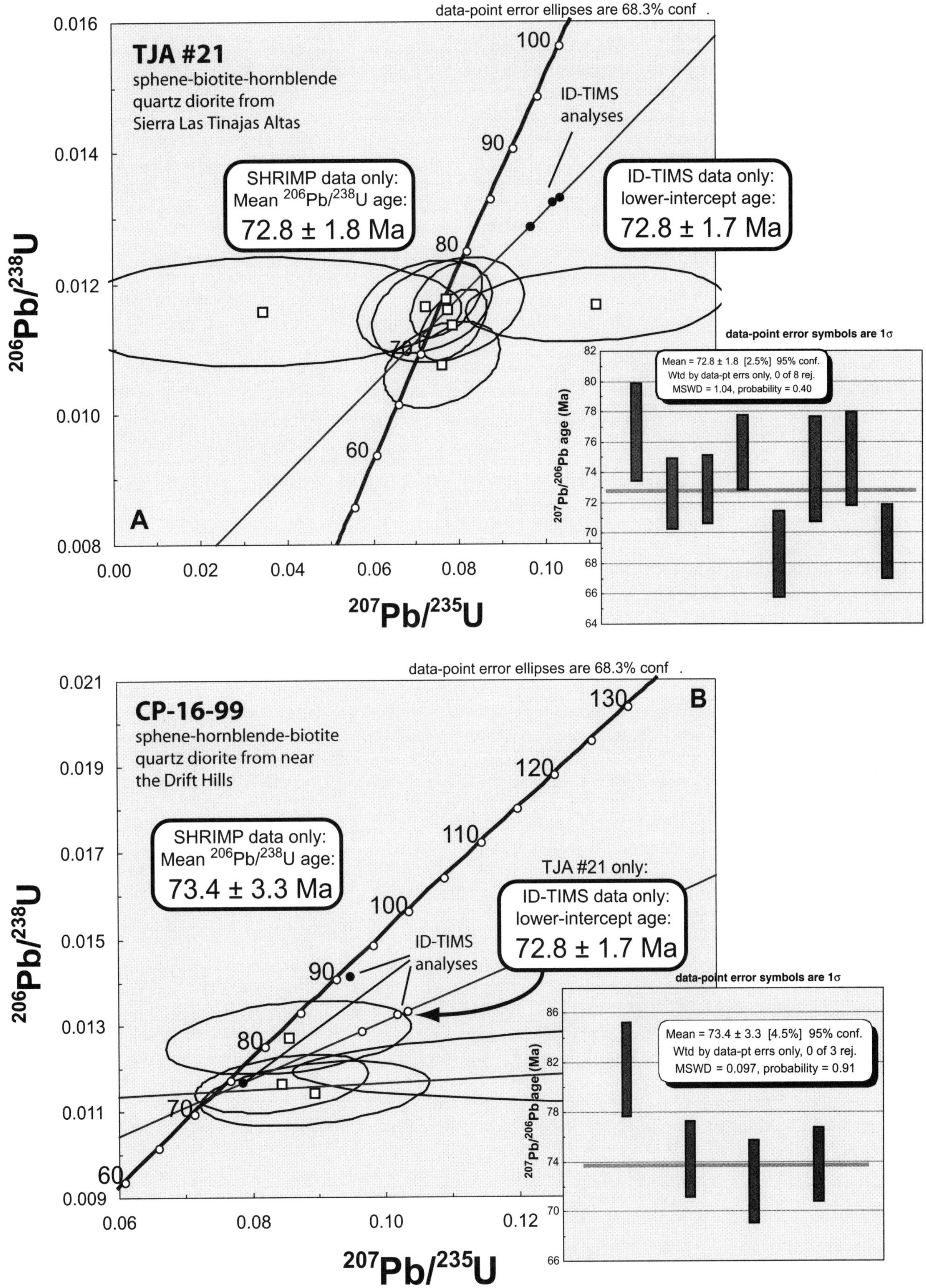

data-point error ellipses are 68.3% conf.
TJA #21
sphene-biotite-hornblende quartz diorite from Sierra Las Tinajas Altas
ID-TIMS analyses
SHRIMP data only: Mean 206Pb/238U age: 72.8 ± 1.8 Ma
ID-TIMS data only: lower-intercept age: 72.8 ± 1.7 Ma
A
206Pb/238U
207Pb/235U
data-point error symbols are 1σ
Mean = 72.8 ± 1.8 [2.5%] 95% conf.
Wtd by data-pt errs only, 0 of 8 rej.
MSWD = 1.04, probability = 0.40
207Pb/206Pb age (Ma)
data-point error ellipses are 68.3% conf.
CP-16-99
sphene-hornblende-biotite quartz diorite from near the Drift Hills
B
SHRIMP data only: Mean 206Pb/238U age: 73.4 ± 3.3 Ma
TJA #21 only:
ID-TIMS data only: lower-intercept age: 72.8 ± 1.7 Ma
ID-TIMS analyses
206Pb/238U
207Pb/235U
data-point error symbols are 1σ
Mean = 73.4 ± 3.3 [4.5%] 95% conf.
Wtd by data-pt errs only, 0 of 3 rej.
MSWD = 0.097, probability = 0.91
207Pb/206Pb age (Ma)

indicate a lower-intercept age of 72.8 ± 1.7 Ma, which we interpret as the best estimate of the age of emplacement of this quartz diorite. The analyses lie along a discordia that indicates mixing between zircon formed during crystallization of the quartz diorite melt and zircon assimilated from basement rocks. The concordia upper-intercept age of 1705 ± 160 Ma represents an average age of assimilated crust that is consistent with Paleoproterozoic ages we have determined from nearby basement exposures.

We also analyzed 8 zircons from TJA #21 on the SHRIMP, taking U-Pb isotopic data from the central areas. All but two data were essentially concordant; the results of these analyses yielded a weighted mean $^{206}Pb/^{238}U$ age of 72.8 ± 1.8 Ma (Fig. 18A) that is statistically identical to the lower-intercept age resulting from the ID-TIMS method.

CP-16-99. This sample is a weakly foliated sphene-hornblende-biotite quartz diorite collected from a sill intruded into Paleoproterozoic syenogranite near the Drift Hills within the Eastern Complex. Its zircons exhibit a mixture of characteristics: some long and slender and clear, others stubby and subrounded and tan-colored. Some stubby grains contain inherited cores in the form of rounded, anhedral grains. Only very clear euhedral, prismatic grains were selected by handpicking, and two fractions were analyzed using the ID-TIMS method (Fig. 18B; solid circles). The results are unfortunately difficult to interpret. Both analyses lie slightly off concordia such that a chord through them yields ages that are not meaningful. However, one of the analyses plots on the discordia defined by sample TJA #21 (open circles) and has a $^{206}Pb/^{238}U$ age of 74.7 ± 1.0 Ma (Table 1A), suggesting a similar age and zircon behavior.

Four spot analyses from the centers of zircon from this sample were carried out on the SHRIMP and confirm that the quartz diorite was emplaced during Late Cretaceous time. Three of these data yielded a weighted mean $^{206}Pb/^{238}U$ age of 73.4 ± 3.3 Ma (Fig. 18B), which is our best estimate of the time of magma crystallization and is statistically indistinguishable from the age of 72.8 ± 1.7 Ma for quartz diorite TJA #21 from Sierra Las Tinajas Altas within the Western Complex.

U-Th-Pb, Rb-Sr, AND Sm-Nd WHOLE-ROCK ANALYSES

Whole-rock U-Th-Pb, Rb-Sr, and Sm-Nd isotopic analyses were performed on all the samples dated by U-Pb zircon in hopes of providing petrogenetic information. Two additional undated Paleoproterozoic granite gneiss samples from the Sierra Choclo Duro region were also analyzed. A special multisystematic chemical procedure from a single-dissolution of whole-rock powder provides direct correlation of the isotopic systems (Tatsumoto and Unruh, 1976; Premo et al., 1989; Premo and Tatsumoto, 1991, 1992). The U-Th-Pb analytical data is given in Table 2, and the Rb-Sr and Sm-Nd analytical data is given in Table 3. Analytical methods are described in Appendix A.

The successful use of Pb-Sr-Nd isotopes on old crustal rocks to obtain meaningful petrogenetic information is well documented (e.g., Dickin, 1995), but depends strongly on the metamorphic condition of the samples, such that the U-Th-Pb, Rb-Sr, or Sm-Nd isotopic systematics have not undergone significant disturbance(s) subsequent to emplacement or deposition. If that is the case and adequate age data is available for the samples, then initial Pb-Sr-Nd compositions or signatures may be calculated that represent the isotopic composition of the rock at its formation. These initial signatures may then be compared to model isotopic compositions for major magma-producing reservoirs of the outer regions of Earth (e.g., mantle, lower crust, upper crust) in order to evaluate their probable tectonic origin. In addition, the initial signatures may be compared to other initial compositions from other rock terranes of the southwestern United States to determine their genetic relationship, if any.

However, because we know that most, if not all, of the Paleoproterozoic samples analyzed in this study have been significantly metamorphosed and altered, it is suspected that their whole-rock isotopic systematics may be disturbed, but might preserve some information about that disturbance. Despite the probable difficulties, initial Pb-Sr-Nd signatures for these samples were calculated using their corresponding U-Pb zircon ages, and these values are given in Tables 2 and 3.

Whole-Rock Isochron Ages

Pb isotopic compositions for these samples range from $^{206}Pb/^{204}Pb$ values of ~16.8 to a radiogenic value of 24.8 (Table 2). Good correlation with $^{207}Pb/^{204}Pb$ is shown for almost all of the samples in this study, and a reasonably good $^{206}Pb/^{204}Pb$–$^{207}Pb/^{204}Pb$ isochron can be defined for several groups of rock units (Fig. 19A). Six of eight metagranitoid rocks (including two undated samples) define an isochron age of 1443 ± 44 Ma. Two samples, CD-12 #19A and Mina La Joya #1-98, were excluded from the calculation as they plot slightly above the isochron. In addition, the four metasedimentary samples define an isochron age of 1419 ± 140 Ma. Combining the metagranitoid and metasedimentary analyses (recrystallized rocks), an isochron age of 1441 ± 39 Ma is calculated, matching well with the age of 1433 ± 8 Ma given by secondary zircon overgrowths from the metasedimentary samples (see below). Because we know the actual age of formation for these rocks is older than ca. 1645 Ma, the whole-rock Pb-Pb age results indicates to us that the 1432 Ma magmatic event was pervasive enough to produce zircon overgrowths and reset the Pb isotopic systematics in the older host rocks, particularly in the highly recrystallized Eastern Complex.

The Pb-Pb isochron does intersect the Stacey and Kramers (1975) model Pb evolution curve at an age of 1722 Ma (Fig. 19A) corresponding to $^{206}Pb/^{204}Pb$ and $^{207}Pb/^{204}Pb$ values of ~15.72 and 15.32, respectively, that represent reasonable initial uranogenic Pb values for these rocks. The U-Pb isotopic systematics of several samples yielded initial uranogenic Pb values close to these model values; these include the granite gneiss Alacranes #1, the meta-arkose CD-12 #13, quartzite CD-12 #20, and the Mesoproterozoic granite PZ-23B (Table 2).

TABLE 2. U-Th-Pb ANALYTICAL DATA FOR WHOLE-ROCK SAMPLES OF THE EL PINACATE AND CAZEBA PRIETA REGION

Sample Name	Sample Wgt (mg)	U (ppm)	Th (ppm)	Pb (ppm)	$^{206}Pb/^{204}Pb^a$	$^{207}Pb/^{204}Pb^a$	$^{208}Pb/^{204}Pb^a$	$^{238}U/^{204}Pb^a$	$^{232}Th/^{204}Pb^a$	$^{206}Pb/^{204}Pb^b$ (Initial)	$^{207}Pb/^{204}Pb^b$ (Initial)	$^{208}Pb/^{204}Pb^b$ (Initial)
PALEOPROTEROZOIC SAMPLES												
Metagranitoid Samples												
ALACRANES #1	134	1.99	7.72	9.53	19.251	15.643	39.079	13.51	54.24	15.103	15.205	34.249
					(0.06)[@]	(0.09)	(0.12)	(0.25)	(0.40)			
ALACRANES #5	131	0.19	1.62	8.52	16.966	15.424	37.130	1.327	12.00	16.558	15.381	36.059
					(0.07)	(0.10)	(0.13)	(1.1)	(0.26)			
CP-17-99	206	5.24	41.90	48.30	24.806	16.138	38.332	7.516	62.17	22.608	15.915	33.031
					(0.06)	(0.09)	(0.12)	(0.22)	(0.70)			
MINA LA JOYA #1-98	214	1.78	16.20	7.79	17.811	15.532	38.855	14.49	136.4	(13.451)	(15.079)	(26.909)
					(0.06)	(0.09)	(0.12)	(0.24)	(0.57)			
CD-3 #4	93	2.94	16.10	24.00	18.417	15.561	39.314	7.863	44.68	16.136	15.331	35.534
					(0.06)	(0.09)	(0.12)	(0.39)	(1.5)			
CD-12 #19	129	4.27	20.50	26.60	19.637	15.704	40.448	10.66	52.94	16.516	15.387	35.935
					(0.07)	(0.10)	(0.13)	(0.69)	(2.2)			
Metasedimentary Samples												
CD-12 #5	138	1.10	20.80	37.70	16.794	15.405	38.477	1.810	35.28	16.270	15.352	35.497
meta-arkose					(0.06)	(0.09)	(0.12)	(2.4)	(0.36)			
CD-12 #13	139	2.05	14.40	25.10	17.376	15.472	37.649	5.029	36.60	15.919	15.325	34.554
meta-arkose					(0.06)	(0.09)	(0.12)	(1.5)	(3.4)			
CD-12 #20	126	2.99	18.90	13.80	19.425	15.647	41.580	14.55	95.03	15.225	15.225	33.575
quartzite					(0.06)	(0.09)	(0.12)	(1.5)	(3.2)			
CD-12 #2	141	0.90	17.10	22.90	17.347	15.467	39.170	2.469	48.75	16.632	15.395	35.050
quartzite					(0.06)	(0.09)	(0.12)	(1.3)	(0.38)			
MESOPROTEROZOIC SAMPLE												
PZ-23B	171	2.16	8.81	15.80	17.519	15.463	37.692	8.486	35.77	15.452	15.279	35.114
					(0.06)	(0.09)	(0.12)	(0.49)	(6.6)			
CRETACEOUS SAMPLES												
CP-16-99	193	2.10	5.93	13.10	18.045	15.528	37.572	10.04	29.22	17.930	15.523	37.466
					(0.07)	(0.10)	(0.13)	(0.26)	(0.22)			
TJA #21	181	2.16	4.54	10.80	18.484	15.569	37.812	12.59	27.34	18.341	15.562	37.713
					(0.06)	(0.09)	(0.13)	(0.91)	(0.94)			

[a]Corrected ratios; corrected for mass fractionation (values given in Appendix A) and blank Pb (values given in Appendix A).

[b]Initial ratios, calculated by subtracting the amount of ^{206}Pb, ^{207}Pb, and ^{208}Pb that has accumulated from the decay of U and Th in each sample since the formation of the rock, taken to be the zircon age, if known.

[@]Number in parentheses is uncertainty at 2 sigma, in percent, for the number directly above. Initial values shown in parentheses are interpreted to be inaccurate and therefore unreliable; values are overcorrected probably due to addition of the parent isotope during alteration (i.e., open-system behavior).

TABLE 3. Rb-Sr AND Sm-Nd ANALYTICAL DATA FOR WHOLE-ROCK SAMPLES OF THE EL PINACATE AND CAZEBA PRIETA REGION

Sample Name	Sample Wgt (mg)	Rb* (ppm)	Sr* (ppm)	Sm* (ppm)	Nd* (ppm)	$^{87}Rb/^{86}Sr$†	$^{87}Sr/^{86}Sr$†	$^{147}Sm/^{144}Nd$†	$^{143}Nd/^{144}Nd$†	Initial§ $^{87}Sr/^{86}Sr$	ε_{Nd} (t)
PALEOPROTEROZOIC SAMPLES											
Metagranitoid Samples											
ALACRANES #1	134	118	96.4	5.54	28.5	3.5716 ± 83	0.758748 ± 21	0.11726 ± 160	0.511899 ± 10	(0.6701)	3.23
ALACRANES #5	131	37.8	607	3.04	15.9	0.1799 ± 6	0.706628 ± 17	0.11552 ± 12	0.511919 ± 15	0.70216	4.04
CP-17-99	206	194	140	5.39	32.1	4.0434 ± 237	0.778710 ± 21	0.10129 ± 20	0.511706 ± 10	(0.6826)	2.11
MINA LA JOYA #1-98	214	79.4	97.1	3.85	28.5	2.3742 ± 47	0.746396 ± 16	0.08161 ± 18	0.511546 ± 12	(0.6885)	3.79
CD-3 #4	93	157	141	13.1	69.1	3.2263 ± 66	0.777834 ± 32	0.11402 ± 18	0.511902 ± 14	0.70172	3.11
CD-12 #19	129	265	114	11.7	66.9	6.8303 ± 250	0.854309 ± 30	0.10522 ± 29	0.511745 ± 22	(0.6919)	2.05
Metasedimentary Samples											
CD-12 #5 meta-arkose	138	427	97.6	7.60	46.2	13.023 ± 85	1.00378 ± 2	0.09937 ± 20	0.511734 ± 15	(0.6970)	2.88
CD-12 #13 meta-arkose	139	433	78.9	10.2	51.0	16.454 ± 84	1.07111 ± 5	0.12096 ± 39	0.511937 ± 15	(0.6834)	2.31
CD-12 #20 quartzite	126	109	1620	8.20	45.9	0.19504 ± 43	0.718573 ± 21	0.10793 ± 35	0.511832 ± 14	0.71398	2.29
CD-12 #2 quartzite	141	321	148	7.43	42.9	6.3697± 199	0.846275 ± 21	0.10465 ± 29	0.511776 ± 13	(0.6962)	2.60
MESOPROTEROZOIC SAMPLE											
PZ-23B	171	115	239	8.17	45.7	1.3976 ± 43	0.728157 ± 22	0.10794 ± 25	0.511806 ± 10	(0.6999)	–0.21
CRETACEOUS SAMPLES											
CP-16-99	193	120	618	5.65	31.4	0.5621± 60	0.707505 ± 14	0.10857 ± 25	0.512152 ± 15	0.70692	–8.62
TJA #21	181	69.8	717	3.86	19.7	0.2815 ± 36	0.707141 ± 15	0.11849 ± 12	0.512066 ± 15	0.70685	–10.39

*Concentration uncertainties for Rb and Sr are ~1.0 % and ~0.5 %, respectively; uncertainties for Sm and Nd are ~0.5 % and ~0.1 %, respectively.

†Isotopic ratios corrected for blank and mass fractionation, $^{87}Sr/^{86}Sr$ data are normalized to $^{86}Sr/^{88}Sr$ = 0.1194 and monitored for instrumental bias using NBS SRM 987 standard. The mean value of $^{87}Sr/^{86}Sr$ for 38 analyses of the Sr standard was 0.710258 ± 6. $^{143}Nd/^{144}Nd$ data are normalized to $^{146}Nd/^{144}Nd$ = 0.7219 and monitored for instrumental bias using the La Jolla Nd standard. The mean value of $^{143}Nd/^{144}Nd$ for 28 analyses of the La Jolla Nd standard was 0.511855 ± 4. Uncertainties shown below the ratio value are given at the 2σ level.

§Initial $^{87}Sr/^{86}Sr$ ratios were calculated using U-Pb zircon ages determined in this study ; $\lambda = 1.42 \times 10^{-11}$/yr; present day $(^{87}Sr/^{86}Sr)UR$ = 0.7045, and $(^{87}Rb/^{86}Sr)UR$ = 0.0824, where UR = uniform reservoir.

Initial $^{143}Nd/^{144}Nd$ ratios and ε_{Nd} are calculated using U-Pb zircon ages determined in this study; $\lambda = 6.54 \times 10^{-12}$/yr; present day $(^{143}Nd/^{144}Nd)CHUR$ = 0.512636, and $(^{147}Sm/^{144}Nd)CHUR$ = 0.1967, where CHUR = chondritic uniform reservoir.

Initial values shown in parentheses are interpreted to be inaccurate and therefore unreliable; values are overcorrected probably due to addition of the parent isotope during alteration (i.e. open-system behavior).

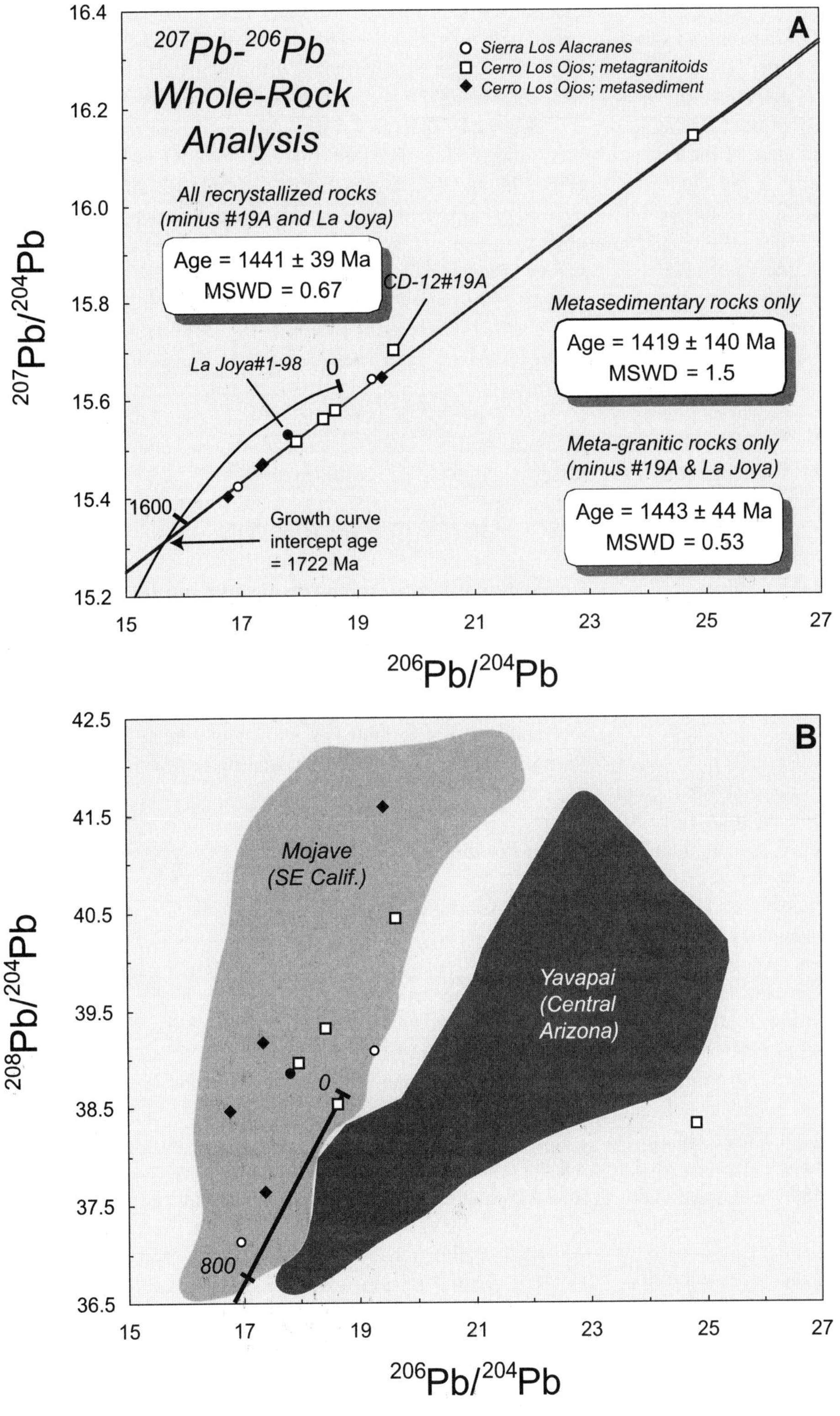

Figure 19. Pb-Pb correlation diagrams illustrating possible age relations between specific suites of whole-rock samples (Eastern Complex metasedimentary and metagranitoid samples, and Western Complex Alacranes; from whole-rock powders of the same samples used for U-Pb zircon geochronology). (A) $^{206}Pb/^{204}Pb$ vs. $^{207}Pb/^{204}Pb$ correlation diagram showing an array that represents the growth of the Pb isotopic system within these whole-rocks, and corresponding to an age of 1441 ± 39 Ma (mean square of weighted deviates [MSWD] = 0.67) for all recrystallized samples except for metagranitoids CD-12 #19A and Mina La Joya #1. Metagranitoids from the Eastern Complex yield a four-point isochron age of 1443 ± 44 Ma (MSWD = 0.53; minus CD-12 #19A). Pb data from two metagranitoid samples not dated in this study were added for this diagram. In addition, the four metasedimentary samples define an isochron age of 1419 ± 140 Ma (MSWD = 1.5). The age of 1441 ± 39 Ma defined by recrystallized samples matches well with the age given by secondary zircon overgrowths at 1433 ± 8 Ma. We interpret these results to indicate that the 1.43 Ga event was strong enough to reset the U-Pb system. (B) $^{206}Pb/^{204}Pb$ vs. $^{208}Pb/^{204}Pb$ correlation diagram with Western and Eastern Complex sample data compared to data from both the Mojave and Yavapai Provinces, indicating the likelihood that U and Th have been fractionated in these rocks (see text for explanation).

However, for the whole suite of recrystallized samples, an ill-defined U-Pb correlation yielded an isochron age of 1142 ± 190 Ma with an initial $^{206}Pb/^{204}Pb$ value of 16.65 ± 0.28 (not shown), indicating the likelihood that the U-Pb isotopic systematics were disturbed at least twice in most of these rocks subsequent to their formation.

The $^{206}Pb/^{204}Pb$–$^{208}Pb/^{204}Pb$ isotopic systematics for these disturbed samples are plotted against fields representing the bulk of the data from Mojave Province rocks and those of the Yavapai Province in central Arizona (Fig. 19B), because the thorogenic Pb isotopic parameter was found to be useful in distinguishing rocks of these two provinces (Wooden and Miller, 1990). As can be seen, all samples save one, Drift Hills granitoid CP-17-99, have thorogenic Pb signatures more similar to Mojave signatures than Yavapai values from central Arizona. However, this correlation does not hold true for the initial Nd isotopic signatures.

The whole-rock Rb-Sr isotopic systematics (Table 3) are variable for these samples but do produce a poorly defined isochron age of 1567 ± 88 Ma with an initial $^{87}Sr/^{86}Sr$ value of ~0.703 for the recrystallized Eastern Complex rocks. We believe this age may be of geologic importance as discussed below, although it is also within error of their crystallization (zircon) ages.

This same suite of Eastern Complex samples yielded a Sm-Nd isochron age of ca. 1635 Ma and initial ε_{Nd} value of ~+2.5 (not shown), indicating that the Sm-Nd isotopic systematics are probably not significantly altered or disturbed, and calculated initial Nd values (Table 3) are therefore accurate.

Initial Nd Signatures

Initial ε_{Nd} values for the whole-rock suite that are plotted against U-Pb zircon age in an Nd evolution diagram (Fig. 20) clearly illustrate the depleted nature (positive ε_{Nd}) of both the Western and Eastern Complex rocks. All three of the western samples lie within the field of Nd Province 3 (Bennett and DePaolo, 1987) but are redefined by data with Nd model ages between ca. 1650 and 1800 Ma, and ε_{Nd} values between +2.5 and +5 at ca. 1.65 Ga. The Western Complex samples also plot near the field that defines the North American block (NA; Fig. 20B) from data for Quitovac in Iriondo (2001) and Iriondo et al. (2004). Some of the eastern samples also plot within the field for Nd Province 3, although several straddle the vaguely defined boundary between Nd Provinces 2 and 3. In either case, the samples from our study area are similar in both age and Nd signature to rocks of the Yavapai and/or Mazatzal Provinces of the southwest United States (Bennett and DePaolo, 1987). None of the Nd isotopic data from our sample suite lie within either the Caborca field, as defined by sampling from Quitovac (Iriondo, 2001; Iriondo et al., 2004) that lies within Nd Province 2, or within Nd Province 1 with Nd model ages between 2.0 and 2.3 Ga, indicative of the Mojave Province (Bennett and DePaolo, 1987). These results have implications for the configuration of Proterozoic basement terranes.

Nd isotopic data from the Cochise block (CB; Fig. 20B) of southeastern Arizona (Eisele and Isachsen, 2001) is shown for comparison; their granitoid data (upper field shown) is of the same age, but slightly more depleted than the Eastern Complex (more positive Nd signatures). The lower, smaller field is from metasedimentary rocks in the Pinal Basin, and they plot between Iriondo's North American block samples and the metasedimentary rocks from Cerro Los Ojos, also slightly within the Nd Province 2 field.

Sample PZ-23B lies at the juncture of Nd Provinces 2 and 3 along a direct line of descent from the "North American" samples (Fig. 20A and B), suggesting that sample PZ-23B was largely derived from the older, Paleoproterozoic depleted crust; again consistent with the initial Pb isotopic results and the U-Pb zircon results (Table 2; Fig. 17). The 1.43 Ga magmatic event therefore appears to be mainly one of crustal melting in this region.

The two Cretaceous samples, CP-16-99 and TJA #21, plot with negative ε_{Nd} values between about −8 and −10 at their age of crystallization (73 Ma; Fig. 20A), indicating that their melts, on average, consisted of more than 50% older, probably Proterozoic basement and less than 50% subcontinental lithospheric mantle (assuming $\varepsilon_{Nd} = 0 \pm 2$ for this mantle).

DISCUSSION AND IMPLICATIONS

Discussion of U-Pb Zircon Results of Samples from the Eastern Complex

A comparison of least discordant (<10%) $^{207}Pb/^{206}Pb$ ages from the four metasedimentary samples as well as the three recrystallized metagranites is shown in Figure 21. They vary somewhat at the extremities, but otherwise appear to be surprisingly similar.

Age of Plutonism

As stated previously, SHRIMP U-Pb zircon data from two of the metagranite samples of the Eastern Complex (CD-3 #4 and CD-12 #19A) scatter along concordia with $^{207}Pb/^{206}Pb$ ages between ca. 1550 and 1750 Ma, resulting in weighted mean $^{207}Pb/^{206}Pb$ ages with expanded errors (±19 m.y. and ±15 m.y.) and limited confidence as to their true crystallization ages within those error limits. However, because field evidence strongly suggests that these samples were comagmatic, we would expect that their true crystallization or emplacement ages are within a 5 m.y. time span. And again, the least deformed metagranite is CP-17-99 from the Drift Hills that is farther away from the 1432 Ma granite, and yielded the weighted mean $^{207}Pb/^{206}Pb$ age of 1646 ± 10 Ma (Figs. 12 and 21) that we believe best represents the true crystallization and/or emplacement age for these plutons. This age for the Eastern Complex granitoids is supported by other U-Pb zircon ages from basement ranges to the south and east in Sonora. Three granitic gneiss samples from Sierra Hornaday, 10–15 km to the south, yielded weighted mean $^{207}Pb/^{206}Pb$ ages between 1644 and 1650 Ma (with smaller errors) that are similar to a weighted mean $^{207}Pb/^{206}Pb$ age of 1644 Ma from a gneissic granite sample 30 km east of Cerro Los Ojos (Nourse

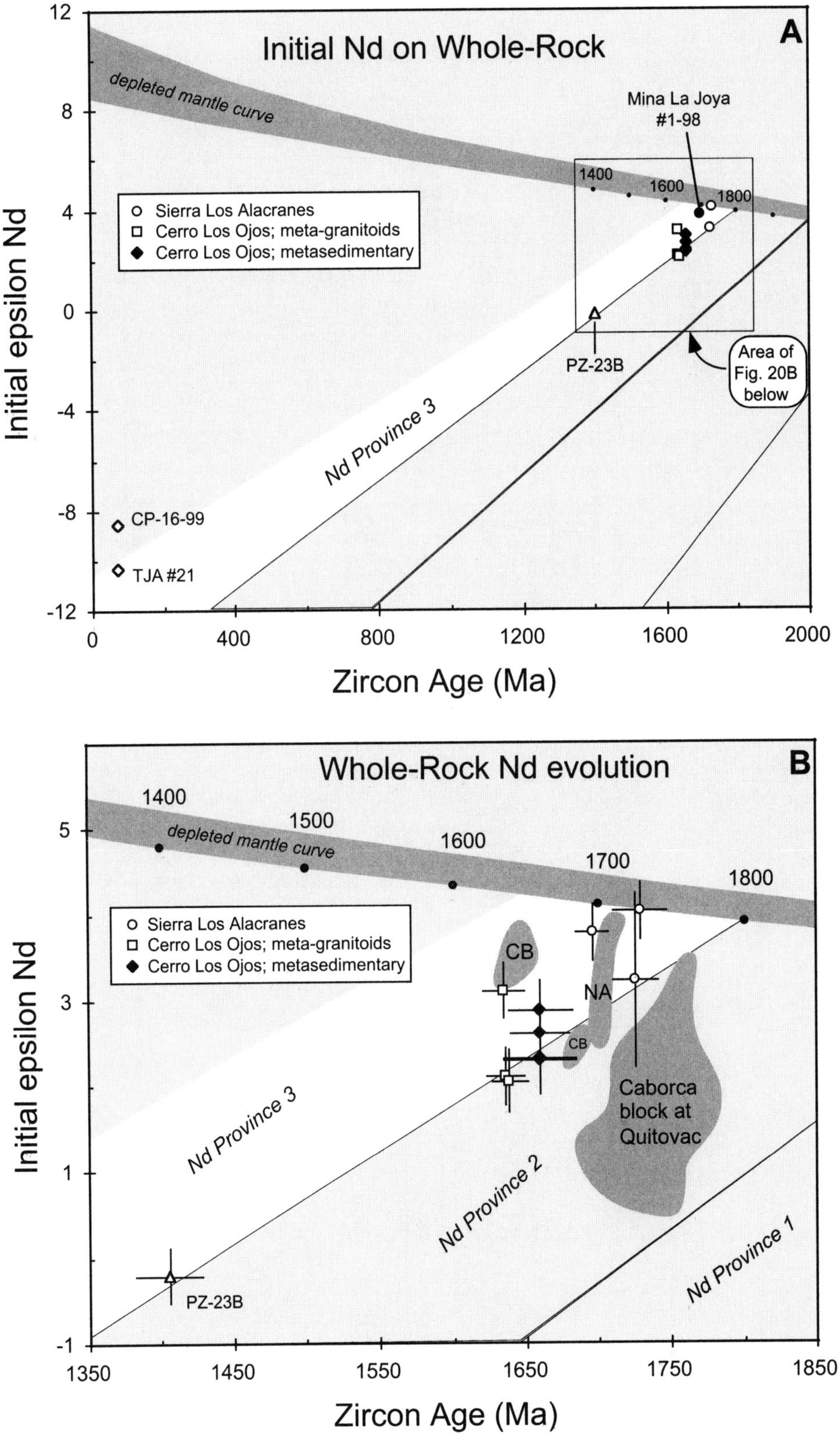

Figure 20. Nd evolution diagrams, illustrating possible age and origin relations between specific suites of whole-rock samples (Eastern Complex metasedimentary, metagranitoid, and Western Complex Alacranes; from whole-rock powders of the same samples used for U-Pb zircon geochronology). (A) Nd evolution diagram from 0 to 2000 Ma; initial Nd (in ε_{Nd} units) vs. zircon age in Ma, showing sample data from this study. All Paleoproterozoic data shown exhibit positive initial Nd values indicative of mainly a juvenile (mantle) origin. Mesoproterozoic sample PZ-23B (open triangle) lies on a path indicating probable derivation by partial melting of the Paleoproterozoic basement, and Cretaceous samples CP-16-99 and TJA #21 (open diamonds) appear to be derived mainly from Proterozoic crustal sources but include a minor mantle (more positive Nd) component. (B) Nd evolution diagram from 1350 to 1850 Ma; initial Nd (in ε_{Nd} units) vs. zircon age in Ma (expanded area of data in Fig. 20A), showing comparison of samples from this study with Paleoproterozoic samples from nearby Proterozoic crustal provinces or blocks. Data from Iriondo (2001), Iriondo et al. (2004), Eisele and Isachsen (2001), and Premo et al. (unpublished data). Data from this study plot around previously reported data for the Caborca and North America (NA) blocks of northern Sonora as well as the Cochise block (CB) of southeastern Arizona, but either within Nd Province 3 or straddling the boundary between Nd Provinces 2 and 3 (Bennett and DePaolo, 1987). In either case, these data are similar to values reported for rocks from the Yavapai and Mazatzal Provinces of central Arizona to New Mexico. The data are unlike those reported from the Caborca block at Quitovac.

Comparison of SHRIMP spot analyses

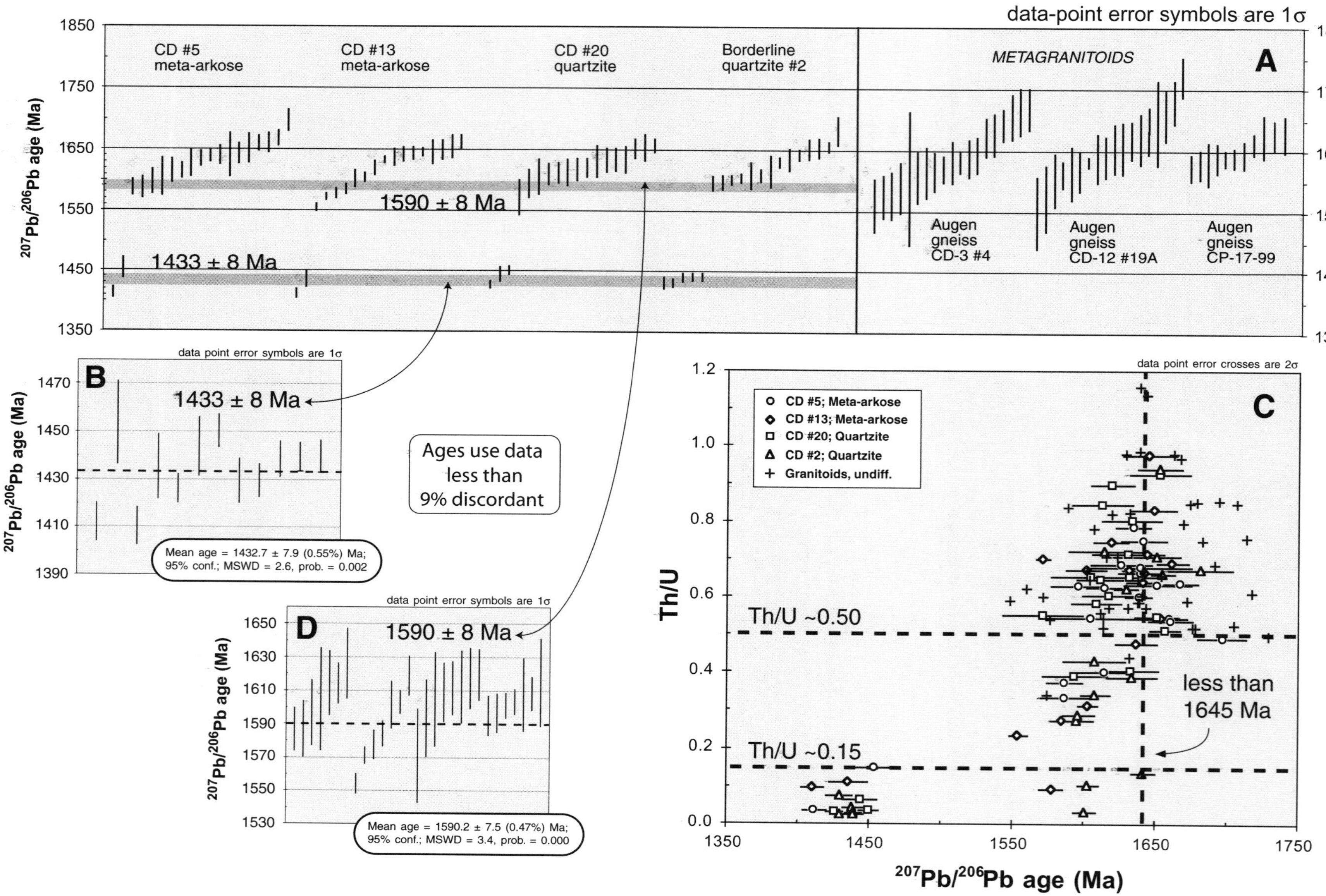

and others, unpublished data). A magmatic age of ca. 1645 Ma is therefore interpreted to represent the time of the most pervasive plutonic event in the basement north of Sierra El Pinacate and east of Sierra Los Alacranes.

Ages of Provenance

Least-discordant $^{207}Pb/^{206}Pb$ age ranges for the four metasedimentary samples are only slightly different, however all have ages between 1550 and 1600 Ma, and none greater than 1700 Ma (Fig. 21; Table 1B). With the exception of only one spot age (1697 ± 17 Ma; sample CD-12 #5), all metasedimentary sample age data do not exceed ca. 1665 Ma. Samples CD-12 #13 and CD-12 #20 have essentially identical maximum provenance ages of 1662 ± 11 and 1657 ± 11 Ma, respectively, whereas samples CD-12 #2 and CD-12 #5 record a slightly older provenance, 1681 ± 23 Ma and 1697 ± 17 Ma, respectively, although the former result is within error of the other samples maximum age of ca. 1665 Ma. The midsection of each pattern is marked by a series of recurring $^{207}Pb/^{206}Pb$ ages at ca. 1640–1660 Ma, suggesting a probable magmatic source age or provenance. Although these least-discordant $^{207}Pb/^{206}Pb$ ages for the four metasedimentary samples include ones that are younger than ca. 1645 Ma, an estimate of the time that plutons are interpreted to have intruded the metasediment protolith(s), we believe the data should be interpreted as indicating a probable provenance age range between ca. 1645 and 1665 Ma, with a possible, yet minor component between ca. 1680 and 1710 Ma. We speculate here that these sediments were shed from a restricted, nearby source not more than five million years prior to being intruded.

Similar provenance age ranges are not unexpected for two of the metasedimentary samples (CD-12 #13 and #20), as these were collected from stratigraphic horizons separated by less than 50 m (Fig. 6), implying that they were exposed to the same source(s) of detritus. The two remaining samples (CD-12 #2 and #5) were collected at much deeper and shallower levels, respectively, of a sandstone section that may have originally exceeded 2 km in thickness. These samples yielded only two grains with ages older than 1665 Ma. Thus, it appears that a very thick sequence of arkose and quartzose sandstone accumulated between ca. 1665 Ma and 1650 Ma. Shortly thereafter (ca. 1645 Ma), these sediments were extensively intruded by granite. The fact that none of the samples preserve sources older than 1700 Ma is quite striking and rather unexpected, because such sources are known directly west (Sierra Los Alacranes) and to the north in Arizona. Furthermore, we have yet to identify a nearby source for the 1665 Ma to 1650 Ma detrital zircons. At the present time, plutonic rocks with these ages are known only in southeastern California and northwestern Arizona (e.g., Silver, 1971; Wooden and Miller, 1990; Hawkins et al., 1996), southeastern Arizona (e.g., Silver, 1965; Conway, 1976; Erickson and Bowring, 1990; Powicki et al., 1993; Eisele and Isachsen, 2001), as well as parts of the Australian basement (e.g., Page et al., 2000; Black et al., 1997; Daly et al., 1996).

Figure 21. Comparison of data from sensitive high-resolution ion microprobe (SHRIMP) spot analyses for Paleoproterozoic metagranitoids and metasedimentary samples from the Eastern Complex. (A) Composite plot comparing SHRIMP $^{207}Pb/^{206}Pb$ ages from zircon. Weighted mean ages for "disturbances" of detrital zircons for metasedimentary samples are 1590 ± 8 Ma and 1433 ± 8 Ma, shown as gray horizontal bars. (B) Diagram illustrating variation of SHRIMP spot ages about the weighted mean age of 1433 ± 8 Ma obtained from mainly high-U rims of detrital zircon from the four metasedimentary samples. (C) Graph of $^{207}Pb/^{206}Pb$ age vs. Th/U for 85 detrital and 50+ magmatic zircons from the Eastern Complex, illustrating that nearly all of the detrital analyses with $^{207}Pb/^{206}Pb$ ages over ca. 1645 Ma (mean age for metagranitoids) have Th/U values of ~0.5 or greater. Nearly all of the Th/U values for the metagranitoids (crosses) also plot above the ~0.5 level, suggesting that detrital analyses with Th/U values above ~0.5 come from magmatic grains. Because the true ages for the detrital grains cannot be younger than the 1645 Ma mean age for the metagranitoids that intrude them, detrital spot ages younger than 1645 Ma must have suffered a disturbance to their U-Pb systematics. Data from the literature (e.g., Hoskin and Black, 2000) document the "low" Th/U of metamorphic zircon overgrowths. For detrital analyses with Th/U values ~0.15 or less, $^{207}Pb/^{206}Pb$ ages are interpreted to represent the timing for metamorphic zircon growth, which occurred at 1433 ± 8 Ma (Fig. 21B) and 1590 ± 8 Ma (Fig. 18D). Representative detrital analyses are found at both mean ages. Detrital analyses with Th/U values between ~0.15 and 0.5 all have $^{207}Pb/^{206}Pb$ ages between 1555 and 1645 Ma, and are interpreted to have suffered either Pb loss or metamorphic overgrowths during that time period. Detrital analyses with Th/U values greater than ~0.5 and $^{207}Pb/^{206}Pb$ ages younger than 1645 Ma are interpreted to have suffered Pb loss during metamorphism between ca. 1555 and 1610 Ma. (D) Diagram illustrating variation of SHRIMP spot ages about the weighted mean age of 1590 ± 8 Ma, obtained from both cores and rims of detrital zircon from the four metasedimentary samples.

Ages of Recrystallization and Metamorphism

The timing of profound dynamic-thermal metamorphism is ultimately constrained between the age of emplacement of the metagranitoids at ca. 1645 Ma and intrusion of a 1432 Ma granite that sharply crosscuts the strong fabric developed in these rocks. However, several lines of evidence suggest to us that there are at least two distinct metamorphic events recorded in the zircon data of the metasedimentary samples.

The younger and more definitive event is recorded by twelve SHRIMP analyses from the four metasedimentary samples that were taken on rims with dark-CL imaging (high U); these analyses yielded significantly younger $^{207}Pb/^{206}Pb$ ages between 1410 Ma and 1454 Ma (Figs. 16, 21A and 21B). The weighted mean of these twelve $^{207}Pb/^{206}Pb$ ages is 1433 ± 8 Ma (MSWD = 2.6; Fig. 21B). We interpret the dark-CL rims to be new zircon (overgrowths) grown during metamorphism immediately preceding emplacement of the Cerro Los Ojos granite at 1432 Ma.

Another piece of evidence that this interpretation is indeed the case comes from evaluating Th/U versus $^{207}Pb/^{206}Pb$ age (Fig. 21C). Recrystallized or metamorphosed zircons typically exhibit low Th/U values (<0.10), and are distinct from primary magmatic zircons that normally have Th/U values greater than

~0.2 (Williams and Claesson, 1987; Rubatto and Gebauer, 2000; Rubatto, 2002; Hoskin and Schaltegger, 2003). Th/U values for our samples can be calculated from the SHRIMP analytical data (Table 1B) and all of the 1.43 Ga analyses from the four metasedimentary samples have Th/U values less than 0.15 and most are less than 0.1 (Fig. 21C).

The older metamorphic event recorded in these samples is more cryptic. A possible age for this suspected isotopic disturbance can be ascertained from spot ages younger than 1630 Ma and extending to ca. 1555 Ma (Figs. 21A and 21D). A change of slope in each of the age patterns for the metasedimentary samples is produced by a significant 15–20 m.y. drop in $^{207}Pb/^{206}Pb$ ages from ca. 1640 to 1620 Ma. Spot ages younger than 1630 Ma and 9% or less discordant were used to calculate a weighted mean $^{207}Pb/^{206}Pb$ age of 1590 ± 8 Ma for metamorphism/recrystallization (Figs. 21A and 21D). In addition to the slope change, the 1630 Ma cutoff is based on the fact that the depositional age of these sediments must be the same or older than the granites that intruded them at ca. 1645 Ma (±~15 m.y.). Zircon analyses that yielded spot ages younger than ca. 1630 Ma and were only slightly discordant must have been isotopically disturbed in some way. Although their weighted mean age is 1590 ± 8 Ma, visually a median age might be closer to ca. 1600 Ma (Fig. 21D). The weighted mean age is heavily influenced by the youngest ages from sample CD-12 #13 that happen to have small errors. Excluding those, all metasedimentary analyses have a broad band of "affected" ages around ca. 1600 Ma. Similarly, the youngest spot ages of ca. 1555 Ma for the metagranitoids from Cerro Los Ojos are suggestive that similar effects are also found in zircons of at least two of those samples.

Whereas this data may be cryptic and our attempt to quantify it into a meaningful metamorphic event a bit skeptical, we believe it to be real for the following reasons:

1. None of the "intermediate" spot ages are younger than ca. 1555 Ma (Figs. 21A and 21C), as might be expected if these analyses represented partial resetting or secondary overgrowth due to the 1.43 Ga event (i.e., there aren't any ages between 1450 and 1550 Ma). The ages do not recur at any one age for any sample, but rather occur as a range of ages between ca. 1570 and 1620 Ma for all samples.

2. Several analyses with Th/U values less than 0.15 are found at $^{207}Pb/^{206}Pb$ ages of ca. 1575 and 1600 Ma (Fig. 21C). Th/U values for our Paleoproterozoic samples help to distinguish magmatic zircon from metamorphic growth, and therefore help us to differentiate between Pb loss and secondary zircon mixtures in disturbed grains.

Most of our data from detrital zircon cores show Th/U values between 0.25 and 0.75, and have $^{207}Pb/^{206}Pb$ ages between ca. 1580 Ma and 1670 Ma (Fig. 21C). Knowing that the age for the crosscutting metagranites is ca. 1645 Ma, then more than half of the data must represent disturbed ages. Data with Th/U values less than 0.5 are all younger than ca. 1645 Ma. This is important because all except two of nearly 50 analyses from the metagranites (crosses in Fig. 21C) have Th/U greater than 0.5, indicating the likely range of Th/U values in regionally available magmatic zircons. Applying this characteristic to the metasedimentary analyses with $^{207}Pb/^{206}Pb$ ages younger than 1630 Ma, some have Th/U greater than 0.5 and their "affected" ages are likely the result of Pb loss between ca. 1575 and 1600 Ma. Analyses with Th/U values less than 0.5 have probably been affected by metamorphism, producing secondary zircon material, although Pb loss cannot be totally ruled out. Analyses with Th/U values less than ~0.15 are interpreted here to indicate zones of metamorphically produced secondary zircon growth. This value is in agreement with some of the data from the literature, but obviously there is a gray zone about a definitive upper Th/U value for secondary zircon overgrowths and no doubt is dependent on other thermodynamic considerations. Several analyses with Th/U values less than 0.15 are found at $^{207}Pb/^{206}Pb$ ages of ca. 1575 and 1600 Ma, and we believe these data are supporting evidence for a distinct age of metamorphism/recrystallization within rocks of the Eastern Complex and particularly at Cerro Los Ojos. Therefore, we infer that most of the zircons were isotopically disturbed or annealed to varying degrees by a metamorphic event at ca. 1575–1600 Ma.

3. Tectonism and associated metamorphism in the age range 1560–1620 Ma has been documented in other Paleoproterozoic terranes of the southwestern United States and appears to be associated with large-scale crustal tectonism along pre-existing shear zones thought to be accretionary structures (e.g., Premo and Van Schmus, 1989; Bickford et al., 1989; Shaw et al., 1999; Premo and Fanning, 2000; Duebendorfer and Chamberlain, 2002; Strickland et al., 2004; Duebendorfer et al., 2004). Furthermore, similar isotopic results have been documented in zircons from recrystallized granitic gneisses in northeast Queensland, Australia (Hoskin and Black, 2000), interpreted as the result of high-grade metamorphism during the Jana orogeny between 1550 and 1600 Ma.

The Paleoproterozoic granites and their metasandstone hosts share a strongly recrystallized tectonic fabric that is sharply intruded by the 1432 Ma Cerro Los Ojos granite. We suspect that these "intermediate" ages (ca. 1550–1620 Ma) record an isotopic disturbance, accompanied by ductile deformation, that is distinct from the 1432 Ma event marked by zircon overgrowths. The age data suggest that thermal and/or fluid effects persisted until ca. 1575 Ma and possibly to ca. 1550 Ma. Thus, an important characteristic feature of the Eastern Complex is evidence of a cryptic metamorphic event at ca. 1575–1600 Ma. These age data are summarized in an age comparison diagram (Fig. 22) and illustrate that the probable age ranges for provenance of the metasedimentary rocks, the age of plutonism defined by the overlapping ages of the three orthogneisses at 1646 ± 10 Ma that is distinguishable from the two suspected ages of metamorphism—one at ca. 1600 Ma and the other at 1433 ± 8 Ma (see also Table 4).

Implications for Reconstruction Models of the Laurentian Western Margin

The northwest Sonora–southwest Arizona border region preserves a rich assemblage of Proterozoic rocks and structures

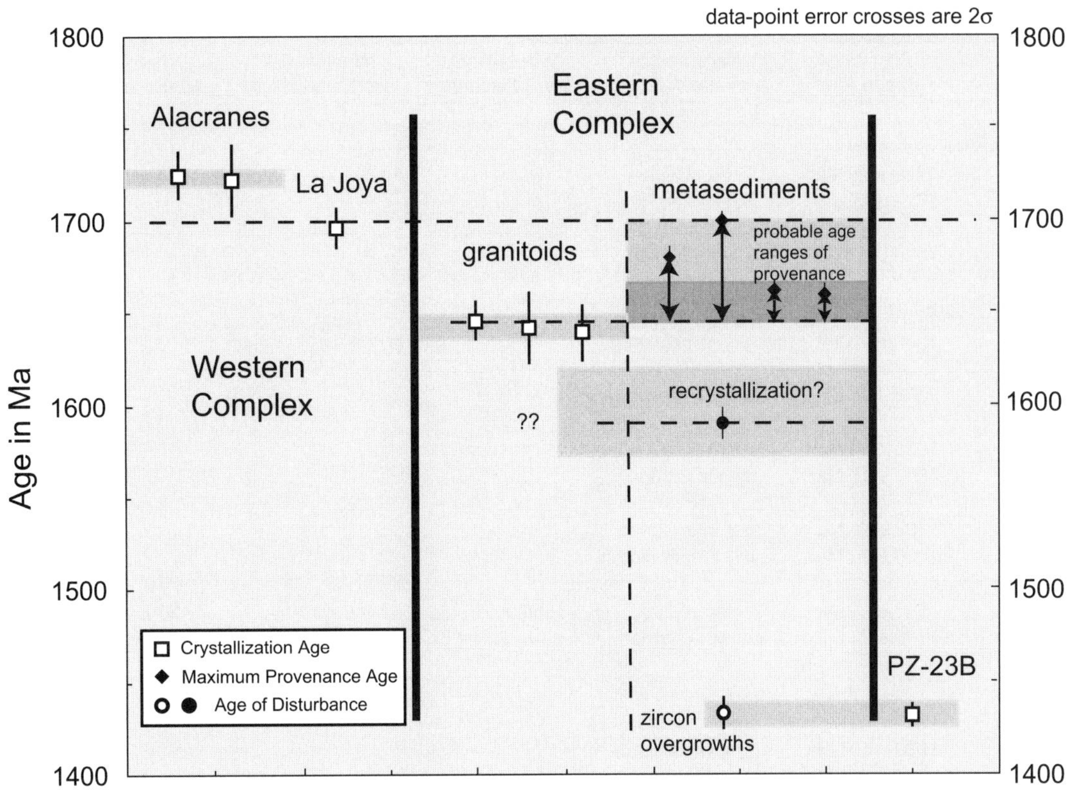

Figure 22. Age summary diagram comparing the timing of plutonism within the Western (ca. 1722 Ma) and Eastern (ca. 1645 Ma) Complexes, probable provenance age ranges for metasedimentary samples (ca. 1645–1700 Ma), and inferred ages of metamorphism (see text for explanation). We speculate that rocks of the Eastern Complex were recrystallized at 1590 ± 8 Ma, but later overprinted at 1433 ± 8 Ma. The age of emplacement of a plutonic unit (1432 ± 6 Ma) in the Eastern Complex is indistinguishable from the age of the younger metamorphic event.

in a key geographic location. Situated at the southwest edge of the Laurentian craton, the study area records a geologic history that predates the breakup of Rodinia at ca. 750 Ma (Stewart, 1972; Ross et al., 1989; Karlstrom et al., 2000; Timmons et al., 2001). Certain elements of this history should be shared by whichever continent was once attached to western Sonora. A major goal of the following discussion is to define an assemblage of lithologic and structural fingerprints in southwesternmost Laurentia that may ultimately provide a unique template in Rodinia reconstructions.

Unlike many Precambrian exposures of southern Arizona and southern California, the border region lacks the ductile deformation overprint of Late Cretaceous–early Tertiary Laramide compression (Davis, 1979; Ehlig, 1981; Haxel et al., 1984; Reynolds et al., 1986a) or mid-Tertiary core complex-style extension (Davis, 1980; Spencer and Reynolds, 1986; Reynolds et al., 1986b; Richard, 1994). Phanerozoic disturbances are limited to thermal effects of the Gunnery Range batholith that have reset $^{40}Ar/^{39}Ar$ biotite ages of Proterozoic basement to 55 Ma (Nourse et al., 2000), tilting of Neogene sedimentary and volcanic strata that overlie Late Cretaceous and Proterozoic crystalline rocks, and dextral displacement on San Andreas–type strike-slip faults. Although our analysis does not restore late Cenozoic extension and distributed dextral shear, the general continuity of Proterozoic geology and structure persists throughout the area mapped. Therefore, the study area is ideally suited for addressing questions regarding configuration of Proterozoic crust near the southwest fringe of Laurentia. Do our basement complexes represent southern projections of the Mojave, Yavapai, or Mazatzal crustal provinces, or something different? Did the Paleoproterozoic tectonic fabrics result from localized arc accretion or the suturing of a major continent during assembly of Rodinia? Which continent was attached to southwest Laurentia prior to 750 Ma? To what degree has Proterozoic basement along the Sonoran margin of Laurentia been modified by Phanerozoic strike-slip faults?

Relationships to Nearby Proterozoic Crustal Provinces of Laurentia

Our work in the Sonora-Arizona border region defines two contrasting Proterozoic basement complexes that have ages and radiogenic isotope signatures that generally overlap with the nearby Yavapai and Mazatzal crustal provinces. However, peculiar regional map patterns, discordant structural geometries, and differences in deformational chronology cause us to ponder the definition of the "Yavapai" and "Mazatzal" orogenic belts in Sonora.

TABLE 4. A COMPARISON OF ISOTOPIC AGES FROM SAMPLES IN THIS STUDY

	Accepted Zircon Age	Interpretation of Zircon Age	Whole-Rock Sm-Nd Isochron	Whole-Rock Pb-Pb Isochron	Whole-Rock Rb-Sr Isochron	Whole-Rock U-Pb Isochron
Western Complex Metagranitoids						
Sierra los Alacranes	1725 ± 25 Ma	crystallization		1545 ± 160 Ma	1074 ± 2.5 Ma	1108.7 ± 7.5 Ma
	1722 ± 19 Ma	crystallization				
			ca. 1585 Ma?			
Mina La Joya	1696 ± 11 Ma	crystallization				
Eastern Complex Metagranitoids						
Drift Hills	1636 ± 13 Ma	crystallization				
	1668 ± 18 Ma	avg. inheritance				
Cerro los Ojos	1635 ± 14 Ma	crystallization				
	1639 + 12 Ma	crystallization				
	1702 ± 28 Ma	avg. inheritance				
	1725 ± 33 Ma	avg. inheritance				
	1601 ± 22 Ma	metamorphism/	1560 ± 110 Ma	1440 ± 48 Ma	1530 ± 240 Ma	968 ± 18 Ma ?
	1605 ± 27 Ma	recrystallization				
Metasediments						
Cerro los Ojos	1642–1681 Ma	provenance				
	1638–1697 Ma	provenance				
	1637–1662 Ma	provenance				
	1645–1657 Ma	provenance				
	1590 ± 8 Ma	metamorphism/ recrystallization			(1530 ± 150 Ma)	
	1433 ± 8 Ma	zircon overgrowths	1451 ± 140 Ma	1419 ± 140 Ma	1441.6 ± 4.5 Ma	1071 ± 35 Ma
					1369 ± 46 Ma	1023 ± 18 Ma
Mesoproterozoic						
Granitoid	1432 ± 6 Ma	crystallization				

Although numerous Proterozoic exposures in Sonora have yet to be studied in detail, some intriguing map relationships result when our data set is compared with recent work from the Proterozoic gneisses near Quitovac (Iriondo, 2001; Iriondo et al., 2004) and Caborca (Premo et al., 2003; Iriondo and Premo, 2003).

The gneisses of Sierra Los Alacranes are part of a northwest-elongate belt of Paleoproterozoic basement, extending from Yuma to a point 100 km southeast of Caborca, that marks the truncated edge of Laurentia (Fig. 1; see also Anderson and Silver, this volume). Our work shows that three deformed granitoids in Sierra Los Alacranes were emplaced between 1725 Ma and 1696 Ma. About 100 km to the southeast, fourteen granitic gneiss samples of comparable depleted isotopic character near Quitovac have yielded upper-intercept concordia ages between 1777 Ma and 1693 Ma (Iriondo, 2001). Granitoids from the Caborca area yield ages within this same range (Premo et al., 2003; Anderson and Silver, this volume), but several preserve strongly enriched Nd isotopic signatures ($\varepsilon_{Nd} = -0.5$ to -2.7) reminiscent of the Mojave Province (Iriondo and Premo, 2003). Inboard of this western gneiss belt lies the younger Choclo Duro–Cabeza Prieta assemblage (Fig. 1), composed of sediments derived from sources no older than 1700 Ma, intruded by granites at ca. 1645 Ma.

The Western and Eastern Complexes both display fold trends and outcrop patterns (Figs. 3 and 6) oriented at high angles to the general northeasterly structural grain of the Yavapai-Mazatzal orogenic belt in central Arizona and New Mexico. At the scale of Figure 1, the older western gneiss belt cuts across the southwest projection of the Yavapai-Mazatzal Province boundary (Karlstrom et al., 1987). Furthermore, the geometry and timing of deformation, e.g., southwest- or northwest-vergent shear and pervasive recrystallization accompanied by isotopic disturbance at ca. 1.6 Ga, appear to distinguish our study area from the Arizona provinces. We concur with Karlstrom and Bowring (1988) that growth of Paleoproterozoic crust in southwest Laurentia probably did not involve a simple accretion mechanism such as that proposed by Condie (1982), in which northeast-trending juvenile arcs were progressively sutured to the craton from north to south.

Age ranges and initial Nd isotopic ratios from our dated Paleoproterozoic samples are compared on Figure 20 with published data from nearby Proterozoic crustal provinces or basement blocks. All of our samples plot in the depleted or primitive field, with ε_{Nd} values between +2.0 to +4.0, and differ from isotopically enriched Paleoproterozoic rocks of the Mojave Province, which has ε_{Nd} values that range from –6.3 to –1.6 (Bennett and DePaolo, 1987; Rämö and Calzia, 1998). Our samples also lack the Archean inheritance that characterizes gneisses of the Mojave Province (Wooden and Miller, 1990; Wooden, 2000). However, both Proterozoic complexes preserve tectonic fabrics of similar age and style to those that overprint the western Mojave Province (Barth et al., 2000).

The three samples from the Western Complex fall within the age and isotopic ranges (i.e., Nd Province 3 or 2) for "Yavapai-type" crust in central Arizona (Fig. 20). However, major structures of Sierra Los Alacranes are oriented oblique to those associated with the ca. 1.7 Ga Yavapai orogeny (Karlstrom and Bowring, 1988). Also, regional metamorphism and deformation postdated emplacement of the Mina La Joya granite at 1696 ± 11 Ma and is probably younger than the Yavapai orogeny. In terms of nearby localities in Sonora, the Western Complex may correlate with some Proterozoic exposures near Quitovac (Iriondo et al., 2004). Our Joya 1–98 sample compares well with Iriondo's "North America" block, whereas samples Alacranes #1 and Alacranes #5 have similar ages but are somewhat more depleted than the "Caborca block" west of Quitovac (Fig. 20B). Southeast of Caborca (Fig. 1), several granitic rocks with similar ages exhibit enriched "Mojave-type" Nd signatures (Iriondo and Premo, 2003; Iriondo, unpublished data), suggesting something special about this part of the Caborca block. We postulate that Paleoproterozoic granitoids situated between Sierra Los Alacranes and Quitovac collectively represent a depleted magmatic arc emplaced between 1777 Ma and 1693 Ma. The southeasterly trend of this arc may reflect an original oroclinal bend in the Yavapai Province or the Yavapai-Mojave transition zone. Further studies are needed to constrain the age(s) of deformational fabrics preserved in these rocks.

Granite gneiss and metasandstone samples from the Eastern Complex are distinctly younger than most rocks from Yavapai, Mojave, or Quitovac, but fall within the general age range (1.71 Ga to 1.63 Ga) of the Mazatzal crustal province (Karlstrom and Bowring, 1988), the southern part of which contains the Pinal block and the Cochise block (Eisele and Isachsen, 2001; Anderson and Silver, this volume). The ages of all three granite gneiss samples (ca. 1645 Ma) overlap with granite and quartz porphyry from the Cochise block of southeastern Arizona (Eisele and Isachsen, 2001; Fig. 20), and sample CD-3 #4 has a comparable depleted ε_{Nd} value of +3.2 (Fig. 20B). Like the Cochise block, the Eastern Complex also contains 1.4 Ga granite. However, metasedimentary host strata in Pinal block and Cochise block contain a significantly older detrital component (ca. 1730 Ma; Eisele and Isachsen, 2001). The source of detrital zircons (predominantly 1665 Ma to 1650 Ma) in our four metasandstone samples remains enigmatic. We have yet to identify nearby outcrops of granite or felsic volcanic rock that could have supplied the great volumes of arkose and quartzose sandstone. The Western Complex granitoids are too old; likewise, we see no north-central Arizona ("Yavapai") detrital signature. None of the Eastern Complex granites could have supplied detritus to the Pinal block, but they could represent one detrital source for sandstones of the Cochise block.

Although the Eastern Complex has an appropriate location and primary age/isotopic signature to be part of the Mazatzal Province, the timing of deformation and metamorphism appears to be younger. Granites that intrude tectonic fabrics of the "Mazatzal" orogeny in the Pinal block and the Cochise block have been dated at 1657 ± 13 Ma and 1643 ± 3 Ma, respectively (Eisele and Isachsen, 2001). In the Eastern Complex, three granites with a composite weighted mean age of 1644 ± 8 Ma are demonstratively pretectonic. As discussed earlier, profound isotopic disturbance of zircons from this region suggests that penetrative deformation and recrystallization that affected these granites culminated at ca. 1.6 Ga.

We favor an interpretation in which the Eastern Complex is an extension of the Mazatzal Province and the Western Complex is part of a southeasterly arcuate protrusion connected to the Yavapai Province. Here we utilize the Karlstrom and Bowring (1988, p. 562) definition of "province" to be "a large tract composed of several distinct tectonostratigraphic terranes or blocks that were assembled during one major pulse of convergent tectonism." The granitoids of Sierra Los Alacranes and Quitovac record depleted arc magmatism between 1777 Ma and 1693 Ma. Subduction polarity of this arc is speculative, as are paleogeographic and genetic relationships to coeval granitoids of the Mojave crustal province (Wooden and Miller, 1990; Barth et al., 2000). Present-day map relations suggest a marked discordance between the Alacranes-Quitovac portion of the Yavapai Province and depleted arcs of the Mazatzal Province that formed between 1.69 Ga and 1.63 Ga (Eisele and Isachsen, 2001). Structural and geochronological relationships suggest a possible collision of the Eastern Complex with the western fringing arc at ca. 1.6 Ga. If the ca. 1645 Ma Eastern Complex granites represent the southwestward continuation of Cochise block arc magmatism, a younger tectonic mechanism (suturing or accretion) is needed to explain the deformational fabrics that penetrate these rocks.

Further work is needed to identify Paleoproterozoic rocks and structures in adjacent areas that share similar histories and isotopic signatures. We are especially interested in exploring possible structural and temporal correlations to basement rocks of the Joshua Tree terrane (Bender et al., 1993) and San Gabriel terrane (Barth et al., 2001), both of which occur along the southwestern edge of the Mojave Province (Fig. 1). Also, the minimum age of deformation in the Western Complex is poorly constrained. Similarities in metamorphic fabric and structural vergence between the Western and Eastern Complexes suggest a common origin, but the precise time of deformation has yet to be established.

In light of the above discussion, we postulate that tectonism at ca. 1.6 Ga, associated with generally west-vergent noncoaxial deformation, is a distinguishing feature of Proterozoic basement situated along the Sonoran margin of Laurentia. This regionally important tectonic episode is distinctly younger than the 1710–1700 Ma Ivanpah orogeny (Wooden and Dewitt, 1991), the ca. 1.7 Ga Yavapai orogeny (Karlstrom et al., 1987), and the ca. 1.65 Ga Mazatzal orogeny (Karlstrom and Bowring, 1988; Karlstrom et al., 1990). Such an event may also be recorded in the southwestern part of the Mojave Province (Barth et al., 2000), where deformational fabrics of comparable orientation and style developed between ca. 1.65 Ga and ca. 1.4 Ga. Tectonism and associated metamorphism in the age range 1560–1620 Ma have been documented in other Paleoproterozoic terranes of Colorado and Wyoming, and indicate regional reactivation of shear zones thought to be accretionary structures (e.g., Premo and Van Schmus, 1989; Bickford et al., 1989; Shaw et al., 1999; Premo and Fanning, 2000; Duebendorfer and Chamberlain, 2002; Strickland et al., 2004; Duebendorfer et al., 2004). Perhaps fabrics formed at ca. 1.6 Ga provide a new fingerprint that can facilitate Rodinia reconstructions involving southwestern Laurentia.

Bearing on Hypothetical Strike-Slip Modifications

The Mojave-Sonora megashear controversy. Proterozoic basement exposures in our study area crop out directly west of the proposed trace of the Mojave-Sonora megashear, a hypothetical left-lateral transform fault that presumably trimmed the southwestern edge of North America during Late Jurassic time (Silver and Anderson, 1974; Anderson and Silver, this volume), and translated Proterozoic basement rocks of the Caborca block (Anderson and Silver, 1981) to the position indicated on Figure 1. As shown in three publications (Anderson and Silver, 1979; Anderson and Silver, 1981; Campbell and Anderson, 2003) the megashear trace crosses Mexican Highway 2 east of El Pinacate, transects the Proterozoic outcrops at Quitovac, and continues southeastward past Caborca. The megashear model as interpreted by various authors (see below) invokes sinistral translations of 800–1000 km. If the transform existed at the location originally proposed, both the Eastern and Western Complexes should be considered part of the Caborca allochthon, and would have been positioned adjacent to western Nevada prior to Late Jurassic time. Thus questions bearing on location of our study area in pre–750 Ma Rodinia reconstructions demand assessment of possible Phanerozoic strike-slip disruptions.

According to the original model (Silver and Anderson, 1974), the Mojave-Sonora megashear represents one of several sinistral transform faults that facilitated opening of the Gulf of Mexico during Middle-Late Jurassic rifting phases of Pangea (see also Anderson and Schmidt, 1983). One of several arguments for its existence is the apparent juxtaposition of "Yavapai-type" Proterozoic basement (loosely defined as containing 1.8–1.72 Ga crystalline rocks) against "Mazatzal-type" basement (containing 1.71–1.63 Ga rocks) along a northwest line extending from Caborca, Sonora, to the Inyo Mountains of California (Silver and Anderson, 1974; Anderson and Silver, this volume; Fig. 1). Similarities between unconformably overlying Neoproterozoic and Paleozoic sections in these two areas (see summary in Stewart et al., 2002) suggested total post-Permian lateral displacement of ~800 km. Other work (Jones et al., 1995; Anderson et al., this volume) argues that the Middle Jurassic magmatic arc has been trimmed and sinistrally displaced a comparable distance along the southeastern projection of the megashear. Recent studies of distinctive Triassic fossils in the El Antimonio Formation (west of Caborca) indicate possible biostratigraphic ties to rocks of northwestern Nevada (distance ~1000 km; Stanley and Gonzalez-Leon, 1995) or southeastern California (distance ~800 km; Gonzalez et al., this volume). At the type locality of the megashear southwest of Sonoita, Campbell and Anderson (2003) have mapped mylonites with left-lateral shear indicators that are developed in Triassic intrusive and Middle Jurassic volcanic rocks but are crosscut by Late Cretaceous plutons.

The megashear hypothesis continues to incite controversy, in particular regarding where and how to position such a major structure across the Mojave Desert of California. Presumably the Mojave Province was sutured to the western Yavapai Province at ca. 1730 Ma (Barth et al., 2000; Duebendorfer et al., 2001), so it is difficult to project a large-displacement Jurassic strike-slip fault through this region. If the megashear model is correct, one or both of our Paleoproterozoic basement complexes have been translated around the Mojave Province from an original position adjacent to western Nevada, or the megashear must wrap around them to the west in some complex fashion.

The California-Coahuila transform. The California-Coahuila transform, recently proposed by Dickinson (2000), offers an alternative solution to problems involving apparent left-lateral truncation of southwest North America and juxtaposition of Mesozoic oceanic basement against eastern California. According to Dickinson's hypothesis, the Caborca block was translated 950 km southeastward into Mexico along a transform fault that linked the Sonoman orogen of northwestern Nevada with a Permian magmatic arc in eastern Mexico. This transform was presumably active between Early Permian and Middle Triassic time (Dickinson and Lawton, 2001), creating similar piercing line offsets of pre-Permian rocks that are attributed by others to activity of the Late Jurassic Mojave-Sonora megashear. The model provides one explanation for the "Permian truncation event" in southeastern California, an observed juxtaposition of Paleozoic eugeoclinal strata against miogeoclinal strata that may have begun as early as mid-Pennsylvanian time (Stone and Stevens, 1988; Stevens et al., this volume).

As shown in Figures 3, 4, and 5 of Dickinson and Lawton (2001), the California-Coahuila transform is located directly southwest of our Western Complex. This arrangement implies that our study area lies entirely within autochthonous North America, and that oceanic basement farther west was created during Permian-Triassic seafloor spreading that accompanied transform faulting.

Limits on location of a hypothetical transform near the Sonora-Arizona border. The Mojave-Sonora megashear and California-Coahuila transform hypotheses are similar in their geometric treatment of the Caborca block as a displaced fragment of Proterozoic basement and overlapping Cordilleran miogeocline. Our study area cannot resolve differences in timing between the Permian-Triassic and Late Jurassic transform models, because the ages of rocks near the hypothetical shear zones are either too old (Proterozoic) or too young (Late Cretaceous) to provide useful constraints. However, the distinct ages, compositions, and isotopic characteristics of our two Proterozoic complexes offer a means to test whether or not a major transform has affected the basement of northwestern Sonora. If such a fault exists near our study area, we can place limitations on its possible location and speculate on where certain blocks should reconstruct. Three alternative scenarios showing hypothetical pre–Late Jurassic paleogeography are explored in Figure 23A–C. These may be compared to Figure 1, a "no transform" scenario that assumes the basement configuration in Sonora resulted from complicated Paleoproterozoic accretion mechanisms.

Figure 23A shows a transform fault at the location proposed in the original Silver and Anderson (1974) model. In this perspective, both of our Proterozoic complexes are treated as displaced basement fragments attached to the north end of the Caborca block. The main implication of this model is restoration of a composite Paleoproterozoic arc to a position adjacent to northwestern Nevada. Because no corresponding basement has been recognized in Nevada or northern California, this model requires existence of an exotic arc west of the rifted margin of Laurentia before Permian or Late Jurassic time. One cumbersome feature is the Mojave Province ("Mojavia"), which appears to have been accreted to the Yavapai Province at ca. 1730 Ma (Duebendorfer et al., 2001). In order to preserve Mojavia in its present-day location, the hypothetical transform fault must skirt around the western edge of Mojavia in a complex fashion such that the displaced Paleoproterozoic arcs and Caborca block restore adjacent to western Nevada and the Inyo Mountains, respectively.

Figure 23B positions a transform fault between the Eastern and Western Complexes where its surface trace is presumably obscured by the northwest-trending Late Cretaceous–Paleogene batholith. In this scenario, the Eastern Complex is treated as an autochthonous continuation of the Mazatzal Province, and the Western Complex has been displaced with the Caborca block. The resulting configuration is similar to the original Silver-Anderson model in that a western 1.8–1.7 Ga basement block is juxtaposed against an eastern 1.7–1.6 Ga basement block. South-trending folds in the Eastern Complex appear to be deflected from northeasterly structural trends preserved in the Mazatzal Province of central Arizona. This deflection could have resulted from drag associated with sinistral displacement on the transform. Alternatively, the Eastern Complex is an isolated block that has rotated counterclockwise between a minor eastern strand and major western strand of the transform system. Figure 23B implies that the Western Complex is part of a depleted 1.77–1.69 Ga magmatic arc that originated 800+ km to the northwest somewhere outboard of Nevada.

In Figure 23C, the transform fault is located southwest of the Western Complex (similar to the position proposed by Dickinson and Lawton, 2001), and the boundary between the Western and Eastern complexes represents an undisturbed Proterozoic suture. In this model autochthonous Quitovac basement is attached to the Sierra Los Alacranes arc rocks. The resulting configuration implies that the Caborca block is a trimmed-off fragment of western Laurentia that skirted around the Mojave Province and our study area during its translation to the southeast.

The models shown in Figures 23B and 23C both result in substantial misalignment with surface exposures of the Mojave-Sonora megashear zone mapped by Campbell and Anderson (2003) near Sonoita, Sonora. Part of this misalignment could be an artifact of superimposed mid-Tertiary extension. As described earlier, the entire study area may be part of a crustal panel that was translated southwestward in the upper plate of a major detachment fault. If a large magnitude of extension (50 km) is restored, the megashear trace still makes a pronounced left step or jog in the vicinity of the El Pinacate volcanic field.

Implications for Rodinia Reconstruction

We return to one of our original questions: How do the two Proterozoic basement complexes fit into a pre–750 Ma reconstruction of Rodinia? It should be clear from the preceding discussion that a variety of configurations is possible depending on the degree to which our study area has been affected by Phanerozoic strike-slip faulting. Figures 1 and 23A–C offer alternative pre-Permian paleogeographic reconstructions for southwestern Laurentia. Achieving a unique match between one of these scenarios and some other continent (or continents) is beyond the scope of this article. However, we summarize below some key characteristics that should prove useful in developing ties to parts of continents that might include Antarctica (Moores, 1991; Dalziel, 1991), Australia (Brookfield, 1993; Karlstrom et al., 1999; Burrett and Berry, 2000), Siberia (Sears and Price, 2000), or south China (Li et al., 1995, 2002).

Essential features of the Western Complex include 1725–1696 Ma granitoids with depleted initial Nd isotopic signatures intruded into previously deformed, layered magmatic arc rocks. Dominant structures are southwest-vergent noncoaxial fabrics and northwest trending map-scale folds that probably developed between 1696 Ma and 1.1 Ga.

Noteworthy Paleoproterozoic lithologic and structural elements of the Eastern Complex include a thick package of arkosic and quartzose sandstone deposited after ca. 1665 Ma and intruded by depleted granites at ca. 1645 Ma. South-trending folds and regional metamorphic fabrics developed at ca. 1.6 Ga are intruded by 1432 Ma granite and 1.1 Ga(?) diabase. Orientations of the late Paleoproterozoic folds and the Mesoproterozoic diabase dike swarm may provide important structural markers for achieving a match to another continent.

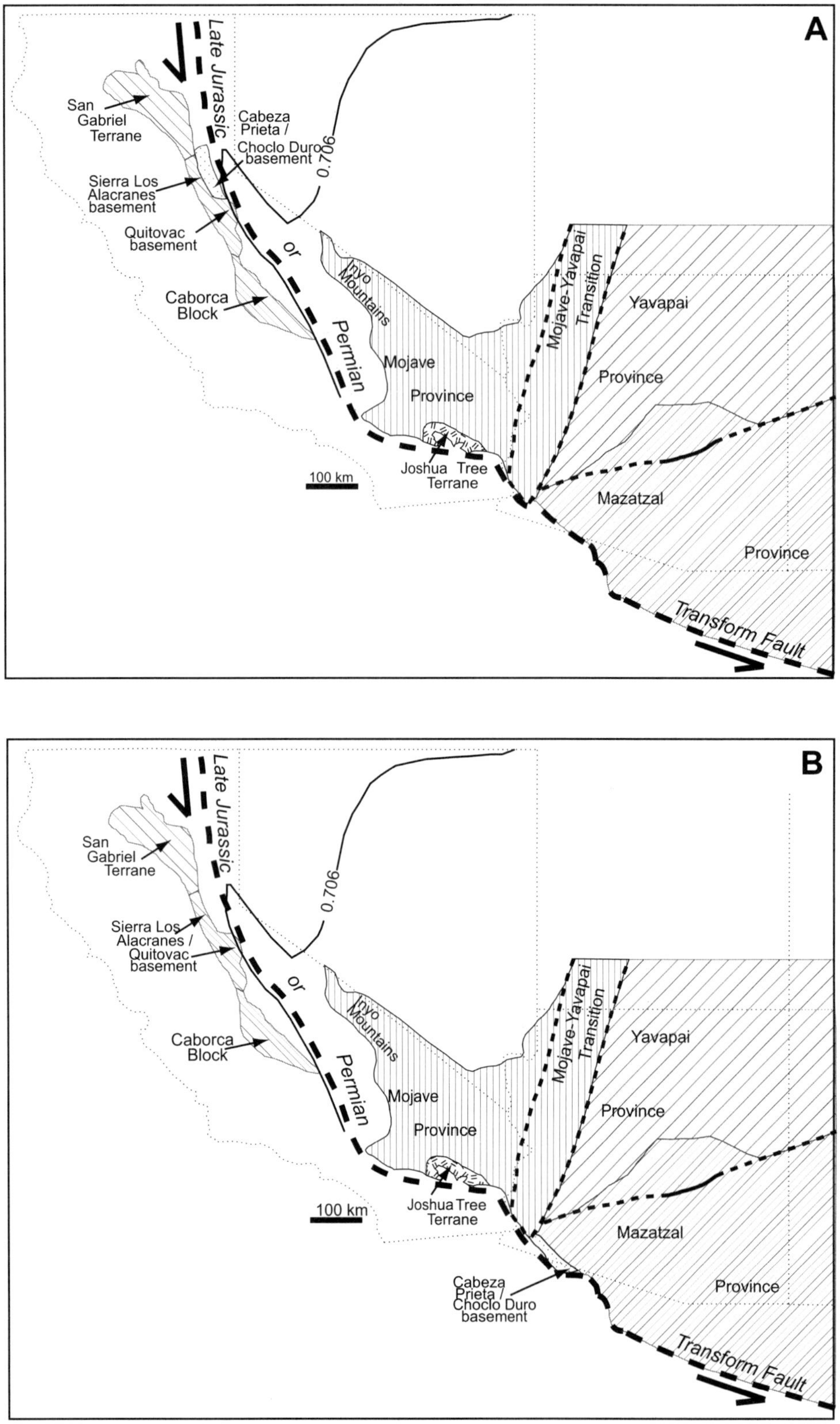

Figure 23 (*on this and following page*). Neoproterozoic paleogeographic maps showing possible basement configurations of western Laurentia that incorporate variable effects of hypothetical Permian or Late Jurassic sinistral transform faults on the Caborca block, the Eastern Complex, and Western Complex. Figure 1 constitutes the template for palinspastic reconstruction. Location of the 0.706 initial Sr isopleth is from Kistler and Peterman (1973). (A) Paleogeographic model showing a megashear in the position originally proposed by Silver and Anderson (1974). In this reconstruction, the Western Complex (Sierra Los Alacranes) and the Eastern Complex (Cabeza Prieta/Choclo Duro) are treated as allochthonous terranes attached to the Caborca block and the San Gabriel terrane. (B) Paleogeographic model showing a megashear positioned between the Western Complex (Sierra Los Alacranes) and the Eastern Complex (Cabeza Prieta/Choclo Duro). In this reconstruction, Sierra Los Alacranes and Quitovac basement are treated as allochthonous terranes attached to the Caborca block and the San Gabriel terrane.

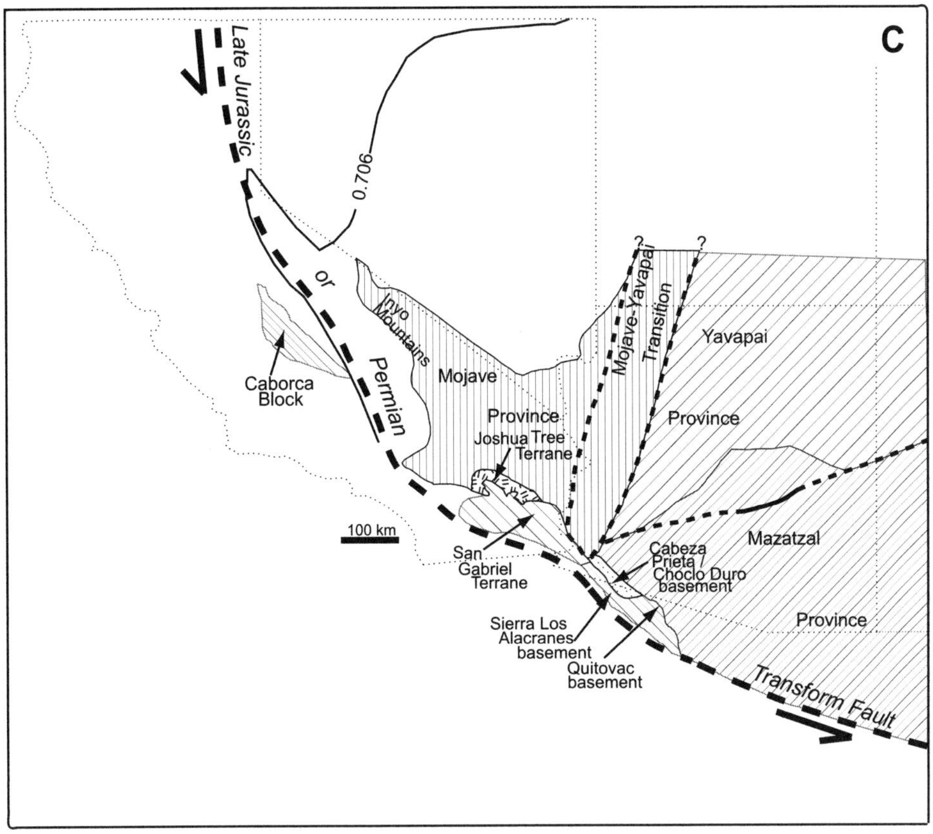

Figure 23 *(continued)*. (C) Paleogeographic model showing a megashear positioned west of both basement complexes. In this reconstruction, the Caborca block is the only allochthonous terrane, restoring to a location ~950 km northwest of its present-day location.

Earlier, we alluded that northeastern Australia may be a good place to look for rocks comparable to the Western and Eastern Complexes. Perusal of Australian literature illuminates some striking congruences in SHRIMP U-Pb zircon ages, for example: (1) a regionally important felsic volcanic event occurred ca. 1725 Ma adjacent to the Murphy inlier (Page et al., 2000), (2) rhyolitic tuffs from separate horizons in the Isa Superbasin were erupted in seven discrete pulses between 1668 Ma and 1585 Ma (Page et al., 2000), (3) two samples of granitic gneiss from the Georgetown region of Queensland display rims of recrystallized zircon (ca. 1560 Ma) that surround magmatic cores dated at 1696 ± 2 and 1684 ± 2 Ma (Hoskin and Black, 2000), and (4) a single occurrence of 1433 Ma granite is reported from the Savannah Province of north Queensland (Withnell et al., 1997). We are particularly intrigued by the evidence for a widespread amphibolite- to granulite-grade metamorphic event between ca. 1590 and ca. 1550 Ma, locally accompanied by plutonism. Numerous concordia plots from the Mt. Isa–McArthur River region suggesting early Mesoproterozoic recrystallization or metamorphism of late Paleoproterozoic primary zircons (Page et al., 2000) bear a marked resemblance to our plots from the Eastern Complex (Figs. 12–15). These are but a few of the similarities that call for more detailed comparison between northwest Mexico and northeastern Australia.

CONCLUSIONS

Proterozoic basement of the northwest Sonora–southwest Arizona international border region is composed of an older Western Complex and a younger Eastern Complex. The Western Complex contains 1725–1696 Ma granitoids with depleted Nd isotopic signatures intruded into banded gneisses derived from magmatic arc protoliths of uncertain age. The granitoids share southwest-vergent shear fabrics with their host rocks and are folded at map scale about northwesterly hinges. The Eastern Complex is composed of predominantly arkosic and quartzose metasandstones containing predominantly 1665–1650 Ma detritus, intruded by sheets of granite dated at ca. 1645 Ma. These rocks were folded about southerly hinges and regionally metamorphosed at ca. 1.6 Ga. U-Pb zircon ages on high-U rims record new zircon that accompanied emplacement of granite at 1432 Ma. Conspicuous 1.1 Ga(?) diabase dikes with northwesterly and northerly strikes intrude all of these Eastern Complex rocks.

The two Proterozoic basement complexes are "stitched together" by a Late Cretaceous batholith composed of 73 Ma quartz diorite or granodiorite and Paleogene leucocratic monzogranite. The region experienced northeast-southwest extension of uncertain magnitude during Miocene(?) time, resulting in the present-day distribution of northwest-trending tilt blocks.

Northwest-striking late Cenozoic strike-slip faults record ~50 km distributed dextral displacement.

The study area preserves a variety of distinct Proterozoic rock assemblages and structures to provide piercing lines useful in reconstructions of the Rodinia supercontinent. Comparisons with previously established crustal provinces of southwestern Laurentia suggest that the Eastern Complex is a continuation of the Mazatzal Province, and the Western Complex is part of an enigmatic southwestern protuberance of the Yavapai Province. The original Proterozoic position of the Western Complex along the southwestern edge of Laurentia remains uncertain because of questions regarding possible Jurassic or Permian disruption by a major sinistral strike-slip fault system.

ACKNOWLEDGMENTS

The principal author is grateful to Lee Silver and Tom Anderson for introducing him to the Proterozoic basement along Highway 2, and sharing results of geochronological work from the late 1960's. Silver contributed to the petrographic analyses and Anderson suggested important focus areas. Anderson supported the field studies of E. Stahl and J. Dembosky in the Cabeza Prieta National Wildlife Refuge. Dembosky provided color-enhanced Landsat images for reconnaissance work. Special thanks to Isabelle Brownfield and Heather Lowers of the U.S. Geological Survey (USGS) for their assistance in obtaining imaging of zircons using the scanning electron microscope in the USGS Denver Microbeam Laboratory. Joe Wooden of the USGS graciously provided access to the SHRIMP Laboratory at Stanford and shared his perspectives on the southwestern Cordilleran geology. The isotopic work was partially funded by the Southern California Areal Mapping Project (SCAMP). Numerous Cal Poly Pomona undergraduate students assisted with the mapping during 16 trips between 1997 and 2004: R. Acosta, T. Watkins, M. Espinoza, V. Vathanasin, J. DeLand, A. Wingfield, R. Burns, S. Gibson, S. Marusich, J. Navarro, M. Magener, C. Sanford, S. Wilkins, C. Horsley, D. Curtis, K. Ross, D. Hashim, J. De Loera, J. Strand, L. Annis, L. Michalka, A. Varnell, J. Utick, E. Fromboise, P. Cortez, and J. Beal. The Cal Poly Geological Sciences Department and several undergraduate students provided field vehicles for many of the mapping excursions. Samples were prepared and minerals separated with the assistance of Dan Miggins at the USGS facility in Denver. Formal reviews by Tom Anderson and an anonymous person helped us to improve the manuscript and clarify issues with the geochronology presentation.

APPENDIX A: ANALYTICAL METHODS

Sample Preparation

Mineral separates, including zircon, from all samples were separated using conventional methods. Samples weighing between 1 and 5 kg were crushed, pulverized, and an aliquot taken for whole-rock analysis. The remainder was sieved to less than 100 mesh (150 μm), heavy minerals were concentrated using a Wilfley table, and the concentrate was then put through a heavy liquid (MeI; $\rho = 3.33$). This zircon-enriched concentrate was then magnetically split to obtain a nonmagnetic fraction from which analyzed zircon fractions were handpicked.

U-Pb Zircon Geochronology: Isotope Dilution–Thermal Ionization Mass Spectrometry (ID-TIMS)

Prior to dissolution, individual fractions of a few grains to tens of grains were weighed into PFA-Teflon microvials (Ludwig design), digitally imaged, then cleaned very briefly with cold, distilled 6*N* HNO_3, and finally dissolved with suprapure, distilled concentrated HF + HNO_3 in a large (6.5-cm-diameter) Parr-type, TFE-Teflon, dissolution vessel at 210 °C for ~3–7 d using the HF-vapor technique of Krogh (1978). The fractions were then spiked with a ^{205}Pb-^{233}U-^{236}U-^{230}Th dilute tracer solution and reheated to achieve isotopic equilibration. Pb was extracted from the dissolved zircon fractions using AG 1-X8 anion exchange resin in Teflon microcolumns using a very dilute HBr medium. Pb residues were then redissolved in H_3PO_4 and loaded onto single Re filaments. The Pb laboratory contamination (blank) varied between 8 and 40 pg total Pb. U and Th were then extracted from the Pb effluent using AG 1-X8 anion exchange resin in a different, slightly larger, Teflon microcolumn using a 7*N* HNO_3 medium, and residues were loaded onto Re filaments using dilute HNO_3. U and Th blank levels were between 15 and 25, and 3–6 pg, respectively.

U-Th-Pb isotopic ratios were measured using a fully automated (using the programming of Ludwig, 1993), multisample, single-collector, VG Isomass 54R mass spectrometer. U-Th-Pb isotopic ratios were corrected using the algorithms and programming of Ludwig (1980, 1985). The Pb data was corrected for a mass fractionation of 0.08% ± 0.03% per a.m.u. from Faraday cup runs and 0.30% ± 0.05% per a.m.u. from Daly collector runs, as determined from multiple runs of NBS Pb standards SRM-981 and 982, and corrected to the values of Todt et al. (1993). The Pb data were further corrected for laboratory contamination (blank) with measured composition of $^{206}Pb/^{204}Pb = 19.34 \pm 0.53$, $^{207}Pb/^{204}Pb = 15.53 \pm 0.08$, and $^{208}Pb/^{204}Pb = 38.11 \pm 0.20$ from multiple determinations, and initial common Pb using the values of Stacey and Kramers (1975) for an approximate age of the sample, and $^{238}U/^{204}Pb = 9.74$ for a second-stage evolution. U and Th ratios were measured in the triple filament mode and corrected for a mass fractionation of 0.10 ± 0.03% per a.m.u. and laboratory blank.

Concordia intercept ages were calculated using decay constants from Steiger and Jäger (1977) and the algorithms of Ludwig (1980, 2001a) that use the regression approach of York (1969); uncertainties are reported at the 95% confidence level. All isotopic diagrams were plotted using Isoplot/Ex (Ludwig, 2001a).

U-Pb Zircon Geochronology: Sensitive High-Resolution Ion MicroProbe (SHRIMP)

SHRIMP procedures used in this study are similar to those reported in Williams (1998). Zircons handpicked from the same population separated for ID-TIMS work and chips of zircon standard R33 were mounted in epoxy, ground to nearly half their thickness using 1500 grit, wet-dry sandpaper, and polished with 6- and 1-µm-grit diamond suspension abrasive. Transmitted- and reflected-light photos were taken of all mounted grains. In addition, CL (cathodoluminescence) images of all zircons were prepared prior to analysis and used to reveal internal zoning related to chemical composition in order to avoid possible problematic areas within grains. The mounts were cleaned in 1*N* HCl and gold-coated for maximum surface conductivity.

The U-Th-Pb analyses were made using the SHRIMP-RG housed in Green Hall at Stanford University, California, and co-owned by the U.S. Geological Survey. The primary oxygen ion beam operated at ~2–4 nA and excavated an area of ~25–30 µm in diameter to a depth of ~1 µm, and sensitivity ranged from 5 to 30 cps per ppm Pb. Data for each spot were collected in sets of either five or six scans through the mass range. The reduced $^{206}Pb/^{238}U$ ratios were normalized to zircon standard R33 (419 Ma; from monzodiorite, Braintree Complex, Vermont; R. Mundil, Berkeley Geochronology Center, 1999, personal commun.; S.L. Kamo, Jack Satterly Geochronology Laboratory, Royal Ontario Museum, 2001, personal commun.; J.N. Aleinikoff, 2003, personal commun.), and either SL13 (238 ppm U) or CZ3 (550 ppm U) depending on the mount; these standard values are based on conventional U-Pb dating of replicate isotope dilution analyses of milligram-sized fragments. Analyses of samples and standard were alternated for the closest control of Pb/U ratios. U and Pb concentrations are accurate to ~10%–20%. SHRIMP isotopic data were reduced and plotted using the Isoplot/Ex and Squid programs of Ludwig (2001a, 2001b).

U-Th-Pb, Rb-Sr, and Sm-Nd Whole-Rock Isotope Geochemistry: ID-TIMS

The analytical techniques used for the simultaneous, single-dissolution of U-Th-Pb, Rb-Sr, and Sm-Nd analysis on whole rocks and mineral separates were similar to those reported in more detail by Tatsumoto and Unruh (1976), and Premo and Loucks (2000). The whole rocks were dissolved in 7 mL PFA Teflon vials with ultrapure concentrated HF + HNO_3 and then spiked with a dilute mixed tracer of ^{205}Pb-^{233}U-^{236}U-^{230}Th as well as dilute mixed tracers of ^{84}Sr-^{87}Rb and ^{150}Nd-^{149}Sm. The procedures for the extraction of Pb and U, and treatment of the data were the same as those listed above, except that the Pb laboratory contamination (blank) varied between 50 and 90 pg total Pb (avg. = 67 ± 8 pg), and had a measured composition of $^{206}Pb/^{204}Pb$ = 18.681 ± 0.064, $^{207}Pb/^{204}Pb$ = 15.432 ± 0.033, and $^{208}Pb/^{204}Pb$ = 37.720 ± 0.120 from multiple determinations. The effluent was then passed through a large (30 mL resin-volume) column with AG 50W-X8 cation exchange resin, separating Rb, Sr, and the rare earth elements (REE). Sm was separated from Nd using AG50W-X8 cation exchange resin and the α-isobutyric method of Lugmair et al. (1975), and then they were loaded with very dilute H_3PO_4 acid onto tantalum filaments in the triple filament mode, and run on a fully automated, multisample, single-collector, VG Isomass 54R mass spectrometer. Laboratory contamination levels of total Sr typically ranged between 0.05 and 0.3 ng, and total Nd ranged between 0.03 and 0.25 ng.

Rb and Sr concentration uncertainties are ~1.0% and ~0.5%, respectively; and Sm and Nd concentration uncertainties are ~0.5% and ~0.1%, respectively. All isotopic ratios were corrected for blank and mass fractionation, $^{87}Sr/^{86}Sr$ data were normalized to $^{86}Sr/^{88}Sr$ = 0.1194, and monitored for instrumental bias using NBS SRM 987 standard; the mean value of $^{87}Sr/^{86}Sr$ for 38 analyses the Sr standard was 0.710258 ± 6 over the period of this study. $^{143}Nd/^{144}Nd$ data were normalized to $^{146}Nd/^{144}Nd$ = 0.7219 and monitored for instrumental bias using the La Jolla Nd standard; the mean value of $^{143}Nd/^{144}Nd$ for 28 analyses of the La Jolla Nd standard was 0.511855 ± 4 over the period of this study.

Initial $^{87}Sr/^{86}Sr$ ratios were calculated using U-Pb zircon ages determined in this study; $\lambda = 1.42 \times 10^{-11}$/yr; present-day $(^{87}Sr/^{86}Sr)UR$ = 0.7045, and $(^{87}Rb/^{86}Sr)UR$ = 0.0824, where UR = uniform reservoir. Initial $^{143}Nd/^{144}Nd$ ratios and ε_{Nd} are calculated using U-Pb zircon ages determined in this study; $\lambda = 6.54 \times 10^{-12}$/yr; present-day $(^{143}Nd/^{144}Nd)CHUR$ = 0.512636, and $(^{147}Sm/^{144}Nd)CHUR$ = 0.1967, where CHUR = chondritic uniform reservoir.

Uncertainties on isotopic ratios are given in Tables 2 and 3 and are reported at the 2σ level. See footnotes of Tables 2 and 3 for measured ratio correction values.

REFERENCES CITED

Anderson, J.L., 1989, Proterozoic anorogenic granites of the southwestern United States, *in* Jenny, J.P., and Reynolds, S.J., eds., Geologic evolution of Arizona: Arizona Geological Society Digest, v. 17, p. 211–238.

Anderson, T.H., and Schmidt, V.A., 1983, A model of the evolution of Middle America and the Gulf of Mexico–Caribbean Sea region during Mesozoic time: Geological Society of America Bulletin, v. 94, p. 941–966, doi: 10.1130/0016-7606(1983)94<941:TEOMAA>2.0.CO;2.

Anderson, T.H., and Silver, L.T., 1979, The role of the Mojave-Sonora megashear in the tectonic evolution of northern Sonora, *in* Anderson, T.H., and Roldán-Quintana, J., eds., Geology of northern Sonora: The Geological Society of America Field Trip Guidebook, p. 59–68.

Anderson, T.H., and Silver, L.T., 1981, An overview of Precambrian rocks in Sonora, Mexico: Universidad Nacional Autónoma de México, Instituto de Geología, Revista, v. 5, no. 2, p. 131–139.

Barth, A.P., Wooden, J.L., Coleman, D.S., and Fanning, C.M., 2000, Geochronology of the Proterozoic basement of southwesternmost North America, and the origin and evolution of the Mojave crustal province: Tectonics, v. 19, no. 4, p. 616–629, doi: 10.1029/1999TC001145.

Barth, A.P., Wooden, J.L., and Coleman, D.S., 2001, SHRIMP-RG U-Pb zircon geochronology of Mesoproterozoic metamorphism and plutonism in the southwesternmost United States: Journal of Geology, v. 109, p. 319–329, doi: 10.1086/319975.

Bender, E.E., Morrison, J., Anderson, J.L., and Wooden, J.L., 1993, Early Proterozoic ties between two suspect terranes and the Mojave crustal block of the southwestern U.S.: Journal of Geology, v. 101, p. 715–728.

Bennett, V.C., and DePaulo, D.J., 1987, Proterozoic crustal history of the western United Sates as determined by neodymium isotopic mapping: Geological Society of America Bulletin, v. 99, p. 674–685, doi: 10.1130/0016-7606(1987)99<674:PCHOTW>2.0.CO;2.

Bickford, M.E., and Anderson, J.L., 1993, Middle Proterozoic magmatism, *in* Reed, J.C., Jr., et al., eds., Precambrian: Conterminous United States: Geological Society of America, Decade of North American Geology, v. C-2, p. 281–292.

Bickford, M.E., Cullers, R.L., Shuster, R.D., Premo, W.R., and Van Schmus, W.R., 1989, U-Pb zircon geochronology of Proterozoic and Cambrian plutons in the Wet Mountains and southern Front Range, Colorado, *in* Grambling, J.A., and Tewksbury, B.J., eds., Proterozoic geology of the southern Rocky Mountains: Boulder, Colorado, Geological Society of America Special Paper 235, p. 49–64.

Black, L.P., Carr, G.R., and Sun, S.S., 1997, Applied isotope geochronology and geochemistry, Chapter 13, *in* Bain, J.H.C., and Draper, J.J., eds., North Queensland geology: Australian Geological Survey Organisation (AGSO) Bulletin 240, Queensland Geology 9, p. 429–447.

Brookfield, M.E., 1993, Neoproterozoic Laurentia-Australia fit: Geology, v. 21, p. 683–686.

Burrett, C., and Berry, R., 2000, Proterozoic Australia–western Unites States (AUSWUS) fit between Laurentia and Australia: Geology, v. 28, p. 103–106, doi: 10.1130/0091-7613(2000)028<0103:PAWUSA>2.3.CO;2.

Campbell, P.L., and Anderson, T.H., 2003, Structure and kinematics along a segment of the Mojave-Sonora megashear: A strike-slip fault that truncates the Jurassic continental magmatic arc of southwestern North America: Tectonics, v. 22, no. 6, p. 16-1–16-21.

Condie, K.C., 1982, Plate tectonics model for Proterozoic continental accretion in the southwestern United States: Geology, v. 10, p. 37–42, doi: 10.1130/0091-7613(1982)10<37:PMFPCA>2.0.CO;2.

Conway, C.M., 1976, Petrology, structure and evolution of a Precambrian volcanic and plutonic complex, Tonto Basin, Gila County, Arizona [Ph.D. dissertation]: Pasadena, California, California Institute of Technology, 460 p.

Daly, S.J., Fanning, C.M., and Fairclough, M.C., 1996, Tectonic evolution and exploration potential of the Gawler craton: AGSO Journal of Australian Geology and Geophysics, v. 17, p. 145–168.

Dalziel, I.W.D., 1991, Pacific margins of Laurentia and East Antarctica–Australia as a conjugate rift pair: Evidence and implications for an Eocambrian supercontinent: Geology, v. 19, p. 598–601, doi: 10.1130/0091-7613(1991)019<0598:PMOLAE>2.3.CO;2.

Davis, G.H., 1979, Laramide folding and faulting in southeastern Arizona: American Journal of Science, v. 279, p. 543–569.

Davis, G.H., 1980, Structural characteristics of metamorphic core complexes, southern Arizona: Geological Society of America Memoir 153, p. 35–77.

Dickin, A., 1995, Radiogenic isotope geology: Cambridge, Cambridge University Press, 490 p.

Dickinson, W.R., 2000, Geodynamic interpretation of Paleozoic tectonic trends oriented oblique to the Mesozoic Klamath-Sierran continental margin in California, *in* Gehrels, G.E., and Soreghan, M J., eds., Paleozoic and Triassic paleogeography and tectonics of western Nevada and northern California: Geological Society of America Special Paper 347, p. 209–245.

Dickinson, W.R., and Lawton, T.F., 2001, Carboniferous to Cretaceous assembly and fragmentation of Mexico: Geological Society of America Bulletin, v. 113, no. 9, p. 1142–1160, doi: 10.1130/0016-7606(2001)113<1142:CTCAAF>2.0.CO;2.

Dillon, J.T., and Ehlig, P.L., 1993, Displacement on the southern San Andreas fault, *in* Powell, R.E., Weldon, R.J., II, and Matti, J.C., eds., The San Andreas Fault System: Displacement, Palinspastic Reconstruction, and Geologic Evolution: Geological Society of America Memoir 178, p. 199–216.

Duebendorfer, E.M., and Chamberlain, K.R., 2002, The Cheyenne belt in the Sierra Madre, Wyoming, is not the Cheyenne belt: It is a ca. 1.4-Ga thrust-tear fault system: Geological Society of America Abstracts with Programs, v. 34, no. 6, abs. 106-3, p. 242.

Duebendorfer, E.M., Chamberlain, K.R., and Jones, C.S., 2001, Proterozoic tectonic history of the Cerbat Mountains, northwestern Arizona: Implications for crustal assembly in the southwestern United States: Geological Society of America Bulletin, v. 113, no. 5, p. 575–590, doi: 10.1130/0016-7606(2001)113<0575:PTHOTC>2.0.CO;2.

Duebendorfer, E., Chamberlain, K., Heizler, M., and Harper, K., 2004, Structural and thermochronologic evidence for a ca. 1.6 Ga contractional event in southern Wyoming: Geological Society of America Abstracts with Programs, v. 36, no. 5, p. 405.

Ehlig, P.L., 1981, Origin and tectonic history of the basement terrane of the San Gabriel Mountains, central Transverse Ranges, *in* Ernst, W.G., ed., The geotectonic development of California: Rubey Volume 1: Englewood Cliffs, New Jersey, Prentice-Hall, p. 253–283.

Eisele, J., and Isachsen, C.E., 2001, Crustal growth in southern Arizona: U-Pb geochronological and Sm-Nd isotopic evidence for addition of the Paleoproterozoic Cochise block to the Mazatzal Province: American Journal of Science, v. 301, p. 773–797.

Erickson, R.C., and Bowring, S.A., 1990, Petrology and geochemistry of the Sommer Gneiss, Dos Cabezas Mountains, Arizona: Dating the Mazatzal orogeny: Geological Society of America Abstracts with Programs, v. 22, no. 3, p. 21.

Hawkins, D.P., Bowring, S.A., Ilig, B.R., Karlstrom, K.E., and Williams, M.L., 1996, U-Pb geochronologic constraints on the Paleoproterozoic crustal evolution of the Upper Granite Gorge, Grand Canyon, Arizona: Geological Society of America Bulletin, v. 108, p. 1167–1181, doi: 10.1130/0016-7606(1996)108<1167:UPGCOT>2.3.CO;2.

Haxel, G.B., Tosdal, R.M., May, D.J., and Wright, J.E., 1984, Latest Cretaceous and early Tertiary orogenesis in south-central Arizona: Thrust faulting, regional metamorphism, and granitic plutonism: Geological Society of America Bulletin, v. 95, p. 631–653, doi: 10.1130/0016-7606(1984)95<631:LCAETO>2.0.CO;2.

Heaman, L.M., and Grotzinger, J.P., 1992, 1.08 Ga diabase sills in the Pahrump Group, California: Implications for development of the Cordilleran margin: Geology, v. 20, p. 637–640, doi: 10.1130/0091-7613(1992)020<0637:GDSITP>2.3.CO;2.

Hoskin, P.W.O., and Black, L.P., 2000, Metamorphic zircon formation by solid-state recrystallization of protolith igneous zircon: Journal of Metamorphic Geology, v. 18, p. 423–439.

Hoskin, P.W.O., and Schaltegger, U., 2003, Chapter 2: The composition of zircon and igneous and metamorphic petrogenesis, *in* Hanchar, J.M., and Hoskin, P.W.O., eds., Zircon: Reviews in Mineralogy and Geochemistry, v. 53, p. 27–62.

Howard, K.E., 1991, Intrusion of horizontal dikes: Tectonic significance of Middle Proterozoic diabase sheets widespread in the upper crust of the southwestern United States: Journal of Geophysical Research, v. 96, no. B7, p. 12,461–12,478.

Iriondo, A., 2001, Proterozoic basements and their Laramide juxtaposition in northwestern Sonora, Mexico: Tectonic constraints on the southwestern margin of Laurentia [Ph.D. thesis]: Boulder, Colorado, University of Colorado, 222 p.

Iriondo, A., and Premo, W.R., 2003, The Caborca block: An inhomogeneous piece of Paleoproterozoic crust in Sonora: Geological Society of America Abstracts with Programs, v. 35, no. 4, p. 67.

Iriondo, A., Premo, W.R., Martinez-Torres, L.M., Budahn, J.R., Atkinson, W.W., Siems, D.F., and Guaras-Gonzalez, B., 2004, Isotopic, geochemical, and temporal characterization of Proterozoic basement rocks in the Quitovac region, northwestern Sonora, Mexico: Implications for reconstruction of the southwestern margin of Laurentia: Geological Society of America Bulletin, v. 116, no. 1–2, p. 154–170.

Jones, N.W., McKee, J.W., Anderson, T.H., and Silver, L.T., 1995, Jurassic volcanic rocks and northeastern Mexico: A possible remnant of a Cordilleran magmatic arc, *in* Jacques-Ayala, C., González-León, C., and Roldán-Quintana, J., eds., Studies of Mesozoic geology in Sonora, Mexico and adjacent regions: Geological Society of America Special Paper 301, p. 179–190.

Karlstrom, K.E., and Bowring, S.A., 1988, Early Proterozoic assembly of tectonostratigraphic terranes in southwestern North America: Journal of Geology, v. 96, p. 561–576.

Karlstrom, K.E., Bowring, S.A., and Conway, C.M., 1987, Tectonic significance of an early Proterozoic two-province boundary in central Arizona: Geological Society of America Bulletin, v. 99, p. 529–538, doi: 10.1130/0016-7606(1987)99<529:TSOAEP>2.0.CO;2.

Karlstrom, K.E., Doe, M.F., Wessels, R.L., Bowring, S.A., Dann, J.C., and Williams, M.L., 1990, Juxtaposition of Proterozoic crustal blocks: 1.65–1.60 Ga Mazatzal orogeny, *in* Gehrels, G.E., and Spencer, J.E., eds., Geologic excursions through the Sonora desert region, Arizona and Sonora: Arizona Geological Survey Special Paper 7, p. 114–123.

Karlstrom, K.E., Harlan, S.S., Williams, M.L., McLelland, J., Geissman, J.W., and Åhäll, K.-I., 1999, Refining Rodinia: Geologic evidence for the Australian–Western U.S. connection in the Proterozoic: GSA Today, v. 9, no. 5, p. 1–7.

Karlstrom, K.E., Bowring, S.A., Dehler, C.M., Knoll, A.H., Porter, S.M., Sharp, Z., Des Marais, D.J., Weil, A.B., Geissman, J.W., Elrick, M., Timmons, M.J., Keefe, K., and Crossey, L.J., 2000, Chuar Group of the Grand Canyon: Record of breakup of Rodinia, associated change in the global carbon cycle, and ecosystem expansion by 740 Ma: Geology, v. 28, p. 619–622, doi: 10.1130/0091-7613(2000)028<0619:CGOTGC>2.3.CO;2.

Kistler, R.W., and Peterman, Z.E., 1973, Variations in Sr, Rb, K, Na and initial $^{87}Sr/^{86}Sr$ in Mesozoic granitic rocks and intruded wall rocks in central

California: Geological Society of America Bulletin, v. 84, p. 3489–3512, doi: 10.1130/0016-7606(1973)84<3489:VISRKN>2.0.CO;2.

Krogh, T.E., 1978, Vapour transfer for the dissolution of zircons in a multi-sample capsule at high-pressure, *in* Zartman, R.E., ed., Short papers of the Fourth International Conference, Geochronology, Cosmochronology, Isotope Geology: U.S. Geological Survey Open-File Report 78–701, p. 233–234.

Li, Z.X., Zhang, L., and Powell, C.McA., 1995, South China in Rodinia: Part of the missing link between Australia–East Antarctica and Laurentia?: Geology, v. 23, p. 407–410, doi: 10.1130/0091-7613(1995)023<0407:SCIRPO>2.3.CO;2.

Li, Zheng-Xiang, Li, Xian-hua, Zhou, Hanwen, and Kinny, P.D., 2002, Grenvillian continental collision in south China: New SHRIMP U-Pb zircon results and implications for the configuration of Rodinia: Geology, v. 30, p. 163–166, doi: 10.1130/0091-7613(2002)030<0163:GCCISC>2.0.CO;2.

Ludwig, K.R., 1980, Calculation of uncertainties of U-Pb data: Earth and Planetary Science Letters, v. 46, p. 212–220, doi: 10.1016/0012-821X(80)90007-2.

Ludwig, K.R., 1985, PBDAT200: A computer program for processing raw Pb-U-Th isotope data: U.S. Geological Survey Open-File Report 85–547, 54 p.

Ludwig, K.R., 1993, ANALYST: VER. 2.20: A computer program for control of a thermal-ionization, single-collector mass spectrometer, user's manual: U.S. Geological Survey Open-File Report 92-543, revised Dec. 7, 1993, 94 p.

Ludwig, K.R., 2001a, ISOPLOT/Ex, version 2.49: A geochronological toolkit for Microsoft Excel: Berkeley Geochronology Center, Special Publication 1a, 56 p.

Ludwig, K.R., 2001b, SQUID, version 1.00: A user's manual: Berkeley Geochronology Center, Special Publication 2, 17 p.

Lugmair, G.W., Scheinin, N.B., and Marti, K., 1975, Sm-Nd age and history of Apollo 17 basalt 75075: Evidence for early differentiation of the lunar exterior: Proceedings of the 6th Lunar Science Conference: New York, Pergamon Press, p. 1419–1429.

Lynch, D.J., and Gutman, J.T., 1987, Volcanic structures and alkaline rocks in the Pinacate volcanic field of Sonora, Mexico, *in* Davis, G.H., and Vanden Dolder, E.M., eds., Geological diversity of Arizona and its margins: Excursions to choice areas: Arizona Bureau of Geology and Mineral Technology, Geological Survey Branch, Special Paper 5, p. 309–322.

Moores, E.M., 1991, Southwest U.S.–East Antarctic (SWEAT) connection: A hypothesis: Geology, v. 19, p. 425–428, doi: 10.1130/0091-7613(1991)019<0425:SUSEAS>2.3.CO;2.

Nourse, J.A., Iriondo, A., Premo, W.R., Kunk, M.J., and Stahl, E.R., 2000, New constraints on the southwestern margin of Laurentia from the Pinacate–Cabeza Prieta basement complex: Sonora and Arizona: Geological Society of America Abstracts with Programs, v. 32, no. 7, p. A-318.

Page, R.W., Jackson, M.J., and Krassay, A.A., 2000, Constraining sequence stratigraphy in north Australian basins: SHRIMP U-Pb zircon geochronology between Mt. Isa and McArthur River: Australian Journal of Earth Sciences, v. 47, p. 431–459, doi: 10.1046/j.1440-0952.2000.00797.x.

Powell, R.E., 1982, Crystalline basement terranes in the southern eastern Transverse ranges, California, *in* Cooper, J.D., compiler, Geologic Excursions in the Transverse Ranges, Field Trip Guidebook for the 78th Cordilleran section meeting of GSA, Anahaeim, California, p. 109–151.

Powicki, D.A., Williams, M.L., and Bowring, S.A., 1993, The four peaks Shear Zone: Timing constraints on the early Proterozoic Mazatzal orogeny in central Arizona: Geological Society of America Abstracts with Programs, v. 25, no. 6, p. A-48.

Premo, W.R., and Fanning, C.M., 2000, SHRIMP U-Pb zircon ages for Big Creek gneiss, Wyoming and Boulder Creek batholith, Colorado: Implications for timing of Paleoproterozoic accretion of the northern Colorado province: Rocky Mountain Geology, v. 35, no. 1, p. 31–50, doi: 10.2113/35.1.31.

Premo, W.R., and Loucks, R.R., 2000, Age and Pb-Sr-Nd isotopic systematics of plutonic rocks from the Green Mountain magmatic arc, southeastern Wyoming: Isotopic characterization of a Paleoproterozoic island arc system: Rocky Mountain Geology, v. 35, no. 1, p. 51–70, doi: 10.2113/35.1.51.

Premo, W.R., and Tatsumoto, M., 1991, U-Th-Pb isotopic systematics of Lunar Norite 78235: Proceedings of Lunar and Planetary Science Conference, v. 21, Lunar and Planetary Institute, p. 89–100.

Premo, W.R., and Tatsumoto, M., 1992, U-Th-Pb, Rb-Sr, and Sm-Nd isotopic systematics of Lunar Troctolitic Cumulate 76535: Implications on the age and origin of this early lunar, deep-seated cumulate: Proceedings of Lunar and Planetary Science Conference, v. 22, p. 381–397.

Premo, W.R., and Van Schmus, W.R., 1989, Zircon geochronology of Precambrian rocks in southeastern Wyoming and northern Colorado, *in* Grambling, J.A., and Tewksbury, B.J., eds., Proterozoic geology of the southern Rocky Mountains: Geological Society of America Special Paper 235, p. 13–32.

Premo, W.R., Tatsumoto, M., and Wang, J.-W., 1989, Pb isotopes in anorthositic breccias 67075 and 62237: A search for primitive lunar lead: Proceedings of the Nineteenth Lunar and Planetary Science Conference, Lunar and Planetary Institute Houston, Texas, p. 61–71.

Premo, W.R., Iriondo, A., and Nourse, J.A., 2003, U-Pb zircon geochronology of Paleoproterozoic basement in Northwestern Sonora, Mexico: Evidence for affinity to the SW U.S. provinces: Geological Society of America Abstracts with Programs, v. 35, no. 4, p. 67.

Rämö, O.T., and Calzia, J.P., 1998, Nd isotopic composition of cratonic rocks in the southern Death Valley region: Evidence for a substantial Archean source component in Mojavia: Geology, v. 26, p. 891–894, doi: 10.1130/0091-7613(1998)026<0891:NICOCR>2.3.CO;2.

Reynolds, S.J., 1998, Geologic map of Arizona: Arizona Geological Survey Map 26, scale 1:1,000,000.

Reynolds, S.J., Spencer, J.E., Richard, S.M., and Laubach, S.E., 1986a, Mesozoic structures in west central Arizona: Arizona Geological Society Digest, v. 16, p. 35–51.

Reynolds, S.J., Shaqfiqullah, M., Damon, P.E., and DeWitt, E., 1986b, Early Miocene mylonitization and detachment faulting, South Mountains, central Arizona: Geology, v. 14, p. 283–286, doi: 10.1130/0091-7613(1986)14<283:EMMADF>2.0.CO;2.

Richard, S.M., 1994, Reconstruction of Miocene extension in the Arizona Basin and Range and adjacent areas: Arizona Geological Survey Open-File Report 94-5, 9 p., 2 figs., 1 plate, scale 1:500,000.

Ross, G.M., McMechan, M.E., and Hein, F.J., 1989, Proterozoic history: The birth of the miogeocline, *in* Ricketts, B.D., ed., Western Canada sedimentary basin: A case history: Calgary, Alberta, Canadian Society of Petroleum Geologists, p. 79–104.

Rubatto, D., 2002, Zircon trace element geochemistry; partitioning with garnet and the link between U-Pb ages and metamorphism: Chemical Geology, v. 184, no. 1–2, p. 123–138, doi: 10.1016/S0009-2541(01)00355-2.

Rubatto, D., and Gebauer, D., 2000, Use of cathodoluminescence for U-Pb zircon dating by ion microprobe; some examples from the Western Alps, *in* Pagel, M., Barbin, V., Blanc, P., and Ohnenstetter, D., eds., Cathodoluminescence in geosciences: Berlin, Springer, p. 373–400.

Sears, J.W., and Price, R.A., 2000, New look at the Siberian connection: No SWEAT: Geology, v. 28, no. 5, p. 423–426, doi: 10.1130/0091-7613(2000)028<0423:NLATSC>2.3.CO;2.

Shafiqullah, M., Damon, P.E., Lynch, D.J., Reynolds, S.J., Rehrig, W.A., and Raymond, R.H., 1980, K-Ar geochronology and geologic history of southwestern Arizona and adjacent areas: Arizona Geological Society Digest, v. XII, p. 201–260.

Shaw, C.A., Snee, L.W., Selverstone, J., and Reed, J.C., Jr., 1999, $^{40}Ar/^{39}Ar$ thermochronology of Mesoproterozoic metamorphism in the Colorado Front Range: Journal of Geology, v. 107, p. 49–67, doi: 10.1086/314335.

Silver, L.T., 1965, Mazatzal orogeny and tectonic episodicity: Geological Society of America Special Paper 82, p. 185–186.

Silver, L.T., 1971, Problems of crystalline rocks of the Transverse Ranges: Geological Society of America Abstracts with Programs, v. 3, p. 193.

Silver, L.T., 1978, Precambrian formations and Precambrian history in Cochise County, southeastern Arizona: 29th Field Conference Guidebook, New Mexico Geological Society, p. 157–163.

Silver, L.T., and Anderson, T.H., 1974, Possible left-lateral early to middle Mesozoic truncation of the southwestern North American craton: Geological Society of America Abstracts with Programs, v. 6, p. 955.

Spencer, J.E., and Reynolds, S.J., 1986, Some aspects of middle Tertiary tectonics of Arizona and southeastern California, *in* Beatty, B., and Wilkinson, P.A.K., eds., Frontiers in the geology and ore deposits of Arizona and the Southwest: Arizona Geological Society Digest, v. 16, p. 102–107.

Stacey, J.S., and Kramers, J.D., 1975, Approximation of terrestrial lead isotope evolution by a two-stage model: Earth and Planetary Science Letters, v. 26, p. 207–221, doi: 10.1016/0012-821X(75)90088-6.

Stanley, G.D., and González-León, C.M., 1995, Paleogeographic and tectonic implications of Triassic fossils and strata from the Antimonio Formation, northwestern Sonora, *in* Jacques-Ayala, C., González-León, C. M., and Roldán-Quintana, J., eds., Studies on the Mesozoic of Sonora and adjacent areas: Geological Society of America Special Paper 301, p. 1–17.

Steiger, R.H., and Jäger, E., 1977, Subcommission on geochronology: Convention on the use of decay constants in geo- and cosmochronology: Earth and Planetary Science Letters, v. 36, p. 359–362, doi: 10.1016/0012-821X(77)90060-7.

Stewart, J.H., 1972, Initial deposits of the Cordilleran geosyncline: Evidence of a Late Precambrian (<850 Ma) continental separation: Geological Society of America Bulletin, v. 83, p. 1345–1360.

Stewart, J.H., Amaya-Martinez, R., and Palmer, A.R., 2002, Neoproterozoic and Cambrian strata of Sonora, Mexico: Rodinian supercontinent to Laurentian Cordilleran margin, *in* Barth, A.P., ed., Contributions to crustal evolution of the Southwestern United States: Geological Society of America Special Paper 365, p. 5–48.

Stone, P., and Stevens, C.H., 1988, Pennsylvanian and Early Permian paleogeography of east-central California: Implications for the shape of the continental margin and timing of continental truncation: Geology, v. 16, p. 330–333, doi: 10.1130/0091-7613(1988)016<0330:PAEPPO>2.3.CO;2.

Strickland, D., Chamberlain, K., and Duebendorfer, E., 2004, New U-Pb dates of syn-deformational minerals that directly date multiple tectonic events at 1.75 and 1.62 Ga along the Cheyenne belt suture zone, southeastern Wyoming: Geological Society of America Abstracts with Programs, v. 36, no. 5, p. 404.

Tatsumoto, M., and Unruh, D.M., 1976, KREEP basalt age: Grain by grain U-Th-Pb systematics study of the quartz monzodiorite clast 15405, 88: Proceedings of the 7th Lunar and Planetary Science Conference, p. 2107–2129.

Timmons, M.J., Karlstrom, K.E., Dehler, C.M., Geissman, J.W., and Heizler, M.T., 2001, Proterozoic multistage (~1.1 and ~0.8 Ga) extension in the Grand Canyon Supergroup and establishment of northwest and north-south tectonic grains in the southwestern United States: Geological Society of America Bulletin, v. 113, p. 163–181, doi: 10.1130/0016-7606(2001)113<0163:PMCAGE>2.0.CO;2.

Todt, W., Cliff, R.A., Hanser, A., and Hofmann, A.W., 1993, Recalibration of NBS lead standards using a ^{202}Pb + ^{205}Pb double spike: 7th Meeting of the European Union of Geosciences Abstracts, April 4–8th, 1993, Strasbourg, p. 96.

Watkins, T., 2003, Geologic history of Proterozoic basement in the area of Desierto de Altar, Sonora, Mexico, near Sierra Los Ojos; with possible relations to the Rodinia supercontinent [senior thesis]: California State Polytechnic University, Pomona, California, 42 p.

Williams, I.S., 1998, U-Th-Pb geochronology by ion microprobe, *in* McKibben, M.A., and Shanks, W.C., eds., Applications of microanalytical techniques to understanding mineralizing processes: Reviews in Economic Geology, v. 7, p. 1–35.

Williams, I.S., and Claesson, S., 1987, Isotopic evidence for the Precambrian provenance and Caledonian metamorphism of high-grade paragneisses from the Seve Nappes, Scandinavian Caledonides. II. Ion microprobe zircon U-Th-Pb: Contributions to Mineralogy and Petrology, v. 97, p. 205–217, doi: 10.1007/BF00371240.

Withnell, I.W., Draper, J.J., Mackenzie, D.E., Knutson, J., Blewett, R.S., Hutton, L.J., Bultitude, R.J., Wellman, P., McConachie, B.A., Bain, J.H.C., Donchak, P.J.T., Lang, S.C., Domaglia, J., Symonds, P.A., and Rienks, I.P., 1997, Review of geologic provinces and basins of north Queensland, Chapter 14, *in* Bain, J.H.C., and Draper, J.J., eds., North Queensland geology: Australian Geological Survey Organisation (AGSO) Bulletin 240, Queensland Geology 9, p. 449–470.

Wooden, J.L., 2000, What is the Mojave crustal province: Geological Society of America Abstracts with Programs, v. 32, no. 7, p. A-318.

Wooden, J.L., and DeWitt, E., 1991, Pb isotopic evidence for the boundary between the rarely Proterozoic Mojave and Central Arizona crustal provinces in western Arizona, *in* Karlstrom, K.E., ed., Proterozoic geology and ore deposits of Arizona: Arizona Geological Society Digest, v. 19, p. 27–50.

Wooden, J.L., and Miller, D.M., 1990, Chronologic and isotopic framework for early Proterozoic crustal evolution in the eastern Mojave Desert region, SE California: Journal of Geophysical Research, v. 95, no. B12, p. 20,133–20,146.

York, D., 1969, Least-squares fitting of a straight line with correlated errors: Earth and Planetary Science Letters, v. 5, p. 320–324.

Manuscript Accepted by the Society 24 March 2005

Geological Society of America
Special Paper 393
2005

Paleoproterozoic Mojave province in northwestern Mexico? Isotopic and U-Pb zircon geochronologic studies of Precambrian and Cambrian crystalline and sedimentary rocks, Caborca, Sonora

G. Lang Farmer*
CIRES and Department of Geological Sciences, University of Colorado, Boulder, Colorado 80309, USA

Samuel A. Bowring
Jennifer Matzel
Department of Earth, Atmospheric, and Planetary Sciences, Massachusetts Institute of Technology, Cambridge, Massachusetts 03319, USA

Guadalupe Espinosa Maldonado
Departamento de Geología, Universidad de Sonora, Hermosillo, Sonora, México

Christopher Fedo
Department of Geology, George Washington University, Washington, D.C. 20052, USA

Joe Wooden
U.S. Geological Survey, 345 Middlefield Road, Menlo Park, California 90425, USA

ABSTRACT

Whole-rock Nd isotopic data and U-Pb zircon geochronology from Precambrian crystalline rocks in the Caborca area, northern Sonora, reveal that these rocks are most likely a segment of the Paleoproterozoic Mojave province. Supporting this conclusion are the observations that paragneiss from the ≥1.75 Ga Bamori Complex has a 2.4 Ga Nd model age and contains detrital zircons ranging in age from Paleoproterozoic (1.75 Ga) to Archean (3.2 Ga). Paragneisses with similar age and isotopic characteristics occur in the Mojave province in southern California. In addition, "A-type" granite exposed at the southern end of Cerro Rajon has ca 2.0 Ga Nd model age and a U-Pb zircon age of 1.71 Ga, which are similar to those of Paleoproterozoic granites in the Mojave province. Unlike the U.S. Mojave province, the Caborcan crust contains ca. 1.1 Ga granite (Aibo Granite), which our new Nd isotopic data suggest is largely the product of anatexis of the local Precambrian basement. Detrital zircons from Neoproterozoic to early Cambrian miogeoclinal arenites at Caborca show dominant populations ca. 1.7 Ga, ca. 1.4 Ga, and ca. 1.1 Ga, with subordinate Early

*farmer@cires.colorado.edu.

Farmer, G.L., Bowring, S.A., Matzel, J., Espinosa Maldonado, G., Fedo, C., and Wooden, J., 2005, Paleoproterozoic Mojave province in northwestern Mexico? Isotopic and U-Pb zircon geochronologic studies of Precambrian and Cambrian crystalline and sedimentary rocks, Caborca, Sonora, *in* Anderson, T.H., Nourse, J.A., McKee, J.W., and Steiner, M.B., eds., The Mojave-Sonora megashear hypothesis: Development, assessment, and alternatives: Geological Society of America Special Paper 393, p. 183–198. doi: 10.1130/2005.2393(05). For permission to copy, contact editing@geosociety.org.

Cambrian and Archean zircons. These zircons were likely derived predominately from North American crust to the east and northeast, and not from the underlying Caborcan basement. The general age and isotopic similarities between Mojave province basement and overlying miogeoclinal sedimentary rocks in Sonora and southern California is necessary, but not sufficient, proof of the hypothesis that Sonoran crust is allochthonous and was transported to its current position during the Mesozoic along the proposed Mojave-Sonora megashear. One viable alternative model is that the Caborcan Precambrian crust is an isolated, autochthonous segment of Mojave province crust that shares a similar, but not identical, Proterozoic geological history with Mojave province crust found in the southwest United States

Keywords: Precambrian geology, Nd isotopes, zircon geochronology, siliciclastic sedimentary rocks, Sonora.

INTRODUCTION

Paleoproterozoic crystalline basement and overlying Neoproterozoic and Paleozoic miogeoclinal sedimentary rocks in northern Sonora have long been recognized (Cooper and Arellano, 1946; Giletti and Damon, 1961; Damon et al., 1962; Anderson and Silver, 1970; Merrian and Eells, 1979; Stewart et al., 1984), but the relationship, if any, between these and similar rocks in the southwestern United States remains unresolved. For example, Paleoproterozoic basement rocks in the southwest United States have been subdivided into three major provinces (Mojave, Yavapai, and Mazatzal) on the basis of their age and Nd and Pb isotopic compositions (Silver, 1965; Bennett and DePaolo, 1987; Karlstrom and Bowring, 1988; Wooden and Miller, 1990). However, a paucity of isotopic data from basement rocks in Sonora has prevented the correlation to Paleoproterozoic rocks in western North America (the exception being the Quitovac area in northern Sonora, Figure 1; Iriondo et al., 2004). In addition, it is not clear that lithologic similarities between Neoproterozoic and early Paleozoic miogeoclinal sedimentary rocks in Sonora and the southwest United States necessarily requires that they have been deposited in close proximity. These similarities could instead simply reflect accumulation of the original sediments in similar depositional environments (Stewart et al., 1984). This issue remains unresolved despite detailed U-Pb studies of detrital zircons and whole-rock Nd studies of miogeoclinal sedimentary rocks in both Sonora and the southwest United States (Gehrels and Stewart, 1998; Farmer et al., 2001).

The primary motivation for further assessing the nature of both the Sonoran basement rocks and miogeoclinal sedimentary rocks is the possibility that both may have been displaced as much as 1000 km during the Phanerozoic (either in late Paleozoic or in Jurassic) along the proposed left-lateral Mojave-Sonora megashear (Silver and Anderson, 1974; Stewart, 1988). Silver and Anderson (1974) proposed that the Sonoran Precambrian rocks formed directly adjacent to rocks currently exposed west of Death Valley, California. However, other workers have argued that Precambrian and early Paleozoic rocks in Sonora are autochthonous and simply represent the intact continuation of the western margin of Laurentia into northern Mexico (Taylor et al., 1991; Molina-Garza and Geissman, 1999).

In this study, we attempt to further assess these hypotheses. We obtained Nd isotopic data and U-Pb zircon ages from several basement rocks exposed in the Caborca area in northern Sonora (Fig. 1) and detrital U-Pb zircon age spectra from a set of Neoproterozoic to early Cambrian miogeoclinal sedimentary rocks for which whole-rock Nd isotopic data were already available (Farmer et al., 2001). Our data reveal that Caborca basement rocks may represent a segment of the Mojave province crust, equivalent to Paleoproterozoic rocks in westernmost Arizona and southeastern California, and that these basement rocks could represent the source rocks of the oldest portions of the miogeocline exposed at Caborca. The occurrence of a segment of the Mojave province in the Caborca area supports but does not prove the existence of the Mojave-Sonora megashear.

Geologic Setting

The Precambrian basement in northwestern Sonora (hereafter referred to as the Caborca "block" after Campa and Coney, 1983) consists of Paleoproterozoic gneiss intruded by ca. 1.66 Ga to 1.75 Ga intermediate to silicic composition igneous rocks and abundant ca. 1.4 Ga and rare 1.1 Ga granitic rocks (Anderson and Silver, 1981). In the Caborca area, the oldest exposed rocks are variably metamorphosed supracrustal rocks and small-volume mafic to silicic composition intrusive rocks of the Bamori Complex. An example of the latter intrusive rock yielded a U-Pb zircon age of ca. 1.78 Ga (Anderson and Silver, 1971). Younger granitic rocks exist in the area (Fig. 1) and have been presumed to be ≥1. 7 Ga and/or 1.4 Ga in age (Anderson and Silver, 1981; Stewart et al., 1984), but the only published U-Pb zircon age for these granitic rocks is from the 1.14 Ga micrographic Aibo Granite (Anderson and Silver, 1971). The Aibo Granite belongs to a suite of ca. 1.1 Ga intrusive igneous rocks found in northwestern Sonora (Anderson and Silver, 1997; Iriondo et al., 2004). No published chemical or whole-rock Nd or Pb isotopic data exist for any of the Caborca area Paleoproterozoic or Mesoproterozoic rocks.

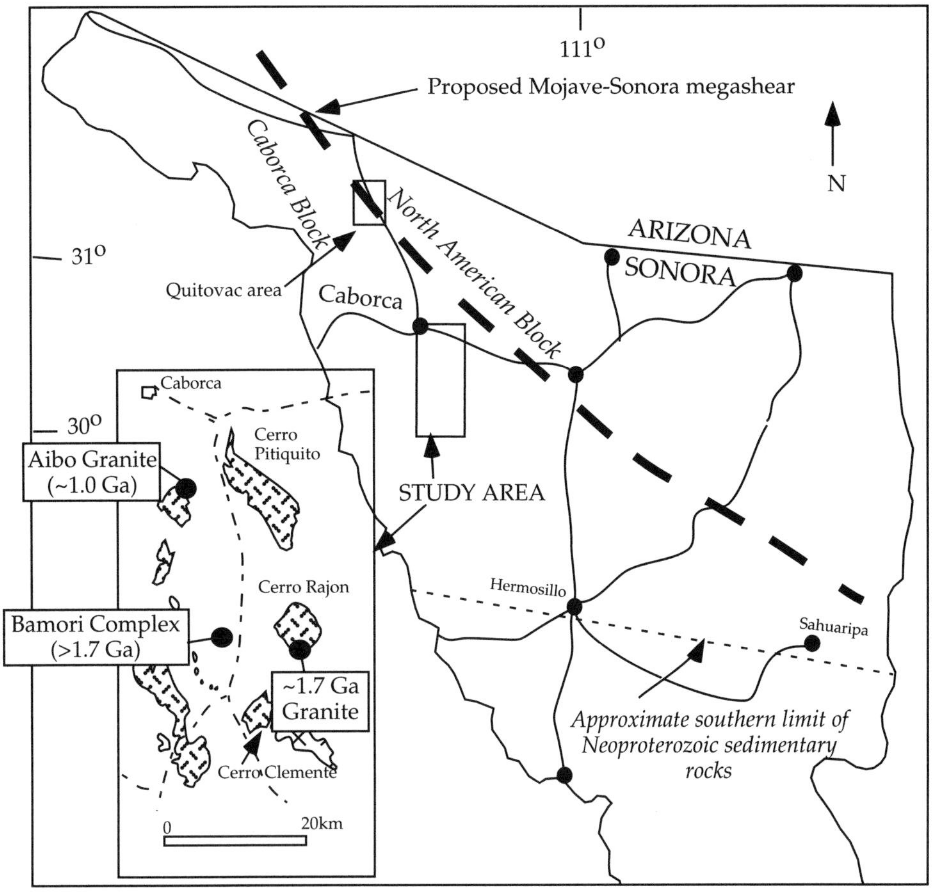

Figure 1. Location of study area, northern Sonora, Mexico. Inset shows locations of Caborca Precambrian basement samples. Modified after Stewart et al. (1984).

Neoproterozoic and Lower Cambrian sedimentary rocks that crop out in the northern half of Sonora are interpreted to be part of a fluvial to shallow-marine, miogeoclinal succession originally deposited on the rifted southwestern margin of Laurentia (Stewart et al., 1984). An ~3 km thick section of this succession exposed just south of Caborca within and near Cerro Rajon (Figs. 1 and 2; Stewart et al., 1984) consists primarily of alternating units of dolomite, dolomitic limestone, quartz arenite, feldspathic arenite, and siltstone. Sandstone of Unit 1 (Fig. 2) of the Puerto Blanco Formation, however, also contains mafic volcaniclastic material and an altered greenstone interpreted as an Early Cambrian basalt flow (Stewart et al., 1984). Paleocurrent directions for miogeoclinal sediments in the Caborca area are variable and reveal no clear indication of the predominant direction of sediment transport, unlike their counterparts in the southwest United States, where paleocurrent directions are well defined (Stewart, 1970). Nevertheless, it is generally considered that the siliciclastic sediments were derived from the erosion of Precambrian rocks to the north and/or east (Stewart et al., 1984). Such a conclusion is consistent with the Nd isotopic compositions of siliciclastic sedimentary rocks in the Caborca sequence, which generally match those of Paleoproterozoic crust in southwestern North America (Farmer et al., 2001).

Stewart et al. (1984) recognized that the miogeoclinal sedimentary sequence in Sonora is similar lithologically to that of Neoproterozoic to Middle Cambrian miogeoclinal sedimentary rocks in the southern Great Basin. In addition, the Nd isotopic compositions of Neoproterozoic siliciclastic sedimentary rocks in the Death Valley and Caborca regions vary in a similar fashion with stratigraphic position (Farmer et al., 2001). In particular, the siliciclastic sedimentary successions in both areas include a prominent Early Cambrian excursion toward higher ε_{Nd} values (>–5), represented in the Death Valley area by the arkosic middle member of the Wood Canyon Formation in Death Valley and at Caborca by the correlative Unit 1 sandstone of the Puerto Blanco Formation (Fig. 2; Farmer et al., 2001). These data are consistent with the concept that the miogeoclinal sequence and underlying Precambrian basement at Caborca are allochthonous and were transported ~800 km to the southeast during the Early Jurassic as a result of left-lateral movement along the proposed Mojave-Sonora megashear (Fig. 1; Anderson and Schmidt, 1983).

SAMPLES

For the purposes of further characterizing the age, composition, and Nd isotopic compositions of Precambrian basement

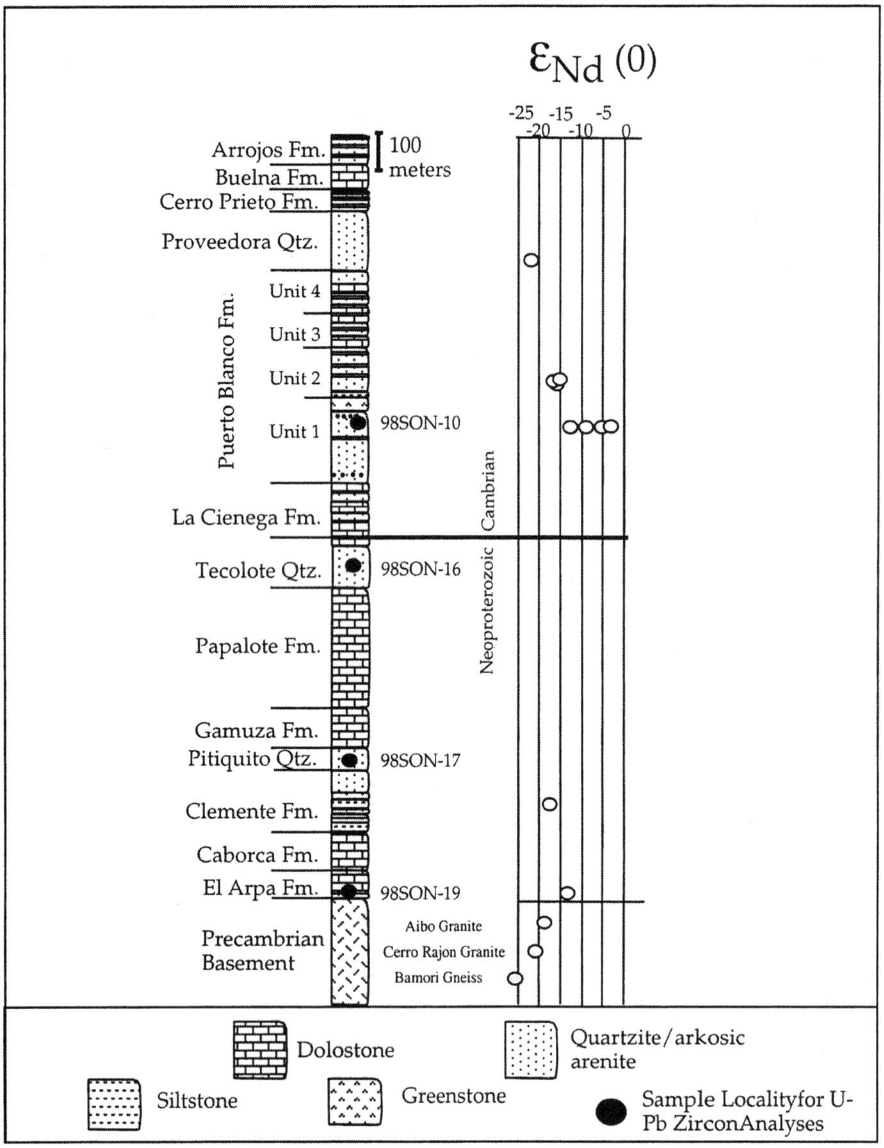

Figure 2. Stratigraphic columns for Neoproterozoic to Cambrian miogeoclinal sedimentary rocks at Caborca (Stewart et al., 1984), ε_{Nd} (0) values for sedimentary rocks (modified from Farmer et al., 2001) and for Precambrian crystalline rocks, and stratigraphic positions of sedimentary rock samples used for zircon U-Pb analyses.

rocks in the Caborca area, we sampled a quartzo-feldspathic gneiss from the Bamori Complex and two granitic bodies; the previously undated granite at the southern end of Cerro Rajon (hereafter referred to as the Cerro Rajon Granite) and the Mesoproterozoic Aibo Granite. The Cerro Rajon Granite sample consists of potassium feldspar phenocrysts set in a finer-grained matrix of quartz and biotite, as well as smaller amounts of Fe-oxides and titanite. The Aibo Granite sample is a micrographic granite typical of this pluton (Anderson and Silver, 1971). For these rocks, we obtained major and trace element compositions, whole-rock Nd and Pb isotopic data, and high-precision (TIMS), single-zircon U-Pb age determinations.

To complement our previous chemical and Nd isotope studies of Neoproterozoic and Cambrian miogeoclinal sedimentary rocks in the Caborca area and to further compare the provenance of miogeoclinal sediments rocks here and in the Death Valley region, we selected four quartz to feldspathic arenites for detrital zircon age determinations (Fig. 2). Low-precision age spectra were obtained using the SHRIMP-RG at Stanford University for Unit 1 (sandstone) of the Puerto Blanco Formation, the Tecolote Quartzite, the Pitiquito Quartzite, and the El Arpa Formation (feldspathic arenite), which rest unconformably on the Cerro Rajon Granite. High-precision U-Pb age determinations were also obtained from five to ten grains from each of these units.

ANALYTICAL TECHNIQUES

Whole-rock samples from the Precambrian basement rocks were submitted to a commercial laboratory (Activation Laboratories) for major and trace element analyses (Table 1). All of the Sm

TABLE 1. CHEMICAL COMPOSITIONS OF PRECAMBRIAN GNEISS AND GRANITIC ROCKS, CABORCA AREA*

Sample:	DIC 98-1	98-SON-20	98-SON-1	Reproducibilities***
Location	Aibo	Cerro Rajon	Bamori Complex	
UTM(E)†	392385	408588	399805	
UTM(N)	3374830	3363085	3357366	
Rock Type	Alkalic granite	Porphyritic Granite	Paragneiss	
Age§	1.1	1.7	>1.7	
SiO_2	77.1	65.9	85.8	
TiO_2	0.2	1.0	0.2	
Al_2O_3	11.7	15.1	7.65	
MnO	0.01	0.06	0.01	
MgO	0.28	2.05	0.53	
Fe_2O_3(t)	1.57	6.04	1.34	
CaO	0.08	1.24	0.17	
Na_2O	1.4	2.9	1.8	
K_2O	7.6	5.2	2.5	
P_2O_5	0.03	0.39	0.07	
Total	98.9	97.2	98.9	
mol Al_2O_3/ ($CaO+Na_2O+K_2O$)	1.09	1.18	1.28	
FeOt/(FeOt+MgO)	0.83	0.73	0.70	
Trace elements**				
Ba	271	1420	681	5
Cs	3.55	2.26	0.75	5
Co	b.d.††	10.5	3.39	10
Hf	8.39	13.5	5.24	7
Nb	27.3	25.0	8.71	4
Rb	348	204	75.1	2
Sc	2.00	15.00	2.00	6
Sr	29.2	175	47.3	15
Ta	2.26	1.74	0.28	9
Th	49.7	14.0	5.64	1
U	8.99	3.07	1.01	2
Y	89.8	58.8	12.0	19
Zn	b.d.	91.1	b.d.	23
Zr	215	482	176	10
Pb	19.3	19.2	5.61	17
La	139	93.6	16.8	7
Ce	242	180	30.1	3
Pr	28.6	20.9	3.46	3
Nd	104	80.2	12.9	3
Sm	19.2	14.9	2.55	3
Eu	1.19	2.86	0.68	4
Gd	15.3	11.8	2.07	7
Tb	2.64	2.01	0.35	1
Dy	15.4	11.1	2.05	3
Ho	3.01	2.04	0.40	3
Er	8.56	5.73	1.15	5
Tm	1.34	0.86	0.19	2
Yb	8.41	5.31	1.25	3
Lu	1.20	0.78	0.19	5
$(La/Yb)_N$	10.7	11.4	8.66	
Eu/Eu* §§	0.21	0.67	0.92	

*Major element data by ICP-MS through Activation Laboratories. Oxide percentages renormalized to 100% (anhydrous). Actual analyses totals reported in final column. Total iron reported as Fe_2O_3.

†Sample locations given in UTM coordinates, zone 12, using World Geodetic System of 1984 (WGS 84) as map datum.

§Ages are in b.y. and are from this study.

**Trace element concentrations in parts per million (ppm).

††Below detection

§§Eu/Eu*= [Eu]/([Gd]*[Sm])^0.5

***External reproducibilities estimated from multiple standard analyses during study period (presented as % of mean standard concentrations) and Nd isotopic dilution determinations and the Nd and Pb whole-rock isotopic analyses were conducted at the University of Colorado, Boulder, using analytical procedures described in Table 2.

Lower-precision U-Pb zircon dates were obtained using the SHRIMP-RG (sensitive high-resolution ion microprobe with reverse geometry) at the Stanford–U.S. Geological Survey Microisotopic Analytical Center. For each of the four samples analyzed, an epoxy grain mount of at least 60 individual zircon grains was prepared. No attempt was made to sort zircons by color or morphology and each mount is considered to represent a random sampling of the zircons present in the sample. Details of the SHRIMP-RG analytical techniques are described elsewhere (DeGraaff-Surpless et al., 2002). The analytical data are presented as Supplementary Data. The detrital zircon age data are plotted on cumulative probability curves calculated using Isoplot/Ex (Fig. 3; Ludwig, 2000). In these diagrams, each age measurement plots as a normal distribution with a 2σ uncertainty, and the cumulative curves represent a summation of each individual normal distribution. For most of the samples, two cumulative curves are shown, one for all grains analyzed and one for those grains with <10% discordancy, as calculated through $^{207}Pb/^{206}Pb$, $^{206}Pb/^{238}U$, and $^{207}Pb/^{238}U$ ages as plotted on a Weatherill concordia.

Techniques for zircon separation and dissolution in preparation for high-precision TIMS U-Pb, and the details of the TIMS analyses themselves, are given in the Appendix and in Table 3.

RESULTS

Precambrian Crystalline Basement

The gneiss sample from the Bamori Complex (SON98-1) is a siliceous (86 wt% SiO_2; Fig. 4) highly peraluminous rock (molecular $Al_2O_3/(CaO+Na_2O+K_2O)$, or A/CNK = 1.3; Table 1). It has low overall rare earth element (REE) and large ion lithophile (LIL) abundances and a low $(La/Yb)_N$ (8.7; Table 1) relative to the other basement samples. The whole-rock ε_{Nd} value for this sample is low (–25; Fig. 5) and yields a Paleoproterozoic Nd model age (T_{DM} = 2.4 Ga; Table 2). Whole-rock Pb isotopic compositions are unradiogenic ($^{208}Pb/^{204}Pb$ = 37.95, $^{207}Pb/^{204}Pb$ = 15.625, and $^{206}Pb/^{204}Pb$ = 18.259; Fig. 6).

Zircon from the Bamori gneiss defines two populations based on morphology. One population includes elongate, euhdral to subhedral prisms, and the other population consists of equant, multifaceted prisms; however, neither grain shape corresponds to a distinct age population. Analyses of eight zircons yielded a wide range of Pb-Pb dates from 1748.9 ± 1.3 Ma to 3220.8 ± 0.7 Ma with varying degrees of discordance (0.1–4.5%) (Table 3; Fig. 7A). Only one error ellipse overlaps concordia at 1845.1 ± 0.8 Ma.

The Cerro Rajon Granite (SON98-20) is potassic (Fig. 4) and moderately peraluminous, with a high FeO/FeO+MgO ratio (0.73), high light rare earth element (LREE) abundances, and moderate negative Eu anomaly (Eu*/Eu = 0.7; Table 1). All of these chemical characteristics are those attributed to "A-type" granitic rocks (Collins et al., 1982). The measured whole-rock ε_{Nd} value is low (–21; T_{DM} = 2.0 Ga; Fig. 5) and this granite had the least radiogenic Pb of any of the analyzed basement rock samples ($^{208}Pb/^{204}Pb$ = 37.424, $^{207}Pb/^{204}Pb$ = 15.56, and $^{206}Pb/^{204}Pb$ = 17.652; Fig. 6).

Zircon from the Cerro Rajon Granite forms a single population of elongate, euhedral to subhedral prisms. Nine zircon analyses yield a restricted range of Pb-Pb dates from 1687.8 ± 1.8 Ma to 1701.5 ± 1.0 Ma (Table 3; Fig. 7B). Discordance varies from 6.4% to –0.2%; however, it is not possible to obtain a statistically significant discordia through these analyses, suggesting multiple Pb loss events. The weighted mean $^{207}Pb/^{206}Pb$ date of the three most concordant analyses is 1699 ± 1.8 Ma (mean square of weighted deviates [MSWD] = 1.46), which we interpret as the best estimate of the crystallization age of this granite.

The Aibo Granite (DIC 98-1) has significantly higher wt% SiO_2 (77) and higher wt% K_2O (7.6) than the Cerro Rajon Granite (Fig. 4). Both the Aibo and Cerro Rajon Granites are

TABLE 2. Nd AND Pd ISOTOPIC COMPOSITIONS OF PRECAMBRIAN GNEISS AND GRANITIC ROCKS, CABORCA AREA

Sample	Sm (ppm)*	Nd (ppm)	$^{147}Sm/^{144}Nd$	$^{143}Nd/^{144}Nd$ (m)†	e_{Nd} (0)§	T_{DM} (Ga)	m**	w**	$^{208}Pb/^{204}Pb$†† (m)	$^{207}Pb/^{204}Pb$ (m)	$^{206}Pb/^{204}Pb$ (m)	DJ§§
DIC 98-1	19.1	98.2	0.118	0.511699 ± 10	–18.3	2.1	32	182	41.515	15.760	20.686	–2.7
98-SON-20	24.5	149	0.099	0.511553 ± 7	–21.2	2.0	9.9	0.47	37.424	15.560	17.652	8.9
98-SON-1	2.5	14.1	0.107	0.511364 ± 6	–24.9	2.4	11	0.51	37.947	15.625	18.259	9.1

*Isotope dilution concentration determinations accurate to ~1% for Rb and Sr, and 0.5% for Sm and Nd. Total procedural blanks averaged ~1ng for Pb and Sr, and 100pg for Nd, during study period.

†Measured $^{143}Nd/^{144}Nd$ normalized to $^{146}Nd/^{144}Nd$ = 0.7219. Analyses were dynamic mode, three-collector measurements. Thirty-three measurements of the La Jolla Nd standard during the study period yielded a mean $^{143}Nd/^{144}Nd$ = 0.511838 ± 8 (2σ mean).

§εNd values calculated using a present-day $^{143}Nd/^{144}Nd$ (CHUR) = 0.512638.

**μ = $^{238}U/^{204}Pb$, ω = $^{232}Th/^{204}Pb$. U, Th, and Pb concentrations from Table 1.

†† Pb isotopic analyses were four-collector static mode measurements. Sixteen measurements of SRM-981 during the study period yielded $^{208}Pb/^{204}Pb$ = 36.56 ± 0.03, $^{207}Pb/^{204}Pb$=15.449 ± 0.008, $^{206}Pb/^{204}Pb$ = 16.905 ± 0.007 (2σ mean). Measured Pb isotope ratios were corrected to SRM-981 values ($^{208}Pb/^{204}Pb$ = 36.721, $^{207}Pb/^{204}Pb$ = 15.491, $^{206}Pb/^{204}Pb$ = 16.937).

§§ΔJ is "Delta Jerome" calculated for galena ore at Jerome, Arizona, using method of Wooden and DeWitt (1991).

LREE enriched ($(La/Yb)_N$ = 10.7 and 11.4, respectively), but Aibo Granite has notably lower Ba and Sr concentrations and a more pronounced negative Eu anomaly (Eu*/Eu = 0.2; Table 1). Measured ε_{Nd} values for the Aibo Granite are somewhat higher than determined for the other basement samples (−18) and the sample contains considerably more radiogenic Pb, reflecting its high μ and ω values (Table 2; Figs. 5 and 6)

Zircons from the Aibo Granite also constitute a single population of elongate, euhedral prisms. Four zircon analyses range from concordant to 0.9% discordant with Pb-Pb dates of 1075.3 ± 1.3 Ma to 1082.3 ± 1.6 Ma (Table 3; Fig. 7C). The two concordant zircons yield ages of 1075.3 ± 1.3 Ma and 1082.3 ± 1.6 Ma. Our preferred interpretation of these results is that the younger concordant analysis represents the crystallization age of the granite and the older concordant analysis is a grain that contains some inherited Pb.

Sedimentary Rocks

TIMS U-Pb Age Determinations

Two zircon populations were analyzed from the Tecolote Quartzite (SON98-16). The first consisted of large (200–300 μm in length) rounded prisms that yielded Pb-Pb dates of 1765.1 ± 7.3 Ma and 1717.9 ± 1.2 Ma (Table 3). The second population consisted of smaller (50–100 μm in length) euhedral prisms, one of which yielded a Pb-Pb date of 1414.8 ± 3.6 Ma. Zircons from the Pitiquito Quartzite (98SON-17) also have rounded grain edges. All analyses from this sample are discordant, but two equant grains yielded Pb-Pb dates of 1774.2 ± 1.7 Ma and 1831.1 ± 1.3 Ma, whereas a third elongate grain yielded a Pb-Pb date of 1315.6 ± 1.3 Ma (Table 3). Zircons from the El Arpa Formation are typically flat, euhedral, elongate prisms that show little evidence of abrasion due to sedimentary transport. Four zircon analyses yielded two concordant results and two slightly discordant (0.5% and 0.8%) results with Pb-Pb dates between 1081.7 ± 1.5 Ma and 1083.0 ± 5.4 Ma (Table 3). The weighted mean of all four $^{206}Pb/^{207}Pb$ dates is 1082.1 ± 1.0 Ma (MSWD = 0.23) and suggests that detrital zircons in this sample were derived from a source terrane of uniform age. Zircons analyzed from Unit 1 of the Puerto Blanco Formation (SON98-10) consisted of elongate prisms and fragments with rounded grain edges. Four of these grains yielded Pb-Pb dates from 1075.3 ± 1.8 Ma to 1096.5 ± 1.9 Ma, two of which were concordant at 1075.3 ± 1.8 Ma and 1091.6 ± 2.3 Ma (Table 3). A fifth zircon from this sample had a Pb-Pb age of 1329.3 ± 1.3 Ma.

Detrital Zircon U-Pb Age Spectra

The two quartz arenite samples, the Tecolote and Pitiquito Quartzites, have similar detrital zircon age spectra, characterized by a dominant peak ca. 1.71 Ga, with smaller peaks ca. 1.4 Ga, and multiple peaks in the range from 1.1 Ga to 1.2 Ga (Fig. 3). These samples also contain several Archean grains, ranging from 2.4 Ga to ca. 3.0 Ga. The zircon age populations ca. 1.7 Ga and ca. 1.4 Ga are those identified by the TIMS analyses, but the

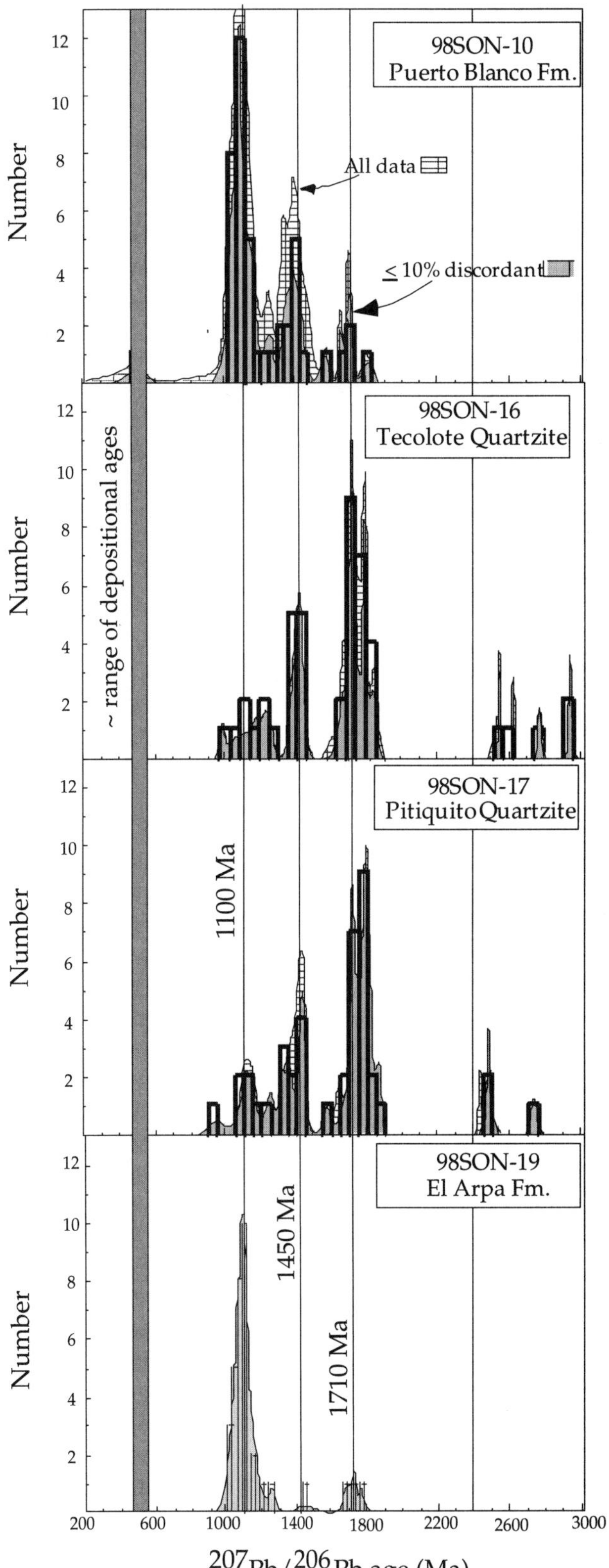

Figure 3. Cumulative probability and histogram plot of SHRIMP-RG detrital zircon ages for selected Caborca miogeoclinal sedimentary rocks (Unit 1, Puerto Blanco Formation; Tecolote Quartzite; Pitiquito Quartzite; El Arpa Formation).

TABLE 3. ZIRCON U-Th-Pb, ISOTOPIC RATIOS, AND AGE DETERMINATIONS FOR CABORCA AREA PRECAMBRIAN AND CAMBRIAN ROCKS

Frac[a]	Composition						Isotopic Ratios								Dates (Ma)					
	Wt (mg)[b]	U (ppm)	Pb (ppm)	$\frac{Th^{c}}{U}$	$\frac{Pb^{*d}}{Pbc}$	Pbc[d] (pg)	$\frac{^{206}Pb^{e}}{^{204}Pb}$	$\frac{^{208}Pb^{f}}{^{206}Pb}$	$\frac{^{206}Pb^{f}}{^{238}U}$	% err[g]	$\frac{^{207}Pb^{f}}{^{235}U}$	% err[g]	$\frac{^{207}Pb^{f}}{^{206}Pb}$	% err[g]	$\frac{^{206}Pb^{h}}{^{238}U}$	$\frac{^{207}Pb^{h}}{^{235}U}$	$\frac{^{207}Pb^{h}}{^{206}Pb}$	±	corr. coef.	Discord-ance
Basement Samples																				
SON98-1																				
z1	1.6	430.3	166	0.380	22.6	11.4	1278	0.110	0.3542	(.15)	5.889	(.19)	0.1206	(.12)	1954.6	1959.6	1964.9	2.1	0.798	0.6
z2	1.5	221.5	73.9	0.393	39.3	2.8	2420	0.116	0.3165	(.16)	4.780	(.18)	0.1095	(.07)	1772.8	1781.4	1791.4	1.3	0.920	1.2
z3	2.0	406.3	142	0.478	225.1	1.3	13447	0.141	0.3244	(.07)	5.023	(.10)	0.1123	(.06)	1811.1	1823.2	1837.1	1.1	0.791	1.6
z5	1.0	333.2	109	0.537	70.0	1.6	4138	0.164	0.2983	(.10)	4.400	(.13)	0.1070	(.07)	1682.7	1712.4	1748.9	1.3	0.832	4.3
z6	5.0	332.0	162	0.343	292.8	2.8	17350	0.101	0.4491	(.11)	10.085	(.12)	0.1629	(.05)	2391.4	2442.6	2485.6	0.8	0.918	4.5
z7	5.0	290.6	101	0.358	845.2	0.6	51963	0.104	0.3312	(.06)	5.152	(.07)	0.1128	(.04)	1844.2	1844.7	1845.1	0.8	0.805	0.1
z9	5.7	358.6	136	0.386	1330.2	0.6	80714	0.112	0.3572	(.07)	5.967	(.08)	0.1211	(.04)	1969.1	1971.0	1973.1	0.7	0.847	0.2
z10	4.6	199.9	152	0.582	1203.3	0.6	63408	0.163	0.6208	(.10)	21.893	(.11)	0.2558	(.04)	3113.1	3179.0	3220.8	0.7	0.927	4.2
SON98-20																				
z2	1.3	284.5	89.2	0.633	35.5	3.2	2058	0.198	0.2800	(.17)	3.996	(.20)	0.1035	(.10)	1591.5	1633.4	1687.8	1.8	0.873	6.4
z3	1.6	119.5	42.1	0.853	28.5	2.4	1591	0.249	0.3026	(.16)	4.348	(.19)	0.1042	(.09)	1704.0	1702.5	1700.5	1.7	0.869	–0.2
z4	2.0	73.4	24.9	0.803	15.2	3.3	861	0.242	0.2924	(.24)	4.187	(.28)	0.1039	(.12)	1653.3	1671.6	1694.5	2.3	0.895	2.8
z5	1.1	267.0	89.7	0.666	57.9	1.6	3345	0.196	0.3001	(.12)	4.310	(.15)	0.1042	(.08)	1691.7	1695.3	1699.8	1.5	0.843	0.5
z6	2.0	251.7	84.0	0.664	185.5	0.9	10675	0.197	0.2979	(.07)	4.281	(.09)	0.1042	(.06)	1680.9	1689.8	1700.8	1.1	0.757	1.3
z7	1.0	376.6	123	0.564	191.8	0.6	11302	0.166	0.2996	(.06)	4.307	(.08)	0.1043	(.06)	1689.1	1694.6	1701.5	1.0	0.718	0.8
z8	1.3	218.7	72.7	0.633	117.3	0.8	6811	0.187	0.2990	(.10)	4.290	(.12)	0.1041	(.06)	1686.3	1691.5	1697.9	1.1	0.869	0.8
z9	1.1	147.0	48.9	0.599	87.5	0.6	5134	0.175	0.3019	(.10)	4.334	(.12)	0.1041	(.06)	1700.5	1699.8	1699.0	1.0	0.877	–0.1
z10	0.6	403.5	130	0.489	108.7	0.7	6529	0.145	0.2988	(.08)	4.293	(.09)	0.1042	(.05)	1685.2	1692.1	1700.7	0.9	0.840	1.0
DIC98-1																				
z1	0.7	427.1	88.1	0.777	63.5	1.0	3639	0.236	0.1828	(.10)	1.904	(.13)	0.0755	(.08)	1082.4	1082.4	1082.3	1.6	0.788	0.0
z2	0.9	210.1	43.6	0.836	68.8	0.6	3887	0.254	0.1813	(.09)	1.882	(.12)	0.0753	(.07)	1074.3	1074.6	1075.3	1.3	0.818	0.1
z3	0.8	547.4	113	0.828	113.4	0.8	6401	0.253	0.1807	(.07)	1.877	(.09)	0.0753	(.05)	1071.0	1073.1	1077.6	2.3	0.787	0.7
z4	0.5	263.4	53.3	0.755	24.7	1.0	1431	0.231	0.1800	(.16)	1.868	(.20)	0.0753	(.12)	1066.9	1069.9	1076.0	1.1	0.817	0.9

(*continued*)

TABLE 3. ZIRCON U-Th-Pb, ISOTOPIC RATIOS, AND AGE DETERMINATIONS FOR CABORCA AREA PRECAMBRIAN AND CAMBRIAN ROCKS (*continued*)

Frac[a]	Composition						Isotopic Ratios								Dates (Ma)					
	Wt (mg)[b]	U (ppm)	Pb (ppm)	Th[c]/U	Pb*[d]/Pbc	Pbc[d] (pg)	$^{206}Pb/^{204}Pb$[e]	$^{208}Pb/^{206}Pb$[f]	$^{206}Pb/^{238}U$[f]	% err[g]	$^{207}Pb/^{235}U$[f]	% err[g]	$^{207}Pb/^{206}Pb$[f]	% err[g]	$^{206}Pb/^{238}U$[h]	$^{207}Pb/^{235}U$[h]	$^{207}Pb/^{206}Pb$[h]	±	corr. coef.	Discord-ance
Sedimentary Samples																				
SON98-10																				
z1	2.3	209.1	39.5	0.377	45.7	2.0	2891	0.114	0.1845	(.10)	1.930	(.15)	0.0759	(.11)	1091.4	1091.4	1091.6	2.3	0.674	0.0
z2	2.8	134.6	31.2	0.343	68.0	1.3	4296	0.104	0.2265	(.11)	2.673	(.12)	0.0856	(.06)	1316.0	1321.0	1329.3	1.3	0.854	1.1
z3	3.4	460.0	85.6	0.336	164.5	1.7	10465	0.102	0.1836	(.07)	1.919	(.10)	0.0758	(.07)	1086.7	1087.7	1089.6	1.5	0.693	0.3
z4	2.0	156.1	28.6	0.322	47.0	1.2	3018	0.098	0.1814	(.09)	1.883	(.13)	0.0753	(.09)	1074.9	1075.0	1075.3	1.8	0.723	0.0
z5	3.3	86.0	18.9	0.991	47.5	1.3	2599	0.300	0.1858	(.10)	1.949	(.14)	0.0761	(.09)	1098.8	1098.0	1096.5	1.9	0.748	–0.2
SON98-16																				
z1	5.0	139.2	49.5	0.821	11.5	20.1	588	0.270	0.2797	(.43)	4.163	(.61)	0.1079	(.40)	1589.8	1666.8	1765.1	7.3	0.753	11.2
z2	8.7	86.9	27.1	0.332	133.9	1.8	8338	0.099	0.3007	(.07)	4.361	(.09)	0.1052	(.07)	1694.6	1705.1	1717.9	1.2	0.699	1.5
z4	1.1	495.0	113	0.197	13.4	8.7	842	0.065	0.2209	(.24)	2.726	(.31)	0.0895	(.19)	1286.9	1335.7	1414.8	3.6	0.800	10.0
SON98-17																				
z3	4.8	28.3	9.7	0.708	33.4	1.4	1901	0.216	0.3022	(.16)	4.520	(.19)	0.1085	(.09)	1702.2	1734.7	1774.2	1.7	0.866	4.6
z4	3.7	216.8	46.8	0.374	121.0	1.4	7517	0.122	0.2077	(.07)	2.434	(.10)	0.0850	(.07)	1216.5	1252.8	1315.6	1.3	0.752	8.3
z5	6.0	166.4	59.3	0.566	226.6	1.6	13268	0.167	0.3241	(.09)	5.002	(.11)	0.1119	(.07)	1809.7	1819.7	1831.1	1.3	0.780	1.3
SON98-19																				
z1	4.8	67.7	14.0	0.777	53.0	1.3	3042	0.236	0.1827	(.10)	1.902	(.14)	0.0755	(.10)	1081.8	1081.9	1082.0	2.1	0.703	0.0
z2	4.9	208.7	45.7	1.067	186.3	1.2	9953	0.325	0.1817	(.10)	1.892	(.13)	0.0755	(.08)	1076.3	1078.1	1081.7	1.5	0.802	0.5
z3	2.1	439.7	95.3	1.020	106.4	1.8	5749	0.312	0.1815	(.07)	1.890	(.12)	0.0755	(.09)	1075.1	1077.6	1082.6	1.9	0.621	0.8
z4	4.6	19.2	4.07	0.891	10.7	1.7	616	0.270	0.1829	(.35)	1.905	(.45)	0.0755	(.27)	1082.7	1082.8	1083.0	5.4	0.804	0.0

[a] All fractions composed of single zircons.

[b] Sample weights were estimated to within 40% using measured grain dimensions and a nominal density of 4.5 g/cm^3 for zircon.

[c] Th contents calculated from radiogenic ^{208}Pb and the $^{207}Pb/^{206}Pb$ date of the sample, assuming concordance between U-Th-Pb systems.

[d] Pb* and Pbc represent radiogenic Pb and common Pb respectively.

[e] Measured ratio corrected for fractionation and spike contribution; Pb fractionation was 0.12 ± 0.04%/a.m.u. for Faraday detector or 0.20 ± 0.04%/a.m.u. for Daly detector analysis, based on daily analysis of NBS-981.

[f] Measured ratios corrected for fractionation, spike, blank, and initial common Pb; nominal U blank = 0.1 pg ± 50% (2σ) for U measured as UO_2 or 1.0 pg ± 50% (2σ) for U measured as U metal; nominal Pb blank = 3.5 pg ± 50% (2σ) or where lower, the total common Pb of the analysis ±10% (2σ); measured laboratory Pb composition: $^{206}Pb/^{204}Pb$ = 19.10, $^{207}Pb/^{204}Pb$ = 15.72, $^{208}Pb/^{204}Pb$ = 38.65 ± 0.01 (2σ); initial Pb composition from model of Stacey and Kramers (1975) at the nominal age of the fraction (i.e., 1.6 or 1.7 Ga).

[g] Numbers in parentheses are the % errors reported at the 2σ confidence interval, propagated using the algorithms of Ludwig (1980).

[h] Isotopic ages calculated using the decay constants of Jaffey et al. (1971): l(235U) = 9.8485 × 10^{-10} yr^{-1} and l(238U) = 1.55125 × 10^{-10} yr^{-1}; error in $^{207}Pb/^{206}Pb$

SHRIMP analyses clearly demonstrate that less abundant zircon grains derived from both 1.1 Ga, and various Archean sources also occur in these samples.

The two feldspathic arenite samples (El Arpa and Unit 1 of the Puerto Blanco Formation) have age spectra that are similar to one another, but distinct from those of the quartz arenites. The feldspathic arenites have a dominant peak ca. 1.1 Ga and minor peaks ca. 1.4 Ga and ca. 1.71 Ga (Fig. 3), consistent with the TIMS age determinations. Both samples lack Archean detrital zircons, and Unit 1 of the Puerto Blanco Formation contains several ca. 500 Ma grains that were not observed in any other sample.

DISCUSSION

Classifying the Caborcan Crystalline Basement

Although we have analyzed only a small number of the Precambrian crystalline basement rocks from the Caborca area, all of these samples are characterized by low measured ε_{Nd} values and Nd model ages between 2.0 and 2.4 Ga. In addition, the Bamori gneiss and Cerro Rajon Granite have Pb isotopic compositions characterized by relatively high $^{207}Pb/^{204}Pb$ ratios for a given $^{206}Pb/^{204}Pb$ (Fig. 6). Wooden and DeWitt (1991) quantified this feature of the Pb isotopic compositions of western U.S. Proterozoic rocks through the use of a ΔJ (Delta Jerome) value (Table 2), with the Mojave province being characterized by ΔJ > 6. Both the Bamori gneiss and Cerro Rajon Granite have high ΔJ values from 8.9 to 9.1 (Table 2). These Nd and Pb isotopic characteristics are exactly those that characterize rocks of the Mojave province (Bennett and DePaolo, 1987; Wooden and DeWitt, 1991; Rämö and Calzia, 1998) and are inferred to reflect the mixture of juvenile Proterozoic crust with existing, low ε_{Nd}, high U/Pb, Th/Pb Archean crustal material. Therefore, it is possible that the Caborcan crystalline basement represents a segment of the Mojave province.

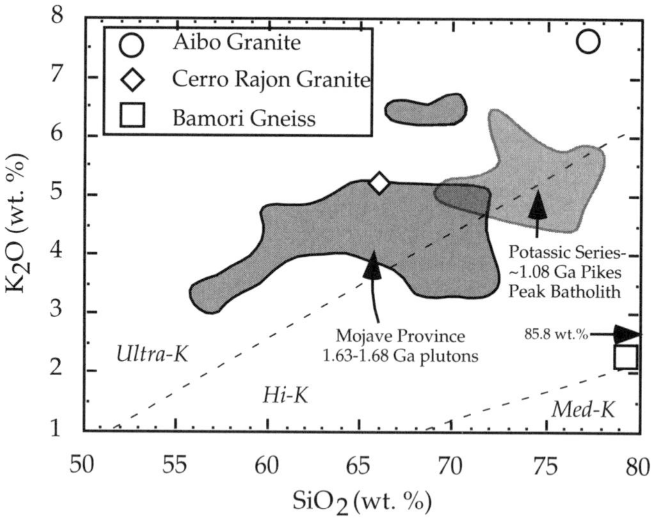

Figure 4. wt% K_20 versus wt% SiO_2 for Caborca area Precambrian basement rocks. Data for Mojave province 1.63–1.68 Ga intrusive igneous rocks from (Bender et al., 1993). Pikes Peak batholith data from Smith et al. (1999).

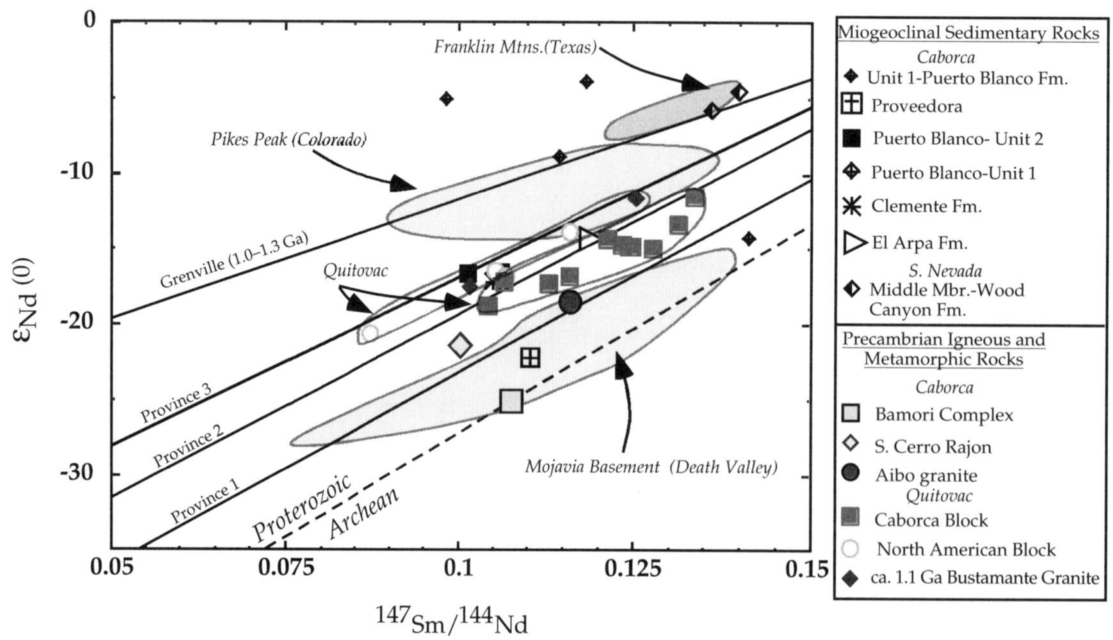

Figure 5. ε_{Nd} (0) versus $^{147}Sm/^{144}Nd$ for Caborcan Precambrian basement and miogeoclinal sedimentary rocks. Data from Patchett and Ruiz (1989), Rämö and Calzia (1998), Smith et al. (1999), Farmer et al. (2001), and Iriondo et al. (2004).

This conclusion is supported by the zircon dates (1.75–3.2 Ga) from the Bamori gneiss. The high wt% SiO_2 (86%) of the Bamori gneiss sample, combined with its peraluminous bulk compositions and low REE abundances (Table 1), clearly indicate that this is metasedimentary rock. Biotite-muscovite paragneisses from the Mojave province, at least in the central Transverse ranges of California (Baldwin gneisses; Barth et al., 2000), have complex zircon U-Th-Pb systematics characterized by $^{207}Pb/^{206}Pb$ dates that vary from 1.76 Ga to 2.7 Ga. Detrital zircons from the Bamori gneiss are also variably discordant and show a range of Paleoproterozoic to Archean $^{207}Pb/^{206}Pb$ dates similar to that of the Baldwin gneisses, although the detrital Archean zircons are as old as 3.2 Ga (Table 3) and so are distinctly older than those found in Mojave province paragneisses in the southwest United States (Wooden and Miller, 1990; Duebendorfer et al., 2001). The age of the youngest Bamori zircon (1.75 Ga) also provides an estimate of the maximum depositional age for the sedimentary protolith that is similar to that estimated for the Baldwin paragneiss (1.76 Ga). The similarity in the detrital zircon ages between the Bamori and Baldwin paragneisses further emphasizes the similarities between the Caborca block basement and the Mojave province.

The Cerro Rajon Granite is 1.7 Ga in age, despite having A-type chemical characteristics usually attributed to "anorogenic" 1.4 Ga granitic rocks in the western United States (Anderson and Bender, 1989). However, highly potassic, moderately peraluminous (A/CNK = 1.0–1.2 at ~65 wt% SiO_2) 1.63–1.68 Ga granitic rocks are common in the Mojave province (Fig. 4; Bender et al., 1993), and we speculate that the Cerro Rajon Granite represents a northern Sonora example of this potassic granitic suite. The low measured ε_{Nd} values for the Cerro Rajon Granites, and its high ΔJ value, are also similar to the values determined for Paleoproterozoic intrusive igneous rocks in the Mojave province (Figs. 5 and 6).

In contrast to the similarities between Caborca and Mojave crustal rocks, highly potassic 1.1 Ga granitic rocks have not been previously reported from the Mojave province, although magmatism and associated metamorphism of this general age are known in the California Transverse Ranges (the 1.19 Ga San Gabriel anorthosite; Barth et al., 1995; Barth et al., 2001) and as 1.1 Ga diabase dikes found throughout much of the southwest United States (Fig. 8) (Heaman and Grotzinger, 1992). Highly potassic Mesoproterozoic granitic rocks are found, however, in the Yavapai province in Colorado (the ca. 1.08 Ga Pikes Peak Granite) (Smith et al., 1999) and at the northern edge of the Mesoproterozoic Grenville province in western Texas (the 1.12 Ga Red Bluff granitic suite, Fig. 8; Shannon et al., 1997). Interestingly, the Aibo Granite has the lowest measured ε_{Nd} value yet determined for any ca. 1.1 Ga potassic granitic rock in western North America, significantly lower than either Pikes Peak Granite, or Red Bluff Granite in north Texas (Fig. 5). We note that the Aibo and Cerro Rajon Granites have essentially identical Nd isotopic compositions at the time of crystallization of the Aibo Granite ($\varepsilon_{Nd\,(1.075\,Ga)}$ ~−7.5) and so we suggest that the Aibo Granite likely represents the product of anatectic melting of the local Caborca block basement. The low ε_{Nd} value of the Aibo Granite relative to other ca. 1.1 Ga granitic rocks therefore also reflects it derivation from the low ε_{Nd} Caborca block crust. The highly silicic compositions of the Aibo Granite, and its prominent negative Eu anomaly (and low Ba and Sr contents, Table 3) may be the result of fractional anatectic melting and removal of plagioclase during subsequent magmatic differentiation, but more detailed conclusions regarding the petrogenesis of the Aibo Granite await further study.

The Nd isotopic compositions of the Caborca basement are distinct from those described for Precambrian basement farther north in Sonora in the Quitovac area. Here, no Paleoproterozoic crystalline rocks, whether belonging to the Caborca block or to the North America block (Iriondo et al., 2004), have Nd isotopic compositions equivalent to those of the Mojave province (Fig. 5). Instead, the Caborca block at Quitovac consists of Yavapai province (or potentially a portion of the Mojave-Yavapai transition zone, as exposed in western Arizona; Duebendorfer et

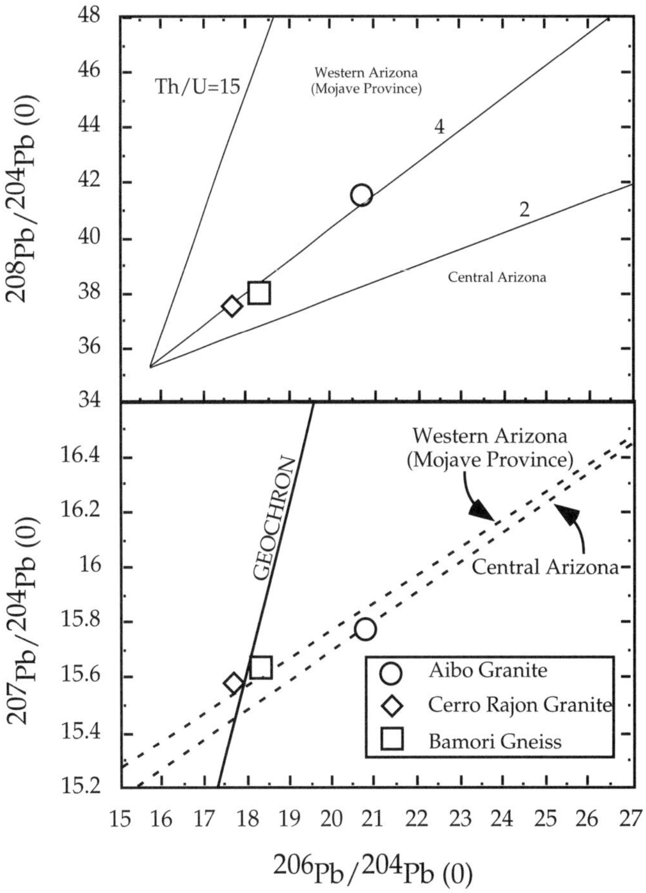

Figure 6. Whole-rock Pb isotopic data for Caborcan basement rocks. Ranges of Pb isotopic compositions for central and western Arizona from Wooden and DeWitt (1991).

al., 2001) that is juxtaposed over Mazatzal province (North American block) to the east along Laramide thrust faults (Iriondo et al., 2004). Therefore, our data demonstrate that the Caborca block may contain a boundary between rocks with Yavapai and Mojave characteristics.

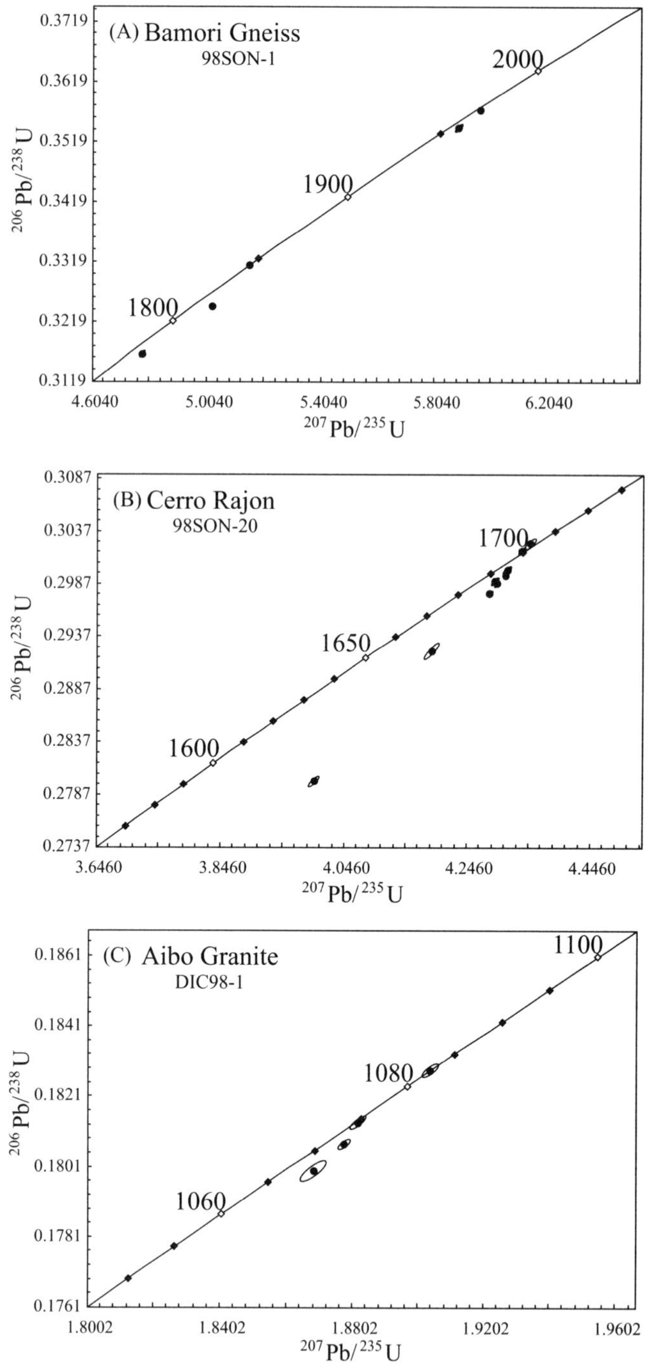

Figure 7. TIMS U-Pb zircons Concordia plots for Caborca Precambrian basement rocks. (A) Bamori gneiss. (B) Cerro Rajon Granite (C) Aibo Granite.

Provenance of Miogeoclinal Sedimentary Rocks

The detrital zircon age spectra of the Caborca miogeoclinal sedimentary rocks are broadly similar to those determined for other miogeoclinal arenites in western North America, implying a similar suite of crustal sources (Stewart et al., 2001). The zircon age peaks at 1.7–1.8 Ga and ca. 1.4 Ga found in all of the analyzed sedimentary rocks (Fig. 3) correspond to known intrusive igneous rock ages within the Paleoproterozoic Yavapai and Mazatzal provinces that make up much of Arizona, New Mexico, Colorado, and Utah (Fig. 8). The Archean grains in the Pitiquito and Tecolote Quartzites, in contrast, were most likely derived from the Wyoming craton although erosion of Mojave province basement may also have provided some grains of this age, given the occurrence of Archean zircons in the latter (including the Bamori Complex). The 500–600 Ma zircons found in Puerto Blanco Formation samples are also present in Cambrian cratonal sedimentary rocks in western North America (Bolsa Quartzite in Sonora; Coronado Sandstone, Arizona) and have been attributed to the erosion of Cambrian igneous rocks found sporadically in the area (Stewart et al., 2001).

The source(s) of zircons in the age range from 1.0 Ga to 1.1 Ga, however, is more enigmatic. This population has been increasingly recognized as a common component of the detrital zircon age spectra of both Cambrian miogeoclinal and cratonal sedimentary rocks in western North America, having been previously noted as the dominant zircon population in the Proveedora Quartzite in Sonora and in the Wood Canyon Formation in southern California and vicinity (Fig. 9; Farmer and Ball, 1997; Stewart et al., 2001). Our data demonstrate that sedimentary rocks with a prominent 1.1 Ga zircon population also occur in the Neoproterozoic portions of the miogeocline in Caborca. We emphasize that because of complicated marine sediment dispersal processes, the Caborca paleocurrent data do not resolve source directions for those sediments containing the 1.1 Ga zircons. This contrasts with lithologically similar fluvial sediments of the Wood Canyon Formation in southern California that shows a strong unimodal paleocurrent pattern indicating that the sediment was derived from the east. One possibility is that the Caborca section could represent a more basinward equivalent, more like rocks exposed in the White-Inyo Mountains in California (Stewart, 1970) in which sediments were transported both normal and parallel to the continental margin.

Stewart et al. (2001) suggested three possible sources of 1.1 Ga zircons in the Cordilleran miogeoclinal sedimentary rocks: the Mesoproterozoic Grenville province exposed in Texas and areas farther east; 1.1 Ga igneous rocks equivalent to the Pikes Peak or Aibo Granites, and a hypothetical, and now completely eroded, volcanoplutonic complex complementary to the 1.1 Ga diabase dikes found throughout the southwest United States (Fig. 8; Farmer and Ball, 1997). Distinguishing between these possibilities is difficult on the basis of detrital zircon ages alone. However, erosion of Grenville age rocks in Texas and

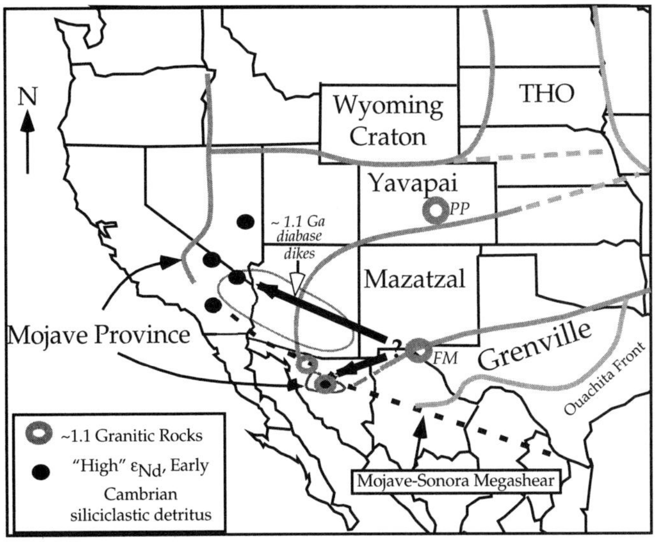

Figure 8. Disposition of Precambrian crust age provinces (Van Schmus et al., 1993), Mojave province (Bennett and DePaolo, 1987), and potential sources of 1.1 Ga detrital zircons found in Neoproterozoic to early Cambrian miogeoclinal sedimentary rocks. THO—Trans-Hudson orogen; FM—Franklin Mountains; PP—Pikes Peak.

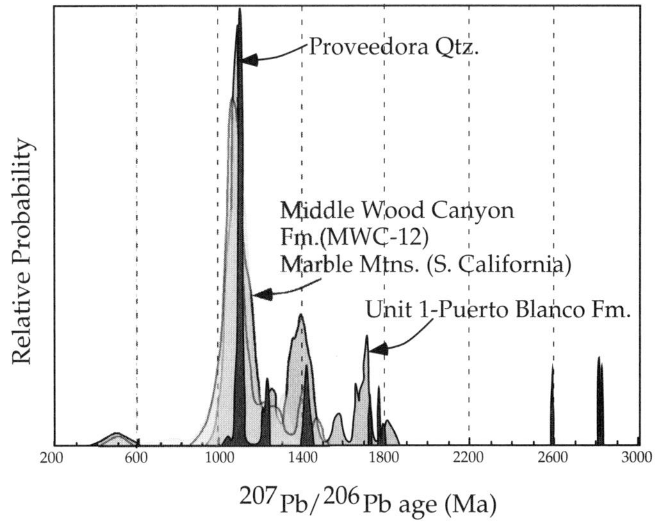

Figure 9. Comparison of cumulative probability distributions for detrital zircon ages from Unit 1, Puerto Blanco Formation, Proveedora Quartzite (Gehrels and Stewart, 1998) and middle member of Wood Canyon Formation, Marble Mountains, southern California (Fedo, unpublished data).

vicinity should yield a range of detrital zircon ages from 1.0 to 1.2 Ga, and not a dominant spike ca. 1.1 Ga (Stewart et al., 2001). If so, then the relatively broad age spectrum for Mesoproterozoic detrital zircon in the Tecolote and Pitiquito quartz arenites (Fig. 3) could be the result of derivation of these zircons from distal sources within the Grenville province to the east. Such a source would not, however, account for the preponderance of 1.1 Ga detrital zircons in the El Arpa Formation and in Unit 1 of the Puerto Blanco Formation (Fig. 3).

For the El Arpa Formation, our single zircon, high-precision U-Pb ages refine the age of the zircon source region in this sample to 1080 ± 2 Ma. This age is essentially equivalent to that of the nearby Aibo Granite. Although the Nd isotopic compositions of the El Arpa Formation and the Aibo Granite are not identical (Fig. 5), the whole-rock ε_{Nd} values of the El Arpa Formation are also lower than expected for detritus derived from other 1.1 Ga granitic rocks in the western United States, including the Pikes Peak and Red Bluff Granites (Fig. 5). As a result, we suggest that the El Arpa Formation represents detritus derived largely from erosion of local Mesoproterozoic igneous rocks, with the somewhat higher ε_{Nd} values of the El Arpa Formation, relative to the Aibo Granite, reflecting the incorporation of detritus derived from higher ε_{Nd} Paleoproterozoic or Mesoproterozoic crust found farther to the east.

In contrast to the El Arpa Formation, concordant, Mesoproterozoic zircons in Unit 1 of the Puerto Blanco Formation, vary from 1073 to 1098 Ma in age and seem to require a crustal source containing Mesoproterozoic igneous rocks that span a larger age range than has been thus far defined for rocks of this age in the Caborca region. Furthermore, the generally high ε_{Nd} values for Unit 1 of the Puerto Blanco Formation are inconsistent with the derivation of the bulk of the detritus in this unit from the local Mojave basement rocks, including the Aibo Granite. Instead, the whole-rock Nd isotopic compositions of Unit 1 of the Puerto Blanco Formation overlap those of more distal 1.1 Ga terranes, including those of the Red Bluff Granite (Fig. 5). However, rather than reflecting the derivation of this unit from a distal high-ε_{Nd} source, the high ε_{Nd} may instead reflect incorporation of local, high-ε_{Nd} volcaniclastic material in this sediment (Farmer et al., 2001). As a result, we cannot yet determine with any confidence the source of the 1.1 Ga zircons in the Puerto Blanco Formation. We do suggest, however, that it may be possible to further resolve this issue through Nd isotopic analyses of the 1.1 Ga zircons themselves, given that Pikes Peak, Aibo, and Red Bluff granitic rocks are distinguishable on the basis of their Nd isotopic compositions (Fig. 5). We currently have such a study under way.

Finally, our isotopic data from Precambrian crystalline rocks in the Caborca area allow a further assessment of the provenance of other miogeoclinal siliciclastic sediments from this region that was not possible in our previous work (Farmer et al., 2001). With the exception of the Proveedora Quartzite and one low-ε_{Nd} sample from Unit 1 of the Puerto Blanco Formation, no siliciclastic sediment has the Nd isotopic compositions expected for material derived exclusively from the underlying Mojave province basement (Figs. 2 and 5). All of these sediments, therefore, must contain detrital components derived from more distal Paleoproterozoic and/or Mesoproterozoic sediment presumably derived from Precambrian crustal sources to the east (Fig. 8).

Implications for Existence of Mojave-Sonora Megashear

One prediction of the Mojave-Sonora megashear hypothesis is that Precambrian and Paleozoic rocks in the Caborca block in Sonora must match the ages, lithologies and isotopic characteristics of rocks of similar age in southern California, the inferred original position of at least some of the Caborca block rocks (Anderson and Schmidt, 1983). Our data do, in fact, demonstrate that the Paleoproterozoic Caborca crystalline basement rocks are similar in age and Nd isotopic compositions to Mojave crystalline basement in southern California, including the occurrence of Archean zircon in Paleoproterozoic paragneisses (Bennett and DePaolo, 1987; Rämö and Calzia, 1998). Furthermore, Neoproterozoic to Cambrian miogeoclinal sedimentary rocks in both the Mojave and Caborca area have similar Nd isotope stratigraphies (Fig. 2; Farmer et al., 2001) and similar detrital zircon age spectra. In particular, the middle member of Wood Canyon Formation in the Marble Mountains (California) and Unit 1 of the Puerto Blanco Formation both have age spectra that are dominated by a peak at 1.1 Ga, subordinate peaks at 1.4 Ga and 1.7 Ga, and an absence of Archean zircon grains (Fig. 9). All of the above observations support the Mojave-Sonora megashear hypothesis. If extant, the megashear presumably bisects the region between the Quitovac and Caborca regions (Fig. 10A), given the lack of Mojave province crust at Quitovac.

It should be emphasized that the above observations are consistent with, but not definitive proof of, the Mojave-Sonora megashear hypothesis. It remains possible that the Caborca block is an autochthonous segment of the Mojave province that is not even contiguous with the Mojave province in southern California, a possibility consistent with the lack of evidence for Mojave province crust in any other portion of northwestern Sonora (Figs. 8 and 10). We also note that Mesoproterozoic granitic rocks similar to the Aibo Granite have not been identified in southern California. The only igneous rock of this age known in the Death Valley region is a ca. 1.08 Ga diabase dike (Heaman and Grotzinger, 1992). This observation suggests that the two regions contain distinctly different Mesoproterozoic igneous lithologies and so may not share identical geological histories prior to the Mesozoic, as required by the Mojave-Sonora megashear hypothesis. Detailed structural and geologic studies of the Precambrian rocks at Caborca might help resolve this issue, given that Mojave province crust in the southwest United States shows distinct evidence for two discrete Paleoproterozoic deformational events, one at 1.75 Ga (the Ivanpah orogeny, interpreted as the suturing of Mojave and Yavapai provinces) and the other coincident with the Yavapai orogeny (interpreted as docking of the Mazatzal province to the combined Mojave and Yavapai crust). If autochthonous, Caborca may have docked directly against Mazatzal province (Fig. 10B) and so lack evidence for the Ivanapah orogenic event.

CONCLUSIONS

Our whole-rock age and isotopic data suggest that the Caborca area Precambrian basement is a previously unrecognized segment of the Paleoproterozoic Mojave province. As such, the Caborca area represents the only known exposure of Mojave province crust in Mexico, and defines the southernmost extent of the Mojave province in North America. What remains

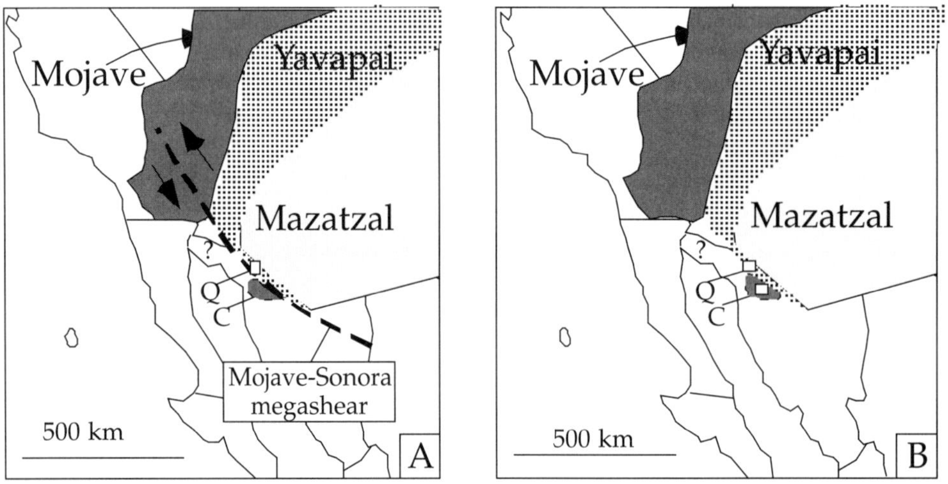

Figure 10. Cartoons depicting possible modes of origin of Caborcan segment of Mojave province. (A) Caborca block is displaced from original position via left-lateral strike slip motion during Mesozoic along proposed Mojave-Sonora megashear. As shown, the Sonoran segment of the megashear juxtaposes Mojave province against the Yavapai/Mazatzal provinces. (B) Caborca block is an autochthonous segment of the Mojave province, against which Yavapai and/or Mazatzal provinces were accreted during the Paleoproterozoic. In this scenario, the Mojave-Sonora megashear need not exist. Because of lack of information on Precambrian basement in northwestern Sonora and northeastern Baja California (question mark), it is not known whether or not Caborca Mojave Province is contiguous with similar Precambrian crust in the southwest United States or instead represents an isolated segment of the Mojave Province. Q—Quitovac; C—Caborca.

unresolved, however, is the relationship of the Caborca crust both to the remainder of the Mojave province and to adjacent Paleoproterozoic crustal terranes to the east.

ACKNOWLEDGMENTS

This research was supported by National Science Foundation grant EAR-9725829 (Farmer) and EAR-0003551 (Bowring). Reviews by Drew Coleman, Dick Tosdal, and Tom Anderson greatly improved the manuscript.

APPENDIX

Mineral separation for the TIMS zircon analyses for all samples was carried out according to standard crushing, heavy liquid, and magnetic separation techniques. Zircon grains were picked in ethanol under a binocular microscope and sorted by their morphology, color, clarity, and inclusion characteristics. All zircon grains selected for U-Pb analysis were air-abraded, washed in $3M$ HNO_3 at 50 °C for 12 h, and ultrasonicated for 1 h. Each zircon was photographed to estimate sample weight, pipetted with acetone into 300 mL Teflon FEP capsules, washed again in $3M$ HNO_3 at 75 °C for 3 h, and finally rinsed three times with $3M$ HNO_3. Zircons were dissolved in 120 mL HF with a mixed ^{205}Pb-^{233}U-^{235}U spike and trace HNO_3 at 220 °C for 48 h, dried to salts, and redissolved in 120 mL 6M HCl at 180 °C for at least 12 h. Pb and U were separated from the sample using an HCl-based anion exchange procedure modified after Krogh (1973). For chemical separations carried out prior to December 2000 (all sedimentary samples), Pb and U were collected in separate beakers and analyzed on separate Re filaments. For chemical separations carried out after December 2000, Pb and U were collected in a single beaker for isotope analysis.

Pb and U were analyzed by conventional thermal ionization mass spectrometry on the MIT VG Sector 54 multicollector mass spectrometer. The separate Pb samples and mixed Pb and U samples were loaded onto previously degassed single Re filaments with a silica gel-H_3PO_4 mixture. Pb isotopes were measured either (1) for ^{207}Pb ion beams $>5 \times 10^{-14}$ A, in a two-cycle dynamic routine with ^{204}Pb in the axial Daly detector and ^{205}Pb through ^{208}Pb in H1-H4 Faraday detectors during the first cycle and ^{205}Pb in the Daly detector and ^{206}Pb through ^{208}Pb in the H1-H3 Faraday detectors during the second cycle, providing real-time Daly gain measurement; or (2) for ^{207}Pb ion beams $<5 \times 10^{-14}$ A, by peak-jumping all ion beams into the axial Daly detector in ion-counting mode. In the case of mixed Pb and U samples, after the Pb analysis was complete, the filament current was increased until U ionization began. U isotopes were then measured as UO_2^+ in static mode with masses 270, 267, and 265 in the axial, L1, and L2 Faraday collectors, respectively. For separate U samples, the sample was mixed with colloidal graphite and $0.1M$ H_3PO_4, loaded onto previously degassed single Re filaments, and U isotopes were measured in static mode with masses 238, 235, and 233 in the axial, L1, and L2 collectors, respectively. Pb isotope fractionation was monitored throughout the study by daily analysis of the NBS-981 common Pb standard, whereas U fractionation was monitored and corrected by use of the double spike.

REFERENCES CITED

Anderson, J.L., and Bender, E.E., 1989, Nature and origin of Proterozoic A-type granitic magmatism in the southwestern United States of America: Lithos, v. 23, p. 19–52, doi: 10.1016/0024-4937(89)90021-2.

Anderson, T.H., and Schmidt, V.A., 1983, The evolution of Middle America and the Gulf of Mexico-Caribbean Sea region during Mesozoic time: Geological Society of America Bulletin, v. 94, p. 941–966, doi: 10.1130/0016-7606(1983)94<941:TEOMAA>2.0.CO;2.

Anderson, T.H., and Silver, L.T., 1970, Reconnaissance survey of Precambrian rocks, northwestern Sonora, Mexico: Geological Society of America Abstracts with Programs, v. 2, p. 484.

Anderson, T.H., and Silver, L.T., 1971, Preliminary history for Precambrian rocks, Bamori region: Sonora: Geological Society of America Abstracts with Programs, v. 3, p. 72–73.

Anderson, T.H., and Silver, L.T., 1981, An overview of Precambrian rocks in Sonora: Revista, Instituto de Geología, v. 5, p. 131–139.

Anderson, T.H., and Silver, L.T., 1997, 1.1 Ga plutons in Sonora, Mexico: Regional tectonic implications: Geological Society of America Abstracts with Programs, v. 29, no. 6, p. 348.

Barth, A.P., Wooden, J.L., and Coleman, D.S., 2001, SHRIMP-RG U-Pb zircon geochronology of Mesoproterozoic metamorphism and plutonism in the southwesternmost United States: Journal of Geology, v. 109, p. 319–327, doi: 10.1086/319975.

Barth, A.P., Wooden, J.L., Coleman, D.S., and Fanning, M.C., 2000, Geochronology of the Proterozoic basement of southwesternmost North America, and the origin and evolution of the Mojave crustal province: Tectonics, v. 19, p. 616–629, doi: 10.1029/1999TC001145.

Barth, A.P., Wooden, J.L., Tosdal, R.M., Morrison, J., Dawson, D.L., and Hernly, B.M., 1995, Origin of gneisses in the aureole of the San Gabriel anorthosite complex and implications for the Proterozoic crustal evolution of southern California: Tectonics, v. 14, p. 736–752, doi: 10.1029/94TC02901.

Bender, E.E., Morrison, J., Anderson, J.L., and Wooden, J.L., 1993, Early Proterozoic ties between two suspect terranes and the Mojave crustal block of the southwestern U.S: Journal of Geology, v. 101, p. 715–728.

Bennett, V.C., and DePaolo, D.J., 1987, Proterozoic crustal history of the western United States as determined by neodymium isotopic mapping: Geological Society of America Bulletin, v. 99, p. 674–685, doi: 10.1130/0016-7606(1987)99<674:PCHOTW>2.0.CO;2.

Campa, M.F., and Coney, P.J., 1983, Tectono-stratigraphic terranes and mineral-resource distributions in Mexico: Canadian Journal of Earth Sciences, v. 20, no. 6, p. 1040–1051.

Collins, W.J., Beams, S.D., White, A.J.R., and Chappell, B.W., 1982, Nature and origin of A-type granites with particular reference to southeastern Australia: Contributions to Mineralogy and Petrology, v. 80, p. 189–200.

Cooper, G.A., and Arellano, A.R.V., 1946, Stratigraphy near Caborca, northwest Sonora, Mexico: American Association of Petroleum Geology Bulletin, v. 30, p. 606–611.

Damon, P.E., Livingston, D.E., and Giletti, B.J., 1962, Extension of the older Precambrian of the southwest into Sonora, Mexico, *in* Abstracts for 1961: Geological Society of America Special Paper 68, p. 158–159.

DeGraaff-Surpless, K., Graham, S.A., Wooden, J.L., and McWilliams, M.O., 2002, Detrital zircon provenance analysis of the Great Valley Group, California: Evolution of an arc-forearc system: Geological Society of America Bulletin, v. 114, p. 1564–1580, doi: 10.1130/0016-7606(2002)114<1564:DZPAOT>2.0.CO;2.

Duebendorfer, E.M., Chamberlain, K.R., and Jones, C.S., 2001, Paleoproterozoic tectonic history of the Cerbat Mountains, northwestern Arizona: Implications for crustal assembly in the southwestern United States: Geological Society of America Bulletin, v. 113, p. 575–590, doi: 10.1130/0016-7606(2001)113<0575:PTHOTC>2.0.CO;2.

Farmer, G.L., and Ball, T.T., 1997, Sources of Middle Proterozoic to Early Cambrian siliciclastic sedimentary rocks in the Great Basin: a Nd isotope study: Geological Society of America Bulletin, v. 109, p. 1193–1205, doi: 10.1130/0016-7606(1997)109<1193:SOMPTE>2.3.CO;2.

Farmer, G.L., Espinoza, G., Morales, M., Martin, M.W., and Bowring, S.A., 2001, Nd isotope constraints on sources of Neoproterozoic to Early Cambrian siliciclastic sedimentary rocks in northern Sonora: Journal of South American Earth Sciences, v. 14, p. 437–446, doi: 10.1016/S0895-9811(01)00053-0.

Gehrels, G.E., and Stewart, J.H., 1998, Detrital zircon U-Pb geochronology of Cambrian to Triassic miogeoclinal and eugeoclinal strata of Sonora, Mexico: Journal of Geophysical Research, v. 103, p. 2471–2487, doi: 10.1029/97JB03251.

Giletti, B.J., and Damon, P.E., 1961, Rubidium-strontium ages of some basement rocks from Arizona and northwestern Mexico: Geological Society of America Bulletin, v. 72, p. 639–643.

Heaman, L.M., and Grotzinger, J.P., 1992, 1.08 Ga diabase sills in the Pahrump Group, California: Implications for development of the Cordilleran miogeocline: Geology, v. 20, p. 637–640, doi: 10.1130/0091-7613(1992)020<0637:GDSITP>2.3.CO;2.

Iriondo, A., Premo, W.R., Martinez-Torres, L.M., Budahn, J.R., Atkinson, W.W.J., Siems, D.F., and Guaras-Gonzalez, B., 2004, Isotopic, geochemical, and temporal characterization of Proterozoic basement rocks in the Quitovac region, northwestern Sonora, Mexico: Implications for the reconstruction of the southwestern margin of Laurentia: Geological Society of America Bulletin, v. 116, p. 154–170, doi: 10.1130/B25138.1.

Jaffey, A.H., Flynn, K.F., Glendeni, L.E., Bentley, W.C., and Essling, A.M., 1971, Precision measurement of half-lives and specific activities of U-235 and U-238: Physical Review C, v. 4, p. 1889–1906.

Karlstrom, K.E., and Bowring, S.A., 1988, Early Proterozoic assembly of tectonostratigraphic terranes in southwestern North America: Journal of Geology, v. 96, p. 561–576.

Krogh, T.E., 1973, A low-contamination method for hydrothermal decomposition of zircon and extraction of U and Pb for isotopic age determinations: Geochimica et Cosmochimica Acta, v. 37, p. 485–494, doi: 10.1016/0016-7037(73)90213-5.

Ludwig, K.R., 1980, Calculation of uncertainties of U-Pb isotope data: Earth and Planetary Science Letters, v. 46, p. 212–220, doi: 10.1016/0012-821X(80)90007-2.

Ludwig, K.R., 2000, User's manual for Isoplot/Ex, Berkeley Geochronology Center Special Publication: Berkeley, California, 56 p.

Merrian, R., and Eells, J.L., 1979, Reconnaissance geologic map of the Caborca quadrangle, Sonora, Mexico: Boletin de Departamento de Geologia Universidad de Sonora, v. 1, p. 87–94.

Molina-Garza, R.S., and Geissman, J.W., 1999, Paleomagnetic data from the Caborca terrane Mexico: Implications for Cordilleran tectonics and the Mojave-Sonora megashear hypothesis: Tectonics, v. 18, p. 293–325, doi: 10.1029/1998TC900030.

Patchett, P.J., and Ruiz, J., 1989, Nd isotopes and the origin of Grenville-age rocks in Texas: Implications for Proterozoic evolution of the United States Mid-continent region: Journal of Geology, v. 97, p. 685–695.

Rämö, O.T., and Calzia, J.P., 1998, Nd isotopic composition of cratonic rocks in the southern Death Valley region: Evidence for a substantial Archean source component in Mojavia: Geology, v. 26, no. 10, p. 891–894, doi: 10.1130/0091-7613(1998)026<0891:NICOCR>2.3.CO;2.

Shannon, W.M., Barnes, C.G., and Bickford, M.E., 1997, Grenville magmatism in west Texas: Petrology and geochemistry of the Red Bluff Granitic Suite: Journal of Petrology, v. 38, p. 1279–1305, doi: 10.1093/petrology/38.10.1279.

Silver, L.T., 1965, Mazatzal orogeny and tectonic episodes, *in* Abstracts for 1964: Geological Society of America Special Paper 82, p. 185–186.

Silver, L.T., and Anderson, T.H., 1974, Possible left-lateral early to middle Mesozoic disruption of the southwestern North American craton margin: Geological Society of America Abstracts with Programs, v. 6, p. 955.

Smith, D.R., Noblett, J., Wobus, R.A., Unruh, D., Douglass, J., Beane, R., Davis, C., Goldman, S., Kay, G., Bustavson, B., Saltoun, B., and Stewart, J.H., 1999, Petrology and geochemistry of late-stage intrusions of the A-type, mid-Proterozoic Pikes Peak batholith (Central Colorado, USA): Implications for petrogenic models: Precambrian Research, v. 98, p. 271–305, doi: 10.1016/S0301-9268(99)00049-2.

Stacey, J.S., and Kramers, J.D., 1975, Approximation of terrestrial lead isotope evolution by a two-stage model: Earth Planetary Science Letters, v. 26, p. 207–221, doi: 10.1016/0012-821X(75)90088-6.

Stewart, J.H., 1970, Upper Precambrian and Lower Cambrian strata in the southern Great Basin, California and Nevada: U.S. Geological Survey Professional Paper 620, 206 p.

Stewart, J.H., 1988, Latest Proterozoic and Paleozoic southern margin of North America and the accretion of Mexico: Geology, v. 16, p. 186–189, doi: 10.1130/0091-7613(1988)016<0186:LPAPSM>2.3.CO;2.

Stewart, J.H., Gehrels, G.E., Barth, A.P., Link, P.K., Christie-Blick, N., and Wrucke, C.T., 2001, Detrital zircon provenance of Mesoproterozoic to Cambrian arenites in the western United States and northwestern Mexico: Geological Society of America Bulletin, v. 113, p. 1343–1356, doi: 10.1130/0016-7606(2001)113<1343:DZPOMT>2.0.CO;2.

Stewart, J.H., McMenamin, M.A.S., and Morales-Ramirez, J.M., 1984, Upper Proterozoic and Cambrian rocks in the Caborca region, Sonora, Mexico: Physical stratigraphy, biostratigraphy, paleocurrent studies, and regional relations, U.S. Geological Survey Professional Paper 1309, 36 p.

Taylor, M.C., Rector, R.R., Carollo, G.F., Colchagoff, B.C., and Gastil, R.G., 1991, How does the paleogeography of Paleozoic Mexico relate to western United States? *in* Cooper, J. D., and Stevens, C. H., eds., Paleozoic paleogeography of the western United States-II: Pacific Section SEPM, p. 829–838.

Van Schmus, W.R., and Bickford, M.E., editors, et al. 1993, Transcontinental Proterozoic Provinces, *in* Reed, J. C., et al., eds., Precambrian: Conterminous U.S.: Geological Society of America, Geology of North America, v. C-2, p. 171–334.

Wooden, J.L., and DeWitt, E., 1991, Pb isotopic evidence for the boundary between the Early Proterozoic Mojave and Central Arizona crustal provinces in western Arizona, *in* Karlstrom, K. E., ed., Proterozoic geology and ore deposits of Arizona: Arizona Geology Society Digest, p. 27–50.

Wooden, J.L., and Miller, D.M., 1990, Chronologic and isotopic framework for Early Proterozoic crustal evolution in the eastern Mojave Desert region: SE California: Journal of Geophysical Research, v. 95, p. 20,133–20,146.

Manuscript Accepted by the Society 24 March 2005

Geological Society of America
Special Paper 393
2005

The crust at northwestern Mexico interpreted from Magsat anomalies: Implications for the existence of the Mojave-Sonora megashear

J.O. Campos-Enríquez
E. Hernández-Quintero
Instituto de Geofísica, Universidad Nacional Autónoma de México, Circuito Institutos s/n Deleg. Coyoacán, 04510 México, D.F., México

M. Lozada-Zumaeta
Instituto Mexicano del Petróleo, Eje Central Lázaro Cárdenas 152, Col. San Bartolo Atepehuacán, 07730 México, D.F., México

ABSTRACT

In an effort to characterize the crustal structure of northwestern Mexico (and constrain the Mojave-Sonora megashear) we studied the Magsat magnetic anomalies from that area. Published anomaly maps covering this area include an extensive positive anomaly covering the southern United States, a positive anomaly over the southern half of the Baja California peninsula, and a magnetic low in between. We interpreted a magnetic profile over these anomalies, focusing on its tectonostratigraphic terrane nature. The profile was further constrained by crustal thicknesses from seismological studies and heat flow data. In our model the Cochimi terrane and the North American craton (Colorado Plateau and southern Basin and Range) are characterized by high magnetic susceptibility in agreement with the mafic nature of their corresponding crusts. The Yuma and the Seri terranes have lower magnetic susceptibilities as expected from their felsic to basic crustal nature. The oceanic crust from the Gulf of California is modeled with a low magnetic susceptibility value due to the high heat flow observed at the extensional basins. Our model satisfies the presence of a subvertical contact between crusts with contrasting magnetic signatures of the Seri terrane (comprising the Caborca subterrane) and the southwestern sector of the North American craton. The Mojave-Sonora megashear itself is below the resolution of the Magsat data. Nevertheless, our model implies that because of their different magnetic signatures, the crystalline basements from the southern United States and northern Mexico (Seri terrane) are of different nature, which does not support the continuity of the North American craton into northwestern Mexico.

Keywords: crust, satellite, magnetic anomalies, Mojave-Sonora megashear.

Campos-Enríquez, J.O., Hernández-Quintero, E., and Lozada-Zumaeta, M., 2005, The crust at northwestern Mexico interpreted from Magsat anomalies: Implications for the existence of the Mojave-Sonora megashear, *in* Anderson, T.H., Nourse, J.A., McKee, J.W., and Steiner, M.B., eds., The Mojave-Sonora megashear hypothesis: Development, assessment, and alternatives: Geological Society of America Special Paper 393, p. 199–208, doi: 10.1130/2005.2393(06). For permission to copy, contact editing@geosociety.org.

INTRODUCTION

Paleo-reconstructions of Pangea result in overlap of Mexico and Middle America with the continent of South America. This problem was evident in the early paleo-reconstructions of Pangea (e.g., Bullard et al., 1965). Subsequent paleo-reconstructions have offered variations of the assemblages of continental masses accommodating the Mexican and Middle American landmasses with little or no overlap with South America (Malfait and Dinkelman, 1972; Van der Voo and French, 1974; Iturralde-Vinent, 1975; Irving, 1977; Scotese et al., 1979; White, 1980; Urrutia-Fucugauchi, 1981; Gose et al., 1982; Pindell and Dewey, 1982; Anderson and Schmidt, 1983). These models propose motions and/or rotations, or simple, large strike-slip movements of landmasses of Mexico relative to cratonic North America along major NW-trending faults (Silver and Anderson, 1974; Scotese et al., 1979; Urrutia-Fucugauchi, 1981; Anderson and Schmidt, 1983).

The Anderson and Schmidt model (1983) includes three large faults separating North America and South America and intervening microplates. Restoration of left-lateral displacement along these faults avoids the overlap encountered in the first paleo-reconstructions. The northernmost microplate of the Anderson and Schmidt model (1983) includes part of present-day northern Mexico. Its southern boundary is postulated to have acted as a strike-slip fault now obscured by the Cenozoic volcanism associated with the Trans-Mexican Volcanic Belt. The northernmost limit corresponds to a transform fault active during the Late Jurassic: the Mojave-Sonora megashear (Silver and Anderson, 1974) (Fig. 1). Northwest Mexico comprises a parautochthonous terrane (displaced 800 km southward from its original position along the Mojave-Sonora megashear). Indeed, the Mojave-Sonora megashear explains an unexpected pattern of Proterozoic crystalline terranes with distinct chronologic histories (Silver and Anderson, 1974).

Recent geochemical and isotopic studies (Valencia-Moreno et al., 2001) reveal 1.0–1.2 Ga North American Precambrian basement well beyond the southern limit of the North American craton as proposed by Sedlock et al. (1993). (e.g., of the trace of the Mojave-Sonora megashear). The fault trace is obscured by later deformation, intrusions, and sedimentary cover. However, several geologic studies have well-documented portions of the Mojave-Sonora megashear in northwest Mexico (Caudillo-Sosa et al., 1996; Campbell and Anderson, 1998; Rodríguez-Castañeda, 2002).

However, geophysically, the Mojave-Sonora megashear has not been delineated, although Aiken et al. (1997) described its possible existence in Chihuahua, and Schellhorn et al. (1991) attributed minor features of the Bouguer anomaly map of northwestern Mexico to the Mojave-Sonora megashear.

Paleomagnetic studies do not unequivocally support the existence of the Mojave-Sonora megashear (Urrutia-Fucugauchi, 1984; Kluger Cohen et al., 1986; Molina-Garza and Geissman, 1996, 1999).

Magnetic anomalies have been interpreted to reveal crustal provinces. Proterozoic structural provinces have been delineated using gravity and aeromagnetic data in several areas, such as Australia (Wellman, 1985) and the United States (Dutch, 1983). Counil et al. (1989) used magnetic satellite data to study Caribbean plate boundaries, in particular the subduction zone along southern Mexico. Information about Proterozoic provinces in northwestern Mexico may also be obtained by the analysis of regional magnetic data, and in particular from magnetic satellite data.

Satellite magnetic anomalies may help to distinguish the crustal structure of the North America terrane (craton) from the Seri terrane separated by the Mojave-Sonora megashear. In this way, satellite magnetic anomalies may be used to bear upon the existence of the Mojave-Sonora megashear.

Satellite magnetic surveys provide the only globally consistent information about crustal magnetization, but their interpretation is not straightforward. The sources of satellite magnetic anomalies are complex combinations of lithospheric characteristics. It is thought that for continental crust, the Moho is the lower boundary of the magnetized crust in areas where the Curie isotherm lies below the Moho (Wasilewski et al., 1979; Mayhew et al., 1985; Wasilewski and Mayhew, 1992). It is also claimed that the lower crust is the origin of long-wavelength anomalies (Hall, 1974; Wasilewski and Mayhew, 1992), and that the upper mantle contributes significantly to these anomalies (Arkani-Hamed and Strangway, 1986; Arkani-Hamed and Strangway, 1987).

However, a drawback of Magsat data is that the spatial resolution ranges from 150 to 300 km (i.e., on the order of the dimensions of many geologic provinces). Despite this resolution satellite magnetic anomalies can still be helpful in studying certain geologic provinces.

Aware of these factors, we undertook an analysis of Magsat anomalies in the neighborhood of the Mojave-Sonora megashear in northwest Mexico. We processed the magnetic information and interpreted a 2-D magnetic profile running with a SW-NE orientation from southern Baja California into Arizona and New Mexico (Fig. 1). The magnetic modeling was constrained by geological, geophysical, and heat flow data.

The objective of this work is to characterize the present crustal structure in northwestern Mexico and eventually constrain the mere existence of the Mojave-Sonora megashear.

SATELLITE MAGNETIC ANOMALIES

Three principal anomalies characterize northwestern Mexico and the southwestern United States as depicted in the Magsat anomaly map (NOAA, 1994) from Mexico and the southern United States (Fig. 2). These three anomalies appear in nine POGO and Magsat magnetic anomaly maps (Langel et al., 1982a, 1982b; Langel and Hinze, 1998; Ravat and Purucker, 1999; Ivers et al., 2000). Aeromagnetic and superficial maps (Campos-Enriquez et al., 1994; Cañon et al., 1993; Aiken et al., 1997; Tanner and the Decade of North American Geology Committee, 1998; Ornelas-Valdés, 1999) of this area show anomalies that may be correlated with the three magnetic satellite anomalies. These correlations indicate that the three anomalies within

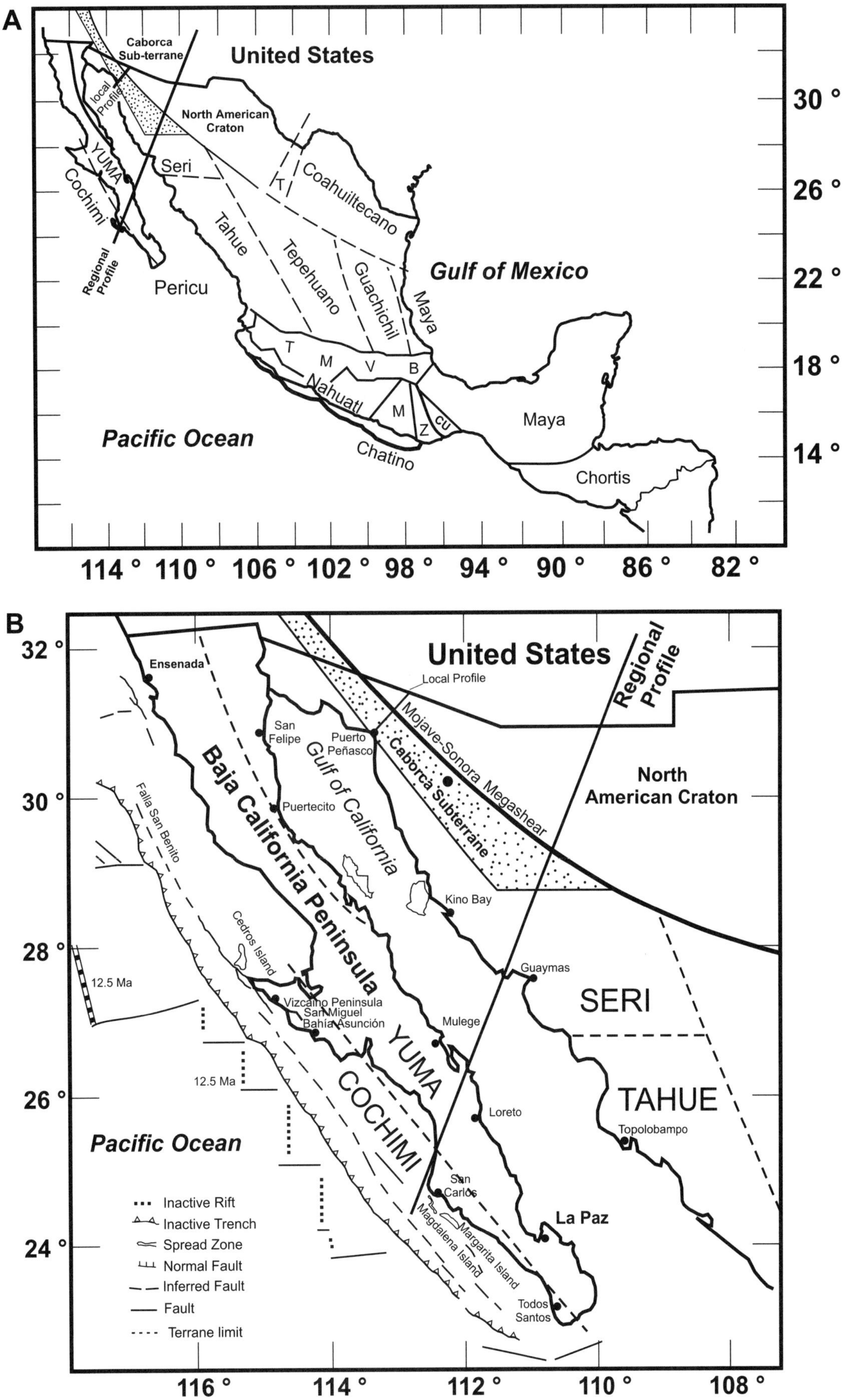

Figure 1. (A) Tectonostratigraphic terranes from Mexico and northern Central America after Sedlock et al. (1993). The dotted area represents the Caborca subterrane. The Mojave-Sonora megashear and the location of the modeled profile are indicated. CU—Cuicateco; M—Mixteco; T—Tarahumara; Z—Zapoteco; TMVB—Trans-Mexican Volcanic Belt. (B) Detailed map of the studied area. The location of the interpreted magnetic profile and of the main geologic elements is indicated. The dotted area represents the Caborca subterrane.

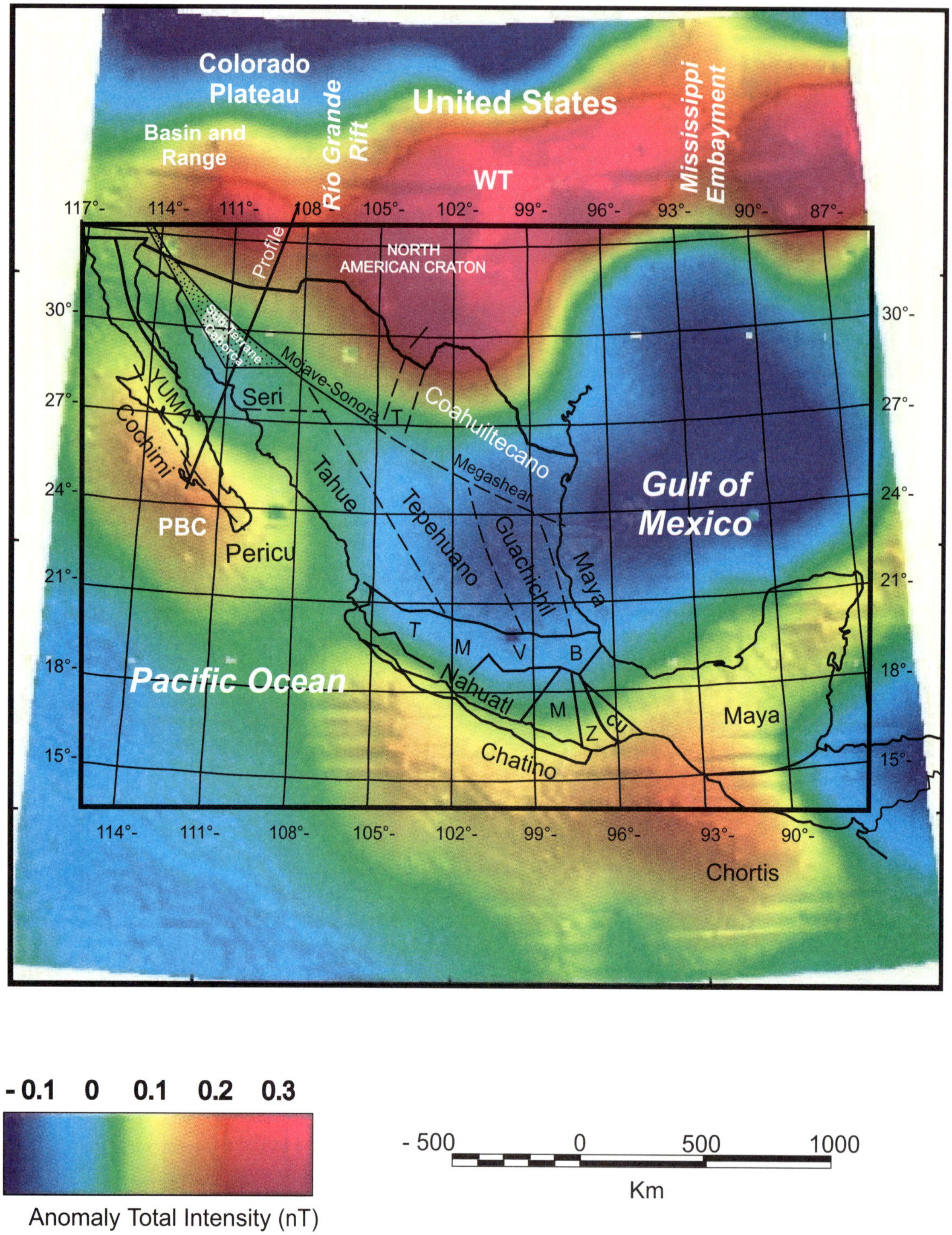

Figure 2. Magsat total intensity magnetic anomaly map of Mexico and southern United States from NOAA (1994). The location of the Mojave-Sonora megashear and of the interpreted magnetic profile is indicated. The dotted area represents the Caborca subterrane. WT and PBC indicate the magnetic anomalies of the southern United States and Baja California, respectively. In the background are indicated for reference the limits of the tectonostratigraphic terranes according to Sedlock et al. (1993).

our study zone are temporally persistent and consistent features of the geomagnetic field (i.e., from the crustal anomaly field).

A conspicuous anomaly, here referred to as WT (Fig. 2), covers a large portion of the United States and is composed of the three highs associated with: (1) the southern portion of the Colorado plateau and neighboring southern Basin and Range, (2) the granite-rhyolite province in Texas and Oklahoma, and (3) the region of the Kentucky-Tennessee boundary. These highs are separated by relative lows (at the Rio Grande rift and over the Mississippi embayment).

A positive anomaly, termed here PBC, is observed over southern Baja California and off the Baja coast over the Pacific. Its maximum is centered just east of the Magdalena and Margarita islands area (approximately at 26° N and 90° W; see Figures 2 and 3). This anomaly reaches continental Mexico at the level of southern Sinaloa.

Between these two anomalies (WT and PBC) we observe a relative low covering the southern two-thirds of Sonora. This relative low constitutes the third anomaly.

INTERPRETATION

The map of satellite anomalies was reduced to the pole in order to locate the anomalies above their sources (Blakely, 1995). We observe that the correlations between anomalies and geologic or geophysical features remain valid. The second vertical derivative (Blakely, 1995) of the reduced to the pole satellite magnetic anomalies is shown in Figure 3. It helps to infer the horizontal limits from the sources of these three anomalies. The WT anomaly is well decomposed into its three components subdivided by the Rio Grande rift and the Mississippi embayment. In particular the relative low over southern Sonora is enhanced.

A preliminary magnetic model orthogonal to the three mentioned anomalies was generated (Figs. 1–4). Its interpretation was based on a forward 2 ½-D model. The profile extends from SW to NE across the Cochimi and Yuma terranes, the Gulf of California, and continental Mexico (Seri terrane), and ends in the North American craton (southern Basin and Range, and southern Colorado plateau) (Figs. 1 and 2). Figure 4 shows the profile and its respective interpretation.

The susceptibility values are assigned according to the nature of the rock units constituting the terranes as discussed below and taking into account corresponding values reported in the geophysical literature (e.g., Langel and Hinze, 1998; Sheriff, 1991; Shive et al., 1992).

The Cochimi terrane that underlies the most conspicuous portion of the PBC anomaly corresponds with the broad marine platform along Baja California's west coast. Structurally (in both the Cedros-Benitos-Vizcaino and Magdalena-Margarita regions) this platform is composed of an imbricate stack of three plates. In the upper plate we have ophiolite, arc, and forearc basin rocks. In the lower plate we have regionally metamorphosed blueschist, and in the middle plate we have a serpentine-matrix mélange. The protolith of the lower plate blueschist is ocean floor basalt, metapillow basalt and massive metabasite weakly to moderately foliated are additional notable elements of the subduction complex (Sedlock, 1988, 1993).

The ophiolitic complex comprises metamorphosed ultramafic rocks, gabbro (layered and massive), diabase, and mafic volcanic rocks including mafic to intermediate dikes and sills and pillow. Garnetiferous amphibolite also crops out within the Magdalena-Margarita region. The subduction complex and the serpentine-matrix mélange occur on the islands of Santa Margarita, Cedros, and San Benito. Pre-arc and arc rocks, ophiolites, and garnetiferous ophiolite crop out on the islands of Santa Margarita, Magdalena, Cedros, and the Vizcaino peninsula (Fig. 1) (Crouch, 1979; Normak et al., 1987; Legg et al., 1991; Sedlock and Hamilton, 1991).

Geologic and geophysical studies conducted near San Miguel in the Vizcaino peninsula confirm the mafic and ultramafic character of the crust as shown by magnetic susceptibilities of ~5×10^{-4} (Castro-Leyva et al., 2001). Because of these results we assign a magnetic susceptibility of 5.2×10^{-4} SI to body 1 (Table 1).

The Yuma terrane is partially associated with the PBC anomaly. Of interest to the study of this anomaly is the chain of plutons of the Peninsular Ranges batholite intruding the Alisitos volcanic arc. The western plutons are of I type (Silver, 1979; Hill et al., 1986; Gromet and Silver, 1987), which are more magnetic (Chappell and White, 1974). This terrane is largely mafic and correspondingly we assign magnetic susceptibility values of 3.9 $\times 10^{-4}$ SI to body 2 (Table 1). The oceanic crust from the Gulf of California was assigned a lesser magnetic susceptibility (3.0×10^{-4} SI, body 3, Table 1) because of the high heat flow observed in the extensional basins.

Body 5 represents the crystalline basement of the Seri terrane. This terrane comprises a crystalline basement, shelfal and basinal rocks deposited on- and outboard of Proterozoic North American continental crust (e.g., Sedlock et al., 1993). Proterozoic crystalline rocks crop out in northwestern Sonora and are inferred to underlie the entire Seri terrane. The basement includes deformed quartz-rich schist and gneiss, quartzite, metarhyodacite, and amphibolite. These rocks are cut by calc-alkaline plutons (1750–1710 Ma) (Anderson and Silver, 1979, 1981). Quartzofeldspathic and amphibolite gneiss were deformed, metamorphosed, and cut by pegmatites ca. 1685–1645 Ma (Anderson and Silver, 1977, 1981). All these rocks are intruded by anorogenic granites (1450 Ma) and micrographic granites (1110 Ma) (Anderson and Silver, 1979; Anderson and Silver, 1981). Because of the silicic nature of body 5, it is assigned a magnetic susceptibility of 4.5×10^{-4} SI (Table 1). Proterozoic crystalline basement rocks of the North American craton are inferred to deepen southward (e.g., figure 16 of Sedlock et al., 1993; Valencia-Moreno et al., 2001).

The magnetic high at the northeastern tip of the profile corresponds to the southern Basin and Range and southern Colorado plateau (e.g., the North American craton). For this area, Purucker et al. (1998) obtained high values of the product of susceptibility

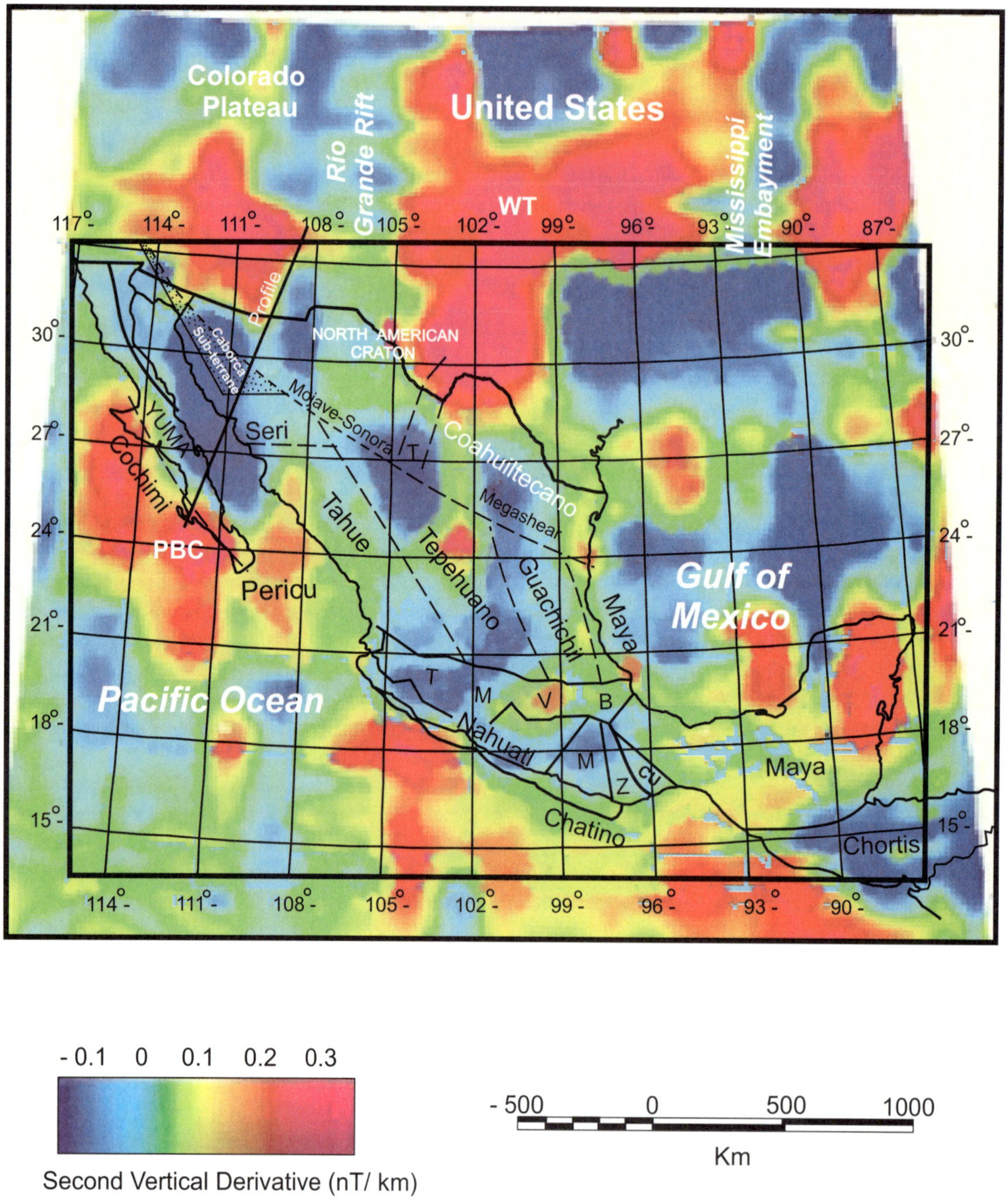

Figure 3. Second vertical derivative of the reduction-to-the-pole of the Magsat total intensity magnetic anomaly map (Fig. 2). The location of the Mojave-Sonora megashear and of the interpreted magnetic profile is indicated. The dotted area represents the Caborca subterrane. WT and PBC indicate the magnetic anomalies of the southern United States and Baja California. In the background are indicated for reference the limits of the tectonostratigraphic terranes according to Sedlock et al. (1993).

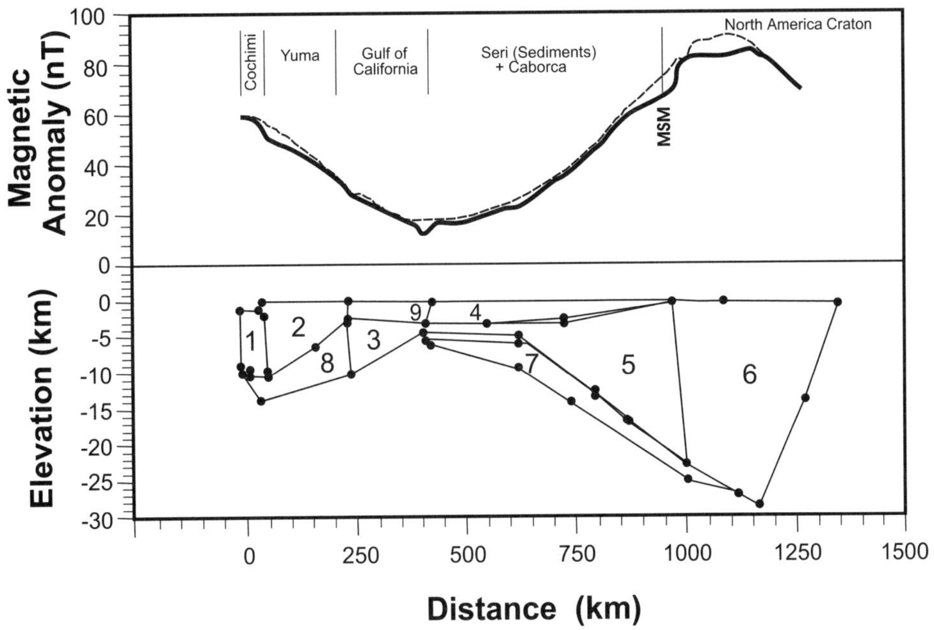

Figure 4. Interpreted magnetic profile. (A) Magnetic profile (location is indicated in Figures 1–3). (B) Magnetic model. Numbers refer to bodies in Table 1. Figures represent magnetic susceptibilities. The limits of the tectonostratigraphic terranes crossed and the location of the proposed Mojave-Sonora megashear are indicated.

times thickness for the magnetized layer beneath the Colorado plateau. The Colorado plateau consists of 35–45 km thick crust (Braile et al., 1989) and the heat flow is less than the regional mean (Mayhew, 1985). It is also characterized by Vp velocities >6.2 km/s (Braile et al., 1989). High average seismic velocities are indicative of a mafic composition (i.e., Smithson et al., 1981). Xenolites from localities at the Colorado Plateau margin include eclogite and piroxenite, indicating that the lower crust (upper mantle) is of a mafic nature (i.e., Arculus and Smith, 1979; Esperanca and Holloway, 1984). In agreement with these data body 6 is assigned a magnetic susceptibility of 5.3×10^{-4} SI (Table 1).

It is thought that at the base of the crust at the Basin and Range province there is a interlayering of mafic-ultramafic rocks formed during repeated episodes of partial melting, underplating of the crust by gabbroic magma, crystal settling, and subhorizontal shearing near the crust mantle boundary (Thompson et al., 1989).

DISCUSSION

As with any geophysical model, this model is non-unique, but the extension and location within the profile of the different bodies are in agreement with the available geological information of the area. In particular, the known limits of the tectonostratigraphic terranes (Sedlock et al., 1993) are assumed to constrain the horizontal extensions and location of the various bodies within the model (Fig. 4).

However, we recognize that a contribution to the magnetic satellite anomalies may come from the lower crust (i.e., Schnetzler, 1985). Bodies 7 and 8 represent the magnetic contribution from the lower crust. The thicknesses of bodies 7 and 8 take into account the available information about the crustal thickness and the heat flow data. For example, preliminary results based upon gravity and magnetotellurics from a traverse across the central portion of the Baja California peninsula (Romo et al., 2001) indicate that the continental crust is ~33 km thick south of the Vizcaino peninsula. According to seismic studies, the crustal thickness in the Baja California peninsula is ~30 km (Thatcher et al., 1971). The oceanic crust west of this peninsula is 10 km thick (beneath the Cedros Island and north of Vizcaino peninsula) (Fischer and Hess, 1963).

According to Thatcher and Brune (1973) the thickness of the oceanic crust in the Gulf of California is ~9 km (and 20 km in its central portion). The Gulf of California is characterized by values of heat flow higher than the mean for oceanic areas (Von Herzen, 1963).

TABLE 1. SUSCEPTIBILITY PROPERTIES OF BODIES PROPOSED IN THE MAGNETIC PROFILE

Body	$\sigma \times 10^{-4}$ (SI)	Geologic correspondence
1	5.2	Cochimi terrane
2	3.9	Yuma terrane
3	3.0	Gulf of California (Oceanic Crust)
4	0.0	Seri terrane (sediments)
5	4.5	Seri terrane (crystalline basement)
6	5.3	North American craton
7	1.2	Lower crust
8	0.9	Lower crust
9	0.0	Gulf of California (water)

Note: Magnetic susceptibilities in u.m. (cgs) units of the bodies constituting the magnetic model of Figure 4.

According to Thatcher and Brune (1973) the crustal thickness in Sonora (Mexico mainland) is similar to that of the Baja California peninsula (~25 km). The thickness of body 6 is assumed to be ~25 km, taking into account these data.

Calderon-Riveroll (1978) and Schellhorn et al. (1991) elaborated crustal models for our study area. These models show a two layer crust, mantle, and upper lithosphere, and were based on gravity data and seismic constraints. Our model is consistent with the crustal structure inferred by these authors. In particular the crustal thicknesses are similar. According to Schellhorn et al. (1991), the Mojave-Sonora megashear is not recognizable in the gravity data, although a crustal boundary is indicated by a change in the trend or truncation of various features in northwestern Sonora. A more or less continuous gravity gradient from the Pacific Ocean toward the continent is interrupted by a relative high at the extrapolated position of the Mojave-Sonora megashear. This feature is modeled by a body in the middle of the crust. The models from Calderon-Riveroll (1978) do not intersect the possible trace of the Mojave-Sonora megashear. However, these profiles incorporate sea magnetic measurements in Baja California and corroborate the mafic nature of the rock outcrops at the Vizcaino peninsula. In contrast to these gravity models, our model shows the presence of a suture zone at the place of the proposed Mojave-Sonora megashear.

In our magnetic model, bodies 5 and 6 are separated by a subvertical contact. The contact between these two bodies was modeled with several dips and positions. In all of our models the contact remains subvertical. As already mentioned, the Mojave-Sonora megashear itself is not resolvable in the Magsat data. However, our model shows the presence of a contact between two bodies with contrasting magnetic susceptibility. This contact may be interpreted as a suture zone between two different geologic units (terranes): the North American craton (Matzazal Archean province) and the Caborca subterrane (Seri terrane). This supports evidence for the existence of the Mojave-Sonora megashear.

Global magnetic satellite studies (e.g., Arkani-Hamed and Strangway, 1985) or regional studies of the United States (e.g., Mayhew, 1982; Schnetzler, 1985; Purucker et al., 1998) report crustal magnetization or susceptibility of the crust for the study area. In these studies the PBC and WT anomalies are associated with units of higher magnetizations, whereas the relative magnetic low over Sonora is associated with a relatively lower crustal magnetization. Our magnetic model agrees with these studies: bodies 1 and 6, associated with the Cochimi terrane and the North American craton (Colorado Plateau), respectively, are assigned high magnetic susceptibility values, in contrast to the Seri terrane (including the Caborca subterrane, e.g., body 5).

ACKNOWLEDGMENTS

Critical reviews and comments by J. Arkani-Hamed, W. Harbut, Thomas H. Anderson, Roberto Molina, and an anonymous reviewer helped to improve the manuscript.

REFERENCES CITED

Aiken, C.L.V., Balde, M., Xueming, X., Abdel-Salam, M.G., De la Fuente, M., and Mena, M., 1997, Integrated studies of Mexico with gravity, magnetic, and GIS database: The Leading Edge, v. 16, p. 1779–1785, doi: 10.1190/1.1437580.

Anderson, T.H., and Schmidt, V.A., 1983, The evolution of Middle America and the Gulf of Mexico-Caribbean Sea during Mesozoic time: Geological Society of America Bulletin, v. 94, p. 941–966, doi: 10.1130/0016-7606(1983)94<941:TEOMAA>2.0.CO;2.

Anderson, T.H., and Silver, L.T., 1977, Geochronometric and stratigraphic outlines of the Precambrian rocks of northwestern Mexico: Geological Society of America Abstracts with Programs, v. 9, p. 880.

Anderson, T.H., and Silver, L.T., 1979, The role of the Mojave-Sonora megashear, the evolution of northern Sonora, *in* Anderson, T.H., and Roldán-Quintana, J., eds., Geology of northern Sonora, Guidebook, Field Trip # 27 prepared for the 1979 Geological Society of America Annual Meeting, University of Pittsburgh, Pennsylvania, p. 59–68.

Anderson, T.H., and Silver, L.T., 1981, An overview of Precambrian rocks in Sonora, Mexico: Revista del Instituto de Geología, Universidad Nacional Autónoma de México, v. 5, p. 131–139.

Arculus, R.J., and Smith, D., 1979, Eclogite, piroxenite, and amphibolite in inclusions in the Sullivan Buttes latite, Chino Valley, Yavapai County, Arizona, *in* Boyd, F.R. and Meyer, H.O.A., eds., The mantle sample; Inclusions in kimberlites and other volcanics: Washington, D.C., American Geophysical Union, p. 309–317.

Arkani-Hamed, J., and Strangway, D.W., 1985, Lateral variations of apparent magnetic susceptibility of lithosphere deduced from magsat data: Journal of Geophysical Research, v. 90, p. 2655–2664.

Arkani-Hamed, J., and Strangway, D.W., 1986, Effective magnetic susceptibility of the oceanic upper mantle derived from Magsat data: Geophysical Research Letters, v. 13, p. 999–1002.

Arkani-Hamed, J., and Strangway, D.W., 1987, An interpretation of magnetic signatures of subduction zones detected by Magsat: Tectonophysics, v. 133, p. 45–55, doi: 10.1016/0040-1951(87)90279-4.

Blakely, R.G., 1995, Potential theory in gravity and magnetic applications: Cambridge University Press, 437 p.

Braile, L.W., Hinze, H.J., von Fese, R.R.B., and Keller, G.R., 1989, Seismic properties of the crust and uppermost mantle of the conterminous United States and adjacent Canada, *in* Pakiser, L.C., and Mooney, W.D., eds., Geophysical framework of the United States, Geological Society of America Memoir 172, p. 177–203.

Bullard, E.C., Everett, J., and Smith, A.G., 1965, A symposium on continental drift: Philosophical Transactions of the Royal Astronomical Society of London, Ser. A, v. 258, p. 41–51.

Calderon-Riveroll, G., 1978, A marine geophysical study of Vizcaino Bay and the continental margin of western Mexico between 27° and 30 °N [Ph.D. dissertation]: Corvallis, Oregon State University, 95 p.

Campbell, P.A., and Anderson, T.H., 1998, Structure and kinematics along a Jurassic plate boundary transform, the Mojave-Sonora megashear, northwestern Sonora, Mexico, *in* Gold Deposits of Northern Sonora, Mexico, Guidebook Series, Society of Economic Geologists, v. 30, p. 177–186.

Campos-Enriquez, J.O., Hernández-Quintero, E., Nolasco-Chávez, H., Torres-Orozco, A., Cañon-Amaro, G., and Urrutia-Fucugauchi, J., 1994, A preliminary assessment of IGRF-1990 for Mexico: Physics of the Earth and Planetary Interiors, v. 82, p. 105–111, doi: 10.1016/0031-9201(94)90083-3.

Cañon, A., Hernández-Quintero, J.E., Nolasco, H., and Orozco, A., 1993, Cartas magnéticas de la República Mexicana para la época 1990.0 escala 1: 5,000,000, Instituto de Geofisica, UNAM, Instituto Nacional de Estadistica Geografía e Informática, INEGI, Mexico.

Castro-Leyva, T. de J., Delgado-Argote, L.A., and García-Abdeslem, J., 2001, Geología y magnetometría del complejo máfico-ultramáfico Puerto Nuevo en el área de San Miguel, Peninsula de Vizcaino, Baja California Sur: GEOS, Transactions of the Mexican Geophysical Union, v. 21, p. 3–21.

Caudillo-Sosa, G., Oviedo-Lucero, L.F., and Rodríguez-Castañeda, J.L., 1996, Falla Quitovac-Resultado de un evento de transpresión del "Mojave-Sonora megashear", Noroeste de Sonora, México: Revista Mexicana de Ciencias Geológicas, v. 13, p. 140–151.

Chappell, B.W., and White, A.J.R., 1974, Two contrasting granite types: Pacific Geology, v. 8, p. 173–174.

Counil, J.-L., Achache, J., and Galdeano, A., 1989, Long-wavelength magnetic anomalies in the Caribbean: Plate boundaries and allochthonous continental blocks: Journal of Geophysical Research, v. 94, p. 7419–7431.

Crouch, J.K., 1979, Neogene tectonic evolution of the western Transverse Ranges and the California Continental Borderland: Geological Society of America Bulletin, v. 90, p. 338–345, doi: 10.1130/0016-7606(1979)90<338:NTEOTC>2.0.CO;2.

Dutch, S.I., 1983, Proterozoic structural provinces in the north-central United States: Geology, v. 11, p. 478–481, doi: 10.1130/0091-7613(1983)11<478:PSPITN>2.0.CO;2.

Esperanca, S., and Holloway, J.R., 1984, Lower crustal nodules from the Camp Creek latite, Carefree, Arizona, *in* Kornprobst, J., ed., Kimberlite; 2, The mantle and crust-mantle relationship: Amsterdam, Elsevier, p. 214–227.

Fisher, R.L., and Hess, H.H., 1963, Trenches, *in* The Sea-Ideas and observations in progress, No. 3, The earth beneath the sea (and) history: New York, Interscience Publishers, p. 411–436.

Gose, W.A., Belcher, R.C., and Scott, G.R., 1982, Paleomagnetic results from northeastern Mexico: Evidence for large Mesozoic rotations: Geology, v. 10, p. 50–54, doi: 10.1130/0091-7613(1982)10<50:PRFNME>2.0.CO;2.

Gromet, L.P., and Silver, L.T., 1987, REE variations across the Peninsular Ranges batholith: Implications for batholithic petrogenesis and crustal growth in magmatic arcs: Journal of Petrology, v. 28, p. 75–125.

Hall, D.H., 1974, Long-wavelength aeromagnetic anomalies and deep crustal magnetization in Manitoba and northwestern Ontario, Canada: Journal of Geophysical Research, v. 40, p. 403–430.

Hill, R.I., Silver, L.T., and Taylor, H.P., Jr., 1986, Coupled Sr-O isotope variations as an indicator of source heterogeneity for the northern Peninsular Ranges batholith: Contributions to Mineralogy and Petrology, v. 92, p. 351–361, doi: 10.1007/BF00572164.

Irving, E., 1977, Drift of the major continental blocks since the Devonian: Nature, v. 270, p. 304–309.

Iturralde-Vinent, M.A., 1975, Problems in applications of modern tectonics hypothesis to Cuba and Caribbean region: American Association of Petroleum Geologists Bulletin, v. 59, p. 838–855.

Ivers, D.J., Stenning, R.J., Turner, J., and Winch, D.E., 2000, Oersted and Magsat scalar anomaly fields: Earth, Planets and Space, v. 52, p. 1213–1225.

Kluger Cohen, K., Anderson, T.H., and Schmidt, V.A., 1986, A paleomagnetic test of the proposed Mojave-Sonora megashear in northwestern Mexico: Tectonophysics, v. 131, p. 23–51, doi: 10.1016/0040-1951(86)90266-0.

Langel, R.A., and Hinze, W.J., 1998, The magnetic field of the earth's lithosphere: the satellite perspective: Cambridge, UK, Cambridge University Press, 429 p.

Langel, R.A., Schnetzler, C.C., Phillips, J.D., and Horner, R.J., 1982a, Initial scalar magnetic anomaly map from Magsat: Geophysical Research Letters, v. 9, p. 265–268.

Langel, R.A., Schnetzler, C.C., Phillips, J.D., and Horner, R.J., 1982b, Initial vector magnetic anomaly map from Magsat: Geophysical Research Letters, v. 9, p. 273–276.

Legg, M.R., Wong, O.V., and Suárez-Vidal, F., 1991, Geologic structure and tectonics of the inner continental borderland of northern Baja California, *in* Dauphin, J.P., and Simoneit, B.R.T., eds., The Gulf and Peninsular Province of the Californias: American Association of Petroleum Geologists Memoir 47, p. 145–177.

Malfait, B., and Dinkelman, M., 1972, Circum-Caribbean tectonic and igneous activity and the evolution of the Caribbean plate: Geological Society of America Bulletin, v. 83, p. 251–272.

Mayhew, M.A., 1982, An equivalent layer magnetization model for the United States derived from satellite altitude magnetic anomalies: Journal of Geophysical Research, v. 87, p. 4837–4845.

Mayhew, M.A., 1985, Curie isotherm surfaces inferred from high altitude magnetic anomaly data: Journal of Geophysical Research, v. 90, p. 2647–2654.

Mayhew, M.A., Johnson, B.D., and Wasilewski, O.J., 1985, A review of problems and progress in studies of satellite magnetic anomalies: Journal of Geophysical Research, v. 90, p. 2511–2522.

Molina-Garza, R.S., and Geissman, J.W., 1996, Timing of deformation and accretion of the Antimonio terrane, Sonora, from paleomagnetic data: Geology, v. 24, p. 1131–1134, doi: 10.1130/0091-7613(1996)024<1131:TODAAO>2.3.CO;2.

Molina-Garza, R.S., and Geissman, J.W., 1999, Paleomagnetic data from the Caborca terrane, Mexico: Implications for Cordilleran tectonics and the Mojave-Sonora megashear hypothesis: Tectonics, v. 18, p. 293–325, doi: 10.1029/1998TC900030.

National Oceanic and Atmospheric Administration (NOAA), 1994, Geophysics of North America, CD-ROM. Boulder, Colorado, National Oceanic and Atmospheric Administration, U.S. Department of Commerce, National Data Center.

Normak, W.R., Spencer, J.E., and Ingle, J.C., Jr., 1987, Geology and Neogene history of the Pacific continental margin of Baja California Sur, Mexico, *in* Scholl, D.W., Grantz, A., and Vedder, J.G., eds., Geology and resource potential of the continental margin of western North America and adjacent ocean basins—Beaufort Sea to Baja California, Vol. 6, Circum Pacific Council for Energy and Mineral Resources, Earth Sciences Series, p. 449–472.

Ornelas-Valdés, R., 1999, Elaboración de la carta aeromagnética de la República Mexicana [Master's thesis]: México, D.F., Universidad Nacional Autónoma de México, 73 p.

Pindell, J., and Dewey, J., 1982, Permo-Triassic reconstruction of western Pangea and the evolution of the Gulf of Mexico/Caribbean region: Tectonics, v. 1, p. 179–211.

Purucker, M.E., Langel, R.A., Rajaram, M., and Raymond, C., 1998, Global magnetization models with a priori information: Journal of Geophysical Research, v. 103, p. 2563–2584, doi: 10.1029/97JB02935.

Ravat, D., and Purucker, M., 1999, The future of satellite magnetic anomaly studies is bright: The Leading Edge, v. 18, p. 326–329, doi: 10.1190/1.1438283.

Rodríguez-Castañeda, J.L., 2002, Tectónica cretácica y terciaria en la margen suroeste del Alto de Cananea, Sonora norte-central [Ph.D. thesis]: México, D.F., Universidad Nacional Autónoma de México, 217 p.

Romo, J.M., García-Abdeslem, J., Gómez-Treviño, E., Esparza, F., and Flores-Luna, C., 2001, Resultados preliminares de un perfil geofísico a través de la región central de la península de Baja California, México: GEOS, Transactions of the Mexican Geophysical Union, v. 21, p. 96–107.

Schellhorn, R.W., Aiken, C.L.V., and de la Fuente, M.F., 1991, Bouguer gravity anomalies and crustal structure in northwestern Mexico, *in* Dauphin, J.P., and Simoneit, B.R.T., eds., The gulf and peninsular province of the Californias: American Association of Petroleum Geologists Memoir, 47, p. 197–215.

Schnetzler, C.C., 1985, An estimation of continental crust magnetization and susceptibility from Magsat data for the conterminous United States: Journal of Geophysical Research, v. 90, p. 2617–2620.

Scotese, C.R., Bambach, R.K., Barton, C., Van der Voo, R., and Ziegler, A.M., 1979, Paleozoic base maps: Journal of Geology, v. 87, p. 217–277.

Sedlock, R.L., 1988, Tectonic setting of blueschists and island-arc terranes of west-central Baja California, Mexico: Geology, v. 16, p. 623–626, doi: 10.1130/0091-7613(1988)016<0623:TSOBAI>2.3.CO;2.

Sedlock, R.L., 1993, Mesozoic geology and tectonics of blueschist and associated oceanic terranes in the Cedros-Vizcaino-San Benito and Magdalena-Margarita region, Baja California, Mexico, *in* Dunne, G.C., and McDougall, K.A., eds., Mesozoic paleography of western United States: Pacific Section, Society of Economic Paleontologists and Mineralogists, v. 71, p. 113–125.

Sedlock, R.L., and Hamilton, D.H., 1991, Late Cenozoic tectonic evolution of southwestern California: Journal of Geophysical Research, v. 96, p. 2325–2352.

Sedlock, R.L., Ortega-Gutiérrez, F., and Speed, R.C., 1993, Tectonostratigraphic terranes and tectonic evolution of Mexico: Geological Society of America Special Paper 278, 153 p.

Sheriff, R.E., 1991, Encyclopedic dictionary of exploration geophysics: Geophysical Reference Series, 3rd ed. Society of Exploration Geophysicists. 376 p.

Shive, P.N., Blakely, R.J., Frost, B.R., and Fountain, D.M., 1992, Magnetic properties of the lower continental crust, *in* Fountain, D.M., Arculus, R., and Kay, R.W., eds., Continental lower crust: Developments in Geotectonics, v. 23, p. 145–177.

Silver, L.T., 1979, Peninsular Ranges batholith: A case study in continental margin magmatic arc setting, characteristics, and evolution: Geological Society of America Abstracts with Programs, v. 11, p. 517.

Silver, L.T., and Anderson, T.H., 1974, Possible left-lateral early to middle Mesozoic disruption of the southwestern North American craton margin: Geological Society of America Abstracts with Programs, v. 6, p. 995.

Smithson, S.B., Johnson, R.A., and Wong, Y.K., 1981, Mean crustal velocity: A critical parameter for interpreting crustal structure and crustal growth: Earth and Planetary Science Letters, v. 53, p. 323–332, doi: 10.1016/0012-821X(81)90037-6.

Tanner, J.G., and the Decade of North American Geology Committee, 1988, Gravity anomaly map for North America: The Leading Edge, v. 7, no. 11, p. 15–18.

Thatcher, W., and Brune, J.N., 1973, Surface waves and crustal structure in the Gulf of California region: Bulletin of the Seismological Society of America, v. 63, p. 1689–1698.

Thatcher, W.J., Brune, N., and Clay, D.N., 1971, Seismic evidence on the crustal structure of the Imperial valley region: Geological Society of America Abstracts with Programs, v. 3, p. 208.

Thompson, G.A., Catchings, R., Goodwin, E., Holbrook, S., Jarchow, C., Mann, C., McCarthy, J., and Okaya, D., 1989, Geophysics of the western Basin and Range province, *in* Pakiser, L.C., and Mooney, W.D., eds., Geophysical framework of the United States: Geological Society of America Memoir 172, p. 177–203.

Urrutia-Fucugauchi, J., 1981, Paleomagnetic evidence for tectonic rotation of northern Mexico and the continuity of the Cordilleran orogenic belt between Nevada and Chihuahua: Geology, v. 9, p. 178–183, doi: 10.1130/0091-7613(1981)9<178:PEFTRO>2.0.CO;2.

Urrutia-Fucugauchi, J., 1984, On the tectonic evolution of Mexico: Paleomagnetic constraints, *in* Van der Voo, R., Scotese, C.R., and Bonhommet, N., eds., Plate reconstructions from Paleozoic paleomagnetism, American Geophysical Union Geodynamic Series, 12, Washington, D.C., American Geological Union, p. 29–47.

Valencia-Moreno, M., Ruiz, J., Barton, M.D., Patchett, P.J., Zürcher, L., Hodkinson, D.G., and Roldán-Quintana, J., 2001, A chemical and isotopic study of the Laramide granitic belt of northwestern Mexico: Identification of the southern edge of the North American Precambrian basement: Geological Society of America Bulletin, v. 113, p. 1409–1422, doi: 10.1130/0016-7606(2001)113<1409:ACAISO>2.0.CO;2.

Van der Voo, R., and French, R., 1974, Apparent polar wandering for the Atlantic-bordering continents: Earth Science Reviews, v. 10, p. 99–119, doi: 10.1016/0012-8252(74)90082-8.

Von Herzen, R.P., 1963, Geothermal heat flow in the gulfs of California and Aden: Science, v. 40, p. 1207–1208.

Wasilewski, P.J., and Mayhew, M.A., 1992, The Moho as a magnetic boundary revisited: Geophysical Research Letters, v. 19, p. 2259–2262.

Wasilewski, P., Thomas, H.H., and Mayhew, M.A., 1979, The Moho as a magnetic boundary: Geophysical Research Letters, v. 6, p. 541–544.

Wellman, P., 1985, Block structure of continental crust derived from gravity and magnetic maps, with Australian examples, *in* Hinze, W.J., ed., The utility of regional gravity and magnetic anomaly maps, Society of Exploration Geophysicists, p. 102–108.

White, G.W., 1980, Permian-Triassic continental reconstruction of the Gulf of Mexico-Caribbean area: Nature, v. 283, p. 823–826, doi: 10.1038/283823a0.

Manuscript Accepted by the Society 24 March 2005

Printed in the USA

Geological Society of America
Special Paper 393
2005

Evidence for Mojave-Sonora megashear—Systematic left-lateral offset of Neoproterozoic to Lower Jurassic strata and facies, western United States and northwestern Mexico

John H. Stewart
U.S. Geological Survey, 345 Middlefield Road, Menlo Park, California 94025, USA

ABSTRACT

Major successions as well as individual units of Neoproterozoic to Lower Jurassic strata and facies appear to be systematically offset left laterally from eastern California and western Nevada in the western United States to Sonora, Mexico. This pattern is most evident in units such as the "Johnnie oolite," a 1- to 2-m-thick oolite of the Neoproterozoic Rainstorm Member of the Johnnie Formation in the western United States and of the Clemente Formation in Sonora. The pattern is also evident in the Lower Cambrian Zabriskie Quartzite of the western United States and the correlative Proveedora Quartzite in Sonora. Matching of isopach lines of the Zabriskie Quartzite and Proveedora Quartzite suggests ~700–800 km of left-lateral offset. The offset pattern is also apparent in the distribution of distinctive lithologic types, unconformities, and fossil assemblages in other rocks ranging in age from Neoproterozoic to Early Jurassic. In the western United States, the distribution of facies in Neoproterozoic and Paleozoic strata indicates that the Cordilleran miogeocline trends north-south. A north-south trend is also suggested in Sonora, and if so is compatible with offset of the miogeocline but not with the ideas that the miogeocline wrapped around the continental margin and trends east-west in Sonora. An imperfect stratigraphic match of supposed offset segments along the megashear is apparent. Some units, such as the "Johnnie oolite" and Zabriskie-Proveedora, show almost perfect correspondence, but other units are significantly different. The differences seem to indicate that the indigenous succession of the western United States and offset segments in Mexico were not precisely side by side before offset but were separated by an area—now buried, eroded, or destroyed—that contained strata of intermediate facies.

Keywords: Mojave-Sonora megashear, left-lateral fault, Neoproterozoic to Jurassic, Mexico, western United States.

*Email: stewart@usgs.gov

Stewart, J.H., 2005, Evidence for Mojave-Sonora megashear—Systematic left-lateral offset of Neoproterozoic to Lower Jurassic strata and facies, western United States and northwestern Mexico, *in* Anderson, T.H., Nourse, J.A., McKee, J.W., and Steiner, M.B., eds., The Mojave-Sonora megashear hypothesis: Development, assessment, and alternatives: Geological Society of America Special Paper 393, p. 209–231. doi: 10.1130/2005.2393(07). For permission to copy, contact editing@geosociety.org.

INTRODUCTION

The Mojave-Sonora megashear was proposed by Silver and Anderson (1974), Anderson and Silver (1979, 1984), and Anderson and Schmidt (1983) as a Middle or Late Jurassic left-lateral fault that displaced Paleoproterozoic and Mesoproterozoic crystalline basement rocks from southeastern California to Sonora. As a part of their original proposal, Silver and Anderson (1974) noted "intriguing similarities" between Cambrian strata in the Inyo Mountains in eastern California on the northeast side of the fault and in the Caborca area in northern Sonora on the southeast side of the fault. They used this similarity to suggest an offset of 700–800 km along the proposed megashear. Since the original work of Silver and Anderson, much has been learned about the Neoproterozoic and Paleozoic stratigraphy in the western United States and Sonora that bears on the "intriguing similarities" of the rocks possibly offset by the proposed fault. This paper summarizes this new information and interprets the data as supporting the original idea of the Mojave-Sonora megashear.

Two different models of the Neoproterozoic to Triassic continental margin of the western United States and northern Mexico have been proposed (Stewart et al., 1984) (Fig. 1). In one, Neoproterozoic to Triassic rocks wrap around the Laurentian (ancestral North American) continental margin with only minor, if any, lateral structural displacement. In the other, the Laurentian margin is truncated by major left-lateral offset along the proposed Mojave-Sonora megashear (Stewart et al., 1984). Debate between proponents of these two ideas has been lively, and no consensus has been reached concerning which of the two models is correct. Discussions emphasizing one or the other of the two ideas include Silver and Anderson (1974, 1983), Anderson and Silver (1979, 1984), Dickinson (1981) Anderson and Schmidt (1983), Stewart et al. (1984), Cohen et al. (1986), Ketner (1986), Jacques-Ayala et al. (1991), Anderson et al. (1991, 1995), Stevens et al. (1992), Sedlock et al. (1993), DeJong and Jacques-Ayala (1993), Bartolini (1993), Stanley and González-León (1995), Poole et al. (1995a, 1995b, 1991, 2000a), Marzolf and Anderson (1996, 2000), Anderson (1997), Marzolf (1997), Caudillo-Sosa et al. (1996), González-León (1997a, 1997b), Campbell and Anderson (1998), Molina-Garza and Geissman (1998, 1999), Iriondo et al. (1998), Iriondo (2001a, 2001b), Stewart et al. (1990, 1997, 1999), Dembosky et al. (1999), Poole and Amaya-Martínez (2000), and Dickinson and Lawton (2001).

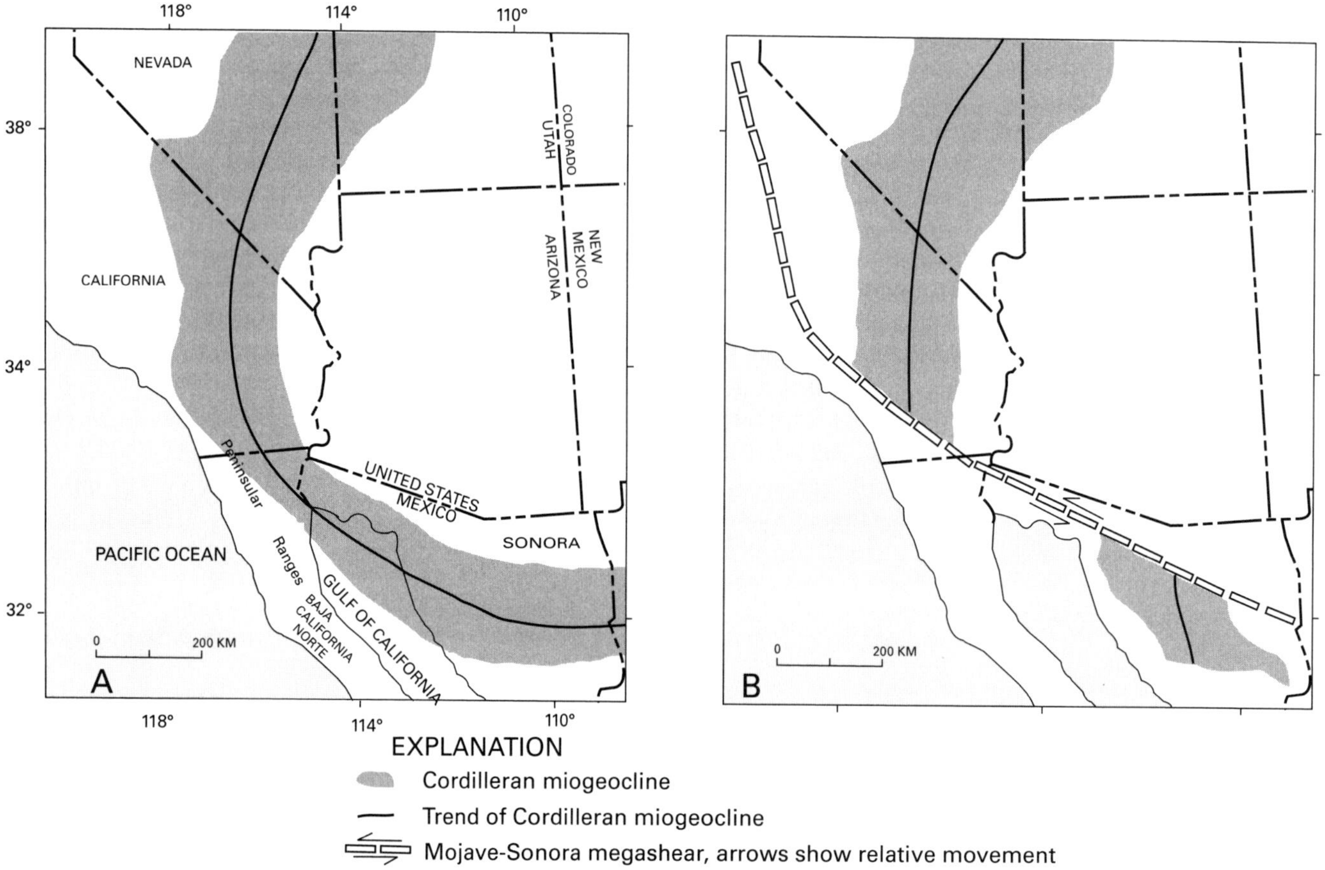

Figure 1. Wraparound and Mojave-Sonora megashear models of Neoproterozoic and Paleozoic rocks in the western United States and northwest Mexico.

This report describes the distribution, lithofacies, and depositional trends of Neoproterozoic to Triassic strata in the western United States and northwest Mexico, information that is interpreted here to support the concept of major left-lateral offset on the Mojave-Sonora megashear. In a series of accompanying maps (Figs. 2–14), outcrops of Neoproterozoic to Lower Jurassic rocks are shown in their present positions—no attempt is made of a palinspastic reconstruction. A thick double-dashed line on these figures separates outcrops that are considered to be displaced. This line is an attempt to approximate the Middle Jurassic position of the megashear relative to present-day outcrops. However, the line differs in important details, as will be discussed below, from the original proposed position of the megashear by Silver and Anderson (1974), Anderson and Silver (1979, 1984), and Anderson and Schmidt (1983). The difference between the position of the megashear shown here and the position originally proposed may be due in part to post–Late Jurassic deformation that has disrupted the original positions of the megashear and of Neoproterozoic to Lower Jurassic strata. The complexities of such deformation in the western United States are summarized by Schweickert and Lahren (1990), Burchfiel et al. (1992), Wernicke (1992), Allmendinger (1992), Stewart and Crowell (1992), and Saleeby (1992).

Outcrops of Paleozoic and Lower Mesozoic rocks west of the San Andreas fault (Burchfiel et al., 1992; Stewart and Crowell, 1992) in the Peninsular Ranges of California and Baja California Norte (Gastil, 1993) are not considered here. If reconstructed palinspastically for 300 km of right lateral movement on the San Andreas fault, these strata would lie west to northwest of Paleozoic to Lower Mesozoic strata in Sonora. The Paleozoic and Lower Mesozoic strata in the Peninsular Ranges and nearby areas are difficult to evaluate palinspastically. They have been considered by Gastil (1993) as a southern continuation of the Cordilleran miogeoclinal and eugeoclinal belts, whereas Madrid et al. (1992a) and Stewart (unpublished data) have noted that at least some of the rocks, including the Coyote terrane of Madrid et al. (1992a), are unusual and consider them to be possibly a separate terrane outboard and west of the Cordilleran miogeocline. For the purposes of this paper, I adopt the latter interpretation and do not show Paleozoic or Lower Mesozoic strata in the Peninsular Ranges of southernmost California and Baja California Norte.

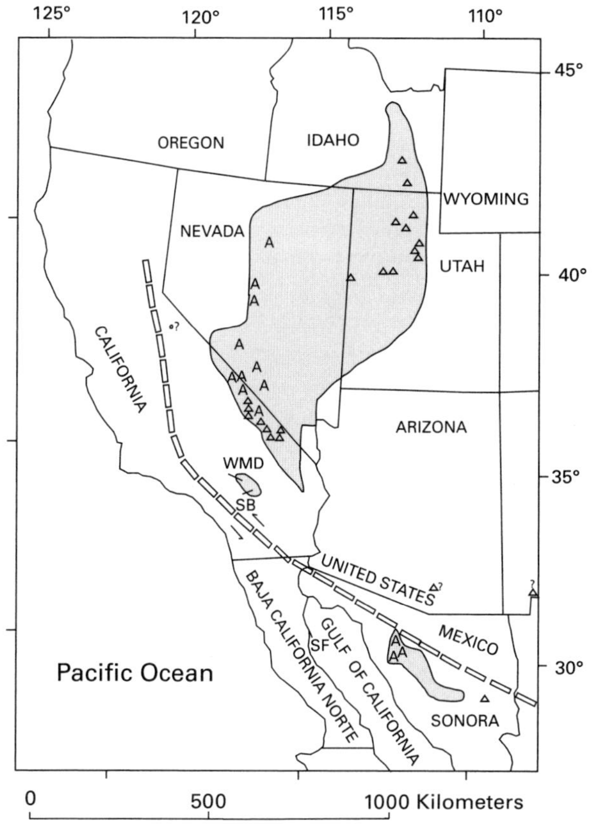

EXPLANATION

Distribution of Neoproterozoic and Lower Cambrian miogeoclinal strata

Δ Diamictite

A Archeocyathids

Mojave-Sonora megashear. Postion of megashear is not corrected palinspastically for post-Late Jurassic structural dislocations. See text for discussion of location of megashear. Arrows show relative motion.

Localities: SF, San Felipe; SB; San Bernardino Mountains; WMD, western Mojave Desert

Figure 2. Distribution of Neoproterozoic and Lower Cambrian miogeoclinal strata in the western United States and northwest Mexico and of Neoproterozoic diamictite (tillite) and Early Cambrian archeocyathids. Based on Stewart (1970, 1991, 1992), Stewart et al. (2002), and Stewart and Poole (2002). Archeocyathids after Stewart (1970), Stewart and McKee (1977), Stewart et al. (1984), Gangloff (1976), and Debrenne et al. (1989). Diamictite after Stewart (1991) and Corseti et al. (2001).

NEOPROTEROZOIC AND LOWER CAMBRIAN MIOGEOCLINAL STRATA

Neoproterozoic and Lower Cambrian miogeoclinal strata are extensively exposed in the western United States and northeastern Mexico and are the most useful strata in evaluating the concept of the megashear. The term "miogeocline" is used here as originally defined by Dietz and Holden (1967) as a wedge-shaped continental shelf (continental terrace) deposit, similar to Cretaceous and younger deposits along the Atlantic margin of the United States. In the western United States, these deposits are referred to as the Cordilleran miogeocline.

Neoproterozoic and Lower Cambrian strata in the Cordilleran miogeocline (Fig. 2) are widely exposed in eastern California, eastern Nevada, northwestern Utah, and southern Idaho (Stewart and Suczek, 1977; Christie-Blick and Levy, 1989; Stewart, 1991, 1992). They thin toward the east and reach a depositional limit in easternmost California and Nevada, central Utah, and eastern Idaho. Toward the south they extend into the San Bernardino Mountains and the western Mojave Desert (Fig. 2) (Stewart and Poole, 1975; Cameron, 1981, 1982; Miller, 1981; Brown, 1991; Martin and Walker, 1991). South-southeast of the San Bernardino Mountains, Neoproterozoic and Lower Cambrian strata do not crop out for ~600 km. Outcrops in Sonora, Mexico (Stewart et al., 1984; Stewart et al., 2002), lie southeast of this gap in outcrops. Strata originally considered to be Neoproterozoic at San Felipe along the east coast of Baja California (Anderson, 1993) at San Felipe are now considered to be Ordovician (Fig. 2) (Gehrels et al., 2002).

In detail, individual Neoproterozoic and Cambrian units show an even better defined separation of outcrops between the eastern California–southern Nevada region and northwest Sonora. The distribution of the Neoproterozoic "Johnnie oolite" is particularly striking (Fig. 3). This oolite is only 1 or 2 m thick and forms a distinctive marker unit in eastern California and southern Nevada (Stewart, 1970), where it lies in the lower part of the Rainstorm Member of the Johnnie Formation. Apparently the same oolite bed is present in northwestern Sonora, where it forms part of the Clemente Formation (Stewart et al., 1984, 2002). In both the California–southern Nevada area and northwest Sonora, the oolite is overlain by distinctive pale red siltstone, limy siltstone, and silty limestone that commonly contain drag marks and flute casts. These detailed lithostratigraphic similarities provide strong support to the idea that the oolite in the California-Nevada area and in Sonora was originally continuous and close to one another.

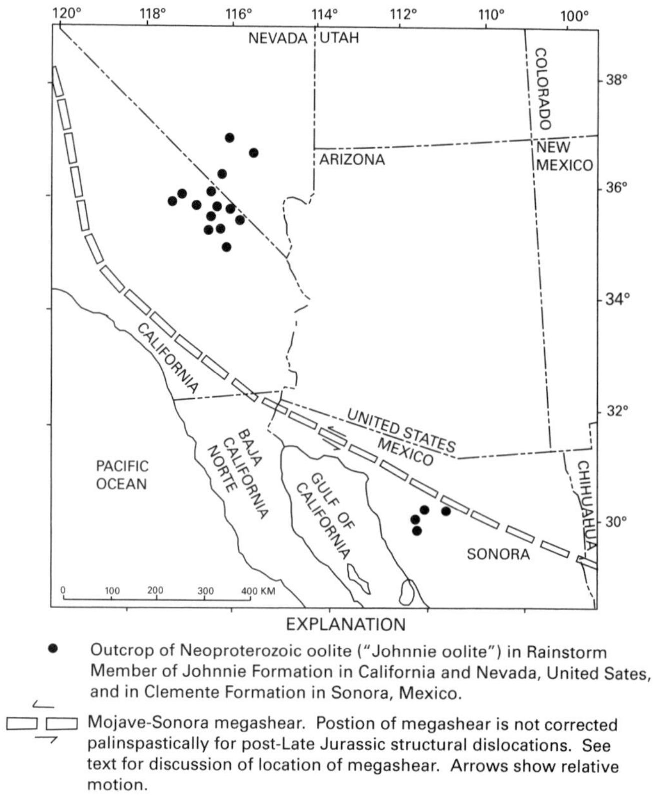

Figure 3. Distribution of Neoproterozoic "Johnnie oolite" of the Rainstorm Member of the Johnnie Formation in California and Nevada and of the Clemente Formation in Sonora. Based on Stewart (1970) and Stewart et al. (1984, 2002).

The Lower Cambrian Zabriskie Quartzite and the correlative Proveedora Quartzite in Sonora show a similar outcrop distribution to the "Johnnie oolite" (Fig. 4). The correlation of the Zabriskie Quartzite and the Proveedora Quartzite is clear. Both units are bracketed by similar trilobite faunal zones and both are composed almost entirely of quartz grains, in contrast to other Neoproterozoic and Lower Cambrian quartzites that contain small to significant amounts of feldspar. Both the Zabriskie Quartzite and Proveedora Quartzite contain abundant *Skolithos* (vertical burrows) in the lower part and both thicken to the west, suggesting that in each case the outer part of the miogeocline lay to the west (Fig. 4). Apparent offset of the 100- and 200-m isopachs of the Zabriskie Quartzite and Proveedora Quartzite indicates 700–800 km of left-lateral displacement.

The similarity of strata in the western United States and Sonora is also illustrated by the distribution of Neoproterozoic diamictite (tillite) and early Cambrian archeocythids (Fig. 2). The diamictite occupies an inner to middle shelf position (see summary of distribution in Christie-Blick and Levy [1989] and Stewart [1991, 1992]) and in the western United States extends from Idaho to southern Nevada and eastern California. An outcrop of diamictite in eastern Sonora has recently been described by Corsetti et al. (2001). Archeocyathids are present in Lower Cambrian strata in central and southern Nevada and eastern California (Stewart, 1970; Gangloff, 1976) and southeast of a gap in outcrops, in northwest Sonora (Cooper et al., 1952; Stewart et al., 1984; Debrenne, 1987; Debrenne et al., 1989).

The overall distribution of Neoproterozoic and Lower Cambrian miogeoclinal strata and of individual units of this age is considered to best fit with the megashear model rather than the wraparound model (Fig. 1). The south-southwest trend of Neoproterozoic and Lower Cambrian miogeoclinal strata across southern Idaho, western Utah, Nevada, and eastern California is well established (Stewart, 1991; Christie Blick and Levy, 1989). The abrupt change in position of these strata southeastward to Sonora (Fig. 2) is striking and compatible with the megashear model. More importantly, the original distribution of the Johnnie oolite (Fig. 3) and of the Zabriskie-Proveedora quartzites (Fig. 4) must have been relatively restricted, and their present widely spaced distribution between the western United States and Sonora is suggestive of offset. The distribution pattern of diamictite (tillite) deposits and of archeocyathids also supports the megashear model. The position of the archeocyathids (outer shelf) northwest of diamictite deposits (inner or middle shelf) in the western United States as well as in northwest Sonora is consistent with offset of one area from the other.

Neoproterozoic and Lower Cambrian miogeoclinal strata in Sonora (Stewart et al., 2002) consist of thin Lower Cambrian inner shelf deposits on the east and of thicker Neoproterozoic and Lower Cambrian outer shelf deposits on the west or northwest (Fig. 5). This pattern is consistent with the offset of a north-

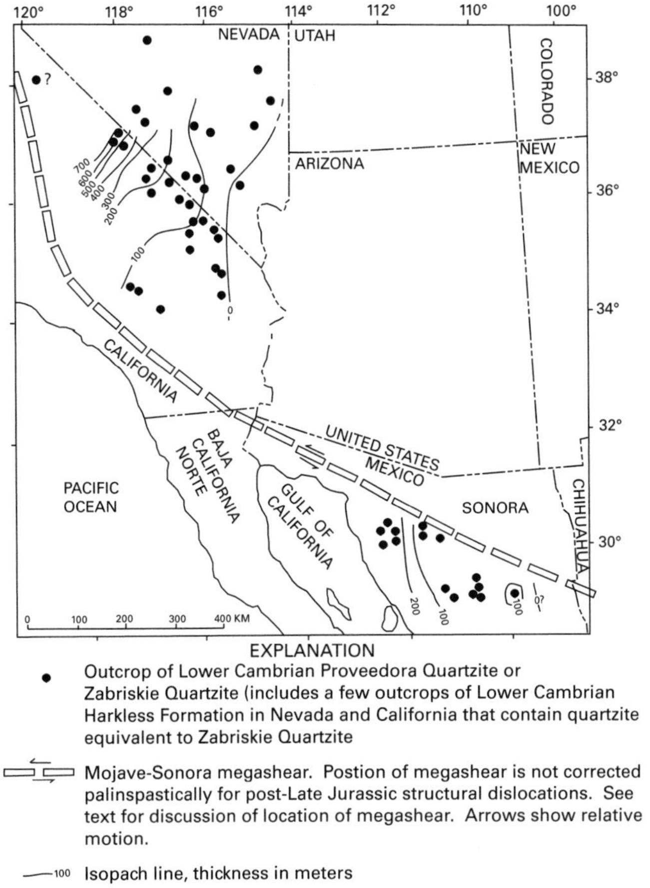

Figure 4. Distribution and thickness of Lower Cambrian Zabriskie Quartzite in California and Nevada and of the correlative Proveedora Quartzite in Sonora. Based on Stewart (1970) and Stewart et al. (1984).

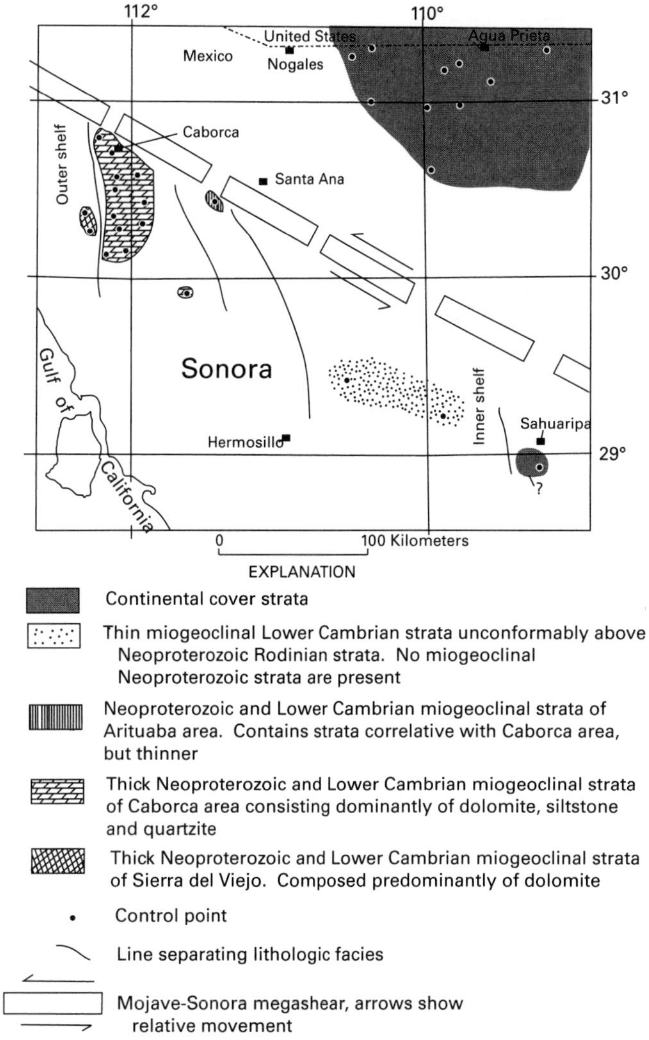

Figure 5. Facies trends of Neoproterozoic and Lower Cambrian strata in Sonora. Based on data in Stewart et al. (2002).

trending miogeoclinal belt, but not with a wraparound margin in which the inner shelf deposits are on the north and the outer shelf deposits are on the south.

MIDDLE AND UPPER CAMBRIAN MIOGEOCLINAL STRATA

Middle and Upper Cambrian miogeoclinal strata consist mostly of thick carbonate successions (Stewart and Suczek, 1977; Stewart, 1991; Palmer and Hintze, 1992; Stewart et al., 2002). They extend from Idaho, south to eastern California, into the San Bernardino Mountains, and then, south of a 600-km gap in outcrops, into northwest Sonora (Fig. 6). Correlation of specific units is restricted to the lowermost part of the Middle and Upper Cambrian succession including the upper part of the Lower Cambrian succession. These rocks consist of the Carrara Formation in eastern California and southern Nevada (Palmer and Halley, 1979) and of the correlative combined Buelna, Cerro Prieto, and Arrojos Formations in northeast Sonora (Cooper et al., 1952; Palmer, 1981; Stewart et al., 1984, 2002). The Carrara Formation consists of shale and siltstone, lime mudstone, oolitic limestone, algal-bound limestone, and oncoid limestone. The Buelna, Cerro Prieto Formation, and Arrojos are lithostratigraphically similar to the Carrara Formation and contain a trilobite fauna correlative to that in the Carrara Formation. The Carrara and the Cerro Prieto Formations both contain distinctive oncoid limestone. The distinctive stratigraphy of the Carrara Formation is found only in eastern California and southern Nevada, and in the correlative Buelna, Cerro Prieto, and Arrojos Formations in Sonora. This similar stratigraphy is suggestive of offset of one to the other. Strata above the Carrara Formation in eastern California and southern Nevada are assigned to the thick carbonate succession of the Bonanza King Formation. The incompletely exposed Tren Formation in Sonora is correlated with the Bonanza King Formation. Correlations of Cambrian strata younger than the Tren Formation in Sonora are sketchy and uncertain (Stewart et al., 1990, 2002).

The distribution pattern of outcrops of Middle and Upper Cambrian miogeoclinal strata is similar to that for Neoproterozoic and Lower Cambrian miogeoclinal strata, suggesting offset along the megashear.

ORDOVICIAN MIOGEOCLINAL STRATA

Ordovician miogeoclinal strata (Ross, 1977; Ross et al., 1991, 1992; Poole and Sandberg, 1992a, 1992b) consist mostly of carbonate rock and widespread Middle Ordovician quartzite, locally containing interstratified limy sandstone. The Middle Ordovician quartzites are referred to as the Eureka Quartzite in most areas of the western United States (Ketner, 1968) and as the Peña Blanca Quartzite in Mexico (Ketner, 1986; Ketner and Noll, 1987; Poole et al., 2000a). The Ordovician miogeoclinal strata have the same left-lateral change in distribution pattern as other Neoproterozoic and Paleozoic strata in the western United States and Mexico (Fig. 7). Ketner (1986) has indicated that the Eureka Quartzite and the Peña Blanca Quartzite are correlative and favors the idea that the Peña Blanca is offset from the Eureka. Ketner indicates that the Eureka Quartzite and Peña Blanca Quartzite are similar in internal stratigraphy, thickness, grain size, sorting, and mineral composition. Ketner's study was in the Cerro Cobachi area, and subsequent work in the Sierra Lopez area (Stewart et al., 1990; Poole et al., 1995a) and near Rancho Placeritos (Poole et al., 2000a) revealed additional outcrops of the Peña Blanca Formation and a more complete section of Ordovician strata. In the Sierra Lopez area and near Rancho Placeritos, gastropods and brachiopods of the *Palliseria* subzone and *Anomalorthis* zone are recognized below the Peña Blanca Quartzite (Stewart et al., 1990; Poole et al., 2000a) in a position similar to the position of these fossil zones below the Eureka Quartzite in the western United States (Ross, 1970; Stevens, 1986).

Ordovician strata are thicker and of a greater age span to the west in Sonora. Strata at Rancho Placeritos (Poole et al., 2000a), Sierra Lopez (Stewart et al., 1990), and Rancho Bízani (Poole et al., 1995a) contain the largest age span of Ordovician strata in Sonora, indicative of a middle to outer shelf position in comparison to localities to the east or southeast (Poole et al., 1995a, 2000a). This relation is consistent with the megashear model.

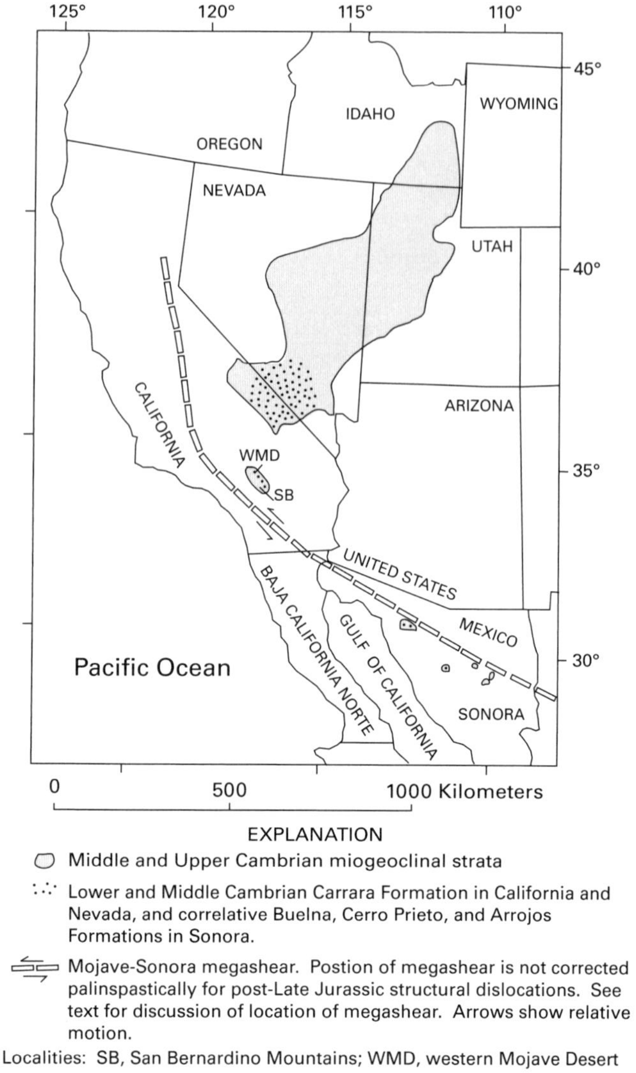

Figure 6. Distribution of Middle and Upper Cambrian miogeoclinal strata in the western United States and northwestern Mexico. Based on Palmer (1971), Stewart and Suczek (1977), Palmer and Halley (1979), Stewart (1991, 1992), Stewart et al. (2002), and Stewart and Poole (2002)

SILURIAN AND DEVONIAN MIOGEOCLINAL STRATA

Silurian and Devonian strata consist largely of carbonate and sandy carbonate rocks in the western United States (Poole et al., 1977; Poole and Sandberg, 1977, 1992a, 1992b; Johnson et al., 1991; Sheehan and Boucot, 1991; Sandberg and Poole, 1992a, 1992b) and in Sonora (Poole et al., 1997, 1998, 2000a; Stewart et al., 1997; Stewart and Poole, 2002). These rocks show the same apparent left-lateral offset as other Neoproterozoic to Lower Jurassic strata (Fig. 8). Silurian strata in Sonora crop out only in the Rancho Placeritos, Cerro Carnero (southern Rancho Barajita area), and Rancho Bízani area (Poole et al., 1997, 1998; Stewart and Poole, 2002). These three localities are considered to indicate a Silurian outer shelf position (Poole et al., 1997, 1998), and if so, suggest that the outer shelf was to the west, compatible with the megashear model.

To the east in Sonora, Devonian strata truncate underlying Silurian and Ordovician strata. At Rancho Placeritos in central Sonora, Devonian strata rest on Silurian strata. Eastward, Devonian, or Devonian (?), strata rest unconformably on the Middle Ordovician Peña Blanca Quartzite or possible equivalents (Poole et al., 1995a). At Sierra Agua Verde, the easternmost continuous section of Ordovician and Devonian strata in Sonora, Upper Devonian strata rest with an angular unconformably on Lower Ordovician strata (Stewart et al., 1999). This truncation of under-

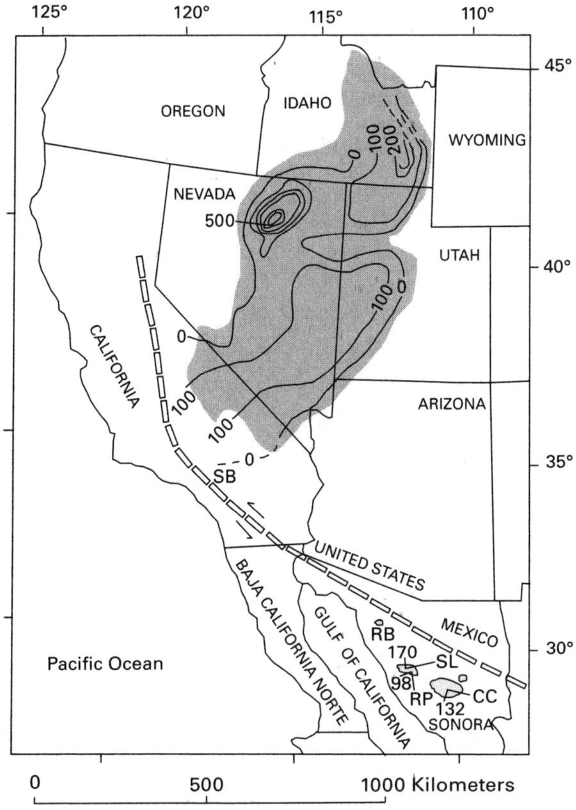

EXPLANATION

Ordovician miogeoclinal strata

100 — Isopach of Middle Ordovician Eureka Quartzite and equivalent strata in western United States, thicknesses in meters. The Eureka and other Ordovician strata are absent in the Paleozoic in the Paleozoic succession of the San Bernardino Mountains (SBM)

98 Thickness in meters of Middle Ordovician Peña Blanca Quartzite in Sonora

Mojave-Sonora megashear. Position of megashear is not corrected palinspastically for post-Late Jurassic structural dislocations. See text for discussion of location of megashear. Arrows show relative motion.

Localities: CC, Cerro Cobachi; RB, Rancho Bízani; RP, Rancho Placeritos; SB, San Bernardino Mountains; SL, Sierra Lopez

Figure 7. Distribution and thickness of Ordovician miogeoclinal strata in the western United States and northwestern Mexico. Based on Ross (1977), Ross et al. (1991, 1992), Poole and Sandberg (1992a), Poole et al. (1995a), and Stewart and Poole (2002).

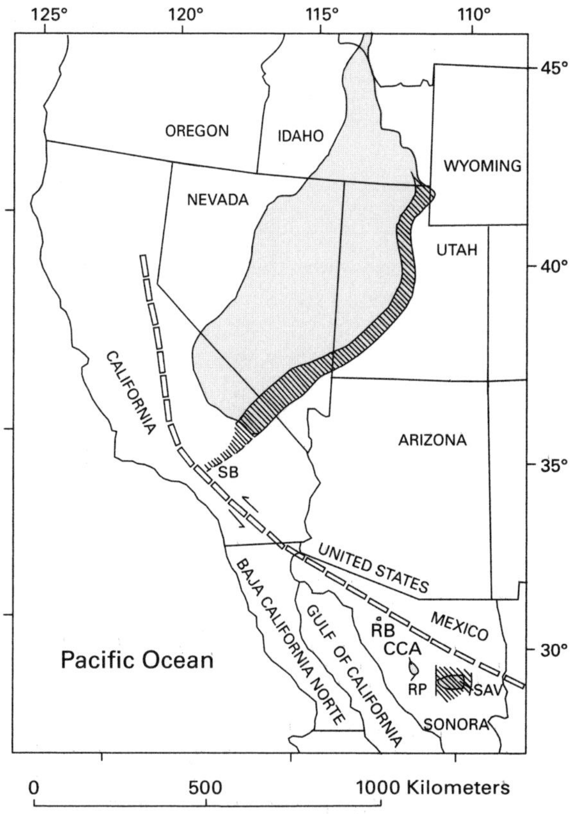

EXPLANATION

Zone of unconformable overlap of Devonian strata over Silurian and Ordovician strata. Dashed where inferred.

Silurian and Devonian miogeoclinal strata

Mojave-Sonora megashear. Position of megashear is not corrected palinspastically for post-Late Jurassic structural dislocations. See text for discussion of location of megashear. Arrows show relative motion.

Localities: CCA, Cerro Carnero; RB, Rancho Bízani; RP, Rancho Placeritas; SAV, Sierra Agua Verde; SB, San Bernardino Mountains

Figure 8. Distribution of Silurian and Devonian miogeoclinal strata in the western United States and northwestern Mexico. Based on Poole et al. (1977), Poole and Sandberg (1977, 1992a, 1992b), Sandberg and Poole (1992a, 1992b), Sheehan and Boucot (1991), Johnson et al. (1991), Poole et al. (1997, 1998, 2000a, 2000b), Stewart et al. (1997), and Stewart and Poole (2002).

lying rocks and overlap of Devonian strata is similar to relations in the western United States where Devonian strata along the eastern margin of the Cordilleran miogeocline unconformably overlap Ordovician and Silurian strata and rest on Cambrian strata in areas to the east (Gans, 1974; Hills and Kottlowski, 1983; Hintze, 1988). Outcrops in Sonora are not complete enough to expose the presumed depositional relations of Devonian strata unconformably on Cambrian strata. Outcrops of Paleozoic strata in thc San Bcrnardino Mountains provide control on the trend of the zone of Devonian truncation. In the San Bernardino Mountains, no outcrops of Ordovician or Silurian strata are known, and Devonian strata rest unconformably on Cambrian strata, a relation corresponding to areas east of the zone of truncation.

MISSISSIPPIAN MIOGEOCLINAL STRATA AND THE ANTLER OROGENIC BELT

The onset of the Late Devonian and Early Mississippian Antler orogeny led to a marked change in the sedimentary and structural setting of the western United States (Poole, 1974;

Poole and Sandberg, 1977, 1991; Miller et al., 1992). The Antler allochthon (Fig. 9) is composed of ocean-basin siliceous strata and was emplaced eastward above Devonian and older Paleozoic miogeoclinal strata producing the Antler orogenic highland. This highland fed coarse to fine clastic material into the Antler trough to the east. East of this trough is a basin of largely shale, and to the east of this is limestone along the eastern part of the miogeocline.

The Mississipian stratigraphy and structure of miogeoclinal strata in Sonora is significantly different from that in the western United States. In Sonora, Mississipian miogeoclinal strata are mostly limestone (Stewart et al., 1997, 1999; Poole et al., 2000b; Stewart and Poole, 2002). No Antler orogenic highland, no coarse to fine Antler trough strata, and no shale-basin strata are recognized. The known record of the Antler orogeny is largely confined to easternmost California, Nevada, and southern Idaho.

Outcrops of Mississippian strata in Sonora are sparse, and conceivably, strata related to the Antler orogeny are covered—or if they are present, they have not been discovered. Another possibility is that the strata related to the Antler orogeny originally lay west of known outcrops of Mississippian strata in Sonora. In this case, the Mississippian carbonate strata of Sonora could be an offset segment of Mississippian strata that lay east of the Antler belt in eastern California.

A further possibility, an idea favored here, is that strata related to the Antler orogeny died out southward in eastern California and never existed in Sonora (Fig. 2) (Poole et al., 2000a). Such a possibility is supported by the apparent absence of Antler strata in Sonora. A fairly complete section of Devonian and Mississippian strata close to the presumed western margin of the Mississippian shelf at Rancho Placeritos in Sonora (Fig. 9) (Poole et al., 2000b) does not contain coarse clastic strata that could have been derived from a continuation of the Antler highland.

The location of the change from areas containing indications of the Antler orogeny to those that do not is problematic. The southward track of the belt through Nevada is well defined from structural and sedimentary data (Poole and Sandberg, 1991). Where the belt enters California, its outcrop width lessens and recognition of structures related to the belt are less certain, although Schweickert and Lahren (1987) and Greene et al. (1997) have described what they consider to be the Roberts Mountain allochthon emplaced during the Antler orogeny in the eastern part of the Sierra Nevada in easternmost California. South of these outcrops, the only possible record of the Antler orogeny is in the El Paso Mountains, the Pilot Knob Valley area, and the Shadow Mountains in the Mojave Desert region of southern California (Fig. 9). In the El Paso Mountains, no definitive structural record of the Antler orogeny has been recognized, and pre-Mississippian and Mississippian and younger Paleozoic strata contain deformational fabrics similar to each other (Carr et al., 1997). However, Poole (1974) and Carr et al. (1992, 1997) interpret clastic strata of Early and Late Mississippian age in the El Paso Mountains as foreland-basin deposits derived from the Antler orogenic highland. In the Pilot Knob Valley area, Carr et al. (1992) interpret metaconglomerate and metaargillite as

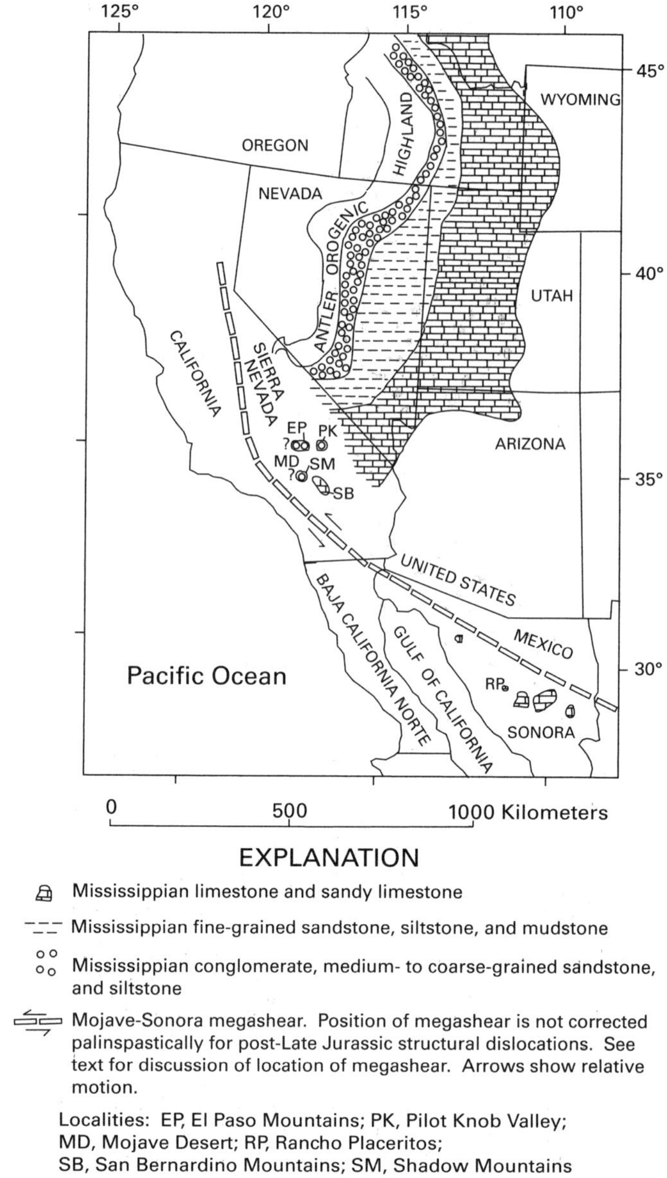

Figure 9. Distribution of Mississippian miogeoclinal and foreland-basin strata in the western United States and northwestern Mexico. Based on Poole (1974), Poole and Sandberg (1977, 1991), Miller et al. (1992), and Stewart and Poole (2002). In the western United States, distribution and facies patterns are of Upper Mississippian strata, modified from Figures 9 and 10 of Poole and Sandberg (1991).

syngenetic deposits related to the Antler orogeny. In the Shadow Mountains, Poole (1974) similarly has interpreted clastic strata as Antler foreland basin deposits, although Martin and Walker (1991) have argued against this interpretation and doubt that Antler belt rocks exist in the Shadow Mountains.

The Antler orogenic belt may die out somewhere south of outcrops in the eastern Sierra Nevada. However, if the interpretation of Antler foreland deposits in the El Paso Mountains and Shadow Mountains is correct, then the Antler orogenic belt extends into the central part of the Mojave Desert region, but perhaps not into the westernmost part. The exact distribution of the Antler orogenic belt is critically important for the interpretation presented here of offset on the Mojave-Sonora megashear. If the Antler orogenic belt does extend to the megashear and is offset left-laterally, it should reappear in Sonora, and so far it has not been recognized there.

PENNSYLVANIAN AND PERMIAN MIOGEOCLINAL STRATA

Pennsylvanian and Permian miogeoclinal strata (Fig. 10) in the western United States (Rich, 1977; Stevens, 1977, 1991; Ross, 1991) and Sonora (Pérez-Ramos, 1992; Stewart et al., 1997, 1999; Stewart and Poole, 2002) consist predominantly of thick successions of shallow-water carbonate strata containing troughs of deeper-water turbiditic silty to sandy carbonate and sandstone and local conglomerate. The strata show the same general distribution pattern as other Neoproterozoic to Jurassic rocks, a pattern that is consistent with an interpretation of offset on the Mojave-Sonora megashear. Fusulinids of the McCloud fauna, or of a McCloud-like fauna, are present in the Inyo Mountains of eastern California (Magginetti et al., 1988), in the Mojave Desert (Ross and Ross, 1983), in central Sonora (Nestell, 2000), and at scattered localities elsewhere in the western United States (Fig. 10). This fauna is generally considered to have formed in a continental borderland outboard of the Cordilleran miogeocline, or perhaps at the edge of the miogeocline. The presence of the McCloud-like fauna indicates a similarity of the eastern California and Sonora strata, suggestive that one was offset from the other. However, the widespread distribution of this fauna precludes the precise matching of one area to another. Arguing against possible offset are Mississippian to Permian strata at Sierra Santa Teresa in Sonora that have a different stratigraphy and history than comparable age strata in eastern California (Stewart et al., 1997).

The Upper Permian Monos Formation in the El Antimonio area of northwest Sonora (Cooper et al., 1953; Stewart et al., 1990, 1997) may lie near or outboard of the Cordilleran margin (Fig. 10). The Monos Formation consists of two different lithologic units, a lower turbiditic succession of siltstone, limy siltstone, and detrital fossiliferous limestone and an upper unit of resistant fossiliferous limestone locally containing giant fusulinids (Dunbar, 1953). The lower turbiditic unit has been interpreted as a relatively deep-water deposit (Stewart et al., 1990, 1997) near or outboard of the miogeocline, and the upper unit could be either shallow-water deposits or slide blocks into a deep-water environment. The gigantic fusulinids are unusual. Similar fusulinids are not known in strata of the Cordilleran miogeocline and are only known in the western United States in accreted terranes of the eastern Klamath Mountains in northern California and in northeastern Washington (Ross and Ross,

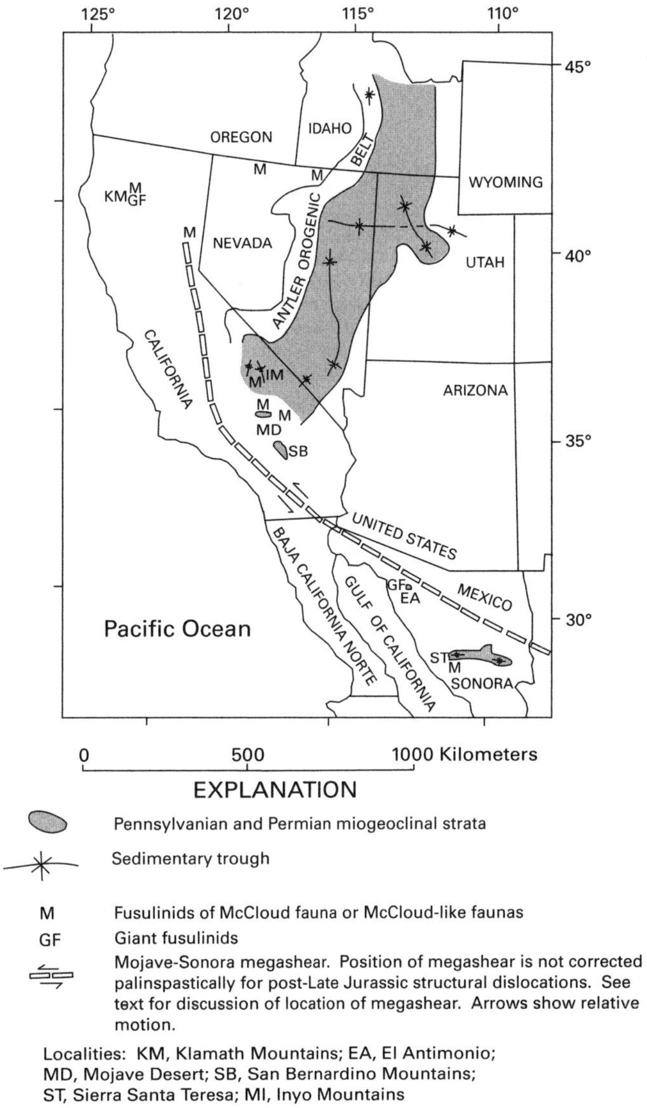

Figure 10. Distribution of Pennsylvanian and Permian miogeoclinal rocks in the western United States and northwestern Mexico. Based on Rich (1977), Ross (1991), Stevens (1991), Stewart et al. (1997), and Stewart and Poole (2002). Distribution of McCloud or McCloud-like fauna after Ross and Ross (1983), Magginetti et al. (1988), and Nestell (2000).

1983). The evidence from the Monos Formation suggests that it may be an accreted terrane or in an unusual setting near the edge of the miogeocline. In either case, the edge of the Upper Permian miogeocline appears to lie in a westward position, compatible with the megashear model and not with the wraparound model.

MARINE UPPER TRIASSIC AND LOWER JURASSIC ROCKS

Marine and locally nonmarine rocks of Upper Triassic and Early Jurassic age (Fig. 11) are widespread in central and western

Nevada, eastern California, and Sonora. Only strata considered to have been part of ancestral North America are considered here. Outboard terranes such as the Taylorsville area in the northern Sierra Nevada, the Klamath Mountains in northern California, and the Blue Mountains in southeastern Oregon are excluded. In central and western Nevada, marine Upper Triassic and Lower Jurassic rocks (Silberling and Roberts, 1962; Nichols and Silberling, 1977; Stewart, 1980, 1997) consist of limestone, shale, sandstone, and conglomerate, largely of Late Triassic age, overlain by a thick succession of generally fine-grained clastic strata of Early Jurassic age. Interstratified volcanic rocks are locally present in Upper Triassic strata. The exact southward continuation of these rocks is uncertain, but they appear to connect with strata in the northern Sierra Nevada (Schweickert, 1978; Stewart, 1997) and then southward to form the Kings sequence in the central and southern Sierra Nevada. The Kings sequence (Saleeby et al., 1978; Saleeby and Busby-Spera, 1992; Saleeby and Busby, 1993) is a complex assemblage of quartzite, marble, calc-silicate rocks, schist, chert-argillite, and slate, silicic metavolcanic rocks, and basalt-andesite metavolcanic rocks. Several Late Triassic and Early Jurassic fossil localities are known in the Kings sequence. The correlation with the Nevada and northern Nevada rocks with the Kings sequence has been proposed previously by Stewart (1988) and Saleeby and Busby-Spera (1992). Saleeby and Busby-Spera (1992) show a belt of widespread shallow-water Upper Triassic carbonate rocks extending from western Nevada southward to form the lower part of the Kings sequence. The amount of igneous rock in this belt is variable, and they classify the eastern part of the belt in Nevada as a backarc basin and the western part of the belt in westernmost Nevada and the entire Kings sequence in California as a magmatic arc containing abundant sedimentary rocks.

South of the southern Sierra Nevada, marine Upper Triassic and Lower Jurassic strata are not exposed for a distance of ~700 km. Exposures of Upper Triassic and Lower Jurassic strata reappear in Sonora on the southwest side of the megashear. These strata vary from marine limestone, mudstone, sandstone, and lenticular conglomerate in northwest Sonora (Stanley and González-León, 1995; González-León et al., 1996; Lucas and Estep, 1999) to interstratified marine to nonmarine successions of siltstone, carbonaceous shale, coal (graphite), sandstone, pebble to boulder conglomerate, and sparse tuff (Cojan and Potter, 1991; Stewart and Roldán-Quintana, 1991; Stewart et al., 1997) in east-central Sonora.

Stanley and González-León (1995) have suggested that strata that contain a distinctive Late Triassic sponge and coral assemblage in the Luning Formation in western Nevada and the Antimonio Formation (equivalent to the combined Antimonio, Rio Asunción, and Sierra de Santa Rosa Formations of Lucas and Estep, 1999) in the El Antimonio area in Sonora were offset from western Nevada to Sonora (Fig. 11). However, the evidence of apparent offset is questionable because the coralline part of the Luning Formation is lowest Norian and the presumably correlative coralline part of the Antimonio Formation is uppermost

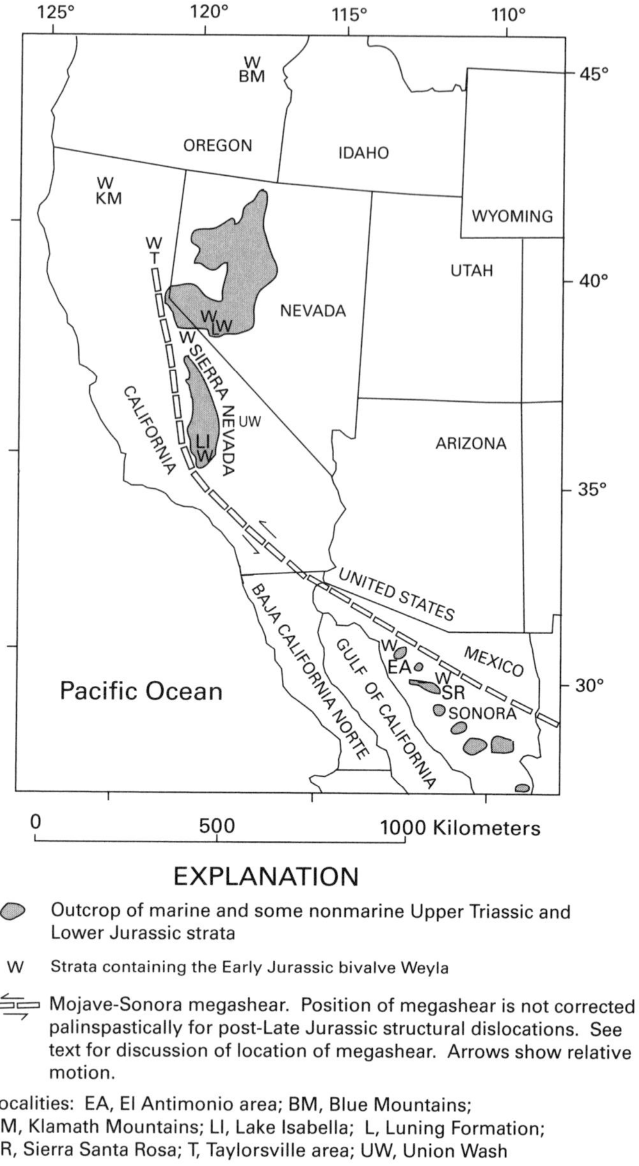

Figure 11. Distribution of Upper Triassic and Lower Jurassic marine and nonmarine continental margin strata. Based on Stewart (1980, 1997), Nokleberg (1983), Saleeby et al. (1978), and Stewart et al. (1997). Distribution of the Early Jurassic bivalve *Weyla* after Saleeby and Busby (1993), Lucas and Estep (1997, 1999), Damborenea and González-León (1997), and N.J. Silberling (2002, personal commun.).

Middle Norian (Silberling, 1984; Lucas and Estep, 1999; N.J. Silberling, personal commun., 2002). Marzolf (1997), Anderson (1997), and Marzolf and Anderson (2000) also have suggested offset of lower Mesozoic strata from western Nevada and easternmost California to Sonora on the basis of facies distribution, hiatal geometry, and cross-cutting relations. On the basis of Lower Triassic marine sedimentary strata, Lucas et al. (1997; Lucas and Estep, 1999) suggest a correlation of Lower and Middle(?)

Triassic marine sedimentary strata from the Union Wash area in eastern California to the El Antimonio area in Sonora, suggesting the possibility that the two localities may have once been close together and later offset along the megashear. However, strata above the Lower and Middle(?) Triassic strata at Union Wash consist of undated volcanic rocks, interstratified with volcanic sandstone and conglomerate, unlike the stratigraphic sequence in the El Antimonio area. Either the two sections are not correlative, or Upper Triassic and Lower Jurassic marine strata comparable to that in the El Antimonio area have been eroded, presumably by pre–Late Jurassic erosion.

An alternative interpretation, following in part a proposal by González-León (1997a), is that the Antimonio Formation is offset from the Kings sequence in the southern Sierra Nevada of California, rather than from eastern California areas such as Union Wash (Fig. 11). The Kings sequence and Antimonio Formation are similar in that they both contain shallow-marine carbonate, marine siltstone and sandstone, and fluvial conglomerate (Jones and Moore, 1973; Girty, 1985; Saleeby and Busby, 1993; Moore et al., 1994; González-León et al., 1996; González-León, 1997a, 1997b; Lucas and Estep, 1999), as well as deep-water turbiditic siliciclastic strata (Saleeby et al., 1978; Nokleberg, 1983; Saleeby and Busby, 1993; González-León, 1997a, 1997b). The Kings sequence in the Lake Isabella area of the southern Sierra Nevada contains the Early Jurassic bivalve *Weyla* (Saleeby and Busby, 1993), as does the Antimonio Formation in the El Antimonio area, Sonora, and the Sierra de Santa Rosa Formation in the Sierra de Santa Rosa, Sonora (Damborenea and González-León, 1997; Lucas and Estep, 1997, 1999). The presence of *Weyla* in the southern Sierra Nevada and northern Sonora indicates a similarity between the two areas that is compatible with offset of one area to the other, although the widespread distribution of *Weyla* in the western United States precludes precise matching of sections on the basis of this fossil alone.

The Kings sequence and Antimonio Formation, although similar, are in some respects, as outlined above, different in that the Kings sequence contains an apparently greater amount of turbiditic strata, as well as local olistoliths and common volcanic units. However, deep-water turbiditic strata form a part of the Antimonio Formation (González-León, 1997a, 1997b). In addition, volcaniclastic sandstone and conglomerate and sparse tuff is present in both the Antimonio Formation (Antimonio, Rio Asunción, Sierra de Santa Rosa Formations of Lucas and Estep [1999]) in the El Antimonio area (Stanley and González-León, 1995; González-León, 1997a, 1997b) and the Sierra de Santa Rosa Formation in Sierra Santa Rosa (Hardy, 1981). Although differences between the Upper Triassic and Lower Jurassic rocks in the southern Sierra Nevada and those in northern Sonora are apparent, the differences appear no greater than changes in facies in time-equivalent rocks in western Nevada and easternmost California (Stewart, 1997). The rocks in western Nevada and easternmost California range from deep-water turbiditic clastic strata, to mixed deep-water clastics and shallow-marine carbonate, to successions with thick volcanic rocks, and finally to thick succession of mostly shallow-water carbonate strata with no deep-water interbeds or volcanic strata (Stewart, 1997). Nevertheless, the correlation of the Kings sequence and the Antimonio Formation is uncertain because of the greater amount of volcanic rock in the Kings sequence.

The amount of apparent displacement of marine Upper Triassic and Lower Jurassic strata from the southern Sierra Nevada to northern Sonora is similar in position and distance to suggested displacements on the megashear of the Neoproterozoic and Paleozoic strata described above. By Middle Jurassic time, this pattern of apparent displacement is no longer recognized, suggesting that displacement on the supposed Mojave Sonora megashear is younger than Early Jurassic. By Middle Jurassic time, a continuous magmatic arc that started at least locally in the Late Permian and Triassic had developed along the continental margin (Saleeby and Busby-Spera, 1992).

DISTRIBUTION OF CORDILLERAN MIOGEOCLINE

The Cordilleran miogeocline is continuous from Idaho, through western Utah, eastern Nevada, and into eastern California (Fig. 12). Outcrops of thick Proterozoic and Cambrian strata in the San Bernardino Mountains and western Mojave Desert region of southern California were considered by Cameron (1981, 1982) and Stewart et al. (1984) to be a continuation of the Cordilleran miogeocline, but a reappraisal (Stewart et al., 2005) suggests that perhaps only the uppermost part of the succession is Neoproterozoic and part of the Cordilleran miogeocline. The San Bernardino Mountains section appears to lie near the eastern margin of the miogeocline. South of the San Bernardino Mountains, the miogeocline does not crop out for ~600 km. South of this gap in outcrops, the miogeocline reappears in Sonora. The trend of the miogeocline is southerly in the United States and Sonora. The apparent offset and the trend of the miogeocline in the western United States and Sonora are consistent with the megashear model. The miogeocline has been proposed (Stewart et al., 2002) to be truncated on the south by the Central Sonora fault, a possible left-lateral transpressional fault thought to be of Permian age. The nature of this fault is controversial. It usually has been considered to be a thrust, rather than a left-lateral transpressional fault, and to have emplaced Paleozoic offshelf deep-water strata northward onto the continental margin (Poole et al. 1995b; Poole and Amaya-Martínez, 2000).

ORDOVICIAN TO PERMIAN OFFSHELF DEEP-WATER SILICEOUS ROCKS

Ordovician to Permian siliceous rocks in the western United States and northwest Mexico are referred to here as offshelf deep-water siliceous rock. These rocks have commonly been called either eugeoclinal western facies, siliceous and volcanic assemblage, or ocean-basin rocks. Lithologically, they are composed dominantly of chert, shale, siltstone, carbonate, quartzite, turbiditic clastic strata, and local volcanics. They are

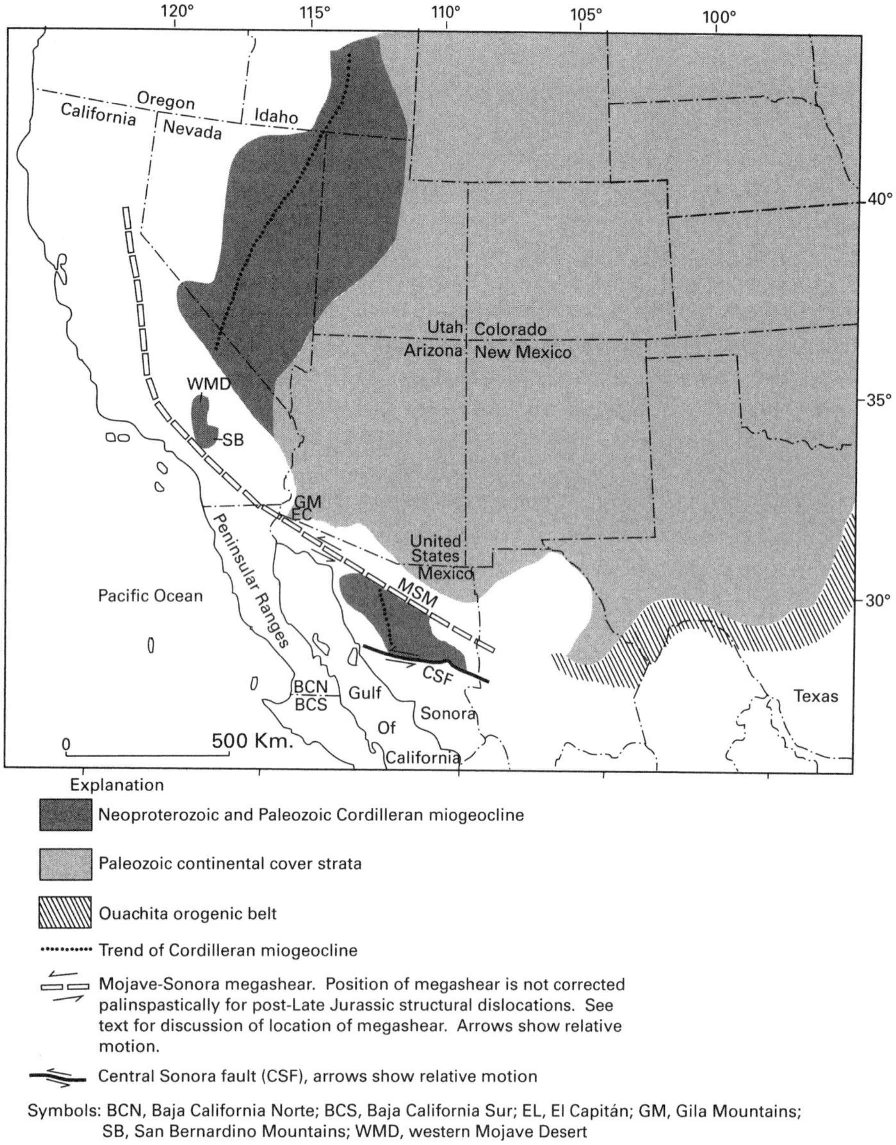

Figure 12. Distribution of Cordilleran miogeocline in the western United States and northwest Mexico.

commonly considered to have been deposited in deep water west of the miogeoclinal belt and to have fringed the western margin of ancestral North America. Transport along thrust faults emplaced these strata over miogeoclinal strata of ancestral North America. The history of emplacement is complex. In Nevada, the first emplacement took place during the Late Devonian to Early Mississippian Antler orogeny, when Ordovician to Devonian strata were transported along the Roberts Mountains thrust over miogeoclinal strata. Renewed offshelf deep-water deposition took place in Nevada in the Pennsylvanian and Permian, and these rocks were in turn emplaced over miogeoclinal strata during the Late Permian to Early Triassic Sonoma orogeny. The history of emplacement in Mexico is different. Here Ordovician to Permian deep-water siliceous strata are in a continuous sedimentary succession that was emplaced over, or transpressionally against, miogeoclinal strata in the Permian or Early Triassic. As described previously, no evidence of the Antler orogeny has been described in Mexico, and the relation of a continuous succession of siliceous strata unbroken by a thrust comparable to the Roberts Mountains thrust is indicative of the difference in tectonic history between Nevada and Mexico.

Ordovician to Devonian deep-water siliceous rocks of the Roberts Mountains allochthon and comparable age strata in the Sonora allochthons (Fig. 13) consist of relatively deep-water marine strata originally deposited along the Laurentian continental margin outboard of the Cordilleran miogeocline (Madrid et al., 1992b). The rocks consist of complexly deformed radiolarian chert, graptolitic shale, quartzite, feldspathic sandstone, siltstone, minor mafic lava flows, bedded tuff, limestone, and bedded barite (Poole et al. 1991, 1995b; Madrid et al., 1992b;

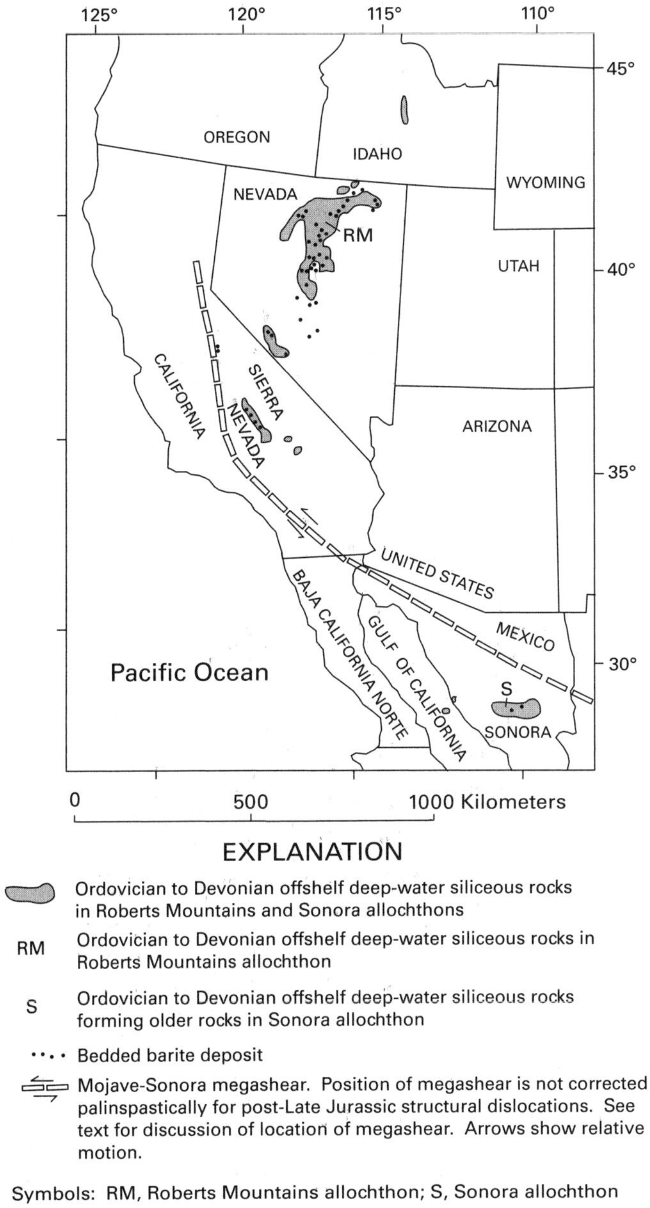

Figure 13. Distribution of Ordovician to Devonian offshelf deep-water rocks in the Roberts Mountains allochthon in the western United States and in the Sonora allochthon in Sonora. Based on Madrid et al. (1992b), Poole (1992), Poole et al. (1995b), and Stewart and Poole (2002).

Poole and Amaya-Martínez, 2000). These rocks show the same apparent left-lateral offset as do other Neoproterozoic to Lower Jurassic strata. Poole and Madrid (1988) and Poole et al. (1991) indicate that the exhalative stratiform barite deposits present in Devonian strata in Sonora are comparable to those in Nevada, although they emphasize that the strata in the Roberts Mountains and Sonora allochthons that contain the barite are composed of different lithofacies and underwent different styles of deformation at different times. Clearly, as in the case of Mississippian miogeoclinal and orogenic strata, lithofacies and structural events changed along the continental margin, but where these changes took place and how the Sonora rocks relate to those in the western United States remain unresolved. In central Sonora, different facies of Paleozoic eugeoclinal strata are present in apparently different structural plates within 18 km of one another (Bartolini et al., 1989, 1991; Stewart et al., 1990), suggesting that along-margin changes in lithofacies could be large and that contraction along thrusts may have contributed to the juxtaposition of different facies.

Pennsylvanian and Permian offshelf deep-water siliceous strata, including local upper Mississippian strata, comprise the Golconda allochthon of the Sonoma orogeny in northern Nevada and the Sierra Nevada of California, as well as strata in the Mojave Desert and comparable age strata in the Sonora allochthon in Sonora (Fig. 14). In northern Nevada, these strata are assigned to the Havallah sequence and consist of radiolarian chert, argillite, siltstone, silty limestone, and greenstone (metabasalt) (Silberling and Roberts, 1962; Stewart et al., 1986; Miller et al., 1992). In west-central Nevada, comparable strata in part of the Mina Formation consist of fine- and coarse-grained volcanogenic turbidite, and lesser amounts of chert, pelitic rocks, and quartz sandstone associated locally with ultramatic rocks (Speed, 1977). In the Sierra Nevada (Fig. 14), small outcrops of ultramafic rocks, basaltic pillow lava, and chert-argillite breccia are included by Schweickert and Lahren (1987) in the Golconda allochthon. In the Mojave Desert, Pennsylvanian and Permian rocks, or suspected Pennsylvanian to Permian rocks, consist of deep-water siliceous successions in the El Paso Mountains (Martin and Walker, 1991; Carr et al., 1997) and the Goldstone, Paradise Range, and northern Lane Mountain area (Martin and Walker, 1991). These strata consist of metamorphosed pelitic strata, chert, limestone, silty limestone, quartzite, conglomerate, basalt, and andesite. In Sonora, Pennsylvanian and Permian deep-water siliceous strata consist of radiolarian chert, siltstone, turbiditic sandstone, conglomerate, barite, limestone, and minor dolomite (Ketner and Noll, 1987; Stewart et al., 1990; Poole and Amaya-Martínez, 2000; Stewart and Poole, 2002). Permian deep-water turbiditic siltstone and fine-grained sandstone comprising the Mina México Formation in east-central Sonora have been considered allochthonous by Radelli et al. (1987) and part of the Sonora allochthon by F.G. Poole (*in* Stewart and Poole, 2002). Alternatively, the Mina México Formation is a deep-water deposit that formed in basins on the shelf and is not allochthonous (Stewart et al., 1990, 1997). The latter interpretation is adopted here because the Mina México Formation may rest depositionally on Permian miogeoclinal strata. The Mina México Formation is not shown on Figure 14. The Pennsylvanian and Permian deep-water siliceous rocks are more variable in composition along the Cordilleran continental margin than Ordovician to Devonian deep-water siliceous strata but show the same general distribution pattern and apparent left-lateral offset as do other Neoproterozoic to Lower Jurassic strata.

IMPERFECT STRATIGRAPHIC MATCH OF OFFSET MIOGEOCLINAL SEGMENTS ACROSS THE MEGASHEAR

A persistent problem in evaluating the megashear model is whether strata on one side of the proposed fault are close enough in lithology and stratigraphy to match supposed offset strata on the other side of the fault. If strata in the western United States are similar to those of the proposed offset segment in Sonora, then the concept of the megashear is supported. If, on the other hand, the strata are significantly different, then the wraparound model in which facies changes could take place parallel to the continental margin is more likely.

Comparison of the Neoproterozoic and Cambrian strata of the Nopah Range in the Death Valley area of eastern California with that in the Caborca area (Fig. 15) illustrates the problem. These two sections are the closest match in Neoproterozoic and Cambrian stratigraphy between the western United States and Sonora, and if the megashear model is valid, the two sections have been offset from one another by hundreds of kilometers of left-lateral displacement.

Before supposed offset on the megashear, however, the two sections were not necessarily closely adjacent to one another. As shown on Figure 15, the Nopah Range section lies nearly 200 km northeast of the inferred position of the megashear. If this position relative to the megashear is valid, then the Nopah Range and Caborca sections were originally 200 km apart. This 200 km of separation may be increased or decreased by structural displacements (thrust faults, extensional detachment faults, strike-slip faults) that disrupted the relative geographic positions of the Nopah Range and Caborca areas. As outlined below, the supposed 200 km of original separation between the Nopah Range and Caborca sections appears sufficient to allow for necessary facies changes between the two sections. In this case, the areas containing strata of intermediate facies have either been buried, eroded, or destroyed by intrusions or tectonic denudation.

In this discussion, the focus is mainly on shelf-parallel (sedimentary strike) changes in facies. On the basis of the thickness and lithology of the successions, the Nopah Range and Caborca sections probably are at similar positions on the shelf, and thus comparisons of facies should be parallel to the shelf. Across-shelf changes in facies are larger than shelf-parallel changes, but are not described here.

Comparison of the Nopah Range and Caborca sections follows:

1. The Noonday Dolomite and Johnnie Formation in the lower part of the section in the Nopah Range are lithologically similar to the El Arpa and Caborca Formations of the Caborca area, but the thickness and stratigraphic successions are considerably different. However, along shelf, changes in facies are large in the lower part of the Neoproterozoic succession of the western United States and Sonora. The Noonday Dolomite in the Nopah section is a platform facies (Williams et al., 1974). This facies is replaced abruptly to the south by a basinal facies composed of coarse breccia, arkosic sandstone, siltstone, shale, dolomite, and limestone, probably associated with Neoproterozoic faulting. In the Death Valley area, the Johnnie Formation is divided into six members that, except for the Rainstorm Member at the top, cannot be recognized along sedimentary strike in areas adjacent to this region (Stewart, 1970). In Sonora, the El Arpa and Caborca

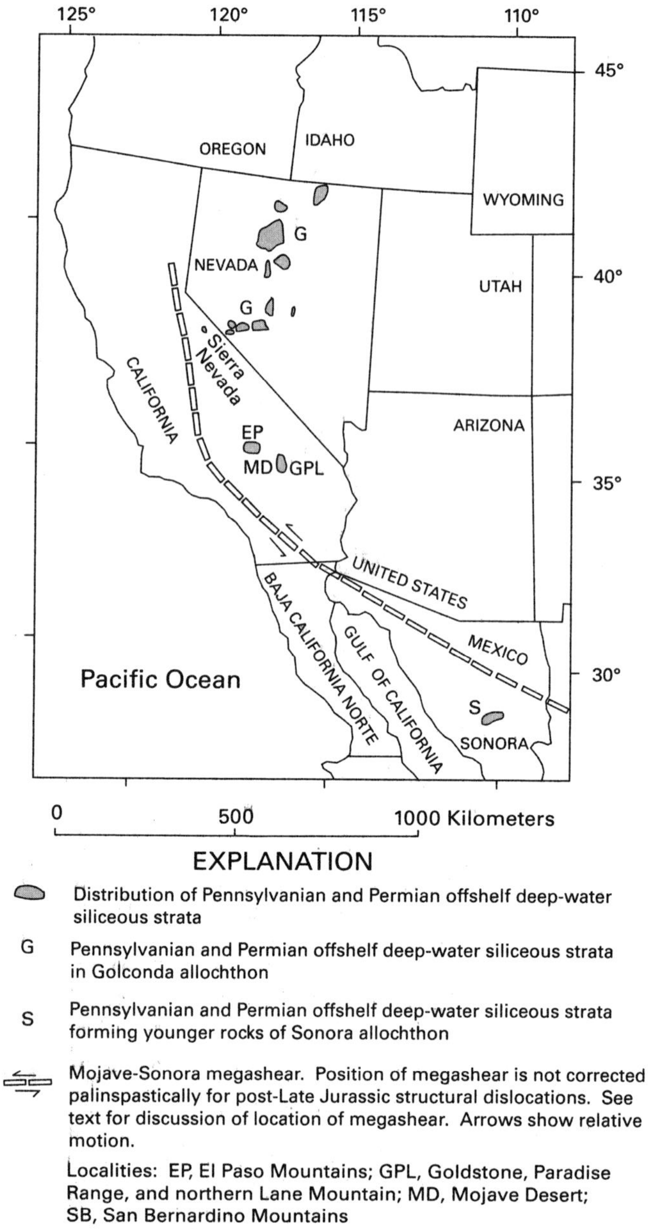

Figure 14. Distribution Pennsylvanian to Permian offshelf deep-water rocks in the Golconda allochthon in Nevada and in the Sonora allochthon in Sonora. Based on Stewart (1980), Schweickert and Lahren (1990), Carr et al. (1997), Martin and Walker (1991), Ketner and Noll (1987), Stewart et al. (1990), Poole and Amaya-Martínez (2000), and Stewart and Poole (2002).

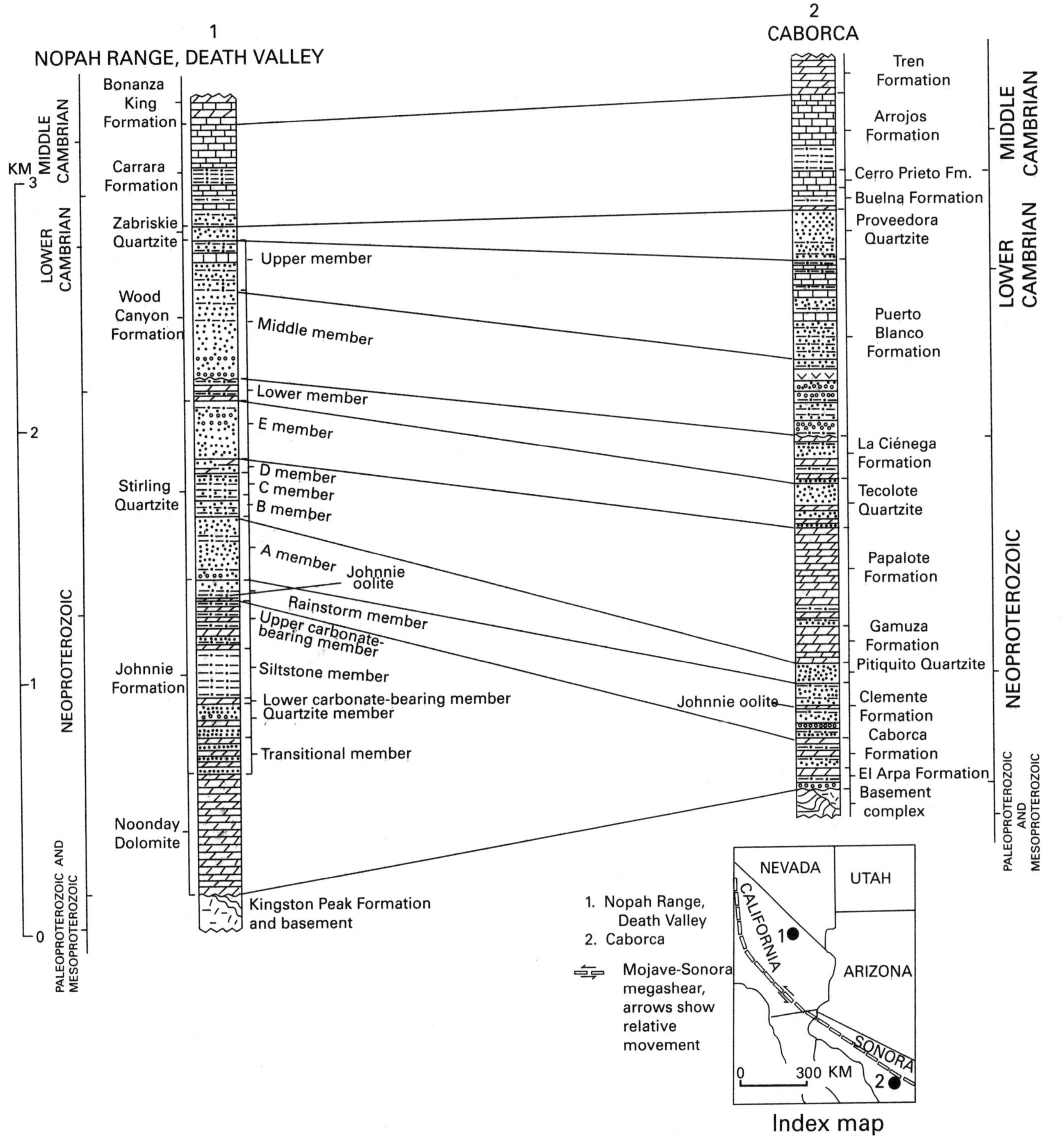

Figure 15. Comparison of stratigraphic succession in the Death Valley region, eastern California and Caborca, northwest Sonora. Lithologic columns based on Stewart (1970) and Stewart et al. (2002). Fm—Formation.

Formations are at the base of the Neoproterozoic succession at Caborca and consist of dolomite, sandy dolomite, and minor amounts of quartzite and sandstone. These formations are not recognized at Rancho Curiel ~100 km to the south-southeast where the lower part of the section is composed of sandstone, quartzite, conglomerate, and dolomite (Stewart et al., 2002).

2. The Rainstorm Member of the Johnnie Formation, a widespread and lithologically consistent unit in eastern California and southern Nevada (Stewart, 1970), and the correlative Clemente Formation in Sonora are the lowest strata that clearly are lithologically similar, and in the case of the Johnnie oolite, are essentially identical.

3. The A member of the Stirling Quartzite and the Pitiquito Quartzite are lithologically similar, although they differ in thickness.

4. The B, C, and D members of the Stirling Quartzite are considered to be related to the Gamuza and Papalote Formations, although these parts of the Nopah and Caborca sections are considerably different. The B, C, and D members are largely siliciclastic strata in the Nopah Range, and the Gamuza and Papalote Formations are largely dolomite, some with large conical stromatolites. However, the D member of the Stirling Quartzite contains dolomite and sandy dolomite in the Nopah section, and 100 km northwest, the D member is 137 m thick and composed largely of dolomite (Stewart, 1970).

5. The E member of the Stirling Quartzite consists largely of quartzite and is correlated with the Tecolote Formation, which is composed of quartzite and minor dolomite.

6. The lower member of the Wood Canyon Formation is correlated with the La Ciénega Formation. Both these units characteristically consist of siltstone and three dolomite layers, the lower one of which contains latest Neoproterozoic cloudiniid fossils in the Caborca area and cloudiniid, or possible cloudiniid, in the Death Valley area (Hagadorn and Fedo, 2000).

7. The Middle member of the Wood Canyon Formation is correlated with the lower part of the Puerto Blanco Formation. Both units are above a significant regional unconformity (Fedo and Cooper, 2001; Stewart et al., 2002) that is correlated from the western United States to northwest Mexico. Strata above the unconformity differ considerably between the Nopah Range and Caborca. In the Nopah Range, the Middle Member of the Wood Canyon Formation consists of feldspathic arenite and minor interstratified siltstone. In the Caborca area, the lower part of the Puerto Blanco Formation consists of volcaniclastic sandstone to boulder conglomerate, greenstone, quartzite, siltstone, and sparse dolomite (Stewart et al., 1984). In the Caborca area, the stratigraphy of the Puerto Blanco Formation varies significantly (Stewart et al., 1984), apparently in response to structural movements associated with the eruption of basalt (now greenstone). The difference between the Nopah section and the Caborca section is probably related to these apparently local structural and igneous events in the Caborca area.

8. The upper member of the Wood Canyon Formation and the correlative upper part of the Puerto Blanco Formation consist of fine-grained quartzite and siltstone and interstratified limestone. Both units are characterized by Early Cambrian trilobites, trace fossils, and, in the Caborca area, archeocyathids. Archeocyathids are not present in the Nopah section but are present in the Upper Member elsewhere in the Death Valley area (Stewart, 1970).

9. The Zabriskie Quartzite in the Nopah Range section and correlative Proveedora Quartzite in the Caborca section are remarkably similar. Both are supermature quartzite composed almost entirely of quartz. *Skolithos* tubes are abundant in the lower part of both units.

10. The Carrara Formation in the Nopah Range section and the correlative Buelna, Cerro Prieto, and Arrojos Formations in the Caborca section are also remarkably similar. Both are characterized by interstratified siltstone, limestone, and silty limestone. Early and Middle Cambrian trilobites are abundant in both successions. Both successions contain distinctive oncoid-bearing limestone.

11. The Bonanza King Formation of the Nopah Range section and the correlative incompletely exposed Tren Formation of the Caborca Formation are both thick successions of carbonate rock.

In summary, the Nopah Range and Caborca areas contain Neoproterozoic and Cambrian successions that are similar, but far from identical. If the two successions were offset directly, then they should be identical. However, as indicated above, some units in these successions change facies rapidly and the sections may originally have been 200 km apart. If the Mojave-Sonora megashear model is correct, the imperfect match between the Nopah Range and Caborca areas must be explained by facies changes in an area between the two sections where rocks of appropriate facies are now buried, eroded, or destroyed by intrusion or tectonic denudation. A possibility is that the destruction of strata may have taken place by subduction erosion (von Huene and Scholl, 1991) related to a Jurassic magmatic arc in western North America (Saleeby and Busby-Spera, 1992).

OFFSET OF PROTEROZOIC BASEMENT ROCKS

The apparent offset of Neoproterozoic to Lower Jurassic strata from the western United States to Sonora is similar to the distribution pattern and apparent offset of Paleoproterozoic to Mesoproterozoic crystalline basement rocks. Anderson and coauthors (Silver and Anderson, 1974; Anderson and Silver, 1979, 1984, this volume; Anderson and Schmidt, 1983) have outlined two juxtaposed basement terranes, one to the northeast containing igneous rocks as old as 1.7–1.6 Ga and one to the south containing igneous rocks as old as 1.8–1.7 Ga. The 1.8–1.7-Ga igneous terrane is considered by Silver and Anderson to have been displaced left-laterally into Sonora from a broad west-southwest–trending band of rocks of that age extending across the western United States. The supposedly displaced terrane containing 1.8–1.7-Ga igneous rocks lies in the same region as the supposedly displaced Cordilleran miogeocline in Sonora. Such a correspondence is expected because the miogeoclinal strata lie unconformably on the Proterozoic basement rocks, and displacement of Proterozoic igneous terranes should be matched by a similar displacement of Neoproterozoic to Lower Jurassic strata. Farmer (2003) and Iriondo (2003), on the basis of Nd isotope and U-Pb zircon data, indicate that some basement rocks in the Caborca region are similar to those in the Mojave crustal province of southeastern California, a relationship seemingly supportive of the concept of the Mojave-Sonora megashear.

POSITION OF THE MEGASHEAR

In Figures 1–14, the megashear is shown as a thick double-dashed line that separates outcrops on either side of the supposed fault. This dashed line is considered to be an approximation of where the fault may have been relative to present-day outcrops of Neoproterozoic to Lower Jurassic strata. Middle Jurassic, Cretaceous, and early Tertiary structural activity may have disrupted the trace of the fault by thrusting, strike-slip movement, extensional detachment, or igneous intrusion, and strata exposed on both sides of the fault may also have been displaced. The present-day trace of the fault may be highly discontinuous and irregular and in a different position relative to present-day outcrops than shown on the figures. A possible surface exposure of the fault has been described in northwestern Sonora (Campbell and Anderson, 1998). Alternatively, displacement of rocks from the western United States to northern Mexico may have taken place on a number of left-lateral strike-slip faults as well on low-angle, mainly detachment, faults. Such a possibility might account for the difficulty of locating a single thoroughgoing trace of the presumed megashear.

The trace of the Mojave-Sonora megashear in Sonora shown here is similar in most places to that originally proposed by Silver and Anderson (1974), Anderson and Silver (1979, 1984), and Anderson and Schmidt (1983) on the basis of the ages of Proterozoic basement rocks, but in northeasternmost Sonora and the western United States, the trace shown here is significantly different from what they show. The problem involves two regions, the El Capitán (Ejido Aquiles Serdán) and Gila Mountains region of northwesternmost Sonora and southwesternmost Arizona (Leveille, 1984; Fitts, 1989; Hamilton, 1987) and the San Bernardino Mountains–western Mojave Desert region of California (Stewart and Poole, 1975; Cameron, 1981, 1982; Miller, 1981; Martin and Walker, 1991) (Fig. 12). In the El Capitán and Gila Mountains region, thin Paleozoic cratonal cover strata are present and are shown here north of the megashear. Such cratonal cover strata are present throughout southern Arizona and southern California and everywhere lie north of the megashear. However, on the basis of ages of Proterozoic basement terranes, these two areas are considered to be south of the megashear of Silver and Anderson (1974), Anderson and Silver (1979), and Anderson and Schmidt (1983). L.T. Silver (*in* Stewart et al., 1984; J.H. Stewart, 1995, personal commun.) reconciles this apparent conflict by proposing southward transport of large slices of rock, including the El Capitán and Gila Mountains region, across the megashear.

Outcrops of uppermost Neoproterozoic and lower Paleozoic strata in the San Bernardino Mountains–western Mojave Desert region are clearly part of the Cordilleran miogeocline and appear to be a southwestern continuation of the miogeoclinal from eastern California (Brown, 1991). However, Silver and Anderson (1974), Anderson and Silver (1979, 1984), and Anderson and Schmidt (1983), on the basis of ages of Precambrian basement rocks, place the megashear ~150 km to the east in the eastern Mojave Desert region. If the megashear exists, then either positioning it in the eastern Mojave Desert is incorrect, or the entire western Mojave Desert–San Bernardino block has been transported 150 km westward across the megashear to its present position as proposed by L.T. Silver (*in* Stewart et al., 1984).

North of the Mojave Desert–San Bernaradino Mountains, the trace of the megashear has been placed east of the Sierra Nevada region, or in the easternmost Sierra Nevada, by Stanley and González-León (1995), Marzolf (1997), Anderson (1997), and Marzolf and Anderson (2000). This position is based on the interpreted offset of Lower Mesozoic strata, sequences, structures, and igneous plutons and clasts. The inferred position of the megashear is drawn differently here. The Upper Triassic and Lower Jurassic Kings sequence of the Sierra Nevada is considered the southern continuation, as described above, of marine Upper Triassic and Lower Jurassic strata of central and western Nevada and eastern California; Upper Triassic and Lower Jurassic strata in northern Sonora are interpreted to have been offset from the Kings sequence in the southern Sierra Nevada (Fig. 11). If this interpretation is correct, the megashear lies west of the King sequence and truncates it south of the southern Sierra Nevada (Fig. 12). As mentioned previously, the correlation of the King sequence and the Antimonio Formation is speculative, and thus placement of the megashear west of the Kings sequence is also speculative.

AMOUNT OF OFFSET

The amount of left-lateral offset on the Mojave-Sonora megashear was originally estimated as 700–800 km by Silver and Anderson (1974), Anderson and Silver (1979, 1984), and Anderson and Schmidt (1983) on the basis of possible offset of Paleozoic strata. Ketner (1986), on the basis of suggested offset of the Middle Ordovician Eureka Quartzite in the western United States and the Peña Blanca Quartzite in Sonora, implies an offset of 1100 km. Stanley and González-León (1995) indicate 800–1000 km on the basis of proposed offset of Triassic strata. An estimate of 700–800 km is indicated in this report on the basis of the proposed offset of isopach lines in the Zabriskie Quartzite and Proveedora Quartzite (Fig. 4). The latter estimate is perhaps the most realistic because it is based on detailed information of a proposed tie across the megashear. However, larger apparent offsets are possible if major structural displacements unrelated to the megashear have relocated the positions of large regions.

AGE OF OFFSET

The Mojave-Sonora megashear was originally proposed to be of Early to Middle Mesozoic age (Silver and Anderson,

1974) and was later defined as mainly Late Jurassic (Anderson and Schmidt, 1983; Anderson and Silver, 1979). A Jurassic age for the megashear is compatible with the information presented in this article that Upper Triassic and Lower Jurassic strata are offset by the megashear, but Upper Jurassic rocks are not. On the other hand, as discussed above, the correlations of possible offset segments of Mesozoic strata across the megashear is uncertain and allows speculation that the megashear is older than Lower Jurassic. If the megashear is not Late Jurassic, it most likely is late Paleozoic, as proposed by Stevens et al. (1992) and, in a modified interpretation, by Dickinson and Lawton (2001).

CONCLUSIONS

No unequivocal evidence has been presented here or elsewhere about whether Neoproterozoic to Lower Jurassic strata wrapped around the continental margin of Laurenia from the western United States to Sonora, or whether these strata were offset by left-lateral motion along the proposed Mojave-Sonora megashear or along related faults (Fig. 1). However, the conclusion of this article is that the weight of evidence is in favor of major left-lateral offset. The evidence for offset is as follows:

1. The distribution patterns of outcrops and facies of Neoproterozoic to Permian miogeoclinal strata, Upper Triassic to Lower Jurassic marine strata, and Orovician to Permian offshelf deep-water siliceous rocks show a systematic left-lateral change in position from the western United States to Sonora (Figs. 2–14).

2. The distribution of individual units such as the Neoproterozoic "Johnnie oolite" and the Lower Cambrian Zabriskie-Proveedora quartzites (Figs. 3 and 4) show a distinct left-lateral separation of outcrops from the western United States to Sonora. Both these units are lithologically distinctive and relatively restricted in distribution, and their two-part distribution pattern is highly suggestive of structural offset. The apparent offset of the isopach trends of the Zabriskie Quartzite supports this conclusion.

3. Distinctive units, unconformities, and fossil assemblages show a consistent left-lateral change in position. Such features consist of Neoproterozoic diamictite (tillite) in the Neoproterozoic (Fig. 2); archeocyathids in the Lower Cambrian (Fig. 2); the Carrara Formation in the western United States and the combined Buelna, Cerro Prieto, and Arrojos Formations in Sonora (Fig. 6); the Middle Ordovician Eureka Quartzite of the western United States and the correlative Peña Blanca Quartzite in Sonora (Fig. 7); unconformable overlap of Devonian strata over Silurian and Ordovician strata at the eastern edge of the miogeocline in the western United States and similar overlap of Devonian strata in Sonora; bedded barite deposits in the Roberts Mountains; and Sonora allochthons, Permian faunas of McCloud-like affinities in eastern California and of unnamed Permian strata in Sonora (Fig. 11), and apparent offset of strata containing the Late Jurassic bivalve *Weyla* from the Lake Isabella area in the southern Sierra Nevada to northern Sonora.

4. The trend of the Neoproterozoic Cordilleran miogeocline (Fig. 12) is southerly in the western United States and appears to be southerly in Sonora. If the Cordilleran miogeocline wraps around the Laurentian continental margin, then the trend of the miogeocline in Sonora should be closer to east-west or east-southeast rather than south. No evidence that the trend of the miogeocline curves eastward in Sonora has been presented, and present information indicates that outer shelf deposits in Sonora are to the west and inner shelf deposits are to the east, at odds with the concept of the wraparound model in which outer shelf deposits should be to the south and inner shelf deposits to the north.

5. The compatibility of information on the distribution and facies of Neoproterozoic to Lower Jurassic strata and of independent information on the distribution of dated rocks of Proterozoic basement terranes (Silver and Anderson, 1974; Anderson and Silver, 1979, 1984; Anderson and Schmidt, 1983) enhances the idea of offset on the megashear. Both sets of information show a left-lateral change in outcrop distribution of rocks and facies from the western United States to Sonora. Where exposed, the Neoproterozoic strata lie unconformably on the basement rocks, so that any proposed left-lateral movement of basement rocks should be matched, as it is, by a comparable displacement of Neoproterozoic and younger strata.

The stratigraphy of the proposed displaced segment of Neoproterozoic to Lower Jurassic strata in Sonora is an imperfect match for the presumably indigenous sections in the western United States. Stratigraphic differences between the sections can be ascribed to facies changes parallel to the continental margin that took place in what is now a gap between present-day outcrops and the proposed trace of the megashear. Comparable outcrops in Sonora and the western United States may have originally been as much as 200 km apart before offset on the megashear.

In Sonora, the inferred Middle Jurassic position of the megashear based on the position of Neoproterozoic and Lower Jurassic strata is similar to the position of the megashear based on ages of Proterozoic basement rocks. However, to the northwest in California, the position of the megashear based on the distribution of the Neoproterozoic to Lower Jurassic strata differs significantly from that based on the ages of Proterozoic basement rocks. This difference indicates either that the position of the megashear based on the position of Proterozoic basement rocks needs to be revised, or large structural blocks have been transported across the megashear to sites southwest of the megashear.

Offset isopach lines of the Zabriskie Quartzite in the western United States and the correlative Proveedora Quartzite in Sonora indicate 700–800 km of left-lateral offset on the megashear. Previous estimates of offset range from 600 km to 1100 km.

ACKNOWLEDGMENTS

The report has been greatly improved by helpful comments by Paul Stone, N.J. Silberling, and T.H. Anderson.

REFERENCES CITED

Allmendinger, R.W., 1992, Fold and thrust tectonics in the Cordilleran region, western United States exclusive of the accreted terranes, *in* Burchfiel, B.C., et al., eds., The Cordilleran orogen, conterminous U.S.: Boulder, Colorado, Geological Society of America, The Geology of North America, v. G-3, p. 583–608.

Anderson, P.V., 1993, Prebatholithic stratigraphy of the San Felipe area, Baja California, *in* Gastil, R.G., and Miller, R.H., eds., The prebatholithic stratigraphy of peninsular California: Geological Society of America Special Paper 279, p. 1–10.

Anderson, T.H., 1997, Restoration of the Caborca block—Sedimentologic implications for Triassic and Jurassic strata, *in* González-León, C.M., and Stanley, G.D., Jr., eds., International workshop on the geology of Sonora (Memoir, preliminary publication): Hermosillo, México, Instituto de Geología, Estación Regional del Noroeste, Universidad Nacional Autónoma de México, p. 1–4.

Anderson, T.H., and Schmidt, V.A., 1983, The evolution of Middle America and the Gulf of Mexico–Caribbean Sea region during Mesozoic time: Geological Society of America Bulletin, v. 94, p. 941–966, doi: 10.1130/0016-7606(1983)94<941:TEOMAA>2.0.CO;2.

Anderson, T.H., and Silver, L.T., 1979, The role of the Mojave-Sonora megashear in the tectonic evolution of northern Sonora, *in* Anderson, T.H., and Roldán-Quintana, J. eds., Geology of northern Sonora: Pittsburgh, Pennsylvania, University of Pittsburgh, and Hermosillo, Sonora, Instituto de Geología, Universidad Nacional Autónoma de México, Geological Society of America Annual Meeting, Guidebook Trip 27, p. 59–68.

Anderson, T.H., and Silver, L.T., 1984, An overview of Precambrian rocks in Sonora: Mexico, D.F., Instituto de Geología, Universidad Nacional Autónoma de México, Revista Mexicana de Ciencias Geológica, v. 5, no. 2, p. 131–139.

Anderson, T.H., McKee, J.W., Jones, N.W., and Rodriguez-Castaneda, J.L., 1991, Transpressional structures along the late Jurassic Mojave-Sonora megashear: Geological Society of America Abstracts with Programs, v. 23, no. 5, p. 251.

Anderson, T.H., McKee, J.W., McKee, M.B., and Nourse, J.A., 1995, Coarse clastic deposits of Late Jurassic age in southwestern North America: A record of pull-apart basins: Geological Society of America Abstracts with Programs, v. 27, no. 6, p. 75.

Bartolini, C., 1993, Tectonic emplacement and accretion of Paleozoic eugeoclinal, allochthonous assemblages in central Sonora, Mexico—A new, Cordilleran tectonostratigraphic terrane, *in* Ortega-Gutiérrez, F., et al., eds., Proceedings, First Circum-Pacific and Circum-Atlantic Terrane Conference, Guanajuato, Mexico: Mexico, D.F., Instituto de Geología, Universidad Nacional Autónoma de México, Revista, p. 12–16.

Bartolini, C., Stewart, J.H., Carter, C., Murchey, B.L., and Repetski, J.E., 1989, Stratigraphy of Paleozoic eugeoclinal strata in Sierra El Aliso, central Sonora, Mexico: Hermosillo, Sonora, México, Universidad de Sonora, Boletín Segunda Epoca, v. 6, nos. 1-2, p. 11–21.

Bartolini, C., Stewart, J.H., Carter, C., Murchey, B.L., and Repetski, J.E., 1991, Stratigraphy of Paleozoic eugeoclinal strata in Sierra El Aliso, central Sonora, Mexico [correction of incomplete publication of 1989 article]: Hermosillo, Sonora, México, Universidad de Sonora, Boletín Segunda Epoca, v. 8, no. 2, p. 73–78.

Brown, H.J., 1991, Stratigraphy and paleogeographic setting of Paleozoic rocks in the San Bernardino Mountains, California, *in* Cooper, J.D., and Stevens, C.H., eds., Paleozoic paleogeography of the western United States II, v. 1: Los Angeles, California, Pacific Section, Society of Economic Paleontologists and Mineralogists, v. 67, p. 193–207.

Burchfiel, B.C., Cowan, D.S., and Davis, G.A., 1992, Tectonic overview of the Cordilleran orogen in the western United States, *in* Burchfiel, B.C., et al., eds., The Cordilleran orogen, conterminous U.S.: Boulder, Colorado, Geological Society of America, The Geology of North America, v. G-3, p. 407–480.

Cameron, C.S., 1981, Geology of the Sugarloaf and Delamar Mountain areas, San Bernardino Mountains, California [Ph.D. dissertation]: Cambridge, Massachusetts Institute of Technology, 399 p.

Cameron, C.S., 1982, Stratigraphy and significance of the Upper Precambrian Big Bear Group, *in* Cooper, J.D., et al., eds., Geology of selected areas in the San Bernardino Mountains, western Mojave Desert, and southern Great Basin, California: Geological Society of America, Cordilleran Section Annual Meeting, Anaheim, California, 1982, Guidebook and Volume, p. 5–20.

Campbell, P.A., and Anderson, T.H., 1998, Structure and kinematics along a Jurassic plate boundary transform, the Mojave-Sonora megashear, northwestern Sonora, Mexico: Guidebook Series, Society of Economic Geologists, v. 30, p. 177–186.

Carr, M.D., Harris, A.G., Poole, F.G., and Fleck, R.J., 1992, Stratigraphy and structure of Paleozoic outer continental-margin rocks in Pilot Knob Valley, north-central Mojave Desert, California: U.S. Geological Survey Bulletin 2015, 33 p.

Carr, M.D., Christiansen, R.L., Poole, F.G., and Goodge, J.W., 1997, Bedrock geologic map of the El Paso Mountains in the Garlock and El Paso Peaks 7-1/2′ quadrangles, Kern County, California: U.S. Geological Survey Miscellaneous Geologic Investigations Map I-2389, scale 1:24,000.

Caudillo-Sosa, G., Oviedo-Lucero, L.F., and Rodriguez-Castaneda, J.L., 1996, Falla Quitovac: Resultado de un evento de transpression del "Mojave-Sonora megashear," noroeste de Sonora, México: Mexico, D.F., Instituto de Geología, Universidad Nacional Autónoma de México, Revista, p. 140–151.

Christie-Blick, N., and Levy, M., 1989, Stratigraphy and tectonic framework of Upper Proterozoic and Cambrian rocks in the western United States, *in* Christie-Blick, N., and Levy, M., eds., Late Proterozoic and Cambrian tectonics, sedimentation, and record of metazoan radiation in the western United States, 28th International Geological Congress, Field Trip Guidebook T331, American Geophysical Union, p. 7–21.

Cohen, K.K., Anderson, T.H., and Schmidt, V.A., 1986, A paleomagnetic test of the proposed Mojave-Sonora megashear in northwestern Mexico: Tectonophysics, v. 131, no. 1–2, p. 23–51, doi: 10.1016/0040-1951(86)90266-0.

Cojan, I., and Potter, P., 1991, Depositional environment, petrology, and provenance of the Santa Clara Formation, Upper Triassic Barranca Group, eastern Sonora, Mexico, *in* Pérez-Segura, E., and Jacques-Ayala, C., eds., Studies of Sonoran geology: Geological Society of America Special Paper 254, p. 37–50.

Cooper, G.A., Arellano, A.R.V., Johnson, J.H., and Okulitch, V.J., Stoyanow, A., and Lochman, C., 1952, Cambrian stratigraphy and paleontology near Caborca, northwest Sonora, Mexico: Smithsonian Miscellaneous Collections, v. 119, no. 1, 184 p.

Cooper, G.A., Dunbar, C.O., Duncan, H., Miller, A.K., and Knight, J.B., 1953, Permian fauna at El Antimonio, western Sonora, Mexico: Smithsonian Miscellaneous Collections, v. 119, no. 2, 111 p.

Corsetti, F.A., Stewart, J.H., and Hagadorn, J.W., 2001, Neoproterozoic diamictite-cap carbonate succession from eastern Sonora, Mexico: Geological Society of America Abstracts with Programs, v. 33, no. 5, p. A21.

Damborenea, S.E., and González-León, C., 1997, Late Triassic and Early Jurassic bivalves from Sonora, Mexico: Mexico, D.F., Instituto de Geología, Universidad Nacional Autónoma de México, Revista Mexicana de Ciencias Geológicas, v. 14, no. 1, p. 178–201.

Debrenne, F., 1987, Archaeocyatha from Mexico in the Smithsonian Institution—New data from recent collectings: Lyon, France: Geobios, v. 20, no. 2, p. 267–273.

Debrenne, F., Gandin, A., and Rowland, S.M., 1989, Lower Cambrian bioconstructions in northwestern Mexico (Sonora)—Depositional setting, paleoecology and systematics of archaeocyaths: Lyon, France, Geobios, v. 22, no. 2, p. 137–195.

DeJong, K.A., and Jacques-Ayala, C., 1993, Tectonic evolution of northwest Sonora, Mexico: Geological Society of America Abstracts with Programs, v. 25, no. 6, p. 284.

Dembosky, J.A., Jr., Anderson, T.H., Stahl, E.R., and Nourse, J.A., 1999, Field constraints on the trace of the Mojave-Sonora megashear in southwestern Arizona: Geological Society of America Abstracts with Programs, v. 31, no. 7, p. 115.

Dickinson, W.R., 1981, Plate tectonic evolution of the southern Cordillera, *in* Dickinson, W.R., and Payne, W.D., eds., Relations of tectonics to ore deposits in the southern Cordillera: Tucson, Arizona, Arizona Geological Society Digest, v. 14, p. 113–135.

Dickinson, W.R., and Lawton, T.F., 2001, Carboniferous to Cretaceous assembly and fragmentation of Mexico: Geological Society of America Bulletin, v. 113, p. 1142–1160, doi: 10.1130/0016-7606(2001)113<1142:CTCAAF>2.0.CO;2.

Dietz, R.S., and Holden, J.C., 1967, Miogeoclines (miogeosynclines) in space and time: Journal of Geology, v. 74, p. 566–583.

Dunbar, C.O., 1953, A giant Permian fusuline from Sonora, *in* Cooper, G.A, et al., eds., Permian fauna at El Antimonio, western Sonora, Mexico: Smithsonian Miscellaneous Collections, v. 119, no. 2, p. 14–19.

Farmer, G.L., 2003, Isotopic evidence on the nature of Precambrian basement and overlying sedimentary rocks, Caborca, Sonora: Geological Society of America Abstracts with Programs, v. 35, no. 4, p. 67.

Fedo, C.M., and Cooper, J.D., 2001, Sedimentary and sequence stratigraphy of Neoproterozoic and Cambrian units across a craton-margin hinge zone, southeastern California, and implications for the early evolution of the Cordilleran margin: Sedimentary Geology, v. 141–142, p. 501–522, doi: 10.1016/S0037-0738(01)00088-4.

Fitts, D.E., 1989, Metasedimentary strata of Ejido Serdan, Mexico [M.S. thesis]: San Diego, California, San Diego State University, 147 p.

Gangloff, R.A., 1976, Archaeocyathids of eastern California and western Nevada, *in* Moore, J.N., and Fritsche, A.E., eds., Depositional environments of lower Paleozoic rocks in the White-Inyo Mountains, Inyo County, California: Los Angeles, California, Pacific Section, Society of Economic Paleontologists and Mineralogists, p. 16–30.

Gans, W.T., 1974, Correlation and redefinition of the Goodsprings Dolomite, southern Nevada and eastern California: Geological Society of America Bulletin, v. 85, p. 189–200, doi: 10.1130/0016-7606(1974)85<189:CAROTG>2.0.CO;2.

Gastil, R.G., 1993, Prebatholithic history of peninsular California, *in* Gastil, R.G., and Miller, R.H., eds., The prebatholithic stratigraphy of Peninsular California: Geological Society of America Special Paper 279, p. 145–156.

Gehrels, G.E., Stewart, J.H., and Ketner, K.B., 2002, Cordilleran-margin quartzites in Baja California—Implications for tectonic transport: Earth and Planetary Science Letters, v. 199, p. 201–210, doi: 10.1016/S0012-821X(02)00542-3.

Girty, G.H., 1985, Boyden Cave Roof Pendant, west-central Sierra Nevada, Madera and Tulare Counties: Sacramento, California: California Division of Mines and Geology, California Geology, v. 38, p. 51–55.

González-León, C.M., 1997a, Sequence stratigraphy and paleogeographic setting of the Antimonio Formation (Late Permian–Early Jurassic) Sonora, Mexico, *in* Stanley, G.D., Jr., and González-León, C.M., eds., Special issue dedicated to the International Workshop on the Geology of Northwestern Sonora, Mexico: Hermosillo, México, Instituto de Geología, Universidad Nacional Autónoma de México, p. 136–148.

González-León, C.M., 1997b, Stratigraphy and paleogeographic setting of the Antimonio Formation, Sonora, Mexico, *in* González-León, C.M., and Stanley, G.D., Jr., eds., International Workshop on the Geology of Sonora: Hermosillo, México, Instituto de Geología, Estación Regional del Noroeste, Universidad Nacional Autónoma de México, p. 29–40.

González-León, C.M., Taylor, D.G., and Stanley, G.D., Jr., 1996, The Antimonio Formation in Sonora, Mexico, and the Triassic-Jurassic boundary: Canadian Journal of Earth Sciences, v. 33, no. 3, p. 418–428.

Greene, D.C., Schweickert, R.A., and Stevens, C.H., 1997, Roberts Mountains allochthon and the western margin of the Cordilleran miogeocline in the northern Ritter Range pendant, eastern Sierra Nevada: Geological Society of America Bulletin, v. 109, p. 1294, 1305.

Hagadorn, J.W., and Fedo, C.W., 2000, Terminal Proterozoic cloudiniids from southwestern North America: Geological Society of America Abstracts with Programs, v. 32, no. 7, p. A300.

Hamilton, W., 1987, Mesozoic geology and tectonics of the Big Maria Mountains region, southeastern Arizona, *in* Dickinson, W.R., and Klute, M.A., eds., Mesozoic rocks of southern Arizona and adjacent areas: Arizona Geological Society Digest v. 18, p. 33–47.

Hardy, L.R., 1981, Geology of central Sierra de Santa Rosa, Mexico, *in* Ortlieb, L., and Roldán-Quintana, J., eds., Geology of northwestern Mexico and southern Arizona: Hermosillo, México, Instituto de Geología, Universidad Nacional Autónoma de México, p. 73–98.

Hills, J.M., and Kottlowski, F.E., 1983, coordinators, Correlation of stratigraphic units of North America (COSUNA) Project; Southwest/Southwest mid-continent region: Tulsa, Oklahoma, American Association of Petroleum Geologists.

Hintze, L.F., 1988, Geologic history of Utah: Brigham Young University Studies, Special Publication 7, 202 p.

Iriondo, A., 2001a, A testable hypothesis for the distribution of Paleoproterozoic provinces along the SW margin of Laurentia: Insights from NW Sonora, Mexico: Geological Society of America Abstracts with Programs, v. 33, no. 5, p. A3.

Iriondo, A., 2001b, Proterozoic basements and their Laramide juxtaposition in NW Sonora, Mexico: Tectonic constraints on the WS margin of Laurentia [Ph.D. dissertation]: Boulder, University of Colorado, 222 p.

Iriondo, A., 2003, The Caborca block: An inhomogeneous piece of Paleoproterozoic crust in Sonora: Geological Society of America, Abstracts with Programs, v. 35, no. 4, p. 67.

Iriondo, A., Premo, W.R., Budahn, J.R., Martinez-Torres, L.M., and Atkinson, W.W., Jr., 1998, Age and origin of Precambrian basement in the Quitovac region, Sonora, Mexico: Implications for extension of the U.S. mid-continent and the existence of the Mojave-Sonora megashear: Geological Society of America Abstracts with Programs, v. 30, no. 7, p. 170.

Jacques-Ayala, C., Garcia y Barragan, J.C., and DeJong, K.A., 1991, The interpreted trace of the Mojave-Sonora megashear in northwest Sonora: A Laramide thrust front and middle Tertiary detachment zone, *in* Kogan, R.R., and Carreno, A.L., eds., Convencion sobre la evolucion geologíca de México: Primer congreso mexicano de mineralogia, Pachuca, México: Coyoacan, México, Universidad Nacional Autónoma de México, p. 78–80.

Johnson, J.G., Sandberg, C.A., and Poole, F.G., 1991, Devonian lithofacies of western United States, *in* Cooper, J.D., and Stevens, C.H., eds., Paleogeography of the western United States—II, v. 1: Los Angeles, California, Pacific Section, Society of Economic Paleontologists and Mineralogists, p. 83–105.

Jones, D.I., and Moore, J.G., 1973, Lower Jurassic ammonite from the south-central Sierra Nevada, California: U.S. Geological Survey Journal of Research, v. 1, p. 453–458.

Ketner, K.B., 1968, Origin of Ordovician quartzite in the Cordilleran miogeosyncline: U.S. Geological Survey Professional Paper 600-B, p. B169–B177.

Ketner, K.B., 1986, Eureka quartzite in Mexico? Tectonic implications: Geology, v. 14, p. 1027–1030, doi: 10.1130/0091-7613(1986)14<1027:EQIMI>2.0.CO;2.

Ketner, K.B., and Noll, J.H., Jr., 1987, Preliminary geologic map of the Cerro Cobachi area, Sonora, Mexico: U.S. Geological Survey Miscellaneous Field Studies Map MF-1980, scale 1:20,000.

Leveille, G.P., 1984, Geology of El Capitan, Sonora, Mexico [M.Sc. thesis]: San Diego, California, San Diego State University, 122 p.

Lucas, S.G., and Estep, J.W., 1997, The Liassic bivalve *Weyla* from Sonora, Mexico, *in* González-León, C.M., and Stanley, G.D., Jr., eds., International Workshop on the Geology of Sonora: Hermosillo, México, Estación Regional del Noroeste, Instituto de Geología, Universidad Nacional Autónoma de México, p. 57.

Lucas, S.G., and Estep, J.W., 1999, Permian, Triassic, and Jurassic stratigraphy, biostratigraphy, and sequence stratigraphy in the Sierra del Alamo Muerto, Sonora, Mexico, *in* Bartolini, C., et al., eds., Mesozoic sedimentary and tectonic history of north-central Mexico: Geological Society of America Special Paper 340, p. 271–286.

Lucas, S.P., Kues, B.S., Estep, J.W., and González-León, C.M., 1997, Permian-Triassic boundary at El Antimonio, Sonora, Mexico: Universidad Nacional Autónoma de México, Revista Mexicana de Ciencias Geológicas, v. 14, p. 149–154.

Madrid, R.J., Poole, F.G., and Miller, M.M., 1992a, Devonian and older rocks in accreted terranes, *in* Burchfiel, B.C., et al., eds., The Cordilleran orogen, conterminous U.S.: Boulder, Colorado, Geological Society of America, The Geology of North America, v. G-3, p. 36–43.

Madrid, R.J., Poole, F.G., and Wrucke, C.T., 1992b, Rocks of the Antler orogen—The Roberts Mountains allochthon, *in* Burchfiel, B.C., et al., eds., The Cordilleran orogen, conterminous U.S.: Boulder, Colorado, Geological Society of America, The Geology of North America, v. G-3, p. 28–34.

Magginetti, R.T., Stevens, C.H., and Stone, P., 1988, Early Permian fusulinids from the Owens Valley Group, east-central California: Geological Society of America Special Paper 217, 61 p.

Martin, M.W., and Walker, J.D., 1991, Upper Precambrian to Paleozoic paleogeographic reconstruction of the Mojave Desert, California, *in* Cooper, J.D., and Stevens, C.H., eds., Paleozoic paleogeography of the western United States II, v. 1: Los Angeles, California, Pacific Section, Society of Economic Paleontologists and Mineralogists, v. 67, p. 167–192.

Marzolf, J.E., 1997, Sequence stratigraphy of the Colorado Plateau, early Mesozoic paleogeography, and the Sonoran megashear, *in* González-León, C.M., and Stanley, G.D., Jr., eds., International Workshop on the Geology of Sonora (Memoir, preliminary publication): Hermosillo, México, Estación Regional del Noroeste, Instituto de Geología, Universidad Nacional Autónoma de México, p. 60–61.

Marzolf, J.E., and Anderson, T.H., 1996, Early Mesozoic paleogeography and the Mojave-Sonora megashear: Geological Society of America Abstracts with Programs, v. 28, no. 7, p. 114–115.

Marzolf, J.E., and Anderson, T.H., 2000, Facies distribution, hiatal geometry, and cross-cutting relations of Lower Mesozoic tectonosequence boundaries: Constraints on Jurassic displacement of the Caborca terrane, *in* Calmus, T., and Pérez-Segura, E., eds., Cuarta Reunion sobre la geología del noroeste de México y areas adyacentes: Hermosillo, México, Estación Regional del Noroeste, Instituto de Geología, Universidad Nacional Autónoma de México, and Departamento de Geología, Universidad de Sonora, México, Libro de Resumenes, Publicaciones Occasionales 2, p. 60–61.

Miller, E.L., 1981, Geology of the Victorville region, California: Part I and II: Geological Society of America Bulletin, v. 92, p. 160–163, 554–163.

Miller, E.L., Miller, M.M., Stevens, C.H., Wright, J.E., and Madrid, R., 1992, Late Paleozoic paleogeographic and tectonic evolution of the western U.S. Cordillera, *in* Burchfiel, B.C., et al., eds., The Cordilleran orogen, conterminous U.S.: Boulder, Colorado, Geological Society of America, The Geology of North America, v. G-3, p. 57–106.

Molina-Garza, R.S., and Geissman, J.W., 1998, Paleomagnetic evidence against Jurassic left-lateral (southeastward) displacement of the Caborca Terrane: Guidebook Series, Society of Economic Geologists, v. 30, p. 187–201.

Molina-Garza, R.S., and Geissman, J.W., 1999, Paleomagnetic data from the Caborca Terrane, Mexico: Implications for Cordilleran tectonics and the Mojave-Sonora megashear hypothesis: Tectonics, v. 18, no. 2, p. 293–325, doi: 10.1029/1998TC900030.

Moore, J.G., Nokleberg, W.J., and Sisson, T.W., 1994, Geologic road guide to Kings Canyon and Sequoia National Parks, Central Sierra Nevada, California: U.S. Geological Survey Open-File Report, 52 p.

Nestell, M., 2000, Fusulinacean biostratigraphy of Mexico, *in* Calmus, T., and Pérez-Segura, E., eds., Cuarta reunion sobre la geología del noroeste de México y áreas adyacentes, Libro de Resumenes: Hermosillo, Sonora, Instituto de Geología, Universidad Nacional Autónoma de México, and Universidad de Sonora, Publicaciones Ocasionales 2, p. 71–72.

Nichols, K.M., and Silberling, N.J., 1977, Stratigraphy and depositional history of the Star Peak Group (Triassic), northwestern Nevada: Geological Society of America Special Paper 178, 73 p.

Nokleberg, W.J., 1983, Wallrocks of the central Nevada batholith, California—A collage of accreted tectono-stratigraphic terranes: U.S. Geological Survey Professional Paper 1255, 28 p.

Palmer, A.R., 1971, The Cambrian of the Great Basin and adjacent areas, western United States, *in* Holland, C.H., ed., Cambrian of the New World: London, Wiley-Interscience, p. 1–78.

Palmer, A.R., 1981, On the correlation of Grand Cycle tops, *in* Taylor, M.E., ed., Short papers for the Second International Symposium on the Cambrian System: U.S. Geological Survey Open-File Report 81-743, p. 156–159.

Palmer, A.R., and Halley, R.B., 1979, Physical stratigraphy and trilobite biostratigraphy of the Carrara Formation (Lower and Middle Cambrian) in the southern Great Basin: U.S. Geological Survey Professional Paper 1047, 131 p.

Palmer, A.R., and Hintze, L.F., 1992, Middle Cambrian to Lower Ordovician rocks, *in* Burchfiel, B.C., et al., eds., The Cordilleran orogen, conterminous U.S.: Boulder, Colorado, Geological Society of America, The Geology of North America, v. G-3, p. 20–22.

Pérez-Ramos, O., 1992, Permian biostratigraphy and correlation between southeastern Arizona and Sonora: Hermosillo, Sonora, México, Universidad de Sonora, Boletín del Departamento de Geología, v. 9, no. 2, p. 1–74.

Poole, F.G., 1974, Flysch deposits of the Antler foreland basin, western United States, *in* W.R. Dickinson, ed., Tectonics and sedimentation: Society of Economic Paleontologists and Mineralogists Special Publication 22, p. 58–82.

Poole, F.G., 1992, Sedimentary mineral deposits, *in* Burchfiel, B.C., et al., eds., The Cordilleran orogen, conterminous U.S.: Boulder, Colorado, Geological Society of America, The Geology of North America, v. G-3, p. 43–46.

Poole, F.G., and Amaya-Martínez, R., 2000, The Sonora orogen in the Barita de Sonora mine area east of Mazatán, central Sonora: Guide to Field Trip 3 (March 5, 2000) for the Fourth Symposium on the Geology of northwest Mexico and adjacent areas (March 6–8, 2000): Hermosillo, Sonora, México, University of Sonora, Center for the Arts, 27 p.

Poole, F.G., and Madrid, R.J., 1988, Comparison of allochthonous Paleozoic eugeoclinal rocks in the Sonoran, Marathon, and Antler orogens: Geological Society of America Abstracts with Programs, v. 20, no. 7, p. 267.

Poole, F.G., and Sandberg, C.A., 1977, Mississippian paleogeography and tectonics of the western United States, *in* Stewart, J.H., et al., eds., Paleozoic paleogeography of the western United States, Pacific Coast Paleogeography Symposium 1: Los Angeles, California, Pacific Section, Society of Economic Paleontologists and Mineralogists, p. 67–86.

Poole, F.G., and Sandberg, C.A., 1991, Mississippian paleogeography and conodont biostratigraphy of the western United States, *in* Cooper, J.D., and Stevens, C.H., eds., Paleogeography of the western United States—II, v. 1: Los Angeles, California, Pacific Section, Society of Economic Paleontologists and Mineralogists, p. 107–136.

Poole, F.G., and Sandberg, C.A., 1992a, Upper Ordovician—Cincinnatian rocks, *in* Burchfiel, B.C., et al., eds., The Cordilleran orogen, conterminous U.S.: Boulder, Colorado, Geological Society of America, The Geology of North America, v. G-3, p. 24–25.

Poole, F.G., and Sandberg, C.A., 1992b, Lower Silurian to Middle Devonian—Llandoverian to Givetian rocks, *in* Burchfiel, B.C., et al., eds., The Cordilleran orogen, conterminous U.S.: Boulder, Colorado, Geological Society of America, The Geology of North America, v. G-3, p. 24–25.

Poole, F.G., Sandberg, C.A., and Boucot, A.J., 1977, Silurian and Devonian paleogeography of the western United States, *in* Stewart, J.H., et al., eds., Paleozoic paleogeography of the western United States, Pacific Coast Paleogeography Symposium 1: Los Angeles, California, Pacific Section, Society of Economic Paleontologists and Mineralogists, p. 39–66.

Poole, F.G., Madrid, R.J., and Oliva-Becerril, J.F., 1991, Geological setting and origin of stratiform barite in central Sonora, Mexico, *in* Raines, G.L., et al., eds., Geology and ore deposits of the Great Basin: Reno, Geological Society of Nevada, v. 1, p. 517–522.

Poole, F.G., Stewart, J.H., Repetski, J.E., Harris, A.G., Ross, R.J., Jr., Ketner, K.B., Amaya-Martínez, R., and Morales-Ramirez, J.M., 1995a, Ordovician carbonate-shelf rocks of Sonora, Mexico, *in* Cooper, J.D., et al., eds., Ordovician odyssey: Short Papers for the Seventh International Symposium on the Ordovician System: Fullerton, California, Pacific Section, Society for Sedimentary Geology (SEPM), book 77, p. 267–275.

Poole, F.G., Stewart, J.H., Berry, W.B.N., Harris, A.G., Repetski, J.E., Madrid, R.J., Ketner, K.B., Carter, C., and Morales-Ramirez, J.M., 1995b, Ordovician ocean-basin rocks of Sonora, Mexico, *in* Cooper, J.D., et al., eds., Ordovician odyssey: Short Papers for the Seventh International Symposium on the Ordovician System, Las Vegas: Fullerton, California, Pacific Section, Society for Sedimentary Geology (SEPM), book 77, p. 277–284.

Poole, F.G., Page, W.R., and Amaya-Martínez, R., 1997, Newly discovered Silurian carbonate-shelf rocks in west-central Sonora, Mexico: Geological Society of America Abstracts with Programs, v. 29, no. 6, p. 483.

Poole, F.G., Page, W.R., and Amaya-Martínez, R., 1998, Rocas Silúricas de plataforma en Sonora: Mundo Minero (México), Enero 1998, año 2, no. 18, p. 6, 9, 24.

Poole, F.G., Amaya-Martínez, R., and Page, W.R., 2000a, Silurian and Devonian carbonate-shelf rocks and Lower Jurassic sequence near Rancho Placeritos, west-central Sonora: Guide to Field Trip 2 (March 4, 2000) for the Fourth Symposium on the Geology of Northwest Mexico and adjacent areas (March 6–8, 2000): Hermosillo, Sonora, México, University of Sonora, Center for the Arts, 24 p.

Poole, F.G., Amaya-Martínez, R., and Harris, A.G., 2000b, Early Mississippian North American carbonate-platform margin in west-central Sonora, *in* Cuarta Reunion sobre la Geología del Northoeste de México y áreas adyacentes, Libro de Resumenes: Hermosillo, Sonora, Publicaciones Ocasionales 2, p. 95–96.

Radelli, L., Menicucci, S., Mesnier, H.Ph., Araux, E., Calmus, T., Amaya-M., R., Barrera-M., E., Dominguez-P., E., Navarro-M., L., and Soto-C., L., 1987, Allochthonous Paleozoic bodies of central Sonora: Universidad de Sonora, Boletin del Departmento de Geología Segunda Epoca, v. 4, nos. 1-2, p. 1–15.

Rich, M., 1977, Pennsylvanian paleogeograpic patterns in the western United States, *in* Stewart, J.H., et al., eds., Paleozoic paleogeography of the western United States, Pacific Coast Paleogeography Symposium 1: Los Angeles, California, Pacific Section, Society of Economic Paleontologists and Mineralogists, p. 87–111.

Ross, C.A., 1991, Pennsylvanian paleogeography of the western United States, *in* Cooper, J.D., and Stevens, C.H., eds., Paleogeography of the western United States—II, v. 1: Los Angeles, California, Pacific Section, Society of Economic Paleontologists and Mineralogists, p. 137–148.

Ross, C.A., and Ross, J.R.P., 1983, Late Paleozoic accreted terranes of western North America, *in* Stevens, C.H., ed., Pre-Jurassic rocks in western North American suspect terranes: Los Angeles, California, Pacific Section, Society of Economic Paleontologists and Mineralogists, p. 7–22.

Ross, R.J., Jr., 1970, Ordovician brachiopods, trilobites, and stratigraphy in eastern and central Nevada: U.S. Geological Survey Professional Paper 639, 103 p.

Ross, R.J., Jr., 1977, Ordovician paleogeography of the western United States, *in* Stewart, J.H., et al., eds., Paleozoic paleogeography of the western United States, Pacific Coast Paleogeography Symposium 1: Los Angeles,

California, Pacific Section, Society of Economic Paleontologists and Mineralogists, p. 19–38.

Ross, R.J., Jr., James, N.P., Hintze, L.F., and Ketner, K.B., 1991, Early Middle Ordovician (Whiterock) paleogeography of Basin Ranges, *in* Cooper, J.D., and Stevens, C.H., eds., Paleogeography of the western United States—II, v. 1: Los Angeles, California, Pacific Section, Society of Economic Paleontologists and Mineralogists, p. 39–50.

Ross, R.J., Jr., Hintze, L.F., and Poole, F.G., 1992, Middle Ordovician—Whiterockian and Mohawkian, *in* Burchfiel, B.C., et al., eds., The Cordilleran orogen, conterminous U.S.: Boulder, Colorado, Geological Society of America, The Geology of North America, v. G-3, p. 22–24.

Saleeby, J.B., 1992, Petrotectonic and paleogeographic settings of U.S. Cordilleran ophiolites, *in* Burchfiel, B.C., et al., eds., The Cordilleran orogen, conterminous U.S.: Boulder, Colorado, Geological Society of America, The Geology of North America, v. G-3, p. 653–682.

Saleeby, J.B., and Busby-Spera, C., 1992, Early Mesozoic tectonic evolution of the western U.S. Cordillera, *in* Burchfiel, B.C., et al., eds., The Cordilleran orogen, conterminous U.S.: Boulder, Colorado, Geological Society of America, The Geology of North America, v. G-3, p. 107–168.

Saleeby, J.B., and Busby, C., 1993, Paleogeographic and tectonic setting of axial and western metamorphic framework rocks of the southern Sierra Nevada, California, *in* Dunn, G.C., and McDougall, K.A., eds., Mesozoic paleogeography of the western United States—II: Los Angeles, California, Pacific Section, Society of Paleontologists and Mineralogists, Book 71, p. 197–226.

Saleeby, J.B., Goodin, S.E., Sharp, W.D., and Busby, C.J., 1978, Early Mesozoic paleotectonic-paleogeographic reconstruction of the southern Sierra Nevada region, *in* Howell, D.G., and McDougall, K.A., eds., Mesozoic paleogeography of the western United States, Pacific Coast Paleogeography Symposium 2: Los Angeles, California, Pacific Section, Society of Economic Paleontologists and Mineralogists, p. 311–336.

Sandberg, C.A., and Poole, F.G., 1992a, Lower Upper Devonian—Frasnian and lower Famennian rocks, *in* Burchfiel, B.C., et al., eds., The Cordilleran orogen, conterminous U.S.: Boulder, Colorado, Geological Society of America, The Geology of North America, v. G-3, p. 26–27.

Sandberg, C.A., and Poole, F.G., 1992b, upper Upper Devonian—Middle and upper Famennian rocks, *in* Burchfiel, B.C., et al., eds., The Cordilleran orogen, conterminous U.S.: Boulder, Colorado, Geological Society of America, The Geology of North America, v. G-3, p. 27–28.

Schweickert, R.A., 1978, Triassic and Jurassic paleogeography of the Sierra Nevada and adjacent regions, California and western Nevada, *in* Howell, D.G., and McDougall, K.A., eds., Mesozoic paleogeography of the western United States, Pacific Coast Paleogeography Symposium 2: Los Angeles, California, Pacific Section, Society of Economic Paleontologists and Mineralogists, p. 361–384.

Schweickert, R.A., and Lahren, M.M., 1987, Continuation of the Antler and Sonoma orogenic belts to the eastern Sierra Nevada, California, and Late Triassic thrusting in a compressional arc: Geology, v. 15, p. 270–273, doi: 10.1130/0091-7613(1987)15<270:COAASO>2.0.CO;2.

Schweickert, R.A., and Lahren, M.M., 1990, Speculative reconstruction of the Mojave-Snow Lake fault: Implications for Paleozoic and Mesozoic orogenesis in the western United States: Tectonics, v. 9, no. 6, p. 1609–1629.

Sedlock, R.L., Ortega-Gutiérrez, F., and Speed, R.C., 1993, Tectonostratigraphic terranes and tectonic evolution of Mexico: Geological Society of America Special Paper 278, 153 p.

Sheehan, P.M., and Boucot, A.J., 1991, Silurian paleogeography of the western United States, *in* Cooper, J.D., and Stevens, C.H., eds., Paleogeography of the western United States—II, v. 1: Los Angeles, California, Pacific Section, Society of Economic Paleontologists and Mineralogists, p. 51–82.

Speed, R.C., 1977, Excelsior Formation, west central Nevada: Stratigraphic appraisal, new divisions, and paleogeographic interpretations, *in* Stewart, J.H., et al., eds., Paleozoic paleogeography of the western United States, Pacific Coast Paleogeography Symposium 1: Los Angeles, California, Pacific Section, Society of Economic Paleontologists and Mineralogists, p. 325–336.

Silberling, N.J., 1984, Map showing localities and correlation of age-diagnostic lower Mesozoic megafossils, Walker Lake 1° × 2° quadrangle, Nevada and California: U.S. Geological Survey Miscellaneous Field Studies Map MF-1382-O.

Silberling, N.J., and Roberts, R.J., 1962, Pre-Tertiary stratigraphy and structure of northwestern Nevada: Geological Society of America Special Paper 72, 58 p.

Silver, L.T., and Anderson, T.H., 1974, Possible left-lateral early to middle Mesozoic disruption of the southwestern North America craton margin: Geological Society of America Abstracts with Programs, v. 6, p. 955–956.

Silver, L.T., and Anderson, T.H., 1983, Further evidence and analysis of the role of the Mojave-Sonora megashear(s) in Mesozoic Cordilleran tectonics: Geological Society of America Abstracts with Programs, v. 15, no. 5, p. 273.

Stanley, G.D., Jr., and González-León, C.M., 1995, Paleogeographic and tectonic implications of Triassic fossils and strata from the Antimonio Formation, northwestern Sonora, *in* Jacques-Ayala, D., ed., Studies on the Mesozoic of Sonora and adjacent areas: Geological Society of America Special Paper 301, p. 1–16.

Stevens, C.H., 1977, Permian depositional provinces and tectonics, western United States, *in* Stewart, J.H., et al., eds., Paleozoic paleogeography of the western United States, Pacific Coast Paleogeography Symposium 1: Los Angeles, California, Pacific Section, Society of Economic Paleontologists and Mineralogists, p. 113–136.

Stevens, C.H., 1986, Evolution of the Ordovician through Middle Pennsylvanian carbonate shelf in east-central California: Geological Society of America Bulletin, v. 97, p. 11–25, doi: 10.1130/0016-7606(1986)97<11:EOTOTM>2.0.CO;2.

Stevens, C.H., 1991, Permian paleogeography of the western United States, *in* Cooper, J.D., and Stevens, C.H., eds., Paleogeography of the western United States—II, v. 1: Los Angeles, California, Pacific Section, Society of Economic Paleontologists and Mineralogists, p. 149–166.

Stevens, C.H., Stone, P., and Kistler, R.W., 1992, A speculative reconstruction of the middle Paleozoic continental margin of southwestern North America: Tectonics, v. 11, no. 2, p. 405–419.

Stewart, J.H., 1970, Upper Precambrian and Lower Cambrian strata in the southern Great Basin, California and Nevada: U.S. Geological Survey Professional Paper 620, 206 p.

Stewart, J.H., 1980, Geology of Nevada: Nevada Bureau of Mines and Geology Special Publication 4, 136 p.

Stewart, J.H., 1988, Tectonics of the Walker Lane Belt, western Great Basin—Mesozoic and Cenozoic deformation in a zone of shear, *in* Ernst, W.G., ed., Metamorphism and crustal evolution of the western United States, Rubey v. 7: Englewood Cliffs, New Jersey, Prentice-Hall, p. 683–713.

Stewart, J.H., 1991, Latest Precambrian and Cambrian rocks of the western United States—An overview, *in* Cooper, J.D., and Stevens, C.H., eds., Paleozoic paleogeography of the western United States: Pacific Coast Paleogeography Symposium 2, Los Angeles, California, Pacific Section, Society of Economic Paleontologists and Mineralogists, p. 13–38.

Stewart, J.H., 1992, Late Proterozoic and Lower Cambrian rocks, *in* Burchfiel, B.C., et al., eds., The Cordilleran orogen, conterminous U.S.: Boulder, Colorado, Geological Society of America, The Geology of North America, v. G-3, p. 16–20.

Stewart, J.H., 1997, Triassic and Jurassic stratigraphy and paleogeography of west-central Nevada and eastern California: U.S. Geological Survey Open-File Report 97-495, 57 p.

Stewart, J.H., 2005, Eolian deposits in the Neoproterozoic Big Bear Group in the San Bernardino Mountains, California, USA, *in* Calzia, J.P., ed., Fifty years of Death Valley research: A volume in honor of Lauren A. Wright and Bennie Troxel: Earth-Science Reviews Special Volume (in press).

Stewart, J.H., and Crowell, J.C., 1992, Strike-slip tectonics in the Cordilleran region, western United States, *in* Burchfiel, B.C., et al., eds., The Cordilleran orogen, conterminous U.S.: Boulder, Colorado, Geological Society of America, The Geology of North America, v. G-3, p. 609–628.

Stewart, J.H., and Poole, F.G., 1975, Extension of the Cordilleran miogeosynclinal belt to the San Andreas fault, southern California: Geological Society of America Bulletin, v. 86, p. 205–212, doi: 10.1130/0016-7606(1975)86<205:EOTCMB>2.0.CO;2.

Stewart, J.H., and Poole, F.G., 2002, Inventory of Neoproterozoic and Paleozoic strata in Sonora, Mexico: U.S. Geological Survey Open-File Report 02-97, http://geopubs.wr.usgs.gov/open-file/of02-97/ (Accessed April 2005).

Stewart, J.H., and McKee, E.H., 1977, Geology and mineral deposits of Lander County, Nevada: Nevada Bureau of Mines and Geology Bulletin 88, pt. 1, 59 p.

Stewart, J.H., and Roldán-Quintana, J., 1991, Upper Triassic Barranca Group; Nonmarine and shallow-marine rift-basin deposits of northwestern Mexico, *in* Pérez-Segura, E., and Jacques-Ayala, C., eds., Studies of Sonoran geology: Geological Society of America Special Paper 254, p. 19–36.

Stewart, J.H., and Suczek, C.A., 1977, Cambrian and latest Precambrian paleogeography and tectonics of the western United States, *in* Stewart, J.H., et al., eds., Paleozoic paleogeography of the western United States, Pacific Coast Paleogeography Symposium 1: Los Angeles, California, Pacific Section, Society of Economic Paleontologists and Mineralogists, p. 1–18.

Stewart, J.H., McMenamin, M.A.S., and Morales-Ramirez, J.M., 1984, Upper Proterozoic and Cambrian rocks in the Caborca region, Sonora, Mexico—Physical stratigraphy, biostratigraphy, paleocurrent studies, and regional relations: U.S. Geological Survey Professional Paper 1309, 36 p.

Stewart, J.H., Murchey, B., Jones, D.L., and Wardlaw, B.R., 1986, Paleontologic evidence for complex tectonic interlayering of Mississippian to Permian deep-water rocks of the Golconda allochthon in Tobin Range, north-central Nevada: Geological Society of America Bulletin, v. 97, p. 1122–1132, doi: 10.1130/0016-7606(1986)97<1122:PEFCTI>2.0.CO;2.

Stewart, J.H., Poole, F.G., Ketner, K.B., and Madrid, R.J., Roldán-Quintana, J., Amaya-Martínez, R., 1990, Tectonics and stratigraphy of the Paleozoic and Triassic southern margin of North America, Sonora, Mexico, *in* Gehrels, G.E. and Spencer, J.E., eds., Geologic excursions through the Sonoran Desert region, Arizona and Sonora: Arizona Geological Survey Special Paper 7, p. 183–202.

Stewart, J.H., Amaya-Martínez, R., Stamm, R.G., Wardlaw, B.R., Stanley, G.D., and Stevens, C.H., 1997, Stratigraphy and regional significance of Mississippian to Jurassic rocks in Sierra Santa Teresa, Sonora, Mexico: Mexico, D.F., Universidad Nacional Autónoma de México, Revista Mexicana de Ciencias Geológicas, v. 14, no. 2, p. 115–135.

Stewart, J.H., Poole, F.G., Harris, A.G., Repetski, J.E., Wardlaw, B.R., Mamet, B.L., and Morales-Ramirez, J.M., 1999, Neoproterozoic (?) to Pennsylvanian inner-shelf, miogeoclinal strata in Sierra Agua Verde, Sonora, Mexico: Mexico, D.F., Universidad Nacional Autónoma de México, Revista Mexicana de Ciencias Geológicas, v. 16, no. 1, p. 35–62.

Stewart, J.H., Amaya-Martínez, R., and Palmer, A.R., 2002, Neoproterozoic and Cambrian strata of Sonora, Mexico: Rodinian supercontinent to Laurentian Cordilleran margin, *in* Barth, A., ed., Contributions to crustal evolution of the southwestern United States: Geological Society of America Special Paper 365, p. 5–47.

von Huene, R., and Scholl, D.W., 1991, Observations at convergent margins concerning sediment subduction, subduction erosion, and growth of continental crust: Reviews of Geophysics, v. 29, p. 279–316.

Wernicke, B., 1992, Cenozoic extensional tectonics of the U.S. Cordilleran, *in* Burchfiel, B.C., et al., eds., The Cordilleran orogen, conterminous U.S.: Boulder, Colorado, Geological Society of America, The Geology of North America, v. G-3, p. 553–582.

Williams, E.G., Wright, L.A., and Troxel, B.W., 1974, The Noonday Dolomite and equivalent stratigraphic units, southern Death Valley region, California, *in* Guidebook, Death Valley region, California and Nevada: Shoshone, California, Shoshone Valley Publishing, p. 73–77.

Manuscript Accepted by the Society 24 March 2005

Geological Society of America
Special Paper 393
2005

Review of Upper Paleozoic and Lower Mesozoic stratigraphy and depositional environments of central and west Mexico: Constraints on terrane analysis and paleogeography

Elena Centeno-García
Instituto de Geologia, Universidad Nacional Autónoma de México, Ciudad Universitaria, México D.F., 04510 México

ABSTRACT

Reconstructing the geological evolution of central and western Mexico during the end of the Paleozoic and the beginning of the Mesozoic is very difficult because of a lack of exposures. The few outcrops available, and indirect information obtained from geophysical and geochemical data suggests that Central and Western Mexico are made up of a mosaic of pre-Jurassic terranes, and that previously defined terranes are mostly composites of basements of different origins. Most of those terranes are allochthonous with respect to North America, but some developed not far from their present position. It has been suggested that the Coahuila and Sierra Madre terranes (Oaxaquia block), part of Gondwana during Early Paleozoic, collided with North America by Late Paleozoic time. However, their Mississippian faunas of North American affinity suggest that the collision might have occurred earlier. The nature of the basement of the Central terrane is unknown, but it is inferred to be allochthonous because there is an accretionary prism at its NE boundary. The basement of the Parral and Tahue terranes is formed by a deformed volcano-sedimentary complex of Early Paleozoic age, whose origin and paleogeographic evolution remains unknown. The Caborca and Cortes terranes are formed by Proterozoic metamorphic complexes and an accreted eugeoclinal Paleozoic sedimentary wedge. The basement of the Zihuatanejo terrane is made up of Triassic ocean-floor continental-rise assemblages accreted in Early Jurassic time.

An overview of new stratigraphic and geochronologic data indicates that a number of tectonic events occurred during Late Paleozoic to Early Mesozoic time. A continental arc with a paleo-Pacific, east-dipping subduction zone evolved from Carboniferous to Early Permian time in eastern Mexico (Oaxaquia), and it was in part contemporaneous to deformation in the Ouachita belt. This was followed by a period of volcanic quiescence during middle Permian. A more felsic arc, with a different distribution of the volcanic axis, developed along all the paleo-Pacific margin in the Permo-Triassic. Terranes in northwestern Mexico show a completely different geological evolution during the Carboniferous and Permian time. They were characterized by passive margin sedimentation and by folding and thrusting of eugeoclinal rocks in the Mississippian and Late Permian. By Late Triassic, a passive or rifting

Centeno-García, E., 2005, Review of Upper Paleozoic and Lower Mesozoic stratigraphy and depositional environments of central and west Mexico: Constraints on terrane analysis and paleogeography, *in* Anderson, T.H., Nourse, J.A., McKee, J.W., and Steiner, M.B., eds., The Mojave-Sonora megashear hypothesis: Development, assessment, and alternatives: Geological Society of America Special Paper 393, p. 233–258. doi: 10.1130/2005.2393(08). For permission to copy, contact editing@geosociety.org.

margin developed along the western margin of Oaxaquia, and thick successions of continent-derived sediments were accumulated on the paleocontinental shelf and slope (Potosi Fan) and in a marginal active oceanic basin (Arteaga Basin). Those rocks were deformed and accreted to nuclear Mexico by Late Triassic–Early Jurassic time, before the development of the Late Jurassic continental arc that was widespread along western and central Mexico.

Keywords: Paleozoic, Triassic, Mexico, tectonic, stratigraphy.

INTRODUCTION

The region of central Mexico broadly corresponds to the southern end of the North America Craton, where a complex tectonic scenario took place at least during the Paleozoic and most of the Mesozoic. During the end of the Paleozoic and Early Mesozoic, the main tectonic processes that occurred in the Atlantic margin of the North America Craton were the assemblage and breakup of Pangea and the opening of the Gulf of Mexico. Thus, it has been proposed that parts of eastern Mexico should have been involved first in a collisional event during the end of the Paleozoic, followed by rifting and extensional tectonics beginning in Late Triassic and ending about Middle Jurassic time (e.g., Pindell, 1985; Ross and Scotese, 1988; Rowley and Pindell, 1989; Molina-Garza et al., 1992, Dickinson and Lawton, 2001).

In contrast, the Pacific margin of North America was characterized mostly by subduction related processes, strike-slip faulting and accretion of terranes. Therefore, it has been suggested that parts of western Mexico were also involved in subduction, strike-slip faulting, and/or accretion of terranes (e.g., Coney, 1983; Saleeby and Busby-Spera, 1992; Centeno-García et al., 1993b, 2003; Sedlock et al., 1993; Dickinson and Lawton, 2001).

Recorded in Mexico are the interactions between the tectonics of the Pacific and Atlantic Margins, making it an important region for understanding the tectonic evolution of North America as a whole, and for reconstructing the processes that can occur in a complex transitional zone. Campa and Coney (1983), Coney and Campa (1987), Sedlock et al. (1993), and Dickinson and Lawton (2001), among others, have proposed several models for the terrane configuration and tectonic evolution of Mexico. Most of the terranes were defined on the basis of differences in Jurassic-Cretaceous stratigraphy. However, detailed studies of the few exposures of pre-Jurassic rocks suggest a more complex terrane configuration that recorded accretions and major displacements before Late Jurassic–Cretaceous time. In other words, looking at the pre-Jurassic stratigraphy, the terranes as defined up to date are in fact composite, with pre-Jurassic, strongly deformed assemblages. This includes the Guerrero terrane, which was previously considered an allochthonous, Cretaceous intraoceanic arc (Tardy et al., 1994; Dickinson and Lawton, 2001).

Overall, lithological, geochemical, and isotopic data suggest that the basement of western Mexico is made up of more juvenile material than the east and the north (Patchett and Ruíz, 1987; Ruíz et al., 1988a, 1988b; Centeno-García et al., 1993b; Talavera-Mendoza et al., 1995; McDowell et al., 1999; Mendoza and Suástegui, 2000; Valencia-Moreno et al., 1999, 2001). Considering this major difference in composition of the basement, the terranes of central and western Mexico can be classified in two general groups: first, terranes that have a Precambrian/Lower Paleozoic? basement and/or have old crustal signatures in the isotopic composition of their younger igneous rocks; and second, terranes whose oldest rocks are Uppermost Paleozoic to Mesozoic and/or show juvenile isotopic signatures in their igneous and metamorphic rocks. Older terranes are located in the east and northwest, and younger terranes are mostly in the central and west of Mexico, except for the Maya terrane of eastern Mexico, which is characterized by Paleozoic metamorphic rocks. Figure 1 shows the possible distribution of pre–Late Jurassic terranes, some of which are proposed in this paper.

The rocks of all these terranes are not well exposed, and most of the contacts are inferred. The only contacts that can be

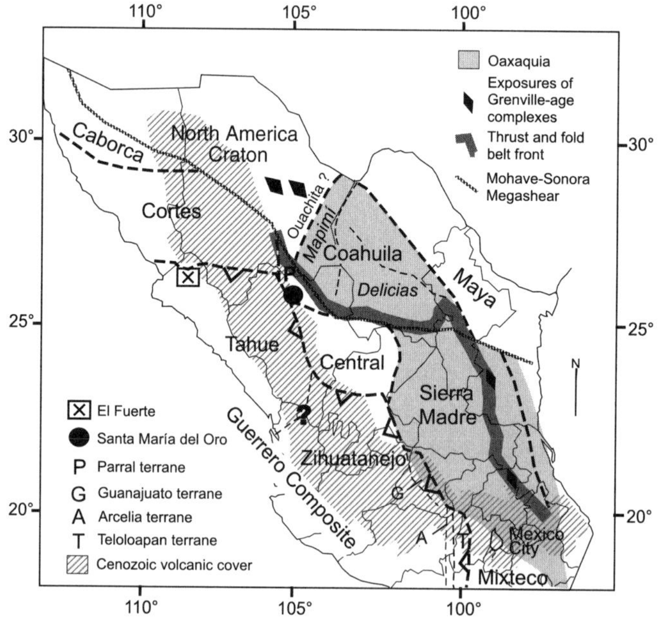

Figure 1. Terrane configuration of North Central Mexico, modified from Coney and Campa (1987) and Sedlock et al. (1993). Thick dashed line corresponds to boundary of Guerrero composite terrane. Thin dashed line shows possible location of other terrane boundaries, based on indirect evidence.

observed are between the Guerrero and surrounding terranes, exposed southeast of Zacatecas City, and in Santa Maria del Oro, Durango. All other contacts are inferred on the basis of four criteria: (1) isotopic signatures of granitic intrusions, felsic volcanic rocks and xenoliths (Cameron and Cameron, 1985; Patchett and Ruíz, 1987; Ruíz et al., 1988a, 1988b; McDowell et al., 1999; Torres et al., 1999; Valencia-Moreno et al., 1999, among others); (2) geophysical data (Mickus and Montana, 1999; Moreno et al., 2000); (3) subsurface data from oil exploration drilling; and (4) structural styles and trends of the Sierra Madre Oriental fold and thrust belt, which may be reflecting major changes in crustal composition and/or previously formed major structures.

The main objective of this paper is to discuss recent data on the Upper Paleozoic Lower Mesozoic stratigraphy, depositional environments, paleontology, structural geology, and provenance of sedimentary rocks of central and western Mexico, and to review the terrane configuration of the area. Also, some of the existing tectonic models for the evolution of the western paleocontinental margin of Mexico are discussed and new models presented.

TERRANES OF EAST AND NORTHWESTERN MEXICO

Most of eastern and northwestern Mexico has Proterozoic basement, but only a small portion of it is considered to be part of the autochthonous Craton of North America. The rest of the Proterozoic basement is interpreted to be an assortment of allochthonous terranes (Campa and Coney, 1983; Coney and Campa, 1987; Sedlock et al., 1993), which includes the Caborca, Coahuila (Delicias, Mapimi), and Sierra Madre terranes (Fig. 1). The Coahuila (Delicias, Mapimi) and Sierra Madre terranes seem to share a common origin, thus they have been defined as the Oaxaquia block or microcontinent, which was disrupted by Late Paleozoic–Mesozoic deformational events (Ortega-Gutíerrez et al., 1995). The Maya and Cortes terranes are made up of deformed Paleozoic rocks (Coney and Campa, 1987). This section contains a short discussion of the nature of the basement of these seven terranes and the North America Craton in Mexico, as well as a description of their Upper Paleozoic–Early Mesozoic stratigraphy.

North America Craton

The basement of northern Chihuahua and part of northeastern Sonora has been interpreted to be the southern continuation of the North America craton but is heterogeneous in composition (Cameron and Cameron, 1985; McDowell et al., 1999). It is made up of pelitic and volcanic schists that are intruded by 1.7–1.6-Ga granites (Anderson and Silver, 1977). Also, an allochthonous block of Grenvillian rocks that are exposed in Sierra del Cuervo (Los Filtros) and a small outcrop of metamorphic basement rocks at Carrizalillo, in north-central Chihuahua, have been considered part of the North America Craton (Coney and Campa, 1987; Ruíz et al., 1988b; Blount et al., 1988) (Fig. 1). Boundaries of the North America Craton in Chihuahua on Figure 1 are all inferred, and they have been traced on the basis of data by Moreno et al. (2000), Cameron and Cameron (1985), McDowell et al. (1999), and Ruíz et al. (1988b), among others.

Sedimentation during the Paleozoic in northern Chihuahua and Sonora states has been correlated to the basins that developed in Southern Texas, New Mexico, and Arizona (Bridges, 1964). Interpreting from the published stratigraphic columns (Bridges, 1964), subsidence was important throughout that time and led to the deposition of up to 3000 m in thickness of Paleozoic marine sediments. Sandstone layers contained within the calcareous succession seem to be quartz-rich arenites (Díaz and Navarro, 1964), which suggests a cratonal provenance.

Permian rocks deposited on the North America craton are mostly calcareous and occur near the U.S.-Mexico border (Sierra Las Palomas, Los Chinos and Moyotes Wells, Sierra de Teras-Bavispe) (12 and 12a in Fig. 2) (Díaz and Navarro, 1964; Tovar, 1968; Reynolds, 1972; Patterson, 1978). They are made up of thick to thin bedded limestone and dolostone that contain chert lenses and nodules. They are interbedded with variable amounts of shale, quartz-rich sandstone, and sandy limestone. Conglomerates formed of calcareous clasts and layers of evaporites (gypsum) have been reported from the base of the succession, suggesting periods of subareal exposure (Díaz and Navarro, 1964; Tovar, 1968; Reynolds, 1972; Patterson, 1978). Horizons with abundant fossils can be found at all localities and include a wide range of corals, brachiopods, fusulinids, crinoids, and bivalves. Fossil associations, and pelletoid and oolithic limestone suggest shallow marine environments. There might have been periods of subaereal exposure, as suggested by the evaporites. The ages of the Permian rocks range from Wolfcampian to Leonardian. Their basal contact with Pennsylvanian rocks is transitional (Díaz and Navarro, 1964; Tovar, 1968; Reynolds, 1972; Patterson, 1978).

Paleozoic sedimentary rocks in central Chihuahua State are well exposed in the Carrizalillo, Placer de Guadalupe, and Sierra del Cuervo areas (location 1 and 2 in Figs. 2 and 3). They range from Proterozoic(?)-Cambrian to Permian, and are strongly deformed. No major angular unconformities within the succession have been reported, except for an unconformity described by Bridges (1964), from Placer de Guadalupe, where Permian rock unconformably overlie Early Pennsylvanian–Early Permian limestone. Permian rocks of central Chihuahua are more clastic than rocks of similar ages in northwest Chihuahua (location 1 and 2 in Fig. 2). They are made up of siltstone and sandstone, some conglomerate as channel-fill deposits, and lenses of fossiliferous, calcareous debris flows. Permian rocks overlie Early Pennsylvanian–Early Permian limestone, mostly in a parallel, slightly erosional contact in Carrizalillo. Fauna associations from Placer de Guadalupe suggest a shallow marine depositional environment, however most of the sediments are turbidites, thus they might be submarine fan deposits (Hadschy and Dyer, 1987).

Pennsylvanian-Permian rocks of the Sierra del Cuervo (location 2 in Figs. 2 and 3) are mostly thin-bedded sandstone and shale, and basinal limestone, with sparse conglomerate, bentonite, and bedded chert. Sandstone becomes more quartz-rich

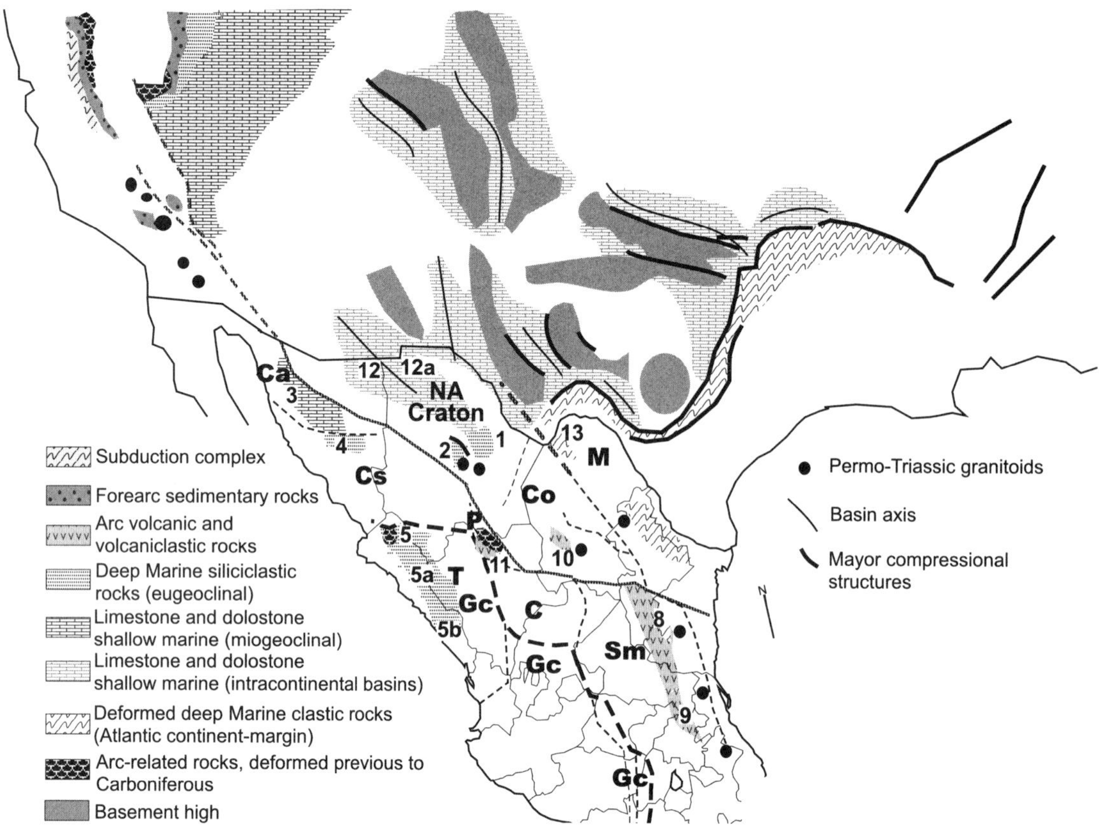

Figure 2. Location of outcrops of Carboniferous-Permian rocks. Patterns indicate depositional environments and tectonic setting. Regional extension of such settings is unknown because of the lack of exposures. Tectonic elements of North America are also included for comparison (from Miller et al., 1999). Black dots indicate the distribution of Permo-Triassic granitoids. 1—Plomosas-Placer de Guadalupe; 2—Sierra del Cuervo (1 and 2 represent the southern extension into Chihuahua of the North America Craton); 3—Caborca region; 4—Mazatan region and southern Sonora; 5—El Fuerte, 5a—San Jose de Gracia, and 5b—Mazatlán, Sinaloa; 8—western Ciudad Victoria, Tamaulipas; 9—Tuzancoa Formation in Huayacocotla-Molango region, Hidalgo; 10—Las Delicias, Coahuila; 11—Santa Maria del Oro, Durango; 12 and 12a—Sierra Los Chinos, Coahuila and NW Sonora; 13—Sierra del Carmen, Coahuila. Thick dashed line corresponds to the boundary of the Guerrero composite terrane (Gc), which includes the Tahue terrane (T). Other terranes are: Sierra Madre (Sm), Coahuila (Co), Maya (M), Parral (P), Caborca (Ca), Cortes (Cs), and Central (C).

arenite upward, with more abundant lithic clasts at the base (carbonate and argillite grains); feldspar grains increase upward as well (Handschy and Dyer, 1987). Sedimentary structures suggest that the succession was deposited by turbidites in a submarine fan environment (Handschy and Dyer, 1987). These rocks were strongly deformed before the deposition of Cretaceous units, but the age of deformation has not been constrained (Handschy and Dyer, 1987).

There has not been a systematic sedimentological and provenance study of the Upper Paleozoic assemblages. Likewise, with no systematic study on reconstructing depositional environments and paleowater depths, there can only be loose constraints on the tectonic setting of these sedimentary basins. However, some important information can be obtained from published data.

Rock and fossil associations in Placer de Guadalupe indicate a change from mostly calcareous shallow marine environments to deeper quartz-rich submarine fan facies. Bridges (1964) suggests that the sedimentary characteristics of the Placer de Guadalupe turbidites are more similar to foreland basins than to deep marine eugeoclinal deposits of the Marathon collisional front. More detailed studies could be done to characterize the sedimentation and provenance of these rocks. But from the evidence described in the literature, it might be possible that the Placer de Guadalupe succession evolved from calcareous platform to foreland sedimentation in Pennsylvanian-Permian time. Those rocks have also been interpreted as basins related to strike-slip motion active during late Pennsylvanian–earliest Permian time (Lawton and Giles, 2000).

Sedimentary rocks at the Sierra del Cuervo seem to be similar to the deposits of Placer de Guadalupe. They have been interpreted as part of eugeoclinal deposits of the paleocontinental margin of North America (continental slope-rise) (Handschy and Dyer, 1987). Their relationship with coeval rocks from Caborca and Cortes terranes has not been determined. Two important differences between these rocks and coeval units from Caborca and Cortes terranes are as follows: apparently they contain some volcaniclastics (altered ash) (Bridges, 1964); and the basement is involved in the deformation, which is not seen in accreted eugeoclinal successions throughout the Cordillera (Miller et al., 1992).

There are no outcrops of Uppermost Permian or Triassic sedimentary rocks in Chihuahua or northeast Sonora. Only a few granitic intrusions of Upper Permian age (250 ± 20 and 267 ± 21 Ma) have been reported (Torres et al., 1999). Paleozoic rocks are unconformably covered by calcareous marine deposits (La Casita Formation) that contain abundant ammonites of Late Jurassic age (Kimeridgian and Tithonian) (Bridges, 1964).

Maya Terrane

The Maya terrane (Fig. 1) was defined on the basis of the report of Paleozoic metamorphic rocks in the subsurface along the Gulf of Mexico (Campa and Coney, 1983, Coney and Campa, 1987, Murillo and Torres, 1987; Sedlock et al., 1993) Whether the Maya terrane extends further north is unknown. However, information from oil exploration boreholes in the states of Coahuila and Nuevo León, as well as gravity data from Northeastern Mexico, suggest that Paleozoic metamorphic rocks might extend north along the northern part of Tamaulipas, Nuevo León, and Coahuila states (Schurbet and Cebull, 1987; Ramirez-Ramirez, 1992; Moreno et al., 2000). The only exposure of the Maya terrane's basement is located near the U.S.-Mexico border. These rocks are made of strongly deformed sandstone and shale that yielded one Rb/Sr age of 240 Ma (Carpenter, 1997) (location 13 in Fig. 2). Scarce information on Paleozoic rocks is found in unpublished reports of the Mexican national oil company Pemex. Core samples of schist yielded K/Ar ages between 277 and 204 Ma, and granite samples of 358–144 Ma (Ramírez-Ramírez, 1992; Torres et al., 1999). Discussion on the origin and extension of the Maya terrane is outside the scope of this paper.

Caborca Terrane

The oldest rocks of Mexico form the basement of the Caborca terrane (Fig. 1). They are orthogneisses and high-grade schist complexes intruded by meta plutonic rocks 1.8–1.7 Ga in age (Anderson and Silver, 1971). The oldest sialic basement of Mexico is exposed in the Caborca terrane (Bamori complex and Aibo granite) (Anderson and Silver, 1977, 1981). There are different opinions on the origin and correlation of its basement with respect to the metamorphic complexes of North America (Anderson and Silver, 1979; Stewart et al., 1990; McDowell et al., 1999; Iriondo et al., 2000).

The northern and eastern limits of the Caborca terrane (Fig. 1) have been placed at slightly different locations (Anderson and Silver, 1979; Coney and Campa, 1987; Mc Dowell et al., 1999; Dickinson and Lawton, 2001). Its southern limit is considered to be placed where continental slope-rise sedimentary rocks (eugeoclinal) are thrust over shallow marine successions at the Mina La Barita, Sonora (Gastil and Miller, 1983; Stewart et al., 1990; Poole et al., 1992).

The sedimentary cover of the Caborca terrane ranges in age from Upper Proterozoic to Permian. Permian rocks are mostly calcareous (location 3 in Figs. 2 and 3). The succession is made up of fossiliferous limestone and dolostone, with abundant chert nodules, and extends from Carboniferous to Lower Permian (Cooper and Arellano, 1946). Shale and sandstone are rare. In the Sierra del Alamo, there is a succession of limestone and dolomite (Los Monos Formation) of mid-Permian age (Guadalupian), but its contact relationships with the other described units are unknown. This unit is covered disconformably by the El Antimonio Formation (Cooper and Arellano, 1946; González-León, 1996, 1997; González-León et al., this volume). The Antimonio stratigraphy is described in detail in this volume; this unit is formed by interbedded thin beds of calcareous siltstone with calcareous nodules, limestone, sandstone, and some beds of conglomerate. Clasts from the conglomerate are mostly chert, limestone, quartzite, rhyolite, quartz porphyry, and granite. It contains, at its base, fossils that are similar to those in underlying Guadalupian rocks and extends into the Triassic and Jurassic (location 3 in Figs. 2 and 3) (González-León, 1996, 1997; González-León et al., this volume). These rocks were deposited in shallow marine environments. The calcareous successions have been interpreted as the southern continuation of the Western U.S. miogeoclinal deposits (Stewart, 1988). González- León et al. (this volume) and other authors consider the Antimonio rocks to be mostly deposited in a forearc setting, but its basal part might have been deposited in a marine platform environment.

Triassic stratigraphy of the Caborca terrane is discussed in detail by González-León et al. in this volume. It is a succession of sandstone, siltstone, and limestone with some conglomerate horizons (Antimonio Group, location 3 in Figs. 2 and 3) that were deposited mostly in shallow marine, but also some in continental environments. Sediments toward the top of the succession were deposited in deeper water. Ages range from Middle to Late Triassic (Carnian) (González-León, 1997). The succession contains abundant fossils and has been interpreted as deposited in a forearc setting and changes upward to Jurassic marine rocks (González-León et al., this volume).

Cortes Terrane

The Cortes terrane (Fig. 1) was originally defined as allochthonous, Early Paleozoic, deep-marine rocks (eugeoclinal) emplaced on thinned continental crust (Poole and Madrid, 1986, Coney and Campa, 1987; Stewart et al., 1990). Pb and Nd isotopic composition from Cenozoic granitoids and volcanic rocks emplaced in the Cortes terrane suggest that its basement might be Proterozoic (McDowell et al., 1999; Valencia-Moreno et al., 1999, 2001). However, there is a difference in the isotopic signatures of the Cenozoic magmatism between the Caborca and Cortes terranes, suggesting that the basement of the Cortes terrane

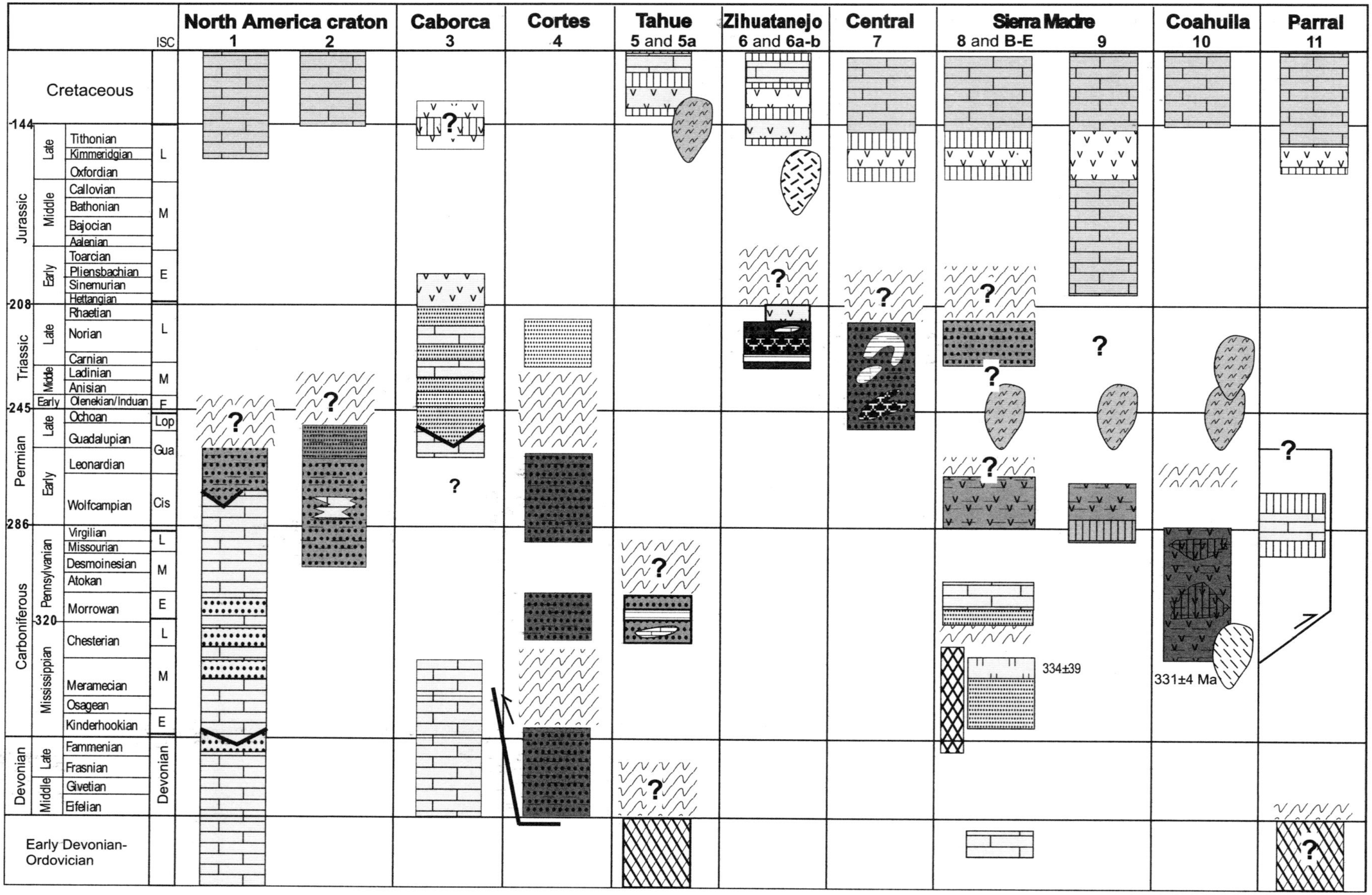
North America craton
1
2
Caborca
3
Cortes
4
Tahue
5 and 5a
Zihuatanejo
6 and 6a-b
Central
7
Sierra Madre
8 and B-E
9
Coahuila
10
Parral
11
ISC
Cretaceous
144
Jurassic
Late
Middle
Early
Tithonian
Kimmeridgian
Oxfordian
Callovian
Bathonian
Bajocian
Aalenian
Toarcian
Pliensbachian
Sinemurian
Hettangian
208
Triassic
Rhaetian
Norian
Carnian
Ladinian
Anisian
Olenekian/Induan
245
Permian
Ochoan
Guadalupian
Leonardian
Wolfcampian
Lop
Gua
Cis
286
Carboniferous
Pennsylvanian
Mississippian
320
Virgilian
Missourian
Desmoinesian
Atokan
Morrowan
Chesterian
Meramecian
Osagean
Kinderhookian
Devonian
Fammenian
Frasnian
Givetian
Eifelian
L
M
E
Early Devonian-
Ordovician
334±39
331±4 Ma

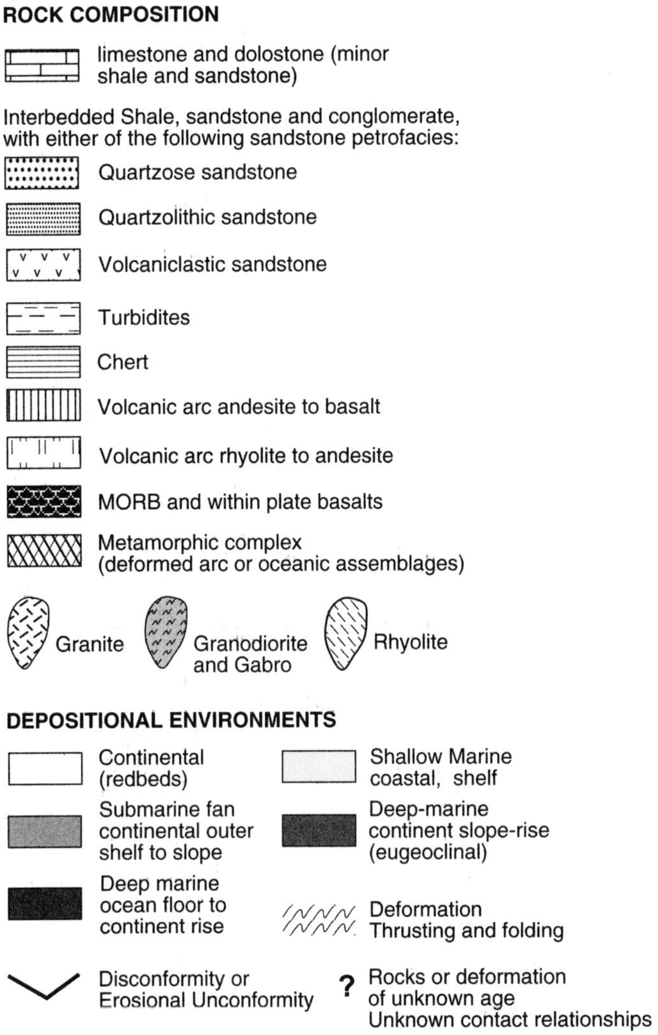

Figure 3 (*on this and previous page*). Correlation table of key stratigraphic columns (source references in text). 1—Carrizalillo and Placer de Guadalupe; 2—Sierra del Cuervo (1 and 2 represent southern extension into Chihuahua of North America Craton); 3—Caborca region; 4—Mazatan region and southern Sonora; 5—El Fuerte; 5a—San José de Gracia (5 and 5a both in northern Sinaloa); 6—Arteaga region, Michoacán; 6a—Zacatecas City; 7—Pico de Teyra region, Zacatecas; 8—western Ciudad Victoria, Tamaulipas; 9—Tuzancoa Formation in Huayacocotla-Molango region, Hidalgo; 10—Las Delicias, Coahuila; 11—Santa María del Oro, Durango.

could be thinned Proterozoic rocks, perhaps the same as in the Caborca terrane, or a Proterozoic basement different from that of the Caborca terrane (McDowell et al., 1999; Valencia-Moreno et al., 1999, 2001).

The northern limit of the Cortes terrane is placed where the Paleozoic deep-marine sedimentary rocks are thrust over calcareous shelf facies of the Caborca terrane (Fig. 1) (Stewart et al., 1990). The southern limit has not been well defined, and it is inferred to be north of El Fuerte, Sinaloa, on the basis of the last northern occurrence of Cretaceous marine volcanic rocks of the Guerrero terrane (Fig. 1) (Servais et al., 1982; Henry and Fredrikson, 1987; Roldán-Quintana et al., 1993; Freydier et al., 1995).

A thrusting event during the Mississippian deformed deep marine rocks of Ordovician and Devonian–Lower Mississippian age (location 4 in Figs. 2 and 3) (Poole and Madrid, 1988; Stewart et al., 1990; Poole et al., 1992). Overlying this event, the Upper Mississippian to Lower Pennsylvanian succession is characterized by submarine turbidites made up of fine-grain sandstone, shale and siltstone, and channel-fill conglomerate (Poole and Madrid, 1988; Stewart et al., 1990; Poole et al., 1992). Upper Pennsylvanian and Lower Permian rocks are a thin-bedded rhythmic succession of mudstone and siltstone that contains chert and detrital limestone beds all deposited by turbiditic flows (Poole and Madrid, 1988; Stewart et al., 1990; Poole et al., 1992). They were deposited in a deep marine environment and are interpreted to be part of the Paleozoic eugeoclinal deposits (continental slope-rise) of western North America (Poole and Madrid, 1988; Stewart et al., 1990; Poole et al., 1992). The change in sedimentation to deeper facies during Early Permian time has been interpreted as a change from a stable continental margin to an active margin (subduction or strike-slip) (Stewart et al., 1997). Apparently a second deformational event originated folding and trusting of these units by Late Permian to Early Triassic time, because continental rocks of Upper Triassic age lie unconformably over the deep-marine strata (location 4 in Figs. 2 and 3) (Stewart et al., 1990).

Upper Triassic rocks that overlie in an angular unconformity the Upper Paleozoic rocks are exposed in southern Sonora State (Barranca Group, column 4 in Fig. 3) (Stewart and Roldán-Quintana, 1991). They form a succession of continental and marine deposits and are made up of sandstone, shale and conglomerate with abundant coal beds, and some tuffaceous layers (Stewart and Roldán-Quintana, 1991). They contain abundant fossil plants and marine fossils such as ammonites, pelecypods, and brachiopods of Late Triassic (Carnian to Norian) age (Stewart et al., 1990). These rocks have been interpreted as deposited in a rift-type basin (Stewart and Roldán-Quintana, 1991). They are overlain by Upper Cretaceous continental sedimentary and volcanic rocks (Stewart and Roldán-Quintana, 1991).

Oaxaquia Block

Ortega-Gutíerrez et al. (1995) proposed that the Proterozoic basements of the Sierra Madre and Coahuila terranes, together with southern Grenvillian terranes (Zapoteco terrane, parts of the Juarez and Maya terranes), might have evolved together as a large piece of continent, which they called Oaxaquia (Fig. 1). Other authors extend Grenvillian basement to areas with Paleozoic metamorphic rocks (e.g., Coahuila, Tampico, and Del Sur blocks of Dickinson and Lawton [2001]). However, regional geology indicates that large parts of those areas have metamorphic Paleozoic basements and are considered to be different terranes (e.g., Ramírez-Ramírez, 1978; Yañez et al., 1991; Ortega-Gutíerrez et al., 1999). In this paper, the subdivision of the northern Oaxaquia block into the Coahuila and Sierra Madre terranes will be used.

Paleogeographic evolution of the Oaxaquia Block (includes Coahuila and Sierra Madre terranes) seems to be quite complex. Its main metamorphic event has been related to collision between north Amazonia and northeast Laurentia in the Proterozoic (Grenville Orogeny) (Keppie and Ortega-Gutíerrez, 1999; Ruíz et al., 1999). Most of the models agree that Oaxaquia stay in Gondwana during the end of the Proterozoic and most of the Paleozoic. A collisional event between Oaxaquia and Laurentia seems to have occurred during the mid-Paleozoic (Restrepo-Pace et al., 1994, 1997; Keppie et al., 1996; Keppie and Ortega-Gutíerrez, 1999; Ortega-Gutíerrez et al., 1999). Silurian faunas from sedimentary units located in northern Oaxaquia, as well as Ordovician faunas of southern Oaxaquia, are distinctly different from fauna of comparable age in adjacent parts of North America and have Old World realm/Gondwana affinity (Robison and Pantoja-Alor, 1968; Boucot et al., 1997; Stewart et al., 1999). This suggests that Oaxaquia remained allochthonous with respect to North America and was part of Gondwana at least up to the Silurian (Stewart et al., 1999). Late Paleozoic tectonic evolution of Oaxaquia will be discussed later.

Coahuila Terrane

There are no exposures of the basement of the Coahuila terrane (Fig. 1). Moreno et al. (1993, 2000), among other authors, suggest that the Ouachita belt may extend into Mexico, between the Chihuahua and Coahuila terranes (Fig. 1). The Coahuila terrane has been subdivided into smaller terranes, such as the Mapimi terrane (Moreno et al., 1993, 2000) and the Delicias terrane (McKee et al., 1999), both apparently floored by Grenvillean-age crust (Cameron and Cameron, 1985; Lopez, 1997; Lopez et al., 2001). In this paper, the Delicias and Mapimi terranes are grouped with the rest of the Coahuila terrane and are referred herein only as the Coahuila terrane. Xenoliths have been collected near the inferred contact between the North America craton in Chihuahua and the Coahuila terrane; they yielded Grenville isotopic signatures (La Olivina) (Rudnick and Cameron, 1991; Ruíz et al., 1988a). These xenoliths have been interpreted as part of the North America craton (Ruíz et al., 1988b; Sedlock et al., 1993; Dickinson and Lawton, 2001) or as the basement of the Mapimi terrane (Moreno et al., 2000).

With no exposures of the north-western contact between the Coahuila terrane and the North America craton, its location and nature remain uncertain. The eastern limit of the terrane is not exposed, and it has been located at different positions by different authors (Coney and Campa, 1987; Sedlock et al., 1993; Ortega-Gutíerrez et al., 1995; Dickinson and Lawton, 2001). It is located in Figure 1 on the basis of reports of Paleozoic metamorphic rocks in the subsurface that are considered part of the Maya terrane (Ramírez-Ramírez, 1992).

Paleozoic rocks are exposed only in the Delicias region (Delicias Formation) (location 10 in Figs. 2 and 3). These rocks have been described in detail by McKee et al. (1999). They are made up of an Upper Mississippian to Upper Permian (Guadalupian) succession of turbidites and debris flows that contain large blocks, from several to tens of meters, of limestone, volcanic, and volcaniclastic rocks and some conglomerate (King, 1944; Wardlaw et al., 1979; McKee et al., 1988, 1999; Lopez, 1997). A peperite intrusive with a U/Pb age of 331 Ma was emplaced in the lower parts of the unit, and two granodiorite intrusions that cut the succession have K/Ar ages of 256 and 266 (Lopez, 1997; Torres et al., 1999; Lopez et al., 2001). Most volcanic rocks range in composition from rhyolite to andesite, but there are a few basalts (Lopez, 1997). Rare conglomerates within the Paleozoic succession contain metamorphic clasts, including Precambrian gneisses, granite, quartz, schist, and clasts of limestone that suggest a mixed provenance of volcanic arc and uplifted basement (Lopez, 1997; Lopez et al., 2001). These rocks have been interpreted as deposited on the margin of a continental arc (McKee et al., 1999). They were deformed after Ouachita and apparently before Late Triassic (McKee et al., 1999), but surely before deposition of Cretaceous rocks (McKee et al., 1999), and are overlain, in a regional unconformity, by Cretaceous rocks.

Sierra Madre Terrane

Basement rocks of northern Sierra Madre terrane (part of the Oaxaquia block) are exposed in Ciudad Victoria and Molango areas (Huiznópala and Novillo Gneisses) (Fig. 1). Both have metamorphic ages between 911 and 1080 Ma, and Nd model ages from 1.4 to 1.8 Ga that are similar to the those of Grenville Belt (Patchett and Ruíz, 1987; Ruíz et al., 1988b; Lawlor et al., 1999).

Upper Paleozoic rocks of northern Sierra Madre terrane are exposed east of Ciudad Victoria, Tamaulipas State (location 8 in Figs. 2 and 3) (Carrillo-Bravo, 1961). An update on the stratigraphy and paleontology of this locality was made by Stewart et al. (1999). The Lower Mississippian (Vicente Guerrero Formation) succession is made up of fine-grained sandstone, siltstone, and shale and contains a few conglomerate lenses at the base (Stewart et al., 1999). To the top, there are interbedded rhyolite flows that yielded an Early Mississippian U/Pb age (334 ± 39 Ma) (Boucot et al., 1997; Stewart et al., 1999). Shale and sandstone contain abundant brachiopods, gastropods, fusulinids, and corals (Stewart et al., 1999). Primary structures suggest that they were deposited by turbiditic and other gravity flows, maybe in a deltaic and/or submarine fan depositional environment.

In the same area, near Ciudad Victoria, where Upper Paleozoic rocks are exposed, metamorphic rocks (Granjeno Schist) of apparently similar age are also exposed (location 8 in Figs. 2 and 3). They are made up of metamorphosed shale, sandstone, volcanic, and ultramafic rocks that yielded Mississippian K/Ar and Rb/Sr ages (ca. 330 Ma) (Ramírez-Ramírez, 1978; Garrison, 1978). Those rocks are in fault contact with the rest of the Proterozoic and Early Paleozoic units (Ramírez-Ramírez, 1978).

Lower Pennsylvanian rocks (Del Monte Formation) rest unconformably on Silurian, Mississippian, and even the metamorphic rocks of the Granjeno Schist (Fig. 3) (Ortiz-Ubilla et al., 1988; Centeno-García et al., 1998; Stewart et al., 1999). They are mostly turbidites made up of alternating sandstone, shale, siltstone, and some conglomerate. The conglomerate contains clasts of sandstone, siltstone, felsic volcanic, gneiss, and schist fragments.

The Lower Permian (Guacamaya Formation) is a thick turbiditic succession made up of fine-grained sandstone interbedded with shale and siltstone (location 8 in Figs. 2 and 3). It contains a few conglomerates made up of volcanic clasts. Some andesitic breccia was observed as clasts in the present river wash, but outcrops of such rocks were not found. Sandstone petrography shows a change in grain composition from quartz/metamorphic/sedimentary>volcanic arenites in the Mississippian beds to volcanic>quartz/metamorphic/sedimentary arenites in Pennsylvanian to Permian layers (Centeno-García et al., 1998). Felsic grains (rhyolite, felsite) are more abundant in the Mississippian; in contrast, intermediate and mafic grains (andesite, trachyte) are more abundant in the Permian sandstone (Centeno-García et al., 1998). The Proterozoic and Paleozoic units are unconformably overlain by Mesozoic red beds (Jurassic?).

To the south, in the central part of the Sierra Madre terrane, Upper Pennsylvanian to Lower Permian rocks (Tuzancoa Formation) are exposed in northern Hidalgo state (location 9 in Figs. 2 and 3; Rosales-Lagarde et al., 1997; Rosales-Lagarde, 2002). At its base, this unit is made of quartz-rich sandstone and shale. This changes upward to andesitic brecciated and massive lava flows, volcanic conglomerate and tuff interbedded with volcaniclastic turbidites (alternating shale and sandstone), and some calcareous debris flows containing abundant crinoids stems, solitary corals, and fusulinids. Volcaniclastic rocks at the top contain conglomerate lenses, deposited as channel fill, made up of volcanic, granite, gneiss, and limestone clasts (Rosales-Lagarde et al., 1997; Rosales-Lagarde, 2002). Fossil crinoids, fusulinids, and brachiopods are dated as Early Permian (Wolfcampian-Leonardian), but some taxa have ranges that go down to Virgilian (Upper Pennsylvanian) (Carrillo-Bravo, 1961; Arellano-Gil et al., 1998; Rosales-Lagarde, 2002). The chemical composition of the volcanic rocks indicates that they are mostly andesitic to basaltic-andesites and were formed in an arc setting (Rosales-Lagarde, 2002). Geochemical and isotopic signatures of the volcanic rocks of the Tuzancoa Formation are very similar to those reported by Lopez (1997) from the Mississippian-Permian rocks from the Coahuila terrane (Rosales-Lagarde, 2002). Apparently Jurassic rocks rest on the Permian succession without an evident angular unconformity, because bedding from both units are parallel (Rosales-Lagarde, 2002).

A belt of Permo-Triassic granitoids cuts all previously described units of the Sierra Madre terrane and extends into the Coahuila terrane and farther north into the North America Craton. This belt has a general northwest-southeast trend (Fig. 2) and is made up of diorite to granodiorite intrusives. Their geochemical and isotopic signatures suggest a continental arc origin, and their ages range from 287 to 232 Ma (Torres et al., 1999). Torres et al. (1999) considered that this magmatic belt corresponds to the shift from Atlantic collisional to Pacific subduction-related tectonics. There are compositional differences between the Permo-Triassic granitoids and the Pennsylvanian-Permian volcanic rocks of Delicias and Tuzancoa that suggest they were derived from different magma sources (Lopez, 1997; Rosales-Lagarde, 2002). In addition, the granitoids cut the volcano-sedimentary rocks of Delicias (Lopez, 1997; McKee et al., 1999) and Tuzancoa (Ochoa-Camarillo, 1997, personal commun.). Therefore, this belt suggests that arc magmatism in eastern Mexico continued up to the Earliest Triassic time, but its axis probably migrated, since it is placed eastward from previous Pennsylvanian-Permian arc and cuts other units.

A thick Triassic sedimentary succession is exposed along the western margin of the Sierra Madre terrane. It consists almost exclusively of siliciclastic rocks deposited by turbiditic and other gravity flows in a submarine fan setting. These rocks are exposed in Peñon Blanco and Charcas (column 8 in Fig. 3; location B and C in Fig. 4) (Labarthe et al., 1982; Silva-Romo, 1993; Tristán-González and Torres-Hernández; 1994; Centeno-García and Silva-

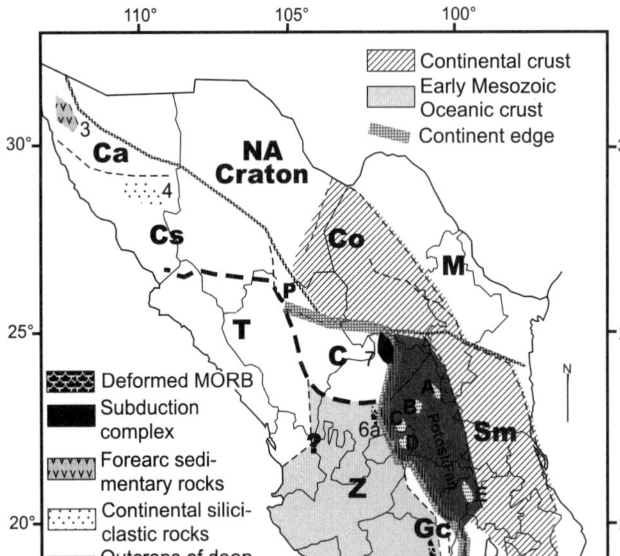

Figure 4. Location of Upper Triassic units of Mexico. Patterns indicate their depositional environment and tectonic setting. Units of unconstrained age are the accretionary complex of the Taray Formation (7) which may be Permian-Triassic or Upper Triassic in age; and rocks of Real de Catorce (A) that might be Pennsylvanian or Carnian in age. 3—Antimonio; 4—Barranca Group; 6 and 6b—Arteaga Complex; 6a—Zacatecas Formation. Main exposures of the Potosi fan: B—Charchas, C—Peñón Blanco, D—Ojo Caliente, E—Tolimán. Thick gray line indicates present location of the Early Mesozoic continent edge. All terranes to the west of the line are younger oceanic fragments or displaced Paleozoic terranes. Thick dashed line is the boundary of the Guerrero composite terrane (Gc), which includes the Tahue (T) and Zihuatanejo (Z) terranes. Other terranes are: Sierra Madre (Sm), Coahuila (Co), Maya (M), Parral (P), Caborca (Ca), Cortes (Cs), and Central (C).

Romo, 1997; Barboza-Gudiño et al., 1998; Bartolini et al., 2002). Although they have been considered part of the Zacatecas Formation, they were redefined as La Ballena Formation by Silva-Romo (1993), Centeno-García and Silva-Romo (1997), and Silva-Romo et al. (2000) on the basis of major stratigraphic differences with respect to the type locality of the Zacatecas Formation. Rocks of La Ballena Formation are made up of quartz-rich sandstone and shale, and they rarely contain channel-fill conglomerate. Clasts in conglomeratic beds are made up of quartz, chert, and a few felsic volcanics (Silva-Romo, 1993; Centeno-García and Silva-Romo, 1997). Ammonites and bivalves of Late Triassic (Carnian) age have been reported from both the Peñon Blanco and Charcas areas and are similar to those reported from the Zacatecas Formation in Zacatecas city (Cantu-Chapa, 1969; Silva-Romo, 1987; Bartolini et al., 2002). Primary structures are indicative of gravity flows (turbidites and submarine slumps) deposited probably in a high-energy environment. The lack of chert and deep marine faunas suggest that the deposition of this unit might have occurred in a distal platform or near the continental slope. Original thickness is uncertain because of its tight folding, but up to 4640 m of the unit has been cut by exploration drilling without reaching the base of the sequence (Lopez-Infanzon, 1986). The unit was deformed and locally metamorphosed before deposition of Upper Jurassic continental volcanic and clastic formations (column 8, Fig. 3) (Silva-Romo, 1993; Tristán-González and Torres-Hernández, 1994). Similar rocks are exposed in Ojo Caliente and Toliman areas (locations D and E in Fig. 4). The name of Potosi Fan is used for this thick succession of marine siliciclastics that suffered major folding and thrusting before Middle-Late Jurassic time.

A similar though undated succession of pre-Jurassic siliciclastic rocks is exposed in San Luis Potosi State (location A in Fig. 4). They have been considered part of the Upper Triassic units located to the west (La Ballena Formation, locations B–D in Fig. 4) (Bartolini et al., 2002). However, the stratigraphies of the two have some differences. The oldest part of the succession in Real de Catorce is made up of turbiditic, fine-grained, quartz-rich sandstone and shale, and thick layers of massive shale that contain abundant limestone nodules and limestone intraclasts, the last containing a few crinoid stems (Member A). Member A contains several fine-grained andesite-basaltic dikes and massive shallow intrusives. There are also layers of laminar green beds that might be strongly deformed diabase or volcaniclastic deposits. The unit shows evidence of at least two phases of deformation. The first developed closely spaced cleavage and some low-grade green schist zones; the second produced tight folding and open cleavage. To the top of the succession, there is a thick unit of interbedded sandstone and shale, deposited from turbidite currents as well, and thick channel-fill lenses of fine-grained conglomerate (Member B).

There is no evidence of a major unconformity between Member A and Member B, but their depositional environments seem to be different. Volcanic rocks within Member A were originally considered part of the volcanic rocks of Jurassic age that are exposed in the area (Maher et al., 1991; Barboza-Gudiño et al., 1998; Franco Rubio, 1999), but field relationships indicate that some of those volcanic rocks are part of the old succession and are not related to overlying Jurassic continental volcanic-sedimentary units. The age of the marine siliciclastic rocks has not been well constrained. There are reports of Pennsylvanian palinomorph grains and pieces of Calamites (Paleozoic) (Franco-Rubio, 1999; Bartolini et al., 2002). However, possible bivalve molds of Late Triassic (Carnian) age that are similar to those from the Zacatecas Formation in Zacatecas city have been reported as well (Barboza-Gudiño et al., 1998).

It might be possible that two different successions (one Upper Paleozoic and other Upper Triassic) are exposed in Real de Catorce, but further studies are needed to constrain the origin and age of these rocks. Those rocks, as well as the La Ballena Formation at Peñón Blanco and Charcas, were deformed before the deposition of Middle-Upper Jurassic continental arc successions (Bartolini et al., 2002).

TERRANES OF CENTRAL AND WESTERN MEXICO

There are very scarce exposures of pre-Jurassic rocks in central and western Mexico. Thus, it is difficult to reconstruct the pre-Jurassic terrane configuration and tectonic evolution of this vast area. Overall, the pre-Jurassic units of western and central Mexico are characterized by large amounts of deep-marine siliciclastic successions. Two types of assemblages can be differentiated: siliciclastic units that are associated with island arc volcanic and volcaniclastic rocks with Early Paleozoic ages (El Fuerte and Santa María del Oro) (Fig. 1, location 5 and 11 in Fig. 2), and siliciclastic rocks that are associated with mid-ocean ridge basalts (MORB) and oceanic-island basalt (OIB) volcanic rocks and scarce or no island-arc volcaniclastics (Pico de Teyra, Zacatecas, Arteaga), and contain Upper Paleozoic(?) and/or Upper Triassic fossils (Fig. 3). On the basis of this difference, and considering the diversity in composition of their overlapping units, a new terrane configuration for central and western Mexico is proposed in Figure 1. This section contains a short discussion of the nature of the basement of these terranes, as well as a description of their Upper Paleozoic–Early Mesozoic stratigraphy.

Central Terrane

The basement of the Central terrane (Zacatecas state) remains unknown (Fig. 1), but it is inferred to be different from the Sierra Madre and Coahuila basements because the Central contains an accretionary complex (Taray Formation) as its northeastern boundary and only surface exposure (Anderson et al., 1990, 1991, this volume; Jones et al., 1995). The Jurassic-Cretaceous succession was deposited unconformably on the Taray subduction complex of the Central terrane (column 7, Fig. 3) and is in tectonic contact (thrust by) to the south and to the west with the Guerrero Composite terrane. The subduction zone was probably constructed along the Oaxaquia continental margin sometime between Late Permian–Early Jurassic time. To the north, the contact with the Coahuila terrane is inferred to be fault, as suggested by the contrast in the thickness of Cretaceous units and contrast in the deformation between the two terranes. The relationship of the Central terrane with the Parral terrane is unknown.

The Taray accretionary complex is described in detail by Anderson et al. in this volume. The assemblage consists of a highly disrupted rhythmic succession of quartz-rich sandstone and shale, interbedded with scarce thin layers of black chert. There are some beds of detrital limestone that contain fragments of crinoids, gastropods, corals, bivalves, and bryozoa, and some beds of conglomerates. Both detrital limestone and conglomerates are channel-fill deposits (Díaz-Salgado et al., 2003). Primary structures in undisturbed areas suggest that the sandstone and shale, as well as the limestone and conglomerate, are mostly turbidite flow deposits in a deep marine setting.

The Taray sedimentary rocks constitutes a matrix within which blocks of black and green chert, pillowed basalts, serpentinite, and scarce crystallized limestone can be found (Fig. 3) (Díaz-Salgado et al., 2003). The age of this unit remains undetermined. There are reports of fusulinids from a limestone block (Upper Paleozoic?) (Anderson et al., 1990), and detrital zircons do not exceed Late Permian ages (Díaz-Salgado et al., 2003). However, there is a report of molds of pelecypoda that resemble those from Upper Triassic (Carnian) rocks of Zacatecas (Barboza-Gudiño et al., 1998; Bartolini et al., 2002). The complex is overlain unconformably by Oxfordian volcano-sedimentary rocks (Anderson et al., 1990; Díaz-Salgado et al., 2003), and thus deposition and deformation should have occurred sometime between latest Permian and Early-Middle Jurassic.

Parral Terrane

The Parral terrane was originally defined as a thick succession of Upper Mesozoic turbiditic sandstones (Pacheco et al., 1984, Coney and Campa, 1987). However, exposures of older rocks suggest that there was accretion of terranes before the deposition of the turbidites, and that those turbidites are in fact an overlapping assemblage (Fig. 1).

The basement of the Parral terrane is considered in this paper to be represented by the metamorphic rocks (Pescadito Schist) of Santa Maria del Oro, Durango (location 11 in Figs. 2 and 3). They are made up of a muscovite schist and a chlorite schist. The protolith of the muscovite schist was quartz-rich sandstone and black shale, and of the chlorite schist deformed lava flows, dikes, and volcaniclastics (Araujo and Arenas, 1986). The Pescadito Schist yielded K/Ar ages around 326 ± 26 Ma (Zaldivar and Garduño, 1984) and 350 Ma (Eguiluz and Campa, 1982). Ar/Ar dates from a syntectonic granitic dike yielded ages as old as Devonian (360 Ma, A. Iriondo, 2003, personal commun.). This metamorphic unit is in tectonic contact with a succession of pillowed basalts interbedded with volcaniclastic rocks (Fig. 3). The volcaniclastic rocks contain large blocks of limestone and interbedded layers of limestone that contain brachiopods and abundant crinoid stems of Late Paleozoic age (Carboniferous?) (Zaldivar and Garduño, 1984).

The basaltic pillowed lavas and volcaniclastic succession have been interpreted by some authors to be younger than Late Paleozoic (Cretaceous?). They have been considered part of the Guerrero terrane, based only on the observation that there are no volcanic rocks in Paleozoic units west of Santa Maria del Oro (Tahue and Cortes terranes) (Egiluz and Campa, 1982; Pacheco et al., 1984; Aranda et al., 1988). The same authors suggest that the limestone blocks containing crinoid stems of Late Paleozoic age might be exotic. Alternatively, these volcanic, sedimentary rocks may be part of the Delicias arc assemblage, because their lava composition and fossil ages resemble rocks of volcanic origin in the Coahuila terrane (Delicias) located a few kilometers to the east, but this should be tested with further studies.

Northeast of Santa María del Oro, red beds that transitionally change to limestone rest unconformably on the metamorphic complex (Fig. 3) (Pescadito Schist) (Araujo and Arenas, 1986), but their relationship with the marine volcanic and volcaniclastics has not been determined. The limestone beds contain Titonian ammonites (Contreras-Montero et al., 1988). Contact relationships among the Parral, Coahuila, Cortes, and Central terranes are unknown because the contacts are covered by overlapping Jurassic-Cretaceous continental and marine successions, and Cenozoic volcanics.

Guerrero Composite Terrane

Some authors have interpreted the Guerrero terrane to be an allochthonous oceanic arc that developed in the paleo-Pacific and collided with Mexico by Late Cretaceous time (Lapierre et al., 1992; Tardy et al., 1994; Freydier et al., 1996; Dickinson and Lawton, 2001). However, there is strong evidence that this terrane had an older history of accretions and that its basement is made up of several strongly deformed pre-Cretaceous rocks that form a heterogeneous basement upon which the arc was built (Centeno-García et al., 1993a, 1993b; Elías-Herrera et al., 2000). Those older rocks were interpreted as part of an under thrust subduction complex by Dickinson and Lawton (2001). However, field evidence shows that deformation of those rocks occurred before the development of the arc. This evidence includes the following: Cretaceous arc-related dikes and intrusives cutting the basement; Jurassic granitoids intruding the basement rocks; regional erosion and a major regional nonconformity between arc rocks and deformed basement; and clasts of basement rocks and clasts of Jurassic granitoids in conglomerate layers at the base and within the arc succession (Vidal-Serratos, 1991; Centeno-García et al., 1993a, 1993b, 2003). This evidence indicates that the arc, at least in western Mexico, was not tectonically emplaced onto the older rocks but was built on the prearc Paleozoic–Early Mesozoic rocks that form its basement (Centeno-García et al., 1993a, 1993b; Elías-Herrera et al., 2000; Mendoza and Suastegui, 2000; Centeno-García et al., 2003). Evolution of these prearc rocks is discussed in this paper.

On the bases of variations in the arc stratigraphy and in the composition of basement rocks, the Guerrero composite terrane has been divided into the Tahue, Zihuatanejo, Arcelia-Guanajuato, and Teloloapan terranes (Fig. 1) (Ramírez-Espinosa et al., 1991; Centeno-García et al., 1993b, 2003; Talavera-Mendoza et al., 1995). Exposures of prearc rocks can be found in the Tahue, Zihuatanejo, and Teloloapan terranes. The basement of the Arcelia-Guanajuato terrane remains unknown.

Tahue Terrane

The Tahue terrane was first defined by Sedlock et al. (1993) as a different terrane from the Guerrero terrane. However, because it contains large volumes of volcanic-sedimentary rocks of the Cretaceous Guerrero arc (Roldán-Quintana et al., 1993; Freydier et al., 1995), the Tahue terrane has been included within the Guerrero composite terrane in this paper (Fig. 1). This northwestern part of the Guerrero composite terrane is the least known, and contact relationships among its units have not been studied in detail. Its oldest known rocks are metamorphosed volcanic-sedimentary rocks of the El Fuerte Complex (location 5 in Figs. 2 and 3) (Mullan, 1978; Roldán-Quintana et al., 1993; Poole and Perry, 1998). This complex is made up of green schist, pelitic schist, quartzite, and marble. Protoliths were made up of volcanic rocks, rhyolitic to andesitic in composition, quartz-rich sandstone, shale, limestone, and volcaniclastics that contain Ordovician conodonts (Poole and Perry, 1998).

Exposures of Upper Paleozoic rocks are located south of El Fuerte, in northern Sinaloa state (San José de Gracia town) (location 5a in Figs. 2 and 3) (Carrillo-Martínez, 1971; Gastil et al., 1991). They are made up of strongly deformed siliciclastic turbidites, thin-bedded calcareous shale (slumps), and chert. The turbidites contain blocks of limestone with chert nodules. This unit contains fossils of mid-Pennsylvanian to Early Permian age (Carrillo-Martinez, 1971; Gastil et al., 1991). They have been interpreted as deposited on a deep marine environment (Gastil et al., 1991). Deformed and partially metamorphosed turbidites of Mazatlán city, made up of interbedded shale and quartz-rich sandstone, might be the southern extension of the San Jose de Gracia rocks (location 5b in Fig. 2). However, those rocks do not contain chert and have only exotic blocks of recrystallized limestone (marble) (Arredondo-Guerrero and Centeno-García, 2003). The age of this unit is unknown. It was apparently deposited on a submarine fan environment (Arredondo and Centeno-García, 2003).

Mafic and ultramafic intrusions that are part of the Cretaceous basic magmatism of the Guerrero arc cut the El Fuerte Complex and deformed sedimentary rocks of Mazatlán (Henry and Fredrikson, 1987; Roldán-Quintana et al., 1993). Thus, these metamorphic rocks are considered to be the basement of the arc in this part of the Guerrero terrane (Valencia-Moreno, 1998). Upper Paleozoic rocks of the San José de Gracia area are overlain by Cretaceous volcanic and sedimentary rocks of Guerrero arc (Henry and Fredrikson, 1987; Roldán-Quintana et al., 1993).

Relationships between Paleozoic rocks of the Tahue terrane and other coeval rocks in surrounding terranes have not been constrained. Both Lower and Upper Paleozoic rocks of the Tahue terrane seem to have similarities with rocks of the Cortes and Parral terranes. However, direct correlation might not be feasible because the Jurassic-Cretaceous Guerrero arc, which was constructed on the Tahue Paleozoic rocks, has suffered considerable displacement during its Jurassic-Cretaceous evolution.

Zihuatanejo Terrane

The basement of parts of the Zihuatanejo terrane (Guerrero composite terrane) crop out in several localities (locations 6 and 6a–b in Figs. 3 and 4). The northernmost exposure of these rocks is located in the surroundings of Zacatecas city (Zacatecas Formation, location 6a in Fig. 4), where they are mostly characterized by interbedded quartz-rich sandstone, shale, and a few layers of pillowed basalts (Ranson et al., 1982; Cuevas-Pérez, 1983; and Monod and Calvet, 1991; Centeno-García and Silva-Romo, 1997). Pillow lavas are MORB geochemically and are different from lavas in the Cretaceous arc. Ammonoids and pelecypods from the Zacatecas Formation are Upper Triassic (Carnian) in age (Burckhardt and Scalia, 1906).

In a tectonic thrust contact, the Zacatecas Formation is overlain by a succession of unknown age. This succession, whose role in the geologic evolution of the area remains uncertain, is made up of fine-grained volcaniclastic rocks, volcanic breccias, and thin-bedded limestone. In a similar thrust contact, these undated volcaniclastic rocks are overlain by Cretaceous pillow lavas and volcaniclastic rocks of island arc affinity (Lapierre et al., 1992; Centeno-García and Silva-Romo, 1997). The Triassic rocks and undated volcaniclastic rocks were deformed before their tectonic contact with the Cretaceous rocks; they have foliation, tight folding, and some mylonitic zones that are not found in the Cretaceous rocks. The Zacatecas Formation has strong similarities with Triassic successions in southern Zihuatanejo terrane (Arteaga Complex) and has been interpreted as accreted open ocean-floor assemblages (back-arc basin?) that received sediments from a continental margin (Centeno-García and Silva-Romo, 1997).

To the south, the most complete exposure of Early Mesozoic rocks is located near the coast in Michoacán State (Arteaga complex, location 6 in Figs. 3 and 4). They are composed of several lithologic units, including siliciclastics, green volcaniclastics, pillow basalts, diabase, gabbros, black and green chert, and exotic blocks of recrystallized limestone (Fig. 3). Approximately 60% of the exposures are of siliciclastic sediments such as black shale, quartz-rich sandstone, and some black chert. Conglomerates are rare, and their clasts are made up of quartz, black and white chert, and black siltstone; rarely, felsic and granitic(?) clasts are found. The siliciclastics are occasionally interbedded with scarce packets (some up to 200 m thick) of light green metamorphosed shale and very fine-grained sandstone, interbedded with very thin layers of recrystallized limestone. They are interpreted as metavolcaniclastics derived from a MORB and/or primitive arc source (Centeno-García et al., 1993, 2003). Basaltic pillow lava flows and massive blocks of basalt have geochemical and isotopic composition that are similar to those in mid-ocean ridges (MORB) (Centeno-García et al., 1993, 2003). Light green–aquamarine–color chert is interbedded with the siliciclastics, but sometimes it forms blocks originated by strong shearing and deformation. It is distinct from the black thin-bedded chert of the siliciclastics. Large limestone blocks (recrystallized), up to tens of meters in diameter, are sporadically found within the siliciclastics sediments.

Considering that the siliciclastic rocks of the Arteaga Complex do not contain interbedded limestone or calcareous fragments, the limestone blocks are interpreted as olistoliths (Centeno-García, 1994; Centeno-García et al., 2003). Original thickness is unknown, but the minimum structural thickness, observed between two thrust planes that contain pillow lavas at the sole, is ~1500 m. The age of deposition and deformation of the Arteaga Complex has not been well constrained. There is one report of radiolarian fossils of Late Triassic (Ladinian-Carnian) age (Campa et al., 1982). Detrital zircons are not younger than around 260 Ma (Centeno-García et al., 2003). Thus, deposition is considered to be Late Permian to Late Triassic or younger.

The Arteaga complex is strongly deformed and, in some areas, metamorphosed to greenschist facies, forming a "broken formation" type of structure. Contacts between the siliciclastics (that constitutes the matrix) and the pillow lavas, green chert, volcaniclastics, and limestone blocks are sheared, forming large lenses of tens to hundreds of meters in size, giving the complex a "block in matrix" aspect, characteristic of an accretionary complex. Jurassic granitic intrusions that cut the deformation and metamorphism of the Arteaga complex are Oxfordian-Kimerigdian in age and set some constraints on the age of deformation of the complex (Centeno-García et al., 2003). Scarce geochronological data indicates that sedimentation and deformation of the Arteaga complex have occurred sometime between the Norian and the Oxfordian (Centeno-García et al., 2003). Volcanic and volcaniclastic rocks of the Cretaceous arc-assemblage rest unconformably on the Arteaga complex (Centeno-García et al., 1993b, 2003). Localities 6b on Figure 4 are other exposures of the Arteaga Complex (Centeno-García et al., 1993a).

Sedimentary structures, lithofacies and fossil associations suggest that the Arteaga sediments were deposited in a deep ocean environment, probably contemporaneous with part of the rift magmatic activity (Centeno-García et al., 2003). The basin received volcaniclastics that may have been either deep-marine deposits derived from the erosion and eruption of the oceanic ridge basalts or air-fall ashes erupted from some oceanic island arc. The erratic blocks of limestone might be olistoliths derived from platform deposits and carried down from the continental slope. The rocks of the complex originated in a marginal back-arc basin or in an open ocean environment (Fig. 4) (ocean-floor/continental rise setting). The structures of the complex suggest that it was deformed by subduction processes and corresponds to the upper levels of an accretionary complex (Centeno-García et al., 2003). Rocks of the Arteaga Complex are covered by Aptian-Albian marine volcanic and volcaniclastic rocks (Centeno-García et al., 2003).

DEPOSITIONAL ENVIRONMENTS, SANDSTONE PROVENANCE, AND REGIONAL CORRELATIONS

Paleozoic

The Paleozoic units previously described can be divided into four different groups or assemblages that have major differences in their stratigraphic and structural characteristics: (1) the Paleozoic assemblage of the North America craton (Chihuahua and Northeast Sonora); (2) rocks of the Sierra Madre and Coahuila terranes; (3) Paleozoic units of Caborca and Cortes terranes; and (4) Paleozoic rocks of the Parral and Tahue terranes.

The Paleozoic rocks of North America craton in Mexico (Chihuahua and northeast Sonora states) have the following distinctive features: (1) Throughout most of the Paleozoic, the sedimentation occurred in shallow marine environments, except for the Lower Permian Formations at Sierra del Cuervo and Placer de Guadalupe, which show an increment in clastic sedimentation and a deepening in the depositional environment. (2) There are no major unconformities among the succession, and thus pre–Late Permian orogenies were not apparently recorded in this area. (3) There are no reports of significant volcanic activity. (4) The crystalline basement rocks are involved in the deformation. (5) A major angular unconformity with Kimeridgian rocks that suggest at least one pre–Late Jurassic post–Early Permian phase of deformation (Bridges, 1964). Paleozoic rocks of Chihuahua and Northeastern Sonora have been correlated with the stratigraphy of Southwestern Texas and New Mexico (Flawn et al., 1961; Bridges, 1964; Pearson, 1964). Provenance studies on the Paleozoic rocks of Chihuahua have not been done.

Several publications discuss the Late Paleozoic correlation of the Caborca and Cortes terranes with other areas of the western North America margin (e.g., Stewart et al., 1990; Poole et al., 1992). Whether they were transported toward the south, and when, are still under debate, but their correlation with Cordilleran tectonics seems to be well constrained. Some of their main features are as follows: (1) shallow marine depositional environments throughout most of the Paleozoic in the Caborca terrane; (2) deep marine sedimentation during most of the Paleozoic in the Cortes terrane; (3) a Mississippian deformational event, followed by a Permo-Triassic compressive event, followed by continental sedimentation in the Triassic in the Cortes terrane; (4) a continuous sedimentation, without major angular unconformities, from Permian to the Jurassic in the Caborca terrane; and (5) no evidence of sedimentary influence from volcanic activity up to the Early Jurassic. These features are common with southern parts of the U.S. Cordilleran belt (Miller et al., 1992). Studies of zircon provenance from Paleozoic rocks from both Caborca and Cortes terranes cannot conclusively constrain the paleogeography of those terranes (Gehrels and Stewart, 1998).

The stratigraphies of the Coahuila and Sierra Madre terranes seem to have strong correlations, at least by Late Paleozoic–Early Mesozoic time. The two main common features that are not found in the other assemblages are as follows: both terranes have recorded Mississippian to Early Permian submarine arc magmatism; and they contain a belt of Permo-Triassic granitoids. One of the differences between the Sierra Madre and Coahuila terranes is that there is an angular unconformity between Mississippian and Pennsylvanian rocks to the east of the Sierra Madre terrane (Ciudad Victoria) that was not observed in the Coahuila terrane (Delicias) (Ortiz-Ubilla et al., 1988; Centeno-García et al., 1998; McKee et al., 1999; Stewart et al., 1999). Both Coahuila and at least northern Sierra Madre show evidence of a major folding event before the deposition of Jurassic or Cretaceous marine and continental rocks.

The Parral terrane and rocks at the El Fuerte complex of the Tahue terrane recorded a pre-Devonian magmatic event that is not found elsewhere in northern Mexico. Both localities show similarities in their stratigraphy, containing Ordovician? to Devonian? arc magmatism. However, they show major differences in their Jurassic-Cretaceous cover. Magmatism of similar age has not been identified in other parts of Northern Mexico. Early Paleozoic arc magmatism has been recorded in eastern areas of the Cordilleran Belt in the USA and Canada (Burchfiel et al., 1992), suggesting a possible Cordilleran origin. Alternatively, metamorphic rocks of the Parral terrane could be the southern continuation of the Oaxaquia–North America collisional belt.

Units with Unconstrained Ages

The age of the succession at Real de Catorce (Sierra Madre terrane) has not been well constrained, and there is not enough information for reconstructing its tectonic and depositional environment. The rocks are mostly marine turbidites that share three characteristics in common with Pennsylvanian-Permian rocks of the Sierra Madre terrane: they contain limestone with crinoid stems as intraclasts; they are associated with volcanic rocks of arc affinity; and they were deformed before deposition of mid-Jurassic rocks. If the Paleozoic age is confirmed, they could be correlated with the basal part of the Tuzancoa arc, which contains siliciclastic rocks and basic volcanics.

Rocks of the Central terrane (Taray Formation) were deformed by subduction processes, and it has not been determined whether they were first deposited on a passive margin or whether they represent a trench-filling succession. Similarities with Triassic sediments of the Sierra Madre terrane have led several authors to suggest that this unit is part of the Triassic submarine fan (Potosi Fan) (Silva-Romo et al., 2000; Bartolini et al., 2002) because turbidites of the Taray Formation have a Sm/Nd and U/Pb zircon provenance that is similar to those in the Potosi Fan (Centeno-García et al., 2003). If a Triassic age for sedimentation is proven, then the rocks might belong to an Early Jurassic subduction zone.

Triassic

The data available suggest that the continental margin during Triassic time was located west of the Potosi Fan along the limit between the Sierra Madre, Central, and Guerrero terranes (Fig. 4). Facies and fossil associations of the turbiditic siliciclastics of the Potosi Fan suggest that they were deposited on a continental shelf-slope environment. The Potosi Fan does not seem to have any correlation with rocks of the Antimonio area (Fig. 4) because the last were deposited on a forearc setting (González-León et al., this volume). Continental to shallow marine Triassic sedimentary units in Cortes terrane are interpreted as deposited on rift basins, but they contain volcanic rocks (tuffs); thus, an arc-related environment is also plausible (Stewart and Roldán-Quintana, 1991).

Lithologic assemblages of the Triassic rocks of the Zihuatanejo terrane (Guerrero composite) suggest an ocean-basin environment of deposition. Their siliciclastic sediments were derived from continental areas and transported to the ocean floor by turbiditic flows. It is possible that deposition of these sediments was contemporaneous with at least part of the rift magmatic activity (MORB signatures in the lavas). Whether the Zacatecas and Arteaga Complexes originated in an active back-arc basin or an open-ocean environment is still uncertain. The only evidence of association with island arc magmatism is the volcaniclastic sediment at the Arteaga Complex, but their geochemical signatures are not specific for discriminating between MORB-derived or primitive oceanic island arc magmatism. The abundance of cratonic-derived siliciclastic rocks indicate that rocks of the Arteaga Complex were deposited near a continent. Pb ages of detrital zircons suggest that the siliciclastics of the Arteaga Complex might have been deeper facies (ocean floor) of the Potosi Fan (Centeno-García et al., 2003). This evidence suggests that the Guerrero Composite terrane did not evolve away from the continental margin, as previously proposed, but as a marginal terrane.

PALEOGEOGRAPHIC MODELS AND TECTONIC EVOLUTION

Several models have been proposed for the sedimentation and tectonic settings of the western margin of Mexico during Paleozoic-Mesozoic time (e.g., Coney, 1978, 1983; Dickinson and Coney, 1980; Damon et al., 1981; Pindell and Dewey, 1982; Cuevas-Pérez, 1983; Sedlock et al., 1993; Ortega-Gutíerrez et al., 1994; Jones et al., 1995, Dickinson and Lawton, 2001). Most of them show generalized tectonic models for large time spans. However, field evidence shows that the end of the Paleozoic and the Early Mesozoic were characterized by a series of tectonic events that occurred over very short time periods. In this section, a more detailed tectonic evolution on shorter time slices is proposed.

Paleozoic Paleogeographic Models

There are a number of models interpreting the assembling of Pangea, but they basically fall into three groups (only samples referenced included):

1. Models in which the Coahuila and Sierra Madre terranes (part of the Oaxaquia Block) remain attached to Gondwana until both collide with North America during Carboniferous-Permian time (Fig. 5A). The collision with North America is via consump-

tion of an oceanic basin, via a subduction zone placed on the northeastern Gondwana margin, that existed along the Marathon Ouachita–Chihuahua belt (e.g., Lopez, 1997; Dickinson and Lawton, 2001). In this model, Carboniferous-Permian volcanism of the Coahuila and Sierra Madre terranes is associated to the consumption of this oceanic basin placed between the continents.

2. Models with a collisional front between Coahuila-Sierra Madre terranes and North America at the Ouachita-Marathon-Chihuahua belt, changing toward the south into a continental arc (Fig. 5b). In this model, Carboniferous-Permian volcanism of the Coahuila and Sierra Madre terranes is related to a subduction zone along the Pacific margin of those terranes (e.g., Sedlock et al., 1993; Ortega-Gutíerrez et al., 1994).

3. Models with the Coahuila–Sierra Madre–North America collisional zone to the east of Coahuila-Sierra Madre terranes, where Permian rocks of the Sierra Madre terrane are interpreted as orogenic flysch (e.g., Pindell, 1985).

There is not enough data to strongly support any one of these three groups of models over the other two.

The model in Figure 5A (group 1) illustrates a Carboniferous-Permian collision of the Coahuila-Sierra Madre terranes (Oaxaquia block) with North America. It seems to explain the difference in stratigraphy of the Coahuila-Sierra Madre terranes with that of the south-central part of the North America craton (northern Chihuahua state). However, it has several shortcomings. It does not explain why the Paleozoic rocks at Placer de Guadalupe did not experience deformation during Mississippian to Early Permian time. This model does not explain the strong affinities of Mississippian faunas of the sedimentary cover of Coahuila and Sierra Madre terranes with the midcontinent province in North America (Stewart et al., 1999; Navarro-Santillan et al., 2002). This model also does not give an explanation to the occurrence of Mississippian metamorphic rocks at the subsurface on the Maya terrane, and the eastern margin of Sierra Madre terrane in Ciudad Victoria (Granjeno Schist) (Ramírez-Ramírez, 1978, 1992).

Tectonic models of group 2 (Fig. 5B) can better explain the time overlapping between the magmatism of the Coahuila and Sierra Madre terranes and deformation on the Ouachita-Marathon belt. However, they do not explain the Mississippian faunal affinities between Coahuila and Sierra Madre terranes and North America.

The model in Figure 5C (modified from Pindell, 1985) (group 3) seems to fit all the evidence available up to date. The main difference between this model and those in Figures 5A and 5B (groups 1 and 2) is that the Coahuila and Sierra Madre terranes (part of the Oaxaquia Block) detached from Gondwana *before* Gondwana collided with North America and was accreted to North America between the Silurian and the Carboniferous. In this model, the belt of metamorphic rocks located to the east of Coahuila and Sierra Madre terranes (Maya terrane) is in a position that strongly suggests it might be the continuation of deep marine successions of the collisional front between Gondwana and North America. Faunal affinities suggest that collision of Coahuila, Sierra Madre, terranes (and other southern terranes that form Oaxaquia) is more likely to have occurred between Silurian and Mississippian time. Gravity and magnetic anomalies recorded along Chihuahua state, which have been considered to be the prolongation of the Ouachita belt, might be the pre-Mississippian accretionary belt between the Coahuila-Sierra Madre terranes and North America or a belt of structures perhaps related to the evolution of the Ouachita Marathon belt but developed into the continent, originating foreland basins toward the west (Chihuahua).

Mississippian-Permian Paleogeography

After the collision, a Mississippian to Early-Late Permian (Leonardian-Guadalupian) arc developed along the Coahuila-Sierra Madre terranes (Oaxaquia) (Delicias and Tuzancoa) (Fig. 5C). It is inferred that this arc was related to an east-dipping Pacific subduction zone, placed on the western side of these terranes (Fig. 5). However, field evidence of subduction complexes of this age has not been found. In this scenario, rocks at central Sierra Madre terrane (Tuzancoa) would have been formed in an intra-arc setting, with nearby volcanic centers of mafic composition. Thus, deposition at Delicias (Coahuila terrane) seems to have occurred in a back-arc basin because it contains less amount of volcanic rocks, more felsic magmatism, and more influence from a cratonal source in the sedimentation. Clast derived from Pan-African and Grenville complexes contained in conglomerates of the Delicias succession might have been derived from uplifted zones in east-southern Mexico and South America (Fig. 5) (Lopez et al., 2001).

No evidence of subduction-related facies has been found in the Carboniferous–Lower Permian sedimentary units of Chihuahua (Rara Formation in Sierra del Cuervo, and Plomosas) (Fig. 3), except for a slight increment in feldspar in sandstones in the Rara Formation toward the top (reported in Handchy and Dyer, 1987). There is one report of a possible rhyolitic flow within Plomosas formation (Grajales-Nishimura et al., 1992), but no abundant volcanism or volcanic detritus within the sediments has been found. Also, there is no evidence of arc activity in the Carboniferous to Permian sedimentation in the Caborca, Cortes, and Tahue successions. Instead, thick successions of siliciclastic rocks were deposited, probably in a passive and/or trailing margin environment. The Cortes terrane experienced deformation during the Early Mississippian, but such a deformation event was not recorded in Chihuahua. The deepening of the sedimentation during the Early Permian in the Cortes terrane has been interpreted as related to a change from a stable continental margin to an active margin (subduction or strike-slip) (Stewart et al., 1997).

Lower Permian volcanic rocks of the Mojave Desert (Walker, 1988) suggest that subduction might have extended toward the north along the paleo-Pacific margin. Contemporaneous strike slip occurred in the Mojave Desert (Stone and Stevens, 1988; Walker, 1988). Whether this Permian strike-slip system extends to Chihuahua or Northeastern Sonora remains unknown. Dickinson and Lawton (2001) suggested that this strike-slip system originated the translation of the Caborca and Cortes terranes to

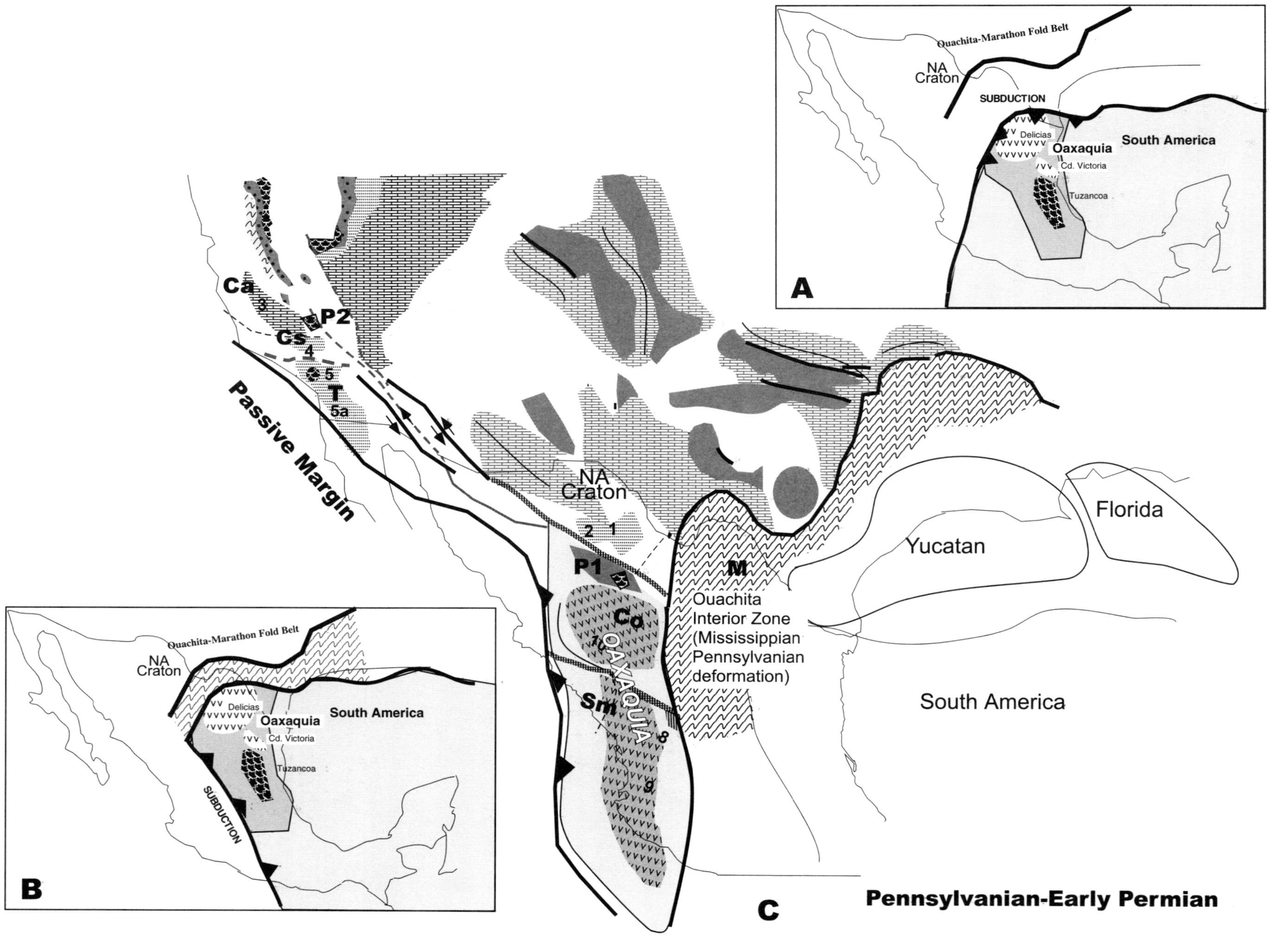
A
Quachita-Marathon Fold Belt
NA Craton
SUBDUCTION
Delicias
Oaxaquia
Cd. Victoria
Tuzancoa
South America
B
Quachita-Marathon Fold Belt
NA Craton
Delicias
Oaxaquia
Cd. Victoria
Tuzancoa
South America
SUBDUCTION
C
Ca
3
P2
Cs
4
5
T
5a
Passive Margin
NA Craton
2
1
P1
Co
10
OAXAQUIA
Sm
8
9
M
Ouachita Interior Zone (Mississippian Pennsylvanian deformation)
Yucatan
Florida
South America
Pennsylvanian-Early Permian

Figure 5. (A) Model that proposes collision of Coahuila and Sierra Madre terranes (Oaxaquia) in Carboniferous-Permian time (models of group 1). (B) Model that proposes development of continental arc in Oaxaquia, with Pacific subduction zone, contemporaneous to collision along southern extension of Ouachita belt (models of group 2). (C) Paleogeographic reconstruction for Pennsylvanian to Early Permian time proposed in this paper (modified from Pindell, 1985). Patterns are same as Figure 2. Tectonic elements of North America are also included for comparison (from Smith and Miller, 1990; Miller et al., 1992). Black lines are active tectonic elements (thrusts and faults). Gray lines indicate location of younger shear zones that were used to move blocks to their possible Paleozoic position. Terranes: Ca—Caborca; Cs—Cortes; T—Tahue; Co—Coahuila; Ch—Chihuahua; Sm—Sierra Madre; M—Maya. Other symbols and abbreviations from Figure 2.

their present position in the Carboniferous–Early Permian. However, there is presently no field evidence to support this aspect of their model. Translation of the Caborca and Cortes terranes will be discussed latter.

Permo-Triassic Paleogeography

Apparently there was a gap in the magmatic activity between Leonardian and Ochoan in eastern Mexico. Then arc magmatism was reestablished by Latest Permian–Early Triassic time (Fig. 6). There was not only a gap in the volcanism, but also a change in the composition and distribution of the magmatic centers. Upper volcanic-sedimentary levels of this arc were not preserved in Mexico, only a belt of granitoids along eastern and northern Mexico. These granitoids cut and tie together the Pennsylvanian–Lower Permian units of the North America Craton (Chihuahua and Sonora) to the Coahuila, Maya, and Sierra Madre terranes and suggest that a subduction zone extended along the paleo-Pacific continental margin of Mexico. If the accretionary prism of Central terrane (Taray accretionary complex in Figs. 4 and 6) is demonstrated to be Permo-Triassic in age, it would be an important piece of evidence for the location of the continental margin for that time. Permo-Triassic granitoids have been reported from the Mojave Desert, as well as in other parts of the western United States (Miller et al., 1992, 1995); thus, this continental arc had extended all along the southwestern margin of the North American Craton.

Paleozoic Paleogeography of the Caborca and Cortes Terranes

The Caborca and Cortes terranes seem to have a different Paleozoic tectonic evolution compared with eastern and central terranes of Mexico (Fig. 5). First, there is no evidence of magmatic activity in the Caborca or Cortes terranes during this time. Second, the Cortes terrane has recorded at least two deformational events with thrusting and folding that have not been identified in either the western Sierra Madre and Coahuila terranes, or in the Craton in Chihuahua. The first event occurred in the Mississippian and is younger than the Antler Orogeny (Poole and Madrid, 1988; Stewart et al., 1990; Burchfiel et al., 1992). The second event occurred in Late Permian to Early Triassic time and has been considered the main deformational event that placed deep-marine continental slope-rise sediments (eugeoclinal) onto the shallow marine platform facies (Sonoran Orogeny) (Fig. 5) (Poole and Madrid, 1988; Stewart et al., 1990; Poole et al., 1995). This second event of deformation has been related to the Sonoma Orogeny and was not recorded in the Caborca terrane (Stewart et al., 1990; Poole et al., 1992).

Whether Late Permian to Early Triassic deformation had affected other parts of Mexico is still uncertain. Upper Paleozoic sedimentary rocks of Chihuahua, as well as those of Coahuila and northern Sierra Madre terranes, were deformed at least before the deposition of the Upper Jurassic–Cretaceous units. The contact between Paleozoic and Triassic sedimentary rocks of Sierra Madre terrane is not exposed. Thus, the age of deformation of the Paleozoic rocks remains unknown. The only evidence of pre-Triassic deformation in eastern-central Mexico is a post-deformation pluton that cuts folded rocks of the Delicias arc in the Coahuila terrane. This intrusive yielded a Late Triassic age (Lopez, 1997; McKee et al., 1999).

Middle-Late Triassic Paleogeographic Reconstruction

Magmatic activity along eastern Mexico seem to be not younger than ca. 232 Ma (Torres et al., 1999), except for the one granitoid at Delicias, Coahuila, dated at 218 ± 4 Ma (Lopez, 1997). Also, no detrital zircons younger than 232 Ma have been found in Upper Triassic units of central and southwestern Mexico (Centeno-García et al., 2003). Apparently, magmatism did not restart until Early(?) to Middle Jurassic time in all Mexico (Jones et al., 1995; Grajales et al., 1992; Centeno-García et al., 2003; Fastovsky, et al., this volume). Thus, there is no evidence of continuous magmatism along the paleo-Pacific coast from Permian to Jurassic, as proposed by Dickinson and Lawton (2001). Instead, the Permo-Triassic volcanic arc was followed by a period of rifting? or passive margin from Ladinian(?) to Norian time, as evidenced by the stratigraphic record of central and western Mexico.

Figure 7 shows a possible tectonic scenario for Ladinian(?) to Norian time. As mentioned before, evidence of well-spread arc-related magmatic activity during this period of time has not been found in Mexico. Although there is no direct evidence, it seems likely that this period of time was characterized by continuous uplift of the eastern and north-central parts of Mexico. This uplift might have been related to the first rifting stages of Pangea, and it seems to have produced massive erosion of the Paleozoic-Precambrian rocks. This erosional event originated the large volumes of siliciclastic turbidites that formed the submarine Potosi Fan, deposited in the western continental platform and slope/rise of the Sierra Madre terrane.

Geochemical and isotopic signatures of younger volcanic rocks, as well as the regional distribution of the siliciclastic rocks of the Triassic Potosi Fan, suggest that the continent margin of

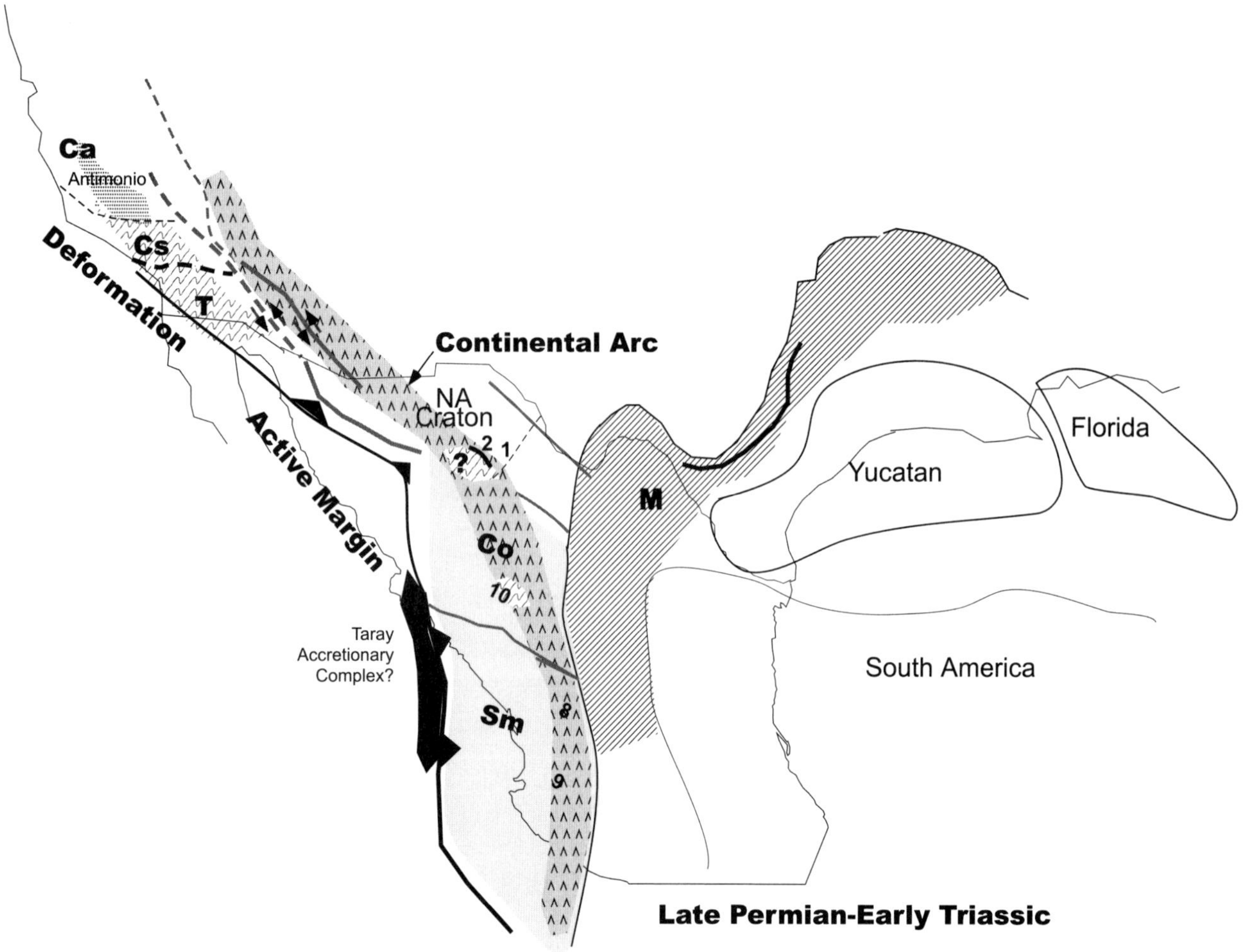

Figure 6. Permo-Triassic paleogeographic reconstruction. Granitoid belt of eastern Mexico is interpreted to be part of continental arc that developed during that time. East-dipping active subduction was probably placed along continental margin, which was in part contemporaneous to Sonoma Orogeny in western USA (Miller et al., 1992). Deformation of rocks in Cortes and Tahue terranes occurred in Late Permian–Early Triassic (Sonoran orogeny). If contemporaneous deformation occurred in Chihuahua (1 and 2) and Delicias (10), it has not been well constrained.

eastern and northern Mexico extended along the western boundary of the Sierra Madre terrane before thrusting of the Zihuatanejo terrane (pre–latest Triassic–Early Jurassic). This margin was located approximately along the present limit between Sierra Madre and Guerrero terrane (Fig. 7) (e.g., Ruíz et al., 1988b; Anderson et al., 1990; Yañez et al., 1991; Torres et al., 1999; Centeno-García and Silva-Romo, 1997; Bartolini et al., 2002). If the accretionary complex of the Central terrane is post-Triassic, the continental margin would have extend along the limit between the Central and Sierra Madre and Coahuila terranes (Figs. 4 and 7).

An oceanic basin located to the west of this paleocontinental margin received sediments derived from the Potosi submarine fan, as indicated by their provenance (Fig. 7) (Centeno-García et al., 2003). Evidence of contemporaneous rift-related volcanism is found in the Arteaga Complex (Zihuatanejo terrane). This basin may have evolved as a back arc basin, as suggested by the primitive geochemical signatures of its volcaniclastic rocks, or as an open ocean-floor basin, where the mid-ocean ridge received sediments from the continent (Centeno-García et al., 1993; Centeno-García et al., 2003).

Early Jurassic

The exact age and number of compressional events that deformed the Triassic rocks in the Zihuatanejo and the Sierra Madre terranes (and maybe the Central terrane?) are still uncertain. However, evidence suggests that rocks of Taray, Zacatecas, and Arteaga were deformed in a subduction complex. Preliminary isotopic data suggest that the Arteaga oceanic basin might have collided against nuclear Mexico sometime between the Rhaetian and the Callovian (Fig. 8A) (Centeno-García, 1994; Centeno-García and Silva-Romo, 1997; Centeno-García et al., 2003). During this deformational event, the thick siliciclastic succession of the Potosi Fan was thrust eastward, over the Sierra Madre terrane.

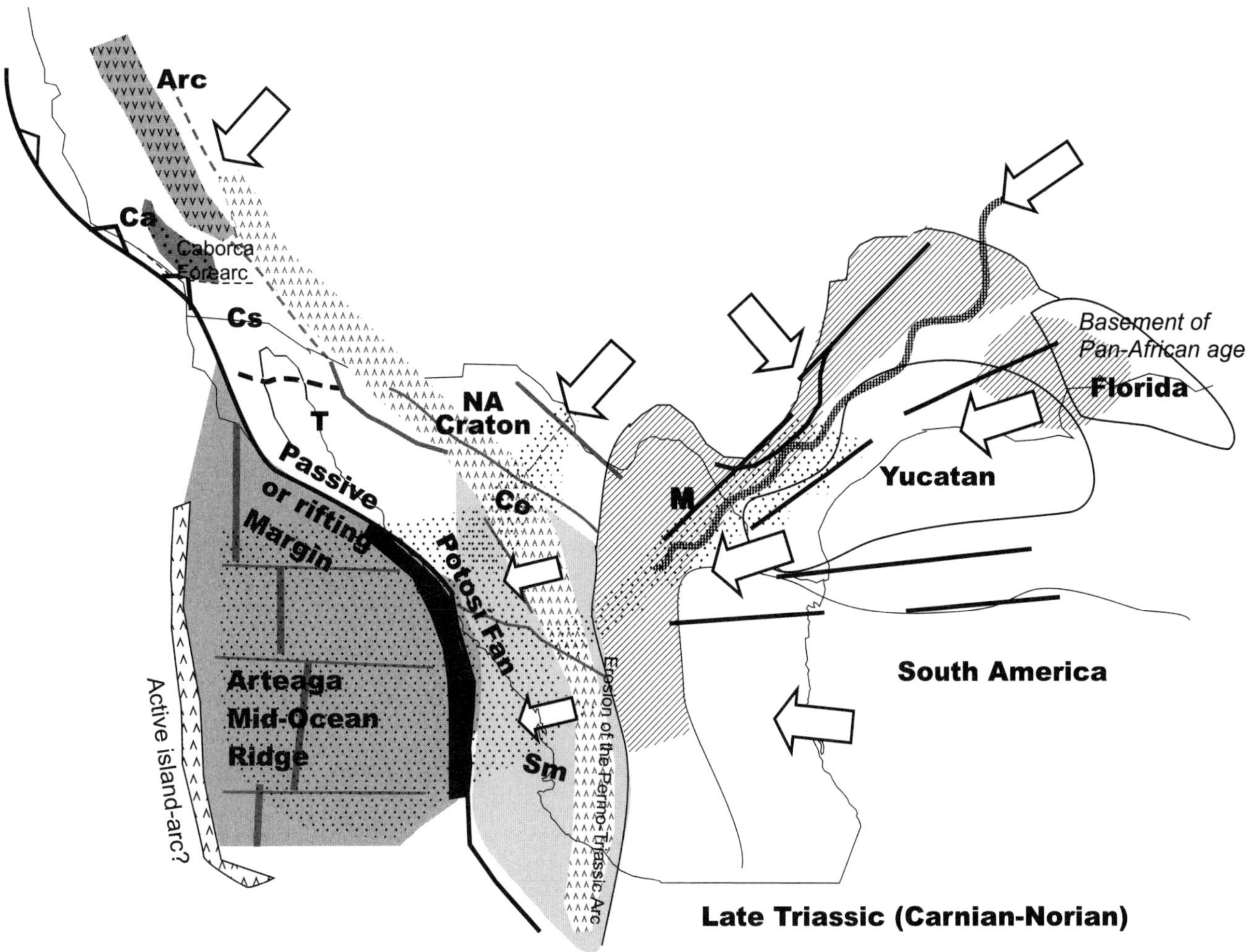

Figure 7. Paleogeographic reconstruction of north central Mexico by Late Triassic time. Arrows indicate possible provenance of sediments and areas with major erosion. Large submarine fan extended along central Mexico and spread in ocean-floor. This oceanic basin had an active ridge and might have been a piece of large oceanic basin or a rift-basin in a marginal sea (back arc?). Abbreviations of names of terranes as in other figures.

The regional angular unconformity that places Oxfordian-Kimmeridgian volcanic and sedimentary rocks on the deformed Triassic rocks of the Sierra Madre and Central terranes, and the granitic intrusions of the same age that cut the Arteaga Complex in the Zihuatanejo terrane (Centeno-García et al., 2003), are evidence of the regional extension of this orogenic event. This deformation was followed by widespread continental arc magmatism along all western Mexico by Middle-Late Jurassic (Fig. 8B), and by the rifting of the Zihuatanejo terrane during the Late Jurassic–Early Cretaceous time, to form the Early Cretaceous Arperos Basin.

Early Mesozoic Paleogeography of the Caborca and Cortes Terranes

Early Mesozoic evolution of the Caborca terrane has major differences with the evolution of terranes of central and eastern Mexico. In Caborca, sedimentation was continuous from Triassic to Early Jurassic (González-León et al., this volume), whereas a passive margin followed by a subduction-related deformational event was occurring in the Sierra Madre, Central(?), and Zihuatanejo terranes (Figs. 7 and 8). The sedimentation in Caborca was related to an active arc during Early Jurassic time, as evidenced by the detrital zircons collected from the Antimonio Group (González-León et al., this volume). Early Jurassic magmatism has been documented in Nevada (Riggs et al., 1993), but it seems to be absent from the Mojave Desert to the Sierra Madre terrane. Thus, provenance supports a northern location of the Caborca terrane for Early Jurassic time. Triassic sedimentary rocks of the Cortes terrane are interpreted as deposited in rift-related basins (Stewart and Roldán-Quintana, 1991). Discussion on the relationships between Triassic rocks of the Caborca and Cortes terranes is presented by González-León et al. (this volume).

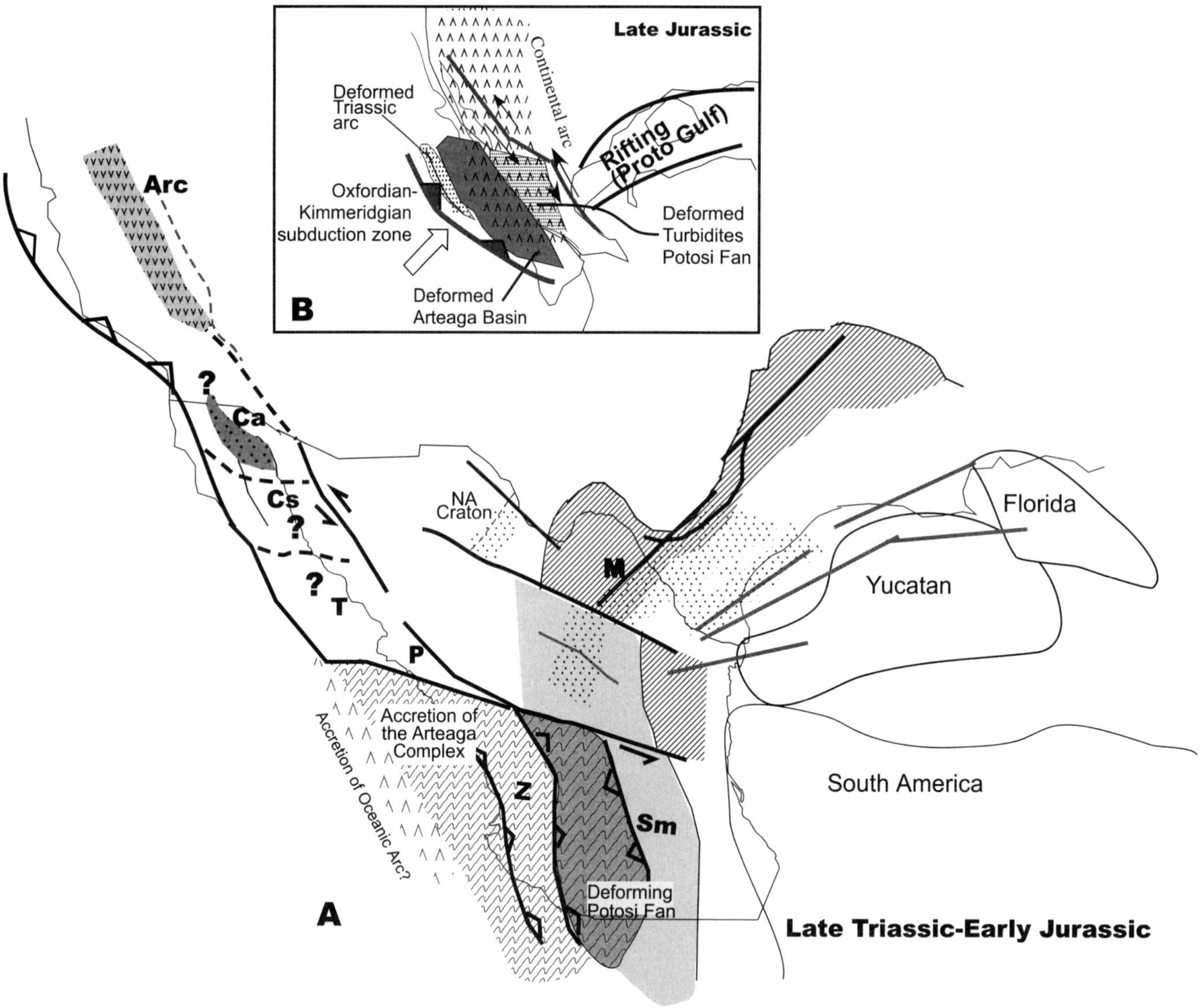

Figure 8. (A) Tectonic reconstruction by Early Jurassic time. Rocks of Triassic submarine fan and Triassic ocean basin were accreted toward continental margin. Whether Caborca, Cortes, Tahue, and Parral terranes were displaced southward during this deformational event is still uncertain. (B) After collision of Arteaga basin, arc magmatism was widespread in Zihuatanejo, Central, and Sierra Madre terranes. At least Parral terrane should be near Central terrane by this time because it contains rocks related to this arc. Abbreviations of names of terranes as in other figures.

Paleogeography of the Tahue and Parral Terranes

The metamorphic rocks of the Tahue and Parral terranes are similar. Both are made up of metavolcanic rocks (lava flows and volcaniclastics) interlayered with metamorphosed quartz-rich sandstone and shale. Their origin remains uncertain. The rocks at Tahue terrane (El Fuerte) contain Ordovician conodonts (Poole and Perry, 1998), but the age of deformation is unknown. Basement rocks of the Parral terrane (Pescadito Schist) were deformed and metamorphosed in the Late Devonian, but the age of deposition is unknown. Whether both metamorphic units had originally been one unit or not has not been determined. They are considered in this paper as different complexes because their overlapping units are different.

Volcanic-sedimentary successions of the Cretaceous Guerrero arc are deposited on the El Fuerte metamorphic rocks. In contrast, pre-Titonian continental red beds and volcanic rocks, as well as Titonian-Cenomanian marine calcareous rocks, were deposited on the metamorphic units of Parral Terrane.

Understanding the origin of these metamorphic units will set constraints on regional displacements and paleogeography. Preliminarily, there are two plausible options. One is that metamorphic rocks of the Parral terrane could be the southern continuation of the accretionary zone between North America and

Oaxaquia (southern continuation of the Ouachita) (P1 in Fig. 5). An alternative origin for the metamorphic rocks of Parral terrane is that they were displaced from a northwestern latitude in the Cordillera (P2 on Fig. 5), because metamorphic volcanic-sedimentary rocks of similar age have been reported from Nevada and California (Miller et al., 1992).

CONCLUDING REMARKS

One of the main problems reconstructing the Late Paleozoic–Early Mesozoic paleogeographic position of the northwestern terranes with respect to central and eastern terranes is that differences in the stratigraphy and structure can be explained by either lateral changes in the tectonic settings, or major translations associated to strike-slip faulting. There is not enough evidence collected to date to strongly favor any of the models proposed in this paper or by other authors; however, any model should take into consideration the following field evidence.

1. The Coahuila and Sierra Madre terranes must have been attached to North America (Chihuahua) by Late Permian–Early Triassic time because they are both intruded by the Permo-Triassic granitic belt. Looking at the present distribution of those granitoids, a minimum sinistral strike-slip displacement between the Coahuila and Sierra Madre terranes of ~200 can be calculated (Fig. 2). Age of this displacement remains unknown.

2. Whether the Caborca and Cortes terranes were displaced southward in the Late Paleozoic is still uncertain. The evidence favoring this displacement is the possible Late Paleozoic truncation of the continental margin in the Mojave Desert region. The evidence against this early movement is that Lower Jurassic arc-related zircons were shed into the Caborca terrane Antimonio Group, because volcanism of this age has not been documented south of the Mojave Desert region.

3. There is no evidence of subduction-related magmatism in central and eastern Mexico from Middle Triassic to Early(?)-Middle Jurassic time. Composition and thickness of the Upper Triassic turbidites of the Potosi Fan suggest that the western margin of the Sierra Madre terrane was a passive margin or rifting margin during the Late Triassic. The continental edge was located approximately along the limit of the Sierra Madre terrane with the Central and Guerrero Composite terranes (Fig. 4).

4. Field evidence suggest that a major regional compressional event occurred sometime in between Latest Triassic and Early Jurassic time in the Sierra Madre and Zihuatanejo terranes. During this event, the basement of the Zihuatanejo terrane (Arteaga Complex and Zacatecas Formation) probably collided in a subduction zone with the western margin of Sierra Madre terrane (Oaxaquia). Then it probably was rifted apart during Late Jurassic–Cretaceous time.

5. Whether the Taray subduction complex was part of the deformational event described in (4) is still uncertain, but field relationships suggest that it was accreted before the development of the Middle-Late Jurassic continental arc.

6. The present position of the Early Mesozoic continental edge is evidence of strike-slip displacements in central Mexico. This edge is defined by the Taray accretionary prism, which has no continuation to the north in the Coahuila terrane. All the rocks exposed in Coahuila and Parral terranes are much older than Taray (Figs. 3 and 4). The margin might have been transported toward the east, with a minimum sinistral displacement of 300 km. Age of this displacement is unknown, but it should be post–Late Permian on the basis of the maximum age of the Taray Formation.

7. The Parral, Central, and Sierra Madre terranes must have been together before latest Middle to early Late Jurassic time, because they have a common overlapping volcano-sedimentary cover of Middle-Late Jurassic age. Volcanic xenoliths of the same age found in La Popa (NW Monterrey City), located in the Coahuila terrane, open the possibility that those three terranes might not have been far from the Coahuila terrane. Middle-Upper Jurassic granitoids that cut the Arteaga complex suggest that the subduction zone of the Middle-Late Jurassic continental arc might not be the Taray Formation, as proposed by Anderson et al. (1990), but might have been much farther toward the west of the Arteaga complex (Fig. 8B).

8. The Paleozoic basement of the Tahue terrane (El Fuerte, San José de Gracia) and the Triassic basement of the Zihuatanejo terrane (Arteaga Complex and Zacatecas Formation) should have been together before Late Jurassic, because the Jurassic-Cretaceous volcanic arc that defines the Guerrero terrane was built on both units. Thus, if the El Fuerte and San José de Gracia rocks were displaced from a northern position, they should have been in a southern latitude before the Late Jurassic.

9. So far, direct stratigraphic or provenance correlation between Paleozoic–Early Mesozoic rocks in west (Caborca, Cortes and Tahue terranes) and east-central (Central and Zihuatanejo, Coahuila and Sierra Madre terranes, and the craton in Chihuahua) Mexico has not been done. In contrast, much evidence has been collected on the stratigraphic similarities between western Mexico and the southeastern United States (Stanley and González-León, 1995; Marzolf, 2000, among others). Until the correlation studies within Mexico are done, tectonic modeling will be based on only one body of evidence.

Looking at the overall present distribution of the Permo-Triassic units, as well as at the stratigraphy and tectonic evolution of their overlapping units, major lateral displacements could have happened (1) before the Late Jurassic volcanic arc (contemporaneous to the Early Jurassic orogeny?), (2) during the development of this volcanic arc and the rifting of the basements of the Guerrero composite terrane, and/or (3) during the collision of the Guerrero arc and formation of the Sierra Madre fold and thrust belt (Late Cretaceous–Early Cenozoic) of eastern Mexico.

ACKNOWLEDGMENTS

This paper is a contribution to the PAPIIT projects IN116599 and IN101095, funded by the National Autonomous University of Mexico (UNAM), and by the Institute of Geology, UNAM. Thanks to Gilberto Silva, Carlos González, Martin Valencia, Francisco Sour, Pedro Arredondo, Ciro Díaz, Martin Guerrero, and Pedro Corona; their comments and discussions were fundamental to this paper. Special thanks to Ronald C. Blakey and Norris W. Jones for their reviews and comments, which greatly improved the paper.

REFERENCES CITED

Anderson, T.H., and Silver, L.T., 1971, Preliminary history for Precambrian rocks, Bamori región, Sonora, México: Geological Society of America Abstracts with Programs, v. 3, p. 72–73.

Anderson, T.H., and Silver, L.T., 1977, U-Pb isotope ages of granitic plutons near Cananea: Sonora, Economic Geology and the Bulletin of the Society of Economic Geologists, v. 72, no. 5, p. 827–836.

Anderson, T.H., and Silver, L.T., 1979, The role of the Mojave-Sonora megashear in the tectonic evolution of northern Sonora, *in* Anderson, T.H. and Roldán-Quintana, J., eds, Geology of northern Sonora: Geological Society of America Annual Meeting, San Diego, California, Guidebook, Field Trip 27, p. 59–68.

Anderson, T.H., and Silver L.T., 1981, An overview of Precambrian rocks in Sonora, México: Universidad Nacional Autónoma de México, Instituto de Geología, Revista, v. 5, p. 131–139.

Anderson, T.H., McKee, J.W., and Jones, N.W., 1990, Jurassic (?) melange in north-central Mexico, Geological Society of America, Cordilleran Section, 86th annual meeting: Geological Society of America Abstracts with Programs, v. 22, no. 3, p. 3.

Anderson, T.H., McKee, J.W., Jones, N.W., 1991, A northwest trending, Jurassic fold nappe, northernmost Zacatecas, Mexico, Tectonics, v. 10, n. 2, p. 383–401.

Aranda, M., Quintero-Legorreta, O., and Martinez, E., 1988, Palinomorfos del Jurasico temprano de la Formacion Gran Tesoro, Santa Maria del Oro, Durango: Revista Instituto de Geologia, v. 7, no. 1, p. 112–115.

Araujo M.J., Arenas P.R., 1986, Estudio tectonico-sedimentario, en el mar mexicano Estados de Chihuahua y Durango: Boletín Sociedad Geológica Mexicana, v. 47-2, p. 43–71.

Arellano-Gil, J., Vachard, D., Yussim, S., and Flores de Dios-González, L., 1998, Aspectos estratigráficos, estructurales y paleogeográficos del Pérmico inferior al Jurásico inferior en Pemuxco, Estado de Hidalgo, México: Revista Mexicana de Ciencias Geológicas, v. 15, p. 9–13.

Arredondo-Guerrero, P., and Centeno-García, E., 2003, Geology of the Mazatlan region, southern Sinaloa state, Mexico: GSA 99th Cordilleran Section Annual Meeting, Abstracts with Programs, v. 35, no. 4, p. 71.

Barboza-Gudiño, J.R., Tristán González, M., and Torres Hernández, J.R., 1998, The Late Triassic–Early Jurassic active continental margin of western North America in northeastern Mexico: Geofisica Internacional, v. 37, no. 4, p. 283–292.

Bartolini, C., Cantú-Chapa, A., Lang, H., and Barboza-Gudiño, R., 2002, The Triassic Zacatecas Formation in Central Mexico: Paleotectonic, paleogeographic and paleobiogeographic implications, *in* Bartolini C., et al., eds, The western Gulf of Mexico Basin: Tectonics, sedimentary basins and petroleum systems, AAPG Memoir 75, p. 295–315.

Blount, J.G., Walker, N.W., and Carlson, W.D., 1988, Geochemistry and U-Pb zircon ages of mid-Proterozoic metaigneous rocks from Chihuahua, Mexico: Geological Society of America, v. 20, no. 7, p. 205.

Boucot, A.J., Blodgett, R.B., and Stewart, J.H., 1997, European Province Late Silurian brachiopods from the Ciudad Victoria area, Tamaulipas, northeastern Mexico, *in* Klapper, G., et al., eds., Paleozoic sequence stratigraphy, biostratigraphy, and biogeography: Studies in Honor of J. Grenville (Jess) Johnson: Geological Society of America Special Paper 321, p. 273–293.

Bridges, L.W., 1964, Stratigraphy of Mina Plomosas–Placer de Guadalupe Area, *in* Geology of Mina Plomosas–Placer de Guadalupe Area, Chihuahua, México: Field guidebook: West Texas Geological Society Publication 64-50, pp. 50–93.

Burchfiel, B.C., Cowan, D.S., and Davis, G.A., 1992, Tectonic overview of the Cordilleran orogen in the western United States, *in* Burchfiel, B.C., et al., eds., The Cordilleran orogen, Conterminous U.S.: Geological Society of America, The Geology of North America, v. G-3, p. 407–479.

Burckhardt, C., and Scalia, S., 1906, Géologie des environs de Zacatecas: International Geological Congress 10th México Excursion Guidebook 16, 26 p.

Cameron K.L., and Cameron M., 1985, Rare earth element, $^{87}Sr/^{86}Sr$, and $^{143}Nd/^{144}Nd$ composition of Cenozoic orogenic dacites from Baja California, northwestern Mexico, and adjacent west Texas: Evidence of predominance of a subcrustal component: Contributions to Mineralogy and Petrology, v. 91, p. 1–11.

Campa, M.F., and Coney, P.J., 1983, Tectono-stratigraphic terranes and mineral resource distributions in México: Canadian Journal of Earth Sciences, v. 20, p. 1040–1051.

Campa, M.F., Ramírez, J., and Bloome, C., 1982, La secuencia volcanico-sedimentaria metamorfizada del Triasico (Ladiniano-Carnico) de la region de Tumbiscatio, Michoacán: Sociedad Geológica Mexicana, VI Convención Geológica Nacional, 6 Resúmenes, 48 p.

Cantu-Chapa, A., 1969, Una nueva localidad Triásico Superior en México: Revista Instituto Mexicano del Petróleo, v. 1, no. 2, p. 71–72.

Carrillo-Bravo, B.J., 1961, Geología del Anticlinorio Huizachal-Peregrina al NW de Ciudad Victoria, Tamps: Asociación Mexicana de Geólogos Petroleros, Boletín, v. 13, p. 1–98.

Carrillo-Martínez, 1971, Geología de la Hoja San José de Gracia, Sinaloa [tesis profesional]: Mexico City, Universidad Nacional Autónoma de México, Facultad de Ingeniería, 154 p.

Carpenter, D.L., 1997, Tectonic history of the metamorphic basement rocks of the Sierra del Carmen, Coahuila, Mexico: Geological Society of America Bulletin, v. 109, no. 10, p. 1321–1332, doi: 10.1130/0016-7606(1997)109<1321:THOTMB>2.3.CO;2.

Centeno-García, E., 1994, Tectonic evolution of the Guerrero Terrane, western Mexico [Ph.D. thesis]: Tucson, University of Arizona, 220 p.

Centeno-García, E., and Silva-Romo, G., 1997, Geology, geochemistry, and tectonic evolution of central Mexico during Mesozoic time: Revista de Ciencias Geológicas, Instituto de Geología UNAM, v. 14, no. 2, p. 244–260.

Centeno-García, E., García, J.L., Guerrero-Suástegui, M., Ramírez-Espinosa, J., Salinas-Prieto, J.C., and Talavera-Mendoza, O., 1993a, Geology of the southern part of the Guerrero terrane, Ciudad Altamirano-Teloloapan Area, *in* Ortega-Gutiérrez F., Coney P., Centeno-García E., Gómez-Caballero A., eds., Terrane Geology of Southern México, First Circum-Pacific and Circum-Atlantic Terrane Conference, Guidebook of Field Trip B: Guanajuato Mexico, UNAM, Instituto de Geología, p. 22–33.

Centeno-García, E., Ruíz, J., Coney, P.J., Patchett, P.J., and Ortega-Gutíerrez, F., 1993b, Geology of the Guerrero terrane and its role in the tectonic evolution of the Southern North America Cordillera from new geochemical data: Geology, v. 21, p. 419–422, doi: 10.1130/0091-7613(1993)021<0419:GTOMIR>2.3.CO;2.

Centeno-García, E., Sánchez-Zavala, J.L., and Ortega-Gutíerrez, F., 1998, Late Proterozoic– Paleozoic stratigraphy of Mexico and its role in Gondwana-Laurentia interactions: Internacional Geological Correlation Programme, Project 376 meeting, Laurentian-Gondwanan connections before Pangea, Oaxaca, México, p. 13.

Centeno-García, E., Corona-Chavez, P., Talavera-Mendoza, O., and Iriondo, A., 2003, Geology and tectonic evolution of the Western Guerrero terrane—A transect from Puerto Vallarta to Zihuatanejo, México, *in* Geologic Transects across Cordilleran México, guidebook for field trips of the 99th GSA Cordilleran Section Meeting, UNAM Instituto de Geologia Publicación Especial 1, p. 201–228.

Coney, P.J., 1978, Mesozoic-Cenozoic Cordilleran plate tectonics: Geological Society of America Memoir 152, p. 33–49.

Coney, P.J., 1983, Un modelo tectónico de México y sus relaciones con América del Norte: América del Sur y el Caribe, Revista del Instituto Mexicano del Petróleo, v. 15, no. 1, p. 6–15.

Coney, P.J., and Campa, M.F., 1987, Lithotectonic Terrane Map of México (west of the 91st meridian): U.S. Geological Survey Miscellaneous Field Studies Map and Report MF-1874-D, scale: 1:10,000,000, 1 sheet.

Contreras-Montero, B., Martínez-Cortes, A., and Gómez-Luna, M.A., 1988, Bioestratigrafía y sedimentología del Jurásico Superior en San Pedro del Gallo, Durango, México: Revista del Instituto Mexicano del Petróleo, v. 20, no. 3, p. 5–29.

Cooper, G.A., and Arellano, A.R.V., 1946, Stratigraphy near Caborca, Northwest Sonora, Mexico: American Association Petroleum Geologists, Bulletin, v. 30, p. 606–619.

Cuevas-Perez, E., 1983, Evolución geológica mesozoica del Estado de Zacatecas, México. Mesozoic geologic evolution of Zacatecas, Mexico, Zentralblatt für Geologie und Palaeontologie, Teil I: Allgemeine, Angewandte, Regionale und Historische Geologie, v. 1, no. 3-4, p. 190–201.

Damon, P.E., Shafiqullah, M., Clark, K.F., 1981, Evolución de los arcos magmáticos en México y su relación con la metalogénesis: Revista, Instituto de Geologia, Universidad Nacional Autónoma de México, v. 5, no. 2, p. 223–238.

Díaz, T.G., and Navarro, G., 1964, Litología y correlación estratigráfica del Paleozoico Superior en la region de Palomas, Chihuahua, México: Asociación Mexicana de Geólogos Petroleros Boletín, v. 16, p. 107–120.

Díaz-Salgado, C., Centeno-García, E., and Gehrels, G., 2003, Stratigraphy, depositional environments, and tectonic significance of the Taray Formation, northern Zacatecas state, Mexico: GSA 99th Cordilleran Section Annual Meeting, Abstracts with Programs, v. 35, n. 4, p. 71.

Dickinson, W.R., and Coney, P.J., 1980, Plate tectonic constrains on the origin of the Gulf of México, *in* Pilger, R.H., ed., The origin of the Gulf of México and the early opening of central north Atlantic Ocean: Louisiana State University Symposium, p. 27–36.

Dickinson, W.R., and Lawton, T.F., 2001, Carboniferous to Cretaceous assembly and fragmentation of Mexico: Geological Society of America Bulletin, v. 113, no. 9, p. 1142–1160, doi: 10.1130/0016-7606(2001)113<1142:CTCAAF>2.0.CO;2.

Eguiluz, de A.S., and Campa, M.F., 1982, Problemas tectónicos del sector de San Pedro El Gallo, en los Estados de Chihuahua y Durango: Boletín Asociación Mexicana de Geólogos Petroleros, v. 34, n. 2, p. 5–42.

Elías-Herrera, M., Sánchez-Zavala, J.L., and Macias-Romo, C., 2000, Geologic and geochronologic data of the Guerrero terrane in the Tejupilco area, southern Mexico: new constrains on its tectonic interpretation: Journal of South American Earth Sciences, v. 13, p. 355–375, doi: 10.1016/S0895-9811(00)00029-8.

Flawn, P.T., Goldstein, A., Jr., King, P.B., and Weaver, C.E., 1961, The Ouachita system: University of Texas Publications 6120, 401 p.

Franco-Rubio, M., 1999, Geology of the basement below the decollement surface, Sierra de Catorce, San Luis Potosi, Mexico, *in* Bartolini, C., Wilson, J.L., Lawton, T.F., Mesozoic sedimentary and tectonic history of north-central Mexico: Geological Society of America Special Paper 340, p. 211–227.

Freydier, C., Lapierre, H., Tardy, M., Coulon, C., Martinez-Reyes, J., and Orsini, J.B., 1995, Les formations magmatiques de Porohui (Sinaloa); temoins de l'evolution geodynamique mesozoique et tertiaire des Cordilleres mexicaines, Comptes Rendus de l'Academie des Sciences, Serie II: Sciences de la Terre et des Planetes, v. 321, no. 6, p. 529–536.

Freydier, C., Martinez, R.J., Lapierre, H., Tardy, M., and Coulon, C., 1996, The Early Cretaceous Arperos oceanic basin (western Mexico): Geochemical evidence for an aseismic ridge formed near a spreading center: Tectonophysics, v. 259, no. 4, p. 343–367.

Garrison, J.R., 1978, Reinterpretation of isotopic age data from the Granjeno Schist: Ciudad Victoria, Tamaulipas, Universidad Nacional Autonoma de México, Instituto de Geología, Revista, v. 2, no. 1, p. 87–89.

Gastil, G., and Miller, R.H., 1983, Prebatholithic terranes of southern and peninsular California, USA, and México: Status report, *in* Stevens, C.H., ed., Pre-Jurassic rocks, western North American suspect terranes: Society of Economic Mineralogists and Paleontologists, Pacific Section, p. 49–61.

Gastil, G., Miller, R., Anderson, P., Crocker, J., Campbell, M., Buch, P., Lothringer, C., Leier-Engelhardt, P., DeLattre, M., Hoobs, J., and Roldán-Quintana, J., 1991, The relation between the Paleozoic strata an opposite sides of the Gulf of California, *in* Pérez-Segura, E., and Jacques-Ayala, C., eds., Studies of Sonoran geology: Geological Society of America Special Paper 254, p. 1–7.

Gehrels, G.E., and Stewart, J.H., 1998, Detrital zircon U-Pb geochronology of Cambrian to Triassic miogeoclinal and eugeoclinal strata of Sonora, Mexico: Journal of Geophysical Research, v. 103, no. B2, p. 2471–2487, doi: 10.1029/97JB03251.

González-León, C., 1996, Stratigraphy and paleogeographic setting of the Antimonio Formation, *in* González-León and Stanley G, ed., U.S.-Mexico cooperative research: International Workshop on the Geology of Sonora, Estacion Regional del Noroeste, Publicaciones Ocasionales 1, Instituto de Geología UNAM, Memoir, p. 23–32.

González-León, C., 1997, Sequence stratigraphy and paleogeographic setting of the Antimonio Formation (Late Permian–Early Jurassic), Sonora, Mexico: Revista Mexicana de Ciencias Geológicas, v. 14, p. 136–148.

Grajales-Nishimura, M., Terrell, D.J., and Damon, P.E., 1992, Evidencias de la prolongación del Arco Magmático Cordillerano del Triásico Tardío-Jurásico en Chihuahua, Durango y Coahuila: Asociación Mexicana de Geólogos Petroleros Boletín, v. 42, no. 2, p. 1–18.

Handschy, J.W., and Dyer, R., 1987, Polyphase deformation in Sierra del Cuervo, Chihuahua, Mexico: Evidence for Ancestral Rocky Mountain tectonics in the Ouachita foreland of northern Mexico: Geological Society of America Bulletin, v. 99, p. 618–632, doi: 10.1130/0016-7606(1987)99<618:PDISDC>2.0.CO;2.

Henry, C.D., and Fredrikson, G., 1987, Geology of part of southern Sinaloa, Mexico adjacent to the Gulf of California: Geological Society of America. Map and Chart Series, v. MCH063, p. 1–14.

Iriondo, A., Martinez Torres, L.M., and Kunk, M.J., 2000, Late Cretaceous–early Tertiary juxtaposition of Paleoproterozoic basement blocks in NW Sonora, Mexico: Testing the Mojave-Sonora megashear hypothesis: Abstracts of the 31st International Geological Congress, v. 31, n. 8, unpaginated.

Jones, N.W., McKee, J.W., Anderson, T.H., and Silver, L.T., 1995, Jurassic volcanic rocks in northeastern Mexico: A possible remnant of a Cordilleran magmatic arc, *in* Jacques-Ayala, C., González-León, C.M., and Roldán-Quintana, J., eds., 1996, Studies on the Mesozoic of Sonora and adjacent areas: Geological Society of America Special Paper 301, p. 179–190.

Keppie, J.D., and Ortega-Gutíerrez, F., 1999, Middle American Precambrian basement: A missing piece of the reconstructed 1 Ga orogen, *in* Ramos, V.A., and Keppie, J.D., eds., Laurentia-Gondwana connections before Pangea: Geological Society of America Special Paper 336, p. 199–210.

Keppie, J.D., Dostal, J., Murphy, J.B., and Nance, R.D., 1996, Terrane transfer between eastern Laurentia and western Gondwana in the early Paleozoic: Constrains on global reconstructions, *in* Nance, R.D., and Thompson, M.D., eds., Avalonian and related peri-Gondwanan terranes of the circum–North Atlantic: Geological Society of America Special Paper 304, p. 369–380.

King, R.E., 1944, Geology and paleontology of the Permian area northwest of Las Delicias, southwestern Coahuila, Mexico: Geological Society of America Special Paper, 174 p.

Labarthe, G., Tristán, M., and Aguillón, R.A., 1982, Estudio geológico-minero del área de Peñón Blanco, estados de San Luis Potosí y Zacatecas: Instituto de Geología y Metalurgia, Universidad Autónoma de San Luis Potosí, Folleto Técnico 76, 80 p.

Lapierre, H., Ortiz, L.E., Abouchami, W., Monod, O., Coulon, C., and Zimmermann, J.L., 1992, A crustal section of an intra-oceanic island arc: The Late Jurassic–Early Cretaceous Guanajuato magmatic sequence, central Mexico: Earth and Planetary Science Letters, v. 108, p. 61–77, doi: 10.1016/0012-821X(92)90060-9.

Lawlor, P.J., Ortega Gutíerrez, F., Cameron, K.L., Ochoa Camarillo, H., Lopez, R., and Sampson, D.E., 1999, U-Pb geochronology, geochemistry, and provenance of the Grenvillian Huiznopala Gneiss of eastern Mexico: Precambrian Research, v. 94, no. 1–2, p. 73–99, doi: 10.1016/S0301-9268(98)00108-9.

Lawton, T.F., and Giles, K.A., 2000, Southwestern ancestral Rocky Mountains province, Arizona, New Mexico, Sonora and Chihuahua: Broken foreland or continental borderland?: Geological Society of America Abstracts with Programs, v. 32, no. 7, p. 466–467.

Lopez Infanzon, M., 1986, Petrologia y radiometria de rocas igneas y metamorficas de Mexico: Boletin de la Asociacion Mexicana de Geologos Petroleros, v. 38, no. 2, p. 59–98.

Lopez, R., 1997, High-Mg andesites from the Gila Bend Mountains, southwestern Arizona; evidence for hydrous melting of lithosphere during Miocene extension; the pre-Jurassic geotectonic evolution of the Coahuila Terrane, northwestern Mexico; Grenville basement, a late Paleozoic arc, Triassic plutonism, and the events south of the Ouachita suture [Ph.D. thesis]: Santa Cruz, University of California, 147 p.

Lopez, R., Cameron, K.L., and Jones, N.W., 2001, Evidence for Paleoproterozoic, Grenvillian, and Pan-African age Gondwanan crust beneath northeastern Mexico: Precambrian Research, v. 107, p. 195–214, doi: 10.1016/S0301-9268(00)00140-6.

Maher, D.J., Jones, N.W., McKee, J.W., and Anderson, T.H., 1991, Volcanic rocks at Sierra de Catorce, San Luis Potosi, Mexico: A new piece for the Jurassic-arc puzzle: Geological Society of America Abstracts with Programs, v. 23, no. 5, p. 133.

Marzolf, J.E., 2000, Triassic paleogeography of the Panthalassan margin of southwestern North America: A regional synthesis based on sequence stratigraphy: Zentralblatt für Geologie und Paläontologie, Teil 1: Heft, v. 11–12, p. 1497–1538.

McDowell, F.W., Housh, T.B., and Wark, D.A., 1999, Nature of the crust beneath west-central Chihuahua, Mexico, based upon Sr, Nd, and Pb isotopic compositions at the Tomochic volcanic center: Geological Society of America Bulletin, v. 111, no. 6, p. 823–830, doi: 10.1130/0016-7606(1999)111<0823:NOTCBW>2.3.CO;2.

McKee, J.W., Jones, N.W., and Anderson, T.H., 1988, Las Delicias Basin: A record of late Paleozoic arc volcanism in northeastern Mexico: Geology, v. 16, no. 1, p. 37–40, doi: 10.1130/0091-7613(1988)016<0037:LDBARO>2.3.CO;2.

McKee, J.W., Jones, N.W., and Anderson, T.H., 1999, Late Paleozoic and early Mesozoic history of the Las Delicias terrane, Coahuila, Mexico, *in* Bartolini, C., et al., eds, Mesozoic sedimentary and tectonic history of north-central Mexico: Geological Society of America, Special Paper 340, p. 161–189.

Mendoza, O.T., and Suástegui, M.G., 2000, Geochemistry and isotopic composition of the Guerrero terrane (western México): Implications for the tectonomagmatic evolution of southwestern North America during the Late Mesozoic: Journal of South American Earth Sciences, v. 13, no. 4, p. 297–324, doi: 10.1016/S0895-9811(00)00026-2.

Mickus, K.L., and Montana, C., 1999, Crustal structure of northeastern Mexico revealed through the analysis of gravity data: Geological Society of America Special Paper 340, p. 357–371.

Miller, E.L., Miller, M.M., Stevens, C.H., Wright, J.E., and Madrid, R., 1992, Late Paleozoic paleogeographic and tectonic evolution of the western U.S. Cordillera, *in* Burchfiel, B.C., et al., eds., The Cordilleran Orogen, Conterminous U.S.: Geological Society of America, The Geology of North America, v. G-3, p. 57–106.

Miller, J.S., Glazner, A.F., Walker, J.D., and Martin, M.W., 1995, Geochronologic and isotopic evidence for Triassic-Jurassic emplacement of the eugeoclinal allochthon in the Mojave Desert region, California: Geological Society of America Bulletin, v. 107, p. 1441–1457, doi: 10.1130/0016-7606(1995)107<1441:GAIEFT>2.3.CO;2.

Molina-Garza, R.S., Van der Voo, R., and Urrutia-Fucugauchi, J., 1992, Paleomagnetism of the Chiapas massif, southern México: Evidence for rotation of the Maya Block and implications for the opening of the Gulf of México: Geological Society of America Bulletin, v. 104, p. 1156–1168, doi: 10.1130/0016-7606(1992)104<1156:POTCMS>2.3.CO;2.

Monod, O., and Calvet, P., 1991, Structural and Stratigraphic reinterpretation of the Triassic units near Zacatecas, Zac., Central México: Evidence of a Laramide nappe pile: Zentralblatt fuer Geologie und Palaeontologie, Teil I: Allgemeine, Angewandte, Regionale und Historische Geologie, v. 1, no. 6, p. 1533–1544.

Moreno, F.A., Keller, G.R., and Mickus, K.L., 1993, The extent of the Ouachita continental magin in northern Mexico, *in* Ortega-Gutíerrez F., et al., eds., Proceedings of the First Circum-Pacific and Circum-Atlantic Terrane Conference, Guanajuato, Mexico, p. 100–103.

Moreno, F.A., Mickus, K.L., and Keller, G.R., 2000, Crustal structure and location of the Ouachita orogenic belt in northern Mexico: Geofisica Internacional, v. 39, no. 3, p. 229–246.

Mullan, H.S., 1978, Evolution of the Nevadan orogen in northwestern México: Geological Society of America Bulletin, v. 89, p. 1175–1188, doi: 10.1130/0016-7606(1978)89<1175:EOPOTN>2.0.CO;2.

Murillo, M.G., and Torres V.R., 1987, Mapa Petrogenético y Radiométrico de la República Mexicana: Instituto Mexicano del Petroleo, Subdirección de Tecnología de Exploración, Informe Inédito, Proyecto c-2010, 78 p.

Navarro-Santillan D., Sour-Tovar F., and Centeno-García E., 2002, Lower Mississippian (Osagean) Brachiopods from the Santiago Formation, Oaxaca, Mexico: Stratigraphic and Tectonic Implications: Journal of South American Earth Sciences, v. 15, p. 327–336.

Ortega-Gutíerrez, F., Sedlock, R.L., and Speed, R.C., 1994, Phanerozoic tectonic evolution of Mexico, *in* Speed R.C., ed., Phanerozoic evolution of North American continent-ocean transitions, Geological Society of America, DNAG Continent-Ocean Transect volume, p. 265–306.

Ortega-Guitérrez, F., Ruíz, J., and Centeno-García, E., 1995, Oaxaquia—A Proterozoic microcontinent accreted to North America during the Late Paleozoic: Geology, v. 23, p. 1127–1130, doi: 10.1130/0091-7613(1995)023<1127:OAPMAT>2.3.CO;2.

Ortega-Gutíerrez, F., Elías-Herrera, M., Reyes-Salas, M., Macias-Romo, C., and Lopez, R., 1999, Late Ordovician–Early Silurian continental collisional orogeny in southern Mexico and its bearing on Gondwana-Laurentia connections: Geology, v. 27, no. 8, p. 719–722, doi: 10.1130/0091-7613(1999)027<0719:LOESCC>2.3.CO;2.

Ortiz-Ubilla, A., Lopez-Ocampo, E., and Gutíerrez-Briones, F.J., 1988, Interpretación Geológica Regional en el Prospecto Cd. Victoria-Villagran: Tamaulipas, Instituto Mexicano del Petroleo, Report C-4014, 141 p.

Pacheco, G.C., Castro, M.R., and Gómez, M.A., 1984, Confluencia de terrenos estratotectónicos en Santa Maria del Oro, Durango, Mexico, Revista del Instituto Mexicano del Petroleo, v. 16, no. 1, p. 7–20.

Patchett, P.J., and Ruíz, J., 1987, Nd isotopic ages of crust formation and metamorphism in the Precambrian of eastern and southern México: Contributions to Mineralogy and Petrology, v. 96, p. 523–528, doi: 10.1007/BF01166697.

Patterson, W.D., 1978, Geology of Permian rocks near Ascencion norhern Chihuahua, México [M.S. thesis]: El Paso, Texas, El Paso University, 70 p.

Pearson, B., 1964, Some regional relationships, Trans-Pecos Texas and Northeastern Chihuahua: Field guidebook: West Texas Geological Society Publication, v. 64-50, p. 109–112.

Pindell, J., Dewey, J.F, 1982, Permo-Triassic reconstruction of western Pangea and the evolution of the Gulf of Mexico/Caribbean region: Tectonics, v. 1, no. 2, p. 179–211.

Pindell, J.L., 1985, Alleghanian reconstruction and subsequente evolution of the Gulf of México, Bahamas, and Proto-Caribean: Tectonics, v. 4, p. 1–39.

Poole, F.G., and Madrid, R.J., 1986, Paleozoic rocks in Sonora (Mexico) and their relation to the southwestern continental margin of North America: Geological Society of America Abstracts with Programs, v. 18, no. 6, p. 720–721.

Poole, F.G., and Madrid, R.J., 1988, Allochthonous Paleozoic eugeoclinal rocks of Barita de Sonora mine area, central Sonora, Mexico, *in* Rodriguez-Torres, R., ed., El Paleozoico de la Región Central del Estado de Sonora: Segundo Simposio sobre la Geología y Minería del Estado de Sonora, Excursiones de Campo, Instituto de Geología, Universidad Nacional Autónoma de México, Hermosillo, Sonora, p. 32–41.

Poole, F.G., Stewart, J.H., Palmer, A.R., Sanberg, C.A., Madrid, R.J., and Ketner, K.B., Carter, C., and Morales-Ramírez, J.M., 1992, Latest Precambrian to latest Devonian time-development of a continental margin, *in* Burchfiel, B.C., et al., eds., The Cordilleran orogen, conterminous U.S.: Boulder, Colorado, Geological Society of America, The Geology of North America, v. G-3, p. 9–56.

Poole, F.G., Stewart, J.H., Berry, W.B.N., Harris, A.G., Repetski, J.E., Madrid, R.J., Ketner, K.B., Carter, C., Morales-Ramirez, J.M., 1995, Ordovician ocean-basin rocks of Sonora, Mexico: Pacific Section, Society of Economic Paleontologists and Mineralogists, Field Trip Guidebook v. 77, p. 277–284.

Poole, F.G., and Perry, W.J., Jr., 1998, Laurentia-Gondwana continental margins in Northern Mexcio, and their late Paleozoic collision: IGCP Project 376 annual meeting: Laurentia-Gondwanan Connections before Pangea, Programs and Abstracts, p. 27.

Ramírez-Espinosa, J., Campa, M.F., Talavera, O., and Guerrero, M., 1991, Caracterización de los arcos insulares de la Sierra Madre del Sur y sus implicaciones tectónicas: Convención sobre la evolución Geológica Mexicana, 1er Congreso Mexicano de Mineralogía, Pachuca, Memoir, p. 163–166.

Ramírez-Ramírez, R.C., 1978, Reinterpretación Tectónica del Esquisto Granjeno de Ciudad Victoria, Tamaulipas: Universidad Nacional Autónoma de México, Instituto de Geología, Revista, v. 2, p. 31–36.

Ramírez-Ramírez, C., 1992, Pre-Mesozoic geology of Huizachal-Peregrina Anticlinorium, Ciudad Victoria, Tamaulipas, and adjacent parts of eastern Mexico [Ph.D. dissertation]: University of Texas at Austin, 450 p.

Ranson, W.A., Fernandez, L.A., Simmons, W.B., Jr., and de la Vega, E.S., 1982, Petrology of the Metamorphic rocks of Zacatecas, México: Sociedad Geológica Mexicana, v. 43, no. 1, p. 37–59.

Restrepo-Pace, P., Ruíz, J., and Cosca, M., 1994, Geochronology of basement uplifts in the Colombian Andes and evidence for terrane transfer from NW South America to southern Mexico in mid-Paleozoic time: Eos (Transactions, American Geophysical Union) v. 75, no. 44, p. 199.

Restrepo-Pace, P.A., Ruíz, J., Gehrels, G.E., and Cosca, M., 1997, Geochronology and Nd isotopic data of Grenville-age rocks in the Colombian Andes: New constraints for late Proterozoic–early Paleozoic paleocontinental reconstructions of the Americas: Earth and Planetary Science Letters, v. 150, no. 3–4, p. 427–441.

Reynolds, M.S., 1972, Informe Final Pozo Los Chinos No. 1, Petróleos Mexicanos, Superintendencia General de Exploración, Zona Noreste, Informe Geológico, NE-M-1194, 105 p.

Riggs, N.R., Mattinson, J.M., and Busby, C.J., 1993, Correlation of Jurassic eolian strata between the magmatic arc and the Colorado Plateau: New U-Pb geochronologic data from southern Arizona: Geological

Society of America Bulletin, v. 105, p. 1231–1246, doi: 10.1130/0016-7606(1993)105<1231:COJESB>2.3.CO;2.

Robison, R., and Pantoja-Alor, J., 1968, Tremadocian trilobites from the Nochixtlán region, Oaxaca, México: Journal of Paleontology, v. 42, no. 3, p. 767–800.

Roldán-Quintana, J., González-León, C.M., and Amaya-Martinez, R., 1993, Geologic constraints on the northern limit of the Guerrero terrane in northwestern Mexico, *in* Ortega Gutíerrez, F., et al., eds., Proceedings of the 1st Circum-Pacific and Circum-Atlantic Terrane Conference, Guanajuato México, p. 124–127.

Rosales-Lagarde, L., 2002, Estratigrafía y geoquímica de la secuencia volcanosedimentaria Paleozoica del noreste del Estado de Hidalgo, México [M.Sc. thesis]: Posgrado en Ciencias de la Tierra, Universidad Nacional Autónoma de México, 81 p.

Rosales-Lagarde, L., Centeno-García, E., Ochoa-Camarillo, H., and Sour-Tovar, F., 1997, Permian volcanism in eastern Mexico—Preliminary report: Universidad Autónoma de Hidalgo, Instituto de Investigaciones en Ciencias de la Tierra y Universidad Nacional Autónoma de México, Instituto de Geología, II Convención Geológica Sobre la Evolución Geológica de México y Recursos Asociados, Pachuca, Hidalgo, Libro Guía de las excursiones geológicas: Excursión, v. 1, p. 27–32.

Ross, M.I., and Scotese, C.R., 1988, A hierarchical tectonic model of the Gulf of México and Caribean Region: Tectonophysics, v. 155, p. 139–168, doi: 10.1016/0040-1951(88)90263-6.

Rowley, D.B., and Pindell, J.L., 1989, End Paleozoic–Early Mesozoic western Pangean reconstruction and its implications for the distribution of Precambrian and Paleozoic rocks around Meso-America: Precambrian Research, v. 42, p. 411–444, doi: 10.1016/0301-9268(89)90022-3.

Rudnick, R.L., and Cameron, K.L., 1991, Age diversity of the deep crust in northern Mexico: Geology, v. 19, no. 12, p. 1197–1200, doi: 10.1130/0091-7613(1991)019<1197:ADOTDC>2.3.CO;2.

Ruíz, J., Patchett, P.J., and Arculus, R.J., 1988a, Nd-Sr isotope composition of lower crustal xenoliths—Evidence for the origin of mid-Tertiary felsic volcanics in Mexico: Contributions to Mineralogy and Petrology, v. 99, p. 36–43, doi: 10.1007/BF00399363.

Ruíz, J., Patchett, P.J., and Ortega-Gutíerrez, F., 1988b, Proterozoic and Phanerozoic basement terranes of México from Nd isotopic studies: Geological Society of America Bulletin, v. 100, p. 274–281, doi: 10.1130/0016-7606(1988)100<0274:PAPBTO>2.3.CO;2.

Ruíz, J., Tosdal, R.M., Restrepo, P.A., and Murillo-Muneton, G., 1999, Pb isotope evidence for Colombia–southern Mexico connections in the Proterozoic, *in* Ramos, V.A., and Keppie, J.D., eds., Laurentia-Gondwana connections before Pangea: Geological Society of America Special Paper 336, p. 199–210.

Saleeby, J.B., and Busby-Spera, C., 1992, Early Mesozoic tectonic evolution of western U.S. Cordillera, *in* Burchfiel, B.C., et al., eds., The Cordilleran orogen, conterminous U.S.: Boulder, Colorado, Geological Society of America, The Geology of North America, v. G-3,p. 107–168.

Sedlock, R.L., Ortega-Gutíerrez, F., and Speed, R.C., 1993. Tectonostratigraphic terranes and tectonic evolution of Mexico: Geological Society of America Special Paper 278, 153 p.

Servais, M., Rojo, Y.R., and Colorado, L.D., 1982, Estudio de las rocas basicas y ultrabasicas de Sinaloa y Guanajuato: Postulacion de un paleogolfo de Baja California y de una digitacion tethysiana en Mexico central: Geomimet, v. 3, no. 115, p. 53–71.

Shurbet, D.H., Cebull, S.E., 1987, Tectonic interpretation of the westernmost part of the Ouachita-Marathon (Hercynian) orogenic belt, West Texas-Mexico, Geology: v. 15, p. 458–461.

Silva-Romo, G., 1987, Amonoideos del area Moctezuma-Salinas, estados de San Luis Potosi y Zacatecas. Ammonoids from the Moctezuma-Salinas area, San Luis Potosi and Zacatecas states, Revista de la Sociedad Mexicana de Paleontologia, v. 1, no. 1, p. 347–363.

Silva-Romo, G., 1993, Estudio de la Estratigrafía y Estructuras Tectónicas de la Sierra de Salinas, Edos. de San Luis Potosí y Zacatecas [M.S. thesis]: Universidad Nacional Autónoma de México, Facultad de Ciencias, México, 111 p.

Silva-Romo, G., Arellano Gil, J., Mendoza Rosales, C., Nieto Obregon, J., 2000, A submarine fan in the Mesa Central, Mexico: Journal of South American Earth Sciences, v. 13, n. 4–5, p. 429–442.

Smith, D.L., and Miller, E.L., 1990, Late Paleozoic extension in the Great Basin, western United States: Geology, v. 18, p. 712–715, doi: 10.1130/0091-7613(1990)018<0712:LPEITG>2.3.CO;2.

Stanley, G.D., Jr., and González-León, C.M., 1995, Paleogeographic and tectonic implications of Triassic fossils and strata from the Antimonio Formation, northwestern Sonora: Geological Society of America Special Paper 301, p. 17–20.

Stewart, J.H., 1988, Latest Proterozoic and Paleozoic southern margin of North America and the accretion of Mexico: Geology, v. 16, p. 186–189.

Stewart, J.H., and Roldán-Quintana, J., 1991, Upper Triassic Barranca Group: Nonmarine and shallow-marine rift-basin deposits of northwestern Mexico, *in* Pérez-Segura, E., and Jacques-Ayala, C., eds., Studies of Sonoran geology: Geological Society of America Special Paper 254, p. 19–36.

Stewart, J.H., Poole, F.G., Ketner, K.B., Madrid, R.J., Roldán-Quintana, J., and Amaya-Martínez, R., 1990, Tectonics and stratigraphy of the Paleozoic and Triassic southern margin of North America, Sonora, México, *in* Gehrels, G.E., and Spencer, J.E., eds., Geologic excursions through the Sonoran Desert region, Arizona and Sonora: Arizona Geological Survey Special Paper 7, p. 183–202.

Stewart, J.H., Amaya Martinez, R., Stamm, R.G., Wardlaw, B.R., Stanley, G.D., Jr., and Stevens, C.H., 1997, Stratigraphy and regional significance of Mississippian to Jurassic rocks in Sierra Santa Teresa, Sonora, Mexico: Revista Mexicana de Ciencias Geologicas, v. 14, no. 2, p. 115–135.

Stewart, J.H., Blodgett, R.B., Boucot, A.J., Carter, J.L., and Lopez, R., 1999, Exotic Paleozoic strata of Gondwanan provenance near Ciudad Victoria, Tamaulipas, Mexico, *in* Ramos, V.A., and Keppie, J.D., eds., Laurentia-Gondwana connections before Pangea: Geological Society of America Special Paper 336, p. 227–252.

Stone, P., and Stevens, C.H., 1988, Pennsylvanian and Early Permian paleogeography of east-central California: Implications for the shape of the continental margin and the timing of continental truncation: Geology, v. 16, p. 330–333, doi: 10.1130/0091-7613(1988)016<0330:PAEPPO>2.3.CO;2.

Talavera-Mendoza, O., Ramírez, E.J., and Guerrero, S.M., 1995, Petrology and geochemistry of the Teloloapan subterrane, a Lower Cretaceous evolved intra-oceanic island-arc: Geofísica Internacional, v. 34, p. 3–22.

Tardy, M., Lapierre, H., Freydier, C., Coulon, C., Gill, J.B., Mercier de Lepinay, B., Beck, C., Martinez, J., Talavera, M., Ortiz, E., Stein, G., Bourdier, J.L., and Yta, M., 1994, The Guerrero suspect terrane (western Mexico) and coeval arc terranes (the Greater Antilles and the Western Cordillera of Colombia): A late Mesozoic intra-oceanic arc accreted to cratonal America during the Creteaceous: Tectonophysics, v. 230, no. 1-2, p. 49–73, doi: 10.1016/0040-1951(94)90146-5.

Torres, R., Ruíz, J., Patchett, P.J., and Grajales, J.M., 1999, Permo-Triassic continental arc in eastern Mexico: Tectonic implications for reconstructions of southern North America, *in* Bartolini, C., et al., eds., Mesozoic sedimentary and tectonic history of north-central Mexico: Geological Society of America Special Paper 340, p. 191–196.

Tovar, R.J., 1968, Medición a detalle de la sección paleozoica expuesta en la Sierra de Teras Bavispe, Sonora: Petróleos Mexicanos, Superintendencia General de Exploración, Distrito Chihuahua, Informe Geológico, NE-M-1073, s/p.

Tristán-González, M., and Torres-Hernández, J.R., Geología del área de Charcas, Estado de San Luis Potosí, 1994, Universidad Nacional Autónoma de México: Instituto de Geología, Revista Mexicana de Ciencias Geológicas, v. 11, n. 2, p. 117–138.

Valencia-Moreno, M., 1998, Geochemistry of Laramide granitoids and associated porphyry copper mineralization in NW Mexico [Ph.D. thesis]: Tucson, University of Arizona,164 p.

Valencia-Moreno, M., Ruíz, J., and Roldán-Quintana, J., 1999, Geochemistry of Laramide granitic rocks across the southern margin of the Paleozoic North American continent, Central Sonora, Mexico: International Geology Review, v. 41, p. 845–857.

Valencia-Moreno, M., Ruíz, J., Barton, M.D., Patchett, P.J., Zürcher, L., Hodkinson, D., and Roldán-Quintana, J., 2001, A chemical and isotopic study of the Laramide granitic belt of northwestern Mexico: Identification of the southern edge of the North American Precambrian basement: Geological Society of America Bulletin, v. 113, p. 1409–1422, doi: 10.1130/0016-7606(2001)113<1409:ACAISO>2.0.CO;2.

Vidal-Serratos, R., 1991, Estratigrafía y tectónica de la región de Zihuatanejo, Estado de Guerrero, Sierra Madre del Sur, Convención sobre la evolución Geológica Mexicana: 1er Congreso Mexicano de Mineralogía, Pachuca, Memoir, p. 231–233.

Walker, J.D., 1988, Permian and Triassic rocks of the Mojave Desert and their implications for the timing and mechanisms of continental truncation: Tectonics, v. 7, p. 685–709.

Wardlaw, B.R., Furnish, W.M., and Nestell, M.K., 1979, Geology and paleontology of the Permian beds near Las Delicias, Coahuila, Mexico: Geological Society of America Bulletin, v. 90, no. 1, p. 111–116, doi: 10.1130/0016-7606(1979)90<111:GAPOTP>2.0.CO;2.

Yañez, P., Ruíz, J., Patchett, P.J., Ortega-Gutíerrez, F., and Gehrels, G., 1991, Isotopic studies of the Acatlan Complex, southern México: Implications for Paleozoic North American Tectonics: Geological Society of America Bulletin, v. 103, no. 6, p. 817–828, doi: 10.1130/0016-7606(1991)103<0817:ISOTAC>2.3.CO;2.

Zaldivar, R.J., and Garduño, M.V.H., 1984, Estudio estratigrafico y estructural de las rocas del Paleozoico Superior de Santa Maria del Oro, Durango, y sus implicaciones tectónicas [abs.]: Annual Meeting, Geologocal Society of Mexico, p. 37–38.

Manuscript Accepted by the Society 24 March 2005

Printed in the USA

Geological Society of America
Special Paper 393
2005

New data on the lithostratigraphy, detrital zircon and Nd isotope provenance, and paleogeographic setting of the El Antimonio Group, Sonora, Mexico

Carlos M. González-León
Instituto de Geología, ERNO, Universidad Nacional Autónoma de México, Apartado Postal 1039, Hermosillo, Sonora, México 83000

George D. Stanley Jr.
Department of Geology, University of Montana, Missoula, Montana 59812, USA

George E. Gehrels
Department of Geosciences, University of Arizona, Tucson, Arizona, 85721, USA

Elena Centeno-García
Instituto de Geología, Universidad Nacional Autónoma de México, México D.F. 04510, México

ABSTRACT

The El Antimonio Group is herein proposed as a new lithostratigraphic unit that encompasses the Antimonio, Río Asunción, and Sierra de Santa Rosa Formations in a revised nomenclature from Lucas and Estep (1999b). The type section for the Antimonio, Río Asunción, and the lower part of the Sierra de Santa Rosa Formations is located in the Sierra del Álamo, whereas the representative upper part of the Sierra de Santa Rosa Formation is located in the mountains of same name in northwestern Sonora. The ~4.5-km-thick sedimentary succession of this group is abundantly fossiliferous, and its biostratigraphic age is constrained between the Late Permian and Early Jurassic. The 3.4-km-thick section that crops out in the Sierra del Álamo is divided into 14 unconformity-bounded sequences that are tens to hundreds of meters thick and grade from the base upward from a fluvial to shallow marine conglomerate to open marine shale. The El Antimonio succession is correlated with several other Triassic and Jurassic sections that are known in Sonora, all of which are located south of the proposed trace of the Mojave-Sonora megashear. The closest Triassic and Lower Jurassic sections that are located north of the Mojave-Sonora megashear that we correlate with the El Antimonio are known in southern Nevada and southeastern California and include the Moenkopi, Virgin Limestone, Union Wash, Silverlake, and Fairview Valley Formations and the Kings sequence. On the basis of these proposed correlations, we suggest that the El Antimonio Group was deposited in an evolving

cmgleon@servidor.unam.mx

González-León, C.M., Stanley, Jr., G.D., Gehrels, G.E., and Centeno-García, E., 2005, New data on the lithostratigraphy, detrital zircon and Nd isotope provenance, and paleogeographic setting of the El Antimonio Group, Sonora, Mexico, *in* Anderson, T.H., Nourse, J.A., McKee, J.W., and Steiner, M.B., eds., The Mojave-Sonora megashear hypothesis: Development, assessment, and alternatives: Geological Society of America Special Paper 393, p. 259–282. doi: 10.1130/2005.2393(09). For permission to copy, contact editing@geosociety.org.

shallow shelf (Upper Permian–Triassic) to fore-arc basin (Lower Jurassic) that was originally positioned adjacent to southern California and later translated to its present position, along with the Caborca block, by left-lateral Jurassic displacement of the Mojave-Sonora megashear. In this proposed paleogeography, a lower Mesozoic magmatic arc that accumulated volcanic, volcaniclastic, and shallow marine sedimentation in the Mojave Desert and along the California-Nevada border separated the El Antimonio basin from a shallow shelf that developed to the north.

New U-Pb geochronology on detrital zircon and Sm/Nd isotope and petrographic data from terrigenous samples of the El Antimonio Group may help to elucidate its provenance and to support this paleogeography. Zircon grains from samples of the lower, middle, and upper parts of the El Antimonio Group yielded ages that cluster around 1.8, 1.6–1.7, 1.4, and 1.00–1.18 Ga and 340, 270–240, and 190 Ma. The Proterozoic zircons are interpreted to indicate provenance from the basement provinces of the southwestern United States, although a reworked source for these grains is also possible as they are present in the Cordilleran miogeocline and off-shelf assemblages of Nevada and California and in Proterozoic and Paleozoic strata in Sonora. The closest known sources for the Permian and Lower Triassic zircons are plutons and volcanic rocks that formed a lower Mesozoic magmatic arc extending from southeastern to northern California and western Nevada. Probable sources for the single zircon grain dated at 340 Ma are the Sierra-Klamath terranes, according to interpretation by other authors of grains of similar age in rocks of Nevada. Grains dated around 190 Ma in the youngest sample most probably reveal provenance from the Lower Jurassic magmatic arc of southeastern California or southern Arizona.

The Sm/Nd isotopic data from three samples of the lower, middle, and upper parts of the El Antimonio Group indicate a progressive decrease in model ages, from the base upward (T_{DM} = 1.9–1.8 to T_{DM} = 1.13 Ga) of this succession, indicating a most probable derivation from the Yavapai and Grenville provinces in the southwestern United States. Sandstone and conglomerate clast composition in the El Antimonio Group indicate mixed sources of provenance from sedimentary and vulcanoplutonic origin. These most probably correspond to the Proterozoic and Paleozoic sedimentary successions of southwestern North America and to the Triassic-Jurassic magmatic arc of this same region, respectively.

Keywords: Upper Permian, Triassic–Lower Jurassic, El Antimonio Group, Sonora, Mexico.

INTRODUCTION

An Upper Permian–Triassic and Lower Jurassic sedimentary succession previously referred to as the Antimonio Formation (González-León, 1980, 1997a) in northwestern Sonora, Mexico, may represent the most complete lower Mesozoic stratigraphic record of the North America Cordillera. This section crops out around the old El Antimonio mine, in the northern foothills of the Sierra del Álamo mountain range, near Caborca in northwestern Sonora (Fig. 1). The Antimonio rocks were considered part of the allochthonous Antimonio terrane (González-León, 1989), and because of its thickness and relative completeness, this succession stands as an important comparative section for both the craton and for other outboard terranes. It is important also in our goals of achieving a better understanding of the depositional and tectonic processes operating along the southwestern margin of North America from the late Permian to the Jurassic.

The succession is well exposed in the Sierra del Álamo (Fig. 1), although other less complete and correlative sections crop out in several ranges that extend to the south and into central Sonora. In the Sierra del Álamo, the Antimonio succession unconformably overlies the Permian (Guadalupian) Monos Formation (Cooper and Arellano, 1946), but in other localities it is either in fault contact or its relationships with other rocks is unclear. Based on previous work of González-León (1997a), the 3.4-km-thick stratigraphic succession is composed of 14 fining-upward, unconformity-bounded sequences (Fig. 2).

This report summarizes the stratigraphy of the Upper Permian–lower Mesozoic Antimonio strata, proposes a revision to its lithostratigraphic nomenclature, and presents new data from detrital zircon U/Pb geochronology, petrographic studies, and whole-rock Nd isotope analysis of sandstone and shale samples. We also include new observations from the lithostratigraphy and paleontology of other Triassic and Juras-

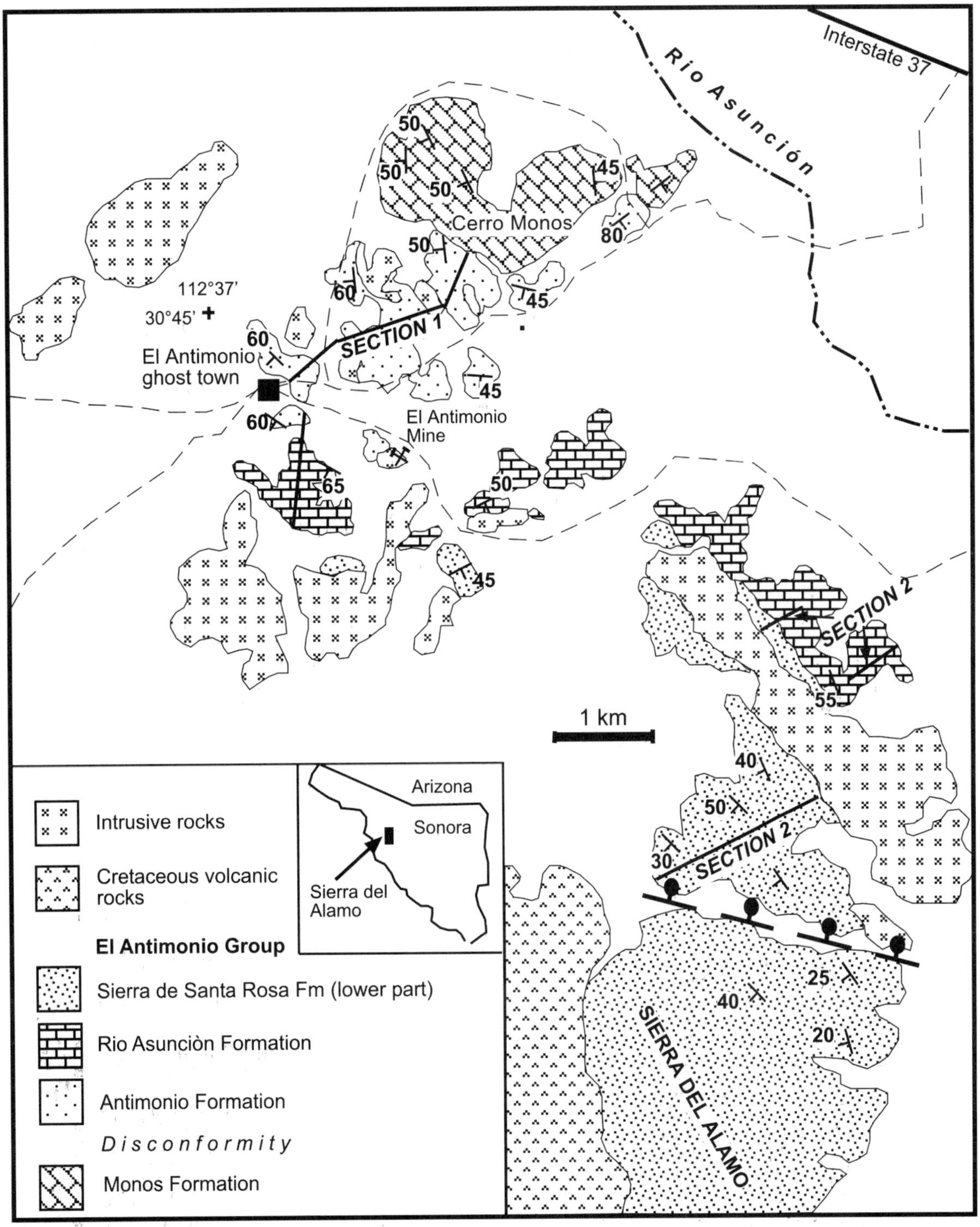

Figure 1. Geologic map of the northern part of the Sierra del Álamo, northwestern Sonora (modified from González-León, 1997a), showing formations of the El Antimonio Group and relationships to older and younger rocks. Section lines indicate location of measured type sections detailed in Figure 2.

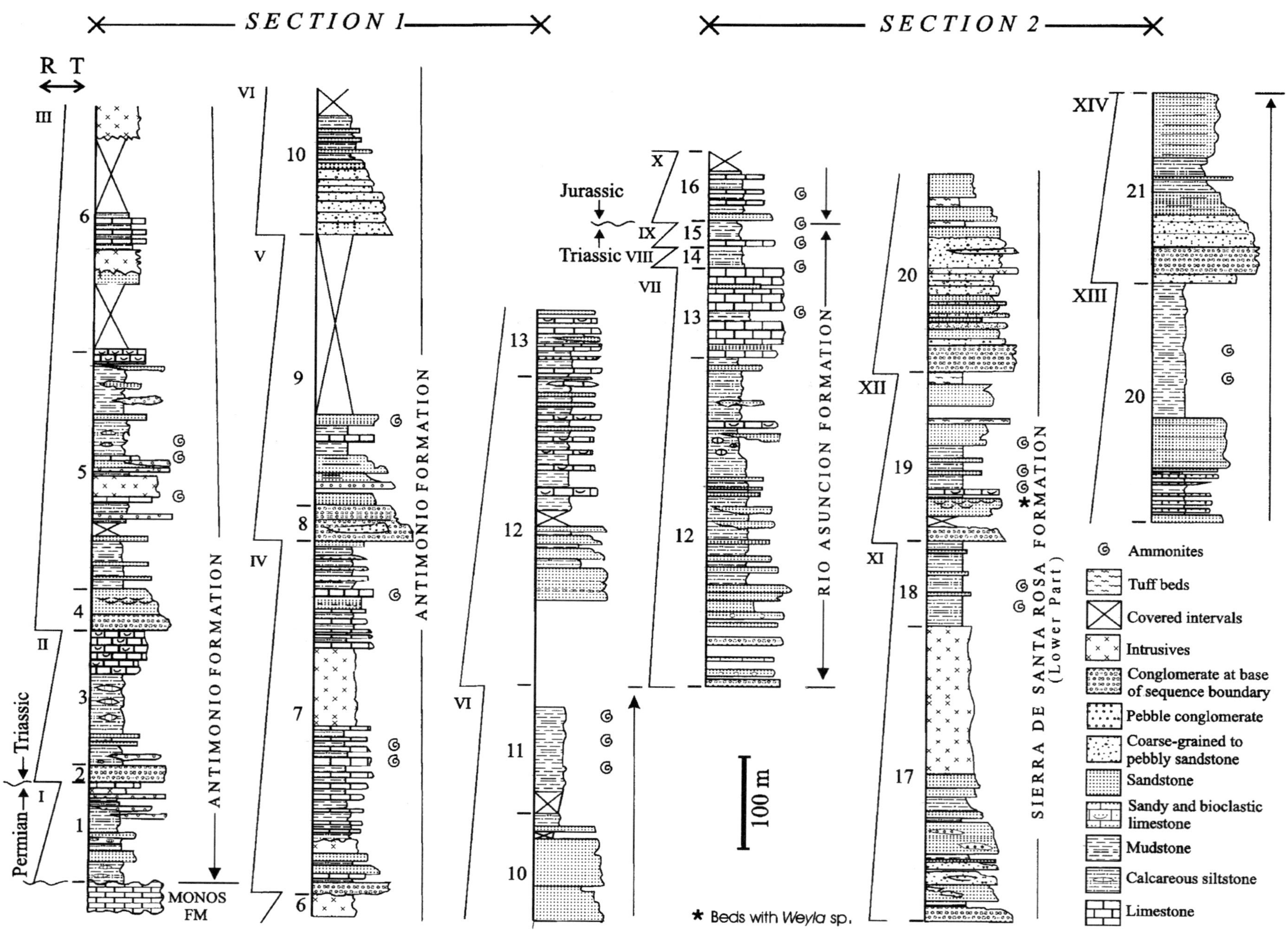

Figure 2. Stratigraphic type section of the El Antimonio Group in the northern part of the Sierra del Álamo, Sonora. See Figure 1 for location of measured sections 1 and 2. R-T—regression-trangression. Roman numerals indicate sequences; numbers designate units.

sic sections of Sonora and Baja California to propose regional correlations. We argue that these new data enable a better understanding of the original early Mesozoic paleogeographic position of these rocks and allow reexamination of the role that the Mojave-Sonora megashear may have played in the tectonic evolution of the whole succession.

STRATIGRAPHY

A detailed stratigraphic analysis of the thick Antimonio succession was undertaken by careful measurement of the whole 3.4-km-thick section at the Sierra del Álamo locality (González-León, 1997a). Fourteen unconformity-bounded sequences divide the section, and these have been numbered from the base upward (Fig. 2). Most of these sequences start with a basal, fluvial, or marginal marine conglomerate and are succeeded upward by shallow marine fine-grained sandstone, siltstone, and limestone. Some of them contain in their upper part finer-grained lithologies representing deeper marine environments. In the uppermost Jurassic part of the section, the Antimonio succession undergoes a transition from carbonate and terrigenous fine-grained lithologies to predominantly terrigenous deposition.

Sequences I through VII were measured along section 1 (Fig. 1) in the northern part of the Sierra del Álamo, in the low hills between Cerro Monos and the old El Antimonio mine. Although section 1 is poorly exposed and cut by younger intrusive rocks, a section where strata consistently dip to the southwest and where the age of the sequences are in temporal succession was measured. Sequences VII through XIV were measured along section 2 in the northeastern part of the mountain range where outcrops are well exposed.

The 100-m-thick sequence I consists of thin-bedded calcareous red siltstone with calcareous nodules, minor thin-bedded limestone, subordinate medium- to thick-bedded, fine-grained sandstone, and occasional thin, lenticular beds of pebble conglomerate. Clasts in the conglomerate are of sedimentary (chert, limestone, and quartzite) and igneous (quartz porphyry and granitic) rocks. Brachiopod fossils identified by Lucas et al. (1997a) from the red siltstone in the lower part of this sequence belong to *Orbiculoidea* sp., *Composita* sp., and *Spiriferellina sonorensis* Cooper. The same fossils are also found in the underlying Monos Formation (Cooper, 1953), which is assigned to the Guadalupian in the Upper Permian. On this basis, Lucas et al. (1997a) assigned sequence I to the Guadalupian.

The 115-m-thick sequence II contains in its lowermost part a 15-m-thick, massive, clast-supported, pebble conglomerate of interpreted fluvial origin that is composed of poorly sorted and subangular clasts of limestone, chert, rhyolite, and granite. It is overlain by a shallow-marine succession of calcareous red siltstone with rare shell fragments and intercalated thin beds of fine-grained sandstone, whereas its upper part is composed of thin- to medium-bedded, muddy to sandy limestone with occasional shell remains. The Permian-Triassic boundary was placed tentatively at the base of this sequence by Lucas et al. (1997a).

Sequence III is 500 m thick and its base is formed by a 15-m-thick, poorly sorted, clast-supported, pebble to boulder conglomerate of fluvial origin. This is an upward-fining conglomerate with subrounded to subangular clasts of limestone, granitic rocks, chert, quartzite, and sandstone. Paleozoic limestone clasts with brachiopods, bryozoans, and the large fusulinid *Parafusulina antimonioensis* [whose type species was described by Dunbar (1953) from the underlying Monos Formation] are abundant. The sequence continues upward with coarse- to fine-grained, trough cross-stratified sandstone, red siltstone, and sandy to muddy limestone intervals that become dominant toward the upper part of the sequence. Poorly preserved ammonoids and conodonts from the limestone intervals indicate this sequence dates from the Spathian (Lucas et al., 1997b).

Sequence IV is 330 m thick and contains at its base a clast-supported, chert-pebble conglomerate of fluvial origin overlain by a succession of interbedded, fine-grained, red sandstone, siltstone, and limestone beds of shallow to deep marine origin. Ammonoids from this sequence indicate it is from the Anisian in the Middle Triassic (Estep et al., 1997a). The basal unit of the overlying sequence V is a fining-upward, poorly sorted, clast-supported conglomerate that grades upward to a succession of red sandstone, siltstone, and limestone with aulacoceratids, ammonoids, and ichthyosaur bones. Although the upper part of this sequence is covered, its measured thickness is 323 m, and it dates probably from the Ladinian or Carnian (González-León, 1997a).

The 440-m-thick sequence VI is poorly exposed. Its lower part is defined by a shallow marine succession of pebble- to coarse-grained pebble sandstone that grades upward to fine- and medium-grained sandstone and siltstone. In its upper part, this sequence contains a fossiliferous unit of interbedded light gray to reddish, massive to finely laminated mudstone, encrinitic limestone, fossiliferous packstone, and minor fine-grained sandstone. Based on previous reports (Keller, 1928; Burckhardt, 1930; Callaway and Massare, 1989; Lucas and González-León, 1995; Estep at al., 1997b) of its abundant fossils, mostly ammonoids, this sequence is assigned to the Carnian.

The basal part (unit 12, Fig. 2) of the 390-m-thick sequence VII is a fluvial succession of fine- to coarse-pebble conglomerate and coarse-grained to pebble sandstone and siltstone with fossil logs, which grades upward to a succession of calcareous sandstone and coquinoid limestone of shallow marine origin. The limestone beds have yielded the bivalves *Mysidioptera mexicana, Palaeolopha* cf. *haidingeriana* (Emmrich), and *Myophoriopsis sonorensis* (Damborenea and González-León, 1997). The upper part of the sequence (unit 13, Fig. 2) is composed of thick-bedded richly fossiliferous biostromal limestone. It contains interbedded mudstone, fine-grained sandstone, and dolomitic limestone. Its abundant fauna include scleractinian corals, chambered sponges, ammonoids, bivalves (*Mysidioptera mexicana,* among others), and other reefal organisms (Stanley et al, 1994; González-León et al., 1996; McRoberts, 1997; Damborenea and González-León, 1997; Goodwin and Stanley, 1997; Stanley and González-León, 1997; Goodwin; 1999). Unit 13 contains ammonoids, *Heterastridium* and other fossils that date from the upper Norian (Columbianus and Cordilleranus zones). Because of the presence of *Mysidioptera* (a Norian genus that lived in the Tethyan region but also is known from several Upper Triassic localities in North America, according to Damborenea and González-León, 1997) in both units 12 and 13, we consider unit 12 to date also from the Norian.

Sequence VIII is composed of a 17-m-thick unit of massive calcareous siltstone with bivalves, crinoids, and ammonoids from the Rhaetian (Amoenum Zone). The overlying sequence IX ranges in thickness from 27 to 43 m. Its base is erosional and characterized by a lenticular interval 2 m to 8 m thick of distinctive reddish, bioclastic to sandy limestone with an abundant, diverse, reworked fauna that includes ammonoids, bivalves, colonial corals, nautiloids, crinoids, and bones and teeth most likely attributed to ichthyosaurs and fish. This interval grades upward to calcareous siltstone, fine-grained sandstone, and laminated to massive mudstone with a fauna characterized by Rhaetian ammonoids that include *Choristoceras* (González-León et al., 1996).

Sequence X is 60 m thick and contains the first Jurassic strata. It is composed in its basal part of a laterally continuous, base-erosive, and coarse-grained to pebble sandstone that grades upward to mudstone/siltstone and interbedded fossiliferous limestone of shallow marine origin. This sequence contains abundant ammonoids from the uppermost Hettangian to the lowermost Sinemurian (Canadensis and Trigonatum Zones) (González-León et al., 1996; Taylor et al., 2001). The boundary between sequences IX and X is an erosive, disconformable surface along which most of the upper Rhaetian and lower Hettangian strata are missing. The Triassic-Jurassic boundary was proposed to fall at this erosional surface (González-León, 1997b; Lucas and Estep, 1999a; González-León et al., 2000).

The 280-m-thick sequence XI presents in its lower part a 20-m-thick, fining-upward, clast-supported, pebble conglomerate with clasts of quartzite, volcanic, and intrusive rocks, and chert. This conglomerate grades upward to interbedded coarse-grained sandstone and siltstone of fluvial origin. Most of the middle part of the sequence is cut out by a dioritic intrusive, and its upper part consists of massive to finely laminated calcareous mudstone and siltstone with ammonoids from the lower Sinemurian (Leslei Zone) (Taylor et al., 2001).

Sequence XII is 115 m thick. Its base is a 5-m-thick unit of massive, lenticular, pebble-conglomerate that is overlain gradationally by coarse-grained, thick-bedded, cross-stratified sandstone and sandy coquina. The upper part of this sequence is composed of massive mudstone/siltstone with intercalations of thin-bedded bioclastic limestone, fine-grained sandstone, and thin volcanic ash tuffs beds. The bioclastic limestone contains bivalves (*Weyla* sp., among others) and ammonoids from the lower part of the upper Sinemurian (Carinatum and Jamesi Zones; Pálfy and González-León, 2000; Taylor et al., 2001).

Sequence XIII is 465 m thick. Its lower part is composed of an amalgamated succession of thick beds of pebble conglomerate, pebbly sandstone, coarse-grained sandstone, and minor interbeds of mudstone/siltstone and tuff beds. The middle and upper parts of the sequence are dominated by fine-grained sandstone and thin-bedded to laminated mudstone and siltstone with sedimentary and biogenic structures that indicate sedimentation in basinal settings with turbidite currents. Some turbidite beds contain poorly preserved ammonites that may represent the upper Sinemurian (Pálfy and González-León, 2000).

The lower part of sequence XIV is formed of amalgamated, very thick beds of clast-supported, pebble conglomerate that grades upward to a succession of sandstone and interbedded siltstone. Clasts in the conglomerate that is interpreted as of fluvial origin include quartzite, chert, limestone, and volcanic rocks. The upper part of this 150-m-thick sequence is a succession of medium- to coarse-grained, amalgamated sandstone beds of probable shallow marine origin. This sequence is not dated because fossils were not found, but we consider it date from the Lower Jurassic as are subjacent units of the same succession.

PROPOSED LITHOSTRATIGRAPHIC REVISIONS TO THE ANTIMONIO SUCCESSION

While the stratigraphic succession of the Sierra del Álamo was first assigned to the Antimonio Formation by González-León (1980), some previous authors had described it partially (Cooper and Arellano, 1946; White and Guiza, 1949). González-León (1980) divided the Antimonio Formation into a lower member from the Upper Triassic and an upper member from the Lower Jurassic. Further lithostratigraphic analysis and paleontologic refinement indicated that the Antimonio Formation actually ranges in age from Upper Permian to Early Jurassic (González-León, 1997a; Lucas et al., 1997a). More recently, Lucas and Estep (1999b) proposed to divide the Antimonio into three formations: the Antimonio Formation, the Río Asunción Formation, and the Sierra de Santa Rosa Formation. Additionally, these authors recognized the so-called "*Mysidioptera* beds" as an informal unit.

The Antimonio Formation of Lucas and Estep (1999b) comprises sequence I through sequence VI as described herein, encompassing Upper Permian (Guadalupian) to Upper Triassic (Carnian) strata. The Río Asunción Formation of Lucas and Estep (1999b) begins at the base of the biostromal limestone (unit 13, Fig. 2) in the upper part of sequence VII and continues up through sequence X, whereas their Sierra de Santa Rosa Formation comprises sequences XI to XIV (Fig. 2). The Lucas and Estep scheme for subdivision of the Antimonio succession is reasonable in that it distinguishes three units, each of which can be recognized easily on the basis of a unique lithostratigraphy. We make some revisions herein to the scheme of these authors, however.

Lucas and Estep (1999b) introduced the term "*Mysidioptera* beds" to refer to the sedimentary succession of unit 12 in sequence VII described herein. Although these authors recognized that "...these strata are younger than the upper Carnian strata..." (p. 276) of sequence VI and that a "...general consistency of dip and correct temporal succession suggest that the *Mysidioptera*-bearing strata overlie the upper Carnian strata..." (p. 276), they did not assign these strata to a formation, because

"...lower and upper depositional contacts are unknown and exposures limited..." (p. 276). Previously, Lucas et al. (1996) had proposed a thrust(?) fault contact between the afore mentioned "*Mysidioptera* beds" and the underlying Carnian strata, based mostly on the misidentification of the bivalve *Mysidioptera* as the Lower Jurassic bivalve *Weyla* (Lucas and Estep, 1997, Figs. 1C and 1D, p. 45; see corrections by Damborenea and González-León, 1997). This misleading identification as *Weyla,* however, still persisted in the work of Lucas and Estep (1999b) where "*Weyla* beds" were identified in strata corresponding to unit 13 (p. 278, Fig. 6 therein), and also in their Table 1 (p. 277) where they listed *Mysidioptera* as a Lower Jurassic fossil. Lucas and Estep (1999b) also interpreted the top of the "*Mysidioptera* beds" as a major, continuous thrust fault contact with the overlaying biostromal limestone of unit 13.

Although the contact between sequences VI and VII is poorly exposed, we interpret both sequences as part of the same stratigraphic succession, based on their continuous consistency of dip and correct temporal succession. Similarly, our field investigations fail to confirm the presence of any major tectonic dislocation in the otherwise continuous stratigraphic succession between unit 12 (the "*Mysidioptera* beds") and unit 13. The bivalve assemblage identified in unit 12 of sequence VII is not sufficient to assign an age more precise than Upper Triassic, but we consider it to be Norian based on the fact that *Mysidioptera mexicana,* present in unit 12, also occurs in the overlying unit 13 along with ammonoids from the Norian. Therefore, we recognize that units 12 and 13 are part of the same continuous, sedimentary succession of sequence VII.

Such conclusions led us to make the following modifications to the stratigraphic nomenclature. We follow the concept of a redefined Antimonio Formation of Lucas and Estep (1999b), as the sedimentary succession that includes sequences I (Guadalupian) thorough VI (Carnian). However we abandon Lucas and Estep's "*Mysidioptera* beds," recognize it as unit 12, and include it along with unit 13 in sequence VII (see also González-León, 1997a). We modify the Río Asunción Formation of Lucas and Estep (1999b) to include sequences VII, VIII, and IX (Norian–Rhaetian) in its succession but not sequence X, contrary to Lucas and Estep. We believe that the disconformity that separates sequences IX and X, which is also the boundary between the Triassic and Jurassic strata in this section, is a distinctive feature that places the boundary between the Río Asunción and Sierra de Santa Rosa Formations (Fig. 2).

The uppermost part of the Sierra del Álamo succession that includes sequences X to XIV (uppermost Hettangian to lower upper Sinemurian) does not correlate lithologically with the stratotype of the Sierra de Santa Rosa Formation exposed in the locality of the Sierra de Santa Rosa Mountains (Fig. 3). As originally defined by Hardy (1981) and later reported by Damborenea and González-León (1997), the stratotype of the Sierra de Santa Rosa Formation is a mostly terrigenous marine succession with ammonites that indicate it ranges from the uppermost Sinemurian to the lower Pliensbachian (see discussion below). The bivalve content of the two sections is also different, but both have in common the presence of *Weyla (W.) alata* (von Buch) (Damborenea and González-León, 1997). Because both sections apparently straddle the upper Sinemurian and have in common the presence of *Weyla* (an index fossil from the Hettangian(?)-Sinemurian to lower Toarcian in the Americas—see Aberhan, 1994, 1998, and Damborenea and González-León, 1997), we recognize the Lower Jurassic rocks at the Sierra del Álamo as the lowermost part of the Sierra de Santa Rosa Formation, following the proposal of Lucas and Estep (1999b).

Furthermore, we propose a new larger lithostratigraphic unit composed of the Antimonio, Río Asunción, and Sierra de Santa Rosa Formations that we name herein the El Antimonio Group. The stratotype of this group is located in the northern foothills of the Sierra del Álamo where the type sections of the Antimonio, Río Asunción, and lower part of the Sierra de Santa Rosa Formations were measured (Fig. 1), and in the central part of the Sierra de Santa Rosa where the type section of the upper part of the Sierra de Santa Rosa Formation crops out [following the lithostratigraphy of Hardy (1981) and Damborenea and González-León (1997)].

OTHER CORRELATIVE TRIASSIC AND JURASSIC STRATA

Several successions of marine Triassic and Lower Jurassic strata crop out at other localities in northwestern and central Sonora, including the Barra Los Tanques, Cerro La Flojera, Sierra Santa Teresa, Pozos de Serna, Sierra de Santa Rosa, Cerro Basura, Cerro Los Chinos, and Sierra de López localities (Fig. 3). Possible correlations of these successions, as well as two localities from Baja California (El Volcán and Vizcaino Peninsula areas), to parts of the El Antimonio Group are described below (Fig. 4).

Barra Los Tanques is located 35 km southwest of the Sierra del Álamo. Goodwin (1999) reported from that area a 45-m-thick section of bedded limestone, minor dolomite, lenticular reefoid carbonate buildups, and limy sandstone beds with scleractinian corals, sponges [including a new sphinctozooid sponge reported by Senowbari-Daryan et al. (2001)], and conodonts from the Norian. According to its lithostratigraphy and coral fauna, this section is similar to unit 13 of sequence VII herein described (Goodwin, 1999) (Fig. 4). The Barra Los Tanques Norian section has been interpreted as tectonically overlying Paleozoic(?) strata (Stanley and González-León, 1995) and is tectonically overlain by a thick section of Jurassic(?) conglomerate.

Upper Triassic marine strata are present in small outcrops just southeast of Hermosillo (Fig. 3). In Cerro La Flojera, an 80-m-thick section of fine-grained sandstone, siltstone, and limestone and minor pebble-conglomerate contains ammonites from the upper Carnian (Lucas and González-León, 1994). In Sierra Santa Teresa a 544-m-thick terrigenous and carbonate section contains a fossiliferous assemblage (sponges, corals, and ammonoids, among others) from the Upper Triassic (Norian) and Lower Jurassic(?) (Stewart et al., 1997a; Goodwin and Stanley, 1997; Goodwin, 1999) (Figs. 3 and 4).

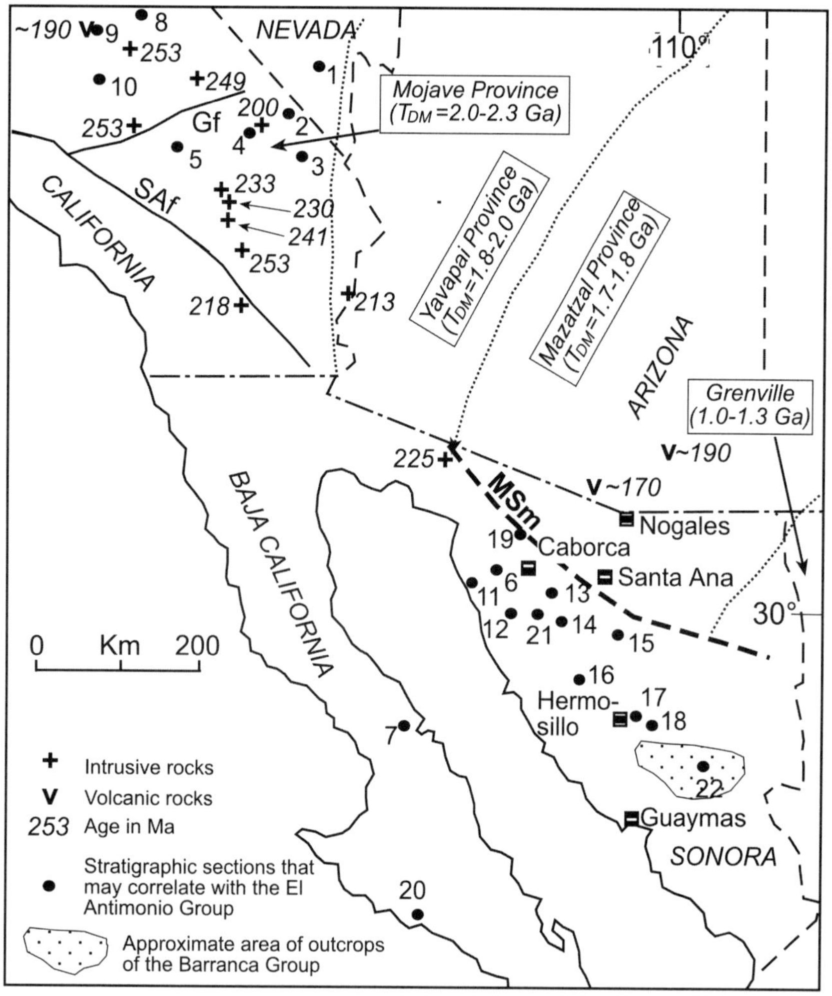

Figure 3. Map showing location of outcrops of the El Antimonio Group in the Sierra del Álamo, and other possible correlative Triassic and Jurassic sections in Sonora, Baja California, and the Mojave Desert. 1—Spring Mountains, 2—Clark Mountains, 3—Providence Mountains, 4—Soda Mountains, 5—Black Mountains, 6—Sierra del Álamo, 7—El Volcán area, 8—Inyo Mountains, 9—Mineral King, 10—Southern Sierra, 11—Barra Los Tanques, 12—Pozos de Serna, 13—Cerro Los Chinos, 14—Sierra de Santa Rosa, 15—Sierra Caracáhui, 16—Sierra de López, 17—Cerro La Flojera, 18—Sierra Santa Teresa, 19—Cerro Basura–Puerto El Álamo, 20—Vizcaino, 21—Sierra La Jojoba, 22—Sierra San Javier. Sections from southwestern United States are redrawn from Walker (1987) and Saleeby and Busby-Spera (1992). Also indicated are Permian, Triassic, and Lower Jurassic igneous rocks and their ages in Ma (modified from Fig. 1 in Barth et al., 1990; Riggs et al., 1993; Barth et al., 1997, 2001). Boundaries and ages of Proterozoic provinces taken from Farmer et al. (2001). MSm—inferred trace of the Mojave-Sonora megashear, SAf—San Andreas fault, Gf—Garlock fault. T_{DM}—Nd model ages.

Other Triassic localities in northwestern Mexico that correlate with the El Antimonio succession are the Del Indio Formation and the San Hipólito Formation in Baja California. The Del Indio Formation (Philip, 1984) is defined as a small outcrop from the Lower Triassic (Smithian) located in the eastern part of the northern Baja California Peninsula (El Volcán area, Fig. 3), ~250 km southwest of the Sierra del Álamo (Fig. 3). It is composed of chert pebble conglomerate, sandstone, limestone, and argillite that disconformably rest on Upper Permian turbiditic strata (Delattre, 1984; Buch, 1984). Outcrops of the 2.4-km-thick San Hipólito Formation (Finch and Abbott, 1977) occur in the Vizcaino Peninsula on the Pacific coast of Baja California (Fig. 3). The lowermost "chert member" of the San Hipólito Formation conformably (?) overlies interbedded pillow lavas and limestone beds that contain fossils from the Upper Triassic. The "chert member" and the overlying "limestone member" contain radiolarians from the Norian (Whalen and Pessagno, 1984). The "breccia member" is formed from olistolithic blocks of shallow water limestone up to tens of meters in diameter and interbedded basaltic lava flows. The breccia limestone clasts contain a representative fauna from the Norian. Most of these limestone blocks were derived from a carbonate platform whose exact paleogeographic location remains unknown, although according to our observations (C.M.G.-L. and G.D.S, Jr.), scleractinian corals, sponges, and *Heterastridium conglobatum* found in these clasts are the same species as those found in unit 13 of the Río Asunción Formation. The overlying "sandstone member" contains radiolarians from the Lower Jurassic (Whalen and Pessagno, 1984). Considering that the Baja California peninsula was part of Sonora prior to the Miocene and that the basaltic lava flows of the San Hipólito Formation are of oceanic-arc affinity (Centeno-García and Valencia-Moreno, 2001), we speculate that this region might have originated relatively close to the El Antimonio basin.

An incomplete Lower Jurassic sedimentary succession crops out at the Pozos de Serna area (Figs. 3 and 4). The lowermost part of this section is composed of interbedded coarse-grained sandstone and pebble conglomerate, ~100 m thick, that is separated from the upper part of the section by a covered interval ~100 m thick. The lowermost 50 m of this upper section is composed of thin-bedded, fine-grained, calcareous sandstone that is overlain by

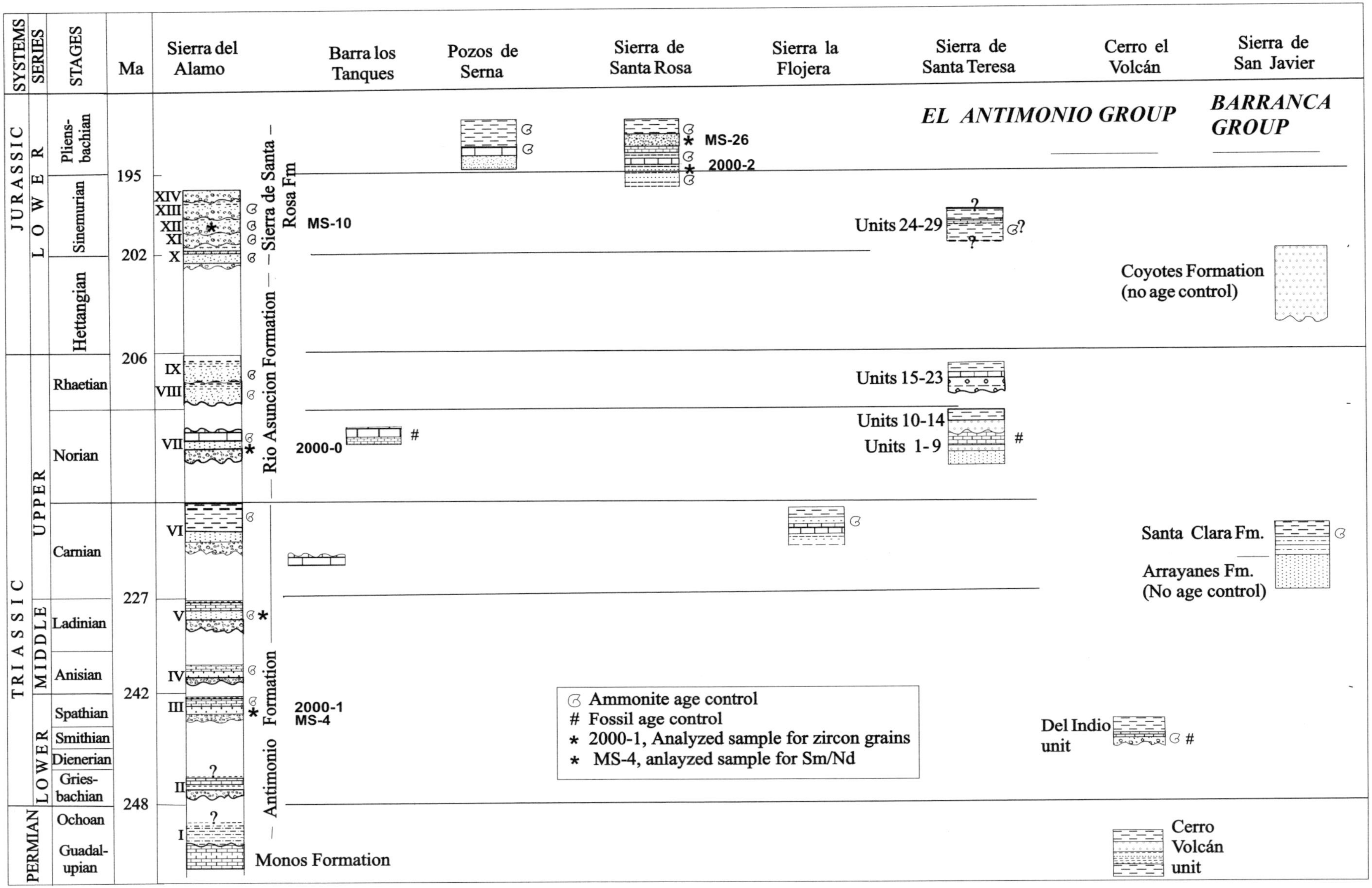

Figure 4. Regional chronocorrelation of Permian–Triassic–Lower Jurassic strata of the El Antimonio Group in Sonora and Baja California. Also shown is the stratigraphic position of sandstone and shale samples analyzed for U/Pb detrital zircons and Sm/Nd isotopic composition. Ages according to the 1999 Geologic Time Scale of The Geological Society of America (compiled by Palmer and Geissman, 1999). Patterns are same as in Figure 2.

75 m of calcareous shale with ammonoids and bivalves. The section is followed upward by an interval 30 m thick of reddish brown, cross-stratified, sandy limestone beds with interbedded siltstone. The middle part of the upper section is dominated by thin-bedded to laminated calcareous shale with rare thin interbeds of dark-colored, micritic limestone where an interval with ammonoids is present. The uppermost 150 m of the section is composed of fine- to medium-grained, thin- to medium-bedded calcareous sandstone with interbedded thick beds of sandy limestone and minor intervals of blue shale. According to Dowlen and Gastil (1981) and Calmus et al. (1997), the Pozos de Serna section is faulted against Proterozoic sedimentary rocks, but according to Lucas et al. (1999), its lower part is in depositional contact with Proterozoic limestone. Our field observations suggest that the contact is a normal fault (González-León, 1997a). Preliminary identifications of an assemblage of ammonites from the lowermost fossiliferous interval indicate it is dominated by several species of *Tropidoceras* indicative of the lower Pliensbachian (Pálfy and González-León, 2000). On the basis of ammonoid identifications, other authors have assigned this section to the Upper Jurassic (Beauvais and Stump, 1976), Lower Jurassic (lower Pliensbachian) (Linares et al., 1997), Sinemurian to Bajocian-Bathonian (Calmus et al., 1997), and uppermost Sinemurian (Lucas et al., 1999).

A Lower Jurassic section in the Sierra de Santa Rosa Mountains (Fig. 4) was first studied by Hardy (1981), who named it the Sierra de Santa Rosa Formation and divided it into lower, middle, and upper members. Hardy (1981) reported from this locality a 1460-m-thick section of fossiliferous, shallow- to deep-marine sedimentary strata, the base of which is not exposed, that tectonically overlies a Precambrian metamorphic basement. Its rich invertebrate faunas are dominated by ammonoids and bivalves that have been previously studied by Keller (1928), Jaworski (1929), Flores (1929), Hardy (1981), and Lucas and Estep (1997). Damborenea and González-León (1997) reported the details of the stratigraphy in a taxonomic study of the diverse fauna of this section with 16 bivalve taxa (including *Weyla, Falcimytilus, Pinna, Modiolus, Pholadomya, Myoconcha, Ceratomya, Gresslya,* and *Platymyoidea,* among others). Rare, solitary corals from the middle member of this unit, previously known from the Hettangian and Sinemurian of Europe, are herein identified as *Montlivaultia haimei* (Fig. 5). Two distinctive and as yet unstudied reeflike carbonate buildups (A and B, Fig. 6) were recently discovered by two of us (C.M.G.-L. and G.D.S, Jr.) in the middle member of this formation. Owing to the nearly world-wide, post–mass-extinction eclipse of reef faunas during much of the Early Jurassic, this discovery seems to be significant. The stratigraphically lower carbonate buildup A is 30 m thick and composed of massive limestone with abundant colonial and branching corals and subordinate bivalves. The uppermost carbonate interval is 25 m thick and in its middle part contains a 4-m-thick boundstone formed by branching and platy coral colonies that reach up to 2 m high. These carbonate bodies are separated by a 15-m-thick interval of medium- to coarse-grained sandstone with interbedded mudstone. Colonial scleractinian corals dominate the fauna, and some of the fossils are partly silicified. The larger corals and other reefal fossils are presently part of an ongoing study to understand more about the taxonomic composition and paleogeography of the fauna.

Ammonites collected from this succession in the Sierra de Santa Rosa locality indicate that it ranges in age from the uppermost Sinemurian to the lower Pliensbachian. According to Pálfy and González-León (2000), the lower part of the unit yielded *Paltechioceras* from the uppermost Sinemurian Harbledownense Zone, the middle part of the unit has yielded *Metaderoceras* that most likely belongs to the lower Pliensbachian, whereas the upper part of the formation yielded an assemblage of *Metaderoceras*, *Dubariceras,* and early species of *Protogrammoceras* that represent the highest part of the lower Pliensbachian. Other outcrops of possible Lower Jurassic rocks that are close to the Sierra de Santa Rosa occur in Sierra la Jojoba (Fig. 3) from where Pálfy and González-León (2000) reported a preliminary identification of *Dactylioceras* from the lower Toarcian.

The Sierra de Santa Rosa Formation is conformably overlain by the mostly volcanic succession of the Cerro San Luis Formation that is at least 900 m thick (Hardy, 1981). It has at its base a conglomerate up to 70 m thick with clasts of quartzarenite and dolomite. The upper part is composed of a heterogeneous assemblage of andesitic agglomerate, lava flows, lapilli tuffs, volcanic breccia, and volcanic arenite. Hardy (1981) speculated that the Cerro San Luis Formation probably dates from the Lower to Middle Jurassic, but no positive proofs of its age are yet available.

Small outcrops of Lower Jurassic rocks were reported from the localities of Cerro Basura, Sierra de Caracáhui, and Sierra

Figure 5. Illustration of the coral *Montlivaultia haimei* collected from the middle member of the Sierra de Santa Rosa Formation in Sierra de Santa Rosa

de López (Fig. 3). In the Cerro Basura (Fig. 3) it crops out as intensely deformed volcanic and volcaniclastic rocks of probable Jurassic age that are arranged in a complex of allochthonous sheets and folds (Corona, 1980). In one locality at Cerro Basura within this deformed complex, a Lower Jurassic (Sinemurian) ammonite (*Vermiceras* sp.) has been found (cited in Jacques-Ayala, 1993). From Sierra de Caracáhui (Fig. 3), Flores (1929) reported Lower Jurassic rocks with the bivalve *Weyla*. At Sierra de López (Rancho Placeritos area), a Lower Jurassic section consists of a succession of siltstone, sandstone, conglomerate, limestone, and dolostone with Sinemurian ammonites (Poole et al., 2000) that unconformably overlie Mississippian strata. Similarly, according to Stewart et al. (1997a), probable Upper Triassic and Jurassic(?) strata with poorly preserved ammonites appear to rest depositionally on Devonian carbonate rocks at Sierra El Carnero located ~15 km north of Sierra de López. Also, close to this locality and north of Hermosillo, Cambrian strata are thrust over Lower Jurassic sedimentary rocks (Morales-Montaño et al., 1990).

Other Triassic rocks of mixed continental and restricted marine character crop out in several sections in central Sonora and are best exposed in Sierra San Javier (Figs. 3 and 4). This sedimentary succession belongs to the 3-km-thick Barranca Group that comprises the Arrayanes, Santa Clara, and Coyotes Formations (Alencaster, 1961a, 1961b; Stewart and Roldán-Quintana, 1991). The Arrayanes Formation is a reddish terrigenous unit formed of sandstone and interbedded siltstone and pebble to cobble conglomerate of probably fluvial origin (Stewart and Roldán-Quintana, 1991). The Santa Clara Formation gradationally overlies the Arrayanes and is composed of interstratified siltstone, shale, sandstone, conglomerate, carbonaceous shale, coal beds, and rare thin tuff layers. Plant fossils that are common in the shale and siltstone and a local marine fauna of brachiopods, bivalves, and ammonites indicate this unit is a deltaic to marginal marine deposit from the Upper Triassic (Carnian and probably Norian) (Alencaster, 1961a, 1961b; Stewart and Roldán-Quintana, 1991; Weber, 1997). The Coyotes Formation is a 600-m-thick, dominantly pebble-and-boulder clast-supported conglomerate of probable fluvial and alluvial origin that overlies the Santa Clara Formation in a suggested unconformable contact (Stewart and Roldán-Quintana, 1991). The east to west regional outcrop pattern characterizing the Barranca Group in central Sonora suggested to Stewart and Roldán-Quintana (1991) that this unit might have accumulated in a rift-type basin.

An important point to be considered here is the question of whether the Antimonio rocks or stratigraphic successions similar to those in Sonora occur north of the Caborca region. From a regional point of view, the closest strata to the north of this region that are correlative with the Antimonio rocks are only present in the Mojave Desert region. There, coeval Lower and Middle Triassic rocks are represented in part by the sedimentary successions of the Moenkopi, Virgin Limestone, Union Wash, Silverlake, and Fairview Valley Formations (Fig. 3). They are known from the Spring, Clark, Providence, New York, Soda, and Black Mountains (Fig. 7), among other localities (Walker, 1987, 1988). These Triassic units unconformably overlie Upper Paleozoic carbonate successions and locally intrusive rocks of the lowermost Triassic (Saleeby and Busby-Spera, 1992; Walker, 1987). In that region, the Upper Triassic and Lower Jurassic are represented by both sedimentary and volcanic rocks of the Kings sequence in localities of the central and southern Sierra Nevada. It is interesting to note that the Kings sequence in the southern Sierra is represented by Middle and Upper Triassic clastic and carbonate strata and that the Lower Jurassic is represented by clastic and carbonate strata that, like the El Antimonio, also contain the distinctive bivalve *Weyla* (Saleeby and Busby-Spera, 1992; see also expanded discussion below).

PETROGRAPHIC DATA

Twenty-five thin sections of sandstone samples and clasts from conglomerates were studied. Fourteen samples of sandstone were point-counted and results are shown in a Qm-F-Lt (quartz-feldspar-lithics) standard ternary diagram (after Dickinson, 1985) in Figure 8. Sandstone composition varies from litharenite to subarkose throughout the succession. Sandstones from both Sierra del Álamo and Sierra de Santa Rosa localities

Figure 6. Field view looking northwest showing the Lower Jurassic reeflike limestone bodies, A and B, that are present in the middle member of the Sierra de Santa Rosa Formation in the locality of the same name. Limestone body A is ~30 m thick. Dip of strata is 28° to the west.

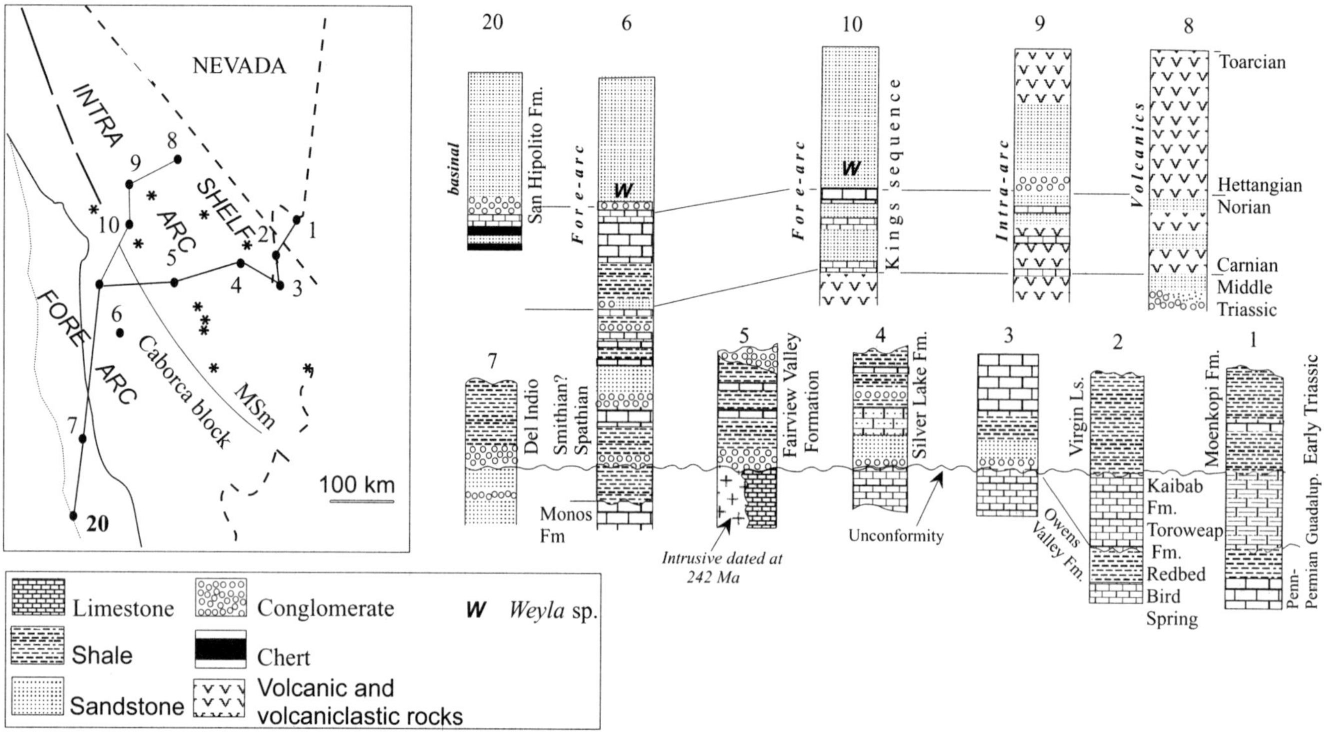

Figure 7. Proposed correlation of Triassic and Lower Jurassic strata of Sonora, southwestern United States, and Baja California. Sections and localities 1–5 taken from Walker (1987, 1988). 1—Spring Mountains, 2—Clark Mountains, 3—Providence Mountains, 4—Soda Mountains, 5—Black Mountains, 6—this work. Section 7 (El Volcán area), from Philip (1984). Sections 8 (South Inyo Range), 9 (Mineral King), and 10 (southern Sierra) redrawn from Saleeby and Busby-Spera (1992). Section 20 (Vizcaino Peninsula) from Finch and Abbott (1977) and Whalen and Pessagno (1984). Patterns are same as in Figure 2. Inset map shows location of correlated sections and speculative palinspastic restoration prior to southeastward translation (~800 km) of the Caborca block by the Mojave-Sonora megashear (MSm). Localities 7 and 20 are restored to their approximate location before the opening of the Gulf of California (following palinspastic reconstruction in Gastil and Miller, 1984, Fig. 3). Star symbol indicates location of igneous rocks shown on Figure 3. ***W*** indicates the presence of the bivalve *Weyla* sp.

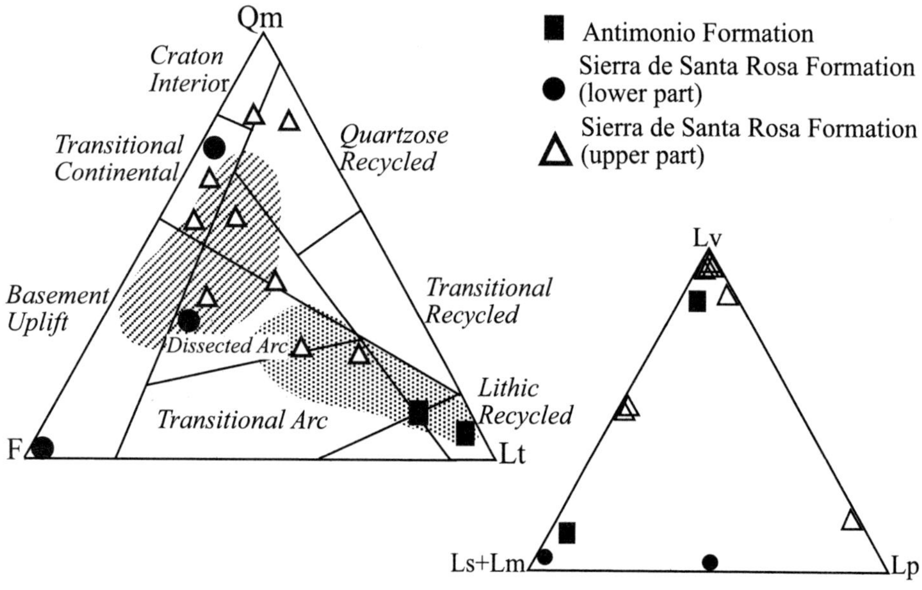

Figure 8. Point-counting data of fourteen sandstone samples from the El Antimonio Group. Shaded areas in the Qm-F-Lt triangle show data previously reported by Stanley and González-León (1995) from the Antimonio Formation (stippled) and Sierra de Santa Rosa Formation (striped). Qm—monocrystalline quartz; F—total feldspar grains; Lt—total lithic fragments (including polycrystalline quartz); Lv—volcanic lithic fragments; Ls + Lm—sedimentary and metamorphic lithic fragments; Lp—plutonic lithic fragments.

have subangular to angular grains and are typically grain supported. Grains are mostly made up of quartz and feldspar, both plagioclase and potassic feldspar, but lithics are abundant in the Antimonio Formation. Myrmekitic and perthitic textures are common in feldspar, suggesting a possible plutonic origin. Lithic grains are mostly of igneous origin and include microcrystalline quartz of probable volcanic origin, quartz/feldspar mosaics with holocrystalline texture (granites?), volcanic glass, felsitic grains, and scarce grains of sandstone and quartz/mica schist. Volcanic grains with trachytic textures were only found in sandstone samples from the upper part of the Sierra de Santa Rosa Formation. Overall, an increase in the percentage of volcanic fragments of rhyolitic composition toward the upper part of the El Antimonio Group succession was observed. There is also an increase in the percentage of volcanic grains of intermediate composition (dacite-andesite) toward the upper part of the Sierra de Santa Rosa Formation, in the locality of the same name.

Clasts of conglomerate from the Triassic Antimonio Formation are mostly igneous (>60%), made up of rhyolite (with large phenocrysts of feldspar in a fine-grained groundmass), granite (holocrystalline, coarse-grained quartz, and feldspar), quartz porphyry, and syenite. Some altered volcanic glass and cherty clasts of probable volcanic origin are also found. Sedimentary clasts are made up of limestone, quartzite (with well-rounded quartz grains), and clean chert that might be of volcanic origin. Clasts of conglomerates and pebbles from coarse-grained sandstone from the lower part of the Sierra de Santa Rosa Formation have a composition similar to the conglomerates from the Antimonio Formation, but in addition, some quartz/mica schist clasts (<2%) and detrital mica flakes were found. Conglomerate from the upper part of the Sierra de Santa Rosa Formation, in the Sierra de Santa Rosa area, consists of clasts of rhyolite and ignimbrite (60%), quartzite and chert (volcanic?) (30%), andesite to dacite (~10%), and subordinate sandstone.

The lack of rounded grains and the high percentage of feldspar grains suggest that sandstone did not suffer high abrasion or chemical weathering during transportation, or the sources were not located far from the depositional area.

ZIRCON PROVENANCE

Three coarse-grained sandstone samples from the El Antimonio Group were analyzed for detrital zircon provenance (see Fig. 4). Sample 2000–1 was collected from sequence III (Spathian), in the lower part of the Antimonio Formation. Sample 2000–0 comes from sequence VII (Norian) of the Río Asunción Formation in the Sierra del Álamo, and sample 2000–2 is from the middle member of the Sierra de Santa Rosa Formation in the Sierra de Santa Rosa. A total of 62 grains were dated (Fig. 9), which yielded ages that cluster around 190 Ma, 250 Ma, 270 Ma, 340 Ma, 1.00–1.18 Ga, 1.4 Ga, 1.6–1.7 Ga, and 1.8 Ga.

The younger zircon grains from sample 2000–1 (Figs. 9A and 9B, Table 1) date from the Permian. One grain is concordant at 274 ± 10 Ma, and seven grains are interpreted to be ca. 274 Ma, the latter containing inherited components from ca. 1.6–1.7 Ga. There are also two grains dated ca. 1.08 Ga, three grains dated ca. 1.43 Ga, eight grains dated between 1.56 Ga and 1.73 Ga, and one grain dated ca. 1.87 Ga. Sample 2000–0 (Fig. 9C, Table 1) yielded two zircon grains with apparent concordant ages at 244 and 256 Ma (Lower Triassic to Upper Permian), one grain with apparent concordant age at 340 Ma, five grains dated 1.08–1.18 Ga, two grains dated 1.42 Ga, and ten grains dated between 1.63 Ga and 1.73 Ga. Sample 2000–2 (Figs. 9D and 9E, Table 1) yielded the younger zircon grains, including eleven grains with a similar age of ca. 190 Ma (Lower Jurassic) and seven grains dated ca. 1.04–1.09 Ga, one grain dated ca. 1.44 Ga, and one grain dated ca. 1.6 Ga.

Zircon grains dated at ca. 1.1, 1.43, and 1.6–1.8 Ga are common in all three samples and most probably reflect provenance from the Precambrian basement in the southwestern United States and/or northwestern Sonora. Those dated between 1.6 and 1.8 Ga could have been derived from the metamorphic and igneous basements of the western United States (Mojave, Yavapai, and Mazatzal provinces) and the Caborca terrane (Fig. 3). Similarly, zircons dated at ca. 1.43 Ga were most probably derived from granitoids of that age that are common in the southern United States and also locally present in the basement of the Caborca terrane in Sonora (Anderson and Silver, 1979). As discussed by Gross et al. (2000), possible local sources for the 1.1 Ga grains include granitic intrusives with similar ages that are present or are possibly covered or eroded in western North America, as well as a few intrusives [the Aibo granite dated at 1110 ± 10 Ma (U/Pb, zircon) by Anderson et al., 1979] in the Caborca terrane. Zircon grains of all of these ages are abundant however in upper Proterozoic to Permian strata in Sonora (Gehrels and Stewart, 1998; Gross et al., 2000; Stewart et al., 2001) and in the miogeocline of Nevada and off-shelf assemblages of Nevada and California (Robert Mountains and Golconda allochthons and the northern Sierra Terrane; Manuszak et al., 2000), for which a reworked source from these rocks is also possible.

The most probable sources of provenance for the detrital zircons with ages between 240 and 270 Ma are Upper Permian and Lower Triassic igneous rocks in southeastern California. For instance, plutons dated between 288 and 240 Ma are present in the El Paso Mountains and Mojave Desert region (see Miller et al., 1995; Barth et al., 1997), and volcanic rocks dated ca. 283 Ma (Walker, 1988) are reported from the central Mojave Desert. Further to the north, in west-central Nevada, equivalent rocks to the McConnell Canyon Volcanics date to ca. 242 Ma (Stewart et al., 1997b; Schweickert and Lahren, 1993). A significant population of zircon grains dated ca. 190 Ma that were collected from the middle member of the Sierra de Santa Rosa Formation correlate with the Pliensbachian age (according to Geological Society of America Time Scale, Palmer and Geissman, compilers, 1999), indicated by ammonites identified from this member. Although igneous rocks of this age are rare in the region, these zircons most probably reflect provenance from the oldest parts of the Jurassic continental volcanic arc that developed in southwestern

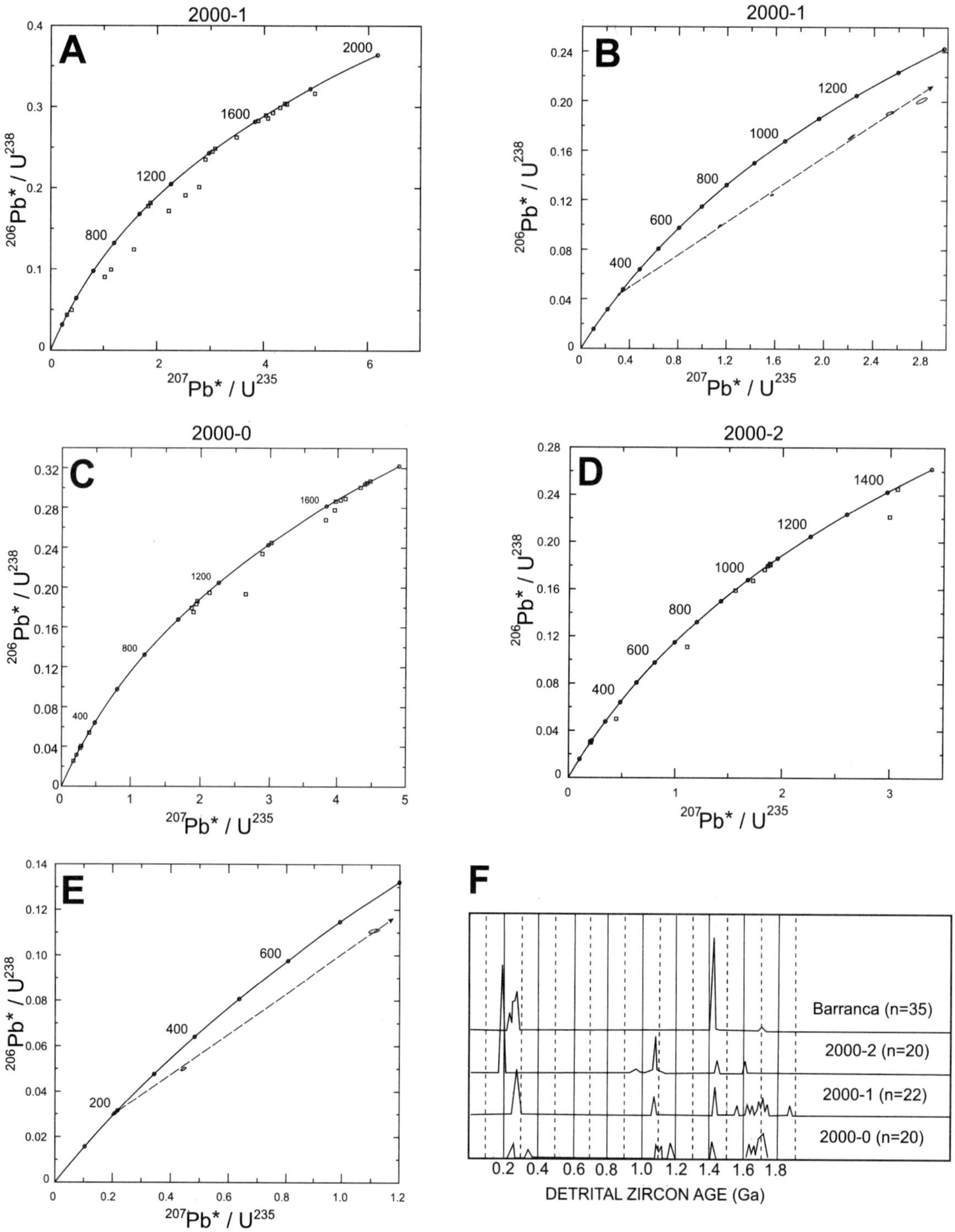

Figure 9. Concordia curves of analyzed detrital zircon samples. A–B, Sample 2000–1; C, Sample 2000–0; D–E, Sample 2000–2; F, plot of relative age probability for the three samples analyzed, together with the available ages from Triassic strata of the Barranca Group (Gehrels and Stewart, 1998). Each curve includes the interpreted ages from Table 1, with each age plotted as a probability distribution centered on the mean age. The individual distributions are then added into a single curve and divided by the number of constituent analyses, such that each curve contains the same area.

North America, for example, the Mount Wrightson Formation (Asmerom et al., 1990; Riggs and Busby-Spera, 1990; Riggs et al., 1993) and arc-volcanic rocks in the Mineral King pendant of the Sierra Nevada (reported in Riggs et al., 1993, p. 1242) (Fig. 3). One of the grains dated yielded an age of 340 Ma. It is a rare contribution and its possible provenance is discussed below.

Zircon grains dated by Gehrels and Stewart (1998) from the Upper Triassic, Santa Clara Formation of the Barranca Group in central Sonora (Fig. 9F), yielded age clusters of 225–235 Ma and 250–280 Ma, which may indicate possible provenance from the Permian and Triassic igneous rocks reported above. This sample also yielded grains with ages around 1.42 Ga and 1.70 Ga, which similarly might indicate provenance from the Precambrian basement of southwestern United States or Sonora (Anderson and Silver, 1979; Gehrels and Stewart, 1998).

Sm/Nd GEOCHEMICAL DATA

One sandstone sample from the Antimonio Formation (sample MS4 from sequence III, Fig. 4) and one from the lower part of the Sierra de Santa Rosa Formation (sample MS10 from sequence XII) in the Sierra del Álamo and one siltstone sample from the upper part of this latter unit collected from the Sierra de Santa Rosa locality (sample MS26) were analyzed for whole-rock Sm/Nd isotopic composition. Sandstone samples from the Antimonio and lower part of the Sierra de Santa Rosa Formations have initial epsilon neodymium (ε_{Ndi}) values of −13.81 and −10.55 (Table 2), and depleted mantle (DePaolo, 1981) Nd model ages (T_{DM}) that range from 1.92 to 1.82 Ga suggesting derivation almost entirely from an old crustal source. The siltstone sample collected from the Sierra de Santa Rosa area yielded an epsilon neodymium (ε_{Ndi}) value of −4.52 (Table 2) and a T_{DM} model age of 1.13 Ga suggesting addition of sediments from more juvenile sources than those of samples collected in lower parts of the column.

DISCUSSION

Stratigraphic Correlation of the El Antimonio Group

The most complete succession of the El Antimonio Group occurs in localities of the Sierra del Álamo and Sierra de Santa Rosa, where its age is constrained between Late Permian–Triassic and Early Jurassic (late Hettangian–Pliensbachian and possibly early Toarcian). The Antimonio and Río Asunción Formations are well represented in the northern foothills of the Sierra del Álamo, whereas the Lower Jurassic strata (upper Hettangian–upper Sinemurian) that crop out at this same locality are assigned to the lower part of the Sierra de Santa Rosa Formation (modified from Lucas and Estep, 1999b). The upper Sinemurian to lower Pliensbachian sedimentary section that crops out in the Sierra de Santa Rosa is a younger succession that we consider as the upper part of the Sierra de Santa Rosa Formation. At this latter locality, the Sierra de Santa Rosa Formation tectonically overlies Precambrian metamorphic rocks (Hardy, 1981), such that its lower part has been cut out by faulting. Although a continuous and more complete section of this formation has not yet been found, we infer that the Lower Jurassic strata of the Sierra del Álamo and Sierra de Santa Rosa are part of the same stratigraphic succession (as did Lucas and Estep, 1999b), but some strata from its middle part in between these sections could be missing. Work in progress on bivalve and ammonoid identifications by M. Aberhan (Germany) and J. Pálfy (Hungary) might help to better clarify the stratigraphic correlation between these two localities.

Other sections that correlate with and are considered to be part of the El Antimonio Group are widely scattered in other areas of Sonora and Baja California (Figs. 3 and 4). For instance, a correlative distinctive interval that crops out from central (Sierra de Santa Teresa) to northwestern Sonora (Sierra del Álamo and Barra Los Tanques) is the biostromal limestone of sequence VII that is also recognized because of its typical reefal fauna. Other sections that are partly correlated on the basis of fossil and age similarities are the upper Carnian section at Sierra La Flojera and the lower Pliensbachian section of Pozos de Serna. Correlation of the El Antimonio Group with other sections in Sonora is made on the basis of age and fossil (mostly ammonoids) similarities, although the stratigraphy of some of them is still poorly known. The stratigraphy of the Del Indio and San Hipólito Formations in Baja California is well known, and both formations are considered to represent basinal facies that correlate with the El Antimonio. The fact that the lower contact of these successions in most of the areas in Sonora is tectonic or doubtful led González-León (1980) to propose that the El Antimonio succession, along with the Upper Permian Monos Formation, was a tectonic allochthon designated as the Antimonio terrane.

Proposed Paleogeography of the El Antimonio Basin

It has been proposed that the Antimonio rocks represent deposits of a fore-arc basin that originally was adjacent to the region of southern California in the southwestern United States (Stanley and González-León, 1995; González-León, 1997a; Marzolf, 2000) and that was later translated during the Jurassic, along with the Caborca block, to its present position. One of the most likely mechanisms would have been the left-lateral movements postulated along the Mojave-Sonora megashear (MSm, Fig. 3) by Silver and Anderson (1974; see also Anderson et al., 1979; Anderson and Schmidt, 1983). Although the Mojave-Sonora megashear remains controversial and its existence has been challenged (see discussion below), we present here some geologic arguments that can support these ideas.

The hypothesis that the El Antimonio succession was deposited in a basin located adjacent to the southwestern margin of the United States and later translated to its present position by the Mojave-Sonora megashear is based on the regional geologic observation that to the north of the trace of the proposed Mojave-Sonora megashear, the closest sections that most likely correlate with the El Antimonio Group are not located in northern Sonora but are represented by Triassic and Lower Jurassic strata of

TABLE 1. U-Pb ISOTOPIC DATA AND AGES

Grain type	wt. (µg)	Pb_c (pg)	U (ppm)	Isotopic ratios				Apparent ages (Ma)			Interpreted age (Ma)
				$^{206}Pb_m/^{204}Pb$	$^{206}Pb_c/^{208}Pb$	$^{206}Pb^*/^{238}Pb$	$^{207}Pb^*/^{235}Pb$	$^{206}Pb^*/^{238}Pb$	$^{207}Pb/^{235}Pb$	$^{207}Pb^*/^{206}Pb$	
2000-1											
CE	24	14	103	482	3.3	0.043470 ± 1.46	0.31050 ± 2.61	274	275	277 ± 46	274 ± 10
CE	25	18	405	1656	6.3	0.049465 ± 0.49	0.39584 ± 0.99	311	339	531 ± 18	274 ± 20
CE	25	7	438	8595	12.5	0.089925 ± 0.41	1.02186 ± 0.51	555	715	1255 ± 6	274 ± 20
CE	22	16	458	3646	30.3	0.099334 ± 0.67	1.14323 ± 0.87	611	774	1280 ± 10	274 ± 20
CE	25	24	300	2320	7.2	0.124263 ± 0.47	1.57254 ± 0.64	755	959	1462 ± 10	274 ± 20
CE	21	12	717	12575	25.4	0.171582 ± 0.94	2.22024 ± 0.97	1021	1187	1515 ± 10	274 ± 20
ME	23	235	991	1052	13.3	0.190846 ± 0.59	2.53146 ± 1.03	1126	1281	1552 ± 15	274 ± 20
ME	22	715	3013	1086	12.4	0.201020 ± 1.02	2.78882 ± 1.28	1181	1352	1635 ± 13	274 ± 20
CE	26	5	84	4145	7.3	0.181419 ± 0.58	1.87953 ± 0.79	1075	1074	1072 ± 10	1072 ± 11
ME	19	9	480	10940	8.4	0.177391 ± 0.60	1.83893 ± 0.70	1053	1059	1074 ± 7	1075 ± 10
LE	21	9	206	6767	7.6	0.234647 ± 1.10	2.90454 ± 1.15	1359	1383	1421 ± 6	1424 ± 10
LE	15	11	800	16080	51.2	0.244688 ± 0.52	3.03471 ± 0.57	1411	1416	1425 ± 5	1425 ± 10
LE	18	48	269	1463	4.9	0.248407 ± 1.13	3.08659 ± 1.31	1430	1429	1428 ± 12	1428 ± 12
ME	20	41	1268	19610	43.3	0.262404 ± 0.77	3.48877 ± 0.80	1502	1525	1556 ± 5	1559 ± 10
LE	16	10	1020	27674	221.1	0.282870 ± 0.51	3.89303 ± 0.55	1606	1612	1621 ± 5	1621± 10
ME	18	11	719	19715	73.4	0.289654 ± 0.59	4.03892 ± 0.64	1640	1642	1645 ± 5	1645 ± 10
LE	19	15	1165	25300	36.8	0.292644 ± 0.39	4.17039 ± 0.44	1655	1668	1685 ± 5	1686 ± 10
ME	20	845	1101	440	8.8	0.285863 ± 1.22	4.07790 ± 2.22	1621	1650	1687 ± 31	1690 ± 32
ME	16	11	852	22440	10.1	0.299116 ± 0.44	4.30598 ± 0.49	1687	1695	1704 ± 5	1705 ± 10
LE	18	11	264	7710	17.0	0.303981 ± 0.39	4.39528 ± 0.49	1711	1712	1712 ± 10	1712 ± 10
ME	15	19	2056	29600	31.8	0.303290 ± 0.72	4.43240 ± 0.75	1708	1718	1732 ± 4	1733 ± 10
LE	20	8	322	16550	16.9	0.316466 ± 0.61	4.97765 ± 0.65	1772	1816	1865 ± 4	1869 ± 10
2000-0											
LE	14	47	437	317	4.5	0.038494 ± 0.81	0.27260 ± 2.83	244	245	257 ± 59	244 ±15
CE	31	12	162	1042	7.6	0.040431 ± 0.88	0.28687 ± 1.33	256	256	262 ± 22	256 ± 10
LE	10	10	912	3140	10.6	0.054143 ± 0.05	0.40072 ± 0.79	340	342	358 ± 13	340 ± 13
LR	11	65	1523	2820	7.4	0.179420 ± 0.71	1.87081 ± 0.87	1064	1071	1085 ± 10	1087 ± 10
LE	14	9	101	1830	8.1	0.186205 ± 0.72	1.95480 ± 0.95	1101	1100	1099 ± 12	1099 ± 12
LE	9	10	163	1680	7.9	0.183212 ± 0.88	1.93749 ± 1.07	1085	1094	1113 ± 18	1115 ± 12
LE	12	11	261	3130	7.9	0.175299 ± 0.05	1.89893 ± 0.65	1041	1081	1161 ± 7	1170 ± 12
LE	18	23	173	1620	6.7	0.194707 ± 0.69	2.12742 ± 0.99	1147	1158	1178 ± 13	1180 ± 14
LE	9	12	338	3950	5.7	0.244824 ± 0.53	3.01682 ± 0.62	1412	1412	1412 ± 6	1412 ± 10
LR	8	9	223	2810	4.4	0.233770 ± 0.63	2.88746 ± 0.69	1354	1379	1417 ± 6	1420 ± 10
LR	10	7	91	2280	18.2	0.286784 ± 0.72	3.97157 ± 0.83	1625	1628	1632 ± 10	1633 ± 10
LE	9	79	1467	1925	7.4	0.193423 ± 051	2.64898 ± 0.63	1140	1314	1612 ± 6	1639 ± 32
LE	7	3	52	18800	8.3	0.287905 ± 0.50	4.04313 ± 0.59	1631	1643	1658 ± 6	1659 ± 10
LE	6	7	228	3600	11.7	0.289444 ± 0.62	4.11364 ± 0.73	1639	1657	1680 ± 7	1682 ± 10
LE	16	61	507	2220	23.8	0.277766 ± 0.45	3.95666 ± 0.55	1580	1625	1685 ± 6	1689 ± 10
LR	12	58	1234	4190	17.1	0.267751 ± 0.52	3.82480 ± 0.64	1529	1598	1690 ± 7	1696 ± 10
ME	10	4	722	28500	13.9	0.300599 ± 0.59	4.33013 ± 0.63	1694	1699	1705 ± 4	1706 ± 10
LE	14	9	113	2880	10.7	0.303987 ± 0.71	4.39482 ± 0.79	1711	1711	1712 ± 6	1712 ± 10
LE	9	5	147	4750	6.6	0.305281 ± 0.62	4.42372 ± 0.73	1717	1717	1716 ± 7	1716 ± 10
LE	12	9	279	6800	11.5	0.306992 ± 0.52	4.47103 ± 0.64	1726	1726	1725 ± 7	1725 ± 10

(*continued*)

TABLE 1. U-Pb ISOTOPIC DATA AND AGES (*continued*)

Grain type	wt. (μg)	Pb_c (pg)	U (ppm)	Isotopic ratios				Apparent ages (Ma)			Interpreted age (Ma)
				$^{206}Pb_m/^{204}Pb$	$^{206}Pb_c/^{208}Pb$	$^{206}Pb^*/^{238}Pb$	$^{207}Pb^*/^{235}Pb$	$^{206}Pb^*/^{238}Pb$	$^{207}Pb/^{235}Pb$	$^{207}Pb^*/^{206}Pb$	
2002-2											
CE	29	19	240	696	4.6	0.029728 ± 0.84	0.20433 ± 2.57	189	189	188 ± 54	189 ± 10
CE	24	10	204	898	3.7	0.029918 ± 1.12	0.20477 ± 2.19	190	189	178 ± 42	190 ± 10
CE	26	8	157	990	5.8	0.029926 ± 1.35	0.20771 ± 2.23	190	192	211 ± 39	190 ± 10
CE	25	9	207	1166	4.6	0.030205 ± 1.30	0.20810 ± 1.98	192	192	193 ± 33	192 ± 10
CE	26	12	163	639	4.2	0.030254 ± 1.25	0.20883 ± 2.91	192	193	198 ± 58	192 ± 10
CE	24	9	247	1251	4.6	0.030513 ± 1.36	0.21120 ± 1.92	194	195	204 ± 30	194 ± 10
CE	27	5	189	1760	5.3	0.030635 ± 1.05	0.20863 ± 1.45	195	192	167 ± 22	195 ± 10
LE	20	10	244	933	4.9	0.030675 ± 1.06	0.20948 ± 2.09	195	193	173 ± 40	195 ± 10
CE	27	18	206	593	4.3	0.030761 ± 1.39	0.21226 ± 3.23	195	195	197 ± 65	197 ± 10
CE	31	10	195	1152	6.4	0.030733 ± 0.91	0.21111 ± 1.72	195	195	187 ± 32	195 ± 10
CE	30	9	114	1197	4.5	0.049856 ± 1.24	0.44458 ± 1.70	314	373	764 ± 23	194 ± 20
ME	15	35	473	1990	7.4	0.158876 ± 2.13	1.56242 ± 2.90	951	955	967 ± 40	968 ± 42
LE	15	20	190	960	9.3	0.111008 ± 0.65	1.11356 ± 1.35	679	760	1007 ± 22	1044 ± 55
ME	15	16	592	5472	13.9	0.167209 ± 0.60	1.72708 ± 0.74	997	1019	1066 ± 9	1071 ± 11
LE	19	17	153	1803	9.9	0.181706 ± 0.53	1.88656 ± 0.82	1076	1076	1077 ± 12	1077 ± 12
LE	18	14	173	2455	9.3	0.179787 ± 1.02	1.86615 ± 1.13	1066	1069	1076 ± 9	1077 ± 10
LE	18	11	1857	31750	9.5	0.176450 ± 0.86	1.83491 ± 0.89	1048	1058	1080 ± 5	1082 ± 10
ME	13	205	273	207	3.8	0.180578 ± 1.65	1.88801 ± 3.37	1070	1077	1091 ± 55	1092 ± 58
LE	16	9	463	12465	16.1	0.245160 ± 0.52	3.06591 ± 0.58	1414	1424	1440 ± 5	1442 ± 10
ME	18	50	1084	4970	4.3	0.221419 ± 0.67	2.99540 ± 0.77	1289	1407	1589 ± 7	1605 ± 18

Notes: Grain types: ME—medium pink euhedral; LE—light pink euhedral; CE—colorless euhedral; LR—light pink rounded. All grains abraded in air abrasion device. $^{206}Pb/^{204}Pb$ is measured ratio, uncorrected for blank, spike, or fractionation $^{206}Pb/^{208}Pb$ is corrected for blank, spike, and fractionation. All uncertainties are at the 95% confidence level. Uncertainties in isotope ratios are in percent. Uncertainties in ages are in millions of years. Most concentrations have an uncertainty of 25% due to uncertainty in weight of grain. Constants used: $^{238}U/^{235}U$ = 137.88. Decay constant for $^{235}U = 9.8485 \times 10^{-10}$. Decay constant for $^{238}U = 1.55125 \times 10^{-10}$. Isotope ratios are adjusted as follows: (1) mass dependent corrections factors of: 0.14 ± 0.06% amu for Pb and 0.04 ± 0.04 % amu for UO_2; (2) Pb ratios corrected for 0.005 ± 0.003 ng blank with $^{206}Pb/^{204}Pb$ = 18.6 ± 0.3, $^{207}Pb/^{204}Pb$ = 15.5 ± 0.3, and $^{208}Pb/^{204}Pb$ = 38.0 ± 0.8; (3) U has been adjusted for 0.001 ± 0.001 ng blank; (4) initial Pb from Stacey and Kramers (1975), with uncertainties of 1.0 for $^{296}Pb/^{204}Pb$, 0.3 for $^{207}Pb/^{204}Pb$, and 2.0 for $^{208}Pb/^{204}Pb$. Interpreted ages for grains less than 1.0 Ga are from $^{206}Pb^*/^{238}U$ ages for concordant grains and from lower concordia intercepts for discordant grains. Interpreted ages for >1.0 Ga grains are from upper concordia intercepts when projected from 100 ± 100 Ma, which is a reasonable age of isotopic disturbance in this region. All analyses conducted using conventional isotope dilution and thermal ionization mass spectrometry, as described by Gehrels (2000).

TABLE 2. WHOLE-ROCK Sm/Nd ISOTOPIC ANALYSIS FROM SEDIMENTARY ROCKS OF THE EL ANTIMONIO GROUP

Sample	Lithology	Age (Ma)	Sm (ppm)	Nd (ppm)	$^{147}Sm/^{144}Nd$	$^{143}Nd/^{144}Nd$	1σ	ε_{Ndo}	1σ	ε_{Ndi}	T_{DM} (Ma)
MS4. Antimonio Formation	Sandstone	242	4.05	21.13	0.116	0.511802	19	−16.31	0.37	−13.81	1923
MS10. Sierra de Santa Rosa Formation (lower part)	Sandstone	198	6.86	32.53	0.128	0.512008	17	−12.29	0.33	−10.55	1826
MS26 Sierra de Santa Rosa Formation (upper part)	Siltstone	196	6.91	36.55	0.114	0.512300	17	−6.59	0.33	−4.52	1132

Note: Analyses were performed at Laboratorio Universitario de Geoquímica Isotópica at UNAM, on a Finnigan MAT 262 mass spectrometer. Blanks were Nd<300 pg, duplicates <+0.3 εNd units, and LaJolla standard is 0.511880 ± 22 × 10^{-6}.

southern Nevada and southeastern California (Figs. 3 and 7). For example, according to Lucas et al. (1997b), sequences II and III of the El Antimonio are comparable to sequences of that region from the Smithian and Spathian. Some of the mostly marine Triassic and Lower Jurassic sections of the Mojave Desert and southern Nevada have lithologies and fossils that are common also in some sequences of the El Antimonio Group (Stanley and González-León, 1995; Estep et al., 1997b). Based on similar ammonite and conodont faunas, Philip (1984) correlated rocks of the Del Indio unit in Baja California with Lower Triassic strata of the Union Wash and Moenkopi Formations in east-central California and southern Nevada. Marzolf (2000), however, considered the Triassic section of the El Antimonio Group to be part of the Moenkopi, Holbrook, Chinle, and Dinosaur Canyon tectonosequences of western United States and proposed that its original position was adjacent to western California.

Another case of lithologic and fossil similarities of the El Antimonio succession with nearby regions of southwestern United States are Norian biostromes of unit 13 whose biota consist of calcareous algae and at least 9 species of scleractinian corals, four species of sphinctozooid sponges, several encrusting hydrozoans and disjectoporoids, and a diverse molluscan assemblage (Stanley et al., 1994). Unit 13 is remarkably similar in thickness, faunal composition, biotic relationships, and depositional environments to counterparts in a Norian succession exposed in various allochthons in mountain ranges of west-central Nevada (Stanley, 1979; Goodwin, 1999). Similarly, Carnian and Norian carbonates with interbedded local clastic and volcanic rocks that are widespread in parts of eastern California (northern Sierra Nevada) and western Nevada (Schweickert and Lahren, 1993; Stewart et al., 1997b), along with the Norian carbonates of the El Antimonio, may be part of a widespread carbonate shelf that formed during the Late Triassic in that region. In the same way, Blodgett and Frýda (2001) noted that Late Triassic (middle Norian) gastropods from the Clan Alpine Mountains (Nevada) and the Río Asunción Formation (Sierra del Álamo) share many species in common and proposed that they belong to a single biogeographic province termed the Nevada-Sonora Province.

The stratigraphic sequences of the Antimonio and Río Asunción Formations are interpreted to record marginal marine to shallow marine sedimentation on a tectonically subsiding platform that, in part, was probably influenced by third-order sea level cycles. These rocks started to deposit at the time when a NW-SE–trending subduction zone began to develop beneath the continental margin of the southwestern United States (Saleeby and Busby-Spera, 1992), giving origin to "early magmatic activity" (245 Ma to ca. 210 Ma according to Walker et al., 2002, p. 8) along the California-Nevada border region and regions to the north (see also Fig. 3 and Barth et al., 1990, 1997 for rock localities distribution). Rocks of this arc include Upper Permian and Lower Triassic granites in the San Bernardino Mountains (Barth et al., 2001), Lower Triassic granitic plutons that are scattered in the southeastern Sierra Nevada and El Paso Mountains (Miller et al., 1995), granitic and alkaline intrusive rocks of probable Middle Triassic age that are sparse in the Mojave region (Saleeby and Busby-Spera, 1992), and Upper Triassic plutonic rocks that occur along the east-central Sierra Nevada, White-Inyo Mountains, and the Mojave Desert region (Saleeby and Busby-Spera, 1992, p. 132). A single Lower Triassic pluton (225 Ma, reported by Stewart et al., 1986) from Sonora, located south of the Mojave-Sonora megashear (Fig. 3), is clearly a part of this arc.

We consider the Antimonio and Río Asunción Formations of Sonora to be correlative with Triassic strata of the southwestern United States and propose that both regions lay originally together, forming part of the same basin (Fig. 7). The Antimonio and Río Asunción Formations would have been deposited south of the continental magmatic arc (Fig. 7) and in an intermediate position between the Triassic strata to the north and deeper marine Triassic deposits to the south, like the Del Indio unit and San Hipólito Formation in Baja California. The South Inyo Range and Mineral King sections of southeastern California (Fig. 7) that are dominated by Middle Triassic to Lower Jurassic arc volcanism may be part of the volcanic arc located north of the original position of the El Antimonio sections. The Lower Jurassic [Hettangian–Toarcian (?)] strata of the Sierra de Santa Rosa Formation that records deeper-water depositional environments and increased rates of sedimentation (Fig. 10) suggest a sudden deepening of the basin by the time the Middle Jurassic continental volcanic arc began to be established. Therefore, deposition of the Sierra de Santa Rosa Formation is interpreted to have occurred in a fore-arc setting located south of the continental magmatic arc (Stanley and González-León, 1995; González-

León, 1997a; Lucas et al., 1999), similar to the Jurassic part of the Kings sequence in southern California. The Kings sequence in the southern Sierra Nevada contains Upper Triassic–Lower Jurassic fossils (including *Weyla* sp.) in shelf quartzite-carbonate that grades upward into a thick turbidite succession of probable fore-arc basin strata (Saleeby et al., 1978; Saleeby and Busby-Spera, 1986, 1992). *Weyla* is a Lower Jurassic index fossil that is also present in strata of the Sierra de Santa Rosa Formation.

New data of sandstone composition, plus those reported by Stanley and González-León (1995), when plotted in a Qm-F-Lt triangle, indicate a change in source of provenance from a recycled orogen for the Antimonio Formation to a transitional continental block for the Sierra de Santa Rosa Formation (Fig. 8). The diverse composition of lithic grains in sandstone and of clasts in conglomerate of the Antimonio Formation indicates mixed volcanoplutonic and sedimentary sources of provenance, this last most probably representing the cratonic to miogeoclinal Proterozoic and Paleozoic successions of the southwestern United States. The closest probable source for the volcanoplutonic components are Permian and Triassic plutons in the southwestern United States (Mojave Desert region and El Paso Mountains as discussed above) as are for the zircon grains with ages between 240 and 270 Ma found in samples 2000–0 and 2000–1. A single zircon grain from the Río Asunción Formation that yielded an age of 340 Ma is a rare contribution, although scarce grains of that age are present in Triassic rocks of Nevada, like in the Lovelock assemblage and the Bishop Canyon Formation of the Black Rock terrane (Manuszak et al., 2000; Darby et al., 2000). Sources for grains of that age have been attributed by these authors to igneous rocks of the Sierra-Klamath terranes, which is also a probable source for the grain in the El Antimonio Group. Similarly, lithic grains of volcanic, plutonic, and, less commonly, of sedimentary origin in sandstone of the Sierra de Santa Rosa strata indicate a mixed influence of provenance (Fig. 8). However, tuff beds and common volcanic and plutonic clasts in that formation, along with the zircon grain population with ages around 190 Ma in sample 2000–2, most probably indicate a marked provenance of early stages of the Jurassic continental magmatic arc that developed in southeastern California and southern Arizona (Busby-Spera, 1988; Tosdal et al., 1989).

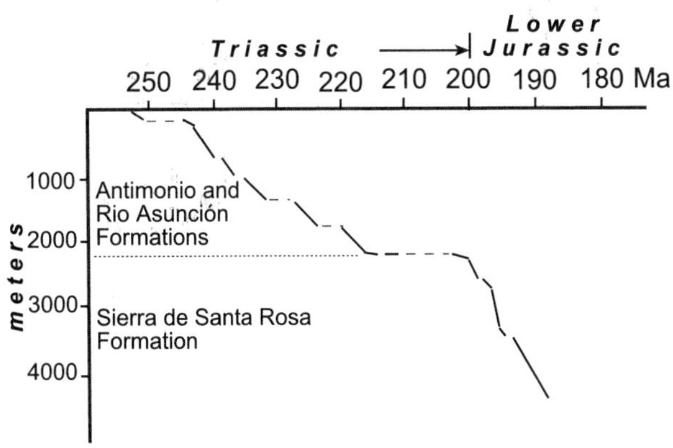

Figure 10. Speculative subsidence curve for the not decompacted sedimentary stratigraphic column of the El Antimonio Group. Numerical ages are estimated from biostratigraphic data.

The Sm/Nd isotopic composition (Table 2) of sandstones from the Antimonio and Sierra de Santa Rosa Formations show a decrease in model ages upward in the section of the El Antimonio Group and agree with provenance interpretations based on the zircon analysis. Nd model ages of 1.9 Ga and 1.8 Ga obtained from samples of the Antimonio and lower part of the Sierra de Santa Rosa Formations respectively, indicate crustal signatures of the Yavapai Province that show T_{DM} between 1.8 and 2.0 Ga (Farmer et al., 2001). Sample MS26 (Fig. 4) collected from the upper Sierra de Santa Rosa Formation yielded a Nd model age of 1.13 Ga, most probably indicating influence of crust of Grenville age (Farmer et al., 2001).

The isotopic composition of sedimentary rocks from the El Antimonio Group is significantly different than values obtained in other Triassic (Norian) siliciclastic units in central and southwestern Mexico. These units located in the states of San Luis Potosi, Zacatecas, and Michoacán have Nd model ages ranging from 1.2 to 1.5 Ga (Centeno-García et al., 1993; Centeno-García and Silva-Romo, 1997). Therefore, the provenance of sedimentary rocks of the Antimonio Group seems to have more correlation with northwestern formations than with southern localities.

We believe that the data presented herein of the El Antimonio Group provide support for the idea that these rocks were originally deposited in a fore-arc basin (Stanley and González-León, 1995) that might have been located adjacent to the southwestern margin of the United States (Fig. 7). We also infer that, following its deposition, the El Antimonio succession was translated southeastward to its present position by the postulated Jurassic movements along the Mojave-Sonora megashear. A paleomagnetically documented (Molina-Garza and Geissman, 1996) tectonic phase of Jurassic compressive deformation that affected the El Antimonio Group rocks could have occurred just before or was contemporaneous with movement of the Mojave-Sonora megashear (Stanley and González-León, 1995).

Ongoing Debate Regarding the Mojave-Sonora Megashear

The Mojave-Sonora megashear may be regarded as a controversial, hypothetical, NW-SE, left-lateral strike-slip fault, originally proposed to explain apparent truncation and juxtaposition of Precambrian basement provinces to the north and south of its trace in southwestern North America (Silver and Anderson, 1974), but it goes much further in explaining other observations. It explains lithocorrelations between similar Neoproterozoic to lower Paleozoic miogeoclinal strata of Sonora and the Mojave Desert region (Stewart et al., 1984), even though some workers have favored the idea that the Paleozoic miogeoclinal rocks of Sonora are a wrap-around continuation of similar rocks in southwestern United States (Peiffer-Rangin, 1979; Poole et al., 1991). Other authors state that there may not be enough conclusive evidence to either support or refute the concept (Stewart et al., 1997a; Stewart et al., 1999), and still others have proposed that truncation of the Neoproterozoic and Paleozoic margin of the southwestern United States was through a Paleozoic rather than a Jurassic fault (Stevens et al., 1992; Dickinson and Lawton, 2001). These later hypotheses, however, do not attempt to explain or reconcile the lower Mesozoic rocks discussed herein.

Some workers have refuted the Mojave-Sonora megashear idea by citing the lack of conclusive evidence that could support a Jurassic, southeastward displacement of ~800 km of the Caborca block along this fault. Among these last contributions, perhaps the most notable is the work by Molina-Garza and Geissman (1999). In a comprehensive paleomagnetic study of the rocks of the Caborca region, they sampled several outcrops of Neoproterozoic, Jurassic, and Cretaceous strata, including the Jurassic rocks of Sierra de Santa Rosa, Pozo de Serna, and Cerro Basura areas (Fig. 3) that yielded primary Jurassic and secondary Cretaceous magnetizations, respectively. On the basis of their results, these authors rejected the hypothesis of the Mojave-Sonora megashear based on the following arguments that, according to them, contradict the southward displacement of the Caborca block with respect to North America. (1) Rocks of the (displaced) Caborca block located south of the Mojave-Sonora megashear and those of the (in situ) Cordilleran Jurassic volcanic arc located to the north of the Mojave-Sonora megashear have similar Jurassic magnetization; and (2) the Caborca block was significantly rotated in a clockwise direction during the Jurassic. We now present the following observations that do not support such an important and far-reaching conclusion.

Rocks that Molina-Garza and Geissman (1999) sampled in the localities of Cerro Basura and Puerto El Álamo (Fig. 3) were considered by these authors to be positioned north of the trace of the Mojave-Sonora megashear and to be part of the Jurassic volcanic arc. These rocks yielded primary Jurassic magnetizations similar to a Jurassic magnetization found in rocks of the El Antimonio Group, located south of the trace of the Mojave-Sonora megashear. Consequently, Molina-Garza and Geissman (1999) concluded that this observation did not support the left-lateral displacement of the Caborca block along the Mojave-Sonora megashear. However, as indicated by González-León (1989) and Stanley and González-León (1995), rocks of the Cerro Basura are intensely deformed sedimentary, volcanic and volcaniclastic rocks (Corona, 1980). They have lithologies similar to rocks found in the Sierra de Santa Rosa Formation in the Sierra del Álamo (González-León, 1980) and contain similar Sinemurian ammonites (Jacques-Ayala, 1993). Thus, we argue that Cerro Basura is not part of the autochthonous volcanic arc but is part of the displaced El Antimonio succession. Likewise, rocks sampled by Molina-Garza and Geissman (1999) in Puerto El Álamo are clearly not part of the Jurassic volcanic arc as these authors assumed; rather, as indicated by Nourse (1995, 2001), they correspond to the younger tectonostratigraphic sequences of the Upper Jurassic–Lower Cretaceous Bisbee Group. The significant Jurassic clockwise rotation noted by Molina-Garza and Geissman (1999) in the Caborca block that they considered proof of inconsistency for the left-lateral displacement during the Jurassic is not a very firm argument with which to refute the postulated movement of the megashear. Clockwise rotation of crustal blocks in response to left-lateral fault displacements is not an unexpected phenomenon, and such occurrences have been documented from several regions (Taylor et al., 1998) and in the Andean Cordillera south of Peru (Beck, 1998; Taylor et al., 1998; Abels and Bischoff, 1999).

Relationship to the Triassic Barranca Group

The El Antimonio Group is in part chronocorrelative with the Barranca Group, but they record different tectonic and depositional settings. Although it has been proposed that the El Antimonio succession was deposited in a fore-arc basin whereas the Barranca Group was part of a rift-basin fill (Stewart and Roldán-Quintana, 1991), both successions would have been deposited further to the northwest if they were transported by the Mojave-Sonora megashear. As discussed above, the Santa Clara Formation of the Barranca Group presents zircon data that could indicate, at least in part, provenance from ca. 1.4 Ga and 1.7 Ga plutonic rocks of the southwestern United States. Stanley and González-León (1995) proposed that the El Antimonio and Barranca Groups represent separate successions and included the El Antimonio (and correlative sections) in the Antimonio terrane. However, as argued by Stewart et al. (1997a), a depositional link between both depositional basins could be indicated by the Norian–Lower Jurassic(?) strata of Sierra de Santa Teresa, which offers lithologic and paleontologic resemblances between both entities. Lucas (1996) proposed correlation of the 600-m-thick, pebble-and-boulder clast-supported conglomerate Coyotes Formation of the Barranca Group with the Sierra de Santa Rosa Formation based on supposed similar lithostratigraphies, but as described herein, there are no lithologic similarities between the nonmarine deposits of the Coyotes Formation and the fore-arc deposits of the Sierra de Santa Rosa.

CONCLUSIONS

The El Antimonio Group is herein proposed as a new lithostratigraphic unit that encompasses the Antimonio, Río Asunción, and Sierra de Santa Rosa Formations following a revision of the formational nomenclature proposed by Lucas and Estep (1999b). The most complete section of the Upper Permian–Triassic and Lower Jurassic sedimentary succession of the El Antimonio Group crops out in the Sierra del Álamo and Sierra de Santa Rosa Mountains in northwestern Sonora and is interpreted to represent marginal marine to shallow marine environments in its lower part and deeper marine environments in its upper part. The El Antimonio succession unconformably overlies the Upper Permian Monos Formation in the Sierra del Álamo, whereas its upper part is thrust on Proterozoic metamorphic rocks in the Sierra de Santa Rosa. Several other incomplete Triassic and Lower Jurassic sections that crop out in neighboring localities and to the south in Sonora and Baja California are considered to be correlative with the El Antimonio Group. To the north of these areas and north of the Mojave-Sonora megashear, the only probable correlative sections are shallow marine Triassic and Lower Jurassic strata of the Mojave Desert region in southern Nevada and southeastern California. Lower Triassic successions of that region that may correlate with the Antimonio Formation include the Moenkopi, Virgin Limestone, Union Wash, Silverlake, and Fairview Valley Formations. Upper Triassic and Lower Jurassic sections of the Kings sequence in the central and southern Sierra Nevada in California, which include sedimentary and volcanic rocks, are believed to have been formed in an intra-arc setting (the lower Mesozoic volcanic arc) with the influence of marine sedimentation and represent contemporaneous sedimentation with the El Antimonio Group.

On the basis of these proposed correlations and because we consider the El Antimonio to be a part of the Caborca block, we infer that this succession was deposited in a fore-arc basin originally located adjacent to southern California, which later translated to its present position by left-lateral Jurassic displacement along the Mojave-Sonora megashear. In this proposed paleogeography, the El Antimonio basin was separated from a shallow shelf located to the north (Fig. 3) by the lower Mesozoic volcanic arc that developed in the Mojave Desert and along the California-Nevada border. A single Triassic pluton located on the Caborca block may represent a record of that arc.

The new U-Pb geochronology on detrital zircon and Sm/Nd isotope and petrographic data obtained from terrigenous samples of the El Antimonio Group elucidate its provenance and support the proposed paleogeography. Sixty-two zircon grains from three sandstone samples of the lower, middle, and upper parts of the El Antimonio Group that were dated yielded ages that cluster around 1.8, 1.6–1.7, 1.4, and 1.00–1.18 Ga and 340, 270–240, and 190 Ma. The Proterozoic zircons are interpreted to reflect provenance from the igneous and metamorphic basement provinces of the southwestern United States, although they could also be reworked from the Cordilleran miogeocline and off-shelf assemblages of Nevada and California (such as the Robert Mountains and Golconda allochthons and northern Sierra terrane), where they are also present. Zircon grains of these ages are similarly abundant in Proterozoic and Paleozoic strata in Sonora, so a reworked source from these rocks is also possible.

The closest known sources for the Permian and Lower Triassic zircons are plutons and less abundant volcanic rocks, which during the early Mesozoic formed a volcanic arc that extended from southeastern California (Mojave Desert) to northern California and western Nevada. Probable sources for the single zircon grain dated at 340 Ma are the Sierra-Klamath terranes, according to interpretations by other authors of grains of similar age in rocks of Nevada. Grains from the youngest sample, dated around 190 Ma, most probably reveal provenance from early magmatic activity of the volcanic arc that resumed activity later, during the Early Jurassic, in southwestern North America.

The Sm/Nd isotopic data indicate a progressive decrease in model ages (1.9–1.8–1.13 Ga) from the base upward of the El Antimonio succession, indicating a most probable connection to the Yavapai and Grenville provinces in southwestern United States.

Sandstone and conglomerate clast compositions in the El Antimonio Group indicate an important influence of sedimentary and igneous sources. Lithic grains in sandstone are dominated by sedimentary, volcanic, and plutonic compositions as are clasts in conglomerate. These may indicate provenance from the Proterozoic and Paleozoic cratonic and miogeoclinal cover and from the Triassic-Jurassic magmatic arc of southwestern North America, respectively.

ACKNOWLEDGMENTS

C. González-León and E. Centeno-García acknowledge support from project PAPIIT # IN116599, Universidad Nacional Autónoma de México. G.D. Stanley Jr. acknowledges support from the National Science Foundation (INT-9603171). Thorough and helpful reviews of this manuscript were provided by Jonathan Nourse, Calvin H. Stevens, J. Douglas Walker, and Martin Aberhan though not all of them share our views regarding paleogeographic interpretations. We thank personnel of the Laboratorio Universitario de Geoquímica Isotópica, UNAM, for running the Nd isotopic analyses and Sr. Ruben Romero for drafting the figures.

REFERENCES CITED

Abels, A., and Bischoff, L., 1999, Clockwise block rotations in northern Chile: Indicators for a large-scale domino mechanism during the middle–late Eocene: Geology, v. 27, p. 751–754, doi: 10.1130/0091-7613(1999)027<0751:CBRINC>2.3.CO;2.

Aberhan, M., 1994, Early Jurassic Bivalvia of northern Chile. Part I. Subclasses Palaeotaxodonta, Pteriomorphia, and Isofilibranchia: Beringeria, v. 13, p. 3–115.

Aberhan, M., 1998, Early Jurassic Bivalvia of western Canada. Part I. Subclasses Palaeotaxodonta, Pteriomorphia, and Isofilibranchia: Beringeria, v. 21, p. 57–150.

Alencaster, G., 1961a, Estratigrafía del Triásico Superior de la parte central del Estado de Sonora: Universidad Nacional Autónoma de México, Instituto de Geología, Paleontología Mexicana, No. 11, p. 1–18.

Alencaster, G., 1961b, Fauna fósil de la Formación Santa Clara (Cárnico) del Estado de Sonora: Universidad Nacional Autónoma de México, Instituto de Geología, Paleontología Mexicana, No. 11, p. 19–45.

Anderson, T.H., and Schmidt, V.A., 1983, The evolution of middle America and the Gulf of Mexico–Caribbean Sea region during Mesozoic time: Geological Society of America Bulletin, v. 94, p. 941–966, doi: 10.1130/0016-7606(1983)94<941:TEOMAA>2.0.CO;2.

Anderson, T.H., and Silver, L.T., 1979, The role of the Mojave-Sonora megashear in the tectonic evolution of northern Sonora, *in* Anderson, T.H., and Roldán-Quintana, J., eds., Geology of northern Sonora: Geological Society of America Annual Meeting, San Diego, California, Guidebook-Field Trip 27, p. 59–68.

Anderson, T.H., Eells, J.L., and Silver, L.T., 1979, Precambrian and Paleozoic rocks of the Caborca region, Sonora, Mexico, *in* Anderson, T.H., and Roldán-Quintana, J., eds., Geology of northern Sonora: Geological Society of America Annual Meeting, San Diego, California, Guidebook-Field Trip 27, p. 1–22.

Asmerom, Y., Zartman, R.E., Damon, P.E., and Shafiqullah, M., 1990, Zircon U-Th-Pb and whole-rock Rb-Sr age patterns of lower Mesozoic igneous rocks in the Santa Rita Mountains, southeast Arizona: Implications for Mesozoic magmatism and tectonics in the southern Cordillera: Geological Society of America Bulletin, v. 102, p. 961–968, doi: 10.1130/0016-7606(1990)102<0961:ZUTPAW>2.3.CO;2.

Barth, A.P., Tosdal, R.M., and Wooden, J.L., 1990, A petrologic comparison of Triassic plutonism in the San Gabriel and Mule Mountains, southern California: Journal of Geophysical Research, v. 95, p. 20,075–20,096.

Barth, A.P., Tosdal, R.M., Wooden, J.L., and Howard, K.A., 1997, Triassic plutonism in southern California: Southward younging of arc initiation along a truncated continental margin: Tectonics, v. 16, p. 290–304, doi: 10.1029/96TC03596.

Barth, A.P., Jacobson, C.E., Coleman, D.S., and Wooden, J.L., 2001, Construction and tectonic evolution of Cordilleran continental crust: Examples from the San Gabriel and San Bernardino Mountains, *in* Dunne, G., and Cooper, J., compilers, Geologic Excursions in the California Deserts and Adjacent Transverse Ranges: Pacific Section, SEPM (Society for Sedimentary Geology), Book 88, p. 17–53.

Beauvais, L., and Stump, T.E., 1976, Corals, molluscs, and paleogeography of Late Jurassic strata of the Cerro Pozo Serna, Sonora, Mexico: Palaeogeography, Palaeoclimatology, Palaeoecology, v. 19, p. 275–301, doi: 10.1016/0031-0182(76)90030-4.

Beck, M.E., Jr., 1998, On the mechanism of crustal block rotations in the central Andes: Tectonophysics, v. 299, p. 75–92, doi: 10.1016/S0040-1951(98)00199-1.

Blodgett, R.B., and Frýda, J., 2001, Upper Triassic gastropod biogeography of western North America: Geological Society of America Abstracts with Programs, v. 33, no. 3, p. A-53.

Buch, P.I., 1984, Upper Permian(?) and Lower Triassic metasedimentary rocks, northeastern Baja California, Mexico, *in* Frizzell, V.A., Jr., ed., Geology of the Baja California Peninsula: Pacific Section, Society of Economic Paleontologists and Mineralogists, v. 39, p. 31–36.

Burckhardt, C., 1930, Étude synthétique sur le Mésozoïque mexicain: Mémoire de la Société Paléontologique Suisse, v. 49, part I, 123 p.

Busby-Spera, C.J., 1988, Speculative tectonic model for the early Mesozoic arc of the southwest Cordilleran United States: Geology, v. 16, p. 1121–1125, doi: 10.1130/0091-7613(1988)016<1121:STMFTE>2.3.CO;2.

Callaway, J.M., and Massare, J.A., 1989, *Shastasaurus altispinus* (Ichthyosauria, Shastasauridae) from the Upper Triassic of the El Antimonio district, northwestern Sonora, Mexico: Journal of Paleontology, v. 63, p. 930–939.

Calmus, T., Pérez-Segura, E., and Stinnesbeck, W., 1997, New Middle and Early Jurassic fauna in the Pozos de Serna sequence (Sonora, Mexico)—Consequences for the identity and position of the Caborca terrane: Comptes Rendus de l'Académie des Sciences, Earth and Planetary Science, v. 325, p. 257–263, doi: 10.1016/S1251-8050(97)88298-5.

Centeno-García, E., and Silva-Romo, G., 1997, Geology, geochemistry, and tectonic evolution of central Mexico during Mesozoic time: Revista Mexicana de Ciencias Geológicas, v. 14, p. 244–260.

Centeno-García, E., and Valencia-Moreno, M., 2001, Geoquímica de las rocas volcánicas y volcaniclásticas de la Formación San Hipólito (Triásico-Jurásico), Peninsula de Vizcaíno, Baja California: Actas INAGEQ, v. 7, no. 1, p. 61.

Centeno-García, E., Ruíz, J., Coney, P., Patchett, J.P., and Ortega-Gutiérrez, F., 1993, Guerrero Terrane of Mexico: Its role in the Southern Cordillera from new geochemical data: Geology, v. 21, p. 419–422, doi: 10.1130/0091-7613(1993)021<0419:GTOMIR>2.3.CO;2.

Cooper, G.A., 1953, Permian fauna at El Antimonio, western Sonora: Smithsonian Miscellaneous Collections, v. 119, no. 2, p. 1–13.

Cooper, G.A., and Arellano, R.V., 1946, Stratigraphy near Caborca, northwest Sonora, Mexico: American Association of Petroleum Geologists Bulletin, v. 30, p. 606–619.

Corona, F.V., 1980, Reconnaissance geology of Sierra La Gloria and Cerro Basura, northwestern Sonora, Mexico [M.S. thesis]: Pittsburgh, University of Pittsburg, 232 p.

Damborenea, S.E., and González-León, C.M., 1997, Late Triassic and Early Jurassic bivalves from Sonora, Mexico: Revista Mexicana de Ciencias Geológicas, v. 14, p. 178–201.

Darby, J.B., Wyld, S.J., and Gehrels, G.E., 2000, Provenance and paleogeography of the Black Rock terrane, northwestern Nevada: Implications of U-Pb detrital zircon geochronology, *in* Soreghan, M.J., and Gehrels, G.E., eds., Paleozoic and Triassic paleogeography and tectonics of western Nevada and northern California: Geological Society of America Special Paper 347, p. 77–87.

Delattre, M., 1984, Permian miogeoclinal strata at El Volcán, Baja California, Mexico, *in* Frizzell, V.A., Jr., ed., Geology of the Baja California Peninsula: Pacific Section, Society of Economic Paleontologists and Mineralogists, v. 39, p. 23–29.

DePaolo, D.J., 1981, Radiogenic isotopes and crustal evolution, *in* O'Connell, R.J., and Fyfe, W.S., eds., Evolution of the Earth: American Geophysical Union, Geodynamics Series, v. 5, p. 59–68.

Dickinson, W.R., 1985, Interpreting provenance relations from detrital modes of sandstones, *in* Zuffa, G.G., ed., Provenance of Arenites: Dordrecht, Holland, Reidel, p. 333–361.

Dickinson, W.R., and Lawton, T.F., 2001, Carboniferous to Cretaceous assembly and fragmentation of Mexico: Geological Society of America Bulletin, v. 113, p. 1142–1160, doi: 10.1130/0016-7606(2001)113<1142:CTCAAF>2.0.CO;2.

Dowlen, R.J., and Gastil, G.R., 1981, Reconnaissance geology of Cerro Pozo Serna, western Sonora, Mexico, *in* Ortlieb, L., and Roldán-Quintana, J., eds., Geology of northwestern Mexico and southern Arizona: Geological Society of America Cordilleran Section Annual Meeting, Field Guides and Papers: Hermosillo, Instituto de Geología, Universidad Nacional Autónoma de México, p. 431–435.

Dunbar, C.O., 1953, A giant Permian fusuline from Sonora: Smithsonian Miscellaneous Collections, v. 119, no. 2, p. 14–19.

Estep, J.W., Lucas, S.G., and González-León, C., 1997a, Middle Triassic ammonites from Sonora, México: Revista Mexicana de Ciencias Geológicas, v. 14, p. 155–159.

Estep, J.W., Lucas, S.G., and González-León, C., 1997b, Late Triassic (late Carnian) ammonites at El Antimonio, Sonora: Estación Regional del Noroeste, Instituto de Geología, Universidad Nacional Autónoma de México Publicaciones Ocasionales, v. 1, p. 16–18.

Farmer, G.L., Espinoza, G., Morales, M., Martin, M.W., and Bowring, S.A., 2001, Nd isotope constraints on sources of Neoproterozoic to Early Cambrian siliciclastic sedimentary rocks in northern Sonora: Journal of South American Earth Sciences, v. 14, p. 437–446, doi: 10.1016/S0895-9811(01)00053-0.

Finch, J.W., and Abbott, P.L., 1977, Petrology of a Triassic marine section, Vizcaino Peninsula, Baja California, Mexico: Sedimentary Geology, v. 19, p. 253–273, doi: 10.1016/0037-0738(77)90039-2.

Flores, T., 1929, Reconocimientos geológicos en la región central del estado de Sonora: Instituto Geológico de México, Boletín 49, 267 p.

Gastil, G., and Miller, R., 1984, Prebatholithic paleogeography of Peninsular California and adjacent Mexico, *in* Frizzell, V.A., Jr., ed., Geology of the Baja California Peninsula: Pacific Section, Society of Economic Paleontologists and Mineralogists, v. 39, p. 9–16.

Gehrels, G.E., 2000, Introduction to detrital zircon studies of Paleozoic and Triassic strata in western Nevada and northern California, *in* Soreghan, M.J., and Gehrels, G.E., eds., Paleozoic and Triassic paleogeography and tectonics of western Nevada and northern California: Geological Society of America Special Paper 347, p. 1–17.

Gehrels, G.E., and Stewart, J.H., 1998, Detrital zircon U-Pb geochronology of Cambrian to Triassic miogeoclinal strata of Sonora, Mexico: Journal of Geophysical Research, v. 103, p. 2471–2488, doi: 10.1029/97JB03251.

González-León, C., 1980, La Formación Antimonio (Triásico Superior–Jurásico Inferior) en la Sierra del Álamo, Estado de Sonora: Universidad Nacional Autónoma de México, Instituto de Geología, Revista, v. 4, p. 13–18.

González-León, C., 1989, Evolución de terrenos mesozoicos en el noroeste de México: Universidad de Sonora, Departamento de Geología, Boletín, v. 6, p. 39–54.

González-León, C., 1997a, Sequence stratigraphy and paleogeographic setting of the Antimonio Formation (late Permian–Early Jurassic), Sonora, Mexico: Revista Mexicana de Ciencias Geológicas, v. 14, p. 136–148.

González-León, C., 1997b, The Triassic-Jurassic boundary in the Antimonio Formation from new evidences: Estación Regional del Noroeste, Instituto de Geología, Universidad Nacional Autónoma de México Publicaciones Ocasionales, v. 1, p. 33–36.

González-León, C., Taylor, D.G., and Stanley, G.D., Jr., 1996, The Antimonio Formation in Sonora, Mexico and the Triassic-Jurassic boundary: Canadian Journal of Earth Sciences, v. 33, p. 418–428.

González-León, C., Stanley, G.D., Jr., and Taylor, D.G., 2000, Ammonoid discoveries in the Antimonio Formation, Sonora, Mexico: New constraints on the Triassic-Jurassic boundary: Journal of South American Earth Sciences, v. 13, p. 491–497, doi: 10.1016/S0895-9811(00)00038-9.

Goodwin, D.G., 1999, Paleontology, paleoecology, and depositional environments within the Upper Triassic (Norian) carbonate strata of the Antimonio Formation, northwest Sonora, Mexico [M.S. thesis]: Missoula, University of Montana, 209 p.

Goodwin, D.G., and Stanley, G.D., Jr., 1997, Norian sponge and coral biostromes in the Antimonio Formation, northwestern Sonora, Mexico: Revista Mexicana de Ciencias Geológicas, v. 14, p. 160–166.

Gross, E.L., Stewart, J.H., and Gehrels, G.H., 2000, Detrital zircon geochronology of Neoproterozoic to Middle Cambrian miogeoclinal and platformal strata: Northwestern Sonora: Geofísica Internacional, v. 39, p. 295–308.

Hardy, L.R., 1981, Geology of the central Sierra de Santa Rosa, Sonora, Mexico, *in* Ortlieb, L., and Roldán-Quintana, J., eds., Geology of northwestern Mexico and southern Arizona: Geological Society of America Cordilleran Section Annual Meeting, Field Guides and Papers: Hermosillo, Instituto de Geología, Universidad Nacional Autónoma de México, p. 73–98.

Jacques-Ayala, C., 1993, The Cretaceous in the Caborca-Santa Ana region, northern Sonora, Mexico [Ph.D. dissert.]: Cincinnati, University of Cincinnati, 145 p.

Jaworski, E., 1929, Eine Liasfauna aus Nordwest-Mexiko: Abhandlungen der Schweizerischen Palaeontologischen Gesellschaft, v. 48, no. 4, 12 p.

Keller, W.T., 1928, Stratigraphische beobachtungen in Sonora (nordwest Mexico): Eclogae Geologicae Helvetiae, v. 21, p. 327–335.

Linares, A., Olóriz, F., and Villaseñor, A.B., 1997, Presencia de *Tropidoceras flandrini* (Dumortier) en Pozo de Serna, Sonora (México): Revista Espanola de Paleontologia, v. 12, p. 257–264.

Lucas, S.G., 1996, Correlation and tectonic significance of Lower Jurassic conglomerates in Sonora Mexico, *in* Morales, M., ed., The Continental Jurassic: Museum of Northern Arizona Bulletin 60, p. 497–501.

Lucas, S.G., and Estep, J.W., 1997, The Liassic bivalve *Weyla* from Sonora, Mexico: Estación Regional del Noroeste, Instituto de Geología, Universidad Nacional Autónoma de México Publicaciones Ocasionales, v. 1, p. 45.

Lucas, S.G., and Estep, J.W., 1999a, Triassic-Jurassic boundary in the Sierra del Álamo Muerto, Sonora, Mexico: Albertiana, v. 23, p. 36–41.

Lucas, S.G., and Estep, J.W., 1999b, Permian, Triassic, and Jurassic stratigraphy, biostratigraphy, and sequence stratigraphy in the Sierra del Álamo Muerto, Sonora, Mexico, *in* Bartolini, C., Wilson, J.L., and Lawton, T.F., eds., Mesozoic sedimentary and tectonic history of north-central Mexico: Geological Society of America Special Paper 340, p. 271–286.

Lucas, S.G., and González-León, C., 1994, Marine Upper Triassic strata at Sierra la Flojera, Sonora, Mexico: Neues Jahrbuch für Geologie und Paläontologie Monatshefte, v. 1994, no. 1, p. 34–40.

Lucas, S.G., and González-León, C., 1995, Ichthyosaurs from the Upper Triassic of Sonora and the biochronology of Triassic ichthyosaurs, *in* Jacques-Ayala, C., González-León, C., and Roldán-Quintana, J., eds., Studies on the Mesozoic of Sonora and adjacent areas: Geological Society of America Special Paper 301, p. 17–20.

Lucas, S.G., Marzolf, J.E., Silberling, N.J., and Steiner, M.B., 1996, Lower Mesozoic strata near El Antimonio, Caborca block, Sonora, Mexico: Reevaluation of structure and stratigraphy: Geological Society of America Abstracts with Programs, v. 28, no. 7, p. A-184.

Lucas, S.G., Kues, B.S., Estep, J.W., and González-León, C.M., 1997a, Permian-Triassic boundary at El Antimonio, Sonora, Mexico: Revista Mexicana de Ciencias Geológicas, v. 14, p. 149–154.

Lucas, S.G., Estep, J.W., González-León, C., Paull, R.K., Silberling, N.J., Steiner, M.B., and Marzolf, J.E., 1997b, Early Triassic ammonites and conodonts from Sonora, northwestern Mexico: Neues Jahrbuch für Geologie und Paläontologie Abhandlungen, v. 9, p. 562–574.

Lucas, S.G., Estep, J.W., and Molina-Garza, R.S., 1999, Early Jurassic stratigraphy and ammonites at Cerro Pozos de Serna, Sonora, Mexico, and their tectonic significance: Neues Jahrbuch für Geologie und Paläontologie Monatshefte, v. 1999, no. 6, p. 357–371.

Manuszak, J.D., Satterfield, J.I., and Gehrels, G.E., 2000, Detrital zircon geochronology of Upper Triassic strata in western Nevada, *in* Soreghan, M.J., and Gehrels, G.E., eds., Paleozoic and Triassic paleogeography and tectonics of western Nevada and northern California: Geological Society of America Special Paper 347, p. 109–118.

Marzolf, J.E., 2000, Triassic paleogeography of the Panthalassan margin of southwestern North America: A regional synthesis based on sequence stratigraphy: Zentralblatt für Geologie und Paläontologie, Teil I, v. 11–12, p. 1497–1538.

McRoberts, C.A., 1997, Late Triassic (Norian-Rhaetian) bivalves from the Antimonio Formation, northwestern Sonora, Mexico: Revista Mexicana de Ciencias Geológicas, v. 14, p. 167–177.

Miller, J.S., Glazner, A.F., Walker, J.D., and Martin, M.W., 1995, Geochronologic and isotopic evidence for Triassic-Jurassic emplacement of the eugeoclinal allochthon in the Mojave Desert region, California: Geological Society of America Bulletin, v. 107, p. 1441–1457, doi: 10.1130/0016-7606(1995)107<1441:GAIEFT>2.3.CO;2.

Molina-Garza, R.S., and Geissman, J.W., 1996, Timing of deformation and accretion of the Antimonio terrane, Sonora, from paleomagnetic data: Geology, v. 24, p. 1131–1134, doi: 10.1130/0091-7613(1996)024<1131:TODAAO>2.3.CO;2.

Molina-Garza, R.S., and Geissman, J.W., 1999, Paleomagnetic data from the Caborca terrane, Mexico: Implications for the Cordilleran tectonics and the Mojave-Sonora megashear hypothesis: Tectonics, v. 18, p. 293–325, doi: 10.1029/1998TC900030.

Morales-Montaño, M., Cota-Reyna, J., and López-Soto, R., 1990, Cambrian-Jurassic relations in Sonora, México: Geological Society of America Abstracts with Programs, v. 22, no. 7, p. A-114.

Nourse, J.A., 1995, Jurassic-Cretaceous paleogeography of the Magdalena region, northern Sonora, and its influence on the positioning of Tertiary metamorphic core complexes, *in* Jacques-Ayala, C., González-León, C.M., and Roldán-Quintana, J., eds., Studies on the Mesozoic of Sonora and adjacent areas: Geological Society of America Special Paper 301, p. 59–78.

Nourse, J.A., 2001, Tectonic insights from an Upper Jurassic–Lower Cretaceous stretched-clast conglomerate, Caborca-Altar region, Sonora, Mexico: Journal of South American Earth Sciences, v. 14, p. 453–474, doi: 10.1016/S0895-9811(01)00051-7.

Pálfy, J., and González-León, C., 2000, Lower Jurassic ammonoid biostratigraphy of the Antimonio terrane, Sonora, northwestern Mexico, *in* Calmus, T., and Pérez-Segura, E., eds., Cuarta reunión sobre la geología del Noroeste de México y áreas adyacentes, Libro de Resúmenes: Estación Regional del Noroeste, Instituto de Geología, Universidad Nacional Autónoma de México Publicaciones Ocasionales No. 2, p. 87–88.

Palmer, A.R., and Geissman, J., compilers, 1999, Geologic Time Scale: Geological Society of America, one sheet.

Peiffer-Rangin, F., 1979, Les zones isopiques du Paleozoique inferieur du nord-ouest mexicain, temois du relais entre les Appalaches et la cordillere ouest-americaine: Comptes Rendus de l'Académie des Sciences de Paris, series D, v. 288, no. 20, p. 1517–1519.

Philip, B.I., 1984, Upper Permian (?) and Lower Triassic metasedimentary rocks, northeastern Baja California, Mexico, *in* Frizzell, V.A., Jr., ed., Geology of the Baja California Peninsula: Pacific Section, Society of Economic Paleontologists and Mineralogists, v. 39, p. 31–36.

Poole, F.G., Madrid, R.J., and Oliva-Becerril, J.F., 1991, Geological setting and origin of stratiform barite in central Sonora, Mexico, *in* Raines, G.L., Lisle, R.E., Schafer, R.W., and Wilkinson, W.H., eds., Geology and ore deposits of the Great Basin, v. 1: Reno, NV, Geological Society of Nevada, p. 517–522.

Poole, F.G., Amaya-Martínez, R., and Page, W.R., 2000, Silurian and Devonian carbonate-shelf rocks and Lower Jurassic sequence near Rancho Placeritos, west-central Sonora: 4th Meeting on the Geology of northwest Mexico and adjacent areas, Field Trip 2: Hermosillo, Instituto de Geología, Universidad Nacional Autónoma de México, 24 p.

Riggs, N.R., and Busby-Spera, C.J., 1990, Evolution of a multi-vent volcanic complex within a subsiding arc graben depression: Mount Wrightson Formation, Arizona: Geological Society of America Bulletin, v. 102, p. 1114–1135, doi: 10.1130/0016-7606(1990)102<1114:EOAMVV>2.3.CO;2.

Riggs, N.R., Mattinson, J.M., and Busby, C.J., 1993, Correlation of Jurassic eolian strata between the magmatic arc and the Colorado Plateau: New U-Pb geochronologic data from southern Arizona: Geological Society of America Bulletin, v. 105, p. 1231–1246, doi: 10.1130/0016-7606(1993)105<1231:COJESB>2.3.CO;2.

Saleeby, J.B., and Busby-Spera, C.J., 1986, Fieldtrip guide to metamorphic framework rocks of the Lake Isabella area, southern Sierra Nevada, California, *in* Dunne, G.C., ed., Mesozoic and Cenozoic structural evolution of selected areas, east-central California: Guidebook and Volume, Geological Society of America Cordilleran Section Meeting, p. 81–94.

Saleeby, J.B., and Busby-Spera, C., 1992, Early Mesozoic tectonic evolution of the western United States Cordillera, *in* Burchfield, B.C., Lipman, P.W., and Zoback, M.L., eds., The Cordilleran Orogen: Conterminous US: Boulder, Colorado, Geological Society of America, The Geology of North America, v. G-3, p. 107–168.

Saleeby, J.B., Goodin, S.E., Sharp, W.D., and Busby, C.J., 1978, Early Mesozoic paleotectonic-paleogeographic reconstruction of the southern Sierra Nevada region, *in* Howell, D.G., and McDougall, K.A., eds., Mesozoic paleogeography of the western United States, Pacific Coast Paleogeography Symposium 2, Los Angeles, California, Pacific Section, Society of Economic Paleontologists and Mineralogists, p. 311–336.

Schweickert, R.A., and Lahren, M.M., 1993, Triassic-Jurassic magmatic arc in eastern California and western Nevada: Arc evolution, cryptic tectonic breaks, and significance of the Mojave-Snow Lake fault, *in* Dunn, G., and McDougall, K., eds., Mesozoic Paleogeography of the Western United States-II: Los Angeles, California, Pacific Section, Society of Economic Paleontologists and Mineralogists, Book 71, p. 227–246.

Senowbari-Daryan, B., Stanley, G.D., Jr., and González-León, C., 2001, A new Triassic sponge from the Antimonio terrane, Sonora, Mexico: Journal of South American Earth Sciences, v. 14, p. 447–452, doi: 10.1016/S0895-9811(01)00048-7.

Silver, L.T., and Anderson, T.H., 1974, Possible left-lateral early to middle Mesozoic disruption of the southwestern North America craton: Geological Society of America Abstracts with Program, v. 6, p. 955.

Stacey, J.S., and Kramers, J.D., 1975, Approximation of terrestrial lead isotope evolution by a two-stage model: Earth and Planetary Science Letters, v. 26, p. 207–221, doi: 10.1016/0012-821X(75)90088-6.

Stanley, G.D., Jr., 1979, Paleoecology, structure and distribution of Triassic coral buildups in western North America: University of Kansas Paleontological Contributions, v. 65, p. 1–58.

Stanley, G.D., Jr., and González-León, C.M., 1995, Paleogeographic and tectonic implications of Triassic fossils and strata from the Antimonio Formation, northwestern Sonora: Geological Society of America Special Paper 301, p. 1–16.

Stanley, G.D., Jr., and González-León, C., 1997, New Late Triassic scleractinian corals from the Antimonio Formation, northwestern Sonora, Mexico: Revista Mexicana de Ciencias Geológicas, v. 14, p. 202–207.

Stanley, G.D., Jr., González-León, C., Sandy, M.R., Senowbari-Daryan, B., Doyle, P., Tamura, M., and Erwin, D.H., 1994, Upper Triassic (Carnian-Norian) invertebrates from the Antimonio Formation, Sonora, Mexico: The Paleontological Society Memoir 36, 33 p.

Stevens, C.H., Stone, P., and Kistler, R.W., 1992, A speculative reconstruction of the middle Paleozoic continental margin of southwestern North America: Tectonics, v. 11, p. 405–419.

Stewart, J.H., and Roldán-Quintana, J., 1991, Upper Triassic Barranca Group; nonmarine and shallow-marine rift-basin deposits of northwestern Mexico: Geological Society of America Special Paper 254, p. 19–36.

Stewart, J.H., McMenamin, M.A.S., and Morales-Ramírez, J.M., 1984, Upper Proterozoic and Cambrian rocks in the Caborca region, Sonora, Mexico—Physical stratigraphy, biostratigraphy, paleocurrent studies, and regional relations: U.S. Geological Survey Professional Paper 1309, 36 p.

Stewart, J.H., Anderson, T.H., Haxel, G.B., Silver, L.T., and Wright, J.E., 1986, Late Triassic paleogeography of the southern Cordillera: The problem of a source for voluminous volcanic detritus in the Chinle Formation of the Colorado Plateau: Geology, v. 14, p. 567–570, doi: 10.1130/0091-7613(1986)14<567:LTPOTS>2.0.CO;2.

Stewart, J.H., Amaya-Martínez, R., Stamm, R.G., Wardlaw, B.R., Stanley, G.D., and Stevens, C.H., 1997a, Stratigraphy and regional significance of Mississippian to Jurassic rocks in Sierra Santa Teresa, Sonora, Mexico: Revista Mexicana de Ciencias Geológicas, v. 14, p. 115–135.

Stewart, J.H., Silberling, N.J., and Harwood, D.S., 1997b, Triassic and Jurassic stratigraphy and paleogeography of west-central Nevada and eastern California: U.S. Geological Survey, Open File Report 97–495, 57 p.

Stewart, J.H., Poole, F.G., Harris, A.G., Repetski, J.E., Wardlaw, B.R., Mamet, B.L., and Morales-Ramírez, J.M., 1999, Neoproterozoic to Pennsylvanian inner-shelf miogeoclinal strata in Sierra Agua Verde, Sonora, Mexico: Revista Mexicana de Ciencias Geológicas, v. 16, p. 35–62.

Stewart, J.H., Gehrels, G.E., Barth, A.P., Link, P.K., Christie-Blick, N., and Wrucke, C.T., 2001, Detrital zircon provenance of Mesoproterozoic to Cambrian arenites in the western United States and northwestern Mexico: Geological Society of America Bulletin, v. 113, p. 1343–1356, doi: 10.1130/0016-7606(2001)113<1343:DZPOMT>2.0.CO;2.

Taylor, D.G., Guex, J., and Rakus, M., 2001, Hettangian and Sinemurian ammonoid zonation for the western Cordillera of North America: Bulletin de Géologie Lausanne, v. 350, p. 381–421.

Taylor, G.K., Grocott, J., Pope, A., and Randall, D.E., 1998, Mesozoic fault systems, deformation and fault block rotation in the Andean forearc: A crustal scale strike-slip duplex in the Coastal Cordillera of northern Chile: Tectonophysics, v. 299, p. 93–109, doi: 10.1016/S0040-1951(98)00200-5.

Tosdal, R.M., Haxel, G.B., and Wright, J.E., 1989, Jurassic geology of the Sonoran Desert region, southern Arizona, southeastern California, and northernmost Sonora: Construction of a continental-margin magmatic arc, *in* Jenny, J.P., and Reynolds, S.J., eds., Geologic evolution of Arizona: Arizona Geological Society Digest 17, p. 397–434.

Walker, J.D., 1987, Permian and Middle Triassic rocks of the Mojave Desert, *in* Dickinson, W.R., and Klute, M.A., eds., Mesozoic rocks of southern Arizona and adjacent areas: Arizona Geological Society Digest, v. 18, p. 1–14.

Walker, J.D., 1988, Permian and Triassic rocks of the Mojave Desert and their implications for timing and mechanisms of continental truncation: Tectonics, v. 7, no. 3, p. 685–709.

Walker, J.D., Martin, M.K., and Glazner, A.F., 2002, Late Paleozoic to Mesozoic development of the Mojave Desert and environs, California, *in* Glazner, A.F., Walker, J.D., and Bartley, J.M., eds., Geologic evolution of the Mojave Desert and Southwestern Basin and Range: Geological Society of America Memoir 195, p. 1–18.

Weber, R., 1997, How old is the Triassic flora of Sonora and Tamaulipas and news on Leonardian floras in Puebla and Hidalgo: Revista Mexicana de Ciencias Geológicas, v. 14, p. 225–243.

Whalen, P.A., and Pessagno, E.A., Jr., 1984, Lower Jurassic radiolaria, San Hipólito Formation, Vizcaino Peninsula, Baja California Sur, *in* Frizzell, V.A., Jr., ed., Geology of the Baja California Peninsula: Pacific Section, Society of Economic Paleontologists and Mineralogists, v. 39, p. 53–66.

White, D., and Guiza, R., 1949, Los yacimientos antimoniales de la región de El Antimonio, Estado de Sonora: Instituto Nacional de Investigación Recursos Minerales (México), v. 23, 48 p.

Manuscript Accepted by the Society 24 March 2005

Geological Society of America
Special Paper 393
2005

Lower Mesozoic facies and crosscutting sequence boundaries: Constraints on displacement of the Caborca terrane

John E. Marzolf
Department of Geology, Southern Illinois University, Carbondale, Illinois 62901, USA

Thomas H. Anderson
Department of Geology and Planetary Science, University of Pittsburgh, Pittsburgh, Pennsylvania 15260, USA

ABSTRACT

Five Triassic and Jurassic tectonosequences recognized on the Colorado Plateau have been identified within the Caborca terrane in Sonora, Mexico. Piercing lines defined by truncation of tectonosequence boundaries constrain the pre-offset position of the terrane. Restoration of 1040 ± 290 km of Middle to Late Jurassic left slip along the Mojave-Sonora megashear places the Caborca block in a paleogeographic position that satisfies the constraints provided by truncated tectonosequences and yields predictable regional facies distributions for all tectonosequences. The restoration is consistent with previous estimates of displacements based upon offset of correlative (1) terrains of crystalline basement and (2) pre-Oxfordian cover.

Keywords: Jurassic, megashear, displacement, sequence boundaries, piercing points.

INTRODUCTION

Since proposed by Silver and Anderson (1974), the Mojave-Sonora megashear has remained controversial. Resolution of the controversy over the existence of the megashear is of fundamental importance in understanding not only the tectonic evolution of the Cordilleran margin, but also the opening of the Gulf of Mexico (Anderson and Schmidt, 1983) and the acceleration of Atlantic opening (Marzolf, 2002) associated with Pangean breakup. Fundamental questions about the hypothetical megashear are: (1) where the megashear is located within the United States; (2) what the magnitude of displacement is; and (3) when the displacement occurred.

Initially, the Mojave-Sonora megashear was proposed to explain: (1) an unexpected pattern of Proterozoic crystalline terranes with distinct chronologic histories (Anderson and Silver, 1981); (2) the distribution of cover rocks overlying these basement terranes; and (3) the distribution of Jurassic magmatic units (Anderson and Silver, 1979; Jones et al., 1995). The Mojave-Sonora megashear was postulated to accommodate ~800 km of left-lateral displacement of the Caborca terrane during the Jurassic (Silver and Anderson, 1974). The terrane is named after Caborca, a town near the northern edge of the terrane where exposures of Proterozoic basement and sedimentary cover contrast strongly with rocks to the northeast (e.g., Cooper and Arellano, 1946; Arellano, 1956; Damon et al., 1962; Fries, 1962; Eells, 1972; Anderson and Silver, 1981; and Stewart et al., 1984).

Early attempts to confirm the postulated displacement on the Mojave-Sonora megashear focused upon similarity of lithofacies and biofacies among Neoproterozoic and lower Paleozoic strata covering the crystalline basement of the Caborca terrane and correlative strata in southeastern California and northwestern Nevada (e.g., Anderson and Silver, 1981; Silver and Anderson, 1974; and Stewart et al., 1984). Continuing studies of the stratigraphic

Marzolf, J.E., and Anderson, T.H., 2005, Lower Mesozoic facies and crosscutting sequence boundaries: Constraints on displacement of the Caborca terrane, *in* Anderson, T.H., Nourse, J.A., McKee, J.W., and Steiner, M.B., eds., The Mojave-Sonora megashear hypothesis: Development, assessment, and alternatives: Geological Society of America Special Paper 393, p. 283–308. doi: 10.1130/2005.2393(10). For permission to copy, contact editing@geosociety.org.

cover (e.g., Stewart et al., 1990, 1997a, 1999, 2002, this volume) have expanded the bases upon which comparisons may be made without weakening the displacement hypothesis. Studies of stratigraphically higher units in the Caborca-terrane section also have led to correlations with units far to the north. Ketner (1986) proposed 1000 km of offset of Ordovician Eureka Quartzite from rocks that he considered to be correlative in Sonora. Stanley and González-León (1995), who compared Triassic strata from El Antimonio in the Caborca terrane with Triassic strata in western Nevada, suggest similar displacement. Although these studies suggest similar displacement along the megashear, the low precision inherent in definition of facies boundaries has not provided a compelling argument in favor of megashear displacement.

Alternative tectonic and paleogeographic models (González-León, 1997; Poole and Madrid, 1988; Dickinson and Lawton, 2001) have been proposed to explain the stratigraphy observed in the Caborca terrane. Molina Garza and Geissman (1999) present paleomagnetic data from Triassic and Jurassic rocks in Sonora, which they conclude do not support significant left-lateral displacement of the Caborca terrane in the Middle to Late Jurassic. Stewart et al. (1984) present a summary of tectonic and paleogeographic options to explain the stratigraphy of the Caborca terrane.

We argue that unconformity-bounded stratigraphic sequences within the Triassic and Jurassic stratigraphy of the southwestern United States and northwestern Mexico delineate a pattern of erosional truncation that: (1) is incompatible with the present location of Triassic and Lower Jurassic strata on the Caborca terrane, and (2) provides piercing points that may be used to restore the Caborca terrane to its pre-displacement location in northwest Nevada. We begin by reviewing the most recent revisions to the definition of Triassic and Jurassic stratigraphic sequences on the Colorado Plateau and biostratigraphic- and isotopic-age constraints on sequences and their bounding unconformities. Using these data, we follow with regional correlation of Triassic and Lower and Middle Jurassic stratigraphic sequences among terranes that lie to the south, southwest, west, and northwest of the Colorado Plateau. Regional correlation reveals: (1) a spatial distribution of lithofacies within sequences and an hiatal geometry of sequence boundaries that are not consistent with the orientation of depositional dip predicted by tectonic and paleogeographic models that postulate that the Caborca terrane is autochthonous; (2) lines and piercing points defined by crosscutting sequence boundaries that support ~1000 km of left-lateral displacement along the megashear; and (3) a maximum age of displacement of middle Early Jurassic.

REGIONAL UNCONFORMITIES AND SEQUENCES

Pipiringos and O'Sullivan (1978) and Peterson and Pipiringos (1979) delineate eight regional unconformities that subdivide the Triassic and Jurassic stratigraphy of the Western Interior of the United States. The bodies of sedimentary rocks bounded by these regional unconformities are, by the definition of Sloss (1963), sequences. As currently defined, sequences are nearly conformable bodies of genetically related strata bounded above and below by unconformities or their correlative conformities (Van Wagoner et al., 1988). Although Pipiringos and O'Sullivan (1978) and Peterson and Pipiringos (1979) incorrectly presume that unconformities must lie at system, series, and stage boundaries, they rightly emphasize the utility of the unconformities for regional correlation within the Triassic and Jurassic nonmarine strata of the Western Interior.

Refinements, additions, and deletions (Blakey, 1994; Peterson, 1994; Steiner et al., 1993; Marzolf, 1993b, 1994; Marzolf and Lucas, 1996) to the stratigraphic subdivisions proposed by Pipiringos and O'Sullivan (1978) and Peterson and Pipiringos (1979) currently yield seven sequences (Marzolf, 1994, 1996, 2000) within the Triassic and Jurassic stratigraphy of the Colorado Plateau. The seven sequences are bounded by regional unconformities that are coeval with changes in sedimentary basin geometry or tectonic setting. Therefore, these sequences are tectonosequences in the sense of Winker (1982), Galloway (1989) and Embry (1988, 1993). From oldest to youngest, these tectonosequences are: the Moenkopi, Holbrook, Chinle, Dinosaur Canyon, Glen Canyon, San Rafael, and Brushy Basin (Fig. 1). Higher order sequences have been recognized within the Moenkopi and Chinle tectonosequences (Lucas, 1993; Marzolf, 1993a, 1993b, 2000). Sequence boundaries of these higher-order sequences are not correlative with changes in sedimentary basin geometry or tectonic setting (Embry, 1988, 1993; Marzolf, 2000). Higher-order sequences recognized within the Moenkopi tectonosequence include the Timpoweap, Virgin, and Shnabkaib (Marzolf, 1993a); within the Chinle tectonosequence, the Shinarump–Blue Mesa and Moss Back–Owl Rock.

AGES OF SEQUENCES AND BOUNDING UNCONFORMITIES

Colorado Plateau

Of the seven tectonosequences recognized on the Colorado Plateau, the lower five, Moenkopi, Holbrook, Chinle, Dinosaur Canyon, and Glen Canyon, may be recognized throughout the southwestern United States and northwestern Mexico, the area shown on Figure 2. We restrict the present discussion to these five tectonosequences and the higher-order Timpoweap, Virgin, Shnabkaib, and Shinarump–Blue Mesa and Moss Back–Owl Rock sequences (Fig. 1). With the exception of marine transgressive strata within the Timpoweap, Virgin, and Shnabkaib sequences of the Moenkopi tectonosequence, paralic shales of the Temple Cap Sandstone, and limestone and shale of the Carmel Formation, the five tectonosequences and included higher-order sequences are exclusively of nonmarine origin.

The Moenkopi tectonosequence is separated from the underlying Lower Permian (Leonardian) Kaibab Limestone by the unconformity U-1 at the base of the Moenkopi Formation (Tr-1 of Pipiringos and O'Sullivan, 1978). Two additional unconformities recognized by McKee (1954) further subdivide the Moenkopi

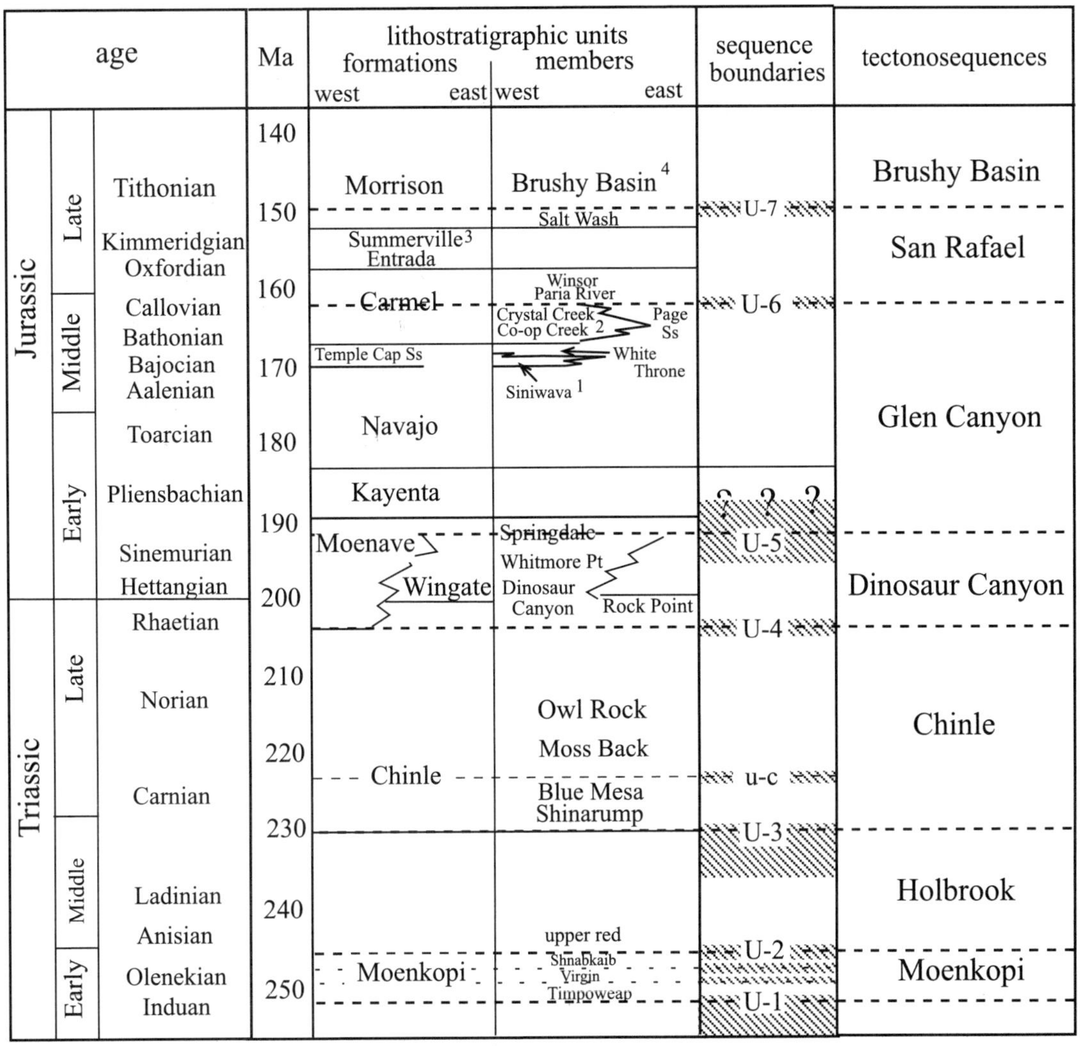

estimated lacuna 1. 170 Ma 2. 166 to 168 Ma (Kowallis and others, 2001) 3. 154.8 +/- 0.5 Ma (Kowallis and others., 1996) 4. 148.1 +/- 0.5 Ma to 150.33 +/- 0.27 Ma (Kowallis and others, 1996)

Figure 1. Early Mesozoic formations, sequence boundaries, and tectonosequences as defined on the southern Colorado plateau. Geochronologic ages for the Triassic from Kozur (2003); for the Jurassic from Pálfy et al. (2000). Ss—sandstone.

tectonosequence into three higher-order sequences, the Timpoweap, Virgin, and Shnabkaib sequences (Marzolf, 1994). The tectonosequence boundary U-2, separating the Moenkopi tectonosequence from the overlying Holbrook tectonosequence, is at the base of the lower massive sandstone near the base of the Moqui Member of the Moenkopi Formation in north-central Arizona and near the base of the upper red member in southwestern Utah (Steiner et al., 1993). The tectonosequence boundary truncates the Moenkopi tectonosequence southward and the Holbrook tectonosequence oversteps the Moenkopi tectonosequence in the same direction.

The Holbrook tectonosequence includes most of the Moqui Member and all of the Holbrook Member of the Moenkopi Formation in northern Arizona, the upper red member of the Moenkopi Formation in southwestern Utah, and the Moody Canyon Member of the Moenkopi Formation in east-central Utah. The upper boundary of the Holbrook tectonosequence is U-3 (Tr-3 of Pipiringos and O'Sullivan, 1978) at the base of the Chinle Formation.

Ammonites (McKee, 1954) and conodonts (Paull and Paull, 1994) collected from marine units within the Timpoweap and Virgin sequences establish respectively a Smithian and early Spathian age for these sequences within the Moenkopi tectonosequence in southwestern Utah. In north-central Arizona, southeastern Utah, and northwestern New Mexico, nonmarine redbeds of the Moenkopi and Holbrook tectonosequences preserve a tetrapod faunal record that indicates an early to late Spathian age for the Moenkopi tectonosequence and an early Anisian age for the Holbrook tectonosequence (Morales, 1993; Lucas, 1999). These age assignments for the Moenkopi and Holbrook tectonosequences are supported by magnetostratigraphy (Steiner et al., 1993). Ladinian rocks are not known on the Colorado Plateau. If deposited, strata of this age were removed prior to deposition of the Shinarump Conglomerate at the base Chinle Formation.

The Chinle tectonosequence is bounded below by U-3. The upper boundary of the Chinle tectonosequence is controversial. Because the age of the Rock Point Member of the Chinle Formation was known to be of late Triassic age, Pipiringos and O'Sullivan (1978) concluded that the Triassic-Jurassic boundary must lie at the contact between the Rock Point Member and the overlying Wingate Sandstone in northeastern Arizona and southeastern Utah. According to them, this required the contact to be an unconformity, which they labeled J-0. As interpreted by Pipiringos and O'Sullivan (1978), J-0 truncated the Rock Point Member westward placing the Dinosaur Canyon Member of the Moenave Formation on the Petrified Forest Member of the Chinle Formation. As a result of these interpretations, the Chinle tectonosequence as originally defined (Marzolf, 1993a) included three higher-order sequences within the Chinle Formation: the Shinarump-Blue Mesa, Moss Back-Owl Rock, and Rock Point (Lucas, 1993). However, the contact of the Rock Point Member of the Chinle Formation with the overlying Wingate Sandstone is conformable (Harshbarger et al., 1957; Lucas et al., 1996; Marzolf et al., 1996). The Rock Point Member unconformably overlies the Owl Rock Member of the Chinle Formation (Lucas, 1993). This unconformity between the Rock Point and Owl Rock members in northeastern Arizona is interpreted to be correlative with the unconformity between the Moenave Formation and underlying Chinle Formation (J-0 of Pipiringos and O'Sullivan, 1978) in southwestern Utah (Lucas et al., 1996; Marzolf, 1994; Marzolf et al., 1996) and is here labeled U-4. Therefore, the Chinle tectonosequence, as now defined, includes only the Shinarump–Blue Mesa and Moss Back–Owl Rock higher order sequences. The boundary between these two sequences is the unconformity u-c, at the base of the Moss Back Member.

The ages of both the Shinarump–Blue Mesa and Moss Back–Owl Rock sequences are well established from extensive collections of vertebrate tetrapods. Based on analysis of these collections, the Shinarump–Blue Mesa sequence is early Carnian in age and the Moss Back–Owl Rock sequence is early to middle Norian in age (Hunt and Lucas, 1991; Lucas and Hunt, 1993, 1989; Lucas, 1999). Magnetostratigraphy of the Chinle Formation supports these age assignments (Steiner and Lucas, 2000).

The Dinosaur Canyon tectonosequence rests on the U-4 unconformity and includes the Rock Point Member of the Chinle Formation and the Wingate Sandstone in northeastern Arizona, the Dinosaur Canyon, Whitmore Point, and lower part of the Springdale Sandstone members of the Moenave Formation in south-central and southwestern Utah. The upper boundary of the Dinosaur Canyon tectonosequence corresponds to the unconformity U-5 at the base of the Kayenta Formation in southeastern Utah and northeastern Arizona and within the Springdale Sandstone in south-central and southwestern Utah. The existence of an unconformity at this stratigraphic horizon is supported by the vertebrate tetrapod record (Clark and Fastovsky, 1986) and magnetostratigraphy (Hutny and Steiner, 2001).

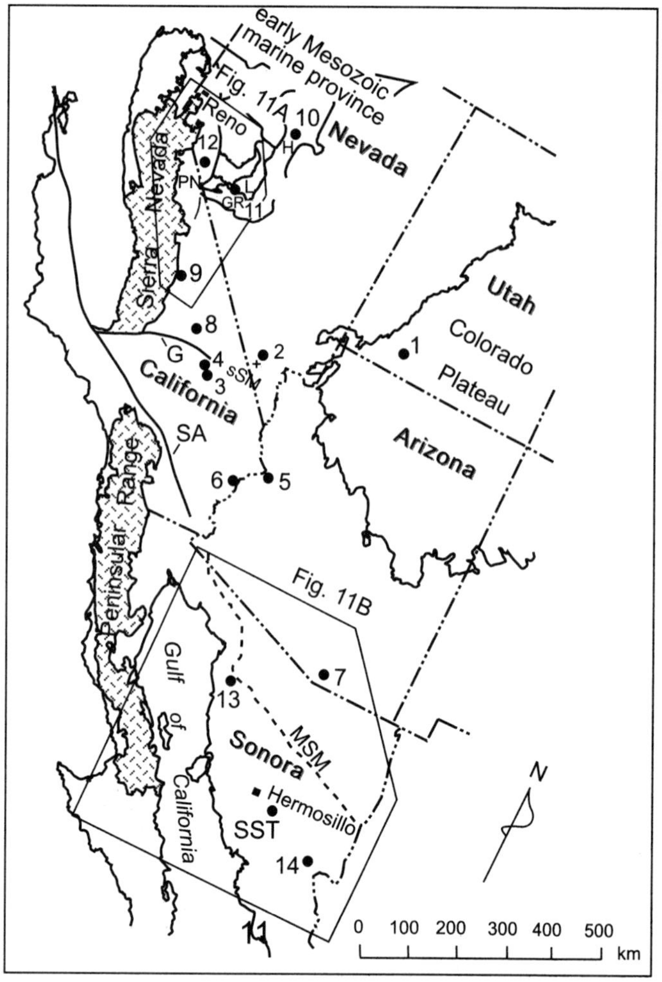

Figure 2. (A) Location of stratigraphic columns with numbers referred to in text. Numbered locations correspond to numbered columns in Figures 3, 4, 5A, and 5B. G—Garlock fault; GR—Gold Range terrane; H—Humboldt terrane; L—Luning terrane; MSM—Mojave-Sonora megashear (postulated trace); PN—Pine Nut terrane; SA—San Andreas fault; sSM—southern Spring Mountains; SST—Sierra de Santa Teresa.

Based on comparison of palynomorphs in the Whitmore Point Member of the Moenave Formation with the Portland Formation of the Newark Series in the eastern United States, Peterson and Pipiringos (1979) assigned a middle Early Jurassic age to the Whitmore Point Member. Recent reevaluation of the palynoflora of the Portland Formation (Fowell and Olsen, 1993; Olsen, 2002) suggests it is earliest Jurassic (early Hettangian) in age. Correlation with the Moenave Formation implies an early Hettangian age for the Whitmore Point Member as well. Existing biostratigraphic data coupled with new magnetostratigaphic data (Hutny and Steiner, 2001) and regional stratigraphic relations (Marzolf et al., 1996) suggests the Triassic-Jurassic boundary lies within the Dinosaur Canyon tectonosequence (Marzolf, 1994, 2000). The base of the tectonosequence is probably Rhaetian. The top of the sequence may be Jurassic but probably no younger than early Sinemurian or early Pliensbachian.

Above the Dinosaur Canyon tectonosequence, the U-5 unconformity separates the Wingate Sandstone from the overlying Kayenta Formation in northeastern Arizona and the Whitmore Point Member and lower part of the Springdale Sandstone from the upper part of the Springdale Sandstone Member of the Moenave Formation to the west. In southwestern Utah, the Glen Canyon tectonosequence includes, in ascending order, the upper part of the Springdale Sandstone Member of the Moenave Formation, the Kayenta Formation, the Navajo Sandstone, the Temple Cap Sandstone, and the lower two members, the Co-op Creek and Crystal Creek members, of the Carmel Formation. The Co-op Creek and Crystal Creek members interfinger eastward with and are replaced by the Page Sandstone in south-central Utah (Fig. 1).

The age of the base of the Glen Canyon tectonosequence is uncertain. The age of the Kayenta Formation is estimated from stratigraphic position, limited vertebrate skeletal fragments (Padian, 1989), and ichnofossil preservation coupled with magnetostratigraphic data (Hutny and Steiner, 2001). All available data suggest a middle Early Jurassic [Sinnemurian to Pliensbachian (ca. 192 to ca. 185 Ma)] age. The age of the upper Glen Canyon tectonosequence is established by marine invertebrates and isotopes. Bivalves from the Co-op Creek Member of the Carmel Formation are correlated with the more fossiliferous Sliderock and Rich members of the Twin Creek Limestone in Wyoming. The Sliderock and Rich members contain ammonites of late early to late Bajocian age (Imlay, 1967a, 1967b, 1980; Pálfy et al., 2000). Recent analyses of ashfall tuffs in the Co-op Creek Member yield U-Pb isotopic ages ranging from 168 to 166 Ma (Kowallis et al., 2001). Tuffs in the underlying Temple Cap Sandstone yield ages of 169 Ma near the top to 170 Ma at its base, directly above the Navajo Sandstone. As the presumption that an unconformity must exist between Lower and Middle Jurassic rocks (Pipiringos and O'Sullivan, 1978; Peterson and Pipiringos, 1979) is unjustified, and as there are no data, neither paleontologic nor isotopic, to support restricting the age of the Navajo Sandstone to the Early Jurassic (Toarcian), the age of the top of the Navajo Sandstone probably is as young as Bajocian (ca. 170 Ma).

Southern Nevada

Except for the overprint of Tertiary extensional deformation, which breaks the stratigraphic continuity, the distinctive lithologies of the Moenkopi, Chinle, and Navajo formations have long been recognized in southern Nevada east of the Mesozoic overthrust belt (Longwell, 1925; Hewett, 1931; Stewart et al., 1972a, 1972b). As on the Colorado Plateau, the Moenkopi Formation unconformably overlies the Lower Permian Kaibab Formation. The Moenkopi Formation records the westward thinning of nonmarine and paralic redbeds and westward thickening of transgressive marine shale and limestone. Marine-invertebrate faunas of these transgressive units indicate Smithian and Spathian ages establishing correlation with the Timpoweap and Virgin Limestone members (Larson, 1966). Stratigraphically higher evaporite beds are correlated with the Shnabkaib Member and overlying redbeds with the upper red member.

Distinctive quartzite- and chert-clast conglomerate overlain by variegated shale identify the Chinle Formation (Stewart et al., 1972a; Wilson and Stewart, 1967; Blakey and Middleton, 1983; Marzolf, 1990). Wilson and Stewart (1967) assigned redbeds unconformably overlying the Chinle Formation to the Moenave and Kayenta formations undivided. Marzolf (1994) restricted these redbeds to the Kayenta Formation, concluding that the Moenave Formation is erosionally absent in southern Nevada. This conclusion is supported by magnetostratigraphic data (Calderone and Beard, 1994). The Kayenta Formation intertongues with the overlying Aztec Sandstone (Wilson and Stewart, 1967; Blakey and Middleton, 1983; Marzolf, 1988b, 1990)

The type locality of the Aztec Sandstone is Aztec Tank near Goodsprings, Nevada, (Hewett, 1931) near the southern end of the nearly continuous sandstone cliffs in the southern Spring Mountains. The Aztec Sandstone is unequivocally the lithostratigraphic equivalent of the Navajo Sandstone. Strata correlative with the Temple Cap Sandstone and Carmel Formation are absent in southern Nevada. The Aztec Sandstone is unconformably overlain by Cretaceous or Tertiary strata or lies structurally below Paleozoic carbonates.

Lower Mesozoic strata of southern Nevada are assigned to the Moenkopi, Holbrook, Chinle, and Glen Canyon tectonosequences. The unconformity that separates the Moenkopi Formation from the underlying Kaibab Formation is correlative with U-1. The Moenkopi Formation, from the base of the Timpoweap Member to the top of the Shnabkaib Member, is assigned to the Moenkopi tectonosequence; the overlying upper red member to the Holbrook tectonosequence. The unconformity near the base of the upper red member is correlative with U-2. A caliche-clast, pebble conglomerate at the base of the Shinarump Conglomerate Member of the Chinle Formation is assigned to the Chinle tectonoseqeunce (Marzolf, 1990). Both the Shinarump–Blue Mesa and Moss Back–Owl Rock sequences within the Chinle tectonosequence and the unconformity, u-c, that separates them, have been recognized in southern Nevada (Lucas and Marzolf, 1993). The overlying Kayenta Formation and Aztec Sandstone are assigned to the Glen Canyon tectonosequence. The unconformity at the base of the Kayenta Formation is correlative with U-5. The Dinosaur Canyon tectonosequence and its basal sequence boundary, U-4, are erosionally truncated near the southwestern corner of the Colorado Plateau.

Within the Mesozoic overthrust belt of southern Nevada, outcrops of the Moenkopi Formation in the Spring Mountains west of Las Vegas are predominantly marine shale and limestone of Smithian and early Spathian age (Larson, 1966; Larson and Lane, 1964). The upper Moenkopi Formation and younger Triassic and Jurassic rocks have been removed below the present erosion surface. In the Meadow Valley Mountains northeast of Las Vegas, a nearly complete section of the Moenkopi Formation from the base of the Timpoweap Member to the base of the upper red member is exposed (Shorb; 1983, Marzolf, 1993a, 1994, 2000). At several locations within the Spring and Meadow Valley mountains, the unconformity U-1 at the base of the Moenkopi Formation has been incised stratigraphically downward through the Kaibab Limestone into underlying Permian redbeds. The incised valleys are filled with limestone pebble, cobble, and boulder conglomerate derived from the Kaibab Limestone (Longwell, 1925; Shorb, 1983; Marzolf, 1993a, 1994).

REGIONAL CORRELATION

Triassic and Jurassic tectonosequences and higher-order sequences defined here are genetically related stratal successions within the chronostratigraphic framework provided by their bounding unconformities (Wheeler, 1964; Van Wagoner et al., 1990). We emphasize correlation of tectonosequences and their included higher-order sequences. Regional correlation of sequences is established by placing maximum and minimum age constraints on sequence-bounding unconformities. To avoid uncertainties inherent in geochronologic scales, age constraints are established by correlating paleontologic ages with paleontologic ages and isotopic ages with isotopic ages. Age constraints on the sequence boundaries of the Moenkopi, Holbrook, Chinle, Dinosaur Canyon, and Glen Canyon tectonosequences on the Colorado Plateau (Fig. 1) permit reliable regional correlation of the these five tectonosequences and their included higher-order sequences.

Colorado Plateau and Southern Nevada to Southern and Western Arizona, the Mojave Desert, and the Southwestern Great Basin

From the Colorado Plateau and southern Nevada to the Mojave Desert and southwestern Great Basin, strata correlated with the Moenkopi Formation include the Buckskin Formation in central western Arizona (Reynolds et al., 1987); the Silver Lake Formation in the northeastern Mojave Desert (Walker and Wardlaw, 1989); the Butte Valley Formation at the southern end of Death Valley (Johnson, 1957), and the Union Wash Formation in the Argus and Slate ranges and Inyo Mountains in east-central California (Lewis et al., 1983; Stone et al., 1991). Correlation of the Buckskin Formation (Fig. 3, col. 5) with the Moenkopi Formation is based upon lithologic similarity, particularly anhydrite schist with evaporates of the Shnakaib Member, and stratigraphic position. Correlation of the other formations with the Moenkopi Formation rely, in addition, on marine invertebrate fossils. Conodonts and ammonites establish the Spathian age of the Silver

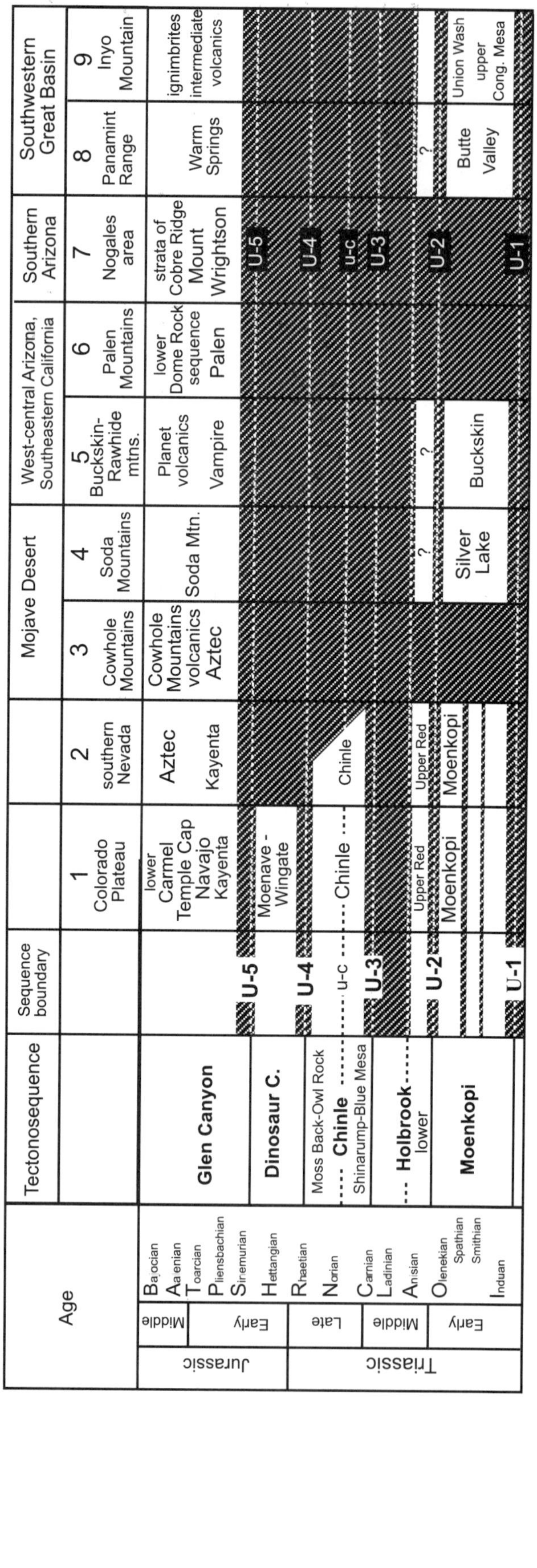

Figure 3. Correlation of tectonosequences from the Colorado Plateau and southern Nevada to the Mojave Desert, south-central Arizona, and the southwestern Great Basin. Location of column numbers given in Figure 2.

Lake Formation (Walker and Wardlaw, 1989; S.G. Lucas, 1995, personal commun.) (Fig. 3, col. 4) and the Smithian (*Meekoceras* zone) and Spathian (*Subcolumbites/tirolites* and *Prohungarites* zones) ages of the Union Wash Formation (Stone et al., 1991) (Fig. 3, col. 9). An Early Triassic (Spathian) age assignment for the Butte Valley Formation is based on lithologic similarity to the Lower Triassic strata in the overthrust belt of southern Nevada and the presence of poorly preserved Spathian ammonites (*Subcolumbites/tirolites* zone) (Johnson, 1957) (Fig. 3, col. 8).

These Lower Triassic formations correlated with the Moenkopi Formation are assigned to the Moenkopi tectonosequence (Fig. 3). With the possible exception of the Union Wash Formation in the Inyo Mountains, Middle Triassic Anisian strata assignable to the Holbrook tectonosequence may be present, but have not been documented. Tentative identification of Anisian ammonites (*Caurus* Zone) in the transition from marine to nonmarine strata near the top of the Union Wash Formation (Stone et al., 1991) support assignment of these strata to the Holbrook tectonosequence.

To the west, southwest, and south of the Colorado Plateau, nonmarine and marine Jurassic strata of the Colorado Plateau are laterally replaced by an assemblage of volcanic rocks that characterizes the Andean-style Jurassic province of the Laurentian cratonal margin. The lower part of the volcanic assemblage comprises interbedded predominantly intermediate volcanic rocks and eolian sandstone; the upper part, widespread, ignimbrite-dominated silicic volcanic rocks. Throughout the Mojave Desert and southern Arizona, the volcanic assemblage overlies Lower Triassic or older rocks.

Since the early 1950s, most geologists who have studied the interbedded eolian sandstone and volcanic rocks in the Mojave Desert have correlated them with the Navajo and Aztec sandstones of the Colorado Plateau and southern Nevada, respectively (Hewett, 1956; Grose, 1959; Novitsky and Burchfiel, 1973; Hamilton, 1982, 1987; Marzolf, 1983, 1988a, 1988b, 1990, 1991, 1993b, 1994; Bilodeau and Keith, 1986; Porter, 1986; Reynolds et al., 1989). Marzolf (1990, 1993b) presented largely stratigraphic arguments supplemented by available U-Pb zircon ages from the arc province and K/Ar ages from the Colorado Plateau in support of this correlation.

More recently, the more silicic volcanic rocks have been isotopically dated in the Mojave Desert (Busby-Spera, 1988; Riggs and Busby-Spera, 1990; Fackler-Adams et al., 1997; Busby et al., 2002; Schermer et al., 2002), western and southern Arizona (Reynolds et al., 1987; Riggs et al., 1993), and southwestern Great Basin (Dunne and Walker, 1993). In summary, interpreted U-Pb ages on zircons from volcanic rocks interbedded with eolian sandstone range from ca. 185–170 Ma. Interpreted ages from the overlying, ignimbrite-dominated succession range from ca. 170–165 Ma (Fig. 4).

Relying on these isotopic ages, Busby-Spera (1988), Riggs and Busby-Spera (1990), Riggs and Blakey, (1993), Riggs et al. (1993), Schermer and Busby (1994), Busby et al., (2002), and

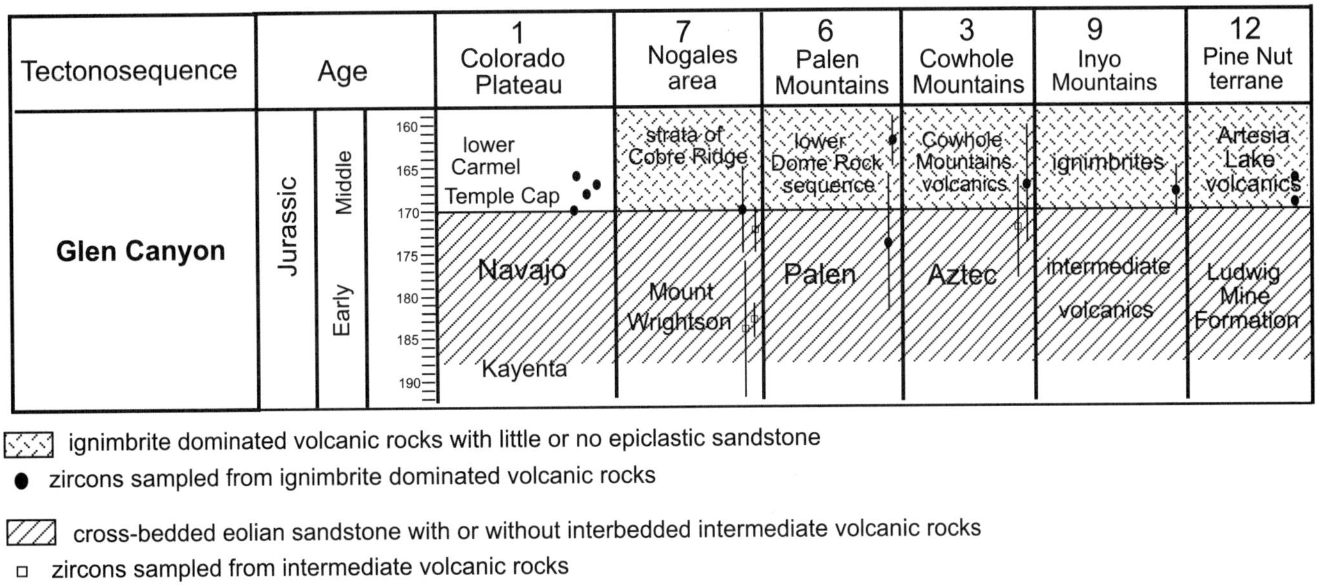

Figure 4. Uranium-lead ages on zircons obtained from volcanic rocks interbedded with and overlying crossbedded, eolian sandstone in the Mojave Desert, southern Arizona, east-central California, and northwestern Nevada. Published ages were obtained from ash beds in the Temple Cap and lower Carmel formations on the southern Colorado Plateau (Kowallis et al., 2001), and ignimbrite-dominated volcanic rocks in: the Cowhole Mountain Volcanics in the Cowhole Mountains (Busby-Spera, 1988); the lower Dome Rock sequence in the Palen Mountains (Fackler-Adams et al., 1997); strata of Cobre Ridge in the Nogales area (Riggs et al., 1993); ignimbrites in the Inyo Mountains (Dunne and Walker, 1993); the Artesia Lake Volcanics in the Pine Nut terrane (Dilles and Wright, 1988). Published ages were obtained from intermediate volcanic rocks interbedded with eolian sandstone in the Mount Wrightson Formation in the Nogales area (Riggs et al., 1993) and the Aztec Sandstone in the Cowhole Mountains (Busby-Spera, 1988).

Schermer et al. (2002) have suggested that the interbedded eolian sandstone and volcanic rocks in the Cowhole Mountains, Palen Mountains, and the Nogales area are correlative with the Entrada, Page, or Temple Cap sandstones. This interpretation is based on the widely accepted assignment of a Toarcian age to the Navajo Sandstone (Pipiringos and O'Sullivan, 1978; Peterson and Pipiringos, 1979), compounded by geochronologic scales that place the minimum age of the Toarcian at 180 Ma or greater (Palmer, 1983; Harland et al., 1990; Gradstein et al., 1994). Recent U-Pb or Ar/Ar ages obtained from the Temple Cap Sandstone and lower Carmel Formation (Kowallis et al., 2001) make correlation of the arc-assemblage eolianites with the Entrada Sandstone untenable. Busby et al. (2002) now conclude that the Middle Jurassic arc-assemblage eolianites in the Mojave Desert and southern Arizona are age equivalents of the Temple Cap and Page sandstones. This interpretation requires westward truncation of the Navajo (Aztec) Sandstone beneath an unconformity at the base of Temple Cap Sandstone or Carmel Formation, presumably either the J-1 or J-2 unconformities of Pipiringos and O'Sullivan (1978) and Peterson and Pipiringos (1979). This interpretation obscures inter-regional correlations of Jurassic sedimentary and volcanic rocks. Furthermore, these correlations are fundamental to recognition of the pattern of cross-cutting sequence boundaries, upon which much of the discussion and conclusions set forth in the following pages, rely.

As stated previously, we argue that presumption of an unconformity at the Early Jurassic–Middle Jurassic boundary is unwarranted, and restriction of the Navajo Sandstone to the Early Jurassic (Toarcian) cannot be justified on the basis of available paleontologic or isotopic ages. Instead, we conclude that the Navajo and Aztec sandstones accumulated over a period of about 15 m.y. (Marzolf, 1988b) and are late Early to early Middle Jurassic in age (ca. 185 Ma to 170 Ma). Therefore, the lower assemblage of interbedded volcanic rocks, having ages of 185 Ma to 170 Ma, and eolian sandstone of the Jurassic arc province are correlative with the Navajo and Aztec sandstones. The upper assemblage of ignimbrites having ages ranging from 170 Ma to ca. 162 Ma is correlative with the Temple Cap, lower Carmel, and Page formations (Marzolf, 1993b). We assign both assemblages to the Glen Canyon tectonosequence (Fig. 3). We further conclude that the stratigraphic relationship of the Jurassic volcanic arc assemblage to underlying Lower Triassic and older rocks resulted from regional truncation below the basal Glen Canyon tectonosequence boundary.

Mojave Desert to Southwestern Great Basin

Volcanic rocks interbedded with cross-stratified eolian sandstone characteristic of the Glen Canyon tectonosequence in the Mojave Desert and southern Arizona are not present in the southwestern Great Basin. Correlation of andesitic volcanic rocks in the southern Panamint Range (Fig. 3, col. 8), Argus and Slate ranges, and the Inyo Mountains (Fig. 3, col. 9) with the Glen Canyon tectonosequence relies on occurrence of andesitic rocks interbedded with eolian sandstone and analogy with stratigraphic relations in the Mojave Desert and southern Arizona. In the latter region, the Glen Canyon tectonosequence rests unconformably on Lower and Middle(?) Triassic or older rocks. Rocks here assigned to the Glen Canyon tectonosequence lie unconformably on Permian rocks in southern Arizona, (Haxel et al., 1988) and southeastern California (Hamilton, 1982, 1987; LeVeque, 1982; Fackler-Adams et al., 1997); on rocks as old as Precambrian in thrust slices in west-central Arizona (Reynolds et al., 1989); and on middle Paleozoic rocks in the Cowhole Mountains (Novitsky-Evans, 1978; Marzolf, 1991) Elsewhere in west-central Arizona (Reynolds et al., 1989) and in the western Soda Mountains (Walker and Wardlaw, 1989) and Mescal Range (Marzolf, 1991), eolian sandstone or eolian sandstone interbedded with andesitic volcanic rocks of the Glen Canyon tectonosequence unconformably overlies Lower Triassic rocks of the Moenkopi tectonosequence. Thus, interbedded intermediate volcanic rocks and eolian sandstone of the lower Glen Canyon tectonosequence overlie rocks no younger than the Lower Triassic Moenkopi tectonosequence (or possibly the Holbrook tectonosequence) and are overlain by isotopically dated 170–165 Ma ignimbrites of the upper Glen Canyon tectonosequence.

In the southern Panamint Range, intermediate volcanic rocks of the Warm Spring Formation unconformably overlie the Lower Triassic Butte Valley Formation, which we assign to the Moenkopi tectonosequence. The unconformity is marked by limestone-cobble conglomerate at the base of the Warm Spring volcanic rocks. Ignimbrites are not present above the andesitic rocks and the Warm Spring Formation remains undated. In the southern Inyo Mountains in east-central California, 200–400 m of unnamed and undated volcanogenic conglomerate, breccia, sandstone and siltstone, and thin basaltic lava flows unconformably overlie the Lower and Middle(?) Triassic Union Wash Formation (Dunne and Walker, 1993) of the Moenkopi and Holbrook(?) tectonosequences. As in the Panamint Range, a limestone-clast conglomerate at the base of the volcanogenic succession marks the unconformity. Silicic volcanic rocks dominated by welded tuffs overlie the volcanogenic assemblage. The tuffs have yielded U-Pb ages of 168 ± 3 to 169 ± 4 Ma (Dunne and Walker, 1993). These rocks are therefore correlative with ignimbrites in the upper part of the Glen Canyon tectonosequence of the Mojave Desert and the Temple Cap Sandstone of the Colorado Plateau.

The unnamed volcanogenic assemblage in the Inyo Mountains lies stratigraphically between the Moenkopi tectonosequence below and ignimbrites of the upper Glen Canyon tectonosequence. We infer that the Warm Spring Formation in the southern Panamint Range occupies the same stratigraphic position. Therefore, we interpret the Warm Spring Formation and the unnamed volcanogenic assemblage in the Inyo Mountains to be Early to early Middle Jurassic in age, correlative with the interbedded eolian sandstone and intermediate volcanic rocks of the Aztec sandstone. The eolian sandstone has pinched out westward. The intermediate volcanogenic assemblage and overlying 168–169 Ma ignimbrites in the Inyo Mountains, similar rocks in the Argus and Slate ranges, and the Warm Spring Formation in the southern Panamint Range are assigned to the Glen Canyon tectonosequence (Marzolf, 1993b, 1994) (Fig. 3).

Colorado Plateau to Northwestern Nevada

Triassic and Lower and Middle Jurassic paralic, shallow-marine, and basinal sedimentary rocks and volcanic rocks crop out extensively in basin-bounded mountain ranges in the western Great Basin of west-central Nevada (Muller and Ferguson, 1939; Silberling, 1959; Speed and Silberling, 1989; Oldow et al., 1993; Stewart et al., 1997b). These lower Mesozoic rocks are distributed among eight allochthonous and parautochthonous terranes (Oldow, 1984; Silberling et al., 1987) of the early Mesozoic marine province of Speed (1978) (Fig. 2). Although lithofacies vary considerably, four lithologically distinct chronostratigraphic assemblages have been recognized within and among terranes.

Although postulated displacements among terranes are on the order of 10s to 100s of kilometers and crustal shortening within terranes is of the same order of magnitude (Oldow et al., 1993), marine invertebrate fossils collected from unconformity-bounded sequences, supplemented by isotopic ages on volcanic rocks, permits reliable correlation of the unconformities among mountain ranges within and among terranes (Silberling and Wallace, 1969; Nichols and Silberling, 1977; Speed, 1978; Speed and Silberling, 1989; Oldow et al., 1993; Stewart et al., 1997b). Biochronostratigraphic and isotopic age control on these unconformities establishes correlation with sequence boundaries of the Moenkopi, Holbrook, Chinle, Dinosaur Canyon, and Glen Canyon tectonosequences and the Shinarump–Blue Mesa and Moss Back–Owl Rock higher-order sequences defined on the Colorado Plateau (Marzolf, 1994, 2000) (Fig. 5).

No single terrane contains a complete Triassic and Jurassic stratigraphic column that includes all five tectonosequences, Moenkopi through Glen Canyon. The youngest, pre-Cretaceous Mesozoic rocks in any of the allochthonous Lovelock, Sand Springs, Pamlico, Luning, and Pine Nut terranes are Bajocian sedimentary rocks (Oldow et al., 1993) and 165.8 Ma volcanic rocks (Dilles and Wright, 1988). Lower Triassic rocks crop out in the parautochthonous Gold Range terrane and Lower, Middle, and Upper Triassic rocks crop out in the parautochthonous Humboldt terrane. Isolated outcrops of Lower Triassic and Lower Jurassic rocks are known in the Gold Range terrane but the lower Mesozoic stratigraphy is not sufficiently well known to identify and correlate sequences and sequence boundaries.

The most complete stratigraphic section of Lower and Middle Triassic rocks in the early Mesozoic marine province is in the parautochthonous Humboldt terrane. Lower Triassic rocks are included in the Koipato Group and Lower, Middle, and Upper Triassic rocks are included in the Star Peak Group (Silberling and Wallace, 1969; Nichols and Silberling, 1977). The Star Peak Group is overlain by the Upper Triassic Grass Valley, Dun Glen, and Winnemucca formations. From oldest to youngest, the Star Peak Group comprises the Tobin, Dixie Valley, Favret, and

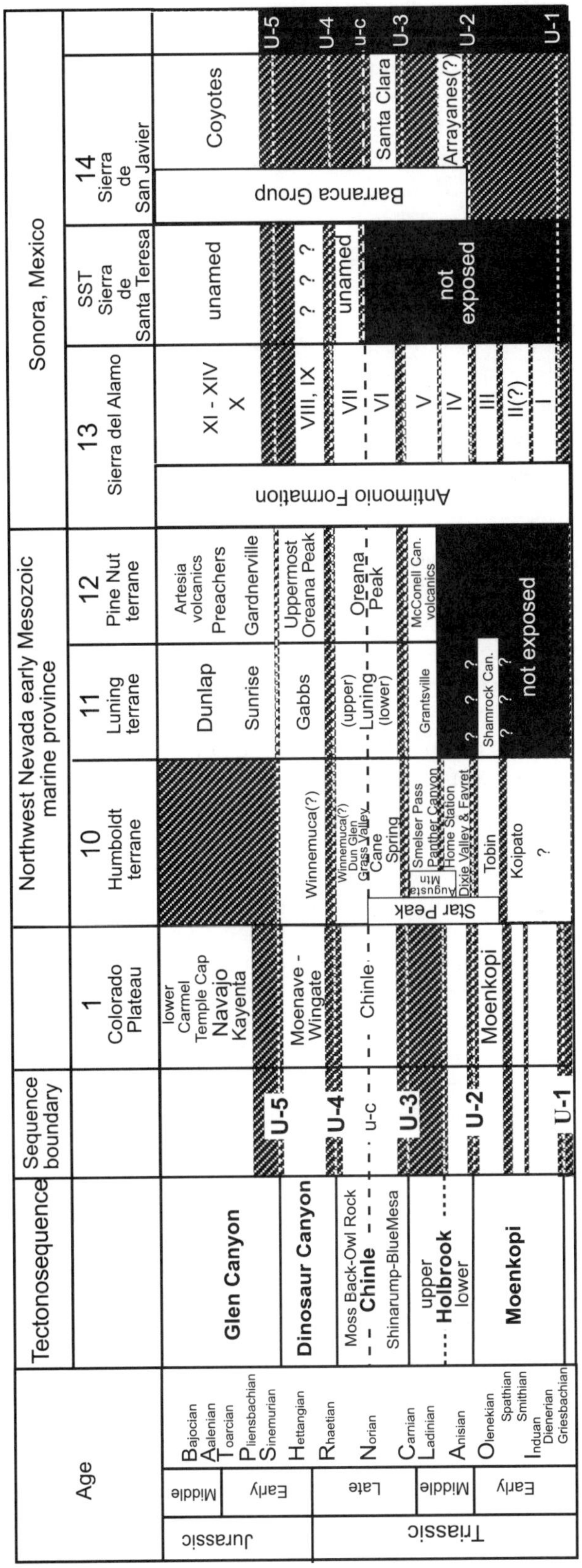

Figure 5. Correlation of tectonosequences from the Colorado Plateau to the early Mesozoic marine province in western Nevada and Sonora, Mexico. Location of column numbers given in Figure 2. Can.—Canyon.

Augusta Mountain formations. The Augusta Mountain Formation is further subdivided into the Home Station, Panther Canyon, and Smelser Pass members.

Widespread unconformities subdivide the Triassic stratigraphic succession of the Humboldt terrane into six sequences. The six sequences include: (1) the Koipato Group; (2) the Tobin Formation; (3) the Dixie Valley and Favret formations and Home Station member of the Augusta Mountain Formation; (4) the Panther Canyon and Smelser Pass members of the Augusta Mountain Formation; (5) the Cane spring Formation; and (6) the Grass Valley, Dun Glen, and Winnemucca formations (Fig. 6B) (Marzolf, 1993a, 2000). Age of the sequences and their bounding unconformities is provided by ammonites and other age-diagnostic marine invertebrate megafossils (Silberling and Wallace, 1969; Nichols and Silberling, 1977) (Fig. 6A, 6B; col. 10).

Only two of the unconformities included within the interval from the Koipato Group to the Winnemucca Formation are correlative with tectonosequence boundaries defined on the Colorado Plateau. The unconformity separating the Tobin and Dixie Valley formations is bracketed between late Spathian (*Subcolumbites zone*) strata below and early Anisian (*Shoshonensis* zone) strata above and is therefore correlative with the boundary between the Moenkopi and Holbrook tectonosequences, U-2. The base of the Cane Spring Formation is a Shinarump-like, chert-pebble to cobble conglomerate and crossbedded, conglomeratic sandstone up to 100 m thick (Nichols and Silberling, 1977). The basal conglomerate fills a broad paleovalley incised across the subaerially exposed underlying Smelser Pass Member of the Augusta Mountain Formation (Marzolf, 1993a, 1994, 2000). The unconformity between the Augusta Mountain and the Cane Spring formations is bracketed between strata of early (*Desatoyense zone*) and mid Carnian (*Dilleri* zone) ages. It is correlative with the unconformity at the base of the Shinarump Conglomerate, U-3.

Other unconformities included within the interval from the Koipato Group to the Winnemucca Formation are either of local extent, boundaries of higher-order sequences, or lie within the Middle Triassic interval removed by erosion on the Colorado Plateau. The unconformity between the Koipato and Star Peak groups is potentially a higher-order sequence boundary possibly correlative with the boundary between the Virgin and Shnabkaib (*Subcolumbites* zone) sequences in the Moenkopi tectonosequence. The unconformity between the Cane Spring and Grass Valley formations lies between late Carnian rocks below and early Norian (*Kerri* zone) strata above. It is correlative with the u-c sequence boundary, which divides the Chinle tectonosequence into the Shinarump–Blue Mesa and Moss Back–Owl Rock sequences, on the Colorado Plateau. Finally, the intra-formational unconformity at the base of the Panther Canyon member of the Augusta Mountain Formation is bracketed between rocks of early Ladinian and early Carnian (*Desatoyense zone*) ages. It divides the stratigraphic interval between the Dixie Valley Formation and the Smelser Pass Member of the Augusta Mountain Formation into a lower and upper Holbrook tectonosequence (Marzolf, 2000). Rocks of these ages and their included unconformity were removed by erosion on the Colorado Plateau prior deposition of the Chinle tectonosequence. Latest Triassic and Early to Middle Jurassic rocks equivalent to the Dinosaur Canyon and Glen Canyon tectonosequences of the Colorado Plateau have not been identified in the parautochthonous Humboldt terrane.

The most complete stratigraphic sections of Upper Triassic and Early to Middle Jurassic rocks crop out in allochthonous terranes of the early Mesozoic marine province. Sequence-stratigraphic analysis of 15 stratigraphic sections representing lithotectonic assemblages from each of four allochthonous terranes in the southern part of the province delineates three inter-terrane unconformities (sequence boundaries 1, 2, and 3 of Oldow et al., 1993) bounding four correlative, complete, or partial sequences. Ammonites, encompassing 50 ammonite zones, distributed among all four terranes (Silberling, 1984) provide high-resolution age control on sequences and sequence boundaries (Fig. 6A and 6B). Age constraints placed on sequence boundaries 1, 2, and 3 of Oldow et al. (1993) establish correlation with U-3, U-4, and U-5 of the Colorado Plateau.

Stratigraphic sections in the Luning and Pine Nut terranes (Fig. 6B; columns 11 and 12) are particularly germane to discussion of displacement and timing of offset on the Mojave-Sonora megashear. From bottom to top, sequences defined from a composite stratigraphic section from the Luning terrane include: (1) the greenstone of Shamrock Canyon and Grantsville Formation; (2) the Luning Formation; (3) the Gabbs Formation; and (4) the Sunrise Formation and overlying Dunlap Formation or lithostratigraphically equivalent Westgate Formation. Lithostratigraphic units included within correlative sequences in the Pine Nut Mountains in the Pine Nut terrane include, from oldest to youngest: (2) most of the Oreana Peak Formation; (3) the uppermost units of the Oreana peak Formation; and (4) the Gardnerville and Preachers formations and overlying Veta Granda and Gold Bug formations. In the Singatse Range, correlative sequences include: (2) the Malachite Mine Formation, overlying the volcanics of McConnell Canyon, and the lower member of the overlying Mason Valley Limestone; (3) the upper member of the Mason Valley Limestone; and (4) the Gardnerville and Ludwig Mine formations and Artesia Lake Volcanics (Stewart et al., 1997b). U-Pb ages obtained from hypabyssal and volcanic rocks within stratigraphic sections in the Pine Nut terrane add isotopic age control to paleontologically dated sequences (Dilles and Wright, 1988; Wyld and Wright, 1993).

Ammonites from the upper Grantsville Formation indicate a late middle Ladinian age. The sequence, which includes the Grantsville Formation, is correlative with the Smelser Pass Member of the Augusta Mountain Formation assigned to the upper Holbrook tectonosequence. Unconformity 1 of Oldow et al. (1993) between the Grantsville and Luning formations correlates with U-3. The overlying sequence, the Luning Formation, ranges in age from middle late Carnian (*Welleri* zone) to late early Norian (*Magnus* zone). In the Union District and Shoshone Mountains, this sequence comprises two shallowing-upward cycles ranging in age from middle late Carnian (*Welleri* zone)

A

System	Series	Stage	Substage	No.	Zone
Jurassic	Lower	Toarcian	Upper	50	Levesquei
				49	Thouarsense
				48	Variabilis
			Lower	47	Bifrons
				46	Falcifer
				45	Tenuicostatum
		Pliensbachian	Upper	44	Spinatum
				43	Margaritatus
			Lower	42	Davoei
				41	Ibex
				40	Jamesoni
		Sinemurian	Upper	39	Raricostatum
				38	Oxinotum
				37	Obtusum
			Lower	36	Turneri
				35	Semicostatum
				34	Bucklandi
		Hettangian		33	Angulata
				32	Liasicus
				31	Planorbis
Triassic	Upper	Rhaetian		30	Crickmayi
				29	Amoenum
		Norian	Upper	28	Cordilleranus
			Middle	27	Columbianus - upper
				26	Columbianus - lower
				25	Rutherfordi
			Lower	24	Magnus
				23	Dawsoni
				22	Kerri
		Carnian	Upper	21	Macrolobatus
				20	Welleri
				19	Dilleri
			Lower	18	Nanseni
				17	Desatoyense
	Middle	Ladinian	Upper	16	Sutherlandi
				15	Maclearni
				14	Meginae
			Lower	13	Poseidon
				12	Subasperum
		Anisian		11	Occidentalus
				10	Meeki
				9	Rotelliformis
				8	Shoshonensis
				7	Anagymnotoceras
				6	Caurus
	Lower	Spathian		5	Haugi
				4	Prohungarites/subcolumbites beds
				3	Columbites/tirolites beds
		Smithian		2	Anasibirites beds
				1	Meekoceras beds

EXPLANATION: FIGURES 6B & 6C

- coal
- evaporites
- carbonate rocks; limestone & dolostone
- fine-grained, marine siliciclastid rocks
- sandstone with large-scale cross-stratification
- chert-clast, nonmarine red beds
- fluvial-deltaic sandstone and finer siliciclastic rocks
- sandstone
- carbonate-clast conglomerate
- chert-clast orthoconglomerate
- oligomict orthoconglomerate and conglomeratic sandstone
- volcanic-clast conglomerate
- volcanic & volcaniclastic rocks
- hypabyssal igneous rocks
- tectonosequence boundary
- higher-order sequence boundary
- 22● Dot indicates stratigraphic position of age-significant fossils. Numbers refer to ammonite zones in Figure 6A.

Figure 6 (*on this and following two pages*). (A) Ammonite zonation for the Triassic and Early to Middle Jurassic after Silberling and Tozer (1968) and Oldow et al. (1993); Kozur (2003). (B) Definition of sequences and biochronostratigraphic correlation of sequence boundaries within stratigraphic columns of the Humboldt, Luning, and Pine Nut terranes. Closed circles give stratigraphic location of ammonites and other marine invertebrate fossils after Silberling (1984) and Oldow et al. (1993); numbers refer to ammonite zones of Figure 6A. Numbers in parentheses on sequence boundaries are designations of boundaries of Oldow et al. (1993). Humboldt stratigraphic column and fossil locations constructed from Nichols and Silberling (1977). (C) Definition of sequences and biochronostratigraphic correlation of sequence boundaries from the Humboldt, Luning, and Pine Nut terranes to stratigraphic columns of the Sierra del Alamo and the Sierra de San Javier, Sonora. Stratigraphic column, sequences, fossil location and age designation for the Sierra del Alamo from González-León (1997) and Taylor et al. (1996); stratigraphic column, fossil location and age designation for the Sierra de San Javier from Stewart and Roldán-Quintana (1991) and Alencaster (1961). Fm—Formation; M—Member; Grp—Group.

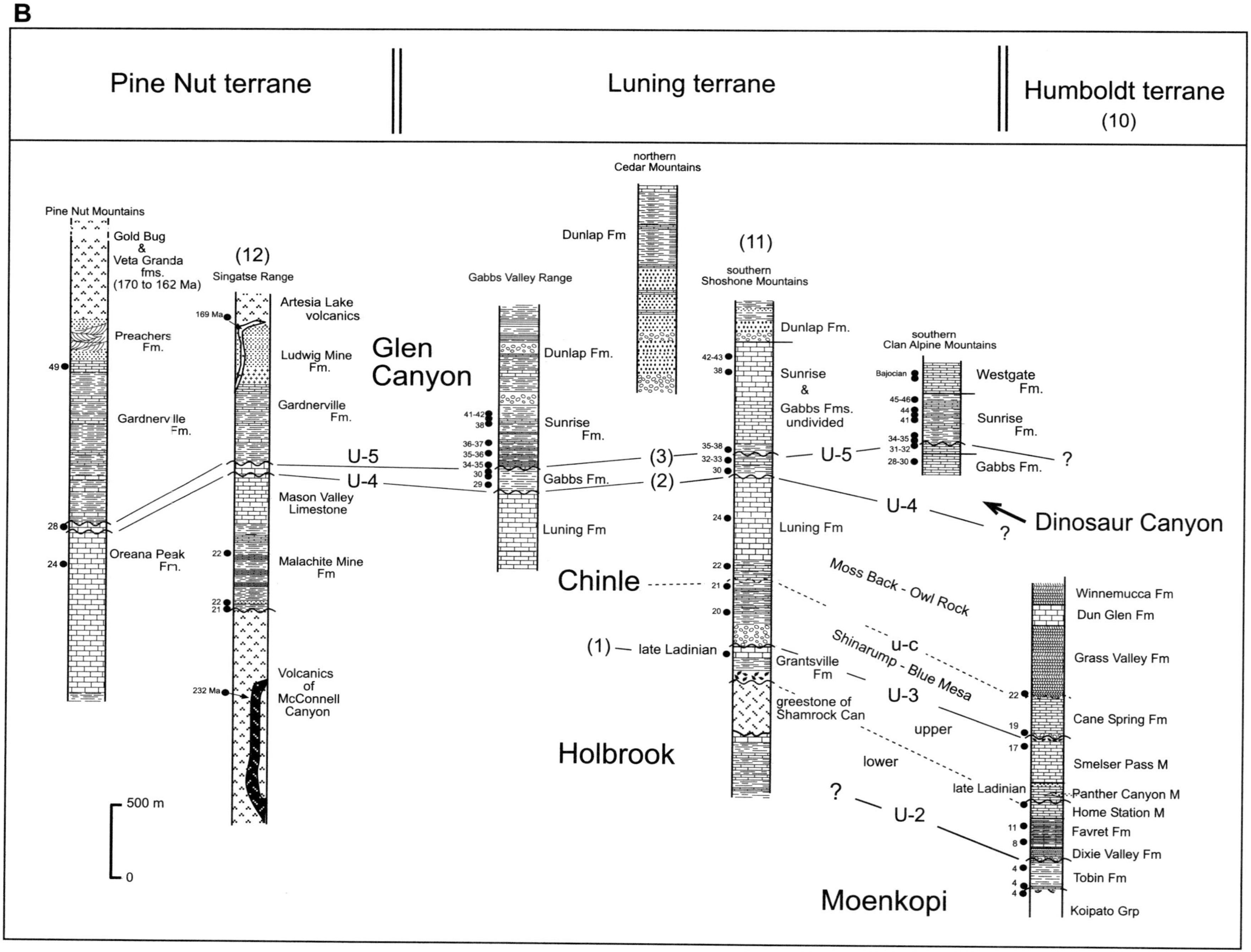

B
Pine Nut terrane
Luning terrane
Humboldt terrane
(10)
Pine Nut Mountains
Gold Bug & Veta Granda fms. (170 to 162 Ma)
Preachers Fm.
Gardnerville Fm.
Oreana Peak Fm.
(12)
Singatse Range
Artesia Lake volcanics
Ludwig Mine Fm.
Gardnerville Fm.
Mason Valley Limestone
Malachite Mine Fm
Volcanics of McConnell Canyon
169 Ma
232 Ma
Glen Canyon
U-5
U-4
Gabbs Valley Range
Dunlap Fm.
Sunrise Fm.
Gabbs Fm.
Luning Fm
northern Cedar Mountains
Dunlap Fm
(3)
(2)
(11)
southern Shoshone Mountains
Dunlap Fm.
Sunrise & Gabbs Fms. undivided
Luning Fm
Grantsville Fm
greestone of Shamrock Can
Chinle
(1) — late Ladinian
Holbrook
southern Clan Alpine Mountains
Bajocian
Westgate Fm.
Sunrise Fm.
Gabbs Fm.
?
Dinosaur Canyon
Moss Back - Owl Rock
u-c
Shinarump - Blue Mesa
U-3
upper
lower
late Ladinian
U-2
Moenkopi
Winnemucca Fm
Dun Glen Fm
Grass Valley Fm
Cane Spring Fm
Smelser Pass M
Panther Canyon M
Home Station M
Favret Fm
Dixie Valley Fm
Tobin Fm
Koipato Grp
500 m
0

C

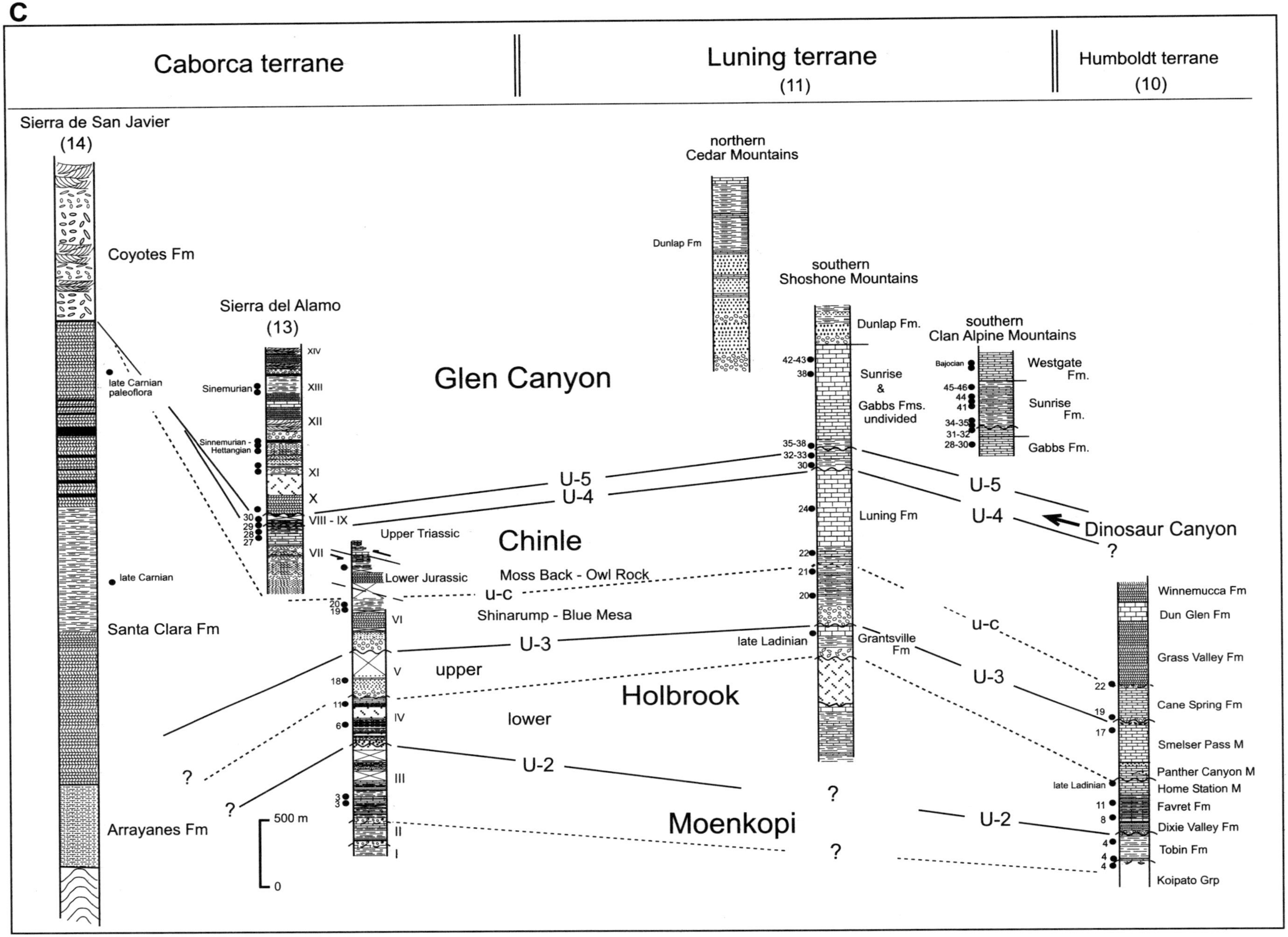

Caborca terrane
Luning terrane
(11)
Humboldt terrane
(10)
Sierra de San Javier
(14)
Coyotes Fm
late Carnian paleoflora
late Carnian
Santa Clara Fm
Arrayanes Fm
Sierra del Alamo
(13)
Sinemurian
Sinnemurian - Hettangian
XIV
XIII
XII
XI
X
VIII - IX
VII
30
29
28
27
Upper Triassic
Lower Jurassic
20
19
18
11
6
3
3
VI
V
IV
III
II
I
500 m
0
Glen Canyon
Chinle
Moss Back - Owl Rock
u-c
Shinarump - Blue Mesa
U-5
U-4
U-3
upper
lower
Holbrook
U-2
Moenkopi
?
northern
Cedar Mountains
Dunlap Fm
southern
Shoshone Mountains
Dunlap Fm.
Sunrise
&
Gabbs Fms.
undivided
Luning Fm
Grantsville
Fm
late Ladinian
42-43
38
35-38
32-33
30
24
22
21
20
southern
Clan Alpine Mountains
Bajocian
45-46
44
41
34-35
31-32
28-30
Westgate
Fm.
Sunrise
Fm.
Gabbs Fm.
Dinosaur Canyon
Winnemucca Fm
Dun Glen Fm
Grass Valley Fm
Cane Spring Fm
Smelser Pass M
Panther Canyon M
Home Station M
Favret Fm
Dixie Valley Fm
Tobin Fm
Koipato Grp
22
19
17
late Ladinian
11
8
4
4
4

through late late Carnian (*Macrolobatus* zone) and, early early Norian (*Kerri* zone) through late early Norian (*Magnus* zone). The thick, primary dolostone at the top of the Luning Formation is unfossiliferous and could be as young as late Norian (*Cordilleranus* Zone) (N.J. Silberling, 2004, personal commun.). These age constraints imply correlation with the Shinarump–Blue Mesa and Moss Back–Owl Rock higher-order sequences, respectively, within the Chinle tectonosequence. The transgressive surface separating the cycles is correlative with u-c.

The Luning Formation is overlain by the sequence defined by the Gabbs Formation. Unconformity 2 of Oldow et al. (1993), between the Luning and Gabbs formations, correlates with U-4. The stratigraphic position of the upper sequence boundary near the Gabbs-Sunrise contact (unconformity 3 of Oldow et al. (1993)) relative to ammonite zones is ambiguous. Thus, sequence 2 ranges in age from early Rhaetian (*Amoenum* zone) through late Hettangian (*Angulata* zone) or early early Sinemurian (*Bucklandi* zone). The Gabbs Formation, the Gabbs and lowermost Sunrise formations, and correlative strata of the uppermost Oreana Peak Formation in the Pine Nut terrane are assigned to the Dinosaur Canyon tectonosequence, which contains the Triassic-Jurassic boundary (Taylor et al., 1992, 2000).

Strata correlative with the Glen Canyon tectonosequence are included in the Sunrise, Dunlap, and Westgate formations of the Luning terrane and the Gardnerville, Preachers, and Ludwig Mine formations and the Artesia Lake, Fulstone, Veta Grande, and Gold Bug volcanic rocks, of the Pine Nut terrane. The Sunrise Formation ranges from basal Jurassic (lower Hettangian) through early Toarcian (*Tenuicostatum* and *Falcifer* zones). The basal Glen Canyon tectonosequence boundary lies within the lower Sunrise Formation. The Dunlap Formation is sparsely fossiliferous, but contains Sinemurian and Pliensbachian fossils (Silberling, 1984; Stewart et al., 1997b). In the southernmost Clan Alpine Mountains of the Luning terrane, the upper Sunrise Formation contains ammonites of the *Tenuicostatum* zone indicating an early early Toarcian age. The Westgate Formation, the lithostratigraphic equivalent of the Dunlap Formation, contains ammonites of Bajocian age. In the Pine Nut terrane, the age of the uppermost Gardnerville Formation just below the Preachers Formation contains Sinemurian (Stewart et al., 1997b) and late Toarcian ammonites (*Thouarsense* zone). In the Singatse Range, the Artesia Lake Volcanics unconformably(?) overlie the Ludwig Mine Formation (Oldow et al., 1993). The unconformity and overlying Artesia Lake Volcanics are intruded by granitoid rocks which yield U-Pb ages of 168.5 and 169.4 Ma (Dilles and Wright, 1988). The Artesia Lake Volcanics are overlain by the Fulstone Volcanics which have yielded a U-Pb age of 166.5 Ma and are intruded by 165 Ma granitoids (Proffett and Dilles, 1984; Dilles and Wright, 1988). In the Pine Nut Mountains, the lithostratigraphic equivalent of the Ludwig Mine Formation, the Preachers Formation, is overlain successively by volcanic rocks of the Veta Grande and Gold Bug formations. Interpreted U-Pb maximum and minimum ages of the Veta Grande and Gold Bug formations are 170 Ma and 162 Ma, respectively (Wyld and Wright, 1993). The Artesia Lake and Fulstone volcanics are correlative with tuffs in the Temple Cap Sandstone and ignimbrites in southern Arizona, the Mojave Desert, and the Inyo Mountains in east-central California. The Veta Grande and Gold Bug formations are correlative with the Temple Cap Sandstone and lower Carmel Formation. All of these volcanic rocks are therefore assigned to the upper Glen Canyon tectonosequence. The unconformity 3 of Oldow and others near the base of the Sunrise Formation and near the base of the Gardnerville Formation correlates with U-5.

Ammonites within the predominantly marine Triassic and Lower to Middle Jurassic stratigraphy, supplemented by limited isotopic dates on volcanic and cross-cutting intrusive rocks, place very tight age constraints on widespread regional unconformities recognized within the lower Mesozoic marine province. These age constraints permit reliable correlation with sequence boundaries of the Colorado Plateau. Although age control on sequence boundaries of the Colorado Plateau is moderately well established by a variety of nonmarine fossil taxa, correlation with sequence boundaries of the lower Mesozoic marine province greatly improves the resolution of age control and ties these boundaries to the ammonite marine standard.

In summary, U-2 lies near the Spathian-Anisian boundary between the *Subcolumbites* and *Shoshonensis* ammonite zones; U-3 near the middle Carnian between the *Desatoyense* and *Dilleri* zones; and u-c near the Carnian boundary between the *Macrolobatus* and *Kerri* zones. U-4 lies near the Norian-Rhaetian boundary between the *Cordilleranus* and *Amoenum* zones. Finally, U-5 at the base of the Glen Canyon tectonosequence lies near the Hettangian-Sinnemurian boundary either between the *Angulata* and *Bucklandi* zones or between the *Bucklandi* and *Semicostatum* zones. This ambiguity is not easily resolved in northwestern Nevada.

Northwestern Nevada to Sonora, Mexico

In Sonora, Mexico, Triassic rocks are unknown and Lower Jurassic rocks are exclusively volcanic northeast of the postulated trace of the Mojave-Sonora megashear. Southwest of the postulated megashear, both marine and nonmarine rocks crop out. In the Sierra del Alamo, west of the town of Caborca in northwestern Sonora, Triassic and Jurassic marine sedimentary rocks are included in the Antimonio Formation (González-León, 1980) (Fig. 6C, col. 13). The Antimonio Formation is a succession of predominantly shallow-marine strata nearly 4 km thick, which overlies the Upper Permian (Guadalupian) Monos Formation. Originally described as Upper Triassic and Lower Jurassic (González-León, 1980), the Antimonio Formation is now recognized to include Upper Permian, and Lower and Middle Triassic, as well as Upper Triassic and Lower Jurassic strata (Lucas et al., 1997a).

González-León (1997) constructed a composite stratigraphic section of the Antimonio Formation from two measured sections. He subdivided the composite section into 14 unconformity-bounded sequences, which he labeled, from oldest to youngest, I through XIV. Five of the unconformities are correlative with tectonosequence boundaries U-1, U-2, U-3, U-4, and U-5 (Fig. 6C, col.13). A sixth unconformity is correlative with the boundary between the lower and upper Holbrook sequences. All other unconformities are correlative with higher order sequences within the Moenkopi and Chinle tectonosequences, a transgressive surface within the Dinosaur Canyon tectonosequence, and local, autocyclic unconformities within the Glen Canyon tectonosequence.

The unconformity between sequences I and II is tentatively correlated with U-1. Sequence I contains fragmental brachiopods (Lucas et al., 1997b) and silicified fusulinids (Marzolf, unpublished) that indicate a Late Permian (Guadalupian) age. Sequence III has yielded ammonites of the *Subcolumbites/Tirolites* zone and is, therefore, correlative with the Spathian Virgin or Shnabkaib sequence of the Moenkopi tectonosequence. Although fossils have not been recovered from sequence II, it is tentatively correlated with the Smithian Timpoweap sequence. The unconformity between sequences III and IV is correlated with U-2. Sequence IV contains ammonoids identified as *Paracrochordiceras* and *Paranevadites* indicating a late Anisian age (*Occidentalis* zone) and assignment to the lower Holbrook tectonosequence. The unconformity between sequences IV and V is correlated with the sequence boundary between the upper and lower Holbrook sequences. Vertebrae of shastasaurid ichthyosaurs and the pectinoid bivalve *Halobia* suggest sequence V is early Carnian in age (Estep et al., 1997). However, a Ladinian age cannot be ruled out (González-León, 1997). Assignment of a Ladinian or early Carnian age suggest correlation with the Panther Canyon and Smelser Pass members of the Augusta Mountain Formation of the Star Peak Group and assignment of sequence V to the upper Holbrook tectonosequence.

The unconformity between sequences V and VI is correlated with U-3. The lower part of sequence VI contains ammonites indicating a middle to late Carnian age (*Sirenites nanseni, Dilleri, and Welleri* zones). A covered interval separates the lower and upper part of sequence VI. The age of strata in the upper part of sequence VI is controversial (Lucas et al., 1997a; Damborena and González-León, 1997). We interpret a fault in the covered interval places Lower Jurassic rocks of the upper part of sequence VI on the upper Carnian rocks of the lower part. Sequence VII is in thrust contact over the upper part of sequence VI. Sequence VII contains an abundant shallow-water marine fauna (Stanley et al., 1994; McRoberts, 1997, Damborena and González-León, 1997). The late Carnian and Norian ages of sequences VI and VII place both sequences within the Chinle tectonosequence. The well-established late Carnian age of sequence VI and the middle to late Norian age of sequence VII establishes correlation with the Shinarump–Blue Mesa and Moss Back–Owl Rock sequences, respectively. Although complicated by structure, the inferred contact between these two sequences is correlated with u-c.

Sequences VIII and IX contain ammonites indicating a Rhaetian age (*Amoenum* and *Crickmayi* zones, respectively). We suggest the boundary between VIII and IX is a transgressive surface and assign both to the Dinosaur Canyon tectonosequence. The unconformity between sequences VII and VIII is correlated with U-4 and the unconformity between sequences IX and X is correlated with U-5. Sequence X contains ammonites and bivalves indicating late Hettangian to early Sinemurian ages (Taylor et al., 1996; 2000). The erosional absence of latest Triassic (uppermost *Crickmayi* zone) and lower and middle Hettangian (*Planorbis* and *Liasicus* zones) strata below U-5 indicates a greater hiatus in Sonora than in western Nevada.

Sequences X through XIV are assigned to the Glen Canyon tectonosequence. The shallow-marine strata of sequence X are lithologically similar to the Sunrise Formation. Sequences XI, XII, XIII, and XI have a combined thickness of over 800 m dominated by coarse-grained siliciclastic rocks including conspicuous pebble to cobble conglomerate and cross-stratified sandstone. Unconformities bounding these sequences are at the bases of conglomerates, conglomeratic sandstones, and sandstones and are interpreted to be of autocyclic origin and local in extent. Ammonites in sequences XI and XII and the bivalve Weyla in sequence XII indicate a Sinemurian age. Ammonites abundant in a few beds of sequence XIII indicate a Sinemurian to Pliensbachian age (Taylor et al., 1996, 2000). Fossils have not been recovered from sequence XIV, but it is probably of Pliensbachian age or younger (González-León, 1997).

In central eastern Sonora in the Sierra de San Javier, Stewart and Roldán-Quintana (1991) assign lower Mesozoic strata to the Barranca Group (Fig. 6C, col. 14). The Barranca Group rests with angular unconformity on lower Paleozoic strata and is overlain by volcanic rocks of Cretaceous age (McDowell et al., 2001). The Barranca Group comprises three formations. From oldest to youngest, they are the Arrayanes, Santa Clara, and Coyotes formations. Fossils have not been recovered from the Arrayanes Formation. Its age, therefore, is unknown. It may be Middle or Late Triassic in age. A shallow-marine fauna including brachiopods, pelecypods, and a single ammonite in strata near the middle of the Santa Clara Formation (Alencaster, 1961; Stewart and Roldán-Quintana, 1991) indicate a late Carnian age. The Santa Clara becomes deltaic upward. The fine-grained deltaic sediments preserve an abundant paleoflora. On the basis of this paleoflora, Weber (1980; 1997) concluded that the upper Santa Clara Formation is also Carnian in age. The overlying Coyotes Formation represents an abrupt change in lithology from the fine-grained, organic-rich siliciclastic sediments of the Santa Clara Formation to sandy metaquartzite-cobble to boulder conglomerate and conglomeratic sandstone. The stratigraphically upward progression from shallow-marine to deltaic strata displayed by the Santa Clara suggests transition from the transgressive and low highstand systems tracts to progradational highstand delta. Neither an overlying sequence boundary, nor a lowstand, nor transgressive systems tract is evident. The Santa Clara Formation is therefore considered to be the record of a single late Carnian sequence.

The nonmarine conglomerates and conglomeratic sandstone of the Coyotes Formation are incompatible with the deltaic depositional environments of the upper Santa Clara Formation. The contact of the Coyotes Formation with the underlying Santa Clara

Formation is abrupt, suggesting an unconformity. Furthermore, elsewhere to the northwest in Sonora, petrologically similar sandy conglomerates and conglomeratic sandstones overlie or contain interbeds of finer-grained ammonite-bearing rocks. The ammonites are exclusively middle Early Jurassic, principally Sinemurian in age, e.g., sequences XI through XIV in the upper Antimonio Formation. Petrologically similar lower Mesozoic conglomerates or sandstones of a documented age other than Early Jurassic are unknown in Sonora. For these reasons, Lucas and Marzolf (1994) and Lucas (1996) concluded that the unfossiliferous Coyotes Formation is unconformable on the Santa Clara Formation and is Early Jurassic in age. The unconformity at the base of the Coyotes Formation is U-5 and the Coyotes Formation is assigned to the Lower Jurassic Glen Canyon tectonosequence. The Arrayanes Formation is tentatively assigned to the upper Holbrook tectonosequence; the Santa Clara Formation to the Shinarump–Blue Mesa sequence within the Chinle tectonosequence. The contact between the Arrayanes and Santa Clara formations is the unconformity U-3 (Fig. 6C, col. 14).

FACIES

In the Western Interior of the United States, facies distributions and sedimentological data record a dramatic change in depositional strike from north-south during Moenkopi tectonosequence deposition to east-west from the beginning of Holbrook tectonosequence deposition through the remainder of the Triassic into the earliest Jurassic. Upper Triassic to Lower Jurassic tectonosequences record a change from terrestrial to marine depositional environments from southeast to northwest (Fig. 7). Nonmarine, fluvial and eolian strata predominate on the Colorado Plateau. Deltaic, shallow-marine, and basinal strata predominate in northwestern Nevada. In both regions, facies boundaries trend northeast to southwest.

During the interval of erosion responsible for the unconformity U-3 at the base of the Chinle tectonosequence, broad valleys were incised from southeast to northwest across the central Colorado Plateau and from south to north along the plateau's southwestern margin and adjacent southern Nevada (Blakey and Gubitosa, 1983; Marzolf, 1990). Sedimentological studies and paleocurrent indicators document southeast to northwest and south to north sediment transport by an extensive fluvial system throughout Chinle deposition (Stewart et al., 1972b; Dubiel, 1994, Lucas, 1993). The sediment source was volcanically active (Stewart et al., 1986) and exposed crystalline basement. Paleoslope indicators suggest that the source terrain extended from eastern Chihuahua, Mexico, to the present Gulf of California. The nonmarine, fluvial sediments of the Chinle tectonosequence on the Colorado Plateau are correlative with deltaic strata of the Humboldt terrane, deltaic to shallow-marine strata of the Luning terrane, and basinal strata of the Sand Springs and Lovelock terranes of northwestern Nevada (Elison and Speed, 1988; Oldow et al., 1990; Marzolf, 1993a, 2000). The Late Triassic shoreline lay along the eastern and southern margins of the early Mesozoic

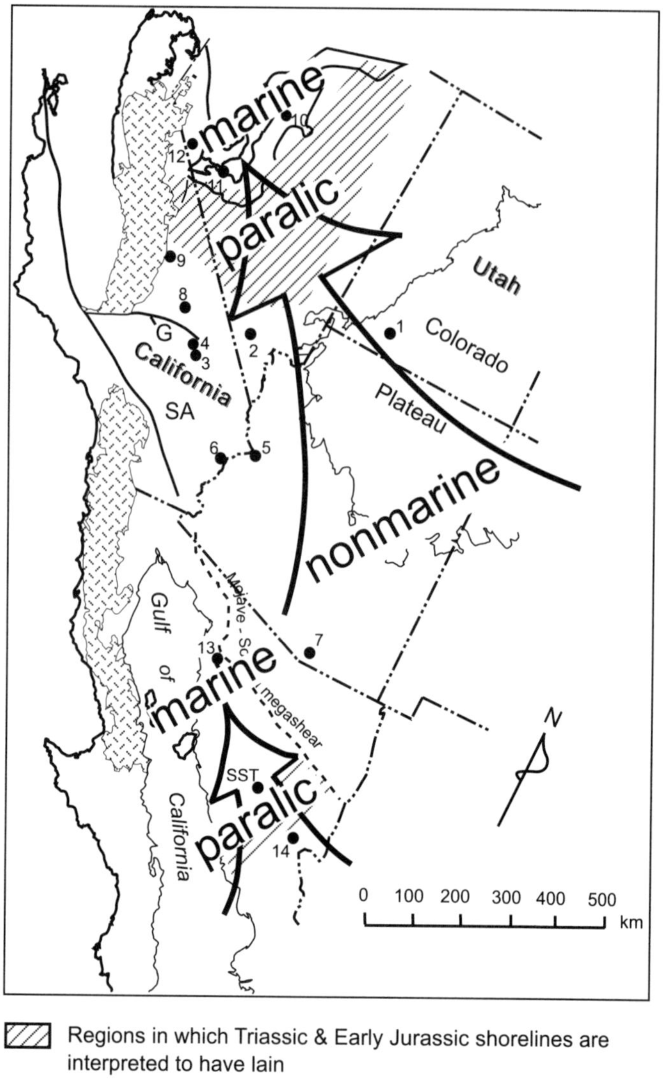

Figure 7. Facies gradient of nonmarine to marine strata from the Colorado Plateau to the early Mesozoic marine province of western Nevada compared to facies gradients of coeval strata in the Caborca terrane in Sonora. SA—San Andreas fault; SST—Sierra de Santa Teresa

marine province in northwestern Nevada. Shifting shoreline position is recorded by transgressive-regressive cycles developed in paralic facies of the Luning Formation (Oldow, 1981) and the Cane Spring, Grass Valley, Dun Glen, and Winnemucca formations of the Humboldt terrane (Silberling and Wallace, 1969; Elison and Speed, 1988; Marzolf, 1993a, 1994, 2000).

On the Colorado Plateau, the Dinosaur Canyon tectonosequence includes nonmarine sandstone, siltstone, and mudstone deposited in ephemeral streams and lakes of the lower Moenave Formation and eolian sandstone of the Wingate Sandstone. In northwestern Nevada, the sequence comprises shelfal impure limestone, siltstone and mudstone of the Gabbs Formation in the Luning terrane and the uppermost Oreana Peak Formation in the Pine

Nut terrane (Oldow et al., 1993; Marzolf, 1994). The shoreline during Dinosaur Canyon tectonosequences has not been identified but, must have lain between the Colorado Plateau on the southeast and the early Mesozoic marine province on the northwest.

In the middle Early Jurassic, the Colorado Plateau and Western Interior became a back-arc basin. The basin filled with nonmarine fluvial and eolian sediments of the lower Glen Canyon tectonosequence over the entire Colorado Plateau and southern Nevada. To the south of the Colorado Plateau and the southwest of southern Nevada, the basin was closed by the middle Early to early Middle Jurassic volcanic arc. In these directions, eolian sandstone interfingers with and is laterally replaced by subaerial intermediate volcanic rocks (Marzolf, 1983, 1990, 1994). Westward from the central to northern Colorado Plateau, nonmarine eolian deposition extended as far west as northeastern Nevada. Farther west in the early Mesozoic marine province of northwestern Nevada, the lower Glen Canyon tectonosequence includes shallow-marine strata of the Sunrise and upper Gardnerville formations overlain by normal-fault bounded interbedded nonmarine and marine conglomerate, sandstone, siltstone, and mudstone of the Dunlap Formation (Oldow and Bartel, 1987; Oldow et al., 1993). On the south, the shoreline lay between the Luning terrane and the nonmarine volcanic rocks of the lower Glen Canyon Group in the southern Inyo Mountains. To the east, the shoreline lay between the eastern margin of the early Mesozoic marine province and northeastern Nevada.

In central eastern Sonora, the Chinle tectonosequence comprises deltaic and fluvial sedimentary rocks of the Santa Clara Formation (Cojan and Potter, 1991; Stewart and Roldán-Quintana, 1991) and passes northwestward to marine-shelf strata in northwestern Sonora (González-León, 1980, 1997; Marzolf, 1997). Basinal rocks, containing detrital zircons with an age distribution nearly identical to those of the Chinle tectonosequence in northwestern Nevada (Gehrels et al., 1993), crop out in northern Baja California (Kimbrough, D.L., 2001, personal commun.). The Dinosaur Canyon tectonosequence is absent in central eastern Sonora, but in northwestern Sonora, it comprises shallow-marine strata similar to the Gabbs Formation in northwestern Nevada (Stanley and González-León, 1995). The lower Glen Canyon tectonosequence was deposited in fluvial and alluvial fan environments and is composed of nonmarine conglomerate and conglomeratic sandstone in central eastern Sonora. Marine strata are absent. In northwestern Sonora, the Glen Canyon tectonosequence is lithologically similar to correlative rocks in northwestern Nevada including shallow-marine strata of the Sunrise Formation overlain by marginal-marine sandstone and conglomerate of the Dunlap Formation (Marzolf and Lucas, 1994). We interpret perceptible differences in lithology between marine rocks of the Luning and Pine Nut terranes and the Caborca terrane to be the result of differences in the depth of water in which the marine sediments were deposited. In general, the marine strata of the Luning and Pine Nut terranes were deposited in deeper water, farther from shore than shallow-water marine strata of correlative sequences in the Caborca terrane.

DISCUSSION

The absence of the Upper Triassic Chinle tectonosequence from the southwestern corner of the Great Basin and Mojave Desert, coupled with the apparent absence of all Triassic strata from southern Arizona and northern Sonora northeast of the postulated trace of the Mojave-Sonora megashear, is the result of regional truncation of Triassic and Jurassic strata below the basal Chinle, U-3, and basal Glen Canyon, U-5, unconformities (Fig. 8). Marine and marginal marine strata of the Triassic and Lower Jurassic Moenkopi, Holbrook, Chinle, Dinosaur Canyon and Glen Canyon tectonosequences of the northern Caborca terrane lie west and southwest of a broad region of northern Sonora, southern Arizona, and southeastern California in which Triassic rocks are unknown and Lower Jurassic rocks comprise volcanic rocks and volcanic rocks interbedded with eolian sandstone. These same sequences lie south of a broad region extending from southern Nevada and east-central California across the Mojave Desert over which the Upper Triassic Chinle and Upper Triassic–Lower Jurassic Dinosaur Canyon tectonosequences have been removed below the basal Glen Canyon tectonosequence boundary U-5 prior to deposition of the Lower Jurassic lower Glen Canyon tectonosequence (Fig. 8). This regional pattern of erosional truncations appears to preclude the lower Mesozoic strata of Sonora from having occupied their present location in the early Mesozoic.

Comparison of facies between Sonora and northwestern Nevada lead Stanley and González-León (1995) to suggest displacement of the Caborca terrane from a location in proximity to the Luning terrane. We agree. Similarities are recorded in the sequence stratigraphy of the early Mesozoic marine province and the Caborca terrane. These two regions are the only locations in all of southwestern North America where the Moenkopi, Holbrook, Chinle, Dinosaur Canyon, and Glen Canyon tectonosequences are developed in predominantly marine-shelf sedimentary rocks. The most striking similarity is that in all of southwestern North America, the early Mesozoic marine province and the Caborca terrane are the only two localities where the late Ladinian to early Carnian strata of the upper Holbrook sequence are present below U-3. In both regions, this sequence comprises shallow-marine rocks. In southwestern North America, the conspicuous unconformity near the Ladinian-Carnian boundary at the base of the upper Holbrook sequence has only been identified in the early Mesozoic marine province and the Caborca terrane. In addition, basal unconformities of the Chinle, Dinosaur Canyon, and Glen Canyon tectonosequences of the early Mesozoic marine province and the northwestern end of the Caborca terrane record comparable hiati approaching conformity in shallow-marine rocks.

Facies within the Chinle and lower Glen Canyon tectonosequences suggest that the entire Caborca terrane in Sonora was displaced from the western extension of the subsequently shortened Gold Range and Luning terranes. In Sonora, the Late Triassic and Early Jurassic shoreline lay between the Sierra de Santa Teresa and the Sierra de San Javier. In east-central California and northwestern Nevada, the shoreline lay between the Mesozoic marine province and the Inyo Mountains. Lithofacies and biofacies suggest that, prior to displacement, the marine strata at the northwestern end of the Caborca terrane lay in close proximity but south, landward, of the southern boundary of the Luning terrane. The predominantly nonmarine rocks of the southeastern end of the Caborca terrane lay still farther south of the Luning terrane.

If the nonmarine, marginal-marine, and shallow-marine Triassic and Lower Jurassic strata of the Caborca terrane are postulated to be autochthonous, then the present distribution of lower Mesozoic sedimentary and volcanic rocks in southwestern North America requires that the early Mesozoic shoreline between the Colorado Plateau and northwestern Nevada must have extended to central eastern Sonora. If the right-angle bend in the Golconda allochthon, the Mina deflection of Wetterauer, is interpreted as a post-Jurassic oroclinal bend (Albers, 1967; Wetterauer, 1977; Stewart et al., 1968), then reconstruction of pre-Cretaceous paleogeography requires straightening the bend producing a Triassic and Early Jurassic shoreline trending from northwest to southeast. The shoreline must have lain east of the present location of the Caborca terrane (Fig. 9). The alternative interpretation—that the Mina deflection is a primary feature in the western margin of North America (Ferguson and Muller, 1949; Geissman et al., 1984; Oldow, 1984; Speed and Silberling, 1989)—requires a reconstruction that produces a shoreline trending eastward north of the present location of the Caborca terrane. Regardless of which of these two reconstructions is chosen, both interpretations predict that depositional dip of Triassic and Lower Jurassic tectonosequences in the Caborca terrane should trend southward or southwestward (Fig. 9) producing: (1) landward to seaward facies gradients also trending from north or northeast to the south or southwest; (2) facies boundaries parallel to a northwest-southeast–trending or east-west–trending shoreline; and (3) an increase in the hiatus of coastally onlapping sequence boundaries in the landward direction, i.e., northeastward or northward.

Observed facies gradients and orientation of facies boundaries within the Chinle and Glen Canyon tectonosequences on the Caborca terrane are not compatible with the southward-trending or southwestward-trending depositional dip predicted for an autochthonous Caborca terrane. For example, if the late Triassic shoreline lay north of the Caborca terrane during Chinle tectonosequence deposition, then the volcanically active source terrain exposing Precambrian basement must have lain within the 400 km between the Palen and Big Maria mountains and the northern end of the Caborca terrane near Caborca. Relatively coarse fluvial volcanogenic sediment carried southward would be expected near Caborca. Facies should become increasingly offshore southward. This is not what is observed. Instead, Late Triassic landward to seaward lithofacies gradients in Sonora show the same trend as that of the southwestern United States—southeast to northwest, at a right angle to that predicted for an autochthonous Caborca terrane (Fig. 7). Deltaic facies are present in central eastern Sonora, marine-shelf facies in northwest Sonora, and basinal facies in northern Baja California.

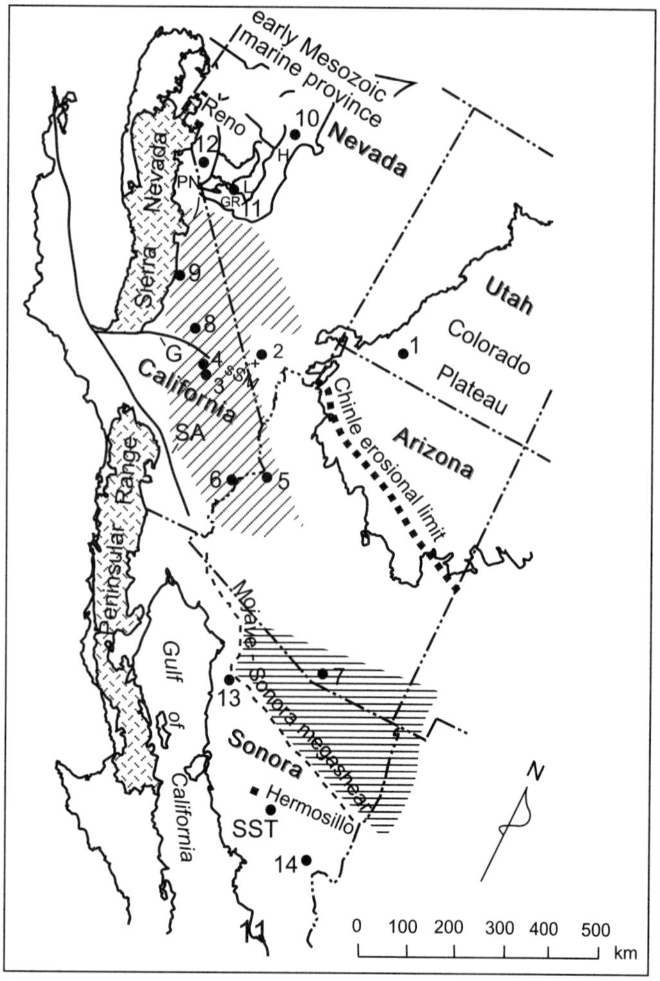

Dinosaur Canyon & Chinle tectonosequences absent
all Triassic sedimentary rocks absent

Figure 8. Regions in south-central Arizona and northern Sonora where all Triassic strata are missing as a result of erosional truncation below the Upper Triassic sequence boundary, U-3, and the Lower Jurassic sequence boundary, U-5; and region in the southern Great Basin and Mojave Desert where the Dinosaur Canyon and Chinle tectonosequences are missing as a result of erosional truncation below the Lower Jurassic sequence boundary, U-5. G—Garlock fault; GR—Gold Range terrane; H—Humboldt terrane; L—Luning terrane; PN—Pine Nut terrane; SA—San Andreas fault; sSM—southern Spring Mountains; SST—Sierra de Santa Teresa.

The presence of deltaic rocks of the Chinle tectonosequence in central eastern Sonora and correlative marine strata of the same sequence in the Sierra de Santa Teresa indicates the Late Triassic shoreline lay in eastern Sonora.

In the Jurassic, the Andean-style arc closed the back-arc basin at its southern end. The Caborca terrane is presently located southwest of the arc. Therefore, an autochthonous Caborca terrane requires that the arc lay between the Caborca terrane and the southern end of the back-arc basin. These geographic relations lead Molina Garza and Geissman (1999) to suggest that

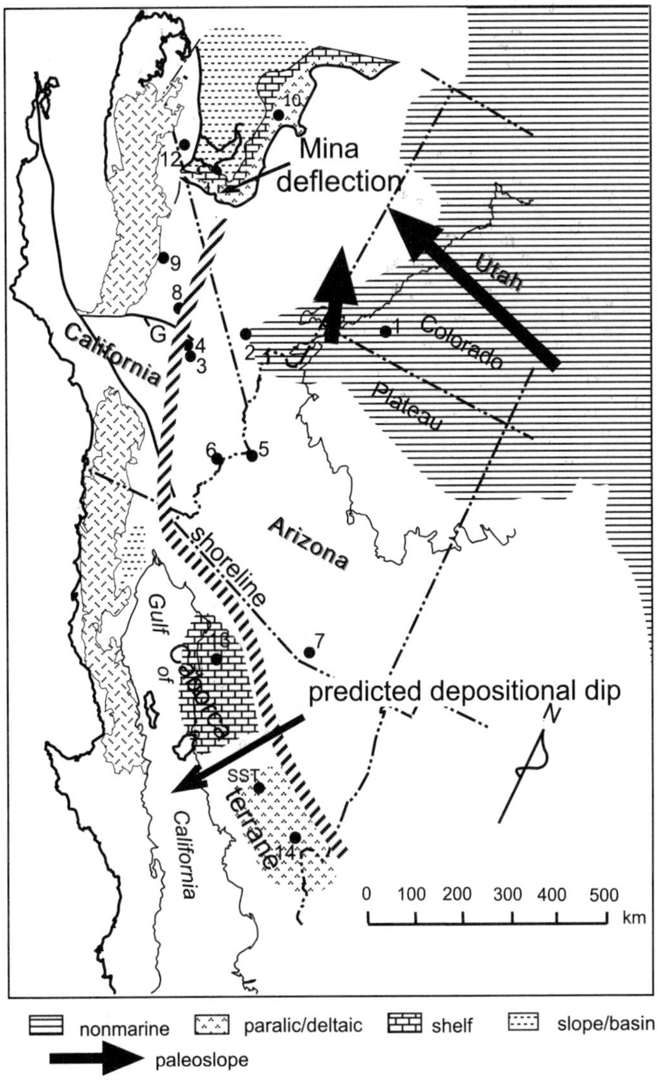

Figure 9. Location of the Upper Triassic Chinle tectonosequence shoreline as interpreted for an autochthonous Caborca terrane. The predicted depositional dip is shown in relation to the observed facies distribution. Compare with observed facies gradients of Figure 7. G—Garlock fault; SST—Sierra de Santa Teresa.

the Lower Jurassic marine rocks of the Caborca terrane were deposited in place in a fore-arc basin. This interpretation predicts northeast to southwest depositional dip. The facies gradient from nonmarine to marine should trend southwestward in rocks with a significant volcanic component. Facies boundaries should trend northwest to southeast parallel to the postulated shoreline. The hiatus at the basal sequence boundary, U-5, should increase landward, i.e., from southwest to northeast.

As with the Chinle tectonosequence, observations do not support the predictions. The Lower Jurassic rocks of the Glen Canyon tectonosequence are nonmarine fluvial and alluvial-fan deposits in central eastern Sonora. Where present in central and northwestern Sonora, the Glen Canyon tectonosequence is composed of shallow-marine siliciclastic rocks overlain by mature quartzose sandstones and quartzite-clast conglomerate similar to shallow-marine strata of the Sunrise Formation and overlying interbedded nonmarine and marine strata of the Dunlap Formation in northwestern Nevada. The observed southeast to northwest, marine to nonmarine facies gradient is developed along the axis of the postulated fore-arc basin parallel to shoreline. Both the shallow-marine siliciclastic rocks and overlying quartzose, coarser-grained clastics contain a paucity of volcanic detritus.

The northwest to southeast orientation of facies boundaries predicted for an autochthonous Caborca terrane is not consistent with observed facies gradients. The southeast to northwest facies gradients of both the Chinle and Glen Canyon tectonosequences on the Caborca terrane imply that the boundaries have a northeast to southwest trend. If this interpretation for the orientation of facies boundaries is correct, then these boundaries either (1) intersect the postulated fore-arc-basin shoreline at a high angle or (2) are abruptly truncated along the postulated trace of the Mojave-Sonora megashear. Facies boundaries intersecting the shoreline at a high angle are not sedimentologically plausible; whereas, the absence of lower Mesozoic marine and nonmarine rocks to the east of the postulated trace of the megashear is consistent with truncation.

The Glen Canyon tectonosequence in Sonora displays a hiatal geometry that is consistent with a location on a north-facing coastline and incompatible with a coastline facing westward or southward (Fig. 10). In the Sierra del Alamo, the Glen canyon tectonosequence rests unconformably on latest Triassic (Rhaetian) rocks of the Dinosaur tectonosequence. In the Sierra de Santa Teresa, the Glen Canyon tectonosequence rests on the Late Triassic (Norian) Moss Back–Owl Rock sequence of the Chinle tectonosequence. In the Sierra de San Javier, the Glen Canyon tectonosequence rests on the Shinarump–Blue Mesa sequence of the Chinle tectonosequence. U-5 has cut down section from the Dinosaur Canyon tectonosequence in the Sierra del Alamo in northwestern Sonora to the Moss Back–Owl Rock sequence of the Chinle tectonosequence in the Sierra de Santa Teresa in central Sonora to the Shinarump–Blue Mesa sequence of the Chinle tectonosequence in the Sierra de San Javier in central eastern Sonora. The U-5 hiatus increases from northwest to southeast (Fig. 10), not from southwest to northeast as predicted for an autochthonous Caborca terrane. In addition, the hiatus in northwestern Sonora is greater than at the top of the Dinosaur Canyon tectonosequence in the Luning terrane, where late Hettangian or early Sinemurian strata of the Glen Canyon tectonosequence rest on middle to lower Hettangian strata of the Dinosaur Canyon tectonosequence. This observation implies that the Antimonio sequences were deposited landward of the Luning sequences.

Stratigraphic relations described here are not consistent with accumulation of early Mesozoic strata upon an autochthonous Caborca terrane as concluded by Molina Garza and Geissman (1999). Rather than compatibility with a southward-, southwestward-, or westward-trending depositional dip

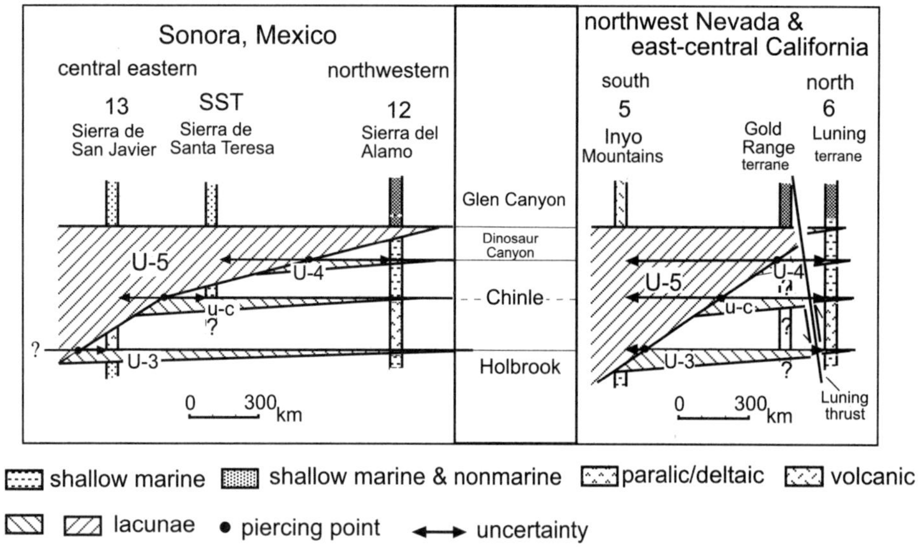

Figure 10. Comparison of hiatal geometries in space-time cross sections along line of section from columns 11–9 in northwestern Nevada and east-central California (Fig. 11B) to line of section from columns 13–14 in Sonora (Fig. 11C). Cross-cutting relations among Late Triassic and Early Jurassic regional unconformities U-3 through U-5 define piercing points (closed dots). Arrows indicate uncertainty in location of piercing points. SST—Sierra de Santa Teresa.

predicted for an autochthonous Caborca terrane, the observed facies gradients, facies boundaries, and hiatal geometries in Sonora are consistent with location on an east-west–trending coastline with northward depositional dip. Similarities in Triassic and Lower Jurassic sequence stratigraphy, particularly the Middle Triassic Holbrook tectonosequence, between the Caborca terrane in Sonora and the early Mesozoic marine province in northwest Nevada suggest the coastline lay along the Mina deflection on the southern margin of the early Mesozoic marine province. Although facies within the five Triassic and Lower Jurassic tectonosequences recognized in Sonora apparently lay landward of those in northwestern Nevada, similarities of Sonoran facies to correlative facies in the Humboldt and Luning terranes are consistent with this interpretation.

REGIONAL UNCONFORMITIES: CROSS-CUTTING RELATIONS

Although outcrops of Triassic and Lower and Middle Jurassic rocks are widely scattered, particularly in Sonora, correlation of tectonosequences across southwestern United States and northwestern Mexico reveals a regional pattern of sequence distribution (Fig. 8). The absence of the Chinle and Dinosaur tectonosequences over large regions implies truncation of sequence boundaries U-3, u-c, and U-4 below U-5.

Because the present erosion surface truncates progressively older unconformities (U-5, U-4, u-c, and U-3) southward from southern Utah and northern Arizona to the present southern erosional limit of the Chinle tectonosequence, locations of intersections formed by truncation of U-4, U-c, and U-3 below U-5 are uncertain. However, the absence of the Dinosaur Canyon and older tectonosequences below the Glen Canyon tectonosequence in southern Arizona implies truncation of U-4, u-c, and U-3 by U-5 between the southern erosional limit of the Dinosaur Canyon and Chinle tectonosequences, respectively, and southern Arizona (Fig. 11A).

To the west and southwest of the Colorado Plateau, locations of intersections formed by truncation of U-4 and U-3 by U-5 are well defined (Marzolf, 1991). U-5 cuts down section southwestward from the Colorado Plateau, truncating the Dinosaur Canyon tectonosequence and its basal sequence boundary U-4 west of the southwestern corner of the Plateau in southeastern Nevada. Farther west, in the southern Spring Mountains, sSM, (Fig. 11A), U-5 truncates the Chinle tectonosequence and its basal boundary, U-3. As a result, interbedded eolian sandstone and volcanic rocks of the Glen Canyon tectonosequence rest on the Moenkopi, or possibly the Holbrook, tectonosequence or older rocks throughout the Mojave Desert. Farther west and northwest, volcanic and volcaniclastic rocks of the Glen Canyon tectonosequence rest on the Moenkopi tectonosequence, or possibly the lower Holbrook tectonosequence, in the southern Panamint Range (Fig. 3, col. 8) and the Inyo Mountains (Fig. 3, col. 9) (Marzolf, 1994).

Along the Mina deflection, Triassic and Lower and Middle Jurassic strata exposed in the Luning assemblage comprise a nearly conformable stratigraphy including the upper part of the Holbrook tectonosequence, and the Chinle, Dinosaur Canyon, and Glen Canyon tectonosequences (Marzolf, 2000). The Chinle tectonosequence includes both Shinarump–Blue Mesa and Moss Back–Owl Rock sequences separated by the unconformity u-c. Because the Glen Canyon tectonosequence rests on the Moenkopi, or possibly the lower Holbrook, tectonosequence in the southern Inyo Mountains, U-5, must truncate U-4, u-c, and U-3 between the Luning assemblage and the Inyo Mountains (Fig. 11B; Fig. 2, columns 11 and 9). Stratigraphic relations among lower Mesozoic strata of the Gold Range, Golconda, and Roberts Mountains allochthons, lying between the Luning assemblage and the Inyo Mountains are not sufficiently well known to further constrain the truncations.

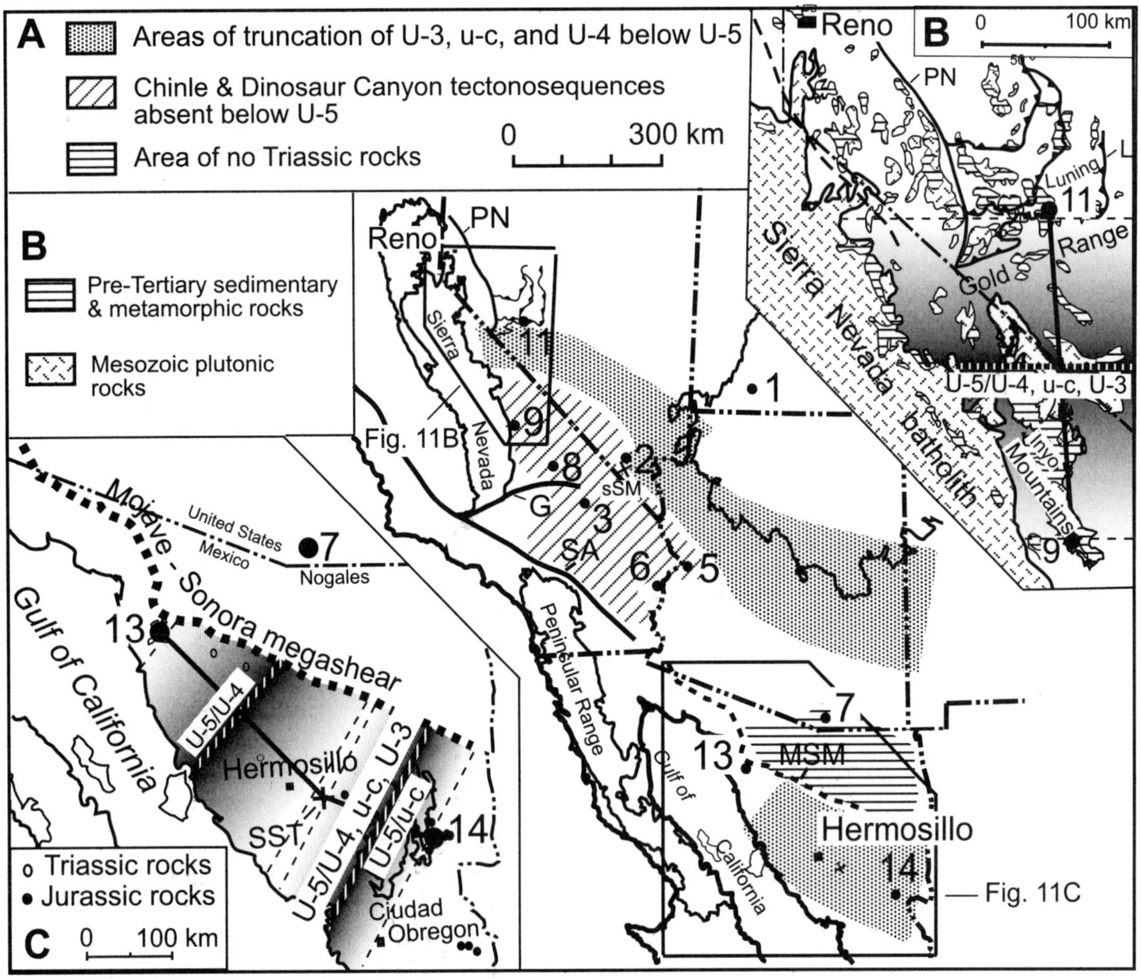

Figure 11. Location of lines defined by truncation of Upper Triassic sequence boundaries below the Lower Jurassic sequence boundary, U-5. Lines define piercing points on the Mojave-Sonora megashear. (A) Regions in the southwestern United States and in Sonora in which the truncations of U-4, u-c, and U-3 are interpreted to lie. (B) Location in northwestern Nevada and east-central California of composite mean of lines defined by truncation of U-3, u-c, and U-4 below U-5; PN—Pine Nut fault; L—Luning thrust fault. (C) Locations in Sonora of lines defined by truncation of U-4 below U-5, u-c below U-5, and composite mean of truncation of U-3, u-c, and U-4 below U-5. SST—Sierra de Santa Teresa.

In the Sierra del Alamo in northwestern Sonora, all five Triassic and Lower and Middle Jurassic tectonosequences, Moenkopi, Holbrook, Chinle, Dinosaur Canyon, and Glen Canyon, are included in the nearly 3-km-thick Antimonio Formation (Fig. 5, col. 13). The late Hettangian or early Sinemurian marine strata of the Glen Canyon tectonosequence overlie Rhaetian strata of the Dinosaur Canyon tectonosequence. The Dinosaur Canyon tectonosequence overlies the Chinle tectonosequence, which includes both the Shinarump–Blue Mesa and Moss Back–Owl Rock sequences of the Chinle tectonosequence separated by the u-c unconformity. In central eastern Sonora, the Glen Canyon tectonosequence rests unconformably on the Shinarump–Blue Mesa sequence (Fig. 5, col. 14). The Dinosaur Canyon tectonosequence and the Moss Back–Owl rock sequence are absent.

In the Sierra de Santa Teresa near Hermosillo, lower Mesozoic strata are exposed in the lower plate of a post–Early Jurassic thrust fault (Stewart et al., 1997a). Interbedded limestone, argillite, siltstone, sandstone, and conglomerate are more than 540 m thick. Conodonts in limestone ~85 m above the base of the section indicate a middle Norian age. Poorly preserved ammonites collected from interbedded siltstone and lime mudstone ~100 m below the top of the section suggest a late Hettangian (*Angulata* zone) age. The Norian strata near the base of the section are correlative with the Moss Back–Owl Rock sequence in the Chinle tectonosequence. The ammonite-bearing strata near the top of the section are correlative with the basal Glen Canyon tectonosequence. Preliminary observations (Marzolf, 1995, unpublished) suggest that strata correlative with the Dinosaur Canyon tectonosequence in the Antimonio section are not present in the more than 300 m between Norian and Hettangian strata. If this interpretation proves correct, then the Glen Canyon tectonosequence rests on marine strata of the Moss Back–Owl Rock sequence. The Dinosaur Canyon tectonosequence is absent. The stratigraphic relations in the Sierra del Alamo, Sierra de Santa Teresa, and the Sierra de San Javier (Fig. 5, col. 13, SST, and 14) indicate that U-5 at the base of the Glen Canyon tectonosequence truncates U-4 between the Sierra del Alamo and the Sierra de Santa Teresa. U-5 truncates u-c between the Sierra de Santa Teresa and the Sierra de San Javier (Figs. 10 and 11C).

Erosional truncations of sequence boundaries by stratigraphically higher sequence boundaries form lines. Uncertainties in the geographic location of the stratigraphically defined lines expand the lines into linear belts. The centers of these belts are taken as the best estimates of the lines of truncation (Fig. 11B and 11C). These lines define piercing points where they intersect the Mojave-Sonora megashear. These points constrain the magnitude of displacement on the megashear. The lines are defined by truncation of U-3 by U-5 (U-5/U-3), truncation of u-c by U-5 (U-5/u-c), and truncation of U-4 by U-5 (U-5/U-4). In northwestern Nevada and east-central California, U-5/U-3, U-5/u-c, and U-5/U-4 can only be constrained to lie between the Inyo Mountains and the Luning assemblage. The mid-point between these two locations is taken as the best estimate for the composite mean of the three truncations (Fig. 11B, columns 9 and 11). In Sonora, U-5/U-4 lies between the Sierra del Alamo (Fig. 11C, col. 13) in northwestern Sonora and the Sierra de Santa Teresa (Fig. 11C, col. SST) and U-5/u-c lies between the Sierra de Santa Teresa and central eastern Sonora (Fig. 11C, col. 14). U-5/U-3 must lie an unknown distance east of central eastern Sonora. Because of the inability to separate piercing points in northwestern Nevada and eastern California, no advantage is gained by their discrimination in Sonora. U-5/u-c is taken as the best estimate for the composite-mean location of U-5/U-3, U-5/u-c, and U-5/U-4 (Fig. 11C). When the composite mean of U-5/U-4, u-c, and U-3 in Sonora is restored to its corresponding piercing point in northwestern Nevada, the resulting estimate of left-lateral displacement is 1040 km ± 290 km. Although uncertainty over the precise locations of U-5/U-3, U-5/u-c, and U-5/U-4 are great, the minimum estimate of displacement is 750 km.

Locations of piercing points in northwestern Nevada and east-central California do not take into account south-directed crustal shortening in the Gold Range and Luning terranes. The true magnitude of crustal shortening may be significantly larger than the estimates of uncertainty in the location of U-5/U-3, U-5/u-c, and U-5/U-4. The palinspastically restored Caborca terrane may have lain at the western end of the Gold Range and Luning terranes entirely within the area of structural overlap between the two terranes. Displacement may have been on the order of 1500 km.

CONCLUSIONS

Correlative offset piercing points defined by truncations of successive Triassic and Lower Jurassic sequence boundaries in Sonora and in east-central California and northwestern Nevada indicate more than 1000 km of left-lateral displacement. Restoration of offsets places the Caborca terrane west of the Luning terrane along the Mina deflection. Facies, facies gradients, facies boundaries, and hiatal geometries in Sonora are consistent with this restoration.

The timing of displacement along the megashear is constrained by the age of the youngest crosscutting sequence boundary. Therefore, displacement had to occur after the early Sinemurian (ca. 195 Ma) as indicated by the age of the unconformity, U-5, at the base of the lower Glen Canyon tectonosequence. An upper age limit near the Oxfordian-Kimmeridgian boundary (ca. 150 Ma) is probable based on several lines of structural and stratigraphic evidence. Clearly, displacement could not have occurred in the Permian or Early Triassic (Stone and Stevens, 1988; Dickinson and Lawton, 2001). Permian faulting outboard of the Caborca terrane is not precluded.

ACKNOWLEDGMENTS

We gratefully acknowledge reviews of the original manuscript by N.J. Silberling, J.I. Satterfield, and D.G. Taylor. Their critical comments made significant contributions to this final version.

REFERENCES CITED

Albers, J.P., 1967, Belt of sigmoidal bending and right-lateral faulting in the western Great Basin: Geological Society of America Bulletin, v. 78, p. 143–176.

Alencaster, G. de Cserna, 1961, Fauna fósil de la Formación Santa Clara (Cárnico) del Estado de Sonora, parte I *in* Paleontología del Triássico Superior de Sonora: Universidad Nacional Autónoma de México, Instituto de Geología, Paleontología Mexicana número 11, 18 p.

Anderson, T.H., and Schmidt, V.A., 1983, The evolution of middle America and the Gulf of Mexico–Caribbean Sea region during Mesozoic time: Geological Society of America Bulletin, v. 94, p. 941–966, doi: 10.1130/0016-7606(1983)94<941:TEOMAA>2.0.CO;2.

Anderson, T.H., and Silver, L.T., 1979, The role of the Mojave Sonora megashear in the tectonic evolution of northern Sonora, *in* Anderson, T.H., and Roldán-Quintana, J., eds., Geology of northern Sonora, Guidebook-fieldtrip #27, Annual meeting of the Geological Society of America: Instituto de Geologia, Universidad Nacional Autónoma de México, Hermosillo, Sonora, Mexico and the University of Pittsburgh, Pittsburgh, Pennsylvania, p. 59–68.

Anderson, T.H., and Silver, L.T., 1981, An Overview of Precambrian Rocks in Sonora: Universidad Nacional Autónoma de México, Instituto de Geología, Revista, v. 5, no. 2, p. 131–139.

Arellano, A.R.V., 1956, Relaciónes del Cámbrico de Caborca, especialmente con la base del Paleozoico: Congreso Geológico Internacional, 20th, México, 1956; El Sistema Cambrico, su Paleogeografía y el Problema de su Base–Symposium, v. 2, p. 509–527.

Bilodeau, W.L., and Keith, S.B., 1986, Lower Jurassic Navajo-Aztec–equivalent sandstones in southern Arizona and their paleogeographic significance: Bulletin of the American Association of Petroleum Geologists, v. 70, p. 690–701.

Blakey, R.C., 1994, Paleogeographic and tectonic controls on some Lower and Middle Jurassic erg deposits, Colorado plateau, *in* Caputo, M., Peterson, J.E., and Franczyk, K.J., eds., Mesozoic systems of the Rocky Mountain Region, USA: Rocky Mountain Section of the Society of Sedimentary Geology, Denver, Colorado, p. 273–298.

Blakey, R.C., and Gubitosa, R., 1983, Late Triassic paleogeography and depositional history of the Chinle Formation, southern Utah and northern Arizona, *in* Reynolds, M.W., and Dolly, E.D., eds., Mesozoic paleogeography of the west-central United States: Rocky Mountain Section, Society of Economic Paleontologists and Mineralogists, Denver, Colorado, p. 57–76.

Blakey, R.C., and Middleton, L.T., 1983, Lower Mesozoic stratigraphy and depositional systems, southwest Colorado Plateau, *in* Marzolf, J.E., and Dunne, G.C., eds., Evolution of Early Mesozoic Tectonostratigraphic Environments—Southwestern Utah to Southern Inyo Mountains: Utah Geological and Mineral Survey Special Studies 60, Guidebook Part 2, p. 33–39.

Busby-Spera, C.J., 1988, Speculative tectonic model for the early Mesozoic arc of the southwest Cordilleran United States: Geology, v. 16, p. 1121–1125, doi: 10.1130/0091-7613(1988)016<1121:STMFTE>2.3.CO;2.

Busby, C.J., Schermer, E.R., and Mattinson, J.M., 2002, Extensional arc setting and ages of Middle Jurassic eolianites, Cowhole Mountains (eastern Mojave Desert block, California) *in* Glazner, A.F., Walker, J.D., and Bartley, J.M., eds., Geologic Evolution of the Mojave Desert and Southwestern Basin and Range: Geological Society of America Memoir 195, p. 79–91.

Calderone, G.J., and Beard, L.S., 1994, New paleomagnetic results from Mesozoic red beds in the South Virgin Mountains area, southern Nevada: Geological Society of America Abstracts with Programs, v. 26, no. 2, p. 42–43.

Clark, J.M., and Fastovsky, D.E., 1986, Vertebrate biostratigraphy of the Glen Canyon Group in northern Arizona, *in* Padian, K., ed., The beginning of the age of dinosaurs: faunal changes across the Triassic-Jurassic boundary: New York, Cambridge University Press, p. 285–301.

Cojan, I., and Potter, P.E., 1991, Depositional environment, petrology, and provenance of the Santa Clara Formation, Upper Triassic Barranca Group, eastern Sonora, Mexico, *in* Pérez-Segura, E., and Jacques-Ayala, C., eds., Studies of Sonoran Geology: Geological Society of America Special Paper 254, p. 37–50.

Cooper, G.A., and Arellano, A.R.V., 1946, Stratigraphy near Caborca, northwest Sonora, Mexico: American Association of Petroleum Geologists Bulletin, v. 30, no. 4, p. 606–611.

Damborena, S.E., and González-León, C.M., 1997, Late Triassic and Early Jurassic bivalves from Sonora, Mexico: Revista Mexicana de Ciencias Geológicas, v. 14, no. 2, p. 178–201.

Damon, P.E., Livingston, D.E., Mauger, R.L., Giletti, B.J., and Alor, J.P., 1962, Edad del Precambrico "Anterior" y de otras rocas del zócalo de la region de Caborca-Altar de la parte noroccidental del Estado de Sonora, México: Universidad Nacional Autónoma de México, Instituto de Geología Boletín, v. 64, p. 11–44.

Dickinson, W.R., and Lawton, T.F., 2001, Carboniferous to Cretaceous assembly and fragmentation of Mexico: Geological Society of America Bulletin, v. 113, p. 1142–1160, doi: 10.1130/0016-7606(2001)113<1142:CTCAAF>2.0.CO;2.

Dilles, J.H., and Wright, J.E., 1988, The chronology of early Mesozoic arc magmatism in the Yerington district of western Nevada and its regional implications: Geological Society of America Bulletin, v. 100, p. 644–652.

Dubiel, R.F., 1994, Triassic deposystems, paleogeography, and paleoclimate of the Western Interior, *in* Caputo, M., Peterson, J.E., and Franczyk, K.J., eds., Mesozoic systems of the Rocky Mountain Region: USA, Rocky Mountain Section of the Society of Sedimentary Geology, Denver, Colorado, p. 133–168.

Dunne, G.C., and Walker, D., 1993, Age of Jurassic volcanism and tectonism, southern Owens valley region, east-central California: Geological Society of America Bulletin, v. 105, p. 1223–1230, doi: 10.1130/0016-7606(1993)105<1223:AOJVAT>2.3.CO;2.

Eells, J.L., 1972, Geology of the Sierra de la Berruga, northwestern Sonora, Mexico [M.S. thesis]: San Diego, California State University, 77 p.

Elison, M.W., and Speed, R.C., 1988, Triassic flysch of the Fencemaker allochthon, East Range, Nevada: Fan facies and provenance: Geological Society of America Bulletin, v. 100, p. 185–199, doi: 10.1130/0016-7606(1988)100<0185:TFOTFA>2.3.CO;2.

Embry, A.F., 1988, Triassic Sea-level changes: Evidence from the Canadian archipelago, *in* Wilgus, C.K., et al., eds., Sea-level changes: An integrated approach: Society of Economic Paleontologists and Mineralogists Special Publication 42, p. 249–259.

Embry, A.F., 1993, Triassic T-R sequence analysis, Sverdrup basin, Canadian Arctic archipelago: Canadian Society of Petroleum Geologists, Carboniferous to Jurassic Pangea, Program and Abstracts, Calgary, Alberta, p. 89.

Estep, J.W., Lucas, S.G., and González-León, C.M., 1997, Middle Triassic ammonites from Sonora, Mexico: Revista Mexicana de Ciencias Geologicas, v. 14, no. 2, p. 155–159.

Fackler-Adams, B.N., Busby, C.J., and Mattinson, J.M., 1997, Jurassic magmatism and sedimentation in the Palen Mountains, southeastern California: Implications for regional tectonic controls on the Mesozoic continental arc: Geological Society of America Bulletin, v. 109, p. 1464–1484, doi: 10.1130/0016-7606(1997)109<1464:JMASIT>2.3.CO;2.

Ferguson, H.G., and Muller, S.W., 1949, Structural geology of the Hawthorne and Tonopah quadrangles, Nevada: U.S. Geological Survey Professional Paper 216, 55 p.

Fowell, S.J., and Olsen, P.E., 1993, Time calibration of Triassic/Jurassic microfloral turnover, eastern North America: Tectonophysics, v. 2s2, p. 361–369, doi: 10.1016/0040-1951(93)90359-R.

Fries, C., Jr., 1962, Reseña de la geología del Estado de Sonora, con énfasis en el Paleozoico: Asociación Mexicana de Geólogos Petroleros Boletín, v. 14, p. 257–273.

Galloway, W.E., 1989, Genetic stratigraphic sequences in basin analysis II: Application to northwest Gulf of Mexico Cenozoic basin: American Association of Petroleum Geologists Bulletin, v. 73, p. 143–154.

Gehrels, G.E., Dickinson, W.R., and Riggs, N., 1993, Detrital zircon provenance link of the Chinle Formation and the Auld Lang Syne Group in Nevada: Geological Society of America Abstracts with Programs, v. 25, no. 5, p. 41.

Geissman, J.W., Callian, J.T., Oldow, J.S., and Humphries, S.E., 1984, Paleomagnetic assessment of oroflexural deformation in west-central Nevada and significance for emplacement of allochthonous assemblages: Tectonics, v. 3, p. 179–200.

González-León, C.M., 1980, La Formación Antimonio (Triássico Superior-Jurásico Inferior) en la Sierra del Alamo, Sonora: Universidad Nacional Autónoma de México, Instituto de Geología Revista, v. 4, p. 13–18.

González-León, C.M., 1997, Sequence stratigraphy and paleogeographic setting of the Antimonio Formation (Late Permian–Early Jurassic, Sonora, Mexico: Revista Mexicana de Ciencias Geologicas, v. 14, no. 2, p. 136–148.

Gradstein, F.M., Agterberg, F.P., Ogg, J.G., Hardenbol, J., van Veen, P., Thierry, J., and Huang, Z., 1994, A Mesozoic time scale: Journal of Geophysical Research, v. 99, p. 24051–24074, doi: 10.1029/94JB01889.

Grose, L.T., 1959, Structure and petrology of the northeast part of the Soda Mountains, San Bernardino County, California: Geological Society of America Bulletin, v. 70, p. 1509–1548.

Hamilton, W., 1982, Structural evolution of the Big Maria Mountains, northeastern Riverside County, southeastern California, *in* Frost, E.G., and Martin, D.L., eds., Mesozoic-Cenozoic evolution of the Colorado River region, California, Arizona, and Nevada, San Diego, Cordilleran Publishers, p. 1–27.

Hamilton, W., 1987, Mesozoic geology and tectonics of the Big Maria Mountains Region, southeastern California, *in* Dickinson, W.R., and Klute, M.A., eds., Mesozoic Rocks of Southern Arizona and Adjacent Areas: Arizona Geological Digest, v. 18, p. 33–47.

Harland, W.B., Armstrong, R.L., Cox, A.V., Craig, L.E., Smith, A.G., and Smith, D.G., 1990, A geologic time scale, 1989: Cambridge University Press, Cambridge, 263 p.

Harshbarger, J.W., Repenning, C.A., and Irwin, J.H., 1957, Stratigraphy of the uppermost Triassic and the Jurassic rocks of the Navajo country [Colorado Plateau]: U.S. Geological Survey Professional Paper 291, 74 p.

Haxel, G.B., Anderson, T.H., Riggs, N.R., and Goodwin, L.B., 1988, The Papago terrane: a crustal anomaly in south-central Arizona and north-central Sonora: Geological Society of America Abstracts with Programs, v. 20, p. 168.

Hewett, D.F., 1931, Geology and ore deposits of the Goodsprings Quadrangle, Nevada: U. S. Geological Survey Professional Paper 162, 172 p.

Hewett, D.F., 1956, Geology and mineral resources of the Ivanpah Quadrangle, Nevada and California: U.S. Geological Survey Professional Paper 275, 172 p.

Hunt, A.P., and Lucas, S.G., 1991, The Paleorhinus biochron and the correlation of the non-marine Upper Triassic of Pangaea: Palaeontology, v. 34, p. 487–501.

Hutny, M.K., and Steiner, M.B., 2001, Abrupt change in North America Plate Motion: Magnetostratigraphy and paleopoles of the Early Jurassic Moenave Formation: Eos (Transactions, American Geophysical Union), Fall Meeting, v. 82, p. F319.

Imlay, R.W., 1967a, Twin Creek Limestone (Jurassic) in the Western Interior of the United States: U.S. Geological Survey Professional Paper 540.

Imlay, R.W., 1967b, Jurassic paleobiogeography of the conterminous United States in its continental setting: U.S. Geological Survey Professional Paper 1062.

Imlay, R.W., 1980, Jurassic paleobiogeography of the conterminous United States in its continental setting: U.S. Geological Survey Professional Paper 1062.

Johnson, B.K., 1957, Geology of part of the Manly Peak Quadrangle, southern Panamint Range, California: University of California, Publications in Geological Sciences, v. 30, p. 353–423.

Jones, N.W., McKee, J.W., Anderson, T.H., and Silver, L. T., 1995, Jurassic volcanic rocks in northeastern Mexico: A possible remnant of a Cordilleran magmatic arc, *in* Jacques-Ayala, C., González-León, C.M., and Roldán-Quintana, J., eds., Studies of the Mesozoic of Sonora and Adjacent Areas: Boulder, Colorado, Geological Society of America Special Paper 301, p. 179–190.

Ketner, K.B., 1986, Eureka Quartzite in Mexico?—Tectonic implications: Geology, v. 14, no. 12, p. 1,027–1030, doi: 10.1130/0091-7613(1986)14<1027:EQIMI>2.0.CO;2.

Kowallis, B.J., Christiansen, E.H., Deino, A.L., Peterson, F., Turner, C.E., and Blakey, R.C., 1996, Implications of a new high precision ^{40}Ar/^{39}Ar Laser-Microprobe ages for the Jurassic time scale, *in* Morales, M., ed., The Continental Jurassic: Museum of Northern Arizona Bulletin 60, p. 7–8.

Kowallis, B.J., Christiansen, E.H., Deino, A.L., Zhang, C., and Everett, B., 2001, The record of Middle Jurassic volcanism in the Carmel and Temple Cap formations of southwestern Utah: Geological Society of America Bulletin, v. 113, p. 373–387, doi: 10.1130/0016-7606(2001)113<0373:TROMJV>2.0.CO;2.

Kozur, H.W., 2003, Integrated ammonoid, connodont and radiolarian zonation of the Triassic and some remarks to Stage/Substage subdivision and the numeric age of the Triassic stages: Albertiana, no. 28, p. 57–74.

Larson, A.R., 1966, Stratigraphy and paleontology of the Moenkopi Formation in southern Nevada [Ph. D. Thesis]: Los Angeles, California, University of California, 256 p.

Larson, A.R., and Lane, N.G., 1964, Repetitive bedding in Triassic sediments in Clark County, Nevada: Kansas Geological Survey Bulletin, v. 169, p. 265–274.

LeVeque, R.A., 1982, Stratigraphy and structure of the Palen Formation, Palen Mountains, southeastern California, *in* Frost, E.G., Martin, D.L., eds., Mesozoic-Cenozoic evolution of the Colorado River region, California, Arizona, and Nevada: San Diego, Cordilleran Publishers, p. 267–273.

Lewis, M., Whitman, C., and Steven, C.H., 1983, Lower Triassic marine sedimentary rocks in east-central California, *in* Marzolf, J.E., and Dunne, G.C., eds., Evolution of early Mesozoic tectonostratigraphic environments—Southwestern Colorado Plateau to Southern Inyo Mountains: Geological Society of America, Annual Meeting of Cordilleran and Rocky Mountain Sections, Guide to field trip no. 7., p. 50–54.

Longwell, C.R., 1925, The pre-Triassic unconformity in southern Nevada: American Journal of Science, v. 10, p. 93–106.

Lucas, S.G., 1993, The Chinle Group: Revised stratigraphy and biochronology of Upper Triassic nonmarine strata in the western United States: Museum of Northern Arizona Bulletin, v. 59, p. 27–50.

Lucas, S.G., 1996, Correlation and tectonic significance of Lower Jurassic conglomerates in Sonora, Mexico, *in* Morales, M., ed., The Continental Jurassic: Museum of Northern Arizona Bulletin 60, p. 497–501.

Lucas, S.G., 1999, Tetrapod correlation of the non-marine Triassic: Zentralblatt für Geologie und Paläontologie, v. 3, p. 498–521.

Lucas, S.G., and Hunt, A.P., 1989, Vertebrate biochronology of the Late Triassic: 28th International Geological Congress (Washington, D. C.), Abstracts, v. 2, p. 335–336.

Lucas, S.G., and Hunt, A.P., 1993, Tetrapod biochronology of the Chinle Group (Upper Triassic), western United States, *in* Lucas, S.G., and Morales, M., eds., The nonmarine Triassic: Bulletin 3, New Mexico Museum of Natural History and Science, Albuquerque, New Mexico, p. 327–330.

Lucas, S.G., and Marzolf, J.E., 1993, Stratigraphy and sequence stratigraphic interpretations of Upper Triassic strata in Nevada, *in* Dunne, G. and McDougall, K., eds., Mesozoic paleogeography of the western United States-II: Pacific Section SEPM Book 71, p. 375–388.

Lucas, S.G., and Marzolf, J.E., 1994, Upper Triassic–Lower Jurassic Barranca Group of Sonora, Mexico: Depositional setting, age constraints, and stratigraphic subdivision: Geological Society of America Abstracts with Programs, v. 26, no. 2, p. 68.

Lucas, S.G., Anderson, O.J., Marzolf, J.E., and Morales, M., 1996, The Triassic-Jurassic transition on the Colorado Plateau: Is J-0 a Triassic unconformity?: Geological Society of America Abstracts with Programs, v. 28, no. 7, p. A-308.

Lucas, S.G., Silberling, N.J., and Marzolf, J.E., 1997a, Triassic sequence stratigraphy and structural complexity in the Sierra del Alamo Muerto, Sonora, Mexico, *in* González-León, C.M., and Stanley, G.D., eds., U.S.–Mexico Cooperative Research: International Workshop on the Geology of Sonora, Publicaciones Occasionales No. 1, Estación Regional del Noreste, Instituto de Geólogia, Universidad Nacional Autónoma de Mexico, p. 46–47.

Lucas, S.G., Kues, B.S., Estep, J.W., and González-León, C.M., 1997b, Permian-Triassic boundary at El Antimonio, Sonora, Mexico, *in* González-León, C.M. and Stanley, G.D., eds., U.S.–Mexico Cooperative Research: International Workshop on the Geology of Sonora, Publicaciones Occasionales No. 1, Estación Regional del Noreste, Instituto de Geología, Universidad Nacional Autonóma de México, p.149–154.

Marzolf, J.E., 1983, Early Mesozoic eolian transition from cratonal margin to orogenic-volcanic arc, *in* Marzolf, J.E., and Dunne, G.C., eds., Evolution of early Mesozoic tectonostratigraphic environments—Southwestern Colorado Plateau to Southern Inyo Mountains: Geological Society of America, Annual Meeting of Cordilleran and Rocky Mountain Sections, Guide to field trip no. 7., p. 39–46.

Marzolf, J.E., 1988a, Reconstruction of Late Triassic and Early and Middle Jurassic sedimentary basins: southwestern Colorado Plateau to the eastern Mojave Desert, *in* Weide, D.L., and Faber, M.L., eds., This Extended Land, Geological Journeys in the southern Basin and Range: Geological Society of America, Cordilleran Section, Field Trip Guidebook, p. 177–200.

Marzolf, J.E., 1988b, Controls on late Paleozoic and early Mesozoic eolian deposition of the western United States: Sedimentary Geology, v. 56, p. 167–191, doi: 10.1016/0037-0738(88)90053-X.

Marzolf, J.E., 1990, Reconstruction of extensionally dismembered early Mesozoic sedimentary basins; South-western Colorado Plateau to the eastern Mojave Desert, *in* Wernicke, B.P., ed., Basin and Range extensional tectonics near the latitude of Las Vegas, Nevada: Geological Society of America Memoir 176, p. 477–500.

Marzolf, J.E., 1991, Lower Jurassic unconformity (J-0) from the Colorado Plateau to the eastern Mojave Desert: Evidence of a major tectonic event at the close of the Triassic: Geology, v. 19, p. 320–323, doi: 10.1130/0091-7613(1991)019<0320:LJUJFT>2.3.CO;2.

Marzolf, J.E., 1993a, Sequence-stratigraphic relationships across the palinspastically reconstructed Cordilleran Triassic cratonal margin, *in* Lucas, S.G., and Morales, M., eds., The nonmarine Triassic: New Mexico Museum of Natural History, Bulletin 3, p. 331–343.

Marzolf, J.E., 1993b, Palinspastic reconstruction of early Mesozoic sedimentary basins near the latitude of Las Vegas: Implications for the early Mesozoic Cordilleran cratonal margin, *in* Dunne, G.C., and McDougall, K., eds., Mesozoic paleogeography of the western United States-II: SEPM (Society for Sedimentary Geology), Book 71, p. 433–462.

Marzolf, J.E., 1994, Reconstruction of the early Mesozoic Cordilleran cratonal margin adjacent to the Colorado Plateau, *in* Caputo, M., Peterson, J.E., and Franczyk, K.J., eds., Mesozoic systems of the Rocky Mountain Region, USA: Rocky Mountain Section of the Society of Sedimentary Geology, Denver, Colorado, p. 181–216.

Marzolf, J.E., 1996, Sequence stratigraphy of the Colorado Plateau and early Mesozoic Paleogeography, *in* Morales, M., ed., The Continental Jurassic: Museum of Northern Arizona Bulletin 60, p. 441–442.

Marzolf, J.E., 1997, Sequence stratigraphy of the Colorado Plateau, early Mesozoic paleogeography, and the Sonoran megashear, *in* González-León, C.M., and Stanley, G.D., eds., U.S.–Mexico Cooperative Research: International Workshop on the Geology of Sonora, Publicaciones Occasionales No. 1, Estación Regional del Noreste, Instituto de Geología, Universidad Nacional Autonóma de México, p.48–52.

Marzolf, J.E., 2000, Triassic paleogeography of the Panthalassan margin of Southwestern North America: A sequence stratigraphy–based regional synthesis: Zentralblatt für Geologie und Paläontologie, v. 3, p. 1497–1538.

Marzolf, J.E., 2002, Jurassic Tectonosequences, tectono-structural events, and apparent polar wander of the North American Cordillera: Implications for North American Evolution during Pangean Breakup: 6th International Symposium on the Jurassic System, Abstracts and Program, p. 119.

Marzolf, J.E., and Lucas, S.G., 1994, Lower Mesozoic tectonosequences and the Mojave-Sonoran puzzle: Geological Society of America Abstracts with Programs, v. 26, no. 2, p. 70.

Marzolf, J.E., and Lucas, S.G., 1996, Lower Mesozoic sequences of the Colorado Plateau *in* Morales, M., ed., The Continental Jurassic: Museum of Northern Arizona Bulletin 60, p. 437–438.

Marzolf, J.E., Lucas, S.G., Satterfield, J.I., and Silberling, N.J., 1996, Cordilleran Triassic-Jurassic transition: The Dinosaur Canyon sequence: Geological Society of America Annual Meeting, Abstracts with Programs, v. 28, no. 7, p. A-308.

McDowell, F.W., Roldán-Quintana, J., and Connelly, J.N., 2001, Duration of Late Cretaceous–early Tertiary magmatism in east-central Sonora, Mexico: Geological Society of America Bulletin, v. 113, p. 521–531.

McKee, E.D., 1954, Stratigraphy and history of the Moenkopi Formation of Triassic age: Geological Society of America Memoir 61, 133 p.

McRoberts, C.A., 1997, Late Triassic (Norian-Rhaetian) bivalves from the Antimonio Formation northwestern Sonora, Mexico: Revista Mexicana de Ciencias Geológicas, v. 14, no. 2, p. 167–177.

Molina-Garza, R.S., and Geissman, J.W., 1999, Paleomagnetic data from the Caborca Terrane, Mexico: Implications for Cordilleran tectonics and the Mojave-Sonora Megashear hypothesis: Tectonics, v. 18, p. 293–295, doi: 10.1029/1998TC900030.

Morales, M., 1993, Tetrapod biostratigraphy of the Lower-Middle Triassic Moenkopi Formation, *in* Lucas, S.G., and Morales, M., eds., The nonmarine Triassic: New Mexico Museum of Natural History & Science Bulletin No. 3, p. 355–356

Muller, S.W., and Ferguson, H.G., 1939, Mesozoic stratigraphy of the Hawthorne and Tonopah quadrangles, Nevada: Geological Society of America Bulletin, v. 50, p. 1573–1624.

Nichols, K.M., and Silberling, N.J., 1977, Stratigraphy and depositional history of the Star Peak Group (Triassic), northwestern Nevada: Geological Society of America Special Paper 178, 73 p.

Novitsky-Evans, J.M., 1978, Geology of the Cowhole Mountains, southeastern California: structural, stratigraphic, and geochemical studies [Ph.D. thesis] Houston, Texas, Rice University, 95 p.

Novitsky, J.M., and Burchfiel, B.C., 1973, Pre-Aztec (Upper Triassic (?)–Lower Jurassic) thrusting, Cowhole Mountains, southeastern California: Geological Society of America Abstracts with Programs, v. 5, no. 7, p. 755.

Oldow, J.S., 1981, Structure and stratigraphy of the Luning allochthon and the kinematics of allochthon emplacement, pilot Mountains, west-central Nevada: Geological Society of America Bulletin, Part I, v. 92, p. 888–911, Part II, p. 1647–1669.

Oldow, J.S., 1984, Evolution of a late Mesozoic back-arc fold and thrust belt, northwestern Great Basin, USA: Tectonophysics, v. 102, p. 245–274, doi: 10.1016/0040-1951(84)90016-7.

Oldow, J.S., and Bartel, R.L., 1987, Early to Middle(?) Jurassic extensional tectonism in the western Great Basin: Growth faulting and synorogenic deposition of the Dunlap Formation: Geology, v. 15, no. 8, p. 740–743, doi: 10.1130/0091-7613(1987)15<740:ETMJET>2.0.CO;2.

Oldow, J.S., Bartel, R.L., and Gelber, A.W., 1990, Depositional setting and regional relationships of basinal assemblages: Pershing Ridge Group and Fencemaker Canyon sequence in northwestern Nevada: Geological Society of America Bulletin, v. 102, p. 193–222, doi: 10.1130/0016-7606(1990)102<0193:DSARRO>2.3.CO;2.

Oldow, J.S., Satterfield, J.I., and Silberling, N.J., 1993, Jurassic to Cretaceous transpressional deformation in the Mesozoic marine province of the northwestern Great Basin, *in* Lahren, M.M., Trexler, J.H., Jr., and Spinosa, C., eds., Crustal evolution of the Great Basin and the Sierra Nevada: Cordilleran and Rocky Mountain Sections, Geological Society of America Guidebook, Department of Geological Sciences, University of Nevada, Reno, p. 129–147.

Olsen, P.E., 2002, Field guide for non-marine boundary events in the Newark Basin (New Jersey, Pennsylvania, and Connecticut), eastern United States and their litho-, chrono-, and biostratigraphic context: Field Workshop IGCP 458, June 8–11, 2002, Lamont-Doherty Earth Observatory of Columbia University, Palisades New York, 102 p.

Padian, K., 1989, Presence of the dinosaur Scelidosaurus indicates Jurassic age for the Kayenta Formation (Glen Canyon Group), northern Arizona: Geology, v. 17, p. 438–441, doi: 10.1130/0091-7613(1989)017<0438:POTDSI>2.3.CO;2.

Pálfy, J., Smith, P.L., and Mortensen, J.K., 2000, A U-Pb and ^{40}Ar/^{39}Ar time scale for the Jurassic: Canadian Journal of Earth Sciences, v. 37, p. 923–944, doi: 10.1139/cjes-37-6-923.

Palmer, A.R., 1983, The decade of North American geology 1983 geologic time scale: Geology, v. 11, p. 503–504, doi: 10.1130/0091-7613(1983)11<503:TDONAG>2.0.CO;2.

Paull, R.K., and Paull, R.A., 1994, Lower Triassic transgressive-regressive sequences in the Rocky Mountains, eastern Great Basin, and Colorado Plateau, USA, *in* Caputo, M., Peterson, J.E., and Franczyk, K.J., eds., Mesozoic systems of the Rocky Mountain Region, USA: Rocky Mountain Section of the Society of Sedimentary Geology, Denver, Colorado, p. 1169–180.

Peterson, F., 1994, Sand dunes, sabkhas, streams, and shallow seas: Jurassic paleogeography in the southern part of the Western Interior basin, *in* Caputo, M., Peterson, J.E., and Franczyk, K.J., eds., Mesozoic systems of the Rocky Mountain Region, USA: Rocky Mountain Section of the Society of Sedimentary Geology, Denver, Colorado, p. 233–272.

Peterson, F., and Pipiringos, G.N., 1979, Stratigraphic relations of the Navajo Sandstone to Middle Jurassic formations, southern Utah and northern Arizona: U.S. Geological Survey Professional Paper 1035F, 43 p.

Pipiringos, G.N., and O'Sullivan, R.B., 1978, Principal unconformities in Triassic and Jurassic rocks, western interior United States—A preliminary survey: U.S. Geological Survey Professional Paper 1035A, 29 p.

Poole, F.G., and Madrid, R.J., 1988, Comparison of allochthonous Paleozoic eugeoclinal rocks in the Sonoran, Marathon, and Antler orogens: Geological Society of America Abstracts with Programs, v. 20, no. 7, p. A267.

Porter, M.L., 1986, Sedimentary record of erg migration: Geology, v. 14, p. 497–500, doi: 10.1130/0091-7613(1986)14<497:SROEM>2.0.CO;2.

Proffett, J.M., Jr., and Dilles, J.H., 1984, Geologic map of the Yerington district, Nevada: Nevada Bureau of Mines and Geology Map 77, scale 1:24,000.

Reynolds, S.E., Spencer, J.E., and DeWitt, E., 1987, Stratigraphy U-Th-Pb geochronology of Triassic and Jurassic rocks in west-central Arizona, *in* Dickinson, W.R., and Klute, M.A., eds., Mesozoic rocks of southern Arizona and adjacent areas: Arizona Geological Digest, v. 18, p. 65–80.

Reynolds, S.E., Spencer, J.E., Asmerom, Y., Dewitt, E., and Laubach, S.E., 1989, Early Mesozoic uplift in west-central Arizona and southeastern California: Geology, v. 17, p. 207–211, doi: 10.1130/0091-7613(1989)017<0207:EMUIWC>2.3.CO;2.

Riggs, N.R., and Blakey, R.C., 1993, Early and Middle Jurassic paleogeography of Arizona and adjacent areas, *in* Dunne, G., and McDougall, K., eds., Mesozoic paleogeography of the western United States-II: Pacific Section SEPM (Society for Sedimentary Geology) Book 71, p. 347–373.

Riggs, N.R., and Busby-Spera, C.J., 1990, Evolution of a multi-vent volcanic complex within a subsiding arc graben depression, Mount Wrightson Formation, Arizona: Geological Society of America Bulletin, v. 102, p. 1114–1135, doi: 10.1130/0016-7606(1990)102<1114:EOAMVV>2.3.CO;2.

Riggs, N.R., Mattinson, J.M., and Busby-Spera, C.J., 1993, Correlation of Jurassic eolian strata between the magmatic arc and the Colorado Plateau: New U-Pb geochronologic data from southern Arizona: Geological Society of America Bulletin, v. 105, p. 1231–1246, doi: 10.1130/0016-7606(1993)105<1231:COJESB>2.3.CO;2.

Schermer, E.R., and Busby, C.J., 1994, Jurassic magmatism in the central Mojave Desert: Implications for arc paleogeography and preservation of continental volcanic sequences: Geological Society of America Bulletin, v. 106, p. 767–790, doi: 10.1130/0016-7606(1994)106<0767:JMITCM>2.3.CO;2.

Schermer, E.R., Busby, C.J., and Mattinson, J.M., 2002, Paleogeographic and tectonic implications of Jurassic sedimentary and volcanic sequences, *in* Glazner, A.F., Walker, J.D., and Bartley, J.M., eds., Geologic Evolution of the Mojave Desert and Southwestern Basin and Range: Geological Society of America Memoir 195, p. 93–115.

Shorb, W.M., 1983, Stratigraphy, facies analysis and depositional environments of the Moenkopi Formation (Lower Triassic), Washington County, Utah and Clark and Lincoln Counties, Nevada [M.S. thesis]: Durham, North Carolina, Duke University, 205 p.

Silberling, N.J., 1959, Pre-Tertiary stratigraphy and Upper Triassic paleontology of the Union district, Shoshone Mountains, Nevada: U.S. Geological Survey Professional Paper 322, 67 p.

Silberling, N.J., 1984, Map showing localities and correlations of age-diagnostic lower Mesozoic megafossils, Walker Lake 1° × 2° quadrangle.

Silberling, N.J., and Tozer, E.T., 1968, Biostratigraphic classification of the marine Triassic in North America: Geological Society of America Special Paper 110, 63 p.

Silberling, N.J., and Wallace, R.E., 1969, Stratigraphy of the Star Peak Group (Triassic) and overlying rocks, Humboldt Range, Nevada: U.S. Geological Survey Professional Paper 592, 50 p.

Silberling, N.J., Jones, D.L., Blake, M.C., Jr., and Howell, D.G., 1987, Lithotectonic terrane map of the western coterminous United States: U. S. Geological Survey Miscellaneous Field Studies Map Mf-1874-C, scale 1:2,500,000.

Silver, L.T., and Anderson, T.H., 1974, Possible left-lateral early to middle Mesozoic disruption of the southwestern North American craton margin: Geological Society of America Abstracts with Programs, v. 4, p. 955–956.

Sloss, L.L., 1963, Sequences in the cratonic interior of North America: Geological Society of North America Bulletin, v. 74, p. 93–114.

Speed, R.C., 1978, Paleogeography and plate tectonic evolution of the early Mesozoic marine province of the western Great Basin, *in* Howell, D.G., and McDougall, K., eds., Mesozoic paleogeography of the western United States: Society of Economic Paleontologists and Mineralogists, Pacific Section, Pacific Coast Paleogeography Symposium 2, p. 253–270.

Speed, R.C., and Silberling, N.J., 1989, Early Mesozoic tectonics of the western Great Basin, Nevada: 28th International Geological Congress, Field Trip Guidebook T122, American Geophysical Union, Washington, D.C., p. 1–54.

Stanley, G.D., Jr., and González-León, C.M., 1995, Paleogeographic and tectonic implications of Triassic fossils and strata from the Antimonio Formation, northwestern Sonora, *in* Jacques-Ayala, C., Gonzaléz-León, C.M., and Roldán-Quintana, J., eds., Studies on the Mesozoic of Sonora and adjacent areas: Geological Society of America Special Paper 301, p. 1–16.

Stanley, G.D., Jr., González-León, C.M., Sandy, M.R., Senowbari-Daryan, B., Doyle, P., Tamura, M., and Erwin, D.H., 1994, Upper Triassic invertebrates from the Antimonio Formation, Sonora, Mexico: Paleontological Society Memoir 36, 33 p.

Steiner, M.B., and Lucas, S.G., 2000, Paleomagnetism of the Late Triassic Petrified Forest Formation, Chinle Group, western United States: Further Evidence of "Large" Rotation of the Colorado Plateau: Journal of Geophysical Research, v. 105, p. 25,791–25,808, doi: 10.1029/2000JB900093.

Steiner, M.B., Morales, M., and Shoemaker, E.M., 1993, Magnetostratigraphic, biostratigraphic, and lithologic correlations in Triassic strata of the western U.S., *in* McNeill, D., Aissaoui, D., and Hurley, N., eds., Applications of paleomagnetism to sedimentary geology: Society of Economic Paleontologists and Mineralogists Special Publication 50, p. 41–57.

Stewart, J.H., and Roldán-Quintana, J., 1991, Upper Triassic Barranca Group: Non-marine and shallow-marine rift-basin deposits of northwestern Mexico, *in* Dickinson, W.R., ed., Tectonics and sedimentation: Society of Economic Paleontologists and Mineralogists Special Publication 22, p. 28–57.

Stewart, J.H., Albers, J.P., and Poole, F.G., 1968, Summary of regional evidence for right-lateral displacement in the western Great Basin: Geological Society of America Bulletin, v. 79, p. 1407–1413.

Stewart, J.H., Poole, F.G., and Wilson, R.F., 1972a, Stratigraphy and origin of the Chinle Formation and related Upper Triassic strata in the Colorado Plateau region: U.S. Geological Survey Professional Paper 690, 336 p.

Stewart, J.H., Poole, F.G., and Wilson, R.F., 1972b, Stratigraphy and origin of the Triassic Moenkopi Formation and related strata in the Colorado Plateau region: U.S. Geological Survey Professional Paper 691, 105 p.

Stewart, J.H., McMenamin, M.A.S., and Morales-Ramirez, J.M., 1984, Upper Proterozoic and Cambrian rocks in the Caborca region, Sonora, Mexico—physical stratigraphy, biostratigraphy, paleocurrent studies, and regional relations: U.S. Geological Survey Professional Paper 1309, 36 p.

Stewart, J.H., Anderson, T.H., Haxel, G.B., Silver, L.T., and Wright, J.E., 1986, Late Triassic paleotopography of the southern Cordillera: The problem of a source area for voluminous volcanic detritus in the Chinle Formation of the Colorado Plateau region: Geology, v. 14, p. 567–570, doi: 10.1130/0091-7613(1986)14<567:LTPOTS>2.0.CO;2.

Stewart, J.H., Poole, F.G., Ketner, K.B., and Madrid, R.J., Roldán-Quintana, Jaime, and Amaya-Martinez, Ricardo, 1990, Tectonics of the Paleozoic and Triassic southern margin of North America, Sonora, Mexico, *in* Gehrels, G.E., and Spencer, J.E., eds., Geologic Excursions through the Sonoran Desert Region, Arizona and Sonora: Arizona Geological Survey Special Paper 7, p. 183–202.

Stewart, J.H., Amaya-Martinez, R., Stamm, R.G., Wardlaw, B.R., Stanley, G.D., and Stevens, C.H., 1997a, Stratigraphy and regional significance of Mississippian to Jurassic rocks in Sierra Santa Teresa, Sonora, Mexico: Revista Mexicana de Ciencias Geologicas, v. 14, no. 2, p. 115–135.

Stewart, J.H., Silberling, N.J., and Harwood, N.S., 1997b, Triassic and Jurassic stratigraphy and paleogeography of west-central Nevada and eastern California with a correlation diagram of Triassic and Jurassic rocks: U.S. Geological Survey Open-File Report 97-495.

Stewart, J.H., Poole, F.G., Harris, A.G., Repetski, J.E., Wardlaw, B.R., Mamet, B.L., and Morales-Ramirez, J.M., 1999, Neoproterozoic(?) to Pennsylvanian inner-shelf, miogeoclinal strata in Sierra Agua Verde, Sonora, Mexico: Revista Mexicana de Ciencias Geológicas, v. 16, no. 1, p. 35–62.

Stewart, J.H., Amaya-Martínez, R., and Palmer, A.R., 2002, Neoproterozoic and Cambrian strata of Sonora, Mexico: Rodinian supercontinent to Laurentian Cordilleran margin, *in* Barth, A., ed., Contributions to Crustal Evolution of the Southwestern United States: Boulder, Colorado, Geological Society of America Special Paper 365, p. 5–48.

Stone, P.A., and Stevens, C.H., 1988, Pennsylvanian and Early Permian paleogeography of east-central California: Implications for the shape of the continental margin and the timing of continental truncation: Geology, v. 16, p. 330–333, doi: 10.1130/0091-7613(1988)016<0330:PAEPPO>2.3.CO;2.

Stone, P.A., Stevens, C.H., and Orchard, M.J., 1991, Stratigraphy of the Lower and Middle(?) Triassic Union Wash Formation, east-central California: U.S. Geological Survey Bulletin 1928, 26 p.

Taylor, D.G., Boelling, K.E., Holser, W.T., Margaritz, M., and Guex, J., 1992, Ammonite biostratigraphy and geochemistry of latest Triassic and earliest Jurassic strata from the Gabbs and Sunrise formations, Nevada: Geological Society of America Abstracts with Programs, v. 24, no. 5, p. 85.

Taylor, D.G., González-León, C.M., and Lucas, S.G., 1996, Sinemurian (Early Jurassic) ammonites from the Antimonio Formation, Sonora, Mexico: Geological Society of America Abstracts with Programs, v. 28, no. 5, p. 116.

Taylor, D.G., Boelling, K., and Guex, J., 2000, The Triassic/Jurassic system boundary in the Gabbs Formation, Nevada: GeoResearch Forum, v. 6, p. 225–236.

Van Wagoner, J.C., Posamentier, H.W., Mitchum, R.M., Fail, P.R., and Sarg, J.F., Loutit, T.S., and Hardenbol, J., 1988, An overview of the fundamentals of sequence stratigraphy and key definitions, *in* Wilgus, C.K., et al., eds., Sea-level changes: and Integrated Approach, Society of Economic Paleontologists and Mineralogists Special Publication No. 42, p. 39–45.

Van Wagoner, J.C., Mitchum, R.M., Campion, K.M., and Rahmanian, V.D., 1990, Siliciclastic sequence stratigraphy in well logs, cores, and outcrops: Tulsa, Oklahoma, American Association of Petroleum Geologists, Methods in Exploration Series, No. 7, 55 p.

Walker, J.D., and Wardlaw, B.R., 1989, Implications of Paleozoic and Mesozoic rocks in the Soda Mountains, northeastern Mojave Desert, California, for late Paleozoic and Mesozoic Cordilleran orogenesis: Geological Society of America Bulletin, v. 101, p. 1574–1583, doi: 10.1130/0016-7606(1989)101<1574:IOPAMR>2.3.CO;2.

Weber, R., 1980, Megafósiles de coniferas del Triásico y del Cretácio Tardio de México y consideraciones generales sobre las coniferas mesozoicas de México: Universidad Nacional Autónoma de México, Instituto de Geología, Revista, v.4, no. 2, p. 111–124.

Weber, R., 1997, How old is the Triassic flora of Sonora and Tamaulipas and news on Leonardian floras in Puebla and Hidalgo, Mexico: Revista Mexicana de Ciencias Geológicas, v. 14, no. 2, p. 225–243.

Wetterauer, R.H., 1977, The mina deflection—a new interpretation based on the history of the Lower Jurassic Dunlap Formation, western Nevada [Ph.D. Thesis]: Northwestern University, Evanston, Illinois, 155 p.

Wheeler, H.E., 1964, Base level transit cycle: Kansas State Geological Survey Bulletin, v. 2, p. 623–630.

Wilson, R.F., and Stewart, J.H., 1967, Correlation of Upper Triassic and Triassic(?) formations between southwestern Utah and southern Nevada: U.S. Geological Survey Bulletin1244-D, p. D1–D19.

Winker, C.D., 1982, Cenozoic shelf margins, northwestern Gulf of Mexico: Transactions, Gulf Coast Association of Geological Societies, v. 32, p. 427–448.

Wyld, S.J., and Wright, J.E., 1993, Mesozoic stratigraphy and structural history of the southern Pine Nut Range, west-central Nevada, *in* Dunne, G.C., and McDougall, K., eds., Mesozoic paleogeography of the western United States-II: Pacific Section SEPM (Society for Sedimentary Geology, Book 71, p. 289–306.

Manuscript Accepted by the Society 24 March 2005

Printed in the USA

Geological Society of America
Special Paper 393
2005

Possible Early Triassic location of the Caborca Block

Maureen B. Steiner
Department of Geology & Geophysics, University of Wyoming, Laramie, Wyoming 82071, USA

Oleg Pinos
Spencer G. Lucas
New Mexico Museum of Natural History & Science, 1801 Mountain Road NW, Albuquerque, New Mexico 87104, USA

John E. Marzolf
Department of Geology, Southern Illinois University, Carbondale, Illinois 62901, USA

John W. Estep*

ABSTRACT

Shallow-marine Triassic red sedimentary rocks and diabase intrusives were investigated on the Caborca Block in Sonora, Mexico. The lower 212 m half of the sequence was sampled as a magnetostratigraphic section. Samples exhibit exceedingly linearly decaying remanent magnetization and pass a fold test. Unblocking temperatures suggest that the remanence is carried by magnetite. The beds are inferred to be Early Triassic in age because they overlie Permian strata and are overlain by late Early Triassic (Spathian) *Tirolites*-bearing strata. The red bed samples exhibit an apparently reversed polarity (northern-hemisphere) remanence. Comparison of this polarity to a global compilation of Early Triassic magnetostratigraphy, combined with the age of the superposed beds and the sequence stratigraphic framework, suggests that the age of these beds and their magnetization may be middle Early Triassic (Dienerian). The remanence suggests a paleolatitude of magnetization of 21° N (±4°), so that in the Early Triassic, the Caborca Block may have lain off of western North America near the present location of Seattle, Washington. The overlying red sedimentary rocks containing Spathian ammonites have been remagnetized in a recent geomagnetic field direction. The entire sedimentary section has been intruded by diabase sills; yet oddly, diabase samples gave only widely scattered directions.

The sampling site and Caborca Block are bordered by the left-lateral Mojave-Sonora megashear, but the paleopole is rotated clockwise relative to the North America Early Triassic reference pole, compatible with transport of the terrane in conjunction with right-lateral strike-slip faulting. Many terranes along the western North American margin have been shown to exhibit a history of Jurassic left-lateral transport followed by Cretaceous-Tertiary right-lateral movement (Beck, 1991). The current location of Caborca relative to its inferred Early Triassic paleolocation and the clockwise displacement of the Early Triassic paleopole may stem from a Jurassic left-lateral transport as postulated for the Mojave-Sonora megashear, followed by

*Deceased.

Steiner, M.B., Pinos, O., Lucas, S.G., Marzolf, J.E., and Estep, J.W., 2005, Possible Early Triassic location of the Caborca Block, *in* Anderson, T.H., Nourse, J.A., McKee, J.W., and Steiner, M.B., eds., The Mojave-Sonora megashear hypothesis: Development, assessment, and alternatives: Geological Society of America Special Paper 393, p. 309–328. doi: 10.1130/2005.2393(11). For permission to copy, contact editing@geosociety.org.

post-Early Cretaceous right-lateral motion, as observed in numerous other western North American terranes. The important point is that because of the multiplicity of terrane histories, e.g., northward then southward motion relative to cratonic North America, the inference of right-lateral transport for the Caborca Block does not, and cannot, disprove the existence of the left-lateral Mojave-Sonora megashear.

Keywords: Caborca block; Early Triassic, Mojave-Sonora megashear, magnetostratigraphy, ammonites.

INTRODUCTION

Geology and Age

In northwestern Sonora, Mexico, near the town of Caborca (Fig. 1), the Sierra del Alamo Muerto (30.8° N, 248.4° E [112.6° W]) hosts a thick sedimentary sequence of Permian through Early Jurassic strata (González-León, 1980; 1997). The lowest part of the exposed stratigraphic sequence consists of a thick succession of dolomite and limestone, the Monos Formation, of Wordian age (Cooper, 1953). Overlying the Monos beds are 3.4 km of section designated as the Antimonio Formation (González-León, 1980). Diabase has intruded the sequence, most commonly as layer-parallel sills between beds, but occasionally occurring as crosscutting dikes. The thick Antimonio Formation is divided by thick conglomerates into several sequences; the conglomerates all are abundantly intruded by diabase.

The lower 500 m of the Antimonio beds are shown in Figure 1. Overlying the Monos carbonates are 50 m of orange-red, thin-bedded calcareous siltstones (0–50 m, Fig. 1) containing Permian brachipods, possibly of Guadalupian age (Lucas et al., 1997). Diabase dikes and sills extensively intrude the contact between the siltstones and Monos carbonates, obscuring the contact relationship. A thick conglomerate interval overlies the Permian orange siltstones and is extensively intruded by fine- to medium-grained diabase (conglomerate and diabase #1 in Fig. 1). About 23 m (105–128 m) of orange-red siltstone overlie the diabase, which is in turn overlain by a sequence of somewhat thicker-bedded calcareous siltstones of a dark red to red-brown color. The dark red siltstones form a sequence ~93 m thick (128–220 m, Fig. 1). This succession of siltstones is capped by another thick conglomerate sequence, also extensively intruded by diabase (conglomerate #2, Fig. 1). This second conglomerate is distinctive because it contains clasts of the basal Monos Formation, and is referred to as the Monos-clast conglomerate.

More red beds are present above the Monos-clast conglomerate, largely mudstone with lesser sand- and silt-stone, and containing early Spathian ammonites (Lucas et al., 1997). Despite similar appearances of the red strata above and below Monos-clast conglomerate, appreciable differences in the magnetization and demagnetization responses are observed; hence the red strata below are referred to as the "lower red beds" and those above are the "upper red beds." A fault may separate the two red bed sequences; however, appreciable cover makes this uncertain. Above the Spathian ammonite-bearing beds, physically similar strata occur, but contain Anisian (early Middle Triassic) ammonites. The Anisian beds are unconformably overlain by Late Triassic Carnian strata (Estep et al., 1997). The stratigraphic sequence appears to continue to the south for several kilometers, mostly under an extensive cover of recent sedimentation.

The next well-exposed sequence is 68 m of orange-red sandy limestones (Lucas and González-León, 1995) containing a late Carnian ammonite fauna; among other ammonites, the presence of *Tropites dilleri* places these beds in the *dilleri* Zone of the late Carnian (Silberling in Stanley and González-León, 1995). This exposure ("Dilleri Hill") has been explored for antimony ore, found more abundantly in other nearby exposures around El Antimonio. The orange-red strata presently are within a mineralized zone, which supported a small mining exploration at this outcrop. A small, unsuccessful mining operation had been attempted at this dilleri hill outcrop.

Sampling

For the most part, samples were cored on outcrop with a gasoline-powered engine, although some block sampling had been carried out during a previous reconnaissance visit. Sampling was carried out in a magnetostratigraphic manner, beginning in the Permian beds containing the Guadalupian(?) brachiopods (beds 8 and 9 in Lucas et al., 1997) in Figure 1. Eight reconnaissance cores were collected from the lower 13 m of orange Permian siltstones. The next sampling was from the thick diabase intruding the conglomerate at ~78–103 m. Overlying orange-red strata, apparently baked by the underlying diabase, were sampled at 117 through 122 m. At ~128 m, a well-bedded sequence of red calcareous, silicified siltstones, mudstones, and limestones are encountered, in 93 m of apparently stratigraphically continuous exposure. We sampled these beds between 128 and 220 m, at 0.3–1.0 m intervals, collecting 61 samples. Within this red bed sequence, at 160 m, a thin (3.5 m) sill is present. The intrusive rock is fine-grained diabase, similar to the previous diabase encountered, and has well-exposed baked contacts on both margins. Both intrusive and baked contact rocks were sampled.

At 221 m lies a second thick conglomerate (Fig. 1), ~40 m thick and pervasively intruded; no samples were taken. Above the conglomerate sequence, another 120 m of thinner-bedded, identically colored, red siltstones, mudstones, sandstones, and limestones crop out. These beds appear to be a continuation of the

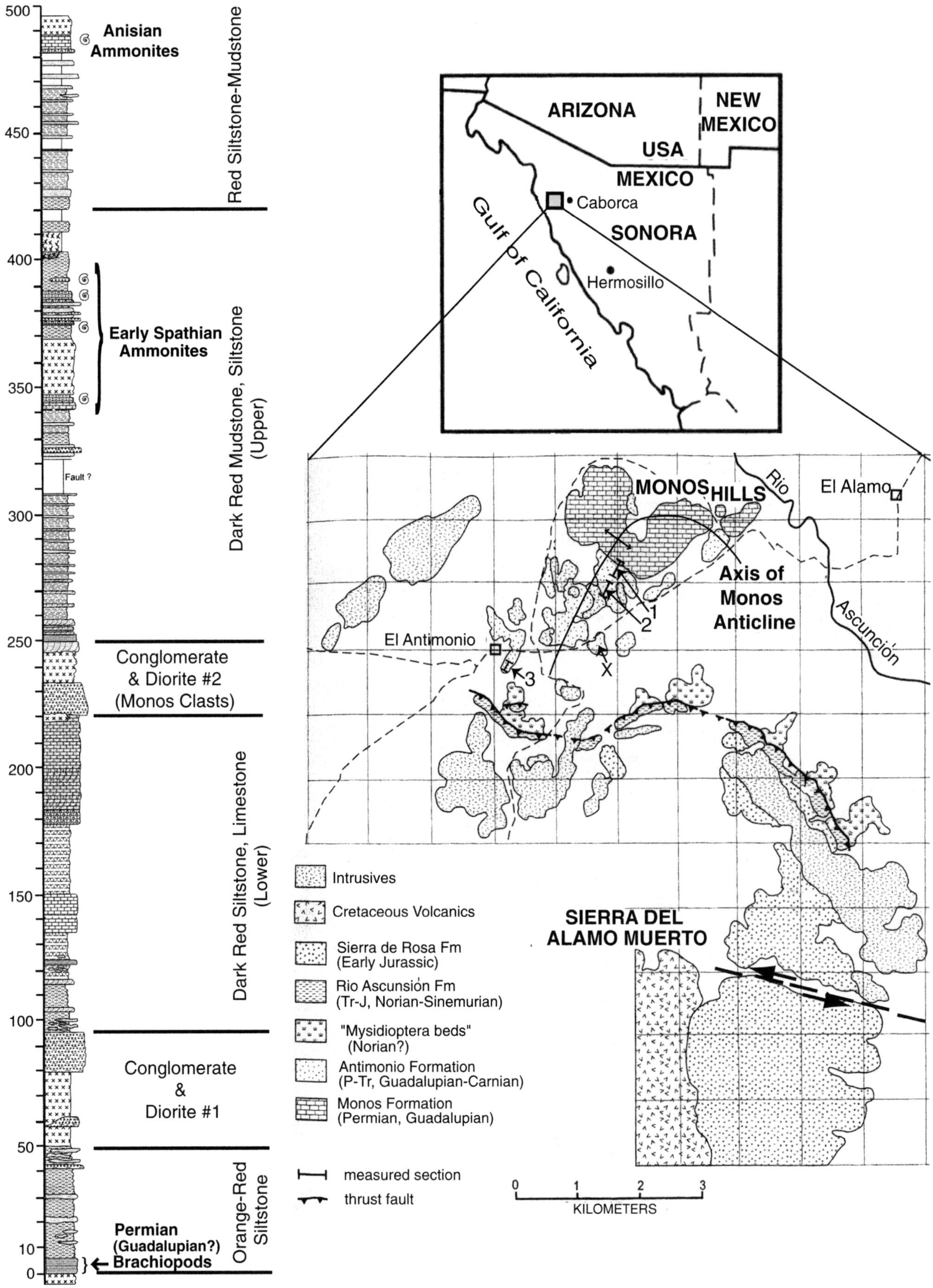

Figure 1. Location map showing the sampling sites (sections 1, 2, 3) and the stratigraphic sequence described in the text (stratigraphy modified from Lucas and Estep, 1999, and González-León, 1997).

stratigraphic sequence below; however, a saddle between exposures and a covered interval may indicate that a fault separates the two red bed sequences (Fig. 1). The exposures of this upper red bed sequence differ from the lower sequence in that the lower red beds are exposed in relatively fresh outcrop along a stream cut, whereas the upper red beds are exposed on a hill slope. Not only is the exposure poorer, but also a larger number of intrusives are present near the upper sequence. Hence, this upper sequence of red beds was sampled more sparsely; samples were taken at intervals of 3–6 m, terminating at the intrusion forming the top of the hill. Twenty-seven samples were cored from these upper red beds, including two baked contacts.

At the exposure containing *dilleri* Zone ammonites, only the upper 50 of the 68 m are well exposed. We block-sampled a preliminary suite of representative strata at 5 m intervals through 35 m of the upper part of the exposed section. Eight blocks yielded 17 cores that were investigated paleomagnetically.

PALEOMAGNETIC DATA

Demagnetization

Distinctly different remanent magnetizations and demagnetization responses were exhibited by the lower and upper parts of the Early Triassic red strata; therefore, their paleomagnetic data are discussed separately, in stratigraphic order. Only thermal demagnetization was conducted on the red bed samples. Both alternating field (AF) and thermal demagnetizations were performed on the igneous rocks. Thermal demagnetization was carried out in 100 °C increments from 150 to 350 °C; thereafter, the rocks were heated at 435, 480, 520, 560, 590, 635, 645, 655, 660, 668, and 674 °C. Intrusive rocks were heated in those same increments to 520 °C, followed by more closely spaced steps: 540, 560, 575, 585, 595, 605, 615, 625, and 635 °C. AF treatments began with 2.5 and 5.0 mT treatments, followed by treatment in 5 mT increments through 60 mT. Characteristic directions were computed for samples with linear demagnetization behavior using the principal component analysis technique (Kirschvink, 1980). Mean directions were computed with Fisherian statistics, and are listed in Table 1. The fold tests of McFadden (1990) were employed on characteristic directions.

The presently observed geomagnetic field direction at this site is 11.0° declination, 56.1° inclination, and the theoretical geocentric axial dipole direction is 0°, 50.0°.

Late Permian Siltstones (0–15 m)

(NRM_0) directions were widely dispersed. Thermal demagnetization showed linear remanence decay between 250 and 560 °C, but directions remain dispersed, although significantly less than the NRM_0 directions. All directions are negative inclinations directed S or SE (in situ) up to 590 °C; between 590 and 674 °C, another magnetization component was revealed, also with negative inclinations, but northerly declinations (in situ), and appreciable dispersion. Despite bedding dips of 28–78° to 240° SW, correction for dip does not reduce the dispersion among either component observed during demagnetization.

Lower Red Bed Sequence (128–220 m)

NRM_0 directions exhibited declinations largely between 180° and 270° and generally shallow negative inclinations. Directions after tilt correction are slightly more tightly grouped

TABLE 1.

EARLY TRIASSIC RED BEDS FOLD

Site	Dec	Inc	n	*k*	α_{95}	γ_c	γ_o
In situ (magnetization not acquired after folding)	206.5	4.0	21	28.1673	6.1	SCOS2=9.520	@ 95%=5.335 @ 99%=7.483
Corrected (magnetization acquired before folding)	191.9	–36.7	21	41.6558	5.0	SCOS2=0.182	

MEAN DIRECTION AND PALEOPOLE

Site		Dec	Inc	n	*k*	α_{95}	E Long	N Lat	d_p	d_m
Sediments										
Early Triassic red beds*	Mean D & I	191.9	–36.7	21	41.6558	5.0				
	Mean VGP			21	56.7924	4.3	16.9	75.6		
Igneous rocks										
Igneous at 78 m (uncorrected)		200.0	–18.4	5	15.8	19.8	87.4	57.3	9.9	19.8
Igneous at 160 m (uncorrected)		150.2	–24.1	3	8.9	44.0	130.5	57.1	25.1	47.0
Contacts at 160 m (uncorrected)		188.4	–39.6	2	299.1	14.5				
Contacts at 160 m (corrected)		139.7	–38.6	2	28.1	49.0				

Note: Fold test used McFadden (1990) Program "Fold2." Dec—declination; Inc—inclination; n—number of samples; D & I—declination and inclination; VGP—Virtual geomagnetic poles.
* Bedding-corrected coordinates.

than in situ directions (Fig. 2A, 2B), and may suggest a northern-hemisphere reversed polarity magnetization. Initial intensities were similar among samples, averaging 1×10^{-5} mA/m.

Approximately one-half of these samples were thermally demagnetized. Little remanence decay occurs between 350 and 560 °C, and the same direction is displayed. At 560 °C, remanence is lost abruptly, decaying quickly toward the origin of orthogonal axes plots (Fig. 3). Generally, a vector with the present geomagnetic field direction is removed first; thereafter, the samples exhibit minor decay toward or to the origin between 350 and 560 °C (Fig. 3). Most remanence is lost between 560 and 620 °C. Demagnetization above 560–590 °C resulted in highly scattered directions (with some tendency to reside in the NE quadrant such as in the Permian siltstones).

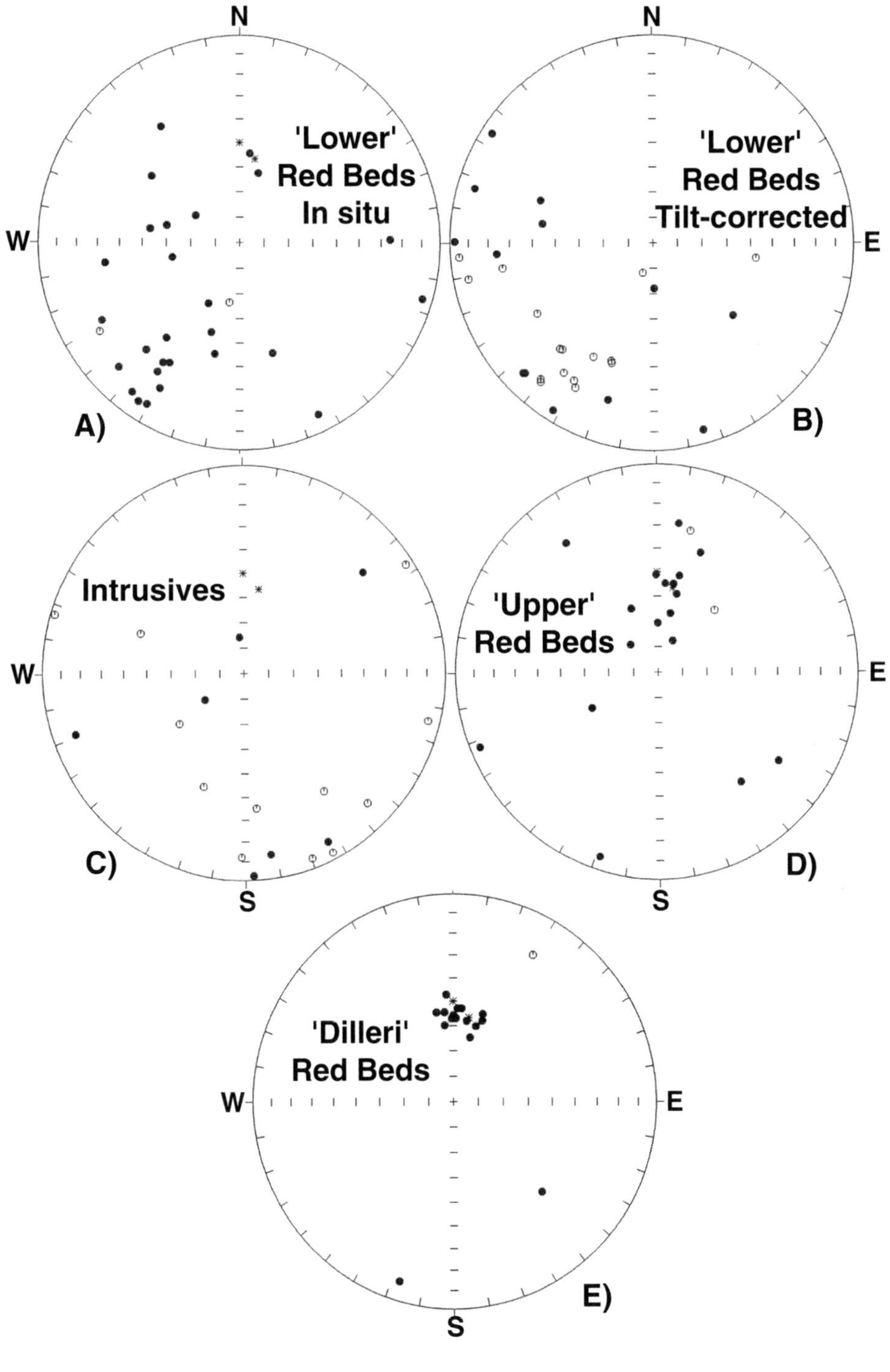

Figure 2. NRM_0 directions. (A) Lower portion of Lower Triassic in in situ and (B) bedding-corrected coordinates; (C) all diabase samples (in situ coordinates); (D) upper portion of Lower Triassic strata in in situ coordinates; (E) Upper Triassic, *dilleri*-bearing strata (in situ coordinates). Equal area projection; closed (open) symbols indicate lower (upper) hemisphere inclinations; asterisks denote the present and geocentric axial dipole field directions for the site.

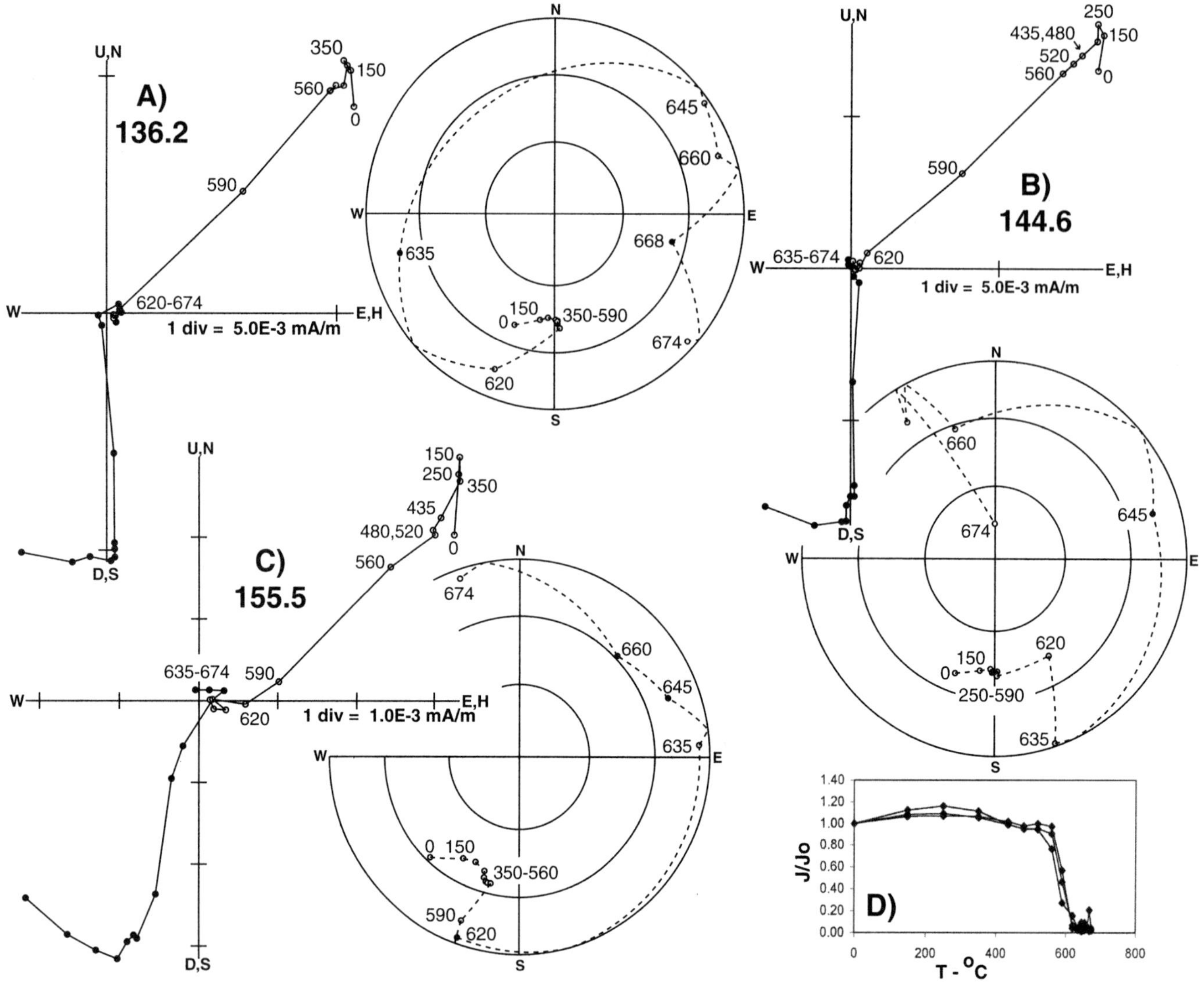

Figure 3. Responses to demagnetization of typical samples from strata between 128 and 160 m (A–C). All are plotted in tilt-corrected coordinates. (D) Normalized intensity during demagnetization for these three samples. Equal-area stereographic projections show lower (upper) hemisphere directions as solid (open) circles. The orthogonal projections display inclination in the plane of declination (Kirschvink, 1980); solid (open) circles represent the horizontal (vertical) components of the magnetization.

The thin sill at 160 m within these red beds forms a division in demagnetization responses of red bed samples. Nearly every sample below the sill is characterized by linear decay toward the origin and "square-shouldered" (see Fig. 3) intensity decay when plotted with respect to the initial remanence intensity (J/J_0). Samples above the thin sill (168–212 m) exhibit similar directions to those of the samples below (Fig. 4), but commonly the decay is not directed to the origin of orthogonal axes plots and unblocking temperatures commonly are more distributed than in the beds below the sill. Moreover, normalized decay of intensity with temperature commonly is not as rectilinear as in the beds below the intrusion (Fig. 4). Also, in contrast to the beds below the intrusive, not all of the red bed samples above the intrusion exhibit the direction characteristic of all lower beds. Three of the upper samples either exhibit that direction at a single temperature step or not at all; the strata between 215 and 220 m do not exhibit that direction or any common direction.

Intrusive Rocks

Six occurrences of igneous rocks were sampled in this study. Sills were sampled in the stratigraphic sequence at 78 m, 160 m, 222 m, 242 m, and 410 m. In addition, a dike, or a sill not entirely restricted to a bedding plane orientation, was sampled in a neighboring sedimentary exposure about a kilometer away; this site is referred to merely as "X" (Fig. 1) because it bears no

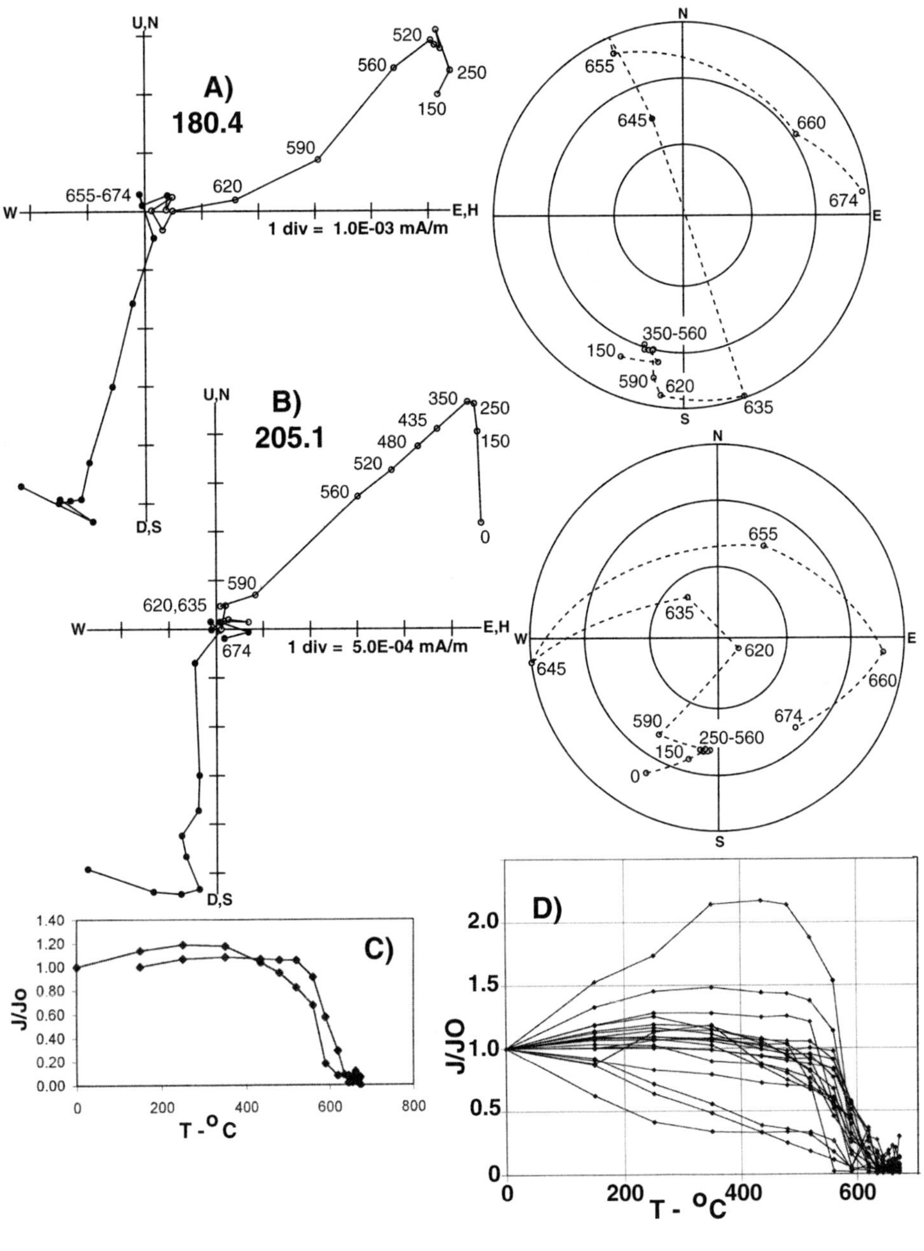

Figure 4. Representative samples from 168 to 210 m showing responses to demagnetization (A, B) and their normalized intensity losses (C). Directions are displayed in tilt-corrected coordinates. (D) Normalized intensity loses of all samples used in paleopole calculation. Projections and symbols as in Figure 3.

stratigraphic relationship to the main sequence of this study and simply presented another opportunity to sample the intrusive rocks. Surprisingly, the various intrusives exhibited some widely varying NRM_0 directions (Fig. 2C), and also a large variation in intensities.

The thin sill at 160 m is most illustrative, and therefore is discussed first. Five regions within this sill and its baked contacts were sampled. The baked margins consist of green calcareous siltstone outlining the sill on either side. Samples from the baked siltstone exhibited demagnetization behavior identical to the non-baked red beds; the baked siltstones show linear decay and square-shouldered demagnetization (Figs. 5A, 5C), and their remanence spectra is the same as the red bed samples distant from the sill. However, the characteristic directions of the baked margin siltstones are distinctly different from the red strata, exhibiting more easterly declinations, as will be discussed.

The diabase of the sill loses an appreciable amount of remanence between 20 and 250 °C, as shown by sample "162.6" in Figure 5B and the lowest intensity sample in the J/Jo plot. Sample "162.6" is from the center of this 3.5 m sill, and the remanence removed up to 250 °C is that of the recent geomagnetic field. Samples on either side, closer to the margins of the sill, show similar appreciable remanence loss, but the direction removed from 0 to 250 °C is the characteristic direction of the sample. This central sample may reflect the response of a coarser grain size, as expected at the more slowly cooled center of an

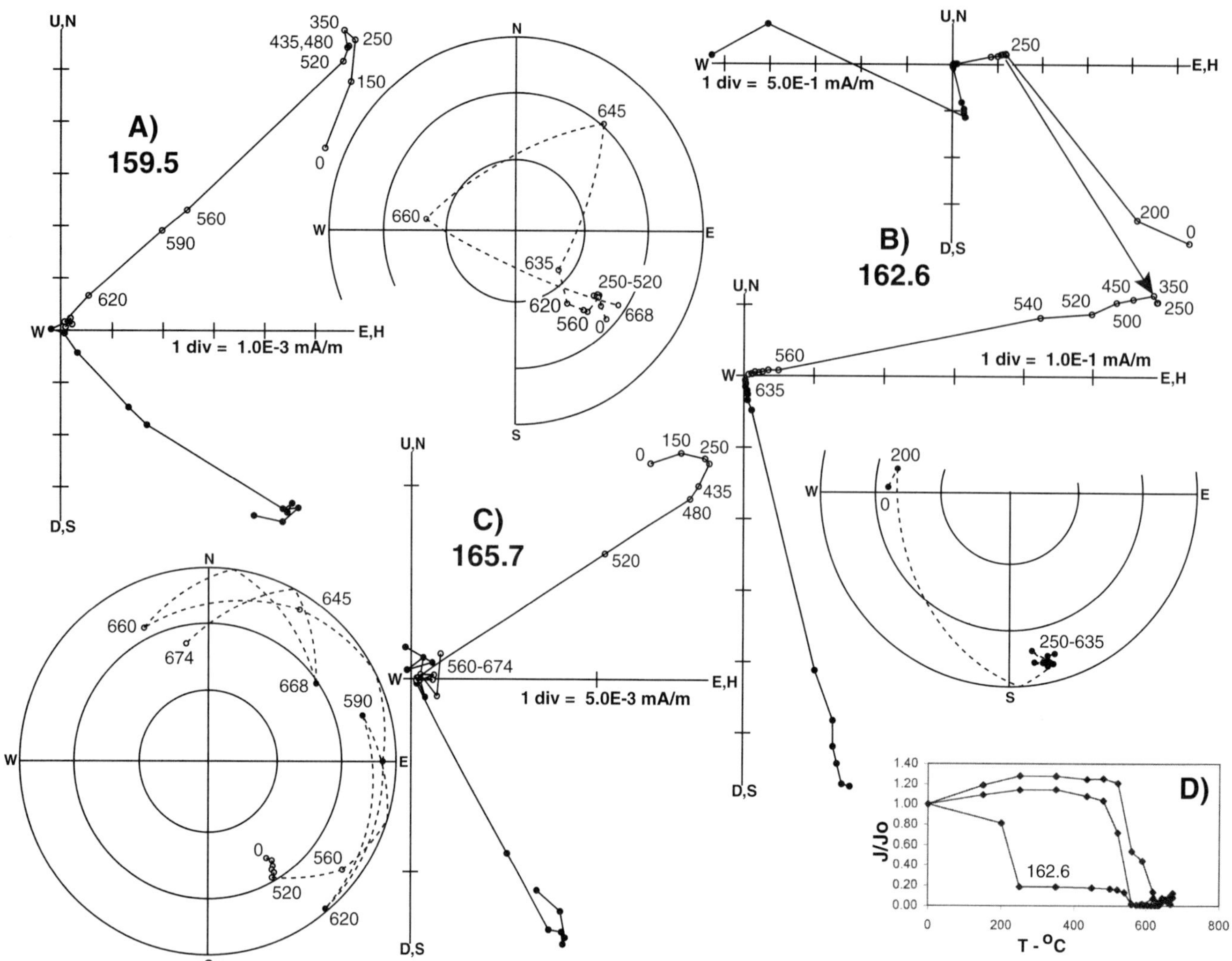

Figure 5. Responses to demagnetization of samples from the baked contacts (A, C) and the sill (B) at 160–166 m. Projections and symbols as in Figure 3; directions displayed in tilt-corrected coordinates.

intrusive; coarser grain size magnetite is prone to acquire remanence in the most recent fields to which it is exposed (viscous remanence acquisition). Unblocking temperatures are fairly discrete: in addition to the appreciable remanence loss by 250 °C, a second significant remanence loss occurs between 540 and 560 °C. Decay is univectorial to the origin above 250 °C.

The NRM_0 and characteristic directions for the thin sill are shown in Figure 6A, labeled "i." The characteristic directions of the baked contacts ("c") are included for comparison. The plot of Figure 6A labeled "Contacts" displays the directions from the two baked contacts both before unfolding (left) and tilt-corrected (right), illustrating the better agreement prior to tilt correction. The baked contacts thus suggest a post-folding timing of intrusion. The "CHR" plot of Figure 6A also suggests a relationship between the characteristic directions of the sill, the contact beds, and the enclosing red strata: the baked strata display directions intermediate between those of the red beds ("Sed Mean," Fig. 6A) and the much more easterly igneous rock characteristic declinations. The means of each directional group, listed in Table 1, further emphasize this apparent relationship. Because of the suggestion of a post-folding intrusion, all other directions illustrated in Figure 6 are shown in in situ attitudes.

The stratigraphically lowest sill sampled lies between 78 m and ~105 m (Fig. 1); this sill is ~37 m of fine-grained diabase, intruded below and into the thick conglomerate beds interpreted to mark the beginning of Early Triassic deposition (Lucas et al., 1997). Six samples were examined. NRM_0 directions were similar and mostly located in the southeast quadrant with near-zero inclinations (Fig. 6B). However, NRM_0 intensities were exceedingly variable, ranging from 8×10^{-5} to 2×10^{-1} mA/m. The NRM_0 direction was stable through AF fields of 8 mT and temperatures of 140–200 °C in the samples. Samples generally

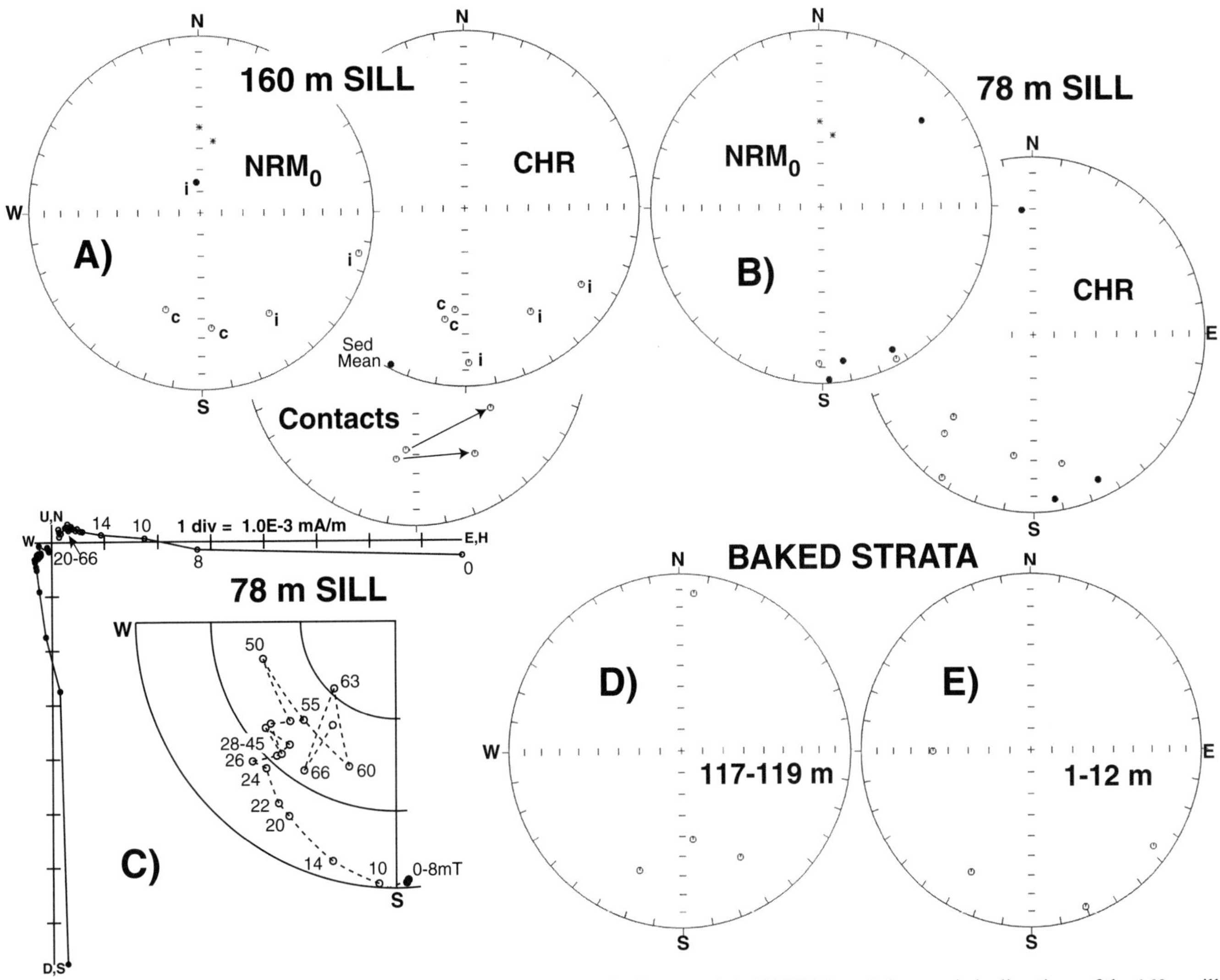

Figure 6. Initial and characteristic directions of the lower two intrusive bodies sampled. (A) NRM_0 and characteristic directions of the 160 m sill igneous (i) and contact (c) rocks in in situ coordinates, and the baked contact samples in in situ and bedding-corrected coordinates. (B) NRM_0 and characteristic directions of the 78 m sill. (C) AF demagnetization of one of the samples from the 78 m sill. (D, E) Characteristic directions from rocks probably baked by nearby intrusives. Directions shown in in situ coordinates; symbols and projections as in Figure 3.

displayed relatively linear decay of magnetization during demagnetization, as shown by the example of AF demagnetization in Figure 6C. However, the characteristic directions are somewhat more poorly grouped than the initial directions (Fig. 6B).

Beds above the 78 m sill (spanning 78 to ~105 m), at 117–119 m, appear to be baked, both from an orange color of the strata and a more indurated nature. Their proximity to the 78 m sill suggests probable baking by this intrusive. Their characteristic directions, shown in Figure 6D (in in situ coordinates), mostly agree with those of the 78 m sill (Fig. 6B).

The Guadalupian(?) beds between 1 and 12 m probably constitute a third zone of baked strata. Again the physical appearance of these beds, orange color and appreciable induration, is consistent with baking, as is their proximity to the extensive intrusive network between these beds and the massive Monos carbonates. The samples were taken from 1 to 12 m away from the intrusives. The sample at 1 m exhibits a characteristic direction essentially the same as the igneous rocks of the 78 m and 160 m sills, and two of the three higher samples that were demagnetized give characteristic directions that also are similar to those igneous rocks (Fig. 6E).

The other four intrusives sampled, at 222 m, 242 m, 410 m, and "X," were sampled only in extreme reconnaissance fashion with 1–3 samples each because it was anticipated that the generally fine-grained igneous rocks would be strongly and stably magnetized and would give the same results; this proved to be entirely incorrect. Directions before and after demagnetization within any one of these intrusives differed wildly, and differed between intrusives, in contrast to the relatively similar within- and between-intrusive directions characteristic of the two intrusives shown in Figures 6A and B. The intrusive rocks sampled at 222 and 242 m are likely the same body; the three samples exhibit univectorial decay to the origin during demagnetization, but the three characteristic directions are scattered over ~270° of arc. The intrusive at 410 m also exhibits univectorial to the origin remanence decay, and an appreciably different direction than the other intrusives. The three samples of the X intrusive do not decay linearly, and are relatively unstable to demagnetization; these too yield highly scattered directions. These directions are distributed over 180° of arc, but it is an arc that is at approximately right angles to that formed by the 222–242 m samples.

Spathian-Ammonite-Bearing Upper Red Beds (300–420 m)

These beds appear to be a same-color, but finer-grained version of the lower red beds. However, although NRM_0 intensities (~1.2×10^{-5} mA/m) were quite similar to those of the lower red beds, their magnetizations differ significantly. Many of the NRM_0 directions clustered near the present-day geomagnetic field direction (Fig. 2D).

Twenty samples were thermally demagnetized. About 50% of the samples decayed linearly in response to thermal demagnetization to 520 °C (Fig. 7). About half of those samples, 25% of the total sample population, exhibited linear decay to the origin, and the characteristic direction was the present-day geomagnetic field direction (e.g., Figure 7D). The other half (another 25%) decayed linearly, but not to the origin, and also exhibited present-day geomagnetic field direction as their characteristic directions (e.g., Fig. 7C). In both groups, remanence directions change abruptly above 520–560 °C. The remaining 50% of the samples exhibited largely overlapping blocking temperature spectra, as shown by in Figure 7A.

In addition, 60% of these samples exhibit a common direction at the 520° or 560 °C temperature steps (e.g., in Fig. 7A), directions with negative inclinations and S-SE declinations (in situ coordinates). The 520–560 °C directions are reasonably similar among samples, but generally are not stable and are displayed for only the single heating step. Thereafter, directions change by large amounts, as shown in Figure 7A. Other samples remain in the southerly direction while exhibiting large changes of direction at demagnetization steps above 520–560 °C (e.g., Fig. 7C). Among these upper red bed samples, a few rare samples low in the sequence exhibit only a southerly, negative direction throughout demagnetization, as shown in Figure 7B. Some of the stratigraphically highest samples did not exhibit negative-inclination southerly directions at 560 °C, but display such directions at higher temperatures—635–674 °C (Fig. 7D). Thus, a common factor among this upper group of red bed samples is that at least one point in their demagnetization responses, all samples display negative inclination directions in a S-SE direction. These directions are distinct from that of the lower red beds, which are directed to the south or southwest.

Upper Triassic Strata

NRM_0 directions almost entirely reflected the present and/or geocentric axial dipole geomagnetic field directions (Fig. 2E). Thermal demagnetization was performed on one or two samples from each block; heating increments were slightly more closely spaced than were used on the Early Triassic strata. Most samples responded with linear decay of magnetization to the origin of orthogonal-axes plots (Fig. 8), and exhibited either normal or reversed polarity magnetization (Figs. 8A, 8B). The two polarities of remanence are displayed sequentially in the stratigraphic sequence (Fig. 9); normal polarity remanence characterizes the lower part of the sampled section, changing to reversed polarity at approximately the level of the small mine. One block sample held both polarities of magnetization within the same sample (Fig. 8C); the two samples from this dual-polarity block exhibited the same demagnetization responses. This block was located physically between a block yielding normal polarity and another yielding reversed polarity magnetizations.

DISCUSSION OF RESULTS

All characteristic directions obtained in this study are shown in Figure 10. Each sampling group is discussed separately and compared to the other data. The main solid result obtained from this study is that from the lower Lower Triassic red beds.

Permian Red Beds

The stratigraphically lowest samples, from Permian calcareous siltstones, possessed directions below 560 °C (Fig. 6D) that resemble those of the intrusive rocks. Also like the intrusives, dispersion among directions is high. The strata sampled lie adjacent to a heavily intruded region between the massive Monos carbonates and the overlying thin-bedded Permian and Triassic strata, and probably were baked by the intrusives. Other than supporting the fact that a SE direction is associated with the intrusives, this small group of samples contributes little else to this study.

Lower Red Beds of the Lower Triassic (128 m and 210 m)

The characteristic directions of these red beds (Fig. 10A) were all defined by linear remanence decay; most commonly that decay occurs between 435 through 560 °C. Some samples decay to the origin of orthogonal axes plots, whereas others exhibit linear decay not including the origin. Greater than 90% of the remanence decays by the 560–590 °C heating. The normalized intensity plot in Figure 4D shows the intensity decay of all samples used in the mean direction and pole calculation. Most of these decay patterns suggest a characteristic remanence that may be carried by magnetite. A remanence held above 590 °C, undoubtedly carried by hematite, is relatively scattered but lies in the NE quadrant; the identity of this direction has not been ascertained at this time.

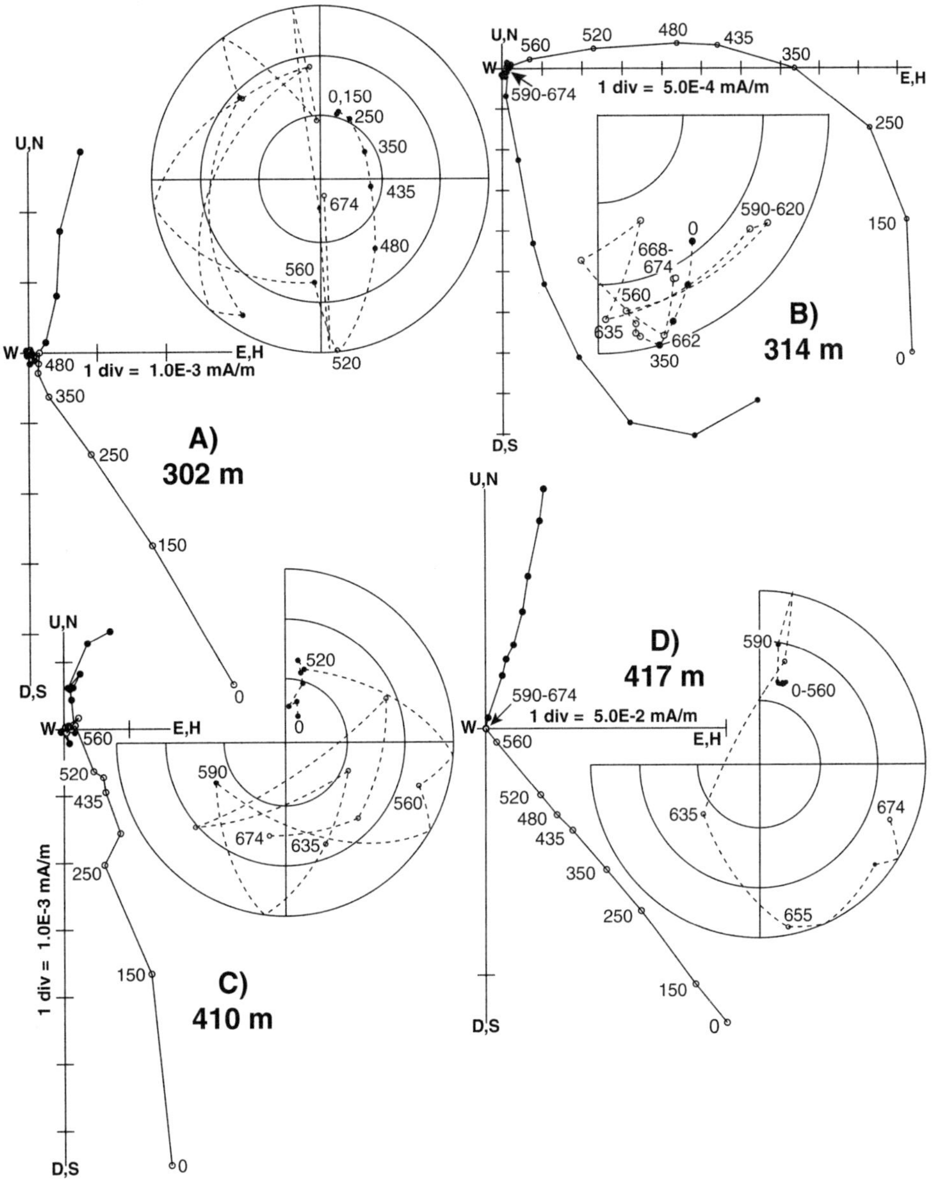

Figure 7. Responses to demagnetization of the upper Lower Triassic red beds (300-420 m). All shown in in situ coordinates. Symbols and projections as in Figure 3.

A true fold test (between two different limbs of a structure) is not possible for the strata sampled in this study because the magnetostratigraphic sequence sampled was located entirely on the west limb of the Monos anticline (see Fig. 1). Thus all beds sampled dip west, but vary appreciably in dip direction (220° to 268°), as well as angle of dip (49° to 75°). The bed attitudes are sufficiently different that conventional Fisher statistics demonstrates a smaller dispersion among directions (k, Table 1) after correction for bed attitudes ($k = 43.9$) than in the in situ coordinates ($k = 28.2$), suggesting that the magnetization may be pre-folding. McFadden's (1990) fold test is sufficiently flexible to test samples from one limb of a fold with appreciably different bedding attitudes. Using his definition 2 of the test statistic (McFadden, 1990, p. 167), a positive fold test was obtained (Table 1), indicating that this remanence is pre-folding.

All of the samples from this sequence yielded characteristic directions suggestive of northern-hemisphere reversed geomagnetic field polarity (Fig. 10A). This section of reversed polarity lies below the sequence containing Spathian *Tirolites* ammonites (Fig. 11). Depending on the length of time represented by the conglomerate and whether or not a fault exists between conglomerate and the ammonite beds, this reversed polarity may indicate a Dienerian age for these beds.

A relatively large number of Early Triassic magnetostratigraphic sections now have been studied (Fig. 12). These sections include the biostratigraphically dated sequences in Figure 12 (Arctic stratotype, Chinese sections, the marine part

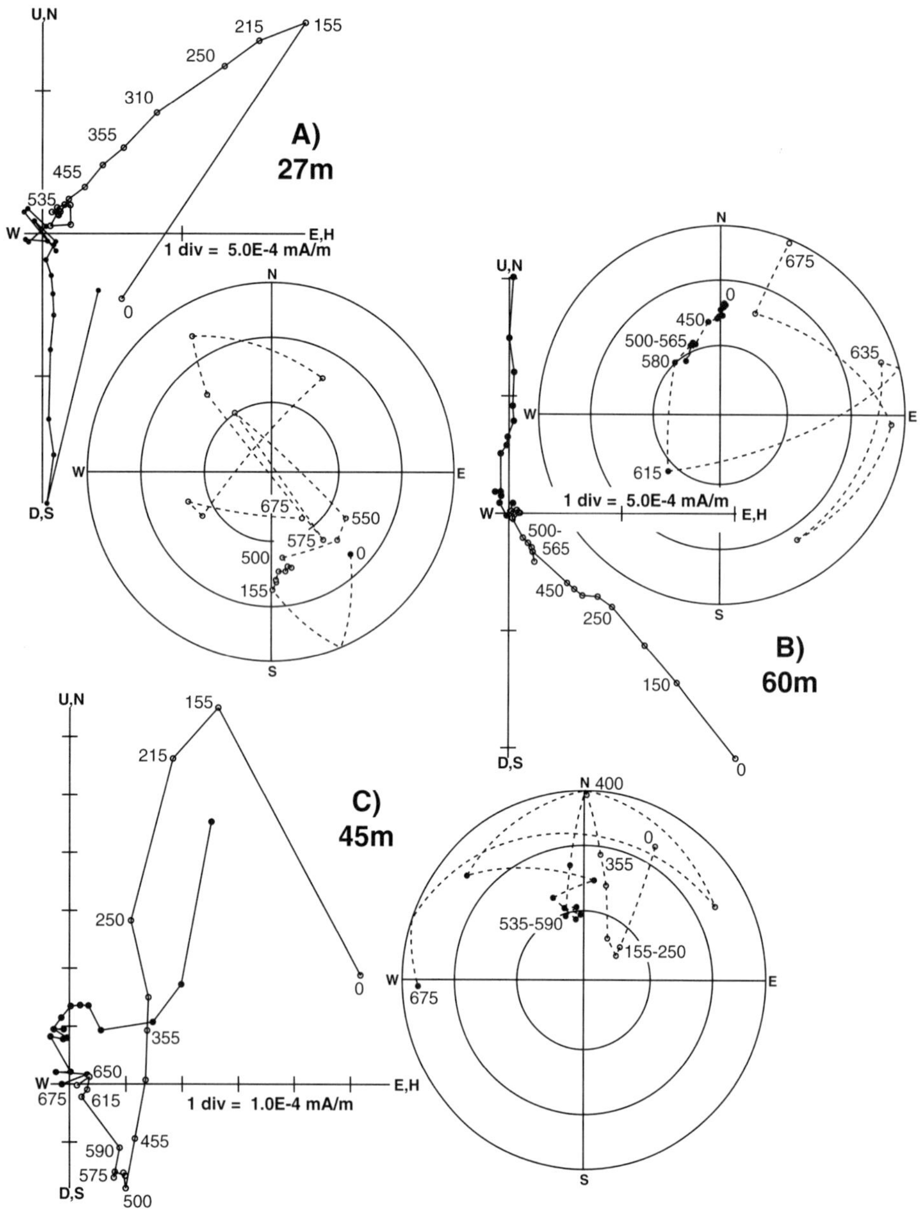

Figure 8. Responses to demagnetization of the Upper Triassic samples. All shown in in situ coordinates. Symbols and projections as in Figure 3.

of the western North American Moenkopi Formation section, the Iranian sequence, and the Italian section); terrestrial red bed sequences include the North American Moenkopi and Chugwater, the Polish Basin, and Russian sequences. All Early Triassic magnetostratigraphic sections indicate essentially the same pattern, thus clearly defining the reversal pattern of the Early Triassic geomagnetic field. A basal short-duration normal polarity interval is followed by a more lengthy interval of dominantly reversed polarity, followed by a normal polarity interval associated with Smithian stage fauna, and Spathian fauna beginning in reversed polarity (Fig. 12).

The reversed polarity of the Antimonio lower red beds is separated from the portion containing the Spathian ammonites by a thick conglomerate; in sequence stratigraphic terms, these beds lie in the sequence below the Spathian sequence. A sequence boundary below the occurrence of *Tirolites* ammonites is very reminiscent of the stratigraphy of the North American Moenkopi Formation. Like the Antimonio Formation, the Moenkopi Formation also contains the *Tirolites* ammonite genus. Moreover, in the Moenkopi Formation, the Smithian stage largely is represented by a conglomerate whose clasts contain *Meekoceras* ammonites. In the context of global magnetostratigraphy of the Early Triassic (Fig. 12), and particularly in relationship to the Moenkopi litho- and magnetostratigraphy, the association in the Antimonio beds of reversed polarity, a conglomeratic sequence and implied

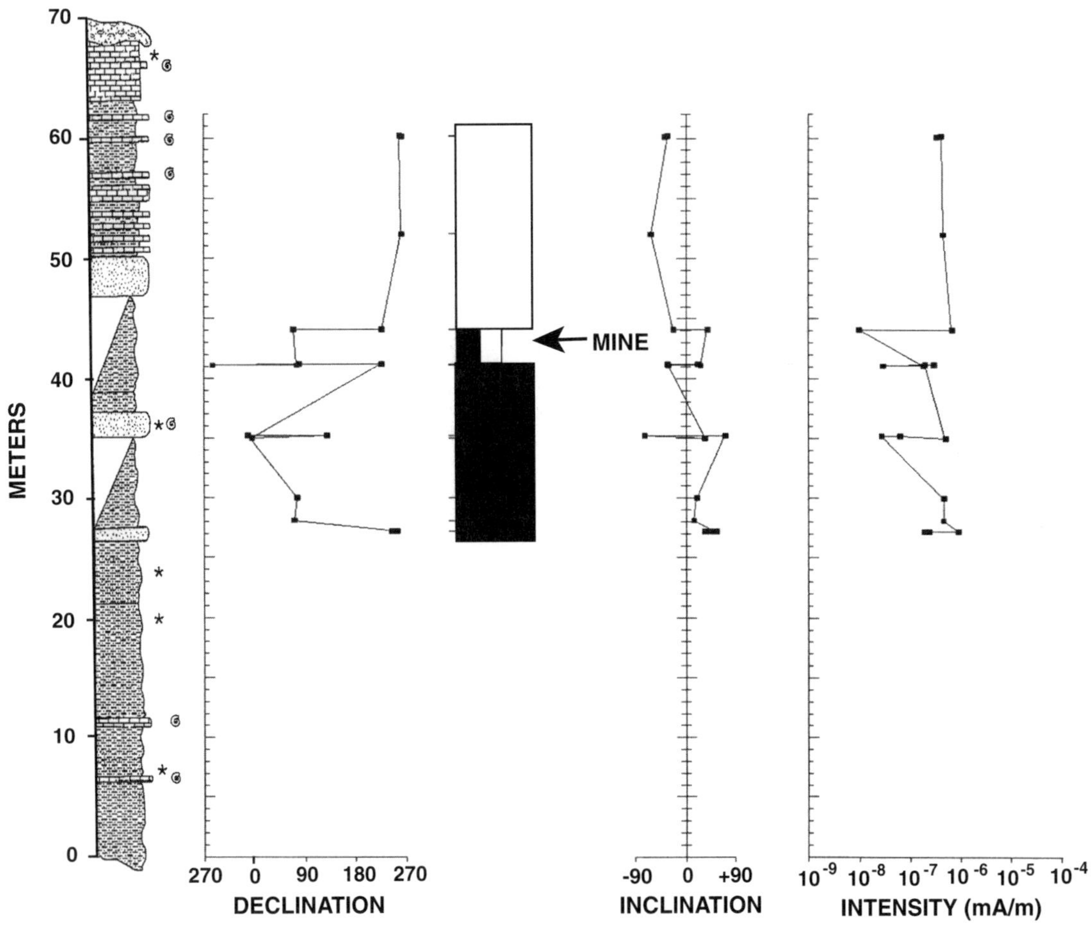

Figure 9. Upper Triassic strata magnetostratigraphy, showing characteristic directions in in situ coordinates. White (black) = reversed (normal) polarity; partial fill in the polarity column indicates less certainty due to poor sample demagnetization behavior. Stratigraphy from Lucas and González-León, 1995; all ammonite symbols indicate Carnian (Late Triassic)-age ammonites; stars indicate ichthyosaur remains.

hiatus, and the presence of *Tirolites* suggests that the lower red bed sequence is Dienerian in age.

Upper Red Beds of the Lower Triassic (300–420 m)

These strata possess characteristic directions that reflect recent normal polarity magnetizations in their in situ attitudes (Fig. 10B). The tendency to exhibit the recent, normal polarity geomagnetic field direction appears to increase up-section in this sample group, which also is the direction approaching the exposed intrusive bodies. These samples also exhibit a high temperature (>520 °C) reversed polarity, S-SE directed component, commonly observed only at a single demagnetization increment. The directions of this component are similar to those of the igneous rocks, and might represent the same field direction. Unfortunately, these beds containing age-diagnostic Spathian ammonites have been remagnetized, apparently more than one time in their history.

Upper Triassic Red Beds

These strata also have been remagnetized in Tertiary geomagnetic fields, both normal and reversed polarities. Their characteristic directions are shown in Figure 10C. The recent normal and reversed geomagnetic field directions occur in a coherent stratigraphic sequence. The mine located in this outcrop is right at the change of polarity. These data indicate that the area was remagnetized in Tertiary time and may indicate that the mineralization of the area is Tertiary in age; they also suggest that mineral emplacement was associated with a change in geomagnetic field polarity (Fig. 9).

Intrusive Rocks in the Lower Triassic Section

The directions from the igneous rocks and their baked contacts, and from rocks suspected of being baked by nearby intrusives, are shown in Figure 10C (in situ coordinates). The disper-

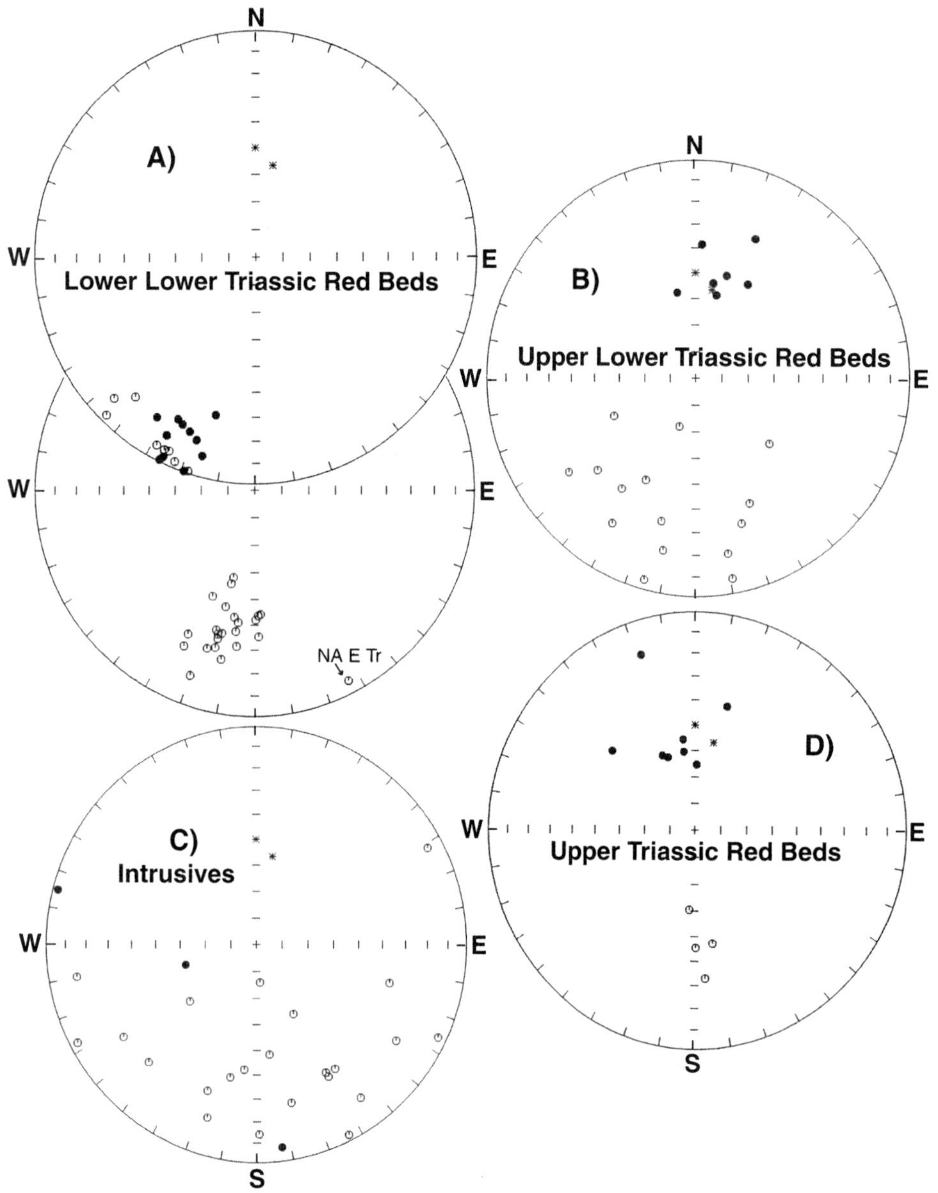

Figure 10. (A) Characteristic directions from the lower red beds of the Lower Triassic (128 and 210 m), plotted both in situ (top), and after correction for the dip of the strata (bottom). (B) Upper red beds of the Spathian Lower Triassic (300–420 m): the ≥560 °C direction discussed in text combined with a few characteristic directions (in situ coordinates). (C) Characteristic directions from all igneous rocks and their contact samples, plotted in in situ coordinates. (D) Characteristic directions from the Late Triassic strata (in situ coordinates). Projections and symbols as in Figure 2. E Tr—Early Triassic; NA—North America.

sion is appreciable and suggests that a single age geomagnetic field direction has not been isolated. The directions and reversed polarity of the upper, Spathian red beds bear notable similarity to the intrusion-related population, in the general direction, polarity, and large dispersion. The relationship suggests some remagnetization of the Spathian rocks by the intrusives, but the abundant evidence of a recent geomagnetic field suggests yet another remagnetization later in time.

Despite collection of igneous rocks from six different intrusive bodies, other than reversed polarity directed southward, no truly coherent magnetization was obtained. Other than the 78 m and 160 m intrusions, intrusives directions varied by 180 to 270° within the same body, but did not exhibit the same patterns of directional variation between bodies. Mean directions were computed for the 78 m and 160 m intrusives, but the means are not similar (Table 1), and the confidence limits are too large for any meaningful conclusion.

The elusive nature of the true igneous rock direction is puzzling; the rocks simply do not appear to have retained coherent magnetizations. Yet, with one exception, the intrusives are quite fine-grained; they border on having been intruded at a hypabyssal, rather than true plutonic, depth. One possible explanation for large dispersion of the igneous rock directions and the failure to record a more coherent representation of the geomagnetic field at the time of emplacement may be the time of the intrusion. If these intrusives were emplaced in the Middle Jurassic, a time of

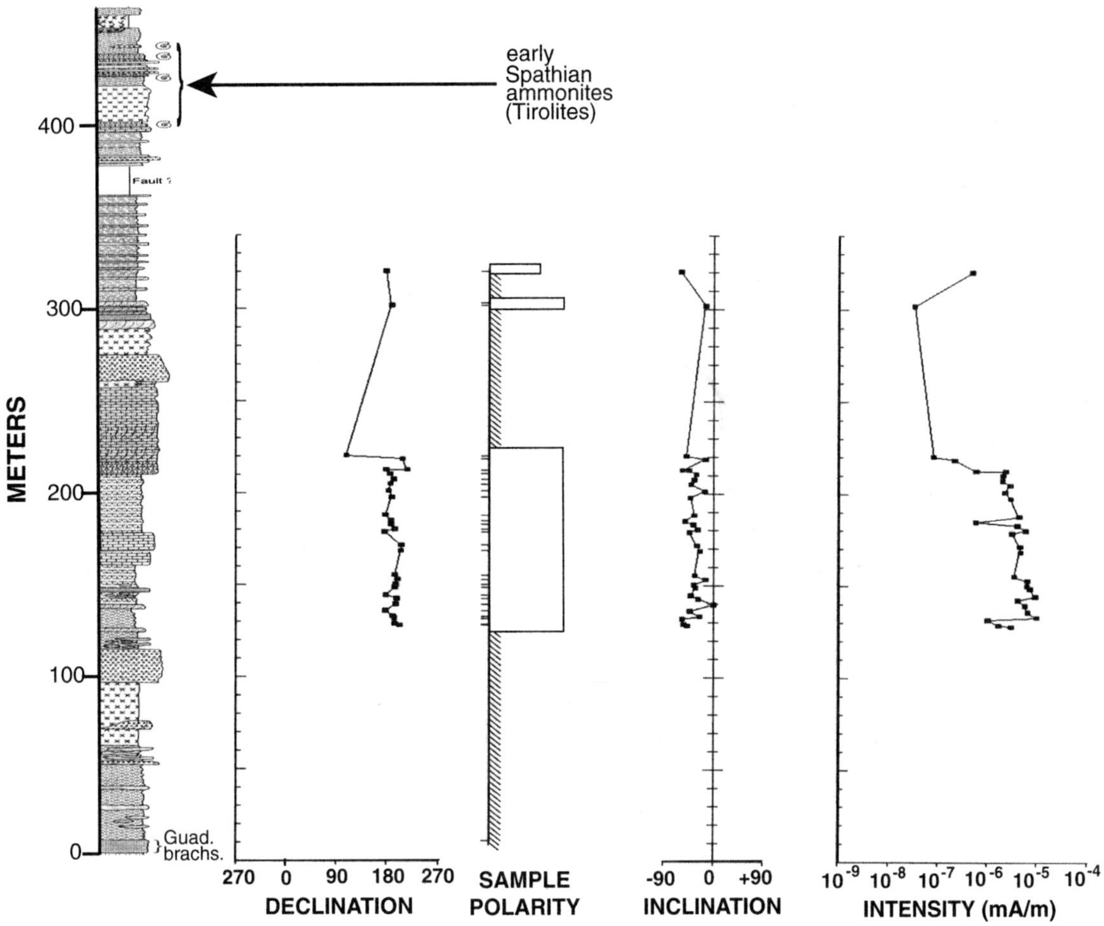

Figure 11. Magnetostratigraphic plot of the characteristic directions of the lower Early Triassic strata (bedding-corrected coordinates). Symbols as in Figure 9.

active tectonics along southern North America, they may also have been intruded during the time in which the geomagnetic field was reversing at an extraordinary rate of ca. 10,000 yr; this geomagnetic field reversal behavior caused the formation of the marine magnetic anomaly Jurassic Quiet Zone (Steiner, 1999, 2001). Because these igneous rocks are not surface eruptions, the cooling rate may have been too slow to allow a magnetization to imprint before the field changed. If this proves to be correct, those igneous rocks would have had to be intruded between 160 and 175 Ma. An alternative explanation is scatter induced by remagnetization during a later intrusive episode.

Comparison with Previous Work

Two earlier studies investigated some of these red beds in the Monos anticline region. Cohen et al. (1986) sampled these strata at their locality 4. The direction they obtained is not the same as was found for the lower red beds of this study, but similar to some of the igneous and contact rock directions observed during this study. Moreover, they obtained a statistically identical direction from Early Jurassic arenaceous limestones and sandy limestones in outcrops ~3.5 km east of the Monos site (their locality 2). The cratonic North American paleopoles for Early Triassic and Early Jurassic times differ by more than 30°; therefore recovery (Cohen et al., 1986) of statistically the same direction from both Early Triassic and Early Jurassic strata probably indicates a remagnetization of both sites. This conclusion is supported by the directional similarity of these sites to the igneous rocks of the present study.

Molina Garza and Geissman (1996) also sampled red bed strata in the Monos Hills anticline region. They sampled both the east and west limbs of the anticlinal structure, and applied a fold test to the site means of the two limbs to test the age of the magnetization. The result was a negative fold test, thus they concluded that the magnetization post-dated folding. Although, they state that the two limbs of the fold shared a common in situ direction, their table 1 shows statistically different in situ site directions for the two limbs. Their sites on the east limb all gave directions that were the same as the presently observed geomagnetic field direction at this locality, whereas the west limb localities exhibited uniformly reversed polarity and directions

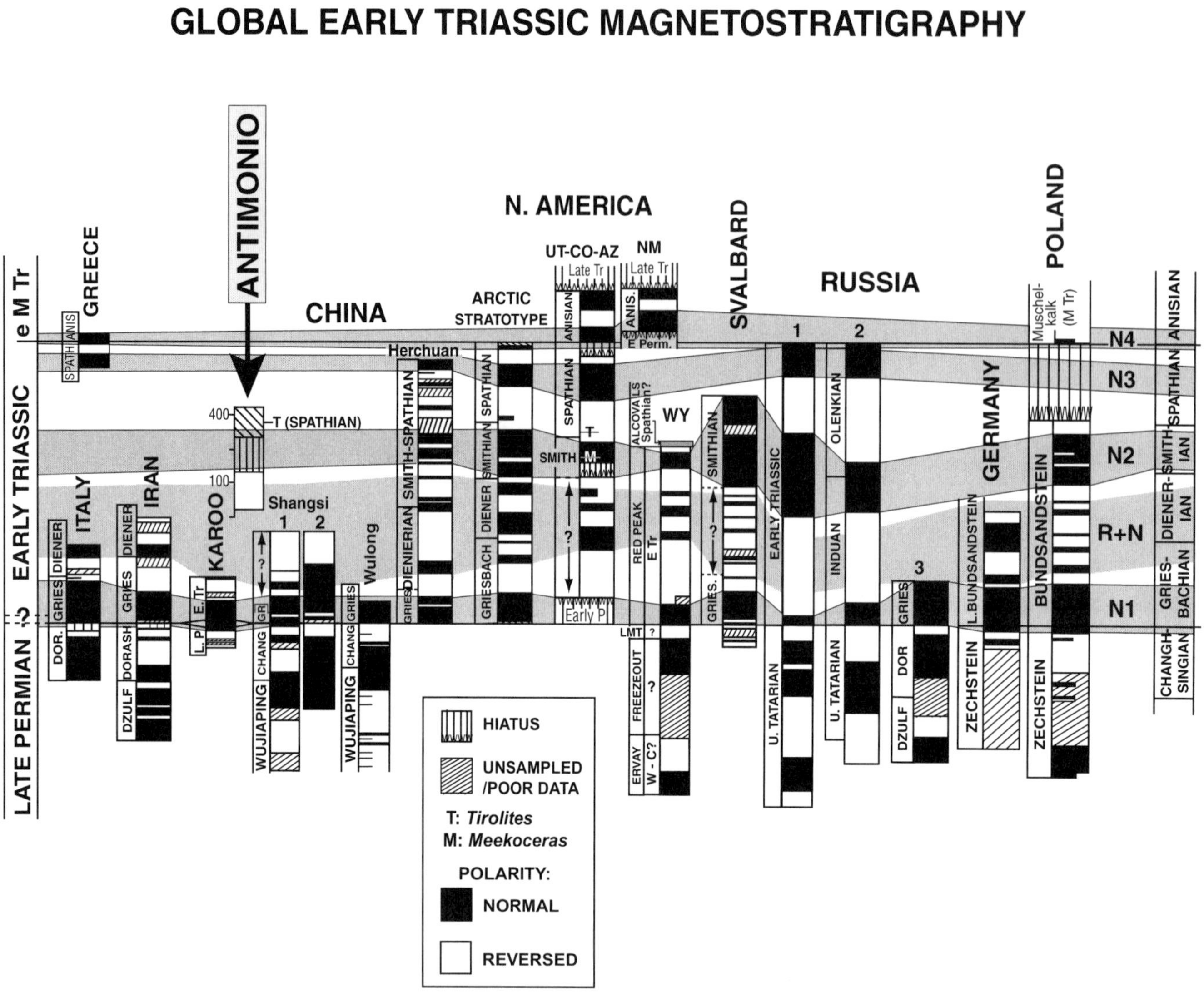

Figure 12. Global correlation of Early Triassic magnetostratigraphy. Early Triassic geomagnetic polarity history is characterized by three normal polarity intervals (labeled "1", "2", "3") that are observed globally in magnetostratigraphic studies. Also distinctive of the Early Triassic geomagnetic polarity history is the lengthy lower reversed polarity interval between "1" and "2," which commonly contains three relatively shorter normal polarity events. Data are from the Arctic stratotypes (Ogg and Steiner, 1991); shallow marine rocks of South China (Steiner et al., 1989; Heller et al., 1995); Iran (Gallet et al., 2000); Greece (Muttoni et al., 1995); Italy (Scholger et al., 2000); a composite of the Moenkopi Formation of western NA (Helsley and Steiner, 1974; Purucker et al., 1980; Steiner et al., 1993); a composite of the Chugwater Group from western NA (Picard, 1964; Shive et al., 1984; Steiner, unpublished data); New Mexico Moenkopi Formation (Steiner and Lucas, 1992); Poland (Nawrocki, 1997); Russia (Molostovsky, 1983; Gurevich and Slautsitays, 1985); and the data of this study ("Sonora"). "T" indicates the location of *Tirolites* ammonites or their equivalent; "M" indicates the presence of *Meekoceras* ammonites.

that were statistically identical to those from the lower red beds obtained in this study. The statistically different directions of the two limbs before unfolding and the coincidence of the directions from one limb with the present field direction suggest that this fold test is inappropriate and that it does not prove a post-folding remanence.

Had unfolding brought the magnetization directions from the two limbs together, it would have been appropriate to consider it merely coincidence that the directions from one limb coincided with the present field direction. Because that was not the result, it is likely that the eastern limb has been remagnetized in a recent geomagnetic field, the same result found in the upper red bed sequence of this study. The presence of recent field directions in some of the red beds investigated by both this study and that of Molina Garza and Geissman (1996) indicates that a recent event, probably intrusive, has remagnetized some of the strata of this area.

Subsequent to concluding that the Monos anticline red beds reflected a post-folding magnetization, but in order to compare their results to equivalent magnetization-age rocks north of the Mojave-Sonora megashear, Molina Garza and Geissman (1996) applied two different tilt corrections to the magnetizations of those strata. However, the basis and appropriateness of these corrections is unclear. First, they corrected for a Cretaceous tilt that was determined from volcanic rocks exposed in the Sierra del Alamo Muerto, an exposure just south of the Monos Hills (Fig. 1). However, the volcanic rocks were described as largely volcanic extrusives containing some volcanogenic sedimentary strata (González-León, 1980, 1997); therefore the dip they exhibit may well be an initial, not a tectonic, dip. Moreover, these rocks do not overlie the strata of the Monos Hills, and crop out no closer than 8 km to the Monos Hills exposures. Further, the Monos Hills and Sierra del Alamo Muerto outcrops are reportedly separated by a thrust fault (Lucas and Estep, 1999) and a strike-slip fault (González-León, 1997). Therefore, even if the dips of the Cretaceous volcanics of the Sierra del Alamo Muerto do indicate a Cretaceous tectonic tilting event, the appropriateness of correcting for that dip in the Sierra del Alamo Muerto is by no means clear.

Molina Garza and Geissman (1996) also applied a second correction for a Tertiary tilt. However, the evidence of this tilting event was not given, nor was the amount and direction. A Tertiary tilt appears contradictory to their claim of stability of the Antimonio terrane since the Cretaceous.

Various ages are shown for the strata and intrusives that Molina Garza and Geissman (1996) studied paleomagnetically, and they constructed a scenario of the geologic history from the combined age and paleomagnetic data. However, the origin of these ages was not clear. Apparently the igneous rock ages were not radiometrically determined, because each "Cretaceous" or "Tertiary" designation carries an appended "(?)," and a literature search did not reveal a radiometric or paleontologic basis for these ages. Yet ages of Cretaceous ("K") and Jurassic ("J") have been attached to the magnetizations of the sedimentary and igneous rocks; the origin of the ages must be clear in order to evaluate their geologic scenario.

Paleopole and Caborca Block Tectonics

The magnetization of the Lower Triassic lower red bed strata yields a well-defined, pre-folding paleopole. Virtual geomagnetic poles (VGPs) were computed from individual characteristic directions and a mean pole was calculated from them (Table 1). Several facts suggest that this may be an Early Triassic remanence. The fold test suggests that the magnetization is pre-folding, although the age of folding is unknown. The underlying beds are Permian and the overlying are late Lower Triassic (Spathian). The unblocking temperatures suggest a magnetite-series remanence carrier, hence a detrital remanence. In a red bed depositional environment, magnetite can only be detrital in origin, because it is not energetically possible to crystallize magnetite under the oxidizing conditions necessary to produce and retain red pigment. Thus, the sum of the evidence suggests that this remanence may date from the Early Triassic.

The Antimonio paleopole is plotted in Figure 13A, along with the cratonic North American apparent polar wander (APW) curve; the Antimonio and North America Early Triassic reference pole are shown with shaded confidence limits. If the Antimonio red bed magnetization is Early Triassic in age, the corresponding paleopole is rotated 42 ± 4.3° clockwise from the cratonic North American Early Triassic reference pole.

The Antimonio sampling locality lies on the Caborca Block, a tectonic element bounded on its northern edge by the Mojave-Sonora megashear. The sampling location is currently ~40 km south of the Mojave-Sonora megashear (Dembosky et al., 1998). Clockwise displacement of the Antimonio region and Caborca Block is inconsistent with left-lateral motion along the Mojave-Sonora megashear. However, the left-lateral motion of the Mojave-Sonora megashear occurred during the Middle to early Late Jurassic, no later than ca. 150–160 Ma (Dembosky et al., 1998). Paleomagnetic studies have demonstrated that since the Early Cretaceous, many terranes bordering the western edge of North America have been translated northward and rotated clockwise, the data for which have been summarized by Beck (1986, 1989a, 1989b, 1991, and 1992). Beck's (1991) summary paleopole plot is reproduced in Figure 13B, to which the North America Late Cretaceous reference pole has been added. The post-Early Cretaceous southward motion of terranes observation has been criticized because some of the evidence is from plutonic rocks that lack paleohorizontal indicators. However, Beck (1991, his fig. 2) subsequently demonstrated that approximately an equal number of bedded sedimentary rocks as plutonic rocks show this displacement.

Beck (1989a, 1989b, 1992) also has shown that prior to the Early Cretaceous, southward terrane motion and counterclockwise rotation occurred relative to cratonic North America. Figure 13C shows Jurassic and older age terrane poles, in which both counterclockwise and clockwise rotations relative to the North America APW curve are evident. North America APW demonstrates that the Late Jurassic was a time of relatively rapid northerly motion of North America, between ca. 160 and ca. 80 Ma (Steiner, 1983; May and Butler, 1986; Steiner, 2003). It would appear that some terranes were, in effect, left behind during this rapid northward motion of the adjacent North America plate. In addition to APW data, structural data also imply a period of notable left-lateral motion along the entire western edge of the North America plate (Ave-Lamont and Oldow, 1988).

Hence, the clockwise rotation of the Antimonio strata might be related to this Late Cretaceous-Tertiary southward transport of terranes. The later Cretaceous-Tertiary terrane motion would overprint any earlier Jurassic motion that might have occurred as a result of the Mojave-Sonora megashear bordering the block. Because opposite motions of some terranes have occurred through time, it cannot be asserted that left-lateral-related motion of a given terrane did not take place prior to the right-lateral activity. Therefore, because of the later elements of terrane geologic histories after the megashear became inactive, clockwise rotations observed in terranes adjacent to the Mojave-Sonora

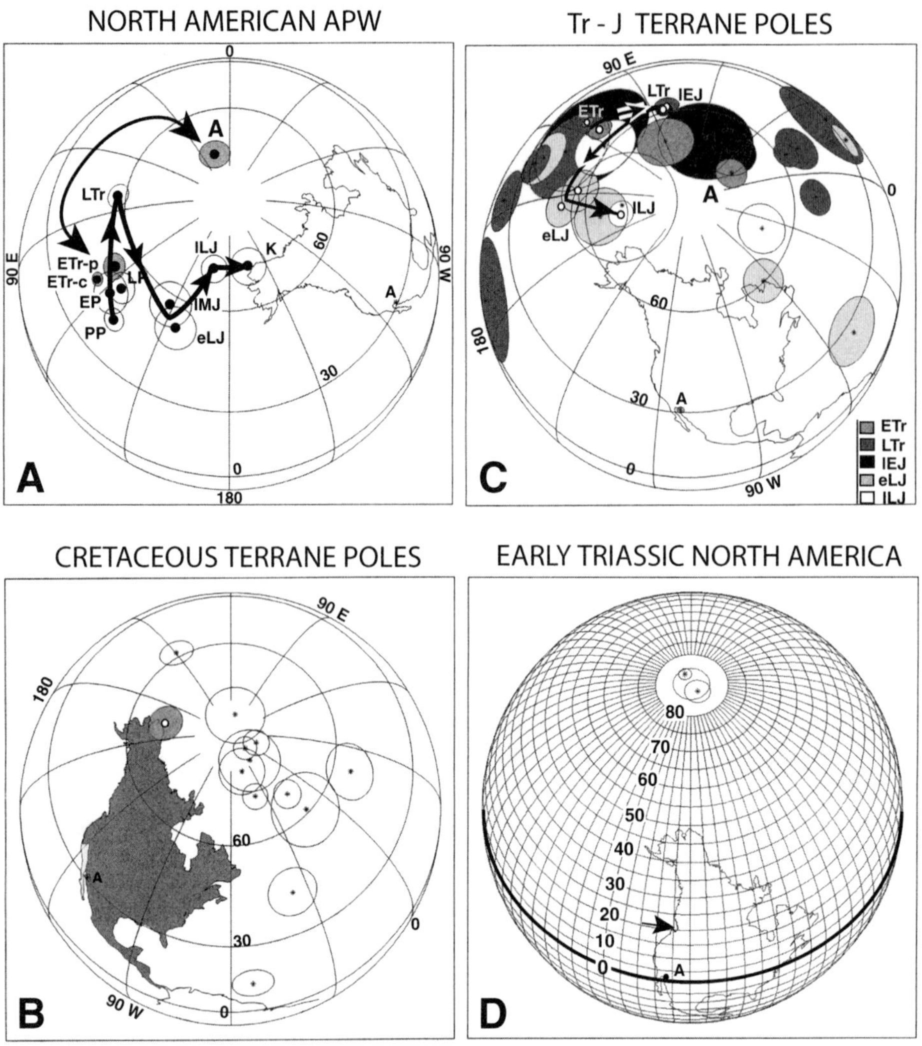

13. (A) Paleopoles illustrating North American apparent polar wander (APW) from Late Pennsylvanian (PP) through Cretaceous (K) (Steiner, 2003) and the pole from the lower portion of the Early Triassic red beds near El Antimonio (A); EP—Early Permian, LP—Late Permian, ETr—Early Triassic, LTr—Late Triassic, lMJ—late Middle Jurassic, eLJ—early Late Jurassic, lLJ—late Late Jurassic. The α_{95} confidence limits are shaded for the pole from this study (A) and the Early Triassic reference poles (both the Early Triassic craton and Colorado Plateau [rotated 9° counterclockwise] mean poles are shown). (B) Displacement of Cretaceous western North American terrane paleopoles (asterisks) (from Beck, 1991) relative to the cratonic North American Cretaceous reference pole (open circle with shaded confidence limit). (C) Jurassic and Triassic western North American terrane paleopoles relative to the Triassic-Jurassic portion of the North American APW curve. Shading indicates similar age terrane (asterisks) and reference poles of the North American APW curve (open circles). (D) Early Triassic paleolatitudes of cratonic North America inferred from the craton and Colorado Plateau mean poles (asterisks). The current location of the El Antimonio site (A), and the location implied by its Early Triassic paleopole (arrow) are illustrated.

megashear cannot be used as evidence that the megashear does not exist (Molina Garza and Geissman, 1996, 1998, 1999). In summary, the Caborca Block Lower Triassic remanence may be explained as a combination of left-lateral Mojave-Sonora megashear and right-lateral Late Cretaceous-Tertiary rotation.

Paleogeographic Location

The paleolatitude of the Lower Triassic Antimonio strata, implied by the mean direction relative to bedding, is 20.9 ± 4.3° N. The Early Triassic paleogeography of North America is shown in Figure 13D, based on cratonic reference data. Had the Caborca locality been part of North America during the Early Triassic, and in its present location, its Lower Triassic strata would indicate an equatorial paleolatitude of ~3° N (Fig. 13D). If the red bed remanence is Early Triassic in age, the 21° N paleolatitude suggests that the Caborca Block resided near Seattle, Washington, in the Early Triassic. If true, this implies that the Caborca Block (or at least its northwestern part) moved southward since the Early Triassic with other terranes adjacent to North America.

Similar bio- and lithostratigraphies may support the location of the Antimonio site and Caborca Block in the Seattle region during the Early Triassic. The *Tirolites* ammonite genus occurs both in the Antimonio strata and in Idaho and Utah, although the genus also occurs in southern Europe from the Alps to Iran and in southern Russia (Tozer, 1984; Silberling, 2002, personal commun.). Late Triassic ichthyosaurs *Shastasaurus Altispinus* and *Toretocnemus,* possibly *Toretocnemus californicus*, occur in the Antimonio Formation (Callaway and Massare, 1989; Lucas and González-León, 1995) in association with *Tropites dilleri* and other ammonites of the *dilleri* Zone. The *Shastasaurus Altispinus* holotype occurs in the *dilleri* Zone in Shasta County, northern California. Whereas *Shastasaurus* also occurs in southern Europe, *Toretocnemus californicus* is only known from Shasta County and from the Caborca Block. In southeastern-most Idaho, another ichthyosaur, *Cymbospondylus* sp., is found in association with a *Tirolites* fauna (Massare and Calloway, 1994; Massare, 2002, personal commun.), and *Tirolites* ammonites reside in the Antimonio red beds. Hence, similar ichthyosaurs and ammonites occur in the Triassic strata of El Antimonio and the western and northwestern United States; however, both are swimming fauna, so other than similarity and permissibility, little can be inferred. Finally, lithostratigraphically similar calcareous fine-grained red bed Early Triassic strata occur in the Antimonio Formation and the Ankarah Formation of Utah and Idaho.

SUMMARY

Paleomagnetic data were obtained from red calcareous siltstones. The magnetizations exhibit a linearly decaying remanence and pass a fold test. Unblocking temperatures suggest that the remanence is carried by magnetite, suggesting that the magnetization may be depositional. The unfossiliferous red beds are inferred to be Early Triassic because they are underlain by Permian strata and overlain by *Tirolites* ammonites-bearing strata (Spathian, i.e., late Early Triassic age). Sequence stratigraphy and global magnetostratigraphy suggest that these beds are Dienerian (middle Early Triassic) in age. The inclination of magnetization indicates a paleolatitude of magnetization of 21 ± 4° N, suggesting that these Caborca Block strata were deposited near the Early Triassic paleolatitude of Seattle, Washington. The left-lateral transport necessary to move the El Antimonio sites from ~20° paleolatitude to the present latitude is a motion consistent with movement along the postulated Mojave-Sonora megashear. However, the paleopole from the data is rotated 42 ± 4° clockwise from the North American Early Triassic pole. Clockwise rotation of the sampling region might have occurred during the Middle Cretaceous to middle Tertiary episode of northward terrane transport and clockwise rotation that affected many terranes on the western margin of North America (Beck, 1991). Significantly, because of the history of both northward and southward transport for numerous western North American terranes, these data cannot provide direct evidence of, nor disprove, the existence of the Mojave-Sonora megashear.

ACKNOWLEDGMENTS

We gratefully acknowledge the assistance of Carlos González-León in accompanying us to the various sites and acquainting us with the stratigraphy. Mryl Beck provided numerous constructive comments, which are gratefully acknowledged.

REFERENCES CITED

Ave-Lamont, H.G., and Oldow, J.S., 1988, Early Mesozoic southward migration of Cordilleran transpressional terranes: Tectonics, v. 7, p. 1057–1075.

Beck, M.E., Jr., 1986, Model for late Mesozoic-early Tertiary Tectonics of costal California and western Mexico and speculations on the origin of the San Andres Fault: Tectonics, v. 5, p. 49–64.

Beck, M.E., Jr., 1989a, Paleomagnetism of continental North America: Implications for displacement of crustal blocks within the Western Cordillera, Baja California to British Columbia, *in* Pakiser, L.C., and Mooney, W.D., eds., Geophysical framework of the continental United States: Geological Society of America Memoir 172, p. 471–492.

Beck, M.E., Jr., 1989b, Block rotations in the continental crust: Examples from western North America, *in* Kissel, C., and Laj, C., eds., Paleomagnetic rotations and continental deformation: Dordrecht/Boston/London, Kluwer Academic Publishers, Nato ASI Series, Series C: Math & Physical Sciences, v. 254, p. 1–16.

Beck, M.E., Jr., 1991, Case for transport of Baja and coastal southern California: Paleomagnetic data, analysis, and alternatives: Geology, v. 19, p. 506–509, doi: 10.1130/0091-7613(1991)019<0506:CFNTOB>2.3.CO;2.

Beck, M.E., Jr., 1992, Tectonic significance of paleomagnetic results from the western conterminous United States, *in* Birchfield, B.C, Lipman, P.W., and Zoback, M.L., eds., The Cordilleran origin: Conterminous U.S.: Boulder, Colorado, Geological Society of America, Geology of North America, v. G-3, p. 683–697.

Callaway, J.M., and Massare, J.A., 1989, *Shastasaurus Altispinus* (Ichthyosauria, Shastasauridae) from the Upper Triassic of the El Antimonio District, northwestern Sonora, Mexico: Journal of Paleontology, v. 63, p. 930–939.

Cohen, K.K., Anderson, T.H., and Schmidt, V.A., 1986, A paleomagnetic test of the proposed Mojave-Sonora megashear in northwest Mexico: Tectonophysics, v. 131, p. 23–51, doi: 10.1016/0040-1951(86)90266-0.

Cooper, G.A., 1953, Stratigraphy and faunal zones, *in* Cooper, G.A., Dunbar, C.O., Duncan, H., Miller, A.K., and Knight, J.B., eds., Permian fauna at El Antimonio, western Sonora, Mexico: Smithsonian Miscellaneous Collections, v. 119, no. 2 (Publication 4108), p. 1–13.

Dembosky, J.A., Anderson, T.H., Nourse, J.A., and Silver, L.T., 1998, Caborca, San Gabriel, and Joshua Tree: Terranes at the southwestern margin of North America: Geological Society of America Abstracts with Programs, v. 30, no. 7, p. 355.

Estep, J.W., Lucas, S.G., and González-León, C.M., 1997, Middle Triassic ammonites from Sonora, Mexico: Revista Mexicana de Ciencias Geologicas, v. 14, no. 2, p. 155–159.

Helsley, C.E., and Steiner, M.B., 1974, Paleomagnetism of the Lower Triassic Moenkopi Formation: Geological Society of America Bulletin, v. 85, p. 457–464, doi: 10.1130/0016-7606(1974)85<457:POTLTM>2.0.CO;2.

Gallet, Y., Krystyn, L., Besse, J., Saidi, A., and Ricou, L.-E., 2000, New constraints on the Upper Permian Lower Triassic geomagnetic polarity timescale from the Abadeh section (central Iran): Journal of Geophysical Research, v. 105, p. 2805–2815, doi: 10.1029/1999JB900218.

González-León, C.M., 1980, La Formacion Antimonio (Triassico Superior-Jurassico Inferior) en la Sierra del Alamo, estado de Sonora: Universidad Nacional Autónoma de Mexico, Instituto de Geologia, Revista, v. 4, no. 1, p. 13–18, 136–148.

González-León, C.M., 1997, Sequence stratigraphy and paleogeographic setting of the Antimonio Formation (Late Permian-Early Jurassic), Sonora, Mexico: Revista Mexicana de Ciencias Geologicas, v. 14, no. 2.

Gurevich, Ye.L., and Slautsitays, I.P., 1985, A paleomagnetic section in the Upper Permian and Triassic deposits of Novaya Zemlya: International Geology Review, v. 27, p. 168–177.

Heller, F., Chen, H., Dobson, J., and Haag, M., 1995, Permian-Triassic magnetostratigraphy–new results from South China: Earth and Planetary Science Letters, v. 89, p. 281–295.

Lucas, S.G., and Estep, J.W., 1999, Permian, Triassic, and Jurassic stratigraphy, biostratigraphy, and sequence stratigraphy in the Sierra del Alamo Muerto, Sonora, Mexico, *in* Bartolini, C., Wilson, J.L., and Lawton, T.F., eds., Mesozoic sedimentary and tectonic history of north-central Mexico: Geological Society of America Special Paper 340, p. 271–286.

Lucas, S.G., and González-León, C.M., 1995, Ichthyosaurs from the Upper Triassic of Sonora and the biochronology of Triassic ichthyosaurs, *in* Jacques-Ayala, C., González-León, C.M., and Roldán-Quintana, J., eds., Studies on the Mesozoic of Sonora and adjacent areas: Geological Society of America Special Paper 301, p. 17–20.

Lucas, S.G., Estep, J.W., González-León, C.M., Paull, R.K., Silberling, N.J., Steiner, M.B., and Marzolf, J.E., 1997, Early Triassic ammonites and conodonts from Sonora, northwestern Mexico: Neues Jahrbuch fur Geologie und Palaontologie, Monatshefte, v. 1997, no. 12, p. 562–574.

Kirschvink, J.L., 1980, The least-squares line and plane and the analysis of paleomagnetic data: Geophysical Journal, v. 62, p. 699–718.

Massare, J.A., and Calloway, J.M., 1994, *Cymbospondylus* (Ichthyosauria: Shastasauridae) from the Lower Triassic Thaynes Formation of southeastern Idaho: Journal of Vertebrate Paleontology, v. 14, p. 139–141.

May, S.R., and Butler, R.F., 1986, North American Jurassic apparent polar wander: Implications for plate motion, paleogeography, and Cordilleran tectonics: Journal of Geophysical Research, v. 91, p. 11545–11555.

McFadden, P.L., 1990, A new fold test for paleomagnetic studies: Geophysical Journal International, v. 103, p. 163–169.

Molina Garza, R.S., and Geissman, J.W., 1996, Timing of deformation and accretion of the Antimonio terrane, Sonora, from paleomagnetic data: Geology, v. 24, p. 1131–1134, doi: 10.1130/0091-7613(1996)024<1131:TODAAO>2.3.CO;2.

Molina Garza, R.S., and Geissman, J.W., 1998, Paleomagnetic evidence against Jurassic left-lateral (southeastward displacement of the Caborca terrane, *in* Gold deposits of northern Sonora, Mexico: Guidebook, Society of Economic Geology, Field Conference Oct 19–24, 1998, p. 187–201.

Molina Garza, R.S., and Geissman, J.W., 1999, Paleomagnetic data from the Caborca terrane, Mexico: Implications for the Cordilleran tectonics and the Mojave-Sonora megashear hypothesis: Tectonics, v. 18, p. 293–325, doi: 10.1029/1998TC900030.

Molostovsky, E.A., 1983, Paleomagnetic stratigraphy of the Eastern European part of the USSR (in Russian): Saratov, Russia, University of Saratov, 167 p.

Muttoni, G., Kent, D.V., and Gaetani, M., 1995, Magnetostratigraphy of a Lower-Middle Triassic boundary section from Chios (Greece): Physics of the Earth and Planetary Interiors, v. 92, p. 245–260, doi: 10.1016/0031-9201(95)03021-4.

Nawrocki, J., 1997, Permian to Early Triassic magnetostratigraphy from the Central European basin in Poland: Implications on regional and worldwide correlations: Earth and Planetary Science Letters, v. 152, p. 37–58, doi: 10.1016/S0012-821X(97)00147-7.

Ogg, J.G., and Steiner, M.B., 1991, Early Triassic magnetic polarity timescale—Interpretation of magnetostratigraphy, ammonite zonation and sea level changes: Earth and Planetary Science Letters, v. 107, p. 69–89, doi: 10.1016/0012-821X(91)90044-I.

Picard, D.M., 1964, Paleomagnetic correlation of units within the Chugwater (Triassic) Formation, west-central Wyoming: American Association Petroleum Geologists Bulletin, v. 48, p. 269–291.

Purucker, M.E., Elston, D.P., and Shoemaker, E.M., 1980, Early acquisition of characteristic magnetization in red beds of the Moenkopi Formation (Triassic), Gray Mountain, Arizona: Journal of Geophysical Research, v. 85, p. 997–1012.

Scholger, R., Mauritsch, H.J., and Brandner, R., 2000, Permian-Triassic magnetostratigraphy from the Southern Alps (Italy): Earth and Planetary Science Letters, v. 176, p. 495–508, doi: 10.1016/S0012-821X(00)00026-1.

Shive, P.N., Steiner, M., and Huycke, D.T., 1984, Magnetostratigraphy and the timing of remanence acquisition in the Triassic Chugwater Formation of Wyoming: Journal of Geophysical Research, v. 89, p. 1801–1815.

Stanley, G.D., and González-León, C.M., 1995, Paleographic and tectonic implications of Triassic fossils and strata from the Antimonio Formation, northwest Sonora, *in* Jacques-Ayala, C., González-León, C.M., and Roldán-Quintana, J., eds., Studies of the Mesozoic of Sonora and adjacent areas: Geological Society of America Special Paper 301, p. 1–16.

Steiner, M.B., 1983, Mesozoic plate motions of North America, *in* Reynolds, M., ed., Mesozoic paleogeography of the West-Central United States: Rocky Mountain Society of Economic Paleontologists, p. 1–11.

Steiner, M.B., 1999, FINALLY!! Origin(s) of the Jurassic Quiet Zone: Eos (Transactions, American Geophysical Union), v. 80, p. F 508.

Steiner, M.B., 2001, Tango in the Mid-Jurassic: 10,000-yr geomagnetic field reversals, Eos (Transactions, American Geophysical Union), v. 82, p. F443.

Steiner, M.B., 2003, A cratonic Middle Jurassic paleopole: Callovian-Oxfordian pole stillstand (J-2 cusp), rotation of the Colorado Plateau, and Jurassic North American apparent polar wander: Tectonics, 22, #3, 1016, 10.1029/2001TC001284, 21 p.

Steiner, M., and Lucas, S., 1992, A Middle Triassic paleomagnetic pole for the North American craton: Geological Society of America Bulletin, v. 104, p. 993–998, doi: 10.1130/0016-7606(1992)104<0993:AMTPPF>2.3.CO;2.

Steiner, M.B., Ogg, J.G., Zhang, Z., and Sun, S., 1989, The Late Permian/Early Triassic magnetic polarity time scale and plate motions of South China: Journal of Geophysical Research, v. 94, p. 7343–7363.

Steiner, M., Morales, M., and Shoemaker, E.M., 1993, Magnetostratigraphic, biostratigraphic and lithologic correlations in Triassic strata of the western U.S., *in* Aissioui, D., McNeill, D., and Hurley, N., eds., Applications of paleomagnetism to sedimentary geology: SEPM Special Publication 49, p. 41–57.

Tozer, E.T., 1984, The Trias and its ammonoids: The evolution of a time scale: Geological Survey of Canada Miscellaneous Report 35, 171 p.

Manuscript Accepted by the Society 24 March 2005

Printed in the USA

Geological Society of America
Special Paper 393
2005

Middle Jurassic Topawa Group, Baboquivari Mountains, south-central Arizona: Volcanic and sedimentary record of deep basins within the Jurassic magmatic arc

Gordon B. Haxel
U.S. Geological Survey, 2255 N. Gemini Dr., Flagstaff, Arizona 86001, USA

James E. Wright
Department of Geology, University of Georgia, Athens, Georgia 30602, USA

Nancy R. Riggs
Department of Geology, Northern Arizona University, Flagstaff, Arizona 86011, USA

Richard M. Tosdal
Mineral Deposit Research Unit, Department of Earth and Ocean Sciences, University of British Columbia, Vancouver, British Columbia, Canada V6T 1Z4

Daniel J. May
Vice President for Academic Affairs, University of Findlay, Findlay, Ohio 45840, USA

ABSTRACT

Among supracrustal sequences of the Jurassic magmatic arc of the southwestern Cordillera, the Middle Jurassic Topawa Group, Baboquivari Mountains, south-central Arizona, is remarkable for its lithologic diversity and substantial stratigraphic thickness, ≈8 km. The Topawa Group comprises four units (in order of decreasing age): (1) Ali Molina Formation—largely pyroclastic rhyolite with interlayered eolian and fluvial arenite, and overlying conglomerate and sandstone; (2) Pitoikam Formation—conglomerate, sedimentary breccia, and sandstone overlain by interbedded siltstone and sandstone; (3) Mulberry Wash Formation—rhyolite lava flows, flow breccias, and mass-flow breccias, with intercalated intraformational conglomerate, sedimentary breccia, and sandstone, plus sparse within-plate alkali basalt and comendite in the upper part; and (4) Tinaja Spring Porphyry—intrusive rhyolite. The Mulberry Wash alkali basalt and comendite are genetically unrelated to the dominant calcalkaline rhyolite. U-Pb isotopic analyses of zircon from volcanic and intrusive rocks indicate the Topawa Group, despite its considerable thickness, represents only several million years of Middle Jurassic time, between approximately 170 and 165 Ma.

Sedimentary rocks of the Topawa Group record mixing of detritus from a minimum of three sources: a dominant local source of porphyritic silicic volcanic and

Haxel, G.B., Wright, J.E., Riggs, N.R., Tosdal, R.M., and May, D.J., 2005, Middle Jurassic Topawa Group, Baboquivari Mountains, south-central Arizona: Volcanic and sedimentary record of deep basins within the Jurassic magmatic arc, *in* Anderson, T.H., Nourse, J.A., McKee, J.W., and Steiner, M.B., eds., The Mojave-Sonora megashear hypothesis: Development, assessment, and alternatives: Geological Society of America Special Paper 393, p. 329–357. doi: 10.1130/2005.2393(12).

subvolcanic rocks, identical or similar to those of the Topawa Group itself; Mesoproterozoic or Cambrian conglomerates in central or southeast Arizona, which contributed well-rounded, highly durable, polycyclic quartzite pebbles; and eolian sand fields, related to Middle Jurassic ergs that lay to the north of the magmatic arc and are now preserved on the Colorado Plateau.

As the Topawa Group evidently represents only a relatively short interval of time, it does not record long-term evolution of the Jurassic magmatic arc, but rather represents a Middle Jurassic "stratigraphic snapshot" of the arc. This particular view of the arc has been preserved primarily because the Topawa Group accumulated in deep intra-arc basins. These nonmarine basins were fundamentally tectonic and extensional, rather than volcano-tectonic, in origin. Evidence from the Topawa Group supports two previous paleogeographic inferences: the Middle Jurassic magmatic arc in southern Arizona was relatively low standing, and externally derived sediment was introduced into the arc from the continent (northeast) side, without appreciable travel along the arc. We speculate that because the Topawa Group intra-arc basins were deep and rapidly subsiding, they became the locus of a major (though probably intermittent) fluvial system, which flowed into the low-standing magmatic arc from its northeast flank.

Keywords: Arizona, Jurassic, magmatic arc, tectonic basin, stratigraphy, U-Pb geochronology, rhyolite, comendite, alkali basalt, conglomerate, sandstone.

INTRODUCTION

Regional Setting of Jurassic Magmatic Arc Rocks, South-Central Arizona

Geologic maps of southern Arizona and adjacent southeast California and northern Sonora published before about 1980 (e.g., Wilson, 1960; Wilson et al., 1960; Cooley, 1967; Wilson et al., 1969; Merriam, 1972; Jennings, 1977) suggested that most of the extensive plutonic, metavolcanic, and metasedimentary rocks of this region were of four ages: Precambrian, Mesozoic undifferentiated, Cretaceous, or "Laramide" (Late Cretaceous to early Tertiary). To the surprise of most geologists involved, geologic mapping and U-Pb geochronology undertaken in the 1970s and 1980s showed that many of these terranes are in fact Jurassic (Anderson and Silver, 1978, 1979; Haxel et al., 1980b; Reynolds et al., 1987; Richard et al., 1987; Fox and Miller, 1990; Riggs et al., 1993; Beikman et al., 1995; Richard et al., 2000). These Jurassic terranes of the Sonoran Desert region are now recognized as one segment of a broad magmatic arc that extended across southwestern and western North America in Early through Late Jurassic time (Coney, 1978; Tosdal et al., 1989; Busby-Spera et al., 1990; Saleeby et al., 1992; Riggs and Blakey, 1993; Jones et al., 1995; Barton, 1996; Schermer et al., 2002).

Within the Sonoran Desert segment of this magmatic arc (Fig. 1; Tosdal et al., 1989), plutonic rocks are extensive but volcanic and related sedimentary rocks less common. In many ranges where Jurassic supracrustal rocks do occur, their stratigraphy and original petrologic character have been disrupted or obscured by Jurassic, Laramide, or middle Tertiary deformation, regional metamorphism, alteration, or plutonism (e.g., Haxel et

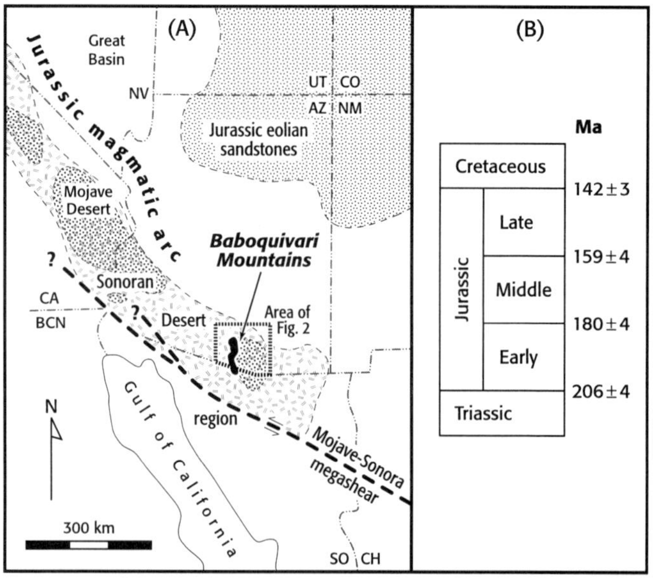

Figure 1. (A) Geologic setting of the Baboquivari Mountains, southern Arizona, in the context of the Early to Late Jurassic magmatic arc (Tosdal et al., 1989), hypothesized Middle to Late Jurassic Mojave-Sonora megashear (Jones et al., 1995; Anderson et al., this volume), and Early to Middle Jurassic eolian sandstones of the Colorado Plateau (Riggs and Blakey, 1993). Stippled areas within the magmatic arc schematically represent regions where quartz arenites—eolian, fluvial, or marine—are interlayered with intermediate to silicic volcanic rocks (after Riggs et al., 1993; Haxel et al., 1985). In the Great Basin, widely scattered Jurassic igneous rocks crop out in a back-arc region as much as a few hundred kilometers east of the main Jurassic arc (Saleeby et al., 1992; Johnson, 2000). In northeast Sonora (SO) and northwest Chihuahua (CH), the Jurassic arc passes beneath middle Tertiary volcanic rocks of the Sierra Madre Oriental. (B) Summary Jurassic time scale (Gradstein and Ogg, 1996).

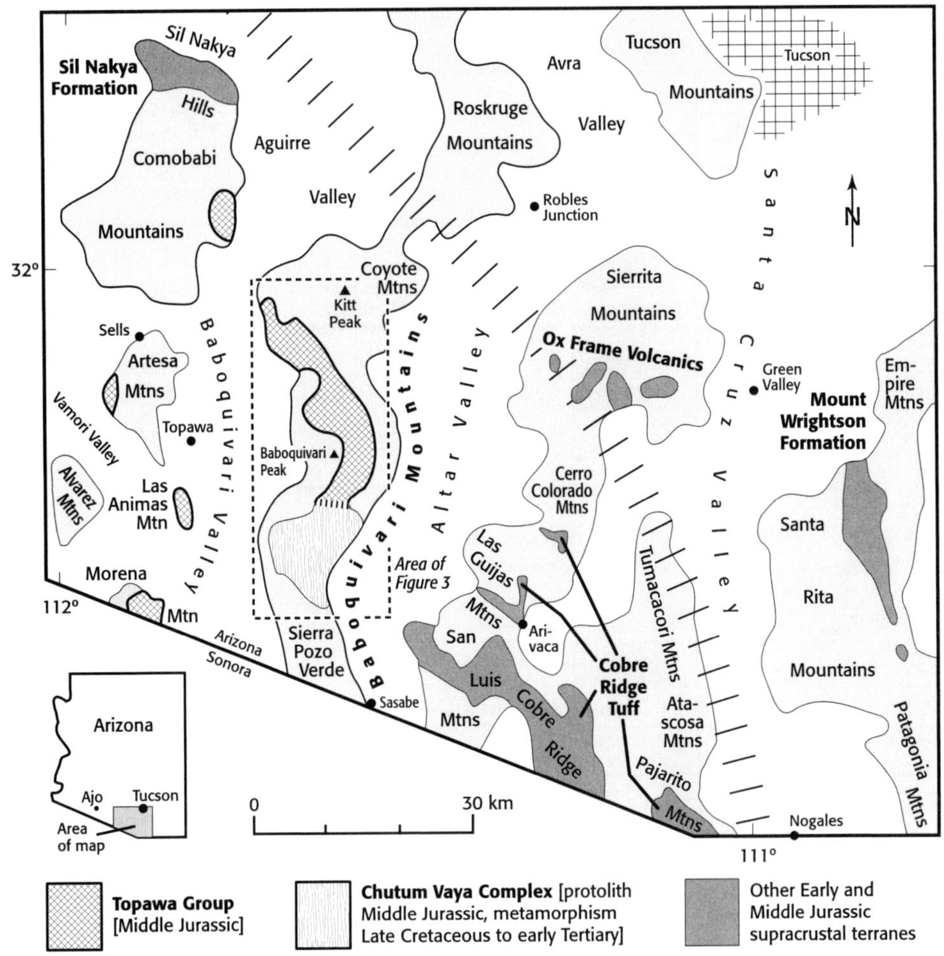

Figure 2. Early and Middle Jurassic supracrustal terranes in south-central Arizona. The principal units described in this paper—Topawa Group and Chutum Vaya Complex—are mapped with patterns. Other Early and Middle Jurassic supracrustal units, discussed briefly in the text, are indicated by gray fill. Light gray indicates mountain ranges and others areas of bedrock outcrop. The Chutum Vaya Complex consists of Late Cretaceous to early Tertiary metamorphic rocks derived largely but not entirely from the Topawa Group. The broad, southwest-concave arc across the central part of the map indicates the gradational northeast boundary of the Papago domain (Haxel et al., 1984; Riggs and Haxel, 1990; Tosdal et al., 1990; Goodwin and Haxel, 1990; Nourse et al., 1994; Anderson et al., this volume). This map is after Beikman et al. (1995) and Richard et al. (2000); its location is shown in Figure 1.

al., 1984, 2002; Richard et al., 1987; Grubensky et al., 1993; Nourse et al., 1994; Anderson et al., this volume). Furthermore, many Jurassic supracrustal terranes are dominated by rather monotonous rhyolite and metarhyolite (collectively known by the field term "the quartz porphyry"). Jurassic plutonic rocks and homogeneous volcanic rocks, though of petrologic interest, can yield only limited information about eruptive and depositional environments. Therefore, those few Jurassic supracrustal sequences that are lithologically diverse, thick, and relatively well preserved are of special interest, as they can provide specific insights into paleogeography, paleotectonics, and paleoclimate.

The Topawa Group

One of the most remarkable of such supracrustal sequences is the Topawa Group, in the Baboquivari Mountains, south-central Arizona (Figs. 1–4). The Topawa Group consists of rhyolitic to dacitic volcanic and subvolcanic rocks, interlayered with several types of conglomerate, sandstone, and siltstone. It also includes minor, distinctive within-plate alkali basalt and trachyte-comendite. Stratigraphic thickness is approximately 8 km. Such a thick sequence might be expected to record long-term evolution of the Jurassic magmatic arc. Ironically, though, stratigraphic and geochronologic data indicate that the Topawa Group represents only a short interval, probably only several million years, of Middle Jurassic time.

In this paper we refine the stratigraphic nomenclature and outline the stratigraphy of the Topawa Group; present new U-Pb isotopic age data; describe the lithology of the volcanic, subvolcanic, and sedimentary rocks; and briefly examine the geochemistry of the igneous rocks. We then use these data to show that the Topawa Group preserves a readily interpretable record of several successive volcanic and sedimentary environments within the Jurassic magmatic arc. We also consider relations of the Topawa Group to other Early to Middle Jurassic magmatic arc terranes in southern Arizona and northern Sonora, Middle Jurassic eolianites on the Colorado Plateau, the hypothesized Middle to Late Jurassic Mojave-Sonora megashear, and distinctive rhyolite clasts found in Paleogene conglomerate in coastal southern California. Finally, we discuss implications of the Topawa Group for understanding the paleogeography of the Jurassic magmatic arc in the southern Arizona region. Specifically, we infer that the

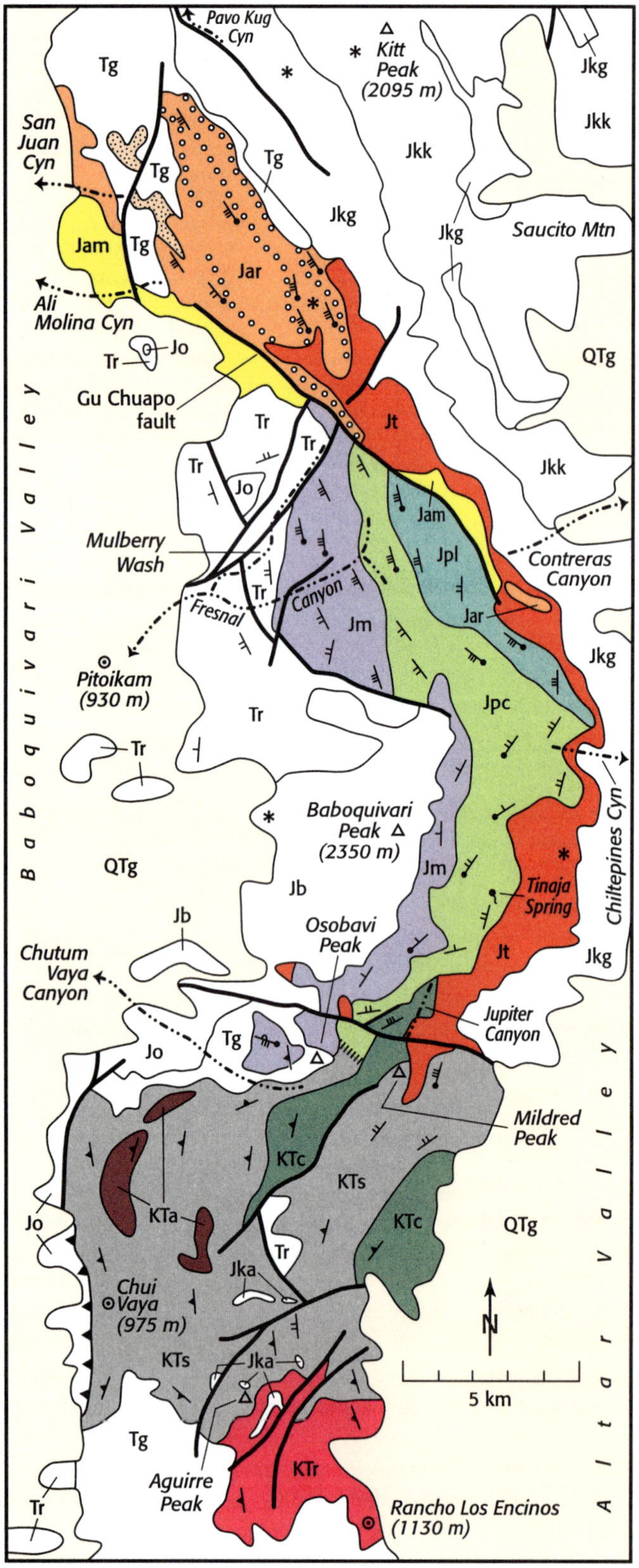

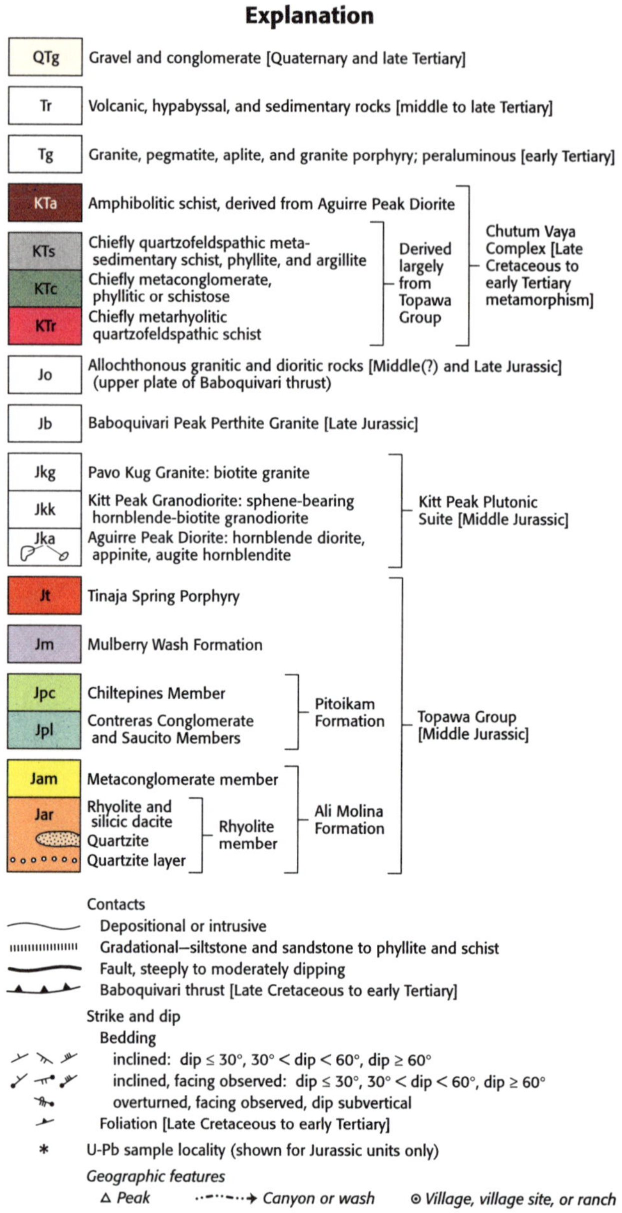

Figure 3. Geologic map of the northern and central Baboquivari Mountains, emphasizing Middle Jurassic supracrustal and hypabyssal rocks and derivative metamorphic rocks (after Haxel et al., 1980a, 1982; Beikman et al., 1995). The location of this map is shown in Figure 2. Geographic features, including those mentioned in Figure 4, are labeled in italic. Strike-and-dip symbols in the rhyolite member of the Ali Molina Formation (unit Jar) schematically represent the attitude of bedding in nearby quartzite layers. Within the Chutum Vaya Complex, each of the units KTs, KTc, and KTr contains subordinate amounts of the lithologic types characteristic of the other two units. For instance unit KTs, chiefly schist and phyllite, also includes some metaconglomerate and metarhyolite. Three major early and middle Tertiary dike swarms and numerous small middle Tertiary rhyolite intrusions are not shown.

Topawa Group was deposited and preserved in two deep, tectonic basins or subbasins within the arc.

TOPAWA GROUP

Definition

Three broadly similar and related sections of Jurassic supracrustal rocks crop out in the Baboquivari Mountains (Fig. 3). Lithologic correlations, described in detail in the following sections, allow reconstruction of a single stratigraphic sequence (Fig. 4, columns 1 and 2). The Jurassic supracrustal sequence in the Baboquivari Mountains is hereby named the Topawa Group. The Topawa Group comprises three formations (in ascending stratigraphic order): Ali Molina, Pitoikam, and Mulberry Wash. All three are dominantly either volcanic and sedimentary or sedimentary; the Ali Molina and Mulberry Wash Formations contain subordinate subvolcanic rocks as well. The Topawa Group also includes a fourth unit, the hypabyssal Tinaja Spring Porphyry, closely related to the volcanic and subvolcanic rocks of the Ali Molina and Mulberry Wash Formations.

Stratigraphic Nomenclature

The three supracrustal formations now incorporated into the Topawa Group were originally named by Heindl and Fair (1965) as Ali Molina Metamorphic Complex, Pitoikam Formation, and Mulberry Wash Volcanic Formation. Subsequent mapping of the Ali Molina unit showed that most of it was stratigraphically coherent, so the name was revised to Ali Molina Formation (Haxel et al., 1980a). Inclusion of "Volcanic" in the name of the Mulberry Wash unit is misleading, as this formation contains a significant proportion of sedimentary rocks, and is no more overwhelmingly volcanic than the associated Ali Molina Formation. We therefore revise the name to Mulberry Wash Formation.

Heindl and Fair (1965) divided the Pitoikam Formation into three members: lower, Contreras Conglomerate, and Chiltepines. We assign to the lower unit the name Saucito Member, and use the other two names as originally defined.

All volcanic and sedimentary rocks stratigraphically above the Pitoikam Formation and below an unconformity at the base of middle Tertiary strata (Figs. 3 and 4) were included in the Mulberry Wash [Volcanic] Formation by Fair (1965). Heindl and Fair (1965) designated the upper one-third of this formation a separate unit, the Chiuli Shaik Formation. Although the Chiuli Shaik Formation of Heindl and Fair (1965) is separated from the underlying Mulberry Wash unit by a minor erosional unconformity, the two formations are stratigraphically and structurally conformable, and they contain very similar volcanic and volcaniclastic rocks. In particular, both formations include distinctive trachyte-comendite volcanic rocks. We believe the original nomenclature of Fair (1965) more accurately represents the geologic history recorded by these strata, and abandon as unnecessary the name Chiuli Shaik Formation.

In the southern part of the central Baboquivari Mountains, strata of the Topawa Group pass gradationally southward into a broad metamorphic terrane (Figs. 2–4), the Chutum Vaya Complex (Haxel et al., 1982, 1984; Goodwin and Haxel, 1990). Although derived largely from the Jurassic Topawa Group, this metamorphic complex is not considered part of the Topawa Group, for three reasons. First, the distribution and lithology of rock units within the Chutum Vaya Complex owes in large part to Late Cretaceous and early Tertiary deformation, regional metamorphism, and granitic plutonism. Second, the Chutum Vaya Complex contains some supracrustal rock types not present in the Topawa Group. Third, the Chutum Vaya Complex includes amphibolitic schist derived from Jurassic dioritic rocks not directly related to the Topawa Group.

Age

Fair (1961) initially suggested that the sequence now included in the Topawa Group spans the Cretaceous-Tertiary boundary. Subsequently, Fair (1965) discussed possible Jurassic(?) and Cretaceous ages, based upon various regional correlations. Heindl and Fair (1965) simply assigned a Mesozoic(?) age. Haxel et al. (1980a; Wright et al., 1981) revised the age to Early Jurassic, based upon interpretation of problematic U-Pb isotopic data. New U-Pb geochronologic data presented in this paper firmly indicate a Middle Jurassic age for the Topawa Group.

STRATIGRAPHY AND INTERNAL CORRELATIONS

Three separate sections of Jurassic volcanic and sedimentary rocks crop out in the Baboquivari Mountains (Figs. 3 and 4). Two of these lie on the west side of the northern and central Baboquivari Mountains; they are separated from one another by the Gu Chuapo fault, a local structure of no great displacement. The section northeast of the Gu Chuapo fault is a single unit, the Ali Molina Formation. The section southwest of the fault consists of two units: the Pitoikam and Mulberry Wash Formations. The third Jurassic section crops out farther south in the Baboquivari Mountains, where the Chutum Vaya Complex spans the entire width of the range. Some of the metasedimentary rocks of the Chutum Vaya Complex can be traced northward into unmetamorphosed strata of the Pitoikam Formation. These three sections are stratigraphically overlapping, and lithologic correlations between them permit reconstruction of a single Middle Jurassic stratigraphic sequence, the Topawa Group (Fig. 4, columns 1 and 2). Stratigraphic thickness of the Topawa Group is approximately 8 km (±1 km). This is a minimum value, as neither the original base nor top of the Topawa Group is preserved.

Ali Molina Formation

The Ali Molina Formation comprises two units, the rhyolite member and the overlying metaconglomerate member (Fig. 3; Fig. 4, column 1). Although these rocks are incipiently to

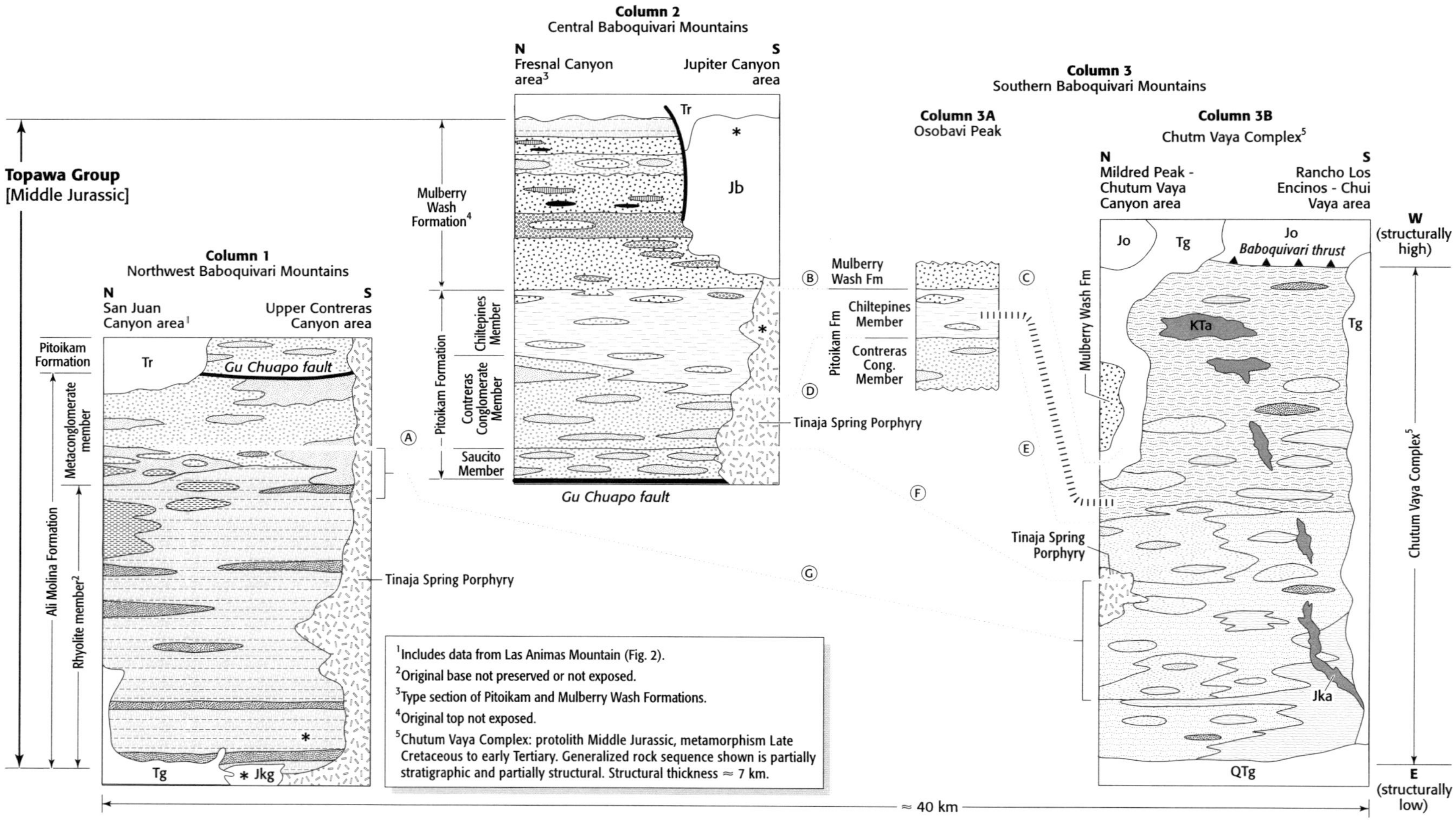

Figure 4 (*continued on following page*). Schematic lithologic columns showing the stratigraphic (columns 1, 2, 3A) and structural-stratigraphic (column 3B) relations of Middle Jurassic supracrustal and hypabyssal rocks in the Baboquivari Mountains. The Topawa Group comprises the Ali Molina, Pitoikam, and Mulberry Wash Formations, plus the related hypabyssal Tinaja Spring Porphyry. Locations of geographic features used to designate the four lithologic columns are shown in Figure 3. Encircled letters label stratigraphic tie lines, discussed in the text. Depiction of lateral variations and intercalations is highly schematic and not to scale. Subordinate Middle Jurassic subvolcanic intrusions within the Ali Molina and Mulberry Wash Formations are not shown. Likewise, widespread and locally abundant early and middle Tertiary intrusions are not represented. Within the Chutum Vaya Complex (Goodwin and Haxel, 1990), the complicated arrangement of lithologic types (summarized in highly schematic fashion in column 3B) is partially stratigraphic and partially structural; internal intrusive relations are also locally important. Minor marble and calcsilicate schist in the Chutum Vaya Complex have no counterparts in the Topawa Group. Unit labels in this Figure correspond to those in Figure 3. A larger version of this figure is provided as a PDF file on the CD-ROM accompanying this volume.

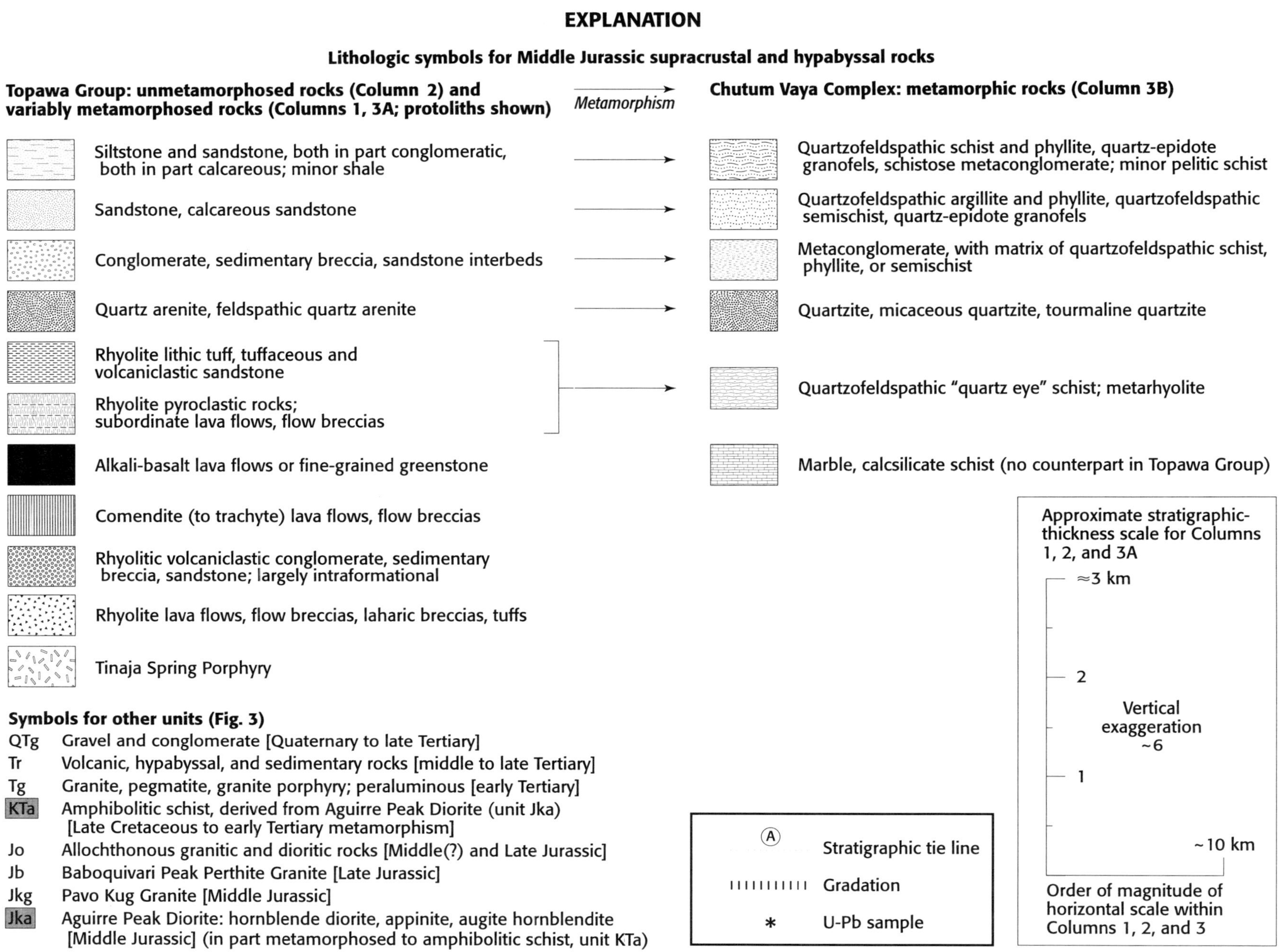

Figure 4 (*continued*).

strongly metamorphosed, large-scale stratigraphic relations are preserved or readily inferred.

The thick rhyolite member consists largely of volcanic rhyolite with interlayered quartzite derived from quartz arenite and feldspathic quartz arenite. Moderately to steeply dipping quartzite layers are laterally continuous over distances of 2–7 km, and numerous crossbeds indicate that these quartzite layers consistently face (young) southwestward. Many of the volcanic rocks have a planar fabric, defined by relict flattened pumice, that consistently parallels bedding in the quartzite layers. The rhyolite member is thus a stratigraphically coherent, homoclinal sequence. Within the rhyolitic rocks, intrusive contacts between volcanic and subvolcanic facies are visible locally. In most places, however, distinctions between extrusive and intrusive rocks are not apparent, owing to mediocre outcrop and obscuring of textures by metamorphic recrystallization. At least three major quartzite layers extend continuously across much or all of the exposed width of the rhyolite member, indicating that sizeable intrusions are not present (Fig. 3). Evidently, the igneous rocks of the rhyolite member are largely volcanic.

The upper rhyolite member and the metaconglomerate member are exposed in continuous sequence only in the area around San Juan Canyon (Fig. 3). Here, rhyolitic metavolcanic rocks are overlain by sericitic metasandstone, in turn overlain by metaconglomerate with subordinate metasandstone (Fig. 4, column 1). Although sedimentary bedding and volcanic layering in the more strongly schistose of these rocks may have been partially folded or transposed into foliation, the gross lithologic sequence here evidently is stratigraphic. A few kilometers to the southeast the upper Ali Molina Formation is disrupted by intrusion and faulting, but the basic sequence metarhyolite-metasandstone-metaconglomerate (in ascending order) still seems to hold.

This stratigraphic sequence inferred for the upper, metamorphosed part of the Ali Molina Formation is corroborated by evidence from nearby Las Animas Mountain, a small outlier in the Baboquivari Valley (Fig. 2). At the north end of this ridge, metavolcanic rocks closely resembling those of the Ali Molina Formation are overlain by metasandstone, in part conglomeratic, in turn overlain by metaconglomerate. A similar sequence also occurs, in more strongly metamorphosed rocks, in a third area—the eastern Comobabi Mountains (Fig. 2; Haxel et al., 1978, 1984).

Pitoikam and Mulberry Wash Formations

A homoclinal, west-facing sequence of virtually unmetamorphosed sedimentary, volcanic, and minor subvolcanic rocks in the Fresnal Canyon area comprises two formations (in ascending stratigraphic order): the sedimentary Pitoikam Formation and the predominantly volcanic and volcaniclastic Mulberry Wash Formation (Fig. 3; Fig. 4, column 2). The Pitoikam Formation is in turn subdivided into three members: the thin Saucito Member, consisting of subequal amounts of interbedded conglomerate and sandstone to siltstone; the massive Contreras Conglomerate Member, composed of conglomerate and sedimentary breccia with subordinate conglomeratic sandstone; and the finer-grained Chiltepines Member, largely siltstone and sandstone. Contacts between the three members of the Pitoikam Formation are gradational.

The contact between the upper Pitoikam Formation and lower Mulberry Wash Formation also is gradational. In the area east of Baboquivari Peak and north of Jupiter Canyon, and to a lesser extent in the Fresnal Canyon area, the upper part of the Chiltepines Member of the Pitoikam Formation contains sparse, thin interlayers of rhyolitic flows and flow breccias identical to some of those of the overlying lower Mulberry Wash Formation. Northeast of Baboquivari Peak, conglomerate lenses in the uppermost Chiltepines Member locally contain clasts identical to some of the volcanic flows of the basal part of the Mulberry Wash Formation in Fresnal Canyon, indicating that Mulberry Wash–like volcanic material was being eroded in latest Pitoikam time. Finally, the middle part of the Mulberry Wash Formation contains thin interbeds of sandstone and conglomeratic sandstone composed entirely of reworked volcanic detritus. Evidently, the Pitoikam and Mulberry Wash Formations represent a single depositional sequence that records an abrupt but transitional change from a dominantly sedimentary environment to a dominantly volcanic environment.

The contact between the Pitoikam Formation and the overlying Mulberry Wash Formation was interpreted as a significant disconformity by Heindl and Fair (1965), but our data indicate the two formations are approximately the same age. Structurally, they are conformable; local appearance of slight discordance apparently results from minor faulting, probably Tertiary, along their contact. More importantly, field observations (just described) clearly show that the Pitoikam–Mulberry Wash contact is stratigraphically gradational.

The principal evidence that led Heindl and Fair (1965) to postulate a temporally significant disconformity between the two formations was misleading. Within the area (around Fresnal Canyon) mapped by Fair (1965) and Heindl and Fair (1965), a swarm of rhyolite dikes that intrudes the Pitoikam Formation appears not to intrude the Mulberry Wash Formation. This observation suggested to Heindl and Fair (1965) that a major event, and therefore an appreciable interval of time, intervened between deposition of the two formation. However, south of Fresnal Canyon numerous dikes of this swarm do intrude the Mulberry Wash Formation (Haxel et al., 1980a, 1982). Moreover, Heindl and Fair did not have the benefit of isotopic ages. We now know that the rhyolite dikes are middle Tertiary (≈24 Ma, U-Pb and K-Ar ages), much younger than the Middle Jurassic Pitoikam and Mulberry Wash Formations.

Stratigraphic Relations of the Ali Molina and Pitoikam Formations

The upper Ali Molina Formation is separated from the lower Pitoikam Formation by the Gu Chuapo fault (Fig. 3). Nevertheless, original stratigraphic relations between the Ali Molina and

Pitoikam Formations can be inferred by direct lithologic correlation of the two formations, and through lithologic correlation of both formations with the Chutum Vaya Complex (Fig. 4).

As the Saucito and Contreras Conglomerate Members of the Pitoikam Formation entirely lack volcanic rocks, they cannot correlate with the rhyolite member of the Ali Molina Formation. Rather, the metasandstone-dominated lower part and the metaconglomerate-dominated upper part of the metaconglomerate member of the Ali Molina Formation evidently correlate with the Saucito and Contreras Conglomerate Members, respectively, of the Pitoikam Formation (Fig. 4, tie line A). The alternative—that the Pitoikam Formation is entirely younger than the Ali Molina Formation—is quite unlikely, considering the close proximity and lithologic correspondence of the upper Ali Molina and lower Pitoikam.

In the low-grade metasedimentary and metavolcanic rocks of the northeast part of the Chutum Vaya Complex, sedimentary bedding and stratigraphic relations are widely preserved. The Chiltepines Member of the Pitoikam Formation extends continuously from its type section in Fresnal Canyon southward into the Chutum Vaya Complex, in the area between Osobavi Peak and Mildred Peak (Fig. 3). Here, siltstone and sandstone of the Chiltepines Member grade into quartzofeldspathic phyllite and schist; calcareous sandstone and siltstone are progressively converted to quartz-epidote granofels and schist. Field observation of this gradation from siltstone and sandstone into low-grade metasedimentary rocks first provided key evidence establishing a firm correlation between the Chiltepines Member of the Pitoikam Formation and part of the Chutum Vaya Complex (Fig. 4, tie lines B, C, D, E). Between Osobavi and Mildred Peaks and in the Jupiter Canyon area (Figs. 3 and 4), the Chiltepines Member and derivative phyllite and schist are gradationally underlain by a low-grade metaconglomerate unit. Given its position, this metaconglomerate must correlate with the Contreras Member of the Pitoikam Formation (Fig. 4, tie lines D, E, F). Below this Contreras-derived metaconglomerate, east and southeast of Mildred Peak, metarhyolite is widely interlayered with metaconglomerate and metasandstone. These interlayered metavolcanic and metasedimentary rocks in the northeast and east part of the Chutum Vaya Complex apparently are broadly correlative (Fig. 4, tie line G) with the upper part of the Ali Molina Formation, where metarhyolite, metaconglomerate, and metasandstone are similarly interlayered.

Direct lithologic correlation of the upper Ali Molina Formation with the lower Pitoikam Formation and independent correlations of both of these formations with the least metamorphosed parts of the Chutum Vaya Complex thus are mutually supportive (Fig. 4).

Tinaja Spring Porphyry

Over a distance of nearly 30 km through the northern and central Baboquivari Mountains (Fig. 3), the Ali Molina and Pitoikam Formations are intruded by a distinctive rhyolite porphyry, commonly strongly altered and characteristically iron-oxide stained in outcrop. This porphyry (or a virtually identical one) also locally intrudes the lower Mulberry Wash Formation (Beikman et al., 1995). This rhyolite porphyry unit at the base of the Topawa Group is hereby named the Tinaja Spring Porphyry.

Its map pattern suggests that the Tinaja Spring Porphyry is related to the supracrustal units of the Topawa Group. Furthermore, the porphyry is similar in appearance, petrography, and geochemistry (see following) to many of the volcanic and subvolcanic rhyolites of the Ali Molina and Mulberry Wash Formations. The porphyry most closely resembles the Ali Molina rhyolites. But, the Tinaja Spring Porphyry must be at least slightly younger than the Ali Molina Formation, as it intrudes the Pitoikam Formation and Mulberry Wash Formations as well. Probably, the Tinaja Spring Porphyry is broadly comagmatic with both the Ali Molina and Mulberry Wash Formations. Alternatively, the Tinaja Spring Porphyry as mapped may be composite unit, comprising two very similar components—an older one related to the Ali Molina Formation and a slightly younger one related to the Mulberry Wash Formation.

URANIUM-LEAD ISOTOPIC AGES

Topawa Group

We first reported inferred U-Pb ages ≈190 Ma for volcanic rocks of the Topawa Group in an abstract in 1981 (Wright et al., 1981; see also Tosdal et al., 1989; Riggs and Haxel, 1990). These geochronologic data (Appendix; see the GSA Data Repository[1] or the CD-ROM accompanying this volume) were determined for multigrain size and magnetic fractions of zircon, analyzed by thermal-ionization mass spectrometry (TIMS). Interpretation of these discordant U-Pb ages, which evidently reflect both inheritance of premagmatic zircon and postmagmatic loss of radiogenic Pb, is not straightforward (cf. Riggs et al., 1993). The most prominent feature of the Topawa Group TIMS data is tight clustering of all $^{207}Pb^{*}/^{206}Pb^{*}$ apparent ages at 190 ± 2 Ma (Pb* denotes radiogenic lead). This inferred age is internally consistent, but seems in conflict with regional data indicating that in many other areas of southern Arizona and adjacent northern Sonora similar Jurassic igneous rocks are significantly younger, ≈180–160 Ma (Anderson and Silver, 1978, 1979; Tosdal et al., 1989; Riggs and Haxel, 1990; Riggs, et al., 1993; Anderson et al., this volume).

In 2003, we determined a new U-Pb age for zircon from the lower Ali Molina Formation (Fig. 3, 4), using the Stanford University–U.S. Geological Survey sensitive high-resolution ion microprobe–reverse geometry mass spectrometer (SHRIMP-RG) (http://shrimprg.stanford.edu/). A single spot on each of twelve

[1]GSA Data Repository item 2005141, Appendix: U-Pb isotopic data determined by thermal-ionization mass spectrometry, and Table A1, TIMS U-Pb isotopic ratios and apparent ages of zircon from the Topawa Group and the Sil Nakya Formation, southern Arizona, is available online at www.geosociety.org/pubs/ft2005.htm, or on request from editing@geosociety.org or Documents Secretary, GSA, P.O. Box 9140, Boulder, CO 80301, USA.

zircon grains was analyzed (Table 1). The corrected, weighted-mean $^{206}Pb^*/^{238}U$ age of these grains is 170 ± 3 Ma, which we infer is the age of the Ali Molina Formation.

A single fraction of zircon from the Tinaja Spring Porphyry, analyzed by TIMS, yielded slightly discordant $^{206}Pb^*/^{238}U$ and $^{207}Pb^*/^{235}U$ apparent ages (Table 2). This single analysis does not provide a geologic age for the Tinaja Spring Porphyry. The $^{206}Pb^*/^{238}U$ apparent age, 174 ± 2 Ma, is marginally older than the age of the Ali Molina Formation (170 ± 3 Ma), which the Tinaja Spring Porphyry intrudes. If discordance is attributed to modest inheritance of Proterozoic radiogenic Pb (Anderson and Silver, 1977), the inferred age of the Tinaja Spring Porphyry is a few million years younger than the $^{206}Pb^*/^{238}U$ apparent age, and is indistinguishable from the age of the Ali Molina Formation. This interpretation, though permissive only, is consistent with the inference, from field relations and petrologic data, that the Ali Molina Formation and Tinaja Spring Porphyry are related.

Younger Granitoids

Most of the area of the northern and central Baboquivari Mountains not underlain by the Topawa Group is made of younger Jurassic plutonic rocks (Fig. 3; Haxel et al., 1980a; Beikman et al., 1995). The most extensive of these plutonic terranes is the calcalkaline Kitt Peak plutonic suite, comprising three units (in order of decreasing age, based upon intrusive relations): Aguirre Peak Diorite, Kitt Peak Granodiorite, and Pavo Kug Granite. The Aguirre Peak Diorite intrudes sedimentary rocks of the Pitoikam Formation. Pitoikam-derived (meta)sedimentary rocks at Morena Mountain (Fig. 2; Beikman et al., 1995) are intruded by Kitt Peak Granodiorite. The Pavo Kug Granite intrudes the Tinaja Spring Porphyry, which in turn intrudes the supracrustal units of the Topawa Group. Thus, the ages of these plutonic rocks constrain the age of the Topawa Group. (Petrogenetic relations of the Topawa Group and the Kitt Peak plutonic suite will be discussed in a paper in preparation.)

A fourth plutonic unit, the Baboquivari Peak Perthite Granite (including subordinate quartz syenite), intrudes the Mulberry Wash Formation and Tinaja Spring Porphyry. This mildly alkaline stock is petrographically and geochemically distinct from the Kitt Peak calcalkaline suite (Tosdal et al., 1989).

We determined U-Pb TIMS isotopic ages for one fraction of zircon from each of three plutonic units: Kitt Peak Granodiorite, Pavo Kug Granite, and Baboquivari Peak Perthite Granite (Table 2; Figs. 3 and 4). Isotopic ratios of zircon from the Kitt Peak Granodiorite are normally concordant ($^{206}Pb^*/^{238}U$ apparent age less than $^{207}Pb^*/^{235}U$ apparent age by only 0.3 m.y.) at 165 Ma. We interpret the age of the Kitt Peak Granodiorite as 165 ± 1 or 2 Ma. Isotopic ratios of zircon from the Pavo Kug and Baboquivari Peak Granites are marginally discordant (0.9 Ma); the best estimate of the age of each pluton is its $^{206}Pb^*/^{238}U$ apparent age. If this marginal discordance is attributed to slight inheritance of Proterozoic radiogenic Pb, the inferred ages are decreased by about 1 or 2 m.y. We estimate the Pavo Kug Granite is 159 ± 2 Ma, and the Baboquivari Peak Perthite Granite is 146 ± 2 Ma.

A body of perthite granite and quartz syenite petrographically and chemically similar to the Baboquivari Peak Perthite Granite crops out near Rancho San Moises in northern Sonora (about

TABLE 1. SHRIMP U-Pb ISOTOPIC AGES OF ZIRCONS FROM THE RHYOLITE MEMBER OF THE MIDDLE JURASSIC ALI MOLINA FORMATION, TOPAWA GROUP, BABOQUIVARI MOUNTAINS, SOUTHERN ARIZONA

Zircon grain[a]	Concentration (μg/g)[c] U	Concentration (μg/g)[c] $^{206}Pb^*$	Atomic ratios $^{238}U/^{206}Pb$	error[d]	$^{207}Pb/^{206}Pb$	error[d]	207Pb-corrected ages (Ma) $^{206}Pb^*/^{238}U$	error[e]
1	287	6.6	37.15	1.9	0.0495	2.8	171.2	3.2
2	126	2.9	36.79	1.8	0.0502	4.2	172.8	3.2
3	615	14.2	37.18	1.6	0.0502	1.9	171.0	2.7
4	477	11.0	37.35	1.6	0.0488	2.2	170.5	2.7
5[b]	527	11.6	39.08	1.6	0.0501	2.1	162.7	2.6
6	252	5.6	38.56	1.7	0.0492	3.0	165.1	2.8
7	230	5.5	35.91	3.6	0.0495	3.1	177.1	6.3
8	176	4.1	36.69	1.8	0.0486	4.0	173.6	3.2
9[b]	131	2.9	39.20	1.8	0.0497	4.1	162.3	3.0
10	655	14.7	38.23	1.7	0.0507	1.8	166.2	2.8
11	179	4.2	36.26	1.8	0.0519	3.4	174.8	3.1
12	183	4.1	37.88	1.7	0.0519	3.4	167.4	2.9
Weighted-average age							170.1	2.8

Note: SHRIMP—sensitive high-resolution ion microprobe.
[a] Rock sample PUP 300.
[b] Data for these two zircon grains are not included in calculation of the weighted-average age.
[c] Pb* denotes radiogenic Pb.
[d] Percent; one sigma.
[e] Two sigma.

20 km southwest of Morena Mountain, Fig. 2). This granitic rock has a U-Pb age of 149 Ma (Anderson et al., this volume).

Summary: Age Span of the Topawa Group

The lower Ali Molina Formation is 170 ± 3 Ma. The Pitoikam Formation is intruded by the Aguirre Peak Diorite, which in turn is intruded by the Kitt Peak Granodiorite. Therefore the Pitoikam Formation is older than 165 ± 2 Ma. The Mulberry Wash Formation is not much younger than the Pitoikam Formation, as the two units are stratigraphically continuous. The lithologic homogeneity of the dominantly volcanic and volcaniclastic Mulberry Wash Formation suggests that it spans a relatively short interval. We infer from these data that the entire Topawa Group represents only a few or several million years of Middle Jurassic time (Fig. 1B), between approximately 170 and 165 Ma.

Other, slightly less restrictive age data support this conclusion. All three supracrustal formations of the Topawa Group are intruded by the Tinaja Spring Porphyry, which is older than the Pavo Kug Granite, inferred age 159 Ma. From this looser (but independent) constraint, the age span of the Topawa Group can be no more than approximately 11 million years.

LITHOLOGY, PETROGRAPHY, AND SEDIMENTARY STRUCTURES

The following descriptions of the rock types of the Topawa Group are summarized from Fair (1965), Heindl and Fair (1965), Haxel et al. (1980a, 1982), and Seaman (1983); and from our heretofore unpublished field and petrographic observations.

Ali Molina Formation, Rhyolite Member

Rhyolite

Igneous rocks of the rhyolite member of the Ali Molina Formation are incipiently to strongly metamorphosed. They range from volcanic rocks to quartzofeldspathic schist with characteristic quartz "eyes" that are relict phenocrysts. The most abundant rock types have relict phenocrysts, including various combinations of plagioclase, quartz, alkali feldspar, and biotite (Table 3), and are obviously silicic volcanic or subvolcanic rocks, but display no other evidence of their origin. In some places, however, one or more of four types of relict primary features are discernable. (1) Eutaxitic foliation is moderately common. It typically is visible on weathered surfaces (Fig. 5A) but not apparent on freshly broken surfaces nor in thin section. We interpret this relict texture as derived from compacted pumice, and the rocks as pyroclastic. (2) Some rocks contain lithic fragments, or both lithic fragments and eutaxitic foliation. In others, the phenocrysts appear broken. We regard these rocks as lithic tuffs or other volcaniclastic rocks. Such rocks are most common around and north of San Juan Canyon, where they are strongly disrupted and their original character obscured by regional metamorphism and intrusion of early Tertiary granite (Fig. 3). (3) Breccia with

TABLE 2. TIMS U-Pb ISOTOPIC AGES OF ZIRCON FROM MIDDLE AND LATE JURASSIC HYPABYSSAL AND PLUTONIC UNITS, BABOQUIVARI MOUNTAINS, SOUTHERN ARIZONA[a]

Unit[b]	Sample	Size fraction[c] (mesh)	Concentration (μg/g)[d]		Observed ratios[e]			Atomic ratios[f]			Apparent ages (Ma)[g,h]		
			U	$^{206}Pb^*$	$^{206}Pb/^{204}Pb$	$^{207}Pb/^{206}Pb$	$^{208}Pb/^{206}Pb$	$^{206}Pb^*/^{238}U$	$^{207}Pb^*/^{235}U$	$^{207}Pb^*/^{206}Pb^*$	$^{206}Pb^*/^{238}U$	$^{207}Pb^*/^{235}U$	$^{207}Pb^*/^{206}Pb^*$
TSP	PUP 39	<325	801.0	18.84	1008	0.06470	0.21776	0.02738	0.18929	0.05015	174.1	176.0	202
KPG	PUP 11	<200	781	17.4	2762	0.05476	0.24018	0.02591	0.17667	0.04945	164.9	165.2	169
PKG	PUP 12	<200	1289	27.8	3906	0.05329	0.16019	0.02509	0.17134	0.04954	159.7	160.6	173
BPPG	PUP 19	<200	371.9	7.35	1316	0.06456	0.15894	0.02300	0.15634	0.04930	146.6	147.5	162

[a] TIMS—thermal-ionization mass spectrometry. Pb* denotes radiogenic Pb. Analytical methods: Wright (1981).

[b] TSP—Tinaja Spring Porphyry; KPG—Kitt Peak Granodiorite; PKG—Pavo Kug Granite; BPPG—Baboquivari Peak Perthite Granite.

[c] Analyzed zircon fractions weighed 10–25 mg.

[d] Estimated precision 0.25%.

[e] Precision: $^{208}Pb/^{206}Pb$ and $^{207}Pb/^{206}Pb$, ≤0.10 %; $^{206}Pb/^{204}Pb$, ≤1.0 %.

[f] As all zircon fractions yielded $^{206}Pb/^{204}Pb$ measured ratios ≥1000, all common Pb has been ascribed to the sample processing blank. Total Pb blanks were 0.2–0.5 ng. Common Pb corrections: $^{206}Pb/^{204}Pb$ = 18.6, $^{207}Pb/^{204}Pb$ = 15.6, $^{208}Pb/^{204}Pb$ = 38.0.

[g] Decay constants (Steiger and Jäger, 1977): ^{235}U, 0.98485 x 10^{-9} yr^{-1}; ^{238}U, 0.155125 x 10^{-9} yr^{-1}; $^{238}U/^{235}U$ = 137.88.

[h] Estimated errors: $^{206}Pb^*/^{238}U$ age, ±0.3 % (~0.5 Ma); $^{207}Pb^*/^{235}U$ age, ±0.7 % (~1 Ma).

TABLE 3. PHENOCRYST ASSEMBLAGES IN VOLCANIC AND HYPABYSSAL ROCKS OF THE MIDDLE JURASSIC TOPAWA GROUP, SOUTH-CENTRAL ARIZONA

		Common phenocryst assemblages[1]	
Unit[2]	Rock type	Feldspar, quartz	Mafic minerals
Ali Molina Formation[3] (15)	Rhyolite[4]	pl, qtz	op
		pl, qtz, af	bio, op
		pl	bio, op, am
		pl, af	
Mulberry Wash Formation (28)	Rhyolite[4]	pl, qtz, af	bio, op, am
		pl, qtz	bio, op, hb
			op
			bio, op, hb, am?
	Comendite[4, 5]	afa	op
		afa, qtz	am
		none	op, am
	Basalt	none	none
		pl	am
			am, op
Tinaja Spring Porphyry[6] (11)	Rhyolite[4]	pl, qtz, af	bio, op, am
		qtz, pl, af	bio, op
			none

[1]Feldspar-quartz phenocryst assemblages and mafic phenocryst assemblages are listed independently. For each rock type, feldspar-quartz and mafic-mineral assemblages are listed in order of decreasing frequency of occurrence. For each phenocryst assemblage, minerals are listed in order of decreasing abundance. Mineral abbreviations: af, alkali feldspar; afa, anorthoclase; am, mafic mineral(s), altered beyond recognition (probably includes some completely altered biotite); bio, biotite; hb, hornblende (or other amphibole); op, opaque mineral; pl, plagioclase; qtz, quartz.

[2]In parentheses: number of thin sections examined.

[3]Excluding strongly metamorphosed rhyolite (quartzofeldspathic schist).

[4]Modal abundances of phenocrysts are typically 20–60 volume percent.

[5]Five of six comendite samples have anorthoclase phenocrysts (Fig. 5C); the sixth is virtually aphyric.

[6]Includes data from Seaman (1983).

matrix similar or identical to angular clasts (autobreccia) occurs locally; we interpret these rocks as flow breccias. They are considerably less common than rocks with eutaxitic foliation. (4) Internal intrusive contacts, though apparently uncommon, show that the rhyolite member locally includes subvolcanic intrusions. However, the continuity of several quartzite layers (see previous) shows that most of the unit consists of extrusive rocks. Overall, pyroclastic and volcaniclastic rocks evidently are more common than flows and flow breccias or subvolcanic rocks, but a large proportion of the Ali Molina (meta)rhyolite can not be confidently assigned to any of these modes of origin.

In thin section, the groundmass of the rhyolites is generally recrystallized to a granoblastic or schistose intergrowth of feldspar and quartz with metamorphic biotite, muscovite, magnetite, epidote, actinolite, and, occasionally, tourmaline. A few of the least-recrystallized rhyolite samples show relict spherulitic devitrification texture; relict glass shards and pumice generally are not apparent.

Quartzite

Variably metamorphosed arenites in the rhyolite member of the Ali Molina Formation range from sandstone or orthoquartzite to quartzite to micaceous quartzite. Many of these rocks appear virtually unmetamorphosed in outcrop and hand specimen, but in thin section are seen to be partially or largely recrystallized (though without much intragranular deformation). Preservation of fine, even, millimeter-scale sedimentary bedding or lamination (Fig. 5B) is common. Quartzite layers range in thickness from several meters to somewhat more than 100 m. Thicker layers contain, at least locally, minor interbedded volcanic or volcaniclastic rocks. Many of the thicker layers, particularly those less metamorphosed, are conspicuously crossbedded. Crossbeds are commonly larger in scale than the 1- to 5-m size of typical outcrops (Bilodeau and Keith, 1986).

The least metamorphosed Ali Molina quartzites are composed predominantly of medium to coarse, fairly well-rounded and well-sorted, rather homogeneous quartz and subordinate feldspar grains. Quartz arenite predominates, but some layers in the upper part of the rhyolite member include appreciable feldspathic arenite as well. Some quartzite layers within the upper part of the member are bimodal in grain-size distribution, due to a subordinate admixed component of subhedral, relatively large (~1 mm), less well-sorted volcanic quartz and feldspar grains. A few quartzite layers contain dark-gray to black beds colored by metamorphic tourmaline, presumably reflecting premetamorphic, hydrothermal introduction of boron from the enclosing volcanic rocks.

Ali Molina Formation, Metaconglomerate Member; Pitoikam Formation, Saucito and Contreras Conglomerate Members

Conglomerate and Sedimentary Breccia

The lower Pitoikam Formation (Saucito and Contreras Conglomerate Members) is well exposed on a high ridge in the upper Fresnal Canyon area. These strata are well indurated but only locally and incipiently metamorphosed. Conglomerate and sedimentary breccia form numerous lenticular beds typically several meters to more than 10 m thick. Internal bedding and sorting range from fair to poor. Channels are visible in many outcrops. Lenticular interbeds of sandstone and pebbly sandstone are common. In the correlative metaconglomerate member of the Ali Molina Formation (Fig. 4), primary lithology of fine-grained rocks is poorly preserved, but clasts in conglomerates are still readily identifiable. Clasts in the conglomerate of the lower Pitoikam Formation and the metaconglomerate member of the Ali Molina Formation (Fig. 4) are pebbles, cobbles, and, less commonly, small boulders, of two distinct types.

The first type, most abundant in the lower part of the Ali Molina metaconglomerate member, is very-well-rounded, typically subspheroidal, mostly pebble-size clasts of white to dark-gray quartzite. These fine-grained, conchoidally fracturing, durable clasts must have been derived from thoroughly metamorphosed quartz arenite or chert, or some possibly from vein quartz. These clasts are not intraformational—they could not have been eroded from the quartzite layers of the rhyolite member of the Ali Molina Formation, for two reasons. First, the rhyolite-member quartzites were metamorphosed in Late Cretaceous to early Tertiary time, so they were unmetamorphosed when the metaconglomerate-member quartzite clasts were deposited. Second, even in their present partially metamorphosed condition the rhyolite-member quartzites widely preserve detrital sand grains visible with a hand lens (Fig. 5B), but detrital grains are rarely apparent in the quartzite clasts.

The second clast type is angular to rounded, mostly pebble- to cobble-size clasts of silicic to intermediate porphyritic volcanic

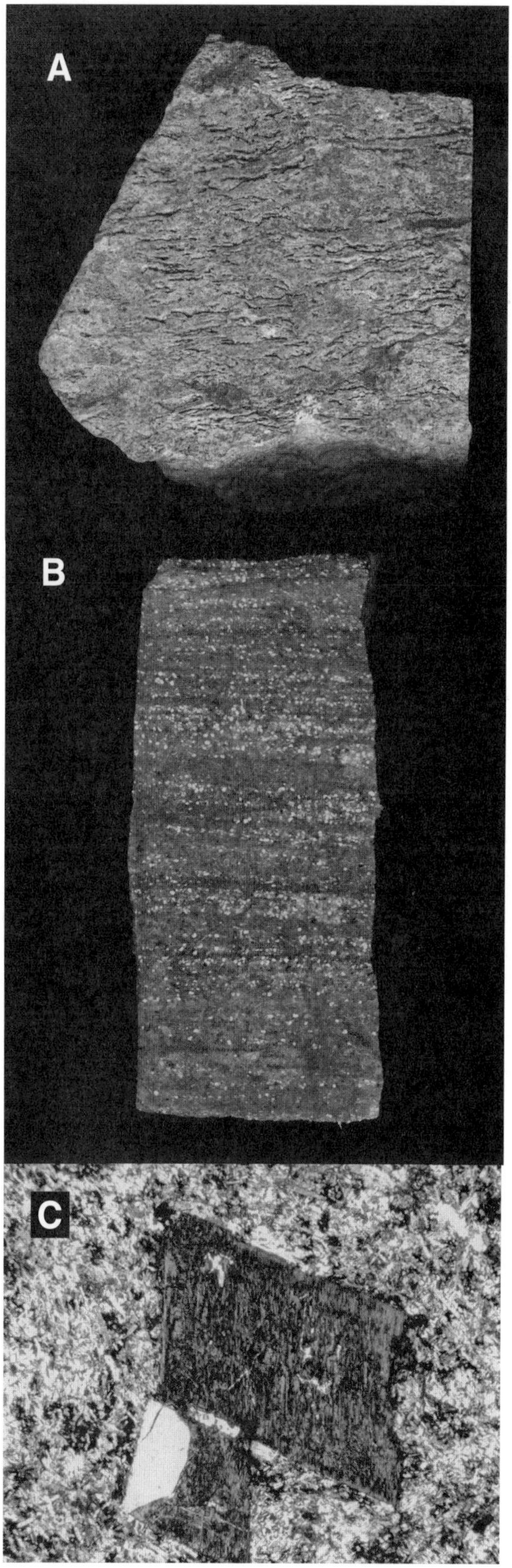

Figure 5. Petrographic features of Topawa Group volcanic and sedimentary rocks. (A) Weathered surface (9 × 8 cm) of weakly metamorphosed rhyolite, Ali Molina Formation, showing relict eutaxitic foliation. (B) Sawn surface (10 × 4.5 cm) of protoquartzite derived from slightly feldspathic quartz arenite, upper rhyolite member, Ali Molina Formation, showing preservation of fine, even sedimentary bedding. In thin section, fine sand and silt grains are slightly or partially recrystallized but coarse sand grains are not. (C) Photomicrograph of euhedral anorthoclase phenocryst (longer diagonal ≈3 mm), comendite, Mulberry Wash Formation. This anorthoclase crystal, shown between crossed polarizers near its mean extinction position, displays the characteristic fine tartan twinning and moiré extinction.

and subvolcanic or hypabyssal rocks. The most common clast varieties are purplish to reddish feldspar porphyry and quartz-feldspar porphyry. This second clast type is increasingly common upward in the metaconglomerate member of the Ali Molina Formation, and predominates in the conglomerates and sedimentary breccias of the lower Pitoikam Formation. All of these clasts are broadly similar to, and many closely resemble, the volcanic and subvolcanic rocks of the Ali Molina Formation. The Pitoikam Formation contains a somewhat higher proportion of these local volcanic and subvolcanic clasts than does the correlative metaconglomerate member of the Ali Molina Formation.

In all the conglomerates and sedimentary breccias of the Topawa Group, clasts of fine-grained granite or aplogranite are rare. Clasts of medium- or coarse-grained granitic rocks, diabase, orthoquartzite, or limestone are very rare or apparently absent.

Pitoikam Formation, Chiltepines Member

Siltstone, Sandstone, and Mudstone

The Chiltepines Member of the Pitoikam Formation is the least resistant to weathering of the Topawa Group units. Where best exposed and least affected by alteration or incipient metamorphism, around its type section in Fresnal Canyon (Fig. 3), the Chiltepines Member consists mostly of dark-colored siltstone; fine- to coarse-grained wacke, locally pebbly; and subordinate mudstone or shale (or, locally, argillite). These strata contain interbeds of brown or red pebble conglomerate and rare interbeds of tan arenite. Some of the siltstone and wacke are calcareous, but limestone beds are rare.

In thin section, the siltstones and sandstones are characterized by mixing of widely variable proportions of angular silicic volcanic detritus—subhedral quartz and feldspar, and rock fragments—and more strongly rounded, presumably nonvolcanic detritus. Field observation suggests that the matrix of the associated conglomerates has a similar character.

Mulberry Wash Formation

Rhyolite and Comendite

The Mulberry Wash Formation contains two types of silicic volcanic and subvolcanic rocks—rhyolite and comendite. Rhyolite is much more abundant. Although the rhyolite and comendite are petrographically and geochemically distinct (see below), they have similar modes of occurrence, look much alike in outcrop, and can be described together.

Within the Mulberry Wash Formation variations in color, mostly various shades of red, brown, gray, and purple, give the impression of greater lithologic or petrographic diversity than is actually present. This range of colors apparently owes largely to variations in type and intensity of alteration, rather than to primary or essential differences. These multihued silicic igneous rocks are chiefly lava flows and flows breccias, accompanied by some derivative monolithologic mass-flow breccia deposits. Thin dikes or small irregular intrusions are locally apparent. In the volcanic flows and subvolcanic intrusions flow banding, often contorted, is common. We found no rocks with obvious eutaxitic foliation. In thin section, the groundmass is generally altered to a turbid, fine-grained, reddish mixture of plagioclase, alkali feldspar, quartz, iron oxides, chlorite, and probably other minerals.

Basalt

Uncommon mafic rocks in the upper Mulberry Wash Formation occur as lava flows, typically a few meters thick, and, less commonly, as small dikes or sills. In thin section these fine-grained, sparsely porphyritic to aphyric rocks are intermediate in character between altered basalt and low-grade greenstone. Thoroughly sericitized plagioclase laths define relict intersertal to seriate, or trachytic, groundmass textures.

Conglomerate, Sedimentary Breccia, and Sandstone

Much of the middle part of the Mulberry Wash Formation in the Fresnal Canyon type area consists of volcanic-clast pebble to boulder conglomerate and sedimentary breccia. Many layers are virtually monolithologic. Some layers are accompanied by thin lenses of volcaniclastic sandstone, composed almost entirely of angular, typically subhedral, poorly to moderately sorted volcanic detritus—largely quartz, feldspar, and biotite grains, and rock fragments. Small-scale bedding or lamination, crossbedding, and channeling are common. Most or all of this conglomerate, sedimentary breccia, and sandstone is intraformational, obviously derived from Mulberry Wash silicic flows and flow breccias. One sedimentary interval in the upper part of the Mulberry Wash Formation, above a minor internal erosional unconformity, is more heterogeneous, containing conglomerate, wacke, mudstone, and a little marl and limestone, interbedded with silicic volcanic rocks similar to the rest of the formation.

Tinaja Spring Porphyry

The Tinaja Spring Porphyry generally is less metamorphosed but more altered than rhyolite in the Ali Molina and Mulberry Wash Formations. Within the Tinaja Spring Porphyry, textural variations from outcrop to outcrop are common. Most of the unit is porphyry characterized by feldspar and quartz phenocrysts, typically 0.5–1 cm across. Some rocks have smaller biotite phenocrysts as well. Aplogranite and aplite, equigranular or less commonly porphyritic, are subordinate but widespread. Outcrops of the Tinaja Spring Porphyry characteristically are iron-oxide stained. In some places the rocks are silicified. In thin section, the groundmass typically is finely crystalline and generally not schistose; a few sections show well-preserved aplitic groundmass texture.

Phenocryst Assemblages of the Ali Molina Formation, Mulberry Wash Formation, and Tinaja Spring Porphyry

In the next section we show that the geochemistry of the Topawa Group volcanic and hypabyssal rocks has been perturbed by pervasive alteration. The best extant evidence as to the original composition of these igneous rocks is their relict phenocryst assemblages (Table 3). Volcanic, subvolcanic, and hypabyssal rocks of the Ali Molina and Mulberry Wash Formations and Tinaja Spring Porphyry typically contain about 20 to 60 volume percent phenocrysts (visual estimates); aphyric rocks are rare. Quartz, plagioclase, and alkali feldspar phenocrysts are readily identifiable (particularly in thin sections stained for the latter). Most mafic silicate phenocrysts are altered beyond certain recognition, except that relict biotite phenocrysts are widely preserved. (Likewise, Wilshire [1957] noted that, in propylitically altered rocks, biotite is the most resistant mafic silicate mineral.) Remnants of highly altered hornblende phenocrysts are discernable in some specimens of Mulberry Wash rhyolite.

Phenocryst assemblages of most of the igneous rocks from the Ali Molina and Mulberry Wash Formations and Tinaja Spring Porphyry typically include various combinations of quartz, alkali feldspar, and biotite, indicating rhyolitic to dacitic compositions. Some uncommon silicic volcanic flows in the Mulberry Wash Formation contain anorthoclase (Fig. 5C), as their only feldspar phenocryst. This distinctive mineralogy, combined with equally distinctive geochemistry, indicates these rocks are comendites.

GEOCHEMISTRY OF VOLCANIC AND HYPABYSSAL ROCKS

Alteration; Geochemical Classification

Thirty-six samples of volcanic and hypabyssal rocks from the Topawa Group have been analyzed for major elements and several trace elements (Table 4; Figs. 6–10). Most geochemical variation diagrams involving alkali or alkali earth elements show little evidence of either definite trends or coherent groupings (Fig. 6). As metamorphosed and unmetamorphosed rocks seem equally affected, much or most of this disarray apparently results from premetamorphic alteration. Particularly telling is the wide and erratic variation of K_2O/Na_2O ratios, which span the thousandfold range from 0.04 to >38. Even if a few extreme values are excluded, this ratio still ranges from ≲0.3 to ≳5 (Fig. 6A). More than one-quarter of the samples have $K_2O/Na_2O \geq 2$. These unusually high K_2O/Na_2O ratios (LeMaitre, 1976; Cox et al., 1979; Bergman, 1987) are far too variable to represent primary values within one or a few suites of related igneous rocks. Further compelling evidence of strong and extensive alteration of the Topawa Group igneous rocks is their consistently highly oxidized condition: median $Fe_2O_3/(Fe_2O_3+FeO)$ is 0.89, and nearly one-quarter of the samples have values ≥0.95.

The data plotted in Figure 6A suggest that alteration of the Topawa Group involved alkali exchange. Rhyolites with high and low K_2O/Na_2O appear to be most common in the upper and lower parts of the stratigraphic section, respectively. This arrangement may represent zonation of potassic and sodic alteration within a Jurassic hydrothermal system (cf. Battles and Barton, 1995; Barton and Johnson, 1996). (Alteration of the Topawa Group will be further discussed in a paper in preparation.)

Because of their pervasive alteration, the igneous rocks of the Topawa Group can not be properly classified using the standard $SiO_2 - (Na_2O+K_2O)$ variation diagram (LeMaitre, 2002). Instead, we have experimented with various diagrams involving elements that are considered relatively immobile or for which moderate postmagmatic changes in concentration are not overly important. The diagrams we found most definitive are two specifically designed to classify altered or metamorphosed rocks (Fig. 7A, B; Winchester and Floyd, 1977; Floyd and Winchester, 1978).

Rhyolite

For the majority silicic volcanic and hypabyssal rocks of the Topawa Group, geochemical classification is merely corroborative. Relict phenocrysts (Table 3) clearly indicate the rocks are rhyolites and/or dacites; the geochemical diagrams support this classification (Fig. 7). Ratios of Si to Al (generally considered the least mobile of the major rock-forming elements) are still entirely normal for rhyolite and dacite (Fig. 7C), so silica contents evidently have not been grossly altered. Concentrations of SiO_2 range from 63 to 78 wt%, with more than two-thirds of the samples in the range of 68% to 77%—more silicic than average dacite (65 % SiO_2), and clustered around average rhyolite (73 % SiO_2)(Fig. 8A). We refer to these rocks collectively as rhyolite, with the understanding that the Topawa Group includes subordinate silicic dacite as well.

Zirconium content is a sensitive indicator of alkalinity in igneous rocks (Erlank et al., 1978; Linthout, 1984; Winchester and Floyd, 1977; Tosdal et al., 1989). Moderate Zr concentrations, ≈150 – 300 µg/g (Fig. 8A), indicate that the Topawa Group rhyolites are ordinary calcalkaline, orogenic igneous rocks (cf. Ewart, 1979, 1982; Saunders et al., 1980). The initial Nd isotopic composition of a representative rhyolite from the Ali Molina Formation, ε_{Nd} = –8.5, is consistent with this conclusion (Asmerom et al., 1991).

Comendite

Six samples from the Mulberry Wash Formation plot separately from the Topawa Group rhyolites on most variation diagrams (e.g., Figs. 7 and 8). High concentrations of La, Ce, Y and, especially, Zr and Fe, compared with the calcalkaline rhyolites, indicate these six special rocks are distinctly alkaline. They are further distinguished from rhyolite by anorthoclase phenocrysts (Fig. 5C), with or without quartz phenocrysts (Table 3). Phenocrysts of anorthoclase ± quartz are characteristic of oversaturated (in silica) peralkaline igneous rocks (Sutherland, 1974). In the classification diagrams, these six rocks tend to plot in the field of comendite and pantellerite, marginal to the field of trachyte (Fig. 7A, B). Relative concentrations of SiO_2, Al_2O_3, and FeO* likewise suggest a range of compositions from comenditic trachyte through comendite (Fig. 9). For simplicity, we refer to these uncommon, anorthoclase-phyric, alkaline volcanic rocks in

TABLE 4. REPRESENTATIVE GEOCHEMICAL ANALYSES OF JURASSIC VOLCANIC AND HYPABYSSAL ROCKS,TOPAWA GROUP, BABOQUIVARI MOUNTAINS, SOUTHERN ARIZONA {1}

Rock type	Rhyolite								Comendite		Basalt	
Unit {2}	AMF	MWF	AMF	TSP	MWF	AMF	AMF	Median	MWF	MWF	MWF	MWF
Sample {3}	BG116	BG134	BG28	BG49	BD120	BG41	PUP17	{9}	SP602	BD118	SP202A	SP234
Major elements (%)												
SiO_2	66.2	69.0	69.4	70.2	71.8	72.9	76.3	71	68.5	71.2	46.6	48.3
Al_2O_3	16.8	14.9	15.0	15.0	14.6	14.4	12.2	14.5	14.1	13.6	16.5	15.4
Fe_2O_3	3.81	3.24	2.02	1.41	2.32	1.23	1.61	2.4	5.66	4.57	10.6	10.0
FeO	0.10	0.17	0.89	1.04	0.19	0.48	0.61	0.48	0.17	0.18	1.26	1.00
MgO	1.11	0.46	0.70	0.75	0.32	0.36	0.34	0.56	0.34	0.12	5.65	6.59
CaO	3.19	0.49	0.84	1.53	0.18	0.33	0.78	0.84	0.14	0.04	6.60	4.68
Na_2O	3.72	3.95	4.27	3.48	3.50	5.26	4.64	3.7	2.76	4.55	4.18	4.35
K_2O	3.35	5.52	4.87	5.07	5.12	3.66	2.50	5.1	6.62	4.47	1.54	2.25
TiO_2	0.54	0.45	0.40	0.36	0.33	0.39	0.12	0.38	0.53	0.49	1.72	1.76
P_2O_5	0.17	0.13	0.09	0.10	0.08	<0.05	<0.05	0.12	0.06	<0.05	0.44	0.62
MnO	0.10	0.08	0.06	0.07	0.05	<0.02	0.11	0.07	0.11	0.06	0.23	0.27
H_2O	0.48	0.63	0.45	0.54	0.88	0.24	0.24	0.53	0.65	0.35	2.75	2.85
CO_2	<0.01	0.27	<0.01	<0.01	0.01	<0.01	<0.01	0.01	0.03	<0.01	1.86	1.61
Total {4}	99.7	99.5	99.3	99.9	99.6	99.5	99.6		99.7	99.7	100.2	99.9
Fe^{3+}/Fe {5}	0.97	0.95	0.69	0.57	0.92	0.72	0.73	0.86	0.97	0.96	0.89	0.91
mg {6}	0.41	0.25	0.36	0.42	0.23	0.33	0.26	0.36	0.12	0.06	0.53	0.59
K_2O/Na_2O	0.90	1.40	1.14	1.46	1.46	0.70	0.54	1.4	2.40	0.98	0.37	0.52
Alkali SI {7}	0.58	0.84	0.82	0.75	0.77	0.88	0.85	0.83	0.83	0.91	0.52	0.62
Alumina SI {8}	1.08	1.11	1.09	1.07	1.26	1.09	1.04	1.09	1.18	1.10	0.80	0.85
Trace elements (μg/g)												
F	700	700	1100	1400	400	600	500	600	200	100	800	1200
Rb	106	232	195	189	118	143	75	180	178	180	74	75
Sr	400	140	140	240	120	82	69	135	150	160	720	360
Ba	1200	1400	1300	1200	1200	1100	460	1200	350	180	550	860
La	57	32	29	38	25	42	36	36	83	53	22	29
Ce	130	<100	<100	<100	<100	<100	<100	<100	180	150	<100	<100
Y	30	26	24	26	26	44	45	26	78	63	31	35
Zr	210	180	240	220	180	290	240	200	800	680	240	390
V	45	29	<10	26	<10	<10	<10		<10	<10	280	210
Cr	<10	<10	<10	<10	<10	<10	<10	<10	<10	<10	150	50
Co	9	5	2	4	3	2	2	4	2	2	47	41
Ni	8	5	4	4	4	4	4	5	9	9	94	69
Cu	3	9	2	29	7	3	1	4	6	8	28	7
Ga	15	16	15	14	10	15	13	15	23	20	18	17

{1} Analytical methods: major elements—wavelength-dispersive x-ray fluorescence (Taggart et al., 1987); FeO, H_2O, CO_2, F—several methods (Jackson et al., 1987); Rb—energy-dispersive x-ray fluorescence (Johnson and King, 1987); other trace elements—direct-current-arc quantitative emission spectrometry (Golightly et al., 1987).

{2} AMF, Ali Molina Formation; MWF, Mulberry Wash Formation; TSP, Tinaja Spring Porphyry.

{3} Representative least-altered samples.

{4} Total includes all Fe as Fe_2O_3, F, correction for F ≙ O, and BaO.

{5} Weight percent $Fe_2O_3/(Fe_2O_3+FeO)$.

{6} Molar MgO/(MgO+FeO), with weight percent $Fe_2O_3/(Fe_2O_3+FeO)$ set to 0.2.

{7} Alkali saturation index, molar $(Na_2O+K_2O)/Al_2O_3$.

{8} Alumina saturation index, molar $Al_2O_3/(CaO+Na_2O+K_2O)$.

{9} Median of 26 analyses, from the Ali Molina Formation (11 analyses; including one from Asmerom et al., 1991), Mulberry Wash Formation (5), Tinaja Spring Porphyry (6), and Chutum Vaya Complex (4).

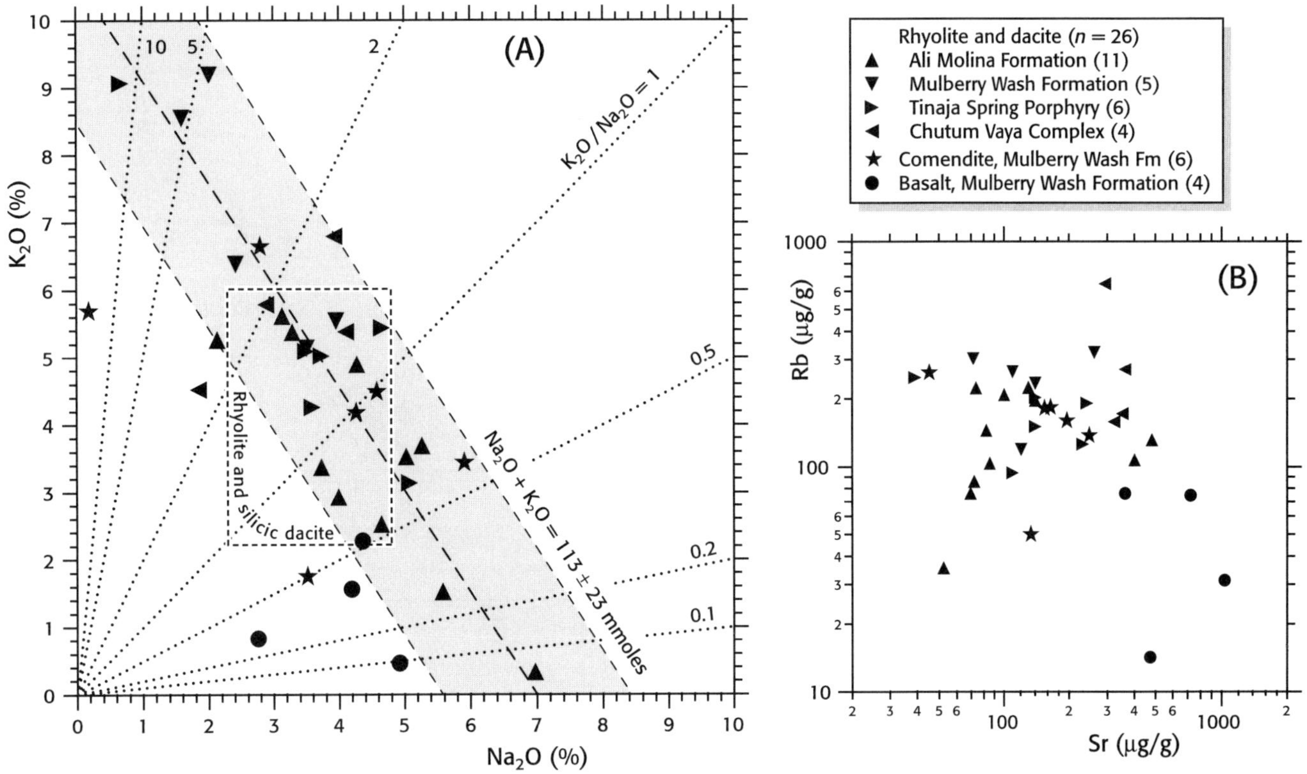

Figure 6. Examples of variation diagrams involving alkali and alkali-earth elements, illustrating the typical high degree of scatter and absence of definite trends or clusters, and the very wide range of K_2O/Na_2O ratios. In A, all of the rhyolites except one have semi-constant total concentration of molar major alkali oxides, $(Na_2O/62.0) + (K_2O/94.2) = (113 \pm 23$ mmoles)/100 g of rock, suggesting that alteration involved alkali exchange. The rectangle labeled "Rhyolite and silicic dacite" represents the most common range of compositions (arithmetic mean ± one standard deviation) of rhyolites and rhyodacites (presumably minimally altered) (LeMaitre, 1976). Two of the Topawa Group analyses (one rhyolite, one basalt) plotted in this and Figures 7, 8, and 10 are from Asmerom et al. (1991).

the Mulberry Wash Formation as comendite. As the six comendite samples all are quite highly altered (generally even more so than the rhyolites), more detailed examination of their igneous geochemistry is probably futile.

Basalt

The four basaltic samples from the Mulberry Wash Formation have high H_2O and CO_2 (Table 4), consistent with their greenstone mineralogy. Normalized volatile-free, their analyses have 49%–52% SiO_2. In the $Zr/TiO_2 - SiO_2$ classification diagram, they plot in or near the alkali basalt and basanite-trachybasalt-nephelinite fields (Fig. 7A). Attempting to further distinguish between these categories of basaltic rocks would be pointless, given our small number of strongly altered samples; we simply describe the Mulberry Wash rocks as alkali basalt. Distinctively high abundances of Ti, Zr, and Y show that the Mulberry Wash mafic rocks are compositionally like within-plate basalts, not magmatic arc basalts (Fig. 10). Their initial Nd isotopic composition, $\varepsilon_{Nd} = +2.0$, further argues that these basalts are not petrogenetically related to arc magmatism (Asmerom et al., 1991).

Petrogenetic Relations of Rhyolite, Basalt, and Comendite

Three factors constrain the relation between rhyolite and basalt. (1) Rhyolite is overwhelmingly dominant; basalt makes up only a tiny fraction, much less than one percent, of the Topawa Group. (2) A pronounced compositional gap separates the basalt, with ≈50% SiO_2, and the rhyolite and silicic dacite, 63%–78% SiO_2. (3) The rhyolites are calcalkaline magmatic arc rocks whereas the mafic rocks are alkaline within-plate basalts, and the two rock types have quite distinct initial Nd isotopic compositions. Considered together, these factors preclude any direct petrogenetic connection between Topawa Group rhyolite and basalt.

More broadly, Nd isotopic compositions indicate the alkali basalt in the Mulberry Wash Formation is regionally unique. Other Jurassic igneous rocks in south-central Arizona, ranging in composition from diorite to granite and rhyolite, have moderately negative ε_{Nd}, –9 to –4 (Farmer and DePaolo, 1984; Asmerom et al., 1991). These are normal values for a continental magmatic arc. Mulberry Wash alkali basalt is the only Jurassic rock in this region known to have positive ε_{Nd}.

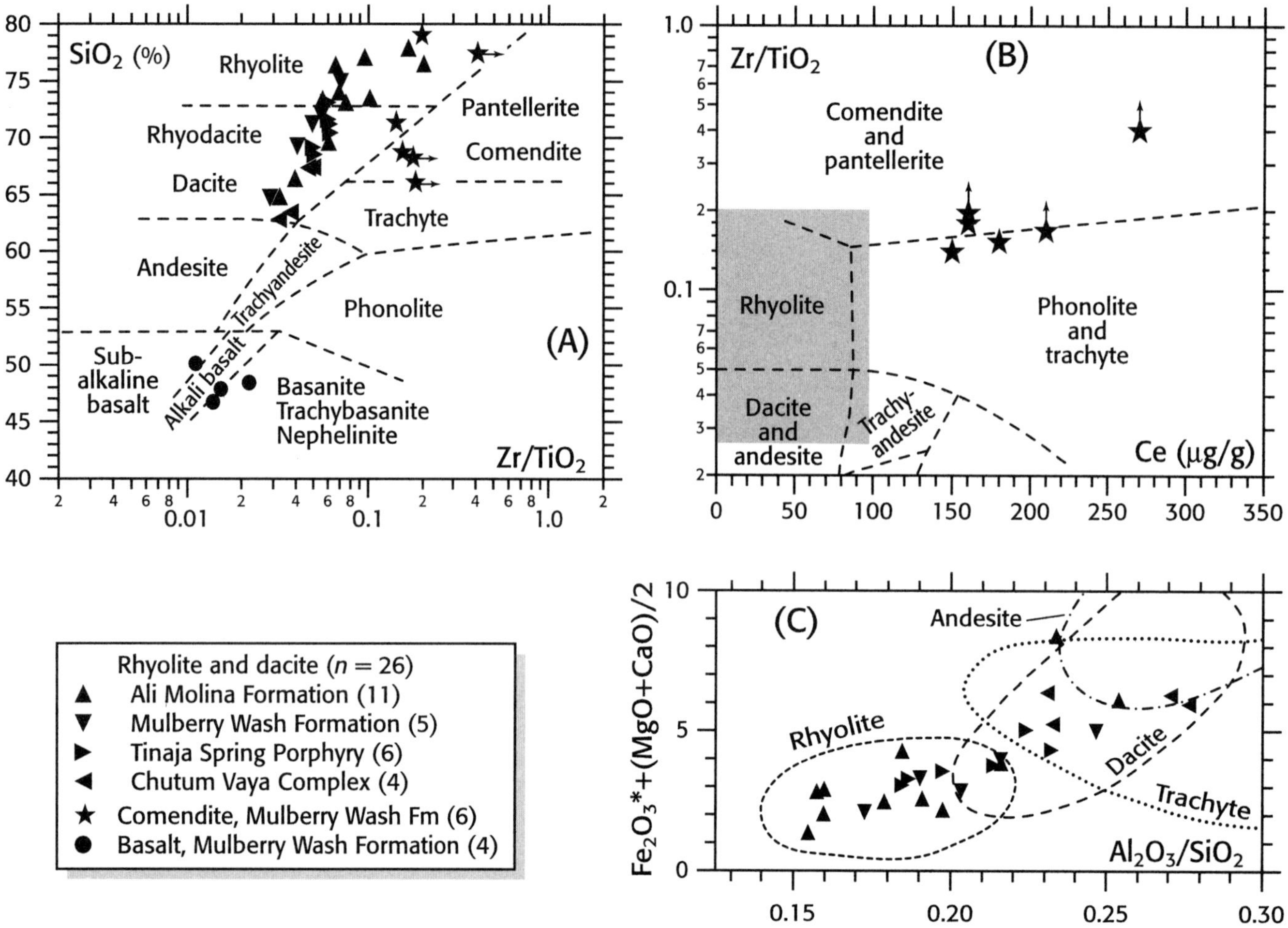

Figure 7. (A, B) Variation diagrams intended for geochemical classification of altered and metamorphosed rocks (Winchester and Floyd, 1977; Floyd and Winchester, 1978). Three comendite samples from the Mulberry Wash Formation (marked with arrows) contain >1000 μg/g Zr; plotted values of Zr/TiO_2 are minimums. In B, all rhyolite samples contain <100 μg/g Ce and thus plot within the shaded rectangle. (C) Classification of common volcanic rocks according to Al_2O_3/SiO_2 and "basicity index" (ordinate), after Church (1975). Fe_2O_3* denotes total iron as Fe_2O_3.

Geochemical data indicate no compositional continuum between Topawa Group rhyolite and comendite (Figs. 7 and 8). Furthermore, rhyolite was erupted and intruded during the entire history of the Topawa Group, whereas comendite is much less abundant, and restricted to the upper one-quarter of the Group (Fig 4). Apparently, the rhyolite and comendite are not directly related. Instead, the comenditic rocks appear to constitute their own series, ranging from trachyte to comendite proper (Fig. 9). Such a connection is to be expected, as trachyte-comendite lineages are well known (e.g., Bevier, 1981; Middlemost, 1981; Parker, 1983).

These two alkaline rock types—within-plate alkali basalt and comendite—are restricted to the upper one-quarter of the Topawa Group, and evidently are both genetically unrelated to the dominant calcalkaline rhyolite. The spatial and temporal association of these two types of alkaline rock surely is not coincidental. Alkali basalt-peralkaline rhyolite associations are characteristic of continental extensional environments, such as the Neogene Basin and Range region of the western United States (e.g., Noble and Parker, 1975; Noble et al., 1978; Rytuba and McKee, 1984; Ngounouno et al., 2000). The trace element and Nd isotopic composition of the Mulberry Wash alkali basalt further suggest an extensional environment.

DEPOSITIONAL ENVIRONMENTS AND PROVENANCE

The Topawa Group records two episodes of rhyolitic volcanism, represented by the rhyolite member of the Ali Molina Formation and the Mulberry Wash Formation, separated by a hiatus in volcanism during which the clastic sediments of the Pitoikam Formation and correlative metaconglomerate member of the Ali Molina Formation were deposited (Figs. 3 and 4).

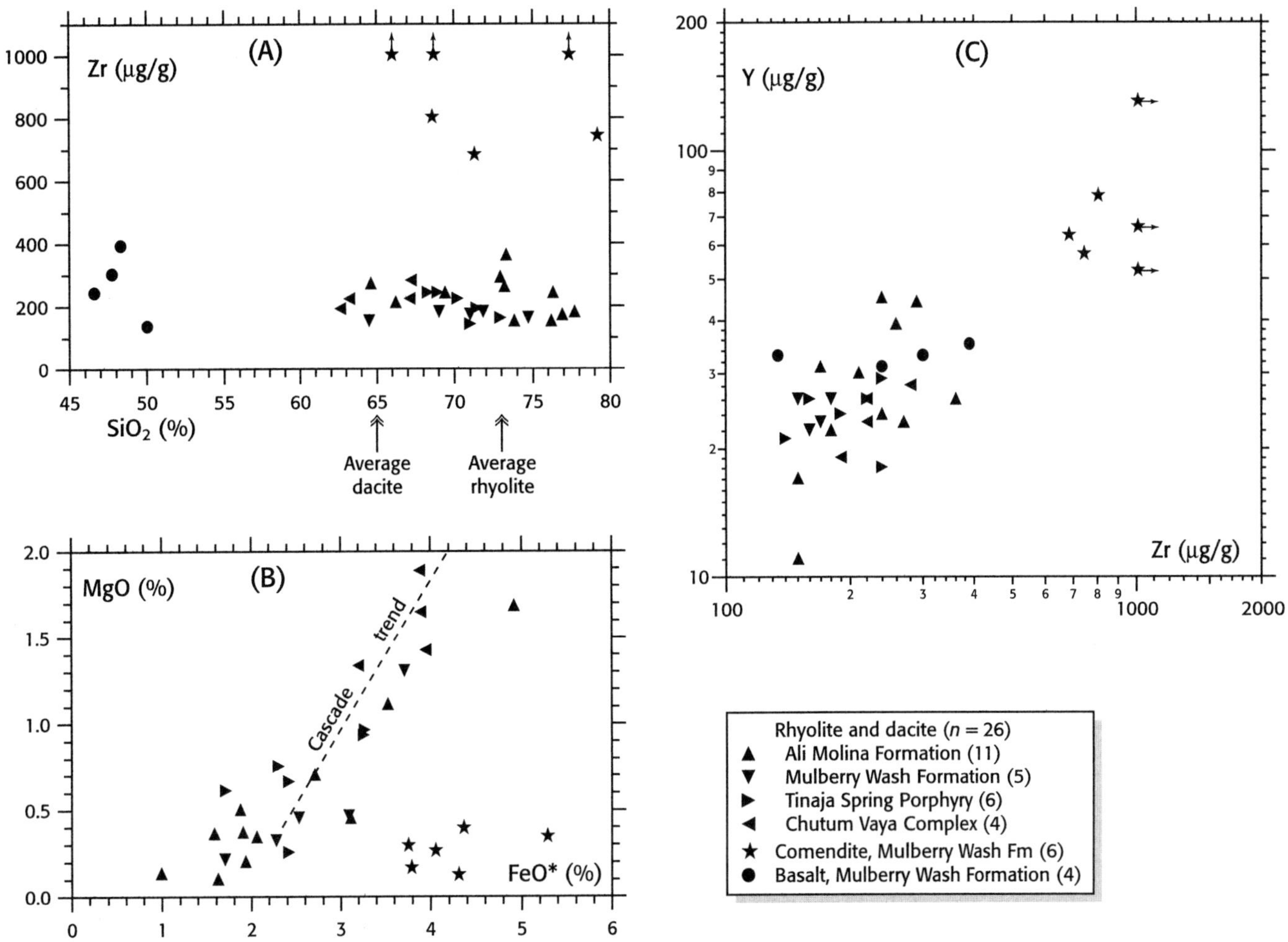

Figure 8. Variation diagrams showing SiO_2 contents of Topawa Group rhyolite (A) and geochemical distinctions between calcalkaline rhyolite of the Topawa Group and comendite of the Mulberry Wash Formation (A, B, C). In A, average SiO_2 content of rhyolite and dacite is from LeMaitre (1976). In A and C, three comendite samples from the Mulberry Wash Formation (marked with arrows) contain >1000 µg/g Zr; plotted values are minimums. In B, the dashed line indicates the low-FeO*, low-MgO part of the Cascade trend of Lowder and Carmichael (1970). FeO* denotes total iron as FeO.

Ali Molina Formation

Despite its considerable thickness, the rhyolite member of the Ali Molina Formation probably was not deposited in a caldera. Caldera-margin facies such as megabreccia are absent. Furthermore, the rhyolite member contains numerous individual thin volcanic units, at least a dozen in the central part of its outcrop area, rather than the few thick units expected as caldera fill. Eruptive centers for the rhyolite-member volcanic rocks remain unknown. The abundance of lithic tuffs in the northwest part of the Ali Molina Formation might suggest a volcanic vent in this direction. If so, the vent probably either was within the area now occupied by a stock of early Tertiary granite or is covered by late Cenozoic gravel (Fig. 3).

The Ali Molina Formation quartzite beds that locally preserve large crossbeds very likely are eolian (Bilodeau and Keith, 1986). Other quartzites probably represent either eolian deposits in which indicative sedimentary structures are not preserved, or fluvially reworked eolian sand. The association of volcanic rocks and sandstone beds indicates deposition in a basin: though outflow facies of pyroclastic flows can form extensive volcanic plateaus, such an environment would not favor accumulation of the thick, laterally continuous sand deposits. Conversion of the quartzite layers from wind-blown sand to eolian sandstone required prolonged interaction with ground water, further implying a basinal setting. Evidently, quartzite layers within the rhyolite member represent several incursions of eolian and fluvial sand into a partially filled noncaldera volcanic or structural depression, during intervals between major eruptions. The absence of detritus coarser than sand size suggests that the margins of the depression were low, well removed from the part of the basin fill now preserved in the rhyolite member, or both. The metaconglomerate member of the Ali Molina Formation appears to represent final sedimentary filling of the rhyolite-

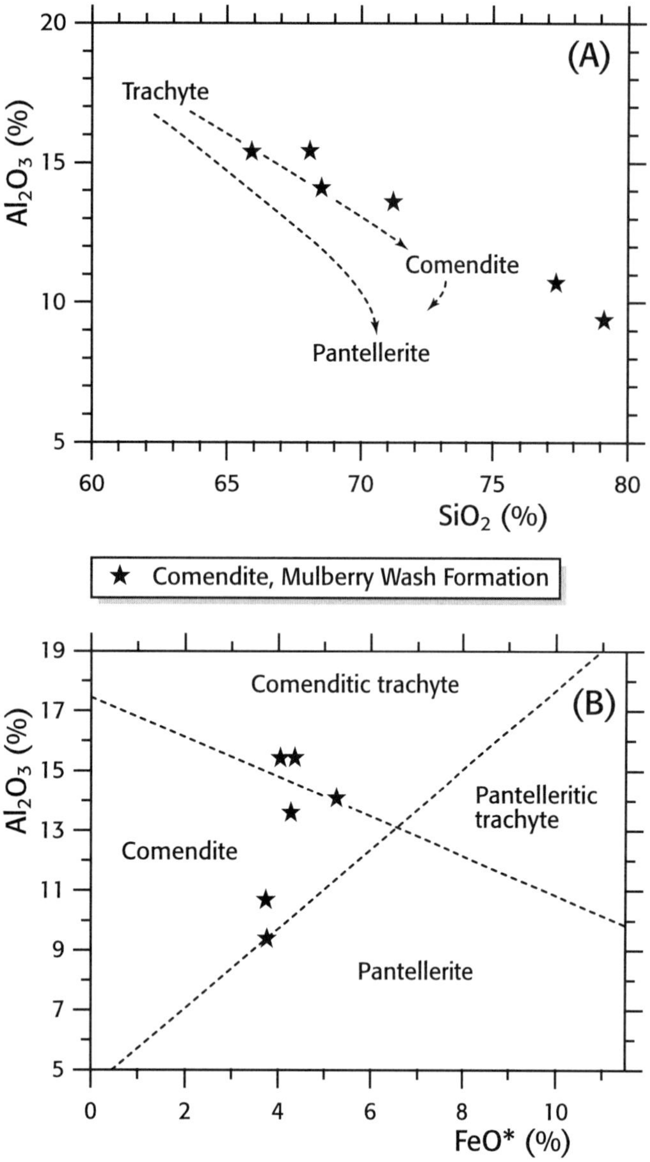

Figure 9. Classification diagrams for oversaturated peralkaline volcanic rocks (A, after Noble, 1968; B, after Macdonald, 1974), showing composition of trachytic to comenditic rocks of the Mulberry Wash Formation. FeO* denotes total iron as FeO.

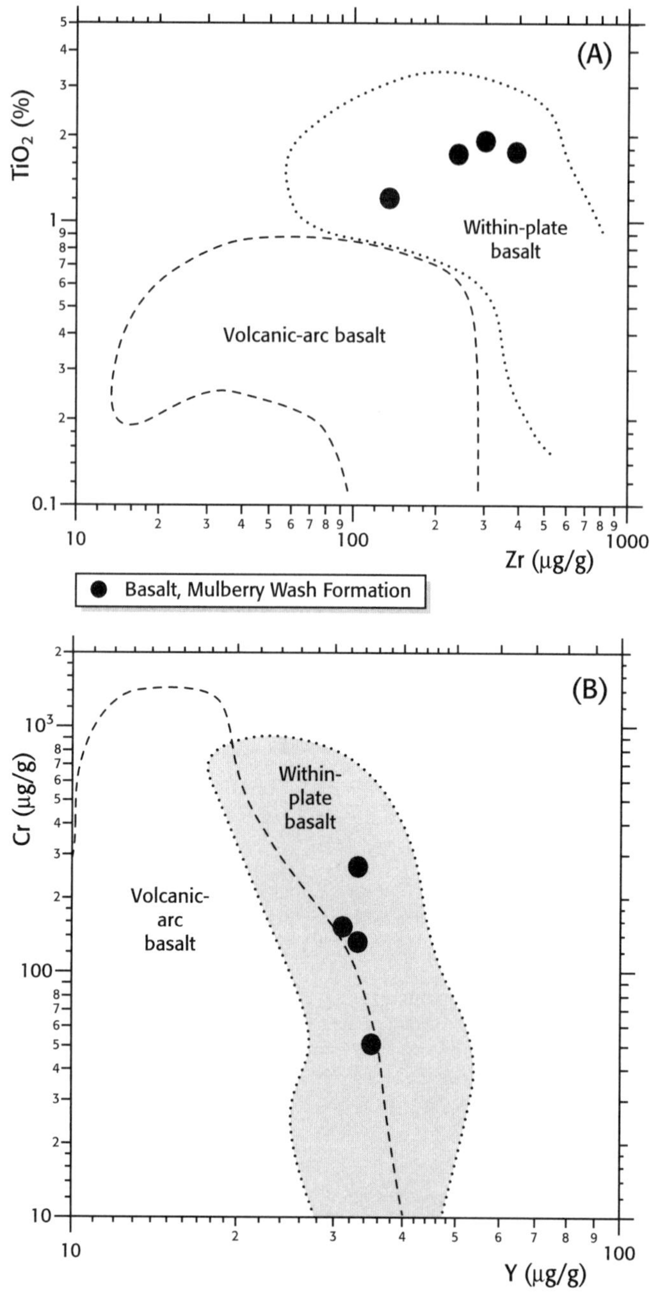

Figure 10. Examples of variation diagrams distinguishing volcanic arc and within-plate basalts (after Pearce, 1982), applied to basalt of the Mulberry Wash Formation. One analysis is from Asmerom et al. (1991).

member depression, probably by fluvial deposits, during the post–rhyolite-member hiatus in volcanism.

Pitoikam Formation

In the Saucito and Contreras Conglomerate Members of the Pitoikam Formation, the preponderance of thickly bedded, channeled conglomerate and sedimentary breccia strongly suggests an alluvial fan environment; subordinate interlayered sandstone and conglomerate presumably represents intra-fan fluvial deposits. These strata are closely associated with the Gu Chuapo fault (Fig. 3). Although its northwest part has undergone minor Tertiary movement, the Gu Chuapo fault evidently is fundamentally a Middle Jurassic structure, older than the Tinaja Spring Porphyry, which terminates the fault at its southeast end (Fig. 3; Haxel et al., 1980a). Strata of the Saucito and Contreras Members evidently were shed from the scarp of the Gu Chuapo fault. This fault was more likely tectonic than volcano-tectonic in origin, as it was active during the essentially nonvolcanic interval represented by the Pitoikam Formation. In support of this interpretation, the lower Pitoikam Formation contains a higher proportion of local volcanic and subvolcanic clasts than does the correlative metaconglomerate member of the Ali Molina Formation, not directly associated with the Gu Chuapo fault.

The Chiltepines Member of the Pitoikam Formation gradationally overlies the Contreras conglomerate Member. Although diagnostic sedimentary structures have not been documented, the sandstone, siltstone, and minor conglomerate of the Chiltepines Member probably represent a fluvial environment that lay basinward of and succeeded the alluvial fan deposits of the lower Pitoikam Formation.

Mulberry Wash Formation

Over the fairly small area in which the full sequence of the Mulberry Wash Formation is exposed, in Fresnal Canyon (Fig. 3), neither we nor Fair (1965) identified any vent facies. The rhyolite flows, flow breccias, and mass-flow breccias that constitute much of the Mulberry Wash Formation here must have accumulated relatively near their sources, so the volcanic center(s) probably were not too far removed from the Fresnal Canyon area. The narrow band of Mulberry Wash Formation extending from upper Chiltepines Canyon southward to Chutum Vaya Canyon (Fig. 3) contains a greater proportion of volcaniclastic conglomerate and lesser proportion of lava flows and flow breccias, suggesting that the volcanic vent(s) lay somewhere to the north. Sedimentary structures indicate that the abundant layers of volcaniclastic conglomerate and sandstone within the Mulberry Wash Formation represent contemporaneous fluvial reworking of the volcanic flows and breccias. Thin lenses of limestone in the upper Mulberry Wash Formation suggest small, localized lakes (Fair, 1965).

Provenance

The Middle Jurassic sedimentary rocks of the Topawa Group record mixing of detritus from a minimum of three distinct and unrelated source terranes. The dominant source was a local or nearby volcanic to hypabyssal terrane generally similar or identical to the igneous rocks of the Topawa Group itself. This source provided the abundant porphyritic, silicic pebbles and cobbles in the Saucito and Contreras Conglomerate Members of the Pitoikam Formation and the metaconglomerate member of the Ali Molina Formation. This local source also contributed the relatively angular volcanic quartz and feldspar grains present, in subordinate amounts, in some quartzite layers in rhyolite member of the Ali Molina Formation. Beds of conglomerate and sedimentary breccia in the Mulberry Wash Formation are derived almost entirely from material just like the flows and flows breccias in which they are intercalated. Very little nonlocal detritus is apparent in most of the Mulberry Wash Formation.

A second source terrane provided the relatively well-rounded and well-sorted, quartz-rich sand in the quartzites of the rhyolite member of the Ali Molina Formation. As explained below, eolian sandstones in the Ali Molina Formation evidently correlate with and are related to those widely preserved in the Middle Jurassic San Rafael Group of the southern Colorado Plateau (Fig. 1). Mature, quartz-rich sand in the Ali Molina Formation, both eolian and fluvial, presumably had sources in the coeval ergs to the north.

The third sedimentary source is indicated by the well-rounded quartzite pebbles that are most common in the metaconglomerate member of the Ali Molina Formation. These thoroughly metamorphosed quartzite pebbles could not have come from the local volcanic-subvolcanic source terrane. They must also have a different source than that of the sand in the Ali Molina quartzites, as the pebbles are much too large to be eolian, and generally do not occur in the fluvial (reworked eolian) quartzite layers. Significance of these quartzite pebbles is discussed in a later section.

REGIONAL EXTENT, CORRELATIONS, AND SIGNIFICANCE OF THE TOPAWA GROUP

In this section, we discuss relations of the Middle Jurassic Topawa Group to other Early to Middle Jurassic sequences in southern Arizona and northern Sonora (Fig. 2); Middle Jurassic eolian sandstones exposed on the Colorado Plateau, ≥400 km to the north (Fig. 1); source areas for quartzite pebbles; the Middle to Late Jurassic Mojave-Sonora megashear (Figs. 1 and 11); and distinctive rhyolite clasts in Eocene conglomerates in coastal southern California (Fig. 12).

By correlating supracrustal units within southern Arizona, we mean that they accumulated in broadly similar environments

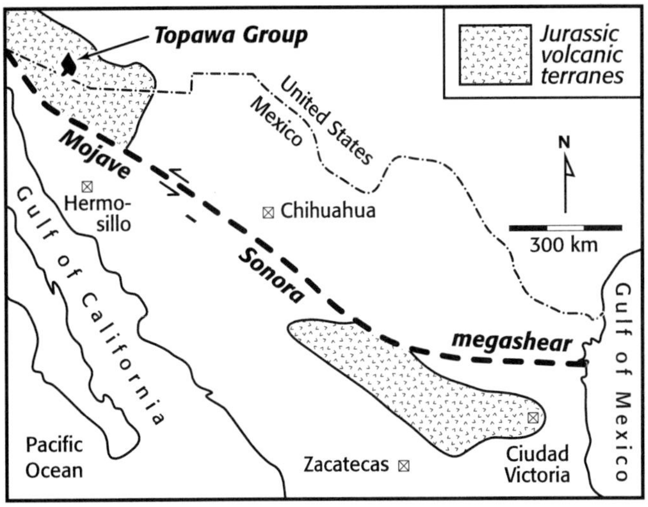

Figure 11. Offset of the Jurassic magmatic arc by the Middle to Late Jurassic Mojave-Sonora megashear, according to Jones et al. (1995). The area mapped in black represents the inferred minimum extent of the greater Topawa Group, including derivative metamorphic rocks (Figs. 2–4).

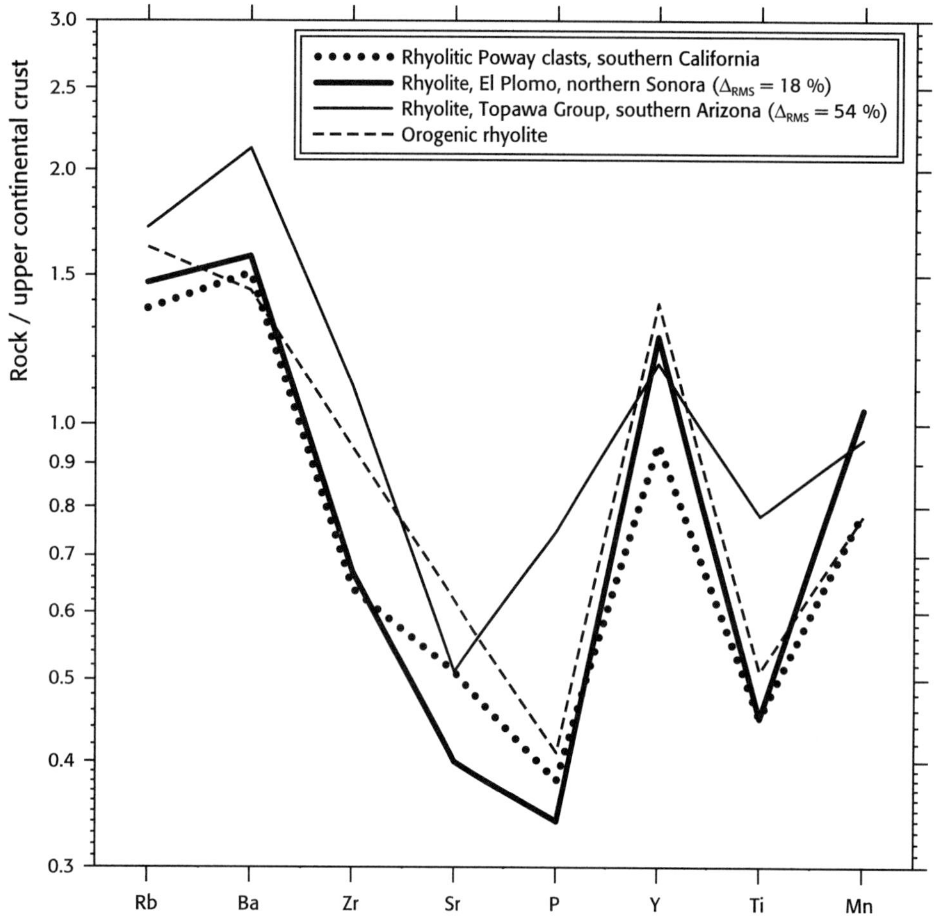

Figure 12. Multielement diagram comparing trace element geochemistry of (1) selected rhyolitic "Poway clasts" from Eocene conglomerates near San Diego, southern California (see text); (2) Jurassic rhyolite west of El Plomo, northern Sonora (about 55 km southwest of Morena Mountain; Fig. 2) identified by Abbott and Smith (1989) as a probable source for this particular subset of the Poway clasts; (3) rhyolite of the Jurassic Topawa Group, Baboquivari Mountains, southern Arizona (Table 4); and (4) orogenic rhyolites, western United States (Ewart, 1979, Table 4-A). All four compositions plotted are arithmetic means, normalized to average upper continental crust (Taylor and McLennan, 1995). Numbers in parentheses are the root-mean-square (RMS) difference (percent) from the average composition of the selected Poway clasts. Poway and El Plomo data are from Abbott and Smith (1989).

at approximately the same time, within the resolution of U-Pb TIMS and SHRIMP isotopic ages for Mesozoic rocks (typically ±2–5 Ma). In this context correlation does not imply that the units are precisely the same age, nor that their volcanic rocks necessarily had the same eruptive source. Correlation of supracrustal sequences in southern Arizona with sedimentary rocks on the Colorado Plateau implies simply that they are approximately the same age.

Regional extent of the Topawa Group

In addition to the main outcrop area of Topawa Group in the Baboquivari Mountains, similar rocks occur in three smaller, nearby areas farther west: eastern Comobabi Mountains, Las Animas Mountain, and Morena Mountain (Figs. 2 and 3; Beikman et al., 1995). Based upon close correspondence in lithology and stratigraphy to the Ali Molina and Pitoikam Formations, these strata can confidently be considered outliers of the Topawa Group. Metasedimentary rocks on the west side of the Artesa Mountains (May and Haxel, 1980) probably also are derived from the upper Ali Molina or lower Pitoikam. These correlations indicate that the Topawa Group must be considerably more extensive, in the subsurface, than its present outcrop areas. Much of the Topawa Group (including derivative metamorphic rocks) evidently is buried, beneath Late Cretaceous to early Tertiary crystalline thrust plates (Haxel et al., 1984) and middle Tertiary volcanic rocks and Neogene valley-fill deposits (Figs. 2 and 3).

Possible Correlation with Other Volcanic and Sedimentary Rocks in South-Central Arizona and North-Central Sonora

(Meta)rhyolites and (meta)sedimentary rocks broadly similar to those of the Ali Molina and Pitoikam Formations and Tinaja Spring Porphyry are widely exposed in ranges west and south of the Baboquivari-Comobabi-Artesa-Morena Mountains region (Fig. 2), north of the Mojave-Sonora megashear, in north-central Sonora and south-central Arizona (e.g., Corona, 1979; Haxel et al., 1984; Tosdal et al., 1989; Goodwin and Haxel, 1990; Beikman et al., 1995; Nourse et al., 1994; Nourse, 1995; Iriondo, 2001; Anderson et al., this volume). Metarhyolite, metaconglomerate, and quartzofeldspathic phyllite, semischist, and schist are particularly common. At least some of these rocks very likely are derived from protoliths correlative with the Topawa Group. Further study is needed to evaluate such possible correlations.

Strata similar to and correlative with the Mulberry Wash Formation have yet to be identified elsewhere in south-central Arizona and north-central Sonora. But, such rocks probably have not been specifically sought. Two characteristics of the Mulberry Wash Formation could aid in recognition of related strata: rhyolite flows and flow breccias with abundant intraformational conglomerate and sandstone, and the association of trachyte-comendite with alkali basalt.

Regional correlations of the Topawa Group face a couple of pitfalls. (1) Without detailed stratigraphic data and U-Pb isotopic ages, conglomerate or metaconglomerate correlative with the Contreras Conglomerate Member of the Pitoikam Formation might be mistaken for the Glance Conglomerate, the basal unit of the Late Jurassic to Early Cretaceous Bisbee Group (Bilodeau et al., 1987; Dickinson et al., 1989; Nourse, 2001). Likewise, metasandstone and metasiltstone correlative with the overlying Chiltepines Member of the Pitoikam Formation might be incorrectly interpreted as derived from the middle Bisbee Group. (2) Rhyolitic lava flows and flow breccias correlative with the Middle Jurassic Mulberry Wash Formation might be confused with somewhat similar rocks belonging to the Late Jurassic Artesa sequence of Tosdal et al. (1989).

Quartz Arenite–Rhyolite Volcanic Sequences in Southern Arizona

The peculiar association of silicic volcanic rocks and interbedded eolian quartz arenite that characterizes the Middle Jurassic Ali Molina Formation is repeated in four other Early to Middle Jurassic units in southern Arizona: Cobre Ridge Tuff, Mount Wrightson Formation, Ox Frame Volcanics, and Sil Nakya Formation (Fig. 2). The unusual nature of this silicic volcanic–quartz arenite association and sparse geochronologic data led to previous interpretations that all of these units are Early Jurassic and approximately coeval (Bilodeau and Keith, 1979, 1986; Haxel et al., 1980b; Busby-Spera, 1988). Subsequent U-Pb geochronologic data have complicated this simple picture.

The Cobre Ridge Tuff represents a Middle Jurassic caldera complex in which eolian sand accumulated during lulls in volcanism (Riggs and Busby-Spera, 1991). The Cobre Ridge Tuff and Topawa Group occur in similar geologic settings, on opposite sides of the Altar Valley (Fig. 2; Riggs, 1987; Riggs and Haxel, 1990). Volcanic rocks of the Cobre Ridge Tuff are predominantly quartz-, feldspar-, and biotite-phyric rhyolites, generally similar to those of the Ali Molina Formation (though no distinctive volcanic units common to both formations have been recognized). Likewise, conglomerate, sandstone, and siltstone that gradationally overlie the Cobre Ridge Tuff are much like the Pitoikam Formation. The U-Pb isotopic age of the Cobre Ridge Tuff is 170 ± 5 Ma (Riggs et al., 1993). Thus, the Topawa Group (170 ± 3 Ma) and Cobre Ridge Tuff are the same age (within the resolution of U-Pb geochronology). Their proximity, lithologic similarity, and correlation suggest they are simply two pieces of the same Middle Jurassic arc terrane. No rocks corresponding to the Mulberry Wash Formation crop out in the Cobre Ridge area. Quite possibly these were originally present, but were eroded prior to deposition of Late Jurassic conglomerate that unconformably overlies the Cobre Ridge Tuff (Knight, 1970, Riggs, 1987).

The lower member of the Mount Wrightson Formation (Drewes, 1971) consists largely of andesitic and dacitic flows, flow breccias, and subvolcanic intrusions, representing multiple small volcanic vents (Riggs and Busby-Spera, 1990). This unit contains only very minor interbedded eolian sandstone. Its U-Pb age is ≈184 ± 8 Ma (Asmerom et al., 1990; Riggs et al., 1993). The middle and upper members of the Mount Wrightson Formation are a thick (≈4 km) succession of dominantly silicic volcanic rocks, chiefly tuffs and ignimbrites derived both from volcanic centers within the multi-vent complex and more distant calderas. Layers of eolian arenite are common and thick, particularly in the upper member. U-Pb geochronology suggests (somewhat uncertain) ages ≈183 to 170 Ma for these strata. The Topawa Group may correlate with some part of the upper Mount Wrightson Formation, which is approximately 175 or 170 Ma.

Lithologically, the Ox Frame Volcanics (Cooper, 1971; Bilodeau and Keith, 1986) and Sil Nakya Formation (Haxel et al., 1978) are more like the Mount Wrightson Formation than like the Topawa Group and Cobre Ridge Tuff. In particular, the Mount Wrightson, Ox Frame, and Sil Nakya all include andesite or trachyandesite as well as rhyolite. These three units may be parts of a distinctive belt that characterizes the northeast margin of the Jurassic magmatic arc in southern Arizona (Figs. 1 and 2). No U-Pb ages are available for the Ox Frame Volcanics and Sil Nakya Formation. The Ox Frame is intruded by a 175 Ma pluton (U-Pb: Spencer et al., 2003). The Sil Nakya is no younger than Late Jurassic, as it is unconformably overlain by conglomerate of that age (Cohen et al., 1986; Bilodeau et al., 1987; Dickinson et al., 1989).

Correlation of Jurassic Arenites in Southern Arizona with Eolian Sandstones of the Colorado Plateau

Drewes (1971) first proposed that quartz arenite in early Mesozoic volcanic and sedimentary sequences in southern Arizona correlates with continental eolianites on the Colorado Plateau. This hypothesis was fully developed and applied specifically to Jurassic sequences in southern Arizona by Bilodeau and Keith (1979, 1986). Generally, thick layers of quartz arenite with high-angle crossbedding in southern Arizona correlate with eolian sandstones in the Early and Middle Jurassic Glen Canyon and San Rafael Groups on the Colorado Plateau (Riggs et al., 1993; Riggs and Blakey, 1993). The 170 Ma age of the Ali Molina Formation indicates that it and the Cobre Ridge Tuff most likely correlate with the lower part of the Middle Jurassic San Rafael Group of northern Arizona and southern Utah.

Quartzite pebbles in the Ali Molina and Pitoikam Formations

Quartzite pebbles in the upper Ali Molina Formation and correlative lower Pitoikam Formation (hereafter, Ali Molina

Formation) are important as the only nonlocal detritus larger than sand size widely preserved in the Topawa Group. The quartzite pebbles are notably more rounded than the accompanying locally derived volcanic pebbles and cobbles, yet very likely were more durable—more difficult to round. These quartzite pebbles could have been transported long distances, possibly in multiple cycles. Their roundness and homogeneity indicates they were eroded from some older, pre–Middle Jurassic conglomerate. Possible sources of the Ali Molina quartzite pebbles include: (1) conglomerate units in the Mesoproterozoic Apache Group and Troy Quartzite (Wrucke, 1989) and in Cambrian strata (Hayes and Cone, 1975; Hereford, 1975), southern and central Arizona; (2) Triassic(?) conglomerate, southeast Arizona (Riggs and Richard, 1994); and (3) conglomerate in the Triassic Chinle Formation, northern Arizona (Thordarson et al., 1972). The Triassic(?) conglomerate and the Chinle Formation are relatively far away, and their pebbles include more chert and are less strongly metamorphosed than the Ali Molina pebbles. The Mesoproterozoic and Cambrian conglomerate units crop out considerably closer, and are lithologically more plausible as sources. These Mesoproterozoic and Cambrian conglomerates are associated with a diversity of other rock types, such as coarse-grained granite, diorite, diabase, arenite or orthoquartzite, and limestone. Clasts of some of these rocks can be readily recognized in Late Jurassic conglomerates in south-central Arizona (for example in the Sheridan and Comobabi Mountains: Briskey et al., 1978; Haxel et al., 1978). However, these other rock types are unrepresented, or at least very poorly represented, among the Ali Molina pebbles. Evidently, the fluvial system(s) that transported and deposited the Ali Molina quartzite pebbles were quite efficient in fractionating them from other, less resistant, clast types.

Relation of the Topawa Group to the Mojave-Sonora Megashear

The hypothesized Middle to Late Jurassic Mojave-Sonora megashear extends from southern California to northeastern Mexico and has approximately 800 km of sinistral offset (Silver and Anderson, 1974; Anderson and Silver, 1979; Anderson et al., this volume; Farmer et al., this volume; Stewart, this volume). Though concealed over much of its length, the Mojave-Sonora megashear is well exposed in northwest Sonora as a subvertical mylonite zone (Tosdal et al., 1990; Campbell and Anderson, 2003). The megashear truncates the Jurassic magmatic arc in northern Sonora (Fig. 1). The offset continuation of the arc evidently occurs north of Zacatecas, northeast Mexico, where exposures of Jurassic volcanic and hypabyssal rocks are widespread (Fig. 11; Jones et al., 1995). Evidence for slip on the western part of the megashear comes from offset of several distinctive Proterozoic and Paleozoic terranes. Displacement on the eastern megashear is less well known, as the Jurassic arc has provided no markers of offset (other than the arc itself). The Topawa Group might be helpful in this regard.

In the absence of geometric piercing points that constrain actual slip, displacement on a major strike-slip fault may be estimated from offset of terranes that share a distinctive geologic history (e.g., Crowell, 1962). Two features of the Topawa Group might be useful in determining displacement on the Mojave-Sonora megashear: the quartz arenite–rhyolite association in the Ali Molina Formation, and trachyte-comendite and alkali basalt in the upper Mulberry Wash Formation. The quartz arenite–rhyolite association is widespread in Jurassic arc rocks in southern Arizona and northern Sonora and also in the Mojave Desert region (Figs. 1 and 2; Riggs and Blakey, 1993; Nourse, 1995; Schermer et al., 2002), and thus could provide only an imprecise estimate of fault offset. On the other hand, the distinctive alkaline igneous rocks of the Mulberry Wash Formation evidently are quite restricted in their distribution in southern Arizona and northern Sonora. If associated trachyte-comendite and alkali basalt were found in northeast Mexico, in Jurassic rocks the same age as the Mulberry Wash Formation (≈170 Ma) and in a similar stratigraphic setting, their discovery could possibly lead to a more precise indication of displacement on the eastern Mojave-Sonora megashear.

Topawa Group as a Possible Source for the Eocene "Poway clasts" of Southern California

A longstanding problem of southern California geology concerns the sources of distinctive tuffaceous rhyolite clasts (the "Poway clasts") that are widespread in Eocene conglomerates of coastal southern California and northern Baja California (Woodford et al., 1968; Kies and Abbott, 1983). Petrographic and major element characteristics of the Poway clasts preclude derivation from several prominent Mesozoic volcanic terranes in southern California. Based on comparisons of trace element abundances and U-Pb zircon apparent ages, Abbott and Smith (1978, 1989) concluded that at least some of the Poway clasts were derived from Jurassic rhyolites in northern Sonora. We have not examined the Poway clasts in any detail, but published descriptions suggest similarity to some of the rhyolites of the Topawa Group. Available, limited trace element data suggest that Topawa Group rhyolite is unlike the deliberately selected Poway clasts analyzed by Abbott and Smith (1989) (Fig. 12). However, this subset is not necessarily representative of the Poway clasts as a whole. The possibility that some of the Poway clasts are derived from the Topawa Group merits further study.

TECTONIC, PALEOGEOGRAPHIC, AND DEPOSITIONAL SETTING

Our stratigraphic and U-Pb geochronologic data indicate that the Topawa Group, though lithologically diverse and ≈8 km thick, represents only a short interval of Middle Jurassic time, probably only several million years. Thus, the Topawa Group does not record long-term evolution of the Jurassic magmatic arc but rather represents a Middle Jurassic "stratigraphic snapshot" of the arc at one place. We believe this particular view of the arc has been preserved primarily because the Topawa Group accumulated in deep intra-arc basins.

As we interpret the stratigraphic record, the Ali Molina Formation was deposited in one basin, and Pitoikam and Mulberry Wash Formations in a slightly younger and adjacent basin, one margin of which was formed by uplands of Ali Molina and Ali Molina–like rhyolitic rocks. Remnants of the Topawa Group now preserved or exposed are not extensive enough to indicate whether these two basins were independent, or were subbasins that constituted parts of some larger structural depression. For simplicity we refer simply to two basins. If this view is correct, neither basin need have been deeper than about one-half of the 8 km total stratigraphic thickness of the Topawa Group (Fig. 4).

Jurassic Arc Setting of the Topawa Group

In south-central Arizona, the northwest-southeast–trending inboard edge of the Early to Middle Jurassic magmatic arc apparently passes just northeast of the Sil Nakya Hills, southern Sierrita Mountains, and northern Santa Rita Mountains (Figs. 1 and 2; Tosdal et al., 1989; Richard et al., 2000). (Scattered Jurassic rocks farther northeast are Middle[?] to Late Jurassic.) The Topawa Group lies some 25–40 km southwest of the inboard edge of the arc, and at least two-thirds of the Topawa Group is volcanic rocks and sedimentary rocks derived from volcanic rocks (Figs. 3 and 4). Clearly, the supracrustal rocks of the Topawa Group were deposited in basins within the magmatic arc.

The substantial stratigraphic thickness of the Topawa Group and the apparently short time interval that it represents imply both deep basins, to allow for accumulation, and a high subsidence rate, to account for preservation of the strata. Evidently, the Topawa Group basins were fundamentally tectonic and extensional, rather than volcano-tectonic, in origin. Stratigraphic data presented earlier indicate that the basin within which the rhyolite and eolian and fluvial arenite of the Ali Molina Formation were deposited was not a caldera. The Pitoikam Formation is closely connected with a Middle Jurassic fault scarp active during a nonvolcanic interval. The association of within-plate alkali basalt ($\varepsilon_{Nd} = +2$) and comendite in the upper Mulberry Wash Formation further indicates an environment of tectonic extension. Some of these extensional characteristics might suggest a back-arc rather than arc environment, but the Topawa Group was deposited within the main arc. (The southern Arizona region evidently lacked Jurassic back-arc magmatism.)

Paleogeography of the Jurassic Arc during Ali Molina Time

Evidence from sandstone and conglomerate of the Ali Molina Formation supports two previous paleogeographic inferences: the Middle Jurassic magmatic arc in southern Arizona was relatively low standing (Busby-Spera, 1988; Busby-Spera et al., 1990), and externally derived sediment was introduced into the arc from the continent (northeast) side, without appreciable travel along the length of the arc (Riggs and Busby-Spera, 1991).

Because the basin in which the Ali Molina Formation, including its eolian sand layers, was deposited lay well within the magmatic arc, the arc must have been low or at least had low gaps through which wind-blown sand could travel. Eolian sand can be carried high into mountain ranges or onto volcanic plateaus, but preservation and lithification of such deposits, particularly as broad sandstone sheets, is quite unlikely. Much of the Ali Molina sand appears virtually uncontaminated by volcanic detritus, arguing against the longer residence time required if the sand had traveled any significant distance along the arc. This conclusion is consistent with the general southwesterly trend of paleowind directions in the Ali Molina Formation and Cobre Ridge Tuff (Bilodeau and Keith, 1986).

Highly durable quartzite pebbles in the upper Ali Molina Formation evidently were derived from Mesoproterozoic or Cambrian sources to the north or northeast. Appreciable travel of these pebbles along the arc is unlikely, as the pebbles are far too large to have been moved by wind, and longitudinal fluvial transport would have been impeded both by low volcanic edifices and by local depressions. Rather, the pebbles must have been fluvially transported into the arc from the side. Therefore, the northeast flank of the magmatic arc, or at least basins within the arc, must have been lower than some adjacent areas farther north. Furthermore, clasts of other characteristic but less resistant rock types from the source area(s) were eliminated during transport. Such efficient removal requires thorough weathering, which implies slow transport, which in turn implies low gradients, and therefore low relief on the northeast flank of the arc.

Deposition of the Pitoikam Formation, in a Deep Tectonic Basin

The most remarkable aspect of the Topawa Group is the Pitoikam Formation—a section more than 2 km thick consisting entirely of noneolian clastic sedimentary strata (Fig. 4), with virtually no intercalated volcanic rocks. In other Jurassic magmatic arc supracrustal sequences in the southern Cordillera, intra-volcanic sedimentary units either are relatively thin or, if thicker, consist largely of eolian sandstone (Busby-Spera et al., 1990; Riggs and Blakey, 1993). No other thick sedimentary section analogous to the Pitoikam Formation has been documented.

We infer that the Pitoikam Formation records formation and filling of a progressively deepening tectonic basin, bounded on the east and northeast by the Middle Jurassic Gu Chuapo fault (Fig. 13). (The western margin of the basin is not exposed.) Conglomerate, sedimentary breccia, and subordinate sandstone of the lower Pitoikam Formation are alluvial fan facies deposited along and below the fault scarp. The finer-grained Saucito Member of the Pitoikam Formation was deposited during initiation of the fault scarp, and the coarser-grained Contreras Conglomerate Member records its full development. The footwall-block source of these coarse clastic deposits was made of volcanic and hypabyssal rocks of the slightly older rhyolite member of the Ali Molina Formation, and probably other, similar terranes not now preserved or exposed. The overlying Chiltepines Member of the Pitoikam Formation, largely sandstone and siltstone, represents fluvial facies deposited nearer the axis of the basin (Leeder, 1995). The thickness and relatively fine grained character of the Chiltepines Member suggest a substantial through-flowing axial drainage. This fluvial system may have been a continuation of, or evolved from, the one that introduced externally derived quartzite pebbles slightly earlier, in late Ali Molina time. Deposition of the Chiltepines fine-grained sediments was abruptly terminated by the resumption of volcanism in earliest Mulberry Wash time (heralded by minor precursors in latest Chiltepines time). However, the abundance within the Mulberry Wash Formation of intraformational conglomerate, sedimentary breccia, and sandstone suggests that this fluvial system persisted well into Mulberry Wash time (Fig. 4), flowing through the Mulberry Wash volcanic field.

We speculate that because the Topawa Group intra-arc basins were rapidly subsiding, they became the locus of a major fluvial system, which flowed into the low-standing magmatic arc from its northeast flank. The fluvial system was probably intermittent, as the Middle Jurassic climate in this region was arid (Parrish, 1993). This hypothesized drainage system accounts for at least three distinctive features of the Topawa Group: the quartzite pebbles in the upper Ali Molina Formation, the considerable volume of fine-grained sediment in the Chiltepines Member of the Pitoikam Formation, and the extensive intraformational sedimentary deposits of the Mulberry Wash Formation.

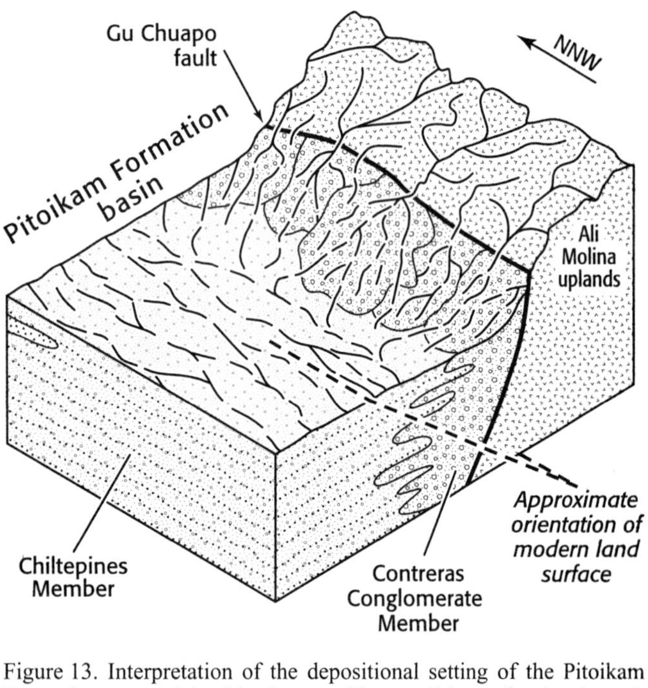

Figure 13. Interpretation of the depositional setting of the Pitoikam Formation, as explained in the text. The small lens of volcanic rock in the northwest corner of the diagram represents a precursor of Mulberry Wash volcanism; see text and Figure 4. The plane labeled as the approximate orientation of the modern land surface dips east ≈55°. When this plane is rotated about its strike to horizontal, the strata of the Pitoikam Formation rotate to their present moderately west-dipping orientation in the upper Fresnal Canyon area (Fig. 3), and the Gu Chuapo fault becomes subvertical, as it is today. This diagram chiefly after Leeder (1995).

SUMMARY: GEOLOGIC HISTORY

The geologic history inferred from the Topawa Group and surrounding rocks can be summarized as a series of events or episodes:

(1) In the Baboquivari Mountains region, arc magmatism commenced in middle Middle Jurassic time, ≈170 Ma. Rhyolitic pyroclastic rocks of the Ali Molina Formation accumulated in a large basin within the arc. During several intervals between eruptions, eolian sand blew into the basin from the northeast.

(2) As volcanism waned, in late Ali Molina time, streams deposited cobbles and pebbles of porphyritic rocks eroded from nearby, contemporaneous or slightly older volcanic and hypabyssal terranes, and durable quartzite pebbles derived from external sources north or northeast of the arc.

(3) At about the same time, another extensional tectonic basin or subbasin began to form, bounded on the east by the Gu Chuapo fault. As faulting progressed, alluvial fans shed from the fault scarp deposited the Saucito and Contreras Conglomerate Members of the Pitoikam Formation.

(4) Basinward from the alluvial fans, a major fluvial system deposited sand, slit, and gravel—the Chiltepines Member. This regime persisted through late Pitoikam time.

(5) Volcanism abruptly resumed, with eruption of rhyolitic flows and flow breccias of the Mulberry Wash Formation. Many of these volcanic rocks were reworked as mass-flow deposits; and by streams, probably a continuation of the Chiltepines fluvial system. In late Mulberry Wash time, as crustal extension continued, flows of alkali basalt and trachyte-comendite erupted sporadically.

Episodes (1) through (5) evidently spanned only several million years of middle Middle Jurassic time.

(6) The Topawa Group was intruded in late Middle Jurassic time (165–159 Ma) by the Kitt Peak plutonic suite, and in late Late Jurassic time (146 Ma) by the Baboquivari Peak Perthite Granite.

(7) Subsequently, the Topawa Group was affected by two regional orogenic episodes: Late Cretaceous to early Tertiary compressional deformation and granitic plutonism (Haxel et al., 1984), and middle Tertiary crustal extension (Spencer and Reynolds, 1989; Goodwin and Haxel, 1990; Tosdal et. al., 1990). During the first episode, some rocks of the Topawa Group were incipiently to strongly metamorphosed. The present moderately west-dipping, west-facing attitude of the Topawa Group (Figs. 3 and 13) represents the combined effects of Middle Jurassic plutonism, Late Cretaceous to early Tertiary deformation, and middle Tertiary tilting.

ACKNOWLEDGMENTS

Our understanding of the Jurassic of southern Arizona has been furthered by some 25 years of field trips, mapping, discussions, and friendly arguments with Tom Anderson. A field trip into the Baboquivari Mountains with Cathy Busby in 1988 helped solidify our interpretation of the Pitoikam Formation. More recently, David Johnson provided insight into alteration of the Topawa Group. Along the way, the ideas and data of a number of other geologists also contributed to this study: Sue Beard, Bill Bilodeau, Peter Coney, Bill Dickinson, Laurel Goodwin, Richard Hereford, Stan Keith, Karen Klugger-Cohen, Juan Manuel Morales, Lynda Nicholson, Wes Peirce, Steve Reynolds, Steve Richard, Jim Rytuba, Sheila Seaman, Lee Silver, Jack Stewart, and Ted Theodore. Terri Lucero, George Tilton, and Joe Wooden facilitated our U-Pb isotopic analyses. Thoughtful reviews by Sue Beard, Bill Bilodeau, David Johnson, and Jon Nourse guided us in refining this paper.

REFERENCES CITED

Abbott, P.L., and Smith, T.E., 1978, Trace-element comparison of clasts in Eocene conglomerates, southwestern California and northwestern Mexico: Journal of Geology, v. 86, p. 753–762.

Abbott, P.L., and Smith, T.E., 1989, Sonora, Mexico, source for the Eocene Poway conglomerate of southern California: Geology, v. 17, p. 329–332, doi: 10.1130/0091-7613(1989)017<0329:SMSFTE>2.3.CO;2.

Anderson, T.H., and Silver, L.T., 1977, U-Pb isotope ages of granitic plutons near Cananea, Sonora: Economic Geology and the Bulletin of the Society of Economic Geologists, v. 72, p. 827–836.

Anderson, T.H., and Silver, L.T., 1978, Jurassic magmatism in Sonora, Mexico: Geological Society of America Abstracts with Programs, v. 10, p. 359.

Anderson, T.H., and Silver, L.T., 1979, The role of the Mojave-Sonora megashear in the tectonic evolution of northern Sonora, *in* Anderson, T.H., and Roldán-Quintana, J., eds., Geology of the northern Sonora, Annual Meeting of the Geological Society of America Field Trip Guidebook, Field Trip no. 27: Hermosillo, Sonora, Instituto de Geologia, Universidad National Autonoma de Mexico, and Pittsburgh, Pennsylvania, University of Pittsburgh, p. 59–68.

Asmerom, Y., Zartman, R.E., Damon, P.E., and Shafiqullah, M., 1990, Zircon U-Th-Pb and whole-rock Rb-Sr age patterns of lower Mesozoic igneous rocks in the Santa Rita Mountains, southeast Arizona: Implications for Mesozoic magmatism and tectonics in the southern Cordillera: Geological Society of America Bulletin, v. 102, p. 961–968, doi: 10.1130/0016-7606(1990)102<0961:ZUTPAW>2.3.CO;2.

Asmerom, Y., Patchett, P.J., and Damon, P.E., 1991, Crust-mantle interaction in continental arcs: Inferences from the Mesozoic arc in the southwestern United States: Contributions to Mineralogy and Petrology, v. 107, p. 124–134.

Barton, M.D., 1996, Granite magmatism and metallogeny of southwestern North America, *in* Brown, M., Candela, P.A., Peck, D.L., Stephens, W.E., Walker, R.J., and Zen, E., eds., Third Hutton Symposium On The Origin Of Granites And Related Rocks: Geological Society of America Special Paper 315, p. 261–280.

Barton, M.D., and Johnson, D.A., 1996, Evaporitic-source model for igneous-related Fe oxide-(REE-Cu-Au-U) mineralization: Geology, v. 24, p. 259–262, doi: 10.1130/0091-7613(1996)024<0259:ESMFIR>2.3.CO;2.

Battles, D.A., and Barton, M.D., 1995, Arc-related sodic hydrothermal alteration in the western United States: Geology, v. 23, p. 913–916, doi: 10.1130/0091-7613(1995)023<0913:ARSHAI>2.3.CO;2.

Beikman, H.M., Haxel, G.B., and Miller, R.J., 1995, Geologic map of the Tohono O'Odham Indian Nation, southern Arizona: U.S. Geological Survey Miscellaneous Geologic Investigations Map I-2017, scale 1:125,000.

Bergman, S.C., 1987, Lamproites and other potassium-rich igneous rocks: A review of their occurrence, mineralogy and geochemistry, *in* Fitton, J.G., and Upton, B.G.J., eds., Alkaline igneous rocks: Geological Society [London] Special Publication 30, p. 103–190.

Bevier, M.L., 1981, The Rainbow Range, British Columbia: A Miocene peralkaline shield volcano: Journal of Volcanology and Geothermal Research, v. 11, p. 225–251, doi: 10.1016/0377-0273(81)90025-1.

Bilodeau, W.L., and Keith, S.B., 1979, Intercalated volcanics and eolian "Aztec-Navajo-like" sandstones in southeast Arizona; another clue to the Jurassic-Triassic paleo-tectonic puzzle of the southwestern U.S: Geological Society of America Abstracts with Programs, v. 11, p. 70.

Bilodeau, W.L., and Keith, S.B., 1986, Lower Jurassic Navajo-Aztec-equivalent sandstones in southern Arizona and their paleogeographic significance: American Association of Petroleum Geologists Bulletin, v. 70, p. 690–704.

Bilodeau, W.L., Kluth, C.F., and Vedder, L.K., 1987, Regional stratigraphic, sedimentologic, and tectonic relationships of the Glance Conglomerate in southeastern Arizona, *in* Dickinson, W.R., and Klute, M.A., eds., Mesozoic rocks of southern Arizona and adjacent areas: Arizona Geological Society Digest, v. 18, p. 229–256.

Briskey, J.A., Haxel, G.B., Peterson, J.A., and Theodore, T.G., 1978, Reconnaissance geologic map of the Gu Achi 15' quadrangle, Arizona: U.S. Geological Survey Miscellaneous Field Studies Map MF-965, scale 1:62,500.

Busby-Spera, C.J., 1988, Speculative tectonic model for the early Mesozoic arc of the southwest Cordilleran United States: Geology, v. 16, p. 1121–1125, doi: 10.1130/0091-7613(1988)016<1121:STMFTE>2.3.CO;2.

Busby-Spera, C.J., Mattinson, J.M., Riggs, N.R., and Schermer, E.R., 1990, The Triassic-Jurassic magmatic arc in the Mojave-Sonoran Deserts and the Sierran-Klamath region; similarities and differences in paleographic evolution, *in* Harwood, D.S., and Miller, M.M., eds., Paleozoic and early Mesozoic paleogeographic relations; Sierra Nevada, Klamath Mountains, and related terranes: Geological Society of America Special Paper 255, p. 325–337.

Campbell, P.A., and Anderson, T.H., 2003, Mylonite along the Mojave-Sonora megashear: Tectonics, v. 22, no. 6, p. 1077, doi: 10.1029/2002TC001367.

Church, B.N., 1975, Quantitative classification and chemical comparison of common volcanic rocks: Geological Society of America Bulletin, v. 86, p. 257–263, doi: 10.1130/0016-7606(1975)86<257:QCACCO>2.0.CO;2.

Cohen, K.K., Anderson, T.H., and Schmidt, V.A., 1986, A paleomagnetic test of the proposed Mojave-Sonora megashear in northwestern Mexico: Tectonophysics, v. 131, p. 23–51, doi: 10.1016/0040-1951(86)90266-0.

Coney, P.J., 1978, The plate tectonic setting of southeastern Arizona, *in* Callender, J.F., Wilt, J.C., Clemons, R.E., and James, H.L., eds., Land of Cochise; southeastern Arizona: New Mexico Geological Society Guidebook, v. 29, p. 285–290.

Cooley, M.E., 1967, Arizona highway geologic map: Cooley, Arizona, Arizona Geological Society, scale 1:1,000,000.

Cooper, J.R., 1971, Mesozoic stratigraphy of the Sierrita Mountains, Pima County, Arizona: U.S. Geological Survey Professional Paper 658-D, 42 p.

Corona, F.V., 1979, Reconnaissance geology of the Sierra La Gloria and Cerro Basura, northwestern Sonora, Mexico, *in* Anderson, T.H., and Roldán-Quintana, J., eds., Geology of the northern Sonora, Annual Meeting of the Geological Society of America Field Trip Guidebook, Field Trip no. 27: Hermosillo, Sonora, Instituto de Geologia, Universidad National Autonoma de Mexico, and Pittsburgh, Pennsylvania, University of Pittsburgh, p. 32–48.

Cox, K.G., Bell, J.D., and Pankhurst, R.J., 1979, The interpretation of igneous rocks: London, Allen and Unwin, 450 p.

Crowell, J.C., 1962, Displacement along the San Andreas fault, California: Geological Society of America Special Paper 71, 61 p.

Dickinson, W.R., Fiorillo, A.R., Hall, D.L., Monreal, R., Potochnik, A.R., and Swift, P.N., 1989, Cretaceous strata of southern Arizona, *in* Jenny, J.P., and Reynolds, S.J., eds., Geologic evolution of Arizona: Arizona Geological Society Digest, v. 17, p. 447–461.

Drewes, H., 1971, Mesozoic stratigraphy of the Santa Rita Mountains, southeast of Tucson, Arizona: U.S. Geological Survey Professional Paper 658-C, 81 p.

Erlank, A.J., Smith, H.S., Marchant, J.W., Cardoso, M.P., and Ahrens, L.H., 1978, Zirconium: Abundance in common igneous rock types; crustal abundance, *in* Wedepohl, K.H., ed., Handbook of geochemistry, v. II/4, Elements Kr (36) to Ba (56): Berlin, Springer-Verlag, p. 40-E-1–40-E-27.

Ewart, A., 1979, A review of the mineralogy and chemistry of Tertiary–Recent dacitic, rhyolitic, and related salic volcanic rocks, *in* Barker, F., ed., Trondhjemites, dacites and related rocks: Amsterdam, Elsevier, p. 13–121.

Ewart, A., 1982, The mineralogy and petrology of Tertiary–Recent orogenic volcanic rocks: With special reference to the andesitic-basaltic compositional range, *in* Thorpe, R.S., ed., Andesites; orogenic andesites and related rocks: New York, Wiley, p. 25–95.

Fair, C.L., 1961, Probable Cretaceous-Tertiary section in Fresnal Canyon, Baboquivari Mountains: Arizona Geological Society Digest, v. 4, p. 93–94.

Fair, C.L., 1965, Geology of the Fresnal Canyon area, Baboquivari Mountains, Pima County, Arizona [Ph.D. dissertation]: Tucson, University of Arizona, 89 p., geologic map scale 1:12,000.

Farmer, G.L., and DePaolo, D.J., 1984, Origin of Mesozoic and Tertiary granite in the western United States and implications of pre-Mesozoic crustal structure. 2. Nd and Sr isotopic studies of unmineralized and Cu- and Mo-mineralized granite in the Precambrian craton: Journal of Geophysical Research, v. 89, no. B12, p. 10,141–10,160.

Floyd, P.A., and Winchester, J.A., 1978, Identification and discrimination of altered and metamorphosed volcanic rocks using immobile elements: Chemical Geology, v. 21, p. 291–306, doi: 10.1016/0009-2541(78)90050-5.

Fox, L.K., and Miller, D.M., 1990, Jurassic granitoids and related rocks of the southern Bristol Mountains, southern Providence Mountains, and Colton Hills, Mojave Desert, California, *in* Anderson, J.L., ed., The nature and origin of Cordilleran magmatism: Geological Society of America Memoir 174 p. 111–132.

Golightly, D.W., Dorrzapf, A.F., Mays, R.E., Fries, T.L., and Conklin, N.M., 1987, Analysis of geological materials by direct-current arc emission spectrography and spectrometry, *in* Baedecker, P.A., ed., Methods for geochemical analysis: U.S. Geological Survey Bulletin 1770, p. A1–A13.

Goodwin, L.B., and Haxel, G.B., 1990, Structural evolution of the southern Baboquivari Mountains, south-central Arizona and north-central Sonora: Tectonics, v. 9, p. 1077–1095.

Gradstein, F.M., and Ogg, J., 1996, A Phanerozoic time scale: Episodes, v. 19, p. 3–5.

Grubensky, M.J., Haxel, G.B., and Koch, R.D., 1993, Geologic map of the Castle Dome Mountains, southwestern Arizona: U.S. Geological Survey Miscellaneous Investigations Series Map I-2138, scale 1:62,500.

Haxel, G.B., Briskey, J.A., Rytuba, J.J., Bergquist, J.R., Blacet, P.M., and Miller, S.T., 1978, Reconnaissance geologic map of the Comobabi Mountains 15' quadrangle, Arizona: U.S. Geological Survey Miscellaneous Field Studies Map MF-964, scale 1:62,500.

Haxel, G.B., May, D.J., Wright, J.E., and Tosdal, R.M., 1980a, Reconnaissance geologic map of the Baboquivari Peak 15' quadrangle, Arizona: U.S. Geological Survey Miscellaneous Field Studies Map MF-1251, scale 1:62,500.

Haxel, G.B., Wright, J.E., May, D.J., and Tosdal, R.M., 1980b, Reconnaissance geology of the Mesozoic and lower Cenozoic rocks of the southern Papago Indian Reservation, Arizona: A preliminary report, *in* Jenny, J.P., and Stone, C., eds., Studies in western Arizona: Arizona Geological Society Digest, v. 12, p. 17–29.

Haxel, G.B., May, D.J., and Tosdal, R.M., 1982, Reconnaissance geologic map of the Presumido Peak 15' quadrangle, Arizona: U.S. Geological Survey Miscellaneous Field Studies Map MF-1378, scale 1:62,500.

Haxel, G.B., Tosdal, R.M., May, D.J., and Wright, J.E., 1984, Late Cretaceous and early Tertiary orogenesis in south-central Arizona: Thrust faulting, regional metamorphism, and granitic plutonism: Geological Society of America Bulletin, v. 95, p. 631–653, doi: 10.1130/0016-7606(1984)95<631:LCAETO>2.0.CO;2.

Haxel, G.B., Tosdal, R.M., and Dillon, J.T., 1985, Tectonic setting and lithology of the Winterhaven Formation, a new Mesozoic stratigraphic unit in southeasternmost California and southwestern Arizona: U.S. Geological Survey Bulletin 1599, 19 p.

Haxel, G.B., Jacobson, C.E., Richard, S.M., Tosdal, R.M., and Grubensky, M.J., 2002, The Orocopia Schist in southwest Arizona: Early Tertiary oceanic rocks trapped or transported far inland, *in* Barth, A., ed., Contributions to the crustal evolution of the southwestern United States: Geological Society of America Special Paper 365, p. 99–128.

Hayes, P.T., and Cone, G.C., 1975, Cambrian and Ordovician rocks of southern Arizona and New Mexico and westernmost Texas: U.S. Geological Survey Professional Paper 873, 98 p.

Heindl, L.A., and Fair, C.L., 1965, Mesozoic (?) rocks in the Baboquivari Mountains, Papago Indian Reservation, southern Arizona: U.S. Geological Survey Bulletin 1194-I, 12 p.

Hereford, R., 1975, Chino Valley Formation (Cambrian?) in northwestern Arizona: Geological Society of America Bulletin, v. 86, p. 677–682, doi: 10.1130/0016-7606(1975)86<677:CVFCIN>2.0.CO;2.

Iriondo, A., 2001, Proterozoic basements and their Laramide juxtaposition in northwest Sonora, Mexico: Tectonic constraints on the southwest margin of Laurentia [Ph.D. dissertation]: Boulder, University of Colorado, 222 p., 3 plates, geologic map scale 1:30,000, PDF documents.

Jackson, L.L., Brown, F.W., and Neil, S.T., 1987, Major and minor elements requiring individual determination, classical whole rock analysis, and rapid rock analysis, *in* Baedecker, P.A., ed., Methods for geochemical analysis: U.S. Geological Survey Bulletin 1770, p. G1–G23.

Jennings, C.R., 1977, Geologic map of California: Sacramento, California, California Division of Mines and Geology, scale 1:750,000.

Johnson, D.A., 2000, Comparative studies of iron-oxide mineralization, Great Basin [Ph.D. dissertation]: Tucson, University of Arizona, 450 p.

Johnson, R.G., and King, B.-S.L., 1987, Energy-dispersive X-ray fluorescence spectrometry, *in* Baedecker, P.A., ed., Methods for geochemical analysis: U.S. Geological Survey Bulletin 1770, p. F1–F5.

Jones, N.W., McKee, J.W., Anderson, T.H., and Silver, L.T., 1995, Jurassic volcanic rocks in northeastern Mexico: A possible remnant of a Cordilleran magmatic arc, *in* Jacques-Ayala, C., González-León, C.M., and Roldán-Quintana, J., eds., Studies on the Mesozoic of Sonora and adjacent areas: Geological Society of America Special Paper 301, p. 179–190.

Kies, R.P., and Abbott, P.L., 1983, Rhyolite clast populations and tectonics in the California continental borderland: Journal of Sedimentary Petrology, v. 53, p. 461–475.

Knight, L.H., 1970, Structure and mineralization of the Oro Blanco mining district, Santa Cruz County, Arizona [Ph.D. dissertation]: Tucson, University of Arizona, 172 p.

Leeder, M.R., 1995, Continental rifts and proto-oceanic rift troughs, *in* Busby, C.J., and Ingersoll, R.V., eds., Tectonics of sedimentary rifts: Oxford, Blackwell, p. 119–148.

Le Maitre, R.W., 1976, The chemical variability of some common igneous rocks: Journal of Petrology, v. 17, p. 589–637.

Le Maitre, R.W., 2002, editor, Igneous rocks: A classification and glossary of terms. Recommendations of the International Union of Geological Sciences Subcommission on the Systematics of Igneous Rocks: Cambridge University Press, 236 p.

Linthout, K., 1984, Alkali-zirconosilicates in peralkaline rocks: Contributions to Mineralogy and Petrology, v. 86, p. 155–158, doi: 10.1007/BF00381842.

Lowder, G.G., and Carmichael, I.S.E., 1970, The volcanoes and caldera of Talasea, New Britain: Geology and petrology: Geological Society of America Bulletin, v. 82, p. 3489–3502.

Macdonald, R., 1974, Nomenclature and petrochemistry of the peralkaline oversaturated extrusive rocks, *in* Bailey, D.K., Barberi, F., and Macdonald, R., eds., Oversaturated peralkaline volcanic rocks: Bulletin Volcanologique, v. 38, p. 498–516.

May, D.J., and Haxel, G.B., 1980, Reconnaissance bedrock geologic map of the Sells 15' quadrangle, Pima County, Arizona: U.S. Geological Survey Miscellaneous Field Studies Map MF-1166, scale 1:62,500.

Merriam, R., 1972, Reconnaissance geologic map of the Sonoita quadrangle, northwestern Sonora, Mexico: Geological Society of America Bulletin, v. 83, p. 3533–3536.

Middlemost, E.A.K., 1981, The Canobolas complex, New South Wales, an alkaline shield volcano: Journal of the Geological Society of Australia, v. 28, p. 33–49.

Ngounouno, I., Déruelle, B., and Demaiffe, D., 2000, Petrology of the bimodal Cenozoic volcanism of the Kapsiki plateau (northernmost Cameroon, central Africa): Journal of Volcanology and Geothermal Research, v. 102, p. 21–44, doi: 10.1016/S0377-0273(00)00180-3.

Noble, D.C., 1968, Systematic variation of major elements in comendite and pantellerite glass: Earth and Planetary Science Letters, v. 4, p. 167–172.

Noble, D.C., and Parker, D.F., 1975, Peralkaline silicic volcanic rocks of the western United States, *in* Bailey, D.K., Barberi, F., and Macdonald, R., eds., Oversaturated peralkaline volcanic rocks: Bulletin Volcanologique, v. 38, p. 803–827.

Noble, D.C., Silberman, M.L., Megard, F., and Bowman, H.R., 1978, Comendite (peralkaline rhyolite) and basalt in the Mitu Group, Peru: Evidence for Permian-Triassic lithospheric extension in the central Andes: U.S. Geological Survey Journal of Research, v. 6, p. 453–457.

Nourse, J.A., 1995, Jurassic-Cretaceous paleogeography of the Magdalena region, northern Sonora, and its influence on the positioning of Tertiary metamorphic core complexes, *in* Jacques-Ayala, C., González-León, C.M., and Roldán-Quintana, J., eds., Studies on the Mesozoic of Sonora and adjacent areas: Geological Society of America Special Paper 301, p. 59–78.

Nourse, J.A., 2001, Tectonic insights from an Upper Jurassic–Lower Cretaceous stretched-clast conglomerate, Caborca-Altar region, Sonora, Mexico: Journal of South American Earth Sciences, v. 14, p. 453–474, doi: 10.1016/S0895-9811(01)00051-7.

Nourse, J.A., Anderson, T.H., and Silver, L.T., 1994, Tertiary metamorphic core complexes in Sonora, northwestern Mexico: Tectonics, v. 13, p. 1161–1182, doi: 10.1029/93TC03324.

Parker, D.F., 1983, Origin of the trachyte-quartz trachyte-peralkalic rhyolite suite of the Oligocene Paisano volcano, Trans-Pecos Texas: Geological Society of America Bulletin, v. 94, p. 614–629, doi: 10.1130/0016-7606(1983)94<614:OOTTTR>2.0.CO;2.

Parrish, J.T., 1993, Mesozoic climates of the Colorado Plateau, *in* Morales, M., ed., Aspects of Mesozoic geology and paleontology of the Colorado Plateau: Museum of Northern Arizona Bulletin 59, p. 1–11.

Pearce, J.A., 1982, Trace element characteristics of lavas from destructive plate boundaries, *in* Thorpe, R.S., ed., Andesites; orogenic andesites and related rocks: New York, Wiley, p. 525–548.

Reynolds, S.J., Spencer, J.E., and DeWitt, E., 1987, Stratigraphy and U-Th-Pb geochronology of Triassic and Jurassic rocks in west-central Arizona, *in* Dickinson, W.R., and Klute, M.A., eds., Mesozoic rocks of southern Arizona and adjacent areas: Arizona Geological Society Digest, v. 18, p. 65–80.

Richard, S.M., Reynolds, S.J., and Spencer, J.E., 1987, Mesozoic stratigraphy of the Little Harquahala and Harquahala Mountains, west-central Arizona, *in* Dickinson, W.R., and Klute, M.A., eds., Mesozoic rocks of southern Arizona and adjacent areas: Arizona Geological Society Digest, v. 18, p. 101–119.

Richard, S.M., Reynolds, S.J., Spencer, J.E., and Pearthree, P.A., 2000, compilers, Geologic map of Arizona: Arizona Geological Survey Map 35, scale 1:1,000,000.

Riggs, N.R., 1987, Stratigraphy, structure, and geochemistry of Mesozoic rocks in the Pajarito Mountains, Santa Cruz County, Arizona, *in* Dickinson, W.R., and Klute, M.A., eds., Mesozoic rocks of southern Arizona and adjacent areas: Arizona Geological Society Digest, v. 18, p. 165–175.

Riggs, N.R., and Blakey, R.C., 1993, Early and Middle Jurassic paleogeography and volcanology of Arizona and adjacent areas, *in* Dunn, G., and McDougall, K.A., eds., Mesozoic paleogeography of the western United States-II: Pacific Section, Society of Economic Paleontologists and Mineralogists, Book 71, p. 347–375.

Riggs, N.R., and Busby-Spera, C.J., 1990, Evolution of a multi-vent volcanic complex within a subsiding arc graben depression: Mount Wrightson Formation, Arizona: Geological Society of America Bulletin, v. 102, p. 1114–1135, doi: 10.1130/0016-7606(1990)102<1114:EOAMVV>2.3.CO;2.

Riggs, N.R., and Busby-Spera, C.J., 1991, Facies analysis of an ancient, dismembered, large caldera complex and implications for intra-arc subsidence: Middle Jurassic strata of Cobre Ridge, southern Arizona, USA: Sedimentary Geology, v. 74, p. 39–68, doi: 10.1016/0037-0738(91)90034-B.

Riggs, N.R., and Haxel, G.B., 1990, The Early to Middle Jurassic magmatic arc in southern Arizona: Plutons to sand dunes, *in* Gehrels, G.E., and Spencer, J.E., eds., Geologic excursions through the Sonoran Desert region, Arizona and Sonora: Arizona Geological Survey Special Paper 7, p. 90–103.

Riggs, N.R., and Richard, S.M., 1994, Timing of post-Permian uplift in southern and western Arizona: Geological Society of America Abstracts with Programs, v. 26, no. 2, p. 84.

Riggs, N.R., Mattinson, J.M., and Busby, C.J., 1993, Correlation of Jurassic eolian strata between the magmatic arc and the Colorado Plateau: New U-Pb geochronologic data from southern Arizona: Geological Society of America Bulletin, v. 105, p. 1231–1246, doi: 10.1130/0016-7606(1993)105<1231:COJESB>2.3.CO;2.

Rytuba, J.J., and McKee, E.H., 1984, Peralkaline ash flow tuffs and calderas of the McDermitt Volcanic Field, southeast Oregon and north-central Nevada: Journal of Geophysical Research, v. 89, no. B10, p. 8616–8628.

Saleeby, J.B., Busby-Spera, C., Oldow, J.S., Dunne, G.C., Wright, J.E., Cowan, D.S., Walker, N.W., and Allmendinger, R.W., 1992, Early Mesozoic tectonic evolution of the western U.S. Cordillera, *in* Burchfiel, B.C., Lipman, P.W., and Zoback, M.L., eds., The Cordilleran orogen: Conterminous U.S.: Boulder, Colorado, Geological Society of America, Decade of North American Geology, v. G-3, p. 107–168.

Saunders, A.D., Tarney, J., and Weaver, S.D., 1980, Transverse geochemical variations across the Antarctic Peninsula: Implications for the genesis of calc-alkaline magmas: Earth and Planetary Science Letters, v. 46, p. 344–360, doi: 10.1016/0012-821X(80)90050-3.

Schermer, E.R., Busby, C.J., and Mattinson, J.M., 2002, Paleogeographic and tectonic implications of Jurassic sedimentary and volcanic sequences in the central Mojave block, *in* Glazner, A.F., Walker, J.D., and Bartley, J.M., eds., Geologic evolution of the Mojave Desert and southwestern Basin and Range: Geological Society of America Memoir 195, p. 93–115.

Seaman, S.J., 1983, Geology and ore potential of the Jupiter Canyon region, Baboquivari Mountains, Arizona [M.S. thesis]: Tucson, University of Arizona, 99 p., geologic map scale 1:24,000.

Silver, L.T., and Anderson, T.A., 1974, Possible left-lateral early to middle Mesozoic disruption of the southwestern North America craton margin: Geological Society of America Abstracts with Programs, v. 6, p. 955–956.

Spencer, J.E., and Reynolds, S.J., 1989, Middle Tertiary tectonics, *in* Jenny, J.P., and Reynolds, S.J., eds., Geologic evolution of Arizona: Arizona Geological Society Digest, v. 17, p. 539–574.

Spencer, J.E., Isachsen, C.E., Ferguson, C.A., Richard, S.M., Skotnicki, S.J., Wooden, J., and Riggs, N.R., 2003, U-Pb isotope geochronologic data from 23 igneous rock units in central and southeastern Arizona: Arizona Geological Survey Open-File Report 03-08, 40 p.

Steiger, R.H., and Jäger, E., 1977, Subcommission on geochronology: Conventions on the use of decay constants in geo- and cosmochronology: Earth and Planetary Science Letters, v. 36, p. 359–371, doi: 10.1016/0012-821X(77)90060-7.

Sutherland, D.S., 1974, Petrography and mineralogy of the peralkaline silicic rocks, *in* Bailey, D.K., Barberi, F., and Macdonald, R., eds., Oversaturated peralkaline volcanic rocks: Bulletin Volcanologique, v. 38, p. 517–547.

Taggart, J.E., Lindsay, J.R., Scott, B.A., Vivit, D.V., Bartel, A.J., and Stewart, K.C., 1987, Analysis of geologic materials by wavelength dispersive x-ray fluorescence spectrometry, *in* Baedecker, P.A., ed., Methods for geochemical analysis: U.S. Geological Survey Bulletin 1770, p. E1–E19.

Taylor, S.R., and McLennan, S.M., 1995, The geochemical evolution of the continental crust: Reviews of Geophysics, v. 33, p. 241–265, doi: 10.1029/95RG00262.

Thordarson, W., Albee, H.F., and Stewart, J.H., 1972, Conglomerate studies, *in* Stewart, J.H., Poole, F.G., Wilson, R.F., eds., Stratigraphy and origin of the Chinle Formation and related Upper Triassic strata in the Colorado Plateau region: U.S. Geological Survey Professional Paper 690, p. 62–75.

Tosdal, R.M., Haxel, G.B., and Wright, J.E., 1989, Jurassic geology of the Sonoran Desert region, southern Arizona, southeastern California, and northernmost Sonora: Construction of a continental-margin magmatic arc, *in* Jenny, J.P., and Reynolds, S.J., eds., Geologic evolution of Arizona: Arizona Geological Society Digest, v. 17, p. 397–434.

Tosdal, R.M., Haxel, G.B., Anderson, T.H., Connors, C.D., May, D.J., and Wright, J.E., 1990, Highlights of Jurassic, Late Cretaceous to early Tertiary, and middle Tertiary tectonics, south-central Arizona and north-central Sonora, *in* Gehrels, G.E., and Spencer, J.E., eds., Geologic excursions through the Sonoran Desert region, Arizona and Sonora: Arizona Geological Survey Special Paper 7, p. 76–88.

Wilshire, H.G., 1957, Propylitization of Tertiary volcanic rocks near Ebbetts Pass, Alpine County, California: University of California Publications in Geological Sciences, v. 32, no. 4, p. 243–272.

Wilson, E.D., 1960, Geologic map of Yuma County, Arizona: Tucson, Arizona, Arizona Bureau of Mines, scale 1:375,000.

Wilson, E.D., Moore, R.T., and O'Haire, R.T., 1960, Geologic map of Pima and Santa Cruz Counties, Arizona: Tucson, Arizona, Arizona Bureau of Mines, scale 1:375,000.

Wilson, E.D., Moore, R.T., and Cooper, J.R., 1969, Geologic map of Arizona: Tucson, Arizona, Arizona Geological Survey and U.S. Geological Survey, scale 1:500,000.

Winchester, J.A., and Floyd, P.A., 1977, Geochemical discrimination of different magma series and their differentiation products using immobile elements: Chemical Geology, v. 20, p. 325–343, doi: 10.1016/0009-2541(77)90057-2.

Woodford, A.O., Welday, E.E., and Merriam, R., 1968, Siliceous tuff clasts in the upper Paleogene of southern California: Geological Society of America Bulletin, v. 79, p. 1461–1486.

Wright, J.E., 1981, Geology and U-Pb geochronology of the western Paleozoic and Triassic subprovince, Klamath Mountains, northern California [Ph.D. dissertation]: Santa Barbara, University of California, 300 p.

Wright, J.E., Haxel, G.B., and May, D.J., 1981, Early Jurassic uranium-lead isotopic ages for Mesozoic supracrustal sequences, Baboquivari Mountains, southern Arizona: Geological Society of America Abstracts with Programs, v. 13, p. 115.

Wrucke, C.T., 1989, The Middle Proterozoic Apache Group, Troy Quartzite, and associated diabase of Arizona, *in* Jenny, J.P., and Reynolds, S.J., eds., Geologic evolution of Arizona: Arizona Geological Society Digest, v. 17, p. 239–258.

Manuscript Accepted by the Society 24 March 2005

Geological Society of America
Special Paper 393
2005

Climatic and tectonic controls on Jurassic intra-arc basins related to northward drift of North America

Cathy J. Busby*
Department of Geological Sciences, University of California, Santa Barbara, California 93106, USA

Kari N. Bassett*
Department of Geological Sciences, University of Canterbury, Christchurch, New Zealand

Maureen B. Steiner*
Department of Geology and Geophysics, University of Wyoming, Laramie, Wyoming 82071, USA

Nancy R. Riggs*
Department of Geology, Northern Arizona University, Flagstaff, Arizona 86011, USA

ABSTRACT

Upper Jurassic strike-slip intra-arc basins formed along the axis of earlier Lower to Middle Jurassic extensional intra-arc basins in Arizona. These strike-slip basins developed along the Sawmill Canyon fault zone, which may represent an inboard strand of the Mojave-Sonora megashear system that did not necessarily produce large-scale translations. Subsidence in the Lower to Middle Jurassic extensional arc was uniformly fast and continuous, whereas at least parts of the Upper Jurassic arc experienced rapidly alternating uplift and subsidence, producing numerous large-scale intrabasinal unconformities. Volcanism occurred only at releasing bends or stepovers in the Upper Jurassic arc, producing more episodic and localized eruptions than in the earlier extensional arc. Sediment sources in the Upper Jurassic strike-slip arc were also more localized, with restraining bends shedding sediment into nearby releasing bends. Normal fault scarps were rapidly buried by voluminous pyroclastic debris in the Lower to Middle Jurassic extensional arc, so epiclastic sedimentary deposits are rare, whereas pop-up structures in the Upper Jurassic strike-slip arc shed abundant epiclastic sediment into the basins. Three Upper Jurassic calderas formed along the Sawmill Canyon fault zone where strands of the fault progressively stepped westward in a releasing geometry relative to paleo-Pacific–North America plate motion.

We hypothesize that strike-slip basins in the Upper Jurassic arc formed in response to changing plate motions that induced northward drift of North America, causing sinistral deformation of the paleo-Pacific margin. Drift out of the northern horse latitudes into northern temperate latitudes brought about wetter climatic conditions, with eolianites replaced by fluvial, debris-flow, and lacustrine sediments. "Dry" eruptions

*E-mails: busby@geol.ucsb.edu; kari.bassett@canterbury.ac.nz; Steiner—magnetic@uwyo.edu; nancy.riggs@nau.edu

Busby, C.J., Bassett, K.N., Steiner, M.B., and Riggs, N., 2005, Climatic and tectonic controls on Jurassic intra-arc basins related to northward drift of North America, *in* Anderson, T.H., Nourse, J.A., McKee, J.W., and Steiner, M.B., eds., The Mojave-Sonora megashear hypothesis: Development, assessment, and alternatives: Geological Society of America Special Paper 393, p. 359–376. doi: 10.1130/2005.2393(13). For permission to copy, contact editing@geosociety.org.

of welded ignimbrite were replaced by "wet" eruptions of nonwelded, easily reworked ignimbrite and phreatoplinian fall. This Late Jurassic transition from hyperarid to more temperate climatic conditions may thus form a superregional "time line" that ties the Cordilleran plate margin to events in the interior of the continent.

Keywords: Jurassic, arc, strike slip, continental drift, climate.

INTRODUCTION

Sedimentologic, paleontologic, and paleomagnetic data from well-studied Jurassic back-arc strata on the present-day Colorado Plateau indicate that very rapid Late Jurassic continental drift northward out of the horse latitudes into temperate latitudes brought about a change from hyperarid conditions to wetter climatic conditions (May et al., 1989; Blakey, 1996; Steiner and Marzolf, 1997). These studies have shown that the eolianites that dominate the Lower to Middle Jurassic section in the western interior (e.g., Navajo and Entrada Sandstones) are largely replaced by fluvial and lacustrine deposits in the Upper Jurassic section (Morrison Formation). Turner and Fishman (1991) infer that the climate may have remained semiarid, as inferred from alkaline lake deposits in part of the Morrison Formation, but shallow lakes, streams, and a water table never far below the surface indicate a wetter climate than envisioned for Early to Middle Jurassic time.

In this paper, we show that intra-arc basins near the plate margin in southern Arizona (Fig. 1) record a similar climate change, from eolian-dominated sedimentation in Early to Middle Jurassic time to fluvially dominated sedimentation in Late Jurassic time. We infer that this climatic change produced a change in eruptive style (Fig. 2) that influenced sedimentation far into the plate interior. We also present here a tectonic model for the Upper Jurassic arc in southern Arizona, in which intra-arc strike-slip basins formed in response to the changing plate motions that induced northward drift of North America.

We present our tectonic and paleoclimate models first, in order to provide a coherent framework for the description and interpretation of structural and stratigraphic data. These data are described from ten structural blocks in southern Arizona that straddle the Sawmill Canyon fault zone. This is followed by a discussion of the controls of climate change on volcanic styles and super-regional stratigraphy. Last, we present a tectonic model for the Upper Jurasssic strike-slip arc, contrasting it with our previously published model for the Lower to Middle Jurassic extensional arc.

REGIONAL TECTONIC SETTING

Throughout much of southern Arizona, Upper Jurassic to lower Cretaceous (?) conglomerates and interstratified volcanic rocks occur upsection from more dominantly volcanic sections that are Lower to Middle Jurassic in age (Figs. 1 and 2; Saleeby and Busby-Spera, 1992). These conglomerates are widely referred to as the Glance Conglomerate and in many places lie at the base of Cretaceous nonmarine to marine sections referred to as the Bisbee Group (Bilodeau, 1979; Jacques-Ayala, 1995; Lawton and Olmstead, 1995). The Glance Conglomerate is the basal unit in the Bisbee basin and thus records the tectonic setting during the basin's initial opening. It occurs as piedmont fan and canyon fill deposits with locally interbedded lava flows and ignimbrites in grabens, half grabens, and calderas (Bilodeau, 1979; Dickinson et al., 1987, 1989; Busby and Kokelaar, 1992; Lipman and Hagstrum, 1992; Bassett and Busby, 1996, 1997; Lawton and McMillan, 1999; Dickinson and Lawton, 2001a).

Some workers have inferred that the Glance Conglomerate forms part of the Jurassic arc sequence, particularly in areas where it has abundant interstratified volcanic rocks (Tosdal et al., 1989). Other workers have suggested it records back-arc extension (Bilodeau, 1979). Still others have inferred that the Glance Conglomerate records continental rifting associated with the opening of the Gulf of Mexico (Dickinson et al., 1986, 1987; Lawton and McMillan, 1999). The main "tug of war" has occurred between workers placing the Glance Conglomerate in an arc environment and those placing it in a continental rift environment. A kind of "fence-sitting" model proposes that the Glance Conglomerate records a transitional phase, wherein Gulf of Mexico–related rifts progressively exploited the thermally weakened, structurally attenuated crust of the Jurassic extensional arc (Busby-Spera et al., 1989; Saleeby and Busby-Spera, 1992; Lawton and McMillan, 1999). Strike-slip tectonics related to the Mojave-Sonora megashear of Silver and Anderson (1974; see also Anderson and Silver, this volume) probably influenced this rifting (Anderson et al., 1995; Bassett and Busby, 1997; Lawton et al., 1997; Anderson and Nourse, this volume). We propose a model wherein the Glance Conglomerate records oblique convergence partitioned between intra-arc strike-slip and back-arc trantensional faulting.

The eastern part of the Glance Conglomerate outcrop belt, which is "inboard" relative to the paleo-Pacific subduction margin, contains few interstratified volcanic rocks, and these are mafic in composition (Lawton and Olmstead, 1995; Lawton and McMillan, 1999). Geochemical data on the mafic volcanic rocks from the eastern Bisbee basin indicate eruption in a rift rather than arc environment (Lawton and McMillan, 1999). This rift was referred to as the "Mexican Borderland rift," which was inferred to have been initiated when arc magmatism was abandoned in Arizona (Lawton and McMillan, 1999; Dickinson and Lawton, 2001a, 2001b).

The Glance Conglomerate on the western, outboard edge of the Bisbee basin (relative to the paleo-Pacific subduction margin)

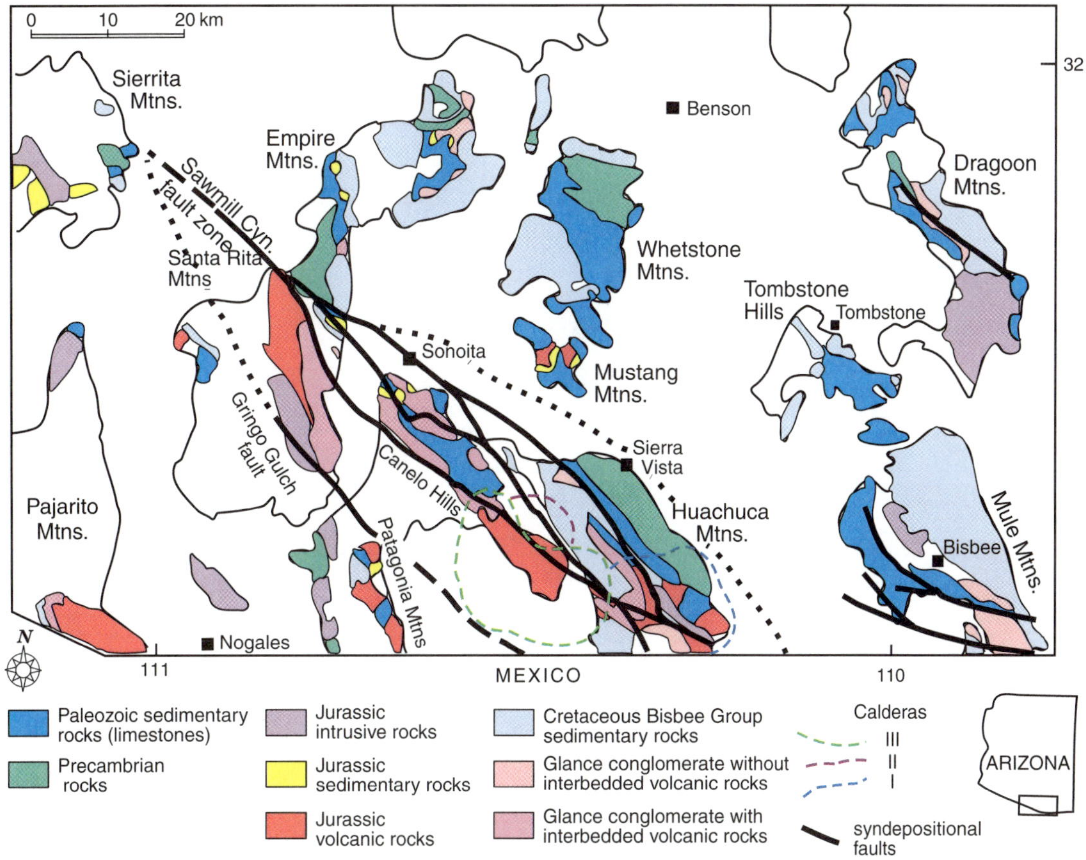

Figure 1. Geologic map of part of southern Arizona, showing Jurassic volcanic, sedimentary, and intrusive rocks, Upper Jurassic–Lower Cretaceous (?) Glance Conglomerate, possible source rocks for the Glance Conglomerate, fault zones that were probably active during Jurassic time (including the Sawmill Canyon fault zone), and Jurassic calderas along the Sawmill Canyon fault zone. Upper Cretaceous and younger rocks in the ranges, as well as Holocene basin fills, are shown in white. Caldera I—Montezuma caldera, caldera II—Turkey canyon caldera, and caldera III—Parker canyon caldera of Lipman and Hagstrum (1992). References for data on Figures 1 and 2: Hayes and Raup (1968), Hayes (1970), Drewes (1971a, 1971b, 1971c, 1972, 1976, 1981), Simons (1972, 1974), Titley (1976), Bilodeau (1979), Kluth (1982,), Vedder (1984), Bilodeau et al. (1987), Dickinson et al. (1986, 1987, 1989), Beatty (1987), Asmerom (1988), Busby-Spera (1988), Tosdal et al. (1989), Riggs and Haxel (1990), Klute (1991), Riggs and Busby-Spera (1990, 1991), Busby and Kokelaar (1992), Lipman and Hagstrum (1992), Riggs et al. (1993), Bassett and Busby (1996, 1997, and this volume), and Busby and Bassett (2005).

contains abundant rhyolitic, dacitic, and andesitic volcanic and volcaniclastic deposits interstratified with boulder breccia-conglomerates. The reported compositions of these volcanic rocks is based largely on phenocryst mineralogy (Hayes and Raup, 1968; Hayes, 1970; Drewes, 1971c; Kluth, 1982; Lipman and Hagstrum, 1992). Geochemical analyses on volcanic rock samples from the Canelo Hills region (Fig. 1) show light rare earth element (LREE) enrichment, Th enrichment, and a strong negative Eu anomaly interpreted to record a variation on continental rift volcanism (Krebs and Ruiz, 1987); however, we believe the compositions indicate a volcanic arc setting (using criteria of Saunders and Tarney, 1984). Geochemical analyses on volcanic rock samples from the Santa Rita Mountains (Basett and Busby, this volume) show large ion lithophile elements (LILE) enrichment, high field strength element (HFSE) depletion, a strong Ta negative anomaly, and shoshonitic composition, all typical of continental arc volcanism (Bassett and Busby, this volume). Volcanic rocks in the western Glance outcrop belt were erupted by a wide variety of mechanisms, including silicic plinian and phreatoplinian fall, andesitic vulcanian explosions, rhyolitic to dacitic lava dome extrusion and collapse into hot block-and-ash flows, dacitic to andesitic flow brecciation, and effusion of coherent (nonbrecciated) basaltic andesite lava flows (Fig. 2; Hayes and Raup, 1968; Hayes, 1970; Drewes, 1971c; Kluth, 1982; Krebs and Ruiz, 1987; Lipman and Hagstrum, 1992; Bassett and Busby, 1996 and this volume; Busby and Bassett, 2005). This wide range in compositions and eruptive behaviors does not support the interpretation of Lawton and McMillan (1999) that volcanic rocks interstratified with Glance Conglomerate are bimodal in composition and restricted to caldera-forming silicic eruptions associated with basalt lava flows, in a manner typical of continental rift settings. Nor do we agree with Dickinson and

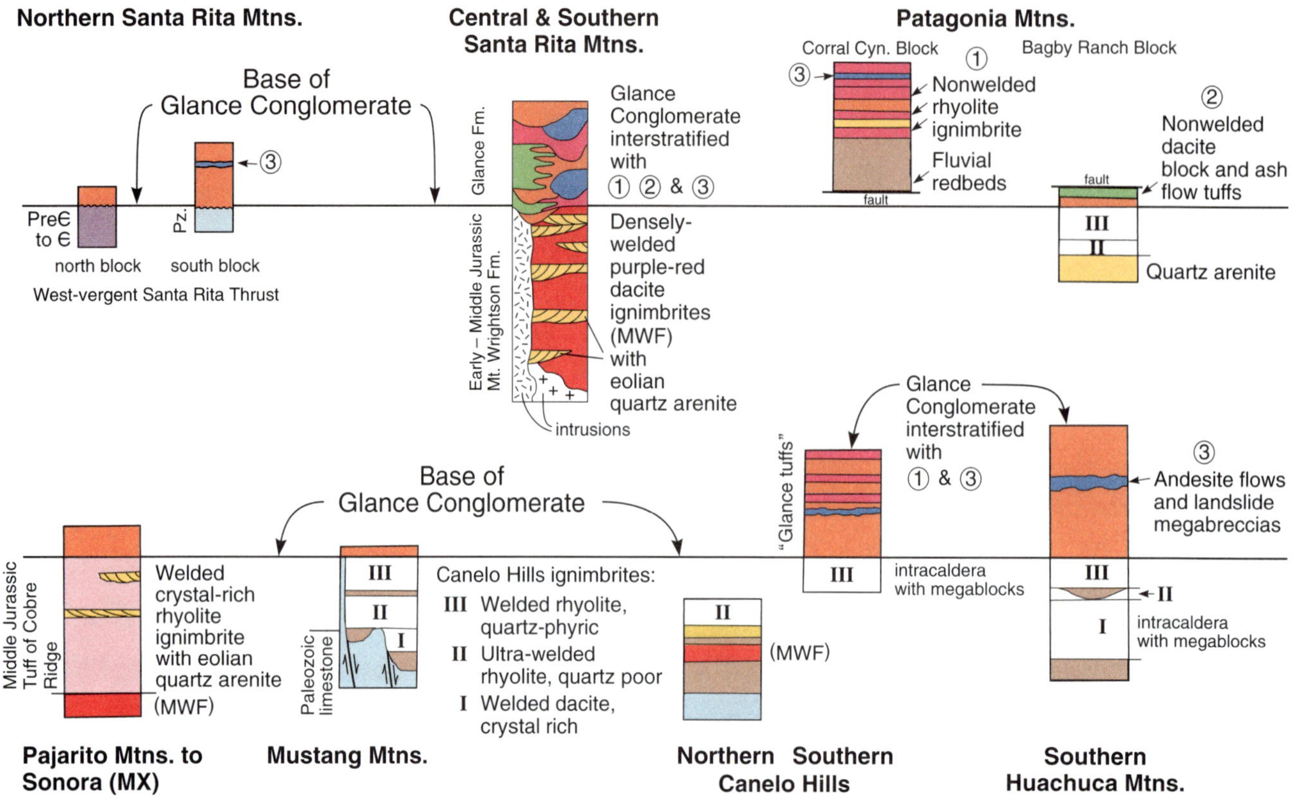

Figure 2. Generalized lithostratigraphic columns showing Lower to Middle Jurassic extensional arc volcanic deposits (Mount Wrightson Formation, tuff of Cobre Ridge, and Canelo Hills volcanics) overlain by Upper Jurassic Glance Conglomerate and interstratified andesitic, dacitic, and rhyolitic volcanic rocks (Glance Conglomerate, "Glance tuffs," and "Glance volcanics"). These columns are hung on the base of the Glance Conglomerate. At most localities the Glance Conglomerate rests on densely welded Middle Jurassic ignimbrites and interstraified eolian quartz arenites (Mount Wrightson Formation—MWF, tuff of Pajarito, and Canelo Hills Volcanics ignimbrites I, II, and II). The Glance Conglomerate contains three types of volcanic units that we correlate from range to range, labeled: (1) nonwelded rhyolite ignimbrites, (2) nonwelded dacite block-and-ash-flow tuffs, and (3) andesite lava flows and flow breccias.

Lawton's (2001a, 2001b) intrepretation that volcanic rocks interstratified with the Glance Conglomerate in the western Bisbee basin represent "post-arc rift volcanics." We interpret the western (outboard) part of the Glance Conglomerate belt to represent intra-arc basins, which formed along strike-slip faults (discussed further below).

USEFULNESS OF IGNIMBRITE STRATIGRAPHY FOR REGIONAL-SCALE CORRELATIONS

The stratigraphic, volcanological, tectonic, and paleoclimatic interpretions presented in this paper take advantage of the widespread nature of ignimbrites. Ignimbrites are defined as pumiceous, ash-rich pyroclastic flow deposits, which may or may not be welded; some other workers use the term "ash-flow tuff" for ignimbrites (Sparks et al., 1973; Freundt et al., 2000). We use distinctive characteristics of ignimbrites for purposes of correlation between ranges, including phenocryst abundance and types, color, degree of welding, lithic contents and compositions, and vertical and lateral organization of depositional structures in the ignimbrites, as well as the nature of distinctive enclosing strata. Using our data and building upon previous work cited in Figure 1, we propose a correlation of Jurassic ignimbrites across a large part of southern Arizona, through ten structural blocks (Fig. 2), described below. This work demonstrates the usefulness of ignimbrite stratigraphy in making correlations between widely separated ranges, despite postdepositional events including Jurassic strike-slip dismemberment, late Mesozoic to early Cenozoic Laramide shortening, and middle to late Cenozoic Basin and Range extension, as well as emplacement of plutons and attendant alteration throughout much of the Mesozoic and Cenozoic.

PROPOSED TECTONIC MODEL

We propose that Upper Jurassic basins of the western belt of Glance formed as intra-arc strike-slip basins along the Sawmill Canyon fault zone (Fig. 1), at the same time that the eastern (inboard) part of the belt experienced back-arc extension, due to strain partitioning along an obliquely convergent margin. The

Sawmill Canyon fault zone is the widest and longest fault zone in the Bisbee basin, and it formed an important conduit for andesitic to rhyolitic magmas throughout the Jurassic (Riggs and Busby-Spera, 1990; Busby and Kokelaar, 1992; Bassett and Busby, 1996 and this volume; Busby and Bassett, 2005). We propose that the Sawmill Canyon fault zone also controlled the siting of at least three calderas identified by Lipman and Hagstrum (1992); these straddle the Sawmill Canyon fault zone (Fig. 1) and erupted ignimbrite units I, II, and III of the Canelo Hills volcanics (Fig. 2; Lipman and Hagstrum, 1992; Busby and Kokelaar, 1992).

The Sawmill Canyon fault zone and related NW-trending, steeply dipping faults form a regional lineament inherited from Precambrian basement and reactivated in Mesozoic and Cenozoic times (Fig. 1; Titley, 1976; Drewes, 1981). The Jurassic movement history of the Sawmill Canyon fault zone has been variably interpreted as normal, dextral, or sinistral (Drewes, 1972; Titley, 1976; Bilodeau, 1979; Drewes, 1981; Busby and Kokelaar, 1992; Lipman and Hagstrum, 1992). This fault marks the northeastern structural boundary of thick Lower to Middle Jurassic arc volcanic successions, interpreted to form the fill in one segment of a 1000-km-long extensional arc graben depression that extended semicontinuously from Sonora, Mexico, to northern California and Nevada (Busby-Spera, 1988). Evidence for movement on the Sawmill Canyon fault zone in Late Jurassic time includes large slabs of shattered rock interbedded with the Glance Conglomerate in the Mustang Mountains, the Huachuca Mountains, and the Canelo Hills (Fig. 1; Hayes and Raup, 1968; Davis et al., 1979; Bilodeau, 1979) as well as along the southern extension of the Sawmill Canyon fault zone into Mexico (McKee and Anderson, 1998). These are widely agreed to represent megabreccias that avalanched off fault scarps. In this paper we present evidence that the Sawmill Canyon fault zone was active as a strike-slip fault during deposition of the Glance Conglomerate and interstratified volcanic rocks.

Upper Jurassic strike-slip intra-arc basin deposits contrast with Lower to Middle Jurassic extensional intra-arc basin deposits in several ways (Table 1). Subsidence in the Lower to Middle Jurassic extensional arc was uniformly fast and continuous along syndepositional normal faults (Busby-Spera, 1988; Busby-Spera et al., 1990; Riggs and Busby-Spera, 1990; Riggs et al., 1993; Haxel et al., this volume). In contrast, at least parts of the Upper Jurassic arc experienced rapidly alternating uplift and subsidence, producing numerous large-scale unconformities, as well as intrabasinal reverse and normal faults (Bassett and Busby, this volume). We interpret this "porpoising" to be the result of temporal alternations between subsiding releasing bends and uplifting restraining bends along the Sawmill Canyon fault zone, and we propose that restraining bends were sites of sediment production. High rates of extension in the Middle to Upper Jurassic arc resulted in high rates of magma production so that basin-bounding fault scarps were buried by pyroclastic deposits, leading to a general lack of talus cones and alluvial fan deposits (Fig. 2), similar to the modern Taupo Volcanic zone. In contrast, Late Jurassic volcanism was more restricted in volume, space, and time (Figs. 1 and 2) so that fault scarps were fringed with breccia-conglomerate units (Glance Conglomerate), in place of being buried by pyroclastic debris. Furthermore, erosion of pop-up structures along restraining bends contributed to the epiclastic sediment supply.

PROPOSED CLIMATIC CONTROLS ON STYLES OF VOLCANISM AND SEDIMENTATION

A super-regional "time line" controlled by climate change may be represented by a contact between: (1) sections with welded ignimbrites, and (2) sections with nonwelded ignimbrites and associated phreatoplinian fall deposits. In southern Arizona, this contact coincides approximately with the base of the Glance Conglomerate (Fig. 2). We do not regard this as a coincidence but rather a fundamental change in tectonic style that accompanied the very rapid northward drift of North America (Table 1).

Lower to Middle Jurassic rocks of the southwestern United States record hyperarid conditions. In the extensional arc of southern Arizona, this is recorded by eolian quartz arenites interstratified with strongly welded ignimbrites with a rarity of fluvial deposits and total lack of lake deposits (e.g., Mount Wrightson Formation and tuff of Cobre Ridge, Fig. 2; Riggs and Busby-Spera, 1990, 1991). The high emplacement temperatures required for strong welding indicates that the eruptions were "dry" (i.e., no interaction with surface water). All of the glass

TABLE 1. CONTRASTING VOLCANIC, STRUCTURAL, AND CLIMATIC CONTROLS ON THE FILL OF EARLY TO MIDDLE JURASSIC VERSUS LATE JURASSIC INTRA-ARC BASINS, ARIZONA

	Early to Middle Jurassic extensional arc	Late Jurassic strike-slip arc
Volcanic controls	More voluminous, widespread, and continuous eruptions => scarps buried by pyroclastic deposits	More restricted eruptions in space and time, episodic => scarps eroded, shedding sediments into the canyons
Structural controls	Uniformly fast and continuous subsidence => unconformities rare	"Porpoising" on all scales, local regions of uplift within basins => numerous big unconformities, intrabasinal highs shed sediment
Climatic controls	Hyperarid: eolianites => "dry" eruptions of welded tuffs, most rheomorphic	Wetter climate: sheetwash and channelized hyperconcentrated flood flow and dilute flow => "wet" eruptions of nonwelded tuffs, plinian, and phreatoplinian fall

shards in these ignimbrites are bubble-wall shards produced by magmatic vesiculation. They lack blocky shards, which are produced by fragmentation upon contact with surface water (Heiken and Wohletz, 1985).

As the climate grew wetter in late Middle to Late Jurassic time, eolian quartz sands were reworked as fluvial deposits (for example, fluvial quartz arenites interstratified with Canelo Hills volcanics and basal "Glance tuffs," Fig. 2). Finally, the climate became wet enough for surface and groundwater to collect in and around vent areas, and the style of volcanism changed dramatically, generating nonwelded to weakly welded ignimbrites and associated phreatoplinian fallout. Nonwelded ignimbrites and phreatoplinian ashes are friable, very easily remobilized pyroclastic deposits. Increased rainfall remobilized this pyroclastic debris by fluvial processes, including channelized stream flow and sheetflood, as well as by debris-flow processes (e.g., Glance Conglomerate, central and southern Santa Rita Mountains, Fig. 2; Bassett and Busby, this volume).

Phreatomagmatic eruptions commonly result in nonwelded ignimbrites, together with widespread and voluminous fall deposits (plinian and phreatoplinian) and surge deposits. They typically occur where lakes or shallow embayments of the sea lie above vents in downsagged areas of caldera evacuation or structurally controlled extension or transtension (Wilson et al., 1995).

Ignimbrites generated from phreatomagmatic eruptions are nonwelded, with blocky shards as well as bubble-wall shards, and commonly lack carbonized plant material, due to cooling of pyroclasts during eruption through water (Self, 1983; Heiken and Wohletz, 1985; McPhie et al., 1993; Bryan et al., 2000; De Rita et al., 2002). Silicic magmatic explosions enhanced by steam explosions produce the highest eruption columns, which provide the momentum to make phreatoplinian ignimbrites the most mobile pyroclastic flows on Earth (Self, 1983; Houghton et al., 2000a).

Phreatoplinian fallout tuffs can be enormously widespread, with thicknesses generally <3–4 m but extending for many hundreds of kilometers in both upwind and downwind directions, with very little change in grain size with distance (Self, 1983). Phreatoplinian ash fall is extremely fine grained and well sorted, with deformational structures typical of water-saturated, cohesive soft sediment, e.g., convolute lamination and plastering of vertical surfaces (Self and Sparks, 1978; Self, 1983; Wilson et al., 1995; De Rita et al., 2002). Phreatoplinian tuffs occur in thin to very thin beds (~1–5 cm) that are tabular to convolute and mantle topography; they are composed largely of blocky glass shards, in addition to bubble-wall shards, and may preserve horizons of accretionary lapilli (Heiken and Wohletz, 1985; McPhie et al., 1993; Morissey et al., 2000; Houghton et al., 2000a; Jurado-Chichay and Walker, 2001). The very thin tabular beds in phreatoplinian ash fall suggest intermittent explosions from highly vesiculated magma interacting with surface water (Houghton et al., 2000a). Phreatoplinian tuffs are commonly interstratified with plinian pumice fall deposits; these are recognized as well-sorted pumice accumulations that mantle topography in thin beds (Fisher and Schminke, 1984; Houghton et al., 2000b). The alternation of plinian and phreatoplinian tuffs results from variability in water influx to the vent (Wohletz, 1986; Jurado-Chichay and Walker, 2001). Phreatoplinian tuffs that lack nonjuvenile rock fragments are more typical of interaction with surface water, rather than aquifers (De Rita et al., 2002).

The pyroclastic rocks interstratified with the Glance Conglomerate are dominated by nonwelded rhyolitic to andesitic phreatomagmatic ignimbrites, with lesser phreatoplinian rhyolitic tuffs and phreatomagmatic andesitic tuffs (Fig. 2). We speculate that the change in eruptive style recorded in the western belt of the Glance Conglomerate was the result of climate change from hyperarid to wetter conditions, induced during rapid northward drift of North America out of the horse latitudes into more temperate latitudes.

To summarize, our new correlations of widespread ignimbrites across ten structural blocks straddling the Sawmill Canyon fault zone (Figs. 1 and 2) permit the following interpretations for Jurassic time: an earlier stage of "dry" ignimbrite eruptions (Lower to Middle Jurassic Mount Wrightson Formation, overlain by Middle Jurassic tuff of Pajarito and Middle Jurassic Canelo Hills volcanics) was succeeded across the region by a later stage of dominantly "wet" ignimbrite eruptions (Upper Jurassic "Glance Conglomerate and volcanics" and "Glance tuffs"). This climatically controlled transition in eruptive style coincides with a tectonically controlled change in sedimentological style, characterized by deposition of compositionally and texturally immature, locally sourced conglomerates (Glance Conglomerate).

PALEOMAGNETIC EVIDENCE FOR VERY RAPID NORTHWARD DRIFT OF NORTH AMERICA IN LATE JURASSIC TIME

Southwestern North America resided in equatorial latitudes from the mid-Pennsylvanian (Supai Group) through the Early Jurassic (~Pliensbachian Kayenta Formation), shown in Figure 3A. For reference, the center of the plateau (e.g., Four Corners area) lay at paleolatitudes of 4°–7° N through the Early Jurassic.

For the end of the Early Jurassic and most of the Middle Jurassic (Toarcian–Bathonian), no paleomagnetic data are available to define the paleolatitude of North American, but by latest Middle Jurassic time (upper Callovian Summerville Formation), the Plateau center lay at 21° N. Paleomagnetic poles with ages within that time gap have been determined (Kluth et al., 1982; May et al., 1986) and lie in sequential relative positions between the Kayenta and Summerville paleopoles, but their reliability is questionable because they come from sites in the block-rotation-prone Basin and Range province. If these sites have not been subject to rotation, a more refined estimate of northward motion versus age can be made. Nevertheless, at this stage of our knowledge, an ~15° northward shift of North America occurred during the Middle Jurassic (Fig. 3A), between ca. 185 Ma and 160 Ma (using the time scale of Gradstein et al., 2004).

The North American plate apparently exhibited little motion with respect to the geomagnetic pole between the latest Callovian and late Oxfordian (ca. 155–150 Ma; Fig. 3B), demonstrated by statistically identical paleopoles from the Summerville and overlying lower Morrison Formations. This "stillstand" of the plate is indicated from data at locations on both the Colorado Plateau and the craton. The J-5 unconformity, which formed in this time frame, may therefore have encompassed little time, consistent with its only intermittent presence and aerially restricted nature (Imlay, 1980).

Paleomagnetic data suggest that a second period of northward drift of North America and the plateau, of more rapid nature, occurred in the Late Jurassic, after deposition of the lower sand-dominated (Salt Wash) member of the Morrison Formation and prior to the deposition of the variegated tuffaceous claystones of the Brushy Basin member. This shift was ~13° northward (Fig. 3). In addition, between deposition of the two members, the Colorado Plateau was rotated 4°–5° clockwise relative to cratonic North America. No field evidence of a hiatus has been observed, but the two very different paleopole positions require one. The change in sedimentary facies across the unconformity may be indicative of the tectonic event that caused rotation of the Colorado Plateau. We infer that abundant ashes above the unconformity record a climatically controlled transition in eruptive style in the arc, i.e., progressively "wetter," as North America drifted into temperate latitudes. The age of the postulated hiatus is between ca. 152 and 150 Ma, estimated from a comparison of the magnetostratigraphy to the seafloor record (Steiner, 1998) and from $^{40}Ar/^{39}Ar$ dating of tuffs in the formation (Kowallis et al., 1998; Gradstein et al., 2004).

In summary, southwestern North America was translated ~15° northward from equatorial latitudes between ca. 185 and 160 Ma and even more rapidly, another ~13° further northward, between ca. 152 Ma and 150 Ma.

LITHOSTRATIGRAPHY OF STRUCTURAL BLOCKS STRADDLING THE SAWMILL CANYON FAULT ZONE

We summarize stratigraphic and structural data that support the tectonic and paleoclimatic models presented above (Table 1 and Fig. 4); for more complete reviews of the Jurassic arc and the Glance Conglomerate in southern Arizona, see Dickinson et al. (1986, 1987, 1989), Bilodeau et al. (1987), Busby-Spera (1988), Tosdal et al. (1989), Saleeby and Busby-Spera (1992), and Dickinson and Lawton (2001a).

We make stratigraphic correlations between ranges (Fig. 2) on the basis of lithologic similarities, using outcrop and thin section characteristics. Lithologic correlations are more convincing where two or more lithologically distinct units occur in stratigraphic succession. The correlations presented here are based largely on our reconnaissance mapping, using published maps and papers cited in the caption for Figure 1; we have not made our own detailed maps, cross sections, and measured sections of most of these ranges, except in the Santa Rita Mountains (Bassett and Busby, this volume; Riggs and Busby-Spera, 1990) and the Pajarito Mountains (Riggs and Busby-Spera, 1991). The lithostratigraphic columns are generalized, composite sections intended to summarize the predominant and lithologically distinctive units used for correlation in this paper, and are by no means detailed or exhaustive. We intend for these correlations and the model that results from them to be tested by future workers using geochemistry, geochronology, and additional field mapping.

Generalized lithostratigraphic columns of Figure 2 are hung on the base of the Glance Conglomerate. This rests on densely welded Middle Jurassic ignimbrites and interstratified eolian quartz arenites at most localities, and correlations between these ignimbrites are summarized in Figure 2, as well as the types of volcanic units correlated from range to range in the Glance Conglomerate. The columns are arranged in a manner that makes it easiest to see the correlation of units between structural blocks of the present-day Basin and Range topography, rather than being arranged in any consistent fashion relative to the various strands of the Sawmill Canyon fault zone. The northern Santa Rita Mountains and the Mustang Mountains lie northwest of all proposed traces of the Sawmill Canyon fault zone, and the Pajarito and Patagonia Mountains lie southwest of all proposed traces. We propose that all the other ranges lie between various strands of the fault zone (Fig. 1), including the central and southern Santa Rita Mountains, the northern and southern Canelo Hills, and the Huachuca Mountains.

Santa Rita Mountains, Northern Block

The northern Santa Rita Mountains structural block is bounded to the south by the Sawmill Canyon fault zone (Fig. 1) and is further subdivided into two structural blocks by a Laramide (Upper Cretaceous–lower Tertiary) thrust fault. Glance Conglomerate in the north sub-block overlies the Cambrian Bolsa Quartzite and the Precambrian Continental Granodiorite, and in the south sub-block it rests on Paleozoic carbonates (Fig. 2; Drewes, 1972, 1976). The Glance Conglomerate passes upward from carbonate clast-dominated conglomerate to granodiorite-clast-dominated conglomerate, recording the stripping of Paleozoic cover down to basement. There are no Lower to Middle Jurassic volcanic rocks below the Glance Conglomerate here, but an andesite lava flow lies within it (Fig. 2).

Precambrian to Paleozoic source rocks for the Glance Conglomerate are abundant northwest of the Sawmill Canyon fault zone and absent south of it. This is probably because they are deeply buried under the Lower to Middle Jurassic extensional arc graben depression of California and Arizona (Busby-Spera, 1988). This is the reason Dickinson and Lawton's (Fig. 7 *in* Dickinson and Lawton, 2001a) NW-SE trending boundary between volcanic-clast-dominated and limestone-clast-dominated Glance Conglomerate coincides with the Sawmill Canyon fault zone of Figure 1.

The lack of Lower to Middle Jurassic volcanic rocks in the northern Santa Rita structural block is consistent with the interpretation that the Sawmill Canyon fault zone formed the northeastern boundary of the extensional arc graben depression shown

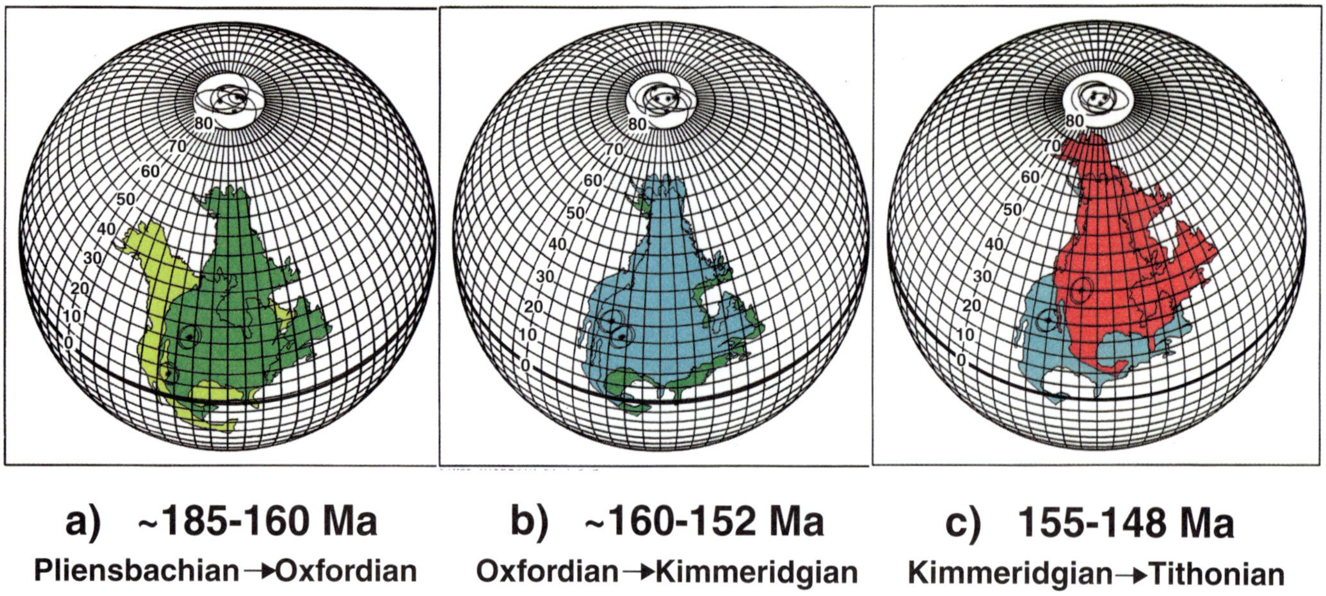

Figure 3. Changes in paleolatitude and paleo-orientation of the North American continent during the Jurassic for the times during which data is available. Chartreuse: Sinemurian-Pliensbachian, or ca. 185 Ma (Kayenta Formation); green: upper Callovian–lower Oxfordian, ca. 160–155 Ma (Summerville Formation); blue: upper Oxfordian–lower Kimmeridgian, ca. 155–152 Ma (lower Morrison Formation); red: upper Kimmeridgian–lower Tithonian, 150–148 Ma (Upper Morrison Formation). The heavy line designates the equator, and paleolatitude is shown in 5° increments.

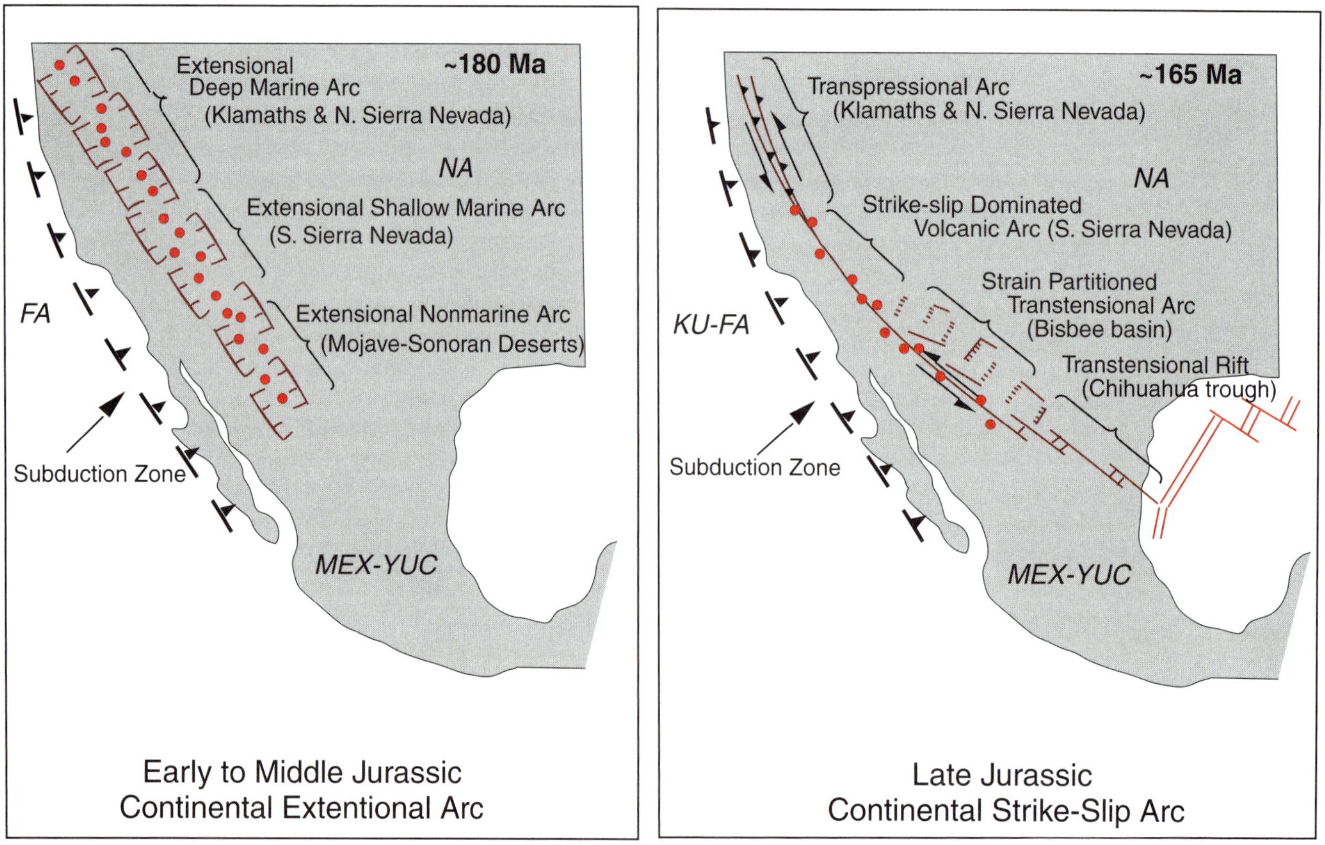

Figure 4. Regional reconstructions of the Jurassic plate boundary in the Cordilleran United States and Mexico. (A) The Lower to Middle Jurassic arc was dominantly extensional, probably due to high rates of slab rollback during subduction of old, cold lithosphere (Busby-Spera, 1988). (B) The Upper Jurassic arc reflects sinistral oblique convergence, with transpression in the north and transtension in the south (Saleeby and Busby-Spera, 1992). We propose here that strain partitioning in the south caused intra-arc strike-slip faulting along the western edge of the Bisbee basin and back-arc transtensional faulting along the eastern edge. Red dots represent arc volcanoes, hachured lines represent grabens, and parallel lines represent rifts. FA—Farallon plate; KU—Kula plate; NA—North America.

in Figure 4, perhaps representing an area of footwall uplift at that time. The Mustang Mountains (Fig. 1; discussed below) is the only range northeast of the Sawmill Canyon fault zone that contains a section of Lower or Middle Jurassic volcanic rocks, and these are relatively thin caldera outflow ignimbrites (described below).

Santa Rita Mountains, Central and Southern Block

Jurassic strata of the central and southern Santa Rita Mountains include (Fig. 2): (1) the Lower to Middle Jurassic Mount Wrightson Formation, described in detail by Riggs and Busby-Spera (1990), and (2) the Upper Jurassic Temporal, Bathtub, and Glance Formations of Drewes (1971a), redefined as the Santa Rita Glance Conglomerate and described in detail by Bassett and Busby (this volume).

The Lower to Middle Jurassic Mount Wrightson Formation records hyperarid conditions, with eolian quartz arenites and a total lack of fluvial or lacustrine deposits, an eruptive style that was "dry," with all of the ignimbrites strongly welded, and a complete lack of phreatoplinian or phreatomagmatic tuff (Riggs and Busby-Spera, 1991). This is in strong contrast with the Upper Jurassic Santa Rita Glance Conglomerate, which has abundant fluvially reworked tuffs and mudflow deposits and an eruptive style that was "wet," with predominantly nonwelded ignimbrites interstratified with phreatoplinian rhyolite tuffs and ignimbrites, as well as phreatomagmatic andesite tuffs and reworked tuffs (described in detail by Busby and Bassett, 2005). Another major difference between the Mount Wrightson Formation and the Santa Rita Glance Conglomerate is that the Mount Wrightson Formation lacks the boulder breccia-conglomerates that dominate the Santa Rita Glance Conglomerate adjacent to the Sawmill Canyon fault zone (Fig. 2). We attribute this to a change in tectonic style along the Sawmill Canyon fault zone, from extensional in the Early to Middle Jurassic (Busby-Spera, 1988; Riggs and Busby-Spera, 1990) to strike slip in the Late Jurassic (Bassett and Busby, this volume).

The most distinctive feature of the Lower to Middle Jurassic Mount Wrightson Formation, densely welded purple-red dacite ignimbrites, is emphasized in Figure 2, and we correlate these with ignimbrites in the Northern Canelo Hills and the Pajarito Mountains of Arizona and Sonora (Fig. 2 and described below). These ignimbrites are distinctive not only for their color and dense welding but also for their lack of quartz phenocrysts, which are commonly present in other Lower to Middle Jurassic continental arc ignimbrites of Arizona and California. However, in addition to the distinctive ignimbrites, in the Santa Rita Mountains the Mount Wrightson Formation contains a diverse array of volcanic lithofacies that represent a complex of multiple small intermediate to silicic eruptive centers, referred to as a "multivent complex" (Riggs and Busby-Spera, 1990). There, rapid intra-arc subsidence resulted in a complex juxtaposition of vent proximal, medial, and distal facies, because vent areas were buried by volcanic/volcaniclastic material derived from adjacent vents in the multivent complex. Rapid intra-arc subsidence also resulted in repeated burial of vent areas (and other environments) by eolian quartz arenites equivalent in age to major ergs on the Colorado Plateau (Riggs et al., 1993). The distinctive purple-red dacite ignimbrites of the Mount Wrightson Formation basin accumulated as outflow ignimbrites presumably erupted from an unidentified caldera/caldera complex elsewhere in the arc (Riggs and Busby-Spera, 1991).

The Upper Jurassic Santa Rita Glance Conglomerate unconformably overlies the underlying Lower to Middle Jurassic Mount Wrightson Formation and is similar in that both are largely composed of silicic to intermediate composition multivent complexes preserved in a basin that also accumulated and preserved outflow ignimbrites erupted elsewhere (Bassett and Busby, this volume; Busby and Bassett, 2005). We consider it very significant that multivent complexes formed in the area of the present-day Santa Rita Mountains throughout the Early, Middle, and Late Jurassic. We interpret this to reflect proximity to a continuously active fault zone (Sawmill Canyon), with strands of the fault frequently plumbing small batches of magma to the surface. We propose that multivent volcanic complexes should be common along major fault zones, be they normal faults (e.g., Mount Wrightson Formation) or strike-slip faults (e.g., Santa Rita Glance Conglomerate).

We interpret the Santa Rita Glance Conglomerate as the fill of an intra-arc strike-slip basin (Bassett and Busby, this volume). Evidence for an intra-arc, rather than rift, setting for this basin includes: (1) a wide compositional range and relative proportions of those compositions (~30% andesitic, ~25% dacitic, and ~45% rhyolitic), (2) a wide range in eruptive styles, and (3) strong LILE enrichment, HFSE depletion, a strong Ta negative anomaly, and shoshonitic composition, all typical of continental arc volcanism (Bassett and Busby, this volume). The Sawmill Canyon fault zone was the master fault for the intra-arc strike-slip basin, as shown by lithofacies distribution, basin asymmetry, vent locations, structural style, and the geometry of intrabasinal unconformities. A less regionally significant fault zone lay along the south end of the basin, and we name this fault the Gringo Gulch fault zone (Fig. 1). The basin also contains numerous intrabasinal syndepositional high-angle faults, which alternated in space and time between normal-slip and reverse-slip separation. The basin fill thickens dramatically away from subordinate basin-bounding faults ("shallow" end) toward the master strike-slip fault ("deep" end) where subsidence was greatest. Talus cone-alluvial fan deposits (Glance Conglomerate *sensu stricto*) are largely restricted to the master fault–proximal (deep) end of the basin. Nearly all of the proximal extrabasinal volcanic rocks were erupted from the master fault zone, where they became trapped in the deep end of the basin, and most of the intrabasinal volcanic rocks were erupted from vents along splays of the master fault and trapped in the deep end. These processes kept the deep (master fault–proximal) end of the basin overfilled, so that it could not provide accommodation for the reworked tuffs and extrabasinally sourced ignimbrites that dominate the shallow (underfilled) end of the basin. This pattern of basin fill contrasts markedly with that of nonvolcanic strike-slip basins on transform margins, where clastic sedimentation commonly cannot keep pace with subsidence in the master fault–proximal end (Busby and Bassett, 2005). Eight large-scale unconformable surfaces occur within the basin. Five are asymmetrical and show extreme vertical relief (460–910 m) with very high paleoslope gradients (40°–71°) that dip away from the master fault. We interpret these as fault scarps or paleoslide scars produced during basin uplift events at restraining bends along the master fault. The other three unconformities are symmetrical and less steep, with vertical relief of 200–600 m and paleoslope gradients of 20°–25°. We interpret these surfaces to be the walls of steep paleocanyons (Bassett and Busby, this volume).

The structural style and the unconformities recorded in the Upper Jurassic Santa Rita Glance Formation are typical of basins affected by both releasing bends and restraining bends along strike-slip faults (Bassett and Busby, this volume). Alternating reactivation of normal faults as reverse faults, and vice versa, is a distinguishing feature of releasing-bend strike-slip basins that have been affected by restraining bends, either simultaneously or successively through time (Crowell, 2003; Link, 2003; Nilsen and Sylvester, 1995). When a strike-slip fault system is slightly transtensional, the sizes of the restraining bends (where reverse faulting occurs) are less than the size of the releasing bends (where normal faulting occurs); therefore, net subsidence over time will result in partial preservation of the basin fill (Cowan et al., 1989; Cowan and Pettinga, 1992; Cowan et al., 1989; Wood et al., 1994). We infer that the deep unconformities in the Santa Rita Glance Formation were created during uplift at small restraining bends, leaving most of the releasing bend basin deposits undeformed (Bassett and Busby, this volume).

Patagonia Mountains, Bagby Ranch Block

Jurassic strata of the Patagonia Mountains occur in two structural blocks, separated by a fault zone of uncertain age and amount and sense of offset (Simons, 1972). The fault zone cuts the top of the Bagby Ranch block and the base of the Corral Canyon block section (Fig. 2). We correlate red beds at the top of the Bagby Ranch block with red beds in basal strata of the Corral Canyon block and interpret them as fault-distal facies of Glance Conglomerate (Fig. 2). This interpretation is supported by the volcanic stratigraphy of the two blocks, which we correlate with units in the Santa Rita Mountains (Fig. 2), discussed here.

Basal strata of the Bagby Ranch block consist of a very thick (~200 m) texturally and compositionally mature quartz arenite with trough cross-lamination, ripple cross-lamination, planar lamination, and cut and fill structures. We interpret this to record fluvial reworking of eolian sands. This is overlain by a distinctive phenocryst-poor rheomorphic rhyolite ignimbrite. This very densely welded ignimbrite is pale tan to gray, containing less than 10% small phenocrysts of quartz and feldspar and abundant distinctive spherulitic and lithophysal crystallization textures. It is identical to the "rhyolite lava" member of the Canelo Hills Volcanics of Hayes (1970), which was reinterpeted to be a "lava-like" or ultrawelded ignimbrite by Lipman and Hagstrum (1992). They correlated this ignimbrite (unit II of Canelo Hills Volcanics, Fig. 2) from the southern Huachuca Mountains through the Canelo Hills to the Mustang Mountains and inferred that it was erupted from the Turkey Canyon caldera (caldera II, Fig. 1) sited along the Sawmill Canyon fault zone. We extend this correlation to the Patagonia Mountains (Fig. 2). Our correlation is supported by the presence of Canelo Hills Volcanics unit III on top of unit II in the Patagonia Mountains (Fig. 2). This is a phenocryst-rich welded ignimbrite characterized by abundant quartz and feldspar, referred to as the "welded tuff member" of the Canelo Hills Volcanics of Hayes (1970); like unit II, it was correlated from the southern Huachuca Mountains through the Canelo Hills to the Mustang Moutains by Lipman and Hagstrum (1992), who identified and named the Parker Canyon caldera (caldera II, Fig. 1) as its source. This in turn is overlain by red sandstones and siltstones whose stratigraphic position directly above the Canelo Hills Volcanics is similar to that of the Glance Conglomerate in three other structural blocks (Hayes and Raup, 1968; Vedder, 1984; Lipman and Hagstrum, 1992). Therefore, we infer that these red beds represent a fault-distal, fluvial facies of the Glance Conglomerate, similar to fluvial redbeds in the master fault–distal part of the Santa Rita Mountains. These are in turn overlain by dacite block-and-ash-flow tuffs (Fig. 2).

Dacite block-and-ash-flow tuffs in the Bagby Ranch block of the Patagonia Mountains (Fig. 2) appear identical to those of the Santa Rita Mountains except that they are more fine grained. They are gray, massive, monomict, poorly sorted matrix-supported tuff breccia and lapilli tuff. They are composed of ~75% nonvesiculated (dense) dacitic blocks and lapilli in a matrix of ash-sized fragments of the same composition and shape. Pumice is absent. In both the Santa Rita and the Patagonia Mountains, the dacite block-and-ash-flow tuffs have ~15%–30% crystals, mostly plagioclase with lesser biotite and quartz, with a higher crystal content in the matrix than the clasts, probably due to loss of vitric component during flow. In the Santa Rita Mountains, these deposits coarsen dramatically toward the Sawmill Canyon fault zone, and they are inferred to have been generated by collapses on lava domes sited along the fault (Busby and Bassett, this volume).

Several features of the dacite block-and-ash-flow tuffs in the Patagonia Mountains indicate they are distal equivalents of those of the Santa Rita Mountains. They are predominately lapilli tuffs rather than tuff breccias; they occur in flow units 1–2 m thick rather than tens of meters thick; there are only a few of them rather than dozens of flow units; and they lack the delicate, plastically deformed "amoeboid" clasts that occur (although rarely) in the Santa Rita Mountains. The present-day distance of the block-and-ash-flow deposits from their proximal equivalents in the Santa Rita Mountains (~25 km) is a rather long one for block-and-ash flows to travel. Perhaps there was enough relief on the Sawmill Canyon fault zone to generate the momentum necessary to drive block-and-ash-flows that far; the enormous scale of buried fault scarps and slide scars in the Santa Rita section suggests there may have been. Alternatively (or in addition), the two basins may have been originally closer to each other, if they were subsequently offset in a sinistral sense by the Gringo Gulch fault (Fig. 1), which was active during deposition of the Santa Rita Glance Conglomerate (decribed above; Bassett and Busby, this volume). This fault reconstruction would also restore Jurassic intrusions of the two ranges (Fig. 1) to closer proximity.

Dacite block-and-ash-flow tuffs occur within unconformity-bounded sequences 4 and 5 (out of 8) in the Santa Rita Mountains, indicating that the top of the Bagby Ranch block section is correlative with the middle of the Santa Rita Glance Conglomerate.

Patagonia Mountains, Corral Canyon Block

We correlate strata of the Corral Canyon block with the upper part of the Santa Rita Glance Conglomerate, using a distinctive pale greenish white nonwelded ignimbrite that occurs near the top of the Corral Canyon section (Fig. 2) and also in unconformity-bounded sequence 6 (out of 8) in the Santa Rita Mountains (Bassett and Busby, this volume).

The basal unit of the Corral Canyon block consists of red siltstone, conglomerate, and sandstone beds that we refer to as Glance Conglomerate. Most show cut and fill structures and stratification/cross stratification typical of fluvial deposits, although some are massive, matrix-supported beds typical of debris flows. Fault talus breccias are absent. We therefore interpret these redbeds to represent fault-distal to fault-medial facies of the Glance Conglomerate. The conglomerate clasts are largely volcanic, and the granitoid clasts that are present in most conglomerate beds in the Santa Rita Mountains are absent, although many conglomerate beds in the Santa Rita Mountains have 100% volcanic clasts. Siltstones in the middle of the red bed unit have a single interbed of algal limestone, less than a meter thick, which may represent the deposit of a shallow lake.

The Glance Conglomerate is overlain by two ignimbrites that we have not yet recognized in other ranges, separated by a few meters of bedded sandstone. These are in turn overlain by another section of Glance Conglomerate with sandstone and siltstone, in turn overlain by a distinctive pale greenish white ignimbrite that we recognize in unconformity-bounded sequence 6 of the Santa Rita Mountains (labeled 1 on Fig. 2; Bassett and Busby, this volume). This is a pale greenish white nonwelded pumice-rich rhyolite tuff. Pumices are abundant and flattened in a white to pink aphyric matrix of vitric tuff, composed of blocky or splintery shards and bubble-wall shards. The matrix is crystal poor, with less than 10% crystals of plagioclase, quartz, biotite, and traces of potassium-feldspar. Similar nonwelded ignimbrite are also present in the southern Canelo Hills block (labeled 1 on Fig. 2 and discussed below). The top of the section consists of a third ignimbrite we have not recognized in other ranges, partly intruded out along an andesite sill.

Pajarito Mountains (Arizona and Sonora)

The volcanic section in the Pajarito Mountains is dominated by the Middle Jurassic tuff of Pajarito, which fills an enormous composite caldera ~50 km long by 25 km wide, described by Riggs and Busby-Spera (1991). The tuff of Pajarito is very crystal rich (50% phenocrysts), with quartz, plagioclase, and potassium feldspar in equal proportions. It is dated at 169 Ma and contains interbeds of eolian quartz arenites (variably contaminated by volcanic detritus) that are age-correlative with the Temple Cap or Page sandstones on the Colorado Plateau (Riggs et al., 1993). The caldera-filling tuff of Pajarito is conformably overlain by caldera moat deposits (Riggs and Busby-Spera, 1991).

The Mount Wrightson Formation lies in the floor of the caldera in the Las Borregas area of Sonora (Fig. 2; Riggs and Busby-Spera, 1991). This forms a >1-km-thick section of crystal-poor igimbrites lacking quartz, with lesser eolian sandstone. The uppermost of these ignimbrites is identical in outcrop and thin section to crystal-poor, densely welded, reddish purple ignimbrites of the Lower to Middle Jurassic Mount Wrightson Formation in the Santa Rita Mountains, which also lack quartz phenocrysts (Riggs and Busby-Spera, 1990, 1991).

The Glance Conglomerate unconformably overlies the Middle Jurassic volcanic section in the Pajarito Mountains (Knight, 1970; Keith and Theodore, 1975). It lacks volcanic interbeds, possibly due to its distance from the Sawmill Canyon fault zone, which appears to have plumbed magmas to the surface in Late Jurassic time (Fig. 1).

Mustang Mountains

We follow Lipman and Hagstrum's (1992) stratigraphy for the Mustang Mountains, consisting of the triparite ignimbrite sequence, Canelo Hills volcanics units I, II, and III, overlain by the Glance Conglomerate (Fig. 2). Unit I consists of reddish brown welded dacite tuff with ~25% phenocrysts of plagioclase, quartz, and biotite (Lipman and Hagstrum, 1992). In places it rests on Permian carbonates (Lipman and Hagstrum, 1992), but we map this contact as a buttress unconformity, where redbeds intervene between the Paleozoic carbonates and the ignimbrite (Fig. 2). The redbeds that underlie unit I contain slide blocks of Paleozoic limestone encased in siltstone. These are interstratified with breccia-conglomerate beds with chert clasts derived from the Paleozoic limestone as well as volcanic sandstone and conglomerate and lesser quartz arenite, all showing sedimentary structures typical of fluvial deposition (e.g., cut and fill structures, cross-lamination, good sorting). We recognize an angular unconformity of up to 45° between ignimbrite unit I and ignimbrite unit II, which we attribute to rotation on a syndepositional fault with a normal component of offset; this fault cuts unit I and is overlapped by unit II (Fig. 2). We also recognize angular discordance between units II and III, possibly due to rotation on a fault that cuts both units but does not cut the Glance Conglomerate (Fig. 2), which overlies unit II with slight angular discordance.

Canelo Hills volcanics unit II, the crystal-poor ultrawelded spherulitic rhyolite ignimbrite (Lipman and Hagstrum, 1992), contains lineations produced during flow that are useful for determining paleotransport directions. Both units II and III lie in buttress unconformity against faulted Paleozoic limestone (Fig. 2), and lineations in unit II are suggestive of transport at a 45° angle to the Sawmill Canyon fault zone; we speculate that this reflects funneling of flows along paleocanyons controlled by faults antithetic to the Sawmill Canyon fault zone.

Canelo Hills, Northern Block

The "rhyolite flow member" of the Canelo Hills volcanics (now referred to as ignimbrite unit II, Fig. 2) was first described in the northern Canelo Hills by Kluth (1982). This crystal-poor spherulitic ultrawelded rhyolite ignimbrite overlies quartz arenite (Fig. 2) with sedimentary structures typical of fluvial deposition (cut and fill structures and ripple and trough cross-lamination). This may record fluvial reworking of eolian sands. Downsection from this, in a section of redbeds, lies a purplish red, densely welded ignimbrite with plagioclase phenocrysts, which we tentatively correlate with the Mount Wrightson Formation (Fig. 2).

Canelo Hills, Southern and Lone Mountain Blocks

Canelo Hills volcanics unit II in the southern Canelo Hills and at nearby Lone Mountain is a thick intracaldera accumulation with shattered megablocks (Lipman and Hagstrum, 1992). Although those megablocks presumably avalanched off caldera walls, we suggest that the caldera was sited along a fault zone, because we recognize shattered megablocks in the overlying carbonate clast-dominated Glance Conglomerate as well. This passes gradationally upward into the volcanic clast-dominated Glance Conglomerate, with an andesite lava flow near its base and three ignimbrites in its upper part (Fig. 2). These ignimbrites also occur at Lone Mountain. These are nonwelded to weakly welded, pale gray crystal-poor ignimbrites that we tentatively correlate with crystal-poor nonwelded ignimbrites of the Patagonia Mountains and Santa Rita Mountains (Fig. 2).

Southern Huachuca Mountains

We follow Lipman and Hagstrum's (1992) stratigraphy for the southern Huachuca Mountains, consisting of the triparite ignimbrite sequence, Canelo Hills volcanics units I, II, and III, overlain by the Glance Conglomerate (Fig. 2). Unit I here fills the Montezuma caldera of Lipman and Hagstrum (1992) and contains abundant megablocks. We map andesitic to dacitic lava flows and flow breccias interstratified with red beds in the floor of the caldera. Unit II is relatively thin and discontinuous and is locally cut out by fluvial channels filled with volcanic lithic sandstone and reworked tuff. The upper contact of unit III is irregular and passes gradationally upward into unit III megablocks encased in basal Glance limestone-clast conglomerate. Above that, cobble-sized clasts of unit II and then unit III appear in the limestone-clast conglomerate. About halfway through the Glance Conglomerate is the "Glance andesite" of Hayes and Raup (1968), which consists of a chaotic mixture of andesite lava flow and flow breccia with shattered limestone and silicic volcanic megablocks. We interpret these as a rock avalanche deposit, synchronous with (perhaps triggered in some way by?) andesite effusive magmatism. Upsection from this, the Glance Conglomerate contains clasts of Canelo Hills volcanics unit I as well as clasts derived from the redbed section in the floor of the unit I caldera. To summarize, the Glance Conglomerate in the Huachuca Mountains records stripping of Canelo Hills volcanics units III, then II, and then I from a source area that also continuously exposed Paleozoic carbonates.

Summary of New Correlations

Our new correlations across the strands of the Sawmill Cayon fault zone include the following.

1. The Mount Wrightson Formation is correlated from the Santa Rita Mountains into the northern Canelo Hills, where it lies downsection from the Canelo Hills volcanics. It was also previously correlated to the southern extension of the Pajarito Mountains in the Borregas area, where it underlies the tuff of Cobre Ridge (Riggs et al., 1993).
2. Welded ignimbrite units II and III of the Canelo Hills volcanics in the Huachuca Mountains, Mustang Mountains, and southern Canelo Hills blocks (correlatons of Lipman and Hagstrum, 1992) are correlated into the northern Canelo Hills and Patagonia Mountains blocks. These all underlie Glance Conglomerate.
3. Nonwelded gray dacite block-and-ash-flow deposits in the Santa Rita Mountains are correlated into the Bagby Ranch block of the Patagonia Mountains. These are all interstratified with Glance Conglomerate.
4. Nonwelded to very weakly welded white rhyolite ignimbrites and interstratified phreatoplinian tuffs in the Santa Rita Glance Conglomerate are correlated with nonwelded to weakly welded pink to white rhyolite ignimbrites in the Canelo Pass block and Lone Mountain block (both in the southern Canelo Hills) and with white nonwelded rhyolite ignimbrites in the Patagonia Mountains block. These are all interstratified with Glance Conglomerate.

IMPLICATIONS OF "WET" ERUPTIONS FOR STRATIGRAPHY OF THE WESTERN UNITED STATES

The great volume of ash in the Upper Jurassic Morrison Formation (Peterson, 1988), relative to older Jurassic strata, is commonly attributed to prevailing winds blowing toward the ENE (Poole, 1962; Peterson, 1988; Turner and Fishman, 1991). We suggest that another factor must be considered, and that is the style of volcanism produced in a more humid climate where transtensional basins may collect water. Kowallis et al. (2001) used new and previously published age data from ash beds in the western interior to infer two periods of Jurassic volcanism: the first in the Middle Jurassic Carmel and Temple Cap Formations and a second "apparently more extensive record" in the Upper Jurassic Morrison Formation (p. 383). They suggested that the section we refer to as the Santa Rita Glance Conglomerate may have been the source for at least some of the Upper Jurassic ash beds (see "Temporal-Bathtub Formation" of their Fig. 14). We agree with this correlation. We further propose that phreatoplinian ("wet") eruptions recorded by the "Glance volcanics" and "Glance tuffs" in southern Arizona were particularly effective at dispersing ash into the back-arc region during accumulation of the Morrison Formation. We predict that evidence of phreatomagmatic eruptions will be found in other segments of the Upper Jurassic arc in Arizona and southern California.

The presence of nonwelded ignimbrites and phreatoplinian tuffs suggests standing water above vent areas, but do Upper Jurassic lake deposits exist in southern Arizona? The Apache Canyon Formation of the Bisbee basin is interpreted to be the deposit of a shallow- to deep-water permanent lake that formed within a half graben (Soreghan, 1998). These lake deposits are presently exposed to the north of the Sawmill Canyon fault zone, in the Empire and Whetstone Mountains (Fig. 2). Klute (1991) considered these lakebeds to be Early Cretaceous in age, but existing fossil age data permit a Late Jurassic age (Soreghan, 1998). Several features of the lakebeds are more consistent with deposition in a Upper Jurassic graben, rather than in a Cretaceous posttectonic thermal downsag, as discussed by Soreghan (1998). The lacustrine facies distribution as well as beveling of older strata indicate deposition in a deep half graben. The lakebeds interfinger with Glance Conglomerate alluvial fan deposits. Blocks of distinctive laminated lacustrine limestone of the Apache Canyon Formation were resedimented into the Glance conglomerate, indicating syntectonic reworking of partially lithified lake deposits onto alluvial fans, presumably along graben boundary faults. Other evidence of an active tectonic environment includes slumps, subaqueous debris-flow deposits, and conglomerate turbidite channels. The lake was "long lived" and deep, with laminated beds indicative of an anoxic environment that would provide an excellent paleoclimate record if the strata were better dated. This is consistent with deposition in a faulted basin rather than a thermal downsag. Plant fragments concentrated on bedding planes suggest a humid environment with riverine input. Soreghan (1998) interprets siliceous layers in the laminated carbonates to represent ashes washed into the lake, and he reports the presence of at least two tuffs, as well as horizons of silicic volcanic rocks fragments. We are not aware of any tuffs in dated Cretaceous strata of the Bisbee basin. Magnetostratigraphic work could potentially resolve the Cretaceous versus Jurassic age of the Apache Canyon Formation (M. Soreghan, 2003, personal commun.) since reversals occurred in the Jurassic but not in the Early Cretaceous. Other poorly dated lacustrine deposits in southern Arizona that may be Upper Jurassic include those in the Amole Arkose near Tucson (Risley, 1987) as well as possible lakebeds in Sonora (T.F. Lawton, 2003, personal commun.).

MODEL FOR THE BISBEE BASIN AS A STRAIN-PARTITIONED ARC–BACK-ARC SYSTEM

Our interpretation of the Glance Conglomerate as intra-arc strike-slip basin deposits appears to be in conflict with recently published models for postvolcanic arc opening of the Bisbee basin (Lawton and McMillan, 1999; Dickinson and Lawton, 2001a, 2001b). Geochemical analyses from mafic lava flows in the Chiricahua Mountains (80 km east of the Santa Rita Mountains, just east of the Dragoon Mountains, Fig. 1) and the Huachuca Mountains (Fig. 1) show no evidence of slab involvement and are interpreted as continental rift volcanics erupted in a post-arc transtensional borderland (Lawton et al., 1997; Lawton and McMillan, 1999; Dickinson and Lawton, 2001a). The mafic lava flows in the Chiracahua Mountains lie above the Glance Conglomerate within a marine section bearing Upper Jurassic (Kimmeridgian) fossils, and in the Huachuca Mountains they are interbedded with the Glance Conglomerate (Lawton and Olmstead, 1995; Olmstead and Young, 2000; Dickinson and Lawton, 2001a). The mafic lava flows are far less voluminous and make up a very small proportion of the eastern Bisbee deposits relative to andesitic to rhyolitic volcanics of the western Bisbee basin (Fig. 1). The tectonic model presented in Dickinson and Lawton (2001b) calls on slab rollback as the driving force for continental extension in the Bisbee basin (their Figs. 5 and 8b). However, slab rollback cannot exist without subduction (Marsaglia, 1995), which implies the presence of a volcanic arc. If the arc existed, then it would be located on the outboard side of the continental extension. We argue that the Upper Jurassic strata along the Sawmill Canyon fault zone (Fig. 1) represent that arc, and we propose here that strain partitioning caused intra-arc strike-slip faulting along the western edge of the Bisbee basin and back-arc transtensional faulting along the eastern edge (Fig. 4). A slightly different model invoking widely distributed sinistral transtension is proposed by Anderson et al. (1995) and Anderson and Nourse (this volume), who argue that Upper Jurassic pull-apart basins developed across a broader region encompassing the Jurassic arc, the Bisbee basin, and Chihuahua Trough. In both models presented in this volume, strike-slip tectonics are important.

Our evidence for intra-arc strike-slip faulting along the Sawmill Canyon fault zone suggests that plate convergence was oblique (Jarrard, 1986). Oblique convergence is far more common than orthogonal convergence, and at most continental arcs, an obliquity of only 10 degrees off orthogonal results in the formation of strike-slip faults in the upper plate (Fitch, 1972; Jarrard, 1986; McCaffrey, 1992). These faults commonly form in the thermally weakened crust of the arc, particularly on continental crust, which is weaker and better coupled to the subducted slab than oceanic-arc crust (Dewey, 1980; Jarrard, 1986; Ryan and Coleman, 1992; Smith and Landis, 1995). Strike-slip faults, transtensional faults, and block rotations play an important role in modern volcanic arcs; examples include the Central American arc (Burkhart and Self, 1985; Jarrard, 1986; Weinberg, 1992); the Trans-Mexican Volcanic belt (Israde-Alcantara and Garduno-Monroy, 1999);

the Andean arc of southern Chile and Patagonia (Cembrano et al., 1996; Thomson, 2002); the Sumatra arc (Bellier and Sebrier, 1994); the Aeolian arc (Gioncada et al., 2003); the Calabrian arc (Van Dijk, 1994); the Aleutian arc (Geist et al., 1988); the Taupo Volcanic Zone (Cole and Lewis, 1981); the central Philippine arc (Sarewitz and Lewis, 1991); and others. Strike-slip basins are the most tectonically active type of basin (Nilsen and Sylvester, 1995), so the effects of strike-slip faulting on the development of volcanic successions within arcs must be profound.

We propose that transtensional stresses in the overriding plate resulted in partitioned strain within the Bisbee basin, where a back-arc continental rift formed behind (northeast of) a continental arc dominated by strike-slip deformation (Fig. 4). In this model, strain partitioning would have allowed coeval back-arc extension and intra-arc strike-slip deformation in the thermally weakened arc axis. In addition, much of the strike-slip motion from the oblique plate convergence could have been taken up by smaller, nonparallel faults (Klute, 1991) in a probably transtensional back-arc Bisbee basin.

Strain partitioning is more commonly discussed for transpressional settings (de Saint Blanquat et al., 1998), but there are examples of strain partitioning in transtensional settings (Acocella et al., 1999; Marra, 2001; Wesnousky and Jones, 1994). Strain partitioning in transtensional settings appears to be controlled by the relative strengths of faults and spatial or temporal changes in the regional stress field (Wesnousky and Jones, 1994). The presence of arc magmatism clearly affects relative fault strengths, especially at depth (de Saint Blanquat et al., 1998), providing the mechanism for strain partitioning in a transtensional convergent margin.

A possible modern analogue for a strain-partitioned, transtensional arc is the syn-arc back-arc rifting of the Andaman Sea at the obliquely convergent Sumatran-Andaman plate boundary (Curray et al., 1978; Maung, 1987). High obliquity reduces the volume of volcanism occurring within the arc itself as total rates of convergence decrease. Oblique convergence is strain-partitioned into intra-arc and accretionary wedge strike-slip faults and back-arc transtensional rifts. The back-arc rifting is highly oblique, with basin-bounding faults and volcanism obscured by a large sediment supply. Volcanism within the arc includes dome chains erupted along the master strike-slip fault, as well as calderas erupted from strike-slip stepover zones (Bellier and Sebrier, 1994; Bellier et al., 1999).

Our detailed studies in the Santa Rita Mountains show that unconformities and sequences developed synchronously with high-angle faults with normal- and reverse-slip separation that alternated in space and time and controlled the positions of grabens and arches. This structural style is typical of alternating restraining-releasing bend strike-slip basins. In this setting, small polygenetic, multivent volcanic complexes form along strands of the master fault.

We speculate that the Montezuma, Turkey Canyon, and Parker Canyon calderas of southern Arizona (from oldest to youngest, Fig. 1) formed at releasing stepovers in a sinistral fault system that progressively stepped left (to the southeast) with time. Our model makes use of three previously mapped strands of the Sawmill Canyon fault zone (Hayes and Raup, 1968), plus two strands hypothesized here (the strands that lie furthest to the northeast and the southwest on Fig. 1). The northeastern of the two hypothesized strands is entirely buried due to Tertiary Basin and Range extension. Sinistral slip on this hypothesized fault would restore the Canelo Hill Volcanics in the Mustang Mountains closer to their source calderas and provide a master fault for antithetic faults that control the orientation of paleocanyons filled by the ignimbrites in the Mustang Mountains. This hypothesized fault would control the northwestern margin of the oldest caldera. The southwestern of the two hypothesized strands is the Gringo Gulch fault (Fig. 1), the syndepositional fault we have mapped in the Santa Rita Mountains (described above; also see Bassett and Busby, this volume). We speculatively extend the Gringo Gulch fault to the southwestern margin of the youngest caldera (Fig. 1).

We agree with Dickinson and Lawton (2001a, p. 485) that "the influence of regional tectonics on local caldera collapse cannot be excluded" for the Cobre Ridge caldera, although it is not sited on the Sawmill Canyon fault zone or any other recognized regional structures (Fig. 1). Regional fault controls on the siting of the caldera sources for the Canelo Hills volcanics (I, II, and II) are much more obvious (Fig. 1). The Canelo Hills volcanics source calderas were not only sited along strands of the Sawmill Canyon fault zone, they were dismembered by it, so their original shapes (circular versus rectangular) are not easily discerned. We suggest that the calderas, which were postulated by Lipman and Hagstrum (1992) to have been circular (Fig. 1), may have instead been more rectilinear features bounded to the northeast and southwest by strike-slip faults with a normal component of slip and to the northwest and southeast by normal faults with a strike-slip component of slip. The geometry of these would be similar to Pliocene transtensional basins on the Baja California margin of the Gulf of California (Umhoefer et al., 2002). Extension in pull-apart basins may control the location and geometry of calderas, making them more elongate and rectangular in shape than the classic circular caldera structure (Ventura, 1994; Bellier et al., 1999). Caldera collapse occurs in the area of maximum extension rather than over the basin as a whole (Bellier and Sebrier, 1994), and the volume of eruptive material isn't enough to account for the volume of the basin, clearly indicating that subsidence is more controlled by fault kinematics than by magma withdrawal (Bellier and Sebrier, 1994; Ventura, 1994). Lipman's (1997) review of the geometries of calderas indicates that no large-volume pyroclastic eruptions have been proven to cause subsidence along a regional graben. Instead, subequant calderas with polygonal fault boundaries may be influenced by regional structures, although "drastic elongation of…subsidence structures along regional structural trends seems rare" (Lipman, 1997, p. 208).

The erosional unconformity between the Canelo Hills volcanics and the overlying Glance Conglomerate in the Huachuca Mountains (described above) has been interpreted to record a change in tectonic regime (Dickinson and Lawton, 2001a). As discussed above, however, angular (and erosional) unconformities occur between each of the Canelo Hills volcanics units (I, II, and II) as well as between the top volcanic unit and the overlying Glance Conglomerate in the Mustang Mountains. Furthermore, angular and erosional unconformities are well developed throughout the Santa Rita Glance Formation. We interpet these unconformities to record ongoing intra-arc deformation along strike-slip faults.

None of the Upper Jurassic rocks in modern ranges along the Sawmill Canyon fault zone have yet been mapped in as much detail as the Santa Rita Glance Formation. Nor have we yet identified distinctive piercing points across the various strands of the fault zone. However, the widespread occurrence of large slide masses in upper Middle to Upper Jurassic strata along the Sawmill Canyon fault zone (absent from Lower to Middle Jurassic strata, except within calderas) suggests a change in tectonic style. Furthermore, the possibility of sinistral strike-slip control on the development of the three upper Middle Jurassic calderas shown on Figure 1 is tantalizing. Therefore, we extend our robust interpretation of Late Jurassic strike-slip tectonics in the Santa Rita Mountains to other ranges along the Sawmill Canyon fault zone and suggest that future detailed basin analysis work will support this interpretation.

CONCLUSIONS

We propose that a transition from dominantly welded (i.e., DRY) ignimbrite eruptions to dominantly nonwelded (i.e., WET) ignimbrite eruptions occurred in late Middle to Late Jurassic time in the southwestern United States, in response to climate change (Table 1, Figs. 2 and 4). At the same time, eolianites were replaced by fluvial, debris-flow, and lacustrine deposits. This was not only the direct result of a wetter climate, due to greater availability of running water, but also the indirect result of a wetter climate, since "wet" eruptive products are more easily resedimented than "dry" eruptive products. "Wet" explosive eruptions produce extremely widespread ash falls and these likely influenced sedimentation far into the plate interior. We further propose that rapid northward drift of North America, shown by paleomagnetic data, produced this change in climate, and was also responsible for a change in tectonic style within the arc, from dominantly extensional in Early to Middle Jurassic time, to sinistral oblique convergence in late Middle to Late Jurassic time. This resulted in the formation of intra-arc strike-slip basins in southern Arizona.

ACKNOWLEDGMENTS

Thorough reviews by Steve Richard and Mike Wolf are gratefully acknowledged. We are indebted to Bill Dickinson, Peter Lipman, Gordon Haxel, Ken Hon, Peter Kokelaar, Richard Fisher, Michael Ort, Robert Hildebrand, and Malcolm Howells for discussions in the field. The field companionship and patience of CJB's three daughters, Claire, Sophia, and Marion, were very much appreciated. Support was provided by NSF-EAR 92-19739, awarded to Busby.

REFERENCES CITED

Acocella, V., Salvini, F., Funiciello, R., and Funiciello, C., 1999, The role of transfer structures on volcanic activity at Campi Flegri (southern Italy): Journal of Volcanology and Geothermal Research, v. 91, no. 2–4, p. 123–139, doi: 10.1016/S0377-0273(99)00032-3.

Anderson, T.H., McKee, J.W., McKee, M.B., and Nourse, J.A., 1995, Coarse clastic deposits of Late Jurassic age in southwestern North America—A record of pull-apart basins: Geological Society of America Abstracts with Programs, v. 27, no. 6, p. A-75.

Asmerom, Y., 1988, Mesozoic igneous activity in the Southern Cordillera of North America: Implication for tectonics and magma genesis [Ph.D. thesis]: Tucson, Arizona, University of Arizona, 232 p.

Bassett, K., and Busby, C.J., 1996, Interstratified rhyolite dome and andesite volcanic deposits in the Late Jurassic(?) Temporal-Bathtub Formations in southern Arizona: Geological Society of America Abstracts with Programs, v. 28, no. 7, p. A-114.

Bassett, K., and Busby, C., 1997, Intra-arc strike-slip basins in the Late Jurassic southern Cordillera: Structural and climatic controls on deposition: Geological Society of America 1997 Annual Meeting, Abstracts with Programs, v. 29, no. 6, p. A-201.

Beatty, B., 1987, Correlation of some mid-Mesozoic redbeds and quartz sandstones in the Santa Rita Mountains, Mustang Mountains, and Canelo Hills, southeastern Arizona [M.S. thesis]: Tucson, Arizona, University of Arizona, 29 p.

Bellier, O., and Sebrier, M., 1994, Relationship between tectonism and volcanism along the Great Sumatran fault zone deduced by SPOT image analyses: Tectonophysics, v. 233, p. 215–231.

Bellier, O., Bellon, H., Sebrier, M., and Sutanto, M.R.C., 1999, K-Ar age of the Ranau tuffs; implications for the Ranau Caldera emplacement and slip-partitioning in Sumatra (Indonesia): Tectonophysics, v. 312, p. 347–359, doi: 10.1016/S0040-1951(99)00198-5.

Bilodeau, W.L., 1979, Early Cretaceous tectonics and deposition of the Glance conglomerate, southeastern Arizona [Ph.D. thesis]: Stanford, California, Stanford University, 145 p.

Bilodeau, W.L., Kluth, C.F., and Vedder, L.K., 1987, Regional stratigraphic, sedimentologic, and tectonic relationships of the Glance Conglomerate in southern Arizona, *in* Dickinson, W.R., and Klute, M.A., eds., Mesozoic rocks of southern Arizona and adjacent areas: Arizona Geological Society Digest 18, p. 229–256.

Blakey, R.C., 1996, Jurassic regional unconformities of the Colorado Plateau: The key to accurate correlation and interpretation of the area's Jurassic geologic history, *in* Morales, M., ed., The continental Jurassic: Museum of Northern Arizona Bulletin 60, p. 439–478.

Bryan, S.E., Cas, R.A.F., and Marti, J., 2000, The 0.57 Ma plinian eruption of the Granadilla Member, Tenerife (Canary Islands): An example of complexity in eruption dynamics and evolution: Journal of Volcanology and Geothermal Research, v. 103, p. 209–238, doi: 10.1016/S0377-0273(00)00225-0.

Burkart, B., and Self, S., 1985, Extension and rotation of crustal blocks in northern Central America and effect on the volcanic arc: Geology, v. 13, p. 22–26, doi: 10.1130/0091-7613(1985)13<22:EAROCB>2.0.CO;2.

Busby, C.J. and Bassett, K., 2005, Volcanic facies architecture of an intra-arc strike slip basin, Santa Rita Mountains, Southern Arizona: Geological Society of America, Abstracts with programs, v. 37, no. 4, p. 64.

Busby, C.J., and Kokelaar, B.P., 1992, Development of calderas in a transtensional fault zone: 29th International Geological Congress Abstracts, Kyoto, Japan, v. 2, p. 482.

Busby-Spera, C.J., 1988, Speculative tectonic model for the Early Mesozoic arc of the southwest Cordilleran United States: Geology, v. 16, p. 1121–1125, doi: 10.1130/0091-7613(1988)016<1121:STMFTE>2.3.CO;2.

Busby-Spera, C.J., Mattinson, J.M., and Adams, B., 1989, The Jurassic transition from continental intra-arc extension to Gulf of Mexico-related rifting in the southwest Cordillera: Eos (Transactions, American Geophysical Union), v. 70, p. 1300.

Busby-Spera, C.J., Mattinson, J.M., Riggs, N.R., and Schermer, E.R., 1990, The Triassic-Jurassic magmatic arc in the Mojave-Sonoran deserts and the Sierran-Klamath region; similarities and differences in paleogeographic evolution, *in* Harwood, D., and Miller, M.M., eds., Paleozoic and Early Mesozoic paleogeographic relations; Sierra Nevada, Klamath Mountains, and related terranes: Geological Society of America Special Paper 255, p. 325–338.

Cembrano, J., Herve, F., and Lavenu, A., 1996, The Liquine Ofqui fault zone: A long lived intra-arc fault system in southern Chile: Tectonophysics, v. 259, p. 55–66, doi: 10.1016/0040-1951(95)00066-6.

Cole, J.W., and Lewis, K.B., 1981, Evolution of the Taupo-Hikurangi subduction system: Tectonophysics, v. 72, p. 1–21, doi: 10.1016/0040-1951(81)90084-6.

Cowan, H., and Pettinga, J.R., 1992, Late Quaternary tectonics and landscape evolution: Glynn Wye, North Canterbury, Field Trip Guides, 1992 Joint Annual Conference of Geological Society of New Zealand and New Zealand Geophysical Society, Geological Society of New Zealand Miscellaneous Publication 63B, p. 51–73.

Cowan, H.A., Pettinga, J.R., and Smith, I.E.M., 1989, Transtension and structural complexity along the Hope Fault, Glynn Wye to Hammer Basin, North Canterbury: Geological Society of New Zealand Miscellaneous Publication, v. 43, p. 31.

Crowell, J.C., 2003, Introduction to Geology of Ridge Basin, Southern California, *in* Crowell, J.C., ed., Evolution of Ridge Basin, southern California: An interplay of sedimentation and tectonics: Geological Society of America Special Paper 367, p. 1–15.

Curray, J.R., Moore, D.G., Lawver, L.A., Emmel, F.J., Raitt, R.W., Henry, M., and Kieckhefer, R., 1978, Tectonics of the Andaman Sea and Burma, *in* Watkins, J.S., Montadert, L., and Dickerson, P.W., eds., Geological and Geophysical Investigations of Continental Margins: Tulsa, Oklahoma, American Association of Petroleum Geologists, p. 189–198.

Davis, G.H., Phillips, M.P., Reynolds, S.J., and Varga, R.J., 1979, Origin and provenance of some exotic blocks in lower Mesozoic red-bed basin deposits, southern Arizona: Geological Society of America Bulletin, v. 90, p. I-376–I-384.

De Rita, D., Giordano, G., Esposito, A., Fabbri, M., and Rodani, S., 2002, Large volume phreatomagmatic ignimbrites from the Colli Albani volcano (middle Pleistocene, Italy): Journal of Volcanology and Geothermal Research, v. 118, p. 77–98, doi: 10.1016/S0377-0273(02)00251-2.

de Saint Blanquat, M., Tikoff, B., Tessier, C., and Vigneresse, J.L., 1998, Transpressional kinematics and magmatic arcs: Geological Society Special Publication 135, p. 327–340.

Dewey, J.F., 1980, Episodicity, sequence, and style at convergent plate boundaries, *in* Strangway, D.W., ed., The Continental Crust and Its Mineral Deposits: Geological Association of Canada, Special Paper 20, p. 533–573.

Dickinson, W.R., and Lawton, T.F., 2001a, Tectonic setting and sandstone petrofacies of the Bisbee basin (USA-Mexico): Journal of South American Earth Sciences, v. 14, p. 475–504, doi: 10.1016/S0895-9811(01)00046-3.

Dickinson, W.R., and Lawton, T.F., 2001b, Carboniferous to Cretaceous assembly and fragmentation of Mexico: Geological Society of America Bulletin, v. 113, p. 1142–1160, doi: 10.1130/0016-7606(2001)113<1142:CTCAAF>2.0.CO;2.

Dickinson, W.R., Klute, M.A., and Swift, P.N., 1986, The Bisbee Basin and its bearing on Late Mesozoic paleogeographic and paleotectonic relations between the Cordilleran and Caribbean regions, *in* Abbott, P.L., ed., Cretaceous Stratigraphy of Western North America: Society of Economic Paleontologists and Mineralogists, Field-Trip Guidebook Pacific Section, p. 51–62.

Dickinson, W.R., Klute, M.A., and Bilodeau, W.L., 1987, Tectonic setting and sedimentological features of upper Mesozoic strata in southeastern Arizona, *in* Davis, G.H., and NandenDolder, E.M., eds., Geologic Diversity of Arizona and Its Margins: Excursions to Choice Areas: Arizona Geological Survey Special Paper 5, p. 266–279.

Dickinson, W.R., Fiorillo, A.R., Hall, D.L., Monreal, R., Potochnik, A.R., and Swift, P.N., 1989, Cretaceous strata of southern Arizona, *in* Jenny, J.P., and Reynolds, S.J., eds., Geologic Evolution of Arizona: Arizona Geological Society Digest, v. 17, p. 397–434.

Drewes, H., 1971a, Geologic map of the Mount Wrightson Quadrangle, southeast of Tucson, Santa Cruz and Pima Counties, Arizona: U.S. Geological Survey Map I-614, scale 1:48,000, 1 sheet.

Drewes, H., 1971b, Geologic map of the Sahuarita Quadrangle, southeast of Tucson, Pima County, Arizona: U.S. Geological Survey Map I-613, scale 1:48,000, 1 sheet.

Drewes, H., 1971c, Mesozoic stratigraphy of the Santa Rita Mountains, southeast of Tucson, Arizona: U.S. Geological Survey Professional Paper 748, 35 p.

Drewes, H., 1972, Structural geology of the Santa Rita Mountains, southeast of Tucson, Arizona: U.S. Geological Survey Professional Paper 658-C, 81 p.

Drewes, H., 1976, Plutonic rocks of the Santa Rita Mountains, southeast of Tucson, Arizona: U.S. Geological Survey Professional Paper 915, 75 p.

Drewes, H., 1981, Tectonics of southeastern Arizona: U.S. Geological Survey Professional Paper 1144, 96 p.

Fisher, R.V., and Schmincke, H.-U., 1984, Pyroclastic Rocks: New York, Springer-Verlag, 472 p.

Fitch, T.J., 1972, Plate convergence, transcurrent faults, and internal deformation adjacent to southeast Asia and the western Pacific: Journal of Geophysical Research, v. 77, p. 4432–4460.

Freundt, A., Wilson, C.J.N., and Carey, S.N., 2000, Ignimbrites and block-and-ash flow deposits, *in* Sigurdsson, H., Houghton, B.F., McNutt, S.R., Rymer, H., and Stix, J., eds., Encyclopoedia of Volcanoes: Academic Press, San Diego, p. 581–600.

Geist, E.L., Childs, J.R., and Scholl, D.W., 1988, The origin of summit basins of the Aleutian Ridge: Implications for block rotation of an arc massif: Tectonics, v. 7, p. 327–341.

Gioncada, A., Mazzuoli, R., Bisson, M., and Pareschi, M.T., 2003, Petrology of volcanic products younger than 42 ka on the Lipari-Vulcano complex (Aeolian Islands, Italy): An example of volcanism controlled by tectonics: Journal of Volcanology and Geothermal Research, v. 122, p. 191–220, doi: 10.1016/S0377-0273(02)00502-4.

Gradstein, F.M., Ogg, J.G., and Smith, A.G., et al., 2004, A Geologic Time Scale 2004: Cambridge University Press.

Hayes, P.T., 1970, Mesozoic stratigraphy of the Mule and Huachuca Mountains, Arizona: U.S. Geological Survey Professional Paper 658-A, 27 p.

Hayes, P.T., and Raup, R.B., 1968, Geologic map of the Huachuca and Mustang Mountains, Arizona: U.S. Geological Survey Miscellaneous Investigations Map I-509, 1:62,500, 1 sheet.

Heiken, G., and Wohletz, G., 1985, Volcanic Ash: University of California Press, Berkeley, 246 p.

Houghton, B.F., Wilson, C.J.N., Smith, R.T., and Gilbert, J.S., 2000a, Phreatoplinian eruptions, *in* Sigurdsson, H., Houghton, B.F., McNutt, S.R., Rymer, H., and Stix, J., eds., Encyclopedia of Volcanoes: Academic Press, New York, p. 513–526.

Houghton, B.F., Wilson, C.J.N., and Pyle, D.M., 2000b, Pyroclastic fall deposits, *in* Sigurdsson, H., Houghton, B.F., McNutt, S.R., Rymer, H., and Stix, J., eds., Encyclopedia of Volcanoes: Academic Press, New York, p. 555–570.

Imlay, R.W., 1980, Jurassic paleobiogeography of the conterminous United States in its continental setting: U.S. Geological Survey Professional Paper 1062, 134 p.

Israde-Alcantara, I., and Garduno-Monroy, V.H., 1999, Lacustrine record in a volcanic intra-arc setting: The evolution of the Late Neogene Cuitzeo basin system (central-western Mexico, Michoacan): Palaeogeography, Palaeoclimatology, Palaeoecology, v. 151, p. 209–227, doi: 10.1016/S0031-0182(99)00024-3.

Jacques-Ayala, C., 1995, Paleogeography and provenance of the Lower Cretaceous Bisbee Group in the Caborca-Santa Ana area, northwestern Sonora, *in* Jacques-Ayala, C., González-León, C.M., and Roldán-Quintana, J., eds., Studies of the Mesozoic of Sonora and adjacent areas: Geological Society of America Special Paper 301, p. 79–98.

Jarrard, R.D., 1986, Relations among subduction parameters: Reviews of Geophysics, v. 24, p. 217–284.

Jurado-Chichay, Z., and Walker, G.P.L., 2001, Variability of plinian fall deposits: Examples from Okataina Volcanic Centre, New Zealand: Journal of Volcanology and Geothermal Research, v. 111, p. 239–263, doi: 10.1016/S0377-0273(01)00229-3.

Keith, W.J., and Theodore, T.G., 1975, Reconnaissance geologic map of the Arivaca quadrangle, Arizona: U.S. Geological Survey Miscellaneous Field Studies Map MF-678, scale 1:62,500.

Klute, M.A., 1991, Sedimentology, sandstone petrofacies, and tectonic setting of the Late Mesozoic Bisbee Basin, southeastern Arizona [Ph.D. thesis]: Tucson, Arizona, University of Arizona, 268 p.

Kluth, C.F., 1982, Geology and mid-Mesozoic tectonics of the northern Canelo Hills, Santa Cruz County, Arizona [Ph.D. thesis]: University of Arizona, Tucson, 245 p.

Kluth, C.F., Butler, R.F., Harding, L.E., Shafiqullah, M., and Damon, P.E., 1982, Paleomagnetism of Late Jurassic rocks in the northern Canelo Hills, southeastern Arizona: Journal of Geophysical Research, v. 87, p. 7079–7086.

Knight, L.H., 1970, Structure and mineralization of the Oro Blanco Mining District, Santa Cruz County, Arizona [Ph.D. dissertation]: Tucson, Arizona, University of Arizona, 172 p.

Kowallis, B.J., Christiansen, E.H., Deino, A.L., Peterson, F., Turner, C.E., Kunk, M.J., and Obradovich, J.D., 1998, The age of the Morrison Formation, *in*

Carpenter, K., Chure, D.J., and Kirkland, J.I., eds., The Morrison Formation: An Interdisciplinary Study: Modern Geology, v. 22, p. 235–260.

Kowallis, B.J., Christiansen, E.H., Deino, A.L., Zhang, C., and Everett, B.H., 2001, The record of Middle Jurassic volcanism in the Carmel and Temple Cap Formations of southwestern Utah: Geological Society of America Bulletin, v. 113, p. 373–387, doi: 10.1130/0016-7606(2001)113<0373:TROMJV>2.0.CO;2.

Krebs, C.K., and Ruiz, J., 1987, Geochemistry of the Canelo Hills volcanics and implications for the Jurassic tectonic setting of southeastern Arizona: Arizona Geological Society Digest, v. 18, p. 139–151.

Lawton, T.F., and McMillan, N.J., 1999, Arc abandonment as a cause for passive continental rifting: Comparison of the Jurassic Mexican Borderland rift and the Cenozoic Rio Grande rift: Geology, v. 27, p. 779–782, doi: 10.1130/0091-7613(1999)027<0779:AAAACF>2.3.CO;2.

Lawton, T.F., and Olmstead, G.A., 1995, Stratigraphy and structure of part of the Bisbee group, northeastern Chiricahua Mountains, Arizona, *in* Jacques-Ayala, C., González-León, C.M., and Roldán-Quintana, J., eds., Studies in the Mesozoic of Sonora and Adjacent Areas: Geological Society of America Special Paper 301, p. 21–39.

Lawton, T.F., Garrison, J.M., and McMillan, N.J., 1997, Late Jurassic transtensional borderland on the southwestern margin of North America: Geological Society of America Abstracts with Programs, v. 29, no. 6, p. A-200.

Link, M., 2003, Depositional systems and sedimentary facies of the Mio-Pliocene Ridge Basin Group, Ridge Basin, California, *in* Crowell, J. C., ed., Evolution of Ridge Basin, southern California: An interplay of sedimentation and tectonics. Geological Society of America Special Paper 367, p. 17–87.

Lipman, P.W., 1997, Subsidence of sh-flow calderas: Relation to caldera size and magma-chamber geometry: Bulletin of Volcanology, v. 59, p. 198–218, doi: 10.1007/s004450050186.

Lipman, P.W., and Hagstrum, J.T., 1992, Jurassic ashflow sheets, calderas, and related intrusions of the Cordilleran volcanic arc in southeastern Arizona: Implications for regional tectonics and ore deposits: Geological Society of America Bulletin, v. 104, p. 32–39, doi: 10.1130/0016-7606(1992)104<0032:JAFSCA>2.3.CO;2.

Marra, F., 2001, Strike-slip faulting and block rotation: A possible trigger mechanism for lava flows in the Alban Hills?: Journal of Structural Geology, v. 23, p. 127–141, doi: 10.1016/S0191-8141(00)00068-7.

Marsaglia, K., 1995, Interarc and backarc basins, *in* Busby, C.J., and Ingersoll, R.V., eds., Tectonics of Sedimentary Basins: Oxford, Blackwell Science, p. 299–330.

Maung, H., 1987, Transcurrent movements in the Burma-Andaman Sea region: Geology, v. 15, p. 911–912,

May, S.R., Butler, R.F., Shafiqullah, M., and Damon, P.E., 1986, Paleomagnetism of Jurassic volcanic rocks in the Patagonia Mountains, southeastern Arizona; implications for the North American 170 Ma reference pole: Journal of Geophysical Research, v. 91, p. 11,545–11,555.

May, S.R., Beck, M.E., Jr., and Butler, R.F., 1989, North American apparent polar wander, plate motion, and left-oblique convergence, Late Jurassic–Early Cretaceous orogenic consequences: Tectonics, v. 8, p. 433–451.

McCaffrey, R., 1992, Oblique plate convergence, slip vectors, and forearc deformation: Journal of Geophysical Research, v. 97, B6, p. 8905–8915.

McKee, M.B., and Anderson, T.H., 1998, Mass-gravity deposits and structures in the Lower Cretaceous of Sonora, Mexico: Geological Society of America Bulletin, v. 110, p. 1516–1529, doi: 10.1130/0016-7606(1998)110<1516:MGDASI>2.3.CO;2.

McPhie, J., Doyle, M., and Allen, R., 1993, Volcanic Textures: A Guide to the Interpretation of textures in volcanic rocks: Hobart, Center for Ore Deposit and Exploration Studies, University of Tasmania, 196 p.

Morissey, M., Zimanowski, B., Wohletz, K., and Buettner, R., 2000, Phreatomagmatic fragmentation, *in* Sigurdsson, H., Houghton, B.F., McNutt, S.R., Rymer, H., and Stix, J., eds., Encyclopedia of Volcanoes: Academic Press, New York, p. 431–446.

Nilsen, T.H., and Sylvester, A.G., 1995, Strike-slip basins, *in* Busby, C.J., and Ingersoll, R.V., eds., Tectonics of Sedimentary Basins: Cambridge, Massachusetts, Blackwell Science, p. 425–457.

Olmstead, G.A., and Young, K., 2000, Late Jurassic ammonites from the northeastern Chiricahua Mountains, southeast Arizona: New Mexico Geology, v. 22, p. 1–7.

Peterson, F., 1988, A synthesis of the Jurassic system in the southern Rocky Mountain region, *in* Sloss, L.L., ed., Sedimentary Cover—North American Craton: U.S.: Geological Society of America, The Geology of North America, v. D-2, p. 65–76.

Poole, F.G., 1962, Wind directions in Late Paleozoic to Middle Mesozoic time on the Colorado Plateau, *in* Short Papers in Geology, Hydrology and Topography: U.S. Geological Survey Professional Paper 450-D, p. D147–D151.

Riggs, N.R., and Busby-Spera, C.J., 1990, Evolution of a multi-vent volcanic complex within a subsiding arc graben depression: Mount Wrightson Formation, southern Arizona: Geological Society of America Bulletin, v. 102, p. 1114–1135, doi: 10.1130/0016-7606(1990)102<1114:EOAMVV>2.3.CO;2.

Riggs, N.R., and Busby-Spera, C.J., 1991, Facies analysis of an ancient large caldera complex and implications for intra-arc subsidence: Middle Jurassic Cobre Ridge Group, southern Arizona, USA, *in* Cas, R., and Busby-Spera, C., eds., Sedimentary Geology, Special Issue on Volcanogenic Sedimentation, v. 74, nos. 1–4, p. 39–70.

Riggs, N.R., and Haxel, G.B., 1990, The Early to Middle Jurassic magmatic arc in southern Arizona: Plutons to sand dunes, *in* Gehrels, G.E., and Spencer, J.E., eds., Geologic Excursions through the Sonoran Desert Region, Arizona and Sonora: Arizona Geological Survey Special Paper 7, p. 90–103.

Riggs, N.R., Mattinson, J.M., and Busby, C.J., 1993, Correlation of Mesozoic eolian strata between the magmatic arc and the Colorado Plateau: New U-Pb geochronologic data from southern Arizona: Geological Society of America Bulletin, v. 105, p. 1231–1246, doi: 10.1130/0016-7606(1993)105<1231:COJESB>2.3.CO;2.

Risley, R., 1987, Sedimentation and stratigraphy of the Lower Cretaceous Amole Arkose, Tucson Mountains, Arizona, *in* Dickinson, W.R., and Klute, M.A., eds., Mesozoic rocks of southern Arizona and adjacent areas: Arizona Geological Society Digest 18, p. 215–228.

Ryan, H.F., and Coleman, P.J., 1992, Composite transform-convergent plate boundaries: Description and discussion: Marine and Petroleum Geology, v. 9, p. 89–97, doi: 10.1016/0264-8172(92)90006-Z.

Saleeby, J.B., and Busby-Spera, C.J., 1992, Early Mesozoic tectonic evolution of the western U.S. Cordillera, *in* Burchfiel, B.C., Lipman, P.W., and Zoback, M.L., eds., The Cordilleran Orogen: Conterminous U.S.: Boulder, Colorado, Geological Society of America, The Geology of North America, v. G-3, p. 107–168.

Sarewitz, D.R., and Lewis, S.D., 1991, The Marinduque intra-arc basin, Philippines: Basin genesis and in situ ophiolite development in a strike-slip setting: Geological Society of America Bulletin, v. 103, p. 597–614, doi: 10.1130/0016-7606(1991)103<0597:TMIABP>2.3.CO;2.

Saunders, A.D., and Tarney, J., 1984, Geochemical characteristics of basaltic volcanism within back-arc basins, *in* Kokelaar, B.P., and Howells, M.F., eds., Marginal Basins: Geological Society of London Special Publication 16, p. 59–76

Self, S., 1983, Large scale phreatomagmatic silicic volcanism: A case study from New Zealand: Journal Volcanology and Geothermal Research, v. 17, p. 433–469.

Self, S., and Sparks, R.S.J., 1978, Characteristics of widespread pyroclastic deposits formed by the interaction of silicic magma and water: Bulletin of Volcanology, v. 41, p. 196–212.

Silver, L.T., and Anderson, T.H., 1974, Possible left-lateral early to middle Mesozoic disruption of the southwestern North American craton margin: Geological Society of America Abstracts with Programs, v. 6, no. 7, p. 955–956.

Simons, F.A., 1972, Mesozoic Stratigraphy of the Patagonia Mountains and adjoining areas, Santa Cruz County, Arizona: U.S. Geological Survey Professional Paper 658-E, 23 p.

Simons, F.S., 1974, Geologic map and sections of the Nogales and Lochiel quadrangles, Santa Cruz County, Arizona: USGS Miscellaneous Investigations Series Map I-762.1 sheet, 1:62,500.

Smith, G.A., and Landis, C.A., 1995, Intra-arc basins, *in* Busby, C.J., and Ingersoll, R.V. eds., Tectonics of Sedimentary Basins: Oxford, Blackwell Science, p. 263–298.

Soreghan, M.J., 1998, Facies distribution within an ancient asymmetric lake basin: The Apache Canyon Formation, Bisbee Basin, Arizona, *in* Pitmna, J.K., and Carrol, A.R., eds., Modern and Ancient Lake Systems: Utah Geological Society Guidebook 26, p. 163–190.

Sparks, R.S.J., Self, S., and Walker, G.P.L., 1973, Products of ignimbrite eruptions: Geology, v. 1, p. 115–118, doi: 10.1130/0091-7613(1973)1<115:POIE>2.0.CO;2.

Steiner, M.B., 1998, Age, correlation, and tectonic implications of Morrison Formation paleomagnetic data, including rotation of the Colorado Plateau, *in* Carpenter, K., Chure, D.J., and Kirkland, J.I., eds., The

Morrison Formation: An Interdisciplinary Study: Modern Geology, v. 22, p. 261–282.

Steiner, M.B., and Marzolf, J.E., 1997, Jurassic tectonics from tectonosequences and paleomagnetism: Geological Society of America Abstracts with Programs, v. 29, no. 6, p. A-203.

Thomson, S.N., 2002, Late Cenozoic geomorphic and tectonic evolution of the Patagonian Andes between latitudes 42°S and 46°S: An appraisal based on fission-track results from the transpressional intra-arc Liquine-Ofqui fault zone: Geological Society of America Bulletin, v. 114, p. 1159–1173.

Titley, S.R., 1976, Evidence for a Mesozoic linear tectonic pattern in southeastern Arizona: Arizona Geological Society Digest, v. 10, p. 71–101.

Tosdal, R.M., Haxel, G.B., and Wright, J.E., 1989, Jurassic geology of the Sonoran Desert region, southern Arizona, southeastern California, and northernmost Sonora: Construction of a continental-margin magmatic arc, *in* Jenny, J.P., and Reynolds, S.J., eds., Geologic Evolution of Arizona: Arizona Geological Society Digest, v. 17, p. 397–434.

Turner, C.E., and Fishman, N.S., 1991, Jurassic lake T'oo'dichi': A large alkaline, saline lake, Morrison Formation, eastern Colorado Plateau: Geological Society of America Bulletin, v. 103, p. 538–558, doi: 10.1130/0016-7606(1991)103<0538:JLTODA>2.3.CO;2.

Umhoefer, P.J., Mayer, L., and Dorsey, R.J., 2002, Evolution of the margin of the Gulf of California near Loreto, Baja California peninsula, Mexico: GSA Bulletin, v. 114, p. 849–868.

Van Dijk, J.P., 1994, Late Neogene kinematics of intra-arc oblique shear zones: The Petilia-Rizzuto fault zone (Calabrian arc, central Mediterranean): Tectonics, v. 13, no. 5, p. 1201–1230, doi: 10.1029/93TC03551.

Vedder, L.K., 1984, Stratigraphic relationship between the Late Jurassic Canelo Hills Volcanics and the Glance Conglomerate, southeastern Arizona [M.S. thesis]: Tucson, Arizona, University of Arizona, 129 p.

Ventura, G., 1994, Tectonics, structural evolution and caldera formation on Vulcano Island (Aeolian Archipelago, southern Tyrrhenian Sea): Journal of Volcanology and Geothermal Research, v. 60, no. 3-4, p. 207–224, doi: 10.1016/0377-0273(94)90052-3.

Weinberg, R.F., 1992, Neotectonic development of western Nicaragua: Tectonics, v. 11, p. 1010–1017.

Wesnousky, S.G., and Jones, C.H., 1994, Oblique slip, slip partitioning, spatial and temporal changes in the stress field, and the relative strength of active faults in the Basin and Range, western United States: Geology, v. 22, p. 1031–1034, doi: 10.1130/0091-7613(1994)022<1031:OSSPSA>2.3.CO;2.

Wilson, C.J.N., Houghton, B.F., McWilliams, M.O., Lanphere, M.A., Weaver, S.D., and Briggs, R.M., 1995, Volcanic and structural evolution of Taupo Volcanic Zone, New Zealand, a review: Journal of Volcanology and Geothermal Research, v. 68, p. 1–28, doi: 10.1016/0377-0273(95)00006-G.

Wohletz, K., 1986, Explosive magma-water interactions: Thermodynamics, explosion mechanisms, and field studies: Bulletin of Volcanology, v. 48, p. 245–264, doi: 10.1007/BF01081754.

Wood, R.A., Pettinga, J.R., Bannister, S., Lamarche, G., and McMorran, T.J., 1994, Structure of the Hanmer strike-slip basin, Hope Fault, New Zealand: Geological Society of America Bulletin, v. 106, p. 1459–1473, doi: 10.1130/0016-7606(1994)106<1459:SOTHSS>2.3.CO;2.

Manuscript Accepted by the Society 24 March 2005

Printed in the USA

Geological Society of America
Special Paper 393
2005

Tectonic setting of the Glance Conglomerate along the Sawmill Canyon fault zone, southern Arizona: A sequence analysis of an intra-arc strike-slip basin

Kari N. Bassett*
Department of Geological Sciences, University of Canterbury, Prvt. Bag 4800, Christchurch, New Zealand

Cathy J. Busby*
Department of Geological Sciences, University of California, Santa Barbara, California 93106, USA

ABSTRACT

In the western Bisbee Basin of southern Arizona, detailed mapping and sequence analysis of the Glance Conglomerate along the largest basin-bounding fault, the Sawmill Canyon fault zone, reveals interbedded clastic, volcanic, and volcaniclastic lithofacies and their relationship to intrabasinal faulting, unconformities, and basin-bounding faults. The basin fill is dominated by small polygenetic, multivent volcanic complexes ranging in composition from rhyolite to andesite typical of continental arc volcanism. Syndepositional basin-bounding faults, the Sawmill Canyon and Gringo Gulch fault zones, controlled subsidence within the basin and plumbed small batches of magma to the surface. Small intrabasinal faults show stratigraphically limited offsets that alternate between normal and reverse separation. Eight unconformable surfaces occur within the basin. Five are asymmetrical, with one very steep wall and one gradually sloping wall. They show extreme vertical relief (460–910 m) with very high paleoslope gradients (40°–71°) that dip away from the master fault. We interpret these as uplifted fault scarps or paleoslide scars. The other three unconformities are symmetrical, V-shaped surfaces that have less steep walls, with vertical relief of 200–600 m and paleoslope gradients of 20°–25°. We interpret the symmetrical surfaces to be walls of deep paleocanyons cut during basin uplift events or following large ignimbrite eruptions.

Analysis of the unconformably bound stratigraphic sequences shows deposition to be related to subsidence along large basin-bounding faults modified by intrabasinal, high-angle, syndepositional normal and reverse faults. Erosion of the sequence-bounding unconformities took place during uplift associated with basin inversion. Alternation of uplift and subsidence and the juxtaposition of intrabasinal reverse and normal faults is typical of strike-slip basins. We interpret the Glance Conglomerate in the Santa Rita Mountains as the fill of an intra-arc strike-slip basin where strike-slip deformation was concentrated along the thermally weakened arc axis. We suggest a model for the Bisbee Basin of a strain-partitioned, obliquely convergent continental arc with backarc extension-transtension.

Keywords: Bisbee Basin, sequence stratigraphy, Glance Conglomerate, volcanology, strike-slip basin, intra-arc basin.

*Bassett, corresponding author—kari.bassett@canterbury.ac.nz; busby@geol.ucsb.edu.

Bassett, K.N., and Busby, C.J., 2005, Tectonic setting of the Glance Conglomerate along the Sawmill Canyon fault zone, southern Arizona: A sequence analysis of an intra-arc strike-slip basin, *in* Anderson, T.H., Nourse, J.A., McKee, J.W., and Steiner, M.B., eds., The Mojave-Sonora megashear hypothesis: Development, assessment, and alternatives: Geological Society of America Special Paper 393, p. 377–400. doi: 10.1130/2005.2393(14). For permission to copy, contact editing@geosociety.org.

INTRODUCTION

The tectonic setting of the Glance Conglomerate in southern Arizona has been interpreted as either a continental rift, particularly in eastern outcrops where conglomerates are interbedded with rare basaltic volcanics, or as a continental arc with backarc extension, particularly in western outcrops where conglomerates are interbedded with voluminous rhyolitic to andesitic volcanics. Throughout much of southern Arizona, Upper Jurassic to Lower Cretaceous(?) conglomerate and interstratified volcanic rocks overlie Lower to Middle Jurassic volcanic sections (Saleeby et al., 1992). The conglomerates, generally called the Glance Conglomerate, commonly lie at the base of Cretaceous nonmarine to marine sections referred to as the Bisbee Group (Fig. 1; Bilodeau, 1979; Jacques-Ayala, 1995; Lawton and Olmstead, 1995). The Glance Conglomerate is the basal unit in the regional Bisbee Basin and thus records the tectonic setting during the basin's initial formation. The conglomerates accumulated as piedmont fan and canyon fill deposits with locally interbedded lava flows and ignimbrites in grabens, half grabens, and calderas (Bassett and Busby, 1997; Bassett and Busby, 1996a, 1996b; Bilodeau, 1979; Busby and Kokelaar, 1992; Dickinson et al., 1987; Lawton and McMillan, 1999; Lipman and Hagstrum, 1992).

The Glance Conglomerate in the northeastern Bisbee Basin of New Mexico and eastern Arizona, which is "inboard" relative to the paleo-Pacific subduction margin, contains few interstratified volcanic rocks, and these are basaltic in composition (Fig. 1) (Lawton and Olmstead, 1995; Lawton and McMillan, 1999). Geochemical data on the basalts from the eastern Bisbee basin indicate eruption in a rift, rather than arc, environment (Lawton and McMillan, 1999). In contrast, the Glance Conglomerate on the southwestern, outboard edge of the Bisbee basin contains abundant rhyolitic, dacitic, and andesitic volcanic and volcaniclastic deposits interstratified with boulder breccia-conglomerates (Fig. 1) (Drewes, 1971b; Hayes, 1970a; Hayes and Raup, 1968; Kluth, 1982; Lipman and Hagstrum, 1992). Geochemical

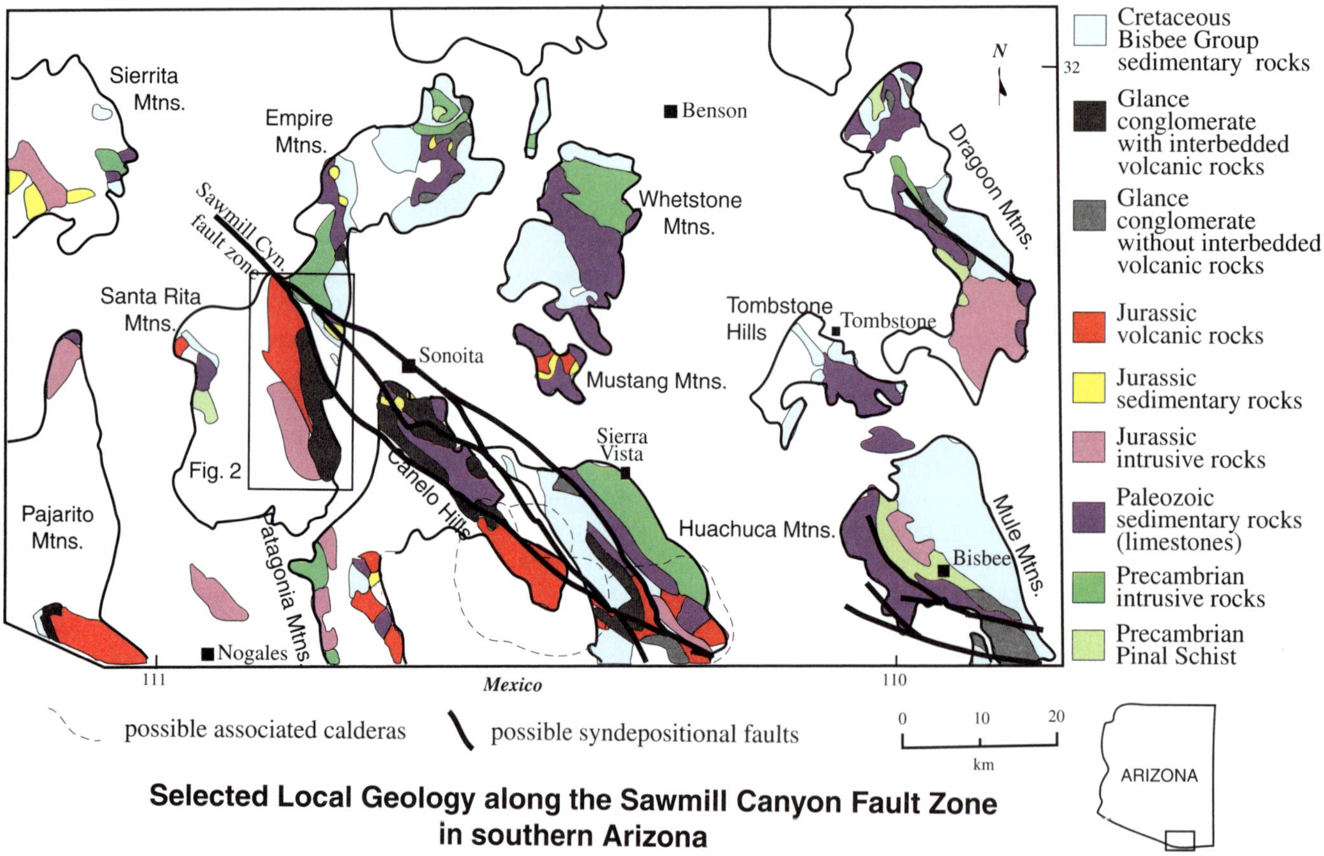

Figure 1. Geologic map of part of southern Arizona showing Upper Jurassic–Lower Cretaceous Bisbee Basin deposits. This map shows the basal Glance Conglomerate and overlying strata of the Bisbee Group, older rocks that acted as sources for the Glance Conglomerate, fault zones that were active in Jurassic time (including the Sawmill Canyon fault zone), and Jurassic calderas along the Sawmill Canyon fault zone. Upper Cretaceous and younger rocks are left blank. The box outlines the field area shown in Figure 2. Compiled from previous researchers (Asmerom, 1988; Bassett and Busby, 1997; Bassett and Busby, 1996a, 1996b; Beatty, 1987; Bilodeau, 1979; Busby and Bassett, 2005; Busby-Spera, 1988; Busby-Spera and Kokelaar, 1991; Dickinson et al., 1989; Dickinson et al., 1987; Dickinson et al., 1986).

analyses of volcanic rock samples from the Canelo Hills region (Fig. 1) show light rare earth element (LREE) enrichment, Th enrichment, and a strong negative Eu anomaly (Krebs and Ruiz, 1987) interpreted to record waning continental arc volcanism (Lawton and McMillan, 1999).

The tectonic setting of the Glance Conglomerate of southern Arizona has been interpreted in several ways. Some suggest that it forms part of the Jurassic arc sequence, particularly in areas where it has abundant interstratified volcanic rocks (Nourse, 1995; Tosdal et al., 1989), with the inboard deposits recording backarc extension (Bilodeau, 1979). Others have inferred that the Glance Conglomerate records continental rifting associated with the opening of the Gulf of Mexico (Dickinson et al., 1986, 1987; Dickinson and Lawton, 2001a, 2001b; Lawton and McMillan, 1999). A model was proposed in which the Gulf of Mexico–related rifts progressively exploited the thermally weakened, structurally attenuated crust of the Jurassic arc (Busby-Spera et al., 1989; Dickinson et al., 1986, 1987; Dickinson and Lawton, 2001a, 2001b; Lawton and McMillan, 1999; Saleeby et al., 1992). Strike-slip tectonics related to the Mojave-Sonora megashear (Silver and Anderson, 1983) may or may not have played a role. Evidence for movement on the Sawmill Canyon fault zone of the western Bisbee Basin in Late Jurassic time includes large slide masses interbedded with the Glance Conglomerate in the Mustang Mountains, the Huachuca Mountains, and the Canelo Hills (Fig. 1; Davis et al., 1979; Hayes and Raup, 1968) as well as along the southern extension of the Sawmill Canyon fault zone into Mexico (McKee and Anderson, 1998).

We have focused our study along the Sawmill Canyon fault zone in the western Bisbee Basin because it is the widest and longest fault zone in the larger basin (Fig. 1) and because it formed an important conduit for andesitic to rhyolitic magmas throughout the Jurassic for the western Glance Conglomerate (Bassett and Busby, 1996b; Busby-Spera and Kokelaar, 1991; Riggs and Busby-Spera, 1990). The Sawmill Canyon fault zone and related NW-striking, steep faults form a regional lineament inherited from Precambrian basement and reactivated in Mesozoic and Cenozoic times (Fig. 1; Drewes, 1981; Titley, 1976). The Jurassic movement history of the Sawmill Canyon fault zone has been variably interpreted as normal or strike slip with either dextral or sinistral movement (Bilodeau, 1979; Busby-Spera and Kokelaar, 1991; Drewes, 1972, 1981; Hagstrum and Lipman, 1991; Lipman and Hagstrum, 1992; Titley, 1976). In this paper, we place volcanic and sedimentary lithofacies into a sequence stratigraphic framework by mapping unconformity-bounded depositional sequences and by examining the relationships of the unconformities and depositional sequences to intrabasinal and extrabasinal faulting and volcanism. This permits reconstruction of the structural and paleogeomorphic evolution of the volcanic Glance Conglomerate along the Sawmill Canyon fault zone in southern Arizona.

Glance Conglomerate in the Santa Rita Mountains

The strata west of the Sawmill Canyon fault zone in the Santa Rita Mountains were originally mapped as the Temporal, Bathtub, and Glance Formations, subdivided into members (Drewes, 1971a). Bilodeau (1979) considered these strata to be part of the basal Glance Conglomerate of the Bisbee Basin, however, and we agree. In the Santa Rita Mountains, the Glance Conglomerate is more volcanic than sedimentary and is correlative with the "Glance tuffs" informally used for ignimbrites interstratified with Glance Conglomerate elsewhere in the western Bisbee Basin (Vedder, 1984). These include volcanic and volcaniclastic rocks of andesitic, dacitic, and rhyolitic composition interstratified with boulder to cobble breccia-conglomerates. Inferred source areas for both volcanic and sedimentary lithofacies range from intrabasinal to proximal extrabasinal to distal extrabasinal sources (Table 1).

Our lithofacies mapping (Fig. 2) shows there is no basis for the distinction of the three formations proposed by Drewes (1971a), let alone individual members. Lithofacies of the three "formations" repeat and interfinger with each other (Fig. 2); furthermore, we map major unconformities that crosscut formational and member boundaries (Fig. 2). For these reasons, we formally place the Temporal and Bathtub Formations into the Glance Conglomerate (Fig. 3). We have also remapped the rhyolitic plinian and dome deposits of the underlying Mount Wrightson Formation and the andesitic volcanic sandstone lithofacies of some of the overlying Gringo Gulch Formation into the Glance Conglomerate due to their lithologic similarity (Fig. 3).

The Glance Conglomerate crops out in a 20 × 6.5 km elongate belt extending southward from the Sawmill Canyon fault zone (Fig. 2). Beds are unfolded and have a regional strike roughly north-south and dip toward the NW-SE–striking Sawmill Canyon fault zone. Thus the outcrop represents an oblique 2-D cross section in map view that lies at ~55° angle to the regional strike of the fault zone. The Glance Conglomerate is cut by splays of the Sawmill Canyon fault zone to the northeast and is buried by Quaternary gravels to the southeast (Fig. 2). It lies unconformably on the Middle Jurassic Mount Wrightson Formation, the Middle Jurassic Piper Gulch monzonite, and the Middle Jurassic Squaw Gulch granite to the southwest and northwest (Fig. 2).

The Santa Rita Glance Conglomerate was deposited in two sub-basins separated by a paleohigh formed by high-angle faults (Fig. 2). Syndepositional intrabasinal faults are small and short lived with stratigraphically limited offsets. The Glance Conglomerate also contains multiple, deeply and steeply incised, intraformational unconformities of varying lateral extent.

The age of the Santa Rita Glance is likely Upper Jurassic (and/or uppermost Middle Jurassic) based on crosscutting relationships and clast compositions. The underlying Mount Wrightson Formation commonly dips ~20° steeper than the Glance Conglomerate and is cut by a deep unconformity that shows more than 0.86 km of vertical relief across the field area (Fig. 2). The conglomerates contain abundant distinctive clasts of red ultrawelded ignimbrite from the underlying Mount Wrightson Formation (Figs. 3 and 4). U/Pb zircon dates on abraded and acid-washed zircons from several samples throughout the Mount Wrightson Formation indicate that it accumulated between

TABLE 1. LITHOFACIES DESCRIPTIONS AND INTERPRETATIONS

ROCK NAME	SEQUENCES	FIELD CHARACTERISTICS	THIN SECTION CHARACTERISTICS	FACIES INTERPRETATION
boulder breccia-conglomerates	in northern and southern sub-basins; sequences 1, 2, 3, 4, 5, 7, and 8; sequence 8 was previously mapped as Glance conglomerate by Drewes (1971)	massive boulder breccia-cgl.; D99=7–8 m, D90=2–3 m, D50=60 cm; polymict, very angular to subrounded clasts of local derivation. ARKOSIC SANDSTONE MATRIX variety is very poorly sorted, crudely stratified with rare sandstone interbeds and has brick red matrix with dominantly rounded granite clasts; DACITIC LITHIC TUFF MATRIX variety is massive, matrix supported, very poorly sorted and has greenish matrix with dominantly angular f.g. volcanic clasts RHYOLITIC PUMICE LAPILLI TUFF MATRIX variety immediately overlies rhyolitic plinian tuffs in sequence 1 in the northern sub-basin; RHYOLITIC QUARTZ PHYRIC PUMICE LAPILLI TUFF MATRIX variety immediately overlies rhyolitic quartz crystal-rich ignimbrite	ARKOSIC SANDSTONE MATRIX VARIETY—clasts dominated by granite lithics and few silicic volcanics; matrix contains free xls. of kspar, qtz., microcline > biotite, plag. DACITIC LITHIC TUFF MATRIX VARIETY—clasts dominated by f.g. dacite or rhyolite volcanic lithics; matrix contains free xls. of plag. > qtz. RHYOLITIC PUMICE LAPILLI TUFF MATRIX and RHYOLITIC QUARTZ PHYRIC PUMICE LAPILLI TUFF MATRIX—clasts dominated by granite and andesite; matrix contains plag., qtz., volcanic lithics, pumice	talus breccia and avalanche deposition within ~100 m of the fault scarp; proximal to medial debris flow deposition within 3 km of fault zone; distal sheetwash and channelized stream flow within 5 km of fault zone. Clasts mostly locally derived from outcrop in Santa Rita Mtns. Differences in the matrix determined by availability of volcanic material.
dacitic block and ash flow tuffs	in northern sub-basin only, sequences 4 and 5, but also occurs as matrix to boulder conglomerates in sequences 4 and 5	greenish-gray, massive, monomict tuff breccias; matrix supported; D90<1 m, D50~10–30 cm or D90=30 cm, D50<5 cm; interfingers gradationally with boulder conglomerates	fragmental textures of 75% dense clasts; matrix has broken, smaller xls. relative to dense clasts and moderate to high xl. content (15–30%)of 20% plag., 5% qtz., 5% biotite with rare flows containing <5% hb; contains rare accidental granite lithics	dacitic block and ash flows probably generated by lava dome collapse at north-eastern basin margin
andesitic lava flows, flow breccias and intrusions	sequences 2, 3, 4, and 5	dark gray to black; generally coherent with large plag. xls., commonly flow aligned; locally crosscuts stratigraphy; locally is flow banded; locally is brecciated and vesicular	varying xl. content 10–50% tot., dominantly of plagioclase, tr.–3% amphibole, <5% Cpx, 3% apatite may be present; locally groundmass is holocrystalline; rare granite accidental lithics introduce Kspar, qtz., titanite	andesitic lava flows, flow breccias and shallow-level intrusions that locally vent to provided surface lava flow and pyroclastic flows
andesitic ignimbrites	sequences 2, 3, and 7	lavender to purple-red, xl. rich, massive, scoriaceous lapilli tuff; scorias are black, commonly flattened; cognate lithics, D90=60 cm at base; one is locally welded and is less xl. rich; rare and small (3-5 cm) accidentals of granite, rhyolite	flattened scoria in matrix of bubble-wall shards; xl. rich 35–40% tot. of 30% plag., 5–10% biotite, tr. qtz.; granite accidentals may introduce Kspar; the welded on is less xl.-rich ~10–15% tot.	andesitic pumice-rich pyroclastic flows, nonwelded
andesitic vitric tuffs and tuffaceous sandstones	sequences 4, 5, and 7; portions of sequences 4–5 were previously mapped as Gringo Gulch volcanics by Drewes (1971)	lavender-gray to purplish black, textures vary depending on the degree of reworking from VITRIC TUFF to VOLCANIC LITHIC SANDSTONE—very poorly sorted, dark colored sandstone/tuff with angular, andesite grains; largely indistinctly stratified in medium beds; no imbrication; rare small channels of clast-supported, crudely stratified, polylithic boulder conglomerate similar to boulder conglomerate lithofacies	VITRIC TUFF: ~90% shards, dominantly bubble wall but also platy, blocky, and splintery, and scoria shreds; 10% andesite lithics; xls. of plag., biotite; VOLCANIC LITHIC SANDSTONE: andesitic lithics show a range of vesicularity from dense to scoria, with dense clasts dominating the most reworked layers; broken to subrounded xls. of plag. >> biotite >> qtz. > hb.; may have pseudomatrix of squashed and altered pumices	tuffaceous material minimally reworked by fluvial unconfined sheetwash, hyperconcentrated flood flow and channelized traction bedload
andesitic vulcanian breccia	sequence 7 only	dark gray to black, monomict clast-supported tuff breccia in multimeter thick beds, some with top few decimeters normally graded; dense andesitic clasts, to 1 m in size, with rare granite or aphyric rhyolite accidental lithics; broken plag. and biotite xls. in matrix	generally xl. rich 30–40% to 15% tot.; dominantly 30% plag., 5% biotite, <3% hb; mafics generally altered to oxides; dense lithics of same composition as matrix	andesitic vulcanian eruptions—moderate to violent ejection of solid fragments of new lava, generating block-rich avalanches
rhyolitic intrusions	probable feeder to sequence 6 dome package and intrudes sequences 1, 2, 3	white, aphyric, coherent, with very indistinct flow banding	xl. poor 5–10% tot. of Kspar or plag.>>qtz.; flow banded	rhyolitic shallow-level intrusion feeding upper rhyolite dome package
rhyolitic dome and dome breccias	sequences 1 and 6; portions were previously mapped as Mount Wrightson Formation by Drewes (1971)	white, flow banded, coherent to brecciated with very large clasts (~1 m); may have remobilized pink porcellanite surrounding blocks of dome breccia	very xl. poor to aphyric ~5% tot. of <5% plag., tr. qtz., tr. biotite; f.g. qtz.-fld. mosaic matrix; flow banding common; rare granite accidentals may introduce kspar	rhyolitic lava domes with flow banding, coherent interiors mantled by breccias

(*continued*)

TABLE 1. LITHOFACIES DESCRIPTIONS AND INTERPRETATIONS (*continued*)

ROCK NAME	SEQUENCES	FIELD CHARACTERISTICS	THIN SECTION CHARACTERISTICS	FACIES INTERPRETATION
rhyolitic block and ash flow tuffs	sequence 6 only	white to pink, monomict tuff breccia; angular dense clasts range in degree of vesicularity from banded rhyolite to rarer pumice; white to pink, aphyric matrix; matrix supported	very xl. poor 5–10% tot. of 5–10% plag., tr. qtz., tr. biotite; dense clast blocks in a vitriclastic matrix	rhyolitic block and ash flow deposited lateral to rhyolitic dome breccias, generated by dome collapse
rhyolitic crystal-poor ignimbrite	sequence 6 only	abundant white pumice, commonly flattened, in a pink to white aphyric matrix of vitric tuff; thin (~1 m) lithic-bearing ignimbrites occur locally interbedded with plinian tuffs	xl. poor to aphyric <5% tot. of 1-2% plag., tr. biotite, tr. qtz., tr. Kspar; >95% pumice shreds and shards, shards dominantly blocky or splintery, with lesser bubble wall, locally occurring lithics 2–3 cm	rhyolitic ignimbrite (pumice-rich pyroclastic flow), blocky shards indicate phreatoplinian ("wet") eruption
rhyolitic, white, high-grade ignimbrite	sequence 6 only	white, aphyric with planar to contorted ultrawelded horizons alternating gradationally with weakly welded horizons showing flattened pumice; ultrawelded horizons show lineations	aphyric to xl. poor ~5% tot. of <5% plag., <1% qtz., <1% biotite; ultrawelded horizons show extreme plastic deformation of shards and stretching of pumice; moderately welded horizons show sintering of shards and moderate flattening of pumice; bubble-wall shards only, no blocky or splintered shards	rhyolitic high-grade ignimbrite (ultrawelded, pumice-rich pyroclastic flow)
rhyolite plinian-phreatoplinian tuffs	sequences 1, 3, 4, 6 and 7; portions were previously mapped as Mount Wrightson Formation by Drewes (1971)	in order of abundance PLINIAN PUMICE FALL DEPOSITS—light greenish-white, pumice lapillistone, commonly thinly tabular-bedded and mantling topography REWORKED PLINIAN FALL DEPOSITS of medium-bedded, erosively based, lenticular, lithic-enriched and vitric-depleted lapilli tuffs PHREATOPLINIAN FALL DEPOSITS—pinkish red, very thinly tabular-bedded porcellanite (extremely fine-grained tuff); may show convolute lamination and other soft-sediment deformation: also occurs as deformed remnants within dome breccia CRYSTAL-RICH (?)PLINIAN FALL DEPOSITS—plag., qtz., biotite xl. rich, thin tabular beds, some mantling topography, xls. are css sized	PLINIAN PUMICE FALL DEPOSITS: ~80–90% pumice shreds (some long tube), 10–20% bubble-wall shard, aphyric; REWORKED: very xl. poor to 15-20% tot. of 10% qtz., 10% plag.; up to 15% lithics of amygdaloidal andesite, granite, aphyric rhyolite, welded tuff; remainder is pumice PHREATOPLINIAN FALL DEPOSITS: ~80-90% of blocky shards; remainder irresolvably fine grained CRYSTAL-RICH (?)PLINIAN FALL DEPOSITS: xl. rich ~40% tot. of 25% plag., 10% qtz., 5% biotite; remainder ~60% is pumice shreds	PLINIAN PUMICE FALL are widely dispersed sheets of stratified pumice derived from high eruption columns REWORKED PLINIAN FALL scoured bases, lenticularity, loss of delicate vitric components and introduction of lithics indicate remobilization by water PHREATOPLINIAN FALL abundant fines, increased sorting, pronounced bedding and blocky shards indicate interaction of vesiculating magma with surface water producing abundant steam and very fine ash CRYSTAL-RICH (?)PLINIAN FALL DEPOSITS crystal-rich indicates proximal plinian fall
rhyolitic crystal-rich ignimbrites	sequence 3 and 6, two occurrences only	white, massive tuff with relict fiamme, xl. rich; qtz. and biotite xls. are large and obvious (css to vcss sized); top locally more pumiceous; rare accidental granite lithics; upper ignimbrite has smaller xls.	xl. rich 30–35% tot. of 15% qtz., 10% plag., 5% biotite, <2% kspar; xls. are large (3–4 mm in sequence 3, 1–3 mm in seq. 6); 70% vitric content of nonwelded bubble wall shards and pumice shreds, but generally altered (vapor phase alteration?); feldspars commonly completely calcitized	silicic crystal and pumice rich pyroclastic flow, nonwelded
rhyolitic lithic-rich ignimbrite	sequence 4 only	white, massive tuff with well-preserved fiamme; lithic rich ~20–30%; D99=2 m, D90=20–50cm, D50=5 cm; lithics of white tuff, red andesite, siltstone, granite; outcrops in southern paleocanyon completely altered by later andesitic intrusion	xl. poor <15% tot. of plag>>qtz.>Kspar, no biotite; well-preserved nonwelded vitriclastic texture of pumice shreds and bubble wall shards; lithics ~20–30% tot. of andesite and granite with microcline; southern samples completely altered	silicic lithic and pumice rich pyroclastic flow, nonwelded
rhyolitic limestone lithic ignimbrite	sequence 4 only; portions were previously mapped as Gringo Gulch volcanics by Drewes (1971)	white to light lavender, massive tuffs with fiamme; xl. poor to aphyric; multiple flow units with welded centers showing more pronounced flattening of pumice; relatively lithic poor but with distinctive compositions including limestone, marble and amphibolite	xl. poor 5–13% of 10% plag., 3% biotite, tr. qtz.; plag. often completely altered to calcite; bubble wall shards and pumice shreds	silicic pumice rich pyroclastic flows bearing distinctive carbonate and metacarbonate lithics, nonwelded to welded in centers of flow units
rhyolitic, red, high-grade ignimbrites	sequences 3 and 5, one in each sequence; upper ignimbrite previously mapped as fault sliver of Mount Wrightson Formation by Drewes (1971)	red-maroon tuff with highly elongate fiamme; xl. poor with plag. only; top is spherlulitic and blocky; passes laterally to N into unit of 3 m blocks in boulder breccia-conglomerate; resembles older Mount Wrightson Formation. rheomorphic ignimbrites of Riggs and Busby-Spera (1991)	xl. poor <5% tot. of ~5% plag., tr. biotite>qtz.; more xl. rich in nonwelded outcrop ~15% tot.; vitriclastic texture of faint fiamme in highly welded samples to bubble wall shards in nonwelded samples; sometimes spherulitic, hematized; rare vesicles	silicic pumice rich pyroclastic flows, largely ultrawelded

Note: Described in full by Busby and Bassett (2005). Lithofacies are grouped by composition (rhyolitic vs. dacitic vs. andesitic) and inferred source (intrabasinal vs. proximal extrabasinal vs. distal extrabasinal).

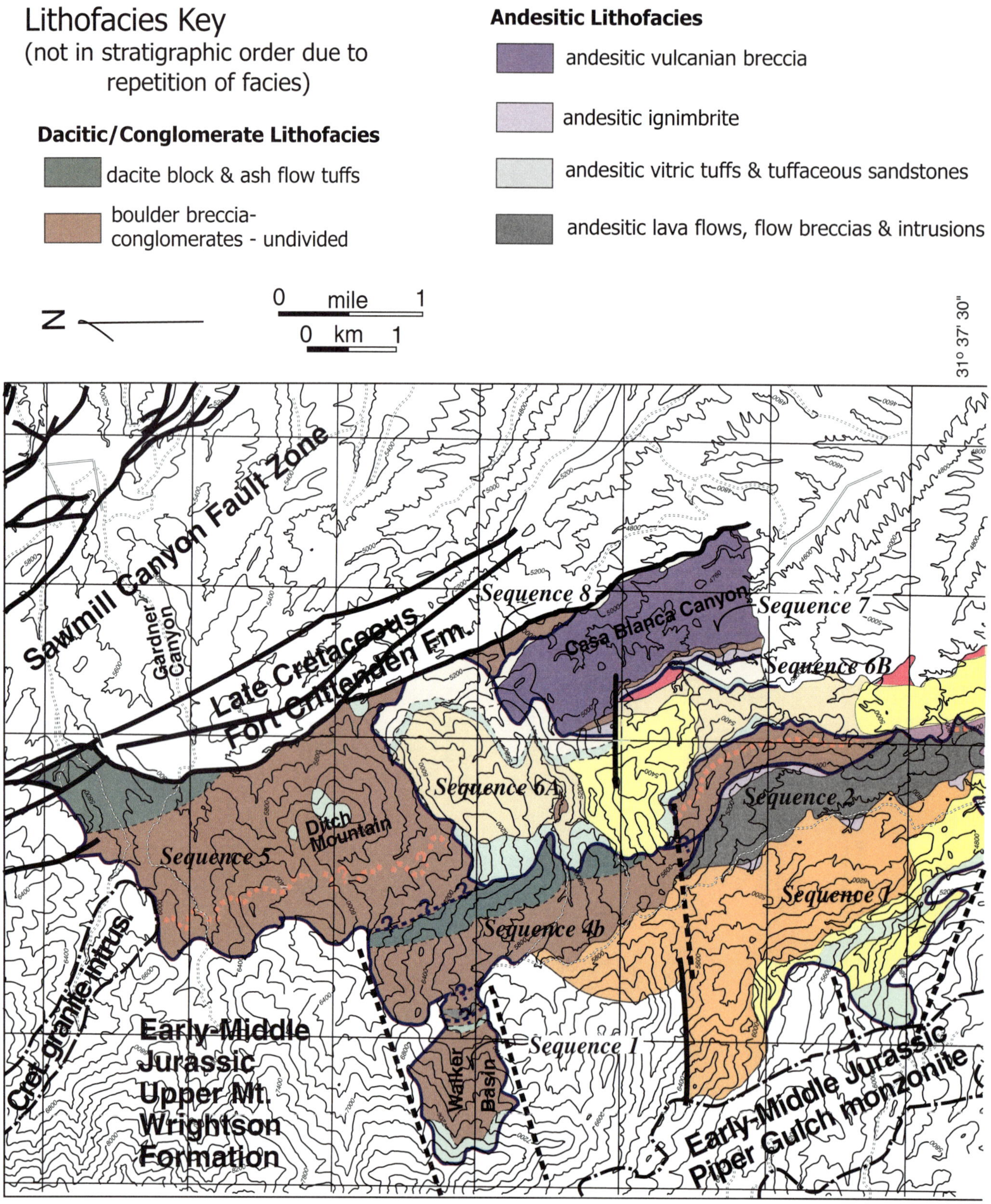
Lithofacies Key
(not in stratigraphic order due to repetition of facies)
Dacitic/Conglomerate Lithofacies
dacite block & ash flow tuffs
boulder breccia-conglomerates - undivided
Andesitic Lithofacies
andesitic vulcanian breccia
andesitic ignimbrite
andesitic vitric tuffs & tuffaceous sandstones
andesitic lava flows, flow breccias & intrusions
N
0 mile 1
0 km 1
31° 37' 30"
Sawmill Canyon Fault Zone
Gardner Canyon
Late Cretaceous Fort Crittenden Fm.
Sequence 8
Casa Blanca Canyon
Sequence 7
Sequence 6B
Sequence 6A
Sequence 2
Ditch Mountain
Sequence 5
Sequence 4b
Sequence 1
Cret. granite intrus.
Early-Middle Jurassic Upper Mt. Wrightson Formation
Walker Basin
Sequence 1
Early-Middle Jurassic Piper Gulch monzonite

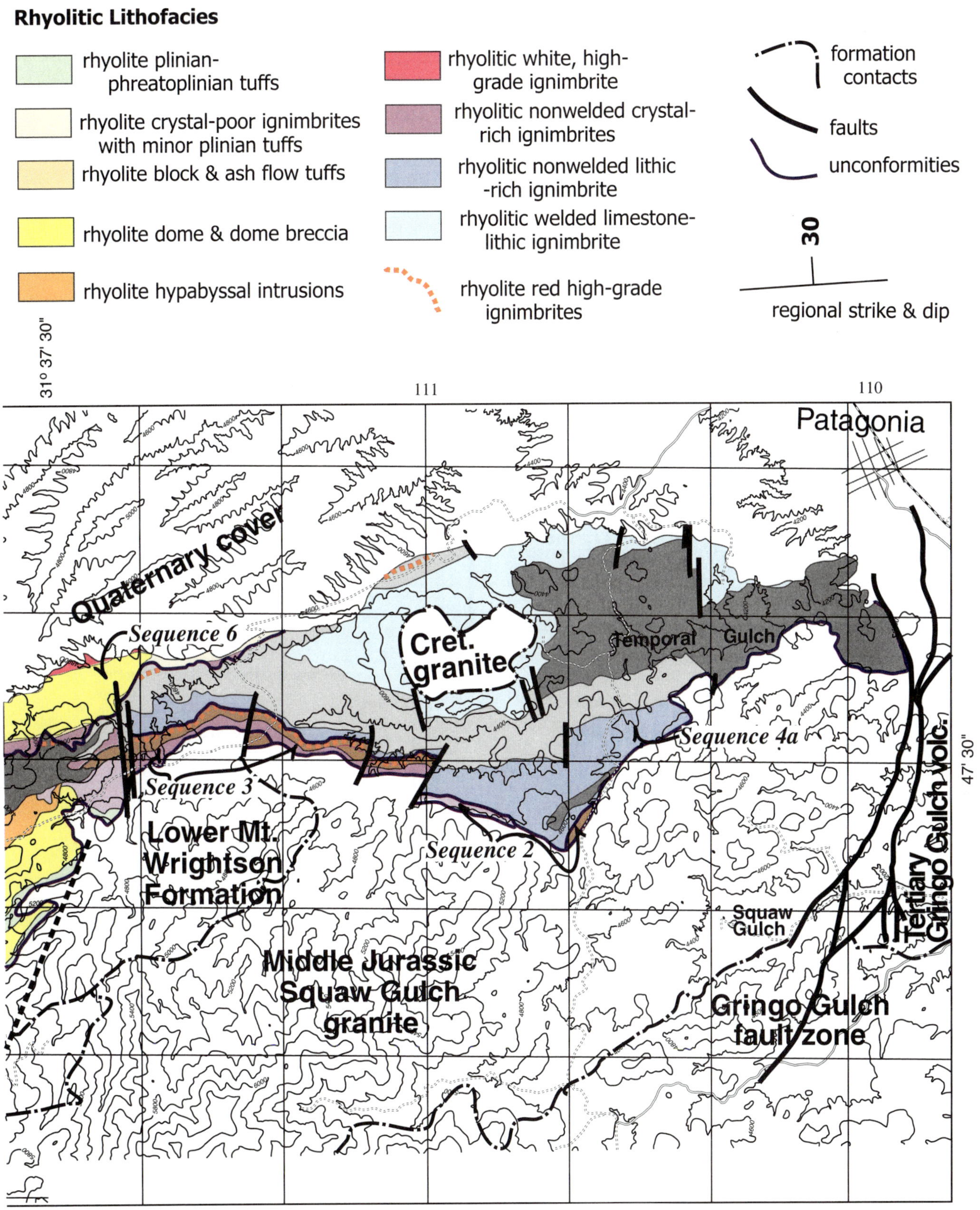

Figure 2 (*on this and previous page*). Lithofacies and sequence stratigraphic map of the newly defined Glance Conglomerate in the Santa Rita Mountains. The strike and dip of the homoclinal section allows an oblique cross-sectional view of the basin fill. The key shows lithofacies, and the sequences are labeled. Descriptions and interpretations of lithofacies are presented in Table 1.

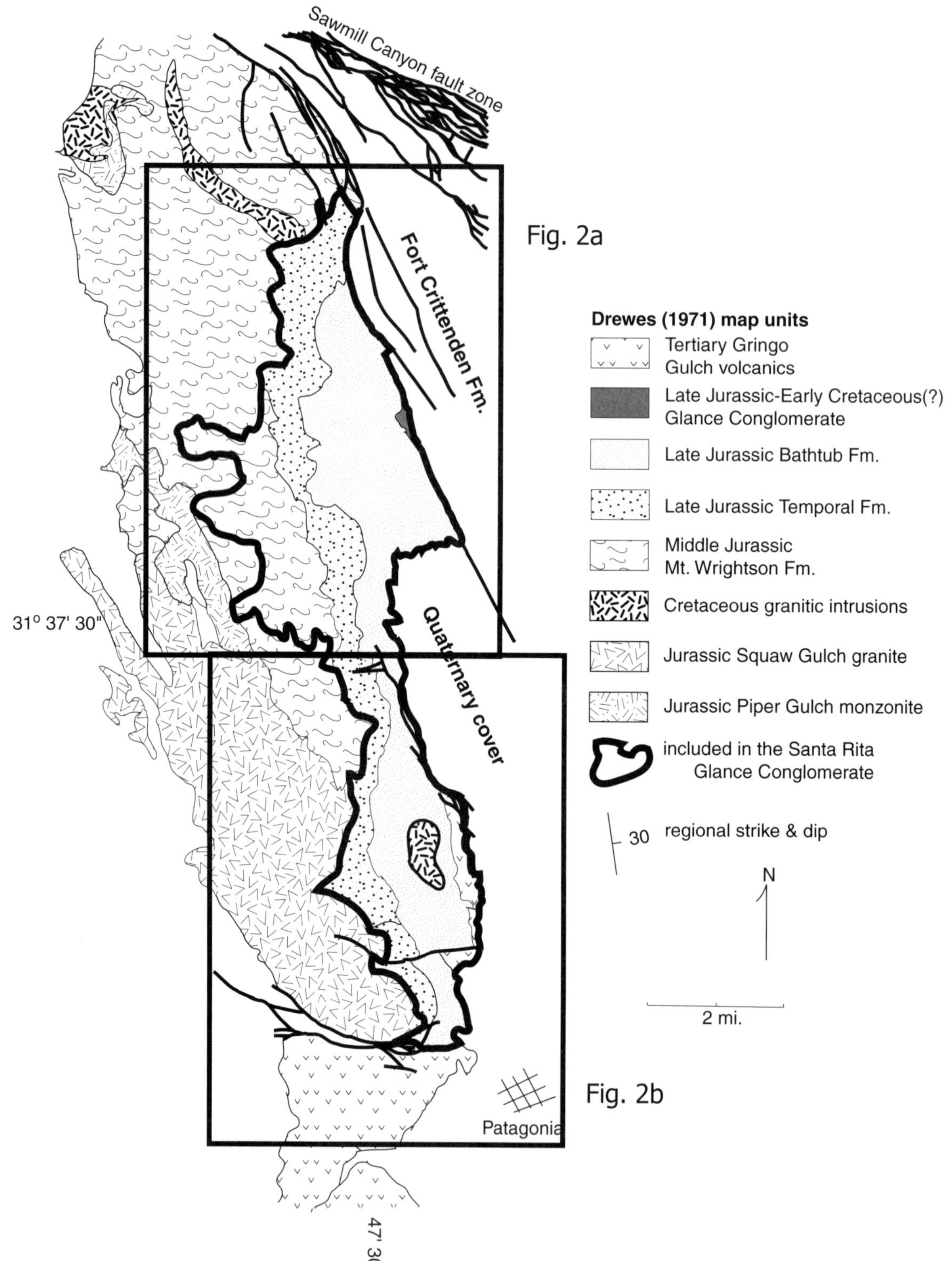

Figure 3. A map showing the remapped Glance Formation in the Santa Rita Mountains incorporating the Temporal, Bathtub, and Glance Formations, and parts of the Mount Wrightson and Gringo Gulch Formations as originally mapped by Drewes (1971a).

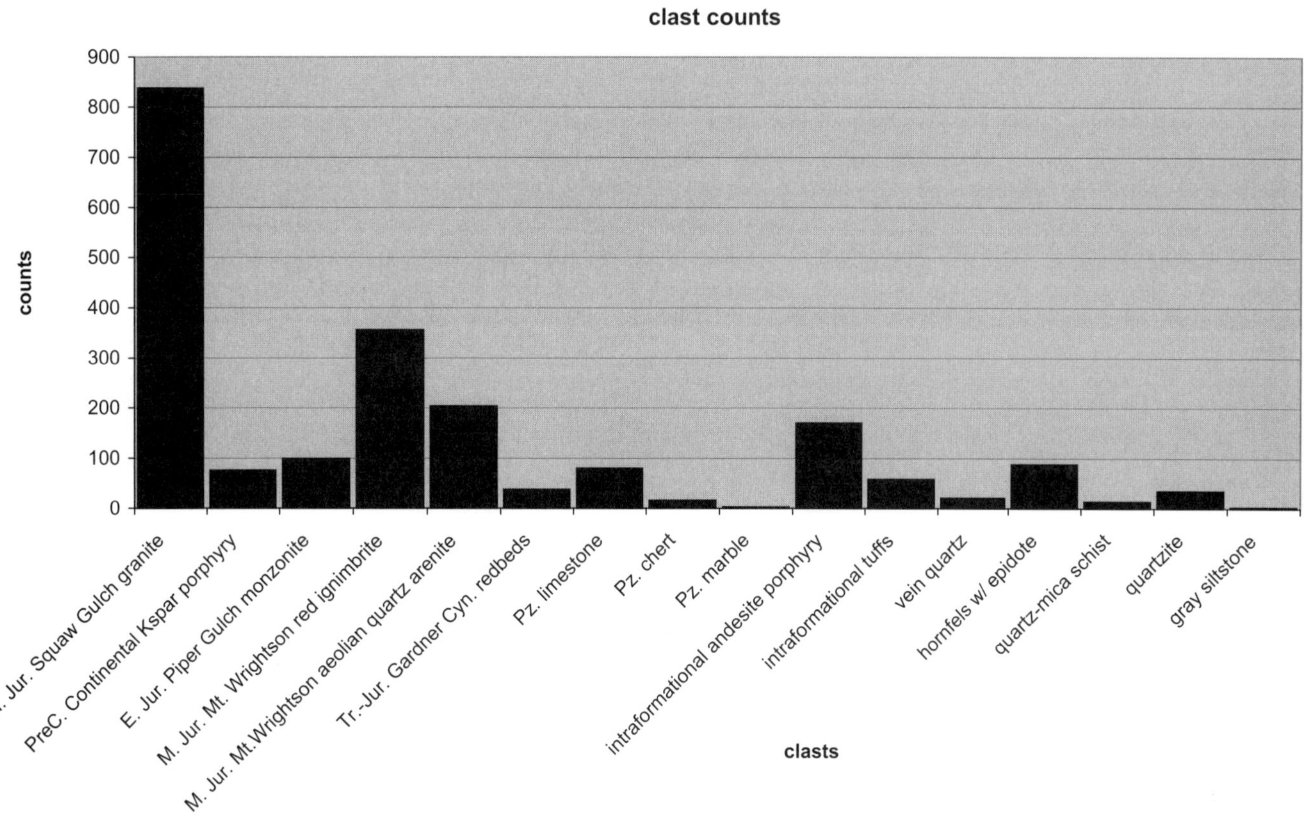

Figure 4. Clast counts from the boulder breccia-conglomerate lithofacies. Data were collected from sequence 5, at five different stratigraphic levels, using 300 counts per site. Nearly all the clasts can be matched to distinctive source rocks that occur in the present-day Santa Rita Mountains. Intraformational clasts indicate cannibalization of older basin fill. Only a small proportion of clasts are too nondistinctive to be correlated to source rocks.

190 Ma and 170 Ma (Riggs et al., 1993), making the Santa Rita Glance Conglomerate younger than 170 Ma. There are no direct constraints on the upper age limit of the Santa Rita Glance, but it is inferred to be pre-Cretaceous, because there are no dated Lower Cretaceous (pre-Laramide) volcanic (or granitic) rocks in southern Arizona. Thus the Santa Rita Glance Conglomerate is fully correlative with the Glance Conglomerate and interbedded basaltic volcanics in the eastern Bisbee Basin, which have yielded Upper Jurassic, Oxfordian ammonites (Lawton and Olmstead, 1995).

LITHOFACIES

The Glance Conglomerate in the Santa Rita Mountains consists of volcanic and volcaniclastic rocks interstratified with boulder to cobble breccia-conglomerates. Units are grouped into lithofacies that are then grouped into associations based on composition (rhyolitic versus dacitic versus andesitic) and on probable source area (intrabasinal versus proximal extrabasinally sourced versus distal extrabasinally sourced; Table 1). Lithologic descriptions are presented in Table 1. Facies interpretations are summarized below and presented in detail in Busby and Bassett (2005).

Boulder Breccia-Conglomerate Lithofacies (Extrabasinally Sourced)

Boulder breccia-conglomerates, typical of the Glance Conglomerate, consist of very coarse-grained, angular, very poorly sorted, massive deposits that become finer grained and more stratified with distance from the Sawmill Canyon fault zone. The matrix is either dacitic lithic tuff, where the breccia-conglomerate is interstratified with the dacitic block-and-ash-flow tuffs, or arkosic sandstone derived from the erosion of the underlying plutons, locally mixed with rhyolitic pumice lapilli tuff, or rhyolitic quartz-phyric pumice lapilli tuff.

Most of the clasts are derived from the underlying formations or from those present in fault slices in the Sawmill Canyon fault zone and include Mount Wrightson red ultrawelded ignimbrite and aeolian quartz arenite, Piper Gulch monzonite, Squaw Gulch granite, Gardener Canyon Formation, Paleozoic limestone and chert, and Proterozoic Continental granite (Fig. 4). Intraformational clasts of rhyolite and andesite are also present.

The boulder breccia-conglomerates are interpreted as talus and debris-flow deposits on alluvial fans. Such fans indicate substantial local relief at the time of deposition. The presence of intraformational clasts indicates cannibalization of older basin fill. The lithofacies has been identified as having a proximal extrabasinal source in the Sawmill Canyon fault zone because clast size, angularity of clasts, and the degree of disorganization of bedding all increase toward the fault zone, and most clast types can be found in fault slivers.

Andesitic Lithofacies Association (Intrabasinally Sourced)

The andesitic lithofacies association consists of andesitic intrusions, block-and-ash-flow tuffs, lava flows and flow breccias, ignimbrite, reworked vitric tuff, and vulcanian breccia (Fig. 2 and Table 1). The association of andesitic intrusions with thick successions of andesitic lava flows and block-and-ash-flow tuffs indicates intrabasinal venting, and therefore all associated lithofacies have been identified as intrabasinally sourced. Location of intrusions is generally controlled by intrabasinal faults (Fig. 2).

Dacitic Lithofacies (Extrabasinally Sourced)

Massive, matrix-supported, poorly sorted, and monomictic dacitic block-and-ash-flow tuff contain clasts up to 1 m, decreasing in size away from the Sawmill Canyon fault zone. Clast sizes, bed thickness, and clast-to-matrix ratio all increase toward the fault zone, indicating a proximal extrabasinal source. The block-and-ash-flow tuffs are interpreted to be the products of lava dome collapses, from domes displaced and therefore no longer preserved on the opposite side of the Sawmill Canyon fault zone.

Rhyolitic Lithofacies Associations (Extrabasinally Sourced and Intrabasinally Sourced)

The rhyolitic lithofacies association includes effusive and explosive subassociations that are both intrabasinally and extrabasinally sourced (Table 1). For the intrabasinally sourced effusive subassociation, rhyolitic intrusions are mapped directly into the rhyolitic dome and dome breccia lithofacies, and the rhyolitic domes are located on syndepositional faults (Fig. 2). The rhyolitic dome–dome breccia lithofacies is fringed by the rhyolitic block-and-ash-flow tuff lithofacies, interpreted to represent pyroclastic flows generated by lava dome collapse (Busby and Bassett, 2005). This assemblage is interstratified with the explosive subassociation consisting of the two rhyolitic ignimbrite lithofacies and the rhyolitic plinian-phreatoplinian lithofacies (Fig. 2). The ultrafine phreatoplinian deposits indicate magma-water interaction, possibly from the groundwater or eruption through surface lakes. Together, the assemblages record alternating effusive and explosive silicic volcanism within the basin, through vents controlled by syndepositional faulting.

Four types of rhyolitic ignimbrite have been identified as extrabasinally sourced, although the source may be relatively proximal, just outside of the Santa Rita Mountains area or across the Sawmill Canyon fault zone. They are identified as extrabasinally sourced, because they differ from the intrabasinal ignimbrites in phenocryst and lithic compositions, and because there are no identified vents, intrusions, or other proximal deposits with similar mineralogy within the Santa Rita Glance Basin. There are two distinctive quartz-crystal-rich ignimbrites, two lithic-rich ignimbrites (one with distinctive limestone lithics), and two red, high-grade ignimbrites (Table 1). The extrabasinally sourced rhyolitic ignimbrites were deposited mainly in the southern sub-basin, but some spill over the paleohigh for a short distance into the northern sub-basin (Fig. 2). Each of these ignimbrites is restricted to one or two horizons; this fact, and their distinctive textures and compositions, make them useful marker horizons that help to tie together the sequences of the northern and southern sub-basins. These were most likely outflow ignimbrites erupted from calderas elsewhere along the arc on the western side of the Bisbee Basin (Busby and Kokelaar, 1992; Lawton and McMillan, 1999; Busby et al., this volume).

Sequence Stratigraphic Methods

The mapping of unconformable and correlative conformable surfaces and the strata between has allowed us to develop a sequence stratigraphic history in a highly complex setting affected by both volcanism and syndepositional faulting (Fig. 5). The stratigraphy of volcanic and volcaniclastic deposits is very difficult to map due to abrupt lateral lithofacies changes, the episodicity of sediment supply controlled by eruptive style and recurrence rate, and the varying erodability of volcanic products controlled by eruptive style and composition (R.C.M. Smith, 1991; G. Smith, 1991). Volcanic constructs also modify topography within the basin independent of changes in accommodation, affecting sequence stratigraphic architecture.

Erosion is commonly discussed in terms of changes in base level, whether by changes in sea level or tectonically controlled subsidence (Blum and Tornqvist, 2000; van Wagoner, 1995). Yet in volcanic systems, aggradation and erosion can be produced by eruption followed by reequilibration and do not require any changes to base level or accommodation in the system (e.g., (R.C.M. Smith, 1991; G. Smith, 1991). The style of volcanism determines whether the episodic sediment supply is friable and easily remobilized (Fisher and Schmincke, 1984), whereas deposition of large ignimbrites often destroys the drainage network (Buesch, 1991; Manville, 2001; R.C.M. Smith, 1991). As the fluvial systems reestablish their gradient, channels are cut (Buesch, 1991; Manville, 2001; R.C.M. Smith, 1991). The reincision forms a sequence-bounding unconformity without any change to base level. Therefore, how can sequence analysis be used to distinguish volcanic from tectonic sequence boundaries and changes in accommodation?

Differentiating between accumulation and preservation space using geomorphic base level provides a clearer framework for distinguishing tectonic and volcanic controls on accommodation. Accumulation space is the "volume of space that can be filled within present process regimes," such as the volcanic setting described above. Preservation space is the space created when

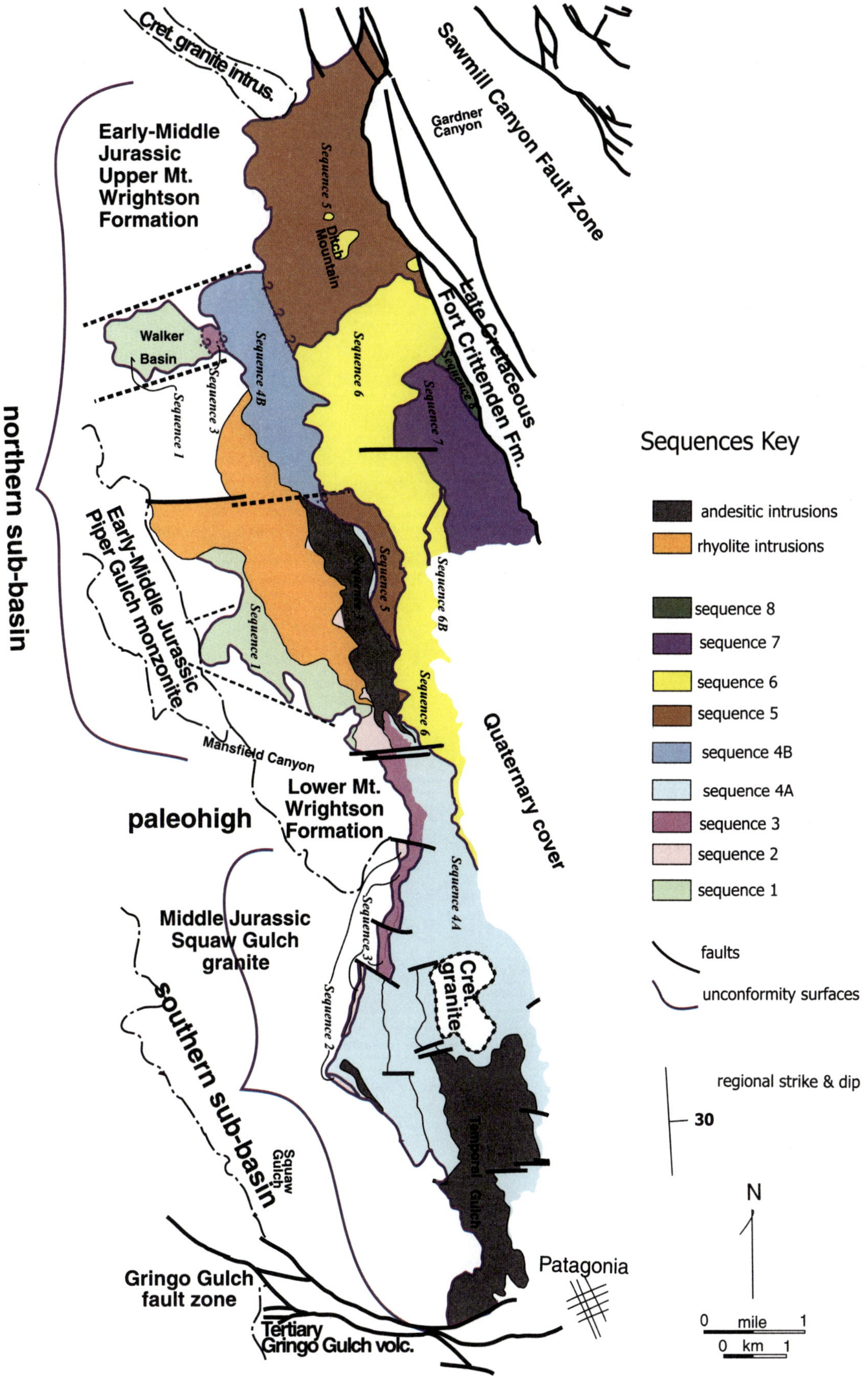

Figure 5. Map showing sequences and sequence boundaries without reference to the lithofacies. Northern and southern sub-basins are labeled. Sequence boundaries take three forms: (1) unconformities, (2) correlative conformities, and (3) syndepositional fault scarps. Unconformities form the majority of all sequence boundaries. Correlation of sequences is based on detailed lithofacies mapping (Fig. 2) and tracing of distinctive marker horizons such as ignimbrites and plinian deposits (Table 1). Sequences are roughly equivalent to time slices in Figure 6.

"subsidence lowers these deposits below possible depths of incision and removal" (Blum and Tornqvist, 2000, p. 20), such as from fault-controlled subsidence. Sequence boundaries in the Santa Rita Member of the Glance Conglomerate illustrate both aspects.

Following the sequence stratigraphic techniques set out by van Wagoner (van Wagoner, 1995; van Wagoner et al., 1988), we use unconformities as sequence boundaries without recourse to models for why they exist (e.g., Pekar et al., 2003). The unconformities in the Santa Rita Mountains are easily identified by their deep incision into the underlying strata (Fig. 2). These can be followed laterally into rare correlative conformities and into syndepositional fault scarps by walking out the surfaces and mapping distinctive marker horizons (Fig. 5). The sequences bounded by the unconformities are made up of the volcanic and volcaniclastic lithofacies presented above (Table 1).

Sequence Analysis of the Santa Rita Glance Conglomerate

Sequence bounding surfaces take three forms: erosional surfaces, fault scarps, and correlative conformable surfaces. A single bounding surface may change along its length through all three forms (Fig. 5). Eight sequence-bounding surfaces have been mapped; five occur in the northern sub-basin only (bounding surfaces 1, 5, 6, 7, and 8), two occur in the southern sub-basin only (bounding surfaces 2 and 3), and one occurs in both (bounding surfaces 4A and 4B; Table 2).

Unconformities within the Santa Rita Glance Conglomerate are recognized where a distinctive lithologic unit is cut by an erosional surface. They are commonly deeply incised into the underlying lithologies, with steeply cut walls. They are commonly, but not always, overlain by a lithologic unit of different composition. Identification requires detailed mapping of surfaces, because of abrupt lateral facies variations typical of volcanic systems. Unconformities are most easily identified where they cut distinctive, widespread, but stratigraphically limited, lithofacies or where they down cut through multiple horizons. Distinctive and widespread units include the rhyolitic plinian-phreatoplinian tuff, the andesitic ignimbrite, the rhyolitic white high-grade ignimbrite, the rhyolitic crystal-rich ignimbrite, the rhyolitic lithic-rich ignimbrite, and the rhyolitic red high-grade ignimbrite (Fig. 2 and Table 1).

Some of the unconformities can be traced laterally into correlative conformities (e.g., (Pekar et al., 2003). Again, correlation is easiest where distinctive, widespread lithofacies form marker horizons. Intrusions pose more difficulties where depositional sequences are "condensed" than where depositional sequences are "expanded." The boulder breccia-conglomerate lithofacies also creates difficulties because it forms amalgamated mono-

TABLE 2A. SURFACE RELIEF, MAXIMUM SURFACE GRADIENTS, AND LATERAL EXTENTS OF UNCONFORMITIES IN THE SANTA RITA GLANCE FORMATION

Unconformities	Minimum relief of surface	Maximum slope gradient	Lateral extent	Comments
unconformity 1	0.86	71 deg.	3.0 km	in N sub-basin
unconformity 2	0.21	20 deg.	2.8 km	unconf. in S sub-basin, conformable in N
unconformity 3	0.21	20 deg.	2.8 km	unconf. in S sub-basin, not present in N
unconformity 4A	0.60	24 deg.	8.7 km	unconf. in S sub-basin
unconformity 4B	0.51	68 deg.	3.9 km	unconf. in N sub-basin
unconformity 5	0.91	48.5 deg.	5.4 km	in northern sub-basin only
unconformity 6	0.77	55 deg.	5.4 km	in northern sub-basin only
unconformity 7	0.46	40 deg.	1.5 km	in northern sub-basin only
unconformity 8	0.13	43 deg.	0.8 km	in northern sub-basin only

Note: The scales of these require tectonic uplift, by (preserved) intrabasinal and (inferred) extrabasinal reverse faults.

TABLE 2B. MAXIMUM PRESERVED THICKNESSES OF THE EIGHT UNCONFORMITY-BOUNDED DEPOSITIONAL SEQUENCES IN THE SANTA RITA GLANCE FORMATION AND THEIR PRESERVED LATERAL EXTENTS

Sequences	Maximum thickness	Minimum thickness	Lateral extent	Comments
sequence 1	0.72	0.46	6.9 km	max in paleohigh, min in N
sequence 2	0.21	0.05	8 km	max in S, min in paleohigh and N
sequence 3	0.13	0.08	6.7 km	max in paleohigh, min in N & S (poor preservation)
sequence 4A	1.08	0.04	12.8 km	max in S, min in paleohigh
sequence 4B	0.51	0.21	3.9 km	max in N, min in paleohigh
sequence 5	0.78	0.01	9.1 km	max in N, min in S
sequence 6	0.77	0.03	5.8 km	max in N, min in S
sequence 7	0.36	0.36	1.9 km	max in N only
sequence 8	0.13	0.13	0.8 km	max in N only

Note: The great thicknesses of these depositional sequences require tectonic downdropping, by both (preserved) intrabasinal and (inferred) extrabasinal normal faults.

lithologic sections up to 1.1 km thick close to the Sawmill Canyon fault zone (Fig. 2 and Table 1). Careful attention to matrix composition and intervening mappable units allows division into at least four parts (Fig. 2).

Some of the unconformity surfaces are cut by syndepositional faults, and the fault scarps make up part of the mappable bounding surface forming buttress unconformities. Intrabasinal faults were clearly active during deposition with many locally ponding ignimbrites (Fig. 2). The faults are all high-angle faults, some with normal separation (sequences 1, 2, 6, and 7) and some with reverse separation (sequences 2 and 4). Some faults that exhibit normal separation were reactivated as faults exhibiting reverse separation (compare sequences 2 and 4), and some faults with normal separation were active synchronously with faults showing reverse separation elsewhere in the basin (sequence 2).

Unconformities in the southern sub-basin have a symmetrical V-shape and are recut to the same level causing successive unconformity surfaces to merge (Fig. 5, sequences 2, 3, and 4). These unconformities generally have slopes of ~20°–25° and vertical relief, measured from the deepest point of incision to the top of the shoulder, ranging from 200 m to 860 m (Table 2). Unconformities in the northern sub-basin are more deeply incised, are significantly steeper with canyon wall slopes ranging from 40° to 71°, and are asymmetric with one steep wall and one shallow wall (Table 2). All of the northern unconformities, except unconformity 1, are steepest on the side near the Sawmill Canyon fault zone (Fig. 5) and all are deeply incised (Table 2).

SEQUENCE STRATIGRAPHIC HISTORY OF THE BASIN

Sequence 1

Sequence boundary 1 is the oldest and forms the basal unconformity for the Glance Conglomerate in the Santa Rita Mountains. The unconformity is one of the most deeply incised bounding surfaces, has a vertical relief of 0.86 km, a lateral extent of at least 3 km, and dips as much as 71° (Figs. 2, 5, and Table 2). The scale of incision into the underlying formations indicates that erosion was controlled by tectonic uplift prior to the deposition of the Santa Rita Glance Conglomerate. This uplift must have been greatest in the south, progressively decreasing toward the north, because Middle Jurassic plutons and the Lower Jurassic lower member of the Mount Wrightson Formation were brought to the surface in the south, whereas the Middle Jurassic middle member of the Mount Wrightson Formation forms the basin floor in the north (Fig. 2; Riggs and Busby-Spera, 1990). The steepness of the unconformity dip is probably partly due to syndepositional faulting and erosion of the fault scarp, making it a combination unconformity and fault plane surface. We treat it as a sequence boundary that formed on an incrementally evolving paleotopographic surface.

Above the basal unconformity, deposition of sequence 1 begins with a widespread rhyolitic plinian-phreatoplinian tuff, which allows us to correlate the basal strata between two troughs whose steep-sided, narrow geometry suggests they were fault bounded (Fig. 6). A rhyolitic lava dome grew in the southern fault-bounded trough while 0.7-km-thick, boulder breccia-conglomerates accumulated in the northern fault-bounded trough. This was followed by a second rhyolitic plinian-phreatoplinian eruption, with a more marked phreatoplinian signature than the first plinian tuff, covering the dome and breccia-conglomerate and providing another marker horizon between the northern and southern fault-bounded troughs. This was followed by construction of a second rhyolitic lava dome in the southern fault-bounded trough. Meanwhile the boulder breccia-conglomerate continued to accumulate in the northern fault-bounded trough, although it is much thinner than the lower one (0.1 km, Fig. 6), suggesting that accommodation was being filled, possibly due to slowing subsidence in the trough. The boulder breccia-conglomerates in the northern fault-bounded trough are dominantly arkosic sandstone matrix, except where they overlie the rhyolitic plinian-phreatoplinian tuffs; there they have the rhyolitic pumice lapilli tuff matrix (Table 1). The boulder breccia-conglomerates represent proximal debris-flow deposits on a highly aggradational fan system, so we do not believe the tuff matrix material was derived from intrabasinal erosion but rather was washed in from surrounding regions that it had mantled.

Sequence 1 records the incremental creation of accommodation in the northern half of the field area (Fig. 6) accomplished by the progressive subsidence of two fault-bounded troughs, each ~0.7 km deep in total. The faults bounding these two troughs are not exposed due to forest cover, but the units comprising the fill of the two troughs are well exposed. We infer that they represent fault-bounded troughs because of their extremely narrow and deep geometry with steeply dipping sides. The map of the underlying Mount Wrightson Formation allows this interpretation (Fig. 2, Riggs and Busby-Spera, 1990).

Sequence 2

Sequence boundary 2 is a symmetrical V-shape with slopes less than 20° dip, 0.21 km vertical relief between shoulders and deepest point of the V, and a preserved width of 2.8 km (Figs. 2, 5, and Table 2). The unconformity is restricted to the southern sub-basin where it forms the basal erosional surface but is offset by small intrabasinal normal faults. Its size and V-shape suggest a paleovalley cut into the Squaw Gulch granite pluton. In the northern sub-basin, sequence 2 is conformable on sequence 1.

Deposition of sequence 2 begins with a granite-clast, boulder breccia-conglomerate with a red, arkosic matrix, which immediately overlies the sequence boundary in the southern sub-basin (Fig. 6). The basal breccia-conglomerate deposits are preserved in downdropped, small, high-angle normal faults on the northern side of the southern sub-basin (Figs. 2 and 5). These faults are clearly syndepositional high-angle faults because the boulder breccia-conglomerate rests depositionally against them (Figs. 2 and 6). At the same time in the northern fault-bounded trough of the northern sub-basin, boulder breccia-conglomerate

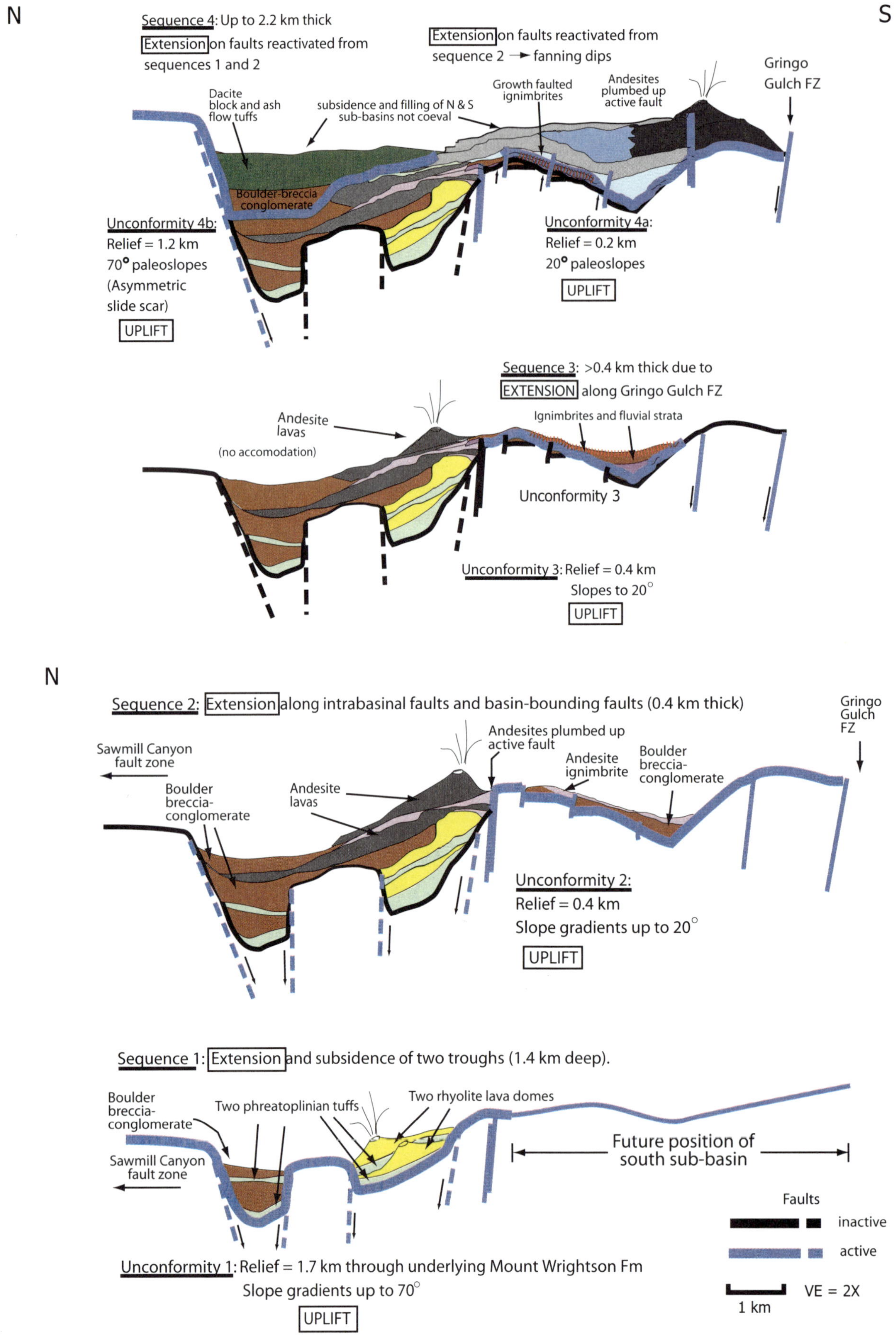

N
S
Sequence 4: Up to 2.2 km thick
Extension on faults reactivated from sequences 1 and 2
Extension on faults reactivated from sequence 2 → fanning dips
Gringo Gulch FZ
Dacite block and ash flow tuffs
subsidence and filling of N & S sub-basins not coeval
Growth faulted ignimbrites
Andesites plumbed up active fault
Boulder-breccia conglomerate
Unconformity 4b:
Relief = 1.2 km
70° paleoslopes
(Asymmetric slide scar)
UPLIFT
Unconformity 4a:
Relief = 0.2 km
20° paleoslopes
UPLIFT
Sequence 3: >0.4 km thick due to
EXTENSION along Gringo Gulch FZ
Andesite lavas
(no accomodation)
Ignimbrites and fluvial strata
Unconformity 3
Unconformity 3: Relief = 0.4 km
Slopes to 20°
UPLIFT
N
S
Sequence 2: Extension along intrabasinal faults and basin-bounding faults (0.4 km thick)
Gringo Gulch FZ
Sawmill Canyon fault zone
Andesites plumbed up active fault
Andesite ignimbrite
Boulder breccia-conglomerate
Boulder breccia-conglomerate
Andesite lavas
Unconformity 2:
Relief = 0.4 km
Slope gradients up to 20°
UPLIFT
Sequence 1: Extension and subsidence of two troughs (1.4 km deep).
Boulder breccia-conglomerate
Two phreatoplinian tuffs
Two rhyolite lava domes
Sawmill Canyon fault zone
Future position of south sub-basin
Faults
inactive
active
1 km
VE = 2X
Unconformity 1: Relief = 1.7 km through underlying Mount Wrightson Fm
Slope gradients up to 70°
UPLIFT

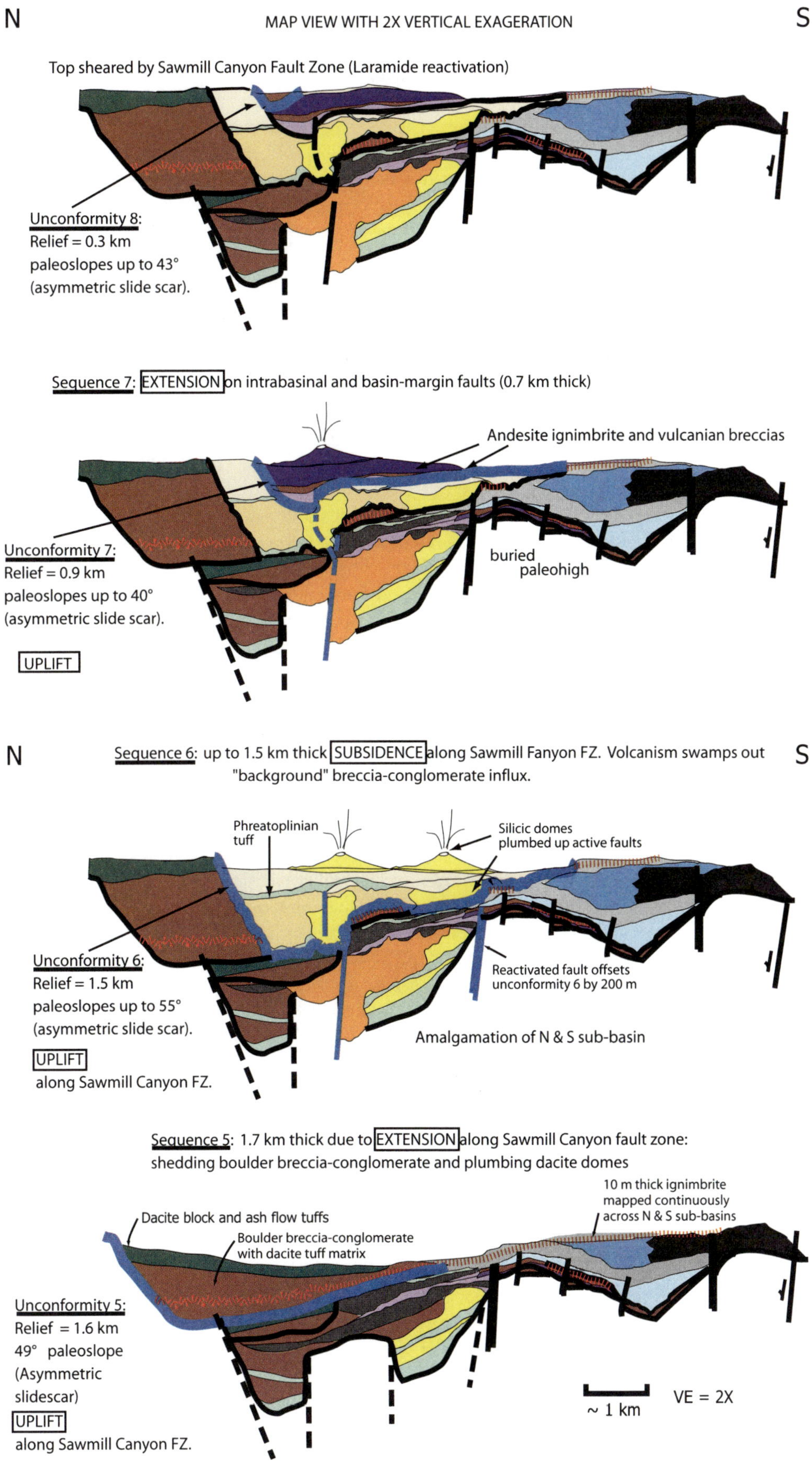

Figure 6 (*on this and previous page*). Evolution of the Santa Rita Glance Basin, divided into eight time slices defined by unconformity-bounded depositional sequences. The cross-sectional view is the same as the 2-D cross section that comprises the outcrop of the basin. Colors for lithofacies map units are the same as those used in Figure 2. Note 2× vertical exaggeration; time slice 8 is identical to our map, which is an oblique cut through bedding. Each time slice includes two major events: (1) cutting of an unconformity (highlighted in blue at the base of the relevant sequence), and (2) deposition of an overlying volcanic and/or sedimentary sequence. Faults that were active during cutting of an unconformity and/or deposition of its overlying sequence are highlighted in blue and thickened for that time slice; if the fault(s) became inactive in succeeding time slices, they are shown in black, but revert to blue color in any later time slices in which they were reactivated, often in the reverse direction. Note the controls of syndepositional faults on siting of vents.

continued to be deposited conformably on sequence 1. Breccia-conglomerate in both the northern and southern sub-basins are overlain by andesitic ignimbrite erupted from a center that developed on high-angle faults forming the paleohigh between the two sub-basins. The faults ponded the andesitic ignimbrites proving that they were active at the time of the eruption (Fig. 2). The fault-controlled andesitic center includes lava flows and sills or small laccoliths (Table 1).

Sequence 2 records continued subsidence in the northern sub-basin and the creation of new accommodation in the southern sub-basin. New accommodation may have been created by activity on the Gringo Gulch fault zone on the southern side of the southern sub-basin (Fig. 2). Thickening of basal sequence 2 strata into the two grabens in the northern sub-basin suggests that the faults there were still active, although the strata could have filled relict basins.

Sequence 3

Sequence boundary 3 is a symmetrical V-shaped unconformity that is 0.21 km deep with slopes up to 20° and a width of ~3 km (Figs. 2, 5, and Table 2). It is restricted to the southern sub-basin and is centered over the paleocanyon defined by unconformity 2 with which it merges. Nearly all of the sequence 2 boulder breccia-conglomerate and andesite in the southern sub-basin were removed under unconformity 3; they are preserved only along the sidewalls of the paleocanyon and in the grabens on the southern flank of the paleohigh of sequence 2. These grabens were inactive during deposition of sequence 3.

Basal strata of sequence 3 consists of a distinctive rhyolitic crystal-rich ignimbrite (Fig. 6 and Table 1), which filled the southern sub-basin and just barely spilled over the paleohigh into the southernmost side of the northern sub-basin. The top of the rhyolitic crystal-rich ignimbrite in the southern sub-basin was fluvially reworked and contains channels with granite boulders. Deposition of the lower of the two distinctive rhyolitic red, high-grade ignimbrites followed.

The maximum preserved thickness of sequence 3 (~130 m, Table 2) is probably a great underestimate of its original thickness because the greatest preserved thickness is on the paleohigh, and ignimbrites normally thicken into lows. There is no sequence 3 in the northern sub-basin. Accumulation space created by the paleocanyon was filled with the ignimbrites, but no preservation space was created from subsidence and faulting.

Sequence 4

Sequence boundary 4 has an overall lateral extent of ~17 km but two segments of this basin-wide surface are recognized (Figs. 2, 5, and Table 2). In the southern sub-basin, sequence boundary 4A is a symmetrical, V-shaped unconformity that has 0.6 km of relief with up to 24° slopes and locally merges with unconformities 2 and 3. It is probable that it was formed by a paleocanyon cutting into sequences 2 and 3. In the northern sub-basin, sequence boundary 4B is an asymmetrical unconformity with 0.51 km relief and steeply dipping slopes. The unconformity completely removes sequence 3 and cuts through sequence 2 andesites. The 68°-sloping northern wall of the unconformity in the northern sub-basin was likely controlled by the same fault that was active during deposition of sequences 1 and 2. This makes it likely that contemporaneous faulting led to progressive development of the sequence boundary and that a 0.51-km-high scarp never existed at any one time. An unconformity with vertical relief and slope gradients of this scale can only have been produced by tectonic activity along basin-bounding faults.

The basal unit of sequence 4A in the southern sub-basin is a distinctive, externally sourced, rhyolitic, lithic-rich ignimbrite (Fig. 6 and Table 1), which ponded in the southern sub-basin and thinned over the paleohigh, ending abruptly at a high-angle reverse fault on the northern margin of the paleohigh. The rhyolitic, lithic-rich ignimbrite is overlain by thick deposits of andesitic vitric tuff and tuffaceous sandstone that are restricted to the southern sub-basin and show paleocurrent directions from the south. The basal andesitic tuff was initially reworked into poorly sorted, indistinctly stratified deposits by overland flow and hyperconcentrated flood flow but became more stratified up section with granitic boulder conglomerates occurring in small 2–3-m-wide channels. The andesitic vitric tuff and tuffaceous sandstone are overlain by the externally sourced, rhyolitic limestone-lithic ignimbrite (Table 1). This ignimbrite appears to interfinger with the andesitic vitric tuff and tuffaceous sandstone, which also overlies it. We infer that the andesitic pyroclastic debris was supplied from a growing laccolith/cryptodome plumbed along the Gringo Gulch fault zone at the southern margin of the basin, because this cryptodome/laccolith intruded late in the deposition of sequence 4, and dikes intrude the basal part of the sequence along basin-bounding and intrabasinal faults.

Sequence 4B in the northern sub-basin consists of boulder breccia-conglomerate ~600 m thick. The matrix is commonly dacitic lithic tuff with rare arkosic sandstone matrix (Fig. 6 and Table 1). The dacitic matrix indicates eruption of proximal extrabasinal dacitic lava domes on the far side of the Sawmill Canyon fault zone, which supplied debris to the basin. These boulder breccia-conglomerates are overlain by primary dacitic block-and-ash-flow deposits ~100 m thick (Figs. 2 and 5).

We cannot correlate any lithologic units, including generally widespread and distinctive ignimbrites, between the southern and northern sub-basins, suggesting to us that the sequences were deposited at two different times. If the two sub-basins did subside at different times, as seems likely, we have no way of determining which one subsided first.

Accommodation must have been created by tectonic subsidence along basin-bounding faults because of the great thickness of strata deposited in both sub-basins. The preserved thickness of sequence 4 is up to 1.1 km in the southern sub-basin and up to 0.51 km in the northern sub-basin (Table 2). The northern sub-basin is filled with nonstratified proximal talus and debris-flow deposits, so fanning dips cannot be demonstrated, but it contains the greatest thickness of strata in a strongly asymmetrically shaped basin. In the southern sub-basin, rotation of strata (by ~10°) supports an interpretation of asymmetric subsidence along a curved normal fault. In addition, the two ignimbrites thin northward over a short distance suggesting that asymmetric subsidence effectively ponded them against the Gringo Gulch fault zone on the southern margin. A minor amount of accommodation was also created within the southern sub-basin by reactivation of four faults. These are clearly syndepositional because basal ignimbrite of sequence 3 is offset by the faults, and the upper ignimbrite of sequence 3 overlaps the faults (Figs. 2 and 5). These four faults were high-angle normal faults during sequence 2, but they were all reactivated as high-angle reverse faults during deposition of sequence 4 and show opposing offsets along their strike.

Sequence 5

Sequence boundary 5 has 0.9 km of relief, with a lateral extent of 5.5 km that widened the basin by ~1.5 km (Figs. 2, 5, and Table 2). It consists of an unconformity along its northern surface that extends across the northern sub-basin to the paleohigh and can be traced into correlative conformity in the southern sub-basin by correlating the distinctive externally sourced, red high-grade ignimbrite. The northern unconformity cuts through the upper member of the Mount Wrightson Formation (Riggs and Busby-Spera, 1990) along a very steep and high erosional surface. This surface is well exposed and does not trace downward into a fault cutting the Mount Wrightson Formation. However, while it does not represent a fault surface, at ~50° slope and 0.9 km depth it seems remarkably steep and deep for a canyon wall (Table 2). It is possible that it represents the scarp of a fault that splays out of the plane of view afforded by the outcrops; alternatively, it may represent a landslide scarp produced by uplift of the basin along the Sawmill Canyon fault zone. In either case, an unconformity of this scale must record tectonic activity.

Sequence 5 in the northern sub-basin consists largely of boulder breccia-conglomerates with a matrix of dacitic lithic tuff and dacitic block-and-ash-flow tuffs (Fig. 6 and Table 1). Less than 10% of the section consists of arkosic sandstone matrix boulder breccia-conglomerate, indicating a nearly constant influx of dacitic debris. The presence of block-and-ash-flow tuffs, formed from collapsing domes, suggests relative proximity to dacite sources, with the most probable location being along the Sawmill Canyon fault zone. These block-and-ash-flow tuffs are lithologically identical to the dacitic block-and-ash-flow tuffs of sequence 4. The block-and-ash-flow tuffs of both sequences are a distinctive part of the basin fill in terms of their relatively high crystal content and color, and they are uniform in terms of textural characteristics and bedding styles (Table 1). This would suggest that they record the growth of a dome complex along the fault zone.

To the south, the paleohigh was buried, and sequence 5 lies conformably upon sequence 4 in the southern sub-basin where sequence 5 consists entirely of the distinctive rhyolitic red high-grade ignimbrite (Fig. 6 and Table 1). It thus provides a stratigraphic tie between the northern and southern sub-basins. The upper rhyolitic red high-grade ignimbrite extends across the paleohigh and thickens northward, where it is more strongly welded. It extends about a kilometer further into the northern sub-basin as an ignimbrite, and beyond that the map unit extends another 4.5 km into the northern sub-basin as a layer of 2–3 m ignimbrite blocks in boulder breccia-conglomerate (Fig. 2). This layer of ignimbrite megablocks is mapped all the way to the northernmost end of the basin, at a stratigraphic position about a third of the way from the base of depositional sequence 5, indicating that this part of the massive boulder breccia-conglomerate section has no unconformities hidden within it.

Accommodation in the northern sub-basin must have been created by tectonic subsidence, probably along the basin-bounding Sawmill Canyon fault zone, because sequence 5 is 0.78 km thick there. Accommodation in the southern sub-basin region during deposition of sequence 5 appears to have been very small, since the preserved thickness is less than 10 m. The intrabasinal faults controlling the location of the paleohigh were inactive since the sequence-bounding surface was unaffected.

Sequence 6

Sequence boundary 6A is an unconformity with 0.77 km relief, slope gradients up to 55°, and a lateral extent of 5.5 km (Figs. 2, 5, and Table 2). It crosses the former paleohigh and partially overlies the southern sub-basin, recording amalgamation of the two sub-basins into one. This is the most deeply incised unconformity with the steepest walls, yet it can be identified as an unconformity surface and not a fault plane. It is very well exposed and there is no fault cutting underlying sequences 1–5, which are not offset. Therefore, the unconformity does not coincide with a fault plane, but it also seems remarkably steep and deep for a canyon wall (~50° slope, Table 2). Again, perhaps it represents either a degraded scarp of a fault that splays out of the plane of view afforded by the outcrops or a headwall scarp of a large run-out landslide created during basin inversion prior to deposition of sequence 6.

Deposition of sequence 6 began with the upper extrabasinally sourced, crystal-rich ignimbrite (Fig. 5 and Table 1). This is very thin and discontinuous because it is largely cut out by later dome intrusions of sequence 6. The crystal-rich ignimbrite was followed by deposition of intrabasinally sourced rhyolitic plinian-phreatoplinian tuffs across the entire basin (Fig. 2). The plinian tuffs are only missing in the areas where they are obscured by the intrusion of succeeding lava domes and also at the southernmost end of the basin where accommodation was low, and friable tuffs could have easily been eroded away. The plinian tuffs locally contain a thin (1 m thick) lithic ignimbrite at their base, which may record vent clearing at the start of the explosive eruption phase. The tuffs become progressively more reworked up section.

Deposition of the plinian tuffs was followed by emplacement of two fault-controlled rhyolitic lava domes (Figs. 2 and 6). This records a switch from explosive to effusive eruptive styles within the basin. The two lava domes lie at approximately the same stratigraphic level, and the block-and-ash-flow tuffs that fringe them interfinger with each other (Figs. 2 and 6). The southern of the two lava domes was bounded by and overlies a fault that displaces unconformity 6, indicating that faulting was active during deposition of sequence 6 and probably plumbed the rhyolite to the surface. There is no direct connection of dome to underlying rhyolitic intrusion in the cross-sectional view afforded by present-day outcrops, but they lie within a couple of hundred meters of each other. The northern of the two lava domes also does not map directly into a rhyolitic intrusion, but is connected to it by a syndepositional fault. This is the same fault that formed the northern boundary of the southern fault-bounded trough during deposition of sequence 1, with ~0.8 km throw down to the south. This reactivated fault offsets unconformity 6 by ~200 m, in the opposite sense (down to the north), and the plinian-phreatoplinian tuff at the base of sequence 6 mantles this fault scarp. Thus the northern fault boundary of the rhyolitic intrusion does not represent rhyolite intrusion downdropped into basement rock; it merely follows the preexisting fault.

The uppermost part of sequence 6 began with deposition of the single stratigraphic occurrence of the distinctive rhyolitic white high-grade ignimbrite (Fig. 6 and Table 1). This was deposited across the tops of both rhyolitic lava domes and their fringing block-and-ash-flow tuffs (Fig. 2). The high-grade ignimbrite is up to 40 m thick and fills basal scours less than a meter deep.

The high-grade ignimbrite is cut out by an erosional surface up to 100 m deep (Figs. 2 and 5). It has far less relief than the rest of the unconformities in the basin and partly because it, and the sequence above it, is poorly exposed north of the northern rhyolite dome, it has been kept as part of sequence 6. Sequence 6B consists of rhyolitic plinian-phreatoplinian tuffs interstratified with 1-m- to 20-m-thick rhyolitic crystal-poor ignimbrites intermittently showing ripples and stratification from fluvial reworking between emplacements. The plinian tuffs dominate the base of the section with ignimbrites increasing in number upward.

Accommodation for sequence 6 must have been created by tectonic subsidence along the Sawmill Canyon fault zone, because it is very thick (up to 0.77 km), and it thickens toward the fault zone. A minor amount of subsidence may have been accommodated by intrabasinal high-angle normal faults that were reactivated from those in sequence 1 (Fig. 2).

Depositional sequence 6 is the only depositional sequence in the entire Santa Rita Glance Formation that has no boulder breccia-conglomerate. This suggests that the rate of volcanic deposition was extremely high, completely swamping out the background influx of extrabasinally sourced coarse-grained detritus.

Sequence 7

Sequence boundary 7 is an unconformity along its entire length, shows a maximum vertical relief of 0.46 km with slopes up to 40°, and has a lateral extent of 1.5 km before it disappears under the cover of Quaternary gravel (Figs. 2, 5, and Table 2). It reappears from beneath the gravel in a very small area of outcrop in the southern area where most of sequence 6 is missing, and the basal andesite ignimbrite of sequence 7 rests unconformably on the upper extrabasinally sourced, crystal-rich ignimbrite of basal sequence 6. Sequence boundary 7 is the third intrabasinal unconformity that is steeply asymmetric and south facing. Like unconformities 5 and 6, unconformity 7 does not trace downward into a fault, and underlying strata are not offset by faults. Therefore, it may also represent the scarp of a fault that splays out of the plane of view afforded by the outcrops or the head scarp of a large runout landslide generated during basin inversion events.

Deposition of sequence 7 began with an andesitic ignimbrite that was deposited across the basin and ponded against a reactivated high-angle fault associated with the northern rhyolitic dome in sequence 6 (Fig. 6). This is overlain by a continuous andesitic vitric tuff-tuffaceous sandstone unit, only 5 m thick. This is overlain in turn by ~300 m of andesitic vulcanian breccias. Sequence 7 is capped by a rhyolitic crystal-poor ignimbrite that is sheared where it lies along a strand of reactivated Sawmill Canyon fault zone.

Accommodation for deposition of sequence 7 (0.36 km thick, Table 2) must have been largely created by basin-bounding faults, but local intrabasinal accommodation was created by an ~50 m offset along a high-angle fault reactivated from sequence 6.

Sequence 8

Sequence boundary 8 and depositional sequence 8 are truncated by the Sawmill Canyon fault zone (Figs. 2 and 5). The sequence boundary is an unconformity that merges locally with unconformity 7. It also appears to be steep and asymmetric, although very little is preserved. It is likely that sequence boundary 8 is another problematic surface that is too steep and deep to have formed exclusively by erosion of a paleocanyon, similar to sequence boundaries 5, 6, and 7.

Sequence 8 is composed entirely of boulder breccia-conglomerate with arkosic sandstone matrix; there are no volcanics associated with sequence 8 (Fig. 6). This is the original Glance Conglomerate unit mapped by Drewes (1971a), yet it is indistinguishable from underlying breccia-conglomerates in both clast composition and depositional facies.

Accommodation was probably created by subsidence along the Sawmill Canyon fault zone.

DISCUSSION

In the southern sub-basin, successive unconformities cut down to the same level indicating that there was no change in the geomorphic base level of the region through time. Thus there were large amounts of accumulation space within the paleocanyon but low amounts of preservation space in the southern sub-basin. The relatively low slopes (20°–25°) of the symmetrical unconformities in the southern sub-basin (unconformities 2, 3,

4A) suggest they may have been carved in response to eruption of ignimbrites filling an older paleocanyon. The paleocanyon was repeatedly filled by ignimbrites and recut to the same geomorphic base level. This changed during sequence 4A; its preservation suggests the creation of preservation space by subsidence along the Gringo Gulch fault zone (Figs. 2 and 5).

In the northern sub-basin, the high relief and great thicknesses of the sequences indicates a large amount of preservation space that must have been created by fault-controlled subsidence. The erosional surfaces in the northern sub-basin (unconformities 1, 5, 6, 7, 8) show 450–900 m of vertical relief on highly asymmetrical surfaces (Figs. 2 and 5). Syndepositional fault scarps are permissible for unconformities 1 and 4B but the steep walls of unconformities 5, 6, 7, and 8 do not merge with faults, and underlying beds are not offset. We present three possible interpretations for the deep, asymmetrical unconformities. (1) They represent extremely deep paleocanyons. This seems unlikely given their depth and asymmetry; we consider this the least likely explanation. (2) They represent faults that all coincidentally splay out of the map view. This is possible if the faults were all reverse faults in a flower structure along the Sawmill Canyon fault zone. (3) These features represent headwall scarps from long run-out landslides (debris avalanches; Rightmer et al., 1995). The lack of debris avalanche deposits is consistent with the interpretation that they were generated at times of uplift of the Santa Rita Glance Basin when there was no preservation space. Debris-avalanche deposits are abundant in correlative strata of adjacent ranges (Davis et al., 1979; Hayes and Raup, 1968; McKee and Anderson, 1998). We prefer interpretations 2 and 3, although whatever interpretation is correct, the depth of erosion indicates that tectonic uplift was the control on the creation of the unconformities in the northern sub-basin.

Basin-bounding faults control accommodation, facies architecture, and volcanism in both the northern and southern sub-basins, although more strongly in the northern sub-basin. Thus basin geometry and subsidence is more from movement on faults than from magma evacuation during eruption of large ignimbrites. The most active basin-bounding fault was the Sawmill Canyon fault zone controlling subsidence in the northern sub-basin.

The Sawmill Canyon fault zone is inferred to be the dominant basin-bounding fault because (1) boulder clast size and angularity increase toward it, (2) boulder breccia-conglomerate facies become more proximal toward it, ranging from distal fluvial facies to medial debris-flow facies to proximal talus facies, and (3) intrabasinal, syndepositional faults show the greatest displacement near it. The Sawmill Canyon fault zone was clearly reactivated during the Upper Cretaceous to lower Tertiary Laramide Orogeny, because it cuts the top of the Glance Conglomerate and juxtaposes fault slivers of younger and older strata against it. Because the Sawmill Canyon fault zone trends at a 55° angle to the strike of the basin fill, it does not coincide with the boundary to the basin margin in the cross sectional view afforded by present-day exposures, although a strand defines the eastern outcrop boundary (Fig. 2). In the third dimension, however, it must have been the basin-bounding fault on the northern side of the Santa Rita Glance basin.

A less regionally significant fault zone with less offset lies along the southern end of the basin; we name this fault the Gringo Gulch fault zone (Fig. 2). This E-W–trending, subvertical fault zone is cut by the Upper Cretaceous Josephine Canyon diorite, making it older than that, and it cuts the Squaw Gulch granite, making it younger than that (Drewes, 1971a). Thus, the Gringo Gulch fault zone is a pre-Upper Cretaceous and therefore pre-Laramide fault that bounds the Santa Rita Glance Conglomerate to the south. The subvertical fault is approximately perpendicular to the N-S strike of the Santa Rita Glance basin fill and is en echelon to the Sawmill Canyon fault zone, suggesting it might be structurally related. Our stratigraphic data suggests that the Gringo Gulch fault zone controlled the southern basin margin and plumbed andesitic volcanic rocks to the surface (Fig. 2). Therefore it forms the southern basin-bounding fault for the local Glance basin.

Intrabasinal faults are common in the basin, commonly offsetting sequence boundaries. Many fault offsets are stratigraphically limited, offsetting beds in only one sequence. Some faults are repeatedly reactivated, often in the reverse sense of direction. These larger intrabasinal faults also often control the site of dome emplacement and plumb magma to the surface. They also control the presence and geometry of the intrabasinal high separating the northern and southern sub-basins.

TECTONIC SETTING

We interpret the Santa Rita Glance Formation as recording continental arc volcanism, based on: (1) the large volume of volcanic deposits relative to sedimentary deposits, (2) the wide range in eruptive styles, and (3) the wide compositional range and relative proportions of those compositions (~30% andesitic, ~25% dacitic, and ~45% rhyolitic volcanic deposits; Busby and Bassett, 2003). This variety of compositions indicates arc volcanism rather than the less voluminous, more bimodal compositions of continental rifting (Wilson, 1989). Unpublished geochemical data from the Santa Rita Glance Formation shows strong large ion lithophile element (LILE) enrichment, high field strength element (HFSE) depletion, a strong Ta negative anomaly, and shoshonitic composition, all typical of continental arc volcanism (see Data Repository[1]). We interpret the Santa Rita Glance Conglomerate as the fill of an intra-arc basin (Busby and Bassett, 2005).

Given that the rest of the Bisbee Basin has continental rift characteristics, the simplest interpretation of the Santa Rita Glance Basin would be of an extensional basin bounded by normal faults subsiding in a half-graben geometry similar to the Rio Grande Rift (Lawton and McMillan, 1999; Mack et al., 1994; Smith et al., 2001). The volcanism could have occurred on a rift

[1]GSA Data Repository item 2005142, Preliminary major and trace element geochemical data, is available online at www.geosociety.org/pubs/ft2005.htm, or on request from editing@geosociety.org or Documents Secretary, GSA, P.O. Box 9140, Boulder, CO 80301, USA.

margin master fault or transform zone (Justet and Spell, 2001; McMillan et al., 2000). However, the Santa Rita Glance Basin contains multiple, very deep unconformities, basin-bounding faults that were syndepositional and steep, and intrabasinal normal faults that were coeval with or reactivated as reverse faults (Fig. 2). The shallower unconformities probably resulted from cutting and refilling of paleocanyons by episodic volcanism, but the deep asymmetrical unconformities in the boulder breccia-conglomerates and the alternation of normal fault and reverse fault movements cannot be explained by a rift model. Instead, we interpret the structural style as that of a strike-slip basin.

The Santa Rita Glance basin contains features that are typical of strike-slip basins. Ancient strike-slip basins are most commonly recognized by the distinctive stratigraphic style that is the result of the tectonic "porpoising," first described by Crowell (1974). Very close spatial and temporal association of releasing bends and restraining bends in the strike-slip fault results in basin subsidence followed by basin inversion and destruction, on a time scale of 105 to 106 years (Barnes et al., 2001; Wood et al., 1994). This produces large-scale intrabasinal unconformities, cannibalization of older basin fill, and deformation of the lower layers (Barnes et al., 2001; Wood et al., 1994). The Santa Rita Glance strata contain numerous intrabasinal unconformities and evidence for cannibalization of older basin fill as determined by clast compositions.

The structural style of ancient strike-slip basins consists of basin-bounding strike-slip faults associated with intrabasinal faults with reverse and normal components of slip that develop simultaneously with grabens and arches, in positions that vary rapidly through time (Barnes and Audru, 1999; Barnes et al., 2001; Crowell, 1982, 2003; Link, 2003; Nilsen and Sylvester, 1995; Wood et al., 1994). Minor intrabasinal faults create "structural arches" that form paleohighs separating the strike-slip basin into sub-basins at any one time (Crowell, 2003; Link, 2003; Nilsen and Sylvester, 1995). Alternating reactivation of normal faults as reverse faults, and vice versa, is a distinguishing feature of releasing-bend strike-slip basins that have been affected by restraining bends, either simultaneously or successively through time (Crowell, 2003; Link, 2003; Nilsen and Sylvester, 1995). All of these structural features are evident in the Santa Rita Glance basin. It is bounded by the strike-slip Sawmill Canyon fault zone. It contains numerous intrabasinal faults that are reactivated, commonly with the opposite sense of movement from reverse to normal or normal to reverse. These intrabasinal faults formed a structural arch between the northern and southern sub-basins that manifested as a paleohigh against which ignimbrites ponded.

We compare the Santa Rita Glance Basin with modern releasing- and restraining-bend strike-slip basins in New Zealand. When a strike-slip fault system is slightly transtensional, the sizes of the restraining bends are less than the size of the releasing bends (Cowan, 1989; Cowan and Pettinga, 1992; Cowan et al., 1989). This can be clearly seen in the modern Hope Fault in New Zealand, which has an overall releasing curvature. Releasing and restraining bends tend to occur in closely spaced couplets along the Hope Fault, with the basins being larger than the adjacent restraining-bend pop-up structures (Fig. 7A; Cowan, 1989; Cowan and Pettinga, 1992; Cowan et al., 1989). The Glynn Wye Basin is a perfect example of a modern releasing-bend basin with an adjacent restraining bend (Fig. 7B). The larger Hanmer Basin along the Hope Fault also has the same couplet structure (Wood et al., 1994). If the restraining bends are of the same scale as the releasing bends, all of the basin fill created at a releasing bend should be inverted and eroded during uplift at the succeeding restraining bend. If the restraining bends are smaller than the releasing bends, however, net subsidence over time will result in partial preservation of the basin fill. In the Glynn Wye Basin the basin is >500 m wide but the restraining bend pop-up is only 150 m wide (Fig. 7B). The difference in size is quite large, meaning that only 30% of the Glynn Wye Basin deposits will be deformed by the subsequent restraining bend thrusting. We suggest that the deep unconformities in the Santa Rita Glance were created during uplift at small restraining bends, leaving most of the releasing bend basin deposits undeformed. Note also the presence of a structural intrabasinal high in the Glynn Wye Basin. We believe this corresponds with the intrabasinal high separating the two sub-basins in the Santa Rita Glance Basin.

Our interpretation of Glance deposits in the western Bisbee basin as an intra-arc strike-slip basin appears to be in conflict with recently published models for postvolcanic arc opening of the Bisbee basin (Dickinson and Lawton, 2001a; Lawton and McMillan, 1999). The tectonic model presented in Dickinson and Lawton (2001b) calls on slab rollback as the driving force for continental extension in the Bisbee basin (their Figs. 5 and 8B). However, slab rollback implies subduction, which implies the presence of a volcanic arc. If the arc existed, then it would be located on the outboard side of the continental extension. We argue that the western Glance Conglomerate forms the continental arc and that the Bisbee basin was the result of slab rollback and backarc extension.

Our evidence for intra-arc strike-slip faulting in the Santa Rita Glance basin suggests that plate convergence was oblique (Jarrard, 1986) with strain partitioning (McCaffrey, 1992) into backarc extension and intra-arc strike slip (Fig. 8). Strain partitioning is more commonly discussed for transpressional settings (de Saint Blanquat et al., 1998), but there are examples of strain partitioning in transtensional settings (Acocella et al., 1999; Marra, 2001; Wesnousky and Jones, 1994) where it appears to be controlled by the relative strengths of faults and spatial or temporal changes in the regional stress field (Wesnousky and Jones, 1994). The presence of arc magmatism clearly affects relative fault strengths, especially at depth (de Saint Blanquat et al., 1998), providing the mechanism for strain partitioning in a transtensional convergent margin. We suggest that the Bisbee basin is partitioned into a backarc extensional domain and an intra-arc strike-slip domain represented by the Santa Rita Glance Conglomerate. In addition, much of the strike-slip motion from the oblique plate convergence could have been taken up by

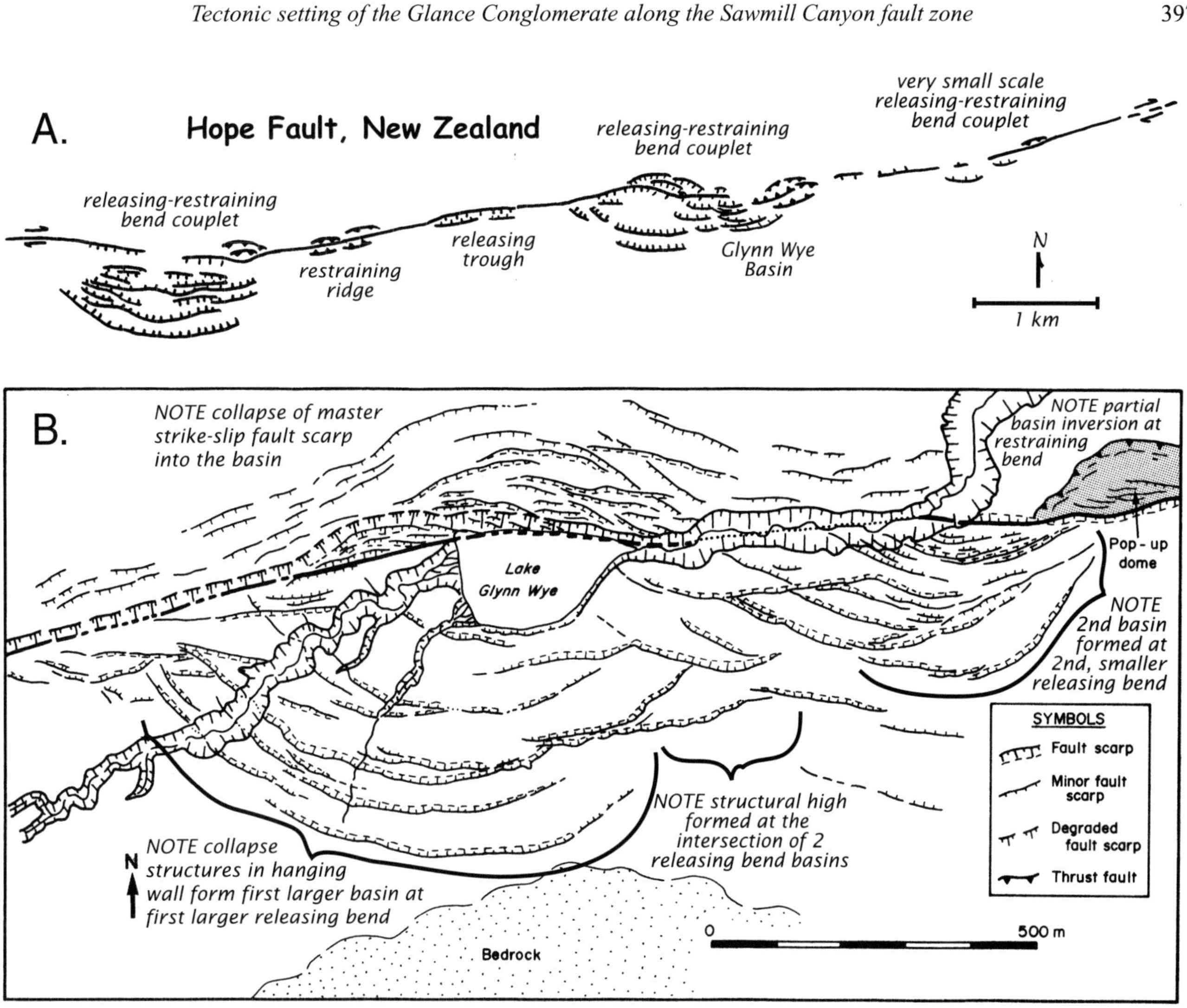

Figure 7. (A) Alternating releasing and restraining bends along the dextral strike-slip Hope Fault in New Zealand as a modern analogue for the Santa Rita Glance Basin along the Sawmill Canyon fault zone. Releasing and restraining bends produce alternating basins and pop-up structures (Cowan, 1989; Cowan and Pettinga, 1992). As the releasing and restraining bends develop along the strike of the strike-slip fault, any single location may undergo repeated subsidence and uplift (basin inversion) events, producing basin fill and unconformities. Along the Hope Fault, releasing and restraining bends occur in pairs, separated by long stretches of lesser activity. The Hope Fault is slightly releasing so the restraining bends are smaller than the releasing bends. (B) Close up of the Glynn Wye releasing-bend basin and associated restraining bend pop-up structure. Note the difference in width suggesting that 70% of the basin will be preserved undeformed during a basin inversion event.

smaller, nonparallel faults (Klute, 1991) in a probably transtensional back-arc Bisbee basin.

A possible modern analogue for a strain-partitioned, transtensional arc is the syn-arc back-arc rifting of the northern Sumatran Arc and Andaman Sea (Curray et al., 1978; Maung, 1987; Mukhopadhyay, 1984). High obliquity reduces the volume of volcanism occurring within the arc itself as total rates of convergence decrease. Oblique convergence is strain partitioned into intra-arc and accretionary wedge strike-slip faults and back-arc transtensional rifting. The back-arc rifting is highly oblique with basin-bounding faults and volcanism obscured by a large sediment supply.

CONCLUSIONS

Our sequence analysis of a volcanically dominated strike-slip basin allowed us to reconstruct the stratigraphic and tectonic history of a complex basin. The cogenetic lithofacies associations of the Santa Rita Glance Conglomerate occur between eight unconformity and correlative conformity sequence boundaries across the basin. These sequence boundaries record two aspects of accommodation: accumulation space, which requires no change in base level, and preservation space, which requires net subsidence. The lower relief, symmetrical, V-shaped unconformities in the southern sub-basin are interpreted as a paleocanyon repeatedly

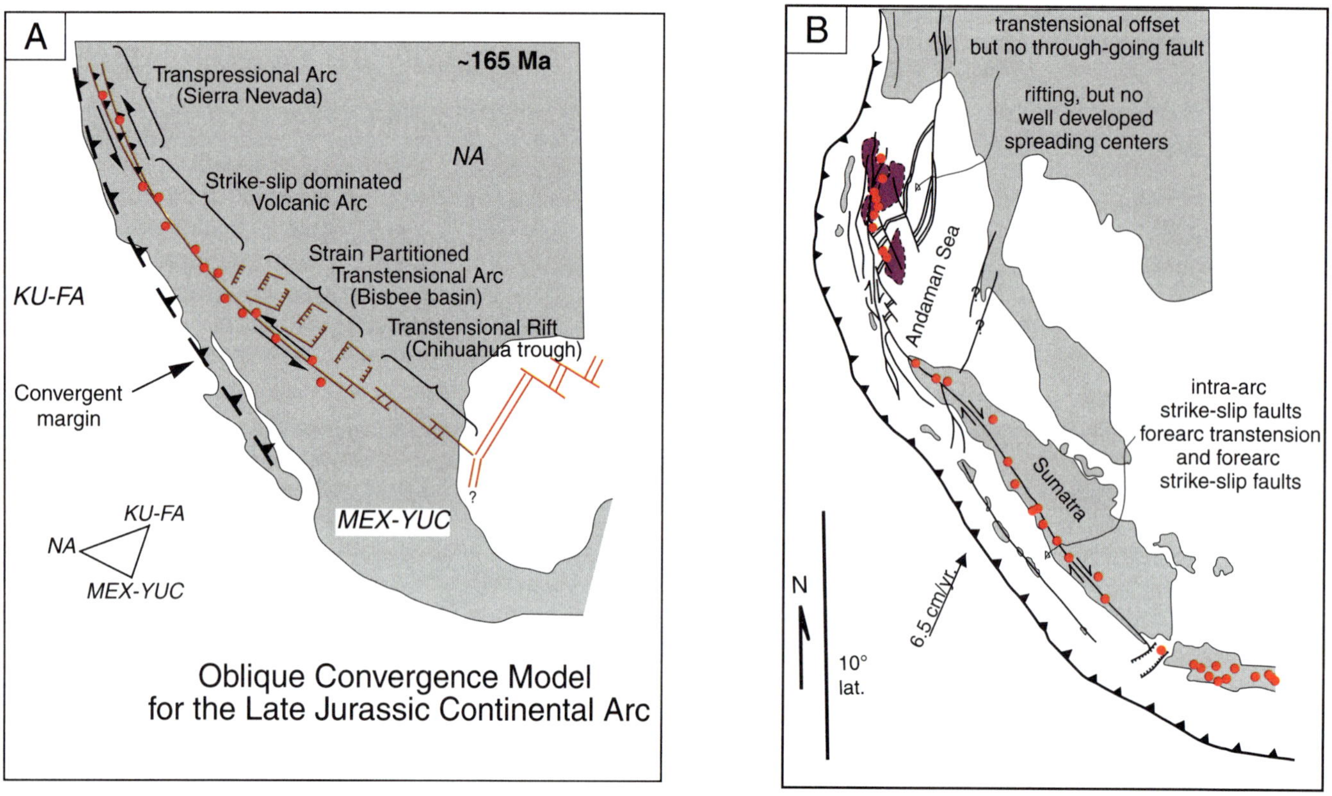

Figure 8. (A) Map of southwestern North America showing inferred tectonic settings and regional reconstruction along the Upper Jurassic plate boundary. Strain partitioning along the continental arc caused intra-arc strike-slip faulting along the western edge of the Bisbee basin and back-arc transtensional faulting along the eastern edge. (B) A modern analog of the strain-partitioned obliquely convergent plate boundary may lie in Sumatra and the Andaman Sea.

filled by volcanic debris and recut by reestablished fluvial systems. This requires no change in base level but is an example of abundant accumulation space. In contrast, unconformities in the northern sub-basin cut deeply into underlying strata, yet thick sequences were preserved, indicating the creation of preservation space from net tectonic subsidence, probably along the Sawmill Canyon fault zone. Many of the unconformity surfaces show extreme vertical relief and are highly asymmetrical; these are interpreted to represent slide scars and fault scarps proximal to the master fault, the Sawmill Canyon fault zone. The unconformities and sequences developed synchronously with high-angle, intrabasinal faults with normal and reverse separation that alternated in space and time and controlled the positions of grabens and arches. This structural style is typical of basins affected by both releasing bends and restraining bends along strike-slip faults.

The intra-arc strike-slip basin fill is dominated by small polygenetic, multivent volcanic complexes that we consider to be typical of basins sited on a major fault zone, where strands of the fault frequently plumb small batches of magma to the surface. Because of this, individual volcanic constructs do not grow large enough to provide significant accommodation in volcano-bounded basins, although they do modify basin floor topography. The use of sequence analysis has allowed greater resolution of basin subsidence and uplift events than would otherwise have been possible, underpinning our intra-arc strike-slip basin tectonic interpretation. We reconcile our tectonic interpretation with other recently published interpretations for the Glance Conglomerate by presenting a model for strain partitioning in an obliquely convergent plate boundary. Strike-slip motion was concentrated along the thermally weakened arc axis, with extension partitioned into the backarc or eastern region of the Bisbee Basin.

ACKNOWLEDGMENTS

We are indebted to Nancy Riggs, Bill Dickinson, Peter Lipman, Ken Hon, Gordon Haxel, Jarg Pettinga, and the late and wonderful Richard Fisher for discussions in the field in southern Arizona and elsewhere. We thank Tim Lawton, Nicholas Christie-Blick, Gary Smith, Jim McKee, and Tom Anderson for their thorough and insightful reviews. Discussions with Doug Walker and Allen Glazner are also acknowledged. Robert Bothman and Senta provided invaluable field assistance. We sincerely thank Robert and Sandy Bowman for allowing us to stay at their guest house. Support was provided by NSF-EAR 92-19739, awarded to C. Busby.

REFERENCES CITED

Acocella, V., Salvini, F., Funiciello, R., and Faccenna, C., 1999, The role of transfer structures on volcanic activity at Campi Flegrei (southern Italy): Journal of Volcanology and Geothermal Research, v. 91, no. 2–4, p. 123 139, doi: 10.1016/S0377-0273(99)00032-3.

Asmerom, Y., 1988, Mesozoic igneous activity in the southern Cordillera of North America: Implication for tectonics and magma genesis [Ph.D. thesis]: Tucson, Arizona, University of Arizona, 232 p.

Barnes, P.M., and Audru, J.-C., 1999, Recognition of active strike-slip faulting from high-resolution marine seismic reflection profiles: Eastern Marlborough fault system, New Zealand: Geological Society of America Bulletin, v. 111, p. 538–559, doi: 10.1130/0016-7606(1999)111<0538:ROASSF>2.3.CO;2.

Barnes, P.M., Sutherland, R., Davy, B., and Delteil, J., 2001, Rapid creation and destruction of sedimentary basins on mature strike-slip faults: An example from the offshore Alpine Fault, New Zealand: Journal of Structural Geology, v. 23, p. 1727–1739, doi: 10.1016/S0191-8141(01)00044-X.

Bassett, K.N., and Busby, C.J., 1996a, Use of ignimbrite stratigraphy for correlation across a regional Mesozoic shear zone; new mapping from S.E. Arizona: Geological Society of America Abstracts with Programs, v. 28, no. 5, p. 47.

Bassett, K.N., and Busby, C.J., 1996b, Interstratified rhyolite dome and andesite volcanic deposits in the Late Jurassic(?) Temporal-Bathtub formations in southern Arizona: Geological Society of America Abstracts with Programs, v. 28, no. 7, p. 114.

Bassett, K., and Busby, C.J., 1997, Intra-arc strike-slip basins in the Late Jurassic Southern Cordillera; structural and climatic controls on deposition: Geological Society of America Abstracts with Programs, v. 29, no. 6, p. 201.

Beatty, B., 1987, Correlation of some mid-Mesozoic redbeds and quartz sandstones in the Santa Rita Mountains, Mustang Mountains, and Canelo Hills, southeastern Arizona [M.S. thesis]: Tucson, Arizona, University of Arizona, 29 p.

Bilodeau, W.L., 1979, Early Cretaceous tectonics and deposition of the Glance conglomerate, southeastern Arizona [Ph.D. thesis]: Stanford, California, Stanford University, 176 p.

Blum, M.D., and Tornqvist, T.E., 2000, Fluvial responses to climate and sea-level change; a review and look forward: Sedimentology, v. 47, Suppl. 1, p. 2–48, doi: 10.1046/j.1365-3091.2000.00008.x.

Buesch, D.C., 1991, Changes in depositional environments resulting from emplacement of a large-volume ignimbrite: Society of Economic Paleontologists and Mineralogists Special Publication 45, p. 139–153.

Busby, C.J. and Bassett, K., 2005, Volcanic facies architecture of an intra-arc strike-slip basin, Santa Rita Mountains, Southern Arizona: Geological Society of America, Abstracts with programs, v. 37, no. 4, p. 64.

Busby-Spera, C.J., 1988, Speculative tectonic model for the early Mesozoic arc of the southwest Cordilleran United States: Geology, v. 16, p. 1121–1125, doi: 10.1130/0091-7613(1988)016<1121:STMFTE>2.3.CO;2.

Busby-Spera, C.J., and Kokelaar, B.P., 1991, Controls of the Sawmill Canyon fault zone on Jurassic magmatism and extension/transtension in southern Arizona: Geological Society of America Abstracts with Programs, v. 23, no. 5, p. 250.

Busby-Spera, C.J., and Kokelaar, B.P., 1992, Development of calderas in a transtensional fault zone [abs.]: 29th International Geological Congress, Kyoto, Japan, Resumes, v. 29, p. 482.

Busby-Spera, C.J., Mattinson, J.M., and Adams, B., 1989, The Jurassic transition for continental intra-arc extension to Gulf of Mexico–related rifting in the southwest Cordillera: Eos (Transactions, American Geophysical Union), v. 70, p. 1300.

Cowan, H., 1989, An evaluation of the late Quaternary displacements and seismic hazard associated with the Hope and Kakapo Faults, Amuri District, North Canterbury [M.Sc. thesis]: Christchurch, New Zealand, University of Canterbury, 213 p.

Cowan, H., and Pettinga, J.R., 1992, Late Quaternary tectonics and landscape evolution Glynn Wye, North Canterbury, Field Trip Guides, 1992 Joint Annual Conference of Geological Society of New Zealand and New Zealand Geophysical Society: Geological Society of New Zealand Miscellaneous Publication 63B, p. 51–73.

Cowan, H.A., Pettinga, J.R., and Smith, I.E.M., 1989, Transtension and structural complexity along the Hope Fault, Glynn Wye to Hammer Basin, North Canterbury: Geological Society of New Zealand Miscellaneous Publication, v. 43, p. 31.

Crowell, J.C., 1974, Origin of late Cenozoic basins in southern California, *in* Dickinson, W.R., ed., Tectonics and Sedimentation: Tulsa, Oklahoma, Society for Economic Paleontologists and Mineralogists Special Publication 22, p. 190–204.

Crowell, J.C., 1982, The tectonics of Ridge basin, southern California, *in* Crowell, J. C., and Link, M.H., eds., Geologic history of Ridge Basin, southern California, SEPM Pacific Section: Los Angeles, California, Society for Economic Paleontologists and Mineralogists, p. 25–41.

Crowell, J.C., 2003, Introduction to Geology of Ridge Basin, Southern California, *in* Crowell, J.C., ed., Evolution of Ridge Basin, southern California: An interplay of sedimentation and tectonics: Geological Society of America Special Paper 367, p. 1–15.

Curray, J.R., Moore, D.G., Lawver, L.A., Emmel, F.J., Raitt, R.W., Henry, M., and Kieckhefer, R., 1978, Tectonics of the Andaman Sea and Burma, *in* Watkins, J. S., Montadert, L., and Dickerson, P. W., eds., Geological and Geophysical Investigations of Continental Margins, AAPG Memoir 29, p. 189–198.

Davis, G.H., Phillips, M.P., Reynolds, S.J., and Varga, R.J., 1979, Origin and provenance of some exotic blocks in lower Mesozoic red-bed basin deposits, southern Arizona: Geological Society of America Bulletin, v. 90, no. 4, p. I-376–I-384.

de Saint Blanquat, M., Tikoff, B., Teyssier, C., and Vigneresse, J.-L., 1998, Transpressional kinematics and magmatic arcs: Geological Society Special Publication 135, p. 327–340.

Dickinson, W.R., and Lawton, T.F., 2001a, Carboniferous to Cretaceous assembly and fragmentation of Mexico: Geological Society of America Bulletin, v. 113, no. 9, p. 1142–1160, doi: 10.1130/0016-7606(2001)113<1142:CTCAAF>2.0.CO;2.

Dickinson, W.R., and Lawton, T.F., 2001b, Tectonic setting and sandstone petrofacies of the Bisbee Basin (USA–Mexico): Journal of South American Earth Sciences, v. 14, no. 5, p. 475–504, doi: 10.1016/S0895-9811(01)00046-3.

Dickinson, W.R., Klute, M.A., and Swift, P.N., 1986, The Bisbee Basin and its bearing on late Mesozoic paleogeographic and paleotectonic relations between the Cordilleran and Caribbean regions, *in* Abbott, P. L., ed., Cretaceous Stratigraphy of Western North America, SEPM Pacific Section, Field Trip Guidebook, v. 46, p. 51–62.

Dickinson, W.R., Klute, M.A., and Bilodeau, W.L., 1987, Tectonic setting and sedimentological features of upper Mesozoic strata in southeastern Arizona, *in* Davis, G.H., Van der Dolder, E.M., eds., Field Trip Guidebook, Geological Society of America 100th Annual Meeting: Arizona Bureau of Geology and Mineral Technology Special Paper 5, p. 266–279.

Dickinson, W.R., Fiorillo, A.R., Hall, D.L., Monreal, R., Potochnik, A.R., and Swift, P.N., 1989, Cretaceous strata of southern Arizona, *in* Jenney, J. P., and Reynolds, S. J., eds., Geologic evolution of Arizona: Tucson, Arizona, Arizona Geological Society, p. 397–434.

Drewes, H., 1971a, Geologic Map of the Mount Wrightson Quadrangle, Southeast of Tucson, Santa Cruz and Pima Counties, Arizona: U.S. Geological Survey. Miscellaneous Geologic Investigations Map, Report I-0614, scale 1:48,000.

Drewes, H., 1971b, Mesozoic stratigraphy of the Santa Rita Mountains, southeast of Tucson, Arizona: U.S. Geological Survey Professional Paper, v. 748, p. 1–35.

Drewes, H., 1972, Structural geology of the Santa Rita Mountains, southeast of Tucson, Arizona: U.S. Geological Survey Professional Paper, v. 658-C, p. 1–81.

Drewes, H., 1981, Tectonics of southeastern Arizona: U.S. Geological Survey Professional Paper, v. 1144, p. 1–96.

Fisher, R.V., and Schmincke, H.U., 1984, Pyroclastic Rocks: New York, Springer-Verlag, 472 p.

Hagstrum, J.T., and Lipman, P.W., 1991, Paleomagnetism of three Upper Jurassic ash-flow sheets in southeastern Arizona: Implications for regional deformation: Geophysical Research Letters, v. 18, p. 1413–1416.

Hayes, P.T., 1970, Mesozoic stratigraphy of the Mule and Huachuca Mountains, Arizona: U.S. Geological Survey Professional Paper, v. 658-A, p. 1–27.

Hayes, P.T., and Raup, R.B., 1968, Geologic map of the Huachuca and Mustang Mountains, Arizona: U.S. Geological Survey, Miscellaneous Geologic Investigations Map, Report I-0509, scale 1:48,000.

Jacques-Ayala, C., 1995, Paleogeography and provenance of the Lower Cretaceous Bisbee Group in the Caborca–Santa Ana area, northwestern Sonora: Geological Society of America Special Paper 301, p. 79–98.

Jarrard, R.D., 1986, Relations among subduction parameters: Reviews of Geophysics, v. 24, p. 217–284.

Justet, L., and Spell, T.L., 2001, Effusive eruptions from a large silicic magma chamber; the Bearhead Rhyolite, Jemez volcanic field, NM: Journal of Volcanology and Geothermal Research, v. 107, no. 4, p. 241–264, doi: 10.1016/S0377-0273(00)00296-1.

Klute, M.A., 1991, Sedimentology, sandstone petrofacies, and tectonic setting of the late Mesozoic Bisbee Basin, southeastern Arizona [Ph.D. thesis]: Tucson, Arizona, University of Arizona, 272 p.

Kluth, C.F., 1982, Geology and mid-Mesozoic tectonics of the northern Canelo Hills, Santa Cruz County, Arizona [Ph.D. thesis]: Tucson, Arizona, University of Arizona, 364 p.

Krebs, C.K., and Ruiz, J., 1987, Geochemistry of the Canelo Hills volcanics and implications for the Jurassic tectonic setting of southeastern Arizona: Arizona Geological Society Digest, v. 18, p. 139–151.

Lawton, J.F., and Olmstead, G.A., 1995, Stratigraphy and structure of the lower part of the Bisbee Group, northeastern Chiricahua Mountains, Arizona, *in* Jacques-Ayala, C., Gonzalez-Leon, C.M., and Roldan-Quintana, J., eds., Studies on the Mesozoic of Sonora and Adjacent Areas: Geological Society of America Special Paper 301, p. 21–39.

Lawton, T.F., and McMillan, N.J., 1999, Arc abandonment as a cause for passive continental rifting; comparison of the Jurassic Mexican Borderland Rift and the Cenozoic Rio Grande Rift: Geology, v. 27, no. 9, p. 779–782, doi: 10.1130/0091-7613(1999)027<0779:AAAACF>2.3.CO;2.

Link, M., 2003, Depositional systems and sedimentary facies of the Miocene-Pliocene Ridge Basin Group, Ridge Basin, California., *in* Crowell, J.C., ed., Evolution of Ridge Basin, southern California: An interplay of sedimentation and tectonics: Geological Society of America Special Paper 367, p. 17–87.

Lipman, P.W., and Hagstrum, J.T., 1992, Jurassic ash-flow sheets, calderas, and related intrusions of the Cordilleran volcanic arc in southeastern Arizona: Implications for regional tectonics and ore deposits: Geological Society of America Bulletin, v. 104, p. 32–39, doi: 10.1130/0016-7606(1992)104<0032:JAFSCA>2.3.CO;2.

Mack, G.H., Seager, W.R., and Kieling, J., 1994, Late Oligocene and Miocene faulting and sedimentation, and evolution of the southern Rio Grande Rift, New Mexico, USA: Sedimentary Geology, v. 92, no. 1–2, p. 79–96, doi: 10.1016/0037-0738(94)90055-8.

Manville, V., 2001, Sedimentology and history of Lake Reporoa; an ephemeral supra-ignimbrite lake, Taupo volcanic zone, New Zealand: Special Publication of the International Association of Sedimentologists, v. 30, p. 109–140.

Marra, F., 2001, Strike-slip faulting and block rotation; a possible trigger mechanism for lava flows in the Alban Hills?: Journal of Structural Geology, v. 23, p. 127–141, doi: 10.1016/S0191-8141(00)00068-7.

Maung, H., 1987, Transcurrent movements in the Burma-Andaman Sea region: Geology, v. 15, p. 911–912, doi: 10.1130/0091-7613(1987)15<911:TMITBS>2.0.CO;2.

McCaffrey, R., 1992, Oblique plate convergence, slip vectors, and forearc deformation: Journal of Geophysical Research, v. 97, B6, p. 8905–8915.

McKee, M.B., and Anderson, T.H., 1998, Mass-gravity deposits and structures in the Lower Cretaceous of Sonora, Mexico: Geological Society of America Bulletin, v. 110, no. 12, p. 1516–1529, doi: 10.1130/0016-7606(1998)110<1516:MGDASI>2.3.CO;2.

McMillan, N.J., Dickin, A.P., and Haag, D., 2000, Evolution of magma source regions in the Rio Grande Rift, southern New Mexico: Geological Society of America Bulletin, v. 112, no. 10, p. 1582–1593, doi: 10.1130/0016-7606(2000)112<1582:EOMSRI>2.0.CO;2.

Mukhopadhyay, M., 1984, Seismotectonics of subduction and back-arc rifting under the Andaman Sea: Tectonophysics, v. 108, p. 229–239, doi: 10.1016/0040-1951(84)90237-3.

Nilsen, T.H., and Sylvester, A.G., 1995, Strike-slip basins, *in* Busby, C.J., and Ingersoll, R.V., eds., Tectonics in Sedimentary Basins: Cambridge, Massachusetts, Blackwell Science, p. 425–457.

Nourse, J.A., 1995, Jurassic-Cretaceous paleogeography of the Magdalena region, northern Sonora, and its influence on the positioning of Tertiary metamorphic core complexes, *in* Jacques-Ayala, C., Gonzalez-Leon, C.M., and Roldan-Quintana, J., eds., Studies on the Mesozoic of Sonora and Adjacent Areas: Geological Society of America Special Paper 301, p. 59–78.

Pekar, S.F., Christie-Blick, N., Miller, K.G., and Kominz, M.A., 2003, Quantitative constraints on the origin of stratigraphic architecture at passive continental margins; Oligocene sedimentation in New Jersey, USA: Journal of Sedimentary Research, v. 72, p. 227–245.

Riggs, N.R., and Busby-Spera, C.J., 1990, Evolution of a multi-vent volcanic complex within a subsiding arc graben depression: Mount Wrightson Formation, Arizona: Geological Society of America Bulletin, v. 102, no. 8, p. 1114–1135.

Riggs, N.R., Mattinson, J.M., and Busby, C.J., 1993, Correlation of Jurassic eolian strata between the magmatic arc and the Colorado Plateau: New U-Pb geochronologic data from southern Arizona: Geological Society of America Bulletin, v. 105, no. 9, p. 1231–1246, doi: 10.1130/0016-7606(1993)105<1231:COJESB>2.3.CO;2.

Rightmer, D. A., Abbott, P. L., and Anonymous, 1995, The Fish Creek sturzstrom, Anza-Borrego Desert State Park, California: Annual Meeting, Association of Engineering Geologists, v. 38, abstract volume, p. 79–80.

Saleeby, J.B., and Busby-Spera, C., with contributions by Oldow, J.S., Dunne, G.C., Wright, J.E., Cowan, D.S., Walker, N.W., and Allmendinger, R.W., 1992, Early Mesozoic tectonic evolution of the western U.S. Cordillera, *in* Burchfiel, B. C., Lipman, P. W., and Zoback, M. L., eds., The Cordilleran Orogen: Conterminous U.S.: Boulder, Colorado, Geological Society of America, The Geology of North America, v. G-3, p. 107–168.

Silver, L.T., and Anderson, T.H., 1983, Further evidence and analysis of the role of the Mojave-Sonora megashear(s) in Mesozoic Cordilleran tectonics: Geological Society of America Abstracts with Programs, v. 15, no. 5, p. 273.

Smith, G., 1991, Facies sequences and geometries in continental volcaniclastic sequences, *in* Fisher, R.V., and Smith, G.A., eds., Sedimentation in Volcanic Settings: Tulsa, Oklahoma, SEPM (Society for Sedimentary Geology), p. 109–122.

Smith, G.A., McIntosh, W.C., and Kuhle, A.J., 2001, Sedimentologic and geomorphic evidence for seesaw subsidence of the Santo Domingo accommodation-zone basin, Rio Grande Rift, New Mexico: Geological Society of America Bulletin, v. 113, no. 5, p. 561–574.

Smith, R.C.M., 1991, Landscape response to a major ignimbrite eruption, Taupo Volcanic Center, New Zealand, *in* Fisher, R.V., and Smith, G. A., eds., Sedimentation in Volcanic Settings: Tulsa, Oklahoma, SEPM (Society for Sedimentary Geology), p. 123–138.

Titley, S.R., 1976, Evidence for a Mesozoic linear tectonic pattern in southeastern Arizona: Arizona Geological Society Digest, v. 10, p. 71–101.

Tosdal, R.M., Haxel, G.B., and Wright, J.E., 1989, Jurassic geology of the Sonoran Desert Region, southern Arizona, southeastern California, and northernmost Sonora: Construction of a continental-margin magmatic arc, *in* Jenney, J.P., and Reynolds, S.J., eds., Geologic Evolution of Arizona: Tucson, Arizona, Arizona Geological Society, p. 397–434.

van Wagoner, J.C., 1995, Sequence stratigraphy and marine to nonmarine facies architecture of foreland basin strata, Book Cliffs, Utah, USA: American Association of Petroleum Geologists Memoir 64, p. 137–223.

van Wagoner, J.C., Posamentier, H.W., Mitchum, R.M.J., Vail, P.R., Sraq, J.F., Loutit, T.S., and Hardenbol, J., 1988, An overview of the fundamentals of sequence stratigraphy and key definitions, *in* Wilgus, C.K., et al., eds., Sea-level changes; an integrated approach: Tulsa, Oklahoma, SEPM (Society for Sedimentary Geology), p. 39–45.

Vedder, L.K., 1984, Stratigraphic relationship between the Late Jurassic Canelo Hills volcanics and the Glance Conglomerate, southeastern Arizona [M.S. thesis]: Tucson, Arizona, University of Arizona, 129 p.

Wesnousky, S.G., and Jones, C.H., 1994, Oblique slip, slip partitioning, spatial and temporal changes in the regional stress field, and the relative strength of active faults in the Basin and Range, Western United States: Geology, v. 22, no. 11, p. 1031–1034, doi: 10.1130/0091-7613(1994)022<1031:OSSPSA>2.3.CO;2.

Wilson, M., 1989, Igneous Petrogenesis: Boston, Unwin Hyman, 466 p.

Wood, R.A., Pettinga, J.R., Bannister, S., Lamarche, G., and McMorran, T.J., 1994, Structure of the Hanmer strike-slip basin, Hope Fault, New Zealand: Geological Society of America Bulletin, v. 106, p. 1459–1473, doi: 10.1130/0016-7606(1994)106<1459:SOTHSS>2.3.CO;2.

Manuscript Accepted by the Society 24 March 2005

Printed in the USA

Geological Society of America
Special Paper 393
2005

Pre–Late Jurassic, fossil-bearing volcanic and sedimentary red beds of Huizachal Canyon, Tamaulipas, Mexico

David E. Fastovsky
O. Don Hermes
Nicholas H. Strater
Department of Geosciences, University of Rhode Island, Kingston, Rhode Island 02881, USA

Samuel A. Bowring
Department of Earth, Atmospheric, and Planetary Sciences, Massachusetts Institute of Technology, Cambridge, Massachusetts 02139, USA

James M. Clark
Department of Biological Sciences, George Washington University, Washington, D.C. 20052, USA

Marisol Montellano
Rene Hernandez R.
Instituto de Geologia, Universidad Nacional Autónoma de México, México 20 D.F., México

ABSTRACT

Huizachal Canyon, one of a series of generally east-west trending canyons in Tamaulipas, northeastern Mexico, exposes a pre–Late Jurassic sequence of fossiliferous pyroclastic and epiclastic rocks, the Huizachal Group. Heretofore considered a sedimentary package associated with either metamorphic or intrusive rocks, structural relationships and petrographic studies presented here indicate that the Huizachal Group was in fact deposited unconformably upon an older, undescribed, sequence of pyroclastic rocks.

Four igneous units are recognized in the steeply dipping older volcaniclastic sequence (respectively, units A–D): a complex suite of pyroclastic flows, accretionary lapilli tuff(s), and lava flows (unit A); a homogeneous, fine-grained felsitic rock (unit B); a sequence of conglomerates (unit C); and a mixed assemblage of rocks including mafic-to-intermediate composition lava flows and intercalated conglomerate and tuff (unit D). Most of these rocks have undergone extensive late-stage or postdepositional silicification, but relatively immobile trace elements demonstrate that these rocks range from subalkaline basalt to rhyolite.

The younger, relatively flat-lying Huizachal Group overlies these rocks in angular unconformity. The fossil assemblage comes from a <10-m-thick sequence in the lower part of the Huizachal Group, which also is the result of pyroclastic volcanic deposition. Some of the organisms entrained within these tuff(s) were reworked by volcanic processes; others appear to have been actively trapped in a manner analogous to Pompeii.

New U-Pb isotopic data from zircon in a volcaniclastic rock from the lowest part of the Huizachal Group (La Boca Formation) yields an age of 189 ± 0.2 Ma (analytical error). The

Fastovsky, D.E., Hermes, O.D., Strater, N.H., Bowring, S.A., Clark, J.M., Montellano, M., and Hernandez R., R., 2005, Pre–Late Jurassic, fossil-bearing volcanic and sedimentary red beds of Huizachal Canyon, Tamaulipas, Mexico, *in* Anderson, T.H., Nourse, J.A., McKee, J.W., and Steiner, M.B., eds., The Mojave-Sonora megashear hypothesis: Development, assessment, and alternatives: Geological Society of America Special Paper 393, p. 401–426. doi: 10.1130/2005.2393(15).

sedimentary rocks immediately above this unit contain fossils considered to be Early Jurassic in age. Thus, the zircon isotopic age agrees with, and is supportive of, the age estimates based upon fossil vertebrates.

Trace element geochemistry of the volcanic units is strongly suggestive of subalkalic ocean-continent Andean volcanism. Thus, the volcanic and sedimentary rocks of Huizachal Canyon were most likely deposited in a convergent plate margin setting instead of an extensional rift system as previously proposed.

RESUMEN

El Cañón del Huizachal, es uno de una serie de cañones con orientación este-oeste en Tamaulipas, en el Noreste de México; ahí aflora el Grupo Huizachal, una secuencia de rocas fosilíferas piroclásticas y epiclásticas pre-jurásicas tardías. Hasta ahora considerado como un paquete asociado con rocas ígneas intrusivas o metamórficas, las relaciones estructurales y los estudios petrográficos presentados aquí sugieren que el Grupo Huizachal estuvo, de hecho, depositado de manera discordante por encima de una secuencia de rocas volcánicas piroclásticas más vieja, no descrita.

Se reconocen cuatro unidades ígneas en la secuencia vulcanoclástica más antigua (respectivamente unidades A–D): un conjunto complejo de flujos piroclásticos, tobas de lapili acrecional, y flujos de lava (unidad A); una roca homogénea félsica de grano fino (unidad B); una secuencia de conglomerados (unidad C); y un conjunto mixto de rocas que incluye flujos de lava de composición máfica a intermedia, así como una toba y un conglomerado intercalados (unidad D). Casi todas estas rocas han sufrido una silicificación extensiva en su fase tardía, o post-deposicional, pero elementos traza relativamente inmóviles demuestran que estas rocas comprenden desde basalto subalcalino a riolita.

El conjunto de fósiles proviene de una secuencia de <10 m de espesor de la parte inferior del Grupo Huizachal. Esta secuencia es también el resultado de un depósito volcánico piroclástico. Algunos de los organismos arrastrados dentro de estas tobas fueron retrabajados por procesos volcánicos; otros parecen haber sido atrapados enérgicamente de una manera análoga a la de Pompeya.

Nuevos datos de U-Pb isotópicos de zircón en una roca vulcanoclástica de la parte más inferior del grupo Huizachal (Formación La Boca) dio una edad de 189 ± 0.2 Ma (error analítico). Las rocas sedimentarias inmediatamente arriba de esta unidad contienen los fósiles considerados como de edad jurásica temprana. De este modo, la edad isotópica del zircón concuerda y apoya la edad estimada basada en los vertebrados fósiles.

La geoquímica de elementos traza de las unidades volcánicas sugiere fuertemente un vulcanismo andino subalcalino océano-continente. Así, las rocas volcánicas y sedimentarias del Cañón Huizachal fueron depositadas probablemente en un ambiente de margen de placa convergente en vez de un sistema de rift extensional como se había propuesto anteriormente.

Keywords: pyroclastic, Huizachal, Early Jurassic, Mexico, geochemistry.

INTRODUCTION

A window into the pre–Late Jurassic of northeastern Mexico is provided by a series of broadly east-west–trending canyons, the bases of which contain exposures of volcanic and/or sedimentary rocks from this time interval. One of these canyons, Huizachal Canyon[1], also preserves a distinctive Early Jurassic vertebrate fauna—the oldest terrestrial vertebrates in Mexico—that bears upon the radiation of Jurassic tetrapods in North America (Clark et al., 1994; Fastovsky et al., 1995). Here we describe and reinterpret pre–Late Jurassic strata within Huizachal Canyon. In the process, we describe and characterize previously unrecognized volcaniclastic rocks in the canyon, and publish the first radiometric date for pre–Late Jurassic rocks in the region, which better constrains the age of the Huizachal Canyon vertebrate fauna.

[1]Huizachal Canyon is commonly referred to as "Huizachal" on Mexican maps. It is widely termed "Huizachal Canyon" in English-language publications; here, therefore, we refer to it by its English name.

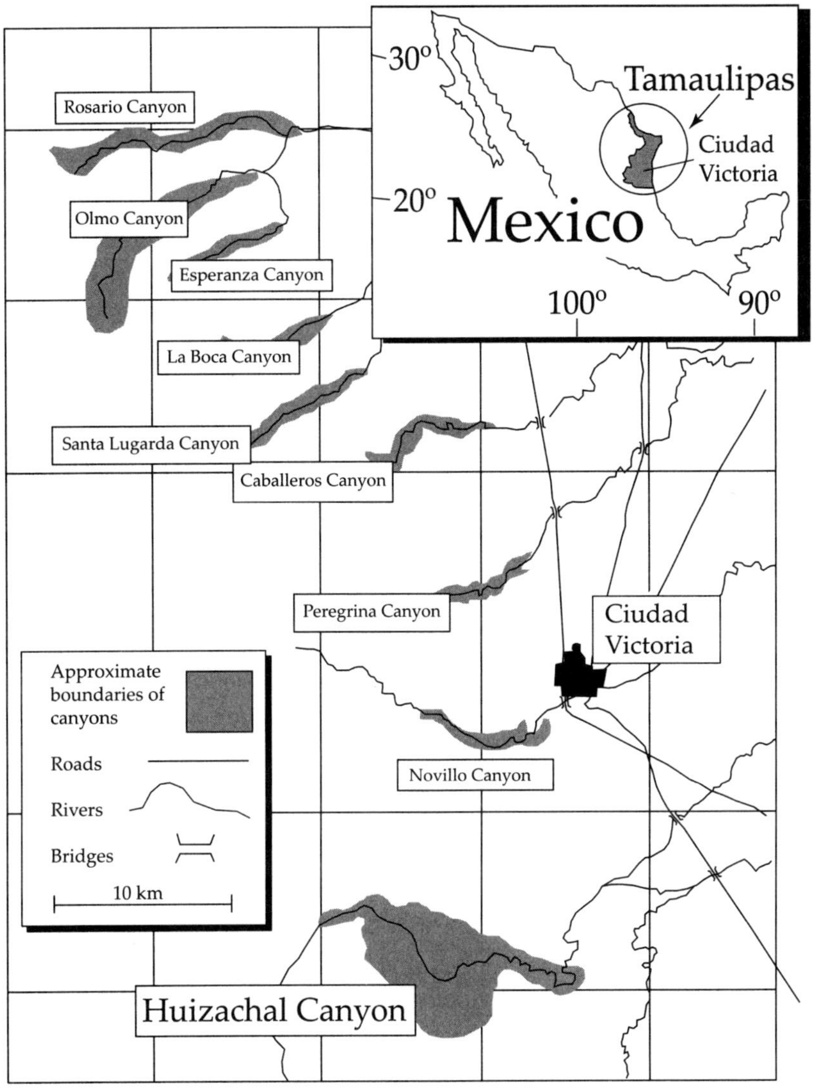

Figure 1. East-west–trending canyons in Ciudad Victoria, Tamaulipas, northeastern Mexico. Inset: Mexico. After Carrillo-Bravo (1961).

Our interpretation of the pre–Late Jurassic rock sequence in Huizachal Canyon leads to the recognition of a distinctly older succession of steeply dipping volcanic and epiclastic rocks (VES) occurring unconformably beneath the subhorizontally oriented vertebrate-bearing and superjacent red bed suite (SS). The pre–Late Jurassic rock succession in Huizachal Canyon appears to have been deposited in an arc-trench setting along a continental margin, rather than in an extensional rift setting as previously interpreted. Moreover, the evidence is mounting that at least some of the vertebrate assemblage was actively trapped by explosive volcanic processes.

Huizachal Canyon

Huizachal Canyon is located ~15 km southwest of Ciudad Victoria, Tamaulipas, northeastern Mexico (Fig. 1). The topographically higher parts of Huizachal Canyon are composed of Oxfordian (Upper Jurassic) and younger marine limestones (Imlay et al., 1948). Toward the base of the canyon, however, are exposed red beds of a volcaniclastic sedimentary succession that constitutes the type locality of the tetrapod-fossil-bearing Huizachal Group (Mixon et al., 1959). This unit is composed of the lower, fossiliferous La Boca Formation and the upper non-fossiliferous La Joya Formation (Imlay et al., 1948; Mixon et al., 1959). Also exposed at the base of the canyon is a succession of older volcanic and epiclastic rocks reflecting episodes of pyroclastic volcanism (Fig. 2). Historically, this latter succession was described either as igneous intrusives into the Huizachal Group, or as basement metamorphic rocks (Mixon 1963a, Carrillo-Bravo, 1963, Belcher, 1979). In our interpretation, however, this weakly metamorphosed volcanic and epiclastic suite of rocks predates the sedimentary suite, and is separated from it by an angular unconformity. Figure 3 is a geological map of Huizachal Canyon, showing the units described in this study.

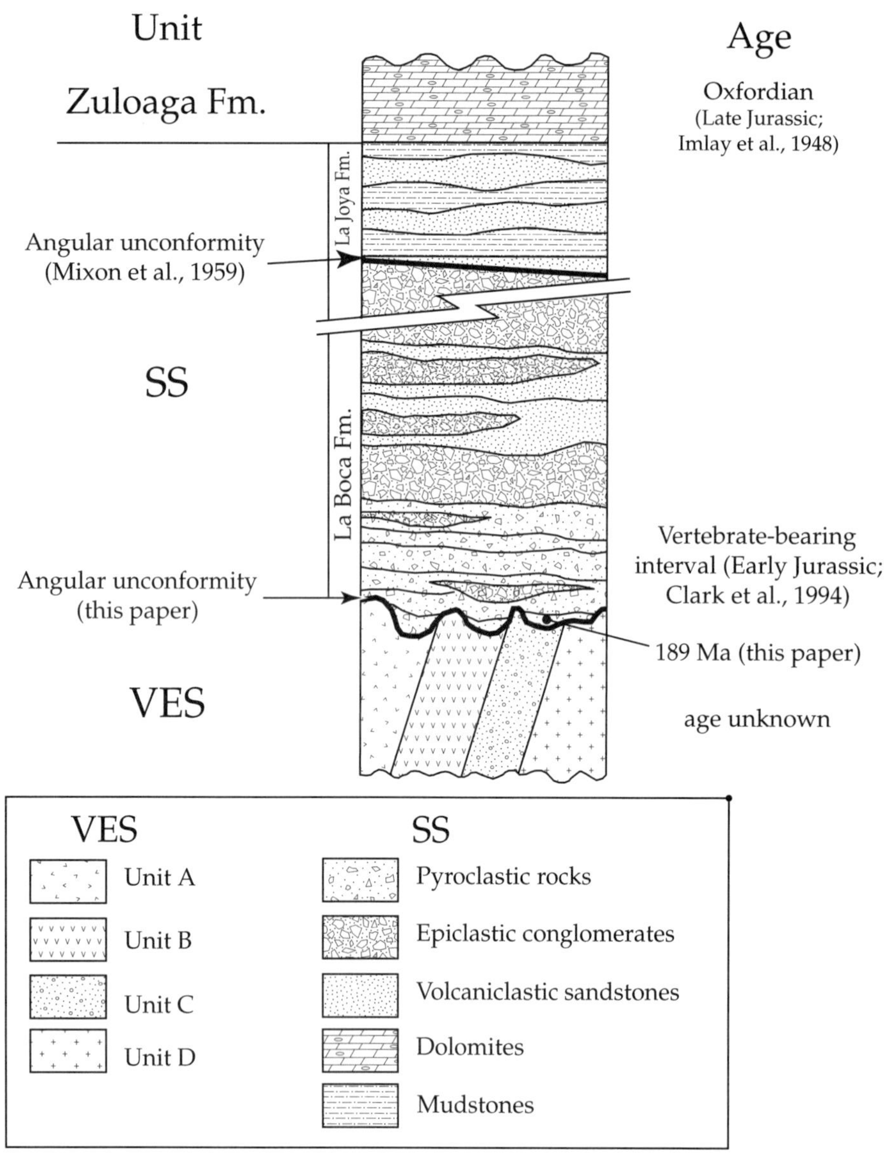

Figure 2. Generalized, schematic stratigraphic section of pre–Late Jurassic red beds, Huizachal Canyon, Tamaulipas, Mexico. The figure emphasizes the VES and the fossil-bearing component of SS. Thickness of VES is unknown; thickness of fossil-bearing interval <10 m. Total thickness of La Boca Formation is ~54 m. Units not drawn to scale.

Previous Studies

Red beds throughout northeastern Mexico were described by Imlay (1938, 1943a, 1943b). He noted that all such units are unconformably overlain by limestones of the Upper Jurassic Zuloaga Formation, and assigned them an Early-Late Jurassic age based on stratigraphic position (Imlay, 1938, 1943a, 1943b). Imlay was thus the first of many successive workers to infer that red beds in the sub-Zuloaga interval throughout the region are stratigraphically equivalent.

Vertebrate-Bearing and Superjacent Red Bed Suite (SS)

The red beds within Huizachal Canyon were initially called the "Huizachal Formation" in technical reports in 1921 (Carrillo-Bravo, 1961) and again in 1925 (Mixon et al., 1959). Finally, Imlay et al. (1948) formally defined the red beds in Huizachal Canyon as the Huizachal Formation. Mixon et al. (1959) and Mixon (1963a, 1963b) elevated the Huizachal Formation to Group status based on a subtle angular unconformity they observed within the SS. They identified a lower La Boca Formation and an upper La Joya Formation[2], and assigned a Late Triassic–Early Jurassic age to the La Boca based upon fossil plants from Novillo Canyon (one canyon to the north; Fig. 1), and an Early, Middle, or Late Jurassic age to the La Joya, based upon stratigraphic position.

Reinforcing the assumption that the red beds were penecontemporaneous, Belcher (1979) and Gose et al. (1982) postulated

[2]The siliciclastic red beds of the La Joya Formation were not studied during the course of this research, other than to confirm the unconformity of Mixon et al. (1959).

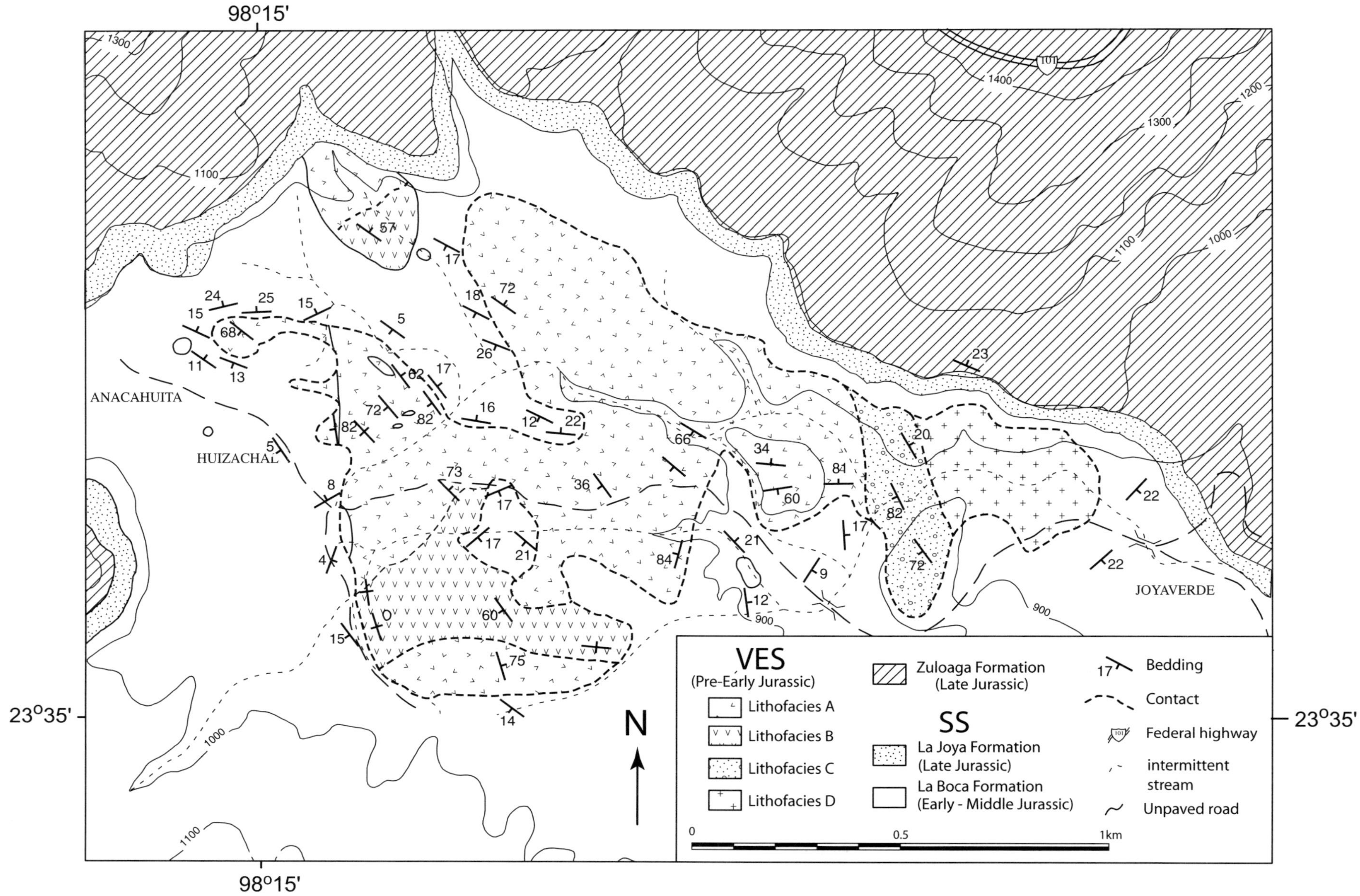

Figure 3. Geological map of Huizachal Canyon, showing exposure limits of units in the VES, the Huizachal Group (SS), and the Zuloaga Limestone. Representative strikes and dips taken from Strater (1993).

that the La Boca Formation was deposited in graben lows produced by the opening of the Gulf of Mexico during the breakup of Pangea (see also Humphrey, 1949; Salvador, 1987; Kesler et al., 1988; Michalzik, 1991). Salvador (1987) asserted that any volcanic material interlayered with the red beds was produced in this rift environment, a point that our study contradicts.

Rueda et al. (1991a, 1991b) separated the La Boca Formation into two "alloformations" (defined and identified on the basis of bounding discontinuities); the Huizachal Alloformation, discordantly overlain by the La Boca Alloformation (though the type section was not specified). Rueda et al. (1991a, 1991b) inferred the La Boca Alloformation to be the product of marginal marine deposition in a local rift basin, based upon palynomorph evidence, algae, and the presence of glauconite.

Clark et al. (1994) revised the age of the La Boca Formation in Huizachal Canyon. Using vertebrate biostratigraphy—in particular, mammals—they estimated an Early-to-Middle Jurassic age for the La Boca Formation in Huizachal Canyon.

Fastovsky et al. (1988, 1995) reported an unconformity between a previously unreported "upturned volcaniclastic sequence" that they identified at the base of the canyon, and the overlying, subhorizontal Huizachal Group. They also estimated the age of the La Boca Formation as Middle Jurassic, based upon a preliminary U-Pb date from zircon. Our study builds on the stratigraphic interpretations of Fastovsky et al. (1988, 1995), and supports the age determination of Clark et al. (1994) for the La Boca Formation. Our conclusion will be that the lower part of the SS—the fossiliferous part of the La Boca Formation—was produced by pyroclastic flows and is as volcaniclastic as anything found in the VES below.

Volcanic and Epiclastic Suite (VES)

The earliest mention of nonsedimentary rocks in Huizachal Canyon comes from a San Antonio Geologic Society road log (Tatum and Dawson, 1936), which notes the presence of "Precambrian igneous rocks," but does not provide a description (see also Murray, 1961). The type section of the Huizachal Formation by Imlay et al. (1948) indicates an erosional contact at the base of the red beds and characterizes the underlying rocks as "metamorphic." In contrast, Carrillo-Bravo (1963), Mixon (1963a), and Belcher (1979) noted the presence of "felsic intrusive rocks," which they claimed crosscut the La Boca; Mixon (1963a) characterized the contact as erosional in some places. Fastovsky et al. (1988; 1995) suggested an angular unconformity between the SS and the VES, an observation reaffirmed in some places by Jones et al. (1995).

PETROLOGY

Two structurally and petrologically distinct groups of rocks comprise the red beds of Huizachal Canyon. The first, VES, consists of a series of steeply dipping pre-Jurassic volcanic rocks (the "upturned volcaniclastic sequence" of Fastovsky et al., 1995). Four lithologically distinct mappable units, here labeled A–D occur within the VES. The second group in Huizachal Canyon, SS, is the gently dipping series of Early Jurassic-aged red beds generally known as the Huizachal Group (following Mixon et al., 1959). Within the SS, structural and stratigraphic relationships support the conclusion of Mixon et al. (1959) and Mixon (1963a, 1963b) that the SS should be divided into the La Boca and La Joya Formations (Strater, 1993).

The steeply dipping rocks of the VES comprise a wide variety of volcanic and volcaniclastic rocks. Using the terminology of Fisher and Schmincke (1984), recognizable units include pyroclastic flows, lava flows, lahars, and pyroclastic fallout deposits. Additional water-reworked pyroclastic and epiclastic rocks also are present. Most units have been extensively altered by a combination of devitrification and late-stage vapor phase or hydrothermal processes that have partly to completely obliterated primary mineralogy, texture, and structure.

Lithology and Petrography

VES—Unit A

Unit A constitutes the majority of the west-central portion of the VES (Fig. 3). This unit consists mainly of pyroclastic volcanic rocks, including pyroclastic debris flows, air-fall lapilli tuffs, lava flows, and some flows or ignimbrites that exhibit rheomorphic features. These rocks are interbedded with lithic tuffs and epiclastic conglomerates. This relatively heterogeneous package of rocks is white, gray, or purple, and generally is poorly stratified except for the presence of local, discontinuous layering.

The pyroclastic debris flows typically contain abundant poorly sorted volcanic clasts, some up to 1–2 m (Fig. 4). Other layers are volcanic tuffs that exhibit more restrictive grain-size but are poorly sorted. These pyroclastic rocks are crystal- and lithic-rich tuff (Fig. 5). Lithic clasts exclusively exhibit volcanic textures. Phenocrysts consist mainly of relict euhedral-subhedral

Figure 4. Field photo of one type of pyroclastic flow from unit A. Unsorted volcanic clasts range in size from centimeter-size to angular and rounded blocks up to a meter.

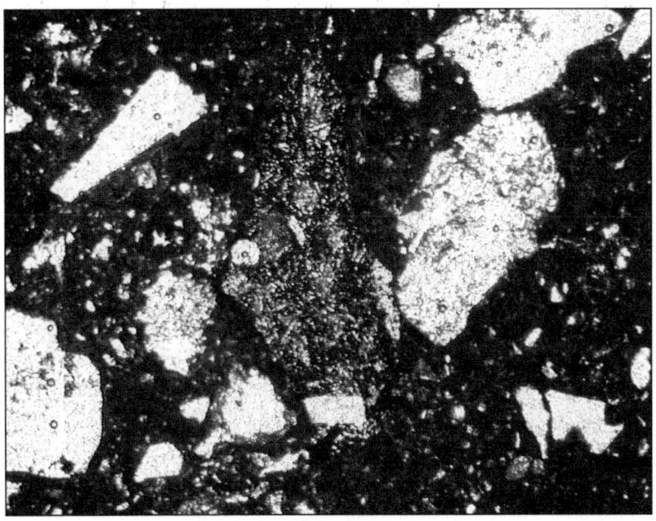

Figure 5. Photomicrograph of tuff from unit A. A volcanic clast is in center of photo, surrounded by relict subhedral phenocrysts of altered plagioclase and angular broken crystal of feldspar. Sample DH031388-8, X-polars, width of field is ~2 mm.

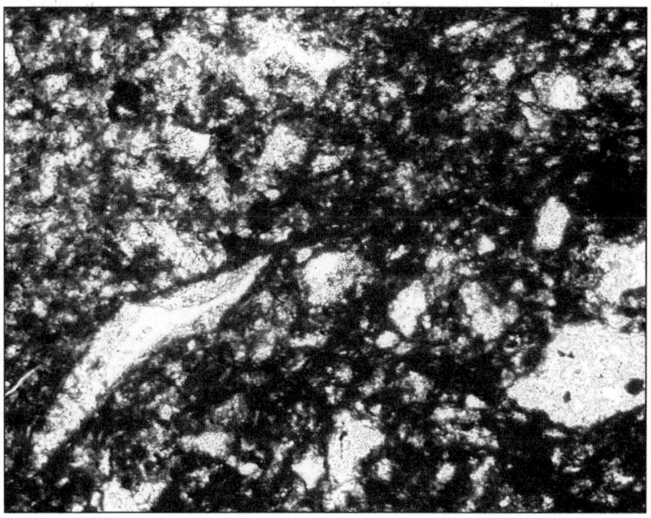

Figure 6. Photomicrograph of tuff from unit A, showing angular and euhedral phenocrysts of altered feldspar, and sparse curvilinear relict devitrified glass shards. Sample DH031388-8, X-polars, width of field is ~2 mm.

plagioclase crystals (pseudomorphed by saussaurite) and sparse microphenocrysts of quartz. Many of these crystals are angular and broken, implying explosive eruption. Some samples contain sparse grains of opaque material that may be pseudomorphs of amphibole or pyroxene. The matrix mainly consists of a fine-grained intergrowth of feldspar, quartz, and opaque minerals, much of which are likely devitrification products of a glassy matrix. Some samples preserve relict well-formed microlites of feldspar, sometimes with an aligned fabric. In certain cases, the microlites exhibit barbed, swallow tail–shaped terminations and skeletal cores consistent with rapid crystal growth upon quenching (Hermes and Schilling, 1976). Some samples contain cuspate-shaped fragments that we interpret to be relict glass shards (Fig. 6) that formed from burst vesicle walls during eruption. Nearly all lithic clasts in these tuffs exhibit volcanic textures and generally are of felsic composition. Some dark-colored, flattened clasts may be either mafic volcanic clasts or devitrified fiamme-like pumice fragments. Lithic clasts generally constitute less than 10% of the tuffs.

Some tuff units contain weakly layered horizons that exhibit distinctive spherulitic textures (Fig. 7). Spherule diameters vary in size from 0.1 to 5 cm; the abundance of these layers varies along and across strike. Locally, some horizons may contain over 50% spherulites up to 5 cm in diameter. In thin section, the

Figure 7. (A) Field photo of weakly layered accretionary lapilli tuff from unit A. Also note the presence of volcanic rock clasts in the tuff, some of which contain sparse spherulites (e.g., at the hammer head). (B) Close-up field photo showing faintly layered lapilli rock with abundant spherulites up to 5 cm in diameter.

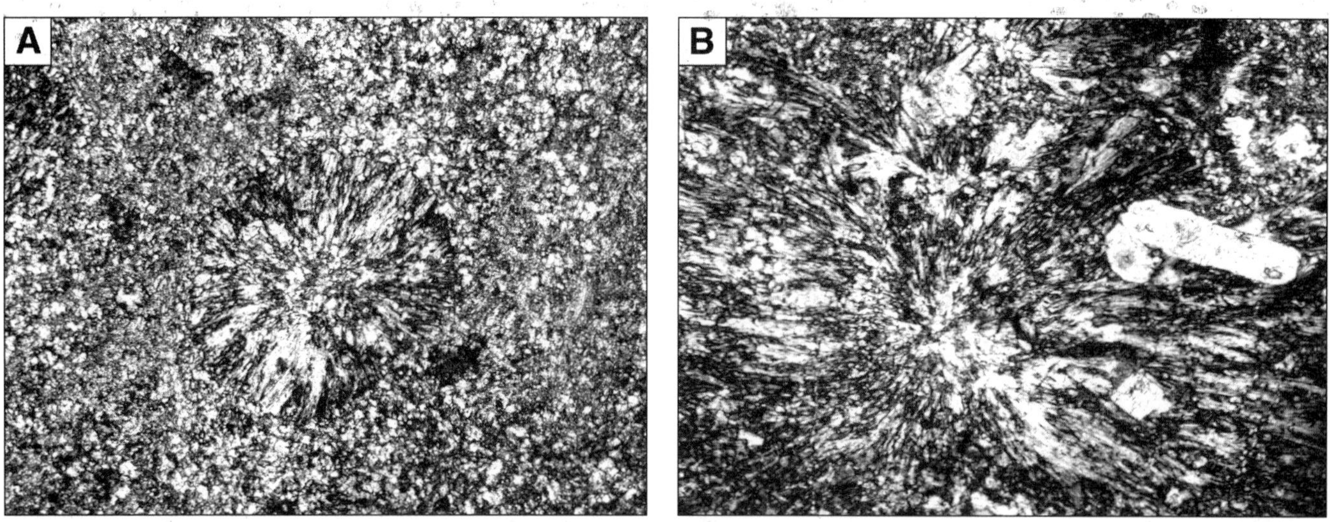

Figure 8. Photomicrograph of accretionary lapilli tuff (Sample DH031388-11; see text). (A) The spherulites consist of radial intergrowths of chalcedony and albite, often nucleated about a central small crystal or clast. X-polars, width of field is ~2 mm. (B) Photomicrograph of a spherulite that encloses several euhedral laths of plagioclase incorporated during accretion; X-polars, width of field = 1 mm.

spherulites are typically cryptic because devitrification and/or alteration processes have obscured much of the original texture; Ross and Smith (1961) point out that spherulite development tends to severely destroy tuff structures. When visible, the spherulites typically exhibit a radiating crystal structure composed of albite and microcrystalline quartz (Fig. 8). Relict euhedral laths of plagioclase commonly are included in the spherulite matrix. All of these textures suggest that these features are devitrified, recrystallized accretionary lapilli; if correct, these strata are volcanic air-fall deposits. An alternate (but less preferred) interpretation is that the spherulites represent late-stage in-filling by circulating fluids of either vesiculated lava flows or rheomorphically deformed welded tuff. However, the striking spherical shapes argue against them being filled vesicles (amygdules), which would most likely be more irregular in shape. Regardless, these kinds of volcanic textures would be expected within a terrain dominated by silica-rich volcanic rocks. Some spherulitic units are green due to the presence of secondary malachite. Other occurrences of malachite are common locally throughout the formation (including in several exploration prospect pits), and reflect the presence of copper carbonate as a component of the silica-rich fluids during the vapor phase stage. Ordinary calcite also is widespread as a late-stage mineral in the matrix of most of the tuffaceous rocks.

Some other rocks within Unit A appear to be devitrified felsic lava flows and/or ignimbrites that contain rheomorphic features. Figure 9 shows abundant contorted banded volcanic clasts within a fine-grained matrix. The deformed banded clasts represent brecciated parts of the flow that have been incorporated into the still-hot, plastically deformed interior parts of the flow.

Most of the volcanic layers in unit A likely formed as pyroclastic flows. Features consistent with a pyroclastic flow origin include the common presence of devitrified glass shards, angular and broken phenocrysts, poorly developed sorting and stratification, the presence of lithic clasts, and fiamme (Ross and Smith, 1961; Fisher and Schmincke, 1984). Other layers within unit A are intercalated lava flows or ignimbrites and air-fall tuffs. Bonnichsen (1982) and Bonnichsen and Kauffman (1987), as well as other workers, have pointed out the difficulty of distinguishing rhyolite sheets that have been erupted as lava flows from ignimbritic pyroclastic flow deposits.

Figure 9. Field photo of a lava flow or ignimbrite from unit A that contains abundant contorted clasts of banded volcanic rock (probably devitrified banded obsidian). We interpret these features as rheomorphic clasts formed by brecciation of flow tops or sides which were enclosed by hot, plastic parts of the flow interior.

The interbedded conglomerates typically contain more than 50% of rounded to angular clasts up to 1 m in diameter. Nearly all clasts are of volcanic origin, and are dominated by felsic compositions. The matrix also is of volcanic origin, typically consisting of crystals of feldspar and quartz. Relict glass shards and quenched microlites are present in the matrix of some samples. We interpret some of the conglomerates as lahars initiated directly by volcanic activity. Of particular interest are the fresher, better-preserved volcanic textures and mineralogy typically preserved in the clasts (compared to the host tuff). These largely unaltered features may reflect rapid incorporation of clasts into a relatively cold matrix that was not subjected to subsequent vapor phase alteration, as was the case for the hot tuffs, especially the lower and interior layers of the tuffs. If the pyroclastic flows were emplaced as hot masses, subsequently welded, and rheomorphically deformed, the accompanying deuteric mineralization and devitrification would have been considerably harsher on the original textures and mineralogy of the clasts.

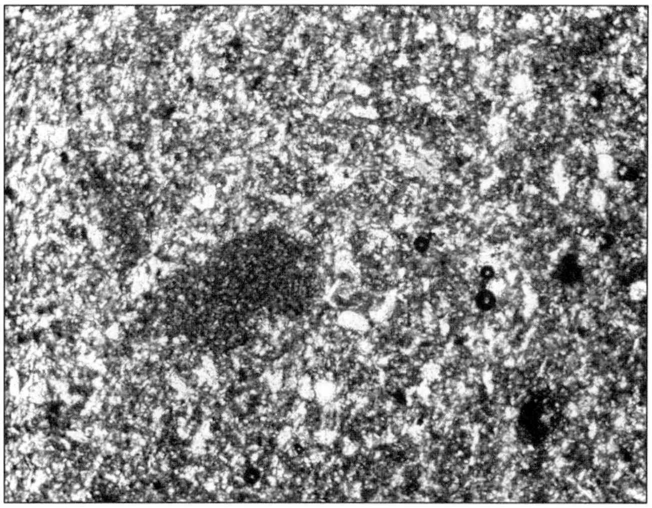

Figure 10. Microphotograph of representative tuff from unit B. Most rocks are massive and fine-grained, containing small dark volcanic clasts and small angular broken crystals of altered feldspar. Sample NHS-21B, X-polars, width of field = 1mm.

VES—Unit B

Unit B consists of generally massive and homogeneous fine-grained felsite that is confined to the west-southwestern portion of the VES (Fig. 3). Color is variably white, pink, purple, green, blue, and black. In places, the rock exhibits faint layering, curvilinear in places. We are uncertain if this layering represents syn-igneous rheomorphic deformation or post-consolidation deformational folds. Epiclastic rocks, or other rocks of sedimentary affinity, are notably absent. Rock clasts are either absent or very sparse and small. In thin section, some rocks of unit B contain euhedral, angular phenocrysts of altered feldspar and small fine-grained darker-colored clasts (Fig. 10). Some microphenocrysts of plagioclase are skeletal and exhibit swallow-tail morphology indicative of quenching. The presence of microphenocrysts embedded in a very fine-grained matrix, and quenched crystals demonstrate that this unit is of extrusive origin. The feldspars commonly are preferentially oriented, which suggests that these rocks were produced by flow rather than air-fall processes. This is consistent with the poor sorting exhibited in these rocks. The matrix consists of a fine-grained intergrowth of quartz, feldspar, and opaque minerals; secondary chlorite and calcite are common in most samples. Some samples are homogeneous fine-grained felsite that exhibit few relict textural or mineralogical features indicative of a volcanic origin. The rocks of this unit are thus interpreted to represent felsic lava flows. The faint folded layering and the aligned microphenocrysts of plagioclase may represent a relict flow structure in a lava flow or rheomorphic deformation developed in a welded tuff.

Figure 11. Unit C in outcrop. Dark clasts are volcanic rock fragments. Hammer is 31 cm in length.

VES—Unit C

Unit C consists of a series of conglomerate beds found in the eastern outcrop of the VES (Fig. 11). Clasts are purple and green, and are generally supported by a red-purple matrix. Ranging in size from coarse sand to small cobbles, these clasts are variably angular; over 95% of the clasts are of volcanic origin. Some clasts exhibit holocrystalline textures, whereas others contain a fine-grained intergrowth of quartz and feldspar that may be devitrified products from an original hypocrystalline mixture of glass and crystals. The matrix consists of small, imbricated feldspar laths and opaque materials, which reflect flow texture. Some clasts exhibit bleached rims that might represent disequilibrium reaction rinds; they are suggestive of clasts incorporated into a hot lahar that induced alteration of the rims of some clasts. At least one hematite-rich mudstone is associated with this unit.

Based on the features and textures listed above, we interpret the rocks of unit C to be the result of volcano-induced debris flows, some of which may have occurred at elevated temperatures. Some parts of unit C may have been deposited by aqueous processes, whereas others may represent pyroclastic volcanic debris flows. Some clasts of this unit appear to have been derived from unit D, which in turn implies that unit D is older than C. Most clasts from these conglomerates show less extensive alteration than do rocks of units A, B, or D.

VES—Unit D

Unit D consists of dark gray-blue-purple, aphanitic-to-slightly porphyritic, mafic-to-intermediate, igneous rocks (Fig. 12). This unit is found at the easternmost limit of the VES, where its exposure is confined to the riverbed (Fig. 3). The most common and best-preserved rocks of unit D are porphyritic basalt to basaltic andesites. Phenocrysts include clinopyroxene and sparse olivine, mostly altered to secondary minerals (Fig. 13). The matrix consists of a felty mass of plagioclase microlites intergrown with opaque minerals. Groundmass plagioclase in some rocks is aligned subparallel to impart a subtrachytic texture. Amygdules of chlorite, chalcedony, and carbonate are common in some rocks. Much of the unit is largely massive, homogenous, and unlayered. Locally, some of these rocks are highly brecciated, resulting in volcanic clasts being immersed in a matrix of secondary chlorite and calcite. We suspect that these breccias represent fragmented cooled tops or sides of flows that were incorporated into the still advancing lava flows. Some samples contain irregular, but elongated slivers of red, hematite-rich material that may represent inclusions of soft-sediment material during emplacement of the flows.

The porphyritic and highly vesicular textures of these rocks and the lack of rock clasts indicate that most rocks of this unit represent a series of lava flows. None of these rocks shows cross-cutting relationships with adjacent rocks, and hence do not appear to be intrusive. Although less prominent than the mafic rocks, we have also found within unit D several intercalated more felsic volcanic units containing fragmental crystals of feldspar and quartz.

SS—Lower Portion of La Boca Formation

The La Boca Formation contains an Early Jurassic terrestrial vertebrate fossil assemblage that includes the oldest terrestrial vertebrates known in Mexico (Fastovsky et al. 1988, 1995; Clark et al., 1994, 1998). Fastovsky et al. (1988, 1995) interpreted the lowest one-third of the La Boca Formation to be the product of subaerial debris flow deposition.

At the base of the La Boca Formation—and conformable with it—is a non-fossiliferous, 2-m-thick, massive, poorly stratified pyroclastic tuff (Fastovsky et al., 1995). The radiometric age reported below was measured on zircon from this rock. The tuff consists of abundant angular and broken phenocrysts of saussauritized plagioclase, with sparse clear-colored phenocrysts of quartz. Fine-grained volcanic rock clasts up to 1 cm are locally abundant. Some clasts are lenticular in shape and may represent flattened clasts of pumice that formed fiamme in the original rock. The groundmass consists of fine-grained angular grains of feldspar and quartz with secondary chlorite, opaque minerals, and calcite. Significant features in most samples are prominent, relict (now devitrified), glass shards (Fig. 14, as well as fig. 6 in

Figure 12. Unit D in outcrop. Unit is weakly bedded, and suggestive of viscous flow. Lens cover is 2.3 cm in diameter.

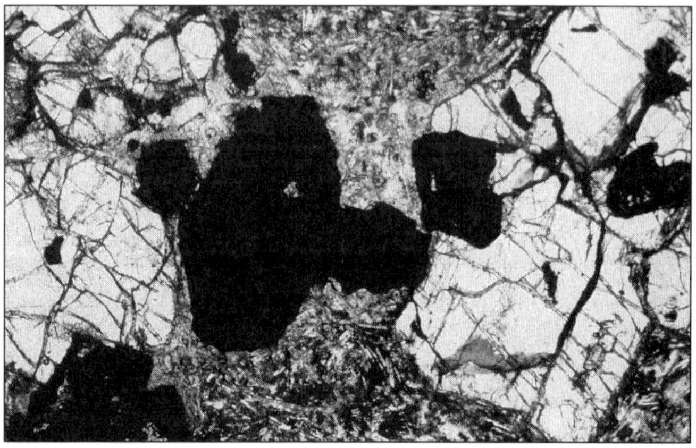

Figure 13. Subhedral phenocrysts of clinopyroxene, olivine, and opaque minerals enclosed in fine-grained matrix of plagioclase microlites. In some rocks, the microlites exhibit a subtrachytic texture. X-polars, width of field is ~1.3 mm.

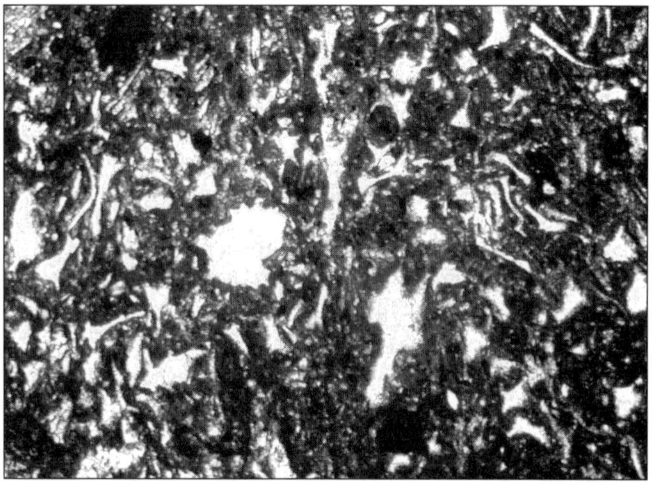

Figure 14. Microphotograph of the La Boca basal tuff. The pyroclastic nature of the rock is demonstrated by the devitrified relict glass shards. Sample Tr082291-1b, plane light, width of field is ~2 mm.

Figure 15. Fossiliferous portion of the La Boca Formation. Note volcaniclastic intraclasts in mottled matrix support. Discoloration rinds around intraclasts are clearly visible. Hammer is 31 cm.

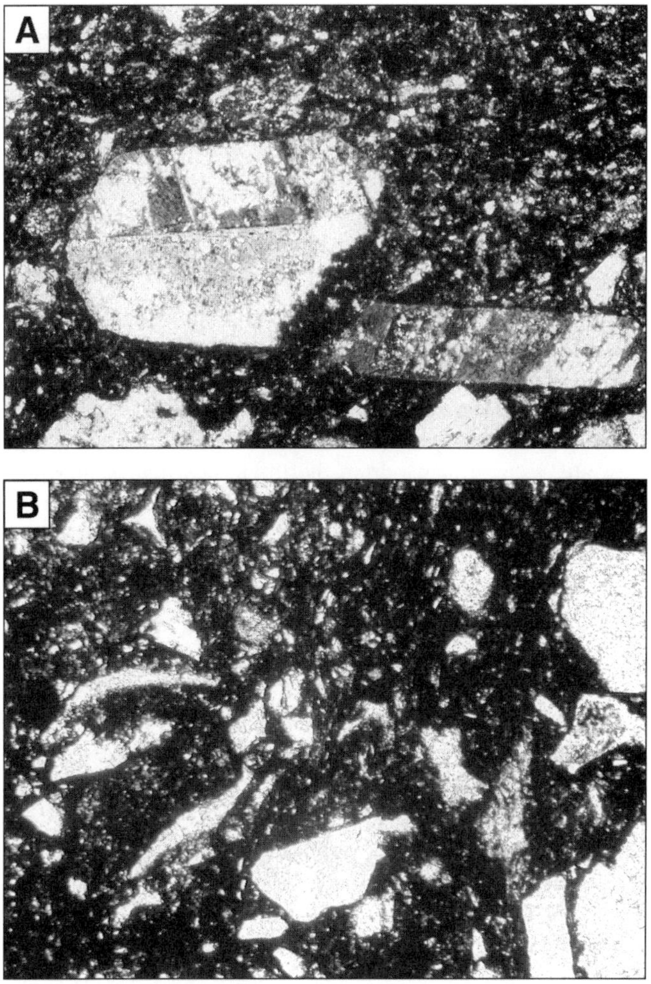

Figure 16. Photomicrographs of volcanic tuff from the horizons in the La Boca Formation that host the fossils. (A) Relict euhedral to subhedral laths of altered feldspar in a devitrified glass matrix that contains smaller microphenocrysts and broken crystals. X-polars, width of field is ~1.3mm. (B) Note the abundant broken phenocrysts of feldspar crystals, and the curvilinear relict glass shards (note the triple-armed shard at top left). These features demonstrate that this tuff formed from a highly explosive eruption. Sample MexBeast, plane light, width of field is ~2 mm.

Fastovsky et al., 1995). We interpret the shards to be remnants of glass-lined vesicle walls that fragmented upon eruption. The shards, as well as the broken phenocrysts and prominent clasts, constitute the main evidence for a pyroclastic origin for this unit. The rock is thus properly considered a tuff. The lack of sorting and stratification in this unit is more consistent with a pyroclastic flow origin, as opposed to air-fall processes.

Stratigraphically above, and conformable with this basal tuff is the fossiliferous part of the La Boca Formation (Figs. 2, 15). The fossils occur just above the basal tuff and, although the unit(s) containing the fossils span less than 10 m of section, the complete sequence consists of ~20 m of poorly bedded (beds 0.5–1.5m thick; see also fig. 7 of Fastovsky et al., 1995) fine-grained rocks intercalated with uncommon, thin (<1-m-thick) lenticular, clast-supported conglomerates. In outcrop, the fossiliferous units generally appear without internal structure. Colors range from a deep red (10R 3/4; Munsell, 1992) to light pink (10YR 8/4). Generally, however, the matrix is mottled to a greater or lesser degree, reflecting the amount of iron oxide cement that has been leached from the rock.

In a previous report (Fastovsky et al., 1995), we characterized the fossiliferous series of the La Boca Formation as "red, silty mudstones," emphasizing the sedimentary component of their origin. In thin section, however, the pyroclastic character of these rocks is apparent and striking. An abundance of small (0.05 mm) angular, doubly terminated plagioclase and potassium feldspar laths, and angular intraclasts of volcanic origin is clearly preserved in a matrix of devitrified fine-grained to very fine-grained material.

The rocks have undergone extensive diagenetic alteration and postdepositional lithification. The dominant cements are calcite, hematite, illite, and silica. Primary ash textures are commonly obscured by extensive cementation, and the illite reported by Fastovsky et al. (1995) appears diagenetic in origin. Volcanic lithic clasts ranging from ash-sized to cobble-sized occur irregularly within the tuffaceous sequence. The clasts are chaotically organized, unstratified, and unsorted. These clasts commonly bear a thin (<1 cm) discoloration rind around their edges (Fig. 15).

Many of the larger matrix fragments show the sharp-edged shapes characteristic of fragmented glassy vesicle walls (Fig. 16; e.g., Fisher and Schmincke, 1984), but are now devitrified. The fossiliferous series of the La Boca Formation shows little to no evidence of reworking and, on the basis of its distinctive petrographic texture, should be termed a tuff.

Volcanic lithic clasts also appear within lenticular, laterally pinching, concave-upward units, suggestive of channelized deposition. These units range from 1 m wide and 20 cm deep to several meters deep and tens of meters wide (Fig. 17). Sedimentary structures suggestive of traction transport (e.g., cross-stratification) were almost never observed (Fastovsky et al., 1995); however, in a few rare instances, a hint of clast imbrication may be present.

In rare cases, pebble-sized (0.5–1 cm in diameter) rounded intraclasts of lighter colored material (2.5YR 5/4; Munsell, 1992) appear interstratified within the bedding . As noted by Fastovsky et al. (1995) they commonly are chaotically organized; however, rarely they are found aligned along bedding planes (Fig. 18). These intraclasts are always well rounded but of variable sphericity, implying some malleability at deposition. In thin section, they are distinguishable from the tuffaceous matrix in which they are embedded only by color, suggesting that they may have originated as ash.

The fossil-bearing sequence clearly represents a series of episodic pyroclastic deposits, interstratified with epiclastic volcanic-lithic conglomerates of probable water-lain origin. The interpretation by Fastovsky et al. (1995), that these deposits formed as the result of sediment gravity flows (e.g., Costa and Wieczorek, 1987), is consistent with the observations and interpretations presented here. While the channelized conglomerates show considerable clast sorting and may thus be the product of traction transport, the fossil-bearing tuff is generally characterized by chaotic textures, extremely poor sorting, and matrix support, features typical of debris flow deposition. Likewise, the alignment of clasts along bedding planes is known from certain kinds of pyroclastic deposits ("alignment bedding" of Fisher and Schmincke, 1984).

The recognition that the fossil-bearing series is of pyroclastic origin has consequences for understanding the origin of the fossils within the sequence. Fastovsky et al. (1995) interpreted the fossil assemblage as a thanatocoenose consisting of reworked vertebrate materials entrained by a series of muddy sediment gravity flow pulses. While some specimens show evidence of water-reworking (Fastovsky et al., 1995), it now appears that at least some of the fossil assemblage was trapped by catastrophic volcanism while still alive. The most startling fossil in the assemblage, an articulated, nearly complete pterosaur (Clark et al., 1998) is, by the delicacy of its bones and the completeness of preservation, a good candidate for having been trapped by a pyroclastic flow such as a nuée ardente. Certainly complete and articulated specimens are well known from nuée ardente deposits (e.g., Sigurdsson et al., 1985).

Figure 17. Channelized conglomerate deposit with fine-grained La Boca Formation tuff deposits. Hammer marks are 1 cm increments.

Figure 18. Well-rounded, pink fine-grained intraclasts aligned along bedding planes in fossiliferous portion of the La Boca Formation. Hammer is 31 cm.

SS—Upper Portion of La Boca Formation

Interbedded sandstones, mudstones, and conglomerates characterize the upper two-thirds of the La Boca Formation. The beds typically thicken up-section, reaching several meters. Hence, cm-scale interbedded mudstones and sandstones are preserved in the basal portion of the upper part, and upward the beds thicken considerably into meter-scale sandstone units with interbedded volcaniclastic conglomerates. The framework grains typical of this interval are highly feldspathic and show little evidence of rounding or abrasion. In some cases, because of secondary alteration, the original feldspar mineralogy is inferred only by shape. Feldspar clasts, when preserved, contain subequal amounts of potassium feldspar and plagioclase. Lesser amounts of volcanic and sedimentary rock fragments are also present, as are small amounts of quartz.

With increasing vertical distance from its basal contact, the La Boca Formation becomes markedly coarser grained. Meter-scale, distinctively bimodal-sized, volcaniclastic conglomerates are a much more dominant component of the depositional setting. These conglomerates are in clast support, with the majority of the clasts ranging between 4 and 7 cm in diameter. The matrix in these units is a well-sorted, coarse sand–sized feldspathic arenite. All of the sandstone bodies show concave bases and sharp erosional contacts with underlying sedimentary rocks. Few

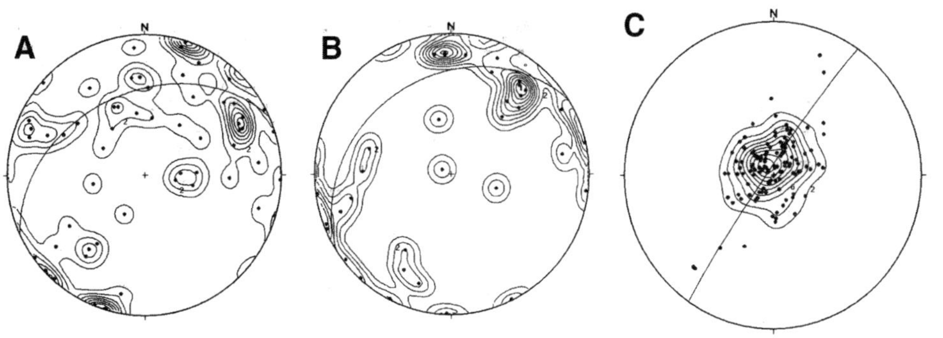

Figure 19. Schmidt equal area projections with density contours of poles to measured layering or bedding. (A) unit A from the VES, n = 62. (B) unit B from the VES, n = 35. (C) La Boca Formation, n = 126.

primary sedimentary structures are preserved, although uncommon imbricated clasts suggest that paleocurrents may have been to the southwest.

The sedimentary environments represented by this unit were not the focus of this study and remain enigmatic. The La Boca Formation has been interpreted by Belcher (1979) to have been deposited by fluvial processes operating in a rift-basin setting. He interpreted the conglomerates as channel fillings in high-velocity channels, and the finer-grained sedimentary rocks as "distributary channels" or "interchannel areas" (i.e., floodplains).

We were unable to identify coherent patterns of lithologies and sedimentary structures that unambiguously allow the inference that a fluvial system produced these rocks. The overwhelming signal from field observations as well as petrographic analysis, particularly for the lower portion of the La Boca formation, is one of deposition by pyroclastic volcanic processes.

STRATIGRAPHY

A focal contribution of this manuscript is the demonstration that the SS sits unconformably above the VES (Figs. 2, 3). In the field, this relationship is most convincingly seen in the west-central part of the mapped area (Fig. 3), where steeply dipping VES is in near contact with more gently dipping SS. For example, the small round-shaped area of La Boca Formation on the map shows gentle dips compared to the steeply dipping VES units A and B that surround the La Boca. In addition, the northwest part of the map shows an erosional window that exposes steeply dipping VES enclosed by more shallow-dipping La Boca Formation rocks (located near the contact with the La Joya Formation). The structural and stratigraphic relationships of VES units A–D are cryptic. Nowhere were we able to find contacts between these units in the field (although, as noted above, the evidence suggests that unit D is older than C). Strikes and dips of bedding and/or layering in these units are locally nearly conformable, but it is uncertain if the units represent a coherent uninterrupted stratigraphy. It is possible that the units may be separated by unconformities or by faults, although we have no direct evidence of such. However, all the units have characteristics that permit us to infer that the rocks form a coherent suite. All rocks of the VES are either volcanic or contain volcanogenic features, thus demonstrating their origin within a volcanic setting. At most locations, the bedding and/or layering is steeply dipping, exhibiting structural attitudes quite distinct from the younger overlying subhorizontal strata (see below).

Structural Relationships

General

Structural data demonstrate that while bedding in the SS is subhorizontal, the layering in rocks of the VES generally exhibit much steeper attitudes and locally are nearly vertical. Figure 19 shows Schmidt equal-area projections of the bedding poles of units A and B from the VES, and the SS, respectively. Table 1 contains the orientation data used in the construction of this figure.

VES

Layering in rocks of unit A show considerable scatter, but overall exhibit a preferential northwest strike with steep dips to the northeast and southwest. A less prominent second population exhibits a northeast strike with moderate to steep dips to the northeast. The average great circle azimuth is 245° with an average plunge of 40° (Fig. 19A). VES unit B likewise exhibits prominent northwest strikes with steep dips to the northeast and southwest. The average great circle azimuth is 238° and average plunge is 31° (Fig. 19B). Collectively, rocks of units A and B are nearly concordant and range in dip from 30 to 90° (average = 66°), with a dominant northwest strike. Subtle, small-scale folding is apparent in some outcrops. The variability in orientation of this sequence suggests that large-scale folds or rotation along faults, although not observable in outcrop, may be present. Insufficient measurements from units C and D disallow the construction of meaningful Schmidt projections. Thus, we do not know whether the contacts between units C and D are concordant, discordant, or faulted; nor do we know their relationships with units A and B.

SS

Structural data demonstrate that bedding planes in the SS are predominantly shallow dipping (an average of 18°), with generally variable strike directions. The average great circle azimuth is 215°, with an average plunge of 86° (Fig. 19C). In Huizachal Canyon, the subhorizontal SS shows no evidence of

TABLE 1: STRIKE AND DIP MEASUREMENTS ON STEEPLY DIPPING VES UNITS A AND B, AND RELATIVELY SUBHORIZONTAL LA BOCA ROCKS

VES UNIT A			VES UNIT B			La Boca					
Station	Strike	Dip	Station	Strike	Dip	Station	Strike	Dip	Station	Strike	Dip
7	42	Vert	2	339	89NE	2	248	8NW(?)	180	162	15E
8	354	73SE	3	187	89NE(?)	17	354	11SE	181	202	4SE
15	16	76NW(?)	42	337	Vert	18	55	9SE	182	332	22NE
26	296	79SW	66	312	46SW	21	11	13SE	184	307	16NE
30	265	81SW	81	350	71SW	24	267	23NE	192	182	11SW
40	357	31SW	82	352	Vert	25	54	12SE	193	322	17NE
43	287	Vert	95	330	82SW	34	272	32SW	194	222	6SE
46	286	85SE	111	297	66NE	41	262	15SW	198	142	8SE(?)
51	335	66SW	112	312	70SW	48	335	28NE	199	202	19SE(?)
54	308	57NE	115	287	71NE	65	20	22SE	200	227	14SE
57	306	51NE(?)	125	277	75SW	67	35	4SE	203	272	21SW
61	26	56SE	127	307	72SW	68	42	4SE	204	207	9SE
87	162	Vert	128	312	Vert	69	317	10SW	205	207	13SE
90	152	Vert	129	292	68SW	70	292	74SW	206	182	27SE
98	312	Vert	130	292	79SW	71	294	16SW	208	182	25SE
100	202	79SE	131	262	76SW	73	307	32SW	209	171	21SE
121	12	29NW	132	267	76SW	77	302	21SW	211	232	26SE
124	292	Vert	133	297	Vert	78	292	24SW	212	182	20SE
135	351	32NW	148	197	28NE	79	282	HOR	213	177	12SE
141	299	26NE	149	142	89NE	80	332	10SW	214	192	19SE
152	307	62NE	150	257	33SW	81	336	15SW	216	147	21SE
154	177	21SE	153	197	53SE	82	332	4SW(?)	217	147	27NE
155	197	Vert	156	167	Vert	83	224	26SW	218	222	7SE
159	217	52SE	157	182	55SE	84	297	14SW	219	147	11SE
160	312	82SW	158	244	Vert	85	312	70NE8	226	277	4SW
169	282	76NE	254	308	56SW	86	352	15SW	227	224	13SE
170	292	Vert	255	316	58SW	88	172	28SW	228	192	8SE
172	317	Vert	256	308	70SW	90	287	26SW	229	229	17SE
177	322	72NE	257	272	Vert(?)	91	307	19SW	230	202	13SW
183	332	67SW	6	267	75SW	93	317	14SW	231	183	17SE
184	332	67SW	7	207	6SW	94	315	40SW	234	208	26SE
185	267	60SE	8	132	65SW	95	322	36SW	235	205	7SE
186	322	72SE	12	302	54NW	101	227	11SE	242	258	17SW
189	202	66SE	18	342	75E	102	268	42SW	243	328	25SW
190	287	Vert	19	347	65NE	106	291	11SW	244	346	19SE

(*continued*)

TABLE 1: STRIKE AND DIP MEASUREMENTS ON STEEPLY DIPPING VES UNITS A AND B, AND RELATIVELY SUBHORIZONTAL LA BOCA ROCKS (*continued*)

VES UNIT A			VES UNIT B			La Boca					
Station	Strike	Dip	Station	Strike	Dip	Station	Strike	Dip	Station	Strike	Dip
191	155	68SE				107	337	5SW	245	352	14SE
195	182	52NW				108	312	9SW	246	347	26SE
196	147	56NW				110	297	65SW	247	16	11SE
197	162	58W				116	294	13SW	248	107	21SW
201	277	34SW				117	307	11NE	250	311	70
202	262	60SW				118	307	15NE	258	225	17SE
207	189	18NW				119	212	21NW	259	252	13S
220	248	44SE				135	60	24NW	260	285	28SW
221	237	64SE				136	332	4SW	261	272	23SW
222	122	56SW				137	307	5NE	264	272	22NE
223	262	37SW				138	307	4NE	265	278	12SW
224	234	34SE				139	297	5NE	266	265	20NW
225	305	36SW				140	296	10NE	267	132	12NW(?)
232	318	67SW				142	252	3NE	268	302	3NE
233	324	Vert				143	212	4NW	269	227	13SE
236	147	40SW				144	325	22NE	270	225	15SE
237	245	45SW				145	327	15SW	302	293	42NE
238	202	66SE				146	48	16NW	307	222	22SE
239	198	78SE				147	268	25NW	308	223	6SE
240	205	83SE				148	222	15NW	11	292	20SW
249	330	72SW				149	242	15NW	113	247	30SW
251	312	Vert				150	252	17N	55	271	12NE
252	32	30SE				151	287	16SW	62	22	10SE
253	291	66SW				162	310	9NE	63	65	10SE
15	272	53SW				166	282	18SW	64	35	23SE
17	107	Vert				168	292	5SE	74	287	35SW
20	182	Vert				169	307	50NE	75	258	32SW
						171	271	5SW	76	262	28SW
						172	307	17NE	89	282	25SW
						173	234	12SE	92	272	14SW
						174	242	10SE	9	55	14NW
						175	247	12SW	10	252	6NW
						176	287	17SE	56	262	10NE
						178	292	18SW	303	292	23NE
						179	292	26SW			

Note: Locations of all stations are shown in Strater (1993).

extensive folding, although locally the units show minor, small-scale deformation such as the development of pencil cleavage. Comparison of the structural attitudes of the VES and SS units demonstrate a high degree of discordancy, which we interpret as an angular unconformity separating the contact of the VES and the SS. This discordancy is directly observable at a number of localities in the field, two of which are depicted in Figure 20. Throughout the field area, the SS can be found in contact with each of the four units of the VES (Fig. 3), and in each case, it is in angular unconformity with the respective unit of VES (see Strater, 1993, for details).

We confirmed the observation of Mixon et al. (1959) that the contact between the La Boca and overlying La Joya Formation is a low-angle angular unconformity that is marked by a conglomerate at the base of La Joya. However, in the northwestern portion of the canyon the La Boca Formation is absent, and the La Joya Formation lies unconformably upon unit B of the VES (Fig. 3). This contact requires either extensive erosion of La Boca sediments before deposition of the La Joya, or substantial topographic relief of an exposed erosion surface upon the rocks of the VES that was not completely blanketed by La Boca sedimentation.

Both the La Boca and La Joya Formations display a pervasive "pencil" cleavage throughout Huizachal Canyon. The average trend and plunge of this cleavage (305.8°, plunging 4.1NW; see Strater, 1993) generally are similar to the axial trend of the Huizachal-Peregrina anticlinorium (~325°, designated by the *Secretaria de Programacción y Presupuesto*, #F14A29, 1978), and it probably resulted from deformation that produced the structure of the anticlinorium.

Our interpretation is that the SS was deposited in the topographic lows of an ancient erosional surface developed on the upturned VES (Figs. 2, 3; see also Fastovsky et al., 1995), a contact that may in turn have been distorted by faulting and/or folding during the formation of the Huizachal-Peregrina anticlinorium. This idea is reinforced by the interpretation that "thick-skinned" basement decoupling was involved in this region, as suggested by Carrillo-Bravo (1961), de Cserna et al. (1977), Ortega-Gutiérrez (1978), and Suter (1987).

GEOCHEMISTRY

VES—Unit A and B

All rocks of unit A and B exhibit very high concentrations of SiO_2, ranging from 70%–85%. Only 6 of 18 samples yield SiO_2 contents of less than 78% (Table 2), the approximate maximum in silicate-rich magmas (Le Maitre, 1976; Best, 1982). Samples containing higher levels of SiO_2 have been significantly silicified by late-stage vapor phase reactions as described in the previous section. Features that demonstrate silica enrichment include the common occurrence of quartz spherulites and lithophysae, devitrification features, a lack of primary minerals and abundant pseudomorphic features, and numerous quartz veinlets and in-filled occurrences. Na_2O values are also anomalously low, demonstrating that sodium was mobilized during secondary alteration and was largely flushed from the system. K, Ca, and Ti also may have been partly lost as well, but concentrations are more variable indicating that mobilization of these elements was not as uniform as for sodium. These alterations make it difficult to accurately classify these rocks based upon major elements, although the petrography clearly demonstrates that they are dominated by felsic-rich volcanic compositions, with more mafic compositions in unit D.

In many cases, however, relatively immobile trace elements such as Y, Zr, and Nb, have proven useful in the classification of altered or metamorphosed volcanic rocks (Pearce and Cann, 1973; Winchester and Floyd, 1977; Floyd and Winchester, 1978; Garcia, 1978; Pearce and Norry, 1979; Meschede, 1986). On a plot of Nb/Y versus Zr/TiO_2 (Fig. 21), most rocks of unit A and B fall within the rhyodacite-dacite field, with several extending into the rhyolite field. Unit A volcanics tend to plot with higher Zr/TiO_2 ratios than unit B, and may have more evolved compositions. Y and Nb exhibit low variability in these rocks and the values here are typical of volcanic rocks (Jakes and White, 1972) and are considered to approximate primary concentrations. Likewise, we consider Zr to be immobile, and thus consider these trace element concentrations to be little changed since emplacement (Winchester and Floyd, 1977). If Ti has been substantially lost during alteration, those few samples that plot as rhyolites may have originally been more dacitic in composition. This diagram is particularly useful because the Zr/TiO2 ratio acts as an index of differentiation, whereas the Nb/Y ratio serves as an alkalinity index. Note that all samples exhibit Nb/Y values that are significantly less than one, and are representative of subalkaline suites rather than alkaline suites of rocks.

VES—Unit D

Rocks of unit D range in SiO_2 from 47% to 57%. Based on major elements, they range from basalt to andesite. On the basis of relatively immobile trace elements, these rocks plot as subalkaline basalt and basaltic andesite (Fig. 21). Thus, the VES, as a group, exhibits a continuum of compositions that range from basalt to rhyolite, typical of many calcalkaline suites of rocks.

SS—La Boca

On a plot of relatively immobile trace elements (Fig. 21), the basal tuff of the La Boca Formation plots as intermediate composition rocks, close to the rhyodacite/dacite–andesite boundary. Thus, their compositions plot as subalkaline andesites to rhyodacites. They clearly are subalkalic in nature based on their low Nb/Y ratio. Compositions are quite similar to the volcanics of unit B in the VES.

TECTONIC SETTING

The rocks of the VES exhibit a range of compositions from subalkaline basalt to rhyolite. Trace-element plots of the mafic

Figure 20. Angular discordance between the VES and the La Boca Formation in outcrop. (A) steeply dipping VES overlain by gently dipping SS. (B) moderately dipping VES overlain in shallow-angled discordance with SS.

TABLE 2. MAJOR ELEMENT AND TRACE ELEMENT COMPOSITIONS OF VOLCANIC ROCKS FROM HUIZACHAL CANYON, MEXICO

	VES unit A											
Sample ID	2	3	5	7	8	9	10	11	12	13	14	15
Major: wt%												
SiO_2	70.83	72.94	83.03	84.74	70.64	84.63	77.37	80.67	84.80	79.61	83.13	78.56
TiO_2	0.02	0.79	0.33	0.09	0.77	0.25	bd	0.16	0.16	0.35	0.33	0.23
Al_2O_3	14.76	14.49	10.10	9.59	16.44	11.54	13.91	11.37	10.05	12.68	10.94	11.94
Fe_2O_3	8.31	8.24	3.26	3.18	7.77	2.59	3.74	2.47	2.43	1.51	0.47	2.77
MnO	0.03	0.03	0.02	0.02	0.02	0.04	0.02	0.03	0.02	0.01	0.04	0.04
MgO	0.11	0.09	1.67	0.80	bd	0.07	0.24	0.17	0.18	0.39	0.55	0.61
CaO	0.56	0.64	0.29	0.29	0.28	0.31	0.26	1.40	0.29	0.28	1.97	1.24
Na_2O	0.10	0.67	bd	0.14	0.20	0.11	bd	0.03	0.26	1.22	bd	0.07
K_2O	4.01	2.62	1.84	1.75	4.14	0.89	3.73	2.79	2.02	3.22	2.86	3.11
P_2O_5	0.20	0.24	0.03	0.05	0.60	0.06	0.06	0.04	0.05	0.02	0.08	0.04
TOTAL	98.93	100.75	100.57	100.65	100.86	100.49	99.33	99.13	100.26	99.29	100.37	98.61
Trace (ppm)												
Rb	89	67	44	45	93	19	90	74	49	73	59	73
Sr	92	90	25	29	36	66	40	34	33	31	32	30
Y	29	29	32	22	33	26	32	36	24	27	31	27
Zr	201	204	157	145	213	170	202	168	153	171	172	179
Nb	10	10	8	7	11	8	9	7	7	10	8	9
Ba	3943	795	296	758	1098	272	2339	566	448	451	277	974
La	2	48	23	18	29	40	8	25	22	33	46	17
Ce	38	99	38	34	64	70	34	56	43	47	68	34
Zn	26	20	46	40	9	19	21	15	13	6	26	3

(*continued*)

rocks from unit D show that the compositions fall in fields of calcalkaline, arc-related basalts (Fig. 22), and are notably distinct from within-plate or alkalic basalts typical of extensional tectonic settings. Spider diagram plots of the mafic rocks shown in Strater (1993) reveal a rather spiked pattern and a distinct Nb trough typical of arc basalts, and the more compatible elements exhibit relatively high normalized values compared to ocean-ocean convergent margins.

Discrimination diagrams for the felsic-rich rocks of units A and B show compositions that are consistent with felsic rocks from volcanic-arc settings (Fig. 23). A spider diagram figured in Strater (1993) shows that these rocks exhibit a negative Nb anomaly and an overall pattern more similar to arc rocks than rift or extensional settings.

As a group, the VES rocks exhibit subalkaline chemistry and a wide range of rock types from basalt to rhyolite. The compositions are more typical of magmas generated at convergent plate tectonic settings, compared to the more characteristic bimodal suites typical of extensional rifting (Wilson, 1989). Moreover, the dominance of felsic volcanic compositions and pyroclastic features is indicative of ocean-continent convergence, as opposed to an ocean-island arc setting. The felsic-rich compositions suggest a continental component in the magmatic source region, a conclusion in accord with the interpretation that these units are floored by Grenville-aged basement (Carrillo-Bravo, 1961, de Cserna, 1971, Ramirez-Ramirez, 1992, Ortega-Gutiérrez, 1978, Anderson and Schmidt, 1983). For these reasons, we believe that the petrologic characteristics point strongly to a convergent tectonic setting of Andean-type which most likely involved subduction of ocean crust beneath continental crust.

The basal tuff unit of the La Boca Formation also plots in fields characteristic of volcanic arc magmatism (Fig. 23). A spider diagram in Strater (1993) shows a pattern similar to felsic volcanic arc rocks with a characteristic trough at Nb. The pattern is similar to the felsic rocks of the VES. Based on geochemistry, the tuff units in the La Boca formation seem to represent igneous activity in an Andean arc setting, coeval with the volcanic and epiclastic sedimentary setting that hosted the vertebrate fossil assemblage.

AGE OF LOWER LA BOCA FORMATION

The presence of volcanic tuff at the base of the fossil-bearing rocks of the La Boca Formation provides the potential to better establish the age of the stratigraphic sequence. Approximately 60 kg of this tuff layer were collected and zircon was extracted using standard heavy liquid and magnetic separation techniques. The zircon crystals are typically clear, doubly terminated prisms that average ~100 μm in longest dimension. Eleven groups of zircon that contained between 8 and 20 grains each were analyzed using standard methods of Isotope Dilution Thermal Ionization Mass Spectrometry. Most of the zircon analyses spread along and subparallel to the concordia curve between 186 and 194 Ma

TABLE 2. MAJOR ELEMENT AND TRACE ELEMENT COMPOSITIONS OF VOLCANIC ROCKS FROM HUIZACHAL CANYON, MEXICO (*continued*)

	VES unit B						VES unit D				La Boca Volcanics				
Sample ID	19	20	21	22	24	25	29	30	31	32	50	51	52	53	59
Major: wt%															
SiO_2	83.75	75.26	79.06	79.96	80.79	79.09	56.84	49.72	46.90	56.20	61.88	75.66	74.96	74.62	84.01
TiO_2	0.21	0.05	0.23	0.65	0.13	0.54	0.87	0.90	0.73	0.77	0.90	0.71	0.76	0.73	0.39
Al_2O_3	9.28	12.47	13.18	12.78	12.98	12.39	18.67	17.73	16.96	15.56	15.41	12.25	13.15	13.43	10.07
Fe_2O_3	2.42	3.71	1.37	2.21	1.37	2.66	9.09	9.73	9.00	9.29	8.88	6.07	7.27	6.78	2.56
MnO	0.02	0.05	0.01	0.01	0.05	0.08	0.10	0.16	0.20	0.16	0.13	0.01	0.06	0.02	0.01
MgO	0.77	1.16	0.70	1.27	1.32	1.60	2.12	5.21	5.64	4.36	2.52	0.45	0.61	0.40	0.45
CaO	0.64	2.05	0.31	0.31	0.32	1.27	5.03	11.68	16.33	8.66	5.94	0.31	0.35	0.30	0.28
Na_2O	0.63	2.61	1.01	bd	0.10	bd	3.85	3.12	bd	2.55	0.63	bd	bd	bd	0.09
K_2O	1.91	2.21	3.69	2.89	2.55	2.58	3.48	1.37	3.59	1.20	3.74	3.02	3.21	3.24	2.74
P_2O_5	0.01	0.04	0.01	0.09	0.06	0.09	0.15	0.12	0.15	0.21	0.16	0.02	0.05	0.03	0.04
TOTAL	99.64	99.61	99.57	100.17	99.67	100.30	100.20	99.74	99.50	98.96	100.19	98.50	100.42	99.55	100.64
Trace (ppm)															
Rb	42	74	52	102	57.8	89	72	28	70	20	100	71	101	70	108
Sr	25	27	54	32	33	31	153	76	86	76	42	25	36	24	32
Y	21	22	22	37	21	27	20	18	16	18	30	38	25	39	26
Zr	151	103	90	187	106	141	100	52	47	90	129	212	128	213	135
Nb	7	8	7	14	9	11	6	4	4	6	10	13	9	13	9
Ba	370	199	1062	551	490	1144	457	249	426	241	1067	301	856	297	660
La	26	47	24	31	47	27	19	6	14	36	22	27	39	32	26
Ce	39	62	49	49	80	47	31	23	22	47	36	48	70	52	50
Zn	27	20	50	20	34	41	81	102	100	109	60	11	25	9	18

(Table 3; Fig. 24). One fraction of zircon has a Pb-Pb date of 306 Ma which indicates a significant, inherited (older) component of Pb. Several data points suggest a small component of slightly older inherited zircon with evidence of superimposed Pb loss. At present we cannot tell if the older zircons are present as distinct, slightly older, detrital grains or as inherited cores within volcanic zircons.

Our preferred interpretation of the data is that the age of the pyroclastic layer is 189 ± 0.2 Ma, which is the lower intercept age calculated from the regression of five zircon fractions (Fig. 24). The small ± value is analytical uncertainty. The true geologic age, especially considering the inherited Pb component, probably is not so tightly constrained. In this interpretation, the younger data points represent Pb loss. In any case, 189 Ma is the maximum age of the pyroclastic layer, an age that is Middle Toarican (late Early Jurassic) according to the time scales of Harland et al. (1989) and Gradstein et al. (1995). The upper intercept is ca. 1.9 Ga, which has no significance other than indicating an inherited component. The only constraint on the age of the inherited component is that it is older than 306 Ma, the oldest Pb-Pb date reported in our data. The Huizachal Canyon fossil assemblage lies directly above the zircon-bearing pyroclastic flow that we sampled.

The age of the Huizachal Canyon fauna is thus somewhat constrained. Clark et al. (1994, p. 300) observed that "the taxa preserved in the Huizachal Group are generally typical of those occurring in assemblages that are considered to be Early Jurassic in age." This statement is contrary to that of Fastovsky et al. (1995), who assigned an early Middle Jurassic age to the assemblage, based upon "preliminary" U-Pb dates from the same tuff dated here. In view of the geochronology data presented here as well as the biostratigraphic evidence presented by Clark et al. (1994), we conclude that the Huizachal Canyon assemblage is best considered late Early Jurassic (Toarcian) in age.

The 189 Ma age of the base of the La Boca Formation also constrains the age of the underlying VES volcaniclastic units. The rocks of the VES must be older than 189 Ma because they occur unconformably below the La Boca. We were unsuccessful in separating zircon or other materials from the VES units that could be used to establish a radiometric age. The angular unconformity that separates the VES from La Boca argues that the VES rocks were deformed to steepen the bedding, followed by erosional unroofing of this basement complex before deposition of the La Boca at 189 Ma.

SUMMARY OF KEY GEOLOGIC OBSERVATIONS IN HUIZACHAL CANYON

Huizachal Canyon is small relative to northeastern Mexico, and the area studied within it covers less than 10 km^2. Nonetheless, observations that emerge from our study of the rocks in Huizachal Canyon should be considered in understanding not only the geologic history of the immediate area, and but also the

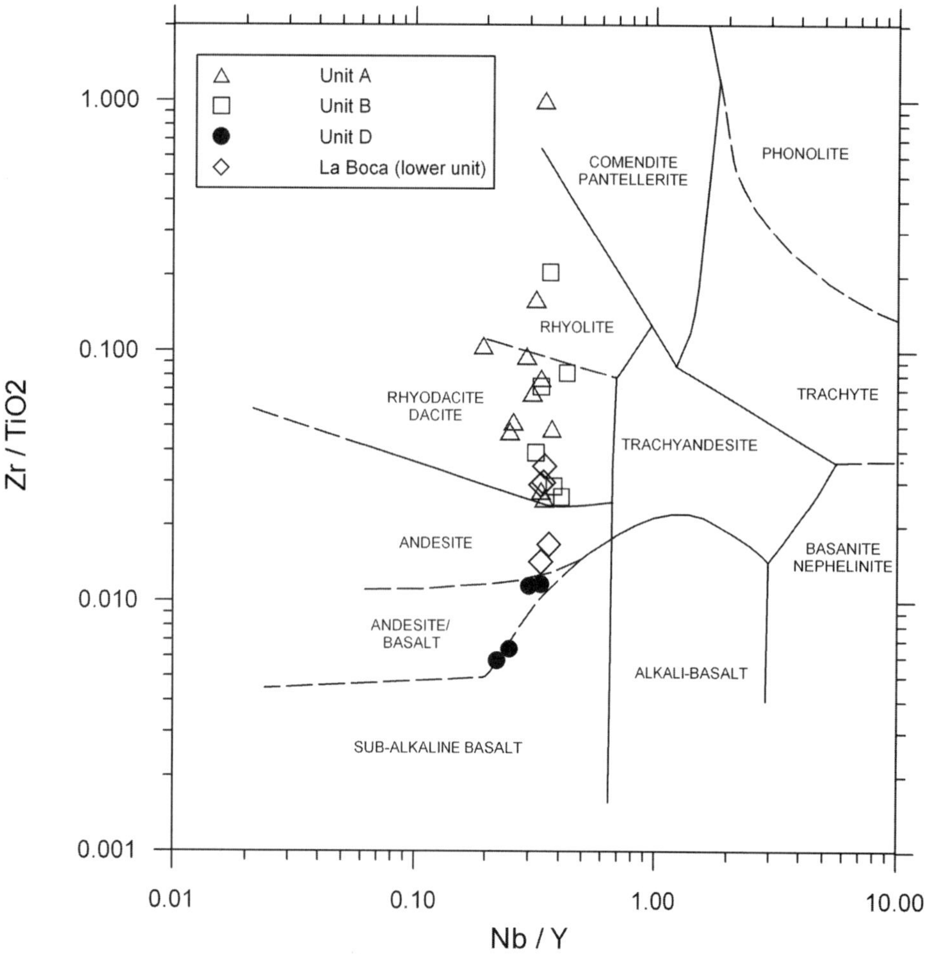

Figure 21. Plot showing distribution of volcanic rocks from the VES (Facies A, B, and D) and the La Boca formation on a plot of Zr/TiO_2 versus Nb/Y (after Winchester and Floyd (1977). Felsic members of the VES plot as rhyolite or rhyolite/dacite, whereas mafic rocks plot as andesite/basalt. Volcanic rocks of the La Boca Formation plot in the fields of andesite to rhyodacite/dacite, and partly overlap compositions of some of the more felsic VES volcanics. The Nb/Y ratio of all samples is less than 0.7, consistent with the subalkaline character of the suites.

regional tectonic history of this part of Mexico. These observations include:

(1) The basement rocks (VES) in Huizachal Canyon consist of a series of volcanic, volcaniclastic, and epiclastic rocks that range from basaltic to rhyolitic-dacitic compositions. Although substantially altered and silicified, relatively immobile trace elements indicate that the suite was subalkalic and generally consistent with those derived at an Andean margin involving ocean-continent subduction. These compositions are not typical of volcanic rocks associated with continental rifting.

(2) We have found no evidence for the presence of intrusive igneous rocks, as indicated by previous workers.

(3) The VES rocks were deformed to form steeply dipping beds that were subsequently eroded producing the angular unconformity that separates the younger, overlying subhorizontal La Boca Formation.

(4) The La Boca Formation (SS) consists of a series of pyroclastic and epiclastic deposits.

(5) Zircon from the basal unit of the La Boca Formation yield a radiometric age of 189 Ma based on U-Pb. These data constrain the age of the underlying VES units to an age >189 Ma. Because the fossiliferous section of the La Boca lies stratigraphically immediately above the basal tuff, its age is similar to the tuff, or slightly younger.

(6) The geochemistry of the La Boca basal tuff also exhibits compositions similar to subduction-related volcanics, quite distinct from those associated with areas of continental rifting.

(7) The units that host the fossils in the La Boca Formation are also volcanic tuffs. The fossil assemblage recovered from the La Boca Formation consists not only of reworked fossil material, but also of organisms trapped by the volcanism.

DISCUSSION

Interpreted Geologic History of the VES

The basement rocks of the Huizachal-Peregrina anticlinorium are generally thought to be Precambrian amphibolite-granulite grade metamorphic rocks associated with the Grenville orogeny (Carrillo-Bravo, 1961; de Cserna, 1971; Ramirez-Ramirez, 1992; Ortega-Gutiérrez, 1978; Anderson and Schmidt, 1983). Such rocks can be found in Peregrina Canyon (Girty,

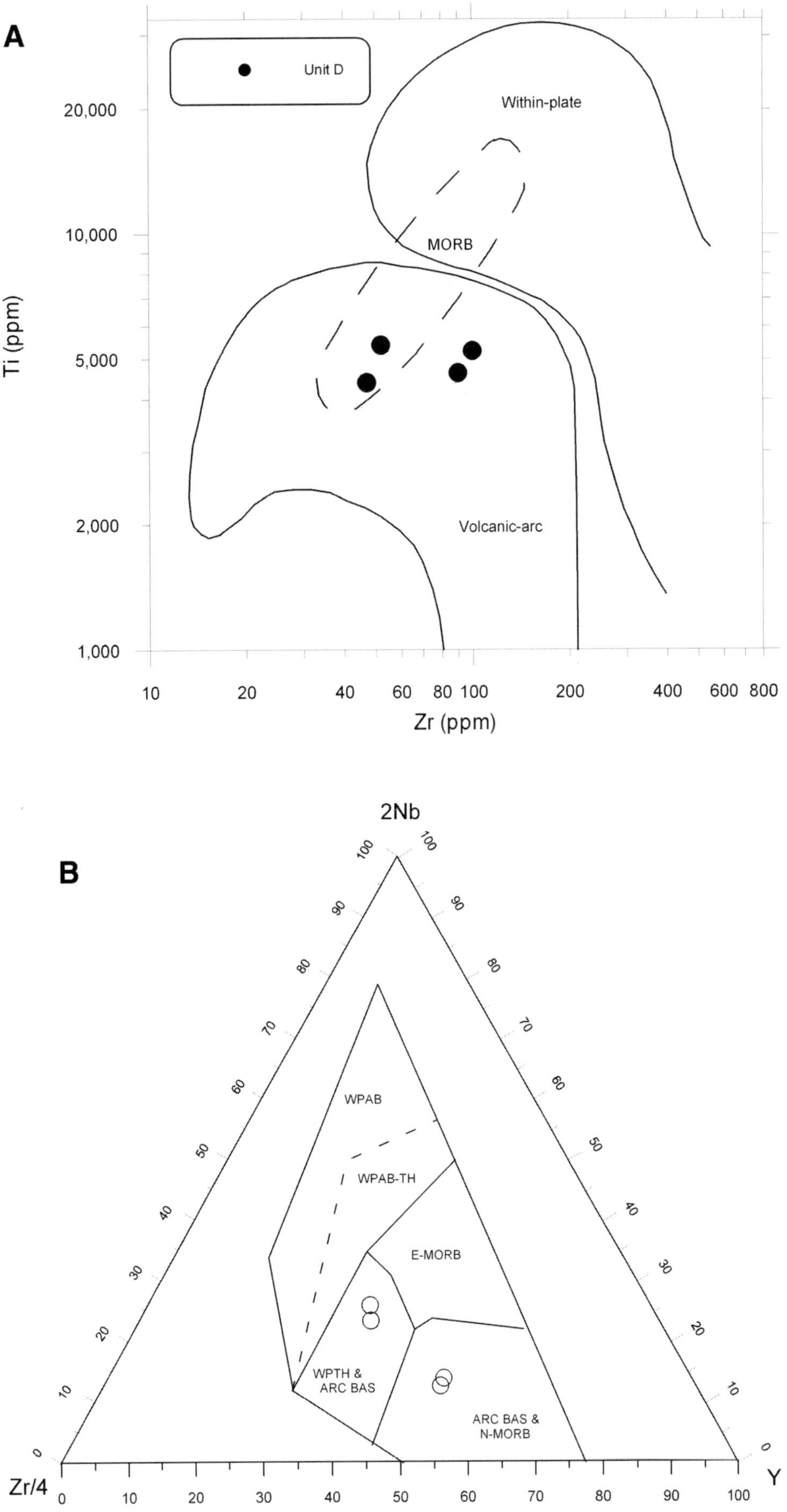

Figure 22. Mafic volcanic rocks from Facies D of the VES plotted on discrimination diagrams. (A) On a Zr versus Ti diagram (after Pearce and Cann, 1973), unit D rocks plot in the volcanic arc field. (B) On the Zr/4-2Nb-Y discrimination diagram (after Meschede, 1986), the rocks plot in the two fields typical of arc basalts. WPAB—with plate basalts; TH—within plate tholeiites; E-MORB—enriched-type mid-ocean ridge basalts; N-MORB—normal-type mid-ocean ridge basalts.

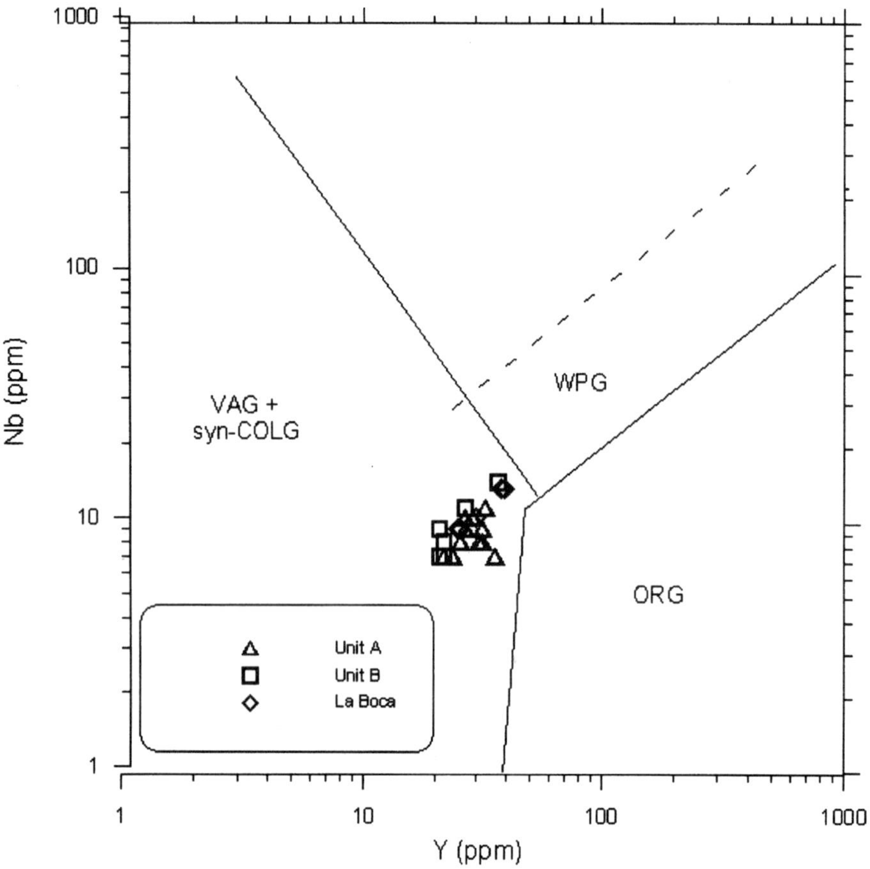

Figure 23. Felsic volcanic rocks from the VES and La Boca Formations plotted on a Y versus Nb tectonic discrimination diagram after Pearce et al. (1984). In using this diagram, we make the assumption that felsic volcanic rocks exhibit similar correlations of trace elements with tectonic setting, to those of plutonic rocks. All samples plot in the field of VAG + syn-COLG. VAG—volcanic-arc granites; syn-COLG—syn-collisional granites; WPG + within-plate granites; ORG—ocean-ridge granites.

1926; Carrillo-Bravo, 1961; Ramirez-Ramirez, 1992), Caballeros Canyon (Carrillo-Bravo, 1961; Ramirez-Ramirez, 1992), and Novillo Canyon (Carrillo-Bravo, 1961; Ramirez-Ramirez, 1978; Ortega-Gutiérrez, 1978). A commonly cited example is the Novillo Gneiss, a gabbroid-feldspathic "orthogneiss" found in Novillo Canyon (Ortega-Gutiérrez, 1978), which has yielded an age range of 912–740 Ma with K/Ar (de Cserna et al., 1977). Because the VES bears none of the high-grade metamorphism typical of these Grenville-aged rocks, it is most likely younger and unrelated to this orogeny.

De Cserna (1970 and 1971) suggested that some rocks in northeastern Mexico were emplaced as thrust sheets during Paleozoic deformation correlative to the North American Taconic Orogeny (see below). Included is the Granjeno Schist of Novillo Canyon (Fig. 1), believed by de Cserna (1971), de Cserna et al. (1977), and de Cserna and Ortega-Gutiérrez (1978) to have been tectonically emplaced during a southern extension of the Late Ordovician Taconic orogeny, and subsequently metamorphosed during the Pennsylvanian-Permian Ouachita-Marathon Orogeny. Because the majority of the rocks in northeastern Mexico associated with Taconic events exhibit considerable evidence of deformation and high-grade metamorphism (de Cserna, 1971), Taconic time probably predated the formation of the VES rocks, which are hydrothermally altered, but only weakly metamorphosed.

The southernmost demonstrable exposure of the Ouachita-Marathon foldbelt is the Marathon Uplift of southern Texas (Stewart, 1988). However, various studies suggest that it actually extends into Mexico (Carrillo-Bravo, 1961; Murray, 1961; Rowett and Hawkins, 1975; Ross, 1979, 1986; Anderson and Schmidt, 1983; Handschy et al., 1987; Shurbet and Cebull, 1987; Ruiz et al., 1987; Ramirez-Ramirez, 1992). Paleozoic rocks in Novillo Canyon, as well as in the other canyons of the Huizachal-Peregrina anticlinorium, have been inferred to be associated with the Ouachita-Marathon system (Carrillo-Bravo, 1961; Lopez-Ramos, 1969; Anderson and Schmidt, 1983; Ewing, 1985; Ramirez-Ramirez, 1992). Devonian strata found in Caballeros Canyon had historically been linked to novaculites typical of the Marathon region (King, 1951; Carrillo-Bravo, 1961; Murray, 1961; Lopez-Ramos, 1969; Ramirez-Ramirez, 1978; Boggs, 1992). However, recent study of this material has instead suggested that it is a series of rhyolitic flows of post-Mississippian age, possibly related to Permian felsic volcanism (Ramirez-Ramirez, 1992), although Ramirez-Ramirez (1992) believes these rocks are nonetheless associated with the Ouachita-Marathon system. He also notes that they may be

TABLE 3. U-PB ISOTOPIC DATA FROM ZIRCON WITHIN THE BASAL TUFF OF THE LA BOCA FORMATION

Sample	Weight (mg)†	Concentration				Errors 2σ (%)						Age (Ma)			Corr. coef.	Pb (pg)††
		U (ppm)	Pb (ppm)	$^{206}Pb^*/^{204}Pb$‡	$^{208}Pb/^{206}Pb$§	$^{206}Pb/^{238}U$#	% err	$^{207}Pb/^{235}U$#	% err	$^{207}Pb/^{206}Pb$#	% err	$^{206}Pb/^{238}U$	$^{207}Pb/^{235}U$	$^{207}Pb/^{206}Pb$		
Hu z5	0.0498	315.8	11.2	1185.2	0.225	0.03086	(.11)	0.22331	(.35)	0.05248	(.32)	**195.9**	**204.7**	**306.3**	0.42	26.6
Hu z6	0.0235	348.9	11.8	6281.4	0.240	0.03046	(.11)	0.21134	(.13)	0.05033	(.06)	**193.4**	**194.7**	**210.1**	0.86	2.6
Hu z7	0.0475	435.2	15.0	2114.4	0.262	0.02991	(.13)	0.20771	(.15)	0.05037	(.08)	**190.0**	**191.6**	**212.2**	0.87	18.9
Hu z2	0.0100	633.1	21.5	1657.7	0.243	0.02992	(.29)	0.20693	(.34)	0.05017	(.17)	**190.0**	**191.0**	**202.7**	0.87	7.4
Hu z2	0.0233	272.2	9.2	1657.7	0.243	0.02986	(.18)	0.20655	(.25)	0.05017	(.17)	**189.7**	**190.7**	**202.7**	0.74	7.4
Hu z1	0.0158	469.5	16.2	1101.0	0.243	0.02989	(.15)	0.20647	(.22)	0.05009	(.16)	**189.9**	**190.6**	**199.2**	0.70	13.1
Hu z3	0.0093	620.4	20.5	1958.2	0.224	0.02977	(.17)	0.20511	(.22)	0.04997	(.14)	**189.1**	**189.4**	**193.7**	0.77	5.7
Z2	0.0135	360.8	12.1	3561.3	0.256	0.02974	(.09)	0.20460	(.18)	0.04989	(.16)	**188.9**	**189.0**	**189.9**	0.51	2.6
Hu z4	0.0388	462.6	15.7	2246.9	0.258	0.02948	(.16)	0.20317	(.19)	0.04999	(.09)	**187.3**	**187.8**	**194.6**	0.86	15.2
Hu z8	0.0190	570.0	19.0	3840.9	0.266	0.02930	(.11)	0.20193	(.17)	0.04998	(.13)	**186.2**	**186.8**	**194.2**	0.65	5.3
Z3	0.0180	297.3	10.0	924.0	0.226	0.02933	(.23)	0.20121	(.47)	0.04976	(.40)	**186.3**	**186.1**	**183.6**	0.54	11.1
Z1	0.0112	295.1	10.6	376.6	0.266	0.02831	(.27)	0.19450	(.61)	0.04983	(.51)	**180.0**	**180.5**	**187.1**	0.55	16.7

Note: Total procedural blank for Pb ranged from 0.65 to 3.7 pg and <1.0 pg for U. Blank isotopic composition: $^{206}Pb/^{204}Pb = 19.10 \pm 0.1$, $^{207}Pb/^{204}Pb = 15.71 \pm 0.1$, $^{208}Pb/^{204}Pb = 38.65 \pm 0.1$. Corr. coef.—correlation coefficient. Age calculations are based on the decay constants of Steiger and Jäger (1977). Common-Pb corrections were calculated by using the model of Stacey and Kramers (1975) and the interpreted age of the sample.

†Sample weights are estimated by using a video monitor and are known to within 40%.

‡Measured ratio corrected for spike and fractionation only.

§Radiogenic Pb.

#Corrected for fractionation, spike, blank, and initial common Pb. Faraday-Daly analyses.

††Total common-Pb in analyses.

correlative to Permian felsic volcanic rocks in Placer de Guadalupe, Chihuahua. However, Stewart (1988) argued that the rocks of Caballeros Canyon bear no resemblance to those of the Ouachita-Marathon system.

Some authors propose that the Ouachita-Marathon orogeny was associated with an arc-trench system in northern Mexico that was responsible for the emplacement of numerous igneous bodies during the late Paleozoic (Walper and Rowett, 1972; Rowett and Hawkins, 1975; Jones et al., 1984). It is possible that the VES rocks were produced in association with this inferred arc-trench system and then deformed during the final formation or breakup of Pangea. If so, then these volcanics might be correlative with the Permian rhyolites of Caballeros Canyon noted by Ramirez-Ramirez (1992). However, we have been unable to recognize any of the marine strata typical of the Ouachita-Marathon system within the VES.

The most promising idea is that the VES might instead be related to a separate arc system associated with the formation of Pangea. Such is thought to be the case for the Pennsylvanian-Permian volcanic rocks of the Las Delicias Basin in Coahuila, north-central Mexico, whose character and timing appear to be different from those typical of the Ouachita-Marathon system (Handschy et al., 1987; Jones et al., 1986; McKee et al., 1988, 1999; Ross, 1986). The Las Delicias rocks are believed by McKee et al. (1988, 1999) to be the products of the Las Delicias arc, an inferred continental marginal feature located south and east of the current location of the Las Delicias Basin (McKee et al., 1999). Las Delicias arc-originated deposition includes volcaniclastic sediments deposited chaotically in arc-proximal, submarine settings. There is nothing in either the geochemistry or the sedimentology of the VES to rule out this setting and mode of emplacement. If the VES rocks are indeed from this arc, then they would be Late Paleozoic in age, and would represent the first evidence in this region for the Las Delicias arc (McKee et al., 1999).

Constraints on the Origin of the La Boca Formation

According to Belcher (1979), Gose et al. (1982), Michalzik (1991), Salvador (1987), and Rueda et al. (1991a and 1991b), the La Boca Formation was deposited in a tensional rifting environment. However, this study indicates that deposition of at least the lower pyroclastic, volcanic portion of the La Boca Formation in Huizachal Canyon likely was coincident with arc-related volcanism during the Early Jurassic. The debris flows observed were accompanied by syndepositional volcanic activity, whose geochemistry suggests subduction tectonics.

Volcanic material observed in the Nazas Formation (and presumed correlative rocks) in San Luis Potosi, northeastern Durango, Chihuahua, and northern Zacatacas is believed to have been produced by the subduction of the Pacific plate along the western margin of Mexico during the Jurassic (McKee and Jones, 1981; Campa and Coney, 1983, Jones et al., 1986, Cohen et al., 1986, Maher et al., 1991, Anderson et al., 1991; Jones et

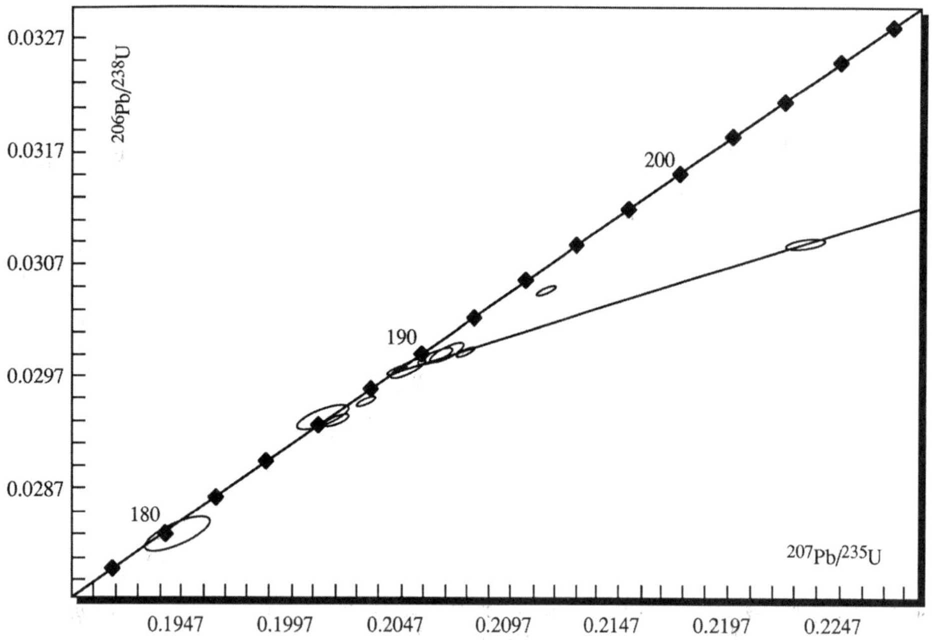

Figure 24. $^{206}Pb/^{238}U$ vs. $^{207}Pb/^{235}U$ concordia plot indicating intercepts at 1885 ± 79 Ma and 189.0 ± 0.2 Ma (MSWD—1.64)

al., 1995). As the La Boca Formation is thought to be correlative to the Nazas Formation (Imlay et al., 1948; Belcher, 1979; Gose et al., 1982; Anderson and Schmidt, 1983; Cohen et al., 1986; Salvador, 1987; Maher et al., 1991; Anderson et al., 1991; Jones et al., 1995; McKee et al., 1997), a similar origin is likely for any volcanic material found therein. If the VES also originated from Pacific subduction, then these two volcanic rock sequences might be from the same arc system or its earlier analog.

INFERRED GEOLOGIC HISTORY OF HUIZACHAL CANYON

This study suggests that the pre–Early Jurassic VES was deposited in a continental arc of Andean type. This arc setting could have been associated either with the collision of North and South America and the formation of Pangea (late Pennsylvanian–early Permian), or Pacific subduction during the Triassic. Much of the silicification and hydrothermal alteration appears to have been imparted as a late-stage vapor mineralization process accompanying pyroclastic and lava flow emplacement. If it originated during the formation of Pangea, deformation of the VES might have occurred in association with the resulting orogeny. If the VES instead originated due to Pacific subduction, deformation probably was caused by the breakup of Pangea, accompanied by eastward transport to its present position.

After erosion of the VES, deposition of the lower portion of the La Boca Formation in the Early Jurassic occurred in erosional topographic lows developed upon the VES and was accompanied by subalkalic volcanism produced through Pacific subduction. This volcanism was concentrated at the base of the formation, where it was deposited atop an irregular topographic erosion surface developed on the upturned VES. The sedimentary rocks of the lower portion of La Boca were deposited through debris-flow processes associated with pyroclastic volcanism. Volcanic events apparently re-exhumed the Early Jurassic vertebrate fauna, as well as entrapping new organisms of approximately similar age. Fluvial processes may have become a factor as deposition continued, resulting in deposition of the upper portion of the La Boca Formation. The La Boca Formation was then deformed slightly and eroded, possibly due to Cordilleran tectonics, and fluvial deposition of the La Joya Formation to form the low-angle unconformity at the top of the La Boca began during the Late Jurassic. Rueda et al. (1991a) suggest increasing marine sedimentation as Mexico was inundated during the latest Jurassic, and the Zuloaga Formation indicates full marine conditions existed in the region during this time. Sometime during the transition from terrestrial to marine sedimentation, the La Joya Formation was slightly tilted and eroded, to produce the unconformity surface on which the Zuloaga was deposited.

During the Paleocene-Eocene transition, folding produced the Huizachal-Peregrina Anticlinorium which was formed in association with the Cordilleran Orogeny. This probably involved "thick-skinned" deformation (intrabasement decoupling), which further deformed the pre-Jurassic VES rocks, as well as the overlying red beds. The cleavage in the red beds most likely was produced by the formation of this anticline.

ACKNOWLEDGMENTS

We sincerely thank the Instituto de Geología, Universidad Nacional Autónoma de Mexico and the Instituto Tamaulipeco de Cultura for their support of this work. F. Ortega, E. Martinez,

G. Lavin Santos del Prado, E. Yepes, and O. Carranza have been instrumental in facilitating this research. The work could not have been accomplished without the support and assistance of the Ejido Huizachal, and we especially thank Fidencio Moreno and his family for their extraordinary assistance, advice, help, and general hospitality; this project could not have happened without them. Assistance in the field was rendered by D.P. Murray and V. Hugo-Reynosa. N. Hamidzada helped produce the figures. Thoughtful comments by N. Riggs, T. Lawton, A. Glazner, W. Muehlberger, J. Roldán Q., and N. Jones greatly improved this manuscript. The work was supported in part by National Geographic Society grants 3017-85 and 3695-87 to J.A. Hopson and Clark, in part by National Science Foundation grants EAR 8917386 and EAR 9218971 to Fastovsky, and EAR 9218871 to Clark, in part by a Dinosaur Society grant to Fastovsky, and in part by the Department of Geosciences, University of Rhode Island.

REFERENCES CITED

Anderson, T.H., McKee, J.W., and Jones, N.W., 1991, A northwest trending, Jurassic fold nappe, northeastern Zacatecas, Mexico: Tectonics, v. 10, p. 383–401.

Anderson, T.H., and Schmidt, V.A., 1983, The evolution of Middle America and the Gulf of Mexico-Caribbean Sea during Mesozoic time: Geological Society of America Bulletin, v. 94, p. 941–966, doi: 10.1130/0016-7606(1983)94<941:TEOMAA>2.0.CO;2.

Belcher, R.C., 1979, Depositional environments, paleomagnetism, and tectonic significance of Huizachal red beds (lower Mesozoic), northeastern Mexico [Ph.D. thesis]: Austin, University of Texas, 276 p.

Best, M.G., 1982, Igneous and Metamorphic Petrology: New York, W.H. Freeman and Company, 630 p.

Boggs, S., Jr., 1992, Petrology of Sedimentary Rocks: New York, Macmillan Publishing Co., 707 p.

Bonnichsen, B., 1982, Rhyolite lava flows in the Bruneau-Jarbidge eruptive center, southwestern Idaho, *in* Bonnichsen, B., and Breckenridge, R.M., eds., Cenozoic Geology of Idaho: Idaho Bureau of Mines and Geology, Bulletin 26, p. 283–320.

Bonnichsen, B., and Kauffman, D.F., 1987, Physical features of rhyolite lava flows in the Snake River Plain volcanic province, southwestern Idaho, *in* Fink, J.H., ed., The Emplacement of Silicic Domes and Lava Flows: Geological Society of America Special Paper 212, p. 119–145.

Campa, M.F., and Coney, P.J., 1983, Tectono-stratigraphic terranes and mineral resource distributions in Mexico: Canadian Journal of Earth Sciences, v. 20, p. 1040–1051.

Carrillo-Bravo, J., 1963, Geology of the Huizachal-Peregrina anticlinorium northwest of Ciudad Victoria, Tamaulipas, *in* Corpus Christi Geological Society Annual Field Trip Guide Book: Geology of Peregrina Canyon and Sierra de el Abra, Mexico: Corpus Christi Geologic Society, Texas, p. 11–23.

Carrillo-Bravo, J., 1961, Geologia del anticlinorio Huizachal-Peregrina al N-W de Ciudad Victoria, Tamaulipas: Asociacion Mexicana de Geologos Petroleos Boletin, v. 13, p. 1–98.

Clark, J.M., Montellano, M., Hopson, J.A., and Hernandez R., R., and Fastovsky, D.E., 1994, An Early or Middle Jurassic tetrapod assemblage from the La Boca formation, northeastern Mexico, *in* Fraser, N.C., and Sues, H.-D., eds., In the Shadow of the Dinosaurs: Cambridge, Cambridge University Press, p. 294–302.

Clark, J.M., Hopson, J.A., Hernandez R., R., Fastovsky, D.E., and Montellano, M., 1998, Foot posture in a primitive pterosaur: Nature, v. 391, p. 886–889, doi: 10.1038/36092.

Cohen, K.K., Anderson, T.H., and Schmidt, V.A., 1986, A paleomagnetic test of the proposed Mojave-Sonora Megashear in northeastern Mexico: Tectonophysics, v. 131, p. 23–51, doi: 10.1016/0040-1951(86)90266-0.

Costa, J.E., and Wieczorek, G.F., eds., 1987, Debris Flows/Avalanches: Process, Recognition, and Mitigation: Geological Society of America Reviews in Engineering VII, 239 p.

de Cserna, Z., 1970, Tectonics of northern Mexico, *in* Seewald, K., and Sundeen, D., eds., Symposium in honor of Professor Ronald K. DeFord, The geologic framework of the Chihuahua tectonic belt: Midland, West Texas Geological Society and University of Texas at Austin, p. 99–118.

de Cserna, Z., and Ortega-Gutiérrez, F., 1978, Reinterpretation of isotopic age data from the Granjero Schist, Ciudad Victoria: Argument: Universidad Nacional Autónoma de Mexico, Instituto de Geologia, Revista, v. 2, p. 212–215.

de Cserna, Z., Graf., J.L., and Ortega-Gutiérrez, F., 1977, Aloctono del Paleozoico Inferior en la region de Ciudad Victoria, Estado de Tamaulipas: Universidad Nacional Autónoma de Mexico, Instituto de Geologia, Revista, v. 1, no, 1, p. 33–43.

de Cserna, Z., 1971, Taconian (Early Caledonian) deformation in the Huasteca Structural Belt of eastern Mexico: American Journal of Science, v. 271, p. 544–550.

Ewing, T.E., 1985, Westward extension of the Devils River uplift-implications for the Paleozoic evolution of the southern margin of North America: Geology, v. 13, p. 433–436, doi: 10.1130/0091-7613(1985)13<433:WEOTDR>2.0.CO;2.

Fastovsky, D.E., Hermes, O.D., Clark, J.M., and Hopson, J.A., 1988, Volcanoes, debris flows, and Mesozoic mammals: Huizachal Group (Early or Middle Jurassic), Tamaulipas, Mexico: Geological Society of America Abstracts with Programs, v. 20, p. A317.

Fastovsky, D.E., Clark, J.M., Strater, N.H., Montellano, M., Hernandez R., R., and Hopson, J.A., 1995, Depositional environments of a Middle Jurassic vertebrate assemblage, Cañon Huizachal, Mexico: Journal of Vertebrate Paleontology, v. 15, p. 561–575.

Fisher, R.V., and Schmincke, H.-U., 1984, Pyroclastic rocks; Springer-Verlag, New York, 472 p.

Floyd, P.A., and Winchester, J.A., 1978, Identification and discrimination of altered and metamorphosed volcanic rocks using immobile elements: Chemical Geology, v. 21, p. 291–306, doi: 10.1016/0009-2541(78)90050-5.

Garcia, M.O., 1978, Criteria for the identification of ancient volcanic arcs: Earth-Science Reviews, v. 14, p. 147–165, doi: 10.1016/0012-8252(78)90002-8.

Girty, G.H., 1926, A new area of Carboniferous rocks in Mexico: Science, v. 63, p. 286–287.

Gose, W.A., Belcher, R.C., and Scott, G.R., 1982, Paleomagnetic results from northeastern Mexico: Evidence for large Mesozoic rotations: Geology, v. 10, p. 50–54, doi: 10.1130/0091-7613(1982)10<50:PRFNME>2.0.CO;2.

Gradstein, F.M., Agterberg, F.P., Ogg, J.G., Hardenbol, J., van Veen, P., Thierry, J., and Huang, Z., 1995, A Triassic, Jurassic, and Cretaceous time scale, *in* Berggren, W.A., Kent, D.V., Aubry, M.-P., and Hardenbol, J., eds., Geochronology, Timescales, and Global Stratigraphic Correlation: SEPM Special Publication no. 54, p. 95–126.

Handschy, J.W., Keller, G.R., and Smith, K.J., 1987, The Ouachita system in northern Mexico: Tectonics, v. 6, p. 323–330.

Harland, W.B., Armstrong, R.L., Cox, A.V., Craig, L.E., Smith, A.G., and Smith, D.G., 1989, A Geologic Time Scale: Cambridge University Press, Cambridge, 263 p.

Hermes, O.D., and Schilling, J.C., 1976, Olivine from Reykjanes—Ridge and Iceland tholeiites, and its significance to the two-mantle source model: Earth and Planetary Science Letters, v. 28, p. 345–355.

Humphrey, W.E., 1949, Geology of the Sierra de los Muertas area, Mexico (with descriptions of Aptian cephalopods from the La Pena Formation): Geological Society of America Bulletin, v. 60, p. 89–176.

Imlay, R.W., 1938, Studies of the Mexican geosyncline: Bulletin of the American Association of Petroleum Geologists, v. 49, p. 1651–1694.

Imlay, R.W., 1943a, Jurassic formations of the Gulf region: Bulletin of the American Association of Petroleum Geologists, v. 27, p. 1407–1533.

Imlay, R.W., 1943b, Evidence for upper Jurassic landmass in eastern Mexico: Bulletin of the American Association of Petroleum Geologists, v. 27, p. 524–538.

Imlay, R.W., Cepeda, E., Alvarez, M., and Diaz, T., 1948, Stratigraphic relations of certain Jurassic formations of eastern Mexico: American Association of Petroleum Geologists Bulletin, v. 32, p. 1750–1761.

Jakes, P., and White, A.J.R., 1972, Major and trace element abundances in volcanic rocks of orogenic areas: Geological Society of America Bulletin, v. 83, p. 29–40.

Jones, N.W., McKee, J.W., Anderson, T.H., and Silver, L.T., 1995, Jurassic volcanic rocks in northeastern Mexico: a possible remnant of a Cordilleran magmatic arc, *in* Jacques-Ayala, C., González-Léon, and Roldán-Quintana, J., eds., Studies on the Mesozoic of Sonora and Adjacent Areas: Geological Society of America Special Paper 301, p. 179–190.

Jones, N.W., McKee, J.W., and Anderson, T.H., 1986, Pre-Cretaceous igneous rocks in north central Mexico: Geological Society of America Abstracts with Programs, v. 18, p. 649.

Jones, N.W., McKee, J.W., Márquez D., B., Tover, J., Long, L.E., and Laudon, T.S., 1984, The Mesozoic La Mula Island, Coahuila, Mexico: Geological Society of America Bulletin, v. 95, p. 1226–1241, doi: 10.1130/0016-7606(1984)95<1226:TMLMIC>2.0.CO;2.

Kesler, M.E., Jones, L.M., and Ruiz, J., 1988, Strontium and sulfur isotope geochemistry of the Galaena Barite District, Nuevo Leon, Mexico: Economic Geology and the Bulletin of the Society of Economic Geologists, v. 83, p. 1907–1917.

King, P.B., 1951, The Tectonics of Middle North America East of the Cordilleran System. Princeton, New Jersey, Princeton University Press, 203 p.

Le Maitre, R.W., 1976, The chemical variability of some common igneous rocks: Journal of Petrology, v. 17, p. 589–598.

Lopez-Ramos, E., 1969, Marine Paleozoic rocks of Mexico: American Association of Petroleum Geologists Bulletin, v. 53, p. 2399–2417.

Maher, D.J., Jones, N.W., McKee, J.W., and Anderson, T.H., 1991, Volcanic rocks at Sierra de Catorce, San Luis Potosi, Mexico: A new piece for the Jurassic-arc puzzle: Geological Society of America Abstracts with Program, v. 23, no. 5, p. A133.

McKee, J.W., Jones, N.W., and Anderson, T.H., 1999, Late Paleozoic and early Mesozoic history of the Las Delicias terrane, Coahuila, Mexico, *in*, Bartolini, C., Wilson, J.L., and Lawton, T.F., eds., Mesozoic Sedimentary and Tectonic History of North-Central Mexico: Geological Society of America Special Paper 340, p. 161–189.

McKee, J.W., Jones, N.W., and Anderson, T.H., 1997, Is "Nazas" a junior synonym of "Huizachal" and does it matter?: Geological Society of America Abstracts with Programs, v. 29, no. 2, p. 39.

McKee, J.W., Jones, N.W., and Anderson, T.H., 1988, Las Delicias basin: A record of late Paleozoic arc volcanism in northeastern Mexico: Geology, v. 16, p. 37–40, doi: 10.1130/0091-7613(1988)016<0037:LDBARO>2.3.CO;2.

McKee, J.W., and Jones, N.W., 1981, Coahuila Peninsula at Sierra Diablo, Chihuahua, Mexico: Geological Society of America Abstracts with Programs, v. 13, p. 242.

Meschede, M., 1986, A method of discriminating between different types of mid-ocean ridge basalts and continental tholeiites with the Nb-Zr-Y diagram: Chemical Geology, v. 56, p. 207–218, doi: 10.1016/0009-2541(86)90004-5.

Michalzik, D., 1991, Facies sequence of Triassic-Jurassic red beds in the Sierra Madre Oriental (NE Mexico) and its relation to the opening of the Gulf of Mexico: Sedimentary Geology, v. 71, p. 243–259, doi: 10.1016/0037-0738(91)90105-M.

Mixon, R.B., 1963a, Geology of the Huizachal redbeds, Sierra Madre Oriental, Mexico [Ph.D. thesis]: Louisiana State University, 98 p.

Mixon, R.B., 1963b, Geology of the Huizachal redbeds, Ciudad Victoria, southwestern Tamaulipas, in Corpus Christi Geological Society Annual Field Trip Guide Book: Geology of Peregrina Canyon and Sierra de el Abra, Mexico: Corpus Christi Geological Society, Texas, p. 24–35.

Mixon, R.B., Grover, E.M., and Teodoro Diaz, G., 1959, Age and correlation of Huizachal Group (Mesozoic), State of Tamaulipas, Mexico: Bulletin of the American Association of Petroleum Geologists. v. 43, p. 757–771.

Munsell Color, 1992, Munsell color charts (1992 edition): MacBeth Division of the Kollmorgen Corporation, Baltimore, Maryland, USA.

Murray, G.E., 1961, Geology of the Atlantic and Gulf Coastal Province of North America: New York, Harper & Brothers Publishers, 692 p.

Ortega-Gutiérrez, F., 1978, El gneis Novillo y rocas metamorficas asociadas en los Canones del Novillo y de la Peregrina, area de Ciudad Victoria, Tamaulipas: Universidad Nacional Autónoma de Mexico, Instituto de Geologia, Revista, v. 2, p. 19–30.

Pearce, J.A., Harris, N.B.W., and Tindle, A.G., 1984, Trace element discrimination diagrams for the tectonic interpretation of granitic rocks: Journal of Petrology, v. 25, p. 956–983.

Pearce, J.A., and Norry, M.J., 1979, Petrogenetic implications of Ti, Zr, Y, and Nb variations in volcanic rocks: Contributions to Mineralogy and Petrology, v. 69, p. 33–47, doi: 10.1007/BF00375192.

Pearce, J.A., and Cann, J.R., 1973, Tectonic setting of basic volcanic rocks determined using trace element analysis: Earth and Planetary Science Letters, v. 19, p. 290–300, doi: 10.1016/0012-821X(73)90129-5.

Ramirez-Ramirez, C., 1992, Pre-Mesozoic geology of Huizachal-Peregrina anticlinorium, Ciudad Victoria, Tamaulipas, and adjacent parts of eastern Mexico [Ph.D. thesis]: University of Texas, Austin, 317 p.

Ramirez-Ramirez, C., 1978, Reinterpretacion tectonica del Esquisto Granjero de Universidad Nacional Autónoma de Mexico: Instituto de Geologia, Revista, v. 2, p. 31–36.

Ross, C.A., 1986, Paleozoic evolution of southern margin of Permian basin: Geological Society of America Bulletin, v. 97, p. 536–554, doi: 10.1130/0016-7606(1986)97<536:PEOSMO>2.0.CO;2.

Ross, C.A., 1979, Late Paleozoic collision of North and South America: Geology, v. 7, p. 41–44, doi: 10.1130/0091-7613(1979)7<41:LPCONA>2.0.CO;2.

Ross, C.S., and Smith, R.L., 1961, Ash-flow tuffs: Their origin, geologic relations, and identification: U.S. Geological Survey Professional Paper 366, 81 p.

Rowett, C.L., and Hawkins, C.M., 1975, Late Paleozoic volcanic island arc-trench system in northern Mexico: Geological Society of America Abstracts with Programs, v. 7, no. 2, p. 230–231.

Rueda G., R., Ocampo, E.L., Duenas, M.A., and Benitez, J.L.R., 1991a, Palinoestratigrafia, petrologia, tectonica y potencial generador de hidrocarburos del alogrupo La Boca en el anticlinorio de Huizachal-Peregrina, Tamaulipas, *in* Verma, S.P., Ramirez F., J.A., Rodriguez De B., C.O., Barbarin C., J.M., Izquierdo M., G., Armienta H., M.A., and Terrel, D.J., eds., Actas fac. ciencias tierra: Linares, UANL, v. 6, p. 203–205.

Rueda G., R., Minero, M., and Uribe, G., 1991b, El analisis por rayos X del residuo palinologico confirma que su color y abundancia son parametros basicos para identificar unidades litoestratigraficas, *in* Verma, S.P., Ramirez F., J.A., Rodriguez De B., C.O., Barbarin C., J.M., Izquierdo M., G., Armienta H., M.A., and Terrel, D.J., eds., Actas fac. ciencias tierra Linares, UANL, v. 6, p. 179–182.

Ruiz, J., Patchett, J.P., and Ortega-Gutiérrez, F., 1987, Proterozoic and Phanerozoic basement terranes in Mexico: Geological Society of America Abstracts with Programs, v. 19, no. 7, p. 826.

Salvador, A., 1987, Late Triassic-Jurassic Paleogeography and origin of the Gulf of Mexico Basin: American Association of Petroleum Geologists Bulletin, v. 71, p. 419–451.

Secretaria de Programacion y Presupuesto, 1978, Carta geologica F14A29: scale 1:50,000.

Shurbet, D.H., and Cebull, S.E., 1987, Tectonic interpretation of the westernmost part of the Ouachita-Marathon (Hercynian) orogenic belt, west Texas-Mexico: Geology, v. 15, p. 458–461, doi: 10.1130/0091-7613(1987)15<458:TIOTWP>2.0.CO;2.

Sigurdsson, H., Carey, S., Cornell, W., and Prescatore, T., 1985, The eruption of Vesuvius in A.D. 79: National Geographic Research, v. 1, p. 332–387.

Stacey, J.S., and Kramers, J.D., 1975, Approximation of terrestrial Lead isotope evolution by a two-stage model: Earth and Planetary Science Letters, v. 26, p. 207–221, doi: 10.1016/0012-821X(75)90088-6.

Steiger, R.H., and Jäger, E., 1977, Subcommission on geochronology: Convention on the use of decay constants in geo- and cosmochronology: Earth and Planetary Science Letters, v. 36, p. 359–362, doi: 10.1016/0012-821X(77)90060-7.

Stewart, J.H., 1988, Latest Proterozoic and Paleozoic southern margin of North America and the accretion of Mexico: Geology, v. 16, p. 186–189, doi: 10.1130/0091-7613(1988)016<0186:LPAPSM>2.3.CO;2.

Strater, N.H., 1993, Origin of the Pre-Jurassic Strata of Cañon Huizachal (Tamaulipas) and their relationship to the Tectonic Evolution of Northeastern Mexico [M.S. thesis]: Kingston, University of Rhode Island, 169 p.

Suter, M., 1987, Structural traverse across the Sierra Madre Oriental fold thrust belt in east-central Mexico: Geological Society of America Bulletin, v. 98, p. 249–264, doi: 10.1130/0016-7606(1987)98<249:STATSM>2.0.CO;2.

Tatum, J.L., and Dawson, J.M. (San Antonio Geological Society), 1936, Geologic road log of Pan-American highway, Laredo to Mexico City: Bulletin of the American Association of Petroleum Geologists, v. 20, p. 457–466.

Walper, J.L., and Rowett, C.L., 1972, Plate tectonics and the origin of the Caribbean Sea and the Gulf of Mexico: Gulf Coast Association of Geological Societies Transactions, v. 22, p. 105–116.

Wilson, M., 1989, Igneous petrogenesis: London, Unwin Hyman, 466 p.

Winchester, J.A., and Floyd, P.A., 1977, Geochemical discrimination of different magma series and their differentiation products using immobile elements: Chemical Geology, v. 20, p. 325–343, doi: 10.1016/0009-2541(77)90057-2.

Manuscript Accepted by the Society 24 March 2005

Geological Society of America
Special Paper 393
2005

The Taray Formation: Jurassic(?) mélange in northern Mexico—Tectonic implications

Thomas H. Anderson
Department of Geology and Planetary Sciences, University of Pittsburgh, Pittsburgh, Pennsylvania 15260, USA

Norris W. Jones
James W. McKee
Department of Geology, University of Wisconsin, Oshkosh, Oshkosh, Wisconsin 54901, USA

ABSTRACT

The Taray Formation is a mudstone-rich body of rock characterized by dismembered beds of sandstone and shale and fragments of a great range of sizes in a finer matrix. The formation is known only from exposures in northern Zacatecas, in the southern part of the San Julian uplift. Structurally, it underlies a thick Jurassic sequence that includes the volcanogenic Caopas, Rodeo, and Nazas formations and younger rocks of La Joya (volcaniclastic) and Zuloaga (limestone) formations. The base of the Taray is unknown. The texture and structures of Taray and its tectonostratigraphic association with Jurassic volcanic rocks suggest that it is formed of mélange of probable Jurassic age.

Sections of interbedded sandstone and mudstone commonly show progressive dismemberment. Initial subtabular beds commonly pinch and swell. Further extension was accommodated by boudinage or by faults, which cut gently across bedding and dissect sandy layers into lenses and slivers. Progressive stratal disruption formed isolated, deformed inclusions of sandstone, commonly mixed with volcanic and cherty debris, encased in dark-gray, fine-grained, foliated, crenulated matrix. Some large boulders of quartzose sand that slid downslope, crumpling sediments at their bows, were subsequently encased by matrix and both were distinctly extended so that the boulders were fractured, whereas the weaker, pliant matrix responded ductilely.

Taray also contains blocks (as much as hundreds of meters in maximum dimension) of diverse composition including: laminated and massively bedded, light-colored chert; fragmented volcanic flow rock; fine, quartzose sandstone; and carbonate beds, some of which contain fusulinids and crinoid debris. Some of these rocks show tight folds that formed, in some cases, while they were semiconsolidated; later they were incorporated in muddy matrix as lithified fragments.

The unexposed contact between mudstone-rich Taray Formation and the volcanic and volcaniclastic beds of the Nazas Formation also marks changes in the attitude of planar structures. We interpret Taray as part of a Jurassic(?) accretionary prism. The tectonostratigraphic sequence at this latitude (west to east) of accretionary prism (Taray), arc (Nazas, Rodeo, Caopas Formations), and overlying clastic cover (La Joya Formation), which developed against and upon a margin of Paleozoic and Precam-

Anderson, T.H., Jones, N.W., and McKee, J.W., 2005, The Taray Formation: Jurassic(?) mélange in northern Mexico—Tectonic implications, *in* Anderson, T.H., Nourse, J.A., McKee, J.W., and Steiner, M.B., eds., The Mojave-Sonora megashear hypothesis: Development, assessment, and alternatives: Geological Society of America Special Paper 393, p. 427–455. doi: 10.1130/2005.2393(16). For permission to copy, contact editing@geosociety.org.

brian crust, cannot be extended northward and must either turn abruptly westward along the southern edge of the Mesozoic Coahuila Island or be truncated and offset, probably along the Mojave-Sonora megashear.

Keywords: Taray Formation, mélange, accretionary prism, Jurassic, volcanic arc.

INTRODUCTION AND REGIONAL SETTING

The Mojave-Sonora megashear is postulated to curve from Sonora east-southeastward across northern Mexico (Anderson and Schmidt, 1983). The inferred trace is constrained by the distribution of pre–Late Jurassic rocks, especially those in the vicinity of the great Monterrey salient of the Sierra Madre Oriental, which is defined principally by resistant ridges of Oxfordian and younger carbonates. The stratigraphic discontinuity cited as an indication of displacement along the fault occurs between Paleozoic rocks (King, 1934, 1944) that underlie the southern part of the Coahuila block (also described as a peninsula or island) and pre-Cretaceous Mesozoic rocks exposed southward in the San Julian uplift (Córdoba-M., 1964), across the inferred megashear (Fig. 1).

During the late 1970s Anderson learned of regional mapping of the relevant pre-Cretaceous rocks being conducted by McKee and Jones in northeastern Mexico. An outcome of their work in the state of Coahuila is the recognition that initial activity along the regional San Marcos fault, a west-northwest–striking structure bounding the northern margin of the Mesozoic Coahuila Island, is contemporaneous with the age proposed for the megashear (McKee and Jones, 1979; 1990). Soon thereafter we joined to conduct regional studies with the objective of assessing the postulated stratigraphic discontinuity between the Coahuila block and the San Julian uplift to the south (Fig. 1).

The initial cooperative study focused upon late Paleozoic, arc-proximal marine strata and related intrusive rocks, named the Las Delicias Formation, that underlie at least 6000 km^2 of the Coahuila block (McKee et al., 1988, 1999). The folded stratified section consists mainly of thousands of meters of sand- to megaboulder-size volcaniclastic debris (mostly andesitic and dacitic) with conspicuous, although volumetrically minor, thick beds composed of allochthonous basin-margin limestone that is intercalated with fine-grained marine pelagic sediment. Coherent or semicoherent allochthonous bodies of carbonate and volcanic rocks that moved down the flanks of the sedimentary basin under the influence of gravity may include blocks with dimensions of hundreds or thousands of meters. One slab may encompass a few hundred square kilometers. Neither the bottom nor the top of the section is exposed. The unit includes strata with ages that fall within the interval from late Mississippian(?) through most of the Permian (McKee et al., 1999).

South of the Delicias terrane across the inferred megashear, which is covered by Cretaceous strata preserved in the Parras basin, major exposures of pre-Cretaceous rocks crop out within the San Julian uplift. Among the hills of the uplift, thousands of meters of flows, breccia, laharic conglomerate, air-fall tuff, ash-flow tuff and tuffaceous siltstone, which are locally strongly deformed, are included within Caopas, Nazas, and Rodeo Formations (Jones et al., 1995). Outcrops of these pre-Oxfordian, Mesozoic, commonly subaerial rocks occur within a discrete region south of the inferred trace of the megashear (McKee et al., 1989).

In the southern part of the San Julian uplift a mudstone-rich body of rock (Taray Formation), which is characterized by dismembered beds of sandstone and shale and by fragments of a great range of sizes in a finer matrix, underlies the volcanogenic rocks of the Nazas Formation (Fig. 2). The unexposed contact between these two units is interrupted by low-dipping faults. Southwestward-directed tectonic transport of the subaerial rocks of the magmatic arc onto fine-grained marine rocks is compatible with the asymmetry of folds recorded by units high in the volcanic section and overlying Oxfordian(?) strata (Anderson et al., 1991).

Our purpose is to: (1) describe the lithologic, compositional, and structural features of Taray that led us to conclude it is part of an accretionary prism (López-Infanzón, 1986), and (2) relate Taray rocks within the San Julian uplift to older and younger units, as well as to the Mojave-Sonora megashear along which the formation has been displaced (Jones et al., 1995).

TARAY FORMATION

We interpret the dismembered muddy rocks composing Taray Formation as mélange. They were first mapped as a formation by Córdoba-M. (1964) during his geologic investigations of the San Julian uplift in northern Zacatecas, Mexico (Figs. 1 and 2). He proposed the name "Taray Formation" for gray and brown phyllite, graywacke, and conglomerate with scattered, bedded masses of dolomite and chert exposed along Arroyo El Taray north of the prominent Pico de Teyra.

The principal exposure of the Taray Formation occupies ~60 km^2, west of Sierra El Solitario de Teyra and around Pico de Teyra (Figs. 2 and 3). Access to the area is by very poor roads and arroyos and most of our field work was done on foot. Taray-like rocks also are exposed in an area of a few hundred square meters in extent along the road just east of Caopas (Fig. 2); these outcrops, which are 15 km north of the main exposure, are surrounded by volcanic rocks of the Jurassic Caopas-Rodeo-Nazas Formations (Jones et al., 1995). López-Infanzón (1986) reported Taray Formation to the east of Cerro Europa, a hill ~4 km north of Caopas and halfway between Caopas and Rodeo, but we were unable to verify this occurrence.

The Taray Formation is a mudstone-rich body of rock characterized by: (1) dismembered beds of sandstone and shale, and (2) fragments or blocks up to hundreds of meters long, in a fine-

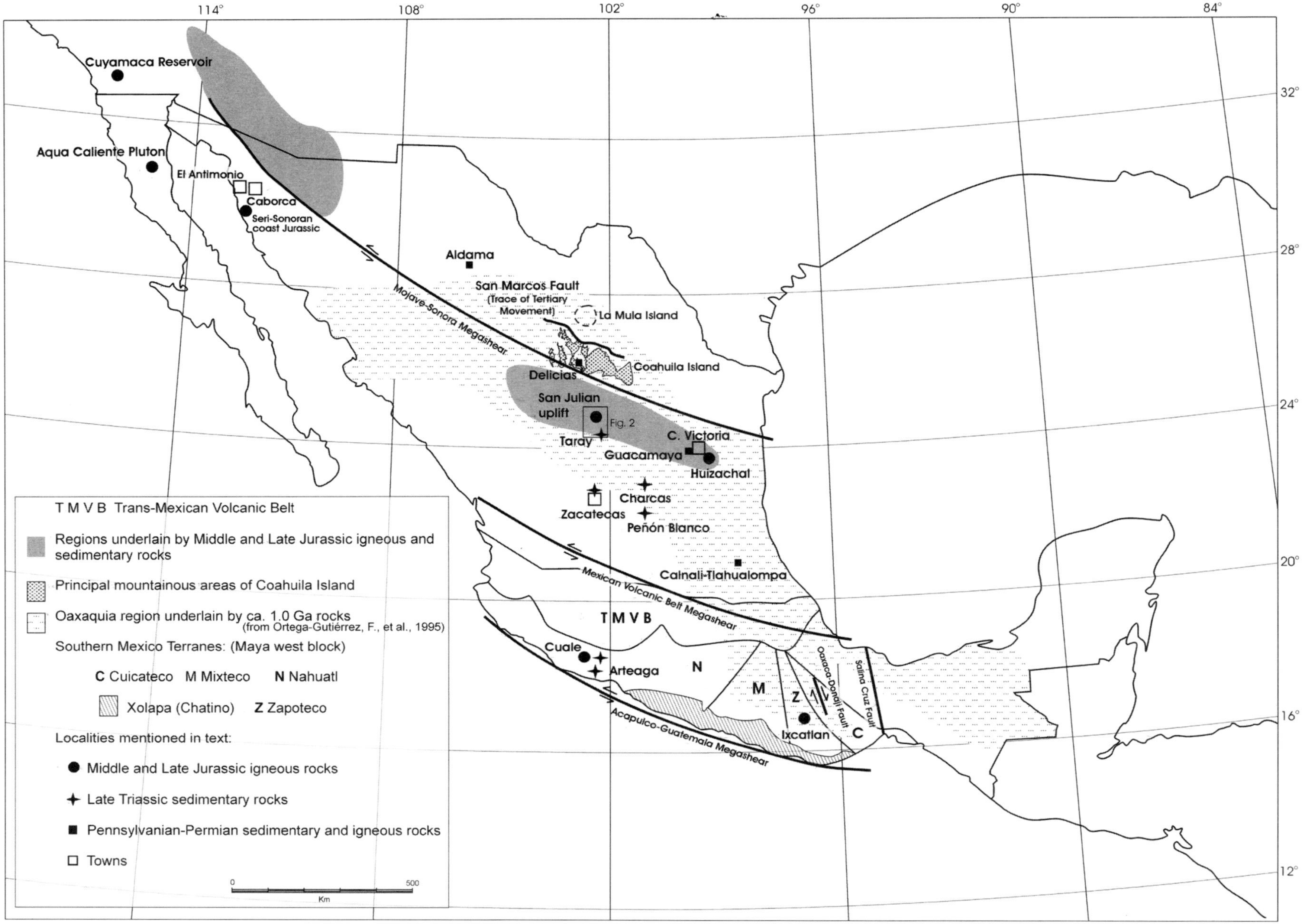

Figure 1. Map of Mexico showing postulated Jurassic tectonic discontinuities and locations of: (1) towns, (2) exposures of Jurassic, Triassic, and late Paleozoic rocks, (3) terranes, and (4) tectonostratigraphic domains mentioned in text.

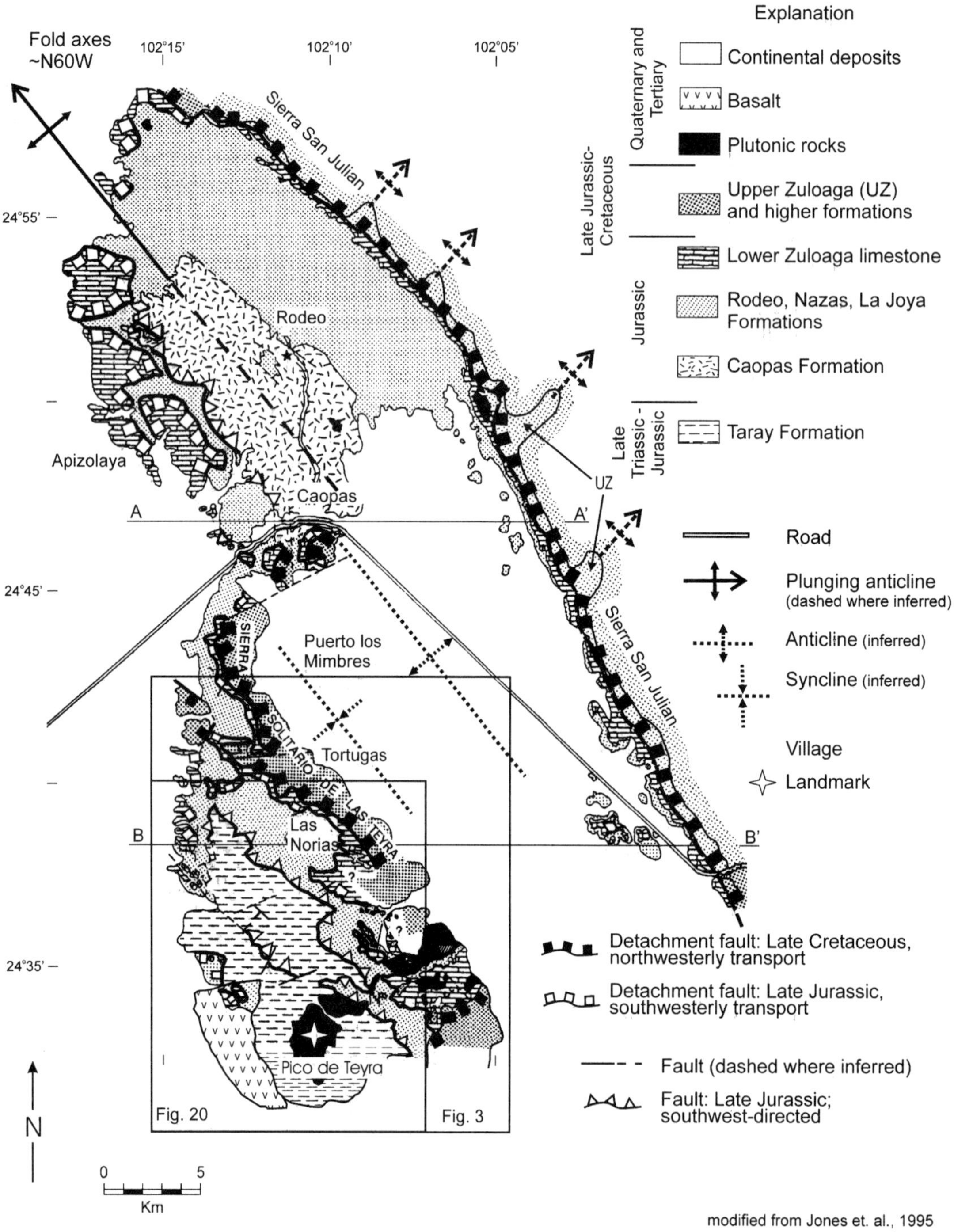

Figure 2. Geologic map of the San Julian Uplift. The Upper Zuloaga Formation was not mapped in the Sierra Solitario de Las Teyra and is included there under the symbol for post-Zuloaga formations, undifferentiated.

grained matrix. Many exposures of matrix composed of dark-gray, fine-grained platy rocks are characterized by closely spaced cleavage that may be planar (poorly to well developed) or phacoidal. Locally the foliated rocks record crenulations. In places we were tempted to label the curved anastomosing cleavage planes in matrix as scaly cleavage (e.g., Lundberg and Moore, 1986; Pini, 1999, fig. 12A therein) equivalent to that which characterizes the tectonosomes of the argille scagliose of the Bologna Apennines of Italy (Pini, 1999). However, we are not confident that the cleavage in the Taray is restricted to bodies comparable to tectonosomes nor does it exhibit typical splintery character.

Parts of the Taray contain tuffaceous horizons. Although these bodies of rock may be compared to progressively extended sequences of mudstone, tuff, chert and sandstone of Cowan's (1985) type II mélange, their stratigraphic position, close to the base of the Nazas, led us to the conclusion that these parts of the Taray had been within reach of a volcanic source other than that related to oceanic volcanism.

We discuss the enigmatic Taray-Nazas contact in a separate section of this report that is devoted to the structural and stratigraphic relationship between those units. The base of the Taray is not exposed.

The age of the principal components of the Taray Formation, such as the common gray, fine-grained foliated rocks, is unknown because (1) the stratigraphic position is not well constrained, (2) the only fossils are in clasts, and (3) no isotopic dates are available. Córdoba-M. (1964) tentatively suggested an Early Devonian to Early Pennsylvanian age based on its perceived lithologic similarity to rocks of that age in the Marathon region of West Texas. Rare clasts (up to boulder size) of crinoidal

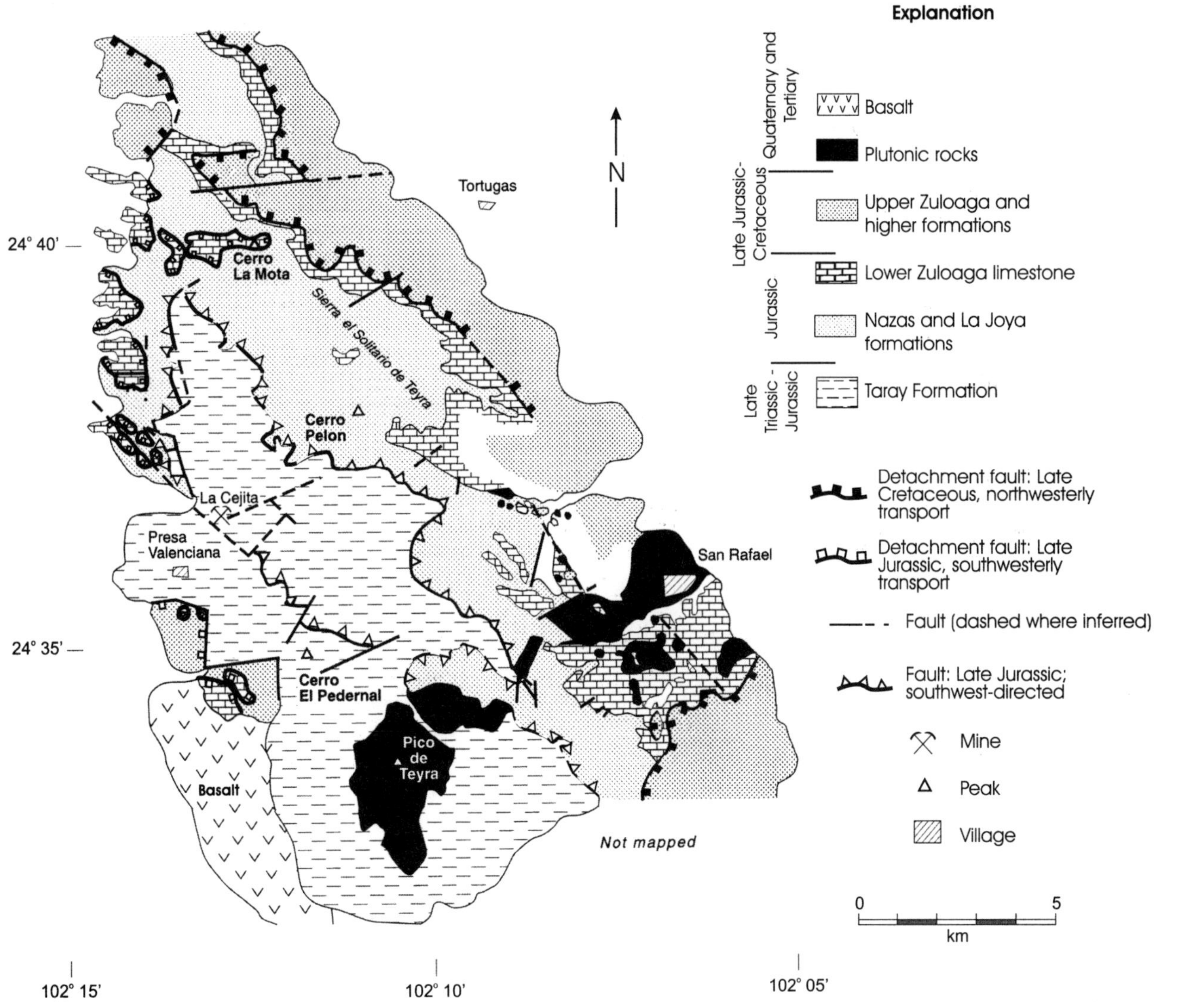

Figure 3. Geologic map of the principal area of exposure of the Taray Formation.

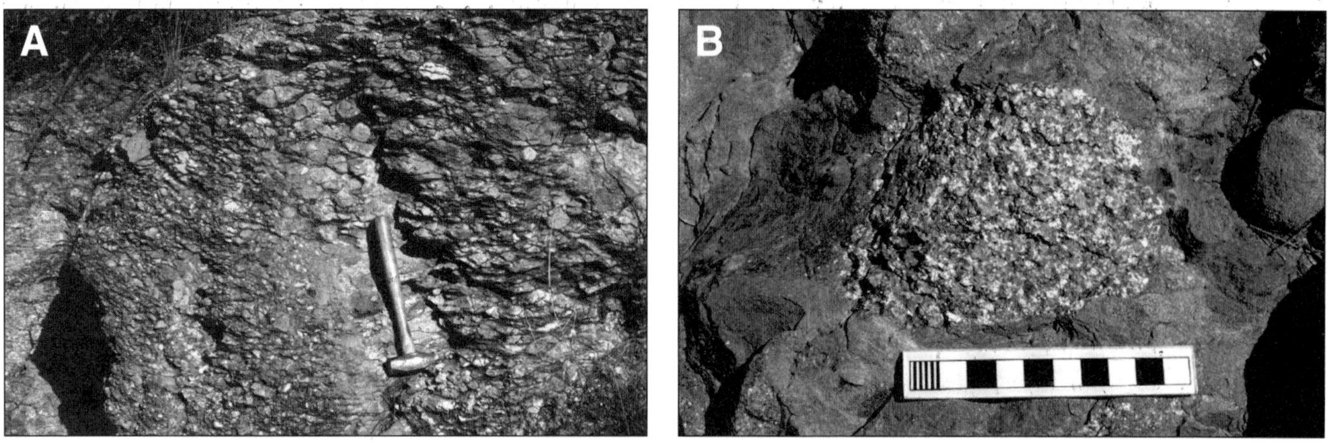

Figure 4. Conglomerate in Taray Formation. (A) Clasts in near-vertical beds are mainly quartz-rich sandstone and chert; note the gentle dip of cleavage. Hammer length = 36 cm. (B) Rare conglomerate with rounded cobble of muscovite granite. Scale in cm.

limestone are assumed to be Paleozoic. Some limestone cobbles contain fusulinids that were identified as Pennsylvanian by Merlynd Nestell (1983, personal commun.). Detrital zircons from sandstone in Taray yield ages as young as 270 Ma, suggesting that the sandstone formed between Late Permian and Nazas time (Middle Jurassic) (Diaz-Salgado et al., 2003). López-Infanzón (1986) and Barboza-Gudino et al. (1999) assigned a Late Triassic age on the basis of their correlation of Taray with the fossiliferous Zacatecas Formation (Burckhardt and Scalia, 1906).

The Zacatecas Formation (Carrillo-Bravo, 1968), exposed in the vicinity of the city of Zacatecas, links Taray Formation to other lithologically similar bodies of fine-grained clastic rocks (e.g., Varales lithofacies of the Arteaga Complex; Centeno-García et al., 2003) and locally mafic marine rocks to the southwest. Together the widespread exposures of comparable strata at Taray (Zacatecas), Charcas (San Luis Potosí), Peñon Blanco (San Luis Potosí and Zacatecas), Zacatecas (Zacatecas) and Arteaga (Michoacan) distinguish a domain of marine clastic rocks, commonly strongly disrupted by layer-parallel extension. These strata occur beneath Jurassic volcanic units (Nazas Formation and others; Blickwede, 1981) of the Sierra Madre terrane, as well as Early Cretaceous volcanic rocks of the Guerrero terrane (Centeno-García and Silva-Romo, 1997). We propose that the mélange-like structures of Taray formed during east-northeast–directed, Middle Jurassic convergence and that the uppermost units of the Taray may be transitional into the Nazas Formation.

Figure 5. Quartzite boulder in Taray Formation with person for scale. Note crumpled sediments to right of boulder.

LITHOLOGIC COMPONENTS OF THE TARAY FORMATION

Overview

Much of the matrix of Taray Formation is gray to dark-gray, light-gray-weathering argillite, slate, or phyllite that shows scant evidence of bedding. However, in most outcrops some sandstone or metasandstone is present, either in lenses or in rare continuous beds. Distinctly bedded sections of sandstone, mudstone, conglomerate, and pebbly mudstone that occur within the Taray are generally within blocks.

Conglomerate, though a relatively minor component, is widespread both as beds and within blocks. Some conglomerate consists of lenses of granules or pebbles of sandstone or slate that were probably derived internally. Other conglomerate bodies contain subrounded to rounded clasts that suggest a history of subaerial transport. Among these latter conglomerates, components include: (1) subangular to rounded granules of metaquartzite or vein quartz, (2) rounded pebbles or cobbles of chert, (3) subrounded cobbles to boulders of sandstone (Fig. 4A), (4) pebbles and cobbles of carbonate, and (5) rounded cobbles of granitoids (one exposure only) (Fig. 4B). A conglomerate

Figure 6. Thin-bedded chert in block in Taray Formation. Note isoclinal fold near centimeter scale.

exposed near the western side of the Taray exposures contains cobbles of crinoid- and fusulinid-bearing limestone.

Blocks up to several hundred meters long, characterized by coherent bedded strata, and other large clasts are distinctive components of the Taray. Clasts, commonly lensoid, of quartzose sandstone that range from pebble to house-size boulders are most common (Fig. 5). The blocks or inclusions up to 300 m long consist of thin- to medium-bedded, light-gray to black chert, some of which was isoclinally folded before incorporation in the Taray (Fig. 6). A complex block, or blocks, containing chert, diabase, scoriaceous material, and turbiditic wacke forms Cerro El Pedernal (Fig. 3; Klein et al., 1990). Inclusions of carbonate are of several types: some are crinoid-bearing limestone; some contain dolomite intimately mixed with volcanic material (near La Cejita mine, Fig. 3); and, on Cerro El Pedernal, ankerite occurs interbedded with magnetite in a laminated to thin-bedded metasediment. Some masses of intermediate to mafic volcanic rock occur near the southeastern edge of the Taray exposure, but they may be fault-bounded slivers of the overlying Nazas Formation.

Matrix

Exposures of massive, fine-grained matrix of the Taray consist of black, greenish-black, gray or silvery gray phyllitic metashale or metasiltstone that is commonly foliated, locally crenulated, and metamorphosed to low greenschist facies. Bedding is difficult to discern in many outcrops.

Thin sections show that grain size is typically coarse silt to clay, but scattered sand-size grains or lensoid granule-size intraclasts are common (Fig. 7). Optical microscopy and X-ray diffraction (XRD) show that subangular quartz is invariably present, 2MIIb clinochlore occurs in nearly all samples, and fine-grained, $2M_1$ white mica (illite, sericite, or muscovite) is a common, but not invariable, component. Albite is less common, and biotite is generally absent or a minor component. Rounded zircon was identified in some.

Foliation, generally evident in thin section, is shown (Fig. 8) by: (1) common extinction of sheet silicates, (2) the presence of wisps of white mica and chlorite, (3) the presence of aligned iron oxide, or (4) the orientation of lensoid intraclasts. Thin seams of opaque minerals (Fig. 7) that separate lensoid domains are interpreted as the residuum of dissolution and define foliation.

Figure 7. Photomicrograph (plane light) of matrix and a few clasts of a conglomeratic phyllite. Matrix and siltstone and sandstone clasts are lensoid as a result of the development of slip surfaces, commonly parallel to compositional layers, and small normal faults that transect the layers and intersect the slip surfaces at acute angles; opaques accentuate foliation. Minerals present include quartz, albite, muscovite, and clinochlore.

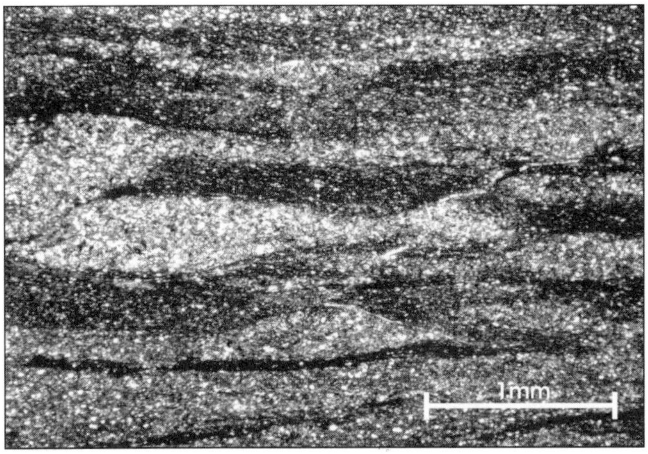

Figure 8. Photomicrograph (crossed polars) of metasiltstone composed of quartz, muscovite, albite, and clinochlore. Extension resulting in the formation of slip surfaces parallel to compositional layers and small normal faults that transect layers has juxtaposed slightly different lithologies and produced lenses. Lenses, common orientation of sheet silicates, and iron oxides on cleavage surfaces define the foliation.

Layers, or zones that resemble beds, are evident in some samples. Dark beds contain more iron oxide and/or chlorite with a strong preferred orientation. Lens-shaped fragments also define crude bedding. Siliceous lenses that resemble chert in hand specimen contain very fine grains of quartz. Some grains may be derived from metamorphosed rock as suggested by irregular domains with sutured boundaries and elongate crystal aggregates. In some samples, fractures filled with iron oxide cut the foliation and bedding. Quartz veins both cut and parallel foliation. Quartz veins in some samples have very irregular shapes and appear to have been deformed.

Locally in the topographically higher parts of the Taray, light-colored clasts and matrix appear to be tuffaceous. We tried to separate these rocks from the rest of the Taray, but the transitions are so subtle that they are not mappable. A thin section of what we interpreted to be light-gray, very fine-grained tuff (Fig. 9) reveals aligned folia of sericite (composing ~50% of the rock), quartz and plagioclase crystals, less than ~0.03 mm in diameter, and chert-like material. Some quartz has shard-like shapes; fragments tend to be elongate, curved and angular. The chert-like material does not clearly occur as grains, but seems to be matrix; it could be devitrified glass.

Figure 9. Photomicrograph (crossed polars) of fine-grained meta-ash-tuff, which consists of quartz, sericite, plagioclase, and particles of probable devitrified volcanic glass. Some quartz fragments are elongate, curved, and angular.

Sandstone

The most prominent masses of sandstone occur as lensoid masses with visible dimensions up to 5 m thick and 10 m long. The sandstone beds vary in composition, but most are quartz-rich. Color varies from dark gray to buff to nearly white, but most are gray. Grain size varies from fine- to coarse-sand and angularity from angular to rounded. Slightly flattened grains and wisps of dark material, probably iron oxides, define weak foliation in some sandstone beds (Fig. 10). Sorting ranges from poor to moderately well, and sandstones are both grain and matrix supported.

Distinct beds (versus lenses) of sandstone are rare. Where present, they show some graded bedding and channeling and may be part of an exotic block. Beds are mostly thin, 2–5 cm, but there are rare beds of sandstone up to 1 m thick.

Some sandstone bodies contain inclusions. In one exposure, massive sandstone contains clasts of quartz sandstone ~0.6 m long, a bed or elongate fragment of chert, and small, 1–10 cm, variously colored wisps of fine sandstone that probably were formed before lithification.

In the topographically upper part of the Taray Formation, some of the sandstone becomes feldspathic. These beds are dark-gray to moderate brown, rusty weathering, fine- to coarse-grained quartz wackes. Some possess a slight foliation and may even be somewhat schistose. Muscovite occurs in some as both detrital and metamorphic grains; biotite is rare. Some have scattered clasts of shale fragments, milky quartz, feldspars, and even carbonate, both as grains and as matrix. In thin section the principal difference from the other sandstones is that lithic fragments, especially volcanic rock fragments, are somewhat more abundant (Fig. 11). Unfortunately, we were unable to map these sandstones as a distinct unit.

Texturally, all of the sandstones are wackes, but the grain mineralogy is variable (Table 1, Figs. 11, 12, 13, 14). The principal grains are quartz (composing from ~18–70% of the sample), plagioclase (trace to 18%), and lithic fragments (0–25%); matrix ranges from 18% to 65%. Sedimentary chert, volcanic chert, and fine-grained polycrystalline quartz are difficult to subdivide.

Differences among individual quartz grains suggest the existence of multiple provenances. Many grains show undulatory extinction or polygonization, and some show deformation lamellae typical of highly sheared rocks suggestive of derivation from metamorphic rocks (Fig. 12). Some grains show very straight extinction, are slightly embayed (Fig. 11), or have inclusions indicating a volcanic source. In one wacke, quartz is angular and has pressure fringes of sericite and chlorite that parallel the plane of foliation. Plagioclase varies from fresh grains with good sharp twins to grains that are slightly to strongly altered (Fig. 13). In a few samples, twin lamellae are bent, but most are undistorted. Much feldspar is untwinned, but staining for K indicated there is very little K-feldspar. Other minerals include clinochlore (which may have formed from detrital Fe-Mg minerals), muscovite, and iron oxides. Zircon and rare hornblende are accessory minerals. Rock fragments include volcanic rocks (ranging from very fine-grained fragments that are unidentifiable to chert-like siliceous rocks to plagioclase-phyric andesite), shale (some with fine quartz grains), chert, chlorite (probably derived from clasts), sandstone, and quartzite (or recrystallized portions of vein quartz).

The matrix of the sandstones is generally composed of very fine-grained quartz and a fine-grained white mica and clinochlore. In some rocks it is difficult to distinguish matrix from distorted rock fragments.

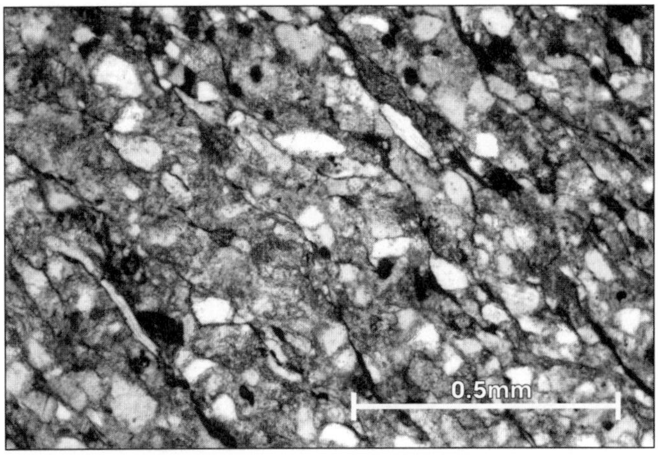

Figure 10. Photomicrograph (plane light) of fine-grained sandstone from lensoid body in Taray Formation. Angular to subangular quartz, plagioclase, and detrital muscovite (lower left) are contained in a matrix of very small crystals of silicates, including quartz and clinochlore, that may be the remains of highly altered volcanic rock fragments. Note the foliation.

Figure 11. Photomicrograph (crossed polars) of medium-grained, grain-supported wacke with grains of quartz, plagioclase, and volcanic rock fragments. Note embayed volcanic quartz in rock fragment. Sample from lensoid body in Taray.

Broken Formation and Pebbly Mudstone

Discontinuous layers of sandstone, recording layer-parallel extension, that occur in a matrix of metashale/siltstone are comparable to broken formation (Hsü, 1974; Kusky and Bradley, 1999) (Fig. 15). Sand horizons in less-disrupted zones are segmented and the broken aspect is clear. Sand bodies in some of these zones range from tabular blocks, up to 2 m by 30 cm, to 1 cm lenticular clasts. In some outcrops, beds pinch and swell in a manner suggesting soft-sediment deformation. In thin section, most clasts are lensoid (Fig. 16); in some, fine-grained phyllitic material interfingers with sandstone so that the term clast is inappropriate. Entirely disrupted sand bodies that are mixed in an irregular fashion within argillaceous matrix are similar to pebbly mudstone as proposed by Hsü (1974) and discussed by Kusky and Bradley (1999). Most common are pebble- to cobble-size, rounded or elongate lenses of

TABLE 1. POINTS COUNTED ON TARAY SANDS

A. No associated volcanic rocks*

Sample:	3	4	5	6	7A	7B	10A	10B	11A	11B	19B	29	30B	38	40B	40C
Qm	19.7	22.7	26.7	46.7	42.0	39.7	53.7	59.7	18.3	26.0	33.3	50.7	19.7	34.0	45.3	33.3
Qp	–	0.7	24.7	6.0	23.0	27.3	4.7	10.3	1.0	2.3	1.3	5.7	2.0	3.7	5.0	25.7
P	18.3	18.0	13.3	12.7	0.3	9.0	4.3	1.7	13.7	10.3	17.3	15.0	7.3	18.0	16.7	8.3
K	4.7	2.3	1.7	–	–	–	–	–	1.7	–	–					
Lv	7.3	8.7	–	3.3	1.0	1.7	–	–	4.3	5.3	12.3	5.3	2.7	18.3	1.3	3.3
Ls	1.3	–	–	0.7	–	0.7	–	–	–	–	6.3	0.3	3.3	7.7	0.3	10.7
M	48.7	47.7	33.7	30.7	33.7	21.7	37.3	28.3	61.0	56.0	29.3	22.3	65.0	18.3	31.3	18.7

B. Volcanic rocks associated*

Sample:	3	8A	8B	8C	8D	9C	9D	9E	10A	10B	10C	11A	11C	14B
Qm	40.3	28.0	25.3	24.0	32.3	44.3	27.0	18.3	23.0	24.0	20.3	17.3	16.7	28.0
Qp	13.7	2.0	1.3		3.3	8.7	13.0	4.3	5.7	2.3	3.7	1.0	42.0	18.0
P	7.3	7.7	13.7	13.0	9.0	2.0	4.3	10.7	11.0	18.7	10.3	18.3	10.7	7.0
K	0.3													
Lv	4.0	9.7	6.3	4.7	8.3	3.7	2.3	5.7	7.0	11.0	22.3	3.0	11.7	1.7
Ls	4.0	8.3	6.3	0.3	0.7	2.0		4.0	3.0	3.0	7.7	2.0		2.7
M	30.3	44.3	47.3	58.0	46.3	39.3	53.3	53.3	50.3	37.7	35.7	58.3	16.3	45.3

Note: 300 points counted, values in percent. Explanation of symbols: Qm—monocrystalline quartz grains; Qp—polycrystalline quartz grains; P—plagioclase grains; K—K-feldspar grains; Lv—volcanic lithic fragments; Ls—sedimentary lithic fragments; M—matrix and other grains.

*No rocks associated with the sandstones in Table 1A are thought to have a volcanic component vs. those in Table 1B. Comparison of Tables 1A and 1B shows that the sands themselves do not reflect this difference. Rocks associated with the sandstones in Table 1B are thought to have a volcanic component vs. those in Table 1A.

Figure 12. Photomicrograph (crossed polars) of a coarser-grained sandstone lense. Quartz in this sample is variable; extinction varies from undulatory to straight, and some lithic grains are polygonized. Plagioclase, sandstone, and volcanic fragments also are present.

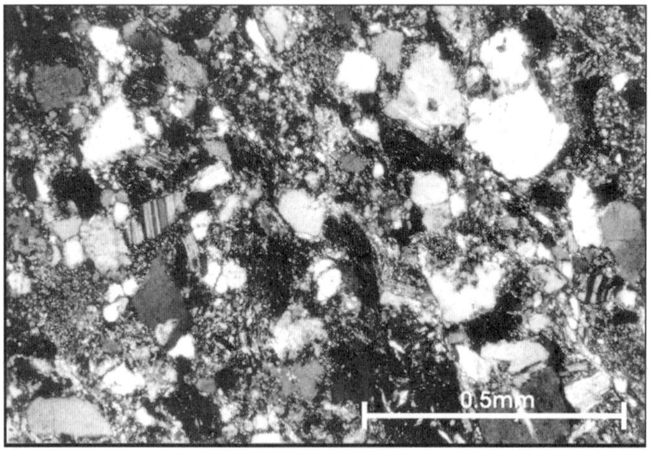

Figure 13. Photomicrograph (crossed polars) of wacke with grains of plagioclase (note albite twinning), quartz, both volcanic and sedimentary lithic fragments, and a few grains of K-feldspar.

Figure 14. Photomicrograph (crossed polars) of wacke with abundant sedimentary and volcanic(?) lithic fragments.

Figure 15. Example of "broken formation" in which what probably were once-continuous layers of fine-grained sandstone in a shale matrix have been dismembered and pinched out. Note overturned slump(?) fold in detached sandstone layer, upper left corner. Field of view ~3.5 m.

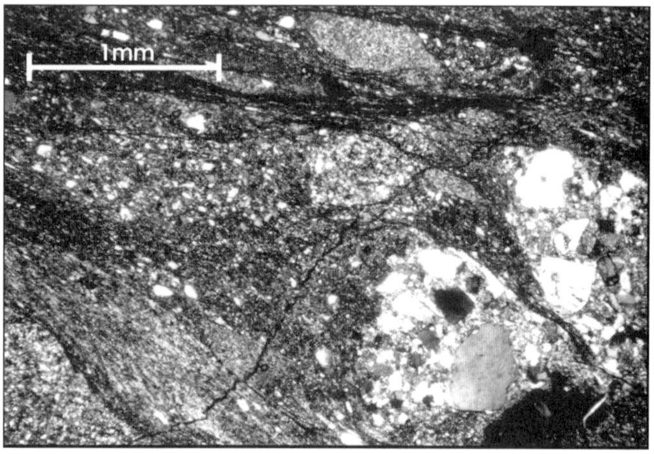

Figure 16. Photomicrograph (crossed polars) of "broken formation" with grains and lenses of siltstone and sandstone contained within a foliated, fine-grained matrix.

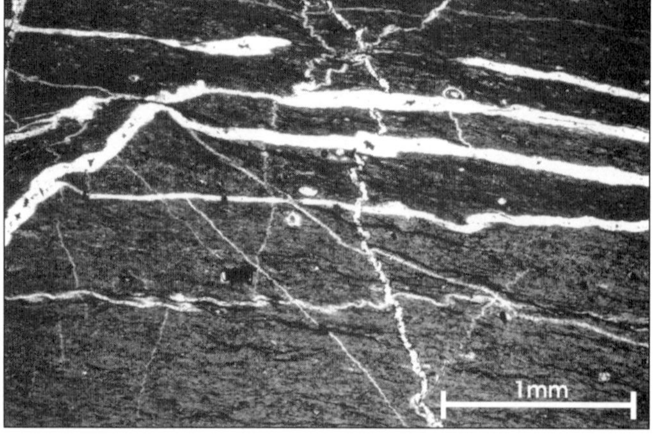

Figure 17. Photomicrograph (plane light) of black chert cut by quartz veins, some of which display ptygmatic folds. Chert contains small amount of clinochlore.

sandstone that occur as inclusions within finer-grained matrix. The clasts of sandstone typically are composed of quartz with some trace amounts of clinochlore and white mica between grains. Generally the fragments are grain supported although some are matrix supported. Grain shapes range from angular (most common) to subrounded. Almost all quartz grains show undulatory extinction. Also present are irregular-shaped, angular grains of plagioclase commonly with bent twin lamellae. One clast has layers of fine- and very fine-grained sand. Many sandstone fragments have a slight foliation indicated by common extinction of sheet silicates and by thin seams of iron oxide (Fig. 10).

Sandstone and shale clasts clearly were derived from the rock that contains them, but other rare clasts of quartzite, granite, and chert are exotic. Exotic clasts include quartzite with coarse quartz grains showing undulatory extinction and sutured boundaries and a clast consisting of quartz and perthitic feldspar (granite). There are a few fragments of very irregularly shaped chert that are difficult to distinguish from the matrix.

Separating discontinuous layers of sandstone and other disrupted sand bodies is matrix of metashale/siltstone. Typically the foliated fine-grained matrix contains a very fine-grained mixture of aligned, finely crystalline flakes of white mica and very fine grains of quartz. Individual grains are very poorly defined with indistinct borders. Aligned iron oxides or concentrations of organic material also help to define the foliation or cleavage. As in some of the other rock types, there are microcrenulations that provide evidence for a second deformation.

Shale clasts also occur within sand, sometimes as pebble-like fragments, sometimes as irregular sheets. Some shale contains irregular fragments and stringers of sandstone, as though they had been mixed and smeared, probably during soft-sediment deformation of sand beds.

Blocks (Or Inclusions)

Chert

Blocks of well-bedded chert may be more than 100 m in maximum exposed dimension. Such masses of chert may form isolated hills; chert is the most prominent and resistant element of Cerro El Pedernal (Fig. 3). Isoclinal and chevron folds (Fig. 6) are conspicuous in these exotic cherts. These folds were formed prior to deposition in the Taray, as shown by the strong discordance among bedding attitudes. Bedded chert is white to light gray to black and weathers brownish-gray to pale buff. Beds typically are 1–5 cm thick but may be up to 1 m. Black chert occurs as thin beds in folded metashale at a few locations. At one, chert occurs as discontinuous, 10–200-cm-long, 1–3-cm-thick bed-like clasts that pinch out or end abruptly. Some of these clasts are isoclinally folded. In thin section, chert shows typical texture with zones or veins of quartz (some of which are ptygmatically folded and/or offset on small faults; Fig. 17). Thin beds of slaty shale that may occur between chert beds range from ~0.5–5 cm thick. The contacts between the shale and chert range from very sharp to gradational. Sharp contacts have iron oxide concentrated along them, whereas iron oxides, though concentrated relative to the rest of the rock, are more disseminated near gradational contacts.

Clinochlore and muscovite are detectable with XRD in all chert samples, and finely disseminated iron oxide is common in some chert samples. Small, somewhat rounded particles of iron oxides are less common, and fractures may contain iron oxide.

Some chert has a volcanic affinity as indicated by possible relic shards and rare angular fragments of volcanic rock. Textural properties suggestive of clastic origin in some cherts are consistent with the volcanic affinity. For example, lamination in at least one sample is due to variation in grain size. Some laminae show fine cross-bed–like texture as though some laminae were eroded before the next was deposited. This is especially noticeable where there are small concentrations of slightly coarser chert contained within a finer-grained matrix. In some cases, these coarse ovoids are truncated by the next bed. In some layers, shreds of white mica are abundant and define a bedding-parallel foliation. We found probable radiolaria in only one thin section, but did not search exhaustively.

Composite Blocks

Some large, matrix-poor blocks are lithologically and structurally heterogeneous, containing volcanic, carbonate, chert, or siliciclastic sedimentary rocks that bear no structural relation to one another. The most conspicuous of these forms Cerro El Pedernal (Fig. 3; Klein et al., 1990), which is capped by a block of chert several hundred meters long. On the north side of that hill, there are a few poor outcrops of highly altered volcanic and volcaniclastic rocks including a black, laminated, water-laid (?) tuff, a coarse clastic rock that we interpret to be a submarine flow of debris, and an enigmatic laminated ankerite-magnetite rock.

Layering in the tuff is due to differences in grain size and relative amounts of sheet silicates and quartz. The sheet silicates are clinochlore and a yellowish biotite; they occur as masses of very small, diversely oriented, indistinct crystals (Fig. 9).

The debris-rich unit is a dark-gray to black rock consisting of a variety of dark fragments. The mass is so altered that it is difficult to identify rock types, even in thin section. Zones or domains with different textures are evident; we infer the presence of sedimentary rocks, including chert and mafic slate, based upon grain size and the abundance of mafic material. Chlorite is locally abundant. Some clasts are replaced by very small, poorly defined, yellowish-green grains of a sheet silicate. In these same zones, there are curved, angular pieces of quartz that are reminiscent of volcanic shards.

Another rock in the Pedernal assemblage is a black, thin-bedded magnetite-ankerite rock. Beds are 0.5–2 cm thick and are graded in terms of mineral percentage. In thin section, the contact between nearly pure ankerite and nearly pure magnetite is sharp. The magnetite near the contact consists of poorly defined grains that are ~0.02 mm in diameter (the same size as the ankerite in this iron-rich part of the sample). In reflected light the iron oxide is milky, suggesting the presence of titanium; XRD indicates it is magnetite. Upward(?) there is a gradational change with a decrease in magnetite and an increase in amount and grain size of carbonate. Near the presumed top of a bed, ankerite comprises ~95% of the rock and consists of crystals or zones of relatively homogenous carbonate in which the individual grains or particles are ~0.2–0.3 mm across, and in which there are some magnetite grains between the carbonates. In some parts of the sample there are also small, poorly defined grains of quartz.

On the south side of Cerro El Pedernal, there are a few small outcrops of diabase, which we interpret as another block in the mélange. The diabase is highly altered, but the original diabasic texture is easily recognizable. Plagioclase is altered to sericite and some calcite and Fe-Mg silicate minerals have been completely replaced by chlorite and calcite. On the southeast corner of the hill, there is an outcrop of proximal turbidite wacke with 15–30-cm-thick beds of sand interbedded with 2-cm-thick beds of shale. The beds are overturned, as evidenced by burrows and sole markings. Because it is unlike other parts of the Taray, we consider it to be part of an exotic block.

Another large composite block contains La Cejita talc mine (Fig. 3). The mine consists of a number of small pits dug into light greenish-yellow to dark greenish-gray zones. Optical and XRD analysis of five samples from these pits indicate that clinochlore is abundant and talc is not; presumably the talc has been mined out. Also in the mine areas are scattered, rusty, siliceous (veined) dolomite blocks up to 10 m or so in exposed diameter. In thin section, abundant dolomite occurs as closely packed crystals or as crystals that are completely surrounded by magnesite (Fig. 18). Dolomite crystals are randomly distributed or concentrated along certain zones. The magnesite is finely crystalline and massive and was identified by XRD methods. Scattered throughout are grains of quartz that appears to be detrital. Talc and clinochlore are minor constituents.

Figure 18. Photomicrograph (crossed polars) of dolomite-magnesite-talc rock from the La Cejita talc mine. Photo shows subhedral crystals of dolomite in matrix of finely crystalline magnesite and talc. Also present are quartz and clinochlore.

Figure 19. Photomicrograph (crossed polars) of mixed volcanic and carbonate rock from ~150 m east of the La Cejita talc mine. Dark areas are mainly clinochlore, light areas are dolomite-magnesite; scattered quartz grains are also present.

Northeast of the mine and exposed along a ridge, dolomitic rocks have a variety of textures, some of which, mesoscopically, appear as though the dolomite contains volcanic clasts. In these rocks, dolomite and magnesite (not distinguishable in thin section) occur as very irregular-shaped patches and aggregates of anhedral grains, intimately mixed with small amounts of chert-like material of variable grain size (Fig. 19). There are also less common patches of clinochlore and fractured, equant grains up to 0.5 mm across of an opaque mineral. The texture and mineralogy of this rock are consistent with our field interpretation of a mixture of volcanic material and carbonate, and it is possible that metasomatism of a volcanogenic parent produced this rock.

Blueschist and other rocks containing other high-pressure minerals, and peridotite and related ultramafic rocks, are not known from Taray Formation. Despite the absence of these lithologies common in occurrences of mélange, some blocks in the Taray are exotic and indicate contributions from multiple sources. One source contains fossiliferous limestone of late Paleozoic age; another provided chert that probably was deformed as soft sediment prior to inclusion in the Taray. The few blocks containing volcanic rocks among siliceous strata are consistent with submarine volcanism. At Cerro El Pedernal, the association of chert with volcanic rocks and a laminated ankerite-magnetite rock suggests a submarine environment that may have included active hydrothermal vents. The turbidite near Cerro El Pedernal likely accumulated offshore, below wave base, perhaps in deep water. The mixed volcanic-carbonate rock near La Cejita mine may have been deposited in shallow water with an active volcano nearby.

METAMORPHISM

Several small plutons of Tertiary(?) granite intrude the Taray and nearby Nazas Formations near the southwest limit of the San Julian uplift, including the one that forms Pico de Teyra

(Fig. 3). Contact metamorphism in the vicinity of Pico de Teyra has formed fine-grained spotted schist; the contact metamorphic aureole is tens to hundreds of meters wide. The schist has a lepidoblastic texture due to white mica and porphyroblastic texture due to small (0.03 mm) porphyroblasts of andalusite and biotitic mica. In places, as the contact is approached, quartz veins and dikes are present and the Taray appears to be somewhat silicified and hydrothermally altered. Elsewhere, the degree of metamorphism is less and does not seem to be related to known plutons.

Regionally, the shale-siltstone protolith is of low greenschist facies. Away from intrusive rocks, the only metamorphic minerals noted in thin section are chlorite and white mica. Most chlorite is clinochlore IIb, as indicated by brown anomalous interference colors, length fast character, and XRD, but in a few thin sections, two chlorites were noted, one with brown and the other with blue anomalous colors. The fine-grained white mica is muscovite, based on XRD. Some larger grains of muscovite and biotite probably are detrital.

Metamorphic grade near the La Cejita mine is somewhat higher than in surrounding rocks, and it is assumed that the mineral assemblages are products of contact metamorphism (there are small intrusions in the area, although no plutonic rocks are exposed at the mine). Microscopic and X-ray studies reveal the following mineral assemblages: clinochlore-Fe-oxide, quartz-clinochlore, quartz-clinochlore-albite, dolomite-talc-Fe-oxide-clinochlore, and dolomite-magnesite-quartz-talc-(clinochlore). The talc-bearing rocks are interpreted as siliceous magnesite-dolomite rocks that are partially replaced by talc.

STRUCTURE

Overview

Planar structures are the principal structural elements recorded in outcrops of Taray Formation (Fig. 20). Planar features include the following: (1) S_0 surfaces of sandy layers (Fig. 21A). (The term "surfaces" is used rather than bedding to indicate our uncertainty about them as primary depositional features; many parallel planes [see C foliation below and Fig. 21C] record movement.) (2) S_1 and C, principal foliations in argillaceous matrix (Fig. 21B–C). (Indications of slippage distinguish C planes that may be parallel to fragment surfaces or that steepen and ramp gently across layers, which commonly show offset.) (3) Axial surfaces of folds (AP_1 and AP_2). (4) Locally developed cleavage (S_2) that is axial planar to folds defined by S_1 (Fig. 21D). (5) Faults along which gouge and/or breccia occur. (These faults record brittle deformation after lithification and are separated from structures that may have formed when the rocks were ductile, perhaps not fully lithified.) (Fig. 21E). (6) Steeply dipping fractures within sandy rocks, commonly filled with quartz. Most planar features strike easterly to northeasterly and dip south. Filled fractures (tension gashes) strike northwesterly.

Linear structures include fold hinges (F_1, F_2), slickenlines (L_f), and intersection lineations (L_{1x0}) (Fig. 21F).

Fabric elements measured in blocks of cherty rock that are encased in matrix are generally distinguished on the data plots, because we were unable to determine unambiguously how they related to deformation common to matrix and aligned fragments. Structures, such as isoclinal folds, within blocks of cherty strata are generally discordant with foliation and folds in matrix, however, exceptions exist.

Planar Structures

Bedding

In general, the measurements of bedding are orientations of the bounding surfaces of tabular fragments of sandy beds against matrix (Fig. 21A). Some beds clearly occur within blocks surrounded by matrix; other beds form seemingly coherent sections but are very likely part of a large block, the boundaries of which are not exposed or recognized. Most sections of interbedded sandstone and mudstone record progressive dismemberment and are probably not part of a block (Fig. 15). Initial attenuation is suggested by subtabular beds that pinch and swell or by those that show boudinage. Numerous small faults that generally show normal separation further dissect the beds into lenses and slivers (Fig. 22). Stratal disruption yields an end product of isolated, aligned inclusions of sandstone, commonly mixed with volcanic and other clasts that are encased in a matrix of somewhat scaly, cleaved, crenulated mudstone (Fig. 23). The oriented slabs of bedding define a "fragment foliation" (Cowan, 1985) consisting of subparallel surfaces, although they are not as well oriented as foliation in matrix. In the northernmost part of the principal Taray exposure, where foliation is not pronounced, well-developed fragment foliation strikes northeast (Fig. 21A).

Some features of deformed beds suggest penecontemporaneous deformation. This type of deformation is recorded by rare, isolated, large boulders of well-cemented sandstone that slid downslope, crumpling sediments at their bows (Fig. 4). An exposure of one of these large blocks, ~20 m long, clearly shows a convex upward, striated base distinguished by sheared and broken shale with thin interbeds of sandstone. Above the striated surface, the base of the coherent block is underplated by pieces broken and corraded from the block. Beneath the underplate, a mixture of pebbles and cobbles of sandstone and black shale with lenticular inclusions of thin-bedded gray shale were overridden. Some of the thin beds are crumpled, and crenulations with hinges parallel to striations at the base of the slide block are observable. Irregular fractures in this horizon are filled with fine-grained sediment. At the southwest bow of the boulder, a sandstone bed is imbricated and folded. A set of quartz-filled fractures in the sandstone boulder is more-or-less orthogonal to slickenlines on the slide surfaces and to the direction of transport (Fig. 21F). The fractures indicate that the boulders were sufficiently coherent during or after sliding for brittle deformation in response to extension or stretching to occur .

Penecontemporaneous intraformational deformation is also inferred by the small, strongly overturned folded layer detached

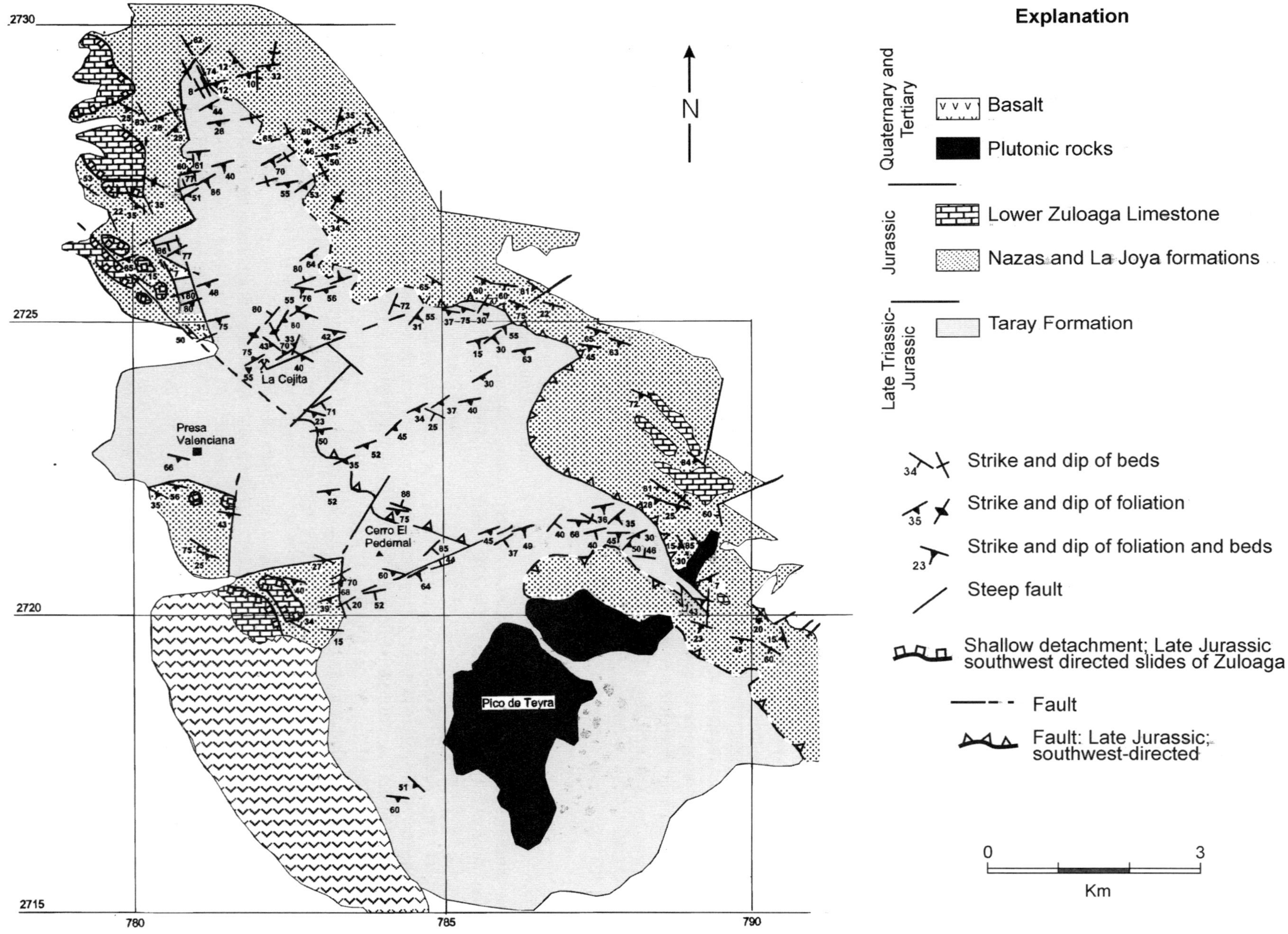

Figure 20. Map of Taray Formation showing bedding and foliation.

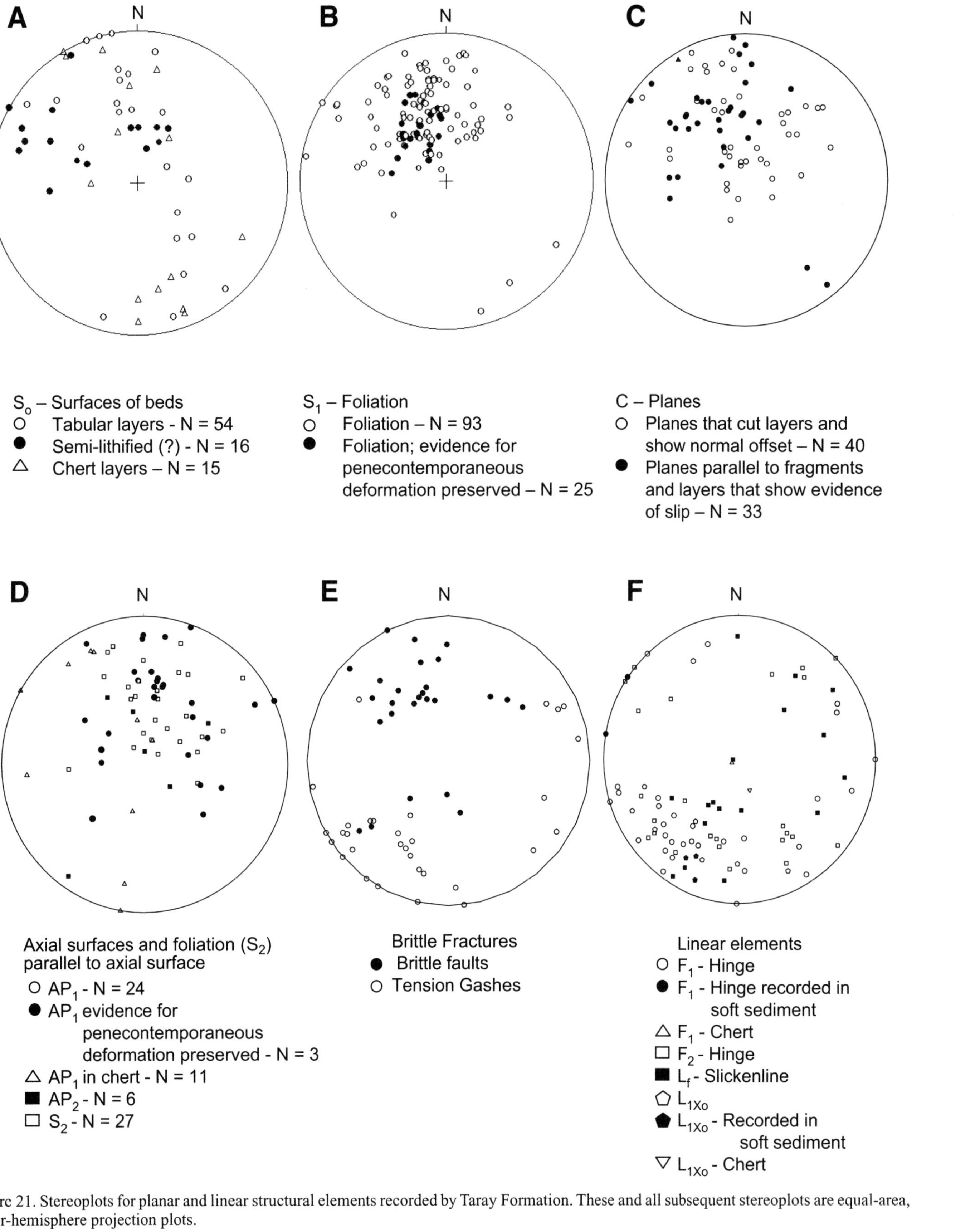

Figurc 21. Stereoplots for planar and linear structural elements recorded by Taray Formation. These and all subsequent stereoplots are equal-area, lower-hemisphere projection plots.

Figure 22. Low angle fault cutting layers below at a high angle and nearly parallel to bedding above. Rock appears to be altered ash-tuff and chert. Hammer length = 36 cm.

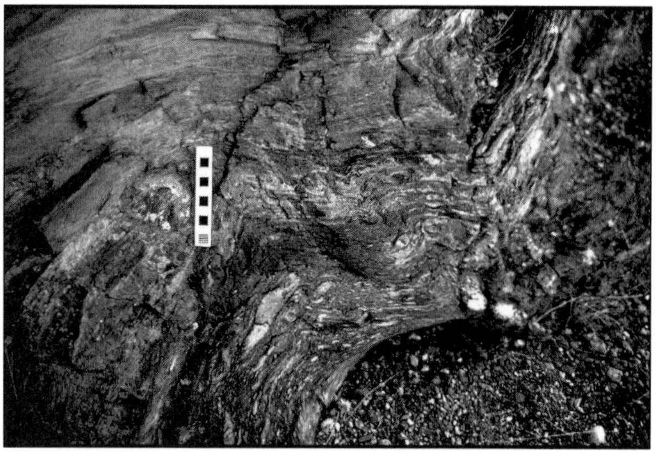

Figure 23. Dismembered, attenuated sandstone in dark, muddy matrix exposed in the crest of a small antiform. Scale length = 10 cm.

from unfolded lower beds (Fig. 15). The thinner, underlying beds are segmented at ramps where faults parallel to layering cut across them. The exposures suggest unconstrained downslope movement followed by similarly directed segmentation during subsequent stretching.

Foliation and Other Spaced Planar Features

Foliation is most conspicuous in the fine-grained matrix but is also evident in some coarser sandy layers that may record spaced cleavage with dips steeper than typical S_1 (Fig. 21B). Strike is parallel to that of bedding. The general dip is gentler than that of bedding; however the dip may vary from nearly vertical to 15°SE (Fig. 21B). Within the matrix, S_1 fabric is accentuated by strong alignment of tabular, lenticular and oblate fragments (Fig. 15). Although prominent, S_1 foliation is not strongly planar and may exhibit common deflections around clasts and inclusions (Fig. 4). S_1 is also commonly tangential or subparallel to gently dipping surfaces that accommodate slip (C planes) as shown by slickenlines or offsets, principally of bed surfaces. Together the foliations resemble phacoidal foliation (Figs. 15, 23; Kusky and Bradley, 1999) composed of cleavage (S_1) and slip surfaces (C planes) (Fig. 21C). Where beds are preserved, C planes commonly accommodate normal displacements of individual layers that indicate extension in the matrix. In places, gash fractures filled with quartz cut the foliated matrix. Distinguishing among anastomosing planar elements with respect to their origins is difficult.

The planarity of foliation may be further modified and obscured by folds (F_2) (Fig. 23) and/or crenulations defined by S_1 foliation. An axial planar S_2 cleavage is developed in the hinge regions of some F_2 folds.

Brittle Faults and Fractures

Brittle faults, marked by breccia and/or gouge, are less well oriented than striated surfaces grouped with the C surfaces (Fig. 21E). In general they strike somewhat more easterly and are slightly steeper. However, some faults fall within the loose cluster of foliation planes and shear surfaces. Where faults cut sandy layers, normal displacements are recorded.

Fractures, commonly lenticular and filled with quartz, are best developed in resistant sandstone blocks and boulders, although they may occur in fine-grained matrix. They are orthogonal to the trend of striations, slickenlines, and hinges of crenulations, and to the dip of bedding and S_1 and C foliations.

Linear Features

Linear features include fold hinges recorded by beds (F_1), hinges of crenulations (F_2), intersection lineation (L_{s1xso}), striations and slickenlines (L_f) (Fig. 21F). Most lineations and fold hinges are co-linear with gentle plunge.

Mesoscopic and Macroscopic Folds

Folds are evident in matrix (Figs. 23 and 24) and large blocks; folds in the latter are commonly isoclinal (Fig. 6). Rare exposures show folds in detached layers (Fig. 15). The distinction between structures formed before sediment was lithified and those formed later is commonly not clear. Confusion arises because many planar structures, i.e., bedding and foliations, are subparallel, and large blocks of fractured (lithified?) material are entrained in foliated shale that we interpret to have been relatively soft, pliable and wet (?) during some part of deformation. We conclude that some folds formed in unlithified sediment.

Environment of Deformation as Suggested by Structural Style

Rocks of Taray Formation that are: (1) disrupted, fragmented bodies of interbedded sandstone and shale that lack coherent stratigraphy, and (2) argillaceous masses containing

Figure 24. Chevron folds with gently dipping axial planes. Plier length = 20 cm.

exotic blocks are comparable to types I and III mélange (Cowan, 1985). Type I mélange is distinguished by sequences of sandstone and mudstone that show variable disruption and fragmentation, principally in response to layer-parallel extension. Although interbedded mudstone and sandstone are the dominant lithologies of Taray Formation, continuous layers of sandstone are restricted to outcrop-scale occurrences and are not preserved as mappable units. Sandstone beds show progressive dismemberment, ranging from tabular layers preserved on the scale of an outcrop, to beds that pinch and swell, to beds dismembered by boudinage or low-angle faults. Some isolated lenses of sand 10 m or more long are interpreted to record boudinage during gravity-driven, downslope sliding of partially lithified sediment followed by further stretching during tectonic extension. These rock masses contain mesoscopic features comparable to those of tectonosomes as recently described by Pini (1999). Similarities include boudinage and disruption of competent beds of sandstone, siltstone, and calcareous layers by fragmentation into blocks with sharp outlines. The fragments show a common trend and strong preferred orientation parallel to the plane of major extension, which coincides with the surfaces of beds.

According to Cowan (1985), masses of rock that contain inclusions: (1) the origins of which are obscure, (2) the structures of which are discordant and the lithologies of which are exotic to the basin, and (3) that are entrained in foliated matrix, are interpreted to belong to Cowan's (1985) type III mélange. The formation of this type requires a multistage process involving: (1) acquisition of rock fragments from diverse, possibly widely separated sources, (2) dispersion of the inclusions during mixing with matrix, and (3) common bulk deformation involving flattening and development of scaly foliation. The common characteristics of type III mélange suggest emplacement as mud-rich, submarine debris flows or olistostromes coupled with migration and intrusion of these materials during upwelling or diapirism within a shallow, nonmetamorphic part of an accretionary prism; we have no evidence of the latter.

Taray contains blocks of many sizes in fine-grained matrix. The most conspicuous are exotic blocks of chert, several hundred meters long that form isolated hills. One of these hills, Cerro El Pedernal, is shown in Figure 3. Other exotic blocks are composed of carbonate, some of which is fossiliferous, quartzose sandstone, and volcanogenic rocks. Sandstone, siltstone, and calcareous layers interbedded with shale also form blocks, as do meter- to centimeter-sized fragments of the layers, which float in matrix. These bodies of disordered rock composed of diverse blocks with definite shapes and sharp outlines randomly distributed in shaly matrix also fit well with Pini's (1999) description of olistostromes.

We interpret the mesoscopic fabric and lithologic composition of Taray Formation as recording progressive layer-parallel extension of interbedded shale and sandstone leading to disruption and fragmentation compatible with sliding and fragmentation during gravity-driven, downslope movement that culminated in mixing of muddy debris. The processes responsible for the formation of the rocks probably occurred within the high part of an accretionary wedge at a convergent margin (Fig. 25, after Cowan, 1985), or near the thrust front (e.g., Pini, 1999, their fig. 48), although deeper settings are not precluded. We were unable to test for the record of coaxial strain, however, some sliding clearly occurred at depths sufficiently shallow so that strongly noncoaxial shear was unlikely.

The processes that we infer to be recorded by structures contained within Taray are comparable to those occurring within actively developing accretionary prisms such as the Nankai (Chamot-Rooke et al., 1992). Near the toe of the eastern part of the Nankai wedge, layers above gentle thrusts bend downward conformably across topographic slopes as steep as 20° to 30°. Locally, bed dips are steeper than the slope and may reach 50–60°. In the same area, a slump scar, the face of which extends for more than 1.5 km, cuts steeply (in places 45°) across almost 1000 m of section composed of subhorizontal mudstone, interpreted as silty turbidite with interbeds of conglomerate contained within stacked thrusts (Lallemand et al., 1992). Fluid is expelled along shallow detachments that are commonly parallel to bedding, and possibly along steep joints. We believe that these conditions must be conducive to the formation of slides, slumps, and debris-rich matrix characteristic of Taray.

STRUCTURAL AND STRATIGRAPHIC RELATIONSHIP BETWEEN TARAY AND JURASSIC VOLCANIC FORMATIONS

The principal planar structures (beds and foliation) in the Taray Formation strike northeast. Beds and foliation in the overlying Jurassic volcanic section, including Nazas, Rodeo, and Caopas Formations, and higher Zuloaga carbonate strike northwest, placing the strikes of the Taray and overlying Jurassic units at right angles to one another. Foliation and bedding in the Jurassic rocks generally conform to the northwest trend of the Late

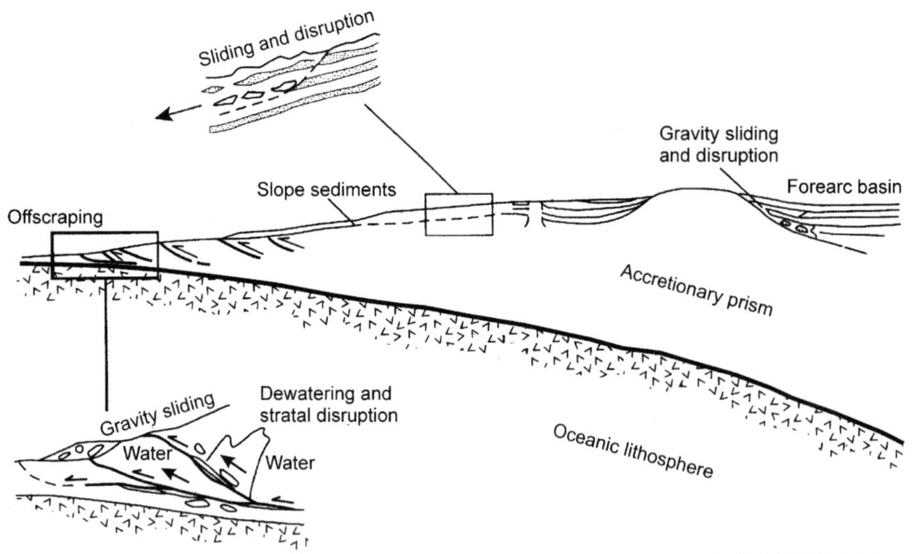

Figure 25. Generalized cross section of an accretionary prism showing possible locations and geologic settings compatible with the interpreted origins of rocks mapped as Taray Formation (after Cowan, 1985).

Jurassic fold that forms the principal structure within the San Julian uplift (Anderson et al., 1991). Well-developed multiple foliations in the Taray indicate a more complex structural history than that recorded by the overlying volcanic rocks.

Despite wide and diligent search, we never found an observable contact between Taray Formation and overlying volcanogenic rock, only covered zones. Córdoba-M. (1964) and López-Infanzón (1986) described the contact as an angular unconformity and our regional structural data (Fig. 21) appear to agree with that interpretation. But other observations indicate to us that the relationship cannot be so succinctly characterized.

The San Julian uplift contains both an early to middle Mesozoic mélange and the rocks of a Jurassic arc. This temporal compatibility of arc and mélange and their geographical and stratigraphic proximity suggest that the two were contemporary members of the same stratigraphic regime. The composition of the mélange adds support to the idea that arc and mélange are related; in the various arroyos that cross the Taray-Nazas contact we saw repeatedly that the parts of the Taray closest to the contact showed an increase in tuffaceous volcanic component, thereby predicting a gradational contact. Because of these relationships (compositional facies, temporal compatibility, and both geographical and stratigraphical proximity), we hesitate to consider the contact to be an unconformity that represents a significant temporal hiatus. We hypothesize: (1) that the zone of contact between forearc and accretionary prism rocks was complex and may have involved accumulation of forearc strata on deformed rock of the prism as portrayed by Cowan (1985) (Fig. 25), (2) that shortening during formation of a Jurassic fold (Anderson et al., 1991) accommodated by low-angle faults resulted in transport of the arc rocks across the forearc and onto the accretionary prism (Figs. 2 and 3), and (3) that the mélange in the footwall of the thrust(s) was largely unaffected by the Jurassic deformation.

Metamorphism is higher grade in the Taray, but the difference may be more apparent than real, because Taray offered a more reactive protolith and it underwent local Tertiary contact metamorphism. Further, the Taray metamorphism is no higher than chlorite-grade (except in the vicinity of Pico de Teyra pluton) and Nazas rocks contain chlorite and white mica in their matrix, suggesting that Nazas also is low greenschist facies.

RESTORATION OF SOME EFFECTS OF SUPERPOSED DEFORMATION: IMPLICATIONS FOR THE PROPOSED TECTONIC SETTING OF TARAY FORMATION

Deformations that occurred after the formation of Taray syndepositional structures include: (1) Late Jurassic folding about northwest-trending hinges during southwest-directed contraction, and accompanying detachment, recorded by beds of the Oxfordian(?) lower Zuloaga limestone and facilitated by interbeds of gypsum (Anderson et al., 1991); (2) north-directed, Late Cretaceous folding above a second detachment also within evaporite-rich strata of the Zuloaga (Tardy, 1980); and (3) Tertiary normal faulting and associated tilting related to northeast-southwest extension (Anonymous, 1981) (Figs. 1 and 2). The faults and detachments and the directions of tectonic transport that we infer from the mapped geological relationships described above are shown on schematic structure sections (Fig. 26). Tertiary tilting is recorded by the 20° eastward plunge of fold hinges in Cretaceous strata along the northeast flank of the San Julian uplift (Fig. 2). The relationships are shown by Tardy (1980, Fig. I_{3-9}), who notes, as have others, the pronounced detachment of Late Jurassic and Cretaceous strata above Oxfordian gypsum. Regionally, east-trending Cretaceous folds with subhorizontal hinges that formed above the detachment are conspicuous topographic and structural features along the great westward bend of

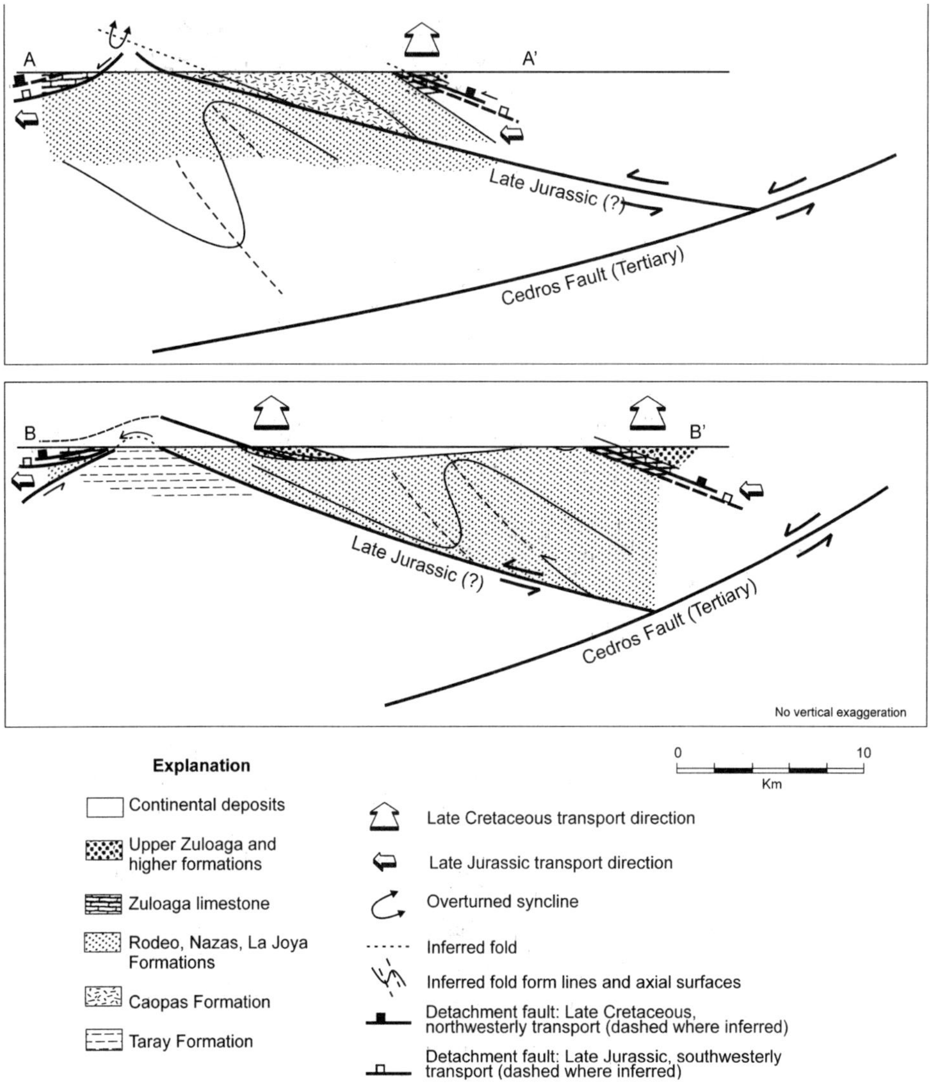

Figure 26. Schematic structure sections (along lines shown in Fig. 2) that illustrate the faults and detachments and the directions of tectonic transport that we infer from the mapped geological relationships. Tertiary tilting is recorded by the 20° dip of postulated planar surfaces toward the inferred Cedros fault. Three deformational events post-formation of the Taray mélange are emphasized in the sections: (1) Late Jurassic folding about northwest-trending hinges during southwest-directed contraction, and accompanying detachment, recorded by beds of the Oxfordian(?) Lower Zuloaga limestone and facilitated by interbeds of gypsum (Anderson et al., 1991); (2) north-directed, Late Cretaceous folding above a second detachment also within evaporite-rich strata of the Zuloaga (Tardy, 1980); and (3) Tertiary normal faulting and associated tilting related to northeast-southwest extension accommodated by the inferred Cedros fault (Anonymous, 1981) (Fig. 2).

the Sierra Madre Oriental. Tardy attributes the eastward plunge of the folds exposed along the northeast flank of the San Julian uplift to tilting during northeast-directed Miocene contraction. Tilting also produced the large northwest-trending anticlinorium manifest as the core of the uplift.

We have argued that deformation of the northwest-trending core of the San Julian uplift is pre-Cretaceous, as indicated by the pronounced deformation of Late Jurassic strata that is not recorded in Cretaceous rocks (Anderson et al., 1991). We agree that the eastward tilt likely occurred during the Tertiary but attribute it to rotation in the hanging wall above a major northwest-striking listric normal fault that occurs in the valley northeast of the uplift (Tardy, 1980; Anonymous, 1981; Fig. 27). The fault disrupts the continuity of the Cretaceous and younger folds as shown on the regional geologic map.

So that we may further assess the relationship between rocks of the Nazas-Rodeo-Caopas volcanic suite (Jones et al., 1995) and the underlying Taray Formation, and to better interpret the kinematic significance of the structures recorded by Taray Formation, we restore the rotation of the hanging wall above the inferred southwest-dipping Tertiary normal fault (Fig. 27). We assume that the plunge shown by Cretaceous folds is the result of Tertiary tilting, and measure the plunge angle of 20° as shown by Tardy (1980). The plunge is restored to horizontal by counterclockwise rotation of 22° about an axis plunging 0°, 330°, parallel to the inferred fault trace. The sense and magnitude of rotation are based upon: (1) down-to-the-southwest normal offset along the fault northeast of the San Julian uplift and (2) sufficient rotation (22°) about the axis trending 330°, which is not orthogonal to the easterly trending Cretaceous fold hinges, to restore the 20° plunge to horizontal.

The effect of the rotation upon the structural elements of the Taray Formation is marked (Fig. 28). Bedding, S_0, and foliations, S_1 and C, are parallel and strike northeast. Each group of planes

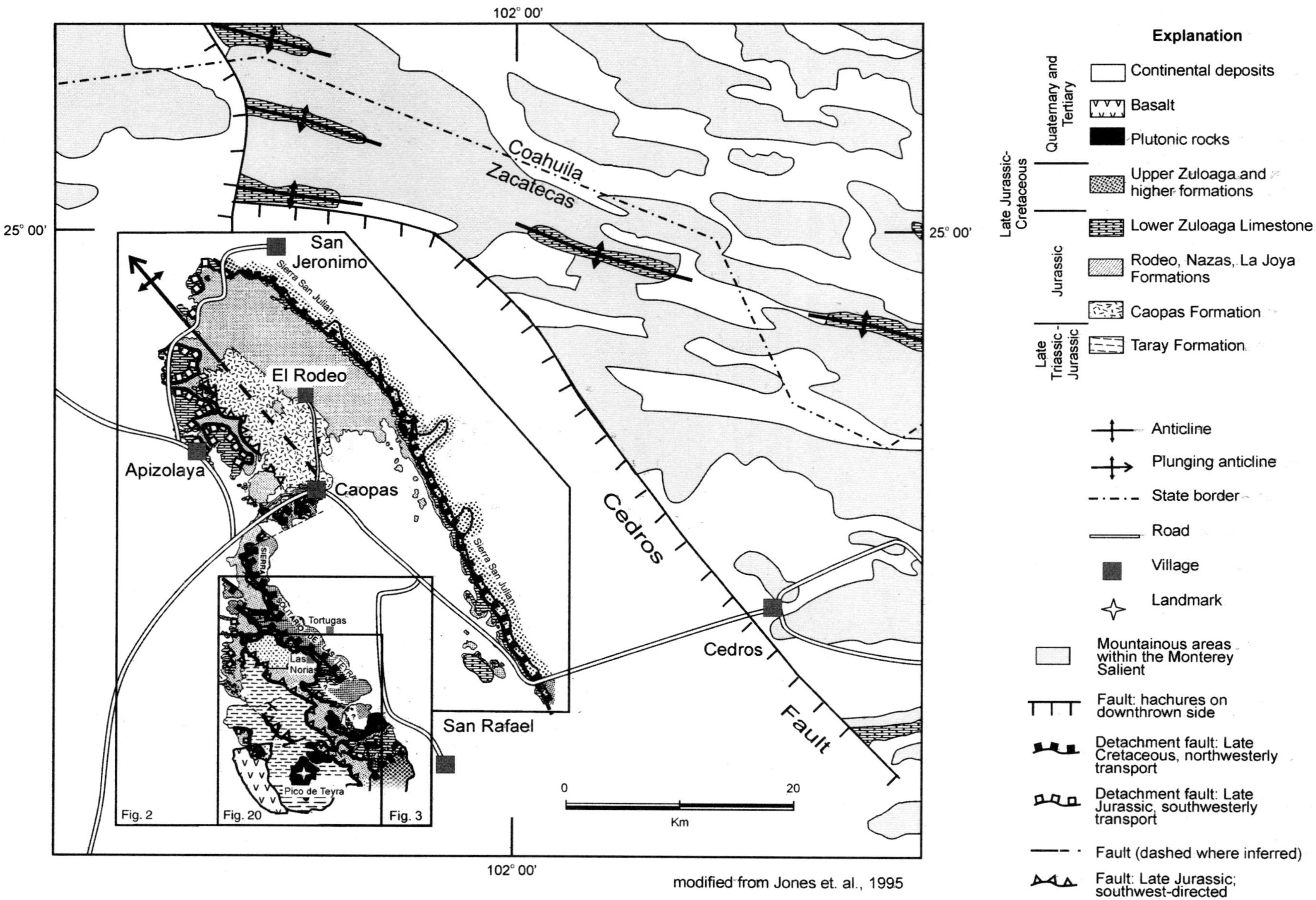

Figure 27. Regional map showing the relationship of San Julian Uplift to Cedros normal fault.

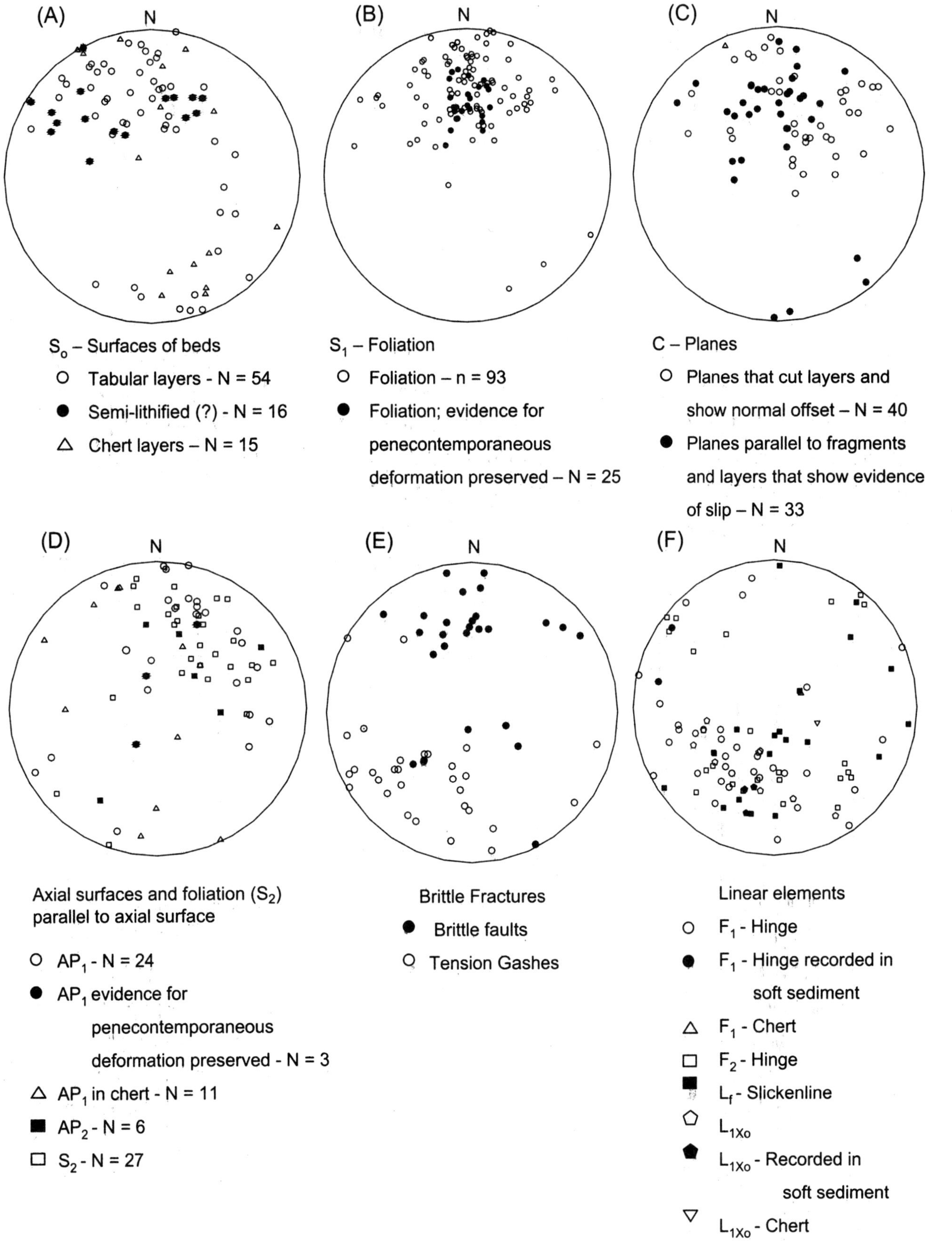

Figure 28. Stereo plots of restored planar and linear structural elements recorded by Taray Formation. All structures have been rotated 22° counterclockwise, as viewed northwest, about an axis plunging 0°, 330°.

dips southeast: S_0, 58°; S_1, 44°; and C, 37°. Many axial planes lie among these planes and they dip the same direction although only 26°. Most linear elements, including fold hinges, intersection lineations, and fault-related striations plunge shallowly southwestward downdip, orthogonal to the strike of planar elements. Extensional fractures (tension gashes) filled with quartz are almost vertical and are mutually perpendicular to the strike of bedding and foliations and the subhorizontal linear structures. The orientations of planar and linear structural elements, restored to positions that we speculate may be close to initial strikes and dips, could accommodate downslope movement of semilithified sediment.

We also corrected northwest-trending folds recorded by overlying Nazas and lower Zuloaga Formation for the effects of Tertiary tilting (Fig. 29). The modest effect shows that the Late Jurassic folds, beneath the detached, east-trending Cretaceous, were asymmetric and verged southwest. We did not restore the effects of Late Jurassic folding upon Taray Formation. We suspect that low-angle faults at the contact between Taray and overlying volcanic rocks accommodated Late Jurassic detachment, related telescoping, and emplacement of the volcanic section across an assumed forearc basin and onto the accretionary prism. Our interpretation also assumes that: (1) the texture and structure of Taray developed as part of a mélange within an accretionary prism, and (2) the mélange most likely formed during Middle Jurassic time (Anderson et al., 1990). Unfortunately the age of Taray may be bracketed only between the late Paleozoic, based upon the age of the youngest detrital zircons reported from sandstone in the formation (Diaz-Salgado et al., 2003), and the age of the Nazas, which is probably no younger than 158 ± 4 Ma (Oxfordian), the age of the Caopas Formation (Jones et al., 1995). This idea remains as a hypothesis in need of further testing.

CORRELATION

Taray Formation shares similarities of composition, lithology, and structure with other northern Mexico, early Mesozoic, marine clastic units: (1) the La Ballena Formation, exposed at Peñon Blanco, San Luis Potosí and Zacatecas (Fig. 1; Labarthe-Hernández et al., 1982), and at Charcas, San Luis Potosí (Fig. 1; Tristán-González and Torres-Hernández, 1994); and (2) the Zacatecas Formation, which is exposed at Zacatecas, Zacatecas (Fig. 1; Centeno-García and Silva-Romo, 1997, and references therein). The La Ballena consists of quartzose sandstone, black shale and conglomerate interpreted as parts of deep marine fans formed from deposits carried by turbidity currents (Silva-Romo, 1993). At Charcas, La Ballena beds record slump-folds and brecciation suggestive of downslope gravity-driven movement. Foliation is commonly parallel to bedding. Late Carnian ammonites in the La Ballena (Cantú-Chapa, 1969; Silva-Romo, 1993) are correlative with those of the Zacatecas Formation at Zacatecas (Burckhardt and Scalia, 1906) that occur within a section of black shale interbedded with quartzose sandstone as well as less common pillow lava and limestone (Ranson et al., 1982; Cuevas-Pérez, 1983; Monod and Calvet, 1991).

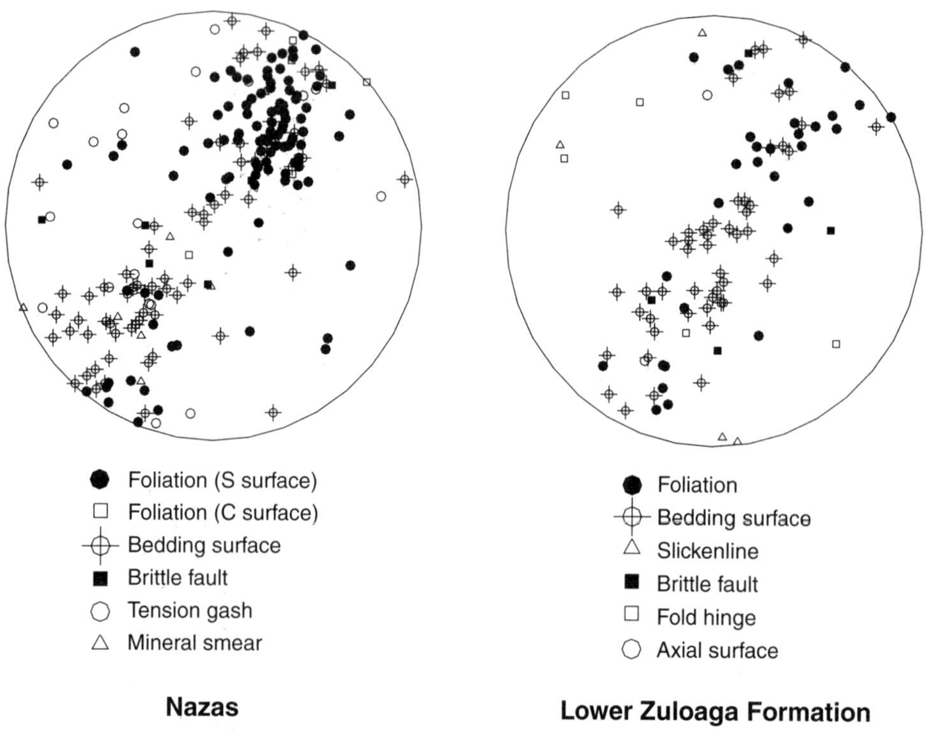

Figure 29. Stereo plots for planar and linear structural elements recorded by Nazas (left) and Zuloaga (right) Formations. All structures have been rotated 22° counterclockwise, as viewed northwest, about an axis plunging 0°, 330°.

Most of the sand in Taray beds is recycled from an orogen, although some sand beds record input from cratonic or dissected arc sources (Table 1; Fig. 30). In the field, we thought we could see an increase in volcanic aspect of the Taray as a whole in some of the topographically higher exposures. However, increased volcanic input is not reflected in the compositions of the sandstones. They indicate a mixed source without a significant component reflecting an active, undissected arc, though we did not sample systematically. Similar to Taray (Fig. 30), La Ballena and Zacatecas Formations (Centeno-García and Silva-Romo, 1997, their fig. 4) plot within the recycled orogen provenance of Dickinson (1985).

These three units (Taray, La Ballena, and Zacatecas) all bear resemblance to certain facies of the Arteaga Complex, which comprises a thick (>1 km) sequence of Triassic-Jurassic rocks exposed near Arteaga, Michoacan (Fig. 1; Centeno-García et al., 2003). In particular, the Varales lithofacies includes siliciclastic units composed of black shale, quartzose sandstone and less common black chert and rare conglomerate, interpreted as channel fill. Predominantly black, dark-gray or dark bluish-gray, 2-cm- to 2-m-thick beds of shale and fine- to medium-grained (rarely coarse-grained) sandstone may be interbedded with thin-bedded black chert or thicker sections of green chert. The sandstone and shale beds are commonly broken and disrupted by the effects of boudinage and may be imbricated with lens-shaped masses, tens to hundreds of meters long in dimension, of pillow lava, green chert, limestone blocks and volcaniclastic rocks. The dark beds of the Varales lithofacies may be interbedded with fine-grained, light-green sandstone and shale of the Jaltomate lithofacies and light-green chert of the Bocana lithofacies. Within the Arteaga Complex, isolated bodies of mafic volcanic rocks with geochemical signatures similar to mid-ocean-ridge basalt (Charapo lithofacies), foliated diorite, gabbro and plagiogranite (Las Juntas lithofacies), and limestone containing chert and crinoid fragments, form blocks and lenses incorporated within the complex during thrusting and shearing. The deformation suggested to Centeno-García et al. (2003) that Arteaga complex may be characterized by broken formation and block-in-matrix texture—deformation styles associated with mélange.

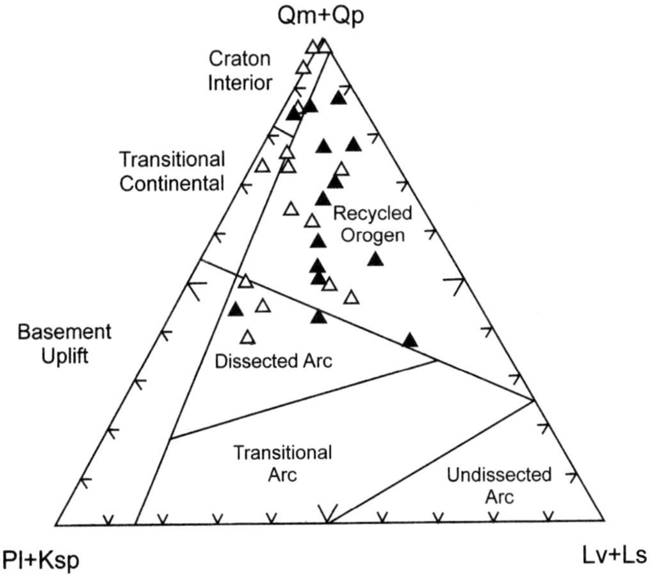

Figure 30. QFL (Quartz-Feldspar-Lithic fragments) diagram after Dickinson (1985) showing provenance categories for sandstones. See Table 1 for meanings of Qm, Qp, P, K, Lv, Ls, and M. Open triangles are samples from Table 1A, those that are not associated with rocks with a volcanic component; solid triangles are samples from Table 1B, those that are associated with rocks with a volcanic component.

In addition to lithological similarity, age constraints at Arteaga, based upon the presence of Late Triassic radiolaria (Campa et al., 1982) and detrital zircons (G. Gehrels, personal commun. to E. Centeno-García, as cited in Centeno-García et al., 2003), are also comparable to Taray. Stratigraphic and structural characteristics of the Arteaga Complex are sufficiently similar to those of Taray that it may be interpreted as part of a correlative accretionary prism, although somewhat more oceanward.

Centeno-García and Silva-Romo (1997) recognized the similarities among the mélange-like units but chose to place them in separate terranes: Taray is included with La Ballena Formation as part of Sierra Madre terrane of northern Mexico, distinguished in part by the postulated presence of sialic basement; Zacatecas Formation and rocks of the Arteaga Complex are grouped together as part of the extensive Guerrero composite terrane, distinguished by oceanic lithosphere, that underlies much of western Mexico.

Recently reported studies are the basis for a less straightforward interpretation by Elias-Herrera et al. (2003). They consider Arteaga to have formed as fill in a basin behind an east-facing offshore arc with sialic roots that was intruded by plutons during the late Early Jurassic ~187 m.y. ago. The arc collided against southern Mexico during the Middle Jurassic.

PLATE TECTONIC IMPLICATIONS

The widely separated exposures of Late Triassic(?) marine strata composed of dark shale, quartzose sandstone, chert and matrix-rich debris enclosing blocks of fossiliferous Paleozoic carbonate and mafic, oceanic igneous rocks, suggest that parts of a Jurassic(?) accretionary wedge underlie more than one region of Mexico.

Jones et al. (1995) proposed that Jurassic volcanic rocks, and accompanying Taray mélange, in northeastern Mexico are displaced from correlative units in northwestern Mexico (Fig. 1). We assess the effect of this proposed displacement upon Taray Formation and its probable equivalents, La Ballena, Zacatecas, and Varales, by means of restoration of displacement along the hypothetical Mojave-Sonora megashear, as well as other parallel faults to the south as shown by Anderson and Schmidt (1983). The maps (Figs. 31 and 32) employ oblique Mercator projections derived from the pole of rotation of the Mojave-Sonora

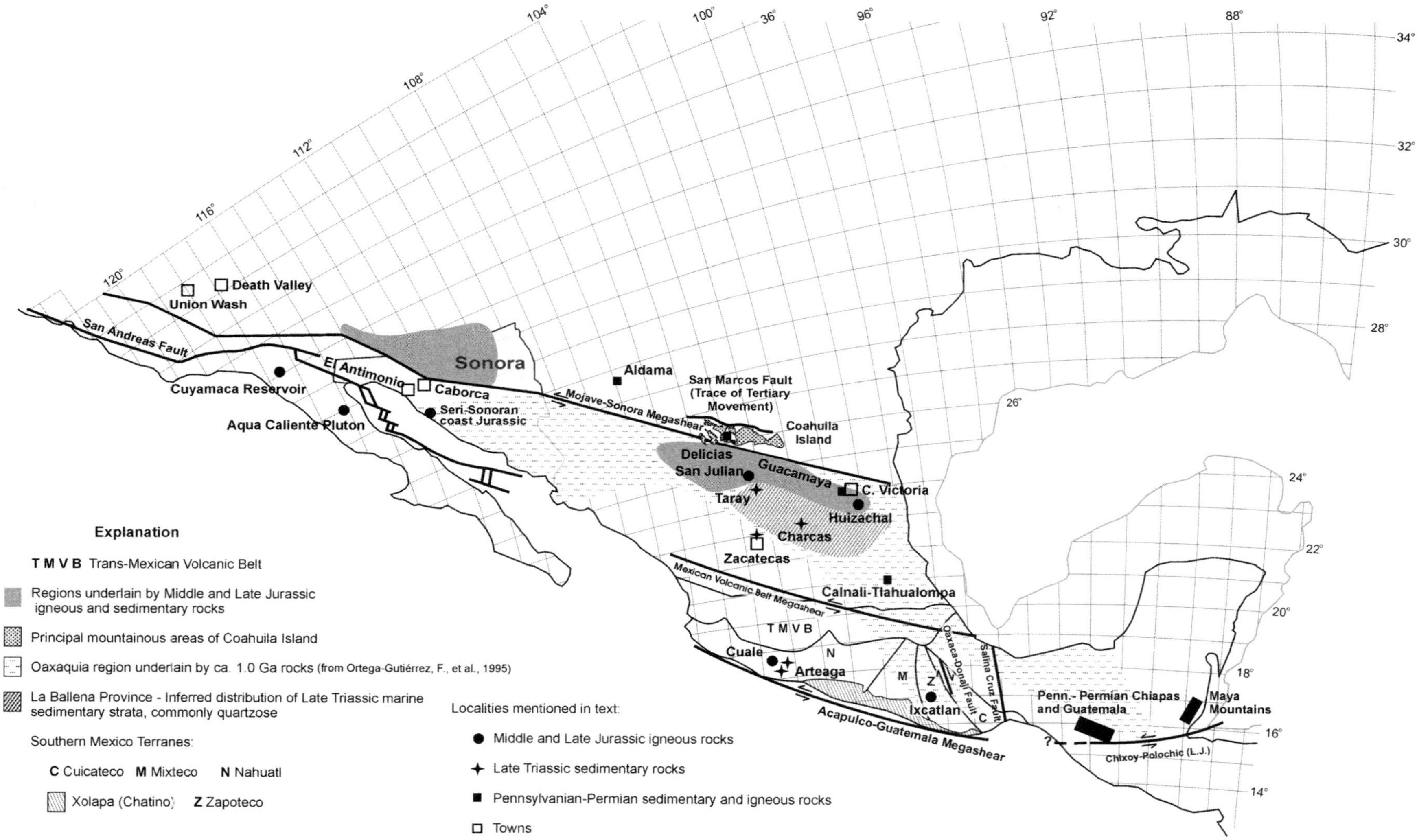

Figure 31. Oblique Mercator projection of map of Mexico showing postulated Jurassic tectonic discontinuities and location of exposures of Jurassic, Triassic, and late Paleozoic rocks.

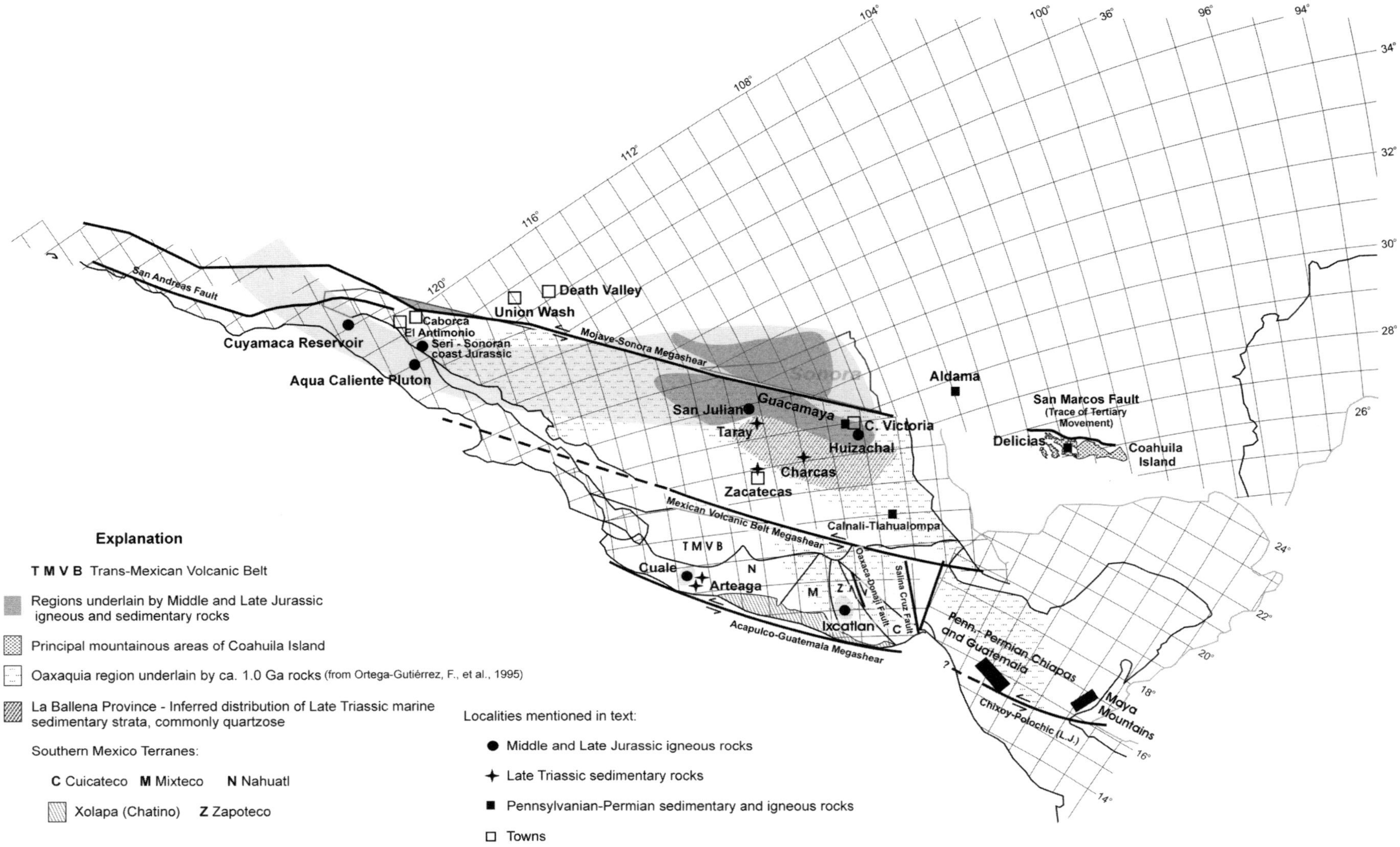

Figure 32. Oblique Mercator projection of Mexico for Middle Jurassic time showing exposures of Jurassic, Triassic, and late Paleozoic rocks restored along principal, inferred Jurassic tectonic discontinuities. See text for discussion.

megashear. As pointed out by Silver (Silver and Anderson, 1974), restoration removes late Paleozoic overlap of South America across much of Mesoamerica.

Restoration of Jurassic igneous rocks of northeastern Mexico (Grajales-Nishimura et al., 1992) to their pre-fault position opposite correlative units in Sonora maintains the oceanward position of the Late Triassic marine rocks relative to the subduction-related, Andean-style, Jurassic igneous province. The Jurassic igneous belt may extend eastward from the San Julian Uplift to Aramberi, Nuevo Leon (Fig. 1), where Middle Jurassic(?) volcanic rocks pinch out against underlying paragneiss. From this point the belt of "arc" rocks must either bend southward or step to the south along a trench-trench transform, perhaps related to the Oaxaca fault (Alaniz-Alvarez et al., 1996) (Figs. 1, 31, and 32). In either case the eastern limit of the Middle Jurassic "arc" likely coincides with exposures of Paleozoic strata above older crystalline rocks exposed in the uplifted core of the Sierra Madre Oriental (e.g., Sedlock et al., 1993).

Exposures of Jurassic igneous rocks that may mark the inboard edge of the accretionary wedge (opposite the Middle Jurassic "arc") are sparse in central Mexico. Chert, rhyolitic and rhyodacitic tuff, conglomerate, breccia, sandstone, and siltstone crop out in the core of El Chilar anticline, east of Guanajuato (Carrillo-Martinez, 1997). Interbeds of siliceous shale contain radiolarians of Middle(?) to Late Jurassic age (Martinez-Hernández, 1979; Chauve et al., 1985). Near Guanajuato (Fig. 1), some samples from mafic and ultramafic rocks (Quintero-Legorreta, 1992) of the basal sequence comprising the Guanajuato arc (Ortiz-Hernández and Martinez-Reyes, 1993) yield Late Jurassic K-Ar dates interpreted as cooling ages of early Mesozoic oceanic lithosphere. However, these results are considered spurious, because excess argon masks the true Early Cretaceous age of this crust (Lapierre et al., 1992).

Farther south is the trans-Mexico volcanic belt, a linear array of active volcanoes that may reflect the presence of a major fault, as suggested by Gastil and Jensky (1973). The volcanic belt and fault (Mexican volcanic belt megashear, Figs. 1, 31, and 32; Anderson and Schmidt, 1983) mark a fundamental terrane boundary. Anderson and Schmidt (1983) treated the rocks south of the Mexican volcanic belt as a single block (Maya West) that they restored northwestward based solely upon geometric constraints stemming from the Bullard reconstruction. Since 1983 few additional correlations have been identified that better constrain the position of the Maya West block. Detrital zircons from Cambrian quartzose sandstone north of the Mojave-Sonora megashear in Sonora show a strong peak (14 of 22 grains) at 1.1 Ga (Stewart et al., 2001). The suggested proximity to Late Proterozoic rocks, such as those that probably form the basement of the Maya West block, may imply further northwestward restoration than inferred by Anderson and Schmidt (1983).

Additional work since 1983 has resulted in recognition of multiple terranes within the Maya West block (Fig. 1; e.g., Campa and Coney, 1983; Sedlock et al., 1993; Centeno-García et al., 2003). On Figure 1, we employ the broader classification scheme of Sedlock et al. (1993), who recognized five terranes in southwestern Mexico. The terranes roughly correspond to principal exposures of the Arteaga Complex (Nahuatl terrane), metamorphosed Paleozoic rocks, e.g., Acatlan Complex (Mixteco terrane), Precambrian Oaxaca Complex (Zapoteco terrane), Jurassic arc (Xolapa Complex within Chatino terrane), and mainly unmetamorphosed, fossiliferous, Upper Paleozoic strata (Maya terrane). The Nahuatl terrane is the southeastern part of the composite Guerrero terrane of Campa and Coney (1983).

The previously discussed Arteaga Complex (Fig. 1; Centeno-García and Silva-Romo, 1997; Campa and Ramirez, 1979; Campa et al., 1982; Campa and Coney, 1983; Centeno-García and Silva-Romo, 1997) is cut by granite plutons, the ages of which fall between ca. 163 Ma (Bathonian) and ca. 158 Ma (Callovian) (Centeno-García et al., 2003, and references therein). Other exposures of Triassic-Jurassic rocks have been discovered northwest of Arteaga, where aluminous schist and chlorite schist, which are probably metamorphosed Late Triassic rocks, comprise the basement at the Cuale mining district (Schaff et al., 2003). There, north-striking faults bound a graben containing rhyolite flows, pyroclastic units, breccia, and conglomerate. Higher in the section, the volcanic units are interbedded with and ultimately buried by black shale, fine sandstone, granule conglomerate, and coarser conglomerate. Dikes and sills that intrude the basin fill are the youngest units. U-Pb interpreted ages from rhyolite low in the section and the intrusive bodies indicate that the basin probably formed shortly before ca. 162 Ma and received sediments and volcanogenic products until at least ca. 156 Ma (Mortensen et al., 2003). Bissig et al. (2003) interpreted the results of trace element geochemistry to indicate that the rhyolites have an "oceanic arc" origin. We offer an alternative interpretation to the idea that the rhyolitic rocks are arc related. The ages and tectonic setting of the Jurassic igneous rocks within the Arteaga Complex and equivalent units are the same as those for pull-apart basins along the Mojave-Sonora megashear (Anderson and Nourse, this volume). Anderson speculates that some of the Jurassic magmatism may not be arc related but records igneous activity through Arteaga crust thinned and broken during transtension related to displacement along the nearby Acapulco-Guatemala megashear (Figs. 1, 31, and 32; Anderson and Schmidt, 1983).

Samples of granitic orthogneiss from the Xolapa Complex within the Chatino terrane yield U-Pb ages between ca. 170 and 158 Ma (Guerrero-García et al., 1978; Ducea et al., 2004). The granitic plutons were emplaced into "basement" composed of Permian and Grenvillian crystalline rocks (Ducea et al., 2003, 2004) and possibly Middle Jurassic sedimentary rocks (Ortega-Gutiérrez and Elias-Herrera, 2003).

A strong tectono-thermal event, recorded in some deep parts of the Acatlan and Oaxacan Complexes at ca. 175 Ma may reflect the effects of an underlying hot spot (Keppie et al., 2003) or be related to arc-related volcanic units that crop out northwest of Oaxaca (Ortega-Gutiérrez et al., 1997). Unmetamorphosed arc-related Jurassic rocks resting on continental crust occur northwest of Oaxaca, where mafic and intermediate tuff and lava interbedded with limestone, conglomeratic sandstone, and limestone conglomerate comprise a section designated as the Ixcatlán arc, which is interpreted as near-shore (Ortega-Gutiérrez et al., 1997). A probable Middle Jurassic age is assigned based upon plant fossils.

In summary, no simple relationship between oceanic rocks, such as Taray, and a Middle Jurassic arc formed in response to subduction of the oceanic rocks is recognized south of the Mexican volcanic belt. Furthermore the Maya East block, composed of Yucatan Peninsula and Guatemala, is awkwardly positioned with respect to Coahuila Island even after restoration of displacements along the inferred Late Jurassic sinistral faults. A determination of a position for this block away from the late Paleozoic collision zone between North and South America awaits further study (cf. Steiner et al., this volume).

CONCLUSIONS

The Taray Formation consists principally of marine sandstone and mudstone or shale that record progressive disruption by layer-parallel extension. Locally, disruption is pervasive. Abundant, large, exotic, predeformed inclusions of diverse shapes, sizes, and compositions contained in argillaceous matrix are an additional characteristic feature. Multiple sources must exist for the blocks, many of which are up to several hundred meters long and are apparently of different ages. The commonly folded inclusions, enclosed in slightly to complexly foliated matrix, cannot have been derived by layer-parallel disruption of bedded sandstone that forms lens-shaped inclusions generally parallel to foliation. Taray lacks inclusions of blueschist facies rocks or ophiolitic rocks.

The lithology, texture, and deformation of these rocks are similar to Cowan's (1985) type I and III mélanges. The principal (penetrative) deformation of the Taray is of the kind most likely to have been imposed on sediment of this kind at a convergent plate boundary along the line of subduction. We interpret the Taray as a mélange and part of an accretionary prism.

The maximum age of the Taray, based on fossils in clasts and on detrital zircons in sandstone, is late Paleozoic. It is either older than the volcanogenic suite (Nazas, Rodeo, and Caopas Formations) that lies above it, or a contemporary of that suite. Thus the Taray may be as young as 158 Ma (latest Middle Jurassic), which is an interpreted U-Pb date obtained on zircons from the Caopas (Jones et al., 1995).

The present contact of Taray mélange with the volcanic rocks of the Nazas Formation marks the discordance of structures within the adjacent units. Nevertheless, the juxtaposition of Taray beneath the volcanic pile permits the hypothesis that this mélange and arc are contemporary facies within the same convergent plate boundary system. The age of the Taray allows that hypothesis. The relatively scarce evidence of volcanism found in the Taray does not support the hypothesis, but does not disallow it. We postulate that the absence of an intervening arc-trench gap, or arc-proximal facies, richer in volcanic rocks and volcaniclastics, is the result of shortening, accommodated by low-angle faults concentrated along the enigmatic contact zone, which brought the volcanic section onto the mélange.

The Taray is lithostratigraphically correlative with the Zacatecas and La Ballena Formations, and with the Varales lithofacies of the Arteaga Complex. These Late Triassic deep-marine units, which may be contemporary with the Taray, are composed of dark shale, chert, and quartzose wacke that may enclose blocks of Paleozoic carbonate and mafic igneous rocks. Each contains some structures commonly associated with rocks formed within accretionary prisms.

Restoration of the mélange-arc suite along the Mojave-Sonora megashear sufficient to remove ~1000 km of left-lateral displacement places it in northwestern Mexico against correlative igneous rocks (Jones et al., 1995). Restoration of ~300 km of left-lateral displacement along a fault coincident with the trans-Mexico volcanic belt (Anderson and Schmidt, 1983) places the Arteaga Complex and associated Precambrian and Paleozoic sialic basement rocks west of postulated Grenvillian rocks of Oaxaquia (Ortega-Gutiérrez et al., 1995).

ACKNOWLEDGMENTS

This work was supported by National Science Foundation grant EAR-8705717 (JWM, NWJ, THA) and by the Faculty Development Fund of the University of Wisconsin–Oshkosh (JWM). R.T. Klein and S.R. Hannula assisted during study of Cerro El Pedernal. Discussions with Jon Blickwede, Fernando Ortega-Gutiérrez, and Elena Centeno-García provided a regional perspective. Scott Davidson carefully drafted the figures and Paul Coyle collated and presented the data shown on the stereo plots. The paper benefited greatly from reviews by Darrel Cowan and Mariano Elias-Herrera.

REFERENCES CITED

Alaniz-Alvarez, S.A., van der Heyden, P., Nieto-Samaniego, A.F., and Ortega-Gutiérrez, F., 1996, Radiometric and kinematic evidence for Middle Jurassic strike-slip faulting in southern Mexico related to the opening of the Gulf of Mexico: Geology, v. 24, p. 443–446, doi: 10.1130/0091-7613(1996)024<0443:RAKEFM>2.3.CO;2.

Anderson, T.H., and Schmidt, V.A., 1983, The evolution of middle America and the Gulf of Mexico–Caribbean Sea region during Mesozoic time: Geological Society of America Bulletin, v. 94, p. 941–966, doi: 10.1130/0016-7606(1983)94<941:TEOMAA>2.0.CO;2.

Anderson, T.H., McKee, J.W., and Jones, N.W., 1990, Jurassic (?) mélange in north-central Mexico: Geological Society of America Abstracts with Programs, v. 22, no. 3, p. 3.

Anderson, T.H., McKee, J.W., and Jones, N.W., 1991, A northwest-trending, Jurassic fold nappe, northernmost Zacatecas, Mexico: Tectonics, v. 10, no. 2, p. 383–401.

Anonymous, 1981, Hoja Monterrey carta geológica: Secretaria de Programacion y Presupuesto, Direccion General de Geografía del Territorio Nacional, México, escala 1:1,000,000.

Barboza-Gudino, J.R., Tristán-González, M., and Torres-Hernández, J.R., 1999, Tectonic setting of pre-Oxfordian units from central and northeastern Mexico: A review, *in* Bartolini, C., Wilson, J.L., and Lawton, T.F., eds., Mesozoic sedimentary and tectonic history of north-central Mexico: Geological Society of America Special Paper 340, p. 197–220.

Bissig, T., Mortensen, J.K., and Hall, B.V., 2003, The volcano-sedimentary setting of the Kuroko type VHMS district of Cuale, Jalisco, Mexico: Geological Society of America Abstracts with Programs, v. 99, no. 5, p. 61.

Blickwede, J.F., 1981, Stratigraphy and petrology of Triassic (?) "Nazas Formation," Sierra de San Julián, Mexico [M.S. thesis]: New Orleans, University of New Orleans, 100 p.

Burckhardt, C., and Scalia, S., 1906, Geologie des environs de Zacatecas: Mexico, D.F., Instituto Geológica de México, 10th International Congress, Excursion Guidebook 16, 26 p.

Campa, M.F., and Ramirez, J., 1979, La evolución geológica y la metalogénesis de Guerrero: Universidad Autónoma de Guerrero, Serie Técnico-Cientifica, v. 1, 84 p.

Campa, M.F., and Coney, P.J., 1983, Tectono-stratigraphic terranes and mineral resource distributions in Mexico: Canadian Journal of Earth Sciences, v. 20, p. 1040–1051.

Campa, M.F., Ramirez, J., and Bloome, C., 1982, La secuencia volcánica-sedimentaria metamorfizada del Triásico (Ladiniano-Cárnico) de la region de Tumbiscatio, Michoacán: Sociedad Geológica Mexicana, 6a Convención Geológica Nacional, Resúmenes, 48 p.

Cantú-Chapa, A., 1969, Una nueva localidad Triásico Superior en México: Revista del Instituto Mexicano del Petroleo, v. 1, no. 2, p. 71–72.

Carrillo-Bravo, J., 1968, Reconocimiento geológico preliminar de la porción central del Altiplano Mexicano: Informe 637-ZN, Pemex, México, Inédito, 31 p.

Carrillo-Martinez, M., 1997, Structural geometry of the Sierra Gorda between Jalpan and Tequisquiapan, State of Queretero, Mexico: Libro-guia de las excursiones geológicas, II Convención sobre la Evolución Geológica de México y Recursos Asociados, Pachuca, Hidalgo, p. 65–71.

Centeno-García, E., and Silva-Romo, G., 1997, Petrogenesis and tectonic evolution of central Mexico during Triassic-Jurassic time, *in* Stanley, G.D., Jr., and González-León, C.M., eds., Special issue dedicated to the International Workshop on the Geology of Northwestern Sonora, Mexico: Revista Mexicana de Ciencias Geológicas, v. 14, no. 2, p. 244–260.

Centeno-García, E., Corona-Chávez, P., Talavera-Mendoza, Ó., and Iriondo, A., 2003, Geology and tectonic evolution of the western Guerrero terrane—A transect from Puerto Vallarta to Zihuatanejo, Mexico, *in* Geological transects across Cordilleran Mexico: 99th Geological Society of America Cordilleran Section Annual Meeting, Field Trip Guidebook, Instituto de Geología, Universidad Nacional Autónoma de México, Special Publication 1, p. 201–228.

Chamot-Rooke, N., Lallemand, S.J., Le Pichon, X., Henry, P., Sibuet, M., Boulègue, J., Foucher, J.-P., Furuta, T., Gamo, T., Glaçon, G., Kobayashi, K., Kuramoto, S., Ogawa, Y., Schultheiss, P., Segawa, J., Takeuchi, A., Tarits, P., and Tokuyama, H., 1992, Tectonic context of fluid venting at the toe of the eastern Nankai accretionary prism: Evidence for a shallow detachment fault: Earth and Planetary Science Letters, v. 109, p. 319–332, doi: 10.1016/0012-821X(92)90095-D.

Chauve, P., Fourcade, E., and Carrillo-Martinez, M., 1985, Les rapports structuraux entre les domaines cordillérain y mésogéen dans la partie centrale du Mexique: Comptes Rendus de l'Académie des Sciences, sér. 2., v. 301, p. 335–340.

Córdoba-M., D.A., 1964, Geology of Apizolaya Quadrangle (east half), northern Zacatecas, Mexico [M.A. thesis]: Austin, The University of Texas, 111 p.

Cowan, D.S., 1985, Structural styles in Mesozoic and Cenozoic mélanges in the western Cordillera of North America: Geological Society of America Bulletin, v. 96, p. 451–462, doi: 10.1130/0016-7606(1985)96<451:SSIMAC>2.0.CO;2.

Cuevas-Pérez, E., 1983, Evolución geológica Mesozoic del Estrada de Zacatecas, México: Zentralblatt für Geologie und Paleontologie, Teil I (3/4), p. 190–201.

Diaz-Salgado, C., Centeno-García, E., and Gehrels, G., 2003, Stratigraphy, depositional environments and tectonic significance of the Taray Formation, northern Zacatecas state, Mexico: Geological Society of America Abstracts with Programs, v. 35, no. 4, p. 71.

Dickinson, W.R., 1985, Interpreting provenance relations from detrital modes of sandstones, *in* Zuffa, G.G., ed., Provenance of arenites: Proceedings of the NATO Advanced Study Institute on Reading Provenance from Arenites: Bingham, Maryland, D. Reidel Publishing, p. 333–361.

Ducea, M., Shoemaker, S., Gehrels, G., Vervoort, J., and Ruiz, J., 2003, Zircon U-Pb geochronology constraints on the magmatic and tectonic evolution of the Xolapa Complex, southern Mexico: Geological Society of America Abstracts with Programs, v. 99, no. 5, p. 66.

Ducea, M., Gehrels, G.F., Shoemaker, S., Ruiz, J., and Valencia, V.A., 2004, Geologic evolution of the Xolapa Complex, southern Mexico: Evidence from U-Pb zircon geochronology: Geological Society of America Bulletin, v. 116, no. 7/8, p. 1016–1025.

Elias-Herrera, M., Ortega-Gutiérrez, F., Sánchez-Zavala, J.L., and Macías-Romo, C., 2003, The real Guerrero terrane, southern Mexico: New insights from recent studies: Geological Society of America Abstracts with Programs, v. 99, no. 5, p. 66.

Gastil, R.G., and Jensky, W.A., 1973, Evidence for strike-slip displacement beneath the trans-Mexican volcanic belt, *in* San Andreas fault symposium: Stanford University Publications in the Geological Sciences, v. 13, p. 171–180.

Grajales-Nishimura, J.M., Terrell, D.J., and Damon, P.E., 1992, Evidencias de la prolongación del arco magmático Cordillerano de Triásico Tardío-Jurásico en Chihuahua, Durango y Coahuila: Asociación Mexicana de Geólogos Petroleros Boletín, v. 42, no. 2, p. 1–18.

Guerrero-García, J.C., Silver, L.T., and Anderson, T.H., 1978, Estudios geocronológicos en el Complejo Xolapa: Boletín de la Sociedad Geológia Mexicana, v. 39, resúmenes, p. 22–23.

Hsü, K.J., 1974, Mélanges and their distinction from olistostromes, *in* Dott, R.H., Jr., and Shaver, R.H., eds., Modern and ancient geosynclinal sedimentation: Society of Economic Paleontologists and Mineralogists Special Publication 19, p. 321–333.

Jones, N.W., McKee, J.W., Anderson, T.H., and Silver, L.T., 1995, Jurassic volcanic rocks in northeastern Mexico: A possible remnant of a Cordilleran magmatic arc, *in* Jacques-Ayala, C., González-León, C.M., and Roldán-Quintana, J., eds., Studies on the Mesozoic of Sonora and adjacent areas: Geological Society of America Special Paper 301, p. 179–190.

Keppie, J.D., Solari, L.A., Ortega-Gutiérrez, F., Elias-Herrera, M., and Nance, R.D., 2003, Paleozoic and Precambrian rocks of southern Mexico—Acatlan and Oaxacan Complexes, *in* Geological transects across Cordilleran Mexico: 99th Geological Society of America Cordilleran Section Annual Meeting Field Trip Guidebook, Instituto de Geología, Universidad Nacional Autónoma de México, Special Publication 1, p. 281–314.

King, R.E., 1934, The Permian of southwestern Coahuila, Mexico: American Journal of Science, v. 27, p. 98–112.

King, R.E., 1944, Geology and paleontology of the Permian area northwest of Las Delicias, southwestern Coahuila, Mexico: Part 1. Geology: Geological Society of America Special Paper 52, p. 3–35.

Klein, R.T., Hannula, S.R., Jones, N.W., McKee, J.W., and Anderson, T.H., 1990, Cerro El Pedernal: Block-in-mélange features in northern Zacatecas, Mexico: Geological Society of America Abstracts with Programs, v. 22, no. 3, p. 35.

Kusky, T.M., and Bradley, D.C., 1999, Kinematic analysis of mélange fabrics: Examples and applications from the McHugh Complex, Kenai Peninsula, Alaska: Journal of Structural Geology, v. 21, p. 1773–1796, doi: 10.1016/S0191-8141(99)00105-4.

Labarthe-Hernández, G., Tristán-González, M., and Aguillon-Robles, A., 1982, Estudio geológico-minero del area de Peñon Blanco estados de San Luis Potosi y Zacatecas: Universidad Autónoma de San Luis Potosí, Instituto de Geología y Metalurgia Folleto Tecnica 76, 80 p.

Lallemand, S.E., Glaçon, G., Lauriat-Rage, A., Fiala-Médioni, A., Cadet, J.-P., Beck, C., Sibuet, M., Iiyama, J.T., Sakai, H., and Taira, A., 1992, Seafloor manifestations of fluid seepage at the top of a 2000-metre-deep ridge in the eastern Nankai accretionary wedge: Long-lived venting and tectonic implications: Earth and Planetary Science Letters, v. 109, p. 333–346, doi: 10.1016/0012-821X(92)90096-E.

Lapierre, H., Ortiz, L.E., Abouchami, W., Monod, O., Coulon, C., and Zimmermann, J.-L., 1992, A crustal section of an intra-oceanic island arc: The Late Jurassic–Early Cretaceous Guanajuato magmatic sequence, central Mexico: Earth and Planetary Science Letters, v. 108, p. 61–77, doi: 10.1016/0012-821X(92)90060-9.

López-Infanzón, M., 1986, Estudio petrogenético de las rocas igneas en las formaciones Huizachal y Nazas: Sociedad Geológica Mexicana Boletin, v. 47, no. 2, p. 1–42.

Lundberg, N., and Moore, J.C., 1986, Macroscopic structural features in Deep Sea Drilling Project cores from forearc regions: Geological Society of America Memoir 166, p. 13–44.

Martinez-Hernández, S., 1979, Contribución al estudio geológico de una porción del sector Vizarrón-Tolimán, Estado de Queretero, México [Bachelor thesis]: Mexico, D.F., Universidad Nacional Autónoma de México, Facultad de Ingenieria, 82 p.

McKee, J.W., and Jones, N.W., 1979, A large Mesozoic fault in Coahuila, Mexico: Geological Society of America Abstracts with Programs, v. 11, no. 9, p. 476.

McKee, J.W., Jones, N.W., and Anderson, T.H., 1988, Las Delicias basin: A record of late Paleozoic are volcanism in northeastern Mexico: Geology, v. 16, p. 37–40, doi: 10.1130/0091-7613(1988)016<0037:LDBARO>2.3.CO;2.

McKee, J.W., Jones, N.W., and Anderson, T.H., 1989, The Jurassic margin of southern North America: Geological Society of America Abstracts with Programs, v. 21, no. 1, p. 34.

McKee, J.W., Jones, N.W., and Long, L.L., 1990, Stratigraphy and provenance of strata along the San Marcos fault, central Coahuila, Mexico: Geological Society of America Bulletin, v. 102, p. 593–614, doi: 10.1130/0016-7606(1990)102<0593:SAPOSA>2.3.CO;2.

McKee, J.W., Jones, N.W., and Anderson, T.H., 1999, Late Paleozoic and early Mesozoic history of the Las Delicias terrane, Coahuila, Mexico, *in* Bartolini, C., Wilson, J.L., and Lawton, T.F., eds., Mesozoic sedimentary and tectonic history of north-central Mexico: Geological Society of America Special Paper 340, p. 161–189.

Monod, O., and Calvet, P., 1991, Structural and stratigraphic re-interpretation of the Triassic units near Zacatecas (Zac.), central Mexico—Evidence of a Laramide nappe pile, *in* Miller, H., Rosenfeld, U., and Weber-Diefenbach, K., eds., Symposium on Latin American Geosciences: Zentralblatt für Geologie und Palaontologie, Teil 1, p. 1533–1542.

Mortensen, J.K., Hall, B.V., Bissig, T., Friedman, R.M., Danielson, T., Oliver, J., Rhys, D.A., and Ross, K.V., 2003, U-Pb zircon age and Pb constraints on the age and origin of volcanogenic massive sulfide deposits in the Guerrero terrane of central Mexico: Geological Society of America Abstracts with Programs, v. 99, no. 5, p. 61–62.

Ortega-Gutiérrez, F., and Elias-Herrera, M., 2003, Wholesale melting of the southern Mixteco terrane and origin of the Xolapa Complex: Geological Society of America Abstracts with Programs, v. 99, no. 5, p. 66.

Ortega-Gutiérrez, F., Ruiz, J., and Centeno-García, E., 1995, Oaxaquia, a Proterozoic microcontinent accreted to North America during the late Paleozoic: Geology, v. 23, p. 1127–1130, doi: 10.1130/0091-7613(1995)023<1127:OAPMAT>2.3.CO;2.

Ortega-Gutiérrez, F., Elias-Herrera, M., and Vega-Carillo, J. de J., 1997, A new Middle Jurassic-Triassic (?) arc-related stratigraphic succession (the Ixcatlán arc) in the southern Mixteco terrane: II Convención Sobre la Evolución Geológica de México Y Recursos Asociados, Pachuca, Hidalgo, Universidad Nacional Autónoma de México, Resúmenes, p. 51–52.

Ortiz-Hernández, L.E., and Martinez-Reyes, J., 1993, Geological structure, petrological and geochemical constraints for the centralmost segment of the Guerrero terrane (Sierra de Guanajuato, central Mexico): First Circum-Pacific and Circum-Atlantic Terrane Conference, Guanajuato, Mexico, November 5–22, Guidebook of Field Trip C, 25 p.

Pini, G.A., 1999, Tectonosomes and olistostromes in the argille scagliose of the northern Apennines, Italy: Geological Society of America Special Paper 335, 70 p.

Quintero-Legorreta, O., 1992, Geología de la region de Comanja, estados de Guanajuato y Jalisco: Universidad Nacional Autónoma de México, Instituto de Geología, Revista, v. 10, no. 1, p. 6–25.

Ranson, W.A., Fernandez, L.A., Simmons, W.B., Jr., and Enciso de la Vega, S.E., 1982, Petrology of the metamorphic rocks of Zacatecas, Zac., México: Boletín de La Sociedad Geológica Mexicana, v. 43, no. 1, p. 37–59.

Schaff, P., Hall, B.V., and Bissig, T., 2003, The Puerto Vallarta batholith and Cuale mining district, Jalisco, Mexico—High diversity parenthood of continental arc magmas and Kuroko-type volcanogenic massive sulphide deposits, *in* Geological transects across Cordilleran Mexico: 99th Geological Society of America Cordilleran Section Annual Meeting Field Trip Guidebook, Instituto de Geología, Universidad Nacional Autónoma de México, Special Publication 1, p. 183–199.

Sedlock, R.L., Ortega-Gutiérrez, F., and Speed, R.C., 1993, Tectonostratigraphic terranes and tectonic evolution of Mexico: Geological Society of America Special Paper 278, 153 p.

Silva-Romo, G., 1993, Estudio de la estratigrafia y estructuras tectónicas de la Sierra de Salinas, Estados de San Luis Potosí y Zacatecas [M.Sc. thesis]: México, D.F., Universidad Nacional Autónoma de México, Facultad de Ciencias, 111 p.

Silver, L.T., and Anderson, T.H., 1974, Possible left-lateral early to middle Mesozoic disruption of the southwestern North American craton margin: Geological Society of America Abstracts with Programs, v. 6, no. 7, p. 955–956.

Stewart, J.H., Gehrels, G.E., Barth, A.P., Link, P.K., Christie-Blick, N., and Wrucke, C.T., 2001, Detrital zircon provenance of Mesoproterozoic to Cambrian arenites in the western United States and northwestern Mexico: Geological Society of America Bulletin, v. 113, p. 1343–1356, doi: 10.1130/0016-7606(2001)113<1343:DZPOMT>2.0.CO;2.

Tardy, M., 1980, Contribution a L'etude geologique de La Sierra Madre Orientale du Mexique: [Ph.D. thesis]: Paris, L'Universite Pierre et Marie Curie, 459 p., 6 plates.

Tristán-González, M., and Torres-Hernández, J.R., 1994, Geología de la sierra de Charcas, Estado de San Luis Potosí, México: Revista Mexicana de Ciencias Geológicas, v. 11, no. 2, p. 117–138.

Manuscript Accepted by the Society 24 March 2005

Printed in the USA

Geological Society of America
Special Paper 393
2005

Pangean reconstruction of the Yucatan Block: Its Permian, Triassic, and Jurassic geologic and tectonic history

Maureen B. Steiner
Department of Geology and Geophysics, University of Wyoming, Laramie, Wyoming 82071, USA

ABSTRACT

Paleomagnetic studies of Paleozoic sedimentary and plutonic rocks demonstrate that the Yucatan Block did not lie between the North and South American plates in the Pangean assembly during the Permian. In the Middle Permian, the Yucatan Block lay in an inverted orientation on the western margin of Pangea at 6–10°S, probably forming part of the NW coast of South America. Subsequently the block rotated in a series of counterclockwise motions as the North and South American plates separated. By 230 Ma, Yucatan had rotated ~47° counterclockwise and moved slightly northward, to the equator. Counterclockwise rotation continued through the Jurassic: ~41° between 230 Ma and ca. Oxfordian time, and another ~47° between the Oxfordian and Tithonian, at which time, the approximate present orientation with respect to North America was achieved. Passage of the Yucatan Block from NW South America into the gap created by the separation of North and South America is a motion consistent with the left-lateral translations along the Mojave-Sonora or similar megashear. The fact that Yucatan has exhibited a counterclockwise motion throughout its Mesozoic history suggests that the microplate may have acted in a ball-bearing fashion between the larger North and South American plates.

The Permian sedimentary remanence appears to be carried dominantly by magnetite and maghemite in red bed strata deposited on the margins of Silurian plutons within the marine Santa Rosa basin of the Maya Mountains. This remanence decays exceedingly linearly to the origin of orthogonal-axes plots. Biostratigraphy indicates a late Pennsylvanian to Middle Permian sedimentary age for the Santa Rosa strata, and the presence of both polarities of remanence in a 110 m magnetostratigraphic sequence of four polarity intervals demonstrates a post-Kiaman, i.e., Middle Permian, magnetization age. Another remanence is exhibited by Silurian plutons, also dual polarity, but corresponding to a paleopole ~60° counterclockwise of that recorded in the Middle Permian sedimentary rocks. Exceedingly uniform K/Ar ages of 231 ± 7 Ma characterize all Maya Mountains plutons and a volcanic complex, indicating a 230 Ma resetting of the K/Ar radiometric systems in plutons dated as Late Silurian by U/Pb (Steiner and Walker, 1996). Furthermore, metamorphic aureoles are developed in the Permian Santa Rosa strata that border the Silurian plutons, suggesting that the 230 Ma resetting was a postintrusion event that involved the margins of the plutons. The spatial relationship between the pluton and sedimentary poles is reminiscent of that between the North American Late Triassic and Permian reference paleopoles.

Steiner, M.B., 2005, Pangean reconstruction of the Yucatan Block: Its Permian, Triassic, and Jurassic geologic and tectonic history, *in* Anderson, T.H., Nourse, J.A., McKee, J.W., and Steiner, M.B., eds., The Mojave-Sonora megashear hypothesis: Development, assessment, and alternatives: Geological Society of America Special Paper 393, p. 457–480. doi: 10.1130/2005.2393(17). For permission to copy, contact editing@geosociety.org.

Therefore the combination of 230 Ma reset K/Ar systems, metamorphic aureoles in strata younger than the plutons, and a magnetization resembling a Late Triassic remanence all suggest that a 230 Ma hydrothermal event remagnetized the igneous rocks and reset their radiometric systems. This probably was an event related to the initial breakup of Pangea.

The Maya Mountains plutons yield a paleopole that is statistically identical to that of the Chiapas Massif to the south. Both plutonic complexes exhibit somewhat dispersed dual polarity magnetization populations, suggesting that both were remagnetized at ca. 230 Ma. Importantly, the identical remanences in these widely separated plutonic complexes indicate that the Yucatan Block (including the Chiapas Massif) has been a structural entity since at least 230 Ma.

Keywords: Pangea, Yucatan, Permian, Late Triassic.

INTRODUCTION

Early reconstructions of the pre-Atlantic continental assembly (e.g., Bullard et al., 1965) result in overlap between northern South America and nuclear Central America, including the Yucatan Block. Subsequent reconstructions involving different types and directions of motion for Yucatan have been proposed to avoid the overlap, scenarios involving clockwise rotation, counterclockwise rotation, or no rotation. The Yucatan Block during Pangea generally is believed to have been situated adjacent to the Texas-Louisiana coastline of southern North America, and hypothesized to have rotated 50°–60° counterclockwise to its present orientation subsequent to Pangean fragmentation (White, 1980; Pindell and Dewey, 1982; Pindell, 1985; Buffler and Sawyer, 1985; Dunbar and Sawyer, 1987; Ross and Scotese, 1988; Rowley and Pindell, 1989; Pindell and Barrett, 1990; Sawyer et al., 1991; Buffler, 1991; Hall and Najmuddin, 1994; Schouten and Klitgord, 1994). Marine geophysical data from the Gulf of Mexico support this reconstruction and motion of Yucatan from Jurassic to present (e.g., Dunbar and Sawyer, 1987; Buffler, 1991; Sawyer et al., 1991; Hall and Najmuddin, 1994; Buffler and Thomas, 1994). The present paleomagnetic study of the Yucatan Block strata was undertaken in 1990, at a time when this hypothesis had yet to be tested. The present study was only a reconnaissance one, aimed at assessing whether stable paleomagnetic recorders existed that could be used to determine the location of the microplate during and subsequent to the Pangean assembly.

Much of the Yucatan Block is buried beneath Cretaceous and Cenozoic sedimentary cover; however, the Maya Mountains of Belize in southeastern Yucatan expose Paleozoic metasedimentary and igneous rocks (Fig. 1). The exposed units include the following:

1. The Late Pennsylvanian through Middle Permian Santa Rosa Group, deposited as mudstone, shale, siltstone, and sandstone, with minor interbedded limestone and conglomerate (Bateson, 1972). Fusulinids, mollusks including ammonites, and brachiopods are the basis of the Late Pennsylvanian to Middle Permian age (Bateson, 1972; Ross, 1979). Acidic volcanic rocks (possibly ca. 300 Ma; Bateson, 1972) interfinger with the sedimentary strata, largely in the southern (but also in the northern) Maya Mountains.

2. Three dominantly granitic intrusive complexes, surrounded by the Santa Rosa strata. Two plutonic complexes occur in the northern Maya Mountains, the Mountain Pine Ridge in the west and Hummingbird–Mullins River complex in the east; the Cockscomb-Sapote complex resides in the southeastern Maya Mountains (Fig. 1). The three igneous complexes yield exceedingly similar K/Ar ages of ca. 230 Ma, as do the biostratigraphically dated Late Pennsylvanian volcanic rocks (Fig. 1) of the southern Maya Mountains (Table 1). Rb/Sr isotopic analysis of the various plutons of the Mountain Pine Ridge complex (those exhibiting K/Ar ages of ca. 230 Ma) yielded a disturbed isochron and an age of 336 ± 10 Ma (Bateson and Hall, 1977). A single Rb/Sr determination also was obtained from the Cockscomb-Sapote plutonic complex, which yielded a 428 ± 40 Ma age (Bateson and Hall, 1977). Subsequently, samples from three plutons of the Mountain Pine Ridge, and one from the Hummingbird–Mullins River complex, were dated with the U/Pb method applied to zircons (Steiner and Walker, 1996). The three plutons of the five compositional-phase Mountain Pine Ridge complex all yield interpreted ages of Late Silurian; the pluton yielding the best-defined intersection with the concordia gave the age of 418 ± 3.6 Ma. A pluton from the Hummingbird–Mullins River complex had suffered lead-loss, but the interpreted results suggest an age of 410–420 Ma (Steiner and Walker, 1996).

The Santa Rosa Group has been variably folded along a generally E-W strike, in folds varying from tight in eastern Belize (55° to 90°) to more open in the west and south. In the reconnaissance studies of the Maya Mountains, the strata are described as regionally metamorphosed to the lower greenschist grade (Dixon, 1956; Bateson and Hall, 1977). In addition, a higher grade of metamorphism is locally superposed over the regional metamorphism of the sedimentary rocks at the margins of the three plutonic complexes (Ower, 1928; Dixon, 1956; Bateson and Hall, 1977). The higher-temperature metamorphism at the margins of the plutonic complexes was described at a time when the dominant age information available for the intrusives was that

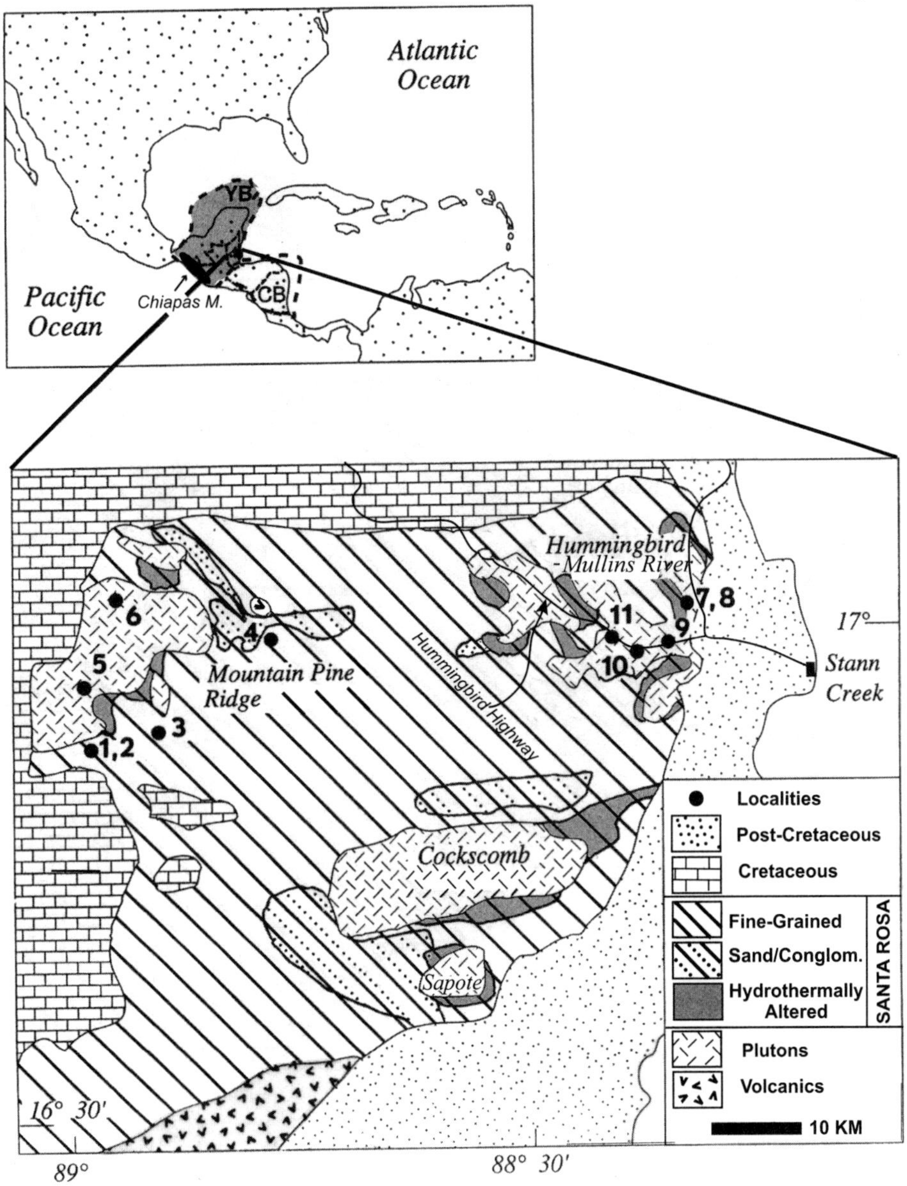

Figure 1. Present plate tectonic setting of the Central America, Yucatan (YB), and Chortis (CB) blocks; the location of the Maya Mountains in Belize is shown by a solid square. The generalized geology of the Maya Mountains is shown in enlargement; numbers identify the paleomagnetic sampling localities.

derived by the K/Ar method. The K/Ar dates were interpreted to indicate that the eastern plutons formed ca. 230 Ma, hence, the marginal alteration was attributed to contact metamorphism related to pluton intrusion (Ower, 1928; Dixon, 1956; Bateson and Hall, 1977).

Sericite and tourmaline characterize the higher-grade metamorphic aureoles around all three plutonic complexes (Dawe, 1984; Ower, 1928; Dixon, 1956; Bateson and Hall, 1977), but the eastern two complexes (Hummingbird–Mullins River and Cockscomb-Sapote) also contain chiastolite (Ower, 1928; Dixon, 1956; Bateson and Hall, 1977). The presence of chiastolite, a high-temperature form of andalusite, indicates higher temperatures around the eastern plutons. In addition, greater aerial extent of the marginal metamorphism exists around the eastern two plutonic complexes. The marginal metamorphism of the western (Mountain Pine Ridge) igneous complex was studied in greater detail. In addition to the presence of tourmaline and sericite in the bordering sedimentary strata, these minerals also exist along joints and fractures within the plutonic rocks (Dawe, 1984). The presence of these minerals in both the adjacent sedimentary strata and in pluton fractures led Dawe to conclude that the western plutonic complex had experienced hydrothermal alteration. The presence or absence of chiastolite is the main mineralogic difference between the margin alteration around the eastern and western plutons (Ower, 1928; Dixon, 1956; Bateson and Hall, 1977; Dawe, 1984).

The relationship between the sedimentary and igneous rocks of the Maya Mountains has long been interpreted as intrusion of the Permian-Carboniferous Santa Rosa Group by all of the eastern, and many of the western plutons, after those strata had been deformed by an episode of folding and regional metamorphism; the Triassic (230 Ma) K/Ar ages were responsible for this interpretation (Bateson and Hall, 1971; Bateson, 1972; Kesler, et al., 1971; Bateson and Hall, 1977; Cole and Andrews-Jones, 1979). Moreover, Late Permian–Early Triassic folding had affected similar sedimentary rocks in Guatemala and elsewhere in Central America (e.g., Dengo and Bohnenberger, 1969), and was considered as supporting evidence of the folding and intrusion scenario for the Maya Mountains. This scenario was promulgated despite (1) the 336 Ma Rb/Sr age of the Mountain Pine Ridge complex, which showed that at least a portion of that western plutonic complex had been emplaced prior to Santa Rosa sedimentation (Bateson and Hall, 1977), and (2) conglomeratic facies adjacent to the eastern margin of the Mountain Pine Ridge complex that indicated probable emergence during Pennsylvanian-Permian sedimentation (Simmons, 1972; Dawe, 1984). The perception that the Santa Rosa Group had been intruded by most of the plutonic rocks around 230 Ma has persisted in even fairly recent literature (e.g., Donnelly, 1990; Woods et al., 1991; Salvador, 1991).

The U/Pb ages have revealed the true geologic history. Samples from the Hummingbird–Mullins River pluton and the three Mountain Pine Ridge plutons were obtained from the same sites as those from which the K/Ar ages were obtained; the U/Pb ages demonstrated that these two plutonic complexes predated Santa Rosa Group sedimentation (Steiner and Walker, 1996). Although the third, Cockscomb-Sapote complex, was not dated by U/Pb methods, sedimentary facies coarser than the typical fine-grained Santa Rosa strata also lie adjacent to its margins (see map of Bateson and Hall, 1977), similar to those around the Mountain Pine Ridge complex. Also similar to the Mountain Pine Ridge plutons, Muncaster (1976) noted some red strata marginal to the Cockscomb complex. Hence, the sedimentary facies around the Cockscomb-Sapote complex indicate that it, too, predated Santa Rosa deposition. Moreover, the accuracy of the single, Late Silurian (428 ± 40 Ma) Rb/Sr age obtained from the Cockscomb-Sapote complex (Bateson and Hall, 1977) is supported by the U/Pb results from the Mountain Pine Ridge and Hummingbird–Mullins River complexes. Thus, the sum of the evidence indicates that all plutonic complexes predated Santa Rosa sedimentation and were emergent at the time that the Santa Rosa Group was deposited around them.

The nonintrusive relationship between the plutons and the Santa Rosa strata indicates that the metamorphic aureoles cannot be the result of contact metamorphism. The uniform 230 Ma K/Ar ages in all plutonic complexes and the volcanic complex (Bateson and Hall, 1977) indicate that a nonmagmatic thermal event occurred at ca. 230 Ma. Like Dawe's (1984) conclusion that the thermal alterations adjacent to and within the western igneous complex were of hydrothermal origin, it is probable that the metamorphic aureoles around all three igneous complexes and the resetting of the K/Ar systems occurred in response to a Triassic hydrothermal event, most likely in relationship to the early rifting between North and South America during initial Pangean breakup. Paleomagnetic results from this study show that the plutonic magnetizations have been reset. Since the plutons are enclosed by sedimentary strata, the hydrothermal fluids probably flowed along the contacts between sedimentary strata and igneous rocks, in effect those contacts acted as fluid guides.

TABLE 1. RADIOMETRIC DETERMINATIONS IN THE MAYA MOUNTAINS, BELIZE, CENTRAL AMERICA

IGNEOUS UNIT	AGE	MATERIAL DATED	REFERENCE
K/Ar			
MPR	205 ± 8	*	2
MPR	~214 ± 8	*	2
MPR	~232 ± 8	*	2
MPR	~231 ± 8	*	2
HMR	231 ± 7	biotite	1,3
HMR	227 ± 7	biotite	1,3
C/S	226 ± 7	biotite	1,3
C/S	237 ± 5	biotite	1,3
C/S	~230 ± 8	*	2,3
Bladen volcanics	~236 ± 8	*	2
K/Ar MEAN of 8	231.2 ±7.0		
K/Ar MEAN of 10	226.9 ± 8.0		
Rb/Sr			
MPR	336 ± 10	10 pt isochron	2
C/S	428 ± 41	single determination	2
U/Pb			
MPR	418 ± 3.6	concordia plot	4
MPR	404 ± 3.3	concordia plot	4
HMR	410–420	lead loss	

Note: MPR—Mountain Pine Ridge complex; HMR—Hummingbird—Mullins River complex; C/S—Cockscomb-Sapote complex. References: 1—Bateson and Hall (1977); 2—Cole and Andrews-Jones (1979); 3—recalculations by Dawe (1984); 4—Steiner and Walker (1996). All ages have been corrected for the new decay constants

PALEOMAGNETISM

Sample Collection

Because this paleomagnetic inquiry was aimed mainly at assessing the presence of rock types suitable for testing Pangean reconstructions, collecting was done in reconnaissance style designed to sample all exposed magnetic recorders that might have long-term stability. The regionally metamorphosed Santa Rosa strata comprise the bulk of the Maya Mountains exposures and were the most extensively sampled material. Hydrothermally altered Santa Rosa strata (previously attributed to contact metamorphism) also were sampled. A number of plutonic rocks were sampled: a diorite pluton, a series of mafic inclusions, the granodiorite hosts, and the granite in direct contact with the hydrothermally altered Santa Rosa strata.

Sedimentary Rocks, Santa Rosa Group

Exposures of these strata were sampled in five areas. Localities #1–4 are adjacent to the Mountain Pine Ridge igneous complex (Fig. 1); locality #7 is adjacent to the Hummingbird–Mullins River complex, in strata mapped as "contact metamorphosed," but which are actually hydrothermally altered.

Localities #1 and #2 abut the southern edge of the Mountain Pine Ridge complex (16.9°N, 271.0°E). These strata are bedded red mudstones, which are only weakly metamorphosed. The strata at localities #1 and #2 dip between 25° and 58° toward 91–130°E. A large stratigraphic coverage was achieved at locality #1, where an old logging road traverses downward from the southern edge of the Mountain Pine Ridge complex to a former rubber plantation on the Macal River, exposing a several-hundred-meter continuous stratigraphic sequence in road cuts. The section is well-bedded, red mudstone-shale; bedding is well developed and easily distinguished from fracture patterns. The upper 110 m of the exposure were sampled with about three samples each in 13 stratigraphically discrete sites.

Locality #2 is from a small road cut (along Cree Road), just a few hundred meters north of locality #1. A 3 m section of strata is exposed, consisting of red mudstone-shale with irregular bleaching of some portions to a white color. The beds dip ~38° toward 91°E. Five stratigraphically distributed sites of 2 samples each were taken.

Locality #3 lies off the SE end of the Mountain Pine Ridge plutonic complex, consisting of four very small road cuts along Kinloch's Line road. Two of the sites are gray to dark-gray mudstones, representing the dominantly marine deposition that characterizes most of the Santa Rosa Group of Belize; the other two consist of reddish-gray mudstone and faintly reddish, dark-gray mudstone. The beds sampled have variable attitudes, dipping due south by 24° to 58°. Two to four samples were collected at each site.

Locality #4 lies due east of the Mountain Pine Ridge complex, along the Baldy Sibun Road. A creek bed intersected by the road (16.9°N, 271.1°E) exposes strata that are essentially pure white phyllitic mudstone with thick (0.15–2.0 mm) bedding-parallel planar layers of blood-red hematite. The beds dip 19° toward N76°. Sixteen samples were taken in the white phyllitic mudstone and four in the hematite layers.

Adjacent to the Mullins River pluton of the Hummingbird–Mullins River igneous complex (locality #7, Fig. 1), some highly indurated bedded strata reside in physical contact with a granite pluton (17.0°N, 271.6°E). The attitude of the beds is vertical. The strata are brown with a slight reddish cast, micaceous, and very fine grained, consistent with having formerly been a shale. Their bedded nature, color, and micaceous nature are similar to the Santa Rosa strata sampled at locality #1, but they are much more indurated. These are strata that had been mapped and described as contact metamorphosed Santa Rosa strata (Ower, 1928; Dixon, 1956; Bateson and Hall, 1971; Bateson, 1972; Bateson and Hall, 1975; Bateson and Hall, 1977). The adjacent pluton gave a 410–420 Ma age (Steiner and Walker, 1996), thus older than the "contact metamorphosed" Santa Rosa strata at its margin. These altered contact strata were collected to inquire into the origin of this alteration, and to thereby gain some insight into the geologic events that have affected the Maya Mountains. Eighteen samples were taken in a stratigraphic sequence of ~12 m of the strata. The adjacent radiometrically dated pluton was also sampled for paleomagnetism, which is described in the igneous results section.

The Santa Rosa strata were demagnetized by alternating field (AF) and thermal demagnetization. AF demagnetization to 60 mT was performed on a few specimens from each locality, but minimal demagnetization was achieved. Most samples were stepwise thermally demagnetized, in 16 incremental heatings from 200 °C to 660 °C. Characteristic directions for each specimen were derived by principal component analysis (Kirschvink, 1980).

Localities #1 and #2

Nearly antipodal directions were displayed in the initial measurements of the natural remanent magnetization (NRM_0). NRM_0 intensities were between 2×10^{-1} and 3×10^{-2} mA/m. Thermal demagnetization first removed a low stability component that either had steep negative inclinations or was a recent geomagnetic field component. Thereafter, an exceptionally linear decay of the remanence to the origin of orthogonal axes plots between 200 °C and 620 °C was observed in all samples (Fig. 2).

Most samples exhibited abrupt declines in remanence intensity above 570 °C (Fig. 3A); 60% to 80% of the remanent intensity had been lost by 570–590 °C, and only 10% or less remained at 620 °C. Loss of remanence at these temperatures and in a narrow temperature range suggests that magnetite and/or an oxidized, cation-deficient magnetite (maghemite) may carry much of the remanent magnetization.

Acquisition of isothermal remanent magnetization (IRM) in representative samples spaced throughout the stratigraphic section is also shown in Figure 3. All samples gained most of their IRM by just above 200 mT (Fig. 3B); relatively minor amounts of remanence continued to be acquired in very high fields. The 200 mT IRM acquisition also may suggest the presence of magnetite-maghemite. The continued minor acquisition between 200 and 1000 mT is due to the hematite component of these red sedimentary rocks.

Characteristic directions of locality #1 samples (Fig. 4A) defined four stratigraphically superposed intervals of opposite polarity magnetizations (Fig. 4B). Moreover above 620 °C, most samples revealed the presence of an additional remanence vector having a polarity opposite, sometimes antipodal, to the remanence held below 620 °C (e.g., Fig. 2D); ~40% of the samples

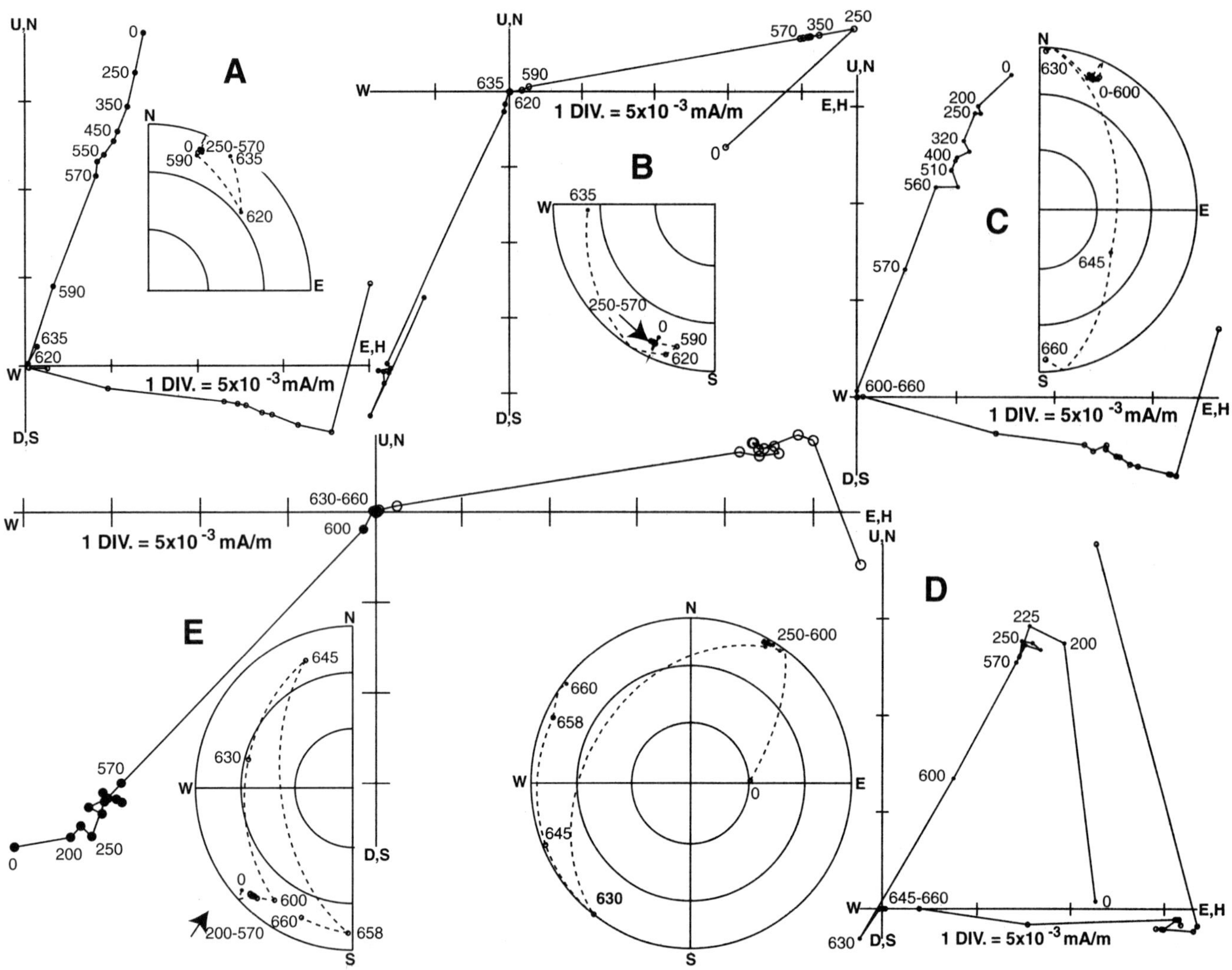

Figure 2. Typical demagnetization behavior of the regionally metamorphosed Santa Rosa Group samples of localities #1 and #2. (A and B) Samples heated to 635 °C. (C–E) Samples heated to 660 °C and examples of the high-temperature display of an opposite polarity remanence. Orthogonal projections display inclination in the plane of declination (Kirschvink, 1980); solid (open) circles represent the horizontal (vertical) components of magnetization. Equal-area stereographic projections show lower (upper) hemisphere directions as solid (open) circles. All plots show data after tectonic correction.

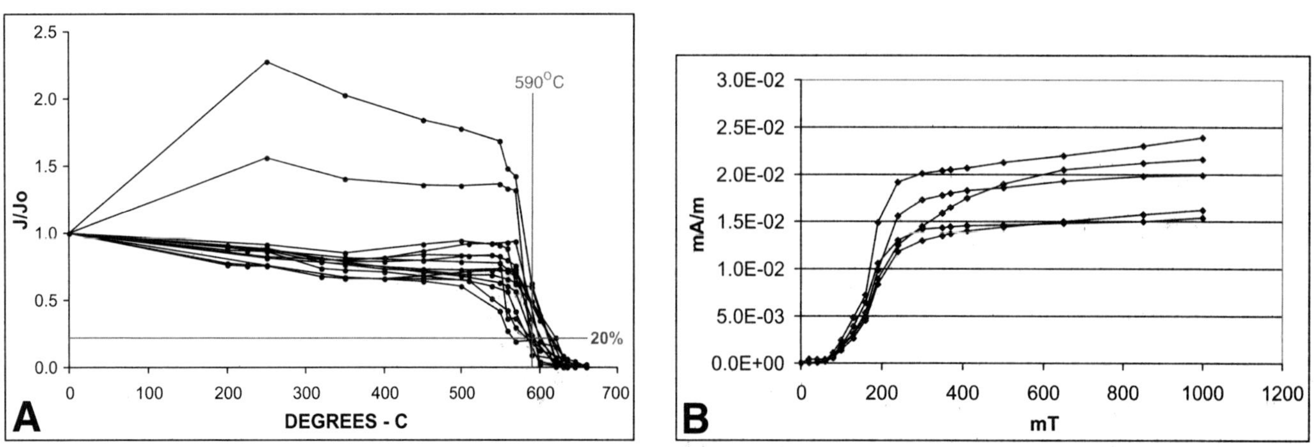

Figure 3. (A) Intensity loss with increasing temperature during thermal demagnetization of the Santa Rosa locality #1 samples. (B) Isothermal remanent magnetization (IRM) acquisition of five locality #1 samples.

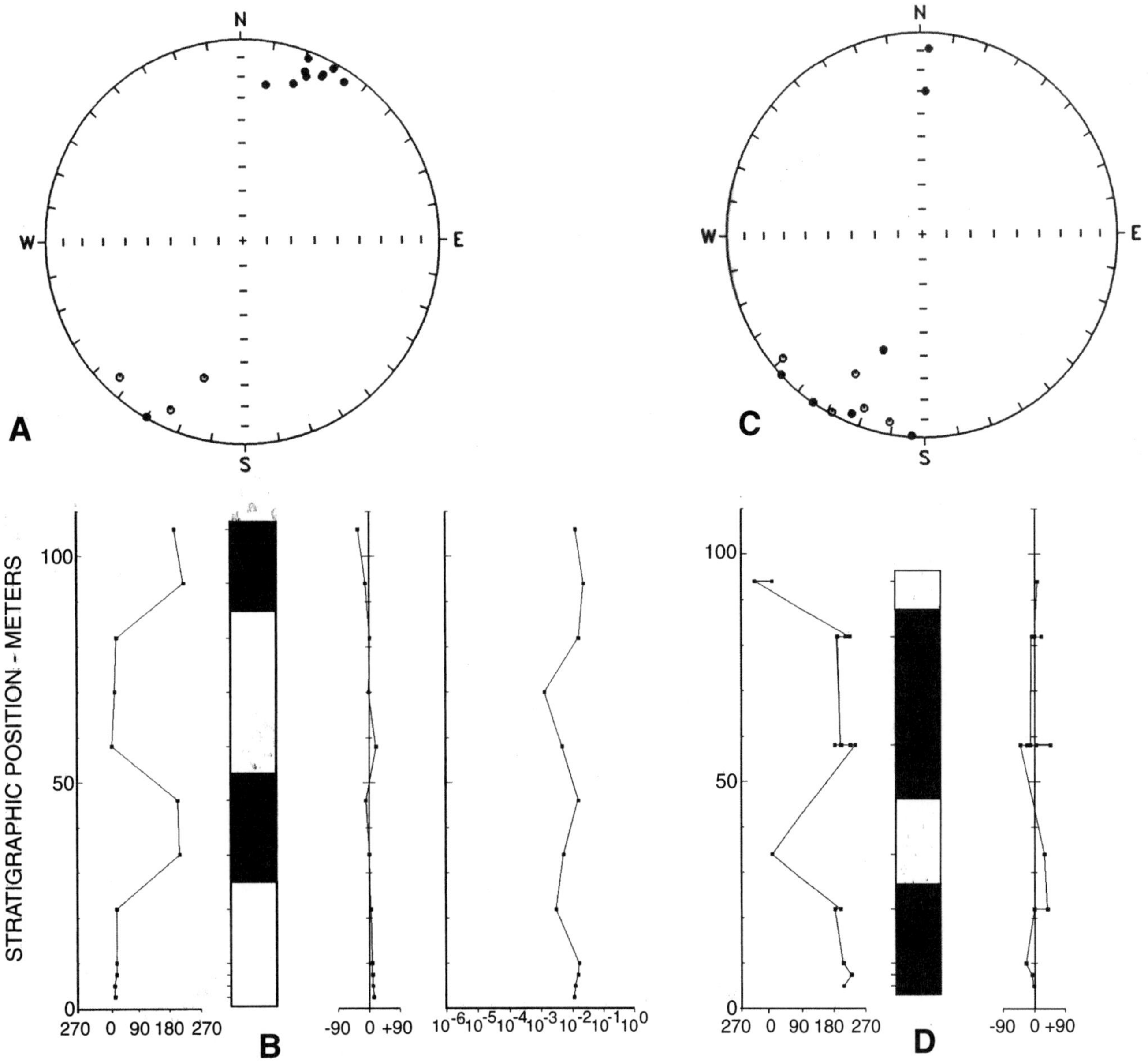

Figure 4. Santa Rosa samples at locality #1. Characteristic remanence directions after tectonic correction are shown in stereographic (A) and stratigraphic (B) projections. Removed vectors during thermal demagnetization are displayed in C and D. Projections and symbols as in Figure 2.

displayed this behavior (Table 2). Other samples exhibited approximately antipodal directions as removed vectors; ~70% of the samples displayed antipodal removed remanence directions (Fig. 4B), commonly between 400–500 °C, 570–590 °C, and/or 620–660 °C. Of particular interest is the fact that this display of approximately antipodal remanence forms an approximate mirror image magnetostratigraphy to that of the characteristic magnetization (Fig. 4D).

Locality #2 strata differ from those of locality #1 in appearing to be partially leached of iron oxide. Nevertheless, samples from this locality behaved identically to those from locality #1. Samples from all five sites displayed exceedingly linear-to-the-origin demagnetization behavior, and exhibited characteristic directions through ~560–620 °C. Moreover, a polarity stratigraphy also appears in these data. Samples from the lowest two sites, spanning ~1 m of this 3 m stratigraphic sequence, exhibited a southwesterly polarity direction like that of locality #1, whereas samples from the three sites in the upper 2 m displayed a northeasterly direction. Tilt-corrected characteristic directions are indistinguishable from those of locality #1.

TABLE 2. SANTA ROSA GROUP MAGNETIZATION

Reversal test	Dec	Inc	*n*	*k*	*k*-ratio	γ_c	γ_o	Classif.
N	208.0	−10.0	23	19.36				
R	23.5	11.6	23	48.34				
					2.5	8.4	4.7	B
Fold test	Dec	Inc	*n*	*k*	α_{95}	Classification		
In situ	197.6	−11.0	46	20.74	4.7	Magnetization not acquired in situ		
Tilt corrected	203.9	−11.2	46	27.28	4.1	Magnetization acquired pre-tilting		
Mean direction	Dec	Inc	*n*	*k*	α_{95}	Lon/E	Lat	Paleolat.
All sample sites*	205.7	−10.8	46	27.60	4.1			
	205.7	−11.1	7	446.68	2.9	202.9	−62.5	5.5 S

Note: Calculations in bedding-corrected coordinates; *n*—number of specimens or sites; *k*—precision parameter (kappa); α_{95}—the half-angle of the cone of confidence about the mean; *k*-ratio—ratio of kappas; γ_c, the critical angle between the means above which the hypothesis of a common population would be rejected with 95% confidence; γ_o is the angle between mean normal and reverse direction; * is the pole plotted in Fig. 10.

Locality #3

Among these mostly gray mudstones, samples of the one reddish-gray outcrop exhibited a southerly declination resembling those of locality #1, but a steeper inclination. Further, this sample displayed the same exceedingly linear demagnetization behavior as observed at localities #1 and #2. Samples from the exposures of gray strata held only the recent geomagnetic field directions prior to bedding correction. The faintly reddish, dark-gray outcrop yielded samples displaying multicomponent demagnetization behavior and overlapping remanence stabilities.

Locality #4

The white phyllitic mudstones samples of this locality were weakly magnetized, generally ~1×10^{-4} mA/m, in sharp contrast to the red mudstones of locality #1. The NRM_0 directions of the samples, prior to bedding correction, had declinations of 260–290° and inclinations of 50–80°. AF demagnetization induced neither direction nor intensity changes. Thermal demagnetization at 250 °C resulted in two-thirds of the sample directions moving to the direction of the present axial geocentric dipole for this site. Through thermal demagnetization to 620 °C, these samples maintained this direction. The other one-third of the samples displayed wildly varying directional behavior and entirely incoherent directional paths during demagnetization. By 620 °C, samples had intensities near the sensitivity limit of the cryogenic magnetometer. The characteristic magnetization directions for the two-thirds of coherently behaved samples are shown in Figure 5A.

The interbedded layers of red hematite were quite strongly magnetized (1×10^{-1} mA/m). Moreover, the samples exhibited a reversed polarity NRM_0 that changed little upon demagnetization through 620 °C. Remanences decayed univectorially to the origin of orthogonal axes vector plots during thermal demagnetization. The reversed polarity characteristic remanence exhibited by these hematitic samples (Fig. 5B) is exactly antipodal to the characteristic remanence direction of the white phyllitic host strata. Thus, the hematite layers record a direction coincident with a reversed present axial dipole geomagnetic field direction at this site.

Locality #7: "Contact Metamorphosed" Strata

An unusual NRM_0 direction is recorded in these strata. In in situ coordinates, the NRM_0 directions are easterly directed with negative inclinations (Fig. 6A). The adjacent granite samples also exhibit a similar magnetization direction (Fig. 6B), as do the igneous rocks of locality #1, both discussed in the next section. The direction is peculiar, both because it certainly is not a recent magnetization (present field = 2.0, 51.3, axial dipole field = 0.0, 31.3), and because of its presence in both sedimentary rocks and adjacent igneous rocks. Upon demagnetization of the locality #7 sedimentary strata, the unusual initial direction was immediately erased.

Other than removal of the initial remanence direction, AF treatment had little effect on the remanence of these strata. Thermal demagnetization resulted in linear remanence decay and abrupt loss of remanent intensity above 560–570 °C in 10 of the 18 samples. The behavior is quite similar to that observed among locality #1 samples, but the resultant directions in in situ coordinates are similar to the axial dipole field directions for this site. Of the remaining eight samples, two exhibited the present field after noisy demagnetization decay, whereas the others exhibited seemingly stable random directions. Some samples also exhibited an opposite polarity remanence in narrow temperature ranges. Characteristic directions and opposite polarity components isolated during demagnetization are shown in Figure 6C; this figure is similar to Figure 5.

Locality #8: Granite in Contact with Locality #7 Strata

The Mullins River pluton lies in direct contact with the "contact metamorphosed" strata of locality #7; the pluton is a gray, muscovite granite. The granite is not fine grained and no mafic inclusions were exposed, but the granite was sampled nevertheless, because the event which caused the contact alteration of the adjacent sediments might have generated a remanence in the granite. Twenty samples were cored from the pluton here, constituting locality #8; 38 specimens were measured. As just mentioned, NRM_0 directions mainly were the same negative, easterly directions observed in the adjacent sediments (Fig. 6B). NRM_0 intensities are relatively weak, commonly between 1×10^{-4} and 5×10^{-3} mA/m. Seven samples were demagnetized, five by thermal demagnetization to 250 °C, and two by AF treatment to 60 mT. The negative easterly NRM_0 direction was removed by the lowest step of either AF or thermal treatment. Demagnetization did not produce coherent remanence directions. Two samples exhibited linear demagnetization and shallow northeasterly directions. Two others displayed shallow northwesterly directions, and two exhibited the presently observed geomagnetic field direction. The last gave widely scattered directions in response to demagnetization.

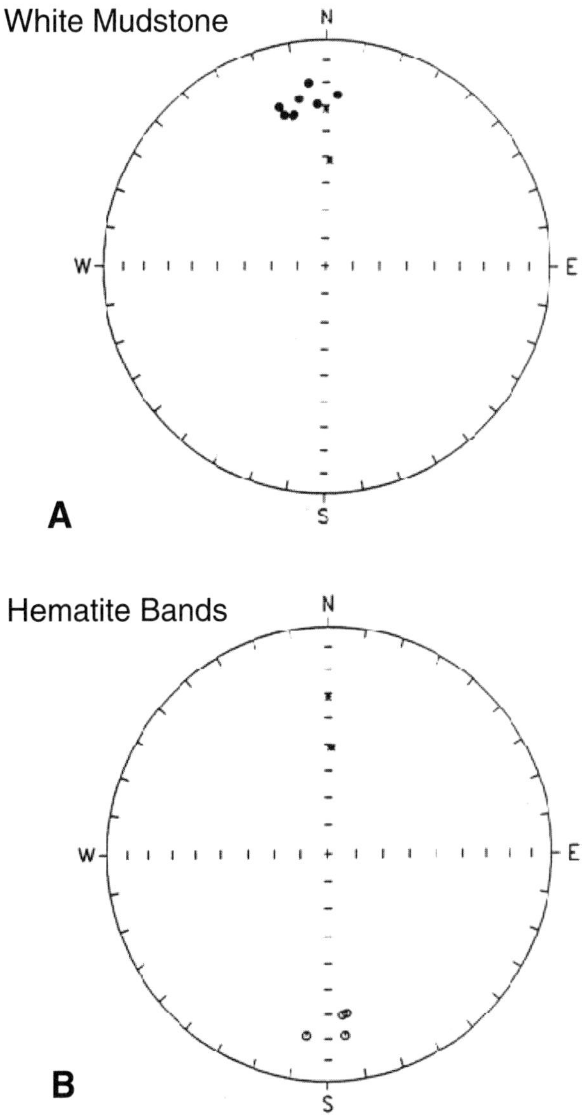

Figure 5. Locality #4 characteristic directions: (A) white phyllite, and (B) hematite beds. Present field and theoretical axial dipole directions are shown as asterisks; other symbols as in Figure 2.

Plutonic Paleomagnetism

Sampling

The majority of the igneous rocks exposed in the Maya Mountains are granites and granodiorites, lithologies that generally are not paleomagnetically reliable. But, the plutonic complex exposures were examined for lithologies that might yield stable paleomagnetic directions. Mafic compositions were sought, either as plutons, intrusions, or inclusions, and granitic exposures were examined for fine-grained granitic rocks.

In the northeast plutonic complex, the Hummingbird pluton abuts the Mullins River pluton that was sampled at locality #8. The Hummingbird pluton is a large pluton of mainly gray granodiorite, which contains numerous small mafic inclusions, 1–2 cm size. Mafic inclusions and the host granodiorite were collected at three adjacent localities along the Hummingbird Highway that traverses the area, localities #9, #10, and #11 (Fig. 1). Locality #9 displayed the largest abundance of mafic inclusions. However, of the 10 cores taken, only four proved to contain wholly mafic inclusion material; 14 specimens were cut from the 10 cores. Ten samples were taken at locality #10, from which 16 specimens were obtained; three are mafic inclusions and seven are granodiorites. At locality #11, 15 samples were cored, five mafic inclusions and ten granodiorites; two additional samples of granodiorite were obtained from the mafic inclusion cores, for a total of 29 specimens. A total of 37 samples were collected at the three localities of the Hummingbird pluton. Although 18 mafic inclusions were cored at the localities, only 12 of the cores contained adequate amounts of the mafic inclusion for study; 25 samples were granodiorite compositions.

In the northwest, five plutons comprise the Mountain Pine Ridge plutonic complex; four are granites, but one is a diorite named the Windward Road pluton. Fifteen diorite samples were cored from the Windward Road pluton at locality #5 (Fig. 1); during coring, the coring engine threw a rod, so a few block samples were collected thereafter. Twenty-three specimens were studied from 17 samples.

Mafic inclusions generally were not observed in the granite plutons of the northwest complex, with one exception. One coarse-grained pink granite pluton hosted a single, large mafic inclusion; the exposure was along Pinol Creek (locality #6, Fig. 1). The inclusion was large enough for six core samples to be fitted within its boundaries.

NRM_0 Directions

A distinct pattern of southwest and northeast directions with shallow inclinations was observed from the initial measurements

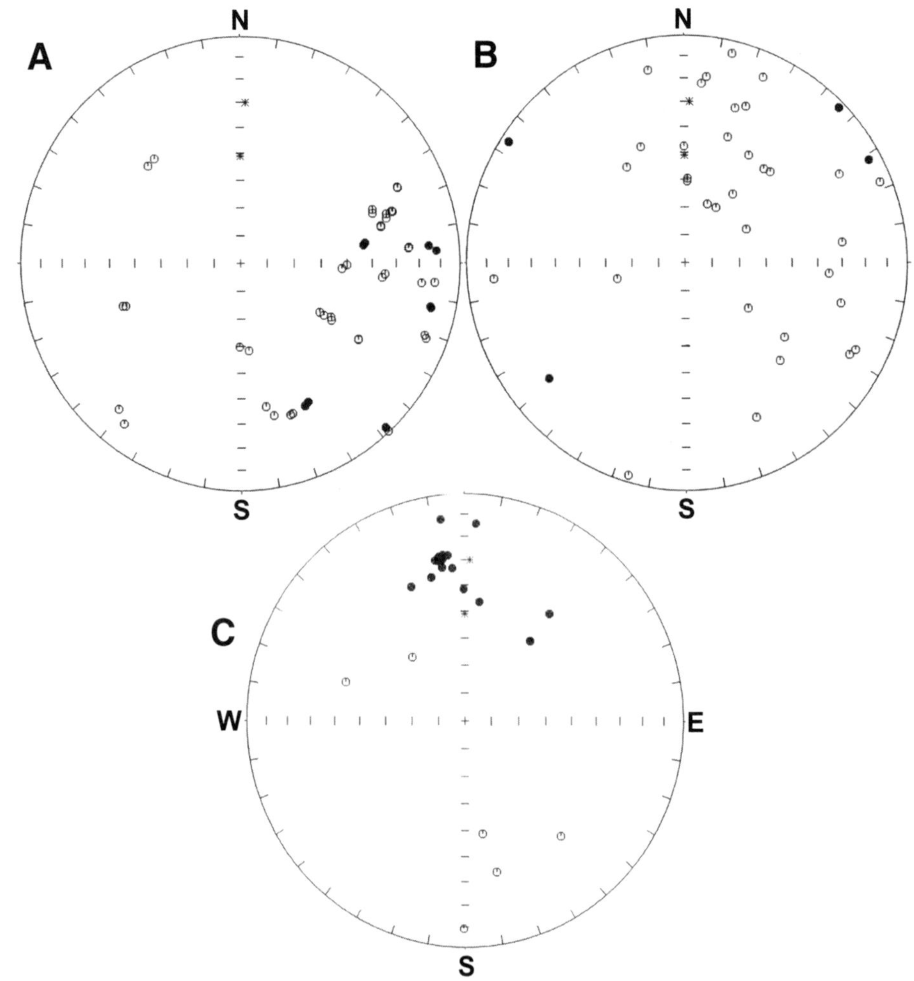

Figure 6. Locality # 7 altered sedimentary strata adjacent to a pluton. (A) Initially measured directions. (B) Initially measured directions of adjacent granites. (C) Stereographic plot of characteristic directions of altered sediments (in situ coordinates). Symbols and projections as in Figure 2.

of the samples of locality #9; the pattern suggests a dual polarity, antipodal population (Fig. 7). NRM_0 directions of all but four are directed to either the southwest or northeast; moreover, a semblance of this pattern is repeated in the other plutons. To emphasize the pattern, a light-gray oval was drawn to encompass the SW directions of locality #9 in Figure 7, and then reflected to the NE. Eight of the ten locality #9 samples exhibited the SW/NE direction in at least one specimen; steep negative directions like those of localities #7 and #8 were displayed by two other specimens, and a third displays a shallow NW direction.

Locality #11 samples displayed some NRM_0 directions like those of locality #9 (Fig. 7). Regardless of composition, eight of the 15 samples exhibited SW/NE NRM_0 directions in at least one specimen, but the population dispersion was greater than at locality #9 (a larger oval was required to encompass SW/NE directions). The majority of NRM_0 directions were either steep negative, steep positive, or shallow directions in the NW (and SE) quadrant.

Locality #10 samples showed few (3) NRM_0 directions in the SW/NE quadrants (Fig. 7); seven of the samples (10 specimens) exhibited only steep negative or steep positive NRM_0 directions, and a few others, NW (and SE) directions.

Among the three localities, no compositional preference, mafic inclusion versus granodiorite, was associated with any of the four NRM_0 directions, SW/NE, steep negative, steep positive, NW (SE). NRM_0 intensities of mafic inclusion samples were between 3×10^{-4} and 6×10^{-2} mA/m and were similar among the granodiorites, 3×10^{-4} to 1×10^{-2} mA/m.

The diorite of locality #5 also showed many NRM_0 directions that were notably similar to locality #9 (Fig. 7). The majority of the population suggests the same SW/NE dual polarity, quasi-antipodal population. The similarity of the two NRM_0 populations of localities #5 and #9 is sufficiently striking that mean directions were calculated for comparison (shown in Table 3). Diorite NRM_0 intensities ranged between 1×10^{-3} to 7×10^{-4} mA/m.

Demagnetization

The most coherent demagnetization responses, and similar final directions revealed by demagnetization, were obtained from the diorite samples of locality #5, hence these samples were demagnetized most intensively. Neither AF nor thermal demagnetization was effective alone, and a variety of AF and thermal combinations were tried. The most effective method was thermal

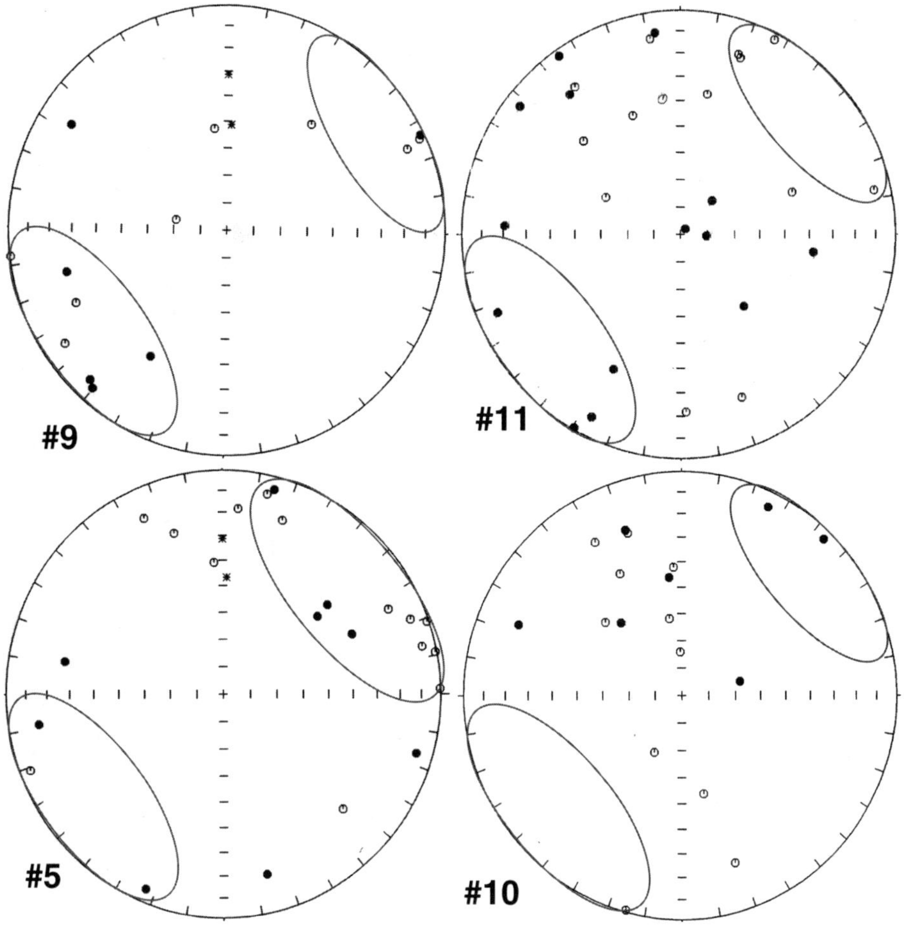

Figure 7. Natural remanent magnetization (NRM_0) of all specimens of mafic inclusions and host granodiorites from localities #9, #10, and #11, and of the diorite pluton of locality #5. Equal area projections; solid (open) circles indicate lower (upper) hemispheres; asterisks indicate the present axial geocentric dipole field and presently observed geomagnetic field directions. Light ovals draw attention to NRM directions that repeat at some localities

TABLE 3. MAGNETIC DIRECTIONS OF PLUTONIC ROCKS

	Dec	Inc	*n/N*	*k*	α_{95}	Lon/E	Lat	α_{95}	*k*	PLat
NRM_0 MEANS										
SITE #9	240.0	8.5	14/1	4.3	21.6					
SITE #5	230.7	7.4	23/1	3.1	20.6					
MEAN NRM_0	234.5	7.9	37/2	3.6	14.7					
DEMAGNETIZED MEAN / PALEOPOLE										
SITE #5 ALL	238.4	-0.8	14/1	8.7	14.3					
SITE #5 ALL-SR+	250.9	–1.4	10/1	16.8	12.2	186.6	18.1 S	9.0	29.5	0.7 S
SITES #9-11 CHR ALL	227.7	–3.8	19/3	7.0	14.1	193.6	40.7 S	9.2	12.0	1.9 S
SITES #9-11 VEC ALL	227.9	0.0	52/3	6.5	8.4	195.8	40.0 S	7.7	7.5	0.0
SITES #9-11 CV ALL	228.1	–1.0	69/3	6.6	7.2	195.0	39.9 S	6.5	7.9	0.5 S
SITES #9-11 CV SR	205.0	–1.3	33/3	15.4	6.7	214.1	62.1 S	5.1	25.1	0.7 N
SITES #9-11 CV 230+	254.3	0.1	32/3	15.4	6.7	185.8	15.1 S	5.1	26.3	0.1 N
PLUTON POLE*			42/2			186.2	16.8 S	4.3	27.4	0.3 S

Note: All poles calculated from virtual geomagnetic poles; + = populations with probable Santa Rosa remanence contamination excluded; NRM_0 = initial measurement of the natural remanent magnetization; CHR = characteristic sample directions; VEC = additional vectors (see text); CV = combined characteristic and vector populations; "SR" and "230" = divisions of the "ALL" populations into Santa Rosa–like directions ("SR") and remaining directions ("230"); n/N = number of samples/sites; *k* = precision parameter; α_{95} = half-angle of the cone of confidence about the mean; * = pole plotted in Fig. 10 (see text).

demagnetization first, in three to seven steps between 54 °C and 275 °C; thereafter samples were subjected to AF demagnetization of 10–40 mT. Finally, the samples were again heated, in 20–30° steps, beginning at 300 °C and continuing through 435 °C, at which point, directions appeared random and intensities weak.

Seventeen specimens from locality #5 were demagnetized, one each from 11 samples and two each from three samples. In general, overlapping stabilities of several magnetizations appear to characterize many of the samples. The majority of the samples (10) exhibited linear components with SW/NE directions. Demagnetization responses trended toward the origin of orthogonal axes plots for some samples, but not for many others. Five specimens exhibited primarily NE directions throughout demagnetization (e.g., Figs. 8A–B). The SW direction was exhibited by four specimens during demagnetization, but these also each displayed an approximately antipodal direction (Figs. 8C and 8E). Four specimens exhibited the present axial dipole field direction; two exhibited it throughout demagnetization, and two exhibited it as one of two components revealed by demagnetization.

Characteristic directions were obtained by analysis of linear segments within the demagnetization trends. Most locality #5 samples exhibited linear components with directions in the NE quadrant with shallow inclinations, and three samples exhibited southwesterly-directed magnetizations (Fig. 9A). Two other samples displayed only the present axial dipole field direction, and one, a shallow SE direction; these three were excluded from Figure 9A. Hence, 14 specimens (11 samples) yielded directions in the NE or SW quadrants as a result of demagnetization. However, among the directions of Figure 9A, two populations appear to exist, one group with declinations between 45° and 90° (and its

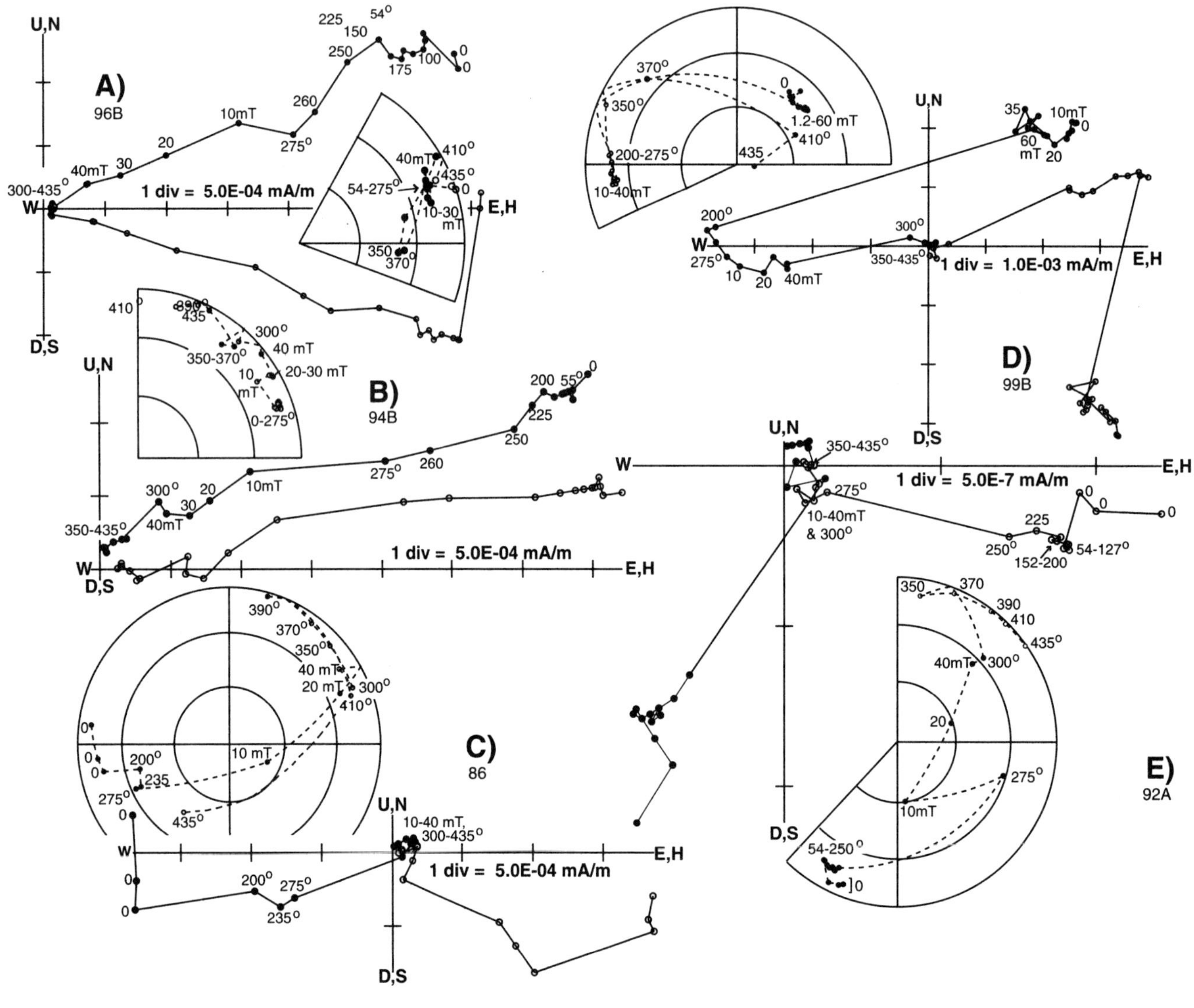

Figure 8. Locality #5 representative demagnetization responses from. Symbols as in Figure 2.

antipode) and another with ~30° declinations (and the antipode). The latter group appears similar to the Santa Rosa direction, so the mean Santa Rosa direction (Table 2) is shown in Figure 9A for comparison.

Few of the mafic inclusion or granodiorite samples of localities #9–#11 responded as well to demagnetization as did those of the locality #5 diorite. A total of 36 specimens from 23 samples, 24 granodiorites and 12 mafic inclusions, were demagnetized. Approximately half were treated with AF, and half with thermal demagnetization. Paired specimens from each of 11 samples were treated, one specimen by AF demagnetization and one by thermal demagnetization. A few specimens were subjected to a combination of treatments, one method followed by the other; in most cases these dual treatments produced no convergence of directions among samples. Specimens representing all of the NRM_0 directional groups observed were chosen for demagnetization.

Despite the variety of demagnetization treatments applied, the result was frequently widely scattered, meaningless directions. AF demagnetization alone generally caused appreciable remanence decay but widely scattered directions. Most NRM_0 directions were not changed in intensity or direction by thermal demagnetization to 250 °C. At locality #9, most SW/NE NRM_0 directions were retained through demagnetization, whereas samples from the other the localities commonly lost their SW/NE NRM_0 directions. As a result of demagnetization, 17 of the 36 specimens exhibited SW/NE linear segments (Fig. 9B). An additional 15 of the 36 specimens did not display the SW/NE directions, either initially or as the result of demagnetization; these yielded either directions in the

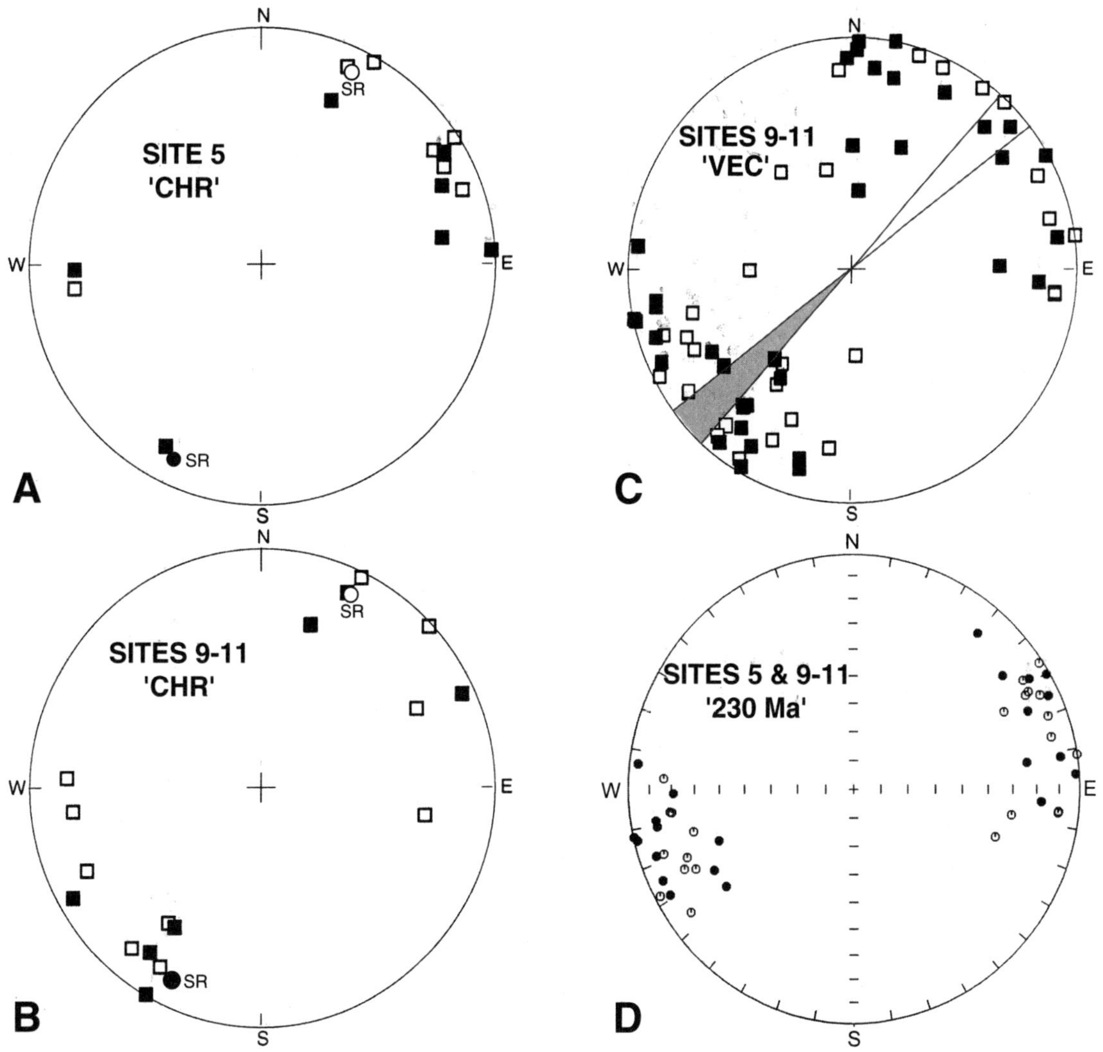

Figure 9. (A) Locality #5 directions from demagnetization. SR—mean direction of the Santa Rosa Group, shown for comparison. (B) Linear components from localities #9–#11, which lay in the SW or NE quadrants. (C) Localities #9–#11 and removed vectors lying in the SW and NE quadrants. (D) Directions remaining (CHR and VEC populations) of sites #5, #9, #10, and #11 after removal of presumed Santa Rosa directions. Equal area projection; solid/open symbols indicate lower/upper hemisphere.

NW or SE quadrants or steep negative directions. Additionally four of the 15 granodiorite samples exhibited the present axial dipole direction as a result of demagnetization.

Among those samples not exhibiting linear segments with directions in SW/NE quadrants, those directions were exhibited as either removed vectors or single-step results during demagnetization. Commonly, these directions accompanied large directional or intensity changes during demagnetization. Because of their frequent appearance in the demagnetization experiments, these vectors were compiled in Figure 9C.

The large mafic inclusion at locality #6 yielded six samples, which all displayed the same NRM_0 direction, with relatively strong magnetizations (~10^{-2} mA/m). The directions were quite stable to both thermal and AF demagnetization. However, the direction, at 165° declination and 50° inclination, is not seen elsewhere, hence its origin is unknown.

INTERPRETATIONS OF RESULTS

Santa Rosa Group

Magnetization Direction

Reversal Test. The antipodal characteristic of the two polarities of magnetization observed in the red Santa Rosa strata was evaluated with McFadden and McElhinny's (1990) reversal test statistics; the test was applied to the sample directions from localities #1, #2, and #3. A positive reversal test was obtained (Table 2), with a classification of "B" (of options A–C), which indicates that the magnetization red strata was acquired prior to folding. The "B" classification implies that the two polarities of magnetization represent antipodal directions of the same age geomagnetic field.

Fold Test. A fold test was also applied to these directions, using McFadden's (1990) fold test; that test indicates a pre-folding magnetization (Table 2), again with a classification of "B" (of options A–C). This result suggests, with relatively strong confidence, that the magnetization was acquired pre-folding.

Mean Direction. Fisherian statistics, applied to the population of all samples from localities #1 and #2 and one sample from locality #3, yielded a mean of 205.7° declination, −10.8° inclination (Table 2). However, to better estimate the confidence limits of the mean, individual directions were converted to virtual geomagnetic poles; the virtual geomagnetic poles (VGPs) then were grouped by stratigraphic proximities of the sites within the section. Seven groups were so defined and a mean VGP computed for each. A paleopole representative of the Santa Rosa Group was calculated from these seven mean VGPs, 202.9, −62.5 (k = 446.68, α_{95} = 2.9; Table 2).

Magnetization Origin, Timing, and Age

The reconnaissance mapping of the Maya Mountains observed the color of the Santa Rosa strata to be gray to dark-gray (or black) (Ower, 1928; Dixon, 1956; Bateson and Hall, 1977). However, the strata encountered in this investigation were almost exclusively red in color. Red strata also were observed by Simmons (1972), who reported an abundance of red Santa Rosa strata around the southern and eastern portions of the northwestern Mountain Pine Ridge plutonic complex. Additionally, red strata were reported adjacent to the C/S plutonic complex (Muncaster, 1976). These observations suggest that red colored strata occur adjacent to the plutons. The age relationship between the Silurian plutons and the Late Pennsylvanian to Middle Permian sedimentary rocks allows for exposure of the plutons during deposition of the Santa Rosa Group. Furthermore, the increase in grain size in proximity to the plutons (e.g., conglomerates and sandstones; Bateson and Hall, 1977) indicates that the plutons were exposed during Santa Rosa sedimentation. Red colored around the intrusive complexes is consistent with weathering of exposed granitic rock in tropical latitudes; significant oxidation would occur, thus yielding oxidized, red sediment. Thus, localization of the red Santa Rosa strata around the plutonic rocks is predictable consequence of the conditions of deposition; presumably, a red bed type of environment of deposition existed around the exposed plutons.

The fold and reversal tests implied an early magnetization of the red Santa Rosa strata, and the presence of reversals indicates that magnetization was acquired near the time of deposition. In addition, several rock-magnetic characteristics also indicate a depositional magnetization. A large portion of the magnetization is unblocked at temperatures between 570 and 620 °C; temperatures below 580 °C suggest a magnetite carrier. The 590–620 °C unblocking temperatures are similar to Curie temperatures reported for maghemite with small amounts of impurities (e.g., 591 and 645 °C; Michel et al., 1951, and Ozdemir and Banerjee, 1984, respectively). It is probable that magnetite deposited in a red bed setting would oxidize partially or wholly to maghemite. Furthermore, most of the IRM was acquired by 250 mT, field strengths indicative of IRM acquisition by magnetite or maghemite, and certainly not characteristic of hematite. The IRM acquisition data resemble those of very fine grained, non-interacting magnetite grains, which were observed to saturate in fields between 300 and 400 mT (see Dunlop and Ozdemir, 1997). Although the 570–620 °C unblocking temperatures could be explained by the presence of a narrow grain size range of hematite, such hematite would not acquire most of its IRM in fields below 250 mT. In addition to the apparent presence of magnetite-maghemite minerals, the presence of hematite is indicated by the relatively small amount of remanence that is unblocked above 620 °C and the continued minor acquisition of IRM above 250 mT. Thus, the reversal test, the unblocking temperatures, and the IRM acquisition characteristics suggest a magnetization near deposition, and the unblocking temperatures and IRM acquisition suggest that a significant magnetization carrier is magnetite and maghemite. The presence of both magnetite and hematite in red bed strata is not uncommon, having been observed a number of times (e.g., Helsley, 1969; Steiner and Helsley, 1974; Hillhouse et al., 1982; Shive et al., 1984; Steiner, 1988; and Steiner and Lucas, 2000).

The presence of magnetite and maghemite magnetization carriers in red strata argues for a partially detrital magnetization, because magnetite and maghemite cannot be crystallized by precipitation under the oxidizing conditions necessary to generate hematitic red pigment. Hence, if magnetite is present, it must have been transported to the depositional site as detrital grains; as mentioned, maghemite could be created by oxidation of the magnetite. In the sedimentary environment, such oxidation may have occurred either during the granitic weathering, sediment transport, or after deposition of the magnetite. Magnetite in red calcareous sediments has been shown to carry a stable magnetization from near the time of deposition (e.g., Channell et. al., 1982; Steiner et al., 1987), suggesting a perhaps enhanced stability of magnetite under somewhat oxidizing conditions of deposition; the co-occurrence of magnetite and hematite in the red bed environment also may argue for the stability of some of the magnetite deposited in an oxidizing environment.

Early acquisition of magnetization also is indicated by the sequence of multiple polarity intervals in stratigraphic succession. The presence of reversals in a stratigraphic sequence implies acquisition of magnetization close enough in time to sediment deposition to be able to record successive changes of polarity of the geomagnetic field. Additionally, the observation that most samples carry an additional, weak remanence with polarity opposite to the characteristic magnetization supports the concept that the characteristic magnetization was acquired early. Because the weak opposite polarity magnetization alternates exactly out of phase stratigraphically with the characteristic remanence, this suggests a scenario in which the characteristic remanence was recorded during deposition and the weak antipodal magnetization was acquired as an early chemical remanence. In other words, magnetite was deposited and perhaps partially oxidized, and slightly later hematite crystallized, but the growth of hematite continued into the polarity interval following the one in which the magnetite was deposited. Observations of such dual polarity remanences within a single stratigraphic horizon in red sedimentary rocks have been documented previously, in both terrestrial red beds (e.g., Roy and Park, 1974; Purucker et al., 1980) and red marine limestones (Channell et al., 1982). In both depositional environments, acquisition of a depositionally related magnetization was followed by a second remanence imprinted during the subsequent polarity interval, as hematite continued to crystallize at or below the depositional interface. Thus, the presence in the Santa Rosa red beds of the additional magnetization antipodal to the characteristic remanence, observed in an out-of-phase magnetostratigraphy further indicates that the characteristic Santa Rosa remanence dates from the time of sediment deposition.

Although the literature describes the Santa Rosa Group as having been subjected to a lower greenschist facies metamorphism (see introduction), the Santa Rosa strata on the flanks of the Mountain Pine Ridge pluton have not been metamorphosed much at all, probably not above zeolite facies. Therefore, the survival of a magnetite remanence within a metamorphic rock does not require explanation, although magnetite-borne remanences have been observed to survive greenschist facies metamorphism in the red, volcanoclastic strata and basalts within the Blue Mountains of Oregon (Hillhouse et al., 1982). Those strata are red sandstones and siltstones containing Triassic faunal elements, and although the input was volcanoclastic rather than plutonic detritus, those strata also had magnetizations held in both hematite and magnetite, and also recorded both polarities of the Triassic geomagnetic field.

The age of the Santa Rosa magnetization is indicated by consideration of the fauna and the presence of multiple reversals. The fauna of the Santa Rosa Group shows that these largely marine strata were deposited in Late Pennsylvanian through Middle Permian times (e.g., Bateson and Hall, 1977; Ross, 1979). However, the acquisition of a dual polarity remanence could not have been acquired in the Late Pennsylvanian or the Early Permian, because the geomagnetic field remained in one polarity from the Early Carboniferous through the Early Permian (the Kiaman Reversed Superchron). Although the age of the end of this lengthy period of constant reversed polarity of the geomagnetic field is not known precisely, the oldest biostratigraphically dated Permian magnetizations of normal polarity are Wordian to early Capitanian, that is, early Middle Permian in age (Heller, et al., 1995; Steiner, 2001). Thus, the magnetization of the sediments around the Mountain Pine Ridge complex can be no older than early Middle Permian. The nearby fauna agrees with this age; the youngest diagnostic faunal element in the Santa Rosa Group is the ammonite *Perrinites hilli*, which also occurs in the Middle Permian stratotype in west Texas and is diagnostic of the Roadian Stage, i.e., earliest Middle Permian (Wilde, 1986). *Perrinites hilli* has been found (e.g., Bateson and Hall, 1977) in strata near the locality yielding the Santa Rosa magnetostratigraphic sequence, only about 8 km to the south. Hence, the presence of polarity reversals in the Santa Rosa strata agrees well with the biostratigraphic data, and indicates that sedimentation in the northern Maya Mountains is at least as young as early Middle Permian.

Other Santa Rosa Magnetizations

Magnetizations at localities #4 and #7 reflect Tertiary geomagnetic fields. The white phyllitic mudstones and their bedding plane–bound iron oxide bands at locality #4 exhibited opposite polarities of the present axial geomagnetic field direction magnetization. The presence of red iron oxide bands along bedding planes in the white phyllitic mudstones suggests that these phyllites were originally red as at site #1, but have been leached and the iron oxides segregated along the bedding planes and fractures. The reversed polarity of the iron oxide bands indicates that segregation of these bands took place during a reversed polarity Tertiary geomagnetic field. The low stability of the normal polarity remanence in the white mudstones may suggest a recent viscous remanence.

The hydrothermally altered Santa Rosa beds of locality #7 also exhibited a recent geomagnetic field remanence. This recent remagnetization is attributed to the intensely weathered condition of this exposure of sedimentary strata; the outcrop, in part hosts a spring.

The steep negative initial direction observed in both the metasedimentary strata (locality #7) and in the adjacent granitic pluton (locality #8), as well as occasionally in the neighboring granodiorite pluton (localities #9, #10, #11) is a peculiar direction. It also was observed as the first removed vector in a few of the metasedimentary rocks of locality #1, and in one sample at locality #2. This relatively widespread remanence direction is surprising because its origin is not recognized as any known post-Permian direction. While its origin may be simply the superposition of two oppositely directed components of remanence, it is interesting that similar directions were observed in the Chiapas granites and in the Todos Santos red beds (Molina Garza et al., 1992, their Figures 5 and 8).

Plutonic Remanence

Characteristic Directions

The plutonic complexes of both the eastern and western Maya Mountains exhibit similar dual polarity magnetizations lying in the SW and NE quadrants. These directions are clearly observed in the initial measurements of the plutons of the western locality (#5) and one eastern locality (#9). These NRM_0 directions survive demagnetization in most of the western samples, and in one-third of all eastern samples. After demagnetization, the characteristic directions remain similar to the NRM_0 directions of localities #5 and #9 (Table 3). The term "characteristic" magnetization is defined for the plutonic rocks as the vector observed from the linear progression in the largest number of demagnetization steps during demagnetization, hence being most "characteristic" of that sample or specimen.

At both localities #5 and #9, the characteristic directions actually appear to comprise two populations, one of which mimics the Santa Rosa Group remanence; to illustrate this, the Santa Rosa mean direction is added to Figures 9A and 9B. At locality #5, in which the NRM_0 directions of most samples survive demagnetiztion, only four samples appear to exhibit Santa Rosa directions (Fig. 9A), and the remainder exhibit directions which form a distinctly different dual polarity population.

The characteristic (as defined above) magnetization directions of the one-third of the eastern pluton samples (localities #9, #10, and #11) with directions similar to those of locality #5 are shown in Figure 9B; the other two-thirds of these samples display multiple magnetization components with largely overlapping stabilities. However, many of those samples exhibited directions of one of their multiple components that were similar to locality #5. That component was observed either as the vectors removed at one or several demagnetization steps and/or as a resultant direction at one (or two) demagnetization step(s); these directions are plotted in Figure 9C. The exhibition of directions similar to locality #5 frequently accompanies large directional or intensity changes in these multi-component samples. To distinguish these brief observations of the locality #5 direction from their observation as "characteristic" remanences, the term "vectors" is used in Table 3 and Figure 9.

A mean direction for the western, locality #5 characteristic directions was calculated (Table 3) after excluding two samples with recent geomagnetic field directions and a single sample with a SE direction that is considerably statistically distinct (>2 standard deviations) from the rest of the population. A second mean calculated after exclusion of the four samples with Santa Rosa–like directions; this resulted in a doubling of the *k* value (Table 3), and lends support to the concept of two directional populations, one of which may be a Santa Rosa overprint. An overprint created by weathering in Santa Rosa time is reasonable given the tropical weathering environment in which exposed plutons were eroded to produce the Santa Rosa sediment. The directions of this second, more restricted group then were converted to virtual geomagnetic poles (VGPs), and a mean paleopole was calculated for locality #5 (Table 3).

Both groups of directions from localities #9–#11, "characteristic" and "vectors," also appear to comprise two populations, one of which appears to be the Santa Rosa direction. The other direction is that exhibited by the majority of the locality #5 samples. Mean directions for localities #9–#11 were calculated for each of the groups (CHR and VEC; Table 3). Notably, the "vectors" group displays a gap within the SW quadrant directions, in which no directions lie (shaded triangle of Fig. 9C), possibly because of the presence of two separate populations. This gap (and the reciprocal arc segment in the NE quadrant) was used to separate the directions into the hypothesized two populations, a Santa Rosa–like group and the remainder; NE quadrant sample directions which lay within the "gap" segment were excluded from consideration. Also, directions with steep inclinations (~>±40° inclination) were excluded from both groups, because of the possibility that these represent the peculiar, unidentified steep negative-inclination overprint of localities #7 and #8, and the likelihood that those directions of steep positive inclination may represent the axial dipole field direction at these sites. Means each were calculated for the "characteristic" and "vector" groups (Table 3). Next, the two populations of directions from the "characteristic" group were combined with the corresponding two of the "vectors" group, and two additional means were calculated, one for the Santa Rosa–like directions and one for the remainder (Table 3). Directions of each of the groups were converted to VGPs to calculate the mean paleopoles shown in Table 3. Finally, the non–Santa Rosa VGPs of localities #9–#11 and locality #5 were combined to derive a mean plutonic paleopole for the Maya Mountains (Table 3).

Remanence Origin

The remanence of the igneous rocks is significantly more complex than that of the Santa Rosa Group; the plutonic rocks commonly possess multi-component magnetizations, and only infrequently exhibit univectorial-to-the-origin demagnetization behavior.

The geologic event which reset the K/Ar radiometric systems of all of these Silurian plutons at ca. 230 Ma may have played a role in the remanence complexity. The resetting of K/Ar systems occurred throughout the entire Maya Mountains region (as was illustrated by the K/Ar data listed in Table 1), and the K/Ar systems of all igneous rocks (including volcanics) were reset; further the sedimentary rocks at the pluton margins also were altered. A mean age for the radiometric resetting (from the existing 10 K/Ar determinations of Table 1) is 227 ± 8 Ma, corresponding to the early Late Triassic (Gradstein et al., 1995). However, because two of the ten determinations lie more than a standard deviation away from the other eight; the true age of the event might be 231 ± 8 Ma, or late Middle Triassic (Gradstein et al., 1995). A hydrothermal episode associated with the opening of the Atlantic and Gulf of Mexico would fit this timing.

In addition, the Rb/Sr system also was partially reset in the western plutons, which is evidenced the disturbed isochron which gave an age of 336 ± 10 Ma from the same plutons which presented the ca. 230 Ma K/Ar ages (Bateson and Hall, 1977) and 420 Ma U/Pb ages. The U/Pb system of that plutonic complex also shows evidence of having been disturbed, possibly by the 230 Ma event, because some of the plutons exhibited lead loss (Steiner and Walker, 1996). In the eastern (HMR) plutons, the same reset K/Ar ages and evidence of lead loss in the U/Pb determinations was observed.

Hence, the magnetization complexity of the plutons and the observed direction logically might be the consequence of an overprinting of a magnetization during the tectonic event which reset or disturbed the radiometric systems. The regional extent of the resetting of K/Ar systems in the Maya Mountains and the occurrence of the same magnetization direction in both eastern and western plutonic complexes suggest that the same tectonic event may have been reset both magnetic and radiometric systems. If this reasoning is correct, the age of the dominant igneous magnetization dates from the Middle–Late Triassic boundary, ca. 230 Ma. This postulated Triassic age for the remanence is tested in the next section.

PLATE TECTONIC INTERPRETATION

Yucatan Paleopoles

Four paleopoles now exist to describe the history of the Yucatan Block (Table 4). The Permian Santa Rosa and 230 Ma plutonic poles are considerably displaced from North American Permian and Late Triassic poles. Depending on which polarity represents the normal state of the geomagnetic field, the Santa Rosa Group pole is located either 128° counterclockwise or 52° clockwise from North American Permian paleopoles; likewise, the ca. 230 Ma plutonic paleopole lies either 105° or 75° from the mean North American Late Triassic paleopole. No specific evidence obtained from this study allows determination of which of the dual polarities of each bi-directional population represents the normal polarity state of the geomagnetic field at the time of remanence acquisition. However, by working backward in time from published results, and hypothesizing the minimum amount of total apparent polar motion, this ambiguity is resolved.

Three previous studies have investigated rocks of the Yucatan Block. Gose and Sanchez-Barreda (1981) studied Early Permian limestones just north of the eastern end of the Chiapas Massif, a large granitic batholith at the southern edge of the Yucatan Block (Fig. 1). Their data suggested a 22° counterclockwise rotation of the sampling site with respect to cratonic North America.

The paleomagnetism of the Late Jurassic (Kimmeridgian-Tithonian) San Ricardo Formation was studied by Guerrero et al. (1990); the paleopole was coincident with the same age North American Late Jurassic reference pole, the Morrison Formation. The age of the Morrison Formation is well known, defined broadly by Kimmeridgian-Portlandian vertebrate and invertebrate faunas, but more precisely by direct correlation to the marine magnetic anomaly time scale (Steiner, 1998) and by $^{40}Ar/^{39}Ar$ dating as 150 through 148 Ma (lower to upper beds; Kowallis et al., 1998).

More recently, the granites of the Chiapas Massif itself and the overlying Jurassic Todos Santos Formation strata were investigated (Molina Garza et al., 1992). The characteristic directions of the Chiapas granites were relatively scattered, but formed a two-polarity population of directions with nearly east-west declinations and shallow inclinations (Molina Garza et al., 1992). The Maya Mountains plutonic rocks directions are remarkably similar, in dispersion, direction, and dual polarity to the Chiapas plutonic directions. In fact, the mean directions and paleopoles of the plutonic bodies in the two locations are statistically indistinguishable, as shown in Table 4 and Figure 10A.

TABLE 4. PALEOPOLES FROM THE YUCATAN BLOCK

	AGE	D	I	*n/N*	*k*	α_{95}	Lon/E	Lat	Paleolatitude
SANTA ROSA GROUP	M-L Permian	205.7	−11.1	20/7	449.4	2.9	202.6	62.5 S	5.6 ± 2 S
MAYA MTNS PLUTONS	ca. 230 Ma	252.6	−0.7	42/2	27.4	4.3	186.2	16.8 S	0.3 ± 2 S
CHIAPAS MASSIF*	ca. 230 Ma	261.3	−9.0	56/12	15.4	11.4	176.1	13.4 S	4.5 ± 6 S
TODOS SANTOS1*	M-L Jurassic	293.7	+2.7	16/4	56.8	12.3	170.6	22.6 N	1.3 ± 6 N
"TODOS SANTOS"2*	M-L Permian	187.7	−4.8	16/4	23.7	25.9	180.6	75.0 S	2.4 ± 14S
SAN RICARDO†	L Jurassic	340.3	+19.9	89/1	25.4	3.0	160.0	69.8 N	10.2 ± 2 N

Note: D—declination; I—inclination; *n/N*—number of samples/sites; *k*—precision parameter; α_{95}—the half-angle of the cone of confidence about the mean
*Molina Garza et al. (1992)
†Guerrero et al. (1990).

The Chiapas granites were dated by the Rb/Sr whole-rock isochron method (Damon et al., 1981), which indicated an age of 256 ± 10 Ma. However, the Rb/Sr method has been demonstrated to be unreliable for granites that have undergone any metamorphism (Krogh and Davis, 1973). Metamorphism has affected the Chiapas Massif region; the surrounding sedimentary rocks are very similar to those of the Maya Mountains in metamorphic grade, as well as general composition (McBirney, 1963). Furthermore, Rb/Sr dating of the Maya Mountains (Mountain Pine Ridge) plutons yielded a disturbed isochron, and a resulting date (Bateson and Hall, 1977) which is nearly 100 m.y. younger than the U/Pb zircon age (Steiner and Walker, 1996). This result again demonstrates the fallibility of the Rb/Sr method in metamorphosed plutons. On this basis alone, the Chiapas Massif age is probably unreliable. But the fact that the Chiapas Massif exhibits the same magnetic direction as held by the ca. 420 Ma Maya Mountains plutons also indicates that the Chiapas Massif age probably is not accurate; the Chiapas Massif should be re-dated by the U/Pb method. In addition, the presence of the same remanence in plutonic complexes of differing ages also suggests that these remanences are not crystallization magnetizations in either.

The common remanence direction in the Chiapas and Maya Mountains plutons has three exceedingly important implications.

1. The Chiapas magnetization must be the same age as the Maya Mountains igneous rock remanence, inferred to be 230 Ma. Supporting this argument is the fact that the Chiapas Massif also yielded a few K/Ar ages near 230 Ma (Damon et al., 1981). The presence of the same remanence direction and the 230 Ma

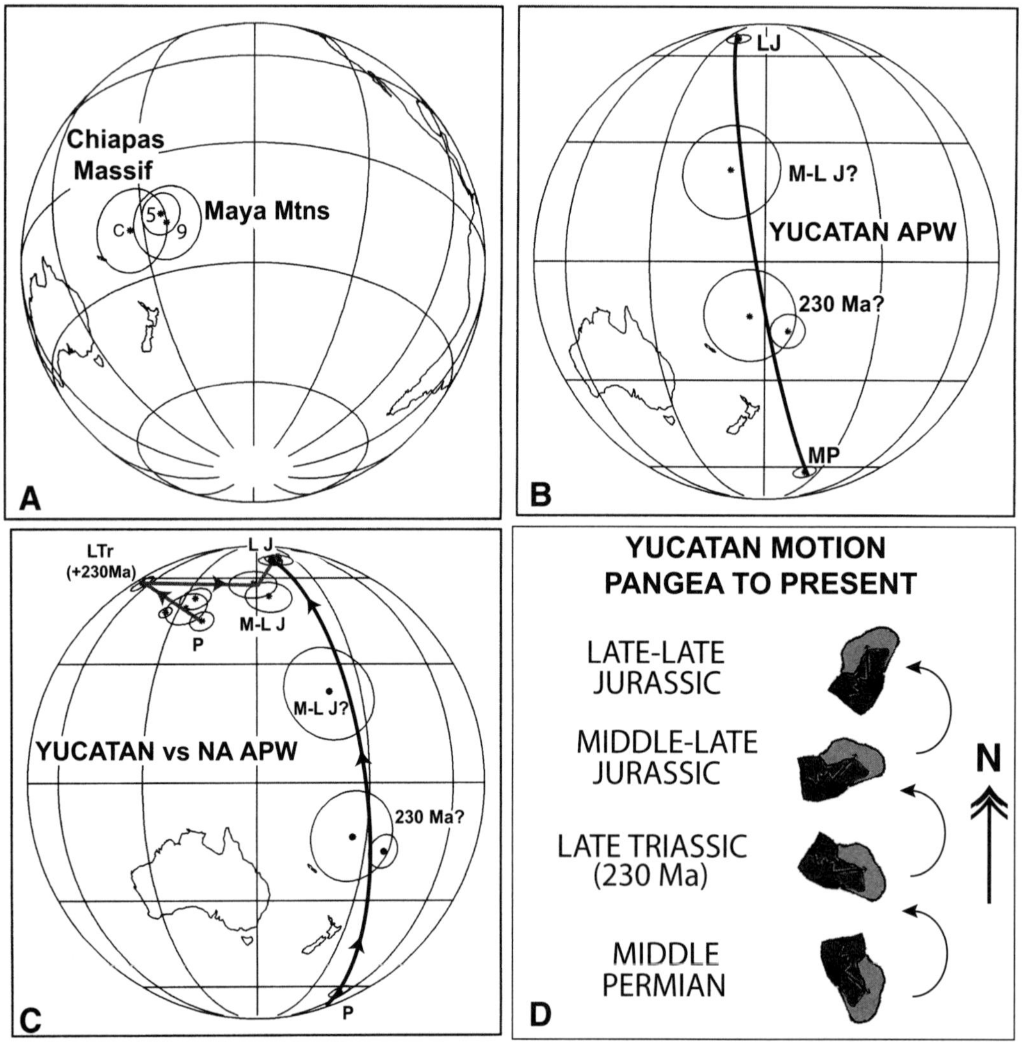

Figure 10. (A) Paleopoles for the plutonic rocks at localities #5 (western Maya Mountains) and #9 (eastern Maya Mountains) compared to that of the Chiapas Massif ("C") of Molina Garza et al. (1992). (B) Paleopoles and inferred apparent polar wander path (APW) for the Yucatan Block from Middle Permian through Late Jurassic. (C) Comparison of Yucatan and North American (NA) APW paths. (D) Successive orientations and rotations of the Yucatan Block from Middle Permian to Late Jurassic. LTr—Late Triassic; M-L J—Middle-Late Jurassic; LJ—Late Jurassic; P—Permian.

ages indicate that a widespread tectonic heating event affected the southern Yucatan Block; the most significant tectonic event occurring at about this time was the initial rifting of Pangea. Initial rifting of the proto–Central Atlantic Ocean is approximately this age, because the oldest infilling sediments in nearly all of the Triassic rift basins of NE North America are late early Carnian (Huber et al., 1993; Lucas and Huber, 1993), or 226–227 Ma (Gradstein et al., 1995). Therefore, the Central Atlantic and presumably initial Gulf of Mexico rifting must have occurred prior to the oldest sedimentation, which was prior to 226–227 Ma, a timing that agrees well with the ca. 230 Ma resetting of K/Ar systems in southern Yucatan, and presumed associated hydrothermal event and remagnetization of the plutons.

2. A second important implication of the presence of the same magnetization in both the Chiapas Massif and the Maya Mountains plutons is that it indicates that the Chiapas Massif has been an integral part of the Yucatan Block since at least ca. 230 Ma. The issue of whether the batholith was a part of the Yucatan Block had been debated, and a number of reconstructions have excluded the Chiapas Massif and immediately adjacent terrane from the Yucatan Block (e.g., Pindell, 1985; Klitgord and Schouten, 1986; Dunbar and Sawyer, 1987; Ross and Scotese, 1988; Pindell and Barrett, 1990). However, as argued by Molina Garza et al. (1992), no geologic evidence exists for the belief that the Chiapas Massif was not integral to the Yucatan Block. The evidence presented here support their argument and demonstrates that the massif has been part of the Yucatan Block since at least the Late Triassic.

3. A third implication of the common remanence direction in both the Maya Mountains and Chiapas plutons is that it demonstrates that the Maya Mountains have not rotated relative to the remainder of the Yucatan Block, a possibility that occasionally has been speculated (e.g., Pindell, 1985).

The Triassic-Jurassic Todos Santos strata studied by Molina Garza et al. (1992) onlap the Chiapas Massif. A dual polarity population of magnetization directions also was obtained from these strata. The age of the Todos Santos strata is poorly known, constrained only as being younger than the granites they onlap (younger than 256 Ma) and older than the overlying San Ricardo Formation. The San Ricardo beds disconformably overlie the Todos Santos Formation in the sampled area and contain a Late Jurassic, Kimmeridgian-Portlandian fauna (Richards, 1963; Alencaster, 1977; Molina Garza et al., 1992). Richards (1963) believed that the Todos Santos Formation was Jurassic because Late Jurassic fossils are found in pebbles in its lower conglomerates. The granites underlying the Todos Santos beds exhibit the ca. 230 Ma remanence, but that remanence was not observed in the Todos Santos beds (Molina Garza et al., 1992), thus the age of the Todos Santos beds probably are post–230 Ma and pre-Kimmeridgian (younger than ca. 156 Ma). Hence, the Todos Santos probably is Jurassic in age.

Molina Garza et al. (1992) studied the Todos Santos Formation at two localities, Matias Romero and Motozintla. The formation did not yield the same mean direction at the two localities. The Matias Romero locality yielded the direction Molina Garza et al. (1992) considered characteristic for the Todos Santos Formation (Table 4), whereas the Motozintla locality yielded a mean direction that is similar to the Santa Rosa direction of this study. Since both the Todos Santos and the Santa Rosa strata in proximity to the granites are red beds, it is possible that the strata sampled at the Motozintla locality were really red Santa Rosa Group strata. Santa Rosa strata overlie granitic intrusives in nearby Guatemala and incorporate abundant granitic detritus into the lower beds (McBirney, 1963; Dengo and Bohnenberger, 1969).

Yucatan Apparent Polar Wander

Although dual polarity characterizes most of the Yucatan magnetizations studied, and hence presents the possibility of polarity ambiguity, no polarity ambiguity exists for the youngest strata studied, the Upper Jurassic San Ricardo Formation, because the San Ricardo paleopole indicates that the Yucatan Block has not moved with respect to North America since San Ricardo deposition. Hence, extrapolating from the San Ricardo/North America Late Jurassic normal polarity sense, and assuming a path of least apparent polar wander, the westerly directed Todos Santos remanence is the Middle–Late Jurassic normal polarity vector. Similarly, the west-southwesterly vector of the ca. 230 Ma igneous rock remanences of the Maya Mountains and Chiapas Massif plutons represents normal polarity, as does the south-southwesterly vector of the Maya Mountains Santa Rosa Group. Defining the vector polarity in this way results in a smooth path describing Yucatan Block motion since the Middle Permian (Fig. 10B). This interpretation implies multiple counterclockwise rotations of the Yucatan Block since Permian times (Fig. 11), the same sense of rotation as was inferred for the Jurassic motion of the block to form the Gulf of Mexico (e.g., White, 1980; Pindell and Dewey, 1982; Dunbar and Sawyer, 1987).

The Santa Rosa paleopole lies 128 ± 5° counterclockwise from the North American Permian paleopoles (Fig. 10C). North American paleopoles are very similar throughout the Permian Period, suggesting that North America was almost stationary from Early through Late Permian time. An Early Permian mean pole for North America, calculated from the average of poles from the upper Casper (Diehl and Shive, 1979), Ingleside (Diehl and Shive, 1981), Abo (Steiner, 1988), and Pictou (Symons, 1990) Formations, and the Cutler Formation pole (Gose and Helsley, 1972, rotated 9° counterclockwise to correct for Colorado Plateau rotation [Steiner, 2003]), lies at 47.1°N, 122.7°E (k = 350.5, α_{95} = 4.1, n = 5). Averaging North American poles from the stratigraphically sequential formations spanning the Late and Middle Permian: the Elk City, Doxy, Seven Rivers, and Marlow Formations of Peterson and Nairn (1971) and the Dewey Lake–Alibates Formations pole (Molina Garza et al., 1989; Steiner, 2001), implies a mean Middle–Late Permian pole at 51.2°N, 124.7°E (k = 351.6, α_{95} = 4.1, n = 5). However, the Santa Rosa paleopole is probably equivalent in age to the earliest Late or latest Early Permian North American poles, so the two mean

Permian poles just calculated were averaged to a single Permian pole for Figure 10C.

North American Late Triassic poles indicate a complete still-stand of the continent during approximately the entire Late Triassic (see Steiner, 2003). The ca. 230 Ma Maya Mountains pluton and the Chiapas Massif paleopoles lay 105 ± 11° counterclockwise from the North American Late Triassic poles (Fig. 10C). The Jurassic Todos Santos pole was estimated by Molina Garza et al. (1992) to be 63 ± 11° counterclockwise of presumed comparable age North American poles. However, the age of the Todos Santos is uncertain, and the present study proposes in the next section that the Todos Santos beds may be somewhat younger; thus its pole may be rotated only 35 ± 12° from comparable North American early Late Jurassic paleopoles.

Yucatan: Pangea to Present

The location of the Yucatan Block during the existence of the Pangean supercontinent has heretofore been unknown. Permian Pangean reconstructions all have shown North and South America separated by a sufficient gap to allow the Yucatan Block to reside between these two plates. That reconstruction was based on the evidence that during the Jurassic, the Yucatan Block rotated 40–50° counterclockwise relative to North America from a location in the proto–Gulf of Mexico region to create the Gulf of Mexico (e.g., Pindell, 1985; Dunbar and Sawyer, 1987; Buffler and Thomas, 1994; Hall and Najmuddin, 1994). Lacking any information to the contrary, this location of Yucatan was carried backward in time into the Permian and the Pangean assembly. The present study

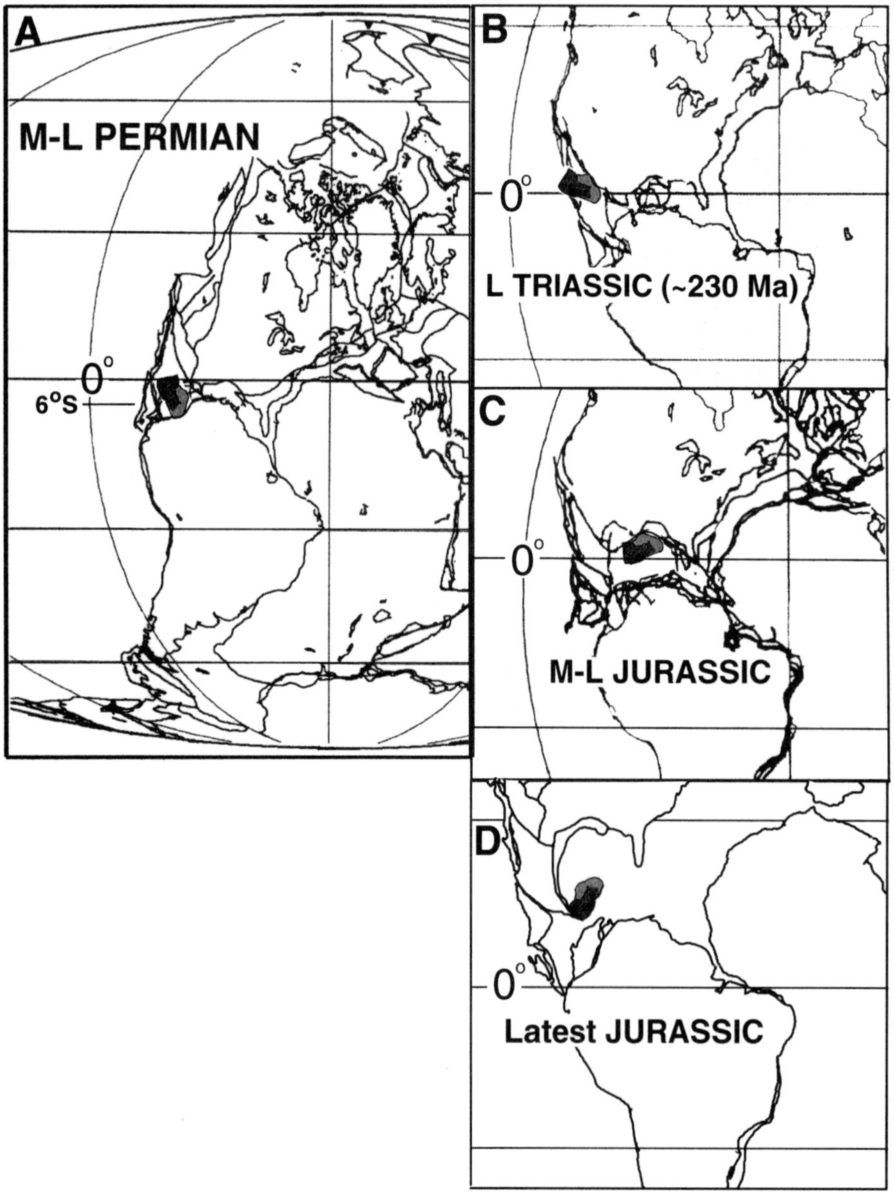

Figure 11. Approximate paleolocations of the Yucatan Block shown on reconstructions modified from Scotese and Langley (1995). Subaerial Yucatan is shaded in dark gray and its surrounding platform in light gray. Present day north on Yucatan is indicated by a tiny arrow. (A) The Yucatan Block in Permian Pangea. (No attempt has been made to reconstruct the many small crustal fragments of western margin of North American, which probably were located further northward along the North American margin.) (B) Yucatan in the Late Triassic. (C) The approximate late Middle–early Late Jurassic (Callovian-Oxfordian) reconstruction. (D) Latest Jurassic, Tithonian, reconstruction, by which time the Yucatan Block had achieved its current location and orientation with respect to North America.

was undertaken in the early 1990's to test this model, which had become dogma in the literature, but had never been tested.

Moreover, this model could not be reconciled with Permian paleomagnetic poles from North and South America; many have argued against Pangean reconstructions with Yucatan lying between the North and South American plates, because of the discordance imposed on Permian paleomagnetic poles from the major plates (e.g., Irving, 1977; Morel and Irving, 1981; Van der Voo et al., 1976, 1984; Van der Voo, 1981; Ziegler, 1981). The Middle Permian data obtained in this study prove that those early arguments of Irving and of Van der Voo were correct, and that the Yucatan Block did not lie between the North American and South American plates during the Pangean assembly. The Middle Permian paleopole of this study demonstrates that the present-day southern end of the block (Maya Mountains) resided in the Southern Hemisphere at 5.5 ± 3°S paleolatitude at that time (Tables 1 and 4). The data further indicate that the Yucatan Block was essentially inverted in orientation relative to the present day (Fig. 11A); this orientation is supported by the subsequently acquired ca. 230 Ma plutonic and Todos Santos remanences. This orientation and paleolatitude allow two possible Southern Hemisphere locations for the Yucatan Block in Permian Pangea, one off northwestern South America and possibly another between northeastern South America and Africa, or more precisely, between South America and the Florida block of Scotese and Langley (1995). It is not possible that Yucatan resided off the western margin of North America because southern North America lay across the equator in the Permian, thus its present western coast was entirely in the Northern Hemisphere at that time (Fig. 11A).

A location adjacent to northwestern South America (Fig. 10A) entails the least complex block motion of Yucatan from Permian through Jurassic times. The Yucatan Block is situated in a position similar to the other crustal blocks that make up Mexico today. The smooth Yucatan apparent polar wander path (Fig. 10B) indicates continuously counterclockwise rotation of Yucatan from Permian through Late Jurassic (Tithonian) times. The reconstruction of Yucatan adjacent to the northwestern South American margin brings its Silurian plutons into the proximity of those in northwest South America. Furthermore, the inverted position of the Yucatan Block required by the Middle Permian remanence places the Late Pennsylvanian–Middle Permian Santa Rosa basin with its andesitic and rhyolitic volcanism (see Bateson and Hall, 1977) on the outer edge of the aggregated Pangean continental mass, in a position which readily allows that volcanism to have been arc-trench related activity along the western margin of Pangea.

The Yucatan Middle Permian paleopole lying ~128° counterclockwise of North American Permian paleopoles, and the ca. 230 Ma Maya and Chiapas paleopoles lying ~105° counterclockwise of North American Late Triassic paleopoles imply that the Yucatan Block rotated ~47° counterclockwise relative to North America between Middle Permian and ca. 230 Ma (Figs. 10D, 11A, and 11B). The ca. 230 Ma orientation of Yucatan derived from this study and that of the Chiapas Massif granites is essentially the same as the paleo-orientation of the Yucatan Block relative to North America proposed by Anderson and Schmidt (1983). Thus the results of this study and that of Molina Garza et al. (1992) validate that part of the Anderson and Schmidt (1983) reconstruction.

Because the time of deposition of the Todos Santos Formation is not well constrained, a means was sought to attempt to more narrowly constrain its age. The conglomeratic and coarse-grained nature of Todos Santos strata imply tectonism during deposition. If the plate tectonic interpretation from this study is correct, that the Yucatan Block lay outside the Gulf of Mexico region during Pangean assembly, then the block must move into that region sometime during the Pangean breakup, but prior to the time when it appears to have become fixed in the North American framework, in the middle Late Jurassic (~Kimmeridgian). Late Jurassic fossils (stage nonspecific) in conglomerates of the basal Todos Santos beds (Richards, 1963) suggest that the tectonism occurred in the Late Jurassic; the superjacent San Ricardo Formation of Kimmeridgian-Tithonian age further narrows the time of Santa Rosa deposition. Together these facts suggest that the Todos Santos strata probably were deposited in the Oxfordian.

The North American plate was stationary relative to the geographic pole during the Callovian and Oxfordian (Steiner, 2003). Some data suggest that the plate experienced a change in plate motion in or after the Oxfordian, at the "J2 cusp" of the North American apparent polar wander path curve. Tectonism might have occurred with a significant change in plate motion, hence Todos Santos deposition may be an expression of plate rearrangement in the Oxfordian.

On this basis, the Todos Santos pole has been compared to the North American Callovian-Oxfordian stillstand poles (from the Summerville and Lower Morrison Formations; Steiner, 2003). This interpretation implies that the Todos Santos pole lay ~35° counterclockwise of the North American poles (Fig. 10C), and that the Yucatan Block underwent an additional ~41° of counterclockwise rotation relative to North America between 230 Ma and the Oxfordian (Figs. 10D, 11B, and 11C). Finally, the paleopole of Guerrero et al. (1990) indicates ~47° of further counterclockwise rotation of the Yucatan Block between deposition of the Todos Santos and San Ricardo Formations (between the Oxfordian and the Kimmeridgian or Tithonian), to bring the microplate to approximately its present orientation.

Initial breakup of Pangea involved a Late Triassic and Jurassic west-northwest motion of the North American plate relative to Europe/Africa (clockwise rotation of the North American plate) to create incipient opening of the Atlantic Ocean and the Gulf of Mexico. Conversely, South America–Africa–Europe moved counterclockwise with respect to North America. The Yucatan Block probably was a fragment of the western margin of South America. However, the counterclockwise rotations of Yucatan after Pangean fragmentation were far greater than that of South America–Africa, suggesting very early separation from the South American plate. The multiple counterclockwise rotations of Yucatan and the large magnitude of the rotations suggest that the small microplate may have acted in a ball-bearing fashion between the large North American, South American, and Pacific plates. For Yucatan to move from its southern latitude Middle Permian location in Pangea to its Late Jurassic proto–Gulf of Mexico orientation requires that Yucatan was translated left-laterally between the North American and South American plates sometime during the succession of counterclockwise rotations it experienced; this left-lateral translation while undergoing counterclockwise rotations further suggests the ball-bearing mode for the microplate during Pangean breakup.

SUMMARY

A depositional remanence of Middle to Late Permian age resides in red metasedimentary strata of the Santa Rosa Group in the Maya Mountains of Belize. The sedimentary rocks are a red bed facies deposited adjacent to exposed Silurian plutons, subsequently metamorphosed to a zeolite facies. Magnetic vectors in the red strata exhibit exceptionally linear demagnetization behavior. Unblocking temperatures and IRM studies demonstrate that a large portion of the remanence is carried by magnetite and maghemite. Hematite also carries a remanence in the same direction as the magnetite/maghemite. A 100 m stratigraphic sequence displays four magnetostratigraphic intervals. A positive reversal test shows that the remanence directions are antipodal, suggesting a deposition-related origin. The biostratigraphy of the Santa Rosa Group is Late Pennsylvanian to early Middle Permian (including *Perrinites hilli*); the presence of reversals demonstrates that the age of the magnetization is early Middle Permian or younger, because the late Paleozoic reversed superchron terminated in early Middle Permian time.

The plutonic rocks of the Maya Mountains display a distinctly different remanence direction. These plutons, dated by U/Pb as Late Silurian (Steiner and Walker, 1996), also display a dual polarity remanence. The same magnetization direction is observed in both eastern and western Maya Mountains plutons, both of which exhibit ca. 420 Ma ages and ca. 230 Ma K-Ar ages. Reset igneous rock K-Ar systems throughout the Maya Mountains and metamorphic aureoles developed in the Pennsylvanian-Permian Santa Rosa strata bordering the plutons suggest the occurrence of a postintrusion hydrothermal event at ca. 230 Ma. The remanence in these igneous rocks, in relationship to the Permian Santa Rosa remanence, resembles the spatial association between the North American Permian and Late Triassic remanences, suggesting that the igneous rocks were remagnetized in the late Triassic (230 Ma) in an event probably related to the initial fragmentation of Pangea.

The presence of this same dual polarity remanence in the Chiapas Massif suggests a widespread tectonic event, but also demonstrates that the Chiapas Massif was a part of the Yucatan Block since at least 230 Ma and that the Maya Mountains have not been rotated relative to the rest of the block since that time.

Paleopoles from four time periods, Permian through latest Jurassic determine the relative location of the Yucatan Block during the Pangean continental assembly and its subsequent motion relative to the North America plate. The Permian paleopole indicates that the Yucatan Block was not situated between the North and South American plates in the Pangean assembly. Yucatan's motion from Pangean fragmentation to Late Jurassic is a series of counterclockwise rotations. The Permian magnetization indicates Yucatan was inverted in orientation and residing adjacent to northwestern South America. By approximately Late Triassic (230 Ma), Yucatan had rotated ~47° counterclockwise. Between the 230 Ma remagnetization of the Maya Mountains plutons and Chiapas Massif and the deposition of the Todos Santos Formation, Yucatan was rotated by another 41° counterclockwise. Based on the abrupt change in North American apparent polar wander path at about Oxfordian time, it is speculated that the Todos Santos Formation and its paleopole (Molina Garza et al., 1992) are around Oxfordian in age, that is, that the conglomeratic Todos Santos beds were deposited in response to the marked change in the direction of plate motions. Between deposition of the Todos Santos and the overlying Kimmeridgian-Tithonian San Ricardo Formations, the Yucatan microplate underwent further counterclockwise rotation of ~47°, achieving its present orientation relative to North America by the end of the Jurassic.

ACKNOWLEDGMENTS

National Science Foundation grant EAR-9019094 supported the sample collection for this study. I thank Minister Evadine Garcia and R. Ramadan of the Ministry of Geology of Belize for their support and advice during sample collection in Belize. I thank Doug Walker for geochronologic advice and assistance during this study, and gratefully acknowledge valuable discussions with Ted Irving and Rob Van der Voo. I am very appreciative of the detailed, constructive reviews provided by Rob Van der Voo, William McDonald, and Thomas Anderson.

REFERENCES CITED

Alencaster, G., 1977, Moluscos y braquiopodos del Jurasico Superior de Chiapas: Universidad Nacional Autónoma de México, Instituto de Geología, Revista, v. 1, p. 151–166.

Anderson, T.H., and Schmidt, V.A., 1983, The evolution of middle America and the Gulf of Mexico–Caribbean Sea region during Mesozoic time: Geological Society of America Bulletin, v. 94, p. 941–966, doi: 10.1130/0016-7606(1983)94<941:TEOMAA>2.0.CO;2.

Bateson, J.H., 1972, New interpretation of geology of Maya Mountains, British Honduras: American Association of Petroleum Geologists Bulletin, v. 56, p. 956–963.

Bateson, J.H., and Hall, I.H.S., 1971, Revised geologic nomenclature for pre-Cretaceous rocks of British Honduras: American Association of Petroleum Geologists Bulletin, v. 55, p. 529–530.

Bateson, J.H., and Hall, I.H.S., 1975, Geological map of the Maya Mountains, Belize: London, Natural Environmental Research Council, HM Ordnance Survey, one sheet, scale 1:130,000 approx.

Bateson, J.H., and Hall, I.H.S., 1977, The geology of the Maya Mountains, Belize: London, HM Stationary Office, Institute of Geological Sciences Overseas Memoir 3, 42 p. and regional map.

Buffler, R.T., 1991, Seismic stratigraphy of the deep Gulf of Mexico basin and adjacent margins, *in* Salvador, A., ed., The Gulf of Mexico Basin: Boulder, Colorado, The Geological Society of America, Decade of North American Geology, v. J, p. 353–387.

Buffler, R.T., and Sawyer, D.S., 1985, Distribution of crust and early history of the Gulf of Mexico Basin: Gulf Coast Association Geological Society Transactions, v. 35, p. 333–344.

Buffler, R.T., and Thomas, W.A., 1994, Crustal structure and evolution of the southeast margin of North America and the Gulf of Mexico Basin, *in* Speed, R.C., ed., Phanerozoic evolution of North American continent-ocean transitions: Boulder, Colorado, Geological Society of America, Decade of North American Geology, Continent-Ocean Transect Volume, p. 219–264.

Bullard, E., Everett, J.E., and Smith, A.G., 1965, The fit of continents around the Atlantic: Philosophical Transactions of the Royal Society of London, v. 258A, p. 41–51.

Channell, J.E.T., Freeman, R., Heller, F., and Lowrie, W., 1982, Timing of diagenetic hematite growth in red pelagic limestones from Gubbio (Italy): Earth and Planetary Science Letters, v. 58, p. 189–201, doi: 10.1016/0012-821X(82)90193-5.

Cole, G.L., and Andrews-Jones, D.A., 1979, Geological and economic evaluation including exploration results of the 1978 field season: Belmopan, Central America, Belize Ministry of Geology, Belize Annual Report #4, 34 p.

Damon, P.E., Shafiqulliah, M., and Clark, K.F., 1981, Evolucion de los arcos magmaticos en Mexico: Universidad Nacional Autónoma de México, v. 5, p. 223–238.

Dawe, S.E., 1984, The geology of the Mountain Pine Ridge area and the relation of the Mountain Pine Ridge granite to the late Paleozoic and early Mesozoic geological history of Belize, Central America [M.S. thesis]: Binghamton, New York, State University of New York–Binghamton, 52 p.

Dengo, G., and Bohnenberger, O., 1969, Structural development of northern Central America: American Association of Petroleum Geologists Memoir 11, p. 203–220.

Diehl, J.F., and Shive, P.N., 1979, Paleomagnetic studies of the Early Permian Ingleside Formation of northern Colorado: Geophysical Journal Royal Astronomical Society, v. 56, p. 271–282.

Diehl, J.F., and Shive, P.N., 1981, Paleomagnetic results from the Late Carboniferous/Early Permian Casper Formation: Implications for northern Appalachian tectonics: Earth and Planetary Science Letters, v. 54, p. 281–292, doi: 10.1016/0012-821X(81)90011-X.

Dixon, C.G., 1956, Geology of southern British Honduras, with notes on adjacent areas: Belize, Belize Government Printer, 85 p.

Donnelly, T.W., Horne, G.S., Finch, R.C., and Lopez-Ramos, E., 1990, Northern Central America: The Maya and Chortis blocks, *in* Dengo, G., and Case, J.E., eds., The Caribbean region: Boulder, Colorado, Geological Society of America, Decade of North American Geology, v. H, p. 37–76.

Dunbar, J.A., and Sawyer, D.S., 1987, Implications of continental crust extension for plate reconstruction: An example from the Gulf of Mexico: Tectonics, v. 6, p. 739–755.

Dunlop, D.J., and Özdemir, Ö., 1997, Rock magnetism: Cambridge, UK, Cambridge University Press, p. 294–312.

Gose, W.A., and Helsley, C.E., 1972, Paleomagnetic and rock-magnetic studies of the Permian Cutler and Elephant Canyon Formations in Utah: Journal of Geophysical Research, v. 77, p. 1534–1548.

Gose, W.A., and Sanchez-Barreda, L.A., 1981, Paleomagnetic results from southern Mexico: Geofisica Internacional, v. 20–3, p. 163–175.

Gradstein, F.M., Agterberg, F.P., Ogg, J.G., Hardenbol, J., Van Veen, P., Thierry, J., and Huang, Z., 1995, A Triassic, Jurassic, and Cretaceous timescale, *in* Geochronology time scales and global stratigraphic correlation: Society of Sedimentary Geology (SEPM) Special Publication 54, p. 95–126.

Guerrero, J.C., Herrero-Brevera, E., and Helsley, C.E., 1990, Paleomagnetic evidence of post-Jurassic stability of southeastern Mexico: Maya terrane: Journal of Geophysical Research, v. 95, p. 7091–7100.

Hall, S.A., and Najmuddin, I.J., 1994, Constraints on the tectonic development of the eastern Gulf of Mexico provided by magnetic anomaly data: Journal of Geophysical Research, v. 99, p. 7161–7175, doi: 10.1029/93JB02570.

Heller, F., Chen, H., Dobson, J., and Haag, M., 1995, Permian-Triassic magnetostratigraphy—New results from South China: Physics of the Earth and Planetary Interiors, v. 89, p. 281–295, doi: 10.1016/0031-9201(94)02993-L.

Helsey, C.E., 1969, Magnetic reversal stratigraphy in the Lower Triassic Moenkopi Formation of western Colorado: Geological Society of America Bulletin, v. 80, p. 2431–2450.

Hillhouse, J.W., Gromme, S.C., and Vallier, T.L., 1982, Paleomagnetism and Mesozoic tectonics of the Seven Devils Volcanic Arc in northeastern Oregon: Journal of Geophysical Research, v. 87, p. 3777–3794.

Huber, P., Lucas, S.G., and Hunt, A.P., 1993, Vertebrate biochronology of the Newark Supergroup, Triassic, eastern North America, *in* Lucas, S.G., and Morales, M., eds., The nonmarine Triassic: New Mexico Museum of Natural History and Science Bulletin, v. 3, p. 179–186.

Irving, E., 1977, Drift of the major continental blocks since the Devonian: Nature, v. 270, p. 304–309.

Irving, E., 1983, Fragmentation of the continents: Surveys in Geophysics, v. 5, p. 299–333, doi: 10.1007/BF01453985.

Kesler, S.E., Bateson, J.H., Josey, W.L., Cramer, G.H., and Simmons, W.A., 1971, Mesoscopic structural homogeneity of Maya Series and Macal Series, Mountain Pine Ridge, British Honduras: American Association of Petroleum Geologists Bulletin, v. 55, p. 97–103.

Kirschvink, J.L., 1980, The least-squares line and plane and the analysis of paleomagnetic data: Geophysical Journal, v. 62, p. 699–718.

Klitgord, K.D., and Schouten, H., 1986, Plate kinematics of the central Atlantic, *in* Vogt, P.R., and Tucholke, B.E., eds., Plate kinematics of the central Atlantic: The western North Atlantic region: Boulder, Colorado, Geological Society of America, Decade of North American Geology, v. M, p. 351–378.

Kowallis, B.J., Christiansen, E.H., Deino, A.L., Peterson, F., Turner, C.E., Kunk, M.J., and Obradovich, J.D., 1998, The age of the Morrison Formation, *in* Carpenter, K., Chure, D.J., and Kirkland, J.I., eds., The Upper Jurassic Morrison Formation: An interdisciplinary study: Modern Geology, v. 22, p. 235–260.

Krogh, T.E., and Davis, G.L., 1973, The effect of regional metamorphism on U-Pb systems in zircons and a comparison with the Rb-Sr systems in the same whole rock and its constituent minerals: Carnegie Institution Yearbook, v. 72, p. 601–610.

Lucas, S.G., and Huber, P., 1993, Revised internal correlation of the Newark Supergroup, Triassic, eastern United States and Canada, *in* Lucas, S.G., and Morales, M., eds., The nonmarine Triassic: New Mexico Museum of Natural History & Science Bulletin, v. 3, p. 311–319.

McBirney, A.R., 1963, Geology of a part of the central Guatemalan Cordillera: University of California Publications in Geological Sciences, v. 39, p. 177–242.

McFadden, P.L., 1990, A new fold test for paleomagnetic studies Geophysical Journal International, v. 103, p. 163–169.

McFadden, P.L., and McElhinny, M.W., 1990, Classification of the reversal test in paleomagnetism: Geophysical Journal International, v. 103, p. 725–729.

Michel, A., Chaudron, G., and Benard, J., 1951, Propiertes des composes ferromagnetiques non metalliques: Journal of Physics E, Scientific Instruments, v. 12, p. 189–201.

Molina Garza, R.S., Geissman, J.W., and Van der Voo, R., 1989, Paleomagnetism of the Dewey Lake Formation (Late Permian), northwest Texas: The end of the Kiaman Superchron in North America: Journal of Geophysical Research, v. 94, p. 17,881–17,888.

Molina Garza, R.S., Van der Voo, R., and Urrutia, J., 1992, Paleomagnetism of the Chiapas Massif, southern Mexico: Evidence for rotation of the Maya block and implications for the opening of the Gulf of Mexico: Geological Society of America Bulletin, v. 104, p. 1156–1168, doi: 10.1130/0016-7606(1992)104<1156:POTCMS>2.3.CO;2.

Morel, P, and Irving, E, 1981, Paleomagnetism and the evolution of Pangea: Journal of Geophysical Research, B, v. 86, p. 1858–1872.

Muncaster, N., 1976, Mineral reconnaissance of granites in the Maya Mountains, Belize, Central America: Belmopan, Belize, Ministry of Natural Resources Report, Petroleum Office, 27 p.

Ower, L.H., 1928, Geology of British Honduras: The Journal of Geology, v. 36, p. 494–509.

Ozdemir, O., and Banerjee, S.K., 1984, High temperature stability of maghemite: Geophysical Research Letters, v. 11, p. 161–164.

Peterson, D.N., and Nairn, A.E.M., 1971, Paleomagnetism of Permian redbeds from the south-western United States: Geophysical Journal, v. 23, p. 191–205.

Pindell, J., 1985, Alleghenian reconstruction and subsequent evolution of the Gulf of Mexico, Bahamas and proto-Caribbean: Tectonics, v. 4, p. 1–39.

Pindell, J., and Barrett, S.F., 1990, Geological perspective of the Caribbean region: A plate-tectonic perspective, *in* Dengo, G., and Case, J.E., eds., The Caribbean region: Boulder, Colorado, Geological Society of America, The Decade of North American Geology, vol. H, p. 405–432.

Pindell, J., and Dewey, J.F., 1982, Permo-Triassic reconstruction of western Pangea and the evolution of the Gulf of Mexico–Caribbean region: Tectonics, v. 1, p. 179–211.

Purucker, M.E., Elston, D.P., and Shoemaker, E.M., 1980, Early acquisition of characteristic magnetization in red beds of the Moenkopi Formation (Triassic), Gray Mountain, Arizona: Journal of Geophysical Research, v. 85, p. 997–1012.

Richards, H.G., 1963, Stratigraphy of earliest Mesozoic sediments in southeastern Mexico and western Guatemala: American Association of Petroleum Geologists Bulletin, v. 47, p. 1861–1870.

Ross, C.A., 1979, Late Paleozoic collision of North and South America: Geology, v. 7, p. 41–44, doi: 10.1130/0091-7613(1979)7<41:LPCONA>2.0.CO;2.

Ross, M.I., and Scotese, C., 1988, A hierarchical tectonic model of the Gulf of Mexico and Caribbean region: Tectonophysics, v. 155, p. 139–168, doi: 10.1016/0040-1951(88)90263-6.

Rowley, D.B., and Pindell, J.L., 1989, End Paleozoic–Early Mesozoic western Pangean reconstruction and its implications for the distribution of Precambrian and Paleozoic rocks around Meso-America: Precambrian Research, v. 42, p. 411–444, doi: 10.1016/0301-9268(89)90022-3.

Roy, J.L., and Park, J.K., 1974, The magnetization process of certain red beds: Vector analysis of chemical and thermal results: Canadian Journal of Earth Sciences, v. 11, p. 437–471.

Salvador, A., 1991, Origin and development of the Gulf of Mexico Basin, *in* Salvador, A., ed., The Gulf of Mexico Basin: Boulder, Colorado, Geological Society of America, The Decade of North American Geology, v. J, p. 389–444.

Sawyer, D.S., Buffler, R.T., and Pilger, R.H., 1991, The crust under the Gulf of Mexico, *in* Salvador, ed., The Gulf of Mexico Basin: Boulder, Colorado, Geological Society of America, The Decade of North American Geology, v. J, p. 53–72.

Schouten, H., and Klitgord, K.D., 1994, Mechanistic solutions to the opening of the Gulf of Mexico: Geology, v. 22, p. 507–510, doi: 10.1130/0091-7613(1994)022<0507:MSTTOO>2.3.CO;2.

Scotese, C.R., and Langley, R.P., 1995, Pangea and the palaeogeography of the Permian, *in* Scholle, P.A., Peryt, T.M., and Ulmer-Scholle, D.S., eds., The Permian of northern Pangea: Berlin, Springer-Verlag, v. 1, p. 3–19.

Shipley, W.E., III, 1978, Geology, petrology and geochemistry of the Mountain Pine Ridge Batholith, Belize, Central America [M.S. thesis]: Golden, Colorado, Colorado School of Mines, 90 p.

Shive, P.N., Steiner, M., and Huycke, D.T., 1982, Magnetostratigraphy and the timing of remanence acquisition in the Triassic Chugwater Formation of Wyoming: Journal of Geophysical Research, v. 89, p. 1801–1815.

Simmons, W.A., 1972, Stratigraphy and sedimentation of the Paleozoic rocks in the Maya Mountains, British Honduras [M.S. thesis]: Baton Rouge, Louisiana, Louisiana State University, 87 p.

Steiner, M.B., 1988, Paleomagnetism of the late Pennsylvanian and Early Permian: A test of the rotation of the Colorado Plateau: Journal of Geophysical Research, v. 93, p. 2201–2215.

Steiner, M.B., 1998, Age, correlation, and tectonic implications of Morrison Formation paleomagnetic data, including rotation of the Colorado Plateau, *in* Carpenter, K., Chure, D.J., and Kirkland, J.I., eds., The Upper Jurassic Morrison Formation: An interdisciplinary study: Modern Geology, v. 22, p. 261–282.

Steiner, M.B., 2001, Magnetostratigraphic correlation and dating of west Texas and New Mexico Late Permian strata, *in* Lucas, S.G., and Ulmer-Scholle, D., eds., Geology of the Llano Estacado: New Mexico Geological Society, 52nd Field Conference Guide, p. 59–68.

Steiner, M.B., 2003, A cratonic Middle Jurassic paleopole: Callovian-Oxfordian pole stillstand (J-2 cusp), rotation of the Colorado Plateau, and Jurassic North American apparent polar wander: Tectonics, v. 22, no. 3, 1016, 21 p., doi: 10.1029/2001TC001284.

Steiner, M.B., and Helsley, C.E., 1974, Magnetic polarity sequence of the Upper Triassic Kayenta Formation: Geology, v. 2, p. 191–194, doi: 10.1130/0091-7613(1974)2<191:MPSOTU>2.0.CO;2.

Steiner, M.B., and Lucas, S.G., 2000, Paleomagnetism of the Late Triassic Petrified Forest Formation, Chinle Group, western USA: Further evidence of "large" rotation of the Colorado Plateau: Journal of Geophysical Research, v. 105, p. 25,791–25,808, doi: 10.1029/2000JB900093.

Steiner, M.B., and Walker, J.D., 1996, Late Silurian U-Pb ages from the Yucatan Block: Journal of Geophysical Research, v. 101, p. 17,727–17,735, doi: 10.1029/96JB00174.

Steiner, M.B., Ogg, J.G., and Sandoval, J., 1987, Jurassic magnetostratigraphy: Part 3. Bajocian-Bathonian of the Subbetics, Southern Spain: Earth and Planetary Science Letters, v. 82, p. 357–372, doi: 10.1016/0012-821X(87)90209-3.

Symons, D.T.A., 1990, Early Permian pole, evidence from the Pictou red beds, Prince Edward Island, Canada: Geology, v. 18, p. 234–237, doi: 10.1130/0091-7613(1990)018<0234:EPPEFT>2.3.CO;2.

Van der Voo, R., 1981, A paleomagnetic comparison of various reconstructions of the Atlantic-bordering continents which have implications for the origin of the Gulf of Mexico, *in* Pilger, R.H., ed., The origin of the Gulf of Mexico and the early opening of the central-North Atlantic Ocean: Houston, Houston Geological Society, Continuing Education Series, p. II-1–II-5.

Van der Voo, R., Mauk, F.J., and French, R.B., 1976, Permian-Triassic continental configurations and the origin of the Gulf of Mexico: Geology, v. 4, p. 177–180, doi: 10.1130/0091-7613(1976)4<177:PCCATO>2.0.CO;2.

Van der Voo, R., Peinado, J., and Scotese, C.R., 1984, A paleomagnetic reevaluation of Pangea reconstructions, *in* Van der Voo, R., Scotese, C.R., and Bonhommet, N., eds., Plate reconstruction from Paleozoic paleomagnetism: Washington, D.C., American Geophysical Union, Geodynamics Series, v. 12, p. 11–26.

White, G.W., 1980, Permian-Triassic continental reconstruction of the Gulf of Mexico–Caribbean area: Nature, v. 283, p. 823–826, doi: 10.1038/283823a0.

Wilde, G.L., 1986, Stratigraphic relationships of the San Andres and Cutoff Formations, Northern Guadalupe Mountains, New Mexico and Texas, *in* Moore, G.E., and Wilde, G.L., eds., Lower and Middle Guadalupian facies, stratigraphy, and reservoir geometries, San Andres/Grayburg Formations, Guadalupe Mountains, New Mexico and Texas: Permian Basin Section, Society of Sedimentary Geology (SEPM) Publication 86-25, p. 63.

Woods, R.D., Salvador, A., and Miles, A.E., 1991, Pre-Triassic, *in* Salvador, A., ed., The Gulf of Mexico Basin: Boulder, Colorado, Geological Society of America, Decade of North American Geology, v. J, p. 109–129.

Ziegler, A.M., 1981, Paleozoic paleogeography, *in* McElhinny, M.W., and Valencio, D.A., eds., Paleoreconstruction of the continents: American Geophysical Union and Geological Society of America, Geodynamics Series, v. 2, p. 31–37.

Manuscript Accepted by the Society 24 March 2005

Printed in the USA

Geological Society of America
Special Paper 393
2005

Mesozoic basin formation, mass-gravity sedimentation, and inversion in northeastern Sonora and southeastern Arizona

James W. McKee*
Department of Geology, University of Wisconsin Oshkosh, Oshkosh, Wisconsin 54901, USA

Mary Beth McKee
7118 Clow Road, Winneconne, Wisconsin 54986, USA

Thomas H. Anderson
Department of Geology and Geophysics, University of Pittsburgh, Pittsburgh, Pennsylvania 15260, USA

ABSTRACT

The El Tule fault extends southeastward from the Huachuca Mountains area of Arizona to run along the base of the Sierra San José in northern Sonora. This fault, with the Jurassic and Cretaceous strata that lie on either side of it, helps to link the geologic histories of the Sierra San José, the Huachuca Mountains, and the Mule Mountains. This fault was active during the Jurassic and formed part of the southern and southwestern limit of a small Jurassic basin that accumulated Glance Conglomerate and contains the type locality of that unit; it was also active during the time of deposition of the upper part of the Bisbee Group, i.e., the Morita, Mural, and Cintura formations. This fault is an important marker for the northern margin of the Cananea high.

Three evidences suggest that a genetic distinction should be made between the basin that collected the Glance Conglomerate and the younger, wider, superposing basin that collected upper Bisbee rocks: (1) There was a depositional pause after active deposition of the Glance (and therefore after active opening of the Glance basin). (2) The subsequent, finer-grained upper Bisbee deposits do not extend entirely over the Cananea high, but are otherwise regional in extent, forming a thick blanket over the inactive fault basin and other similar basins elsewhere in the region. (3) The axial trends of the Jurassic fault basin(s) and the upper Bisbee basin do not coincide.

We offer the hypothesis that the Jurassic fault basins were caused by the same crustal stresses that caused the Mojave-Sonora megashear (their ages of activity and orientations are similar), and that the upper Bisbee basin was the probable result of crustal thinning associated with formation of the extensional Jurassic basins, and of eustasy. We offer this in contrast to the current hypothesis that the Bisbee basin is a rift basin, that is, the product of a single, ongoing Mesozoic process.

Large-scale mass-gravity deposits are signal features both of the Jurassic fault basin (a result of crustal extension) and of the overlying local Laramide basin (which

*Current address: 7118 Clow Road, Winneconne, Wisconsin 54986, USA.

McKee, J.W., McKee, M.B., and Anderson, T.H., 2005, Mesozoic basin formation, mass-gravity sedimentation, and inversion in northeastern Sonora and southeastern Arizona, *in* Anderson, T.H., Nourse, J.A., McKee, J.W., and Steiner, M.B., eds., The Mojave-Sonora megashear hypothesis: Development, assessment, and alternatives: Geological Society of America Special Paper 393, p. 481–507. doi: 10.1130/2005.2393(18). For permission to copy, contact editing@geosociety.org.

followed crustal contraction). Historically, these deposits have been confusing. A mountainous mass of megaclasts (La Negrita Hills) within the Glance Conglomerate has been mistaken for the hanging wall of a Laramide thrust (a contractional feature) instead of the product of Jurassic crustal extension that it is. The 91 Hills have been considered an in situ array of mid-Cretaceous patch reefs instead of the Late Cretaceous convulsive mass-gravity deposits (a product of Laramide contraction and inversion) that they are.

Keywords: Bisbee Group, Cabullona Group, mass-gravity deposition, Sierra San José.

INTRODUCTION

This paper presents the results of geological exploration and mapping (Fig. 1) in northern Sonora, immediately south of the Mule and Huachuca Mountains of Arizona. Our work helps to integrate the Jurassic and Cretaceous history of those mountains and that of the Sierra San José in Sonora; it deals principally with the Upper Jurassic–Lower Cretaceous Bisbee Group, and the Upper Cretaceous Cabullona Group and Fort Crittenden Formation.

The El Tule fault extends between the Huachuca Mountain area and the Sierra San José and enables correlation of geologic events between the two areas. The fault is well exposed, as is the stratigraphic evidence of the strong influence it had on Jurassic and Cretaceous sedimentary patterns.

Cananea high is the name that has been given to an area in northern Sonora that contains no known exposures of the Bisbee Group, and on that basis has been a speculative barrier between northern and southern areas of Bisbee Group sedimentation. The El Tule fault and associated facies give good evidence that the Cananea high was a fault-elevated upland during Bisbee deposition. They also enable the inference of another fault, the Cananea fault, along which additional elevation was generated.

A Jurassic-Neocomian fault-bounded depositional basin, marked by the presence of Glance Conglomerate, lay immediately to the north of the Cananea high and may have extended to the Mule Mountains of Arizona. We hypothesize that this Glance basin (or basins) was formed by transtensional stresses that elsewhere were manifest as the Mojave-Sonora megashear, and that the basin(s) is a pull-apart. Anderson and Nourse (this volume) specify the pull-apart geometry. Our work indicates that younger rocks of the Bisbee Group (Morita, Mural, and Cintura formations) do not rest conformably upon the Glance; instead, they are deposits of a younger, broader basin of different origin. That is, we suggest that the Bisbee basin is more accurately perceived as an array of relatively small fault basins of one origin plus a superposing regional basin of another origin. This is an alternative hypothesis to that of the Bisbee basin as a rift, that is, a single basin, whose origin was a single ongoing process.

Megaclasts, mostly of Paleozoic limestone, are common components of Glance Conglomerate where volcanism was concomitant with Glance deposition. South of the Mule Mountains, however, where the Glance contains no contemporary volcanic rocks, we identified a mountainous mass of megaclasts (La Negrita Hills) that are components of the Glance Conglomerate. These megaclasts are in contact with a younger mass-gravity deposit (91 Hills slide mass), which we include in the Upper Cretaceous Cabullona Group. We consider such large-scale mass-gravity sedimentation to be a characteristic feature of both the Bisbee and Cabullona groups, and point out that failure to recognize this can cause (and has caused) significant errors of interpretation of geological history.

The faults that created the narrow Jurassic basin that we studied here had a broad and enduring effect on the geology of the area. They engendered the conglomerate that partially filled the basin and rendered that conglomerate distinctive by the megaclasts it contains. Later, Laramide rejuvenation and inversion of the Jurassic faults caused uplift and rapid erosion of poorly lithified Bisbee rocks, thereby producing steep-walled drainageways on a Late Cretaceous landscape. Landslides (including 91 Hills) were a result of that. So, the La Negrita Hills deposit was a product of the basin's opening, and the 91 Hills slide was a product of the basin's closing. Ultimately, both were caused by Jurassic faults that we believe share their origins with the Mojave-Sonora megashear.

For convenience of description, we follow Ransome's (1904) method of partitioning the study area according to structural blocks. Stratigraphic units considered in this study are shown in Figure 2.

BASINS

Ransome (1904) and Bilodeau (1979) reported the step-like, down-to-the-south arrangement of fault blocks in the Mule Mountains and pointed out that the Glance Conglomerate (Fig. 3) was thinnest on the highest-standing block (Bisbee block) and progressively thicker on the lower blocks (Copper Queen and Gold Hill blocks). These faults in the Mule Mountains mark one side of a "Glance basin" that was implied by their work. But because the Glance Conglomerate is part of Ransome's Bisbee Group, Bisbee basin became the term for the depositional basin that contains all Bisbee Group rocks, Glance and upper Bisbee, wherever they may be.

Our mapping (Fig. 1) extends the local Glance basin for ~30 km into Mexico and identifies a fault (El Tule fault) that is exposed both at the base of Sierra San José and along Arroyo El Tule southwest of the Huachuca Mountains. This fault extends

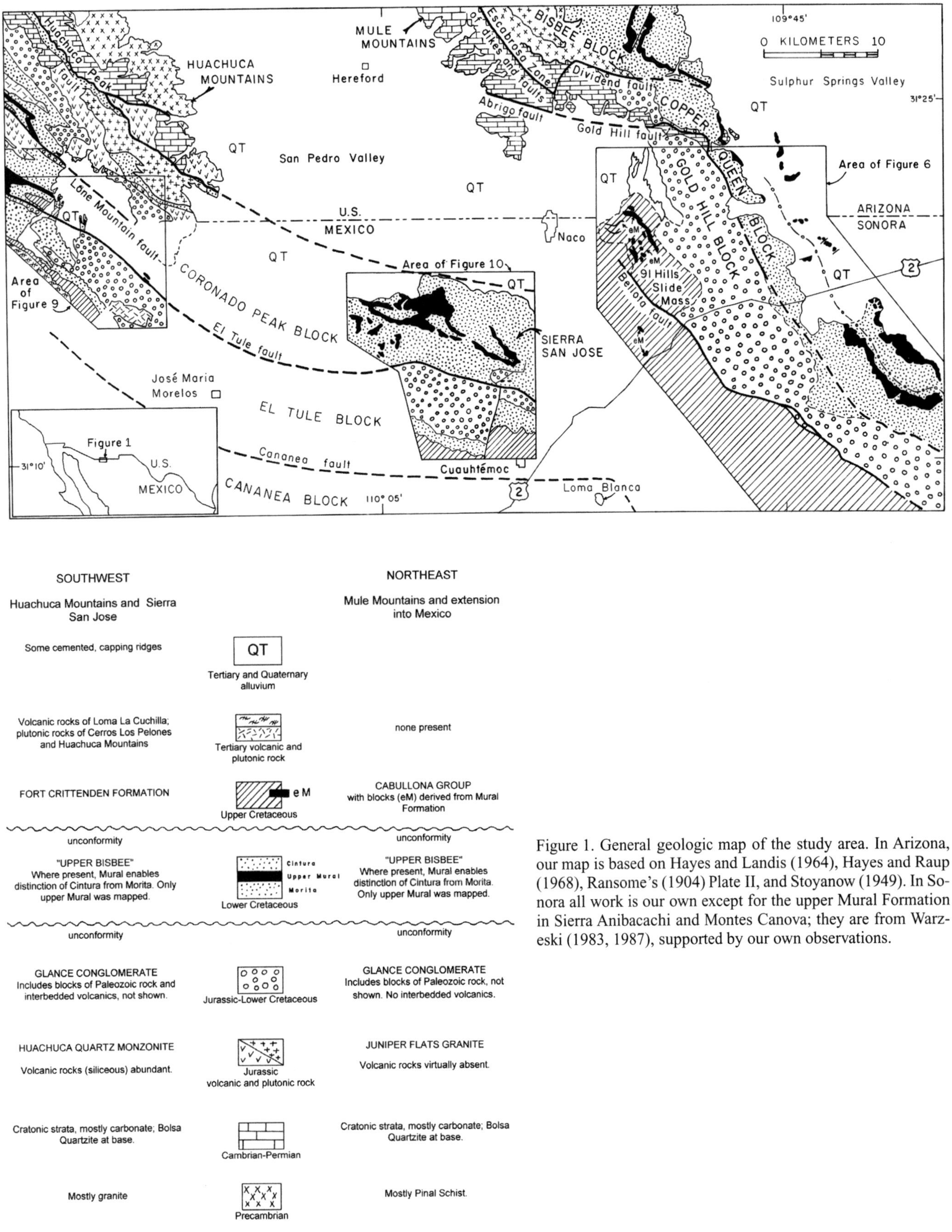

Figure 1. General geologic map of the study area. In Arizona, our map is based on Hayes and Landis (1964), Hayes and Raup (1968), Ransome's (1904) Plate II, and Stoyanow (1949). In Sonora all work is our own except for the upper Mural Formation in Sierra Anibacachi and Montes Canova; they are from Warzeski (1983, 1987), supported by our own observations.

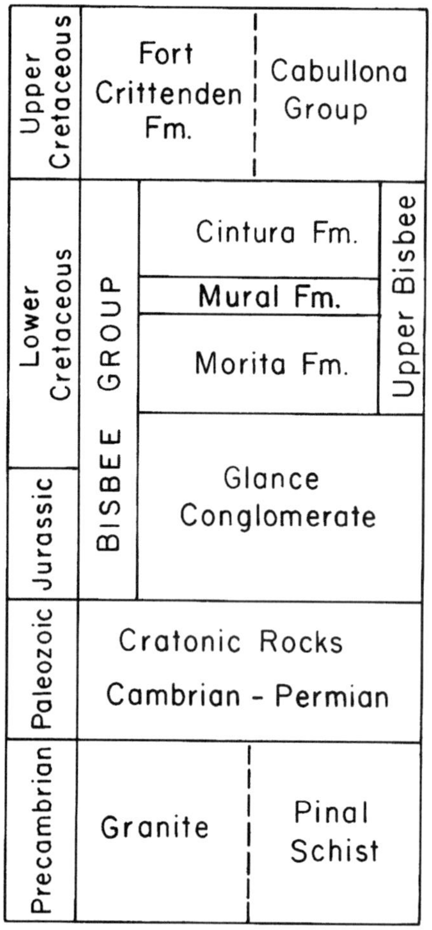

Fort Crittenden Formation, Stoyanow (1949). A thick terrestrial deposit; complex facies variations mostly alluvial/lacustrine.

Cabullona Group, (Taliaferro, 1933). Principally lacustrine. Consists of five formations (González-León, 1994) but two (Corral de Enmedio and Lomas Coloradas) are not exposed in our map areas. The El Cemento Conglomerate is a littoral/sublittoral deposit that interfingers with basinal Camas Sandstone and Packard Shale. We add a basal unit, the Agua Amarga beds.

"upper Bisbee." Collective term for Bisbee Group above the Glance; useful also when absence of Mural prohibits distinction between Morita and Cintura.

Cintura Formation, Ransome (1904). Sandstone and shale, a "crude mirror image of the Morita" (Hayes, 1970). Lithologically indistinguishable from Morita but sandier parts are higher in the Morita, lower in the Cintura.

Mural Formation, Ransome (1904). Named for the limestone of the "light-gray cliff crowning Mural Hill" east of Bisbee, Arizona. Ransome recognized the genetic relationship between the cliff and the largely siliciclastic slope beneath it, so he wrote of the lower and upper Mural. That informal distinction has been retained. Stoyanow (1949) described geographic variability within the lower Mural, but the resulting nomenclature has not been popular. The upper Mural consists almost entirely of thick beds of limestone. Rudists, distinctive scleractinians (*Microsolena*), and the foraminiferan *Orbitolina* are characteristic. The lower Mural is distinguished from the Morita by the relative abundance of marine fossils, mostly molluscan. Thickness estimates of the Mural vary; Ransome's estimate (650 feet or 198 m) is near the middle of the range; the upper Mural forms about 40% of the total thickness of the formation. The Mural Formation is marine.

Morita Formation, Ransome (1904). A "monotonous repetition of sandstones and shales" (Ransome, 1904), marine in its upper parts. The sandstone is feldspathic, occurring in beds that typically have scoured bases and become finer upward until they grade into the overlying mudstone. Mudstone composes more than half the formation, but sandstone is more abundant near the top of the section. 855 meters thick (Hayes, 1970).

Glance Conglomerate, Ransome (1904). Mule Mountains: alluvial, poorly sorted boulder conglomerate and breccia composed mostly of fragments of Precambrian schist and granite, Paleozoic limestone and quartzite, Jurassic granite, and Mesozoic volcanic rocks. Huachuca Mountains and Sierra San José: contemporary volcanic rocks form part of the Glance (Hayes, 1970). Maximum thickness estimates vary, 1000–2200 m.

Figure 2. General stratigraphic column of rocks exposed in the study area.

into Arizona as an unnamed fault shown by Hayes and Raup (1968). The El Tule fault separates the Coronado Peak block from the El Tule block. An inferred fault, the Cananea fault, separates the El Tule block from the Cananea block, where an unconformity represents all of the Bisbee Group. These fault blocks mirror the Bisbee, Copper Queen, and Gold Hill blocks, and form the western and southern sides of this Glance basin. Figure 4A shows the principal faults associated with the basin and the exposures of Glance Conglomerate that mark and verify the basin's presence; Figure 4B shows the fault-bounded blocks.

We name this local Glance basin the Hereford basin (Fig. 4C). Our field work does not identify this basin's northern limit. Provisionally, we place this limit along a westward extension of the Abrigo fault (Figs 1 and 4A–C), which crosses the San Pedro River between the village of Hereford (Fig. 1) and Hereford Ranch. The Hereford basin is a depositional basin that we identify by the thick sequence of coarse sedimentary rocks contained within it, by the faults that created it, and by its temporal separation (discussed below) from immediately overlying rocks.

The Hereford basin is the oldest of four successively younger, superposed, depositional basins: Hereford, "upper Bisbee," Cabullona /Fort Crittenden, and the Tertiary/Quaternary basin beneath the modern San Pedro River. We believe it is useful to think of the Bisbee Group in the study area as having accumulated in two genetically different basins: (1) the Hereford fault basin that collected Glance conglomerate, and (2) a temporally distinct depositional basin of complex origin in which the upper Bisbee lithosome was deposited.

Recent geophysical work in the San Pedro Valley, coupled with well data, (Bultman et al., 1999; Gettings and Houser, 2000) has shown that along the central axis of the valley, pre-Oligocene "basement" lies between 500 and 2000 m beneath the present valley floor. In the area of our study, information from wells along the flanks of the valley does not usefully augment the information from outcrops. Consequently, we are without direct information regarding the Jurassic and Cretaceous rocks of the central area of the Hereford basin. The simplest assumption is that the fanglomerates on the Gold Hill, Coronado Peak, and El Tule blocks (Figs. 1 and 4) are an accurate reflection of Jurassic paleogeography, that is, that coarse alluvial aprons faced one another across the Hereford basin (Fig. 4C), and that finer-grained facies occupied the central part of the basin. Alternatively, the entire central part of the Hereford basin area may have stood somewhat higher than either of the Gold Hill or Coronado Peak blocks, thereby creating two narrow fault basins (Fig. 4D). In that case (Fig. 4D), the Gold Hill basin is marked on three sides by known faults—La Bellota, Abrigo, and Gold Hill. The Coronado Peak basin is almost as well defined by the Cananea and Huachuca Peak faults.

Hereford Basin

In this presentation we follow the single (Hereford) basin model (Figs. 4C and 5A); it is the simpler model and therefore the best vehicle for presenting our field observations.

Northeastern Part of the Hereford Basin

We interpreted the Abrigo and Gold Hill faults (Ransome, 1904) (Fig. 1) as a single reverse fault, that is, the product of Laramide inversion (Lawton, 2000) of the Jurassic Gold Hill and Abrigo faults. These faults, with La Bellota fault (McKee et al., 1994) bound the Gold Hill block, which contains major exposures of Glance Conglomerate, including its type locality.

In Sonora, the Gold Hill fault is exposed just west of Cerro La Muela (Fig. 6), where it separates Morita and Glance formations. Farther southeast, however, the fault is not exposed. Along the base of Sierra Anibacachi, we mapped the trace of the buried fault on the basis of abrupt dip change that is compatible with a drape fold in which the underlying forcing member is the Gold Hill block. The fault roughly parallels the Glance-Morita contact, except near the head of Arroyo Agua Amarga (Fig. 6), where the contact extends farther to the southwest.

Geographical and stratigraphical variation in the Glance Conglomerate on the Gold Hill block is shown in Figure 6. Glance composition is as described in the Mule Mountains (Ransome, 1904; Hayes, 1970; Bilodeau, 1979), except for some components that bear separate mention: (1) La Negrita Hills (our name for a mountainous area that includes Cerro La Negrita, Loma La Morita, both Cerros La Bruja, and "little" Sierra Anibacachi) is a large mass-gravity deposit consisting of megaclasts of Paleozoic limestone; it is part of the Glance Conglomerate. We discuss this deposit in the section "Mass-Gravity Deposition." (2) A large exposure of rhyolite tuff and flow rock south of the southern Cerro La Bruja is probably a megaclast; a pyroclastic rock would be more widespread if it were an in situ deposit contemporary with Glance deposition. (3) Pinal Schist and overlying lower Paleozoic strata are exposed in a narrow arroyo at the southern margin of the La Negrita Hills; these may be a megaclast. (4) Farther southeast, individually mappable clasts (horses?) are exposed along the Bellota fault: Paleozoic limestone lies between the La Bellota fault and the Agua Amarga beds (therefore slightly off the Gold Hill block), and Precambrian Pinal Schist is present still farther southeast. (5) A rounded cobble-boulder conglomerate predominantly of granite porphyry with feldspar crystals up to 13 cm long and 2–3 cm wide is present in Arroyo El Fresnal directly south of Cerro Canoitas. (6) Monolithologic sedimentary breccia and conglomerate composed of schist (Pinal) clasts are abundant around the southwestern and southern margin of La Negrita Hills. Boulders with diameters greater than one meter are common in the breccia (Fig. 7). (7) Thin, red, tabular beds of fine sandstone and shale (Figs. 6 and 8) occur extensively at the top of the Glance section. We interpret these as lacustrine and depict them (Fig. 5A) as having transgressed the toe of the Glance alluvial apron.

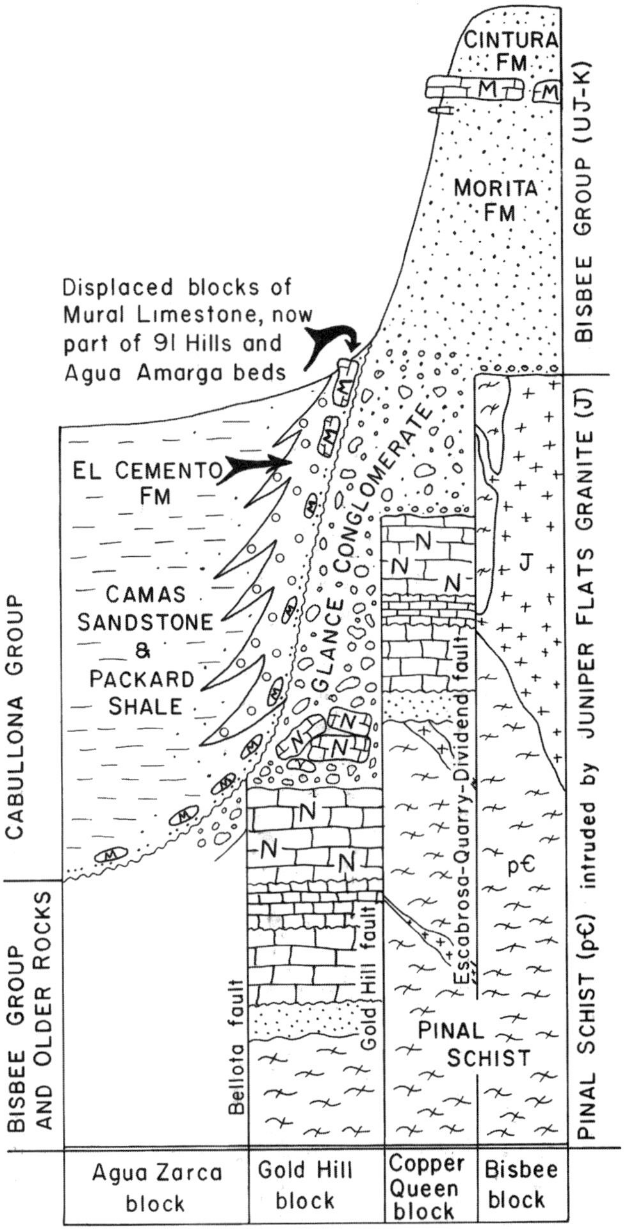

Figure 3. Cartoon representation of stratigraphic and structural relationships in the Mule Mountains and adjacent parts of Sonora. N—Naco Formation; megaclasts on the Gold Hill block represent the rocks of La Negrita Hills.

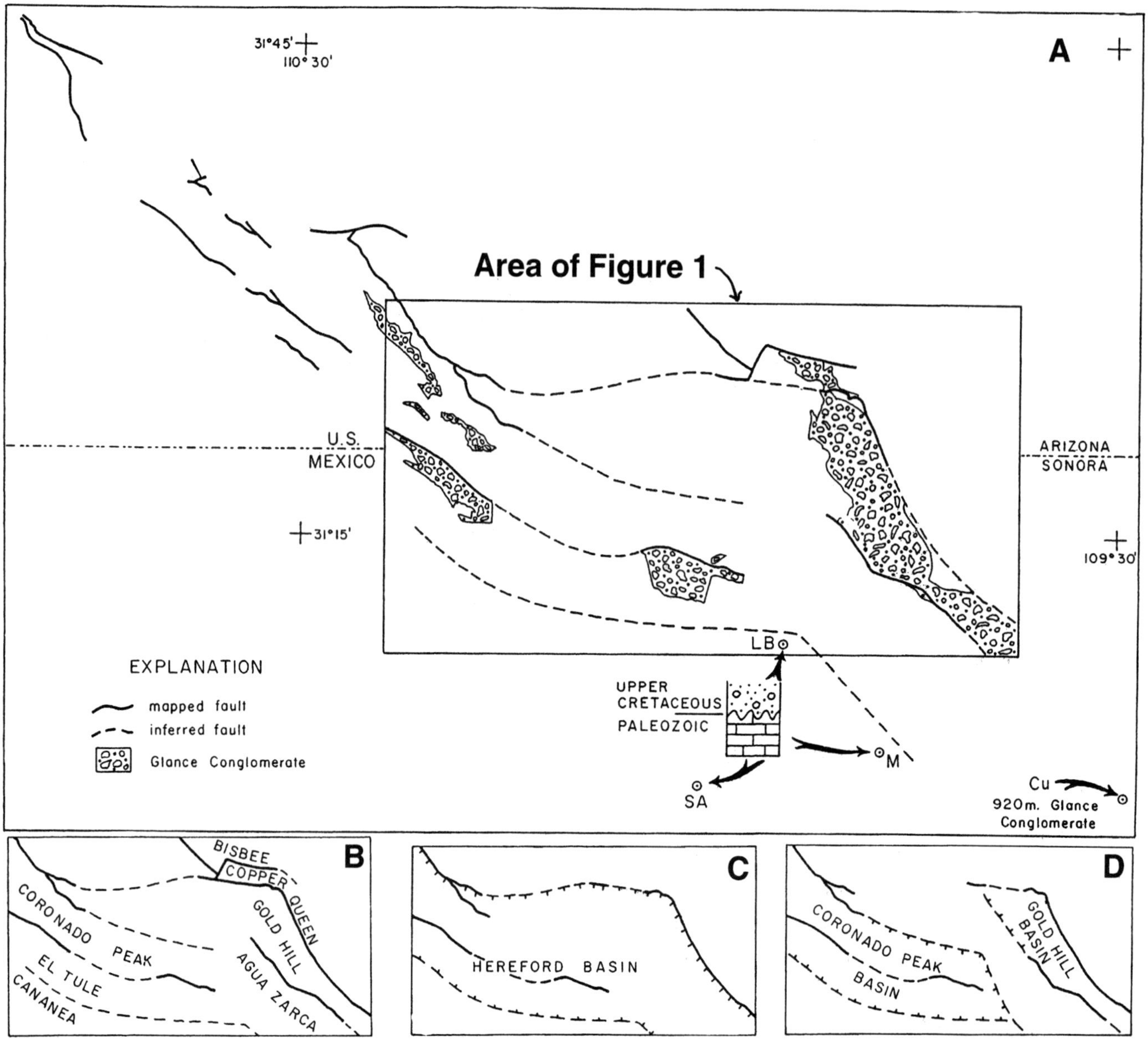

Figure 4. (A) Principal Jurassic faults in and near the study area and exposures of Glance Conglomerate. These are the basic data for delineation of Jurassic fault basin(s). (B) Names and locations of fault blocks bounded by faults active in the Jurassic. Our Agua Zarca block is ill-defined; we use it to refer only to an area west of the La Bellota fault. (C) The Jurassic Hereford basin. (D) A two-basin alternative to the Hereford basin. Anderson and Nourse (this volume) utilize a refinement of this two-basin alternative. LB—Loma Blanca; SA—Sierra Los Ajos; M—Musteñas; Cu—Rancho Culantrillo

In the Mule Mountains of Arizona, the Copper Queen and Bisbee blocks stood higher than the Gold Hill block during Glance deposition (Fig. 3), thereby forming the northeastern flank and rim of the Hereford basin. We carry that interpretation south into Sonora, but there the Glance is buried beneath the Morita on the Copper Queen block. The basin rim is well exposed on the Bisbee block in the Mule Mountains, but is not exposed in Sonora.

Western Part of the Hereford Basin

In the Huachuca Mountains of Arizona (Fig. 1), Hayes and Raup (1968) offered no name for the main Laramide fault that raises Precambrian and lower Paleozoic strata. We call it the Huachuca Peak fault after the peak on its hanging wall. Drewes (1980, 1981) considered this fault as a segment of his regional Cochise fault, but we prefer a local name.

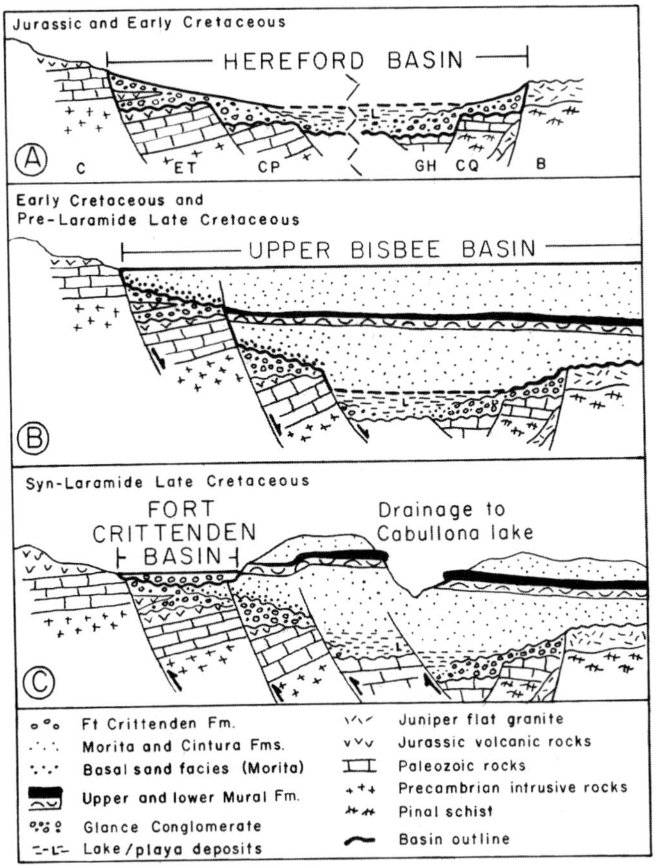

Figure 5. Stratigraphic diagram illustrating basin succession in the study area. This cartoon representation is not a structure section; all faults are shown as normal merely to depict basins. See text for discussion. (A) Hereford basin. Faulting has engendered coarse deposits (Glance Conglomerate), and fine-grained basin-center deposits have transgressed the toe of the alluvial aprons. The El Tule block shows the Glance with contemporaneous volcanic rock. (B) Upper Bisbee basin. Upper Bisbee rocks (Morita, Mural, and Cintura formations) have covered the Hereford basin and Bisbee block; they do not reach across the Cananea block. (C) Fort Crittenden/Cabullona basin. Late Cretaceous inversion of El Tule and Huachuca Peak faults has provided a source for Fort Crittenden sediment, which has just begun to cover the Cananea block. The Fort Crittenden received no Mural blocks because the Mural was in its less competent marginal marine facies there. To the east (area of the modern San Pedro Valley), uplift of the upper Bisbee has caused rapid erosion. Continued inversion and contraction will create, west of the Huachuca Peak fault, the Laramide belt of faults and cylindrical folds. C—Cananea block; ET—El Tule block; CP—Coronado Peak block; GH—Gold Hill block; CQ—Copper Queen block; B—Bisbee block.

We divide this area into three components: (1) The hanging wall of the Huachuca Peak fault (Fig. 1) contains Precambrian and Paleozoic rocks that were raised during Laramide compression. This fault block may also have stood high during the Jurassic and contributed detritus to the Coronado Peak block. (2) The Coronado Peak block and (3) the El Tule block, which are separated by the El Tule fault (Figs. 1, 4, and 9).

On the Coronado Peak block (Huachuca Mountains) and nearby areas there has been some difficulty in separating coarse rocks associated with Jurassic arc magmatism from younger conglomerates with interbedded volcanic rocks that are properly considered the equivalent in age and origin to the coarse basal unit of the Bisbee Group in its type area. Anderson and Nourse (this volume) summarize this problem. There does seem now to be agreement that a substantial thickness of Bisbee Glance is present in the Huachucas (Lipman and Hagstrum, 1992). Vedder (1984) measured two sections of Glance Conglomerate on the Coronado Peak block. Her central Huachuca section is 800 m thick, excluding andesite. Her southern Huachuca section is 960 m thick.

The El Tule block's most striking feature here is the mountainous array of megaclasts that we refer to as the El Caloso Hills (Fig. 9). Three of the highest points in this array are Cerros El Caloso, Agua Dulce, and El Tule. González-León (1986) found that they are composed of Devonian, Mississippian, Pennsylvanian, and Permian limestone. He also identified two Cambrian formations, the Bolsa Quartzite and the Abrigo Limestone. In our map (Fig. 9), we do not distinguish among the different limestones, but we show Bolsa Quartzite and granite. Our interpretation that the El Caloso Hills are formed of megaclasts is discussed in another section of this report, "Mass-Gravity Deposition." The Glance Conglomerate consists of the megaclasts together with the conglomerate, sandstone, and dark flow rock upon and within which they rest. We found no contact evidence of whether the granite and the Bolsa Quartzite are in place, are clasts within the Glance, or are part of a fault sliver that we could not identify. For simplicity, we have mapped them as Glance.

West of the El Caloso Hills, the low-lying terrain contains many scattered exposures of Glance with compositions noted in Figure 9. In the valley surrounded by Cerro El Tule, Cerro Caloso, and the Cerros Los Pelones the oldest Glance rocks are feldspar porphyry flow rock (andesite), tuff, and volcaniclastic sandstone that are exposed along the base of Cerro El Tule. These are overlain by a well-exposed, rounded, Paleozoic-limestone-boulder conglomerate that dips to the southwest and interfingers with reddish sandstone and shale toward the southeast. These two facies (conglomerate, and sandstone and shale) are overlain by andesite flows that show definable bases and flow-brecciated tops.

Southern Part of the Hereford Basin

In the Sierra San José, the Coronado Peak block (Figs. 1, 4B, and 10) contains only one exposure of Glance Conglomerate. It consists of pebble-cobble conglomerate, sandstone, and mudstone that occur in a truncated anticline (Fig. 10) near the El Tule fault. The clasts in the conglomerate are principally granite, feldspar porphyry, and quartzite. This gives us little direct knowledge of the Glance here on the Coronado Peak block, but the Glance exposed on the adjacent part of the El Tule block probably offers a sample of Coronado Peak Glance.

On the El Tule block, the El Granizo fault (Fig. 10) cuts a body of volcanic rock (dark-brown and purple flows and tuffs)

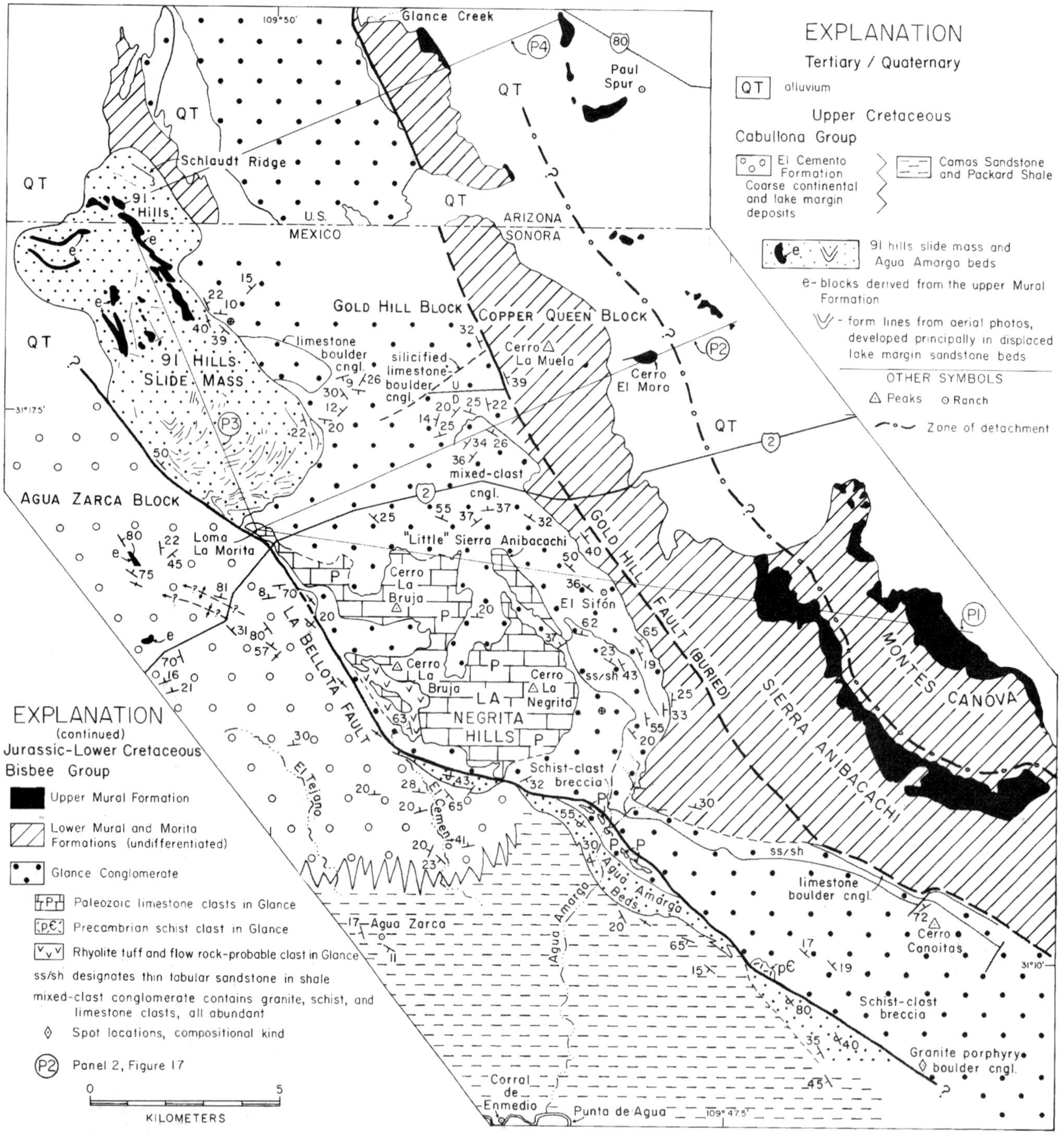

Figure 6. Geologic map of the southern part of the Mule Mountains, and adjacent area in Sonora. The Gold Hill and Copper Queen blocks, and the Copper Queen fault are from Ransome (1904), extended into Sonora. In Arizona, the geology is principally from Hayes and Landis (1964).

Figure 7. Schist-clast sedimentary breccia near the La Bellota fault on the southwestern margin of La Negrita Hills. Hat provides scale.

Figure 8. Thin-bedded fine sandstone and shale of the Glance Conglomerate. Prominent bed across the middle of the photo is 4 cm thick.

that contains megaclasts of Paleozoic limestone (Cerro El Granizo) and one apparent megaclast of Bolsa Quartzite (Cambrian). These rocks dip gently westward and are overlain by the conglomerate, whose composition is shown in Figure 10. In general, the clasts consist of granite, limestone, and volcanic rock; there are no schist-clast conglomerates or breccias.

We interpret a large exposure of granite shown in Figure 10 on the northwestern part of the exposed El Tule block, and the lightly cemented grus and immature arkose next to it, as a relatively high part of the basement terrain upon which Glance Conglomerate was deposited. The body of volcanic rocks that contains Cerro El Granizo could be Glance basement as well, or a part of the Glance.

We could not measure the thickness of the Glance between granite basement and the base of the upper Bisbee; small folds, inherent limited lateral continuity of strata, initial dips, and cover prevented that, but we offer 500 m as a crude estimate.

The Upper Bisbee Basin

The upper Bisbee basin (Fig. 5B) is the depositional basin that received upper Bisbee rocks (Morita, Mural, and Cintura Formations) and is defined by their presence. These rocks cover the Hereford and other local basins. In the area of Figure 1, the upper Bisbee rocks reach a maximum exposed thickness of about 2000 m; they are much thinner (less than 600 m) over the El Tule block and were not deposited on the Cananea block—at least in the area containing Loma Blanca, Musteñas, and Sierra Los Ajos (Fig. 4A).

Upper Bisbee Rocks on the Copper Queen and Bisbee Blocks

In the Mule Mountains (Fig. 1), the upper Bisbee rocks are thick and well exposed. In Sonora the good exposures continue in Sierra Anibacachi (Fig. 6), except that there are no confirmed exposures of the Cintura Formation. The upper part of the Anibacachi sequence was duplicated to form Montes Canova, which is a ridge capped by upper Mural. The valley between the two ridges consists mostly of lower Mural, although some Cintura from the Anibacachi section (or Morita from the Montes Canova section) may be present. The Mural Formation was studied in detail by Warzeski (1983, 1987). Warzeski also reported Glance(?) at the base of Montes Canova. We do not know the origin of this conglomerate, but speculate that it is an Upper Cretaceous deposit. We have not found the name Montes Canova on modern maps, although Rangin (1977b) and Warzeski (1987) both use the term.

Upper Bisbee Rocks on the Coronado Peak and El Tule Blocks

Huachuca Mountains Area. On the Coronado Peak block (Figs. 1 and 4A), the upper Bisbee section differs in three ways from that exposed to the east on the Copper Queen and Bisbee blocks. (1) The Mural Formation, where present, lacks the ledge-forming facies of upper Mural found elsewhere. (2) Sedimentological detail is better preserved, permitting better interpretations of depositional environments. (3) An unconformity can be inferred at the base of the upper Bisbee.

This unconformity is present on both the Coronado Peak and El Tule blocks, and the distinctive facies immediately above and below the unconformity are the same on both blocks. But on the El Tule block, the upper Bisbee section (1) is less than one-third as thick as on the Coronado Peak block, (2) contains a claystone (lacustrine) facies, and (3) contains no Mural Formation. On both blocks, the upper Bisbee is capped by the basal conglomerate of the Fort Crittenden Formation.

These relationships are shown in Figure 11 and section locations are shown (circled letters) in Figure 9. Section A is known in the literature as the Bear Creek section (Hayes, 1970; Bilodeau, 1979; Vedder, 1984; Klute, 1991). We studied section A at the base of the upper Bisbee and observed the remainder. Section B is exposed along a tributary stream that joins Bear Creek at the

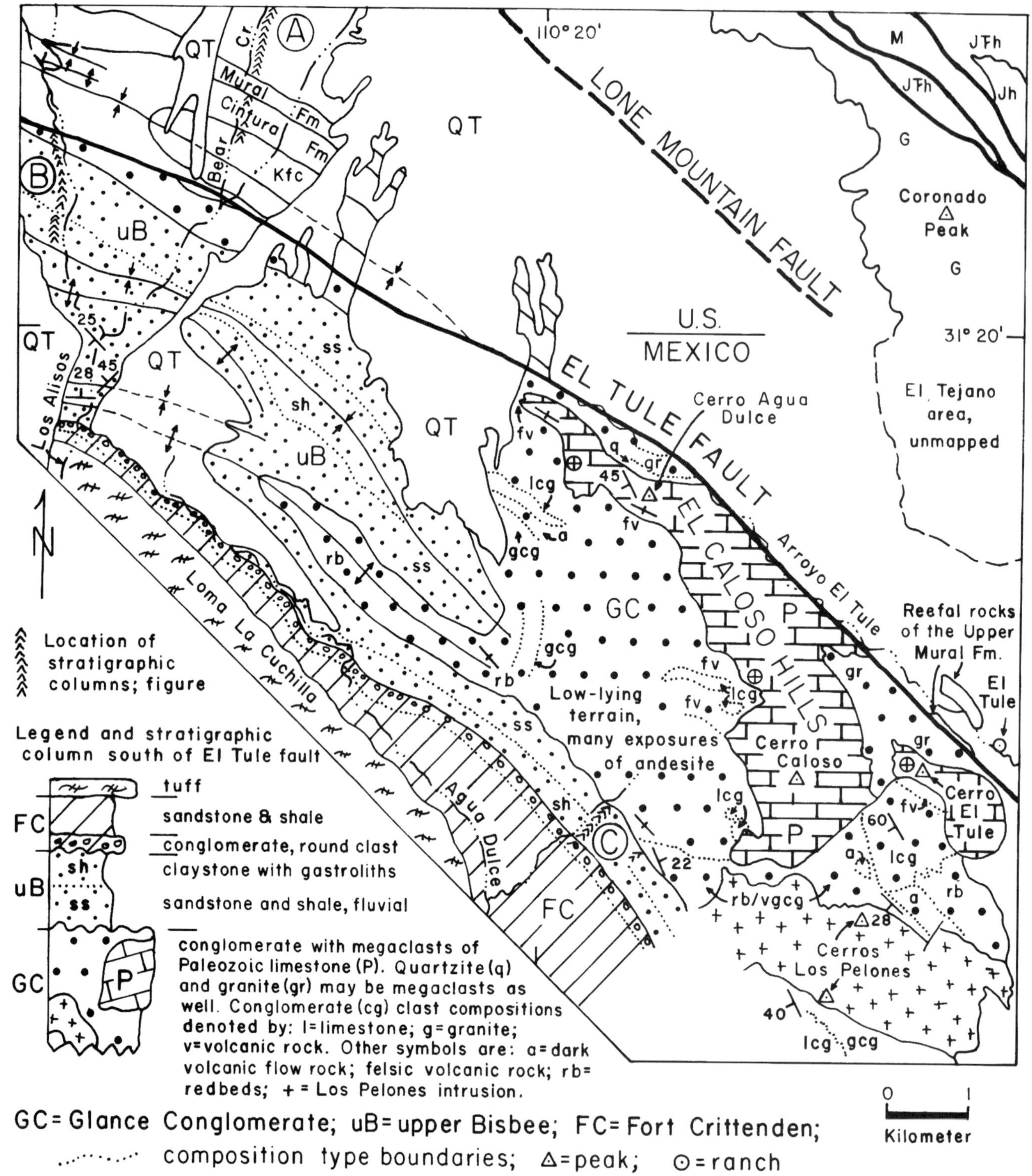

Figure 9. Geologic map of the southernmost part of the Huachuca Mountains and adjacent parts of Sonora. The U.S. portion of this map is from Hayes and Raup (1968).

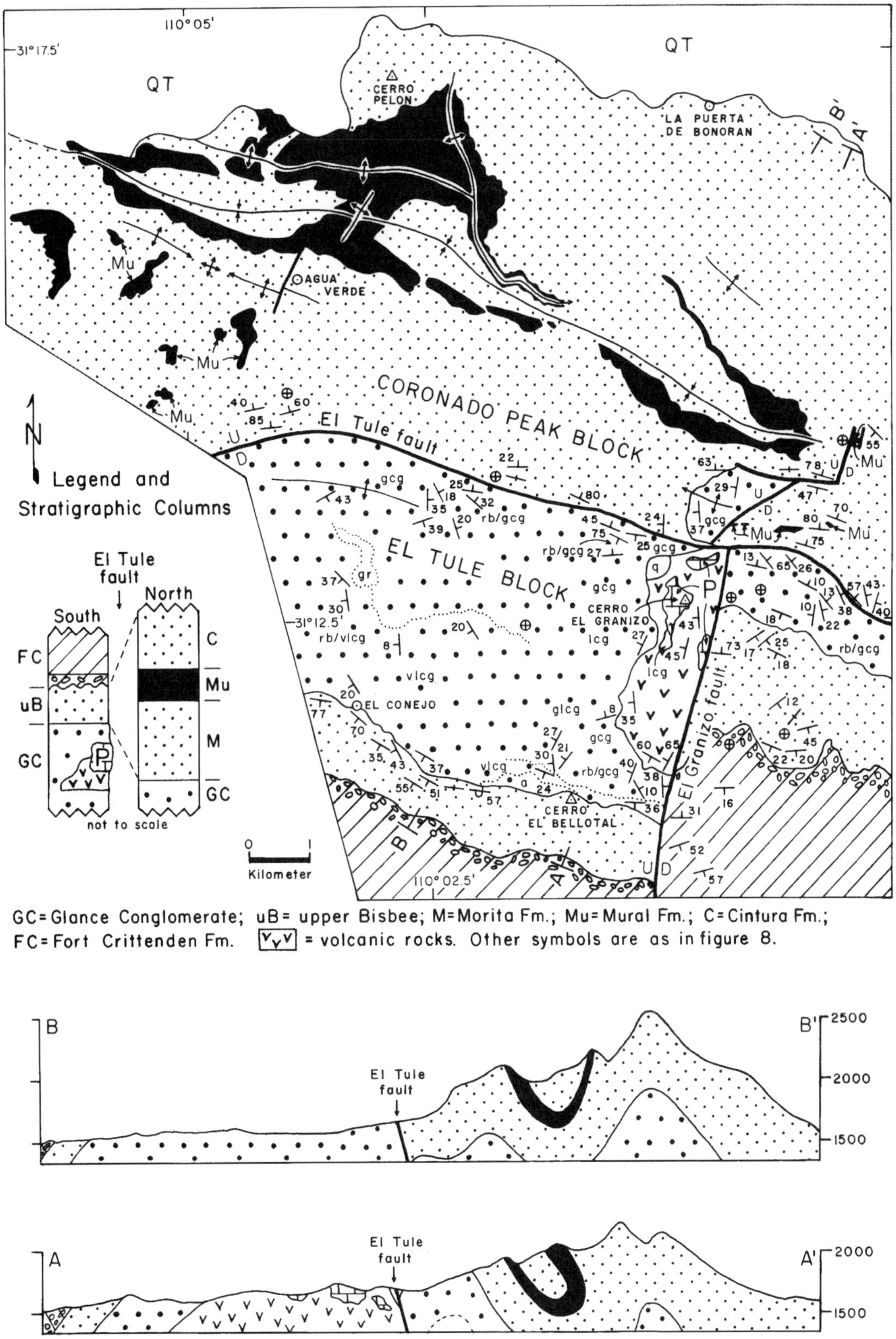

Figure 10. Geologic map and cross sections, Sierra San José.

international boundary; in Mexico the common flowageway is called Los Alisos. Section C begins just below a graded ranch road near the Cerros Los Pelones intrusion.

Regolith on Glance Conglomerate. In section B, which begins at the El Tule fault, the Glance Conglomerate is represented by a rhyolite porphyry breccia that becomes finer and less angular upsection, turning sandy then shaly, with beds of granule-pebble conglomerate 1–15 cm thick. Pedogenic limestone nodules are scattered in the upper parts of the section and are also principal components of the granule-pebble conglomerates.

We interpret the beds above the breccia as a record of the protracted close of Glance deposition, and therefore part of the Glance. During this close of Glance deposition, the finer components of the Glance and the products of Glance weathering were moved about and redeposited by surface waters. Soils were formed, then eroded, and the authigenic nodules they contained were deposited as thin beds of granule-pebble conglomerate. These processes formed a regolith that is a condensed succession of uppermost Glance Conglomerate.

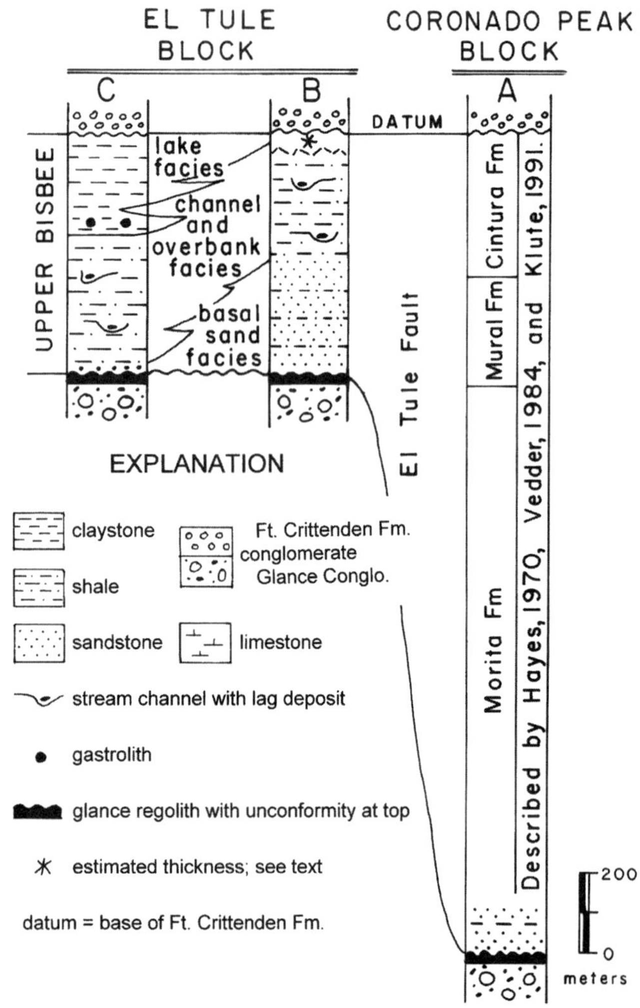

Figure 11. Comparison of upper Bisbee stratigraphy on Coronado Peak and El Tule blocks. Locations of lettered sections are shown in Figure 9. Datum is base of Fort Crittenden Formation.

In the Bear Creek section (section A) both Vedder (1984) and Klute (1991) also interpreted the top of the Glance as a condensed stratigraphic sequence, the product of soil formation and reworking, although they placed it in the Morita Formation. In either case, the Glance regolith (or transition zone of those authors), the unconformity, and the overlying sandstone are virtually the same in both sections A and B. In section C, the regolith is thinner, coarser, contains more micrite, and is largely silicified.

Basal sandstone facies. In section B (Fig. 11), the regolith is covered, in sharp contact, by a thick (15 m) unit of clean, well-sorted, light-colored (pinkish) medium-grained sandstone. This unit consists of stacked beds that are up to 1 m thick. Horizontal laminae are common in these beds, as are cross-beds in sets up to ½ m thick. In some places, very thin, poorly exposed laminae of shale separate beds. Mud rip-ups are common, preserved near the bases of sandstone beds. Some of the rips-ups measure 5 or 6 cm, but are generally much smaller. The large rip-ups (part of the traction load), the height of the cross-bed sets, and the laminar bedding, all in abundance, indicate currents of water with substantial competence. The stacked beds, separated by shale where it is preserved, suggest that the currents were intermittent, and that the mud that accumulated after each episode of current provided the material for the abundant rip-ups at the time of the next swift current. We infer that the mud was relatively dry when it was eroded, else it would have been more likely to disperse, becoming part of the suspended load rather than the bed load. The absence of the larger grain sizes that the current had the competence to transport (and deposit) indicates that they were not available—that the high-relief terrain associated with the Glance was either gone or at some distance.

These rocks permit us to envision, at low water, a broad area of sand with mud-filled, or mud-coated, depressions. The shape of the surface, imposed by running water, was probably modified by dunes as wind moved over the loose sand. Water to erode and transport the sediment of this oddly grain-size–restricted flowageway came intermittently, either because of flooding, or channel shifting, or probably both.

The section above this 15 m unit at the base of the upper Bisbee consists of a repetition of similar sandstone bodies that are generally less than 6 m thick. Many of these contain deep burrows ~1 cm in diameter and rare thin beds of micrite granules. The sandstone bodies are separated by sections (mostly covered) of finer sediment, including minor, thin beds of sandstone. We interpret that these finer-grained sediments are interfluve deposits, and that the micrite granules came from soils developed on the interfluves. The proportion of sandstone bodies in the section diminishes upsection. We mark the top of this portion of the upper Bisbee at a thick, subvertical sandstone that forms a prominent waterfall. This is an arbitrary mark, but we use it to separate this basal sandstone facies of the upper Bisbee from the overlying channel and overbank facies. About 300 m of this basal sandstone facies lie between the unconformity and the waterfall.

In section A, the 15-m-thick basal unit of the sandstone facies is well represented. In section C it consists only of two 1½-m-thick upward-fining sequences of sandstone. Each sequence has a granule–small pebble base and fines upward to a tabular-bedded, medium-grained, lithic wackestone. These beds are separated from the channel and overbank facies by a covered interval of ~15 m.

Channel and Overbank Facies. In the upper part of section B, the proportion of sandstone is further diminished and siltstone (with abundant limestone nodules) dominates the section. Sandstone bodies are principally medium grained, typically contain lag deposits, are of inconstant thickness, and have sharp but irregular basal contacts. They are channel sandstones, part of a meandering stream system. This change in kind of fluvial deposits appears to have taken place gradually in the section. We measured 213 m of section above the waterfall, then stopped because of shallowing of dip, shift of strike, cover, and the Mexican border. But from aerial photos, structural projection, and a subsequent visit to the Sonoran side, we do not think there is much additional section beneath the Fort Crittenden Formation that is exposed there (Fig. 9).

In section C, the channel and overbank facies is indistinguishable from the same facies in section B and is consistent with most of section A. The sand bodies display festoon cross-bedding in sets up to 40 cm thick, erosional bases, lenticular shapes, and contain lag deposits principally of calcareous concretions (detrital) and woody plant debris. Half-inch diameter burrows are common in the sandstone, some concentrated on bed tops, but others deep. Gray and red shales contain calcareous concretions that are authigenic.

Lake Facies. In section C, 323 m above the Glance, there is a 47-m-thick unit composed entirely of purplish claystone and siltstone. Rhythmic "bedding" is due to changes in fissility, probably caused by grain-size variations too fine to detect in the field. We collected several large (up to 6 cm) polished pebbles, probable gastroliths, which were suspended in this very fine-grained rock. Fourteen meters of sandstone overlie this unit, then 187 m more of this same unusual purplish claystone and siltstone. We believe that a continental deposit of this thickness and uniformly fine grain size is most likely to have been deposited in a lake. It is overlain by the basal (cobble) conglomerate of the Fort Crittenden Formation (Fig. 12).

Unconformity. The deposits of Glance regolith must have taken a significant period of time to form. Vedder (1984) reported ~27 m of regolith (her transition zone) in the Bear Creek section (section A, Fig. 11). On the El Tule block, the Glance regolith is more than 100 m thick in section B, and less than 10 m thick in section C. The environment that engendered the regolith was greatly different from the environment (described above) that engendered the basal sandstone facies of the upper Bisbee. Thus there is evidence for three separate—and addible—passages of time: (1) that required for the formation of the regolith; (2) that of the unknown geological events that superposed (at a sharp contact) the rocks of two depositional environments (regolith and basal upper Bisbee) that were not part of the same depositional suite; and (3) that of an inferred degradation vacuity created by the swift currents that deposited the basal sandstone. Together, these mark passage of a significant span of time between the active deposition of the Glance Conglomerate and the onset of upper Bisbee sedimentation. It is incorrect to think of Glance and Morita here as conformable formations. Consequently, we consider the base of the upper Bisbee to be an unconformity within which we include the time value of the regolith.

These considerations recalled our attention to the Glance-Morita contact on the Gold Hill and Copper Queen blocks. The thin (0.5–12 cm), red, tabular beds of well-sorted, fine-grained sandstone and shale (Figs. 6 and 8) are in the same stratigraphic position as the Glance regolith; they lie between coarse, typical Glance Conglomerate and the upper Bisbee. We believe they are lacustrine deposits because they are aqueous, nonmarine, and the sandstones show, by their shape, no evidence of channels. Because lakes do not cling to the slopes of coarse alluvial aprons, we hypothesize that the alluvial apron on the Gold Hill block had by then been transgressed by basin-center

Figure 12. Basal conglomerate of the Fort Crittenden Formation on the El Tule fault block south of the Huachuca Mountains. Hat in center of photo provides scale.

facies (Figs. 5A and 5B). Thus the time value of the surface of transgression plus that of overlying lacustrine rocks is the time separating deposition of typical, coarse Glance and onset of upper Bisbee deposition.

We do not believe it is accurate to refer to the Glance–upper Bisbee contact as simply conformable in this part of Sonora.

Thickness and Facies Change Across the El Tule Fault. The Bear Creek section (section A, Fig. 11) extends to the base of the Fort Crittenden Formation, where it is exposed near the headquarters of the Lone Mountain Ranch. Adjusted for the removed 25 m transition zone, its thickness is 1996 m (Klute, 1991). The upper Bisbee south of the El Tule fault is less than 30% of the thickness measured north of the El Tule fault (Fig. 11). The presence of both the Glance regolith and the distinctive basal sandstones of the upper Bisbee on both sides of the El Tule fault (sections A and B, Figs. 9 and 11) indicates that there was little topographic relief along the fault at the time of their formation and deposition. Subsequent movement along the fault provided space for three times as much sediment on the Coronado Peak block as on the El Tule block, and the basal Fort Crittenden conglomerate is continuous across the fault. Movement along the El Tule fault was syndepositional.

On the Coronado Peak block the Mural Formation abuts the El Tule fault, but there is no Mural Formation on the El Tule block. It is likely that tongues of Mural extended across the fault and were removed by erosion that alternated with deposition on the low-lying but tectonically positive El Tule block.

Sierra San José Area. The Sierra San José is entirely within the Coronado Peak block. It consists of folded Morita, upper and lower Mural, and Cintura formations (Fig. 10). These appear to be comparable in thickness to the upper Bisbee in the Mule and Huachuca Mountains.

South of the El Tule fault (on the El Tule block), a thick unit of clean sandstone above the Glance Conglomerate makes a strike ridge that extends from Cerro El Bellotal (Fig. 10) westward to Rancho El Conejo. This sand body and the beds of fluvial sandstone and shale that lie conformably above it are very similar to rocks in the same stratigraphic position in sections A and B of Figure 11. At the base of Cerro El Bellotal, the sandstone rests on andesite of the Glance Conglomerate, which is similar to the relationship in section C of Figure 11. The sandstone dips to the south, is not conformable with the Glance, and is clearly distinct from it. This section is capped by a rounded-cobble conglomerate that marks the base of the Fort Crittenden—just as in sections A–C of Figure 11. No Mural Formation is present in the section, so we map it is as upper Bisbee. In general, the best exposures of the upper Bisbee are east of the El Granizo fault. A map-based thickness estimate indicates that the upper Bisbee in the El Conejo area (south of the El Tule fault) is about the same thickness as in sections B and C, Figure 11.

Near the La Bellota Fault. Upper Bisbee rocks are absent along the La Bellota fault south of the La Negrita Hills (Fig. 6). Upper Cretaceous rocks there lie unconformably on Glance Conglomerate.

The Fort Crittenden–Cabullona Basin

Laramide contraction reactivated the faults that bounded the Hereford basin and reversed the vertical (normal) component of their movement. This inversion caused uplift (Fig. 5C) and folding of Bisbee rocks of the Coronado Peak and El Tule blocks (Fig. 1). These blocks contain a Laramide belt of folds of "cylindrical elegance" (Davis, 1979, p. 560) that lies entirely southwest of the Huachuca Peak fault (Fig. 1). The folds are well displayed in the Sierra San José, and farther east there are apparent east-west folds in the Cabullona Group (Fig. 6). On that basis, the belt of folds (products of Laramide inversion) may be extended to the La Bellota fault.

The Huachuca Peak fault's elevated hanging wall shed abundant detritus mostly to the west and southwest of the Huachuca Mountains, forming the continental deposits of the Fort Crittenden Formation (Stoyanow, 1949; M.J. Hayes, 1987). In the El Tule area (Fig. 9), the basal part of the Fort Crittenden Formation is 100 m of rounded-cobble and small boulder conglomerate (Fig. 12). Fluvial sandstone and shale overlie this conglomerate, then the tuff of the Loma La Cuchilla. South of Sierra San José, the same rounded-cobble conglomerate is present, exposed in a strike ridge formed of rounded, elongate hills (Fig. 10). Above the conglomerate, the Fort Crittenden consists of some sandstone and shale, then a thick shale unit that is probably a deposit of Lake Cabullona. We have not studied it.

East of the Huachuca Mountains and north and east of Sierra San José, a wide valley was cut into the uplifted Bisbee Group (Fig. 5C). The upper Bisbee, which formed the walls of the valley, was relatively undeformed and easily eroded, except for the ledge-forming limestone of the upper Mural. The stream flowing in this valley was competent to carry cobbles and small boulders from the north. The stream was dammed, Cabullona Lake was formed, and those cobbles and small boulders formed the lake's northern littoral.

The Cabullona Group (Taliaferro, 1933; González-León, 1994; González-León and Lawton, 1995) consists principally of the deposits of this lake. The coarse rocks of the littoral and associated facies are the El Cemento Formation. Where conglomerates of the El Cemento and Glance are in proximity, three factors distinguish them. In the El Cemento: (1) clasts are rounded and support one another so that there is relatively little matrix; (2) light-colored clasts of felsic volcanic rocks are characteristic; and (3) scattered light-colored cobbles of limestone bear Mural fossils, most commonly *Orbitolina*. The El Cemento interfingers with strata of basinal facies, the Camas Sandstone and Packard shale. This interfingering relationship is observable in Arroyo El Cemento and is well displayed in Arroyo El Tejano (Fig. 6). The gross distribution of basinal lacustrine (Packard and Camas formations) and littoral-sublittoral facies (El Cemento Formation) indicates that in the area of Figure 6, Cabullona lake was fed from the north. There are three evidences that the Gold Hill block (part of the Sierra Anibacachi uplift of González-León and Lawton [1995]) was not the major sediment contributor.

(1) Interfingering of El Cemento with basinal strata occurs in a zone that trends roughly east-west, and basinal sediment lies south of the El Cemento. (2) The adjacent Gold Hill block is an unlikely source for the abundant rounded cobbles of felsic volcanic rock that are common in the El Cemento. (3) There are no littoral Cabullona facies along the southern part of the La Bellota fault. Neither are there littoral facies farther basinward of the La Bellota fault, where González-León and Lawton (1995) placed their Cabullona thrust.

We add two additional units to the Cabullona Group. (1) The Agua Amarga beds lie beneath the Camas and Packard formations and record two mass-gravity events. (2) The 91 Hills slide mass rests on both Gold Hill and Agua Zarca blocks and is equivalent to (and probably continuous with) the Agua Amarga beds. Together, these units are a record of mass-gravity events that can be traced for 25 km along the La Bellota fault. We describe them under the section "Mass-Gravity Deposition."

ORIGINS OF THE HEREFORD AND BISBEE BASINS

The Hereford basin is defined by (1) the coarse deposits that it contains, (2) the temporal separation of those deposits (Glance Conglomerate) from upper Bisbee rocks, and (3) the faults that formed the basin's boundaries (Fig. 4). We have no direct information regarding the origin of the faults that bound the Hereford basin, but only two regional tectonic events that might have affected the study area at that time have been hypothesized: rifting (Bilodeau, 1982; Dickinson and Lawton, 2001a) and continental transform faulting (Mojave-Sonora megashear, Silver and Anderson, 1974).

Rifting

The Hereford basin does not fit comfortably with the rift hypothesis. (1) Lawton (2000) hypothesized that in southeastern Arizona and southwestern New Mexico most or all of the Laramide faults are reactivated Jurassic–Early Cretaceous normal faults that had been active during formation of the Bisbee basin, and pointed out that the northwest-southeast trends of these faults form an en echelon pattern (Fig. 13). The northwest trend is about the same as the Hereford basin, diverging at ~20° from the trend of the northern structural margin of the Bisbee basin (Fig. 13) and from the axial trend of the Bisbee core basin of Dickinson and Lawton (2001b). This en echelon pattern is not compatible with simple orthogonal separation. Under the rifting hypothesis, a lateral component of separation, control by older faults, or some other explanation must be offered. (2) The opening of the Hereford basin, and thus the tectonism that formed it, ended before initial deposition of the upper Bisbee rocks; this does not support the idea that both Glance and upper Bisbee are products of a continuing tectonic phenomenon—rifting. (3) The Hereford basin is narrow (Fig. 4A) and therefore not the core of a rift that continued to open, ultimately forming the wide Bisbee basin.

The upper Bisbee rocks also offer only weak support for the rift hypothesis. The northern border of the upper Bisbee lithosome is marked by coastal onlap in southern Arizona and New Mexico and is consistent with the idea of a rift. But to the south and east, the upper Bisbee rocks are merely part of a widespread assemblage of terrestrial, littoral, shelfal, and basinal strata that formed in a variety of contemporaneous environments within, along the margins of, and somewhat inland of a broad Cretaceous sea. Delineation of the post-Glance Bisbee basin is therefore difficult, thereby diminishing the evidence of shape, that is, the geographical evidence of rifting. Lawton and Olmstead (1995) offered one solution to the problem of basin delineation. In their Figure 13 they define the Aptian–early Albian portion of the Bisbee basin as the part of a broad sea whose sedimentary record now happens to be isolated by the younger volcanic mass of the Sierra Madre Occidental. As a practical matter, that is a good definition. But a depositional basin with limits that must be defined operationally does not offer very good geographic evidence of rifting. The older, post-Glance (Neocomian) rocks could offer more useful support of rifting. Correlation of those Bisbee rocks with units in New Mexico and in the Chihuahua Trough is consistent with the rift hypothesis. But (1) the Chihuahua Trough has its roots in the Paleozoic, and (2) Haenggi and Muehlberger (this volume) consider the Chihuahua Trough to be a right-lateral pull-apart.

Continental Transform

Two circumstances favor the idea that the Hereford basin and the Mojave-Sonora megashear are genetically linked. (1) Northwest-striking, Jurassic basin-bounding faults in southeastern Arizona and southwestern New Mexico are parallel to the Mojave-Sonora megashear, suggesting that their origin might lie with the same subcrustal movements that generated the megashear. The en echelon pattern of these faults is compatible with a transform origin. (2) The megashear and the Hereford basin, both Jurassic, became inactive at about the same time; movement along the megashear had ceased by the end of the Oxfordian (Anderson and Schmidt, 1983) and the formation of the Hereford basin predates earliest upper Bisbee. Thus, both orientation of the faults and their timing are consistent with the idea that the Hereford basin and the Mojave-Sonora megashear are related features.

Discussion

Regionally, Glance Conglomerate was deposited "...along the margins of isolated basins, bounded on at least one side by normal faults" (Bilodeau, 1979, p. 122). The Hereford basin is one basin of the array that Bilodeau wrote of and that Lawton (2000) depicted. We believe that the array's origin is linked to the Mojave-Sonora megashear, and that the Glance Conglomerate in these basins was covered by the finer-grained deposits of a much larger depositional system of different origin. We suggest that aggregate crustal extension associated with the formation of the array of Jurassic fault basins caused regional subsidence within

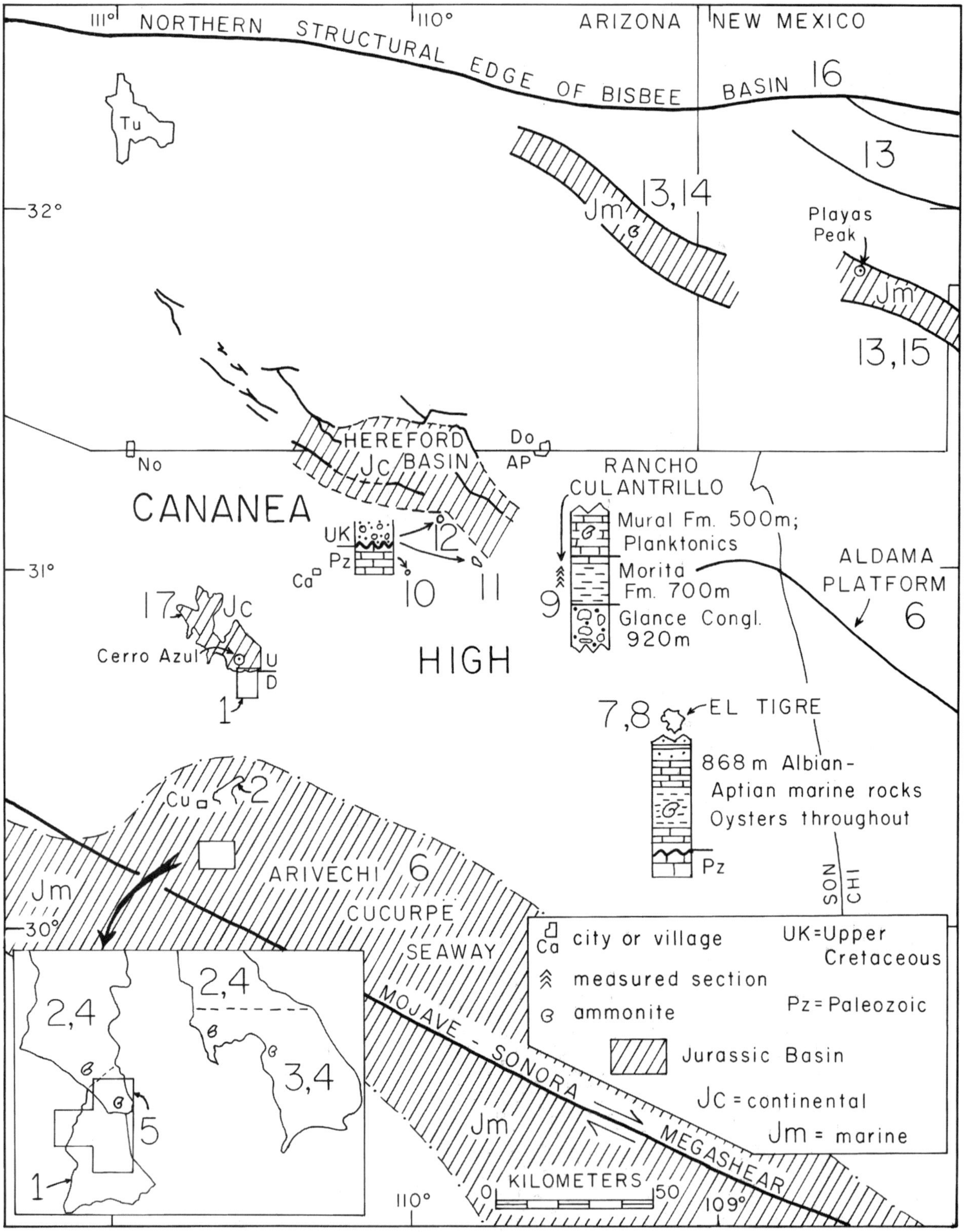

Figure 13. The Cananea high and evidence for its existence. Thick Glance Conglomerate in the Hereford basin and at Rancho Culantrillo mark Jurassic uplift of the northeastern side of the high. Thick exposures of Glance at and northwest of Cerro Azul mark uplift on the southwest. Exposures of Paleozoic–Upper Cretaceous unconformity east of Cananea city support the interpretation of Jurassic and younger uplift. On the north, uplift was along El Tule and Cananea faults (see Figs. 1 and 4B). Jurassic marine waters were on both sides of the high. The El Tigre area was part of a Cananea peninsula before Aptian inundation. Near Cerro Azul, at location 1, Late Cretaceous mass-gravity displacements suggest paleoslope away from the high. Numbers are locations and references to literature. 1—McKee and Anderson, 1998; 2—M.B. McKee unpublished reconnaissance map; 3—Rodríguez-Castañeda, 1984; 4—Rangin, 1977a; 5—Lawton et al., 2004; 6—Haenggi and Muehlberger, this volume; 7—Imlay, 1939; 8—Cerecero-Luna and Castro-Escarrega, 1996; 9—Rosales-Domínguez et al., 1995; 10—Sierra Las Ajos, Page et al., 2003; 11—Cerro Musteñas, Rangin, 1977b; 12—Loma Blanca, Rangin, 1977b; 13—Lawton, 2000; 14—Lawton and Olmstead, 1995; 15—Lucas and Lawton, 2000; 16—Dickinson and Lawton, 2001b; 17—Nourse, 1995.

the zone occupied by those basins. This subsidence, and eustasy, were responsible for deposition of the upper Bisbee rocks. Along the northeastern flank of the Cananea high, this subsidence can be seen in the syndepositional lowering of the Coronado Peak block relative to the Cananea high (Figs. 5B and 11). This lowering of crust beneath the upper Bisbee basin was along the El Tule fault, which had been a Jurassic fault that was active in formation of the Hereford basin.

Our outline and description of the Hereford basin is a simplism, made so by shortage of information. The "lazy S" shape (Mann et al., 1983) of this basin and the en echelon pattern created by Jurassic basins in the region, suggest that this linear fault basin is a strike-slip basin, probably a pull-apart, none of which are as simple as our representation. One of us (Anderson) points out that the Hereford basin as shown here is not kinematically valid as a pull-apart basin. Accordingly, Anderson and Nourse (this volume) show a kinematically valid solution to the opening of the basin as a pull-apart.

We believe that we are offering a useful alternative to the rift hypothesis and we hope that both hypotheses receive continuing evaluation.

THE CANANEA HIGH

The idea of a paleotectonic high around the mining community of Cananea was first suggested by Emmons (1910). McKee (1991), and McKee and Anderson (1998) concluded that reefal upper Mural and associated facies exposed in their study area south of Cerro Azul (Fig. 13, location 1) first formed in shallow waters then slid southward into a contemporary or younger basin. They invoked the southern margin of Emmons's paleotectonic high to provide the shallow waters in which the slide masses had originated and used the name "Cananea high" to identify the feature. This hypothesis of a paleogeographic high is supported by recent work (Grijalva-Noriega, 1995; Rodriguez-Castañeda, 2002) and by additional evidence that is summarized in Figure 13.

At Sierra Los Ajos (Figs. 4A and 13, location 10), Page et al. (2003) described Cabullona-equivalent rocks lying unconformably above a Proterozoic-Paleozoic sequence. At Cerro Musteñas and Loma Blanca (localities 11 and 12, Fig. 13), Upper Cretaceous rocks overlie Paleozoic limestone. The unconformity exposed at these three localities does not prove whether the Bisbee Group's absence is by nondeposition or by erosion, but additional evidence supports absence by nondeposition. (1) Thick deposits of Glance Conglomerate on the El Tule and Coronado Peak blocks (Fig. 4) and at Rancho Culantrillo (Figs. 4A and 13) were certainly derived from some high-standing source very nearby. (2) Upper Bisbee rocks on both sides of the El Tule fault (Fig. 11) demonstrate that the fault was syndepositional and that the El Tule (and Cananea) blocks were high-standing during the time of deposition of the upper Bisbee. We infer a second fault (Cananea fault) that would have prevented upper Bisbee deposition on the Cananea block. Whether or not a thin upper Bisbee section was deposited (then removed), the presence of the Cananea high is equally well indicated. Glance Conglomerate as much as 1 km thick (Nourse, 1995) exposed at and northwest of Cerro Azul (Fig. 13, location 17) completes the picture of a narrow, high-standing, fault-bounded block flanked on both sides by its own coarse Jurassic detritus. These observations validate Emmons's inference.

The evidence of a seafloor sloping southward away from the high (McKee and Anderson 1998) has been disputed. Jacques-Ayala (1995) disagreed with their interpretation of deeper water into which slide masses moved; he thought all water depths there to have been shallow, which would not have allowed a slope. Lawton and others' (2004) established a five-member reference section for the Mural Formation in northern Sonora. Two of the members, the Tuape Shale and Cerro La Puerta Members, were interpreted as products of "deeper water, euxinic deposition recorded by thick intervals of black shale with a pelagic ammonite fauna" that (because of facies relationships) was not part of a "truly deep, bathyal environment" (Lawton et al., 2004, p. 54). Lawton and others recognized these two members in their Cerro La Puerta section, which coincides with McKee and Anderson's (1998) stratigraphic section of the same name. The relatively deep water interpreted by Lawton and others allows a slope down which sliding away from the Cananea high could have occurred. Nevertheless, neither Jacques-Ayala nor Lawton and others agreed that sliding took place. We regret that these authors did not deal with the principal evidence upon which the interpretation of sliding was based. This was geometrical evidence, which is perhaps the most powerful tool of geological interpretation and exposition. This geometrical evidence is a precise, entirely field-based geological map drawn on a 1:10,000 aerial photo base supplemented by the reliable cross sections such a map permits. We maintain that basinward sliding (on two discrete low-angle surfaces) did occur, and offer it again as evidence of the Cananea high.

The Jurassic Cananea high was a peninsular rib of continental crust, a horst flanked by coarse detritus on its northeastern and southwestern sides. Measured through the city of Cananea, its width was ~60 km. On its northeastern side, continental deposits and inferred highlands separated it from Jurassic marine waters farther north (Fig. 13). On the southwestern side, the Jurassic marine basin reached northward nearly to Cerro Azul. Although Cretaceous transgression of the high was probably never completed, the El Tigre area (Fig. 13) was inundated by Aptian time.

MASS-GRAVITY DEPOSITION

The Hereford basin contains assemblages of megaclasts (principally of Paleozoic limestone) that are manifest as mountainous areas on both its eastern side (La Negrita Hills) and the western side (El Caloso Hills). The La Negrita Hills are 10 km long and up to 360 m thick measured from the base of the megaclasts (Fig. 14) to the top of Cerro La Negrita. The El Caloso Hills are 8 km long and up to 200 m thick.

The Cabullona basin (Late Cretaceous) contains abundant clasts and megaclasts of Lower Cretaceous rock. The most

Figure 14. The northern part of La Negrita Hills (middleground) viewed westward, from "little" Sierra Anibacachi. The highest peak in the middleground is the northern of the two Cerros La Bruja (Fig. 6). The quasi-planar, subhorizontal base of the assemblage of megaclasts that form La Negrita Hills is well marked by the vegetation change low on the hillsides. The prominent peak on the horizon, center, is Sierra San José. The knob on the left (southern) flank of Sierra San José is Cerro El Granizo (Fig. 10) which is another megaclast consisting of Paleozoic limestone. The truck on the road at left of photo provides scale.

distinctive of these clasts are composed of reefal upper Mural Limestone that occurs within a thin stratigraphic interval that can be traced for 25 km.

Jurassic–Early Cretaceous Mass-Gravity Deposits

South of the Mule Mountains, Taliaferro (1933) interpreted Cerro La Negrita (he called it La Morita Mountain) as part of a southwest-vergent overturned anticline of Paleozoic rocks that was thrust (Cabullona thrust) over the Cabullona Group. Rangin (1977b) followed that interpretation as did González-León and Lawton (1995), who added the opinion that the Cabullona basin was the product of lithospheric loading imposed by the hanging wall of the Cabullona thrust, which they positioned well basinward (southwest) of the exposures of Agua Amarga beds.

Two observations seem to support this interpretation of a thrust: (1) One view of Cerro La Negrita from the south gives the illusion of an overturned anticline (perhaps the result of a happenstance arrangement of the megaclasts of limestone). (2) The Glance has undeniably been raised, along the steep La Bellota fault, with respect to the Cabullona Group.

Nevertheless, the Paleozoic limestone blocks (Permian Naco Group) are megaclasts of the Glance Conglomerate. We conclude this for four reasons: (1) There is no anticline; the blocks of tilted Paleozoic strata show diverse dips but no organization that suggests a fold. (2) Beneath the megaclasts (that is, beneath the supposed thrust), the Glance contains smaller clasts of Paleozoic limestone; these range from boulders to one clast with exposed dimensions of 30 × 10 m that was imbedded in schist-clast conglomerate. (3) The isolated ridge ("little" Sierra Anibacachi) west of Rancho El Sifón (Fig. 6) dips into the Glance. (4) The base of the mass of Paleozoic blocks is irregular (Fig. 15) but overall it is virtually horizontal (Fig. 14). If that horizontal contact is a thrust, (a) it is devoid of characteristic structural features such as gouge or fault surfaces, (b) La Negrita Hills would be an outlier with the mother thrust nowhere to be found or imagined, and (c) the La Bellota fault would then still be required to raise the Gold Hill block against the Agua Zarca block.

In the Huachuca area, we evaluated whether the El Caloso Hills (Fig. 9) are: (1) in situ, (2) in the hanging wall of a thrust, or (3) megaclasts within the Glance. In some places along the western margin of the limestone mass, reentrants in the mass provide evidence that the Glance conglomerate and andesite lie beneath the limestone. This argues against the first hypothesis, but permits the other two. We found no place where the contact between the Paleozoic rocks and the Glance could be directly examined for features of either thrusting or deposition. Nevertheless, we chose the third hypothesis as the best for four reasons. (1) It would be remarkable for such a thrust to be present without showing itself somewhere along the perimeter of the hanging wall. (2) The irregular shape of the contact is consistent with the base of an "exotic" (unrooted) mass, appearing very much like the one at the base of La Negrita Hills (for example, Fig. 15). (3) Limestone blocks within the mass have various structural attitudes; we see no structural coherence. (4) Blocks of Paleozoic limestone are well known within the Glance (Simons et al., 1966), and in the Huachucas they are associated with andesite. At Montezuma Pass, about 5 km from the El Caloso Hills, Bilodeau et al. (1987, p. 244) reported a middle volcanic member of the Glance that "contains an assortment of exotic blocks of Permian Concha Limestone, andesite, and volcanic flow breccia." The limestone blocks "commonly [occupy] the entire 60 m width of the member, and reach up to 400 m in length."

Figure 15. Blocks of Paleozoic limestone forming a small waterfall at the base of La Negrita Hills, southwestern part. Smaller brecciated fragments of the limestone are in lower right of photo. All clasts rest here on schist-clast breccia. Downstream from this waterfall, the schist-clast breccia contains a megaclast of Paleozoic limestone that measured ~10 × 30 m. Person for scale.

In the Sierra San José area, west of the El Granizo fault (Fig. 10), the isolated blocks of Paleozoic limestone (Cerro El Granizo) that are on and within a suite of volcanic rocks (rhyolite, agglomerate, tuff) are certainly clasts.

Late Cretaceous Mass-Gravity Deposits

91 Hills Slide Mass

The Mural Limestone exposed on the Gold Hills block in both Arizona (91 Hills) and Sonora (Figs. 1, 6, and 16) is megaclasts within a Late Cretaceous mass-gravity deposit that came to rest near and in Lake Cabullona on both sides of the La Bellota fault. We reached this conclusion for the reasons that follow.

1. Reefal Mural rocks within the slide mass are otherwise stratigraphically out of place. In Sonora, the slide mass rests directly on Glance Conglomerate; in Arizona it rests on a thin section of Morita Formation (Fig. 17).

2. The slide mass forms a discrete, mappable unit with an internal structure that is unrelated to rocks beneath it (Fig. 17, panel 3). In its distal (southern) part, that internal structure consists of ridges of sandstone whose strikes and shapes reflect the lobe-like shape of the mass. These can be seen in Figures 16 and 18 and in the form lines (Fig. 6) that were taken from 1: 20,000 and 1:50,000 aerial photos and verified in the field. We tentatively interpret these sandstones as nearshore sands of Lake Cabullona. They slid into contact with Loma La Morita (Fig. 17, panel 3), which is the northernmost part of the La Negrita Hills. Thus at that very unusual contact, an Upper Cretaceous mass-gravity deposit that contains Lower Cretaceous (Aptian-Albian) megaclasts lies against a Paleozoic megaclast in a Jurassic or Lower Cretaceous (Neocomian) mass-gravity deposit.

3. This sliding explains the otherwise puzzling presence of two isolated exposures of upper Mural that are surrounded by Cabullona Group on the Agua Zarca block just north of Highway 2 (Figs. 6 and 19). They are megaclasts of reefal Mural Formation from the 91 Hills slide mass. Their partial cover of Late Cretaceous rock was preserved because of their structurally lower position west of the La Bellota fault.

4. In the northern part of the mass, upper Mural limestone (the light-colored ridges in Fig. 18) does not occupy a single stratigraphic position. One explanation for this was offered by Bergman et al. (1993). They reported an angular discordance of ~15° at the normally concordant contact between beds of the lower Mural (the Lowell Formation of Stoyanow, 1949) and the base of the upper Mural. They inferred Late Cretaceous submarine displacement by gravity, during which the relatively brittle limestone broke into blocks and overrode the lower Mural. We explain this discordance by movement during a Late Cretaceous, terrestrial event even though both events may have occurred.

Figure 16. View, to the north, of the 91 Hills slide mass from Loma La Morita (the northernmost part of La Negrita Hills, north of Highway 2). Foreground: low ridges of folded sandstone in the toe of the slide mass. Middleground: blocks of reefal limestone derived from the upper Mural Limestone; these are the 91 Hills. One of these blocks (Monument Ridge, at arrow) is crossed by the international border and bears border monument number 91. Background: Mule Mountains.

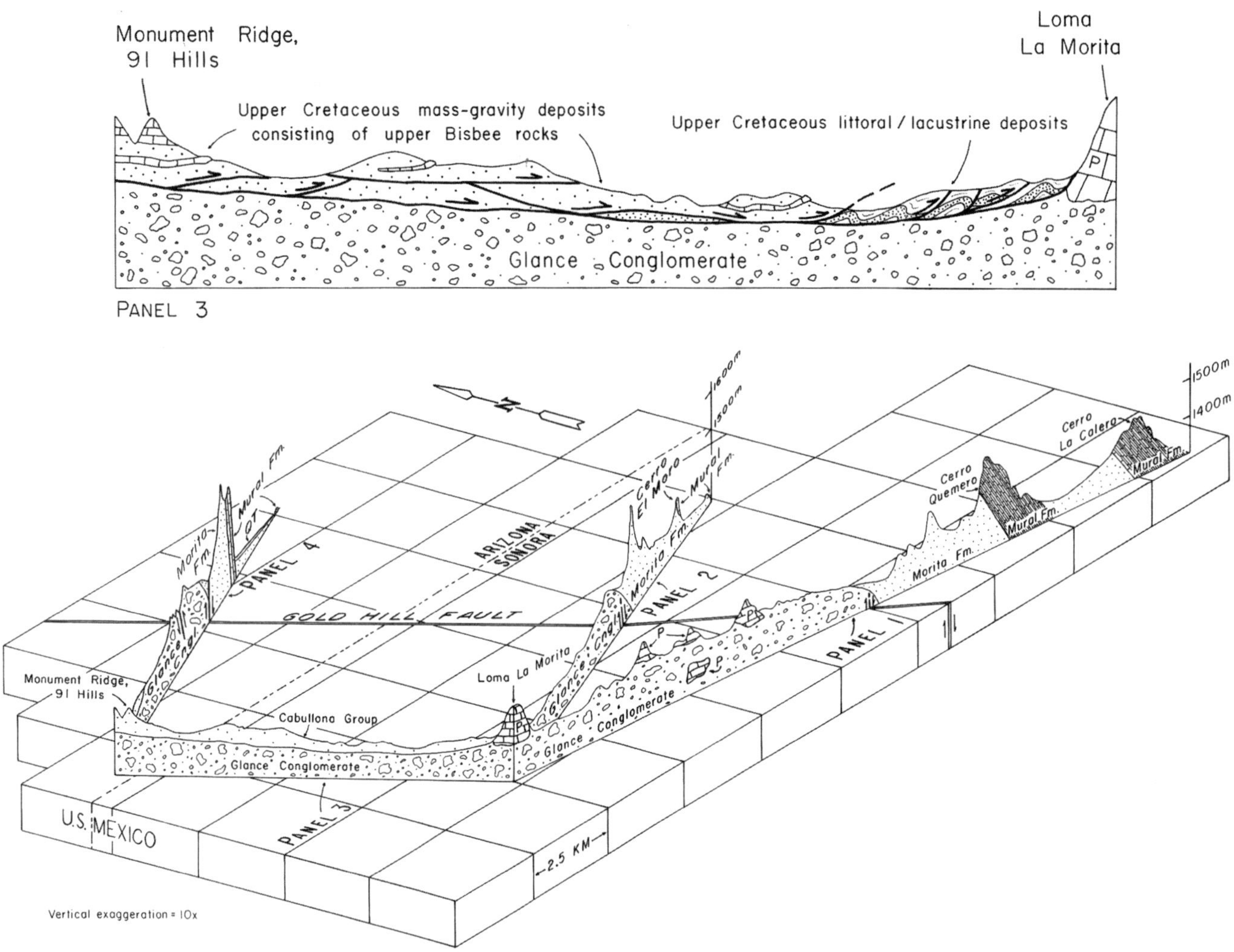

Figure 17. Fence diagram illustrating the relationship of the 91 Hills slide mass to the Bisbee Group. Panel locations are shown in Fig. 6. The structures shown in panel 3 are illustrative, based on field examination of the rocks and on the shapes as seen in Figures 16 and 18. Cerros Quemero and La Calera are peaks on Sierra Anibacachi and Montes Canova, respectively.

The anomalous distribution of blocks of Mural in the slide mass might also have been caused by tears within the sliding mass that permitted one section containing rotated upper Mural to slide farther than another. Bergman and others mapped one such tear. Such internal disruption is evident in Figure 18.

5. We correlate the 91 Hills slide mass and the Agua Amarga mass-gravity deposits that we identified on the basis of entirely different evidence. These two deposits are almost in physical continuity and together extend for about 25 km.

Agua Amarga Beds

Agua Amarga (Toscalita Creek of Taliaferro, 1933) crosses Taliaferro's Snake Ridge and runs out into the Punta de Agua basin (Fig. 6). In doing so, the creek drains southwestward, *downsection*, through Glance Conglomerate. It then crosses three mappable rock units (units A, B, and C, of Fig. 20, I and II). Unit A is red mudstone (in a narrow, obscure exposure) that contains distinctive white, angular crystals of feldspar. Unit B is shale containing beds of well-sorted sandstone with excellent top-and-bottom evidence proving downstream and southwest to be *upsection*. The creek runs within this rock unit for a kilometer or less, then passes though a gap in a ridge and out into the openness of the modern Punta de Agua drainage basin. That ridge consists principally of unit C, which is vertical-to-overturned beds of Mural limestone debris. The clasts forming this sheet of Mural debris are cobbles and boulders up to several meters in maximum exposed dimension. Typical upper Mural fossils, rudists for example, are abundant. We refer to these rock units (A, B, and C) as the Agua Amarga beds, which we place in the Cabullona Group.

The upper Mural debris (unit C) is ~12 km in exposed length and was probably derived from the Gold Hill block where Sierra Anibacachi was in a structurally lower and pre-retreat position. The deposition of unit C was a convulsive mass-gravity event, because it consists of very coarse debris resting on relatively

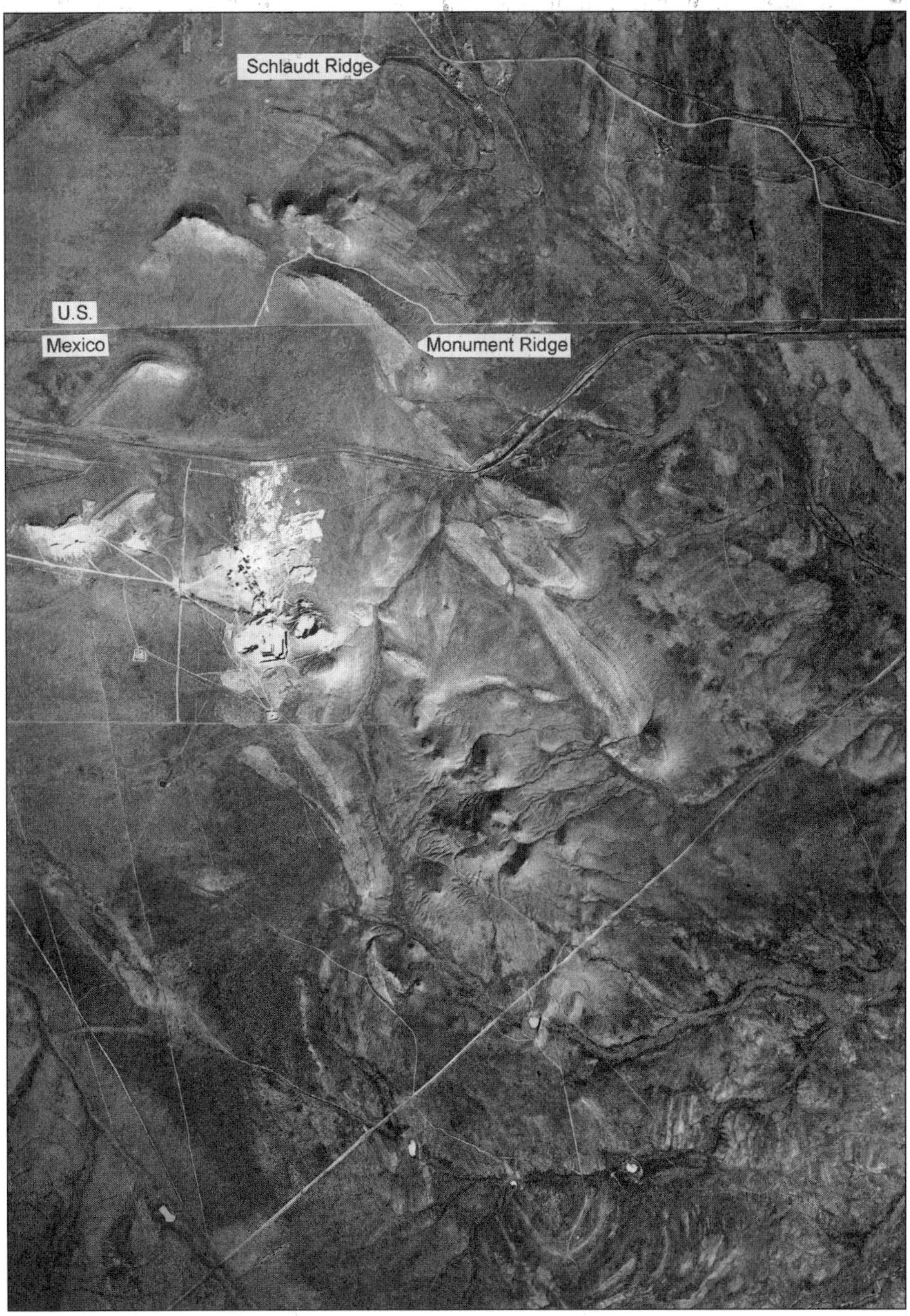

Figure 18. Aerial photo of part of the 91 Hills slide mass. Schlaudt Ridge is part of the slide mass and includes Morita Formation (Stoyanow, 1949). The lighter-toned ridges and knobs are formed of reefal Mural; some have been largely removed by quarrying. Folded sandstone of the slide toe is in the southeastern part of the photo and in the foreground of Figure 16. Notice that there is some consistency of dip in the larger Mural fragments, and that if in situ, they would necessarily be scattered within a thick stratigraphic zone, which the upper Mural Formation is not. Scale: The segment of the international boundary shown here is about 5 km long.

Figure 19. Block of limestone from the reefal upper Mural Limestone surrounded by Cabullona Group west of the La Bellota fault. In Figure 6, this is the northern of the two limestone blocks north of Highway 2. Naco Hills (part of the Mule Mountains) are in the background.

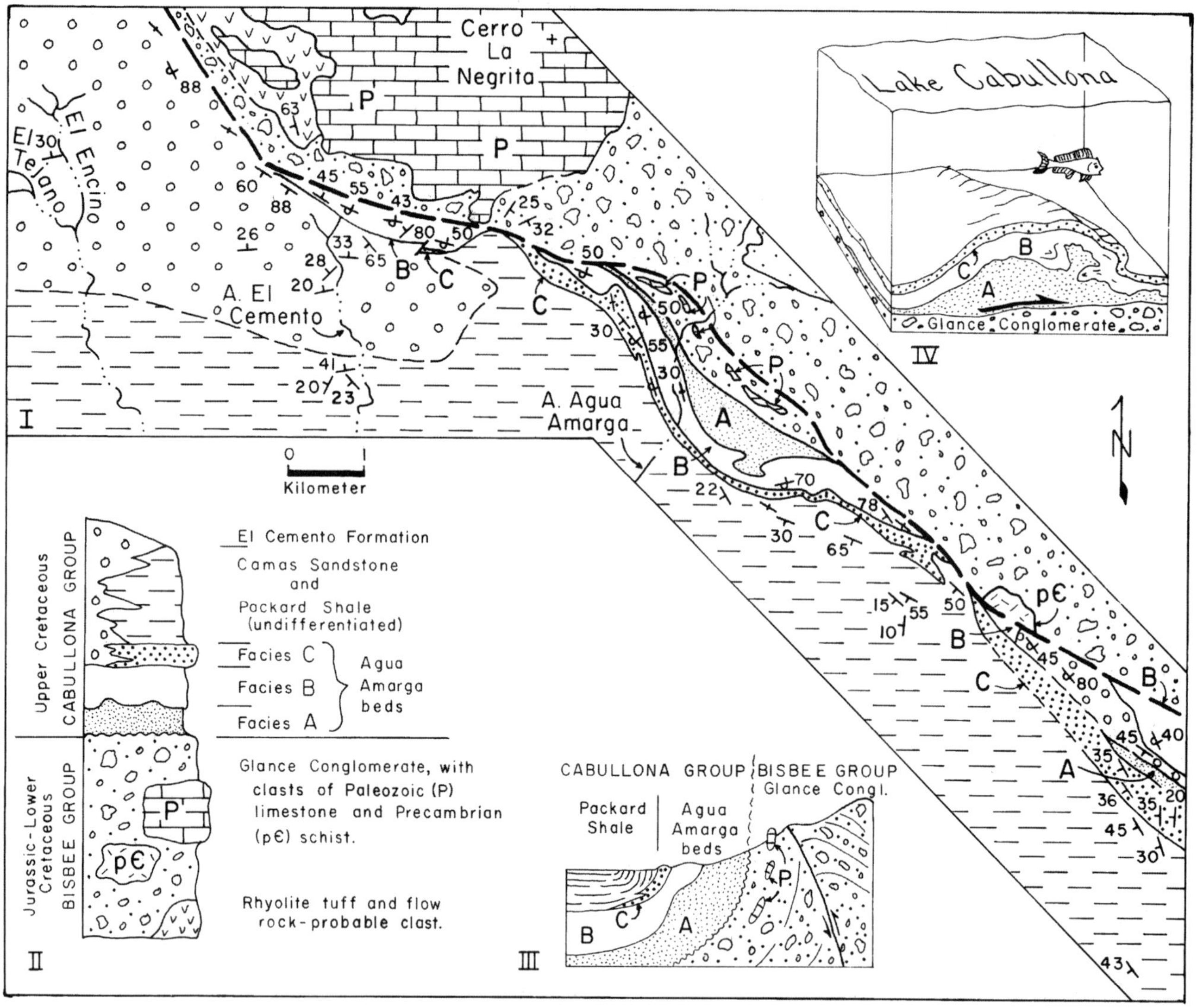

Figure 20. Geologic relationships along the La Bellota fault and the contact between the Bisbee and Cabullona groups. (I) Geologic map. (II) Stratigraphic column and symbol explanation. (III) Cross section, roughly along the course of Agua Amarga, where it crosses the Agua Amarga anticline. (IV) Cartoon depiction of the Agua Amarga anticline on the floor of Lake Cabullona; we assume that overlying Packard Shale, not shown, would have been present to contain the fold as it formed during sliding; fish for scale is ~500 m long.

fine-grained sediment and it is covered by finer, basinal sediment (Packard Shale) of the Late Cretaceous Lake Cabullona.

Following the deposition of unit C and some unknown thickness of basinal lake sediment, the Agua Amarga beds slipped farther basinward. This second episode of mass-gravity movement formed: (1) an anticlinal bulge (Fig. 20, I and IV, and Fig. 21), (2) the anvil-shaped diapir of red mud (unit A) that pierces unit B (Fig. 20, IV, and Fig. 21), and (3) the overturned beds of unit C on the southern limb of the anticline (Fig. 21). Detachment must have been at or near the base of unit A, although we found no movement indicators.

Basal Cabullona Group, including the Agua Amarga beds, was rotated upward during Laramide elevation of the Gold Hill block along the La Bellota fault. Subsequent erosion created the cross section of the Agua Amarga beds (Fig. 21) and revealed the upturned unconformity between Glance Conglomerate and Agua Amarga beds.

The sandstone and shale (facies B) can be mapped up to the southern part of the La Negrita Hills. North of there, only scattered exposures of the sand can be found, but we have the opinion that they correlate with the nearshore(?) sands of the 91 Hills slide mass. The sheet of Mural debris interfingers with the El Cemento, but cannot be traced beyond the area mapped in Figure 20. However, the isolated upper Mural blocks north of Highway 2 and west of Loma La Morita (Figs. 6 and 20), and small pieces of upper Mural (*Orbitolina*-bearing) limestone that we found on the southwestern flank of Loma La Morita show that upper Mural debris was widespread on the floor of Lake Cabullona.

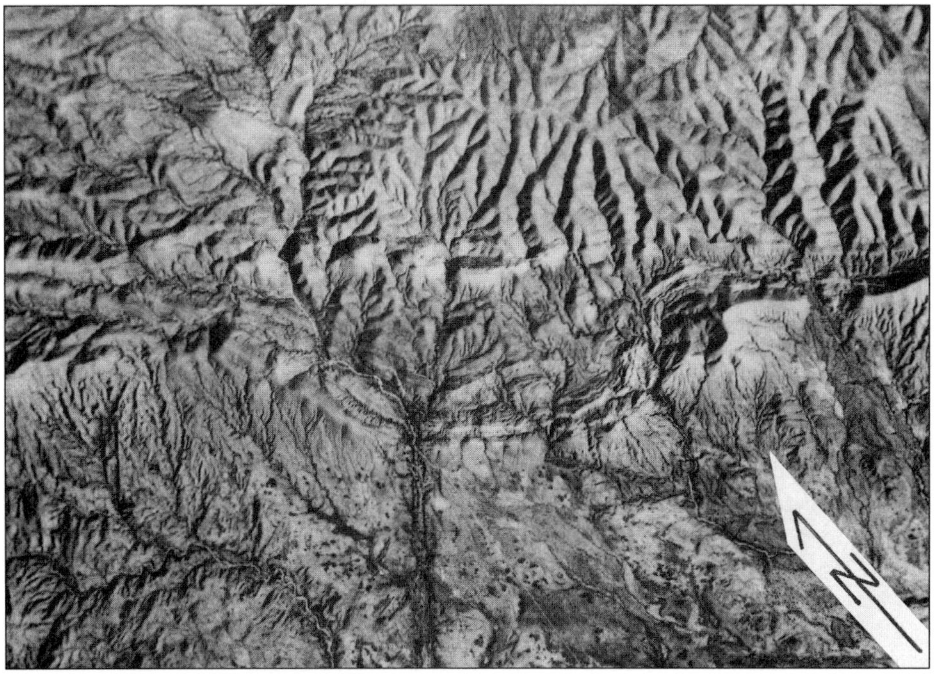

Figure 21. Anticlinal bulge in basal Cabullona rocks (Agua Amarga beds) formed by sliding of these lake sediments. Compare with Figure 20, I and IV. The headwaters of Arroyo Agua Amarga are in the top of the photo, the stream flows through the anticlinal bulge, then it leaves the photo at the bottom. The ridge that best marks the bulge is composed of facies C, the sheet of upper Mural limestone debris. The anvil-shaped diapir of facies A (see Figure 20, I and IV) is visible as well. See Figures 6 and 20 for scale.

In any case, there is a body of debris (Fig. 6) deposited by mass-gravity processes that can be traced from Schlaudt Ridge in Arizona to exposures in Sonora that extend a few kilometers south of Arroyo Agua Amarga—a total distance of more than 25 km.

González-León and Lawton (1995) show that the Cabullona rocks are early Santonian to late Maastrichtian. Lucas et al. (1995) narrow the age somewhat to late Campanian and Maastrichtian. Thus, uplift of the Gold Hill block and removal of the Morita Formation are pre–late Campanian or perhaps pre-Santonian events.

Propensity for Mass-Gravity Movement of Rocks of the Bisbee Group

Glance Conglomerate

We have no special knowledge of the reason for the abundance of megaclasts in the Glance, but pull-apart basins, by the manner of their opening, place fault-bounded blocks in company with topographic relief and earthquakes; megaclasts are a likely result. Volcanism in the western part of the Hereford basin would also contribute to clast size, both by its explosive effects and by "plucking" (Simons et al., 1966). There is little evidence of volcanism in the eastern part of the study area, where the Bisbee Group rests on continental crust, so the origin of the La Negrita Hills may be very different from that of the El Caloso Hills.

Upper Bisbee

By the time of Laramide contraction and uplift, the sand and mud of the upper Bisbee were still relatively young and probably poorly lithified, much like the Tertiary strata of the Gulf of Mexico coastal plain are today. Mid-Late Cretaceous (Santonian) Cabullona Group deposited on Glance Conglomerate from which Morita, Mural, and Cintura formations had been stripped (Fig. 5C) verifies that Laramide uplift caused very rapid removal of the upper Bisbee. This created valley walls of sandstone and mudstone that contained thick ledges of lithified and brittle limestone, the upper Mural. The walls of poorly lithified sand and mud would have retreated at a steady rate compared to the ledges, whose retreat would have been intermittent, by rockfall and slide. Laramide earthquakes may have triggered calving of large portions of the slopes, thereby moving great volumes of upper Bisbee rock downslope, some more-or-less intact, like the 91 Hills slide mass.

This potential for sliding existed during and immediately after deposition of the upper Mural limestone when marine conditions still existed. Then, the water-charged lower Mural and Morita were even poorer supports for the Mural ledge (shelf) than they were later, when they were components of a Late Cretaceous landscape. With a gentle slope, mass-gravity transportation of Mural blocks and debris would be likely. This could be the origin of some of the isolated blocks of upper Mural that are now identified as patch reefs. Patch reefs are common today and in the geologic record, but mostly (special exceptions exist) when seated on carbonate sediment. This is because the conditions of salinity and water temperature that are required for the growth of a patch reef will induce carbonate deposition regionally. For that reason, a patch reef is very unlikely to form while sitting on, and being surrounded by siliciclastic sediment. On the other hand, it is not difficult to imagine a block of reefal Mural complete with scleractinians, having slid into such a setting. Such sliding would have been more likely on the south side of the Cananea high (where a contemporary seaward-facing slope was probably present) than on the north side.

For these reasons, mass-gravity movement should be a signature feature of upper Bisbee rocks. We think it is. We have already mentioned the displaced Mural and associated rocks at Sierra Azul (McKee and Anderson, 1998). In the following paragraphs we explain other occurrences of upper Bisbee rocks that were (or may have been) displaced by mass-gravity processes.

Little Hatchet Mountains. In the northern part of the Little Hatchet Mountains, just south of Playas Peak (Fig. 13), a large exposure (1200 × 200 m) of reefal U-Bar Formation (Mural equivalent) lies amid Upper Cretaceous terrestrial deposits. These terrestrial deposits (equivalent to the Fort Crittenden and Cabullona) rest in normal succession upon older Upper Cretaceous deposits that are equivalent to the Cintura Formation. The anomalous presence of the reefal Lower Cretaceous rock so high in the section has been explained by one or more thrusts (Zeller, 1970; Hodgson, 2000). During our recent examination of these rocks, we concluded that the reefal U-Bar (along with many smaller U-Bar fragments) is a megaclast within a Late Cretaceous terrestrial mass-gravity deposit. The 91 Hills slide mass is a model.

Montes Canova. The structural duplication of section that produced Montes Canova (Fig. 6) may have been by gravity gliding. The idea is brought to mind by the following observations. (1) The upper Mural that forms the northeastern flank (dip slope) of Montes Canova has an unusual undulatory shape (Fig. 22), and the discontinuous string of upper Mural exposures between Montes Canova and Paul Spur (Figs. 1 and 6) do not appear to lie within the same stratigraphic horizon. (2) Features of apparent extensional strain are consistent with sliding. The clearest of these features are fractures and a large array of graceful half-grabens (Fig. 23) in upper Mural on the dip slope of Sierra Anibacachi. Also, walnut-size balls of limestone coated with fine debris and arranged in irregular beds might have been caused by attenuation, pinching, and rolling during movement of thin beds of micrite, which can be plastic prior to lithification. The duplication of section may have been caused by thrusting, but duplication by gravity gliding cannot yet be dismissed.

Sierra San José. The principal folds in the Sierra San José are Laramide cylindrical folds, but there is some structural irregularity within them that may reflect a previous deformation. Northeast of Rancho Agua Verde (Fig. 10), a northeast-trending anticlinal axis cuts almost directly across the axis of a major syncline. The east-west anticline immediately to the north is trun-

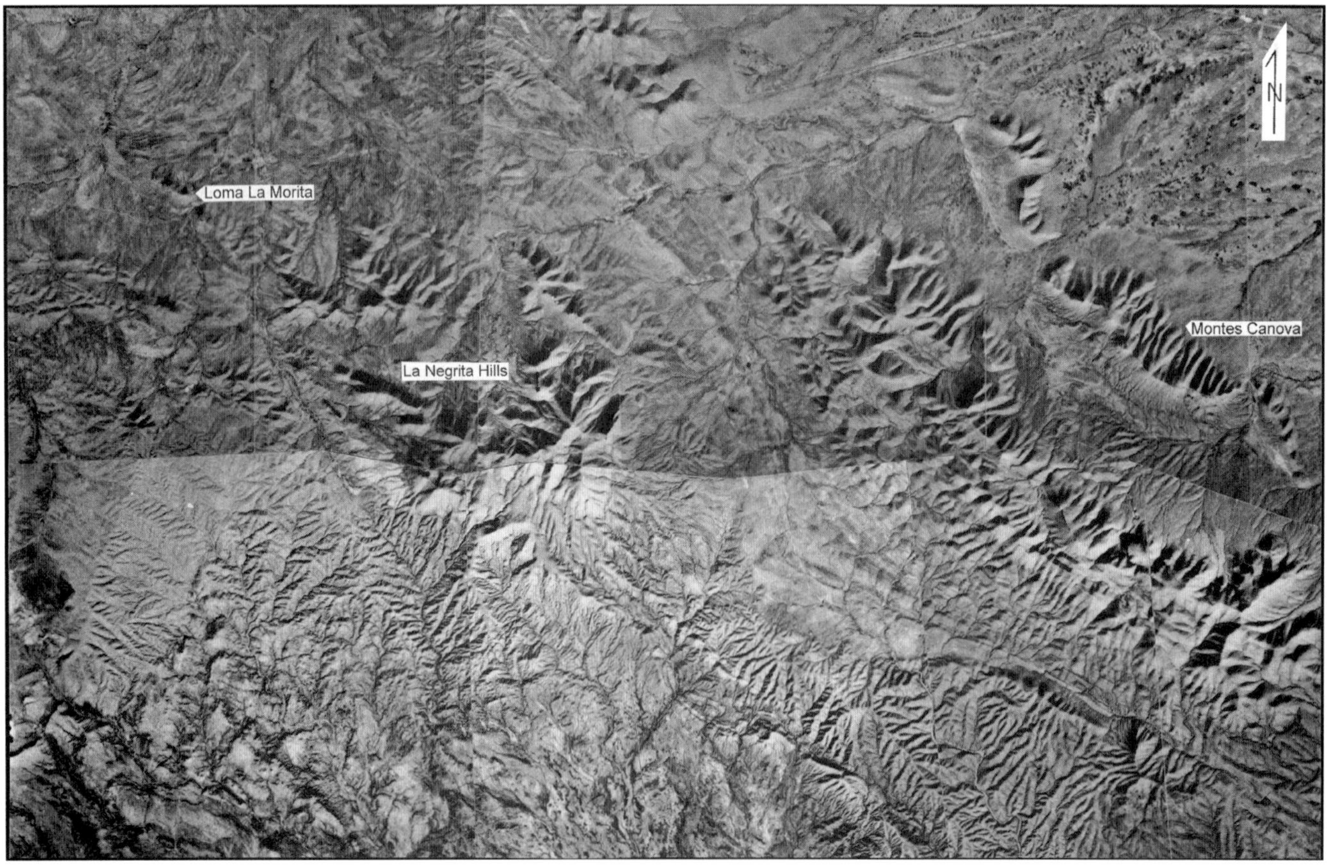

Figure 22. This aerial photograph shows most of the Sonoran part of Figure 6. The mountainous area between Montes Canova and La Negrita Hills is Sierra Anibacachi. The northeasternmost part of La Negrita Hills is also labeled Sierra Anibacachi on modern maps; we refer to it as "little" Sierra Anibacachi.

cated by another anticline, whose trends range from nearly north to southeast. It appears that the upper Mural marked a simple curviplanar surface of the kind that might have formed in rocks with no great overburden while moving and flexing under the force of gravity. Subsequent Laramide deformation of the curviplanar surface would yield the pattern we see today.

Arroyo El Tule. In the Huachuca Mountains, including the Bear Creek section (Fig. 11, section A), the Mural Formation is in its marginal marine facies. Nearby, in Arroyo El Tule (Fig. 9), there are scattered but prominent exposures of reefal upper Mural limestone. (Just two of these outcrops are shown in Figures 1 and 9.) This may mark a very abrupt facies change, or perhaps the reefal Mural represents exposed parts of another 91 Hills–type mass-gravity deposit.

CONCLUSIONS

The Hereford basin (or the alternative Coronado Peak and Gold Hill basins, Fig. 4D) is a small fault basin that contains the type area of the Glance Conglomerate. The basin is defined by faults that bound it, the Glance Conglomerate that partially filled it, and by a depositional pause that separated the Glance from overlying upper Bisbee rocks.

The northeastern boundary of the basin was created principally by the Abrigo and Gold Hill faults. The El Tule and Cananea faults bound the Hereford basin on the southwest and south and are the northern limit of the Cananea high. We place the Jurassic southern boundary of the Cananea high just to the northeast of the thick Glance Conglomerate near Cerro Azul, well north of the Late Jurassic ammonite-bearing shale that is widely exposed near the city of Cucurpe, Sonora. The Cretaceous southern boundary lay north of the Mural Limestone deposits of the Sierra Azul area.

Figure 23. Upper Mural Limestone on the dip slope of Sierra Anibacachi, showing one of many arcuate half-grabens from NE-SW attenuation of the limestone. View is to the northwest.

We hypothesize that the Hereford basin was one of an array of strike-slip (pull-apart) basins that was created by the same transtensional stresses that created the Mojave-Sonora megashear. We base this principally on (1) times of fault activity, the close of which is marked by the depositional pause at the end of coarse Glance deposition, and (2) orientation of bounding faults, including those specified by Lawton (2000).

The Hereford basin was succeeded by a regional depositional basin that we refer to as the upper Bisbee basin. That larger basin is also defined by its deposits, which are the upper part of the Bisbee Group—the Morita, Mural, and Cintura formations. These deposits accumulated in a zone of crustal subsidence caused by crustal thinning associated with the creation of the array of Hereford-like pull-apart basins, and by eustasy.

We offer the concept of the Bisbee Group as the contents of superposed basins of different origins. The Glance Conglomerate accumulated in megashear-related strike-slip fault basins, then regional subsidence and eustatic sea-level rise created the upper Bisbee depositional basin. This is in contrast to the concept of the Bisbee Group as the contents of one basin of one origin (rifting).

The coarse sediment that filled the Hereford basin contains megaclasts in masses of such prominence that they appear on 1:250,000 topographic maps. One of these products of mass-gravity sedimentation (La Negrita Hills) has been incorrectly interpreted as part of the hanging wall of a southwest-vergent Laramide thrust. In that way, products of Jurassic crustal extension (large-scale mass-gravity deposits) were erroneously interpreted as products of Laramide crustal contraction.

The principal effect of Laramide contraction was inversion of the faults that had engendered both the Jurassic basin and the large-scale mass-gravity sedimentation. This inversion elevated the Bisbee Group rocks that filled the basin, folded them (principally west and south of the Huachuca Peak fault), and initiated very rapid erosion of upper Bisbee rocks, forming a steep-walled drainage system. Our evidence of these steep walls is the Late Cretaceous 91 Hills slide mass and the apparently contiguous mass-gravity deposits of the Agua Amarga beds that together can be traced for 25 km.

Mass-gravity deposits are an important feature of both Jurassic and Late Cretaceous basins in the area, and recognition of them is essential to correct interpretation of structural and depositional history.

ACKNOWLEDGMENTS

This work was supported by grants from the National Science Foundation (EAR-9117824) (JWM, MBM, THA) and from the Faculty Development Program of the University of Wisconsin–Oshkosh (JWM). This work would not have been possible without the permission and kind help of ranchers in Sonora and Arizona. We are greatly indebted to them. We also appreciate the courtesy and informative exchanges given us by colleagues. Among them are: William R. Dickinson, Juan Carlos García y Barragán, Carlos González-León, César Jacques-Ayala, Norris

W. Jones, Daniel Lehrmann, Rogelio Monreal, William Page, José Luís Rodríguez-Castañeda, Jaime Roldán-Quintana, and James Lee Wilson. We are particularly grateful for the help and discussions given to us where they are the most helpful—in the field—by Timothy Lawton and Nancy McMillan. The friendship and guidance given to us by Alberto Mueller are much appreciated. Thoughtful reviews by George Davis and Charles Kluth caused great improvements in the manuscript. As in all the years since 1978, Roberto Bernal-Valadez provided invaluable field assistance and companionship.

REFERENCES CITED

Anderson, T.A., and Schmidt, V.A., 1983, The evolution of middle America and the Gulf of Mexico–Caribbean Sea region during Mesozoic time: Geological Society of America Bulletin, v. 94, p. 941–966, doi: 10.1130/0016-7606(1983)94<941:TEOMAA>2.0.CO;2.

Bergman, B.J., McKee, J.W., McKee, M.B., and Anderson, T.H., 1993, Mass-gravity sedimentation and sliding in the Cretaceous Bisbee Group at 91 Hills, southern Arizona: Geological Society of America Abstracts with Programs, v. 25, p. A-68.

Bilodeau, W.L., 1979, Early Cretaceous tectonics and deposition of the Glance Conglomerate, southeastern Arizona [Ph.D. dissertation]: Stanford, Stanford University, 145 p.

Bilodeau, W.L., 1982, Tectonic models for Early Cretaceous rifting in southeastern Arizona: Geology, v. 10, p. 466–470, doi: 10.1130/0091-7613(1982)10<466:TMFECR>2.0.CO;2.

Bilodeau, W.L., Kluth, C.F., and Vedder, L.K., 1987, Regional stratigraphic, sedimentologic, and tectonic relationships of the Glance Conglomerate in southeastern Arizona, *in* Dickinson, W.R., and Klute, M.F., eds., Mesozoic rocks of southern Arizona, and adjacent areas: Arizona Geological Society Digest, v. 18, p. 229–256.

Bultman, M.W., Gettings, M.E., and Wynn, J., 1999, An interpretation of the 1997 airborne electromagnetic (AEM) survey, Fort Huachuca vicinity, Arizona, with digital data: U.S. Geological Survey Open-File Report 99-7-A.

Cerecero-Luna, M., and Castro-Escarrega, J., 1996, Carta geológico-minera, Colonia Oaxaca H-12–B66 Sonora: Pachuca, Hidalgo, Consejo de Recursos Minerales, Secretaría de Comercio y Fomento Industrial, scale 1:50,000.

Davis, G.H., 1979, Laramide folding and faulting in southeastern Arizona: American Journal of Science, v. 279, p. 543–569.

Dickinson, W.L., and Lawton, T.F., 2001a, Carboniferous to Cretaceous assembly and fragmentation of Mexico: Geological Society of America Bulletin, v. 113, no. 9, p. 1142–1160, doi: 10.1130/0016-7606(2001)113<1142:CTCAAF>2.0.CO;2.

Dickinson, W.L., and Lawton, T.F., 2001b, Tectonic setting and sandstone petrofacies of the Bisbee basin (USA-Mexico): Journal of South American Earth Sciences, v. 14, no. 5, p. 475–504.

Drewes, H., 1980, Tectonic map of southeast Arizona: U.S. Geological Survey Miscellaneous Investigation Series Map I-1109, scale 1:125,000.

Drewes, H., 1981, Tectonics of southeastern Arizona: U.S. Geological Survey Professional Paper 1144, 96 p.

Emmons, S.F., 1910, Cananea mining district of Sonora, Mexico: Economic Geology, v. 5, p. 312–356.

Gettings, M.E., and Houser, B.B., 2000, Depth to bedrock of the upper San Pedro valley, Cochise County, southeastern Arizona: U S. Geological Survey Open-File Report 00-138.

González-León, C.M., 1986, Estratigrafia del Paleozoico de la Sierra del Tule, noresta de Sonora: Universidad Nacional Autónoma de México, Instituto de Geología, Revista, v. 6, no. 2, p. 117–135.

González-León, C.M., 1994, Stratigraphy, depositional environments, and origin of the Cabullona basin, northeastern Sonora, Mexico [Ph.D. dissertation]: Tucson, University of Arizona, 144 p.

González-León, C.M., and Lawton, T.F., 1995, Stratigraphy, depositional environments, and origin of the Cabullona basin, northeastern Sonora, *in* Jacques-Ayala, C., González-León, C.M., and Roldán-Quintana, J., eds., Studies on the Mesozoic of Sonora and adjacent areas: Geological Society of America Special Paper 301, p. 121–142.

Grijalva-Noriega, F.G., 1995, El alto Cananea: Un elemento paleogeográphico del Jurassico Tardio: Universidad de Sonora Boletín del Departamento de Geología, v. 12 (2).

Hayes, M.J., 1987, Depositional history of Upper Cretaceous Fort Crittenden Formation in southeastern Arizona, *in* Dickinson, W.R., and Klute, M A., eds., Mesozoic rocks of southern Arizona and adjacent areas: Tucson, Arizona, Arizona Geological Society Digest, v. 18, p. 315–325.

Hayes, P.T., 1970, Mesozoic stratigraphy of the Mule and Huachuca Mountains, Arizona: U.S. Geological Survey Professional Paper 658-A, 28 p.

Hayes, P.T., and Landis, E.R., 1964, Geologic map of the southern part of the Mule Mountains, Cochise County, Arizona: U.S. Geological Survey Miscellaneous Investigations Map I-418, scale 1:48,000.

Hayes, P.T., and Raup, R.B., 1968, Geologic map of the Huachuca and Mustang Mountains, southeastern Arizona: U.S. Geological Survey Miscellaneous Investigations Map I-509, scale 1:48,000.

Hodgson, S.A., 2000, Structural geology and Laramide tectonics of the Little Hatchet Mountains, southwestern New Mexico, *in* Lawton, T.F., McMillan, N.J., and McLemore, V.T., eds., Southwest passage–A trip through the Phanerozoic: New Mexico Geological Society, 51st Field Conference Guidebook, p. 109–116.

Imlay, R.W., 1939, Paleogeographic studies in northeastern Sonora: Geological Society of America Bulletin, v. 50, p. 1723–1744.

Jacques-Ayala, C., 1995, Paleogeography and provenance of Lower Cretaceous Bisbee Group in the Caborca–Santa Ana area, northwestern Sonora, *in* Jacques-Ayala, C., González-León, C.M., and Roldán-Quintana, J., eds., Studies on the Mesozoic of Sonora and adjacent areas: Geological Society of America Special Paper 301, p. 79–98.

Klute, M.A., 1991, Sedimentology, sandstone petrofacies, and tectonic setting of the late Mesozoic Bisbee basin, southeastern Arizona [Ph.D. dissertation]: Tucson, University of Arizona, 266 p.

Lawton, T.F., 2000, Inversion of Late Jurassic–Early Cretaceous extensional faults of the Bisbee basin, southeastern Arizona and southwestern New Mexico, *in* Lawton, T.F., McMillan, N.J., and McLemore, V.T., eds., Southwest passage–A trip through the Phanerozoic: New Mexico Geological Society, 51st Field Conference Guidebook, p. 95–108.

Lawton, T.F., and Olmstead, G.A., 1995, Stratigraphy and structure of the lower part of the Bisbee Group, northeastern Chiricahua Mountains, Arizona, *in* Jacques-Ayala, C., González-León, C.M., and Roldán-Quintana, J., eds., Studies on the Mesozoic of Sonora and adjacent areas: Geological Society of America Special Paper 301, p. 21–39.

Lawton, T.F., González-León, C.M., Lucas, S.G., and Scott, R.W., 2004, Stratigraphy and sedimentology of the Upper Aptian–Upper Albian Mural Limestone (Bisbee Group) in northern Sonora, Mexico: Cretaceous Research, v. 25, p. 43–60, doi: 10.1016/j.cretres.2003.09.003.

Lipman, P.W., and Hagstrum, J.T., 1992, Jurassic ash-flow sheets, calderas, and related intrusions of the Cordilleran volcanic arc in southeastern Arizona: Implications for regional tectonics and ore deposits: Geological Society of America Bulletin, v. 104, p. 32–39, doi: 10.1130/0016-7606(1992)104<0032:JAFSCA>2.3.CO;2.

Lucas, S.G., and Lawton, T.F., 2000, Stratigraphy of the Bisbee Group (Jurassic-Cretaceous), Little Hatchet Mountains, New Mexico, *in* Lawton, T.F., McMillan, N.J., and McLemore, V.T., eds., Southwest Passage–A trip through the Phanerozoic: New Mexico Geological Society, 51st Field Conference Guidebook, p. 175–194.

Lucas, S.G., Kues, B.S., and González-León, C.M., 1995, Paleontology of the Upper Cretaceous Cabullona Group, northeastern Sonora, *in* Jacques-Ayala, C., González-León, C.M., and Roldán-Quintana, J., eds., Studies on the Mesozoic of Sonora and adjacent areas: Geological Society of America Special Paper 301, p. 143–165.

Mann, P., Hempton, M.R., Bradley, D.C., and Burke, K., 1983, Development of pull-apart basins: Journal of Geology, v. 91, no. 5, p. 529–554.

McKee, J.W., McKee, M.B., and Anderson, T.H., 1994, Mountain-forming megablocks in the Jurassic(?) of northernmost Sonora, Mexico,: Tectonic significance: Geological Society of America Abstracts with Programs, v. 26, no. 7, p. 386.

McKee, M.B., 1991, Deformation and stratigraphic relationships of mid-Cretaceous to early Tertiary mass gravity slides in a marine basin in Sonora [Ph.D. dissertation]: Pittsburgh, University of Pittsburgh, 286 p.

McKee, M.B., and Anderson, T.H., 1998, Mass-gravity deposits and structures in the Lower Cretaceous of Sonora, Mexico: Geological Society of America Bulletin, v. 110, no. 12, p. 1516–1529, doi: 10.1130/0016-7606(1998)110<1516:MGDASI>2.3.CO;2.

Nourse, J.A., 1995, Jurassic-Cretaceous paleogeography of the Magdelena region, northern Sonora, and its influence on the positioning of Tertiary metamorphic core complexes, *in* Jacques-Ayala, C., González-León, C.M., and Roldán-Quintana, J., eds., Studies on the Mesozoic of Sonora and adjacent areas: Geological Society of America Special Paper 301, p. 59–78.

Page, W.R., Gray, F., Blodgett, R.B., Maldonado, F., Miller, R.J., and Jamie-Ruiz, R., 2003, Mesozoic-Cenozoic evolution of the northern Sierra Los Ajos, Sonora, Mexico: Geological Society of America Abstracts with Programs, v. 35, no. 4, p. 70.

Rangin, C., 1977a, Sobre la presencia de Jurassico Superior con amonitas en Sonora septentrional: Universidad Nacional Autónoma de México, Instituto de Geología, Revista, v. 1, no. 1, p. 1–4.

Rangin, C., 1977b, Tectonicas sobrepuestos en Sonora septentrional: Universidad Nacional Autónoma de México, Instituto de Geología, Revista, v. 1, no. 1, p. 44–47.

Ransome, F.L., 1904, Geology and mineral deposits of the Bisbee Quadrangle, Arizona: U.S. Geological Survey Professional Paper 21, 167 p.

Rodríguez-Castañeda, J.L., 1984, Geology of the Tuape region, north-central Sonora [M.S. thesis]: Pittsburgh, University of Pittsburgh, 156 p.

Rodríguez-Castañeda, J.L., 2002, Tectónica Cretácica y Terciaria en la margen suroeste del alto Cananea, Sonora norte central [Ph.D. dissertation]: México, D.F., Universidad Nacional Autónoma de México, 217 p.

Rosales-Domínguez, M.C., Grajales-Nashimura, J.M., Sánchez-Ríos, M.A., Gómez-Luna, M.E., and Duenas, M.A., 1995, Biostratigraphy of the Lower Cretaceous Bisbee Group, Rancho Culantrillo area, northeastern Sonora, *in* Jacques-Ayala, C., González-León, C.M., and Roldán-Quintana, J., eds., Studies on the Mesozoic of Sonora and adjacent areas: Geological Society of America Special Paper 301, p. 49–57.

Silver, L.T., and Anderson, T.H., 1974, Possible left-lateral early to middle Mesozoic disruption of the southwestern North American craton margin: Geological Society of America Abstracts with Programs, v. 6, no. 7, p. 955–956.

Simons, F.S., Raup, R.B., and Hayes, P.T., and Drewes, H., 1966, Exotic blocks and coarse breccias in Mesozoic volcanic rocks of southeastern Arizona, *in* Geological Survey research 1966: U.S. Geological Survey Professional Paper 550-D, p. D12–D22.

Stoyanow, A., 1949, Lower Cretaceous stratigraphy in southeastern Arizona: Geological Society of America Memoir 38, 169 p.

Taliaferro, N.L., 1933, An occurrence of Upper Cretaceous sediments in northern Sonora, Mexico: Journal of Geology, v. 41, p. 12–37.

Vedder, L.K., 1984, Stratigraphic relationships between the Late Jurassic Canelo Hills volcanics and the Glance Conglomerate, southeastern Arizona [M.S. thesis]: Tucson, University of Arizona, 129 p.

Warzeski, E.R., 1983, Facies patterns and diagenesis of a Lower Cretaceous carbonate shelf: Northeastern Sonora and southeastern Arizona [Ph.D. dissertation]: Binghamton, State University of New York at Binghamton, 401 p.

Warzeski, E.R., 1987, Revised stratigraphy of the Mural Limestone: A Lower Cretaceous carbonate shelf in Arizona and Sonora, *in* Dickinson, W.R., and Klute, M.A., eds., Mesozoic rocks of southern Arizona: Tucson, Arizona, Arizona Geological Society Digest, v. 18, p. 335–370.

Zeller, R.R., Jr., 1970, Geology of the Little Hatchet Mountains, Hidalgo and Grant Counties, New Mexico: New Mexico Bureau of Mines and Mineral Resources Bulletin 96, 26 p.

Manuscript Accepted by the Society 24 March 2005

Geological Society of America
Special Paper 393
2005

Reactivation of the San Marcos fault during mid-to-late Tertiary extension, Chihuahua, Mexico

José Jorge Aranda-Gómez*
Departamento de Geología Económica, Instituto Potosino de Investigación Científica y Tecnológica, A.C., P.O. Box 3-74, San Luis Potosí, SLP, 78216, México

Todd B. Housh
Department of Geological Sciences, University of Texas at Austin, Austin, Texas 78712, USA

James F. Luhr
Department of Mineral Sciences, Smithsonian Institution, Washington, D.C. 20013-7012, USA

Christopher D. Henry
Nevada Bureau of Mines and Geology, University of Nevada, Reno, Nevada 89557-0088, USA

Tim Becker
Berkeley Geochronology Center, Berkeley, California 94709, USA

Gabriel Chávez-Cabello
Posgrado en Ciencias de la Tierra, Centro de Geociencias, Universidad Nacional Autónoma de México, Campus Juriquilla, P.O. Box 1-742, Querétaro, Qro., 76001, México

ABSTRACT

The San Marcos fault is a >300-km-long, WNW- to NW-trending structure that has been repeatedly reactivated during Mesozoic and Tertiary pulses of deformation. The San Marcos fault has previously been demonstrated to extend across Coahuila in a trend roughly parallel to the proposed Mojave-Sonora megashear. Exposures in this broad region are dominated by Mesozoic marine sedimentary rocks. Northwest of the border between Coahuila and Chihuahua a thick cover of Tertiary volcanic rocks masks the fault trace. We believe that the San Marcos fault trace projects across the eastern branch of the Bolsón de Mapimí, a major Basin and Range structure, into the northern portion of the Sierra del Diablo and into the central part of the Pliocene-Quaternary Camargo volcanic field ($^{40}Ar/^{39}Ar$ ages are 4.7–0.09 Ma). In the same area, upper Tertiary normal faults that bound both branches of the Bolsón de Mapimí show a marked change in orientation. South of the projected San Marcos fault the extensional structures strike ~north-south. Northwest of the San Marcos fault, a conspicuous northwest-trending graben (ca. 4.7–2.1 Ma) crosses the Camargo volcanic

*E-mail address: jaranda@ipicyt.edu.mx.

Aranda-Gómez, J.J., Housh, T.B., Luhr, J.F., Henry, C.D., Becker, T., and Chávez-Cabello, G., 2005, Reactivation of the San Marcos fault during mid-to-late Tertiary extension, Chihuahua, Mexico, *in* Anderson, T.H., Nourse, J.A., McKee, J.W., and Steiner, M.B., eds., The Mojave-Sonora megashear hypothesis: Development, assessment, and alternatives: Geological Society of America Special Paper 393, p. 509–521. doi: 10.1130/2005.2393(19). For permission to copy, contact editing@geosociety.org.

field. A lineament identified in digital elevation models suggests that the San Marcos fault extends beneath the Camargo volcanic field, passes west of Sierra Aguachile, and can be traced at least an additional 300 km to the northwest, which effectively doubles the known or inferred length of the San Marcos fault.

Mid-Tertiary volcanic rocks (ca. 32–14 Ma) are folded into NNW-trending, SSE-plunging synclines in areas immediately northwest and southeast of the Camargo volcanic field. A short distance (~5 km) west of Sierra El Morrión in central Chihuahua another NNW-trending syncline, observable in satellite images and in regional geologic and topographic maps, is developed in mid-Tertiary volcanic rocks. Miocene folding in the volcanic sequence at Peña Blanca has been reported in the literature near the NNW projection of the San Marcos fault.

The central graben of the Camargo volcanic field, the upper Tertiary synclines, and the abrupt change in trend of the Tertiary normal faults are all closely associated with a distinct bend in the San Marcos fault, marking a change in trend from WNW to NW. The Camargo volcanic field central graben and the adjacent synclines were developed in the same northwest-trending segment of the fault. We interpret the central graben of the Camargo volcanic field as a pull-apart basin, whereas the synclines were formed in areas where a small left-lateral component in the fault movement produced shortening. We believe this deformation was accomplished in at least two pulses of late Tertiary extension. An older event that occurred between ca. 14 Ma and 5 Ma produced the left-lateral movement and associated folding. This pulse corresponds to a short episode of regional northwestern extension documented in northern Mexico and the southern United States. The younger event (≤5 Ma) was in part synchronous with the activity in the Camargo volcanic field and was caused by NE to ENE extension.

Keywords: Basin and Range, transtension, transpression, intraplate volcanism.

INTRODUCTION

Several major NW- to WNW-trending, left-lateral megafaults in northern Mexico and the southwestern United States (Fig. 1) have been proposed to explain: (1) the distribution of Precambrian and Paleozoic rocks from various ~north-south–trending structural belts (e.g., De Cserna, 1970; Silver and Anderson, 1974; Anderson and Schmidt, 1983; Stewart, 2003); (2) the stratigraphy, facies changes, provenance, and faunal content of Jurassic and Lower Cretaceous rocks in Chihuahua and Coahuila (McKee and Jones, 1979; McKee et al., 1984; Longoria, 1984; Pessagno et al., 1993); and (3) changes in style of Laramide folding in the Sierra Madre Oriental of northeastern Mexico (e.g., Charleston, 1981; Longoria, 1985; Padilla y Sánchez, 1986). These megafaults are interpreted to have been active during the late Jurassic (e.g., Caudillo-Sosa et al., 1996; Marzolf, 2003) and early Cretaceous. They are probably related to the opening of the Gulf of Mexico (Fig. 1). Structural patterns in the vicinity of the faults suggest they were reactivated during the early Tertiary Laramide Orogeny (e.g., Charleston, 1981; Padilla y Sánchez, 1986; Hennings, 1994). In this paper we address reactivation of the San Marcos fault during mid-to-late Cenozoic extension and the local folding of Tertiary volcanic rocks near the Chihuahua-Coahuila state limit.

The San Marcos fault was described by McKee and Jones (1979) and McKee et al. (1984) as a WNW-trending, ~300-km-long structure (Fig. 1). The trace of the fault is conspicuous in areas where outcrops are dominated by folded marine sedimentary rocks, as in the region between Potrero de La Gavia (Fig. 2) and the eastern branch of the Bolsón de Mapimí (Fig. 3). Farther to the northwest the fault trace is obscured by: (1) a thick sequence of Tertiary volcanic rocks that extends between the Sierra Madre Occidental volcanic field and Trans Pecos Texas (Fig. 1); (2) Basin and Range structures that have approximately the same trend as the San Marcos fault; and (3) Pliocene-Pleistocene mafic alkalic volcanic rocks. Based on changes in the deformation style observed in Mesozoic rocks, Padilla y Sánchez (1986, p. 184) speculated that this structure extends farther northwest to the area between Aldama and Plomosas (Fig. 1), ~300 km from the Bolsón de Mapimí. Stratigraphic and structural evidence along the San Marcos fault indicates several pulses of activity during the late Jurassic and early Cretaceous and during the early Tertiary Laramide Orogeny (McKee et al., 1984, 1990). The youngest left-lateral movement along the fault recognized by McKee et al. (1984) caused thrusting of Cretaceous marine sediments on top of Paleogene (Eocene?) red beds at Sierra Mojada (see inset, Fig. 3). However, the precise timing of this deformation has not been established.

La Babia fault, as proposed by Charleston (1981), is a NW-SE lineament that probably reflects a Jurassic basement structure. This fault separates the Coahuila-Texas Craton to the north from the Coahuila folded belt to the south (inset, Fig. 1). Laramide folds in the region located between the San Marcos fault and La Babia fault are interpreted by Charleston (1981) as products of

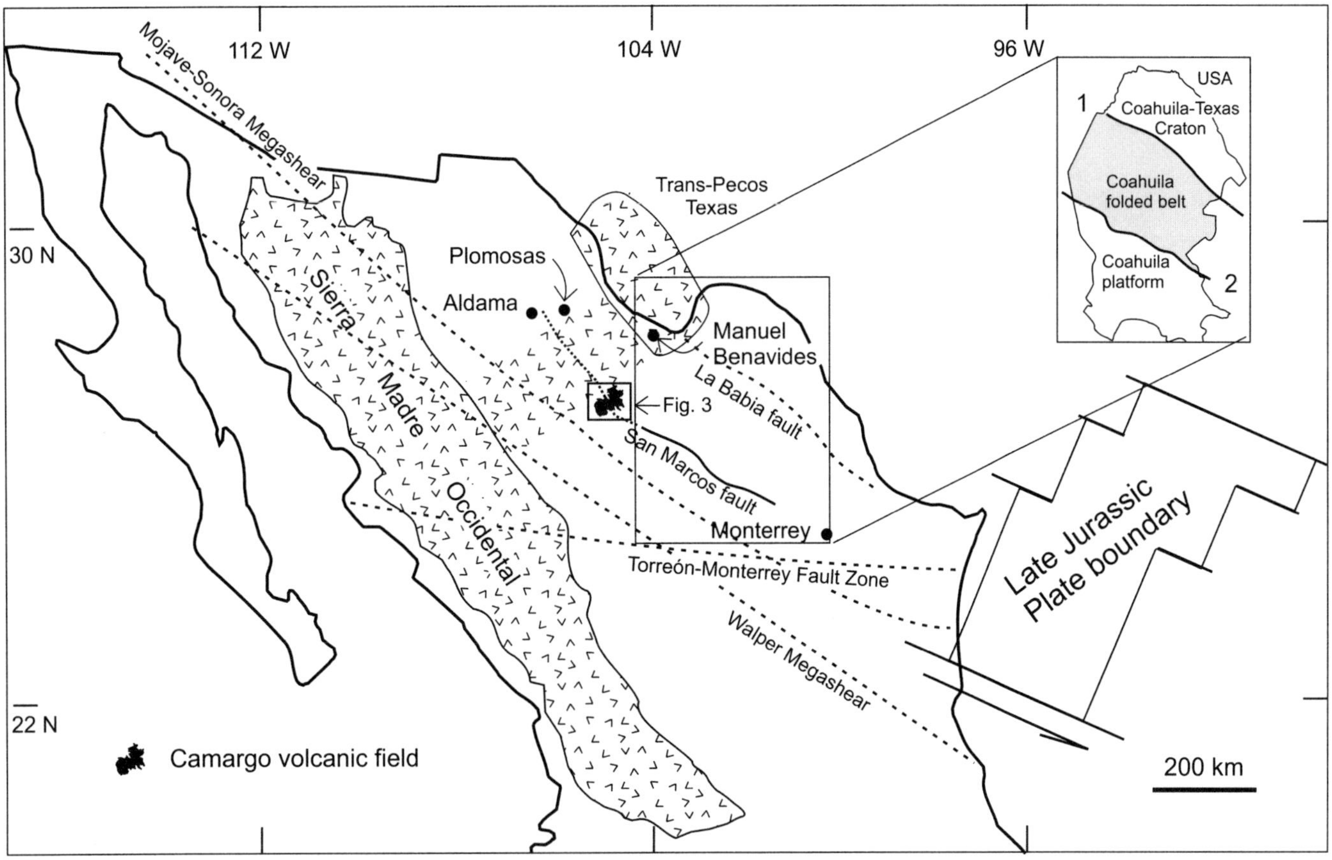

Figure 1. Distribution of Cenozoic volcanic rocks and major faults in central and northern Mexico: Torreón-Monterrey fault zone (De Cserna, 1970), La Babia fault (Charleston, 1974, 1981), Mojave-Sonora Megashear (Anderson and Schmidt, 1983), Walper Megashear (Longoria, 1984), and San Marcos fault (McKee et al., 1984, 1990; Padilla y Sánchez, 1986). Modified after McKee et al. (1984) and Aranda-Gómez et al. (2003). Inset: location of major tectonic features in Coahuila: 1—La Babia fault, 2—San Marcos fault.

shearing caused by transpression between the Coahuila-Texas Craton and the Coahuila Platform (inset, Fig. 1). In this model, both megafaults—San Marcos and La Babia—acted as wrench faults during the early Tertiary Laramide deformation. Folds in mid-Tertiary volcanic rocks similar to those discussed in this paper have been reported in the Manuel Benavides area (Gregory, 1981; Gunderson et al., 1986), and their origin has been tied to mid-to-late Tertiary reactivation of the La Babia fault (Flotte et al., 2004).

Our work has been centered on the Pliocene-Pleistocene (4.7–0.09 Ma) Camargo volcanic field, which is located immediately northwest of a major Basin and Range structure (Bolsón de Mapimí) near the border between Chihuahua and Coahuila (Aranda-Gómez et al., 2003). In this region we documented two large folds developed in mid-Tertiary volcanic rocks, a pronounced change in the trend of the Basin and Range structures, and a conspicuous graben that cuts across the Camargo volcanic field. We believe that all these structures are associated with a concealed segment of the San Marcos fault, which was reactivated during two distinct periods of late Tertiary extension. The purpose of this paper is to describe the late Tertiary folds near the Camargo volcanic field, establish time brackets for this deformation, analyze its origin, and provide additional data in support of the hypothesis that the San Marcos fault extends at least to the Aldama region as a basement structure concealed by volcanic rocks.

REGIONAL SETTING

The immediate area surrounding the Camargo volcanic field is characterized by isolated mountain ranges cored by folded Mesozoic marine sedimentary rocks that were buried by a thick sequence of Tertiary and/or Pliocene-Quaternary volcanic rocks (Fig. 3). The intervening valleys are tectonic basins partially filled with Tertiary to Quaternary continental sediments (Aranda-Gómez et al., 2001). Scattered through the central and northeastern part of the state of Chihuahua are a few outcrops of crystalline basement of Grenvillian age. These, in turn, are in places covered by Paleozoic sequences that might be part of the Ouachita Belt, which disappears south of the United States–Mexico border under a thick cover of younger sedimentary and volcanic rocks.

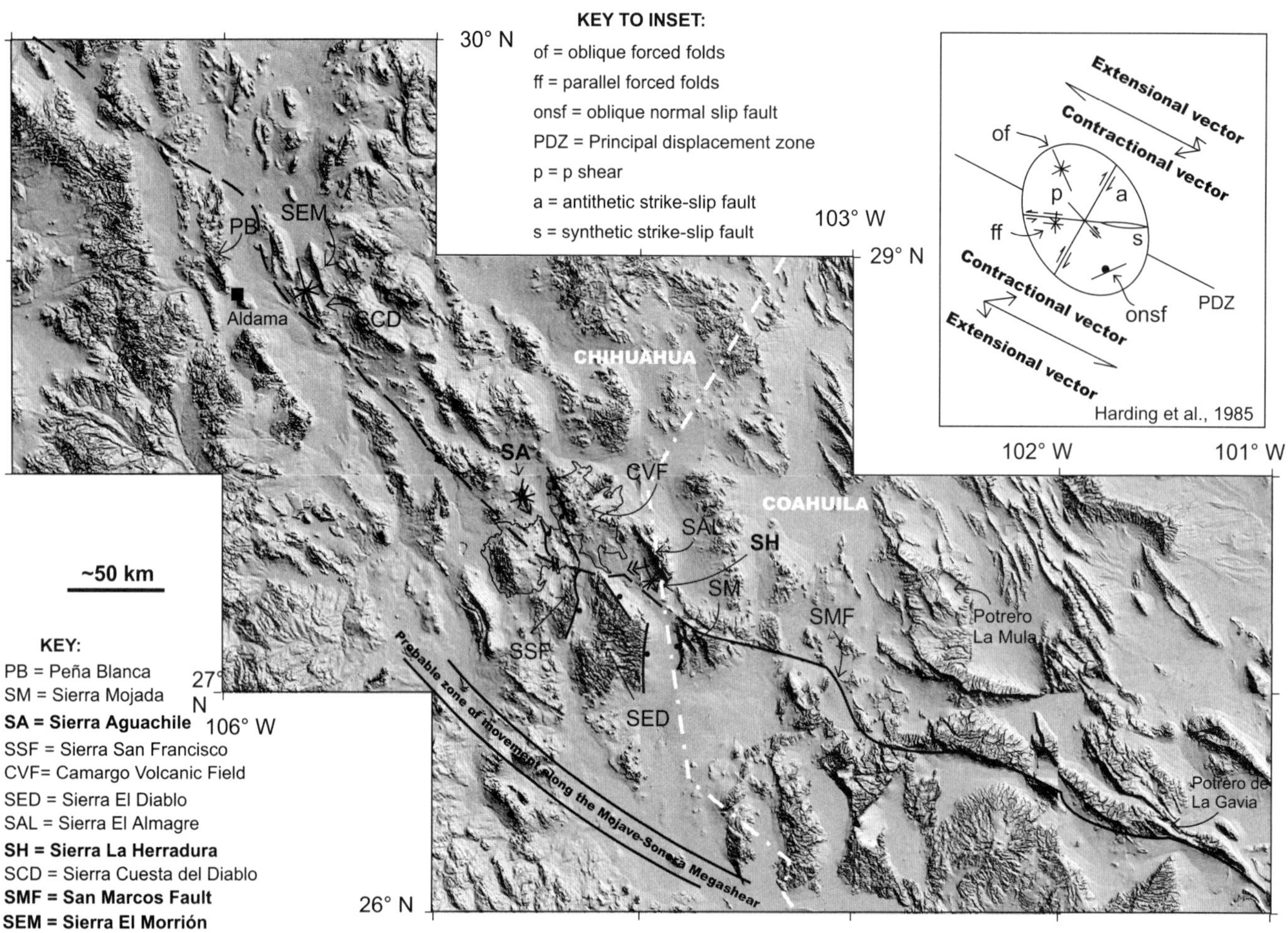

Figure 2. Digital elevation model (DEM) of the southeastern portion of Chihuahua and Coahuila. The image was generated with SURFER commercial software simulating illumination from N45E with a sun elevation of 45°. GEMA data (INEGI, 1994) were transformed to binary SURFER format with the program published by García-Estrada (1997). The solid line labeled SMF is the trace of the San Marcos fault documented by McKee et al. (1984). The heavy dashed line is the extension of the fault proposed in this paper. The location of the hidden portion of the fault trace is interpreted from the location of the isolated folds in the volcanic cover and a lineament interpreted from the DEM. The outline of the Camargo volcanic field and major normal faults in its surroundings are also shown. Inset: Angular relations between structures that tend to form in left-lateral simple shear under ideal conditions (simplified after Harding et al., 1985). Location of the Mojave-Sonora Megashear after McKee et al. (1990).

The extrapolated Ouachita Thrust Front lies west of the Camargo volcanic field (McDowell et al., 1999; Moreno et al., 2000).

The oldest rocks exposed in the area of the Camargo volcanic field are part of the Paleozoic basement of the Coahuila Island (Upper Jurassic–Lower Cretaceous), exposed at the southern end of Sierra El Diablo (McKee et al., 1984, 1990), ~50 km south of the field (Fig. 3). Triassic crystalline rocks are exposed at Potrero de La Mula, ~150 km east of the Camargo volcanic field (Fig. 2), and are interpreted as the basement of the lower Cretaceous La Mula Island (Jones et al., 1984). The inferred boundary between the Jurassic Coahuila Peninsula and the Mexican Sea lies close to the location of the Camargo volcanic field (Jones et al., 1984). This boundary was disrupted by late Jurassic to early Cretaceous activity along the San Marcos fault, which formed the northern boundary of the Coahuila Island (Upper Jurassic–Lower Cretaceous).

The San Marcos fault is parallel to the proposed trend of the Mojave-Sonora megashear (Silver and Anderson, 1974; Anderson and Schmidt, 1983). At the Chihuahua-Coahuila state line, the San Marcos fault trace appears to end abruptly (Figs. 2 and 3). The WNW–trending San Marcos fault has a complex history: several episodes of normal and/or strike-slip faulting with a large south-side-up component occurred while some of the Jurassic and Cretaceous sediments accumulated (McKee et al., 1990). The Mesozoic sedimentary sequence was folded and faulted during the early Tertiary Laramide Orogeny with elevation of the block located north of the San Marcos fault. Post-Eocene left-lateral movement caused thrusting at a restraining bend in the Sierra Mojada (see inset, Fig. 3). Whether this reactivation is related to a late stage of the Laramide Orogeny or is younger remains to be established. This event is recorded by the involvement of the

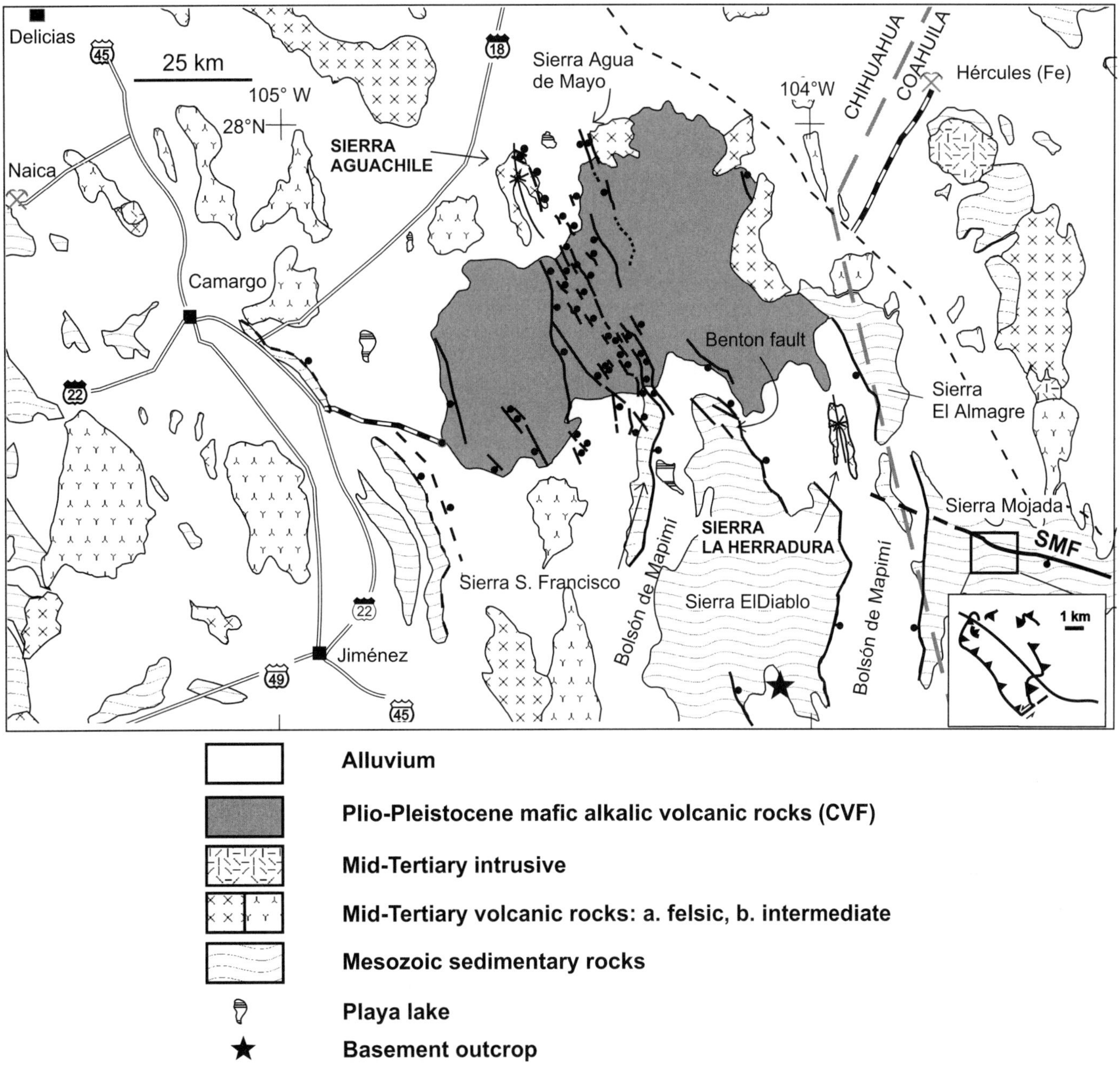

Figure 3. Regional setting of the Camargo volcanic field and the Aguachile and La Herradura synclines (modified after Aranda-Gómez et al., 2003). Paleozoic basement is exposed near the southern end of Sierra El Diablo. The San Marcos fault (SMF), as documented by McKee et al. (1984, 1990) ends at the eastern branch of Bolsón de Mapimí. Inset shows a simplified map of post-Ahuichila thrust faults mapped by McKee and collaborators (1984) in Sierra Mojada. See area of consideration in Figure 1.

Ahuichila formation (late Eocene to early Oligocene: de Cserna, 1956) in the thrusting (McKee et al., 1990).

LOCAL SETTING

The San Marcos fault trace appears to project across the eastern branch of the Bolsón de Mapimí, into the northern portion of the Sierra del Diablo and into the central part of the Pliocene-Pleistocene Camargo volcanic field (Fig. 2). In the same area the trend of upper Tertiary normal faults that bound both branches of the Bolsón de Mapimí changes markedly (Fig. 2). South of the projected San Marcos fault the extensional structures are oriented ~north-south. North of the San Marcos fault trace, a conspicuous northwest-trending graben crosses the Camargo volcanic field

(Aranda-Gómez et al., 2003), and range front fault systems of sierras El Almagre and Agua de Mayo strike northwest. A lineament interpreted from digital-elevation models suggest that the San Marcos fault trace extends across the Camargo volcanic field, passes west of Sierra Aguachile, and can be traced as far to the northwest as the Sierra El Morrión, east of Aldama (Fig. 2).

Aranda-Gómez et al. (2003) have shown that the extensive Camargo volcanic field (~3000 km^2) is covered by intraplate-type mafic alkalic volcanic rocks erupted from more than 300 vents. Twenty-six $^{40}Ar/^{39}Ar$ ages show that the field was formed between 4.7 and 0.09 Ma. (Fig. 4). Volcanic activity began in the southwestern portion of the field and shifted toward the northeast at a rate of ~15 mm/yr. The lava fields are cut by a complex N30°W–trending central graben (Fig. 3), with scarps up to 100 m high. Normal faulting in the central graben occurred between 4.7 and 2.1 Ma. Older faults have been recognized in the western portion of the Camargo volcanic field (i.e., Lagunetas fault: Fig. 4), and there is evidence of Quaternary faulting in the northeastern portion of the Camargo volcanic field (Honorato fault: Fig. 4). This suggests that faulting may have also migrated from southwest to northeast. The central graben of the Camargo volcanic field is an antithetic transfer zone between the north-south–trending, down-to-the-east San Francisco and Sierra del Diablo fault zones and northwest-trending, down-to-the-southwest Sierra Agua de Mayo fault system (Figs. 3 and 4). These opposing normal fault systems meet in the Camargo volcanic field and produce its central graben. The inferred vertical displacement in the central graben, derived from the height of the degraded fault scarps and the subtle tilting of the lava fields suggested by the location of small playa lakes, indicates low-magnitude extension during the Pliocene-Quaternary pulse of extension in the region.

Upper Cenozoic Folding of Tertiary Volcanic Rocks

The Aguachile and La Herradura synclines are two large and prominent structures located at short distances northwest and southeast, respectively, from the Camargo volcanic field (Fig. 4). The Aguachile syncline (Fig. 5) was developed in a volcano-sedimentary sequence similar to that described in the nearby Agua de Mayo range (Smith, 1993). The most distinctive units in the sequence are mid-Tertiary andesitic lava flows and breccias (K-Ar: 32.4 ± 0.8 Ma: Smith, 1993; Smith et al., 1996) characterized by large (1–2 cm long) phenocrysts of plagioclase and a crystal-poor, welded, rhyolitic ignimbrite (K-Ar: 31.4 ± 0.7 Ma: Smith, 1993; Smith et al., 1996). La Herradura syncline (Fig. 6) affects a series of andesitic lava flows and interlayered breccias. The youngest lava flow, exposed in the innermost part of the syncline, is remarkably similar to the 32.4 Ma porphyritic andesites documented by Smith (1993) in the Aguachile and Agua de Mayo ranges. We dated ($^{40}Ar/^{39}Ar$) the groundmass of an underlying aphyric andesite at 33.5 ± 0.3 Ma (Fig. 7). Other smaller folds were developed in a 40–41 Ma mafic sill (Cerro Prieto), a 13.97 ± 0.08 Ma calc-alkaline andesite sill (Cerro La Tortuga: both in Aranda-Gómez et al., 2003), and in Miocene gravel deposits, all in the area immediately west of Sierra Aguachile (Fig. 5). However, the curved axial plane of the Cerro La Tortuga sill is nearly east-west, and the structure has been modified by a small northeast-trending graben. These anomalies may indicate that the Aguachile and La Herradura synclines and the folded sill may have been produced under different deformation regimes. The available data do not provide a solution to this problem, and for the sake of simplicity we assume that all folds involving mid-Tertiary volcanic and younger rocks were formed during a single period of compression. Drape folding is common and produces nearly perpendicular fold axes in plan view in the vicinity of the San Marcos fault (McKee et al., 1984). Whether this applies only to the lower Tertiary Laramide folds or also to the younger deformation remains to be established.

Both Sierra Aguachile and Sierra La Herradura are NNW–trending synclines that were slightly modified by subsequent faulting (Figs. 5 and 6). The Aguachile syncline is an open symmetrical fold ~15 km long with a minimum fold width of 5 km and with a curved axis that plunges to the southeast. Near the nose of the structure there is a set of WNW-trending, down-to-the-NNE normal faults. A NNW-trending, down-to-the-ENE normal fault duplicates the southern part of the eastern limb. This fault is parallel to the central graben of the Camargo volcanic field and aligns with it (Fig. 4).

La Herradura syncline is asymmetrical with a steeper eastern limb, and the fold axis is not curved as in the Aguachile syncline (Figs. 5 and 6). Mesozoic limestone is exposed in the eastern limb immediately beneath the mid-Tertiary volcanic rocks. The structural nose of La Herradura is repeated at least four times by ENE-trending, down-to-the-NNW normal faults. Right-lateral (see inset, Fig. 6) displacement of the axis of the structure indicates that these normal faults have an oblique component. The northern end of the La Herradura structure is a southwest-dipping monocline that was modified by small displacements along an ENE set of left-lateral faults and a northeastern set of right-lateral faults. The total combined length of the syncline and monocline is 25 km, and the minimum fold width of the syncline at its southern end is 3.5 km (Fig. 6).

DISCUSSION

Both the Aguachile and La Herradura ranges are NNW-trending, SE-plunging synclines (Fig. 2) younger than 31 Ma. These structures are symmetrical to the graben that crosses the central part of the Camargo volcanic field, and they lie to the east of the inferred and concealed trace of the San Marcos fault in the area. Interpretation of satellite images (INEGI, 1995) and regional geologic and topographic maps (INEGI, 1981, 1983, 1999) suggests the occurrence of another fold that developed in mid-Tertiary volcanic rocks at Sierra La Cuesta del Infierno, 5 km west of Sierra El Morrión (Fig. 2) close to the lineament interpreted as the projection of the San Marcos fault near Aldama. Miocene folding of Cenozoic volcanic rocks has been reported by Calas (1977) and Tardy (1980) at Sierra Peña Blanca,

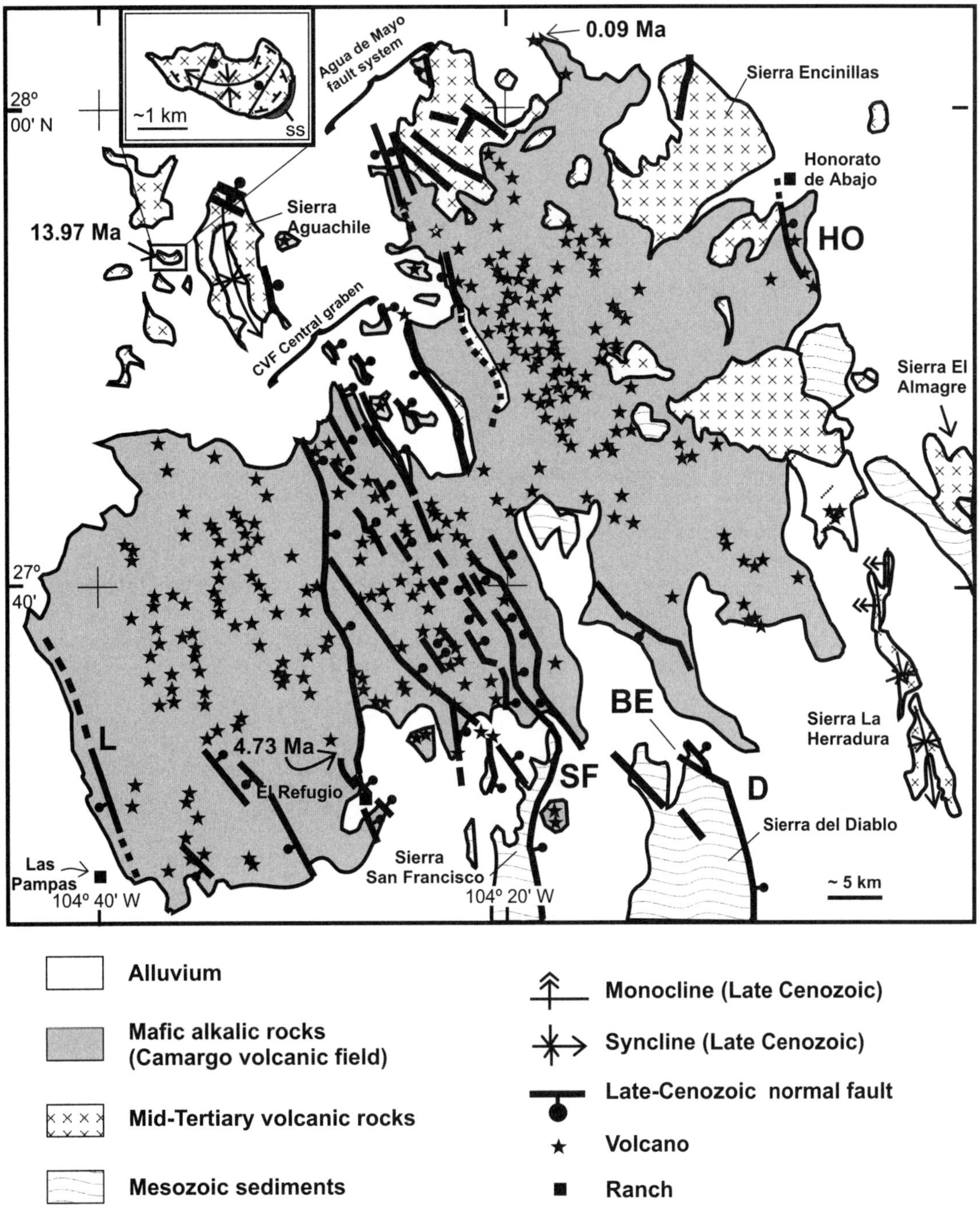

Figure 4. Generalized geologic map of the Camargo volcanic field (simplified from Aranda-Gómez et al., 2003). Inset shows the WNW–trending fold located west of Sierra Aguachile. The fold was formed in a 13.97 Ma ($^{40}Ar/^{39}Ar$: Aranda-Gómez et al., 2003) andesitic sill emplaced in a poorly consolidated clastic sequence made of sandstone (ss) interbedded with conglomerate. Faults mentioned in the text: L—Lagunetas; HO—Honorato; SF—San Francisco; BE—Benton; D—El Diablo.

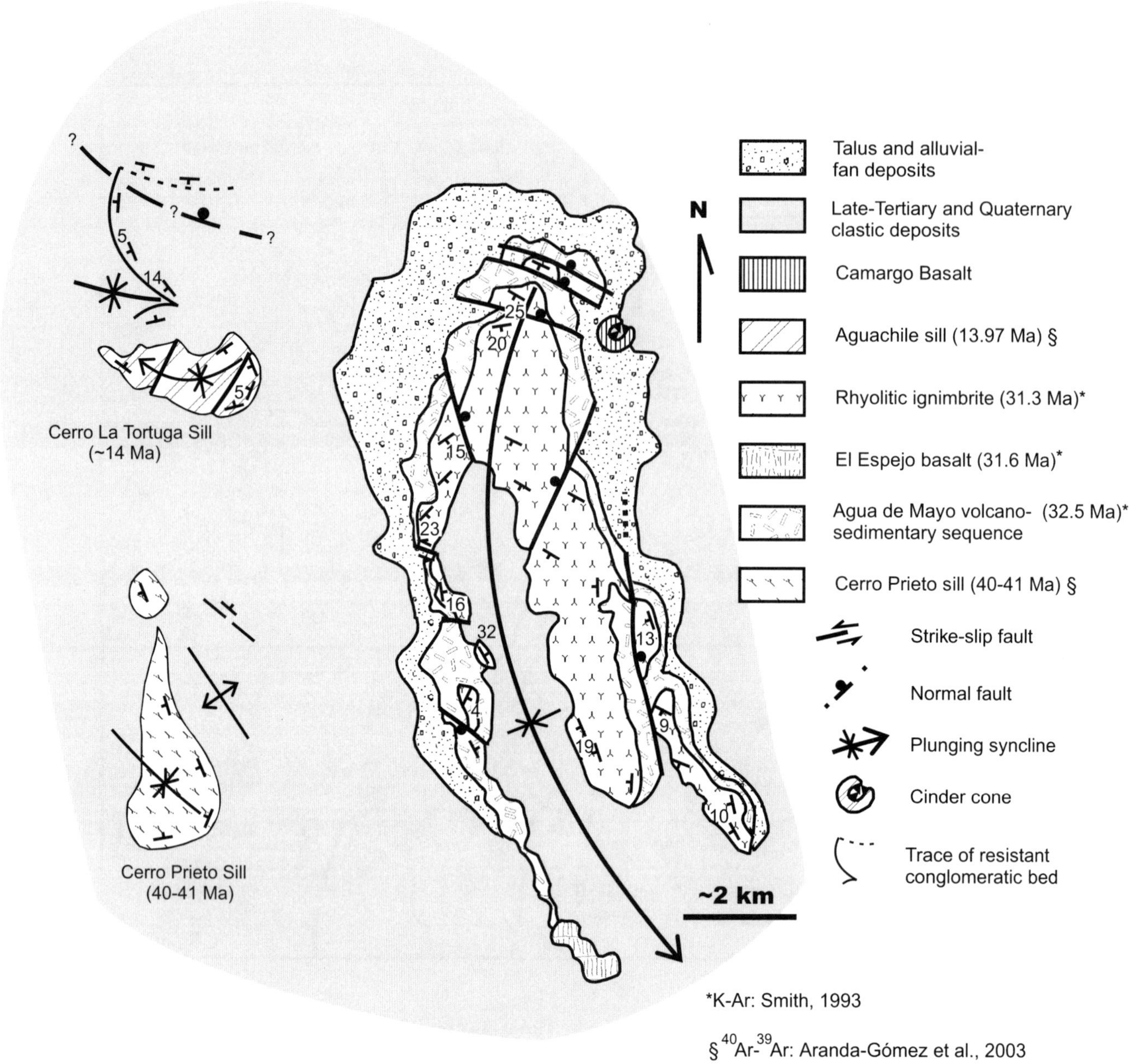

Figure 5. Photogeologic map of the Aguachile syncline.

north of Aldama (Fig. 1). Our data indicate that folding in the Camargo volcanic field area certainly occurred after deposition of a 31.4 Ma felsic ignimbrite exposed in the core of the Sierra Aguachile syncline. If the synclines and the fold in the Cerro La Tortuga sill (Fig. 5) formed at the same time, the local shortening is bracketed by the emplacement of the sill at 13.97 Ma and initiation of the Camargo volcanic field at ca. 4.73 Ma, whose lavas are not folded.

Somewhat similar folds of debated origin have been recognized in the Mexican part of the Trans-Pecos volcanic field, near the northwestern end of the proposed La Babia fault (Fig. 1; Charleston, 1981). Initial mapping identified at least two distinct types of folds (Gregory, 1981; Gunderson et al., 1986). A large anticline that developed only in Cretaceous rocks has an axial trace that curves more than 90°, from northeast to northwest. A 30 Ma granite intruded the core of the anticline. This fold is interpreted to be a Laramide fold that was modified by the intrusion around a caldera (Gregory, 1981; Gunderson et al., 1986). Two northwest-striking synclines that developed in mid-Tertiary (ca. 33 Ma) ash-flow tuffs a few km east of the caldera margin were attributed alternatively to late, post-Laramide contraction (Gunderson et al., 1986) or to magmatic doming related

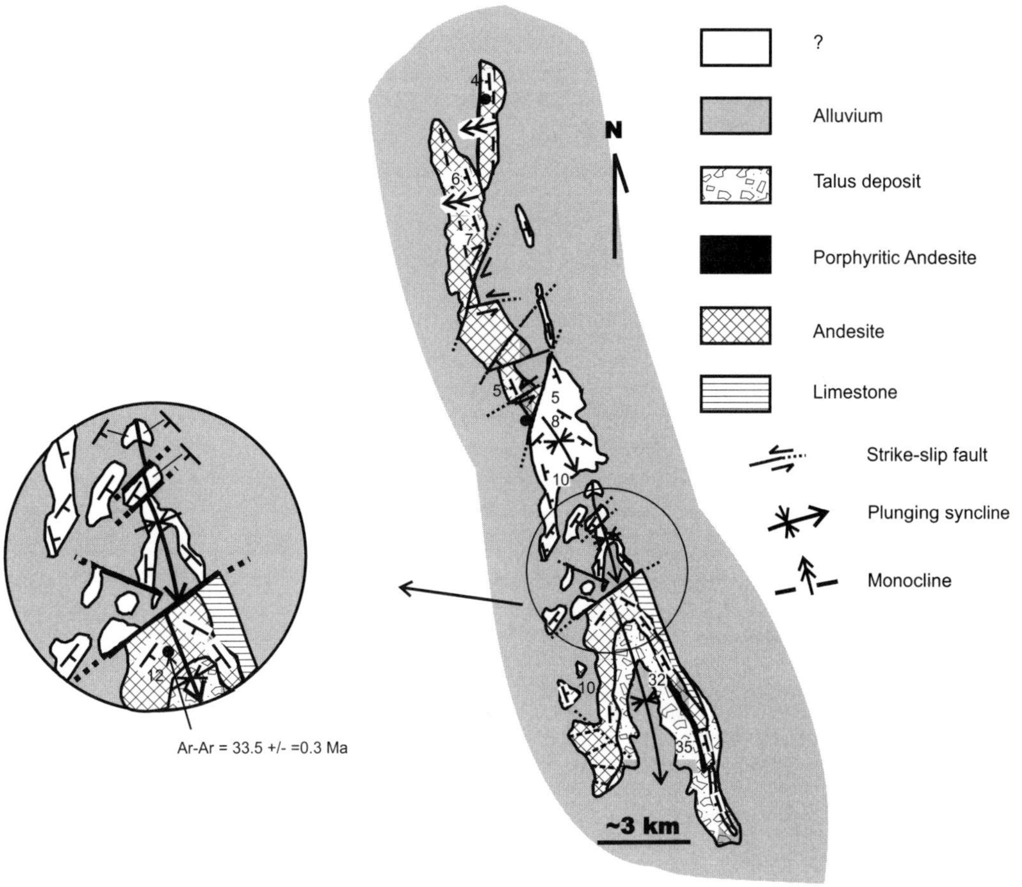

Figure 6. Photogeologic map of La Herradura syncline.

to caldera activity (Gregory, 1981). More recently, Flotte et al. (2004) interpreted the synclines that developed in mid-Tertiary volcanic rocks to have resulted from left-lateral displacement along a northwestern continuation of the La Babia fault, similar to our proposed origin of folds adjacent to the San Marcos fault.

The close spatial association of dilational structures and the late Tertiary synclines near the Camargo volcanic field can be interpreted in several different ways. Our favored view is a two-stage reactivation of the San Marcos fault during late Tertiary Basin and Range extension (Fig. 8). Alternatively, the synclines can be viewed as folds related to extension, formed either as basement-cored structures (e.g., Howard and John, 1997), or as longitudinal folds due to normal fault propagation, or as folds developed in the hanging-wall blocks above ramps in irregular fault planes (Janecke et al., 1998). Local folding in mid-Tertiary volcanic rocks may also be related to magmatic activity, as proposed for the structures in the Manuel Benavides area. In the particular case of a fold that developed solely in a medium- to high-grade welded ignimbrite, it might be related to differential compaction over an irregular pre-eruptive surface.

We favor the first model based on the following arguments. (1) All the upper Cenozoic folds developed in competent units, and both the rocks and the structures were formed at the surface. Field relations indicate it is unlikely that these volcanic rocks were deeply buried at the time of the deformation. Therefore, it is very probable that the volcano-sedimentary sequence at Sierra Aguachile and the overlying ignimbrite would fail in a brittle way in the case of hanging-wall collapse over sharp concave fault bends. (2) Pliocene-Pleistocene vertical displacements in the master faults of the Camargo volcanic field central graben, suggested by the maximum height of the degraded fault scarps and the calculated long-term rate of vertical displacement (Aranda-Gómez et al., 2003), are on the order of a couple of hundred meters. We believe that the scale of the Aguachile syncline, with a structural relief on the order of 300 m, is not consistent with the relatively small vertical displacement in the graben and conclude that the syncline is not a longitudinal fault propagation fold, such as those described by Janecke et al. (1998). (3) At Sierra Mojada (inset, Fig. 3) there is clear evidence of left-lateral reactivation of the San Marcos fault sometime after the Eocene (McKee et al., 1984, 1990). This event caused thrusting of Cretaceous limestones over the Eocene(?) Ahuichila Formation. (4) Upper Cenozoic normal faults markedly change orientation in the area (Figs. 2 and 3). This suggests that a discontinuity in the basement may have influenced the orientation of younger normal faults. (5) Near the northern

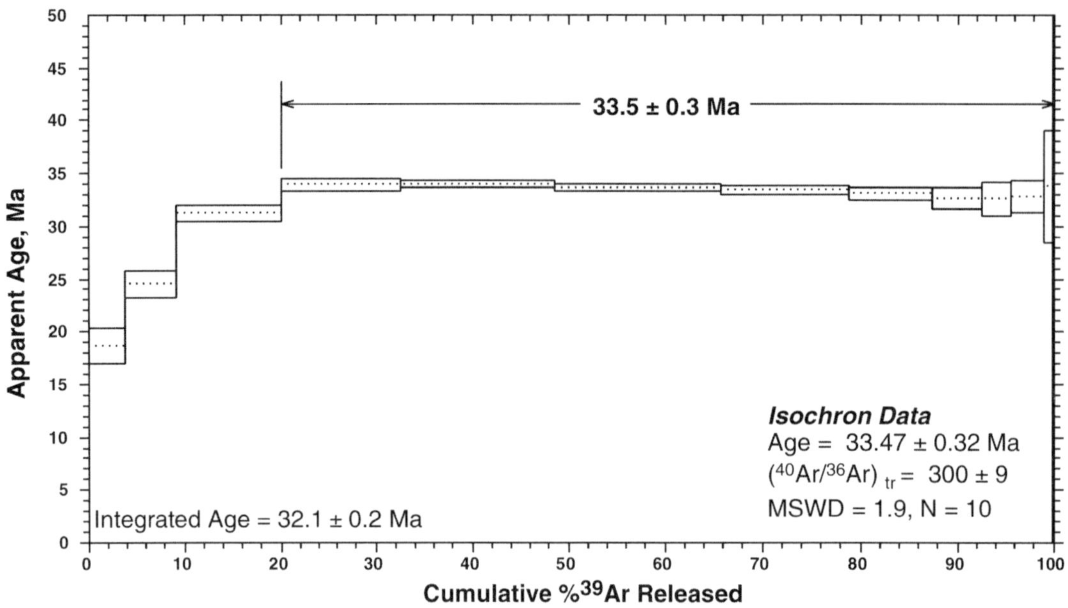

Figure 7. $^{40}Ar/^{39}Ar$ apparent age spectrum and inverse isochron data for the lava sample collected at La Herradura syncline (see Figure 6 for location). The material analyzed was the 250–400 µm size fraction of the groundmass. Uncertainties for age and ratios are ± 2 standard deviations. Vertical height of each increment in spectra is 2 standard deviations. The preferred age is the plateau age in boldface type. Subscript tr denotes the initial or "trapped" component of the $^{40}Ar/^{39}Ar$ ratio. The first three discordant steps may reflect postemplacement heating or alteration after ca. 20 Ma or poor Ar retention in some alteration phase in the matrix. Analytical procedures are described in Aranda-Gómez et al. (2003). Fish Canyon sanidine was used as a neutron-fluence monitor (28.02 Ma; Renne et al., 1998) as in Aranda-Gómez et al. (2003). MSWD—mean square of weighted deviates.

end of Sierra del Diablo, the WNW- to NW-trending Benton fault system (BE, Fig. 4) is parallel to the known trace of the San Marcos fault at nearby Sierra Mojada. The Benton system offsets for a short distance, in an apparent left-lateral sense, the range front down-to-the-east normal fault that bounds Sierra del Diablo (Fig. 3). (6) The complex array of faults that modified La Herradura syncline, together with the fold, are consistent with a left-lateral wrench system (Fig. 6). (7) We are not aware of extensive intrusive bodies near the Aguachile and La Herradura synclines, such as those identified in the Manuel Benavides area by Gregory (1981). Two sills are exposed a short distance west of Sierra Aguachile, and both of them are folded (Fig. 5). The Cerro Prieto mafic sill (40–41 Ma: Aranda-Gómez et al., 2003) is older than the Aguachile volcano-sedimentary sequence (32–31 Ma: Smith, 1993), therefore it cannot be the cause of the deformation. The other sill (Cerro La Tortuga, ca. 14 Ma: Aranda-Gómez et al., 2003) is younger, but the exposed part of the sill is considerably smaller than the Aguachile syncline. If we suppose that a large portion of La Tortuga sill is concealed near the Aguachile syncline, it is hard to explain why the heat of this shallow andesitic intrusion did not disrupt the K-Ar systems in the volcanic rocks exposed in Sierra Aguachile. (8) The differential-compaction hypothesis is not sufficient as an explanation because we observed in the Aguachile syncline that all the volcano-sedimentary sequence is deformed, and in La Herradura there are no ignimbrites in the sequence.

The concealed trace of San Marcos fault that we propose shows distinct bends in the region between Sierra Mojada and the Camargo volcanic field, changing trend from WNW to NW (Fig. 2). Both the central graben in the Camargo volcanic field and the Aguachile and La Herradura synclines were developed north of the northwest-trending parts of the San Marcos fault. We believe this was accomplished in at least two pulses of late Tertiary extension. An older event (<13.97 Ma, >4.73 Ma), which may correspond to a brief episode of regional northwestern extension documented in northern Mexico (<12 Ma: Aranda-Gómez et al., 1997) and southwestern United States (ca. 9 Ma: Zoback et al., 1981; Minor, 1995; Aldrich et al., 1986; Hentz and Henry, 1989), produced left-lateral movement along the San Marcos fault and local compression and associated folding (Fig. 8A). The younger event occurred after a change in the paleostress direction that caused ENE extension (Fig. 8B) and was in part synchronous with the activity in the Camargo volcanic field (4.73–0.09 Ma). Seen this way, the central graben of the Camargo volcanic field may be considered a small pull-apart basin, where both volcanism and faulting migrated as this feature was being formed. The unusually large volume of the Camargo volcanic field (120 km^3; Aranda-Gomez et al., 2003), compared to other intraplate-type

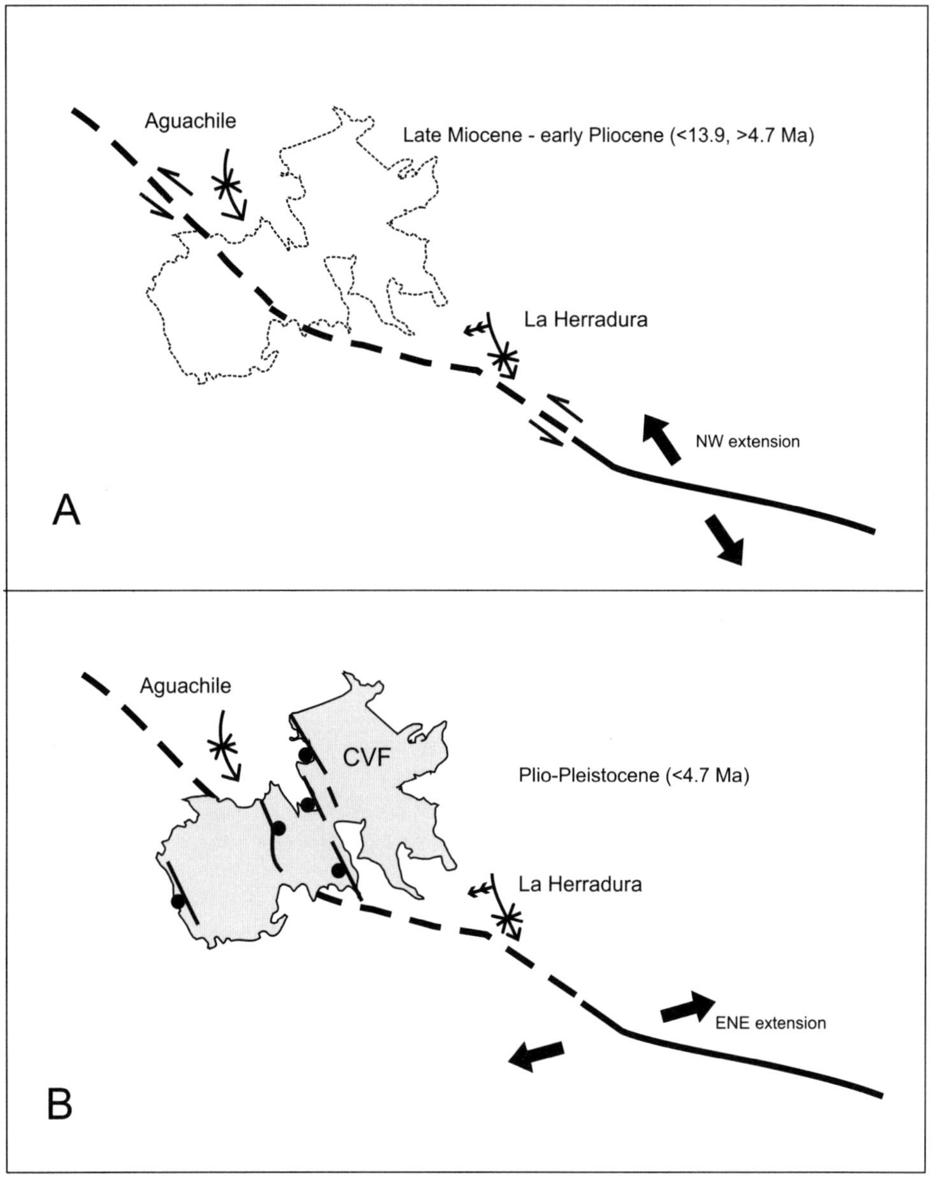

Figure 8. Two-stage model of reactivation of the San Marcos fault during the late Tertiary. (A) Folding associated with northwestern extension during the late Miocene. The future location of the Camargo volcanic field is shown by the dotted outline. (B) Volcanism and normal faulting during the Pliocene-Pleistocene. A small pull-apart basin was formed in the area now occupied by the Camargo volcanic field (CVF). The precise orientation of the stress tensor is unknown.

volcanic fields in central and northern Mexico, may be related to this pull-apart along a large, presumably lithospheric-scale structure that originally formed as a transform fault during formation of the Gulf of Mexico (Fig. 1).

SUMMARY AND CONCLUSIONS

1. Folding of Oligocene volcanic rocks around the Camargo volcanic field occurred sometime between 31 and 4.7 Ma. We believe that the 13.97 Ma sill located west of Sierra Aguachile was also deformed during this event. Therefore, local compression may be restricted between 13.97 and 4.77 Ma. We interpret this deformation as a product of a minor left-lateral movement along a concealed segment of the San Marcos fault, which acted as a restraining bend during a short period of regional northwestern extension (Fig. 8A).

2. We speculate that the post-Eocene thrust fault described by McKee et al. (1984) may have been formed during the late Miocene pulse of northwestern extension. This structure is also located on a northwest-trending fault segment that could act as a restraining bend during a left-lateral reactivation of the San Marcos fault (see inset, Fig. 3).

3. Intraplate-type volcanism and normal faulting occurred along the same northwestern extension of the San Marcos fault. During the Pliocene-Pleistocene, regional ENE extension caused the formation of the central graben as a pull-apart basin (Fig. 8B). The unusually large volume of the Camargo volcanic field and its obvious association with synvolcanic normal faulting, compared

to other intraplate-type volcanic fields in the Mexican Basin and Range, may be explained by this dilational structure developed on a large lithospheric-scale fault.

4. The presence of other folded volcanic units west of Sierra El Morrión supports the continuation of the San Marcos fault as a concealed basement structure for an additional 300 km, as previously proposed by Padilla y Sánchez (1986). Although there is folding in the Peña Blanca volcanic sequence and the inferred age of the deformation (Miocene: Calas, 1977) is similar to that in Sierra de Aguachile, the location of this fold is not entirely consistent with the model proposed here for the synclines near the Camargo volcanic field because the former syncline lies west of the fault trace (Fig. 2).

ACKNOWLEDGMENTS

Our research has been financed by CONACYT grants 3657PT and 37429-T (to J. Aranda-Gómez) and the Smithsonian Institution's Scholarly Studies Program (to J. Luhr). We gratefully acknowledge this support.

REFERENCES CITED

Aldrich, M.J., Jr., Chapin, C.E., and Laughlin, A.W., 1986, Stress history and tectonic development of the Rio Grande rift, New Mexico: Journal of Geophysical Research, v. 91, p. 6199–6211.

Anderson, T.H., and Schmidt, V.A., 1983, The evolution of Middle America and the Gulf of Mexico-Caribbean Sea region during Mesozoic time: Geological Society of America Bulletin, v. 94, p. 941–966, doi: 10.1130/0016-7606(1983)94<941:TEOMAA>2.0.CO;2.

Aranda-Gómez, J.J., Henry, C.D., Luhr, J.F., and McDowell, F.W., 1997, Cenozoic volcanism and tectonics in NW Mexico: A transect across the Sierra Madre Occidental volcanic field and observations on extension-related magmatism in the southern Basin and Range and Gulf of California tectonic provinces, *in* Aguirre-Díaz, G. J., Aranda-Gómez, J. J., Carrasco-Nuñez, G., and Ferrari, L., eds., Magmatism and tectonics in central and northwestern Mexico, a selection of the 1997 IAVCEI General Assembly excursions: México, D.F., Universidad Nacional Autónoma de México, Instituto de Geología, p. 41–84.

Aranda-Gómez, J.J., Housh, T.B., Luhr, J.F., Becker, T., Solorio-Munguía, J.G., and Martínez, E., 2001, Timing of multiepisodic deformation based on the study of continental clastic deposits and volcanic rocks, east-central Chihuahua, México: GEOS, v. 21, no. 3, p. 204.

Aranda-Gómez, J.J., Luhr, J.F., Housh, T.B., Connor, C.B., Becker, T., and Henry, C.D., 2003, Synextensional Pliocene-Pleistocene eruptive activity in the Camargo volcanic field, Chihuahua, México: Geological Society of America Bulletin, v. 115, no. 3, p. 298–313, doi: 10.1130/0016-7606(2003)115<0298:SPPEAI>2.0.CO;2.

Calas, G., 1977, Les phénomenes d'alteration hydrothermale et leur relation avec les minéralizations uraniferes en milieu volcanique: Le cas des ignimbrites tertiaires de la Sierra de Peña Blanca. Chihuahua (Mexique): Bulletin des Sciences Géologiques (Strasbourg), v. 30, no. 1, p. 3–18.

Charleston, S., 1974, Stratigraphy, tectonics and hydrocarbon potential of the Lower Cretaceous, Coahuila Series, Coahuila, Mexico [Ph.D. thesis]: Ann Arbor, University of Michigan, 268 p.

Charleston, S., 1981, A summary of the structural geology and tectonics of the state of Coahuila, Mexico: West Texas Geological Society Publications, v. 81-74, p. 28–36.

Caudillo-Sosa, G., Oviedo-Lucero, L.F., and Rodríguez-Castañeda, J.L., 1996, Falla Quitovac—Resultado de un evento de transpresión del "Mojave-Sonora Megashear," noroeste de Sonora, México: Revista Mexicana de Ciencias Geológicas, v. 13, no. 2, p. 140–151.

De Cserna, Z., 1956, Tectónica de la Sierra Madre Oriental de México entre Torreón y Monterrey, *in* 20th Congreso Geológico Internacional, México: Instituto Nacional para la Investigación de Recursos Minerales de México., 89 p.

De Cserna, Z., 1970, Mesozoic sedimentation, magmatic activity and deformation in northern Mexico, *in* Seewald, K., and Sundeen, D., eds., The geologic framework of the Chihuahua tectonic belt: West Texas Geological Society, Midland, Texas, p. 99–117.

Flotte, N., Martinez, J., Rangin, C., Tardy, M., Husson, L., and Le Pichon, X., 2004, The Rio Grande fault, a major mid-Tertiary left-lateral shear zone [abs.]: GEOS, v. 24, no. 2, p. 280.

García-Estrada, G.H., 1997, Programa en FORTRAN para el manejo de modelos digitales de elevación del terreno del Instituto Nacional de Estadística, Geografía e Informática (INEGI): Geotermia, Revista Mexicana de Geoenergía, v. 12, no. 3, p. 141–162.

Gregory, J.L., 1981, Volcanic stratigraphy and K-Ar ages of the Manuel Benavides area, northeastern Chihuahua, Mexico, and correlations with the Trans-Pecos Texas volcanic province [M.S. thesis]: Austin, Texas, The University of Texas at Austin, 78 p.

Gunderson, R., Cameron, K., and Cameron, M., 1986, Mid-Cenozoic high-K calc-alkalic and alkalic volcanism in eastern Chihuahua, Mexico: Geology and geochemistry of the Benavides-Pozos area: Geological Society of America Bulletin, v. 97, p. 737–753, doi: 10.1130/0016-7606(1986)97<737:MHCAAV>2.0.CO;2.

Harding, T.P., Vierbuchen, R.C., and Christie-Blick, N., 1985, Structural styles, plate tectonic settings, and hydrocarbon traps of divergent (transtensional) wrench faults, *in* Biddle, K., and Christie-Blick, N., eds., Strike-slip deformation, basin formation, and sedimentation: Tulsa, Oklahoma, Society of Economic Paleontologists and Mineralogists Special Publication 31, p. 51–75.

Hennings, P.H., 1994, Structural transect of the southern Chihuahua Fold Belt between Ojinaga and Aldama, Chihuahua, Mexico: Tectonics, v. 13, no. 6, p. 1445–1460, doi: 10.1029/94TC00800.

Hentz, T.F., and Henry, C.D., 1989, Evaporite-hosted native sulfur in Trans-Pecos: Relations to late-phase Basin and Range deformation: Geology, v. 17, p. 400–403, doi: 10.1130/0091-7613(1989)017<0400:EHNSIT>2.3.CO;2.

Howard, K.A., and John, B.E., 1997, Fault-related folding during extension: Plunging basement-cored folds in the Basin and Range: Geology, v. 25, no. 3, p. 223–226, doi: 10.1130/0091-7613(1997)025<0223:FRFDEP>2.3.CO;2.

INEGI, 1981, Carta geológica Chihuahua: México City, Secretaria de Programación y Presupuesto, 1 sheet, scale 1:1,000,000.

INEGI, 1983, Carta geológica Ciudad Delicias H13–11: México City, Instituto Nacional de Estadística, Geografía e Informática, 1 sheet, scale 1:250,000.

INEGI, 1994, Geomodelos de altimetría del Territorio Nacional: Aguascalientes, Ags., Instituto Nacional de Estadística, Geografía e Informática (CD-ROM).

INEGI, 1995, Espaciomapa Ciudad Delicias, Hoja H13–11: México City, Instituto Nacional de Estadística, Geografía e Informática, 1 sheet, scale 1:250,000.

INEGI, 1999, Carta topográfica Trancas, H13C58: México City, Instituto Nacional de Estadística, Geografía e Informática, 1 sheet, scale 1:50,000.

Janecke, S.U., Vandenburg, C.J., and Blankenau, J.J., 1998, Geometry, mechanisms and significance of extensional folds from examples in the Rocky Mountain Basin and Range province, U.S.A.: Journal of Structural Geology, v. 20, no. 7, p. 841–856, doi: 10.1016/S0191-8141(98)00016-9.

Jones, N.W., McKee, J.W., Márquez, D.B., Tovar, J., Long, L.E., and Laudon, T.S., 1984, The Mesozoic La Mula Island, Coahuila, Mexico: Geological Society of America Bulletin, v. 95, p. 1226–1241.

Longoria, J.F., 1984, Stratigraphic studies in the Jurassic of northwestern Mexico: Evidence for the origin of the Sabinas basin, *in* Ventress, W.W., Bebout, D.G, Perkins, B.E., and Moores, C.H., eds., The Jurassic of the Gulf Rim: Austin, Texas, Society of Economic Paleontologists and Mineralogists, Gulf Coast Section, Annual Research Conference Proceedings, p. 171–193.

Longoria, J.F., 1985, Tectonic transpression in the Sierra Madre Oriental, northeastern Mexico: An alternative model: Geology, v. 13, p. 453–456, doi: 10.1130/0091-7613(1985)13<453:TTITSM>2.0.CO;2.

Marzolf, J.E., 2003, The Mojave-Sonoran megashear and reassembly of Cordilleran Triassic-Jurassic tectonostratigraphic assemblages: Geological Society of America Abstracts with Programs, v. 35, no. 4, p. 78.

McDowell, F.W., Housh, T.B., and Wark, D.A., 1999, The crust beneath the Sierra Madre Occidental volcanic field as inferred from isotopic studies at the Tomochic volcanic center, western Chihuahua, Mexico: Geological Society of America Bulletin, v. 111, p. 823–830, doi: 10.1130/0016-7606(1999)111<0823:NOTCBW>2.3.CO;2.

McKee, J.W., and Jones, N.W., 1979, A large Mesozoic fault in Coahuila, Mexico: Geological Society of America Abstracts with Programs, v. 11, p. 476.

McKee, J.W., Jones, N.W., and Long, L.E., 1984, History of recurrent activity along a major fault in northeastern Mexico: Geology, v. 12, p. 103–107, doi: 10.1130/0091-7613(1984)12<103:HORAAA>2.0.CO;2.

McKee, J.W., Jones, N.W., and Long, L.E., 1990, Stratigraphy and provenance of strata along the San Marcos fault, central Coahuila, Mexico: Geological Society of America Bulletin, v. 102, p. 593–614, doi: 10.1130/0016-7606(1990)102<0593:SAPOSA>2.3.CO;2.

Minor, S.A., 1995, Superposed local and regional paleostress: Fault-slip analysis of Neogene extensional faulting near coeval caldera complexes, Yucca Flat, Nevada: Journal of Geophysical Research, v. 100, p. 10,507–10,528.

Moreno, F.A., Mikus, K.L., and Keller, G.R., 2000, Crustal structure and location of the Ouachita orogenic belt in northern Mexico: Geofisica Internacional, v. 39, p. 229–246.

Padilla y Sánchez, R.J., 1986, Post-Paleozoic tectonics of northeast Mexico and its role in the evolution of the Gulf of Mexico: Geofisica Internacional, v. 25, no. 1, p. 157–206.

Pessagno, E.A., Jr., Longoria, J.F., Hull, D.M., and Kelldorf, M.E., 1993, Tectonostratigraphic significance of the San Pedro del Gallo area, Durango, Mexico, *in* Dunne, G.C., and McDougall, K.A., eds., Mesozoic paleogeography of the Western United States–II, p. 141–156.

Renne, P.R., Swisher, C.C., Deino, A.L., Karner, D.B., Owens, T., and DePaolo, D.J., 1998, Intercalibration of standards, absolute ages and uncertainties in $^{40}Ar/^{39}Ar$ dating: Isotope Geoscience, v. 145, no. 1–2, p. 117–152.

Silver, L.T., and Anderson, T.H., 1974, Possible left-lateral early to middle Mesozoic disruption of the southwestern North American craton margin: Geological Society of America Abstracts with Programs, v. 6, p. 955.

Smith, R.D., 1993, The Agua de Mayo mid-Cenozoic volcanic group and related xenoliths from La Olivina SE Chihuahua, Mexico [M.S. thesis]: Sanata Cruz, University of California, 112 p.

Smith, R.D., Cameron, K.L., McDowell, F.W., Niemeyer, S., and Sampson, D.E., 1996, Generation of voluminous silicic magmas and formation of mid-Cenozoic crust beneath north-central Mexico: Evidence from ignimbrites, associated lavas, deep crustal granulites and mantle pyroxenites: Contributions to Mineralogy and Petrology, v. 123, p. 375–389, doi: 10.1007/s004100050163.

Stewart, J.H., 2003, Mojave-Sonora megashear: Evidence from Neoproterozoic to lower Jurassic strata: Geological Society of America Abstracts with Programs, v. 35, no. 4, p. 13.

Tardy, M., 1980, Contribution à l'étude géologique de la Sierra Madre Orientale du Mexique [Doctorat d'Etat thesis]: Département de Géotectonique de l'Université Pierre et Marie Curie à Paris et Département des Sciences de la Terre de l'Université de Savoie à Chambéry, 459 p.

Zoback, M., Anderson, R.E., and Thompson, G.A., 1981, Cainozoic evolution of the state of stress and style of tectonism of the Basin and Range province of the western United States: Philosophical Transactions of the Royal Society of London, Series A, v. 300, p. 407–434.

Manuscript Accepted by the Society 24 March 2005

Geological Society of America
Special Paper 393
2005

A reconstruction of late Miocene to Recent transtension in the southwestern United States and northwesternmost Mexico

John A. Dembosky Jr.*
Thomas H. Anderson
Department of Geology and Planetary Science, University of Pittsburgh, 200 Space Research Coordination Center, Pittsburgh, Pennsylvania 15260, USA

ABSTRACT

A retrodeformable reconstruction of late Miocene and younger deformation based on a transtensional model, which combines extension and strike-slip motion, is proposed herein. The reconstruction incorporates and expands upon previous restorations of late Miocene (Basin and Range) strain east of the San Andreas transform and is supported by Pacific–North America plate kinematics since 15.1 Ma. The study region extends south from the Garlock fault across the Mojave and Sonoran Deserts into southwestern Arizona and Sonora, Mexico. In the reconstruction, proposed late Miocene north-trending physiographic basins are treated as pull-apart structures at releasing steps within a system of linked northwest-striking, right-lateral strike-slip faults. Pull-apart structures are closed by restoring strike-slip fault displacements obtained from published fault offsets where available. In areas where published data are unavailable, restoration is based upon consideration of geologic markers, seismic interpretation, depth to basement studies, and studies of basin geometry. Rotations of crustal blocks in California are modeled after Powell (1993) and have not been recognized in southwestern Arizona and northwestern Sonora, Mexico. The late Miocene reconstruction shows total east-west elongation strain to be 9% (50 km). About 43 km of right-lateral strike-slip displacement is accommodated by northwest-trending faults east of the San Andreas fault between the Gila River and the Garlock fault. Total right-lateral separation of 77 km results from the incorporation of clockwise block rotations in the model.

The reconstruction of late Miocene to Recent transtension: (1) reveals the partial effect of extensional disruption of the surficial trace of the Mojave-Sonora megashear, and (2) partially resolves some of the late Miocene "slip discrepancy" of dextral offsets between the Pacific and North American plates (~635 km), and that recorded in the Gulf of California (~300 km). Furthermore, the reconstruction assesses the relationship between transtensional strain in the desert region of the southwest and faults comprising the San Andreas transform plate boundary.

Keywords: reconstruction, Miocene, basin and range, Mojave, Sonora.

*Present address: Department of Chemistry & Physical Science, Methodist College, Fayetteville, North Carolina 28331, USA; jdembosky@methodist.edu.

Dembosky, J.A., Jr., and Anderson, T.H., 2005, A reconstruction of late Miocene to Recent transtension in the southwestern United States and northwesternmost Mexico, *in* Anderson, T.H., Nourse, J.A., McKee, J.W., and Steiner, M.B., eds., The Mojave-Sonora megashear hypothesis: Development, assessment, and alternatives: Geological Society of America Special Paper 393, p. 521–542. doi: 10.1130/2005.2393(20). For permission to copy, contact editing@geosociety.org.

INTRODUCTION

Paleotectonic reconstructions have become increasingly important as geologists strive to unravel the full history of tectonism along the western margin of North America. Late Miocene reconstructions of the southwestern United States are primarily confined to southern California and include those that restore displacements and rotations: (1) west of the San Andreas transform (e.g., May and Walker, 1989; Matti and Morton, 1993; Dickinson, 1997; Atwater and Stock, 1998; Fritsche, 1998; Sonder and Jones, 1999); (2) east of the San Andreas transform (e.g., Luyendyk et al., 1980, 1985; Luyendyk, 1991; Richard, 1993, 1994; Dickinson, 1996; Schermer et al., 1996; Dokka et al., 1998); and (3) on both sides of the transform (e.g., Powell, 1993).

Some models that incorporate rotated blocks into reconstruction rely heavily upon paleomagnetic data (e.g., Luyendyk et al., 1980, 1985; Carter et al., 1987; Luyendyk, 1991; Dokka et al., 1998). However, the effectiveness of paleomagnetic data to determine block rotation is debated (e.g., Powell, 1993; Dickinson, 1996; Dickinson and Butler, 1998). Restorations by Richard (1993, 1994) and Powell (1993) are based primarily upon mapping and/or correlations derived from structural analyses, but the two models disagree on how to account for or incorporate paleomagnetic data. Restorations of Dickinson (1996, 1997) utilize previously proposed tectonic models for rotation and disagree with paleomagnetic data models of Luyendyk (1991).

Atwater and Stock (1998) estimate that ~635 km of dextral displacement has occurred between the North American and Pacific plates since ca. 12.3 Ma. Of this amount, ~300 km of displacement was accommodated by the southern San Andreas transform in the Gulf of California since 6 Ma (Oskin and Stock, 2003). Much of the large magnitude of extension and right-lateral shear that remains, referred to as "slip discrepancy," must have taken place between 12 and 6 Ma outside the Gulf of California and east of the San Andreas fault (Gans, 1997; Oskin and Stock, 2003). In the model presented herein, several tens of kilometers of the "slip discrepancy" was accommodated in southeastern California, southwestern Arizona, and northwestern Mexico.

Our proposed reconstruction encompasses the region east of the San Andreas transform and south of the Garlock fault, extending southeastward across the Mojave Desert into the Sonoran Desert of Arizona and Sonora, Mexico (Plate 1[1]). The purpose of this work is: (1) to present a new model of late Miocene restoration based upon the working hypothesis that transtension related to the San Andreas transform is the primary mechanism of extension east of the transform, (2) to successfully incorporate previously published field observations and other data into the models, (3) to extend the boundaries of the previously studied regions in California (e.g., Powell, 1993, Richard, 1993, Schermer et al., 1996, and Dokka et al., 1998) southeastward into northwestern Mexico, and (4) to reveal the effects of late Miocene transtension on the paleogeography of major disruptive structures, such as the Mojave-Sonora megashear (Silver and Anderson, 1974).

DISTINCTIONS BETWEEN EARLY MIOCENE AND LATE MIOCENE AND YOUNGER EXTENSION

The present physiography of southwesternmost North America is primarily characterized by basin and range topography (Plate 1). The topography was created largely during Miocene and later extensional episodes in southern California, southwestern Arizona, and northwestern Sonora, Mexico. An early Miocene extensional episode between ca. 27 and 15–13 Ma (Crittenden et al., 1980; Spencer and Reynolds, 1989; Lucchitta and Suneson, 1996) was followed by development of the San Andreas transform system after 12 Ma (e.g., Bohannon et al., 1993).

The early Miocene episode is characterized by core complex structures south and west of the Colorado Plateau (e.g., Spencer and Reynolds, 1991), where stretching lineations in crystalline rocks consistently trend 060° and basins bounded by normal faults trend 330° (parallel to the assumed maximum principal stress direction during this period). Both brittle and ductile (core complex) structures related to early Miocene deformation are presently exposed.

A younger episode of extension distinguished by north-striking, range-bounding faults began after ca. 15 Ma. The north-trending long axes of resulting late Miocene and younger transtensional basins are consistent and differ from the northwest orientation of the early Miocene basin axes (e.g., Burchfiel and Stewart, 1966; Bohannon et al., 1993). Extension was contemporaneous with transforms within the Gulf of California, which became active after 12 Ma, as shown by offsets recorded by ignimbrites (e.g., Oskin et al., 1999; Oskin and Stock, 2003). The amount of extension of this younger period is much less than that of the early Miocene (e.g., Gans, 1997).

PREVIOUS LATE MIOCENE ROTATIONAL MODELS

Previous models of late Miocene extension incorporate rotation as an important component. A significant contribution of Luyendyk et al. (1985) is the proposal that a left bend existed in the pre-Miocene continental margin that acted as an asperity to the northward slippage of the Pacific plate past southwestern North America. The movement of the plates past this asperity is proposed as a mechanism that may have helped initiate the rotation of upper lithospheric blocks in southern California. In a later paper, the amount of clockwise rotation of the eastern Transverse Ranges (Plate 1) was determined by paleomagnetic data to be 41° (Carter et al., 1987). Middle Miocene and younger rotation east of the San Andreas was proposed by Luyendyk (1991) to have been a response to shear associated with transform plate motion after the cessation of early Miocene subduction. Luyendyk also

[1]Plates 1–4 are on the CD-ROM accompanying this volume and are in the GSA Data Repository as item 2005143, available online at www.geosociety.org/pubs/ft2005.htm, or on request from editing@geosociety.org or Documents Secretary, GSA, P.O. Box 9140, Boulder, CO 80301, USA.

suggested that an early Miocene phase of extension accompanied an older phase of block rotations. Our model for reconstruction of the region east of the San Andreas fault does not incorporate this earlier rotation.

Dickinson (1996) offers a comprehensive model of late Miocene deformation based upon the kinematics of block rotation in the Transverse Ranges of southern California that expands upon the work of Luyendyk (1991). The locations of axes of rotation are specifically addressed. A "pinned" model of rotation is preferred by Dickinson (1996) in which the width of the shear zone increases as deformation progresses. Dickinson (1996) also presents an alternative "decoupled" model in which the width of the shear zone is constant within the region of rotation between bounding strike-slip faults. Lastly, Dickinson (1996) demonstrates that holding the widths of both the shear-zone and rotating panel domain constant is not viable.

PREVIOUS LATE MIOCENE TECTONIC RECONSTRUCTIONS

Previous reconstructions of extended crust in southeastern California and west-central Arizona were developed by Powell (1993), Richard (1993, 1994), Schermer et al. (1996), and Dokka et al. (1998). These incorporate both geologic piercing points and block rotations. Powell (1993) describes one of the most comprehensive reconstructions of southern California that focuses on deformation along the San Andreas fault. This reconstruction restored mostly strike-slip fault motion based on the correlation of geologic features. He emphasizes both the eastern Transverse Ranges and the region west of the San Andreas fault.

East of the San Andreas fault, Richard (1993) proposed a restoration of southeasternmost California and southwestern Arizona and recognized the disagreement between fault offsets and paleomagnetic data. He accounted for rotations as part of two stages of late Miocene deformation in the eastern Transverse Ranges, which enabled his restoration to better agree with paleomagnetic data than Powell's (1993). In order to eliminate overlap of the southern California Chocolate Mountains from the rotation of the eastern Transverse Ranges in his reconstruction, Richard (1993) postulated faults and basins with orientations that are difficult to explain as being kinematically active with reference to the generally north-south–oriented maximum principal stress (Prescott et al., 1979) (e.g., the northeast-trending basin northeast of the McCoy Mountains, and the east-trending basin south of the Palen Mountains; Plate 1).

Schermer et al. (1996) presented the first detailed attempt to restore the northeast Mojave block or domain. They concluded that "a simple plane strain model of crustal blocks bounded by discrete, single, strike-slip faults is clearly too simple" to explain rotations in this region. Half of the 65 km of dextral shear in the Eastern California shear zone, a shear zone that accommodates ~15% of the strain between the Pacific–North American plate boundary east of the San Andreas transform (Dokka and Travis, 1990), is estimated by Schermer et al. (1996) to have been accommodated within this domain since 10 Ma. Observed fault slip within the domain suggests that it has rotated 23° clockwise and is not great enough to explain the 63.5° ± 7.6° results from paleomagnetic data. Discrepancies between fault-slip and paleomagnetic results are attributed to clockwise bending of faults bounding fault blocks in the domain (Schermer et al., 1996).

Dokka et al. (1998) address extensional domains in the central Mojave Desert region of southern California. These authors utilize paleomagnetic data to constrain a reconstruction of the Mojave block to ca. 21 Ma, thus including both early and late Miocene extension. In this reconstruction and earlier ones (see Dokka et al., 1998, and references therein), zones of overlap (contraction) and extension are identified that are kinematically difficult to reconcile with a transtensional model.

Restoration of fault offsets west of the San Andreas fault is not revised in this paper. We accept the minimum displacements that Powell (1993) used between the San Andreas fault and Pacific Ocean. Other authors addressing the region west of the San Andreas faults include Dickinson (1983), May and Walker (1989), Harbert (1991), Sedlock and Hamilton (1991), Matti and Morton (1993), Dickinson (1997), Bohannon and Geist (1998), and Fritsche (1998). In general, our conclusions incorporate the Powell (1993) reconstruction.

ASSUMPTIONS OF THE NEW MODEL

In this paper, north-trending basins are treated as pull-apart structures at releasing steps within a system of linked northwest-oriented, right-lateral strike-slip faults striking 320°–330° (Fig. 1A), similar to the basin geometry of Death Valley (e.g., Burchfiel and Stewart, 1966). A complementary set of northeast-striking, left-lateral faults is predicted by this model under conditions of pure shear; however, we see deficient evidence for them within the study region.

We do not incorporate block rotations suggested by paleomagnetic studies (e.g., Luyendyk et al., 1980, 1985; Carter et al., 1987; Luyendyk, 1991) where they are not supported by changes in the orientation of geologic features, such as the Late Jurassic Independence dikes, which generally strike northwest (e.g., Glazner et al., 1994; Ron and Nur, 1996; Plate 1), or early Miocene stretching lineations (060°) and associated orthogonal fractures (330°). We use the modest rotations of Powell (1993) and Schermer et al. (1996) in the eastern Transverse Ranges and northeastern Mojave, respectively. We assume the rotations are related to distributed shear along northwest-striking faults that are compatible with the strain calculations illustrated in Figure 2.

We assume that between late Miocene and present time the plate motion between the Pacific and North American plates east of the San Andreas transform was accommodated by regional transtension (Figs. 1A and 1B) recorded by north-trending basins (e.g., the Growler Valley, Santa Rosa Valley, and La Posa Plain; Plates 1 and 2) and right-lateral strike-slip faults parallel to the trend of the San Andreas, which strikes ~324°. These fault orientations are significantly clockwise of the RM2, NUVEL, and

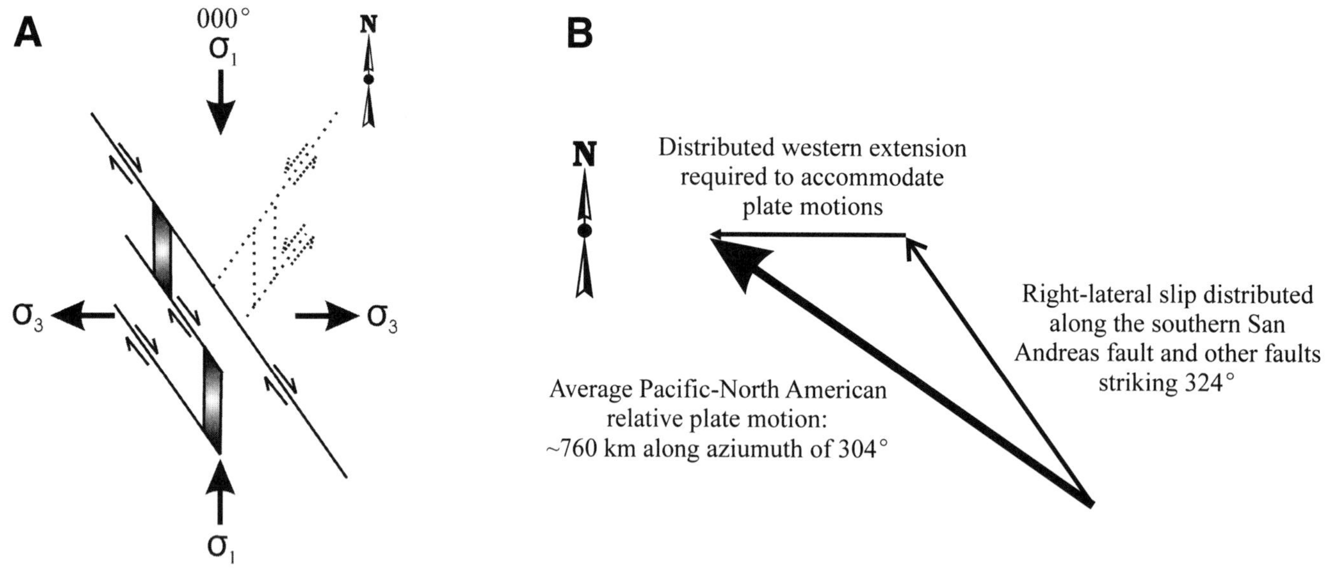

Figure 1. (A) Model for the formation of basins at right releasing steps among northwest-striking, right-lateral faults. Assumed maximum principal stress orientation during the late Miocene is shown for reference. Complementary left-lateral strike-slip faults and related basins are uncommon in the study area and are shown dashed. (B) Vector diagram illustrating the plate tectonic driving mechanism for late Miocene to Recent transtension in southwestern North America. The average Pacific–North America relative plate motion since 15.1 Ma (Atwater and Stock, 1998; Oskin and Stock, 2003) is partitioned into right-lateral strike-slip faulting and east-west extension. The net effect is to produce the system of right-stepping, right-lateral faults illustrated in A, which is recorded within large regions east of the San Andreas fault.

other estimates of the Pacific–North American relative motion direction (e.g., Jordan and Minster, 1988; DeMets et al., 1990; Humphreys and Weldon, 1994; Atwater and Stock, 1998; Oskin and Stock, 2003; Fig. 1B). Further we assume that the orientation of the right-lateral strike-slip faults parallel to the San Andreas and north-trending normal faults is kinematically compatible with a maximum principal stress oriented 000°–010° since the late Miocene (Wright, 1976; Prescott et al., 1979; Rundle, 1986; Ekström and England, 1989).

Basins in the reconstruction are closed primarily by restoring bounding strike-slip fault displacements obtained from published fault offsets (e.g., Richard, 1993; Powell, 1993; Dokka et al., 1998). In areas where published information is unavailable or insufficient to constrain restoration, basin geometries (Kruger et al., 1998), combined with consideration of regional geologic strain markers (such as boundaries between provinces of Proterozoic rocks and terranes), seismic interpretation (Eberly and Stanley, 1978), and depth to basement studies (Oppenheimer and Sumner, 1980), are utilized. Closure of north-trending basins in the study area is based on the following assumptions:

1. Basin width does not accurately record structural extension. The correct amount of extension is almost always less, because of the existence of erosional pediments that extend from flanking mountains.

2. The widths of basins are related to the magnitude of strike-slip displacement along bounding northwest-striking faults (Fig. 3). The extension of each basin (idealized for clarity on Plate 3) is always less than displacement along the bounding strike-slip faults.

3. If the basin geometry is manifest, but the amount of strike-slip offset is unknown, comparisons with basins that have geometry established from seismic reflection and other data provide the basis for an estimate of width and related extension. Miller and John (1999) make the point that the width of a basin and magnitude of extension are partially controlled by the dip angle of bounding normal faults (Fig. 4A and 4B). The steeper the normal fault, the narrower the basin. The calculated extension may be used to estimate the lateral slip on inferred, bounding, northwest-striking faults.

4. Late Miocene fault patterns may be influenced by early Miocene faults. Early Miocene normal faults striking northwest (330°) may have thinned and weakened the crust (upper lithosphere) significantly enough to allow the easy propagation of right-lateral strike-slip faults oriented the same direction in the late Miocene. Early Miocene steep and gentle faults, as well as basin sediments, are exposed in the Copper, Cabeza Prieta, Muggins, Mohawk, and Laguna Mountains, and Baker Peaks in Arizona (Plate 1), indicating uplift following early Miocene extension (Mueller et al., 1982; Pridmore and Craig, 1982). We cannot preclude that changes in base level, especially within the Gila River drainage basin, were in part responsible for this physiography (S. Richard, 2003, personal commun.).

5. Some basins that formed during the early Miocene persisted through the late Miocene to the present time. Late

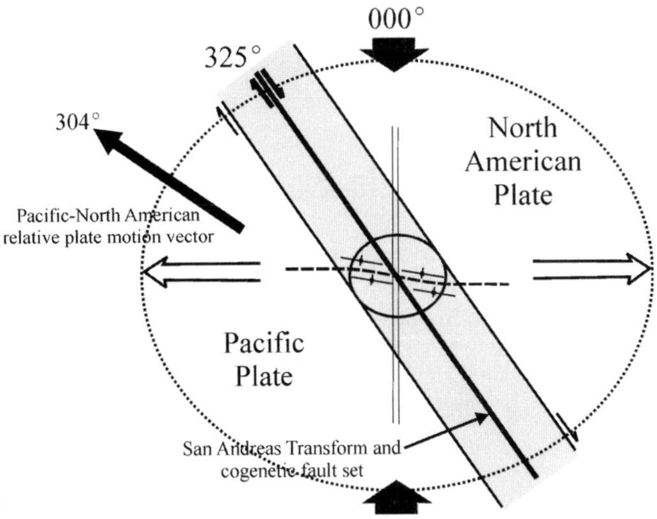

Figure 2. Strain ellipses for late Miocene and younger pure and simple shear in western North America (maximum principal stress shown as black arrows; minimum principal stress shown as white arrows). Simple shear accumulates in response to right-lateral strike-slip faults within a broad zone of faulting along the San Andreas transform and cogenetic faults (shaded). The ellipse for simple shear within the zone of faulting (solid) records 9% east-west extension. Simple shear within the zone may rotate the long axis of the local strain ellipse clockwise. Regional strain accommodates pure shear (dotted ellipse; not to scale). Orientation of cogenetic folds is shown as dashed line and fold symbols and pull-aparts as double bar (modified from Sylvester, 1988). Calculated values for the local strain ellipse: east-west elongation strain of 9%, $\gamma = 0.19$, $\psi = 11°$.

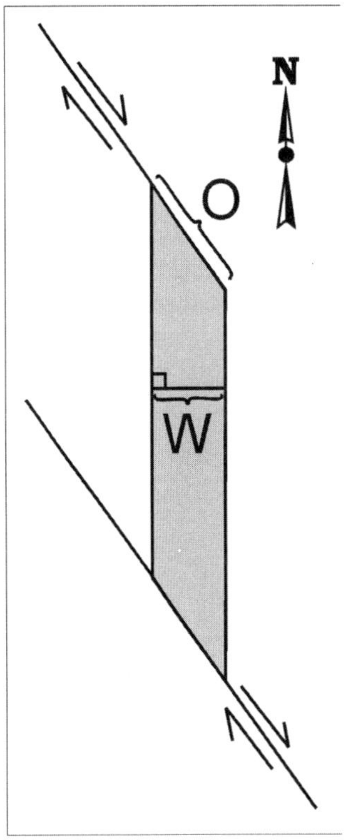

Figure 3. Extensional strain (basin width, W) is always less than the strike-slip offset of bounding faults (O). O is the hypotenuse of a right triangle in this idealized illustration.

Miocene basins are postulated to have been able to form within some extensive early Miocene basins where the crust (upper lithosphere) was strongly attenuated. Many large, deep basins in southwestern Arizona (and Sonora?), such as the Mohawk Valley (Plate 1), were initiated during the early Miocene as inferred from the discovery of buried early Miocene volcanics during drilling and subsurface studies in the region (e.g., Eberly and Stanley, 1978; Mattick et al., 1973).

6. Seismic studies (e.g., Eberly and Stanley, 1978), depth to basement studies (Oppenheimer and Sumner, 1980), and gravity data (Rotstein et al., 1976; Jachens and Griscom, 1985; Mariano et al., 1986; Klein and Kucks, 1991; Richard, 1993) may be used to determine the depth of basins and constrain the location of basins in the reconstruction (Plate 2).

NEW POST–12 MA PALEOTECTONIC RECONSTRUCTION OF THE SOUTHWESTERN UNITED STATES AND NORTHWESTERNMOST MEXICO

We present a reconstruction that includes the southwesternmost United States and northwesternmost Mexico (Plate 1), which is divided into several geographic regions for reference below. The regions are identified in Table 1 of Appendix B,

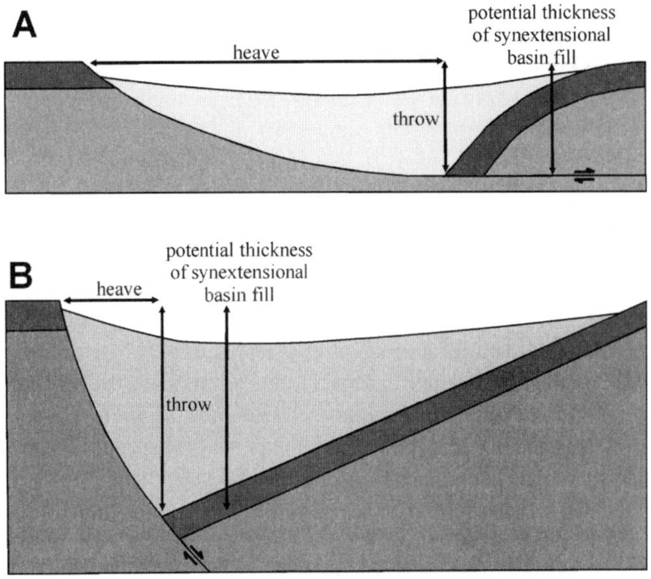

Figure 4. (A) Cross section of a basin formed by a low-angle normal fault. The hanging wall is shown coherent, but may break apart during slip. Due to the listric nature of the fault, the basin's true heave and throw may only be estimated (after Miller and John, 1999). (B) Cross section of a basin formed by a high-angle normal fault. The basin's depth, dependent on fault throw, is greater than the fault's heave (after Miller and John, 1999).

with reference to Plate 3. Many of the geographic and structural boundaries of the regions restored in this study were delineated in Luyendyk et al. (1980).

Our reconstruction is intended to restore brittle deformation in the crust of the study area to ca. 12 Ma (Plate 4). Appendix A includes references used in the paleotectonic reconstruction presented in this report (Plates 1–4). We incorporate Powell's (1993) ~265 km southeastward translation of ranges west of the San Andreas fault at 10–13 Ma in our reconstruction (Plate 4).

Throughout the study area, restoration is primarily by the closure of north-trending basins. Quantitative control is provided by published northwest-striking, right-lateral fault offsets in California and a combination of published and inferred faults and/or fault zones in Arizona and Sonora. Each block of crust (upper lithosphere) in the reconstruction is assigned a reference number that relates to the order in which blocks are restored. Groups of crustal blocks of uniform strain are designated as strain panels. Table 1 of Appendix B identifies each panel, its constituent blocks, and the total amount of strain within each panel that is restored.

The Sonoran Desert Region

The reconstruction of this region, which lies south of the proposed Gila River line, is the most speculative in the model (Plate 3). Transtensional structures similar to those in California are postulated to exist in this region, though little field data can be currently presented to verify exact locations. Information for the restoration of the Sonoran Desert region is presented in Table 2 of Appendix B. The values in Table 2 are based on basin closure and the inferred resultant strike-slip offset. Errors in the magnitude of restored displacements are difficult to determine without additional field study; however, most of the restored displacements are considered conservative. Offsets along postulated northwest-striking faults in the Lechuguilla and Tule Deserts (Plate 1) may be as much as 21 km and 17.5–22.5 km, respectively (Nourse et al., this volume, their Fig. 2). However, some of these structures may record early or middle Miocene displacements.

The ranges of southwesternmost Arizona (e.g., Gila, Tinajas Altas, Copper, Cabeza Prieta, and Mohawk Mountains; Plate 1) are separated by wide, deep basins (e.g., Mattick et al., 1973; Oppenheimer and Sumner, 1980) that suggest significant early Miocene and subsequent extension. The Yuma Hills are restored to align exposures and the proposed extent of the Bouse Formation with those along the western Gila Mountains (Plate 2), as suggested by Mattick et al. (1973). The Hickiwan Valley (Plate 1) may have formed during the late Miocene, but is not restored due to insufficient data.

The time in which most of the basins in the Sonoran Desert region were actively expanding is believed to be very early in the history of the shear system, except those that are still active in the extreme western portion of the region. Most of the proposed faults in the region do not clearly have a surficial outcrop, with the exception of scarps associated with the Algodones fault identified on the Arizona State Geologic Map (Reynolds, 1988). In addition, portions of the proposed faults and basins are covered by the late Tertiary–Quaternary Sentinel and Pinacate volcanic fields, which may have initiated or continued activity in part due to the shear system.

The Castle Dome–Dome Rock Region

Restoration of the Castle Dome–Dome Rock region requires closure of the basin under the La Posa Plain (Appendix B, Table 3). However, northwest-striking faults that crop out within the Dome Rock Mountains were not active as right-lateral structures during the late Cenozoic (Balderman and La Violette, 1977; Marshak, 1980; Sherrod and Tosdal, 1991; Richard, 1993). Therefore, left-lateral strike-slip faults conjugate to the right-lateral strike-slip faults (Fig. 1, dashed) are proposed to accommodate the closure of the north-trending basin under the La Posa Plain (Fig. 5A). The conjugate fault set implies that pure shear coexists with simple shear in deformation of the surface rocks, such as in Nevada (Beard, 1996; Deubendorfer et al., 1998). Only 5 km of displacement has been restored to close the King Valley and La Posa Plain to keep volcanics in the northern Castle Dome and western Kofa Mountains in close proximity.

An alternative fault configuration for the Castle Dome–Dome Rock region that does not include northeast-striking faults can also be made (Fig. 5B). The alternative configuration restores the Dome Rock Mountains and the Moon Mountains to the southeast like all the other blocks and panels in the region.

Eastern Mojave Desert Region

In the reconstruction, the eastern Mojave Desert region is divided into three strain panels (Appendix B, Table 3; Plate 3). In the cyan panel, the timing and magnitude of offset along the hypothesized All-American fault in the southernmost part of the panel are unknown, therefore, the Cargo Muchacho Mountains are not separated from the southernmost California Chocolate Mountains in the reconstruction. A significant difference in our model from previous restorations is the large magnitude of late Miocene and later-age closure of the McCoy Wash Valley and northern Palo Verde Valley of California (Appendix B, Table 3; block 28).

The magenta panel records a total of 7 km of displacement along northwest-trending strike-slip faults (Appendix B, Table 1). The portion of the panel northwest of the Palo Verde Valley is restored by the closure of north-trending basins and the portion southwest of the valley is restored by strike-slip offset on the Vinagre Wash fault (Plate 3). The southern Palo Verde Valley and Cibola Valley are closed more than previous models and the restoration is partitioned with the western Mojave Desert and eastern Transverse Ranges region (Appendix B; Table 1). The total basin width closed between the Mule and Palo Verde Mountains in California and the Trigo Mountains and Trigo Peaks of Arizona totals ~20 km (Plate 3).

The closure of north-trending basins in the northwestern portion of the magenta panel brings outcrops of Paleozoic rock

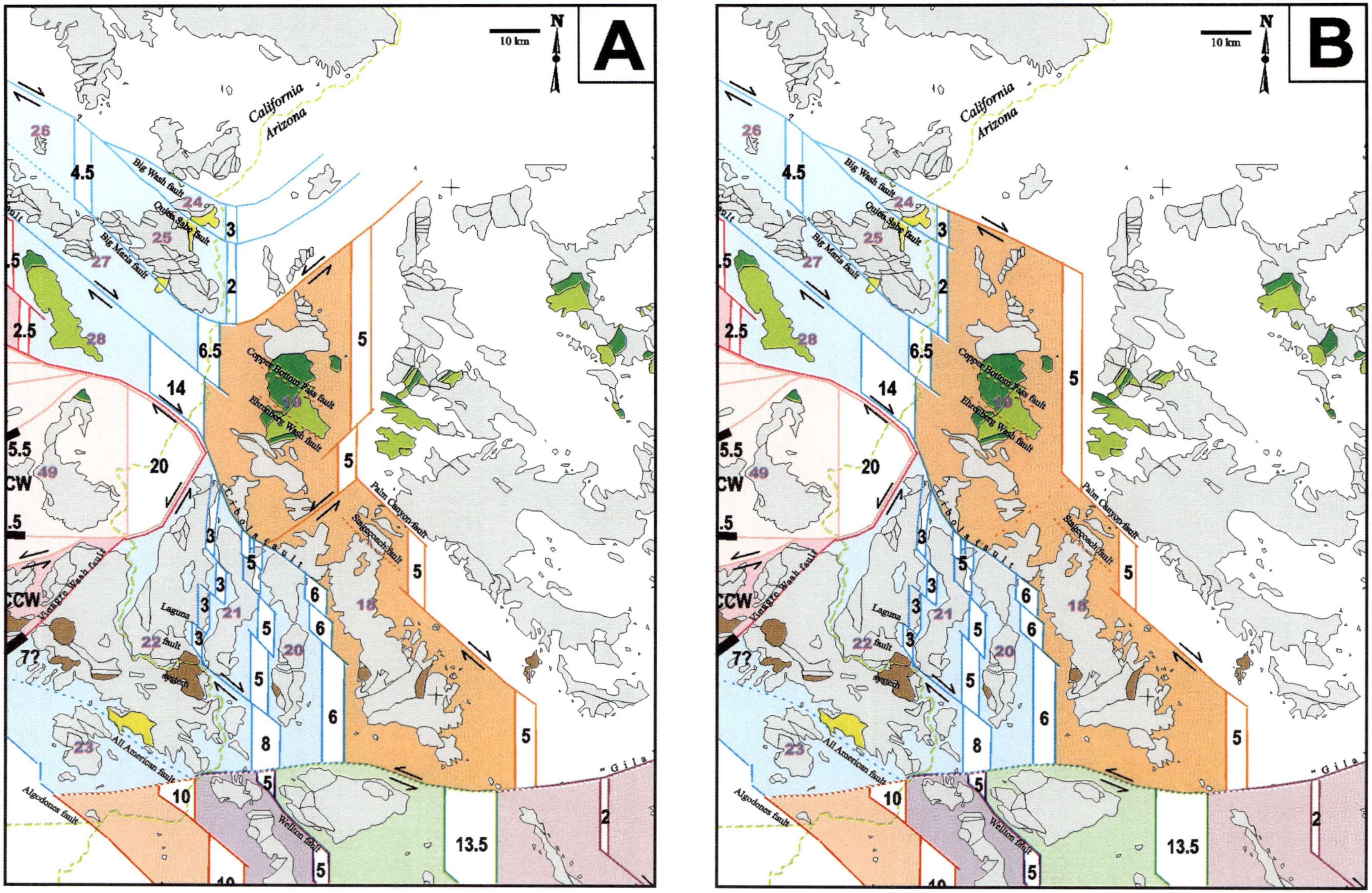

Figure 5. Models for the reconstruction of the Castle Dome–Dome Rock region, which exemplifies a special region. White areas represent Quaternary alluvium; white areas specifically within strain "panels" (outlined purple, green, blue, red, orange, cyan, and pink) represent basins with surficial Quaternary alluvium that are restored on Plate 4. Black numbers represent the restored strike-slip offset in kilometers required to close the surrounding basin in the reconstruction. Blue numbers represent restoration block identifiers and faults are color-coded to correspond with strain "panels" (Appendix B; Plate 3). Gray areas are undifferentiated range outcrop; yellow areas are Pliocene-Pleistocene sediment; brown areas are outcrops of the Orocopia Schist; light- and dark-green areas are sediments associated with the McCoy Mountains Formation. Map base shows present-day geography and geology. (A) Original model for restoration of the Castle Dome–Dome Rock region (highlighted orange) used in the reconstruction (Plates 3 and 4). The illustrated fault configuration includes northeast-striking, left-lateral faults that are conjugate to the northwest-striking, right-lateral faults dominating the region and incorporates some pure shear into the proposed model (Plate 3). (B) Alternate model for restoration of the Castle Dome–Dome Rock region (highlighted orange). The illustrated fault configuration would restore the Dome Rock Mountains (including the Moon Mountains to the north) to the southeast, like all of the other blocks and panels.

in the Palen and Little Maria Mountains closer together than previous reconstructions (Powell, 1993; Richard, 1993), as well as outcrops of the McCoy Mountains Formation (Plate 4). It also helps accommodate east-west shortening in the region when 23° of clockwise rotation of the eastern Transverse Ranges is restored in the western Mojave Desert and eastern Transverse Ranges region. Even after reconstruction, the width of the restored basin(s) under the McCoy Wash, Palo Verde, and Cibola Valleys remains large and is postulated to completely close only after the restoration of early Miocene extension (Plate 4).

Restoration of blocks 34 and 35 (diagonal magenta pattern in Plate 3) follows that of block 33, but the 3 km offset along the Ludlow fault (Dokka et al., 1998) is restored separately from blocks 31–33 and 52 (solid magenta in Plate 3). The displacement of the Ludlow fault is considered to translate northwestward into the Cady or Manix faults and/or dip-slip deformation within the Northeast Mojave block. The minimal restoration of these blocks accommodates the rotation and slip among the northernmost blocks of the eastern Mojave region (and blocks of the Northeast Mojave block) so that generally east-trending faults, such as the Manix and Cady faults (Plate 4), rotate to a more northeasterly trend (Luyendyk, 1991) in the reconstruction.

Northeast Mojave Block

The Northeast Mojave block (Luyendyk, 1991), shown as blocks 30A–30M, is translated with block 30 and internally restored (diagonal cyan pattern in Plate 3; Table 4, Appendix B). Restoration of blocks 30B and 31–35 accommodate the rotation of the Northeast Mojave block, which is restored using Schermer et al. (1996) and Dokka et al. (1998) as guides; however, some faults are extended or postulated to especially separate rotated from nonrotated blocks and accommodate rotation (dashed medium blue faults; Plate 3). In the restoration, several blocks in the Northeast Mojave block include the restoration of 23° of clockwise rotation, close to the values indicated by paleomagnetic studies (Schermer et al., 1996).

Western Mojave Desert and Eastern Transverse Ranges Region

Within the Mojave Desert, reconstruction consists of the restoration of northwest-striking, right-lateral strike-slip offsets reported by Powell (1993) and Dokka et al. (1998) (Appendix B, Table 5). Restoration of crustal blocks between the Pinto Mountain and Garlock faults in the domain result in minor overlaps because of curves in the bounding northwest-striking strike-slip faults. The irregularities of the faults and overlaps in the reconstruction may be the result of east-west shortening during the late Miocene, as suggested by Dickinson (1996). The trace of the Bristol Mountains fault (Plate 4) coincides with the shear zone inferred by Brady (1992).

In order to restore the outcrops of early Paleozoic strata southeast and northwest of Harper Lake (Walker et al., 2002), an additional 7 km of right-lateral offset (Appendix B, Table 6) in the westernmost Mojave is added to that proposed in previous reconstructions (Powell, 1993; Dokka et al., 1998). The additional 7 km of offset is distributed among mapped fault segments. In the westernmost Mojave and along the Garlock fault, contractional features crop out as discussed by Bartley et al. (1990) and where the traces of right-lateral faults bend left near the left-lateral fault.

The Rand and Lava Mountains, Red Mountain, and Almond Mountain located just south of the Garlock fault are not offset along northwest faults in the western Mojave Desert (Appendix B, Table 6). Right-lateral displacement may be accommodated by uplift near the Garlock fault.

Restoration of the eastern Transverse Ranges in the reconstruction follows that of Powell (1993), except in the southern California Chocolate Mountains south of the Mammoth Wash fault (Appendix B, Table 7). Paleomagnetic data for the eastern Transverse Ranges suggests a clockwise rotation of ~41° during the late Miocene (Carter et al., 1987; Luyendyk, 1991); however, this magnitude of rotation is not supported by structural offsets observed in the field along bounding left-lateral strike-slip faults (Powell, 1993; Dickinson, 1996). Therefore, we use geologic offset values and 23° of clockwise rotation suggested by Powell (1993) for the reconstruction. This restoration yields modest overlap along the westerly trending faults of the eastern Transverse Ranges as discussed below.

The more conservative 23° rotations of the eastern Transverse Ranges of California from Powell (1993) are utilized in the proposed reconstruction because we are influenced by the rotation of the Independent dike swarm and early Miocene(?) faults. Twenty-three degrees of rotation is also applied to the Northeast Mojave block, which closely agrees with rotations of 25° presented by Schermer et al. (1996).

The left-lateral Vinagre Wash fault is used as the southern boundary of the eastern Transverse Ranges restoration. The fault is complementary to the northwest-striking faults and is considered to be active only early in the late Miocene.

DISCUSSION OF THE MODEL AND ITS IMPLICATIONS

Overlap

Overlap of restored blocks in the eastern Transverse Ranges and California Chocolate Mountains (among blocks 43–52 in Plate 3) is difficult to avoid in the reconstruction due to incomplete knowledge of all fault offset constraints and geometries, as well as block rotations and deformation along unrecognized faults (Powell, 1993; Richard, 1993). An attempt has been made to minimize overlap in four specific areas of this region in the proposed reconstruction (Plate 4).

1. The southernmost part of Copper Mountain (20 km east of the San Bernardino Mountains) overlaps with the Pinto Mountains across the Pinto Mountain fault. This overlap suggests a more complex fault system east of the San Bernardino Mountains

than the one shown in Plate 3. Copper Mountain is likely a small splay separated from the San Bernardino Mountains by late (or early?) Miocene faults.

2. Using nonuniform block rotations in the eastern Transverse Ranges of California may improve the late Miocene reconstruction. For example, if blocks including the Pinto, Hexie, Eagle, and Cottonwood Mountains are rotated less than 23° (Powell, 1993), then overlap between the Salton Creek and Vinagre Wash faults in the southern California Chocolate Mountains is reduced. If the San Bernardino Mountains are not rotated or rotated slightly, then overlap of the southern boundary of the range across the Pinto Mountain fault with the Pinto Mountains is reduced or eliminated.

Scotti et al. (1991) argued that the disagreement between geologic field observations and the paleomagnetic data may be resolved by oblique slip along the faults surrounding the blocks of the western Transverse Ranges. Powell (1993) pointed out that vertical fault offsets may accommodate overlap in the reconstruction, as suggested in his reconstruction of the eastern Transverse Ranges. In light of oblique-slip instead of strictly strike-slip motion along faults within the eastern Transverse Ranges and California Chocolate Mountains, the removal of all overlap in the model was not considered essential to the proposed reconstruction (Plate 4).

3. As noted above, 7 km of additional dextral restoration in the western Mojave Desert also relieves overlap in the southern California Chocolate Mountains in the vicinity of the Mammoth Wash and Black Eagle faults because contraction is accommodated further north (Appendix B, Table 6).

4. Less closure of the Palo Verde Valley of California would eliminate some overlap in the California Chocolate Mountains around the Black Eagle and Mammoth Wash faults. However, the width of the Palo Verde Valley in the reconstruction is considered a minimum and additional closure of the underlying basin may be warranted.

Overlap in the Trigo, Arizona Chocolate, and Middle Mountains results from restoration of 14 km of displacement along the Cibola fault. This is double the displacement suggested by Richard (1993). The additional offset, in part added from the Laguna fault system (Richard, 1993), is used to transfer the strain restored in the northern Palo Verde Valley into Arizona. The restoration of the Castle Dome Mountains along the proposed Palm Canyon fault (Plate 3), in combination with the Cibola fault and Laguna fault system, reduces the amount of overlap in the late Miocene reconstruction in the southern California Chocolate Mountains, despite the absence of indications of offset greater than 7 km along the Cibola fault (S. Richard, 2003, personal commun.).

Upon restoration, the northern Sierra Cubabi, near the town of Sonoita, Mexico, overlap slightly along north-striking faults in Sonora. The overlap is minor and does not greatly affect the reconstruction because a series of imbricate north-striking normal faults in the area may account for the overlap.

Apparent overlap is locally evident at the eastern boundary of the restoration. The overlap occurs between restored and unrestored domains, especially at the margin of the Northeast Mojave block. Displacement within the Furnace Creek fault zone may contribute to this overlap, however, treatment of that structure is outside the scope of this paper.

Rotation Constraints

The orientation of the Late Jurassic Independence dike swarm may be used to constrain block rotations in the Mojave and eastern Transverse Ranges (e.g., Ron and Nur, 1996). If Independence dikes in rocks within the study area are aligned with those of the Sierra Nevada and form a continuous zone southeastward to the Pinto Mountains with a similar strike (James, 1989; Ron and Nur, 1996), then (1) the Pinto Mountains record 27° of clockwise rotation, close to the 23° used by Powell (1993), (2) dikes in the Granite Mountains record ~42° clockwise rotation, (3) dikes in the Soda Mountains and in Cowhole Mountain record ~23° clockwise rotation, and (4) dikes in the region between Stoddard and Hidalgo Mountain and in the Bristol Mountains record no rotation.

Rotation of the San Bernardino Mountains from paleomagnetic data is inconclusive (Powell, 1993). The late Miocene rotation of the San Bernardino Mountains is not required in the reconstruction herein.

Gila River Line

The Gila River line (Plate 3) emerges as a prominent structure in Arizona as revealed by the reconstruction. Discontinuity between the proposed fault systems on either side of the line is suggested by the pattern of structures and proposed offsets along them. It is possible that the discontinuity was active early in the late Miocene. The trace of the Gila River line, which does not coincide with the Gila River or Gila River Valley, is not well constrained, but corresponds in part with the northern margin of uplifted early Miocene clastic beds near Yuma and Gila Bend (Spencer et al., 1995). The enigmatic line has been recognized by Sherrod and Tosdal (1991) as part of the southern boundary of the structurally extended terrane from the Kofa and Tank Mountains of Arizona west to the California Chocolate and Palo Verde Mountains. Early Miocene volcanic rocks, abundant in ranges north of the line (in the Trigo, Arizona Chocolate, Middle, and Castle Dome Mountains), are absent to the south (Sherrod and Tosdal, 1991). In our model, the Gila River line separates regions recording slightly different magnitudes of extensional strain, thereby serving as a transfer structure similar to the Garlock fault in California (Davis and Burchfiel, 1973).

Differences in the magnitudes of strain north and south of the Gila River line suggest that strain is partitioned across this boundary. Our reconstruction accommodates ~40 km of extension south of the line. In contrast, north of the line the reconstruction cumulatively accommodates ~20 km of strain that is constrained by offsets postulated along the Cibola and/or Palm Canyon faults. The Gila River line also coincides with an east-northeast–oriented, wedge-shaped gap in the reconstruction (Plate 4), ~6 km wide (maximum) and 160 km long that tapers in width to the east in southwestern Arizona. Two explanations are proposed for the gap: (1) The gap may be interpreted as evidence of late Miocene and younger north-south shortening (Fig. 2). Uplift of Tertiary sediments in the Laguna and Muggins Mountains and, perhaps, Signal Butte and Texas Hill (Plate 1), suggests late Miocene contraction along the south side of the Gila River line perpendicular to the maximum principal stress. (2) The gap may be the result of overestimation of strain restored south of the Gila River. For example, in the Yuma Desert and northwesternmost Sonora, geometries of the restored basins among the Yuma Hills, Monument Hills, El Capitan, and Sierra Rosario are poorly constrained.

Late Miocene and Younger Strike-Slip Faults in Southwestern Arizona

Early Miocene low-angle extensional structures in southwesternmost Arizona crop out in the Mohawk and Copper Mountains (Mueller et al., 1982; Pridmore and Craig, 1982). The low-angle structures that are assumed to have formed near the early Miocene brittle-ductile transition have been uplifted and are presumed to be crosscut by younger northwest-oriented faults analogous to right-lateral faults in southern California predicted by the north-south maximum principal stress east of the San Andreas transform (Fig. 1). Some of the northwest-trending faults in southwestern Arizona may have formed along fracture zones that were originally early Miocene normal faults.

Right-lateral faults in southwestern Arizona are postulated to open north-trending pull-apart basins that are schematically defined by aeromagnetic, gravity, and depth to basement studies (e.g., Oppenheimer and Sumner, 1980; Klein and Kucks, 1991; Plate 3). Further evidence for north-trending pull-apart basins in southwestern Arizona includes the north-trending outlines of ranges and segments of ranges, such as the eastern boundaries of the Copper and northern Gila Mountains, respectively.

Chocolate Mountains Anticlinorium

The Chocolate Mountains anticlinorium (Haxel et al., 1985; Fig. 6) is a structure defined by exhumed Orocopia Schist and outcrops of the Winterhaven and Slumgullion Formations (Sherrod and Tosdal, 1991; Jacobson, 1990). Before the restoration of late Miocene deformation, the trace of the anticlinorium trends easterly from Neversweat Ridge to Peter Kane Mountain, where it curves northwest (Plate 1). This curve may record simple shear close to the San Andreas transform that is unrestored in the proposed reconstruction (Plate 4). In an effort to remove at least part of the curvature, displacements on the Cibola fault and Laguna fault system (a right-lateral fault zone primarily through the Trigo Mountains of Arizona) of Richard (1993) are increased (Appendix B, Table 3).

The eastern segment of the Chocolate Mountains anticlinorium hinge in Arizona restores to an orientation near 060° (Plate 4), which is parallel to the early Miocene extension direction (Spencer and Reynolds, 1991). According to Sherrod and Tosdal (1991) the anticlinorium was not present at 26 Ma and

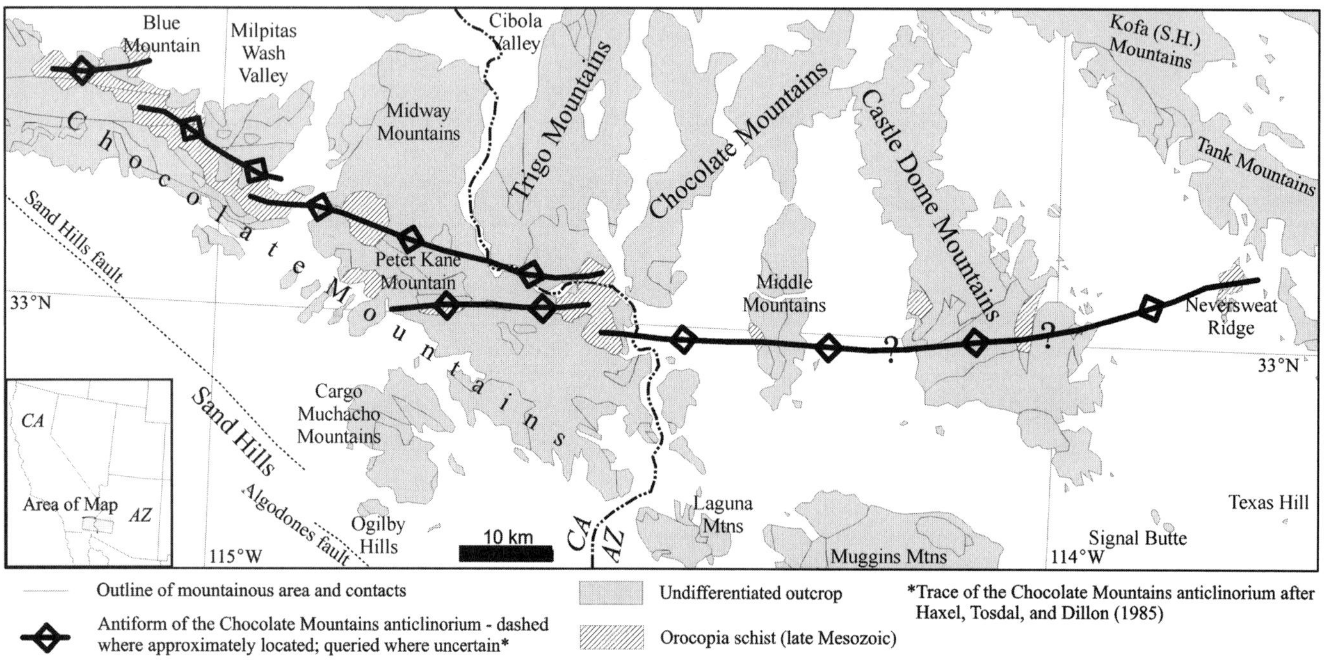

Figure 6. Map illustrating the present location of the Chocolate Mountains anticlinorium.

was likely exhumed ca. 22 Ma. The structure may, therefore, be analogous to the anticlines (corrugations) in the Whipple, Buckskin, Harcuvar, and Harquahala Mountains (e.g., Teel and Frost, 1982; Spencer and Reynolds, 1991) that formed parallel to the extension direction (Teel and Frost, 1982) and may be used as a strain marker for late Miocene deformation (Dembosky et al., 1998). The restored southwestern end of the anticlinorium in Arizona (Plate 4) lies 40–50 km south of its present location. This is in agreement with the cumulative strike-slip displacement between the San Andreas fault and Eastern California shear zone (45–75 km) as stated by Dickinson (1997).

The Chocolate Mountains anticlinorium in Arizona, which is assumed to be unbroken by any singular large offset, roughly follows the same arcuate trend as the outcrop of the McCoy Mountains Formation from the Coxcomb Mountains eastward to the Plomosa Mountains prior to reconstruction (Plate 1). The outcrop of the McCoy Mountains Formation is postulated to record strain in a manner similar to the hinge line of the Chocolate Mountains anticlinorium in Arizona.

Eastern California Shear Zone

The precise boundaries of the Eastern California shear zone are not well constrained (Dokka and Travis, 1990; Dokka et al., 1998). The strain restored along the Granite Mountains fault is transferred north of the Garlock fault beyond the boundary of the reconstruction. Strain along the Bristol Mountains fault may transfer into the Owens Valley (or Panamint Valley?) fault zone. The western Garlock fault has an offset of ~12 km (Powell, 1993), which is similar to that of the Blackwater-Pisgah-Bullion-Delta faults (14 km), which may be in the Eastern California shear zone. As a result, the strain within the Blackwater-Pisgah-Bullion-Delta faults is transferred to the western Garlock fault, and then to the San Andreas transform. Displacement in the Eastern California shear zone may transfer to the San Andreas via the rotation of the eastern Transverse Ranges.

Strain Markers

In addition to the hinge line of the Chocolate Mountains anticlinorium (Fig. 6), we recognize several other pre–late Miocene linear geologic features that may be useful as strain markers for restoring extension and translation. These include the Kane Springs fault (Plate 1), the Independence dike swarm (Plate 1), and general shapes of plutons or batholiths, and boundaries between Paleozoic and Proterozoic provinces in Arizona (Plate 2). The Kane Springs fault extends easterly across the northern Bullion, Rodman, and Ord Mountains near Barstow, California (Dokka, 1989). It is interpreted as an early Miocene right-lateral fault or accommodation zone in the central Mojave Desert. Restoration of displacements of northwest-striking, right-lateral strike-slip faults in the Mojave by Powell (1993) and Dokka et al. (1998) generally aligns proposed segments of the structure so that a continuous trace results (Plate 4). The importance of this fault during early Miocene extension is controversial. For example, Glazner et al. (2000) stated that the amount of early Miocene displacement on this fault may be small and that the proposed fault segments of the Kane Springs fault, especially a section called the Silver Bell fault, may never have been continuous. Even if it is not a fault with significant offset, the Kane Springs fault serves as a significant boundary between Tertiary sedimentary rocks and Mesozoic plutonic rocks that can serve as a strain marker (Dokka, 1983).

Unrestored pre–late Miocene deformation is evident in the reconstruction by the misalignment of the Independence dike swarm (Plate 4). Additional unrestored late Miocene or pre–late Miocene deformation is also present in the restoration since the hinge line of the Chocolate Mountains anticlinorium is not straightened, nor are the outcrops of the Orocopia Schist geographically grouped very close together.

The deformed shape of plutonic masses as a mappable unit may be used as imprecise strain markers. The Gunnery Range batholith in the Tinajas Altas and Cabeza Prieta Mountains and Sierra Pinta of southwestern Arizona restores to an ellipse, the long axis of which trends northeast (pink unit in Plates 1 and 4). If plutons such as the Gunnery Range batholith had a generally circular original shape, then this result suggests that additional restoration of early Miocene extension is required.

Offset Precambrian terrane and province boundaries (Plate 2) are potentially useful markers for reconstructing southwestern North America. Boundaries have recently been defined and are still being studied throughout the region (Silver, 1964, 1968; Wooden et al., 1988; DeWitt, 1989; Anderson, 1989a, 1989b; Bowring and Karlstrom, 1990; Wooden and Miller, 1990; Chamberlain and Bowring, 1990; Nourse et al., this volume; Anderson and Silver, this volume). The fault boundary that separates the composite Hexie Mountains–Eagle Mountains assemblages (also referred to as the San Gabriel–Joshua Tree terrane) and Caborca terrane from the autochthonous Mazatzal, Yavapai, and Mojave Provinces (Anderson and Silver, 1981; Dembosky et al., 1998, Dembosky, 2001) is postulated to be a steeply dipping paleotransform plate boundary called the Mojave-Sonora megashear (Anderson and Silver, this volume).

Although the trace of the Mojave-Sonora megashear has been obscured by subsequent tectonism (Dembosky et al., 1999), its position is constrained by outcrops of the parautochthonous provinces and terranes that it separates (Plate 2; Anderson and Silver, this volume; Nourse et al., this volume). North of the Pinto Mountain fault, west-directed thrusting (latest Cretaceous and earliest Tertiary) and detachment (early Miocene) have carried Mojave Province crust and cover rocks west (e.g., Silver, 1983; May, 1989; Silver and Nourse, 1991; Martin and Walker, 1992; Fletcher et al., 2002; Anderson et al., this volume). We postulate the surficial trace of the megashear would lie west of the San Andreas, whereas the root would extend through footwall rocks beneath the allochthonous cover of the Mojave Desert (Plates 2 and 4).

Magnitude of Extension and Right-Lateral Displacement

The roughly north-south orientation of the effective maximum principal stress within the region east of the San Andreas transform (Fig. 2) has resulted in primarily westward extension since the late Miocene. The amount of east-west elongation extension from the longitude of the Santa Rosa Valley of Arizona to the longitude of Fairmont Butte in the western Mojave Desert (southeast of the junction of the San Andreas and Garlock faults; Plate 1) is ~9% or 50 km (l_o = 537 km; l_f = 587 km; l is length when used to measure elongation; o is original length (Miocene); f is final length).

The total right-lateral displacements may be estimated from sums of the restored strains in Appendix B, Table 1. The total displacement in the Sonoran Desert region, which is the area south of the Gila River line, is 39.5 km. North of the Gila River line to the Garlock fault, the total restored strain is 43 km. Both regions have approximately the same amount of offset. However, Fairmont Butte in the northwestern Mojave records 77 km total right-lateral separation with respect to a point at the southeast corner of the study area on the Mexican border (Plate 4). This larger value is due to rotation of the eastern Transverse Ranges (i.e., displacements are accommodated along east-west and northeast faults in addition to northwest-trending faults).

FUTURE WORK

The southeasternmost part of California, including the southern Chocolate Mountains of California, the Palo Verde and Little Mule Mountains, and the Black Hills, is an area that has proved difficult to restore in this model, as well as in others (e.g., Powell, 1993; Richard, 1993). Further study of this area may resolve the problem of overlap in restorations in this region. Information needed to achieve better results includes determining if and how blocks in this region have rotated and by how much, the amount and timing of offset of the Vinagre Wash, Coon Hollow, and, perhaps, undiscovered faults. The roles of the Black Eagle and Mammoth Wash faults (Dillon, 1975), as well as the Clemens Well fault (Powell, 1993), remain in question in this region.

Structural analyses in southwestern Arizona and northwesternmost Mexico are needed to verify or refute the boundaries of basins and fault offsets proposed in this reconstruction. Additional geologic mapping in western Sonora near the eastern margin of the Gulf of California would provide a particularly useful test of the ideas presented above.

CONCLUSIONS

The reconstruction of late Miocene to recent transtension: (1) reveals the partial effect of extensional disruption of the surficial trace of the Mojave-Sonora megashear, (2) partially resolves some of the late Miocene "slip discrepancy" of dextral offsets between the Pacific and North American plates (~635 km), and that recorded in the Gulf of California (~300 km), and (3) assesses the relationship between transtensional strain in the desert region of the southwest and faults comprising the San Andreas transform plate boundary.

Late Miocene to Recent extension east of the San Andreas is explained by a systematic pattern of north-trending basins at steps in northwest-striking, right-lateral strike-slip faults in a transtensional stress regime. The releasing bend steps that form pull-apart basins are employed to create a retrodeformable model of extension in southwesternmost North America that is consistent with published constraints on Pacific–North American relative plate motion. The model for late Tertiary extension reveals that: (1) 9% east-west extensional strain (50 km) occurred during late Miocene and later deformation, accompanied by 43 km of right-lateral displacement parallel with the San Andreas between the Gila River line and Garlock fault, (2) total right-lateral separation of 77 km results from the incorporation of clockwise block rotations in the model. Early Miocene detachment structures in southwesternmost Arizona crop out in uplifted ranges and are crosscut and displaced by the late Tertiary younger structures. The Gila River line is a discontinuity across which panels of consistent late Miocene extensional strain do not extend, however, extension north and south of this discontinuity is similar.

ACKNOWLEDGMENTS

Staff at Organ Pipe Cactus National Monument and Cabeza Prieta National Wildlife Refuge provided assistance and access to the monument and refuge. The United States Marine Corps, Yuma, granted permission to visit the Barry M. Goldwater Range. Field assistance and/or mapping information provided by Gordon Haxel, Jonathan A. Nourse, Erin Stahl, and Leon T. Silver is greatly appreciated. Stephen Richard and Jonathan A. Nourse generously provided useful advice and constructive criticism. John M. Fletcher, Carl Jacobson, and Nadine McQuarrie performed constructive reviews on a previous version. Funding for this project has in part been supplied by the University of Pittsburgh Provost Central Research Development Fund. Thanks also to John A. Dembosky Sr. for support in the field and tolerating the desert.

APPENDIX A. REFERENCES FOR THE PLATES OF SOUTHWESTERN NORTH AMERICA

The following references for the digital map of southwestern North America composed to illustrate the paleotectonic reconstruction have been categorized below. The map datum for the plates is NAD27 with a Lambert conformal conic projection (standard parallels of 33°N [and 45°N]).

Geology and Outcrop Patterns

- Anonymous (1981) [Mexican Los Vidrios quadrangle]
- Anonymous (1982) [Mexican Puerto Peñasco quadrangle]
- Anonymous (1983) [Mexican Mexicali quadrangle]
- Beikman, Haxel, and Miller (1995)

- Gray, Miller, Grubensky, Tosdal, Haxel, Peterson, May, and Silver (1987)
- Gray, Miller, Peterson, May, Tosdal, and Kahle (1985)
- Grubensky, Haxel, and Demsey (1995)
- Grubensky, Haxel, and Kock (1993)
- Jennings (1977)
- Olmstead (1972)
- Oppenheimer and Sumner (1980)
- Powell (1993)
- Reynolds (1988)
- Richard (1992a, 1992b, 1992c)
- Richard (1993)
- Unpublished mapping and reconnaissance by John A. Dembosky Jr., Thomas H. Anderson, Jonathan A. Nourse, Erin Stahl, and Leon T. Silver

Faults

- Beikman, Haxel, and Miller (1995)
- Dokka, Ross, and Lu (1998)
- Dokka (1989)
- Gray, Miller, Peterson, May, Tosdal, and Kahle (1985)
- Grubensky, Haxel, and Demsey (1995)
- Grubensky, Haxel, and Kock (1993)
- Jennings (1977)
- Kruger, Faulds, Reynolds, and Okaya (1998)
- Mattick, Olmstead, and Zohdy (1973)
- Olmstead (1972)
- Powell (1993)
- Reynolds (1988)
- Richard (1993)
- Richard (1992a, 1992b, 1992c)
- Sherrod and Tosdal (1991)
- Spencer and Reynolds (1991)
- Stewart (1998)
- Walker, Black, Berry, Davis, Andrew, and Mitsdarfer (2002)
- Wells and Hillhouse (1989)

Depth to Basement

- Oppenheimer and Sumner (1980)

Terrane Boundaries

- Anderson and Silver (1981)
- Bowring and Karlstrom (1990)
- Chamberlain and Bowring (1990)
- Gray, Miller, Grubensky, Tosdal, Haxel, Peterson, May, and Silver (1987)
- Silver (1964)
- Silver (1968)
- Wooden, Miller, and Howard (1988)
- Wooden and Miller (1990)
- Unpublished mapping and reconnaissance by John A. Dembosky Jr., Thomas H. Anderson, Jonathan A. Nourse, Erin Stahl, and Leon T. Silver

Outcrop Pattern of the Independence Dike Swarm

- Glazner, Walker, Bartley, Fletcher, Martin, Schermer, Boettcher, Miller, Fillmore, and Linn (1994)
- Ron and Nur (1996)

APPENDIX B. RECONSTRUCTION DATA TABLES

TABLE 1. STRAIN "PANELS" OF RESTORATION OUTLINED ON PLATE 3

Geographic region	Strain "panel" (color in Plate 3)	Constituent block numbers	Total amount of restored strain within "panel" (km)
Sonoran Desert region	Gray	1	2
	Yellow	2	2
	Purple	3–12	7
	Green	13–15	13.5
	Blue	16	5
	Red	17	10
Castle Dome–Dome Rock region	Orange	18–19	5
Eastern Mojave Desert region	Cyan	20–30	14
	Magenta	31–33, 52	7
	Magenta, diagonal line pattern	34 and 35	3
Northeast Mojave Desert region	Cyan, diagonal line pattern*	30A	0
Western Mojave Desert and eastern Transverse Ranges region	Pink	36–51	~14

Note: All strain is restored along strike-slip faults. Data concerning constituent block numbers are found in Tables 2–6.
*Diagonal line pattern is used to clarify block outline.

TABLE 2. RESTORATION OF BLOCKS IN THE SONORA DESERT REGION SOUTH OF THE GILA RIVER LINE

Block number	Amount restored (km)	Comments
1	2	Unnamed fault(s); constrained by basin width between Vaiva Hills and Tai Momoli Mountains; closes part of Rainbow Valley and Santa Rosa Valley; the great depth of the Santa Rosa Valley (Oppenheimer and Sumner, 1980) suggests additional unrestored extension; eastern boundary of reconstruction
2	2	Unnamed fault(s); constrained by basin width between the Sand Tank Mountains and Copperosity Hills; closes unnamed valley between the Gila Bend and Maricopa Mountains, southern Vekol Valley, and part of Quijotoa Valley
3	1.5	Kuakatch fault*; constrained by the width of the Barajita Valley basin; closes Barajita Valley and parts of La Quituni Valley and The Great Plain
4	3	Crater Range fault*; constrained primarily by the width of northernmost valley of the Ajo basin; closes the Valley of the Ajo and Sonoyta Valley; displacement partitioned with block 3
5	4	Cipriano fault*[1]; constrained by cumulative extension in the hills southeast of Sonoyta, Sonora; partially closes the La Abra Plain
6	1	Unnamed fault(s); constrained by basin width in the Cipriano Hills
7	1	Unnamed fault(s); constrained by postulated strike-slip offset in the Agua Dulce Mountains; partially closes the southernmost Growler Valley
8	2	Extension of Saguaro Gap fault* [no outcrop in block]; constrained by offset in block 7; partially closes the southern Growler Valley
9	1	Saguaro Gap fault*; constrained by postulated strike-slip offset in the southern Granite Mountains; partially closes the southern Growler Valley
10	2	Unnamed fault(s); constrained by widths of the northern Growler Valley and southern Cristobal Valley basins closed in the model; potentially underestimated
11	2	Unnamed fault(s); constrained by the width of the basin between the Aguila Mountains and the Crater Range; postulated basin north of the Crater Range Fault is presently partially obscured by the Sentinel Plain volcanic field
12	2	Unnamed fault(s); constrained by the width of the basin between the Bryan Mountains and Antelope Hills; partially closes the Cristobal Valley in the model
13	1	Northern extension of Cipriano fault*[1]; constrained by basin width west of the Antelope Hills; partially closes the unnamed valley between the Agua Dulce and Bryan Mountains in the model
14	3.5	Unnamed fault(s); constrained by the width of the unnamed basin immediately east of the southern Sierra Pinta that is partially closed in the model
15	10	Unnamed fault(s) and Outlier fault*; constrained by width of the northern Mohawk Valley and Pinacate volcanic field basins that are partially closed in the model
16	5	Wellton*, Puente Cuates*, and other unnamed faults; constrained by basin width between the Laguna and Muggins Mountains; partially closes the Lechuguilla Desert basin in the model
17	10	Unnamed fault(s); constrained by Bouse Formation with respect to blocks 17 and 23; possible extension of Algodones fault; restoration of El Capitan unsure; partially closes the Yuma Desert basin southward into Sonora in the model

Note: The magnitude of restoration along northwest-trending, right-lateral strike-slip faults is measured relative to the adjacent block to the southeast. "Block number" is the number assigned to the fault-bounded area of crust shown in Plate 3. Some valleys are referred to as partially closed by the indicated restoration because they were initiated during the early Miocene and would only be completely restored in a reconstruction of core complex-age faulting. References for faults are shown as footnotes and contain additional information about the indicated faults.

*Newly postulated faults with informal names

[1]Extension of fault trace introduced in Haxel et al. (1984) and Tosdal et al. (1990).

TABLE 3. RECONSTRUCTION OF BLOCKS IN THE EASTERN MOJAVE DESERT REGION IN ARIZONA, NORTH OF THE GILA RIVER AND CALIFORNIA NORTH OF THE SHEEP HOLE FAULT

Block number	Amount restored (km)	Comments
18	5	Palm Canyon fault* [& Stagecoach fault*]; constrained by the offset across the fault (unknown) and the width of the King Valley basin, which is partially closed in the model
19	5	Unnamed fault(s); constrained by the width of the La Posa Plain basin that is partially closed in the model; restores northeast
20	6	Cibola fault[1]; constrained by the offset on the Cibola fault (Richard, 1993) and width of Castle Dome Plain basin
21	5	Cibola fault[1]; constrained by the offset on the Cibola fault (Richard, 1993) and width of the northern Indian Wash Valley basin
22	8	Cibola fault[1]; constrained by the width of the southern Indian Wash Valley basin and incorporates the Laguna fault system[1] (8 km restored at southern end) [overestimated?]
23	0	All-American fault*; constraints are indefinite, but partly controlled by location of the Bouse Formation; the Cargo Muchacho Mountains are not separated from the southern California Chocolate Mountains in the model
24	3	Big Wash fault[1]; partially closes Parker Valley
25	2	Quien Sabe fault[1]; accommodates offset of 1 km between the Riverside Mountains and northern Big Maria Mountains; partially closes the southern Parker Valley in the model
26	4.5	Extension of Big Wash fault[1]; partially closes the valley between the Arica and Big Maria Mountains
27	6.5	Big Maria fault[1]; results in 4.5 km offset between the Big Maria and Little Maria Mountains, offset strictly with reference to block 25
28	14	Calumet–Packard Well fault[1]; partially closes McCoy Wash Valley and northern Palo Verde Valley; offset strictly with reference to block 27
29	2	Extension of Big Wash fault?; constrained by width of basin between the Calumet and Granite Mountains; partially closes the Cadiz Valley
30/30A-30M	4.5	Granite Mountains fault[2]; includes translation of the Northeast Mojave block (30C-30M)
31	2.5	Calumet-Packard Well[1] fault; partially or completely closes the Ford Lake Wash Valley
32	4.5	Calumet-Packard Well[1] fault; partially closes the Palen Valley
33/33A	7	Bristol Mountains fault[2]; constrained by the >6 km offset (7 km) reported by Dokka et al. (1998); partially closes the valley between the Bullion and Calumet Mountains; includes the 23° counterclockwise rotation of block 33A accommodated by the Manix[2] (12 km offset) and Cady[2] (1 km offset) faults
34/34A	<1	Broadwell Lake fault[2]; constrained by the observed <1 km offset along the Broadwell Lake fault (Dokka et al., 1998); includes the 23° counterclockwise rotation of block 34A accommodated by the Cady[2] (1 km offset) and Sleeping Beauty[2] (12 km offset) faults
35	3	Ludlow[2]–Cleghorn Lakes–Sheep Hole[1,3] fault; recorded displacement is along the Ludlow fault portion of the fault strand (Dokka et al., 1998); partitioned with a southern extension of the Broadwell Lake fault (dashed fault on Plate 3) to sustain the <1 km offset on the northern Broadwell Lake fault and open the valley south of the Lava Hills

Note: The magnitude of restoration along right-lateral strike-slip faults and/or fault zones is measured relative to the adjacent block to the southeast. "Block number" is the number assigned to the fault-bounded area of crust shown in Plate 3. Some valleys are referred to as partially closed by the indicated restoration because they were initiated during the early Miocene and would only be completely restored in a reconstruction of core complex-age faulting. References for faults are shown as footnotes and include much more information about individual faults, including restraints.

*Newly postulated faults, informal names.

[1]Richard (1993).

[2]Dokka et al. (1998).

[3]Powell (1993).

TABLE 4. RESTORATIONS OF LEFT-LATERAL STRIKE-SLIP AND ROTATIONS OF BLOCKS WITHIN THE NORTHEAST MOJAVE BLOCK

Block number	Amount restored (km) and rotation (degrees counter-clockwise)	Comments
30A	0; 0°	Granite Mountains fault[1]; translated directly with block 30
30B	12; 23°	Manix fault[1]; offset with respect to blocks 30K, 30L, and 30M; requires ~4 km offset along a proposed fault bordering blocks 30, 30A, and 33, and ~5 km offset along a fault between the Cady Mountains and Cave Mountain that may be a northern extension of the Ludlow fault
30C	3; 0°	Goldstone Lake fault[1,2]; offset with respect to block 30I
30D	4; 23°	Desert Spring King fault[2]; offset with respect to block 30E; block is assumed to be rotated
30E	0; 23°	Unnamed accommodating fault between the Granite and Avawatz Mountains
30F	1.5; 23°	McLean–Alpine Valley–Drinkwater faults[2]; offset with respect to block 30D (and 30E)
30G	4; 23°	Nelson Lake fault and Fort Irwin fault zone[1,2]; offset with respect to block 30F
30H	3.5; 23°	Coyote Canyon fault and Tiefort Mountain faults[1,2]; offset with respect to block 30G
30I	5; 23°	Bicycle Lake fault[1,2]; offset with respect to block 30H
30J	0; 23°	Pin of rotation is along northernmost trace of the Garlock fault
30K	10; 23°	Red Pass fault[1] on western side of block; offset with respect to blocks 30H, 30I, and 30M; estimated as a result of restoration of blocks 30D–30I
30L	0; 0°	No net slip along western extension of the Coyote Lake fault[1,2]; accommodating slip on south side of block results in an offset of 12 km along the Manix fault[1] as blocks 30B, 33A, and 34A rotate
30M	11; 23°	Coyote Lake fault[1,2]; offset with respect to blocks 30C and 30I; opening of the valley between the Calico Mountains and Alvord Mountain results, but may be partitioned with or instead be the opening of the valley between Alvord Mountain and the Cronese Mountains, as suggested by the intervening Pliocene/Pleistocene sediments

Note: "Block number" is the number assigned to the fault-bounded area of crust shown in Plate 3. Amount restored is concurrently restored during the 4.5 km translation with block 30. References for faults are shown as footnotes and include much more information about individual faults, including restraints.

[1]Dokka et al. (1998).

[2]Schermer et al. (1996).

TABLE 5. RESTORATION OF THE WESTERN MOJAVE DESERT OF CALIFORNIA BETWEEN THE PINTO MOUNTAIN AND GARLOCK FAULTS

Block number	Amount restored (km)	Comments
36/36A	14	Blackwater-Pisgah-Bullion-Delta(-Valley Mountain?) fault[1,2]
37	9	(Blackwater–)Calico–West Calico–Mesquite Lake fault[1,2]
38	4	(Harper–Harper Lake–)Camp Rock–Emerson fault[1,2]
39	4	Southern extension of Emerson fault[1] (and Copper Mountain fault?); restored offset along this compensating feature results in the overlap of Copper Mountain and the Pinto Mountains and complements the restoration of block 38
40	3	Lockhart-Lenwood fault[1,2]
41	3	Spring-Helendale fault[1,2]
42	3	Mirage Valley fault[1]

Note: The magnitude of restoration along northwest-trending, right-lateral strike-slip faults is measured relative to the adjacent block to the northeast. "Block number" is the number assigned to the fault-bounded area of crust shown in Plate 3. References for faults are shown as footnotes and include much more information about individual faults, including restraints.

[1]Powell (1993).

[2]Dokka et al. (1998).

TABLE 6. RESTORATIONS MADE IN THE WESTERNMOST MOJAVE DESERT OF CALIFORNIA IN ADDITION TO THOSE OF POWELL (1993) AND DOKKA ET AL. (1998)

Block number	Amount restored (km)	Comments
38A	7	Additional restoration of Harper–Harper Lake fault
40A	7	Additional restoration of Lockhart-Lenwood fault
41A	7	Additional restoration of Spring-Helendale fault
42A	7	Additional restoration of Mirage Valley fault

Note: The magnitude of restoration along northwest-trending, right-lateral strike-slip faults is measured relative to the adjacent block to the southeast. "Block number" is the number assigned to the fault-bounded area of crust shown in Plate 3.

TABLE 7. RESTORATION OF THE EASTERN TRANSVERSE RANGES OF SOUTHEASTERN CALIFORNIA

Block number	Amount restored (km) and rotation (degrees counter-clockwise)	Comments
43	0; 8°	San Bernardino Mountains; restored with block 38 then rotated with the pin in the northeastern corner of the block[1]
44	16; 23°	Pinto Mountain fault[1,2,3]
45	5; 23°	Blue Cut fault[1,2,3]; internal deformation not considered
46	11; 23°	Chiriaco fault[1,2,3]
47	3; 23°	Corn Springs Wash fault[1,2]; offset relative to block 46
48	2; 23°	Ship Creek fault[1,2]; offset relative to block 46
49	5.5; 0°	Wiley-Indian fault*; 5.5 km slip is restored relative to block 50; in the model, block 49 is shifted 20 km due east, then 5 km due south relative to its original position; estimated 5.5 km basin in the Indian Well Valley is postulated
50	8; 23°	Salton Creek fault[1,2]; block rotation also includes slip along the Wiley-Indian fault* and hypothesized fault south of the Black Hills and Palo Verde Mountains to "close" Indian Well Valley
51	5; 23°	Mammoth Wash fault[1,2]; offset relative to block 50 after its restoration; difficult to restore due to rotation of blocks 50 and 51
52 relative to 51	8; 0°	Black Eagle fault[1,2]; rotation of block 51 results in overlap of block 51 onto block 52
52 relative to 22	7?; 0°	Vinagre Wash fault[2]; hypothesized fault in southern California Chocolate Mountains; this block may rotate; restoration of offset removes slight left bend in original outcrop of the Orocopia Schist

Note: The magnitude of restoration along left-lateral, strike-slip faults is measured relative to the adjacent block to the north, except block 47. "Block number" is the number assigned to the fault-bounded area of crust shown in Plate 3. References for faults shown as footnotes include much more information about individual faults, including restraints.

*Proposed name of inferred fault by Powell (1993) between the Little Chuckwalla and Mule Mountains.

[1]Powell (1993).

[2]Richard (1993).

[3]Dokka et al. (1998).

REFERENCES CITED

Anderson, P., 1989a, Proterozoic plate tectonic evolution of Arizona, *in* Jenney, J.P., and Reynolds, S.J., eds., Geologic evolution of Arizona: Tucson, Arizona, Arizona Geological Society Digest, v. 17, p. 17–55.

Anderson, P., 1989b, Stratigraphic framework, volcanic-plutonic evolution, and vertical deformation of the Proterozoic volcanic belts of central Arizona, *in* Jenney, J.P., and Reynolds, S.J., eds., Geologic evolution of Arizona: Tucson, Arizona, Arizona Geological Society Digest, v. 17, p. 57–147.

Anderson, T.H., and Silver, L.T., 1981, An overview of Precambrian rock in Sonora: Universidad National Autonoma de Mexico, Instituto de Geología, Revista, v. 5, no. 2, p. 131–139.

Anonymous, 1981, Carta geologia, Los Vidrios, Secretaría de Programación y Presupuesto, Estados Unidos Mexicanos, I12–10, scale 1:250,000.

Anonymous, 1982, Carta geologia, Puerto Peñasco, Secretaría de Programación y Presupuesto, Estados Unidos Mexicanos, H12–1, scale 1:250,000.

Anonymous, 1983, Carta geologia, Mexicali, Secretaría de Programación y Presupuesto, Estados Unidos Mexicanos, I11–12, scale 1:250,000.

Atwater, T., and Stock, J., 1998, Pacific–North America plate tectonics of the Neogene southwestern United States: An update, *in* Ernst, W.G., and Nelson, C.A., eds., Integrated earth and environmental evolution of the southwestern United States: The Clarence A. Hall, Jr. Volume: Columbia, Maryland, Bellwether Publishing, Ltd., for the Geological Society of America, p. 393–420.

Balderman, M.A., and La Violette, J.W., 1977, Northwest-trending, left-separation faults in the southern Dome Rock Mountains, Yuma County, Arizona: Geological Society of America Abstracts with Programs, v. 10, p. 384–385.

Bartley, J.M., Glazner, A.F., and Schermer, E.R., 1990, North-south contraction of the Mojave block and strike-slip tectonics in southern California: Science, v. 248, p. 1398–1401.

Beard, L.S., 1996, Paleogeography of the Horse Spring Formation in relation to the Lake Mead fault system, Virgin Mountains, Nevada and Arizona, *in* Beratan, K.K., ed., Reconstructing the history of Basin and Range extension using sedimentology and stratigraphy: Geological Society of America Special Paper 303, p. 27–60.

Beikman, H.M., Haxel, G.B., and Miller, R.J., 1995, Geologic map of the Tohono O'Odham Indian Nation, southern Arizona: U.S. Geological Survey Miscellaneous Investigations Map I-2017, scale 1:125,000.

Bohannon, R.G., and Geist, E., 1998, Upper crustal structure and Neogene tectonic development of the California continental borderland: Geological Society of America Bulletin, v. 110, no. 6, p. 779–800, doi: 10.1130/0016-7606(1998)110<0779:UCSANT>2.3.CO;2.

Bohannon, R.G., Grow, J.A., Miller, J.J., and Blank, R.H., Jr., 1993, Seismic stratigraphy and tectonic development of Virgin River depression and associated basins, southeastern Nevada and northwestern Arizona: Geological Society of America Bulletin, v. 105, p. 501–520, doi: 10.1130/0016-7606(1993)105<0501:SSATDO>2.3.CO;2.

Bowring, S.A., and Karlstrom, K.E., 1990, Growth, stabilization, and reactivation of Proterozoic lithosphere in the southwestern United States: Geology, v. 18, p. 1203–1206, doi: 10.1130/0091-7613(1990)018<1203:GSAROP>2.3.CO;2.

Brady, R.H., III, 1992, The Eastern California shear zone in the northern Bristol Mountains, southeastern California, *in* Richard, S.M., ed., Deformation associated with the Neogene Eastern California shear zone, southwestern Arizona and southeastern California: Redlands, California, San Bernardino County Museum Special Publication 92–1, p. 6–10.

Burchfiel, B.C., and Stewart, J.H., 1966, "Pull-apart" origin of the central segment of Death Valley, California: Geological Society of America Bulletin, v. 77, p. 439–442.

Carter, J.N., Luyendyk, B.P., and Terres, R.R., 1987, Neogene clockwise tectonic rotation of the eastern Transverse Ranges, California, suggested by paleomagnetic vectors: Geological Society of America Bulletin, v. 98, p. 199–206, doi: 10.1130/0016-7606(1987)98<199:NCTROT>2.0.CO;2.

Chamberlain, K.R., and Bowring, S.A., 1990, Proterozoic geochronologic and isotopic boundary in NW Arizona: Journal of Geology, v. 98, p. 399–416.

Crittenden, M.D., Jr., Coney, P.J., and Davis, G.H., eds., 1980, Cordilleran metamorphic core complexes: Geological Society of America Memoir 153, 490 p.

Davis, G.A., and Burchfiel, B.C., 1973, Garlock fault, an intracoastal transform structure, southern California: Geological Society of America Bulletin, v. 84, p. 1407–1422, doi: 10.1130/0016-7606(1973)84<1407:GFAITS>2.0.CO;2.

Dembosky, J.A., Jr., 2001, Paleotectonic restoration of Miocene extensional strain of crystalline rock provinces and terranes in southwestern North America [Ph.D. dissertation]: Pittsburgh, Pennsylvania, University of Pittsburgh, 153 p., 12 plates.

Dembosky, J.A., Anderson, T.H., Nourse, J.A., and Silver, L.T., 1998, Caborca, San Gabriel, and Joshua Tree: Terranes at the southwestern margin of North America [abs.]: The Geological Society of America Abstracts with Programs, v 30, no. 7, p. A-355.

Dembosky, J.A., Anderson, T.H., Stahl, E.R., and Nourse, J.A., 1999, Field constraints on the trace of the Mojave-Sonora megashear in southwestern Arizona [abs.]: The Geological Society of America Abstracts with Programs, v. 31, no. 7, p. A-115.

DeMets, C., Gordon, R.G., Argus, D.F., and Stein, S., 1990, Current plate motions: Geophysical Journal International, v. 101, p. 425–478.

DeWitt, E., 1989, Geochemistry and tectonic polarity of Early Proterozoic (1700–1750 Ma) plutonic rocks, north-central Arizona, *in* Jenney, J.P., and Reynolds, S.J., eds., Geologic evolution of Arizona: Tucson, Arizona, Arizona Geological Society Digest, v. 17, p. 149–163.

Dickinson, W.R., 1983, Cretaceous sinistral strike slip along Nacimiento fault in coastal California: American Association of Petroleum Geologists Bulletin, v. 67, no. 4, p. 624–645.

Dickinson, W.R., 1996, Kinematics of transrotational tectonism in the California Transverse Ranges and its contribution to cumulative slip along the San Andreas transform fault system: Geological Society of America Special Paper 305, 46 p.

Dickinson, W.R., 1997, Tectonic implications of Cenozoic volcanism in coastal California: Geological Society of America Bulletin, v. 109, no. 8, p. 936–954, doi: 10.1130/0016-7606(1997)109<0936:OTIOCV>2.3.CO;2.

Dickinson, W.R., and Butler, R. F., 1998, Coastal and Baja California paleomagnetism reconsidered: Geological Society of America Bulletin, v. 110, p. 1268–1280.

Dillon, J.T., 1975, Geology of the Chocolate and Cargo Muchacho Mountains, southeasternmost California [Ph.D. dissertation]: Santa Barbara, California, University of California, 575 p.

Dokka, R.K., 1983, Displacements on late Cenozoic strike-slip faults of the central Mojave Desert, California: Geology, v. 11, p. 305–308, doi: 10.1130/0091-7613(1983)11<305:DOLCSF>2.0.CO;2.

Dokka, R.K., 1989, The Mojave extensional belt of southern California: Tectonics, v. 8, no. 2, p. 363–390.

Dokka, R.K., and Travis, C.J., 1990, Role of the Eastern California shear zone in accommodating Pacific–North American plate motion: Geophysical Research Letters, v. 17, no. 9, p. 1323–1326.

Dokka, R.K., Ross, T.M., and Lu, G., 1998, The trans Mojave-Sierran shear zone and its role in early Miocene collapse of southwestern North America, *in* Holdsworth, R.E., Strachan, R.A., and Dewey, J.F., eds., Continental transpressional and transtensional tectonics: Geological Society [London] Special Publication 135, p. 183–202.

Duebendorfer, E.M., Beard, L.S., and Smith, E.I., 1998, Restoration of Tertiary deformation in the Lake Mead region, southern Nevada: The role of strike-slip transfer faults, *in* Faulds, J.E., and Stewart, J.H., eds., Accommodation zones and transfer zones: The regional segmentation of the Basin and Range Province: Geological Society of America Special Paper 323, p. 127–148.

Eberly, L.D., and Stanley, T.B., Jr., 1978, Cenozoic stratigraphy and geologic history of southwestern Arizona: Geological Society of America Bulletin, v. 89, p. 921–940, doi: 10.1130/0016-7606(1978)89<921:CSAGHO>2.0.CO;2.

Ekström, G., and England, P., 1989, Seismic strain rates in regions of distributed continental deformation: Journal of Geophysical Research, v. 94, no. B8, p. 10,231–10,257.

Fletcher, J.M., Miller, J.S., Martin, M.W., Boettcher, S.S., Glazner, A.F., and Bartley, J.M., 2002, Cretaceous arc tectonism in the Mojave block: Profound crustal modification that controlled subsequent tectonic regimes, *in* Glazner, A.F., Walker, J.D., and Bartley, J.M., eds., Geologic evolution of the Mojave Desert and southwestern Basin and Range: Geological Society of America Memoir 195, p. 131–149.

Fritsche, A.E., 1998, Miocene paleogeography of southwestern California and its implications regarding basin terminology *in* Ernst, W.G., and Nelson, C.A., eds., Integrated earth and environmental evolution of the southwestern United States: The Clarence A. Hall Jr. Volume: Columbia, Maryland, Bellwether Publishing, Ltd., for the Geological Society of America, p. 484–502.

Gans, P.B., 1997, Large-magnitude Oligo-Miocene extension in southern Sonora: Implications for the tectonic evolution of northwest Mexico: Tectonics, v. 16, no. 3, p. 388–408, doi: 10.1029/97TC00496.

Glazner, A.F., Walker, J.D., Bartley, J.M., Fletcher, J.M., Martin, M.W., Schermer, E.R., Boettcher, S.S., Miller, J.S., Fillmore, R.P., and Linn, J.K., 1994, Reconstruction of the Mojave block, *in* McGill, S.F., and Ross, T.M., eds., Geological investigations of an active margin: San Bernardino, California, Cordilleran Section, Geological Society of America Field Trip Guidebook, Field Trip 1, p. 4–30.

Glazner, A.F., Bartley, J.M., and Sanner, W.K., 2000, Nature of the southwestern boundary of the central Mojave Tertiary province, Rodman Mountains, California: Geological Society of America Bulletin, v. 112, no. 1, p. 34–44, doi: 10.1130/0016-7606(2000)112<0034:NOTSBO>2.3.CO;2.

Gray, F., Miller, R.J., Peterson, D.W., May, D.J., Tosdal, R.M., and Kahle, K., 1985, Geologic map of the Growler Mountains, Pima and Maricopa Counties, Arizona: U.S. Geological Survey Miscellaneous Field Studies Map MF-1681, scale 1:62,500.

Gray, F., Miller, R.J., Grubensky, M.J., Tosdal, R.M., Haxel, G.B., Peterson, D.W., May, D.J., and Silver, L.T., 1987, Geologic map of the Ajo and Lukeville 1° × 2° quadrangle, southwest Arizona: U.S. Geological Survey Open-File Report 87-347, scale 1:250,000, 23 p.

Grubensky, M.J., Haxel, G.B., and Kock, R.D., 1993, Geologic map of the Castle Dome Mountains, southwestern Arizona: U.S. Geological Survey Miscellaneous Investigations Map I-2138, scale 1:62,500.

Grubensky, M.J., Haxel, G.B., and Demsey, K.A., 1995, Geologic map of the southeastern Kofa Mountains and western Tank Mountains, southwestern Arizona: U.S. Geological Survey Miscellaneous Investigations Map I-2454, scale 1:62,500.

Harbert, W., 1991, Late Neogene relative motions of the Pacific and North American plates: Tectonics, v. 10, no. 1, p. 1–15.

Haxel, G.B., Tosdal, R.M., May, D.M., and Wright, J.E., 1984, Latest Cretaceous and early Tertiary orogenesis in south-central Arizona: Thrust faulting, regional metamorphism, and granitic plutonism: Geological Society of America Bulletin, v. 95, p. 631–653, doi: 10.1130/0016-7606(1984)95<631:LCAETO>2.0.CO;2.

Haxel, G.B., Tosdal, R.M., and Dillon, J.T., 1985, Tectonic setting and lithology of the Winterhaven Formation: A new Mesozoic stratigraphic unit in southeasternmost California and southwestern Arizona: U.S. Geological Survey Bulletin 1599, 19 p.

Humphreys, E.D., and Weldon, R.J., 1994, Deformation across the western United States: A local estimate of Pacific–North America transform deformation: Journal of Geophysical Research, v. 99, no. B10, p. 19,975–20,010, doi: 10.1029/94JB00899.

Jachens, R.C., and Griscom, A., 1985, An isostatic residual gravity map of California—A residual map for interpretation of anomalies from intracrustal sources, *in* Hinze, W.J., ed., The utility of regional gravity and magnetic anomaly maps: Tulsa, Oklahoma, Society of Exploration Geophysicists, p. 347–360.

Jacobson, C.E., 1990, The ^{40}Ar/^{39}Ar geochronology of the Pelona Schist and related rocks, southern California: Journal of Geophysical Research, v. 95, no. B9, p. 509–528.

James, E.W., 1989, Southern extension of the Independence dike swarm of eastern California: Geology, v. 17, p. 587–590, doi: 10.1130/0091-7613(1989)017<0587:SEOTID>2.3.CO;2.

Jennings, C.W., 1977, Geologic map of California: Sacramento, California, California Department of Conservation, Division of Mines and Geology, scale 1:750,000.

Jordan, T.H., and Minster, J.B., 1988, Measuring crustal deformation in the American West: Scientific American, v. 259, no. 2, p. 48–58.

Klein, D.P., and Kucks, R.P., 1991, Gravity and aeromagnetic anomaly maps of the Ajo and Lukeville 1° × 2° quadrangles, southwestern Arizona: U.S. Geological Survey Miscellaneous Investigations Map I-2139, scale 1:250,000.

Kruger, J.M., Faulds, J.E., Reynolds, S.J., and Okaya, D.A., 1998, Seismic reflection evidence for detachment polarity beneath a major accommodation zone, west-central Arizona, *in* Faulds, J. E., and Stewart, J. H., eds., Accommodation zones and transfer zones: The regional segmentation of the Basin and Range Province: Geological Society of America Special Paper 323, p. 89–113.

Lucchitta, I., and Suneson, N.H., 1996, Timing and character of deformation along the margin of a metamorphic core complex, west-central Arizona, *in* Beratan, K.K., ed., Reconstructing the history of Basin and Range extension using sedimentology and stratigraphy: Boulder, Colorado, Geological Society of America Special Paper 303, p. 147–170.

Luyendyk, B.P., 1991, A model for Neogene crustal rotations, transtension, and transpression in southern California: Geological Society of America Bulletin, v. 103, p. 1528–1536, doi: 10.1130/0016-7606(1991)103<1528:AMFNCR>2.3.CO;2.

Luyendyk, B.P., Kamerling, M.J., and Terres, R., 1980, Geometric model for Neogene crustal rotations in southern California: Geological Society of America Bulletin, Part I, v. 91, p. 211–217.

Luyendyk, B.P., Kamerling, M.J., Terres, R.R., and Hornafius, J.S., 1985, Simple shear of southern California during Neogene time suggested by paleomagnetic declinations: Journal of Geophysical Research, v. 90, no. B14, p. 12,454–12,466.

Mariano, J., Helferty, M.G., and Gage, T.B., 1986, Bouguer and isostatic residual gravity maps of the Colorado River region, including the Kingman, Needles, Salton Sea, and El Centro quadrangles: U.S. Geological Survey Open-File Report 86-347, scale 1:250,000, 7 sheets.

Marshak, S., 1980, A preliminary study of Mesozoic geology in the southern Dome Rock Mountains, southwestern Arizona, *in* Jenney, J.P., and Stone, C., eds., Studies in western Arizona: Arizona Geological Society Digest, v. XII, p. 123–133.

Martin, M.W., and Walker, J.D., 1992, Extending the western North American Proterozoic and Paleozoic craton through the Mojave Desert: Geology, v. 20, p. 753–756, doi: 10.1130/0091-7613(1992)020<0753:ETWNAP>2.3.CO;2.

Matti, J.C., and Morton, D.M., 1993, Paleogeographic evolution of the San Andreas fault in southern California: A reconstruction based on a new cross-fault correlation, *in* Powell, R.E., Weldon, R.J., II, and Matti, J.C., eds., The San Andreas fault system: Displacement, palinspastic reconstruction, and geologic evolution: Geological Society of America Memoir 178, p. 107–159.

Mattick, R.E., and Olmstead, F.H., and Zohdy, A.A.R., 1973, Geophysical studies in the Yuma Area, Arizona and California: U.S. Geological Survey Professional Paper 726-D, 36 p.

May, D.J., 1989, Late Cretaceous intra-arc thrusting in southern California: Tectonics, v. 8, no. 6, p. 1159–1173.

May, D.J., and Walker, N.W., 1989, Late Cretaceous juxtaposition of metamorphic terranes in the southeastern San Gabriel Mountains, California: Geological Society of America Bulletin, v. 101, p. 1246–1267, doi: 10.1130/0016-7606(1989)101<1246:LCJOMT>2.3.CO;2.

Miller, J.M.G., and John, B.E., 1999, Sedimentation patterns support seismogenic low-angle normal faulting, southeastern California and western Arizona: Geological Society of America Bulletin, v. 111, no. 9, p. 1350–1370, doi: 10.1130/0016-7606(1999)111<1350:SPSSLA>2.3.CO;2.

Mueller, K.J., Frost, E.G., and Haxel, G.B., 1982, Mid-Tertiary detachment faulting in the Mohawk Mountains of southwestern Arizona, *in* Frost, E.G., and Martin, D.L., eds., Mesozoic-Cenozoic tectonic evolution of the Colorado River region, California, Arizona, and Nevada: San Diego, California, Cordilleran Publishers, p. 449–458.

Olmstead, F.H., 1972, Geologic map of the Laguna Dam 7.5-minute quadrangle, Arizona and California: U.S. Geological Survey Geologic Quadrangle Map Series Map GQ-1014, scale 1:24,000.

Oppenheimer, J.M., and Sumner, J.S., 1980, Gravity modeling of the basins in the Basin and Range Province, Arizona, *in* Stone, C.S., and Jenney, J.P., eds., Arizona Geological Society Digest, v. XIII, p. 111–115.

Oskin, M., and Stock, J., 2003, Pacific–North American plate motion and opening of the Upper Delfin basin, northern Gulf of California, Mexico: Geological Society of America Bulletin, v. 115, no. 10, p. 1173–1190, doi: 10.1130/B25154.1.

Oskin, M., Stock, J., Lewis, C., and Martin-Barajas, A., 1999, Separation of Baja California from North America: New constraints on timing and total displacement: Geological Society of America Abstracts with Programs, v. 31, no. 7, p. A-294.

Powell, R.E., 1993, Balanced palinspastic reconstruction of pre–late Cenozoic paleogeology, southern California: Geology and kinematic constraints on evolution of the San Andreas fault system, *in* Powell, R.E., Weldon, R.J., II, and Matti, J.C., eds., The San Andreas fault system: Displacement, palinspastic reconstruction, and geologic evolution: Geological Society of America Memoir 178, p. 1–86.

Prescott, W.H., Savage, J.C., and Kinoshita, W.T., 1979, Strain accumulation rates in the western United States between 1970 and 1978: Journal of Geophysical Research, v. 84, no. B10, p. 5423–5435.

Pridmore, C.L., and Craig, C., 1982, Upper-plate structure and sedimentation of the Baker Peaks area, Yuma County, Arizona, *in* Frost, E.G., and Martin, D.L., eds., Mesozoic-Cenozoic tectonic evolution of the Colorado River region, California, Arizona, and Nevada: San Diego, California, Cordilleran Publishers, p. 357–375.

Reynolds, S.J., 1988, Geologic map of Arizona: Arizona Geological Survey Map 26, scale 1:1,000,000.

Richard, S.M., 1992a, Geologic map of the Imperial Reservoir quadrangle, Yuma County, Arizona, and Imperial County, California: Arizona Geological Survey Open-File Report 92-11, scale 1:24,000.

Richard, S.M., 1992b, Geologic map of the Red Hill NE quadrangle, La Paz and Yuma Counties, Arizona: Arizona Geological Survey Open-File Report 92-12, scale 1:24,000.

Richard, S.M., 1992c, Geologic map of the Red Hill quadrangle, La Paz County, Arizona: Arizona Geological Survey Open-File Report 92-14, scale 1:24,000.

Richard, S.M., 1993, Palinspastic reconstruction of southeastern California and southwestern Arizona for the middle Miocene: Tectonics, v. 12, no. 4, p. 830–854.

Richard, S.M., 1994, Preliminary reconstruction of Miocene extension in the Basin and Range of Arizona and adjacent areas: Arizona Geological Survey Open-File Report 95-4, 11 p.

Ron, H., and Nur, A., 1996, Vertical axis rotations in the Mojave: Evidence from the Independence dike swarm: Geology, v. 24, no. 11, p. 973–976, doi: 10.1130/0091-7613(1996)024<0973:VARITM>2.3.CO;2.

Rotstein, Y., Combs, J., and Biehler, S., 1976, Gravity investigation in the southeastern Mojave Desert, California: Geological Society of America Bulletin, v. 87, p. 981–993, doi: 10.1130/0016-7606(1976)87<981:GIITSM>2.0.CO;2.

Rundle, J.B., 1986, An approach to modeling present-day deformation in southern California: Journal of Geophysical Research, v. 91, no. B2, p. 1947–1959.

Schermer, E.R., Luyendyk, B.P., and Cisowski, S., 1996, Late Cenozoic structure and tectonics of the northern Mojave Desert: Tectonics, v. 15, no. 5, p. 905–932, doi: 10.1029/96TC00131.

Scotti, O., Nur, A., and Estevez, R., 1991, Distributed deformation and block rotation in three dimensions: Journal of Geophysical Research, v. 96, no. B7, p. 12,225–12,243.

Sedlock, R.L., and Hamilton, D.H., 1991, Late Cenozoic tectonic evolution of southwestern California: Journal of Geophysical Research, v. 96, no. B2, p. 2325–2351.

Sherrod, D.R., and Tosdal, R.M., 1991, Geologic setting and Tertiary structural evolution of southwestern Arizona and southeastern California: Journal of Geophysical Research, v. 96, no. B7, p. 12,407–12,423.

Silver, L.T., 1964, Mazatzal orogeny and tectonic episodicity [abs.]: Geological Society of America Special Paper 82, p. 185–186.

Silver, L.T., 1968, Precambrian batholiths of Arizona [abs.]: Geological Society of America, Special Paper 121, p. 558–559.

Silver, L.T., 1983, Paleogene overthrusting in the tectonic evolution of the Transverse Ranges, Mojave and Salinian regions, California: Geological Society of America Abstracts with Programs, v. 15, p. 38.

Silver, L.T., and Anderson, T.H., 1974, Possible left-lateral early to middle Mesozoic disruption of the southwestern North American craton margin: Geological Society of America Abstracts with Programs, v. 6, no. 7, p. 955–956.

Silver, L.T., and Nourse, J.A., 1991, Timing of the Rand thrust and implications for late Cretaceous tectonics in southern California: Geological Society of America Abstracts with Programs, v. 23, no. 5, p. A480.

Sonder, L.J., and Jones, C.H., 1999, Western United States extension: How the West was widened: Annual Review of Earth and Planetary Science, v. 27, p. 417–462, doi: 10.1146/annurev.earth.27.1.417.

Spencer, J.E., and Reynolds, S.J., 1989, Middle Tertiary tectonics of Arizona and adjacent areas, *in* Jenney, J.P., and Reynolds, S.J., Geologic evolution of Arizona: Tucson, Arizona Geological Society Digest, v. 17, p. 539–574.

Spencer, J.E., and Reynolds, S.J., 1991, Tectonics of mid-Tertiary extension along a transect through west central Arizona: Tectonics, v. 10, p. 1204–1221.

Spencer, J.E., Richard, S.M., Reynolds, S.J., Miller, R.J., Shafiqullah, M., Gilbert, W.G., and Grubensky, M.J., 1995, Spatial and temporal relationships between mid-Tertiary magmatism and extension in southwestern Arizona: Journal of Geophysical Research, v. 100, no. B7, p. 10,321–10,351, doi: 10.1029/94JB02817.

Stewart, J.H., 1998, Regional characteristics, tilt domains, and extensional history of the Cenozoic Basin and Range Province, western North America, *in* Faulds, J.E., and Stewart, J.H., eds., Accommodation zones and transfer zones: The regional segmentation of the Basin and Range Province: Geological Society of America Special Paper 323, p. 47–74.

Sylvester, A.G., 1988, Strike-slip faults: Geological Society of America Bulletin, v. 100, p. 1666–1703, doi: 10.1130/0016-7606(1988)100<1666:SSF>2.3.CO;2.

Teel, D.B., and Frost, E.G., 1982, Synorogenic evolution of the Copper Basin Formation in the eastern Whipple Mountains, San Bernardino County, California, *in* Frost, E.G., and Martin, D.L., eds., Mesozoic-Cenozoic tectonic evolution of the Colorado River region, California, Arizona, and Nevada: San Diego, California, Cordilleran Publishers, p. 275–286.

Tosdal, R.M., Haxel, G.B., Anderson, T.H., Connors, C.D., May, D.J., and Wright, J.E., 1990, Highlights of Jurassic, Late Cretaceous to Early Tertiary, and Middle Tertiary tectonics, south-central Arizona and north-central Sonora, *in* Gehrels, G.E., and Spencer, J.E., eds., Geologic excursions through the Sonoran Desert region, Arizona and Sonora: Arizona Geological Survey Paper 7, p. 76–88.

Walker, J.D., Black, R.A., Berry, A.K., Davis, P.J., Andrew, J.E., and Mitsdarfer, J.M., 2002, Geologic maps of the northern Mojave Desert and southwestern Basin and Range Province, California: Explanation of maps on CD-ROM, *in* Glazner, A.F., Walker, J.D., and Bartley, J.M., eds., Geologic evolution of the Mojave Desert and southwestern Basin and Range: Geological Society of America Memoir 195, p. 297–299.

Wells, R.E., and Hillhouse, J.W., 1989, Paleomagnetism and tectonic rotation of the lower Miocene Peach Springs Tuff: Colorado Plateau, Arizona, to Barstow, California: Geological Society of America Bulletin, v. 101, p. 846–863, doi: 10.1130/0016-7606(1989)101<0846:PATROT>2.3.CO;2.

Wooden, J.L., and Miller, D.M., 1990, Chronologic and isotopic framework for Early Proterozoic crustal evolution in the eastern Mojave Desert region, SE California: Journal of Geophysical Research, v. 95, no. B12, p. 20,133–20,146.

Wooden, J.L., Miller, D.M., and Howard, K.A., 1988, Early Proterozoic chronology of the eastern Mojave Desert: Geological Society of America Abstracts with Programs, v. 21, p. 243.

Wright, L., 1976, Late Cenozoic fault patterns and stress fields in the Great Basin and westward displacement of the Sierra Nevada block: Geology, v. 4, no. 8, p. 489–494, doi: 10.1130/0091-7613(1976)4<489:LCFPAS>2.0.CO;2.

Manuscript Accepted by the Society 24 March 2005

Printed in the USA

Geological Society of America
Special Paper 393
2005

Tectonic synthesis of the Ouachita-Marathon-Sonora orogenic margin of southern Laurentia: Stratigraphic and structural implications for timing of deformational events and plate-tectonic model

Forrest G. Poole
U.S. Geological Survey, MS-973, Box 25046, Federal Center, Denver, Colorado 80225, USA

William J. Perry Jr.
U.S. Geological Survey, MS-939, Box 25046, Federal Center, Denver, Colorado 80225, USA

Raul J. Madrid
Independent Scientist, 155 Kaanapali Drive, Napa, California 94558, USA

Ricardo Amaya-Martínez
Departamento de Geología, Universidad de Sonora, Hermosillo, Sonora 83000, México

ABSTRACT

The Ouachita-Marathon-Sonora orogen is a 3000-km-long belt of deformed Paleozoic rocks bordering the southern margin of the Laurentian (North American) craton. Extending from Mississippi and Arkansas (Ouachita) southwestward through Texas (Marathon) and westward through Chihuahua and Sonora (Sonora), the orogenic system formed during a late Paleozoic collisional-subductional event. This event resulted in closure of the Rheic ocean and the development of the orogen as the southern edge of the Laurentia plate was subducted beneath a northward-advancing Gondwanan (South American) continental-margin arc. Foredeeps and foreland basins and uplifts were created on the Laurentian plate continentward of the orogen. Regional stratigraphic and structural relations indicate original physical continuity of the Ouachita-Marathon-Sonora orogenic belt along the entire southern margin of Laurentia.

In the Neoproterozoic and Early Cambrian, the supercontinent Rodinia rifted along trends later followed by the Ouachita-Marathon-Sonora orogenic belt. During the breakup of Rodinia, the promontories and embayments that developed along the south margin of Laurentia are related to a northeast-striking rift system (oceanward of the continental edge) deformed by northwest-striking synrift transform faults that offset the rift-parallel cratonal margin. Initial deposits of the rifted margin were sediments deposited in tectonic sags and basins, which opened oceanward, and in depressions adjacent to transform faults. Shelf deposition began with Middle Cambrian clastic and carbonate sediments and continued with deposition of mostly shallow-marine carbonate sediments. Offshelf deposition began with Upper Cambrian (oldest

Poole, F.G., Perry, W.J., Jr., Madrid, R.J., and Amaya-Martínez, R., 2005, Tectonic synthesis of the Ouachita-Marathon-Sonora orogenic margin of southern Laurentia: Stratigraphic and structural implications for timing of deformational events and plate-tectonic model, *in* Anderson, T.H., Nourse, J.A., McKee, J.W., and Steiner, M.B., eds., The Mojave-Sonora megashear hypothesis: Development, assessment, and alternatives: Geological Society of America Special Paper 393, p. 543–596. doi: 10.1130/2005.2393(21). For permission to copy, contact editing@geosociety.org.

rocks recognized in thrust sheets) clastic sediments and continued into the Early Mississippian with deposition of deep-marine clastic and subordinate carbonate sediments in continental-rise and ocean-basin settings.

Deformation of the southern margin of Laurentia resulted from its late Paleozoic diachronous oblique collision with the South American part of Gondwana and development of volcanic-plutonic arc and associated fore-arc and back-arc assemblages along the northern margin of Gondwana. Westward younging of foredeep and foreland-basin depocenters and decreasing age of basin-fill sediments are in accord with the westward migratory closure of the Laurentia–Gondwana suture. Cambrian to Lower Mississippian preorogenic sediments were deposited in offshelf settings along the margins of the two continents, and Upper Mississippian to Permian synorogenic sediments were deposited in the deep-water ocean basin between the two converging continents and associated arcs, in troughs within the evolving allochthons, and in foredeeps and foreland basins on the craton margins. Ocean-basin sediments were scraped off the ocean floor and transported as allochthons formed during continental collision, in advance of synchronous Pennsylvanian and Permian foredeeps and foreland basins and uplifts. These allochthons were part of a large accretionary wedge formed above a south-dipping subduction zone and thrust northwestward 50–200 km above Laurentian continental-shelf and foredeep rocks. The late Paleozoic synorogenic foreland basins and uplifts developed cratonward as far north as the Transcontinental arch and Ancestral Rocky Mountains. The timing and sense of movement of these intracratonic structures have complex relationships to the collisional margin.

Deformation in all three segments of the Ouachita-Marathon-Sonora orogenic belt began in mid-Mississippian time and ended diachronously in the Late Pennsylvanian in the Ouachita Mountains, Early Permian in the Marathon region, and Late Permian in Sonora. This represents predominantly north to northwest contraction. The westward migratory termination of orogenesis and the related development of foredeeps along the Ouachita-Marathon-Sonora belt are consistent with oblique convergence of Gondwana (Africa and South America) with Laurentia and require some clockwise rotation of South America following initial collision in the Ouachita Mountains region. The Sonora segment of the Ouachita-Marathon-Sonora orogenic system shows that northwestern Gondwana and its associated volcanic-arc terrane once lay south of western Laurentia (south of central Sonora) and was not restricted to areas to the east of Sonora as often shown in Pangaean reconstructions. Data south of the Laurentia continental margin indicate that the Gondwanan crust was extremely variable. Ashfall tuffs in Upper Mississippian through lowermost Pennsylvanian hemipelagites and turbidites in the Ouachita Mountains and Marathon region, and rhyolite flows and ashfall tuffs (bentonites) in Permian flysch in the southern Pedregosa basin of north-central Mexico indicate explosive volcanism along the northern margin of Gondwana. Geochemistry of the upper Paleozoic extrusive Sabine Rhyolite in the flysch sequence of the Sabine uplift area indicates a continental-arc origin. Several inferred remnants of Gondwana crust and volcanic-arc rocks in the southern United States and northwestern Gulf of Mexico are collectively referred to as the Sabine block. Other inferred remnants of Gondwana crust and volcanic-arc rocks in northern Mexico include the Coahuila block in Coahuila and southeast Chihuahua, and the El Fuerte block in Sinaloa, Sonora, and adjacent Chihuahua.

Several lines of critical evidence contradict a late Paleozoic or Mesozoic megashear through Sonora. Both the Rodinia rift system and the Ouachita-Marathon-Sonora orogen represent a continuous southern continental margin westward to Baja and Alta California, across the supposed trace of the hypothetical Mojave-Sonora megashear. Although structures are present that offset the orogen and its associated rift system, their sense of offset is right-stepping rather than left-stepping as required by the megashear. Minor displacements by postorogenic faulting together

with right-stepping transform faulting cumulatively represent ~300 km of displacement in contrast to 600–1100 km of left-lateral offset postulated for the megashear. Stratigraphic facies changes around the southwest end of Laurentia do not support offset by such a megashear. Paleozoic biostratigraphic faunal groups persist throughout the length of the orogen to Baja California and are distinct from those in the Cordilleran margin. These, together with later Mesozoic faunal groups, continue across the hypothetical trace of the megashear with no offset. Finally, paleomagnetic data by other workers are not supportive of the megashear. The megashear concept is not compatible with the information presented in this chapter.

Keywords: Laurentia-Gondwana, Paleozoic-Proterozoic, tectonics, stratigraphy, fold-thrust belt, foreland basins and uplifts.

INTRODUCTION

This chapter summarizes the stratigraphy, structure, and our reconstruction of the southern margin of Proterozoic and Paleozoic North America. Figure 1 shows some of our data sources and location of the Ouachita, Marathon, and Sonora segments of the Ouachita-Marathon-Sonora orogen, which display similar sedimentary facies and structural characteristics. Although Figure 1 shows a continuous thrust-belt line for the orogen (shaded area), Cenozoic plutonic and volcanic rocks of the Sierra Madre Occidental and Late Cretaceous–early Tertiary fold and thrust belts of the Sierra Madre Oriental cut across the structural and sedimentary facies trends of the older orogenic system in northern Mexico (Córdoba and Ortega-Gutiérrez, 1990; Ortega-Gutiérrez et al., 1992). These cross-cutting younger features obscure physical evidence for the linkage between the Marathon and Sonora segments. However, limited geophysical data indicate that northwest and subordinate east-west orogenic trends are present in northern Chihuahua and northeastern Sonora (Pilkington et al., 1992a, 1992b; North American Magnetic Anomaly Group, 2002; G.R. Keller, 2002, written commun.). These geophysical data thus provide supportive evidence of linkage between the two segments.

The structural grain of Paleozoic and younger tectonic activity was established during Neoproterozoic and early Paleozoic rifting and transform faulting (Figs. 2 and 3; Thomas, 1991). The data sources shown on Figure 1 and described in Table 1 help define the position and limits of the Ouachita-Marathon-Sonora orogen and Laurentian continental margin in northern Mexico. Most of this margin in the southern United States is well known from many surface and subsurface geologic studies, but the margin in northern Mexico is not known in detail because Neoproterozoic and Paleozoic rocks are mostly concealed by Mesozoic and Cenozoic rocks and outcrops are widely scattered.

Following the breakup of supercontinent Rodinia, Laurentia (North America) and Gondwana (South America portion) were discrete continental masses until the end of the Permian, when assembly of the supercontinent Pangaea became complete. In this report, we use the terms Laurentia and Gondwana (includes volcanic arc and associated fore-arc and back-arc settings along the northwest margin of Gondwana) in discussing Paleozoic geologic history of the two continents.

This study focuses on northern Mexico, particularly the State of Sonora (Fig. 4), where we have been collecting field data for the past 20 years. For clarity we have restored Baja California and Alta California (west of the Cenozoic San Andreas and Golfo de California transform fault system) southeastward ~400 km to their pre-Tertiary position (Figs. 1–4). The amount of restoration is based on ocean-floor topography and coastline configuration (U.S. Geological Survey, 1977; Mammerickx and Smith, 1981, 1984), and on correlation of rock units in Baja California with those in mainland Mexico (López-Ramos and Sánchez-Mejorada, 1976; Ortega-Gutiérrez et al., 1992). This pre-Tertiary restoration is necessary for proper reconstruction of Precambrian–Mesozoic stratigraphic and structural trends along this portion of the southwestern margin of North America.

Our reconstruction and discussion include: (1) Neoproterozoic–Early Cambrian rifting of Rodinia and initial deposits along the newly developing continental margin (Fig. 2); (2) preorogenic Cambrian–Mississippian continental-shelf and offshelf facies (Fig. 5); (3) late Paleozoic orogenesis along the southern margin of Laurentia (area between frontal thrust symbol and continental-shelf edge hachured line on Fig. 3); (4) synorogenic Mississippian–Permian flysch facies (stipple pattern on Fig. 5); (5) foredeeps and foreland basins (stippled areas) and uplifts (+ pattern areas) that developed in front of the Ouachita-Marathon-Sonora orogenic belt (Fig. 3); (6) stratigraphic and structural similarities of the three exposed segments of the orogenic belt (Figs. 5–10); (7) geology of the Minas de Barita area in central Sonora (Figs. 11–13); (8) timing of deformation and plate-tectonic model (Fig. 14); (9) inferred remnants in the southern United States and northern Mexico of the northwestern margin of Gondwana and associated volcanic-arc terranes (Fig. 3); and (10) consequence of large displacement of Cordilleran miogeocline and associated crustal features along the hypothetical Mojave-Sonora megashear.

The Sonora allochthon of this report and earlier reports by Poole and coworkers (e.g., Poole and Madrid, 1988; Poole and Madrid *in* Stewart et al., 1990; Poole et al., 1995a; Poole and Perry, 1997) was considered by Coney (1981, fig. 1) to be part of the Roberts Mountains allochthon displaced from the southwest-

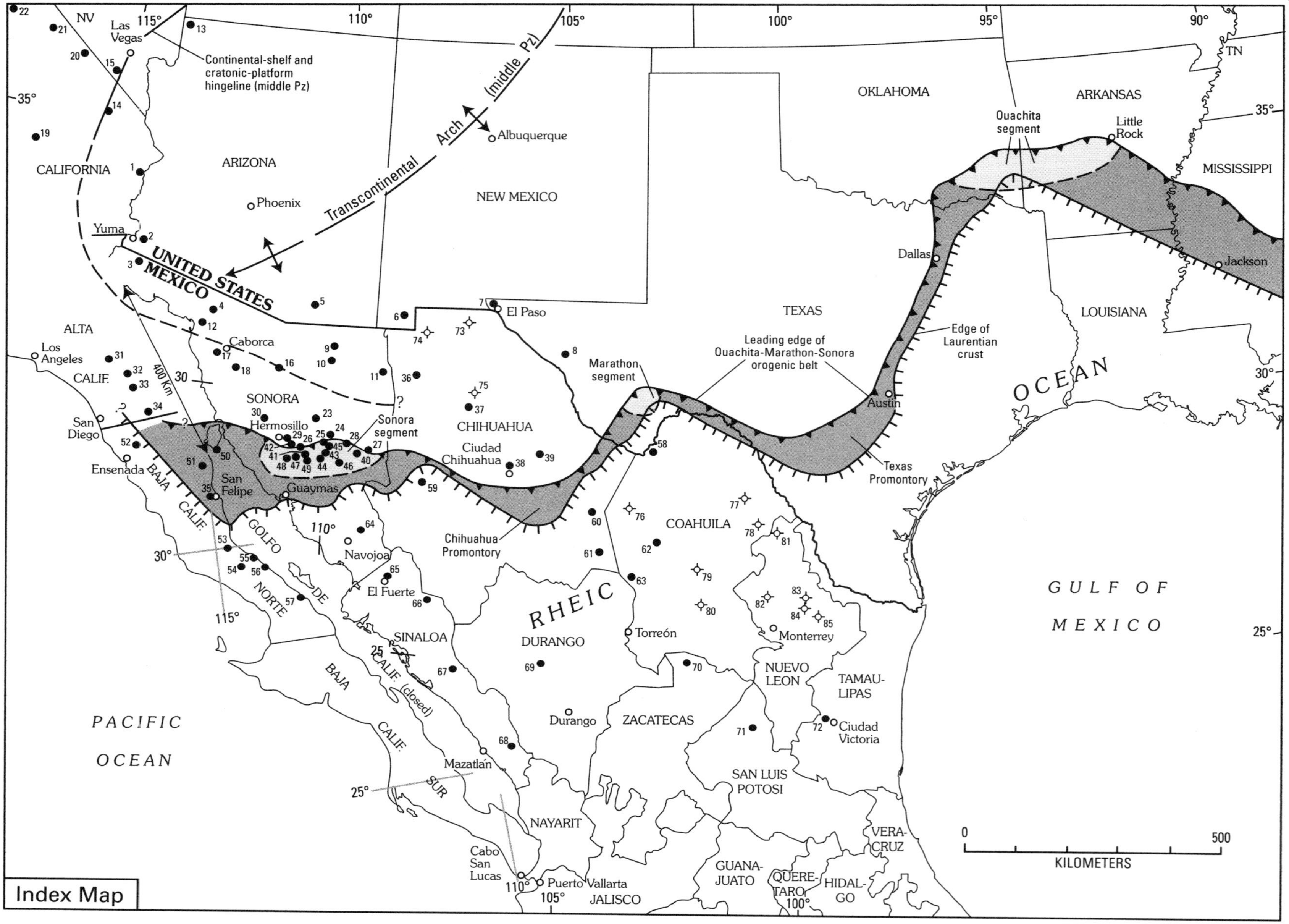

Figure 1. Index map of southern North America (present-day position) and palinspastically restored Baja California and adjacent Alta California (400 km to pre-Tertiary position). Numbers refer to localities mentioned in text. See Table 1 for explanation of data sources. Also shown, are the middle Paleozoic Transcontinental arch, shelf-platform hingeline, edge of Laurentian crust, and leading edge of the late Paleozoic Ouachita-Marathon-Sonora orogenic belt.

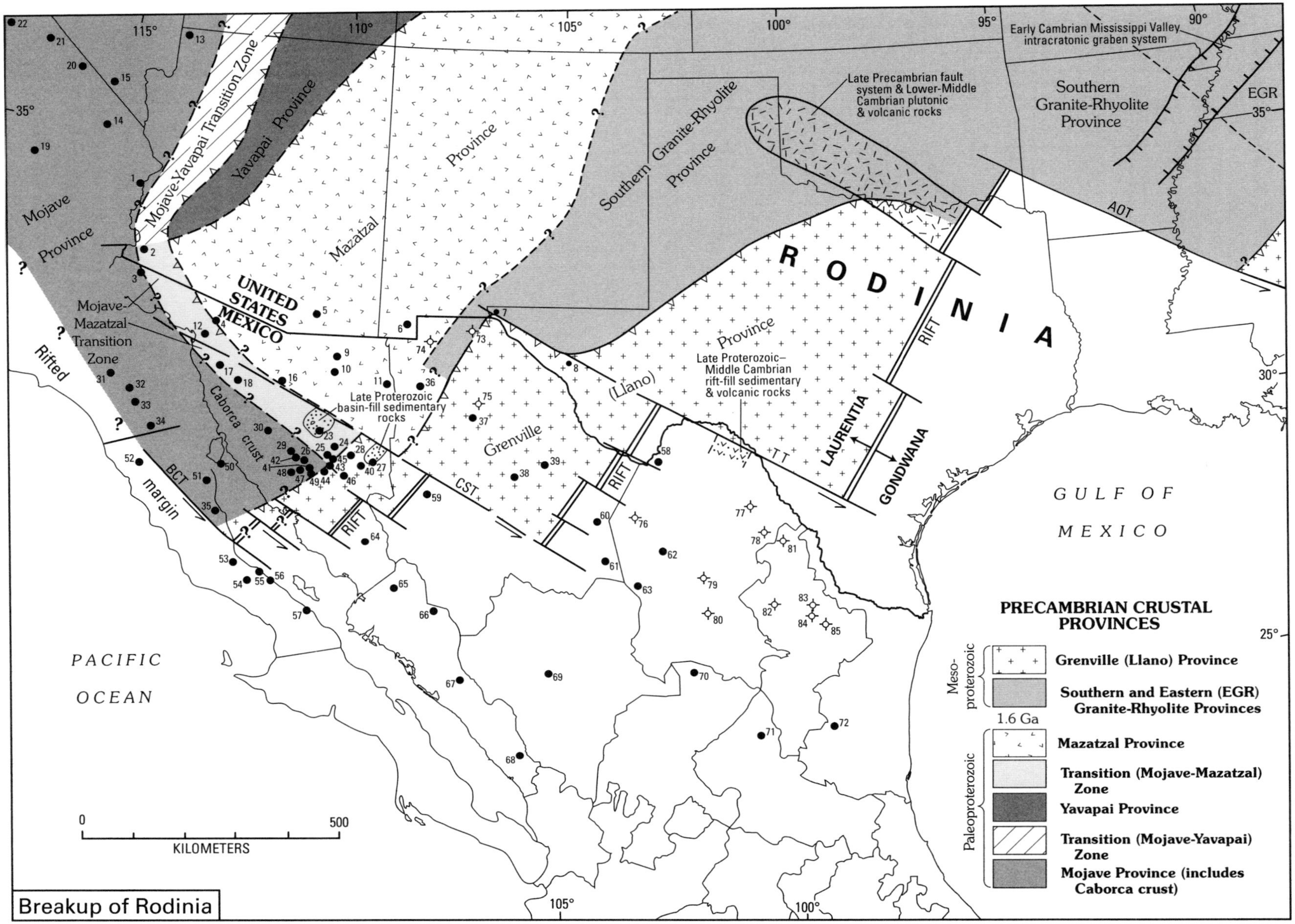

Figure 2. Postulated edge of Laurentian continent at the end of Proterozoic time due to the breakup of Rodinia along a series of rift (double line) and transform (single heavy line) segments. Modified from Speed (1994), Thomas (1991), and Hatcher et al. (1989). In our reconstruction, the original southwest edge of Proterozoic–Paleozoic Laurentian crust extended into Baja California Norte and adjacent Alta California. Precambrian crustal provinces are from Wooden and DeWitt (1991), Anderson et al. (1993), Van Schmus et al. (1993), Karlstrom and Bowring (1993), Karlstrom et al. (2002), Iriondo et al. (2004), and Farmer et al. (this volume). AOT—Alabama-Oklahoma Transform; BCT—Baja California Transform; CST—Chihuahua-Sonora Transform; EGR—Eastern Granite–Rhyolite Province; TT—Texas Transform.

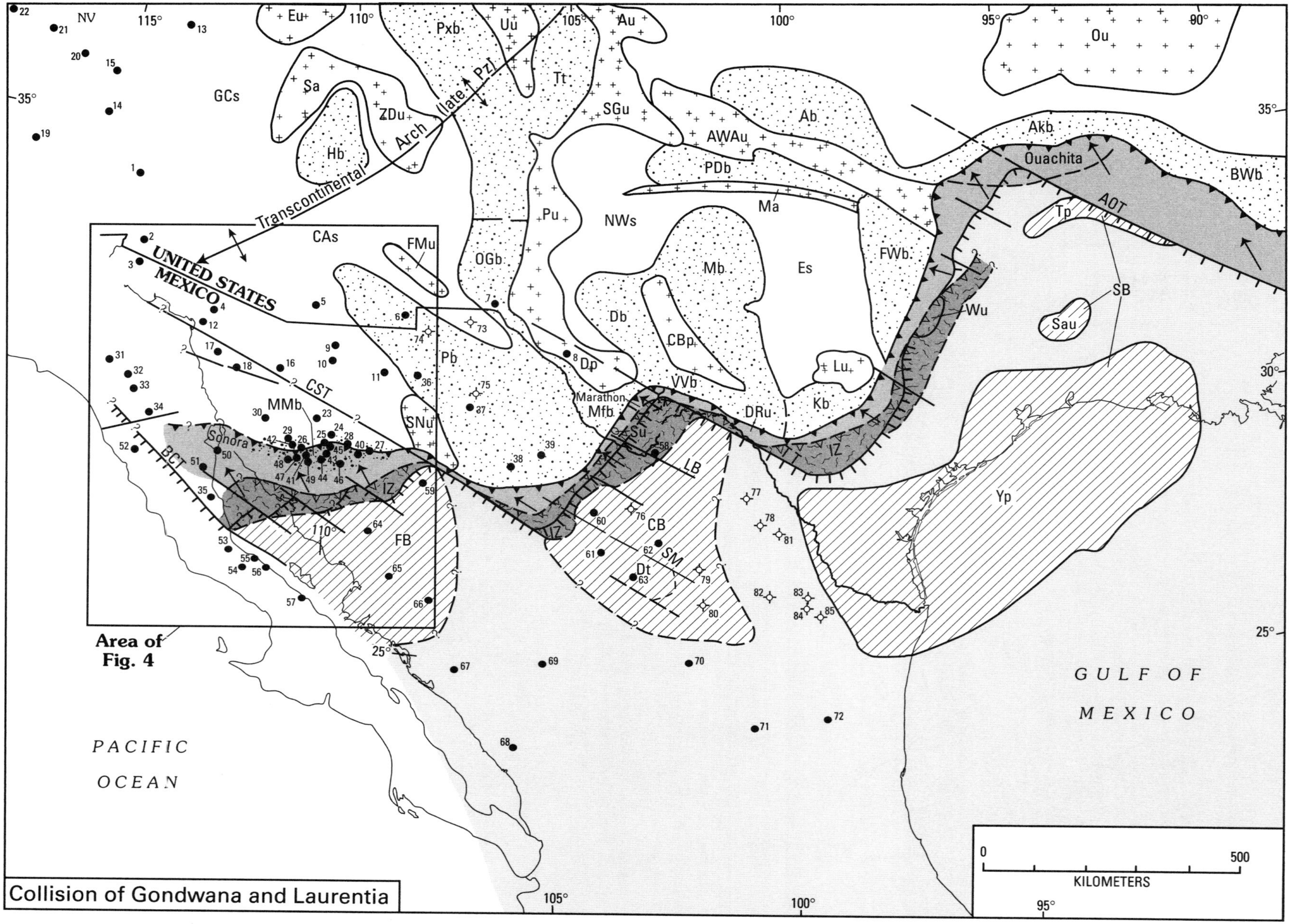

Figure 3. Regional map of southern Laurentia showing late Paleozoic geologic features and localities mentioned in text and described in Table 1, and location of Figure 4.

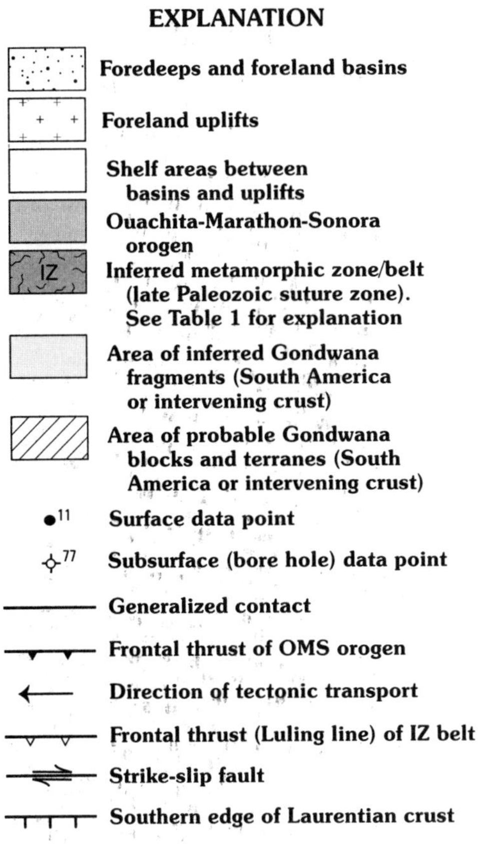

Figure 3 (*continued*).

ern United States, but later was considered by Coney and Campa (1987) as part of their Cortés terrane. Subsequently, Roldán-Quintana et al. (1993) restricted the Cortés terrane to include allochthonous Paleozoic eugeoclinal rocks (Sonora allochthon of this report) and possible Gondwana fragments south of the Laurentian continental edge as inferred in this report (Figs. 3 and 4). Sánchez-Zavala et al. (1999) assigned allochthonous ocean-basin rocks of the Sonora allochthon to the autochthonous-parautochthonous carbonate-shelf rocks of the Caborca terrane of Coney (1981), Coney and Campa (1987), and Roldán-Quintana et al. (1993).

We believe that the Sonora allochthon was emplaced in central Sonora during the Late Permian following deposition of the youngest flysch (Middle Permian) in the accretionary wedge and deposition of the youngest flysch (Middle Permian) in the Mina México foredeep. It seems likely that major thrusting ended before the Triassic, and then in the Early and Middle Triassic, transtensional and/or extensional faulting formed rift basins that subsequently were filled with Upper Triassic sediments (e.g., Barranca Group described by Stewart and Roldán-Quintana, 1991). We are not aware of any documented Lower or Middle Triassic strata in central Sonora; however, the undated Arrayanes Formation of the Barranca Group could be Early or Middle Triassic in age.

The term "foredeep" in this report is applied specifically to an elongate depression (trough) adjacent and subparallel to the Ouachita-Marathon-Sonora orogenic front as a result of crustal flexing and isostatic loading associated with the collision of the Gondwanan (South American) and Laurentian (North American) continents. "Foreland basin" is a more general term applied to an elongate depression more distant from the orogenic front. Synorogenic foreland basins are located adjacent to basement uplifts and arches that developed in the continental crust during orogenesis.

Work of Miser (1929), Hendricks et al. (1937), Flawn et al. (1961), and King (1975) on the Ouachita and Marathon segments and by King (1937, 1975) on the Marathon segment provide background information for this report. Our discussions of stratigraphy and structure in the Ouachita and Marathon segments of the Ouachita-Marathon-Sonora orogen are based largely on papers published in the Geological Society of America "Decade of North American Geology" Volume F-2 (Hatcher et al., 1989) and subsequent biostratigraphic papers by Ellison and Powell (1989), Noble (1990, 1993, 1994), Barrick and Haywa-Branch (1994), and Barrick and Noble (1995). The westernmost part of the Ouachita-Marathon orogenic belt in Mexico has been the subject of diverse interpretations (Shurbet and Cebull, 1987). However, the summary papers by Thompson et al. (1978) and Haenggi (2001) on surface and subsurface data for the Pedregosa basin and Chihuahua trough, and Flawn et al. (1961), Handschy et al. (1987), and McKee et al. (1999) on geologic data for parts of Chihuahua, Coahuila, and Nuevo León have been our main sources of information linking segments of the orogen in northeastern Mexico. Papers published in Geological Society of America Special Paper 279 (edited by Gastil and Miller, 1993) were our main source for pre-batholithic stratigraphy of Baja California. We will focus on northwestern Mexico (Figs. 3 and 4) where our stratigraphic and structural data on the Sonora segment (Sonora allochthon) and associated rocks indicate that they represent the western part of the Ouachita-Marathon orogenic belt, and suggest that the Coahuila block (CB, central area of diagonal-line pattern on Fig. 3), which includes the Las Delicias terrane, is a remnant of the northwestern Gondwana volcanic arc, as is the inferred El Fuerte block (FB, western area of diagonal-line pattern) that attached to North America after they collided with southwestern Laurentia in late Paleozoic time. These inferred fragments of Gondwana and associated volcanic arcs remained attached to North America during subsequent Mesozoic and Cenozoic orogenies.

In Coahuila and in southern Sonora, the collision of cratonic Laurentia and volcanic-arc terranes of northwest Gondwana was the driving force that formed the Marathon and Sonora segments, respectively, of the Ouachita-Marathon-Sonora orogen (Fig. 3). Importantly, the Sonora segment of the orogen indicates that northwestern Gondwana once lay south of western Laurentia (south of central Sonora), and was not confined to areas to the east as often shown in Pangaean reconstructions (e.g., Sedlock et al., 1993; Dickinson and Lawton, 2001). We do not subscribe to the view that the Marathon segment of the orogen extends due south from West Texas through eastern Mexico, where it

TABLE 1. KEY AND EXPLANATIVE GLOSSARY TO FIGURES 1–4

Abbreviation	Explanation
Ab	Anadarko basin. Tectonism (thrust loading) initiated development of the asymmetric Anadarko basin in Late Mississippian (probably Chesterian) time with structural inversion of southern Oklahoma aulacogen (Perry, 1989). Carboniferous transpressional basin, further compressed and overprinted by left-lateral strike-slip faulting in the Late Pennsylvanian and/or Early Permian (Perry, 1989).
AWAu	Amarillo–Wichita–Arbuckle uplift. Sinistral shear zone in North American foreland. Active during the Pennsylvanian (Pindell, 1985; Perry, 1989; Denison, 1989). Basement rocks of the region are known from outcrops in the Wichita and Arbuckle Mountains and from subcrop samples in deep boreholes. Mesoproterozoic basement rocks in the eastern Arbuckle area include granites and diorites dated at 1350–1050 Ma (Ham et al., 1964). Three Cambrian basement-rock groups in the Wichita Mountains include the Carlton Rhyolite, Wichita Granite, and Raggedy Mountains Gabbro dated at 550–500 Ma (Ham et al., 1964). Isotopic ages were determined by U-Pb, Rb-Sr, and K-Ar methods on zircon, feldspar, and biotite, respectively.
Akb	Arkoma basin. Carboniferous foredeep resulting from load of Ouachita thrust sheets to south (Pindell, 1985; Thomas, 1988, 1989; Arbenz, 1989; Meckel et al., 1992). Synorogenic clastic wedge prograding northward and cratonic delta prograding southward to fill basin above older deep-water deposits (Houseknecht, 1986; Link and Roberts, 1986; Thomas, 1989). Foredeep contains Pennsylvanian deltaic, shallow marine, and flysch deposits (Harry and Mickus, 1998). Foredeep defined as a homocline dipping away from the craton and extending beneath the cratonward-directed frontal structures of the Ouachita fold-thrust belt (Thomas, 1988, 1989; Roberts, 1994). Basin consists of a mildly compressed thin-skin fold belt detached along lower Atokan shales from a previously block-faulted substrate (Arbenz, 1989).
AOT	Alabama–Oklahoma transform. Neoproterozoic–Early Cambrian northwest-striking transform fault that offsets northeast-striking rift (Thomas, 1991; North American Magnetic Anomaly Group, 2002). Location based primarily on magnetic anomaly map of North America.
Au	Apishapa uplift. Pennsylvanian-Permian intracratonic uplift between the Sierra Grande and Front Range uplifts.
BCT	Baja California transform. Inferred Neoproterozoic–Early Cambrian northwest-striking transform that formed the southwest edge of early Paleozoic Laurentia (Figs. 2–4). The present southwest edge of Proterozoic basement in northwest Mexico is confined to the area east of this postulated transform fault because the original southwest margin of Proterozoic North America (west of Golfo de California) was fragmented and rifted away.
BWb	Black Warrior basin. Carboniferous foredeep with a triangular outline formed in response to two separate thrust loads: (1) the Ouachita accretionary prism along the southwest side of the basin, and (2) thrust-imbricated passive-margin rocks in the Appalachian thrust belt along the southeast side of the basin (Thomas, 1988, 1989; Meckel et al., 1992; Whiting and Thomas, 1994). Most of the foredeep fill was sediment derived from the Appalachian orogen (Meckel et al., 1992).
CAs	Central Arizona shelf (Ross, 1973). Shelf area south of the Zuni–Defiance uplift and north and west of the Pedregosa basin. Pedregosa shelf of Armin (1987).
CBp	Central Basin platform. Carbonate platform related to north-northwest-trending uplifted basement block of the North American craton separating Midland and Delaware–Val Verde basins. A deep borehole (#1 Nellie) penetrated 1400 m of Phanerozoic sedimentary rocks and ~4460 m of crystalline basement composed of mafic and ultramafic rocks (Keller et al., 1989b). A large, mafic to ultramafic layered (Nellie) intrusion yielded a U-Pb age of 1163 ± 4 Ma and was interpreted to have formed during Keweenawan rifting (Keller et al., 1989b).
CST	Chihuahua–Sonora transform. Inferred Neoproterozoic northwest-striking transform fault offsetting northeast-striking rift formed during breakup of supercontinent Rodinia. Reactivated during late Paleozoic collision of North and South America, early Mesozoic rifting, and late Mesozoic Laramide compression. Location based partly on magnetic anomaly map of North America (North American Magnetic Anomaly Group, 2002).
CB	Coahuila block of this report (part of Coahuiltecano terrane of Sedlock et al.,1993). Consists of the Coahuila or Las Delicias terrane of earlier reports. Includes "Coahuila Peninsula" area of Jones et al. (1984). McKee et al. (1999) named and defined the Delicias terrane as a terrane south of the San Marcos fault. Previous workers used the term Coahuila terrane or block for Gondwana fragments north and south of the San Marcos fault. Tectonic element in northeastern Mexico that was considered part of the Sabine block by Viele and Thomas (1989). The Coahuila/Las Delicias terrane contains flysch deposits with clasts of Paleozoic arc rocks and a basal Mississippian conglomerate with Precambrian granitoid boulders (López et al., 1998). The Precambrian clasts indicate a tectonic history that includes ca. 1232–1217 Ma arc plutonism on ca. 1.8 Ga basement, ca. 1222–1214 Ma late to post-orogenic magmatism, and 579 Ma Pan–African plutonism (López et al., 1998). López et al. (1998) concluded that the Precambrian tectonic history of the Coahuila block precludes a Laurentian origin, and that the Coahuila Grenville-age rocks may be the equivalents of the granulite-grade Grenville rocks in eastern and southern Mexico. Probably Grenville-age crust belonging to Gondwana. We consider the Coahuila block to be a fragment of northwestern Gondwana. See Appendix 2 in Stewart et al. (1999a) for additional information on Coahuila block.
Db	Delaware basin. Early Permian transpressional basin bounded by thrust margin to east (King, 1942)
Dp	Diablo platform. Northwest-trending Early Permian uplifted basement block (Handschy et al., 1987) bounded on the south and southwest by the Marfa foredeep and on the east and northeast by the Delaware foreland basin.
DRu	Devils River uplift. A major subsurface structural feature trending northwest along the boundary between the Marathon segment of the Ouachita–Marathon–Sonora orogenic system and the Val Verde foredeep basin (Nicholas and Rozendal, 1975; Nicholas, 1983; Nicholas and Waddell, 1989). The uplift is ~125 km long and 50 km wide. Nicholas and Rozendal (1975), Nicholas (1983), and Nicholas and Waddell (1989) concluded that the Devils River uplift is part of the chain of external massifs localized along or near the cratonic margin but differs in that it is structurally very shallow and rocks of the interior zone are absent on its crest, probably owing to erosion. Pindell (1985) suggested that the uplift resulted from a right-lateral shear zone created by northwest migration of Marathon segment of Ouachita–Marathon–Sonora orogenic belt along southern side of Llano uplift during ocean closure. A metasedimentary sequence with some interbedded metavolcanic rocks of possible Middle to Early Cambrian age was penetrated from 4000–6800 feet in the Shell Stewart No. 1 well in southern Val Verde County, Texas (Nicholas and Rozendal, 1975). They concluded that Rb-Sr dates on these low-grade metasedimentary and metavolcanic rocks indicate deposition ca. 529 ± 31 Ma. Denison et al. (1977) determined an isochron age of 711 ± 23 Ma (Neoproterozoic) on two whole-rock metarhyolites in this same interval. These rocks may represent Neoproterozoic-early Paleozoic basinal deposits along the Texas transform during the breakup of Rodinia (Fig. 2). The uplift is interpreted by Haenggi et al. (2000) as an accreted block that was originally part of Gondwana.

(*continued*)

TABLE 1. KEY AND EXPLANATIVE GLOSSARY TO FIGURES 1–4 (*continued*)

Abbreviation	Explanation
Dt	Las Delicias terrane (part of Coahuila block). Accreted to North America during Pangaea assembly, probably originally part of Gondwana but stranded during Mesozoic rifting, includes late Paleozoic volcanic arc created by closure between Laurentia and Gondwana plates (Pindell, 1985; McKee et al., 1988, 1999; López et al., 1996). The Las Delicias terrane of McKee et al. (1999) lies south of the Marathon suture, and contains a variety of igneous rocks including Grenville-age boulders in Carboniferous flysch, a small Mississippian granite pluton, Permian arc volcanics (basalt to rhyolite), and Permian and Triassic granitoid plutons (López et al.,1996; McKee et al., 1999). According to McKee et al. (1999), the Las Delicias terrane is bounded on the north by the San Marcos fault (SM on Fig. 3) and on the south by the hypothetical Mojave-Sonora megashear. We consider the Las Delicias terrane to be part of the Coahuila block (labeled CB on Fig. 3).
Es	Eastern shelf. North-trending carbonate shelf separating Fort Worth foredeep and Midland foreland basin that forms the Texas promontory. Kerr foredeep on south side of shelf and Llano uplift.
Eu	Emery uplift/positive area, south-central Utah. Pennsylvanian uplift northwest of the Zuni–Defiance uplift (Blakey, 1980). Also, referred to as the Piute positive area.
FB	El Fuerte block. Accreted to North America during Pangaea assembly, probably part of Gondwana but stranded during Mesozoic rifting, may represent late Paleozoic arc created by closure between Laurentia and Gondwana. A biotite gneiss sample from the Sonobári Complex (Sonobári terrane of Campa and Coney, 1983; part of El Fuerte block of this report) in northern Sinaloa yielded a Late Triassic zircon age (Anderson and Schmidt, 1983). Two other samples (gneiss and amphibolite) from the Complex yielded Nd(t)DM model ages close to 1.0 Ga, which suggests derivation from reworked Proterozoic crust (Valencia-Moreno et al., 2001). A eugeoclinal limestone turbidite bed in the nearby Río Fuerte Group (part of the El Fuerte block) yielded Middle to Late Ordovician conodonts (A.G. Harris, 1996, written commun.). The Rusias terrane of Campa and Coney (1983), which includes upper Paleozoic rocks (loc. 66, Fig. 3), is part of our El Fuerte block. Therefore, the El Fuerte block contains Paleozoic rocks in addition to Upper Triassic metamorphic rocks. The block may represent Gondwana volcanic-arc crust that attached to Laurentia during late Paleozoic collision or it was a separate terrane between North and South America. Sedlock et al. (1993) introduced the term Tahué terrane for mid-Paleozoic metasedimentary rocks of unknown origin in Sinaloa and adjacent areas that were accreted to North America by Late Jurassic time. The Tahué terrane includes the El Fuerte block and part of the extensive Guerrero terrane of Campa and Coney (1983) and Coney and Campa (1987). See Appendix 2 in Stewart et al. (1999a) for additional information on El Fuerte block.
FMu	Florida-Moyotes uplift. Northwest-trending Early Permian uplifted basement block (Handschy et al.,1987).
FWb	Fort Worth basin. Pennsylvanian foredeep related to loading by thrust sheets of Ouachita segment of Ouachita–Marathon–Sonora orogenic belt (Pindell, 1985). Basinal facies and sediment dispersal pattern described by Meckel et al. (1992).
GCs	Grand Canyon shelf. Pennsylvanian and Permian shelf in northwestern Arizona (Blakey and Knapp, 1989).
Hb	Holbrook basin. Early Permian (Leonardian) basin in east-central Arizona that received several hundred meters of fine-grained clastic, evaporite, and carbonate sediment (Blakey, 1980).
IZ	Interior metamorphic zone/belt. Zone of sheared, low-grade, metasedimentary and minor volcanic rocks along the southern margin of the Ouachita–Marathon–Sonora orogenic belt (Fig. 3; Flawn, 1961), consisting of accreted Gondwana crust (James and Henry, 1993a, 1993b). The interior metamorphic belt formed as a result of the late Paleozoic collision of western Gondwana and Laurentia. The metamorphic zone has been identified in several boreholes in Texas, but crops out only at Sierra del Carmen in northwestern Coahuila (loc. 58, Figures 1–3; Flawn et al., 1961; Carpenter, 1997). In Texas, an overthrust contact (Luling front) separates the frontal belt from the interior metamorphic belt that is marked by stronger deformation and metamorphism (Flawn, 1961). However, Nicholas and Waddell (1989) considered the boundary in many places to be transitional and a disturbed belt several kilometers wide. Because of sparse data, the contact is poorly defined in many areas, but may be marked by a general change in metamorphism and structural grain over a wide area (Flawn, 1961). The Luling front and interior metamorphic belt recognized in Texas have not been positively identified in the Ouachita Mountains and Sonora. We presume that the interior belt of accreted Gondwana crust continues westward from Coahuila to Sonora, following the trend of the Ouachita–Marathon–Sonora orogenic belt and southern edge of Laurentia (Fig. 3). The regional extent of the interior belt is unknown, but we believe that it is present along the entire length of the Ouachita–Marathon–Sonora orogenic belt from Mississippi to Sonora. Deformation in the interior belt probably involved basement rocks, whereas deformation in the frontal belt was superficial (Flawn, 1961). Width of the interior belt and its degree of metamorphism and deformation should be extremely variable in this zone. See Appendix 2 in Stewart et al. (1999a) for additional information on the interior metamorphic belt.
Kb	Kerr basin. Middle Pennsylvanian to Early Permian foredeep related to loading by thrust sheets of the Marathon segment of the Ouachita–Marathon–Sonora orogenic belt (Pindell, 1985; Wuellner et al., 1986). Considered a peripheral basin by Wuellner et al. (1986). Sediment source and dispersal pattern described by Meckel et al. (1992).
LB	La Babia fault. Northwest-trending transform fault in northwestern Coahuila and northeastern Chihuahua (Charleston, 1981; Haenggi, 2001). Haenggi (2001) considered the La Babia fault to be the southern limit of the Texas lineament zone of Muehlberger (1980).
Lu	Llano uplift. Structural dome in central Texas that includes the largest exposed area of Grenville-age crystalline rocks in the southern United States (Walker, 1992; Mosher, 1998). The Llano uplift consists of ca. 1360–1232 ± 4 Ma metavolcanic, metaplutonic, and metasedimentary rocks (Walker, 1992; Mosher, 1998). Stable buttress in southern Laurentia during collision with Gondwana; interpreted as a forebulge structure related to the Ouachita segment of Ouachita–Marathon–Sonora orogenic belt (Pindell, 1985).
Ma	Matador arch. Arch east of the Northwest shelf and between the Midland and Palo Duro basins.
Mfb	Marfa basin. Late Pennsylvanian and Early Permian foredeep related to loading by thrust sheets of the Marathon segment of the Ouachita–Marathon–Sonora orogenic belt (Pindell, 1985; Meckel et al., 1992) located south and west of the Diablo platform (King, 1948). Haenggi (2001) suggested a pull-apart origin for the Marfa foredeep caused by Pennsylvanian–Permian left- and right-lateral movements along northwest-trending faults of the Texas "zone" or lineament of Muehlberger (1980).

(*continued*)

TABLE 1. KEY AND EXPLANATIVE GLOSSARY TO FIGURES 1–4 (*continued*)

Abbreviation	Explanation
Mb	Midland basin. Shallow cratonic basin adjacent to Delaware basin, probably had similar structural origin to Delaware basin.
MMb	Mina México basin. Permian foredeep developed on the carbonate shelf in central and east-central Sonora that trends west, subparallel to the leading edge of the Sonora allochthon. Named after Mina México (mine), which was the name (Mina México Formation) applied to this turbidite sequence by Hewett (1978) and Schmidt (1978) in the vicinity of the mine. Foredeep is filled with a thick flysch sequence (>1000 m) of deep-water siliciclastic and minor calciclastic turbidites. Foredeep formed in response to crustal flexing and isostatic loading caused by the north-directed Sonora accretionary wedge during collision of northwest Gondwana and southwest Laurentia. The Sonora allochthon overrode part of the foredeep-basin sequence. Maximum extent of the foredeep is unknown at this time because of sparse exposures and cover by Mesozoic and Cenozoic rocks. See locality 40 discussion for additional information.
NWs	Northwest shelf. Shelf area north of the Delaware foreland basin and east of the Pedernal uplift.
OGb	Orogrande basin. North-trending basin west of Pedernal uplift. Pennsylvanian to Early Permian foreland basin, polycyclic subsidence and depositional history influenced by strike-slip faulting caused by progressive westward suturing of Gondwana and Laurentia (T.F. Lawton and K.A. Giles, 2002, written commun.).
Ou	Ozark uplift. Structural basement high, forebulge related to loading by thrust sheets of Ouachita segment of Ouachita–Marathon–Sonora orogenic belt (Pindell, 1985; Link and Roberts, 1986; Roberts, 1994). Basement ages (U-Pb zircon) are 1480 Ma (Bickford and Mose, 1975), and a granite on the Nashville dome is 1120 ± 30 Ma (Rb-Sr) (Wasserburg et al.,1962).
PDb	Palo Duro basin. West-trending foreland basin south of the Amarillo–Wichita–Arbuckle uplift and north of the Matador arch.
Pu	Pedernal uplift. North-trending basement high, related to development of Delaware and Orogrande foreland basins.
Pb	Pedregosa basin. Foreland basin related to emplacement of thrust sheets of Marathon segment of the Ouachita–Marathon–Sonora orogenic belt in north-central Mexico. Extent and setting of the Late Mississippian–Early Permian Pedregosa basin were discussed by Wilson (1987), Armin (1987), and Armstrong and Mamet (1988). The deeper southern Pedregosa basin may be a foredeep adjacent to the accretionary wedge front.
Pxb	Paradox basin. Pennsylvanian–Permian basin filled with sediment derived from erosion of the Uncompaghre highland. Site of thick carbonate and evaporite sedimentation during the Pennsylvanian and thick clastic sedimentation during the Early Permian.
Sa	Sedona arch. Pennsylvanian and Permian arch in central Arizona west of the Permian Holbrook basin (Blakey, 1980).
Sau	Sabine uplift, Texas–Louisiana border. South of the Ouachita segment of the Ouachita–Marathon–Sonora orogenic belt is a remnant of upper Paleozoic volcanic-arc rocks that were thrust over and sutured to the North American plate (Viele and Thomas, 1989). Nicholas and Waddell (1982, 1989) reported on four boreholes drilled into a complex of pre-Mesozoic subcrop of Carboniferous sediments and crystalline rocks. Two of the boreholes bottomed in rhyolitic flow rock (Sabine Rhyolite) on the south flank of the Sabine uplift. A Rb-Sr date of 255 ± 15 Ma (Late Permian) on the devitrified perlitic rhyolite porphyry was considered too young and most likely pre-Desmoinesian (pre-late Middle Pennsylvanian) by R.E. Denison (*in* Nicholas and Waddell, 1989). They considered the upper Paleozoic felsic volcanic rocks in the Sabine uplift area to be part of a Carboniferous volcanic island arc of unknown extent related to subduction of southern Laurentia, and concluded that deformation in the interior of the Ouachita system had ceased prior to late Middle Pennsylvanian (Desmoinesian) time. Sabine uplift is part of our Sabine block.
SB	Sabine block. Elements of Gondwana accreted to Laurentia during Pangaea assembly, includes Texarkana platform, Sabine uplift, and Yucatán platform, considered fragments of Gondwana continental-margin volcanic arc stranded during Mesozoic rifting. May represent late Paleozoic island arc created by closure between Laurentia and northwest Gondwana (King, 1975; Nicholas and Waddell, 1982; Handschy et al., 1987; Viele and Thomas, 1989). Fragments of our inferred Sabine block are labeled on Figure 3. The Coahuila block of this report was included in the Sabine plate or block by some earlier workers. See Appendix 2 in Stewart et al. (1999a) for additional information on the Sabine block.
Su	Solitario uplift, West Texas. Eroded roof of a radially symmetric late Eocene laccolith exposing a complete section of Upper Cambrian to Lower Pennsylvanian rocks of the Marathon segment (Corry et al., 1990; Henry et al., 1994,1996) of the Ouachita–Marathon–Sonora orogen.
SGu	Sierra Grande uplift. Pennsylvanian-Permian intracratonic uplift in northeastern New Mexico.
SM	San Marcos fault. Northwest-trending transform fault in central Coahuila. Described by McKee and Jones (1979), Charleston (1981), and McKee et al. (1984) from exposures in the Valle de San Marcos. Evidence for Mesozoic displacement on the San Marcos fault is a widespread Upper Jurassic–Lower Cretaceous detrital lithosome north of the fault, composed mostly of scarp-derived sediment (McKee et al., 1984). The San Marcos fault parallels our inferred Neoproterozoic Chihuahua–Sonora transform that formed during Rodinia rifting.
SNu	Sierra del Nido uplift. Defined by a large, negative Bouguer-gravity anomaly interpreted by Moreno et al. (1992) to be a Tertiary batholitic mass, and by Goodell (1993) to represent an uplifted block of Precambrian cratonic crust. Alternatively, the Sierra del Nido block may represent a late Paleozoic foreland uplift in front of the Marathon and Sonora segments of the Ouachita–Marathon–Sonora orogenic belt. The uplift is probably part of the Aldama platform, which is interpreted by Haenggi et al. (2000) as an accreted block that was originally part of western Gondwana. It is conceivable that the major gravity 'low' may represent a flexed depression in front of stacked thrust sheets of the Ouachita–Marathon–Sonora accretionary wedge.
Tp	Texarkana platform, southern Arkansas and adjacent states. Tectonic element south of the Ouachita segment of the Ouachita–Marathon–Sonora orogenic belt is a remnant of an upper plate consisting of upper Paleozoic rocks that was thrust over and sutured to the North American plate (Viele and Thomas, 1989). Part of Sabine block.
TT	Texas transform. Neoproterozoic–Early Cambrian northwest-striking transform fault offsetting northeast-striking rift (Thomas, 1991; North American Magnetic Anomaly Group, 2002). Location based primarily on magnetic anomaly map of North America.
Tt	Taos trough (southern extension of Central Colorado trough). Pennsylvanian–Permian trough filled with sediment derived from erosion of adjacent highlands (Apishapa–Sierra Grande and Uncompahgre uplifts).
Uu	Uncompahgre uplift. Northwest-trending basement high. Active during the Pennsylvanian and Permian. Uplift bounded by late Paleozoic fault systems.

(*continued*)

TABLE 1. KEY AND EXPLANATIVE GLOSSARY TO FIGURES 1–4 (*continued*)

Abbreviation	Explanation
VVb	Val Verde basin. Middle Pennsylvanian to Early Permian foredeep related to loading by thrust sheets of Marathon segment of the Ouachita–Marathon–Sonora orogenic belt (Pindell, 1985; Wuellner et al., 1986). Foredeep fill as much as 6200 m thick accumulated in orogenic clastic wedges, starved basin, submarine fan, and shelf/platform settings (Wuellner et al., 1986). Sediment dispersal pattern described by Meckel et al. (1992).
Yp	Yucatán platform, Gulf Coast region. Tectonic element south of the Ouachita segment of the Ouachita–Marathon–Sonora orogenic belt that separated the Laurentian and Gondwanan plates (Viele and Thomas, 1989). Part of Sabine block.
Wu	Waco uplift. External massif within the interior metamorphic zone/belt of the Ouachita segment in east-central Texas. Seismic and sparse drill-hole data indicate that the Waco uplift is an asymmetrical, complexly faulted, Grenville basement-cored antiform 145 km long and ~16 km wide (Nicholas and Waddell, 1989). Uplift of this basement-cored structure occurred after emplacement of Ouachita ocean-basin rocks over coeval carbonate-shelf rocks (Nicholas and Waddell, 1989).
ZDu	Zuni–Defiance uplift. Pennsylvanian–Permian uplift north of the Central Arizona shelf and south of the Paradox basin.
Cratonic-platform rocks of Laurentia	
1	Big Maria and Little Maria Mountains, easternmost Riverside County, California. Paleozoic cratonic-platform strata of Laurentia. Consists of a complexly deformed and metamorphosed sequence of quartzite, schist, marble, metacherty marble, and calc-silicate rock. Proterozoic basement rocks include gneiss and granite. U-Pb zircon dates indicate a magmatic age of 1.4 Ga (L.T. Silver *in* Hamilton, 1984). The metasedimentary rocks are similar to the Cambrian Tapeats Sandstone, Bright Angel Shale, Muav Limestone, Mississippian Redwall Limestone, Pennsylvanian and Permian Supai and Hermit Shales, Permian Coconino Sandstone and Kaibab and Toroweap Limestones (Hamilton, 1984, 1987; Brown, 1991; Poole, unpublished data) of westernmost Grand Canyon and southeastern Nevada. Upper Cambrian, Ordovician, and Lower Devonian formations are present to the north, but could not be discriminated by Hamilton (1984) in the metadolomite in the Big Maria Mountains. Brown (1991) illustrated the Paleozoic section exposed in the Little Maria Mountains. North American Precambrian basement underlies the Paleozoic section throughout this area. The Paleozoic sequence is intruded by Jurassic granitic rocks.
2	Gila Mountains, southwesternmost Yuma County, Arizona. Paleoproterozoic crystalline basement rocks consist of gneiss, schist, and granite. Premo et al. (2003) obtained a U-Pb zircon age of 1702 ± 14 Ma on a granodioritic gneiss in the Gila Mountains (Interstate 8 roadcut), and a U-Pb zircon age of 1697 ± 11.5 Ma on a granite at Cerro El Aguila (10 km west of Yuma, Arizona). L.T. Silver (*in* Olmsted et al., 1973) obtained a U-Pb zircon age of 1.44 Ga on a porphyritic quartz monzonite collected in Yuma, Arizona, whereas Wasserburg and Lanphere (1965) obtained a Rb-Sr biotite age of 73 Ma on the same rock. The Precambrian date indicates the time of crystallization, and the Late Cretaceous date indicates a metamorphic event related to Laramide tectonism. In the northwestern part of the Gila Mountains, Paleozoic cratonic-platform rocks are probably represented in the metasedimentary sequence that overlies the Precambrian basement rocks. Poole (unpublished data) examined a section in the northern Gila Mountains that consists of bedded, medium- to coarsely crystalline, calcite to dolomite marble with some metachert lenses. The marble was sampled for conodonts, but yielded none. Poole provisionally correlated the marble with the Mississippian Redwall Limestone. Lower Tertiary sedimentary and volcanic rocks unconformably overlie the older metamorphosed sequence.
3	El Capitán (Ejido Serdan), northwesternmost Sonora. Upper Paleozoic platform strata of North America. Consists of a complexly deformed sequence of marbles, quartzites, metamorphosed clastic rocks, and metaplutonic rocks. The metasedimentary rocks are equivalent to the upper Paleozoic Supai Group, Coconino Sandstone, and Kaibab Limestone, and lower Mesozoic Moenkopi Formation and Aztec Sandstone of the southeastern Mojave Desert and Colorado Plateau; they are intruded by Middle Jurassic adamellite plutons (Leveille and Frost, 1984; Poole, unpublished data). The cratonic-platform sequence implies Laurentian Proterozoic basement below. The hypothetical northwest-striking Jurassic Mojave–Sonora megashear, as drawn by Silver and Anderson (1974) and Anderson and Silver (1979), passes northeast of the area.
4	Quitovac area, northwestern Sonora. Two distinct Paleoproterozoic granitoid blocks dated as 1.78–1.69 Ga and 1.71–1.66 Ga by U-Pb zircon geochronology (Iriondo et al., 2004). One block is referred to as the Mojave-Yavapai Transition zone of southwestern Arizona, whereas rocks of the other block (1.71–1.66 Ga) have characteristics similar to those of the Mazatzal province of southern Arizona (Iriondo, 2001). Structural and geochronologic studies indicate that the juxtaposition of the two basement blocks resulted from a north-directed Laramide thrust and not a Proterozoic suture (Iriondo, 2001). Paleozoic rocks have not been positively identified in the Quitovac area, but the La Toñita marble, quartzite, and muscovite schist may represent metamorphosed Paleozoic strata (Iriondo, 2001). Based on regional facies distribution, Paleozoic cratonic-platform rocks should be present in this area. Data do not support the existence of a Jurassic Mojave–Sonora megashear at Quitovac. The trace of the hypothetical northwest-striking megashear of Silver and Anderson (1974) and Anderson and Silver (1979) passes through or very close to the village of Quitovac.
5	Dry Canyon–French Joe Canyon area, Whetstone Mountains, Cochise County, southeastern Arizona. Paleozoic carbonate-platform of southern Laurentia. The sequence includes: Proterozoic Pinal Schist; Cambrian Bolsa Quartzite and Abrigo Limestone; Devonian Martin Formation; Mississippian Escabrosa Limestone; Pennsylvanian Black Prince and Horquilla Limestones and lower Earp Formation; Permian upper Earp Formation, Colina Limestone, Epitaph Formation, Scherrer Formation, Concha Limestone, and Rainvalley Formation (Wrucke and Armstrong, 1987; Schreiber et al., 1990). Paleozoic sequence is unconformably overlain by the Lower Cretaceous Bisbee Group.
6	Big Hatchet Mountains, southwesternmost New Mexico. Paleozoic cratonic platform of southern Laurentia (Zeller, 1965; Hayes and Cone, 1975). The Paleozoic sequence includes: the Cambrian–Ordovician Bliss Sandstone; Ordovician El Paso Formation and Montoya Dolomite; Devonian Percha Shale; Mississippian Escabrosa Limestone and Paradise Formation; Pennsylvanian Horquilla Limestone; and Permian Earp Formation, Colina Limestone, Epitaph Dolomite, Scherrer Formation, and Concha Limestone (Zeller, 1965). The Concha is unconformably overlain by Lower Cretaceous clastic rocks. The Bliss is unconformably underlain by Proterozoic coarse-grained porphyritic granite and quartzite.
7	Franklin Mountains (near El Paso, Texas), westernmost Texas. Paleozoic carbonate platform of southern Laurentia (LaMone, 1969; Bridges, 1970; Hayes and Cone, 1975). The Paleozoic sequence includes the Cambrian–Ordovician Bliss Sandstone; Ordovician El Paso and Montoya Groups; Silurian Fusselman Formation; Devonian Canutillo Formation and Percha Shale; Mississippian Las Cruces, Rancheria, and Helms Formations; Pennsylvanian Magdalena Group; and Permian Hueco Group. The Bliss Sandstone is unconformably underlain by the Mesoproterozoic Red Bluff Granite that includes Caster Limestone, Mundy Breccia, Llanoria Quartzite, and rhyolite porphyry. The Red Bluff Granite yielded zircon U-Pb dates of 1120 ± 35–1086 ± 5 Ma (Shannon et al., 1997; Mosher, 1998); they interpreted the granite to have been emplaced in a continental-rift setting. The Permian Hueco Group is unconformably overlain by the Cretaceous Trinity Group and younger strata.

(*continued*)

TABLE 1. KEY AND EXPLANATIVE GLOSSARY TO FIGURES 1–4 (*continued*)

Abbreviation	Explanation
8	Van Horn Mountains, Trans-Pecos region, West Texas. A thick section of Mesoproterozoic middle-grade metamorphic rocks intruded by pegmatites is exposed in the Mica Mine area (Flawn, 1951, 1956). These rocks are unconformably overlain by Permian (Wolfcampian) strata. The metamorphic section consists of feldspathic quartzite, feldspathic muscovite and biotite schist, and amphibolite. The pegmatites conform to and cut across the foliation of the host rock (Wasserburg et al., 1962). K-Ar and Rb-Sr age determinations on muscovite and biotite schist and hornblende in amphibolite ranged from 1030 ± 30–940 ± 20 Ma, and on muscovite, biotite, and feldspar in pegmatite ranged from 1090 ± 20–970 ± 20 Ma (Wasserburg et al., 1962). Metarhyolites in the Carrizo Mountain Group yielded U-Pb zircon dates of 1380 ± 20–1327 ± 28 Ma (Mosher, 1998). These dates represent the Mesoproterozoic Grenville and Southern Granite-Rhyolite crustal provinces of eastern and southern United States (Fig. 2). Ordovician dolostone and subordinate sandstone of the El Paso, Simpson, and Montoya Groups are exposed in the Baylor and Beach Mountains, 13–15 km north of Van Horn, Texas (Suhm and Ethington, 1975). This sequence was deposited in a shallow-marine depression (Tobosa basin) on the Laurentian cratonic platform.
9	Northern Sierra Los Ajos, northeastern Sonora. Late Paleozoic cratonic platform of Laurentia (Bridges, 1970; Page et al., 2003a). Page et al. (2003a) report that the northern part of the range is cored by Proterozoic, Cambrian, Devonian, Mississippian, and Pennsylvanian strata, equivalent respectively, to: the Pinal Schist, Bolsa Quartzite and Abrigo Limestone, Martin Formation, Escabrosa Limestone, and Horquilla Limestone. These Paleozoic rocks are unconformably overlain by strata equivalent to the Upper Cretaceous Cabullona Group (Page et al., 2003a).
10	Bacoachi (Cerro La Cal), northeastern Sonora. Paleozoic cratonic-platform strata of Laurentia. Proterozoic porphyritic granite unconformably overlain, in ascending order, by: Bolsa Quartzite, Abrigo and Martin Formations, and Escabrosa Limestone (Stewart and Poole, 2002; Stewart et al., 2002; Poole, unpublished data). U-Pb dates on detrital zircons from the Middle Cambrian Bolsa Quartzite range from 1.632 Ga to 525 Ma with a main age peak at 1.108 Ga (Gross et al., 2000; Stewart et al., 2001).
11	Sierra de Teras (Bavispe–El Tigre area), northeastern Sonora. Paleozoic carbonate platform of Laurentia (Imlay, 1939; Bridges, 1970; Holcomb, 1979; Poole et al., 1995b). Lower Ordovician–Middle Permian shallow-marine strata exposed in Santa Rosa Canyon consist of Lower Ordovician limestone unconformably overlain by Lower Mississippian (Osagean) limestone followed by Upper Mississippian, Pennsylvanian, and Lower Permian strata unconformably overlain by Cretaceous rocks (Holcomb, 1979). Lower Ordovician rocks are in fault contact with a granitic pluton of unknown age. The Bavispe shelf, located west of the Pedregosa basin, formed during the Permian.
12	Estación Sahuaro area, western part of Altar Desert, northwestern Sonora. Neoproterozoic(?) and Paleozoic outer cratonic-platform to possibly inner continental-shelf strata of North America. Two prominent limestone ridges, 12–19 km north of railway Estación Sahuaro, consist of complexly deformed Neoproterozoic(?) and Paleozoic limestone, cherty limestone, dolostone, marble, argillite-phyllite, and sandstone-quartzite. Sequence is mostly intertidal nonfossiliferous carbonate rocks except for sparse shallow-subtidal upper Paleozoic limestone units containing crinoids, brachiopods, conodonts, etc. (Poole and Amaya-Martínez, work in progress). Several limestone samples yielded conodonts that were studied by C.A. Sandberg. Conodonts indicate the presence of Lower Mississippian, Lower Pennsylvanian, and Lower and Middle Permian carbonate rocks that are associated with subordinate cherty, sandy, and silty carbonate rocks, quartzite, argillite, and phyllite. Several dolomitic silty limestone beds are burrowed. This area was first studied by Calmus (1993) and Calmus and Sosson (1995). Pre-Neoproterozoic metamorphic basement rocks in the area are part of the Mesozoic Papago terrane (Calmus and Sosson, 1995). Contact relations between the pre-Neoproterozoic, Neoproterozoic(?), and Paleozoic rocks are unknown. Proterozoic and Paleozoic sequence in the Estación Sahuaro area indicates Laurentian continental crust.
13	Virgin Ridge area, northern Virgin Mountains, northwesternmost Mohave County, Arizona. Paleozoic outer cratonic-platform strata of western Laurentia. The sequence consists of Paleoproterozoic crystalline rocks; Cambrian Tapeats Sandstone, Bright Angel Shale, Muav Limestone, unnamed dolomite unit, Nopah Formation; Devonian Sultan Limestone; Mississippian Monte Cristo Limestone; Pennsylvanian and Permian Bird Spring Formation; Permian Esplanade Formation, Coconino Sandstone, Toroweap Formation, and Kaibab Limestone (Bohannon and Lucchitta, 1991). Triassic and Jurassic strata overlie the Paleozoic sequence. The basement rocks are part of the Paleoproterozoic Mojave crustal province (K.E. Karlstrom, 2003, personal commun.). Paleoproterozoic to Mississippian rocks exposed ~50 km to the northwest in the East Mormon Mountains, southeasternmost Lincoln County, Nevada (Poole and B.P. Wernicke, unpublished measured section), are similar to innermost continental-shelf rocks of southern Nevada and southeastern California.
14	Northern Providence Mountains, eastern San Bernardino County, California. Proterozoic basement (gneiss and schist), Neoproterozoic and Cambrian sedimentary rocks (Hazzard, 1954; Stewart, 1970; Stewart and Poole, 1975; Brown, 1991), and Devonian to Triassic sedimentary rocks (Hazzard, 1954; Poole et al., 1977; Brown, 1991). Sequence represents an outermost cratonic-platform setting underlain by Laurentian continental crust.
<u>Continental-shelf rocks of Laurentia</u>	
15	Goodsprings area, southern Spring Mountains, southwestern Clark County, Nevada. Paleozoic carbonate-shelf strata of western Laurentia. Consists of Middle Cambrian to Lower Permian carbonate rock and subordinate shale and sandstone (Hewett, 1931; Carr and Pinkston, 1987; Poole and Sandberg, 1991; Cooper and Edwards, 1991; Poole, unpublished data). Section represents an inner carbonate-shelf setting.
16	Santa Ana–Estación Llano area (includes area of San Francisco gold deposit), north-central Sonora. Proterozoic metamorphic rocks representing at least two Proterozoic complexes (Salas, 1968; Calmus et al., 1992). Of the three Proterozoic metamorphic units (Gauna Rhyolite, San Francisco, and Coyotillo), the San Francisco unit has been dated as 911 and 770 Ma by the $^{40}Ar/^{39}Ar$ method (Calmus et al., 1992). Proterozoic rocks are unconformably overlain by Jurassic strata (Calmus et al., 1992). No Paleozoic rocks are exposed. The hypothetical northwest-striking Jurassic Mojave–Sonora megashear of Silver and Anderson (1974) and Anderson and Silver (1979) passes 5–10 km south of railway Estación Llano (Nourse, 1995).

(*continued*)

TABLE 1. KEY AND EXPLANATIVE GLOSSARY TO FIGURES 1–4 (*continued*)

Abbreviation	Explanation
17	Rancho El Bísani, northwestern Sonora. Paleozoic carbonate-shelf strata of southwest Laurentia. Consists of complexly deformed dolostones, cherty and sandy dolostones, sandstones, and limestones exposed in isolated hills. Outcropping rocks range in age from Late Ordovician to Late Mississippian (Cooper and Arellano, 1946; Easton et al., 1958; Brunner, 1975, 1976; Poole et al., 1995b; Poole and coworkers, unpublished data). Paleozoic shelf sequence implies the presence of Laurentian Proterozoic crust below. Northwest and southeast of Rancho El Bísani and Caborca, Proterozoic rocks are exposed in widely scattered outcrops (Anderson and Silver, 1981; Iriondo and Premo, 2003; Premo et al., 2003). In the northern foothills of Sierra del Álamo (El Antimonio area), 20 km northwest of Rancho El Bísani, the Middle Permian (Guadalupian) Monos Formation crops out in a limited area (Monos Hills). The base of the Monos is not exposed in the Sierra del Álamo. The Upper Permian, Triassic, and Lower Jurassic marine Antimonio Formation is 3400 m thick and disconformably overlies the Monos Formation (González-León, 1980, 1997). The marine Monos Formation consists of 300 m of pale-red siltstone, fine-grained sandstone, and minor clayey limestone in its lower part, and of 200 m of thin- to thick-bedded, medium-gray fossiliferous limestone in its upper part (González-León, 1980, 1997; Stanley and González-León, 1995). The lower part of the Monos Formation is poorly exposed and has yielded no fossils, but the upper part has yielded a varied fauna of fusulinids, brachiopods, pelecypods, scaphopods, corals, cephalopods, gastropods, and sponges studied by Cooper et al. (1953). The paleotectonic setting of the Monos and Antimonio Formations has been interpreted by González-León (1980, 1989, 1997), Stanley and González-León (1995), and most of their co-workers, to be part of an exotic or accreted terrane (Antimonio terrane) resting on the North American Caborca terrane, and by a few other workers (e.g., Gómez-Luna and Martínez-Cortés, 1997) to be near its original position and part of the North American margin. Proponents of an exotic or accreted terrane consider the section to be part of the Cortés terrane (Coney and Campa, 1987) thrust onto the Caborca terrane (Coney and Campa, 1987; González-León, 1997). González-León (1980, 1989, 1997) and Stanley and González-León (1995) correlated part of the Upper Permian–Lower Jurassic Antimonio Formation with the coeval Luning Formation or assemblage in west-central Nevada on the basis of interpreted stratigraphic and paleontologic similarities, and proposed that the Nevada section was transported southeastward to Sonora along the hypothetical left-lateral Mojave–Sonora megashear (MSM) during Middle to Late Jurassic time. Alternatively, Stewart (this volume) proposed that the Antimonio Formation is offset 700–800 km along the MSM from the coeval Upper Triassic and Lower Jurassic Kings sequence exposed in the southern Sierra Nevada of California. In contrast, Gómez-Luna and Martínez-Cortés (1997) reported many stratigraphic, structural, and paleontologic dissimilarities between the Antimonio and Luning Formations, and concluded that the data do not support a tectonic model of transcurrent strike-slip displacement from Nevada to Sonora. Part of the Upper Permian–Lower Jurassic Antimonio Formation in northwest Sonora has some lithologic similarities to the Upper Triassic, Jurassic, and Lower Cretaceous El Gameño Formation/Group in west-central Sonora (Amaya-Martínez, Poole, and D.A. Amaya-Zepeda, work in progress) and to the Upper Triassic Barranca Group in central Sonora (Stewart and Roldán-Quintana, 1991). Both the El Gameño and Barranca unconformably overlie Paleozoic strata that are attached to, and are part of, the North American continent. Similarly, the Jurassic Pozos de Serna sequence unconformably overlies Neoproterozoic or Cambrian strata (Calmus et al., 1997), and is attached to the North American continent. Clearly, lower and middle Mesozoic rocks in Sonora are attached to subjacent Neoproterozoic and Paleozoic rocks of the North American continent. Therefore, they may reside near their original depositional site with minimal subsequent displacement. More studies need to be done on late Paleozoic and Mesozoic rocks in western Sonora to determine their paleogeographic and paleotectonic settings. In the Pozos de Serna area, 65 km south-southeast of Rancho El Bísani, Lower and Middle Jurassic rocks unconformably overlie Neoproterozoic or Cambrian strata of the so-called Caborca terrane (Calmus et al., 1997). They correlated the Jurassic Pozos de Serna rocks with the 2000-m-thick upper member of the Antimonio Formation in Sierra del Álamo (Antimonio terrane of González-León, 1989), and suggested that the attached Caborca and Antimonio terranes are allochthonous and accreted as a unified block to the North American craton. We agree that rocks of the so-called Antimonio and Caborca terranes are part of the same block, but our data indicate that rather than being exotic to North America or displaced along the continental margin, the rocks may be located near their original depositional site on the southwest margin of the North American craton. If so, use of the lithotectonic terrane terms Caborca and Antimonio are probably invalid and should be reevaluated.
18	Sierra El Rajón–Sierra de La Berruga (Cerro Calaveras), northwestern Sonora. In Sierra El Rajón, Neoproterozoic, Lower and Middle Cambrian strata are ~2560 m thick and unconformably overlie Proterozoic porphyritic granite (Stewart et al., 1984; Stewart and Poole, 2002; Stewart et al., 2002). The granite yielded U-Pb zircon ages of 1400 ± 20 Ma (L.T. Silver *in* Stewart et al., 1984) and 1703 ± 12 Ma (Premo et al., 2003). Cambrian strata are unconformably overlain by the Jurassic Rajón Group (Stewart et al., 1984). At Cerro Calaveras, Neoproterozoic and Cambrian sedimentary rocks are unconformably overlain by Mesozoic rocks (Eells, 1972; Anderson, et al., 1978). A foliated granite northwest of Cerro Calaveras yielded a U-Pb zircon age of 1737 ± 17 Ma (Premo et al., 2003). Sedimentary strata represent a continental-shelf setting, and imply Laurentian crust below.
19	Big Bear area, San Bernardino Mountains, southwestern San Bernardino County, California. Paleozoic carbonate-shelf strata of North America. Consists of Paleoproterozoic orthogneisses (1.78–1.65 Ga, Coleman et al., 2003) intruded by plutons (ca. 1750 ± 15 Ma, Silver, 1971), Neoproterozoic–Cambrian metasedimentary rocks, and Devonian-Permian metasedimentary rocks intruded by Permian or Triassic monzonite (Stewart and Poole, 1975; Cameron, 1982; Brown, 1991). Devonian–Permian strata are similar to those in the southern Spring Mountains (Poole, unpublished data). Sedimentary section represents a Laurentian carbonate-shelf setting.
20	Nopah Range, southeasternmost Inyo County, California. Southern part of range exposes Paleoproterozoic crystalline basement unconformably overlain by Neoproterozoic and Cambrian sedimentary rocks (Hazzard, 1937; Stewart, 1970). Northern part of range exposes Ordovician to Pennsylvanian sedimentary rocks (Hazzard, 1937; Cooper and Edwards, 1991; Poole, unpublished data). Sequence represents inner carbonate-shelf setting of western Laurentia.
21	Tucki Mountain area, Death Valley, Inyo County, California. Neoproterozoic and Paleozoic continental-shelf strata of western Laurentia. Consists of a thick sequence of Proterozoic–Permian metamorphic and sedimentary rocks (Hunt and Mabey, 1966) that represent a medial shelf setting. Nearby Proterozoic crystalline rocks represent Laurentian basement.
22	Mazourka Canyon area, southern Inyo Mountains, west-central Inyo County, California. Paleozoic outer continental-shelf and -slope strata of western Laurentia. Consists of Cambrian, Ordovician, Silurian, Devonian, and Mississippian sedimentary rocks (Ross, 1966; Stevens, 1986; Stevens et al., 1996). Pennsylvanian, Permian, and Triassic strata also are present in nearby areas of the southern Inyo Mountains (Stone et al., 1989).

(*continued*)

TABLE 1. KEY AND EXPLANATIVE GLOSSARY TO FIGURES 1–4 (*continued*)

Abbreviation	Explanation
23	Cerro de Oro–El Sotol area, north-central Sonora. Neoproterozoic epicontinental strata described by Stewart et al. (2002) include Las Víboras Group (<1200 m thick) and El Aguila Group (>1000 m thick) deposited near the inferred northwest-striking Chihuahua–Sonora transform of this report, and the similarly located hypothetical Mojave–Sonora megashear of Silver and Anderson (1974) and Anderson and Silver (1979). Sediments may have been deposited in transform basins formed during initial movement along our postulated Neoproterozoic Chihuahua–Sonora transform fault. Strata presumed to overlie Laurentian Proterozoic crust.
24	Sierra Agua Verde, east-central Sonora. Paleozoic carbonate-shelf strata of North America. Over 3700 m of inner-shelf Neoproterozoic(?), Cambrian, Lower Ordovician, Upper Devonian, Mississippian, and Lower and Middle Pennsylvanian strata are exposed (Stewart et al., 1999b). Neoproterozoic and Paleozoic sequence implies the presence of Laurentian continental crust below.
25	Sierra Martínez, central Sonora. Paleozoic carbonate-shelf strata of southern Laurentia. Thick sequence of Ordovician, Devonian, Mississippian, Pennsylvanian, and Permian carbonate-shelf strata (Poole, W.R. Page, and A.G. Harris, unpublished data). The uppermost beds of the carbonate-shelf consist of Lower Permian fusulinid-bearing crinoidal limestone that is conformably overlain by Lower and Middle Permian foredeep flysch (Mina México Formation) through a transition interval 5–15 m thick. On the southeast side of Sierra Martínez, the exposed Mina México turbidite sequence is at least 500 m thick, and is structurally overlain by the Sonora allochthon, which consists of deep-marine Ordovician-Permian rocks (Figs. 10 and 11). Presence of Paleozoic-shelf strata implies Laurentian continental crust below.
26	Rancho Las Norias–Rancho El Albañil area, central Sonora. Paleozoic carbonate-shelf strata of southern Laurentia. Sequence of shelfal Cambrian, Lower and Middle Ordovician, Middle and Upper Devonian, Mississippian, Pennsylvanian, and Lower Permian strata (Page et al., 2003b). Although Proterozoic rocks are not exposed in this area, the Paleozoic shelfal section implies presence of Laurentian continental crust below.
27	Arivechi–Sahuaripa area, east-central Sonora. Neoproterozoic epicontinental strata and Paleozoic carbonate-shelf strata of southern Laurentia. Incomplete sequence of shelfal Neoproterozoic, Cambrian, Ordovician, Mississippian, and Permian strata unconformably overlain by Mesozoic strata (Almazán-Vázquez, 1989; Fernández-Aguirre and Almazán-Vázquez, 1991). In the Sahuaripa area, Neoproterozoic epicontinental strata and Cambrian inner-shelf to outer-platform deposits were described by Stewart et al. (2002) as the Cerro Las Bolas Group (2600 m thick) and Cambrian cratonal-cover strata. Neoproterozoic and Paleozoic sequence implies the presence of Laurentian continental crust below. Area is southwest of the inferred northwest-striking Chihuahua–Sonora transform of this report, and the similarly located hypothetical Jurassic Mojave–Sonora megashear of Silver and Anderson (1974) and Anderson and Silver (1979). Neoproterozoic sediments may be deposited in a transform basin formed during initial movement along our postulated Neoproterozoic Chihuahua–Sonora transform fault.
28	Sierra Batamote, east-central Sonora. Unstudied Paleozoic carbonate-shelf strata of southern Laurentia. The thick sequence is exposed in an uplifted mountain block north of a major east-west-trending high-angle fault that downdrops rocks of the Mina México foredeep on the south (Poole and Amaya-Martínez, work in progress). Both the carbonate-shelf and superjacent foredeep flysch are exposed in roadcuts along Sonora Highway 104 between 4 and 10 km east of the Río Yaqui. The Mina México Formation consists of metamorphosed turbiditic very fine grained quartzite and siltite, and sparse coarse-grained lime grainstone. We interpret the contact between the shelfal limestone (marble) and overlying foredeep quartzite and siltite to be conformable and intertonguing. In addition, we postulate that rocks of the Sonora allochthon will be found in this area. Paleozoic shelfal rocks indicate Laurentian continental crust below.
29	Hermosillo, Sonora. Paleozoic carbonate-shelf strata of southern Laurentia. Outcrops of complexly deformed calcite marbles and calcsilicates exposed in numerous hills and cement quarries within the city limits of Hermosillo. Poole and Amaya-Martínez provisionally correlate these undated metasedimentary carbonate rocks with upper Paleozoic shelfal limestones of the region. Several samples were collected for conodonts, but all were barren. Sequence implies the presence of Laurentian continental crust below. These metamorphosed carbonate rocks occur as roof pendants in the batholithic belt.
30	Rancho Placeritos–Cerro El Carnero, west-central Sonora. Paleozoic carbonate-shelf strata of southwest Laurentia. At Rancho Placeritos strata consist of limestones, cherty and sandy limestones, dolostones, and quartzites intruded by Paleocene granite plutons. Most Paleozoic strata at Cerro El Carnero are metamorphosed. The Paleozoic rocks range in age from Early Ordovician to Early Mississippian; they are unconformably overlain by Mesozoic continental clastic deposits with several interbedded marine deposits (Poole et al., 2000, 2003a, 2003b; Poole, Amaya-Martínez, and D.A. Amaya-Zepeda, unpublished data). The El Gameño Formation/Group consists of Upper Triassic, Jurassic, and Lower Cretaceous rocks that contain abundant volcanic detritus (Poole and Amaya-Martínez, work in progress). Paleozoic shelf sequence implies the presence of Laurentian Proterozoic continental crust below. Paleozoic and Mesozoic strata occur as large roof pendants in batholithic belt.
31	San Jacinto Mountains, Riverside County, southern California, U.S.A. Autochthonous metasedimentary rocks include the Windy Point metamorphic rocks of Sydnor (1975), and the Desert Divide Group of Brown (1980, 1981). The Windy Point metamorphic rocks consist of amphibolite and plagioclase-rich schist and quartz- and carbonate-rich metasedimentary rocks (Todd et al., 1988). The Desert Divide Group consists of a large body of medium- to coarse-grained quartzite on the east called the Ken Quartzite, and a diverse sequence of metacarbonate rocks, schist, and gneiss on the west called Bull Canyon Formation (Todd et al., 1988). Todd et al. (1988) suggested that the sediments were derived largely from a Precambrian source terrane and deposited in a Paleozoic continental-shelf setting on Laurentian cratonic crust.
32	Santa Rosa Mountains, Riverside and San Diego Counties, southern California, U.S.A. Autochthonous metasedimentary rocks include the Palm Canyon Complex or Formation and Santa Rosa Formation (Todd et al., 1988; B.G. Erskine *in* Gastil et al., 1991). The Palm Canyon Complex in the vicinity of Palm Springs, California, is a sequence of intensely deformed metasedimentary rocks, anatexites, and orthogneiss (Todd et al., 1988). B.G. Erskine (*in* Gastil et al., 1991) described the Santa Rosa Formation as ranging from nearly pure quartzite to quartz-rich garnet-sillimanite-cordierite schists, and gneiss, interbedded with marble, and a variety of calc-silicate rocks. According to Erskine, the Palm Canyon Formation is present throughout the extent of the Santa Rosa Mountains, and is similar to the Santa Rosa Formation except that they are strongly migmatitic and generally lack pure metaquartzite. Conodonts of Early Ordovician age were reported from carbonate-quartzite strata in the Santa Rosa Mountains (A.E.J. Engel *in* Gastil, 1985). The Santa Rosa and Palm Canyon Formations were probably part of a Paleozoic miogeoclinal quartz-rich clastic and carbonate-bearing sequence deposited on Laurentian cratonic crust.
33	Borrego Valley, San Diego County, southern California, U.S.A. Ranges west of Borrego Valley contain autochthonous metasedimentary rocks similar to those in the San Jacinto and Santa Rosa Mountains (Todd et al., 1988). Conodonts of Early Ordovician age were reported from carbonate-quartzite strata in the Borrego Valley area (A.E..J. Engel *in* Gastil, 1985). The rocks were probably deposited in a Paleozoic miogeoclinal setting similar to localities 31, 32, and 34.

(*continued*)

TABLE 1. KEY AND EXPLANATIVE GLOSSARY TO FIGURES 1–4 (*continued*)

Abbreviation	Explanation
34	Carrizo Mountain, Coyote Mountains, western Imperial County, southern California, USA. Magnesium-rich carbonate rock (marble), containing Early Ordovician conodonts, structurally overlies metaquartzite and conglomeratic arkose (Dockum, 1982; Gastil and Miller, 1983; Miller and Dockum, 1983; Gastil et al., 1991; Madrid et al., 1992). The carbonate rocks correlate with Lower Ordovician shelf strata of the Pozo Nuevo Limestone in the Sierra López–Rancho Placeritos area (loc. 30) on mainland Mexico. The strata apparently represent a Paleozoic miogeoclinal sequence deposited along the southwestern margin of Laurentia, and not part of an accreted terrane as was postulated by Madrid et al. (1992).
35	San Felipe area, northeastern Baja California Norte. Anderson (1982) reported at least 3800 m of strata that he subdivided into eight map units, consisting of orthoquartzite, marble, meta-argillite, and micaceous quartzite. He interpreted the sequence to represent rapidly deposited blanket sands and sublittoral sheet sands in a shallow-water marine environment. Anderson (1982) and Gastil et al. (1991) correlated the sequence with Neoproterozoic and Lower Cambrian rocks in the Caborca region (loc. 17). Their correlation is unlikely because the section in the Caborca area is dissimilar to the section at San Felipe. Gehrels et al. (2002) reported that detrital zircon-age spectra of San Felipe (loc. 35) and Rancho San Marcos (loc. 52) quartzites are similar to Ordovician arenites of western Laurentia, and suggest that Baja California Norte strata formed in proximity to strata in northwest Mexico and southwest United States. This sequence of questionable early Paleozoic age at San Felipe is anomalous. If the San Felipe section is Ordovician in age, it is probably part of the carbonate-shelf sequence deposited on the southwest margin of Laurentia.
Foredeep and foreland-basin rocks on continental shelf of Laurentia	
36	La Vinata–Sierra de Azcárate area, near Nuevo Casas Grandes in northwestern Chihuahua. Wolfcampian part of late Paleozoic Pedregosa foreland-basin fill consisting of deep-water flysch at La Vinata and basin-margin limestone facies at Sierra de Azcárate. The flysch correlates with the Rara and Plomosas Formations to the southeast in central Chihuahua. Proterozoic basement of southern Laurentia underlies the Pedregosa basin.
37	Sierra de La Mojina, north-central Chihuahua. Lower Cretaceous conglomerate composed of clasts of Neoproterozoic metamorphic and volcanic rocks and Lower Permian rhyolite, implying that the Paleozoic sedimentary sequence was absent in the source area (Denison et al., 1970; Bridges, 1970). Denison et al. (1970) reported metamorphic dates of 255–231 Ma (Permian–Triassic) on muscovite schist (K-Ar, muscovite) boulders in the conglomerate. King (1975) reported 350–330 Ma (Mississippian) metasedimentary clasts in the Lower Cretaceous conglomerate. Although the exact source area for the conglomerate is unknown, Handschy et al. (1987) reported that the freshness of clast lithology and preliminary study of clast imbrication indicate a source immediately to the west. At the base of a graphic section, Ramírez-M. and Acevedo (1957, fig. 32) show a unit composed of reheated green mudstones with isolated calcareous nodules and subordinate sandstone beds that underlies Lower Cretaceous conglomerate and sandstone. Apparently they included this undated unit in the Mesozoic. W.T. Haenggi (2002, written commun.) speculated that the unit could be a facies related to the Marathon segment of the Ouachita–Marathon–Sonora orogen. Other speculations are that this pre-Mesozoic? unit correlates with rocks of the interior metamorphic zone or a Gondwana terrane unit, such as, the flysch-like green unit at Sierra de Catorce (loc. 71) described by Franco-Rubio (1999).
38	Sierra El Cuervo (30 km north of Ciudad Chihuahua), central Chihuahua. Southern part of Pedregosa foreland basin. Upper Pennsylvanian–Lower Permian Rara Formation consisting of a thick sequence of weakly metamorphosed, thinly bedded turbiditic sandstones, shales, and carbonate rocks, and minor conglomerates, altered bentonites, and bedded chert unconformably overlain by Cretaceous shallow-marine and marginal-marine strata (Blount, 1983; Handschy and Dyer, 1987). Basement rocks consist primarily of metagranites with subordinate amounts of granitic gneisses, metagabbro, and metadiorite (Mosher, 1998). U-Pb zircon dates of 1333 +10/–8 Ma on metagrabbo, 1274 +6/–5 Ma on metagranite, and 1080 ± 5 Ma on crosscutting pegmatitic trondhjemites indicate that these rocks are Mesoproterozoic in age (Blount, 1993; Mosher, 1998). In the Rancho Los Filtros area (south-central Sierra El Cuervo), intensely deformed Rara Formation flysch contains tectonic blocks of Mesoproterozoic igneous and metamorphic rocks, including metagranites, amphibolites (metamorphosed mafic dikes), and minor gneisses; two K-Ar dates on hornblende from amphibolites are 1037 and 1024 Ma (Blount, 1983; Mauger et al., 1983). These dates imply that Laurentian Grenville crust underlies the Pedregosa foreland basin in central Chihuahua. Alternatively, this Grenville-age crust may be allochthonous blocks derived from the northwestern margin of Gondwana (South America) during its late Paleozoic collision with southern Laurentia. Easterly paleocurrent directions and clast compositions in the Rara Formation indicate crystalline rock exposed in a source terrane to the west, possibly a basement-cored (Sierra del Nido block?) uplift (Handschy and Dyer, 1987). Late Paleozoic deformation in Sierra El Cuervo includes tight to isoclinal folding of Rara and imbricate thrusting of Rara and Proterozoic rocks, indicating tectonic transport toward the east-southeast, and suggests that a western basement-cored continental block overrode the Pedregosa basin to the east during Ouachita–Marathon–Sonora orogenesis (Handschy and Dyer, 1987). The apparent unroofing of a western basement-cored block, followed by east-southeast-directed thrusting, is inconsistent with the northwest tectonic transport direction of the Ouachita–Marathon–Sonora frontal zone, and suggests that the late Paleozoic structures in Sierra El Cuervo are related to Ancestral Rocky Mountains deformation (Handschy and Dyer, 1987).
39	Mina Plomosas–Placer de Guadalupe–Cerro El Carrizalillo area, east-central Chihuahua. Upper Paleozoic foreland-basin strata and lower Paleozoic carbonate-shelf strata of southern Laurentia are unconformably underlain by Proterozoic schist similar to schist in the Mesoproterozoic Carrizo Mountain Group near Van Horn, Texas (L.W. Bridges, 2002, personal commun.). An eastern outcrop at Cerro El Carrizalillo contains multiply deformed mafic schists intruded by trondhjemite and granite dikes and sills (Blount, 1993; Mosher, 1998). Ages of the mafic schists and granitoids are unknown, but are presumed to be Mesoproterozoic. Lower Paleozoic rocks represent outer-shelf facies whereas upper Paleozoic rocks represent a foreland-basin setting. Fossils in the Paleozoic sequence are described by Bridges (1962) and Benedetti (1976). Late Ordovician, Silurian, and Early Devonian brachiopods in the Solis Limestone are described by Sheehan (1975a and 1975b). Post-middle Wolfcampian to pre-Jurassic tectonic events are recognized in the area (Bridges, 1962; Dyer and Reyes, 1987; Roberts, 1989). Conglomerate in the upper part of the Lower Permian Plomosas Formation indicates tectonism during deposition (Bridges, 1962). Below the conglomerate within the basal Plomosas, are Early Permian fusulinids (Bridges, 1962) and a Leonardian/Guadalupian extrusive rhyolite (270 ± 30 Ma by Pb-alpha, De Cserna et al., 1968); clasts of this rhyolite occur locally in the overlying conglomerate (Bridges, 1962). We interpret the Proterozoic basement, Paleozoic shelf and foreland-basin sequence in this area, to be north of the Ouachita–Marathon–Sonora orogenic belt (see Figures 1 and 3). The rhyolite probably indicates proximity to a volcanic arc along the northern margin of Gondwana.

(*continued*)

TABLE 1. KEY AND EXPLANATIVE GLOSSARY TO FIGURES 1–4 (*continued*)

Abbreviation	Explanation
40	Mina México–Cañon de Río Bacanora–Cerro Santo Domingo area (11–18 km south of the village of Bacanora, Sonora), east-central Sonora. A sequence of Lower to Middle Permian thin-bedded, fine-grained turbiditic quartzite and siltite and sparse bioclastic lime grainstone was named Mina México Formation by Hewett (1978) and Schmidt (1978). The Mina México was interpreted by them to be depositional on Permian shallow-water carbonate rocks that they named La Cueva Member of El Tigre Formation, and by Radelli et al. (1987) as part of a regional allochthon in thrust contact with the underlying Permian shelfal limestones. The contact is mostly covered throughout the area, but where exposed in moderately deformed areas, it is gradational and intertonguing through a transitional interval as thick as 50 m. We consider the Mina México Formation as a thick (>1000 m) flysch sequence deposited in a foredeep (Mina México foredeep) that developed on the subsiding Lower Permian carbonate shelf in front of the evolving Sonora accretionary wedge. Stewart and Poole (2002) reported conodonts of Pennsylvanian to Early Permian age from two different calcareous siltstone layers in the Mina México Formation (collected in 1984 by K.B. Ketner and J.H. Stewart and studied by B.R. Wardlaw and R.G. Stamm). During the past four years, Amaya-Martínez and Poole (work in progress) collected Early to Middle Permian fusulinids (studied by C.H. Stevens) and conodonts (studied by A.G. Harris and C.A. Sandberg) from several calciclastic turbidite beds in the Mina México Formation. Amaya-Martínez and Poole (work in progress) propose to abandon the name El Tigre Formation and apply the name Cerro Santo Domingo Group for the multiple formations (yet to be defined) that constitute the Paleozoic carbonate shelf in this area. We plan to retain the name La Cueva Limestone, but restrict it to the Lower Permian fossiliferous (crinoids, fusulinids, etc.) shallow-water limestone at the top of the carbonate shelf. A probable upper Middle Ordovician quartzite unit (Peña Blanca Quartzite) and overlying thick sequence of Paleozoic carbonate-shelf rocks crop out on the west side of a sierra in the vicinity of Rancho Los Alisitos (Poole and Amaya-Martínez, unpublished data). A thick section of Paleozoic carbonate-shelf strata crop out on the west side of Cerro Santo Domingo (Amaya-Martínez, work in progress). Three fossil collections from the southeast flank of Cerro Santo Domingo were sent to C.H. Stevens by E. Almazán-Vázquez. Stevens studied the collections and reported Early Permian, and probable Pennsylvanian, Mississippian, and Devonian fossils, including fusulinids, brachiopods, and corals (Stewart and Poole, 2002). Cerro Santo Domingo is a major upfaulted mountain block, and the deep-water Permian Mina México Formation, which stratigraphically overlies the Cerro Santo Domingo shallow-water carbonate-shelf section, is exposed on the downfaulted west block. Although not confirmed, we suspect that rocks of the Sonora allochthon will be found in the Mina México area. The Paleozoic shelfal sequence and depositionally attached overlying foredeep deposits imply Laurentian continental crust below.
41	Cerro San Francisco–El Subiate–Rancho La Garita area (12–14 km southeast of Rancho Las Norias), central Sonora. Outcrops of turbiditic, thin-bedded, fine-grained quartzite and silltite, and medium- to thick-bedded conglomeratic lime-grainstone flysch facies were interpreted by Radelli et al. (1987) as allochthonous. Poole correlates this quartzose and lime-grainstone facies with the Permian Mina México Formation in the Minas de Barita-Sierra El Máviro area (loc. 43) and near Ranchos La Mano and Las Trancas in the vicinity of Mina México (loc. 40), all of which are part of the Mina México foredeep fill. Detrital fusulinids in bioclastic lime grainstones are late Leonardian to early Guadalupian in age (C.H. Stevens, 2003, written commun.). Subjacent Permian shelfal limestones in the area also contain late Leonardian to possibly early Guadalupian fusulinids (C.H. Stevens, 2001 and 2003, written commun.), indicating rapid transition between the shallow-water shelf and overlying deep-water foredeep. Contact between the carbonate-shelf and overlying foredeep rocks is mostly covered, but considered by Poole to be transitional and similar to units 6 and 7 at Sierra Santa Teresa (Stewart et al., 1997). The shelfal carbonate sequence and attached overlying foredeep deposits imply Laurentian Proterozoic basement below.
42	Sierra Santa Teresa–Sierra La Flojera, central Sonora. Sierra Santa Teresa consists of ~2300 m of Mississippian, Pennsylvanian, and Permian strata in the upper plate of a post-Late Triassic thrust fault (Stewart et al., 1997), which we believe to be latest Cretaceous in age. Lower-plate rocks consist of 544 m of Upper Triassic and Lower Jurassic strata, and ~600 m of Mesozoic andesitic rock (Stewart et al., 1997). We consider the andesite unit to be part of the Late Cretaceous Tarahumara Formation related to the Tarahumara volcanic arc (Roldán-Quintana, 2002). Upper Lower (Leonardian) Permian and possibly lowermost Middle (Guadalupian) Permian deep-water calcareous siliciclastic rocks (unit 7) overlie Lower Permian (Leonardian) shallow-water fossiliferous limestone (unit 6). Stewart et al. (1997) interpreted the contact between units 6 and 7 as transitional, and indicative of abrupt subsidence of the shallow-water carbonate shelf to accommodate the sudden influx of deep-water turbiditic calcareous siliciclastic sediments. Stewart et al. (1997) correlated unit 7 at Sierra Santa Teresa with the Mina México Formation of east-central Sonora. We make the same correlation and consider unit 7 of Stewart et al. (1997) as flysch deposited in the Mina México foredeep. In the Sierra La Flojera, 12 km west-northwest of Sierra Santa Teresa, Lower Permian shelfal limestone is overlain by >300 m of Lower and possibly lower Middle Permian deep-water turbiditic siltstone, silty limestone, bioclastic limestone, and conglomerate (Stewart et al., 1990). We consider this Permian flysch to be Mina México foredeep fill, and probably includes Leonardian and Guadalupian strata. Although the contact between the shelfal limestone and superjacent foredeep Mina México is not exposed at Sierra La Flojera, stratigraphic relations are similar to that in Sierra Santa Teresa, where the contact between the two formations is transitional and the formations are part of the Laurentian continental shelf. An isolated low hill of Upper Triassic marine strata is present at the northern end of Sierra La Flojera (Stewart et al., 1990; Lucas and González-León, 1994). Although the contact between the Permian Mina México Formation and Upper Triassic strata is not exposed, the close proximity of the two units at Sierra La Flojera suggests that the Triassic deposits may be in stratigraphic continuity with the Permian foredeep flysch. If so, Upper Triassic strata may depositionally overlap Permian strata, which would suggest that the Triassic rocks also are part of the Laurentian continental margin, and not part of an accreted terrane. The Paleozoic sequence here implies the presence of Laurentian continental crust below.
Ocean-basin rocks in Sonora allochthon and subjacent foredeep flysch	
43	Minas de Barita–Sierra El Máviro, central Sonora. Complexly deformed eugeoclinal Ordovician, Silurian?, Upper Devonian, Mississippian, Pennsylvanian, and Permian deep-water facies characterize the Sonora allochthon. Rocks no older than Early Ordovician have been recognized in the eugeoclinal sequence (Fig. 5). Footwall rocks consist of Lower and Middle Permian Mina México foredeep flysch and subjacent Lower Permian and older Paleozoic shelfal limestones. No Proterozoic basement is exposed below the shelf sequence in this area. The Mina México flysch depositionally overlies the shelfal limestone, and the foredeep flysch is overridden by the Sonora allochthon. At Cerro La Ciénega in the Sierra El Máviro, a very fine grained quartzite unit 100–200 m thick (A. Fort-Martínez, 2001, personal commun.) lies between allochthonous Paleozoic eugeoclinal strata (Sonora allochthon) above and Permian carbonate-shelf strata below. We provisionally correlate the quartzite unit with the Mina México Formation of central and east-central Sonora. A strongly deformed thick sequence of turbiditic quartzite, siltite, argillite, and subordinate chert and limestone is correlated with the Los Pozos Formation, Cerro Tasajo Limestone, and Rancho Nuevo Formation exposed in the canyon of arroyo Máviro between 0.3 and 2 km south of Rancho El Salto, which is located several kilometers southeast of Sierra El Máviro (Poole and Amaya-Martínez, work in progress). The folded and faulted Los Pozos, Cerro Tasajo, and Rancho Nuevo forms most of the Sonora allochthon south of Rancho El Salto. Here the thickness of the allochthon is at least 500 m (estimated from topographic relief in the canyon of arroyo Máviro). Sonora allochthon rocks are in fault contact (thrust or reverse fault) with Upper Cretaceous Tarahumara volcanic rocks 300 m south of Rancho El Salto. Paleozoic shelfal sequence implies the presence of Laurentian Proterozoic basement below.

(*continued*)

TABLE 1. KEY AND EXPLANATIVE GLOSSARY TO FIGURES 1–4 (*continued*)

Abbreviation	Explanation
44	Cerro Cobachi (~10 km southeast of village of Cobachi), central Sonora. Outcrops of complexly deformed eugeoclinal Ordovician, Silurian?, Upper Devonian, Mississippian, Pennsylvanian, and Permian deep-water facies of the Sonora allochthon. Footwall rocks consist of shelfal Cambrian? to Lower Permian strata assigned by Ketner and Noll (1987) to the Tinajas Group (lower part of section) and Picacho Colorado Formation (upper part of section). The La Vuelta Colorada Formation, which is the youngest Paleozoic unit recognized in the Cerro Cobachi area, consists of deep-water, fine-grained siliciclastic and sparse coarse-grained calciclastic turbidites containing fusulinids, conodonts, and radiolarians of Early Permian and Pennsylvanian age (Ketner and Noll, 1987). The La Vuelta Colorada probably depositionally overlies the shelfal limestone sequence and is overridden by the Sonora allochthon (Poole and Amaya-Martínez, work in progress). We correlate the La Vuelta Colorada Formation with both the upper Rancho Nuevo Formation and the Mina México Formation of central and east-central Sonora. More work is needed to establish the affinity of the La Vuelta Colorada Formation with similar-age deep-water facies in Sonora. However, we provisionally consider the formation to represent foredeep flysch equivalent to the Mina México Formation. Stratigraphy and structure in the Cobachi area are similar to that in the Minas de Barita area 20 km to the north-northeast. Paleozoic shelfal sequence implies the presence of Laurentian continental crust below.
45	Rebeiquito area, central Sonora. Outcrops of Ordovician black shale, black and gray chert, and turbiditic quartzite beds of the Sonora allochthon. Black shale units in arroyos El Carrizo and El Zacatón, 1.7–2.4 km northeast of the village of Rebeiquito, yielded graptolites of Early, Middle, and Late Ordovician age (Poole et al., 1995a; Bartolini et al., 1995). Bartolini et al. (1990) reported radiolarians of Mississippian age in this area. A strongly deformed sequence, which consists of very thin to thin beds of turbiditic fine- to coarse-grained quartzite, conglomeratic quartzite, siltite, argillite, limestone, and sparse chert, is provisionally assigned to the Rancho Nuevo Formation. These Rancho Nuevo–like strata structurally overlie graptolitic shale of the Ordovician El Mezquite Shale in a roadcut on the north side of Sonora Highway 104 ~2 km northeast of Rebeiquito (Poole and Amaya-Martínez, work in progress). The Rancho Nuevo–like turbiditic sequence is exposed in roadcuts between 2 and 4 km northeast of Rebeiquito. We consider the strata to be part of the Sonora allochthon. Four kilometers northeast of Rebeiquito, the Rancho Nuevo–like strata are faulted against Upper Cretaceous Tarahumara volcanic rocks.
46	Sierra El Aliso–Sierra El Encinal area (northwest and northeast of Tónichi), central Sonora. Outcrops of shale interbedded with thick-bedded quartzites yielded Early, Middle, and Late Ordovician graptolites in Sierra El Aliso (Bartolini et al., 1989). Sparse limestone beds yielded Late Ordovician conodonts (Bartolini et al., 1989). Similar outcrops of shale interbedded with thick-bedded quartzites in the Sierra El Encinal area yielded Middle Ordovician graptolites (Barrera-Moreno and Domínguez-Perla, 1987; Bartolini et al., 1990, 1995). A small outcrop of Lower Permian fusulinid-bearing conglomeratic lime grainstone in Sierra El Aliso (Bartolini et al., 1989) is provisionally correlated with the Rancho Nuevo Formation. Sequence is part of the Sonora allochthon.
47	La Colorada area, central Sonora. Outcrops of allochthonous offshelf turbiditic quartzite, siltite, argillite, and minor limestone/marble contain Middle Ordovician graptolites. Sequence, which is part of the Sonora allochthon, is unconformably overlain by redbeds of probable Triassic age (Poole et al., 1995a). Ordovician sequence is similar to allochthonous Middle Ordovician quartzite-bearing strata in Sierra El Aliso (Bartolini et al., 1989, 1990, 1995).
48	Estación Torres, central Sonora. Outcrops 4 km north of railway Estación Torres in the southern part of the Cerros El Aguaje (18 km west of La Colorada and 32 km southeast of Hermosillo) contain poorly preserved graptolites in shales between thick turbiditic quartzite beds that indicate a Middle to Late Ordovician age (S.C. Finney *in* Bartolini et al., 1995). Claire Carter identified poorly preserved biserial graptolites of Middle Ordovician to Early Silurian age collected by J.H. Stewart and others (Stewart and Poole, 2002) from the same outcrop area 4 km north of Estación Torres. Sequence is part of the Sonora allochthon.
49	Cerros La Herradura (~21 km southeast of La Colorada and ~7 km southwest of the village of San José de Pimas), central Sonora. Bartolini et al. (1990, 1995) reported Late Ordovician graptolites from the Cerros La Herradura. Sequence is part of the Sonora allochthon.
50	Isla Turner/Turner Island (eastern Golfo de California). Outcrops of complexly deformed eugeoclinal argillite and chert in the northern part of the island yielded late Middle to early Late Ordovician graptolites (Poole et al., 1993). Facies are similar to coeval strata in the Sonora allochthon at Minas de Barita and Cerro Cobachi.
51	Sierra Las Pintas, northern Baja California Norte. Leier-Engelhardt (1986) reported that rocks in the northern part of the sierra (estimated thickness 520 m) consist of calcareous siltstone with interbedded sparse crinoidal grainstone and granule conglomerate; normally graded sandy crinoidal grainstone; massive and graded beds of coarse sandstone and clast-supported granule to cobble conglomerate; massive basalt flows and sparse pillow breccia and hyalotuff; bedded chert and calcareous argillite. She reported that rocks in the southern area (720 m) consist of a sequence of bedded chert and argillite, fine sandstone, coarse siltstone, and conglomeratic debris flows; limestone and shale; pillow basalt and massive basalt flows. McEldowney (1970) found corals, crinoids, and brachiopods in the northern area that suggested a Mississippian or Pennsylvanian age. Leier-Engelhardt (1986) found conodont fragments in chert and argillite in the southern area that suggested an Early Devonian to Early Mississippian age. These rocks were probably deposited in a continental-rise or ocean-basin environment along the southwest margin of Laurentia. They correlate with similar age rocks in the Sonora allochthon on the mainland.
<u>Continental-slope and -rise rocks of Laurentia</u>	
52	Rancho San Marcos, northern Baja California Norte. Outcrops of quartzite, carbonate rock, and bedded chert. Age-diagnostic conodonts were recovered from two samples of altered limestone beds in large olistoliths within Mesozoic mélange or flysch (Lothringer, 1993; Gastil and Miller, 1983). Conodonts identified by R.H. Miller (*in* Lothringer, 1993) from a sample collected by Lothringer contains late Early (Ibexian) Ordovician conodonts, and conodonts identified by J.E. Repetski from a sample collected by K.B. Ketner (1.6 km NW of the Lothringer sample) yielded early Early (Ibexian) Ordovician conodonts (Poole et al., 1995b). The Lower Ordovician facies at Rancho San Marcos indicate continental-slope deposits near the carbonate-shelf margin of southwest Laurentia (Fig. 4).
53	Zamora area, central Baja California Norte. On the 30th parallel, Delattre (1984), Buch (1984), Phillips (1984), and Buch and Delattre (1993) described a well-preserved sequence of rocks at least as old as Early Permian. The Lower Permian strata are disconformably overlain by rocks of Early Triassic age, which are unconformably overlain by rocks of mid-Cretaceous age (Delattre, 1984). The Lower Permian sequence is more than 2500 m thick. Paleontologic control was limited to the base of the middle unit, which contains large crinoids (Webster et al., 1985; Webster and Delattre, 1993) and fusulinids tentatively identified as *Parafusulina kummeli* Roberts (C.H. Stevens *in* Delattre, 1984), indicating an Early Permian (Leonardian) age. This sequence represents offshelf deep-water turbiditic deposits.

(*continued*)

TABLE 1. KEY AND EXPLANATIVE GLOSSARY TO FIGURES 1–4 (*continued*)

Abbreviation	Explanation
54	Cañon Calamajué, central Baja California Norte. Sequence of Mississippian bedded chert, flysch, and sandy carbonate rock overlain by Jurassic and Cretaceous volcanic rocks, and a carbonate/clastic sequence of unknown age. Hoobs (1985) described this sequence as consisting of 1000 m of chert and shale, and interbedded shale and coarsening upward lithic sandstone with the uppermost 30 m consisting of chert litharenite, pebbly conglomerate, and limestone. The limestone contains crinoids and conodonts identified as *Gnathodus bilineatus* (Roundy) representing the early Chesterian *bilineatus*-Upper *Cavusgnathus* Zone. This Mississippian section is similar to facies we would expect to occur in the southern part of the Sonora allochthon in central Sonora. Gastil et al. (1991) noted that the apparent base of the Cretaceous section is marked by a quartz-boulder conglomerate with a siliceous volcanic matrix similar to the section on the 30th parallel (loc. 53).
55	Coastal area south of Puerto Calamajué, central Baja California Norte. Nearly continuous exposure of Paleozoic rocks along the coast have been reported in Master's theses by Campbell (1985) and Crocker (1987). Crocker (1987) reported that the northern area consists of a 2000-m sequence of coarse-grained bioclastic packstone, black chert, shale, and quartz arenite. And in the southern part of the northern area, he reported a 1000-m section of mostly thin-bedded lime mudstone, coarse-grained bioclastic packstone, and boulder and cobble conglomerate with minor bedded chert and argillite. Some beds of intraclast granule to boulder conglomerate contain bioclastic debris. Conodont fragments (Spathognathodiform and Gnathodiform elements) collected in both parts of the northern area indicate a Devonian to Mississippian age (Crocker, 1987; Gastil et al. 1991).
56	Several kilometers farther down the coast from locality 55, Campbell (1985) mapped four sequences of strongly deformed metamorphosed strata of upper greenschist to lower amphibolite grade. He described a lower unit (1060 m thick) consisting of thick amalgamated chert and slump-folded lime mudstone overlain by a thick interval of argillite; unit 2 (1170 m thick) consisting of interbedded impure carbonate rock, chert, argillite, and minor quartz arenite with thick-bedded lime mudstone in the middle part, and 85 m of pillowed alkaline basalt in the lower part; unit 3 (1100 m thick) consisting of fine-grained clastic rocks interbedded with abundant bedded chert and impure carbonate rock with a pillowed alkali basalt as thick as 400 m in the lower part; and unit 4 (at least 800 m thick) consisting of impure carbonate rock and intrabasinal pebble to cobble debris-flow beds composed of carbonate-rock and chert clasts with the middle part containing impure quartz sandstone and the upper part composed of interbedded chert, impure carbonate rock, and minor argillite. Crocker (1987) and Gastil et al. (1991) interpreted the northern sequence (loc. 55) argillite and chert as ocean-basin deposits and the carbonate rock with associated turbiditic carbonate flysch as shelf-margin deposits. Campbell (1985) and Gastil et al. (1991) interpreted the southern sequence (loc. 56) as slope-to-basin deposits. We agree with their interpretation of the depositional environment. Both of these sequences are similar, and are facies we would expect to occur along the Laurentian continental slope in central Sonora. Isla Angel de la Guardia, Golfo de California (island southeast of loc. 56). Phillips (1966) and Gastil (1993) reported that the metasedimentary rocks exposed on the southwestern coast of the island were similar to those on the opposite peninsular coast mapped by Campbell (1985). Therefore, it seems likely that Isla de la Guardia was part of Baja California Norte during the Paleozoic. In addition to the Paleozoic sequences at localities 53–56, we would expect this deep-water offshelf sequence also to occur on mainland Mexico along the southern margin of Paleozoic Laurentia.
57	Arroyo El Toro, Bahía Los Angeles area, southern Baja California Norte. Outcrops, including pillow basalt, are similar to deep-water strata of Devonian and Mississippian age at localities 51 and 56 (Gastil, 1993).
Interior metamorphic zone of Marathon segment of Ouachita–Marathon–Sonora orogen in northeastern Mexico	
58	Sierra del Carmen, northwesternmost Coahuila. Only known outcrop of interior zone/belt (IZ on Fig. 3) in northern Mexico is a sequence of fine-grained metasedimentary rocks consisting of thinly interlayered very fine grained graphitic and chloritic sericite schist or phyllite, fine-grained calcite marble, and fine-grained metaquartzite grading into calcareous schist, schistose or phyllitic marble, and phyllitic metaquartzite (Flawn, 1961). Age of metamorphism 263 ± 5 Ma by K/Ar and 275 ± 20 Ma by Rb/Sr (Denison et al., 1969, 1977) and 277 ± 10 Ma by Rb/Sr muscovite-whole rock (Carpenter, 1997) is Early to Middle Permian. Carpenter (1997) considered the interior zone at Sierra del Carmen to be allochthonous and exotic with respect to Laurentia. Flawn (1961) reported that similar rocks were encountered in several wells in southwest Texas. We infer that the interior zone/belt metasedimentary rocks consist of Gondwana crust in the suture zone along the northern margin of Gondwana, and are likely present along the entire southern margin of Laurentia and Ouachita–Marathon–Sonora orogen (Fig. 3).
Gondwana crustal rocks and associated volcanic-arc terranes	
59	Tomóchic volcanic center, west-central Chihuahua. Pb isotopic compositions of the volcanic rocks closely resemble volcanic rocks in eastern Chihuahua and West Texas that were erupted through basement similar to that of the Coahuila block outboard of the Marathon segment of the Ouachita–Marathon–Sonora orogenic belt (McDowell et al.,1999). They interpreted their data to indicate that Coahuila-like basement (El Fuerte block of this report) underlies the volcanic center rather than North American Laurentian crust.
60	La Olivina xenolith locality, southeastern Chihuahua. Deep crustal granulite xenoliths in a basanitic cinder cone in the late Cenozoic Camargo basalt field. They are interpreted in terms of two age groups, pre-Cenozoic and mid-Tertiary, based on ion-probe dating of zircons and on isotopic comparisons of xenoliths to rocks of known age (Ruiz et al., 1988; Cameron et al., 1992). They consider the pre-Cenozoic xenoliths as Proterozoic basement, sedimentary and magmatic-arc rocks related to the late Paleozoic convergence of Laurentia and Gondwana, and perhaps post-orogenic Triassic plutons. Cameron et al. (1992, Fig. 1) interpreted data on the pre-Cenozoic orthogneisses to have isotopic and age affinities with rocks from the southern plate (Gondwana). In contrast, Ruiz et al. (1988, Fig. 1) interpreted Nd isotopic data on the xenoliths to indicate basement belonging to the Grenville (1.0 Ga) tectonothermal event along the southern edge of the northern plate (Laurentia). On Figure 3, we show locality 60 close to the suture zone between Gondwana (Coahuila block of this report) and Laurentia.
Fragments of Gondwana or interposed volcanic-arc terrane	
61	Sierra del Diablo, southeast Chihuahua. Two units of felsic pyroclastic rocks (devitrified and recrystallized ashflow and ashfall tuffs) of unknown age underlie Lower Cretaceous (Albian) limestone (McKee et al., 1990, 1999). Unit 1 is a folded densely welded rhyolitic tuff, and unit 2 is less deformed layered pyroclastic breccias, lapilli tuffs, and ashfall tuffs (Jones and McKee, 1987). Considered basement rocks (Pennsylvanian?, Permian or Triassic) of Coahuila island of McKee et al. (1984).
62	Potrero de La Mula–Sierra del Fuste, west-central Coahuila. Oldest basement rocks exposed are Upper Triassic granitic rocks that make up part of a single large pluton (Jones et al., 1984). Denison et al. (1969) obtained a K-Ar hornblende age of 206 ± 4 Ma from a granodiorite, and Jones et al. (1984) obtained a Rb-Sr whole-rock isochron age of 213 ± 14 Ma from several igneous rock types within the pluton. Jones et al. (1984) concluded that the pluton was emplaced at a shallow depth above a subduction zone, which is consistent with the hypothesis that a late Paleozoic–earliest Mesozoic convergent plate boundary existed in northern Mexico. Lower Cretaceous strata overlap the Late Triassic pluton.

(*continued*)

TABLE 1. KEY AND EXPLANATIVE GLOSSARY TO FIGURES 1–4 (*continued*)

Abbreviation	Explanation
63	Las Delicias area, west-central Coahuila and southeast Chihuahua. Outcrops immediately south of the San Marcos fault (SM, Fig. 3) consist of upper Paleozoic, arc-proximal, deep-water, fossiliferous marine strata that were chiefly deposited during multiple episodes of mass-gravity movement (McKee et al., 1999). Early and Middle Permian fossils were described by Wardlaw et al. (1979) from beds near Las Delicias, Coahuila. The strata, which were named the Las Delicias Formation, are subdivided into 11 units; 7 chaotic lithosomes mark episodes of convulsive sedimentation and 4 bedded lithosomes form the general sedimentary matrix for the chaotic lithosomes (McKee et al., 1999). The formation consists mostly of volcanic rock fragments as large as block size, with minor limestone clasts derived from the basin margin and nearby Las Delicias arc (McKee et al.,1988; López et al., 1996; McKee et al., 1999). It contains large folds that formed during syndepositional sliding. The older parts of the Las Delicias Formation were intruded during the medial Mississippian by the La Pesuña pluton (rhyolite peperite). López et al. (1996) obtained concordant U-Pb zircon dates on several igneous boulders that ranged in age from 331 ± 4 Ma (Late Mississippian) to 268 ± 6 Ma (Middle Permian). The arc volcanics range from basalt to rhyolite, and their isotopic character (eNd(T) = +1.5 to +4 and $^{87}Sr/^{86}Sr(i)$ = 0.7048–0.7068) implies a significant mantle component (López et al., 1996). The Permian granites isotopically overlap with the arc volcanics and may have been feeder sources (López et al., 1996). In the Early or Middle Triassic, the Las Delicias Formation was folded, then intruded again (El Coyote pluton) in the Late Triassic (McKee et al., 1999). McKee et al. (1999) named this tectonically bounded stratigraphic terrane, the Las Delicias terrane. The formation is part of the "Coahuila block" which is part of Gondwana that impacted and attached to Laurentia. A granitoid boulder in the Las Delicias Formation yielded a near concordant U-Pb zircon age of 1215 ± 10 Ma that indicates the Paleozoic rocks in the Las Delicias area are underlain by Grenvillian continental crust (López et al., 1996) of the Gondwanan craton.
64	Presa Aldolfo Ruiz Cortinez (Mocuzari) region (Río Mayo and Arroyo Cedros valleys), southern Sonora. Several outcrops 25–65 km northeast of Navojoa, Sonora, consist of a variety of metasedimentary rocks, including limestone, quartz sandy limestone, quartzite, graphitic siltite and argillite, marble, phyllite, slate, schist, and gneiss. The abundance of plutonic rocks in the region suggest that the metasedimentary rocks are roof pendants. All of the metasedimentary rocks are strongly folded and faulted. Fossils are rare in the sedimentary rocks, and no isotopic ages have been reported on the metamorphic rocks. In 1993, Poole and Amaya-Martínez examined several of these small outcrops, and collected samples of limestone and marble for conodonts, but all samples were barren. Age of the metasedimentary rocks is unknown, but probably includes both Paleozoic and Mesozoic protoliths. Dumble (1900) noted fossiliferous limestones interbedded with quartzites in the vicinity of the old Trigo mine. Some of the limestones contained poorly preserved crinoids, brachiopods, and solitary corals. He considered these fossils to be Silurian in age. Angermann (1904) reported crinoids, corals, and brachiopods in a limestone 8 km south of the village of La Casita (outcrop now under water), and a poorly preserved brachiopod in a limestone near Mina El Trigo. The brachiopod may be a *Spirifer*, which would suggest a late Paleozoic age (King, 1939). King (1939) observed crinoids and other unidentifiable fossils in boulders (as much as 1 m in diameter) in Arroyo Arenoso. The fossiliferous limestones are part of a sedimentary sequence composed of interbedded quartzite, phyllite, slate, and sparse limestone. We consider these outcrops to be fragments of Gondwana and related to the El Fuerte block.
65	El Fuerte area, northern Sinaloa. Low-grade metasedimentary and metavolcanic rocks aggregating 5000 m in thickness comprise the Río Fuerte Group (Mullan, 1978) in the vicinity of Presa Miguel Hidalgo, north of El Fuerte. The Río Fuerte Group includes three formations, in ascending order: Río Fuerte, Topaco, and Corral Falso (Mullan, 1978). The Paleozoic Río Fuerte Formation consists of thin-bedded chert, quartzite, carbonate rock, and argillite; the undated Topaco Formation consists mostly of volcanic rocks with the distinctive Nodular Rhyolite Member near its base; the undated Corral Falso Formation is lithologically similar to the Río Fuerte Formation. A thin turbiditic marbleized limestone bed in the Río Fuerte Formation yielded Middle to Late Ordovician conodonts (collected in 1996 by Poole and W.R. Page, and identified by A.G. Harris). The Río Fuerte Group and the Francisco Gneiss (gneiss dated as Triassic by Anderson and Schmidt, 1983) are part of our El Fuerte block, and herein considered to be fragments of Gondwana that attached to Laurentia during late Paleozoic ocean closure and were stranded during Mesozoic rifting. Triassic metamorphic rocks in Sierra de Sonobári may be related to a post-orogenic event. Some of the offshelf deep-water facies in the El Fuerte block may correlate with similar facies in southern Baja California Norte (Figs. 3 and 4). This area was referred to as the Sonobári terrane by Campa and Coney (1983), but was included in the more extensive Tahué terrane by Sedlock et al. (1993).
66	San José de Gracia–Real de San Javier area, northern Sinaloa. Paleozoic rocks consisting of a lower sequence of chert, shale, lime mudstone, and quartzite, and an upper sequence of shale and quartzite (Carrillo-Martínez, 1971). We consider this area to be part of our El Fuerte block (Fig. 3). A calcarenite in the lower sequence yielded the fusulinid *Millerella*, which ranges from Early Mississippian to Late Pennsylvanian (Carrillo-Martínez, 1971). Near San José de Gracia, G. Gastil and J. Roldán-Quintana (*in* Gastil et al., 1991) found a thick sequence of bedded chert characterized by 5-cm-thick beds with argillite partings, and local beds of quartz arenite, cobble conglomerate, fine-grained clastic rock, and sparse carbonate rock. They collected a conodont sample from a small lens of calcarenite that yielded *Streptognathodus*? and *Gondolella*, indicating a Middle Pennsylvanian to Early Permian age (Gastil et al., 1991). In the vicinity of Real de San Javier, slates, quartzites, and limestones were considered to be Paleozoic in age by the compilers of the Monograph on the State of Sinaloa (Consejo de Recursos Minerales, 1992). We interpret the sequence in the vicinity of San José de Gracia and Real de San Javier to be part of northwest Gondwana that attached to North America during late Paleozoic collision. Alternatively, the Paleozoic rocks in this area may correlate with Laurentian offshelf deep-water facies in southern Baja California Norte. The rocks in this area were referred to as the Rusias terrane by Campa and Coney (1983), but included in the more extensive Tahué terrane by Sedlock et al. (1993).
67	La Prieta mine area, central Sinaloa. Metamorphic rocks consisting of argillites, phyllites, and slates may be Paleozoic in age (Consejo de Recursos Minerales, 1992). The metasedimentary sequence is intruded by granitoids of the Sinaloa batholith. Area included in the Tahué terrane of Sedlock et al. (1993).
68	Copala area, southern Sinaloa. Slates, quartzites, and graywackes of possible Paleozoic age occur as roof pendants in granodiorite of the Sinaloa batholith (Consejo de Recursos Minerales, 1992). Area is within the Tahué terrane of Sedlock et al. (1993).
69	Arroyos El Picacho and Pescaditos, northeast of Santa María del Oro, north-central Durango. The Gran Tesoro Formation consists of folded muscovite schist, crenulated micaceous phyllite, and foliated and fractured slate, metatuff and metalimestone (Main, 1950; Córdoba and Silva-Mora, 1989). Near the top of the formation is a zone of metavolcanic rocks consisting of mafic pillow lavas, and blocks of fossiliferous limestone that may be olistostromes. Fossils reported by Zaldivar-R. and Garduño-M. (1984) and Córdoba and Silva-Mora (1989) include *Linoproductus* aff. *L. prattenianus*, *Rugosochonetes* sp., cf. *Kitakamithyris* sp., cf. *Alispirifer* sp., *Fenestella* aff. *F. modesta*, and numerous fenestellid bryozoans, echinoderms, crinoids, and minor solitary corals. The fossiliferous limestones are probably late Paleozoic in age. The Gran Tesoro Formation is overlain by Mesozoic rocks. Aranda et al. (1988) reported Early Jurassic palynomorphs from the Gran Tesoro Formation. It is unlikely that the palynomorphs came from the Gran Tesoro Formation because the greenschist grade of metamorphism of the formation would destroy or significantly alter any pollen or spores in the metsedimentary rocks. We suspect that the palynomorphs came from overlying Mesozoic rocks. Muscovite schist in the Gran Tesoro Formation yielded an age of 326 ± 26 Ma (Zaldívar-Ruiz and Garduño-M., 1984), which is near the Mississippian-Pennsylvanian systemic boundary. The earliest recorded deformation in the area was at the close of the Paleozoic (Main, 1950), which coincides with the time of collision between northwestern Gondwana (South America) and Laurentia. We consider the Paleozoic rocks in Durango to be part of Gondwana.

(*continued*)

TABLE 1. KEY AND EXPLANATIVE GLOSSARY TO FIGURES 1–4 (*continued*)

Abbreviation	Explanation
70	Caopas–Rodeo area, northern Zacatecas. Outcrops in Valle del Rodeo (Sierra de San Julián) in the vicinity of Ranchos Caopas and El Rodeo consist of upper Paleozoic (Permian?) Caopas Schist and El Rodeo Formation unconformably overlain by Jurassic rocks (De Cserna, 1956; Rogers et al., 1961; Flawn, 1961). De Cserna (1956) and Rogers et al. (1961) described the older Caopas sequence as mica schist with thin beds of quartzite and conglomerate and the younger El Rodeo sequence as interlayered green and violet phyllite and altered basic volcanic rocks, whereas Anderson et al. (1991) and Jones et al. (1995) reported that the Caopas Formation consists of rhyolite porphyry and the Rodeo Formation consists of volcanic and volcaniclastic rocks that are less silicic than the Caopas. Anderson et al. (1991) suggested a pre-Triassic age on the basis of the overlying Nazas Formation presumed by many workers to be Triassic and/or Jurassic in age, and time correlative with part of the Huizachal Group. A relatively undeformed rhyolite porphyry of the Caopas yielded a zircon U-Pb age of 158 ± 4 Ma (Jones et al., 1995). This Late Jurassic isotopic age is in conflict with earlier estimates of a late Paleozoic or pre-Triassic age. Also, the lithologies of the Caopas and El Rodeo Formations described in the above reports do not agree. If the pre-Nazas low-grade metamorphic rocks are late Paleozoic in age, they may be part of western Gondwana that attached to North America. If the pre-Nazas volcanic rocks are Late Jurassic in age, they probably represent a magmatic arc that post-dates the North America-South America collisional event.
71	Sierra de Catorce, northern San Luis Potosí. Outcrops in the canyon near the town of Catorce consist of an older sequence of metamorphosed green sericitic and talcose schists, greenish rocks (originally porphyries?), some thin layers of andesitic tuff breccias, and less metamorphosed green and purple sandy and shaly rocks (partly tuffaceous?) and interbedded brown conglomeratic quartzites composed of rounded quartz pebbles in a matrix of clear quartz sand (Baker, 1922; De Cserna, 1956; Flawn, 1961). Franco-Rubio (1999) correlated a unit of flysch-like green siltstones and sandstones (graywackes and subgraywackes) and minor thin conglomerates to the upper Paleozoic Guacamaya Formation in the Ciudad Victoria area (loc. 72). We consider these rocks to be part of Gondwana that attached to North America.
72	Ciudad Victoria area, west-central Tamaulipas. Outcrops in Cañónes de la Peregrina and de Caballeros near Ciudad Victoria consist of Grenville-age (1018 Ma) Novillo gneiss overlain, in ascending order, by: (1) Silurian Cañón de Caballeros Formation composed of conglomerate, silicic volcanic arenite, limestone, and sandstone and siltstone; (2) Lower Mississippian Vicente Guerrero Formation composed of quartz and lithic arenite, siltstone, and shale; (3) Upper Mississippian El Aserradero Rhyolite (334 ± 39 Ma by U-Pb zircon); (4) unconformably overlying Lower and Middle Pennsylvanian Del Monte Formation composed of bioclastic grainstone, sandstone, and limy sandstone; (5) Lower Permian Guacamaya Formation composed of turbiditic siltstone and sandstone rich in volcanic detritus; and (6) unconformably overlying Jurassic and Cretaceous conglomerate, limestone, siltstone, and sandstone (Stewart et al., 1999a). Boucot et al. (1997) interpreted the brachiopods in the Cañón de Caballeros Formation as having close affinities to northern South America, whereas the overlying Mississippian-Permian strata have North American biogeographic affinities. These relations suggest that the Ciudad Victoria area was originally part of Gondwana (South America), but by Early Mississippian time was close enough to Laurentia (North America) for faunal interchange (Boucot et al., 1997).
Pre-Mesozoic rocks identified in PEMEX (PEtróleos MEXicanos) boreholes in north-central and northeast Mexico	
73	No. 1 Moyotes (Chihuahua). Bottomed in Precambrian granite (Thompson et al., 1978). Granite 890 ± 32 Ma by Rb-Sr from PEMEX files (Haenggi, 2001, table 2). Represents Laurentian continental crust.
74	No. 1 Chinos (Chihuahua). Bottomed in Precambrian rocks (Thompson et al., 1978). Granite 1327 ± 242 Ma by Rb-Sr from PEMEX files (Haenggi, 2001, table 2). Represents Laurentian continental crust.
75	No. 1 Villa Ahumada (Chihuahua). Bottomed in Lower Permian (Wolfcampian) rocks (Thompson, et al., 1978). Penetrated 1115 m of Leonardian and 800 m of Wolfcampian strata (Thompson et al., 1978). Sequence is part of Pedregosa basin fill.
76	Concordia-1 (Coahuila). Penetrated 400 Ma phyllite (Haenggi, 2001). May be part of Coahuila block that attached to Laurentia.
77	No. 2A Peyotes (Coahuila). Bottomed in medium- to light-gray schist that may be part of the interior zone/belt, according to Wilson et al. (1984). Borehole is ~100 km south of our inferred interior belt. Probably part of western Gondwana that attached to North America.
78	No. 1 La Perla (Coahuila). Bottomed in granitic gneiss dated as 358 ± 60 Ma by Rb/Sr method (Denison et al., 1969). Considered to be an exotic terrane by Handschy et al. (1987), and to represent the interior zone/belt by Wilson et al. (1984). Borehole is ~125 km south of our inferred interior belt. Probably part of Gondwana that attached to North America.
79	No. 1 Barril Viejo (Coahuila). Bottomed in pre-Cretaceous igneous rocks. Considered to be an exotic terrane by Handschy et al. (1987). May be part of Coahuila block that attached to Laurentia.
80	No. 1 Paila (Coahuila). Bottomed in pre-Cretaceous igneous rocks. Considered to be an exotic terrane by Handschy et al. (1987). May be part of Coahuila block that attached to Laurentia.
81	No. 1 Garza (Nuevo León). Bottomed in pre-Cretaceous metasedimentary and igneous rocks (Wilson et al., 1984). Considered to be an exotic terrane by Handschy et al. (1987). Borehole is ~125 km south of our inferred interior belt. Probably part of Gondwana that attached to North America.
82	Minas Viejas (Nuevo León). Bottomed in gray schist and phyllite presumed to represent the interior zone/belt by Wilson et al. (1984). Borehole is ~250 km south of our inferred interior belt. Probably part of Gondwana that attached to North America.
83	No. 101 Chapa (Nuevo León). Bottomed in metasediments consisting of metagraywacke and sericite-chlorite schist presumed to represent the interior zone/belt by Wilson et al. (1984). Borehole is ~250 km south of our inferred interior belt. Probably part of Gondwana that attached to North America.
84	No. 1 Ceralvo (Nuevo León). Bottomed in 20 m of metagraywacke (Wilson et al., 1984). Borehole is ~275 km south of our inferred interior belt. Probably part of Gondwana that attached to North America.
85	Carbajal (Nuevo León). Bottomed in Paleozoic slate and sericite-chlorite phyllite (Wilson et al., 1984). Borehole is ~300 km south of our inferred interior belt. Probably part of Gondwana that attached to North America.

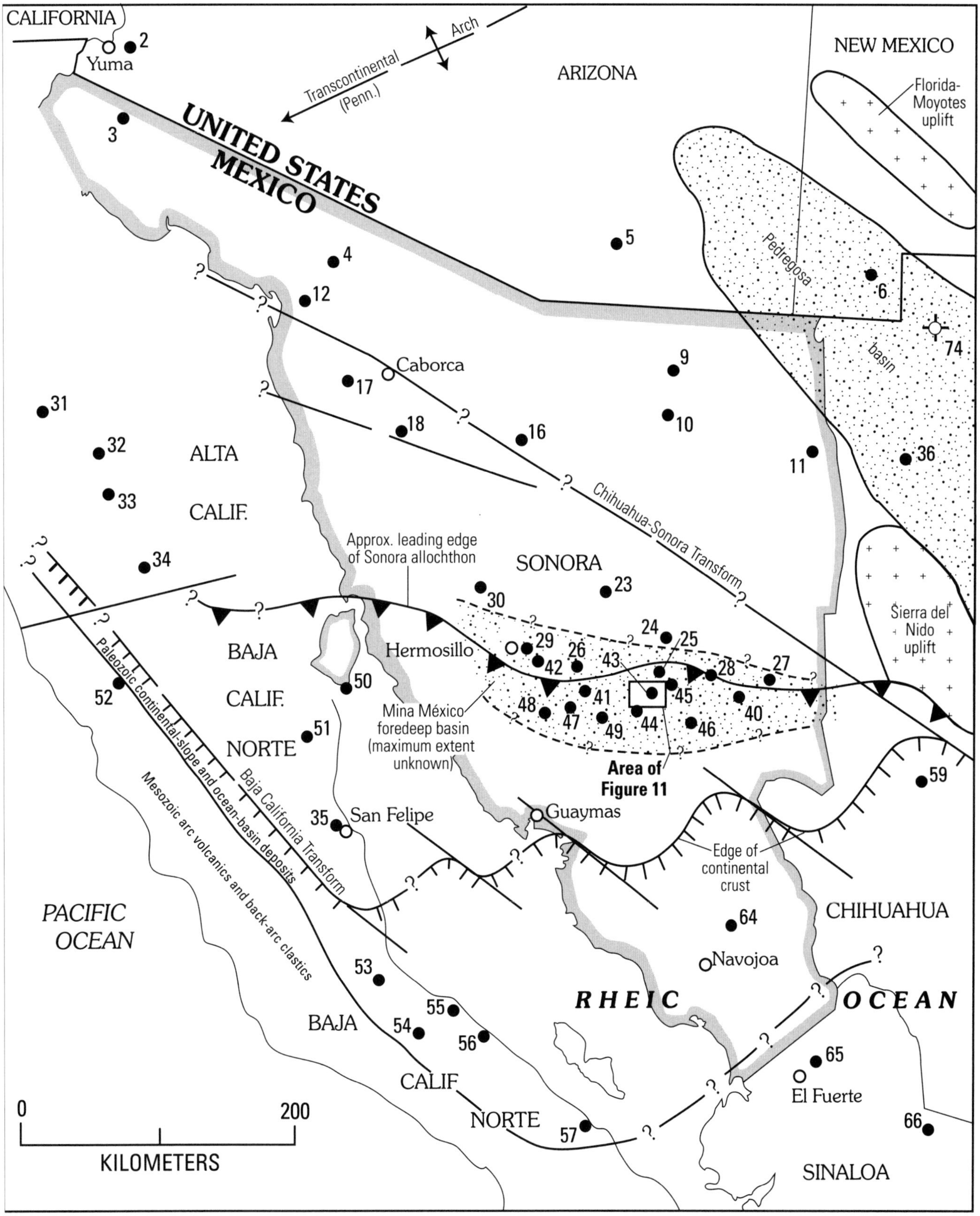

Figure 4. Enlarged map of Sonora and Baja California Norte showing data points and geologic features. Outline of map shown on Figure 3.

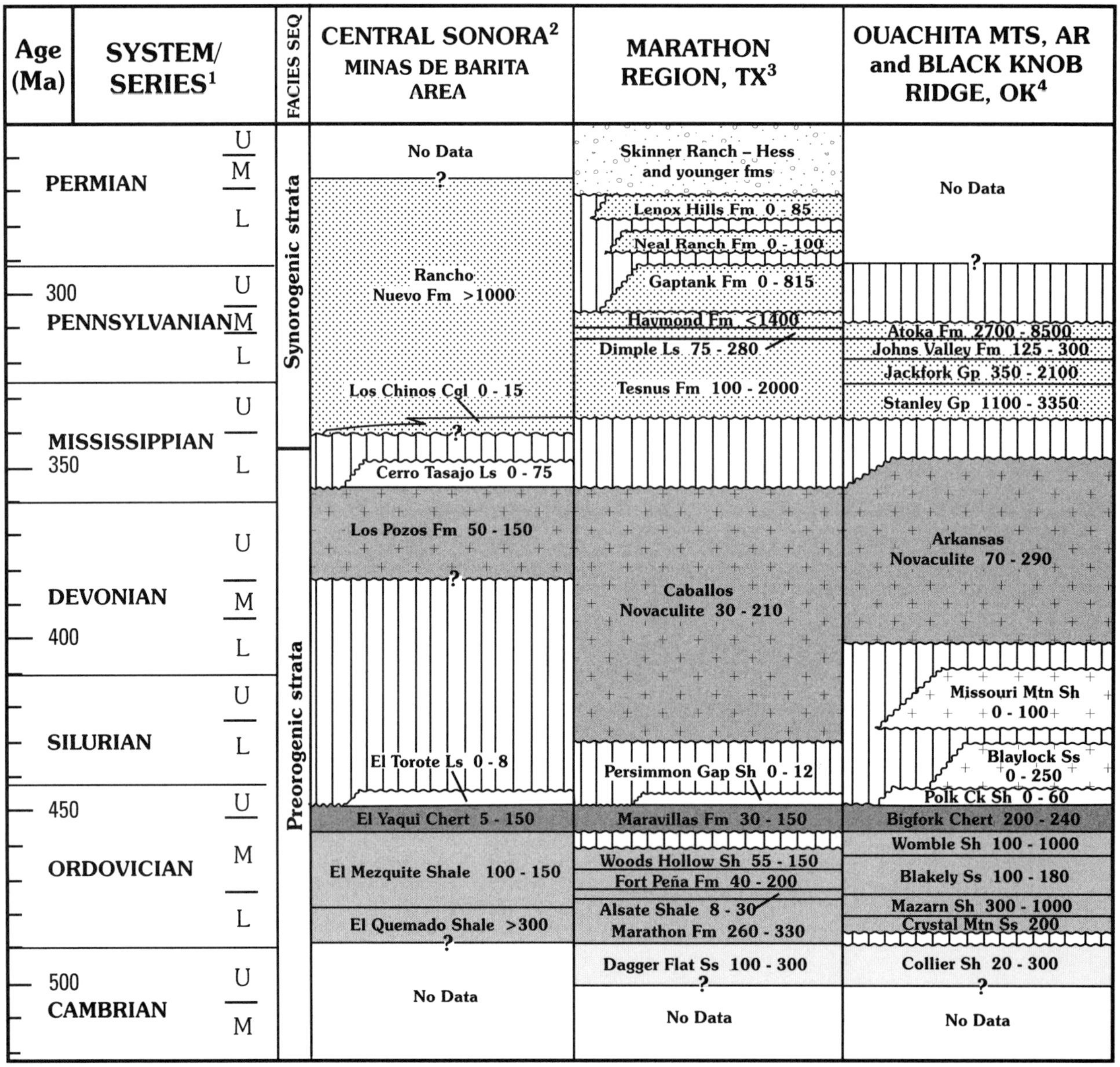

[1]**System/Series boundaries:** Claoué-Long et al. (1992), Cooper (1999), Harland et al. (1990), Jin et al. (1997), Landing et al. (2000), Menning et al. (2000), Sandberg and Ziegler (1996), Sando (1985), Stone et al. (2000), Tucker et al. (1998).

[2]**Sonora:** Poole et al. (in progress).

[3]**Marathon:** Barrick and Noble (1995), Ellison and Powell (1989), Ethington et al. (1989), Finney (1986), McBride (1989), Noble (1990, 1994), Ross (1978, 1979, 1986), Ross and Ross (1995).

[4]**Ouachita:** Barrick and Haywa-Branch (1994), Ethington et al. (1989), Finney (1986), Hass (1951), Lowe (1989), Morris (1989).

Figure 5. Generalized stratigraphic sections characteristic of the Sonora, Marathon, and Ouachita segments of the Ouachita-Marathon-Sonora orogenic system. Thickness ranges in meters. Similar patterns represent correlative strata.

ends, as interpreted by Ross (1986), Handschy et al. (1987), and Dickinson and Lawton (2001). Instead we show that abundant and detailed stratigraphic and structural data indicate a continuous ~3000-km-long Paleozoic carbonate shelf extending across the southern United States and northern Mexico (Poole et al., 2003a), and that late Paleozoic physical connections existed between the Ouachita, Marathon, and Sonora segments of the Ouachita-Marathon-Sonora orogen. Our reconstruction of the southwestern margin of Laurentia infers that exposed Paleozoic carbonate-shelf-like facies in Baja and Alta California and central Sonora are part of the continental shelf, which is underlain by Precambrian basement rocks (Figs. 2–4).

NEOPROTEROZOIC–EARLY CAMBRIAN RIFTING OF RODINIA AND SEDIMENTATION ALONG THE NEW CONTINENTAL MARGIN

The paleo-continent Rodinia was rifted in the Neoproterozoic and Early Cambrian along trends that set the early Paleozoic tectonic framework for the southern margin of the Laurentian craton. Rodinia broke apart between 800 and 550 Ma, according to new paleomagnetic data (Meert and Torsvik, 2003) and between 625 and 555 Ma based on tectonic subsidence curves in early Paleozoic miogeoclines (Bond et al., 1984). The breakup resulted in two separate continents—Laurentia and Gondwana (Fig. 2). The rifting event that produced the Laurentian and west Gondwanan (South America) cratons was probably distributed among multiple rift systems (Cawood et al., 2001). The inferred new margin of Laurentia is an orthogonal northeast-trending rift system offset by a series of northwest-trending synrift transform faults (Fig. 2), resulting in offsets of the continental margin. These major transform faults included the right-stepping Alabama-Oklahoma (545–525 Ma) and Texas (600–570 Ma) transforms in the southern United States (Thomas, 1976, 1977, 1983, 1989, 1991; Thomas et al., 2001, 2002), and the Chihuahua-Sonora (800–700 Ma) transform in northern Mexico (Fig. 2). Assuming these ages of faulting are valid, an eastward younging is indicated. The age differences suggest that rifting occurred over an extended period of time. Apparently, movements along the transform faults were episodic. Transform movement is indicated by the ages of igneous rocks emplaced along the fault systems and by the ages of synrift sedimentary rocks in intracratonic graben systems (Buffler and Thomas, 1994) or transform basins. These Neoproterozoic–Early Cambrian northwest-trending synrift transform fault systems established a tectonic fabric in the Laurentian continental crust that persisted throughout the Paleozoic.

Serpentinites and associated alkalic metagabbros (1025 ± 48 Ma by whole-rock K-Ar) occur as exotic blocks enclosed in the Middle Ordovician upper Womble Shale and as remoblized pods in late Paleozoic thrust zones in the core of the Ouachita Mountains (Cox, 1986; Morris and Stone, 1986; Nielsen et al., 1989). These ultramafic-rock fragments probably represent a Grenville-age upper-mantle or lower-crust ophiolite assemblage that was tectonically dismembered and transported to the upper crust or seafloor during Neoproterozoic–early Paleozoic rifting and transform faulting. The ophiolite fragments were incorporated in Middle Ordovician ocean-basin and continental-rise sediments that later were thrust onto the Laurentian continental shelf during the late Paleozoic collisional-subductional event.

During the rifting phase, the new continental margin included rifted-margin sediment wedges flanking areas of shelf deposition, and grabens and basins opening oceanward (Viele and Thomas, 1989). It is also likely that grabens, sags, and depressions formed adjacent to active synrift transform faults (transform basins) that penetrated and offset continental crust during the rifting phase. Strata of the Neoproterozoic Las Víboras and El Aguila Groups in north-central Sonora (loc. 23 on Fig. 2) and Cerro Las Bolas Group in east-central Sonora (loc. 27 on Fig. 2) may represent this type of deposition in grabens, sags, or basins formed during Neoproterozoic transform faulting in northern Mexico (i.e., Chihuahua-Sonora transform). The Las Víboras Group (~2400 m thick) and El Aguila Group (>1000 m thick) are present in an area of ~5000 km^2 in a region centered 50–70 km north-northeast of Hermosillo, Sonora (Stewart et al., 2002). The strata consist of quartzite, arkosic sandstone, siltstone and stromatolitic dolostone (Stewart et al., 2002). The Cerro Las Bolas Group (2600 m thick) is exposed near Sahuaripa, Sonora, ~160 km east of Hermosillo. The Cerro Las Bolas Group consists of Sturtian glacial deposits (ca. 685 Ma, according to Lund et al., 2003), carbonate rocks, siltstone, sandstone, and conglomerate (Stewart et al., 2002). It seems likely that many of these Neoproterozoic strata, which are near the trace of the postulated northwest-striking Chihuahua-Sonora transform (CST, Fig. 2), were deposited in depressions (transform basins) developed along the transform fault.

The early Paleozoic configuration of the southern Laurentian continental-shelf margin controlled the geometry and development of the late Paleozoic Ouachita-Marathon-Sonora orogenic belt. The postrift early Paleozoic passive shelf margin was continuous along the eastern, southern, and western margins of Laurentia. Figure 2 also shows the distribution of pre-shelf Precambrian crustal provinces that formed the southwestern part of the Laurentian craton (Anderson and Silver, 1981; Wooden et al., 1988; Wooden and DeWitt, 1991; Van Schmus et al., 1993; Mosher, 1998; Karlstrom et al., 2002; Farmer et al., 2003, this volume; Iriondo et al., 2004; A. Iriondo and W.R. Premo, 2004, personal commun.). We have attempted to show the boundaries and relative ages of primary Precambrian crust rather than include later dates on metamorphic and intrusive rocks related to overprinted younger orogenic events (Anderson et al., 1993; Karlstrom and Bowring, 1993).

PALEOZOIC STRATIGRAPHY AND TECTONICS ALONG SOUTHERN MARGIN OF LAURENTIA

The tectonic and depositional framework of the southern margin of Paleozoic North America is shown on Figures 1–4, and abbreviations and numeric labels are defined in Table 1. The sequence of data-point numbers on Figures 1–4 is generally oceanward, beginning with the cratonic platform, then continental shelf and Ouachita-Marathon-Sonora orogen, and ending with the Gondwana megaterrane.

Cratonic Platform and Continental Shelf

The boundary between the cratonic platform and continental shelf is well defined by thickness changes and lithofacies distribution in the western United States and northwestern Mexico, but is gradational in the southern United States and north-central Mexico. In northwestern Mexico (Figs. 3 and 4), cratonic-platform sections occur at El Capitán (loc. 3; Leveille and Frost, 1984; Poole, unpublished data), in the Estación Sahuaro area (loc. 12; Calmus and Sosson, 1995; Poole and Amaya-Martínez, work in progress), in northern Sierra Los Ajos (loc. 9; Page et al., 2003a), and at Bacoachi (loc. 10; Stewart et al., 2002; Stewart and Poole, 2002; Poole, unpublished data). The section at El Capitán is a metamorphosed upper Paleozoic and lower Mesozoic sequence, including recognizable Supai Group, Coconino Sandstone, Kaibab Limestone, Moenkopi Formation, and Aztec Sandstone, similar to correlative formations in western Arizona and southeastern California (Leveille and Frost, 1984). North of Estación Sahuaro, the section includes Neoproterozoic(?) quartzite, argillite, and carbonate rock and Paleozoic limestone and dolostone (locally marble), cherty and sandy carbonate, and subordinate argillite-phyllite and sandstone-quartzite. The section at Bacoachi consists of Precambrian porphyritic granite unconformably overlain by Middle Cambrian Bolsa Quartzite through Mississippian Escabrosa Limestone. Other cratonic-platform outcrops in northeastern Sonora mentioned by Stewart et al. (2002) and Stewart and Poole (2002) are not detailed in this report.

Continental-shelf sections just south of the cratonic platform and southwest of inferred traces of our postulated Chihuahua-Sonora transform (CST, Fig. 2) and the hypothetical Jurassic Mojave-Sonora megashear, are documented at Rancho El Bísani (loc. 17; Poole and coworkers, work in progress), Rancho Placeritos (loc. 30; Poole et al., 1997, 1998, 2000, 2003a, 2003b), Hermosillo (loc. 29; Poole and Amaya-Martínez, work in progress), Sierra Agua Verde (loc. 24; Stewart et al., 1999b), Sierra Martínez (loc. 25; Poole et al., work in progress), Arivechi (loc. 27; Almazán-Vázquez et al., 1986; Almazán-Vázquez and Fernández-Aguirre, 1987; Almazán-Vázquez, 1989; Poole, unpublished data), and Mina Plomosas–Placer de Guadalupe–Cerro El Carrizalillo area (loc. 39; Bridges, 1964; Dyer and Reyes, 1987; Haenggi, 2001; Poole, unpublished data).

Preorogenic Cambrian–Mississippian Continental Shelf

The Laurentian continental shelf is beneath and north of the orogen and south of the Laurentian cratonic platform. Cambrian to Early Mississippian stratigraphic data indicate that a depositionally continuous carbonate shelf bordered the southern and western margins of the Laurentian craton. In the southwestern United States and northwestern Mexico, similar facies are present in some parts of the Neoproterozoic to Ordovician, whereas dissimilar facies are present in the Silurian to Permian. Apparently, continental-shelf depositional environments were similar along the western and southern margins of Laurentia. However, Early Devonian brachiopods in west-central Sonora (loc. 30) are similar to those of correlative Lower Devonian strata in north-central Mexico and the southern and eastern United States, but have not been recognized in the western United States (Poole et al., 2003a). In the Early Devonian, the southwest-trending Transcontinental arch (Fig. 1) may have formed a barrier to migration and mixing of many genera and species of benthonic marine life from the southern shelf of Laurentia in northern Mexico to the western shelf (Cordilleran miogeocline) in the western United States (Poole et al., 2003a). In addition, a combination of sea-level and climatic fluctuations were important factors in faunal distribution on the Paleozoic continental shelf that wrapped around the southwest nose of the Transcontinental arch.

Ouachita-Marathon-Sonora Orogen

The Paleozoic Ouachita-Marathon-Sonora orogen extends continuously ~3000 km from Mississippi to the Golfo de California along the southern margin of Laurentia (shaded area on Figures 1 and 3). The three principal exposures of this late Paleozoic orogen are labeled on Figures 1 and 3. These exposures represent 15% or less of the entire belt that is mostly concealed by younger rocks. Rocks of the offshelf facies were thrust onto the southern margin of the Laurentian craton as South America and associated arc terrane impinged on North America during the final stages of Rheic ocean closing (Scotese, 1997). Geologic data indicate that the late Paleozoic Ouachita-Marathon-Sonora orogenic system follows the shape of the Neoproterozoic–early Paleozoic rifted margin of the Laurentian craton (Thomas, 1976, 1977; Poole and Perry, 1997, 1998).

Figure 5 shows generalized stratigraphic columns for each of the three segments of the Ouachita-Marathon-Sonora orogen. Several erosional hiatuses characterize the stratigraphic sequences in all three segments. Although the sole faults of the orogenic belt are generally not exposed, they are believed to be regional décollement surfaces. Most of our stratigraphic summary of the Ouachita and Marathon segments of the Ouachita-Marathon-Sonora orogenic system is extracted from Ross (1979, 1986), Finney (1986), Lowe (1989), McBride (1989), Morris (1989), Ellison and Powell (1989), Ethington et al. (1989), Noble (1990, 1993), Barrick and Haywa-Branch (1994), and Barrick and Noble (1995); however, we have examined many of the stratigraphic units in the Ouachita Mountains and Marathon region.

Paleozoic rocks in the Ouachita Mountains, Marathon region, and central Sonora region consist primarily of deep-water deposits with sparse megafossils and many age calls must make use of microfossils. Radiolarians are common in most Paleozoic chert and siliceous mudstone units. Conodonts occur sparingly throughout the Paleozoic sequence in limestone and turbiditic limestone beds. Graptolites, phyllocarids, orbiculoids, and sponge spicules are common in lower Paleozoic mudstones and siltstones, and fusulinids are common in upper Paleozoic continental-shelf limestones and synorogenic turbiditic lime grainstones. In central Sonora, brachiopods and worm tubes associated with seafloor vents or seeps occur locally in deep-marine Upper Devonian barite deposits at localities 43 and 44 (Fig. 4).

Preorogenic Upper Cambrian to Lower Mississippian offshelf chert, siliceous shale, sandstone, and minor limestone, 600–3000 m thick within the orogen, represent slow accumulating continental-rise and -slope deposits (Fig. 5).

Foreland and Transform Basins

Pedregosa foreland-basin sections are present in front of the Marathon segment in Chihuahua near Nuevo Casas Grandes in the La Vinata–Sierra de Azcárate area (loc. 36) on the cratonic platform, and near Ciudad Chihuahua (loc. 38 and 39) on the continental shelf. A foredeep (Mina México foredeep) is present on the continental shelf in the vicinity of the frontal thrust of the Sonora segment of the Ouachita-Marathon-Sonora orogen in central and east-central Sonora (Figs. 3 and 4).

Foredeeps and foreland basins and uplifts developed in front of the advancing accretionary wedge (Fig. 3) due to crustal flexing and isostatic loading of the continental crust. The allochthons formed tectonic highlands beginning in Mississippian time, as indicated by the thick Late Mississippian to Permian flysch deposited in the foredeeps. Thick sedimentary sequences derived from erosion of the tectonic highlands and cratonic-platform uplifts accumulated in the rapidly subsiding foredeeps and foreland basins (stippled areas on Fig. 3).

Transform basins likely developed adjacent to evolving transform faults that penetrated the Laurentian crust during the breakup of Rodinia (Fig. 2).

The Texas and Chihuahua promontories project southward from the Laurentian craton between the exposed segments of the orogen (Figs. 1 and 3). The Texas promontory and the southwest-trending segment of the Ouachita-Marathon-Sonora orogenic system between southeast Oklahoma and West Texas are interpreted to be related to northwest-trending transform faults (Alabama–Oklahoma [AOT] and Texas [TT] transforms) offsetting the Neoproterozoic–early Paleozoic northeast-striking rift system (Thomas, 1991; Thomas and Astini, 1999; North American Magnetic Anomaly Group, 2002). The Chihuahua promontory may be similarly related to offsets in the Neoproterozoic–early Paleozoic rifted margin caused by northwest-trending transform faults (Goetz and Dickerson, 1985; Dickerson 1987), including the postulated Neoproterozoic right-stepping Chihuahua-Sonora transform (CST, this report).

Fragments of Gondwana Megaterrane

Several inferred remnants of Gondwana continental-margin volcanic-arc and related fore-arc and back-arc rocks south of the Ouachita-Marathon-Sonora orogen in the southern United States have been collectively designated as the Sabine block (SB, eastern areas of diagonal-line pattern on Figure 3, including Tp, Sau, and Yp; Nicholas and Waddell, 1989; Viele and Thomas, 1989, fig. 19). Boreholes drilled in the Gulf Coastal Plain of eastern Texas (Sabine uplift area, Sau on Fig. 3), ~300 km south of the Ouachita Mountains, penetrated pre-Desmoinesian (pre–upper Middle Pennsylvanian) volcanic rocks and flysch (Nicholas and Waddell, 1982. 1989). They speculated that the volcanic rocks may represent either a local volcanic center or an extensive volcanic-arc system associated with a subduction zone related to the Ouachita segment of the Ouachita-Marathon-Sonora belt. Two inferred remnants of Gondwana volcanic-arc and related fore-arc and back-arc rocks south of the Ouachita-Marathon-Sonora belt in northern Mexico are first, the Coahuila block (collectively the Coahuila terrane of López et al., 1996, and R. López, 1997, written commun., and the Las Delicias terrane of McKee et al., 1999; central area of diagonal-line pattern labeled CB on Fig. 3), and second, the El Fuerte block of this paper (western area of diagonal-line pattern labeled FB on Fig. 3). The Las Delicias terrane (part of the Coahuila block south of the San Marcos fault) exposed in northeast Mexico consists of upper Paleozoic flysch and volcanic-arc rocks, including gravity-flow deposits containing clasts of basalt, andesite, and dacite volcanic rock and volcaniclastic sandstone and conglomerate. The El Fuerte block (Poole and Perry, 1998, and this report) exposed in northwest Mexico consists of polydeformed and metamorphosed ocean-basin and arc-magmatic andesitic to rhyolitic rocks related to a subduction zone. We have not determined all of the components and full extent of our interpreted El Fuerte block.

LATE PALEOZOIC OROGENESIS ALONG THE SOUTHERN MARGIN OF LAURENTIA AND DESCRIPTION OF THE THREE SEGMENTS OF THE OUACHITA-MARATHON-SONORA OROGEN

Segments of the Ouachita-Marathon-Sonora orogen resulted from the growth of an accretionary prism and its thrusting onto the southern margin of the Laurentian craton. The overall geometry and location of the Ouachita-Marathon-Sonora orogenic belt was controlled by the configuration of the southern edge of the Laurentian cratonic basement or continental shelf (hachured line on Figs. 1 and 3) along the southern margin of the Ouachita-Marathon-Sonora orogenic belt, and by the direction of late Paleozoic plate motion based on perpendicularity to fold trends (bold arrows on Fig. 3). The trace of the orogenic belt was inherited from the location and shape of the earlier rifted margin. This is manifested by the belts of compressional structures, which assume a curved trace conforming to the shape of the older rifted continental margin (Dewey and Burke, 1974; Thomas, 1977; Kluth and Coney, 1981). The principal compressional stress caused by plate convergence was northwest directed, which resulted in development of the Ouachita-Marathon-Sonora orogen (Fig. 14).

Embayments and promontories of a continental margin are commonly a consequence of rifting and transform faulting (King,

1975; Thomas, 1976, 1977, 1983, 1991). Figures 2 and 3 show three major Neoproterozoic or early Paleozoic transform faults interpreted to offset the continental margin, and an inferred transform fault (Baja California, BCT) in Baja California Norte that formed the southwesternmost edge of Laurentia.

We have projected the Ouachita-Marathon-Sonora orogenic front and southern edge of continental crust westward from central Chihuahua to central Sonora and northern Baja California Norte on the basis of stratigraphic and structural similarities of the Marathon and Sonora segments, and the location of the Pedregosa foreland basin (Pb) and Mina México foredeep (MMb) and Sierra del Nido (SNu) uplift (Fig. 3), which may be part of the accreted Aldama platform (Haenggi et al., 2000). This latter is conjectural because no boreholes have penetrated continental or oceanic crust beneath the Sierra Madre Occidental in west-central Chihuahua and adjacent Sonora. To the south and southeast of the orogen, our inferred southern edge of the Laurentian craton (hachured line on Figures 1 and 3) is the boundary between continental-shelf strata dominated by shallow-marine carbonate-rock facies and offshelf continental-rise strata dominated by deep-marine mudstone and chert facies. These deep-water siliceous deposits, in part, originated as opaline-silica oozes and include some stratiform barite. Although the margin between shallow-marine and offshelf facies was a fundamental tectonic and stratigraphic boundary, it is not exposed and its original position can be inferred only from indirect surface and subsurface data (Nicholas and Rozendal, 1975; Lillie, 1985; Keller et al., 1989a; Mickus and Keller, 1992; Roberts, 1994; Harry and Londono, 2004). Keller et. al (1989a), for example, presented a geophysical model for the Ouachita Mountains that depicts a preserved early Paleozoic continental margin buried beneath allochthonous Paleozoic strata and younger sedimentary rocks. The southern part of their model indicates the presence of oceanic or highly extended continental crust overlain by ~15 km of mostly Paleozoic sedimentary rock. Our inferred southern limit of Laurentian continental crust (hachured line on Figures 1, 3, and 4) in southern Sonora approximates the location of Proterozoic basement inferred by Valencia-Moreno et al. (2001) using initial Sr ratios. The southern limit of Proterozoic basement in Chihuahua is supported by data presented by McDowell et al. (1999) using Sr, Nd, and Pb isotopic compositions at the Tomóchic volcanic center (loc. 59 on Figures 1–4); Cameron et al. (1992) and Ruiz et al. (1988) using isotopic evidence from orthogneiss xenoliths at La Olivina (loc. 60 on Figures 1 and 3); and James and Henry (1993a, 1993b) using Pb isotopic composition of Tertiary igneous rocks in Texas and northern Chihuahua. At Ciudad Chihuahua, the northerly bend in the late Neoproterozoic–Early Cambrian edge of Laurentia, as shown by Haenggi (2001, figures 1–3), may represent offset of the continental edge by faulting during the rifting phase or later.

In northern Mexico and the southern United States, the mid-Mississippian collision of Laurentia with northwest Gondwana and associated volcanic arc initiated syntectonic deposition along the southern margin of Laurentia that continued into the Permian. The resulting synorogenic Upper Mississippian (Meramecian) to Middle Permian (Guadalupian) rocks record rapid deposition of turbidites, the maximum thickness of which may be as much as 10,000 m in the eastern part of the Ouachita-Marathon-Sonora orogenic system (Morris, 1989). Most of the early synorogenic deposits accumulated rapidly in the deep-water ocean basin that separated Laurentia and the South American part of Gondwana and subsequently in deep-water foredeeps and foreland basins (stippled areas on Fig. 3), whereas in the Ouachita and Marathon segments, some of the late synorogenic and postorogenic upper Paleozoic sediments were deposited slowly in shallow-marine settings.

According to C.H. Stevens (2003, personal commun.), most Pennsylvanian and Permian fusulinids in Sonora have a Texas affiliation, and are not like those of California and Nevada. Ross (1967b, 1979) concluded that biogeographic separation of fusulinid faunas in North America was complete at certain times during the Pennsylvanian and Permian when only a few genera filtered between the faunal realms of south-central United States and northeastern Mexico (Midcontinent–Andean realm) and those of the western United States and western Canada (Eurasian–Arctic realm). The middle to late Paleozoic Transcontinental arch (Figs. 1 and 3) may have been a major barrier to integration of some fusulinid faunas. A close approach of western Gondwana (South America) and southern Laurentia (North America) in the Pennsylvanian is indicated by the migration and mingling of fusulinids on both continental margins (Ross, 1979). The geographic distribution of the Early Permian Midcontinent–Andean fusulinid faunal realm and the regional histories with similar depositional and tectonic patterns indicate that the Ouachita-Marathon-Sonora facies was adjacent to Colombia and Venezuela during the collision of North and South America (Ross, 1979). The Early Mississippian (early Osagean) fauna in the Vicente Guerrero Formation, near Ciudad Victoria (loc. 72), has strong affinities to North American (Ohio) faunas, and suggests that what is now the Ciudad Victoria area may have originally been part of South America but by Mississippian time had been close enough to North America for faunal interchange (Boucot et al., 1997).

The late Paleozoic structural and depositional history of the convergence of Gondwana and Euramerica has many similarities to the Cenozoic history of the southern Lesser Antilles arc platform described by R.C. Speed and coworkers (Torrini and Speed, 1989; Russo and Speed, 1992; Speed et al., 1993).

Central Sonora Region

Paleozoic rocks within the Sonoran segment of the orogen in central Sonora (Fig. 4) consist of Lower Ordovician to Middle Permian ocean-basin and continental-rise strata (Fig. 5). Total thickness of the sequence is unknown because of structural complications, but it probably exceeds two thousand meters on the basis of a composite of partial stratigraphic sections. Preorogenic (pre–Late Mississippian) strata are 500–900 m thick and synorogenic (post-Early Mississippian) strata are estimated to be significantly thicker than 1000 m.

The best exposed Paleozoic offshelf (eugeoclinal) sequence in central Sonora is in the vicinity of Minas de Barita ~100 km east of Hermosillo (loc. 43 on Fig. 4). Peiffer-Rangin et al. (1980) were the first to recognize graptolites and corals of Ordovician age in the Minas de Barita (loc. 43) and Cerro Cobachi (loc. 44) areas. Since 1982, Poole and coworkers have been preparing a detailed geologic map (1:15,000-scale) of part of the Minas de Barita area. Radelli et al. (1987) published a generalized geologic map (~1:60,000-scale) of the Minas de Barita area.

The oldest part of the Paleozoic allochthonous sequence in the Minas de Barita area is the Lower Ordovician El Quemado Shale, which consists of more than 300 m of interbedded mudstone, siltstone, sandstone, chert, and sparse limestone. The formation, which is mostly shale, locally is indurated to argillite, and commonly is silicified and intensely deformed (internally folded and sheared). The El Quemado contains Early Ordovician graptolites, phyllocarids, and conodonts.

The overlying slope-forming Lower and Middle Ordovician El Mezquite Shale consists of 100–150 m of graptolitic silty shale and minor sandstone. In addition to graptolites, orbiculoid brachiopods and siliceous sponge spicules are common. In Sierra La Ciénega (15 km southeast of the Minas de Barita area) and in the sierras east, south and west of Sierra La Ciénega (e.g., Sierra El Aliso, loc. 46, and La Colorada, loc. 47), Middle Ordovician facies containing thick quartzites (deep-sea channel sands) interbedded with metamorphosed shales are equivalent to the quartz-sand deficient El Mezquite Shale in the Minas de Barita (loc. 43) and Cerro Cobachi (loc. 44) areas, indicating a pronounced southerly increase in quartz-sand content. This significant increase in quartz-sand content is believed to be a normal facies transition. However, the apparent closeness of the sand-rich and sand-poor facies suggests a possibility of two separate thrust sheets within the Sonora allochthon. These coeval Ordovician facies are believed to be continental-slope and -rise deposits along the Laurentian margin. The quartzite-bearing eugeoclinal facies correlates with the upper Middle Ordovician Peña Blanca Quartzite of the Laurentian continental shelf to the north, and it sedimentologically ties the offshelf and shelf facies.

The uppermost Middle and Upper Ordovician El Yaqui Chert is a prominent ridge-forming unit consisting of 5–150 m of chert (generally in beds 10–30 cm thick) containing scattered cannon-ball-size chert concretions and sparse limestone beds. The chert has graptolitic siltstone partings (Fig. 6) and minor interbedded siltstone, dolostone, and sparse tuff. Radiolarians are visible in most chert beds. The contact between the El Mezquite Shale and El Yaqui Chert is gradational and locally intertonguing. On Turner Island in the eastern Gulf of California (loc. 50 on Fig. 4), late Middle to early Late Ordovician graptolites were collected from a phyllitic shale unit within an interbedded chert and siliceous shale/argillite sequence (Poole et al., 1993, 1995a) that correlates with parts of the El Mezquite Shale and El Yaqui Chert.

The overlying El Torote Limestone is a thin silty and dolomitic limestone to limy dolomite that marks the top of the Upper Ordovician. It is as much as 8 m thick and locally contains Late Ordovician graptolites in the platy lower part. Locally, the upper El Torote consists of intraclast conglomerate and sedimentary flow folds that have yielded conodonts of Ordovician to Silurian age; therefore, the uppermost El Torote locally may contain some Silurian strata.

A major unconformity separates Upper Ordovician–Silurian? strata and Upper Devonian strata, and the contact zone commonly is mineralized with iron and manganese oxides.

The Upper Devonian and Lower Mississippian Los Pozos Formation is a prominent ridge-forming unit consisting of 50–150 m of interbedded mudstone, siltstone, chert, novaculitic chert, turbiditic sandstone, turbiditic and exhalative barite, and minor limestone and sparse tuff. It contains Late Devonian and early Early Mississippian radiolarians and conodonts. Stratiform barite is present in both the Devonian and Mississippian parts of the Los Pozos. Barite was deposited in multiple turbidite-fan systems throughout the Minas de Barita and Cerro Cobachi areas (loc. 43 and 44). The original source of barite was numerous seafloor hydrothermal vents (Poole et al., 1991). No Lower or Middle Devonian strata have been recognized in the Minas de Barita area; however, Bartolini et al. (1989) reported upper Middle Devonian (Givetian) strata 45 km to the southeast, in the Sierra El Aliso area (loc. 46).

The Lower Mississippian Cerro Tasajo Limestone consists of 0–75 m of turbiditic limestone with minor chert, mudstone, siltstone, sandstone, and conglomerate. It contains late Early Mississippian (middle Osagean) conodonts. In many places, the Cerro Tasajo is absent. A regional angular unconformity between the Early and Late Mississippian records a late Osagean defor-

Figure 6. Outcrop of upper Middle to Upper Ordovician El Yaqui Chert in central Sonora (near mine office of Minas de Barita) showing cyclic thin-bedded chert with shaly siltstone partings. Formation is dated by graptolites in the siltstone partings. Benita Murchey and Juan-Manuel Morales in foreground (February 1983).

mational event (Los Chinos Event) and partial erosion of the pre–Late Mississippian sequence.

The ledge-forming medial Mississippian Los Chinos Conglomerate consists of 0–15 m of chert-gravel conglomerate with a dark-gray chert matrix that rests on the folded and faulted pre–Late Mississippian sequence. This submarine conglomerate ranges from debris-flow deposit to turbidite. The Los Chinos Conglomerate is absent locally due to nondeposition, erosion, or faulting. Most of the clasts are dark-gray chert derived from the El Yaqui Chert and Los Pozos Formation; some of the chert clasts are internally folded. At one locality (north of Rancho Los Chinos), a limestone clast yielded early Early Devonian (Lochkovian) conodonts (C.A. Sandberg, 1994, written commun.). Probable source of the limestone clast was the carbonate shelf to the north and it sedimentologically ties the offshelf to shelf facies.

This basal conglomerate is overlain by the synorogenic Rancho Nuevo Formation that consists of a thick sequence (>1000 m) of Upper Mississippian mudstone, siltstone, chert, and barite, and Lower to Upper Pennsylvanian and Lower to Middle Permian interbedded turbiditic mudstone, siltstone, sandstone, conglomerate, chert, barite, limestone, and minor dolostone (Fig. 5). The Rancho Nuevo contains radiolarians, conodonts, and fusulinids. Stratiform barite is restricted to the Upper Mississippian and Lower Pennsylvanian part of the Rancho Nuevo. Barite was deposited in multiple turbidite-fan systems (Poole et al., 1991). Barite sediment was derived primarily from Upper Devonian–Lower Mississippian barite beds that were eroded and redeposited following mid-Mississippian folding and faulting (Los Chinos Event). An additional barite source may have been syndepositional seafloor vents and seeps. The formation may be more than a thousand meters thick and contains many distinctive lithologic units and presumed disconformities, but member subdivision has been hampered by structural complexities. The formation records deposition in steadily subsiding troughs being filled with flysch sediment shed from structural highs in the orogenic belt and the continental shelf. The Rancho Nuevo Formation contains some beds of pebble to boulder conglomerate suggestive of proximal flysch; however, most of the formation consists of fine-grained clastic turbidites indicative of distal flysch (Fig. 7). Several hiatuses likely occur within the Rancho Nuevo, judging from the many facies changes within the formation. These facies changes probably indicate syntectonic unconformities that record contemporaneous tectonic activity and sedimentation.

The youngest limestones (late Early to early Middle Permian) of the Paleozoic carbonate shelf are conformably overlain by or intertongue with upper Lower to Middle Permian Mina México foredeep flysch (>1000 m thick), which is structurally overlain by rocks of the Sonora allochthon (Figs. 10 and 11). The same sequence of Ordovician to Permian ocean-basin strata is recognized in the Cerro Cobachi area (loc. 44) 20 km southwest of the Minas de Barita area (Noll, 1981; Ketner and Noll, 1987; Riva and Ketner, 1989; Poole, unpublished data).

Upper Permian and Triassic strata have not been recognized in the Minas de Barita and Cerro Cobachi areas; however, 40 km southeast of the Minas de Barita area in Sierra El Aliso (loc. 46) and Sierra de San Javier (Bartolini et al., 1989, 1990; Stewart et al., 1990), and 30 km west-southwest of the Minas de Barita area in the vicinity of La Colorada (loc. 47), undeformed redbeds of probable Triassic age (Barranca Group?) depositionally overlie the Sonora allochthon.

Marathon Region, West Texas

Paleozoic rocks within the Marathon segment of the orogen in West Texas comprise Upper Cambrian to Permian strata (Fig. 5). Paleozoic rocks exposed in the Solitario area (Su on Fig. 3), ~60 km southwest of the Marathon outcrops, are closely related to the former (Corry et al., 1990; Henry et al., 1996). According to Thomson and McBride (1964) and McBride (1989), Paleozoic rocks in the Marathon geosyncline have a composite thickness of 4800 m that can be divided into preorogenic (pre-flysch) and synorogenic (flysch) stages (Fig. 5).

Preorogenic deposits, which are ~950 m thick, include Upper Cambrian to Lower Mississippian rocks. The Upper Cambrian and Lower Ordovician Dagger Flat Sandstone consists of 100–300 m of sandstone and subordinate mudstone, calcarenite, and conglomerate. The Lower Ordovician Marathon Formation is composed of 260–330 m of interbedded sandstone, siltstone, mudstone, limestone, and subordinate calcarenite, limestone-pebble conglomerate, and gravelly mudstone. The Alsate Shale (8–30 m thick) is considered to be the upper member of the Marathon Formation by some workers (Young, 1970). The Middle Ordovician Fort Peña Formation is a ridge-forming unit consisting of 40–200 m of interbedded mudstone, calcarenite, conglomerate, chert, and subordinate gravelly mudstone. Wilson

Figure 7. Outcrop of Pennsylvanian part of Rancho Nuevo Formation in central Sonora (west bench of Las Papas barite pit, Minas de Barita) is part of a thick sequence of turbiditic flysch deposited in a synorogenic basin. This thin-bedded turbiditic distal flysch elsewhere contains radiolarians and conodonts that date the unit.

(1954) and Corry et al. (1990) reported a higher content of quartz sand in the Middle Ordovician Fort Peña and associated strata in the Solitario (Su on Fig. 3) relative to coeval strata to the northeast in the northern and central parts of the Marathon region. Quartzites in the Fort Peña and upper Marathon (Rodrigues Tank Sandstone Member of Corry et al., 1990) Formations are as thick as 10 m in the Solitario and in the Old Jones Ranch section in the southern part of the Marathon region (Wilson, 1954), indicating a pronounced southerly increase in quartz sand. The Middle Ordovician Woods Hollow Shale is composed of 55–150 m of mudstone, siltstone, sandstone, calcarenite, and sparse gravelly mudstone. The Middle and Upper Ordovician Maravillas Formation is a conspicuous ridge-forming unit 30–150 m thick of interbedded black limestone and chert and subordinate mudstone, siltstone, conglomerate, and dolostone. The Upper Ordovician Persimmon Gap Shale (Wilson, 1954) is 0–12 m thick and consists of deep-water shale that lies conformably above the Maravillas Formation in the southern part of the region (Ross, 1986). The Lower Silurian to lowermost Mississippian Caballos Novaculite is a prominent ridge-forming unit 30–210 m thick consisting of white chert or novaculite, varicolored chert, subordinate mudstone, and sparse calcarenite, sandstone, and granule conglomerate. Several disconformities have been recognized within the Caballos. Early Silurian (late Llandoverian) to earliest Mississippian conodonts and radiolarians have been found in the Caballos Novaculite (Barrick, 1987; Noble, 1990, 1994; Barrick and Noble, 1995), which is lithologically similar to the Lower Devonian–Lower Mississippian Arkansas Novaculite of the Ouachita segment, and to the Upper Devonian and Lower Mississippian Los Pozos Formation of the Sonora segment. A major unconformity probably separates the Upper Ordovician Maravillas Formation and the Caballos Novaculite.

Unconformably overlying the Caballos Novaculite is a synorogenic (flysch) sequence ~3800 m thick that consists of interbedded mudstone, siltstone, and sandstone, subordinate conglomerate and chert. Noble (1990) inferred an Early Mississippian basinwide hiatus between them (Fig. 5). Principal stratigraphic units in the synorogenic sequence are the Upper Mississippian and Lower Pennsylvanian Tesnus Formation, Middle Pennsylvanian Dimple Limestone, and Middle Pennsylvanian Haymond Formation. The Tesnus Formation is a 100–2000-m-thick sequence of interbedded turbiditic sandstone, siltstone, mudstone, tuff, and sparse barite concretions that unconformably overlies the Caballos Novaculite. Ellison and Powell (1989) reported Late Mississippian (earliest Chesterian) conodonts from shales just above the base of the Tesnus, and J.E. Barrick (2001, written commun.) recovered Early Pennsylvanian conodonts from chert beds near the top of the Tesnus. Argillized tuff beds are present in the Tesnus (Imoto and McBride, 1990). The Dimple Limestone is composed of 75–280 m of turbiditic calcarenite and limestone conglomerate and subordinate mudstone, siltstone, and chert. The Haymond Formation (Fig. 8) consists of as much as 1400 m of interbedded sandstone, siltstone, mudstone, and an interval of coarse sandstone and gravelly mudstone containing exotic blocks of sedimentary, igneous, and metamorphic rock types. Sparse limestone boulders contain Middle Cambrian fossils (Palmer et al., 1984), and igneous and metamorphic boulders yielded Devonian radiometric Rb-Sr ages of 410–370 Ma (Denison et al., 1969). The Upper Pennsylvanian Gaptank Formation consists of 0–815 m of interbedded mudstone, quartzose sandstone, calcarenite, limestone, and conglomerate. Ross (1967a, 1978) and McBride (1989) considered the Gaptank Formation to represent latest orogenic and post-orogenic (molasse-like) shallow-marine and fluvial facies. The overlying Lower Permian Neal Ranch (0–100 m) and Lenox Hills (0–85 m) Formations were considered by Ross (1967a, 1978) to be shallow-marine and marginal marine deposits in the western part of the Glass Mountains. According to Ross (1978), the Lenox Hills Formation lies on the Dugout Creek thrust sheet, which in turn rests on folded and faulted beds of the Neal Ranch and Gaptank Formations in the western part of the Marathon region. The overlying Skinner Ranch–Hess Formations (Leonardian Series) and younger Permian strata represent latest synorogenic and post-orogenic marine and marginal marine deposits. Stratigraphic and structural evidence in the Glass Mountains indicates that collisional deformation in the Marathon region ended in the middle part of the Early Permian (Ross, 1978, 1986).

Ouachita Mountains, West-Central Arkansas and Black Knob Ridge, Southeastern Oklahoma

The Ouachita segment of the orogen consists of a thick Paleozoic sequence of Upper Cambrian–Middle Pennsylvanian strata (Fig. 5). According to Viele and Thomas (1989), the composite thickness is ~19,500 m. We base most of our description and dating of Ouachita rocks on reports by Hendricks et al.

Figure 8. Outcrop of Middle Pennsylvanian Haymond Formation in West Texas (near roadside geologic marker east of Marathon, Texas) is part of a thick sequence of distal turbiditic flysch deposited in a synorogenic basin. The formation is dated by a fusulinid and constrained by Middle Pennsylvanian conodonts and radiolarians in adjacent formations.

(1937), Hass (1951), Finney, (1986), Ethington et al. (1989), Lowe (1989), Morris (1989), Viele (1989), and Barrick and Haywa-Branch (1994); however, we have also examined many of the Paleozoic formations in the Ouachita Mountains.

Preorogenic rocks of the Ouachita Mountains include nearly 3500 m of strata ranging in age from Late Cambrian to late Early Mississippian (Osagean). The Upper Cambrian and Lower Ordovician Collier Shale consists of 20–300 m of black to dark-gray, commonly graphitic, fissile shale, and slate, and locally, thin beds of limestone and calcarenite, sandstone, and conglomerate. Overlying Lower Ordovician Crystal Mountain Sandstone (~200 m thick) is dominated by turbiditic fine- to medium-grained quartzose, commonly carbonate-cemented sandstone, local conglomerate, and minor interbedded shale. The Lower Ordovician Mazarn Shale (300–1000 m thick) consists of dark-gray to black shale with minor interbedded greenish-weathering shale, laminated siltstone, quartzose sandstone, and finely crystalline gray limestone. The Middle Ordovician Blakely Sandstone (100–180 m thick) is a series of turbiditic thinly to thickly bedded quartzite interbedded with shale. The Middle Ordovician Womble Shale (100–1000 m thick) is mostly graptolitic black fissile shale with minor thin interbeds of quartzose siltstone, sandstone, and calcarenite. The upper Middle and Upper Ordovician Bigfork Chert (200–240 m thick) consists of evenly bedded, light to medium gray-weathering chert (generally in beds 5–15 cm thick) and interbedded carbonate rock and shale (Fig. 9). At Black Knob Ridge, Oklahoma, Finney (1986) reported graptolites from the Bigfork that range in age from late Middle to early Late Ordovician (early Mohawkian to early Cincinnatian). The Bigfork Chert grades upward into the Upper Ordovician Polk Creek Shale (0–60 m thick) that is made up of graptolitic black, fissile, very carbonaceous shale. The Silurian Blaylock Sandstone (0–250 m thick) consists of brownish-weathering, olive- to dark-gray, turbiditic, very fine to medium-grained quartzose sandstone and gray shale. The Silurian Missouri Mountain Shale (0–100 m thick) contains black, gray, pink, green, and buff clayey shale and thin interbeds of turbiditic quartzose siltstone and sandstone. At Black Knob Ridge, Barrick and Haywa-Branch (1994) reported conodonts of Early (late Llandoverian) to Late (Pridolian) Silurian age, and speculated that the Missouri Mountain may be as young as earliest (early Lochkovian) Devonian. The Devonian and Lower Mississippian Arkansas Novaculite (70–290 m thick) is subdivided into three members: a lower novaculite member, a middle chert and shale member, and an upper novaculite member (Miser, 1917). Hass (1951) reported Early Mississippian conodonts in the upper part of the middle division of the Arkansas Novaculite. At Black Knob Ridge, Oklahoma, Barrick and Haywa-Branch (1994) reported Early (late Emsian) to Late (Famennian) Devonian conodonts in the Arkansas Novaculite.

Overlying the Arkansas Novaculite is a thick succession of synorogenic flysch that, according to Morris (1989), attained a maximum thickness of ~7000 m during the Late Mississippian Meramecian through Early Pennsylvanian Morrowan. Another 9000 m of sediments were deposited during the early Middle Pennsylvanian Atokan in the axial parts of a few foreland basins. The Mississippian and Pennsylvanian detritus was deposited mainly by various types of gravity flows (Morris, 1989). The Upper Mississippian Stanley Group, which is as much as 3350 m thick, has been subdivided into three formations (Morris, 1989). Major lithologies, in order of decreasing abundance, include shale, sandstone, tuff, impure chert, and siliceous shale (Morris, 1989). Late Mississippian (Chesterian) conodonts were reported in the lower Stanley Shale by Hass (1950). The uppermost Mississippian and Lower Pennsylvanian Jackfork Group consists of turbiditic light-gray sandstone and subordinate dark-gray shale that has a maximum thickness of 2100 m in the axial part of the Arkoma foredeep (Morris, 1989). The Lower Pennsylvanian Johns Valley Formation overlies the Jackfork Group (Fig. 5). In Oklahoma, the Johns Valley Formation is predominantly a shale unit 125–270 m in thickness, and the lower part is black, laminated, and contorted, whereas the middle and upper parts are light-gray mudstone with intercalated sandstone. The Johns Valley Formation in Oklahoma contains abundant boulders of diverse composition set in contorted shale. In Arkansas, the Johns Valley Formation is ~300 m thick and consists mostly of thin turbiditic sandstone and shale. The lower Middle Pennsylvanian Atoka Formation, which is as thick as 8500 m in the Ouachita Mountains, consists of turbiditic sandstone and interbedded shale. Post-Atoka Formation Pennsylvanian and Permian strata are not exposed in the Ouachita Mountains.

Figure 9. Outcrop of upper Middle to Upper Ordovician Bigfork Chert in Ouachita Mountains (Weyerhaeuser office area, Hot Springs, Arkansas) showing cyclic thin-bedded chert with shaly siltstone partings. The formation is dated by graptolites and conodonts.

Combined neodymium and U-Pb detrital zircon data on turbidites from the preorogenic Blakely and Blaylock Sandstones and the synorogenic Jackfork Group sandstones in Arkansas and Oklahoma (Fig. 5) indicate tectonic recycling of zircons from the Laurentian craton (Gleason et al., 2001).

Stratigraphic and Structural Similarities in the Ouachita-Marathon-Sonora Orogen

Stratigraphic and structural similarities within the three segments of the Ouachita-Marathon-Sonora orogen strongly suggest a physical continuity along the southern margin of Paleozoic North America. This continuity is indicated by stratigraphic correlations, similarities in sedimentary facies, faunal evidence, and periods of depositional hiatuses.

Regional Stratigraphic, Faunal, and Nondepositional Correlations

The total thickness of Paleozoic preorogenic and synorogenic deposits decreases westward from the Ouachita Mountains through the Marathon region to central Sonora (Fig. 5). Several correlative lower and upper Paleozoic units illustrate stratigraphic similarities within the Ouachita-Marathon-Sonora orogenic system (Fig. 5). For example, similar southerly increases in quartz-sand content occur in the Middle Ordovician El Mezquite Shale and correlative strata of the Sonora segment, in the Middle Ordovician Fort Peña and upper Marathon Formations and associated strata in the Marathon-Solitario area (Wilson, 1954; Corry et al., 1990) of the Marathon segment, and in correlative Ordovician strata in McCurtain County, southeasternmost Oklahoma (Wilson, 1954), in the western part of the Ouachita segment of the Ouachita-Marathon-Sonora orogen. Another example is the similarity of the uppermost Middle and Upper Ordovician Bigfork Chert in Arkansas and southeast Oklahoma, which consists of cyclic thin-bedded chert with siltstone partings deposited in a continental-rise setting, and the correlative uppermost Middle and Upper Ordovician El Yaqui Chert in Sonora (cf. Figs. 6 and 9). Similar graptolite faunas in siltstone partings are present in both formations. Novaculitic chert is present in the Upper Devonian part of the Los Pozos Formation in Sonora and forms a major part of the correlative Caballos and Arkansas Novaculites in West Texas and Arkansas. The Upper Mississippian and Lower Pennsylvanian Tesnus Formation and Middle Pennsylvanian Haymond Formation in West Texas are part of a thick sequence of distal turbiditic flysch deposited in a steadily subsiding basin during orogenesis. These sediments were derived from erosion of the evolving Marathon allochthon (McBride, 1989) and are correlative with the Upper Mississippian to Middle Pennsylvanian part of the Rancho Nuevo Formation in central Sonora that also is characterized by a thick sequence of turbiditic flysch (cf. Figs. 7 and 8). The Rancho Nuevo Formation contains age-diagnostic Upper Mississippian–Middle Permian radiolarians, conodonts, and fusulinids. Fusulinids in the Mina México and upper Rancho Nuevo Formations are very similar to those in Wolfcampian, Leonardian, and Guadalupian (Neal Ranch, Lenox Hills, Skinner Ranch–Hess, and Word Formations) strata of the Marathon region, but are dissimilar to those in Lower and Middle Permian strata of Nevada and California (C.H. Stevens, 2003, personal commun.). The similarity of the Sonora and the Chihuahua and Texas fusulinids provides additional evidence supporting a physical connection between these two segments of the Ouachita-Marathon-Sonora orogenic belt. Early and Middle Permian fusulinids on the margins of southern and western Laurentia apparently were partially segregated due to paleogeography and biogeography on opposite sides of the Transcontinental arch.

A late Early Mississippian hiatus is recognized throughout the Ouachita-Marathon-Sonora orogenic system. Noble (1993) discussed a regional late Kinderhookian to late Meramecian hiatus in the Ouachita and Marathon regions. Our stratigraphic data indicate a late Osagean hiatus in the Sonora segment related to the Los Chinos Event. Synorogenic flysch sedimentation in the Sonora segment and Mina México foredeep continued at least into the Middle Permian (middle Guadalupian/Wordian).

Structural Style

Most fold axes and thrust faults exposed in the Ouachita-Marathon-Sonora orogenic system strike northeast, with most thrusting directed toward the northwest, and many folds and thrust faults are overturned to the northwest (Fig. 10). Maximum northward displacement of allochthonous ocean-basin rocks in the Ouachita-Marathon-Sonora orogen is estimated to be 50–200 km (Figs. 1, 3, and 4). This estimate is based on the distance between the autochthonous carbonate-shelf edge and the frontal thrusts of the Ouachita-Marathon-Sonora belt. Similar structural style characterizes the three segments of the Ouachita-Marathon-Sonora orogen (Fig. 10). Structure section A is from Arbenz (1989, plate 11, E–E) and represents the Ouachita segment in southeastern Oklahoma (also see cross section of Wickham et al., 1976, fig. 1). Section B is from Muehlberger and Tauvers (1989, plate 11, combined sections I–I and J–J ; W.R. Muehlberger, 2002, personal commun.) and represents the Marathon segment in West Texas. Lastly, section C is from mapping in progress by Poole and coworkers, which represents the Sonora segment in central Sonora (Fig. 11). Note the much larger scale of section C of the Sonora segment in contrast to sections A and B. Section C (line of section shown on Fig. 11) represents a more limited area of exposed rocks in the Minas de Barita area than in Oklahoma and Texas. The thrust faults in all three sections exhibit typical décollement-style (sled-runner) profiles characteristic of thin-skinned deformation. The thrusts are stacked in multilevel duplexes and strata within the thrust sheets are complexly folded. Two structural-stratigraphic levels characterize all three segments. Tightly folded, preorogenic, offshelf (Ordovician–Lower Mississippian) strata (gray pattern on Figs. 10 and 11) compose the lower level. Synorogenic, deep-water (Upper Mississippian–Permian) flysch strata (compact stipple pattern on Figs. 10 and 11) compose the upper level, which is deformed by broad folds and thrusts. Shallow-marine and marginal-marine strata (labeled

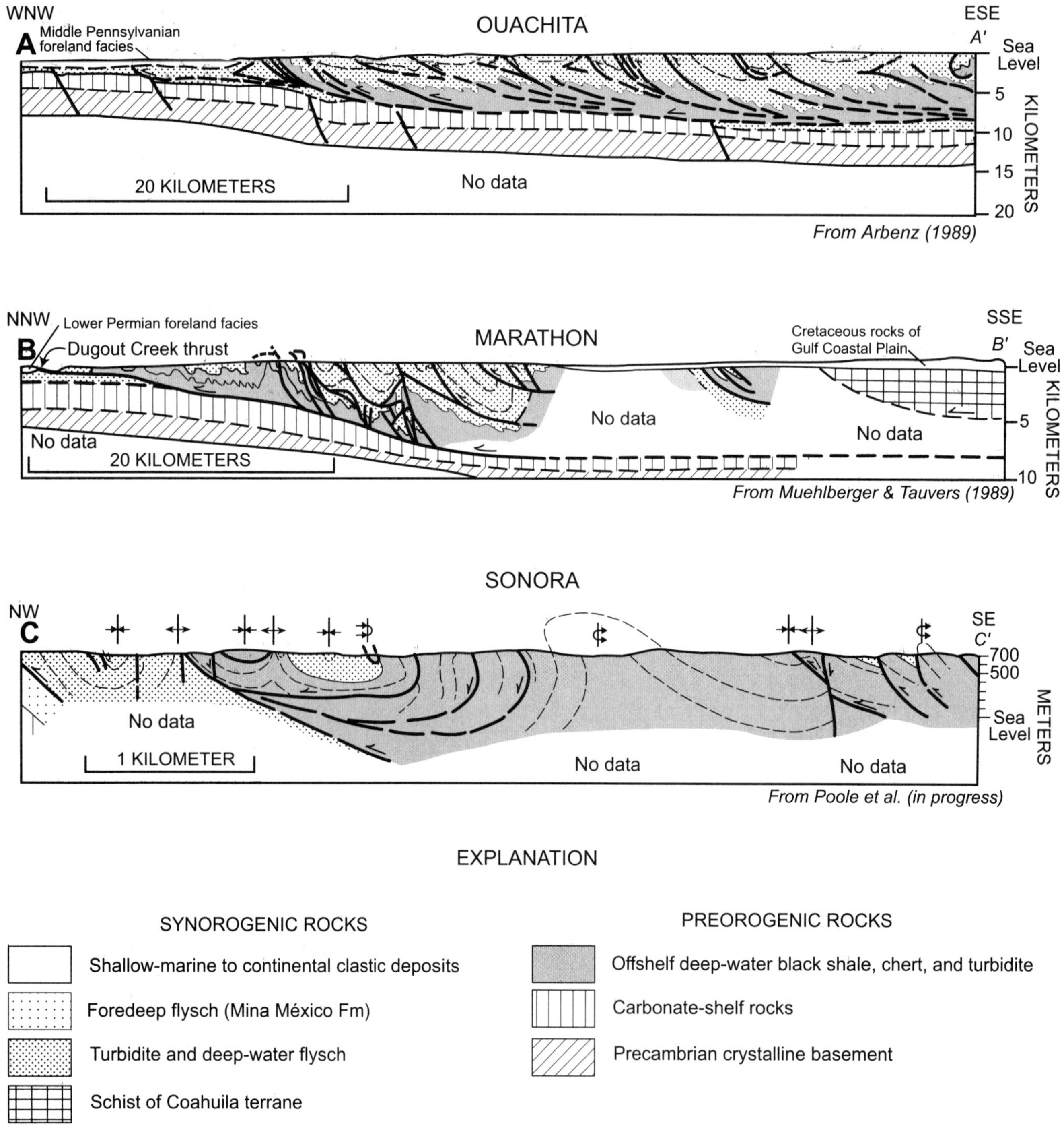

Figure 10. Structural cross sections representative of the three segments of the Ouachita-Marathon-Sonora orogenic system. Location of Sonora segment cross section C–C′ shown on Figure 11.

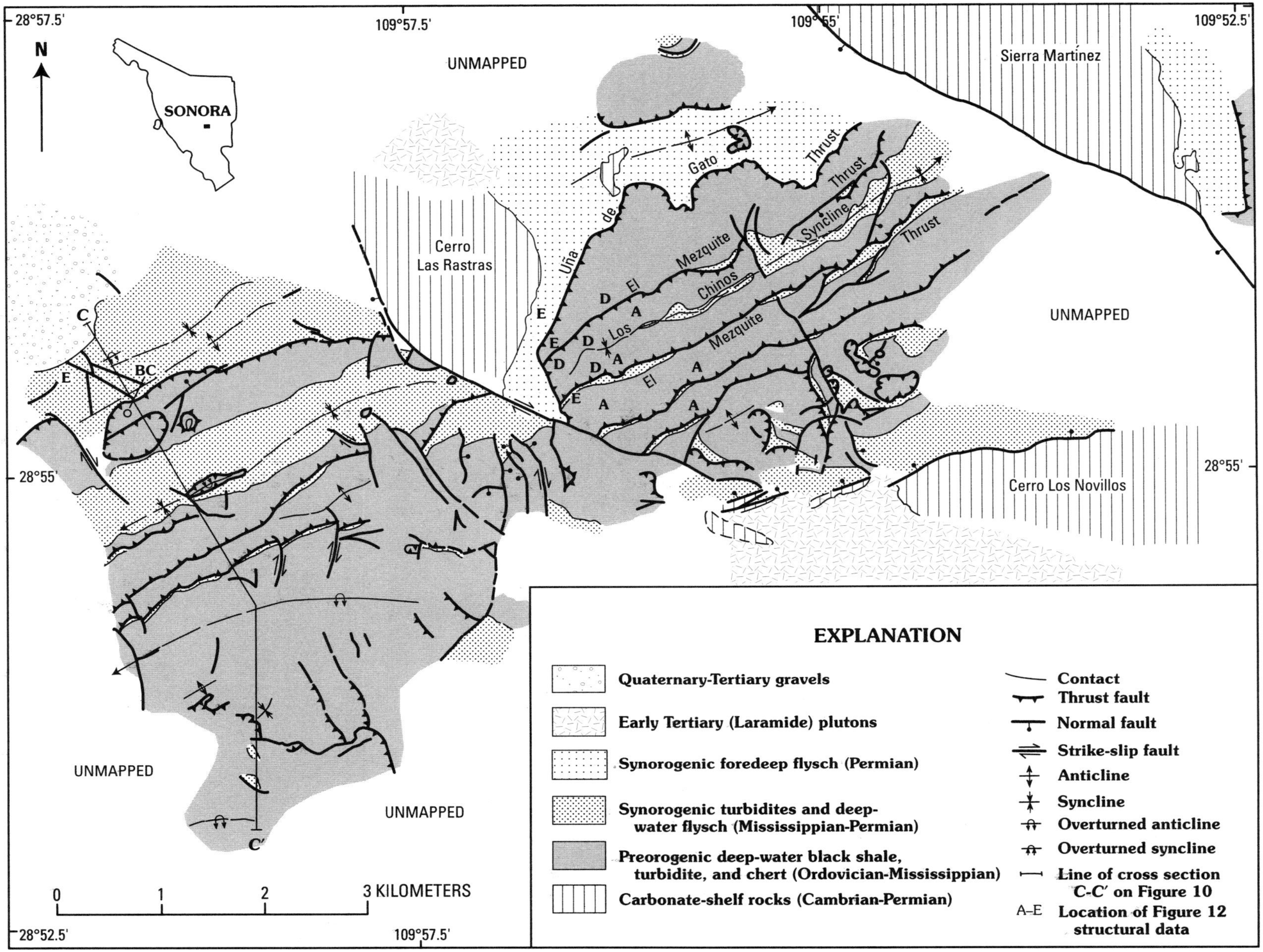

Figure 11. Simplified geologic map of the Minas de Barita area, central Sonora, showing distribution of allochthonous Paleozoic preorogenic and synorogenic rocks, autochthonous to parautochthonous carbonate-shelf Paleozoic rocks and foredeep Permian flysch, major faults and folds, and location of fold-orientation studies of Figure 12. Localities A–E on map denote areas of structural measurements for stereo plots A-E on Figure 12. Outline of map shown on Figure 4.

open areas on Fig. 10, sections A and B) are generally the youngest deposits in the foreland basins.

Bedding measurements in the Sonora allochthon and subjacent foredeep deposits of the Mina México Formation in the Minas de Barita area define a consistent (58–65° azimuth) northeast-southwest fold trend (Figs. 11 and 12A, B, D, and E). Broad folds related to the Late Permian Sonora deformation overprint tight folds that apparently formed during the mid-Mississippian deformational event (Los Chinos Event). Folded thrusts are also present. For example, the older El Mezquite Thrust is exposed on both limbs of the younger Los Chinos Syncline (Fig. 11). These younger broad folds verge N25–35°W, and are displaced by northwest-trending right-lateral faults of less than 5 km displacement (Fig. 11). Most of these strike-slip faults appear to be early Cenozoic structures that accommodate oblique-slip movement. Flow folds within the El Yaqui Chert (Fig. 13) locally indicate a southeasterly direction of tectonic transport, and a tightly defined S37°W axis of folding, more than 20° off the regional fold trend (Fig. 12C compared to 12A, B, D, and E). Major southeast-directed thrusts have not been recognized in the Minas de Barita area (Figs. 10–12). The southeast-verging folds in the El Yaqui Chert (Fig. 13) may be gravity-induced slump folds formed on the south-facing Ordovician continental slope. Alternatively, southeast-verging folds may be related to pre–Late Mississippian northwest-southeast–directed contraction. Younger-over-older attenuation faulting may have affected the northwest margin of the Minas de Barita area, based on data from a shallow borehole.

The Sonora segment consists of several thrust sheets that have discrete deep-water facies (Figs. 10–11). Two different Middle Ordovician facies of the same age characterize the Sonora segment. They include a northerly facies of shale/argillite, and a southerly facies of interbedded argillite and quartzite. Although we believe these two Ordovician facies are laterally gradational, it is possible they occur in separate thrust sheets. The Sonora allochthon rests directly on footwall Lower to Middle Permian Mina México foredeep flysch in the Minas de Barita area (loc. 43 on Fig. 4; Figs. 10 and 11).

Foredeeps and Foreland Basins and Uplifts

A variety of major basins and uplifts characterize the foreland adjacent to the Ouachita-Marathon-Sonora orogen (Fig. 3). The Ouachita-Marathon-Sonora nappes and evolving thrust belts to the south supplied much of the Upper Mississippian to Permian flysch to asymmetrical foredeeps that developed subparallel and adjacent to the orogenic front (i.e., Black Warrior, Arkoma, Fort Worth, Kerr, Val Verde, Marfa, and Mina México foredeeps). Farther cratonward foreland basins and uplifts tend to be elongate northwest. The Neoproterozoic–Early Cambrian transform-fault systems not only influenced the structure and geometry of the late Paleozoic Ouachita-Marathon-Sonora orogenic belt but also appear to have influenced the location of the cratonic foreland basins (i.e., Anadarko, Palo Duro, Midland, Delaware, Orogrande, and Pedregosa basins). In Colorado, 800–1200 km northwest of the Ouachita-Marathon orogenic front, the late Paleozoic fault-bounded Front Range–Apishapa and Uncompahgre uplifts and associated Central Colorado and Paradox basins, all have northwest trends. The Pedernal uplift in central New Mexico and adjacent western Texas has a northerly trend. The foreland basins and platform uplifts (Fig. 3) are closely associated with the development of the orogenic system (Kluth and Coney, 1981; Ross and Ross, 1985, 1986, 1995; Ross, 1986; Denison, 1989; Atchley et al., 1999). Pennsylvanian and Early Permian tectonics and sedimentation were approximately synchronous in the Ouachita-Marathon-Sonora orogenic system (Houseknecht, 1986; Wuellner et al., 1986; Meckel et al., 1992; Roberts, 1994; Harry and Mickus, 1998). Fault-bounded uplifts and depressions during this period controlled facies distribution and sediment thickness (Atchley et al., 1999). Similarly, the southern Ancestral Rocky Mountains are intracratonic uplifts that formed during Pennsylvanian time (Kluth and Coney, 1981). Location of the Sierra del Nido uplift (SNu, Fig. 3) is based on a large, negative Bouguer-gravity anomaly in west-central Chihuahua (Goodell et al., 1986; Moreno et al., 1992). This gravity "low" may represent an uplifted block of Precambrian crust (Goodell, 1993) or a Tertiary batholithic mass near the eastern boundary of the Sierra Madre Occidental volcanic field (Moreno et al., 1992). Another possible explanation for the Sierra del Nido uplift is that it represents a late Paleozoic foreland uplift in front of the Marathon and Sonora segments of the Ouachita-Marathon-Sonora orogen. Haenggi et al. (2000) interpreted the Aldama platform, which includes the Sierra del Nido uplift, as an accreted block that originated apart from North America, a less likely explanation. It is conceivable that the major gravity "low" may represent a flexed depression in front of stacked thrust sheets of the Ouachita-Marathon-Sonora accretionary wedge.

Foreland uplifts resulted from the arc-continent collision of Gondwana and Laurentia that produced the Ouachita-Marathon-Sonora orogenic system. The compressive stresses were transmitted far into the continental interior along transform faults resulting in both vertical and strike-slip movement on those faults, giving rise to a complex system of paired uplifts and fault basins (Walper, 1977). Many foreland basins, elongate parallel to the northwest structural grain, were probably inherited from transform faults formed during the Neoproterozoic–early Paleozoic rifting phase. The asymmetric geometry of the foredeeps is consistent with a downward flexure under an isostatic load of thrust sheets (Beaumont, 1981) that made up the Ouachita-Marathon-Sonora orogen. The thickest sediment accumulation is generally near the frontal margin of imbricately stacked thrust sheets where deep troughs (foredeeps) formed and subsided as they filled with turbiditic sediment. Depending on geologic settings within the Ouachita-Marathon-Sonora orogenic system, source of sediment in the foredeeps and foreland basins could be from both the uplifted orogen and the uplifted North American craton or continental shelf. Both sources are known from the sedimentary record, including paleocurrent directions and clast compositions (McBride, 1989; Morris, 1989; Meckel et al., 1992). Crustal

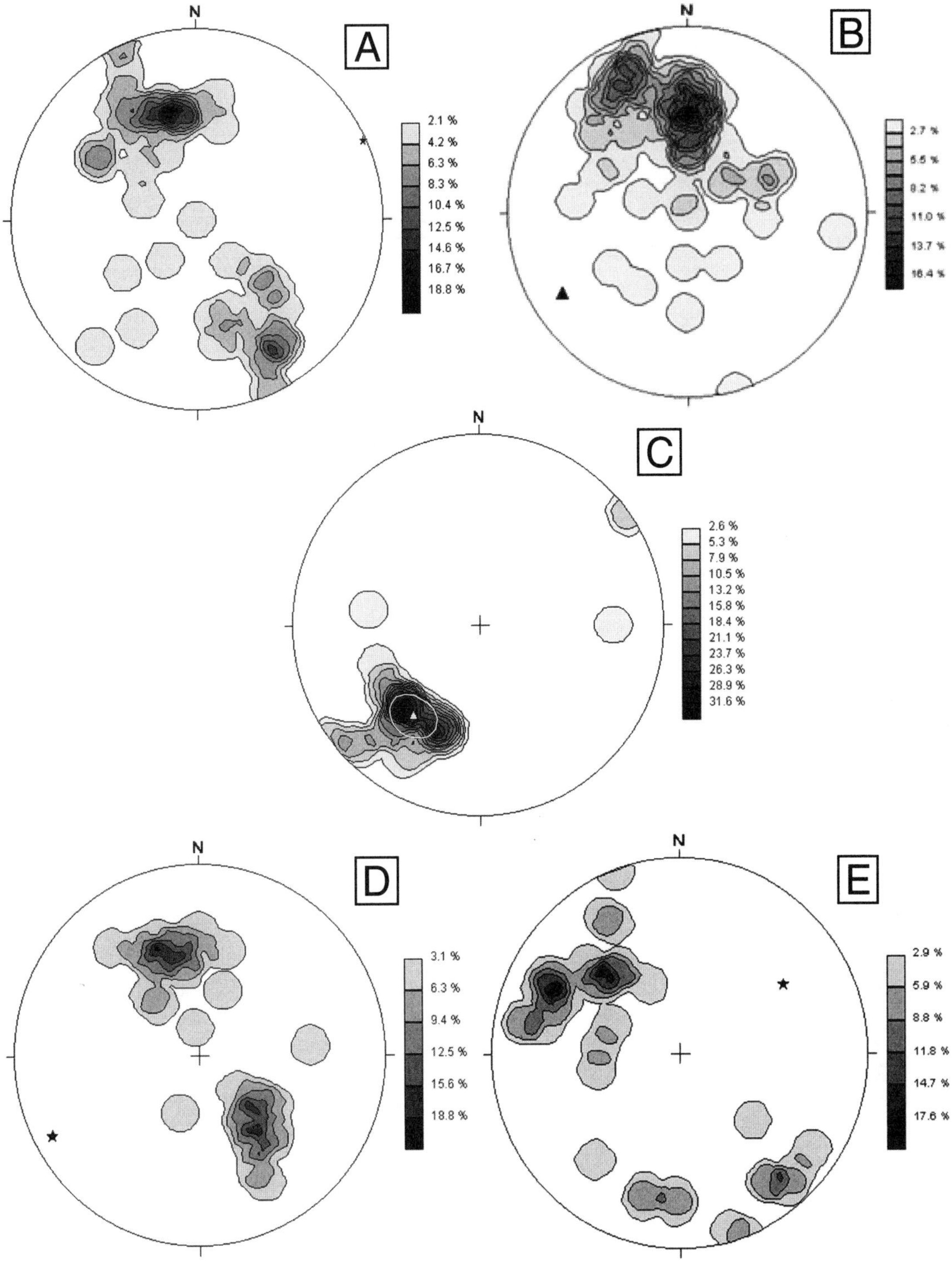

Figure 12. Lower hemisphere, equal-area contour stereo plots of poles-to-bedding and fold and crenulation axes in folded Paleozoic strata of the Sonora allochthon and subjacent Permian foredeep strata in the Minas de Barita area, central Sonora (Perry compilation). (A) Contour plot of 47 poles-to-bedding in folded Ordovician El Yaqui Chert and El Mezquite Shale in areas A on Figure 11; contours represent density of poles-to-bedding (% of poles per 1% search area); fold axis (★) calculated as pole-to-girdle distribution, trends N65°E and plunges 2.5° NE, assuming cylindrical folding. (B) Contour plot of 73 poles-to-bedding in tightly folded El Yaqui Chert along haul road just east of grinding plant; contours same as 12A; fold axis (▲), calculated as in 12A, trends S58°W and plunges 18° SW; data from outcrop shown in Figure 13. (C) Contour plot of 38 fold and crenulation axes in tightly folded El Yaqui Chert along haul road just east of grinding plant; (Δ) marks Fisher mean fold axis which trends S37°W and plunges 40° SW; white oval around Fisher mean represents 95% cone of confidence; data from outcrop shown in Figure 13. (D) Stereo plot of 32 poles-to-bedding in Devonian-Mississippian Los Pozos Formation 1–2.3 km north of Rancho Los Chinos; contours as in 12A; calculated pole-to-girdle fold axis (★) trends S64°W and plunges 11° SW. (E) Stereo plot of 21 poles-to-bedding in Mississippian Los Chinos Conglomerate, Mississippian-Permian Rancho Nuevo Formation, and subjacent Permian Mina México Formation; contours as in 12A; pole-to-girdle fold axis (★) trends N61°E and plunges 32° NE. Areas of structural measurements shown on Figure 11.

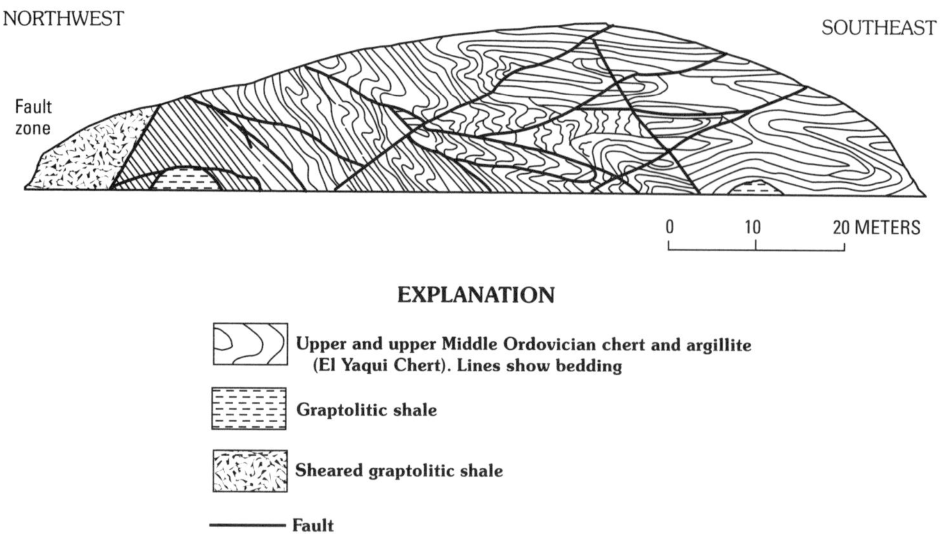

Figure 13. Sketch diagram of tightly folded El Yaqui Chert exposed in the haul-road cut east of grinding plant at Minas de Barita (Madrid field sketch, ca. 1985). See Figure 11 for location of outcrop (BC) and see Figures 12B and 12C for equal-area plots of poles-to-bedding and fold and crenulation axes measured in this roadcut exposure.

flexing and synorogenic sedimentation in many of the foredeeps produced upflexed areas, or forebulges. Foredeeps and foreland basins, which extend southwest from Mississippi to Sonora, are directly related to late Paleozoic closure of the ocean between the Gondwana volcanic arc and Laurentia craton (Figs. 3 and 14). The foredeeps within and in front of the advancing orogenic front were progressively deformed until the forward motion of the orogenic belt ceased (Ross, 1986). Orogenic foredeeps and foreland basins (stipple pattern on Fig. 3) from northeast to southwest include the Black Warrior (BWb), Arkoma (Akb), Anadarko (Ab), Fort Worth (FWb), Palo Duro (PDb), Kerr (Kb), Val Verde (VVb), Midland (Mb), Delaware (Db), Marfa (Mfb), Orogrande (OGb), Pedregosa (Pb), and Mina México (MMb).

The Pedregosa foreland basin in northern Chihuahua and adjacent Sonora is related to the Marathon and Sonora segments of the Ouachita-Marathon-Sonora orogen in northern Mexico (Fig. 3). The Pedregosa basin, as shown on Figure 3, may consist of several sub-basins or separate foreland basins. Armin (1987) presented a comprehensive summary on the sedimentology and tectonic significance of Lower Permian strata in the Pedregosa basin of southeastern Arizona, southwestern New Mexico, and northern Chihuahua. Thompson et al. (1978) compiled information on oil and gas exploration wells in the Pedregosa basin. About 12 km west of Nuevo Casas Grandes in the area of La Vinata (loc. 36 on Fig. 3), Lower Permian turbiditic flysch deposits of the Rara/Villa Aldama Formation are ~200 m thick, and contain conodonts and fusulinids of late Wolfcampian to early Leonardian age. A.G. Harris (1997, written commun.) identified the conodonts *Sweetognathus behnkeni* Kozur, *Neostreptognathodus prayi* Behnken, and *Mesogondolella idahoensis* (Younquist, Hawley, and Miller), and C.H. Stevens (1996, written commun.) identified the fusulinid *Schwagerina* cf. *S. crebrisepta*. Interbedded black shale and very thin limestone beds contain the deep-water trace fossils *Cosmorhaphe* and *Scalarituba*(?). This flysch sequence defines the deep-marine basinal part of the southeast-trending Pedregosa basin. About 35 km northwest of Nuevo Casas Grandes in the Sierra de Azcárate (near loc. 36 on Fig. 1), detrital limestones correlative with the Permian Plomosas Formation consist of bioclastic limestone turbidites containing conodonts and fusulinids of latest Wolfcampian to earliest Leonardian age. A.G. Harris (1997, written commun.) identified the conodonts *Mesogondolella bisselli* (Clark and Behnken), *Sweetognathus inornatus* Ritter or *Homeoiranognathus* sp., *Sweetognathus whitei* (Rhodes) transitional to *S. behnkeni* Kozur and *S. behnkeni* Kozur, and C.H. Stevens (1996, written commun.) identified the fusulinid *Schwagerina guembeli* from a 12-cm-thick turbiditic lime wackestone to grainstone estimated to be 50 m below the top or ~265 m above the base of the formation; this turbidite unit was correlated with the Horquilla Formation by Torres-Espinoza et al. (1987). This detrital limestone facies is believed to represent the Lower Permian depositional-slope margin of the Pedregosa basin in northern Chihuahua. In the Sierra El Cuervo (loc. 38 on Fig. 3), intensely folded and faulted sandstones and shales of the Upper Pennsylvanian and Lower Permian Rara Formation (or Villa Aldama Formation) are unconformably overlain by Cretaceous strata (Handschy and Dyer, 1987). We interpret the Rara/Villa Aldama Formation in north-central Mexico to be a foredeep or foreland-basin deposit in front of the northward advancing accretionary wedge between West Texas and central Sonora.

The Mina México foredeep in central and east-central Sonora is related to the Sonora segment of the Ouachita-Marathon-Sonora orogen in northwestern Mexico. The Mina México foredeep (named after the Mina México gold-silver vein deposit hosted by Permian turbiditic quartzites, 13 km south of the village of Bacanora, Sonora), as shown on Figures 3 and 4, trends westward, subparallel to the frontal thrust of the Sonora allochthon. The northward advancing thrust sheets probably restricted

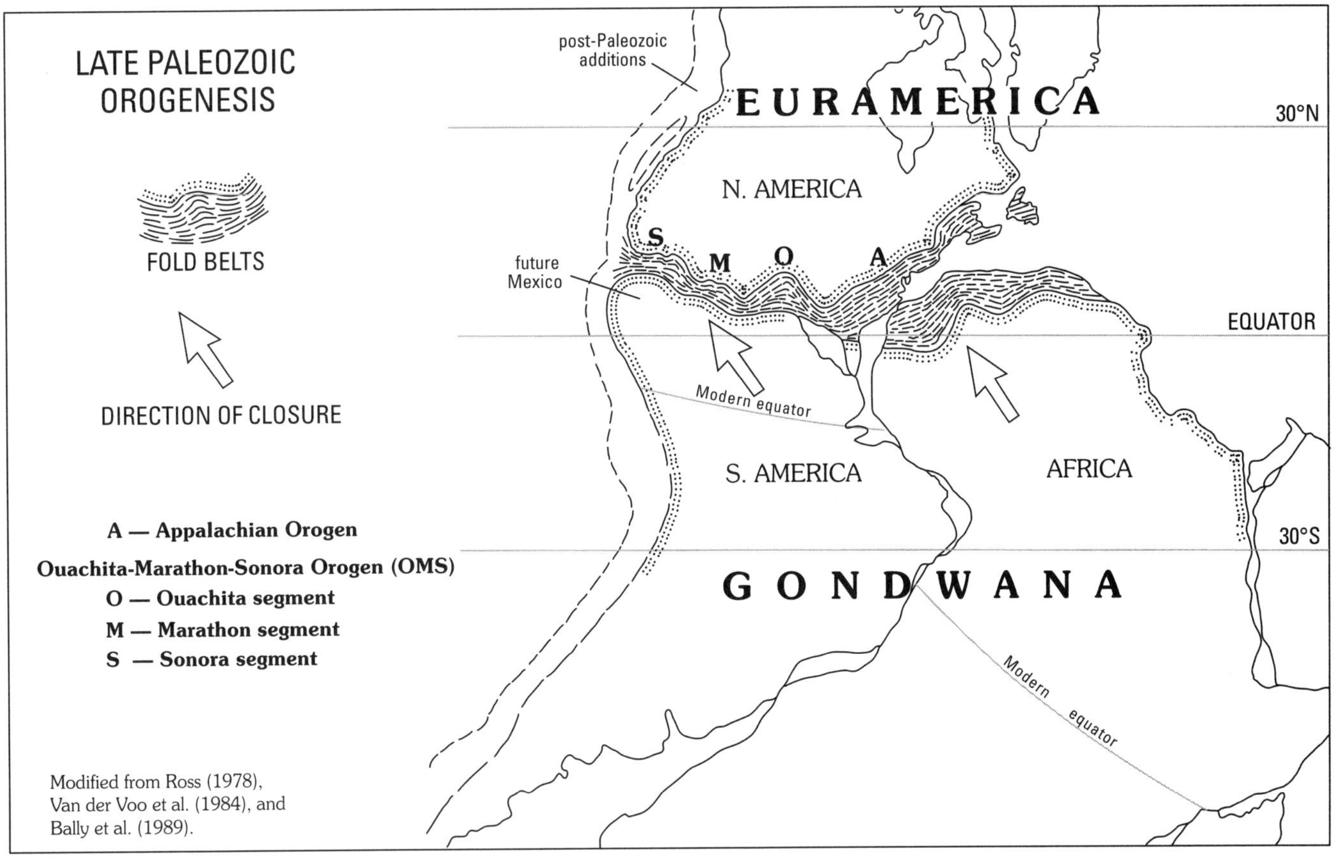

Figure 14. Plate-tectonic reconstruction of Laurentia and Gondwana showing an oblique collisional origin and diachronous suturing of the late Paleozoic fold and thrust belt along the margins of Euramerica and Gondwana.

and narrowed the foredeep, and following collision, much of the foredeep was uplifted and denuded. Minimum extent of the foredeep, based on a small number of data points (Fig. 4, locs. 40–42), appears to be ~160 km in length and 30 km in width. The foredeep probably consists of several structural sub-basins. Basin-fill sediments, which are represented by the Mina México Formation (named in theses by Hewett, 1978, and Schmidt, 1978), consist of a thick sequence (>1000 m) of thin-bedded, very fine to fine-grained siliciclastic and sparse coarse-grained calciclastic turbidites. Siliciclastic turbidites (quartzites) that we examined are composed primarily of angular to subrounded monocrystalline quartz and subordinate angular chert grains. Hewett (1978) noted sparse feldspar, zircon, mica, and tourmaline grains (loc. 40). Gehrels and Stewart (1998) reported a feldspathic quartzite in the Mina México Formation in the northern Sierra Santa Teresa (loc. 42). Sandstone composition primarily reflects tectonic setting and provenance terranes (Dickinson and Suczek, 1979; Dickinson, 1985; Schwab, 1986). Composition of the quartzose and feldspathic sandstones/quartzites in the Mina México foredeep indicates two major source terranes—the Laurentian craton to the north and the orogenic belt to the south. Pyrite is ubiquitous throughout the Mina México Formation, which accounts for its pervasive olive, brown, and orange weathering colors. Interturbidite siltstone layers contain bathyal zone (water depth 200–2000 m) ichnofossils, including *Lophoctenium*, *Scalarituba*, *Cosmorhaphe* and other *Nereites*-facies trace fossils. The calciclastic turbidites are composed of bioclastic debris, including abundant fragments of crinoids and brachiopods, sparse conodonts, and locally common fusulinids of Early to Middle Permian age. C.H. Stevens (2003, written commun.) identified the fusulinids *Parafusulina*? sp., *Parafusulina*? aff. *P. empirensis*, *Parafusulina sullivanensis* Ross?, *Parafusulina wildei*, and *Schubertella* sp. from several calciclastic turbidites south of Bacanora (loc. 40) and west of Cerro San Francisco (loc. 41). In addition, Stevens (*in* Stewart et al., 1990) identified the fusulinids *Schwagerina diversiformis*, *Schwagerina youngquisti*?, and *Schwagerina grandensis* of middle and late Wolfcampian age from calciclastic turbidites at Sierra La Flojera (loc. 42), and Stewart et al. (1997) reported *Parafusulina* sp. aff. *P. boesei* of late Leonardian or early Guadalupian age from their unit 7 at Sierra Santa Teresa (loc. 42). Most of the fossils were transported into the flysch basin from their shallow-water habitat on the continental shelf and possibly from outer-shelf limestones incorporated in the accretionary wedge of the Sonora segment of the orogen. Most of the calciclastic turbidites contain some angular chert and rounded quartz grains that were probably derived from erosion of both the carbonate shelf and the evolving Sonora allochthon. Lithofacies and fossils in the Mina México Formation indicate Early and Middle Permian sedimentation related to erosion of both the evolving Sonora allochthon and flexed Laurentian shelf.

The contact between shallow-water shelfal limestones and superjacent deep-water Mina México flysch is covered in most places, and has been interpreted as both structural and depositional. The contact is exposed in Sierra Santa Teresa (loc. 42, Fig. 4) where Stewart et al. (1997) considered it to be conformable. In one exposure, Amaya-Martínez and Poole (field work in 2004) found the contact to be transitional through an interval of 8 m. The contact is locally exposed in Cañon de Río Bacanora (loc. 40, Fig. 4) where Hewett (1978) interpreted it to be conformable. Amaya-Martínez and Poole (work in progress) studied several exposures of the contact in the Mina México and Minas de Barita areas (locs. 40 and 43, Fig. 4), and found it to be transitional and intertonguing through an interval of 5–50 m. Poole (*in* Stewart and Poole, 2002) tentatively sided with the Radelli et al. (1987) interpretation that the contact was structural, but Amaya-Martínez and Poole have since determined that the regional contact is depositional. The depositional contact is disrupted where deformation is intense, but is preserved where deformation is moderate.

The late Paleozoic La Vuelta Colorada Formation in the Cerro Cobachi area (loc. 44) correlates with the Mina México and Rancho Nuevo Formations of central Sonora (loc. 43) and the Mina México Formation of east-central Sonora (loc. 40). We consider it to be synorogenic flysch related to the evolving Sonora allochthon or foredeep. An unnamed very fine grained quartzite unit 100–200 m thick (A. Fort-Martínez, 2001, personal commun.), which conformably overlies Permian shallow-water shelfal limestone at Cerro La Ciénega in the Sierra El Máviro (loc. 43), is correlated with the Mina México Formation.

The thick Rancho Nuevo Formation in the Minas de Barita area (Figs. 5 and 10) is part of the Sonora allochthon and interpreted as synorogenic flysch deposited in an offshelf setting or within the evolving submarine accretionary wedge. Detrital fusulinids and conodonts in the Rancho Nuevo range in age from Early Pennsylvanian to early Middle Permian. The Permian part is coeval with foredeep flysch of the Mina México Formation. The Mina México, which is >1000 m thick in the Minas de Barita area, is overridden by the Sonora allochthon (Figs. 10 and 11).

The geometry and full extent of the Mina México foredeep and other possible foreland basins in Sonora are unknown at this time owing to sparse upper Paleozoic-rock outcrops and widespread cover by Mesozoic and Cenozoic sedimentary, plutonic, and volcanic rocks.

On the basis of fusulinid and conodont faunas, the shallow-water shelfal limestones that underlie the deep-water flysch of the Mina México Formation are Early to early Middle Permian in age, whereas the youngest fusulinids and conodonts in calciclastic turbidites of the Mina México are Middle Permian in age. The relatively thin transitional contact indicates rapid subsidence of the carbonate shelf as the Mina México foredeep developed by crustal flexing and thrust-sheet loading. Where exposed in areas of moderate deformation, a transitional/intertonguing interval 5–50 m in thickness is preserved between the subjacent shelfal limestone and foredeep flysch. In Sierra Santa Teresa (loc. 42, Fig. 4), marine orbiculoid brachiopods and transported land plants occur together 9–10 m above the base of the Mina México Formation (unit 7 of Stewart et. al., 1997), indicating the presence of Permian land plants nearby when the foredeep began forming in this area (Amaya-Martínez and Poole, field work in 2004). U-Pb zircon data on a sample of turbiditic feldspathic sandstone from unit 7 (Mina México Formation) indicate zircon sources on the Laurentian craton (Gehrels and Stewart, 1998).

PALEOZOIC ROCKS IN BAJA CALIFORNIA NORTE AND ALTA CALIFORNIA

By restoring present Baja and Alta California southeastward ~400 km to their pre-Tertiary position (Figs. 1–4), Paleozoic rocks exposed in Baja California Norte and Alta California can be more confidently correlated with Paleozoic facies exposed in mainland Mexico. Distribution of Paleozoic rocks and facies in northern Baja California Norte and adjacent southern Alta California (Fig. 4) indicates the presence of Laurentian continental shelf and underlying Precambrian basement as far west as the hachured line and Baja California transform fault shown on Figures 1–4. Immediately west of the continental edge is a narrow belt of North America-related Paleozoic continental-slope and ocean-basin rocks. We infer that the narrowness of the Paleozoic facies belt is the result of compression when the Pacific oceanic plate impinged on North America in early Mesozoic time. West of this belt is a complex assemblage of Mesozoic sedimentary rocks, arc, fore-arc, and intra-arc volcanic rocks, and back-arc clastic rocks reflecting this impingement (Fig. 4; Gastil et al., 1975, 1991; Gastil and Miller, 1984; Gastil, 1993).

PLATE-TECTONIC MODEL

Evolutionary models of the margins of North and South America must integrate the geology of the region (Fig. 3) into a kinematic framework defined by a late Paleozoic continental reconstruction of western Pangaea and an accurate subsequent relative motion history between North and South America (Pindell, 1985). Furthermore, the way in which closure is achieved in the reconstruction must provide an internally consistent rearrangement of interpreted tectonic provinces and metamorphic zones from the various continental blocks, as attempted by Pindell (1985).

Our plate-tectonic reconstruction of the Laurentian portion of Euramerica (northern Pangaea) and the South America and Africa portions of Gondwana shows oblique collision of the late Paleozoic Ouachita-Marathon-Sonora fold and thrust belt along the adjacent margins of Laurentia and Gondwana (Fig. 14) that is modified from Ross (1978), Van der Voo et al. (1984), and Bally et al. (1989). Locations of the Appalachian orogen and segments of the Ouachita-Marathon-Sonora orogen are respectively labeled A, O, M, and S (Fig. 14). Strong rotational motion between North and South America allowed close approach and grazing of the continents over a long period of time (Johnson and Tarling, 1985). The initial close approach occurred in the Early Mississippian (early Osagean) when faunal interchange is indicated by strong affinities between North and South America faunas (Boucot et al., 1997). In addition, proximity of Laurentia and Gondwana (South America portion) in Late Mississippian to Permian time is recorded by ashfall tuffs and volcaniclastic deposits on the southern margin of Laurentia in the Stanley Group (Niem, 1977; Decatur and Rosenfeld, 1982; Loomis et al., 1994), Tesnus Formation (Imoto and McBride, 1990), and Rara Formation (Handschy and Dyer, 1987). Also, a rhyolite flow and conglomeratic deposits in the Lower Permian Plomosas Formation in the southern Pedregosa foreland basin (Bridges, 1962, 1970; Armin, 1987) indicate nearby volcanism and tectonic activity. On the northern margin of Gondwana, arc volcanics and feeder plutons of Late Mississippian through Permian age in the Coahuila–Las Delicias terrane (López et al., 1996) and extrusive rhyolite porphyry (Sabine Rhyolite) of Late Permian (Rb-Sr date of 255 ± 15 Ma) or pre–Late Pennsylvanian age in the Sabine uplift area (Nicholas and Waddell, 1989) indicate that Gondwana was close to Laurentia (Euramerica) in the late Paleozoic. Diachronous westward younging of late-stage orogenic activity (i.e., Late Pennsylvanian in the Ouachita Mountains, Early Permian in the Marathon region, and Late Permian in Sonora) is compatible with diachronous westward final closure of Gondwana against Laurentia during plate collision. Deformation began during the medial Mississippian in all three segments. Neoproterozoic–early Paleozoic history records the rifting of the southern margin of the North American craton and the opening of the Iapetus and Rheic oceans (Scotese, 1997; Timmons et al., 2001). Late Paleozoic history records closing of the ocean basins by the south-directed subduction of Laurentia and the northward thrusting of its offshelf oceanic facies. The offshelf, preorogenic rocks and the deep-water synorogenic rocks are remnants of oceanic facies deposited south of the Laurentian continental shelf. Detrital zircons in preorogenic and synorogenic sandstones along the southern margin of Laurentia indicate multiple source areas on the Laurentian craton (Gehrels and Stewart, 1998; Gleason et al., 2001). Foreland basins and uplifts extend inboard beyond the boundaries of the orogenic belt (Fig. 3), and provide sedimentological data relevant to timing of deformation in the Ouachita-Marathon-Sonora fold and thrust belt. Development of foredeeps progressed westward during oblique convergence and suturing of Gondwana and Laurentia (Dickinson and Lawton, 2003).

Accreted Blocks and Terranes

Several remnants of the northern margin of Gondwana and associated volcanic arcs or interposed volcanic-arc terranes have been postulated south of the Ouachita-Marathon-Sonora orogenic system in the southern United States and northern Gulf of Mexico (collectively referred to as the Sabine block), and two other major remnants have been postulated in northern Mexico referred to as the Coahuila–Las Delicias terrane of the Coahuila block, and the Sonobári and Rusias terranes of the El Fuerte block (Fig. 3). The collision zone between Laurentia (Euramerica) and Gondwana or associated volcanic arc was an accumulation of variably deformed and metamorphosed rocks now exemplified by the interior metamorphic zone or belt (IZ on Fig. 3). The interior zone consists primarily of accreted Gondwana crust. The only known outcrop of the interior zone consists of fine-grained metasedimentary rocks in Sierra del Carmen in northwesternmost Coahuila (loc. 58 on Figs. 1 and 3; Flawn et al., 1961; Handschy et al., 1987; Carpenter, 1997). In West Texas and adjacent Chihuahua and Coahuila, common Pb isotopic data suggest that the interior zone is isotopically distinct from Laurentia and is allochthonous (James and Henry, 1993a, 1993b; Carpenter, 1997). Elsewhere, the interior belt has been recognized only in boreholes penetrating Paleozoic and Precambrian metamorphic rocks (Fig. 1 and Table 1). No deep boreholes have been drilled in Sonora. The lateral extent of this late Paleozoic suture zone or belt can only be inferred because most of the belt is concealed by Mesozoic and Cenozoic rocks. Mesozoic rifting left fragments of Gondwana attached to North America. A series of microcontinent-sized fragments of Gondwana (South America) make up most of Mesoamerica to the south of the Marathon and Sonora segments of the Ouachita-Marathon-Sonora orogen (Yañez et al., 1991). These fragments are extremely variable in stratigraphy, metamorphism, igneous activity, and structure.

Sabine block (SB). This block includes the subcrop Texarkana platform (Tp), Sabine uplift (Sau), and Yucatán platform (Yp), designated collectively as the Sabine plate or block (Viele and Thomas, 1989, fig. 19; Fig. 3). The original Sabine block may have been a large element connected to Gondwana that was dismembered during oblique collision with Laurentia, or it may have been many small plates or microcontinents, including isolated volcanic-arc terranes.

Coahuila block (CB). This is mostly a subcrop composite element (López et al., 1996; R. López, 1997, written commun.) that includes the Las Delicias terrane (McKee et al., 1999), labeled Dt in northeast Mexico (loc. 63 on Figs. 1 and 3). The terrane consists of Pennsylvanian and Permian flysch and volcanic-arc rocks in the Las Delicias basin area that has been studied and described by Robert López, James McKee, Norris Jones, and Thomas Anderson. A granitoid boulder in the Carboniferous flysch yielded a U-Pb zircon age of 1215 ± 10 Ma, indicating that basement rock of the terrane is Grenville-age continental crust (López et al., 1996). We consider this Mesoproterozoic crust to be part of the Gondwana craton. The flysch also contains Permian basalt, andesite, and dacite volcanic debris and volcaniclastic sandstone and conglomerate. López et al. (1996) reported three other important U-Pb zircon dates on rocks in the Las Delicias terrane including: (1) an Upper Mississippian granite intruding terrane rocks yielded a concordant U-Pb age of 331 ± 4 Ma, which may record the first magmatism of the incipient Las Delicias arc; (2) a Lower Permian dacite clast from the youngest volcanic-debris-bearing flysch yielded a concordant U-Pb age of 268 ± 6 Ma; and (3) the final pulse of magmatism was the intrusion of Permian and Triassic granitoid plutons yielding ages of 286 Ma and 236–220 Ma. Rock distribution and geologic setting suggest a magmatic arc associated with a south-dipping subduction zone. We interpret the Las Delicias terrane to be a remnant of the marginal magmatic and/or volcanic arc of northwest Gondwana that formed as it impacted Laurentia in the late Paleozoic to create the Marathon segment of the orogen.

El Fuerte block (FB). This is mostly a subcrop and poorly known element that includes the Francisco Gneiss and the Río Fuerte Group composed of polydeformed and metamorphosed ocean-basin and arc-magmatic andesitic to rhyolitic rocks related to a subduction zone and magmatic arc (loc. 65 on Figs. 1, 3, and 4). The Francisco Gneiss and Río Fuerte Group rocks are not in contact and their geologic relationship is unknown at this time. Rocks in the El Fuerte area were described by De Cserna and Kent (1961) and by Mullan (1978). A biotite gneiss sample from Sierra de Sonobári yielded a Late Triassic U-Pb zircon age of ca. 220 Ma (Anderson and Schmidt, 1983). Two other samples of the Francisco Gneiss (a quartzofeldspathic gneiss and an amphibolite) from Sierra de Sonobári yielded $Nd_{(t)DM}$ model ages close to 1.0 Ga, which suggests that they were probably derived from reworked Proterozoic crust (Valencia-Moreno et al., 2001). The lower part of the Río Fuerte Group is composed of chert, argillite, siltite, quartzite, and minor limestone. A thin turbiditic limestone in the lower part of this Group yielded late Middle to Late Ordovician conodonts (A.G. Harris, 1997, written commun.). It is likely that the Group contains other Paleozoic-age rocks. We consider the Paleozoic rocks in the vicinity of San José de Gracia and Real de San Javier (loc. 66, Fig. 3) to be part of our El Fuerte block. The rocks consist of chert, shale, limestone, quartzite, and local conglomerate (Carrillo-Martínez, 1971; Gastil et al., 1991; Consejo de Recursos Minerales, 1992). Fusulinids and conodonts of late Paleozoic age have been found in the San José de Gracia area (Carrillo-Martínez, 1971; Gastil et al., 1991). Although this faunal and stratigraphic information provides no direct tie to Gondwana, the position of these pre-Mesozoic rocks outboard of the southern edge of continental Laurentia is indicative of an exotic origin. The regional lithofacies distribution suggests that the Laurentia-Gondwana suture zone lies between latitudes 27°30′ and 28°30′ in southern Sonora. Our El Fuerte block includes the Sonobári and Rusias terranes of Campa and Coney (1983). The El Fuerte block (Sonobári Complex of De Cserna and Kent, 1961; Anderson and Schmidt, 1983; Campa and Coney, 1983) was interpreted as part of the Mesozoic Guerrero terrane of western Mexico by Coney (1981) and Roldán-Quintana et al. (1993), whereas Coney and Campa (1987) and Sánchez-Zavala et al. (1999) considered it part of the Cortés terrane (consisting mainly of Paleozoic rocks) of northwest Mexico. Sedlock et al. (1993) introduced the term Tahué terrane for mid-Paleozoic metasedimentary rocks of unknown origin in Sinaloa and adjacent areas that were accreted to North America by Late Jurassic time. Their newly defined Tahué terrane includes part of the extensive Guerrero terrane. We interpret the El Fuerte block to be a volcanic-arc remnant of northwest Gondwana that impacted Laurentia during the late Paleozoic to create the Sonora segment, which was the result of convergence of the northwest South American part of Gondwana and southwest Laurentia (Figs. 3, 4, and 14).

Future Studies

Future research must include multidiscipline approaches to confirm or refute the affiliation of the Sabine block and Coahuila and El Fuerte blocks to northwestern Gondwana (South America). It will be necessary to conduct detailed studies of the stratigraphy, igneous geology, and isotopes of the rocks that constitute the various terranes and blocks, and correlative rocks on the northern margin of present-day South America. For example, the preorogenic lower Paleozoic passive-margin deposits and synorogenic upper Paleozoic foredeep deposits in the Mérida Andes of Venezuela (Audemard and Serrano, 2001) need to be studied in detail, and compared to similar-age deposits of the Ouachita-Marathon-Sonora orogenic system. Of particular interest for study in Venezuela, will be the 2000 m of Permian flysch (X. Márquez, 2002, personal commun.) that was deposited during the late Paleozoic collision of northwestern Gondwana and southern Laurentia. Paleozoic rocks on the southern margin of the Rheic ocean (northwestern Gondwana) should be analogous to those on the northern margin of the Rheic ocean (southern Laurentia).

Mojave-Sonora Megashear

The concept of the Mojave-Sonora megashear proposed by Silver and Anderson (1974) and Anderson and Silver (1979) and expanded by Anderson and Schmidt (1983) is one of the most controversial ideas regarding the Mesozoic evolution of mainland Mexico and the Gulf of Mexico. These and later workers have postulated that the Mojave-Sonora megashear was a left-lateral megafault that truncated the western edge of the North American craton and offset Proterozoic, Paleozoic, and lower Mesozoic rocks and structures 600–1100 km southward from California to Sonora during early Late Jurassic time. The Mojave-Sonora megashear was referred to by Dickinson and Lawton (2001) as the California-Coahuila transform of Permian–Triassic age. Most geologists working in northern Mexico have accepted and utilized the hypothetical Mojave-Sonora megashear to explain their interpretation of Precambrian basement, Neoproterozoic, Paleozoic, and lower Mesozoic rock distributions in southwestern North America (e.g., Coney, 1981; Dickinson, 1981, 1989, 2000; Campa and Coney, 1983; Stewart et al., 1984, 1990, 1999b, 2001, 2002; Pindell, 1985; Ketner, 1986; Coney and Campa, 1987; McKee et al., 1988, 1990; Stewart, 1988, this volume; Stevens et al., 1992; Bartolini, 1993; Sedlock et al., 1993; Speed, 1994; Ortega-Gutiérrez et al., 1994; Gehrels et al., 1995;

Jones et al., 1995; Stanley and González-León, 1995; Marzolf and Anderson, 1996, this volume; Ye et al., 1996; González-León, 1997; McDowell et al., 1999; Sánchez-Zavala et al., 1999; Dickinson and Lawton, 2001, 2003; Valencia-Moreno et al., 2001; Campbell and Anderson, 2003; Anderson and Silver, this volume), and most authors in this volume. Since its proposal in 1974, many stratigraphic and structural complexities and contradictions have developed with the concept of the Mojave-Sonora megashear. Some authors have attempted to accommodate these complexities and contradictions by repositioning the postulated Mojave-Sonora megashear in northwestern Mexico and southwestern United States. Because there is no evidence for a Late Jurassic megashear passing through southeastern California or southwestern Arizona (e.g., Leveille and Frost, 1984; Hamilton, 1987), some proponents of the Mojave-Sonora megashear have reoriented the megashear and restricted it to Sonora. Some workers have restricted the Mojave-Sonora megashear to northwesternmost Sonora on the basis of the stratigraphy and structure of Proterozoic, Paleozoic, and Mesozoic rocks in the Caborca-Altar area (Jacques-Ayala et al., 1990) and northwest of Caborca on the basis of the southwest extension of the Mesozoic Papago terrane of southwestern Arizona and adjacent Sonora (Calmus and Sosson, 1995; Calmus et al., 1997).

Rather than being a first-order Permian–Triassic or early Late Jurassic structure, we find that our studies in Sonora and other segments of the Ouachita-Marathon-Sonora orogen dispel the Mojave-Sonora megashear (Figs. 2–4) as a viable interpretation for offsets of this throughgoing and continuous orogen and its subjacent continental shelf. Instead, limited offset of the late Paleozoic orogen and Paleozoic shelf may indicate only minor reactivation of an inferred northwest-striking, right-stepping Neoproterozoic transform fault in Sonora and Chihuahua (i.e., Chihuahua-Sonora transform of this report). These transform faults offset a northeast-striking rift system (of Rodinia) that broke open the North American continental crust beginning in the Neoproterozoic (850–742 Ma, Timmons et al., 2001). Our Chihuahua-Sonora transform therefore differs in age and origin from the California-Coahuila transform of Dickinson and Lawton (2001). The ca. 1.1 Ga Aibo Granite (Anderson and Silver, 1971; Farmer et al., this volume) and other igneous and metamorphic rocks of Grenville age (1.2–1.0 Ga) in northwestern Mexico are most likely related to accretion of the Mesoproterozoic Grenville crust to southern Laurentia (Fig. 2). Farmer et al. (this volume) suggested that the Aibo Granite likely represents the product of anatectic melting of the local Caborca basement. Farther northeast this Neoproterozoic–early Paleozoic rifting and opening of the Iapetus–Rheic ocean produced a Laurentian margin along which the late Paleozoic Appalachian orogen and Ouachita-Marathon segments of the Ouachita-Marathon-Sonora orogen subsequently formed (Thomas, 1991; AOT, Alabama-Oklahoma transform and TT, Texas transform, Fig. 2). Further, our studies indicate that this rifting event extended west-northwest and included the Laurentian continental crust in Sonora, Baja California Norte, and adjacent Alta California. The breakup of Rodinia probably resulted in multiple rift episodes (Timmons et al., 2001). It is important to note that changes in Paleozoic paleogeography and biogeography have important implications for the non-viability of the megashear (Poole et al., 2003a, 2003b).

Stratigraphic and Structural Arguments Against the Mojave-Sonora Megashear

Stewart et al. (1990) discussed arguments for and against the Mojave-Sonora megashear model as originally proposed by Silver and Anderson (1974), Anderson and Silver (1979), and Anderson and Schmidt (1983). Stewart (this volume) presents his stratigraphic interpretations to support the Mojave-Sonora megashear model as it was originally proposed. Most geologists have applied that concept of the Mojave-Sonora megashear to their interpretation of the geology of Sonora; however, many recent studies of Sonoran geology do not support the Mojave-Sonora megashear model. Geologic findings, which do not support the Mojave-Sonora megashear model, are summarized below.

We recognize Neoproterozoic and Paleozoic strata thickness and facies changes within the continental-shelf and ocean-basin rocks along the entire western North American margin, but coeval rocks in Sonora and Baja California Norte exhibit different facies and tectonic style from those in the southwestern United States. Although there is stratigraphic evidence for a connection of Paleozoic shelf facies around the southwest end of Laurentia, most facies are different enough to negate large magnitude offset along a late Paleozoic or Mesozoic megashear. In Sonora and adjacent Arizona and California, we recognize fault displacements of relatively small magnitude that offset Paleozoic facies. These faults are probably Late Cretaceous to early Tertiary (Laramide) in age. These include several west-northwest–trending right-lateral faults of less than 5 km displacement in the Minas de Barita area (Fig. 11).

Polarity problems with "Caborca-Hermosillo block." Proponents of the Mojave-Sonora megashear argue that the so-called "Caborca-Hermosillo (also referred to as Serí) block" in Sonora is part of the Cordilleran shelf that was displaced southward 600–800 km from California. If so, the Roberts Mountains allochthon (Antler orogen or eugeoclinal assemblage) of southwestern Nevada and east-central California should be present in northwesternmost Sonora along the north margin of the "Caborca-Hermosillo block." No rocks of the Devonian–Mississippian Antler (Roberts Mountains allochthon) or Permian–Triassic Sonoma (Golconda allochthon) orogens have been found in Sonora or Baja California Norte. In addition, rocks of the Sonora allochthon in central Sonora are notably different in stratigraphy, structural style, and timing of deformation than the orogens in the southwestern United States. Furthermore, Cordilleran cratonic-platform facies should be present at the southern end of the "Caborca-Hermosillo block" in the vicinity of Hermosillo, Sonora. Instead, Paleozoic facies in the vicinity of Hermosillo represent carbonate-shelf facies of the southern margin of Laurentia, which is depositionally overlain by Permian foredeep rocks (Mina México flysch). East of Hermosillo, Permian foredeep rocks are overridden by Ordovician–Permian oceanic rocks of the north-directed Sonora allochthon. The Mina México foredeep and related Sonora accretionary wedge are unique to Sonora, and their content and setting are unlike coeval rocks in California and Nevada. The Sonora miogeoclinal section, like the Cordilleran miogeocline in southwestern United States, is probably underlain by Paleoproterozoic Mojave province basement (A. Iriondo and W.R. Premo, 2004, personal commun.; Farmer et al., this volume). It will be important to obtain more dates on the Precambrian basement in the "Caborca-Hermosillo block" (informal Caborca crust of this report, Fig. 2) to determine whether or not it is a continuation of the Mojave province. Isotopic evidence indicates that most of the Caborca crust (Fig. 2) ranges in age from 1.8 to 1.6 Ga, and is related to the Mojave province (Anderson and Silver, 1981; Coleman et al., 2003; Iriondo and Premo, 2003; Farmer et al., 2003, this volume; Nourse et al., 2003; Premo et al., 2003; Iriondo et al., 2004). Two younger tectonomagmatic events are recognized in the Mojave province of southwestern United States and northwestern Mexico. Following protracted assembly of Paleoproterozoic lithosphere from 1.8 to 1.6 Ga, major Mesoproterozoic contractional-magmatic events occurred from 1.5 to 1.3 Ga and from 1.1 to 1.0 Ga (Nyman et al., 1994; Karlstrom et al., 2002). The 1.5–1.3 Ga event coincides with development of the Southern and Eastern Granite-Rhyolite provinces, and the 1.1–1.0 Ga event coincides with development of the Grenville (Llano) province (Fig. 2). We suggest that western Sonora and northern Baja California Norte and Alta California are underlain by crust similar in age to the Mojave province, and that the Mojave province is continuous from California into northeastern Baja California Norte and western Sonora as shown in Figure 2. The Paleozoic paleogeography of southwestern United States and northwestern Mexico, in our view, negates the concept that the so-called "Caborca-Hermosillo block" is a displaced Cordilleran shelf block from California.

The Neoproterozoic and lower Paleozoic rocks have some similarities in both the Cordilleran miogeocline and the Caborca-Hermosillo miogeoclinal section (Stewart et al., 1984; Link et al., 1993); however, there are significant differences that hamper one-to-one physical correlation. It seems likely that the continental shelf between California and Sonora was continuous and had similar environments of deposition that account for the persistence of many Neoproterozoic, Cambrian, and Ordovician stratigraphic units from California to Sonora. However, the shelf distance between the two areas probably accounts for many of the variations in lithofacies as well as missing units. The middle and upper Paleozoic rocks of Sonora are significantly different from coeval rocks in California. Regional tectonics may have altered the original shelf setting and the emergence of the southwest-trending Transcontinental arch may have influenced water depth and sediment and faunal distribution between California and Sonora. The Transcontinental arch in Arizona and New Mexico migrated northwest in the middle and late Paleozoic in response to northwest-directed convergence along the southwest margin of Laurentia (see Figs. 1 and 3). Poole et al. (2003a) studied an Early Devonian brachiopod fauna from a carbonate-shelf section in west-central Sonora (loc. 30, Figs. 1, 3, and 4) ~100 km southwest of the hypothetical trace of the Mojave-Sonora megashear. They concluded that the fauna belongs to the Appohimchi Subprovince of the Eastern Americas Realm (Laurentian southern margin), and that many of the genera and species in Sonora have not been recognized in the western United States (Laurentian western margin). Biogeographic distribution of the Appohimchi brachiopod fauna in west-central Sonora indicates continuous Early Devonian shelf deposition along the southern margin of Laurentia without any significant offset. The largely emergent southwest-trending Early Devonian Transcontinental arch and associated offshore currents probably formed a barrier to migration and mixing of many genera and species of brachiopods from the southern shelf of Laurentia in northern Mexico to the western shelf (Cordilleran miogeocline) in the western United States (Poole et al., 2003a). These data argue against the hypothesis of large magnitude (600–1100 km) left-lateral offset of the Cordilleran carbonate-shelf rocks from California to Sonora along a northwest-striking late Paleozoic or early to middle Mesozoic Mojave-Sonora megashear. Gómez-Luna and Martínez-Cortés (1997) compared Triassic ammonoid successions in northwestern Sonora and west-central Nevada, and concluded that the large number of dissimilarities between the Luning assemblage and the El Antimonio sequence do not support offset along a Mojave-Sonora megashear as proposed by Stanley and González-León (1995). Iriondo (2001) and Iriondo et al. (2004) collected radioisotopic (Pb, Sr, Nd) and trace-element geochemical data on many Proterozoic granitic basement rocks in the Quitovac region (loc. 4) of northwestern Sonora, and concluded that the bulk of the data and their implications do not support the existence of a Mojave-Sonora megashear in the Quitovac region. Molina-Garza and Geissman (1999) collected paleomagnetic data from Neoproterozoic and Paleozoic continental-shelf strata, arc-derived Lower Jurassic marine strata, and Jurassic volcanic and volcaniclastic rocks in northwest Mexico (Caborca terrane of Coney, 1981), and concluded that paleomagnetic evidence is not consistent with the Mojave-Sonora megashear hypothesis of early Late Jurassic left-lateral strike-slip motion of the continental crust of northern Mexico. Megashear proponents must explain these diverse data sets, which evidently disprove the hypothesis of large magnitude displacement, if the megashear is to remain a credible concept in Cordilleran tectonics.

SUMMARY AND CONCLUSIONS

The supercontinent Rodinia broke apart in Neoproterozoic and early Paleozoic time (800–525 Ma) to form Laurentia and Gondwana. Figures 1–4 show a continuous rifted continental margin of southern Laurentia, whose development and geometry were modified by mostly right-stepping transform faulting. Shelf and offshelf deposits accumulated on the newly formed continental margin. Distribution of the contrasting early Paleozoic carbonate-shelf and ocean-basin facies indicates a shelf edge oceanward of the Ouachita-Marathon-Sonora orogen. Euramerica began to form in Late Silurian–Devonian time by the collision of Baltica and Laurentia (Scotese, 1997). The Ouachita-Marathon-Sonora orogen evolved in late Paleozoic time by oblique collision of northwest Gondwana and associated volcanic-arc terrane and southern Euramerica (Fig. 14). The Ouachita-Marathon-Sonora orogen developed as an overthrust belt along the entire 3000-km-long southern continental margin of Laurentia. Three prominent segments of the orogen still exist, documenting the location and shape of the rifted southern margin of Laurentia (Figs. 1–4). Sedimentary facies and structural style are similar in the three exposed segments of the Ouachita-Marathon-Sonora orogen (Figs. 5–10). Continuity of the margin is also indicated by many genera and species of invertebrate fossils that occur throughout the extent of the southern margin of Laurentia. The Ouachita-Marathon-Sonora orogen and its subjacent continental shelf probably extend westward to Baja and Alta California. This relation indicates that the margin, along with its superjacent Ouachita-Marathon-Sonora orogen, was continuous with little cumulative offsets. Maximum northward displacement of ocean-basin rocks onto continental-shelf and foredeep rocks was 50–200 km. Thin-skinned transport to the northwest is virtually coaxial in the Marathon and Sonora segments. Deformation began in mid-Mississippian time in all three segments of the orogen, but ended diachronously in Late Pennsylvanian time in the Ouachita Mountains, Early Permian time in the Marathon region, and Late Permian time in Sonora.

The widespread foreland basin and uplift development north of the orogen were coincident with accretionary wedge tectonics caused by the late Paleozoic collision of Laurentia with Gondwana (proto–South America). Rapidly subsiding foredeeps and foreland basins formed on the continental shelf where synorogenic sediments, eroded from the accretionary wedge to the south and craton to the north, were rapidly deposited as flysch in asymmetric foredeeps near the thrust front. At the same time, flysch deposits probably accumulated in depressions within the evolving accretionary wedge, and were incorporated in the allochthons as they moved northward. In central Sonora, Lower and Middle Permian foredeep deposits of the Mina México Formation were overridden by the Sonora allochthon in the Late Permian (Figs. 10 and 11).

The above summary of stratigraphic, faunal, and structural continuities argues against the idea that the Mojave-Sonora megashear existed in Sonora. Instead, lesser offset (~300 km) occurred along transform faults of Neoproterozoic to at least Early Mississippian age. Our postulated northwest-striking Chihuahua-Sonora transform is confined to Chihuahua and Sonora and possibly southern California, and does not project into Arizona. Furthermore, lithofacies and biofacies of most Paleozoic strata in California and Nevada are significantly different from age-equivalent strata in Sonora. At times during the Paleozoic, the Transcontinental arch in Arizona and New Mexico formed a barrier to facies continuity and faunal mixing between the southern and western shelves of Laurentia.

We suspect that the Caborca crust (1.8–1.6 Ga) beneath the miogeoclinal rocks in Sonora and adjacent Baja California Norte and Alta California is a continuation of the Paleoproterozoic Mojave crustal province of southern California, southern Nevada, and westernmost Arizona (Fig. 2), and is not offset by the hypothetical Mojave-Sonora megashear.

Elements of the Sabine block and the Coahuila–Las Delicias terrane of the Coahuila block are interpreted to be remnants of northwest Gondwana volcanic-arc terranes that impacted southern Laurentia to form the Ouachita and Marathon segments of the Ouachita-Marathon-Sonora orogen. Likewise, the El Fuerte block may be a remnant of a northwest Gondwana volcanic-arc terrane that impacted southwestern Laurentia to form the Sonora segment of the Ouachita-Marathon-Sonora orogen.

An ocean of inferred narrow width separated Laurentia and Gondwana during pre-Mississippian time. Early synorogenic sediments were deposited in the ocean basin that separated northwest Gondwana (South America) and southern Laurentia (North America), and late synorogenic sediments were deposited in foredeeps and foreland basins and within evolving accretionary wedges. The ocean basin closed during the late Paleozoic, beginning in the east and ending in the west. Metamorphosed ocean-basin deposits and volcanic-arc rocks lie along the northern margin of Gondwana in southern Sonora and northern Sinaloa and adjacent Chihuahua (El Fuerte block of this report), the Coahuila block in southwestern Coahuila and southeastern Chihuahua (Coahuila terrane of López et al., 1996, or Las Delicias terrane of McKee et al., 1999), and the Sabine block in the southern United States and Gulf of Mexico. Except for exposures in Sierra del Carmen (loc. 58), the inferred collision zone (IZ belt on Fig. 3) in northern Mexico is concealed by Mesozoic and Cenozoic rocks. The Sonora segment of the orogen, including related foredeep and foreland basins and uplifts, are overprinted by extensive Late Cretaceous and early Tertiary (Laramide) folding and faulting, Jurassic to Paleogene magmatism and volcanism, and Neogene (<25 Ma) Basin-and-Range faulting (see maps by Córdoba and Ortega-Gutiérrez, 1990, and Ortega-Gutiérrez et al., 1992). The leading edge of Gondwana and associated arc rocks in northern Mexico likely extends from Coahuila westward to south-central Sonora near Guaymas, and probably as far west as Baja California Norte (Figs. 3 and 4). We consider most of the Precambrian and Paleozoic rocks in Mesoamerica south of the Ouachita-Marathon-Sonora orogenic belt to be displaced fragments of Gondwana (Figs. 3 and 14).

ACKNOWLEDGMENTS

Many USGS and Mexican colleagues contributed to our field work in Sonora through assistance in mapping and discussions of the geology. Important contributions in the field by J. Fernando Oliva-Becerril, Juan Manuel Morales-Ramírez, Guillermo A. Salas-Pizá, Leonard C. Bolich, William R. Page, John H. Stewart, Martín Valencia-Moreno, Luis Fernando Estrella-F., Pablo Kirschbaum, Angel Fort-Martínez, Alejandro Verdugo-Harita (deceased), Manuel de Moises González-Carlos, Elena Centeno-García, Daniel A. Amaya-Zepeda, and many others helped expedite the mapping. Daniel H. Knepper Jr. kindly processed Landsat 7 Plus data of several areas in Sonora, which were useful in map compilation. James A. Messerich and Jerry L. Bevins of the U.S. Geological Survey (USGS) GeoPhotogrammetry Facility patiently set stereo models for map compilation of many areas in Sonora. We wish to thank Sonora ranch owners for permission to work on their ranch properties. We are grateful to personnel of Minas de Barita S.A. de R.L. de C.V. Unidad Sonora (Chemical Products Corporation), especially present mine manager David B. Croas and exploration manager Angel Fort-Martínez, for providing accommodations at the mine during mapping sessions.

Fossil determinations have been important in deciphering the Paleozoic stratigraphy and structure of northern Mexico. We acknowledge the following paleontologists for their invaluable support: Anita G. Harris (USGS, Reston, Virginia, and Deerfield Beach, Florida) identified conodonts from Ordovician-Permian units, Bruce R. Wardlaw (USGS, Reston, Virginia) identified conodonts from some upper Paleozoic units, and Charles A. Sandberg (USGS, Denver, Colorado) identified conodonts from several Paleozoic units; William B.N. Berry (University of California, Berkeley) identified graptolites from Ordovician units; Benita L. Murchey (USGS, Menlo Park, California) identified radiolarians from Devonian-Pennsylvanian units; Reuben J. Ross Jr. (deceased, USGS, Denver, Colorado, and Colorado School of Mines, Golden, Colorado) identified brachiopods and trilobites from Ordovician units; Robert J. Elias (University of Manitoba, Winnipeg, Canada) identified corals from Upper Ordovician units; Peter M. Sheehan (Milwaukee Public Museum, Milwaukee, Wisconsin) identified brachiopods from Upper Ordovician units; Arthur J. Boucot (Oregon State University, Corvallis, Oregon) identified brachiopods from Devonian units; Raymond C. Douglass (USGS, Washington, D.C.) and Calvin H. Stevens (San José State University, San José, California) identified fusulinids from Pennsylvanian and Permian units.

The senior author has greatly benefited from discussions with Patricia W. Dickerson and William R. Muehlberger (University of Texas, Austin, Texas), G. Randy Keller, Jr. (University of Texas, El Paso, Texas), Walter T. Haenggi (Missouri City, Texas), and L.W. Dan Bridges (Aurora, Colorado) on the geology of Texas and Chihuahua; George W. Viele (deceased, University of Missouri, Columbia, Missouri) and William A. Thomas (University of Kentucky, Lexington, Kentucky) on the geology of Arkansas, Oklahoma, and Texas; Charles G. Stone (Arkansas Geological Commission) on the geology of Arkansas; Alexander Iriondo (University of Colorado, Boulder, Colorado, and USGS, Denver, Colorado), Wayne R. Premo (USGS, Denver, Colorado), G. Lang Farmer (CIRES and University of Colorado, Boulder, Colorado) and Karl E. Karlstrom (University of New Mexico, Albuquerque, New Mexico) on Precambrian crustal provinces of southwestern United States and northwestern Mexico; R. Gordon Gastil (San Diego State University, San Diego, California) on the geology of Baja and Alta California; James E. Barrick (Texas Tech University, Lubbock, Texas) on conodont biostratigraphy of the Marathon and Ouachita segments; and Calvin H. Stevens (San José State University, San José, California) on affinities of Sonora fusulinids to those in West Texas, Chihuahua, and California.

We are grateful to USGS managers Floyd Gray, Mark E. Gettings, Edward A. du Bray, Poul Emsbo, and Albert H. Hofstra for their financial support for field and laboratory expenses during the past several years. We thank Arthur J. Boucot, L.W. Dan Bridges, Warren B. Hamilton, Alexander Iriondo, Timothy F. Lawton, James W. McKee, William R. Muehlberger, William R. Page, Dwight L. Schmidt, Calvin H. Stevens, John H. Stewart, and William A. Thomas for their helpful reviews of different versions of the manuscript, and to Barbara J. Ramsey and Norma J. Maes for computer preparation of the figures.

We end our paper with two quotes from the Nobel Prize-winning philosopher, scientist and social critic Bertrand Russell. "Nobody should be CERTAIN of anything. If you're certain, you're certainly wrong." "Even when the experts all agree, they may be mistaken."

REFERENCED CITED

Almazán-Vázquez, E., 1989, El Cámbrico-Ordovícico de Arivechi, en la región centrooriental del estado de Sonora: Universidad Nacional Autónoma de México, Instituto de Geología, Revista, v. 8, núm. 1, p. 58–66.

Almazán-Vázquez, E., and Fernández-Aguirre, M.A., 1987, Los terrenos Paleozoicos de la región serrana de Sonora, *in* Paleozoico de Chihuahua (Excursión Geológica No. 2, Libreto Guía), Gaceta Geológica, v. 1, no. 1, p. 97–104.

Almazán-Vázquez, E., Palmer, A.R., and Robison, R.A., 1986, Constraints on the southern margin of North America during the early Paleozoic: Geological Society of America Abstracts with Programs, v. 18, no. 6, p. 525.

Anderson, J.L., Wooden, J.L., and Bender, E.E., 1993, Mojave province of southern California and vicinity, *in* Van Schmus, W.R., and Bickford, M.E., co-eds., Transcontinental Proterozoic provinces, *in* Reed, J.C., Jr., Bickford, M.E., Houston, R.S., Link, P.K., Rankin, D.W., Sims, P.K., and Van Schmus, W.R., eds., Precambrian: Conterminous U.S.: Boulder, Colorado, Geological Society of America, The Geology of North America, v. C-2, chapter 4, p. 176–188.

Anderson, P.V., 1982, Prebatholithic stratigraphy of the San Felipe area, Baja California, Mexico [M.S. thesis]: San Diego, California, San Diego State University, 100 p.

Anderson, T.H., and Schmidt, V.A., 1983, The evolution of Middle America and the Gulf of Mexico–Caribbean Sea region during Mesozoic time: Geological Society of America Bulletin, v. 94, no. 8, p. 941–966, doi: 10.1130/0016-7606(1983)94<941:TEOMAA>2.0.CO;2.

Anderson, T.H., and Silver, L.T., 1971, Preliminary history for Precambrian rocks, Bamori region, Sonora, Mexico: Geological Society of America Abstracts with Programs, v. 3, no. 2, p. 72–73.

Anderson, T.H., and Silver, L.T., 1979, The role of the Mojave-Sonora megashear in the tectonic evolution of northern Sonora, *in* Anderson, T.H., and Roldán-Quintana, J., eds., Geology of northern Sonora: University of Pittsburgh and Universidad Nacional Autónoma de México, Instituto de Geología, Geological Society of America Annual Meeting, Hermosillo, Sonora, Guidebook trip 27, p. 59–68.

Anderson, T.H., and Silver, L.T., 1981, An overview of Precambrian rocks of Sonora: Universidad Nacional Autónoma de México, Revista, v. 5, núm. 2, p. 131–139.

Anderson, T.H., Eells, J.L., and Silver, L.T., 1978, Rocas Precámbricas y Paleozoicas de la región de Caborca, Sonora, México, *in* Roldán-Quintana, J., and Salas-Pizá, G.A., eds., Primer Simposio sobre la Geología y Potencial Minero en el Estado de Sonora (Libreto Guía), Instituto de Geología, U.N.A.M., Hermosillo, Sonora, México, p. 5–34.

Anderson, T.H., McKee, J.W., and Jones, N.W., 1991, A northwest trending Jurassic fold nappe, northernmost Zacatecas, Mexico: Tectonics, v. 10, no. 2, p. 383–401.

Angermann, E., 1904, Apuntes sobre el Paleozoico en Sonora: Instituto Geológico de México, Parergones, tomo 1, núm. 3, p. 81–90.

Aranda, M., Quintero, O., and Martínez-Hernández, E., 1988, Palinomorfos del Jurásico Temprano de la Formación Gran Tesoro, Santa María del Oro, Durango: Universidad Nacional Autónoma de México, Revista, v. 7, núm. 1, p. 112–115.

Arbenz, J.K., 1989, Ouachita thrust belt and Arkoma basin, *in* Hatcher, R.D., Jr., Thomas, W.A., and Viele, G.W., eds., The Appalachian-Ouachita orogen in the United States: Boulder, Colorado, Geological Society of America, The Geology of North America, v. F-2, chapter 21, p. 621–634.

Armin, R.A., 1987, Sedimentology and tectonic significance of Wolfcampian (Lower Permian) conglomerates in the Pedregosa basin: Southeastern Arizona, southwestern New Mexico, and northern Mexico: Geological Society of America Bulletin, v. 99, no. 1, p. 42–65, doi: 10.1130/0016-7606(1987)99<42:SATSOW>2.0.CO;2.

Armstrong, A.K., and Mamet, B.L., 1988, Mississippian (Lower Carboniferous) biostratigraphy, facies, and microfossils, Pedregosa basin, southeastern Arizona and southwestern New Mexico: U.S. Geological Survey Bulletin 1826, 40 p.

Atchley, S.C., Kozar, M.G., and Yose, L.A., 1999, A predictive model for reservoir distribution in the Permian (Leonardian) Clear Fork and Glorieta Formations, Robertson field area, West Texas: American Association of Petroleum Geologists Bulletin, v. 83, no. 7, p. 1031–1056.

Audemard, F.E., and Serrano, I.C., 2001, Future petroliferous provinces of Venezuela, *in* Downey, M.W., Threet, J.C., and Morgan, W.A., eds., Petroleum provinces of the twenty-first century: American Association of Petroleum Geologists Memoir 74, chapter 18, p. 353–372.

Baker, C.L., 1922, General geology of Catorce mining district: Transactions of the American Institute of Mining and Metallurgical Engineers, v. 66, p. 42–48.

Bally, A.W., Scotese, C.R., and Ross, M.I., 1989, North America; Plate-tectonic setting and tectonic elements, *in* Bally, A.W., and Palmer, A.R., eds., The Geology of North America—An overview: Boulder, Colorado, Geological Society of America, The Geology of North America, v. A, chapter 1, p. 1–15.

Barrera-Moreno, E., and Domínguez-Perla, J.E., 1987, Geología de la carta Tónichi (H12D65) con énfasis en el Paleozoico porción central del Estado de Sonora [tesis]: Hermosillo, Sonora, México, Universidad de Sonora, 93 p.

Barrick, J.E., 1987, Conodont biostratigraphy of the Caballos Novaculite (Early Devonian–Early Mississippian), northwestern Marathon uplift, west Texas, *in* Austin, R.L., ed., Conodonts: Investigative techniques and applications: British Micropalaeontological Society Series, Ellis Horwood Ltd., Chichester, England, p. 120–135.

Barrick, J.E., and Haywa-Branch, J.N., 1994, Conodont biostratigraphy of the Missouri Mountain Shale (Silurian–Early Devonian?) and the Arkansas Novaculite (Devonian), Black Knob Ridge, Atoka County, Oklahoma, *in* Suneson, N.H., and Hemish, L.A., eds., Geology and resources of the eastern Ouachita Mountains frontal belt and southeastern Arkoma basin, Oklahoma: Oklahoma Geological Survey Guidebook 29, p. 161–177.

Barrick, J.E., and Noble, P.J., 1995, Early Devonian conodonts from a limestone horizon in the Caballos Novaculite, Marathon uplift, west Texas: Journal of Paleontology, v. 69, no. 6, p. 1112–1122.

Bartolini, C., 1993, Tectonic emplacement and accretion of Paleozoic eugeoclinal, allochthonous assemblages in central Sonora, Mexico—a new Cordilleran tectonostratigraphic terrane, *in* Ortega-Gutiérrez, F., Coney, P.J., Centeno-García, E., and Gómez-Caballero, A., eds., Proceedings of the First circum-Pacific and circum-Atlantic terrane conference: Guanajuato, México, Universidad Nacional Autónoma de México and Instituto de Geología, p. 12–16.

Bartolini, C., Stewart, J.H., Carter, C., Murchey, B.L., and Repetski, J.E., 1989, Stratigraphy of Paleozoic eugeoclinal strata in Sierra El Aliso, central Sonora, Mexico: Universidad de Sonora, Boletín del Departamento de Geología, Boletín, Segunda Epoca, v. 6, nos. 1 and 2, p. 11–21.

Bartolini, C., Morales-Montaño, M., Spinosa, C., and Finney, S.C., 1990, Paleozoic off-shelf successions unconformably overlain by Triassic strata in central Sonora, Mexico—evidence for a Permo–Triassic tectonic event?: Geological Society of America Abstracts with Programs, v. 22, no. 3, p. A114.

Bartolini, C., Morales-Montaño, M., Barrera-Moreno, E., and Domínguez-Perla, J.E., Navarro-Martínez, L.A., Soto-Contreras, L.A., Finney, S.C., and Carter, C., 1995, Geological reconnaissance of Ordovician deep-marine sequences in central Sonora, Mexico, *in* Cooper, J.D., Droser, M.L., and Finney, S.C., eds., Ordovician Odyssey: Short Papers for the Seventh International Symposium on the Ordovician System: Fullerton, California, Pacific Section Society for Sedimentary Geology (SEPM), book no. 77, p. 285–289.

Beaumont, C., 1981, Foreland basins: Geophysical Journal of Royal Astronomical Society, v. 65, no. 2, p. 291–329.

Benedetti, S.J., 1976, Paleozoic conodonts from the Placer de Guadalupe area, east-central Chihuahua, Mexico [M.S. thesis]: Fort Worth, Texas, Texas Christian University, 141 p.

Bickford, M.E., and Mose, D.G., 1975, Geochronology of Precambrian rocks in the St. Francois Mountains, southeastern Missouri: Geological Society of America Special Paper 165, 48 p.

Blakey, R.C., 1980, Pennsylvanian and Early Permian paleogeography, southern Colorado Plateau and vicinity, *in* Fouch, T.D., and Magathan, E.R., eds., Paleozoic paleogeography of the west-central United States: Rocky Mountain paleogeography symposium 1: Denver, Colorado, Rocky Mountain Section, Society of Economic Paleontologists and Mineralogists, p. 239–257.

Blakey, R.C., and Knapp, R., 1989, Pennsylvanian and Permian geology of Arizona, *in* Jenney, J.P., and Reynolds, S.J., eds., Geologic evolution of Arizona: Tucson, Arizona, Arizona Geological Society Digest 17, p. 313–347.

Blount, J.G., 1983, The geology of the Rancho Los Filtros area, Chihuahua, Mexico, *in* Clark, K.F., and Goodell, P.C., eds., Geology and mineral resources of north-central Chihuahua: El Paso Geological Society, Field Conference Guidebook, p. 157–164.

Blount, J.G., 1993, The geochemistry, petrogenesis, and geochronology of the Precambrian meta-igneous rocks of Sierra Del Cuervo and Cerro El Carrizalillo, Chihuahua, Mexico [Ph.D. thesis]: Austin, Texas, University of Texas at Austin, 242 p.

Bohannon, R.G., and Lucchitta, I., 1991, Geologic map of the Mount Bangs quadrangle, Mohave County, Arizona: U.S. Geological Survey Miscellaneous Investigations Series, Map I-2166, scale 1:24,000.

Bond, G.C., Nickeson, P.A., and Kominz, M.A., 1984, Breakup of a supercontinent between 625 Ma and 555 Ma: New evidence and implications for continental histories: Earth and Planetary Science Letters, v. 70, no. 2, p. 325–345, doi: 10.1016/0012-821X(84)90017-7.

Boucot, A.J., Blodgett, R.B., and Stewart, J.H., 1997, European Province Late Silurian brachiopods from the Ciudad Victoria area, Tamaulipas, northeast Mexico, *in* Klapper, G., Murphy, M.A., and Talent, J.A., eds., Paleozoic sequence stratigraphy, biostratigraphy, and biogeography; Studies in honor of J. Granville ("Jess") Johnson: Geological Society of America Special Paper 321, p. 273–293.

Bridges, L.W., II, 1962, Geology of Mina Plomosas area, Chihuahua, Mexico [Ph.D. thesis]: Austin, Texas, University of Texas at Austin, 241 p.

Bridges, L.W., II, 1964, Stratigraphy of Mina Plomosas-Placer de Guadalupe area, *in* Geology of Mina Plomosas-Placer de Guadalupe area, Chihuahua, Mexico: West Texas Geological Society Publication 64-50, p. 50–59.

Bridges, L.W., II, 1970, Paleozoic history of the southern Chihuahua tectonic belt, *in* Seewald, K. and Sundeen, D., eds., The geologic framework of the Chihuahua tectonic belt; Symposium in honor of Professor Ronald K. DeFord: Midland, Texas, West Texas Geological Society, p. 67–74.

Brown, A.R., 1980, Limestone deposits of the Desert Divide, San Jacinto Mountains, California, *in* Fife, D.L., and Brown, A.R., eds., Geology and mineral wealth of the California desert: Santa Ana, California, South Coast Geological Society, Dibblee Volume, p. 284–293.

Brown, A.R., 1981, Structural history of the metamorphic, granitic and cataclastic rocks in southeastern San Jacinto Mountains, *in* Brown, A.R., and Ruff, R.W., eds., Geology of the San Jacinto Mountains: Santa Ana, California, South Coast Geological Society, Annual Fieldtrip Guidebook, no. 9, p. 100–138.

Brown, H.J., 1991, Stratigraphy and paleogeographic setting of Paleozoic rocks in the San Bernardino Mountains, California, *in* Cooper, J.D., and Stevens, C.H., eds., Paleozoic paleogeography of the western United States—II: Los Angeles, California, Pacific Section, Society of Economic Paleontologists and Mineralogists, v. 1, book 67, p. 193– 207.

Brunner, P., 1975, Estudio estratigráfico del Devónico en el área de El Bísani, Caborca, Sonora: Revista del Instituto Mexicano del Petróleo, v. 7, núm. 1, p. 16–45.

Brunner, P., 1976, Litología y bioestratigrafía del Misisípico en el área de El Bísani, Caborca, Sonora: Revista del Instituto Mexicano del Petróleo, v. 8, núm. 3, p. 7–41.

Buch, I.P., 1984, Upper Permian(?) and Lower Triassic metasedimentary rocks, northeastern Baja California, Mexico, *in* Frizzell, V.A., Jr., ed., Geology of the Baja California peninsula: Los Angeles, California, Pacific Section, Society of Economic Paleontologists and Mineralogists, v. 39, p. 31–36.

Buch, I.P., and Delattre, M.P., 1993, Permian and Lower Triassic stratigraphy along the 30th parallel, eastern Baja California Norte, Mexico, *in* Gastil, R.G., and Miller, R.H., eds., The prebatholithic stratigraphy of peninsular California: Geological Society of America Special Paper 279, p. 77–90.

Buffler, R.T., and Thomas, W.A., 1994, Crustal structure and evolution of the southeastern margin of North America and the Gulf of Mexico basin, *in* Speed, R.C., ed., Phanerozoic evolution of North American continent-ocean transitions: Boulder, Colorado, Geological Society of America, Geology of North America Continent-Ocean Transects, chapter 4, p. 219–264.

Calmus, T., 1993, Reconnaissance geologic map of the Estación Sahuaro area (Altar Desert, Sonora): Tectonic implications, *in* González-León, C., and Vega-Granillo, E.L., eds., Libro de Resúmenes, III Simposio de la Geología de Sonora y áreas adyacentes, p. 19–22.

Calmus, T., and Sosson, M., 1995, Southwestern extension of the Papago terrane into the Altar Desert region, northwestern Sonora, and its implications, *in* Jacques-Ayala, C., González-León, C.M., and Roldán-Quintana, J., eds., Studies on the Mesozoic of Sonora and adjacent areas: Geological Society of America Special Paper 301, p. 99–109.

Calmus, T., Pérez-Segura, E., and Herrera-Urbina, S., 1992, Geology of Estación Llano (Sonora): A new Proterozoic basement and the San Francisco gold deposit, *in* Clark, K.F., Roldán-Quintana, J., and Schmidt, R.H., eds., Geology and mineral resources of northern Sierra Madre Occidental, Mexico: Guidebook for the 1992 Field Conference, El Paso Geological Society, p. 313–321.

Calmus, T., Pérez-Segura, E., and Stinnesbeck, W., 1997, La structuration de la marge pacifique nord-américaine et du <<terrane>> Caborca: apports de la découverte d'une faune du Jurassique inférieur et moyen dans la série de Pozos de Serna (Sonora, Mexique)/New Early and Middle Jurassic fauna in the Pozos de Serna sequence (Sonora, Mexico): Consequences for the identity and the position of the Caborca terrane: Sciences de la terre et des planètes/Earth and Planetary Sciences, v. 325, p. 257–263, C.R. Académie des sciences, Elsevier, Paris, France.

Cameron, C.S., 1982, Stratigraphy and significance of the upper Precambrian Big Bear Group, *in* Cooper, J.D., Troxel, B.W., and Wright, L.A., eds., Geology of selected areas in the San Bernardino Mountains, western Mojave Desert, and southern Great Basin, California, Field Trip Number 9: Shoshone, California, The Death Valley Publishing Company, p. 5–20.

Cameron, K.L., Robinson, J.V., Niemeyer, S., Nimz, G.J., Kuentz, D.C., Harmon, R.S., Bohlen, S.R., and Collerson, K.D., 1992, Contrasting styles of pre-Cenozoic and mid-Tertiary crustal evolution in northern Mexico: Evidence from deep crustal xenoliths from La Olivina: Journal of Geophysical Research, v. 97, no. B12, p. 17353–17376.

Campa, M.F., and Coney, P.J., 1983, Tectono-stratigraphic terranes and mineral resource distributions of Mexico: Canadian Journal of Earth Sciences, v. 20, p. 1040–1051.

Campbell, M.J., 1985, Prebatholithic stratigraphy of the northeastern Sierra La Asamblea, Baja California, Mexico [M.S. thesis]: San Diego, California, San Diego State University, 127 p.

Campbell, P.A., and Anderson, T.H., 2003, Structure and kinematics along a segment of the Mojave-Sonora megashear: A strike-slip fault that truncates the Jurassic continental magmatic arc of southwestern North America: Tectonics, v. 22, no. 6, p. 16-1–16-21.

Carrillo-Martínez, M., 1971, Geología de la hoja San José de Gracia, Sinaloa: Universidad Nacional Autónoma de México, Facultad de Ingeniería, 25 p.

Carpenter, D.L., 1997, Tectonic history of the metamorphic basement rocks of the Sierra del Carmen, Coahuila, Mexico: Geological Society of America Bulletin, v. 109, no. 10, p. 1321–1332, doi: 10.1130/0016-7606(1997)109<1321:THOTMB>2.3.CO;2.

Carr, M.D., and Pinkston, J.C., 1987, Geologic map of the Goodsprings district, southern Spring Mountains, Clark County, Nevada: U.S. Geological Survey Miscellaneous Field Studies, Map MF-1514, scale 1:24,000.

Cawood, P.A., McCausland, P.J.A., and Dunning, G.R., 2001, Opening Iapetus: Constraints from the Laurentian margin in Newfoundland: Geological Society of America Bulletin, v. 113, no. 4, p. 443–453, doi: 10.1130/0016-7606(2001)113<0443:OICFTL>2.0.CO;2.

Charleston, S., 1981, A summary of the structural geology and tectonics of the State of Coahuila, Mexico, *in* Charleston, S., Smith, Ch.I., and Brown, J.B., eds., Lower Cretaceous stratigraphy and structure, northern Mexico; Field trip guidebook: West Texas Geological Society Publication 81-74, p. 28–36.

Claoué-Long, J.C., Jones, P.J., Roberts, J., and Maxwell, S., 1992, The numerical age of the Devonian–Carboniferous boundary: Geological Magazine, v. 129, no. 3, p. 281–291.

Coleman, D.S., Barth, A.P., and Wooden, J.L., 2003, Metamorphism and orogenesis of Proterozoic rocks in the southwestern Mojave province: Geological Society of America Abstracts with Programs, v. 35, no. 4, p. 67.

Coney, P.J., 1981, Accretionary tectonics in western North America, *in* Dickinson, W.R., and Payne, W.D., eds., Relations of tectonics to ore deposits in the southern Cordillera: Arizona Geological Society Digest v. 14, p. 23–37.

Coney, P.J., and Campa, M.F., 1987, Lithotectonic terrane map of Mexico (west of the 91st Meridian): U.S. Geological Survey Miscellaneous Field Studies Map MF-1874-D, scale 1:2,500,000.

Consejo de Recursos Minerales, 1992, Geological-mining monograph of the State of Sinaloa, 159 p.

Cooper, G.A., and Arellano, A.R.V., 1946, Stratigraphy near Caborca, northwest Sonora, Mexico: American Association of Petroleum Geologists Bulletin, v. 40, no. 4, p. 606–619.

Cooper, G.A., Dunbar, C.O., Duncan, H., Miller, A.K., and Knight, J.B., 1953, Permian fauna at El Antimonio, western Sonora, Mexico: Smithsonian Miscellaneous Collections, v. 119, no. 2, 111 p.

Cooper, J.D., and Edwards, J.C., 1991, Cambro–Ordovician craton-margin carbonate section, southern Great Basin: A sequence-stratigraphic perspective, *in* Cooper, J.D., and Stevens, C.H., eds., Paleozoic paleogeography of the western United States—II: Los Angeles, California, Pacific Section, Society of Economic Paleontologists and Mineralogists, v. 1, book 67, p. 237–252.

Cooper, R.A., 1999, The Ordovician time scale—calibration of graptolite and conodont zones, *in* Kraft, P., and Fatka, O., eds., *Quo vadis* Ordovician?: Short Papers of the Eighth International Symposium on the Ordovician System: Prague, Czech Republic, Acta Universitatis Carolinae Geologica, v. 43, no. 1/2, p. 1–4.

Córdoba, D., and Ortega-Gutiérrez, F., 1990, Tectónica IV.2.1, Atlas Nacional de México: Universidad Nacional Autónoma de México, Instituto de Geografía, map scale 1:4,000,000.

Córdoba, D.A., and Silva-Mora, L., 1989, Marco geológico del área de Revolución-Puerta de Cabrera, Estado de Durango: Universidad Nacional Autónoma de México, Instituto de Geología, Revista, v. 8, núm. 2, p. 111–122.

Corry, C.E., Herrin, E., McDowell, F.W., and Phillips, K.A., 1990, Geology of the Solitario, Trans-Pecos Texas: Geological Society of America Special Paper 250, 122 p.

Cox, T.L., 1986, The Warner serpentine–talc (soapstone) pits, *in* Stone, C.G., and Haley, B.R., eds., Sedimentary and igneous rocks of the Ouachita Mountains of Arkansas; A guidebook with contributed papers, part 2: Little Rock, Arkansas, Arkansas Geological Commission Guidebook 86-3, p. 16–18.

Crocker, J., 1987, Stratigraphy and structure of Paleozoic metasediments south of Bahía Calamajué, Baja California, Mexico [M.S. thesis]: San Diego, California, San Diego State University, 129 p.

Decatur, S.H., and Rosenfeld, J.H., 1982, Mississippian radioactive marker beds in West Texas and their possible relationship to the Ouachita orogeny: Geological Society of America Abstracts with Programs, v. 14, no. 3, p. 109.

De Cserna, Z., 1956, Tectónica de la Sierra Madre Oriental de México, entre Torreón y Monterrey: XX Congreso Geológico Internacional, 87 p.

De Cserna, Z., and Kent, B.H., 1961, Mapa geológico de reconocimiento y secciones estructurales de la región de San Blas y El Fuerte, Estados de Sinaloa y Sonora: Universidad Nacional Autónoma de México e Instituto de Geología, Cartas Geológicas y Mineras No. 4, escala 1:100,000.

De Cserna, Z., Rincón-Orta, C., Solorio-Munguía, J., and Schmitter-Villada, E., 1968, Una edad radiométrica Pérmica temprana de la región de Placer de Guadalupe, noreste de Chihuahua: Boletín de la Sociedad Geología Mexicana, v. 31, núm. 1, p. 65–73.

Delattre, M.P., 1984, Permian miogeoclinal strata at El Volcan, Baja California, Mexico, *in* Frizzel, V.R., Jr., ed., Geology of the Baja California peninsula: Los Angeles, California, Pacific Section, Society of Economic Paleontologists and Mineralogists, p. 23–29.

Denison, R.E., 1989, Foreland structure adjacent to the Ouachita foldbelt, *in* Hatcher, R.D., Jr., Thomas, W.A., and Viele, G.W., eds., The Appalachian-Ouachita orogen in the United States: Boulder, Colorado, Geological Society of America, The Geology of North America, v. F-2, chapter 25, p. 681–688.

Denison, R.E., Kenney, G.S., Burke, W.H., Jr., and Hetherington, E.A., Jr., 1969, Isotope ages of igneous and metamorphic boulders from the Haymond Formation (Pennsylvanian), Marathon Basin, Texas, and their significance: Geological Society of America Bulletin, v. 80, no. 2, p. 245–256.

Denison, R.E., Burke, W.H., Jr., Hetherington, E.A., Jr., and Otto, J.B., 1970, Basement rock framework of parts of Texas, southern New Mexico and northern Mexico, *in* Seewald, K., and Sundeen, D., eds., The geologic framework of the Chihuahua tectonic belt; Symposium in honor of Professor Ronald K. DeFord: Midland, Texas, West Texas Geological Society, p. 3–14.

Denison, R.E., Burke, W.H., Jr., Otto, J.B., and Hetherington, E.A., Jr., 1977, Age of igneous and metamorphic activity affecting the Ouachita foldbelt, *in* Stone, C.G., ed., Symposium on the geology of the Ouachita Mountains, v. 1: Little Rock, Arkansas, Arkansas Geological Commission, p. 25–40.

Dewey, J.F., and Burke, K., 1974, Hot spots and continental break-up: Implications for collisional orogeny: Geology, v. 2, no. 2, p. 57–60, doi: 10.1130/0091-7613(1974)2<57:HSACBI>2.0.CO;2.

Dickerson, P.W., 1987, Structural and depositional setting of the SW U.S. and northern Mexico along a Paleozoic transform plate margin, *in* Paleozoico de Chihuahua (Excursión Geológica No. 2, Libreto Guía), Gaceta Geológica, v. 1, no. 1, p. 129–159.

Dickinson, W.R., 1981, Plate tectonic evolution of the southern Cordillera, *in* Dickinson, W.R., and Payne, W.D., eds., Relations of tectonics to ore deposits in the southern Cordillera: Tucson, Arizona, Arizona Geological Society Digest v. 14, p. 113–135.

Dickinson, W.R., 1985, Interpreting provenance relations from detrital modes of sandstones, *in* Zuffa, G.G., ed., Provenance of arenites: NATO ASI Series C: Mathematical and Physical Sciences v. 148: D. Reidel Publishing Company, Dordrecht/Boston/Lancaster, p. 333–361.

Dickinson, W.R., 1989, Tectonic setting of Arizona through geologic time, *in* Jenney, J.P., and Reynolds, S.J., eds., Geologic evolution of Arizona: Arizona Geological Society Digest 17, p. 1–16.

Dickinson, W.R., 2000, Geodynamic interpretation of Paleozoic tectonic trends oriented oblique to the Mesozoic Klamath–Sierran continental margin in California, *in* Soreghan, M.J., and Gehrels, G.E., eds., Paleozoic and Triassic paleogeography and tectonics of western Nevada and northern California: Geological Society of America Special Paper 347, p. 209–245.

Dickinson, W.R., and Lawton, T.F., 2001, Carboniferous to Cretaceous assembly and fragmentation of Mexico: Geological Society of America Bulletin, v. 113, no. 9, p. 1142–1160, doi: 10.1130/0016-7606(2001)113<1142:CTCAAF>2.0.CO;2.

Dickinson, W.R., and Lawton, T.F., 2003, Sequential intercontinental suturing as the ultimate control for Pennsylvanian Ancestral Rocky Mountains deformation: Geology, v. 31, no. 7, p. 609–612, doi: 10.1130/0091-7613(2003)031<0609:SISATU>2.0.CO;2.

Dickinson, W.R., and Suczek, C.A., 1979, Plate tectonics and sandstone compositions: American Association of Petroleum Geologists Bulletin, v. 63, no. 12, p. 2164–2182.

Dockum, M.S., 1982, Greenschist-facies carbonates, eastern Coyote Mountains, western Imperial County, California [M.S. thesis]: San Diego, California, San Diego State University, 89 p.

Dumble, E.T., 1900, Notes on the geology of Sonora, Mexico: American Institute of Mining Engineers Transactions, v. 29, p. 122–152.

Dyer, R., and Reyes, I.A., 1987, The Geology of Cerro El Carrizalillo, Chihuahua, Mexico: Preliminary findings, *in* Paleozoico de Chihuahua (Excursión Geológica No. 2, Libreto Guía), Gaceta Geológica, v. 1, no. 1, p. 108–128.

Easton, W.H., Sanders, J.E., Knight, J.B., and Miller, A.K., 1958, Mississippian fauna in northwestern Sonora, Mexico: Smithsonian Miscellaneous Collections, v. 119, no. 3, 87 p., 9 pl.

Eells, J.L., 1972, Geology of the Sierra de La Berruga, northwestern Sonora, Mexico [M.S. thesis]: San Diego, California, San Diego State University, 77 p.

Ellison, S.P., Jr., and Powell, J.D., 1989, Tesnus (Mississippian) conodonts, Brewster County, Texas, *in* Ziegler, W., ed., Papers on Ordovician to Triassic conodonts: 1st International Senckenberg Conference and 5th European Conodont Symposium (ECOS V), Contribution III, Courier Forschungsinstitut Senckenberg, v. 117, p. 359–373.

Ethington, R.L., Finney, S.C., and Repetski, J.E., 1989, Biostratigraphy of the Paleozoic rocks of the Ouachita orogen, Arkansas, Oklahoma, west Texas, *in* Hatcher, R.D., Jr., Thomas, W.A., and Viele, G.W., eds., The Appalachian-Ouachita orogen in the United States: Boulder, Colorado, Geological Society of America, The Geology of North America, v. F-2, chapter 17, p. 563–574.

Farmer, G.L., Bowring, S.A., Espinoza, I., and Matzel, J.P., 2003, Isotopic evidence of the nature of Precambrian basement and overlying sedimentary rocks, Caborca, Sonora: Geological Society of America Abstracts with Programs, v. 35, no. 4, p. 67.

Fernández-Aguirre, M.A., and Almazán-Vázquez, E., 1991, Carta Geológica Arivechi: Gobierno del Estado de Sonora, Dirección General de Fomento Minero, Universidad Nacional Autónoma de México, Instituto de Geología, escala 1:50,000.

Finney, S.C., 1986, Graptolite biofacies and correlation of eustatic, subsidence, and tectonic events in the Middle to Upper Ordovician of North America: Palaios, v. 1, no. 5, p. 435–461.

Flawn, P.T., 1951, Pegmatites of the Van Horn Mountains, Texas: Economic Geology and the Bulletin of the Society of Economic Geologists, v. 46, no. 2, p. 163–192.

Flawn, P.T., 1956, Basement rocks of Texas and southeast New Mexico: University of Texas Publication 5605, 261 p.

Flawn, P.T., 1961, The Ouachita structural belt in Mexico, *in* Flawn, P.T., Goldstein, A., Jr., King, P.B., and Weaver, C.E., eds., The Ouachita system: Austin, Texas, University of Texas, Bureau of Economic Geology Publication No. 6120, p. 99–106.

Flawn, P.T., Goldstein, A., Jr., King, P.B., and Weaver, C.E., 1961, The Ouachita system: Austin, Texas, University of Texas, Bureau of Economic Geology Publication No. 6120, 401 p.

Franco-Rubio, M., 1999, Geology of the basement below the décollement surface, Sierra de Catorce, San Luis Potosí, Mexico, *in* Bartolini, C., Wilson, J.L., and Lawton, T.F., eds., Mesozoic sedimentary and tectonic history of north-central Mexico: Geological Society of America Special Paper 340, p. 211–227.

Gastil, G., 1985, Terranes of peninsular California and adjacent Sonora, *in* Howell, D.G., ed., Tectonostratigraphic terranes of the circum-Pacific region: Houston, Texas, Circum-Pacific Council for Energy and Mineral Resources Earth Science Series, no. 1, p. 273–283.

Gastil, R.G., 1993, Prebatholithic history of peninsular California, *in* Gastil, R.G., and Miller, R.H., eds., The prebatholithic stratigraphy of peninsular California: Geological Society of America Special Paper 279, p. 145–156.

Gastil, R.G., and Miller, R.H., 1983, Prebatholithic terranes of southern and peninsular California, U.S.A. and Mexico; Status report, *in* Stevens, C.H., ed., Pre-Jurassic rocks in western North America suspect terranes: Los Angeles, California, Pacific Section, Society of Economic Paleontologists and Mineralogists, p. 49–61.

Gastil, G., and Miller, R., 1984, Prebatholithic paleogeography of peninsular California and adjacent Mexico, *in* Frizzell, V.A., Jr., ed., Geology of the Baja California peninsula: Los Angeles, California, Pacific Section, Society of Economic Paleontologists and Mineralogists, v. 39, p. 9–16.

Gastil, R.G., and Miller, R.H., eds., 1993, The prebatholithic stratigraphy of peninsular California: Geological Society of America Special Paper 279, 163 p.

Gastil, R.G., Phillips, R.P., and Allison, E.C., 1975, Reconnaissance geology of the State of Baja California: Geological Society of America Memoir 140, 170 p.

Gastil, G., Miller, R., Anderson, P., Crocker, J., Campbell, M., Buch, P., Lothringer, C., Leier-Engelhardt, P., Delattre, M., Hoobs, J., and Roldán-Quintana, J., 1991, The relation between the Paleozoic strata on opposite sides of the Gulf of California, *in* Pérez-Segura, E., and Jacques-Ayala, C., eds., Studies of Sonoran geology: Geological Society of America Special Paper 254, p. 7–17.

Gehrels, G.E., and Stewart, J.H., 1998, Detrital zircon U-Pb geochronology of Cambrian to Triassic miogeoclinal and eugeoclinal strata of Sonora, Mexico: Journal of Geophysical Research, v. 103, no. B2, p. 2471–2487, doi: 10.1029/97JB03251.

Gehrels, G.E., Dickinson, W.R., Ross, G.M., Stewart, J.H., and Howell, D.G., 1995, Detrital zircon reference for Cambrian to Triassic miogeoclinal strata of western North America: Geology, v. 23, no. 9, p. 831–834, doi: 10.1130/0091-7613(1995)023<0831:DZRFCT>2.3.CO;2.

Gehrels, G.E., Stewart, J.H., and Ketner, K.B., 2002, Cordilleran-margin quartzites in Baja California—implications for tectonic transport: Earth and Planetary Science Letters, v. 199, p. 201–210, doi: 10.1016/S0012-821X(02)00542-3.

Gleason, J.D., Gehrels, G.E., and Finney, S.C., 2001, Tectonic recycling in the Paleozoic Ouachita assemblage from U-Pb detrital zircon studies: Supplement to Eos (Transactions, American Geophysical Union), v. 82, no. 20, p. S435.

Goetz, L.K., and Dickerson, P.W., 1985, A Paleozoic transform margin in Arizona, New Mexico, West Texas, and northern Mexico, *in* Dickerson, P.W., and Muehlberger, W.R., eds., Structure and tectonics of Trans-Pecos Texas: West Texas Geological Society Publication 85-81, p. 173–184.

Gómez-Luna, M.E., and Martínez-Cortés, A., 1997, Relationships and differences between the Triassic ammonoid successions of northwestern Sonora, Mexico, and west-central Nevada, U.S.A: Revista Mexicana de Ciencias Geológicas, v. 14, núm. 2, p. 208–218.

González-León, C.M., 1980, La Formación Antimonio (Triásico Superior–Jurásico Inferior) en la Sierra del Álamo, Estado de Sonora: Universidad Nacional Autónoma de México, Instituto de Geología, Revista, v. 4, núm. 1, p. 13–18.

González-León, C.M., 1989, Evolución de terrenos Mesozoicos en el noroeste de México: Universidad de Sonora, Boletín del Departamento de Geología, Boletín, Segunda Epoca, v. 6, nos. 1 y 2, p. 39–54.

González-León, C.M., 1997, Sequence stratigraphy and paleogeographic setting of the Antimonio Formation (Late Permian–Early Jurassic), Sonora, Mexico: Revista Mexicana de Ciencias Geológicas, v. 14, núm. 2, p. 136–148.

Goodell, P.C., 1993, Precambrian cratonic crust of west-central Chihuahua—The Sierra del Nido block, *in* González-León, C., and Vega-Granillo, E.L., eds., Libro de Resúmenes, III Simposio de la Geología de Sonora y áreas adyacentes, p. 45.

Goodell, P.C., Keller, G.R., and Dyer, J.R., 1986, The Sierra del Nido tectonic block—A newly recognized cratonic feature in northern Mexico, *in* Precambrian de Chihuahua, Chihuahua–Cerro Carrizalillo, Chihuahua–Sierra del Cuervo: Sociedad Geológica Mexicana A.C. Excursión Geológica, Universidad Autónoma de Chihuahua, p. 36.

Gross, E.L., Stewart, J.H., and Gehrels, G.E., 2000, Detrital zircon geochronology of Neoproterozoic to Middle Cambrian miogeoclinal and platformal strata: Northwest Sonora, Mexico: Geofísica Internacional, v. 39, núm. 4, p. 295–308.

Haenggi, W.T., 2001, Tectonic history of the Chihuahua trough, Mexico and adjacent USA, Part 1: The pre-Mesozoic setting: Boletín de la Sociedad Geológica Mexicana, tomo LIV, no. 1, p. 28–66.

Haenggi, W.T., Dickerson, P.W., and Muehlberger, W.R., 2000, Ancestral Rocky Mountain uplifts and accreted blocks at the southern margin of N. America (Chihuahua and W. Texas): Geological Society of America Abstracts with Programs, v. 32, no. 7, p. A-317.

Ham, W.E., Denison, R.E., and Merritt, C.A., 1964, Basement rocks and structural evolution of southern Oklahoma: Oklahoma Geological Survey Bulletin 95, 302 p.

Hamilton, W.B., 1984, Generalized geologic map of the Big Maria Mountains region, northeastern Riverside County, southeastern California: U.S. Geological Survey Open-File Report 84-407, scale 1:48,000.

Hamilton, W.B., 1987, Mesozoic geology and tectonics of the Big Maria Mountains region, southeastern California, *in* Dickinson, W.R., and Klute, M.A., eds., Mesozoic rocks of southern Arizona and adjacent areas: Tucson, Arizona, Arizona Geological Society Digest 18, p. 33–47.

Handschy, J.W., and Dyer, R., 1987, Polyphase deformation in Sierra del Cuervo, Chihuahua, Mexico: Evidence for Ancestral Rocky Mountain tectonics in the Ouachita foreland of northern Mexico: Geological Society of America Bulletin, v. 99, no. 5, p. 618–632, doi: 10.1130/0016-7606(1987)99<618:PDISDC>2.0.CO;2.

Handschy, J.W., Keller, G.R., and Smith, K.J., 1987, The Ouachita system in northern Mexico: Tectonics, v. 6, no. 3, p. 323–330.

Harland, W.B., Armstrong, R.L., Cox, A.V., Craig, L.E., Smith, A.G., and Smith, D.G., 1990, A geologic time scale, 263 p. and wall chart (=GTS 89): Cambridge, New York, Port Chester, Melbourne, Sydney (Cambridge University Press).

Harry, D.L., and Londono, J., 2004, Structure and evolution of the central Gulf of Mexico continental margin and coastal plain, southeast United States: Geological Society of America Bulletin, v. 116, no. 1/2, p. 188–199.

Harry, D.L., and Mickus, K.L., 1998, Gravity constraints on lithosphere flexure and the structure of the late Paleozoic Ouachita orogen in Arkansas and Oklahoma, south central North America: Tectonics, v. 17, no. 2, p. 187–202, doi: 10.1029/97TC03786.

Hass, W.H., 1950, Age of lower part of Stanley Shale: American Association of Petroleum Geologists Bulletin, v. 34, no. 7, p. 1578–1584.

Hass, W.H., 1951, The Arkansas Novaculite: American Association of Petroleum Geologists Bulletin, v. 35, no. 12, p. 2526–2541.

Hatcher, R.D., Jr., Thomas, W.A., and Viele, G.W., eds., 1989, The Appalachian-Ouachita orogen in the United States: Boulder, Colorado, Geological Society of America, The Geology of North America, v. F-2, 30 chapters, 12 plates, 767 p.

Hayes, P.T., and Cone, G.C., 1975, Cambrian and Ordovician rocks of southern Arizona and New Mexico and westernmost Texas: U.S. Geological Survey Professional Paper 873, 98 p.

Hazzard, J.C., 1937, Paleozoic section in the Nopah and Resting Springs Mountains, Inyo County, California: California Journal of Mines and Geology, v. 33, no. 4, p. 273–339.

Hazzard, J.C., 1954, Rocks and structures of the northern Providence Mountains, San Bernardino County, California, *in* Jahns, R.H., ed., Geology of southern California: California Division of Mines and Geology Bulletin 170, v. 1, chapter 4, part 4, p. 27–35.

Hendricks, T.A., Knechtel, M.M., and Bridge, J., 1937, Geology of Black Knob Ridge, Oklahoma: American Association of Petroleum Geologists Bulletin, v. 21, no. 1, p. 1–29.

Henry, C.D., Muehlberger, W.R., Erdlac, R.J., Jr., Price, J.G., and Dickerson, P.W., 1994, Geology of the Solitario, Trans-Pecos Texas (Geological Society of America Special Paper 250): Discussion and reply: Geological Society of America Bulletin, v. 106, no. 4, p. 560–569.

Henry, C.D., Muehlberger, W.R., McCormick, C.L., Davis, L.L., Erdlac, R.J., Jr., Kunk, M.J., and Smith, C.I., 1996, Geology of the Solitario Dome, Trans-Pecos Texas: Paleozoic, Mesozoic, and Cenozoic sedimentation, tectonism, and magmatism: Bureau of Economic Geology, The University of Texas at Austin, Report of Investigations No. 240, 182 p.

Hewett, D.F., 1931, Geology and ore deposits of the Goodsprings quadrangle, Nevada: U.S. Geological Survey Professional Paper 162, 172 p.

Hewett, R.L., 1978, Geology of the Cerro La Zacatera area, Sonora, Mexico [M.S. thesis]: Flagstaff, Arizona, Northern Arizona University, 99 p.

Holcomb, R.A., 1979, Conodont biostratigraphy of the Paleozoic carbonates near Bavispe, Sonora, Mexico [M.S. thesis]: Fort Worth, Texas, Texas Christian University, 99 p.

Hoobs, J.H., 1985, Carboniferous island-arc and associated rocks from the Mission Calamajué area, Baja California, Mexico [M.S. thesis]: San Diego, California, San Diego State University, 122 p.

Houseknecht, D.W., 1986, Evolution from passive margin to foreland basin: The Atoka Formation of the Arkoma basin, south-central U.S.A., *in* Allen, P.A., and Homewood, P., eds., Foreland basins: International Association of Sedimentologists Special Publication Number 8, p. 327–345.

Hunt, C.B., and Mabey, D.R., 1966, Stratigraphy and structure Death Valley, California: U.S. Geological Survey Professional Paper 494-A, 162 p.

Imlay, R.W., 1939, Paleogeographic studies in northeastern Sonora: Geological Society of America Bulletin, v. 50, no. 11, p. 1723–1744.

Imoto, N., and McBride, E.F., 1990, Volcanism recorded in Tesnus Formation, Marathon uplift, Texas, *in* Laroche, T.M., and Higgins, L., eds., Marathon thrust belt: Structure, stratigraphy, and hydrocarbon potential: 1990 Field Seminar, West Texas Geological Society, Permian Basin Section, Society of Economic Paleontologists Society, p. 93–98.

Iriondo, A., 2001, Proterozoic basements and their Laramide juxtaposition in NW Sonora, Mexico: Tectonic constraints on the SW margin of Laurentia [Ph.D. thesis]: Boulder, Colorado, University of Colorado, 222 p.

Iriondo, A., and Premo, W.R., 2003, The Caborca block: An inhomogeneous piece of Paleoproterozoic crust in Sonora: Geological Society of America Abstracts with Programs, v. 35, no. 4, p. 67.

Iriondo, A., Premo, W.R., Martínez-Torres, L.M., Budahn, J.R., Atkinson, W.W., Jr., Siems, D.F., and Guarás-González, B., 2004, Isotopic, geochemical, and temporal characterization of Proterozoic basement rocks in the Quitovac region, northwestern Sonora, Mexico: Implications for the reconstruction of the southwestern margin of Laurentia: Geological Society of America Bulletin, v. 116, no. 1/2, p. 154–170; Data Repository item 2004027.

Jacques-Ayala, C., García y Barragán, J.C., and De Jong, K.A., 1990, Caborca-Altar geology: Cretaceous sedimentation and compression, Tertiary uplift and extension, *in* Gehrels, G.E., and Spencer, J.E., eds., Geologic excursions through the Sonoran Desert region, Arizona and Sonora: Tucson, Arizona, Arizona Geological Survey Special Paper 7, p. 165–182.

James, E.W., and Henry, C.D., 1993a, Southeastern extent of the North American craton in Texas and northern Chihuahua as revealed by Pb isotopes: Geological Society of America Bulletin, v. 105, no. 1, p. 116–126, doi: 10.1130/0016-7606(1993)105<0116:SEOTNA>2.3.CO;2.

James, E.W., and Henry, C.D., 1993b, Pb isotopes of ore deposits in Trans-Pecos Texas and northeastern Chihuahua, Mexico: Economic Geology and the Bulletin of the Society of Economic Geologists, v. 88, no. 4, p. 934–947.

Jin, Y., Wardlaw, B.R., Glenister, B.F., and Kotlyar, G.V., 1997, Permian chronostratigraphic subdivisions: Episodes, International Union of Geological Sciences, v. 20, no. 1, p. 10–15.

Johnson, G.A.L., and Tarling, D.H., 1985, Continental convergence and closing seas during the Carboniferous, *in* Tarling, D.H., chairman, Symposium no. 7, Continental plate reasemblies for the Carboniferous: Compte Rendu—10th Congrès International de Stratigraphie et de Géologie du Carbonifère (Madrid, Spain, 1983), v. 4, p. 163–168.

Jones, N.W., and McKee, J.W., 1987, Pre-Cretaceous volcanic rocks at Sierra Diablo, Chihuahua, *in* Paleozoico de Chihuahua (Excursión Geológica No. 2, Libreto Guía), Gaceta Geológica, v. 1, no. 1, p. 82–96.

Jones, N.W., McKee, J.W., Márquez-D., B., Tovar, J., Long, L.E., and Laudon, T.S., 1984, The Mesozoic La Mula Island, Coahuila, Mexico: Geological Society of America Bulletin, v. 95, no. 10, p. 1226–1241, doi: 10.1130/0016-7606(1984)95<1226:TMLMIC>2.0.CO;2.

Jones, N.W., McKee, J.W., Anderson, T.H., and Silver, L.T., 1995, Jurassic volcanic rocks in northeastern Mexico: A possible remnant of a Cordilleran magmatic arc, *in* Jacques-Ayala, C., González-León, C.M., and Roldán-Quintana, J., eds., Studies on the Mesozoic of Sonora and adjacent areas: Geological Society of America Special Paper 301, p. 179–190.

Karlstrom, K.E., and Bowring, S.A., 1993, Proterozoic orogenic history of Arizona, *in* Van Schmus, W.R., and Bickford, M.E., co-eds., Transcontinental Proterozoic provinces, *in* Reed, J.C., Jr., Bickford, M.E., Houston, R.S., Link, P.K., Rankin, D.W., Sims, P.K., and Van Schmus, W.R., eds., Precambrian: Conterminous U.S.: Boulder, Colorado, Geological Society of America, The Geology of North America, v. C-2, chapter 4, p. 188–211.

Karlstrom, K.E., and 28 authors of CD-ROM Working Group, 2002, Structure and evolution of the lithosphere beneath the Rocky Mountains: Initial results from the CD-ROM experiment: GSA Today, v. 12, no. 3, p. 4–10.

Keller, G.R., Braile, L.W., McMechan, G.A., Thomas, W.A., and Harder, S.H., Wen-Fong Chang, and Jardine, W.G., 1989a, Paleozoic continent-ocean transition in the Ouachita Mountains imaged from PASSCAL wide-angle seismic reflection-refraction data: Geology, v. 17, no. 2, p. 119–122.

Keller, G.R., Hills, J.M., Baker, M.R., and Wallin, E.T., 1989b, Geophysical and geochronological constraints on the extent and age of mafic intrusions in the basement of west Texas and eastern New Mexico: Geology, v. 17, no. 11, p. 1049–1052, doi: 10.1130/0091-7613(1989)017<1049:GAGCOT>2.3.CO;2.

Ketner, K.B., 1986, Eureka Quartzite in Mexico?—tectonic implications: Geology, v. 14, no. 12, p. 1027–1030, doi: 10.1130/0091-7613(1986)14<1027:EQIMI>2.0.CO;2.

Ketner, K.B., and Noll, J.H., Jr., 1987, Preliminary geologic map of the Cerro Cobachi area, Sonora, Mexico: U.S. Geological Survey Miscellaneous Field Studies Map MF-1980, scale 1:20,000.

King, P.B., 1937, Geology of the Marathon region, Texas: U.S. Geological Survey Professional Paper 187, 148 p.

King, P.B., 1942, Permian of west Texas and southeastern New Mexico: American Association of Petroleum Geologists Bulletin, v. 26, no. 4, p. 535–763.

King, P.B., 1948, Geology of the southern Guadalupe Mountains, Texas: U.S. Geological Survey Professional Paper 215, 183 p.

King, P.B., 1975, Ancient southern margin of North America: Geology, v. 3, no. 12, p. 732–734, doi: 10.1130/0091-7613(1975)3<732:ASMONA>2.0.CO;2.

King, R.E., 1939, Geological reconnaissance in northern Sierra Madre Occidental of Mexico: Geological Society of America Bulletin, v. 50, no. 11, p. 1625–1722.

Kluth, C.F., and Coney, P.J., 1981, Plate tectonics of the Ancestral Rocky Mountains: Geology, v. 9, no. 1, p. 10–15, doi: 10.1130/0091-7613(1981)9<10:PTOTAR>2.0.CO;2.

LaMone, D.V., 1969, Lower Paleozoic rocks in the El Paso area, *in* Córdoba, D.A., Wengerd, S.A., and Shomaker, J., eds., Guidebook of the Border Region: New Mexico Geological Society, Twentieth Field Conference, p. 68–79.

Landing, E., Bowring, S.A., Davidek, K.L., Rushton, A.W.A., Fortey, R.A., and Wimbledon, W.A.P., 2000, Cambrian–Ordovician boundary age and duration of the lowest Ordovician Tremadoc Series based on U-Pb zircon dates from Avalonian Wales: Geological Magazine, v. 137, no. 5, p. 485–494, doi: 10.1017/S0016756800004507.

Leier-Engelhardt, P.J., 1986, Middle Paleozoic strata of the Sierra Las Pinta, northeastern Baja California Norte, Mexico [M.S. thesis]: San Diego, California, San Diego State University, 169 p.

Leveille, G., and Frost, E.G., 1984, Deformed upper Paleozoic-lower Mesozoic cratonic strata, El Capitán, Sonora, Mexico: Geological Society of America Abstracts with Programs, v. 16, no. 6, p. 575.

Lillie, R.J., 1985, Tectonically buried continent/ocean boundary, Ouachita Mountains, Arkansas: Geology, v. 13, no. 1, p. 18–21, doi: 10.1130/0091-7613(1985)13<18:TBCBOM>2.0.CO;2.

Link, M.H., and Roberts, M.T., 1986, Pennsylvanian paleogeography for the Ozarks, Arkoma, and Ouachita basins in east-central Arkansas, *in* Stone, C.G., and Haley, B.R., eds., Sedimentary and igneous rocks of the Ouachita Mountains of Arkansas; A guidebook with contributed papers, part 2: Little Rock, Arkansas, Arkansas Geological Commission Guidebook 86-3, p. 37–60.

Link, P.K., Christie-Blick, N., Stewart, J.H., Miller, J.M.G., Devlin, W.J., and Levy, M., 1993, Late Proterozoic strata of the United States Cordillera, *in* Link, P.K., ed., Middle and Late Proterozoic stratified rocks of the western U.S. Cordillera, Colorado Plateau, and Basin and Range province, *in* Reed, J.C., Jr., Bickford, M.E., Houston, R.S., Link, P.K., Rankin, D.W., Sims, P.K., and Van Schmus, W.R., eds., Precambrian: Conterminous U.S.: Boulder, Colorado, Geological Society of America, The Geology of North America, v. C-2, chapter 6, p. 536–558.

Loomis, J., Weaver, B., and Blatt, H., 1994, Geochemistry of Mississippian tuffs from the Ouachita Mountains, and implications for the tectonics of the Ouachita orogen, Oklahoma and Arkansas: Geological Society of America Bulletin, v. 106, no. 9, p. 1158–1171, doi: 10.1130/0016-7606(1994)106<1158:GOMTFT>2.3.CO;2.

López, R., Jones, N.W., Cameron, K.L., and Hernandez, C., 1996, The pre-Jurassic evolution of the Coahuila terrane, Mexico: No evidence of a major change in magmatic source during the course of the Ouachita orogeny: Eos (Transactions, American Geophysical Union), v. 77, no. 46, p. F759–F760.

López, R., Cameron, K.L., and Jones, N.W., 1998, New paleo-Proterozoic, Grenvillian, and Pan–African U-Pb zircon ages of boulders from the Coahuila terrane: Implications for the Pre-Cambrian tectonic history of northern Mexico: Geological Society of America Abstracts with Programs, v. 30, no. 7, p. A-354.

López-Ramos, E., and Sánchez-Mejorada, S.H., 1976, Carta geológica de la República Mexicana [edición 4]: Comité de la Carta Geológica de México, escala 1:2,000,000. Map is available from Instituto de Geología, U.N.A.M., Ciudad Universitaria.

Lothringer, C.J., 1993, Allochthonous Ordovician strata of Rancho San Marcos, Baja California Norte, Mexico, *in* Gastil, R.G., and Miller, R.H., eds., The prebatholithic stratigraphy of peninsular California: Geological Society of America Special Paper 279, p. 11–22.

Lowe, D.R., 1989, Stratigraphy, sedimentology, and depositional setting of pre-orogenic rocks of the Ouachita Mountains, Arkansas and Oklahoma, *in* Hatcher, R.D., Jr., Thomas, W.A., and Viele, G.W., eds., The Appalachian-Ouachita orogen in the United States: Boulder, Colorado, Geological Society of America, The Geology of North America, v. F-2, chapter 18, p. 575–590.

Lucas, S.G., and González-León, C., 1994, Marine Upper Triassic strata at Sierra la Flojera, Sonora, Mexico: Neues Jahrbuch für Geologie und Paläontologie Monatshefte, v. 1, p. 34–40.

Lund, K., Aleinikoff, J.N., Evans, K.V., and Fanning, C.M., 2003, SHRIMP U-Pb geochronology of Neoproterozoic Windermere Supergroup, central Idaho: Implications for rifting of western Laurentia and synchroneity of Sturtian glacial deposits: Geological Society of America Bulletin, v. 115, no. 3, p. 349–372, doi: 10.1130/0016-7606(2003)115<0349:SUPGON>2.0.CO;2.

Madrid, R.J., Poole, F.G., and Miller, M.M., 1992, Devonian and older rocks in accreted terranes, *in* Poole, F.G., and 8 others, Latest Precambrian to latest Devonian time; Development of a continental margin, *in* Burchfiel, B.C., Lipman, P.W., and Zoback, M.L., eds., The Cordilleran orogen: Conterminous U.S.: Boulder, Colorado, Geological Society of America, The Geology of North America, v. G-3, chapter 2, p. 36–43.

Main, F.H., 1950, Structure and stratigraphy of the Inde-Cieneguillas district, Durango, Mexico: Geological Society of America Bulletin, v. 61, no. 12, p. 1483.

Mammerickx, J., and Smith, S.M., 1981, Bathymetry of the northeast Pacific: Geological Society of America Map and Chart Series MC-43, scale at equator 1:6,442,194.

Mammerickx, J., and Smith, S.M., 1984, General bathymetric chart of the oceans (GEBCO): Ottawa, Canada, Canadian Hydrographic Service, sheet 5-07, scale at equator 10,000,000.

Marzolf, J.E., and Anderson, T.H., 1996, Early Mesozoic paleogeography and the Mojave-Sonora megashear: Geological Society of America Abstracts with Programs, v. 28, no. 7, p. A-114.

Mauger, R.L., McDowell, F.W., and Blount, J.G., 1983, Grenville-age Precambrian rocks of the Los Filtros area near Aldama, Chihuahua, Mexico, *in* Clark, K.F., and Goodell, P.C., eds., Geology and mineral resources of north-central Chihuahua: El Paso Geological Society, Field Conference Guidebook, p. 165–168.

McBride, E.F., 1989, Stratigraphy and sedimentary history of pre-Permian Paleozoic rocks of the Marathon uplift, *in* Hatcher, R.D., Jr., Thomas, W.A., and Viele, G.W., eds., The Appalachian-Ouachita orogen in the United States: Boulder, Colorado, Geological Society of America, The Geology of North America, v. F-2, chapter 20, p. 603–620.

McDowell, F.W., Housh, T.B., and Wark, D.A., 1999, Nature of the crust beneath west-central Chihuahua, Mexico, based upon Sr, Nd, and Pb isotopic compositions at the Tomóchic volcanic center: Geological Society of America Bulletin, v. 111, no. 6, p. 823–830, doi: 10.1130/0016-7606(1999)111<0823:NOTCBW>2.3.CO;2.

McEldowney, R.C., 1970, Geology of the northern Sierra Pinta, Baja California, Mexico [M.S. thesis]: San Diego, California, San Diego State University, 78 p.

McKee, J.W., and Jones, N.W., 1979, A large Mesozoic fault in Coahuila, Mexico: Geological Society of America Abstracts with Programs, v. 11, no. 7, p. 476.

McKee, J.W., Jones, N.W., and Long, L.E., 1984, History of recurrent activity along a major fault in northeastern Mexico: Geology, v. 12, no. 2, p. 103–107, doi: 10.1130/0091-7613(1984)12<103:HORAAA>2.0.CO;2.

McKee, J.W., Jones, N.W., and Anderson, T.H., 1988, Las Delicias basin–A record of late Paleozoic arc volcanism in northeastern Mexico: Geology, v. 16, no. 1, p. 37–40, doi: 10.1130/0091-7613(1988)016<0037:LDBARO>2.3.CO;2.

McKee, J.W., Jones, N.W., and Long, L.E., 1990, Stratigraphy and provenance of strata along the San Marcos fault, central Coahuila, Mexico: Geological Society of America Bulletin, v. 102, no. 5, p. 593–614, doi: 10.1130/0016-7606(1990)102<0593:SAPOSA>2.3.CO;2.

McKee, J.W., Jones, N.W., and Anderson, T.H., 1999, Late Paleozoic and early Mesozoic history of the Las Delicias terrane, Coahuila, Mexico, *in* Bartolini, C., Wilson, J.L., and Lawton, T.F., eds., Mesozoic sedimentary and tectonic history of north-central Mexico: Geological Society of America Special Paper 340, p. 161–189.

Meckel, L.D., Jr., Smith, D.G., and Wells, L.A., 1992, Ouachita foredeep basins: Regional paleogeography and habitat of hydrocarbons, *in* Macqueen, R.W., and Leckie, D.A., eds., Foreland basins and fold belts: American Association of Petroleum Geologists Memoir 55, chapter 15, p. 427–444.

Meert, J.G., and Torsvik, T.H., 2003, The making and unmaking of a supercontinent: Rodinia revisited: Tectonophysics, v. 375, nos. 1–4, Special issue, p. 261–288.

Menning, M., Weyer, D., Drozdzewski, G., van Amerom, H.W.J., and Wendt, I., 2000, A Carboniferous time scale 2000: discussion and use of geological parameters as time indicators from Central and Western Europe: Geologisches Jahrbuch, A 156, p. 3–44.

Mickus, K.L., and Keller, G.R., 1992, Lithospheric structure of the south-central United States: Geology, v. 20, no. 4, p. 335–338, doi: 10.1130/0091-7613(1992)020<0335:LSOTSC>2.3.CO;2.

Miller, R.H., and Dockum, M.S., 1983, Ordovician conodonts from metamorphosed carbonates of the Salton trough, California: Geology, v. 11, no. 7, p. 410–412, doi: 10.1130/0091-7613(1983)11<410:OCFMCO>2.0.CO;2.

Miser, H.D., 1917, Manganese deposits of the Caddo Gap and DeQueen quadrangles, Arkansas: U.S. Geological Survey Bulletin 660-C, p. 59–122.

Miser, H.D., 1929, Structure of the Ouachita Mountains of Oklahoma and Arkansas: Oklahoma Geological Survey Bulletin 50, 30 p.

Molina-Garza, R.S., and Geissman, J.W., 1999, Paleomagnetic data from the Caborca terrane, Mexico: Implications for Cordilleran tectonics and the Mojave-Sonora megashear hypothesis: Tectonics, v. 18, no. 2, p. 293–325, doi: 10.1029/1998TC900030.

Moreno, F.A., Montana, C.J., and Keller, G.R., 1992, Gravity anomalies in northern Mexico, *in* Clark, K.F., Roldán-Quintana, J., and Schmidt, R.H., eds., Geology and mineral resources of northern Sierra Madre Occidental, Mexico: Guidebook for the 1992 Field Conference, El Paso Geological Society, p. 203–209.

Morris, E.M., and Stone, C.G., 1986, A preliminary report on the metagabbros of the Ouachita core, *in* Stone, C.G., and Haley, B.R., eds., Sedimentary and igneous rocks of the Ouachita Mountains of Arkansas; A guidebook with contributed papers, part 2: Little Rock, Arkansas, Arkansas Geological Commission Guidebook 86-3, p. 87–90.

Morris, R.C., 1989, Stratigraphy and sedimentary history of post-Arkansas Novaculite Carboniferous rocks of the Ouachita Mountains, *in* Hatcher, R.D., Jr., Thomas, W.A., and Viele, G.W., eds., The Appalachian-Ouachita orogen in the United States: Boulder, Colorado, Geological Society of America, The Geology of North America, v. F-2, chapter 19, p. 591–602.

Mosher, S., 1998, Tectonic evolution of the southern Laurentian Grenville orogenic belt: Geological Society of America Bulletin, v. 110, no. 11, p. 1357–1375, doi: 10.1130/0016-7606(1998)110<1357:TEOTSL>2.3.CO;2.

Muehlberger, W.R., 1980, Texas lineament revisited, *in* Dickerson, P.W., Hoffer, J.M., and Callender, J.F., eds., Trans-Pecos region, southeastern New Mexico and West Texas: New Mexico Geological Society, Guidebook 31, p. 113–121.

Muehlberger, W.R., and Tauvers, P.R., 1989, Marathon fold-thrust belt, west Texas, *in* Hatcher, R.D., Jr., Thomas, W.A., and Viele, G.W., eds., The Appalachian-Ouachita orogen in the United States: Boulder, Colorado, Geological Society of America, The Geology of North America, v. F-2, chapter 24, p. 673–680.

Mullan, H.S., 1978, Evolution of part of the Nevadan orogen in northwestern Mexico: Geological Society of America Bulletin, v. 89, no. 8, p. 1175–1188, doi: 10.1130/0016-7606(1978)89<1175:EOPOTN>2.0.CO;2.

Nicholas, R.L., 1983, Devils River uplift, *in* Bolden, G.P., ed., Structure and stratigraphy of the Val Verde Basin–Devils River Uplift, Texas: West Texas Geological Society Publication 83–77, p. 125–137.

Nicholas, R.L., and Rozendal, R.A., 1975, Subsurface positive elements within Ouachita foldbelt in Texas and their relation to Paleozoic cratonic margin: American Association of Petroleum Geologists Bulletin, v. 59, no. 2, p. 193–216.

Nicholas, R.L., and Waddell, D.E., 1982, New Paleozoic subsurface data from the north-central Gulf Coast: Geological Society of America Abstracts with Programs, v. 14, no. 7, p. 576.

Nicholas, R.L., and Waddell, D.E., 1989, The Ouachita system in the subsurface of Texas, Arkansas, and Louisiana, *in* Hatcher, R.D., Jr., Thomas, W.A., and Viele, G.W., eds., The Appalachian-Ouachita orogen in the United States: Boulder, Colorado, Geological Society of America, The Geology of North America, v. F-2, chapter 23, p. 661–672.

Nielsen, K.C., Viele, G.W., and Zimmerman, J., 1989, Structural setting of the Benton–Broken Bow uplifts, *in* Hatcher, R.D., Jr., Thomas, W.A., and Viele, G.W., eds., The Appalachian-Ouachita orogen in the United States: Boulder, Colorado, Geological Society of America, The Geology of North America, v. F-2, chapter 22, p. 635–660.

Niem, A.R., 1977, Mississippian pyroclastic flow and ash-fall deposits in the deep-marine Ouachita flysch basin, Oklahoma and Arkansas: Geological Society of America Bulletin, v. 88, no. 1, p. 49–61, doi: 10.1130/0016-7606(1977)88<49:MPFAAD>2.0.CO;2.

Noble, P.J., 1990, Radiolarian biostratigraphy: Evidence for an Early Mississippian basin-wide hiatus at the Caballos–Tesnus boundary, Marathon basin, Texas, *in* Laroche, T.M., and Higgins, L., eds., Marathon thrust belt: Structure, stratigraphy, and hydrocarbon potential: Midland, Texas, West Texas Geological Society and Permian Basin Section, Society of Economic Paleontologists and Mineralogists Field Seminar, p. 83–91.

Noble, P.J., 1993, Paleoceanographic and tectonic implications of a regionally extensive Early Mississippian hiatus in the Ouachita system, southern mid-continental United States: Geology, v. 21, no. 4, p. 315–318, doi: 10.1130/0091-7613(1993)021<0315:PATIOA>2.3.CO;2.

Noble, P.J., 1994, Silurian radiolarian zonation for the Caballos Novaculite, Marathon uplift, west Texas: Bulletin of American Paleontology, v. 106, no. 345, 55 p.

Noll, J.H., Jr., 1981, Geology of the Picacho Colorado area, northern Sierra de Cobachi, central Sonora, Mexico [M.S. thesis]: Flagstaff, Arizona, Northern Arizona University, 165 p.

North American Magnetic Anomaly Group (NAMAG) [Geological Survey of Canada, United States Geological Survey, and Consejo de Recursos Minerales de México], 2002, Magnetic Anomaly Map of North America: U.S. Geological Survey Special Map, scale 1:10,000,000, accompanied by 31 p. text.

Nourse, J.A., 1995, Jurassic–Cretaceous paleogeography of the Magdalena region, northern Sonora, and its influence on the positioning of Tertiary metamorphic core complexes, *in* Jacques-Ayala, C., González-León, C.M., and Roldán-Quintana, J., eds., Studies on the Mesozoic of Sonora and adjacent areas: Geological Society of America Special Paper 301, p. 59–78.

Nourse, J.A., Premo, W.R., and Iriondo, A., 2003, Gneisses of the NW Sonora–SW Arizona border region: Two distinct Paleoproterozoic complexes at the southwest edge of Laurentia: Geological Society of America Abstracts with Programs, v. 35, no. 4, p. 67.

Nyman, M.W., Karlstrom, K.E., Kirby, E., and Graubard, C.M., 1994, Mesoproterozoic contractional orogeny in western North America: Evidence from ca. 1.4 Ga plutons: Geology, v. 22, no. 10, p. 901–904, doi: 10.1130/0091-7613(1994)022<0901:MCOIWN>2.3.CO;2.

Olmsted, F.H., Loeltz, O.J., and Irelan, B., 1973, Geohydrology of the Yuma area, Arizona and California: U.S. Geological Survey Professional Paper 486-H, 227 p.

Ortega-Gutiérrez, F., Mitre-Salazar, L.M., Roldán-Quintana, J., Aranda-Gómez, J., Moran-Zenteno, D., Alaniz-Alvarez, S., y Nieto-Samaniego, A., 1992, Carta geológica de la República Mexicana: Consejo de Recursos Minerales e Instituto de Geología, escala 1:2,000,000.

Ortega-Gutiérrez, F., Sedlock, R.L., and Speed, R.C., 1994, Phanerozoic tectonic evolution of Mexico, *in* Speed, R.C., ed., Phanerozoic evolution of North American continent-ocean transitions: Boulder, Colorado, Geological Society of America, Geology of North America, Continent-Ocean Transects, chapter 5, p. 265–306.

Page, W.R., Gray, F., Blodgett, R.M., Maldonado, F., Miller, R.J., and Jaime-Ruiz, R., 2003a, Mesozoic-Cenozoic tectonic evolution of the northern Sierra Los Ajos, Sonora, Mexico: Geological Society of America Abstracts with Programs, v. 35, no. 4, p. 70.

Page, W.R., Harris, A.G., Poole, F.G., and Repetski, J.E., 2003b, Reinterpretation of the stratigraphy and structure of the Rancho Las Norias area, central Sonora, Mexico: Journal of South American Earth Sciences, v. 16, p. 523–540, doi: 10.1016/S0895-9811(03)00135-4.

Palmer, A.R., DeMis, W.D., Muehlberger, W.R., and Robison, R.A., 1984, Geological implications of Middle Cambrian boulders from the Haymond Formation (Pennsylvanian) in the Marathon basin, west Texas: Geology, v. 12, no. 2, p. 91–94, doi: 10.1130/0091-7613(1984)12<91:GIOMCB>2.0.CO;2.

Peiffer-Rangin, F., Echávarri-Pérez, A., Salas-Pizá, G., and Rangin, C., 1980, Sur la présence de l'Ordovicien à graptolites dans le nord-ouest de Mexique: C.R. Academy of Science, Paris: Series D, v. 290, p. 13–16.

Perry, W.J., Jr., 1989, Tectonic evolution of the Anadarko basin region, Oklahoma: U.S. Geological Survey Bulletin 1866-A, 19 p.

Phillips, J.R., 1984, "Middle" Cretaceous metasedimentary rocks of La Olvidada, northeastern Baja California, Mexico, *in* Frizzell, V.A., Jr., ed., Geology of the Baja California peninsula: Los Angeles, California, Pacific Section, Society of Economic Paleontologists and Mineralogists, v. 39, p. 37–41.

Phillips, R.P., 1966, Reconnaissance geology of some of the northwestern islands in the Gulf of California: Geological Society of America Special Paper 101, Abstracts for 1966, p. 327.

Pilkington, M., Grieve, R.A.F., Rupert, J.D., and Halpenny, J.F., 1992a, Gravity anomaly map with shaded relief of gradient of North America: Geological Survey of Canada, Map 1807A, scale 1:10,000,000.

Pilkington, M., Grieve, R.A.F., Rupert, J.D., and Halpenny, J.F., 1992b, Isostatic gravity anomaly map of North America: Geological Survey of Canada, Map 1808A, scale 1:10,000,000.

Pindell, J.L., 1985, Alleghenian reconstruction and subsequent evolution of the Gulf of Mexico, Bahamas, and proto-Caribbean: Tectonics, v. 4, no. 1, p. 1–39.

Poole, F.G., and Madrid, R.J., 1988, Comparison of allochthonous Paleozoic eugeoclinal rocks in the Sonoran, Marathon, and Antler orogens: Geological Society of America Abstracts with Programs, v. 20, no. 7, p. A267.

Poole, F.G., and Perry, W.J., Jr., 1997, The late Paleozoic Ouachita-Marathon-Sonora orogenic system along the southern margin of North America: Geological Society of America Abstracts with Programs, v. 29, no. 2, p. 43.

Poole, F.G., and Perry, W.J., Jr., 1998, Laurentia–Gondwana continental margins in northern Mexico, and their late Paleozoic collision, *in* Program and Abstracts for Conference on Laurentia–Gondwana connections before Pangea (IGCP Project 376): Oaxaca City, Oaxaca, Mexico, October 1998, p. 27.

Poole, F.G., and Sandberg, C.A., 1991, Mississippian paleogeography and conodont biostratigraphy of the Western United States, *in* Cooper, J.D., and Stevens, C.H., eds., Paleozoic paleogeography of the Western United States—II: Los Angeles, California, Society of Economic Paleontologists and Mineralogists, Pacific Section, book 67, p. 107–136.

Poole, F.G., Sandberg, C.A., and Boucot, A.J., 1977, Silurian and Devonian paleogeography of the Western United States, *in* Stewart, J.H., Stevens, C.H., and Fritsche, A.E., eds., Paleozoic paleogeography of the Western United States: Los Angeles, California, Society of Economic Paleontologists and Mineralogists, Pacific Section, Pacific Coast Paleogeography Symposium 1, p. 39–65.

Poole, F.G., Madrid, R.J., and Oliva-Becerril, J.F., 1991, Geological setting and origin of stratiform barite in central Sonora, Mexico, *in* Raines, G.L., Lisle, R.E., Schafer, R.W., and Wilkinson, W.H., eds., Geology and ore deposits of the Great Basin: Reno, Nevada, Geological Society of Nevada Symposium Proceedings, v. I, p. 517–522.

Poole, F.G., Berry, W.B.N., and Madrid, R.J., 1993, Allochthonous Ordovician eugeoclinal rocks on Turner Island, eastern Gulf of California, and their paleotectonic significance: Geological Society of America Abstracts with Programs, v. 25, no. 5, p. 134–135.

Poole, F.G., Stewart, J.H., Berry, W.B.N., Harris, A.G., Repetski, J.E., Madrid, R.J., Ketner, K.B., Carter, C., and Morales-Ramirez, J.M., 1995a, Ordovician ocean-basin rocks of Sonora, Mexico, *in* Cooper, J.D., Droser, M.L., and Finney, S.C., eds., Ordovician Odyssey: Short Papers for the Seventh International Symposium on the Ordovician System: Fullerton, California, Pacific Section Society for Sedimentary Geology (SEPM), book no. 77, p. 277–284.

Poole, F.G., Stewart, J.H., Repetski, J.E., Harris, A.G., Ross, R.J., Jr., and Ketner, K.B., Amaya-Martínez, R., and Morales-Ramirez, J.M., 1995b, Ordovician carbonate-shelf rocks of Sonora, Mexico, *in* Cooper, J.D., Droser, M.L., and Finney, S.C., eds., Ordovician Odyssey: Short Papers for the Seventh International Symposium on the Ordovician System: Fullerton, California, Pacific Section Society for Sedimentary Geology (SEPM), book no. 77, p. 267–275.

Poole, F.G., Page, W.R., and Amaya-Martínez, R., 1997, Newly discovered Silurian carbonate-shelf rocks in west-central Sonora, Mexico: Geological Society of America Abstracts with Programs, v. 29, no. 6, p. 438.

Poole, F.G., Page, W.R., and Amaya-Martínez, R., 1998, Rocas Silúricas de plataforma en Sonora: Mundo Minero (México), Enero 1998, año 2, núm. 18, p. 6, 9, and 24.

Poole, F.G., Amaya-Martínez, R., and Page, W.R., 2000, Silurian and Devonian carbonate-shelf rocks and Lower Jurassic sequence near Rancho Placeritos, west-central Sonora: Field guide for Field Trip 2 (March 4, 2000), Fourth symposium on the geology of northwest Mexico and adjacent areas (March 6–8, 2000), Center for the Arts, University of Sonora, Hermosillo, Mexico, 24 p.

Poole, F.G., Boucot, A.J., Amaya-Martínez, R., Sandberg, C.A., Harris, A.G., and Page, W.R., 2003a, Early Devonian brachiopods in west-central Sonora indicate depositional continuity along southern margin of Laurentia: Geological Society of America Abstracts with Programs, v. 35, no. 4, p. 13–14.

Poole, F.G., Amaya-Martínez, R., and Harris, A.G., 2003b, Early Mississippian North American carbonate-shelf margin in west-central Sonora, Mexico: Geological Society of America Abstracts with Programs, v. 35, no. 4, p. 14.

Premo, W.R., Iriondo, A., and Nourse, J.A., 2003, U-Pb geochronology of Paleoproterozoic basement in northwestern Sonora, Mexico: Evidence for affinity to SW US provinces: Geological Society of America Abstracts with Programs, v. 35, no. 4, p. 67.

Radelli, L., Menicucci, S., Mesnier, H.Ph., Araux-Sánchez, E., Calmus, T., Amaya-Martínez, R., Barrera-Moreno, E., Domínguez-Perla, J.E., Navarro-Martínez, L.A., and Soto-Contreras, L.A., 1987, Allochthonous Paleozoic bodies of central Sonora: Universidad de Sonora, Boletín del Departamento de Geología, Boletín, Segunda Epoca, v. 4, nos. 1 and 2, p. 1–15.

Ramírez-M., J.C., and Acevedo-C., F., 1957, Notas sobre la geología de Chihuahua: Boletín de la Asociación Mexicana de Geólogos Petroleros, v. 9, nos. 9 y 10, p. 583–770.

Riva, J.F., and Ketner, K.B., 1989, Ordovician graptolites from the northern Sierra de Cobachi, Sonora, Mexico: Transactions of the Royal Society of Edinburgh: Earth Sciences, v. 80, p. 71–90.

Roberts, D.C., 1989, Sedimentation and tectonics of the Cerro Los Panales area, east-central Chihuahua, Mexico [M.S. thesis]: El Paso, Texas, University of Texas at El Paso, 188 p.

Roberts, M.T., 1994, Geologic relations along a regional cross section from Spavinaw to Broken Bow, eastern Oklahoma, *in* Suneson, N.H., and Hemish, L.A., eds., Geology and resources of the eastern Ouachita Mountains frontal belt and southeastern Arkoma basin, Oklahoma: Oklahoma Geological Survey Guidebook 29, p. 137–159.

Rogers, C.L., de Cserna, Z., Van Vloten, R., Tavera-Amezcua, E., and Ojeda-Rivera, J., 1961, Reconocimiento geológico y depósitos de fosfatos del norte de Zacatecas y áreas adyacentes en Coahuila, Nuevo León y San Luis Potosí: Consejo de Recursos Naturales no Renovables, Boletín 56, 322 p.

Roldán-Quintana, J.S.F., 2002, Caracterización geológico-geoquímica y evolución del arco magmático Mesozoico-Terciario entre San Carlos y Maycoba, sur de Sonora [D.C. tesis]: Ciudad México, Universidad Nacional Autónoma de México, 185 p.

Roldán-Quintana, J., González-León, C., and Amaya-Martínez, R., 1993, Geologic constraints on the northern limit of the Guerrero terrane in northwestern Mexico, *in* Ortega-Gutiérrez, F., Coney, P.J., Centeno-García, E., and Gómez-Caballero, A., eds., Proceedings of the First circum-Pacific and circum-Atlantic terrane conference (Guanajuato, México): Universidad Nacional Autónoma de México and Instituto de Geología, p. 124–127.

Ross, C.A., 1967a, Stratigraphy and depositional history of the Gaptank Formation (Pennsylvanian), west Texas: Geological Society of America Bulletin, v. 78, no. 3, p. 369–384.

Ross, C.A., 1967b, Development of fusulinid (Foraminiferida) faunal realms: Journal of Paleontology, v. 41, no. 6, p. 1341–1354.

Ross, C.A., 1973, Pennsylvanian and Early Permian depositional history, southeastern Arizona: American Association of Petroleum Geologists Bulletin, v. 57, no. 5, p. 887–912.

Ross, C.A., 1978, Late Pennsylvanian and Early Permian sedimentary rocks and tectonic setting of the Marathon geosyncline, *in* Mazzullo, S.J., ed., Tectonics and Paleozoic facies of the Marathon geosyncline, West Texas: Permian Basin Section, Society of Economic Paleontologists and Mineralogists Publication 78-17, p. 89–93.

Ross, C.A., 1979, Late Paleozoic collision of North and South America: Geology, v. 7, no. 1, p. 41–44, doi: 10.1130/0091-7613(1979)7<41:LPCONA>2.0.CO;2.

Ross, C.A., 1986, Paleozoic evolution of southern margin of Permian basin: Geological Society of America Bulletin, v. 97, no. 5, p. 536–554, doi: 10.1130/0016-7606(1986)97<536:PEOSMO>2.0.CO;2.

Ross, C.A., and Ross, J.R.P., 1985, Paleozoic tectonics and sedimentation in West Texas, southern New Mexico, and southern Arizona, *in* Dickerson, P.W., and Muehlberger, W.R., eds., Structure and tectonics of Trans-Pecos Texas: West Texas Geological Society Publication 85–81, p. 221–230.

Ross, C.A., and Ross, J.R.P., 1986, Paleozoic paleotectonics and sedimentation in Arizona and New Mexico, *in* Peterson J.A., ed., Paleotectonics and sedimentation in the Rocky Mountain region, United States: American Association of Petroleum Geologists Memoir 41, p. 653–668.

Ross, C.A., and Ross, J.R.P., 1995, Permian sequence stratigraphy, *in* Scholle, P.A., Peryt, T.M., and Ulmer-Scholle, D.S., eds., The Permian of northern Pangea: Springer-Verlag, Germany, Paleogeography, paleoclimates, stratigraphy, v. 1, p. 98–123.

Ross, D.C., 1966, Stratigraphy of some Paleozoic formations in the Independence quadrangle, Inyo County, California: U.S. Geological Survey Professional Paper 396, 64 p.

Ruiz, J., Patchett, P.J., and Ortega-Gutiérrez, F., 1988, Proterozoic and Phanerozoic basement terranes of Mexico from Nd isotopic studies: Geological Society of America Bulletin, v. 100, no. 2, p. 274–281, doi: 10.1130/0016-7606(1988)100<0274:PAPBTO>2.3.CO;2.

Russo, R.M., and Speed, R.C., 1992, Oblique collision and tectonic wedging of the South American continent and Caribbean terranes: Geology, v. 20, no. 5, p. 447–450, doi: 10.1130/0091-7613(1992)020<0447:OCATWO>2.3.CO;2.

Salas, G.A., 1968, Areal geology and petrology of the igneous rocks of the Santa Ana region, northwest Sonora: Boletín de la Sociedad Geológica Mexicana, v. 31, núm. 1, p. 11–63.

Sánchez-Zavala, J.L., Centeno-García, E., and Ortega-Gutiérrez, F., 1999, Review of Paleozoic stratigraphy of México and its role in the Gondwana–Laurentia connections, *in* Ramos, V.A., and Keppie, J.D., eds., Laurentia–Gondwana connections before Pangea: Geological Society of America Special Paper 336, p. 211–226.

Sandberg, C.A., and Ziegler, W., 1996, Devonian conodont biochronology in geologic time calibration: Senckenbergiana lethaea 76, p. 259–265; Frankfurt/Main, Germany.

Sando, W.J., 1985, Revised Mississippian time scale, Western Interior Region, conterminous United States: U.S. Geological Survey Bulletin 1605-A, p. A15–A26.

Schmidt, G.T., 1978, Geology of the northern Sierra El Encinals, Sonora, Mexico [M.S. thesis]: Flagstaff, Arizona, Northern Arizona University, 80 p.

Schreiber, J.F., Jr., Armin, R.A., Connolly, W.M., Stanton, R.J., Jr., Armstrong, A.K., Lyons, T.W., and Wrucke, C.T., 1990, Upper Paleozoic stratigraphy of the Whetstone Mountains, Cochise and Pima Counties, Arizona, *in* Gehrels, G.E., and Spencer, J.E., eds., Geologic excursions through the Sonoran Desert region, Arizona and Sonora: Tucson, Arizona, Arizona Geological Survey Special Paper 7, p. 104–113.

Schwab, F.L., 1986, Sedimentary 'signatures' of foreland basin assemblages: real or counterfeit? *in* Allen, P.A., and Homewood, P., eds., Foreland basins: International Association of Sedimentologists Special Publication Number 8, p. 395–410.

Scotese, C.R., 1997, Paleogeographic atlas, PALEOMAP Progress Report 90-0497: University of Texas at Arlington, Arlington, Texas, 45 p.

Sedlock, R.L., Ortega-Gutiérrez, F., and Speed, R.C., 1993, Tectonostratigraphic terranes and tectonic evolution of Mexico: Geological Society of America Special Paper 278, 153 p.

Shannon, W.M., Barnes, C.G., and Bickford, M.E., 1997, Grenville magmatism in West Texas: Petrology and geochemistry of the Red Bluff granitic suite: Journal of Petrology, v. 38, no. 10, p. 1279–1305, doi: 10.1093/petrology 38.10.1279.

Sheehan, P.M., 1975a, Upper Ordovician and Silurian brachiopods from the Solis Limestone, Chihuahua, Mexico: Journal of Paleontology, v. 49, no. 1, p. 200–211.

Sheehan, P.M., 1975b, Lower Devonian brachiopods from the Solis Limestone, Chihuahua, Mexico: Journal of Paleontology, v. 49, no. 3, p. 445–471.

Shurbet, D.H., and Cebull, S.E., 1987, Tectonic interpretation of the westernmost part of the Ouachita–Marathon (Hercynian) orogenic belt, west Texas–Mexico: Geology, v. 15, no. 5, p. 458–461, doi: 10.1130/0091-7613(1987)15<458:TIOTWP>2.0.CO;2.

Silver, L.T., 1971, Problems of crystalline rocks of the Transverse Ranges: Geological Society of America Abstracts with Programs, v. 3, no. 2, p. 193–194.

Silver, L.T., and Anderson, T.H., 1974, Possible left-lateral early to middle Mesozoic disruption of the southwestern North America craton margin: Geological Society of America Abstracts with Programs, v. 6, no. 7, p. 955–956.

Speed, R.C., 1994, North American continent-ocean transitions over Phanerozoic time, *in* Speed, R.C., ed., Phanerozoic evolution of North American continent-ocean transitions: Boulder, Colorado, Geological Society of America, Geology of North America Continent-Ocean Transects, chapter 1, p. 1–86.

Speed, R.C., Smith-Horowitz, P.L., Perch-Nielsen, K.v.S., Saunders, J.B., and Sanfilippo, A.B., 1993, Southern Lesser Antilles arc platform: Pre-late Miocene stratigraphy, structure, and tectonic evolution: Geological Society of America Special Paper 277, 98 p.

Stanley, G.D., Jr., and González-León, C.M., 1995, Paleogeographic and tectonic implications of Triassic fossils and strata from the Antimonio Formation, northwestern Sonora, *in* Jacques-Ayala, C., González-León, C.M., and Roldán-Quintana, J., eds., Studies on the Mesozoic of Sonora and adjacent areas: Geological Society of America Special Paper 301, p. 1–16.

Stevens, C.H., 1986, Evolution of the Ordovician through Middle Pennsylvanian carbonate shelf in east-central California: Geological Society of America Bulletin, v. 97, no. 1, p. 11–25, doi: 10.1130/0016-7606(1986)97<11:EOTOTM>2.0.CO;2.

Stevens, C.H., Stone, P., and Kistler, R.W., 1992, A speculative reconstruction of the middle Paleozoic continental margin of southwestern North America: Tectonics, v. 11, no. 2, p. 405–419.

Stevens, C.H., Klingman, D.S., Sandberg, C.A., Stone, P., Belasky, P., Poole, F.G., and Snow, J.K., 1996, Mississippian stratigraphic framework of east-central California and southern Nevada with revision of Upper Devonian and Mississippian stratigraphic units in Inyo County, California: U.S. Geological Survey Bulletin 1988-J, 39 p.

Stewart, J.H., 1970, Upper Precambrian and Lower Cambrian strata in the southern Great Basin, California and Nevada: U.S. Geological Survey Professional Paper 620, 206 p.

Stewart, J.H., 1988, Latest Proterozoic and Paleozoic southern margin of North America and the accretion of Mexico: Geology, v. 16, no. 2, p. 186–189, doi: 10.1130/0091-7613(1988)016<0186:LPAPSM>2.3.CO;2.

Stewart, J.H., and Poole, F.G., 1975, Extension of the Cordilleran miogeosynclinal belt to the San Andreas fault, southern California: Geological Society of America Bulletin, v. 86, no. 2, p. 205–212, doi: 10.1130/0016-7606(1975)86<205:EOTCMB>2.0.CO;2.

Stewart, J.H., and Poole, F.G., 2002, Inventory of Neoproterozoic and Paleozoic strata in Sonora, Mexico: U.S. Geological Survey Open-File Report 02-97, 50 p.

Stewart, J.H., and Roldán-Quintana, J., 1991, Upper Triassic Barranca Group; nonmarine and shallow-marine rift-basin deposits of northwestern Mexico, *in* Pérez-Segura, E., and Jacques-Ayala, C., eds., Studies of Sonoran geology: Geological Society of America Special Paper 254, p. 19–36.

Stewart, J.H., McMenamin, M.A.S., and Morales-Ramírez, J.M., 1984, Upper Proterozoic and Cambrian rocks in the Caborca region, Sonora, Mexico––physical stratigraphy, biostratigraphy, paleocurrent studies, and regional relations: U.S. Geological Survey Professional Paper 1309, 36 p.

Stewart, J.H., Poole, F.G., Ketner, K.B., Madrid, R.J., Roldán-Quintana, J., and Amaya-Martínez, R., 1990, Tectonics and stratigraphy of the Paleozoic and Triassic southern margin of North America, Sonora, Mexico, *in* Gehrels, G.E., and Spencer, J.E., eds., Geologic excursions through the Sonoran Desert region, Arizona and Sonora: Tucson, Arizona, Arizona Geological Survey Special Paper 7, p. 183–202.

Stewart, J.H., Amaya-Martínez, R., Stamm, R.G., Wardlaw, B.R., Stanley, G.D., Jr., and Stevens, C.H., 1997, Stratigraphy and regional significance of Mississippian to Jurassic rocks in Sierra Santa Teresa, Sonora, Mexico: Universidad Nacional Autónoma de México, Instituto de Geología, Revista Mexicana de Ciencias Geológicas, v. 14, núm. 2, p. 115–135.

Stewart, J.H., Blodgett, R.B., Boucot, A.J., Carter, J.L., and López, R., 1999a, Exotic Paleozoic strata of Gondwanan provenance near Ciudad Victoria, Tamaulipas, México, *in* Ramos, V.A., and Keppie, J.D., eds., Laurentia–Gondwana connections before Pangea: Geological Society of America Special Paper 336, p. 227–252.

Stewart, J.H., Poole, F.G., Harris, A.G., Repetski, J.E., Wardlaw, B.R., Mamet, B.L., and Morales-Ramírez, J.M., 1999b, Neoproterozoic(?) to Pennsylvanian inner-shelf, miogeoclinal strata in Sierra Agua Verde, Sonora, Mexico: Universidad Nacional Autónoma de México, Instituto de Geología, y Sociedad Geológica Mexicana, Revista Mexicana de Ciencias Geológicas, v. 16, núm. 1, p. 35–62.

Stewart, J.H., Gehrels, G.E., Barth, A.P., Link, P.K., Christie-Blick, N., and Wrucke, C.T., 2001, Detrital zircon provenance of Mesoproterozoic to Cambrian arenites in the western United States and northwestern Mexico: Geological Society of America Bulletin, v. 113, no. 10, p. 1343–1356, doi: 10.1130/0016-7606(2001)113<1343:DZPOMT>2.0.CO;2.

Stewart, J.H., Amaya-Martínez, R., and Palmer, A.R., 2002, Neoproterozoic and Cambrian strata of Sonora, Mexico: Rodinian supercontinent to Laurentian Cordilleran margin, *in* Barth, A., ed., Contributions to crustal evolution of the southwestern United States: Geological Society of America Special Paper 365, p. 5–48.

Stone, P., Dunne, G.C., Stevens, C.H., and Gulliver, R.M., 1989, Geologic map of Paleozoic and Mesozoic rocks in parts of the Darwin and adjacent quadrangles, Inyo County, California: U.S. Geological Survey Miscellaneous Investigation Series Map I-1932, scale 1:31,250.

Stone, P., Stevens, C.H., Spinosa, C., Furnish, W.M., Glenister, B.F., and Wardlaw, B.R., 2000, Stratigraphic relations and tectonic significance of rocks near the Permian–Triassic boundary, southern Inyo Mountains, California: Boulder, Colorado, Geological Society of America Map and Chart Series MCH086, 32 p.

Suhm, R.W., and Ethington, R.L., 1975, Stratigraphy and conodonts of Simpson Group (Middle Ordovician), Beach and Baylor Mountains, West Texas: American Association of Petroleum Geologists Bulletin, v. 39, no. 7, p. 1126–1135.

Sydnor, R.H., 1975, Geology of the northeast border zone of the San Jacinto pluton, Palm Springs, California [M.S. thesis]: Riverside, California, University of California, 121 p.

Thomas, W.A., 1976, Evolution of Ouachita–Applachian continental margin: Journal of Geology, v. 84, no. 3, p. 323–342.

Thomas, W.A., 1977, Evolution of Appalachian-Ouachita salients and recesses from reentrants and promontories in the continental margin: American Journal of Science, v. 277, no. 10, p. 1233–1278.

Thomas, W.A., 1983, Continental margins, orogenic belts, and intracratonic structures: Geology, v. 11, no. 5, p. 270–272, doi: 10.1130/0091-7613(1983)11<270:CMOBAI>2.0.CO;2.

Thomas, W.A., 1988, The Black Warrior basin, *in* Sloss, L.L., ed., Sedimentary cover—North American craton: U.S.: Boulder, Colorado, Geological Society of America, The Geology of North America, v. D-2, chapter 16, p. 471–492.

Thomas, W.A., 1989, The Appalachian-Ouachita orogen beneath the Gulf Coastal Plain between the outcrops in the Appalachian and Ouachita Mountains, *in* Hatcher, R.D., Jr., Thomas, W.A., and Viele, G.W., eds., The Appalachian-Ouachita orogen in the United States: Boulder, Colorado, Geological Society of America, The Geology of North America, v. F- 2, chapter 15, p. 537–553.

Thomas, W.A., 1991, The Appalachian-Ouachita rifted margin of southeastern North America: Geological Society of America Bulletin, v. 103, no. 3, p. 415–431, doi: 10.1130/0016-7606(1991)103<0415:TAORMO>2.3.CO;2.

Thomas, W.A., and Astini, R.A., 1999, Simple-shear conjugate rift margins of the Argentine Precordillera and the Ouachita embayment of Laurentia: Geological Society of America Bulletin, v. 111, no. 7, p. 1069–1079, doi: 10.1130/0016-7606(1999)111<1069:SSCRMO>2.3.CO;2.

Thomas, W.A., Astini, R.A., and Denison, R.E., 2001, Strontium isotopes, age, tectonic setting of Cambrian salinas along the rift and transform margins of the Argentine Precordillera and southern Laurentia: Journal of Geology, v. 109, no. 2, p. 231–246, doi: 10.1086/319241.

Thomas, W.A., Astini, R.A., and Bayona, G., 2002, Ordovician collision of the Argentine Precordillera with Gondwana, independent of Laurentian Taconic orogeny: Tectonophysics, v. 345, p. 131–152, doi: 10.1016/S0040-1951(01)00210-4.

Thompson, S., III, Tovar-R., J.C., and Conley, J.N., 1978, Oil and gas exploration wells in the Pedregosa basin, *in* Callender, J.F., Wilt, J.C., and R.E. Clemons, eds., Land of Cochise, southeastern Arizona: New Mexico Geological Society Guidebook, 29th Field Conference, p. 331–342.

Thomson, A., and McBride, E.F., 1964, Summary of the geologic history of the Marathon geosyncline: Permian Basin Section, Society of Economic Paleontologists and Mineralogists, Field Trip Symposium and Guidebook, Publication 64-9, p. 52–60.

Timmons, J.M., Karlstrom, K.E., Dehler, C.M., Geissman, J.W., and Heizler, M.T., 2001, Proterozoic multistage (ca. 1.1 and 0.8 Ga) extension recorded in the Grand Canyon Supergroup and establishment of northwest- and north-trending tectonic grains in the southwestern United States: Geological Society of America Bulletin, v. 113, no. 2, p. 163–180, doi: 10.1130/0016-7606(2001)113<0163:PMCAGE>2.0.CO;2.

Todd, V.R., Erskine, B.G., and Morton, D.M., 1988, Metamorphic and tectonic evolution of the northern Peninsular Ranges batholith, southern California, *in* Ernst, W.G., ed., Metamorphism and crustal evolution of the Western United States: Prentice- Hall, Englewood Cliffs, New Jersey, Rubey Volume VII, chapter 32, p. 894–937.

Torres-Espinoza, J.A., Salinas-Hinojosa, C.R., and López-Flores, A., 1987, Estratigrafía preliminar del Paleozoico en las áreas de la Vinata y Sierra Azcárate, Noroeste de Chihuahua, México, *in* Paleozoico de Chihuahua (Excursión Geológica No. 2, Libreto Guía), Gaceta Geológica, v. 1, no. 1, p. 203–217.

Torrini, R., Jr., and Speed, R.C., 1989, Tectonic wedging in the forearc basin—accretionary prism transition, Lesser Antilles forearc: Journal of Geophysical Research, v. 94, no. B8, p. 10549–10584.

Tucker, R.D., Bradley, D.C., Ver Straeten, C.A., Harris, A.G., Ebert, J.R., and McCutcheon, S.R., 1998, New U-Pb zircon ages and the duration and division of Devonian time: Earth and Planetary Science Letters, v. 158, p. 175–186, doi: 10.1016/S0012-821X(98)00050-8.

U.S. Geological Survey, 1977, Geographic map of the circum-Pacific region, northeast quadrant: Circum-Pacific map project, American Association of Petroleum Geologists, scale 1:10,000,000.

Valencia-Moreno, M., Ruiz, J., Barton, M.D., Patchett, P.J., Zürcher, L., Hodkinson, D.G., and Roldán-Quintana, J., 2001, A chemical and isotopic study of the Laramide granitic belt of northwestern Mexico: Identification of the southern edge of the North American Precambrian basement: Geological Society of America Bulletin, v. 113, no. 11, p. 1409–1422, doi: 10.1130/0016-7606(2001)113<1409:ACAISO>2.0.CO;2.

Van der Voo, R., Peinado, J., and Scotese, C.R., 1984, A paleomagnetic reevaluation of Pangea reconstructions, *in* Van der Voo, R., Scotese, C.R., and Bonhommet, N., eds., Plate reconstruction from Paleozoic paleomagnetism: American Geophysical Union Geodynamics Series, v. 12, p. 11–26.

Van Schmus, W.R., and Bickford, M.E., co-editors, and 23 authors, 1993, Transcontinental Proterozoic provinces, *in* Reed, J.C., Jr., Bickford, M.E., Houston, R.S., Link, P.K., Rankin, D.W., Sims, P.K., and Van Schmus, W.R., eds., Precambrian: Conterminous U.S.: Boulder, Colorado, Geological Society of America, The Geology of North America, v. C-2, chapter 4, p. 171–334.

Viele, G.W., 1989, The Ouachita orogenic belt, *in* Hatcher, R.D., Jr., Thomas, W.A., and Viele, G.W., eds., The Appalachian-Ouachita orogen in the United States: Boulder, Colorado, Geological Society of America, The Geology of North America, v. F-2, chapter 16, p. 555–561.

Viele, G.W., and Thomas, W.A., 1989, Tectonic synthesis of the Ouachita orogenic belt, *in* Hatcher, R.D., Jr., Thomas, W.A., and Viele, G.W., eds., The Appalachian-Ouachita orogen in the United States: Boulder, Colorado,

Geological Society of America, The Geology of North America, v. F-2, chapter 27, p. 695–728.

Walker, N., 1992, Middle Proterozoic geologic evolution of Llano uplift, Texas: Evidence from U-Pb zircon geochronometry: Geological Society of America Bulletin, v. 104, no. 4, p. 494–504, doi: 10.1130/0016-7606(1992)104<0494:MPGEOL>2.3.CO;2.

Walper, J.L., 1977, Paleozoic tectonics of the southern margin of North America: Transactions of the Gulf Coast Association of Geological Societies, v. 27, p. 230–241.

Wardlaw, B.R., Furnish, W.M., and Nestell, M.K., 1979, Geology and paleontology of the Permian beds near Las Delicias, Coahuila, Mexico: Geological Society of America Bulletin, v. 90, no. 1, p. 111–116, doi: 10.1130/0016-7606(1979)90<111:GAPOTP>2.0.CO;2.

Wasserburg, G.J., and Lanphere, M.A., 1965, Age determinations in the Precambrian of Arizona and Nevada: Geological Society of America Bulletin, v. 76, no. 7, p. 735–758.

Wasserburg, G.J., Wetherill, G.W., Silver, L.T., and Flawn, P.T., 1962, A study of the ages of the Precambrian of Texas: Journal of Geophysical Research, v. 67, p. 4021–4047.

Webster, G.D., and Delattre, M.P., 1993, Permian crinoid columnals from the Zamora Formation near El Volcan, Baja California Norte, Mexico, *in* Gastil, R.G., and Miller, R.H., eds., The prebatholithic stratigraphy of peninsular California: Geological Society of America Special Paper 279, p. 91–95.

Webster, G.D., Gastil, G., and Delattre, M., 1985, World's largest crinoid columnals from Baja California Norte: Geological Society of America Abstracts with Programs, v. 17, no. 6, p. 417.

Whiting, B.M., and Thomas, W.A., 1994, Three-dimensional controls on subsidence of a foreland basin associated with a thrust-belt recess: Black Warrior basin, Alabama and Mississippi: Geology, v. 22, no. 8, p. 727–730, doi: 10.1130/0091-7613(1994)022<0727:TDCOSO>2.3.CO;2.

Wickham, J., Roeder, D., and Briggs, G., 1976, Plate tectonics models for the Ouachita foldbelt: Geology, v. 4, no. 3, p. 173–176, doi: 10.1130/0091-7613(1976)4<173:PTMFTO>2.0.CO;2.

Wilson, J.L., 1954, Ordovician stratigraphy in the Marathon folded belt, West Texas: American Association of Petroleum Geologists Bulletin, v. 38, no. 12, p. 2455–2475.

Wilson, J.L., 1987, The late Paleozoic geologic history of southern New Mexico and Chihuahua, *in* Paleozoico de Chihuahua (Excursión Geológica No. 2, Libreto Guía), Gaceta Geológica, v. 1, no. 1, p. 36–53.

Wilson, J.L., Ward, W.C., and Finneran, J.M., 1984, Upper Jurassic and Lower Cretaceous platform and basin systems, Monterrey-Saltillo area, northeast Mexico: Gulf Coast Section, Society of Economic Paleontologists and Mineralogists, Field Guide, 76 p.

Wooden, J.L., and DeWitt, E., 1991, Pb isotopic evidence for the boundary between the Early Proterozoic Mojave and Central Arizona crustal provinces in western Arizona, *in* Karlstrom, K.E., ed., Proterozoic geology and ore deposits of Arizona: Arizona Geological Society Digest 19, p. 27–50.

Wooden, J.L., Stacey, J.S., Howard, K.A., Doe, B.R., and Miller, D.M., 1988, Pb isotopic evidence for the formation of Proterozoic crust in the southwestern United States, *in* Ernst, W.G., ed., Metamorphism and crustal evolution of the Western United States: Prentice-Hall, Englewood Cliffs, New Jersey, Rubey Volume VII, chapter 3, p. 69–86.

Wrucke, C.T., and Armstrong, A.K., 1987, Paleozoic stratigraphic section in Dry Canyon, Whetstone Mountains, Cochise County, Arizona, *in* Hill, M.L., ed., Cordilleran Section: Geological Society of America Centennial Field Guide Volume 1, p. 29–34.

Wuellner, D.E., Lehtonen, L.R., and James, W.C., 1986, Sedimentary-tectonic development of the Marathon and Val Verde basins, West Texas, U.S.A.: A Permo–Carboniferous migrating foredeep, *in* Allen, P.A., and Homewood, P., eds., Foreland basins: International Association of Sedimentologists Special Publication Number 8, p. 347–368.

Yañez, P., Ruiz, J., Patchett, P.J., Ortega-Gutiérrez, F., and Gehrels, G.E., 1991, Isotopic studies of the Acatlán complex, southern Mexico: Implications for Paleozoic North American tectonics: Geological Society of America Bulletin, v. 103, no. 6, p. 817–828, doi: 10.1130/0016-7606(1991)103<0817:ISOTAC>2.3.CO;2.

Ye, H., Royden, L., Burchfiel, C., and Schuepbach, M., 1996, Late Paleozoic deformation of interior North America: The Greater Ancestral Rocky Mountains: American Association of Petroleum Geologists Bulletin, v. 80, no. 9, p. 1397–1432.

Young, L.M., 1970, Early Ordovician sedimentary history of the Marathon geosyncline, Trans-Pecos Texas: American Association of Petroleum Geologists Bulletin, v. 54, no. 12, p. 2303–2316.

Zaldívar-Ruiz, J., and Garduño-M., V.H., 1984, Estudio estratigráfico y estructural de las rocas del Paleozoico superior de Santa María del Oro, Durango, y sus implicaciones tectónicas [resumen]: México, D.F., Sociedad Geológica Mexicana, VII Convención Nacional, Resúmenes, p. 37–38.

Zeller, R.A., Jr., 1965, Stratigraphy of the Big Hatchet Mountains area, New Mexico: New Mexico Bureau of Mines and Mineral Resources Memoir 16, 128 p.

Manuscript Accepted by the Society 24 March 2005

Printed in the USA

Geological Society of America
Special Paper 393
2005

A new reconstruction of the Paleozoic continental margin of southwestern North America: Implications for the nature and timing of continental truncation and the possible role of the Mojave-Sonora megashear

Calvin H. Stevens*
San Jose State University, Department of Geology, San Jose, California 95192, USA

Paul Stone*
U.S. Geological Survey, 345 Middlefield Road, Menlo Park, California 94025, USA

Jonathan S. Miller*
San Jose State University, Department of Geology, San Jose, California 95192, USA

ABSTRACT

Data bearing on interpretations of the Paleozoic and Mesozoic paleogeography of southwestern North America are important for testing the hypothesis that the Paleozoic miogeocline in this region has been tectonically truncated, and if so, for ascertaining the time of the event and the possible role of the Mojave-Sonora megashear. Here, we present an analysis of existing and new data permitting reconstruction of the Paleozoic continental margin of southwestern North America. Significant new and recent information incorporated into this reconstruction includes (1) spatial distribution of Middle to Upper Devonian continental-margin facies belts, (2) positions of other paleogeographically significant sedimentary boundaries on the Paleozoic continental shelf, (3) distribution of Upper Permian through Upper Triassic plutonic rocks, and (4) evidence that the southern Sierra Nevada and western Mojave Desert are underlain by continental crust.

After restoring the geology of western Nevada and California along known and inferred strike-slip faults, we find that the Devonian facies belts and pre-Pennsylvanian sedimentary boundaries define an arcuate, generally south-trending continental margin that appears to be truncated on the southwest. A Pennsylvanian basin, a Permian coral belt, and a belt of Upper Permian to Upper Triassic plutons stretching from Sonora, Mexico, into westernmost central Nevada, cut across the older facies belts, suggesting that truncation of the continental margin occurred in the Pennsylvanian. We postulate that the main truncating structure was a left-lateral transform fault zone that extended from the Mojave-Sonora megashear in northwestern Mexico to

*E-mails: Stevens@geosun.sjsu.edu; pastone@usgs.gov; jsmiller@email.sjsu.edu

Stevens, C.H., Stone, P., and Miller, J.S., 2005, A new reconstruction of the Paleozoic continental margin of southwestern North America: Implications for the nature and timing of continental truncation and the possible role of the Mojave-Sonora megashear, *in* Anderson, T.H., Nourse, J.A., McKee, J.W., and Steiner, M.B., eds., The Mojave-Sonora megashear hypothesis: Development, assessment, and alternatives: Geological Society of America Special Paper 393, p. 597–618. doi: 10.1130/2005.2393(22). For permission to copy, contact editing@geosociety.org.

the Foothills Suture in California. The Caborca block of northwestern Mexico, where Devonian facies belts and pre-Pennsylvanian sedimentary boundaries like those in California have been identified, is interpreted to represent a missing fragment of the continental margin that underwent ~400 km of left-lateral displacement along this fault zone. If this model is correct, the Mojave-Sonora megashear played a direct role in the Pennsylvanian truncation of the continental margin, and any younger displacement on this fault has been relatively small.

Keywords: Paleozoic, paleogeography, southwestern North America, Mojave-Sonora megashear, continental truncation.

INTRODUCTION

The apparent termination of the Neoproterozoic to middle Paleozoic Cordilleran miogeocline in southern California has led many geologists to hypothesize that the miogeocline and underlying continental margin were tectonically truncated in late Paleozoic to early Mesozoic time (e.g., Hamilton, 1969; Burchfiel and Davis, 1972; Davis et al., 1978; Walker, 1988; Stone and Stevens, 1988; Dickinson, 2000; Dickinson and Lawton, 2001). The precise age and location of the truncating fault, however, have been subjects of considerable controversy, and some investigators have argued that the miogeocline was not truncated but that it simply wraps around the southwestern margin of the North American continent (e.g., Snow, 1992; Poole et al., 2000b).

Precambrian basement and miogeoclinal rocks in the Caborca region of northern Mexico, here referred to as the Caborca block (Fig. 1), are lithologically similar to those in California and were interpreted by Silver and Anderson (1974) to have been displaced several hundred kilometers southeastward on a fault they called the Mojave-Sonora megashear. As pointed out by Davis et al. (1978), this proposed displacement of the Caborca block is consistent with the general concept of continental truncation. Based on stratigraphic and structural evidence in Mexico, however, the age of displacement on the Mojave-Sonora megashear is most commonly regarded as Late Jurassic (Anderson and Silver, 1979; Anderson and Schmidt, 1983), an age incompatible with the idea that the truncation event occurred during the late Paleozoic or early Mesozoic. For this and other reasons, there has been little consensus on the nature of continental truncation and the possible role of the Mojave-Sonora megashear in such an event.

Recently, Dickinson (2000) made a case for late Paleozoic to earliest Mesozoic continental truncation and left-lateral displacement of the Caborca block and proposed the name California-Coahuila transform for the truncating fault to avoid confusion with the Mojave-Sonora megashear. His interpretation, based on multiple lines of evidence and reasoning, implies that any Late Jurassic movement on the megashear was relatively small and may have overprinted the California-Coahuila transform upon which much greater displacement had occurred. This conclusion is supported by paleomagnetic data that are inconsistent with major Late Jurassic or younger left-lateral displacement of the Caborca block but are permissive of earlier displacement (Molina Garza and Geissman, 1999). Additional data, however, are still needed to fully test this and other possible alternative concepts of continental truncation.

Geologic investigations during the past decade have produced new data that provide a basis for reexamining the paleogeographic evolution of the southern California continental margin and existing concepts of continental truncation. Primary among these are the results of studies of (1) continental margin rocks in the El Paso Mountains and northern Mojave Desert (Carr et al., 1997; Martin and Walker, 1991, 1995; Schermer et al., 2001; Walker et al., 2002) and in the east-central Sierra Nevada (Stevens and Greene, 1999); (2) the geochemistry of igneous rocks in the Mojave Desert region (Miller et al., 1996, 2000; Ducea, 2001); and (3) Upper Permian to Upper Triassic plutons in southeastern California (Miller et al., 1995; Barth et al., 1997). These new data lead us to propose an alternative to previous reconstructions of the southern California continental margin, which in turn leads to new insights into the existence, location, and age of a continental truncation boundary in this region.

The reconstruction presented here is based on a number of paleogeographic features of different ages. These are (1) Middle to Upper Devonian continental-margin facies belts, (2) sedimentary boundaries on the Paleozoic continental shelf, and (3) position of the margin of Precambrian continental crust as marked by geochemical characteristics. These features (Fig. 2) are discussed below and are later reconstructed by restoring displacements on the known and inferred Cenozoic and upper Mesozoic faults that have displaced them. Special emphasis is placed on newly recognized Middle to Upper Devonian facies belts and their relationship to the distribution of Upper Permian to Upper Triassic plutons.

MIDDLE TO UPPER DEVONIAN FACIES BELTS

The western North American continental margin is characterized by structural complexities that have inhibited detailed interpretation of the depositional settings and original paleogeographic positions of many Paleozoic facies belts. Therefore, in most previous studies of lower and middle Paleozoic facies in west-central Nevada and east-central California, rocks have been considered to be eugeoclinal or miogeoclinal, commonly with recognition of a more or less transitional facies. These rather vague terms are useful for broad generalizations, but are not

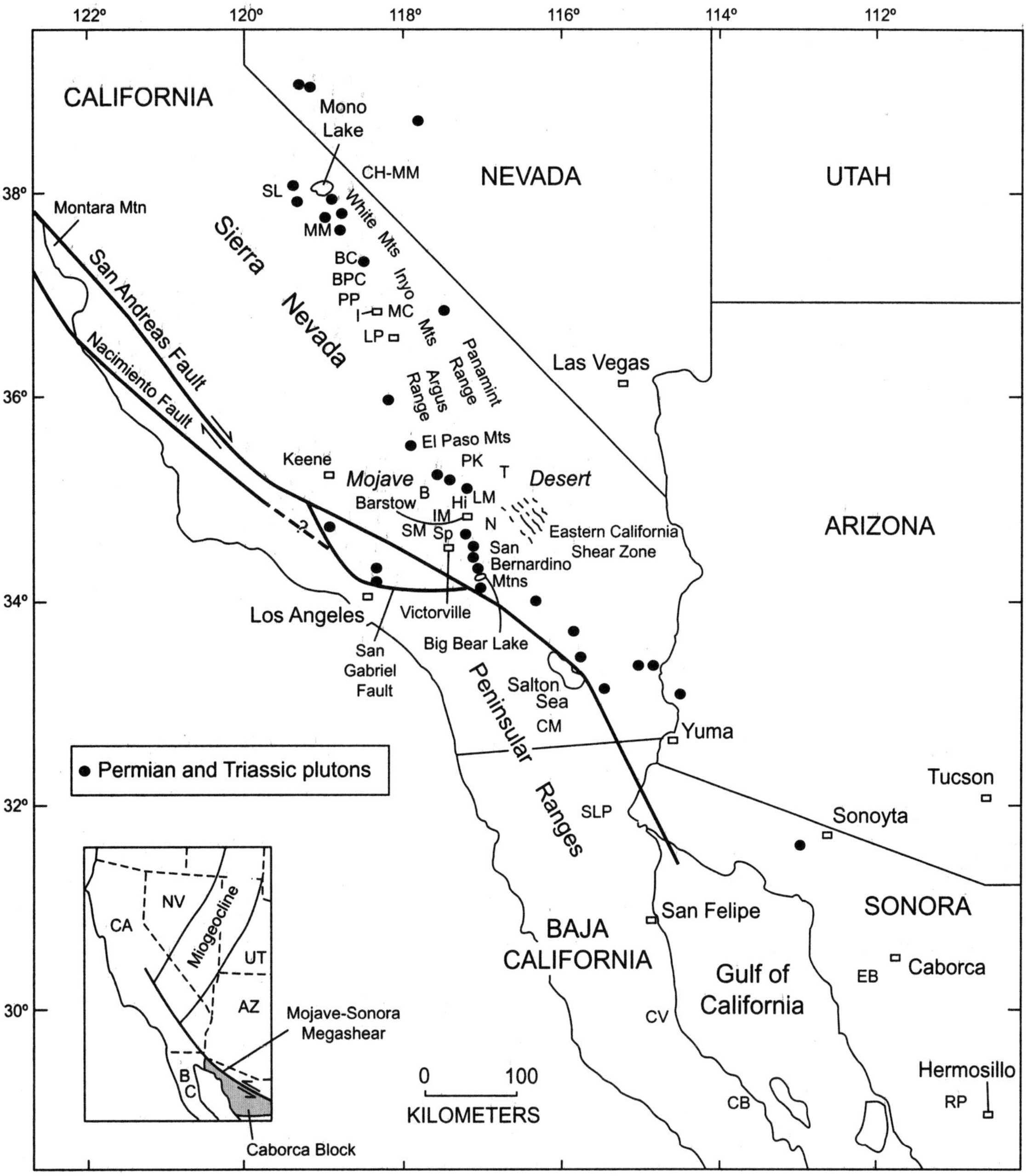

Figure 1. Map of southwestern North America showing localities referred to in the text and locations of Permian and Triassic plutons in the region. B—The Buttes; BC—Bishop Creek pendant; BPC—Big Pine Creek pendant; CB—Canal Ballenas area; CH-MM—Candelaria Hills–Miller Mountain area; CM—Coyote Mountains; CV—Cerro El Volcán; EB—El Bisani; Hi—Hinkley Hills; I—Independence; IM—Iron Mountain; LM—Lane Mountain; LP—Lone Pine; MC—Mazourka Canyon; MM—Mount Morrison pendant; N—Newberry Mountains; PK—Pilot Knob Valley; PP—Pinchot Pass pendant; RP—Rancho Placeritos; SL—Saddlebag Lake pendant; SM—Shadow Mountains; Sp—Sparkhule Mountain; SLP—Sierra Las Pinta; T—Tiefort Mountains. Locations of Permian and Triassic plutons from Snow et al. (1991), Bateman (1992), Dunne and Saleeby (1993), Schweickert and Lahren (1993a), Miller et al. (1995), Barth et al. (1997), and A.P. Barth (2003, personal commun.).

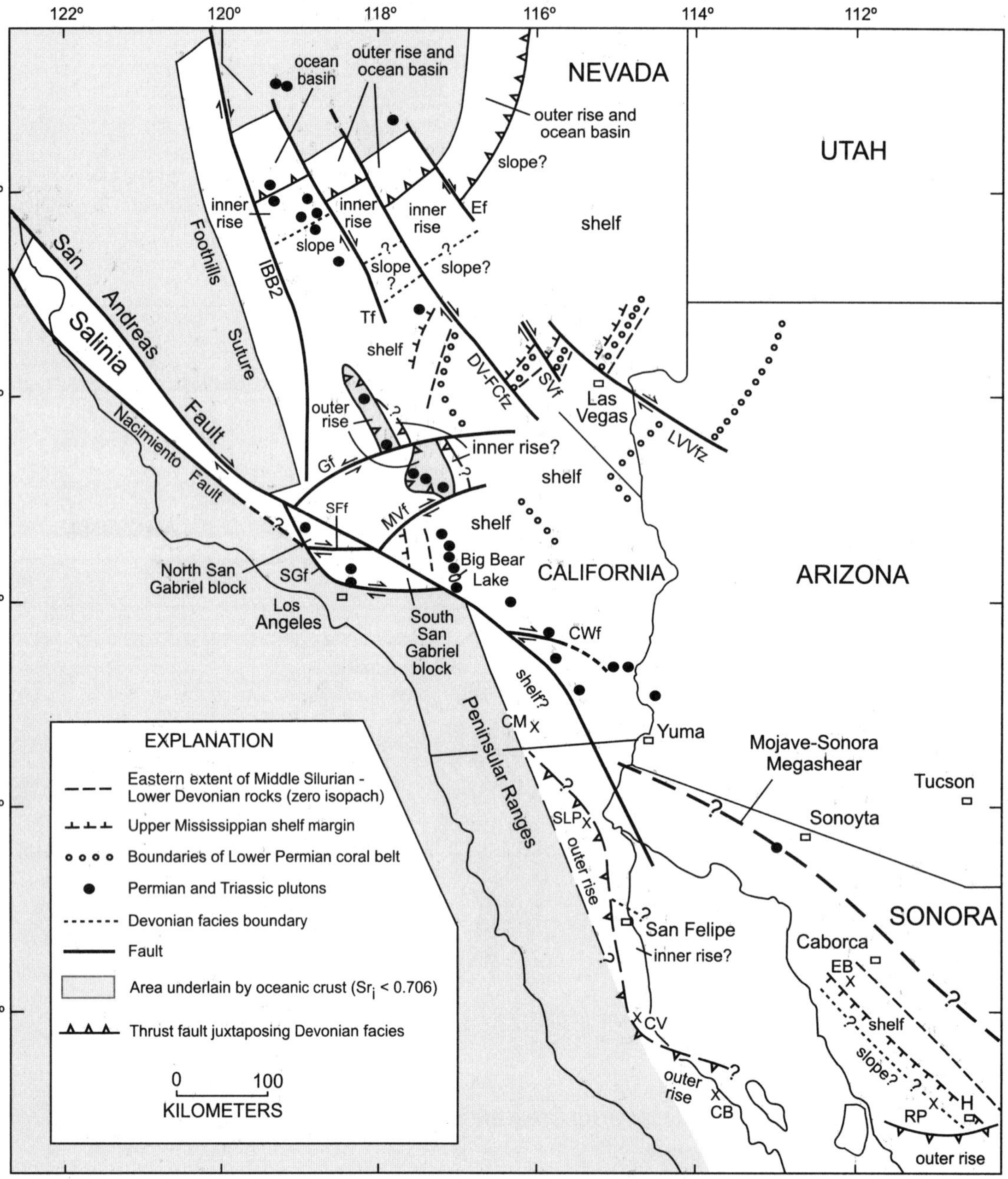

Figure 2. Index map showing Devonian facies belts, selected Paleozoic sedimentary boundaries on the continental shelf, areas interpreted to be underlain by oceanic crust, and major faults that have displaced facies belts and sedimentary boundaries. Note that in several areas rocks of the outer rise and ocean basin now are underlain by continental crust because of thrusting of those facies over the continental margin. In addition, outer-rise rocks not now exposed in the Saddlebag Lake area are inferred to underlie the ocean basin facies in that area. CB—Canal Ballenas area; CM—Coyote Mountains; CV—Cerro El Volcán; CWf—Clemens Well fault; DV-FCfz—Death Valley–Furnace Creek fault zone; EB—El Bisani; Ef—Excelsior fault; Gf—Garlock fault; H—Hermosillo; IBB2—Intrabatholithic break 2; LVVfz—Las Vegas Valley fault zone; MVf—Mojave Valley fault; RP—Rancho Placeritos; SFf—San Francisquito fault; SGf—San Gabriel fault; SLP—Sierra Las Pinta; SVf—Stewart Valley fault; Tf—Tinemaha fault.

adequate for close comparison of rock sequences, especially where only small parts of the original stratigraphic section are preserved. This is especially true in south-central California where generally small outcrops of metamorphosed rocks are all that remain of once broad Paleozoic facies belts.

In the early to middle Paleozoic, prior to the Antler orogeny in latest Devonian to Mississippian time, the western margin of North America was passive (e.g., Dickinson, 1981). As such, this margin should have been geomorphically similar to modern passive margins. Today, breaks in slope can be employed to define broad passive-margin settings; these are the continental shelf, continental slope, continental rise, and ocean basin. In ancient deposits these geomorphic settings can be inferred on the basis of lithologic character, as shown by Poole et al. (1992) for lower Paleozoic rock sequences throughout western North America.

This general categorization of continental-margin settings is useful, but there are problems in application to ancient margins because of changes in their geographic positions through time. For example, the shelf margin of the western United States was marked by backstepping through most of the Paleozoic (e.g., Stevens, 1986; Poole et al., 1992). Thus, areas that at one time were part of the shelf later became part of the slope. In the Inyo Mountains of east-central California (Fig. 1) the Lower Ordovician shelf break occurs at the north end of Mazourka Canyon; by the Devonian the shelf break had retreated at least 30 km southeastward (Stevens, 1986). In the northernmost Inyo Mountains, shelf deposits represent the Upper Cambrian, but in the White Mountains to the northwest (Fig. 1), the coeval Emigrant Formation (Nelson, 1966) may represent the slope. If so, the shelf margin lay in the White Mountains at that time, ~40 km northwest of the position of the Lower Ordovician shelf margin. Farther west in the Mount Morrison and related pendants of the eastern Sierra Nevada (Morrison block of Stevens and Greene, 1999), all lower and middle Paleozoic rocks are of slope and rise facies. In the Big Pine Creek and Pinchot Pass pendants (Fig. 1), however, the shelfal Lower Cambrian Poleta and Campito Formations are exposed (Moore and Foster, 1980; Bartley et al., 2001). Although the relations of these latter units to those in the Morrison block are uncertain, it is likely that the rocks in these pendants represent the strata beneath the units exposed in the Morrison block. If so, the Lower Cambrian shelf margin lay west of these pendants. Thus, use of isolated outcrops of rocks of different ages for identification of facies belts may be misleading. Instead, whenever possible, definitions of depositional settings should be based on coeval rocks.

Because of the strong contrast between various Middle to Upper Devonian facies and their widespread occurrence in southwestern North America, we use rocks of this age to define our facies belts. These belts are assigned to the most appropriate geomorphic settings on the basis of the character of the deposits most common to those settings today. Here we subdivide the continental rise into an inner part and an outer part, which for simplicity are referred to as inner rise and outer rise. Thus, we distinguish the following Middle to Upper Devonian marine facies belts in west-central Nevada and eastern California: (1) continental shelf, (2) continental slope, (3) inner rise, (4) outer rise, and (5) ocean basin. The present locations of these facies belts are shown in Figure 2.

Thick sequences consisting exclusively of rocks deposited on a shallow-water platform comprise the continental shelf facies. Thin sequences consisting primarily of hemipelagic and pelagic rocks, commonly with a few sediment–gravity-flow deposits and in places cut by submarine channels, are considered to belong to the continental slope. The inner-rise facies consists of very thick sequences of deep-water, coarse-grained sediment–gravity-flow deposits interbedded with hemipelagic and pelagic sediment. The outer-rise facies is similar to the inner-rise facies except that it may be thinner and basaltic igneous rocks and/or bedded barite deposits may be present. The ocean-basin facies is composed of relatively thin, almost exclusively pelagic and hemipelagic deposits locally with basaltic volcanic rocks and/or bedded barite.

The margin of Precambrian continental crust in western North America has been considered to be marked by the Sr_i = 0.706 isotopic isopleth in Mesozoic plutonic rocks (e.g., Kistler and Peterman, 1973; Kistler, 1978). On this basis, it appears that as in modern passive margins, the Middle to Upper Devonian sediments of the continental shelf, continental rise, and inner rise were deposited on continental crust. Rocks of the outer rise and ocean basin presumably were deposited on oceanic crust, although some of them have since been thrust onto the continent (e.g., Stevens et al., 1992; Fig. 2).

Continental Shelf

In western North America this facies belt, which includes all Middle to Upper Devonian rocks in the southeastern part of the Inyo Mountains and areas to the southeast (e.g., Stevens, 1986; Figs. 1 and 2), consists of thick-bedded carbonates along with minor quartzite. These rocks reach a thickness of 500–550 m in the southern Inyo Mountains region (Hall and MacKevett, 1962; Merriam, 1963). The boundary between the continental shelf and the continental slope, which occurs in the southwestern Inyo Mountains, is concealed, but the transition must occur somewhere between Independence and Lone Pine (Fig. 1). Shallow-water carbonates in the San Bernardino Mountains (Brown, 1991) and in the vicinity of Victorville (Miller, 1981) define the southernmost and westernmost known extent of Middle to Upper Devonian continental shelf deposits in California (Fig. 1).

In the Caborca block of northern Sonora (Fig. 1), Devonian continental shelf sequences characterized by shallow-water carbonate rocks (Stewart and Poole, 2002) are well represented. The shelf-slope break is placed near Rancho Placeritos northwest of Hermosillo (Fig. 2) based on the presence of both platform and debris-flow deposits in the Devonian strata at that locality (Poole et al., 2000a, 2000b).

Paleozoic rocks are poorly represented in the Peninsular Ranges. A very thick carbonate-rich sequence in the Coyote Mountains of southernmost California, however, probably represents shelf deposition (Fig. 2), although slope or inner rise is possible. Ordovician conodonts (Miller and Dockum, 1983) are the only fossils known from this sequence, so the assignment here to the Devonian shelf facies is questionable. Highly metamorphosed carbonate and quartzitic rocks exposed discontinuously from the Coyote Mountains area to the north end of the Peninsular Ranges (Gastil et al., 1991) are undated but are lithologically consistent with the interpretation that this area was part of the Devonian shelf.

Continental Slope

Rocks in Mazourka Canyon in the western Inyo Mountains (Fig. 1) are here considered to represent the continental slope. Middle Devonian rocks in this area are composed of hemipelagic and pelagic sediment commonly with a few sediment–gravity-flow deposits. Black, radiolarian-bearing chert of Late Devonian age fills submarine channels cut into these older slope deposits (Stevens, 1986). In central Mazourka Canyon, Middle to Upper Devonian continental slope deposits, based on data from Ross (1966) and Miller (1976), apparently are ~150 m thick.

We interpret the continental slope facies belt to extend at least as far northwest as the Bishop Creek pendant in the eastern Sierra Nevada (Fig. 1), where Stevens and Greene (1999) reported exposures of the Squares Tunnel Formation, consisting entirely of pelagic and hemipelagic rocks, to be only ~120 m thick. Here, the Squares Tunnel Formation probably represents slow deposition throughout both the Middle and Late Devonian. The absence in this pendant of a thick Devonian submarine-fan sequence similar to that which characterizes the inner rise facies in the Mount Morrison pendant farther to the northwest (Stevens and Greene, 1999) indicates that the boundary between the continental slope and the inner rise lies between these two pendants. Based on this interpretation, the minimum width of the continental slope is ~35 km.

Inner Rise

The inner rise is represented by Middle to Upper Devonian rocks in all of the pendants of the Morrison block except the Bishop Creek pendant, as noted above. These rocks, which comprise the Sevahah Group (Mount Morrison Sandstone and Squares Tunnel Formation), consist of massive calcareous quartzite, sandy limestone, and dark gray argillite and chert and have been interpreted to represent a submarine fan complex encased in pelagic and hemipelagic sediment (Stevens and Greene, 1999). A major channel filled with conglomerate and calcareous quartzose sandstone, considered to be the primary distributary channel for the fan, is exposed in the southeastern part of the Mount Morrison pendant.

Stevens and Greene (1999) indicated that the Devonian rocks of the inner-rise facies belt in the Morrison block have been displaced ~65 km northward relative to the Inyo Mountains. Across this offset, rocks representing the main distributary channel in the southeastern part of the Mount Morrison pendant can be tied to the channeled continental slope facies in the western Inyo Mountains on the basis of their distinctive lithology and age (Stevens and Greene, 1999). The erosional Middle Devonian sequence boundary beneath the channels can then be traced onto the shelf where it is overlain by quartz sandstone similar in age and lithology to that in the submarine fan.

The Middle to Upper Devonian inner-rise section in the Mount Morrison pendant is probably more than 1000 m thick (Stevens and Greene, 1999), much thicker than the coeval continental shelf section and vastly thicker than the rocks composing the continental slope facies. These rocks thin to ~400 m in the Log Cabin Mine pendant 50 km to the north, in a general oceanward direction. Because of the paucity of outcrops, the original width of the inner-rise facies belt cannot be ascertained.

Middle to Upper Devonian inner-rise strata are not known to be exposed anywhere in California or western Nevada outside the Morrison block. In the eastern El Paso Mountains (Fig. 1), however, three units included by Carr et al. (1997) in the metasedimentary rocks of El Paso Peaks are quite similar to coeval Cambrian(?)–Lower Devonian(?) formations that underlie the Middle to Upper Devonian inner-rise strata described by Stevens and Greene (1999) in the Morrison block. The oldest unit (cs of Carr et al., 1997) contains moderately thin-bedded limestone similar to that in the Upper Cambrian(?) Mount Aggie Formation in the Mount Morrison pendant. The next younger unit (Oem of Carr et al., 1997) contains quartzite similar to that in the Ordovician Convict Lake Formation in the Mount Morrison pendant, although it also contains prominent brown silty mudstone resembling the Ordovician Barrel Spring Formation (of Ross, 1966) in the slope facies in the northern Inyo Mountains. The uppermost unit (DOeu of Carr et al., 1997) consists mostly of calc-silicate hornfels that resembles the Silurian(?) to Lower Devonian(?) Aspen Meadow Formation in the Mount Morrison pendant. Based on the close stratigraphic comparison between lower Paleozoic rocks in the eastern El Paso Mountains and the Morrison block, this sequence of units in the El Paso Mountains very likely was overlain by Middle to Upper Devonian rocks similar to those in the Morrison block, even though such rocks are not preserved. Thus, we provisionally place the eastern El Paso Mountains in the inner-rise facies belt (Fig. 2), although an outer slope position is possible. Other isolated outcrops in the northern Mojave Desert correlated by Carr et al. (1992) with the El Paso Peaks section probably also originally underlay Devonian rocks of the inner-rise facies belt. In Figure 2, we show a thrust fault bounding this area because it appears to be out of place with respect to the shelf facies to the east.

Lower Paleozoic rocks exposed on the southern flank of Miller Mountain in west-central Nevada (Fig. 1) also are considered correlative with the section below the Middle to Upper Devonian inner-rise strata in the Morrison block (Stevens and Greene, 1999). These rocks consist of fine-grained limestone, calc-silicate rocks, graptolitic shale, and radiolarian chert, corresponding to the Upper Cambrian(?) and Ordovician Mount

Aggie and Convict Lake formations of the Morrison block. As in the eastern El Paso Mountains, we interpret this correlation as provisional evidence for placing the southern flank of Miller Mountain in the inner-rise facies belt, although the outer slope belt is possible.

In northeastern Baja California near San Felipe (Fig. 2), a very thick sequence of orthoquartzite, marble, and meta-argillite (Gastil et al., 1991) overlies continental crust (Gromet and Silver, 1987). Gastil et al. (1991) and Anderson (1993) considered these rocks to be shallow-water deposits of late Precambrian to Early Cambrian age, in part correlative with rocks near Caborca. However, analysis of detrital zircons in the orthoquartzite suggests that these rocks are actually Ordovician and were deposited in off-shelf environments (Gehrels et al., 2002). If this interpretation is correct, it seems likely that this sequence was deposited on an Ordovician inner rise. Thus, by analogy with the Morrison block in California, where Ordovician and Devonian inner-rise facies are superimposed, we here questionably assign the rocks near San Felipe to the Devonian inner-rise facies belt.

Outer Rise

A belt of Paleozoic rocks (metasedimentary rocks of Colorado Camp) in the western El Paso Mountains includes Cambrian and Ordovician rocks recognized by Carr et al. (1997) to represent a somewhat different facies than the metasedimentary rocks of El Paso Peaks in the eastern El Paso Mountains. This western belt includes Devonian calcareous quartz sandstone essentially identical to the Middle Devonian submarine fan deposits recognized in the Morrison block, dark gray argillite and chert, and both basaltic and andesitic volcanic rocks (Carr et al., 1997). This rock assemblage indicates assignment to the outer-rise facies belt as defined here. Middle Devonian calcareous sandstone similar to that in the western El Paso Mountains and apparently displaced from them by the Garlock fault occurs in Pilot Knob Valley (Carr et al., 1992). Calcareous sandstone associated with bedded barite and basaltic volcanic rocks is present also in the southern Sierra Nevada (Dunne and Suczek, 1991). Upper Permian and Triassic plutonic rocks that intrude this facies belt have isotopic signatures indicative of having risen through oceanic crust (Miller et al., 1995).

On the north side of Miller Mountain and in the Candelaria Hills (Fig. 1), a structurally complicated sequence of Ordovician to Devonian rocks includes amygdaloidal greenstone (Stevens and Greene, 1999), barite deposits (Stewart, 1984), and moderately thick exposures of calcareous sandstone representing a Middle Devonian fan complex. These rocks are very similar to those in the western El Paso Mountains and also probably represent the outer-rise facies.

In Sonora, south of Hermosillo (Fig. 2), a thick sequence of Ordovician to Devonian deep-water rocks is interpreted to have been thrust over the continental slope and onto the margin of the continental shelf in the Late Permian (Stewart, 1988; Stewart et al., 1990; Poole et al., 1995). Most of these deep-water rocks are considered to represent the outer-rise facies because shelf-derived sediment–gravity-flow deposits are associated with barite deposits (Poole et al., 1995).

In north-central Baja California at Sierra Las Pinta and along Canal Ballenas (Fig. 2), rocks of Devonian to Mississippian age consist of shelf-derived sediment, basaltic volcanic rocks, and bedded chert (Gastil et al., 1991). These rocks clearly belong to the outer-rise facies belt as defined in this paper. By analogy with the structural relations near Hermosillo, we speculate that these rocks have been thrust onto the continental margin as shown in Figure 2. We infer the thrust to underlie these rocks and to pass west of the inner-rise(?) strata exposed near San Felipe. The inferred thrust is shown also passing west of a very thick sequence of Permian turbiditic strata described near Cerro El Volcán (Fig. 2) by Buch and Delattre (1993). These strata evidently accumulated in relatively deep water near the shelf margin, and we thus interpret them to be in a structural position below the inferred thrust.

Ocean Basin

The ocean-basin facies is represented by undated lower Paleozoic rocks of the Antler allochthon exposed west of Mono Lake (Figs. 1 and 2). These rocks consist mostly of intensely deformed black chert with light gray weathering phosphatic lenses, stringers, and nodules, siliceous argillite and locally slate (Greene et al., 1997). Some of the phosphatic rocks are similar to the Devonian Squares Tunnel Formation in the Morrison block and western Inyo Mountains. These rocks presumably were deformed during the Antler orogeny in latest Devonian and Early Mississippian time (Greene et al., 1997) and were uplifted and eroded later in the Mississippian. This resulted in transport of coarse clastic sediment eastward onto rocks of the Devonian outer-rise facies belt, as seen in the western El Paso Mountains (Carr et al., 1997) and the southern Sierra Nevada (Dunne and Suczek, 1991). Later, in the Triassic, the ocean-basin rocks west of Mono Lake were thrust against inner-rise rocks of the Morrison belt (Stevens and Greene, 1999).

SEDIMENTARY BOUNDARIES ON THE PALEOZOIC CONTINENTAL SHELF

Many sedimentary boundaries, several of which appear to be truncated in southern California (e.g., Brown, 1991), have been recognized in the Paleozoic continental shelf rocks of the western United States. All of these boundaries are of potential significance for evaluation of the continental truncation hypothesis. Here, however, only three boundaries are considered: (1) the eastern extent (zero isopach) of middle Silurian to Lower Devonian rocks, (2) the lower Meramecian (Upper Mississippian) shelf margin, and (3) the western and eastern margins of the Lower Permian "*Thysanophyllum*" coral belt (Fig. 2). These boundaries were selected because all three are easily recognizable, they bracket a possible time of truncation of the continent, and the first two can be recognized in Mexico, thus providing information useful in evaluating whether or not there has been a major offset of the geology in the region. The location of the zero isopach of middle Silurian–Lower Devonian strata in southern Nevada and eastern California, as shown in Figure 2, is from Snow (1992); the location of the Meramecian shelf margin in California and southern Nevada is from Stevens et al. (1991); and the position and orientation of the Permian "*Thysanophyllum*" coral belt (a coral now referred to as *Protowentzelella*) is from Stevens (1982).

The locations of the middle Silurian–Lower Devonian zero isopach and the Meramecian shelf margin in the western Mojave Desert–San Bernardino Mountains region are of prime importance for our paleogeographic reconstructions because these locations establish regional trends that probably reflect the trend of the coeval continental margins. Silurian and Lower Devonian rocks are absent in this region (Miller, 1981), indicating that the zero isopach lies somewhere to the west. The Bullion Limestone, which represents the Meramecian carbonate shelf, is recognized in the San Bernardino Mountains (Brown, 1991), and pure calcitic marble that almost surely represents the Bullion Limestone is exposed at Sparkhule Mountain just north of Victorville. This latter marble is in contact, probably on a minor fault, with silty limestone from which we recently recovered Lower to Middle Pennsylvanian conodonts (identified by S.M. Ritter, 2002, personal commun.) and which we identify as Bird Spring Formation. Thus, the Meramecian shelf margin also must lie west of the Paleozoic sections exposed in the San Bernardino Mountains and near Victorville.

The placement of these two boundaries can be more closely estimated in the Mojave Desert–San Bernardino Mountains region by comparing the thickness of the various Mississippian formations in the San Bernardino Mountains (Brown, 1991) with those elsewhere in southern California and Nevada. This comparison indicates that the section in the San Bernardino Mountains best correlates with sections exposed a short distance north of Las Vegas (Poole and Sandberg, 1977) and in the Panamint and Argus Ranges (Fig. 1) in east-central California (Stevens et al., 1979). Those sections are close to both the middle Silurian–Lower Devonian zero isopach and the margin of the Meramecian shelf (Fig. 2). Thus, these two facies boundaries probably originally lay only a short distance west of outcrops in the San Bernardino Mountains as shown on Figure 2, approximately due south of the Argus and Panamint Ranges.

In Sonora, the middle Silurian–Lower Devonian zero isopach, the margin of the Meramecian shelf, and the Middle-Upper Devonian shelf-slope boundary (Fig. 2) can be estimated on the basis of scattered outcrops. Two especially important sections are near Rancho Placeritos, northwest of Hermosillo (Poole et al., 2000a, 2000b), and at El Bisani, west of Caborca (Stewart and Poole, 2002; Fig. 2). Silurian dolomite near Rancho Placeritos contains a wide variety of platform fossils including colonial corals and stromatoporoids. The Devonian section is composed of limestone and sandy limestone with one interval of debris-flow deposits, and the Lower Mississippian sequence contains turbiditic limestone. This section shows the typical backstepping of the shelf margin through time as seen elsewhere in western North America (Stevens, 1986; Poole et al., 1992), so it clearly represents a transition from the continental shelf to the continental slope as defined in this paper. The El Bisani section contains Ordovician to Mississippian shelf carbonates (Stewart and Poole, 2002). The presence of shelfal Silurian rocks indicates that this section represents the outer part of the continental shelf. Together these two sections define the approximate positions and trends of the zero middle Silurian–Lower Devonian zero isopach and the Mississippian shelf margin in northwestern Sonora (Fig. 2).

MARGIN OF PRECAMBRIAN CONTINENTAL CRUST

The Sr_i = 0.706 isopleth parallels Devonian facies belts (Fig. 2) in Nevada and easternmost California and probably represents the Proterozoic continental margin along which the Cordilleran miogeocline developed (e.g., Kistler, 1978; Stevens et al., 1992). Near the California border, this isopleth is offset by several Mesozoic and Cenozoic right-lateral faults (Kistler, 1990; Kistler, 1993; Robinson, 1993; A. Robinson, 1994, personal commun.). Farther west in the central Sierra Nevada, the Sr_i = 0.706 isopleth changes from a southwestward to a south-southeastward orientation (Fig. 2) and coincides in part with the Foothills Suture of Saleeby et al. (1978). This part of the isopleth has been interpreted to reflect late Paleozoic or early Mesozoic truncation of the continental margin and subsequent accretion of oceanic terranes (e.g., Saleeby, 1992).

Complications to the significance of the Sr_i = 0.706 isopleth arise in the southern Sierra Nevada and western Mojave Desert, where Mesozoic plutonic rocks with Sr_i >0.706 and $\delta^{18}O$ >+9 have been interpreted to record anatexis of continent-derived sedimentary rocks (e.g., Saleeby et al., 1987; Kistler, 1990; Stevens et al., 1992) rather than to have been derived from melting of crystalline Precambrian continental crust. According to this interpretation, the margin of continental basement in these areas could lie considerably east of the Sr_i = 0.706 isopleth.

Alternative interpretations, however, are possible. Juxtaposition of granitic sources by low-angle faulting prior to intrusion, for instance, has only recently been considered. With this in mind, Miller et al. (1996) suggested that muscovite-garnet granites in the western Mojave Desert with high $\delta^{18}O$ values were derived from underplating and melting of immature forearc sedimentary rocks (e.g., Pelona-Orocopia-Rand schists or similar sources) that were thrust under sialic crust. More recently, in an investigation of geochemical signatures of Miocene volcanic rocks in the Mojave Desert, Miller et al. (2000) pointed out that the most important and abrupt geochemical change in these rocks (high large ion lithophiles [LILEs], light rare earth element [LREE] enrichment, and depletions of high field strength elements [HFSE]) occurs at ~116°W longitude, well east of outcrops assignable to the Paleozoic continental shelf (Fig. 2). Miller et al. (2000) interpreted this isotopic boundary to mark the western limit of Proterozoic subcontinental mantle that farther west had been completely stripped away by Late Cretaceous–early Tertiary shallow subduction and concluded that Precambrian basement extended much farther west into the area considered by Kistler (1990) to be allochthonous Panthalassan lithosphere. Zircons from Jurassic, Cretaceous, and Tertiary plutonic rocks in the central Mojave Desert (e.g., Walker et al., 1995; Schermer et al., 2001; Miller and Walker, 2002) show appreciable Proterozoic inheritance, which is compatible with the presence of Precambrian basement in this region. Most recently Saleeby (2003) has accepted the premise that greatly thinned continental crust underlies the western Mojave Desert.

Although much of the old continental crust in the western Mojave Desert probably has been reworked by the voluminous Mesozoic batholithic magmatism, scattered exposures of Precambrian and Paleozoic rocks in the region also tend to suggest that the underlying crust is sialic in nature. Sialic basement evidently crops out as far west as the Tiefort Mountains (Fig. 1), where it is overlain by rocks provisionally correlated with the Precambrian to Cambrian Johnnie Formation, Stirling Quartzite, and Wood Canyon Formation (Schermer et al., 2001). Proterozoic basement also has been identified in unnamed hills south of Barstow and in the Newberry Mountains (Fig. 1) a short distance to the east (Martin and Walker, 1992). In the Victorville area and in the Shadow Mountains to the west (Fig. 1), the miogeoclinal succession of the Wood Canyon Formation through the Bonanza King Dolomite of Cambrian age (Miller, 1981; Martin and Walker, 1995) also presumably lies on continental crust. This tends to be confirmed in the Shadow Mountains where rocks assigned to the Cambrian Zabriskie Quartzite contain crystalline basement clasts and 1.7 Ga detrital zircons, implying a nearby source of Proterozoic basement rocks (Martin and Walker, 1992). Highly metamorphosed sedimentary rocks, primarily marble and quartzite, that probably represent Neoproterozoic to Paleozoic miogeoclinal strata deposited on continental crust are exposed also in the Hinkley Hills (Kiser, 1981), at Iron Mountain (Boettcher and Walker, 1993), and at the Buttes (Fletcher et al., 1995).

Farther northwest in the Keene area (Fig. 1) of the Tehachapi Mountains, exposed prebatholithic rocks are dominantly calcitic marble with lesser dolomitic marble, pure to impure quartzite, and mica schist (Ross, 1989). In the southwestern United States, dolomite is most common in continental shelf rocks of Neoproterozoic to middle Paleozoic age. Thus, the presence of abundant dolomite tends to suggest that these probable shelf rocks are of middle Paleozoic or older age. Consistent with this interpretation, limestone in the Bean Canyon Formation (Ross, 1989) some 35 km south of Keene has recently yielded a poorly preserved sponge of probable Paleozoic age (J.K. Rigby Sr., 2003, personal commun.). Thus, as indicated by Ducea (2001), Paleozoic shelf rocks underlain by sialic crust probably extend westward to the Foothills Suture (Fig. 2).

The rather simple isotopic picture of apparent decrease westward of continental contributions to younger igneous rocks (Miller et al., 2000; Ducea, 2001) is interrupted in southernmost Sierra Nevada and the northern Mojave Desert. Here, an elongate, northwest-trending area, bisected and offset by the Garlock fault, is characterized by anomalous Upper Permian to Middle Triassic (260–240 Ma) plutons with Sr_i <0.706 (Miller et al., 1995; Fig. 2). This area (western El Paso terrane) commonly has been interpreted to be underlain by oceanic basement (e.g., Kistler, 1978, 1990) and is apparently out of place with respect to the surrounding sialic terrane. As noted earlier, the anomalous plutons intrude Paleozoic rocks here assigned to the outer-rise facies belt. Several previous workers (e.g., Davis et al., 1978; Stevens et al., 1992) have inferred that this anomalous oceanic terrane was derived from the Antler orogenic belt to the northwest and displaced into its present position by late Paleozoic or early Mesozoic left-lateral strike-slip faulting associated with continental truncation. Miller et al. (1995), however, showed that the anomalous plutons are essentially surrounded by Middle Jurassic plutons (as old as 179 Ma) that have continental lithosphere isotopic signatures. This has led to the alternative interpretation that the outer-rise rocks and older plutons lie structurally above continental basement as a result of overthrusting that took place between 240 Ma and 179 Ma (Miller et al., 1995). This interpretation is compatible with our facies analysis of the rocks in the El Paso Mountains and is incorporated into the paleogeographic reconstruction presented later.

West of the San Andreas fault, the western margin of Precambrian continental crust coincides with the Nacimiento fault zone on the west side of Salinia and the San Gabriel fault to the south (Kistler, 1990; Figs. 1 and 2). The Sr_i = 0.706 isopleth crosses Salinia from east to west at Montara Mountain (Figs. 1 and 2), apparently marking the northern margin of sialic basement in this terrane (Kistler and Champion, 2001). In the Peninsular Ranges of southern California, the Sr_i = 0.706 isopleth extends southward to the international border (Kistler, 1990; Fig. 2). From there the isopleth continues into Baja California to the approximate latitude of San Felipe (Gromet and Silver, 1987). The isopleth as shown by the above authors is complicated in detail, and on Figure 2 we have generalized its location.

UPPER PERMIAN TO UPPER TRIASSIC PLUTONS

Recent studies have shown that plutons ranging in age from Upper Permian to Upper Triassic are more widespread in California and adjacent areas than previously thought (Snow et al., 1991; Bateman, 1992; Dunne and Saleeby, 1993; Schweickert and Lahren, 1993a; Miller et al., 1995; Barth et al., 1997). Middle and Upper Triassic plutons are the most widespread, but in the Mojave Desert there are a few Upper Permian and Lower Triassic plutons as well (A.P. Barth, 2002, personal commun.). Because plutons of all of these ages are geographically intermingled, especially in the Mojave Desert, they are grouped together for the purposes of this paper.

In modern geographic coordinates these plutons form a diffuse, northwest-trending belt extending from west-central Nevada through southern California into northern Mexico (Figs. 1 and 2). This belt presumably developed above an east-dipping subduction zone that approximately paralleled a northwest-trending continental margin (e.g., Barth et al., 1997). Significantly, the plutonic belt cuts across all the previously described Devonian facies belts, apparently reflecting a change in the orientation of the continental margin prior to emplacement of the plutons. This relationship, which also was noted by Dickinson (2000), is more clearly revealed through the paleogeographic reconstructions presented later.

Whereas the majority of the plutons are considered to be relatively in place, those that intrude the oceanic western El Paso terrane are presumed to be displaced with respect to the remainder of the plutonic belt as a result of younger thrusting (Miller et al., 1995), as noted previously. The inferred original locations of these plutons will be shown in our later reconstructions.

MOJAVE-SONORA MEGASHEAR

The precise location of the Mojave-Sonora megashear, which bounds the northeastern side of the Caborca block, is controversial, and various authors have placed it in somewhat different locations. Here we generally follow the trace shown by Anderson and Schmidt (1983) in Mexico extending north of Caborca and then intersecting the common corner of Arizona, Sonora, and Baja California before being cut off by the San Andreas fault (Fig. 2). This interpretation differs slightly from that of Anderson and Silver (1979), who implied that the megashear extended in a more northerly direction from the vicinity of Sonoyta and passed east of Yuma. That fault path is consistent with an apparent major discontinuity in Precambrian crystalline rocks. Such a path, however, is difficult to reconcile with the continuity of Paleozoic cratonal and miogeoclinal strata across the Mojave Desert region to the San Andreas fault (e.g., Stewart and Poole, 1975; Stone et al., 1983) if the megashear is, in fact, a major left-lateral fault along which the Caborca block has been displaced southward.

PALEOGEOGRAPHIC RECONSTRUCTION AND IMPLICATIONS FOR CONTINENTAL TRUNCATION

The Middle to Upper Devonian facies belts, the sedimentary boundaries on the continental shelf, the distribution of Permian and Triassic plutons, and the margin of Precambrian continental crust delineated on Figure 2 can be analyzed together to test the hypothesis of continental truncation. In their present configuration, however, the possible relationship between these features and an inferred truncation of the continental margin is not obvious because of numerous fault offsets. Here we attempt to restore these features to their original configurations through a stepwise paleogeographic reconstruction designed to address the question of whether or not the continental margin has been truncated and, if so, when this event took place.

Step 1: Restoration of Cenozoic and Upper Mesozoic Strike-Slip Faults

For our first reconstruction (Fig. 3), we restore displacements on most of the faults shown in Figure 2 that offset the geologic features of interest here. Most of these faults are strike-slip faults of Cenozoic age, although at least one (IBB2 in the Sierra Nevada) is a cryptic strike-slip fault of Mesozoic age (Kistler, 1993). Faults that have resulted primarily in contractional deformation (e.g., Dunne, 1986; Snow, 1992) or extension (Snow and Wernicke, 2000) are not considered here because their combined effects are uncertain, although some of these faults could have caused significant deformation of the features shown in Figure 2.

The amounts of displacement removed for the reconstruction shown in Figure 3 are as follows:

1. Death Valley–Furnace Creek fault zone: 35–80 km dextral, decreasing to the northwest (Stevens et al., 1991).
2. Eastern California shear zone: 65 km dextral (Dokka and Travis, 1990).
3. Excelsior fault zone: 50 km dextral (Stewart, 1985).
4. Garlock fault: 60 km sinistral (Davis and Burchfiel, 1973).
5. Intrabatholithic break 2 (IBB2): 100 km dextral (Kistler, 1993).
6. Las Vegas Valley fault zone: 50 km dextral (Stevens et al., 1991).
7. Mojave Valley fault: 65 km dextral (Martin et al., 1993).
8. San Andreas fault system: 330 km dextral (Stewart and Crowell, 1992). In southern California, this displacement is partitioned as follows: 160 km on the Pliocene to the present San Andreas fault; 60 km on the upper Miocene San Gabriel fault; and 110 km on the lower to middle Miocene Clemens Well–San Francisquito fault (simplified from Powell, 1993). Following Powell (1993, p. 30), we do not extend the Clemens Well fault indefinitely southeastward into the continent but infer that its displacement is taken up by extensional faulting in the vicinity of Yuma. Our restoration of this displacement thus results in a shortening of the distance between the plutons near Yuma and the pluton near Sonoyta (Fig. 3).
9. Proto–San Andreas fault: 100 km dextral. This hypothetical pre-Miocene fault displacement of terranes west of the San Gabriel block is required to line up the $Sr_i = 0.706$ isopleth at Montara Mountain near the north end of Salinia with that in the Tehachapi Mountains near Keene (Fig. 2). This reconstruction is similar to that of Dickinson (1983) but requires less displacement than the 195 km required for that reconstruction because different criteria are used for determining the total offset of Salinia.
10. Stewart Valley fault: 30 km dextral (Stevens et al., 1991).
11. Tinemaha fault: 65 km dextral (Stevens and Greene, 1999). This fault was originally postulated as an Upper Triassic structure, although new evidence suggests that it may be Cretaceous or younger (Glazner et al., 2003; Stevens et al., 2003).

In addition to restoring these offsets, the Gulf of California was closed, considerable liberties were taken in straightening the $Sr_i = 0.706$ isopleth, especially in westernmost Nevada and the northern Sierra Nevada, and many of the fault-bounded blocks were distorted in order to complete the assembly.

The amounts of displacement on some of the faults listed above are uncertain, but most of these uncertainties do not seriously affect the overall shape of the reconstruction. Disagreements

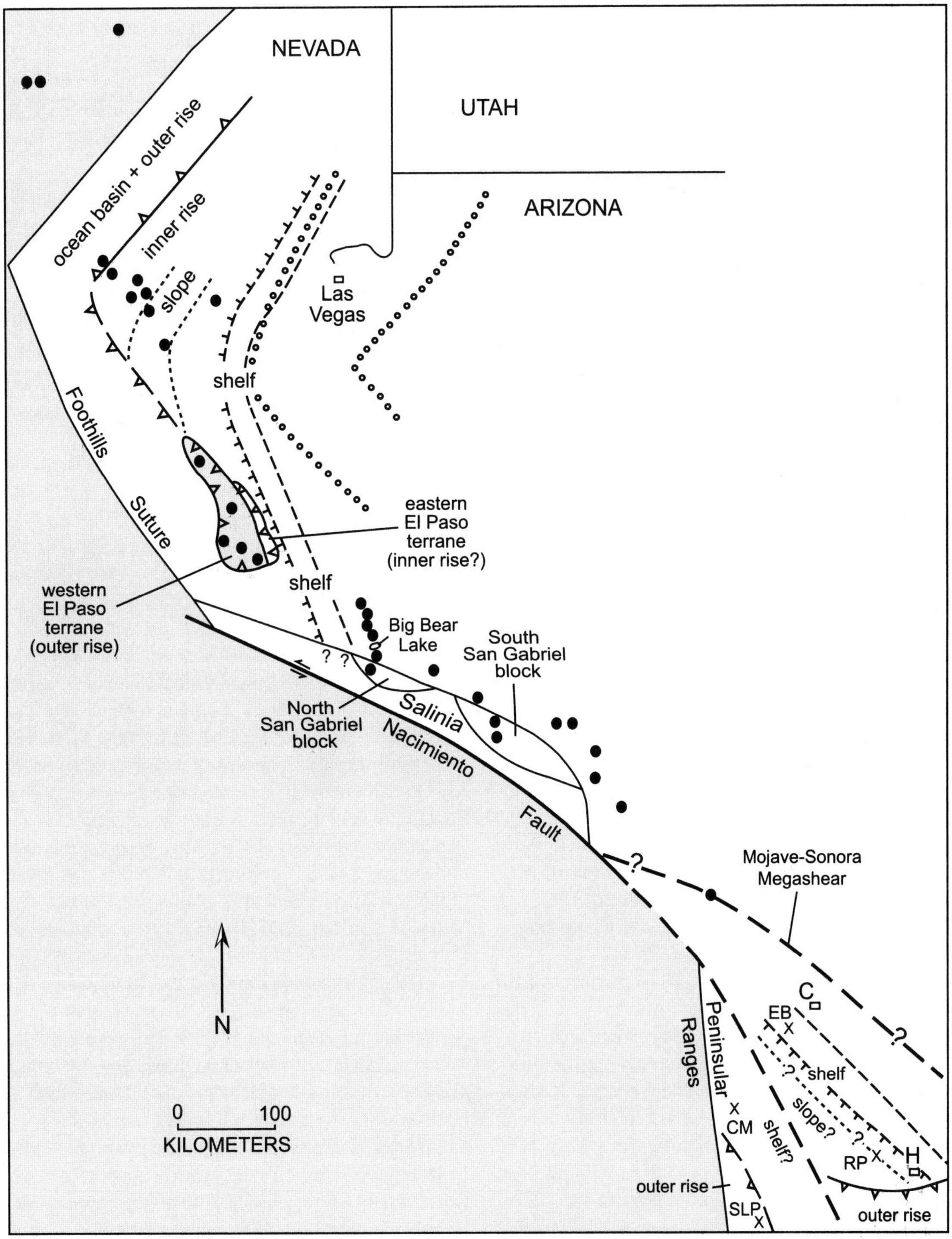

Figure 3. Paleogeographic reconstruction, step 1. Restoration of Cenozoic and Mesozoic strike-slip faults. See discussion in text. Note distortion of San Gabriel blocks and Salinia to allow trace of Nacimiento fault and its inferred southeastern extension to be drawn as a smooth line. C—Caborca. Other abbreviations and symbols as in Figure 2.

on the amount of displacement on Intrabatholithic break 2 (IBB2) of Kistler (1993), however, are very significant. This cryptic fault is equivalent to the axial intrabatholithic break (AIB) of Saleeby and Busby (1993) and much of the Mojave–Snow Lake fault of Schweickert and Lahren (1991). Schweickert and Lahren (1991) considered dextral displacement on the Mojave–Snow Lake fault to be ~400 km, with the Snow Lake pendant, located ~35 km west of Saddlebag Lake (Fig. 1), having originated in the western Mojave Desert where rocks they interpreted to be similar to those at Snow Lake are exposed. Kistler (1993), however, showed only ~100 km of dextral offset of the Sr_i = 0.706 isopleth across IBB2, and Saleeby and Busby (1993) inferred a similar amount of displacement based on several lines of evidence suggesting that the Snow Lake pendant originated west of the Inyo Mountains. Recently, Lewis and Girty (2001) presented petrographic and geochemical data suggesting that displacement on that fault was less than 200 km. Our own reconnaissance of the Mount Pinchot area west of the central Inyo Mountains (Fig. 1) has indicated that clastic rocks including conglomerates of possible Triassic or Jurassic age overlie rocks identified as Cambrian by Bartley et al. (2001). Therefore, the succession near Mount Pinchot may be similar to that in the Snow Lake pendant ~100 km to the north. We therefore follow Kistler (1993) and Saleeby and Busby (1993) in the interpretation that displacement on this fault is ~100 km, the amount necessary to align the inferred margin of Precambrian continental crust.

This first reconstruction (Fig. 3) moves Salinia southeastward to a position west of the Mojave Desert and moves the Peninsular Ranges southeastward to a position west of the Caborca block in northwestern Mexico. The western margin of Precambrian crust is restored to a relatively simple path that trends southwestward from Nevada into California, then turns southeastward to follow the Foothills Suture into the northwestern part of Salinia. From here the margin of Precambrian crust follows the Nacimiento fault, which is not restored in this step of the reconstruction, southeastward to the northern Peninsular Ranges. Following Dickinson (1983), we consider the Nacimiento fault to be a major sinistral strike-slip fault of mid-Cretaceous age.

In southern Nevada and California, the reconstruction of Figure 3 restores the three Paleozoic sedimentary boundaries described earlier to what we interpret as their original configurations. Of particular interest are the middle Silurian–Lower Devonian zero isopach and the Meramecian shelf margin. Both of these boundaries, as reconstructed here, trend south-southwestward across southern Nevada into California, where they bend and extend in a south-southeastward direction to the area west of Big Bear Lake. Both boundaries are at an angle to the margin of continental crust along the southwestern edge of Salinia and appear to be truncated.

Also in this first reconstruction, the oceanic western El Paso terrane (which includes rocks of the outer-rise facies belt) is restored across the Garlock fault but remains in an anomalous position surrounded by areas underlain by continental crust (Fig. 3). The thrust fault or system of thrust faults interpreted to bound this terrane and the inferred inner-rise area to the east (eastern El Paso terrane) are interpreted to extend northwest to the Saddlebag Lake area in the eastern Sierra Nevada and then northeast into Nevada. The age and extent of this thrust system are discussed in more detail later.

Step 2: Restoration of Cretaceous Sinistral Displacement of the Peninsular Ranges on the Nacimiento Fault

The structural evolution of the Nacimiento fault is uncertain and probably complex (e.g., Page, 1981). Dickinson (1983), however, proposed a paleogeographic reconstruction in which the Nacimiento fault is interpreted to have originated as a major sinistral strike-slip fault of mid-Cretaceous age. After removing Cenozoic displacements on the San Andreas and proto–San Andreas fault systems, the reconstruction of Dickinson (1983) restores the Peninsular Ranges to a position west of Salinia by removing 560 km of mid-Cretaceous sinistral slip on the Nacimiento fault, which aligns older Mesozoic lithologic belts from the Peninsular Ranges through Salinia and into the Sierra Nevada. We accept the fundamental premise of this interpretation and use it as the basis for the second step of our reconstruction (Fig. 4).

To perform this restoration, however, it is first necessary to decide where to project the Nacimiento fault to the southeast. One possibility (see Fig. 3) would be to connect the Nacimiento fault with the Mojave-Sonora megashear, with which it is approximately aligned. Reversal of sinistral displacement on a combined Nacimiento fault and Mojave-Sonora megashear would restore the Caborca block together with the Peninsular Ranges to a position southwest of Salinia, resulting in a plausible alignment of some paleogeographic features. Such a fault connection, however, is evidently prohibited by relations in Mexico that limit the age of faulting on the megashear to Late Jurassic or older (Anderson and Silver, 1979). This discrepancy was also noted by Dickinson (1983), who depicted the southeastward extension of the Nacimiento fault to pass southwest of the Mojave-Sonora megashear. In concert with this interpretation, we here place the southeastward extension of the Nacimiento fault between the Peninsular Ranges and the Caborca block as shown in Figure 3.

Restoration of displacement on the Nacimiento fault in our second reconstruction (Fig. 4) therefore puts the Peninsular Ranges southwest of Salinia while leaving the Caborca block in place. We restore a displacement of 600 km, which aligns the Sr_i = 0.706 isopleth in the northern Peninsular Ranges with that in Salinia and is in reasonable agreement with the 560 km inferred by Dickinson (1983). This alignment is not required by any independent data but is consistent with our interpretation of the late Paleozoic evolution of the continental margin, as discussed in subsequent steps of our reconstruction.

Basis for Further Reconstruction

Figure 4 shows a number of features that suggest the southern California continental margin was truncated at some time

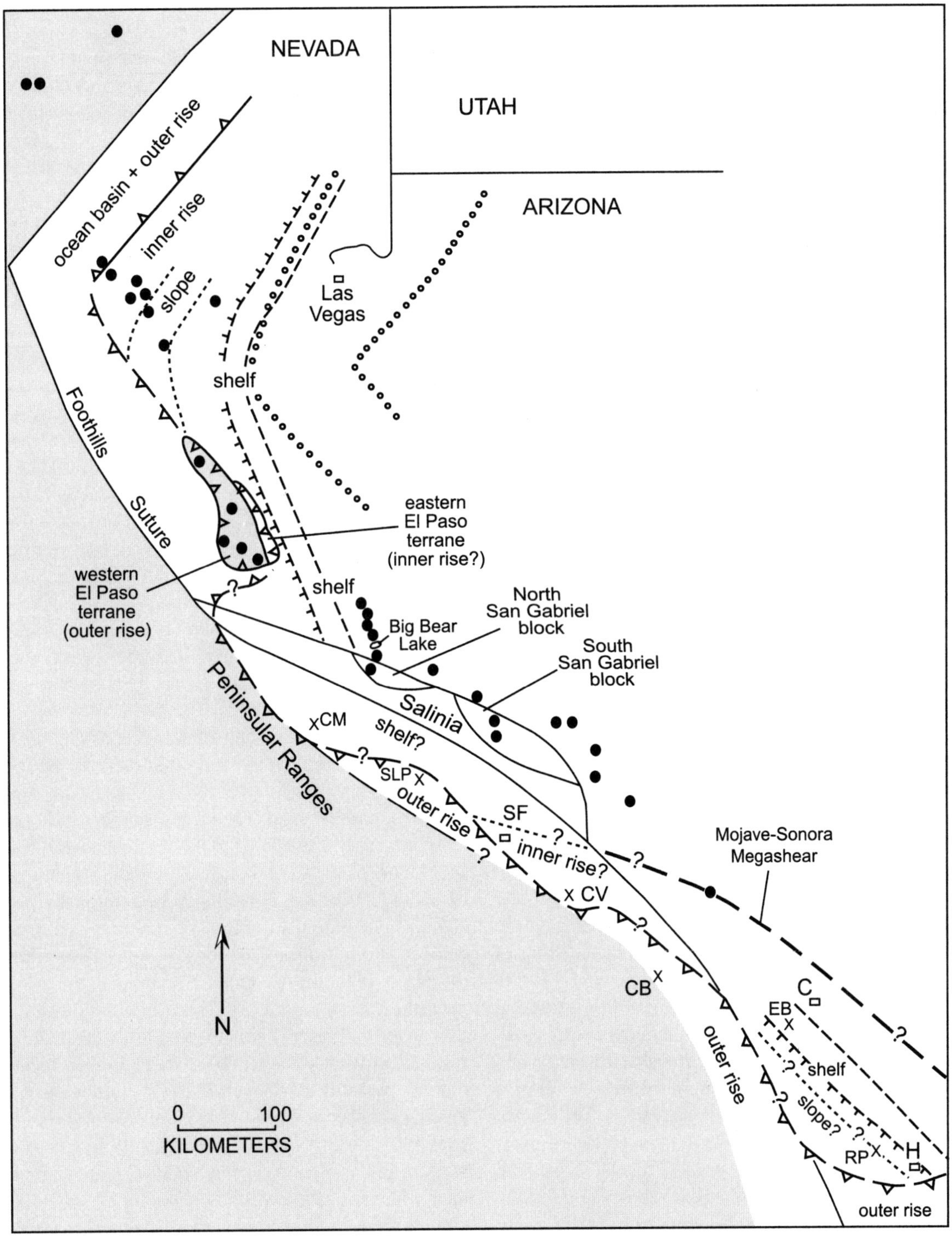

Figure 4. Paleogeographic reconstruction, step 2. Restoration of Cretaceous sinistral displacement of the Peninsular Ranges on the Nacimiento fault. See discussion in text. Abbreviations and symbols as in Figure 2.

prior to sinistral displacement on the Nacimiento fault. Truncation is most strongly indicated by an apparent sinistral offset of Paleozoic facies belts and sedimentary boundaries from California to the Caborca area in Mexico (Fig. 4). Any further reconstruction of the continental margin rests on an interpretation of the timing of the truncational event: whether it was Late Jurassic, late Paleozoic to early Mesozoic, or as early as Pennsylvanian.

Several lines of evidence suggest a pre-Jurassic truncation. First, the Upper Permian to Upper Triassic plutonic belt, as restored in Figure 4, cuts across all of the Devonian facies belts, the middle Silurian–Lower Devonian zero isopach, and the Meramecian shelf margin. This relationship suggests that initiation of subduction beneath a new, southeast-trending continental margin formed prior to the Jurassic. Second, a pre-Permian truncation would explain the marked southeasterly bend of the Permian coral belt in eastern California (Fig. 4). This belt diverges from the Meramecian shelf margin, suggesting that the continental margin changed to a more southeastward orientation between Late Mississippian and Early Permian time. Third, a Pennsylvanian age of truncation would account for the development of a Pennsylvanian to earliest Permian turbidite basin (the Keeler basin of Stevens et al., 2001) that cuts across the Mississippian carbonate shelf immediately southeast of the Inyo Mountains (Stone and Stevens, 1988). Finally, a Carboniferous to Permian truncation along the Foothills Suture explains relations of the Kings-Kaweah ophiolite belt in the southwestern Sierra Nevada (Saleeby, 1992). On the basis of this evidence, we favor the concept that continental truncation was initiated during the Pennsylvanian. This interpretation guides the remaining steps of our paleogeographic reconstruction.

Step 3: Restoration of Triassic Thrust Displacement of the El Paso Terrane

In Figure 4, the eastern and western El Paso terranes, interpreted to represent the inner- and outer-rise facies belts, are still out of place with respect to the Devonian continental shelf and slope. As indicated earlier, Miller et al. (1995) considered the western El Paso terrane to be allochthonous, with thrusting of these rocks onto continental crust between 240 and 179 Ma (early Middle Triassic to early Middle Jurassic). If this is true and if the western Mojave Desert is underlain by continental crust, as argued earlier, the western El Paso terrane must have been displaced from a position at least 100 km to the west, the closest location of rocks underlain by oceanic crust (Fig. 4). The eastern El Paso terrane similarly appears to be ~100 km too far east with respect to the Devonian shelf. Based on these relationships, both the eastern and western El Paso terranes are shown to be in the upper plate of a thrust fault that carried them ~100 km eastward and emplaced them above rocks of the Devonian continental shelf facies (Fig. 4).

We further speculate that this inferred thrust fault extended northwestward to connect with or to be related to a Triassic fault in the Saddlebag Lake pendant in the Sierra Nevada that Schweickert and Lahren (1987, 1993b) correlated with the Golconda thrust in north-central Nevada. Here we refer to this fault as the "Golconda" thrust because of its questionable correlation with the Golconda thrust in its type area. The "Golconda" thrust in the Saddlebag Lake pendant is interpreted to have emplaced lower to middle Paleozoic ocean-basin rocks against inner-rise facies (Stevens and Greene, 1999) between ca. 245 and 220 Ma (Schweickert and Lahren, 1987, 1993b). The amount of displacement on this fault in the Saddlebag Lake pendant is unknown, but considering the great difference in lithology and thickness between upper Paleozoic facies in the two plates (Stevens and Greene, 1999), it must have been large. Thus, the thrusts in both the Saddlebag Lake pendant and the El Paso Mountains could have been active between ca. 240 and 220 Ma, and despite the lack of direct evidence, we suggest that both thrusts may represent the same event. From Saddlebag Lake, the "Golconda" thrust is shown extending northeastward through the Miller Mountain–Candelaria Hills area (e.g., Schweickert and Lahren, 1993b; Fig. 4).

Finally, this thrust system as shown in Figure 4 is extended southeastward to underlie outer-rise strata in the Peninsular Ranges of Baja California and in Sonora near Hermosillo. This depiction of thrust faults should be viewed as highly diagrammatic and speculative; in fact, a direct connection between the Late Permian thrust faulting documented near Hermosillo (Poole et al., 1995) and the Triassic thrust faulting in the Saddlebag Lake area seems improbable. For the purposes of our reconstruction, however, it is reasonable to infer a general continuity of thrust faults that emplaced outer-rise and ocean-basin rocks onto the continental margin at various times between the Late Permian and Late Triassic.

Thus, in the next step of the reconstruction (Fig. 5), we restore the El Paso terrane and other rocks in the upper plate of the "Golconda" thrust system to the west where they are shown overlying oceanic crust. The outer-rise rocks in Sonora and Baja California also are restored to the south and west, respectively.

Figure 5 shows a northwest-trending continental margin extending from the Foothills Suture through the Peninsular Ranges into Baja California. We interpret this margin to have resulted from truncation along a left-lateral transform fault of which both the Foothills Suture and the Mojave-Sonora megashear are a part, as discussed in more detail below. Along this margin, Upper Permian to Upper Triassic plutons are restored to what we infer to be their original locations. These plutons probably developed above a northwest-trending, east-dipping subduction zone located a relatively short distance (perhaps 100–200 km) west of the continental margin. Plutons in the western El Paso terrane appear somewhat anomalous because of their inferred location west of the Foothills Suture. Perhaps these plutons reflect a local irregularity in the strike or dip of the subduction zone or segmentation of the subducting slab.

Plutonism in the El Paso Mountains and the Victorville–San Bernardino Mountains area largely postdated latest Permian to earliest Triassic deformation and metamorphism (Miller, 1981; Miller and Cameron, 1982; Carr et al., 1997). Folds and thrust

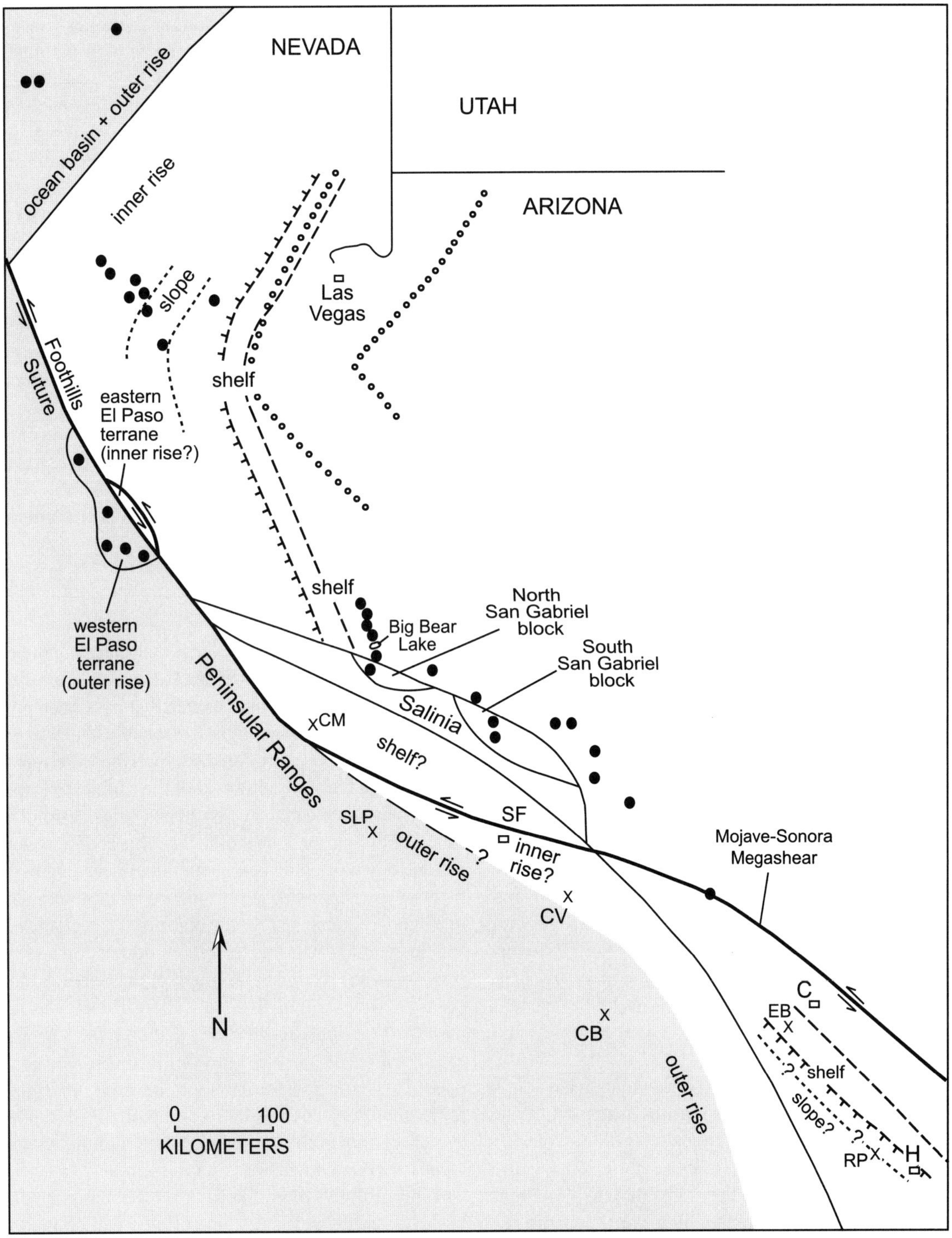

Figure 5. Paleogeographic reconstruction, step 3. Restoration of Triassic thrust displacement of the El Paso Mountains and other areas. Also includes restoration of Late Permian thrust displacement near Hermosillo. See discussion in text. Abbreviations and symbols as in Figure 2.

faults of this age in the El Paso Mountains (Carr et al., 1997) are west-vergent and therefore consistent with deformation above an east-dipping subduction zone.

Steps 4A and 4B: Two-phase Restoration of Pennsylvanian Sinistral Displacements on a Continental Truncation Fault System

Figure 5 depicts the margin of Precambrian continental crust and the associated paleogeographic features as we interpret their configurations at about Middle Triassic time, prior to thrusting of deep-water rocks onto the continental margin but after continental truncation and displacement of the Caborca block on the Mojave-Sonora megashear. In this reconstruction, we interpret the megashear to continue northwestward from northern Mexico across the Peninsular Ranges of California, following the imprecisely located boundary between the shelf(?) and inner-rise(?) facies belts, then to follow the ocean-continent boundary and connect with the Foothills Suture. This restored fault path defines what we interpret as the probable locus of the Pennsylvanian continental truncation that is broadly analogous to the California-Coahuila transform of Dickinson (2000). Across this proposed truncational fault, the Devonian facies boundaries, the middle Silurian–Lower Devonian zero isopach, and the Meramecian shelf margin in the Caborca block appear to be sinistrally offset by ~400 km from the analogous features in California.

The rationale for our proposed path of the truncation fault in the Peninsular Ranges (Fig. 5) is that the rocks in the Coyote Mountains probably are shelfal and therefore in place relative to the shelf rocks near Big Bear Lake, whereas the inner-rise(?) rocks near San Felipe appear to be significantly offset from those in California. This fault path, although plausible, must be regarded as speculative because of the questionable nature of our facies-belt assignments in the Peninsular Ranges.

The western El Paso terrane also appears to be sinistrally offset by ~400 km from the restored locations of ocean-basin and outer-rise rocks at Miller Mountain to the north, consistent with the apparent offset of the Caborca block. The eastern El Paso terrane, however, which we interpret to represent the inner-rise belt, is offset only ~200 km from the inner-rise facies belt in eastern California. This smaller offset requires this terrane to represent a stranded block surrounded by faults.

Restoration of 400 km of sinistral displacement on the continental truncation fault shown in Figure 5 results in step 4A of our reconstruction (Fig. 6).

As shown in Figure 6, the margin of Precambrian crust in the northern Peninsular Ranges has not been brought into alignment with the continent-ocean boundary stretching from westernmost central Nevada to eastern California, as might be expected in a complete reconstruction of the continental margin. Thus, we propose the existence of an earlier left-lateral truncational fault to account for the remaining apparent offset of ~175 km. This proposed fault is shown extending oceanward of the western El Paso terrane, following the Foothills Suture, and separating the Caborca terrane from the southern part of the Peninsular Ranges (Fig. 6). In the final step (4B) of our reconstruction (Fig. 7), displacement on this proposed fault is restored, aligning the margin of Precambrian crust and bringing the inner-rise(?) rocks near San Felipe into reasonable alignment with the inner-rise facies belt of California.

This final reconstruction leads to the following interpretation of the late Paleozoic to early Mesozoic evolution of the continental margin. First, a sliver of continental crust consisting of the southern part of the Peninsular Ranges was carried southward ~175 km, probably in the Middle Pennsylvanian. Later, probably also in the Pennsylvanian, the northern and southern parts of the truncation fault jumped to a more easterly position. At this time the western El Paso terrane and the Caborca block were carried ~400 km southward. Later, in the Late Permian to Late Triassic, convergence led to subduction and emplacement of plutonic rocks. The latter part of this episode was accompanied by emplacement of the "Golconda" thrust over the continental margin. This scenario is consistent with the evolution of the Foothills Suture, which is interpreted to have originated as a Carboniferous to Permian transform margin against which the Kings-Kaweah ophiolite was subsequently emplaced in later Permian to Triassic time (Saleeby, 1992).

DISCUSSION

The idea that the continental margin of southwestern North America was modified by a late Paleozoic tectonic event is supported by several lines of evidence that have been presented here and by many other geologists. Probably the best evidence for this interpretation is provided by the Upper Permian to Triassic plutonic belt that cuts across all of the Middle to Upper Devonian facies belts delineated in this paper (Figs. 3–5). The existence of this plutonic belt suggests that the continental margin was tectonically reoriented before subduction leading to the emplacement of the plutons began. Evidence already summarized points to a Pennsylvanian age for the onset of this tectonic modification. It is less certain that this tectonic modification and reorientation necessarily involved displacement of the Caborca block. It is still possible, for example, that the Caborca block was not displaced until the Late Jurassic. The idea that both the Caborca block and the El Paso terrane were sinistrally displaced along the same upper Paleozoic transform margin, however, is attractive not only for its simplicity but also for its compatibility with the interpreted evolution of both the continental shelf and the orientation of the Upper Permian to Upper Triassic plutonic belt. We therefore consider it much more likely that Late Jurassic activity on the Mojave-Sonora megashear overprinted earlier, major movement on an upper Paleozoic transform fault.

The alternative idea that the Caborca block has never been displaced with respect to California also needs to be considered. One aspect of our new paleogeographic reconstructions (Figs. 3–7) is that the pre-Pennsylvanian continental shelf evidently trended south-southeastward across the Mojave Desert to the Big

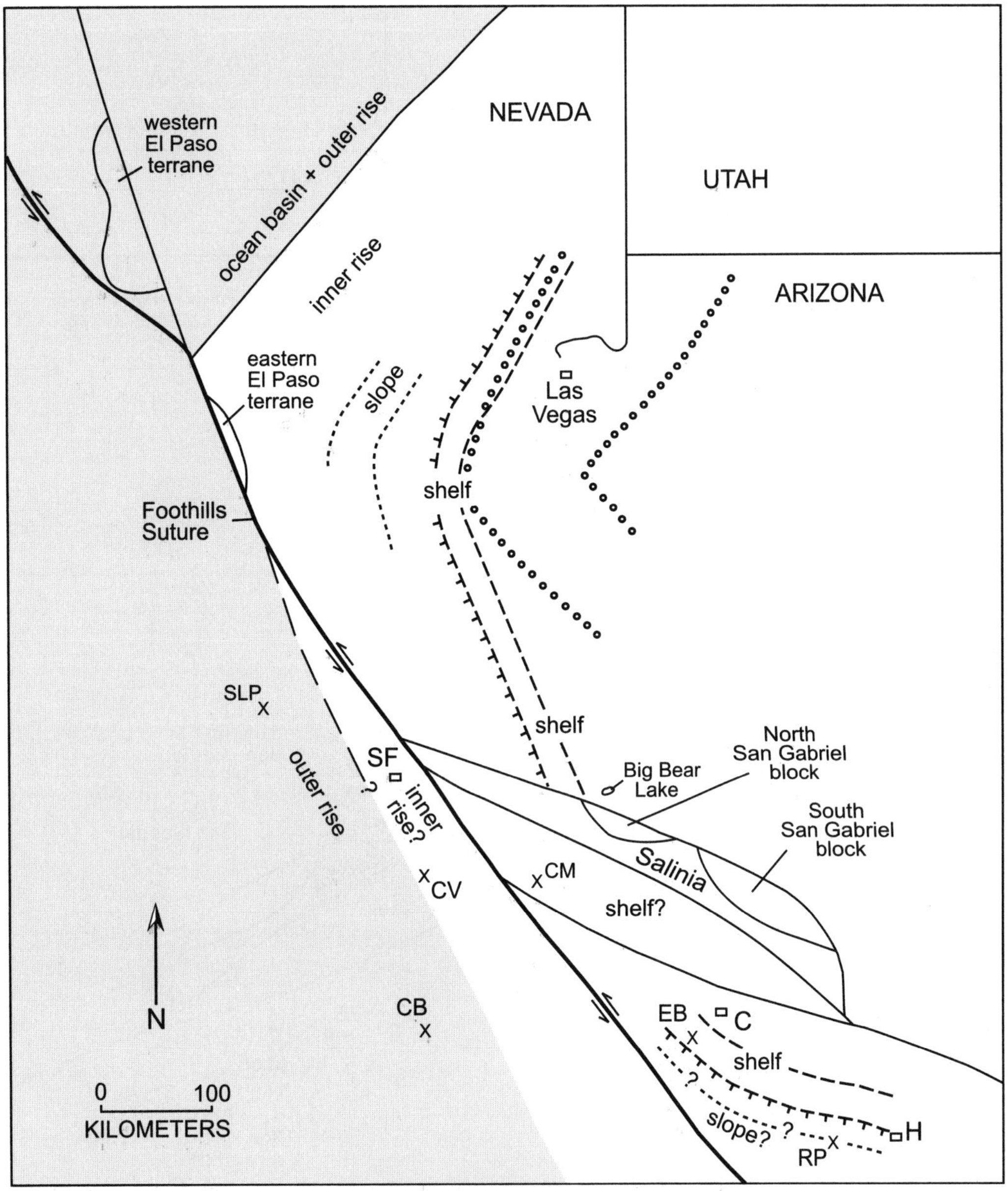

Figure 6. Paleogeographic reconstruction, step 4A. Restoration of sinistral displacement on a proposed Pennsylvanian continental truncation fault including the Mojave-Sonora megashear. See discussion in text. Abbreviations and symbols as in Figure 2.

Bear Lake area, rather than continuing on the south-southwestward trend exhibited in southern Nevada and east-central California. Evidently, the shelf and underlying continental margin were "wrapping around" southeastern California toward Mexico even before the proposed Pennsylvanian truncation. As a result, the angle between the pre-Pennsylvanian continental margin of California and a proposed southeast-trending truncational fault like that shown in Figure 5 is smaller than has been presumed in most previous models of continental truncation (e.g., Stone and Stevens, 1988). This smaller angle decreases the amount of apparent offset of facies belts and sedimentary boundaries between California and the Caborca block, thereby reducing the need for major displacement on a truncating fault to align these features. In the reconstruction of Figure 5, for example, it is possible to envision a direct or nearly direct paleogeographic connection between California and Sonora that would not require much if any displacement of the Caborca block. Most of the evidence, however, still points to the conclusion that the Caborca block has indeed been displaced from California, although the offset may not be as large as previously proposed.

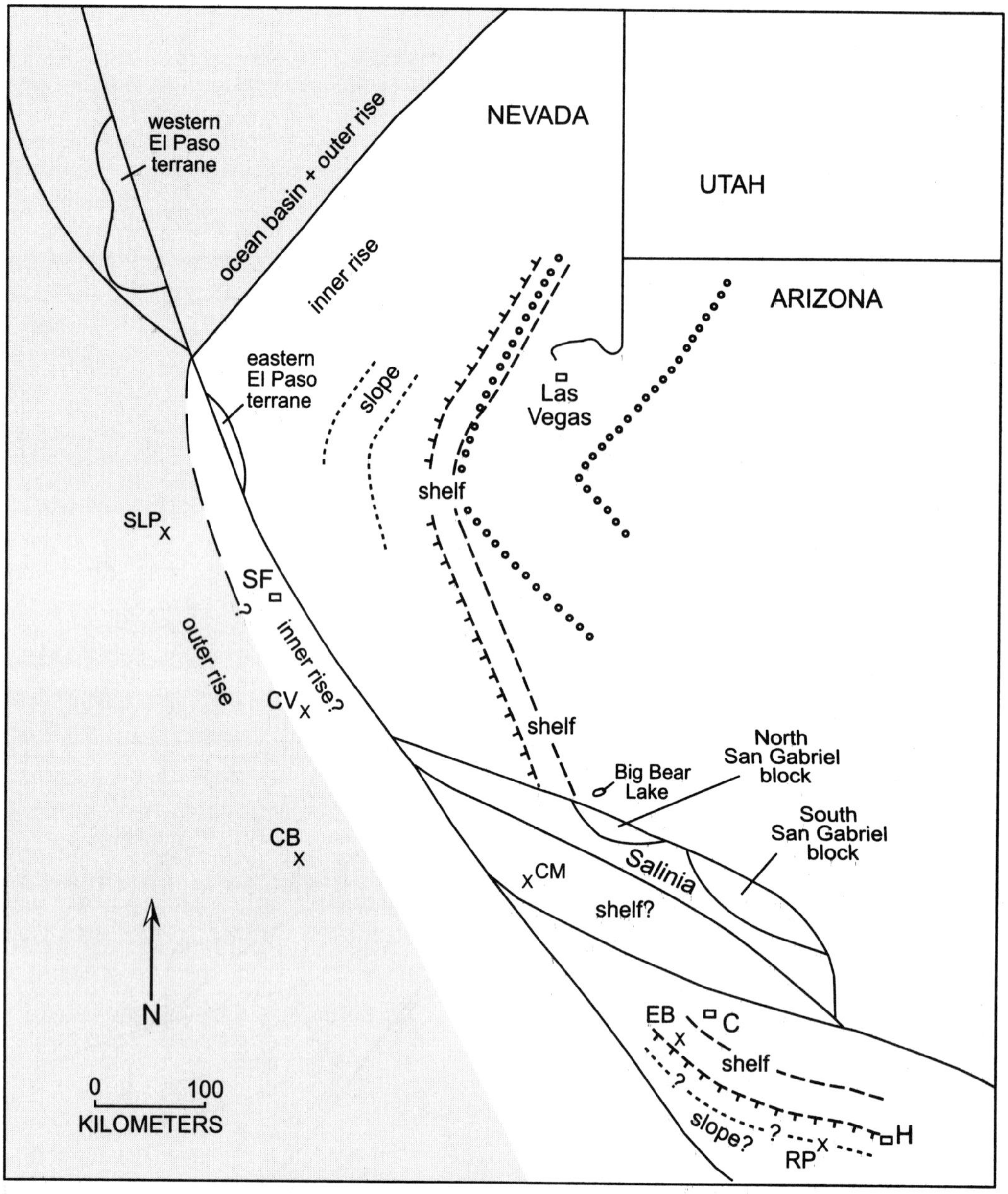

Figure 7. Paleogeographic reconstruction, step 4B. Restoration of sinistral displacement on a proposed Pennsylvanian continental truncation fault predating the Mojave-Sonora megashear. See discussion in text. Abbreviations and symbols as in Figure 2.

The paleogeographic reconstruction presented here is similar in some respects to that of Stevens et al. (1992), who also interpreted the continental margin to have been truncated by late Paleozoic sinistral faulting that caused large southward displacement of both the Caborca block and rocks of the El Paso Mountains. This previous reconstruction, however, placed the truncational fault east of the present position of the El Paso Mountains rather than along the Foothills Suture to the west and did not require the El Paso terrane to have been thrust eastward to its present position in the Late Triassic. This previous reconstruction cannot be completely ruled out; in fact, it would result in a more linear belt of Permian and Triassic plutons than in the present reconstruction after restoration of the inferred Late Triassic thrust faulting (compare Figs. 4 and 5). On the other hand, the previous reconstruction does not account for the presence of Precambrian continental crust west of the El Paso Mountains, as interpreted in this paper, or for the probable role of the Foothills Suture in a late Paleozoic continental truncation event (Saleeby, 1992). Thus, the previous reconstruction would require substantial modification to remain a viable alternative to the one presented here.

CONCLUSIONS

The available geologic data, including the positions and orientations of Devonian facies belts, pre-Pennsylvanian sedimentary boundaries on the continental shelf, and extensive geochemical data, suggest that the Neoproterozoic to Paleozoic miogeocline and underlying continental margin in southern California have been truncated. The trends of the Middle Pennsylvanian–Lower Permian Keeler basin and the belt of Upper Permian to Upper Triassic plutons in western North America, both of which cut across older facies belts, suggest that the truncation was initiated during the Pennsylvanian. This interpretation is supported by the trend of the Lower Permian "*Thysanophyllum*" coral belt, which is more or less parallel to that of the plutonic belts (Figs. 3–5).

Our paleogeographic reconstruction of the Paleozoic continental margin leads us to propose a two-phase truncational event on two different but related sinistral transform faults. First, the southern Peninsular Ranges, which originally lay west of east-central California, were displaced ~175 km southward on a fault that passed west of the Caborca block. Later, the Caborca block and the western El Paso terrane were transported ~400 km southward on a new fault that partially followed the older fault but diverged from it both to the north and south. In Mexico, we interpret this strand of the fault to be equivalent to the Mojave-Sonora megashear, which we therefore consider to have played an integral role in this late Paleozoic (Pennsylvanian) continental truncation event. If this interpretation is correct, Late Jurassic displacement on the megashear was minor and overprinted earlier, major movement on an upper Paleozoic transform fault.

ACKNOWLEDGMENTS

We are indebted to Andy Barth for providing us with information on the Permian and Triassic plutons in the central and southern California area, to Richard Sedlock for his views on the post–early Mesozoic structural development of California, and to Howard Brown for suggestions on correlation of rocks in the Mojave Desert region. We also thank James McKee, Benita Murchey, Allison Palmer, Norm Silberling, Jack Stewart, and Maureen Steiner for their helpful reviews and Scott McPeek for his help in drafting the figures. Finally, we deeply appreciate the invitation from Tom Anderson to write a paper for this volume.

REFERENCES CITED

Anderson, P.V., 1993, Prebatholithic stratigraphy of the San Felipe area, Baja California Norte, Mexico, *in* Gastil, R.G., and Miller, R.H., eds., The prebatholithic stratigraphy of peninsular California: Geological Society of America Special Paper 279, p. 1–10.

Anderson, T.H., and Schmidt, V.A., 1983, The evolution of Middle America and the Gulf of Mexico–Caribbean Sea region during Mesozoic time: Geological Society of America Bulletin, v. 94, p. 941–966, doi: 10.1130/0016-7606(1983)94<941:TEOMAA>2.0.CO;2.

Anderson, T.H., and Silver, L.T., 1979, The role of the Mojave-Sonora megashear in the tectonic evolution of northern Sonora, *in* Anderson, T.H., and Roldan-Quintana, J., eds., Geology of northern Sonora: Hermosillo, Sonora, Sonora Instituto de Geología, Universidad Nacional Autónoma de México, p. 59–68.

Barth, A.P., Tosdal, R.M., Wooden, J.L., and Howard, K.A., 1997, Triassic plutonism in southern California: Southward younging of arc initiation along a truncated continental margin: Tectonics, v. 16, p. 290–304, doi: 10.1029/96TC03596.

Bartley, J.M., Glazner, A.F., and Mahan, K.H., 2001, Structure of the Cardinal Mountain interpluton screen, Sierra Nevada batholith, California: Geological Society of America Abstracts with Programs, v. 33, no. 3, p. A-50.

Bateman, P.C., 1992, Plutonism in the central part of the Sierra Nevada batholith, California: U.S. Geological Survey Professional Paper 1483, 186 p.

Boettcher, S.S., and Walker, J.D., 1993, Geologic evolution of Iron Mountain, central Mojave Desert, California: Tectonics, v. 12, p. 372–386.

Brown, H.J., 1991, Stratigraphy and paleogeographic setting of Paleozoic rocks in the San Bernardino Mountains, California, *in* Cooper, J.D., and Stevens, C.H., eds., Paleozoic paleogeography of the western United States—II: Los Angeles, Pacific Section, Society of Economic Paleontologists and Mineralogists, Book 67, p. 193–207.

Buch, I.P., and Delattre, M.P., 1993, Permian and Lower Triassic stratigraphy along the 30th parallel, eastern Baja California Norte, Mexico, *in* Gastil, R.G., and Miller, R.H., eds., The prebatholithic stratigraphy of peninsular California: Geological Society of America Special Paper 279, p. 77–90.

Burchfiel, B.C., and Davis, G.A., 1972, Structural framework and evolution of the southern part of the Cordilleran orogen, western United States: American Journal of Science, v. 272, p. 97–118.

Carr, M.D., Harris, A.G., Poole, F.G., and Fleck, R.J., 1992, Stratigraphy and structure of Paleozoic outer continental-margin rocks in Pilot Knob Valley, north-central Mojave Desert, California: U.S. Geological Survey Bulletin 2015, 33 p.

Carr, M.D., Christiansen, R.L., Poole, F.G., and Goodge, J.W., 1997, Bedrock geologic map of the El Paso Mountains in the Garlock and El Paso Peaks 7 1/2′ quadrangles, Kern County, California: U.S. Geological Survey Miscellaneous Investigations Series Map I-2389, scale 1:24,000.

Davis, G.A., and Burchfiel, B.C., 1973, Garlock fault: An intracontinental transform structure, southern California: Geological Society of America Bulletin, v. 84, p. 1407–1422, doi: 10.1130/0016-7606(1973)84<1407:GFAITS>2.0.CO;2.

Davis, G.A., Monger, J.W.H., and Burchfiel, B.C., 1978, Mesozoic construction of the Cordilleran "collage," central British Columbia to central California, *in* Howell, D.G., and McDougall, K.A., eds., Mesozoic paleogeography of the western United States: Los Angeles, Pacific Section, Society of Economic Paleontologists and Mineralogists Pacific Coast Paleogeography Symposium 2, p. 1–32.

Dickinson, W.R., 1981, Plate tectonics and the continental margin of California, *in* Ernst, W.G., ed., The geotectonic development of California, Rubey Volume 1: Englewood Cliffs, New Jersey, Prentice-Hall, Inc., p. 2–28.

Dickinson, W.R., 1983, Cretaceous sinistral strike slip along Nacimiento fault in coastal California: American Association of Petroleum Geologists Bulletin, v. 67, p. 624–645.

Dickinson, W.R., 2000, Geodynamic interpretation of Paleozoic tectonic trends oriented oblique to the Mesozoic Klamath-Sierran continental margin in California, *in* Soreghan, M.J., and Gehrels, G.E., eds., Paleozoic and Triassic paleogeography and tectonics of western Nevada and northern California: Geological Society of America Special Paper 347, p. 209–245.

Dickinson, W.R., and Lawton, T.F., 2001, Carboniferous to Cretaceous assembly and fragmentation of Mexico: Geological Society of America Bulletin, v. 113, p. 1142–1160, doi: 10.1130/0016-7606(2001)113<1142:CTCAAF>2.0.CO;2.

Dokka, R.K., and Travis, C.J., 1990, Late Cenozoic strike-slip faulting in the Mojave Desert, California: Tectonics, v. 9, p. 311–340.

Ducea, M., 2001, The California arc: Thick granitic batholiths, eclogitic residues, lithospheric-scale thrusting, and magmatic flare-ups: GSA Today, v. 11, no. 11, p. 4–10, doi: 10.1130/1052-5173(2001)011<0004:TCATGB>2.0.CO;2.

Dunne, G.C., 1986, Geologic evolution of the southern Inyo Range, Darwin Plateau, and Argus and Slate Ranges, east-central California—An overview, *in* Dunne, G.C., ed., Mesozoic and Cenozoic structural evolution of selected areas, east-central California, Guidebook and Volume, Field Trips 2 and 14: Los Angeles, Cordilleran Section, Geological Society of America, p. 3–21.

Dunne, G.C., and Saleeby, J.B., 1993, Kern plateau shear zone, southern Sierra Nevada—New data concerning age and northward continuation: Geological Society of America Abstracts with Programs, v. 25, no. 5, p. 33.

Dunne, G.C., and Suczek, C.A., 1991, Early Paleozoic eugeoclinal strata in the Kern Plateau pendants, southern Sierra Nevada, California, *in* Cooper, J.D., and Stevens, C.H., eds., Paleozoic paleogeography of the Western United States—II: Los Angeles, Pacific Section, Society of Economic Paleontologists and Mineralogists, Book 67, p. 677–692.

Fletcher, J.M., Bartley, J.M., Martin, M.W., Glazner, A.F., and Walker, J.D., 1995, Large-magnitude continental extension: An example from the central Mojave metamorphic core complex: Geological Society of America Bulletin, v. 107, p. 1468–1483, doi: 10.1130/0016-7606(1995)107<1468:LMCEAE>2.3.CO;2.

Gastil, G., Miller, R., Anderson, P., Crocker, J., Campbell, M., Buch, P., Lothringer, C., Leier-Engelhardt, P., DeLattre, M., and Hoobs, J., 1991, The relation between the Paleozoic strata on opposite sides of the Gulf of California, *in* Pérez-Segura, E., and Jacques-Ayala, C., eds., Studies of Sonoran geology: Geological Society of America Special Paper 254, p. 7–17.

Gehrels, G.E., Stewart, J.H., and Ketner, K.B., 2002, Cordilleran-margin quartzites in Baja California—Implications for tectonic transport: Earth and Planetary Science Letters, v. 199, p. 201–210, doi: 10.1016/S0012-821X(02)00542-3.

Glazner, A.F., Bartley, J.M., Coleman, D., and Kylander-Clark, A.R.C., 2003, Evidence from the Independence dike swarm for 65 km or more of post-Jurassic dextral offset across Owens Valley, California: Geological Society of America Abstracts with Programs, v. 34, no. 5, p. 80.

Greene, D.C., Schweickert, R.A., and Stevens, C.H., 1997, The Roberts Mountains allochthon and the western margin of the Cordilleran miogeocline in the Northern Ritter Range pendant, eastern Sierra Nevada, California: Geological Society of America Bulletin, v. 109, p. 1294–1305, doi: 10.1130/0016-7606(1997)109<1294:RMAATW>2.3.CO;2.

Gromet, L.P., and Silver, L.T., 1987, REE variations across the Peninsular Ranges batholith: Implications for batholithic petrogenesis and crustal growth in magmatic arcs: Journal of Petrology, v. 28, p. 75–125.

Hall, W.E., and MacKevett, E.M., Jr., 1962, Geology and ore deposits of the Darwin quadrangle, Inyo County, California: U.S. Geological Survey Professional Paper 368, 87 p.

Hamilton, W., 1969, Mesozoic California and the underflow of Pacific mantle: Geological Society of America Bulletin, v. 80, p. 2409–2430.

Kiser, N.L., 1981, Stratigraphy, structure and metamorphism in the Hinkley Hills, Barstow, California [M.S. thesis]: Palo Alto, California, Stanford University, 70 p.

Kistler, R.W., 1978, Mesozoic paleogeography of California: A viewpoint from isotope geology, *in* Howell, D.G., and McDougall, K.A., eds., Mesozoic paleogeography of the western United States: Los Angeles, Pacific Section, Society of Economic Paleontologists and Mineralogists Pacific Coast Paleogeography Symposium 2, p. 75–84.

Kistler, R.W., 1990, Two different lithosphere types in the Sierra Nevada, California, *in* Anderson, J.L., ed., The Nature and Origin of Cordilleran Magmatism: Geological Society of America Memoir 174, p. 271–281.

Kistler, R.W., 1993, Mesozoic intrabatholithic faulting, Sierra Nevada, California, *in* Dunne, G.C., and McDougall, K.A., eds., Mesozoic paleogeography of the western United States—II: Los Angeles, Pacific Section, Society of Economic Paleontologists and Mineralogists, Book 71, p. 247–261.

Kistler, R.W., and Champion, D.E., 2001, Rb-Sr whole-rock and mineral ages, K-Ar, ^{40}Ar/^{39}Ar, and U-Pb mineral ages, and strontium, lead, neodymium, and oxygen isotopic compositions for granitic rocks from the Salinian composite terrane, California: U.S. Geological Survey Open-File Report 01-453, 83 p.

Kistler, R.W., and Peterman, Z.E., 1973, Variations in Sr, Rb, K, Na, and initial ^{87}Sr/^{86}Sr in Mesozoic granitic rocks and intruded wallrocks in central California: Geological Society of America Bulletin, v. 84, p. 3489–3512, doi: 10.1130/0016-7606(1973)84<3489:VISRKN>2.0.CO;2.

Lewis, J.G., and Girty, G.H., 2001, Tectonic implications of a petrographic and geochemical characterization of the Lower to Middle Jurassic Sailor Canyon Formation, northern Sierra Nevada, California: Geology, v. 29, p. 627–630, doi: 10.1130/0091-7613(2001)029<0627:TIOAPA>2.0.CO;2.

Martin, M.W., and Walker, J.D., 1991, Upper Precambrian to Paleozoic paleogeographic reconstruction of the Mojave Desert, California, *in* Cooper, J.D., and Stevens, C.H., eds., Paleozoic paleogeography of the western United States—II: Los Angeles, Pacific Section, Society Economic Paleontologists and Mineralogists, Book 67, p. 167–192.

Martin, M.W., and Walker, J.D., 1992, Extending the western North American Proterozoic and Paleozoic continental crust through the Mojave Desert: Geology, v. 20, p. 753–756, doi: 10.1130/0091-7613(1992)020<0753:ETWNAP>2.3.CO;2.

Martin, M.W., and Walker, J.D., 1995, Stratigraphy and paleogeographic significance of metamorphic rocks in the Shadow Mountains, western Mojave Desert, California: Geological Society of America Bulletin, v. 107, p. 354–366, doi: 10.1130/0016-7606(1995)107<0354:SAPSOM>2.3.CO;2.

Martin, M.W., Glazner, A.F., Walker, J.D., and Schermer, E.R., 1993, Evidence for right-lateral transfer faulting accommodating en echelon Miocene extension, Mojave Desert, California: Geology, v. 21, p. 355–358, doi: 10.1130/0091-7613(1993)021<0355:EFRLTF>2.3.CO;2.

Merriam, C.W., 1963, Geology of the Cerro Gordo mining district, Inyo County, California: U.S. Geological Survey Professional Paper 408, 83 p.

Miller, E.L., 1981, Geology of the Victorville region, California: Geological Society of America, v. 92, part 1, p. 160–163, part 2, p. 554–608.

Miller, E.L., and Cameron, C.S., 1982, Late Precambrian to Late Cretaceous evolution of the southwestern Mojave Desert, California, *in* Cooper, J.D., Troxel, B.W., and Wright, L.A., eds., Geology of selected areas in the San Bernardino Mountains, western Mojave Desert, and southern Great Basin, California, Guidebook and Volume, Field Trip 9: Los Angeles, Cordilleran Section, Geological Society of America, p. 21–34.

Miller, J.S., and Walker, J.D., 2002, Mesozoic geologic evolution of Alvord Mountain, central Mojave Desert, California, *in* Glazner, A.F., Walker, J.D., and Bartley, J.M., eds., Geologic evolution of the Mojave Desert and southwestern Basin and Range: Geological Society of America Memoir 195, p. 59–77.

Miller, J.S., Walker, J.D., Glazner, A.F., and Martin, M.W., 1995, Geochronologic and isotope evidence for Triassic-Jurassic emplacement of the eugeoclinal allochthon in the Mojave Desert region, California: Geological Society of America Bulletin, v. 107, p. 1441–1457, doi: 10.1130/0016-7606(1995)107<1441:GAIEFT>2.3.CO;2.

Miller, J.S., Glazner, A.F., and Crowe, D.E., 1996, Muscovite-garnet granites in the Mojave Desert: Relation to crustal structure of the Cretaceous arc: Geology, v. 24, p. 335–338, doi: 10.1130/0091-7613(1996)024<0335:MGGITM>2.3.CO;2.

Miller, J.S., Glazner, A.F., Farmer, G.L., Suayah, I.B., and Keith, L.A., 2000, A Sr, Nd, and Pb isotopic study of mantle domains and crustal structure from Miocene volcanic rocks in the Mojave Desert, California: Geological Society of America Bulletin, v. 112, p. 1264–1279, doi: 10.1130/0016-7606(2000)112<1264:ASNAPI>2.3.CO;2.

Miller, R.H., 1976, Revision of Upper Ordovician, Silurian, and Lower Devonian stratigraphy, southwestern Great Basin: Geological Society of America Bulletin, v. 87, p. 961–968, doi: 10.1130/0016-7606(1976)87<961:ROUOSA>2.0.CO;2.

Miller, R.H., and Dockum, M.S., 1983, Ordovician conodonts from metamorphosed carbonates of the Salton Trough, California: Geology, v. 11, p. 410–412, doi: 10.1130/0091-7613(1983)11<410:OCFMCO>2.0.CO;2.

Molina Garza, R.S., and Geissman, J.W., 1999, Paleomagnetic data from the Caborca terrane, Mexico: Implications for Cordilleran tectonics and the Mojave-Sonora megashear hypothesis: Tectonics, v. 18, p. 293–325, doi: 10.1029/1998TC900030.

Moore, J.H., and Foster, C.T., Jr., 1980, Lower Paleozoic metasedimentary rocks in the east-central Sierra Nevada, California: Correlation with Great Basin formations: Geological Society of America Bulletin, v. 91, p. 37–43, doi: 10.1130/0016-7606(1980)91<37:LPMRIT>2.0.CO;2.

Nelson, C.A., 1966, Geologic map of the Blanco Mountain quadrangle, Inyo and Mono Counties, California: U.S. Geological Survey Map GQ-529, scale 1:62,500.

Page, B.M., 1981, The southern Coast Ranges, *in* Ernst, W.G., ed., The geotectonic development of California, Rubey Volume 1: Englewood Cliffs, New Jersey, Prentice-Hall, Inc., p. 329–417.

Poole, F.G., and Sandberg, C.A., 1977, Mississippian paleogeography and tectonics of the western United States, *in* Stewart, J.H., Stevens, C.H., and Fritsche, A.E., eds., Paleozoic paleogeography of the western United States: Los Angeles, Pacific Section, Society of Economic Paleontologists and Mineralogists Pacific Coast Paleogeography Symposium 1, p. 67–85.

Poole, F.G., Stewart, J.H., Palmer, A.R., Sandberg, C.A., Madrid, R.J., Ross, R.J., Jr., Hintze, L.F., Miller, M.M., and Wrucke, C.T., 1992, Latest Precambrian to latest Devonian time; development of a continental margin, *in* Burchfiel, B.C., Lipman, P.W., and Zoback, M.L., eds., The Cordilleran orogen: Conterminous U.S.: Geological Society of America, The Geology of North America, v. G-3, p. 9–56.

Poole, F.G., Stewart, J.H., Berry, W.B.N., Harris, A.G., Repetski, J.E., Madrid, R.J., Ketner, K.B., Carter, C., and Morales-Ramirez, J.M., 1995, Ordovician ocean-basin rocks of Sonora, Mexico, *in* Cooper, J.D., Droser, M.L., and Finney, S.C., eds., Ordovician odyssey: Short papers for the Seventh International Symposium on the Ordovician System: Fullerton, California, Pacific Section Society of Economic Paleontologists and Mineralogists, Book 77, p. 277–284.

Poole, F.G., Amaya-Martinez, R., and Harris, A.G., 2000a, Early Mississippian North American carbonate-platform margin in west-central Sonora, *in* Calmus, T., and Perez-Segura, E., eds., Cuarta reunion sobre la geología del noroeste de Mexico y areas adyacentes: Hermosillo, Sonora, Centro de las Artes, Universidad de Sonora, p. 95–96.

Poole, F.G., Amaya-Martinez, R., and Page, W.R., 2000b, Silurian and Devonian carbonate-shelf rocks and Lower Jurassic sequence near Rancho Placeritos, west-central Sonora: Fourth Symposium on the geology of northwest Mexico and adjacent areas, Field Trip 2: Hermosillo, Sonora, Center for the Arts, University of Sonora, 24 p.

Powell, R.E., 1993, Balanced palinspastic reconstruction of pre–late Cenozoic paleogeology, southern California: Geologic and kinematic constraints on evolution of the San Andreas fault system, *in* Powell, R.E., Weldon, R.J., II, and Matti, J.C., eds., The San Andreas fault system: Displacement, palinspastic reconstruction, and geologic evolution: Geological Society of America Memoir 178, p. 1–106.

Robinson, A.C., 1993, Regionally and compositionally coherent magma sources for Mesozoic and Tertiary granitoids in Nevada and eastern California north of 37 degrees latitude: Geological Society of America Abstracts with Programs, v. 25, p. 139.

Ross, D.C., 1966, Stratigraphy of some Paleozoic formations in the Independence quadrangle, Inyo County, California: U.S. Geological Survey Professional Paper 396, 64 p.

Ross, D.C., 1989, The metamorphic and plutonic rocks of the southernmost Sierra Nevada, California, and their tectonic framework: U.S. Geological Survey Professional Paper 1381, 159 p.

Saleeby, J.B., 1992, Petrotectonic and paleogeographic settings of U.S. Cordilleran ophiolites, *in* Burchfiel, B.C., Lipman, P.W., and Zoback, M.L., eds., The Cordilleran orogen: Conterminous U.S.: Geological Society of America, Geology of North America, v. G-3, p. 653–682.

Saleeby, J.B., 2003, Segmentation of the Laramide slab—Evidence from the southern Sierra Nevada region: Geological Society of America Bulletin, v. 115, p. 655–668, doi: 10.1130/0016-7606(2003)115<0655:SOTLSF>2.0.CO;2.

Saleeby, J.B., and Busby, C., 1993, Paleogeographic and tectonic setting of axial and western metamorphic framework rocks of the southern Sierra Nevada, California, *in* Dunne, G.C., and McDougall, K.A., eds., Mesozoic paleogeography of the western United States—II: Los Angeles, Pacific Section, Society of Economic Paleontologists and Mineralogists, Book 71, p. 197–225.

Saleeby, J.B., Goodin, S.E., Sharp, W.D., and Busby, C.J., 1978, Early Mesozoic paleotectonic-paleogeographic reconstruction of the southern Sierra Nevada region, *in* Howell, D.G., and McDougall, K.A., eds., Mesozoic paleogeography of the western United States: Los Angeles, Pacific Section, Society of Economic Paleontologists and Mineralogists Pacific Coast Paleogeography Symposium 2, p. 311–336.

Saleeby, J.B., Sams, D.B., and Kistler, R.W., 1987, U/Pb zircon, strontium and oxygen isotopic and geochronological study of the southernmost Sierra Nevada batholith, California: Journal of Geophysical Research, v. 92, p. 10,443–10,466.

Schermer, E.R., Stephens, K.A., and Walker, J.D., 2001, Paleogeographic and tectonic implications of the geology of the Tiefort Mountains, northern Mojave Desert, California: Geological Society of America Bulletin, v. 113, p. 920–938, doi: 10.1130/0016-7606(2001)113<0920:PATIOT>2.0.CO;2.

Schweickert, R.A., and Lahren, M.M., 1987, Continuation of the Antler and Sonoma orogenic belts to the eastern Sierra Nevada, California, and Late Triassic thrusting in a compressional arc: Geology, v. 15, p. 270–273, doi: 10.1130/0091-7613(1987)15<270:COAASO>2.0.CO;2.

Schweickert, R.A., and Lahren, M.M., 1991, Age and tectonic significance of metamorphic rocks along the axis of the Sierra Nevada batholith: A critical reappraisal, *in* Cooper, J.D., and Stevens, C.H., eds., Paleozoic paleogeography of the western United States—II: Los Angeles, Pacific Section, Society of Economic Paleontologists and Mineralogists, Book 67, p. 653–676.

Schweickert, R.A., and Lahren, M.M., 1993a, Triassic-Jurassic magmatic arc in eastern California and western Nevada: Arc evolution, cryptic tectonic breaks, and significance of the Mojave-Snow Lake fault, *in* Dunne, G., and McDougall, K., eds., Mesozoic paleogeography of the western United States—II: Los Angeles, Pacific Section, Society of Economic Paleontologists and Mineralogists, Book 71, p. 227–246.

Schweickert, R.A., and Lahren, M.M., 1993b, Tectonics of the east-central Sierra Nevada—Saddlebag Lake and Northern Ritter Range pendants, *in* Lahren, M.M., Trexler, J.H., Jr., and Spinosa, C., eds., Crustal evolution of the Great Basin and Sierra Nevada: Reno, Nevada, Cordilleran/Rocky Mountain Section, Geological Society of America Guidebook, Department of Geological Sciences, University of Nevada, Reno, p. 313–351.

Silver, L.T., and Anderson, T.H., 1974, Possible left-lateral early to middle Mesozoic disruption of the southwestern North America craton margin: Geological Society of America Abstracts with Programs, v. 6, no. 7, p. 955–956.

Snow, J.K., 1992, Large-magnitude Permian shortening and continental-margin tectonics in the southern Cordillera: Geological Society of America Bulletin, v. 104, p. 80–105, doi: 10.1130/0016-7606(1992)104<0080:LMPSAC>2.3.CO;2.

Snow, J.K., and Wernicke, B.P., 2000, Cenozoic tectonism in the central Basin and Range: Magnitude, rate, and distribution of upper crustal strain: American Journal of Science, v. 300, p. 659–719.

Snow, J.K., Asmeron, Y., and Lux, D.R., 1991, Permian-Triassic plutonism and tectonics, Death Valley region, California and Nevada: Geology, v. 19, p. 629–632, doi: 10.1130/0091-7613(1991)019<0629:PTPATD>2.3.CO;2.

Stevens, C.H., 1982, The Early Permian *Thysanophyllum* coral belt: Another clue to Permian plate-tectonic reconstructions: Geological Society of America Bulletin, v. 93, p. 798–803, doi: 10.1130/0016-7606(1982)93<798:TEPTCB>2.0.CO;2.

Stevens, C.H., 1986, Evolution of the Ordovician through Middle Pennsylvanian carbonate shelf in east-central California: Geological Society of America Bulletin, v. 97, p. 11–25, doi: 10.1130/0016-7606(1986)97<11:EOTOTM>2.0.CO;2.

Stevens, C.H., and Greene, D.C., 1999, Stratigraphy, depositional history, and tectonic evolution of Paleozoic continental-margin rocks in roof pendants of the eastern Sierra Nevada, California: Geological Society of America Bulletin, v. 111, p. 919–933, doi: 10.1130/0016-7606(1999)111<0919:SDHATE>2.3.CO;2.

Stevens, C.H., Dunne, G.C., and Randall, R.G., 1979, Carboniferous stratigraphy of part of eastern California, *in* Saul, R.B., et al., eds., The Mississippian and Pennsylvanian (Carboniferous) Systems in the United States—California, Oregon, and Washington: U.S. Geological Survey Professional Paper 1110-CC, p. CC-1–CC-21.

Stevens, C.H., Stone, P., and Belasky, P., 1991, Paleogeographic and structural significance of an Upper Mississippian facies boundary in southern Nevada and east-central California: Geological Society of America Bulletin, v. 103, p. 876–885, doi: 10.1130/0016-7606(1991)103<0876:PASSOA>2.3.CO;2.

Stevens, C.H., Stone, P., and Kistler, R.W., 1992, A speculative reconstruction of the middle Paleozoic continental margin of southwestern North America: Tectonics, v. 11, p. 405–419.

Stevens, C.H., Stone, P., and Ritter, S.M., 2001, Conodont and fusulinid biostratigraphy and history of the Pennsylvanian to Lower Permian Keeler basin, east-central California: Brigham Young University Geology Studies, v. 46, p. 99–142.

Stevens, C.H., Stone, P., and Greene, D.C., 2003, Reply: Correlation of Permian and Triassic deformations in the western Great Basin and eastern Sierra Nevada: Evidence from the northern Inyo Mountains near Tinemaha Reservoir, east-central California: Geological Society of America Bulletin, v. 115, p. 1309–1311, doi: 10.1130/B25385D.1.

Stewart, J.H., 1984, Siliceous Ordovician strata and structurally interleaved Devonian calcareous gravity flow deposits in Miller Mountain area, western Nevada, *in* Lintz, J., Jr., ed., Western geological excursions, v. 4: Boulder, Colorado, Geological Society of America, and Reno, Nevada, University of Nevada, p. 81–93.

Stewart, J.H., 1985, East-trending dextral faults in the western Great Basin: An explanation for anomalous trends of pre-Cenozoic strata and Cenozoic faults: Tectonics, v. 4, p. 547–564.

Stewart, J.H., 1988, Latest Proterozoic and Paleozoic southern margin of North America and the accretion of Mexico: Geology, v. 16, p. 186–189, doi: 10.1130/0091-7613(1988)016<0186:LPAPSM>2.3.CO;2.

Stewart, J.H., and Crowell, J.C., 1992, Strike-slip tectonics in the Cordilleran region, western United States, *in* Burchfiel, B.C., Lipman, P.W., and Zoback, M.L., eds., The Cordilleran Orogen: Conterminous U.S.: Geological Society of America, The Geology of North America, v. G-3, p. 609–628.

Stewart, J.H., and Poole, F.G., 1975, Extension of the Cordilleran miogeosynclinal belt to the San Andreas fault, southern California: Geological Society of America Bulletin, v. 86, p. 205–212, doi: 10.1130/0016-7606(1975)86<205:EOTCMB>2.0.CO;2.

Stewart, J.H., and Poole, F.G., 2002, Inventory of Neoproterozoic and Paleozoic strata in Sonora, Mexico: U.S. Geological Survey Open-File Report 02-97, 50 p.

Stewart, J.H., Poole, F.G., Ketner, K.B., Madrid, R.J., Roldan-Quintana, J., and Amaya-Martinez, R., 1990, Tectonics and stratigraphy of the Paleozoic and Triassic southern margin of North America, Sonora, Mexico, *in* Gehrels, G.E., and Spencer, J.E., eds., Geologic excursions through the Sonoran Desert region, Arizona and Sonora: Arizona Geological Survey Special Paper 7, p. 183–202.

Stone, P., and Stevens, C.H., 1988, An angular unconformity in the Permian section of east-central California: Geological Society of America Bulletin, v. 100, p. 547–551, doi: 10.1130/0016-7606(1988)100<0547:AAUITP>2.3.CO;2.

Stone, P., Howard, K.A., and Hamilton, W., 1983, Correlation of metamorphosed Paleozoic strata of the southeastern Mojave Desert region, California and Arizona: Geological Society of America Bulletin, v. 94, p. 1135–1147, doi: 10.1130/0016-7606(1983)94<1135:COMPSO>2.0.CO;2.

Walker, J.D., 1988, Permian and Triassic rocks of the Mojave Desert and their implications for timing and mechanisms of continental truncation: Tectonics, v. 7, p. 685–709.

Walker, J.D., Fletcher, J.M., Fillmore, R.P., Martin, M.W., Taylor, W.J., Glazner, A.F., and Bartley, J.M., 1995, Connection between igneous activity and extension in the central Mojave metamorphic core complex: Journal of Geophysical Research, v. 100, p. 10,477–10,494, doi: 10.1029/94JB03132.

Walker, J.D., Martin, M.W., and Glazner, A.F., 2002, Late Paleozoic to Mesozoic development of the Mojave Desert and environs, California, *in* Glazner, A.F., Walker, J.D., and Bartley, J.M., eds., Geologic evolution of the Mojave Desert and southwestern Basin and Range: Geological Society of America Memoir 195, p. 1–18.

Manuscript Accepted by the Society 24 March 2005

Printed in the USA

Geological Society of America
Special Paper 393
2005

Chihuahua trough—A Jurassic pull-apart basin

Walter T. Haenggi
2007 Tradewinds Drive, Missouri City, Texas 77459, USA

William R. Muehlberger
Department of Geological Sciences, The University of Texas at Austin, 1 University Station C1100, Austin, Texas 78712-0254, USA

ABSTRACT

The Chihuahua trough is a right-lateral pull-apart basin that began to form ca. 159 to ca. 156 Ma (Oxfordian) during a period of relative counterclockwise rotation of the North American plate. The Chihuahua trough developed as a right step between two right-lateral fault systems—the Texas "zone" and the San Marcos fault zone, which has been interpreted as an element of the Mojave-Sonora megashear. By this interpretation, the Mojave-Sonora fault zone of the Caborca area cannot—and need not—be projected to join the San Marcos fault zone as a continuous system in medial Jurassic time.

Jurassic seas were well established by latest Oxfordian time, and there was little change in basin configuration throughout the remainder of Late Jurassic and Early Cretaceous time. Elements of a broad zone of intersecting preexisting northwest-trending and north-trending structures along the southwest border of the North American craton provide the fabric for development of the pull-apart basin between the Diablo and Aldama platforms. While the presence of Middle to Upper Jurassic marine rocks in Chihuahua and Sonora has received comment in previous articles, this paper provides a regionally consistent tectonic mechanism for creating two arms of the Mar Mexicano.

Keywords: Chihuahua, right-lateral, pull-apart, southwest border North America, lineaments.

INTRODUCTION

This paper is a brief condensation of ideas and data presented in "The tectonic history of the Chihuahua trough, Mexico and adjacent USA, Part II: Mesozoic and Cenozoic" (Haenggi, 2002). Part I, "the pre-Mesozoic setting" (Haenggi, 2001) deals with tectonic events that preceded development of the Chihuahua trough (Fig. 1) including the late Paleozoic Pedregosa basin that was, in effect, a proto-Chihuahua trough. The Chihuahua trough is a Mesozoic depositional basin that includes the area where a Late Jurassic pull-apart basin developed in northeastern Chihuahua and adjacent parts of Texas, New Mexico, and Sonora. In latest Aptian and Albian time there was a general marine transgression onto adjacent emergent areas. The Late Jurassic–Aptian Chihuahua trough joined the Bisbee basin toward the northwest and was connected to the Sabinas basin and Mar Mexicano to the southeast and south (see inset in Fig. 1). The Chihuahua trough of this paper has arbitrary southern and northwestern boundaries separating it from the Sabinas and Bisbee basins (Fig. 1). The southern boundary of the trough is along a portion of the boundary of the North American craton defined on the basis of lead isotope data (James and Henry, 1993) and along the Alamitos lineament, a feature of uncertain tectonic significance observed on space shuttle and satellite images. The Alamitos lineament

Haenggi, W.T., and Muehlberger, W.R. 2005, Chihuahua trough—A Jurassic pull-apart basin, *in* Anderson, T.H., Nourse, J.A., McKee, J.W., and Steiner, M.B., eds., The Mojave-Sonora megashear hypothesis: Development, assessment, and alternatives: Geological Society of America Special Paper 393, p. 619–630, doi: 10.1130/2005.2393(23). For permission to copy, contact editing@geosociety.org.

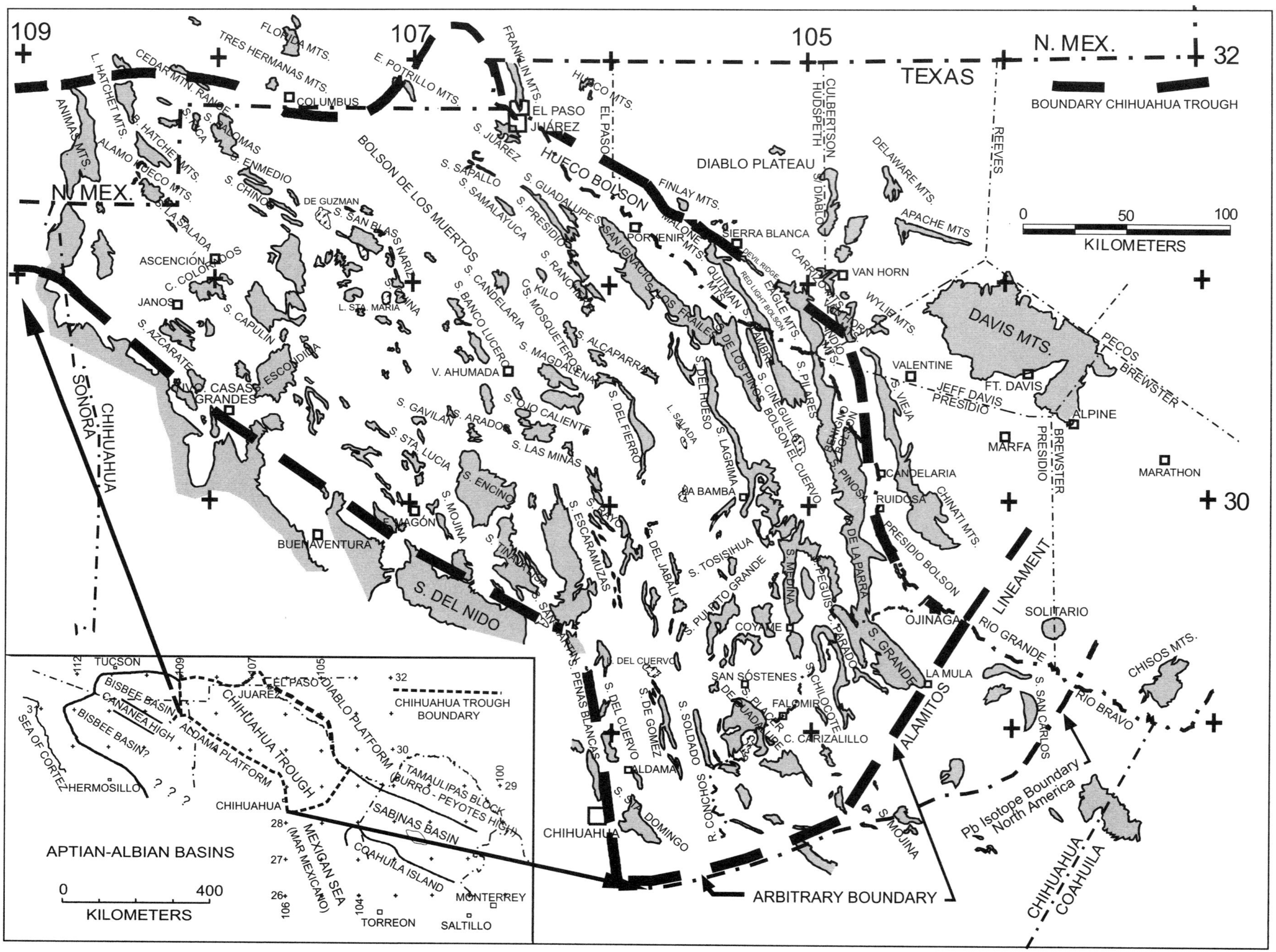
N. MEX.
TEXAS
BOUNDARY CHIHUAHUA TROUGH
KILOMETERS
0
50
100
109
107
105
32
30
FLORIDA MTS.
TRES HERMANAS MTS.
CEDAR MTN. RANGE
L. HATCHET MTS.
ANIMAS MTS.
ALAMO HUECO MTS.
S. HATCHET MTS.
E. POTRILLO MTS.
COLUMBUS
FRANKLIN MTS.
EL PASO
JUÁREZ
HUECO MTS.
S. JUAREZ
HUECO BOLSON
FINLAY MTS.
DIABLO PLATEAU
CULBERTSON
HUDSPETH
S/ DIABLO
REEVES
DELAWARE MTS.
APACHE MTS
SIERRA BLANCA
VAN HORN
PORVENIR
MALONE MTS.
QUITMAN MTS.
DEVIL RIDGE
RED LIGHT BOLSON
EAGLE MTS.
WYLIE MTS.
DAVIS MTS.
PECOS
BREWSTER
FT. DAVIS
VALENTINE
JEFF DAVIS
PRESIDIO
ALPINE
MARFA
MARATHON
S. VIEJA
CANDELARIA
RUIDOSA
CHINATI MTS.
PRESIDIO BOLSON
LINEAMENT
SOLITARIO
OJINAGA
RIO GRANDE
CHISOS MTS.
LA MULA
ALAMITOS
S. SAN CARLOS
RIO BRAVO
Pb Isotope Boundary
North America
CHIHUAHUA
COAHUILA
ARBITRARY BOUNDARY
S. MOJINA
BOLSON DE LOS MUERTOS
S. SAPALLO
S. SAMALAYUCA
S. GUADALUPE
S. PRESIDIO
S. SAN IGNACIO
LOS FRAILES
S. DE LOS PINOS
S. CINEGUILLO
BOLSON EL CUERVO
S. PILARES
S. PINOSA
DE LA PARRA
S. GRANDE
S. PEGUIS
C. PARADO
S. MEDINA
S. CHILOCOTE
FALOMIR
C. CARIZALILLO
S. DEL HUESO
S. LAGRIMA
LA BAMBA
L. SALADA
S. TOSISIHUA
S. PUERTO GRANDE
COYAME
SAN SÓSTENES
S. PLACER
DE GUADALUPE
S. DEL JABALI
S. SOLDADO
R. CONCHOS
S. DE GOMEZ
S. DEL CUERVO
ALDAMA
S. STA. DOMINGO
CHIHUAHUA
S. PEÑAS BLANCAS
S. SAN MARTIN
S. ESCARAMUZAS
S. DEL FIERRO
S. ALCAPARRA
S. MOSQUETEROS
S. MAGDALENA
S. RANCHO
C. KILO
S. CANDELARIA
S. BANCO LUCERO
V. AHUMADA
S. OJO CALIENTE
S. ARADOS
S. LAS MINAS
S. GAVILAN
S. STA. LUCIA
S. ENCINO
S. TINAJA
S. MOJINA
F. MAGÓN
S. DEL NIDO
BUENAVENTURA
DE GUZMAN
S. SAN BLAS
S. NARIZ
L. STA. MARIA
S. ENMEDIO
S. CHINOS
S. PALOMAS
S. RICA
S. LA SALADA
ASCENCIÓN
C. COLORADOS
JANOS
S. CAPULIN
ESCONDIDA
S. AZCARATE
CASAS GRANDES
SONORA
APTIAN-ALBIAN BASINS
KILOMETERS
400
TUCSON
BISBEE BASIN
CANANEA HIGH
BISBEE BASIN?
SEA OF CORTEZ
HERMOSILLO
ALDAMA PLATFORM
CHIHUAHUA TROUGH
DIABLO PLATFORM
CHIHUAHUA TROUGH BOUNDARY
TAMAULIPAS BLOCK
(BURRO - PEYOTES HIGH)
SABINAS BASIN
COAHUILA ISLAND
MEXIGAN SEA
(MAR MEXICANO)
MONTERREY
SALTILLO
TORREON
112
109
107
105
100
106
104
31
30
29
28
27
26

Figure 1. Map (after Haenggi, 2002) showing extent of Chihuahua trough, sierras (shaded) and bolsons (topographic basins), lead isotope boundary of North America and Alamitos lineament. Inset shows location of Chihuahua trough relative to Bisbee and Sabinas basins, the Mar Mexicano, and paleogeographic positive areas. Reproduced with permission of Sociedad Geológica Mexicana.

intersects the lead isotope boundary of North America in Chihuahua, ~30 km west of the Sierra Mojina (see Fig. 1), then is subparallel to and ~60 km inboard (northwest) of the eastern part of the lead isotope boundary. The arbitrary northwestern boundary of the trough is along the 109th Meridian.

Upper Jurassic rocks in the Sabinas basin, Chihuahua trough, and the easternmost part of the Bisbee basin indicate that the Chihuahua trough is part of a Jurassic marine basin that extended from the Gulf of Mexico to southeastern Arizona (Fig. 2A). West of the Chihuahua trough, in Arizona, Oxfordian and/or Kimmeridgian ammonites are described from the northeastern Chiricahua Mountains, Arizona (Lawton and Olmstead, 1995; Olmstead and Young, 2000). In Sonora, Oxfordian ammonites are reported near Cucurpe (Rangin, 1977); east of Arivechi, Almazán-Vázquez, and Palafox-Reyes (2000) describe Kimmeridgian ammonites. Between these areas of Arizona and Sonora, Jurassic rocks are absent in the southern Chiracahua Mountains (Rucker horst, Lindberg, 1987), Peloncillo Mountains (Hayes, 1970), and Mule Mountains (Bilodeau et al., 1987). This leads to the conclusion that the distribution of Jurassic rocks in Chihuahua, Sonora, Arizona, New Mexico, and Texas is the result of deposition in two northwest-trending arms of the Mar Mexicano (Fig. 2A): (1) Chihuahua trough–Bisbee basin, and (2) "Arivechi-Cucurpe" seaway.

The Chihuahua trough is generally interpreted to be the result of a Jurassic extensional event that also developed the contiguous Bisbee and Sabinas basins (Busby-Spera and Kokelaar, 1991; Fackler-Adams et al., 1997; and Lawton et al., 1997). There is general agreement that extension is related to the opening of the Atlantic Ocean and specifically to formation of the Gulf of Mexico. Extension probably commenced sometime during the Middle Jurassic, and by late Oxfordian time seas extended from the Gulf of Mexico to southeastern Arizona and into north-central Sonora (Fig. 2A).

The history of development of the Bisbee and Sabinas basins, while more or less synchronous with the formation of the Chihuahua trough, is complicated by Jurassic igneous activity (including rift-related volcanism). Extensive volcanic and volcaniclastic rocks accumulated in the Bisbee basin as the basin formed—a link between Cordilleran arc magmatism and extension of the Gulf of Mexico. The basal Mesozoic unit in the Sabinas basin is a Middle Jurassic(?) conglomerate with some extrusive volcanic rocks (Santamaría-O. et al., 1991) that may be related to the waning stages of a Permian-Triassic continental arc described by Torres et al. (1999). Garrison and McMillan (1999) interpreted allogenic blocks of greenschist facies metamorphic rocks in the Papalote evaporite diapir in the southeastern Sabinas basin (Laramide La Popa basin; see Fig. 5 for location) as evidence of continental rift basalts being erupted during Jurassic time. In contrast to the Bisbee and Sabinas basins, there is no known Jurassic igneous activity in the Chihuahua trough. Basalt flows do occur in a probable Jurassic formation in the Little Hatchet Mountains, New Mexico (D, Fig. 2B), in the extreme northwestern part of the trough (Lucas and Lawton, 2000), and extensive basal Mesozoic syntectonic conglomerate is apparently absent.

CHIHUAHUA TROUGH

With the possible exception of a thick (up to 360 m; Ward, 1977) conglomerate, mapped as Permian Plomosas Formation, locally underlying the Jurassic La Casita Formation in the vicinity of Placer de Guadalupe (K, Fig. 2B), syntectonic Mesozoic basal conglomerates and coeval volcanic rocks are not recognized in the Chihuahua trough. In northwestern Chihuahua, there is a basal Mesozoic conglomerate (Lower Cretaceous) that has been called "Glance" (Brown and Dyer, 1987). This unit occurs conformably below Lower Cretaceous rocks, is generally less than 10 m thick, and contains clasts derived from the underlying Paleozoic section. Neither this conglomerate nor basal Jurassic conglomerates near Placer de Guadalupe and Cerro Carrizalillo (Roberts and Dyer, 1988) have aspects that can be ascribed to deposition in alluvial fans along faults with high relief—they are similar to the ubiquitous Lower Cretaceous basal conglomerates and sandstones of Texas, which are deposits along shorelines of seas advancing onto a tectonically stable continental margin with islands, peninsulas, etc.

Conclusions regarding depositional setting and paleogeology of Jurassic rocks in the Chihuahua trough are speculations because data are insufficient to resolve details of abrupt lateral facies changes and to define significant variations of thickness. At least 1000 m of predominantly siliciclastic marine sedimentary rocks were deposited in the deeper parts of the Chihuahua trough during Late Jurassic time. In Chihuahua, the oldest observed Jurassic rocks are assigned to the La Casita Formation. Near Carrizalillo (L, Fig. 2B), the La Casita contains a thin (less than 100 m) basal fluvial facies, a very thick shelf to basinal facies, and an uppermost gypsum facies (Roberts and Dyer, 1988). Here and wherever observed in the Chihuahua trough, the La Casita grades upward into a shallow-water regressive facies that persists into Neocomian time (Navarrete, Torcer, and Alcaparra Formations; see Fig. 3 for nomenclature of Jurassic and Cretaceous formations of the Chihuahua trough). Extensive evaporites that form the principal décollement of the Chihuahua tectonic belt occur in this latest Tithonian–Neocomian regressive facies. Thickening of the Jurassic–Lower Cretaceous section from platform areas into the trough is dramatic; e.g., typical Diablo Platform section sections (Jurassic absent) are ~800 m thick and, in the adjacent trough sections are in excess of 4500 m thick (Haenggi, 2002, Figures 6 and 7 therein).

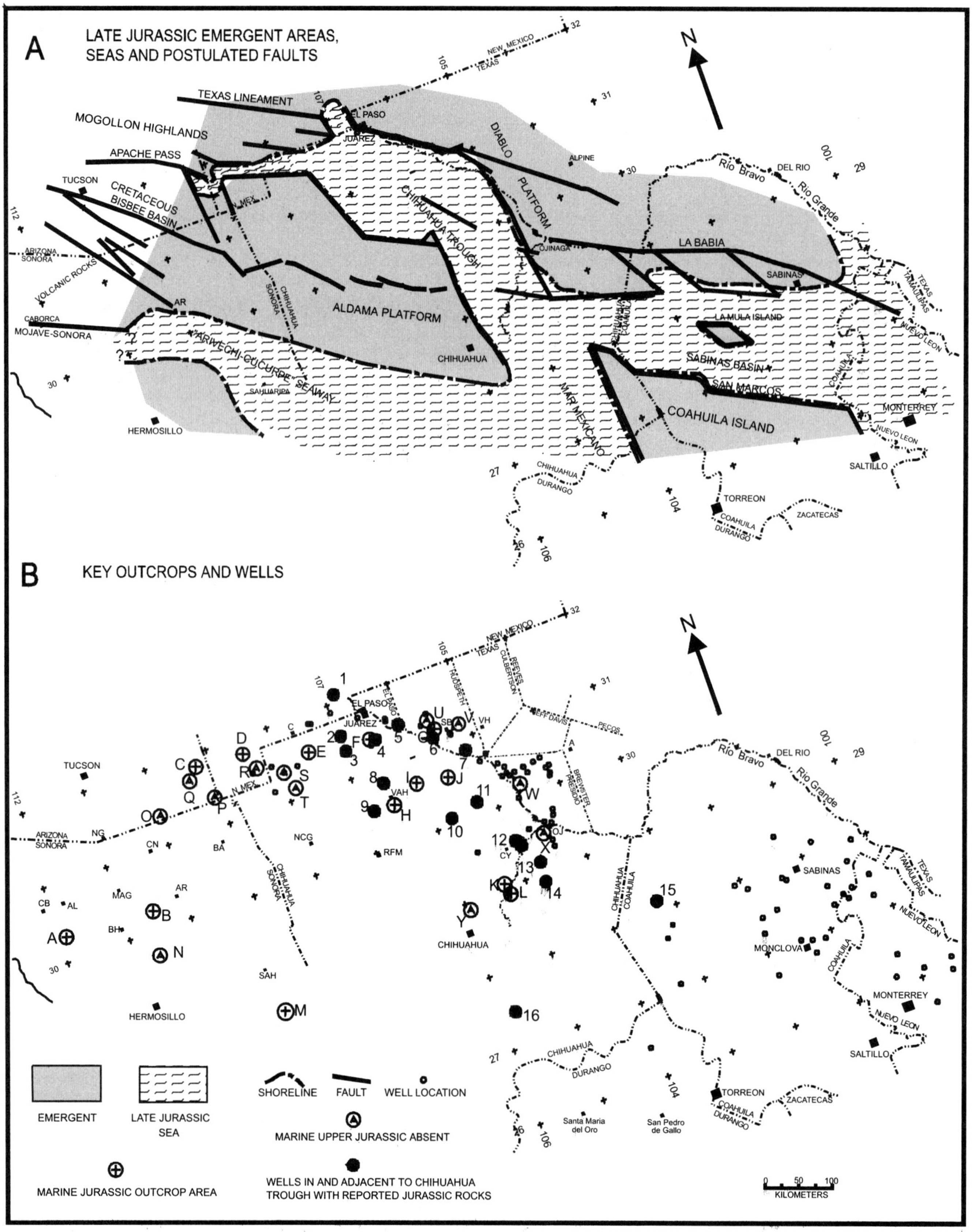
A
LATE JURASSIC EMERGENT AREAS, SEAS AND POSTULATED FAULTS
N
TEXAS LINEAMENT
MOGOLLON HIGHLANDS
APACHE PASS
CRETACEOUS BISBEE BASIN
VOLCANIC ROCKS
MOJAVE-SONORA
ARIVECHI-CUCURPE SEAWAY
CHIHUAHUA TROUGH
DIABLO PLATFORM
ALDAMA PLATFORM
LA BABIA
LA MULA ISLAND
SABINAS BASIN
SAN MARCOS
COAHUILA ISLAND
MAR MEXICANO
EL PASO
JUAREZ
TUCSON
CABORCA
HERMOSILLO
SAHUARIPA
CHIHUAHUA
OJINAGA
ALPINE
DEL RIO
Rio Bravo
Rio Grande
SABINAS
MONTERREY
SALTILLO
TORREON
B
KEY OUTCROPS AND WELLS
MONCLOVA
Santa Maria del Oro
San Pedro de Gallo
EMERGENT
LATE JURASSIC SEA
SHORELINE
FAULT
WELL LOCATION
MARINE UPPER JURASSIC ABSENT
MARINE JURASSIC OUTCROP AREA
WELLS IN AND ADJACENT TO CHIHUAHUA TROUGH WITH REPORTED JURASSIC ROCKS
0 50 100
KILOMETERS

Figure 2. (A) Map (after Haenggi, 2002) of Late Jurassic marine basins, emergent areas and postulated Jurassic fault zones. (B) Map (after Haenggi, 2002) showing key outcrop areas and wells used to construct A (see Table 1). Marine Lower Jurassic outcrop area at **(A)**—Pozo Serna; reported Upper Jurassic marine outcrop areas: **(B)**—Curcupe area; **(C)**—Chiricahua Mts.; **(D)**—Little Hatchet Mts.; **(E)**—S. Aguila; **(F)**—S. Samalayuca; **(G)**—Malone Mts.; **(H)**—S. de la Mina; **(I)**—S. Alcaparra; **(J)**—Cerros Prietos; **(K)**—Placer de Guadalupe; **(L)**—C. Carrizalillo; **(M)**—Arivechi area. Wells with reported Jurassic rocks: **1**—Mobil 32-1; **2**—Moyotes-1; **3**—Sapallo-1; **4**—Presidio-1; **5**—State Univ. DW-1; **6**—Bask State 9-1; **7**—Emmett-1; **8**—Banco Lucero-1; **9**—V. Ahumada-1; **10**—Menonita-1; **11**—El Hueso-1; **12**—C. Parado 1 and 2; **13**—Maijoma-1; **14**—Apache-1; **15**—Concordia-1; **16**—Lago Toronto-1. Wells and surface locations where Jurassic rocks are absent: **(N)**—Cerro de Oro; **(O)**—Mule Mts.; **(P)**—Peloncillo Mts.; **(Q)**—Rucker horst; **(R)**—Big Hatchet Mts.; **(S)**—Espia-1; **(T)**—Ascención-1; **(U)**—Gardner and Moseley-1; **(V)**—State FV-1/State 11-1; **(W)**—S. Vieja area wells; **(X)**—Ojinaga area wells; **(Y)**—S. del Cuervo. Abbreviations for cities and towns: A—Alpine, Texas; AL—Altar; Sonora; AR—Arizpe, Sonora; BH—Benjamin Hill, Sonora; C—Columbus, New Mexico; CB—Caborca, Sonora; CN—Cananea, Sonora; CY—Coyame, Chihuahua; MAG—Magdalena, Sonora; NCG—Nuevo Casas Grandes, Chihuahua; NG—Nogales, Arizona; OJ—Ojinaga, Chihuahua; RFM—R.F. Magón, Chihuahua; SAH—Sahuaripa, Sonora; SB—Sierra Blanca, Texas; VAH—V. Ahumada, Chihuahua; VH—Van Horn, Texas. Reproduced with permission of Sociedad Geológica Mexicana.

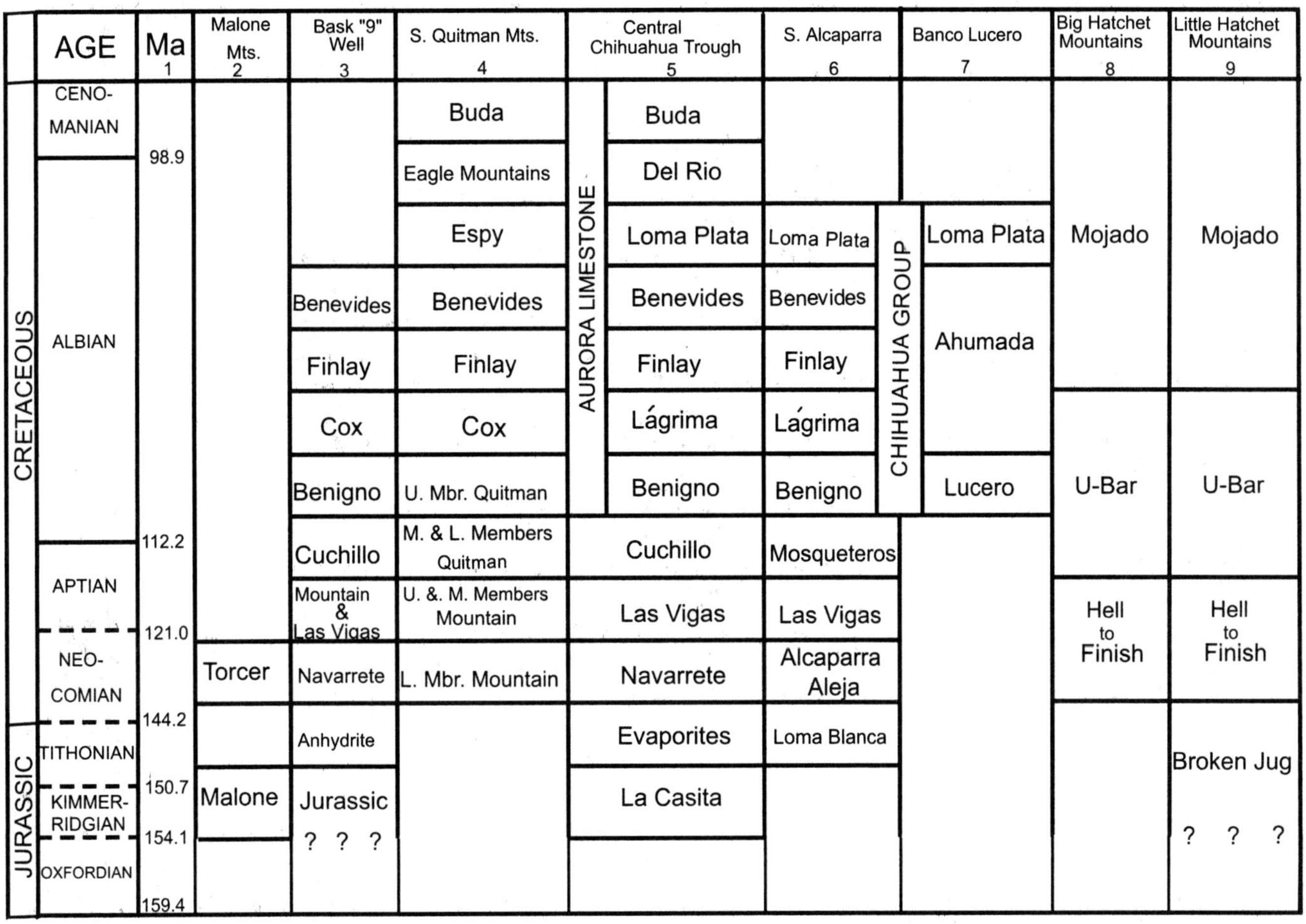

Figure 3. Chart showing Chihuahua trough nomenclature of Jurassic and Lower Cretaceous formations (after Haenggi, 2002). References: 1—Gradstein et al., 1994; 2—Malone Mountains, Albritton and Smith (1965); 3—Bask "9" well, interpretation of well logs; 4—S. Quitman Mountains, Jones and Reaser (1970); 5—Central Chihuahua trough, Haenggi (1966), DeFord and Haenggi (1970); 6—S. Alcaparra, Rodríguez Torres (1969); 7—Banco Lucero, Córdoba (1969); 8—Big Hatchet Mts., Zeller (1965); 9—Little Hatchet Mts., Lucas and Lawton (2000). Reproduced with permission of Sociedad Geológica Mexicana.

TABLE 1. KEY WELLS AND OUTCROP AREAS ON FIGURE 2B

No. on Figure 2B	Well/locality	Data source
Jurassic Marine Rocks Reported		
1	GRIMM/MOBIL 32-1	Thompson (1982)
2	PEMEX/MOYOTES-1	Thompson et al. (1978)
3	PEMEX/SAPALLO-1	Thompson et al. (1978)
4	PEMEX/PRESIDIO-1	Thompson et al. (1978)
5	HUMBLE/ST. UNIV. DW-1	Uphoff (1978)
6	BORDER EX./BASK STATE-9	Well logs (Texaco, 1998, personal commun.)
7	TEXACO/EMMETT-1	Well logs (Texaco, 1998, personal commun.)
8	PEMEX/BANCO LUCERO-1	Thompson et al. (1978)
9	PEMEX/VILLA AHUMADA-1	Thompson et al. (1978)
10	PEMEX/MENONITA-1	Dyer (1987, p. 18).
11	PEMEX/EL HUESO-1	PEMEX files (1973, personal commun.)
12	PEMEX/CUCHILLO PARADO 1&2	Ramírez-M. and Acevedo-C. (1957)
13	PEMEX/MAIJOMA-1	Limón-González (1986); López-Ramos (1988)
14	PEMEX/APACHE-1	Limón-González (1986); López-Ramos (1988)
15	PEMEX/CONCORDIA-1	López-Ramos (1988); Santamaría-O. et al. (1991)
16	PEMEX/LAGO TORONTO-1	López-Ramos (1988)
A	POZO SERNA (LOWER JURASSIC ROCKS)	Pálfy and González-Leon (2000); Calmus et al. (1997)
B	CUCURPE AREA	Rangin (1977)
C	NORTHEASTERN CHIRICAHUA MTS.	Lawton and Olmstead (1995)
D	LITTLE HATCHET MOUNTAINS	Lawton and Olmstead (1995)
E	SIERRA AGUILA	Reyeros de Castillo (1974)
F	SIERRA SAMAYALUCA	Cantu-Chapa (1970)
G	MALONE MOUNTAINS	Albritton and Smith (1965)
H	SIERRA DE LA MINA	Ramírez-M. and Acevedo-C. (1957)
I	SIERRA ALCAPARRA	Rodríguez-Torres (1969)
J	CERROS PRIETOS	Araujo-Mendieta and Casar-González (1987)
K	PLACER DE GUADALUPE	Araujo-Mendieta and Casar-González (1987)
L	CERRO CARRIZALILLO	Roberts and Dyer (1988)
M	ARIVECHI AREA	Almazán-Vázquez and Palafox-Reyes (2000)
Jurassic Marine Rocks Absent		
N	CERRO DE ORO	González-León and Lucas (1995)
O	MULE MOUNTAINS	Bilodeau et al. (1987)
P	PELONCILLO MOUNTAINS	Hayes (1970)
Q	RUCKER HORST (SOUTHERN CHIRICAHUA MTS.)	Lindberg (1987)
R	BIG HATCHET MOUNTAINS	Zeller (1965)
S	PEMEX/ESPIA-1	Thompson et al. (1978)
T	PEMEX/ASCENCION-1	Thompson et al. (1978)
U	WESTERN STS./G & M-1	Albritton and Smith (1965)
V	BORDER EX./STATE 11-1 TEXACO/STATE FV-1	Well logs (Texaco, 1998, personal commun.)
W	Sierra Vieja Area Wells GULF/PRESIDIO TRUST-1 GULF/SWAFFORD-1 GULF/1ST NATL. BANK FT. WORTH-1 HUNT/PRESIDIO TRUST-1	Gulf Oil Co. (pre-1984, personal commun.)
X	Ojinaga Area Wells	
	PEMEX/OJINAGA-1	Cantu-Chapa et al. (1985); Limón-González (1986); López-Ramos (1988)
	ARCO/PRESIDIO STATE-1	DeJong and Addy (1992)
	WEST/PRESIDIO TRUST-1	Gulf Oil Co. (pre-1984, personal commun.)
	WEST/COCKBURN, PRESIDIO TRUST-1	Gulf Oil Co. (pre-1984, personal commun.)
Y	SIERRA DEL CUERVO	Ramírez-M. and Acevedo-C. (1957)

Conclusions from all of this are:

1. The Chihuahua trough developed rapidly as a marine sedimentary basin, probably during latest Middle and earliest Late Jurassic time.

2. Topographic relief was slight; tectonic activity did not create a terrain that supplied conglomerates to extensive alluvial fans.

3. Termination of Jurassic facies and thick sections along the edges of the trough are abrupt and suggest that the basin is fault-bounded—essentially the same basin boundaries persisted throughout Neocomian and Aptian time. Throughout this period, sediment influx and basin subsidence were essentially balanced and some of the boundary faults eventually attained displacements of more than 1000 m.

4. With the possible exception of local basalt flows in the extreme northwestern corner of the trough (Little Hatchet Mountains, New Mexico; Lucas and Lawton, 2000), there is no evidence of igneous activity coeval with basin formation and development.

These conclusions suggest that the origin of the Chihuahua trough is, in some way, different from that of the more-or-less coeval Bisbee and Sabinas basins.

Figure 4A shows a pre-Jurassic regional structural fabric consisting of the broad zone of northwesterly trending lineaments and intersecting north-trending zones of crustal weakness (after Muehlberger and Moustafa, 1984). Elements of the northwest-trending zone, along the southwest margin of the North American craton, have been recurrently active since middle Proterozoic time (Muehlberger, 1980; Haenggi, 2001). Intermittent activity of faults of the northerly trending zones has probably occurred over a similar time period. Figure 4B shows an interpretation of the Chihuahua trough forming as a pull-apart during right-lateral motion along northwest-trending zones of crustal weakness. All of the following discussion involving Jurassic faulting in the Chihuahua trough is speculation—there are no documented Jurassic offsets on faults within or bounding the trough.

Evidence for Jurassic right-lateral motion along the southwestern boundary of North America is from studies of the Kane Fracture Zone in the Central Atlantic Ocean that show significant changes in relative plate motion from orientation and structure of the zone (Tucholke and Schouten, 1988). A pronounced kink in the trace of the Kane and North Kane fracture zones in the Jurassic Magnetic Quiet Zone (between the Blake Spur Magnetic Anomaly and Anomaly M-25; see Muehlberger, 1992) is caused by changes in relative plate motion during Late Jurassic time. Adjusting the Tucholke and Schouten dates from the Decade of North American Geology time scale (Palmer, 1983) to comply with the time scale of Gradstein et al. (1994), interpolated ages of relative plate motion, indicated by the kink, are at ca. 162 Ma (20°–25° clockwise) and at ca. 159 to ca. 156 Ma (45° counterclockwise). Counterclockwise rotation (ca. 159 to ca. 156 Ma) of North America (moving away from a relatively stationary African plate) would have caused right-lateral offsets along susceptible elements of the northwest-trending fabric along its boundary. North America moved toward the east relative to the accreted terranes (see Fig. 5 for boundary) that formed Mexico in those days. (N.B.: the preceding clockwise rotation could cause left-lateral displacements along elements of the fabric.)

During the interval of clockwise rotation, the Chihuahua trough was born as a pull-apart feature with its configuration controlled by a preexisting north-trending fabric (Fig. 4B). The ca. 159 to ca. 156 Ma interval during which the trough first appeared is during Oxfordian time (time scale of Gradstein et. al., 1994). The Chihuahua trough is the result of the interactions between the right-lateral Texas "zone" of Muehlberger (1980) and the San Marcos fault of McKee et al., (1990) with the north-trending fabric.

SABINAS BASIN

A reddish pre–Upper Jurassic "basal" conglomerate is reported from numerous wells in the Sabinas basin (Santamaría-O. et al., 1991). This basal conglomerate is only found in the subsurface and, based on similarity to sequences to the south and southeast that contain Jurassic vertebrates and pollen, Santamaría-O. et al. interpreted it to be a Lower to Middle Jurassic continental deposit. Their interpretation shows conglomerate accumulating in grabens developed between a series of elevated blocks and islands; the prevalent trend of basins is 75° west, and rhomb-shaped islands and basins, bounded by faults trending 15° west, occur throughout the area. Considerable topography is indicated because Jurassic sediments are absent on the highest structural blocks, Oxfordian rocks are found only in the lowest blocks, and Kimmeridgian-Tithonian rocks occur on "intermediate" blocks. McKee et al. (1990) described a Jurassic marine conglomerate from Valle de San Marcos (Fig. 5), which occupies a similar stratigraphic position to the basal conglomerate. Here there are ~1600 m of sandstone and conglomerate (Las Palomas and Sierra El Granizo beds, base not exposed) below a section containing Tithonian ammonites (Tanque Cuatro Palmas beds).

Santamaría-O. et al. (1991) reported that eight wells in the extreme southeastern Sabinas basin encountered extrusive dacite, rhyodacite, andesite, trachyte, and some basalt interbeds in the basal conglomerate (Fig. 5). Garrison and McMillan (1999) interpreted the protolith of allogenic blocks of greenschist facies metavolcanic rocks in the El Papalote evaporite diapir (Fig. 5) in the Laramide La Popa basin (southeastern Sabinas basin) as Jurassic continental rift basalt. Santamaría-O. et al. (1991) reported Jurassic (Callovian) K-Ar dates from granite in PEMEX Pecten-1 (160 ± 6 Ma), ~20 km southeast of Monclova and PEMEX Menchaca-1A (164 ± 7 Ma), ~60 km northwest of Monclova (see Fig. 5 for locations).

DISCUSSION

If the reported dates from Pecten-1 and Menchaca-1 are valid, Jurassic granite was emplaced into a terrane consisting of Late Permian–Early Triassic granite that extends along the Mexi-

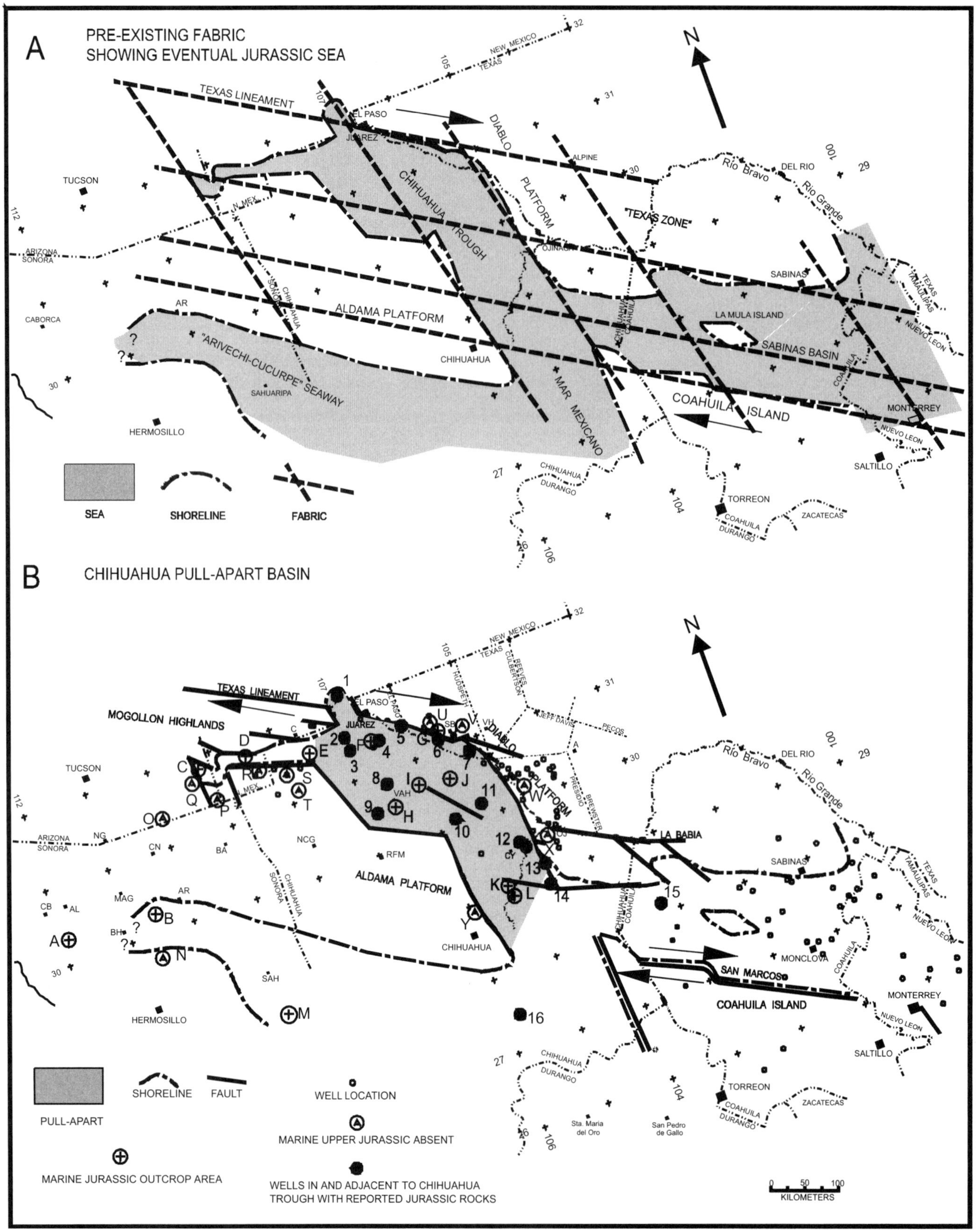

Figure 4. (A) Map (after Haenggi, 2002) showing pre-Jurassic structural fabric (after Muehlberger and Moustafa, 1984) and extent of Jurassic ocean, well and surface control shown by B. (B) Map (after Haenggi, 2002) showing mechanism for formation of Chihuahua trough as a pull-apart basin. A and B show Jurassic emergent areas. B shows locations of marine Jurassic outcrops, wells with reported Jurassic rocks, and surface locations where Jurassic rocks are absent. Symbols for wells, surface locations, and cities and towns same as for Figure 2B. Reproduced with permission of Sociedad Geológica Mexicana.

can coast from the central Sabinas Basin to the latitude of the city of Veracruz (Torres et al., 1999; Salvador, 1991, plate 3 therein). If Permian-Triassic granitic rocks formed a contiguous terrane, as Salvador's map implies, it is unlikely that Mesozoic sinistral (or dextral) displacements of any magnitude have occurred during the past 160 m.y. along the various northwest-trending megashears postulated south of the Sabinas basin because the trend of the Permian-Triassic granites crosses the projected megashears with little or no apparent displacement. See Jones et al. (1995) for a representative argument favoring major displacement. If Jurassic dates for granites within the Permian-Triassic plutonic trend are correct, the rather widespread pre-Oxfordian volcanic rocks reported by Jones et al. (1995) and Santamaría-O. et al. (1991) from northeastern Mexico may be coeval with emplacement of granite in a tectonic setting separate from the well-known Jurassic arc along the western coast of North America (Busby-Spera, 1988).

Perhaps these Jurassic igneous rocks reflect a rejuvenation of whatever caused emplacement of the Permian-Triassic granites (consumption of a late Paleozoic–earliest Triassic ocean behind an advancing arc?). The Permian-Triassic granites formed subsequent to development of a late Paleozoic island arc (Las Delicias arc) located immediately south of the San Marcos fault (McKee et al., 1988).

SUMMARY

During a Late Jurassic (ca. 159 to ca. 156 Ma) period of relative counterclockwise rotation of the North American plate, reactivation of elements of an ancient system of intersecting zones of crustal weakness caused the Chihuahua trough to form as a pull-apart basin. The trough, located between the Diablo and Aldama platforms, occupies the site of the late Paleozoic Pedregosa basin and was an area of tectonic subsidence, at varying rates, from inception to earliest Albian time. The trough was the locus of a marine transgression during Late Jurassic time (Oxfordian?-Kimmeridgian). During Tithonian and Neocomian (Early Cretaceous) time, as relative basin subsidence rates decreased, extensive evaporite deposits were deposited in large areas of the eastern part of the trough and are a key for interpretation of the structural style of the Laramide inversion of the Chihuahua trough. If a pull-apart origin of the Chihuahua trough is accepted, it implies that a Mojave-Sonora Megashear does not cross the area depicted by Figure 5 and that two northwest-trending arms of the Mar Mexicano formed in the study area at the time of the opening of the Gulf of Mexico.

ACKNOWLEDGMENTS

Pat Dickerson provided many thoughts and observations from her extensive work in the general area. She reviewed the manuscript and has made her extensive library and knowledge of the literature available. Data for wells near the Quitman and Eagle mountains were provided by Texaco through the efforts of Susan Longacre and Vic Smith. Pat Dickerson and Jim Wolleben, formerly with Gulf Oil Company, provided data for wells in the Marfa Basin. During the 1960s and early 1970s, Teodoro Díaz-G., Arsenio Navarro-G., Jorge Tovar-R., and J. Valencia-R., provided their interpretations of PEMEX petroleum exploration efforts in Chihuahua. Fred McDowell explained the significance of isotope studies for delineation of the North American craton. We thank the Owen Coates Fund of the Geology Foundation, John A. and Katherine G. Jackson School of Geosciences, The University of Texas at Austin, for funding partial page charges.

REFERENCES CITED

Albritton, C.C., and Smith, J.F., Jr., 1965, Geology of the Sierra Blanca area, Hudspeth County, Texas: U.S. Geological Survey Professional Paper 479, 131 p.

Almazán-Vázquez, E., and Palafox-Reyes, J.J., 2000, Secuencia samitica del Jurasico Tardío con amonitas del genero *Subdichotomoceras* (Kimmeridgiano) expuesta al oriente de Arivechi, Sonora, México: Cuarta Reunión sobre la Geología del Noroeste de México y Areas Adyacentes: Libro de Resúmenes, Universidad Nacional Autónoma de México, Instituto de Geología, Estación Regional del Noroeste, Publicaciones Ocasionales no. 2, p. 1–2.

Araujo-Mendieta, J., and Casar-González, R., 1987, Estratigrafía y sedimentología del Jurasico Superior en la Cuenca de Chihuahua, norte de México: Revista del Instituto Mexicano del Petróleo, v. 19, no. 1, p. 6–29.

Bilodeau, W.L., Kluth, C.F., and Vedder, L.K., 1987, Regional stratigraphic, sedimentologic, and tectonic relationships of the Glance Conglomerate in southeastern Arizona, *in* Dickinson, W.R., and Klute, M.A., eds., Mesozoic rocks of southern Arizona and adjacent Areas: Arizona Geological Society Digest, v. 18, p. 229–256.

Brown, M.L., and Dyer, R., 1987, Mesozoic geology of northwestern Chihuahua, Mexico, *in* Dickinson, W.R., and Klute, M.A., eds., Mesozoic rocks of southern Arizona and adjacent Areas: Arizona Geological Society Digest, v. 18, p. 381–394.

Busby-Spera, C.J., 1988, Speculative tectonic model for the early Mesozoic arc of the southwest Cordilleran United States: Geology, v. 16, p. 1121–1125, doi: 10.1130/0091-7613(1988)016<1121:STMFTE>2.3.CO;2.

Busby-Spera, C.J., and Kokelaar, B.P., 1991, Controls of the Sawmill Canyon fault zone on Jurassic magmatism and extension/transtension in southern Arizona: Geological Society of America Abstracts with Programs, v. 23, no. 5, p. A-250.

Calmus, T., Pérez-Segura, E., and Stinnesbeck, W., 1997, La structuration de la marge Pacifique Nordamericaine et du "terrane" Caborca; apports de la découverte d'une faune du Jurassique Inférieur et Moyen dans la série de Pozo Serna (Sonora Mexique): Comptes rendus de l'Académie des Sciences (Paris), Sciences de la Terre et des planets (Earth and Planetary Sciences), v. 325, p. 257–263.

Cantu-Chapa, A., 1970, El Kimeridgiano Inferior de Samalayuca, Chihuahua: Revista del Instituto Mexicano del Petróleo, v. 11, no. 3, p. 40–44.

Cantu-Chapa, C.M., Sandoval-Silva, R., and Arenas-Partida, R., 1985, Evolución sedimentaria del Cretácico Inferior en el norte de México: Revista Instituto Mexicano del Petróleo, v. 17, no. 2, p. 14–37.

Carpenter, D.L., 1997, Tectonic history of the metamorphic basement rocks of the Sierra del Carmen, Coahuila, Mexico: Geological Society of America Bulletin, v. 109, p. 1321–1332, doi: 10.1130/0016-7606(1997)109<1321:THOTMB>2.3.CO;2.

Córdoba, D.A., 1969, Mesozoic stratigraphy of northeastern Chihuahua, Mexico, *in* Córdoba, D., Wengard, S.A., and Shomaker, J., eds., The border region: New Mexico Geological Society, 20th Field Conference Guidebook, p. 91–101.

De Cserna, Z., De Rincón-Ortega., C., Solorio-Munguia, and Schmitter V., 1968, Una edad radiometrica Permica temprana de la región de Placer de Guadalupe, noreste de Chihuahua: Sociedad Geológica Mexicana, Boletín 31, p. 65–73.

DeFord, R.K., and Haenggi, W.T., 1970, Stratigraphic nomenclature of Cretaceous rocks in northeastern Chihuahua, *in* Seewald, K., and Sundeen, D., eds., The geologic framework of the Chihuahua tectonic belt: Symposium in honor of Professor Ronald K. DeFord: West Texas Geological Society and the University of Texas at Austin, p. 175–196.

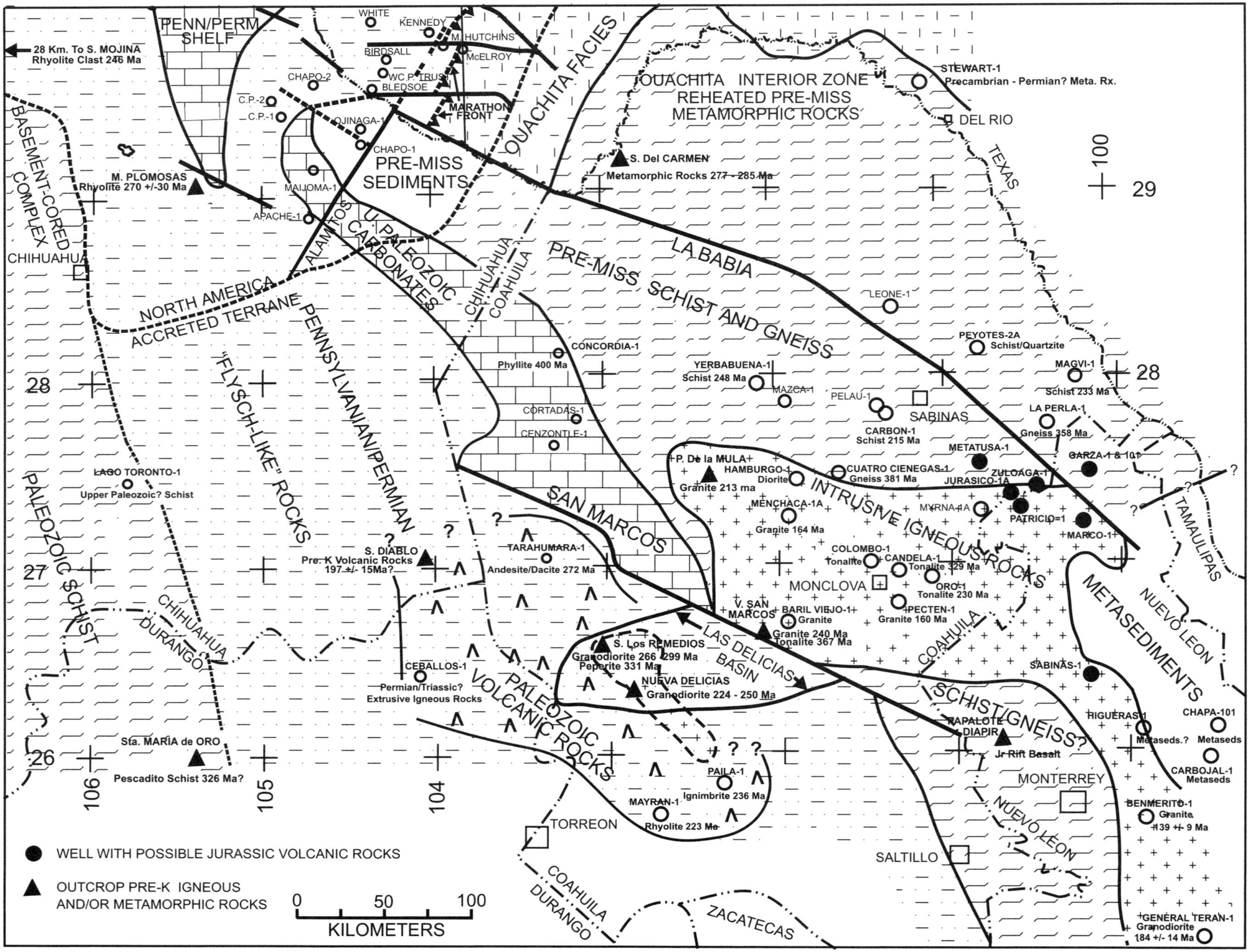
PENN/PERM SHELF
28 Km. To S. MOJINA Rhyolite Clast 246 Ma
WHITE
KENNEDY
M. HUTCHINS
BIRDSALL
McELROY
CHAPO-2
WC.P. TRUST
BLEDSOE
C.P.-2
C.P.-1
MARATHON FRONT
OJINAGA-1
CHAPO-1
PRE-MISS SEDIMENTS
MAIJOMA-1
APACHE-1
M. PLOMOSAS Rhyolite 270 +/-30 Ma
BASEMENT-CORED COMPLEX
CHIHUAHUA
ALAMITOS
U. PALEOZOIC CARBONATES
OUACHITA FACIES
OUACHITA INTERIOR ZONE REHEATED PRE-MISS METAMORPHIC ROCKS
STEWART-1 Precambrian - Permian? Meta. Rx.
DEL RIO
TEXAS
S. Del CARMEN Metamorphic Rocks 277 - 285 Ma
100
29
CHIHUAHUA
COAHUILA
LA BABIA
PRE-MISS SCHIST AND GNEISS
NORTH AMERICA
ACCRETED TERRANE
PENNSYLVANIAN/PERMIAN
"FLYSCH-LIKE" ROCKS
LEONE-1
PEYOTES-2A Schist/Quartzite
MAGVI-1 Schist 233 Ma
28
CONCORDIA-1 Phyllite 400 Ma
YERBABUENA-1 Schist 248 Ma
MAZCA-1
PELAU-1
SABINAS
CARBON-1 Schist 215 Ma
LA PERLA-1 Gneiss 358 Ma
CORTADAS-1
CENZONTLE-1
METATUSA-1
GARZA-1 & 101
LAGO TORONTO-1 Upper Paleozoic? Schist
P. De la MULA
HAMBURGO-1 Diorite
Granite 213 ma
CUATRO CIENEGAS-1 Gneiss 381 Ma
ZULOAGA-1
JURASICO-1A
INTRUSIVE IGNEOUS ROCKS
MENCHACA-1A Granite 164 Ma
MYRNA-1A
PATRICIO-1
MARCO-1
TAMAULIPAS
SAN MARCOS
S. DIABLO Pre-K Volcanic Rocks 197 +/- 15Ma?
TARAHUMARA-1 Andesite/Dacite 272 Ma
COLOMBO-1 Tonalite
CANDELA-1 Tonalite 329 Ma
ORO-1 Tonalite 230 Ma
MONCLOVA
PECTEN-1 Granite 160 Ma
METASEDIMENTS
NUEVO LEON
27
PALEOZOIC SCHIST
CHIHUAHUA
DURANGO
V. SAN MARCOS
BARIL VIEJO-1 Granite
Granite 240 Ma Tonalite 367 Ma
S. Los REMEDIOS Granodiorite 266 - 299 Ma Peperite 331 Ma
LAS DELICIAS BASIN
NUEVA DELICIAS Granodiorite 224 - 250 Ma
COAHUILA
CEBALLOS-1 Permian/Triassic? Extrusive Igneous Rocks
PALEOZOIC VOLCANIC ROCKS
SABINAS-1
SCHIST/GNEISS?
HIGUERAS-1 Metaseds.?
CHAPA-101 Metaseds
CARBOJAL-1 Metaseds
NAPALOTE DIAPIR
Jr Rift Basalt
Sta. MARIA de ORO Pescadito Schist 326 Ma?
26
106
105
104
PAILA-1 Ignimbrite 236 Ma
MAYRAN-1 Rhyolite 223 Ma
MONTERREY
BENMERITO-1 Granite 139 +/- 9 Ma
TORREON
NUEVO LEON
SALTILLO
WELL WITH POSSIBLE JURASSIC VOLCANIC ROCKS
OUTCROP PRE-K IGNEOUS AND/OR METAMORPHIC ROCKS
0 50 100
KILOMETERS
COAHUILA
DURANGO
ZACATECAS
GENERAL TERAN-1 Granodiorite 184 +/- 14 Ma

Figure 5. Map showing pre–Late Jurassic geology of Coahuila, southeastern Chihuahua, and adjacent areas of Durango, Texas, and Nuevo Leon; outcrops of and wells penetrating pre-Cretaceous intrusive igneous, volcanic and metamorphic rocks, wells penetrating presumed Jurassic volcanic rocks and other wells. Data sources: Sierra del Carmen, Carpenter (1997); Stewart-1, Nicholas and Waddell (1989); Sierra Mojina clast, Denison et al. (1970); Minas Plomosas area, De Cserna et al. (1968); Santa Maria del Oro, Benmerito-1 and General Teran-1, Wilson (1990); Sierra Diablo, Jones and McKee, (1987); Peyotes-2A, Flawn (1961); Lago Toronto-1 and Ceballos-1, López-Ramos (1988); Papalote diapir, Garrison and McMillan (1999); all other outcrops and wells in Coahuila, Durango, and Nuevo Leon are cited by and/or discussed by Santamaría-O. et al. (1991); all other wells in Chihuahua and Texas from Haenggi (2001). Paleogeology of central and southeastern Coahuila modified from Santamaría-O. et al., (1991); Las Delicias basin (minimum extent of Las Delicias Formation) from McKee et al. (1999); pre–late Paleozoic boundary of "North America" from Haenggi (2001).

DeJong, H.W., and Addy, S.K., 1992, Broad view indicates hydrocarbon potential low in far West Texas: Trans-Pecos Texas–2 (Conclusion): Oil and Gas Journal, January, v. 27, p. 97–102.

Denison, R.E., Burke, W.H., Jr., Hetherington, E.A., and Otto, J.B., 1970, Basement rock framework of parts of Texas, southern New Mexico and northern Mexico, *in* Seewald, K., and Sundeen, D., eds., The geologic framework of the Chihuahua tectonic belt: Symposium in honor of Professor Ronald K. DeFord: West Texas Geological Society and the University of Texas at Austin, p. 3–14.

Dyer, R., 1987, First day, Part A; Road log from El Paso/Ciudad Juárez via Villa Ahumada, El Sueco to Ciudad Chihuahua, *in* Excursión Geológica, Libreto Guía de Caminos: Sociedad Geológica Mexicana, Universidad Autónoma de Chihuahua, University of Texas at El Paso, p. 2–30.

Fackler-Adams, B.N., Busby, C.J., and Mattinson, J.M., 1997, Jurassic magmatism and sedimentation in the Palen Mountains, southeastern California: Implications for regional tectonic controls on the Mesozoic continental arc: Geological Society of America Bulletin, v. 109, p. 1464–1484, doi: 10.1130/0016-7606(1997)109<1464:JMASIT>2.3.CO;2.

Flawn, P.T., 1961, The Ouachita structural belt in Mexico, *in* The Ouachita System: Texas Bureau of Economic Geology, Publication 6120, p. 99–106.

Garrison, J.M., and McMillan, N.J., 1999, Jurassic continental rift magmatism in northeast Mexico: Allogenic metaigneous blocks in the El Papalote evaporite diapir, La Popa basin, Nuevo Leon, Mexico, *in* Bartolini, C., Wilson, J.L., and Lawton, T.F., eds., Mesozoic sedimentary and tectonic history of north-central Gulf of Mexico: Geological Society of America Special Paper 340, p. 319–332.

González-León, C.M., and Lucas, S.G., 1995, Stratigraphy and paleontology of the Early Cretaceous Cerro de Oro Formation, central Sonora, *in* Jacques-Ayala, C., González-León, C.M., and Roldán-Quintana, J., eds., Studies on the Mesozoic of Sonora and adjacent areas: Geological Society of America Special Paper 301, p. 79–98.

Gradstein, F.M., Agterberg, F.P., Ogg, J.C., Hardenbol, J., van Veen, P., Thierry, J., and Huang, Z., 1994, A Mesozoic time scale: Journal of Geophysical Research, v. 99, p. 24,051–24,074, doi: 10.1029/94JB01889.

Haenggi, W.T., 1966, Geology of El Cuervo area, northeastern Chihuahua, Mexico [Ph.D. dissertation]: University of Texas at Austin, 403 p.

Haenggi, W.T., 2001, Tectonic history of the Chihuahua trough, Mexico and adjacent USA, Part I: the pre-Mesozoic setting: Boletín Sociedad Geológica Mexicana, v. LIV, p. 28–66, http://www.unicit.unam.mx/SGM.html (Accessed 15 Mar. 2003).

Haenggi, W.T., 2002, Tectonic history of the Chihuahua trough, Mexico and adjacent USA, Part II: Mesozoic and Cenozoic: Boletín Sociedad Geológica Mexicana, v. LV, p. 38–94, http://www.unicit.unam.mx/SGM.html (Accessed 15 Mar. 2003).

Hayes, P.T., 1970, Cretaceous paleogeography of southeastern Arizona and adjacent areas: U.S. Geological Survey Professional Paper 658-B, 28 p.

James, E.W., and Henry, C.D., 1993, Pb isotopes of ore deposits in Trans-Pecos Texas and northeastern Chihuahua, Mexico: Basement, igneous and sedimentary sources of metals: Economic Geology and the Bulletin of the Society of Economic Geologists, v. 88, p. 934–947.

Jones, B.R., and Reaser, D.F., 1970, Geology of southern Quitman Mountains, Hudspeth County, Texas: Texas Bureau of Economic Geology, Geologic Quadrangle Map No. 39.

Jones, N.W., and McKee, J.W., 1987, Pre Cretaceous volcanic rocks at Sierra Diablo, Chihuahua: Universidad Autónoma de Chihuahua, Facultad de Ingenieria and Sociedad Geológica Mexicana, Delegación Chihuahua, Gaceta Geológica, v. 1, no. 1, p. 82–96.

Jones, N.W., McKee, J.W., Anderson, T.H., and Silver, L.T., 1995, Jurassic volcanic rocks in northeastern Mexico: A possible remnant of a Cordilleran magmatic arc, *in* Jacques-Ayala, C., González-León, C.M., and Roldán-Quintana, J., eds., Studies on the Mesozoic of Sonora and adjacent areas: Geological Society of America Special Paper 301, p. 179–190.

Lawton, T.F., and Olmstead, G.A., 1995, Stratigraphy and structure of the lower part of the Bisbee Group, northeastern Chiricahua Mountains, Arizona, *in* Jacques-Ayala, C., González-León, C.M., and Roldán-Quintana, J., eds., Studies on the Mesozoic of Sonora and adjacent areas: Geological Society of America Special Paper 301, p. 21–39.

Lawton, T.F., Garrison, J.M., and McMillan, N.J., 1997, Late Jurassic transtensional borderland on the southwestern margin of North America: Geological Society of America Abstracts with Programs, v. 29, no. 6, p. A-200–A-201.

Limón-González, M., 1986, Evaluación geológico-geoquimica de la provincia de Chihuahua: Asociación Mexicana de Geólogos Petroleros, Boletín, v. XXXVIII, p. 3–58.

Lindberg, F.A., 1987, Cretaceous sedimentary geology of the Rucker Canyon Area, Cochise County, Arizona, *in* Dickinson, W.R., and Klute, M.A., eds., Mesozoic rocks of southern Arizona and adjacent areas: Arizona Geological Society Digest, v. 18, p. 283–300.

López-Ramos, E., 1988, Geologia y aprovechamiento integral de las perforaciones en el altiplano Mexicano: Geomimet, no. 151, p. 84–100.

Lucas, S.G., and Lawton, T.F., 2000, Stratigraphy of the Bisbee Group (Jurassic-Cretaceous), Little Hatchet Mountains, New Mexico, *in* Lawton, T.F., McMillan, N.J., and McLemore, V.T., eds., Southwest passage—A trip through the Phanerozoic: New Mexico Geological Society, 51st Field Conference Guidebook, p. 175–194.

McKee, J.W., Jones, N.W., and Anderson, T.H., 1988, Las Delicias basin: A record of late Paleozoic arc volcanism in northeastern Mexico: Geology, v. 16, p. 37–40, doi: 10.1130/0091-7613(1988)016<0037:LDBARO>2.3.CO;2.

McKee, J.W., Jones, N.W., and Anderson, T.H., 1990, Stratigraphy and provenance of strata along the San Marcos fault, central Coahuila, Mexico: Geological Society of America Bulletin, v. 102, p. 593–614, doi: 10.1130/0016-7606(1990)102<0593:SAPOSA>2.3.CO;2.

McKee, J.W., Jones, N.W., and Anderson, T.H., 1999, Late Paleozoic and early Mesozoic history of the Las Delicias terrane, Coahuila, Mexico, *in* Bartolini, C., Wilson, J.L., and Lawton, T.F., eds. Mesozoic sedimentary and tectonic history of north-central Mexico: Geological Society of America Special Paper 340, p. 161–189.

Muehlberger, W.R., 1980, Texas lineament revisited, *in* Dickerson, P.W., and Hoffer, J.M., eds., Trans-Pecos region: New Mexico Geological Society, 31st Field Conference Guidebook, p. 113–121.

Muehlberger, W.R., 1992, Tectonic map of North America, southeast sheet: American Association of Petroleum Geologists, scale 1:5,000,000.

Muehlberger, W.R., and Moustafa, A.R., 1984, Late Cenozoic pull-apart graben development, Big Bend Region, Texas: American Association of Petroleum Geologists Bulletin, v. 64, p. 510.

Nicholas, R.L., and Waddell, D.E., 1989, The Ouachita system in the subsurface of Texas, Arkansas and Louisiana, *in* Hatcher, R.D., Jr., Thomas, W.A., and Viele, G.W., eds., The Appalachian-Ouachita Orogen in the United States: Geological Society of America, Geology of North America, v. F-2, p. 661–672.

Olmstead, G.A., and Young, K., 2000, Late Jurassic ammonites form the northeastern Chiricahua Mountains, southeast Arizona: New Mexico Geology, v. 22, p. 1–7.

Pálfy, J., and González-León, C.M., 2000, Lower Jurassic ammonoid biostratigraphy of the Antimonio terrane, Sonora, northwestern Mexico: Cuarto Reunión sobre la Geología del Noroeste de México y Areas Adyacentes: Libro de Resumes, Universidad Nacional Autónoma de México, Instituto de Geología, Estación Regional del Noroeste, Publicaciones Ocasionales, no. 2, p. 87–88.

Palmer, A.R., compiler, 1983, The Decade of North American Geology 1983 geologic time scale: Geology, v. 11, p. 503–504.

Ramírez-M., J.C., and Acevedo-C., F., 1957, Notas sobre la geólogía de Chihuahua: Asociación Mexicana de Geólogos Petroleros, Boletín, v. IX, p. 583–772.

Rangin, C., 1977, Sobre la presencia del Jurásico Superior con amonitas in Sonora septentrional: Universidad Nacional Autónoma de México, Instituto de Geología, Revista, p. 1–4.

Reyeros de Castillo, M.M., 1974, Corales del Jurásico superior de Chihuahua: Universidad Nacional Autónoma de México, Instituto de Geólogia, Paleontología Mexicana, no. 40, p. 7–43.

Roberts, D.C., and Dyer, R., 1988, A preliminary report on the geology of the Cerro Panales area, east-central Chihuahua, Mexico, *in* Clark, K.F., Goodell, P.C., and Hoffer, J.M., eds., Stratigraphy, tectonics and resources of parts of Sierra Madre Occidental province, Mexico: El Paso Geological Society, Guidebook 1988 Field Conference, p. 159–172.

Rodríguez-Torres, R., 1969, Mesozoic stratigraphy of Sierra de la Alcaparra, northeastern Chihuahua, Mexico, *in* Córdoba, D., Wengard, S.A., and Shomaker, J., eds., The border region: New Mexico Geological Society, 20th Field Conference Guidebook, p. 173–175.

Salvador, A., 1991, Structure at the base and subcrop below Mesozoic marine section, Gulf of Mexico, *in* Salvador, A., ed., The Gulf of Mexico Basin: Boulder, Colorado, Geological Society of America, Geology of North America, v. J, Plate 3, scale 1:5,000,000.

Santamaría-O., D., Ortuño-A., F., Adatte, T., Ortiz-U., A., Riba-R., A., and Franco-N., S., 1991, Evolución Geodinamica de la Cuenca de Sabinas y sus Implicaciones Petroleras Estado de Coahuila: Instituto Mexicano del Petroleo, Subdirección de Technología de Exploración, Gerencia de Investigación Aplicada a la Exploración, tomo I, 209 p.

Thompson, S., III, 1982, Oil and gas exploration wells in southwestern New Mexico, *in* Powers, R.B., ed., Geologic Studies of the Cordilleran Thrust Belt: Rocky Mountain Association of Geologists, v. II, p. 521–536.

Thompson, S., III, and Tovar R., J.C., and Conley, J.N., 1978, Oil and gas exploration wells in the Pedregosa basin, *in* Land of Cochise: New Mexico Geological Society, 29th Field Conference Guidebook, p. 331–342.

Torres, R., Ruiz, J., Patchett, P.J., and Grajales, J.M., 1999, Permo-Triassic continental arc in eastern Mexico: Tectonic implications for reconstruction of southern North America, *in* Bartolini, C., Wilson, J.L., and Lawton, T.F., eds., Mesozoic sedimentary and tectonic history of north-central Gulf of Mexico: Geological Society of America Special Paper 340, p. 191–196.

Tucholke, B.E., and Schouten, H., 1988, Kane Fracture Zone: Marine Geophysical Researches, v. 10, p. 1–39.

Uphoff, T.L., 1978, Subsurface stratigraphy and structure of the Mesilla and Hueco bolsons, El Paso region, Texas and New Mexico [M.S. thesis]: University of Texas at El Paso, 66 p.

Ward, C.A., 1977, Structural geology and tectonic history of Paleozoic rocks in the Sierra de Las Monillas, east-central Chihuahua, Mexico [M.S. thesis]: Fort Worth, Texas Christian University, 54 p.

Wilson, J.L., 1990, Basement structural controls on Mesozoic carbonate facies in northeastern Mexico—a review: Special Publications of the International Association of Sedimentologists, v. 9, p. 235–255.

Zeller, R.A., Jr., 1965, Stratigraphy of the Big Hatchet Mountains area, New Mexico: New Mexico Bureau of Mines and Mineral Resources Memoir 16, 128 p.

Manuscript Accepted by the Society 24 March 2005

Printed in the USA

Geological Society of America
Special Paper 393
2005

Northward Laramide thrusting in the Quitovac region, northwestern Sonora, Mexico: Implications for the juxtaposition of Paleoproterozoic basement blocks and the Mojave-Sonora megashear hypothesis

Alexander Iriondo*
Department of Geological Sciences, University of Colorado, Boulder, Colorado 80309, USA, and *U.S. Geological Survey, MS 974, Box 25046, Denver Federal Center, Denver, Colorado 80225, USA*

Luis M. Martínez-Torres
Departamento de Geodinámica, Universidad del País Vasco, Apartado 644, E-48080 Bilbao, Spain

Michael J. Kunk
U.S. Geological Survey, MS 963, Box 25046, Denver Federal Center, Denver, Colorado 80225, USA

William W. Atkinson Jr.
Department of Geological Sciences, University of Colorado, Boulder, Colorado 80309, USA

Wayne R. Premo
U.S. Geological Survey, MS 963, Box 25046, Denver Federal Center, Denver, Colorado 80225, USA

William C. McIntosh
New Mexico Bureau of Mines and Mineral Resources, Socorro, New Mexico 87801, USA

ABSTRACT

Restoration of 12%–30% Basin and Range extension allows direct interpretation of ductile fabrics associated with a stack of Laramide thrust faults in the Quitovac region in northwestern Sonora. The inferred direction of displacement of these thrusts varies gradually from N63°W to N23°E and is interpreted to represent a clockwise rotation of the direction of Laramide thrusting through time. The thrust faults represent a piggy-back sequence of thrusting propagating north, toward the foreland. The average direction and sense of displacement of the thrusts is N18°W, and the cumulative 45 km of estimated northward-directed displacement corresponds to ~86% of shortening.

*Present address: Centro de Geociencias, UNAM, Campus Juriquilla, C.P. 76230 Juriquilla, Querétaro, México; e-mail: iriondo@geociencias.unam.mx.

Iriondo, A., Martínez-Torres, L.M., Kunk, M.J., Atkinson, W.W., Jr., Premo, W.R., and McIntosh, W.C., 2005, Northward Laramide thrusting in the Quitovac region, northwestern Sonora, Mexico: Implications for the juxtaposition of Paleoproterozoic basement blocks and the Mojave-Sonora megashear hypothesis, *in* Anderson, T.H., Nourse, J.A., McKee, J.W., and Steiner, M.B., eds., The Mojave-Sonora megashear hypothesis: Development, assessment, and alternatives: Geological Society of America Special Paper 393, p. 631–669. doi: 10.1130/2005.2393(24). For permission to copy, contact editing@geosociety.org.

Based on geochronological constraints, onset of thrusting in Quitovac occurred sometime between 75 and 61 Ma, whereas cessation occurred at ca. 39 Ma. The presence of Paleocene-Eocene orogenic gold mineralization, spatially associated with thrusting, strengthens our idea that compressional tectonism associated with the Laramide orogeny is a very important and widespread dynamometamorphic event in the region.

Similarities in age, kinematics, and structural stratigraphy indicate that the thrusting in the Quitovac region may be equivalent to the Laramide Quitobaquito Thrust in southwestern Arizona. In both areas, thrust faults juxtapose the Paleoproterozoic Caborca and "North America" basement blocks. This juxtaposition was previously proposed as exclusively related to movements along the hypothetical Upper Jurassic Mojave-Sonora megashear. The Laramide northward displacements and clockwise rotations recorded in the Caborca block rocks in Quitovac contradict the southward displacements (~800 km) and counterclockwise rotations inherent in the left-lateral Upper Jurassic Mojave-Sonora megashear hypothesis. We conclude that if this megashear exists in northwestern Sonora, its trace should be to the southwest of the Quitovac region.

Keywords: Laramide, thrusting, megashear, Sonora, Mexico.

INTRODUCTION

The term Laramide has been widely used in southeastern California, southern Arizona, and northern Sonora to explain a large-magnitude, tectonomagmatic, metamorphic, and sedimentary event from the Late Cretaceous into early Tertiary time (e.g., Damon et al., 1964; Armstrong, 1974; Coney, 1976; Davis, 1979; Dickinson et al., 1988). The Laramide orogeny involved different styles of deformation along the North American Cordillera (e.g., thick- and thin-skin styles of deformation), but the exact timing and the nature of this compressional event are still a matter of debate (e.g., Coney and Reynolds, 1977; Dickinson and Snyder, 1978; Haxel et al., 1984; Bird, 1998; Dumitru et al., 1991; Maxson and Tikoff, 1996).

In a summary of the Late Cretaceous to early Tertiary evolution of southwestern Arizona and southeastern California, Miller et al. (1992) proposed that widespread magmatism, metamorphism, and ductile thrusting of Precambrian basement and associated supracrustal rocks were characteristic of the region at that time. Similarly, in northwestern Sonora most of the ductile compressional fabrics and associated greenschist-facies regional metamorphism along thrust faults have been interpreted as Upper Cretaceous to lower Tertiary in age (e.g., De Jong et al., 1988; Jacques-Ayala et al., 1990; Sosson and Calmus, 1990; Pubellier et al., 1995; Jacques-Ayala and De Jong, 1996; Nourse, 2001).

However, other hypotheses proposed that the compressional ductile fabrics in Proterozoic, Paleozoic, and Middle Jurassic rocks developed during thrusting in a transpressive regime along the hypothetical Mojave-Sonora megashear (Anderson et al., 1984; Connors, 1990; Connors et al., 1989; Anderson et al., 1992; Caudillo-Sosa et al., 1996; Campbell and Anderson, 1998, 2003). These authors suggested a Late Jurassic age for deformation but did not observe correlative fabrics in the Upper Cretaceous two-mica granites outcropping in the region. In their perspective, much of the Mojave-Sonora megashear has been obscured by intense Late Cretaceous tectonism, Tertiary volcanism, and Mid-Tertiary metamorphic core complexes and Basin and Range extension. Nonetheless, well-preserved exposures with postulated megashear fabrics have been proposed in Sierra Los Tanques and in the Quitovac region (Fig. 1; Connors et al., 1989; Tosdal et al., 1990; Anderson et al., 1992; Campbell and Anderson, 1998, 2003).

Ongoing debate in Sonora centers on whether the age of metamorphic and deformational fabrics associated with thrusting is Upper Jurassic or Laramide. Complicating the issue are ductile fabrics representing other episodes of deformation that predate the hypothetical Upper Jurassic Mojave-Sonora megashear (this paper) or postdate the Laramide event (metamorphic core complexes; e.g., De Jong et al., 1988; Nourse et al., 1994). This added complexity establishes a premise that several episodes of ductile deformation, including reactivation of older fabrics, might have occurred in some areas in Sonora.

The first step to evaluate this premise was to restore Basin and Range extension accumulated in tilted blocks in the Quitovac region. Coupled with geologic mapping and $^{40}Ar/^{39}Ar$ geochronology, this restoration allows for the direct interpretation of the timing and nature of ductile fabrics. The Quitovac region offers an excellent opportunity to test whether these fabrics are associated with the juxtaposition of the Paleoproterozoic Caborca and "North America" basement blocks represent a single event of the postulated Mojave-Sonora megashear or instead represent several episodes of ductile deformation.

THE MOJAVE-SONORA MEGASHEAR HYPOTHESIS IN NORTHWESTERN SONORA

The Mojave-Sonora megashear in northwestern Sonora is postulated to be a northwest-trending, left-lateral strike-slip

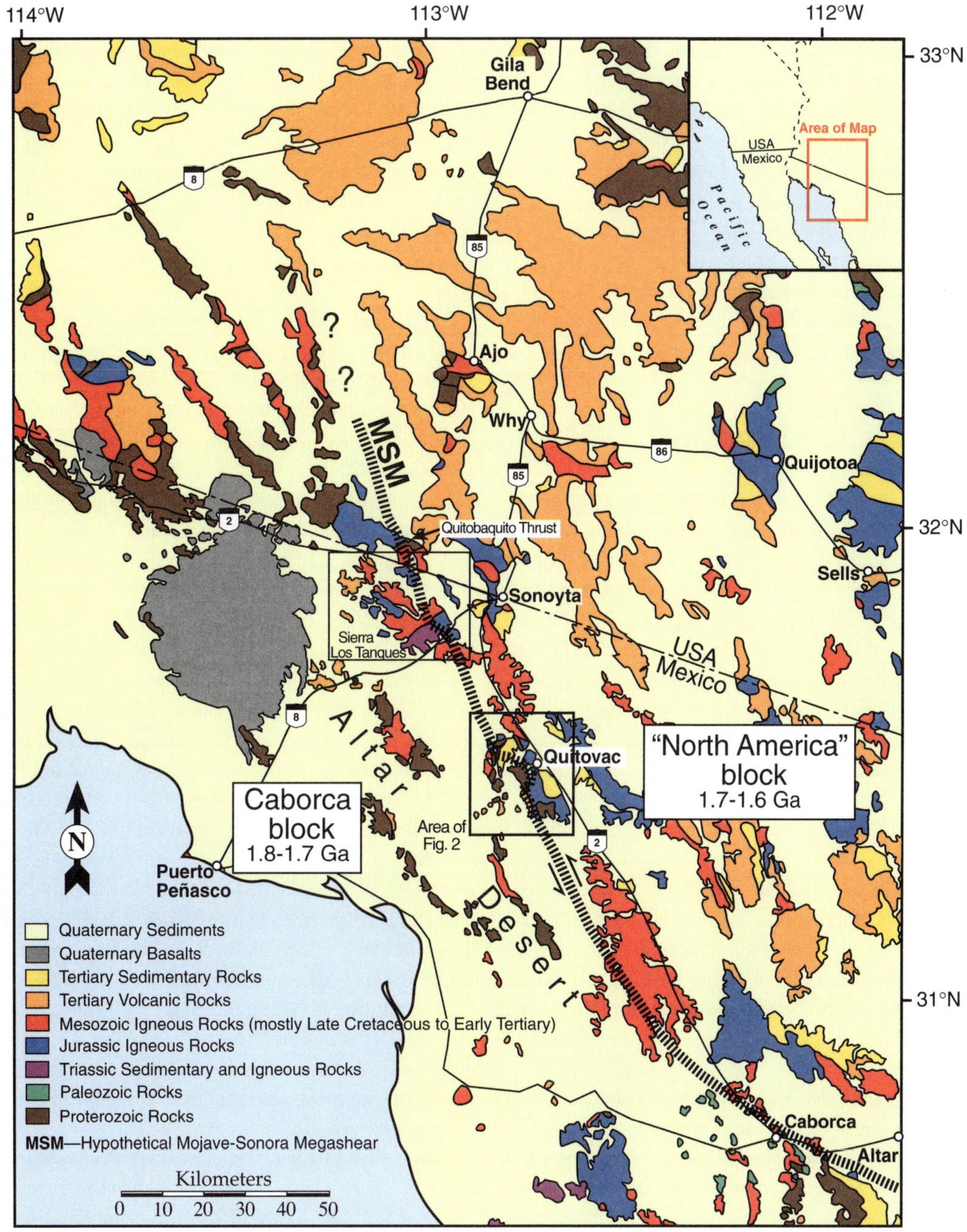

Figure 1. Regional geological map of northwestern Sonora and southwestern Arizona compiled by Iriondo (2001, unpublished map adapted from SSP, 1981; Reynolds, 1988; Tosdal et al., 1990; Fernández-Aguirre et al., 1993). MSM—Mojave-Sonora megashear after T.H. Anderson and C.D. Connors in Tosdal et al. (1990). The Caborca (1.8–1.7 Ga) and "North America" (1.7–1.6 Ga) Proterozoic blocks of Silver and Anderson (1974, 1983) and Anderson and Silver (1978, 1979) are separated by the left-lateral Mojave-Sonora megashear (dashed black line). Outlined boxes represent the Quitovac area (Fig. 2) and Sierra Los Tanques "type area" for the Mojave-Sonora megashear (Campbell and Anderson, 1998, 2003).

megashear of Upper Jurassic age that divides Sonora into two juxtaposed Paleoproterozoic crystalline basement blocks of different ages (Fig. 1; Silver and Anderson, 1974, 1983; Anderson and Silver, 1979). The older Caborca basement (1.8–1.7 Ga) composed of volcanic, granitic, and pegmatitic rocks lies southwest of the proposed megashear (Fig. 1). The younger "North America" basement block, composed of 1.7–1.6 Ga metasedimentary and plutonic rocks, lies northeast of the megashear (Fig. 1). Due to similarities in rock assemblage and geochronology, Anderson and Silver (1981) proposed that the "North America" basement block rocks are correlative with "Pinal Schist" of the Mazatzal crustal province in Arizona. The proposed 700–800 km of left-lateral, Late Jurassic displacement along the Mojave-Sonora megashear (Anderson and Schmidt, 1983) would restore the Caborca block against the westernmost part of the Mojave province, in the vicinity of the Death Valley–Inyo Mountains region. In support of this interpretation, Stewart (1988) suggested that the supracrustal Neoproterozoic and Paleozoic rocks in Death Valley are physically correlative with those in the Caborca region in northwestern Sonora. If this interpretation is correct the basement rocks of the Caborca block in northwestern Sonora should have strong geochemical and isotopic affinities with those in the Mojave crustal province.

GEOLOGY OF THE QUITOVAC REGION

The Quitovac region is located in the Altar Desert in northwestern Sonora, ~45 km south of the United States–Mexico international border at Sonoyta, Sonora (Fig. 1). Geologic mapping in the Quitovac region (Fig. 2), aided by air photography and Landsat TM imagery, has been integrated with geochronologic studies to obtain a basic understanding of the rocks affected by ductile compressional deformation along thrust faults and the subsequent events of Miocene volcanism and extension.

Pre-Tertiary Rocks and Fabrics

This group of pre-Tertiary rocks includes rocks affected by regional metamorphism and compressional ductile deformation(s) associated with thrusting in the Quitovac region. Locally, Proterozoic basement rocks record high-grade metamorphism as indicated by the presence of amphibolite units of Mesoproterozoic age. At this point we do not have a good understanding of this older deformational and metamorphic event. Additional work is needed to document the metamorphic history of these rocks.

Thrusting is the youngest and most widespread ductile deformational event in Quitovac, and it is best recorded in Middle Jurassic and Upper Cretaceous rocks. The thrusting occurred at lowermost greenschist facies as indicated by the presence of epidote and chlorite in the country rocks, away from the ductile fabrics. White mica grew only in the ductile to brittle-ductile shear zones (high-strain rocks) associated with thrusting. We suspect that the white mica was formed during shearing in an environment with high-K fluids traveling along the thrusts. We propose that the white mica grew below its argon closure temperature (350 °C; Purdy and Jäger, 1976) during this thrusting and that it can be used to date the time of thrusting. Other data that supports this hypothesis include (1) a $^{40}Ar/^{39}Ar$ cooling age of 72 Ma (Iriondo, work in progress) for biotite from country rocks adjacent to the shear zones. The closure temperature for biotite is 280 ± 40 °C (Harrison et al., 1985). If the metamorphic temperatures had reached the white mica closure temperature, the biotite would have been reset; and (2) orogenic gold mineralization spatially and temporally linked to regional thrusting formed at ~270 °C based on estimates of homogenization temperatures from fluid inclusions (Albinson, 1989). Although $^{40}Ar/^{39}Ar$ ages of the white mica should represent the time of its crystallization and growth, we would expect that there would be a gradient in the age spectra of such samples if the thrusting occurred over an extended period of time.

Proterozoic and Paleozoic(?) Rocks

The oldest rocks exposed in the Quitovac region are Paleoproterozoic granites and gneisses of the Caborca and "North America" blocks (Iriondo, 2001; Iriondo et al., 2004). The main Paleoproterozoic intrusive rocks in the Caborca block are biotite granites and two-mica granites with ages ranging from 1.78 to 1.69 Ga (unit Pgc; Fig. 2). In the "North America" block, the Paleoproterozoic rocks are granites, quartz monzonite, and associated gneisses with ages between 1.71 and 1.66 Ga (units Pga and Pbga; Fig. 2). We speculate that these basement rocks have suffered metamorphism in the early Mesoproterozoic as reflected by the existence of amphibolites in both basement blocks (units Pac and Paa). However, the only age constraint for the amphibolite in the "North America" block is a $^{40}Ar/^{39}Ar$ hornblende age of 1.4 Ga (Iriondo, 2001). Bodies of presumably early Mesoproterozoic pegmatite (unit Ppa) locally intrude this amphibolite. Amphibolites from the Caborca block have not been dated.

Subsequently, the Caborca block has recorded an event, of so-called anorogenic magmatism, at ca. 1.1 Ga (e.g., Anderson and Silver, 1997). This event is represented in the Quitovac region by granites in the Campo de Bustamante area (included in unit Pgc). We believe that deformational processes accompanied this magmatic event, as indicated by the presence of muscovite schist in the Caborca block (unit Pshc), tentatively dated at 1.1 Ga. Therefore, these granites in the Quitovac region may not represent anorogenic magmatism. The temporal, isotopic and geochemical characteristics of these Proterozoic plutonic and metamorphic rocks are discussed at length in Iriondo (2001) and Iriondo et al. (2004).

In La Toñita area, part of the Caborca block, we have mapped layers of marble (Pzma) and quartzite (Pzqtz) that may be Paleozoic or Mesozoic in age (Fig. 2). We contemplate a Paleozoic age for these units because they occur at an unconformity with the overlying Mid-Jurassic volcanic rocks. However, it is possible that these rocks are as young as Lower or Mid-Jurassic. The greenschist metamorphic grade observed in the marble makes fossil determination difficult. The exception may be the identification of conodonts that are preserved sometimes even at this metamorphic grade. In addition, U-Pb zircon studies in the quartzite unit may help constrain the maximum possible age of these units.

Mesozoic Rocks (Mid-Jurassic and Upper Cretaceous)

Both Proterozoic basement blocks are unconformably overlain by metamorphosed Mid-Jurassic volcanic and volcaniclastic

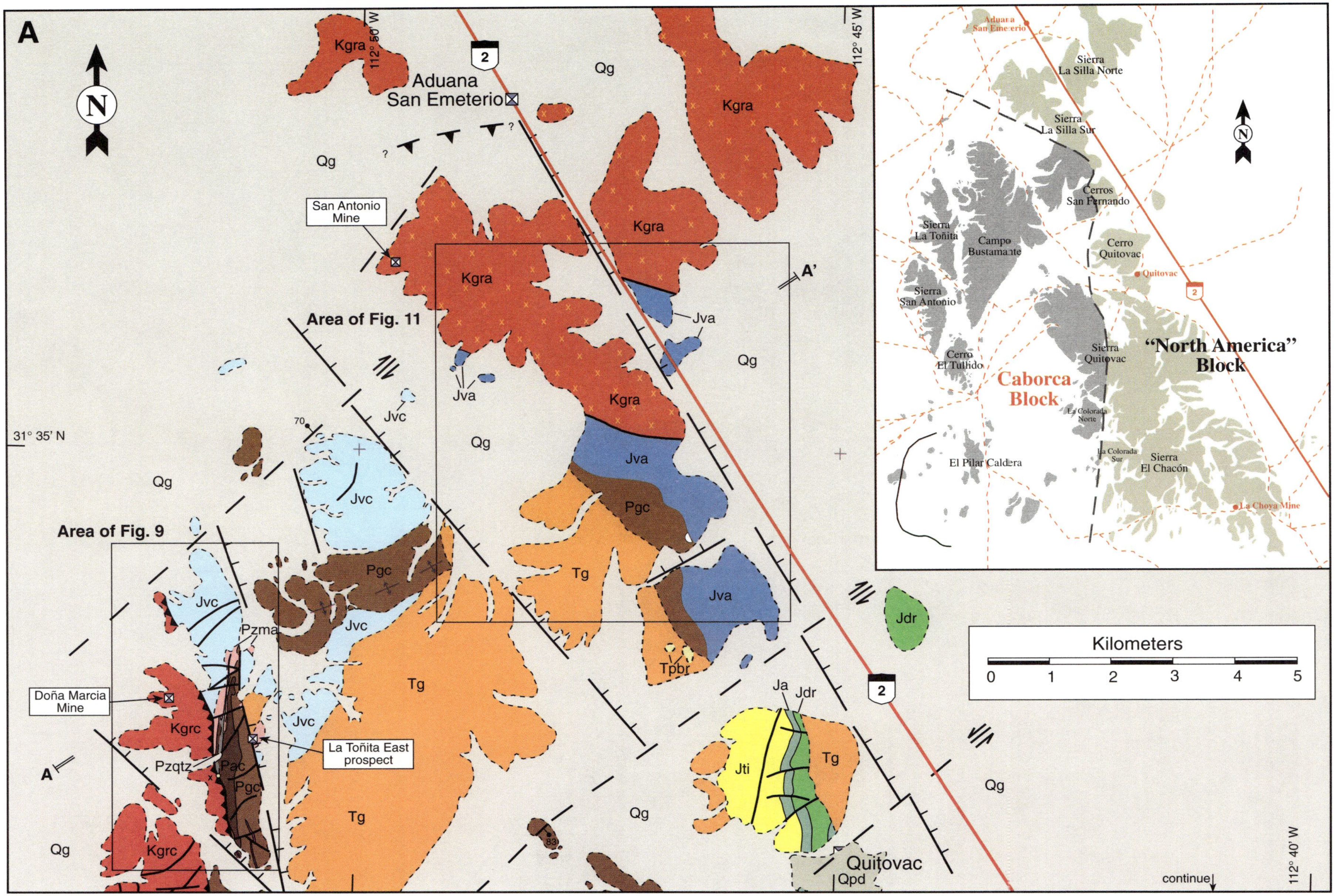

Figure 2. Simplified geologic map of the Quitovac region, northwestern Sonora. See Figure 1 for location of geologic map. A–A′, B–B′, and C–C′ indicate locations of cross sections shown in Figure 4. Outlined boxes represent areas of Figures 8–11. Inset: Caborca and "North America" basement blocks and location of key geographical features in the Quitovac region. (*Continued on following two pages.*)

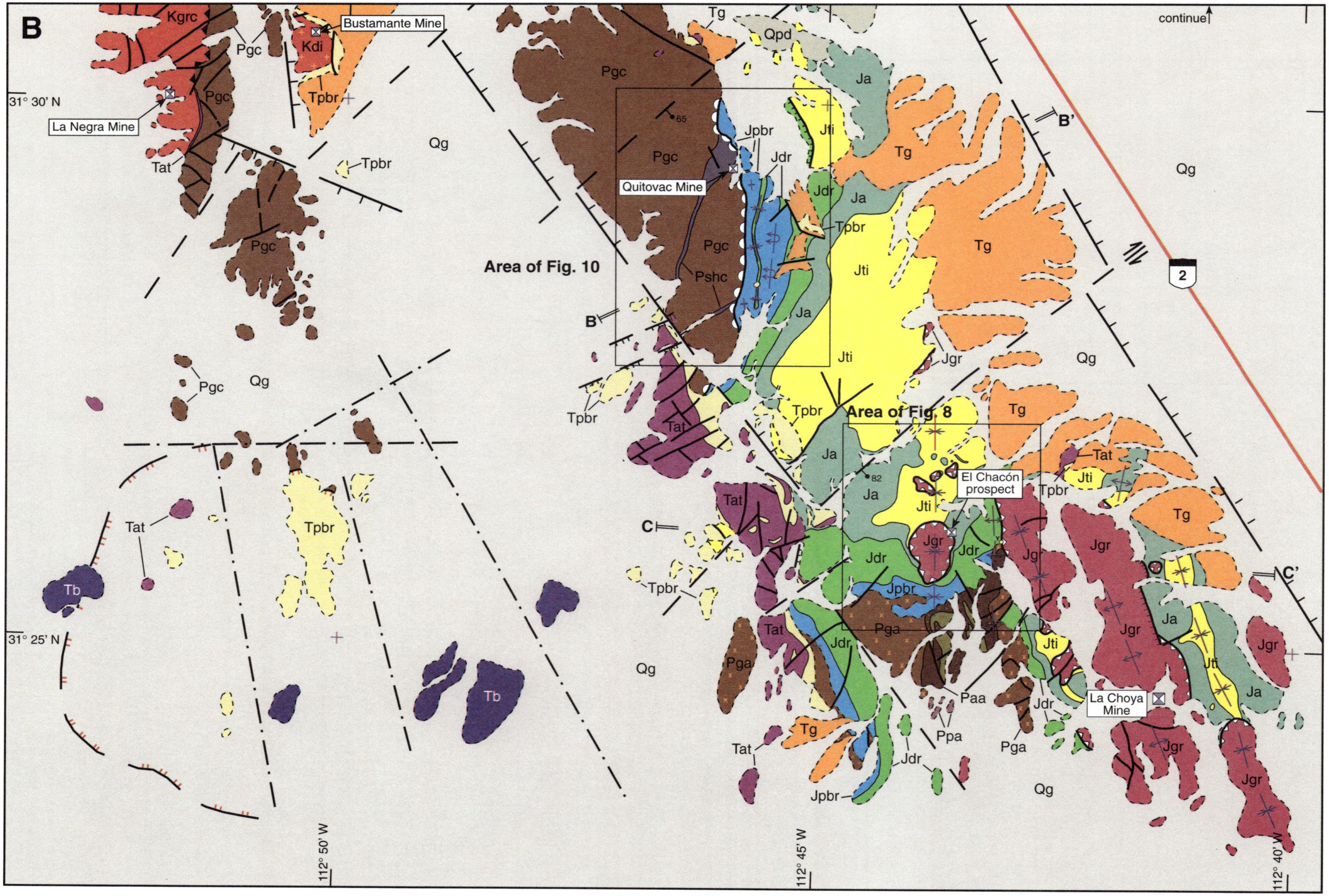

Figure 2 (*continued*).

DESCRIPTION OF MAP UNITS

Qg — Gravel and alluvium (Quaternary to Present)–Alluvium in present-day river and stream channels, flood plains, and locally eolian deposits (sand dunes from El Pilar Caldera).

Unconformity

Qpd — Playa deposits (Pleistocene to Present)–Lacustrine deposits with Mammoth bones at the town of Quitovac.

Unconformity

Tg — Gravels (Post-Early Miocene)–Unconsolidated gravels containing clasts of Early Miocene, Late Cretaceous, Middle Jurassic, and Proterozoic rocks of the Quitovac region. Locally monolithologic.

Unconformity

El Pilar Caldera Volcanic Rocks ($^{40}Ar/^{39}Ar$ Geochronology by Iriondo et al., 1996)

Tb — Basalt lava flows (Early Miocene)–(biotite: 17.46 ± 0.04 Ma).

Tat — Andesite and trachyte lava flows (Early Miocene)–Local basal vitrophyre (sanidine:17.90 ± 0.09 Ma). Overlying andesite flow (biotite: 17.72 ± 0.07 Ma). Also mafic and intermediate dikes intruding two-mica granites at Sierra San Antonio and Sierra La Silla (hornblende: 17.37 ± 0.10 Ma; and whole rock: 16.96 ± 0.16 Ma).

Tpbr — Pyroclastic rocks (Early Mioc–Unit composed of ignimbrite with numerous brecciated lithic fragments of many pre-Earlyene) Miocene lithologies in the Quitovac region (locally blocks of country rock); biotite: 18.03 ± 0.04 Ma; sanidine 17.97 Ma. Also rhyolite flows and water-laid tuff with some small lenses of lacustrine limestone.

Unconformity

Caborca Block / "North America" Block

Rock units from the Caborca and "North America" blocks with different degrees of Late Cretaceous-Early Tertiary metamorphism (greenschist facies) and brittle and ductile deformation (dynamometamorphism). Locally it is possible to define older tectonometamorphic events possibly of Proterozoic age. Laramide (65-48 Ma) orogenic gold mineralization is also present in these rocks.

Caborca Block

Kdi — Campo Bustamante fine-grained diorite (Late Cretaceous: 76.1 ± 3.7 Ma, U-Pb zircon)

Kgrc — Caborca two-mica granite and aplitic dikes/sills (Late Cretaceous: 79.4 ± 1.3 Ma, U-Pb zircon)

Jvc — Undifferentiated Caborca volcanic rocks (Middle Jurassic)– –Mostly deformed pyroclastic rocks with minor amounts of felsic and mafic flows in Campo Bustamante and Sierra La Toñita.

Unconformity

Pzma — La Toñita marble (Paleozoic? or Middle Jurassic?)

Pzqtz — La Toñita quartzite (Paleozoic? or Middle Jurassic?)–Locally becomes a muscovite schist.

Unconformity

Pshc — Caborca muscovite schist (Late Mesoproterozoic?) Locally becomes mylonitic.

Pac — Caborca amphibolite (Early Mesoproterozoic?)

Pgc — Caborca granite and gneiss (Late Paleoproterozoic)–Granite, quartz-rich granite and two-mica granite with U-Pb zircon geochronology from 1.78-1.69 Ga. Locally in Campo Bustamante ~1.1 Ga granite.

"North America" Block

Kgra — La Silla two-mica granite and aplitic dikes/sills (Late Cretaceous: ~80 Ma, U-Pb zircon)

Jgr — El Chacón biotite granite, granite (s.s.), and aplitic granite (Middle Jurassic)–Four different samples of three different lithologies at ~166 Ma (U-Pb zircon).

Jti, Ja, Jdr, Jpbr — Jva — Undifferentiated "North America" volcanic rocks (Middle Jura–Volcanic rocks in Sierra La Silla Sur and Cerrosssic) San Fernando are Jpbr, Jdr, Ja, and Jti.

Jti — Pyroclastic and porphyritic unit (Middle Jurassic)–Ignimbrite and granitic porphyry (175.6 ± 0.38 Ma, U-Pb zircon).

Ja — Andesite and mafic lava flows (Middle Jurassic)

Jdr — Dacite, rhyodacite and latite lava flows (Middle Jurassic)

Jpbr — Pyroclastic rocks (Middle Jurassic)–Unit composed of ignimbrite (locally with large boulders) of andesitic composition. Close to fault (thrust) contacts we observe a phyllite to fine-grained micaceous schist. Local dacite dikes/sills.

Unconformity

Ppa — "North America" pegmatites (Early Mesoproterozoic?)

Paa — "North America" amphibolite (Early Mesoproterozoic)– ($^{40}Ar/^{39}Ar$ hornblende: ~1.4 Ga).

Pga — "North America" granite and gneiss (Late Paleoproterozoic) Granite and quartz monzonite with U-Pb zircon geochronology from 1.71-1.66 Ga.

Pbga — "North America" banded gneiss (Late Paleoproterozoic?)– Granitic protolith possibly Pga.

EXPLANATION OF MAP SYMBOLS

- Normal contact
- Unconformity
- Fault contact
- 50 Fault with dip
- Inferred normal fault
- Inferred normal fault showing dip
- Strike-slip fault
- Magnetic lineament
- Margin of El Pilar volcanic caldera
- La Toñita Thrust (LTT)
- Campo Bustamante Thrust (CBT)
- El Chacón Thrust (CHT)
- Quitovac Thrust (QT)
- Hypothetical La Silla Thrust (LST)
- 35 Strike and dip of bedding
- Horizontal bedding
- Vertical bedding
- 45 Strike and dip of foliation
- 22 50 Strike and dip of foliation, showing bearing and plunge of lineation
- Vertical foliation or cleavage
- Anticline
- Overturned anticline
- Syncline
- A — A' Geological cross-sections (Fig. 4)
- Sample for $^{40}Ar/^{39}Ar$ geochronology (Figs. 9-12)
- Town or gold mine
- 2 Principal paved road (Mexican Highway)

Figure 2 (*continued*).

rocks. The volcanic sequence in the "North America" block is best preserved away from the main thrusts. In some places, unmetamorphosed protoliths were recognized and their stratigraphic relationships identified. This sequence crops out in La Silla Sur, Cerros San Fernando, Cerro Quitovac, Sierra Quitovac, and Sierra El Chacón (Fig. 2). The base of the volcanic sequence is composed of ignimbrites with different textures and degrees of metamorphism (unit Jpbr; Fig. 2). Locally, these rocks are intruded by dacitic dikes and sills that may be part of the overlying unit composed of dacite, rhyodacite, and latite flows (unit Jdr). A mafic unit (unit Ja), consisting mostly of andesite flows, was deposited conformably on top of the Jdr unit. The uppermost volcanic unit (unit Jti) is composed of ignimbrite and granitic porphyry of Mid-Jurassic age. U-Pb zircon geochronology from a single sample of granitic porphyry yielded two concordant $^{206}Pb/^{238}U$ ages at 172.3 ± 1.86 Ma and 175.6 ± 0.38 Ma (Iriondo, 2001) that we cannot interpret as to which is the true crystallization of the volcanic rock. In Sierra La Silla Sur and Cerros San Fernando (Fig. 2) the sequence of Mid-Jurassic rocks has been mapped as undifferentiated "North America" volcanic rocks (Jva) because of the structural and metamorphic complexity of the area. However, these volcanic and volcaniclastic rocks are the same as those mapped to the south in Sierra Quitovac and El Chacón (Jpbr, Jdr, Ja, and Jti).

The Mid-Jurassic volcanic sequence overlying the Caborca Proterozoic basement was also deformed and metamorphosed and has been mapped in Sierra La Toñita and Campo Bustamante as undifferentiated Caborca volcanic rocks (unit Jvc; Fig. 2). These rocks are composed principally of pyroclastic rocks and minor amounts of felsic and mafic flows and are similar to the volcanic rocks mapped in the "North America" block. In other words, the Jurassic volcanic rocks of the Caborca block are not different enough from those in the "North America" block to consider them "exotic" to the region.

We propose that Mid-Jurassic and Upper Cretaceous granites in the Quitovac region were emplaced along northward-directed Laramide thrusts. Perhaps the best example is El Chacón klippe, a Mid-Jurassic granite (unit Jgr) thrust over the Mid-Jurassic volcanic sequence in Sierra El Chacón (Fig. 2). This unit is composed of different igneous phases; the most predominant being a biotite sphene-rich granite dated at 165.9 ± 0.31 Ma (concordant U-Pb zircon age; Iriondo, 2001). The age of El Chacón biotite granite is comparable with intrusions in south-central Arizona (e.g., Kitt Peak) and northern Sonora that are part of the Jurassic magmatic arc in southwestern North America (e.g., Tosdal et al., 1989).

Upper Cretaceous, two-mica, hornblende-free granitic plutons and aplitic dikes and sills crop out in Sierras La Toñita and San Antonio within the Caborca block (unit Kgrc; Fig. 2). The granite presents ductile deformation near a thrust plane that juxtaposes the granite against pre-Cretaceous plutonic and volcanic rocks of the Caborca block. Zircons from this granite yielded a U-Pb discordia with a lower intercept age of 79.4 ± 1.3 Ma (Iriondo, 2001), which we interpret as the age of crystallization of the pluton. The upper intercept with concordia yielded a Paleoproterozoic age that we interpret as an average age of inheritance introduced by melting of the Caborca basement or sediments derived from it. A fine-grained equigranular diorite intrusion (unit Kdi) crops out in the southern part of the Campo Bustamante area (Fig. 2). A concordant U-Pb zircon age of 76.1 ± 3.7 Ma has been interpreted as the emplacement age of the diorite intrusion (Iriondo, 2001). The nature of the geological contact between this unit and older units of the Caborca block is unknown because the contact has been obscured by overlying early Miocene ignimbrites (unit Tpbr) and younger gravels (unit Tg).

The La Silla two-mica granite and associated aplitic dikes and sills (unit Kgra) crop out in the Sierra La Silla Sur (Fig. 2). U-Pb zircon geochronologic age data of the two-mica granite suggests an age of ca. 80 Ma (Iriondo, 2001). Inheritance similar to that in zircons from the two-mica granite of the Caborca block is also observed in this granite. This unit has been tectonically overlain by "North America" Mid-Jurassic volcanic rocks and presents the highest degree of ductile deformation and metamorphism along the thrust contact.

Most of the pre-Tertiary units in the Quitovac region have been locally mineralized by Paleocene-Eocene (65–48 Ma) gold-bearing quartz veins classified by Iriondo and Atkinson (2000) as orogenic gold veins. These authors also indicated that the quartz veins are spatially and temporally linked to the Laramide ductile fabrics in the region. Figure 2 shows the location of some of the gold mines and prospects in which gold mineralization was found. Later in the text we briefly discuss the $^{40}Ar/^{39}Ar$ geochronology of these gold veins and their implications to constrain the Laramide orogeny in the Quitovac region.

Upper Cenozoic Rocks and Extensional Structures

After the Late Cretaceous–early Tertiary Laramide orogeny (75–40 Ma), the Quitovac region was blanketed by calc-alkaline lower Miocene volcanic rocks (Iriondo et al., 1996). The El Pilar Caldera volcanic sequence unconformably overlies pre-Laramide rocks. The lower part of the sequence is composed of ignimbrite with numerous brecciated lithic fragments of many of the pre-lower Miocene rocks present in the Quitovac region (unit Tpbr; Fig. 2). The ignimbrite has been dated at two different localities at 18.03 ± 0.04 Ma (biotite) and 17.97 Ma (sanidine) (Iriondo et al., 1996). Locally, the ignimbrite contains large blocks up to 30 m in diameter of country rock that we interpret as a result of caldera collapse. Outcrops of such ignimbrite help define part of the El Pilar Caldera margin (Fig. 2; Iriondo et al., 1996). Minor amounts of rhyolite flows and water-laid tuffs with small lenses of lacustrine limestone are locally interbedded with the ignimbrites.

Conformably overlying the pyroclastic unit are andesite and trachyte flows (unit Tat; Fig. 2). Sanidine from a trachyte basal vitrophere was dated with $^{40}Ar/^{39}Ar$ geochronology at 17.90 ± 0.09 Ma and biotite from the overlying andesite flow at 17.72 ± 0.07 Ma (Iriondo et al., 1996). Dikes and sills of these lithologies are common in Sierra San Antonio (hornblende: 17.37 ± 0.10 Ma) and Sierra La Silla Sur (whole rock: 16.96 ± 0.16 Ma) and yielded slightly younger $^{40}Ar/^{39}Ar$ ages than their counterpart flows, explained by slower cooling of the hypabyssal injections (Iriondo et al., 1996). These two lower units of lower Miocene volcanic rocks (Tpbr and Tat) were later faulted and tilted during the Basin and Range extensional event.

Subhorizontal basaltic flows (unit Tb; Fig. 2) cap the volcanic sequence in the Quitovac region and were dated with a biotite separate at 17.46 ± 0.04 Ma (Iriondo et al., 1996). The contact of the basalts with older volcanic rocks is not exposed, but their structural character and age suggest that the unit is the upper part of the El Pilar Caldera volcanic sequence. These basalts appear to postdate Basin and Range extensional faulting in the Quitovac region.

Early Miocene uplift resulted in high rates of denudation and deposition as observed in the basin-filling Tertiary gravels (unit Tg). These gravels stand topographically higher than younger Quaternary gravels (unit Qg) in modern river and stream channels (Fig. 2). Locally, in the El Pilar Caldera area, we observe eolian deposits (sand dunes) of Quaternary to present age. The town of Quitovac is underlain by lacustrine deposits (unit Qpd; Fig. 2) of probable Pleistocene age, as indicated by the presence of mammoth bones.

Basin and Range Extensional Faults

The most prominent topographical feature in the Quitovac region is the distribution of northwest-trending, elongated mountain ranges and valleys (Fig. 2). This pattern is the result of widespread Tertiary extension well documented in other parts of southwestern North America and defined as the Basin and Range Province (e.g., Stewart, 1978; Eaton, 1982; Coney, 1987; Wernicke et al., 1987; Henry and Aranda-Gomez, 1992; Stewart and Roldán-Quintana, 1994; Lee et al., 1996; Gans, 1997). The nature of this extensional event along the southwestern margin of the North America plate remains controversial and has been linked to different processes, including changes in plate motion, steepening and/or removal of the subduction slab, change in convergence rates, and gravitational collapse of overthickened crust (see Lee et al., 1996, and references therein).

The Basin and Range event in the Quitovac region is represented by two distinct sets of faults. However, the same or slightly similar distensive stress field could explain both sets. The first set consists of closely spaced (1–8 km) normal faults with a N30°W orientation and dips between 30° and 60° to the ENE and locally to the WSW (Figs. 2 and 3). Displacing these faults, and with an orthogonal orientation of approximately N50°E, is a set of steeply dipping strike-slip faults with predominantly dextral motion (Figs. 2 and 3). These subvertical faults probably represent zones between blocks accommodating differential extension. Traces of both sets of faults are only locally exposed, because they are partially covered by younger basin-fill deposits (Fig. 2). Nonetheless, these faults have topographic expression allowing their subsurface trace to be inferred.

As discussed later, we propose that the first set of normal faults (N30°W) is responsible for the tilting of blocks in the Quitovac region. Based on mapped offsets and orientation of slickenlines we calculated minimum vertical throws of these faults to range from ~300–1,770 m (Fig. 3).

From north to south, the four main strike-slip faults mapped in the Quitovac region dip 70°NW, 83°SE, 65°SE, and 82°SE, respectively. Slickenlines on the southernmost fault plunge consistently at low angles (~10°SW), indicating dextral motion. The fault separating the La Silla and Quitovac blocks (Fig. 3) represents a maximum horizontal displacement of 2.3 km, excluding a potential component of normal slip.

The oldest lower Miocene volcanic rocks affected by the extension in the Quitovac region are ignimbrites dated at ca. 18 Ma (Iriondo et al., 1996). Extensional faulting in the Quitovac region may have been remarkably short lived (half a million years) as indicated by the inferred subhorizontal attitude of the ca. 17.5 Ma capping basalts (Iriondo et al., 1996). Locally, the post–lower Miocene and Quaternary gravels appear to be little affected by the extensional faults, and we have not observed significant tilting of these units that would imply extension at or since those times.

The common fault geometry at the surface and the parallel extension direction of the faults lead us to speculate that the regional structure is caused by a domino- or bookshelf-style rotation of blocks during normal slip on the faults. This interpretation is consistent with other models in the Basin and Range Province of southern Sonora (e.g., Gans, 1997).

Tilting of Basin and Range Fault blocks

The intersection between the system of normal faults (N30°W) and the strike-slip faults (N50°E) has divided the Quitovac region into irregular, westerly tilted fault blocks: La Silla, Campo Bustamante, La Toñita, Quitovac North, Quitovac South, El Chacón, and El Pilar Caldera tilted blocks (Fig. 3).

Most of these blocks have been rotated about horizontal axes subparallel to the system of normal faults. The angle of tilting of the La Silla, Quitovac (North and South), and El Chacón blocks has been determined from the dip of lower Miocene ignimbrites and locally from the dip of interbedded layers of lacustrine limestone. The ignimbrites in the Campo Bustamante, Quitovac North, and El Chacón blocks are subhorizontal and considered to be little affected by Basin and Range normal and strike-slip faults (Fig. 3). In the La Silla block, the ignimbrites dip 35°SW, and in the Quitovac S block the dip is ~45°W (Fig. 3).

The ignimbrite sequence has not been preserved in La Toñita block; therefore, the amount of tilting has been estimated from the geologic map and cross sections. Assuming that the thrust fault mapped in the La Toñita area was subhorizontal prior to the Basin and Range event (this is true for thrust faults in adjacent tilting blocks in the Quitovac region) and knowing the dip of the normal faults bounding the La Toñita block, we have determined an angle of tilting of ~60°WSW (Fig. 3).

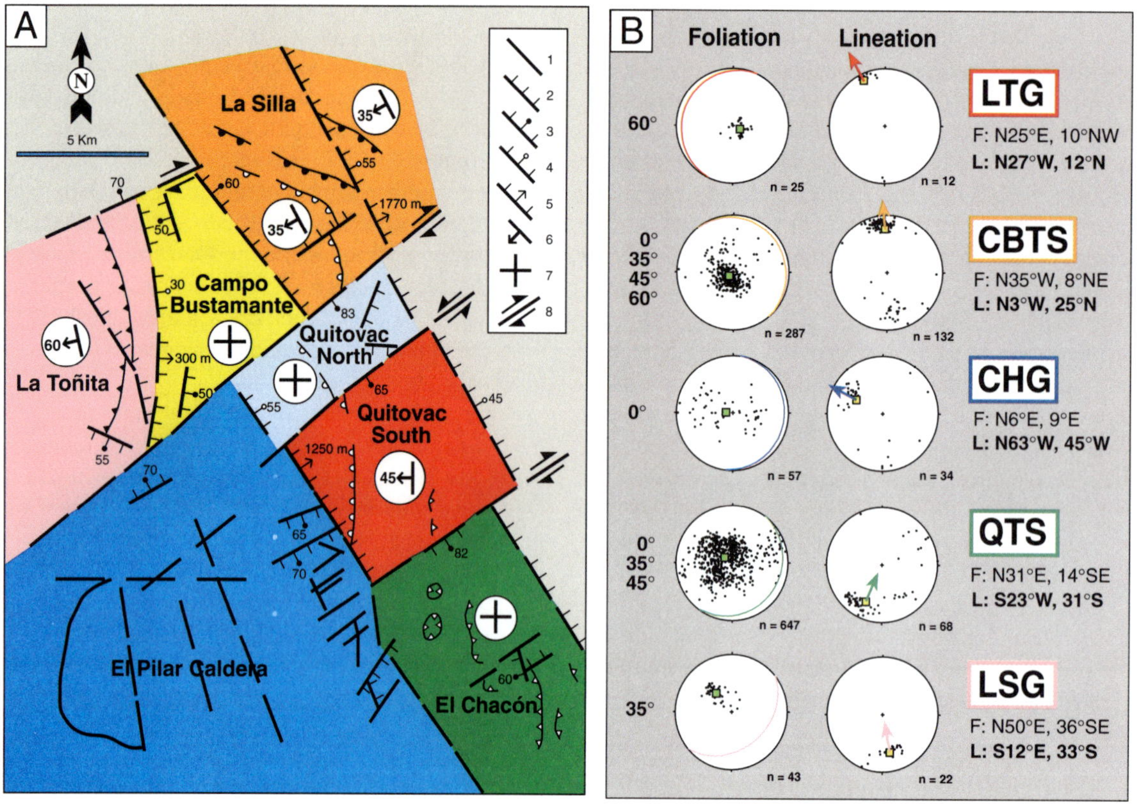

Figure 3. (A) Tilting of Basin and Range fault blocks in the Quitovac region. 1—inferred fault, 2—normal fault, 3—normal fault with measured dip, 4—normal fault with calculated dip, 5—throw of fault, 6—amount of tilting of Basin and Range blocks obtained from geometric analysis and from dip of preextension lower Miocene volcanic rocks, 7—block without tilting, 8—subvertical strike-slip faults. (B) Lower hemisphere stereonet plots of foliation and lineation from the different tectonic sheets in the Quitovac region (Fig. 5). Black circles represent the poles to foliation and lineations corrected for Basin and Range tilting (the amount of correction is shown in the left part of each set of stereonet plots). Green and yellow squares inside stereonet plots symbolize the average foliation and lineation values for the different tectonic sheets. Colored arrows in the lineation plots show the average direction of displacement for the different sheets. LTG—La Toñita Granite, CBTS—Campo Bustamante Tectonic Sheet, CHG—El Chacón Granite, QTS—Quitovac Tectonic Sheet, LSG—La Silla Granite.

The magnitude of the Basin and Range extension has been estimated in the Quitovac region by palinspastically restoring the effects of down-dip displacement along the system of normal faults. The cross sections in Figure 4 (A–A′ and B–B′) have been restored perpendicular to the N30°W normal fault system. The restoration of section A–A′ yields a minimum extension of 3100 m, which corresponds to ~30% extension. For section B–B′, the restoration yields 1600 m, or 12% extension. These are minimum estimates, because there may be other faults concealed beneath the gravels that could account for larger amounts of extension. Section C–C′ has not been restored because the geological section cuts across the untilted El Chacón block.

The tilting of blocks in this fashion exhumes deep structural levels in the footwall of the normal faults. In fact, in the Quitovac region, the average direction of Laramide transport (N18°W–S18°E) is subparallel to the axis of rotation of the blocks (N30°W). Therefore, these tilted blocks provide oriented cross sections parallel to the shear direction of the Laramide thrusts and allow for direct assessment of the sense of shearing of the thrust sheets. However, in blocks without significant tilting, we observe only the upper structural levels, requiring geometrical extrapolation of the lower structural levels from adjacent tilted blocks.

STRUCTURAL STRATIGRAPHY

Geologic mapping (Fig. 2) and geochronologic studies in the Quitovac region have permitted the identification of five distinct tectonic sheets bounded by thrust faults. We have named these tectonic sheets based on their geographical location as follows: La Toñita Granite, Campo Bustamante Tectonic Sheet, El Chacón Granite, Quitovac Tectonic Sheet, and La Silla Granite (Fig. 5A). The bounding thrust faults are named according to the widely adopted hangingwall terminology. The identified thrusts, from high to low structural levels, are La Toñita thrust, Campo Bustamante thrust, El Chacón thrust, Quitovac thrust,

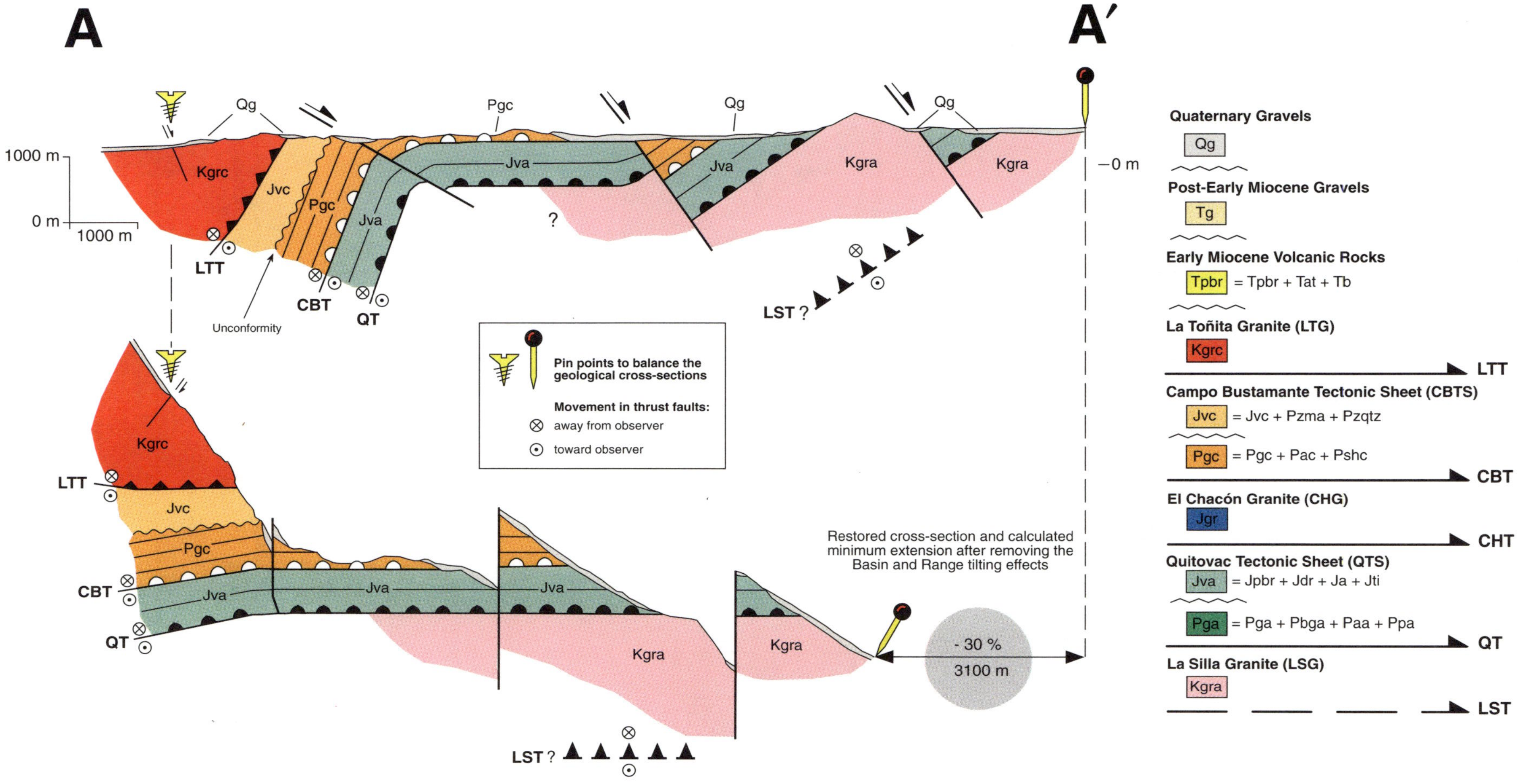

Figure 4. Cross sections of the Quitovac region. Locations of the cross sections are given in both Figures 2 and 3. For sections A–A′ and B–B′ the palinspastically restored section is shown after removing the extensional Basin and Range offsets. Section C–C′ has not been restored because it is an area without significant tilting. (*Continued on following page.*)

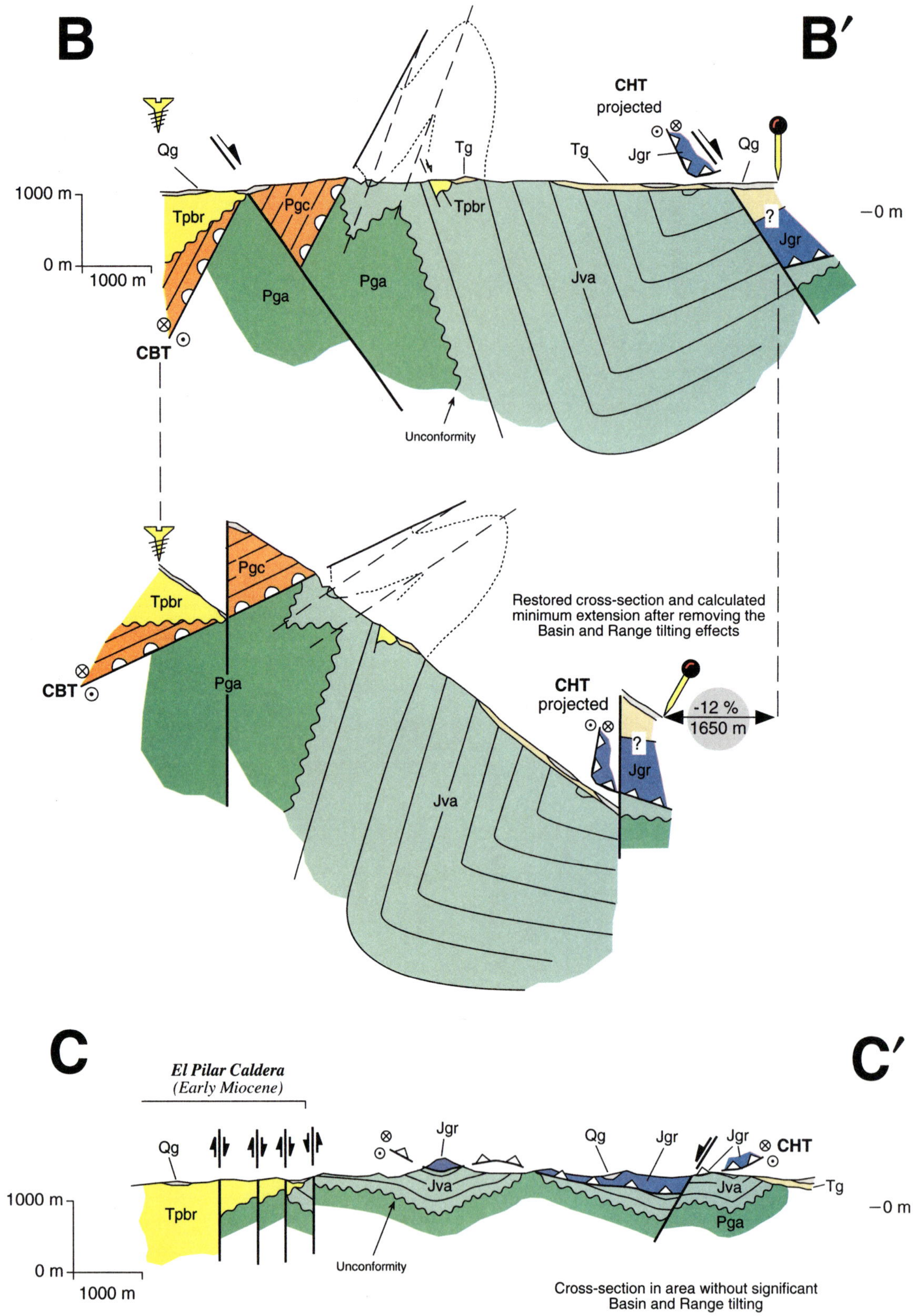

Figure 4 (*continued*).

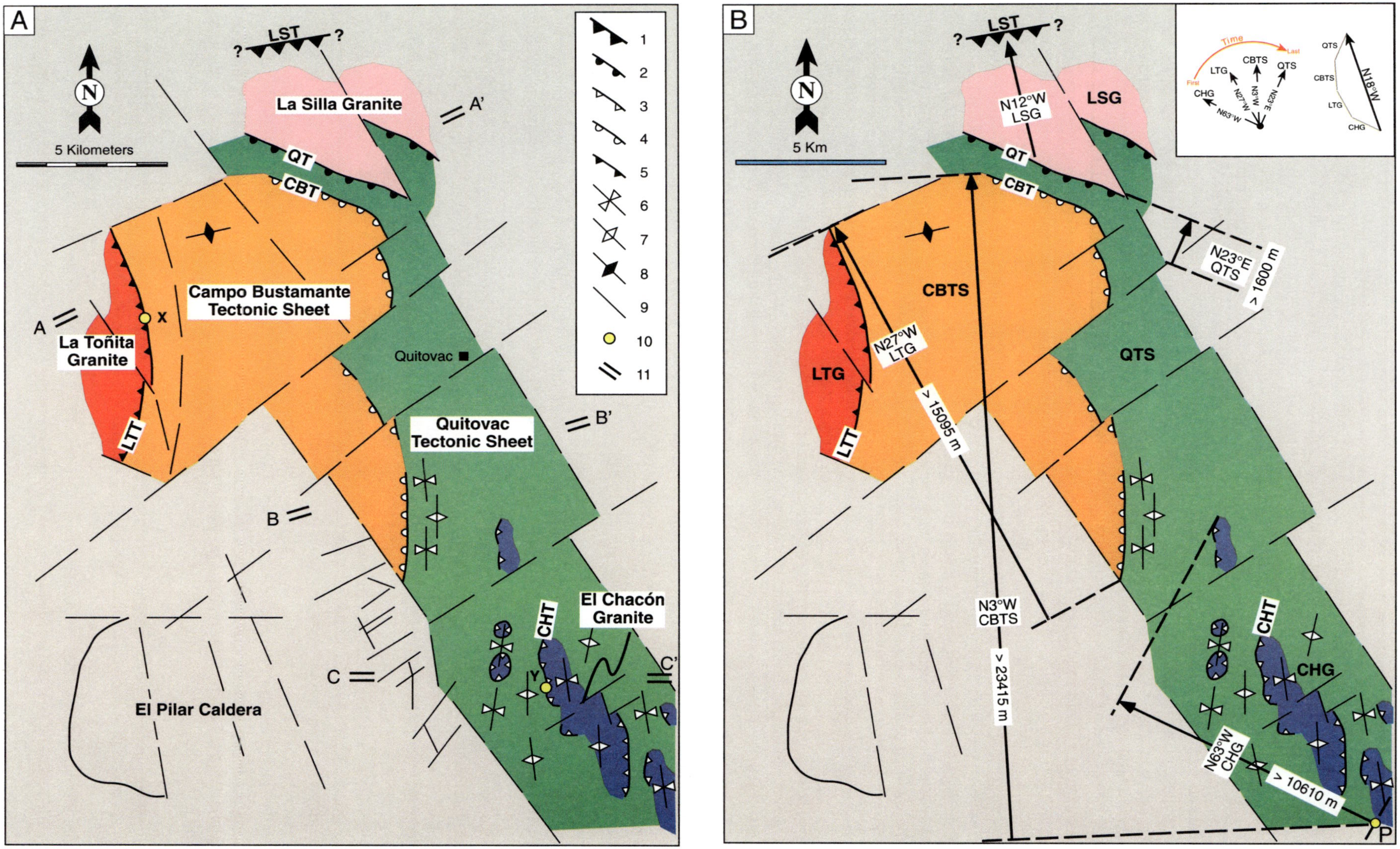

Figure 5. (A) Generalized map of the different tectonic sheets and main structures in the Quitovac region. 1—hypothetical La Silla Thrust (LST), 2—Quitovac Thrust (QT), 3—El Chacón Thrust (CHT), 4—Campo Bustamante Thrust (CBT), 5—La Toñita Thrust (LTT), 6—syncline in the Quitovac Tectonic Sheet, 7—anticline in the Quitovac Tectonic Sheet, 8—anticline in Campo Bustamante Tectonic Sheet (Campo Bustamante fault-bend fold), 9—Tertiary Basin and Range faults, 10—intersection point between basement-cover unconformity and the overlying thrust fault, 11—Cross sections in Figure 4. (B) Direction and amount of Laramide displacements in the Quitovac region. The vectors indicate the direction and sense of displacement of the different tectonic sheets, after removing the Basin and Range effects (see Fig. 3) and using field and microscopic kinematic indicators observed in rocks affected by the deformation. Length of the vector indicates the amount of overlap of one tectonic sheet with respect to its parallochthon. Point P represents the southernmost outcrop of the allochthonous El Chacón Granite (CHG).

and La Silla Thrust (Fig. 5A). A summary of the structural and geochronologic data for the thrust faults in the Quitovac region is presented in Table 1.

La Toñita Granite

The Upper Cretaceous two-mica La Toñita Granite crops out in the Sierra La Toñita and Sierra San Antonio areas (unit Kgrc; Figs. 2 and 5A). The granite is ~1800 m thick and is separated from the underlying parallochthonous Campo Bustamante Tectonic Sheet by the La Toñita thrust. This thrust developed an upper-plate greenschist-facies mylonite zone that is locally 2 m thick. Kinematic indicators in this mylonitized granite suggest top-to-the-NNW displacement for the thrust (Fig. 6A). Older, lower-plate rocks have developed a subparallel foliation to the thrust plane. It is interesting to mention that this thrust presents a younger-over-older relationship, a feature that appears quite common in granitoids involved in ductile deformation in the region. The thrust plane in the southern part of Sierra San Antonio has been obscured by the intrusion of a lower Miocene andesitic dike (Fig. 2). Quaternary gravels cover the upper contact with the La Toñita Granite.

Internally, the granite has local bands of mylonite several centimeters thick subparallel to the La Toñita thrust. However, the most prominent structural feature is a system of conjugate fractures with E-W (35°N) and N60°W (45°S) orientations. The fractures formed contemporaneously with thrusting and were mineralized by Paleocene-Eocene orogenic gold-bearing fluids. Doña Marcia and La Negra Mines (Fig. 2) contain examples of such gold veins within the La Toñita Granite and locally present mylonitization and boudinage. Gold veins subparallel to the thrust plane are also locally present.

Campo Bustamante Tectonic Sheet

The footwall of La Toñita Thrust is the Campo Bustamante Tectonic Sheet. This parallochthonous sheet crops out in the eastern half of Sierra La Toñita and Sierra San Antonio, Cerro El Tullido, Campo Bustamante, the southern part of Sierra La Silla Sur, and in the western part of Sierra Quitovac (Figs. 2 and 5A).

The Campo Bustamante Tectonic Sheet is composed of three different rock sequences. The lower sequence consists of Proterozoic igneous and metamorphic basement rocks of the Caborca block (Pgc, Pac, and Pshc) and has an estimated thickness of 700–1100 m. Overlying the Proterozoic rocks is a thin (120 m) middle sequence composed of Paleozoic(?) marble and quartzite (Pzma and Pzqtz). Mid-Jurassic metamorphosed volcanic and volcaniclastic rocks (Jvc) compose the upper sequence and account for ~450 m of the total thickness of the Campo Bustamante Tectonic Sheet. An Upper Cretaceous diorite (Kdi) is also part of the Campo Bustamante Tectonic Sheet, but its structural relationship with the aforementioned sequences is unknown. All these pre-Laramide rocks are unconformably overlain by Miocene volcanic rocks and younger gravels.

From geologic mapping and systematic measurement of foliations, we infer that the rock sequences in the Campo Bustamante Tectonic Sheet are discordant and their contacts were originally angular unconformities. The angle between the Mid-Jurassic and Paleozoic(?) sequences in the La Toñita area is ~5°, and the angle between the Paleozoic(?) and the Proterozoic Caborca basement is also 5°. In the northern part of Campo Bustamante the Mid-Jurassic volcanic sequence rests directly on the Proterozoic basement rocks, whereas the Paleozoic(?) sequence is absent, possibly due to erosion.

Internally, the Campo Bustamante Tectonic Sheet rock sequences show extensive Laramide foliation, especially in the Paleozoic(?) marble and quartzite in Sierra La Toñita, where we observed tight isoclinal folds verging to the north (Figs. 6B and 6C). Foliation in these rocks resulted from the superposition of the tectonic foliation (F) on the earlier compositional layering (S_0). Locally in the Campo Bustamante area, the Proterozoic basement rocks have rootless fold hinge lines without a predominant direction as well as intersection lineations with the Laramide foliation. Obviously, these sparse structures in the Proterozoic rocks are older than the Laramide ones. However, the intense Laramide deformational overprint obscures a straightforward analysis and/or geochronology of the older deformational event(s). The only exception may be the upper Mesoproterozoic(?) Caborca Muscovite Schist (Pshc) that appears to represent an older structure reactivated during the Laramide orogeny. The schist foliation is subparallel to the Campo Bustamante Thrust and concordant with the regional Laramide fabric for the Campo Bustamante Tectonic Sheet rocks.

The La Toñita Thrust is a frontal footwall ramp that truncates the stratigraphy of the Campo Bustamante Tectonic Sheet and is perpendicular to the average direction of displacement for La Toñita Granite. The angle for the ramp varies from 1° to 40° (average 20°) and was determined from our geologic mapping as well as the stereonet plots for the measured foliations (Fig. 3).

In Sierra La Silla Sur, the Campo Bustamante Thrust develops a thin greenschist-facies mylonite (~0.5 m) from the Proterozoic granite (Pgc; Fig. 6D) and juxtaposes Campo Bustamante Tectonic Sheet rocks over Mid-Jurassic volcanic rocks (Jva) of the parallochthonous Quitovac Tectonic Sheet. The mylonite presents kinematic indicators with top-to-the-north displacement for the thrust (Fig. 6D), corroborating previously defined criteria for displacement of the Campo Bustamante thrust. In this case, the thrust presents the classical older-over-younger relationship. Fabrics in the lower-plate rocks are also subparallel to the thrust.

The presence of Mid-Jurassic rocks (Jvc) of the Campo Bustamante Tectonic Sheet in the northern part of the Campo Bustamante area and the geometric relation of foliation with its parallochthon (the Quitovac Tectonic Sheet) suggest that the Campo Bustamante Thrust forms a hangingwall ramp. This interpretation is supported by the presence of an east-west–trending hangingwall anticline, the Campo Bustamante anticline (CBA; Fig. 5) that affects the Mid-Jurassic volcanic rocks and the underlying Proterozoic basement rocks. The average dip of

TABLE 1. SUMMARY OF STRUCTURAL AND GEOCHRONOLOGIC DATA FOR THE DIFFERENT TECTONIC SHEETS IN THE QUITOVAC REGION, NW SONORA

	LTG La Toñita Granite	CBTS Campo Bustamante Tectonic Sheet (Caborca Block)	CHG El Chacón Granite	QTS Quitovac Tectonic Sheet ("N.A." Block)	LSG La Silla Granite
Geological units*	Kgrc	Jvc = Jvc + Pzma + Pzqtz Pgc = Pgc + Pac + Psch	Jgr	Jva = Jva + Jpbr + Jdr + Ja + Jti Pga = Pga + Pbga + Paa + Ppa	Kgra
Minimum thickness	Kgrc = 1800 m	Jvc = 450 m Pzma – Pzqtz = 120 m Pgc = 700–1100 m	Jgr = 250 m	Jva = 550–4100 m Pga >200 m	Kgra = 1700 m
Bounding thrust faults					
Upper contact	?	LTT	CBT?	CBT and CHT	QT
Lower contact	LTT La Toñita Thrust	CBT Campo Bustamante Thrust	CHT El Chacón Thrust	QT Quitovac Thrust	LST? La Silla Thrust
Amount of B&R tilting removed †	60°	0°, 35°, 45°, and 60°	0°	0°, 35°, and 45°	35°
Restored average§					
foliation	N25°E, 10°NW	N35°W, 8°NE	N6°E, 9°E	N31°E, 14°SE	N50°E, 36°SE
lineation	N27°W, 12°N	N3°W, 25°N	N63°W, 45°W	S23°W, 31°S	S12°E, 33°S
Sense of displacement#	N27°W	N3°W	N63°W	N23°E	N12°W
Amount of displacement					
measured	15,095 m	23,415 m	10,610 m	>1600 m	?
N18°W projected**	15,095 m	19,580 m	8335 m	>975 m	?
Crustal shortening	29%	37%	16%	>4%	?
Time of thrusting††					
onset	≥55 Ma	<55 Ma and >53 Ma	≥61 Ma	≥53 Ma	<53 Ma
cessation	≤49 Ma	?	≤39 Ma	≤40 Ma	?
Order of "piggy-back" sequence	2	3	1	4	5
Occurrence of orogenic gold mineralization	Doña Marcia mine La Negra mine	La Toñita East prosp. C. Bustamante mine Quitovac mine	La Choya mine El Chacón prospect	—	San Antonio mine

*Abbreviations for geological units shown in Figure 2.
†The amount of corrected Basin and Range tilting effects depends on the tilting block from which we restore the tilting (Fig. 3).
§Foliation and lineation after restoring the Basin and Range tilting effects.
#Sense of displacement based on average lineation and shear sense indicators.
**N18°W is the average direction of Laramide transport calculated adding the individual displacement vectors for the different tectonic sheets in the Quitovac region.
††Minimum and maximum $^{40}Ar/^{39}Ar$ ages for onset and cessation of thrusting, respectively.

A
Ms
Ks
B
C
D
C
S
E
C
S
F
G

the east-west foliation in the northern limb of the anticline is 10° to the north whereas the same foliations in the southern limb dip 30° to the south. This anticline has been interpreted as a fault-bend fold perpendicular to the NNW displacement direction of the Campo Bustamante Tectonic Sheet.

In Sierra Quitovac, the Campo Bustamante Thrust becomes a lateral ramp that places Proterozoic rocks of the Caborca block over Mid-Jurassic volcanic rocks of the "North America" block. As we will explain in the following section, this thrust fault also juxtaposes El Chacón Granite.

Paleocene-Eocene orogenic gold mineralization is abundant in this tectonic sheet and is represented by gold veins in La Toñita East prospect and Campo Bustamante and Quitovac Mines (Fig. 2). These gold veins formed contemporaneously with Laramide thrusting as indicated by ductile shearing present in the veins.

El Chacón Granite

The Mid-Jurassic El Chacón Granite crops out in Sierra Quitovac (unit Jgr; Figs. 2 and 5A). The El Chacón Thrust separates the allochthonous granite from the underlying Proterozoic and Mid-Jurassic volcanic sequences of the Quitovac Tectonic Sheet (later described). At Cerro El Chacón, the granite forms a klippe that has developed a 9-m-thick greenschist-facies mylonite along the thrust fault. Shear indicators in the mylonite consistently indicate top-to-the-northwest displacement of the El Chacón Thrust (Fig. 6E). The estimated thickness of the El Chacón Granite in the klippe is ~250 m. Internally, the granite contains a subtle foliation and local bands of several centimeter-thick mylonites away from the El Chacón thrust. We mapped another dozen klippen near Sierra Quitovac (Fig. 2) where the thrust contacts show a basal mylonite zone significantly thinner (0.1–0.5 m) than the one mapped beneath the El Chacón klippe. Lower-plate rocks, mostly Mid-Jurassic volcanic rocks, are deformed in greenschist facies and present foliation subparallel to the thrust plane. However, the upper-plate rocks appear to absorb most of the deformation in this younger-over-older thrust.

We infer that the allochthonous granite was emplaced along a footwall ramp over the Quitovac Tectonic Sheet. The thrust clearly truncates Proterozoic and Mid-Jurassic rocks of the Quitovac Tectonic Sheet. The upper tectonic contact of the granite was eroded during the Basin and Range event. We interpret that the Campo Bustamante Thrust tectonically separates the El Chacón Granite from the overlying Campo Bustamante Tectonic Sheet.

Other prominent structural features in the El Chacón Granite are open NNW-SSE–trending folds that clearly affect the foliation and the thrust plane geometry (Figs. 2 and 4). These are longitudinal lower-plate folds that are kilometer spaced and perhaps associated with the juxtaposition of the overlying Campo Bustamante Tectonic Sheet along a lateral ramp of the Campo Bustamante thrust. Since the folds affect the thrust plane of El Chacón Granite, the folding is younger than the thrust. This folding produced inverted topographic relief in the area; the synclines were preserved (klippen), whereas the anticlines were eroded (Figs. 2 and 4).

Paleocene-Eocene orogenic gold veins are also abundant in the El Chacón Granite. Brittle to brittle-ductile rheological conditions of the granite probably facilitated the development of fractures that host gold-bearing quartz veins. In most cases, these gold veins were deposited subparallel to the Laramide foliation and are deformed in ductile conditions. La Choya Mine and El Chacón Prospect (Fig. 2) are the best examples of orogenic gold mineralization in El Chacón Granite.

Quitovac Tectonic Sheet

The Quitovac Tectonic Sheet crops out at Sierra La Silla Sur, Cerros San Fernando, Cerro Quitovac, the eastern part of Sierra Quitovac, and in Sierra El Chacón (Figs. 2 and 5A). This tectonic sheet makes up the "North America" block in the Quitovac region and is composed of two main rock sequences. The lower sequence is composed of Proterozoic igneous and metamorphic basement rocks (Pbga, Pga, Paa, and Ppa) and is overlain by an upper sequence of Mid-Jurassic volcanic and volcaniclastic rocks (Jpbr, Jdr, Ja, Jti, and Jva). The base of the Proterozoic sequence does not crop out in Sierra Quitovac, and the calculated 200 m thickness represents only a portion of the true thickness of this

Figure 6. Mesoscopic and microscopic kinematic indicators in ductile to brittle-ductile shear zones associated with thrust faults in the Quitovac region. All photographs were taken along planes containing the mineral-stretching lineation and normal to the foliation. (A) Cross-polarized light photomicrograph of mylonite sample SA-12 formed after the Upper Cretaceous La Toñita two-mica granite (field of view ~3 cm looking to the WSW). Rigid K-feldspar grain (Ks) showing antithetic shearing deformed mainly by brittle fracturing and cataclastic flow (below 300 °C). Muscovite porphyroclasts (Ms) forming a "mica fish" with tails of dynamically recrystallized fine-grained mica. S-C fabrics for this mylonite were better shown in outcrop and hand samples. The foliation was defined by preferred orientations of rotated K-feldspar porphyroclasts and white mica depicting right-lateral sense of shearing indicating top-to-the-NNW displacement for the La Toñita thrust. (B, C) Paleozoic (?) quartzite (Pzqtz) from Sierra La Toñita presenting isoclinal folding with NNW vergency (views looking to the west). (D) Outcrop of mylonite sample formed after Proterozoic granite from Sierra La Silla Sur presenting S-C fabrics indicating right-lateral shearing (view to the west) and corroborating the northward displacement of the Campo Bustamante thrust. (E) Polished hand sample of mylonite SA-2 formed after the Upper Jurassic El Chacón biotite granite (field of view ~15 cm looking to the SSW). Specimen showing S-C fabrics as well as K-feldspar porphyroclasts with dynamically recrystallized asymmetric tails (sigma-type), all indicating right-lateral shearing suggesting top-to-the-NW displacement for the El Chacon thrust. (F) En echelon extension gashes (quartz veins) in a more brittle zone with right-lateral shear sense indicators suggesting NNE displacements for the Quitovac Thrust in Sierra La Silla (view to the WNW). (G) Large rotated porphyroclast of a more resistant part of the La Silla two-mica granite (view to the west) showing right-lateral sense of shearing indicating northward displacement for the lower-plate rocks from the Quitovac trust in Sierra La Silla Sur.

sequence. The Mid-Jurassic volcanic sequence varies significantly in thickness from north (4100 m) to south (550 m).

Similar to the Campo Bustamante Tectonic Sheet basement rocks, the Proterozoic basement rocks in the Quitovac Tectonic Sheet locally have structures of older deformational event(s). The best example is the lower Mesoproterozoic amphibolite unit (Paa) that crops out in the southern part of Sierra Quitovac. This unit presents strong Laramide deformation superimposed on complex folding and faulting of possible early Mesoproterozoic age.

The Quitovac Tectonic Sheet is the footwall of both the Campo Bustamante Thrust and El Chacón thrust. In Sierra Quitovac, the lateral ramp of the Campo Bustamante Thrust juxtaposes Caborca basement rocks and associated cover rocks over the Mid-Jurassic volcanic sequence of the "North America" block. The previously described NNW-SSE–longitudinal folding that affected the El Chacón Granite is also present in the rocks of the Quitovac Tectonic Sheet. These passive lower-plate longitudinal folds verge ENE and were produced by the tectonic juxtaposition of the allochthonous Campo Bustamante Tectonic Sheet along the lateral ramp of the Campo Bustamante Thrust. This folding does not affect the rocks in the Campo Bustamante Tectonic Sheet in Sierra Quitovac, but the lack of folding of the Quitovac Thrust appears to indicate that folding predates the emplacement of the Quitovac Tectonic Sheet.

The Quitovac Thrust crops out in Sierra La Silla Sur and juxtaposes undifferentiated Mid-Jurassic volcanic rocks (Jva) of the Quitovac Tectonic Sheet over the Upper Cretaceous La Silla Granite. The thrust developed a 20- to 30-m-thick mylonite zone that can be traced for ~3 km along strike. Most of the mylonite was formed in the Mid-Jurassic hangingwall volcanic rocks (Fig. 6F), but a significant amount of the mylonite has a granitic protolith (Fig. 6G). This band of high-strain deformation records the highest greenschist metamorphic grade for the Laramide orogeny in the Quitovac region. This observation supports our interpretation that the Quitovac Thrust is the deepest fault in the observed stack of thrusts in the Quitovac region (Fig. 7). The absence of significant Paleocene-Eocene gold mineralization in the Quitovac Tectonic Sheet is probably an indication of deep structural levels that would not permit brittle deformation.

La Silla Granite

The Upper Cretaceous La Silla Granite crops out in Sierra La Silla North and South (unit Kgra; Fig. 2). This unit is the lowest Laramide tectonic unit in the sequence of thrust sheets in the Quitovac region, and the structural block has a minimum estimated thickness of 1700 m. As noted earlier, La Silla Granite is the footwall of the Quitovac Thrust. It is difficult to estimate how much of the mylonite was formed in the granite, but estimates based on petrographic studies indicate that at least 8 m of the mylonite had a granitic protolith.

Even though Quaternary gravels cover the base of the sequence, the subtle internal foliation and the existence of thin mylonite bands ~100 m away from the thrust plane suggest that La Silla Granite may be a parallochthonous unit in the region. For that reason, we hypothesize the existence of a buried thrust north of the La Silla area that we named the La Silla Thrust (LST; Fig. 5A).

Paleocene-Eocene orogenic gold mineralization is sparse in La Silla Granite. We speculate that the granite was too deep for brittle deformation. If the strain rate was approximately the same for all the thrust sheets, then the depth could explain the lack of brittle deformation in the lower units. The only exception may be the San Antonio Mine (Fig. 2), where small occurrences of orogenic gold mineralization were found. These gold veins were later reheated by lower Miocene dikes as indicated by the lower Miocene age obtained on a white-mica sample from this deposit (discussed later; Fig. 14).

LARAMIDE THRUSTING

Proterozoic crystalline rocks of the Campo Bustamante and Quitovac Tectonic Sheets represent the Caborca and "North America" basement blocks of Anderson and Silver (1981), respectively. Rocks from both basements have complex fabrics due to the superposition of the Laramide structures onto preexisting structures of probable Proterozoic and possible Jurassic age. Nonetheless, the intensity of the Laramide orogeny recorded in these rocks is very strong and the fabrics are consistent with those developed in the Mid-Jurassic and Upper Cretaceous rocks.

The Laramide orogeny is the youngest compressional event recorded in the Quitovac region and probably in most of northwestern Sonora. This orogeny is characterized by regional foliation and local mylonitization along thrust faults and intense regional metamorphism and plutonism. It is proposed that the Laramide orogeny in the region started some time in the Late Cretaceous (ca. 75 Ma) and ended in the Eocene (ca. 40 Ma) (e.g., Damon et al., 1964; Coney, 1976; Dickinson et al., 1988). During this time, the previously described tectonic sheets in the Quitovac region were emplaced along northward-directed thrusts. In our reconstruction of the Laramide displacements, we utilize a "thin-skin" model in which we do not account for the internal deformation accumulated within the tectonic sheets or along thrust faults.

Direction, Sense, and Quantification of Thrusting

The average direction of transport for the different tectonic sheets in the Quitovac region has been determined by systematically measuring foliations and stretching mineral lineations within the units and along the thrust faults separating them. In addition, we have determined the sense of displacement using shear sense indicators in mylonites (e.g., S-C fabrics, asymmetric porphyroclasts). Some examples of such structures are presented in Figure 6. The best mylonites were formed after granitoids and some porphyritic volcanic rocks. Kinematic indicators were studied at mesoscopic (outcrop) and microscopic level, and all the observations were made in planes containing the mineral-stretching lineation and normal to the foliation. In general, the best sense of motion was determined in rock outcrops where

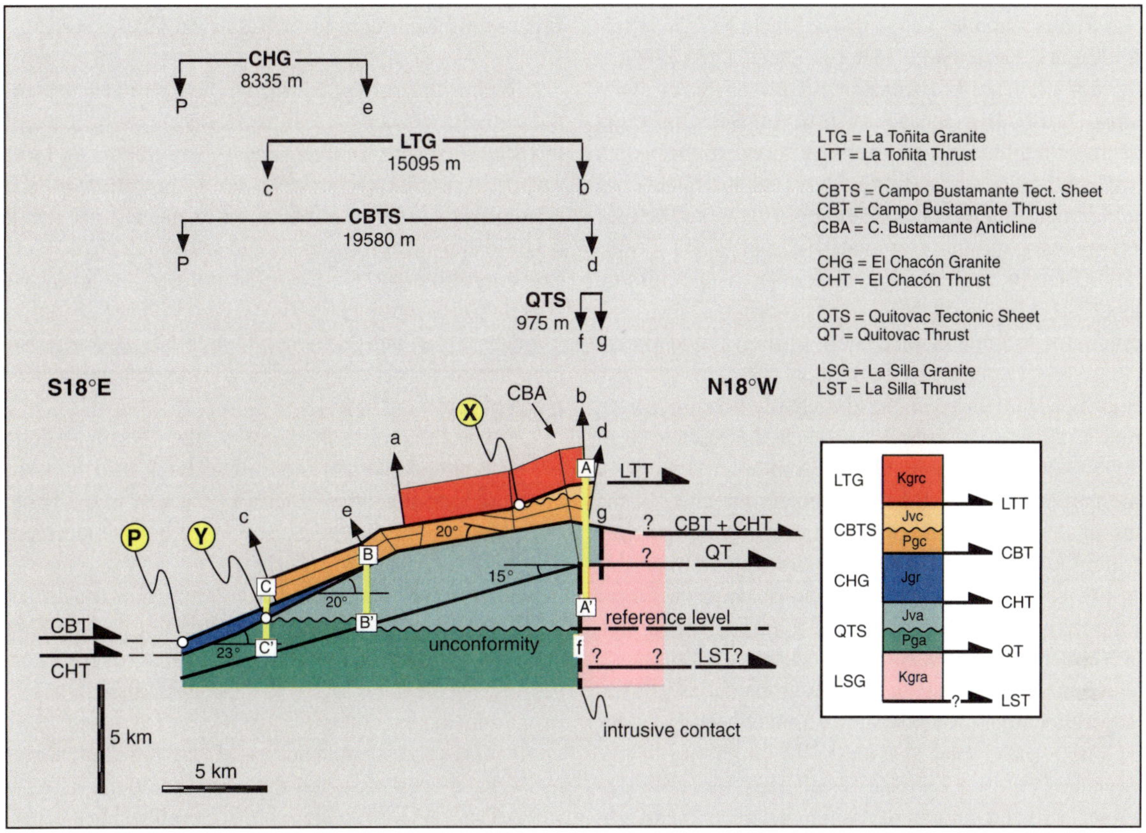

Figure 7. Balanced cross section, including geometric elements used for the structural restoration of Laramide deformation of Figure 12. Orientation for the geologic section S18°E–N18°W has been obtained from the average displacement vector from Figure 5B. The reference level chosen for the balanced section is the unconformity separating basement rocks from supracrustal rocks in the Quitovac Tectonic Sheet. Vertical yellow lines represent the projected cross sections from Figure 4, after eliminating the effects of the Basin and Range extension. Points X and Y are the intersection points represented in Figure 5A. The lower-case letters a–g and point P represent reference points to quantify overlap of each of the tectonic sheets. Upper segments represent the amount of overlap of each tectonic sheet with respect to its parallochthon. Ramp angles have been determined from map measurements and from the stereonet plots in Figure 3. Structural stratigraphy in the lower-right corner corresponds with the one shown in Figure 4.

numerous examples of kinematics indicators were put to a test. Selected oriented samples were also collected to corroborate the sense indicators observed in the field.

In most cases, the upper-plate rocks in these thrusts in the Quitovac region accumulate the highest deformation and metamorphic grade. In this type of ductile environment it is common to observe older-over-younger as well as younger-over-older thrust relationships that complicate the classical perception on thrusting that was developed using rocks from more brittle environments.

The average Laramide foliation and lineation for each of the tectonic sheets was restored to the pre–Basin and Range position by rotating the amount of observed and/or estimated tilting for each particular block about a horizontal axis (discussed earlier; Fig. 3). After restoration of the Basin and Range extensional event, the internal deformation of the allochthonous sheets (more specifically the widespread foliations and lineations) acquires new values for some of the units. The restored lineations present new values that have changed from +4° to –4° with respect to the original measurements.

The lineations and shear sense criteria indicate northward displacements of the tectonic sheets between N63°W and N23°E. The average displacement direction for each of the units (Fig. 5B) has been calculated using the average lineation obtained for each of the sheets after removing the Basin and Range effects (Fig. 3). The values for average lineation are shown in Table 1.

The minimum displacement of the tectonic sheets in the Quitovac region was estimated by graphically determining the amount of overlap between each of the allochthonous sheets with respect to their parallochthon in the direction of Laramide tectonic transport. This is the only method available to quantify thrust displacement in the Quitovac region because of the lack of cut-off lines or other structural markers necessary to facilitate more accurate determinations. Therefore, the following estimates only provide minimum displacements for the thrust faults.

The La Toñita Granite was displaced in a N27°W direction over the Campo Bustamante Tectonic Sheet. The La Toñita Granite overlaps the Campo Bustamante Tectonic Sheet along the La Toñita Thrust for a distance of 9025 m. Since no other outcrops of the allochthonous granite were found further south over its parallochthon (Campo Bustamante Tectonic Sheet), we use the southernmost outcrop of the Campo Bustamante Tectonic Sheet to determine the minimum amount of displacement for this unit (Fig. 5B). Therefore, the La Toñita Granite has a minimum displacement of 15.1 km in the N27°W direction.

The minimum tectonic displacement of the Campo Bustamante Tectonic Sheet over its parallochthon, the Quitovac Tectonic Sheet, is measured based on the amount of overlap between the two units. In Figure 5B point P represents the southernmost outcrop of the Quitovac Tectonic Sheet. From point P, we have traced a line perpendicular to the displacement direction of the Campo Bustamante thrust, and somewhere south of this line we should find the root of the thrust. The distance between this intersection line and the front of the Campo Bustamante Thrust represents the minimum displacement (23.4 km) of the Campo Bustamante Tectonic Sheet in a N3°W direction.

The allochthonous El Chacón Granite crops out as klippen on top of its parallochthonous Quitovac Tectonic Sheet (Fig. 5B). El Chacón Granite was displaced at least 10.6 km in the N63°W direction over the Quitovac Tectonic Sheet. This minimum displacement represents the amount of overlap between the allochthonous El Chacón Granite and its underlying Quitovac Tectonic Sheet parallochthon.

To estimate the displacement of the Quitovac Tectonic Sheet with respect to its parallochthon, La Silla Granite, overlap relationships cannot be used as in the previous tectonic sheets, because here there are only outcrops of the front of the Quitovac Thrust. However, we have measured a minimum displacement of 1600 m in a N23°E direction for the Quitovac Thrust based on geologic mapping and topographic changes (Fig. 5B). The displacement of the Quitovac Tectonic Sheet should be much larger, especially if we consider the size of the mylonite zone developed along the Quitovac Thrust. It is possible that displacement on the Quitovac Tectonic Sheet could be many kilometers, similar to that of the other tectonic sheets in the Quitovac region.

Geometric criteria are not available to quantify the displacement of La Silla Granite because there are no outcrops of the hypothetical La Silla Thrust. It is possible that this unit is autochthonous in the region, but evidence of internal deformation (thin mylonite bands) away from the upper-bounding Quitovac Thrust lead us to propose that La Silla Granite may be a parallochthonous unit displaced in a N12°W direction (Fig. 5B).

The summation of the individual displacement vectors for the different tectonic units in the Quitovac region, excluding the vector for La Silla Granite, has yielded an average direction of Laramide tectonic transport N18°W (Fig. 5B). As discussed later, we estimate 45 km of cumulative northward-directed displacement for the thrusts in the Quitovac area.

Geometric Elements for a Balanced Cross Section

Basin and Range tilting in the Quitovac region hampers the construction of a cross section containing the entire stack of tectonic sheets. To overcome this problem, we have selected a vertical plane of representation in the direction of Laramide transport (N18°W) to which we projected all the geometric relationships from our geologic map and cross sections. The resulting cross section has been used as a reference for a balanced cross section (Fig. 7).

The following geometric elements from our geologic mapping and stereonet diagrams (Fig. 3) have been projected to the plane of representation.

- The unconformity between the "North America" basement block rocks and the overlying Mid-Jurassic volcanic sequence south of Sierra El Chacón in the Quitovac Tectonic Sheet has been used as the horizontal reference level for our reconstruction (Fig. 7).
- Angles of frontal and lateral ramps of the different thrust faults, as well as the measured angles for the unconformities between basement rocks and overlying sequences have been used for the Caborca and "North America" blocks.
- The original thicknesses of the different tectonic sheets have been projected from cross sections A–A′, B–B′, and C–C′ (Fig. 4) and are shown as yellow lines in the projected section. The maximum thickness for El Chacón Granite has been measured directly from the geologic map.
- The intersection between La Toñita Thrust and the basement-cover unconformity of Campo Bustamante Tectonic Sheet is represented by point X. In the same manner, the intersection between El Chacón Granite and the basement-cover unconformity of the Quitovac Tectonic Sheet is represented by point Y.
- The overlap of each of the tectonic sheets, that is, the front and back of each thrust, is indicated by letters a–g and point P. The segments correspond to projected displacements for each of the tectonic sheets and represent only their apparent displacement in the N18°W direction and not the minimum displacement previously estimated from the geologic map (Fig. 5B).
- The dip of the two limbs for the Campo Bustamante Anticline has been preserved because the fold axis is perpendicular to the average direction of Laramide transport.

Time of Thrusting: $^{40}Ar/^{39}Ar$ Geochronology

Because the Upper Cretaceous granites in the region (La Toñita Granite and La Silla Granite) are clearly affected by this intense period of compressional deformation, we propose that the Laramide orogeny would be the most logical event to explain thrusting in the Quitovac region. To further evaluate this supposition and to help determine the sequence of thrusting in the Quitovac region, we have analyzed $^{40}Ar/^{39}Ar$ geochronology on syntectonic muscovites from dynamically metamorphosed rocks associated with the thrust faults.

Analytical results of $^{40}Ar/^{39}Ar$ incremental heating experiments on muscovites from such metamorphosed rocks are presented in Tables 2–5, and corresponding age spectra are shown in Figures 8–11. See Appendix A for $^{40}Ar/^{39}Ar$ analytical procedure.

El Chacón Thrust

To evaluate the time of movement along El Chacón Thrust, we have collected samples affected by the thrust in the Sierra El Chacón area (Fig. 8). A mylonite sample developed along the thrust fault and three schists in the footwall of the thrust were selected to separate metamorphic muscovite for $^{40}Ar/^{39}Ar$ geochronology studies. These muscovites are an unequivocal product of shearing, because none of their protoliths contained muscovite (units Jgr, Jdr, and Jpbr).

The $^{40}Ar/^{39}Ar$ spectra obtained for these muscovite separates provide a range of apparent ages with staircase and sigmoidal patterns (Fig. 8). In the case of mylonite sample SA-2 (Fig. 6E), the age spectrum for the muscovite is highly discordant and characterized by young ages at low extraction temperatures (39 Ma) that systematically climb with higher extraction temperatures up to 61 Ma (Fig. 8). For the schist samples (S-493, S-499, and S-501), $^{40}Ar/^{39}Ar$ spectra display marked sigmoidal patterns between 50 and 61 Ma (Fig. 8). These patterns show climbing ages in the low-temperature portions, followed by a nearly flat, plateau-like segment at intermediate temperatures, and again climbing ages in the high-temperature portions. Age spectra with staircase and sigmoidal patterns are typical for muscovite in mylonites and schists in orogenic belts that contain multiple generations of muscovite and/or superimposed deformation fabrics (e.g., Wintsch et al., 1996; Kirschner et al., 1996). It was not possible to determine more than one generation of muscovite from petrographic studies of these particular rocks. Future microprobe analyses of the muscovites may clarify this situation.

We interpret the range in ages between 39 and 61 Ma (combined age spectra on Fig. 8) as the minimum time period over which deformation and associated neocrystallization and growth of muscovite occurred along the El Chacón Thrust. The oldest age for the muscovites (61 Ma) represents the minimum possible age for the onset of movement along the thrust fault, whereas the youngest age (39 Ma) represents the oldest possible age for cessation of movement. We propose that movement along the El Chacón Thrust started at 61 Ma only for the purpose of determining the sequence of thrusting in the Quitovac region.

La Toñita Thrust

The time of thrusting for La Toñita Thrust has been evaluated using one mylonite sample (SA-12; Fig. 6A) developed along the thrust fault and a schist sample (S-487) from the footwall of the thrust (Fig. 9). The protolith for the mylonite rock is the Upper Cretaceous La Toñita two-mica granite. The schist, on the other hand, was metamorphosed from a Mid-Jurassic pyroclastic rock that did not contain muscovite.

The $^{40}Ar/^{39}Ar$ spectra for these two samples present sigmoidal patterns from 49 to 58 Ma (Fig. 9). We interpret these data similarly to the samples from El Chacón Thrust. The age range between 49 and 58 Ma represents the minimum time period over which deformation took place along La Toñita Thrust. In the case of the mylonite sample (SA-12), we have petrographic evidence to propose two distinct generations of muscovite. The older generation is the original muscovite from the Upper Cretaceous two-mica granite that acts as porphyroclasts in the mylonite (Fig. 6A). The younger generation is the fabric-forming muscovite associated with the thrusting event. Petrographically, the schist appears to have only a fabric-forming generation of muscovite and that may explain the younger ages for the high-temperature steps in the schist. For that reason, we believe that it is more appropriate to use the muscovite from the schist to constrain the onset of thrusting. We propose that the 55 Ma age for muscovite in the schist represents the minimum possible age for the onset of movement along La Toñita Thrust, whereas the 49 Ma represents the oldest possible age for cessation of movement. For the purpose of determining the sequence of thrusting, we propose that movement along La Toñita Thrust started at 55 Ma.

Campo Bustamante Thrust

To constrain the timing of movement along the Campo Bustamante Thrust, we used a mylonite sample from the hangingwall of the thrust (Fig. 10).

The $^{40}Ar/^{39}Ar$ spectrum for a muscovite separate shows a staircase to sigmoidal pattern with a large age gradient between the low- and high-temperature portions (Fig. 10). Obviously, the age gradient between 138 and 1140 Ma cannot be interpreted as the time period for a single deformational event. This is a complex age spectrum that could imply superposition and/or reactivation of deformational fabrics. We suspect that the original rock was a Paleoproterozoic granite of the Caborca block, and we tentatively hypothesize that this granitic rock suffered a first stage of deformation in which muscovite grew in the late Mesoproterozoic (at or before 1140 Ma). Such a deformational event could be associated with the Grenville orogeny, because we have evidence of plutonism at ca. 1.1 Ga in the Campo Bustamante area. We propose that a younger reactivation of the mylonite took place sometime after 138 Ma. Since this age represents the maximum possible age for the newer deformational event recorded in this mylonite, we hypothesize that the compressional Laramide orogeny would cause the reactivation of the mylonite.

Quitovac Thrust

In the Sierra La Silla Sur and Cerros San Fernando area, we collected four mylonite samples in the hangingwall and footwall of the Quitovac Thrust (Fig. 11).

TABLE 2. $^{40}Ar/^{39}Ar$ ANALYTICAL DATA FOR WHITE MICAS IN SIERRA EL CHACÓN AREA

T (°C)	K/Ca**	$^{36}Ar/^{39}Ar$ (× 10^{-3})	$^{40}Ar^*/^{39}Ar_K$	$^{39}Ar_K$ †† (× 10^{-15} mol)	$^{39}Ar_K$ (% of total)	$^{40}Ar^*$ (%)	Age (Ma)	±1σ§§ (Ma)
S-493	Muscovite	J = 0.0015749 ± 0.10%		wt = 1.25 mg		NM-71 Lab# 8096-01		
775	9.7	4.2	19.2657	2.449	7.1	94.0	53.92	± 0.24
825	14.5	2.9	19.5269	2.813	8.2	95.9	54.64	± 0.22
875	30.2	2.6	20.9838	5.326	15.4	96.5	58.66	± 0.17
925	46.5	1.8	20.3702	7.644	22.1	97.5	56.97	± 0.15
975	26.9	1.6	19.9377	5.852	17.0	97.7	55.78	± 0.17
1010	11.6	3.0	19.9880	2.708	7.8	95.8	55.92	± 0.26
1050	13.0	5.2	20.3579	1.549	4.5	93.0	56.93	± 0.36
1200	13.1	3.2	21.7596	6.175	17.9	95.8	60.79	± 0.18
						Total gas age = 57.22		
S-499	Muscovite	J = 0.0015653 ± 0.10%		wt = 1.12 m		NM-71 Lab# 8095-01		
775	26.6	15.5	17.8140	1.007	3.8	79.5	49.62	± 0.54
825	34.8	9.8	18.8132	1.233	4.7	86.6	52.36	± 0.43
875	56.8	7.3	19.8066	2.959	11.2	90.2	55.08	± 0.27
925	91.7	3.3	20.2695	6.530	24.8	95.4	56.35	± 0.16
975	16.6	3.3	20.0653	3.649	13.9	95.4	55.79	± 0.21
1010	8.1	4.6	20.0685	1.824	6.9	93.7	55.80	± 0.30
1050	24.6	4.3	20.0652	2.619	9.9	94.1	55.79	± 0.23
1200	11.7	3.7	20.5412	6.498	24.7	95.0	57.10	± 0.16
						Total gas age = 55.78		
S-501	Muscovite	J = 0.0015743 ± 0.10%		wt = 1.38 mg		NM-71 Lab# 8097-01		
775	34.1	9.6	17.7007	1.159	4.7	86.2	49.58	± 0.41
825	38.3	7.6	18.4342	1.300	5.2	89.2	51.61	± 0.36
875	44.4	8.8	19.3490	2.836	11.4	88.1	54.13	± 0.26
925	62.9	4.0	19.8310	5.696	22.9	94.3	55.46	± 0.17
975	23.9	4.5	19.5877	3.118	12.5	93.6	54.79	± 0.21
1010	12.0	5.8	19.7494	1.946	7.8	92.0	55.24	± 0.29
1050	24.1	3.3	20.0682	3.386	13.6	95.3	56.11	± 0.19
1200	17.8	3.2	20.3347	5.430	21.8	95.6	56.85	± 0.17
						Total fusion age = 55.12		
SA-2	Muscovite	J = 0.0007896 ± 0.09%		wt = 7.95 mg		NM-47 Lab# 6438-01		
600	3.9	144.1	27.9001	2.469	2.1	39.6	39.31	± 0.61
650	3.1	21.1	33.0372	4.280	3.6	84.2	46.46	± 0.22
675	2.7	8.2	35.3187	3.102	2.6	93.6	49.62	± 0.22
700	2.7	6.0	36.1068	3.711	3.1	95.3	50.71	± 0.20
730	3.0	4.1	36.5196	4.563	3.8	96.8	51.29	± 0.16
760	3.7	3.5	36.6445	5.453	4.6	97.3	51.46	± 0.17
800	4.1	3.0	36.9024	6.821	5.7	97.7	51.82	± 0.14
840	4.7	3.0	37.0745	9.629	8.1	97.7	52.05	± 0.15
860	5.4	3.1	37.4613	6.767	5.7	97.6	52.59	± 0.13
900	5.3	3.1	38.4116	8.340	7.0	97.7	53.90	± 0.14
950	6.0	3.5	39.2616	13.088	11.0	97.4	55.08	± 0.14
1000	7.2	3.2	40.1517	18.071	15.2	97.7	56.31	± 0.12
1075	4.2	4.0	40.5063	18.190	15.3	97.2	56.80	± 0.14
1200	3.4	9.2	40.7790	12.962	10.9	93.8	57.17	± 0.14
1650	0.4	66.6	43.6013	1.440	1.2	69.0	61.06	± 0.55
						Total gas age = 53.95		

*Ages calculated assuming an initial $^{40}Ar/^{36}Ar$ = 295.5.
**Apparent K/Ca ratios were calculated using the equation given in Fleck et al. (1977).
†† $^{39}Ar_K$ concentrations were calculated using the measured sensitivity of the mass spectrometer and thus have a precision of ~5%.
§§ Ages of individual steps do not include the error in the irradiation parameter J. No error is calculated for the total fusion age.

TABLE 3. $^{40}Ar/^{39}Ar$ ANALYTICAL DATA FOR WHITE MICAS IN SIERRA LA TOÑITA AREA

T (°C)	K/Ca**	$^{36}Ar/^{39}Ar$ ($\times 10^{-3}$)	$^{40}Ar^*/^{39}Ar_K$	$^{39}Ar_K$†† ($\times 10^{-15}$ mol)	$^{39}Ar_K$ (% of total)	$^{40}Ar^*$ (%)	Age (Ma)	±1σ§§ (Ma)
S-487	Muscovite	J = 0.0015640 ± 0.10%		wt = 2.35 mg		NM-71 Lab# 8098-01		
650	7.7	21.3	17.7149	0.415	0.8	73.8	49.31	± 0.96
700	9.8	10.2	18.5675	0.584	1.1	86.0	51.64	± 0.70
775	13.0	6.1	18.8752	2.096	3.8	91.3	52.49	± 0.31
825	19.7	3.4	19.0968	2.513	4.6	95.0	53.10	± 0.24
875	34.6	4.1	19.0057	5.803	10.6	94.0	52.85	± 0.17
925	55.6	1.5	19.2811	11.972	21.9	97.7	53.60	± 0.15
975	31.0	1.3	19.0783	10.257	18.8	98.1	53.05	± 0.13
1010	7.5	1.5	19.0848	6.369	11.7	97.7	53.06	± 0.14
1050	17.6	1.4	19.5248	6.289	11.5	97.9	54.27	± 0.15
1200	5.9	3.5	19.7760	8.265	15.1	95.0	54.96	± 0.16
						Total gas age = 53.52		
SA-12	Muscovite	J = 0.0007894 ± 0.09%		wt = 7.88 mg		NM-47 Lab# 6437-01		
775	5.2	70.9	34.9853	0.710	0.4	62.5	49.15	± 0.98
825	9.3	27.2	37.4173	1.696	1.0	82.3	52.52	± 0.40
875	41.1	9.8	40.2849	6.071	3.7	93.3	56.48	± 0.17
900	126.6	3.1	41.2234	10.949	6.7	97.8	57.78	± 0.13
930	238.1	1.9	41.1101	17.827	11.0	98.7	57.62	± 0.15
970	243.9	1.3	40.7319	20.401	12.5	99.0	57.10	± 0.13
1010	232.6	1.4	40.5024	24.806	15.2	99.0	56.78	± 0.13
1050	212.8	1.9	40.2015	16.035	9.9	98.6	56.37	± 0.17
1110	111.1	2.4	40.1684	16.236	10.0	98.3	56.32	± 0.19
1200	81.3	2.5	40.1772	31.017	19.1	98.2	56.33	± 0.15
1300	6.2	15.2	40.5653	4.575	2.8	90.0	56.87	± 0.24
1650	11.4	6.2	41.2855	12.415	7.6	95.7	57.86	± 0.18
						Total gas age = 56.80		

*Ages calculated assuming an initial $^{40}Ar/^{36}Ar$ = 295.5.

** Apparent K/Ca ratios were calculated using the equation given in Fleck et al. (1977).

†† $^{39}Ar_K$ concentrations were calculated using the measured sensitivity of the mass spectrometer and thus have a precision of ~5%.

§§ Ages of individual steps do not include the error in the irradiation parameter J. No error is calculated for the total fusion age.

TABLE 4. $^{40}Ar/^{39}Ar$ ANALYTICAL DATA FOR A WHITE MICA SAMPLE IN SIERRA QUITOVAC

T (°C)	K/Ca**	$^{36}Ar/^{39}Ar$ ($\times 10^{-3}$)	$^{40}Ar^*/^{39}Ar_K$	$^{39}Ar_K$†† ($\times 10^{-15}$ mol)	$^{39}Ar_K$ (% of total)	$^{40}Ar^*$ (%)	Age (Ma)	±1σ§§ (Ma)
S-471	Muscovite	J = 0.0151978 ± 0.09%		wt = 1.72 mg		NM-79 Lab# 8449-01		
600	16.31	130.7	5.2423	2.864	0.7	12.0	138.30	± 6.26
650	57.80	7.4	8.3615	3.735	0.9	79.4	215.82	± 1.48
700	109.89	1.9	9.8458	6.851	1.6	94.6	251.57	± 0.87
775	222.22	1.7	11.5916	19.348	4.6	95.8	292.73	± 0.64
825	263.16	1.0	15.3604	21.833	5.2	98.1	378.51	± 0.74
875	384.61	0.9	25.9439	34.390	8.2	99.0	599.65	± 1.19
925	555.56	0.8	37.6697	48.588	11.6	99.4	816.64	± 1.65
975	625.00	1.0	42.8065	35.156	11.6	99.3	904.05	± 1.40
1010	555.56	0.9	44.7168	32.784	7.8	99.4	935.51	± 1.56
1050	555.56	0.3	44.1165	30.164	7.2	99.8	925.68	± 1.74
1200	769.23	0.3	51.8866	154.009	36.8	99.8	1048.91	± 1.89
1300	263.16	3.2	58.6703	13.552	3.2	98.4	1150.02	± 1.72
1650	91.74	44.1	57.9765	2.297	0.5	81.6	1139.93	± 3.68
						Total gas age = 874.30		

**Apparent K/Ca ratios were calculated using the equation given in Fleck et al. (1977).

†† $^{39}Ar_K$ concentrations were calculated using the measured sensitivity of the mass spectrometer and thus have a precision of ~5%.

§§Ages of individual steps do not include the error in the irradiation parameter J. No error is calculated for the total fusion age.

*Ages calculated assuming an initial $^{40}Ar/^{36}Ar$ = 295.5.

TABLE 5. $^{40}Ar/^{39}Ar$ ANALYTICAL DATA FOR WHITE MICAS IN SIERRA LA SILLA SUR AND CERROS SAN FERNANDO AREA

T (°C)	K/Ca**	$^{36}Ar/^{39}Ar$ ($\times 10^{-3}$)	$^{40}Ar^*/^{39}Ar_K$	$^{39}Ar_K$†† ($\times 10^{-15}$ mol)	$^{39}Ar_K$ (% of total)	$^{40}Ar^*$ (%)	Age (Ma)	±1σ§§ (Ma)
S-310	Muscovite	J = 0.0007550 ± 0.25%		wt = 2.47 mg		NM-58 Lab# 7196		
750	b.l.	119.3	21.1522	0.316	3.3	37.5	28.58	± 0.65
800	b.l.	9.1	24.2027	0.320	3.3	90.0	32.67	± 0.31
875	86.21	13.7	26.9781	0.382	4.0	87.0	36.38	± 0.27
925	109.90	11.0	35.7082	2.037	21.3	91.7	47.99	± 0.19
975	133.30	5.9	36.9262	1.787	18.7	95.5	49.61	± 0.18
1000	75.19	6.5	36.9994	0.740	7.7	95.0	49.71	± 0.21
1030	39.84	7.0	37.3552	0.638	6.7	94.7	50.18	± 0.22
1070	312.50	4.2	38.6810	0.802	8.4	96.9	51.93	± 0.22
1170	109.90	2.7	39.4833	2.549	26.6	98.0	52.99	± 0.19
						Total gas age =	48.63	
S-481	Muscovite	J = 0.0015706 ± 0.10%		wt = 1.93 mg		NM-71 Lab# 8101-01		
775	87.72	8.4	14.3079	1.855	3.9	85.2	40.09	± 0.30
825	104.17	8.9	15.2328	2.001	4.2	85.3	42.65	± 0.31
875	714.29	5.5	16.5250	5.097	10.6	91.0	46.23	± 0.19
925	833.34	2.5	16.8045	11.444	23.9	95.8	47.00	± 0.12
975	1666.6	2.5	16.8114	5.770	12.1	95.7	47.02	± 0.14
1010	b.l.	3.4	16.7624	2.867	6.0	94.4	46.88	± 0.26
1050	b.l.	3.2	16.9575	3.692	7.7	94.8	47.42	± 0.18
1200	1000.0	1.3	17.3215	15.133	31.6	97.8	48.42	± 0.12
						Total gas age =	46.95	
S-484	Muscovite	J = 0.0015684 ± 0.10%		wt = 1.69 mg		NM-71 Lab# 8102-01		
650	b.l.	61.2	26.3438	0.028	0.2	59.3	73.04	± 12.88
700	b.l.	49.7	17.1670	0.063	0.4	53.9	47.93	± 5.68
775	b.l.	26.3	16.7277	0.239	1.3	68.2	46.72	± 1.67
825	b.l.	12.4	18.1757	0.628	3.5	83.2	50.71	± 0.66
875	1428.6	2.7	18.3366	4.850	27.2	95.9	51.15	± 0.16
925	625.0	1.3	18.2394	4.853	27.2	97.9	50.88	± 0.16
975	250.0	2.5	17.6696	1.760	9.9	96.0	49.32	± 0.30
1010	b.l.	5.0	17.6465	0.588	3.3	92.2	49.25	± 0.71
1050	181.8	6.7	17.5745	0.741	4.2	89.8	49.05	± 0.52
1200	416.7	3.2	18.2449	4.088	22.9	95.0	50.90	± 0.18
						Total gas age =	50.64	
S-486	Muscovite	J = 0.0015545 ± 0.10%		wt = 2.35 mg		NM-71 Lab# 8099-01		
600	41.1	138.6	9.7609	0.419	0.6	19.2	27.17	± 1.86
650	b.l.	41.3	16.7803	0.340	0.5	57.9	46.46	± 1.37
700	109.9	32.2	16.7032	0.519	0.8	63.7	46.24	± 0.94
775	62.5	9.7	16.5301	1.895	2.9	85.3	45.77	± 0.30
825	119.0	7.2	17.0052	4.073	6.3	88.9	47.07	± 0.21
875	62.5	2.1	16.9884	13.756	21.4	96.5	47.02	± 0.12
925	434.8	1.0	17.0267	15.197	23.6	98.3	47.13	± 0.11
975	3333.3	1.5	16.9538	6.961	10.8	97.4	46.93	± 0.13
1010	b.l.	2.8	16.8225	3.445	5.4	95.3	46.57	± 0.19
1050	714.3	1.8	16.9019	4.878	7.6	97.0	46.79	± 0.14
1200	222.2	1.0	17.2101	11.458	17.8	98.3	47.63	± 0.12
1300	126.6	5.5	17.2621	1.043	1.6	91.4	47.77	± 0.39
1650	1.1	29.3	18.8879	0.389	0.6	68.7	52.21	± 1.15
						Total gas age =	46.98	

*Ages calculated assuming an initial $^{40}Ar/^{36}Ar = 295.5$

**Apparent K/Ca ratios were calculated using the equation given in Fleck et al. (1977); b.l.—below detection limit for ^{37}Ar.

†† $^{39}Ar_K$ concentrations were calculated using the measured sensitivity of the mass spectrometer and thus have a precision of ~5%.

§§ Ages of individual steps do not include the error in the irradiation parameter J. No error is calculated for the total fusion age.

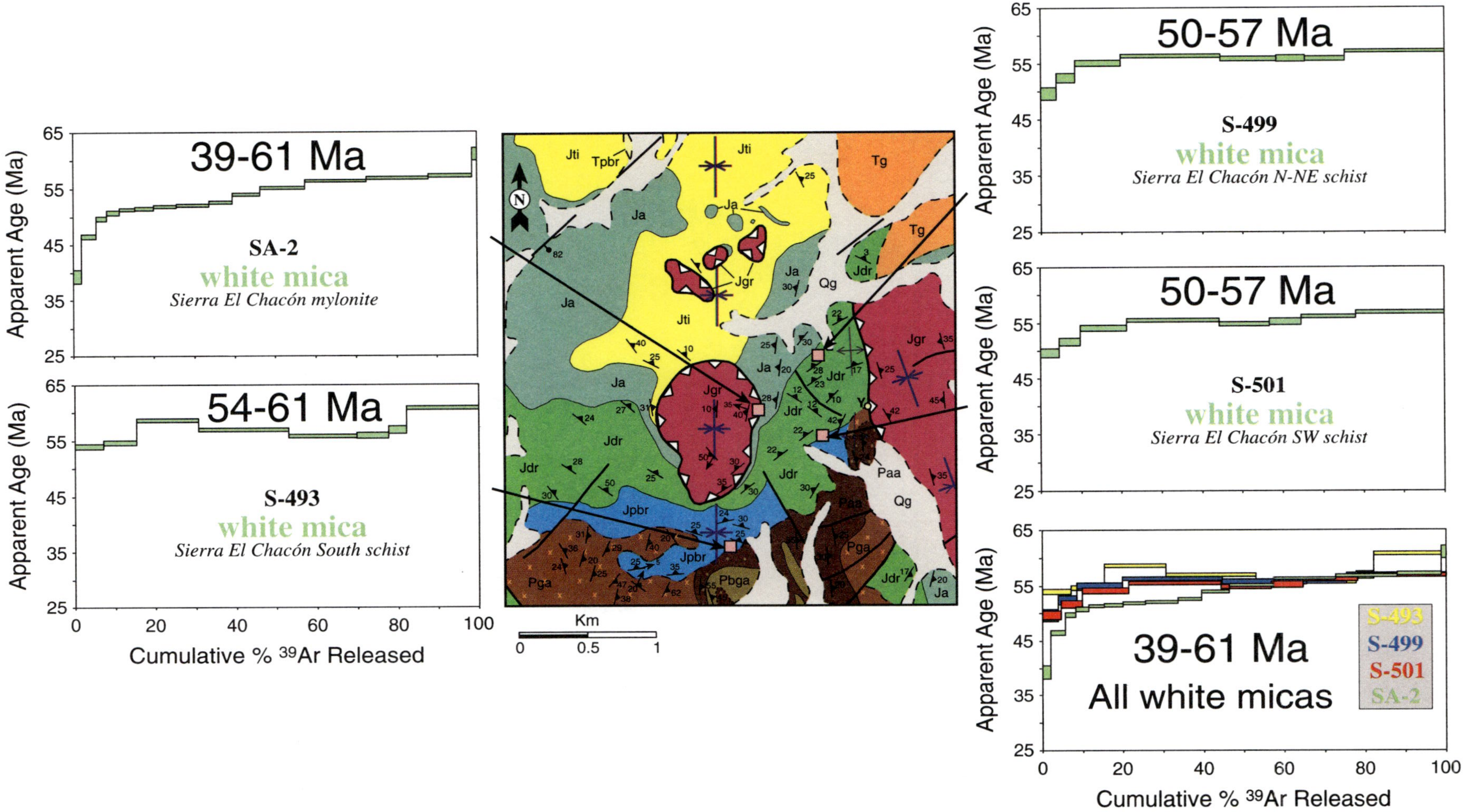

Figure 8. Geologic map of Sierra El Chacón area showing the $^{40}Ar/^{39}Ar$ age spectra and sample locations for dynamometamorphic rocks. See Figure 2 for description of map units and symbols. Analytical results of $^{40}Ar/^{39}Ar$ incremental heating experiments are presented in Table 2.

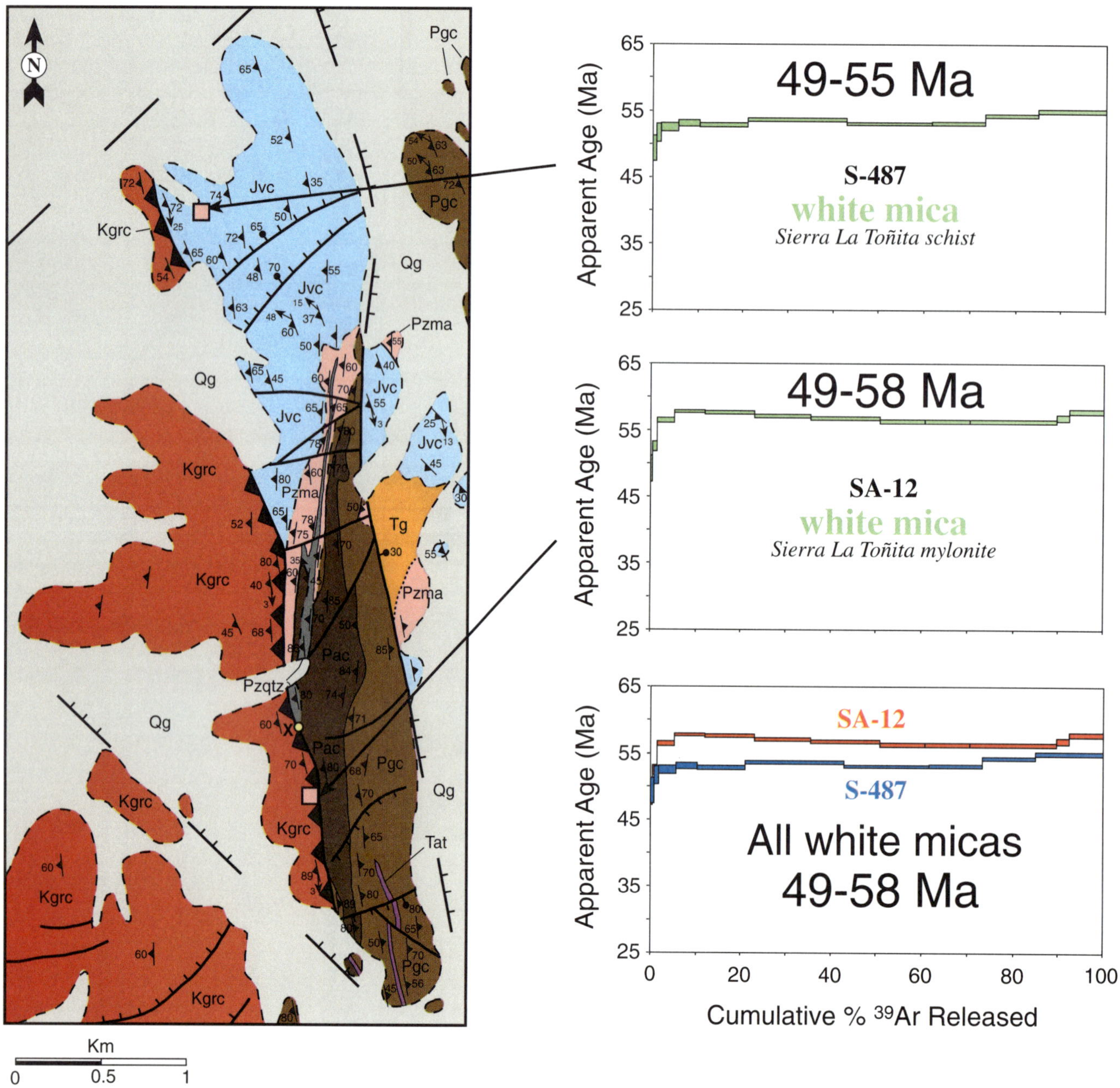

Figure 9. Geologic map of Sierra La Toñita area showing the $^{40}Ar/^{39}Ar$ age spectra and sample locations for dynamometamorphic rocks. See Figure 2 for description of map units and symbols. Analytical results of $^{40}Ar/^{39}Ar$ incremental heating experiments are presented in Table 3.

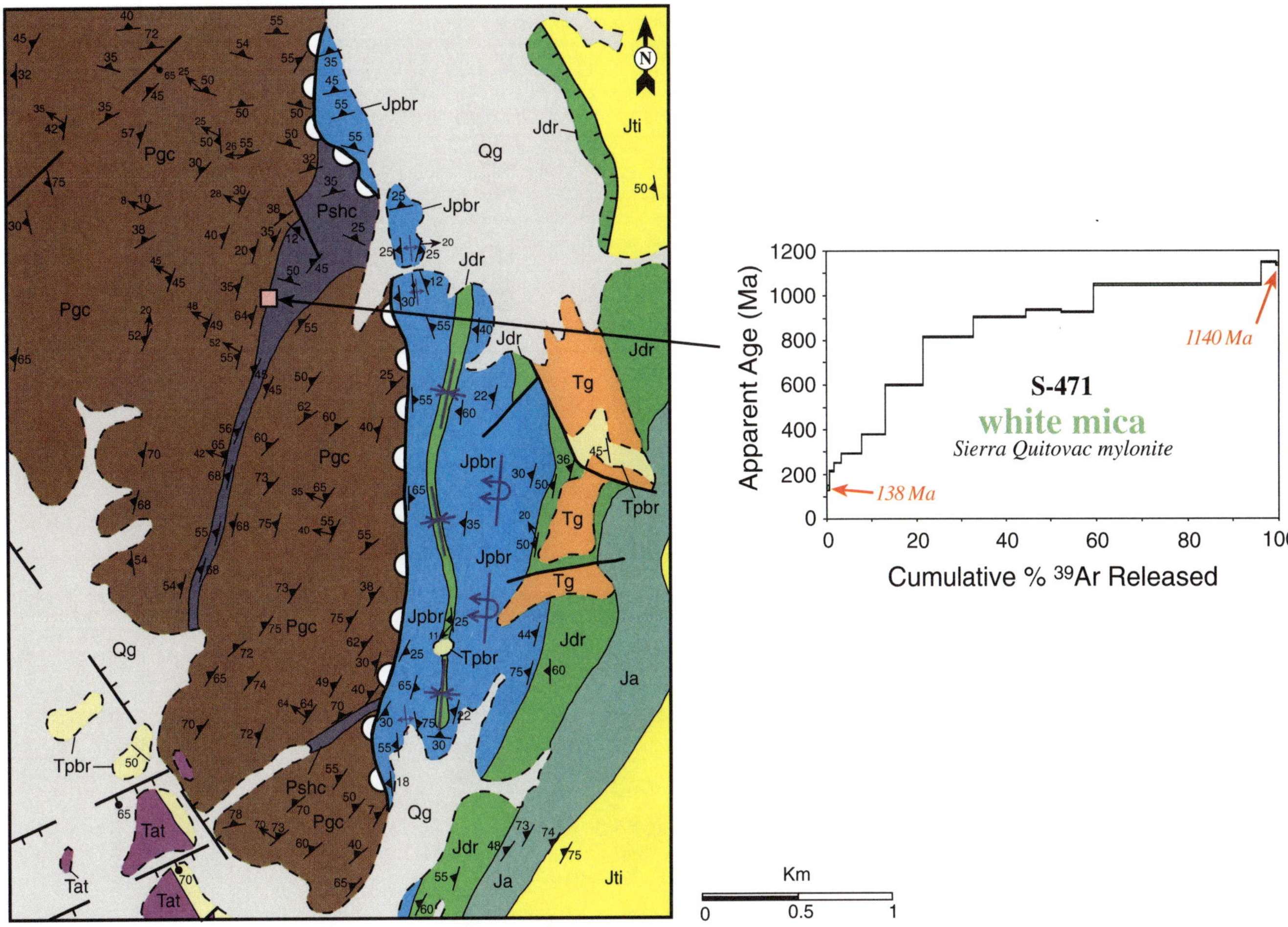

Figure 10. Geologic map of Sierra Quitovac area showing the $^{40}Ar/^{39}Ar$ age spectrum and sample location for dynamometamorphic rocks. See Figure 2 for description of map units and symbols. Analytical results of $^{40}Ar/^{39}Ar$ incremental heating experiments are presented in Table 4.

Muscovites from these samples present $^{40}Ar/^{39}Ar$ spectra with staircase and sigmoidal patterns similar to muscovites from the previously described rocks. However, in this case there may be an additional complication for samples S-486 and S-310 (Fig. 11). We interpret the young ages at the low-temperature portions of the spectra for these two samples (27 and 29 Ma) to represent partial argon loss due to a reheating event related to lower Miocene dikes. In Sierra La Silla Sur, we have observed numerous mafic dikes intruding the Upper Cretaceous two-mica granite and the Mid-Jurassic volcanic rocks. Sample S-486 was collected ~40 m away from one of the mafic dikes that was dated at 16.96 ± 0.16 Ma (Iriondo et al., 1996). If this interpretation of partial isotopic resetting were correct, it would be more appropriate to use the other two mylonite samples to constrain the time of cessation of movement along the Quitovac Thrust.

Sample S-481 was collected in the footwall of the thrust and is a mylonite developed from the La Silla two-mica granite. The $^{40}Ar/^{39}Ar$ spectrum presents a staircase pattern between 40 and 48 Ma. We propose that the 40 Ma age represents the oldest possible age for cessation of movement along the Quitovac Thrust. It is important to mention that this time for cessation of thrusting is in close agreement with the 39 Ma age proposed for the Chacón Thrust. The mylonite sample S-484 was formed in the hanging-wall of the thrust fault from a Mid-Jurassic volcanic protolith. The muscovite from this mylonite shows an age spectrum with a sigmoidal pattern between 48 and 51 Ma, but this age range does not help constrain the initiation or cessation of the thrust fault because it falls within a wider range of ages determined by the other samples in the area.

We interpret the age range between 40 and 53 Ma as the minimum time period over which deformation took place along the Quitovac Thrust. Interpreting the data in the same manner as the previous samples, we propose that the 53 Ma age represents the minimum possible age for the onset of movement along the Quitovac Thrust, and the 40 Ma age represents our interpreted time of cessation of thrusting. For the purpose of determining the sequence of thrusting, we propose that movement along the Quitovac Thrust began at 53 Ma.

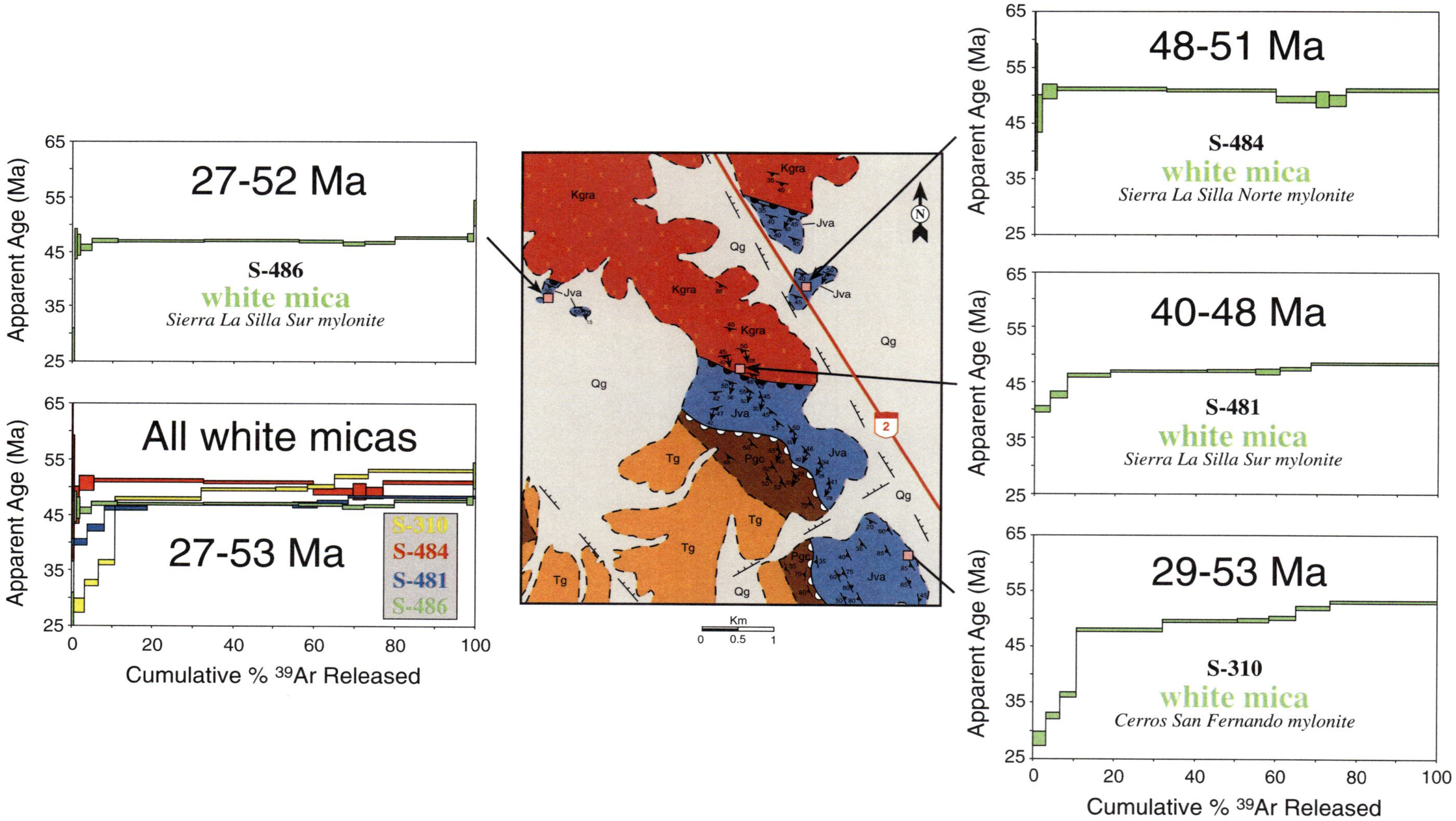

Figure 11. Geologic map of Sierra La Silla Sur and Cerros San Fernando area showing the $^{40}Ar/^{39}Ar$ age spectra and sample locations for dynamometamorphic rocks. See Figure 2 for description of map units and symbols. Analytical results of $^{40}Ar/^{39}Ar$ incremental heating experiments are presented in Table 5.

Tentative Sequence of Laramide Thrusting

A tentative sequence of Laramide thrusting in the Quitovac region has been determined combining the geometric field relationships with the previously described $^{40}Ar/^{39}Ar$ geochronology of muscovites associated with the thrusts. Below is a discussion of observed field relationships that are important in accessing the sequence of thrusting in the Quitovac area.

The longitudinal folding recorded in the El Chacón Thrust, which resulted from the emplacement of the Campo Bustamante Tectonic Sheet along a lateral ramp of the Campo Bustamante Thrust, suggests that the El Chacón Thrust is older than the Campo Bustamante thrust. On the other hand, the absence of folding in the Quitovac Thrust is used to propose that this thrust is younger than the Campo Bustamante Thrust. This interpretation, in which the emplacement of the Campo Bustamante Tectonic Sheet predates and postdates the emplacement of El Chacón Granite and the Quitovac Tectonic Sheet, respectively, is supported by our $^{40}Ar/^{39}Ar$ geochronology. Since the onset of movement along El Chacón Thrust took place at 61 Ma and that of the Quitovac Thrust at 53 Ma, the onset of movement along the Campo Bustamante Thrust is constrained between 61 and 53 Ma.

We do not have definitive field evidence to accurately determine the sequence of thrusting of the La Toñita Granite, but we propose that this granite was thrust over the Campo Bustamante Tectonic Sheet before this tectonic sheet folded El Chacón Thrust. This interpretation further constrains the onset of movement along the Campo Bustamante Thrust between 55 Ma, the age of La Toñita Thrust, and 53 Ma.

Overall, the sequence of thrusting appears to have propagated in a transport direction toward the foreland. This kind of thrusting sequence is termed piggy-back and implies that the highest thrust in the stack is the oldest and the lowest represents the youngest thrust in the sequence. For this reason, we propose that if the lowest, hypothetical La Silla Thrust exists, it should be the latest thrust to move in the stack. This piggy-back sequence of thrusting in the Quitovac region is somewhat irregular, because the El Chacón Thrust is out of sequence with the other sheets. We are not certain why this thrust is out of sequence, but we speculate that a gradual change in the regional stress field produced a clockwise rotation of the direction of thrusting through time (Fig. 5B) that could have forced younger tectonic sheets to overthrust older sheets.

We acknowledge that alternative sequences of thrusting can be proposed for the Quitovac area. However, we believe that our proposed sequence is the most probable, because it is based on integrated geometric relationships and $^{40}Ar/^{39}Ar$ geochronology resulting from comprehensive studies in the region.

Structural Restoration of Laramide Deformation (Evolutionary History)

Using our balanced cross section in Figure 7 and the previously discussed sequence and timing of thrusting, we restored the thrusts in the Quitovac region assuming a model of thin-skin deformation (Fig. 12). We present a sequential history of Laramide thrusts from old to young.

Sometime before 61 Ma, the Caborca and "North America" Paleoproterozoic basement blocks, represented by the Campo Bustamante Tectonic Sheet and the Quitovac Tectonic Sheet, respectively, were tentatively separated by their original Paleoproterozoic suture zone (stage A, Fig. 12). The geometry of this suture zone is unknown in northwestern Sonora at this time. For restoration purposes, we assume that the younger Mesozoic granitoids (La Toñita Granite, El Chacón Granite, and La Silla Granite) have subvertical intrusive contacts with the older basement rocks. This assumption is supported by the small thickness of the El Chacón and La Toñita granite sheets with respect to their lengths.

At 61 Ma, a thin slice of El Chacón Granite was thrust over the Quitovac Tectonic Sheet ("North America" basement block) along El Chacón Thrust (stage B, Fig. 12). The Late Cretaceous cooling history recorded in hornblende, biotite, and K-feldspar separates from this Upper Jurassic granite (Iriondo, work in progress) implies that El Chacón Granite was warm at or near the time of décollement. This isotopic resetting could have been caused by heat released from Late Cretaceous two-mica intrusions. To simplify our reconstruction, we speculate that the Campo Bustamante Tectonic Sheet and La Toñita Granite traveled on top of El Chacón Granite. The displacement for the El Chacón Granite, projected in our plane of representation, accounts for 8.3 km and corresponds to 16% of restored Laramide shortening (stage B, Fig. 12).

At 55 Ma, we infer that the Upper Cretaceous La Toñita Granite was still warm based on its $^{40}Ar/^{39}Ar$ cooling history (Iriondo, work in progress). The competence difference between the warm granite and its relatively cool basement host could have triggered the décollement in the La Toñita Granite. The displacement for La Toñita Granite in the average direction of Laramide transport is 15.1 km and amounts to 29% of restored shortening (stage C, Fig. 12). The cumulative displacement at this stage is 23.3 km and is the equivalent of 45% shortening.

Sometime after 55 Ma, the Paleoproterozoic Caborca (Campo Bustamante Tectonic Sheet) and "North America" (Quitovac Tectonic Sheet) basement blocks were juxtaposed along the Campo Bustamante Thrust (stage D, Fig. 12). During this time the footwall rocks in Sierra El Chacón and Sierra Quitovac (El Chacón Granite and Quitovac Tectonic Sheet) developed longitudinal folding along a lateral ramp of the thrust. The projected displacement of the Campo Bustamante Tectonic Sheet is 19.6 km or 37% of shortening (stage D, Fig. 12). The cumulative Laramide displacement is 42.8 km (82% shortening).

At 53 Ma, the Quitovac Tectonic Sheet was thrust over the La Silla Granite along the Quitovac Thrust (stage E, Fig. 12). The minimum projected displacement for this sheet is over 975 m (4%). We previously underestimated the amount of displacement and shortening for this particular thrust, judging from the thickness of the mylonite developed along it. The total cumulative

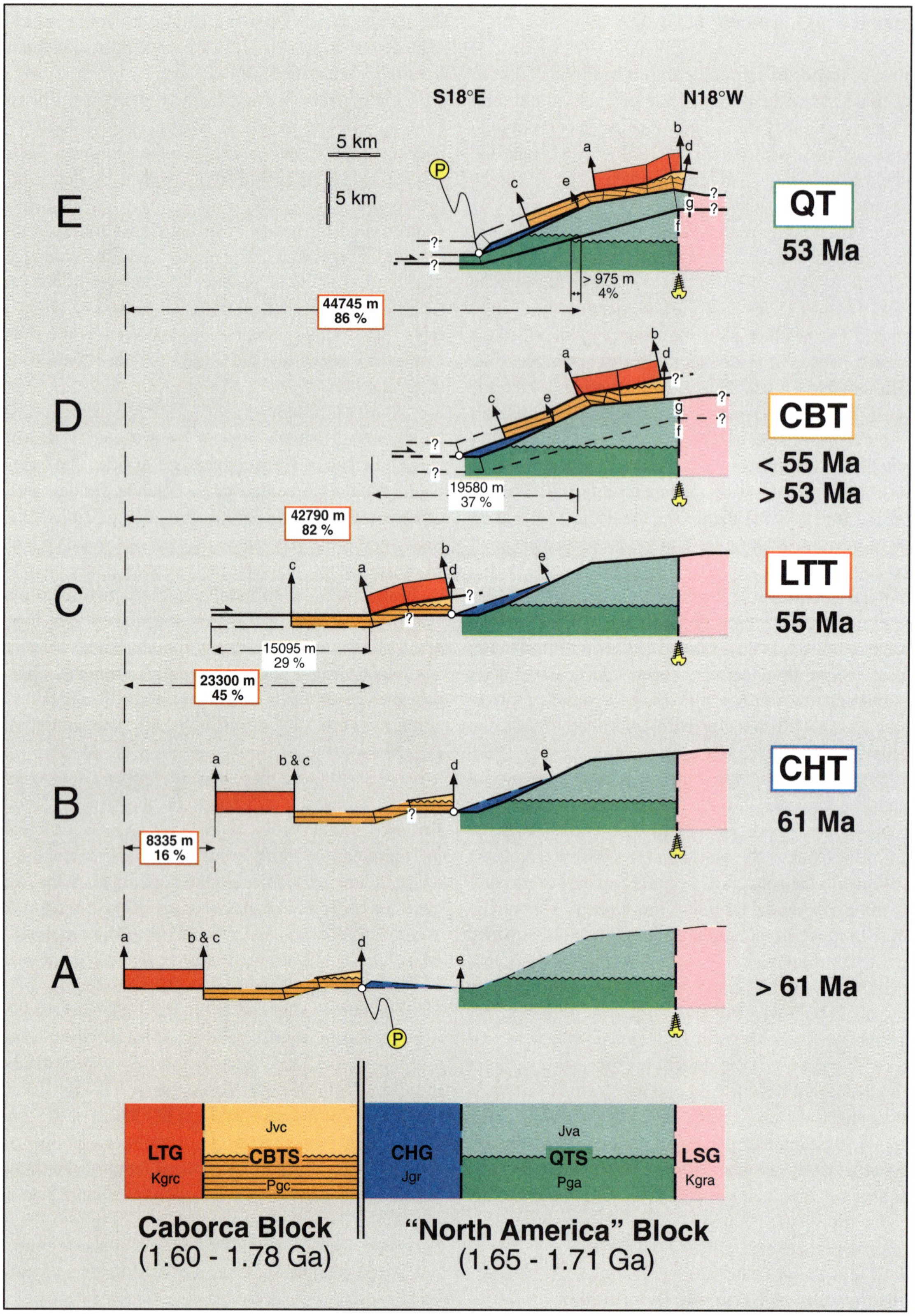

Figure 12. Structural restoration of Laramide deformation in the Quitovac region. Abbreviations and detailed explanation of the geometric elements used for this restoration are shown in Figure 7. Numbers inside the red boxes represent the cumulative amount of displacement and percentage of restored Laramide shortening. Time lapse for the different stages of thrusting (stages A through E) is based on $^{40}Ar/^{39}Ar$ geochronology discussed in the text. Age range for the Caborca and "North America" basement blocks is after Iriondo et al., 2004.

displacement along the stack of thrusts in the Quitovac region, without considering the hypothetical La Silla Thrust, is 44.7 km and corresponds to 86% of restored Laramide crustal shortening (stage E, Fig. 12). Stage E represents the pre–Basin and Range configuration depicted in our projected balanced cross section in Figure 7.

This reconstruction has been used to produce palinspastic maps that show our interpretation of the evolutionary history of the Quitovac region from the beginning of Laramide thrusting, through early Miocene volcanism, to Basin and Range extension (Fig. 13).

DISCUSSION AND REGIONAL TECTONIC IMPLICATIONS

Basin and Range Extensional Tectonics

Based on age bracketing by lower Miocene volcanic rocks, extensional faulting in the Quitovac region may have been remarkably short lived, occurring between 18 Ma and 17.5 Ma. The recorded tilting angles between 0° and 60° and the estimated stress field for extension (σ_3 ~N60°E) are consistent with what has been proposed in adjacent areas in Sonora and Sinaloa (e.g., Henry, 1989; Stewart and Roldán-Quintana, 1994; Gans, 1997). Stewart and Roldán-Quintana (1994) considered extension of this age and style in Sonora as pre–Basin and Range. The estimated 12%–30% extension accumulated in the Quitovac region is significantly smaller than estimates from southern Sonora (~90%; Gans, 1997) but very similar to estimated values in Baja California (~16%; Lee et al., 1996) and Sinaloa (20%–50%; Henry, 1989).

The timing and direction of extension in the Quitovac region partly coincides with extension recorded along low-angle detachment faults associated with a NW-SE belt of metamorphic core complexes in north-central Sonora. These detachment faults crop out 100 km east of the Quitovac region, were active in late Oligocene–early Miocene time (25–15 Ma), and extended in an average direction N40°E (Nourse et al., 1994; Stewart and Roldán-Quintana, 1994).

North-Directed Laramide Compressional Tectonics

Our restoration of Laramide deformation in the Quitovac region represents the first attempt in northwestern Sonora to quantify Laramide shortening using a thin-skin model of deformation in which both the crystalline Proterozoic basement and associated supracrustal rocks are involved.

The estimated direction of displacement for the thrust sheets in the Quitovac region varies through time from N63°W to N23°E. These thrusts represent a piggy-back sequence that migrated north, toward the foreland. A gradual change in the regional stress field could have produced a clockwise rotation in the direction of thrusting. In support of this interpretation, paleomagnetic data for Cretaceous rocks of the Caborca block, south of the Quitovac region, were interpreted to indicate that the kinematics of Late Cretaceous deformation and metamorphism involved small components of clockwise rotation (Molina-Garza and Geissman, 1999). This interpretation is also supported by global plate reconstructions by Stock and Molnar (1988). These authors indicated that during the Laramide orogeny, the Farallon plate was being subducted eastward under North America, but the convergence vector between the plates rotated in a clockwise sense during this time. Assuming a NNW-SSE orientation of the coastline in northwestern Mexico during the Late Cretaceous, we hypothesize that crustal shortening recorded in the Quitovac region (North America plate) was initially caused by NNW compression, the result of plate convergence with a strong right-oblique component. Subsequently, the convergence vector was changing gradually toward the northeast, caused by perpendicular subduction of the Farallon plate under North America.

The ENE vergency of folds below the Campo Bustamante Thrust in Sierra Quitovac was used in previous studies to propose that thrusting in the Quitovac region was caused by ENE-SSW compression (Caudillo-Sosa and Oviedo-Lucero, 1990; Caudillo-Sosa et al., 1996). These authors named this thrust fault the Quitovac Fault and indicated that thrusting was associated with transpression along the Upper Jurassic Mojave-Sonora megashear. Their time constraints for the thrusting were the age of the deformed rocks (Mid-Jurassic) and the age of two-mica granites in the region (Upper Cretaceous) that they assumed represented postdeformation rocks. We interpret this folding in Sierra Quitovac to represent Laramide lower-plate folds formed during the juxtaposition of the Campo Bustamante Tectonic Sheet over the Quitovac Tectonic Sheet along a lateral ramp of the Campo Bustamante Thrust.

The average direction and sense of displacement for the stack of thrusts in the Quitovac region is N18°W and is significantly different from the wide range of directions (north, northeast, and east) proposed for different thrust systems in southeastern California, southwestern Arizona, and central Sonora (e.g., Haxel et al., 1984; De Jong et al., 1988; Tosdal, 1990; Dillon et al., 1990; Sosson and Calmus, 1990; Miller et al., 1992; Calmus and Sosson, 1995; Pubellier et al., 1995; Jacques Ayala and De Jong, 1996; Campbell and Anderson, 1998; Nourse, 2001). Despite these differences, most of the movement recorded in these thrusts systems was toward the continent. Some of the regional discrepancies may be attributed to regional clockwise rotations about the vertical axis, possibly related to post-Laramide tectonism. One good example is the Neogene clockwise rotation related to the San Andreas Fault system in the Chocolate Mountains, southeastern California (Costello, 1985). Here a paleomagnetic study suggested that the original direction of thrusting was northward instead of the currently recorded northeastward direction.

Northward displacement of Caborca block rocks is also supported by sandstone provenance studies in the Lower Cretaceous Bisbee Group in the Caborca–Santa Ana area in northwestern Sonora (Jacques-Ayala, 1995). This study indicated that the absence of sediment derived from Proterozoic/Paleozoic Caborca

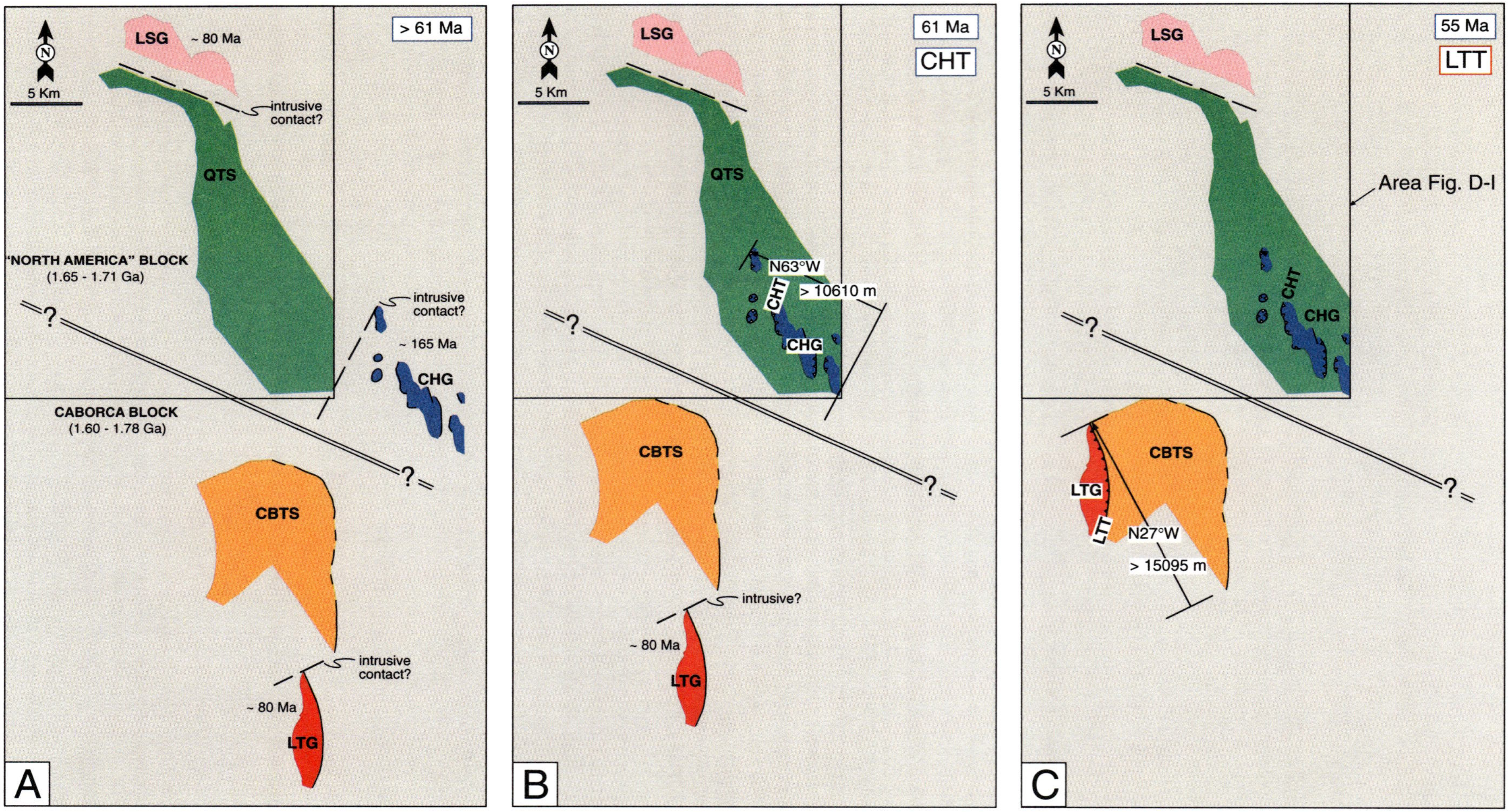

Figure 13 (*on this and following page*). Palinspastic maps of the Quitovac region showing the sequence of Laramide thrusting and subsequent early Miocene magmatism and Basin and Range tectonism. Upper-left box in stages A–C corresponds with the area shown for stages D–I. See Figure 5A for explanation of map units and symbols.

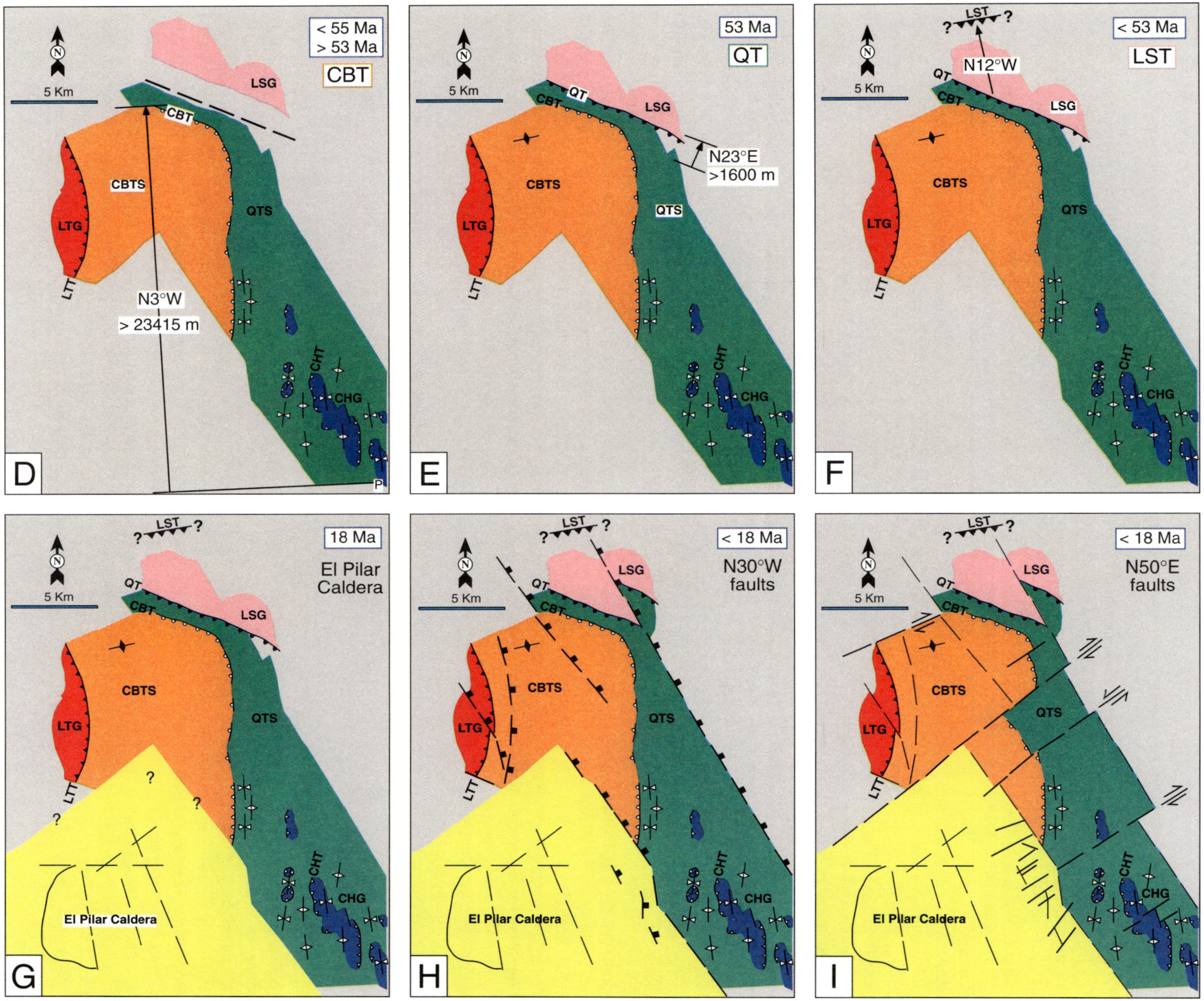

< 55 Ma
> 53 Ma
CBT
5 Km
N
LSG
CBT
CBTS
LTG
LTT
QTS
N3°W
> 23415 m
CHT
CHG
P
D
53 Ma
QT
5 Km
N
QT
CBT
LSG
N23°E
>1600 m
CBTS
LTG
LTT
QTS
CHT
CHG
E
< 53 Ma
LST
5 Km
N
? LST ?
N12°W
QT
CBT
LSG
CBTS
LTG
LTT
QTS
CHT
CHG
F
18 Ma
El Pilar
Caldera
5 Km
N
? LST ?
QT
CBT
LSG
CBTS
LTG
LTT
QTS
?
El Pilar Caldera
CHT
CHG
G
< 18 Ma
N30°W
faults
5 Km
N
? LST ?
QT
CBT
LSG
CBTS
LTG
LTT
QTS
El Pilar Caldera
CHT
CHG
H
< 18 Ma
N50°E
faults
5 Km
N
? LST ?
QT
CBT
LSG
CBTS
LTG
LTT
QTS
El Pilar Caldera
CHT
CHG
I

block rocks in the Bisbee Group suggested that this block was not part of the landscape in the Caborca–Santa Ana area during deposition of Bisbee Group sediments (Jacques-Ayala, 1995). Jacques-Ayala proposed that the Caborca block was thrust northward over the Cretaceous sedimentary sequence sometime during the Laramide orogeny. In the same region, paleomagnetic data for Cretaceous rocks of the Caborca block also support the idea of northward Laramide displacements on the order of a few hundreds of kilometers (<~300 km; Molina-Garza and Geissman, 1999).

The cumulative 45 km of northward-directed displacement estimated along the thrusts in the Quitovac region (Fig. 12) is comparable to displacements of a few tens of kilometers for the Mule Mountains Thrust (Tosdal, 1990) and the estimated 50 km of northward displacement for the Chocolate Mountains Thrust in southeastern California (Haxel and Tosdal, 1986).

The nearest, best-studied Laramide Thrust in the region is the Quitobaquito Thrust in southwestern Arizona (Haxel et al., 1984; Sosson, 1990), located a few kilometers north of Sierra Los Tanques (Fig. 1). This thrust juxtaposes Paleoproterozoic basement rocks of the Caborca block, equivalent to our Campo Bustamante Tectonic Sheet, over a sequence of Mid-Jurassic metamorphosed volcanic and volcaniclastic rocks, similar to our Quitovac Tectonic Sheet (Haxel et al., 1984; Campbell and Anderson, 1998, 2003). Vergence of folds below the Quitobaquito Thrust was used by Haxel et al. (1984) to propose an approximately northeastward displacement of the allochthonous rocks. However, more recent data collected by Sosson (1990) indicate that stretch lineations in pebbles from a metaconglomerate unit are roughly aligned S15°E and that shear criteria indicate a top-to-the-north sense of displacement (N15°W).

Similarities in age, kinematics, and structural stratigraphy indicate that the Quitobaquito Thrust may be equivalent to the Campo Bustamante Thrust in the Quitovac region. The direction of displacement for the Campo Bustamante Thrust is N3°W (Fig. 5B), which is very similar to the N15°W proposed for the Quitobaquito Thrust. Campbell and Anderson (1998) estimated that the displacement of the Quitobaquito Thrust was on the order of 10 km, which is similar to our estimate of 23.4 km for the Campo Bustamante Thrust. This is an important correlation, because both Laramide thrusts involved the juxtaposition of Paleoproterozoic Caborca and "North America" basement blocks.

Based on a structural analysis, Campbell and Anderson (1998, 2003) hypothesized that the Quitobaquito Thrust was correlative with structures near the Mojave-Sonora megashear in Sierra Los Tanques area in northwestern Sonora and formed during upper Jurassic transpression rather than Laramide compression. This is a controversial hypothesis because the Quitobaquito Thrust, a relatively well-constrained Late Cretaceous–early Tertiary ductile thrust, is rooted in Upper Jurassic mylonites of the postulated Mojave-Sonora megashear. It is important to present unequivocal temporal evidence that the ductile fabrics in Sierra Los Tanques are Upper Jurassic or Upper Cretaceous to lower Tertiary. Second, it will be necessary to determine whether the Laramide Quitobaquito Thrust resulted from reactivation of an older structure associated with the Mojave-Sonora megashear. Future mapping and detailed geochronologic studies in both areas should help solve this controversy.

Age of Ductile Fabrics Associated with Thrusting

We have presented mapping and geochronological evidence that the ductile fabrics in Quitovac were the result of Laramide thrusting. In addition, we speculated that the Laramide Campo Bustamante Thrust resulted from the reactivation of an older Proterozoic ductile shear zone. However, no evidence of Upper Jurassic ductile fabrics or evidence of Laramide reactivation of Jurassic fabrics in connection with the Mojave-Sonora megashear was found in the Quitovac region. Ductile fabrics associated with the extensional episode of metamorphic core complexes in north-central Sonora were not identified either.

We propose that the onset of thrusting in the Quitovac region occurred sometime between 75 and 61 Ma, the age of granites affected by deformation and the oldest age recorded by $^{40}Ar/^{39}Ar$ geochronology of the ductile fabrics, respectively. Nonetheless, we acknowledge that the $^{40}Ar/^{39}Ar$ geochronology of muscovites from the deformed rocks does not yield precise age constraints on the onset of thrusting. From these data, we cannot discard the idea that for some of these ductile fabrics, the measured ages may represent regional uplift, or be the result of a thermal resetting event. In fact, in a few outcrops in the Quitovac region we have observed pegmatitic and granitic dikes of supposedly Upper Cretaceous age cutting ductile fabrics. However, many similar dikes are strongly mylonitized and/or boudinaged with ductile fabrics parallel or nearly parallel to that in the enclosing metamorphosed rocks. These observations suggest that the Upper Cretaceous granites in the Quitovac region are pre- to synkinematic.

The proposed Laramide age of ductile thrusting of Proterozoic basement and associated supracrustal rocks in the Quitovac region is similar to ages obtained for thrust faults in the Chocolate and Mule Mountains (e.g., Tosdal, 1990; Miller et al., 1992) and the Orocopia Schist in southeastern California (e.g., Jacobson, 1990); the Quitobaquito Thrust in southwestern Arizona (Haxel et al., 1984); and in the Sierras El Batamote and La Víbora in northern Sonora (De Jong et al., 1988; Jacques-Ayala and De Jong, 1996). Thus, compressional deformation in the Quitovac region is coeval with Laramide thrusts that extend from southeastern California to southwestern Arizona and northern Sonora.

It has been proposed that movement in the Quitobaquito Thrust in southwestern Arizona postdated the Aquajita Spring Granite of Upper Cretaceous age and that thrusting terminated at 55–60 Ma based on K-Ar geochronology of minerals from ductile fabrics (Haxel et al., 1984). Their K-Ar geochronology of whole-rock, muscovite, and biotite for mylonites sampled in the upper and lower plates of the thrust is comparable to our $^{40}Ar/^{39}Ar$ total gas ages (47–57 Ma) listed in Tables 2–5. However, the $^{40}Ar/^{39}Ar$ age spectra reported herein for samples in the Quitovac region better constrains the cessation of thrusting in the region at ca. 39 Ma, the youngest age proposed for the end of the Laramide orogeny in northwestern Sonora (Fig. 14).

Orogenic Gold Mineralization and the Laramide Orogeny

Gold-bearing quartz veins in the Quitovac region (Fig. 2) were hosted in high-strain rocks in brittle to brittle-ductile metamorphic conditions associated with Laramide thrusting. The proposed age range of gold mineralization in the Quitovac region is between 65 and 48 Ma based on $^{40}Ar/^{39}Ar$ geochronology of muscovite separates (Fig. 14; Iriondo, 2001). Some of the muscovite samples have been partially (La Choya, Quitovac, and Doña Marcia Mines and La Toñita Este prospect) or totally reset (San Antonio Mine) by thermal effects of early Miocene volcanism (Fig. 14). However, the majority of these muscovite ages represent "birthdays," because lower closure temperature minerals separated from rocks hosting the gold veins (e.g., biotite and K-feldspar) yielded significantly older $^{40}Ar/^{39}Ar$ ages than the muscovites (Iriondo, 2001). This observation precludes the idea that these ages represent uplift or thermal resetting at or sometime before Paleocene-Eocene time.

Trace element geochemistry, fluid inclusion, and stable isotope studies of these gold-bearing quartz veins were used by Iriondo and Atkinson (2000) to classify them as orogenic gold veins. This type of vein formation requires a contemporaneous orogeny to thicken the crust and trigger metamorphic dehydration of basement rocks to form the mineralizing fluids (e.g., Groves and Phillips, 1987; Phillips and Powell, 1992; Goldfarb et al., 1993). These veins are usually emplaced during the late orogenic stages of deformation and metamorphism. The Laramide orogeny recorded in the Quitovac region, constrained herein between ca. 75 and 39 Ma (Fig. 14), appropriately explains the widespread Paleocene-Eocene gold-mineralizing event in the region. Iriondo and Atkinson (2000) have proposed that the roots of Laramide thrusts and associated structures in the Quitovac region acted as deep fluid conduits for the ascent of hot metamorphic fluids to shallower levels forming the gold-bearing quartz veins.

The age and origin of these gold veins strengthens our idea that the compressional tectonism associated with the Laramide orogeny is the most important dynamometamorphic event recorded in pre-Eocene rocks in the Quitovac region. In addition, we discard our previously discussed idea that ages of muscovites from the ductile fabrics associated with thrusting could have represented regional uplift or were the result of a thermal resetting event. Based on reconnaissance studies in the region, we propose that this type of orogenic gold mineralization forms a NW-SE belt from central Sonora through southwestern Arizona to southeastern California. If all the orogenic gold mineralization present in the region is contemporaneous with that studied in Quitovac, it would suggest that the Laramide orogeny is a widespread event along this belt.

In summary, similarities in age, kinematics, and spatial association between the Quitovac thrusts and thrusts that extend from southeastern California to southwestern Arizona and northern Sonora indicate that they may be related within a context of progressive Late Cretaceous–early Tertiary deformation in the region. We propose that the juxtaposition of the Caborca and "North America" basement blocks in the Quitovac region represents a Laramide Thrust that may be the result of reactivation of an older Proterozoic shear zone.

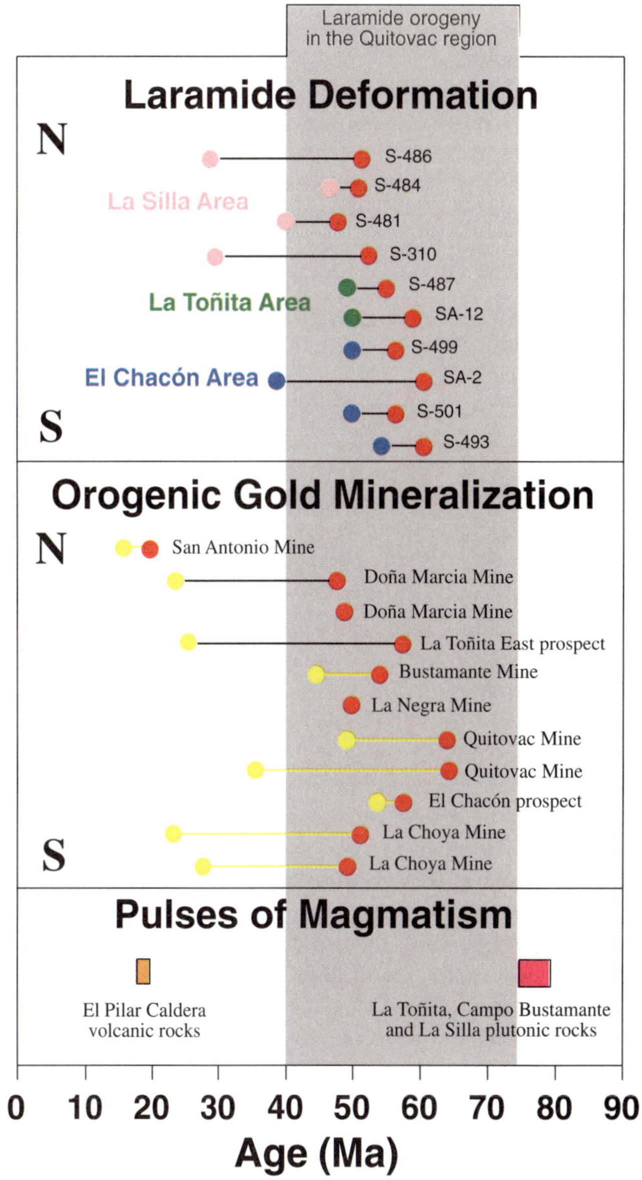

Figure 14. Age bracketing for the Laramide orogeny in the Quitovac region (ca. 75–39 Ma). The pulse of Laramide deformation corresponds to the $^{40}Ar/^{39}Ar$ data discussed in the text. Colored dots represent the maximum and minimum ages obtained for the muscovite samples. Pulse of orogenic gold mineralization represents $^{40}Ar/^{39}Ar$ data for muscovite samples associated with gold-bearing quartz veins (Iriondo, 2001). The colored dots also represent maximum and minimum ages for the gold veins. Post-Laramide El Pilar Caldera volcanic rocks were dated with $^{40}Ar/^{39}Ar$ geochronology (Iriondo et al., 1996). Pre- to syntectonic plutonic rocks were dated with U-Pb zircon geochronology (Iriondo, 2001).

The Hypothetical Mojave-Sonora Megashear in the Quitovac Region

It seems likely to us that many of the ductile thrust faults mapped along the proposed trace of the megashear in northwestern Sonora, including the Quitovac region, can be attributed to the Late Cretaceous–early Tertiary Laramide orogeny. If this conclusion is true, record of the older Mojave-Sonora megashear could have been partially or completely obliterated by superposed episodes of younger magmatism, thrusting, extension, and volcanism.

Our structural interpretation does not support the idea that the ductile fabrics associated with thrusting in the Quitovac region are exclusively related to transpressive movements along the Upper Jurassic Mojave-Sonora megashear (Anderson et al., 1984; Connors et al., 1989; Anderson et al., 1992; Caudillo-Sosa et al., 1996; Campbell and Anderson, 1998, 2003) or are even the result of reactivation of an older Late Jurassic thrust. The northward Laramide displacements and clockwise rotations recorded in the Caborca block rocks in the Quitovac region contradict the southward displacements (~800 km) and counterclockwise rotations inherent in the left-lateral Upper Jurassic Mojave-Sonora megashear hypothesis. Paleomagnetic studies south of Quitovac (Molina-Garza and Geissman, 1999) not only predicted the northward motion and clockwise rotations recorded in Caborca block rocks in the Quitovac region but also indicated that similarities of Jurassic magnetizations in rocks across the proposed megashear were not consistent with the Mojave-Sonora megashear hypothesis of Late Jurassic left-lateral strike-slip motion of the crust in northwestern Sonora.

Basement studies by Iriondo (2001) and Iriondo et al. (2004) have proposed that the Caborca and "North America" blocks in the Quitovac region have different suites of rocks and are not easily distinguished by U-Pb zircon geochronology alone, as predicted by the Mojave-Sonora megashear hypothesis. Instead, these authors undertook additional geochemistry and radiogenic isotope (Pb-Sr-Nd) studies to effectively distinguish them. Iriondo (2001) and Iriondo et al. (2004) also concluded that the Caborca basement block rocks in the Quitovac region do not correlate with the highly evolved geochemical and isotopic signature from rocks of the Mojave crustal province in the Death Valley–Inyo Mountains region—the expected signature if we were to restore the Caborca basement block to its proposed original position after displacement of ~800 km along the postulated Mojave-Sonora megashear. These facts, combined with the proposed Late Cretaceous–early Tertiary age for the juxtaposition of the Caborca and "North America" basement blocks shown in this study, do not support the presence of the postulated Upper Jurassic Mojave-Sonora megashear in the Quitovac region.

Nonetheless, these findings at Quitovac do not eliminate the possibility of a megashear elsewhere in northwestern Sonora, but if the Mojave-Sonora megashear exists, we propose that its trace should be located southwest of the Quitovac region. For example, Calmus and Sosson (1995), using similarities in Mid-Jurassic stratigraphy and structures on both sides of the megashear, proposed that the trace of the Mojave-Sonora megashear in northwestern Sonora should be placed farther to the southwest, approximately along a WNW-ESE line between Caborca and Puerto Peñasco (see Fig. 1 for geographical location of these two towns).

As stated by Silver and Anderson (1974, p. 955), "Confirmation of this inferred structure is vital to continental reconstruction. Complex younger structures, which have been superimposed, must be resolved." We believe that studies such as the one presented here represent an important step in that direction and should help achieve a consensus on the complex Mesozoic-Cenozoic evolutionary history of the region.

CONCLUSIONS

1. Restoration of the short-lived lower Miocene Basin and Range in the Quitovac region indicates a minimum of 12%–30% extension. This extension is accumulated in tilted blocks with angles between 0° and 60°.

2. The estimated direction of movement for the Laramide thrust faults studied in the Quitovac region gradually varies from N63°W to N23°E and is interpreted to represent a clockwise rotation in the direction of Laramide thrusting through time. The faults represent a piggy-back sequence of thrusting migrating in an average direction N18°W, toward the foreland. The cumulative 45 km of estimated northward-directed displacement corresponds to ~86% of Laramide thin-skinned crustal shortening.

3. The onset of thrusting in the Quitovac region occurred sometime between 75 and 61 Ma. Cessation of thrusting occurred at ca. 39 Ma. This age is the youngest proposed for the end of the Laramide orogeny in northwestern Sonora.

4. The occurrence of Paleocene-Eocene orogenic gold mineralization is spatially associated with thrusting. Thus compressional tectonism associated with the Laramide orogeny is, in fact, a very important and widespread dynamometamorphic event in the region.

5. Thrusting in the Quitovac region may correlate with the Laramide Quitobaquito Thrust in southwestern Arizona based on similarities in age, kinematics, and structural stratigraphy. In both areas, Laramide thrust faults are involved in the juxtaposition of Paleoproterozoic Caborca and "North America" basement blocks.

6. Northward Laramide displacements and clockwise rotations recorded in the Caborca block rocks in the Quitovac region contradict the southward Late Jurassic displacements and counterclockwise rotations inherent in the left-lateral Mojave-Sonora megashear hypothesis. If the megashear exists in northwestern Sonora, its trace should be southwest of the Quitovac region.

ACKNOWLEDGMENTS

The senior author wishes to thank the Basque Government in Spain for its support through a predoctoral fellowship from 1992 to 1996 at the University of Colorado at Boulder. Fieldwork for this project was partially supported by the University of Colorado EMARC field fellowship (summer of 1994 and 1995) and by HECLA Mining Company. Financial support for $^{40}Ar/^{39}Ar$ geochronology was in part obtained from HECLA, Santa Cruz Gold, Geomaque, and Peñoles Mining Companies. We thank Daniel Miggins for assistance with mineral preparations for the geochronology studies. We are grateful to Ric Page, George Gehrels, and Jonathan Nourse for their helpful comments to clarify the concepts presented in this paper.

APPENDIX A

$^{40}Ar/^{39}Ar$ Analytical Procedures and Interpretation Methods

Eleven white mica samples were concentrated from crushed metamorphic rocks using conventional heavy liquid and magnetic separation techniques at the U.S. Geological Survey in Denver, Colorado. Greater than 99% pure mineral separates were obtained by removing impurities by handpicking under a binocular microscope.

Argon analyses were performed at the New Mexico Bureau of Mines and Mineral Resources, located at New Mexico Tech (NMT) in Socorro, New Mexico. Ten of the samples and a number of flux monitors were irradiated at the Nuclear Science Center reactor at Texas A&M University and one sample (S-471) at the Ford Reactor at the University of Michigan. At the Texas A&M reactor, the D-3 position was used, whereas for the Ford reactor we used the L67.

Following irradiation, the samples were step heated in a double vacuum Mo resistance furnace. The Mo furnace crucible is heated with a W heating element and the temperature is monitored with a W-Re thermocouple placed in a hole drilled into the bottom of the crucible. Evolved gases were purified during heating with a SAES GP-50-ST101 getter for 7 min, followed by additional cleanup with another hot SAES GP-50-ST101 getter for 6–7 min. Argon was analyzed with a MAP 215-50 mass spectrometer operated in static mode using a Johnston electron multiplier with an overall sensitivity of 1×10^{-6} mol/pA. Fish Canyon Tuff sanidine (NMGRL FC-1 and FC-2) at 27.84 Ma was used as flux monitor. The monitors were fused with a 10W Synrad CO_2 continuous laser, and evolved gases were purified for 2 min using a cold GP-50-ST101 getter. J-factors were determined to a precision of 0.1% (2 sigma) by analyzing 4 single-crystal aliquots from each of 4 radial positions around the irradiation vessel. The extraction system and mass spectrometer blanks and backgrounds were measured throughout the course of the analyses. Correction factors for interfering nuclear reactions during irradiation were determined using K- and Ca-rich glasses and salts.

We used age spectra diagrams to plot the apparent age of each incrementally heated gas fraction versus the cumulative $\%^{39}Ar_K$ released, with steps increasing in temperature from left to right (see Figures 9–12). Each apparent age is calculated assuming that the trapped argon (argon not produced by in situ decay of ^{40}K) has the modern day atmospheric $^{40}Ar/^{36}Ar$ value of 295.5. All errors in Tables 2–5 are reported at 1 sigma. The decay constant and isotopic abundances are those suggested by Steiger and Jäger (1977).

REFERENCES CITED

Albinson, T.F., 1989, Vetas mesotermales auríferas del sector norte del Estado de Sonora: Memorias Técnicas, XVIII Convención Nacional de la Asociación de Ingenieros de Minas, Metalurgistas y Geólogos de México, A.C., p. 20–40.

Anderson, T.H., and Schmidt, V.A., 1983, The evolution of Middle America and the Gulf of Mexico–Caribbean Sea region during Mesozoic time: Geological Society of America Bulletin, v. 94, p. 941–966, doi: 10.1130/0016-7606(1983)94<941:TEOMAA>2.0.CO;2.

Anderson, T.H., and Silver, L.T., 1978, The nature and extent of Precambrian rocks in Sonora, Mexico, *in* Roldán-Quintana, J., and Salas, G.A., eds., Resúmenes Primer Simposio sobre la Geología y Potencial Minero en el Estado de Sonora, Hermosillo, Instituto de Geología, UNAM, p. 9–10.

Anderson, T.H., and Silver, L.T., 1979, The role of the Mojave-Sonora megashear in the tectonic evolution of northern Sonora, *in* Anderson, T.H., and Roldán-Quintana, J., eds., Geology of northern Sonora: Geological Society of America Field Trip Guidebook 7, p. 59–68.

Anderson, T.H., and Silver, L.T., 1981, An overview of Precambrian rocks in Sonora: Revista del Instituto de Geología, UNAM, v. 5, p. 131–139.

Anderson, T.H., and Silver, L.T., 1997, 1.1 Ga plutons in Sonora, Mexico: Regional Tectonic implications: Geological Society of America Abstracts with Programs, v. 29, no. 6, p. 348.

Anderson, T.H., Bajek, D.T., Chepega, J.R., Ichikawa, K.M., Rodríguez-Castañeda, J.L., Stephens, W.W., and Silver, L.T., 1984, Crystalline thrust sheets near the Mojave-Sonora megashear: Geological Society of America Abstracts with Programs, v. 16, p. 430.

Anderson, T.H., Campbell, P.A., and Silver, L.T., 1992, Mylonite at the Mojave-Sonora megashear, northwestern Mexico: Geological Society of America Abstracts with Programs, v. 24, no. 7, p. A147.

Armstrong, R.L., 1974, Magmatism, orogenic timing, and orogenic diachronism in the Cordillera from Mexico to Canada: Nature, v. 247, p. 348–351.

Bird, P., 1998, Kinematic history of the Laramide orogeny in latitudes 35°–49°N, western United States: Tectonics, v. 17, p. 780–801, doi: 10.1029/98TC02698.

Calmus, T., and Sosson, M., 1995, Southwestern extension of the Papago terrane into the Altar Desert region, northwestern Sonora, and its implications, *in* Jacques-Ayala, C., González-León, C.M., and Roldán-Quintana, J., eds., Studies on the Mesozoic of Sonora and Adjacent Areas: Geological Society of America Special Paper 301, p. 99–109.

Campbell, P.A., and Anderson, T.H., 1998, Structure and kinematics along a Jurassic plate boundary transform, the Mojave-Sonora Megashear, northwestern Sonora, Mexico, *in* Clark, K.F., ed., Gold Deposits of Northern Sonora, Mexico: Society of Economic Geologists Guidebook Series 30, p. 177–186.

Campbell, P.A., and Anderson, T.H., 2003, Structure and kinematics along a segment of the Mojave-Sonora megashear: A strike-slip fault that truncates the Jurassic continental magmatic arc of southwestern North America: Tectonics, v. 22, 1077, doi: 10.1029/2002TC001367.

Caudillo-Sosa, G., and Oviedo-Lucero, L.F., 1990, Geología del área Quitovac, Municipio de Puerto Peñasco, Sonora [Tesis de Geólogo]: Hermosillo, Universidad de Sonora, México, 90 p.

Caudillo-Sosa, G., Oviedo-Lucero, L.F., and Rodríguez-Castañeda, J.L., 1996, Falla Quitovac: Resultado de un evento de transpresión del "Mojave-Sonora megashear," noroeste de Sonora, México: Revista Mexicana de Ciencias Geológicas, UNAM, v. 13, p. 140–151.

Coney, P.J., 1976, Plate Tectonics and the Laramide Orogeny, *in* Woodward, L.A., and Northrop, S.A., eds., Tectonics and Mineral Resources of Southwestern North America: New Mexico Society Special Publication 6, p. 5–10.

Coney, P.J., 1987, The regional tectonic setting and possible causes of Cenozoic extension in the North American Cordillera, *in* Coward, M.P., Dewey, J.F., and Hancock, P.L., eds., Continental Extensional Tectonics: Geological Society Special Publication 28, p. 177–186.

Coney, P.J., and Reynolds, S.J., 1977, Cordilleran Benioff zones: Nature, v. 270, p. 403–406.

Connors, C.D., 1990, Geology of the Quitovac/Sierra La Toñita area, northwestern Sonora, Mexico [Master's thesis]: University of Pittsburgh, 74 p.

Connors, C.D., Anderson, T.H., and Silver, L.T., 1989, Expression and structural analysis of the Mojave-Sonora megashear in northwestern Sonora, Mexico: Geological Society of America Abstracts with Programs, v. 21, no. 6, p. A91.

Costello, S.C., 1985, A paleomagnetic investigation of mid-Tertiary volcanic rocks in the lower Colorado River area, Arizona and California [Master's thesis]: San Diego, California, San Diego State University, 109 p.

Damon, P.E., Mauger, R.L., and Bikerman, M., 1964, K-Ar dating of Laramide plutonic and volcanic rocks within the Basin and Range province of Arizona and Sonora, *in* Sastry, M.V.A., and Bhatia, S.B., eds., Cretaceous-Tertiary Boundary Including Volcanic Activity: 22nd International Geological Congress, India, Part 3 Proceedings of Section 3: New Delhi, p. 45–55.

Davis, G.H., 1979, Laramide folding and faulting in southeastern Arizona: American Journal of Science, v. 279, p. 543–569.

De Jong, K.A., Escarcega-Escarcega, J.A., and Damon, P.E., 1988, Eastward thrusting, southwestward folding, and westward backsliding in the Sierra La Víbora, Sonora, Mexico: Geology, v. 16, p. 904–907, doi: 10.1130/0091-7613(1988)016<0904:ETSFAW>2.3.CO;2.

Dickinson, W.R., and Snyder, W.S., 1978, Plate tectonics of the Laramide Orogeny, *in* Matthews, V., III, ed., Laramide Folding Associated with Basement Block Faulting in the Western United States: Geological Society of America Memoir 151, p. 355–366.

Dickinson, W.R., Klute, M.A., Hayes, M.J., Janecke, S.U., Lundin, E.R., McKittrick, M.A., and Olivares, M.D., 1988, Paleogeographic and paleotectonic setting of Laramide sedimentary basins in the central Rocky Mountain region: Geological Society of America Bulletin, v. 100, p. 1023–1039, doi: 10.1130/0016-7606(1988)100<1023:PAPSOL>2.3.CO;2.

Dillon, J.T., Haxel, G.B., and Tosdal, R.M., 1990, Structural evidence for northeastward movement on the Chocolate Mountains thrust, southeasternmost California: Journal of Geophysical Research, v. 95, p. 19,953–19,971.

Dumitru, T.A., Gans, P.B., Foster, D.A., and Miller, E.L., 1991, Refrigeration of the western Cordilleran lithosphere during Laramide shallow-angle subduction: Geology, v. 19, p. 1145–1148, doi: 10.1130/0091-7613(1991)019<1145:ROTWCL>2.3.CO;2.

Eaton, G.P., 1982, The Basin and Range province, origin and tectonic significance: Annual Review of Earth and Planetary Sciences, v. 10, p. 409–440, doi: 10.1146/annurev.ea.10.050182.002205.

Fernández-Aguirre, M.A., Monreal-Saavedra, R., and Grijalva-Haro, A.S., 1993, Carta Geológica Sonora (Escala 1:500,000): Hermosillo, Sonora, México, Centro de Estudios Superiores del Estado de Sonora—CESUES.

Fleck, R.J., Sutter, J.F., and Elliot, D.H., 1977, Interpretation of discordant ^{40}Ar/^{39}Ar age spectra of Mesozoic tholeiites from Antarctica: Geochimica et Cosmochimica Acta, v. 41, p. 15–32, doi: 10.1016/0016-7037(77)90184-3.

Gans, P.B., 1997, Large-magnitude Oligo-Miocene extension in southern Sonora: Implications for the tectonic evolution of northwest Mexico: Tectonics, v. 16, p. 388–408, doi: 10.1029/97TC00496.

Goldfarb, R.J., Snee, L.W., and Pickthorn, W.J., 1993, Orogenesis, high-T thermal events, and gold vein formation within metamorphic rocks of the Alaskan Cordillera: Mineralogical Magazine, v. 57, p. 375–394.

Groves, D.I., and Phillips, G.N., 1987, The genesis and tectonic control on Archean gold deposits: A metamorphic replacement model: Ore Geology Reviews, v. 2, p. 287–322, doi: 10.1016/0169-1368(87)90009-6.

Harrison, T.M., Duncan, I., and McDougall, I., 1985, Diffusion of ^{40}Ar in biotite: Temperature, pressure and compositional effects: Geochimica et Cosmochimica Acta, v. 49, p. 2461–2468, doi: 10.1016/0016-7037(85)90246-7.

Haxel, G.B., and Tosdal, R.M., 1986, Significance of the Orocopia Schist and Chocolate Mountains thrust in the late Mesozoic tectonic evolution of the southeastern California–southwestern Arizona region, *in* Beatty, B., and Wilkinson, P.A.K., eds., Frontiers in Geology and Ore Deposits of Arizona and the Southwest: Arizona Geological Society Digest, v. 16, p. 52–61.

Haxel, G.B., Tosdal, R.M., May, D.J., and Wright, J.E., 1984, Latest Cretaceous and early Tertiary orogenesis in south-central Arizona: Thrust faulting, regional metamorphism, and granitic plutonism: Geological Society of America Bulletin, v. 95, p. 631–653, doi: 10.1130/0016-7606(1984)95<631:LCAETO>2.0.CO;2.

Henry, C.D., 1989, Late Cenozoic Basin and Range structure in western Mexico adjacent to the Gulf of California: Geological Society of America Bulletin, v. 101, p. 1147–1156, doi: 10.1130/0016-7606(1989)101<1147:LCBARS>2.3.CO;2.

Henry, C.D., and Aranda-Gomez, J.J., 1992, The real southern Basin and Range: Mid- to late Cenozoic extension in Mexico: Geology, v. 20, p. 701–704, doi: 10.1130/0091-7613(1992)020<0701:TRSBAR>2.3.CO;2.

Iriondo, A., 2001, Proterozoic basements and their Laramide juxtaposition in NW Sonora, Mexico: Tectonic constraints on the SW margin of Laurentia [Ph.D. thesis]: University of Colorado, Boulder, Colorado, 222 p.

Iriondo, A., and Atkinson, W.W., Jr., 2000, Orogenic gold mineralization along the proposed trace of the Mojave-Sonora megashear: Evidence for the Laramide orogeny in NW Sonora, Mexico: Geological Society of America Abstracts with Programs, v. 32, no. 7, p. A-393.

Iriondo, A., Atkinson, W.W., Jr., and Martinez-Torres, L.M., 1996, El Pilar Caldera: An example of early Miocene calc-alkaline magmatism in the Quitovac area, northwestern Sonora, Mexico: Geological Society of America Abstracts with Programs, v. 28, no. 7, p. A-441–A-442.

Iriondo, A., Premo, W.R., Martínez-Torres, L.M., Budahn, J.R., Atkinson, W.W., Jr., Siems, D.F., and Guarás-González, B., 2004, Isotopic, geochemical and temporal characterization of Proterozoic basement rocks in the Quitovac region, northwestern Sonora, Mexico: Implications for the reconstruction of the southwestern margin of Laurentia: Geological Society of America Bulletin, v. 116, no. 1/2, p. 154–170.

Jacobson, C.E., 1990, The ^{40}Ar/^{39}Ar geochronology of the Pelona schist and related rocks, southern California: Journal of Geophysical Research, v. 95, p. 509–528.

Jacques-Ayala, C., 1995, Paleogeography and provenance of the Lower Cretaceous Bisbee Group in the Caborca-Santa Ana area, northwestern Sonora, *in* Jacques-Ayala, C., González-León, C.M., and Roldán-Quintana, J., eds., Studies on the Mesozoic of Sonora and Adjacent Areas: Geological Society of America Special Paper 301, p. 79–98.

Jacques-Ayala, C., and De Jong, K.A., 1996, Extension of the Orocopia Schist Belt into northwestern Sonora, Mexico: A working hypothesis: Geological Society of America Abstracts with Programs, v. 28, no. 5, p. 78.

Jacques-Ayala, C., García y Barragán, J.C., and De Jong, K.A., 1990, Caborca-Altar Geology: Cretaceous sedimentation and compression, Tertiary uplift and extension, *in* Gehrels, G.E., and Spencer, J.E., eds., Geologic Excursions through the Sonoran Desert Region, Arizona and Sonora: Arizona Geological Survey Special Paper 7, p. 165–182.

Kirschner, D.L., Cosca, M.A., Masson, H., and Hunziker, J.C., 1996, Staircase ^{40}Ar/^{39}Ar spectra of fine-grained white mica: Timing and duration of deformation and empirical constraints on argon diffusion: Geology, v. 24, p. 747–750, doi: 10.1130/0091-7613(1996)024<0747:SAASOF>2.3.CO;2.

Lee, J., Miller, M.M., Crippen, R., Hacker, B., and Ledesma-Vazquez, J., 1996, Middle Miocene extension in the Gulf Extensional Province, Baja California: Evidence from the southern Sierra Juarez: Geological Society of America Bulletin, v. 108, p. 505–525, doi: 10.1130/0016-7606(1996)108<0505:MMEITG>2.3.CO;2.

Maxson, J.A., and Tikoff, B., 1996, Hit-and-run collision model for the Laramide orogeny, western United States: Geology, v. 24, p. 968–972, doi: 10.1130/0091-7613(1996)024<0968:HARCMF>2.3.CO;2.

Miller, D.M., Nilsen, T.H., and Bilodeau, W.L., 1992, Late Cretaceous to Eocene geologic evolution of the U.S. Cordillera, *in* Burchfiel, B.C., Lipman, P.W., and Zoback, M.L., eds., The Cordilleran Orogen: Conterminous U.S.: Geological Society of America, The Geology of North America, v. G-3, p. 205–260.

Molina-Garza, R.S., and Geissman, J.W., 1999, Paleomagnetic data from the Caborca terrane, Mexico: Implications for the Cordilleran tectonics and the Mojave-Sonora megashear hypothesis: Tectonics, v. 18, p. 293–325, doi: 10.1029/1998TC900030.

Nourse, J.A., 2001, Tectonic insights from a Late Jurassic–Early Cretaceous stretched-clast conglomerate, Altar-Caborca region, northwest Sonora, Mexico: Journal of South American Earth Sciences, v. 14, p. 453–474, doi: 10.1016/S0895-9811(01)00051-7.

Nourse, J.A., Anderson, T.H., and Silver, L.T., 1994, Tertiary metamorphic core complexes in Sonora, northwestern Mexico: Tectonics, v. 13, p. 1161–1182, doi: 10.1029/93TC03324.

Phillips, G.N., and Powell, R., 1992, Gold-only provinces and their common features: Economic Geology Research Unit (School of Earth Sciences) Contribution, v. 43, p. 1–27.

Pubellier, M., Rangin, C., Rascon, B., Chorowicz, J., and Bellon, H., 1995, Cenomanian thrust tectonics in the Sahuaripa region, Sonora: Implications about northwestern Mexico megashears, *in* Jacques-Ayala, C., González-León, C.M., and Roldán-Quintana, J., eds., Studies on the Mesozoic of Sonora and Adjacent Areas: Geological Society of America Special Paper 301, p. 111–120.

Purdy, J.W., and Jäger, E., 1976, K-Ar ages on rock-forming minerals from the Central Alps: Memoirs, Institute of Geology and Mineralogy, University of Padova, v. 30, p. 1–31.

Reynolds, S.J., 1988, Geologic Map of Arizona (Scale 1:1,000,000): Tucson, Arizona, Arizona Geological Survey, Map 26.

Silver, L.T., and Anderson, T.H., 1974, Possible left-lateral early to middle Mesozoic disruption of the southwestern North American Craton margin: Geological Society of America Abstracts with Programs, v. 6, p. 955.

Silver, L.T., and Anderson, T.H., 1983, Further evidence and analysis of the role of the Mojave-Sonora megashear(s) in Mesozoic Cordilleran tectonics: Geological Society of America Abstracts with Programs, v. 15, p. 273.

Sosson, M., 1990, Le passage des Cordillères nord-américaines aux Sierras Madres mexicaines, sud de l'Arizona (Etats-Unis) et nord du Sonora [Ph.D. thesis]: Nice, France, Université de Nice, Sophia Antipolis, 410 p.

Sosson, M., and Calmus, T., 1990, Early Late Cretaceous overthrusting in northern Sonora responsible for a large scale displacement of the Nevadian Belt, *in* Aubouin, J., and Bourgois, J., eds., Tectonics of Circum-

Pacific Continental Margins: Proceedings, International Geological Congress, 28th, July 1989: Utrecht, The Netherlands, VSP, p. 37–50.

SSP, 1981, Carta Geológica Hoja Tijuana: Direccíon General de Geografía del Territorio Nacional, México, D.F., scale 1:1,000,000.

Steiger, R.H., and Jäger, E., 1977, Subcommission on geochronology: Convention on the use of decay constants in geo- and cosmochronology: Earth and Planetary Science Letters, v. 36, p. 359–362, doi: 10.1016/0012-821X(77)90060-7.

Stewart, J.H., 1978, Basin-range structure in western North America, *in* Smith, R.B., and Eaton, G.P., eds., A Review, Cenozoic Tectonics and Regional Geophysics of the Western Cordillera: Geological Society of America Memoir 152, p. 1–31.

Stewart, J.H., 1988, Latest Proterozoic and Paleozoic southern margin of North America and the accretion of Mexico: Geology, v. 16, p. 186–189, doi: 10.1130/0091-7613(1988)016<0186:LPAPSM>2.3.CO;2.

Stewart, J.H., and Roldán-Quintana, J., 1994, Map showing late Cenozoic extensional tilt patterns and associated structures in Sonora and adjacent areas, Mexico: U.S. Geological Survey Miscellaneous Field Studies, Map MF-2238.

Stock, J.M., and Molnar, P., 1988, Uncertainties and implications of the Late Cretaceous and Tertiary position of North America relative to the Farallon, Kula, and Pacific plates: Tectonics, v. 7, p. 1339–1384.

Tosdal, R.M., 1990, Constraints on the tectonics of the Mule Mountains thrust system, southeast California and southwest Arizona: Journal of Geophysical Research, v. 95, p. 20,025–20,048.

Tosdal, R.M., Haxel, G.B., and Wright, J.E., 1989, Jurassic geology of the Sonoran Desert region, southern Arizona, southeast California, and northernmost Sonora: Construction of a continental-margin magmatic arc, *in* Jenny, J.P., and Reynolds, S.J., eds., Geologic evolution of Arizona: Arizona Geological Society Digest 17, p. 397–434.

Tosdal, R.M., Haxel, G.B., Anderson, T.H., Connors, C.D., May, D.J., and Wright, J.E., 1990, Highlights of Jurassic, Late Cretaceous to Early Tertiary, and Middle Tertiary tectonics, south-central Arizona and north-central Sonora, *in* Gehrels, G.E., and Spencer, J.E., eds., Geologic Excursions Through the Sonoran Desert Region, Arizona and Sonora: Arizona Geological Survey Special Paper 7, p. 76–89.

Wernicke, B.P., Christiansen, R.L., England, P.C., and Sonder, L.J., 1987, Tectonomagmatic evolution of Cenozoic extension in the North American Cordilleran, *in* Cowan, M.P., Dewey, J.F., and Hankcock, P.L., eds., Continental Extensional Tectonics: Geological Society Special Publication 28, p. 203–221.

Wintsch, R.P., Kunk, M.J., and Epstein, J.B., 1996, ^{40}Ar/^{39}Ar whole-rock data constraints on Acadian diagenesis and Alleghanian cleavage in the Martinsburg Formation, eastern Pennsylvania: American Journal of Science, v. 296, p. 766–788.

Manuscript Accepted by the Society 24 March 2005

Geological Society of America
Special Paper 393
2005

Isotope provinces in Laramide and mid-Tertiary igneous rocks of northwestern Mexico (Chihuahua and Sonora) and their relation to basement configuration

Todd B. Housh
Fred W. McDowell
Department of Geological Sciences, The University of Texas at Austin, Austin, Texas 78712, USA

ABSTRACT

We have identified five distinct isotope provinces in northwestern Mexico in a study of the Sr, Nd, and Pb isotope geochemistry of Laramide and mid-Tertiary igneous rocks in Sonora and Chihuahua as well as adjacent portions of southwestern United States. Province A, in northern Sonora, is characterized by very unradiogenic Nd and Pb isotopic compositions and radiogenic Sr isotopic compositions. Unradiogenic Nd and radiogenic Sr and Pb isotopic compositions characterize Province B, in central Sonora. Provinces A and B are flanked to the east and south by Provinces C, D, and E, which are characterized by more radiogenic Nd and less radiogenic Sr isotopic compositions than observed in Provinces A and B. Rocks from Provinces C, D, and E have similar Sr and Nd isotopic compositions, although the variation observed in rocks from Province E is more restricted than in Provinces C and D. These provinces are unique, however, in their Pb isotopic compositions; rocks from Province C are the least radiogenic whereas those from Province E are the most radiogenic.

The identified isotope provinces correlate well with the constraints imposed by the geology of the southwestern United States and northwestern Mexico. The primary factor controlling the geographic distribution of isotope provinces in northwestern Mexico appears to be the age of the crust that the magmas passed through and with which they interacted. Provinces A and B are interpreted to reflect the area underlain by old, early Proterozoic basement in northwestern Sonora; the significance of the difference in Pb isotopic compositions between rocks in Provinces A and B is not clear. We interpret Province C, located in northern Chihuahua and adjacent regions of Texas, to reflect the distribution of mid-Proterozoic crust of the North American craton that has been overprinted by Grenvillian deformation. Province D is located in a belt running from central Chihuahua into southwestern Texas and is interpreted to reflect the distribution of exotic crust that was accreted to the southern margin of the North American craton during Ouachita convergence; this crustal block consists of a Paleozoic arc sequence developed on Proterozoic basement. Lastly, we interpret Province E, located in southern Sonora and Chihuahua, as exotic Jurassic-Cretaceous arc crust accreted during the late Mesozoic.

The observed distribution of isotope provinces in northwestern Mexico is not consistent with the Mojave-Sonora megashear hypothesis. In particular, the NNE-trend-

Housh, T.B., and McDowell, F.W., 2005, Isotope provinces in Laramide and mid-Tertiary igneous rocks of northwestern Mexico (Chihuahua and Sonora) and their relation to basement configuration, *in* Anderson, T.H., Nourse, J.A., McKee, J.W., and Steiner, M.B., eds., The Mojave-Sonora megashear hypothesis: Development, assessment, and alternatives: Geological Society of America Special Paper 393, p. 671–692. doi: 10.1130/2005.2393(25). For permission to copy, contact editing@geosociety.org.

ing eastern boundary of the two western isotope provinces (Provinces A and B) does not show the 700–800 km of left-lateral offset that would be expected if this boundary was intersected by the Mojave-Sonora megashear.

Keywords: Mexico, Sierra Madre Occidental, radiogenic isotopes, basement.

INTRODUCTION

It has long been recognized that variation in the radiogenic isotopic compositions of magmas is due, in part, to the age and nature of the crust the magmas traverse. Kistler and Peterman (1973) examined $^{87}Sr/^{86}Sr$ in Mesozoic volcanic and plutonic rocks in California and found that the geographical distribution of Sr isotopic compositions of the rocks varied systematically and could be contoured. The contour of $^{87}Sr/^{86}Sr = 0.706$ has been interpreted to reflect the limit of miogeosynclinal sediments at the edge of North American Precambrian crust.

Zartman (1974) examined the Pb isotopic compositions of plutonic and volcanic rocks and ore deposits throughout western North America. He found three distinct Pb isotopic provinces in the southwestern United States: one province corresponds to areas underlain by early Proterozoic basement in New Mexico, Arizona, southern Nevada, and southeastern California. A second province is in central and northern Nevada where the miogeosynclinal strata deposited on the western end of the Precambrian craton can be in excess of 15 km thick. A third Pb province is located in California, west of the first two provinces and the $^{87}Sr/^{86}Sr = 0.706$ line. This province lies outboard of the Precambrian craton of North America.

Further studies of Mesozoic and Tertiary igneous rocks in the southwestern United States have demonstrated, in general, a good correlation between Sr, Nd, and Pb isotopic composition and Precambrian age provinces in the region (Farmer and DePaolo, 1983, 1984; Lang and Titley, 1998; Bouse et al., 1999). These studies have also demonstrated that the isotopic characteristics of magmatism in an area are not solely a function of the age of the underlying crust, but are also dependent upon the flux rate of mantle-derived magma, crustal thickness, and the composition of the underlying crust (e.g., DePaolo and Farmer, 1984; Bouse et al., 1999). The work described below follows upon the methodology of these earlier studies and uses Sr, Nd, and Pb isotopic compositions of Laramide and Tertiary igneous rocks to delineate the distribution of isotope provinces in northwestern Mexico. We then relate this distribution to that of older crustal blocks in northwestern Mexico.

PRE-MESOZOIC BASEMENT GEOLOGY IN NORTHWESTERN MEXICO

Outcrop Distribution

The exposed Precambrian basement of the southwestern United States has been divided into distinct provinces on the basis of U-Pb geochronology and Nd and Pb isotopic studies (e.g., Bennett and DePaolo 1987; Chamberlain and Bowring, 1990; Karlstrom and Bowring, 1988; Karlstrom et al., 1987; Wooden et al., 1988; Wooden and Miller, 1990). These basement provinces and their southwesterly trending boundaries, disappear beneath younger rocks in the states of Sonora and Chihuahua in northern Mexico (Fig. 1). In general, the configuration of the Precambrian basement domains defining the southwestern margin of the North American craton at their boundary with Phanerozoic accreted terranes in northwestern Mexico is poorly understood. Extensive deposits of Phanerozoic sedimentary rocks, especially Tertiary continental and Cretaceous marine sequences, as well as Upper Cretaceous and Tertiary volcanic and plutonic rocks blanket the region, effectively burying older basement. Data that bear directly on the distribution of basement provinces may be derived from scattered basement outcrops and well intercepts, as well as from rare xenoliths included in Tertiary volcanic rocks in a few localities. But these types of data are particularly sparse east and south of northern Sonora and do not effectively constrain the basement framework in northwestern Mexico.

The extent to which the North American craton extends into northern Mexico and the configuration of its domains in the region are issues of some controversy. The North American craton has been envisioned as extending down into central Mexico as a long peninsula (e.g., Dickinson, 1981; Shurbet and Cebull, 1987). Others have argued that the southern margin of the North American craton extends only through central Sonora and Chihuahua in northwestern Mexico and that Precambrian material to the south was accreted in the late Paleozoic during the Ouachita orogeny (e.g., Campa and Coney, 1983; Coney and Campa, 1987; Stewart, 1988; Sedlock et al., 1993). In this study we have addressed the issues of the location of the southern margin of the North American craton in northwestern Mexico as well as the distribution of basement provinces in this area by looking at the radiogenic isotopic compositions of widespread Upper Cretaceous through mid-Tertiary igneous rocks.

Direct data on the age of basement in much of northwestern Mexico, particularly outside of Sonora, are sparse (Fig. 2). Within northern Sonora there are numerous outcrops of Proterozoic basement, many of which have been dated (Anderson and Silver, 1979; Anderson and Silver, 1981). Anderson and Silver found that these rocks can be divided into two belts of distinct age: a northeastern belt comprising rocks with ages between 1.6 and 1.7 Ga, and a southwestern belt (the "Caborca block") comprising rocks with ages between 1.7 and 1.8 Ga. The northeastern belt is believed to be continuous with the Mazatzal province in southern Arizona and New Mexico. The southwestern

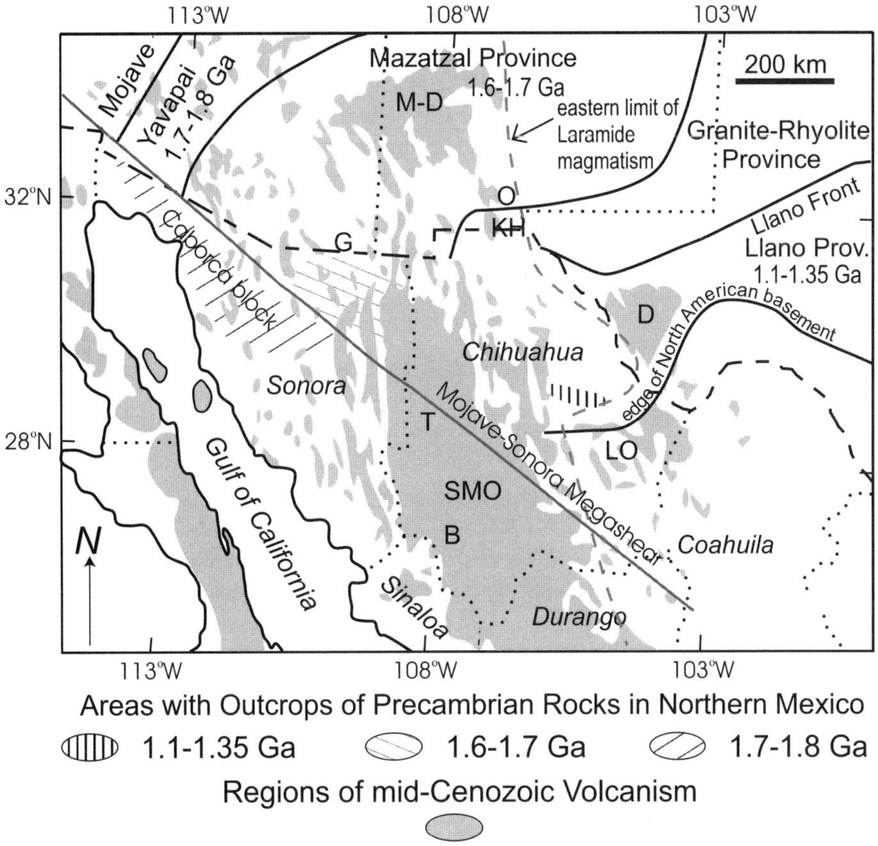

Figure 1. Distribution of mid-Tertiary volcanic rocks in northwestern Mexico and southwestern United States and areas with outcrops of Precambrian rocks in northwestern Mexico. Distribution of Precambrian age provinces in southwestern United States is modified from Hoffman (1988). The position for the edge of North American basement in western Texas and eastern Chihuahua is from James and Henry (1993a). The location of the Mojave-Sonora megashear is from Anderson and Silver (1979) and Anderson and Schmidt (1983). Localities discussed in the text: B—Batopilas; D—Davis Mountains; G—Geronimo volcanic field; KH—Kilbourne Hole; LO—La Olivina (Camargo volcanic field); M-D—Mogollon-Datil volcanic field; O—Organ Mountains; SMO—Sierra Madre Occidental; T—Tomóchic.

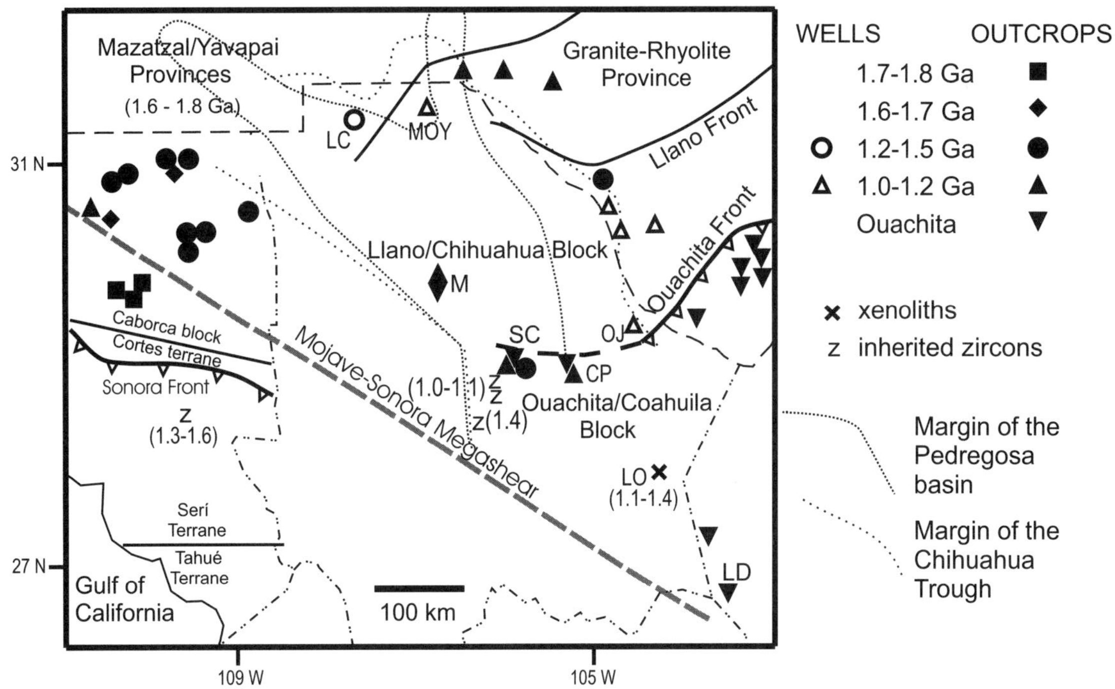

Figure 2. Constraints on the basement framework of northwestern Mexico. These constraints include: (1) the distribution of dated basement, either in outcrop or in wells; (2) localities where lower crustal xenoliths have been studied; (3) areas where U-Pb geochronologic studies of Laramide and mid-Tertiary igneous rocks have documented inherited zircons in the rocks; and (4) the locations of important tectonic boundaries. SC—Sierra de Cuervo; CP—Cerro el Carrizalillo and Placer de Guadalupe; LD—Las Delicias, LC—Los Chinos-1; LO—La Olivina; M—Sierra de la Mojina; MOY—Moyotes; OJ—Ojinaga-1. The margins for the Pedregosa basin and Chihuahua trough have been modified from Haenggi (2001, 2002). See text for discussion and references.

1.7–1.8 Ga belt extends from the southern Inyo Mountain region in California; its NNW trending structural grain apparently truncates the northeast trending belt of 1.6–1.7 Ga Proterozoic rocks in Arizona, New Mexico, and northern Sonora. Silver and Anderson (1974) proposed the Mojave-Sonora megashear to account for this juxtaposition of Proterozoic orogenic belts. Late Precambrian and Paleozoic depositional trends are truncated along the megashear as well, and a left-lateral offset of 700–800 km has been suggested on the basis of stratigraphic similarities between the Inyo Mountains and the Caborca area, on opposite sides of the megashear (Anderson and Silver 1979). The eastern extension of the megashear shown in Figure 2 is from Anderson and Schmidt (1983) who interpreted the Mojave-Sonora megashear as a major left-lateral transform fault associated with the opening of the Gulf of Mexico during Late Jurassic time.

Upper Proterozoic to Paleozoic miogeoclinal strata are exposed in northern Sonora as far south as about latitude 29°N (Stewart et al., 1990). Most of the southern portion of Sonora, on the other hand, is underlain by Ordovician through Permian eugeoclinal strata (Stewart et al., 1990; Poole et al., 1995). These eugeoclinal strata were deformed and thrust northward over the Paleozoic miogeoclinal strata sometime between middle Permian and Late Triassic time (Poole et al., 1991); the frontal thrust of the Sonora orogenic belt is illustrated in Figure 2. The area of northern Sonora south of the Mojave-Sonora megashear comprises the Caborca block, also known as the Serí terrane of Sedlock et al. (1993).

Central and southern Sonora are believed to be underlain by what has variably been called the Cortes terrane (Coney and Campa, 1987), or the Serí and Tahué terranes (Sedlock et al., 1993). Sedlock and others extend Precambrian crystalline basement of the Serí terrane as far south as ~27°N, whereas Coney and Campa implicitly question the presence of Precambrian basement in southern Sonora and place the boundary between the Caborca block and the Cortes terrane, a block of Paleozoic and Mesozoic rocks, at ~29°N. The Tahué terrane of Sedlock et al. (1993) is roughly equivalent to the Guerrero terrane, which is classically believed to have been accreted to North America in Late Cretaceous and to comprise Upper Jurassic to middle Cretaceous oceanic arc-related rocks and associated sedimentary units upon basement of uncertain affinity. A noteworthy feature of the accreted material in southern Sonora and northern Sinaloa is that here it includes Upper Paleozoic eugeoclinal strata in addition to Jurassic to Lower Cretaceous arc rocks (Sedlock et al., 1993). Sedlock et al. (1993) believe that the Tahué terrane was initially accreted to North America during Late Jurassic time.

Exposures of Precambrian basement are rare in Chihuahua, and are limited to two localities: Sierra del Cuervo and possibly Cerro El Carrizalillo, 100 km east of Sierra del Cuervo (Fig. 2). Undated amphibolite facies mafic schists intruded by trondhjemite, tonalite, and granite exposed in Cerro El Carrizalillo have been correlated with Grenvillian rocks in central Texas (Blount, 1993; Mosher, 1998). Metaigneous rocks in Sierra del Cuervo have U-Pb ages between 1080 and 1333 Ma (Blount et al., 1988; Blount, 1993). Blount dated orogenesis in Sierra del Cuervo around 1080 Ma. Finally, a Lower Cretaceous conglomerate at Sierra de la Mojina is dominated by quartzofeldspathic schists and unmetamorphosed rhyolite clasts. Denison et al. (1971) obtained a poorly constrained Late Proterozoic age from the quartzofeldspathic schists (695 ± 10 Ma, Rb-Sr; the isochron is poor, however, and Denison et al. expanded the error to 100 m.y.) and generally Permian ages for the unmetamorphosed rhyolitic clasts (233–287 Ma, K-Ar and Rb-Sr; only one of the samples has an age younger than 245 Ma). While the conglomerate does not qualify as an actual exposure of Proterozoic and Permian rocks, the source area for the clasts is relatively near.

Three wells, all near the U.S. border, penetrate into Precambrian basement in northern Chihuahua (Fig. 2); one well (LC) penetrated a granite with a Rb-Sr age of 1327 ± 242 Ma, while the other two wells (MOY and OJ) penetrated granites with ages (Rb-Sr) between 890 and 977 Ma (Haenggi, 2001). While the uncertainties for these latter two ages are not reported, it is likely that these ages overlap those of >1.0 Ga Grenvillian rocks in western and central Texas. The sparse outcrop and well data suggest that much of northern Chihuahua is underlain by mid-Proterozoic basement, some of which has been affected by Grenvillian orogenesis.

Northern Chihuahua was the site of cratonic shelf sedimentation along a passive margin of the North American craton during the Paleozoic prior to the Mississippian (Haenggi, 2001), as well as of two later basins: the Pennsylvanian-Permian Pedregosa Basin (Haenggi, 2001) and the Jurassic-Cretaceous Chihuahua trough (Haenggi, 2002). The Pedregosa basin (Fig. 2) contains a thick sequence of siliciclastic sediments bounded to the north and east by a carbonate shelf (Haenggi, 2001). The region, however, suffered several phases of deformation during the late Paleozoic (Haenggi, 2001). Haenggi describes three periods of deformation in various localities in northern Chihuahua and adjacent regions between the Mississippian and Pennsylvanian-Permian; these deformations are recognized in numerous localities of the Ancestral Rocky Mountains Province (e.g., Handschy et al., 1987; Handschy and Dyer, 1987). A fourth period of deformation occurred during the Permian (Late Wolfcampian–Leonardian–Guadalupian?). This fourth phase of deformation is best developed in the Placer de Guadalupe, Carrizalillo, and Sierra del Cuervo regions (Fig. 2) and may be related to either Ouachita deformation or the Ancestral Rocky Mountains (e.g., Handschy and Dyer, 1987). Stewart (1988, 1992) and Coney and Campa (1987) have suggested that a Late Proterozoic and Paleozoic continental margin extended across central Chihuahua, connecting the Sonoran and Ouachita continental margins. This interpretation is not compatible with the Mojave-Sonora megashear model which would palinspastically locate the Sonoran margin near Nevada in Paleozoic time.

The configuration of the extension of the Ouachita-Marathon orogenic system into northern Mexico is controversial, mostly due to the lack of appropriate exposures. The orogenic belt has been postulated to turn abruptly southward in northern Mexico (e.g., Handschy et al., 1987), to turn southwestward across Chihuahua (Armin, 1987; Stewart, 1988), or to terminate against a transform margin (Stewart, 1992; see Shurbet and Cebull, 1987, for discussion).

The continental fragment that collided with North America during the Ouachita convergence is known in Mexico as the Coahuila (Coney and Campa, 1987) or Coahuiltecano (Sedlock et al., 1993) block and is best known from exposures in the state of Coahuila. The Coahuila block is believed to be the remnant of a Late Paleozoic continental arc and forearc that formed the leading edge of Gondwana before collision with North America, and comprises upper Paleozoic low-grade metamorphic rocks, arc-derived flysch, and arc rocks cut by postorogenic Triassic plutons. In the Las Delicias area of Coahuila, the arc was active between ca. 331 Ma and 270 Ma (Lopez et al., 1997), while U-Pb geochronologic studies of gneissic and granitoid clasts in a Mississippian conglomerate in the area indicate that crystalline basement with ages ca. 1.85 Ga, 1.2–1.3 Ga, and 0.58 Ga is present in the Coahuila block (Lopez et al., 1997, 1998). McKee et al. (1999) have suggested, however, that the rocks in the Las Delicias area may not have been involved in Ouachita orogenesis, but may have been somewhere else at that time. They were, nonetheless, believed to be in place by the end of the Mesozoic (McKee et al., 1984, 1999).

Xenoliths and Zircon Inheritance

Evidence for basement in southern Chihuahua south of Stewart's proposed continental margin is mostly based upon study of crustal xenoliths (from La Olivina in the Camargo volcanic field; Fig. 2) and from inherited components of zircons identified during geochronologic studies of Laramide and mid-Tertiary volcanic rocks. Lower crustal xenoliths are abundant in the La Olivina cinder cone in the late Tertiary Camargo volcanic field in southeastern Chihuahua and include orthogneisses, paragneisses, and mafic granulites (Cameron et al., 1992). Sensitive high-resolution ion microprobe studies of xenoliths from La Olivina have identified zircons with U-Pb ages of 1, 25–37, ca. 200, 350, 1100 and 1370 Ma (Rudnick and Cameron, 1991). Similarly, Cameron et al. (1992) identified two groups of granulites at La Olivina on the basis of their isotope geochemistry: pre-Cenozoic and mid-Tertiary. The bulk of these xenoliths are interpreted to be pre-Cenozoic and represent samples of Proterozoic basement, sedimentary, and magmatic rocks related to Paleozoic convergence of North America and the Coahuila block, and perhaps postorogenic Triassic plutons. Cameron et al. (1992) associated the lower crustal xenoliths from La Olivina with the Coahuila block, which is believed to be an exotic continental fragment.

U-Pb geochronologic studies of Laramide and mid-Tertiary volcanic rocks from central Sonora and central Chihuahua have identified inherited components in the zircons from these rocks. In central Chihuahua, six of thirteen samples of Upper Cretaceous and Tertiary volcanic and plutonic rocks studied by McDowell and Mauger (1994) had identifiable inherited components. Inheritance ages were poorly constrained, but three samples have upper intercepts of: 1155 ± 20 Ma, 1040 ± 80 Ma, and 1450 ± 200 Ma. These ages are roughly comparable to ca. 1.1–1.3 Ga U-Pb ages for local basement rocks in Sierra del Cuervo (Blount et al., 1988; Blount, 1993), ca. 1.1 Ga and 1.37 Ga ages from lower crustal xenoliths at La Olivina in southeastern Chihuahua (Rudnick and Cameron, 1991), and ca. 1.2 Ga rocks from the Las Delicias area in Coahuila (Lopez et al., 1997, 1998). Crust with Grenvillian ages appears to extend as far west as central Chihuahua.

McDowell et al. (2001) dated six Upper Cretaceous and early Tertiary volcanic rocks in east-central Sonora; four of these samples exhibited significant inheritance. Three of the samples had regressed upper intercept ages of 1329 ± 252 Ma, 1448 ± 10 Ma, and 1575 ± 220 Ma. It is noteworthy that only one of these three ages overlaps (with large uncertainties) with 1.7–1.75 Ga ages reported by Anderson and Silver (1979) for Precambrian rocks in northern Sonora. Anderson and Silver (1977, 1981) do report that porphyritic granites with ages between 1.4 and 1.5 Ga, as well as one granite with an age of 1.1 Ga are exposed in northern Sonora (Anderson and Silver, 1979). The inheritance ages reported by McDowell et al. (2001) are probably best interpreted in terms of sampling these younger, mid-Proterozoic rocks.

LATE CRETACEOUS THROUGH MID-TERTIARY MAGMATISM IN NORTHWESTERN MEXICO

Northwestern Mexico has experienced several episodes of convergent continental margin magmatism since Cretaceous time. In Sonora and Chihuahua, these episodes include Late Cretaceous to early Tertiary batholiths and associated andesites and rhyolites as well as the middle Tertiary ignimbrites of the Sierra Madre Occidental. Exposures of Early Cretaceous to Eocene batholiths associated with Laramide magmatism are common along the Pacific margin of Mexico, including: the Peninsular Ranges batholith of Baja California (Silver and Chappell, 1988); the Sonora batholith and associated volcanic rocks (Damon et al., 1983; McDowell et al., 2001); the Sinaloa batholith (Henry and Fredrikson, 1987); and vast tracts of plutons in southern Mexico in the states of Jalisco, Colima, and Guerrero (Schaaf et al., 1995). Within Sonora, batholithic rocks are primarily calc-alkalic granodiorites and tonalites, although two-mica granites are locally present (Anderson et al., 1980; Roldán-Quintana, 1991). Ages of the batholithic rocks in Sonora range between 91 and 43 Ma (see compilation in McDowell et al., 2001). Upper Cretaceous to Eocene volcanic rocks are common in eastern Sonora and are also found in Chihuahua where not buried by younger volcanic rocks (McDowell and Mauger, 1994; González-León et al., 2000; McDowell et al., 2001). These Upper Cretaceous to Eocene volcanic rocks are predominantly composed of propylitically altered andesites and felsic volcanic rocks and were erupted through broadly the time interval represented by the plutonic rocks in Sonora.

The immense mid-Tertiary volcanic accumulations within the Sierra Madre Occidental and surrounding regions of western Mexico (Fig. 1) constitute a remarkable geologic feature (McDowell and Clabaugh, 1979; McDowell et al., 1990). The thick, dominantly felsic sections represent the surface expression of an intense and short-lived period of magmatism between 46 and 27.5 Ma coinciding with the final stages of subduction of the Farallon plate beneath western North America (McDowell et al., 1990; Wark et al., 1990). Typically, volcanic sections within the Sierra Madre Occidental are more than 1000 m thick and include several ignimbrite units. The base of the section is rarely exposed. In the northern part of the Sierra Madre Occidental, the felsic units are overlain without stratigraphic break by mafic lavas. These basaltic andesites represent the youngest volcanic activity throughout much of the Sierra Madre Occidental, ranging in age from 34 to 18 Ma. Cameron et al. (1989) have referred to these lavas as SCORBA (Southern Cordillera Basaltic Andesite), inferring for them a tectonic setting distinct from the underlying felsic units. In contrast, detailed mapping and petrologic studies at the Tomóchic volcanic complex in the central Sierra Madre Occidental (Fig. 1) have shown that the basaltic andesites are linked to the underlying rhyolites and also have documented a genetic relationship between the rhyolites and other intermediate composition lavas occurring beneath them (Wark, 1991).

The mid-Tertiary volcanic rocks thin markedly both east and west of the Sierra Madre Occidental. To the east in central Chihuahua, basal exposures are common and the onset of ignimbritic volcanism has been well dated at 46 Ma (McDowell and Mauger, 1994). Felsic volcanic activity terminated abruptly in central Chihuahua at 27.5 Ma. To the west of the Sierra Madre Occidental in eastern Sonora, abrupt thinning of the volcanic section marks a distinct western boundary to the volcanic province. Typically, sections are less than 500 m thick, and ages are from 35 to 28 Ma, locally as young as 24 Ma (McDowell, 1993; Bockoven, 1980; Cochemé and Demant, 1991). Flows of basaltic andesite occur just above, or intercalated with, the youngest felsic units.

In south-central Sonora units of the Sierra Madre Occidental are apparently absent. In their place are younger (27–10 Ma) and thinner (<100 m) volcanic sections that were deposited along with coarse clastic sediments in north- to northwest-trending, linear, fault-bounded basins (McDowell et al., 1997). Basin formation and sedimentation were associated with Basin-and-Range faulting. In western Sonora, adjacent to the Gulf of California, Neogene volcanism becomes more prominent. At Sierra Santa Ursula to the north of Guaymas, the volcanic section is ~1000 m thick (Mora-Alvarez, 1992, 1993; Mora-Alvarez and McDowell, 2000). Its main portion is similar in composition to rocks of the Sierra Madre Occidental, although it is younger (23–10 Ma) and pyroclastic units are less prominent. Together with other volcanism of age 24–10 Ma in the Gulf of California region (Sawlan, 1991; Stock, 1989), the Sierra Santa Ursula section represents a magmatic arc associated with subduction of the last fragments of the Farallon plate (Sawlan, 1991; Stock and Lee, 1994).

PROJECT METHODOLOGY

General Approach

Several Precambrian and Phanerozoic crustal provinces trend southward into northern Mexico, including: the 1.8–2.3 Ga Mojave Province, the 1.7–1.8 Ga Yavapai Province, the 1.6–1.7 Ga Mazatzal Province, the mid-Proterozoic Granite-Rhyolite Terrane, the 1.1–1.3 Ga Llano Province, and the late Paleozoic Ouachita belt (Fig. 2). To the south there are Paleozoic and Mesozoic terranes as well, although their distribution is less well known (e.g., see Campa and Coney, 1983; Coney and Campa, 1987; Sedlock et al., 1993). All of these crustal provinces are buried in northern Mexico by thick and extensive deposits of Mesozoic and Tertiary sedimentary rocks as well as by Upper Mesozoic and Tertiary igneous rocks described above. Fortunately, these crustal provinces span a range of ~2 billion years and have, consequently, very diverse radiogenic isotope characteristics. We have found that regional patterns exist in the radiogenic isotopic characteristics of Laramide through mid-Tertiary igneous rocks in northwestern Mexico that relate to the distribution of basement domains in northwestern Mexico.

This study has focused on looking at the radiogenic isotope geochemistry of Upper Mesozoic through mid-Tertiary igneous rocks in Chihuahua, Sonora and adjacent regions of the southwestern United States. We have examined the Sr, Nd, and Pb isotopic systematics of over 90 new samples of mid-Tertiary and Upper Cretaceous igneous rocks from Chihuahua and Sonora. The isotopic data that we have generated, combined with other published isotopic data from this region, as well as from adjacent portions of the southwestern United States, display distinct spatial variations that we use to delineate what we term "isotope provinces." Figure 3 illustrates the locations of samples we have analyzed for this study as well as the locations of the samples with published isotopic data that have been utilized for this study.

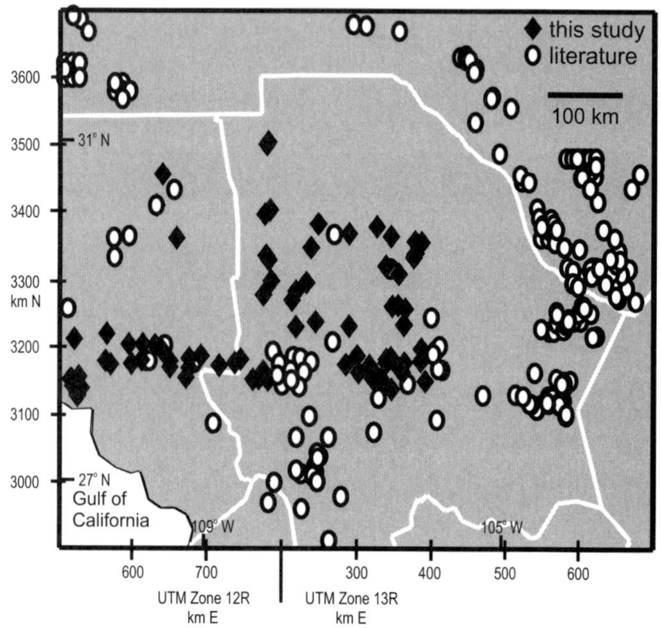

Figure 3. Locations of samples analyzed for this study as well as of samples with published data from the literature.

Our selection of samples was dependent upon our goal of producing a map of isotopic provinces. As a consequence, our sampling was primarily directed at intrusive bodies and small lava flows, as such rocks are not far removed (laterally) from their point of origin. Conversely, we have avoided sampling large ash-flows and related rocks that may have flowed over a large distance unless we specifically knew the location of the caldera or other eruptive center for a unit. We acknowledge that this methodology was particularly discriminatory within the Sierra Madre Occidental itself, where the exposed volcanic sequence is dominated by large ash-flow tuffs and related rocks; however, a lack of detailed mapping in most of this region has precluded pinpointing the sources of most of these rocks. Many of the samples we have analyzed were from suites collected during the progress of several Ph.D. and M.S. mapping studies completed at the University of Texas at Austin; the remainder we collected ourselves during reconnaissance sampling trips in Chihuahua and Sonora.

Previous Isotopic Studies

There is currently a significant amount of published isotopic data from Cretaceous and Tertiary igneous rocks in the southwestern United States, as well as a lesser amount of published data for rocks from northern Mexico. Areas from which there are published analyses include: Laramide and mid-Tertiary plutonic and volcanic rocks in southern Arizona (Farmer and DePaolo, 1984; Lang and Titley, 1998; Anthony and Titley, 1988, Bouse et al., 1999), the mid-Tertiary Mogollon-Datil volcanic field in southern New Mexico (Bikerman et al., 1992; Davis and Hawkesworth, 1993, 1995; Menzies et al., 1983), the Organ Needle pluton in southern New Mexico (Verplanck et al., 1999), mid-Tertiary volcanic rocks from the Uvas volcanic field in southern New Mexico (McMillan and Dickin, 1994), and mid-Tertiary volcanic rocks in western Texas (Cameron et al., 1996; Potter, 1996; James and Henry, 1993a). Published data in northwestern Mexico include mid-Tertiary volcanic rocks at several locations in western Chihuahua (Lanphere et al., 1980; Barreiro et al., 1982; Cameron and Cameron, 1985; Albrecht et al., 1990; McDowell et al., 1999; Albrecht and Goldstein, 2000), a number of locations in eastern Chihuahua (Cameron and Cameron, 1985; Gunderson et al., 1986; Cameron et al., 1989; James and Henry, 1993a; James et al., 1994; Smith et al., 1996), Laramide plutonic rocks in Sonora (Damon et al., 1983; Mead et al., 1988), as well as basaltic andesites throughout northwestern Mexico (Cameron et al., 1989). Late Tertiary basaltic fields for which there are published data include the Geronimo volcanic field in southern Arizona (Menzies et al., 1983; Kempton et al., 1990) and Camargo volcanic field in southeastern Chihuahua (Nimz et al., 1993), as well as basalts from Kilbourne Hole in southern New Mexico (Roden et al., 1988; McMillan et al., 2000). Whereas a few of the published data sets contain data for Sr, Nd, and Pb isotopes, this is not the rule, and many are lacking analyses of one or more of these elements.

Crustal Contamination

We have attempted in this study to indirectly document the distribution of basement provinces in northwestern Mexico through examining the effect of crustal contamination on the isotopic compositions of Upper Cretaceous to mid-Tertiary igneous rocks in the region. The extensive mid-Tertiary Sierra Madre Occidental volcanic province in Chihuahua and Sonora and Laramide igneous rocks in Sonora have concealed much of the basement geology of northwestern Mexico. However, petrographic, geochemical, and isotopic studies have demonstrated that crustal contamination was a factor in the evolution of these magmatic systems, although the amount of contamination has been hotly debated (Cameron and Cameron, 1985; Ruiz et al., 1988; Cameron and Robinson, 1990; Wark, 1991; Smith et al., 1996). This assessment has, for the most part, been based upon modeling of bulk major- and trace-element concentrations of volcanic rocks, and upon comparison of their isotopic compositions with those of crustal xenoliths entrained within much younger basaltic magmas at the La Olivina xenolith locality, within the ca. 2 Ma Camargo volcanic field. To one set of investigators these data suggest that the felsic rocks have evolved from a massive input of mantle-derived magma into the lower crust, followed by assimilation–fractional crystallization processes that typically require ~20% of bulk crustal assimilant (e.g., Cameron and Cameron, 1985; Gunderson et al., 1986; Smith et al., 1996). To others, the data are consistent with the addition of more than 50% lower and middle crustal material during the evolution of these magmas (e.g., Ruiz et al., 1988).

Irrespective of arguments about the amount of crustal contamination, we have clearly demonstrated that contamination has a significant effect upon the isotopic compositions of evolved magmatic rocks in this region. In our study of the Tomóchic volcanic center in the central Sierra Madre Occidental in western Chihuahua, we saw that the evolved lavas are characterized by significantly more enriched Nd and Sr isotopic compositions and slightly more radiogenic Pb isotopic compositions than are the cogenetic basaltic andesites (McDowell et al., 1999). This observation is important because it demonstrates that the isotopic compositions of the magmatic rocks in this region were affected by the nature of the crust they passed through on the way to the surface.

It might appear that resolution of the problem of the amount of crustal contamination has major implications for the premise of our study, namely that the isotopic compositions of these rocks can be used to map out the distribution of basement domains. However, we maintain that the controversy is only of limited relevance and that our approach will be viable in any case. First, the relative amount of a crustal versus mantle contribution of a particular element in a primitive magma that has undergone crustal assimilation is not the same as the bulk proportion of crustal material assimilated, if the concentrations of that element within the primitive magma and the assimilant are not the same. For an element that is concentrated within the crust (e.g., Pb), relatively small amounts of bulk assimilation will result in a modified magma in which the budget of that element is dominated by a crustal contribution. (For example, assume a primitive magma with 1 ppm Pb and a crustal assimilant with 30 ppm Pb; 20% bulk assimilation will result in a magma with 6.8 ppm Pb, 88% of which is derived from the assimilant.) Sr, on the other hand, is a less sensitive indicator of crustal contamination because Sr concentrations in basalts are commonly much higher than in crustal contaminants. Similarly, Nd concentrations can be highly variable in basalt and in the crust. In both cases, however, large differences in Sr or Nd isotopic compositions between a basaltic magma and crustal contaminants will result in significant changes in the isotopic composition of the contaminated magma. Our study at the Tomóchic volcanic center (McDowell et al., 1999) has demonstrated that we can observe crustal effects in the Sr, Nd, and Pb isotopic compositions of these rocks. The difference in the geochemical behavior of these three elements is probably one of the main reasons that the isotope province boundaries are not defined neatly by all of the three isotopic systems.

Second, we will demonstrate that there are systematic geographic variations in radiogenic isotope compositions of mid-Tertiary volcanic rocks within northwestern Mexico and adjacent regions of the southwestern United States that have otherwise similar major-element compositions (cf., Albrecht and Goldstein, 2000). These isotopic differences could reflect regional variations in the amount of crustal contamination experienced by these magmas. However, the overall compositional similarity of these rocks over a large region suggests that the differences more likely reflect differences in the nature of the lithospheric columns beneath them.

Experimental Techniques

We report here the Sr, Nd, and Pb isotopic compositions of ~90 samples of mid-Tertiary and Upper Cretaceous igneous rocks from Chihuahua and Sonora measured at the mass spectrometry facility at The University of Texas at Austin. These include analyses of both whole-rock powders as well as Pb isotope measurements of feldspar separates. A description of the samples analyzed is presented in Table 1, and the isotopic data are presented in Table 2. A detailed discussion of the analytical techniques and measurement protocols are found in Housh and McMahon (2000). The mean of ten analyses of our internal Nd reference standard made during the course of this study is $^{143}Nd/^{144}Nd = 0.512085 \pm 0.000012$ (2σ) ($\varepsilon_{Nd} = 10.79 \pm 0.23$) relative to a value of the La Jolla Nd standard of $\varepsilon_{Nd} = -15.27$. The mean of ten analyses of National Institute of Standards and Technology standard reference material (NIST SRM) 987 made during the course of this study is $^{87}Sr/^{86}Sr = 0.710266 \pm 0.000011$ (2σ). The analyses of 24 Pb isotopic analyses of NIST SRM 981 resulted in a mean uncertainty 0.06%/amu (2σ).

ISOTOPIC VARIATIONS IN NORTHWESTERN MEXICO AND SOUTHWESTERN UNITED STATES

Variations in the radiogenic isotopic compositions of Laramide through mid-Tertiary igneous rocks in northwestern Mexico and the southwestern United States as well as late Tertiary alkaline lavas of the Geronimo and Camargo volcanic fields and Kilbourne Hole are summarized in Figures 4A and 4B. The data summarized in these figures are all from published sources (they do not include our new data) and have been broadly grouped into geographic regions. Late Tertiary mafic alkaline rocks from the Geronimo and Camargo volcanic fields and Kilbourne Hole are characterized by high ε_{Nd} and low $^{87}Sr/^{86}Sr$, as well as Pb isotopic compositions that lie on the Northern Hemisphere Reference Line (NHRL) of Hart (1984). These volcanic centers developed during a period of regional extension and their isotopic compositions are interpreted to reflect those of the asthenospheric mantle underlying this region with little influence from crustal contamination (Menzies et al., 1983; Roden et al., 1988; Nimz, et al., 1993).

In contrast, mid-Tertiary igneous rocks from the Mogollon-Datil volcanic field and Organ Mountains of New Mexico as well as Laramide and mid-Tertiary igneous rocks of southeastern Arizona were emplaced through the one of the oldest crustal blocks in this region, the 1.6–1.7 Ga crust of the Mazatzal Province. These rocks have the lowest values of ε_{Nd} and highest $^{87}Sr/^{86}Sr$ as well as unradiogenic Pb isotopic compositions that are characterized by $^{207}Pb/^{204}Pb$ values that lie well above the NHRL. These radiogenic isotopic characteristics reflect the interaction of mantle derived magmas with a crustal reservoir (characterized by high Rb/Sr and low Sm/Nd and U/Pb) that has long been separated from the mantle.

The next oldest region is the Llano block. This block comprises samples from west Texas that were erupted through 1.1–1.35 Ga crust of the southwestern extension of the Grenville Province. These rocks have moderate values of ε_{Nd} and $^{87}Sr/^{86}Sr$, clustering around bulk earth, as well as very unradiogenic Pb isotopic compositions that overlap the field of samples from New Mexico and Arizona that were erupted through the Mazatzal Province.

Mid-Tertiary rocks from northwestern Mexico likewise have moderate values of ε_{Nd} and $^{87}Sr/^{86}Sr$, clustering around bulk earth and overlapping data from the Llano block. The Pb isotopic compositions of igneous rocks from this region are, however, distinctly more radiogenic than those observed in rocks from Texas, New Mexico, and Arizona. These Pb isotopic compositions also lie the farthest from NHRL. The plotted samples from northwestern Mexico range in composition from basaltic andesite through rhyolite, and the more mafic rocks commonly have slightly higher values of ε_{Nd} and $^{207}Pb/^{204}Pb$, and slightly lower $^{87}Sr/^{86}Sr$ than the felsic rocks. These differences are small, however, and generally lie within the range of isotopic compositions observed in this region (c.f. McDowell et al., 1999).

From the published data summarized in Figures 4A and 4B, we see that there are distinct regional variations in the isotopic compositions of Cenozoic igneous rocks that can, in part, be related to differences in the age and nature of the underlying continental crust. We will next examine, with the addition of our newly obtained data, the geographic distribution of the isotopic compositions of Laramide through mid-Tertiary igneous rocks more closely to determine if coherent, mappable provinces can be delineated.

RESULTS AND DISCUSSION

Isotopic and Geochemical Fractionation Processes Pertinent to Northwestern Mexico

The isotopic character of magmatism in a given locality is dependent on the character and flux rate of the primary mantle-derived magmas, the age and character of the underlying crust, and local petrogenetic processes (e.g., DePaolo and Farmer, 1984; Bouse et al., 1999). There are, unfortunately, few data bearing on the character of primary magmas associated with Laramide and mid-Tertiary arc magmatism in northwestern Mexico. In the only case where this issue has been explicitly studied, Nimz et al. (1993) identified two types of pyroxenites from the suite of xenoliths associated with basanites from the late Tertiary La Olivina cinder cone in the Camargo volcanic field in southeastern Chihuahua. One group of the pyroxenites has isotopic and geochemical compositions consistent with their origin through crystal accumulation in magma bodies associated with the late Tertiary basanites, whereas the isotopic and geochemical compositions of the other group are consistent with crystal accumulation in magma bodies associated with mid-Tertiary magmatism in the region. Nimz and others suggested that the mantle magmas associated with mid-Tertiary magmatism in the region were characterized by $^{87}Sr/^{86}Sr$ = ~0.7043; ε_{Nd} = ~+1; and $^{206}Pb/^{204}Pb$ = ~18.7. There are also a number of rocks in several of the isotopic provinces that have substantially more primitive Nd isotopic compositions (i.e., ε_{Nd} >+1) indicating that the primary magmas were more isotopically primitive than Nimz and others suggest (e.g., Figs. 5, 6A).

Local petrogenetic sources have a strong influence on the isotopic character of magmas in an area. Note, for example, the tendency for the Sr isotopic compositions of samples in southeast Chihuahua (Figs. 5, 6A) to run to very radiogenic compositions ($^{87}Sr/^{86}Sr$ >0.7100) without a concomitant decrease in ε_{Nd}. This feature is limited only to some of the rocks from the Camargo area in southeastern Chihuahua, and is the only place in this region where we suspect that a significant sedimentary component was involved in the evolution of magmas. Smith et al. (1996) suggested that the very radiogenic Sr observed in "upper ignimbrite" unit in this area is a feature resulting from assimilation of a sedimentary component.

McDowell et al. (1999) presented data on volcanic rocks from the Tomóchic volcanic center in western Chihuahua that range in composition from basaltic andesite to high-silica rhyolite. The Sr and Nd isotopic data are well correlated forming an array from $^{87}Sr/^{86}Sr$ = 0.7043–0.7089 (only one value is >0.7070), and ε_{Nd} = +2.3 to –5.2 (only one value <–3.0). The basaltic andesites, furthermore, all have ε_{Nd} >–0.7, whereas the evolved rocks all have ε_{Nd} <+0.5 (and even so, there is only one sample of an evolved rock that overlapped the field of the basaltic andesites). These isotopic variations were interpreted to reflect varying amounts of crustal contamination (McDowell et al., 1999).

Despite the amount of isotopic variability that may be displayed within one volcanic center, the range in isotopic compositions within the whole of northwestern Mexico is greater, and the range of compositions seen within a given isotopic province is, in general, consistent and unique (e.g., Figs. 5 and 6). We interpret this to indicate that although there may be some variability in mantle inputs and flux as well as local petrogenetic processes, the gross isotopic structure of Laramide and mid-Tertiary igneous rocks in this region is determined by the nature of the underlying basement.

Proposed Isotope Provinces in Northwestern Mexico

We have constructed maps that illustrate the initial isotopic compositions of Laramide through mid-Tertiary igneous rocks of northwestern Mexico and contiguous portions of southwestern United States (Figs. 5A, 5B, and 5C). The points represented in these maps do not correspond perfectly, as most of the data sets are incomplete in terms of reporting Sr, Nd, and Pb isotopic data. Five isotope provinces have been identified in this portion of northwestern Mexico based upon the Sr, Nd, and Pb isotopic character of the Laramide and mid-Tertiary igneous rocks found in them. The goal in defining the isotope provinces was to delineate geographic regions with unique isotopic characteristics. How well this has been accomplished may be evaluated in Figure 6A, where initial $^{87}Sr/^{86}Sr$ is plotted versus ε_{Nd} for all of the samples included in Figures 5A and 5B for which there are both Sr and Nd isotopic data, and Figure 6B, where initial $^{206}Pb/^{204}Pb$ is plotted versus $^{207}Pb/^{204}Pb$. The uniqueness of the isotopic characteristics of a given province was in two cases observed in only one isotope system (e.g., Provinces C and D). The remaining provinces have characteristics in multiple isotopic systems that define them as unique from the rest.

Initial Sr and Nd isotopic compositions are in general well correlated throughout the region of northwestern Mexico and southwestern United States, although excursions to high $^{87}Sr/^{86}Sr$ can be seen without a concomitant decrease in ε_{Nd} (e.g., Province D in Fig. 6A). It should be noted that the bulk of these high $^{87}Sr/^{86}Sr$ samples are mid-Tertiary igneous rocks from southeastern Chihuahua; this increase in radiogenic Sr may reflect a peculiarity of the magmatic systems in that area and are probably related to the effect of local contaminants. It can readily be seen upon inspection of the maps in Figures 5A and 5B and the plot in Figure 6A that northern Sonora and southeastern Arizona (Provinces A and B) are generally characterized by high values of initial $^{87}Sr/^{86}Sr$ and very low values of initial ε_{Nd} in contrast to the rest of the region. Although a few rocks with primitive isotopic compositions from these provinces can be seen in Figure 6A, the preponderance of the samples from Provinces A and B have Sr and Nd isotopic compositions unlike that seen elsewhere in northwestern Mexico or west Texas. These two provinces are not,

TABLE 1.

Sample information

Sample	Rock type or unit	Age	Age reference	UTM km N	UTM km E	UTM grid zone	State	Material analyzed
3-25-3	andesite agglomerate	32 Ma	1	3130.1	768.2	12R	Chihuahua	feldspar
C1	lava dome	30 Ma	2	3121.5	762.1	12R	Chihuahua	feldspar
C2	lava dome	30 Ma	2	3120.7	754.5	12R	Chihuahua	feldspar
11-14-1	andesite agglomerate	32 Ma	1	3147.2	738.4	12R	Sonora	feldspar
MA-1	granodiorite	63 Ma	1	3142.4	730.2	12R	Sonora	feldspar
3-21-2	rhyolite lava	23 Ma	1	3138.4	709.6	12R	Sonora	feldspar
SAD-6	Penas Azules formation	68 Ma	3	3192.9	365.3	13R	Chihuahua	feldspar
CM-38	andesite	53 Ma	3	3213.3	367.0	13R	Chihuahua	feldspar
CM-636	diorite	37 Ma	3	3269.1	341.0	13R	Chihuahua	feldspar
SGD 2	Gallego rhyolite	37.9 Ma	4	3285.7	380.6	13R	Chihuahua	feldspar
SGD 5	Mesteno rhyolite	36.5 Ma	4	3300.5	389.3	13R	Chihuahua	feldspar
SGC 4	Mesteno rhyolite	36 Ma	4	3286.7	381.3	13R	Chihuahua	whole rock
SGC 5	Milagro bas. andesite	29 Ma	4	3281.4	378.8	13R	Chihuahua	whole rock
SGD 6	Agate andesite	38–39 Ma	4	3301.7	380.5	13R	Chihuahua	whole rock
SGD 12	Milagro bas. andesite	29.4 Ma	4	3308.7	348.2	13R	Chihuahua	whole rock
CM-44	Cryptic tuff	29.6 Ma	3	3217.9	357.8	13R	Chihuahua	feldspar
MEX-3	Quintas tuff	45.7 Ma	3	3192.0	365.7	13R	Chihuahua	feldspar
CH97-3	Quintas tuff	46 Ma	3	3193.2	366.0	13R	Chihuahua	whole rock
CH97-5	Cryptic tuff	30 Ma	3	3218.7	356.1	13R	Chihuahua	whole rock
CH97-6	basaltic andesite	ca 30 Ma	5	3216.8	349.8	13R	Chihuahua	whole rock
CH97-9	basaltic andesite	ca 30 Ma	5	3259.5	358.5	13R	Chihuahua	whole rock
CH97-11	rhyolite lava	38.3 Ma	5	3311.9	291.7	13R	Chihuahua	whole rock
CH97-11d	rhyolite lava	38.3 Ma	5	3311.9	291.7	13R	Chihuahua	feldspar
CH97-12	basaltic andesite	27.0 Ma	5	3293.6	241.3	13R	Chihuahua	whole rock
CH97-13	basaltic andesite	27.9 Ma	5	3247.8	235.3	13R	Chihuahua	whole rock
CH97-14	dacite dome	27.2 Ma	5	3236.0	221.7	13R	Chihuahua	whole rock
CH97-15	basaltic andesite	ca. 30 Ma	5	3250.3	778.8	12R	Chihuahua	whole rock
CH97-16	rhyolite lava	ca. 30 Ma	5	3277.5	775.2	12R	Chihuahua	whole rock
CH97-17	andesite	ca. 30 Ma	5	3283.6	771.8	12R	Chihuahua	whole rock
CH97-18	basaltic andesite	27.7 Ma	5	3232.2	768.5	12R	Chihuahua	whole rock
CH97-20	basaltic andesite	29.2 Ma	5	3189.3	222.0	13R	Chihuahua	whole rock
CH97-21	basaltic andesite	26.2 Ma	5	3196.6	247.4	13R	Chihuahua	whole rock
M21*	Trincheras basalt	10.1 Ma	6	3108.6	523.4	12R	Sonora	whole rock
M28*	Trincheras basalt	10.3 Ma	6	3123.7	519.6	12R	Sonora	whole rock
M35*	LaEspuela andesite	22.7 Ma	6	3119.9	508.9	12R	Sonora	whole rock
M38*	Ventana felsic lava	12.2 Ma	6	3100.9	517.7	12R	Sonora	wr, feldspar
M48*	Ventana felsic lava	11.5 Ma	6	3111.5	512.2	12R	Sonora	wr, feldspar
M51*	El Popalote granite	14.4 Ma	7	3103.0	516.5	12R	Sonora	wr, feldspar
M53*	Ventana felsic lava	12.3 Ma	6	3108.0	522.1	12R	Sonora	wr, feldspar
M56*	Ventana felsic lava	11.8 Ma	6	3112.3	515.9	12R	Sonora	wr, feldspar
M58*	Mezquite dacite	18.8 Ma	6	3104.7	516.8	12R	Sonora	wr, feldspar
M60*	Mezquite dacite	15.3 Ma	6	3119.8	510.3	12R	Sonora	wr, feldspar
M61*	Trincheras basalt	8.5 Ma	6	3098.6	518.1	12R	Sonora	whole rock
C-16	basaltic andesite	28 Ma	2	3224.7	216.9	13R	Chihuahua	feldspar
D75A	rhyolite	32.0 Ma	8	3129.9	304.6	13R	Chihuahua	feldspar
I521	Buenavista rhyolite	31.4 Ma	3	3142.4	342.2	13R	Chihuahua	feldspar
Per	Perales intermediate lava	42.3 Ma	3	3142.3	366.5	13R	Chihuahua	feldspar
J369	felsic lava	30.6 Ma	3	3128.7	322.1	13R	Chihuahua	feldspar

(*continued*)

TABLE 1. (*continued*)

Sample information								
Sample	Rock type or unit	Age	Age reference	UTM km N	UTM km E	UTM grid zone	State	Material analyzed
W 200	felsic lava	32.6 Ma	3	3121.5	332.2	13R	Chihuahua	feldspar
W 300	interm. lava	35 Ma	3	3121.5	323.7	13R	Chihuahua	feldspar
DZ 31	rhyolite lava	29.8 Ma	3	3108.2	349.1	13R	Chihuahua	feldspar
PLM	trachyandesite flow	31 Ma	3	3135.3	359.5	13R	Chihuahua	feldspar
I 532	andesite flow	30 Ma	3	3146.3	342.5	13R	Chihuahua	feldspar
RND	rhyolite flow	34.7 Ma	9	3118.6	394.6	13R	Chihuahua	feldspar
C10	mafic lava	28 Ma	2	3190.1	291.6	13R	Chihuahua	whole rock
D70D	mafic lava	28 Ma	8	3139.6	287.8	13R	Chihuahua	whole rock
D76A	mafic lava	26 Ma	8	3150.5	302.6	13R	Chihuahua	whole rock
D53D	mafic lava	27 Ma	3	3138.8	320.5	13R	Chihuahua	whole rock
W176	mafic lava	21–31 Ma	10	3121.8	335.5	13R	Chihuahua	whole rock
I133	mafic lava	ca 31 Ma	11	3149.1	348.3	13R	Chihuahua	whole rock
PC-1	mafic lava	32 Ma	3	3130.2	363.6	13R	Chihuahua	whole rock
ECD	mafic lava	30.5 Ma	9	3144.6	387.4	13R	Chihuahua	whole rock
4RSRB	mafic lava	30–33 Ma	9	3160.2	389.2	13R	Chihuahua	whole rock
SO 106	dacite dome	28.7 Ma	12	3145.5	605.6	12R	Sonora	feldspar
SO 35	granodiorite/tonalite	ca. 62 Ma	13	3136.1	643.6	12R	Sonora	feldspar
SO 15	granodiorite	53 Ma	13	3150.9	685.3	12R	Sonora	feldspar
SO 2	granodiorite	65 Ma	13	3166.0	606.9	12R	Sonora	whole rock
SO 7	intrusive	66 Ma	13	3166.3	588.0	12R	Sonora	whole rock
SO 8	granite	55 Ma	13	3179.5	557.8	12R	Sonora	whole rock
SO 26	granodiorite	60 Ma	13	3173.0	513.3	12R	Sonora	whole rock
SO 64	granodiorite	63 Ma	13	3140.5	591.3	12R	Sonora	whole rock
SO 3	granodiorite	67 Ma	13	3164.7	624.1	12R	Sonora	whole rock
SO 5	granodiorite	59 Ma	13	3159.9	631.8	12R	Sonora	whole rock
SO 14	mafic flow	14.4 Ma	12	3148.4	669.4	12R	Sonora	whole rock
SO 25	granodiorite	81 Ma	6	3094.7	487.6	12R	Sonora	whole rock
SO 28	mafic flow	23.5 Ma	12	3173.0	513.3	12R	Sonora	whole rock
SO 40	basalt	17.1 Ma	12	3135.1	670.1	12R	Sonora	whole rock
SO 41	basalt	27.7 Ma	12	3123.0	665.3	12R	Sonora	whole rock
SO 48	basalt	27.5 Ma	12	3144.3	643.7	12R	Sonora	whole rock
SO 60	basalt	8.9 Ma	12	3144.5	556.6	12R	Sonora	whole rock
SO 63	granite	45 Ma	13	3140.1	563.1	12R	Sonora	whole rock
CH98-2d	rhyolite	38.0 Ma	5	3320.6	328.9	13R	Chihuahua	feldspar
CH98-3	basaltic andesite	26.2 Ma	5	3324.6	251.0	13R	Chihuahua	whole rock
CH98-5d	rhyolite	30.9 Ma	5	3342.5	776.8	12R	Chihuahua	feldspar
CH98-5	rhyolite	30.9 Ma	5	3342.5	776.8	12R	Chihuahua	whole rock
CH98-6	basaltic andesite	ca. 30 Ma	5	3337.1	769.8	12R	Chihuahua	whole rock
CH98-7d	rhyolite	26.5 Ma	5	3429.1	772.6	12R	Chihuahua	feldspar
CH98-7	rhyolite	26.5 Ma	5	3429.1	772.6	12R	Chihuahua	whole rock
CH98-8	basaltic andesite	25.8 Ma	5	3432.0	774.6	12R	Chihuahua	whole rock
CH98-10	basaltic andesite	29.4 Ma	5	3390.4	633.3	12R	Sonora	whole rock
CH98-11d	granodiorite	56.9 Ma	5	3348.3	625.5	12R	Sonora	feldspar
CH98-11	granodiorite	56.9 Ma	5	3348.3	625.5	12R	Sonora	whole rock
CH98-17d	granodiorite	63.6 Ma	5	3306.6	652.0	12R	Sonora	feldspar
CH98-17	granodiorite	63.6 Ma	5	3306.6	652.0	12R	Sonora	whole rock

References: 1—Bockoven (1980); 2—Swanson (1977); 3—McDowell and Mauger (1994); 4—Keller et al. (1982); 5—McDowell and Housh (unpublished data); 6—Mora-Alvarez and McDowell (2000); 7—Mora-Klepeis et al. (1997); 8—Duex (1983); 9—Megaw and McDowell (1983); 10—Wark (1983); 11—Ide (1986); 12—McDowell et al. (1997); 13—McDowell and Roldán-Quintana (unpublished data).

TABLE 2.

Measured and age-corrected isotopic ratios

	Measured Ratios						Initial Ratios					
sample	$^{143}Nd/^{144}Nd$	εNd	$^{87}Sr/^{86}Sr$	$^{206}Pb/^{204}Pb$	$^{207}Pb/^{204}Pb$	$^{208}Pb/^{204}Pb$	$^{143}Nd/^{144}Nd$ i	εNd i	$^{87}Sr/^{86}Sr$ i	$^{206}Pb/^{204}Pb$ i	$^{207}Pb/^{204}Pb$ i	$^{208}Pb/^{204}Pb$ i
3-25-3				18.638	15.596	38.607				18.638	15.596	38.607
C1				18.664	15.583	38.558				18.664	15.583	38.558
C2				18.645	15.583	38.543				18.645	15.583	38.543
11-14-1				18.636	15.579	38.506				18.636	15.579	38.506
MA-1				18.608	15.600	38.615				18.608	15.600	38.615
3-21-2				18.728	15.609	38.597				18.728	15.609	38.597
SAD-6				18.587	15.591	38.451				18.587	15.591	38.451
CM-38				18.462	15.546	38.335				18.462	15.546	38.335
CM-636				18.432	15.535	38.426				18.432	15.535	38.426
SGD 2				18.297	15.522	38.369				18.297	15.522	38.369
SGD 5				18.615	15.557	38.570				18.615	15.557	38.570
SGC 4	0.512495	−2.8	0.712171				0.51246	−2.7	0.70848			
SGC 5	0.512579	−1.2	0.704553	18.081	15.511	38.010	0.51255	−0.9	0.70452	18.016	15.508	37.949
SGD 6	0.512518	−2.3	0.706981	18.579	15.653	38.848	0.51248	−2.2	0.70649	18.494	15.649	38.767
SGD 12	0.512531	−2.1	0.704600	17.677	15.501	37.679	0.51251	−1.8	0.70455	17.643	15.500	37.630
CM-44				18.848	15.614	38.728				18.848	15.614	38.728
MEX-3				18.681	15.601	38.560				18.681	15.601	38.560
CH97-3	0.512632	−0.1	0.714163				0.51261	0.1	0.70900			
CH97-5	0.512546	−1.8					0.51252	−1.5				
CH97-6	0.512616	−0.4	0.704624	18.259	15.525	38.209	0.51259	−0.2	0.70458	18.203	15.522	38.149
CH97-9	0.512531	−2.1	0.705109	18.160	15.517	38.060	0.51251	−1.8	0.70504	18.131	15.516	38.029
CH97-11	0.512434	−4.0					0.51240	−3.8				
CH97-11d				18.287	15.615	38.713				18.287	15.615	38.713
CH97-12	0.512513	−2.4	0.706025	18.456	15.550	38.424	0.51249	−2.2	0.70599	18.436	15.549	38.401
CH97-13	0.512577	−1.2	0.705376	18.663	15.627	38.715	0.51255	−1.0	0.70532	18.632	15.626	38.684
CH97-14	0.512513	−2.4	0.706068	18.601	15.564	38.486	0.51249	−2.2	0.70596	18.576	15.563	38.461
CH97-15	0.512658	0.4	0.705425	18.674	15.574	38.426	0.51263	0.6	0.70539	18.641	15.573	38.394
CH97-16	0.512519	−2.3	0.709463	18.659	15.605	38.632	0.51249	−2.0	0.70608	18.596	15.602	38.574
CH97-17	0.512574	−1.2	0.705669	18.593	15.574	38.504	0.51255	−1.0	0.70559	18.573	15.573	38.484
CH97-18	0.512562	−1.5	0.705668	18.545	15.554	38.414	0.51254	−1.2	0.70562	18.515	15.553	38.384
CH97-20	0.512731	1.8	0.704774	18.758	15.591	38.524	0.51270	2.1	0.70475	18.728	15.590	38.489
CH97-21	0.512557	−1.6	0.705977	18.750	15.592	38.542	0.51253	−1.4	0.70594	18.726	15.591	38.518
M21*	0.512767	2.5	0.704587	18.973	15.643	38.718	0.51276	2.6	0.70457	18.951	15.642	38.697
M28*	0.512691	1.0	0.704770	18.989	15.669	38.815	0.51268	1.1	0.70475	18.967	15.668	38.794
M35*	0.512691	1.0	0.703987	18.872	15.633	38.562	0.51267	1.3	0.70395	18.821	15.631	38.514
M38*	0.512663	0.5	0.705468	18.975	15.626	38.696	0.51265	0.6	0.70530	18.949	15.625	38.670
M48*	0.512679	0.8	0.705543	19.026	15.608	38.646	0.51267	0.9	0.70539	19.000	15.606	38.620
M51*	0.512603	−0.7	0.705655	18.972	15.640	38.734	0.51259	−0.4	0.70499	18.928	15.638	38.692
M53*	0.512640	0.0	0.705604	18.988	15.626	38.682	0.51263	0.1	0.70527	18.961	15.625	38.657
M56*	0.512681	0.8	0.705774	19.044	15.621	38.695	0.51267	0.9	0.70545	19.017	15.619	38.670
M58*	0.512633	−0.1	0.705172	18.971	15.647	38.759	0.51261	0.0	0.70492	18.929	15.645	38.719
M60*	0.512576	−1.2	0.705586	18.977	15.618	38.658	0.51256	−1.1	0.70533	18.944	15.616	38.626
M61*	0.512710	1.4	0.704975	18.983	15.655	38.783	0.51270	1.5	0.70485	18.963	15.654	38.764
C-16				18.820	15.621	38.659				18.820	15.621	38.659
D75A				18.843	15.650	38.716				18.843	15.650	38.716
I521				18.810	15.644	38.700				18.810	15.644	38.700
Per				18.728	15.644	38.626				18.728	15.644	38.626

(*continued*)

TABLE 2. (*continued*)

Measured and age-corrected isotopic ratios

sample	Measured Ratios						Initial Ratios					
	$^{143}Nd/^{144}Nd$	εNd	$^{87}Sr/^{86}Sr$	$^{206}Pb/^{204}Pb$	$^{207}Pb/^{204}Pb$	$^{208}Pb/^{204}Pb$	$^{143}Nd/^{144}Nd$ i	εNd i	$^{87}Sr/^{86}Sr$ i	$^{206}Pb/^{204}Pb$ i	$^{207}Pb/^{204}Pb$ i	$^{208}Pb/^{204}Pb$ i
J369				18.769	15.597	38.530				18.769	15.597	38.530
W 200				18.877	15.714	38.912				18.877	15.714	38.912
W 300				18.367	15.571	38.240				18.367	15.571	38.240
DZ 31				18.891	15.683	38.801				18.891	15.683	38.801
PLM				18.764	15.636	38.648				18.764	15.636	38.648
I 532				18.818	15.670	38.772				18.818	15.670	38.772
RND				18.544	15.594	38.402				18.544	15.594	38.402
C10	0.512545	−1.8	0.706335	18.668	15.609	38.601	0.51252	−1.6	0.70630	18.601	15.606	38.538
D70D	0.512541	−1.9	0.705227	18.730	15.605	38.547	0.51251	−1.7	0.70517	18.701	15.604	38.516
D76A	0.512643	0.1	0.705230	18.749	15.597	38.590	0.51262	0.3	0.70520	18.682	15.594	38.527
D53D	0.512676	0.7	0.704678	18.839	15.622	38.610	0.51265	1.0	0.70465	18.772	15.619	38.547
W176			0.704028	18.785	15.596	38.477			0.70393	18.718	15.593	38.415
I133	0.512667	0.6	0.704239	18.565	15.580	38.388	0.51264	0.8	0.70419	18.498	15.577	38.325
PC-1			0.705816	18.868	15.631	38.649			0.70562	18.801	15.628	38.586
ECD	0.512524	−2.2	0.705256	18.316	15.545	38.230	0.51250	−2.0	0.70518	18.249	15.541	38.167
4RSRB	0.512551	−1.7	0.705236	18.618	15.585	38.513	0.51252	−1.4	0.70515	18.577	15.583	38.458
SO 106				18.814	15.631	38.676				18.814	15.631	38.676
SO 35				18.891	15.655	38.783				18.891	15.655	38.783
SO 15				18.872	15.621	38.650				18.872	15.621	38.650
SO 2	0.512315	−6.3	0.707885	19.616	15.693	39.303	0.51226	−5.9	0.70676	19.231	15.675	38.974
SO 7	0.512310	−6.4	0.708040	19.456	15.660	39.150	0.51225	−6.1	0.70673	18.908	15.634	38.739
SO 8	0.512110	−10.3	0.712823	19.481	15.733	39.151	0.51206	−9.8	0.70798	19.339	15.726	38.918
SO 26			0.707435	19.272	15.690	39.007			0.70701	19.109	15.683	38.847
SO 64			0.707357	19.093	15.709	38.994			0.70669	19.007	15.705	38.854
SO 3	0.512217	−8.2	0.707232	19.069	15.678	38.826	0.51216	−7.7	0.70689	18.910	15.670	38.644
SO 5	0.512411	−4.4	0.706299	18.848	15.645	38.728	0.51236	−4.0	0.70617	18.625	15.634	38.655
SO 14	0.512404	−4.6	0.706319	18.927	15.634	38.707	0.51239	−4.5	0.70626	18.905	15.633	38.692
SO 25	0.512434	−4.0	0.706288	19.054	15.673	38.893	0.51236	−3.4	0.70571	18.834	15.663	38.677
SO 28	0.512443	−3.8	0.705707	18.874	15.635	38.645	0.51239	−3.3	0.70559	18.779	15.630	38.567
SO 40	0.512571	−1.3	0.705629	18.928	15.641	38.708	0.51256	−1.2	0.70559	18.848	15.637	38.655
SO 41	0.512548	−1.8	0.705538	18.852	15.629	38.646	0.51252	−1.5	0.70547	18.804	15.627	38.614
SO 48	0.512598	−0.8	0.704636	18.593	15.585	38.464	0.51257	−0.6	0.70461	18.476	15.580	38.406
SO 60	0.512918	5.5	0.703618	18.864	15.609	38.481	0.51291	5.5	0.70362	18.822	15.607	38.467
SO 63	0.512301	−6.6		19.115	15.659	38.794	0.51226	−6.2		19.076	15.658	38.718
CH98-2d				18.179	15.616	38.508				18.179	15.616	38.508
CH98-3	0.512521	−2.3	0.705990	18.613	15.580	38.567	0.51249	−2.1	0.70594	18.585	15.579	38.534
CH98-5d				18.444	15.561	38.518				18.444	15.561	38.518
CH98-5	0.512532	−2.1					0.51251	−1.7				
CH98-6	0.512388	−4.9	0.705598	17.761	15.493	37.954	0.51236	−4.6	0.70550	17.727	15.491	37.896
CH98-7d				18.856	15.592	38.712				18.856	15.592	38.712
CH98-7	0.512372	−5.2					0.51234	−5.1				
CH98-8	0.512355	−5.5	0.707635	17.796	15.498	38.180	0.51233	−5.3	0.70751	17.762	15.497	38.134
CH98-10			0.705512	18.521	15.553	38.397			0.70549	18.484	15.551	38.346
CH98-11d				18.601	15.585	38.546				18.601	15.585	38.546
CH98-11			0.707377						0.70697			
CH98-17d				18.820	15.648	38.702				18.820	15.648	38.702
CH98-17	0.512406	−4.5	0.707557				0.51235	−4.1	0.70701			

* Nd and Sr analyses provided by Gabriela Mora-Klepeis

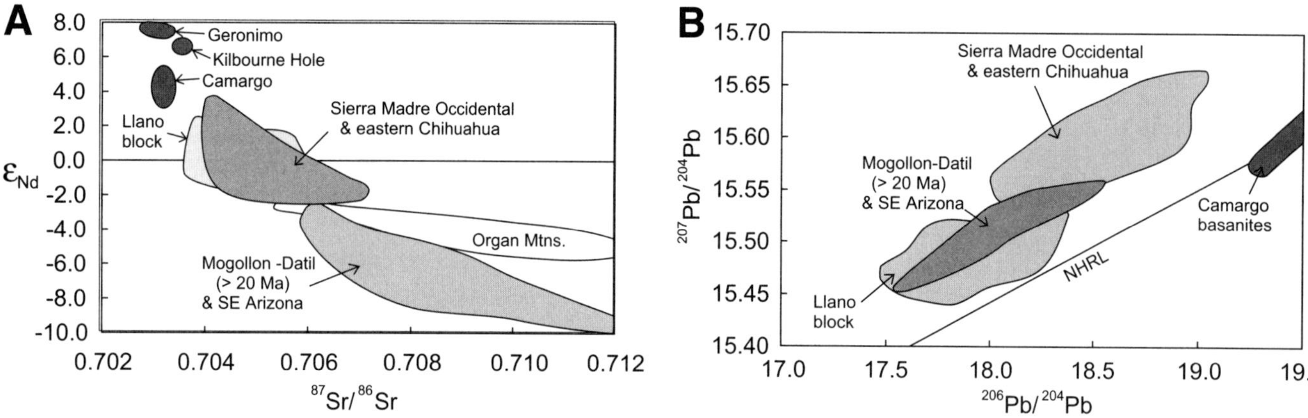

Figure 4. (A) Plots of initial $^{87}Sr/^{86}Sr$ versus ε_{Nd}. (B) Initial $^{206}Pb/^{204}Pb$ versus $^{207}Pb/^{204}Pb$ summarizing the variations observed in published data for Laramide, mid-Tertiary, and late Tertiary igneous rocks from northwestern Mexico and the southwestern United States. Sources of data: Farmer and DePaolo (1984), Lang and Titley (1998, used only data for districts in the "South Pb Province" contiguous with Sonora), Anthony and Titley (1988), Bouse et al. (1999, used only data for districts in the "South Pb Province" contiguous with Sonora), Bikerman et al. (1992), Davis and Hawkesworth (1993, 1995), Menzies et al. (1983), Verplanck et al. (1999), McMillan and Dickin (1994), Cameron et al. (1996), Potter (1996), James and Henry (1993a), McDowell et al. (1999), Lanphere et al. (1980), Barreiro et al. (1982), Cameron and Cameron (1985), Albrecht et al. (1990), Albrecht and Goldstein (2000), Gunderson et al. (1986), Cameron et al. (1989), James et al. (1994), Smith et al. (1996), Damon et al. (1983), Mead et al. (1988), Menzies et al. (1983), Nimz et al. (1993), and Roden et al. (1988). NHRL is the Northern Hemisphere Reference Line from Hart (1984).

however, particularly distinctive from each other in terms of their Sr and Nd isotopic characteristics, although the highest values of initial $^{87}Sr/^{86}Sr$ are observed in Province A.

The Pb isotopic compositions of Laramide and mid-Tertiary igneous rocks in Provinces A and B are different from rocks from the rest of the region as well (Fig. 6B). The most radiogenic Pb isotopic compositions observed in rocks in northwestern Mexico and contiguous areas of the southwestern United States are located in Province B, whereas the least radiogenic Pb isotopic compositions observed in rocks in this area are located in Province A. Pb isotopic data delineate between samples from Provinces A and B (Fig. 6B) in a way not possible with the Sr and Nd isotopic data where there is much compositional overlap. Due to sparse sample coverage in this region, the boundary between Provinces A and B is poorly constrained and could, but for one data point, coincide with the location of the Sonora front. More work clearly needs to be done in this area.

Provinces A and B are bounded to the south and east by regions characterized by generally lower values of $^{87}Sr/^{86}Sr$ and higher ε_{Nd} (Provinces C, D, and E). Within this broad region, Pb isotopic data are the most useful in delineating distinct geographic provinces with unique characteristics (Fig. 6B). Here it can be seen that, in general, Pb isotopic compositions become more radiogenic southward from Province C to D to E; Kuentz et al. (1992) have also made this observation. Sr and Nd data are less unique in this area, although it can be seen that (1) the Sr and Nd isotopic data from Province E are far less variable and, in general, more primitive than the data from Provinces C and D; and (2) that Sr isotopic compositions for samples from Province D are far more likely to extend to high $^{87}Sr/^{86}Sr$ without a concomitant decrease in ε_{Nd} than from any other area (Fig. 6A). Kuentz et al. (1992) argued that Sr isotopic data of mafic to intermediate mid-Cenozoic volcanic rocks defined a major lithospheric boundary "almost precisely at the Rio Grande" between Chihuahua and west Texas, although it should be mentioned that their study was different from ours to the extent that they were looking at mafic samples. While we do not have their data or sample locations, an examination of this region combining both data sets would be useful to resolve the nature of the province boundaries in this area.

James and Henry (1993a, 1993b) and James et al. (1994) delineated a Pb isotopic boundary in west Texas and eastern Chihuahua based upon data from mid-Tertiary volcanic rocks and ore deposits. Their Pb boundary has been adopted in this study as the eastern two-thirds of the boundary between Provinces C and D (Fig. 5C). In the Trans-Pecos area, where the bulk of their samples were located, James and Henry (1993a) found a SW-NE–trending boundary zone separating regions characterized by unradiogenic Pb isotopic compositions to the northwest and more radiogenic Pb isotopic compositions to the southeast (Fig. 5C). Within an intermediate zone ~50 km wide mid-Tertiary igneous rocks have Pb isotopic compositions that overlap those of rocks from both the northwest and southeast regions. James and Henry interpreted this feature as a dipping lithospheric boundary separating Precambrian crust of North America to the northwest from crust accreted during the Ouachita orogeny at the end of the Paleozoic to the southeast. These lithospheric blocks are interpreted to overlap in the boundary zone.

James and Henry extended this boundary southwestward into central Chihuahua in studies of ore deposits (James and Henry, 1993b) and mid-Tertiary volcanic rocks (James et al.,

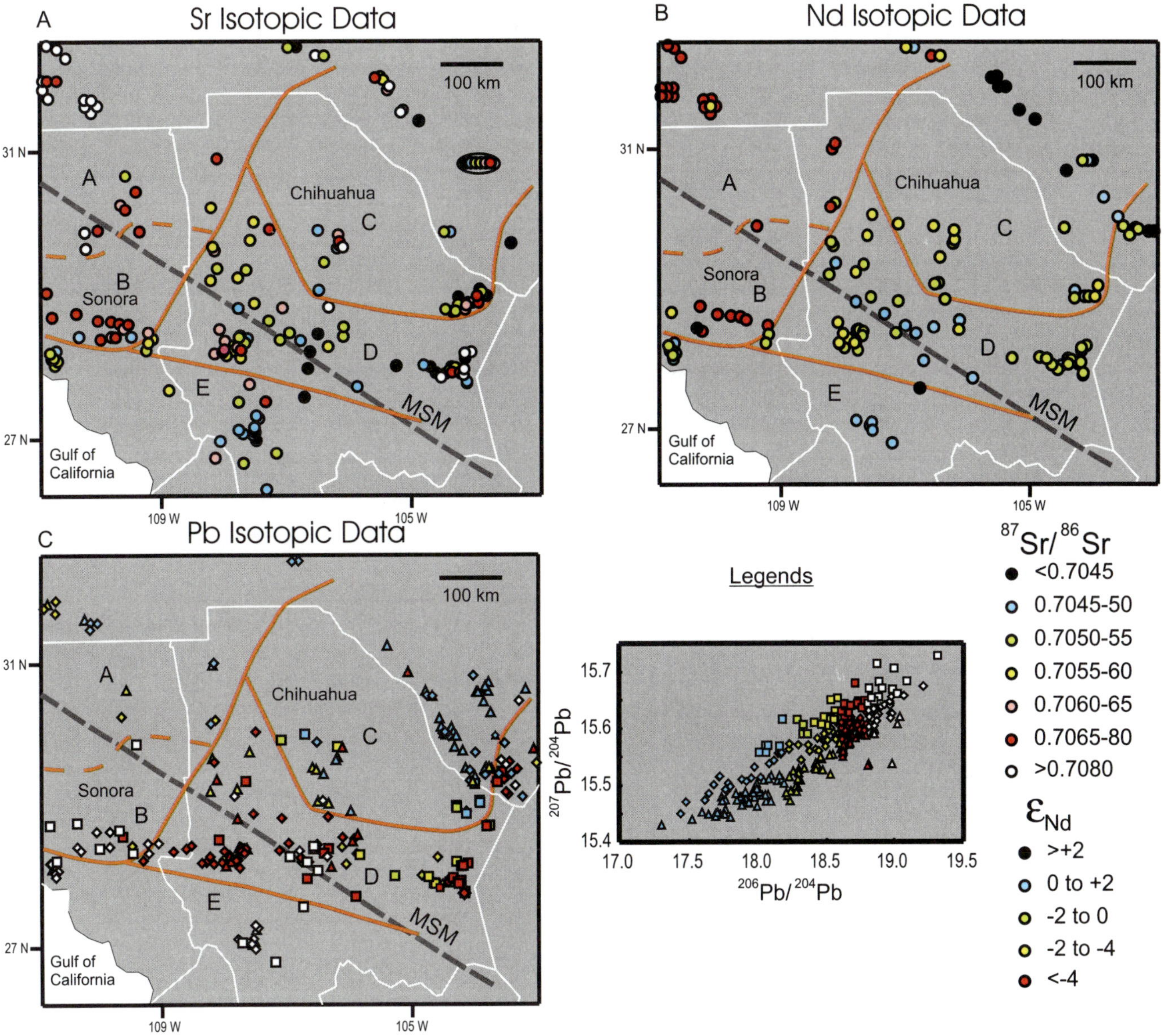

Figure 5. Maps illustrating the isotopic compositions of Laramide through mid-Tertiary igneous rocks of northwestern Mexico and contiguous portions of the southwestern United States. (A) Initial $^{87}Sr/^{86}Sr$. (B) Initial ε_{Nd}. (C) Initial Pb isotopic composition ($^{206}Pb/^{204}Pb$ and $^{207}Pb/^{204}Pb$). The samples have been binned according to their isotopic composition as represented in the legends in the lower right; note that the color of the Pb points correlates to $^{206}Pb/^{204}Pb$, whereas the shape of the symbol correlates to both $^{206}Pb/^{204}Pb$ and $^{207}Pb/^{204}Pb$ (i.e., samples with low $^{207}Pb/^{204}Pb$ at a given $^{206}Pb/^{204}Pb$ are plotted as triangles, samples with high $^{207}Pb/^{204}Pb$ at a given $^{206}Pb/^{204}Pb$ are plotted as squares, and samples with intermediate $^{207}Pb/^{204}Pb$ at a given $^{206}Pb/^{204}Pb$ are plotted as diamonds). The data plotted in these maps are both new data from this study as well as the data represented in Figure 3, less those from the late Tertiary Geronimo and Camargo volcanic fields and Kilbourne Hole (see Figure 4 for sources). The data enclosed in an oval in southwest Texas are data from the Davis Mountains, which are illustrated schematically for convenience (Cameron et al., 1996). Also illustrated in these maps are the boundaries between isotope provinces as determined in this study. MSM—Mojave-Sonora megashear.

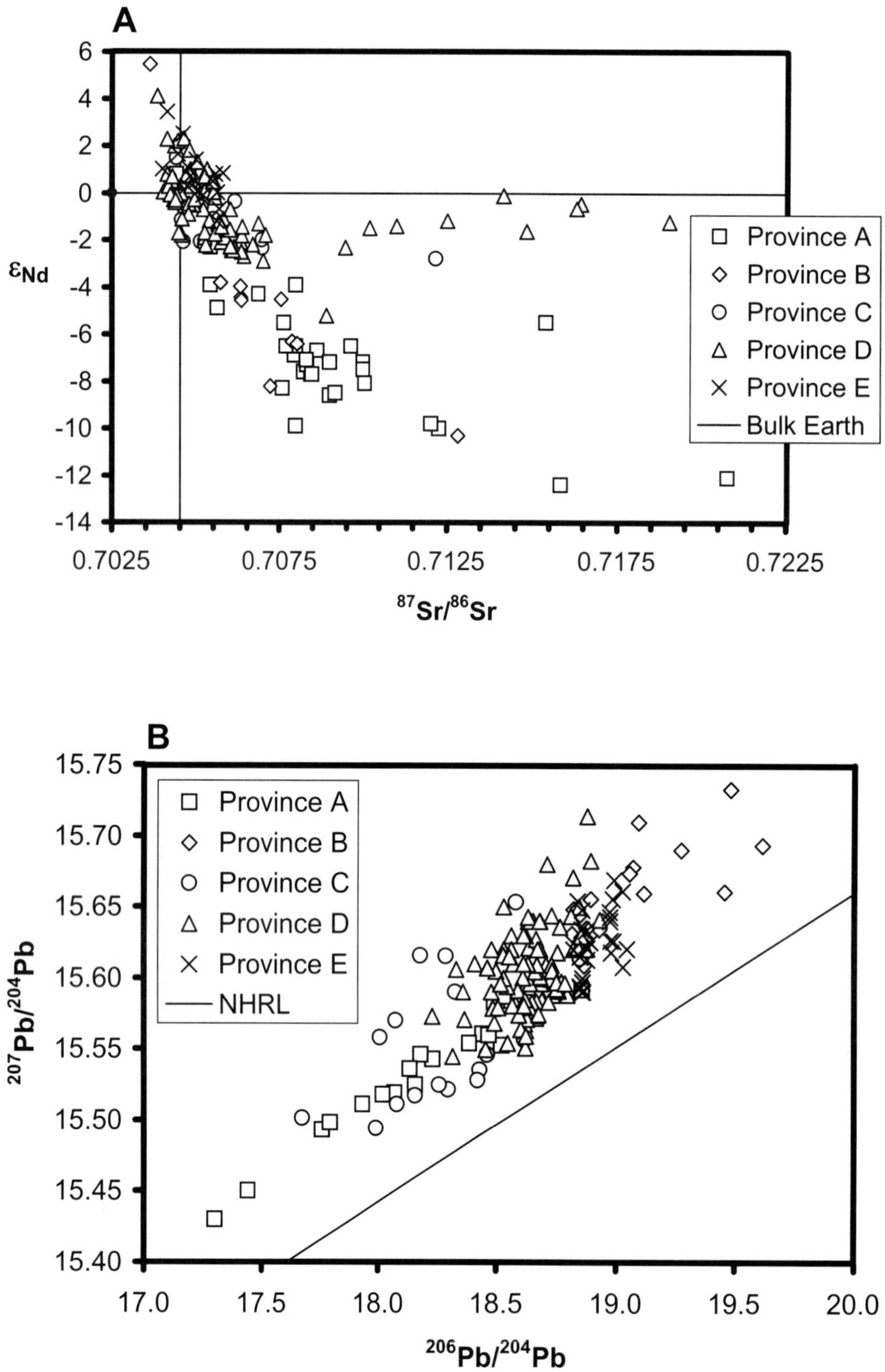

Figure 6. (A) Plot of initial $^{87}Sr/^{86}Sr$ vs. ε_{Nd} for the samples included in Figures 5A and B for which there were both Sr and Nd isotopic data. (B) Plot of initial $^{206}Pb/^{204}Pb$ vs. $^{207}Pb/^{204}Pb$ for samples plotted in Figure 5C. The samples in both of these plots are coded according to the province in which they are located. In both A and B data for rocks in Texas have been omitted for clarity. NHRL is the Northern Hemisphere Reference Line from Hart (1984).

1994). Inspection of Figure 5C, however, indicates that it cannot continue west out of central Chihuahua, but rather heads to the northwest where it is truncated against Province A in northwestern Chihuahua.

South of Province D is a region characterized by radiogenic Pb isotopic compositions with little variability that we term Province E. As noted above, Province E is also characterized by Sr and Nd isotopic compositions that cluster broadly around bulk earth, and while they do overlap with Provinces C and D, the low values of ε_{Nd} and high $^{87}Sr/^{86}Sr$ found in rocks from Provinces C and D are not found there. There is a strong isotopic contrast, however, between Provinces E and B in central Sonora where the low values of ε_{Nd} and high $^{87}Sr/^{86}Sr$ seen in Province B are abruptly terminated against Province E. Damon et al. (1983), Mead et al. (1988), and Valencia-Moreno et al. (2001) have previously noted the strong isotopic contrast between Provinces B and E in southern Sonora and have interpreted the boundary as the southern limit of the North American craton in Sonora.

On the basis of an integrated evaluation of Sr, Nd and Pb isotopic data from Laramide and mid-Tertiary igneous rocks in northwestern Mexico, we have defined five distinct isotopic domains. We feel that the isotope provinces in northwestern Mexico, as drawn, do a reasonably good job in delineating geographically contained regions with unique isotopic characteristics, as is demonstrated in Figures 6A and 6B. On the basis of the good discrimination that we see in the Sr, Nd, or Pb isotopic data, we feel reasonably confident in the locations of our isotope province boundaries.

Comparison of the Proposed Isotope Provinces to Basement Geology

The isotope provinces of northwestern Mexico identified in this study are compared to our current knowledge of basement in the region in Figure 7. Province A in northern Sonora contains numerous outcrops of early Proterozoic rocks, and the Laramide and Tertiary igneous rocks in this province are characterized by radiogenic Sr and unradiogenic Nd isotopic compositions. Similar Sr and Nd isotopic compositions are also observed in Laramide and Tertiary igneous rocks from Province B. Although there are no exposures of early Proterozoic rocks in Province B, miogeoclinal Paleozoic strata underlie the northern portion of the province, with the Sonora front roughly bisecting the province. We interpret Provinces A and B to reflect the distribution of thick sections of early Proterozoic crust in northwestern Mexico. It is not clear why Province A is characterized by unradiogenic Pb isotopic compositions in contrast to Province B, but this feature clearly indicates that the crust of Province A is characterized by lower time-integrated U/Pb ratios than Province B.

The Sr and Nd isotopic contrast between Provinces A and B and the remaining isotope provinces in northwestern Mexico is large and the transition appears to be relatively abrupt (Figs. 5A and 5A). We interpret this to indicate that thick sections of early Proterozoic crust are not present to the south and east of Provinces A and B. This transition is nearly orthogonal to a prominent feature of northwestern Mexico—the southeast-trending Mojave-Sonora megashear. Despite the fact that the boundary between Provinces A and B is not constrained by numerous samples, the presence of a sample with radiogenic Pb isotopic compositions in the northeastern portion of Province B north of the Mojave-Sonora megashear suggests that the megashear does not form the boundary between provinces A and B. Even though up to 800 km of left-lateral movement has been postulated for the Mojave-Sonora megashear (Silver and Anderson, 1974; Anderson and Silver, 1979), there is no indication of that magnitude of offset of the boundary between Province B and Province D to the east. In fact, Province B barely reaches the Chihuahua-Sonora

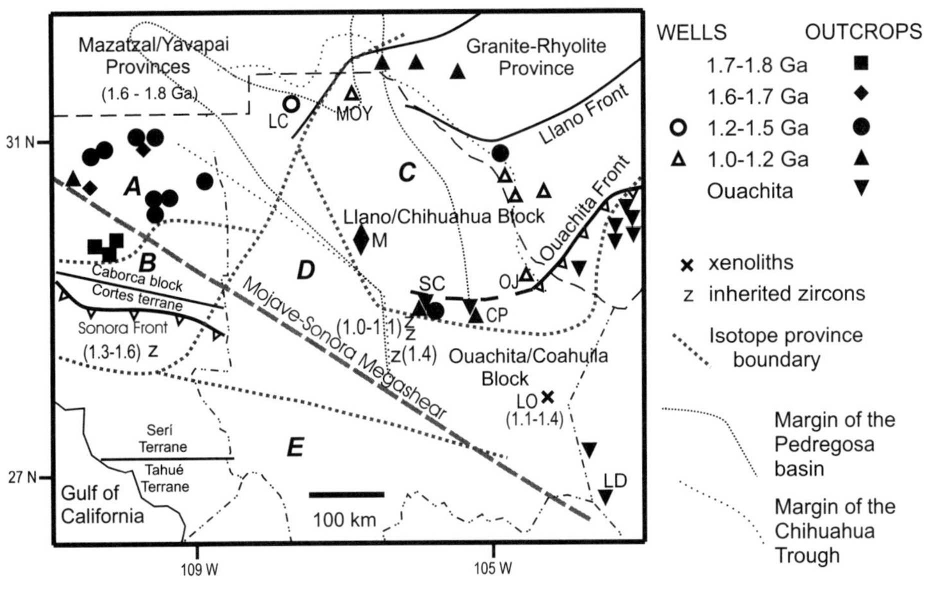

Figure 7. This map compares the isotope province boundaries identified in this study with the constraints on the basement geology of northwestern Mexico from Figure 2 (see Figure 2 and discussion for details). SC—Sierra de Cuervo; CP—Cerro el Carrizalillo and Placer de Guadalupe; LD—Las Delicias; LC—Los Chinos-1; LO—La Olivina; M—Sierra de la Mojina; MOY—Moyotes; OJ—Ojinaga-1. The margins for the Pedregosa basin and Chihuahua trough have been modified from Haenggi (2001, 2002).

border south of the Mojave-Sonora megashear. This observation is difficult to reconcile with the postulated displacement along that tectonic feature.

The area termed the Llano/Chihuahua Block on Figure 7 is best known in west Texas where mid-Proterozoic crust was reworked during Grenvillian orogenic events. South of the Llano/Chihuahua Block is the Ouachita/Coahuila Block that was accreted to North America at the end of the Paleozoic. Geochronologic data discussed above indicate that the Ouachita/Coahuila Block comprises early and mid-Proterozoic rocks as well as Grenvillian components that are somewhat similar to what is observed in the Llano/Chihuahua Block. There is little age difference between these two crustal blocks, and isotopic differences are expected to be small. Mid-Tertiary volcanic rocks from both areas have Sr and Nd isotopic compositions surrounding bulk earth, consistent with the lack of a significant age difference between the crusts of these two blocks. In west Texas and northeastern Chihuahua, however, mid-Tertiary volcanic rocks in the region of the Llano/Chihuahua Block are characterized by relatively unradiogenic Pb isotopic compositions and comprise isotope Province C. In the area of the Ouachita/Coahuila Block mid-Tertiary volcanic rocks have more radiogenic Pb isotopic compositions than those from Province C. James and Henry (1993a, 1993b) and James et al. (1994) were able to map a Pb isotopic boundary between relatively unradiogenic Province C rocks and more radiogenic Province D rocks from west Texas through central Chihuahua. This boundary was interpreted as the extent of North American cratonic crust of the Llano/Chihuahua Block, which has been overthrust by accreted material along the eastern portions of the Province C-D boundary during the Ouachita orogenic event.

Based upon Pb isotopic data, the Province C-D boundary takes a sharp bend to the northwest in central Chihuahua and continues to the point at which these provinces are abutted by Province A. The Province C-D boundary in northwestern Chihuahua approximately follows the southwest margin of the late Paleozoic Pedregosa basin and the Jurassic-Cretaceous Chihuahua trough. Haenggi (2001) speculated that the southwestern and western side of the Pedregosa basin in northwestern Chihuahua was bounded by an accreted continental terrane that was unroofing during the development of the basin in the Permian. This interpretation was based upon: (1) studies of sedimentary rocks in Sierra del Cuervo by Handschy and Dyer (1987) that indicated a basement-cored block was unroofing in west central Chihuahua during the Permian; (2) gravity data; (3) interpretation of space imagery; and (4) an interpretation of the provenance of late Proterozoic quartzofeldspathic and Permian volcanic clasts in Sierra de la Mojina (Denison et al., 1971).

The observation that the Province C-D boundary roughly parallels the edge of a major basin raises the possibility that the isotopic compositions of igneous rocks within Province C are not related to the nature of the basement at all, but rather reflect upper crustal contamination within the thick clastic and carbonate sequences of the Pedregosa basin and the Chihuahua trough. Two points are important to note, however. The first is that the Province C-D boundary actually cuts across the Chihuahua trough in southeastern Chihuahua, and the second is that not all of the Province C rocks occur within the area of the Pedregosa basin and Chihuahua trough. Almost all of the Province C samples from west Texas lie outside of these basins, and yet they have isotopic compositions similar to the Province C samples from Chihuahua that do lie within these basins. It is especially relevant that samples with the characteristic unradiogenic Pb isotopic compositions that are the hallmark of Province C are present throughout the Province, not merely within one portion of it (Fig. 5C). These two observations weigh heavily against the argument that the existence of Province C merely reflects upper crustal interaction within the thick sedimentary sequences of the Pedregosa basin and Chihuahua trough.

Province E is interpreted to reflect what has variably been termed the Cortes terrane (Coney and Campa, 1987) or the southern Serí and Tahué terranes (Sedlock et al., 1993), although its areal extent does not correspond perfectly with the terrane maps published by either group. The relatively primitive Sr and Nd isotopic compositions of Laramide and mid-Tertiary igneous rocks from this province (Fig. 6A) probably reflect the younger and more juvenile nature of the accreted crust comprising this terrane. This is particularly evident in contrast to rocks from Province B, which we believe to be underlain by thick early Proterozoic crust. Abundant Upper Cretaceous and Tertiary igneous rocks and significant isotopic contrasts should allow tracing of this important boundary further to the south and east in Mexico.

CONCLUSIONS

An analysis of published as well as new radiogenic isotopic data for Laramide and mid-Tertiary igneous rocks in northwestern Mexico and the southwestern United States documents the existence of five isotope provinces. The major factor controlling the isotopic variability in the region is the age and nature of the underlying crust, although we do recognize lesser variations in the isotope data reflecting differences in mantle input and local petrogenetic processes.

The two western isotope provinces (Provinces A and B) have very similar Sr and Nd isotopic geochemistry characterized by radiogenic Sr and unradiogenic Nd isotopic compositions. The Pb isotopic compositions of rocks from these two provinces are, however, very dissimilar. We interpret these provinces to reflect the distribution of thick, early Proterozoic crust in northern and central Sonora as well as the northwestern corner of Chihuahua. The eastern boundaries of these provinces do not show the ca. 700–800 km of left-lateral offset proposed along the Mojave-Sonora megashear.

The two eastern isotope provinces also have similar Sr and Nd isotope geochemistry characterized by Sr and Nd isotopic compositions surrounding bulk earth. On the basis of the Pb isotope geochemistry, however, it is possible to map the boundary between mid-Proterozoic crust of the North American craton

(Llano/Chihuahua block) and accreted crust of the Ouachita/Coahuila block. The deflection of this boundary to the northwest in central Chihuahua is, however, not well understood but may indeed be real.

Finally, more primitive Sr and Nd isotopic compositions as well as radiogenic Pb isotopic compositions are observed in Laramide and mid-Tertiary igneous rocks across the southern portion of the study area. We interpret this isotope province to reflect the distribution of younger, more juvenile accreted crust of the Guerrero/Cortes/Tahué terrane.

ACKNOWLEDGMENTS

This study has been supported by National Science Foundation grant EAR-9705773 and the Geology Foundation and Jackson School of Geosciences at the University of Texas at Austin. We thank Jaime Roldán Quintana, Richard Mauger, Kirt Kempter, and Fernando de la Fuente for assistance and consultation in the field. We gratefully acknowledge Gabriela Mora-Klepeis for the use of her Sr and Nd isotopic data on volcanic rocks from Sierra Santa Ursula prior to publication. We are indebted to Timothy Lawton, Nancy J. McMillan, and Jonathan Nourse for their helpful reviews.

REFERENCES CITED

Albrecht, A., Ward, D.W., Bufler, R., Dobbelstein, A., and Brookins, D.G., 1990, Application of the Rb-Sr dating technique to Tertiary siliceous igneous rocks, northern Sierra Madre Occidental, Chihuahua, Mexico: Isochron/West, v. 55, p. 6–9.

Albrecht, A., and Goldstein, S.L., 2000, Effects of basement composition and age on silicic magmas across an accreted terrane-Precambrian crust boundary, Sierra Madre Occidental, Mexico: Journal of South American Earth Sciences, v. 13, p. 255–273, doi: 10.1016/S0895-9811(00)00014-6.

Anderson, T.H., and Schmidt, V.A., 1983, The evolution of Middle America and the Gulf of Mexico–Caribbean Sea region during Mesozoic time: Geological Society of America Bulletin, v. 94, p. 941–966, doi: 10.1130/0016-7606(1983)94<941:TEOMAA>2.0.CO;2.

Anderson, T.H., and Silver, L.T., 1977, U-Pb isotope ages of granitic plutons near Cananea, Sonora: Economic Geology and the Bulletin of the Society of Economic Geologists, v. 72, p. 827–836.

Anderson, T.H., and Silver, L.T., 1979, The role of the Mojave-Sonora Megashear in the tectonic evolution of northern Sonora, *in* Anderson, T.H., and Roldán-Quintana, J., eds., Geology of Northern Sonora, Guidebook, Field Trip #27: Geological Society of America, p. 59–68.

Anderson, T.H., and Silver, L.T., 1981, An overview of Precambrian rocks in Sonora: Universidad Nacional Autónoma de Mexico, Instituto de Geología, Revista, v. 5, p. 131–139.

Anderson, T.H., Silver, L.T., and Salas, G.A., 1980, Distribution and U-Pb isotope ages of some lineated plutons, northwestern Mexico, *in* Crittenden, M.D., Jr., Coney, P.J., and Davis, G.H., eds., Cordilleran metamorphic core complexes: Geological Society of America, p. 269–283.

Anthony, E.Y., and Titley, S.R., 1988, Progressive mixing of isotopic reservoirs during magma genesis at the Sierrita porphyry copper deposit, Arizona; inverse solutions: Geochimica et Cosmochimica Acta, v. 52, p. 2235–2249, doi: 10.1016/0016-7037(88)90126-3.

Armin, R.A., 1987, Sedimentology and tectonic significance of Wolfcampian (Lower Permian) conglomerates in the Pedregosa basin: Southeastern Arizona, southwestern New Mexico, and northern Mexico: Geological Society of America Bulletin, v. 99, p. 42–65, doi: 10.1130/0016-7606(1987)99<42:SATSOW>2.0.CO;2.

Barreiro, B., Cameron, K.L., Cameron, M., and Grunder, A., 1982, A lead isotope study of silicic and intermediate volcanic rocks from south-central Chile and the Sierra Madre Occidental, Mexico: Carnegie Institution of Washington Yearbook, v. 81, p. 494–498.

Bennett, V.C., and DePaolo, D.J., 1987, Proterozoic crustal history of the western United States as determined by neodymium isotopic mapping: Geological Society of America Bulletin, v. 99, p. 674–685, doi: 10.1130/0016-7606(1987)99<674:PCHOTW>2.0.CO;2.

Bikerman, M., Bell, K., and Card, J.W., 1992, Strontium and neodymium isotopic study of the western Mogollon-Datil volcanic region, New Mexico, USA: Contributions to Mineralogy and Petrology, v. 109, p. 459–470, doi: 10.1007/BF00306549.

Blount, J.G., Walker, N.W., and Carlson, W.D., 1988, Geochemistry and U-Pb zircon ages of mid-Proterozoic metaigneous rocks from Chihuahua, Mexico: Geological Society of America Abstracts with Programs, v. 20, p. A-205.

Blount, J.G., 1993, The geochemistry, petrogenesis, and geochronology of the Precambrian meta-igneous rocks of Sierra del Cuervo and Cerro El Carrizalillo, Chihuahua, Mexico [Ph.D. dissertation]: Austin, University of Texas at Austin, 242 p.

Bockoven, N.T., 1980, Reconnaissance geology of the Yécora-Ocampo area, Sonora and Chihuahua, Mexico [Ph.D. dissertation]: Austin, University of Texas at Austin, 241 p.

Bouse, R.M., Ruiz, J., Titley, S.R., Tosdal, R.M., and Wooden, J.L., 1999, Lead isotope compositions of Late Cretaceous and early Tertiary igneous rocks and sulfide minerals in Arizona; implications for the sources of plutons and metals in porphyry copper deposits: Economic Geology and the Bulletin of the Society of Economic Geologists, v. 94, p. 211–244.

Cameron, K.L., and Cameron, M., 1985, Rare earth element, $^{87}Sr/^{86}Sr$, and $^{143}Nd/^{144}Nd$ compositions of Cenozoic orogenic dacites from Baja California, northwestern Mexico, and adjacent west Texas: evidence for predominance of a subcrustal component: Contributions to Mineralogy and Petrology, v. 91, p. 1–11, doi: 10.1007/BF00429422.

Cameron, K.L., Nimz, G.J., Kuentz, D., Niemeyer, S., and Gunn, S., 1989, Southern Cordilleran basaltic andesite suite, southern Chihuahua, Mexico: A link between Tertiary continental arc and flood basalt magmatism in North America: Journal of Geophysical Research, v. 94, p. 7817–7840.

Cameron, K.L., Parker, D.F., and Sampson, D.E., 1996, Testing crustal melting models for the origin of flood rhyolites; a Nd-Pb-Sr isotopic study of the Tertiary Davis Mountains volcanic field, West Texas: Journal of Geophysical Research, v. 101, p. 20,407–20,422, doi: 10.1029/96JB00935.

Cameron, K.L., and Robinson, J.V., 1990, Comments on "Nd-Sr isotopic compositions of lower crustal xenoliths: evidence for the origin of mid-Tertiary felsic volcanics in Mexico" by J. Ruiz, P.J. Patchett, and R.J. Arculus: Contributions to Mineralogy and Petrology, v. 104, p. 609–614, doi: 10.1007/BF00306668.

Cameron, K.L., Robinson, J.V., Niemeyer, S., Nimz, G.J., Kuentz, D.C., Harmon, R.S., Bohlen, S.R., and Collerson, K.D., 1992, Contrasting styles of pre-Cenozoic and mid-Tertiary crustal evolution in northern Mexico: Evidence from deep crustal xenoliths from La Olivina: Journal of Geophysical Research, v. 97, p. 17353–17376.

Campa, M.F., and Coney, P.J., 1983, Tectonostratigraphic terranes and mineral resource distributions in Mexico: Canadian Journal of Earth Sciences, v. 20, p. 1040–1051.

Chamberlain, K.R., and Bowring, S.A., 1990, Proterozoic geochronologic and isotopic boundary in NW Arizona: Journal of Geology, v. 98, p. 399–416.

Cochemé, J.J., and Demant, A., 1991, Geology of the Yécora area, northern Sierra Madre Occidental, Mexico, *in* Pérez-Segura, E., and Jacques-Ayala, C., eds., Studies of Sonoran Geology: Geological Society of America Special Paper 254, p. 81–94.

Coney, P.J., and Campa, M.F., 1987, Lithotectonic terrane map of Mexico (west of the 91st meridian): U.S. Geological Survey Miscellaneous Field Studies Map MF-1874-D, scale 1:2,500,000.

Damon, P.E., Shafiqullah, M., Roldán-Quintana, J., and Cochemé, J.J., 1983, El Batolito Laramide (90–40 Ma) de Sonora: Memoria XV Convencion Nacional, Asociación de Ingenieros de Minas, Metalurgistas y Geólogos de México, A.C., p. 65–95.

Davis, J., and Hawkesworth, C., 1993, The petrogenesis of 30–20 Ma basic and intermediate volcanics from the Mogollon-Datil volcanic field, New Mexico, USA: Contributions to Mineralogy and Petrology, v. 115, p. 165–183, doi: 10.1007/BF00321218.

Davis, J.M., and Hawkesworth, C.J., 1995, Geochemical and tectonic transitions in the evolution of the Mogollon-Datil volcanic field, New Mexico, U.S.A: Chemical Geology, v. 119, p. 31–53, doi: 10.1016/0009-2541(94)00094-O.

Denison, R.E., Burke, W.H., Jr., Hetherinton, E.A., and Otto, J.B., 1971, The basement rock framework of parts of Texas, southern New Mexico and northern Mexico, *in* The geologic framework of the Chihuahua tectonic belt: West Texas Geological Society Publication 71–59, p. 3–14.

DePaolo, D.J., and Farmer, G.L., 1984, Isotopic data bearing on the origin of Mesozoic and Tertiary granitic rocks in the western United States: Philosophical Transactions of the Royal Society of London, Series A: Mathematical and Physical Sciences, v. 310, p. 743–753.

Dickinson, W.R., 1981, Plate tectonic evolution of the southern Cordillera, *in* Dickinson, W.R., and Payne, W.D., eds., Relations of tectonics to ore deposits in the southern Cordillera: Arizona Geological Digest, v. 14, p. 113–135.

Duex, T.W., 1983, Geology, geochemistry, and geochronology of volcanic rocks between Cuauhtemoc and La Junta, central Chihuahua, Mexico [Ph.D. dissertation]: Austin, University of Texas at Austin, 234 p.

Farmer, G.L., and DePaolo, D.J., 1983, Origin of Mesozoic and Tertiary granite in the western United States and implications for pre-Mesozoic crustal structure: 1. Nd and Sr isotopic studies in the geocline of the northern Great Basin: Journal of Geophysical Research, v. 88, p. 3379–3401.

Farmer, G.L., and DePaolo, D.J., 1984, Origin of Mesozoic and Tertiary granite in the western United States and implications for pre-Mesozoic crustal structure: 2. Nd and Sr isotopic studies of unmineralized and Cu- and Mo-mineralized granite in the Precambrian craton: Journal of Geophysical Research, v. 89, p. 10141–10160.

González-León, C.M., McIntosh, W.C., Lozano-Santacruz, R., Valencia-Moreno, M., Amaya-Martínez, R., and Rodríguez-Castañeda, J.L., 2000, Cretaceous and Tertiary sedimentary, magmatic, and tectonic evolution of north-central Sonora (Arizpe and Bacanuchi Quadrangles), northwest Mexico: Geological Society of America Bulletin, v. 112, p. 600–610, doi: 10.1130/0016-7606(2000)112<0600:CATSMA>2.3.CO;2.

Gunderson, R., Cameron, K., and Cameron, M., 1986, Mid-Cenozoic high-K calc-alkalic and alkalic volcanism in eastern Chihuahua, Mexico; geology and geochemistry of the Benavides-Pozos area: Geological Society of America Bulletin, v. 97, p. 737–753, doi: 10.1130/0016-7606(1986)97<737:MHCAAV>2.0.CO;2.

Haenggi, W.T., 2001, Tectonic history of the Chihuahua trough, Mexico and adjacent USA, Part I: the pre-Mesozoic setting: Boletín de la Sociedad Geológica Mexicana, v. 54, p. 28–66.

Haenggi, W.T., 2002, Tectonic history of the Chihuahua trough, Mexico and adjacent USA, Part II: Mesozoic and Cenozoic: Boletín de la Sociedad Geológica Mexicana, v. 55, p. 38–94.

Handschy, J.W., and Dyer, R., 1987, Polyphase deformation in Sierra del Cuervo, Chihuahua, Mexico: Evidence for Ancestral Rocky Mountain tectonics in the Ouachita foreland of northern Mexico: Geological Society of America Bulletin, v. 99, p. 618–632, doi: 10.1130/0016-7606(1987)99<618:PDISDC>2.0.CO;2.

Handschy, J.W., Keller, G.R., and Smith, K.J., 1987, The Ouachita system in northern Mexico: Tectonics, v. 6, p. 323–330.

Hart, S.R., 1984, A large-scale isotopic anomaly in the Southern Hemisphere mantle: Nature, v. 309, p. 753–757, doi: 10.1038/309753a0.

Henry, C.D., and Fredrikson, G., 1987, Geology of part of southern Sinaloa, Mexico, adjacent to the Gulf of California: Geological Society of America Map and Chart Series, MCH067, 1 sheet, scale 1:250,000, 14 p.

Hoffman, P.F., 1988, United Plates of America, The birth of a craton: Early Proterozoic assembly and growth of Laurentia: Annual Reviews of Earth and Planetary Science, v. 16, p. 543–603, doi: 10.1146/annurev.ea.16.050188.002551.

Housh, T., and McMahon, T.P., 2000, Ancient isotopic characteristics of Neogene potassic magmatism in western New Guinea (Irian Jaya, Indonesia): Lithos, v. 50, p. 217–239, doi: 10.1016/S0024-4937(99)00043-2.

Ide, S., 1986, Geology of mid-Tertiary volcanic rocks in the Laboricita-General Trias area, central Chihuahua, Mexico [M.A. thesis]: Austin, Texas, University of Texas at Austin, p. 155.

James, E.W., and Henry, C.D., 1993a, Southeastern extent of the North American craton in Texas and northern Chihuahua as revealed by Pb isotopes: Geological Society of America Bulletin, v. 105, p. 116–126, doi: 10.1130/0016-7606(1993)105<0116:SEOTNA>2.3.CO;2.

James, E.W., and Henry, C.D., 1993b, Pb isotopes of ore deposits in Trans-Pecos Texas and northeastern Chihuahua, Mexico: Basement, igneous, and sedimentary sources of metals: Economic Geology and the Bulletin of the Society of Economic Geologists, v. 88, p. 934–947.

James, E.W., Henry, C.D., and McDowell, F.W., 1994, Defining the southeastern limit of North American middle Proterozoic basement in Texas and Chihuahua with lead isotopes: Geological Society of America Abstracts with Programs, v. 26, no. 6, p. 20.

Karlstrom, K.E., and Bowring, S.A., 1988, Early Proterozoic assembly of tectonostratigraphic terranes in southwestern North America: Journal of Geology, v. 96, p. 561–576.

Karlstrom, K.E., Bowring, S.A., and Conway, C.M., 1987, Tectonic significance of an Early Proterozoic two-province boundary in central Arizona: Geological Society of America Bulletin, v. 99, p. 529–538, doi: 10.1130/0016-7606(1987)99<529:TSOAEP>2.0.CO;2.

Keller, P.C., Bockoven, N.T., and McDowell, F.W., 1982, Tertiary volcanic history of the Sierra del Gallego area, Chihuahua, Mexico: Geological Society of America Bulletin, v. 93, p. 303–314, doi: 10.1130/0016-7606(1982)93<303:TVHOTS>2.0.CO;2.

Kempton, P.D., Harmon, R.S., Hawkesworth, C.J., and Moorbath, S., 1990, Petrology and geochemistry of lower crustal granulites from the Geronimo Volcanic Field, southeastern Arizona: Geochimica et Cosmochimica Acta, v. 54, p. 3401–3426, doi: 10.1016/0016-7037(90)90294-U.

Kistler, R.W., and Peterman, Z.E., 1973, Variations in Sr, Rb, K, Na, and initial Sr^{87}/Sr^{86} in Mesozoic granitic rocks and intruded wall rocks in central California: Geological Society of America Bulletin, v. 84, p. 3489–3512, doi: 10.1130/0016-7606(1973)84<3489:VISRKN>2.0.CO;2.

Kuentz, D.C., Cameron, K.L., and Niemeyer, S., 1992, Isotope provinces of southeastern Chihuahua and adjacent west Texas defined by mid-Cenozoic volcanic rocks, *in* Clark, K.F., Roldán-Quintana, J., and Schmidt, R.H., eds., Guidebook for the 1992 Field Conference: El Paso Geological Society Publication 24, p. 111–118.

Lang, J.R., and Titley, S.R., 1998, Isotopic and geochemical characteristics of Laramide magmatic systems in Arizona and implications for the genesis of porphyry copper deposits: Economic Geology and the Bulletin of the Society of Economic Geologists, v. 93, p. 138–170.

Lanphere, M.A., Cameron, K.L., and Cameron, M., 1980, Sr isotope geochemistry of voluminous rhyolitic ignimbrites and related rocks, western Mexico: Nature, v. 286, p. 594–596, doi: 10.1038/286594a0.

Lopez, R., Cameron, K.L., and Jones, N.W., 1997, Evidence from U-Pb zircon ages for the Grenvillian reworking of 1.8 Ga crust south of the Ouachita suture in the Coahuila terrane, northern Mexico: Eos (Transactions, American Geophysical Union), v. 78, p. F785.

Lopez, R., Cameron, K.L., and Jones, N.W., 1998, New Paleoproterozoic, Grenvillian, and Pan-African U-Pb zircon ages of boulders from the Coahuila terrane; implications for the Precambrian tectonic history of northern Mexico: Geological Society of America Abstracts with Programs, v. 30, no. 7, p. 354.

McDowell, F.W., 1993, The volcanic section near Yécora, Sonora and its regional significance, *in* Rodríguez, J.L. and Calmus, T., eds., Libreto-Guía Excursiónes del III Simpósio de la Geología de Sonora y areas adjacentes, p. 185–194.

McDowell, F.W., and Clabaugh, S.E., 1979, Ignimbrites of the Sierra Madre Occidental and their relation to the tectonic history of western Mexico, *in* Chapin, C.E., and Elson, W.E., eds., Ash-flow tuffs: Geological Society of America Special Paper 180, p. 113–124.

McDowell, F.W., Wark, D.A., and Aguirre-Díaz, G., 1990, The Tertiary ignimbrite flare-up in western Mexico: Geological Society of America Abstracts with Programs, v. 22, no. 3, p. 66.

McDowell, F.W., and Mauger, R.W., 1994, K-Ar and U-Pb zircon chronology of Late Cretaceous and Tertiary magmatism in central Chihuahua State, Mexico: Geological Society of America Bulletin, v. 106, p. 118–132, doi: 10.1130/0016-7606(1994)106<0118:KAAUPZ>2.3.CO;2.

McDowell, F.W., Roldán-Quintana, J., and Amaya-Martínez, R., 1997, Interrelationship of sedimentary and volcanic deposits associated with Tertiary extension in Sonora, Mexico: Geological Society of America Bulletin, v. 109, p. 1349–1360, doi: 10.1130/0016-7606(1997)109<1349:IOSAVD>2.3.CO;2.

McDowell, F.W., Housh, T.B., and Wark, D.A., 1999, Nature of the crust beneath west-central Chihuahua, Mexico, based upon Sr, Nd, and Pb isotopic compositions at the Tomóchic volcanic center: Geological Society of America Bulletin, v. 111, p. 823–830, doi: 10.1130/0016-7606(1999)111<0823:NOTCBW>2.3.CO;2.

McDowell, F.W., Roldán-Quintana, J., and Connelly, J.N., 2001, Duration of Late Cretaceous-early Tertiary magmatism in east-central Sonora, Mexico: Geological Society of America Bulletin, v. 113, p. 521–531, doi: 10.1130/0016-7606(2001)113<0521:DOLCET>2.0.CO;2.

McKee, J.W., Jones, N.W., and Anderson, T.H., 1999, Late Paleozoic and early Mesozoic history of the Las Delicias terrane, Coahuila, Mexico, *in*

Bartolini, C., Wilson, J.L., and Lawton, T., eds., Mesozoic Sedimentary and Tectonic History of North-Central Mexico: Geological Society of America Special Paper 340, p. 161–189.

McKee, J.W., Jones, N.W., and Long, L.E., 1984, History of recurrent activity along a major fault in northeastern Mexico: Geology, v. 12, p. 103–107.

McMillan, N.J., and Dickin, A.P., 1994, The Uvas volcanic field; early Rio Grande Rift lithospheric magmatism in southern New Mexico: Geological Society of America Abstracts with Programs, v. 26, no. 7, p. 367–368.

McMillan, N.J., Dickin, A.P., and Haag, D., 2000, Evolution of magma source regions in the Rio Grande rift, southern New Mexico: Geological Society of America Bulletin, v. 112, p. 1582–1593, doi: 10.1130/0016-7606(2000)112<1582:EOMSRI>2.0.CO;2.

Mead, R.D., Kesler, S.E., Foland, K.A., and Jones, L.M., 1988, Relationship of Sonoran tungsten mineralization to the metallogenic evolution of Mexico, *in* Clark, K.F., Salas P., and Guillermo, A., eds., A special issue devoted to the geology and mineral deposits of Mexico: Economic Geology, v. 83, p. 1943–1965.

Megaw, P.K.M., and McDowell, F.W., 1983, Geology and geochronology of volcanic rocks of the Sierra Pastorias area, Chihuahua, Mexico, *in* Clark, K.F., and Goodell, P.C., eds., Geology and mineral resources of north-central Chihuahua: El Paso Geological Society Publication 15, p. 195–204.

Menzies, M.A., Leeman, W.P., and Hawkesworth, C.J., 1983, Isotope geochemistry of Cenozoic volcanic rocks reveals mantle heterogeneity below western USA: Nature, v. 303, p. 205–209, doi: 10.1038/303205a0.

Mora-Alvarez, G., 1992, History of Cenozoic magmatism in the Sierra Santa Ursula, Sonora, Mexico [M.A. thesis]: Austin, The University of Texas at Austin, 152 p.

Mora-Alvarez, G., 1993, Relaciones Estratigráficas y Geochronológicas entre las Unidades Volcánicas de la Sierra Santa Ursula, en Sonora, y el Magmatismo de la región del Golfo de California, *in* Delgado-Argote, L.A., and Martín-Barajas, A., eds. Contribuciones a la Tectónica del Occidente de México: Unión Geofísica Mexicana Monografía No. 1, p. 123–146.

Mora-Alvarez, G., and McDowell, F.W., 2000, Miocene volcanism during late subduction and early rifting in the Sierra Santa Ursula of western Sonora, Mexico, *in* Delgado-Granados, H., Aguirre-Díaz, G., and Stock, J.M., eds., Cenozoic Tectonics and Volcanism of Mexico: Geological Society of America Special Paper 334, p. 123–141.

Mora-Klepeis, G., McDowell, F.W., and Ortega-Rivera, A., 1997, Late Miocene volcanism in west-central Sonora: an expression of rift or subduction-related magmatism: Geological Society of America Abstracts with Programs, v. 29, no. 6, p. A481.

Mosher, S., 1998, Tectonic evolution of the southern Laurentian Grenville orogenic belt: Geological Society of America Bulletin, v. 110, p. 1357–1375, doi: 10.1130/0016-7606(1998)110<1357:TEOTSL>2.3.CO;2.

Nimz, G.J., Cameron, K.L., and Niemeyer, S., 1993, The La Olivina pyroxenite suite and the isotopic compositions of mantle basalts parental to the mid-Cenozoic arc volcanism of northern Mexico: Journal of Geophysical Research, v. 98, p. 6489–6509.

Poole, F.G., Madrid, R.J., and Oliva-Becerril, F., 1991, Geological setting and origin of the stratiform barite in central Sonora, Mexico, *in* Raines, G.L., et al., eds., Geology and ore deposits of the Great Basin, Volume I: Geological Society of Nevada, p. 517–522.

Poole, F.G., Stewart, J.H., Berry, W.B.N., Harris, A.G., Repetski, J.E., Madrid, R.J., Ketner, K.B., Carter, C., and Morales-Ramirez, J.M., 1995, Ordovician ocean-basin rocks of Sonora, Mexico, *in* Cooper, J.D., Droser, M.L., and Finney, S.C., eds., Ordovician Odyssey: Short Papers for the Seventh International Symposium on the Ordovician System: Pacific Section Society for Sedimentary Geology, p. 277–284.

Potter, L.S., 1996, Chemical and isotopic variation along strike in the Eastern alkalic belt, Trans-Pecos magmatic province, Texas and New Mexico [Ph.D. dissertation]: Austin, University of Texas at Austin, p. 247.

Roden, M.F., Irving, A.J., and Murthy, V.R., 1988, Isotopic and trace element composition of the upper mantle beneath a young continental rift; results from Kilbourne Hole, New Mexico: Geochimica et Cosmochimica Acta, v. 52, p. 461–473, doi: 10.1016/0016-7037(88)90101-9.

Roldán-Quintana, J., 1991, Geology and chemical composition of El Jaralito and Aconchi batholiths in east-central Sonora, *in* Pérez-Segura, E., and Jacques-Ayala, C., eds., Studies of Sonoran Geology: Geological Society of America Special Paper 254, p. 19–36.

Rudnick, R.L., and Cameron, K.L., 1991, Age diversity of the deep crust in northern Mexico: Geology, v. 19, p. 1197–1200, doi: 10.1130/0091-7613(1991)019<1197:ADOTDC>2.3.CO;2.

Ruiz, J., Patchett, P.J., and Arculus, R.J., 1988, Nd-Sr isotopic compositions of lower crustal xenoliths: Evidence for the origin of mid-Tertiary felsic volcanics in Mexico: Contributions to Mineralogy and Petrology, v. 99, p. 36–43, doi: 10.1007/BF00399363.

Sawlan, M.G., 1991, Magmatic evolution of the Gulf of California rift: American Association of Petroleum Geologists Memoir 47, p. 301–369.

Schaaf, P., Morán-Zenteno, D., Hernández-Bernal, M.S., Solís-Pichardo, G., Tolson, G., and Köhler, H., 1995, Paleogene continental margin truncation in southwestern Mexico: Geochronological evidence: Tectonics, v. 14, p. 1339–1350, doi: 10.1029/95TC01928.

Sedlock, R.L., Ortega-Gutierrez, F., and Speed, R.C., 1993, Tectonostratigraphic terranes and tectonic evolution of Mexico: Geological Society of America Special Paper 278, 153 p.

Shurbet, D.H., and Cebull, S.E., 1987, Tectonic interpretation of the westernmost part of the Ouachita-Marathon (Hercynian) orogenic belt, west Texas-Mexico: Geology, v. 15, p. 458–461, doi: 10.1130/0091-7613(1987)15<458:TIOTWP>2.0.CO;2.

Silver, L.T., and Anderson, T.H., 1974, Possible left-lateral early to middle Mesozoic disruption of the southwestern North American craton margin: Geological Society of America Abstracts with Programs, v. 6, p. 955.

Silver, L.T., and Chappell, B.W., 1988, The Peninsular Ranges batholith: An insight into the evolution of the Cordilleran batholiths of southwestern North America: Royal Society of Edinburgh Transactions, v. 79, p. 105–121.

Smith, R.D., Cameron, K.L., McDowell, F.W., Niemeyer, S., and Sampson, D.E., 1996, Generation of voluminous silicic magmas and formation of mid-Cenozoic crust beneath north-central Mexico: evidence from ignimbrites, associated lavas, deep crustal granulites, and mantle pyroxenites: Contributions to Mineralogy and Petrology, v. 123, p. 375–389, doi: 10.1007/s004100050163.

Stewart, J.H., 1988, Latest Proterozoic and Paleozoic southern margin of North America and the accretion of Mexico: Geology, v. 16, p. 186–189, doi: 10.1130/0091-7613(1988)016<0186:LPAPSM>2.3.CO;2.

Stewart, J.H., 1992, Late Proterozoic and Paleozoic southern margin of North America in northern Mexico, *in* Clark, K.F., Roldán-Quintana, J., and Schmidt, R.H., eds., Geology and Mineral Resources of Northern Sierra Madre Occidental, Mexico: El Paso Geological Society Publication 24, p. 291–299.

Stewart, J.H., Poole, F.G., Ketner, K.B., Madrid, R.J., Roldán-Quintana, J., and Amaya-Martínez, R., 1990, Tectonics and stratigraphy of the Paleozoic and Triassic southern margin of North America, Sonora, Mexico, *in* Gehrels, G.E., and Spencer, J.E., eds., Geological excursions through the Sonoran desert region, Arizona and Sonora: Arizona Bureau of Geology and Mineral Technology Special Paper 7, p. 183–202.

Stock, J.M., 1989, Sequence and geochronology of Miocene rocks adjacent to the main Gulf escarpment: Southern Valle Chico, Baja California Norte, Mexico: Geofísica Internacional, v. 28, p. 851–896.

Stock, J.M., and Lee, J., 1994, Do microplates in subduction zones leave a geological record?: Tectonics, v. 13, p. 1472–1487, doi: 10.1029/94TC01808.

Swanson, E.R., 1977, Reconnaissance geology of the Tomochic-Ocampo area, Sierra Madre Occidental, Chihuahua, Mexico [Ph.D. dissertation]: Austin, University of Texas at Austin, 154 p.

Valencia-Moreno, M., Ruiz, J., Barton, M.D., Patchett, P.J., Zürcher, L., Hodkinson, D.G., and Roldán-Quintana, J., 2001, A chemical and isotopic study of the Laramide granitic belt of northwestern Mexico: Identification of the southern edge of the North American Precambrian basement: Geological Society of America Bulletin, v. 113, p. 1409–1422, doi: 10.1130/0016-7606(2001)113<1409:ACAISO>2.0.CO;2.

Verplanck, P.L., Farmer, G.L., McCurry, M., and Mertzman, S.A., 1999, The chemical and isotopic differentiation of an epizonal magma body; Organ Needle Pluton, New Mexico: Journal of Petrology, v. 40, p. 653–678, doi: 10.1093/petrology/40.4.653.

Wark, D.A., 1983, Geology of the mid-Tertiary volcanic terrane at Buenos Aires, Chihuahua, Mexico [M.A. thesis]: Austin, University of Texas at Austin, p. 156.

Wark, D.A., 1991, The role of mafic and intermediate composition magmas in rhyolite genesis, Sierra Madre Occidental, Mexico: Journal of Geophysical Research, v. 96, p. 13389–13411.

Wark, D.A., Kempter, K.A., and McDowell, F.W., 1990, Evolution of waning calcalkalic magmatism, northern Sierra Madre Occidental, Mexico: Geological Society of America Bulletin, v. 102, p. 1555–1564, doi: 10.1130/0016-7606(1990)102<1555:EOWSRM>2.3.CO;2.

Wooden, J.L., and Miller, D.M., 1990, Chronologic and isotopic framework for Early Proterozoic crustal evolution in the eastern Mojave desert region, SE California: Journal of Geophysical Research, v. 95, p. 20133–20146.

Wooden, J.L., Stacey, J.S., Doe, B.R., Howard, K.A., and Miller, D.M., 1988, Pb isotopic evidence for the formation of Proterozoic crust in the southwestern United States, *in* Ernst, W.G., ed., Metamorphism and crustal evolution of the western United States (Rubey Volume VII): Englewood Cliffs; Prentice Hall, p. 68–86.

Zartman, R.E., 1974, Lead isotope provinces in the Cordillera of the western United States and their geologic significance: Economic Geology and the Bulletin of the Society of Economic Geologists, v. 69, p. 792–805.

Manuscript Accepted by the Society 24 March 2005

Printed in the USA

Index

C

D

G

H

I

J

K

O

P